KIRK-OTHMER

# ENCYCLOPEDIA OF
# CHEMICAL
# TECHNOLOGY

## FOURTH EDITION

VOLUME **19**

PIGMENTS
TO
POWDERS, HANDLING

**EXECUTIVE EDITOR**
Jacqueline I. Kroschwitz

**EDITOR**
Mary Howe-Grant

# KIRK-OTHMER

# ENCYCLOPEDIA OF CHEMICAL TECHNOLOGY

## FOURTH EDITION

### VOLUME 19

PIGMENTS
TO
POWDERS, HANDLING

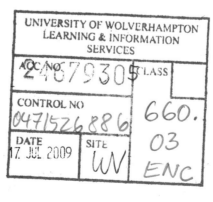

A Wiley-Interscience Publication
**JOHN WILEY & SONS**
New York  •  Chichester  •  Brisbane  •  Toronto  •  Singapore

**Library of Congress Cataloging-in-Publication Data**

Encyclopedia of chemical technology/executive editor, Jacqueline
  I. Kroschwitz; editor, Mary Howe-Grant. — 4th ed.
    p. cm.
  At head of title: Kirk-Othmer.
  "A Wiley-Interscience publication."
  Contents: v. 19, Pigments to Powders, Handling
  ISBN 0-471-52688-6 (v. 19)
  1. Chemistry, Technical—Encyclopedias.  I. Kirk, Raymond E.
(Raymond Eller), 1890–1957.  II. Othmer, Donald F. (Donald
Frederick), 1904–1995.  III. Kroschwitz, Jacqueline I., 1942– .
IV. Howe-Grant, Mary, 1943– .  V. Title: Kirk-Othmer encyclopedia
of chemical technology.
TP9.E685 1992                                                        91-16789
660'.03—dc20

# CONTENTS

# EDITORIAL STAFF
# FOR VOLUME 19

*Executive Editor:* **Jacqueline I. Kroschwitz**
*Editor:* **Mary Howe-Grant**
*Associate Managing Editor:* **Lindy Humphreys**
*Copy Editors:* **Lawrence Altieri**
              **Jonathan Lee**

# CONTRIBUTORS
# TO VOLUME 19

**Anthony Anton,** *E. I. du Pont de Nemours & Co., Inc., Wilmington, Delaware,* Fibers (under Polyamides)

**Darlene M. Back,** *Union Carbide Corporation, Danbury, Connecticut,* Ethylene oxide polymers (under Polyethers)

**J. W. Barlow,** *University of Texas, Austin,* Polymer blends

**Daniel J. Brunelle,** *General Electric, Schenectady, New York,* Polycarbonates

**David F. Cadogan,** *European Council for Plasticizers and Intermediates, Brussels, Belgium,* Plasticizers

**John W. Carson,** *Jenike and Johanson, Inc., Westford, Massachusetts,* Bulk powders (under Powders, handling)

**Kuen-Wai Chiu,** *Callery Chemical Company, Pittsburgh, Pennsylvania,* Potassium

**Elke M. Clark,** *Union Carbide Corporation, Danbury, Connecticut,* Ethylene oxide polymers (under Polyethers)

**M. P. Dreyfuss,** *Consultant, Midland, Michigan,* Tetrahydrofuran and oxetane polymers (under Polyethers)

**P. Dreyfuss,** *Consultant, Midland, Michigan,* Tetrahydrofuran and oxetane polymers (under Polyethers)

**Seán G. Dwyer,** *S. C. Johnson & Sons, Inc., Racine, Wisconsin,* Polishes

**H. W. Earhart,** *Consultant, Wichita, Kansas,* Polymethylbenzenes

**Anthony J. East,** *Hoechst Celanese Corporation, Summit, New Jersey,* Polyesters, thermoplastic

**M. Jamal El-Hibri,** *Amoco Polymers Inc., Alpharetta, Georgia,* Polysulfones (under Polymers containing sulfur)

**Mark B. Freilich,** *The University of Memphis, Tennessee,* Potassium compounds

**Barbara J. Furches,** *The Dow Chemical Company, Midland, Michigan,* Plastics testing

**Steven D. Gagnon,** *BASF Corporation, Geismar, Louisiana,* Propylene oxide polymers (under Polyethers)

**Jon F. Geibel,** *Phillips Petroleum Company, Bartlesville, Oklahoma,* Poly(phenylene sulfide) (under Polymers containing sulfur)

**Christen M. Giandomenico,** *Johnson Matthew, West Chester, Pennsylvania,* Platinum-group metals, compounds

**Michael Golden,** *Hoechst Celanese Corporation, Summit, New Jersey,* Polyesters, thermoplastic

**Stanley E. Handman,** *Consultant, Plainview, New York,* Piping systems

**Christopher J. Howick,** *European Vinyls Corporation, Cheshire, United Kingdom,* Plasticizers

**E. E. Jaffe,** *Ciba-Geigy Corporation, Newport, Delaware,* Organic (under Pigments)

**Alexy D. Kachkovski,** *National Academy of Sciences of the Ukraine, Kiev,* Polymethine dyes

**Henno Keskkula,** *University of Texas, Austin,* Polymer blends

**Andrew P. Komin,** *Koch Chemical Company, Wichita, Kansas,* Polymethylbenzenes

**John Leland,** *Phillips 66, Bartlesville, Oklahoma,* Poly(phenylene sulfide) (under Polymers containing sulfur)

**Gerd Leston,** *Consultant, Pittsburgh, Pennsylvania,* (Polyhydroxy)benzenes

**George M. Long,** *Institute of Gas Technology, Chicago, Illinois,* Pipelines

**Subhash Makhija,** *Consultant, Summit, New Jersey,* Polyesters, thermoplastic

**Joseph Marinelli,** *Peabody Solids Flow, Charlotte, North Carolina,* Bulk powders (under Powders, handling)

**Roland E. Meissner III,** *The Ralph M. Parsons Company, Pasadena, California*, Plant layout; Plant location

**Lester R. Morss,** *Argonne National Laboratory, Argonne, Illinois*, Plutonium and plutonium compounds

**Ralph D. Nelson, Jr.,** *E. I. du Pont de Nemours & Company, Inc., Wilmington, Delaware*, Dispersion of powders in liquids (under Powders, handling)

**Mirek Novotny,** *Cerdec Corporation, Washington, Pennsylvania*, Inorganic (under Pigments)

**Julia I. O'Farrelly,** *Johnson Matthey Technology Center, Reading, United Kingdom*, Platinum-group metals

**Richard P. Palluzi,** *Exxon Research and Engineering Company, Florham Park, New Jersey*, Pilot plants

**Robert J. Palmer,** *Du Pont de Nemours International SA, Geneva, Switzerland*, Plastics (under Polyamides)

**Donald R. Paul,** *University of Texas, Austin*, Polymer blends

**Richard L. Petersen,** *The University of Memphis, Tennessee*, Potassium compounds

**Louise C. Potter,** *Johnson Matthey Technology Center, Reading, United Kingdom*, Platinum-group metals

**Gerfried Pruckmayr,** *Du Pont Specialty Chemicals, Wilmington, Delaware*, Tetrahydrofuran and oxetane polymers (under Polyethers)

**Richard W. Prugh,** *Process Safety Engineering, Inc., Wilmington, Delaware*, Plant safety

**Ramesh Ramachandran,** *Union Carbide Corporation, Danbury, Connecticut*, Ethylene oxide polymers (under Polyethers)

**Francis J. Randall,** *S. C. Johnson & Sons, Inc., Racine, Wisconsin*, Polishes

**Stephen L. Rosen,** *University of Missouri, Rolla*, Polymers

**Michael Scherrer,** *Morton International, Inc., Woodstock, Illinois*, Polysulfides (under Polymers containing sulfur)

**Jeffrey Selley,** *Consultant, Durham, North Carolina*, Polyesters, unsaturated

**Richard J. Seymour,** *Johnson Matthey Technology Center, Reading, United Kingdom*, Platinum-group metals

**Mark D. Smith,** *AlliedSignal Aerospace Company, Kansas City, Missouri*, Plasma technology

**Z. Solc,** *University of Pardúbice, Czech Republic*, Inorganic (under Pigments)

**Graham Swift,** *Rohm and Haas Research Laboratories, Spring House, Pennsylvania*, Polymers, environmentally degradable

**Tohru Takekoshi,** *General Electric, Schenectady, New York*, Polyimides

**David B. Todd,** *Stevens Institute of Technology, Hoboken, New Jersey*, Plastics processing

**M. Trojan,** *University of Pardúbice, Czech Republic*, Inorganic (under Pigments)

**David Vietti,** *Morton International, Inc., Woodstock, Illinois*, Polysulfides (under Polymers containing sulfur)

**Joseph N. Weber,** *Du Pont Nylon, Wilmington, Delaware*, General (under Polyamides)

**Thomas P. Whaley,** *Consultant, Sun City, Arizona*, Pipelines

**Dwain M. White,** *General Electric, Schenectady, New York*, Aromatic (under Polyethers)

**Marino Xanthos,** *Stevens Institute of Technology, Hoboken, New Jersey*, Plastics processing

# NOTE ON CHEMICAL ABSTRACTS SERVICE REGISTRY NUMBERS AND NOMENCLATURE

Chemical Abstracts Service (CAS) Registry Numbers are unique numerical identifiers assigned to substances recorded in the CAS Registry System. They appear in brackets in the *Chemical Abstracts* (CA) substance and formula indexes following the names of compounds. A single compound may have synonyms in the chemical literature. A simple compound like phenethylamine can be named $\beta$-phenylethylamine or, as in *Chemical Abstracts*, benzeneethanamine. The usefulness of the *Encyclopedia* depends on accessibility through the most common correct name of a substance. Because of this diversity in nomenclature careful attention has been given to the problem in order to assist the reader as much as possible, especially in locating the systematic CA index name by means of the Registry Number. For this purpose, the reader may refer to the CAS Registry Handbook—Number Section which lists in numerical order the Registry Number with the *Chemical Abstracts* index name and the molecular formula; eg, **458-88-8**, Piperidine, 2-propyl-, (*S*)-, $C_8H_{17}N$; in the *Encyclopedia* this compound would be found under its common name, coniine [*458-88-8*]. Alternatively, this information can be retrieved electronically from CAS Online. In many cases molecular formulas have also been provided in the *Encyclopedia* text to facilitate electronic searching. The Registry Number is a valuable link for the reader in retrieving additional published information on substances and also as a point of access for on-line data bases.

In all cases, the CAS Registry Numbers have been given for title compounds in articles and for all compounds in the index. All specific substances indexed in *Chemical Abstracts* since 1965 are included in the CAS Registry System as are a large number of substances derived from a variety of reference works. The CAS Registry System identifies a substance on the basis of an unambiguous computer-language description of its molecular structure including stereochemical detail. The Registry Number is a machine-checkable number (like a Social Security number) assigned in sequential order to each substance as it enters the registry system. The value of the number lies in the fact that it is a concise and unique means of substance identification, which is independent of, and therefore bridges, many systems of chemical nomenclature. For polymers, one Registry Number may be used for the entire family; eg, polyoxyethylene (20) sorbitan monolaurate has the same number as all of its polyoxyethylene homologues.

Cross-references are inserted in the index for many common names and for some systematic names. Trademark names appear in the index. Names that are incorrect, misleading, or ambiguous are avoided. Formulas are given very frequently in the text to help in identifying compounds. The spelling and form used, even for industrial names, follow American chemical usage, but not always the usage of *Chemical Abstracts* (eg, *coniine* is used instead of *(S)-2-propylpiperidine*, *aniline* instead of *benzenamine*, and *acrylic acid* instead of *2-propenoic acid*).

There are variations in representation of rings in different disciplines. The dye industry does not designate aromaticity or double bonds in rings. All double bonds and aromaticity are shown in the *Encyclopedia* as a matter of course. For example, tetralin has an aromatic ring and a saturated ring and its structure

appears in the *Encyclopedia* with its common name, Registry Number enclosed in brackets, and parenthetical CA index name, ie, tetralin [*119-64-2*] (1,2,3,4-tetrahydronaphthalene). With names and structural formulas, and especially with CAS Registry Numbers, the aim is to help the reader have a concise means of substance identification.

# CONVERSION FACTORS, ABBREVIATIONS, AND UNIT SYMBOLS

## SI Units (Adopted 1960)

The International System of Units (abbreviated SI), is being implemented throughout the world. This measurement system is a modernized version of the MKSA (meter, kilogram, second, ampere) system, and its details are published and controlled by an international treaty organization (The International Bureau of Weights and Measures) (1).

SI units are divided into three classes:

### BASE UNITS

| | |
|---|---|
| length | meter[†] (m) |
| mass | kilogram (kg) |
| time | second (s) |
| electric current | ampere (A) |
| thermodynamic temperature[‡] | kelvin (K) |
| amount of substance | mole (mol) |
| luminous intensity | candela (cd) |

### SUPPLEMENTARY UNITS

| | |
|---|---|
| plane angle | radian (rad) |
| solid angle | steradian (sr) |

[†]The spellings "metre" and "litre" are preferred by ASTM; however, "-er" is used in the *Encyclopedia*.

[‡]Wide use is made of Celsius temperature ($t$) defined by

$$t = T - T_0$$

where $T$ is the thermodynamic temperature, expressed in kelvin, and $T_0 = 273.15$ K by definition. A temperature interval may be expressed in degrees Celsius as well as in kelvin.

DERIVED UNITS AND OTHER ACCEPTABLE UNITS

These units are formed by combining base units, supplementary units, and other derived units (2–4). Those derived units having special names and symbols are marked with an asterisk in the list below.

| Quantity | Unit | Symbol | Acceptable equivalent |
|---|---|---|---|
| *absorbed dose | gray | Gy | J/kg |
| acceleration | meter per second squared | $m/s^2$ | |
| *activity (of a radionuclide) | becquerel | Bq | 1/s |
| area | square kilometer | $km^2$ | |
| | square hectometer | $hm^2$ | ha (hectare) |
| | square meter | $m^2$ | |
| concentration (of amount of substance) | mole per cubic meter | $mol/m^3$ | |
| current density | ampere per square meter | $A//m^2$ | |
| density, mass density | kilogram per cubic meter | $kg/m^3$ | $g/L$; $mg/cm^3$ |
| dipole moment (quantity) | coulomb meter | C·m | |
| *dose equivalent | sievert | Sv | J/kg |
| *electric capacitance | farad | F | C/V |
| *electric charge, quantity of electricity | coulomb | C | A·s |
| electric charge density | coulomb per cubic meter | $C/m^3$ | |
| *electric conductance | siemens | S | A/V |
| electric field strength | volt per meter | V/m | |
| electric flux density | coulomb per square meter | $C/m^2$ | |
| *electric potential, potential difference, electromotive force | volt | V | W/A |
| *electric resistance | ohm | Ω | V/A |
| *energy, work, quantity of heat | megajoule | MJ | |
| | kilojoule | kJ | |
| | joule | J | N·m |
| | electronvolt[†] | eV[†] | |
| | kilowatt-hour[†] | kW·h[†] | |
| energy density | joule per cubic meter | $J/m^3$ | |
| *force | kilonewton | kN | |
| | newton | N | $kg·m/s^2$ |

[†]This non-SI unit is recognized by the CIPM as having to be retained because of practical importance or use in specialized fields (1).

| Quantity | Unit | Symbol | Acceptable equivalent |
|---|---|---|---|
| *frequency | megahertz | MHz | |
| | hertz | Hz | 1/s |
| heat capacity, entropy | joule per kelvin | J/K | |
| heat capacity (specific), specific entropy | joule per kilogram kelvin | J/(kg·K) | |
| heat-transfer coefficient | watt per square meter kelvin | W/(m²·K) | |
| *illuminance | lux | lx | lm/m² |
| *inductance | henry | H | Wb/A |
| linear density | kilogram per meter | kg/m | |
| luminance | candela per square meter | cd/m² | |
| *luminous flux | lumen | lm | cd·sr |
| magnetic field strength | ampere per meter | A/m | |
| *magnetic flux | weber | Wb | V·s |
| *magnetic flux density | tesla | T | Wb/m² |
| molar energy | joule per mole | J/mol | |
| molar entropy, molar heat capacity | joule per mole kelvin | J/(mol·K) | |
| moment of force, torque | newton meter | N·m | |
| momentum | kilogram meter per second | kg·m/s | |
| permeability | henry per meter | H/m | |
| permittivity | farad per meter | F/m | |
| *power, heat flow rate, radiant flux | kilowatt | kW | |
| | watt | W | J/s |
| power density, heat flux density, irradiance | watt per square meter | W/m² | |
| *pressure, stress | megapascal | MPa | |
| | kilopascal | kPa | |
| | pascal | Pa | N/m² |
| sound level | decibel | dB | |
| specific energy | joule per kilogram | J/kg | |
| specific volume | cubic meter per kilogram | m³/kg | |
| surface tension | newton per meter | N/m | |
| thermal conductivity | watt per meter kelvin | W/(m·K) | |
| velocity | meter per second | m/s | |
| | kilometer per hour | km/h | |
| viscosity, dynamic | pascal second | Pa·s | |
| | millipascal second | mPa·s | |
| viscosity, kinematic | square meter per second | m²/s | |
| | square millimeter per second | mm²/s | |

| Quantity | Unit | Symbol | Acceptable equivalent |
|---|---|---|---|
| volume | cubic meter | $m^3$ | |
| | cubic diameter | $dm^3$ | L (liter) (5) |
| | cubic centimeter | $cm^3$ | mL |
| wave number | 1 per meter | $m^{-1}$ | |
| | 1 per centimeter | $cm^{-1}$ | |

In addition, there are 16 prefixes used to indicate order of magnitude, as follows:

| Multiplication factor | Prefix | Symbol | Note |
|---|---|---|---|
| $10^{18}$ | exa | E | |
| $10^{15}$ | peta | P | |
| $10^{12}$ | tera | T | |
| $10^{9}$ | giga | G | |
| $10^{6}$ | mega | M | |
| $10^{3}$ | kilo | k | |
| $10^{2}$ | hecto | h[a] | [a]Although hecto, deka, deci, and centi |
| 10 | deka | da[a] | are SI prefixes, their use should be |
| $10^{-1}$ | deci | d[a] | avoided except for SI unit-multiples |
| $10^{-2}$ | centi | c[a] | for area and volume and nontech- |
| $10^{-3}$ | milli | m | nical use of centimeter, as for body |
| $10^{-6}$ | micro | $\mu$ | and clothing measurement. |
| $10^{-9}$ | nano | n | |
| $10^{-12}$ | pico | p | |
| $10^{-15}$ | femto | f | |
| $10^{-18}$ | atto | a | |

For a complete description of SI and its use the reader is referred to ASTM E380 (4) and the article UNITS AND CONVERSION FACTORS which appears in Vol. 24.

A representative list of conversion factors from non-SI to SI units is presented herewith. Factors are given to four significant figures. Exact relationships are followed by a dagger. A more complete list is given in the latest editions of ASTM E380 (4) and ANSI Z210.1 (6).

## Conversion Factors to SI Units

| To convert from | To | Multiply by |
|---|---|---|
| acre | square meter ($m^2$) | $4.047 \times 10^3$ |
| angstrom | meter (m) | $1.0 \times 10^{-10\dagger}$ |
| are | square meter ($m^2$) | $1.0 \times 10^{2\dagger}$ |

[†]Exact.

| To convert from | To | Multiply by |
|---|---|---|
| astronomical unit | meter (m) | $1.496 \times 10^{11}$ |
| atmosphere, standard | pascal (Pa) | $1.013 \times 10^{5}$ |
| bar | pascal (Pa) | $1.0 \times 10^{5\dagger}$ |
| barn | square meter (m$^2$) | $1.0 \times 10^{-28\dagger}$ |
| barrel (42 U.S. liquid gallons) | cubic meter (m$^3$) | 0.1590 |
| Bohr magneton ($\mu_B$) | J/T | $9.274 \times 10^{-24}$ |
| Btu (International Table) | joule (J) | $1.055 \times 10^{3}$ |
| Btu (mean) | joule (J) | $1.056 \times 10^{3}$ |
| Btu (thermochemical) | joule (J) | $1.054 \times 10^{3}$ |
| bushel | cubic meter (m$^3$) | $3.524 \times 10^{-2}$ |
| calorie (International Table) | joule (J) | 4.187 |
| calorie (mean) | joule (J) | 4.190 |
| calorie (thermochemical) | joule (J) | $4.184^{\dagger}$ |
| centipoise | pascal second (Pa·s) | $1.0 \times 10^{-3\dagger}$ |
| centistokes | square millimeter per second (mm$^2$/s) | $1.0^{\dagger}$ |
| cfm (cubic foot per minute) | cubic meter per second (m$^3$/s) | $4.72 \times 10^{-4}$ |
| cubic inch | cubic meter (m$^3$) | $1.639 \times 10^{-5}$ |
| cubic foot | cubic meter (m$^3$) | $2.832 \times 10^{-2}$ |
| cubic yard | cubic meter (m$^3$) | 0.7646 |
| curie | becquerel (Bq) | $3.70 \times 10^{10\dagger}$ |
| debye | coulomb meter (C·m) | $3.336 \times 10^{-30}$ |
| degree (angle) | radian (rad) | $1.745 \times 10^{-2}$ |
| denier (international) | kilogram per meter (kg/m) | $1.111 \times 10^{-7}$ |
|  | tex$^{\ddagger}$ | 0.1111 |
| dram (apothecaries') | kilogram (kg) | $3.888 \times 10^{-3}$ |
| dram (avoirdupois) | kilogram (kg) | $1.772 \times 10^{-3}$ |
| dram (U.S. fluid) | cubic meter (m$^3$) | $3.697 \times 10^{-6}$ |
| dyne | newton (N) | $1.0 \times 10^{-5\dagger}$ |
| dyne/cm | newton per meter (N/m) | $1.0 \times 10^{-3\dagger}$ |
| electronvolt | joule (J) | $1.602 \times 10^{-19}$ |
| erg | joule (J) | $1.0 \times 10^{-7\dagger}$ |
| fathom | meter (m) | 1.829 |
| fluid ounce (U.S.) | cubic meter (m$^3$) | $2.957 \times 10^{-5}$ |
| foot | meter (m) | $0.3048^{\dagger}$ |
| footcandle | lux (lx) | 10.76 |
| furlong | meter (m) | $2.012 \times 10^{-2}$ |
| gal | meter per second squared (m/s$^2$) | $1.0 \times 10^{-2\dagger}$ |
| gallon (U.S. dry) | cubic meter (m$^3$) | $4.405 \times 10^{-3}$ |
| gallon (U.S. liquid) | cubic meter (m$^3$) | $3.785 \times 10^{-3}$ |
| gallon per minute (gpm) | cubic meter per second (m$^3$/s) | $6.309 \times 10^{-5}$ |
|  | cubic meter per hour (m$^3$/h) | 0.2271 |

$^{\dagger}$Exact.
$^{\ddagger}$See footnote on p. xiii.

| To convert from | To | Multiply by |
| --- | --- | --- |
| gauss | tesla (T) | $1.0 \times 10^{-4}$ |
| gilbert | ampere (A) | 0.7958 |
| gill (U.S.) | cubic meter ($m^3$) | $1.183 \times 10^{-4}$ |
| grade | radian | $1.571 \times 10^{-2}$ |
| grain | kilogram (kg) | $6.480 \times 10^{-5}$ |
| gram force per denier | newton per tex (N/tex) | $8.826 \times 10^{-2}$ |
| hectare | square meter ($m^2$) | $1.0 \times 10^{4\dagger}$ |
| horsepower (550 ft·lbf/s) | watt (W) | $7.457 \times 10^2$ |
| horsepower (boiler) | watt (W) | $9.810 \times 10^3$ |
| horsepower (electric) | watt (W) | $7.46 \times 10^{2\dagger}$ |
| hundredweight (long) | kilogram (kg) | 50.80 |
| hundredweight (short) | kilogram (kg) | 45.36 |
| inch | meter (m) | $2.54 \times 10^{-2\dagger}$ |
| inch of mercury (32°F) | pascal (Pa) | $3.386 \times 10^3$ |
| inch of water (39.2°F) | pascal (Pa) | $2.491 \times 10^2$ |
| kilogram-force | newton (N) | 9.807 |
| kilowatt hour | megajoule (MJ) | $3.6^{\dagger}$ |
| kip | newton (N) | $4.448 \times 10^3$ |
| knot (international) | meter per second (m/S) | 0.5144 |
| lambert | candela per square meter (cd/$m^3$) | $3.183 \times 10^3$ |
| league (British nautical) | meter (m) | $5.559 \times 10^3$ |
| league (statute) | meter (m) | $4.828 \times 10^3$ |
| light year | meter (m) | $9.461 \times 10^{15}$ |
| liter (for fluids only) | cubic meter ($m^3$) | $1.0 \times 10^{-3\dagger}$ |
| maxwell | weber (Wb) | $1.0 \times 10^{-8\dagger}$ |
| micron | meter (m) | $1.0 \times 10^{-6\dagger}$ |
| mil | meter (m) | $2.54 \times 10^{-5\dagger}$ |
| mile (statute) | meter (m) | $1.609 \times 10^3$ |
| mile (U.S. nautical) | meter (m) | $1.852 \times 10^{3\dagger}$ |
| mile per hour | meter per second (m/s) | 0.4470 |
| millibar | pascal (Pa) | $1.0 \times 10^2$ |
| millimeter of mercury (0°C) | pascal (Pa) | $1.333 \times 10^{2\dagger}$ |
| minute (angular) | radian | $2.909 \times 10^{-4}$ |
| myriagram | kilogram (kg) | 10 |
| myriameter | kilometer (km) | 10 |
| oersted | ampere per meter (A/m) | 79.58 |
| ounce (avoirdupois) | kilogram (kg) | $2.835 \times 10^{-2}$ |
| ounce (troy) | kilogram (kg) | $3.110 \times 10^{-2}$ |
| ounce (U.S. fluid) | cubic meter ($m^3$) | $2.957 \times 10^{-5}$ |
| ounce-force | newton (N) | 0.2780 |
| peck (U.S.) | cubic meter ($m^3$) | $8.810 \times 10^{-3}$ |
| pennyweight | kilogram (kg) | $1.555 \times 10^{-3}$ |
| pint (U.S. dry) | cubic meter ($m^3$) | $5.506 \times 10^{-4}$ |
| pint (U.S. liquid) | cubic meter ($m^3$) | $4.732 \times 10^{-4}$ |

$^{\dagger}$Exact.

| To convert from | To | Multiply by |
|---|---|---|
| poise (absolute viscosity) | pascal second (Pa·s) | $0.10^{\dagger}$ |
| pound (avoirdupois) | kilogram (kg) | 0.4536 |
| pound (troy) | kilogram (kg) | 0.3732 |
| poundal | newton (N) | 0.1383 |
| pound-force | newton (N) | 4.448 |
| pound force per square inch (psi) | pascal (Pa) | $6.895 \times 10^3$ |
| quart (U.S. dry) | cubic meter (m³) | $1.101 \times 10^{-3}$ |
| quart (U.S. liquid) | cubic meter (m³) | $9.464 \times 10^{-4}$ |
| quintal | kilogram (kg) | $1.0 \times 10^{2\dagger}$ |
| rad | gray (Gy) | $1.0 \times 10^{-2\dagger}$ |
| rod | meter (m) | 5.029 |
| roentgen | coulomb per kilogram (C/kg) | $2.58 \times 10^{-4}$ |
| second (angle) | radian (rad) | $4.848 \times 10^{-6\dagger}$ |
| section | square meter (m²) | $2.590 \times 10^6$ |
| slug | kilogram (kg) | 14.59 |
| spherical candle power | lumen (lm) | 12.57 |
| square inch | square meter (m²) | $6.452 \times 10^{-4}$ |
| square foot | square meter (m²) | $9.290 \times 10^{-2}$ |
| square mile | square meter (m²) | $2.590 \times 10^6$ |
| square yard | square meter (m²) | 0.8361 |
| stere | cubic meter (m³) | $1.0^{\dagger}$ |
| stokes (kinematic viscosity) | square meter per second (m²/s) | $1.0 \times 10^{-4\dagger}$ |
| tex | kilogram per meter (kg/m) | $1.0 \times 10^{-6\dagger}$ |
| ton (long, 2240 pounds) | kilogram (kg) | $1.016 \times 10^3$ |
| ton (metric) (tonne) | kilogram (kg) | $1.0 \times 10^{3\dagger}$ |
| ton (short, 2000 pounds) | kilogram (kg) | $9.072 \times 10^2$ |
| torr | pascal (Pa) | $1.333 \times 10^2$ |
| unit pole | weber (Wb) | $1.257 \times 10^{-7}$ |
| yard | meter (m) | $0.9144^{\dagger}$ |

$^{\dagger}$Exact.

## Abbreviations and Unit Symbols

Following is a list of common abbreviations and unit symbols used in the *Encyclopedia*. In general they agree with those listed in *American National Standard Abbreviations for Use on Drawings and in Text* (*ANSI Y1.1*) (6) and *American National Standard Letter Symbols for Units in Science and Technology* (*ANSI Y10*) (6). Also included is a list of acronyms for a number of private and government organizations as well as common industrial solvents, polymers, and other chemicals.

*Rules for Writing Unit Symbols* (4):

1. Unit symbols are printed in upright letters (roman) regardless of the type style used in the surrounding text.
2. Unit symbols are unaltered in the plural.
3. Unit symbols are not followed by a period except when used at the end of a sentence.
4. Letter unit symbols are generally printed lower-case (for example, cd for candela) unless the unit name has been derived from a proper name, in which case the first letter of the symbol is capitalized (W, Pa). Prefixes and unit symbols retain their prescribed form regardless of the surrounding typography.
5. In the complete expression for a quantity, a space should be left between the numerical value and the unit symbol. For example, write 2.37 lm, *not* 2.37lm, and 35 mm, *not* 35mm. When the quantity is used in an adjectival sense, a hyphen is often used, for example, 35-mm film. *Exception:* No space is left between the numerical value and the symbols of degree, minute, and second of plane angle, degree Celsius, and the percent sign.
6. No space is used between the prefix and unit symbol (for example, kg).
7. Symbols, not abbreviations, should be used for units. For example, use "A," not "amp," for ampere.
8. When multiplying unit symbols, use a raised dot:

$$\text{N·m} \quad \text{for} \quad \text{newton meter}$$

In the case of W·h, the dot may be omitted, thus:

$$\text{Wh}$$

An exception to this practice is made for computer printouts, automatic typewriter work, etc, where the raised dot is not possible, and a dot on the line may be used.

9. When dividing unit symbols, use one of the following forms:

$$\text{m/s} \quad or \quad \text{m·s}^{-1} \quad or \quad \frac{\text{m}}{\text{s}}$$

In no case should more than one slash be used in the same expression unless parentheses are inserted to avoid ambiguity. For example, write:

$$\text{J/(mol·K)} \quad or \quad \text{J·mol}^{-1}\text{·K}^{-1} \quad or \quad \text{(J/mol)/K}$$

but *not*

$$\text{J/mol/K}$$

10. Do not mix symbols and unit names in the same expression. Write:

joules per kilogram   *or*   J/kg   *or*   J·kg$^{-1}$

but *not*

joules/kilogram   *nor*   joules/kg   *nor*   joules·kg$^{-1}$

## ABBREVIATIONS AND UNITS

| | | | |
|---|---|---|---|
| A | ampere | AOAC | Association of Official Analytical Chemists |
| A | anion (eg, HA) | | |
| *A* | mass number | AOCS | American Oil Chemists' Society |
| a | atto (prefix for 10$^{-18}$) | | |
| AATCC | American Association of Textile Chemists and Colorists | APHA | American Public Health Association |
| | | API | American Petroleum Institute |
| ABS | acrylonitrile–butadiene–styrene | | |
| | | aq | aqueous |
| abs | absolute | Ar | aryl |
| ac | alternating current, *n.* | *ar-* | aromatic |
| a-c | alternating current, *adj.* | *as-* | asymmetric(al) |
| *ac-* | alicyclic | ASHRAE | American Society of Heating, Refrigerating, and Air Conditioning Engineers |
| acac | acetylacetonate | | |
| ACGIH | American Conference of Governmental Industrial Hygienists | | |
| | | ASM | American Society for Metals |
| ACS | American Chemical Society | | |
| | | ASME | American Society of Mechanical Engineers |
| AGA | American Gas Association | | |
| Ah | ampere hour | ASTM | American Society for Testing and Materials |
| AIChE | American Institute of Chemical Engineers | | |
| | | at no. | atomic number |
| AIME | American Institute of Mining, Metallurgical, and Petroleum Engineers | at wt | atomic weight |
| | | av(g) | average |
| | | AWS | American Welding Society |
| | | *b* | bonding orbital |
| AIP | American Institute of Physics | bbl | barrel |
| | | bcc | body-centered cubic |
| AISI | American Iron and Steel Institute | BCT | body-centered tetragonal |
| | | Bé | Baumé |
| alc | alcohol(ic) | BET | Brunauer-Emmett-Teller (adsorption equation) |
| Alk | alkyl | | |
| alk | alkaline (not alkali) | bid | twice daily |
| amt | amount | Boc | *t*-butyloxycarbonyl |
| amu | atomic mass unit | BOD | biochemical (biological) oxygen demand |
| ANSI | American National Standards Institute | | |
| | | bp | boiling point |
| AO | atomic orbital | Bq | becquerel |

| | | | |
|---|---|---|---|
| C | coulomb | DIN | Deutsche Industrie Normen |
| °C | degree Celsius | | |
| C- | denoting attachment to carbon | dl-; DL- | racemic |
| | | DMA | dimethylacetamide |
| c | centi (prefix for $10^{-2}$) | DMF | dimethylformamide |
| c | critical | DMG | dimethyl glyoxime |
| ca | circa (approximately) | DMSO | dimethyl sulfoxide |
| cd | candela; current density; circular dichroism | DOD | Department of Defense |
| | | DOE | Department of Energy |
| CFR | Code of Federal Regulations | DOT | Department of Transportation |
| cgs | centimeter-gram-second | DP | degree of polymerization |
| CI | Color Index | dp | dew point |
| cis- | isomer in which substituted groups are on same side of double bond between C atoms | DPH | diamond pyramid hardness |
| | | dstl(d) | distill(ed) |
| cl | carload | dta | differential thermal analysis |
| cm | centimeter | (E)- | entgegen; opposed |
| cmil | circular mil | ε | dielectric constant (unitless number) |
| cmpd | compound | | |
| CNS | central nervous system | e | electron |
| CoA | coenzyme A | ECU | electrochemical unit |
| COD | chemical oxygen demand | ed. | edited, edition, editor |
| coml | commercial(ly) | ED | effective dose |
| cp | chemically pure | EDTA | ethylenediaminetetra-acetic acid |
| cph | close-packed hexagonal | | |
| CPSC | Consumer Product Safety Commission | emf | electromotive force |
| | | emu | electromagnetic unit |
| cryst | crystalline | en | ethylene diamine |
| cub | cubic | eng | engineering |
| D | debye | EPA | Environmental Protection Agency |
| D- | denoting configurational relationship | | |
| | | epr | electron paramagnetic resonance |
| **d** | differential operator | | |
| d | day; deci (prefix for $10^{-1}$) | eq. | equation |
| d | density | esca | electron spectroscopy for chemical analysis |
| d- | dextro-, dextrorotatory | | |
| da | deka (prefix for $10^1$) | esp | especially |
| dB | decibel | esr | electron-spin resonance |
| dc | direct current, n. | est(d) | estimate(d) |
| d-c | direct current, adj. | estn | estimation |
| dec | decompose | esu | electrostatic unit |
| detd | determined | exp | experiment, experimental |
| detn | determination | ext(d) | extract(ed) |
| Di | didymium, a mixture of all lanthanons | F | farad (capacitance) |
| | | F | faraday (96,487 C) |
| dia | diameter | f | femto (prefix for $10^{-15}$) |
| dil | dilute | | |

| | | | |
|---|---|---|---|
| FAO | Food and Agriculture Organization (United Nations) | hyd | hydrated, hydrous |
| | | hyg | hygroscopic |
| | | Hz | hertz |
| fcc | face-centered cubic | $i$ (eg, Pr$^i$) | iso (eg, isopropyl) |
| FDA | Food and Drug Administration | $i$- | inactive (eg, $i$-methionine) |
| | | IACS | International Annealed Copper Standard |
| FEA | Federal Energy Administration | ibp | initial boiling point |
| FHSA | Federal Hazardous Substances Act | IC | integrated circuit |
| | | ICC | Interstate Commerce Commission |
| fob | free on board | | |
| fp | freezing point | ICT | International Critical Table |
| FPC | Federal Power Commission | | |
| | | ID | inside diameter; infective dose |
| FRB | Federal Reserve Board | | |
| frz | freezing | ip | intraperitoneal |
| G | giga (prefix for $10^9$) | IPS | iron pipe size |
| $G$ | gravitational constant = $6.67 \times 10^{11}$ N·m$^2$/kg$^2$ | ir | infrared |
| | | IRLG | Interagency Regulatory Liaison Group |
| g | gram | | |
| (g) | gas, only as in $H_2O$(g) | ISO | International Organization Standardization |
| $g$ | gravitational acceleration | | |
| gc | gas chromatography | ITS-90 | International Temperature Scale (NIST) |
| gem- | geminal | | |
| glc | gas–liquid chromatography | IU | International Unit |
| | | IUPAC | International Union of Pure and Applied Chemistry |
| g-mol wt; gmw | gram-molecular weight | | |
| GNP | gross national product | IV | iodine value |
| gpc | gel-permeation chromatography | iv | intravenous |
| | | J | joule |
| GRAS | Generally Recognized as Safe | K | kelvin |
| | | k | kilo (prefix for $10^3$) |
| grd | ground | kg | kilogram |
| Gy | gray | L | denoting configurational relationship |
| H | henry | | |
| h | hour; hecto (prefix for $10^2$) | L | liter (for fluids only) (5) |
| ha | hectare | $l$- | levo-, levorotatory |
| HB | Brinell hardness number | (l) | liquid, only as in $NH_3$(l) |
| Hb | hemoglobin | LC$_{50}$ | conc lethal to 50% of the animals tested |
| hcp | hexagonal close-packed | | |
| hex | hexagonal | LCAO | linear combination of atomic orbitals |
| HK | Knoop hardness number | | |
| hplc | high performance liquid chromatography | lc | liquid chromatography |
| | | LCD | liquid crystal display |
| | | lcl | less than carload lots |
| HRC | Rockwell hardness (C scale) | LD$_{50}$ | dose lethal to 50% of the animals tested |
| HV | Vickers hardness number | | |

| | | | |
|---|---|---|---|
| LED | light-emitting diode | $N$- | denoting attachment to nitrogen |
| liq | liquid | | |
| lm | lumen | $n$ (as $n_D^{20}$) | index of refraction (for 20°C and sodium light) |
| ln | logarithm (natural) | | |
| LNG | liquefied natural gas | $^n$ (as Bu$^n$), | |
| log | logarithm (common) | $n$- | normal (straight-chain structure) |
| LOI | limiting oxygen index | | |
| LPG | liquefied petroleum gas | $n$ | neutron |
| ltl | less than truckload lots | n | nano (prefix for $10^9$) |
| lx | lux | na | not available |
| M | mega (prefix for $10^6$); metal (as in MA) | NAS | National Academy of Sciences |
| $M$ | molar; actual mass | NASA | National Aeronautics and Space Administration |
| $\overline{M}_w$ | weight-average mol wt | | |
| $\overline{M}_n$ | number-average mol wt | nat | natural |
| m | meter; milli (prefix for $10^{-3}$) | ndt | nondestructive testing |
| | | neg | negative |
| $m$ | molal | NF | *National Formulary* |
| $m$- | meta | NIH | National Institutes of Health |
| max | maximum | | |
| MCA | Chemical Manufacturers' Association (was Manufacturing Chemists Association) | NIOSH | National Institute of Occupational Safety and Health |
| | | NIST | National Institute of Standards and Technology (formerly National Bureau of Standards) |
| MEK | methyl ethyl ketone | | |
| meq | milliequivalent | | |
| mfd | manufactured | | |
| mfg | manufacturing | | |
| mfr | manufacturer | nmr | nuclear magnetic resonance |
| MIBC | methyl isobutyl carbinol | | |
| MIBK | methyl isobutyl ketone | NND | New and Nonofficial Drugs (AMA) |
| MIC | minimum inhibiting concentration | | |
| | | no. | number |
| min | minute; minimum | NOI-(BN) | not otherwise indexed (by name) |
| mL | milliliter | | |
| MLD | minimum lethal dose | NOS | not otherwise specified |
| MO | molecular orbital | nqr | nuclear quadruple resonance |
| mo | month | | |
| mol | mole | NRC | Nuclear Regulatory Commission; National Research Council |
| mol wt | molecular weight | | |
| mp | melting point | | |
| MR | molar refraction | NRI | New Ring Index |
| ms | mass spectrometry | NSF | National Science Foundation |
| MSDS | material safety data sheet | | |
| mxt | mixture | NTA | nitrilotriacetic acid |
| $\mu$ | micro (prefix for $10^{-6}$) | NTP | normal temperature and pressure (25°C and 101.3 kPa or 1 atm) |
| N | newton (force) | | |
| $N$ | normal (concentration); neutron number | | |

| | | | |
|---|---|---|---|
| NTSB | National Transportation Safety Board | qv | quod vide (which see) |
| O- | denoting attachment to oxygen | R | univalent hydrocarbon radical |
| o- | ortho | (R)- | rectus (clockwise configuration) |
| OD | outside diameter | r | precision of data |
| OPEC | Organization of Petroleum Exporting Countries | rad | radian; radius |
| o-phen | o-phenanthridine | RCRA | Resource Conservation and Recovery Act |
| OSHA | Occupational Safety and Health Administration | rds | rate-determining step |
| | | ref. | reference |
| owf | on weight of fiber | rf | radio frequency, n. |
| Ω | ohm | r-f | radio frequency, adj. |
| P | peta (prefix for $10^{15}$) | rh | relative humidity |
| p | pico (prefix for $10^{-12}$) | RI | Ring Index |
| p- | para | rms | root-mean square |
| p | proton | rpm | rotations per minute |
| p. | page | rps | revolutions per second |
| Pa | pascal (pressure) | RT | room temperature |
| PEL | personal exposure limit based on an 8-h exposure | RTECS | Registry of Toxic Effects of Chemical Substances |
| pd | potential difference | $^{s}$ (eg, Bu$^{s}$); sec- | secondary (eg, secondary butyl) |
| pH | negative logarithm of the effective hydrogen ion concentration | S | siemens |
| | | (S)- | sinister (counterclockwise configuration) |
| phr | parts per hundred of resin (rubber) | S- | denoting attachment to sulfur |
| p-i-n | positive-intrinsic-negative | s- | symmetric(al) |
| pmr | proton magnetic resonance | s | second |
| p-n | positive-negative | s | solid, only as in $H_2O(s)$ |
| po | per os (oral) | SAE | Society of Automotive Engineers |
| POP | polyoxypropylene | | |
| pos | positive | SAN | styrene-acrylonitrile |
| pp. | pages | sat(d) | saturate(d) |
| ppb | parts per billion ($10^9$) | satn | saturation |
| ppm | parts per million ($10^6$) | SBS | styrene–butadiene–styrene |
| ppmv | parts per million by volume | | |
| ppmwt | parts per million by weight | sc | subcutaneous |
| PPO | poly(phenyl oxide) | SCF | self-consistent field; standard cubic feet |
| ppt(d) | precipitate(d) | | |
| pptn | precipitation | Sch | Schultz number |
| Pr (no.) | foreign prototype (number) | sem | scanning electron microscope(y) |
| pt | point; part | | |
| PVC | poly(vinyl chloride) | SFs | Saybolt Furol seconds |
| pwd | powder | sl sol | slightly soluble |
| py | pyridine | sol | soluble |

| | | | |
|---|---|---|---|
| soln | solution | *trans-* | isomer in which |
| soly | solubility | | substituted groups are |
| sp | specific; species | | on opposite sides of |
| sp gr | specific gravity | | double bond between C |
| sr | steradian | | atoms |
| std | standard | TSCA | Toxic Substances Control |
| STP | standard temperature and | | Act |
| | pressure (0°C and 101.3 | TWA | time-weighted average |
| | kPa) | Twad | Twaddell |
| sub | sublime(s) | UL | Underwriters' Laboratory |
| SUs | Saybolt Universal seconds | USDA | United States Department |
| syn | synthetic | | of Agriculture |
| $^t$ (eg, Bu$^t$), | | USP | *United States* |
| *t-, tert-* | tertiary (eg, tertiary | | *Pharmacopeia* |
| | butyl) | uv | ultraviolet |
| T | tera (prefix for $10^{12}$); tesla | V | volt (emf) |
| | (magnetic flux density) | var | variable |
| t | metric ton (tonne) | *vic-* | vicinal |
| t | temperature | vol | volume (not volatile) |
| TAPPI | Technical Association of | vs | versus |
| | the Pulp and Paper | v sol | very soluble |
| | Industry | W | watt |
| TCC | Tagliabue closed cup | Wb | weber |
| tex | tex (linear density) | Wh | watt hour |
| $T_g$ | glass-transition | WHO | World Health |
| | temperature | | Organization (United |
| tga | thermogravimetric | | Nations) |
| | analysis | wk | week |
| THF | tetrahydrofuran | yr | year |
| tlc | thin layer chromatography | (Z)- | zusammen; together; |
| TLV | threshold limit value | | atomic number |

| Non-SI (Unacceptable and Obsolete) Units | | Use |
|---|---|---|
| Å | angstrom | nm |
| at | atmosphere, technical | Pa |
| atm | atmosphere, standard | Pa |
| b | barn | $cm^2$ |
| bar$^†$ | bar | Pa |
| bbl | barrel | $m^3$ |
| bhp | brake horsepower | W |
| Btu | British thermal unit | J |
| bu | bushel | $m^3$; L |
| cal | calorie | J |
| cfm | cubic foot per minute | $m^3/s$ |
| Ci | curie | Bq |
| cSt | centistokes | $mm^2/s$ |
| c/s | cycle per second | Hz |

$^†$Do not use bar ($10^5$ Pa) or millibar ($10^2$ Pa) because they are not SI units, and are accepted internationally only for a limited time in special fields because of existing usage.

| Non-SI (Unacceptable and Obsolete) Units | | Use |
|---|---|---|
| cu | cubic | exponential form |
| D | debye | $C \cdot m$ |
| den | denier | tex |
| dr | dram | kg |
| dyn | dyne | N |
| dyn/cm | dyne per centimeter | mN/m |
| erg | erg | J |
| eu | entropy unit | J/K |
| °F | degree Fahrenheit | °C; K |
| fc | footcandle | lx |
| fl | footlambert | lx |
| fl oz | fluid ounce | $m^3$; L |
| ft | foot | m |
| ft·lbf | foot pound-force | J |
| gf den | gram-force per denier | N/tex |
| G | gauss | T |
| Gal | gal | $m/s^2$ |
| gal | gallon | $m^3$; L |
| Gb | gilbert | A |
| gpm | gallon per minute | $(m^3/s)$; $(m^3/h)$ |
| gr | grain | kg |
| hp | horsepower | W |
| ihp | indicated horsepower | W |
| in. | inch | m |
| in. Hg | inch of mercury | Pa |
| in. $H_2O$ | inch of water | Pa |
| in.-lbf | inch pound-force | J |
| kcal | kilo-calorie | J |
| kgf | kilogram-force | N |
| kilo | for kilogram | kg |
| L | lambert | lx |
| lb | pound | kg |
| lbf | pound-force | N |
| mho | mho | S |
| mi | mile | m |
| MM | million | M |
| mm Hg | millimeter of mercury | Pa |
| $m\mu$ | millimicron | nm |
| mph | miles per hour | km/h |
| $\mu$ | micron | $\mu$m |
| Oe | oersted | A/m |
| oz | ounce | kg |
| ozf | ounce-force | N |
| $\eta$ | poise | Pa·s |
| P | poise | Pa·s |
| ph | phot | lx |
| psi | pounds-force per square inch | Pa |
| psia | pounds-force per square inch absolute | Pa |
| psig | pounds-force per square inch gage | Pa |
| qt | quart | $m^3$; L |
| °R | degree Rankine | K |
| rd | rad | Gy |
| sb | stilb | lx |
| SCF | standard cubic foot | $m^3$ |
| sq | square | exponential form |
| thm | therm | J |
| yd | yard | m |

## BIBLIOGRAPHY

1. The International Bureau of Weights and Measures, BIPM (Parc de Saint-Cloud, France) is described in Appendix X2 of Ref. 4. This bureau operates under the exclusive supervision of the International Committee for Weights and Measures (CIPM).
2. *Metric Editorial Guide (ANMC-78-1)*, latest ed., American National Metric Council, 5410 Grosvenor Lane, Bethesda, Md. 20814, 1981.
3. *SI Units and Recommendations for the Use of Their Multiples and of Certain Other Units (ISO 1000-1981)*, American National Standards Institute, 1430 Broadway, New York, 10018, 1981.
4. Based on *ASTM E380-89a (Standard Practice for Use of the International System of Units (SI))*, American Society for Testing and Materials, 1916 Race Street, Philadelphia, Pa. 19103, 1989.
5. *Fed. Reg.*, Dec. 10, 1976 (41 FR 36414).
6. For ANSI address, see Ref. 3.

R. P. LUKENS
ASTM Committee E-43 on SI Practice

**P**

*Continued*

# PIGMENTS

## INORGANIC

Inorganic pigments, black, white, or colored inorganic substances produced and marketed as fine powders, are an integral part of many decorative and protective coatings (qv) and are used for the mass coloration of plastics, fibers (qv), paper (qv), rubber, glass (qv), cement (qv), glazes, and porcelain enamels (see ENAMELS, PORCELAIN OR VITREOUS). These materials are colorants in printing inks (qv), cosmetics (qv), and markers, eg, crayons. In all these applications the pigments are dispersed, ie, they do not dissolve, in the media forming a heterogeneous mixture (see PIGMENT DISPERSIONS). In nature, inorganic pigments contribute to the color of some rocks and minerals (see COLORANTS FOR CERAMICS; COLORANTS FOR FOOD, DRUGS, COSMETICS, AND MEDICAL DEVICES; COLORANTS FOR PLASTICS; PAINTS, ARCHITECTURAL).

History is rich with examples of the use of inorganic pigments. Archeological findings document the use of red ochre for aesthetic, religious, and other purposes. Preservation of primitive pictures drawn by prehistoric peoples 30,000 years ago, found on the walls of caves in southern France, northern Spain, and northern Africa, demonstrates one of the strongest aspects of inorganic pigments, ie, their long lasting stability. Inorganic pigments have also been used to color and decorate clayware since ancient times.

Originally, only fine powders used for coloring various media were defined as pigments. This definition has been expanding, and in the 1990s many powdery materials, eg, metallic powders and powders having magnetic or anticorrosive

1

properties, that are intentionally dispersed (not dissolved) into media to increase value and/or impart some special properties, are sometimes also classified as pigments. In some cases, pigments are also used as raw materials to prepare other pigments or chemicals, eg, $Fe_2O_3$, $TiO_2$, etc, or in some other applications, eg, as catalyst supports (carbon).

Chemically, inorganic pigments are quite simple materials and include elements, their oxides, mixed oxides, sulfides, chromates, silicates, phosphates, and carbonates. Laboratory preparation of these materials can be quite simple,

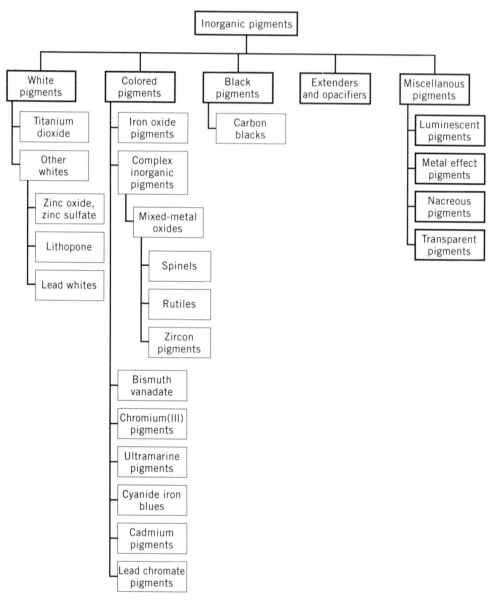

**Fig. 1.**   Classification of inorganic pigments.

but their large-scale production is complex and demands attention to every detail of the manufacturing process. This is because the application usefulness of inorganic pigments is determined by physical as well as chemical properties. Particle size, shape, and surface properties are as important in the pigment performance as chemical composition. For inorganic pigments that can exist in several crystal structures, controls must be exercised to make sure that the proper crystal habit having the optimum coloristic properties is produced.

Historical classification of inorganic pigments into the naturally occurring ones and synthetically produced ones is no longer useful. The majority of pigments is manufactured synthetically. For the purpose of this review, inorganic pigments are classified according to Figure 1.

## Properties

The value of pigments results from their physical–optical properties. These are primarily determined by the pigments' physical characteristics (crystal structure, particle size and distribution, particle shape, agglomeration, etc) and chemical properties (chemical composition, purity, stability, etc). The two most important physical–optical assets of pigments are the ability to color the environment in which they are dispersed and to make it opaque.

The opacity of a pigment lies in its ability to prevent a transmission of light through the medium. By doing so the pigmented medium obscures the subject on which it is applied. When light enters a pigmented medium and hits a pigment particle, the light can either be absorbed or dispersed by the particle. White pigments disperse the whole visible light spectrum more effectively than they absorb it; black pigments do the opposite. Color results when pigment particles absorb only certain portions of the visible light spectrum while dispersing the rest of it. Absorbed light changes into heat energy, which in the long run might have detrimental effects on the medium or pigment.

The opacity of pigments is a function of the pigment particle size and the difference between the pigment's refractive index and that of the media in which pigment particles are dispersed. The multiple light dispersion in the pigment–medium interface results in the appearance that the light is transmitted through a much thicker layer than it actually is. A pigment having a particle size between $0.16-0.28$ $\mu$m gives the maximum dispersion of the visible light. Any agglomeration of pigment particles can affect their opacity. Much effort has been spent to optimize particle size and size distribution and to prevent any particle agglomeration in order to achieve a maximum pigment opacity. The commercial value of pigments can also be enhanced by various surface treatments. The function of these treatments is to improve pigments' coloristic value or to simplify physical handling.

The most commonly measured pigment properties are elemental analysis, impurity content, crystal structure, particle size and shape, particle size distribution, density, and surface area. These parameters are measured so that pigments' producers can better control production, and set up meaningful physical and chemical pigments' specifications. Measurements of these properties are not specific only to pigments. The techniques applied are commonly used to characterize powders and solid materials and the measuring methods have been standardized in various industries.

Coloristic properties of pigments are best evaluated by dispersing them into the media they were developed to color, eg, plastics, glass enamels, glazes, etc. The measured characteristics include color, color strength, opacity, lightfastness, weathering, heat stability, chemical stability, and rheological properties. Many of these characteristics are not determined by the pigments alone. The dispersing media and the processing conditions can strongly influence the results. Some of these characteristics are meaningful only for specific applications. For these reasons many of the specifics of testing are developed between suppliers and their customers. Because pigments, as any fine powders, have a tendency to segregate by size during transportation and handling, the use of proper sampling (qv) methods is critical for getting meaningful physical and chemical data (see POWDERS, HANDLING).

**Chemical Properties.** Elemental analysis, impurity content, and stoichiometry are determined by chemical or instrumental analysis. The use of instrumental analytical methods (qv) is increasing because these are usually faster, can be automated, and can be used to determine very small concentrations of elements (see TRACE AND RESIDUE ANALYSIS). Atomic absorption spectroscopy and x-ray fluorescence methods are the most useful instrumental techniques in determining chemical compositions of inorganic pigments. Chemical analysis of principal components is carried out to determine pigment stoichiometry. Analysis of trace elements is important. The presence of undesirable elements, such as heavy metals, even in small amounts, can make the pigment unusable for environmental reasons.

*Crystal Structure.* Crystal structure, the information about compounds such as impurities or unreacted materials present in the pigment, the presence of various crystal phases, and the degree of the crystallinity, can be resolved using an x-ray diffractometer. Chemically, pigments can contain only one compound, such as single oxides (eg, $Fe_2O_3$, $Cr_2O_3$, $TiO_2$, etc), or several compounds. Inclusion pigments, encapsulated pigments, and lithopones are typical examples of multicompound pigments. Even a single-compound pigment can exist in several crystal structures, not all of which have equal coloristic properties. A good pigment contains a colorant as a principal phase. Secondary phases that might be present only reduce coloristic properties. Mixed-metal oxides represent a special class of pigments where several oxides react and form a new compound, or some of the cations of the host lattice of one of the oxides are replaced by the cations of a chromophore. The structure of the host lattice may be preserved, but it can be distorted. In most cases, analyzed pigments are not completely unknown and the powder diffraction pattern is sufficient in finding present phases or unreacted starting materials in the pigment. The x-ray analysis has become an indispensable tool of the inorganic pigments' development and production (see X-RAY TECHNOLOGY).

Powder diffraction patterns have three main features that can be measured: *d*-spacings, peak intensities, and peak shapes. Because these patterns are a characteristic fingerprint for each crystalline phase, a computer can quickly compare the measured pattern with a standard pattern from its database and recommend the best match. Whereas the measurement of *d*-spacings is quite straightforward, the determination of peak intensities can be influenced by sample preparation. Any preferred orientation, or presence of several larger crystals in the sample, makes the interpretation of the intensity data difficult. The most common structures of inorganic pigments are rutile, anatase, and spinel.

**Physical and Chemical Properties.** *Particle Size.* Particle size and distribution are the most fundamental measured properties of powders (see SIZE MEASUREMENT OF PARTICLES). These properties impact a number of pigment characteristics. Those affected the most are the color (1,2), color strength, hiding power, and rheological properties. Particle size and distribution data can be easily misinterpreted. Only the data for spherical powders are easy to measure and interpret. Actual powders, however, consist of a population of particles of many different shapes. To complicate the matter further, powders are usually not formed from a mixture of single, free-flowing particles. The particles can be interconnected by weak forces, ie, electrostatic forces or liquid bridges, forming agglomerates; or by solid bridges, ie, chemical bonds or sintered necks, resulting in hard aggregates. To prevent powder dusting and make handling easier, some pigments are intentionally agglomerated to granules by the addition of granulating agents.

To permit a good description of powder population, a representative sample of the powder must be collected, measured, and the results interpreted using statistical methods. To simplify the mathematical evaluation it is usually assumed that particles are spherical and particle size is calculated as an average size. Particle size distribution (PSD) can either be presented in a graphical form as a distribution function, a histogram, or in a tabular form. For inorganic pigments to be useful in most applications, they must have an average particle size between 0.1 and 10 $\mu$m. For these reasons only some of the particle size analysis (PSA) techniques have become widely accepted by the inorganic pigments industry.

Sieving analysis is one of the oldest and simplest methods in determining pigment particle size distribution. The analysis can be carried out using dry or wet samples. In some cases, only one sieve is used and the specification is set up as a maximum allowed retain of the sample on that particular sieve. Sieving can also be employed as the last operation in the pigment production to guarantee the absence of particles larger than the selected sieve aperture.

Sedimentation (qv) techniques, whether based on gravitational forces or centrifugation, derive the particle size from the measured travel rates of particles in a liquid. Before the particle analysis is carried out, the sample is usually dispersed in a medium to break down granules, agglomerates, and aggregates. The dispersion process might involve a simple stirring of the powder into a liquid, but the use of an ultrasonic dispersion is preferred (see ULTRASONICS).

Some particle size measuring techniques are more particle shape sensitive than others. Data obtained by different methods can be significantly different, and whenever a particle size is reported, the measuring technique and conditions should always be mentioned. Even using the same equipment, the extremes of the distributions (low and high 10%) are usually not readily reproducible.

A recent trend in particle analysis has been the introduction of personal computer-based automation (3). Sophisticated software packages can be used to automate and speed up the analysis. In some cases these computers can even carry out continuous process control (qv) (see COMPUTER TECHNOLOGY). The latest machines also allow the measurements of smaller particles and can detect a wider range of sizes. Machines based on light-scattering principles are being more widely accepted by the industry because of speed. An average analysis takes from 1–2 minutes, whereas those based on sedimentation principles require from 10–120 minutes.

*Pigment Opacity, Hiding Power, and Tinting Strength.* A character, ie, color or pattern, of a substrate becomes obscured when coated with a pigment containing film such as a paint or a ceramic glaze. The degree of the obscuration, known as opacity, depends on the amount and type of pigment used and the thickness of the applied film. The color of the reflected light is the result of the combination of colors of substrate, pigment, and dispersion medium. In most cases the dispersion medium is clear and does not contribute to the color. At a certain pigment loading and/or film thickness the substrate becomes completely hidden and the pigment determines its color.

The ability of a coating to hide the substrate is called its hiding power. Hiding power of a uniform coating is expressed as the area of substrate that can be hidden by a unit volume of the coating ($ft^2$/gal or $m^2$/L). When the hiding power of a paint is low, several coats must be applied to increase the film thickness to hide the substrate. Some paints have excellent hiding power and only one coating may suffice to hide the surface.

The ability of a pigment to change the color of an opaque film is known as its tinting strength. Both the hiding power and tinting strength are the fundamental pigment properties.

For white pigments, the hiding power can be expressed through the Lorentz-Lorenz equation (4) as a function of pigment, $n_p$, and medium, $n_m$, refractive indexes. For most organic binders having an average refractive index of about 1.5, and for pigments having a refractive index between 1.5 to 2.75, the hiding power (HP) can be approximated, using the Lorentz-Lorenz equation, by the following.

$$HP \approx 0.16(n_p - n_m)^2$$

The larger the difference is between the pigment and binder refractive indexes, the better the hiding power of the coating. For color pigments the relationship between the hiding power and the pigment's physical properties is considerably more complex. Hiding power is also affected by the pigment particle size.

Hiding power and tinting strength of a pigment also depend on concentration in the organic matrix. The concentration is usually expressed as the pigment volume concentration:

$$\text{pigment volume concentration} = V_P/(V_P + V_O)$$

where $V_P$ is the pigment volume and $V_O$ is the volume of the organic matrix.

Hiding power of a pigment increases with the increasing pigment volume concentration. A critical pigment volume concentration (cpvc) is the pigment concentration where individual pigment particles are dispersed in the matrix without directly touching each other. This concentration can be determined from the amount of linseed oil that is necessary to form a pigment paste (5). The knowledge of the critical pigment volume concentration is particularly important for coating formulation. Coatings having a higher pigment volume concentration than cpvc are porous and do not protect substrates well. Coatings having too low pigment volume concentration have a tendency to blister.

Hiding power and tinting strength can be determined visually or instrumentally. Visually, the hiding power is measured by applying successive thin layers of coating until the substrate pattern is not visible. The amount of coating that is necessary to reach that point directly relates to the hiding power. This method is not very precise because the point where the pattern becomes obliterated is not easily reproduced. The measurement of tinting strength is achieved visually by adding a white pigment to the sample until its lightness matches the lightness of the standard.

Instrumentally, both the hiding power and tinting strength can be determined from the amount of the incident light reflectance of coated white and black substrates. Relationships derived from Kubelka and Munk theory (6) are applied in actual calculations.

*Color Matching.* Color matching is a process in which a technician prepares a formulation, ie, a mixture of pigments in a desired medium, that has the color effects desired by the customer. A good color match in one medium, eg, plastic, is not always a good match in another medium, eg, ceramic glaze. Thus the medium as well as the processing conditions have to be identical to those used by the customer.

Experienced color matchers can achieve a good color match by trial and error without using any instrumentation. In some cases, however, this technique can be a lengthy process, and should the desired match be outside the color space defined by the available color standards, the technician might spend too much time just to determine that the match is not possible. To get the most cost-effective match using a low metamerism in the shortest possible time, the use of a computer color matching system is preferable.

As the color space is three-dimensional, a mixture of at least three colorants is needed for color matching; the use of more than six colorants can introduce metamerism into the color. For opaque systems, eg, ceramic glazes, color matching can be carried out using three high chroma pigment standards such as iron zircon coral [68187-13-3], vanadium zircon blue [68186-95-8], and praseodymium zircon yellow [68187-15-5]. These three pigments are commonly referred to in the ceramic tile industry as the triaxial stains. All the colors within the color triangle defined by these three stains can be matched and the desired value, ie, lightness or darkness, can be theoretically adjusted by the addition of a black or white stain. Because the good dispersion of small quantities of black or white stains on a production scale is difficult, the value is usually adjusted by changing the stains' loadings. The addition of white stains also improves the system opacity. A number of color matching philosophies have been developed in the industry. Color matching still falls somewhere between an art and a science (7).

An important property of a pigment is its ability to maintain its color when exposed to light, weather, heat, and chemicals. This property is seldom measured for pigments alone. Rather it is determined for the dispersion of a pigment in a desired medium, eg, paints or plastics, and in many cases it is compared to the performance of a standard pigment. The observed changes are the result of complex pigment and media reactions and their possible interactions. In all evaluations, time of exposure plays a role.

*Lightfastness.* Many pigments, when exposed to high intensity light such as direct sunlight or uv lamp, can get darker, change their shade or lose the color saturation. The color and its saturation change mainly for organic pigments. Inorganic pigments, particularly those containing ions that can exist in several oxidation states, eg, Pb, Hg, Cr, Cu, etc, usually get darker. Some color changes can be reversible; others are permanent.

Light, particularly its uv component, can also attack the organic medium in which the pigments are dispersed. The breakage of C–H, C–C, and C–O bonds in the medium leads to overall coating deterioration. Use of pigments that absorb light in the uv region can therefore improve the coating stability.

Some pigments are photoactive and can accelerate the decomposition of the organic matrix. This decomposition frees pigment particles from the matrix, resulting in a so-called chalking of the paints or plastics. A continuous loss of pigment occurs. Chalking caused by titanium dioxide white, particularly its anatase [1317-70-0] form, is the best known example of a pigment accelerating the decomposition of the organic matrix. The mechanism of the chalking process is quite complicated and requires the presence of both water and oxygen (8,9). The whole process can be accelerated by heat. To reduce chalking of white paints, titanium dioxide pigment particles are surface treated using inorganic and organic coatings.

Lightfastness is measured by exposing pigmented film to an artificial or natural, eg, Florida exposure, light for a predetermined time. It is a relative term where the color of a sample exposed to a known light source is compared to its original color values. To make it meaningful, all conditions of the exposure have to be well defined.

*Weathering.* Weathering is the ability of the colored system, ie, the coating, paint, etc, not the pigment alone, to resist light and environmental conditions. Changes in color and gloss are two main factors that are evaluated in weathering tests. Various accelerated weathering procedures available in color testing laboratories may give a quick assessment of the pigment–medium's long-term performance. However, before a final judgment of the system performance is made, accelerated test results should always be confirmed by outside exposure. Even that is not without problem as there is no standard weather available and to average the changes in weather, long-term exposures are needed. In the United States, Florida weather is considered to be an industry standard for testing colored coatings and plastics.

*Heat Stability and Chemical Resistance.* Heat stability is measured as a change in the hue of the colored system and a degree of yellowing of the white system after exposure to a desired high temperature for a certain time. This property is particularly important in coloring engineering plastics. It can also be expressed as the maximum temperature at which the color of the system does not change. In many cases there is a need to compare the heat stability of several pigments. This can be done by the heat/time exposure of pigmented binders and by comparing, usually in graphic form, the measured color differences.

In determining the chemical resistance, color changes of pigmented binder surfaces are measured after their exposure to various chemicals, such as water–sulfur dioxide or water–sodium chloride systems. These systems imitate the environment to which the colored articles could become exposed.

*Pigments' Aftertreatments.* The surfaces of pigment particles can have different properties and composition than the particle centers. This disparity can be caused by the absorption of ions during wet milling, eg, the –OH groups, on the surface. In some cases, surfaces are modified intentionally to improve the pigments' application properties, interaction with the organic matrix, and weather resistance.

Most inorganic pigments are hydrophylic and therefore can be readily wetted only by polar solvents, eg, water. The wettability and dispersion of inorganic pigments in an organic matrix (polymer, solvent) can be improved by the physical or chemical absorption of surface-active compounds containing polar groups, such as $-NH_2$, $-OH$, or longer aliphatic chains on pigment particles. The absorption of these compounds makes the pigment surface hydrophobic. Compounds that help to form a bridge between inorganic particles and an organic polymeric matrix are called coupling agents. The most common coupling agents are based on tetrafunctional organometallic compounds of titanium, silicon, aluminum, and zirconium (10) (see ORGANOMETALLICS (SUPPLEMENT)).

In some cases, a pigment's thermal and chemical resistance can be improved by the encapsulation of the pigment particles by an insoluble, colorless layer of metal oxide or oxide–hydroxide, eg, silica, $SiO_2$. The function of such a shell is to prevent direct contact and reaction between the pigment surface and the organic matrix in which the pigment is dispersed (11).

## Specifications, Standards, and Quality Control

Whereas the production flow charts of inorganic pigments appear to be simple, the actual processes can be very complicated. Many pigments are not pure chemical compounds, but can be multiphase systems contaminated with various impurities and modifiers. Because pigments are fine powders, the physical properties are as critical to their application performance as are the chemical properties.

Some raw materials used in pigment production are mined minerals in which the amount of minor contaminants cannot be controlled. In some cases, pigment producers are only minor users of these materials, and have little leverage in negotiating specifications. Pigment producers must therefore understand all the variables affecting product quality and know how to compensate for any material and process variations to deliver the inorganic pigment with minimal lot-to-lot variation.

As of this writing (ca 1995) there are no international standards governing the production of inorganic pigments. Pigment producers meeting at least the International Standard ISO 9002 have a quality system established that guarantees some controls over product consistency. The pigment business is highly customer-oriented.

Production and product quality of most pigment producers are controlled through pigments' standards. Whenever a pigment is developed or significantly improved, a new standard is set. Because all future productions of that pigment are compared and adjusted to this standard, it is important that the standard represents the average production results, not necessarily the best ones. If all production processes are under control, the properties of the produced pigment's lots are evenly distributed around those of the standard. The producer can then match the standard by blending pigments with those having better and worse properties. In actuality, the process is much more complicated because pigment properties cannot be characterized by a single number.

## Production

World production of inorganic pigments in 1993 was estimated to be around $5 \times 10^6$ metric tons. Sales were close to $15 billion. This output can be compared with the world production of organic pigments, which for 1994 was assessed at 160,000 metric tons, worth $4 billion (12). The future growth of inorganic pigments' sales is expected to be strongly influenced by environmental policies. The use of environmentally safe pigments is expected to continue to grow. The demand for pigments based on lead or cadmium, however, could eventually disappear. Production volumes of the key inorganic pigments are shown in Figure 2.

About six million metric tons of carbon blacks are produced worldwide, but most of these are not used for pigmentary applications. The United States is the largest producer, ca 1.8 million metric tons, followed by Western Europe and Japan. Consumption in the United States in 1994 was estimated at 1.4 million metric tons. The automotive industry consumed about $1 \times 10^6$ t for tire production. For pigmentary applications the use is estimated to be less than $150 \times 10^3$ t. Automotive tires are getting smaller and lasting longer, thus consumption of carbon blacks in the United States has been steadily decreasing. The Pacific Rim region consumes about $1.7 \times 10^6$ t of carbon black, Western Europe about $1.1 \times 10^6$ t, and the former Eastern Europe about $1.5 \times 10^6$ t. Overall,

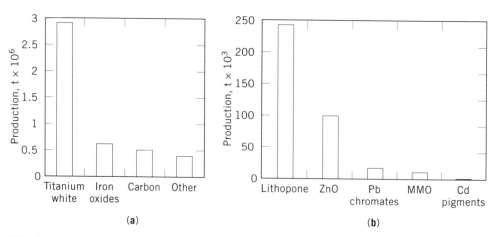

**Fig. 2.** 1993 Production volumes for (**a**) principal inorganic pigments in units of millions of metric tons, and (**b**) some of the other pigments in thousands of metric tons. MMO = mixed metal oxides.

about 90% of the produced carbon blacks are consumed by tire manufacturers (13). The world's leading producers of carbon blacks are Cabot Corp. and J. M. Huber Corp. in the United States, and Degussa AG in Germany. These companies have production facilities in various parts of world (see CARBON, CARBON BLACK).

Titanium dioxide [13463-67-7] is by far the most often used inorganic pigment (14). In 1993 the estimated worldwide plant capacity was around $3.7 \times 10^6$ t. Plant utilization in that year was only about 78%, thus the world demand for $TiO_2$ pigment in 1993 can be estimated to have been about $2.9 \times 10^6$ t. Growth in worldwide production of this pigment has been phenomenal since it was first produced in 1918.

| Year | Production, $t \times 10^3$ | Year | Production, $t \times 10^3$ |
|------|------|------|------|
| 1925 | 4 | 1970 | 1936 |
| 1939 | 100 | 1985 | 2959 |
| 1965 | 1389 | 1994 | 3060 |

U.S. production of titanium dioxide in 1993 was $1.16 \times 10^6$ t, an increase of about 2.2% over 1992 (15). The annual production increased on an average of 5.4% in the 1980s (16). There are approximately 50 world producers of titanium dioxide pigments. The world's leading suppliers are Du Pont ($\sim$ 24% worldwide share), Tioxide (the U.K. Co. owned by ICI), and SCM Corp. Other producers include Kerr McGee, Tiwest, Kemira Oy, Kronos, and Bayer AG (see TITANIUM COMPOUNDS, INORGANIC).

Iron oxide [1309-37-1] is the second most used inorganic pigment having an estimated 1993 worldwide consumption around 600,000–650,000 t. Out of that total production, the United States consumed about 200,000–250,000 t, mainly for pigmenting construction products (34%) and coating applications (23%). Western Europe consumed about 300,000 t, the United Kingdom accounting for about 20% of that. The Japanese demand for iron oxide in 1993 was rather small, only about 30,000 t, but is expected to increase significantly by the twenty-first century (17).

Iron oxide is also used for nonpigmentary applications, eg, ferrites (qv) and foundry sands making total world demand for iron oxide close to $1 \times 10^6$ t. The principal worldwide producers of iron oxide pigments are Bayer AG (ca 300,000 t/yr worldwide), and Harcros Pigments Inc., a subsidiary of Harrisons & Crosfield PLC. In the United States, Bayer produces the Bayferrox line of iron oxide pigments in New Martinsville (see IRON COMPOUNDS).

Other pigments are consumed in considerably smaller amounts. Moreover, the market data for these pigments are not as readily available as those for carbon blacks, titanium dioxide, and iron oxides.

Lithopone [8006-32-4], an important white pigment, is produced mainly in the former Soviet Union, the Czech Republic, and the People's Republic of China. In China, the nation's output for lithopone peaked in 1990 at about 233,000 t. In 1992 it was only 168,000 t. About one-third of the lithopone produced in

China was exported. Annual worldwide production can be estimated to be under 250,000 t as of this writing (ca 1995) (18) (see BARIUM COMPOUNDS).

Whereas the worldwide production of zinc oxide [1314-13-2] is estimated to be around 500,000 t annually, only about 75,000–100,000 t are used for pigmentary applications by the paint industry. About 250,000 t are consumed by the rubber industry, and the rest is used in the production of plastics, paper, cosmetics, pharmaceutical products, ceramics, and glass (see ZINC COMPOUNDS).

Owing to environmental regulations, the consumption of lead-based pigments, lead chromate [7758-97-6], in particular, has been decreasing steadily at an average annual rate of about 5% since the early 1980s. In 1993 production was estimated to be around 20,000 t (see LEAD COMPOUNDS). Cadmium-based pigments have been hit particularly hard by environmental regulations. As the result, consumption in 1993 was estimated to be between 1.5–5 thousand metric tons. The future of this pigment category is particularly questionable in North America.

The worldwide consumption of the mixed-metal oxide (MMO) pigments, sometimes referred to as complex inorganic color pigments, was estimated to be close to 11,000 metric tons in 1993. The principal worldwide suppliers of these pigments are Bayer AG and Cerdec AG in Germany, Ferro Corp. and Engelhard Corp. in the United States, and Cookson Matthey Ceramics in the United Kingdom. Cerdec AG was formed in 1993 as the joint venture between Degussa's Ceramic Division and Drakenfeld Colors of Ciba; Cookson Matthey Ceramics, formed in 1994, is a joint venture between Johnson Matthey's Color & Print Division and Cookson's Ceramic Supplies and Minerals business. All these companies have worldwide manufacturing and/or distribution centers. Other suppliers of mixed-metal oxide pigments include Shepherd Color in the United States, Dr. H. Heubach in Germany, and Ishihara in Japan.

## White Pigments

The most common white pigments are titanium dioxide, zinc oxide, leaded zinc oxide, zinc sulfide [1314-98-3], and lithopone, a mixture of zinc sulfide and barium sulfate [7727-43-7]. The use of lead whites and antimony oxides has been decreasing steadily for environmental reasons.

**Titanium Dioxide.**  Chemically, titanium white is titanium dioxide either in an anatase or rutile form. The history of this pigment is relatively recent compared to other white pigments. Commercial production of this pigment was discovered in the early 1900s during the investigation of ways to convert ilmenite [12168-52-4] to iron or titanium–iron alloys. The first industrial productions of titanium white were in 1918 in Norway, the United States, and Germany. These processes did not produce pure titanium dioxide as a white pigment, but as a mixture with barium sulfate.

The first manufacturing of pure titanium white (anatase form) for pigment use was reported in 1923 in France. However, the real growth of the production and use of titanium white pigments began in the early 1930s. Growth continues in the mid-1990s, but the rate of growth has decreased. In 1994 worldwide production was expected to exceed three million tons.

*Properties.* Crystals of titanium dioxide, $TiO_2$, can exist in one of the three crystal forms: rutile [1317-80-2], anatase [1317-70-0], and brookite [12188-41-9]. Both rutile and anatase crystallize in the tetragonal system, brookite in the rhombic system. Only anatase and rutile forms have good pigmentary properties, and rutile is more thermally stable. Anatase transforms rapidly to rutile at a temperature above 700°C. Compared to other white pigments, titanium dioxide has the highest refractive index, giving white paints formulated with these pigments the highest coverage, ca 38 $m^2/g$. Because of the higher refractive index, rutile-based paints have a slightly higher hiding power than anatase-based ones.

Titanium whites resist various atmospheric contaminants such as sulfur dioxide, carbon dioxide, and hydrogen sulfide. Under normal conditions they are not readily reduced, oxidized, or attacked by weak inorganic and organic acids. Titanium dioxide dissolves slightly in bases, hydrofluoric acid, and hot sulfuric acid. Owing to its chemical inertness, titanium dioxide is a nontoxic, environmentally preferred white pigment.

*Production.* Titanium is the seventh most common metallic element in the earth's crust. Titanium minerals are plentiful in nature (19). The most common mineral/raw materials used for the production of titanium dioxide pigments are shown in Table 1.

Ilmenite is more abundant than rutile. Ilmenite world supplies are estimated to meet the requirements of the $TiO_2$ industry into the twenty-second century. The largest sources of ilmenite are in Australia, Canada, South Africa, Russia, and the United States. Large, unexplored sources also exist in China. About nine million metric tons of ilmenite are mined annually. Long-term atmospheric effects weather ilmenite into leucoxene [1358-95-8], which contains most of its iron as $Fe^{3+}$. The majority of the world's supply of rutile comes from the beach sands of Australia, Florida, India, Brazil, and South Africa. The total worldwide supply is estimated to be about 50 million metric tons. About a half million tons are mined a year. The supply should last at least until the end of the twenty-first century.

Titanium slag and synthetic rutile are also used as raw materials in the production of titanium whites. Titanium slag results from a metallurgical process during which iron (qv) is removed from ilmenite by reduction with coke in an electric arc furnace at 1200–1600°C. Under these conditions, iron oxide is reduced to metal, melts, and separates from the formed titanium slag. Titanium slag contains 70–75% $TiO_2$ and only 5–8% iron.

**Table 1. Mineral/Raw Materials for TiO₂ Production**

| Mineral/raw material | Main composition | $TiO_2$, % |
|---|---|---|
| ilmenite | $FeO-TiO_2$ | 35–65 |
| leucoxene | $Fe_2O_3-TiO_2(+TiO_2)$ | 60–90 |
| rutile | $TiO_2$ | 90–98 |
| rutile, synthetic | $TiO_2$ | 85–96 |
| anatase | $TiO_2$ | 80–90 |
| titanium slag | $TiO_2(Fe)$ | 70–85 |

Synthetic rutile raw material is produced from ilmenite by reducing the iron oxides and leaching out the metallic iron with hydrochloric or sulfuric acids. In both processes, the objective is to increase the amount of $TiO_2$ in the raw materials.

Titanium white pigments are commercially produced either by the older sulfate process (SP), or by the chloride process (CP). The advantages of the sulfate process are low capital investment and low energy consumption. The disadvantage is the formation of a large amount of by-products: 3–4 tons of ferrous sulfate heptahydrate [7782-63-0], $FeSO_4 \cdot 7H_2O$, and 8 tons of diluted sulfuric acid for each ton of titanium dioxide pigment. Difficulties in finding an effective use for these by-products led to the commercialization of the chloride process. Since the mid-1980s the amount of titanium whites produced by the sulfate process has declined from 65–45%. This trend is expected to continue. Only about 35–40% of the titanium whites are expected to be manufactured by the sulfate process by the year 2000. The chloride process is simple, but requires more expensive raw materials and a higher capital investment. For pigmentation of certain products, eg, automotive top coats, coil-coatings, and PVC, pigments produced by the chloride process are needed. These applications, however, represent only about 10% of world demand. For balance, good quality SP and CP pigments are virtually interchangeable.

*Sulfate Process.* The flow chart of the sulfate process is presented in Figure 3. Titanium-bearing raw materials, ilmenite or titanium slag, are beneficiated before digestion in sulfuric acid. The beneficiation consists of the milling, screening, and drying of raw materials to about 40 $\mu$m particles. To prevent formation of hydrogen during the digestion process, any metallic iron must be magnetically separated at this point (see SEPARATION, MAGNETIC SEPARATION). Digestion of the ground raw materials is carried out using a concentrated sulfuric acid. The reaction is started by adding water or oleum to the reactor. During the addition of sulfuric acid, heat is generated that raises the temperature in the reactor to between 170 and 220°C. This starts the ilmenite decomposition which usually lasts about 12 hours and can be described by the following:

$$FeTiO_3 + 2\,H_2SO_4 \longrightarrow TiOSO_4 + FeSO_4 + 2\,H_2O$$

Under normal conditions about 95–97% of $TiO_2$ from ilmenite is solubilized. Most of the iron in the solution is in the $Fe^{2+}$ oxidation state. Any $Fe^{3+}$ present must be reduced to $Fe^{2+}$ because iron can only be removed by crystallization in its divalent form. The reduction is usually done by adding some scrap iron during the digestion step.

The cake produced by the digestion is extracted with cold water and possibly with some diluted acids from the subsequent processes. During the cake dissolution it is necessary to maintain the temperature close to 65°C, the temperature of iron sulfate maximum solubility. To prevent the reoxidation of the $Fe^{2+}$ ions during processing, a small amount of $Ti^{3+}$ is prepared in the system by the $Ti^{4+}$ reduction. The titanium extract, a solution of titanium oxo-sulfate, iron sulfate, and sulfuric acid, is filtered off. Coagulation agents are usually added to the extract to facilitate the separation of insoluble sludge.

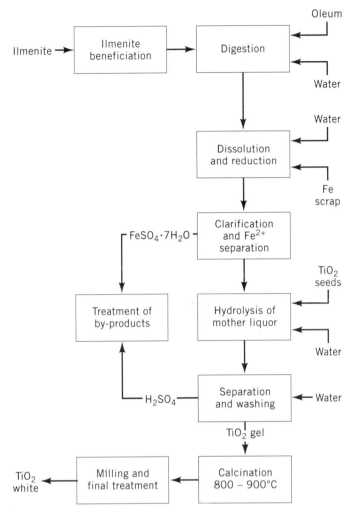

**Fig. 3.** Flow chart for the sulfate process for production of the pigment titanium white.

    The main impurity of the filtrate is the iron(II) sulfate which can be crystallized as the heptahydrate, $FeSO_4 \cdot 7H_2O$, by cooling the solution to a temperature below 15°C. This process is most often carried out in vacuum crystallizers. The crystals of iron sulfate are separated by centrifugation or filtration (qv). To prevent premature hydrolysis of titanium dioxide, the temperature of the above processes should not exceed 70°C.

    The concentrated mother liquor contains a large amount of sulfuric acid in a free form, as titanium oxy-sulfate, and as some metal impurity sulfates. To yield the purest form of hydrated $TiO_2$, the hydrolysis is carried out by adding crystallizing seeds to the filtrate and heating the mixture close to its boiling temperature, ~109°C. The crystal structure of the seeds (anatase or rutile) and their physical properties affect the pigmentary characteristics of the final product.

Chemically, the hydrolysis step can be described by a simple reaction:

$$TiOSO_4 + (n + 1)H_2O \longrightarrow TiO_2 \cdot nH_2O + H_2SO_4$$

Whereas the hydrolysate, $TiO_2 \cdot nH_2O$, does not have any pigmentary properties, its physical characteristics such as particle size eventually affect its value as a pigment. For that reason, the conditions of the hydrolysis step have to be carefully controlled.

To produce the anatase titanium dioxide pigment, the crystallization has to be initiated by anatase microcrystalline seeds. These can be prepared by neutralizing a small portion of the mother liquor, eg, by NaOH addition and heating the formed suspension to 80–90°C. The formed microcrystals have a particle size of about 1 nm and an anatase crystal structure. Adding these microcrystals to the original mother liquor in a concentration of about 0.5–1.0% as related to $TiO_2$ content, and hydrolyzing the mixture for 3–6 hours, yields the pigment having the anatase crystal structure.

To produce the rutile titanium dioxide pigment, hydrolysis of the mother liquor has to be carried out in the presence of a specially prepared hydrosol as a seeding agent. This hydrosol is made by the neutralization of a portion of the mother liquor in the presence of hydrochloric or some other monohydric acid. Because of the large amount of the hydrosol that must be added to the mixture (about 6% concentration), the hydrolysis reaction only takes about one hour.

Because free sulfuric acid is present, the hydrolysate has to be quickly separated from the mother liquor to prevent its possible dissolution. Separation and subsequent water washing is usually carried out by rotary vacuum filters. Even after a good washing of the gel, about 10% of $H_2SO_4$ (in relation to the weight of $TiO_2$) remains in the cake. This remainder is removed during the subsequent calcination step. The hydrolysate is usually doped with small quantities of various chemicals to improve the pigmentary properties of the final product.

Calcination of the washed cake of the hydrated gel of $TiO_2$ is carried out in rotary kilns similar to those used for producing cement (qv). The kilns are directly heated with gas or oil. An excess of air in the kilns is required to prevent the possible reduction of $TiO_2$ to its lower oxides. The presence of such oxides, even in small quantities, gives the final product a bluish gray shade.

During calcination, water is removed at temperatures between 200 and 300°C; sulfur trioxide is removed at temperatures between 480 and 800°C. At about 480°C the crystals of $TiO_2$ are being formed and continue to grow with increasing temperature. To prepare the anatase pigment, the final calcination temperature of the hydrolysate prepared in the presence of anatase seeds should reach about 800–850°C.

To produce the titanium white rutile pigment, the hydrated $TiO_2$ gel prepared in the presence of rutile seeds is calcined at temperatures of 900–930°C. This temperature is quite important because pigments having a particle size of 200–400 nm are produced at these temperatures. When the calcination is carried out at temperatures above 950°C, the particles of $TiO_2$ become considerably larger and do not have optimum pigmentary properties.

The calcined $TiO_2$ exits the rotary kiln in the form of various aggregates and agglomerates. It must be milled to attain optimum pigmentary properties.

The milling can be wet or dry and can be combined with air classification to produce the pigment having its optimal particle size distribution. During the wet milling, soluble salts present in the pigment can be removed.

*Chloride Process.* The flow chart of the chloride process is presented in Figure 4. In the chloride process, finely ground rutile reacts with chlorine in the presence of calcined petroleum coke. At a temperature between 800 and 1200°C, the following reaction occurs:

$$TiO_2 + 2\ Cl_2 + C \longrightarrow TiCl_4 + CO_2$$

The chlorination is mostly carried out in fluidized-bed reactors. Whereas the reaction is slightly exothermic, the heat generated during the reaction is not sufficient to maintain it. Thus, a small amount of oxygen is added to the mixture to react with the coke and to create the necessary amount of heat. To prevent any formation of HCl, all reactants entering the reactor must be completely dry. At the bottom of the chlorination furnace, chlorides of metal impurities present in the titanium source, such as magnesium, calcium, and zircon, accumulate.

Impurities that form volatile chlorides leave as gases at the top of the furnace together with the $TiCl_4$. By cooling those gases, most impurities, with the exception of vanadium and silicon chlorides can be separated from the titanium tetrachloride [7550-45-0]. Vanadium chlorides can be reduced to lower oxidation state chlorides that are solids; highly volatile $SiCl_4$ can be removed from $TiCl_4$ by fractional distillation.

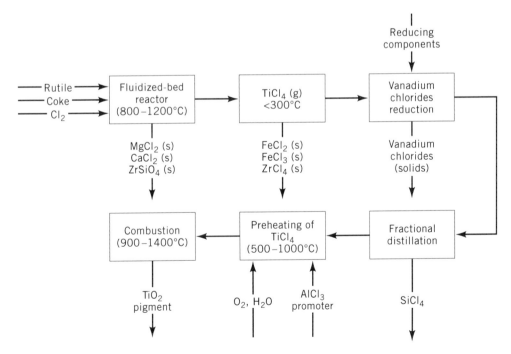

**Fig. 4.** Flow chart for the chloride process for production of the pigment titanium dioxide.

High purity $TiCl_4$ is preheated, mixed with hot oxygen (sometimes also with a small amount of water), and combusted at 900–1400°C to $TiO_2$:

$$TiCl_4 + O_2 \longrightarrow TiO_2 + 2\,Cl_2$$

Factors such as reaction temperature, excess of oxygen, water addition, addition of other minor reactants, eg, $AlCl_3$ to promote the formation of rutile, mixing conditions inside the reactor, and many others influence the quality of $TiO_2$ pigment. In general, titanium white pigments produced by the chloride process exhibit better lightness than those produced by the sulfate process.

For many applications, the individual particles of titanium dioxide pigment have to be encapsulated by the shell of an inorganic oxide, eg, $SiO_2$, and/or coated with an organic surfactant. These coatings fulfill two functions: they prevent a possible reaction between the highly reactive surface of $TiO_2$ and the matrix in which the $TiO_2$ is dispersed; and they improve the dispersability of $TiO_2$ in those matrices. Any reaction between pigments and the matrix can result in poor lightfastness and weathering of the pigmented article. Poor pigment dispersability can have immediate economic impact.

The deposition of inorganic coatings on the surface of pigment particles can be carried out either from a gaseous phase or by precipitation from an aqueous solution. Particles can also be partially coated by absorbing coating agents during milling operations.

*Use.*   Titanium dioxide is mainly used in the production of paints and lacquers (55–60%), plastics (15–20%), and paper (~15%). Other applications include the pigmentation of printing inks, rubber, textiles (qv), leather, synthetic fibers, ceramics, white cement, and cosmetics.

About 100,000 t of titanium dioxide annually are used as formulation components in the production of glass (qv), ceramics, electroceramics, catalysts, and in the production of mixed-metal oxide pigments.

**Other White Pigments.**   *Zinc Oxide.*   By volume, zinc oxide is the second most significant white pigment. It is known as zinc white, Chinese white, or flowers of zinc. Its pigmentary properties are good, providing good coverage. It has a good lightfastness and is well miscible with other pigments. With the increasing popularity of titanium dioxide white pigment in the twentieth century, the pigmentary use of zinc oxide has been declining.

Historically, ZnO is an older pigment than $TiO_2$. In France, the industrial production of zinc oxide started in 1840. In the French process, zinc white was produced by the direct combustion of zinc vapors. Around 1850, the New Jersey Zinc Co. in the United States started producing zinc oxide by a two-step process which became known as the American process. Almost 99% of zinc oxide is produced by one of those two processes (see ZINC COMPOUNDS).

The direct American process is basically a metallurgical operation. Zinc oxide ores are reduced using coal (qv) in a rotary kiln at temperatures of 1000–1200°C to zinc metal. At these temperatures the produced zinc metal evaporates and is oxidized with air to ZnO in the upper part of the kiln.

$$ZnO\ (ore) + C \longrightarrow Zn\ (g) + CO$$

$$Zn\ (g) + 1/2\ O_2 \longrightarrow ZnO$$

Because of the limited availability of high quality zinc oxide ores and the consequent lower quality of produced zinc oxide which can contain oxides of Pb, Cd, etc, the America production process has limited applications.

In the indirect French process, metallurgically refined zinc metal is boiled and the generated vapors combust with air to form zinc oxide. Metallic zinc is the starting material of this latter process, making it more expensive than the American one. The French process is the preferred production method, however, because it yields a purer product. Volatile metals, such as lead and cadmium, are again the main impurities that can be found in zinc white. Complex purification methods, such as fractional distillation, have been devised to reduce the amount of these impurities to acceptable levels.

Different types of furnaces (rotary, muffle) can be used for producing zinc vapors. In most cases these furnaces are heated with gas. The oxidation of zinc vapors is carried out with preheated air in specially designed chambers.

$$Zn \ (g) + 1/2 \ O_2 \longrightarrow ZnO$$

Because the oxidation reaction is highly exothermic, the temperature in the combustion chamber reaches 1200°C and microcrystals of zinc oxide are formed. By controlling the reaction conditions in the combustion chamber, zinc oxide having an optimal particle size (400–700 nm) can be produced. In combustion chambers, an excess of air (30–50%) must be maintained to prevent the undesirable reaction of zinc vapors and zinc oxide with combustion by-products.

$$ZnO + CO_2 \longrightarrow CO + Zn + O_2$$

$$Zn + CO \longrightarrow ZnO + C$$

$$2 \ Zn + CO_2 \longrightarrow 2 \ ZnO + C$$

Formed zinc oxide microcrystals are collected in a series of sedimentation chambers. The largest particles are collected in the first chamber together with nonoxidized zinc, cadmium, and lead oxide. Finer fractions settle in the subsequent chambers. The grades of the produced zinc oxide are determined by its purity, particle size, and shape.

*Use.*  Whereas zinc oxide was originally used as a pigment, its most important application as of this writing (ca 1995) is to aid in vulcanizing synthetic and natural rubber. Up to 5% of ZnO can be present in rubber products. Paint and coating industries are not using zinc white as their main white pigment, but zinc white is used as an additive to improve anticorrosion properties, mildew resistance, and durability of external coatings. Zinc oxide is also used as a chemical in the production of many mixed-metal oxide pigments, particularly spinels.

*Zinc Sulfide.*  Whereas zinc sulfide is mainly important as a component of the composite white pigment lithopone, it also has a limited use as a single pigment. Under the commercial name Sachtolith, pure ZnS is produced by Sachtleben Chemie in Germany.

Zinc sulfide has, after titanium white, the second highest refractive index of all the white pigments. However, its chemical and thermal resistances are inferior to those of $TiO_2$. As a pigment it has a pure, white color shade, reflects uv radiation, and because its particles are spherical the oil number of the pigment is quite low. The sphericity of the particles, as well as their softness, contribute to the low abrasivity of this pigment. This property reduces machinery damage during extrusion of plastics and fibers and therefore in some applications it is preferred over titanium white pigments.

*Production.* Zinc sulfide production started in the United States and in Europe in the 1920s. Starting in the early 1950s, zinc sulfide, like most white pigments, was slowly replaced by the more superior titanium white. Zinc sulfide can be prepared by a process similar to the one used to manufacture lithopone. In the first step, barium sulfide reacts with sodium sulfate to produce sodium sulfide solution:

$$BaS + Na_2SO_4 \longrightarrow Na_2S + BaSO_4$$

In the following step, sodium sulfide reacts with zinc salts to produce the final product:

$$Na_2S + ZnSO_4 \longrightarrow ZnS + Na_2SO_4$$

Most of the heavy-metal impurities present in zinc salt solutions must be removed before the precipitation reaction, or these form insoluble colored sulfides that reduce the whiteness of the zinc sulfide pigment. This end is usually achieved by the addition of zinc metal which reduces most heavy-metal ions to their metallic form. The brightness of zinc sulfide can be improved by the addition of a small amount of cobalt salts (ca 0.04% on a Co/Zn basis) (20). Barium sulfate [7727-43-7] formed in the first step is isolated and can be used as an extender.

Zinc sulfide can also be prepared by the direct reaction of barium sulfide and zinc chloride solution:

$$BaS + ZnCl_2 \longrightarrow ZnS + BaCl_2$$

Commercially, zinc sulfide is available in the standard untreated grade and in several grades having particles treated by surfactants (qv) to improve their dispersability in either aqueous or organic media.

*Use.* Zinc sulfide is used in applications where white color shade and low abrasivity are required. In printing inks and paints it also contributes to stability and good rheological and printing properties. For those reasons, zinc sulfide is mainly utilized for coloring plastics, synthetic fibers, and in the preparation of special coatings, greases, and lubricating oils. When used in color sealing compounds, it also slows down aging and helps to preserve elasticity.

*Lithopone.* Lithopone is a mixture of ZnS and $BaSO_4$. The pigmentary properties of the mixture are determined by zinc sulfide and therefore lithopone pigments are characterized by the amount of ZnS present in the mixture. The amount of ZnS in commercial lithopones varies from 15 to 60%. The most common is a 30% lithopone pigment, containing 28–30% of ZnS and 70–72% of $BaSO_4$.

*Production.* Commercial production of lithopone started in the first half of the nineteenth century, and continued to grow until the middle of the twentieth century when titanium dioxide started to dominate the white, inorganic pigment market. Lithopone is prepared by combining barium sulfide and zinc sulfate solutions at 50–60°C:

$$BaS + ZnSO_4 \longrightarrow ZnS + BaSO_4$$

Both zinc sulfide and barium sulfate are insoluble in water. To improve the stability of lithopone, a small amount of a cobalt salt is added to the precipitated mixture. The mixture has to be filtered off, dried, and calcined. The calcination is carried out in rotary calciners at temperatures between 600 and 700°C. During calcination, the particle size of zinc sulfide grows from its original size (about 0.1 $\mu$m) to the pigmentary optimal size of 0.4–0.6 $\mu$m.

Hot lithopone leaving the calciner has to be cooled quickly to prevent the oxidation of ZnS to ZnO. Rapid cooling is accomplished by its suspension in water that might contain a small amount of sulfuric acid to remove any traces of undesirable zinc oxide. The suspension is then wet milled, filtered off, and dried to the commercial product.

The barium sulfide needed for the reaction is prepared by the reduction of barite [*13462-86-7*], $BaSO_4$, with petroleum coke in a rotary furnace:

$$BaSO_4 + C \longrightarrow BaS + 2\,CO_2$$

The fused product contains about 60–85% barium sulfide, unreacted barium sulfate, and impurities present in barite and ash. The soluble barium sulfide is extracted from the mixture with water and separated from the insoluble impurities by filtration.

The other component of the lithopone precipitation reaction, zinc sulfate, is prepared by the dissolution of various zinc-containing raw materials in sulfuric acid:

$$Zn + H_2SO_4 \longrightarrow ZnSO_4 + H_2$$

The amount of impurities present in the zinc sulfate solution is determined by the local source of zinc-containing material used in the above reaction. To get a good quality lithopone, the impurities must be removed from the solution.

Pure solutions of zinc sulfate can only be used for the preparation of lithopone having a low (up to 30%) percentage of zinc sulfide. Additional zinc compound must be added for the preparation of lithopone having a higher concentration of ZnS. Most often that compound is zinc chloride. One can either use a mixture of sulfuric and hydrochloric acid for the dissolution of zinc-containing raw materials, or directly mix separately prepared solutions of zinc sulfate and zinc chloride. The precipitation of 60% of the lithopone can be approximated by the following reaction:

$$2\,ZnSO_4 + 5\,ZnCl_2 + 7\,BaS \longrightarrow 7\,ZnS + 2\,BaSO_4 + 5\,BaCl_2$$

During this reaction, some barium ions remain in the solution as chlorides and can be utilized for the preparation of other barium compounds.

*Use.*  Lithopones are used in water-based paints because of their excellent alkali resistance, in paper manufacturing as a filler and opacifying pigment, and in rubber and plastics as a whitener and reinforcing agent.

*Lead Whites.*  Basic lead carbonate, sulfate, silicosulfate, and dibasic lead phosphite are commonly referred to as lead whites. Usage is limited because of environmental restrictions placed on the use of lead-containing compounds.

## Colored Pigments

**Iron Oxide Pigments.**  The worldwide consumption of iron oxide pigments represents about 40% of the total production of colored, inorganic pigments. Iron oxides can be produced by the beneficiation of naturally occurring materials or synthetically from iron salts. Natural iron oxide pigments account for only about 30% of total iron oxide production. Some of the natural iron compounds are the oldest pigments known. Chemically, these are oxides or oxide–hydroxide compounds of iron(III). The following oxides and oxide–hydroxides have acceptable pigmentary properties and are components of natural and synthetic iron oxide pigments: the yellow goethite [1310-14-1], $\alpha$-FeO(OH); orange lepidocrocite [12022-37-6], $\gamma$-FeO(OH); red hematite [1317-60-8], $\alpha$-Fe$_2$O$_3$; and brown maghemite [12134-66-6], $\gamma$-Fe$_2$O$_3$.

Black magnetite [1309-38-2], FeO·Fe$_2$O$_3$, a binary iron oxide having a spinel structure, has not gained wide acceptance as a pigment because of its poor tinting strength. Iron oxide is also an important raw material for the production of many mixed-metal oxide pigments.

In general, all iron pigments are characterized by low chroma and excellent lightfastness. They are nontoxic, nonbleeding, and inexpensive. They do not react with weak acids and alkalies, and if they are not contaminated with manganese, do not react with organic solvents. However, properties vary from one oxide to another.

*Natural Iron Oxides.*  The earth's crust contains about 7 wt % iron oxides, but only a few deposits are rich enough in iron to be suitable for mining pigmentary-quality iron oxides. Deposits that are a suitable source of natural iron oxide pigments are usually hydrated aluminum silicates that contain various amounts and forms of iron oxide. Most of these are contaminated with oxides of aluminum, manganese, magnesium, and in some cases carbon and other organic compounds. After a mechanical beneficiation, and in some cases calcination, iron oxides are supplied to the market as red, ocher, sienna, and umber natural pigments. The hue of the natural iron oxide pigments is determined by raw material composition and processing. Material composition varies from one geographic location to another as does the quality of iron pigments. In some cases, iron oxide pigments have a name that reflects the mine location, eg, Persian red which comes from Ormuz Island in the Persian Gulf.

About 60% of the natural iron oxide pigments is used to color cement and other building materials (qv). About 30% is consumed in the production of paints. For coloring plastics and rubber, synthetic iron oxide pigments are preferred. The main advantage of the natural iron oxide pigments, as compared to the synthetic ones, is cost. However, the quality is inferior, and in most cases, they are consumed in close proximity to the mines. As colorants, the natural iron oxides are about 50% weaker than synthetically produced iron oxides.

*Synthetic Iron Oxides.* Iron oxide pigments have been prepared synthetically since the end of the seventeenth century. The first synthetic red iron oxide was obtained as a by-product of the production of sulfuric acid from iron sulfate containing slate. Later, iron oxide pigments were produced directly by the thermal decomposition of iron sulfates. In the 1990s, about 70% of all iron oxide pigments consumed are prepared synthetically.

Advantages of synthetic iron oxides over their natural counterparts include chemical purity, more uniform particle size and size distribution, and in the case of precipitated oxides the ability to prepare the pigment in predispersed vehicle systems by flushing techniques. The popularity of browns, yellows, and earth-tone colors, and environmental regulations that are limiting the use of heavy-metal toxic pigments, are helping to increase the sale of iron oxides worldwide. In Europe, iron oxide pigments dominate such construction materials as ceramics, cement, and roofing granules. In the United States, it is mainly the paint and coating material sectors that are consuming the largest quantities of these pigments. Other areas of synthetic iron oxide applications are plastics, paper, and magnetic recording tapes.

*Iron Oxide Reds.* From a chemical point of view, red iron oxides are based on the structure of hematite, $\alpha$-$Fe_2O_3$, and can be prepared in various shades, from orange through pure red to violet. Different shades are controlled primarily by the oxide's particle size, shape, and surface properties.

*Production.* Four methods are commercially used in the preparation of iron oxide reds: two-stage calcination of $FeSO_4 \cdot 7H_2O$; precipitation from an aqueous solution; thermal dehydration of yellow goethite, $\alpha$-$FeO(OH)$; and oxidation of synthetic black oxide, $Fe_3O_4$.

The final product of all the above processes is iron(III) oxide, $\alpha$-$Fe_2O_3$, but its properties are determined by the method of preparation. Thermal dehydration of goethite yields a pigment of lowest (4.5 g/cm$^3$) density. The highest (5.2 g/cm$^3$) density pigment is one prepared by two-stage calcination. The particle size varies from 0.3 to 4 $\mu$m; the refractive index varies from 2.94 to 3.22.

The largest volume of synthetic red iron oxide is produced by the two-step calcination of iron(II) sulfate. In the first step, the iron(II) sulfate heptahydrate is dehydrated to a monohydrate:

$$FeSO_4 \cdot 7H_2O \longrightarrow FeSO_4 \cdot H_2O + 6\,H_2O$$

The second step involves thermal decomposition of the monohydrate product at a temperature above 650°C in the absence of air (21):

$$6\,FeSO_4 \cdot H_2O \longrightarrow 2\,Fe_2O_3 + Fe_2(SO_4)_3 + 6\,H_2O + 3\,SO_2$$

$$Fe_2(SO_4)_3 \longrightarrow Fe_2O_3 + 3\,SO_3$$

or in the presence of air:

$$6\,FeSO_4 \cdot H_2O + 3/2\,O_2 \longrightarrow 3\,Fe_2O_3 + 6\,SO_3 + 6\,H_2O$$

By-products of these reactions are reclaimed and recycled. The color depends on the size of the particles formed. Size is controlled by regulating the calcination profile, ie, time and temperature. The calcined product is ground, washed, and classified.

Wet preparation of red iron oxides can involve either a hydrothermal process (see HYDROTHERMAL PROCESSING) or a direct precipitation and growth of iron oxide particles on specially prepared nucleating seeds of $Fe_2O_3$. In the hydrothermal process, iron(II) salt is chemically oxidized to iron(III) salt, which is further treated by alkalies to precipitate a hydrated iron(III) oxide gel. The gel can be dehydrated to anhydrous hematite under pressure at a temperature around 150°C.

In the direct precipitation process, the seeds of iron(III) oxide are added to an iron salt solution, most often iron(II) sulfate, which is subsequently oxidized by air. The released sulfuric acid is removed by the addition of metallic iron with which it reacts to iron(II) sulfate. The overall reaction shows that ferrous sulfate is not consumed during the process. It only helps to oxidize metallic iron to ferric oxide:

$$2\ FeSO_4 + 2\ Fe + 3/2\ O_2 \longrightarrow Fe_2O_3 + 2\ FeSO_4$$

The reaction conditions are critical, as hydrated iron oxide, $Fe_2O_3 \cdot H_2O$, can also precipitate. The particles are either spherical or rhombohedral, depending on the nucleating material.

Ferrite reds can also be prepared by calcining synthetic yellow iron oxide and the process parallels the production of reds from natural yellow oxides. Red iron oxide pigments can also be produced by the oxidation of the ferrous oxide component, FeO, of the binary black iron oxide, $FeO \cdot Fe_2O_3$, at 370°C. The original cubic shape of the black iron oxide is retained. A Venetian red iron oxide is prepared by calcining a mixture of iron sulfate and lime. The final product typically contains 40 wt % $Fe_2O_3$, 60 wt % $CaSO_4$. The color range of Venetian red is not as wide as that of the other iron oxides and is limited to light shades.

Synthetic red iron oxides are prepared in a variety of grades from light to dark. These are sold under a variety of names, eg, Indian red, Turkey red, and Venetian red.

*Iron Oxide Yellows.* From a chemical point of view, synthetic iron oxide yellows, also known as iron gelbs, are based on the iron(III) oxide–hydroxide, $\alpha$-FeO(OH), known as goethite. Color varies from light yellows to dark buffs and is primarily determined by particle size, which is usually between 0.1 and 0.8 $\mu$m. Because of their resistance to alkalies, these are used by the building industry to color cement. Thermally, iron oxide yellows are stable up to 177°C; above this temperature they dehydrate to iron(III) oxide;

$$2\ FeO(OH) \longrightarrow Fe_2O_3 + H_2O$$

*Production.* Three commercial processes are used for the production of iron yellows: the Penniman-Zoph process, the precipitation process, and the Laux process.

The Penniman-Zoph process involves the preparation of seeds or nucleating particles by the alkali precipitation of ferrous sulfate. The reaction is carried out at a low temperature using an excess of ferrous ions. The hydroxide is then oxidized to the seeds of hydrated ferric oxide:

$$FeSO_4 + 2\,NaOH \longrightarrow Fe(OH)_2 + Na_2SO_4$$

$$2\,Fe(OH)_2 + 0.5\,O_2 \longrightarrow Fe_2O_3{\cdot}H_2O + H_2O$$

The seeds are transferred to tanks containing scrap iron and a ferrous sulfate solution, and the mixture is heated to a temperature between 70 and 90°C. While the seeds circulate over the scrap iron, air is bubbled through the medium causing the seeds to grow. The process can be described by the following reactions:

$$4\,FeSO_4 + 6\,H_2O + O_2 \longrightarrow 4\,FeO(OH) + 4\,H_2SO_4$$

$$4\,H_2SO_4 + 8\,FeSO_4 + 2\,O_2 \longrightarrow 4\,Fe_2(SO_4)_3 + 4\,H_2O$$

$$4\,Fe_2(SO_4)_3 + 4\,Fe \longrightarrow 12\,FeSO_4$$

The overall reaction can be summarized as,

$$4\,Fe + 3\,O_2 + 2\,H_2O \longrightarrow 4\,FeO(OH)$$

indicating that in this process it is actually the metallic iron that is consumed in the production of yellow gelbs, and ferrous sulfate is the intermediate that is replenished by the reaction of formed sulfuric acid with the scrap iron. Variables that affect production and determine the shade of yellow are the temperature of the reaction, the circulation rate of oxygen, the circulation rate of the ferrous sulfate solution, and the size and shape of the seed particles. As the reaction continues over several days, larger particles develop, which are deeper and redder. The reaction is stopped at the desired hue and the precipitate is washed free of soluble salts, dried carefully, ground, and bagged.

Iron oxide yellows can also be produced by the direct hydrolysis of various ferric solutions with alkalies such as $NaOH$, $Ca(OH)_2$, and $NH_3$. To make this process economical, ferric solutions are prepared by the oxidation of ferrous salts, eg, ferrous chloride and sulfate, that are available as waste from metallurgical operations. The produced precipitate is washed, separated by sedimentation, and dried at about 120°C. Pigments prepared by this method have lower coverage, and because of their high surface area have a high oil absorption.

The Laux process is a modification of the Bechamp reaction that was discovered in 1854. It has been used for the reduction of nitrobenzene to aniline using metallic iron:

$$C_6H_5NO_2 + 2\,Fe + 2\,H_2O \longrightarrow C_6H_5NH_2 + 2\,FeO(OH)$$

Originally, iron oxides were the by-products of this reduction process and did not have any pigmentary properties. Laux (Bayer AG) (22) discovered that when the reduction is carried out in the presence of aluminum or ferrous chlorides, high quality iron pigments can be produced. At the end of the reduction process, aniline is distilled off, and the pigments are separated from the unreacted iron, washed, dried, and ground or micronized. Depending on the reaction conditions, red $\alpha$-$Fe_2O_3$ or yellow $\alpha$-$FeO(OH)$ pigments can be produced by this process.

*Iron Blacks.* Chemically, iron blacks are based on the binary iron oxide, $FeO \cdot Fe_2O_3$. Although the majority is produced in the cubical form, these can also be produced in acicular form. Most of the black iron oxide pigments contain iron(III) oxide impurities, giving a higher ratio of iron(III) than would be expected from the theoretical formula.

*Iron Browns.* Iron browns are often prepared by blending red, yellow, and black synthetic iron oxides to the desired shade. The most effective mixing can be achieved by blending iron oxide pastes, rather than dry powders. After mixing, the paste has to be dried at temperatures around 100°C, as higher temperatures might result in the decomposition of the temperature-sensitive iron yellows and blacks. Iron browns can also be prepared directly by heating hydrated ferric oxides in the presence of phosphoric acid, or alkali phosphates, under atmospheric or increased pressure. The products of precipitation processes, ie, the yellows, blacks, and browns, can also be calcined to reds and browns.

**Complex Inorganic Color Pigments.** Based on the crystal structure, the Color Pigments Manufacturers' Association (CPMA) (formerly Dry Color Manufacturers' Association, DCMA) has classified 53 key inorganic pigments into 14 categories. In 1991 the CPMA decided to call these inorganic colorants complex inorganic color pigments (23). The original name, mixed-metal oxide pigments (MMO), did not accurately describe the chemical nature of all the classified pigments. In particular, several single oxides, eg, $Cr_2O_3$ and $Fe_2O_3$, and some metal salts, eg, cobalt phosphate [13455-36-2], $Co_3(PO_4)_2$, and cobalt lithium phosphate [13824-63-0], $CoLiPO_4$, that were included in the CPMA classification are not mixed-metal oxides.

*Mixed-Metal Oxide Pigments.* Mixed-metal oxide pigments can be considered a subcategory of complex inorganic color pigments. The name, mixed-metal oxides, does not, however, represent the reality as these pigments are not mixtures but rather solid solutions or compounds consisting of two or more metal oxides. Each pigment has a defined crystal structure that is determined by the host lattice. Other oxide(s) interdiffuse at high temperatures into the host lattice structure forming either a solid-state solution or a new compound. Mixed-metal oxide pigments belong to a significant, but by volume, small category of inorganic pigments. Most of them contain metal cations balanced by oxygen anions having structures similar to naturally occurring minerals. Structurally, mixed-metal oxide pigments belong to one of 14 structure types. The most common ones are rutile and spinel [1302-67-6]. The commercial significance is in their thermal, chemical, and light stability, combined with low toxicity. When these are employed for coloring glass enamels and ceramics, they are sometimes referred to as colors or stains; when used to color paints and plastics, they are known as pigments.

The color of mixed-metal oxide pigments results from the incorporation of chromophores, into the structure of stable host oxides. The host can be a single oxide, eg, tin(IV) oxide [18282-10-5], $SnO_2$, or $TiO_2$, or a mixed oxide, eg, zircon

[*10101-52-7*] (zirconium silicate), $ZrSiO_4$, or spinel (magnesium aluminum oxide), $MgAl_2O_4$. Most of the host oxides can be found as minerals in nature, and when pure, are usually colorless. It is the stability of the oxide host lattices that gives these pigments high thermal stability and resistance toward the corrosiveness of mediums in which they are dispersed, eg, molten glass in glazes and enamels. The refractive index of the host oxide also affects the hiding power of the produced pigment. Typical host oxides are baddeleyite [*12036-23-6*], $ZrO_2$; zircon, $ZrSiO_4$; cassiterite, $SnO_2$ sphene, $CaSnSiO_5$; spinels, $MgAl_2O_4$, $TiZn_2O_4$; corund, $Al_2O_3$; rutile, $TiO_2$; garnet, $3CaO \cdot Al_2O_3 \cdot 3SiO_2$; etc. Chromphores are transition-metal ions, such as Fe, Cr, Mn, Ni, Co, Cu, and V; and rare-earth elements, eg, Ce, Pr, Nd, in particular.

When the host is a mixed oxide, the incorporation of the chromophore is best achieved during high temperature formation from single oxides:

$$ZrO_2 + SiO_2 \longrightarrow ZrSiO_4$$

$$CaO + SiO_2 + SnO_2 \longrightarrow CaSnSiO_5$$

When the host is a single oxide, incorporation is best achieved during a high temperature phase transition of the host lattice such as when $TiO_2$ goes from anatase to rutile, or during formation from carbonates or other salts.

The crystal defects of the host lattice structure aid in the incorporation of chromophores. By increasing those defects, reactants can diffuse more easily through the product layers and the pigment is formed faster. The presence of mineralizers can also positively affect the solid-state reaction (24). A mineralizer is a compound that facilitates crystal growth during solid-state reactions by providing a local environment that makes the movement of reactants through the solids' mixture easier. The incorporation of the chromophore into the host lattice usually results in the formation of a substitution, or less often an addition compound.

*Spinel Ceramic Pigments.* Pigments having a spinel structure are widely used by the ceramic and plastic industries (see COLORANTS FOR CERAMICS). They cover a wide range of colors and many are thermally stable up to 1400°C and are resistant to molten glass. Another advantage is their intermiscibility, allowing the user a choice of creating many intermediate colors.

Spinel compounds have a common chemical formula, $AB_2X_4$. Structurally they have a cubic symmetry and are derived from magnesium aluminate, $MgAl_2O_4$, a naturally occurring mineral. More than 100 compounds with the spinel structure are known. Most spinels contain small quantities of other metal oxides, modifiers, to change their color shade without affecting the crystal structure. More than half of the spinel pigments reported by the CPMA are dark (brown and black). Many pigments based on Fe–Cr–Zn (brown), Fe–Cr–Zn–Mn (dark brown), Fe–Cr–Zn–Co (black), and similar compositions, are commercially available for coloring ceramic glazes, enamels, ceramics, and plastics.

*Rutile Ceramic Pigments.* Structurally, all rutile pigments are derived from the most stable titanium dioxide structure, ie, rutile. The crystal structure of rutile is very common for $AX_2$-type compounds such as the oxides of four valent metals, eg, Ti, V, Nb, Mo, W, Mn, Ru, Ge, Sn, Pb, and Te; as well as halides of divalent elements, eg, fluorides of Mg, Mn, Fe, Co, Ni, and Zn.

Rutile pigments, prepared by dissolving chromophoric oxides in an oxidation state different from +4 in the rutile crystal lattice, have been described (25,26). To maintain the proper charge balance of the lattice, additional charge-compensating cations of different metal oxides also have to be dissolved in the rutile structure. Examples of such combinations are $Ni^{2+}$ + $Sb^{5+}$ in 1:2 ratio as $NiO$ + $Sb_2O_5$, $Cr^{3+}$ + $Sb^{5+}$ in 1:1 ratio as $Cr_2O_3$ + $Sb_2O_5$, and $Cr^{3+}$ + $W^{6+}$ in 2:1 ratio as $Cr_2O_3$ + $WO_3$.

Many pigments having such substitutions have been commercialized. The most important one is the Ti–Ni–Sb yellow pigment having nickel oxide [12035-36-8], $NiO$, as the chromophoric component and $Sb^{5+}$ as the charge-compensating cation.

Rutile types of pigments can be prepared by the simple calcination of mixed starting oxides. For example, the Cr–Sb–Ti yellow pigment can be produced by mixing chromium and antimony oxides (10–20% by weight) with titanium dioxide (80–90% by weight). The anatase form of $TiO_2$ is the preferred starting material as it changes to rutile (at temperatures ~900°C) that is defective and highly reactive. The recommended calcination temperature is 1040°C.

Whereas the yellow color of the Ti–Sb–Ni pigment is the most typical color of rutile pigments, other colors such as buffs, Ti–Sb–Cr and Ti–W–Cr; browns, Ti–Sb–Mn; and blacks, Ti–Sb–V, have also been commercialized.

*Zircon Ceramic Pigments.* Zircon pigments are relatively new compared to other complex inorganic color pigments. Blue zircon pigment was discovered in 1948 (27), yellow in the early 1950s, and pink in the 1960s (28,29). The earliest studies describing the effects of reaction conditions on the properties of zircon pigments were published in 1962 (30). Several more recent publications describe the reactions occurring during the formation of zircon pigments (31–35).

Zircon pigments are derived from the tetragonal zirconium silicate, $ZrSiO_4$. The structure is slightly distorted. There are two different Zr–O bonds, one being 0.205 nm and the other 0.241 nm long. This distortion affects the color of the pigment. Because of the high temperature (up to 1600°C) and chemical stability of zirconium silicate, zircon pigments can be used in the formulations of high temperature (1300–1400°C) glazes. Zirconium silicate is also used as an opacifier in porcelain and vitreous enamels.

In pigments, zirconium silicate serves as the host lattice for various chromophores, such as vanadium, praseodymium, iron, etc. Zirconium silicate crystals are usually formed *in situ* during pigment preparation by a high temperature reaction of $ZrO_2$ and $SiO_2$:

$$ZrO_2 + SiO_2 \longrightarrow ZrSiO_4$$

To facilitate this reaction, mineralizers can be added to the mixture. In rare instances, zircon pigments are prepared directly from the zircon mineral, $ZrSiO_4$. Based on the mechanism of color formation, zircon pigments can be divided into two categories, substitution-defect pigments and inclusion pigments.

Defect pigments have more brilliant colors than inclusion pigments, but only a few are known. The blue Zr–Si–V pigment is probably the best known pigment of this type. The pink coral pigment formed by the inclusion of $Fe_2O_3$ into the zirconium silicate lattice is a typical example of an inclusion pigment. The pigment can be described as $(ZrSiO_4)_{1-x}\cdot(Fe_2O_3)_x$. Inclusion of $Cr_2O_3$ results in the well-known green pigment, $(ZrSiO_4)_{1-x}\cdot(Cr_2O_3)_x$.

*Bismuth Vanadate.* The use of lead chromate pigments has been slowly phased out of many applications. Thus a search for more environmentally acceptable relatively inexpensive yellow pigments having excellent coloristic properties has been ongoing.

The use of bismuth vanadate [*14059-33-7*], $BiVO_4$, as a nontoxic, yellow pigment with good hiding strength and lightfastness was patented by Du Pont in 1978 (36). Because of the high cost of bismuth, the pigment could not compete with the inexpensive lead chromate, and for years $BiVO_4$ was only a laboratory curiosity. That situation has changed as the result of environmental pressures, and as of 1995 at least two pigment producers, Ciba and BASF, are marketing this pigment primarily for plastic and paint applications. Some users have already replaced lead chromate in their paint formulations with a combination of organic pigments and bismuth vanadate (see BISMUTH AND BISMUTH COMPOUNDS).

Bismuth vanadate can be produced by chemical precipitation, as well as by high temperature calcination methods. In the wet process, the acidic solution of bismuth nitrate, $Bi(NO_3)_3$, is mixed with the alkaline solution of sodium vanadate, $Na_3VO_4$. The gel formed is filtered off on a filter, pressed, washed, and converted to a crystalline form by calcination at low temperatures of 200–500°C for one hour (37,38).

In the calcination process, a mixture of corresponding oxides and an optional modifier, eg, molybdic acid, are milled together to achieve a homogenous mixture. The mixture is calcined at 750–950°C and milled to a desired particle size. Wet milling in an alkaline medium is recommended to remove any unreacted vanadium salts that are believed to degrade the pigmentary properties of bismuth vanadate (39).

For most applications, bismuth vanadate is not thermally or chemically stable enough, and it has to be encapsulated with a dense, amorphous shell of silica (40,41). It is recommended that bismuth vanadate be granulated to improve its handling and eliminate any dusting problems.

**Chromium(III) Pigments.** There are two green pigments based on chromium in the +3 oxidation state. The first one is chromium oxide [*1308-38-9*], $Cr_2O_3$; the second is hydrated chromium oxide, $Cr_2O_3\cdot xH_2O$. Worldwide production is about 20,000 metric tons. Principal producers are American Chrome and Chemicals in the United States, Bayer in Germany, British Chrome and Chemicals in the United Kingdom, and Nihon Denko in Japan.

*Chromium(III) Green Pigment.* Chromium oxide green is characterized by outstanding lightfastness and has excellent resistance to acids, alkalies, and high temperatures. Because it weathers extremely well, chromium oxide green is applied as a colorant for roofing granules, cement, concrete, and outdoor industrial coatings. It is also used in ceramic applications. Because these pigments can withstand vulcanization conditions and do not degrade, chromium pigments are also used for rubber pigmentation. One unique feature of chromium oxide green is that it reflects infrared radiation similarly to chlorophyll. Therefore, chromium oxide green is used extensively in formulating camouflage coatings for military applications. One drawback for some applications is its abrasiveness.

*Hydrated Chromium(III) Green Pigment.* Hydrated chromium oxide has a brilliant green color and is referred to as Gingnet's green. It exhibits a limited hue range, is semitransparent, and has a low opacity, but provides excellent lightfastness and alkali resistance. Water of hydration limits the heat resistance to application temperatures no higher than 260°C. Thus, the pigment is unsuitable for ceramic use. Transparency permits the formulation of polychromatic finishes. Consumption has diminished since the introduction of phthalocyanine green in the 1940s (see PIGMENTS, ORGANIC).

**Ultramarine Pigments.** Ultramarines are derived from lazurite [*1302-85-8*] (Lapis Lazuli), a semiprecious stone, which was the natural source of ultramarine blue for hundreds of years. Ultramarines can be prepared in many shades. Examples of commercially significant ones are ultramarine blue, ultramarine violet, and ultramarine pink. The ultramarine pigment having a green shade can also be prepared but it is not commercially available. The first German patent issued in 1877 was for the manufacture of ultramarine red (42).

Chemically, ultramarines are complex sodium aluminates having a zeolite structure. Composition varies within certain wt % ranges, ie, $Na_2O$, 19–23; $Al_2O_3$, 23–29; $SiO_2$, 37–50; and S, 8–14.

Ultramarine blues are prepared by a high temperature reaction of intimate mixtures of china clay, sodium carbonate, sulfur, silica, sodium sulfate, and a carbonaceous reducing agent, eg, charcoal, pitch, or rosin.

About 20,000 tons of ultramarines are produced worldwide. The largest manufacturers are Dainichi Seika (Japan), Nubiola (Spain), and Reckitts Colours International (RCI) based in Hull (U.K.), where Isaac Reckitt first began making laundry blue in the 1850s. Holliday Chemical Holdings (HCH of Huddersfield, U.K.) purchased RCI in 1994 (43).

Ultramarine pigments are used in printing inks, textiles, rubber, artists' colors, cosmetics, and laundry bluing. Because of their thermal stability they are also used to color roofing granules.

**Cyanide Iron Blues.** Cyanide iron blue, also known as Prussian blue, is one of the oldest industrially produced, inorganic pigments. Chemically, cyanide iron blues are based on the $\{Fe^{2+}[Fe^{3+}(CN)_6]\}^-$ anion. The charge is balanced by sodium, potassium, or ammonium cations. Modern iron blues are ammonium salts, which are as good or better than the pre-World War I potassium types and were developed in the United States as a consequence of a shortage of potassium during that period. Iron blue pigments were usually named according to their place of production or their original developer. For those reasons they became known as Berlin or Prussian blue, Milori blue, Paris blue, Chinese blue, Toning blue, or Turnbull blue.

Cyanide iron blues can be prepared by several methods. The most common one is the indirect, two-step process. In the first step, a white precipitate (Berlin white), is produced by the reaction of sodium, potassium, or ammonium ferrocyanide and ferrous sulfate:

$$Fe^{2+}SO_4 + M_4^+[Fe^{2+}(CN)_6] \longrightarrow M_2^+ Fe^{2+}[Fe^{2+}(CN)_6]$$

The precipitate is digested with hot sulfuric acid, and following digestion it is oxidized with sodium chlorate or sodium bichromate to iron blue, $M^+\{Fe^{2+}[Fe^{3+}(CN)_6]\}^-$. After oxidation, the iron blue is filtered, washed, dried, and packaged. Iron blues are produced by Degussa AG (Germany) and Dainichiseika (Japan). Iron blues are mainly used by the printing industry for coloring printing inks. In Europe, cyanide blues are used for coloring fungicides (see IRON COMPOUNDS).

**Cadmium Pigments.** Historically, cadmium pigments have been very important, providing a range of clean, bright shades of yellow, orange, red, and maroon colors. This importance, however, has been decreasing continually because of the environmental issues associated with the production and use of Cd-, Se-, and Hg-containing compounds. As of this writing (ca 1995), most users around the world are looking for safer replacements for these pigments. Only a few pigment producers are willing to continue cadmium pigment production (see CADMIUM AND CADMIUM COMPOUNDS).

Pigment color is determined by the ratio of Cd, and Zn or Hg if present, to S and Se in the product and can be changed all the way from primrose to maroon. Mercury substitution for cadmium yields (Cd, Hg)S pigments with red and maroon shades similar to those obtained with selenium substitution. The Cadmium Association provides the simple diagram given in Figure 5 showing the color and composition correlation.

The production volume of cadmium pigments is decreasing steadily as of the mid-1990s. In 1993, production was estimated to be anywhere between 1500 to 5000 metric tons. The reasonable projection for 1994 seems to be around 2000 metric tons representing about one-third of the 1990 production. Producers of cadmium pigments in the United States are Engelhard Corp. and SCM. In Europe, they are Johnson Matthey Colors Ltd., and James M. Brown, Ltd. in the United Kingdom, Cerdec AG in Germany, General Quimica in Spain, and SLMC in France. Dainichiseika and Mitsubishi are two producers of cadmium pigments in Japan.

**Lead Chromate Pigments.** Lead chromate [7758-97-6], $PbCrO_4$, occurs in nature as the orange-red mineral crocoite [14654-05-8]. Synthetically prepared lead chromate and its solid-state solutions with lead sulfate [7446-14-2], $PbSO_4$, or lead molybdate [10190-55-3], $PbMoO_4$, are known to have excellent pigmentary properties. Some lead chromate pigments can also contain basic lead chromate, lead carbonate, and lead phosphate. The usage of these pigments has been steadily decreasing because of environmental regulations restricting the

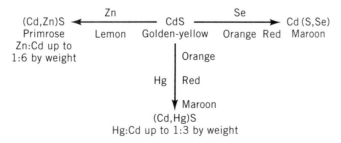

**Fig. 5.** Color and composition correlation of cadmium-based pigments.

production and the use of lead-containing products. The various hues of the lead chrome yellows, chrome oranges, and molybdate oranges depend on the interrelationship of three factors: chemical composition, crystal structure, and particle size.

The basic compositions of the most common commercially available lead chromate pigments are given in Table 2 (44).

Lead chromates are prepared by precipitation techniques from soluble salts in aqueous media. The raw material list includes a number of different lead compounds, eg, litharge, lead nitrate, basic lead acetate, basic lead carbonate, as well as acids, alkalies, sodium bichromate, and sodium chromate. The typical reaction can be represented by the following equation:

$$2\ Pb(NO_3)_2 + Na_2Cr_2O_7 + H_2O \longrightarrow 2\ PbCrO_4 + 2\ NaNO_3 + 2\ HNO_3$$

Insoluble lead chromate can be chemically treated where applicable, then filtered, washed, dried, and ground.

Most lead chromates are surface treated. Inorganic and/or organic compounds may be used. Treatments enhance the working and performance properties of the pigment, eg, wetting, texture, heat stability, and lighfastness. Chrome yellows, oranges, and molybdate oranges are used in a large number of different paint systems, restricted mostly to maintenance and industrial finishes because of the toxicity and potential carcinogenic nature of the lead. Traditional use of these pigments in traffic paint formulations and automotive finishes has been decreasing as the result of environmental regulations. Less toxic, inorganic metal oxide yellow pigments, such as titanium–nickel yellow, praseo yellow, bismuth vanadate, and their combinations with organic pigments, are being used increasingly as a replacement for lead chromate pigments.

The principal producers of lead chromate pigments in North America are Cookson, DCC (owned by Kikuchi), Engelhard, and Wayne Pigment. In Europe, lead chromates are produced by Ciba in Holland, Capelle in France, and Heubach GmGH in Germany.

**Table 2. Composition of Lead Chromate Pigments, wt %[a]**

|  | Lead pigments | | | | |
|---|---|---|---|---|---|
| Constituent | Primrose | Lemon | Medium | Molybdate orange | Chrome orange |
| $PbCrO_4$ | 65–71 | 61–75 | 90–94 | 69–80 | 58.2 |
| $PbMoO_4$ |  |  |  | 9–15 |  |
| $PbSO_4$ | 23–30 | 20–38 | 0–6 | 3–7 |  |
| PbO |  |  |  |  | 39.2–39.6 |
| other | 3–8 | 1–6 | 4–6 | 3–13 | 2.2–2.6 |

[a]Ref. 44.

## Black Pigments

Black pigments can be divided into two basic groups. The first group is represented by carbon blacks. Many other inorganic black pigments, called noncarbon blacks, also are available. These belong chemically to the colored pigment category. Examples are spinel and rutile blacks, iron blacks, and some inclusion zircon pigments.

**Carbon Blacks.** Carbon black is one of the oldest pigments known. It was used by prehistoric peoples for painting pictures on cave walls, and commercially produced by the Chinese as early as 3000 BC.

The annual worldwide production of carbon blacks, which include a large variety of carbonaceous products, was estimated to be around six million metric tons in 1994. More than 90% of this pigment is consumed by the rubber industries, in particular, by the tire industry as a reinforcing agent. The rest (~500,000 t) is used for coloring plastics, printing inks, and paints. Particle size of carbon blacks varies from 5 to 500 $\mu$m and can be controlled by the process conditions and feedstock (see CARBON, CARBON BLACK).

Two processes, partial oxidation and thermal decomposition, are used for the production of carbon blacks. The partial oxidation process can be represented by the following reaction:

$$C_xH_y + y/4\ O_2 \longrightarrow x\ C + y/2\ H_2O$$

Large quantities of carbon blacks, called furnace blacks, are produced by this process in which a highly viscous, heavy aromatic feedstock from petroleum (qv) refining is atomized and dispersed into combustion gases of a secondary fuel that is burned in an excess of air. By controlling the air excess and by the temperature quenching of the product, the partial oxidation of feedstocks can be carried out. The produced carbon black particles are collected, micropulverized, and pelletized by dry or wet methods to provide a low dusting or nondusting product. Dry pelletization is used for the production of pigmentary-grade products.

Thermal decomposition of hydrocarbons is carried out in the absence of oxygen and at a high temperature required to break the carbon–hydrogen bonds.

$$C_xH_y \longrightarrow x\ C + y/2\ H_2$$

When natural gas is used as a feedstock to produce thermal blacks, the reaction is endothermic. In order to maintain the reaction, the reactor has to be kept at about 1300°C. When acetylene is used as the feedstock to produce acetylene blacks, the reaction is exothermic, and the reaction can be run at a temperature between 800 and 1000°C.

The production process or the feedstock is sometimes reflected in the name of the product such as lamp black, acetylene black, bone black, furnace black, or thermal black. The reason for the variety of processes used to produce carbon blacks is that there exists a unique link between the manufacturing process and the performance features of carbon black.

Environmentally, carbon blacks are relatively stable and unreactive. There is no evidence that these materials are toxic to humans or animals. Because they are fine and light, dusting problems might occur and therefore careful housekeeping is necessary. Most carbon blacks are pelletized, which not only reduces dusting problems, but because of the higher bulk density, pelletization also lowers the cost of shipping.

## Extenders and Opacifiers

Extender pigments are low cost, generally colorless or white pigments with a refractive index less than 1.7. Sometimes these pigments are also referred to as fillers (qv). Many extenders are derived from natural sources and display many diverse properties. They are added to various formulations to improve their technical and application properties and to reduce costs. Like pigments, extenders are dispersed in media in which they do not dissolve, but compared to pigments they do not have any significant coloristic properties. They are sometimes referred to as fillers, bulking agents, viscosity modifiers, or reinforcing agents.

In coating applications, extender pigments control gloss, viscosity, texture, suspension, and durability (see COATINGS). Extender pigments also enhance the opacity of white hiding pigments, eg, $TiO_2$. In plastic applications, extenders influence numerous properties of the resin including melt viscosity, thermal conductivity, and electrical properties, tensile strength, and moisture resistance.

With few exceptions, extender pigments can be classified as commodity chemicals. They are manufactured in large quantities with less sophisticated production methods than most pigments. Whereas the majority of pigments is prepared synthetically, extender pigments are mainly manufactured by the beneficiation of natural minerals. Based on chemical composition, extender pigments are carbonate, sulfate, silica-based, and silicate-based compounds.

Opacifiers are fine inorganic powders, usually white, that are used to reduce the transparency of ceramic glazes and porcelain enamels. The coating becomes opaque because the particles of the opacifier scatter and reflect the incident light. When inorganic pigments are combined with white opacifiers, pastel colors are obtained.

To achieve the maximum coating opacity the opacifier particle size should be between 0.2–0.3 $\mu$m. A good opacifier should not be soluble in the vitreous system, should have a refractive index substantially different from the refractive index of the system, should be inexpensive, easily milled to a submicrometer particle size, and thermally stable at the film's firing temperature.

Commercially, the most important opacifiers for glazes are $ZrO_2$, $ZrSiO_4$, and $SnO_2$. Because of low solubility ($<1\%$) in most molten glazes and relatively high refractive index, $SnO_2$ is the most effective opacifier of the group. It is usually added in 4–8 wt % to the glaze, but because it is very expensive, its commercial use is limited. Zirconia is less expensive and has a higher refractive index than tin oxide, but it can be used as an opacifier only in low temperature glazes; at higher temperatures it reacts with $SiO_2$ present in the glaze and forms zircon, $ZrSiO_4$. Because zircon is less expensive, it is added directly to the glazes fired at temperatures above 1000°C. In most glaze mixtures the zircon concentration is between 8–10%. Because of its solubility ~5% at 1200°C in the molten glazes, most glaze coatings contain two zircons, one that did not dissolve during the firing process, and the other that dissolved in the molten glaze, but recrystallized on cooling (45). In porcelain enamels and some glazes that are fired at temperatures considerably lower than 1000°C, titania in anatase form is usually the preferred opacifier.

## Miscellaneous Pigments

**Luminescent Pigments.** Luminescence is the ability of matter to emit light after it absorbs energy (see LUMINESCENT MATERIALS). Materials that have luminescent properties are known as phosphors, or luminescent pigments. If the light emission ceases shortly after the excitation source is removed ($<10^{-8}$ s), the process is fluorescence. The process with longer decay times is referred to as phosphorescence.

Semiconducting sulfides that can be represented by the formula $n\mathrm{ZnS}(1-n)\mathrm{CdS}{:}\mathrm{A}$, where A stands for an activator, and $n = 0.15{-}1$, are typical of fluorescence pigments. Phosphorescence pigments can be expressed by the general formula $n\mathrm{ZnS}(1-n)\mathrm{CdS}{:}\mathrm{Cu}$, where $n = 0.78{-}1.0$ and the amount of the $\mathrm{Cu}^+$ activator is only a few hundredths of a percent. Other phosphorescent sulfide pigments are $n\mathrm{CaS}(1-n)\mathrm{SrS}{:}\mathrm{Bi}$, Cu, where $n = 0{-}1$, and Bi = 0.04 wt % and Cu = 0.01 wt %.

Phosphorescent pigments are used in military applications, plastics, and paints. Zinc sulfide doped with $\mathrm{Ag}^+$ (blue) cations, or with $\mathrm{Cu}^+$ (green) cations are important pigments for the production of color television screens. Phosphorescent sulfide pigments are produced in the United States by Radium Corp. and by Conrad Precision Ind., Inc.

Zinc and calcium silicates, and calcium tungstate doped with activators such an Mn, Pb, Sn, and Eu are just a few examples of a large number of known ionically bonded luminescent pigments. Halide phosphates of the general formula $3\mathrm{Ca}(\mathrm{PO}_4)_2{\cdot}\mathrm{CaX}_2$, where X = F or Cl, and $\mathrm{Sb}^{3+}$ and $\mathrm{Mn}^{2+}$ are the activators, are extensively used in fluorescent lamps. Yttrium vanadate [13566-12-6] doped with europium, $\mathrm{YVO}_4{:}\mathrm{Eu}^{3+}$ is used as red phosphor in color television tubes and a single-crystal $\mathrm{Al}_2\mathrm{O}_3$ doped with about 0.05 wt % of $\mathrm{Cr}^{3+}$ is the basis of the ruby laser. The basis of the neodymium laser is yttrium aluminum garnet (YAG), $\mathrm{Y}_3\mathrm{Al}_5\mathrm{O}_{12}$, activated with $\mathrm{Nd}^{3+}$.

**Metal Effect Pigments.** Some metals, when prepared as small flakes, impart a special metallic appearance to the coatings and plastics in which they are dispersed. Metals most often used in these applications are aluminum (aluminum bronzes), copper and copper–zinc alloys (gold bronzes), and in smaller amounts zinc, tin, nickel, gold, silver, and stainless steel. Metallic pigments were first produced in the Middle Ages when gold leaf was made by hand beating. This was eventually replaced by stamping machines. Metal effect pigments are usually produced in ball mills using dry and wet milling.

**Nacreous Pigments.** Nacreous, ie, pearlescent pigments are used for creating special decorative effects typical of natural pearls. Nacreous pigments are fine, thin, plate-like transparent particles having a high refractive index. Because of these physical characteristics, when dispersed in a transparent film, they produce a silky appearance. When flakes have the right thickness, these pigments can also produce interference colors. Colors result from the interference of two light reflections, one from the upper and the other from the lower surface of a thin film.

Important requirements for achieving pearl essence effects include a plate-like particle shape of pigment coupled with its high refractive index. Manufacture of the most popular nacreous pigments involves coating mica (qv) with

50–300-nm films of $TiO_2$, $Fe_2O_3$, or $Cr_2O_3$. The mica, which alone does not have a high enough refractive index for creating nacreous luster, provides the required transparent platelet base. The oxide coating provides the necessary high refractive index. Because of the controlled thickness of the inorganic oxide films, coated mica can also behave as an interference pigment, and some interesting color shades can be achieved by a combination coating of two oxide films with $TiO_2$ being the one adjacent to mica. Mica flake dimensions are important. The preferred thickness is 0.3–0.6 $\mu$m, the preferred length is 5–110 $\mu$m.

A large number of mica pigment grades were optimized for particular applications, such as industrial grades for plastics, coatings, and printing inks; cosmetic grades for cosmetic use; and exterior grades modified to have good weather resistance, eg, rutile-coated mica, for outside applications. Mica-coated pigments are produced by The Mearl Corp. (U.S.), E. Merck and BASF (Germany), Kemiry OY (Finland), and Tayco (Japan).

**Transparent Pigments.** Pigments having chemical composition corresponding to colored or white opaque pigments can, under certain circumstances, appear transparent in a media. This happens when the particle size of these pigments becomes very small (2–15 nm), and if the particle refractive index is comparable to the refractive index of the media in which the particles are dispersed. Because of the very small particle size, the preparation of these pigments is much more complicated than the preparation of their nontransparent analogues. Their large surface area makes their dispersion difficult and they have a strong tendency to agglomerate.

Transparent iron oxide pigments have excellent weatherability, lightfastness, and chemical resistance, comparable to opaque iron oxides. Transparent yellow pigment is composed of needle particles of FeO(OH) having a thickness of 2–5 nm, a width of 10–20 nm, and a length of 50–100 nm. They are prepared by the precipitation process from a very diluted solution of ferrous salt, eg, 6 wt % ferrous sulfate, followed by the oxidation of the precipitate with atmospheric oxygen at a temperature of less than 25°C. The precipitate is left to mature for about one day, then filtered, dried, and milled.

Transparent red iron oxide is composed mainly of hematite, $\alpha$-$Fe_2O_3$, having primary particles about 10 nm. It is prepared by a precipitation reaction from a dilute solution of an iron salt at a temperature around 30°C, followed by a complete oxidation in the presence of some seeding additives, eg, $MgCl_2$, $CaCl_2$, and $AlCl_3$.

Transparent iron oxides are produced by BASF (Germany), Johnson Matthey (U.K.), and Hilton Davis (U.S.). The mid-1990s annual production is estimated to be around 2000 metric tons.

Two blue pigments can be prepared in transparent form: cyanide iron blue and cobalt aluminum blue. These pigments are used in achieving a blue shade of the metal effect pigments in metallic paints. Transparent cyanide iron blue is prepared by a precipitation reaction similar to the one used for the preparation of the opaque pigment, but considerably lower concentrations of solutions are used. It is produced by Degussa (Germany), Manox (U.K), and Dainichiseika (Japan).

Transparent cobalt aluminum blue is prepared by the precipitation of diluted solutions of cobalt and aluminum with alkalies. The formed precipitate of hydroxides is washed, filtered off, dried, and calcined at about 1000°C.

Titanium dioxide having a primary particle size of 15–30 nm also exhibits a transparency. In general, $TiO_2$ does not absorb light well in the visible region, and because of its small particle size it diffuses light very slightly and does not exhibit its white pigment characteristics. It is used in the production of metallic paints, transparent plastic films (46), and in cosmetics. Pigments having the anatase structure are prepared by burning $TiCl_4$ in the presence of water vapors (flame hydrolysis) at a temperature below 700°C. Rutile transparent pigments can be prepared by a sol–gel process or by precipitation in the presence of protective colloids, (see COLLOIDS; SOL–GEL TECHNOLOGY). Transparent $TiO_2$ is produced by Degussa (Germany), Teikoku Kako Co. (Japan), and Chem America, Inc. (U.S.).

## Environmental Aspects

Some inorganic pigments contain heavy metals. Thus production, use, and disposal are becoming more and more regulated. In the United States there are several federal regulations that control the use and disposal of heavy metals. Those relevant to the inorganic pigments industry are mentioned herein (47).

The Resource Conservation and Recovery Act (RCRA) controls the disposal of hazardous waste. SARA Title III governs the toxic inventory and emission reporting; the Clean Water Act (CWA) sets the limits for metals that can be present in water discharge; and the Clear Air Act (CAA) Amendments of 1990 control the abatements of all materials in the air.

The Occupational Safety and Health Administration (OSHA) regulates the exposure to chemicals in the workplace. From the point of view of the inorganic pigments industry, the limits established for lead and cadmium exposure are particularly important. A comprehensive lead standard adopted by OSHA in 1978 has been successful in reducing the potential for lead contamination in the workplace.

OSHA has enacted the permissible levels for all cadmium compounds, including fumes and dust, to be a maximum of 5 ppm per cubic meter of air over an eight-hour exposure period. As even tighter regulations are anticipated, it is questionable whether the U.S. cadmium pigment producers can manufacture these pigments at competitive prices. About 80% of cadmium-produced pigments are used to color plastics. Research directed toward finding suitable substitutes is ongoing.

The situation of lead chromate and lead molybdate pigments in the United States is similar. These pigments are primarily used for painting traffic signs and automobiles, and for producing industrial paints. As the result of various regulations, usage has been declining about 5–7% a year since the 1970s (48). This trend is likely to continue even though more environmentally acceptable products are being developed. In these newer products the lead bioavailability is lowered through pigment encapsulation. Table 3 lists those metals regulated by federal law that are or might be present in inorganic pigments.

**Packaging.** Products packaging, which constitutes about one-third of the municipal solid waste, is usually decorated with various colors (see PACKAGING). Some of these colors contain heavy metals. As landfill construction and placement became more complicated, the cost of garbage disposal escalated. For

**Table 3. Elements Potentially Present in Inorganic Pigments**

| Element | Federal regulation[a] | | | | |
|---|---|---|---|---|---|
| | RCRA | SARA | OSHA | CWA | CAA |
| Al | | | | + | |
| Ag | + | + | | | |
| As | + | + | | + | + |
| Ba | + | + | | + | |
| Be | | + | | + | |
| Cd | + | + | + | + | + |
| Co | | + | | + | + |
| Cr | + | + | | + | + |
| Cu | | + | | + | |
| Hg | + | + | | + | + |
| Mn | | + | | | + |
| Mo | | | | + | |
| Ni | | + | | + | |
| Pb | + | + | + | + | + |
| Sb | + | + | | + | |
| Se | + | + | | + | |
| Ti | | | | + | |
| Tl | | + | | | |
| Zn | | | | + | |

[a]A + indicates that the element is regulated by the particular act.

example, between 1984 and 1988 the fee for depositing trash in landfills in the northeast United States quadrupled. As a result the Coalition of Northeast Governors (CONEG) developed a model state legislation to regulate cadmium, hexavalent chromium, lead, and mercury in packaging to a gradually decreasing content, not leachability, of 100 ppm of all four metals combined in the four years after adoption. This legislation had a significant effect on the inorganic pigments industry that has relied heavily on the bright red and orange colors of cadmium- and lead-containing pigments. As of this writing (1995), 15 states have enacted this legislation with some limited variations.

Two voluntary incentives being adopted by many U.S. chemical companies are 33–50 reduction in toxic releases and Responsible Care. Using 1988 as a baseline, the 33–50 program seeks a 33% reduction in emission from the plant of the listed toxic chemicals by the end of 1992 and a 50% by 1995. The chemicals that are on the Environmental Protection Agency list and relevant to the inorganic pigments industry are cadmium, chromium, lead, nickel, and their compounds (49). As the U.S. government seeks the absolute reduction of toxic discharge, the production levels are not considered in this incentive, ie, with increased production levels companies have to use more efficient abatement systems to meet this regulation.

Responsible Care is the incentive sponsored by the Chemical Manufacturers' Association (CMA). Any CMA company must embrace the philosophy of continuous improvements of health, safety, and environmental efforts accompanied by an open communication to the public about products and their production. Thus the total impact of any product on the environment, from the extraction of raw materials, their beneficiation, transportation, production of final product, and disposal of the product at the end of its useful life, must be taken into consideration.

Most lead- and cadmium-containing inorganic pigments have very low solubilities in body fluids, therefore toxicity of these materials is extremely low. Additional encapsulation of these pigments into a silica or zircon matrix as practiced by some companies decreases solubility further, making these metals even more environmentally inert (50). Treated pigments are environmentally inert only from the narrow view of immediate application. There is often, however, a weak link in the whole life cycle of the product that can result in a release of a toxic element. Usually that weak link is either at the beginning or the end of the cycle. For example, fly ash particles created by the incineration of products colored with cadmium- or lead-containing inorganic pigments can contain soluble heavy metals that could potentially be leached into the ground water. Mining and beneficiation of heavy metals can also create environmentally hazardous by-products. Even though these potential hazards can be minimized by applying proper industrial practices, the trend to reduce the use of inorganic pigments containing lead and cadmium is expected to continue.

Pigments contribute to the enjoyment and beauty of the world. To assure the future of inorganic pigments, research efforts are directed toward the development of environmentally acceptable pigments, pigments that when produced under well-controlled conditions do not release any toxic materials into the environment whether during production, use, or disposal.

## BIBLIOGRAPHY

"Pigments, Inorganic" in *ECT* 1st ed., Vol. 10, pp. 612–660, by W. A. Gloger, National Lead Co.; in *ECT* 2nd ed., Vol. 15, pp. 495–555, by W. A. Gloger, National Lead Co.; in *ECT* 3rd ed., Vol. 17, pp. 788–838, by R. C. Schiek, Ciba-Geigy Corp.

1. W. P. Hsu and E. Matijevic, *Appl. Opt.* **24**, 1623 (1985).
2. M. Kerker, P. Scheiner, D. D. Cook, J. P. Kratohvil, *J. Colloid Interface Sci.* **71**, 176 (1979).
3. L. M. Sheppard, *Ceramic Bulletin*, **71**(5), 715 (1992).
4. C. Patton, ed., *Pigment Handbook*, Vol. III, John Wiley & Sons, Inc., New York, 1973, p. 289.
5. P. I. Ermilov, E. A. Indejkin, I. A. Tolmacev, *Pigmenty i Pigmentirovannyje lakokrasocnyje materialy*, Chimija, Leningrad, 1987, p. 199.
6. P. Kubelka and F. Munk, *J. Opt. Soc. Amer.* **38**(5), 448 (1948).
7. Ref. 4, p. 272.
8. A. C. Canonico, *Am. Ceram. Soc. Bull.* **64**(10), 1399 (1985).
9. H. G. Volz and co-workers, *Farbe + Lack*, **82**(9), 805 (1976).
10. *Ken-React Reference Manual*, Kenrich Petrochemicals, Inc., Bayonne, N.J., 1993.
11. R. K. IIer, *The Chemistry of Silica*, John Wiley & Sons, Inc., New York, 1979, p. 85.
12. *Eur. Paint Resin News*, **32**(1), 6 (Jan. 1994).
13. *Chem. Mark. Rep.* **245**(1), 7,20 (Jan. 1994).
14. *Chem. Week (Intl. Ed.)*, **153**(14), 15 (Oct. 1993).
15. *Chem. Eng. News*, 13 (Apr. 11, 1994).
16. *Chem. Eng. News*, (June 28, 1993).
17. *Ind. Miner. (London)*, (317), 49 (Feb. 14, 1994).

18. *China Chem. Rep.* 4 (Mar. 10, 1994).
19. J. Barksdale, *Titanium: Its Occurrence, Chemistry and Technology*, 1949.
20. P. A. Lewis, ed., *Pigment Handbook*,Vol. I, John Wiley & Sons, Inc., New York, 1987, p. 44.
21. Z. Solc, M. Sedlak, J. Mracek, *J. Thermal Anal.* **36**, 2103 (1990).
22. Ger. Pat. 463,773 (1925) J. Laux (to I. G. Farbenind).
23. Complex Inorganic Color Pigments Committee, *Dry Color Manufacturer's Asociation*, 3rd ed., DCMA, Alexandria, Va., 1991.
24. Z. Solc and M. Trojan, *Sklar a keramik* **39**, 16 (1989).
25. F. Hund, *Angewante Chemie* **74**, 23 (1962).
26. Ger. Pats. 1,417,246 (1971), 1,417,248 (1971), and 2,223,491 (1973), F. Hund (to Bayer AG).
27. U.S. Pats. 1,441,447, 2,623,833; Brit. Pat. 625,448, C. A. Seabright (to Harshaw Chemical Co.).
28. K. Fuji and K. Sono, *Repts. Gov. Res. Inst. Ceram.* (Kyoto) **6**, 18 (1952).
29. U.S. Pat. 3,166,430 (1965), C. A. Seabright (to Harshaw Chemical Co.).
30. F. T. Booth and G. N. Peel, *Trans. Brit. Ceram. Soc.* **61**(7), 297 (1974).
31. R. W. Batcherol, *Trans. J. Brit. Ceram. Soc.* **73**(8), 297 (1974).
32. R. A. Eppler, *Ceram. Bull.* **56**(2), 213 (1977).
33. S. Pajakoff, A. Vendl, and G. Banik, *Interceram.* (4), 488 (1980).
34. A. Broll, *Keramische Zeitschrift* **30**(6), 324 (1978).
35. M. Trojan, Z. Solc, *Proc. Zirconia '86 (Tokyo II)*, Uchida Rokahubo Publishing Co., Tokyo, 1987, p. 323.
36. U.S. Pat. 4,115,142 (Sept. 19, 1978), R. W. Hess (to E. I. du Pont de Nemours & Co., Inc.).
37. U.S. Pat. 4,115,141 (Sept. 19, 1978), D. H. Piltingsrud (to E. I. du Pont de Nemours & Co., Inc.).
38. U.S. Pat. 4,115,142 (Sept. 19, 1978), R. W. Hess (du Pont de Nemours & Co., Inc.).
39. U.S. Pat. 4,937,063 (June 26, 1990), R. M. Sullivan (to Ciba-Geigy AG).
40. U.S. Pat. 4,063,956 (Dec. 20, 1977), J. F. Higgins (du Pont de Nemours & Co., Inc.).
41. U.S. Pat. 4,316,746 (Feb. 23, 1982), M. Rustioni, C. D'Adda, and L. Balducci (to Montedison Spa).
42. Ger Pat. 1 (1877), J. Zeltner.
43. *Financial Times*, Mar. 25, 1994, p. 22; Mar. 29, 1994, p. 23.
44. A. Zamoyski, presented at *Intertech Environmental Pigments '92 Conference*, National Printing Research Institute.
45. F. T. Booth, G. N. Peel, *Trans. Br. Ceram. Soc.* **58**(9), 532 (1959).
46. EP 270,472 (June 8, 1988), (to BASF Corp., Inmont Div.).
47. M. Novotny, *Environmentally Friendly Pigments*, presented at Industrial Inorganic Chemicals Group, Royal Society of Chemistry, London, Jan. 12, 1994.
48. *Current Industrial Reports, Series M28A*, U.S. Dept. of Commerce, Bureau of Census, Washington, D.C.
49. *Chem. Eng. News*, 8 (July 26, 1993).
50. *Environ. Toxicol. Chem.* **10**, 1247–1253 (1991).

MIREK NOVOTNY
Cerdec Corporation


Z. SOLC
M. TROJAN
University of Pardubice

# ORGANIC

Pigments are colored, colorless, or fluorescent particulate organic or inorganic finely divided solids which are usually insoluble in, and essentially physically and chemically unaffected by, the vehicle or medium in which they are incorporated. They alter appearance either by selective absorption and/or scattering of light. They are usually incorporated by dispersion in a variety of systems and retain their crystal or particulate nature throughout the pigmentation process (1). The large number of systems vary widely from paints to plastics to inks and fibers.

Dyes, on the other hand, are colored substances which are soluble or go into solution during the application process and impart color by selective absorption of light. In contrast to dyes, whose coloristic properties are almost exclusively defined by their chemical structure, the properties of pigments also depend on the physical characteristics of its particles.

In some cases, a single chemical substance can serve both as a dye and as a pigment. For example, indanthrone [81-77-6] (Pigment Blue 60) functions as a blue pigment or as a dye. As a pigment, indanthrone is a particulate, insoluble solid dispersed directly into a vehicle, whereas as a dye it is reduced to a base-soluble hydroquinone derivative and then reoxidized onto a solid substrate.

The description of colored organic pigments excludes consideration of inorganic pigments, as well as black pigments which consist of specially treated forms of carbon and white pigments which are entirely of inorganic origin (see PIGMENTS, INORGANIC). Some commercially significant organic pigments are listed in Table 1.

Pigments are categorized according to their generic name and chemical constitution in the *Color Index* (CI), published by the Society of Dyers and Colourists, and the American Association of Textile Chemists and Colorists. For example, copper phthalocyanine is designated CI Pigment Blue 15, CI number 74160 (2) and is further identified by the Chemical Abstract (CAS) Registry Number [147-14-8]. In practical terms, manufacturers have their own name for a particular pigment, with variations depending on the applications for which it is designed. In this article pigment structures are identified by their abbreviated pigment designations, eg, PB 60, for the convenience of the reader in associating the color hue with the structure. PB, PG, PO, PR, PV, and PY, designate blue, green, orange, red, violet, and yellow pigments, respectively.

Pigments are available in a number of commercial forms including dry powders (either surface treated or untreated), presscakes, flushed colors (thick pastes), fluidized dispersions (pourable pastes), resin predispersed pigments (powders), and plastic color concentrates or master batches (granules) (see PIGMENT DISPERSIONS).

Significant pigment attributes are tinctorial strength, durability (photochemical stability), hiding power, transparency, and heat and solvent resistance. Other properties include brightness (saturation), gloss, rheology, crystal stability, bleed resistance, flocculation resistance, and other properties associated with specialized applications.

In 1856 William H. Perkins at the age of eighteen synthesized for the first time the mauve color mauveine [6373-22-4] by oxidizing aniline containing toluidine with chromic acid. This event ushered in an era of colored synthetic

**Table 1. Selected Commercial Synthetic Organic Pigments**

| Color Index (CI) name | CAS Registry Number | CI constitution number | Pigment class (common name) | Method of preparation |
|---|---|---|---|---|
| Pigment Blue 15 | [147-14-8] | 74160 | copper phthalocyanine | condensation of phthalic anhydride with urea, in presence of copper ions, with or without added chlorophthalic anhydride; subsequent conversion to alpha-phase and stabilization, if necessary |
| Pigment Blue 60 | [81-77-6] | 69800 | indanthrone | intermolecular condensation of 2-amino-anthraquinone in presence of a strong inorganic base and oxidizing agent |
| Pigment Blue 19 | [58569-23-6] | 42750 | triarylcarbonium sulfonate | sulfonation of bis(p-phenylaminophenyl)-3-methyl-4-aminophenylcarbonium sulfate |
| Pigment Blue 1 | [1325-87-7] | 42595:2 | triarylcarbonium PTMA salt (Victoria Blue B) | condensation of 4,4'-bis-N,N-dimethyl aminobenzophenone with N-ethyl-1-naphthylamine, followed by oxidation and salt formation |
| Pigment Blue 24 | [6548-12-5] | 42090:1 | triarylcarbonium Ba salt (Peacock Blue) | condensation of benzaldehyde-o-sulfonic acid with N-ethyl-N-benzylaniline, followed by sulfonation, oxidation, and salt formation |
| Pigment Green 4 | [61725-50-6] | 42000:2 | triarylcarbonium chloride[a] (Malachite Green) | condensation of benzaldehyde with N,N-dimethylaniline, followed by oxidation and salt formation |
| Pigment Green 7 | [1328-53-6] | 74260 | polychloro copper phthalocyanine | chlorination of copper phthalocyanine |
| Pigment Green 36 | [14302-13-7] | 74265 | polybromochloro copper phthalocyanine | bromination/chlorination of copper phthalocyanine |
| Pigment Orange 5 | [3468-63-1] | 12075 | monoazo (Dinitraniline Orange) | coupling of diazotized 2,4-dinitroaniline with 2-naphthol |
| Pigment Orange 13 | [3520-72-7] | 21110 | disazo pyrazolone (Pyrazolone Orange) | coupling of tetrazotized 3,3'-dichloro-benzidine with 1-phenyl-3-methyl-pyrazolone |

42

| Name | Colour Index number | CAS number | Chemical class (common name) | Preparation |
|---|---|---|---|---|
| Pigment Orange 36 | 11780 | [12236-62-3] | monoazo benzimid-azolone | coupling of diazotized 2-nitro-4-chloro-aniline with acetoacetyl-5-amino-benzimidazolone |
| Pigment Orange 43 | 71105 | [4424-06-0] | Perinone | condensation of naphthalene-1,4,5,8-tetracarboxylic acid dianhydride with o-phenylenediamine, and separation of trans isomer |
| Pigment Red 2 | 12310 | [6041-94-7] | Naphthol AS (Naphthol Red) | coupling diazotized 2,5-dichloroaniline with 2-hydroxy-3-naphthanilide |
| Pigment Red 3 | 12120 | [2425-85-6] | Beta-Naphthol (Toluidine Red) | coupling of diazotized 2-nitro-4-methyl-aniline with 2-naphthol |
| Pigment Red 5 | 12490 | [6410-41-9] | Naphthol AS (Carmine) | coupling of diazotized N,N-diethyl-4-methoxymetanilamide with 3-hydroxy-2-naphth-2',4'-dimethoxy-5'-chloro-anilide |
| Pigment Red 17 | 12390 | [6655-84-1] | Naphthol AS | coupling of diazotized 2-methyl-5-nitro-aniline with 3-hydroxy-2-naphth-2'-methylanilide |
| Pigment Red 23 | 12355 | [6471-49-4] | Naphthol AS | coupling of diazotized 2-methoxy-5-nitro-aniline with 3-hydroxy-2'-naphth-3'-nitroanilide |
| Pigment Red 38 | 21210 | [6358-87-8] | disazo pyrazolone (Pyrazolone Red) | coupling of tetrazotized 3,3'-dichloro-benzidine with 1-phenyl-3-carbethoxy-5-pyrazolone |
| Pigment Red 48, barium, calcium, strontium, and manganese salts | 15865:1 15865:2 15865:3 15865:4 | [7585-41-3] [7023-61-2] [15782-05-5] [5280-66-0] | BONA (Ba salt) BONA (Ca salt) BONA (Sr salt) BONA (Mn salt) (Permanent Red 2Bs) | coupling of diazotized 2-amino-4-chloro-5-methylbenzenesulfonic acid with 3-hydroxy-2'-naphthoic acid, followed by salt formation |
| Pigment Red 49, barium and calcium salts | 15630:1 15630:2 | [1103-38-4] [1103-39-5] | Beta Naphthol (Ba salt) Beta Naphthol (Ca salt) | coupling of diazotized 2-amino naphth-alene-1-sulfonic acid with 2-naphthol, followed by salt formation |

**Table 1.** (*Continued*)

| Color Index (CI) name | CAS Registry Number | CI constitution number | Pigment class (common name) | Method of preparation |
|---|---|---|---|---|
| Pigment Red 52, calcium salt | [17852-99-2] | 15860 | BONA (Ca salt) | coupling of diazotized 2-amino-4-methyl-5-chlorobenzenesulfonic acid with 3-hydroxy-2-naphthoic acid, followed by salt formation |
| Pigment Red 53, barium salt | [5160-02-1] | 15585:1 | BONA (Ba salt) (Lake Red C) | coupling of diazotized 2-amino-4-methyl-5-chlorobenzenesulfonic acid with 2-naphthol, followed by salt formation |
| Pigment Red 57, calcium salt | [5281-04-9] | 15850:1 | BONA (Ca salt) (Lithol Rubine B or 4B toner) | coupling of diazotized 2-amino-5-methyl-benzenesulfonic acid with 3-hydroxy-2-naphthoic acid, followed by salt formation |
| Pigment Red 81 | [12224-98-5] | 45160:1 | triarylcarbonium PTMA salt | salt formation between Rhodamine 6G with phosphotungstomolybdic acid (PTMA) |
| Pigment Red 112 | [6535-46-2] | 12370 | Naphthol AS | coupling of diazotized 2,4,5-trichloro-aniline with 3-hydroxy-2-naphth-2'-methylanilide |
| Pigment Red 122 | [980-26-7] | 73915 | Quinacridone | condensation of diakyl succinoylsuccinate with *p*-toluidine, followed by oxidation and hydrolysis, and acid-catalyzed cyclization of 2,5-di(*p*-toluidino)-terephthalic acid |
| Pigment Red 144 | [5280-78-4] | 20735 | disazo condensation | coupling of diazotized 2,5-dichloroaniline with 3-hydroxy-2-napthoic acid, followed by acid chloride formation and reaction with 2-chloro-*p*-phenylene-diamine |
| Pigment Red 170 | [2786-76-7] | 12475 | Naphthol AS | coupling of diazotized *p*-aminobenzamide with 3-hydroxy-2-naphth-2'-ethoxy-anilide |

44

| Pigment | CAS Number | C.I. Number | Class | Synthesis |
|---|---|---|---|---|
| Pigment Red 177 | [4051-63-2] | 65300 | anthraquinone | bimolecular debromination of 1-amino-4-bromoanthraquinone-2-sulfonic acid, followed by desulfonation |
| Pigment Red 179 | [5521-31-3] | 71130 | Perylene | imidation of perylene 1,6,7,12-tetra-carboxylic acid dianhydride with methylamine |
| Pigment Red 202 | [68859-50-7] | 73907 | Quinacridone | condensation of dialkyl succinoyl-succinate with p-chloroaniline, cyclization of resulting diester to 2,9-dichloro-6,13-dihydroquinacri-done, followed by oxidation |
| Pigment Red 254 | [122390-98-1] | 56110 | diketopyrrolopyrrole (DPP) | condensation of diisopropyl succinate with p-chlorobenzonitrile |
| Pigment Violet 1 | [1326-03-0] | 45170:2 | triarylcarbonium PTMA salt | salt formation between Rhodamine B and phosphotungstomolybdic acid |
| Pigment Violet 3 | [1325-82-2] | 42535:2 | triarylcarbonium PTMA salt | salt formation between Methyl Violet and phosphotungstomolyb... acid |
| Pigment Violet 19 | [1047-16-1] | 46500 | Quinacridone | condensation of dialkyl succinoylsuccinate with aniline, cyclization of resulting diester to 6,13-dihydroquinacridone, followed by oxidation |
| Pigment Violet 23 | [6358-30-1] | 51319 | Dioxazine (Carbazole Violet) | condensation of 3-amino-N-ethyl-carbazole with chloranil followed by acid-catalyzed cyclization of the diarylaminodichlorobenzoquinone |
| Pigment Yellow 3 | [6486-23-2] | 11710 | monoazo (Hansa Yellow 10G) | coupling of diazotized 2-nitro-4-chloro-aniline with acetoacet-2-chloroanilide |
| Pigment Yellow 12 | [6358-85-6] | 21090 | diarylide | coupling of tetrazotized 3,3'-dichloro-benzidine with acetoacetanilide |
| Pigment Yellow 13 | [5102-83-0] | 21100 | diarylide | coupling of tetrazotized 3,3'-dichloro-benzidine with acetoacet-2,4-dimethyl-anilide |
| Pigment Yellow 74 | [6358-31-2] | 11741 | monoazo | coupling of diazotized 2-methoxy-4-nitro-aniline with acetoacet-2-methoxyanilide |
| Pigment Yellow 83 | [5567-15-7] | 21108 | diarylide | coupling of diazotized 3,3'-dichlorobenzi-dine with acetoacet-2,5-dimethoxy-4-chloroanilide |

**Table 1. (Continued)**

| Color Index (CI) name | CAS Registry Number | CI constitution number | Pigment class (common name) | Method of preparation |
|---|---|---|---|---|
| Pigment Yellow 93 | [5580-57-4] | 20710 | disazo condensation | coupling of diazotized 3-amino-4-chloro-benzoic acid with 1,4-bis(acetoacetyl-amino)-2-chloro-5-methylbenzene, diacid chloride formation, and amidation with 2-methyl-3-chloroaniline |
| Pigment Yellow 95 | [5280-80-8] | 20034 | disazo condensation | coupling of diazotized 3-amino-4-chloro-benzoic acid with 1,4-bis(acetoacetyl-amino)-2,5-dimethylbenzene, diacid chloride formation, and amidation with 2-methyl-5-chloroaniline |
| Pigment Yellow 110 | [5590-18-1] | 56280 | isoindolinone | reaction of 3,3'-dimethoxy-4,5,6,7-tetra-chloro isoindolenine with p-phenylene-diamine |
| Pigment Yellow 138 | [56731-19-2] | 56300 | quinophthalone | condensation of 2-methyl-8-amino-quinoline with tetrachlorophthalic anhydride |
| Pigment Yellow 139 | [36888-99-0] | 56298 | isoindoline | reaction of 1-amino-3-iminoisoindolenine with barbituric acid |
| Pigment Yellow 154 | [63661-02-9] | 11781 | monoazo benz-imidazolone | coupling of diazotized 2-trifluoro-methylaniline with acetoacetyl-5-aminobenzimidazolone |
| Pigment Yellow 168 | [71832-85-4] | 13960 | monoazo (Ca salt) | coupling diazotized 3-nitro-4-amino-benzenesulfonic acid with acetoacet-2-chloroanilide, followed by salt formation |
| Pigment Yellow 191 | [129423-54-7] | 18795 | monoazo pyrazolone (Ca salt) | coupling of diazotized 2-amino-4-chloro-5-methylbenzenesulfonic acid with 3-methyl-N-(3'-sulfophenyl)-pyrazolone, followed by salt formation |

[a]Or other selected anion.

46

chemistry which continues into the 1990s. Many new organic structures have been discovered and introduced as commercial organic pigments.

The initial synthetic developments were concerned primarily with dyestuffs for the textile industry, and the period up to 1900 was characterized by the discovery and development of many dyes derived from coal-tar intermediates. Rapid advances in color chemistry were initiated after the discovery of diazo compounds and azo derivatives (shown to be largely hydrazone derivatives). The wide color potential of this class of pigments and their relative ease of preparation led to the development of azo colors, which represent the largest fraction of manufactured organic pigments. Commercial development began with the discovery of Lithol Red in 1899 (3). After World War I among the most important pigments produced in the United States were Toluidine Red [2425-85-6], Lake Red C [5160-02-1], and Hansa Yellow. Condensation azo pigments were discovered in 1954. In 1991, on a value basis, all azo pigments represented about 53% of the approximately $4 billion worldwide organic pigment market (4) (see AZO DYES).

The development in 1913 of the brilliant lakes of complex heteropoly acids of phosphorus, molybdenum, and tungsten with basic dyes like Rhodamine B [81-88-9], Victoria Blue [56646-84-5], or Methyl Violet [8004-87-3], led to development of superior fastness pigments relative to the tannin tartar emetic precipitations of the same dyes. Still these pigments were not sufficiently durable for outdoor use. The largest single advance in pigment technology after World War I was the discovery of the relatively complex structure but easily synthesized copper phthalocyanines which were characterized by excellent brightness, strength, bleed resistance, and lightfastness. Copper phthalocyanines share of the world market stands at 27% (4) (see PHTHALOCYANINE COMPOUNDS).

After World War II the most important discovery was the family of red–violet quinacridone pigments, followed by the mostly yellow–orange benzimidazolone and isoindolinone pigments, and the most recent development of the red diaryl pyrrolopyrroles. On a value basis, polycyclic pigments, excluding copper phthalocyanines, represent about 17% of the world organic pigment market (4).

Finding a totally new pigment class or chromophore is a rare event. The latest commercialization of a new family of pigments is represented by the pyrrolopyrroles discovered in 1983. The discovery years of important pigments are compiled in the following.

| Pigment name | Year | Pigment name | Year |
|---|---|---|---|
| Perkin's Mauve | 1856 | Copper Phthalocyanine Blue | 1935 |
| Lithol Red | 1899 | Copper Phthalocyanine | 1938 |
| Indanthrone Blue | 1901 | Green | |
| Toluidine Red | 1905 | Perylenes | 1950 |
| Hansa Yellow | 1909 | Carbazole Violet | 1952 |
| Diarylide Yellow | 1911 | condensation azo pigments | 1954 |
| heteropolyacid complexes with | 1913 | Quinacridones | 1955 |
| basic dyes | | Benzimidazolones | 1960 |
| Red 2B | 1931 | diaryl pyrrolopyrroles | 1983 |

In the discovery of colored pigments the basic synthesis of the pigment is only the beginning of an important development effort. A significant challenge comes in reducing and/or controlling particle size, particle shape, particle size distribution, and conditioning of pigments to achieve desired dispersibility, flocculation resistance, rheology, and other working properties. A large and creative effort goes into the modification of pigments to improve their chemical, photochemical, and physical properties.

## Color and Constitution

As early as 1868 Graebe and Liebermann (5) recognized that color in organic compounds is associated with the presence of multiple bonds. A few years later Witt (6) coined the term chromophore, from the Greek *chroma* meaning color and *phoros* meaning bearer, for groups which give rise to color. The term is used to designate $\pi$-electron-containing moieties (conjugated double bonds) which contribute to the selective absorption of visible light. Generally, organic compounds absorb light in the ultraviolet (210–400 nm) and visible (400–750 nm) region of the spectrum at characteristic wavelengths. The intensities of these absorptions vary due to the excitation of the more loosely held electrons in the molecule. All unsaturated groups have remarkably similar $\pi - \pi^*$ transitions regardless of the atoms contained in the common chromophores.

| ethenyl | nitro | carbonyl | azo | hydrazone |

Isolated, unconjugated chromophoric groups absorb at about 200 nm, at the end of the ultraviolet spectrum. Ethylene absorbs at 162.5 nm. Conjugation shifts absorption to longer wavelengths. Butadiene absorbs at 217 nm, and benzene shows its primary absorption at 254 nm. The $\pi$-electrons in conjugated hydrocarbons tend to be more readily separated from their nuclei, leading to excited states which are more polar than the ground states.

Incorporation of substituents into aromatic systems which are electron donating, eg, hydroxy, methoxy, amino, alkylamino, dialkylamino, etc, tend to shift light absorption to longer wavelengths (bathochromic shifts). Such groups are known as auxochromes, from the Greek *auxo* meaning increase. They are auxiliaries to the chromophores and enhance and modify absorption and color by virtue of their nonbonding electrons. Groups that withdraw electrons have the opposite effect, making electron excitation less facile, and consequently shift absorption to shorter wavelengths (hypsochromic shifts).

The presence of $\pi - \pi$ or $n - \pi$ conjugated systems does not assure absorption of visible light or generation of color. However, all colored organic compounds, including pigments, possess extended conjugated resonance systems. Thus, whereas 1,4-diphenylbutadiene is colorless, 1,6-diphenylhexatriene is colored.

colorless                                           colored

Similarly, 6,13-dihydroquinacridone is virtually colorless, whereas quinacridone is a colored pigment. Generally, the longer the conjugated system the longer the wavelength of the absorbed visible light.

colorless                                       colored

All absorbed light is complementary to reflected light which produces observed color. All colors are a function of the wavelength of absorbed light as shown in Table 2 (see COLOR). Thus, if a pigment absorbs only blue light it imparts an orange color, whereas when it absorbs orange light the observed color is blue.

**Table 2. Colors of Absorbed Light and the Corresponding Complementary Colors as a Function of Wavelength**

| Wavelength, nm | Color of absorbed light | Complementary color |
|---|---|---|
| 400–420 | violet | yellow-green |
| 420–450 | indigo blue | yellow |
| 450–490 | blue | orange |
| 490–510 | blue-green | red |
| 510–530 | green | purple |
| 530–545 | yellow-green | violet |
| 545–580 | yellow | indigo blue |
| 580–630 | orange | blue |
| 630–720 | red | blue-green |

## Properties of Pigments

The physical and chemical characteristics that control and define the performance of a commercial pigment in a vehicle system include its chemical composition, chemical and physical stability, solubility, particle size, shape and particle size distribution, degree of dispersion, crystal morphology including polymorphic forms, refractive index, specific gravity, electronic spectra with particular emphasis on extinction coefficients in the visible spectrum, surface area, and the presence of impurities, extenders, and surface modifying agents. Invariably a pigment is used in a vehicle system, therefore its ultimate performance in use derives from both physical and chemical pigment–vehicle interaction. Performance of most pigments is system dependent.

Unlike inorganic pigments, organic pigments are relatively strong and bright (saturated), but their fastness properties, though adequate for the purposes for which they are used, vary widely from poor to outstanding.

**Strength.** The inherent strength of a pigment depends on its light-absorbing characteristics which are related to its molecular and crystalline structure. In addition, strength is a function of particle size or surface area. The ability of a pigment to absorb light increases with decreasing particle size or increasing surface area, until the particles become entirely translucent or transparent to incident light. Particle size reduction beyond this point does not increase tinctorial strength of a pigmented system. Specific surface areas of organic pigments are typically in the range of $10–100$ $m^2/g$. Whereas most azo pigments are prepared in pigmentary form by aqueous precipitation processes, most other pigments are synthesized in nonpigmentary crude form of relatively large crystal size up to 100 $\mu$m and subsequently reduced to pigmentary size of $0.01–0.5$ $\mu$m either by attrition or precipitation processes. Being finely divided the particles have a great tendency to aggregate and agglomerate into crystal assemblies. To obtain the inherent strength of a pigment the aggregates must be completely broken down to individual crystals by application of work and their reagglomeration or flocculation prevented. Total breakdown to single crystals in practical systems seldom happens.

Pigment strength in a vehicle also depends on the typical character of other components in a pigmented system insofar as they absorb or scatter light. Strength comparisons are usually made with a series of samples featuring varying amounts of a pigment incorporated in a vehicle with a corresponding series in the same vehicle containing a reference pigment. Instrumental comparisons are commonly practiced.

A good absolute theoretical comparison of strength is represented by the area under the absorption bands in the visible spectrum, or less accurately by the molecular extinction coefficients at the maximum wavelength of absorption. On that basis diarylide pigments are about three times stronger than most ordinary monoazo pigments, and copper phthalocyanines are also very strong pigments. With most pigments color strength does not increase continuously until the pigment is molecularly dispersed (7), that is, it is in solution. Color strength passes through a maximum within the particle size range of $0.1–0.01$ $\mu$m. This sets a limit to the potential color strength of a pigment dispersion.

**Brightness or Saturation.** The saturation of a colored pigment is a measure of its brightness or cleanliness as opposed to dullness of hue. Generally, if a pigment absorbs light over a wide range of wavelengths, ie, shows broad absorption bands, or contains more than one chromophore, the pigment is likely to be duller than a pigment with sharp absorption bands due to a single chromophore. Because pigments are frequently used in combinations or blends, the brightness is determined by the selective absorption of the individual pigments and this significantly affects the brightness of the reflected color. A saturated pigment provides flexibility in color blending and its use provides important economic advantages since desaturation can be accomplished by blending with duller and less expensive pigments, like carbon black [1333-86-4] or iron oxide [1309-37-1].

**Fastness.** The fastness of a colored pigment defines its inherent ability to withstand the chemical and physical influences to which it is exposed during and subsequent to its incorporation into a pigmented system. Fastness describes the characteristics of a pigment in terms of its color stability in a pigmented system upon exposure to light, weather, heat, solvents, or various chemical agents. Ideally, a pigment should be insoluble and chemically and photochemically inert. Only a few organic pigments approach such perfection. Fastness properties which are of practical significance are those observed for pigments incorporated into a coating or resin sysem under the exact conditions of incorporation and use.

The development of new resins, plastics, fibers, elastomers, etc, which are processed at progressively higher operating and curing temperatures has created a need for pigments that stand up for relatively long periods of time to a hostile environment. They must remain essentially unaltered when incorporated into plastics such as polypropylene, ABS, or nylon at relatively high temperatures. In reality, in high temperature plastics most organic pigments partially dissolve and undergo particle ripening or growth thus changing color without chemical destruction. Some pigments can change to thermodynamically more stable polymorphic forms with consequent color change, and others simply decompose.

Bleeding properties are also a function of solubility of a pigment in a vehicle, plasticizer, or solvent. Blooming is another manifestation of solubility whereby dissolved pigment migrates from within a pigmented medium to its surface where it is redeposited as pigment crystals and can be readily rubbed off. This is quite prevalent in the coatings (qv) and printing industry.

Once a pigment is incorporated into a system, it is expected to be durable and withstand the combined chemical and physical stresses of weather, solar radiation, heat, water, and industrial pollutants. Because a pigment is totally enveloped by the medium which is itself not inert, various pigments perform differently in different systems. Thus, a pigment may be lightfast or weatherfast in one system and fail in another. In addition, lightfastness or weatherfastness of an organic pigment is dependent on its particle size, since absorbed radiation by pigment particles largely determines the response of a system to light. Relatively little is known about the chemistry responsible for the destruction of organic pigments by solar radiation. However, since destructive light penetrates the pigment surface, the larger the surface area the faster the deterioration. Within a particle, light destroys one layer of molecules after another. Interior pigment molecules are preserved until outer layers are destroyed, therefore relatively large or thick particles resist light longer and are more durable. Thus, some pigments with high surface areas are usable only for indoor applications, and the same pigments with reduced surface areas (larger particle sizes), have found outdoor applications.

The quantitative measurement of pigment or pigmented system deterioration upon exposure to heat or light used to be expressed by visual numerical standards. In modern times color differences are expressed in the CIELAB system which has become the leading method for color characterization (8).

**Dispersibility.** The dispersibility of a pigment is measured by the effort required to develop the full tinctorial potential of a pigment in a vehicle system.

Dispersibility differs from system to system depending on pigment–medium interaction and compatibility.

Small particle size pigments, especially the very small crystals, seldom exist as individual entities, but as strongly coherent aggregates or less firmly bound agglomerates. These are best examined by electron microscopy. Aggregation and agglomeration take place during organic pigment manufacture especially when individual crystals are being produced during attrition or precipitation processes. Aggregate formation is particularly pronounced during drying when water removal from capillaries between crystals causes crystals to be pulled into close contact by capillary attraction (9). These surface forces are a function of particle size and particle size distribution. The larger the particles, the lower the surface forces and the easier it is to disperse the aggregates (10). Small particles cause bridging between large particles and occupy gaps and pores making them less accessible to the vehicle with consequent poor wetting and poorer dispersibility. Similarly, pigments with wide particle size distribution have a greater tendency to reagglomerate or flocculate. Although flocculation is caused by formation of loose units that have been wetted, they do adversely affect the coloristic properties of a pigment–vehicle system by decreasing its tinctorial strength.

A wide variety of additives are being used to reduce aggregation or agglomeration and flocculation to improve dispersibility and color strength of organic pigments. These include resins, especially those related to abietic acid, aliphatic amines, amides, substituted derivatives of pigments themselves, and various combinations thereof. The additives are most effective if they are present when the pigment crystals are being generated.

**Hiding Power and Transparency.**   Hiding power of a pigment is a function of its strength, that is, its absorption coefficients, its particle size, or light-scattering coefficient, and relative refractive indexes of pigment and vehicle. Light scattering has a powerful influence on opacity and, as is seen from Figure 1, goes through a maximum as a function of particle size. The maximum occurs at a particle size which is approximately half the wavelength of absorbed visible light.

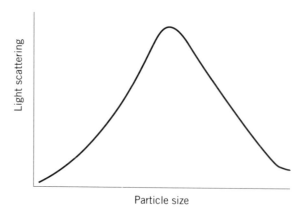

**Fig. 1.**   The effect of particle size on light scattering.

Similarly, a pigment which absorbs much light increases hiding even when light scattering is insufficient, and the higher the refractive index the greater the hiding power of a pigment. In practical systems hiding is determined by the ability of a pigment dispersion to completely hide or cover up a black and white checkered board.

Conversely, to increase transparency light scattering must be minimized by particle size reduction. The smaller the particle size and the better the dispersion the greater the transparency. Upon continued particle size reduction of a pigment, tinctorial strength comes to a maximum but transparency continues to increase. Transparent coatings are important in multicolor printing as well as metallized finishes, particularly in the automotive industry. Transparency is determined by the contrast ratio obtained from reflectance measurements of a pigmented system applied on a black and white background.

**Other Working Properties.** Other properties which facilitate pigment incorporation and use include compatibility with a system, oil absorption, rheological characteristics, gloss, distinctness of image, wettability, migration fastness, plate-out, polymer distortion, etc. Most of these properties are controlled during pigment manufacture or formulation, others require special treatments to overcome undesirable effects. Pigments vary widely in most properties but the requirements for various systems also vary. For example, pigments used in red inks for printing comic books differ vastly from those used in red automotive top-coatings.

## Types of Pigments

All organic pigments have to be synthesized and nearly all have to be conditioned or finished. The physical conditioning is as important as its chemical constitution. This is particularly well illustrated by comparing the appearance of crude copper phthalocyanine with pigmentary material. The difference in tinctorial strength is profound, brought about by the conditioning process which most importantly affects particle size and shape and other characteristics of the particle surface. Pigment conditioning has become an important separate process step in organic pigment manufacture.

### AZO PIGMENTS

Azo pigments provide good examples of materials in which conditioning is an integral part of the synthesis process. The coupling process for azo components is simple. An aromatic amine is diazotized by treatment with nitrous acid under conditions which vary from dilute mineral acid to concentrated sulfuric acid, depending on the basicity of the amine (see AZO DYES). The resulting ice-cold diazo solutions are then mixed with finely divided coupling components which comprise a wide variety, eg, 2-naphthol, 3-hydroxy-2-naphthoic acid and its arylamides, arylamides of acetoacetic acid, pyrazolone derivatives, etc. Cooling is accomplished by adding crushed ice, and heating by introducing live steam. The simplest method, ie, direct coupling, involves running the diazo solution directly into the solution or suspension of the coupling component. In inverse coupling, the coupling component is run into the diazo solution. The most elegant technique, simultaneous coupling, entails running both the diazo and coupling component simultaneously into water or a dilute buffer. This method allows careful control of the pH, as well as other important variables of temperature, rate of coupling, use of auxiliary agents, etc. The latter are of particular importance since they frequently affect rate of nucleation and control particle size, dispersibility, and other pigmentary properties of the resulting products which range in color essentially over the entire visible spectrum.

The monoazo and disazo pigments contain one or more chromophoric groups usually referred to as the azo $-N=N-$ group. However, it has been shown by x-ray diffraction analysis and nuclear magnetic resonance (nmr) techniques that azo pigments, eg, the following, where R = H, Cl, or $CH_3$, exist in the hydrazone rather than the azo tautomeric form (11). The hydrazone form, which has three intramolecular hydrogen bonds, renders the molecule planar (with the exception of the aniline moiety) which is a stabilizing influence. Similar studies in the Naphthol AS series have shown that the hydrazone rather than the azo form is the prevalent tautomer (12).

azo form                          hydrazone form

Azo pigments, one of the oldest and most diverse group of pigments, comprise two types. One type consists of pigments that are insoluble in the aqueous reaction medium in which they are synthesized, eg, Hansa and diarylide yellows and oranges, or arylide reds and maroons. Most of the pigments show poor bleed characteristics, but relatively good acid and alkali resistance. They show acceptable lightfastness in deep shades but poor tint lightfastness. The second type are laked or precipitated azo pigments derived from components substituted with sulfonic and/or carboxylic acid groups. The pigments are rendered insoluble by precipitation as calcium, barium, strontium, or manganese salts. Among the commercially important laked pigments are Lithol Red and Lithol Rubine [5281-04-9], Lake Red C [5160-02-1], Red 2B, and more recently introduced yellows based on pyrazolone sulfonic acid derivatives. These pigments are characterized by good to excellent bleed resistance, poor acid and alkali resistance, fair to good lightfastness in deep shades, and poor tint lightfastness. Also available are special azo pigments which show very good overall properties and therefore find applications in fairly demanding systems.

**Monoazo Pigments.** Monoazo yellow pigments are represented by the following general formula:

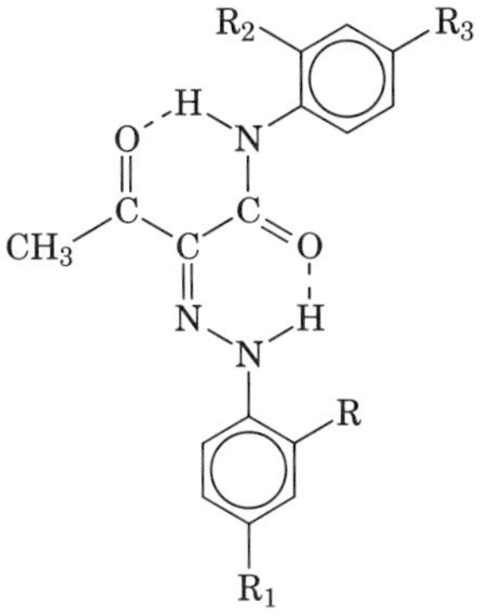

$R_2$ — $R_3$

$CH_3$ — $C$ — $C$ — $C$ = $O$

$R_1$

Many of the pigments carry a nitro group in the diazonium component, usually in the ortho position ($R = NO_2$). Among the acetoacetarylide components the $o$-methoxy derivative ($R_2 = OCH_3$, $R_3 = H$) is one of the most important in the production of azo pigments. The colors of these pigments range from red to green-shade yellows. Commercially important products are shown in Table 3.

**Table 3. Selected Monoazo Pigments**

| Pigment Yellow | CAS Registry Number | Substituents on diazonium component | Substituents on coupling component |
|---|---|---|---|
| 1 | [2512-29-0] | $R = NO_2$, $R_1 = CH_3$ | $R_2 = R_3 = H$ |
| 2 | [6486-26-6] | $R = NO_2$, $R_1 = Cl$ | $R_2 = R_3 = CH_3$ |
| 3 | [6486-23-3] | $R = NO_2$, $R_1 = Cl$ | $R_2 = Cl$, $R_3 = H$ |
| 5 | [4106-67-6] | $R = NO_2$, $R_1 = OCH_3$ | $R_2 = OCH_3$, $R_3 = H$ |
| 65 | [6528-34-3] | $R = NO_2$, $R_1 = OCH_3$ | $R_2 = OCH_3$, $R_3 = H$ |
| 74 | [6358-31-2] | $R = OCH_3$, $R_1 = NO_2$ | $R_2 = OCH_3$, $R_3 = H$ |
| 75 | [52320-66-8] | $R = NO_2$, $R_1 = Cl$ | $R_2 = H$, $R_3 = OC_2H_5$ |

These pigments are sensitive to heat and bleed in most paint solvents. They are, however, resistant to acids and bases. Their tinctorial strength is considerably greater than that of inorganic yellows but they are weaker than the diarylide yellows. They are used extensively in emulsion paints, paper coating compositions, inks (qv), and, depending on particle size, can in some cases be used outdoors because of excellent lightfastness in full shades.

**Benzimidazolones.** This class of pigments derives its name from 5-aminobenzimidazolone [95-23-8] which upon reaction with diketene or 2-hydroxy-3-naphthoyl chloride leads to compounds which can be coupled with a variety of diazotized amines.

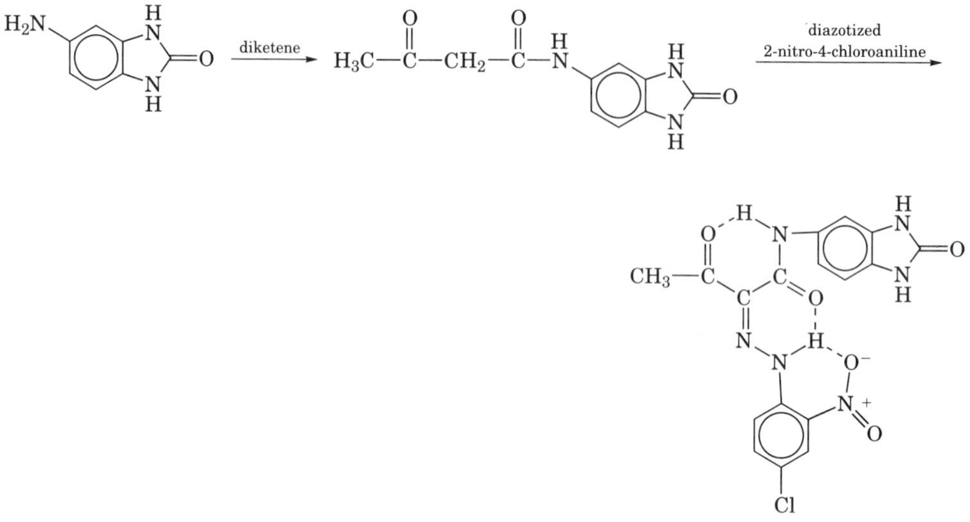

PO 36

The acetoacetarylides yield yellow to orange pigments and the naphthoic acid amides yield red and brown pigments. Because of the considerable insolubility of the generated pigments they usually require a special thermal aftertreatment to grow crystals to a useful pigmentary range. Pigment Yellow 154 [63661-02-9], derived from coupling diazotized 2-trifluoromethylaniline and Pigment Orange 36 [12236-62-3] from 2-nitro-4-chloroaniline with acetoacetyl-5-aminobenzimidazolone, are among a series of pigments which have found application throughout the paint industry for a variety of industrial finishes and in some plastics applications.

**Diarylide Yellows.**    Diarylide or disazo yellow pigments are represented by the following general structural formula:

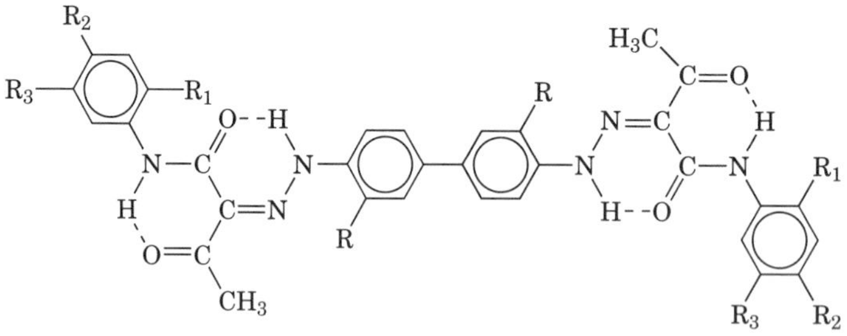

The chemistry and process of manufacture are very similar to the monazo pigments, except a diamine, usually 2,2′-dichlorobenzidine (R = Cl), is tetrazotized and then coupled with two equivalents of an acetoacetarylide. In some cases two different arylides are used, leading to a mixture of three compounds which enhance tinctorial strength and extend the range of available pigments.

Diarylides show significantly greater tinctorial strength and superior bleed and heat resistance than the conventional monoazo pigments. However, they are generally inferior to the monoazo pigments in lightfastness.

Base-soluble resins are frequently coprecipitated with diarylides without proportional diminution in the strength and a significant increase in transparency of the resulting pigments. Based on high strength and versatile transparency most of the arylides are used in a variety of printing inks, and in some plastics where temperature restrictions of 200°C have been imposed (13). Some commercially important products are shown in Table 4.

Unlike most diarylide yellows, Yellow 83, a reddish yellow, possesses very good lightfastness particularly in its opaque form and consequently is used in plastics and paints in addition to inks.

**Table 4. Selected Diarylide Pigments**

| Pigment | CAS Registry Number | Substituents on diazonium component | Substituent on coupling component |
|---------|--------------------|-------------------------------------|-----------------------------------|
| Yellow 12 | [6358-85-6] | R = Cl | $R_1 = R_2 = R_3 = H$ |
| Yellow 13 | [5102-83-0] | R = Cl | $R_1 = R_2 = CH_3$, $R_3 = H$ |
| Yellow 14 | [5468-75-7] | R = Cl | $R_1 = CH_3$, $R_2 = R_3 = H$ |
| Yellow 17 | [4531-49-1] | R = Cl | $R_1 = OCH_3$, $R_2 = R_3 = H$ |
| Yellow 83 | [5567-15-7] | R = Cl | $R_1 = R_3 = OCH_3$, $R_2 = Cl$ |
| Orange 16 | [6505-28-8] | $R = OCH_3$ | $R_1 = R_2 = R_3 = H$ |

**Monoazo Yellow Salts.** Several monoazo yellow salts have gained popularity since the 200°C temperature restriction on the use of diarylide yellows has been imposed. One is Pigment Yellow 168 [71832-85-4], the calcium salt of diazotized 3-nitro-4-aminobenzenesulfonic acid coupled with acetoaceto-2-chloroanilide. It provides a clean, somewhat greenish yellow color which shows good migration resistance but relatively poor tinctorial strength. It is used in polyethylene and inexpensive industrial finishes where the durability requirements are not high.

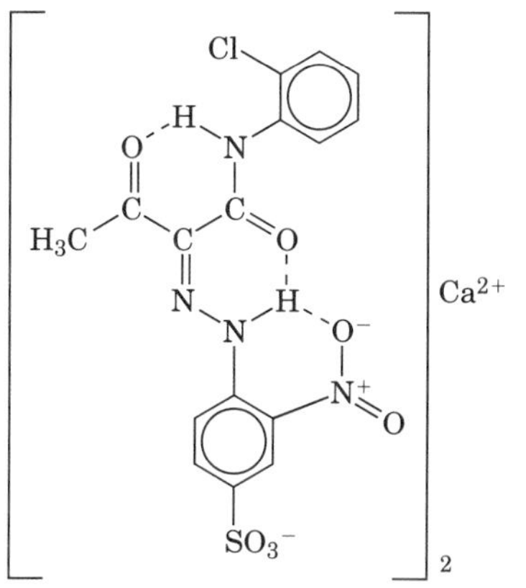

PY 168 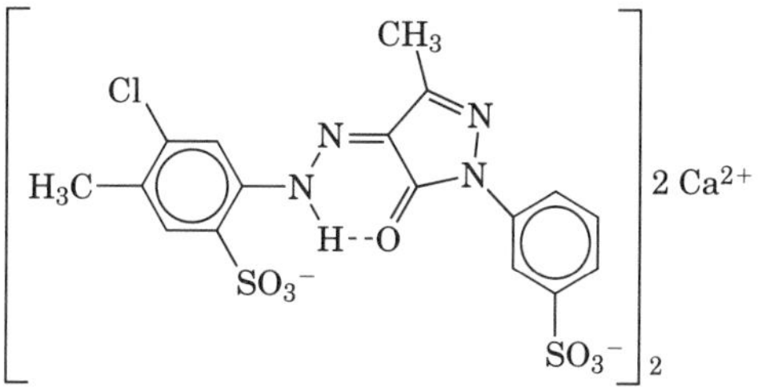                                                PY 191

A second example is the yellow pyrazolone salt, Pigment Yellow 191 [129423-54-7]. It is the calcium salt of diazotized 2-amino-4-chloro-5-methyl-benzenesulfonic acid coupled with 3-methyl-1-[3′–sulfophenyl]-5-pyrazolone and provides a reddish yellow pigment for use in plastics applications. It shows very good heat stability and excellent resistance to nonpolar solvents and commonly used plasticizers (qv). It finds applications in high density polyethylene, polystyrene, and ABS, and shows satisfactory lightfastness.

**Dinitraniline Orange.**  Dinitraniline Orange or Pigment Orange 5 [3468-63-1] is a strong and bright orange pigment with relatively low hiding power and good lightfastness in full shades, but poor tint lightfastness. It shows poor bleed but acceptable base resistance and finds principle application in air-drying systems, including a variety of printing inks.

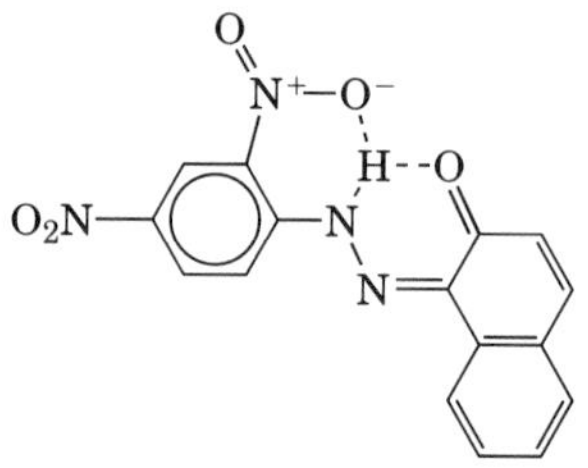

PO 5

**Pyrazolone Orange.**  Pyrazolone Orange or Pigment Orange 13 [3520-72-7] is a disazo pigment of high strength and bright color with good lightfastness in full shades and poor tint lightfastness. It is characterized by fair base and chemical resistance and is used primarily in inks, with limited application in paints and plastics.

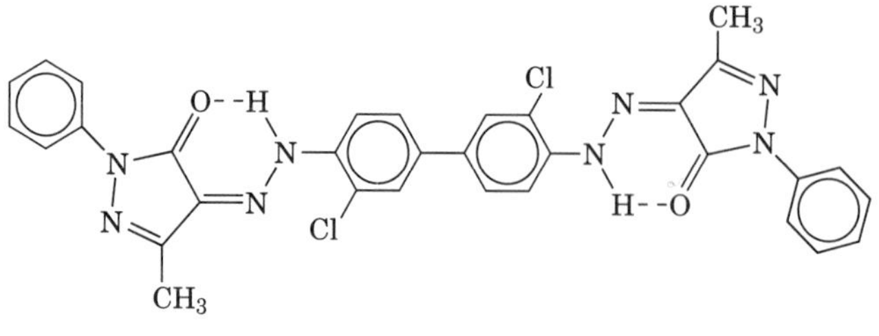

PO 13

**Azo Reds and Maroons.**  *Toluidine Red.*  Pigment Red 3 [2415-85-6] is one of the most popular organic red pigments used in industrial finishes. Its hue varies considerably with particle size and therefore several shades are commercially available. In full shade the pigment is very lightfast and weatherfast but shows poor tint lightfastness. It is characterized by poor fastness toward solvents and is likely to bloom in baking enamels. Its principal application is in air drying paints, and to a limited extent in printing inks.

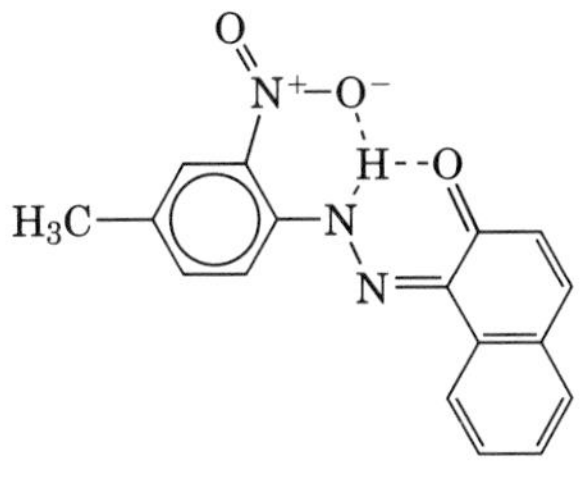

PR 3 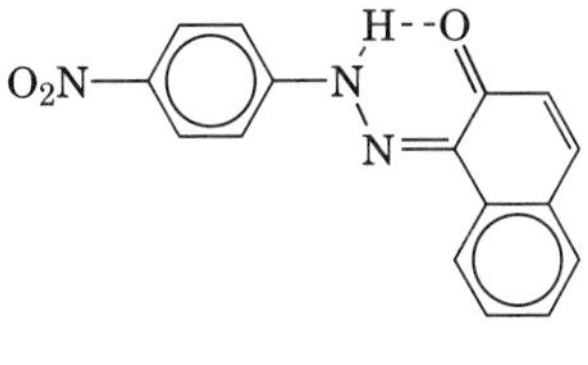                    PR 1

*Para Reds.* Para Red or Pigment Red 1 [*6410-10-2*] is an intense, reasonably opaque red which shows poor lightfastness, particularly in tints. The pigment is bluer than Toluidine Red. The related pigment Parachlor Red (Pigment Red 4 [*2814-77-9*]), derived from 2-chloro-4-nitroaniline, is an intense yellowish red. Both pigments show poor bleed and bake resistance and a tendency to bloom in enamels. The pigments are used in some inks and in low cost articles such as detergents, floor polishes, colored pencils, etc. The use of these pigments has declined markedly as a result of greater quality demands by the coatings industry.

*Lithol Reds.* Lithol Red or Pigment Red 49:1 [*1103-38-4*] is one of the most important of the precipitated salt pigments. They comprise a family of sodium (PR 49), barium (PR 49:1), calcium (PR 49:2), and strontium (PR 49:3) salts of diazotized Tobias acid or 2-naphthylamine-1-sulfonic acid coupled with 2-naphthol. The most popular are the barium and calcium salts, the former being yellower in shade. These reds are used where brightness, bleed resistance, and low cost are of primary importance. They are neither resistant to heat nor chemicals, and are used primarily in printing inks and some inexpensive air-dried industrial paints where good durability is not required.

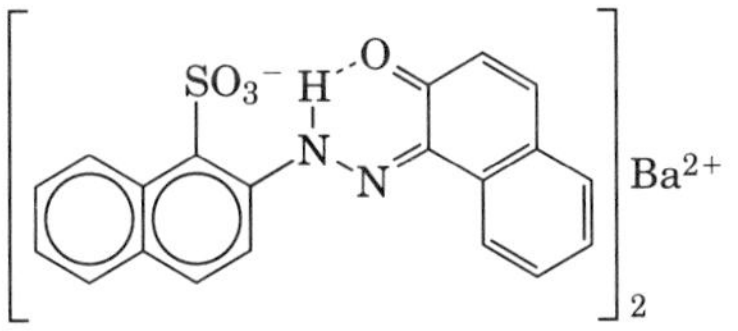

PR 49:1 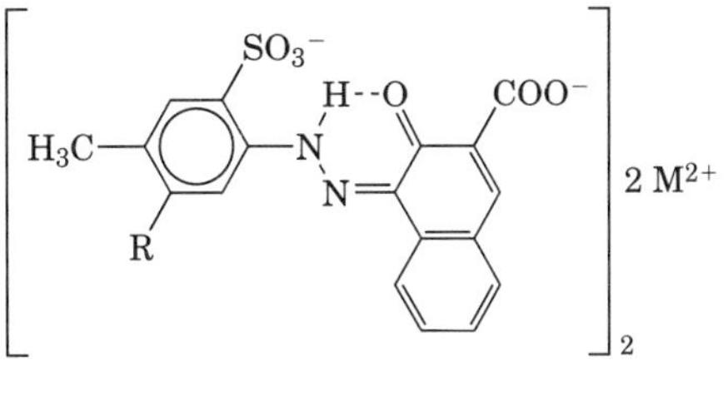             PR 57:1 (R=H; M=Ca)

PR 48 (R=Cl)

*BON Reds and Maroons.* The BON or BONA Reds and Maroons derive their name from β-hydroxynaphthoic acid, also known as 3-hydroxy-2-naphthoic acid [*92-70-6*]. BON is used as a general coupling component for the entire group with various diazotized amines containing salt-forming groups. Insolubilization is effected by precipitating calcium, barium, strontium, or manganese salts. Colors vary from yellowish red to dark maroon.

*Lithol Rubine.* Lithol Rubine (Pigment Red 57 [*5281-04-9*]), also referred to as 4B toner, is the calcium salt of diazotized 2-amino-5-methylbenzenesulfonic acid coupled with 3-hydroxy-2-naphthoic acid. It ranks high among organic pigments in production volume and use.

The probable crystal structure of this salt has been described (14) as a complex of two dianions and two cations forming a large polyatomic ring. By implication this type of molecular packing probably applies to many other similar pigments.

Lithol Rubine is characterized by high tinctorial strength, good bleed, and bake resistance but poor alkali, soap, and acid resistance. Its lightfastness is considered fair and varies within a wide range of shades obtained by inclusion of auxiliary agents. Because the pigment is brighter than Lithol Reds, it is used extensively in oil, gravure, and flexographic inks. It is heat stable up to about 250°C and is therefore a suitable candidate for indoor use in polyethylene and spin-dyed polypropylene.

*Red 2B.* Red 2B defines the important barium (PR 48:1 [7585-41-3]), strontium (PR 48:3), calcium (PR 48:2), and manganese (PR 48:4) salts of diazotized 2-amino-4-chloro-5-methylbenzenesulfonic acid coupled with 3-hydroxy-2-naphthoic acid. The first three are bright red pigments, becoming bluer in that sequence. They exhibit high strength, good bleed, and bake resistance, but poor resistance to alkali, soap, and acids, and fair lightfastness. The main fields of application are printing inks, plastics, and inexpensive industrial paints.

The manganese salt (M = Mn), ie, manganese 2B (PR 48:4) is a bluish red, characterized by superior masstone lightfastness and outdoor durability, and finds use in some automotive and other high quality industrial finishes. It is also used in some plastics, because of reasonably good heat stability, and a variety of printing inks.

*Pyrazolone Reds.* The pyrazolone reds are disazo pigments which provide high color strength and reasonable lightfastness in full shades but poor tint lightfastness, good bake, bleed, and chemical resistance. Some find application in plastics such as poly(vinyl chloride) where they show good dielectric properties, making them useful for cable insulations, in rubber, and specialized printing inks. Pigment Red 38 [6358-87-8] is the product obtained by coupling diazotized 3,3'-dichlorobenzidine with N-phenyl-3-carbethoxy-5-pyrazolone.

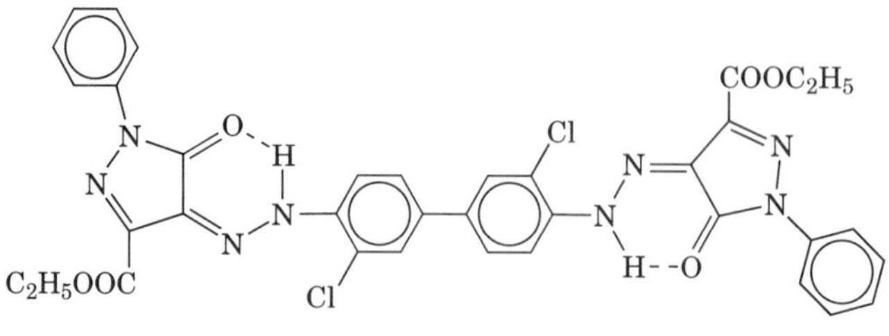

PR 38

**Naphthol Reds and Maroons.** Naphthol Reds and Maroons are monoazo pigments which provide a wide range of colors from yellowish and medium red to bordeaux, maroon, and violet, and are characterized by high strength but marginal migration resistance. Depending on the substitution pattern some are strongly migrating and others are more or less resistant to migration. The introduction of amide functions, in addition to the arylamide of 3-hydroxy-2-naphthoic acid on which all of this type of pigment are based, provides a measure of migration improvement. Lightfastness is generally marginal to good. Pigment Red 112 [6535-46-2] is a brilliant medium red pigment, approaching the shade of Toluidine Red, which is used in a variety of printing inks, air drying, and emulsion paints.

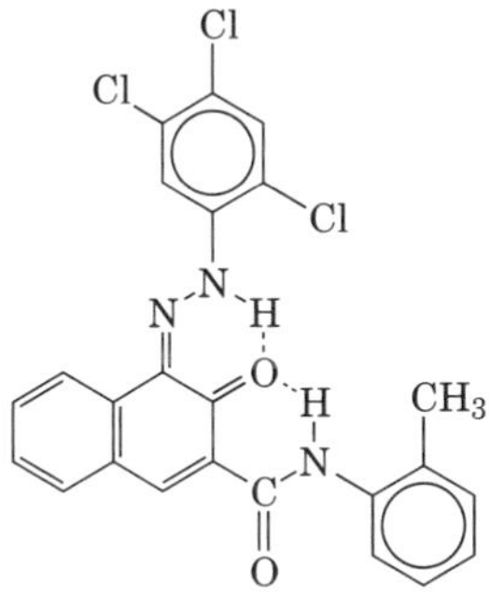

PR 112 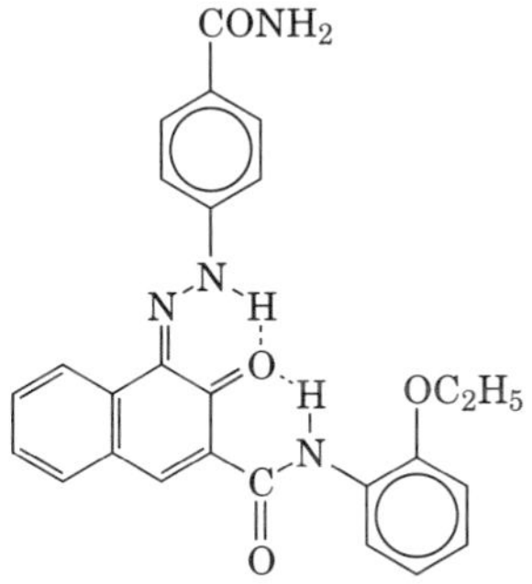                                    PR 170

Another important pigment in this class is Pigment Red 170 [6985-95-1], which provides medium shades of red, and when particle-grown produces an opaque modification which shows improved migration resistance and lightfastness. Like some other Naphthol AS pigments it shows the phenomenon of polymorphism.

It is used in high grade industrial paints and, in combination with high performance pigments, in automotive finishes. The transparent type which is tinctorially strong finds applications in a variety of printing inks.

**Azo Condensation Pigments.** A further improvement in heat stability of azo pigments was achieved by the condensation disazo pigments due to an enlarged molecular framework and higher molecular weight. Formally they are composed of two monoazo or more accurately two monohydrazone units, which are attached to each other by an aromatic dicarbonamide bridge. Both red and yellow pigments are prepared by essentially the same or modified processes. The red group of pigments is synthesized by coupling 3-hydroxy-2-naphthoic acid with a diazotized amine, followed by conversion to the acid chlorides and diamide formation by reaction with aromatic diamines. An example is Pigment Red 144 [5280-78-4] (Fig. 2a).

The yellow members of this pigment class are obtained by coupling a diazotized aminobenzoic acid with a bisacetoacetarylide, followed by conversion to a diacid chloride and reaction with a substituted aromatic amine. An example is Pigment Yellow 93 [5580-57-4] (Fig. 2b).

The pigments are used primarily in plastics, including polypropylene fibers, because of very good bleed resistance, heat stability, and lightfastness. The reds also find use in printing inks, primarily for high quality products.

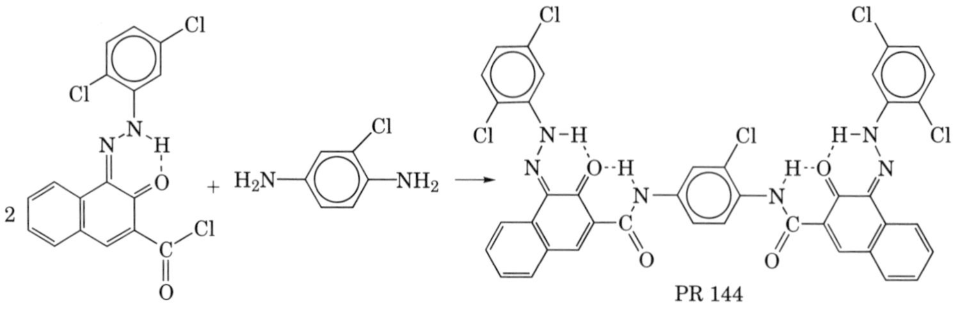

(a)

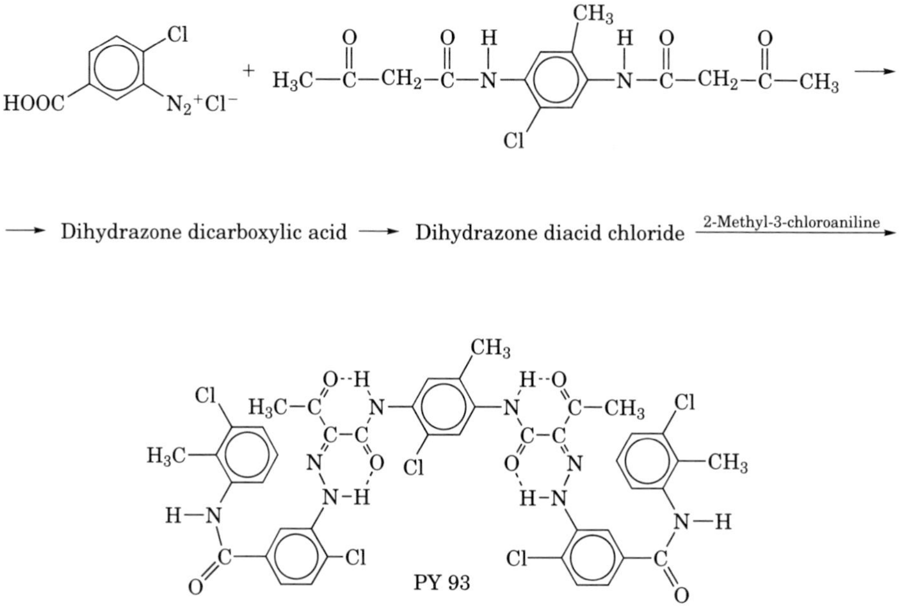

(b)

**Fig. 2.** Azo condensation pigments: (**a**) Pigment Red 144 and (**b**) Pigment Yellow 93.

## LAKES

Lakes are either dry toner pigments that are extended with a solid diluent, or an organic pigment obtained by precipitation of a water-soluble dye, frequently a sulfonic acid, by an inorganic cation or an inorganic substrate such as aluminum hydrate.

Basic dyes are characterized by bright shades and high strength but poor lightfastness. However, when laked by precipitation with soluble salts of organic acids such as tannic acid, or inorganic heteropolyacids like phosphotungstic (PTA; M = W) and phosphomolybdic (PMA; M = Mo), and the combined phos-

$$H_3[P(M_3O_{10})_4] \cdot xH_2O \qquad H_3[P(Mo_3O_{10})_3(W_3O_{10})] \cdot xH_2O$$

PTA; PMA                                PTMA

photungstomolybdic acid (PTMA), the resulting pigments retain the dyes' tinctorial attributes, but become insoluble and show improved lightfastness. Ideally three acidic hydrogens can be replaced by three dye molecules to form pigments which are characterized by uncommonly clean and brilliant red and violet shades. They are, however, not stable to polar solvents or alkali and fail to satisfy stringent fastness requirements, particularly in packaging and special inks. Representative basic dyes which are converted to pigments by complex formation are listed in Table 5.

**Table 5. Typical Basic Dyes Used as Complexes with Heteropolyacids**

| Name of dye | Structure of dye | Anion |
|---|---|---|
| Rhodamine 6G (Pigment Red 81:1) CI 45160:1 | 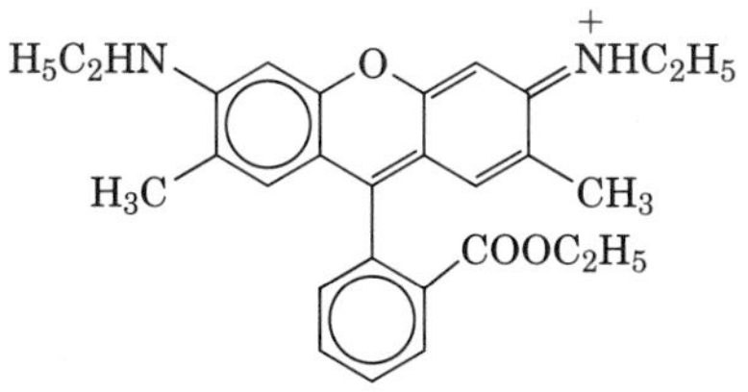 | PTMA |
| Methyl Violet (Pigment Violet 3) CI 42535:2 | 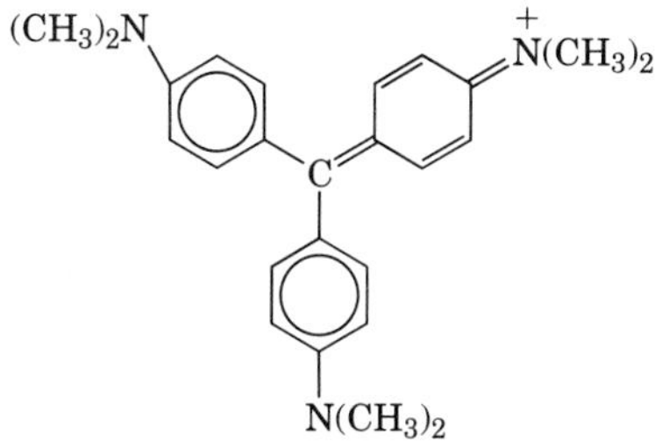 | PTMA PMA |
| Victoria Blue B (Pigment Blue 1) CI 42595:2 | 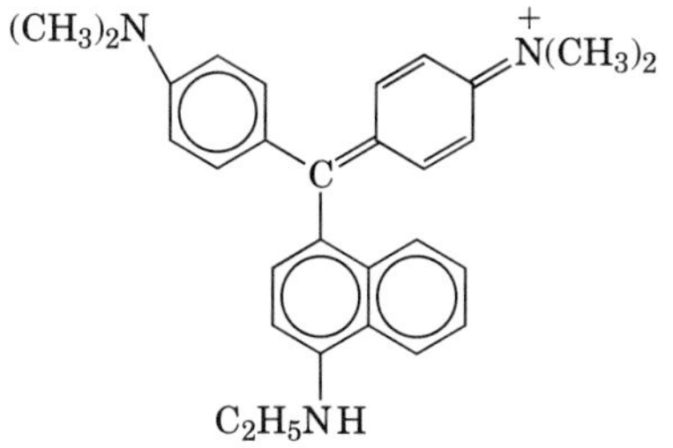 | PTMA PMA |
| Malachite Green (Pigment Green 4) CI 42000:2 | 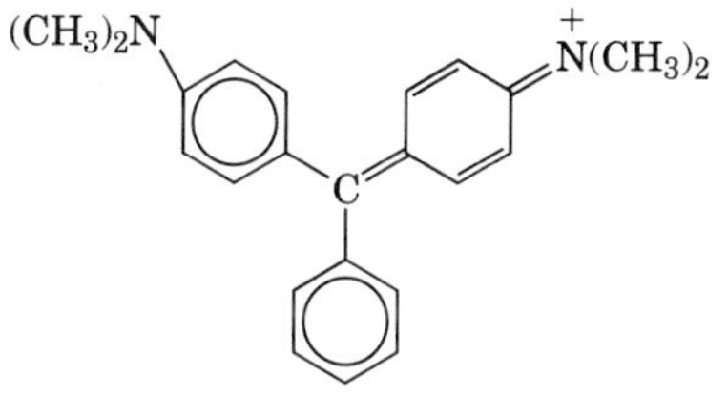 | PTMA PMA |

## COPPER PHTHALOCYANINES

Copper phthalocyanine [147-14-8] (CPC) approximates an ideal pigment (Pigment Blue 15). This class of pigments offers extreme brightness, tinctorial strength, bleed and chemical resistance, stability to heat, and migration. The pigments show excellent weatherfastness but are restricted to the blue and green regions of the spectrum. Phthalocyanine blue and green are among the most important organic pigments on the worldwide market.

**Copper Phthalocyanine Blue.** CPC blue exists in several polymorphic modifications, two of which, the red-shade blue alpha and green-shade blue beta form, are of great commercial significance. Beta is the thermodynamically more stable phase and is the product resulting from manufacture by the two basic processes using either phthalonitrile or phthalic anhydride as starting materials, either in the presence of a solvent or by a dry baking process. The alpha form is usually obtained by conversion from the beta form and has to be stabilized to prevent phase reconversion. The pronounced tendency of the alpha form to flocculate in fluid media is suppressed by special surface treatments and/or the introduction of a small amount of aromatically bound chlorine to form solid solutions between CPC and its chlorinated derivatives. A so-called semichloro-CPC containing an average of about 0.75 chlorine atoms per molecule provides a stable pigment even upon incorporation in high temperature plastics.

Despite the superficially structural complexity of the CPC molecule, it is prepared with comparative ease (15) (see PHTHALOCYANINE COMPOUNDS). The preferred manufacturing process starts with phthalic anhydride and is carried out in a variety of solvents, trichlorobenzene having been the preferred solvent, but it has been replaced with high boiling hydrocarbons or glycols in order to avoid the formation of the hazardous and poorly degradable polychlorinated biphenyls.

The coarse, nonpigmentary crudes are particle size reduced by a variety of processes to obtain pigmentary CPC. Since CPCs are utilized in a variety of industries and a host of applications, the specific property requirements vary widely. A large number of special-purpose types are commercially available designed to provide optimal properties for specific applications.

**Copper Phthalocyanine Green.** CPC green is obtained by electrophilic substitution of CPC blue with chlorine, the degree of chlorination reaching 14–15 chlorines per molecule. The typical polychloro-CPCs are blue-shade green pigments. To provide yellower shades of green, substitution of bromine for chlorine with up to eight or nine bromine atoms per molecule, the rest being chlorine, is carried out. Like CPC blue the green pigments show outstanding pigmentary properties, but are lower in tinting strength with progressive halogen substitution, particularly with bromine.

## QUINACRIDONES

Quinacridone pigments offer generally outstanding fastness properties across the visible spectrum from red-shade yellows to scarlet, maroon, red, magenta, and violet color ranges (16). The pigments are practically insoluble in most common solvents and therefore show excellent migration resistance in most application media. The low solubility is attributed to effective intermolecular hydrogen bonding (17) which is also responsible for the photochemical stability. The various available colors are the result of polymorphism and various substitution patterns. The parent compound Pigment Violet 19 [*1047-16-1*] exists in three polymorphic modifications. The red gamma and violet beta forms are commercial pigments, whereas the red alpha form is metastable. The pigments are synthesized by two principal processes; both involve the common intermediate dialkyl 2,5-diarylamino-3,6-dihydroterephthalate (Fig. 3). Substitution in the 2,9-positions of quinacridone leads to magenta pigments whereas 4,11-substitution causes shifts to shorter wavelengths and scarlet or orange pigments. Oxidation of quinacridone yields quinacridonequinone (QAQ), a yellow pigment which, like the 4,11-disubstituted pigments, is not lightfast. However, in solid solutions with quinacridone, pigments of excellent durability are obtained. Both opaque large particle size and transparent particle size-reduced pigments are commercially available. Table 6 lists some commercial products.

Due to the excellent pigmentary properties, quinacridones are used in many industries but particularly in automotive finishes, emulsion paints, plastics, and fibers.

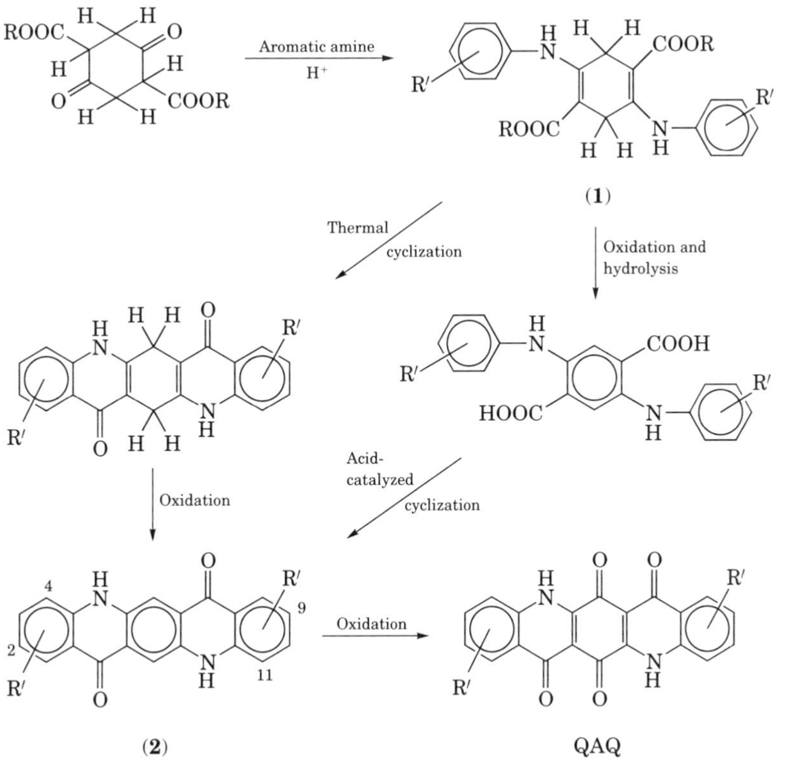

**Fig. 3.** Synthesis of quinacridones where (**1**) is dialkyl 2,5-diarylamino-3,6-dihydro-terephthalate. See Table 6 for representative R′ groups.

**Table 6. Representative Commercial Quinacridone Pigments**

| Pigment composition | CI designation | CAS Registry Number | Color | Structure[a] |
|---|---|---|---|---|
| quinacridone (QA) | PV 19 | [1047-16-1] | opaque yellow-shade red (gamma)[b]; opaque yellow-shade violet (beta)[c] | $R' = H$ |
| 2,9-dimethyl-QA | PR 122 | [980-26-7] | magenta | $R' = 2,9-(CH_3)_2$ |
| 2,9-dichloro-QA | PR 202 | [68859-50-7] | magenta | $R' = 2,9-Cl_2$ |
| quinacridone plus 4,11-dichloroquinacridone | PR 207 | [3089-16-5] | scarlet solid solution | $R' = H + R' = 4, 11-Cl_2$ |
| quinacridone plus quinacridonequinone | PR 206 | [1503-48-6] | maroon solid solution | $R' = H + QAQ$ |

[a]See Fig. 3; especially (**2**) for position numbers.
[b]Also semitransparent blue-shade red (gamma).
[c]Also semitransparent and transparent violet (beta).

### DIARYL PYRROLOPYRROLES

The 1,4-diketo-3,6-diarylpyrrolo(3,4-c)pyrroles are the most recently discovered class of pigments ranging in color from orange to bluish red (18). The pigments are synthesized by base-catalyzed condensation of higher diakyl esters of succinic acid with aromatic nitriles. One important member of this class is Pigment Red 254 [122390-98-1] which is a very opaque yellowish red pigment of outstanding durability, brightness, and chemical resistance. Another is the parent compound Pigment Red 255 [54660-00-3], which is a high performance orange pigment. Both are used in automotive finishes, and a higher strength variation of PR 254 is used in plastics applications.

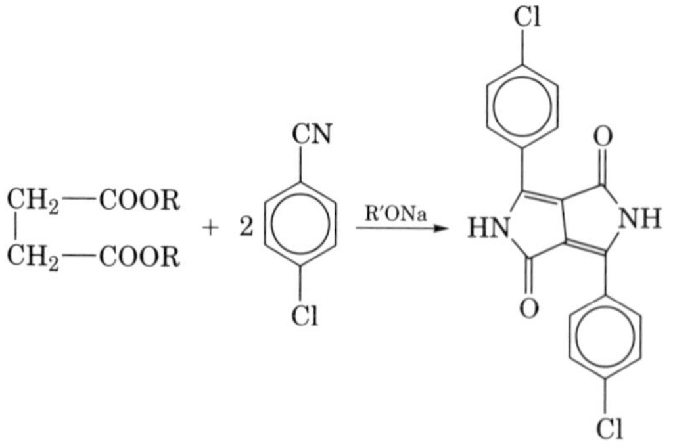

PR 254

VAT DYE PIGMENTS

Vat dyes have been used for a long time for coloring textile fibers. As pigment technology evolved with a variety of physical and chemical methods for conditioning or finishing pigments, they were also successfully applied to largely insoluble dyes. Improved purity of products coupled with optimized particle size and particle size distribution has afforded pigments of considerable commercial interest. Only a few of the very large number of vat dyes have found application in the pigment field.

**Perylenes.** Perylene pigments are either the 3,4,9,10-tetracarboxylic dianhydride or more often N,N'-substituted diimides (Table 7).

**Table 7. Commercial Perylene Pigments**

| Structural formula | R | CI designation | CAS Registry Number | Color |
|---|---|---|---|---|
| 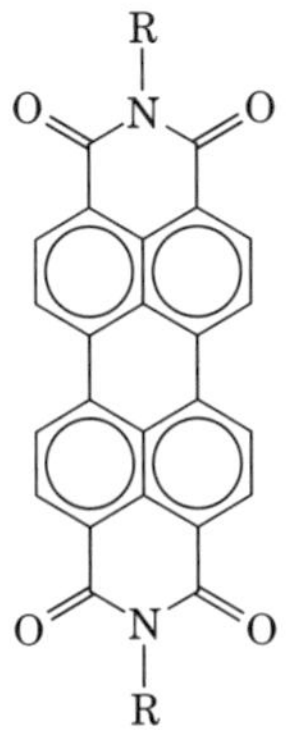 | CH$_3$ | PR 179 | [55231-31-3] | red to maroon |
| | 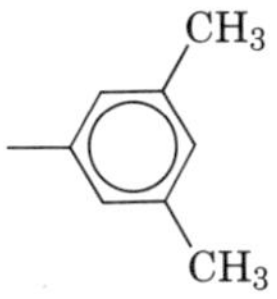 (2,5-dimethylphenyl) | PR 149 | [4948-15-6] | red |
| | 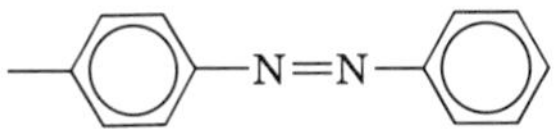 (phenyl-N=N-phenyl) | PR 178 | [3049-71-6] | red |
| | H | PV 29 | [12236-71-4] | bordeaux |

The pigments are manufactured either by reaction of the dianhydride with an amine or N,N'-dialkylation of the diimide. They are characterized by high tinctorial strength, excellent solvent stability, very good weatherfastness, moderate brightness, and range in color from red to violet. An exception is the dianhydride which is not stable to alkali.

Most applications are in high grade industrial paints, especially automotive finishes. Some types (PR 149) are used primarily in plastics and fibers.

**Perinones.** The most important pigment in this family is the orange perinone, Pigment Orange 43 [4424-06-0] which is obtained by reaction of naphthalene-1,4,5,8-tetracarboxylic dianhydride with o-phenylenediamine. The result is a mixture of the cis- and trans-isomers. The commercial product is the orange trans-compound which must be separated from the dull, bluish red cis-isomer, and then conditioned for pigment use. The pigment is fairly weatherfast and heat stable, and is used primarily in plastics and fiber applications.

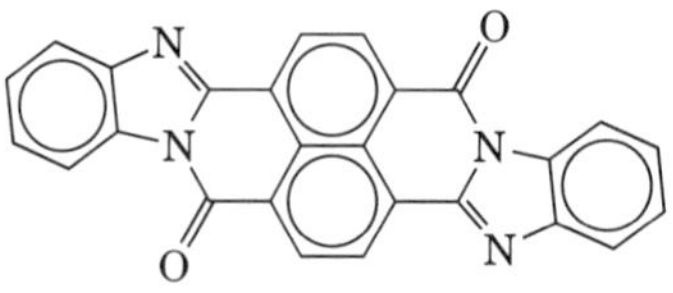

PO 43 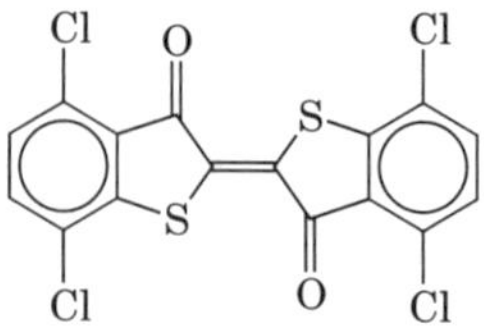                                   PR 88

**Thioindigo.** Thioindigo pigments are produced by oxidative dimerization of variously substituted thioindoxyls or thionaphthenones. The most important pigment in this series is Pigment Red 88 [14295-43-3] a red-violet pigment in which the chlorine substituents are essential to confer marginal to acceptable lightfastness and stability to solvents. Although still used in some paint and plastic systems, it is being replaced by pigments of higher quality.

### AMINOANTHRAQUINONE PIGMENTS

Pigment Red 177 [4051-63-2] has the chemical structure of 4,4'-diamino-1,1'-dianthraquinonyl and is prepared by intermolecular copper-catalyzed debromination of 1-amino-4-bromoanthraquinone-2-sulfonic acid followed by desulfonation. It is the only known pigment with unsubstituted amino groups which are involved in both intra- and intermolecular hydrogen bonding (19). The bluish red pigment is used in plastics, industrial and automotive paints, and specialized inks (see DYES, ANTHRAQUINONE).

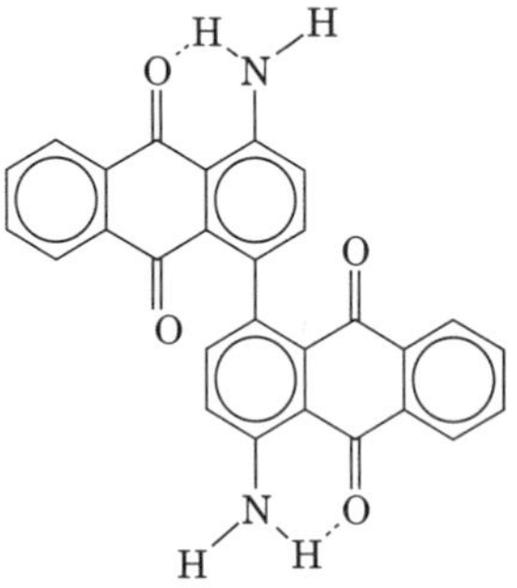

PR 177 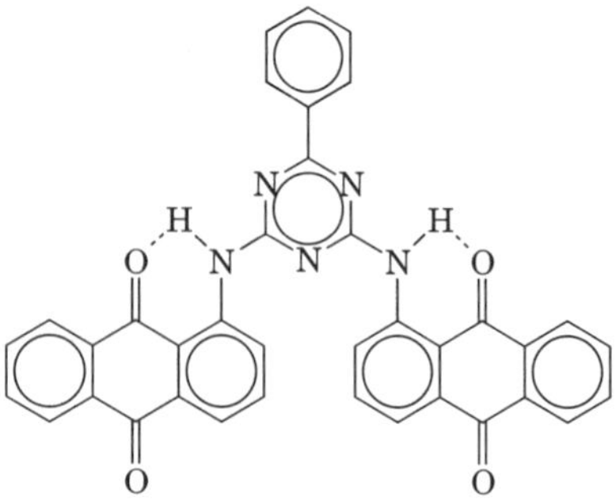                                   PY 147

Pigment Yellow 147 [76168-75-7] is derived from reaction of 1-amino-anthraquinone with 1-phenyl-3,5-dichloro-2,4,6-triazine. It is a reddish shade yellow pigment used primarily in certain plastics and in polyester and polypropylene fibers.

**Indanthrone.**   Pigment Blue 60 [*81-77-6*] is synthesized by oxidative dimerization of 2-aminoanthraquinone in the presence of a strong base, and then particle size reduced by precipitation from very strong sulfuric acid or by milling. The very red-shade blue pigment shows outstanding weatherfastness in full shade as well as in light white reductions. It is used primarily in metallized automotive finishes where it is sometimes more weatherfast than copper phthalocyanine blue.

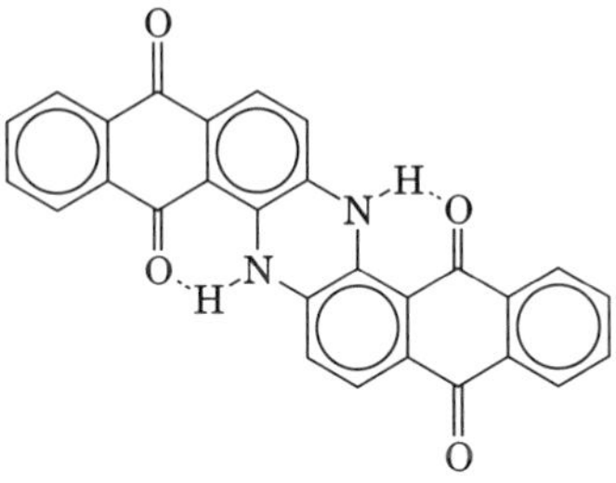

PB 60

## DIOXAZINE

Carbazole Violet (Pigment Violet 23 [*5358-30-1*]) is synthesized by reaction of chloranil with 3-amino-*N*-ethylcarbazole followed by oxidative acid-catalyzed cyclization, and conditioning via milling or acid treatment. Since its discovery in the 1950s the linear structure had been assigned to this pigment. It has now been established that the cyclization goes in an angular fashion, and that the unequivocally synthesized linear compound is different from Pigment Violet 23 (20). The bluish violet pigment is uncommonly strong, resistant to solvents, and shows fair weatherfastness. It is used primarily as a shading pigment with copper phthalocyanines and for toning whites in a variety of systems.

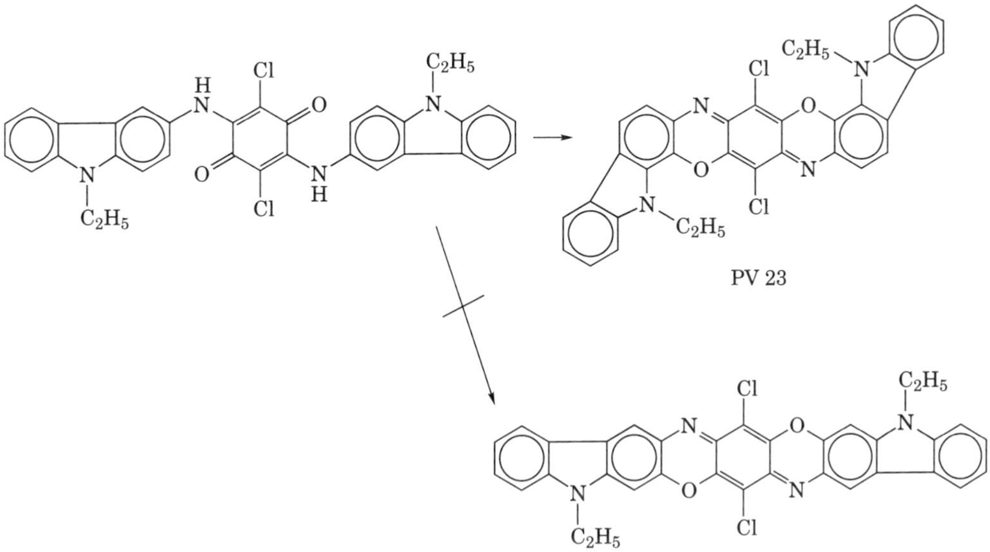

PV 23

### ISOINDOLINONES AND ISOINDOLINES

Tetrachloroisoindolinone pigments are synthesized by condensation of 3,3,
4,5,6,7-hexachloroisoindoline-1-one or the corresponding 3,3-dialkoxy derivatives
with aromatic diamines yielding pigments ranging in color from green-shade yellows to oranges. The pigments are characterized by very good lightfastness, heat
stability, migration resistance, and chemical inertness. Although Pigment Yellow 110 [5590-18-1], a red-shade yellow, is relatively weak, it finds extensive
use in automotive and other high grade finishes and in a variety of plastics
and ink applications.

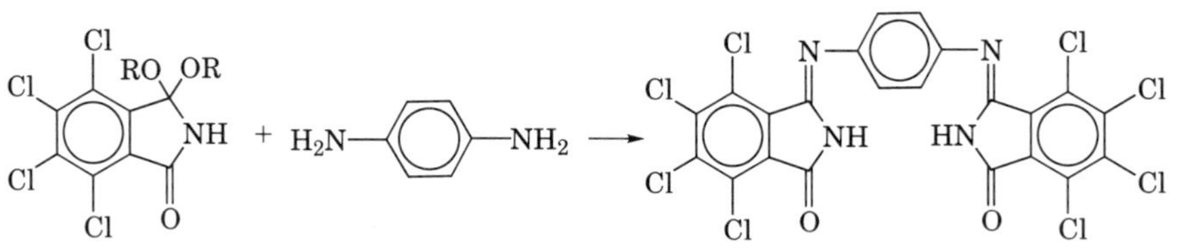

PY 110

Isoindoline pigments are synthesized by reaction of 1-amino-3-
iminoisoindolenine(diiminoisoindoline) with two moles of compounds containing
activated methylene groups. In Pigment Yellow 139 [36888-99-0] the second
component is barbituric acid. PY 139 is a reddish yellow pigment which differs
in color as a function of particle size. The opaque version is the reddest. The
pigment is lightfast but tends to darken in full shades. Although marginal in
chemical resistance the pigment is used in the paint and plastics industries.

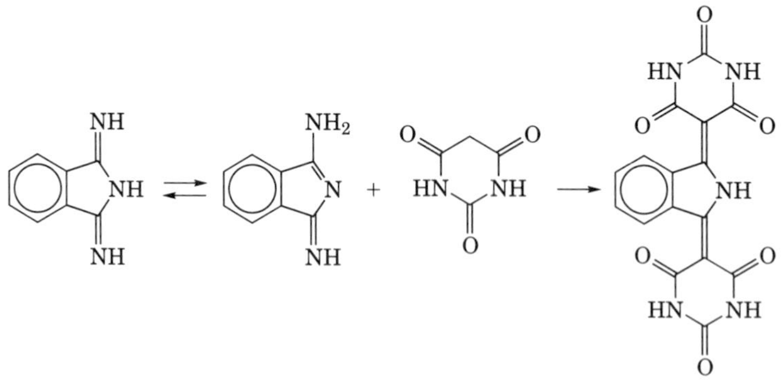

PY139

### QUINOPHTHALONES

The quinophthalone pigments are prepared by condensation of quinaldines with
a variety of aromatic anhydrides. Introduction of a hydroxyl group in the 3-
position of quinaldine yields yellows of improved lightfastness. One pigment

in this series, Pigment Yellow 138 [*56731-19-2*], is a reasonably weatherfast greenish yellow pigment of good heat stability. The main field of application is paints and plastics.

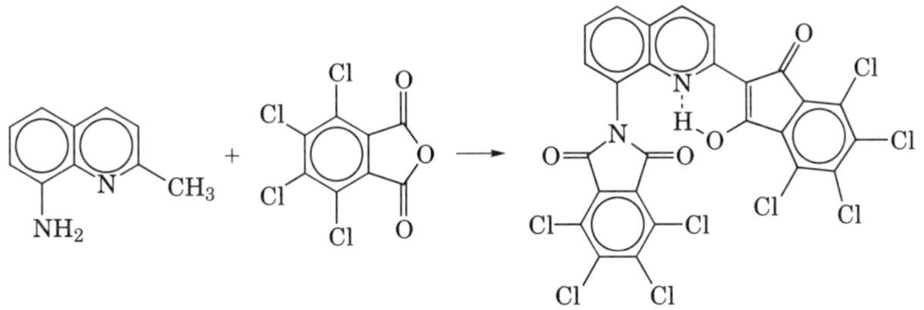

PY 138

## Production and Economic Aspects

In 1993 the five principal producers of organic pigments in the United States were BASF Corp., Ciba-Geigy Corp., Hoechst Celanese Corp., Miles Inc. (Bayer), and Sun Chemical Corp.

　　The following organic pigment manufacturers reported their 1991 production and/or sales to the U.S. International Trade Commission.

Allegheny Chemical Corp.

Apollo Colors, Inc.

BASF Corp.

Baker Fine Color, Inc.

Synalloy Corp., Blackman Uhler Chemical Division

CDR Pigments and Dispersions

Ciba-Geigy Corp.

E. I. du Pont de Nemours, & Co., Inc., Chemicals and Pigments Dept.

C. Lever Co., Inc.

Max Marx Color Corp.

Magruder Color Corp., Inc.

Daicolor-Pope, Inc.

CPS Corp.

PMC, Specialties Group, Inc.

Fabricolor Manufacturing Corp.

Galaxie Chemical Corp.

Hoechst Celanese Corp., Specialty Chemical Group

Cookson Pigment, Inc.

Engelhard Corp.

Industrial Color, Inc.

Indol Color Co., Inc.

Spectrachem Corp.

R-M Industries

Roma Color, Inc.

Sun Chemical Corp., Pigments Division

Paul Uhlich and Co., Inc.

Miles, Inc.

U.S. production and sales of organic pigments during the period of 1983 to 1992 are tabulated in Table 8. More detailed information by color class is provided from 1988 to 1992 in Table 9. In addition, information about export, import, and apparent domestic consumption of organic pigments appears in Tables 10 and 11.

**Table 8. U.S. Production and Sales of Organic Pigments and Lakes, 1983–1992[a]**

| Year | Thousands of metric tons | | Value, $10^3$ \$ | Average unit value,[b] \$/kg |
|------|------------|--------|--------------|--------------|
|      | Production | Sales  |              |              |
| 1983 | 35,446 | 31,576 | 422,434 | 13.38 |
| 1984 | 38,938 | 34,616 | 492,954 | 14.24 |
| 1985 | 36,754 | 31,379 | 447,709 | 14.27 |
| 1986 | 40,237 | 34,869 | 513,132 | 14.72 |
| 1987 | 42,689 | 37,865 | 586,254 | 15.48 |
| 1988 | 52,680 | 39,487 | 594,657 | 15.06 |
| 1989 | 50,360 | 43,236 | 701,552 | 16.33 |
| 1990 | 52,551 | 44,773 | 717,194 | 16.02 |
| 1991 | 51,311 | 39,426 | 643,561 | 16.32 |
| 1992 | 56,944 | 45,789 | 788,697 | 17.23 |

[a]Ref. 21.
[b]Unit values calculated from total sales and values given in Ref. 21.

## Uses

Organic pigments are used for decorative and/or functional effects. In paints, for example, pigments provide color and contribute to exposure durability of the systems, which is particularly true for high performance pigments. Other functional effects include hiding power and high visibility, such as is displayed with daylight fluorescent pigments; the latter are mostly fluorescent dyes dissolved in synthetic resins and are therefore not particularly lightfast. They are used in various printing processes for textiles, plastics, and safety markings of various types (see LUMINESCENT MATERIALS, FLUORESCENT PIGMENTS (DAYLIGHT)).

The most important and established use for pigments is the imparting of color to a variety of materials and compositions. Examples are surface coatings for exteriors and interiors of automobiles and houses with oil- or water-based paints; wood stains, leather and artificial leather finishes, distempers; printing inks for rotogravure, lithographic, and flexographic systems, including inks for metal plates, foil, wallpaper, food wrappers, and general packaging materials; textile printing inks for clothing, awnings, bookcovers, etc; application for paper, rubber, shoe polish, roofing granules, concrete, and cement; soaps, detergents, synthetic resins, and wax compositions; color coating of fertilizers and seeds; and laundry bluing. Pigmentation of textile fibers, nylon, polypropylene, and cellulose acetate is practiced. Organic pigments are also used for coloring various plastics including poly(vinyl chloride) (PVC), polyethylene, polystyrene, acrylonitrile–butadiene–styrene (ABS), polycarbonate, etc. Numerous pigments are also incorporated in artists' material, ie, oils, crayons, chalks, colored pencils, modeling clay, etc.

**Table 9. U.S. Production and Sales of Organic Pigments and Lakes, 1988–1992[a]**

| Year and kind | Production, t | Sales Metric tons | Sales Value, $10^3$ | Average unit value, $/kg |
|---|---|---|---|---|
| **1988** | | | | |
| yellow | 12,591 | 10,593 | 128,939 | 12.17 |
| orange | 1,145 | 1,089 | 16,159 | 14.84 |
| red | 23,074 | 13,416 | 218,508 | 16.29 |
| violet | 1,081 | 1,121 | 45,557 | 40.64 |
| blue | 13,185 | 11,776 | 157,403 | 13.37 |
| green | 1,165 | 1,138 | 22,349 | 19.64 |
| brown/black | 103 | 113 | 1,609 | 14.24 |
| lakes | 337 | 241 | 4,133 | 17.15 |
| **1989** | | | | |
| yellow | 13,122 | 10,750 | 133,756 | 12.44 |
| orange | 1,303 | 1,174 | 17,312 | 14.75 |
| red | 17,158 | 14,927 | 284,131 | 17.69 |
| violet | 2,547 | 2,459 | 85,175 | 34.65 |
| blue | 14.542 | 12,318 | 169,235 | 13.74 |
| green | 1,279 | 1,238 | 25,882 | 20.91 |
| brown/black | 102 | 114 | 1,654 | 14.51 |
| lakes | 307 | 256 | 4,407 | 17.21 |
| **1990** | | | | |
| yellow | 14,176 | 11,126 | 137,826 | 12.39 |
| orange | 1,288 | 1,099 | 15,848 | 14.42 |
| red | 18,448 | 16,754 | 275,982 | 16.47 |
| violet | 808 | 1,362 | 66,939 | 49.15 |
| blue | 16,101 | 12,829 | 186,657 | 14.55 |
| green | 1,421 | 1,378 | 29,940 | 21.73 |
| brown/black | 21 | 21 | 165 | 7.86 |
| lakes | 288 | 204 | 3,837 | 18.81 |
| **1991** | | | | |
| yellow | 14,548 | 10,602 | 135,156 | 12.75 |
| orange | 1,217 | 1,104 | 18,233 | 16.52 |
| red | 15,959 | 12,470 | 224,549 | 18.01 |
| violet | 2,288 | 1,628 | 76,773 | 47.15 |
| blue | 15,569 | 12,070 | 155,348 | 12.87 |
| green | 1,393 | 1,344 | 30,202 | 22.47 |
| lakes | 337 | 208 | 3,300 | 15.87 |
| **1992** | | | | |
| yellow | 15,810 | 13,971 | 169,062 | 12.10 |
| orange | 1,420 | 1,215 | 22,436 | 18.47 |
| red | 19,228 | 14,070 | 275,329 | 19.57 |
| violet | 1,938 | 1,765 | 94,678 | 53.64 |
| blue | 16,969 | 13,195 | 180,753 | 13.70 |
| green | 1,426 | 1,418 | 33,431 | 23.58 |
| brown/black | 43 | 35 | 145 | 4.14 |
| lakes | 110 | 119 | 2,863 | 24.06 |

[a]Ref. 21.

73

**Table 10. U.S. Exports of Pigments and Preparations, by Principle Markets, 1988–1992[a]**

| Market | 1988 | 1989 | 1990 | 1991 | 1992 |
|---|---|---|---|---|---|
| quantity, t | | | | | |
| Belgium | 895 | 3,203 | 3,412 | 2,913 | 4,160 |
| Brazil | 116 | 202 | 199 | 230 | 340 |
| Canada | 975 | 1,403 | 3,029 | 3,608 | 4,605 |
| Germany (West) | 500 | 789 | 919 | 1,577 | 927 |
| Hong Kong | 84 | 320 | 496 | 424 | 638 |
| Italy | 48 | 402 | 224 | 83 | 149 |
| Japan | 330 | 1,140 | 1,032 | 826 | 608 |
| Mexico | 392 | 602 | 460 | 613 | 866 |
| Netherlands | 71 | 489 | 1,014 | 150 | 252 |
| Switzerland | 62 | 126 | 239 | 152 | 402 |
| U.K. | 725 | 284 | 419 | 2,484 | 4,999 |
| other | 1,002 | 2,389 | 3,321 | 2,909 | 3,846 |
| *Total* | *5,200* | *11,349* | *14,764* | *16,019* | *21,793* |
| value, $10^3$ $ | | | | | |
| Belgium | 6,803 | 22,956 | 26,144 | 30,436 | 36,684 |
| Brazil | 1,435 | 3,150 | 3,553 | 3,281 | 4,193 |
| Canada | 9,915 | 13,921 | 27,301 | 30,372 | 38,279 |
| Germany (West) | 5,974 | 6,284 | 9,783 | 8,562 | 7,477 |
| Hong Kong | 671 | 4,318 | 5,851 | 4,324 | 5,838 |
| Italy | 566 | 1,864 | 2,471 | 773 | 2,052 |
| Japan | 5,627 | 17,790 | 19,424 | 11,935 | 9,914 |
| Mexico | 3,046 | 5,829 | 5,761 | 7,080 | 9,196 |
| Netherlands | 1,492 | 10,294 | 14,514 | 3,712 | 4,130 |
| Switzerland | 1,427 | 3,665 | 3,845 | 5,464 | 8,846 |
| U.K. | 5,303 | 2,731 | 2,955 | 7,920 | 17,591 |
| other | 12,748 | 21,321 | 27,815 | 25,461 | 34,300 |
| *Total* | *55,007* | *114,123* | *149,417* | *139,320* | *178,500* |

[a]Ref. 22.

Some pigments or pigment derivatives have found unorthodox experimental applications as photoconductors (23) for copiers and in a soluble form as medical agents for photodynamic therapy (24) by sensitized generation of singlet oxygen against certain animal and human cancers. These newer fields of application are under active investigation.

On a global basis the consumption of organic pigments (4) by various industries in 1991 is detailed in Table 12.

About one-third each worldwide consumption of organic pigments takes place in Western Europe (32%) and North America (29%) (25). Other geographic areas consume as follows: Far East, 18%; China/Russia/Eastern Europe, 12%; Latin America, 4%; Africa and Middle East, 3%; and Australia/New Zealand, 2% (25).

**Testing and Standardization.** Pigments are subjected to a number of tests before they are released to customers. Testing is complicated because of the great diversity of pigment types and uses. A given pigment may be dispersible in one system but poorly dispersible in another, and can exhibit different durability depending on the system. Performance is system dependent. Standardization is

Table 11. U.S. Production, Imports, and Exports of Organic Pigments and Preparations, 1988–1992[a]

| Market | Production | Imports | Exports | Apparent consumption | Imports relative to consumption, % |
|---|---|---|---|---|---|
| quantity, t | | | | | |
| 1988 | 52,680 | 16,410 | 5,200 | 63,890 | 25.7 |
| 1989 | 50,360 | 15,768 | 11,349 | 54,779 | 28.8 |
| 1990 | 52,551 | 21,056 | 14,764 | 58,843 | 35.8 |
| 1991 | 51,311 | 21,236 | 16,019 | 56,528 | 37.6 |
| 1992 | 56,944 | 20,891 | 21,793 | 56,042 | 37.3 |
| value, $10^3$ $ | | | | | |
| 1988 | 594,657 | 161,692 | 55,007 | 701,342 | 23.1 |
| 1989 | 701,552 | 171,806 | 114,123 | 759,235 | 22.6 |
| 1990 | 717,194 | 205,211 | 149,417 | 772,988 | 26.6 |
| 1991 | 643,561 | 237,704 | 139,320 | 741,945 | 32.0 |
| 1992 | 788,697 | 267,722 | 178,500 | 877,919 | 30.5 |

[a]Ref. 22.

Table 12. Worldwide Consumption of Organic Pigments by Industrial Sectors[a]

| Industrial sector | Wt % | Value, % |
|---|---|---|
| printing inks | 54 | 41 |
| paints | 23 | 29 |
| plastics | 18 | 23 |
| special applications | 5 | 7 |

[a]Ref. 4.

carried out against a standard sample for coloristics and a variety of working properties. Among the tests, depending on the pigment type, may be thermal stability, hiding power, rheology, migration, chemical stability, gloss, distinctness of image, durability, etc. Durability or weatherfastness are usually established for a standard and other samples are tested occasionally against the standard by exposure under actual conditions of use. Car paint manufacturers, for example, require exposure in Florida under specific conditions for at least two years. Accelerated exposure under artificial light and simulated weather sequences is also employed, requiring significantly shorter times of exposure. In plastic systems, ink- or fiber-accelerated exposure is generally an acceptable method of testing. The mechanism of photochemical degradation of organic pigments is poorly understood.

In the process of testing, color deviations are expressed in the CIELAB system which projects total color differences either on the axes of the rectangular LAB or the equivalent polar LHC system (26). In either case tested samples must fall within acceptable ranges or limits established versus a standard by the pigment manufacturer and accepted by the pigment user.

In dispersing a pigment by an established method, acceptable pigment strength vs a standard must be achieved, even though an ideal dispersion normally is never realized, that is, not all agglomerates are broken down completely. In effect, there is more color strength built into a pigment than is usually realized commercially, and the dispersion process is simply discontinued for economic reasons.

## Health and Safety Factors

Since pigments are generally insoluble, unlike most dyes, they are usually not bioavailable and consequently are generally not absorbed or metabolized. Nevertheless many health-related studies have been carried out and reported in the literature. A summary (27) of rabbit tests with 192 organic pigments showed that six are skin irritants and 24 cause various degrees of eye irritations, the rest being nonirritating.

Acute toxicity of organic pigments has been studied extensively. The most common measure of toxicity is $LD_{50}$ expressed in mg/kg of body weight which has a lethal effect on 50% of test animals after a single (oral, dermal, etc) administration. These tests assess toxicity vs other known compounds. A large $LD_{50}$ value represents a low degree of toxicity. The Federal Hazardous Substances Act defines a material with an LD above 5000 mg/kg as being nontoxic (28) and the Occupational Safety and Health Administration (OSHA) defines chemicals with an $LD_{50}$ of less than 500 mg/kg as toxic (29). The Ecological and Toxicological Association of Dyestuffs and Organic Pigments Manufacturers (ETAD) published a summary (30) on toxicity testing of 4000 colorants, the National Printing Ink Research Institute (31) tabulated $LD_{50}$ data for 108 organic pigments, and in a NIFAB symposium lecture (32) the toxicology of 194 pigments was reviewed. Most oral $LD_{50}$ values exceeded 5000 mg/kg; only four pigments showed values between 2000 and 5000 mg/kg acute toxicity. By OSHA's definition none of the tested pigments are toxic and an overwhelming number are not as toxic as table salt which has an $LD_{50}$ of 3000 mg/kg. Thus, pigments in general have very low levels of acute toxicity.

Chronic toxicity defines a specific dose or exposure level that will produce measurable, long-term toxic effects, including carcinogenicity. Some cancers are thought to arise due to mutations in body cells. Although the correlation between the Ames mutagenicity test (33) and carcinogenicity is a controversial issue it is nevertheless considered to be a useful tool. Out of 24 pigments given the Ames test (31), only two (PO 5 and PR 1) showed weakly positive results. Eleven pigments (PY 12, 16, 83; PO 5; PR 3, 4, 23, 49, 53:1, 57:1; and PB 60) were tested for carcinogenicity by long-term feeding studies. Results (34) were negative for most and equivocal for three (PO 5 and PR 3 and 23).

One area which requires special comment is a study (13) which showed that certain diarylide pigments processed in polymers above 200°C and particularly above 240°C decompose to give off 3,3'-dichlorobenzidine, an animal carcinogen. As a consequence diarylide pigments (not, however, condensation disazo pigments) are not recommended for use in any applications where they might be exposed to temperatures exceeding 200°C.

The hazards associated with handling pigments is contained in the MSDS mandated by an OSHA Hazards Communication Standard (28) which also requires labeling and employee information and training.

**Ecological Effects.** The starting materials for manufacture of organic pigments are as diverse as the pigments themselves. However, most starting materials are derived from petroleum or natural gas sources. Although many pigments are synthesized in water, a variety of organic solvents are also employed by the industry. The effective utilization of all starting materials and solvents, and reduction of undesirable by-products, is a primary objective of the organic pigment industry because the cost of environmental protection has increased significantly. Notwithstanding the increasing demand for organic pigments, the need for replacement of some heavy-metal inorganic pigments by acceptable organic substitutes, and the continued demand for brighter and more durable pigments in critical applications, the industry strives to accomplish the task with minimum adverse effect on the ecology while continuing its contribution to the aesthetic side of human existence.

## BIBLIOGRAPHY

"Pigments (Organic)" in *ECT* 1st ed., Vol. 10, pp. 660–689, by E. R. Allen, Rutgers University; in *ECT* 2nd ed., Vol. 15, pp. 555–589, by F. F. Ehrich, E. I. du Pont de Nemours & Co., Inc.; in *ECT* 3rd ed., Vol. 17, pp. 838–871, by M. Fytelson, Sandoz Colors & Chemicals.

1. *Safe Handling of Color Pigments*, 1st ed., Color Pigments Manufacturers Association, Inc., 1993.
2. *Color Index*, 3rd ed., Vol. 4, The Society of Dyers and Colourists, Bradford, Yorkshire, England; American Association of Textile Chemists and Colorists, Research Triangle Park, N.C., p. 4618.
3. *Ibid*, p. 4081.
4. A. Zamoyski, "Global Outlook of Organic Pigments from a European Perspective," at *The Intertech Conference on Environmental Pigments*, Herndon, Va., 1993.
5. C. Graebe and C. Liebermann, *Ber. Deut. Chem. Ges.* **1**, 106 (1868).
6. O. N. Witt, *Z. Physik. Chem.* **9**, 522 (1876).
7. A. Hanke, *J. Opt. Soc. Amer.* **56**, 713 (1966).
8. ISO 7724-1-1984, *CIE Standards Colorimetric Observers: Paints and Varnishes—Colorimetry*, Part 1, *Principles*; ISO 7724-2-1984, Part 2, *Colour Measurement*, ISO, Geneva, Switzerland.
9. J. Molliet and D. A. Plant, *J. Oil Colour Chem. Assoc.* **52**, 289 (1969).
10. W. C. Carr, *J. Oil Colour Chem. Assoc.* **50**, 1115 (1967).
11. K. Hunger, E. F. Paulus, and D. Weber, *Farbe + Lack* **88**, 453 (1982); R. K. Harris, P. Johnsen, K. J. Packer, and C. D. Campbell, *J. Chem. Soc., Perkin Trans. II*, 1383 (1987); L. Dongzhi and R. Shengwu, *Dyes Pigments* **22**, 47 (1993).
12. D. Kobelt, E. F. Paulus, and W. Kunstmann, *Acta Crystallogr. Sect.* **B28**, 1319 (1972); *Z. Krystallogr.* **139**, 15 (1974); A. Whitaker, *Z. Kirstallogr.* **146**, 173 (1977).
13. R. Az, B. Dewald, and D. Schnaitman, *Dyes Pigments* **15**, 1 (1991).
14. B. G. Hays, *Amer. Ink Maker* **64**, 13 (1986).
15. F. M. Smith and J. D. Easton, *J. Oil Colour Chem. Assoc.* **49**, 614 (1966).
16. S. S. Labana and L. L. Labana, *Chem. Rev.* **67**, 1 (1967).
17. H. Gaertner, *J. Oil Colour Chem. Assoc.* **46**, 13 (1963).
18. A. Iqbal, L. Cassar, A. C. Rochat, J. Pfenninger, and O. Wallquist, *J. Coat. Technol.* **60**(758), 37 (1988).
19. K. Ogawa, K. J. Scheel, and F. Laves, *Naturwissenschaften* **53**(24), 700 (1966).
20. K. Kozawa and T. Uchida, *J. Heterocyclic Chem.* **27** 1575 (1990); E. Dietz, *11th International Color Symposium*, Montreux, Switzerland, 1991.
21. *Synthetic Organic Chemicals, United States Production and Sales*, USITC Publications 1588 (1983), 1745 (1984), 1892 (1985), 2009 (1986), 2118 (1987), 2219 (1988),

2238 (1989), 2470 (1990), 2607 (1991), 2720 (1992); United States International Trade Commission.

22. *U.S. Imports for Consumption—Harmonized Tariff Schedules*, Publications FT 246 (1988) and FT 247 (1989–1992), U.S. Dept. of Commerce, Washington, D.C.; *U.S. Exports—Harmonized Schedule B*, Publications FT 446 (1988) and FT 447 (1989–1992), U.S. Dept. of Commerce, Washington, D.C.

23. R. O. Loutfy, A. M. Hor, P. M. Kazmaier, R. A. Burt, and G. K. Hamer, *Dyes Pigments* **15**, 139 (1991).

24. W. S. Chan, J. F. Marshall, R. Svensen, J. Bedwell, and I. R. Hart, *Cancer Res.* **50**, 4533 (1990); H. L. van Leengoed and co-workers, *Photochem. Photobiol.* **58**, 233 (1993).

25. Presentation of *The 1992 DCMA Annual Meeting, The World Market for Organic Pigments*; *Amer. Ink Maker* **70**, 38 (1992).

26. ISO 7724-3-1984, 1st ed., *Paints and Varnishes—Colorimetry*, Part 3, *Calculation of Colour Differences*, ISO, Geneva, Switzerland.

27. W. Herbst and K. Hunger, *Industrial Organic Pigments*, VCH Publishers, Inc., New York, 1993, p. 570.

28. *Code of Federal Regulations*, Title 16, part 1500, Federal Hazardous Substances Act Regulation, Consumer Products Safety Commission, Washington, D.C.

29. *Code of Federal Regulations*, Title 29, part 1910.1200, OSHA Hazard Communication Standard, Occupational Safety and Health Administration, Washington, D.C.

30. E. A. Clark and P. Anliker, in O. Hutzinger, ed., *Organic Dyes and Pigments, The Handbook of Environmental Chemistry*, Vol. 3, Part A, Springer-Verlag, Berlin, 1980.

31. *NPIRI Raw Materials Data Handbook*, Vol. 4, *Pigments*, Francis MacDonald Sinclair Memorial Laboratory 7, Lehigh University, Bethlehem, Pa., 1983.

32. K. H. Leist, *Toxicity of Pigments NIFAB Symposium*, Stockholm, 1980.

33. B. N. Ames, J. McCann, and E. Yamasaki, *Mutagen. Res.* **31**, 347 (1975).

34. Ref. 27, p. 572.

*General References*

H. A. Lubs, ed., *The Chemistry of Synthetic Dyes and Pigments*, ACS Monograph Series, Reinhold Publishing Corp., New York, 1955.

D. Patterson, *Pigments, An Introduction to their Physical Chemistry*, Elsevier, New York, 1967.

T. C. Patton, ed., *Pigment Handbook*, Vols. I, II, and III, John Wiley and Sons, Inc., New York, 1973.

K. Venkataraman, *The Chemistry of Synthetic Dyes*, Vol. V, Academic Press, Inc., New York, 1971.

F. H. Moser and A. L. Thomas, *The Phthalocyanines*, 2 Vols., CRC Press, Boca Raton, Fla., 1983.

P. A. Lewis, *Organic Pigments*, Federation of Societies for Coatings Technology, Federation Series on Coatings Technology, 1988.

J. Sanders, *Pigments for Ink makers*, SITA Technology, London, 1989.

E. E. Jaffe, *J. Oil Colour Chem. Assoc.* **75**, 24 (1992).

*Safe Handling of Color Pigments*, Color Pigments Manufacturer Association, Inc., 1993.

H. Zollinger, *Color Chemistry*, 2nd ed., VCH Publishers, Inc., New York, 1993.

W. Herbst and K. Hunger, *Industrial Organic Pigments*, VCH Publishers, Inc., New York, 1993.

E. E. JAFFE
Ciba-Geigy Corporation

# PILOT PLANTS

The design or substantial modification of a new plant or process, its subsequent construction, and start-up represent a tremendous investment of time and money. The rewards are great if a significant improvement is realized; the risks are also great if a costly commercial plant fails to produce as expected. To reduce the degree of risk, lengthy and expensive research programs are often undertaken.

Table 1 shows a sequence of activities that might be followed as part of a research program. Depending on the complexity of the project, some of these steps may be eliminated; a great deal of recycle and feedback is also possible between each step. Each step involves an exponential increase in the resources, time, and money required, as shown in Figure 1. This illustrates the need to minimize the amount of recycling as the project progresses through the various steps to the commercial unit. A substantial amount of time and money has progressively been invested and any desire to change the process to make minor improvements should be resisted. At some point the decision must be made that the remaining unanswered process questions are acceptable risks or that additional, potentially extensive research is required.

A further consideration, which is contrary to minimizing the risks associated with a new product or process, can be the timeliness of the discovery. Timing can be the most important factor in a research program, as in the case of securing a new market with a totally new product or attempting to secure a patent position before a competitor. In this situation, the decision may be made to proceed to commercialization earlier than desirable, prior to satisfactorily resolving all design concerns. This usually requires a more conservative and, hence, more expensive design approach.

A pilot plant is a collection of equipment designed and constructed to investigate some critical aspect(s) of a process operation or perform basic research. It is a tool rather than an end in itself. A pilot plant can range in size from

**Table 1. Typical Research Program Steps**

| Step | Intention |
|---|---|
| origin of concept | initiate ideas or concepts to be investigated |
| basic research | determine basic feasibility of concept |
| preliminary economic evaluation | determine if economic incentive is sufficient to proceed with investigation |
| laboratory research | develop basic data |
| product evaluation | evaluate suitability of product produced |
| process development and preliminary engineering | produce preliminary design of actual commercial plant, resolving process questions as they arise |
| pilot-plant study | prove basic reliability of proposed process or process design |
| demonstration unit | demonstrate feasibility of process on mid-size commercial scale |
| prototype unit | demonstrate feasibility of process on small-size commercial scale |
| commercial plant | sell a product at a profit |

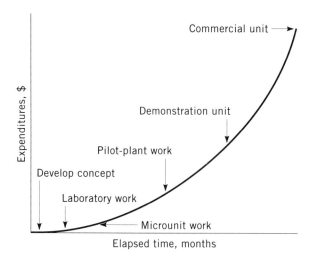

**Fig. 1.** Steps involved in research program for a new pilot plant.

a laboratory bench-top unit to a facility only marginally smaller than a commercial unit. The purposes for its construction and operation can vary widely: confirming feasibility of a proposed process; providing design data; determining the economic feasibility of a new process; determining optimum materials of construction; testing operability of a control scheme; determining the extent of plant maintenance (qv); producing sufficient quantities of product for market evaluation; obtaining kinetic data; screening catalysts; proving areas of advanced technology; providing data for solutions to scale-up problems; providing technical support to an existing process or product; assessing process hazards; determining operating costs; optimizing an existing process; evaluating alternative processes, feedstocks, or operating conditions; and performing basic process research.

The need for a pilot plant is a measure of the degree of uncertainty in developing a process from the research stage to a full commercial plant. A modification to a well-known process may go directly from basic research work to design of a commercial plant; using this approach for a brand new process risks a significant failure. Hence, one or more intermediate size units are usually desirable to demonstrate process feasibility as well as to determine safe scale-up factors.

## Categorizing Pilot Plants

Pilot plants can be categorized by a number of different methods. Size is the most common classification as it is the most uniformly proportional to construction and operating costs; some typical characteristics are shown in Table 2. These characteristics are generalizations and may not be applicable in any specific case. Whereas some microunits are very heavily automated, for example, and some demonstration units are not, the description applies in general when comparing pilot plants as a whole.

The pilot plant's degree of automation is the next most common classification as the instrumentation is usually a large fraction of the initial construction

**Table 2. Typical Classes of Pilot Plants by Size**

| Class | Characteristics | Cost range, $ |
|---|---|---|
| microunit or bench-top unit | small size ($<2$ m$^2$) typically found in hood or on bench-top; small tubing used throughout (3–6 mm dia); limited automation | 10,000–100,000 |
| integrated pilot plant | larger size ($\sim$10–25 m$^2$) found in open bay, walk-in hood, or containment cell; tubing and small (13–25 mm dia) pipe used; fully automated feed and product systems with limited capacity usually included | 50,000–500,000 |
| demonstration or prototype unit | very large size ($>250$ m$^2$) usually in dedicated building or area; pipe used extensively; fully automated usually with dedicated computer system; feed and product systems, including significant storage capacity | $\geq$1,000,000 |

and annual operating costs; some typical characteristics are shown in Table 3. The high cost of operating labor and the difficulties in manually controlling a process while taking accurate data have forced virtually all but the simplest new pilot plants to increase their level of automation so this classification is becoming less useful. These classifications are elastic and other possibilities have been discussed (1).

## Scale-Up

Scale-up is the process of developing a plant design from experimental data obtained from a unit many orders of magnitude smaller. This activity is considered successful if the commercial plant produces the product at planned rates, for planned costs, and of desired quality. This step from pilot plant to full-scale operation is perhaps the most precarious of all the phases of developing a new process because the highest expenses are committed at the stages when the greatest risks occur.

Process plant design has come a long way from the early 1930s when process designers used the rule-of-thumb that a process facility could not be scaled-up more than 10-fold (2). American Oil's Ultracracking unit (Texas City, Texas) for example, was designed from data from a small pilot plant with a scale-up factor of 80,000 (3).

Scale-up problems exist for many reasons including basic equations unsolvable by known mathematical techniques, interconnected physical and chemical aspects of the process resulting in coupled basic equations, solutions used on the pilot plant not suitable on the commercial scale, and unknown equipment performance at sizes never used before. Discussions of scale-up problems and solutions are available (3–10). Although successful pilot-plant operation does not guarantee successful commercial plant operation it considerably increases the

Table 3. Typical Classes of Pilot Plants by Degree of Automation[a]

| Degree of automation | Characteristic | Cost of instrumentation, $[b,c] |
|---|---|---|
| manual | all process conditions set and monitored by operators; poor repeatability; loose process control | <2,000 |
| local control | most important variables monitored by dedicated controllers; tighter process control; possibly some data gathering instrumentation | 10,000–20,000 |
| mixed systems | all variables monitored by controllers (local or distributed) with some computer controlled or monitored; usually data gathering provided by "smart" data logger or computer; usually older system upgraded to more modern standards | 15,000–30,000 |
| automated | all or majority of systems computer controlled; tightest process control; may be supervisory or direct digital control; high repeatability; lowest operating labor; most batch operations may be automated; data gathering integral part of control system | 20,000–60,000 |

[a]A pilot plant typically has significantly more control loops than the average process facility. Caution should be used when applying process correlations in the absence of a detailed design.
[b]Costs include installation of equipment but not field wiring. The figures shown are typical but capable of wide variation.
[c]Based on a typical medium-sized pilot plant of 10–20 control loops. For smaller units (<10 control loops) reduce the above by 25–40%; for larger units (20–40 control loops) increase the costs by 50–100%.

success factor. Care, however, must be taken at all stages of a research program to ensure good scale-up practices are being followed.

## Planned Pilot-Plant Experimentation

Planning the experimental program is an important part of the decision process concerning the type of pilot plant required. This step should be performed as soon as the research program objectives are formulated because the type of experimental program affects the total program cost, the type of pilot plant required, and the way the pilot plant needs to operate. Thus it strongly affects the economic justification of the research program. Planned experimentation also helps reduce the number of tests required, a key factor in controlling annual costs as it is not unusual for the annual operating costs of a pilot plant to be two to three times the initial construction costs. In addition, the experiments planned may force alterations in design or show that a pilot plant is not the best place to obtain some of the desired information. Early planning may also highlight the need for more laboratory work to support the pilot-plant operation. Laboratory work is less costly and frequently gives insight to correlations or models for the process to be studied in the pilot plant.

Pilot-plant design specifications should be established only after careful consideration of the experimental program because decisions on the accuracy of instruments, analyzers, and other equipment should be based on the requirements of the experiments planned for the unit. Flexibility and versatility are important but costly; when provided unnecessarily or too profusely they can result in a unit that is difficult or impossible to operate successfully.

Statistical designs for experiments maximize information and reduce research time and costs. These techniques are less likely to miss synergistic factors affecting performance or product quality, minimize the element of human bias, eliminate less productive avenues of experimentation by taking advantage of previous data, and reduce the number of pilot-plant runs needed to define the effects of variables. Overall, statistical designs increase the confidence level in the experimental results. Additional information is available (10–18) (see DESIGN OF EXPERIMENTS).

## Pilot-Plant Design

The way a pilot plant is designed affects its cost, operability, and effectiveness. All operating ranges may not yet be fully defined at the initial design stage, but a realistic preliminary range is required before the design is commenced, as is a clear definition of the pilot plant's purpose.

The first step in pilot-plant design is to determine whether to design a pilot plant for process modeling or problem investigation. Modeling the process involves reproducing the specific unit operations on a smaller, pilot scale. This promotes safe scale-up, minimizes design time, and reproduces all process operations of interest. It is usually expensive since all operations are reproduced, not just the most important ones. Investigating the problem involves designing a pilot plant to look at a specific area of interest. In this case, the pilot plant may not resemble the commercial operation. Although this approach is usually cheaper and quicker it carries the inherent risk of missing the real problem and producing nothing of value. Several references have addressed each type of approach (1,19–22).

Two approaches are commonly used for pilot-plant design. The first uses conventional design techniques which mimic commercial process design. This usually provides for a safe and operable design as the design methodology is known and the effectiveness of the final results have been proven previously. This approach is not always feasible: the proposed scale of the pilot-plant operation may be completely outside the range of all commercial design techniques, the design suggested may not be economical on the scale envisioned or in the location proposed, or copying the commercial design may carry some inherent limitations that adversely impact operations at ranges or conditions which are of interest to the pilot plant but not to the commercial plant. The alternative approach is to use a design methodology oriented to the pilot plant using many conventional design techniques but trying to maximize the advantages of the pilot-plant operations with regard to scale, technique, and operation. Advantages include economic savings due to reduced equipment, construction, and operating costs, economical and efficient ways around otherwise expensive problems, increased simplicity of operation resulting in reduced maintenance and operating staff, and improved process understanding through use of specialized equipment. The lack of published design information, however, raises the risk factor; hence the quality of the final design is very dependent on the skill level and experience of the design engineer.

## Methods of Estimating Pilot-Plant Costs

Pilot plant costs range from $10,000 to $10,000,000, but the majority are typically in the range of $50,000 to $250,000, assuming an existing facility is available to house them. There are three basic methods for estimating the costs to design and construct a pilot plant: similarity, cost ratios, and detailed labor and materials.

Similarity involves estimating the cost of the pilot plant based on the costs to design and construct a similar unit. This is the fastest method, but it is the least accurate; accuracies of ±100% are not uncommon. This is due to the fact that few similar units are ever identical. Cost ratios develop the cost estimate by relating the overall cost of the pilot plant or a part of the pilot plant to a known factor such as the cost of major process equipment, the number of control loops, the size of the equipment, or a variety of similar factors. The cost estimate is built up by using the ratios to develop the cost of the entire unit, individual equipment, or separate subsystems depending on how detailed a cost ratio estimate is made. Unfortunately, cost ratio information is rarely available for pilot-plant-scale equipment and the lack of a large data pool renders much of what is available suspect regarding its overall accuracy. Depending on the type of cost ratios used and the experience of the personnel making the estimate, accuracies of ±25–50% are typical. Cost ratios can be used more accurately if they are restricted to small subsystems because much information is available in this area. However this usually requires as much or more of an effort as the last method.

Detailed labor and material estimating involves breaking the pilot-plant construction down into a detailed series of small tasks and estimating the labor and materials required for each separate task. This method can produce estimates with accuracies of ±10–20% but requires more effort than either of the previous methods (Table 4) (1,23–28). The difference between a general and detailed cost ratio cost estimate is the amount of detail involved. A general cost ratio estimate might develop the total cost for the pilot-plant control system, or even the pilot plant, from the number of control loops. A detailed cost ratio estimate estimates the cost per loop based on the type of loop. The rest of the estimate must be generated by a large number of similar ratios.

In general, similarity or general cost ratios estimates are used for developing screening estimates when an estimate is required quickly and a limited accuracy is acceptable. Detailed cost ratios or detailed labor and materials estimates are generally developed prior to actual appropriation of funds to develop a better estimate for budgeting and cost control purposes.

**Table 4. Pilot-Plant Cost Estimating Techniques**

| Method (accuracy range, %) | Information required | Time to develop,[a] weeks |
|---|---|---|
| similarity ($\pm50-100$) | costs of similar unit<br>differences from proposed unit | <1 |
| general cost ratios ($\pm25-50$) | all scaling elements, such as control<br>loops, size, or similar factors | 1–2 |
| detailed cost ratios ($\pm15-30$) | detailed P&ID[b]<br>complete equipment list<br>preliminary layout | 4–12 |
| detailed labor and materials ($\pm10-20$) | detailed P&ID[b]<br>preliminary electrical and<br>instrumentation drawings<br>complete equipment and materials list<br>preliminary layout | 4–12 |

[a]After preliminary design is completed.
[b]Piping and instrument diagram.

The costs to operate a pilot plant are a summary of the costs of the feedstock, product disposal, utilities, operating labor, spare parts, maintenance, and support services. Significant efforts on reducing operating costs have focused on reducing the operating labor through automation and unattended operation (1,29–37) as well as controlling waste disposal costs through careful design and planning (1,38).

## Pilot-Plant Space

In general, pilot-plant space can be divided into five basic types: separate buildings, containment cells or barricades, open bays, walk-in hoods, and laboratory areas. A summary of the advantages and disadvantages of each has been given (1).

The space required for a pilot plant varies tremendously with its size and type. A small unit may require only part of a laboratory (perhaps $5-10$ m$^2$), whereas an average pilot plant of $50,000 to $200,000 may require a large room or building (perhaps $500-2000$ m$^2$), excluding extended feed or product storage.

**Scheduling.** A significant concern in all pilot-plant work is minimizing the time involved between project inception and meaningful data generation. Typically, an average pilot plant of $50,000–$250,000 requires from six to eighteen months to progress through this process: two to eight months for design, materials procurement, and scheduling; two to six months for construction; and one to four months for start-up (up to and including first actual run). This lead time can be reduced significantly through careful planning, purchasing long lead time items early, and various construction techniques. Typical lead times for major pieces of equipment for planning purposes have been given (1). Design time can be reduced if standardized designs have been developed in advance for common subsystems. Construction time can be reduced by various methods. Construction can begin on certain subsystems before the final design is complete. The pilot plant can be built a subsection at a time, starting up each subsection as it is completed. Both approaches reduce the lead time but with an added element of risk.

## Pilot-Plant Control Systems

Defining the requirements for a pilot-plant control system is often difficult because process plant experience for comparison and evaluation is commonly lacking and the design is frequently performed by personnel inexperienced in either instrumentation systems or pilot-plant operations. The isolated and often intermittent nature of pilot-plant operations also inhibits evolution and promotes individual unique installations. This complicates the selection process.

Automatic instrumentation systems for pilot plants have historically been based on analogue devices. The trend in these devices has been toward smaller and less expensive units with more flexibility; this increases their suitability for pilot-plant instrumentation systems. Computer-based control systems, once too expensive for anything but larger process plant applications, are also commonly available in sizes and costs suitable for most pilot-plant operations. The additional data gathering and work-up capabilities of these systems make them very attractive. Computer-based systems are growing smaller and less expensive; analogue equipment is being redesigned to use microprocessors. This is blurring the distinction between the two systems. Most organizations are installing computer-based systems on all but the simplest pilot plants (see PROCESS CONTROL).

Three types of computer control systems are commonly used for pilot-plant instrumentation. The first is a centralized system, usually based on a minicomputer or occasionally a mainframe. These systems have large storage capacities, substantial memories, and much associated equipment. They typically control all the pilot plants in an area or facility. Centralized systems are economical if a large number of units are involved but are becoming less common due to their high installation and maintenance costs as well as the limitation that any failure of the central system shuts down all pilot plants involved.

Stand-alone computer systems, usually based on a personal computer (PC) or programmable logic controller (PLC), provide a separate computer system for each pilot plant. This allows for economical expansion for new units, separates pilot plants completely for maintenance and troubleshooting, and often has the lowest initial cost. Standardization can be a problem and software control, data gathering, and storage packages can be limited in size, scope, and capability; these are usually acceptable trade-offs.

Distributed control systems (DCS) are a hybrid of the previous systems, having stand-alone computers for control and data gathering at each pilot plant together with one or more higher level computer(s) for data storage and work-up. Typically more expensive than either central systems or stand-alone computers, distributed control systems have the advantages of both systems. This approach is preferred for pilot plants when it can be economically justified, usually for larger research facilities or very large demonstration units.

Further information on the different systems including a more detailed discussion of their costs, advantages, and disadvantages is available (1,39,40). Information on actual experience is also available (32,34–36,41–47).

The support resources available to service, repair, and calibrate the control system must also be considered when selecting a system; they are considerable and can be estimated from the literature (48–50).

**Instrumentation.**　Pilot plants are usually heavily instrumented compared to commercial plants. It is not uncommon for a pilot plant to have an order of magnitude more control loops and analytical instruments than a commercial plant because of the need for additional information no longer required at the commercial stage. A discussion of all the specific types of instrumentation used on pilot plants is beyond the scope of this article. Further information on some of the more common instrumentation is available (1,51).

Analytical instrumentation for pilot plants is divided into two classes: off-line and on-line. Off-line analysis is a batch operation requiring sample taking, handling, and storage followed by analysis at a later date, frequently in another location. It interferes least with the process, is commonly less expensive, and allows the analysis to be performed under rigorously controlled conditions. Some analyses cannot be made in any other manner because of equipment limitations, although this is becoming less frequent. Disadvantages of off-line analysis include taking, handling, and storing representative samples and the inability to modify process conditions on a real-time basis due to the time lag between sampling and analysis.

On-line analysis is often more expensive and difficult to set up initially but can be more accurate and reliable if performed properly. On-line analyzers can also be used to provide real-time control of a process through a secondary variable such as severity or conversion, as opposed to controlling a primary variable, such as temperature (36,52).

The maintenance of analytical instrumentation requires trained personnel and is a time-consuming task (39,40). An additional problem is the necessity of frequently checking the calibration of the analysis instrumentation and recalibrating if required. Stand-alone data gathering instrumentation, once common in pilot plants, has been virtually replaced in all but the simplest pilot plants by a data gathering computer, usually used for process control as well.

## Feed and Product Handling

One of the most vexing aspects of pilot-plant work can be feed and product handling as a pilot plant is neither designed nor operated as a closed-loop system like a commercial plant. Indeed, the problems involved in handling and storing feed and product materials can sometimes seem to rival the pilot-plant process problems in difficulty.

The toxicity of both feeds and products must be carefully considered during the preliminary design stages (53,54) especially if the feeds or products contain known or suspected carcinogens. Special testing may even be required prior to operation to determine the toxicity of unknown materials.

Provision must be made for sufficient storage of both feeds and products and the storage system should be designed to minimize operator handling time. Sufficient surge time must be available at both ends of the process to minimize the need for close operator attention. High level alarms and cut-offs should be installed on all product vessels that could overflow and low level alarms and cut-offs should be considered for all feed vessels that could be emptied. Vented products must conform to all existing federal, state, and local environmental regulation, as well as good engineering judgment. This may require scrubbing or neutralization, incineration, flaring, or similar measures. Care is required to ensure emergency venting is adequately sized, selected, and installed with products vented to an appropriate height and/or location. If a pilot plant is located in an electrically classified area, flammable feed and product handling becomes more difficult. The same is true if toxic materials are used in a poorly ventilated area.

## Pilot-Plant Start-Up

Pilot-plant start-up is different from principal process plant start because of the smaller scale of the unit, smaller resources committed, lack of advance start-up planning, and limited experience with the pilot-plant process and operation.

The key to a successful pilot-plant start-up is advance planning, at least six to eight weeks before project completion, to allow identification of problem areas and concerns in time for successful resolution. A detailed start-up sequence should be developed listing each task to be performed in chronological order (1). The start-up sequence then allows the development of a list of required resources and a tentative start-up schedule. A successful start-up requires this advance planning as well as adequately trained and experienced personnel with a variety of skills. Many companies have found that a specialized start-up group is a primary asset if pilot-plant work is regularly done. Safety is also a significant concern during start-up because interlock systems are not fully functional and equipment and subsystems are being energized for the first time.

Pilot-plant start-up costs vary widely. The costs booked to the start-up as well as the extent of the commissioning activities labeled as "start-up" vary widely among organizations. Typical costs range from 5 to 50% of construction cost; 10 to 20% is typical. Start-up durations also vary, but a range of one to three months is common depending on the personnel training and expertise.

## Safety in Pilot-Plant Design and Operations

Pilot plants are often more hazardous than process plants, even though they are smaller in size, for many reasons. These include a tendency to relax standard safety review procedures based on the small scale, exceptionally qualified personnel involved, and the experimental nature of the research operations; the lack of established operational practice and experience; lack of information regarding new materials or processes; and lack of effective automatic interlocks due to the frequently changing nature of pilot-plant operations, the desire for wide latitude in operating conditions, and the lack of full-time maintenance personnel.

To minimize these concerns, most organizations require a formal series of safety reviews and hazard analyses for each new or modified pilot plant. At a minimum, this involves analyzing the proposed pilot plant before construction to identify and eliminate any potential hazards, including toxicity or flammability concerns, feed or product handling, disposal problems, relevant government regulations, and potentially harmful reactions. The exact type and timing of these reviews varies between organizations (1,39,40).

The pilot plant must also be carefully designed so that its control and safety systems are "fail-safe" and any unexpected equipment or utility failure brings the unit into a safe and de-energized condition. Unexpected or rapid process changes, if they can herald or lead to dangerous conditions (eg, runaway exothermic reaction), should be continuously monitored by appropriate instrumentation and suitable automatic action provided (1,55–67).

## Future Trends

Bench-top or microunits will continue to be the predominant type of pilot plant constructed due to their lower construction and operating costs. Demonstration and prototype units will become even rarer for the opposite reasons.

The trend toward construction and operation of specialized pilot plants will continue in an effort to reduce construction and installation costs. This will result in an increased need for fast design and construction in order to reduce the lead time for new pilot plants; the lack of existing units will increase an organization's reliance on their ability to develop new units quickly as the need arises. Consulting groups with expertise in these areas will grow, offering their services to smaller companies. Larger companies may find the need to develop and maintain specialized groups with expertise in design, start-up, and operations.

The safety aspects of pilot-plant construction and operation will continue to grow driven by both regulation and an increased awareness of the potential hazards. This will also require specialized skills and training at all levels.

Virtually all new pilot plants will be computer controlled and heavily automated due to the high cost of operating labor, need for high accuracy and repeatability, and ease of data gathering and work-up. Stand-alone computer and programmable logic controller systems will continue to dominate the market because of their low cost and ease of use.

## BIBLIOGRAPHY

"Pilot Plants" in *ECT* 1st ed., Vol. 10, pp. 696–711, by C. F. Montross, General Aniline & Film Corp.; in *ECT* 2nd ed., Vol. 15, pp. 605–637, by D. G. Jordan, Consultant; "Pilot Plants and Microplants" in *ECT* 3rd ed., Vol. 17, pp. 890–905, by R. Van Sweringen, D. J. Cecchini, and R. P. Palluzi, Exxon Research and Engineering Co.

1. R. P. Palluzi, *Pilot Plants: Design, Construction and Operation*, McGraw Hill Book Co., Inc., New York, 1992.
2. R. Katzen, *Chem. Eng.* **95** (Mar. 26, 1968).
3. F. R. Bradbury, *Chem. Tech.* **3**, 532 (1973).
4. H. C. Henry and J. B. Gilbert, *Ind. Chem. Proc. Dev.* **12**, 328 (Mar. 1973).
5. J. M. Smith, *Chem. Eng. Prog.* **64**, 78 (Aug. 1968).
6. R. Katzen, *Chem. Eng.* **95** (Mar. 26, 1968).
7. P. Trambouze, *Chem. Eng.* (Sept. 1979).
8. P. Trambouze, *Chem. Eng.* (Feb. 1990).
9. R. H. Wang and L. T. Fan, *Chem. Eng.* (May 1974).

10. W. L. Gore, *Ind. Eng. Chem.* **43**, 2327 (Oct. 1951).
11. A. C. Ackinson, *Chem. Eng.* **149** (May 9, 1966).
12. W. G. Hunter and A. C. Ackinson, *Chem. Eng.* **159** (June 6, 1966).
13. K. A. Brownlee, *Statistical Theory and Methodology in Science and Engineering*, John Wiley and Sons, Inc., New York, 1965.
14. W. G. Cochran and D. R. Cox, *Experimental Designs*, John Wiley and Sons, Inc., New York, 1965.
15. W. E. Biles and J. J. Swain, *Optimization and Industrial Experimentation*, John Wiley and Sons, Inc., 1980.
16. O. L. Davies, *The Design and Analysis of Industrial Experiments*, Longman, New York, 1978.
17. O. L. Davies and O. L. Goldsmith, *Statistical Methods in Research with Special Reference to the Chemical Industry*, Longman, New York, 1976.
18. C. R. Hicks, *Fundamental Concepts in the Design of Experiments*, Holt, Reinhart and Winston, New York, 1973.
19. L. A. Robbins, *Chem. Eng. Prog.* **75**, 45 (Sept. 1979).
20. J. B. Maerker and J. W. Schall, *Chem. Eng. Prog.* **45**, 1622 (Aug. 1953).
21. D. E. Garrett, *Chem. Eng. Prog.* **55**, 44 (Sept. 1959).
22. J. W. Payne, *Ind. Eng. Chem.* **45**, 1621 (Aug. 1953).
23. R. Perry and co-workers, *Perry's Chemical Engineer's Handbook*, 6th ed., McGraw Hill Book Co., Inc., New York, 1984.
24. J. W. Drew and A. F. Ginder, *Chem. Eng.* (Feb. 9, 1970).
25. C. Shanmugam, *Hydrocarbon Proc.* (Jan. 1983).
26. D. S. Remer and L. H. Chai, *Chem. Eng. Prog.* (Aug. 1990).
27. D. S. Remer and L. H. Chai, *Chem. Eng. Prog.* (Apr. 1990).
28. K. R. Cressman, *Cost Eng.* **25**, 1, 31 (Feb. 1983).
29. C. E. Ellis, *Chem. Eng. Prog.* **64**, 50 (Oct. 1968).
30. J. W. Carr, Jr. and co-workers, *Chem. Eng. Prog.* **61**, 84 (June 1965).
31. C. D. Ackermann, *Chem. Eng. Prog.* **61**, 67 (June 1965).
32. C. D. Ackermann, A. B. Hartmann, and R. A. Wright, *Ind. Eng. Chem. Proc. Des. Dev.* **6**, 476 (Apr. 1967).
33. L. C. Hardison, G. P. Huling, and K. J. Metzger, *Chem. Eng. Prog.* **67**, 50 (Apr. 1971).
34. W. M. Herring and S. E. Shields, *Chem. Eng. Prog.* **61**, 94 (June 1965).
35. L. C. Hardison and R. A. Lengemann, *Chem. Eng. Prog.* (June 1964).
36. R. D. Dean, N. B. Angelo, and R. B. Young, *Chem. Eng. Prog.* (June 1965).
37. C. E. Ellis, *Chem. Eng. Prog.* (Oct. 1968).
38. M. M. Girvotch and J. G. Lowenstein, *Chem. Eng.* (Dec. 1990).
39. R. P. Palluzi, *Chem. Eng.* **97**, 3, 76 (Mar. 1990).
40. R. P. Palluzi, *Chem. Eng. Prog.* **87**, 1, 21 (Jan. 1991).
41. C. D. Ackerman, G. P. Hurling, and K. J. Metzger, *Chem. Eng. Prog.* (Apr. 1971).
42. E. J. Bourne and W. M. Herring, *Chem. Eng. Prog.* (Dec. 1977).
43. M. K. Corrigan, *AICHE Annual Meeting*, San Francisco, Calif., Nov. 1989.
44. W. E. Kaufman, *Chem. Eng. Prog.* (May 1979).
45. L. Uitenham and R. Munjal, *Chem. Eng. Prog.* (Jan. 1991).
46. R. K. Wang, L. T. Fan, and R. A. Van Sweringen, *Chem. Eng. Prog.* **53**, 11, 129 (Nov. 1961).
47. P. Tessier and N. Therien, *Microproc. Microsys.* **11**, 2, 99 (Mar. 1987).
48. A. T. Upfold, *Instru. Technol.* **18**, 46 (Feb. 1971).
49. F. H. Barbin, *Instru. Technol.* **20**, 44 (Nov. 1973).
50. R. W. Purgh, *AIChE National Meeting*, Paper 13 D, Tulsa, Okla., Mar. 10–13, 1974.
51. B. Liptak, ed., *Instrument Engineer's Handbook*, Chilton, Radnor, Pa., 1982.
52. J. E. Osborne, *Chem. Eng. Prog.* **70**, 76 (Nov. 1974).

53. C. H. Powell, *Chem. Eng. Prog.* **67**, 71 (Apr. 1971).
54. F. A. Graf, Jr., *Chem. Eng. Prog.* **63**, 67 (Nov. 1967).
55. R. N. Brummel, *Plant Oper. Prog.* **8**, 4, 228 (Oct. 1989).
56. M. A. Caparo and J. Strickland, *Chem. Eng. Prog.* **13** (Oct. 1989).
57. A. R. Jones and D. J. Hopkins, *Chem. Eng. Prog.* (Dec. 1966).
58. R. W. Prugh, *Chem. Eng. Prog.* **63**, 49 (Nov. 1967).
59. V. Lesins and J. J. Moritz, *Chem. Eng. Prog.* (Jan. 1991).
60. J. C. Dore, *Plant Oper. Prog.* **7**, 4, 223 (Oct. 1988).
61. J. W. Carr, Jr., *Chem. Eng. Prog.* **84**, 9, 52 (Sept. 1988).
62. P. C. Ashbrook and M. M. Renfrew, *Safe Laboratories*, Lewis, Chelsa, Mich., 1991.
63. Center for Chemical Process Safety, *Guidelines for Hazard Evaluation Procedures*, American Institute of Chemical Engineers, New York, 1992.
64. F. A. Graf, *Chem. Eng. Prog.* (Nov. 1967).
65. H. T. Kohlbrand, *Chem. Eng. Prog.* (Apr. 1985).
66. V. Siminski, *Chem. Eng. Prog.* (Jan. 1987).
67. L. V. Beckman, *Chem. Eng.* (Jan. 1992).

*General References*

*Means Mechanical Cost Data*, R. S. Means Co., Kingston, Mass., 1992.
*Building Construction Data*, R. S. Means Co., Kingston, Mass., 1992.
*Means Electrical Cost Data*, R. S. Means Co., Kingston, Mass., 1992.

RICHARD P. PALLUZI
Exxon Research and Engineering Company

## PIPELINE HEATING.  See PIPELINES.

# PIPELINES

Pipelines or pipe lines are continuous large-diameter piping systems, usually buried underground where feasible, through which gases, liquids, or solids suspended in fluids are transported over considerable distances. They are used to move water, wastes, minerals, chemicals, and industrial gases, but primarily crude oil, petroleum products, and natural gas. In the oil and gas business, a pipeline system consists of a trunkline, ie, the large-diameter, high pressure, long-distance portion of the piping system through which crude oil is shipped to refineries, or natural gas and oil products, respectively, are transported to distribution points, and smaller low pressure gathering lines that transport oil or gas from wells to the trunkline. Smaller lines used by natural gas distributors are not considered part of a gas pipeline system (see GAS, NATURAL).

Pipeline transport involves the application of force to the material being moved, either through the use of pumps to transport liquids, compressors to move gases, or flowing water to move solids. In some applications, vacuum may create the pressure differential.

## Pipeline Transport of Gases

Essentially all substances that are gases at standard conditions of temperature and pressure are transported commercially by pipeline, this includes ammonia, carbon dioxide, carbon monoxide, chlorine, ethane, ethylene, helium, hydrogen, methane (natural gas), nitrogen, oxygen, and others. Gases with moderate boiling points can be pipelined in either gaseous or liquid form; liquefied petroleum gases (LPG), carbon dioxide, ammonia, and chlorine are usually shipped as liquids because of the smaller pipeline volume for liquids.

The largest pipeline transport of gas, by far, is the movement of methane (natural gas). Natural gas can be liquefied, but it is not pipelined in liquid form because of cost and safety considerations. For overseas transport, it is shipped as liquefied natural gas (LNG) in insulated tankers, unloaded at special unloading facilities, vaporized, and then transported over land in pipelines as a gas.

**Gas Pipeline Industry.**    In early gas industry days, the U.S. Congress mandated that any pipeline must have an assured 20-year supply of gas before it could receive a permit to operate. The pipeline companies signed long-term purchase contracts with producers to satisfy this mandate and transported the gas to the cities where it was sold and title taken by the gas distribution companies. In the mid-1980s, however, the Federal Energy Regulatory Commission (FERC) began a restructuring of the industry that was to remove the merchant role from the gas pipelines and ultimately establish them as common carriers, comparable to railroads and liquid pipelines. Distribution companies and large gas users, such as corporations, may buy their own gas directly from producers, aggregators, or marketers and contract with pipelines to transport it for a fee.

This industry restructuring, coupled with removal of price controls and a large deliverability–demand inbalance, the so-called gas bubble, led to low gas prices during the late 1980s and early 1990s and resulted in the formation of a spot market for natural gas, a gas futures market, and a need for more sophisticated gas technology. The common carrier status of natural gas pipelines allows many small producers to enter the pipelines, if pipeline capacity is available, and make deliveries on a short-term (spot) basis. Flow computers at entry and exit points along the pipeline provide an instantaneous and accurate account of the gas that enters or leaves the pipeline. The flow computers tie into a large computer at pipeline headquarters to record an accurate account of gas from producers entering the pipeline as well as gas exiting to a gas user or distributor. In addition, FERC Order 636 mandated the establishment of an electronic bulletin board (EBB) for the gas industry, by which anyone with a personal computer can obtain the latest information needed to purchase gas and transport it by pipeline, such as suppliers' prices, available pipeline capacity, etc (1). The EBB will also carry information about capacity release in pipeline operations and permit trading of capacity release options so that elements of pipeline capacity will become a commodity, just as the natural gas itself. The objective is to optimize the efficiency of the national gas pipeline grid (2).

**Methane (Natural Gas).**   Although the first natural gas pipeline was constructed in 1870, most of the gas consumed at that time was manufactured from
coal (qv) and used locally rather than transported by long-distance pipeline. In
the 1990s, natural gas is conveyed in strong, thin-walled, long-distance pipelines
in virtually all principal countries of the world. In the United States, the total
length of gas pipelines is ca $5.82 \times 10^5$ km, which includes ca $4.24 \times 10^5$ km
of transmission pipelines, ca $1.49 \times 10^5$ km of field (gathering) lines, and ca
$0.09 \times 10^5$ km of lines in storage areas (3,4). This represents ca 73% transmission
pipelines, 25.5% gathering lines, and ca 1.5% storage lines. Collectively, long-
distance gas pipelines account for only about one-third the natural gas transport
system in the United States; the other two-thirds are gas distribution mains to
consumers.

Polyethylene (PE), poly(vinyl chloride) (PVC), or polypropylene plastic pipe
is being used in increasingly greater amounts throughout the world in both gas
gathering systems and gas distribution systems. U.S. gas distribution companies
have installed 644,000 km of polyethylene piping since its introduction in 1968
and are adding over 40,000 km each year (5). Plastic pipe is used in sizes greater
than 400-mm dia for certain nonpressurized nongas applications, but is not used
in sizes of greater than 400-mm dia for natural gas and not at all for truly
long-distance natural gas pipelines. When plastic pipe is installed under roads,
casings are often required. Railroads do not allow polyethylene pipe to be used at
crossings; the additional installation costs for crossings can reduce the economic
advantages of using plastic pipe for long pipelines (6).

Transport of natural gas starts with small-diameter gathering lines that
convey the raw gas from individual wells to a collection point and from there to
a gas treating plant, where heavier hydrocarbons (qv), particulates, and water
are separated from the gas. The hydrocarbons known as natural gas liquids
(NGL) are transported as raw mix in liquid pipelines to chemical plants, where
they are fractionated into lighter and heavier fractions that are transported
to market areas in products pipelines. The methane-rich gas, which usually
contains some ethane, is processed to remove particulates, water, and gaseous
impurities such as hydrogen sulfide, carbon dioxide, etc, and then compressed to
the appropriate pressure for transmission by large-diameter pipeline. Pressure
is maintained by large compressors at compressor stations along the length of
the pipeline, usually spaced 80–160 km apart (7). Maximum operating pressures
are 3.45 MPa (500 psi) for older lines and 9.93 MPa (1440 psi) for newer lines.

Of the 125 U.S. gas pipelines listed in 1991, 47 were rated as principal
pipelines, ie, $1.42 \times 10^9$ m$^3$ of gas transported or stored for three consecutive
years (4). The longest gas pipelines originate near the largest U.S. gas producing
areas, primarily the Gulf of Mexico and the mid-continent area of Oklahoma,
Texas, Kansas, New Mexico, etc, and move gas to the more populated areas of
the country, ie, Los Angeles, Chicago, New York, or New Jersey. Maps showing
the locations of natural gas pipelines in the United States (8) and Europe (9)
are available.

Some U.S. natural gas pipeline companies are subsidiaries of gas holding
companies. The largest U.S. natural gas pipeline companies, in terms of overall length of transmission systems are Northern Natural Gas Co., 26,539 km;
Tennessee Gas Pipeline Corp., 23,567 km; Columbia Gas Transmission Co.,

18,481 km; Natural Gas Pipeline Co. of America, 17,200 km; and Transcontinental Gas Pipe Line Corp., 17,071 km. For gas moved in 1994, the four largest pipelines were ANR Pipeline Co., $95,278 \times 10^6$ m$^3$ (3,363,275 MMcf), of which 40.8% was gas moved for others; Transcontinental Gas Pipe Line Corp., 87,050 $\times 10^6$ m$^3$ (3,073,801 MMcf), of which 99.7% was moved for others; Natural Gas Pipeline Co. of America, $83,089 \times 10^6$ m$^3$ (2,933,940 MMcf), of which 87.1% was moved for others; and Northern Natural Gas Co., $56,523 \times 10^6$ m$^3$ (1,995,861 MMcf), with 100% moved for others.

In Europe, Russia's huge natural gas resources are being moved to Western Europe by pipelines from Siberia to the consuming countries. Gas from Algeria's large resources is moved by 2500-km, 1200-mm dia pipeline from the Hassi R'Mel gas field across the Mediterranean Sea and the Strait of Messina to Italy in water depths reaching to 610 m. A 1448-km pipeline will carry Algerian gas to Spain and Portugal via Morocco. Norway's Statoil will build a 1095-km pipeline system (Zeepipe) from the North Sea to Europe. In South America, a 2252-km gas pipeline will transport natural gas from producing fields in Bolivia to Sao Paulo, Brazil, and hook up with a line between Sao Paulo and Rio de Janiero. A gas pipeline that would cover 27,000 km has been proposed by the National Pipeline Research Society of Japan to transport gas from Siberia and Sakhalin Island to Australia by way of Japan, China, Thailand, Malaysia, and Indonesia (9).

*Ammonia Pipelines.*  Ammonia [*7664-41-7*] is a commodity produced from natural gas and may be produced in plants located near its market areas or plants may be located near gas-producing regions and the ammonia transported to the market areas by tank car, barge, or pipeline. The primary ammonia pipelines in the United States are the Mid-America Pipeline Company (MAPCO) and Gulf Central, both connecting ammonia-producing areas in Texas, Oklahoma, and/or the Gulf Coast with farming districts throughout the Midwest. The pipelines are made of high strength pipeline steel, 200–250-mm dia, and operate with an ammonia pressure of 2.07–10 MPa (300–1450 psi) to maintain the ammonia in the liquid state. A limit of 0.2% $H_2O$ and 0% $CO_2$ in the ammonia is specified to minimize stress–corrosion problems. Stress–corrosion cracking of the high strength pipeline steel can occur if the ammonia is contaminated by atmospheric carbon dioxide. Copper-base alloys are corroded by ammonia in the presence of air and water, and equipment containing these metals can be corroded by even small ammonia leaks. The MAPCO ammonia pipeline has resolved these problems by using a special steel containing 0.25% C, 1.0–1.7% Mn, and 0.15% Cu; eliminating copper alloys from valves, fittings, and pumps; stress-relieving fabricated valves or using cast valves and stress-relieving pipe fabrications containing branch connections; and using neoprene for valve trim (10).

A Russian ammonia pipeline of nearly 2400 km extends from Togliatti on the Volga River to the Port of Odessa on the Black Sea, and a 2200-km, 250-mm dia branch line extends from Gorlovka in the Ukraine to Panioutino. The pipeline is constructed of electric-resistance welded steel pipe with 7.9-mm thick walls but uses seamless pipe with 12.7-mm thick walls for river crossings. The pipeline is primed and taped with two layers of polyethylene tape and supplied with a cathodic protection system for the entire pipeline. Mainline operating pressure is 8.15 MPa (1182 psi) and branch-line operating pressure is 9.7 MPa (1406 psi) (11).

*Hydrogen Pipelines.*   The manufactured gas distributed by early gas distribution systems contained up to 50% hydrogen, and at least two large-scale operations in the 1990s have evaluated 10–20% hydrogen mixtures with natural gas, but actual experience with long-distance hydrogen pipelines is rather limited. The oldest hydrogen pipeline (started in 1938) is the Chemische Werke Hüls AG 220-km, 150–300-mm dia system in the German Ruhr Valley that transports $100 \times 10^6$ m$^3$ of hydrogen annually to multiple users at a nominal pressure of 1.55 MPa (225 psi). Following its expansion after 1954, some fires have occurred but no hydrogen embrittlement or explosions (12,13). Other shorter H$_2$ pipelines include a 340-km network in France and Belgium, an 80-km pipeline in South Africa, and two short pipelines in Texas that supply hydrogen to industrial users (14). NASA has piped H$_2$ through short pipelines at their space centers for several years.

Natural gas is roughly eight times more dense than hydrogen (0.72 g/L vs 0.09 g/L). Because the pipeline capacity of a pipeline depends on the square root of the gas density, the pipeline capacity of hydrogen is nearly three times greater than for natural gas. The heating value of hydrogen, however, is only about one-third of the 37.2 MJ/m$^3$ heating value for an equivalent volume of natural gas. Thus, the energy carrying capacity of a given size pipeline is approximately the same when it is carrying either hydrogen or natural gas, provided that it is operating in turbulent flow with the same pressure drop along its length and at the same operating pressure. At higher pressures, the ratio of heating values of natural gas and hydrogen increases and at 5.17 MPa (750 psi); the ratio is 3.8:1, compared with about 3:1 at atmospheric pressure. This also means that hydrogen compressors must handle 3.8 times more gas than natural gas compressors for the same energy throughput, thus indicating higher pipeline transmission costs for hydrogen than for natural gas. One study suggests that regional transport costs for hydrogen may be as much as five times greater than for natural gas at the higher pressures but that long-distance pipeline transport should not show such a large difference (15). Some components currently used for natural gas may be adequate for hydrogen service but not compressors and meters. In view of the different compressor requirements for hydrogen and natural gas, the design of rotary compressors used for natural gas is probably not satisfactory for hydrogen use (12,16).

The question of whether hydrogen embrittlement of pipeline steel is a problem has not been completely resolved. No hydrogen embrittlement problems were reported for hydrogen pipeline transmission at 5.17 MPa (750 psi), and an American Petroleum Institute (API) study of steels operating in hydrogen service indicates no problems at temperatures below 204°C and pressures of 69 MPa (10,000 psi). Other tests of pipeline steels at hydrogen pressures of 3.5–6.9 MPa (500–1000 psi) have shown a loss of tensile ductility and accelerated fatigue-crack growth. Embrittlement is a potential problem in high pressure pipeline transmission of hydrogen (13).

**Industrial Gases.**   Industrial gas (oxygen, nitrogen, etc) pipelines are short compared with long-distance pipelines that transport crude oil, natural gas, or petroleum products; however, more than 80% of the oxygen and more than 60% of the nitrogen produced in the United States by air separation is transported by pipeline. Air-separation plants that are built adjacent to the oxygen or nitrogen users' facilities are usually owned and operated by the user or a second party. However, plants located at more distant locations are usually owned and operated by the oxygen producer and the gases transported by pipeline to multiple users in the area, eg, the 150-km, 406-mm dia Houston Ship Channel pipeline system that supplies oxygen and nitrogen from air-separation plants to chemical plants and oil refineries from Houston to Texas City, Texas (17). Carbon dioxide and hydrogen are also moved by pipeline in the Houston area. Other U.S. industrial gas pipeline systems are located near Gary, Indiana, and along the Mississippi River near the Gulf Coast (18). Multiple users of oxygen (qv), nitrogen (qv), hydrogen (qv), and carbon monoxide (qv) near Rotterdam, the Netherlands, are supplied by pipeline from a nearby industrial gas complex. Oxygen has been transported for many years by a French pipeline connecting Metz with Nancy and extending to Luxembourg and Saarbrucken in Germany.

The application that has led to increased interest in carbon dioxide pipeline transport is enhanced oil recovery (see PETROLEUM). Carbon dioxide flooding is used to liberate oil remaining in nearly depleted petroleum formations and transfer it to the gathering system. An early carbon dioxide pipeline carried by-product $CO_2$ 96 km from a chemical plant in Louisiana to a field in Arkansas, and two other pipelines have shipped $CO_2$ from Colorado to western Texas since the 1980s. Feasibility depends on crude oil prices.

Helium is extracted from natural gas in the southwestern United States and moved by a 685-km, 50-mm dia pipeline to storage in a partially depleted gas field near Amarillo, Texas, as part of the U.S. government's helium conservation program.

**Cryogenic Gases.** Some of the most sophisticated pipeline technology deals with the transport of liquefied gases with very low boiling points, ie, cryogenic gases such as oxygen ($-182.96°C$), argon ($-185.7°C$), nitrogen ($-195.8°C$), hydrogen ($-252.8°C$), and helium ($-268.9°C$). These gases are liquefied by modern cryogenic methods and shipped in cylinders or special storage vessels for bulk liquid transport and storage (see CRYOGENICS). However, development of a system for piping them as liquids has been brought about by the needs of the space program, superconducting magnets, and other high technology areas. The use of liquid hydrogen as a rocket fuel and liquid oxygen as the oxidant has resulted in piping these cryogenic materials to launching and test sites; the use of liquid helium to cool superconducting magnets led to the development of vacuum-jacketed, liquid nitrogen-shielded piping systems (qv). Most of the lines have been relatively small-diameter systems; however, similar but unshielded vacuum-jacketed lines up to 356-mm dia (19) and 8184 kPa (1200 psi), which would qualify as pipeline-sized piping, can be manufactured. CVI, Inc. (Columbus, Ohio) produces pipe for liquid helium consisting of an inner line of stainless steel that is wrapped with aluminized Mylar and glass fiber paper and may be shielded by a patented aluminum extrusion cooled with liquid nitrogen or cold helium gas; the extrusion shield is also wrapped with superinsulating paper. This assembly is completely enclosed, tested, and factory sealed in a vacuum jacket with a stainless steel outer pipe, as shown in Figure 1 (20). Piping systems based on

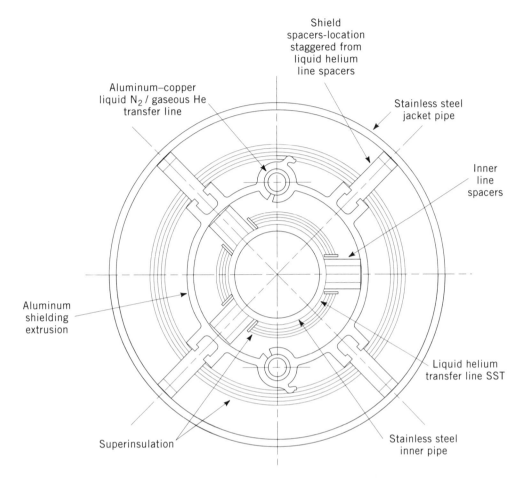

**Fig. 1.**   Vacuum-insulated helium piping.

similar technology are also designed for transfer of liquid oxygen, liquid nitrogen, and liquid hydrogen; liquefied natural gas has also been transferred in vacuum-jacketed piping. Thus far, transfer lines are short by comparison with the longer-range pipelines for other applications, but many thousands of vacuum-jacketed piping installations in the 30–150-m range have been installed, with one system over 8 km in length.

## Pipeline Transport of Liquids

The main technical difference between liquid and gas pipeline transport is the compressibility of the fluid being moved and the use of pumps, rather than compressors, to supply the pressure needed for transport. The primary use for liquids pipelines is the transport of crude oil and petroleum products.

**Crude Oil and Products Pipelines.**   The true pioneering in long-range modern oil pipelines was the construction of two pipelines to move crude oil from Texas to New Jersey and Pennsylvania during World War II. These were named the Big Inch (2156-km, 600-mm dia) and the Little Big Inch (2373-km, 500-mm dia) pipelines.

The crude oil delivery system starts with relatively small-diameter gathering lines from individual producing wells to a main-line pump station, from where it is pumped through a larger transmission trunkline to a refinery or other destination. At the refinery, the crude oil is separated into gasoline, jet fuel, kerosene, distillate fuel oil, etc, and the refined products are transported by products pipelines to markets, storage, shipping terminals, etc. In modern lines, all inputs and outputs are metered, monitored, and remotely controlled by supervisory control and data acquisition (SCADA) computer systems.

Several different refined products are shipped in the same pipeline (batching) by using control methods to minimize intermingling at different interfaces. This is achieved by maintaining high turbulent flow in the pipe. Synthetic rubber spheres have been used as separators at interfaces but are not used often in the 1990s. Intermingling occurs at the interfaces, but it can be estimated and minimized. At the terminal, the comingled fraction is separated from other products and either blended or returned to the refinery. Interfaces between products are delineated with densitometers. This technique is used to ship different types of crude oil in the same pipeline.

The liquids pipeline subsidiaries of principal oil companies with both production and refining facilities generally operate all three types of liquids pipelines: gathering, transmission, and products lines. However, the independent 9000-km, 900-mm dia Colonial Pipeline transports only products. It carries about $105 \times 10^6$ m$^3$ of products annually. Of the nearly 276,000 km of liquids pipelines reported in 1991 by FERC, 47% were products lines, 34% were transmission lines, and only 19% gathering lines. This compares with 35, 34, and 31%, respectively, in the 1970s. The total length of the U.S. operating interstate pipeline system has remained relatively constant since 1982 at about 720,000 km, with roughly 32% represented by liquids pipelines and 62% by natural gas pipelines (4).

At the beginning of 1992, the largest liquids pipelines in the United States, based on pipeline length, were Amoco Pipeline Co., 19,096 km; Mobil Pipe Line Co., 15,026 km; Exxon Pipeline Co., 14,983 km; and Conoco Pipe Line Co., 12,980 km. Distances do not include 1316 km of the Trans-Alaska Pipeline with multiple ownership. In both 1991 and 1992, the product pipeline company with the most product deliveries was Colonial Pipeline with 104,990,000 m$^3$, more than double the amount delivered by Santa Fe Pacific Pipelines, Inc. The top pipeline in terms of crude oil deliveries was the Alyeska Pipeline Service Co., operator of the Trans-Alaska Pipeline System, with movement of 105,735,000 m$^3$ (3).

**Sulfur and Chlorine Pipelines.** Underground sulfur is melted by superheated water and then piped as liquid to the surface with compressed air. At the surface, molten sulfur is transported by heated pipeline to a storage or shipping terminal. One such pipeline, located under 15 m of water in the Gulf of Mexico, is insulated and surrounded by steel casing to which are strapped two 130-mm dia pipelines that carry return water from the deposit. The superheated water is carried from shore to the deposit in a 63.5-mm dia pipe inside the pipeline that carries the molten sulfur (21).

Chlorine is shipped by pipeline in either gaseous or liquid form; however, great care must be taken to ensure that liquid pipelines are operated only in the liquid phase and gaseous pipelines operated only in the gaseous phase. Liquid chlorine has a high coefficient of thermal expansion and high pressure develops with increasing temperature; liquid chlorine trapped between two valves can lead to hydrostatic rupture of the pipeline. Because chlorine is corrosive and has a high vapor pressure, it and the pipeline must be dry before the chlorine enters the pipeline. Operating pressure should not exceed 217 MPa (31,465 psi) and temperature should be no higher than 121°C (22).

**Trans-Alaska Pipeline.** The design and construction of the Trans-Alaska Pipeline was one of the most difficult and ambitious pipeline projects ever attempted. Its 1316-km length extends from the producing fields at Prudhoe Bay across three mountain ranges, where up to 12 m of winter snow and wind chills of −73°C can be expected, to the Valdez Marine Terminal on the southern coast of Alaska. Thirteen bridges were built to carry the pipeline across rivers and difficult terrain. To avoid localized melting of the permafrost by the 32−49°C crude oil and possible instability of the soil, a significant portion of the pipeline was installed above ground, supported by 78,000 vertical support members embedded in permafrost to depths ranging from 4.5 to 18 m. The pipeline is insulated and jacketed to reduce heat losses, and passive refrigeration is used to keep the soil frozen around the vertical support members which can move up or down, thus allowing a 0.6-m vertical movement of the pipeline. The pipeline is laid out in a zigzag fashion to allow a 3.6-m horizontal movement for pipe expansion and contraction and possible seismic disturbances. The buried portion of the pipeline is 611-km long, with 6.4 km of it refrigerated. The aboveground portion of the pipe is at least 1.5 m above the surface to permit caribou migration underneath and at least 3 m above the surface at the 554 special animal crossings. The pipeline is constructed of 1219-mm dia high tensile strength carbon steel, with wall thickness of either 11.7 or 14.3 mm. Although the Prudhoe Bay oil discovery was announced in 1968, the initial pipe of the pipeline was not laid until early 1975 because of many delays. Final weld was made May 31, 1977 (23,24). Final cost, including construction interest, was roughly $11 billion.

## Pipeline Transport of Solids

Pipelines to transport solids are called freight pipelines, of which three different types exist: pneumatic pipelines, the use of which is known as pneumotransport or pneumatic conveying; slurry pipelines, which may also be called hydrotransport or hydraulic conveying; and capsule pipelines. When air or inert gas is used to move the solids in the pipeline, the system is called a pneumatic pipeline and often involves a wheeled vehicle inside the pipeline, propelled by air moving through the pipe (25). Slurry pipelines involve the transport of solid particles suspended in water or another inert liquid. Hydraulic capsule pipelines transport solid material within cylindrical containers, using water flow through the pipeline for propulsion.

**Pneumatic Pipelines.** Pneumatic pipe systems are used to move blood samples, medicine, and supplies between buildings in hospital complexes; cash and receipts in drive-up banks; parts and materials in factories; refuse from apartment complexes; and grain, cement, and many other materials. Most of these are small diameter and usually short; however, a 17-km, 1220-mm dia pneumatic pipeline has been used to transport rock in the former Soviet Union since 1981, and a 3.2-km, 1000-mm dia line has moved limestone from the mine to a cement plant in Japan since 1983 (22).

**Slurry Pipelines.**    Finely divided solids can be transported in pipelines as slurries, using water or another stable liquid as the suspending medium. Flow characteristics of slurries in pipelines depend on the state of subdivision of the solids and their distribution within the fluid system. Although slurry flow and conventional liquid flow are both divided into laminar, transitional, and turbulent flow, the contribution of the solids-to-slurry flow results in a further characterization as either homogeneous or heterogeneous flow. Homogeneous flow of slurries occurs when very finely divided solids are distributed uniformly throughout the suspending medium, eg, bentonite slurries as drilling muds. These slurries are usually quite concentrated and often have viscosities much higher than those of the suspending medium itself. Heterogeneous flow occurs when larger, irregular solid particles are slurried and distribution is not uniform throughout the suspension; these slurries are usually of lower solids concentration and viscosities are comparable to that of the suspending liquid.

The difference between homogeneous and heterogeneous flow is due to the deposition velocity, below which solid particles start to separate from the slurry and build up in the pipeline, and is related to the degree of turbulence and the particle fall rate. Homogeneous slurries behave more like single-fluid systems; deposition velocity is primarily a property of heterogeneous slurries. Deposition velocity increases with increasing particle density, solids concentration, particle size, and pipeline diameter. In the transport of heterogeneous slurries, turbulent flow is an important requirement to prevent solids from building up in the pipeline. Slurries may be classified according to particle size, eg, colloidal ($<1$ $\mu$m), structured (1–50 $\mu$m), and finely dispersed (50–150 $\mu$m, mostly produced by grinding). A polydispersed structured category, often encountered with products of technological processes and produced by dispersion and grinding, is also defined and described as containing a broader ranger of particles, eg, from finer particles to coarser particles, and sometimes lumps (26). A maximum flow velocity to minimize pipe wall erosion must be determined.

Slurry pipeline design is similar to the design of conventional liquid pipelines, except for the slurry preparation stage and, if necessary, an additional step for separating the suspended solids from the suspending liquid at the point of use. In some instances, such as the direct firing of a concentrated homogeneous coal slurry, this final separation step is unnecessary. In the United States, the Coal and Slurry Technology Association has focused its attention on the development of coal–water–fuel (CWF) and its eventual commercialization, rather than on transportation by slurry pipelines (27).

*Coal Slurry Pipelines.*    The only operating U.S. coal slurry pipeline is the 439-km Black Mesa Pipeline that has provided the 1500-MW Mohave power plant of Southern California Edison with coal from the Kayenta Mine in northern Arizona since 1970. It is a 457-mm dia system that annually delivers ~4.5 × $10^6$ t of coal, the plant's only fuel source, as a 48.5–50% slurry. Remote control of slurry and pipeline operations is achieved with a SCADA computer system. In 1992 coal delivery cost from mine to power plant was calculated to be $0.010/t·km ($0.015/t·mi) (28).

Several coal slurry pipelines were planned for the United States during the 1980s, primarily to deliver low sulfur coal from mines in Montana and Wyoming to power plants in Texas and other states in the southern United States. None was built, however, because of vigorous opposition from water conservationists, who opposed using scarce water resources for the slurrying medium, and the railroads, who feared competition for the utility coal markets (29,30). In addition, lower prices for oil and natural gas provided little incentive to develop lower cost competitive fuels; however, a study for the U.S. Department of Energy (Pittsburgh Energy Technology Center (PETC)) indicated that coal–water fuels (CWF) could be produced for less than $1.91/m$^3$ of oil equivalent (31), and it is expected that there will be renewed plans for coal slurry pipelines (32).

The former Soviet Union constructed a 262-km, 508-mm dia experimental coal slurry line between the Belovo open-pit coal mine in Siberia's Kuznets basin to an electric power plant at Novosibirsk, using technology developed by Snam-progetti. Testing began in late 1989 and tentative plans call for construction of two much larger slurry pipelines, each 3000-km long, with capacity to move a total of $33 \times 10^6$ t/yr to industrialized areas near the Ural Mountains (27,33).

*Mineral Slurry Pipelines.*　Bentonite clay slurries, used as drilling muds in oilwells, are probably the most universally used mineral slurries. Although they are not transported by pipelines over great distances, they represent a significant contribution to slurry pipeline technology because they led to the development of powerful pumps needed to force oilwell drilling muds downhole to the rotating bit, where the mud cools the bit and also carries rock chips from the hole back to the surface. The pumps developed for this application contributed greatly to the development of the huge positive-displacement reciprocating pumps used for pumping slurries through pipelines (29). Table 1 gives some examples.

**Table 1. Mineral Slurry Pipelines**

| Pipeline | Location | Slurry | Length, dia (km,mm) | $10^6$ t/yr |
|---|---|---|---|---|
| M. & Chem. Philip | Georgia (U.S.) | kaolin | 25.6, 200 | |
| Trinidad Cement | Australia | clay, limestone | 9, 200 | |
| Calaveras | California (U.S.) | limestone | 27, 150 | 1.36 |
| VALEP | Brazil | 61% phosphate | 120, 250 | 2 |
| SAMARCO | Brazil | 66% iron ore | 400, 500 | 6.35 |
| Da Hong Shan | China | iron ore | | |
| Pena Colorado | Mexico | iron ore | 48, 200 | 1.8 |
| Savage River | Tasmania | iron ore | 85, 250 | 2.3 |
| Waipipi | New Zealand | iron sands | 9.3[a] | |
| | Bougainville | copper ore | 27, 152 | |
| | Bougainville | copper ore | 105, 102 | |
| Hondo | Japan | ore wastes | 71[a] | |
| | South Africa | uranium sands | 19, 150 | |

[a]Length.

*Fiber Slurry Pipelines.* Pipelines to carry suspensions of wood, paper, sludge, etc, have found commercial acceptance. Most of them are less than 15 km long but have diameters of up to 500 mm. These slurries are often concentrated and display viscous plastic properties, although particle sizes may vary; special pumps are used. One such hydrotransport system carries a cellulose slurry by pipeline from the plant to a paper plant near Heidenau, Germany. The 250-mm dia pipeline carries 60 t/d over the 3-km distance to thickeners. In Sweden, a 3.7-km, 500-mm dia pipeline moves cellulose by pressurized hydrotransport from a cellulose plant in Wifstaur to a sulfite plant at Fagervik (26). The former Soviet Union has also been active in pipeline transport of fiber slurries in the cellulose (qv) and paper (qv) industries and the movement of municipal waste from aeration plants. For example, municipal sewage sludge from the Kozhukhovskaya purification plant has been transported by a 2-km, 300-mm dia pipeline to settling areas since 1936, and a 70-km, 400-mm dia pipeline has moved sludge from the Lyublino aeration plant in Moscow since 1954 with no plugging (26).

**Capsule Pipelines.** Capsule pipelines involve the transport of material inside a closed cylindrical container propelled by water flowing through the pipeline. This is called hydrotransport and the diameter of the cylindrical container is approximately 90–95% the diameter of the pipeline. Water is 1000 times more dense than air under standard conditions, so the buoyancy is much greater than in pneumatic pipelines and permits capsules to be suspended at relatively low water flow velocities. First proposed during World War II, the concept was resurrected in the 1960s and has been demonstrated amply in several locations, using various sizes and lengths of pipelines. Considerable research activities on capsule pipelines have been ongoing in several places, including Japan, Australia, South Africa, the Netherlands, and the United States (25). The University of Missouri, Columbia, has the only research center and testing facilities devoted to capsule pipelines in the United States and has a 131-m, 203-mm dia test capsule pipeline devoted to this activity. The concept has been expanded to include a coal log pipeline.

The coal log pipeline involves forming logs of coal by compressing and extruding finely divided coal into cylindrical shapes having a diameter 90–95% the diameter of the pipeline. The coal logs are then injected continuously into a water-filled pipeline for transport by the flowing water. The log length is between 1.5 and 3 times the log diameter and buoyancy is achieved at a water flow rate of 2.4–3.0 m/s. This reduces the energy required for pumping, as well as the friction between coal log and pipeline walls. The coal log pipeline is said to use only one-third as much water for coal transport as coal slurry pipelines, in addition to eliminating the dewatering step. No full-scale coal log pipeline has been built, but short test lines and research activities continue, and a Coal Log Pipeline Consortium has been formed, with both private companies and governmental agencies participating (34).

## Pipeline Technology

Pipeline technology involves design, construction, maintenance (qv), and operation. Although certain aspects of the technology differ under different climatic conditions, whether above or below ground or under water, etc, the basic steps are the same for liquids pipelines as for gas pipelines.

**Design.** Pipeline design begins with a preliminary mapping of the proposed route, noting areas to be avoided and such obstacles as rivers, railroads, and highways. For cross-country pipelines, aerial and ground-control surveys are made to establish the final alignment of the pipeline. Permission to cross all land parcels must be obtained from landowners, and permits obtained from the proper authorities to cross rivers, highways, and railroads; where permission is not obtained, appeal is made to the courts under the right of eminent domain. Approval for gas pipeline construction must be obtained from FERC and/or state public utility commissions, and an environmental impact statement must be filed. Liquid pipelines do not require FERC approval for construction; common carrier pipelines are generally accorded the right of eminent domain and are regulated by FERC or state authority.

While detailed routing is being established, mechanical design begins by establishing pipe size, based on the volume and type of material being transported, design pressures, pipeline length, and spacing between compressor or pumping stations. The strongest steel possible is used and pipe wall thickness is determined by a code-specified design formula that recognizes pipe as an unfired pressure vessel; the design formula for natural gas pipelines is given by the industry code ASME B31.8 and also by federal regulation 49 CFR 12. For steel pipe, the formula is as follows:

$$P = (2St/D) \times F \times E \times T$$

where $P$ = design pressure in kPa (psig), and may be limited per federal standard 192-105 to 75% of the pressure if steel pipe has been subjected to cold expansion to meet the specified minimum yield strength (SMYS) and then heated to 482°C or held above 316°C for more than 1 h, other than by welding; $S$ = yield strength in kPa (psig); $t$ = nominal wall thickness in mm (in.); $F$ = design factor, which is 0.72 in rural areas and 0.40 in urban areas, with intermediate values of 0.60 and 0.50 for gas lines and 0.72 for liquid lines; $E$ = longitudinal joint factor, which depends on the type of pipe weld seam and is 1 if seam strength equals pipe strength; $T$ = temperature derating factor, which is 1 for operating temperatures of 121°C or less but <1 for higher temperatures, such as at compressor discharge; and $D$ = nominal outside diameter in mm (in.).

This formula determines only the primary pipe (hoop) stress caused by the internal fluid pressure perpendicular to the pipe wall; however, it also recognizes secondary hoop stress through the design factor, $F$, which represents a safety factor. The maximum design factor of 0.72 recognizes the secondary stresses caused by the load of the overburden against an imperfect trench bottom, those produced by thermal expansion or contraction against earth constraint on buried pipe, and others. The design factor is lowered in populous areas to provide additional safety and to recognize additional stresses in those areas compared with rural areas; it is also lowered for installations at river crossings, highways, etc. The maximum allowable operating pressure (MAOP) of the pipeline can be affected by the pipe mill test pressure level and/or the hydrostatic strength test after construction is complete; either could underrate the pipeline. Longitudinal primary tensile stresses are half of the primary hoop stress. Specifications for valves, fittings, etc, must be written as part of the design phase. Shut-off valves are required at specific intervals to isolate any pipeline damage; powered valve-closing devices are commonly installed to shut off valves automatically in the event of a line break and isolate the damaged portion of the line. Valves on cross-country pipelines should be of a full round-opening design to permit passage of devices, called pigs, for line cleaning, separating water from gas during hydrostatic testing, corrosion detection, etc. Specifications are made for exterior coatings and cathodic protection to minimize external corrosion, and for interior coatings or additives to minimize interior corrosion, as well as decrease flow resistance and improve transmission of odorants. Interior pipe coatings are usually epoxy paints applied before pipeline assembly.

Spacing between compressor/pumping stations is usually 80–160 km along the pipeline to boost pressure lost to internal friction by the gas or liquid being transported. The compressors/pumps are driven by either reciprocating internal combustion gas engines of up to 4.5 MW (6000 hp), centrifugal gas turbines of up to 11.2 MW (15,000 hp), or electric motors of up to 7.5 MW (10,000 hp). A compressor/pumping station usually consists of several compressors/pumps operating in parallel.

**Construction.** Pipeline construction is a continuous activity required by supply changes due to decreasing production from older fields, opening up new fields, or bringing in new supplies from Canada; obsolete pipelines are decommissioned, additions are made to existing pipelines, or new pipelines are designed and laid to take advantage of new supply sources. Construction of a cross-country pipeline may be divided into segments (spreads) and proceeds by the following steps: right-of-way clearing and grading, trenching, pipe stringing along the right-of-way, pipe bending to fit the trench bottom, welding, cleaning and priming the exterior surface for coating, coating and wrapping for cathodic protection unless it is mill-coated and only the weld areas need coating, lowering pipeline into trench, and backfill of earth over the pipeline.

Alternative construction methods are used for crossing obstacles. For river crossings, floats may be used to carry the pipe across the river with concrete coatings or weights, flooding the floats to sink the pipe to the river bottom where the weights hold it in place. Land under highways is bored with an earth auger followed closely by the pipe or casing, if required, advanced by the same machine. With advances in horizontal directional drilling, bored river or wetlands crossings of up to 762 m of bore have been made.

When construction is complete, the pipeline must be tested for leaks and strength before being put into service; industry code specifies the test procedures. Water is the test fluid of choice for natural gas pipelines, and hydrostatic testing is often carried out beyond the yield strength in order to relieve secondary stresses added during construction or to ensure that all defects are found. Industry code limits on the hoop stress control the test pressures, which are also limited by location classification based on population. Hoop stress is calculated from the formula, $S = PD/2t$, where $S$ is the hoop stress in kPa (psig); $P$ is the internal pressure in kPa (psig), and $D$ and $T$ are the outside pipe diameter and nominal wall thickness, respectively, in mm (in.).

Tests of cross-country pipelines are recommended at 90% of pipe yield strength in order to avoid having insufficient field-test pressure, which can limit the MAOP of the pipeline to less than that calculated by the design formula. According to industry code, this can be achieved only by hydrostatic testing with nonflammable liquid, such as water. Filtered water is used to displace air left in the line during construction; a pipeline pig separates the two phases and eliminates air pockets. The water is pressurized to produce the hoop stress desired and is recommended to be held for a minimum of 8 h to test pipe strength. When the test is finished, the water is displaced by the pressurized fluid for which the line was built, ie, natural gas, crude oil, etc. A pig is used to separate the two phases and remove accumulated water at low points; for natural gas pipelines, multiple pigs also remove water retained on pipe walls that could cause operational problems. The pipeline is tied in to distribution and supply systems, if it passes the tests, and put into service. The final step is a cleanup and posting of required location markers.

Construction of underwater (submarine) pipelines does not take place under water. Pipelines are welded onshore and dragged into position by powerful winches on ships floating on the water surface (for short lines), welded on a specially constructed lay barge, and lowered to the ocean floor by a stinger from one end of the barge or welded onshore, floated on pontoons, and towed to the offshore area where they are lowered into position. For smaller size pipelines, the lines can be welded onshore and spooled onto large reels, placed on special ships, and spooled into the offshore trench. In shallow water, or where endangered by anchors or wave action, submarine pipelines are laid in trenches in the sea bottom (35). Underwater pipelines are concrete-coated or weighted to overcome the buoyancy effect; concrete coatings applied over the primary coating provide additional protection against damage during laying and against corrosion. Submarine pipelines are being used regularly to transport oil, natural gas, and other commodities to shore from offshore locations, such as the Gulf of Mexico, the North Sea, and the Arabian Gulf. One of the longest and technically challenging submarine pipeline systems is a 2599-km, 1200-mm dia pipeline for transporting natural gas across the Mediterranean Sea and the Strait of Messina from gas fields in Algeria to Italy; pipes are laid in water depths to 610 m. Offshore and onshore pipelines require different design factors (36).

*Trenchless Construction.* Increased emphasis on environmental and ecological factors has presented considerable difficulties for normal pipeline trenching, even in unpopulated areas, if considered harmful to wildlife in their native habitats. In densely populated areas, trenching is costly and complex because of airports, commercial buildings, factories, etc. To circumvent them, trenchless procedures are being developed for installing fuel lines, water lines, sewer pipe, etc, or in some cases for steel casings that can serve as conduits for several smaller diameter pipe systems. Horizontal directional drilling is a trenchless construction technique that drills under an obstacle, such as a street, airport runway, or river, and the welded continuous steel pipeline segment is pulled through the opening after horizontal drilling is complete. Another procedure that was used in Berlin to put a large-diameter casing under a railway embankment between a train station and the River Spree (37) involves the use of pneumatic pipe ramming to ram the steel pipe through the soil. Trenchless construction is used commonly for small gas distribution lines.

Microtunneling involves sinking of two vertical shafts, large enough to lower a horizontal drilling machine to the level from which to dig a tunnel between the two vertical shafts to accommodate the pipeline. The technique has proved useful for installing or rehabilitating sewer pipe systems and often includes the sinking of additional shafts to provide access to the moling (horizontal drilling) machines. Microtunneling was originally and principally developed in Europe, but variations are being developed in both the United States and Japan, and costs are expected to drop rapidly with expanded usage (38). Houston and Dallas are among the U.S. cities that have applied this technique to revitalize underground utilities infrastructures (39–41).

**Maintenance.** *Pipeline Pigging.* After construction, a pipeline must be tested, inspected, cleaned, mapped for bends, dents, or ovalties, maintained during operation, and monitored for leaks or corrosion to ensure safety of operation. These operations involve pigging, moving devices called pigs through the pipeline that can carry out these missions. These may be sophisticated electronic devices or something as simple as a rubber ball pumped through the pipeline to displace fluids (42).

Modern pipeline pigging applications begin with the commissioning of a pipeline when pigs are first used to remove air prior to hydrostatic testing and then used to remove any water left after the testing. A gauging (caliper) pig is used to locate any dents or buckles resulting from the laying or backfilling operations. Special cleaning pigs are used to remove scale, dirt, etc, from interior walls. If the pipeline transports multiple products, pigs can be used to separate them at the liquid interfaces. To prevent loss of pipeline efficiency or increased corrosion due to deposits of wax, scale, or bacteria, the deposits are removed with brush pigs or even more sophisticated pin-wheel pigs that remove deposits to a precise depth (43,44). Smart or intelligent pigs involving electronic sensor interfaces with computers and miniature video cameras (45), ultrasonics, magnetic-flux leakage, etc, have been developed to inspect pipeline interiors for integrity, metal loss, corrosion, and other defects; however, some pipelines have physical limitations that do not accommodate smart pigs. The potential of smart pigs to improve pipeline safety is considered to be so great that the Office of Pipelines Safety published a ruling that would require modification of pipelines to permit their use where feasible or practical; however, many petitions for reconsideration were filed and a limited suspension of compliance has been granted while the situation is under study (46). Several pigging operations can be put into a single train (47).

*Corrosion.* Anticorrosion measures have become standard in pipeline design, construction, and maintenance in the oil and gas industries; the principal measures are application of corrosion-preventive coatings and cathodic protection for exterior protection and chemical additives for interior protection. Pipe for pipelines may be bought with a variety of coatings, such as tar, fiber glass, felt and heavy paper, epoxy, polyethylene, etc, either pre-applied or coated and wrapped on the job with special machines as the pipe is lowered into the trench.

An electric detector is used to determine if a coating gap (holiday) exists; bare spots are coated before the pipe is laid (see CORROSION AND CORROSION CONTROL).

Cathodic protection is provided by inducing an electric current on the pipeline to ensure that the electric potential of the metallic pipe is less than the earth surrounding the pipeline after it is laid, by using a sacrificial anode bed with a more electropositive metal than iron, or by thermoelectric protection, using heat to generate a direct current to lower the potential of the pipe. Inducing an electric current involves the use of a rectified alternating current to lower the potential of the pipe relative to the ground, which also provides a way to monitor the system by an aerial pipeline survey. Pulsating the direct current produces a magnetic field with a strength proportional to the current flowing in the pipe; using field coils to measure the magnetic field strength provides a method of monitoring the current, even for underground pipe. If the survey indicates that the magnetic strength is reduced over any segment of the pipeline, it would indicate change of potential and current loss, such as would occur with a hole in the pipeline coating, an electric casing short, a separated coupling, or other event that would permit corrosive action (48).

U.S. Department of Transportation (DOT) statistics on liquids pipelines operated under the *Code of Federal Regulations* (49) indicate that corrosion was the second largest contributor to accidents and failures for the period from 1982 to 1991. These statistics covered an average of 344,575 km of liquids pipelines and were derived from required reports to DOT on all pipeline accidents involving loss of at least 7.95 $m^3$ of liquid, death or bodily harm to any person, fire or explosion, loss of at least 0.8 $m^3$ of highly volatile liquid, or property damage of $5000 or more (50). Similar results were also reported for 1991 in the 1992 DOT/OPS report on both oil and gas pipeline incidents; 62 out of 210 oil pipeline incidents were due to corrosion, of which 74% were due to external corrosion (43). For gas pipelines, 16 of all 71 reported incidents were due to corrosion, of which 63% were reported as due to internal corrosion; however, internal corrosion of gas pipelines is likely only if $CO_2$ and $H_2O$ and/or $H_2S$ are present, as with unprocessed gas in gathering lines.

**Operation.** Operations are controlled from a central computerized office that maintains communication with compressor stations and flow stations along the route; reports are also received from weekly or biweekly aircraft flyovers to inspect for potentially threatening conditions, such as soil erosion, floods, approaching excavation, or any external evidence of leaks. Flow computers along the pipeline route monitor and quantify flow from producers entering the pipeline or flow leaving the line to customers. Pipeline flow, pressure, and other operating variables can be controlled by commands to compressor stations. In systems that have remote control of main-line valves, central control can isolate sections of pipelines with reported or threatened problems and dispatch personnel to the site. If patrols detect indications of significant increases in population density in specific areas, the safety factor in the design formula may be increased in accordance with industry code, and the maximum allowable operating pressure in those areas is reduced. Operations and maintenance of pipelines are also covered in ASME B31.8 and federal regulation 49 CFR 192.

Cross-country gas pipelines generally must odorize the normally odorless, colorless, and tasteless gas in urban and suburban areas, as is required of gas distribution companies. Organosulfur compounds, such as mercaptans, are usually used for this purpose, and code requires that the odor must be strong enough for someone with a normal sense of smell to detect a gas leak into air at one-fifth the lower explosive limit of gas–air mixtures. The latter is about 5%, so the odorant concentration should be about 1%, but most companies odorize more heavily than this as a safety precaution.

## Economic Aspects

Pipelines usually represent the least costly way to move fluid products, including solid slurries, wherever they compete with truck or rail transportation. Comparing the transport costs for removing municipal sludge in the former Soviet Union shows a surprising similarity to a similar comparison of transport costs for moving U.S. crude oil and petroleum (qv) products. The Russian study (26) showed that pipeline transport of sludge is equal to 25–38% the cost of barge transportation, 6–13% the cost of rail transport, and 4–5% the cost of truck hauling. By comparison, pipeline cost/km of moving crude oil and petroleum products in the United States has been only about 17% of rail costs and 3.5% of truck transport costs (26,51).

The cost of building a pipeline is usually divided into four categories: materials (line pipe, fittings, coatings, cathodic protection, etc), right-of-way and damages, labor, and miscellaneous (surveying, engineering, supervision, administration and overhead, interest, contingencies, afudc, and FERC filing fees). The relative contribution of each category depends on the size of pipe, the difficulty of terrain, and delays and added costs due to environmental factors, but generally the costs for materials and labor account for about 75% of the entire cost of the pipeline, with each accounting for about half this amount. The remainder is engineering, overhead, fees, interest, etc. The relative contribution of each category can vary widely in difficult terrain or climate, however, as illustrated by the high labor costs of the Trans-Alaska Pipeline, which were 69.3% of the total, and ca $1.3 \times 10^9$ interest charges during construction or roughly 15% of the total pipeline cost (52).

The variation of average pipeline construction costs with increasing size of line pipe is shown in Figure 2, based on data taken from FERC construction permit applications from July 1991 to July 1992. The cost of a common carrier pipeline project must be reported to the FERC no later than six months after successful hydrostatic testing.

The 10-yr construction costs for land-based gas pipelines show no obvious trend either in the four individual categories or the high–low range, on a $/km basis; however, pipeline land acquisition costs rose steadily, with average $/km cost increasing from $370,881 to $528,355 for proposed U.S. gas pipeline projects over the 1990–1992 period. The $/km costs for 1991–1992 were material, $190,379; labor, $200,127; right-of-way (ROW) and damages, $21,938; and miscellaneous, $115,930 (4). The cost of compressor stations for the same period averaged $1894/kW.

When a natural gas pipeline has been commissioned and is operating, its transmission costs are the operating expenses plus maintenance expenses. In 1989, these two cost elements for gas pipelines were $13,573/$10^6$ m$^3$

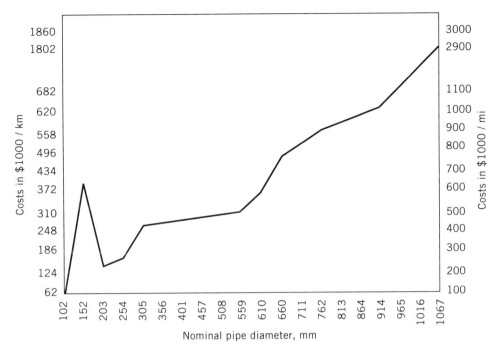

**Fig. 2.**  Gas pipeline construction cost vs pipe size.

and $2872/10^6 m^3$, respectively; for 1990, the corresponding values were $17,669/10^6 m^3$ and $3851/10^6 m^3$, respectively. The primary operating expense for natural gas pipelines is the cost of compression and gas transmission by others, which accounted for ca 40% of the 1989–1992 expenses. The next largest operating expenses were compressor station labor and expenses, and gas for compressor fuel, accounting for 30% of the total (4). Pigging and testing costs depend on pig type, pipeline cleanliness and diameter, extent of corrosion, etc. The Government Accounting Office survey shows a range of \$404–\$1492/km for the cost of smart pig inspections; a similar survey of three foreign smart pig manufacturers indicated a range of \$746–2486/km for a 1609-km, 610-mm dia pipeline. Hydrostatic test costs are even higher; estimates by the U.S. DOT are \$2890/km for a 508-mm dia hazardous liquids pipeline; this cost rises to \$5523–8794/km if costs of transporting, testing, and disposing of the test water and the revenue loss due to the pipeline being out of service are added. The cost of pressure testing and transportation or disposal of test water is about \$6563/km of pipeline, depending on the pipeline diameter, the terrain crossed, and water scarcity in the area (46).

## Safety, Environmental, and Ecological Aspects

Data compiled by the U.S. DOT indicate that pipeline transport is the safest materials transport mode, particularly over long distances. In 1990, fatalities attributed to oil and gas pipelines were significantly lower (8 out of 4679) for all materials transporters in the United States, compared with 599 for rail transport and 3281 for motor transport (trucks) (53).

To ensure the safety of gas pipelines, the Natural Gas Safety Act of 1968 was created to mandate federal regulation of gas storage facilities and pipeline transport of natural gas, with the U.S. Department of Transportation given exclusive authority to regulate safe operation of natural gas pipeline systems that fall within the jurisdiction of the Federal Power Commission under the Natural Gas Act. The Hazardous Liquid Pipeline Safety Act of 1979 and the Hazardous Materials Transportation Act of 1979, and their respective amendments, gave authority for the U.S. DOT to regulate safety for all pipelines operating under the *Code of Federal Regulations* (49). The U.S. DOT continues to regulate safety on pipelines and has established legal standards for all pipeline phases, from design to construction, operation, and maintenance. DOT also issues statistics on pipeline accidents from required reports on accidents involving release of gas or LNG and death or hospitalization or losses in excess of $5000.

Between 1985 and 1991, 1726 natural gas pipeline ruptures and leakages were reported in the United States. These incidents resulted in 634 injuries and 131 fatalities. Third-party damage was the most common cause of these incidents, followed by corrosion. The GAO believes that the corrosion-related incidents can be reduced with the use of smart pigs (46). U.S. DOT 1992 accident statistics showed that 52.5% of U.S. oil spills involving loss of at least 1590 $m^3$ came from pipeline accidents, comparable to the worldwide statistic of 51.5%. The U.S. DOT regulated 344,575 km of liquids pipelines during the 10-yr study period and received reports on 1901 accidents during that time; thus the number of failures per year per 1000 miles was 0.888, of which 27% was due to corrosion and 31% to outside forces (48).

Most ecological issues involve the pipeline's construction phase and the effect that it will have on vegetation, wildlife, topography, population density, and land use. All pipeline design and construction programs must include an environmental impact statement which must be submitted and evaluated before permission to proceed with construction is granted. Construction through ecologically sensitive obstacles, such as rivers and wildlife areas, is often suspended during certain times of the year, such as fish spawning and wildlife migration seasons, unless acceptable alternatives can be found. Archeological concerns must also be addressed.

Greater natural gas use for power generation and vehicular fuel use, to help meet more stringent environmental standards, is expected to expand natural gas pipeline usage. High technology will influence the role of post-Order 636 gas pipelines (54). Pipeline transport of coal could become a reality in the United States, as well as other coal-rich countries, to reduce dependence on imported oil. Greater use of freight pipelines to reduce pollution from trucks and other transportation modes may arise (25). Pipeline research activities are expected to expand through organizations such as Japan's National Pipeline Research Society and the Capsule Pipeline Research Center and Gas Research Institute in the United States.

The concept of transporting heat over great distances with chemical energy pipelines has been considered (55).

## BIBLIOGRAPHY

"Pipelines" in *ECT* 3rd ed., Vol. 17, pp. 906–929, by T. P. Whaley and G. M. Long, Institute of Gas Technology.

1. P. Crow and A. D. Koen, *Oil Gas J.*, 15–20 (Aug. 9, 1993).
2. L. Parent, ed., *The Gas Price Report*, private communication, 1994.
3. J. Watts, ed., *Pipeline Gas J.*, 12–36 (Sept. 1992).
4. W. R. True, *Oil Gas J.*, 41–62 (Nov. 23, 1992).
5. F. W. Griffith, *Pub Utilities Fortnightly* **131**(6), 21 (1993).
6. S. J. Davis, *Pipeline Gas J.*, 44–47 (Dec. 1992).
7. "Transportation of Natural and Other Gas by Pipeline," *Code of Federal Regulations*, Hazardous Materials Regulations Board, U.S. Dept. of Transportation, Minimum Federal Safety Standards, U.S. Printing Office, Washington, D.C., Title 49, Chapt. 1, Part 192.
8. *Map of Major Natural Gas Pipelines*, U.S. Dept. of Energy, for sale by the Superintendent of Documents, U.S. Printing Office, Washington, D.C. Crude oil and products pipeline maps also available.
9. *Pipeline Gas J.*, 14–16 (Aug. 29, 1992); *Gas World Intl.*, 16–24 (May 1993).
10. J. J. Lee, *Oil Gas J.* **66**, 103 (Nov. 4, 1968).
11. N. Hetland, *Pipeline Gas J.* **250**, 38 (Oct. 1978).
12. D. P. Gregory and J. B. Pangborn, *Ann. Rev. Energy* **1**, 279–310 (1976).
13. C. F. Blazek, R. T. Biederman, S. E. Foh, and W. Jasionowski, *Proceedings of the 3rd Annual Hydrogen Meeting, A Blueprint for Hydrogen's Future with Enabling Technologies Workshop*, Mar. 18–29, 1992, pp. 4-203–4-221.
14. J. H. Kelley and R. Hagler, Jr., *Int. J. Hydrogen* **5**, 35–54 (1980).
15. C. Carpetis, in C-J. Winter and J. Nitsch, eds., *Hydrogen as an Energy Carrier*, Springer Verlag, New York, 1988, Chapt. 9.
16. J. Pottier, E. Blondin, and A. Garat, in T. N. Veziroglu and A. N. Protsenko, eds., *Hydrogen Energy Progress VII, Proceedings of the 7th World Hydrogen Energy Conference*, Vol. 2, Moscow, Russia, Sept. 25–29, 1988, pp. 1095–1107.
17. *Oil Gas J.* **6** (Oct. 10, 1983).
18. Technical data, Air Liquide America, Walnut Creek, Calif., 1994.
19. S. Hensley, Vacuum Insulated Pipe, CVI, Inc., Columbus, Ohio, personal communication, 1994.
20. Technical data, *Cryogenic Standard Products*, CVI, Inc., Columbus, Ohio, 1994; U.S. Pat. 4,233,816 (to CVI, Inc.).
21. J. A. Oates, ed., *Pipes and Pipelines Manual and Directory*, 3rd. ed., Scientific Surveys, Ltd., London, 1970, pp. 7–19.
22. *Chlorine Pipelines*, The Chlorine Institute, Inc., Washington, D.C.
23. Technical data, *The Trans Alaska Pipeline*, Alyeska Pipeline Service Co., Anchorage, Alaska, 1992.
24. E. S. Newcomer and P. G. McDevitt, *Oil Gas J.*, 49–57 (Sept. 14, 1992).
25. H. Liu, "Freight Pipelines for Transporting Solids and Packages," presented at *ASCE 1993, Annual Convention and Exposition*, Dallas, Tex., Oct. 26, 1993.
26. A. Y. Smoldyrev and Y. K. Safonov, in W. C. Cooley, ed., *Pipeline Transport of Concentrated Slurries*, Terraspace, Inc., Rockville, Md., trans. by A. L. Peabody, 1979.
27. R. M. Braca, *Pipeline Gas J.*, 32–36 (Jan. 1988).
28. J. K. Anderson, Black Mesa Pipeline, Inc., personal communication, Jan. 5, 1994.
29. E. J. Wasp, *Mech. Eng.* **101**, 38 (Dec. 1979).
30. *Coal* **25**(11), 12–14 (Nov. 1988); **95**(1), 16, 25 (Jan. 1990).
31. *PETC Review* **8**, 24–31 (Spring 1993).

32. S. D. Serkin, Coal and Slurry Technology Association, Washington, D.C., personal communication, Feb. 1994.
33. *Oil Gas J.*, 44 (Sept. 18, 1989).
34. H. Liu, "Coal Log Pipeline; Economics, Water Use, Right-of-Way, and Environmental Impact," presented at *The 10th Annual Pittsburgh Coal Conference*, Sept. 20–24, 1993, Pittsburgh, Pa.
35. R. J. Brown, *Oil Gas J.*, 106–111 (May 2, 1983).
36. M. Hein, *Oil Gas J.*, 146–157 (May 2, 1983).
37. *Pipeline Util. Constr.*, 36 (Oct. 1992).
38. *Pipeline Util. Constr.*, 26–32 (Oct. 1992).
39. *Pipeline Util. Constr.*, 24–26 (Feb. 1993).
40. A. V. Almeida, *Civil Eng.* **62**(9), 71–73; *Fluid Abstr.* **3**(5), 6 (1992).
41. *Pipeline Util. Constr.*, 42–44 (Mar. 1992).
42. P. C. Porter, *Pipeline Gas J.*, 37–47 (Aug. 1993).
43. G. Smith, *Pipeline Gas J.*, 35–40 (Aug. 1992).
44. *Pipes Pipelines Int.*, 5–7 (Nov.–Dec. 1992).
45. *Pipeline Gas J.*, 51 (July 1993).
46. J. Watts, ed., *Pipeline Gas J.*, 55–56 (Dec. 1992).
47. M. S. Keys and R. Evans, *Pipeline Gas J.*, 26–33 (Mar. 1993).
48. R. M. Cameron, *Pipeline Gas J.*, 34–40 (Mar. 1993).
49. *Code of Federal Regulations*, U.S. Dept. of Transportation, U.S. Printing Office, Washington, D.C., Title 49D, Part 195.
50. D. J. Hovey and E. J. Farmer, *Oil Gas J.*, 104–107 (July 7, 1993).
51. *Capsule Pipeline Res. Center NEWS* **2**(1) (Spring 1993).
52. *Oil Gas J.* **76**, 63 (Aug. 13, 1978).
53. *National Transportation Statistics Annual Report*, DOT-UNTSC-RSPA, Washington, D.C., June 1992.
54. L. V. Parent, ed., *Pipe Line Prog.*, 13 (Nov. 1993).
55. N. R. Baker, T. P. Whaley, and co-workers, *Transmission of Energy by Open-Loop Chemical Energy Pipelines*, Institute of Gas Technology Final Report, Sandia Contract No. 87-9181, June 1978, pp. 10–17.

THOMAS P. WHALEY
Consultant

GEORGE M. LONG
Institute of Gas Technology

**PIPERAZINE.**   See DIAMINES AND HIGHER AMINES, ALIPHATIC.

**PIPERIDINE.**   See PYRIDINE AND PYRIDINE DERIVATIVES.

# PIPING SYSTEMS

A piping system provides a conduit for the safe and economical transfer of fluid or fluid-like materials from one location to another, often with provisions for controlling the rate of flow. Piping system design requires specialized knowledge of the selection and application of materials of construction, fluid mechanics (qv), mechanical and structural design, facility safety requirements, and an understanding of the potential hazards and environmental consequences of accidental release of the material being transferred.

Piping materials must be suitable for the temperature, pressure, and corrosivity of the flowing material; the size must be adequate for the flow rate desired; and valving must be provided to enable the flow to be started, stopped, and rate controlled when necessary. In addition, the piping system must meet certain mechanical and structural requirements. The pipe wall thickness must be adequate for anticipated pressures and temperatures and the mechanical loads it carries; thus the supports must be designed to prevent excessive mechanical loads and, when necessary, to avoid pockets owing to bowing of the pipe. Furthermore, the overall piping system must be arranged to provide sufficient flexibility to accommodate the thermal expansion (or contraction) of the pipe and the movements of connected equipment caused by the temperature difference between operating and nonoperating conditions or by foundation settlement. Also, the properties of the fluid must be known, so that appropriate design practices are followed for materials that are hazardous or would adversely affect the environment.

## Materials

Pipe materials are metallic, nonmetallic, or lined metallic. The most common material, carbon steel, is generally the least expensive. In many cases, however, it cannot be used because of the corrosivity or the temperature of the flowing medium. When temperature, rather than corrosivity, is the limiting factor, the lowest grade of steel (qv) that provides the required service life at the design temperature is used. Thus, as the design temperature increases, carbon steel is replaced progressively by carbon–molybdenum steels, chromium–molybdenum steels, and finally, higher alloyed chromium–nickel steels and other high temperature alloys (qv). When the design temperature and pressure require expensive alloy piping, the economics of a higher strength material having a thinner wall should be compared with that of a lower strength material having a thicker wall. When corrosivity of the fluid is of prime importance, because corrosion rate is usually temperature-dependent, perceptive process design and plant layout (qv) can result in significant cost savings by minimizing the amount of expensive piping required.

For moderate temperatures and pressures and if corrosivity eliminates carbon steel, nonmetallic piping offers an alternative. Plastic and reinforced plastic piping and fittings in many different forms are available from a number of manufacturers who also provide extensive mechanical design and corrosion data. The pressure–temperature ratings vary with pipe size and manufacturer. The maximum service temperature and pressure are generally below 150°C and 2.0 MPa (290 psig), respectively. For some materials the pressure rating decreases as the temperature increases. Except for the most straightforward applications of plastic or reinforced plastic piping, the manufacturer should be consulted to ensure that the material is suitable. The environment frequently limits the service rating. Generally, nonmetallic piping is limited to situations where the risk of fire is low. Glass piping can be used to temperatures of about 230°C, limited by the gasket materials rather than the glass (qv), and on pressures dependent on the pipe size but generally below about 700 kPa (100 psig). When product purity is of prime importance or the fluids are highly corrosive, glass piping has no equal.

For pressures above the capabilities of nonmetallic piping and where corrosion is a consideration, metal piping with plastic or glass linings is preferred if shock loading and thermal expansion are not severe. Here the pressure load is carried by the outer metal pipe, which is protected from the flowing fluid by the inner lining. In addition to plastics and glass, other lining materials, eg, metals, can provide corrosion protection for the pressure-carrying pipe. For larger-diameter pipes where the temperature of the flowing fluid requires a high alloy pipe, or is even beyond the suitable temperature range for metals, carbon or alloy steel pipe lined with insulating brick or other heat-resistant ceramic materials, possibly backed up by thermal insulation, may be used. Cost, availability, and ease of installation and repair must be considered in evaluating alternatives.

Design parameters as a function of temperature and design temperature limits are set forth in the ANSI/ASME B31 Piping Codes for a very broad range of materials. These codes, and the additional information available from manufacturers, vendors, and technical societies such as the National Association of Corrosion Engineers provide ample data for the selection of materials for piping systems (1–13).

## Pipe Size

With respect to pipe size, piping systems may be divided into the following three classes.

(1) Connecting equipment where pressure is set by process requirements, such as the pressure in a fractionation tower, and where the pressure loss in the piping system does not determine the pressure-rise requirements of pumps (qv) or compressors or the elevation of towers, drums, or other pieces of equipment. Pipe sizes are generally chosen as the minimum size able to carry the maximum flow rate required at a pressure loss equal to the pressure available. In addition to the friction and velocity-head losses in the piping and fittings, the pressure loss owing to flow meters and valves in the line, particularly those needed for proper operation of control valves, must be included as part of the pressure loss in the piping system. Full consideration must be given to off-design operating conditions that may involve changes in the available pressure differential between the terminals of the piping system and changes in the required flow rate to ensure that the line size is adequate for all anticipated operating conditions. Furthermore, although the available pressure loss may be used to set the pipe size, the velocity in the pipe cannot exceed the velocity of sound in the fluid. If large pressure losses are taken in gas-carrying piping systems, the pressure loss calculations must be based on isothermal or adiabatic flow considerations, including the change in density with pressure, rather than

treating the system as one handling a constant density fluid. Corrosion and noise also affect the commonly applied velocity limitations. Conventional practice results in liquid and vapor velocities generally below 3 and 10 m/s, respectively.

(2) Piping systems where pressure loss determines all or part of the pressure rise developed by pumps or compressors. In these systems the choice of pipe size is strongly influenced by the overall pump (or compressor) and piping economics. Increasing the pipe size reduces the pressure loss, thereby reducing the investment and operating costs of the pump (or compressor), but increasing the piping cost. Reducing the pipe size obviously has the opposite effect (Fig. 1). The best design results in system layout and the minimum total cost, including the cost of fittings, valves, and pipe supports. The latter may change because of flexibility requirements because larger-diameter pipe has lower flexibility. Similarly, the pumping system costs include the cost of the pump and driver, and if electric motor drive is used, the electrical starter and wiring and the effect on overall plant electrical service. If turbine drive is used, the cost of the turbine, steam piping system, and plant steam generation system is included (see STEAM). Depending on the economics of the particular situation, the system design can be selected on investment cost alone, or on a specific payout period where both initial investment and operating costs are considered.

Typical pump-discharge piping in systems with carbon steel pipe is sized for a pressure drop of 0.20–1.0 kPa/m pipe (1.5–7.5 mm Hg/m). With high alloy piping, pressure drops from 0.40 to 2.0 kPa/m (3–15 mm Hg/m) often give the most economical overall design; the lower pressure drop is associated with the higher flow rate and vice versa. For piping systems in gas-compressor or large-capacity pump circuits, attention must be given to the discrete capacity ranges of the equipment. A small increase in the compressor discharge pressure may require a machine with the next larger frame size and involve costs substantially greater than the same pressure increase at a lower pressure level (within the range of the smaller machine). In gas-compressor circuits all the equipment in the circuit should be considered rather than just the piping and the compressor (14,15).

(3) Piping systems where pressure loss contributes to setting the elevations of towers and drums; for example, piping systems connecting tower-bottom

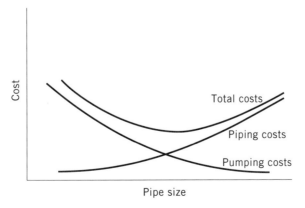

**Fig. 1.** System costs vs pipe size.

draw-offs to pump suctions and piping systems between towers and reboilers. Pumps require a specific minimum net positive suction head (NPSH) at the suction flange, and for liquids at their boiling or bubble-point temperature, the NPSH available consists of the static head of the liquid above the pump suction less the velocity head loss and the friction head loss in the piping. Thus, if a pump requires a 4-m NPSH and the suction piping system has a pressure loss equivalent to 0.5 m liquid, the minimum liquid level in the vessel must be 4.5 m above the pump suction flange. Increasing the tower elevation above grade increases the costs for the tower support, whereas increasing the available NPSH at the pump may enable a less expensive pump to be used. Here again, an economical balance must be found that minimizes the total cost. Pump suction lines are generally sized for pressure drops of less than 0.25 kPa/m (1.9 mm Hg/m) when the available NPSH is critical. The pressure loss to any point in the line should always be less than the static head pressure gain to that point to prevent vaporization in lines handling a liquid at the boiling or bubble point. Pump suction-piping systems in critical NPSH service should always be arranged to minimize the length of piping required, consistent with providing the required piping system flexibility.

The driving force causing circulation in thermosiphon reboilers is derived from the difference in density between the liquid in the piping from the tower to the reboiler and that of the liquid–vapor or vapor in the piping from the reboiler to the tower. Sufficient pressure must be developed to overcome the pressure loss in the reboiler and in the piping. Since the minimum reboiler elevation above grade is determined by access to the reboiler and its piping, the tower draw-off and return nozzles must be sufficiently above the reboiler to develop the required static pressure rise in the draw-off line. If the reboiler circuit is the determining factor in setting the tower elevation, an economic balance between piping cost, reboiler cost, and the cost of increasing the tower elevation must be found. Reboiler piping systems typically are sized for pressure drops of below 0.1 kPa/m pipe (0.75 mm Hg/m). The two-phase flow regimes in the riser should be outside the slug-flow region in horizontal pipe runs with slug flow in the vertical lines at low to moderate kinetic energy levels.

In order to select the pipe size, the pressure loss is calculated and velocity limitations are established. The most important equations for calculation of pressure drop for single-phase (liquid or vapor) Newtonian fluids (viscosity independent of the rate of shear) are those for the determination of the Reynolds number, and the head loss, $h_f$ (16–18).

$$Re = \frac{dv\rho}{\mu} \tag{1}$$

where $Re$ = Reynolds number, nondimensional; $d$ = pipe inside diameter, m; $v$ = fluid velocity, m/s; $\rho$ = fluid density, kg/m$^3$; $\mu$ = fluid dynamic viscosity, Pa·s (1 Pa·s = 10$^3$ cP). Equation 2 is the commonly used Fanning head-loss equation:

$$h_f = 4\left(\frac{fL}{d}\right)\left(\frac{v^2}{2\,g_c}\right) \tag{2}$$

where $h_f$ = friction head loss, m; $f$ = Fanning friction factor, nondimensional; $L$ = length of line, m; $g_c$ = acceleration of gravity, m/s$^2$. The equation is sometimes written as

$$h_f = \left(\frac{f_D L}{d}\right)\left(\frac{v^2}{2\,g_c}\right) \tag{3}$$

and is referred to as the Darcy or Darcy-Weisbach equation. The equations are basically the same, except that the Darcy friction factor, $f_D$, is four times the Fanning friction factor. When using charts of friction factor versus Reynolds number, the proper equation must be used with the proper friction factor.

For laminar flow ($Re$ <2000), generally found only in circuits handling heavy oils or other viscous fluids, $f = 16/Re$. For turbulent flow, the friction factor is dependent on the relative roughness of the pipe and on the Reynolds number. An approximation of the Fanning friction factor for turbulent flow in smooth pipes, reasonably good up to $Re = 150{,}000$, is given by $f = (0.079)/(4 \cdot Re^{1/2})$.

The flow resistance of pipe fittings (elbows, tees, etc) and valves is expressed in terms of either an equivalent length of straight pipe or velocity head loss (head loss = $Kv^2/2\,g_c$). Most handbooks and manufacturers' publications dealing with fluid flow incorporate either tables of equivalent lengths for fittings and valves or $K$ values for velocity head loss. Inasmuch as the velocity in the equipment is generally much lower than in the pipe, a pressure loss equal to at least one velocity head occurs when the fluid is accelerated to the pipe velocity.

There are also many empirical formulas used for calculating the friction head loss in piping systems. These must be used carefully because many are based on the properties of specific fluids and are not applicable over a broad range of fluids, temperatures, and pressures. For example, the Hazen and Williams formula widely used for water flow:

$$h_f = 0.002126L(100/C_{\mathrm{HW}})^{1.852}q^{1.852}/d^{4.8655}$$

where $h_f$, $L$, and $d$ are as previously defined, $q$ = m$^3$/s, and $C_{\mathrm{HW}}$ is the Hazen and Williams constant (essentially a friction factor), based on the properties of water at 15.6°C. For new, clean steel pipe and cast-iron pipe, $C_{\mathrm{HW}}$ has a value of about 130; a value of 100 is commonly used in design to account for the increase in friction as the pipe ages. For fluids with properties significantly different than those of water at 15.6°C, a density correction must be made and a $C_{\mathrm{HW}}$ incorporating a viscosity correction must also be used. Values of $C_{\mathrm{HW}}$ are available in many handbooks covering fluid flow (19).

Occasionally, piping systems are designed to carry multiphase fluids (combinations of gases, liquids, and solids), or non-Newtonian fluids. Sizing piping for such systems is beyond the scope of this article. Publications covering multiphase flow (20) and non-Newtonian flow (21) are available.

## Valves

Valves in piping systems are employed for on–off service (gate, plug, and ball valves), for controlling the fluid-flow rate (globe, needle, angle, butterfly, and diaphragm valves), and for ensuring unidirectional flow (check valves) (22).

Valves for on–off service are often called shutoff or block valves and are required to provide tight (leakproof) shutoff. Under normal plant operating conditions, they are either completely open or completely closed, and usually are situated near or at equipment that might require rapid stoppage or start of flow. On a fractionation tower, block valves usually are installed on all feeding lines and effluent lines, except for the overhead line and the lines to and from the reboiler. These valves have to be readily accessible and be located in the vicinity of the equipment they serve (no further than ca 11.5 m away) in order that an operator located at the valve can see the equipment. Gate valves, as their name implies, have a gate-like disk that moves at right angles to the flowing stream. The disk is wedge shaped and is forced against the valve seats in closing (Fig. 2a). Plug valves have a conical plug, with a through passageway for the fluids, fitted into a conical seat (Fig. 2b). When the passageway is lined up with the connecting piping, the valve is open. A 90° turn of the plug positions the passageway at right angles to the connecting piping and shuts off the flow since the opening of the passageway is closed by the walls of the conical seat. Ball valves are plug valves with spherical plugs, except that the flow passageway is circular instead of being almost rectangular. Ball valves are available with full-size and reduced-size flow passageways. With a full-size passageway, the open ball valve provides a continuous flow path of the same diameter as that of the piping, and hence has a very low pressure loss. In services where absolutely tight shutoff is required for safety reasons, double-block valves with an intermediate valved bleed line are often employed. When both block valves are closed and the bleed line open, leakage fluid can be removed by the bleed line, thereby preventing pressure buildup between the valves and leakage across the downstream valve.

Globe valves have a body configuration (Fig. 2c) that causes the flowing fluid to take an S-shaped path through the valve body. Needle valves generally have a similar internal flow configuration, whereas angle valves cause the flowing fluid to take an L-shaped path. These valves have higher pressure losses than gate, plug, or ball valves; angle valves generally lose less pressure than globe or needle valves. These valves are used for flow control, with the seat and disk or needle configuration selection based on the degree of control required. A flat composition disk and a flat seat or a slightly tapered metal disk and seat are used when close control, particularly near the shutoff position, is not required. When close control is required at all positions, a long taper plug-type disk and long taper seat are used. For fine control, the long taper plug-type disk is replaced by a long tapered cone and the valve is referred to as a needle valve. Butterfly valves (Fig. 3) more closely resemble gate valves than globe valves because flow is controlled by means of a disk positioned across the flow passage for shutoff or parallel to the direction of flow for the wide-open position. Only specially designed butterfly valves provide tight shutoff and their control characteristics are somewhat poorer than desired. However, in situations where flow control of large gas volumes is required and pressure loss must be minimized, the butterfly

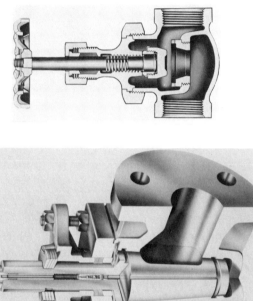

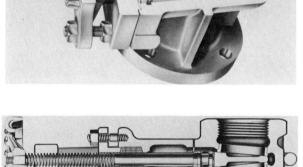

(a)

(b)

(c)

**Fig. 2.** (**a**) Gate valve; (**b**) plug valve; and (**c**) globe valve.

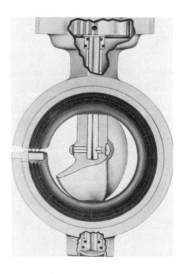

**Fig. 3.**  Butterfly valve.

valve has no competition. The diaphragm or pinch-clamp valve affects closure by a resilient rubber-like diaphragm pressing down on a transverse weir (Fig. 4). A relatively clean sweep of fluid through the valve and over the weir minimizes pressure drop. Crevices and corners are also minimized, which prevents buildup of precipitates or solid deposits. Diaphragm valves are employed where the valve mechanism must be isolated from the fluid or where the fluid contains solids and a tight closure is required. In this type of valve, the diaphragm acts as closing medium as well as sealing gasket. The diaphragm valve is used widely in the food and beverage industry and under corrosive conditions. It is limited to ca 90°C and 1.0 MPa (145 psi), depending on the material of construction. The body can

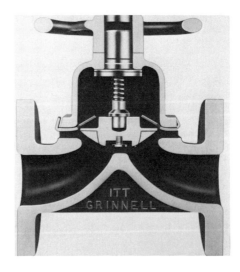

**Fig. 4.**  Diaphragm or pinch-clamp valve.

be made of standard materials; the flexible diaphragm, however, must withstand the limits of temperature and pressure.

Check valves such as the swing check, tilting-disk check, lift check, and ball check prevent flow reversals. The swing-check valve is the most commonly used and has the lowest pressure loss. It is similar to a gate valve with one seat and the disk hinged at one end; thus fluid flowing in one direction lifts the disk off the seat, whereas reverse flow forces the disk against the seat. The tilting-disk check is like a swing check, but with the hinge point close to the center of the disk, partially counterbalancing the disk and thereby reducing the closing speed. The lift check is similar to the globe valve with fluid flow in one direction lifting the disk off its seat and flow reversal forcing the disk against the seat (Fig. 5). These check valves utilize gravity in their operation to keep the valve closed in the nonflowing condition, and hence must be installed in the proper orientation. The ball check valve can have either a straight-through or globe valve configuration; in both, the solid ball is held against the seat by a spring. Sufficient pressure must be available in the fluid to overcome the spring force and permit flow in the flow-through direction. Fluid pressure adds to the spring pressure to keep the ball seated and prevent backflow. Disk-type check valves designed with spring-loaded disks are also available, generally in the lift-check configuration. Spring-loaded check valves can generally be placed in any orientation in the piping system.

Check valves commonly are used in centrifugal pump discharge circuits to prevent damage to the pump and motor resulting from reverse rotation caused by the pump operating as a turbine owing to reverse liquid flow. Spring-loaded check valves and tilting-disk check valves are generally used in systems having irregular or pulsating flow because there is less tendency for these to chatter and slam.

For any given application of any type of valve, temperature, pressure, and corrosivity must be considered in the same manner as for the piping system itself. Valve vendors specify the temperature, pressure, and general service limitation for their valves and these are indicated in the manufacturers' catalogues.

Although it has been common practice to specify the pressure loss in ordinary valves in terms of either equivalent length of straight pipe of the same size or velocity head loss, it is becoming more common to specify flow rate and

**Fig. 5.**  Horizontal lift-check valve.

pressure drop characteristics in the same terms as has been the practice for valves designed specifically for control service, namely, in terms of the valve coefficient, $C_v$. The flow coefficient of a valve is defined as the volume of liquid at a specified density that flows through the fully opened valve with a unit pressure drop, eg, $C_v = 1$ when 3.79 L/min (1 gal/min) pass through the valve with a pressure drop of 6.9 kPa (1 psi) (23,24). In order to use the flow coefficient as a means of valve sizing or to find the valve pressure drop for a given flow rate, the actual flowing conditions are expressed in terms of the defining conditions for the valve flow coefficient. For liquids with kinematic viscosities comparable to that of water (generally <10 mm²/s or 10 cSt):

$$C_v = 11.6\,Q\left(\frac{G_f}{\Delta P}\right)^{1/2} \quad \text{or} \quad \Delta P = G_f\left(\frac{11.6\,Q}{C_v}\right)^2 \tag{4}$$

where $Q$ = flow rate, m³/h; $G_f$ = specific gravity at flowing conditions (water = 1.0 at 15°C); and $\Delta P$ = pressure drop, kPa.

For weight flow, where $W$ = liquid flow rate, 1000 kg/h,

$$C_v = 11.6\,W_l/(G_f\Delta P)^{1/2} \quad \text{or} \quad \Delta P = \frac{1}{G_f}\left(\frac{11.6\,W}{C_v}\right)^2 \tag{5}$$

For gases or vapor (approximately):

$$C_v = W_g/(2.74(\Delta P\cdot\rho)^{1/2}) \quad \text{or} \quad \Delta P = \left(\frac{1}{7.51\,\rho}\right)\left(\frac{W_g}{C_v}\right)^2 \tag{6}$$

where $W_g$ = gas flow rate, kg/h. When the pressure downstream of the valve is less than the critical pressure, the critical pressure is used in calculating the pressure drop across the valve. The density used should correspond to the downstream pressure and temperature, ie, critical pressure and temperature, if the downstream pressure is below the critical.

The valve cannot control if it is at either end of its travel. To ensure controllability, a valve is generally chosen in such a way that at the maximum design flow rate the flow coefficient required is no more than 85% of the wide-open valve flow coefficient, and at the minimum anticipated flow rate requiring control, a flow coefficient of about 10% of the wide-open valve flow coefficient is required. Whenever practical, control valves are located at grade or at platforms, to assure adequate working space for servicing.

The operation of system valves (also starting and shutdown of pumps) has a significant effect on the transient fluid pressures in the piping system because of the acceleration and deceleration of the fluid as it changes its velocity. As a simple example, the maximum head rise caused by the instantaneous closing of a valve is given by

$$h_{\max} = av/g_c \tag{7}$$

where $h_{\max}$ = maximum head rise of fluid, m, and $a$ = pressure wave velocity in the fluid, m/s.

The basic wave velocity is a function of the bulk modulus and density of the fluid, where $K$ = bulk modulus, Pa.

$$a = (K/\rho)^{1/2} \tag{8}$$

Gas entrained in the fluid and the flexibility of the pipe wall both result in lowering of the wave speed. For deaerated water, the wave speed is about 1250 m/s. Detailed methods of analysis and evaluation of hydraulic transients may be found in the literature (25).

## Mechanical and Structural Standards

The design code to which the piping system must usually conform is the American National Standards Institute/American Society of Mechanical Engineers (ANSI/ASME) B31 Code for Pressure Piping (Table 1). The section of the code that must be followed is determined by the general service for which the piping is intended. These sections establish definite rules as to the design formula that must be used. They specify allowable stresses for the piping material, joint efficiencies, and minimum allowances for threading, mechanical strength, and corrosion or minimum thickness. Material specifications are also given that establish underthickness tolerances. The codes also specify how the combined effects of different loading conditions, eg, pressure, thermal expansion, weight, wind, and earthquake, should be evaluated.

Many other codes, standards, specifications, and recommended practices have been developed by various organizations (26). Some apply to specialized piping systems; others, particularly those covering materials and dimensions, are referenced in the ANSI/ASME Code for Pressure Piping.

**Pipe-Wall Thickness.** Once the design pressure and temperature have been established and the pipe material and size selected, the wall thickness is calculated using the appropriate section of the code. In rare cases, a thin pipe must be made thicker to withstand handling. Occasionally the thickness is affected by external loads or vibrations. All codes prescribe essentially the same design formula for metallic hollow circular cylinders under internal pressure:

$$t_{\min} = \frac{PD}{2\,SE + 2\,YP} + C \tag{9}$$

**Table 1. ANSI/ASME B31 Codes for Pressure Piping**[a]

| Section | Piping service | Designation[b] |
|---|---|---|
| 1 | power | B31.1-1992 |
| 3 | chemical plant and petroleum refinery | B31.3-1993 |
| 4 | liquid transportation | B31.4-1992 |
| 5 | refrigeration | B31.5-1992 |
| 8 | gas transmission and distribution | B31.8-1992 |
| 9 | building services | B31.9-1988 |
| 11 | slurry transportation | B31.11-1989 |

[a]Rules for Nuclear Power piping are in Section 3 of the ASME Boiler & Pressure Vessel Code.
[b]Fuel Gas Piping (B31.2) is no longer an ANSI/ASME code. See American Gas Association (AGA) National Fuel Gas Code, Z223.1.

where $t_{\min}$ = minimum wall thickness, m (if pipe is ordered to nominal wall thickness, the manufacturing tolerance on wall thickness must be taken into account); $D$ = outside diameter, m; $E$ = joint efficiency, $P$ = design pressure, Pa (psi); $S$ = allowable tensile stress in material caused by internal pressure, Pa (psi); $C$ = allowance for threading, mechanical strength, or corrosion, m; and $Y$ = coefficient.

This formula is not intended for thick-wall cylinders, and for its limitations the appropriate sections of Code B.31 must be consulted.

The theoretical value of the coefficient $Y$ ranges from 0 to 1. If $Y = 0$, equation 9 reduces to the Barlow or outside diameter formula. The Barlow formula is always conservative and for large $D/t$ the error is small. The value $Y = 0.4$ gives the Boardman approximation of the Lame tangential stress (27) and is used for ferritic steels (482°C or below) and austenitic steels (566°C or below) for almost all sections of the ASME Boiler and Pressure Vessel Code and the ANSI Code for Pressure Piping. At higher temperatures, the influence of creep permits a better distribution of stress and, therefore, a higher value of $Y$ is used. For ferritic steels at 510°C and austenitic steels at 593°C, $Y = 0.5$. At 538°C and above for ferritic steels, and 621°C and above for austenitic steels, $Y = 0.7$.

For most nonmetallic pipes ANSI/ASME B31.3 prescribes a somewhat similar formula for calculating the pipe thickness under internal pressure where the terms are as previously defined:

$$t = \frac{PD}{2S + P} + C \tag{9a}$$

## Supports and Restraints

The design and selection of supports and restraints for piping systems subject to thermal expansion and contraction involves more than simply the support and restraint requirements. In addition to wide ranges of movements, piping systems may be subject to operating conditions involving cyclic or local temperature variations or may be influenced by unusual startup conditions or emergency shutdowns. Therefore, the support design must be reasonably adequate and effective for any set of circumstances. The attachment to the pressure-containing pipe shell should be satisfactory for localized thermal loading and thermal-gradient effects. To accommodate normal or unusual changes in position of the line after several temperature cycles, the supporting and restraining elements are provided with sufficient margin for adjustment (28).

**Supports.**    The spacing of supports is governed by the hot allowable stress of the piping materials; stability, in the case of large-diameter thin-wall pipe; deflection to avoid sagging or pocketing; and the natural frequency of the unsupported length to avoid susceptibility to undesirable vibration.

A particular type of support assembly is selected according to the amount of restraint tolerable by the piping system and the movement to be allowed at each location. Support types are classified as rigid, resilient, and constant-effort; hanging and resting are the two basic arrangements.

Rigid-type resting (or sliding) supports are widely used because these are usually simple in construction, can accommodate substantial horizontal movements, and maintain a constant elevation. Although frictional resistance to movement is inherent in the metal-to-metal sliding contact, thermal expansion forces usually are high enough to overcome this effect. In special cases, to minimize sliding resistance, resting supports are provided with rollers or low friction sliding plates.

Rigid-type hanging supports are employed where frictional resistance is undesirable and overhead structural steel is available (see Fig. 6a). To accommodate large horizontal movements, hanger assemblies must be sufficiently long; the hanger support point is sometimes pre-offset in the direction of line movement to minimize total vertical displacement. Rigid-type hanging supports usually are preferred in high temperature applications for their ease in adjustment during initial load distribution and subsequent readjustment after equilibrium in service temperature is reached.

In general, rigid-type supports are utilized in systems having long horizontal runs where the amount of vertical expression is negligible or the inherent flexibility in the vertical direction is sufficient.

Resilient-type supports employ coil springs either in a hanging or resting application. Springs can be arranged in series or parallel to accommodate more movement or greater load beyond their individual capacity. The resisting effect of these springs varies with the vertical displacement and is proportional to the spring constant.

At locations where variable support reactions are not tolerable over the required movement range, constant-effort springs or counterweights are used. Piping systems supported entirely by constant-effort devices require precise accuracy to counterbalance the total piping load, otherwise the system may be vertically unstable.

Both the variable and constant-effort springs as well as counterweights are provided with adequate movement range and means for load adjustment to allow for unusual line displacements.

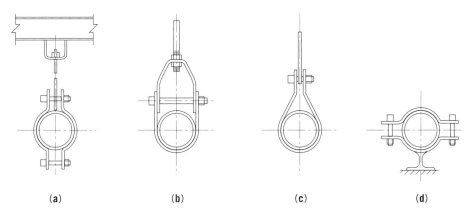

(a)                    (b)                    (c)                    (d)

**Fig. 6.** Nonintegral pipe attachments: (**a**) pipe clamp, rigid-type hanging support; (**b**) Clevis; (**c**) sling; and (**d**) clamped shoe.

**Restraints.**   A restraint limits thermal reactions at equipment and line stresses or expansion movement at specifically desired locations. It may be defined as a device preventing, resisting, or limiting the free thermal movement of a piping system. Because the application of a restraint reduces the inherent flexibility of the piping, its effect on the system is established through calculation.

Restraints are provided to limit movements in any number of directions (Fig. 7). For example, a single-directional arrangement uses a simple tie rod with pin connections (Fig. 7**a**). This type is favored because of low frictional resistance and positive action. Another simple arrangement utilizes a shoe and provides two functions at one point (Fig. 7**b**). Partial restraint along the pipe axis may be accomplished as shown in Figure 7**c**, and Figure 7**d** shows typical restraint perpendicular to the pipe axis.

**Anchors.**   Anchors provide full restraint against the three deflections and three rotations. They usually are subject to large loadings and are often required to develop the full strength of the attached pipe. Their design must, therefore, be sufficiently rigid. Attachment must be integral, the attachment device must be stiff, and the structure to which they are connected must be adequate. The skirt assembly illustrated in Figure 8 is a typical anchor arrangement. Application of anchors is common in piping subject to vibration or in systems using expansion joints to withstand unbalanced pressures.

**Attachments.**   Connections or devices attached directly to the surface of the pipe transmit thermal as well as dynamic or weight loads from the pipe shell to the restraining, bracing, or supporting fixtures. In high temperature service, where the material's strength is low and temperature gradients owing to localized heat loss at the point of attachment to the pipe are high, the most

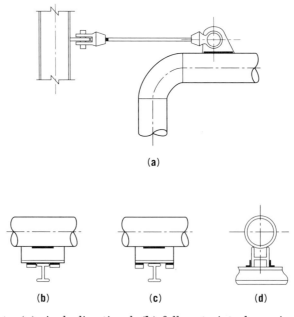

(a)

(b)          (c)          (d)

**Fig. 7.**   Restraints: (**a**) single-directional; (**b**) full restraint along pipe axis; (**c**) partial restraint along pipe axis; and (**d**) guide perpendicular to pipe axis.

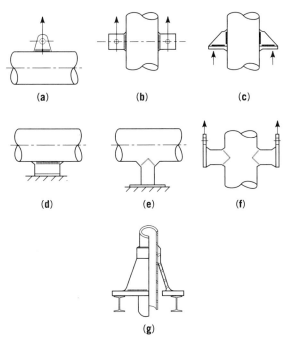

**Fig. 8.** Integral pipe attachments: (**a**) and (**b**), ears; (**c**) lug; (**d**) shoe; (**e**) and (**f**), cylindrical trunnions; and (**g**) skirt.

favorable attachment contour minimizes stress levels and avoids unnecessary stress concentrations.

Pipe attachment devices are either integral or nonintegral with the shell of the pipe. The particular type is selected according to load-carrying capacity, severity of service, and the desirability of welding directly to the pipe.

Nonintegral pipe attachments are used widely for support purposes. They are favored for their simplicity in attachment to the pipe wall, freedom in location, and availability as standardized stock items from many suppliers (see Fig. 6).

Integral pipe attachments are welded directly to the pipe and are usually of special design. They are used in high temperature services, with moderate or severe loads, and in conjunction with supports, braces, and restraints where rigidity with the pipe shell is desired. Integral attachments may be grouped into five types: ears, lugs, shoes, cylindrical trunnions, and skirts (see Fig. 8).

Ears or plates are applied in hanging-type arrangements to horizontal, vertical, or inclined piping.

Lugs are an improvement of the ear type by the addition of an element to provide lateral stiffening which also serves to improve load-carrying capacity and lateral strength. In high temperature services, unless heavily insulated, lugs are susceptible to distortion.

Shoes are attached to horizontal runs of insulating piping. Since they are cut from standard shapes, their length can be varied to accommodate large pipe movements. Although classified as integral in type, they frequently are attached with clamps to fit the pipe contour as illustrated in Figure 6**d**.

Cylindrical trunnions are of more favorable contour for stress distribution and frequently are used in high temperature services. Their load-carrying capacity is high and their structural effect on the pipe shell can easily be determined.

Skirts are used in vessels and towers. They transmit high axial and bending loads and offer favorable geometry for thermal gradients. In piping, when loads are beyond the capacity of lugs and trunnions, skirts are often favored.

Integral-type pipe attachments in elevated temperature service are subjected to the same design, material, fabrication, and inspection requirements as the pipe to which they are attached.

Piping supports, guides, and anchors increase local stresses on the pipe wall at the point of attachment. These stresses derive from continuously acting loads owing to the weight of the piping system carried at these points (pipe, contents, insulation), the pressure in the pipe, and any other loads such as valves, special components, etc. The piping system also must accommodate the additional stresses caused by occasional loads such as wind and earth movements. The code permits higher allowable stresses for the combination of continuous (sustained) loads plus occasional loads than for the continuous loads alone. Higher allowable stresses also are permitted for pressure and temperature variations beyond the design conditions when these are of limited frequency and duration.

Code-allowable stresses are conservative with respect to structural failure that occurs when the limit load is reached, ie, the load that results when component deflections and distortions have destroyed its serviceability. The limit load is generally reached when the stresses throughout a main portion of the component cross section exceed the material yield strength (29).

For piping systems operating within at least 150°C above or below ambient temperature, heat loss or gain, respectively, from the piping to the support system should be evaluated, as well as local thermal stresses due to temperature differences between the pipe and its support attachment. Clamp-type supports insulated from the piping, and extended support connections with the support members covered with insulation at the support junction and for a distance beyond frequently are used for such systems.

## System Flexibility

The need to ensure that the stresses in piping systems meet the appropriate code requirements and the concern that cyclic stresses resulting from events such as periodic heating and cooling of the piping may lead to fatigue failures, make accurate evaluation of the stresses and strains in piping systems a necessity.

**Rigid Systems.**    Literature pertaining to the theoretical analysis of the three-plane rigid piping system is voluminous (30). This literature is expanding steadily and, as it is becoming more abstract, tends to obscure the basic problem which is the analysis of a three-dimensional statically indeterminate structure.

The concepts behind the analysis are not difficult. The piping system is simply a structure composed of numerous straight and curved sections of pipe. Although, for straight pipe, elementary beam theory is sufficient for the solution of the problem, it is not adequate for curved pipe. However, by the introduction of a flexibility factor, $k$, to account for increased flexibility of curved pipe over straight pipe, and a stress intensification factor, $i$, to account for the increase in stress in a curved pipe or any other piping component over that predicted by beam theory, the elementary beam analysis can be used.

The assumptions in the analysis are the usual ones for the theory of linear elasticity, ie, the material is homogeneous and isotropic, stress is proportional to strain, plane sections before bending remain plane after bending, and deflections are small in proportion to the size of the configuration so that the effect of changes in position and shape of the member may be ignored upon flexibility as a whole. Furthermore, axial tension or compression and shear deflection may be ignored, since they are negligible in comparison to the bending and torsional effects. The last assumption is not necessary and most computer programs take these effects into account. It is made herein merely for the purpose of simplification.

A piping system can be evaluated for its displacement and stress either by manual methods (charts, tables, hand calculation) or computerized solution. The latter has become the standard approach; desktop and laptop computers can handle all but the most complicated problems. Manual methods are used only for rough estimates on very simple systems.

*Noncomputerized Calculation.* Design charts or tables are used for hand calculations or simplified formulas such as the guided cantilever equation given in Reference 31, described briefly below.

As shown in Figure 9a, if the system is not constrained, points $B$ and $C$ of the piping expand to $B'$ and $C'$ because of temperature change. The end $C$ moves $\Delta x$ and $\Delta y$ in $x$ and $y$ directions, respectively, but no internal force or stress is generated. However, in the actual case, the ends of the piping are always connected to equipment or other piping as in Figure 9b, and the system is constrained. This is the same as moving the free expanded end $C'$ back to the original point $C$, and the point $B$ to $B''$. The $\Delta x$ is the expansion from leg $AB$, and $\Delta y$ from leg $CB$.

The force and moment in a constrained system can be estimated by the cantilever formula. Leg $AB$ is a cantilever subject to a displacement of $\Delta y$ and leg $CB$ subject to a displacement $\Delta x$. Taking leg $CB$, for example, the task has become the problem of a cantilever beam with length $L$ and displacement of $\Delta x$.

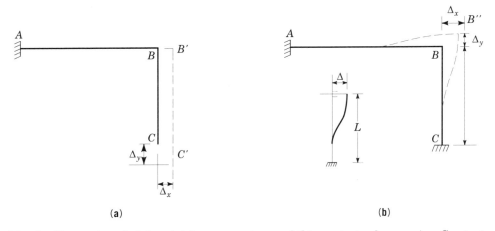

(a)                                                  (b)

**Fig. 9.** Expansion of piping: (**a**) free expansion, and (**b**) constrained expansion. See text.

This problem cannot be readily solved, because the end condition at $B$ is an unknown quantity. However, it can be conservatively solved by assuming there is no rotation at point $B$. This is equivalent to putting a guide at point $B$, and results in higher estimate in force, moment, and stress. The approach is called guided-cantilever method.

From basic beam theory, the moment-displacement and moment-force relationships for a guided cantilever are as follows:

$$M = \frac{6EI\,\Delta}{L^2} \quad \text{and} \quad M = \frac{FL}{2} \tag{10}$$

where $M$ = moment of force, N·m; $F$ = force, N (or kg·m/s$^2$); $L$ = length, m; $E$ = modulus of elasticity, Pa; $I$ = moment of inertia, m$^4$; and $\Delta$ = thermal expansion, m.

For thin-wall pipes:

$$I = \pi \frac{D^3}{8} t \text{ (approx)} \tag{11}$$

$$Z = \pi \frac{D^2}{4} t \text{ (approx)} \tag{12}$$

where $Z$ = section modulus, m$^3$; $D$ = outside diameter of pipe, m; and $t$ = pipe wall thickness, m.

The stress in this pipe resulting from the bending moment caused by the thermal expansion is

$$S = M/Z \tag{13}$$

where $S$ = thermal expansion stress, Pa (psi).

Combining equations 10, 11, 12, and 13 yields the following:

$$S = \frac{3ED\,\Delta}{L^2} \tag{14}$$

Equation 14 can be used to make a quick estimate for the length of leg required to absorb a given thermal expansion, particularly when typical values of $E$ and $S$ are used. For example, using $E = 200$ GPa (ca $29 \times 10^6$ psi), and $S = 172$ MPa (ca 25,000 psi), equation 14 becomes

$$L = 59.069(D\Delta)^{1/2} \cong 60(D\Delta)^{1/2} \tag{15}$$

For systems having multiple legs, the displacement to be absorbed by each leg is distributed proportionally to the cube of the length of the legs as shown in Figure 10.

*Computerized Solutions. Flexibility Matrix Method.* The systematic analysis of a piping system using elbows and straight pipe sections as building

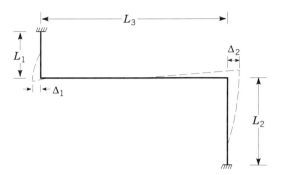

**Fig. 10.** Distribution of displacement, where $\Delta = \Delta_1 + \Delta_2 =$ expansion of $L_3$; $\Delta_1 = (L_1^3/(L_1^3 + L_2^3))\Delta$; and $\Delta_2 = (L_2^3/(L_1^3 + L_2^3))\Delta$. See text.

blocks was developed long before the advent of computer technology (qv). In manual calculations, it is essential that the equations be kept at a minimum. Previously, piping analysis basically was concentrated in the flexibility method because of the smaller number of equations involved. The detailed theory is covered in Reference 30.

In the flexibility method, a basic anchor is assigned and the entire system is worked in reference to this basic anchor. The pipe is left free to expand from the basic anchor and all the other anchors and restraints are considered as released. The forces and moments are then applied to the release anchor and restrained points to force those points back to their installed locations. Therefore, finding those forces and moments is required.

In a three-dimensional system, an anchor involves three forces and three moments; in the direction of each main axis, a directional restraint involves one force, whereas a closed loop involves three forces and three moments. Because each force or moment is counted as one unknown, a system has

$$N = 6 \times (\text{anchors} - 1) + 6 \times \text{loops} + 1 \times \text{restraints} \tag{16}$$

where $N =$ number of unknowns, and $N$ equations are needed for solutions. These equations are formed by summarizing the displacement contributions to each unknown direction that result from each unit force or moment applied at each unknown direction. Then the total displacement at each unknown direction contributed by all forces and moments should be equal but opposite to the free expansion in order to bring the point back to the original location. This should be applicable to all unknown directions.

Therefore, the equations can be written as

$$a_{11}F_1 + a_{12}F_2 + \ldots + a_{1n}F_n = -\Delta_1$$
$$a_{21}F_1 + a_{22}F_2 + \ldots + a_{2n}F_n = -\Delta_2 \tag{17}$$
$$a_{n1}F_1 + a_{n2}F_2 + \ldots + a_{nn}F_n = -\Delta_n$$

where $a_i, j_i =$ displacement at $i$ owing to unit force at $j$, $f_i =$ force or moment required at $i$, and $\Delta i =$ free expansion at $i$.

Equation 17 can be written in matrix form as

$$
\begin{vmatrix}
a_{11}, a_{12}, \ldots a_{1n} \\
a_{21}, a_{22}, \ldots a_{2n} \\
\quad \prime \\
\quad \prime \\
a_{n1}, a_{n2}, \ldots a_{nn}
\end{vmatrix}
\begin{vmatrix}
F_1 \\
F_2 \\
\prime \\
\prime \\
F_n
\end{vmatrix}
=
\begin{vmatrix}
-\Delta_1 \\
-\Delta_2 \\
\prime \\
\prime \\
-\Delta_n
\end{vmatrix}
\tag{18}
$$

or simply as

$$
(A)\{F\} = \{D\} \tag{19}
$$

where $(A)$ is the flexibility matrix, and $\{F\}$ and $\{D\}$ are the force and displacement vectors, respectively; $(A)$ is generated using structural synthesis methods. The displacement vector $\{D\}$ is calculated at all constrained points assuming the constraints are removed. The constraint reactions $\{F\}$ are calculated by solving equation 19. The effect of externally applied loadings and deadweight of the piping is treated by adding an equivalent displacement term in vector $\{D\}$.

The size of the flexibility matrix $(A)$ is relatively small, thus the flexibility matrix method is especially suitable for small computers. Most of the pipe-stress computer programs developed in the early 1960s used this method. The original Kellogg program, the Mare Island (U.S. Navy) program, and programs which originated from the Mare Island program such as TRIFLEX and AUTOFLEX, all used the flexibility method. However, many topological and bookkeeping problems result. For example, the logic used in handling multiconnected loops is very complicated.

*Stiffness Matrix Method.*   The stiffness method can be expressed in matrix form as follows:

$$
(K)\{D\} = \{F\} \tag{20}
$$

where $(K)$ is the stiffness matrix of the piping system, and $\{D\}$ and $\{F\}$ are the displacement and force vectors, respectively. This so-called direct stiffness matrix method has been developed for solutions to general structural–mechanics problems, and its formulation can be found in many textbooks on the matrix methods of structural analysis (32,33).

In contrast to the flexibility method, the stiffness method considers the displacements as unknown quantities in constructing the overall stiffness matrix $(K)$. The force vector $\{F\}$ is first calculated for each load case, then equation 20 is solved for the displacement $\{D\}$. Thermal effects, deadweight, and support displacement loads are converted to an equivalent force vector in $\{F\}$. Internal pipe forces and stresses are then calculated by applying the displacement vector $\{D\}$ to the individual element stiffness matrices.

This method has a simple straightforward logic for even complex systems. Multinested loops are handled like ordinary branched systems, and it can be extended easily to handle dynamic analysis. However, a huge number of equations is involved. The number of unknowns to be solved is roughly equal to six times the number of node points. Therefore, in a simple three-anchor system, the number of equations to be solved in the flexibility method is only 12, whereas the number of equations involved in the direct stiffness method can be substantially larger, depending on the actual number of nodes.

The most recent developments in computational structural analysis are almost all based on the direct stiffness matrix method. As a result, piping stress computer programs such as SIMPLEX, ADLPIPE, NUPIPE, PIPESD, and CAESAR, to name a few, use the stiffness method.

Figure 11 shows a reasonably complex piping system. The number of equations involved in modeling this systems is 19 for the flexibility method, and 144 for the stiffness method.

Modern piping flexibility analysis programs take advantage of the data handling capability of computers and of the physical property and other data needed for the flexibility analysis to calculate the stresses resulting from other loadings, eg, pressure, weight, wind, and seismic effects. Some programs also calculate stresses owing to dynamic effects such as forced vibrations, relief valve discharge, and local pipe failure. Many programs compare the calculated stresses with the code allowable stresses, demonstrating code compliance.

**Flexibility and Stress-Intensification Factors.** The flexibility factor $k$ ($\geq 1.0$) is defined as the ratio between the rotation per unit length of the part in question produced by a given moment to the rotation of a straight pipe (of the same size and schedule) produced by the same moment. A close approximation of the flexibility factor that agrees quite well with theory and experiment for bends is as follows:

$$k = 1.65/h \qquad (21)$$

where $h = tR/r^2$; $t$ = wall thickness, m; $R$ = radius of bend, m; and $r$ = mean pipe radius, m.

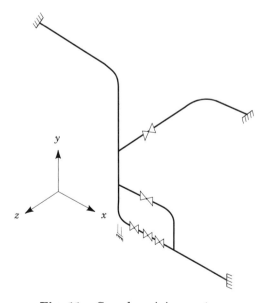

**Fig. 11.** Complex piping system.

Equation 21 applies to the central portion of a large arc under bending, and neglects the effects of internal pressure or end restraints. The effect of internal pressure on the flexibility of large thin-wall pipes can be significant and should be evaluated with the code formula. End restraints also have a considerable effect (34). The ANSI B31.3 code introduces the factors $h^{1/6}$ and $h^{1/3}$ to reduce $k$ for 90° elbows flanged at one and both ends, respectively. For T-sections and flanged connections, no theoretical solution has yet been devised to compute $k$, and for this reason the code assigns these components a $k$ factor of 1.0.

In order to derive values for the theoretical stress concentration factor, a fatigue–stress concentration factor $i$ has been used. The thermal expansion of a pipe is essentially a fatigue problem. This $i$, which is the ANSI B31.3 code value, is defined as the ratio of the bending moment producing fatigue failure in the component in question to that producing fatigue failure in straight pipe. Although this ratio is not strictly valid throughout the range of a large number of cycles, test data reasonably fit the following equation:

$$iSN^{0.2} = C \tag{22}$$

where $i$ is the fatigue–stress concentration factor; $S$, the calculated bending stress in a straight pipe; $N$, the cycles to failure; and $C$ a material constant. The value of $C$ for grade-B carbon steel pipe, with ordinary commercial butt welds, has been found to be 245,000 (35,36). Equation 22, with $C = 245,000$, was the baseline established for $i = 1.0$.

For bends, elbows, miters, and forged and fabricated T-sections, the value of $i$ can be expressed by the following:

$$i = \frac{0.9}{h_e^{2/3}} > 1.0 \tag{23}$$

$$i = \frac{0.75}{h_e^{2/3}} > 1.0 \tag{24}$$

where $h_e$ is the effective flexibility characteristic. Equation 23 applies to in-plane bending in bends, elbows, and miters, and for out-of-plane bending in miters and Ts. Equation 24 applies to out-of-plane bending in bends and elbows. For elbows and bends, $h_e = h$ as previously defined. For miters and Ts, the values can be obtained from Table D1 in Appendix D of B31.3. The end corrections $h^{1/6}$ and $h^{1/3}$ for curved and mitered pipe that were applied to the flexibility factor are again applied to $i$ for curved and mitered pipe. Locations of areas where the stress concentrations occur are available (36).

*The Stress-Range Concept.* The solution of the problem of the rigid system is based on the linear relationship between stress and strain. This relationship allows the superposition of the effects of many individual forces and moments. If the relationship between stress and strain is nonlinear, an elementary problem, such as a single-plane two-member system, can be solved but only with considerable difficulty. Most practical piping systems do, in fact, have stresses that are initially in the nonlinear range. Using linear analysis in an apparently nonlinear problem is justified by the stress-range concept (35).

This concept is explained by Figure 12 which shows the uniaxial stress–strain curve for a ductile material such as carbon steel. If the stress level is at the yield stress $B$ or above, the problem is no longer a linear one.

As a pipeline is heated, strains of such a magnitude are induced into it as to accommodate the thermal expansion of the pipe caused by temperature. In the elastic range, these strains are proportional to the stresses. Above the yield stress, the internal strains still absorb the thermal expansions, but the stress, $\sigma'_e$ computed from strain $\epsilon_2$ by elastic theory, is a fictitious stress. The actual stress is $\sigma_C$ and it depends on the shape of the stress–strain curve. Failure, however, does not occur until $\epsilon_3$ is reached which corresponds to a fictitious stress of many times the yield stress.

As the pipe cools to room temperature, the strain $\epsilon_2$ decreases to zero while a stress, $-\sigma_D$, is produced in the reverse direction. The sum of $\sigma_C + \sigma_D$ is the stress range and is equal to $\sigma'_e$. As long as $\sigma_D$ is no greater than the yield stress in reverse direction any subsequent heating and cooling is in the elastic range. Since $-\sigma_D$ is the stress at room temperature, the modulus of elasticity for computing $\sigma_e$ is the room-temperature modulus $E_c$ rather than the modulus at operating temperature, $E_h$.

*Stresses Allowable by B31.3 Code.* The foregoing discussion of stress range points out the cyclic nature of the piping stress problem and suggests that stress

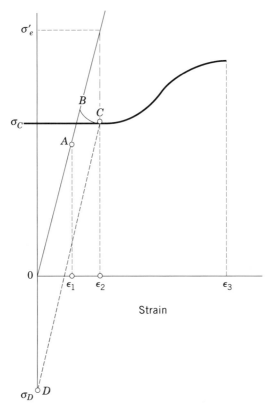

**Fig. 12.** Uniaxial stress–strain curve for an elastic plastic material. See text.

levels should be based on fatigue considerations. The B31.3 code assigns stress levels for the allowable stress range. The allowable stress range for the expansion stress, $\sigma_e'$ is given as follows:

$$S_A = f(1.25\ S_C + 0.25\ S_h) \qquad (25)$$

$S_C$ = basic allowable stress for the material at minimum (cold) metal temperature; $S_h$ = basic allowable stress for the material at maximum (hot) metal temperature; $f$ = stress-range reduction factor for cyclic conditions for the total number of full temperature cycles over the expected life; $f$ = 1.0 for 7000 cycles or less.

The sum of the longitudinal stress (in the corroded condition) caused by pressure weight or other sustained loadings is not allowed to exceed $S_h$. Where the sum of these stresses is less than $S_h'$, the difference between $S_h$ and this sum may be added to the term 0.25 $S_h$ in equation 25.

For the above, the total allowable stress range (for 7000 cycles or less) is as follows:

$$S_A \text{ (total)} = 1.25\ S_C + 1.25\ S_h \qquad (26)$$

If $S_C$ and $S_h$ are two-thirds the yield stress of the materials at room and operating temperatures (which is the case below the creep range), then

$$S_A \text{ (total)} = 5/6(S_{ypc} + S_{yph}) \qquad (27)$$

where $S_{ypc}$ = cold yield stress and $S_{yph}$ = hot yield stress. After initial yielding, the expansion stress, $\sigma_e'$ is in the elastic range.

**Semirigid and Nonrigid Systems.** The calculated results of flexibility analysis with which the designer is concerned are the stress levels and movements at significant points within the piping system, and the magnitude and direction of forces and moments at terminal equipment. When stresses or reactions exceed their allowable limits, modification of the piping arrangement or of the physical cross section of the pipe is necessary.

The simplest method of reducing stresses and reactions is to provide additional pipe in the system in the form of loops or offset-bonds. When physical limitations restrict the use of additional bends, a multiple arrangement of several small-size pipe runs may sometimes be used. Owing to stress intensification, the maximum stress generally occurs at elbows, bends, and Ts. Thus, heavier-walled fittings may reduce the stress without significantly impairing flexibility. Finally, effectively located restraints can reduce thermal effects on the equipment.

If a rigid system is impractical, either for economic or practical reasons, the piping configuration may always be made more flexible with hinge, rotation, or translatory joints. Systems using such devices are commonly classified as semirigid or nonrigid depending on the degree of flexibility.

Figure 13 illustrates two typical semirigid single-plane piping arrangements, one of which is provided with hinge-type joints and the other with a

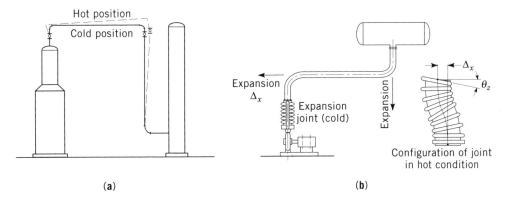

**Fig. 13.**  Semirigid piping systems: (**a**) two-hinge arrangement; (**b**) universal-type arrangement.

universal-type tied joint. The horizontal differential expansion of the system shown in Figure 13a is accommodated by bending the vertical riser, whereas the vertical expansion is absorbed by rotation of the hinges. In the system illustrated in Figure 13b, vertical expansion is accommodated by flexure of the horizontal pipe run and the horizontal expansion is absorbed by angular and lateral displacement of the joint. Arrangements such as these enable the lowering of overall thermal pipe stresses and terminal reactions while maintaining desirable features of self-support and ability to carry longitudinal pressure loads. The computations for flexibility and stress for piping in this category are greatly simplified, especially when the system is contained in only one plane. The fact that the system is only partially restrained should not imply that exact methods of thermal-stress analysis are unnecessary; design conditions or service usually establish the degree of comprehensiveness required. Whatever method of analysis is used, reasonably accurate ranges of movement at the joints are required for proper joint selection.

In the nonrigid piping, all piping members are free of thermal forces and moments. Figure 14 shows a multimember three-hinge system and a single-member free-movement system. As illustrated in Figure 14a, the axial expansion

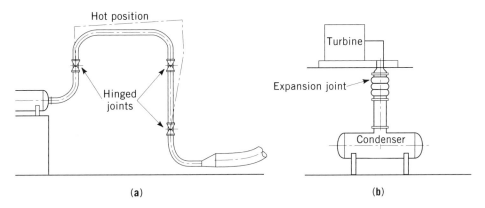

**Fig. 14.**  Nonrigid piping systems: (**a**) three-hinge arrangement; (**b**) free-movement arrangement.

of the pipe members is absorbed by rotary displacement of the hinges. In the free-movement arrangement, the longitudinal expansion of the pipe is absorbed within an axial movement-type device as shown in Figure 14b. For systems such as these, the thermal expansion design involves merely the dimensional evaluation of the joint movements.

The movement-absorbing devices used in semirigid and nonrigid piping systems are usually called expansion joints. Based on the method by which the pressure seal is effected, joints are either of the packed type or the packless or bellows type. Selection depends not only on the required movement but also on the severity of service in terms of pressure and temperature, tolerability of leakage, and the number of service cycles.

There are three varieties of packed joints: the slip type, the swing type, and the ball type. The slip type is used for axial displacement but is also capable of rotation about the pipe axis. Slip types generally are constructed in the form of two telescoping cylinders with a stuffing-box-and-gland arrangement for pressure sealing. The swing, swivel, and revolving types usually permit rotary or hinge motion in only one axis and are sealed by gaskets or packing. The ball type consists of a spherical ball-and-socket arrangement that allows universal movement and is pressure-sealed by means of gaskets.

Packed-type joints are significant for their large-movement capacities and ruggedness in body construction. When properly adapted into a piping system, their performance is satisfactory in both moderate and severe services. However, as with any packed element subject to motion, joints of this type require occasional tightening and repacking.

The packless-type uses a highly flexible pressure-sustaining and movement-absorbing membrane or bellows which is usually either welded to or integral with the body of the joint. A set of bellows elements can be compressed, extended, laterally offset, or angularly rotated (cocked). By means of external superstructure in the form of tie rods, bars, hinges, or gimbal arrangements, various combinations of movements are available. Bellows elements, however, are not suitable for absorbing axial rotation (torsion). For additional pressure capacity, the bellows elements are either provided with external reinforcement in the form of rings or collars, or are specifically designed to provide inherent strength without external support. Usually they are constructed of metal, although not restricted to metallic materials; their application is not limited by temperature. For the types using thin-bellows elements (metal), corrosion-resistant materials are essential. Since flexing of metal bellows subjects them to metal fatigue after a finite number of cycles, their cyclic capacity should be evaluated by means of analysis or tests.

The selection and application of an expansion joint is not as simple as selecting a pipe fitting or a valve and requires a sound understanding of the joint's capabilities and limitations. Improper application of any type of joint can result in serious or damaging effects. However, when properly selected and integrated into the piping system, satisfactory service and safe operation can be expected. Selection and application of bellows expansion joints require special attention to design and installation.

The thin-wall bellows element should be designed for membrane stresses to conform to code-allowable stresses. The sum of membrane and secondary bending stresses should not exceed 1.5 times the yield stress in order to prevent the collapse of the corrugations caused by pressure. Bellows subjected to external pressure can be analyzed in a manner similar to a cylinder, utilizing an equivalent moment of inertia. The fatigue life can be estimated based on the sum of deflections and pressure stresses as compared to $S/N$ curves based on bellows test data or using the curves in B31.3 Appendix X, Metal Bellows Expansion Joints. Formulas for the stress analysis of bellows are available in the Expansion Joints Manufacturing Association (EJMA) Standards (37).

Bellows subjected to internal pressure are subject to a column-buckling phenomenon known as squirm. This can occur in very short bellows as well as in long slender bellows, because of the extreme flexibility of the element. Formulas for elastic squirm are given in the EJMA Standards. Plastic squirm or instability can occur also, but equations to predict such behavior are not well defined. Hence, bellows are best pressure-rated based on extensive testing programs conducted by reputable bellows manufacturers. They also should be shop-hydrotested, if possible. Since squirm is a (fluid) column-buckling phenomenon, end conditions are important. A bellows placed in the center of a flexible pipe can squirm at one-fourth the pressure of a bellows where end rotation is not possible. A fixed-end condition is assumed in the EJMA equation. For reasons of end stability, EJMA recommends that only one bellows be installed between any set of anchors, and that the bellows be placed adjacent to an anchor, with a double set of full lateral restraint pipe guides (not weight-support type) on the other side.

Dual-bellows assemblies, ie, universal-type expansion joints, are particularly vulnerable to squirm, and can experience elastic squirm at one-fourth the pressure of an individual bellows. When large amounts of offset are encountered, as is often the design basis, a pinwheel effect occurs because of unbalanced pressure forces. This effect tends to rotate the center-spool pipe which may lead to bellows rupture. For this reason the center spool should always be stabilized by hinges or tie-rod lugs to prevent such rotation.

Pressurized bellows exert large forces on external anchors or self-restraining hardware. If these forces are not considered in the design, anchor failure can occur leading to bellows extension and rupture. Bellows pressure end load is equal to the product of the cross-sectional area based on the mean diameter times the internal pressure. In externally anchored axial bellows systems, the anchor relieves the piping from the tensile load it would normally carry. In such a case, the pipe itself can buckle because of the inherent instability of the pressurized-fluid column. This fact is not generally realized by piping designers. Column instability of the pipe is prevented by proper spacing of pipe guides, which provide the necessary lateral restraint. Rules for spacing of guides are contained in the EJMA Standards.

Bellows can vibrate, both from internal fluid flow and externally imposed mechanical vibrations. Internal flow liner sleeves prevent flow-induced resonance, which produces bellows fatigue failure in minutes at high flow velocities. Mechanically induced resonant vibration is avoided by a bellows with a natural frequency far away from the forcing frequency, if known. Multiple-ply bellows are less susceptible to vibration failure because of the damping effect of interply friction.

Bellows installations should always be inspected to ensure that temporary shipping bars are removed and that flow liners are installed in the right direction. Bellows generally are not designed to accommodate piping misalignments or to make up length, and hence such practices should be avoided. Inspection should also ensure that all the required anchors and guides are in place before pressurizing the system. Piping systems containing bellows are always pressure tested (hydraulic or pneumatic) without restraining the bellows in order to test main anchors and guides as well.

### Environmental Aspects

The ANSI/ASME B31.3 Code for Pressure Piping recognizes that special provisions are needed for piping conveying toxic materials. Where there is potential for personnel exposure and leakage of a small quantity can result in irreversible harm, the piping is designated in Category M fluid service. The code provides a guide for the determination of toxic fluid services (Appendix M) and additional requirements for this piping (Piping for Category M Fluid Service). The code also has additional requirements for high pressure piping, with high pressure defined as pressure in excess of that allowed by the ANSI B16.5 Class 2500 rating. Piping must be protected against overpressure by either safety relief valves or rupture disks located on the piping or on connected equipment. Piping that can be isolated from protective devices by valves may require protection against overpressure caused by the expansion of trapped materials.

The consequences of failure of components to perform as designed should be considered. With internally lined pipe, safety problems could arise due to lining failure. The system must be so designed that lining failure can be detected and appropriate action taken to prevent a catastrophe. For piping with an internal insulating lining, temperature-sensitive paint (qv) could flag an impending failure. Also, an alloy pressure shell could have sufficient life at the temperature it would attain on insulation failure to provide time for detection and orderly plant shutdown. Water jacketing can be used to provide warning of insulation failure by an increased rate of steaming and provide protection against rapid failure of the piping. B31.3, Appendix G, entitled Safeguarding, outlines some considerations for enhancing safety and reducing releases to the environment. For some services double-pipe systems may be warranted. The outer pipe ensures containment of any leakage from the inner pipe; instrumentation is available to detect leakage into the outer pipe (38).

The Clean Air Act requires control of fugitive emissions from the piping system (flanges, valves, etc) (39,40). This can be accomplished by better design of valve packing and flange gasketing systems, by confining and collecting shields for flanges and valve bonnets, and by good housekeeping and maintenance (qv) practices. Estimates of fugitive emissions, and controlled and uncontrolled process emissions, are generally required by the U.S. EPA in permitting new facilities. Average values of fugitive and process emissions are available from the U.S. EPA (41).

### Piping Industry

In the United States the piping industry, excluding concrete piping products, had sales of approximately $12 billion in 1990 (42). Sales of steel (qv) pipe and tubes were about $5 billion. Cawsl Corp., Bundy Western Hemisphere, and UNR Industries were the sales leaders. Shipments of valves and pipe fittings, flanges, and associated hardware had a value of nearly $3 billion, with Crane Co. and

Cameron Iron Works leading in sales. Gray iron foundries shipped pipe and fittings having a value close to $400 million; American Cast Iron Pipe Co. and U.S. Pipe and Foundry Co. were prominent in this area. Plastic pipe shipments had a value just over $2.6 billion; sales leaders were Lamson and Sessions Co., Phillips Driscopipe Inc., and Advanced Drainage Systems. Fabricated pipe and pipe fittings added slightly more than $1 billion, excluding the value of the purchased materials, to the industry sales, with Valley Industries, Industrial Services, and Stanley G. Flagg Co. leading in sales (see PLASTICS PROCESSING). Among the well-known foreign firms in this industry are Bonney Forge (Italy), Dalmane (Italy), Kawasaki (Japan), Kitz (Japan), Mannesmann (Germany), Nippon Steel (Japan), Raimondi (Italy), and Sumitomo (Japan).

Shipments of concrete pipe in 1990 had a value of >$1.6 billion dollars. Much of this piping was for storm and sanitary sewers, culverts, and irrigation piping. Pressure piping sales were close to $300 million. There are a large number of manufacturers of concrete pipe since much of the pipe is produced reasonably close to the point of use.

## Nomenclature

| Symbol | Definition | Units |
|---|---|---|
| $a$ | pressure wave velocity in the fluid | m/s |
| $(A)$ | flexibility matrix | |
| $a_{i},j_{i}$ | displacement at $i$ owing to unit force at $j$ | |
| $C$ | material constant; allowance for threading, mechanical strength, and corrosion | m |
| $C_{\mathrm{HW}}$ | Hazen and Williams constant | |
| $C_{v}$ | valve flow coefficient | |
| $d$ | inside diameter | m |
| $D$ | outside diameter | m |
| $\{D\}$ | displacement vector | |
| $E$ | joint efficiency; modulus of elasticity | Pa |
| $E_{h}$ | operating temperature modulus | Pa |
| $E_{c}$ | room-temperature modulus | Pa |
| $f$ | Fanning friction factor; nondimensional; stress-range reduction factor | |
| $\{F\}$ | force vector | |
| $f_{D}$ | Darcy friction factor | |
| $f_{t}$ | force or movement required | |
| $g_{c}$ | acceleration of gravity | m/s$^2$ |
| $G_{f}$ | specific gravity at flowing conditions | |
| $h_{e}$ | effective flexibility characteristic | |
| $h_{f}$ | friction head loss | m |
| $h_{\max}$ | maximum head rise of fluid | m |
| $i$ | increase in stress; fatigue-stress concentration factor | |
| $I$ | moment of inertia | m$^4$ |
| $K$ | bulk modulus | Pa |
| $(K)$ | stiffness matrix | |
| $L$ | length of line | m |
| $N$ | cycles to failure; number of unknowns | |
| NPSH | net positive suction head | |

| Symbol | Definition | Units |
|---|---|---|
| $P$ | design pressure | Pa (psi) |
| $q$ | flow rate | $m^3/s$ |
| $Q$ | flow rate | $m^3/h$ |
| $r$ | mean pipe radius | m |
| $R$ | radius of bend | |
| $Re$ | Reynolds number | |
| $S$ | stress | Pa (psi) |
| $S_{ypc}$ | cold yield stress | Pa |
| $S_{yph}$ | hot yield stress | Pa |
| $t$ | wall thickness | m |
| $t_{min}$ | minimum pipe wall thickness | m |
| $v$ | fluid velocity | m/s |
| $W$ | liquid flow rate | 1000 kg/h |
| $W_g$ | gas flow rate | kg/h |
| $Y$ | coefficient | |
| $\Delta$ | expansion to be absorbed | m |
| $\Delta P$ | pressure drop | kPa |
| $\epsilon$ | strain | |
| $\mu$ | fluid dynamic viscosity | Pa·s(=cP) |
| $\rho$ | fluid density | $kg/m^3$ |
| $\sigma$ | expansion stress | Pa |

## BIBLIOGRAPHY

"Pipes and Valves" in *ECT* 1st ed., Vol. 10, pp. 712–722, by R. L. Feder, Allied Chemical & Dye Corp.; "Piping Systems" in *ECT* 2nd ed., Vol. 15, pp. 646–675, by S. E. Handman and J. T. McKeon, The M. W. Kellogg Co.; in *ECT* 3rd ed., Vol. 17, pp. 929–957, by S. E. Handman, The M. W. Kellogg Co.

1. *Metals Handbook*, 10th ed., Vol. 1, ASM International, Materials Park, Ohio, 1990, pp. 617–652, and pp. 930–949.
2. *Haynes High Temperature Alloys*, Haynes International, Inc., Kokomo, Ind., 1984 to 1993.
3. *High-Temperature Characteristics of Stainless Steels, American Iron and Steel Institute*, Nickel Development Institute, Toronto, Ontario, Canada, 1993.
4. L. L. Shreir, ed., *Corrosion*, 3rd ed., Butterworth-Heinemann, Stoneham, Mass., 1993.
5. M. G. Fontana, *Corrosion Engineering*, 3rd ed., McGraw-Hill Book Co., Inc., New York, 1986.
6. R. S. Treseder, *NACE Corrosion Engineer's Reference Book*, 2nd rev. ed., National Association of Corrosion Engineers, Houston, Tex., 1991.
7. *Corrosion Resistance of the Austenitic Chromium–Nickel Stainless Steels in Chemical Environments*, Inco Limited, Nickel Development Institute, Toronto, Ontario, Canada, 1963.
8. *Hastelloy Corrosion Resistant Alloys*, Haynes International, Inc., Kokomo, Ind., 1987 to 1993.
9. *Fiberglass Reinforced Piping Systems for Chemical and Industrial Applications*, Smith Fiberglass Products, Inc., Little Rock, Ark., 1993 to 1994.
10. *Kynar and Kyner Flex*, Elf-Atochem, Philadelphia, Pa., (PL-705), 1993.
11. *Glass Plants Components/Process Pipe Fitting Hardware/Joining and Adaption*, Corning & QVF Pipe, QVF Corning, New York, 1989 to 1992.

12. *Dow Plastic Lined Products*, Dow Chemical USA, Bay City, Mich., 1992 to 1993.
13. M. E. Jones, *Chem. Eng.* **97**(10), 104–111 (Oct. 1990).
14. F. C. Yu, *Hydrocarbon Proc.* **72**(6), 67–74 (June 1993).
15. F. C. Yu, *Hydrocarbon Proc.* **73**(5), 99–106 (May 1994).
16. *Flow of Fluids*, Technical Paper 410, 1991: *Crane Companion*, Crane Co., Jolliey, Ill.
17. *Piping Engineering*, 6th ed., Tube Turns Technologies, Inc., Louisville, Ky., 1968.
18. B. R. Munson, *Fundamentals of Fluid Mechanics*, 2nd ed., John Wiley and Sons, Inc., New York, 1993.
19. *Cameron Hydraulic Data*, 17th ed., Ingersoll-Rand, Woodcliff Lake, N.J., 1992.
20. S. L. Soo, ed., *Multiphase Fluid Dynamics*, Ashgate Publishing Co., Brookfield, Vt., 1990.
21. G. Bohme, *Non-Newtonian Fluid Mechanics*, Elsevier Science Publishing Co., New York, 1987.
22. R. W. Zappe, *Valve Selection Handbook*, Gulf Publishing Co., Houston, Tex., 1991.
23. *Masoneilan Handbook for Control Valve Sizing*, 8th ed., Masoneilan International, Inc., Houston, Tex., 1994.
24. *Flow Equations for Sizing Control Valves*, ISA S 75.01-85, Instrument Society of America, Verona, Pa., 1985.
25. E. B. Wylie and co-workers, *Fluid Transients in Systems*, Prentice Hall, New York, 1993; J. Zarbua, *Water Hammer in Pipe Line Systems*, Elsevier Science Publishing Co., New York, 1993.
26. M. L. Nayyar and co-workers, *Piping Handbook*, 6th ed., McGraw-Hill Book Co., Inc., New York, 1992, pp. A227–A277.
27. S. Timoshenko and J. N. Goodier, *Theory of Elasticity*, McGraw-Hill Book Co., Inc., New York, 1951, p. 60.
28. *Pipe Hangers Catalog*, PH-92R, Grinnell Corp., Exeter, N.H., 1992.
29. I. H. Shames, *Introduction to Solid Mechanics*, 2nd ed., Prentice Hall, New York, 1989, pp. 341–345.
30. The M. W. Kellogg Co., *Design of Piping Systems*, rev. 2nd ed., John Wiley and Sons, Inc., New York, 1964.
31. Ref. 30, p. 97.
32. A. Ghali and A. M. Neville, *Structural Analysis—A Unified Classical and Matrix Approach*, 3rd ed., Chapman Hall, New York, 1990.
33. J. Robinson, *Structural Matrix Analysis for the Engineer*, John Wiley and Sons, Inc., New York, 1966.
34. T. E. Pardue and I. Vigness, *Trans. ASME* **73**, 77 (1951).
35. D. B. Rossheim and A. R. C. Markl, *Trans. ASME* **62**, 443 (1940).
36. A. R. C. Markl, *Trans. ASME* **74**, 287 (1952).
37. *Standards of the Expansion Joint Manufacturers Association*, 6th ed., EJMA Inc., White Plains, N.Y., 1993.
38. *Double-Pipe (DBP 1191)*, PermAlert ESP, a subsidiary of Midwesco, Inc., Niles, Ill., 1991.
39. E. D. Crowley and D. G. Hart, *Hydrocarbon Proc.* **71**(7), 93–96 (July 1992).
40. J. F. Gardner and T. F. Spock, *Hydrocarbon Proc.* **71**(8), 49–52 (Aug. 1992).
41. *Compilation of Air Pollution Emission Factors*, Vol. 1, *Stationary Point and Area Sources*, AP-42, 4th ed., Suppl. F, U.S. Environmental Protection Agency, Washington, D.C., 1993.
42. *Manufacturing USA*, 3rd ed., Gale Research Inc., Detroit, Mich., 1993.

STANLEY E. HANDMAN
Consultant

**PITCH.**   See Tar and pitch.


**PIVALIC ACID.**   See Carboxylic acids.


# PLANT LAYOUT

Plant layout can be the single most important part of the overall design. A good layout can result in significant savings amounting to millions of dollars in erected plant cost. A layout can make the difference between a facility that is easy to operate and maintain, and a plant that is a nightmare. This article presents an overview of the many concepts that go into a good layout and includes a step-by-step procedure to aid the designer in developing the layout. A list of commonly used layout terms and the corresponding definitions is provided. A discussion of the particular equipment layout requirement is also included. Many layout considerations are concerned with the piperack, which is the arterial system of a plant. There are several different modeling concepts, including plastic scale models, three-dimensional computer-aided design (CAD) models, and isometric and orthographic presentations.

## Definitions

*Battery limit* defines the boundary of the unit equipment, which is used in a processing facility, by an imaginary line that completely encompasses the defined site. The term distinguishes areas of responsibility and defines the processing facility for the required scope of work.

*Inside battery limit* (ISBL) is the limit of the processing facility and the equipment contained therein.

*Outside battery limit* (OSBL) defines the work and responsibilities outside the battery limit. The OSBL includes auxiliaries required for the chemical process unit, including tank storage for feedstock and refined products, waste treatment, cooling towers, flares, and utilities that are not included in the battery limit but are required for the unit. Also included are the pipeways, racks, and sleepers, which are used to convey all of the utility's interconnecting process piping to and from the processing facility.

*Plot plan* is the scaled plan drawing of the processing facility.

*Site plan* is a scaled drawing, including the roadway system for adjacent areas and auxiliaries and the space requirements for the processing facility.

*Piperack* is the elevated supporting structure used to convey piping between equipment. This structure is also utilized for cable trays associated with electric power distribution and for instrument trays.

*Sleepers* comprise the grade-level supporting structure for piping between equipment for facilities to tank farms or other remote areas.

*Paving* is the surface preparation in a processing facility which may be concrete, asphalt, gravel, crushed shell, or brick.

*Equipment-handling facilities* include any device used to convey or temporarily support equipment, or provide access for plant maintenance, eg, portable ladders, davits, A-frames, monorails, forklift vehicles, and cranes.

*Alloy piping* comprises all piping that is stainless steel, carbon–molybdenum, or chrome alloys.

*Shutdown or turnaround* is the period when an operating plant or a significant portion of it is out of service, during which replacement of catalyst, demister pads, or sections of piping; repairs to rotating equipment; or cleaning of exchangers is performed. Generally, a shutdown or turnaround is any period during which production is curtailed and maintenance is necessary.


## Importance of a Good Layout

Once the process design optimization has been completed, it is estimated that over 75% of the capital cost has been established. The next significant cost-controlling opportunity is in the selection and purchasing of the equipment, which depends on having a good set of equipment data sheets and general specifications that define the level of equipment quality desired. After purchasing has been initiated, the next area in which substantial plant cost savings can still be realized, or lost, is in the plant layout. The plant layout also has profound implications on both the operability and maintainability of the plant. Obtaining the recommendation of an experienced layout consultant in the early phases of the design can make the difference between a highly successful and a bad project.

Equipment layout and the design of the associated piperack are critical elements of the plant design and require interfacing among many disciplines, including operational and maintenance staffs as well as various design disciplines. Low plant cost is favored by a tightly spaced layout; operations and maintenance (qv) are favored by a more spread-out layout. A compromise between these two extremes needs to be reached early in the process. Some high pressure alloy piping can easily run over $1000 a linear foot. Pipe fittings can be even more expensive. Shutdown time can also be costly, running into hundreds of thousands of dollars per day in lost product revenues when the plant is shut down for maintenance. Providing adequate maintenance space for quick turnarounds can help reduce the down-time cost.

The re-engineering cost for making design modifications as a result of moving equipment after the design layout has been started can be expensive. A firm layout can help keep changes to engineering design to a minimum. It is important to spend a little extra time early in the design to get the equipment located and spaced so that it does not have to be moved in the middle of the design. This keeps engineering on schedule and costs within budget. For this reason it is undesirable to squeeze equipment together too closely in the early part of the design and then to have to spread it out at a later time. A good layout engineer or designer can help provide realistic spacing and select equipment locations that do not need to be changed throughout the design.

A cost-effective layout that provides for easy operability and quick turnarounds is often designed by someone with a background in the piping discipline. This is most likely because of their ability to think in three dimensions and their familiarity with visualizing three-dimensional piping (see PIPING SYSTEMS).

The layout can only be started after a process flow diagram is available. The process flow diagram includes information such as the principal equipment items and order of the process flow. A sized equipment list is also useful from the standpoint of knowing what spaces are required to fit the equipment. Availability of a preliminary piping and instrumentation drawing (P&ID) provides more information to aid in spacing equipment and thinking about piperack requirements.

Development of a preliminary layout requires the following information: (1) process flow diagrams; (2) plot limits; (3) process unit rough area requirements; (4) storage and tankage requirements; (5) expected expansion requirements; (6) waste treating area, usually located at the low point of the plot; (7) rerun and product storage location (preferably close to process unit); (8) locations of roads, rail spurs, and pipeline tie-in points; (9) product and blending area (best if located close to sales loading area); (10) product tankage and loading (should be located near blending); (11) plant roads and access ways for considerations of maintenance access and plant constructability; (12) utility and steam generations (should be adjacent to process plot) if a significant requirement exists; and (13) cooling towers and electrical distribution substations (must be at plot periphery).

## Plot Plans

Figure 1 shows an overall plot plan for a grassroots plant. A grassroots plant design differs from a retrofit design in that it allows the designers full latitude in the plant layout and subjects only them to the overall limitations of the new plant site boundaries and safety-in-design considerations. The process facilities usually occupy a small fraction of the overall plant site area, typically about 25%. However, the process facilities generally cost about half of the overall plant expenditure and the offsites cost the other half. Most petrochemical facilities provide large plot areas for feed and product storage. Feed storage is usually at least 30 days, and product storage can be 60 days or more. Tankage can be quite spread out due to diking requirements for fire control containment of fluids for safety reasons. Space also needs to be provided for all the other facilities that are required to make a plant self-sufficient. These facilities include general office space; parking for employees and contractors; engineering services offices; laboratories; wastewater-treating facilities; product loading racks; shops; warehouses; fire stations; laydown areas for construction; offsites, including boilers, water treating, cooling towers, and electrical distribution; change rooms; railroad spurs; control room; and safety headquarters and equipment. These need to be laid out in a logical manner that considers site terrain, accessibility to roads, soil bearing capability of land, and the climate, including wind directions, winter conditions, hurricanes, and other unusual weather conditions.

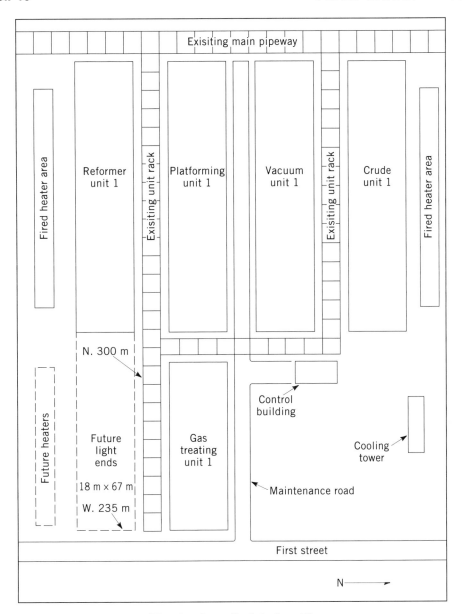

**Fig. 1.** Overall plot plan (1).

Few grassroots facilities, however, are being built in the 1990s. Most companies are adding on to or expanding their existing facilities. These projects are often more challenging because space limitations can frequently be the limiting considerations. Part of a good grassroots layout is to consider what the requirements for future expansion might be, and to provide not only the space anticipated within individual units for expansions and for debottlenecking, but also open areas in the overall site for the addition of new process units. Access to utilities and tie-ins to existing piperacks are critical to avoid moving existing facilities and prevent costly shutdown requirement in order to provide for new additions.

The preliminary plot plan is the first step in the layout process. The plot plan shows the location of the following items which are known or approximated: main equipment, main pipeway, major structures, housed electrical switch, control room, and tie-in connection locations. Individual unit plot plans may also need to be developed. The scale is often increased on the unit plot plans to make it easier to read. By using a computer-aided design (CAD) system, unit plot plans are easily developed from the plot plan file. Unit plot plans are necessary, mainly as background information, for such uses as equipment location plan drawings, foundation locations, excavation drawings, paving plan drawings, fire system, underground drawings, and shop model concept. The overall plot plans are usually drawn to scale one-eighth inch per foot (1 mm/m).

**Initial Sketch.**    Figure 2 shows a process flow diagram for a petrochemical plant (1,2). This drawing shows the feed and products so the designer knows what to allow for these lines in the interunit pipeway routing. The process engineer has indicated with notes which pieces of equipment will be located in elevated structures, such as the overhead condensers, and has also shown which equipment should be located close by other equipment, such as the reboiler next to its column. Primary instrumentation is shown to indicate that room is required for instrument drops to these control valves. All this additional information provides the designer with data to make a good start on the first cut of the layout in a rough sketch form.

The first-cut sketch is usually not drawn to scale but is roughly spaced out. Figure 3 shows a first-cut layout. The utilization of CAD on the first-cut layout can be beneficial when optimization of the plot is necessary. Several important things should be noted in the first-cut layout sketch derived from key information provided by the process engineer. These items are indicated in the legend for Figure 3.

**Scaling the Sketch Example.**    Once the initial sketch is completed, it can be refined to include space requirements and shown to the approximate scale. This is where the designer uses knowledge of spacing and clearances for safety and maintenance considerations. There are several rules of thumb on how much space to leave between each type of equipment. There is also a way of approximating the amount of total space needed in a plot area to provide for all the equipment. For additional assistance, inside as well as outside battery limit equipment spacing charts have been compiled. Figure 4 shows how this information has been converted into a scaled drawing.

Adding the estimated dimensions of equipment and spacing, the designer finds an overall dimension of 65 m (Fig. 4). As more firm information is developed, some of the dimensions may vary slightly, but if so there is no more plot length available; the designer would then have to adjust other dimensions to suit, perhaps by combining some foundations. Alternatively, the air coolers can be located above the piperack to reduce the plot length by over 13 meters. Other considerations include the following: wind direction, which is a key factor in the location of stack and fired heaters; the location of pumps that have autoignition or that create extreme fire hazards, which should be away from under the piperack; soil conditions at plant site, which determine the size and type of foundations;

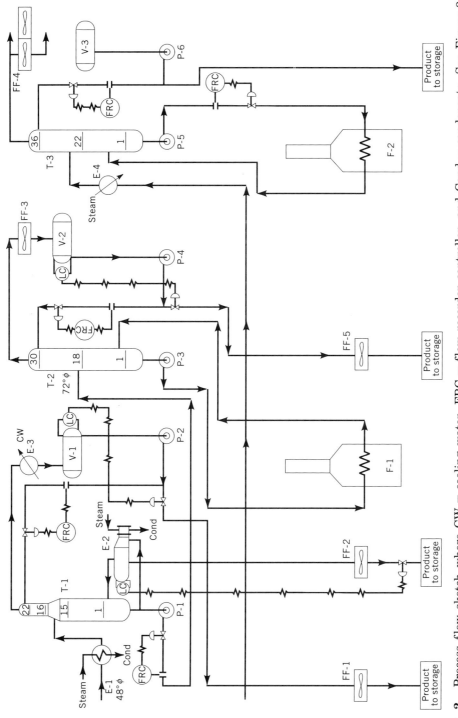

**Fig. 2.** Process flow sketch where CW = cooling water, FRC = flow recorder controller, and Cond = condensate. See Figure 3 for other definitions.

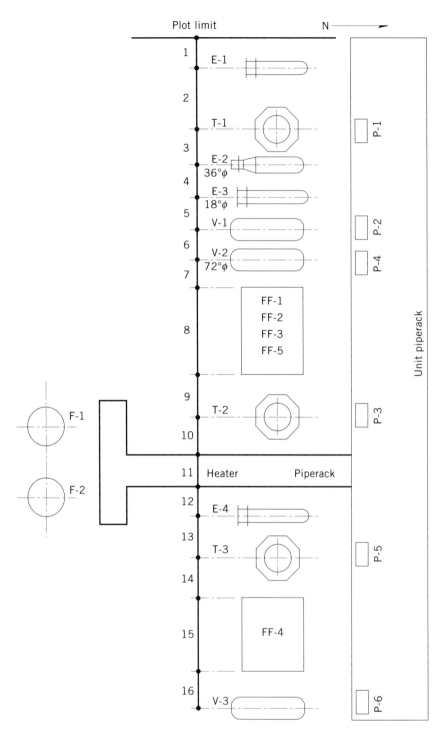

**Fig. 3.** Rough layout sketch: (*1*) the two fired heaters F-1 and F-2 are located together but are separated from the other equipment with a subpipeway connecting the process area to the heater area; (*2*) the reboiler E-2 is located adjacent to its column, T-1. The preheat exchanger E-4 is located adjacent to tower T-3; (*3*) the elevated overhead condenser E-3 is located next to the overhead accumulator V-1. Also, the air condenser FF-3 is located adjacent to its overhead accumulator V-2; (*4*) the rest of the air coolers (FF-1–3, -5) are grouped together in a common fan structure; (*5*) all equipment and related piping is routed to and from the existing piperack saving the addition of a new piperack; (*6*) all pumps (P-1–P-6) are located in a row under the piperack, and each pump and its spare are located close to the respective upstream suction source (*1*).

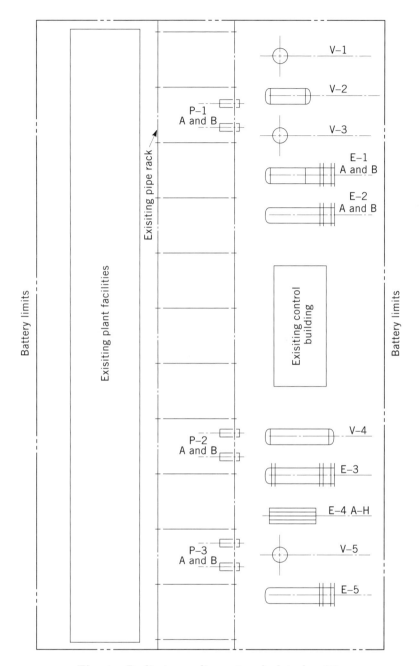

**Fig. 4.** Preliminary dimensioned plot plan (1).

room for loading and unloading catalyst for reactors using solid catalyst; and the use of extremely large pipeline between equipment, which can dictate the location of equipment (piping stress engineers should be involved in the decision).

**Cut-and-Paste Method.** Some experienced layout people bypass all the previously discussed preliminary procedures and go directly to a scaled preliminary plot plan. This is another way of creating a first rough plot plan and is referred to as the cut-and-paste technique. In this method, the background of the plot area is sketched on a blank drawing to scale. This background shows plot limits as well as existing roads and other objects within the facility of the proposed plot. All the equipment items are then cut out to a scale that matches the background plot scale. In order to apply this method, the following information is desirable: process flow diagram for unit; plot size and locations of boundaries; rough equipment sizes, which should be conservative and err on the high side; location of control building and other existing structures can be ignored if off plot; and knowledge of the rough sizes of the lines and which of the piping is critical from a materials standpoint, for example, alloy lines should be kept as short as feasible.

The way the equipment is located on the background is based on the process flow sequence. Again, certain equipment such as fired heaters can be situated first to put them at a safe distance from other equipment. Other large equipment may have to be located where the soil-bearing load is best.

When alloy piping or large bore piping is required, the associated equipment is located together as much as possible to keep the pipe runs short, preferably nozzle-to-nozzle by avoiding the piperack. Items such as elevated overhead condensers are located near the source and destination. Similarly, thermosyphon reboilers need to be placed adjacent to the column they reboil. Where gravity flow is required, these lines must be kept short and sloped. Space allocation for future additions must also be considered.

Once the equipment has been located on the plot, it can be photographed for reference. Because many cases can be evaluated, references to these need to be documented. Client approval is also needed. One way of making a record is to trace over the various arrangements for future reference. Once the plot is finalized, a first pass can then be made at transposing the principal piping lines to determine if any significant piperack problem exists and whether equipment relocation is required. Figure 5 shows a simple flow diagram used to transpose the piping on the preliminary plot plan (2). This is done line by line starting with large bore and critical material piping. A simple example has been used for illustration. The transposition effort can be much more complex with a more detailed unit.

## Equipment Considerations

Every type of equipment requires certain layout considerations. These requirements address both operational and maintenance issues. Practical rules have been provided to guide the designer on the spacing requirements needed to address these concerns. Most of these are rule-of-thumb information based on reviews of existing designs from a maintenance and operation standpoint as well as practical engineering design rules and safety-in-design practices. The following equipment items have been evaluated in detail and are covered by References 3–14. Only summary information is provided here for the most common equipment items: fractionating towers; vessels, including drums, separators, and surge tanks; reactors; heat exchangers, including shell-and-tube, double-pipe,

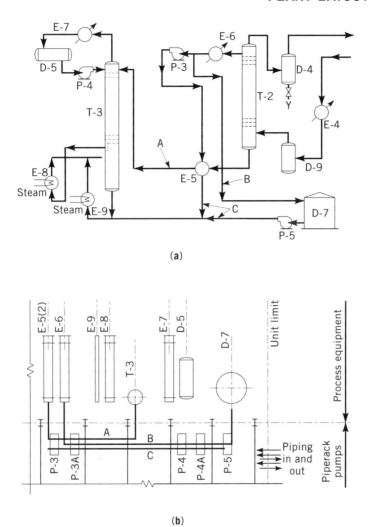

**Fig. 5.** Initial piping transposition: (**a**) process flow diagram and (**b**) plot plan (1).

and plate-type exchangers; fired heaters, both box and vertical cylindrical; compressors, including reciprocating and centrifugal; storage tanks, including fixed roof, floating roof, and open roof; cooling towers, including side- and top-draft fans; pumps, including centrifugal, reciprocating, and in-line; control room; and electrical switch gear.

   **Fractionating Towers.** Most petrochemical facilities include the use of multiple fractionating towers to perform separation of the products by boiling range. Special types of towers include strippers, absorbers, reboiled absorbers, and articulated fractionators. These towers are usually vertical vessels that contain trays or packing to perform the separation. These towers are supported by skirts or other structural means to provide the vertical height required for the bottoms pump net positive suction head (HNSH) as well as the vertical height needed by the reboiler to produce enough hydraulic head to make thermosyphon reboilers circulate. These towers usually include overhead condenser systems and sometimes side strippers.

From a practical layout consideration, the fractionating towers are usually grouped together. In some eastern European countries and in Russia, towers are always grouped together so that personnel access is by platforms with stairs; no ladder access is allowed. Common structures are often used for the overhead condenser systems. These towers need to be spaced so that platforms can be located to provide an entrance area for inspection and tray repair. Towers over 16 meters need to be provided with davits for handling vessel trays and with a clear vertical drop underneath and away from the piperack so that trays can be replaced. Generally, ladder access (if allowed) to the towers is located on the piperack side for easy access by the operators. Towers should all be located on a common center line having about 4–5 m separation away from the piperack column supports for access clearance.

Figure 6 shows a typical plan and tower elevation drawing (6). The piperack is located to the right, and the access to manholes is to the left. Clear access to the pumps is required at the bottom. Piping headroom clearance is provided below the unit pipeway and grade for equipment access. The designer needs to be aware of the various types of tower internals (8). The spacing between these internals is critical from both a process and accessibility standpoint. The feed inlet, the overhead reflux return, the reboiler drawoff nozzle, the reboiler return nozzle, and the intermediate draws all have to be oriented by the layout person and this information must be passed on to the vessel designer so that the vessel nozzles can be oriented properly. A liquid draw nozzle that leaves from a tray with a boiling liquid should immediately turn down in order to allow any vapors to work their way back into the tower. If a horizontal run is required, this can be made at some lower elevation after the vapors have been eliminated. Overhead lines again should be routed so that there are no low spots or pockets in the line where liquid can collect. The layout designer has to be aware of these critical situations and provide the clearances and open spaces in the plot area to allow for these important piping considerations. The process engineer needs to call these out on the process drawing or piping and instrument diagram (P&ID) so that they are not overlooked by the piping designer.

Most towers contain trays or packing required to perform the separation of feed into products. The various internals provide the means of taking material from the column or introducing feeds. Most column problems are a result of improper selection of feed distributors, or using the wrong type of internal in a column to make a transition from one type of tray to another (single pass to double pass). Flooding problems at the transition zone are caused by not providing enough vertical space for gravity flow of liquid between the internals. It is good design practice to provide a manway for access at each important internal. It is also good practice to provide a manway about every 10 trays to minimize the number of trays a person needs to pass through to access all parts of the tower. Manways should be ~60-cm dia where possible. Small-diameter towers may not be big enough to allow such large manways. Sufficient vertical distance should be provided between the trays at the manway to allow access, eg, ~1 m minimum if the internal occupies significant space.

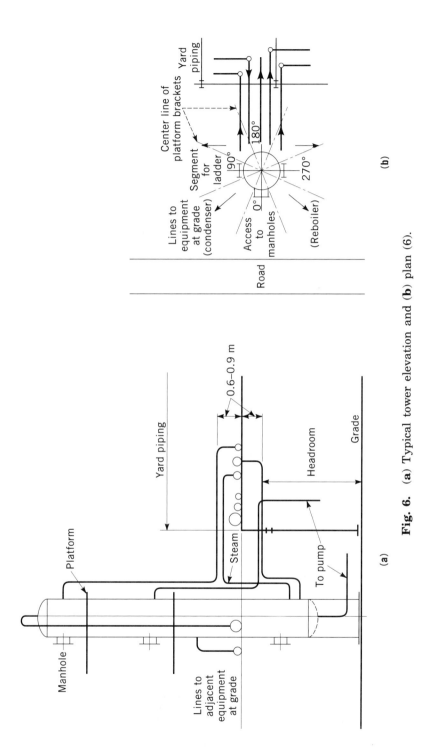

**Fig. 6.** (**a**) Typical tower elevation and (**b**) plan (6).

**155**

**Vessels and Drums.**    Vessels and drums have been discussed in detail (7). Vessels are oriented either in the vertical or horizontal position. The vertical position is preferred if the plot space is tight. Horizontal vessels are easier to support and are preferred when large liquid surge volumes are required. The liquid level displacement height for a unit volume is much less for a horizontal vessel than for a vertical vessel, which makes the control range shorter than for a vertical vessel. The displacement height per unit volume is only approximately linear on horizontal vessel when the level is near the center line, however. This can be a problem if the normal liquid level is too low or too high and the instruments are not tuned for quick response.

Figure 7 shows nozzle locations and support arrangements for a typical horizontal vessel (7). The saddles used for support are sustained by cement pedestals or steel structures. Sufficient clearance between the bottom nozzles and the support saddles needs to be provided for access to the nozzle flange bolts. The manway can be located on the end head of the vessel, the topside of the vessel, or the side of the vessel. The preference is for an end manway wherever possible for accessibility, except when it is limited by the level gauges and controls that are commonly mounted off the heads.

The vessel can be supported off the structure and sometimes off the rack. Some economy may be possible by combining two or more services into a common vessel by using a single vessel that has an internal head. Differential pressure as well as concerns over internal leakage need to be considered for these services. This can be done with vertical vessels as well. A knockout section can be provided below or above the main vessel.

**Reactors.**    Reactors are a special type of vertical vessel. Some reactors are also in horizontal vessels but this is rare. Reference 7 covers reactors in more detail (see also REACTOR TECHNOLOGY). Reactors provide the means by which chemical reactions occur to transform feedstocks into products. Typically, reactors require some type of catalyst. Reactors with catalyst can be of the fixed-bed style for fluid-bed types. Fixed-bed reactors are the most common. The feed often enters the reactor at an elevated temperature and pressure. The reaction mixtures are often corrosive to carbon steel and require some type of stainless steel alloy or an alloy liner for protection. If the vessel wall is less than 6 mm, the vessel is constructed of all alloy if alloy is provided. Thicker reactor walls can be fabricated with a stainless overlay over a carbon steel or other lower alloy base steel at less cost than an all-alloy wall construction.

Reactions are either endothermic and require heating to complete the reaction, or exothermic and raise the temperature, thus requiring some type of cooling such as quenching or an internal heat exchanger to remove reaction heat. The reactors are provided with various types of internals to support the catalyst and distribute the reaction components uniformly across the catalyst area; collection internals remove the products and other distribution.

Figure 8 shows a typical vessel sketch for a vertical reactor with a fixed-bed catalyst (7). The top nozzle is frequently used as an access manway. Below the inlet nozzle, a distributor is used to spread the reactant uniformly across the reactor cross section. Below this is a rough distributor of some other type, such as a bubble cap tray, that provides for more uniform distribution. Whatever is used, a good means of distribution of feed over the catalyst is required

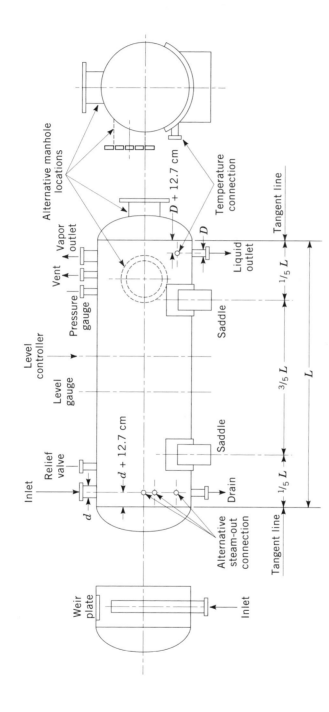

**Fig. 7.** Vessel support and arrangement where $d$ = nozzle diameter, $D$ = vessel diameter, and $L$ = length. Saddles can be straight or tapered. Single supports can be used for small drums. Combining drums saves the number of necessary supports (7).

157

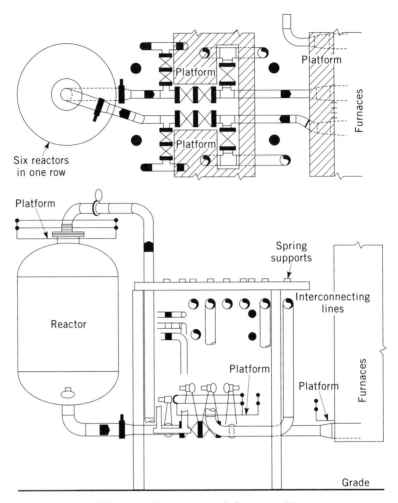

**Fig. 8.**  Reactor vessel drawing (7).

for an effective reaction and efficient use of the catalyst. A good vapor liquid distributor is usually required if the flow is two-phase going into the top of a reactor bed; vapors find the easiest path available. Good mixing requires good uniform distribution of the phases (see MIXING AND BLENDING). The top bed of catalyst often contains some type of trash baskets to localize the collection of feed impurities so that this part of the catalyst bed can be removed and replaced during the run cycle if required without unloading the whole reactor. This reactor can also include an internal quench system to control reactor temperature in the case of exothermic reactions. The quench media is cooler than the fluid in the bed and thus thermal expansion of the distributor should be considered so as to avoid excessive thermal stress. Some part of the internal needs to be free to expand and contract with temperature changes.

Distribution balls are frequently used on the top of each bed; they are often employed to support the catalyst bed from below. The outlet of the reactor needs to include some type of collector in order to retain the catalyst and support balls while allowing the products to leave the reactor.

Figure 8 shows a plan and elevation view of the piping and valving between a typical reactor and furnace in a catalytic reforming unit. In this typical catalytic reforming system there are multiple reactors that have inter-reactor heaters to reheat the reactants in order to increase the reaction conversion. The valving shown allows for each reactor to be taken off line and regenerated while another reactor is put on line in its place. The arrangement of piping between the heater and reactor is made compact because these pipe runs and fittings are constructed of expensive alloy steel. Access platforms are provided to enable the operators to reach the isolation valves. Sufficient clearances are provided under the lowest piping that connects the furnace and reactor to allow adequate headroom for the maintenance of equipment. The reactors are typically supported from a common structure. Either individual heaters or separate coils within a single heater firebox are used for the reheaters, depending on the size of the unit.

**Heat Exchangers.** Heat exchangers are special types of pressure vessel that include internal tubes used to transfer heat between two streams. The hot stream (heat source) is used to heat the cold stream (heat sink). Generally, heat exchangers are of the shell-and-tube type of construction. However, other types of heat exchangers are often used, including double-pipe exchangers, plate heat exchangers, spiral heat exchangers, and air coolers. The other types of exchangers have been described (8) (see HEAT-EXCHANGE TECHNOLOGY).

Typically clustered in groups that shorten the length of the interconnecting piping (3), heat exchangers are located in a row with all of the channels in a line similar to the center-line lineup used for aligning fractionation columns. The channels always face away from the pipeway and an open area is provided to remove the heat exchanger tube bundles periodically for cleaning. There must be enough space in between the exchanger bundles to provide for the clearances and room occupied by insulation and the interconnecting piping. Access has to be provided to the channel flanges for unbolting the tube cover plate for removal during cleaning. The Tubular Exchanger Manufacturers Association (TEMA) head type influences the space required. Whenever a shell cover is present it must be removed as well. The TEMA Type S construction, which uses a floating head attached to the exchanger bundle with a split ring, requires that the rear shell cover plate be removed, the split ring disconnected from the floating head, and the floating head cleared away before the tube bundle can be removed (pulled) from the shell. This operation requires an access area to the rear shell cover and is the main disadvantage of this type of construction. However, it provides for a less costly design than the TEMA T (pull-through design) and also allows for more surface area to be included for a given shell diameter. The U-tube and pull-through construction are often preferred designs for the reason that they do not require additional maintenance to function every time the exchanger is cleaned.

The exchanger piping elevation drawing for a typical exchanger shows the piping connections to and from the channel, the piperack, the shell, and the piperack again (Fig. 9). Also, the double-block valve and bypasses, which are included on certain heat exchangers, must be included on both the tubeside and the shellside in order to isolate the exchanger from the rest of the system. This

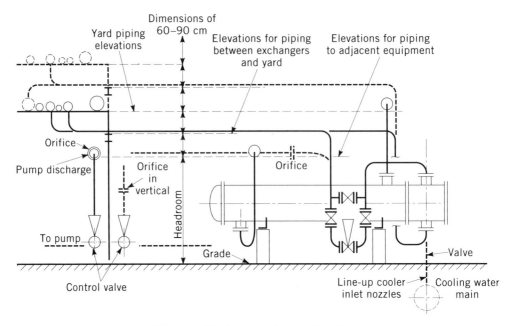

**Fig. 9.**   Exchanger piping elevation.

piping, shown located at the end of the pipeway, is required in order to take the exchanger out of service for cleaning while the rest of the unit is operating. Minimum clearances must be provided for the piping, including the necessary dimensions to make turns (90 degree elbows) and allow room for the isolation valves. Water-cooled heat exchangers are supplied with cooling water from a cooling-water main usually located underground. The cooling-water main is lined up with the channel inlet flange so that a line can be run directly up to the inlet of the tubeside and includes space for a cooling-water supply shut-off valve. The tubeside outlet is also piped back to the cooling-water return header, which is located underground but further away from the channel outlet nozzle.

Where possible the exchanger piping should run nozzle-to-nozzle to minimize the amount of piperack piping and shorten piping runs. Heat exchangers are grouped together and located relatively close to other equipment for this reason. Energy conservation favors maximum heat recovery consistent with the associated savings in fuel and cooling cost reductions.

Locating the heat exchanger bundles in elevated structures is often done where plot space is tight, as shown in Figure 10 (15). These heat exchanger systems are often complex and the piping between the heat exchanger bundles is tightly compacted. Elevated heat exchangers often require special provisions to be able to remove the heat exchanger bundles. Figure 10 shows the complexity of the piping when a heat exchanger system is located in the structure. Heat exchangers can also be stacked vertically on top of each other with no intermediate platforms or supports. In this case, the stacking is done so the exchangers are connected nozzle to nozzle. No valving can be included between exchangers in this type of arrangement. This means that if the exchangers are to be cleaned, they must all be removed from the operating service in banks of stacked ex-

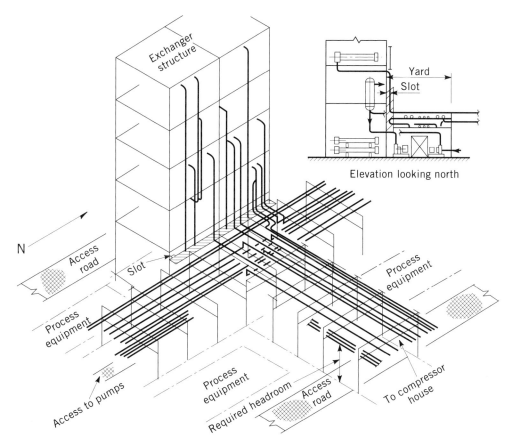

**Fig. 10.**  Exchanger structure piping.

changers. Typically, heat exchangers can only be stacked so that the center line of the top bundle is below five meters from the grade that limits the number of bundles in the stack to three (sometimes two) high.

In high pressure services where 10 MPa (1500 psig) of pressure flange ratings are required, considerable savings in nozzle bolting and head closure space can be achieved by using two types of special connections. The exchanger nozzles can be connected to the process piping by using a Graylock flange (16), a compression type of fitting that requires only two or four bolts rather than the large number of bolts for a conventional flange. When Graylock (Gray Tool Co.) flanges are subjected to lateral stresses due to thermal expansion as the piping heats up, the alignment can change and cause internal displacement and potential leakage from the flange. The piping has to be designed with enough flexibility so that the thermal squirm does not result in undue stresses and cause a flange to become misaligned.

The other space-saving approach is to use the Kobe (Kobe Steel) type of channel enclosure, which does not have all the external head bolts of typical TEMA Type B or Type C head enclosures. These exchangers require special tools to remove the tube bundles and trained maintenance personnel to do the work.

These exchangers should never be located in structures because of the need to be able to access the channel from grade as it is difficult to remove the channel cover plate by using special equipment.

Heat exchangers are usually supported by two pedestals attached to the exchanger saddles. The saddle closest to the channel is the fixed support, which has the saddle bolted tight to the foundation pedestal. About two-thirds of the exchanger weight usually rest on this pedestal because it carries the weight of both the channel and part of the tube bundle as well as half of the shell. The rear end support saddle carries the rest of the weight, which is half the shell weight and part of the tube bundle. The saddle on the end away from the channel end is usually bolted loosely to the pedestal to allow the exchanger shell to grow thermally and expand as the unit heats up. On some heavier exchangers, a Teflon slide plate is required between the top of the pedestal and the support saddle in order to provide reduced friction, so that the saddle can slide on top of the foundation pedestal to reduce the horizontal stress on the pedestal and prevent it from cracking. This is commonly used on exchanger support foundations either in high temperature service, when the exchangers are heavy, or when they are stacked.

Stacking heat exchangers so that the center line is higher than five meters or more than three stacks high can be a problem for maintenance. If more exchangers are required, eg, four, then the exchangers must be stacked in two pair two bundles high, because the surface area exceeds that which can be fabricated into three bundles.

Overhead condensers sometimes need to be located in the structure. Usually, partial condensers need to be elevated above the reflux accumulator. Considerable structure cost reduction can be achieved if the process can use grade-mounted condensers. Mounting the exchangers at grade may require them to be designed with subcooling so that the reflux accumulator can be located above the condenser. This should be considered as part of the process design.

Reboilers need to be located next to the tower they serve, except for the pump-through types, which can be located elsewhere. Fired heater reboilers are always located away from the associated tower and use a pump to circulate the bottoms. Kettle-type reboilers are preferred from an operational and hydraulic standpoint because they can be designed without the worry of having to ensure sufficient head for circulation required by thermosyphon reboilers. However, kettle reboilers require a larger-diameter shell that is more costly, and the reboiler must be supported at a sufficient elevation to get the product to the bottoms pump with adequate NPSH.

Horizontal thermosyphon reboilers are popular because they are less costly than kettle types and because they can be supported close to grade. The piping must be designed to provide enough liquid hydraulic head to overcome the recirculation loop pressure drop. Typically, an elevated draw tray provides the head and a large, low pressure drop inlet line runs from the reboiler draw nozzle to the reboiler inlet nozzle located on the bottom side of the shell. The reboiler outlet nozzle and piping are smaller than the inlet piping to keep the return-line velocity high enough and prevent separation of liquid from vapor and the resultant slug flow from occurring. Space needs to be provided for this piping.

The vertical thermosyphon reboiler has a particular advantage in that the reboiled fluid flows inside the tubes. This permits the use of alloy on the tubeside only if the reboiled fluid is corrosive or requires a noncarbon steel construction. However, the vertical thermosyphon reboiler also has two principal drawbacks. First, the vertical shell must be supported off the column or in some other way. Second, the removal of the tube bundle can be difficult because of the need to remove the outlet piping and provide the access room for pulling the bundle out vertically. The location of the vertical thermosyphon is usually put on the backside of the piperack because of the access room requirements for pulling the bundle vertically for cleaning. Many columns require multiple reboilers because the size of a single reboiler cannot be fabricated in a single shell. This service requires the allocation of space on both sides of the tower and a symmetrical piping design to ensure evenly distributed heat input loads and equal flows to each unit. Side reboilers are often useful inclusions on towers that make separations of mixtures with widely differing boiling points. Side reboiling is sometimes a preferred alternative to preheating the feed if the preheat temperature levels get too high and result in too much vaporization of the heavy component. Side reboilers require special considerations from a plot area and support consideration. They need to be supported in the structure at or near the section of the tower from which they withdraw liquid and return the reboiled mixture of vapor and liquid to the tower. Space needs to be provided for both the piping runs to and from the column and the pipe runs for whatever fluid is used as the heat media.

The location of exchangers is the key to maintenance. Usually the back head is kept at a distance of about three meters from the piperack support columns. Access equipment must be able to get in and remove the shell cover and flange head. Access area must also be provided to handle and remove the shell cover usually located under the piperack. The tube-pulling or rodding-out area must be kept clear to allow access to the channel end. This space should be at least equal to the tube length and about two meters from the tube sheet location. Tube removal space should be allowed for but is not mandatory if grade-mounted heat exchangers are used and mobile maintenance equipment employed to pick up the entire unit and transfer it to the repair shop.

**Fired Heaters.** In many European refineries, the overall plant layout is designed so that all heater exhaust gas is ducted to a centrally located 100-m high stack; in this manner release to the atmosphere is dispersed more broadly. Usually, fired heaters are located a minimum of 15 m from the hydrocarbon-containing equipment. When reactor–heater systems are used, the distance can be shortened if certain safety precautions are included in the design. The piperack can be used to provide part of the separation between the reactor and the heater. Reference 10 covers the various types of heaters and their components. Heaters basically come in two types. The vertical cylindrical heater, which is usually bottom-fired and often has a self supporting stack, is popular from a space-saving standpoint and the capital cost is usually low for this type of design on small units. Vertical heaters use vertical tubes in the radiant section, which can be a process disadvantage in some two-phase systems because of the higher pressure drop and the tendency for vapor and liquid to separate in the upward flow path. The tubes are pulled and removed from the top side of the heater usually by using a pull ring attached to the heater stack. This ring provides a way of removing the tubes without the need for a tall crane. A space of three meters longer than the tube length needs to be provided above the heater to pull the tubes. A lightly traveled road at the access side of the heater is preferred for equipment access. The tube-pulling area must be kept open during maintenance.

Larger-sized heaters are usually horizontal box heaters. The radiant coils can be located either on the side walls so that the units are fired from underneath, or in a center row of tubes in which the heater is fired from both sides to provide a higher heat flux for reducing the radiant surface. An access area at one end of the box is required in order to remove the tubes. Sometimes multiple coils are included in the same box, which may require access to both ends of the box.

Most high efficiency fired heaters include either a convection coil to generate low pressure steam or some other low temperature coil in the convection section. The other option to improve heater thermal efficiency is to include an air preheater. The air preheater makes the layout much more complex for several reasons. First, the stack can no longer be self-supporting. Second, space needs to be provided to accommodate the forced draft fan, the induced draft fan, and the air preheater itself. Some heaters can share a common stack to simplify the addition of convection coils by having all coils in one convection box. A steam coil in the heater convection section requires that a steam drum and the associated piping for natural circulation be provided. This requires a support structure. Environmental considerations may also require a structure to provide a sampling access platform on the stack and the necessary analyzer housings.

Piping for snuffing steam injection into a heater firebox is required to help put out a fire if a tube rupture occurs. The snuffing steam isolation valve needs to be located at an accessible spot remote from the heater. Also, a remote fuel shutoff valve should be located adjacent to the snuffing steam valve so that both valves can be accessed quickly in case of fire.

**Compressors.**    There are two basic types of compressors: centrifugal and reciprocating machines. The centrifugal compressor is usually used for higher volume low head applications. It can be either motor-driven or steam-turbine-driven. In some cases, it may even be gas-turbine-driven. The other type of compressor is the reciprocating machine. It is generally used for lower volumetric flow rates or when higher differential pressures are required. Most reciprocating compressors are multistaged units and equipped with intercooling and knockout drums between compression stages. Because of the nature of the operation, reciprocating compressor piping is often subject to vibrations. If slugs of liquid get into the suction piping, disastrous effects and severe damage to the compressor valves and pistons can occur. Therefore, some type of separation drum is usually located upstream of the compressor unit. The pipe run from the knockout drum to the compressor suction is kept as short as possible, with no low point pockets, and the suction line is often heat-traced to prevent any condensation. Because compressors are expensive equipment items, they should be protected from the elements. Some type of building or housing is usually provided, particularly for reciprocating units. The compressor and its auxiliaries are usually separated from the rest of the process facilities. This isolates them from fire and also moves the noise zone away from the process facility.

Space needs to be provided for the auxiliaries, including the lube oil and seal systems, lube oil cooler, intercoolers, and pulsation dampeners. A control panel or console is usually provided as part of the local console. This panel contains instruments that provide the necessary information for start-up and shutdown, and should also include warning and trouble lights. Access must be provided for motor repair and ultimate replacement needs to be considered. If a steam turbine is used, a surface condenser is probably required with a vacuum system to increase the efficiency. All these additional systems need to be considered in the layout and spacing. In addition, room for pulsation dampeners required between stages has to be included. Aftercoolers may also be required with knockout drums. Reference 8 describes the requirements of compressor layouts and provides many useful piping hints.

**Pumps.** Pumps (qv) are usually located along two pump rows that are used up underneath the piperack. Pumps are oriented with the motor accessible from the aisle way under the piperack. Each pump is located as close to the suction source as possible. Both the main pump and the spare pump are frequently located on the same support foundation. Small in-line pumps may be supported off the process piping. However, most in-house pumps require a pedestal for support. Space has to be provided around the pump for the piping and isolation valves as well as the electric supply. If steam turbines are used, then steam supply and exhaust steam piping are required. Some pumps also require flushing oil piping and cooling water supply to the gland coolers.

Most centrifugal pumps have end-suction and top-discharge nozzles, but both top-suction and top-discharge designs are not uncommon. Multistage pumps require much more space than conventional pumps. The casing is usually vertically split for single-stage pumps and horizontally split for multistage pumps. Many high pressure pumps are designed with a barrel casing.

The layout specialist should be aware of any special space requirements for a pump. Otherwise, pumps are usually fitted into a small area normally considered adequate for a general pump service. Sump pumps and other special types of applications need to be called out.

**Storage Tankage.** Most tankages are located away from the main process area. Occasionally, the process unit includes a day tank. However, it is safer to provide feed surge in a pressure vessel such as a feed surge drum rather than in a tank. On plot process, tankage is usually used only for nonflammable substances and the storage volumes are kept small to minimize the plot space taken up (day tanks are typical of storage volume required). Reference 2 provides some useful information about the required spacing for various types of storage tanks as well as guidance on the minimum required spacing between tanks and between the tank and the facility fence line (property line). It also provides useful information on locating tanks relative to each other, the use of tank dikes, grouping tanks in a common diked area, drainage requirements, and other safety-related information. Each site location must comply with local regulations in addition to general guidelines.

Tank storage of liquids at atmospheric conditions is generally classified into three classes: flammable liquids having a flash point below 100°F (38°C), flammable liquids having a flash point above 100°F but below 140°F (60°C), and combustible liquid having a flash point above 140°F and below 200°F (93°C).

The National Fire Protection Association (NFPA) has developed criteria based on these classifications and the type of tankage used for storage (fixed roof, floating roof, and no roof). These guidelines must be followed in the early layout. It can be costly to move a tank after it has been constructed. Once a tank is located on the plot plan, it tends to be forgotten until construction time. On one project, for instance, local spacing requirements were more conservative than the space provided for tanks already under construction. As a result, it was expedient to provide the community with a dedicated fire truck than to pour a new tank support ring and air-float the half-constructed tank shell to another location to meet the local tank space requirements.

The following are some general minimum spacing requirements. (*1*) Use one-meter minimum space between any two flammable liquid storage tanks. (*2*) The minimum distance between any two adjacent tanks should be at least one-sixth of the sum of their diameters. If one tank is less than one-half the diameter of the other, the tank spacing should be at least one-half the diameter of the smallest tank. (*3*) Crude oil storage tanks in production areas should be spaced a minimum of one meter if under 3000 barrels and at a space of at least one-tank diameter (smallest tank) if over 3000 barrels. (*4*) Unstable, flammable, or combustible liquid tankage should be spaced at one-half the sum of the tank diameters. (*5*) Local regulations, including fire protection, insurance codes, and common practices, may require even greater spacing for grouped tankage of three or more rows of irregularly spaced tanks. Tank pattern and spacing must provide adequate space so firefighting equipment can gain access. (*6*) Liquefied petroleum gas (LPG) requires a minimum vessel spacing of six meters and LPG containers must have dikes that are a minimum of three meters from the side and the center line of the dike.

**Cooling Towers.**    The cooling tower location relative to the prevailing wind direction should be such that the wind hits the short side or the side perpendicular to the inlet louvers. This helps balance the air flow to the two inlet sides.

The direction of the outlet plume relative to local roads should be considered. In foggy weather, the plume can cause the roadway to become a visual problem and a driving safety hazard. Also, the spray carryover from the top of the tower can, in some climates, cause ice to form downwind. The parking areas should never be located downwind of the cooling tower because the chemicals contained in most cooling water can cause severe paint problems. This can cost a plant a large amount of money to repaint the parked cars if a parking area is opened adjacent to the cooling tower.

Air to the cooling tower should be as cool as possible. Equipment that gives off heat should not be located upwind of the tower. The cooling-water pump pit should be located on the side of the tower that minimizes piping. Note that the slope of the bottom of the pit should be such that suction to any of the pumps does not become starved. In addition, water-treating chemical injection equipment and the delivery of these chemicals need to be considered. Also, the water blowdown, which on many towers is regulated by an overflow weir, flows by gravity to the appropriate sewer system, usually the high salts (non-oily) sewer.

Vented risers should be provided on most cooling towers to release only light hydrocarbon leakage from the cooling water before the spray header. No ignition or source of spark should be within 30 m of the vented riser.

**Control Room.** The control room location can be critical to the efficient operation of a facility. One prime concern is to locate it the maximum distance from the most hazardous units. These units are usually the units where LPG or other flammables, eg, hydrocarbons that are heavier than air, can be released and accumulate at grade level. Deadly explosions can occur if a pump seal on a light-ends system fails and the heavier-than-air hydrocarbons collect and are ignited by a flammable source. Also, the sulfur recovery unit area should be kept at a healthy distance away as an upset can cause deadly fumes to accumulate.

A central location where instrument leads are short is preferred. In modern facilities with distributed control systems, all units are controlled from a central control room with few operators. Only a few roving operators are available to spot trouble. It is desirable to deep process equipment a minimum of 8 m away from the control room. Any equipment and hydrocarbon-containing equipment should be separated by at least 15 m if possible. Most control rooms are designed with blastproof construction and have emergency backup power and air conditioning. The room is pressurized to prevent infusion of outside air that may have hydrocarbon content in the explosive range.

## Piperack Considerations

Piperack is considered the arterial system of the process plant (see PIPING SYSTEMS). All of the nonnozzle-to-nozzle pipe runs are made by running pipe to and from the piperack. Utility lines, electric lines, and instrumentation lines are often run on the piperack as well. Figure 11 shows various types of piperack configurations (9). These different types of piperack layouts are referred to by the shape of the piperack: straight-through (which is the most common type), L-shaped, T-shaped, U-shaped, and a combination of these. The individual unit piperacks are ties into the interconnecting plant piperack that provides the link between units. Each of these types of piperack layouts fits a particular need. The designer selects the most appropriate ones depending on the space availability and the equipment layout requirements.

Multielevation piperacks are usually needed to handle all the required services for piping, electrical, utilities, and instrumentation. The two-level rack is one of the most common but three-level ones are also used. The utility lines are usually run in the upper level and the process lines in the lower levels. The larger-diameter lines are located to the outside of the rack to be closest to the column supports. Access platforms are required at the battery limit to provide operators access to the block valves and blinds. If long runs of hot pipe are required, a portion of the pipe rack needs to be dedicated to an expansion loop. A horizontal space in the piperack is provided for a set of lines to be flat-turned into a set of expansion loops with the large pipes located on the outside. All of the pipe turns are in the same horizontal plane, which is an exception to normal

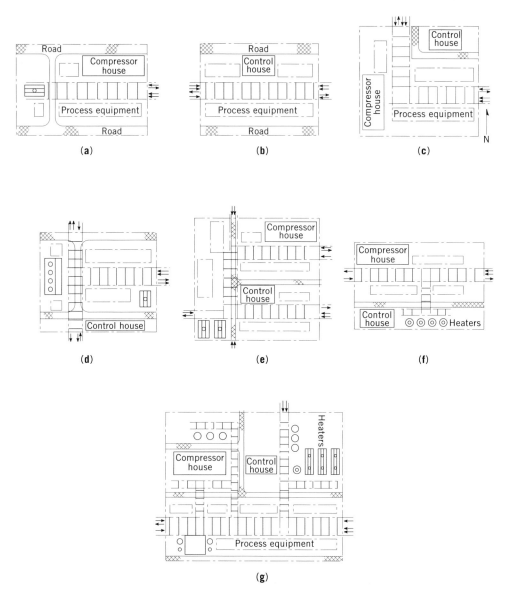

**Fig. 11.** Types of piperack configurations: (**a**) dead-end yard: lines enter and leave one end of yard; (**b**) straight-through yard: lines can enter and leave both ends of the yard; (**c**) L-shaped yard: lines can enter and leave north and east of the plot; (**d**) T-shaped yard: lines can enter and leave on three sides of the plot; (**e**) U-shaped yard: lines can enter and leave all four sides of the plot; (**f**) combination of I- and T-shaped yard; and (**g**) complex yard piping arrangement for a large chemical plant (9).

piping practice. A flat turn takes up and blocks space for other pipes. Flat turns are generally only made from the outside of the rack to minimize this blockage.

High temperature lines that grow due to thermal expansion are supported by shoes welded or strapped to the bottom of the pipe at the pipe support locations. Insulated lines must have insulation breaks at the support or be supported by shoes. Large-diameter lines are often used to support smaller lines at locations in between the supports by running a support beam attached to two larger-sized lines below the smaller lines.

The location of the pipe in the rack is selected to minimize the congestion and eliminate line crossing. If a process line connects two nozzles which are elevated higher than the piperack, then the upper level of the rack is used. Similarly, if the nozzles are both below the piperack, then the lower level rack is used. Other cases in which one nozzle is below and the other above use the least congested part of the piperack. Lines with valves are more easily accessed from the upper level, but require an access platform.

Pumps are usually located under the piperack. An open slot in the piperack has to be provided at the location of the pump suction and pump discharge lines so that the piping can make a straight run down to the pump suction and then the discharge line can be run directly back up to the piperack. Pumps commonly have an end suction and top discharge; but they can also have a top suction and top discharge. The location of these nozzles on the pump can affect the piping configuration. The pump discharge line usually has some type of flow control valve that requires a piping drop from the piperack down to the control valve. The control loop usually has a control valve, a double-block valve around the control valve, and a hand bypass valve so that the valve can be replaced or repaired while the hand bypass is used for temporary control. Double-block and bypass valves should be located at an elevation where the valve and the controller can be accessed without a ladder.

Whenever a change in piping direction occurs, the elevation of the pipe run should also change. If the main piperack is at an elevation of four meters, then the lateral piping can either go up to five meters or drop down to three meters. The piperack can also provide the support for air coolers and other equipment such as elevated drums.

Figure 12 shows the plan and elevation views of a process unit piping (9). A drum is supported off the piperack. Heat exchangers are located far enough back from the support columns so that they are accessible and their shell covers can be removed. Pumps are located underneath the piperack, but sufficient room is provided for maintenance equipment to access the motors and to remove the pump if necessary. The motor is always oriented away from the process equipment and located on that side of the piperack. Instrument valve drops are shown supported from the columns. The instrument trays themselves run on the outside of the support columns. Flat turns are only made from the outside position of the piperack. Nozzle-to-nozzle pipe runs are made whenever possible. Larger lines are located on the outside of the piperack. Connections to nozzles above the rack are made from the top elevation. Recommended spacing of yard pipes on a piperack have been given (15). Minimum equipment spacing requirements for inside as well as outside battery limit equipments, such as offsites, are given in Figures 13 and 14. In Figure 13, where environmental factors control discharge of emissions to the atmosphere, a closed collection header has to be provided in the piperack to flares. Relief valves (above tower) are located above such header and sloped accordingly. Where environmental factors are not of concern, relief valve discharges may discharge as indicated. Steam discharges to the atmosphere at a safe location.

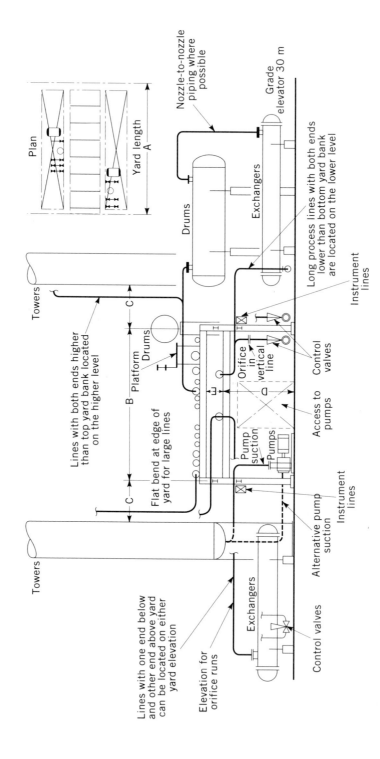

**Fig. 12.** Piperack, piping, and equipment relationships in a petroleum plant. A–E signify dimensions which affect piping cost.

170

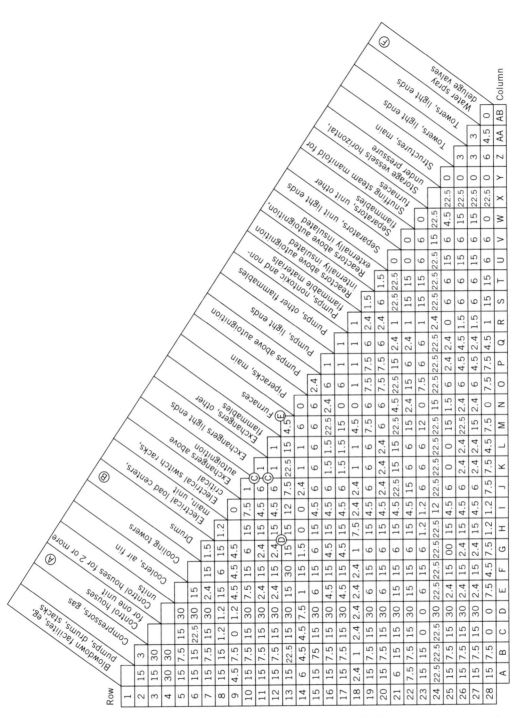

**Fig. 13.** Inside battery limit (ISBL) equipment spacing (distances are in meters): A, spacing of 22.5 m between control houses and equipment containing flammables is preferred to the 15-m spacing shown; B, drums containing nonflammables require minimum space of 1.5 m to other equipment for operating and maintenance access; C, exchangers in the same service as adjacent ones require only minimum spacing of 0.6 m, regardless of operating temperatures; D, 22.5 m required between furnaces and light-ends drums; E, for furnaces operating above 25 kg/(s·m²), spacing to adjacent furnaces is 9 m min; and F, deluge valve is 15 m from equipment being protected.

171

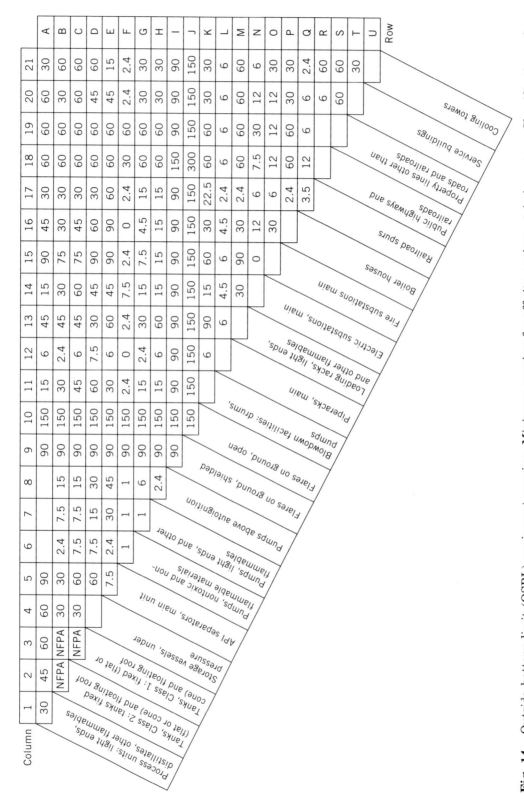

**Fig. 14.** Outside battery limit (OSBL) equipment spacing. Minimum spacing for off-site equipment is in meters. Classifications of tankage are Class 1: high hazard, flash point below 38°C; Class 2: low hazard, flash point above 38°C. NFPA = National Fire Protection Association. Safety standards are calculated by flare stack, height, and sterile-area radius.

## CAD Models

Computer-aided design (CAD) (qv) technology is steadily improving. Output from CAD is readable, and hard-copy drawings are available with consistent line weights, symbols, and dimensions. CAD models still only show two-dimensional representation of three-dimensional models. However, the integration of all equipment, piping, and structural steel makes it easier to detect and interface between two components occupying the same space through interference detection checks. Formerly, plastic models were the only way that this could be checked with any degree of confidence.

Three-dimensional CAD model development still requires some two-dimensional preliminary work such as plot plans. Individual pieces of equipment need to be dimensionally defined by creating an equipment model file with the special and geometric configuration generation. To these models the locations of nozzles, ladders, platforms, and other externals are added, depending on the desired complexity of the model. This information is entered into the computer in various layers of data which permits the CAD designer to display any or all of these layers. Geometry of the structural steel, piperacks, and foundations are entered. Ultimately, piping instrumentation and electrical information is entered to the level of detail desired. The piping file can be extracted and used to generate orthographic and isometric drawings that can include a sized list of materials such as pipe and fittings with piping classifications, materials of construction, and type of alloy identified. These 3-D models can be viewed from almost any perspective by using the zoom-in display and walking through the model. Views can be enlarged by moving the visualization point in closer to the desired reference. Any of the model input levels can be turned off to provide an unobstructed view, ie, by removing instrument, electrical, and structural steel, to see piping only.

**Scale Models.** Replicas of a plant are often prepared as part of the design and can be in several forms: equipment only, equipment and major piping, or complete equipment and piping. Model building is both time-consuming and expensive. Models of whole facilities can easily cost millions of dollars but the money is considered well-spent by many operating companies. The model can be used as a design tool, a construction aid, and an operation and training aid. When the facility has complex piping, 3-D visualization can be depicted on a model that can make for a much more desirable operational layout. Scale models are quickly being replaced by 3-D CAD models.

## Noise Abatement

One of the prime OSHA requirements is control of the noise level, which can usually be met by using proper specification and noise attention devices. Noise can cause hearing impairment and be hazardous to health. It also interferes with work efficiency by inducing stress, hindering conversation of operators, and preventing them from hearing warning signals.

Plant layout and noise suppression material are two general noise abatement methods. Plant layout does not affect noise levels at any given point; however, noise can be abated by screening off a section of the plant. An example of this is to orient cooling towers with their closed faces toward the critical location. This method must also consider wind direction to balance air draft. Tankage can be located to act as a noise screen.

Most rotating equipment includes electric motors or steam dryers that generate noise at a constant frequency. Air cooler fans are a source of noise that can be reduced by lowering the fan speed and increasing the number of blades. Pump motor noise can be reduced by including a shroud or fan cover that is accurately lined. Centrifugal compressor noise reduction can be achieved by blade design and the use of compressor pulsation noise reduction, silencers, and vibration isolation.

Flare noise (roar of combustion) is the most serious because it is elevated and the sound carries. The flare can be located at a remote distance from the operating unit or surrounding community. Noise of steam injection into the burner can be reduced by using multiple nozzles. Furnace noise from air intake, fuel systems, and combustion blower forced draft/induced draft (FD/ID) fans can be reduced by acoustics. The plot plan should be evaluated for noise generation and to find the means of alleviating or moving noise to a less sensitive area (see NOISE POLLUTION AND ABATEMENT METHODS (SUPPLEMENT)).

## BIBLIOGRAPHY

"Plant Layout" in *ECT* 1st ed., Vol. 10, pp. 737–743, by J. Salviani, Chemical Construction Corp.; in *ECT* 2nd ed., Vol. 15, pp. 689–699, by M. M. Goetz, Chemical Construction Corp.; in *ECT* 3rd ed., Vol. 18, pp. 23–43, by F. V. Anderson, Stone & Webster Engineering Corp.

1. R. Weaver, *Process Piping Design*, Vol. 1, Gulf Publishing Co., Houston, Tex., 1973, p. 57 ff.
2. R. Kern, *Petroleum Refiner* **39**(12), 139–146 (Dec. 1960).
3. R. Kern, "Chemical Engineering Refresher: Plant Layout, Part 4: How to Find the Optimum Layout for Heat Exchangers," *Chem. Eng.* **84**(19), 169–178 (Sept. 12, 1977).
4. *Ibid.* "Part 1: Plant Design Activities," **84**(11), 130–136 (May 23, 1977).
5. *Ibid.* "Part. 2: Design Requirements," **84** (July 4, 1977).
6. *Ibid.* "Part 3: Layout Arrangements for Distillation Columns," **84**(17), 153–160 (Aug. 15, 1977).
7. *Ibid.* "Part 5: Arrangement of Process and Storage Vessels," **84**(24), 93–99 (Nov. 7, 1977).
8. *Ibid.* "Part 6: How to Get the Best Process: Plant Layouts for Pumps and Compressors," **84**(26), 131–140 (Dec. 5, 1977).
9. *Ibid.* "Part 7: Piperack Design for Process Plants," **85**(3), 105–112 (Jan. 30, 1978).
10. *Ibid.* "Part 8: Space Requirements and Layout for Process Furnaces," **85**(5), 117–122 (Feb. 27, 1978).
11. *Ibid.* "Part 9: Instrument Arrangements for Ease of Maintenance and Convenient Operation," **85**(9), 127–134 (Apr. 10, 1978).
12. *Ibid.* "Part 10: How to Arrange the Plot Plan for Process Plant," **85**(11), 191–197 (May 8, 1978).
13. *Ibid.* "Part 11: Arranging the Housed Chemical Process Plant," **85**(16), 123–130 (July 17, 1978).
14. *Ibid.* "Part 12: Controlling the Cost Factor in Plant Design," **85**(18), 141–146 (Aug. 14, 1978).

15. R. E. Meissner III and D. C. S. Shelton, *Chem. Eng.* **99**(4), 81–85 (Apr. 1992).
16. Technical data, "Graylock Pipe Fittings and Connectors," Gray Tool Co., 1995.

ROLAND E. MEISSNER III
The Ralph M. Parsons Company

# PLANT LOCATION

Selecting a plant site is critical to the financial success of a plant. Several factors must be considered in selecting a general plant site location. The procedure for choosing a specific plant location can be presented in a series of required steps. After the site is selected, factors that go into getting the facility built, including permitting and the other necessary legal steps, need to be considered.

## Siting Factors

The primary siting factors which influence the selection of a plant location are as follows: environmental considerations, labor availability and productivity, raw material availability, proximity to market, property cost, accessibility to transportation, tax incentives, electric power availability and cost, and living conditions.

Table 1 shows a rating of the heavily industrial state business climate as defined by several criteria. The highest rated state is located to the left. Table 2 shows a similar rating of the least heavily industrial states. Note that the lowest number in the rating scale has the highest rating. These tables can be somewhat misleading in that states like Texas, Louisiana, and California, which have large petrochemical bases, are shown with low ratings.

In the selection of a plant site, it is a good idea to get broad-based input, including information from sales, production, plant engineering, and from the general manager. The first objective is to narrow the range of possible choices. This involves focusing on the most important criteria, which differ widely for each type of facility. Table 3 presents some of the factors to consider for plant siting. An objective rating system needs to be used; a listing of the important criteria and a point system which affords the evaluation committee a means of scoring the pluses and the minuses has been found to be useful. The scoring may need to be weighted so that the factors which most heavily affect the bottom-line profit are more heavily counted. At the end of this phase, the less desirable sites should be discarded so that a more in-depth evaluation of the final candidates can be made.

**Table 1. Which State is Best (Heavily Industrial States)[a,b]**

| Parameter | State | | | | | | | | | | | | | | |
|---|---|---|---|---|---|---|---|---|---|---|---|---|---|---|---|
| | N.C. | Vt. | Wis. | Miss. | Ala. | Va. | Ga. | Ariz. | Tenn. | S.C. | Mo. | Ind. | N.J. | N.H. | R.I. |
| chemical output, $10^9 | 12.9 | 0.28 | 5.11 | 2.03 | 5.28 | 6.23 | 7.96 | 1.85 | 7.41 | 9.25 | 9.04 | 13.9 | 45.4 | 0.5 | 0.83 |
| number of chemical plants | 113 | 8 | 71 | 34 | 57 | 53 | 97 | 31 | 86 | 72 | 117 | 108 | 413 | 14 | 12 |
| employees, 10^3 | 40.9 | 1.1 | 18.6 | 6.9 | 19.2 | 26.3 | 21.7 | 6.8 | 30.9 | 29.8 | 31.1 | 46 | 155.5 | 1.8 | 3.2 |
| energy costs, $/10^6 kJ | 6.93 | 9.03 | 6.26 | 5.50 | 4.66 | 4.76 | 6.03 | 5.75 | 6.03 | 6.20 | 6.12 | 4.43 | 7.66 | 10.8 | 7.79 |
| environmental record | 37 | 2 | 21 | 44 | 47 | 36 | 38 | 40 | 45 | 35 | 33 | 49 | 28 | 8 | 7 |
| environmental policies | 18 | 12 | 6 | 46 | 49 | 22 | 29 | 50 | 40 | 32 | 23 | 27 | 3 | 20 | 10 |
| fiscal policies | 25 | 4 | 30 | 26 | 14 | 31 | 15 | 16 | 12 | 42 | 28 | 22 | 27 | 1 | 8 |
| state-related labor costs | 1 | 24 | 20 | 4 | 12 | 7 | 17 | 31 | 11 | 8 | 21 | 2 | 3 | 27 | 38 |
| labor costs | 11 | 27 | 18 | 5 | 10 | 32 | 15 | 4 | 16 | 7 | 22 | 44 | 41 | 39 | 31 |
| resource productivity | 29 | 18 | 11 | 42 | 40 | 3 | 39 | 36 | 43 | 37 | 26 | 19 | 15 | 44 | 20 |

[a]Ref. 1. Ranks states with the best manufacturing climates in descending order, from left to right, on the basis of four weighted factors.
[b]Heavily industrial states are those accounting for more than 2% of U.S. industrial output, with at least 16% of the work force in manufacturing.

**Table 1.** (*Continued*)

| Parameter | | Minn. | Fla. | Del. | Ill. | Pa. | La. | Calif. | N.Y. | Ohio | Mass. | Maine | Tex. | Conn. | Mich. |
|---|---|---|---|---|---|---|---|---|---|---|---|---|---|---|---|
| | State | | | | | | | | | | | | | | |
| chemical output, $10^9$ | | 6.4 | 9.4 | 6.68 | 22.6 | 26.7 | 13.9 | 20.0 | 34.1 | 20.1 | 6.65 | 0.3 | 33.3 | 8.61 | 21.7 |
| number of chemical plants | | 73 | 100 | 24 | 296 | 240 | 108 | 353 | 239 | 220 | 121 | 5 | 274 | 64 | 176 |
| employees, $10^3$ | | 17.1 | 35.2 | 18.9 | 79 | 89.9 | 46 | 82.9 | 108.4 | 70.2 | 25.9 | 0.9 | 78.4 | 29.5 | 57.4 |
| energy costs, $/10^6$ kJ | | 6.44 | 6.65 | 5.61 | 6.47 | 5.67 | 4.38 | 6.97 | 6.68 | 5.63 | 8.44 | 6.84 | 3.94 | 8.6 | 7.11 |
| environmental record | | 5 | 30 | 27 | 42 | 34 | 49 | 19 | 17 | 46 | 6 | 4 | 48 | 23 | 32 |
| environmental policies | | 7 | 13 | 25 | 17 | 21 | 27 | 1 | 8 | 19 | 9 | 5 | 35 | 4 | 11 |
| fiscal policies | | 35 | 45 | 18 | 19 | 29 | 22 | 33 | 48 | 24 | 6 | 32 | 50 | 20 | 9 |
| state-related labor costs | | 46 | 29 | 15 | 43 | 33 | 2 | 35 | 30 | 36 | 47 | 44 | 45 | 48 | 50 |
| labor costs | | 21 | 12 | 49 | 29 | 43 | 44 | 38 | 45 | 47 | 46 | 36 | 26 | 40 | 50 |
| resource productivity | | 5 | 32 | 7 | 24 | 21 | 19 | 34 | 8 | 31 | 27 | 48 | 12 | 35 | 46 |

Table 2. Which State is Best (Light Industrial States)[a]

| Parameter | State | | | | | | | | | |
|---|---|---|---|---|---|---|---|---|---|---|
| | N. Dak. | Wyo. | S. Dak. | Nebr. | Iowa | Nev. | Utah | W. Va. | Oreg. | Kans. |
| chemical output, billion | 0 | 0.16 | 0.44 | 0 | 1.75 | 0.41 | 1.22 | 6.51 | 0.48 | 3.0 |
| number of chemical plants | 0 | 3 | 2 | 0 | 38 | 13 | 15 | 29 | 18 | 40 |
| employees, thousands | 0 | 0.43 | 0.18 | 0 | 5.8 | 1.6 | 5.6 | 17.3 | 2 | 7.4 |
| energy costs, $/10^6 kJ | 4.21 | 4.76 | 5.81 | 5.51 | 5.24 | 7.67 | 5.39 | 4.19 | 6.17 | 5.63 |
| environmental record | 16 | 25 | 12 | 24 | 29 | 9 | 22 | 41 | 3 | 43 |
| environmental policies | 37 | 44 | 48 | 30 | 16 | 43 | 41 | 45 | 2 | 28 |
| fiscal policies | 44 | 3 | 10 | 17 | 34 | 2 | 40 | 5 | 11 | 37 |
| state-related labor costs | 9 | 13 | 5 | 16 | 18 | 10 | 6 | 25 | 28 | 19 |
| labor costs | 3 | 28 | 2 | 6 | 8 | 19 | 17 | 34 | 20 | 30 |
| resource productivity | 1 | 3 | 17 | 6 | 4 | 50 | 28 | 23 | 38 | 16 |

[a]Ref. 1. Ranks states with the best manufacturing climates in descending order, from left to right, on the basis of four weighted factors.

178

**Table 2. (Continued)**

| | State | | | | | | | | | | |
|---|---|---|---|---|---|---|---|---|---|---|---|
| Parameter | N. Mex. | Alaska | Idaho | Ky. | Ariz. | Md. | Okla. | Colo. | Hawaii | Wash. | Mich. |
| chemical output, billion | 0.61 | 0 | 0.18 | 3.41 | 5.05 | 3.49 | 1.1 | 0.98 | 0.09 | 0.53 | 0.08 |
| number of chemical plants | 8 | 0 | 3 | 58 | 21 | 51 | 28 | 25 | 3 | 19 | 4 |
| employees, thousands | 2.2 | 0 | 0.7 | 16.6 | 15.4 | 11.2 | 4.6 | 3.8 | 0.3 | 2.2 | 0.4 |
| energy costs, $/10^6 kJ | 6.57 | 2.23 | 6.00 | 6.94 | 7.71 | 5.84 | 3.97 | 5.37 | 11.4 | 4.92 | 5.40 |
| environmental record | 20 | 18 | 11 | 39 | 26 | 14 | 31 | 10 | 1 | 13 | 15 |
| environmental policies | 38 | 47 | 36 | 33 | 39 | 15 | 42 | 26 | 24 | 14 | 31 |
| fiscal policies | 41 | 23 | 36 | 13 | 49 | 7 | 39 | 47 | 21 | 38 | 46 |
| state-related labor costs | 37 | 49 | 26 | 34 | 23 | 22 | 32 | 42 | 14 | 39 | 41 |
| labor costs | 1 | 14 | 9 | 23 | 13 | 42 | 24 | 25 | 48 | 33 | 35 |
| resource productivity | 30 | 2 | 41 | 33 | 25 | 14 | 22 | 10 | 45 | 47 | 49 |

**Table 3. Factors to Consider for Plant Siting**

| Plant requirements | Labor | Transportation | Raw materials | Markets | Power |
|---|---|---|---|---|---|
| acreage[a]<br>present/future expansion<br>building[a]<br>drainage, foundation,<br>subsoil, maintenance[b]<br>natural features<br>topography, climate,<br>local supplies<br>public utilities[b]<br>streets, phones, light,<br>heat, gas, power, tele-<br>graph, newspaper,<br>fire protection<br>water supply[a]<br>quantity/quality<br>waste disposal[b]<br>sanitary/industrial<br>advertising value<br>government<br>character, legal permits,<br>police protection, taxes,<br>insurance[b]<br>legal phases<br>laws, status, ordinances,<br>board of health<br>requirements | supply<br>proximity available<br>character<br>wages<br>h/d, overtime, bonus<br>disposition<br>union, content,<br>strikes<br>housing<br>health, sanitation,<br>markets, shopping,<br>schools, churches,<br>recreation | railroads<br>rates, sidings[b],<br>shifting, transfers,<br>belt line<br>waterways<br>boat lines,<br>canals<br>highways<br>motor traffic<br>passenger buses,<br>freight lines<br>street/railways<br>present/future<br>schedules, fares,<br>freight, express | haul<br>quantity<br>permanency<br>quality<br>annual<br>consumption<br>storage<br>requirements | schedule/rates<br>shipping<br>storage<br>requirements<br>(invention)<br>competition<br>shipping costs,<br>custom<br>charges | fuel supply<br>electricity<br>water supply<br>quantity/quality,<br>boiler purposes,<br>condenser water<br>water power<br>(hydroelectric) |

[a]Represents an investment cost.
[b]Represents an operating cost.

## Environmental Considerations

No matter how advantageous a site location may be, if a permit to build cannot be obtained or the uncertainties in getting the necessary permits jeopardize the timing of a project, then it may be necessary to choose another site. Thus, environmental considerations may be, overall, the most important siting factor. California has a reputation for environmentally strict regulations. For example, one chemical company had to pull out of a planned facility because timely permitting on a proposed northern California project was not forthcoming. An oil company has been in an ongoing battle with Santa Barbara, California, and federal regulators on the development of offshore oil fields. After spending hundreds of millions of dollars in court and legal fees, the company is scaling back its planned development.

Some states have become very active in promoting the construction of new plants in order to bolster their local economies. States such as Mississippi, Alabama, Tennessee, and Georgia have started courting the chemical processing industry (CPI) sector. Other states such as Louisiana and Texas had been known for their ease in issuing environmental permits in order to get plants built which would create jobs. However, these states have been tightening up their policies as a result of citizen outcry and the adverse publicity generated by environmental issues that have been brought to public attention.

Many companies are opting to increase the efficiency of existing facilities and add more capacity to selected plants rather than to try to site new facilities, because of the difficulties of the environmental permitting process.

## Labor Availability and Productivity

Plants need to be run by people and the availability of employees can constitute the overriding consideration in certain businesses, in relation to siting. Labor-intensive businesses have to either move to a location where labor is available or move their employees to the new plant site, which can be costly both from the standpoint of the physical move and also with regard to the additional expense of relocating a family from one place to wholly new surroundings. Older and less flexible work forces often choose to accept early retirement, quit, or not to relocate.

Training a new work force can also be costly. Bringing a new employee to the required level of proficiency requires time, and does not always succeed. It is therefore preferred to have access to a work force with developed skills that can be readily converted to suit the special requirements of a new business. The new plant site needs to be accessible to a sufficiently large work force that the incremental new requirements do not cause a shortage to develop that could raise the wage rates above prevailing rates. Usually the factors that make a site desirable also include incentives which have already attracted a skilled work force that can be hired away from other, similar employers. It is advisable to have a labor survey conducted well in advance of site selection in order to avoid surprises. This can be done by hiring local recruiters, by running blind advertisements in local newspapers for key positions that are to be filled first, or by using consultants.

**Living Conditions.**  An investigation of the following community quality-of-life considerations, which are considered important in attracting personnel to relocate, is necessary: living conditions, ie, house availability, costs, and safety; schools, ie, quality, class size, and distance; health care, ie, availability of a major hospital and local doctors; recreational facilities, ie, types and proximity; cultural, ie, musical, plays, and movie theaters; and sports, ie, major or minor league teams.

## Raw Material Availability

For many industries, such as the petroleum (qv) and petrochemical industries, the accessibility of raw materials is the overwhelming factor in selecting a plant site. In the United States, the majority of the refineries are located in three geographic areas either where raw materials are located or where foreign oil can be easily brought in by large tankers. Texas has three significant refining areas which include the Houston Ship Canal, Corpus Christi, and Port Arthur–Beaumont. In Louisiana, the 160-km stretch of the Mississippi River located between New Orleans and Baton Rouge is home to several hundred refining and petrochemical facilities. In California, the refineries are either located near the Long Beach and Los Angeles harbors, in the San Francisco Bay area, or in the San Joaquin Valley. All these areas have either a source of crude oil, access to ports in which foreign oil can be brought by large tankers, or a combination of the two. The East Coast is highly dependent on the importation of foreign crude. The two principal refinery areas are near Philadelphia and in New Jersey, where tankers can bring in much of the crude oil.

The other refining centers in the United States are all located near one of the major pipelines that are supplied either from the Gulf Coast oil fields or from Oklahoma, Texas, or Louisiana fields. The Pacific Northwest refineries were originally supplied mostly by Canadian production. Currently (ca 1995), these refineries are predominantly supplied by Alaskan North Slope crude, which is brought in from Valdez to the refinery ports in tankers.

Most of the petrochemical facilities that depend on a cheap and abundant supply of natural gas are also located in the Gulf Coast, where a surplus of offshore gas is available. These facilities use cracking of natural gas to produce ethylene as the starting point for the refining of their products. This process not only uses natural gas as feedback but also uses the natural gas to fire the ethylene cracking furnaces. About one-fourth of the world's chemical production is located in Texas.

Other industries that are traditionally located close to the source of raw materials include the steel (qv) industry, located close to iron ore; the flour industry, close to wheat fields; the meat-packing industry, close to grazing land for cattle; pulp and paper, close to forests; and the mining industry, close to mines. The inorganic industry, including salts, ash, borax, and gypsum, has always been located near the source of the needed raw material.

## Accessibility to Transportation

For relatively low priced products, the cost of transportation can be a significant deciding factor in plant site selection. The plant needs to be located close to the market. The cost of shipping by tanker is lowest, pipeline is next lowest (see PIPELINES), and truck and railcar shipment is the highest. These last two means are sometimes the only options for some products. Most refineries have the product distribution terminal located outside the plant. Pipelines are used to send blended products to intermediate storage tanks located at the distribution terminal, which could be located hundreds of miles away from the refinery. This minimizes the land haul cost. It also gets the traffic of the truck loading away from the refinery gate.

## Property Cost

Land cost in certain highly desirable petrochemical manufacturing areas can be as high as $25/m^2$ (such as Houston ship canal), whereas in most normal industries land costs are in the $4-12/m^2$ range. Where a lot of land is needed for feed and product storage, the cost of land can be significant. Many of the lots located on the deep-water ship canals are so orientated as to minimize the amount of costly land facing the waterway to only that needed for shipping products and receiving feed with the rest of the plant located farther back, away from the expensive real estate. Plants also tend to locate away from areas where encroaching residential homes drive up the property cost.

## Tax

Several states that have a large number of CPI plants offer various types of tax incentives. Louisiana, for instance, offers a 10-yr tax exemption from property taxes on buildings, equipment, and improvements to land (2). Texas, which has a large petrochemical industry, offers a 7-yr tax abatement program. Neither of these states have a state income tax. Both states offer a tax credit for each job created and provide free worker training.

When considering taxes, all types need to be considered: initial fees, capital value, corporate rate, personal income tax, sales tax, property tax, unemployment insurance, workmen's compensation, and nuisance tax. During the construction phase, several types of taxes may be levied. These include building permits, special fees, assessments, and sewer connection fees.

## Electric Power Availability and Cost

Several industries are highly dependent on cheap electric power. These include the aluminum industry, the Portland cement industry, electrochemical industries such as plating and chlorine production, the glass industry, and the pulp and paper industry. Other industries such as the petrochemical industry, which is highly competitive, depend on low priced power. About two-thirds of the cost of producing ammonia is electrical cost.

These industries try to locate near a source of hydropower (Niagara Falls or Hoover Dam) or near a source of excess nuclear power. They generally work out arrangements to get power at a reduced cost based on being the first one cut off when electric load shedding is required.

Many CPI industries have installed cogeneration facilities so that are carried on a separate balance sheet and are often owned by a third party. The cogeneration unit is usually located next to the CPI facilities so that the latter can take the steam and return condensate while the electric power can either be used or exported to the electric utility grid. This eliminates the disadvantage of the inherently inefficient use of energy by a power producer that generally has to throw away two-thirds of the energy to condense the exhaust steam from the turbines which generate power, rather than consuming the heat for useful purposes.

## European Siting Considerations

European siting considerations are somewhat different than those in the United States. Germany, the Netherlands, France, and Italy were traditionally the favored locations for European CPI industry plants because of their proximity to the markets, cheap energy, and presence of a skilled labor force. However, labor costs, when considering all of the fringe benefits including liberal vacations and retirement policies, make doing business in Europe expensive.

Many firms are finding areas such as Ireland, Scotland, Spain, and southern Italy, where unemployment rates are relatively high and governments offer property tax exemptions, grants toward capital investment, low interest loans, and other incentives, attractive from those standpoints. Some of these areas have developed a good infrastructure based on businesses already attracted to the area.

There is also, however, an increasing resistance by local communities to siting new plants in many of these European areas. Moreover, the environmental regulations have become increasingly stringent in Germany, the Netherlands, and France. It is estimated that the environment and other safety-related costs in Germany have risen to the point where 30% of the total capital investment needed is for environmental and safety measures.

## Site Purchase

Once the site is selected, the land purchase must be made. Usually a third party such as a real estate agent is employed to do this work. It is advisable first to secure an option to buy the land so that all of the potential problems can be evaluated or solved before the purchase is made. Other negotiations besides the purchase or lease of the site also need to be negotiated. Zoning, easement, building costs, access roads, taxes, and the like need to be settled before the land is purchased. It is best to keep the identity of the purchaser secret while all this is being handled. Contact between the plant owner and utilities, railroads, pipelines, and the local community need to be finalized in writing before a final purchase.

**Environmental Permitting.** An environmental baseline is usually required to establish how the plant would affect the surroundings. Air quality issues can be the deciding factor in siting, as air permits often take the most time to obtain. Water pollution also needs to be addressed. Federal, state, and regional regulations are often in conflict with each other. Hiring a consultant to steer a way through this complex myriad of regulations can be useful. What the environmental discharges are and how best to mitigate their effects by providing a plant that is designed to minimize the impact are key issues.

An environmental impact statement (EIS) normally has to be prepared and a public hearing held in order for the community to air its views. This can take up to two years. Getting a public relations effort under way helps in getting these requirements completed in a timely fashion. Efforts should be made to foster good relations with local authorities and neighbors. The opposition to a new plant can often be well organized and articulate. An early assessment of the political climate needs to be made in order to determine if public opinion is in such strong opposition to a new development that no matter what is done, it will be fought.

**Local Site Condition Evaluation.** In addition to visiting the site, drawing up a contour map and geology reports, acquiring soil-bearing information, and a knowledge of boundaries, setbacks, local requirements, utility tie-in locations, sewer connections, access to roadways, pipelines, railroads, etc, may be needed to make a full assessment.

A buffer zone may be required around the plant, and even if it is not required, it is less costly to provide a buffer at the start rather than to buy more property at a later date when the price has gone up.

## Specific Plant Site Considerations

Once a general plant location has been established, a specific plant site location needs to be determined. This requires developing a list of requirements for the plant. One approach is to use the "cornfield" method of plant site assessment, the essence of which is to develop an ideal list of specifications that define what would be required to put the plant in any cornfield. This approach should be carried out by devising a preliminary plant layout (qv). All the requirements for the plant and the supporting infrastructure are put down on a to-scale drawing. This drawing has to be modified to adapt to the features and terrain of the ultimate site. Some of the considerations that should be listed as requirements are as follows: number of employees; size of process units; utility requirements, ie, for cooling water, power, water, fuel, and steam; shipping requirements, ie, road access, and access to railway lines, waterways, and pipelines; disposal requirements for solids waste, chemical sewer loads, and sanitary sewer loads; and permitting factors such as air emission, water emission, and those that are applicable to the particular type of product.

An example of a checklist is provided in Table 4 as a typical item-by-item check for use after a final site has been selected. This example is based on an evaluation for hazards (3).

## Table 4. Facility-Siting Checklist[a]

*Spacing between process components*

have adequate provisions been made
for relieving explosions in process
components?

are operating units and equipment
within units spaced to minimize
potential damage from fires or
explosions in adjacent areas?

are there safe exit routes from
each unit?

has equipment been adequately spaced
and located to safely permit anticipated maintenance (eg, pulling
heat-exchanger bundles, dumping
catalyst, lifting with cranes) and
hot work?

are vessels containing highly hazardous
chemicals located sufficiently
far apart? if not, what
hazards are introduced?

is there adequate access for emergency
vehicles, eg, fire trucks?

can adjacent equipment or facilities
withstand the overpressure
generated by potential explosions?

can adjacent equipment and facilities,
eg, support structures, withstand
flame impingement?

*Location of large inventories*

are large inventories of highly hazardous chemicals located away from
the process area?

is temporary storage provided for raw
materials and finished products
at appropriate locations?

are the inventories for highly hazardous
chemicals held to a minimum?

where applicable, are reflux tanks,
surge drums, and rundown tanks
located in a way that avoids
large-volume concentration of
highly hazardous chemicals
in any one area?

where applicable, has special
consideration been given to
storage and transportation of
explosives?

have the following been considered
in the location of material
handling areas:
 fire hazards

location relative to
 important buildings
safety devices, eg, sprinklers
slope of area (is it level?)

*Location of motor control center*

is the motor control center located
so that it is easily accessible
to operators?

are circuit breakers easy to
identify?

can operators safely open circuit
breakers? have they been trained?

is the motor control center designed
such that it could not be an ignition
source? are the doors always closed?

is a "No Smoking" policy strictly
enforced?

is the motor control center designed
and meant to be a safe haven?

*Location and construction of control
room(s)*

is the control room built to satisfy
current corporate overpressure and
safe-haven standards?

does the construction basis for the
control room satisfy acceptable
criteria

are workers protected in the control
room (or their escape routes) from
all of the following:
 toxic, corrosive, or flammable
  sprays, fumes, mists, or vapors
 thermal radiation from fires
  (including flares)
 overpressure and projectiles from
  explosions
 contamination from spills or runoff
 noise
 contamination of utilities, eg,
  breathing air
 transport of hazardous materials
  from other sites
 possibility of long-term exposure of
  employees to low concentrations
  of process material
 odors
 impacts, eg, from a forklift
 flooding, eg, ruptured storage tank

are vessels containing highly hazardous
chemicals located sufficiently far from
control rooms?

186

Table 4. (*Continued*)

were the following characteristics considered when the control room location was determined:

types of room construction
types and quantities of materials
direction and velocity of prevailing winds
types of reactions and processes
operating pressures and temperatures
ignition sources
fire protection facilities
drainage facilities

if windows are installed, are they of rigid construction with sturdy panes, eg, woven wire-reinforced glass?
is at least one exit located in a direction away from the process area?
do exit doors open outward? are emergency exits provided for multistoried control buildings?
are ends of horizontal vessels facing away from control rooms?
are critical pieces of equipment in the control room well protected? is adequate barricading provided for the control room?
are open pits, trenches, or other pockets where inert, toxic, or flammable vapors could collect located away from control buildings or equipment handling flammable fluids?
where piping, wiring, and conduit enter the building, is the building sealed at the point of entry? have other potential leakage points into the building been adequately sealed?
is the control room located a sufficient distance from excessive vibration sources?
is positive pressure maintained in control rooms located in hazardous areas?
could any structures fail on the control room in an accident?
is the control room roof free from heavy equipment and machinery?

*Location of machine shops, welding shops, electrical substations, and other likely ignition sources*
are likely ignition sources, eg, mainten-

ance shops, roads, and rail spurs, located away from release points for volatile substances (both liquid and vapor)?
are process sewers located away from likely ignition sources?
are all vessels containing highly hazardous chemicals or components containing material above its flash point located away from likely ignition sources?
are the flare and fired-heater systems located so as to minimize hazards to personnel and equipment, with consideration given to normal wind direction and wind velocity and heat potential?

*Location of engineering, lab, administration, or other buildings*
are administration buildings located away from inventories of highly hazardous chemicals?
are administration buildings located away from release points for highly hazardous chemicals?
are workers in administration buildings protected from all of the following:

toxic, corrosive, or flammable sprays, fumes, mists, or vapors
thermal radiation from fires (including flares)
overpressure and projectiles from explosions
contamination of utilities, eg, water
contamination of spills or runoff
noise
transport of hazardous materials from other sites
flooding, eg, ruptured storage tank
odors

*Unit layout and location of facility relative to neighbors*
are large inventories or release points for highly hazardous chemicals located away from public access roads? from vehicular traffic within the plant?
is the unit, or can the unit, be located to minimize the need for offsite or intrasite transportation of hazardous materials?

**Table 4. (*Continued*)**

are workers in adjacent units protected and workers in this unit protected from the effects of all of the following from adjacent units or facilities:

　releases of highly hazardous chemicals

　toxic, corrosive, or flammable sprays, fumes, mists, or vapors

　overpressure from explosions

　contamination from spills or runoff

　odors

　noise

　contamination of utilities, eg, sewers

　transport of hazardous materials from other sites

　impacts, eg, airplane crashes, derailments

　flooding, eg, ruptured storage tank

could specific siting hazards be posed to the site from credible external forces such as high winds, earth movement, utility failure from outside sources, flooding, natural fires, and fog?

is there adequate access for emergency vehicles, eg, fire trucks? are access roads free of the possibility of being blocked by trains, highway congestion, spotting of rail cars, etc?

are access roads well engineered to avoid sharp curves? are traffic signs provided?

is vehicular traffic appropriately restricted from areas where pedestrians could be injured or equipment damaged?

are cooling towers located in such a way that fog generated by them will not be a hazard?

are the ends of horizontal vessels facing away from personnel areas?

is hydrocarbon-handling equipment located outdoors?

are pipe bridges located such that they are not over equipment, including control rooms and administration buildings?

is piping design adequate to withstand potential liquid loads?

*Location of firewater mains and backup, eg, diesel pumps*

are firewater mains easily accessible?

are firewater mains and pumps protected from overpressure and blast debris impact?

is an adequate water supply available for firefighting?

are the firehouse doors pointed away from the process area so that doors will not be damaged by an explosion overpressure?

*Location and adequacy of drains, spills, basins, dikes, and sewers*

are spill containments sloped away from process inventories and potential fire sources?

have precautions been taken to avoid open ditches, pits, sumps, or pockets where inert, toxic, or flammable vapors could collect?

are process sewers that transport hydrocarbons closed systems?

are concrete bulkheads, barricades, or beams installed to protect personnel and adjacent equipment from explosion or fire hazards?

are vehicle barriers installed to prevent impact to critical equipment adjacent to high traffic areas?

do drains empty to areas where material cannot pool?

can dikes hold the largest tank's capacity?

is there access in and out of dikes, pits, etc?

*Location of emergency stations (showers, respirators, personnel protective equipment, etc)*

are emergency stations easily accessible?

are first-aid stations prudently located and adequately equipped?

are safety showers heated, freeze-protected, and wind-protected?

is there a control room alarm for water flow from a safety shower and eyewash station (is there a need for such an alarm)?

*Planning*

what expansion or modification plans are there for the facility?

can the unit be built and maintained without lifting heavy items over operating equipment and piping?

**Table 4. (*Continued*)**

| | |
|---|---|
| are calculations, charts, and other documents available that verify facility siting has been considered in the unit layout? do these documents show that consideration has been given to:<br>  normal direction and velocity of wind<br>  atmospheric dispersion of gases and vapors<br>  estimated radiant heat density that might exist during a fire<br>  estimated overpressure<br>are appropriate security safeguards in place, eg, fences, guard stations?<br>are gates located away from the public roadway, so that the largest trucks can move completely off the roadway while waiting for the gates to be opened?<br>where applicable, are safeguards in place to protect high structures against low flying aircraft? | are adequate safeguards in place to protect employees against exposure to excessive noise, considering the cumulative effect of equipment items located close together?<br>is adequate emergency lighting provided? is there adequate redundant backup power for this lighting?<br>are procedures in place to restrict nonessential or untrained personnel from entering hazardous areas?<br>are indoor safety-control systems such as sprinklers and fire walls provided in buildings where personnel will frequently be located, such as control rooms and administrative buildings?<br>are evacuation plans (from buildings, units, etc) adequate and accessible to personnel?<br>are evacuation drills conducted routinely? |

[a]Ref. 3.

# BIBLIOGRAPHY

"Plant Location" in *ECT* 1st ed., Vol. 10, pp. 744–754, by R. S. Aries, R. S. Aries & Associates; in *ECT* 2nd ed., Vol. 15, pp. 700–720, L. V. Kaltenecker, Chemical Construction Corp.; in *ECT* 3rd ed., Vol. 18, pp. 44–59, by Conway Publications, Inc.

1. J. Charvdhury, *Chem. Eng.* (Feb. 1992).
2. J. E. Granger, *Chem. Eng.* (June 15, 1981).
3. S. J. Wallace and B. L. Hunter, *Hydrocarbon Process.* May 1994.

*General References*

W. B. Speir, "Choosing and Planning Industrial Sites" *Chem. Eng.* (Nov. 30, 1970).
D. V. Bierwert and F. A. Krone, "How to Find Best Site for New Plant" *Chem. Eng.* (Dec. 1995).
G. Ondrey, *Chem. Eng.* (2) (Apr. 1992).
G. Ondrey, *Chem. Eng.* (3) (Apr. 1992).
G. D. L. Morris, "Texas Looks to Building Downstream" *Chem. Week* (May 27–June 3, 1992).
G. D. L. Morris, "Louisiana! Bullish on Chemical Business" *Chem. Week* (May 27–June 3, 1992).
G. D. L. Morris, "Mississippi's Pro-Business Approach" *Chem. Week* (May 27–June 3, 1992).
P. Kemenzis, "Mobile Area is Holding Its Own" *Chem. Week* (May 27–June 3, 1992).

R. E. Maples and M. J. Hyland, "What Is Involved in Major Venture Financing" *Chem. Eng. Progr.* (Jan. 1980).

ROLAND E. MEISSNER III
The Ralph M. Parsons Company

# PLANT SAFETY

Health and safety of personnel and loss prevention are paramount concerns of the chemical industry. Injuries and property damage have high costs, not the least of which are business interruption and loss of trained personnel or equipment. Over the years, worker fatalities have decreased and lost time has leveled, but property losses have increased.

As of the early 1990s, annual worker fatalities ran about 9 per 100,000 employees; annual lost-time disabling injuries ran about 4,000 per 100,000 employees (1). Property losses increased fourfold from the 1970s (2). The trends in fatalities and property losses can probably be ascribed to the increasing complexity and productivity of the highly automated chemical plants, where personnel are isolated from processes. Whereas exposure to health and safety hazards may be reduced, the ability of experienced operating personnel to sense process problems and to correct these problems frequently is decreased. Another aspect of process management which has tended to increase hazards is the effort to reduce the formation of wastes and undesired by-products (see WASTE REDUCTION). This effort requires close approach to temperature and pressure limits, at which points loss of control can be catastrophic (see PROCESS CONTROL). Process and plant safety issues have been discussed (3–8).

Safety assessments of entire processes began with quantification of over-pressure potential and flammability hazards, by measurements of vapor pressure and of flash points and flammability limits, respectively (9–12). Process designers make use of data pertaining to reaction rates and energies for exothermic reactions and unstable chemicals: temperature limits beyond which explosive decompositions or other undesirable behavior can occur; rates of gas or vapor generation for proper design of emergency pressure-relief devices; recommended limits for exposures to toxic materials (13), radiation, noise, and heat; and strengths and corrosion rates of materials of construction (see NOISE POLLUTION AND ABATEMENT (SUPPLEMENT); CORROSION AND CORROSION CONTROL). The application of fault-tree analysis to chemical processes provides a means for quantitatively combining characteristics of process hazards with component and human failure rates to obtain a safety assessment of a process (14) (see HAZARD ANALYSIS AND RISK ASSESSMENT).

Many changes in the requirements for safety in the chemical and petrochemical industries arose in the 1980s and 1990s. Some of these changes were presented as consensus guidelines initiated by industry groups such as the Chemical Manufacturers Association (CMA) and the American Petroleum Institute (API); other changes were legislated by individual states and by the U.S. Government. The objective of these changes is to raise the design, operating, and maintenance standards of all members of these industries to as high a level as is economically possible.

## Pollution Prevention

The responsibility of chemical process managers for preventing air, water, and soil pollution has indirectly influenced plant safety by requiring better control of plant processes to prevent releases of hazardous materials. Regulatory legislation was introduced by the Health, Education, and Welfare Department (Health and Human Services) and the U.S. Environmental Protection Agency (EPA) to require (1) improvements in air quality (1955 Air Pollution Act; 1963 Clean Air Act and its amendments in 1970, 1977, and 1990; 1967 Air Quality Standards and National Air Pollution Acts; and 1970 National Environmental Policy Act); (2) better waste disposal practices (1965 Solid Waste Disposal Act; 1976 Resource Conservation and Recovery Act) (see WASTES, INDUSTRIAL; WASTE TREATMENT, HAZARDOUS WASTES); (3) reduced noise levels (1972 Noise Control Act) (see NOISE POLLUTION AND ABATEMENT (SUPPLEMENT)); (4) improved control of the manufacture and use of toxic materials (1976 Toxic Substances Control Act); and (5) assignment of responsibility to manufacturers for product safety (1972 Consumer Product Safety Act) (15,16).

## Process Safety Management

Several incidents occurring in the latter part of the twentieth century indicated that a significant improvement in the management of process hazards was needed. Those incidents which provoked the greatest industry and legislative response are listed in Table 1. Standards and guidelines, intended to improve the management of process safety, have since been developed and their implementation required. These standards and guidelines are presented herein in chronological order.

**Health and Safety at Work Act.** The Health and Safety Executive (HSE) in the United Kingdom was the leading authority among industrialized nations in establishing standards for process-hazards control (8). Starting in 1974, ie, following the explosion at Flixborough (see Table 1), and empowered by the Health

**Table 1. Process Incidents Resulting in Casualties**

| Description | Location | Date | Fatalities | Serious injury |
|---|---|---|---|---|
| ruptured bellows (vapor cloud explosion) | Flixborough, U.K. | June 1, 1974 | 28 | 89 |
| runaway reaction (toxic cloud) | Bhopal, India | Dec. 3, 1984 | ca 3,000 | ca 200,000 |
| valve opened in error (vapor cloud explosion) | Pasadena, Tex. | Oct. 23, 1989 | 23 | 232 |

and Safety at Work Act (17), the HSE required registration of processes according to the type or quantity of chemical used or produced. The HSE also submits recommendations concerning plant design and maintenance functions, operator training, and methods for evaluating process hazards (see PLANT LAYOUT; PLANT LOCATION).

**Occupational Safety and Health Act.**    Prior to 1985, the U.S. Occupational Safety and Health Administration (OSHA) generally limited its activity to enforcement of codes or consensus standards concerning toxicity and flammability; after 1985, OSHA began applying the "general duty" clause in its enabling legislation more broadly to management of hazardous chemical processes (18).

The Occupational Safety and Health Act (OSHA) enacted in 1970, establishes standards for several types of occupational hazards, including toxicity, noise, equipment guarding and protection against falling, and electrical shock (19). It also promulgates other consensus standards for exit facilities and fire and explosion control. Employers are obliged to keep records concerning occupational injuries and illnesses, and OSHA inspectors are authorized to inspect places of employment for violations of the standards as part of its program for inspection of high risk industries or as the result of an accident or employee complaint. Penalties are assessed for violations, and appeal procedures are established. Employers may apply for variances from the Act, based on equally effective protection, ie, a permanent variance, or inability to comply readily, ie, a temporary variance.

A more recent OSHA development has been the availability of personnel from OSHA and the National Institute for Occupational Safety and Health (NIOSH) for consultation to identify, evaluate, and correct workplace hazards (20).

**State Acts and Regulations.**    The New Jersey Toxic Catastrophe Prevention Act (NJTCPA) was developed following the Bhopal disaster, an incident in Institute, West Virginia in 1985, and several chemical-release incidents in New Jersey during 1986. This Act and its program regulations became effective in 1988 (21,22). A registration quantity is specified for each of the 109 materials listed in the regulations, based on attainment of an acute toxicity concentration at a distance of 100 m from a potential source of a 1-h release (23). This Act was readopted and amended, effective in 1993. This Act does not include the handling or storage of flammable or explosive materials, unless such materials are also toxic and are included in the lists of Extraordinarily Hazardous Substances (EHS).

A California statue requiring hazardous materials management was passed in 1985 (24), but guidance for compliance for industries covered by the Act was not issued until 1988 (25). A revised standard, which became effective in January of 1994 (26), applies to facilities handling any of 128 toxic materials; flammable liquids and gases in quantities of 10,000 lb (4.54 t) or more, except where used as fuel or in atmospheric pressure, ambient temperature tanks; and explosives.

In Delaware, the Regulation for the Management of Extremely Hazardous Substances Act, developed in response to the Bhopal disaster and several chemical-release incidents in Delaware, became effective in 1989 (27,28). The regulations list 88 toxic substances, 32 flammable substances, and 50 explosive substances. A sufficient quantity is specified for each of these materials, based on potential for a catastrophic event at a distance of 100 m from a potential source of a 1-h release.

The Nevada Chemical Catastrophe Prevention Act (NCCPA) was developed in response to the two incidents in the late 1980s which occurred in Nevada. This Act and its program regulations, enacted in 1991, became effective in 1992 (29,30). The legislation applies to facilities handling any of the 128 listed chemicals in quantities above a threshold quantity.

The state of Texas controls chemical process hazards through its statute that established the Texas Air Control Board in 1965. In 1985, guidelines were established for evaluations of community impact of releases of 46 toxic chemicals, if Texas decides that a disaster potential exists (26).

The state of Louisiana amended its air quality regulations (26) to incorporate requirements for chemical accident prevention, and several other states, including Michigan and New York, are considering process safety regulations.

**U.S. Community Right-to-Know Act.** In 1986, the U.S. Congress enacted the Emergency Planning and Community Right-to-Know Act, often called Title III of the Superfund Amendments and Reauthorization Act. This Act requires facilities to notify State Emergency Response Commissions when any of 366 (as of 1988 (31)) extremely hazardous substances are present in quantities at or above the threshold planning quantities. The Act also requires facilities to participate with local emergency planning committees to plan and prepare for chemical emergencies. Moreover, facilities are required to provide critical information on the identities, quantities, and on-site locations to the community.

**Chlorine Institute.** Members of the Chlorine Institute are required to reaffirm their commitment to the Institute's safety pledge, including prevention of chlorine releases. The safety pledge also includes annual safety audits, annual emission and hazard evaluations of chlorine operations, periodic emergency-response test drills, and coordination with local officials for protection of the community (32) (see ALKALI AND CHLORINE PRODUCTS).

**Chemical Manufacturers Association Code.** In 1988, the Chemical Manufacturers Association (CMA) adopted an initiative called Responsible Care: A Public Commitment (33). Members of the CMA commit themselves, as an obligation of membership, to improving performance in response to public concerns about the impact of chemicals on health, safety, and environmental quality.

One of the six elements of Responsible Care is the Codes of Management Practice, and one of the Codes is the Process Safety Code of Management Practices (34). This Code emphasizes management commitment and accountability, information sharing, and community relations, but also includes 11 of the 14 specific elements of the legislated process safety management standards developed later.

**American Petroleum Institute Recommended Practice.** In 1990, the American Petroleum Institute issued a recommended practice on Management of Process Hazards (35). The stated objective was to help prevent the occurrence of, or minimize the consequences of, catastrophic releases of toxic, flammable, or explosive materials.

This recommended practice is intended to apply to facilities that (1) handle or store flammable or explosive substances in such a manner that a release of ca 5 t of gas or vapor could occur in a few minutes and (2) handle toxic substances. The threshold quantity for the toxic materials would be determined using engineering judgment and dispersion modeling, based on a potential for serious danger as a result of exposures of $\leq 1$ h.

**Occupational Safety and Health Act.** *Protection of Employees.* In 1986, shortly after the Bhopal disaster, OSHA contracted to develop a federal standard on process hazards management. A proposed standard was issued in 1990, and the Process Safety Management of Highly Hazardous Chemicals standard was issued and implemented in 1992 (36).

The regulation lists 137 toxic and reactive substances and a threshold quantity for each. The regulation also applies to flammable liquids and gases in quantities of 10,000 lb or more ($\geq 4.5$ metric tons), except hydrocarbon fuels and liquids stored in unpressurized, ambient temperature tanks, as well as to the manufacture of any quantities of explosives (see EXLOSIVES AND PROPELLANTS) and pyrotechnics (qv).

There are 14 aspects required by this OSHA legislation:

| | |
|---|---|
| employee participation | mechanical integrity |
| process safety information | hot work (and other) permits |
| process hazards analysis | management of change |
| operating procedures | incident investigation |
| training | emergency planning and response |
| contractors | compliance audits |
| pre-start-up safety review | trade secrets |

Whereas no quantitative consequence analysis is required by this legislation, the process hazards analysis must include a qualitative evaluation of the possible effects of failure of controls on employees. Details concerning development and implementation of programs for these subjects are available (37–39).

**Clean Air Act Amendments.** *Protection of the Public.* In 1990, the Clean Air Act was amended to include Section 112(r), Prevention of Accidental Releases, and Section 304, Chemical Process Safety Management (26). Section 112(r) states that "operators producing, processing, handling, or storing extremely hazardous substances have a general duty to identify hazards which may result from releases and to design and maintain a safe facility." This section applies to facilities handling any of an initial list of at least 100 chemicals. Section 304 directed OSHA to regulate process safety management and listed the 14 elements of a safety standard required by the EPA legislation:

| | |
|---|---|
| hazard assessment | pre-start-up review |
| management system | management of change |
| process hazard analysis | safety audits |
| process safety information | accident investigation |
| standard operating procedures | emergency response program |
| training | risk management plan (report) |
| mechanical integrity | audits |

In response to the requirements of Section 112(r)(7)(B), the U.S. EPA developed a proposed standard for a Risk Management Program (40). This program was to be established by all stationary sources having a regulated substance present in a process in more than a threshold quantity. The list of regulated toxic substances and threshold quantities (41) contains 77 chemicals; the list of regulated flammable substances contains 63 chemicals. This standard also applies to the handling of explosive materials, by reference to the U.S. Department of Transportation requirements (42), and thus applies to the handling of 43 explosive chemicals, commercial explosives, fireworks, flares, igniters, ammunition, and ordinance having a threshold quantity of 5000 lbs (2270 kg).

The hazard assessment is to include identification of a worst-case scenario and other more likely scenarios for release of a regulated substance, and analyze the off-site consequences of such releases. The release and consequence assessment is to include the rate, duration, and quantity of the release, the distances for exposure or damage (using atmospheric, called "F" stability and a 1.5-m/s wind, and most-often-occurring conditions), populations that could be exposed, and environmental damage that could be expected.

## Process and Production Hazards

There are several steps which individuals and corporations can and should take to minimize the consequences of hazard occurrence, as well as risk to employees and the public. Much of the legislation has adopted a hierarchy of controls that essentially fall into two categories, ie, engineering and administrative (43,44).

Engineering controls may be subdivided into those providing inherent safety and those involving process equipment and conditions. Those providing inherent safety controls include (1) intensification, minimizing the amount of hazardous material or hazardous operations; (2) substitution, using safer materials or safer processing or production methods; and (3) isolation, barricading or distancing to minimize personnel exposure.

Process equipment and conditions controls are (1) containment, designing for plant and process integrity; (2) attenuation, using less severe operating conditions of pressures and temperatures; (3) consequence reduction, designing to minimize accidental release rates; (4) simplification, avoiding complexities in equipment and control systems; (5) use of passive safeguards, eg, relief devices, excess flow valves, and dikes; (6) use of active safeguards, eg, alarm and interlock systems, scrubbers, and remote valves; and (7) risk minimization, eg, ventilation, leak-stopping, spill control, and alerting systems.

The administrative controls are (1) operating procedures for startup, shutdown, response to upsets, and emergencies; (2) maintenance programs: maintaining plant integrity through inspections and testing; (3) process hazards analysis: maintaining and upgrading process integrity; (4) limiting personnel exposure, ie, limiting access and providing personal protective equipment; and (5) emergency procedures for escape and evacuation.

Engineering design of equipment and control of operating conditions is discussed herein.

**Chemical Hazards.** Chemical manufacturers and employees contend with various hazards inherent in production of even commonplace materials. For example, some catalysts used in the manufacture of polyethylene (see OLEFIN POLYMERS) ignite when exposed to air or explode if allowed to become too warm; the

Table 2. Properties of Chemicals[a]

| Chemical | Formula | Boiling point, °C | Toxicity properties, ppm | | |
| --- | --- | --- | --- | --- | --- |
| | | | Odor threshold | Threshold limit | IDLH[b] concentration |
| acetone | $C_3H_6O$ | 56 | 120 | 750 | 20,000 |
| acetylene | $C_2H_2$ | −84 | 620 | 2,500[d] | 2,500 |
| acrylic acid | $C_3H_4O_2$ | 141 | 0.1 | 2 | 750 |
| ammonia | $NH_3$ | −33 | 1.5 | 25 | 500 |
| benzene | $C_6H_6$ | 80 | 11 | 10[g] | 2,000 |
| carbon disulfide | $CS_2$ | 46 | 0.25 | 10 | 500 |
| chlorine | $Cl_2$ | −34 | 0.2 | 0.5 | 25 |
| diethyl ether | $C_4H_{10}O$ | 35 | 0.6 | 400 | 19,000 |
| ethyl alcohol | $C_2H_6O$ | 78 | 30 | 1,000 | |
| ethylene | $C_2H_4$ | −104 | 1,000 | 2,700[d] | |
| hydrogen | $H_2$ | −253 | no | 4,000[d] | |
| hydrogen chloride | HCl | −85 | 12 | 5[i] | 100 |
| hydrogen cyanide | HCN | 26 | 2 | 10[j] | 55 |
| hydrogen sulfide | $H_2S$ | −61 | 0.002 | 10 | 300 |
| methane | $CH_4$ | −161 | no | 5,000[d] | |
| methyl alcohol | $CH_4O$ | 65 | 450 | 200 | 25,000 |
| phosgene | $COCl_2$ | 8 | 0.7 | 0.1 | 2 |
| propane | $C_3H_8$ | −42 | 4,500 | 2,100[d] | 20,000 |
| styrene | $C_8H_8$ | 145 | 3 | 50[m] | 5,000 |
| sulfur dioxide | $SO_2$ | −10 | 1.5 | 2 | 100 |
| vinyl chloride | $C_2H_3Cl$ | −14 | | 5[n] | 5 |

basic ingredient in fluorocarbon polymers, eg, Teflon (see FLUORINE COMPOUNDS, ORGANIC), can become violently self-reactive if overheated or contaminated with caustic substances (45,46); one of the raw materials for the manufacture of acrylic fibers (see FIBERS, ACRYLIC) is the highly toxic hydrogen cyanide (see CYANIDES).

Table 2 lists some of the physical, toxicity, flammability, and reactivity properties of common chemicals (10,13,42,45–51). Also given are some of the quantities specified for reporting spills and for compliance with legislated requirements. The OSHA regulations require that material safety data sheets (MSDS) be developed for all process materials, so that the hazard data can be communicated to employees (52). Characteristics of toxicity, flammability, chemical instability, reactivity and reaction energy, operating conditions, and corrosive properties of construction materials must all be considered in analyzing hazard potentials of chemicals and chemical operations.

*Toxic Materials.* Individuals can come in contact with materials by ingestion, inhalation, skin irritation, skin absorption, and subcutaneous injection (53) (see INDUSTRIAL HYGIENE). If employees are careless about washing before eating, chemical substances on the hands may be ingested after contaminating food, cigarettes, or other materials. For most substances, a single event does not cause serious effects. The continuous daily ingestion of small amounts may, however,

**Table 2. (*Continued*)**

| Flash point, °C | Flammability properties | | NFPA reactivity[c] | Reportable spill, kg |
|---|---|---|---|---|
| | Lower limit, vol % | Autoignition temperature, °C | | |
| −20 | 2.5 | 465 | 0 | 2,270 |
| e | 2.5 | 305 | 3 | |
| 50 | 2.4 | 438 | 2 | 2,270 |
| e | 15 | 651 | 0 | 45[f] |
| −11 | 1.2 | 498 | 0 | 4.5 |
| −30 | 1.3 | 90 | 0 | 45 |
| | | | 0 | 4.5[h] |
| −45 | 1.9 | 180 | 1 | 45 |
| 13 | 3.3 | 363 | 0 | |
| e | 2.7 | 450 | 2 | |
| e | 4.0 | 500 | 0 | |
| | | | 0 | 2,270 |
| −18 | 5.6 | 538 | 2 | 4.5[k] |
| e | 4.0 | 260 | 0 | 45[h] |
| e | 5.0 | 537 | 0 | |
| 11 | 6.0 | 464 | 0 | 2,270 |
| | | | 1 | 4.5[l] |
| e | 2.1 | 450 | 0 | |
| 31 | 0.9 | 490 | 2 | 450 |
| | | | 0 | k |
| −78 | 3.6 | 472 | 1 | 0.45 |

[a]Refs. 10,13,42,45–51.   [b]Immediately Dangerous to Life and Health (IDLH). Not established for acrylic acid; value shown is the emergency response-planning guide-3 (ERPG-3) for exposure up to one hour without life-threatening effects.   [c]National Fire Protection Association (NFPA). The NFPA reactivity stability range is integral from 0 (stable) to 4 (unstable).   [d]Simple asphyxiant; value shown is 10% of the lower flammable limit (LFL).   [e]Gaseous material.   [f]Material has an OSHA threshold of 10,000 lbs (4,500 kg).   [g]Suspected human carcinogen; exposures should be carefully limited to levels as low as reasonably achievable below the TLV.   [h]OSHA threshold of 1500 lbs (680 kg).   [i]Ceiling value; concentration which should not be exceeded; not appropriate to use a time-weighted average.   [j]Ceiling value and skin absorption; special measures to prevent significant cutaneous absorption may be required.   [k]OSHA threshold of 1000 lbs (450 kg).   [l]OSHA threshold of 100 lbs (45 kg).   [m]Skin absorption; special measures to prevent significant cutaneous absorption may be required.   [n]Confirmed human carcinogen; exposures should be carefully limited to levels as low as reasonably achievable below the TLV.

result in harmful accumulation. Inhalation of gases and vapors also may cause toxicological effects. Also, toxic dusts such as beryllium (see BERYLLIUM AND BERYLLIUM ALLOYS), silica (see SILICA COMPOUNDS), and lead compounds (qv), can cause acute or chronic symptoms (54).

Strong acids and strong alkalies can severely burn the skin, chromium compounds can produce skin rashes, and repeated exposure to solvents causes removal of natural oils from the skin. Infection is always a concern for damaged skin. Absorption through the skin is possible for materials that are appreciably soluble in both water and oil, eg, nitrobenzene, aniline, and tetraethyllead. Other materials can be absorbed if first dissolved in extremely good solvents, eg, dimethyl sulfoxide. Subcutaneous injection can occur accidentally by direct exposure of the circulatory system to a chemical by means of a cut or scratch or inadvertent penetration of the skin with a hypodermic needle.

The eyes are particularly susceptible to liquids, gases, and some solids. The conjunctival membrane surrounding the eye is easily irritated. Alkaline materials, eg, hydroxides and amines, destroy the eye tissues rapidly, frequently causing partial or complete loss of vision.

*Physiological Classifications of Contaminants.* The physiological classification of air contaminants is difficult, because the type of action of many gases and vapors depends on concentrations (55). For example, a vapor at one concentration may exert its principal effect as an anesthetic but, at a lower concentration, the same vapor may injure the nervous system, the hematopoietic (blood-forming) system, or some visceral organ (see AESTHETICS; TOXICOLOGY).

*Irritants.* Irritant materials are corrosive or vesicant, ie, cause blisters, and may inflame moist or mucous surfaces. These have essentially the same effect on animals as on humans. The concentration is far more significant than the duration of exposure. Some representative irritants, eg, aldehydes (qv), alkaline dusts and mists, ammonia (qv), hydrogen chloride (qv), hydrogen fluoride, sulfur dioxide, and sulfur trioxide, chiefly affect the upper respiratory tract (51). Other irritants, eg, bromine, chlorine, dimethyl sulfate, fluorine (qv), ozone (qv), sulfur chlorides, and phosphorus chlorides, affect the upper respiratory tract and lung tissues. Irritants that primarily affect terminal respiratory passages and air sacs include arsenic trichloride, nitrogen oxides, and phosgene (qv). Lung irritants are similar to the chemical asphyxiants in that the effects frequently result in asphyxial death.

*Asphyxiants.* Asphyxiants interfere with oxygenation of tissues and may be classified as simple or chemical. Simple asphyxiants are physiologically inert gases that act principally by dilution of atmospheric oxygen below the partial pressure required to maintain an oxygen saturation of the blood sufficient for normal tissue respiration. These include ethane, helium, hydrogen, methane, nitrogen, and nitrous oxide. Chemical asphyxiants either prevent the blood from transporting oxygen from the lungs or prevent normal oxygenation of the tissues even if the blood is well oxygenated. Among the chemical asphyxiants, carbon monoxide combines with hemoglobin; cyanogen, hydrogen cyanide, and nitriles inhibit tissue oxidation by combining with cellular catalysts; and aniline, $N$-methylaniline, $N,N$-dimethylaniline, and toluidine cause formation of methemoglobin. Nitrobenzene, characterized by the nitrite effect, also causes methemoglobin formation, lowers blood pressure, and disturbs breathing. Hydrogen sulfide causes olfactory and respiratory paralysis.

*Anesthetics and Narcotics.* Anesthetics (qv) and narcotics exert their principal action as painkillers without seriously affecting systemic processes. Their depressant action on the central nervous system is governed by their partial pressure in the blood supply to the brain (see PSYCHOPHARMACOLOGICAL AGENTS). In the order of decreasing anesthetic action are acetylenic hydrocarbons, olefins, ethyl ether and isopropyl ether, paraffins, aliphatic ketones, and aliphatic alcohols.

*Systemic Poisons.* Some systemic poisons, the majority of which are halogenated hydrocarbons, cause organic injury to one or more of the visceral organs. Benzene, phenols, and, to some degree, toluene, xylene, and naphthalene, damage the hematopoietic (blood-forming) system. Nerve poisons include carbon disulfide, methanol, and thiophene. Some of the toxic metals are lead, mercury, cadmium, antimony, manganese, and beryllium. Toxic nonmetal inorganics include fluorides and arsenic, phosphorus, selenium, and sulfur compounds.

*Particulate Matter Other Than Systemic Poisons.* Silica and asbestos dust produce fibrosis. Silicon carbide, carbon (other than exhaust emissions), and emery are inert dusts. Many organic dusts, eg, pollen, wood, and resins, cause allergic reactions. Acids, alkalies, fluorides, and chromates are irritants.

*Carcinogens.* Special rules for 17 specific carcinogenic materials have been formulated by OSHA (52). Cancers are thought to be the result of changes in or damage to the deoxyribonucleic acid (DNA) material in the chromosomes (see NUCLEIC ACIDS). Carcinogenic materials include nitrogen mustards and other direct alkylating agents and some forms of electromagnetic and atomic particle radiation. It appears that metabolic activation may be needed before most carcinogens, eg, benzopyrene or vinyl chloride, are effective. Some compounds act as promoters that speed the development of tumors, and other substances, eg, asbestos (qv), appear to act through physical damage to the cells.

Reproductive-hazard materials act by (1) reducing the amount or viability of sperm, eg, as spermatotoxins, eg, dibromochloropropane; (2) by crossing the placenta and thereby injuring the developing embryo, eg, as fetotoxins, or by altering the development of the embryo, eg, as teratogens, such as thalidomide; or (3) by damaging germ cells so that faulty sperm or ova are produced, eg, as mutagens. Several governmental agencies are formulating rules to protect workers against these hazards.

The long latent periods involved in development of cancers make correlation of chemical exposures and disease extremely difficult. This can be countered partly with tests on naturally short-lived animals. Tests on bacteria, eg, the Ames test, may permit rapid detection of cancer potential, although there is no direct relationship between the results of bacterial tests and the effects of the tested chemicals on humans (56).

*Threshold Limit Value.* The American National Standards Institute (ANSI) has published standards regarding the maximum acceptable concentration for certain gases and vapors in the air at work locations. A list of threshold limit values (TLVs), published annually by the American Conference of Governmental Industrial Hygienists (ACGIH), provides the concentrations of dust, mist, or vapor believed to be harmless to most workers when exposed for 5 8-h days per week (13). The 1970 TLVs were adopted by OSHA as a consensus standard for time-weighted averages (TWAs) or ceiling limits. The National Institute for Occupational Safety and Health (NIOSH) has documented concurrence with some of these values or has recommended different and usually lower values in a few cases (57). The American Industrial Hygiene Association (AIHA) has carefully evaluated the effect of several toxic vapors (47 as of 1995 (58)) and has developed Emergency response planning guides (ERPGs).

*Control of Exposure Potential.* Exposure to toxic materials can be controlled by a number of methods, eg, substitution, removal, enclosure, and personal protection. The best method of protecting workers is by substituting a less toxic material for a more toxic substance having equal effectiveness, eg, the use of 1,1,1-trichloroethane for carbon tetrachloride, and toluene for benzene. Ventilation at the work location is much more effective in removing undesirable contaminants than general room ventilation (59). Suitable exhaust hoods or flexible ducts should be utilized to draw off contaminated air as near to the point of chemical release as is feasible. Some operations can be completely enclosed, eg, continuous processing in contrast to batch operation, where process vessels are opened occasionally. Personal protection, the last point of defense, sometimes is the only way in which a worker can be protected from exposure. Such protection includes a hard hat, face shield or goggles, apron, coat, pants, and boots or rubber shoes. Respiratory protection may be provided by a dust respirator, canister gas mask (if sufficient oxygen is always present), self-contained breathing equipment, or airline respirators (53). Recent OSHA requirements for breathing apparatus specify special fit-testing of masks and positive pressure face masks, so that any air contaminants present do not leak inward through gaps at the edges of the mask.

Specially designed impervious suits, eg, Level A suits, are utilized by workers handling some rocket fuels and other highly hazardous compounds (see EXPLOSIVES AND PROPELLANTS). Barrier creams are much less effective than gloves for preventing skin contact.

Where it is necessary to use known cancer-causing substances in industry, OSHA has promulgated rigorous standards for regulated areas, that include and surround the place of use (52). These standards include analysis of processes for adequate engineering controls to prevent releases, strict control of access to potentially hazardous facilities, and high standards for ventilation and personal protective equipment, special work practices, training, health monitoring of employees, and control of used clothing and waste disposal.

*Flammability.* Engineering and operational controls are usually effective in preventing fires involving flammable materials. Modern continuous processing is characterized by retention of such materials in closed systems, thus preventing access to air or ignition sources. However, in batch processing or under some emergency conditions, flammable materials may be released and can be ignited. The basic method of fire prevention is to avoid situations in which flammable materials, air, and ignition sources are in the same place at the same time. The pertinent properties describing the fire hazard of a flammable material have been defined by the National Fire Protection Association (NFPA) (60).

The flash point is a measured temperature at which vapors above the surface of a liquid are just sufficiently concentrated to propagate a flame (10). In practice, materials of concern may be in closed or open containers or may have spilled. Generally, the chosen flash point method should be related to the problem as well as to the type of material; ie, open-cup methods are more significant for open containers or spills, whereas closed-cup methods give more significant information for closed containers, eg, process vessels. A number of commercial flammable liquids contain a moderate amount of noncombustible components, eg, chlorinated hydrocarbons, in order to elevate the closed-cup flash point and thus gain a more favorable classification. When the same material is analyzed by an open-cup method, the flash point is not elevated, ie, after a spill, the noncombustible material would soon be lost and the residue may be highly flammable.

The regulations of the U.S. Department of Transportation (DOT) for the shipping of hazardous materials specify the use of the Tag open-cup test (ASTM D1310) and the Tag closed-cup apparatus (ASTM D56) (9). The Tag closed-cup method is used for liquids with flash points below 80°C, except No. 4 and heavier fuel oils; the Tag open-cup method is for liquids with flash points of −18 to 163°C. Because the flash points of many petroleum products exceed 163°C, the Cleveland open-cup method (ASTM D92) is used for all petroleum products, except fuel oils, with flash points above 80°C. Many fuel oils and other mixtures are excluded from the scope of the above three methods; such materials usually are tested in the Pensky Martin closed-cup test (ASTM D93), which is intended for fuel oils, lubricating oils, viscous materials, and suspensions of solids having flash points of −7 to 370°C. Two other methods that have been developed are for drying oils (ASTM D1393) and for waxes and similar products (ASTM D1437). The test methods are revised frequently by the ASTM.

The ignition temperature or autoignition temperature is the minimum temperature of a flammable mixture that is required to initiate or cause self-sustained combustion without ignition from an external source of energy such as a spark or flame (ASTM D2155).

The lower flammable limit (LFL) or lower explosive limit (LEL) is the minimum concentration of vapor in air below which a flame is not propagated when an ignition source is present (61–64). Below this concentration, the mixture is considered too lean to burn. The lower flammable limit and the flash point of a flammable liquid are closely related by the liquid's vapor pressure characteristics.

The upper flammable limit (UFL) or upper explosive limit (UEL) is the maximum vapor concentration in air at which a flame can propagate. Above this concentration, the mixture is too rich to burn, ie, the oxygen is consumed in the combustion of one particle and there is insufficient oxygen to burn the adjacent particle of fuel. Products of combustion surrounding the first particle tend to quench the flame. The flammable range or explosive range consists of all concentrations between the lower flammable limit and the upper flammable limit. Flammable limits usually refer to flowing materials; explosive limits usually refer to confined or stagnant mixtures. Values for LFL and LEL, or UFL and UEL, are identical.

The stoichiometric concentration is that mixture of fuel and oxidant, usually air, that produces fully oxidized combustion products, chiefly water and carbon dioxide, following ignition. Such mixtures typically are the most easily ignited and produce the highest temperature and pressure at the greatest rates. The limiting oxygen concentration is that concentration below which combustion, usually in air diluted with an inert gas such as nitrogen or carbon dioxide, does not propagate in a mixture of gases or vapors.

The characteristics of flammable and combustible materials can be displayed on a ternary (triangular) graph. The characteristics of several common flammable and combustible materials are presented in Table 2.

Water solubility sometimes is important in determining whether water can be used to dilute or flush away flammable liquids. However, a water solution of flammables can give off sufficient vapors to burn, eg, a 30 vol % solution of ethyl alcohol in water (60 proof) has a flash point which is only 16.6°C above that of pure ethyl alcohol (29.4°C vs 12.8°C).

*Storage of Flammable Materials.* The preferred storage for flammable liquids or gases is in properly designed tanks. Floating roof tanks frequently are used in the petroleum industry for flammable crudes and products (see TANKS AND PRESSURE VESSELS). The vents on cone roof tanks should either be equipped with flame arrestors or the vapor space above the contents should be inerted with a nonflammable gas or vapor, unless the flash point is well above the maximum ambient temperature, the contents are not heated above the flash point, and the tank is not exposed to other tanks containing flammable liquids.

Flammable materials in drums should be stored away from processing and operations buildings and should be protected by sprinkler systems or other automatic fire-extinguishing devices. High vapor pressure materials should not be subjected to summer sunlight temperatures. The transfer of these materials from tanks or plant streams into drums or other containers and from drums into smaller vessels has resulted in many fires. Bonding lines must be used to equalize charges and thus prevent ignition from discharge of static electricity. Also, because static electricity can be generated by free fall of liquid, lines into vessels should discharge below the surface or with no more than a 15-cm free fall to the bottom of the tank.

Flammable liquids used in indoor workplaces and laboratories should be contained in approved safety cans having self-closing spouts and flash arrestors whenever possible. Other flammable or combustible materials may be kept safely in metal containers fitted with fusible-link automatic closures. Ventilation of indoor areas where flammable vapors may be present can greatly reduce the probability of fire.

*Plant Fireproofing.* There is a growing practice in the chemical industry of locating principal equipment out of doors and to enclose only a control room where all instruments and control equipment are centered. The control room should be resistant to potential explosion, fire, and toxicity hazards of processes in the vicinity. Prompt and orderly shutdown of processes following a serious incident is essential in order to minimize personnel-injury and property-loss hazards (65,66).

Steel structures should be protected by approved fireproofing treatment, eg, concrete or insulating, ie, intumescent or ablative materials; untreated steel should be protected by some method of cooling, eg, a water-spray system.

Electrical equipment which is installed or used in areas where there is a fire hazard should be in accordance with the NFPA National Electrical Code, eg, the equipment should be characterized by explosion-proof motors, switchgear, lights, wiring, instrumentation, etc (67). However, it may be more practical to enclose and ventilate such equipment, particularly instrument panels, when a dependable source of clean air is available. Intrinsically safe equipment which uses extremely low voltages and currents may be used in some hazardous locations because any sparks produced do not have enough energy to ignite vapors (68). This equipment is much less expensive than instruments in explosion-proof housings.

Waste facilities should be designed to prevent explosions in sewer systems and typically are comprised of suitable traps, vents, clean-outs, collecting chambers, etc. Flammable gas detectors are installed in sewers to warn of hazardous concentrations, and inert gas blanketing of closed process sumps generally is advisable.

*Vapor Cloud Explosions.* The Flixborough, England, disaster in 1974 demonstrated the potential hazards of flammable vapor releases. In this incident, tons of liquid cyclohexane heated above its atmospheric-pressure boiling point escaped through a damaged expansion bellows. Delayed ignition at a furnace caused a violent explosion that demolished the existing plant and damaged many of the surrounding buildings and residences; 28 fatalities resulted, including 26 in the nearby control building. There were no fatalities away from the plant. More recent flammable fluid releases and vapor cloud explosions occurred in Norco, Louisiana, on May 5, 1988 (7 fatalities); and in Pasadena, Texas, on October 23, 1989 (23 fatalities) (69).

Theories are being developed to account for the high flame speeds that occur in vapor cloud explosions and to explain the blast effects, which differ from those of high explosives in that the far-field damage indicates a higher trinitrotoluene (TNT) equivalent than the damage close to the explosion center (70). Where flammable liquids are processed at temperatures above their boiling points and thus at high pressure, special precautions are required: limiting the storage quantities and flow rates of such materials; providing isolation valves that are operable from the control room, to limit the quantity released in an incident; providing a system that will allow remotely controlled deinventorying to a vent stack or flare; isolating the process by placing it at a distance from occupied buildings; providing blast resistance for occupied buildings; eliminating ignition sources close to potential release points and providing facilities for prompt shutdown of ignition sources downwind from the release; and providing alarms actuated by flammable vapor detectors to alert personnel in local and downwind areas.

*Boiling Liquid Expanding Vapor Explosion.* A phenomenon that has developed upon use of increasingly large storage tanks and railroad tank cars is violent rupture of containers caused by physical expansion of superheated liquid (71). Although this can occur with nonflammable liquids, it is more likely that a container of flammable liquid is involved. Usually a boiling liquid expanding vapor explosion (BLEVE) is caused by heat from surrounding fire or flame from a relief device or a leak at the top of the tank. As the liquid level decreases, direct flame impingement or conduction to the wall area surrounding the vapor phase causes this portion of the vessel to weaken and fail (72). The result is release of the hot, compressed vapor from the vapor space and, in some materials, an explosive flash vaporization of the superheated liquid remaining in the tank. For flammable liquids, the further result is, in addition to the shrapnel and blast effects (73), an extremely dangerous fireball (74). Cooling the container walls with water with an unattended or remote-control water cannon or deluge gun probably is the only practical countermeasure.

*Reactions.* Certain reactions are difficult to control, particularly if there is a failure of instrumentation or cooling water, agitation, or other such systems. The conditions of many chemical reactions often are extreme, eg, temperatures well above 500°C or approaching absolute zero. Changes in metal properties at these temperatures can be dramatic. Ordinarily, steel loses much of its tensile strength with increase in temperature and becomes quite brittle at moderately low temperatures. Stainless steel or other special alloys may be used at elevated pressures, and nonferrous materials usually are needed for very low temperature conditions. Pressures range from high vacuum to megapascals in both laboratory and commercial reactions. Choice of metals and proper design commonly are in accordance with standards (75). Stresses, eg, from temperature cycling, vibrations, load bearing, wear, and earthquakes, also must be considered.

Reaction rates typically are strongly affected by temperature (76,77), usually according to the Arrhenius exponential relationship. However, side reactions, catalytic or equilibrium effects, mass-transfer limitations in heterogeneous (multiphase) reactions, and formation of intermediates may produce unusual behavior (76,77). Proposed or existing reactions should be examined carefully for possible intermediate or side reactions, and the kinetics of these side reactions also should be observed and understood.

Laboratory or pilot plant work does not always provide accurate prediction of a reaction hazard (78). Impure raw materials may replace pure chemicals. Temperature gradients may be quite appreciable in large equipment and local temperatures much higher than indicated by sensing devices. The ratio of cooling surface to reaction mass may be much smaller, and processing times often are longer. The use of metal instead of glass apparatus may exert profound effects upon the course of the reaction. The very mass of material may change the ratio of undesired to desired products.

Certain molecular groups, eg, nitrates; primary and secondary nitramines; aliphatic and aromatic nitro compounds; and organic salts of perchlorates, chlorates, picrates, nitrates, bromates, chlorites, and iodates have explosion potential (79). Less powerful but often more sensitive compounds contain azides, nitroso groups, diazo groups, diazosulfides, peroxides, haloamines, and acetylides. The presence of one or more of these groups in a molecule indicates compound instability. Another safety evaluation is consideration of the oxygen balance if the molecule contains combined oxygen. The closer the molecule approaches self-satisfaction by oxidation, the more powerful the explosive. For example, three nitro groups on an aromatic ring, eg, in trinitrotoluene and trinitrophenol (picric acid), provide a source of readily available oxygen for the rapid burning or explosion of the carbon and hydrogen in the molecule.

It is often difficult to decide when exhaustive and expensive investigations should be undertaken to develop safety data for a new compound. Much depends on the amount of compound available, the forecast for production, and the end use. Several instruments are available to assess quantitatively the stability of chemicals as functions of temperature, time, and pressure (9,63). They are based primarily on heat effects, eg, vent-sizing package (vsp), accelerating rate calorimeters (arc), differential thermal analyzers (dta), and differential scanning calorimeters (dsc), or on weight loss, eg, thermogravimetric analyzers (tga). Some of these devices can be used to analyze the thermal behavior of mixtures, including reaction masses.

A process plant typically consists of a charging system, a reactor system, and an outlet or product system. All reactants must be introduced in the correct order and the reaction must be well understood. Provision should be made for controlling instability or excessive pressure or temperature. Highly viscous materials, or those in which solids are present, may cause fouling, poor agitation, and local overheating, with possible decomposition. Some hazardous reactions must be blanketed with an inert gas for safety and quality control. In any reaction system, relief devices should be installed with well-designed vent systems. For extremely hazardous processes, it may be necessary to provide emergency dumping, dilution, or other emergency controls. Withdrawal or removal of products from closed-system operations can be hazardous. Such locations should be monitored with suitable sensing and alarm devices.

Hazard identification of the contents of in-plant bulk storage tanks, warehouses, etc, may be achieved by a system developed by the NFPA (48). The system makes use of three diamond-shaped areas, which are marked with numbers 0, 1, 2, 3, or 4 indicating increasing hazards of toxicity, flammability, and reactivity, respectively.

*Combustion.* The burning of solid, liquid, and gaseous fuels as a source of energy is very common. Using sufficient and reliable combustion controls, this process seldom causes serious problems. However, some combustion processes are deliberately carried out with an inadequate oxygen supply in order to obtain products of incomplete combustion. Explosive mixtures sometimes occur, and then flashback is a serious problem.

*Oxidation.* There are 10 types of oxidative reactions in use industrially (80). Safe reactions depend on limiting the concentration of oxidizing agents or oxidants, or on low temperature. The following should be used with extreme caution: salts of permanganic acid; hypochlorous acid and salts; sodium chlorite and chlorine dioxide; all chlorates; all peroxides, particularly organic peroxides; nitric acid and nitrogen tetroxide; and ozone (qv).

*Nitration.* All nitration reactions are potentially hazardous because of the explosive nature of the products and the strong oxidizing tendency, characteristic of the nitrating agent. The nitration reaction and the oxidation side reaction are highly exothermic. Therefore, these reactions may be extremely rapid and become uncontrollable. Close temperature control must be maintained (80,81). Sensitivity is enhanced by the presence of impurities, and rapid autocatalytic decompositions, ie, fume-offs, may be violent (see NITRATION).

*Halogenation.* Heats of reaction are highly exothermic for halogens, particularly fluorine (qv), and chain reactions can result in explosions over broad concentration ranges. Halogens also present severely challenging corrosion problems (see CORROSION AND CORROSION CONTROL).

*Hydrogenation.* Except for the difficulties of using hydrogen under very high pressures and at moderately high temperatures, hydrogen reactions are not particularly hazardous. Moderately exothermic, uncontrollable conditions are rarely encountered, except where hydroxylamine intermediates can be formed.

*Polymerization.* Chain reactions proceed quickly following slow initiation. Heat effects can be sudden, especially where catalysts are used, and may become uncontrollable, particularly as the visocosity of the reaction mixture increases (82). In reactors intended to contain exothermic reactions, a maximum differential of 10°C should be maintained between the reactants and the cooling surface. Sufficient surface cooling and auxiliary cooling, eg, refluxing liquid, should be provided. An adequate, dependable supply of coolant is essential for control of exothermic reactions. Instrumentation for control of chemical processes is extremely complex as temperature and pressure limits are approached. However, most instrument-control systems are designed to "fail safe" upon failure of the computer, electric power, or the instrument air supply.

*Corrosion.* Proper attention should be given to the corrosion of all chemical processing equipment, both internal and external. Special alloys that are resistant to stress corrosion induced by chloride ions in stainless steel vessels should be employed where necessary (50,83). Hydrogen can induce similar stress corrosion cracking and blistering. Lining vessels with glass preserves the purity and color of the product and prevents corrosion, which weakens the process vessel. Fumes such as hydrogen chloride can cause stress corrosion on external piping, valves, and other critical pieces of equipment. Periodic inspection and testing of all process vessels is an essential aspect of preventive maintenance (qv).

## Design of Facilities

**Plant Site and Layout.** The choice of a location for a chemical plant depends on a number of factors, including effects on plant personnel and the surrounding community (see PLANT LAYOUT; PLANT LOCATION). The assessment of hazards, based on the flammability of materials, reaction energy, and presence of highly toxic materials, is important (66,84). Consideration is given to possible effects on plant personnel and the community from the worst possible incident (see HAZARD ANALYSIS AND RISK ASSESSMENT). An adequate water supply for process cooling and fire fighting is a vital necessity. Prevailing winds should also be considered.

Open areas around the operating units of a plant act as buffers to the surrounding community. Sufficient clearance should be allowed so that, if tall structures collapse, other on-site buildings or equipment, or off-site properties are not affected. Adequate roadways providing entry to the plant are extremely important, and multiple entries and exits are advisable. An overcrowded plant can lead to damage or shutdown of adjacent units and may impede the movement of vehicles and materials in case of emergency (85). Another consideration is community fire-fighting assistance, first aid, and medical facilities.

Hazard potentials should always be segregated from nonhazardous operations such as offices, laboratories, and warehouses. Operating units having few personnel should be separated from highly populated offices and laboratories to reduce the victim toll in case of emergency. When administrative facilities are located on the periphery of the plant, visitors are less likely to be exposed to operational dangers. Tank-car and tank-truck loading and unloading facilities should be adequately separated from other operating areas as well. Elevated flares or ground-level burning pits must be carefully located and designed to minimize the possibility of igniting flammables in case of spills. Tankage areas should be segregated. Adequate roadways should surround every process unit and principal building for access of maintenance and construction vehicles and fire-protection equipment.

The practice of building single-line processing units as compared to small parallel facilities is growing rapidly. Although the latter may be somewhat more expensive to build and may require somewhat greater manpower to operate, the possibility of a costly total shutdown resulting from a disaster is much less likely to occur. Physical separation of such parallel lines is essential to prevent transmission of explosive or other effects to neighboring units. Some attention should be given to fire and business interruption insurance programs when a choice is being made between a large single-line plant and multiple units.

**Utilities.** *Services and Facilities.* Principal electric power lines run underground reduce the probability of damage from exterior causes. Transformer stations and switchgear need be accessible only to authorized personnel. Repair work on electrical circuits should occur only when circuits are not energized. No work on active lines should be permitted if it is reasonably possible to arrange for shutdown. If shutdown is not possible, work should commence only upon written direction, signed by the top level of management.

Each plant or laboratory should adopt definite rules and procedures for electrical installations and work. All installations should be in accordance with the National Electrical Code (NEC) for the type of hazard, eg, Class I: flammable gas or vapor; Class II: organic, metallic, or conductive dusts; and Class III: combustible fibers; and the degree of process containment, eg, Division 1: open; and Division 2: closed (67). Regardless of the flammability of the materials in the installed operations, changes in procedure involving use of such materials often occur, sometimes without concurrent alteration of the electrical installation.

Generally, it is more economical to prevent explosive atmospheres in rooms than to try to provide explosion-proof electrical equipment. Personnel should never be allowed to work in a hazardous atmosphere. Where such an atmosphere cannot be avoided through control of flammable liquids, gases, and dusts, access to the area involved should be limited and the area segregated by hoods or special ventilation. Electrical equipment on open, outdoor structures more than 8 m above-ground usually is considered free from exposure to more than temporary, local explosive mixtures near leaks (86). Electrical equipment should be grounded to protect personnel from shock hazards and to prevent extraneous sparking, ground currents, and electrical fault heating. Portable tools should be grounded or double-insulated. All electric motors, appliances, lighting fixtures, and other electrical equipment should be similarly grounded by internal wiring or external wire or cables (87).

All steel buildings and outdoor structures and all tanks, drums, open-end hoses, tank cars, trucks, and chemical equipment associated with the handling or use of flammable liquids or gases or in areas where these are handled should be grounded according to the NEC (67). Flexible grounds should be connected to large water pipes or driven grounds but never to electrical conduits, branch sprinkler lines, or gas, steam, or process piping. These grounds should be properly maintained and the electrical resistance to ground should be measured periodically. Lightning protection is achieved through the use of large grounding conductors.

The name, number, and voltage of equipment controlled should be clearly marked on all switch boxes, compensators, and starters. Pins and chains should be required on all butterfly switches. Extension cords should be three-wire and limited to 8-m lengths.

*Water.* Water mains should be connected to plant fire mains at two or more points, so that a sufficient water supply can be delivered in case of emergency. The plant loop and its branches should be adequately valved so that a break can be isolated without affecting a principal part of the system. If there is any question of maintaining adequate pressure, suitable booster pumps should be installed. Any connection made to potable water for process water or cooling water must be made in such a manner that there can be no backflow of possibly contaminated water; check valves alone are not sufficient. The municipal supply should fall freely into a tank from which the water is pumped for process purposes, or commercially available and approved backflow preventers should be used.

For large plants located at natural water sources, special water mains can be used to supply untreated water for emergency use as well as for cooling, process water, or general plant uses other than human consumption and emergency washing facilities. Extreme care must be taken to prohibit any cross-connection of the two systems. Untreated water usually is supplied by pumps (qv) which, preferably, are driven by two unrelated sources of energy, eg, steam (qv) pumps may be installed parallel to diesel pumps. If the equipment is only on standby use, it should be run frequently enough to make sure that it operates when needed.

*Compressed Air.* Explosions have occurred in air compressors as a result of rapid oxidation of oil deposits in the piping between stages of multiple-stage compressors. Use of proper lubricants prevents deposition of oxidizable materials in high pressure piping (see LUBRICATION AND LUBRICANTS). High maintenance standards are required to detect and avoid the hazards associated with broken valves and other sources of hazardous recompression.

Compressor systems for respirators should be completely separate from other uses. Air compressors that are designed to produce breathing-quality air are required. In smaller systems, it may be preferable to use bottled air rather than a compressor. The contents of bottles, however, should be analyzed to ascertain that they do contain air and to assure that the concentrations of any contaminants are acceptably low.

*Safety Showers.* Safety showers and eyewash fountains or hoses should be installed where corrosive or toxic materials are handled. A large-volume, low velocity discharge from directly overhead should effect continuous drenching, ie, a minimum flow of 20 L/min (50 gal/min). Water to outside showers may be heated to a maximum temperature of 27°C by an electric heating cable. The valves for all safety showers should be at the same height and relative position to the shower head, and they should operate in the same way and direction. The shower station should be identified by paint of a bright, contrasting color. In areas where chemicals harmful to the eyes may be encountered, an eyewash fountain or spray should be available in case of splash accidents.

*Ventilation.* When plant equipment is located outdoors, there usually is little need for mechanical ventilation. Many operations must be done indoors, however, and it may be necessary to remove toxic or flammable gases or vapors or process-generated atmospheric heat (53,59). Ventilation and heat-stress standards are intended to avoid hazards associated with high body temperature, heat exhaustion, heat stroke, or discomfort in processes generating high ambient temperatures, eg, glass and steel manufacturing.

The removal of flammable and toxic contaminants is best achieved by exhaust ventilation as near the source of discharge as possible, to minimize the amount of air removed from a building, the energy losses resulting from exhausting conditioned air, and the exposure of personnel to the contaminants (59, 88). General room ventilation, which is expressed as the number of air changes per hour, requires the removal of much air and exposes the entire population in the room to materials that may have been released. Most flammable vapors are heavier than air; thus they tend to flow to the floor or ground level and then travel appreciable distances. Suction should be within a few centimeters of the floor. However, appreciable diffusion takes place, and the entire atmosphere in the building may be contaminated. Local ventilation is accomplished by ductwork, either rigid or flexible, which conveys the vapor to the wall or roof of the building. The blower, preferably a high suction type, should be on the outside of the building so that all ductwork inside the building operates at less than atmospheric pressure; thus leakage does not cause contamination (see FANS AND BLOWERS) (59).

Discharge from blowers is usually directly to the atmosphere unless this presents a pollution problem; if so, scrubbing of vapors or collection of dusts may be required. Atmospheric discharge should be at a sufficient height to obtain the desired amount of dispersion, and the exhaust should be straight up with no weather cap or any other obstruction. Where suction must be at numerous points, either many blowers, ie, one for each point, or a large blower with a manifold system collecting from many points may be used. The latter is very difficult to balance aerodynamically and, although less expensive to install, may be less effective and more costly to operate and maintain.

**Pressure Vessels and Piping.** Some of the most critical components of a chemical plant involve pressure vessels. A thorough knowledge of the American Society of Mechanical Engineers (ASME) Pressure Vessel Code (75) is essential for design and maintenance of chemical plants. Some states have their own codes, which usually conform closely to the ASME version (see HIGH PRESSURE TECHNOLOGY; TANKS AND PRESSURE VESSELS).

*Relief Devices.* Overpressures in process equipment usually can be prevented by automatic or manual controls, in response to pressure sensors (qv) and alarms (see PRESSURE MEASUREMENT). In event of pressure-control failure, overpressure relief can be provided by spring-loaded relief valves or rupture disks. Rules for the sizing and location of pressure-relief devices are described in the ASME Pressure Vessel Code, in API standards (86), and by the NFPA and governmental agencies (60). In general, the set pressure must not exceed the design pressure of the vessel. Special consideration, calculations, and tests may be required for two-phase flow through the relief device, eg, liquid and vapor or gas, or foam or flashing liquid. Study of two-phase relief was sponsored by the American Institute of Chemical Engineers (AIChE) through the Design Institute for Emergency Relief Systems (DIERS).

Relief valves are preferred for use on clean materials, because automatic closure prevents excessive discharge once excessive pressure is relieved. Rupture disks are less susceptible to plugging or other malfunctions but may allow complete emptying of the vessel, thus creating a safety or environmental hazard. Where fluctuating pressures or very corrosive conditions exist, or where polymerizable materials could prevent proper operation of a relief valve, some designers install two safety devices in series, ie, either two rupture disks or an upstream rupture disk followed by a relief valve. With either arrangement, it is imperative that the space between the two relief devices be monitored so that perforation or failure of the relief device closest to the vessel may be detected (86).

Passageways of relief devices must not become obstructed, and discharge piping must be sized to transport the effluent to a safe place and with minimum pressure drop. Blow tanks may be provided for liquid discharges, and gases can be exhausted to stacks or flares. Discharge piping must be anchored to resist discharge reaction and shock effects. Passing inert gas into piping, ducts, blow tanks, and knockout drums or use of flame arrestors in vents is advisable where flammable vapors could form explosive mixtures with air (89) (see PIPING SYS-TEMS). Special precautions, eg, heating devices, may be required if condensation or other forms of pluggage could occur in flame arrestors.

**Materials Handling.** *Liquids.* Liquids usually are moved through pipelines (qv) by pumps. Special alloys, plastic pipe and liners, glass, and ceramics are widely employed in the chemical industry for transport of corrosive liquids. Care is required in making the connections, to prevent exposure of unprotected metal such as flanges and bolts to the corrosive material inside the piping.

Piping design requires consideration of the maximum pressures, temperatures, and flows that might be attained, the corrosive and erosive nature and the viscosities of the materials passing through the piping, the distances between the inlet and discharge points, and the external force and vibrational stresses to which the piping might be subjected. Tests under simulated conditions may be required to define material specifications and maximum flow velocities. The results of such tests may determine the materials of construction, the diameter and length of the piping sections, and the pressure resistance of the piping. Other considerations include the location of the piping, ie, underground or aboveground, and supporting methods; exposure to vehicle traffic; and stresses created by fluctuations in temperature and pressure.

Valves used to isolate sections of piping and to control the flow of materials through piping must be compatible with the materials being handled. Similarly, pumps used to develop motive pressure must be constructed of materials which resist any corrosive or erosive properties of the materials. Centrifugal pumps are widely used in the chemical industry. Precautions are required to prevent operation when suction or discharge valves are closed, particularly when using heat-sensitive materials. Positive-displacement pumps should have relief valves on the discharge or an alternative method of preventing dead-head operation. Pump failures commonly occur at the packing area and can result in the release of toxic or flammable materials.

Drainage valves should be provided at the low points of the plant system, and all piping should slope downward toward them. The specification of safe valving arrangements is required. Double-block valves having intermediate bleed valves should be used for dependable shutoff of hazardous materials flow. By-passes should be provided around control valves only in coolant or diluent lines, not in lines where inadvertent opening could defeat the process-control system or create a hazard. Check valves prevent inadvertent backflow of nuisance materials into feed systems. These should not be relied on alone to maintain isolation of reactive, corrosive, flammable, or toxic materials. For liquids having high coefficients of expansion, such as liquefied gases, pressure-relief valves or expansion chambers must be provided between block valves.

*Solids.* Equipment for transporting and feeding solid materials include belt, flight, and screw conveyors (see CONVEYING), and pneumatic systems. Common hazards associated with solids handling are dust explosions and the escape and dispersion of noxious or combustible dusts (see POWDERS, HANDLING). Two items are of critical importance in designing pneumatic conveying systems. First, if the material is combustible, inert gas should be used as the conveying fluid, or blowout panels or vents should be provided to avoid explosion damage. Pressure systems are preferred over vacuum systems, because these preclude air infiltration. Second, where poisonous or noxious materials are being transported, special attention must be given to recovery of fines from the exit air. A complete recycle of the carrier gas may be desirable. Special guarding of belt and screw conveyors is required to restrict access during start-up, unclogging, and maintenance.

Electric or fuel-powered means of transporting solid materials, such as forklift trucks, should be employed only when full consideration has been given to any hazardous atmospheres in which these might be used. Such transport must be properly maintained to preserve the integrity of built-in safety devices. Operators must be trained to operate transport safely, avoiding even the possibility of puncturing drums or packages containing hazardous materials. Overhead guards should be provided on all riding trucks intended for operation in areas where materials are stored above head height. Lighting not obstructed by the load should be provided if trucks are to operate after dark or in poorly lighted areas (90).

**Plant Construction.** Construction of a chemical plant most often is the responsibility of a construction company, although a few large manufacturers carry out their own construction work. Regulations regarding some construction hazards are described in the OSHA Construction Standards (19,91). Considerable responsibility is placed on the plant owner to evaluate the safety performance and to monitor the operations of construction contractors (36). During construction, the client should gather information that may help in the start-up and operation of the plant. Frequent field checking of dimensions and locations during construction can lead to helpful but relatively inexpensive changes compared to post-construction alterations. Prior to start-up of new facilities, a prestart-up safety review should confirm that (1) construction and equipment is in accordance with design specifications; (2) the equipment is suitable for the process application and has been installed properly; and (3) maintenance materials are on hand. A checklist should be used to evaluate the readiness of new facilities for start-up.

## Operation of Facilities

**Start-Up.** Often key personnel from the design and construction organizations remain at the plant during start-up. Depending on the hazards of the process and materials involved, it may be advisable to use less hazardous materials under working conditions before going ahead with the actual process. For example, a distillation column might be operated on water or high flash-point liquid prior to introduction of a volatile charge stock. In this way, leaks may be detected without serious consequences, and operating and control problems may be identified and corrected. Furthermore, such initial testing provides training for operators under stable conditions.

It frequently is necessary to charge vessels with inert gases before introducing flammable or reactive materials. The hazards of entering vessels that have been filled with inert atmospheres which do not support life must not be overlooked. Start-up procedures should be written in detail well ahead of actual start-up.

**Normal Operation.** The designer of a chemical plant must provide an adequate interface between the process and the operating employees. This is usually accomplished by providing instruments to sense pressures, temperatures, flows, etc, and automatic or remote-operated valves to control the process and utility streams. Alarms and interlock systems provide warnings of process upsets and automatic shutdown for excessive deviations from the desired ranges of control, respectively. Periodic interruption of operations is necessary to ensure that instruments are properly calibrated and that emergency devices would operate if needed (see FLOW MEASUREMENT; TEMPERATURE MEASUREMENT).

**Shutdown.** Written procedures for normal, as well as for emergency, shutdowns should be prepared, rehearsed, and kept up-to-date. Operating supervisors must be responsible for leaving the process equipment in a safe condition or preparing plant equipment for maintenance work.

**Maintenance.** Good plant maintenance (qv) obviates the crash shutdowns that could follow failure of critical components. Maintenance in the chemical industry differs from that in other industries because of the nature of the materials, processes, and types of equipment used. Because much chemical work involves the movement of fluids, gases, and powdered solids from one piece of equipment to another, many pipelines (qv), conveyors, forklift trucks, and other material-handling devices are used. Containers are more likely to be tanks, drums, or some form of closed container than in other industries. Prior to maintenance inside equipment, all lines and equipment containing hazardous materials should be effectively separated, disconnected, blanked, or purged in order to minimize the possibility of release of harmful materials. Maintenance personnel must make sure that all equipment and piping is so prepared.

**Safe Work Practices.** *Locking and Tagging.* Safe maintenance requires that no one work on or be exposed to power-driven equipment without positively disconnecting the source of power beforehand (92). This may be done by locking the electric switch on the power circuit in the off position, disconnecting the motor electrically or mechanically, removing the belt drive, or locking feed valves to prime movers (pneumatic or hydraulic) in the off position, and by blocking the movement of pistons, crank arms, or flywheels. Any exceptions, such as adjusting glands and seals on moving equipment, must be approved by supervisors.

Written procedures should be prepared and thorough training given. In general, these procedures stress that operating supervisors must first identify the equipment and equipment controls. Operating supervisors and each person who is to work on the equipment must individually place their locks on the controls after the controls have been placed in the off position. Each person keeps the key to his or her own lock. The equipment should be rechecked to be certain that it cannot be started; work then can begin. When the work has been completed, the procedure is reversed before control of equipment is relinquished to the operating supervisors.

*Entry into Confined Spaces.* In 1993 OSHA adopted a confined space entry rule (93) requiring employers to evaluate the workplace to (*1*) determine if it contains any confined spaces, (*2*) mark or identify such confined spaces, and (*3*) develop and implement a permit program for entry into such spaces. The program must include a permit system which specifies the steps to be taken to identify, evaluate, control, and monitor possible electrical, mechanical, and chemical hazards; select and use equipment; institute stand-by attendance; and establish communications. The reference standard (93) should be studied for details.

*Hot Work.* The objective of a hot work standard is to prevent fires, explosions, and other causes of injury which might result from workplace ignition sources such as welding (qv), cutting, grinding, and use of electrically powered tools. The OSHA standards have specific requirements (36,94) for fire prevention and protection and a permit system.

*Opening Process Equipment or Piping.* A procedure for opening process equipment or piping which could contain hazardous materials is required by the OSHA Process Safety Management standard (36). The procedure requires appropriate isolation of equipment and flows; use of personal protective equipment; observance of emergency preparedness; obtaining proper authorizations; and employment of specific opening methods.

## Product Handling

**Labeling.** The Federal Hazardous Substance Labeling Act (95) requires that all containers sold to consumers be labeled with appropriate precautionary wording to protect the user and employees from injury resulting from contact with the chemical. The capacities of all packages up to and including 208-L (55-gal) drums customarily are indicated by labels. The information includes identification of the material, notification of principal hazards and precautions for use, and antidotes or first-aid measures if applicable.

**Sampling.** The first consideration in sampling (qv) is protection of the person performing the sampling. Eye and face protection, gloves, and respiratory equipment may be needed. Line sampling usually is carried out at a suitable valve, preferably equipped with a self-closing or fusible-link device. The sampler should be aware of the possibility of a sudden increase in flow or of the flow of high pressure or high temperature material when the valve is opened. Operating personnel should always be told that a sample is being taken.

For toxic materials, it usually is advisable to provide ventilated sampling hoods or breathing-air stations and masks, to assure that the sampler is adequately protected from toxic or flammable vapors and dusts. Special provision for access to and exit from sampling points also may be needed at elevated locations and to avoid tripping or bumping hazards and to ensure that the sampler does not transverse areas not intended as walkways, eg, tank covers or roofs.

Safe sampling facilities are needed at railroad tank car and tank truck unloading areas to avoid falls, particularly in winter or wet weather. Installation of a well-designed loading rack with an adjustable platform reduces this hazard.

Care should be taken so that the car is not jolted by a switching engine or other vehicle during sampling. The same precaution should be taken with tank trucks: the driver should be aware that the sampler is on the truck and be sure not to move it.

**Storage.** Liquid products may be stored in tanks at isolated tank farms, or in drums or cans in warehouses (see TANKS AND PRESSURE VESSELS). Smaller packages, eg, glass bottles, usually are placed in protective cases or carriers. Large quantities of solid materials typically are stored in bins or silos; smaller amounts are packaged in steel or fiber drums, paper boxes, or plastic bags. A source of danger in bulk storage occurs when workers enter tanks or silos for maintenance or other duties. Gases or vapors above the stored materials can cause intoxication, asphyxiation, or explosions, and collapse of the stored materials could cause engulfment and suffocation.

In any warehousing operation, it is essential that incompatible substances be isolated to avoid a reaction in case of a spill or fire. Where highly hazardous materials are stored, it may be advisable for them to be segregated in masonry enclosures protected by suitable fire extinguishing or ventilation equipment. Shock-sensitive or extremely heat-sensitive materials should be stored in separate buildings, and rules for maximum contents should be well-defined and rigidly observed. Scrupulous housekeeping and inventory control helps to minimize storage hazards. For some highly hazardous materials, minimum distances between storage areas are recommended (47,60).

In the design of warehouses in which flammable or combustible materials are to be stored, consideration should be given to the installation of fire walls, fire doors, and duct shut-off dampers. Automatic sprinklers are standard equipment in such locations. Developments in warehousing include high piling of palleted material by computer-controlled handling equipment. However, there is concern about the large size of these warehousing installations and the high value of the contents. For example, fire involving drummed, flammable liquids stacked four or five tiers high would be exceedingly difficult to control using customary sprinkler designs. The use of aqueous film-forming foam has been proposed for control of fire in these warehouses (11). Fire-detection devices such as flame-sensing or ionization-interference types operate much more rapidly than sprinkler heads and are used extensively both as alarms and to activate fixed fire-extinguishing systems.

**Disposal.** Disposal of hazardous waste must be carried out in accordance with precautions against fire and explosion hazards, severe corrosion, severe reactivity with water, toxic effects, and groundwater pollution. Several methods are available, but each has drawbacks. Burning in the open is becoming less acceptable because of the air pollution (qv) that may result from incomplete combustion. Incineration controls the pollution problem somewhat by assuring complete combustion, but care must be taken that the heat release is not so rapid as to damage the incinerator and its auxiliary parts (see INCINERATORS). Disposal through industrial sewers must be in accordance with good waste-disposal practices, and it must observe the restrictions imposed by the receptor and authorities. Volatile, flammable materials may generate explosive vapors in sewers, causing flashbacks that can damage the plant. Suitable separators or treatment facilities usually are necessary to treat waste entering a sewer system.

Reactive wastes should be treated to make these wastes relatively harmless before disposal (96). For example, sodium can be treated with alcohol, and the resulting alkaline solution can be neutralized. Explosive materials usually are taken to safe areas and burned in controlled quantities (see WASTE TREATMENT, HAZARDOUS WASTES).

Wastes sometimes are eliminated by burying; however, landfill operations usually are followed by some constructive use of the area. Therefore, chemicals buried in such a location may handicap the subsequent use of the site. Where there is a possibility that vapors or reactive product gases could rise to the surface or where the chemical might destroy foundations or interfere with well water supplies, this type of disposal cannot be used. Disposal at sea has been practiced along the coastline, but is being discontinued because of effects on marine life and the possibility of materials being washed up along the shore.

Contract disposal agencies offer their services to relieve the chemical industry of unwanted materials; however, the cost of such disposal (primarily incineration) is high. The manufacturer should ascertain that the disposal agency employees are adequately aware of chemical hazards and can responsibly handle and dispose of the waste materials (see WASTES, INDUSTRIAL).

Great efforts have been made by the chemical industry to minimize the generation of waste materials (see WASTE REDUCTION). Of primary importance is minimizing the frequency and quantity of spills and leaks, particularly of solids and liquids. Sometimes this can be accomplished by adjustment of temperatures, pressures, or residence times in reactors; minimizing process upsets by improved control systems; by reclaiming, reworking, or recycling products that do not meet specifications; and by using by-products or waste materials for other uses. However, several of these efforts, such as reducing the safety margin in operating temperature and reducing the frequency of equipment cleaning and area cleanup, can introduce safety hazards, and compensating improvements in process control may be required.

**Transportation of Chemicals.**  Feed materials and finished products are frequently transported by tank truck and railroad tank cars. Design, construction, and movement of these vehicles is regulated by the U.S. Department of Transportation (DOT) (97). The DOT regulations require placarding of material-transport vehicles to alert the public and emergency personnel to the nature of their contents.

Assistance following an accident involving hazardous chemicals during transport can be obtained 24 hours per day from CHEMTREC, which is an industry-supported information network, by telephone at 1-800-424-9300, as of 1995 (see TRANSPORTATION).

When loading or unloading a tank truck or tank car, static bonding lines must be attached between the vehicle and the fixed piping system. Tank truck drivers may be inexperienced in the handling of chemicals and, therefore, it is essential that the plant personnel be alert in checking the driver's operations and in correcting any possible deficiencies (see TRANSPORTATION).

During the loading or unloading of tank truck and tank cars, the brakes should be set and the wheels must be chocked. Warning signs should be in place to prevent unintended movement. During the loading or unloading of tank cars, derails should be locked in place to prevent contact between a switching engine or other cars during the transfer of materials. DOT regulations require that the loading and unloading of tank trucks and tank cars be attended. As an alternative to stationing an employee at the transfer operation, this objective can be accomplished by leak detectors, television monitors, and remote-operated shutoff valves installed on the tank car or tank truck.

In the United States, regulation of barge shipments is overseen by the U.S. Coast Guard. It is essential that the barge be bonded by a cable to the pipe system on shore whenever loading or unloading takes place, in order to minimize the hazards of differences in voltage and static electricity.

## Human Relations

**Personnel Selection and Training.**  The quality of operating personnel is of paramount importance to the safe operation of a chemical plant. Operators must be intelligent and emotionally stable. Excessive use of alcohol and drugs affects reliability and can thereby render workers more susceptible to certain types of toxic exposure. Thorough medical screening is essential to avoid damaging exposures to susceptible individuals, eg, people with respiratory ailments should not be employed in areas where corrosive atmospheres could occur.

Training has assumed significant proportions in some larger companies. A great deal of attention has been given to the development of standard training procedures, including a liberal amount of safety training. A successful method involves the preparation of a manual of standard operating procedures for each unit or plant, and personnel are trained in complete adherence to these procedures. Another method involves job-safety analysis: careful study is made of what an operator is expected to do under normal conditions and how he or she might be trained to react in emergency situations. Discussions assist operators in recognizing hazardous situations, reacting correctly and promptly, and contributing ways of handling such conditions.

Proven steps for on-the-job training include preparing the workers by describing the job and discussing the important points; presenting the operation, encouraging questions, and stressing key points; training in which the operator works under close supervision and errors are corrected as they occur; and then working alone with frequent follow-up by supervisory personnel.

**Medical Programs.**  Large chemical plants have at least one full-time physician who is at the plant five days a week and on call at all other times. Smaller plants either have part-time physicians or take injured employees to a nearby hospital or clinic by arrangement with the company compensation-insurance carrier. When part-time physicians or outside medical services are used, there is little opportunity for medical personnel to become familiar with plant operations or to assist in improving the health aspects of plant work. Therefore, it is essential that chemical-hazards manuals and procedures, which highlight symptoms and methods of treatment, be developed. A full-time industrial physician should devote a substantial amount of time to becoming familiar with the plant, its processes, and the materials employed. Such education enables the physician to be better prepared to treat injuries and illnesses and to advise on preventive measures.

Clinical tests can and should be made prior to employment or work assignment and at frequent intervals thereafter, when employees are exposed to hazardous operations such as handling benzene, mercury, chlorinated solvents, etc. Recordkeeping of injuries and exposures to toxic materials also is important. Disaster planning with other plant personnel and research involving toxicological work on occupational diseases and epidemiology may be included in medical programs.

**First Aid and Rescue.** Immediate treatment is of primary importance in first aid. Thorough knowledge of first aid, as taught in courses by the American Red Cross or the U.S. Bureau of Mines, should be a primary part of chemical plant training programs. Rescue techniques also should be taught and practiced. Many organizations insist that employees use the buddy system in which one fully equipped person remaining outside the hazardous area watches another person and initiates rescue if necessary.

## Fire and Explosion Prevention and Protection

**Fire and Explosion Prevention.** Prevention of fire and explosion takes place in the design of chemical plants. Such prevention involves the study of material characteristics, such as those in Table 1, and processing conditions to determine appropriate hazard avoidance methods. Engineering techniques are available for preventing fires and explosions. Containment of flammable and combustible materials and control of processes which could develop high pressures are also important aspects of fire and explosion prevention.

**Fire and Explosion Protection.** Extinguishment or control of fire is essential. Exposure of personnel to thermal-radiation hazards must be minimized and property protected. Extinguishing fire requires cooling below the flash point, removing the oxidant, or reducing the fuel concentration below the lower flammability limit. For combustible solids and high flash-point liquids, water can be used alone to extinguish fire. Water has an additional benefit as a result of its high specific heat and high latent heat of vaporization: it can be used to cool equipment, structures, and containers of hazardous materials, even where extinguishment is difficult, eg, fires involving low flash-point liquids and flammable gases. Water is the preferred fire-control medium. Designs for automatic sprinkler protection against specific hazards and general area coverage have been well developed and tested (47,60). Such systems may be composed of open sprinkler heads, ie, deluge systems, which are activated by temperature rise or flame detectors, or they may be closed-head systems that are activated by high temperature. One development is an on/off sprinkler head, which limits water damage.

The extinguishing capability of water can be improved, while much of the cooling benefits and low cost can be retained, by including additives, eg, foaming materials, surface-active agents to produce "wet" water, and chemicals, eg, carbon dioxide, to generate inert gases. Foams (qv) generally are formed by adding natural proteins or similar synthetic materials and aerating at nozzles to make a blanket, which floats on flammable materials. Because the foam excludes air and reduces volatilization, it can be used to cover spills and, thus reduce the potential for fire, as well as to extinguish existing fire. Ordinary foams are dissolved by polar solvents, eg, alcohols and ketones, and special alcohol-resistant foams have been developed.

Foam is especially valuable in fighting fires in large storage tanks. The foam usually is discharged remotely onto the surface of liquid from above, but some types of foam are suitable for subsurface injection to avoid explosion-caused disruption of the foam's fire-extinguishing capability and for protection of very large-diameter tanks. Sprinkler systems have been designed to spray foam from nozzles onto materials to be protected, as in warehouses. The effectiveness of water in extinguishing flammable-liquid fires also can be improved by decreasing the size of the droplets, which consequently volatilize rapidly in a fire to form an inert steam atmosphere. Fog also can be used to move or disperse flammable vapors in open areas and to reduce the danger of flash fire or explosion.

In some applications, water can be used in the form of vapor or steam to exclude, dilute, or drive air from enclosed areas. However, extinguishment of fire by means of oxidant reduction can be accomplished more effectively using inert gases, eg, nitrogen (qv), carbon dioxide, halogenated hydrocarbons, or helium-group gases (qv). Carbon dioxide (qv) is particularly useful because of its low cost and absence of residues. The effectiveness of some of these materials, particularly the halogenated hydrocarbons such as Freon or Halon, is greater than could be expected if reduction in oxygen concentration were the only cause for fire extinguishment. For example, a $CO_2$ concentration of 40 vol % and oxygen concentration of 13 vol % is required to extinguish fire in most flammable liquids. An oxygen concentration of 13 vol % would be fatal to any humans in the protected area. In contrast, a concentration of only 4 vol % of Freon 1301, $CBrF_3$, resulting in an oxygen concentration of 20 vol %, provides similar effectiveness. As a result of the Montreal Protocol on Substances that Deplete the Ozone Layer (98), newer fluoro/chloro/hydrocarbon compounds are being developed to replace the bromo/fluorocarbon Freon and Halon chemicals that were installed prior to 1987 (see CHLOROCARBONS AND CHLOROHYDROCARBONS).

Most dry-chemical fire-extinguishing materials also function by inhibiting combustion rather than by cooling or by reducing oxygen concentration. The usual dry-chemical material is a bicarbonate, but some phosphates, eg, ammonium, provide a coating that makes the material suitable for use on fires involving solid combustibles such as rubber tires, wood, and paper.

Portable fire extinguishers are classified according to applicability: Class A for solid combustibles; Class B for flammable liquids; Class C for electrical fires that require a nonconducting agent; and Class D for combustible metals. Water frequently is used for Class A extinguishers; bicarbonates for Class B and Class BC; carbon dioxide or Freon for Class C; ammonium phosphate for Class ABC; and powdered salt, sodium chloride, for Class D.

Protection against explosions is typically provided by explosion-venting, using panels or membranes which vent an incipient explosion before it can develop dangerous pressures (11,60). Protection from explosions can be provided by isolation, either by distance or barricades. Because of the destructive effects of explosions, improvement in explosion-prevention instrumentation, control systems, or overpressure protection should receive high priority.

**Disaster Planning.** Plant managers should recognize the possibility of natural and industrial emergencies and should oversee formulation of a plan of action in case of disaster. The plan should be well documented and be made known to all personnel critical to its implementation. Practice fire and explosion drills should be carried out to make sure that all personnel, ie, employees, visitors, construction workers, contractors, vendors, etc, are accounted for, and that the participants know what to do in a major emergency.

A checklist for total emergency planning has been described (99). In all emergency situations, fire services, safety staff, and medical organization are of paramount importance for the conservation of life and property (100). Plans should be formulated to mobilize off-duty personnel and to bring in outside assistance when necessary. In highly industrial areas, it is usually practical to form a mutual-aid organization. To make such a system work smoothly, it is necessary to have periodic meetings to compare problems and needs, and to discuss new developments. Drills designed to mobilize such assistance are also essential to their smooth functioning (101).

## Control of Process Hazards

**Process Hazards Analysis.** Analysis of processes for unrecognized or inadequately controlled hazards (see HAZARD ANALYSIS AND RISK ASSESSMENT) is required by OSHA (36). The principal methods of analysis, in an approximate ascending order of intensity, are what-if; checklist; failure modes and effects; hazard and operability (HAZOP); and fault-tree analysis. Other complementary methods include human error prediction and cost/benefit analysis. The HAZOP method is the most popular as of 1995 because it can be used to identify hazards, pinpoint their causes and consequences, and disclose the need for protective systems. Fault-tree analysis is the method to be used if a quantitative evaluation of operational safety is needed to justify the implementation of process improvements.

**Human Factors.** Human failings have been found to be responsible for most catastrophic process incidents. Failings range from inadequate commitment of management to safety administration, policies, and programs to inadequate commitment of employees to safety awareness and the observance of safe practices. The OSHA standard (36) requires a formal evaluation of human factors, including training, communications, physiological and psychological stresses, process/human control interfaces, personal protective equipment, exit facilities, etc. Further, OSHA compliance auditors interview employees extensively concerning not only their knowledge of policies and procedures, but also their participation in the development of safety programs and related procedures and in the analysis of processes and production operations for potential hazards.

**Accident Investigation.** A study of all accidents and injuries with the objective of determining the cause or causes can lead to correction of unsafe practices or conditions and prevent recurrence of the accident. Sometimes relatively minor injuries can be the key to disclosing an unsafe condition or an improper operation which has the potential for causing a far worse injury than that which was sustained. Employers are obliged to report to the regional OSHA office the details of any accident that has caused a fatality or hospitalization of five or more personnel. Information needed in accident investigation comes from several sources, eg, examination of the site and of the equipment involved, and temperature, pressure, and flow charts and logbooks (102,103). Eyewitness accounts are necessary but may be contradictory and unreliable; thus caution in interpretation and verification are needed. Consultants or experts from government agencies, eg, the U.S. Bureau of Mines, should be sought to obtain the best interpretation of information involving an explosion.

**Center for Chemical Process Safety.**  In 1985, the American Institute of Chemical Engineers established the Center for Chemical Process Safety (CCPS) (New York). The objective of the CCPS was to help prevent catastrophic chemical accidents by compiling information on the latest scientific and engineering practices, safety programs, and administrative procedures of the larger members of the chemical industry, so that they can be shared with other (and particularly the smaller) members of the chemical and petrochemical industries.

The CCPS has published a substantial library of "Guideline" books, covering many of the chemical-plant safety subjects discussed above, and in much greater detail. As of 1995, the library consisted of the following volumes, titled *Guidelines for...*

Use of Vapor Cloud
  Dispersion Models
Safe Storage and Handling
  of High Toxic
  Hazard Materials
Vapor Release Mitigation
Vapor Cloud Source Dispersion
  Models (Workbook of Test Cases)
Chemical Process Quantitative
  Risk Analysis
Process Equipment Reliability
  Data, with Data Tables
Technical Management of
  Chemical Process Safety (Plant)
Evaluating the Characteristics of
  Vapor Cloud Explosions,
  Flash Fires, and BLEVEs
Technical Management of
  Chemical Process Safety
  (Corporate)
Handling Emergency
  Release Effluents
Safe Automation of
  Chemical Processes
Chemical Reactivity Evaluation and
  Application to Process Design
Preventing Human Error in
  Process Safety
Safety, Health, and Loss Prevention
  in Chemical Processes: Student
  Problems;

Instructor's Guide for
  Undergraduate Engineering
  Curricula
Hazard Evaluation Procedures, with
  Worked Examples, 2nd ed.
Investigating Chemical
  Process Incidents
Auditing Process Safety
  Management Systems
Making Acute Risk Decisions
Process Safety Fundamentals for
  General Plant Operations
Engineering Design for
  Process Safety
Implementing Process Safety
  Management Systems
Evaluating Plant Buildings for
  External Explosions and Fires
Chemical Process Documentation
Chemical Transport Risk Analysis
Safe Process Operations and
  Maintenance
Technical Planning for On-Site
  Emergencies
Writing Effective Operating and
  Maintenance Procedures
Concentration Fluctuations &
  Averaging Time in Vapor Clouds
Expert Systems in Process Safety
Understanding Atmospheric
  Dispersion of Accidental Releases

Other publications include the proceedings of six international conferences and symposia three training courses and a computerized bibliography of *Guideline* references, all available from the Center for Chemical Process Safety (1-800-242-4363).

## BIBLIOGRAPHY

"Safety" in *ECT* 1st ed., Vol. 12. pp. 1–36, by F. A. Van Atta, National Safety Council; in *ECT* 2nd ed., Vol. 17, pp. 694–719, by W. S. Wood, Sun Oil Company, "Plant Safety" in *ECT* 3rd ed., Vol. 18, pp. 60–86, by R. W. Prugh, E. I. du Pont de Nemours & Co., Inc.

1. *Accident Facts*, National Safety Council, 1992, pp. 34, 42. Published annually.
2. *Large Property Damage Losses in the Hydrocarbon—Chemical Industries–A Thirty-Year Review*, 14th ed., Marsh & McLennan, Inc., Chicago, Ill., 1992, p. 2.
3. *Annual Loss Prevention Symposium*, American Institute of Chemical Engineers, New York, 1st, 1967; 29th, 1995.
4. *International Symposium on Runaway Reactions*, American Institute of Chemical Engineers, New York, 1989 and 1995.
5. *Loss Prevention*, Vols. 1–14, American Institute of Chemical Engineers, New York, 1967–1981.
6. *Plant/Operations Progress*, Vols. 1–11, American Institute of Chemical Engineers, 1982–1992; *Process Safety Progress*, Vol. 12, 1993–present.
7. *International Symposium on Loss Prevention and Safety Promotion in the Process Industries*, 1st, European Federation of Chemical Engineering, The Hague/Delft, the Netherlands, 1974; 2nd, Heidelberg, Germany, 1977; 3rd, Basel, Switzerland, 1980; 4th, Harrogate, U.K., 1983; 5th, Cannes, France, 1986; 7th, Taormina, Italy, 1992.
8. Institute of Chemical Engineers (UK), *Chemical Process Hazards*, I, No. 7, 1960; II, No. 15, 1963; III, No. 25, 1967; IV, No. 33, 1972; V, No. 39a, 1974; VI, No. 49, 1977; VII, No. 58, 1980; *Process Industry Hazards*, No. 47, 1976; *Major Loss Prevention in the Process Industries*, No. 34, 1971; *Preventing Major Chemical and Related Process Accidents*, No. 110, 1988.
9. *Flash Point*, D56, D92, D93, D1310, D3143, D3278, D3828, E134, E502; *Autoignition Temperature*, D2155, E659; *Flammability*, D2863, E681; *Impact Sensitivity*, E680; *Minimum Ignition Energy*, E582; *Reactivity*, D2883, E476; *Shock Sensitivity*, D2539; *Thermal Analysis*, E472, E473, E474, E487, E537, American Society for Testing and Materials, Philadelphia, Pa.
10. *Fire Protection Guide to Hazardous Materials*, 11th ed., National Fire Protection Association, Boston, Mass., 1994.
11. *Fire Protection Handbook*, 18th ed., National Fire Protection Association, Boston, Mass., 1995.
12. *The SFPE Handbook of Fire Protection Engineering*, 2nd ed., Society of Fire Protection Engineers, Boston, Mass., 1995.
13. *Threshold Limit Values for Chemical Substances and Physical Agents* (annual); *Documentation of the Threshold Limit Values and the Biological Exposure Indices*, 6th ed, American Conference of Governmental Industrial Hygienists, Cincinnati, Ohio, 1993, updated periodically.
14. R. W. Prugh, *Chem. Eng. Prog.*, **76**(7), 59 (1980); *AIChE* **14**, 1 (1981).
15. *Environmental Reporter Monographs*, Bureau of National Affairs, Washington, D.C., 1970–present.
16. *Occupational Safety and Health Reporter*, Bureau of National Affairs, Washington, D.C., 1970–present.

17. *Health, Safety and Welfare in Connection with Work, and Control of Dangerous Substances and Certain Emissions into the Atmosphere*, U.K. Parliament, U.K., July 31, 1974, Chapt. 37.

18. H. A. Williams and W. Steiger, *Occupational Safety and Health Act of 1970*, Public Law No. 91-596, 29 U.S.C. 651, para. 5(a)(1), the "general duty clause," Dec. 29, 1970.

19. *Occupational Safety and Health Standards*, Title 29, Subtitle B, Chapt. XVII, Part 1910, U.S. Department of Labor, Occupational Safety and Health Administration, Washington, D.C., 1991.

20. *Chem. Proc. Safety Rep.* **4**(7), 1 (May 1994).

21. *Toxic Catastrophe Prevention Act*, New Jersey State Assembly N.J.S.A 13:1B-3, State of New Jersey, Department of Environmental Protection and Energy, Trenton, N.J., Sept. 12, 1985.

22. *Toxic Catastrophe Prevention Act Program*, New Jersey Act of Congress N.J.A.C. 7:31, State of New Jersey, Department of Environmental Protection and Energy, June 20, 1988 and Aug. 1, 1988; readopted with amendments, July 19, 1993.

23. R. Baldini and P. Komosinsky, *J. Loss Prevention Process Ind.*, **1**(2), 147 (July 1988).

24. *Health and Safety Code*, Chap. 6.95, Art. 2, *Hazardous Materials Management*, State of California, Sacramento, Calif., 1986; Sect. 25531, *Risk Management and Prevention Program*, 1987.

25. *Health and Safety Code Risk Management and Prevention Program Guidance*, Chapt. 6.95, Art. 2, Sect. 25534(1), State of California, Governor's Office of Emergency Services, Sacramento, Calif., 1988.

26. *Chemical Process Safety Report*, Sect. 900, 261, 1221, 1421, Thompson Publishing Group, Washington, D.C., 1994.

27. R. A. Barrish, Paper No. 42B, AIChE Spring National Meeting, Orlando, Fla., Mar. 20, 1990.

28. *Regulation for the Management of Extremely Hazardous Substances*, State of Delaware, Department of Natural Resources and Environmental Control, Sept. 25, 1989; *Extremely Hazardous Substances Risk Management Act*, Title 7, Chapt. 77, July 19, 1988.

29. J. A. Johnson, *Proceedings of the 1994 Process Plant Safety Symposium*, American Institute of Chemical Engineers, New York, Mar. 1, 1994, p. 596.

30. *Nevada Chemical Catastrophe Prevention Act*, State of Nevada, Nevada Revised Statutes 459.380 to 459.3874, Reno, Nev., 1991.

31. U.S. Environmental Protection Agency, *Superfund Amendments and Reauthorization Act of 1986*, Title III, *Emergency Planning and Community Right-to-Know Act*, Public Law 99-499, Sect. 302(a), *Emergency Planning—List of Extremely Hazardous Substances*, Title 40, Subchapt. J, Part 355, of the *Code of Federal Regulations* (40 CFR 355), Washington, D.C., Oct. 17, 1986; *Federal Register* (Nov. 17, 1986); **52**(77), 13395 (Apr. 22, 1987); (Feb. 25, 1988).

32. *Member Safety Commitment*, Chlorine Institute, Inc., Washington, D.C., Feb. 5, 1986.

33. L. M. Ramonas, *Ammonia Plant Safety* **31**, 1 (1991).

34. *Process Safety Code of Management Practices*, Chemical Manufacturers Association, Washington, D.C., Sept. 11, 1990; *A Resource Guide for the Process Safety Code of Management Practices* Oct. 1990.

35. *Management of Process Hazards*, Recommended Practice 750, American Petroleum Institute, Washington, D.C., Jan. 1990.

36. *Process Safety Management of Highly Hazardous Chemicals*, Title 29, Subtitle B, Chapt. XVII, Part 1910, Subpart H, Paragraph 119, of the *Code of Federal Regulations* (29 CFR 1910.119), *Federal Register* **57**(36), 6403, U.S. Department of Labor, Occupational Safety and Health Administration (Feb. 24, 1992).

37. J. S. Arendt, *1994 Process Plant Safety Symposium*, American Institute of Chemical Engineers, New York, Feb. 28, 1994, p. 8.

38. *Guidelines for Technical Management of Chemical Process Safety*, Center for Chemical Process Safety, New York, 1989; *Guidelines for Auditing Process Safety Management Systems*, Center for Chemical Process Safety, New York, 1993.

39. *Plant Guidelines for Technical Management of Chemical Process Safety*, Center for Chemical Process Safety, New York, 1992.

40. U.S. Environmental Protection Agency, *Risk Management Programs for Chemical Accidental Release Prevention; Proposed Rule*, Title 40, Part 68, Subpart B, of the Code of Federal Regulations (40 CFR 68), *Federal Register* **58**(201), 54212 (Oct. 20, 1993).

41. U.S. Environmental Protection Agency, *List of Regulated Substances and Thresholds for Accidental Release Prevention and Risk Management Programs for Chemical Accident Release Prevention*, Title 40, Part 68, Subpart C, of the Code of Federal Regulations (40 CFR 68), *Federal Register* **59**(20): 4493 (Jan. 31, 1994).

42. U.S. Department of Transportation, *Tables of Hazardous Materials*, Title 49, Subtitle B, Chapt. I, Subchapt. C, Part 172, Subpart B, Paragraph 101, of the *Code of Federal Regulations* (49 CFR 172.101), 1989.

43. T. A. Kletz, *Plant Design for Safety: A User-Friendly Approach*, Hemisphere Publishing Co., London, 1991.

44. R. W. Prugh, *Loss Prevention* **5**(2), 67 (1992).

45. L. Bretherick, *Handbook of Reactive Chemical Hazards*, 4th ed., CRC Press, Boca Raton, Fla., 1979.

46. *Manual of Hazardous Chemical Reactions*, No. 491M, National Fire Protection Association, Boston, Mass., 1991.

47. *Handbook of Industrial Loss Prevention*, 2nd ed., Factory Mutual Engineering Corp., New York, 1967.

48. *Identification of the Fire Hazards of Materials*, No. 704, National Fire Protection Association, Boston, Mass., 1990.

49. *Pocket Guide to Chemical Hazards*, 2nd ed., National Institute for Occupational Safety and Health, Washington, D.C., 1990.

50. R. H. Perry and D. W. Green, *Perry's Chemical Engineers' Handbook*, 6th ed., McGraw-Hill Book Co., Inc., New York, 1984, pp. 23-3–23-9.

51. N. I. Sax and R. J. Lewis, *Dangerous Properties of Industrial Materials*, 8th ed., Van Nostrand Reinhold Co., Inc., New York, 1992.

52. *Occupational Safety and Health Standards*, Title 29, Subtitle B, Chapt. XVII, Part 1910, Subpart Z, Paragraph 1000, *Air Contaminants*, and Paragraph 1200, *Hazard Communication*, U.S. Department of Labor, Occupational Safety and Health Administration, Washington, D.C., 1991.

53. *Fundamentals of Industrial Hygiene*, 3rd ed., National Safety Council, Chicago, Ill., 1988.

54. A. J. Fleming, C. A. D'Alonzo, and J. A. Zapp, *Modern Occupational Medicine*, Lea & Febiger, Philadelphia, Pa., 1960.

55. G. D. Clayton, F. E. Clayton, L. J. Cralley, and L. V. Cralley, *Patty's Industrial Hygiene and Toxicology*, Wiley-Interscience, New York, 4th ed., Vol. 1, 1991; 3rd ed., Vol. 2, 1993; 2nd ed., Vol. 3, 1993.

56. J. B. Olishifski, *National Safety News* **122**(2), 43 (Aug. 1980).

57. *A Guide to Industrial Respiratory Protection*, National Institute for Occupational Safety and Health, Washington, D.C., 1976.

58. *Emergency Response Planning Guidelines*, American Industrial Hygiene Association, Fairfax, Va., 1995.

59. *Industrial Ventilation*, 20th ed., American Conference of Governmental Industrial Hygienists, Cincinnati, Ohio, 1989.

60. *Fire Codes*, Vols. 1–16, National Fire Protection Association, Boston, Mass., annual.

61. F. T. Bodurtha, *Industrial Explosion Prevention and Protection*, McGraw-Hill Book Co., Inc., New York, 1980.

62. H. F. Coward and G. W. Jones, *Limits of Flammability of Gases and Vapors*, Bulletin 503, U.S. Bureau of Mines, Washington, D.C., 1952.

63. D. R. Stull, *Fundamentals of Fire and Explosion*, AIChE Monograph Series No. 10, Vol. 73, American Institute of Chemical Engineers, New York, 1977.

64. M. G. Zabetakis, *Flammability Characteristics of Combustible Gases and Vapors*, Bulletin 627, U.S. Bureau of Mines, Washington, D.C., 1965.

65. *Process Plant Hazard and Control Building Design*, Chemical Industries Association (U.K.), 1979.

66. *Siting and Construction of New Control Houses for Chemical Manufacturing Plants*, Safety Guide SG-22, Chemical Manufacturers Association (Manufacturing Chemists Association), Washington, D.C., 1978.

67. *National Electrical Code*, No. 70, National Fire Protection Association, Boston, Mass., 1993; *Electrical Installations in Chemical Plants*, No. 497A, 1992.

68. *Electrical Safety Practices*, Monograph Nos. 110–113, Instrument Society of America, Research Triangle Park, N.C., 1965–1972; *Electrical Safety Abstracts*, 4th ed., 1972.

69. *Managing Workplace Safety and Health: The Case of Contract Labor in the U.S. Petrochemical Industry*, Appen. 1-A, John Gray Institute, Beaumont, Tex., July 1991.

70. R. W. Prugh, *International Conference on Vapor Cloud Modeling (Boston)*, Nov. 4, 1987, p. 712.

71. R. C. Reid, *J. Chem. Ed.* **12**(2), 60; **12**(3), 108; **12**(4), 194 (1978).

72. K. Gugan, *Unconfined Vapor Cloud Explosions*, Gulf Publishing, Houston, Tex., 1979.

73. R. W. Prugh, *Chem. Eng. Prog.* **87**(2), 66 (Feb. 1991); *J. Fire Protection Eng.* **3**(1), 9 (Jan. 1991).

74. R. W. Prugh, *Proceedings of the 27th Annual Loss Prevention Symposium*, Houston, Tex., Mar. 30, 1993, Paper No. 8d.

75. *Boiler and Pressure Vessel Code*, Sect. VIII, American Society of Mechanical Engineers, New York, 1995.

76. A. A. Frost and R. G. Pearson, *Kinetics and Mechanisms*, 2nd ed., John Wiley & Sons, Inc., New York, 1961.

77. S. M. Walas, *Reaction Kinetics for Chemical Engineers*, McGraw-Hill Book Co., Inc., New York, 1958.

78. N. V. Steere, *Safety in the Chemical Laboratory*, Vol. 3, Chemical Rubber Company, Cleveland, Ohio, 1973.

79. R. D. Coffee, in H. H. Fawcett and W. S. Wood, *Safety and Accident Prevention in Chemical Operations*, 2nd ed., Wiley-Interscience, New York, 1982, p. 305; *International Symposium on Runaway Reactions*, Center for Chemical Process Safety, New York, 1989, pp. 140, 144, 177, 234.

80. P. H. Groggins, ed., *Unit Processes in Organic Synthesis*, 5th ed., McGraw-Hill Book Co., Inc., New York, 1958.

81. T. L. Davis, *Chemistry of Powder and Explosives*, John Wiley & Sons, Inc., New York, 1943.

82. R. W. Prugh, in J. I. Kroschwitz, ed., *Encylopedia of Polymer Science and Engineering*, Vol. 14, Wiley-Interscience, 1988, p. 805.

83. C. P. Dillon, *Materials Selection for the Chemical Process Industries*, McGraw-Hill Book Co., Inc., 1991.

84. E. E. Ludwig, *Applied Process Design for Chemical and Petrochemical Plants* Gulf Publishing, Houston, Tex., 1979.

85. *Plant Layout and Spacing for Oil and Chemical Plants*, IM.2.5.2, Industrial Risk Insurers, June 3, 1991, IM.17.2.1, Feb. 1, 1990.

86. *Classification of Locations for Electrical Installations in Petroleum Refineries*, RP 500A, 4th ed., American Petroleum Institute, Washington, D.C., 1982; *Guide for Pressure Relieving and Depressuring Systems*, RP 521, 2nd ed., 1982; *Safe Maintenance Practices in Refineries*, RP 2007, 2nd ed., 1983.

87. *Occupational Safety and Health Standards*, Title 29, Subtitle B, Chapt. XVII, Part 1910, Subpart S, Paragraph 304, U.S. Department of Labor, Occupational Safety and Health Administration, Washington, D.C., 1991.

88. *ASHRAE Handbook—Fundamentals*, American Society of Heating, Refrigeration, and Air Conditioning Engineers, New York, 1977.

89. *Flammability of Liquids*, No. 340, Underwriters Laboratories Northbrook, Ill., 1972; *Flame Arresters*, No. 525, 1973; *Flammable Liquid Dispensing*, No. 1238, 1975; *Group Classification of Flammable Liquids and Gases*, Test Report MH8593, 67C2889, 1967.

90. *Forklift Operations*, Safety Guide SG-6, Chemical Manufacturers Association, Washington, D.C., 1960.

91. *Employment Safety and Health Guide*, Commerce Clearing House, Chicago, Ill., 1985.

92. *Control of Hazardous Energy (Lockout/Tagout)*, 29 CFR 1910.147, U.S. Department of Labor, Occupational Safety and Health Administration, Sept. 1, 1989.

93. *Confined Space Entry Program*, 29 CFR 1910.146, U.S. Department of Labor, Occupational Safety and Health Administration, Jan. 14, 1993.

94. *Hot Work Fire Prevention and Protection*, 29 CFR 1910.252, U.S. Department of Labor, Occupational Safety and Health Administration, Apr. 11, 1990.

95. *Identification of Piping Systems*, American National Standards Institute, New York, A13.1, 1981; *Chemical Plant and Refinery Piping*, B31.3, 1990; *Precautionary Labeling of Hazardous Industrial Chemicals*, Z129.1, 1988.

96. G. Lunn and E. B. Sansone, *Destruction of Hazardous Chemicals in the Laboratory*, John Wiley & Sons, Inc., 1990.

97. *Hazardous Materials Regulations—Carriage by Rail*, 49 CFR 174; *Hazardous Materials Regulations—Carriage by Public Highway*, 49 CFR 177, U.S. Department of Transportation, Washington, D.C., Oct. 1, 1989.

98. *Halon 1301 Fire Extinguishing Systems*, NFPA 12A, paragraph 1-5.3, National Fire Protection Association, Boston, Mass., 1992.

99. R. W. Kling and J. Magid, *Industrial Hazard and Safety Handbook*, Butterworths, London, 1979.

100. *Accident Prevention Manual for Industrial Operations—Administration and Programs*, 9th ed., National Safety Council 1988, pp. 20–32, 43–67, 69–99, 127, 149, and 328; *Accident Prevention Manual for Industrial Operations—Engineering and Technology*, 9th ed., pp. 5, 388–397, 399–425, 443–450, 477–481, 489–495, 1988.

101. *Emergency Organization for the Chemical Industry*, Safety Guide SG-4, Chemical Manufacturers Association (Manufacturing Chemists Association), Washington, D.C., 1960.

102. C. Field, *The Study of Missiles Resulting from Accidental Explosion*, U.S. AEC Safety and Fire Protection Bulletin No. 10, Washington, D.C., 1966.

103. *Guidelines for Investigating Chemical Process Incidents*, Center for Chemical Process Safety, New York, 1992.

RICHARD W. PRUGH
Process Safety Engineering, Inc.

# PLASMA TECHNOLOGY

Plasma can be broadly defined as a state of matter in which a significant number of the atoms and/or molecules are electrically charged or ionized. The generally accepted definition is limited to situations wherein the numbers of negative and positive charges are equal, and thus the overall charge of the plasma is neutral. This limitation on charge leaves a fairly extensive subject area. The vast majority of matter in the universe exists in the plasma state. Interstellar space, interplanetary space, and even the stars themselves are plasmas.

Plasma technology, the use of natural or artificially produced plasmas, has grown dramatically in variety and magnitude since the late 1950s. Many different types of plasmas exist or can be naturally or artificially created by various energy sources. Whereas gas plasmas are the most common and the most commonly used, plasmas also exist and find application within solids as well as liquids. Plasmas expected to be of consequence into the twenty-first century include those associated with integrated circuit processing (see INTEGRATED CIRCUITS), polymer production (see POLYMERS), surface modifications and coatings (qv), fusion energy (qv), specialized weapons, and propulsion technology (see EXPLOSIVES AND PROPELLANTS), as well as associated interplanetary space travel (see SPACE PROCESSING).

## Background

**History.** The term plasma was first used by Langmuir in 1928 to describe the main body of a gas discharge (1). Use of the term reflects the Greek origin of the word, to mold. Artificially induced glowing gas discharges mold themselves to the shape of their container and to items being processed within.

Work on plasmas has roots extending back to the Greeks who found that amber rubbed with various materials tended to attract certain objects. The concept of plasma as the fourth state of matter can be traced to Sir William Crookes (2) in 1879. "So distinct are these phenomena from anything which occurs in air or gas at the ordinary tension, that we are led to assume that we are here brought face to face with Matter in a Fourth state or condition, a condition so far removed from the State of gas as a gas is from a liquid." This description has been shown to be accurate over many years of experimentation and application of plasmas.

Many have worked on plasma technology. Table 1 lists a few of the contributors (3–5) and the corresponding dates associated with their concepts.

**Origins.** The theory concerning the origin of the universe, called the Big Bang, indicates that all matter and energy were contained in one unimaginably hot and dense plasma about $15 \times 10^9$ years ago (6). Following the Big Bang explosion came expansion, cooling, and condensing of matter. In some areas, gravitational accumulation of matter progressed into clusters large and dense enough to reform gaseous plasmas by nuclear reaction reheating. Other natural plasmas exist, even some terrestrial types, such as lightning and the aurora borealis. In fact, synthetically formed plasmas are essentially surrounded and outnumbered by naturally occurring types.

**Table 1. Important Concepts in Gaseous Electronics**

| Date | Concept | Originator |
|------|---------|------------|
| 1600 | electricity | Gilbert |
| 1742 | sparks | Desaguliers |
| 1785 | charge losses | Coulomb |
| 1803 | arc (discharge) | Petroff |
| 1808 | diffusion | Dalton |
| 1817 | mobility | Faraday |
| 1821 | arc (name) | Davy |
| 1834 | cathode and anode | Faraday |
| 1834 | ions | Faraday |
| 1848 | striations | Abria |
| 1858 | cathode rays | Plücker |
| 1860 | mean free path | Maxwell |
| 1869 | deflection of cathode rays | Hittorf |
| 1879 | fourth state of matter | Crookes |
| 1880 | Paschen curve | la Rue and Müller |
| 1886 | positive ion rays | Goldstein |
| 1887 | effect of light on spark gaps | Hertz |
| 1888 | emission (photoelectrons) | Hallwachs |
| 1889 | Maxwell-Boltzmann distribution | Nernst |
| 1891 | electron (charge) | Stoney |
| 1895 | x-rays | Röntgen |
| 1897 | (cyclotron) frequency | Lodge |
| 1898 | ionization | Crookes |
| 1899 | transport equations | Townsend |
| 1899 | energy gain equations | Lorentz |
| 1901 | Townsend coefficients | Townsend |
| 1905 | diffusion of charged particles | Einstein |
| 1906 | electron (particle) | Lorentz |
| 1906 | (plasma) frequency | Rayleigh |
| 1914 | ambipolar diffusion | Seeliger |
| 1921 | Ramsauer effect | Ramsauer |
| 1925 | Debye length | Debye and Hückel |
| 1928 | plasma | Langmuir |
| 1935 | velocity distribution functions | Allis |

**Definitions.** When positive charges are fixed in a solid, but the electrons are free to move about, the system is called a solid-state plasma. In a liquid-state plasma, both the positive and negative charges are fully mobile. These solid-state and liquid system are examples of condensed matter plasmas as opposed to gaseous plasmas.

Gaseous plasmas sustained by electric fields at reduced pressure, either under direct or alternating current, are sometimes referred to as glow discharges, because they emit light. There is, in fact, a slight difference between the terms plasma and discharge. There are some regions within discharges which may not strictly adhere to the neutral concept of plasmas (7). However, in general, the two terms are used interchangeably throughout the plasma technology literature.

Central to the categorization of plasmas are electron temperature and electron density. Electrons have a distribution of energies, so it is useful to assume a Maxwellian distribution, in terms of electron energy, $E$, such that

$$f(E) = \frac{2E^{1/2}}{((\pi)^{1/2}(kT_e)^{3/2})\ \exp(-E/kT_e)}$$

where $f(E)$ is proportional to the number of electrons having an energy between $E$ and $E + \Delta E$, $k$ is the Boltzmann constant, and $T_e$ is the electron temperature. The electron energy is given by the following:

$$E = (1/2)mV^2$$

where $m$ is the electron mass and $V$ is the magnitude of the electron velocity. This leads to the relation for the average energy:

$$\int E(f(E))\,dE = (3/2)kT_e$$

and thus for a Maxwellian distribution, $T_e$ is a measure of the average electron energy (7). Some weakly ionized plasmas do not necessarily have Maxwellian distributions of electron energies but it is still common to use the electron temperature, $T_e$, when describing them. Electron temperatures can be expressed in eV or in degrees K, using 1 eV = 11,600 K from $E = kT_e$. Densities of electrons and ions are measured per unit volume, usually as the number of particles per cubic centimeter.

An important characteristic of plasma is that the free charges move in response to an electric field or charge, so as to neutralize or decrease its effect. Reduced to its smallest components, the plasma electrons shield positive ionic charges from the rest of the plasma. The Debye length, $\lambda_D$, given by the following:

$$\lambda_D = \left(\frac{kT}{4\pi ne^2}\right)^{1/2}$$

where $n$ is the electron density and $e$ is the electron charge, is a measure of the extent to which plasma electrons collect in the vicinity of a charge and create a shielded potential. Ordinary, nonplasma gases have an average interparticle spacing of $d = n^{-1/3}$ and a mean free path between collisions of $\lambda = (\pi nD^2)^{-1}$ where $D$ is the atomic or molecular particle size. The number of electrons within a Debye sphere, $N_D$, around an ion is then $N_D = 4\pi\lambda_D^3 n/3$. The volume of a plasma must be significantly larger than $N_D$ or else the shielding would not be complete, a charge imbalance would exist, and the definition of a plasma having overall charge neutrality would be violated.

The primary characteristic frequency of an ordinary gas is the rate of collision $f = \overline{V}/\lambda = \pi \overline{V} n D^2$, where $\overline{V}$ is the mean particle velocity, and $\overline{V} = (8kT/\pi m)^{1/2}$ for particles of mass $m$. Among the special frequencies associated with plasmas, the most notable is the plasma frequency:

$$\omega_P = (4\pi n e^2/m)^{1/2}$$

This frequency is a measure of the vibration rate of the electrons relative to the ions which are considered stationary. For true plasma behavior, plasma frequency, $\omega_P$, must exceed the particle-collision rate, $f$. This plays a central role in the interactions of electromagnetic waves with plasmas. The frequencies of electron plasma waves depend on the plasma frequency and the thermal electron velocity. They propagate in plasmas because the presence of the plasma oscillation at any one point is communicated to nearby regions by the thermal motion. The frequencies of ion plasma waves, also called ion acoustic or plasma sound waves, depend on the electron and ion temperatures as well as on the ion mass. Both electron and ion waves, ie, electrostatic waves, are longitudinal in nature; that is, they consist of compressions and rarefactions (areas of lower density, eg, the area between two compression waves) along the direction of motion.

**Plasma Types.** Figure 1 (7–9) indicates the various types of plasmas according to their electron density and electron temperature. The colder or low electron energy regions contain cold plasmas such as interstellar and interplanetary space; the earth's ionosphere, of which the aurora borealis would be a visible type; alkali-vapor plasmas; some flames; and condensed-state plasmas, including semiconductors (qv).

Gaseous plasmas are sometimes classified as equilibrium or nonequilibrium referring to the electron temperature as compared to the gas temperature. Low pressure glow discharges are generally classified as nonequilibrium plasmas because the electron temperature is significantly greater than the gas temperature. Commonly, glow discharges have electron temperatures in the $10^4$–$10^5$ K (1–10 eV) range, whereas the gas temperature in those discharges is generally less than $5 \times 10^2$ K or near ambient. Low and high pressure arc discharges, also known as plasma jets, have no such large difference and the electron and gas temperatures are both in the $5 \times 10^3$–$10^4$ K range, thus being called equilibrium plasmas. The areas of most interest to plasma chemistry are the glow discharges and arcs.

Controlled thermonuclear fusion experiments and certain types of confined arcs known as pinches have temperatures in the $5 \times 10^5$–$10^7$ K range. However, to be successful, controlled thermonuclear fusion needs to take place from $6 \times 10^7$–$10^9$ K. In fact, the goal of all fusion devices is to produce high ion temperatures in excess of the electron temperature (10).

Also shown in Figure 1 are the Debye screening length and Debye sphere size. For gaseous plasmas, $N_D \gg 1$ (11). Solid-state plasmas or condensed-state plasmas generally exist where $N_D \leq 1$.

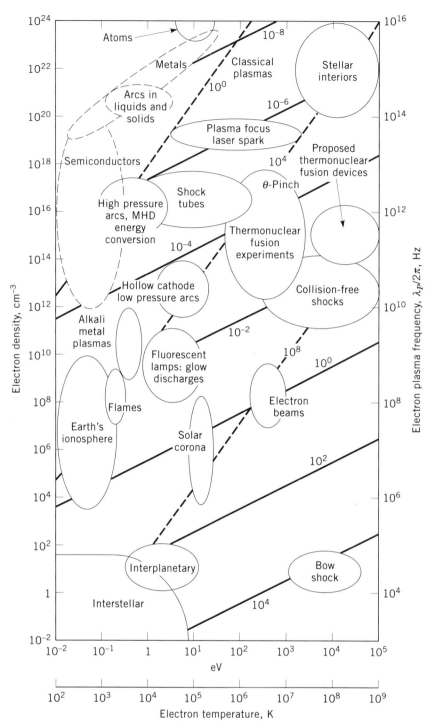

**Fig. 1.** Electron temperature and density regions for plasmas (7–9) where the numbers and the diagonal lines represent (——) the Debye screening length, $\lambda_D$, in centimeters, and (———) Debye sphere volume, $N_D$, in cubic centimeters. The plasma frequency is given on the right-hand axis. Condensed-state plasmas, indicated in dashed line areas, have relatively low temperatures, small Debye spheres, and high densities. Plasmas in metals and semimetals fall along the line separating degenerate quantum plasmas from nondegenerate classical plasmas. Gaseous plasmas, shown in solid line areas, have characteristics that vary widely. MHD = magnetohydrodynamic.

*Field Effects.* Velocities, lengths, and frequencies are intrinsic to gases and plasmas, independent of incident radiation or existing fields. However, some of the more interesting plasma phenomena only appear in the presence of static or dynamic fields. External static electric fields tend to separate and accelerate plasma charges. Such fields and the resultant electron motions can both produce plasmas and heat them. Magnetic field effects in plasmas are so important that at one time plasma physics and magnetohydrodynamics (qv) (MHD) were practically synonymous. However, some plasmas are not magnetized, and some MHD processes do not involve plasmas, eg, the fluid motions in the earth's core which produce the terrestrial magnetic field.

Transverse electromagnetic waves propagate in plasmas if their frequency is greater than the plasma frequency. For a given angular frequency, $\omega$, there is a critical density, $n_c$, above which waves do not penetrate a plasma. The propagation of electromagnetic waves in plasmas has many uses, especially as a probe of plasma conditions.

$$n_c = m\omega^2/4\pi e^2$$

The presence of a static magnetic field within a plasma affects microscopic particle motions and microscopic wave motions. The charged particles execute cyclotron motion and their trajectories are altered into helices along the field lines. The radius of the helix, or the Larmor radius, is given by the following:

$$mcV_\perp/qB$$

for a particle of mass $m$ and charge $q$, and velocity $V_\perp$ normal to the magnetic field of strength $B$, where $c$ is the velocity of light. The cyclotron frequency, $\omega_c$, which is introduced by the presence of the field, is

$$\omega_c = qB/mc$$

The particle–field interaction is the means by which magnetic fields can exert pressures on plasmas and vice versa.

The presence of static magnetic fields does not alter the propagation of longitudinal electrostatic electron or ion waves if the propagation direction is parallel to the field. However, propagation that is orthogonal to the field involves new frequencies that depend on the field strength. In contrast, the propagation of transverse electromagnetic waves in plasmas is altered by a magnetic field regardless of the relative geometry of the direction of motion and the field vector.

Magnetic fields introduce hydromagnetic waves, which are transverse modes of ion motion and wave propagation that do not exist in the absence of an applied $B$ field. The first of these are Alfven, A, waves and their frequency depends on $B$ and $\rho$, the mass density. Such waves move parallel to the applied field having the following velocity:

$$V_A = B/(4\pi\rho)^{1/2}$$

and are similar to the waves that travel along a string. Magnetosonic waves are a second type of hydromagnetic wave, and these propagate perpendicular to the magnetic field. Their frequency depends on the Alfven and the acoustic velocities. Hydromagnetic waves are electromagnetic waves. Even though the applied $B$ field is static or nonoscillatory, the waves are transverse and are characterized by oscillatory electric, $E$, and magnetic, $B$, components.

A good discussion of plasma waves and a tabulation of their characteristics is available (12). Useful plots of the dispersion relations for various frequencies, field conditions, geometries, and detailed mathematical relationships are given in Reference 13.

*Gaseous Plasmas.* Gaseous plasmas are often far from equilibrium and therefore can exhibit microscopic or particle instabilities, and macroscopic or hydromagnetic instabilities (14,15). Microscopic instabilities are caused by departures from the equilibrium Maxwellian distributions for the electrons or ions. Examples of such situations include a plasma expanding while cooling, anisotropies in the velocity distribution caused by applied magnetic fields, or the motion or streaming of a particle beam through a plasma. Macroscopic instabilities produce the motion of the plasma as a whole. Causes include pressure or density gradients or magnetic field curvature. All instabilities represent the tendency of plasmas to reach equilibrium more quickly than is possible by ordinary collisions alone. Instabilities can reduce plasma confinement times by many orders of magnitude, a significant problem in fusion research. The conduction and diffusion of energy through plasmas, and the way in which these processes are influenced by magnetic fields, are described (12,13).

The high energy densities of many gaseous plasmas raise safety concerns. The sources of energy used to produce and heat plasmas, eg, steady-state, high voltage, and high current generators, and capacitors for pulsed electron-discharge heating and laser beams, can be hazardous (16,17). Work with plasmas usually requires careful attention to proper electrical safety precautions and to eye hazards. Even in the absence of lasers (qv), plasmas can pose a threat to vision because the plasmas often are very bright and can emit dangerous levels of uv radiation. X-radiation from plasmas usually is not a safety concern. Most energy from $10^6$ K plasmas is soft and does not escape from the experimental chamber or traverse significant distances in air. The hard x-rays emitted by most plasma sources usually are of very low intensity. However, some low energy, low pressure plasmas, such as electron cyclotron resonance sources, high ($>10^4$ V) voltage, plasma-generating machines and some fusion-energy research devices can emit unsafe x-ray emissions. Shielding, eg, lead or concrete, and distance from the source reduce exposures to acceptable levels.

## Production

Sources of matter and energy are necessary for the production of gaseous plasmas, and such plasmas serve as sources of matter and energy in their applications; ie, gaseous laboratory plasmas can be viewed as transducers of matter and energy. The initial and final forms of the material that enters a plasma and the requisite energy vary widely, depending on the particular plasma source and its utilization.

The molecules that are dissociated and the atoms that are ionized during plasma production can be in any state at the start. Steady-state plasmas are formed most often from gases, although liquids, such as volatile organics, and solids are also used. Gases and solids routinely serve as sources of material in pulsed plasma work.

The energy for plasma formation may be supplied in a variety of ways. The source may be internal, eg, the release of chemical energy in flames. Another energy source involves electrical excitation (18). The externally applied d-c or a-c field excites the plasma, which becomes part of the electric circuit. Electromagnetic fields, eg, in the radio frequency (rf) or microwave range (19), can be used to form plasmas that interact with or feed back into the source. It should be noted that electron motion in an a-c field, especially a r-f one, yields a stable plasma without the secondary electron emission required for d-c plasmas. A fourth kind of energy source includes externally produced beams of photons, eg, laser beams or other energetic particles, that create a plasma by impact and absorption independent of the source. Plasmas may be produced in a fifth manner by strong shock waves. Chemical, discharge, and high frequency sources often produce steady-state plasmas, whereas beam and shock heating usually produces pulsed plasmas.

In the production of plasmas by steady-state electric discharge heating, the current across two electrodes in the gas can vary widely, as indicated in Figure 2 (4,11,20–22). At low current values, externally induced ionization is needed to maintain current flow. Milliampere currents in gases having pressures at about 100 Pa ($10^{-3}$ atm) produce a glow discharge that is sustained by electrons produced by positive-ion bombardment of the cathode. If the current is increased to around 10 A, a self-sustaining arc forms and exists even at pressures above 101 kPa (1 atm).

**Modification and Dissipation.** Changes in the composition or energy of a plasma after it is formed are often desired. For example, materials can be introduced into a plasma and excited, thereby producing information for spectrochemical analysis. Plasma heating is a more common modification. Plasma heating, ie, raising the energies of all the particles comprising the plasma, is of primary importance or interest in fusion science and technology. Many energy sources and coupling mechanisms can be employed to heat plasmas (12). Plasma formation and heating often are driven by the same energy source. However, an entirely separate second source can be used for heating. For example, laser beam and particle beam sources often are used to heat plasmas produced by electric discharges. Some commercial equipment uses simple infrared heat sources to boost the energy of their plasmas. Energetic neutral, ie, atomic beams are employed to heat plasmas that are formed initially by discharges.

Restraining a gaseous plasma from expanding and compressing is also a form of plasma modification. Two reasons for plasma confinement are maintenance of the plasma and exclusion of contaminants. Plasmas may be confined by surrounding material, eg, the technique of wall confinement (23). A second approach to confinement involves the use of magnetic fields. The third class of confinement schemes depends on the inertial tendency of ions and associated electrons to restrain a plasma explosion for a brief but useful length of time, ie, forces active over finite times are required to produce outward particle

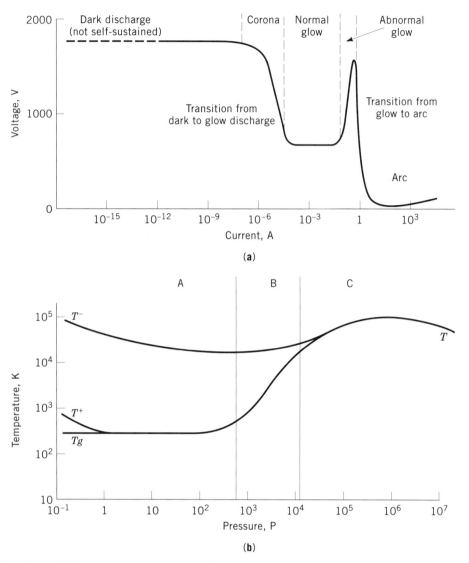

**Fig. 2.** (**a**) Voltage–current relationship for an electrical discharge (4,11,21,22); (**b**) plasma temperature vs pressure. Temperatures, $T$, are indicated for the plasma electrons, $T^-$, and ions, $T^+$, and the neutral gas, $Tg$. Region A is the low pressure nonequilibrium cold-arc regime where $T^+ \approx Tg$, B is the transition area, and C is the high pressure equilibrium hot-arc regime where $T^- \approx T^+ \approx Tg$ (20). To convert Pa to atm, multiply by $9.87 \times 10^{-6}$.

velocities. This inertial confinement is usually, but not necessarily, preceded by inward plasma motion and compression.

Low density plasmas are confined magnetically by a variety of field configurations that are designed to prevent particle losses and overall fluid instabilities (Fig. 3). Fusion-research plasmas, low temperature plasmas used for sputtering and high temperature, can be contained by magnetic confinement which can take several different forms (24). Magnetron sputtering sources include a variety of magnetic field configurations designed to restrain plasma particles near

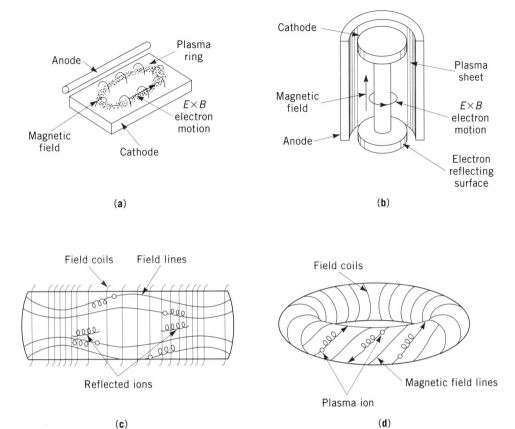

**Fig. 3.** Schematics of magnetic confinement geometries: (**a**) planar and (**b**) cylindrical geometries for magnetron sputtering sources (24); (**c**) open-ended mirror and (**d**) closed toroidal geometries for containing fusion-research plasmas (25).

the cathode from which atoms are sputtered by impact of ions (24). The two basic magnet configurations for fusion-research plasmas are open and closed (25). Increasing strength along magnetic field lines produces magnetic mirrors which exert a retarding force on particles that tend to escape the plasma. However, the plasma in a simple mirror is not stable against overall fluid motion. For stability, the $B$ field must increase in all directions from the plasma outward, ie, the plasma must be in a minimum $B$ location to be magnetohydrodynamically stable against gross translational motion. Curvature of the magnet geometry for mirror systems produces such stability. Nevertheless, sufficiently energetic particles can penetrate the mirror, producing end losses. Closure of the field lines to form loops counteracts such losses, but the field strength decreases radially from the center and particles tend to drift out of the plasma normal to planar closed field lines. Addition of out-of-plane fields, which yields helical fields in a toroidal configuration, produces both particle confinement and MHD stability. The history and physics of magnetic confinement of fusion plasmas are available (26).

High density plasmas can be confined and compressed magnetically by fields produced by strong electric currents flowing in and heating the plasmas, as well as by externally applied fields (27). The radial force produced by the magnetic field, which affects the flowing plasma electrons ($v \times B$ force), produces inward plasma motion and compression. Such confinement is short-lived and not stable, but it produces plasmas with high energy densities. High density plasmas are studied primarily as x-ray and neutron sources.

Inertial confinement depends on plasma heating outpacing plasma expansion. When a target is struck by a short ($<\mu$s) pulse of laser photons, electrons, or ions, it is heated on a time scale comparable to or shorter than that on which the resulting plasma expands and cools. This is true for targets of any geometry. If a cylindrical or spherical target is simultaneously irradiated from two or more sides, the outer surface separates violently, producing a reaction in the opposite direction that accelerates the remainder of the target inward, thereby compressing and heating the central materials (28).

Hot, elevated pressure, fusion-type plasmas lose matter and energy through particle escape and radiation of photons. When matter or energy are no longer supplied to a hot plasma, it expands and cools before interacting with the surroundings and totally dissipating. In low pressure cold plasmas, electron loss to chamber walls and atomic and molecular recombination affects are of more importance. Such processes must be offset during the lifetime of a plasma because the material and energy from a plasma can produce undesirable effects. Energetic particles impinging on solids near plasmas can profoundly alter the material properties. Plasma effects on materials are especially important in proposed thermonuclear fusion reactors. Materials in the high pressure plasma reactors must withstand high temperatures and stresses for an economically useful period (29). Low pressure plasmas generally deliver no such debilitating exposures. It is not uncommon for commercially available plasma units to operate for 20–30 years in production environments with little or no effect on the materials of construction.

## Diagnostics

Plasma diagnostics, the determination of conditions within plasmas, also refers to the broad collection of experimental techniques and associated calibration and analytical methods used to assess the characteristics of plasmas. Noteworthy properties include the identities, concentrations, and energy distributions of the various particle species such as neutrals, electrons, and ions, and their velocity distributions as functions of space and time. Quantities such as plasma flow velocities, turbulence, instabilities, and flow of energy into and out of plasmas by various means also are desired. Most diagnostic methods have limited resolution spatially, temporally, and spectrally. Therefore, plasma characteristics that are derived from measurements generally are averaged. Measured quantities in plasma diagnostics usually are integrated along a line of sight through the plasma to the instrumentation, yielding spatially integrated results. It is possible, however, to measure point temperatures within low temperature plasmas using fluoroptic thermometers (30). No one method for assaying plasmas is universally applicable. A variety of diagnostic tools is needed to characterize a plasma empirically and to compare its empirical and theoretical characteristics.

Diagnostic methods can be categorized most broadly according to those that involve external probes of plasmas or those that rely only on plasma self-emission. Probes of plasmas include solid instruments inserted into gaseous plasmas and beams of photons or charged particles, which are shot through plasmas. Electric voltage and current measurements for discharge-heated plasmas provide useful information, thus the energy source also is a plasma probe. Self-emission can include radiated fields, photons, electrons, ions, and neutrons, ie, essentially any plasma effect or constituent. Gaseous plasma diagnostic methods or techniques and useful compilations of gaseous plasma characteristics measured by them are available (31–43). Plasma diagnostics is, by necessity, a key field of plasma technology.

Electrodes or Langmuir probes may be inserted into plasmas that are large enough (>1 cm) and relatively cool (<$10^4$ K). The net current to the probe is measured as a function of the applied voltage. Electron temperatures, electron and ion densities, and space and wall potentials may be derived from the probe signals. Interaction of plasmas with solid probes tends to perturb plasma conditions.

Monochromatic light from short-pulse lasers may be focused into plasmas that are large enough ($\geq 1$ cm) regardless of their temperatures. These beams cause little alteration of plasma conditions. Elastic Thomson scattering of the light at right angles to the incident direction is measured. Spectral or Doppler broadening of the light, which results from motion of the scattering electrons, yields the electron temperature. The overall scattered intensity is a measure of the electron density. Thomson scattering is a primary diagnostic method for magnetic fusion research plasmas.

Diagnostic techniques that involve natural emissions are applicable to plasmas of all sizes and temperatures and clearly do not perturb the plasma conditions. These are especially useful for the small, high temperature plasmas employed in inertial fusion energy research, but are also finding increased use in understanding the glow discharges so widely used commercially.

The small ($\leq 1$ cm) sizes and brief (<1 $\mu$s) lifetimes of the fusion research plasmas preclude the use of most probe techniques. Laser pulse imaging of such plasmas does yield valuable spatial information, however. Diagnostic methods involving plasma emissions can be used to resolve spatial and temporal nonuniformities in plasma emission (44). For high temperature plasmas, diagnostic methods based on x-ray emission are especially useful, because the x-rays are preferentially emitted from the hottest and densest parts of plasmas. Emission spectrometry is useful in determining relative concentrations of species, reactants, or products in glow discharges. This information can be used to determine process efficiencies and proper end-of-process times to prevent damage to items being plasma treated. Figure 4a is an example of emission spectrometry data for the removal of chlorine contamination from small electronic components using an r-f plasma (45). Figure 4b shows the monitoring of the etching and eventual removal of a polymer layer from a metal substrate (45).

## Types of Plasmas

**Natural Gaseous Plasmas.**  Lightning is the most common atmospheric, plasma related phenomenon (46,47). The separation of charges in clouds, and between clouds and the ground, produces potentials as high as 100 MV. The currents of the discharges are as high as 100 kA. Spectroscopic data have yielded

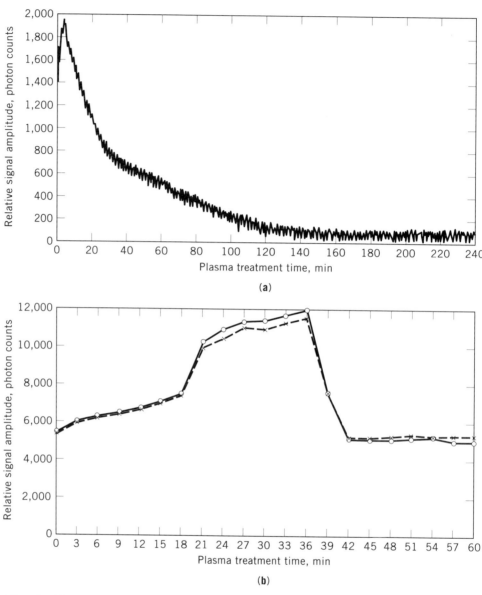

**Fig. 4.** Examples of emission spectrometry as a diagnostic monitoring tool for plasma processing. (**a**) The removal of chlorine contamination from copper diode leads using a hydrogen–nitrogen plasma. Emissions are added together from several wavelengths. (**b**) The etching and eventual removal of a 50-$\mu$m thick polyimide layer from an aluminum substrate, where ($\times$) and ($\circ$) correspond to wavelengths (519.82 and 561.02 nm, respectively) for molecular $CO_2$ (45).

considerable information on plasma conditions within a lightning discharge (48,49).

Meteors produce atmospheric plasmas as their kinetic energy is converted to thermal energy (50). Most particles from space are consumed before they reach an altitude of 50 km. Meteors are of little practical use, although radio waves can be bounced off the plasmas left in their wakes (see EXTRATERRESTRIAL MATERIALS).

Auroras are observed primarily at polar latitudes near the geomagnetic poles and result from impact on the atmosphere of energetic particles that are guided by the earth's electromagnetic field (51,52). The emissions are produced at altitudes of one to several hundred kilometers by electrons having energies up to 10 keV, or by protons having energies as high as 200 keV. Ionization densities that produce auroras reach plasma levels generally following intense solar flare activity.

Absorption of solar uv radiation high in the atmosphere produces a tenuous but important plasma, the ionosphere (53). Many physical processes occur in the ionosphere during the creation and loss of free electrons (54). As shown in Figure 5, the electron density can exceed $10^6$ cm$^{-3}$ and the electron temperature generally is less than 1000 K (0.1 eV) (55). Atmospheric motions below and solar activity above the ionosphere influence its characteristics, which vary markedly with the time of day and the solar cycle and emission. Understanding

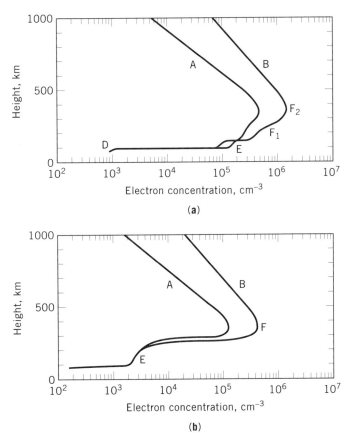

**Fig. 5.** Ionospheric electron density vs height above the earth at the extremes (A = minimum, B = maximum) of the 11-yr sunspot cycle during (**a**) day and (**b**) night (54). D, E, F, $F_1$, and $F_2$ are conventional labels for the indicated regions of the ionosphere.

of the ionosphere has led to the increasingly effective use of it in long-range communication.

The ionosphere is part of the larger magnetosphere, a cavity in the stream of particles from the sun. The cavity is produced by the earth's magnetic field (56,57). The ionosphere and the Van Allen radiation belt lie within the plasmasphere, which extends to a maximum distance of about 15,000 km above the earth's surface.

Magnetospheric plasmas are produced and heavily influenced by solar emissions and activity and by magnetic fields of the planets. Interplanetary plasmas result from solar emission processes alone. Protons in the solar wind have low densities ($10-100/cm^3$) and temperatures below $10^4$ to more than $10^6$ K ($1-10$ eV). Their average outward kinetic energy from the sun is approximately 400 eV (58,59). The various zones and phenomena from the sun's visible surface to the upper atmosphere of the earth have been discussed (60–62).

Classical astronomy is largely concerned with the classification of stars without regard to the details of their constituent plasmas (63). Only more recently have satellite-borne observations begun to yield detailed data from the high temperature regions of other stellar plasmas. Cosmic plasmas of diverse size scales have been discussed (64).

**Plasmas in Condensed Matter.** In contrast to gaseous plasmas, which can be described by kinetic and fluid theories, plasmas in condensed matter are intimately related to the theory of solids (65–67). The formation of a diatomic molecule is a good example of the freedom and high velocities that electrons experience in the condensed state. Electrons that initially are confined to either atom, once the atoms bond, are free to range over the larger molecular volume. In the buildup of larger aggregates of condensed matter, addition of more atoms similarly expands the accessible volume. Bonding electrons increase their velocities when atoms form into molecules and condensed matter.

The energy levels of bonding electrons in a conductor are shown in Figure 6 (68). Because electrons are Fermions, no two can occupy the same spatial and

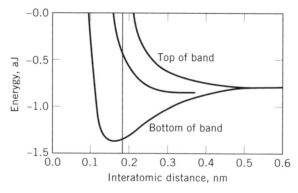

**Fig. 6.**   Schematic energy levels of a solid as a function of interatomic distance where the vertical line represents the equilibrium spacing (68). A band of states obeying Fermi distribution is required by the Pauli principle. High electron velocities and equivalent temperatures exist in conductors even when the lattice is at ordinary temperatures. To convert J to cal, divide by 4.184.

energy states simultaneously, according to the Pauli exclusion principle. Energy bands develop as more atoms agglomerate, forming a solid. These may be continuous, as for metals, or have an energy gap. Such gaps are narrow for semiconductors and wide for insulators. The Fermi distribution or the distribution of electrons in metals is uniform with energy. The bands and the distribution are filled to an energy called the Fermi level, $E_F$, at absolute zero. Above this temperature there are vacancies below $E_F$ and electrons in levels above $E_F$. Electrons at $E_F$ move with velocities near $10^8$ cm/s, which is equivalent to temperatures near 1 eV, even though the lattice is cold ($\sim$300 K). Densities of mobile electrons in metals are approximately $10^{22}$ particles/cm$^3$. The presence of plasmas in metals is readily shown by the excitation of electron plasma waves or plasmons. The characteristic energies of plasmas introduce additional peaks into electron and x-ray spectra from metals (67).

In perfect semiconductors, there are no mobile charges at low temperatures. Temperatures or photon energies high enough to excite electrons across the band gap, leaving mobile holes in the Fermi distribution, produce plasmas in semiconductors. Thermal or photoexcitation produces equal numbers of electrons and holes, similar to the neutral charge situation in gaseous plasmas where the charges of electrons and ions usually balance. Photoexcited plasmas, which are characterized by particle densities of around $10^{17}$–$10^{18}$/cm$^3$, are stable in silicon at temperatures below 20 K and in germanium below 8 K (69,70).

Impurity-produced plasmas in semiconductors do not have to be compensated by charges of the opposite sign. These plasmas can be produced by introduction of either electron donors or electron scavengers, ie, hole producers, into semiconductor lattices. Their densities range from a lower limit set by the ability to produce pure crystals ($\sim$10$^{12}$ particles/cm$^3$) to values in excess of $10^{20}$ particles/cm$^3$. Plasmas in semiconductors generally are dilute, so that the Pauli principle cannot be used to determine their energy distribution. These plasmas are Maxwellian and have the same temperature as the lattice in equilibrium. The uncompensated nature of plasmas in semiconductors, and especially their unique and very wide ranges of density and temperature, make them particularly interesting and useful. Numerous plasma effects, including waves and instabilities, are studied in semiconductor plasmas (71,72). Effects peculiar to solids such as photon scattering of plasma particles and anisotopic effective masses, also are unique to semiconductor plasmas.

Plasmas in gaseous and condensed states are related by more than the principles that govern them. For gaseous matter, there is a continuum of behavior from a low density Maxwellian plasma to a high density Fermion plasma. The boundary region in the density–temperature plane is given by the equality of the thermal energy, $kT_F$, and the Fermi energy, $E_F$, where $h$ is Planck's constant (73).

$$kT_F = E_F = \bar{h}^2/2m_e(3\pi^2 n_F)^{2/3}$$

Densities less than $n_F$ indicate classical, nondegenerate behavior, whereas those above $n_F$ imply quantum, degenerate conditions, such as can be found in condensed-state plasmas (see Fig. 1).

Gaseous and condensed plasmas have been summarized (65,74). Fundamental differences between gaseous and condensed plasmas include their states of excitation and characteristic lengths. Gaseous plasmas are produced by classical collisional effects. Some gaseous plasmas, although clearly not all, are not in equilibrium thermodynamically and are accompanied by photons. In contrast, solid-state metallic plasmas are produced by quantum effects, can be in or near the ground state, and can exist in the absence of photons. Instabilities are common in gaseous plasmas but can be avoided or excited as desired in condensed-state plasmas. Long, mean free paths, compared to the plasma size, are common in gaseous plasmas. However, in solids, scattering usually produces short mean free paths in comparison to the millimeter sizes of ordinary samples. Scattering often interferes with plasma behavior in semiconductor plasma experiments.

## Uses

**Radiation Sources.** Ordinarily, electron beams are produced from solids *in vacuo* by thermal or field-assisted processes. Plasmas also serve as electron sources, but are more uniquely used as ion sources. Whereas ions can be produced by sputtering and field assisted processes in the absence of plasmas, most ion sources involve plasmas (75).

Pulsed plasmas containing hydrogen isotopes can produce bursts of alpha particles and neutrons as a consequence of nuclear reactions. The neutrons are useful for radiation-effects testing and for other materials research. A dense plasma focus filled with deuterium at low pressure has produced $10^{11}$ neutrons in a single pulse (76) (see DEUTERIUM AND TRITIUM). Intense neutron fluxes also are expected from thermonuclear fusion research devices employing either magnetic or inertial confinement.

Plasmas frequently are used as sources of incoherent and coherent electromagnetic radiation. Infrared radiation is emitted by plasmas, but these usually are not employed expressly as ir sources. Visible light sources involving steady-state plasmas, in which the electrons are hot compared to the ions, are common. Fluorescent lights and carbon-arc searchlights are examples. Inert gas plasma displays that are flat and are replacing bulky cathode-ray tubes, also depend on emission of visible light. Direct-current plasmas emit visible and uv radiation for chemical analysis. Pulsed sources of visible and uv light for high speed photography (qv) usually involve plasmas. Plasma uv sources are used commercially for production of microelectronics circuits by lithography. Multimillon-degree ($\times 10^6$ K) plasmas provide uniquely short pulses of incoherent uv and soft x-radiation for spectroscopy, materials analysis, and other applications.

Lasers act as sources and sometimes as amplifiers of coherent ir–uv radiation. Excitation in lasers is provided by external particle or photon pump sources. The high energy densities required to create inverted populations often involve plasma formation. Certain plasmas, eg, cadmium, are produced by small electric discharges, which act as laser sources and amplifiers (77). Efforts that were directed to the improvement of the energy conversion efficiencies at longer wavelengths and the demonstration of an x-ray laser in plasma media were successful (78).

**Chemistry.** The material and energy available in plasmas can be used to excite materials and drive chemical reactions. The unique characteristics of plasmas, especially their abundance of energetic species, have been exploited in plasma chemical applications (79–84).

The analysis of existing materials and the production of new chemicals and materials involve a gamut of gaseous plasma sources. Nonequilibrium or cold plasmas, in which the ion or gas temperature is much less than the electron temperature, are widely used. In these, the electrons provide energy that induces excitations and reactions without excessive heating of the desired products or the surroundings by the heavier particles in the plasmas. Equilibrium or hot plasmas, in which the electrons and ions are characterized by approximately the same temperature, also are used in plasma chemistry, especially for spectrochemical analysis and the processing of refractory materials (see REFRACTORIES) (85) (see Figs. 1 and 2). Plasma-electron temperatures below 50,000 K ($\leq 5$ eV) are most useful for plasma chemistry.

**Chemical Analysis.** Plasma oxidation and other reactions often are used to prepare samples for analysis by either wet or dry methods. Plasma excitation is commonly used with atomic emission or absorption spectroscopy for qualitative and quantitative spectrochemical analysis (86–88).

Samples to be analyzed may be collected on filter papers, either directly from the air, water, or other carriers, or after wet chemical separation or concentration. The filters, composed of organic materials, interfere with subsequent analysis steps. Thus, if the sample is to be heated or excited in steps following filtration, the filter must be removed or destroyed. Ashing to rid the samples of organic materials can be accomplished in oxidizing or nonoxidizing plasmas (89). Excitation is provided by r-f energy coupled either inductively or capacitively to the plasma, although capacitive coupling is the most common commercially available method. Ashing also is used in the preparation of samples for microscopic examination (90).

Many sources of energy are used to excite samples to emit characteristic wavelengths for chemical identification and assay (91,92). Very high temperature sources can be employed but are not necessary. All materials can be vaporized and excited with temperatures of only a few electron volts. The introduction of samples to be analyzed into high temperature or high density plasmas and their uniform excitation often are problematic.

Use of glow-discharge and the related, but geometrically distinct, hollow-cathode sources involves plasma-induced sputtering and excitation (93). Such sources are commonly employed as sources of resonance-line emission in atomic absorption spectroscopy. The analyte is vaporized in a flame at 2000–3400 K. Absorption of the plasma source light in the flame indicates the presence and amount of specific elements (86).

Pulsed spark sources, in which the material to be analyzed is part of one electrode, are used for semiquantitative analyses. The numerous and complex processes involved in spark discharges have been studied in detail by time- and space-resolved spectroscopy (94). The temperature of d-c arcs, into which the analyte is introduced as an aerosol in a flowing carrier gas, eg, argon, is approximately 10,000 K. Numerous experimental and theoretical studies of stabilized plasma arcs are available (79,95).

Plasmas can be produced by radio frequencies that are inductively coupled to aerosol samples in argon streams. Inductively coupled high temperature plasma chemical analysis devices are operated at close to 30 MHz (96).

Temperatures of 10,000–15,000 K are produced having electron densities of about $10^{15}/cm^3$ (91). These conditions tend to ensure vaporization, dissociation, and excitation. Microwaves near 2 GHz also can be used to excite plasmas for analysis (see MICROWAVE TECHNOLOGY). The analyte is carried in an argon gas flow which passes through a cavity excited by a magnetron or other source of microwaves. Temperatures of 6000 K and electron densities of $10^{12}-10^{15}/cm^3$ are normal in microwave-excited analytical plasmas (97). Low temperature analysis is also possible in some situations. A device using low temperature plasma to analyze a surface for organic contamination has been developed. The surface is flushed with low temperature gas plasma species which react with organic molecules and emit light in characteristic wavelengths (98).

High power pulsed lasers are used to produce plasmas and thus to sample and excite the surfaces of solids. Improvements in minimum detectable limits and decreases in background radiation and in interelement interference effects result from the use of two lasers (99) (see SURFACE AND INTERFACE ANALYSIS).

In plasma chromatography, molecular ions of the heavy organic material to be analyzed are produced in an ionizer and pass by means of a shutter electrode into a drift region. The velocity of drift through an inert gas at approximately 101 kPa (1 atm) under the influence of an applied electric field depends on the molecular weight of the sample. The various sonic species are separated and collected every few milliseconds on an electrode. The technique has been employed for studying upper atmosphere ion molecule reactions and for chemical analysis (100).

**Plasma Processing.**  Plasma processing is an extremely broad and growing field. Radio-frequency gas plasma processing, made commercially available in the late 1960s, is an environmentally conscious surface modification and material production technology. The literature covering plasma processing is quite extensive and grows by thousands of references each year. Space permits the listing of only a small number herein (79,101–118).

Plasmas can accelerate reactions that are otherwise slow to the point of impracticality. Moreover, plasmas are often used to accomplish processes not possible by other means, eg, providing atomically clean surfaces of materials that would be damaged by high temperature or wet chemical cleaning. In the plasma processing generally used in industry, no hazardous wastes are generated and, in fact, plasma usage has proven effective at decomposing hazardous waste materials (119). Most plasma processes take a few minutes and after the plasma treatment process is completed, parts or assemblies are immediately ready to be bonded, potted, welded, painted, soldered, or assembled in whatever manner necessary. It is, however, possible to overprocess or damage items being treated. Even low pressure plasmas can result in damage to materials owing to ion, electron, or photon bombardment. Analysis of materials exposed to plasmas is often useful and usually required in both experimental and commercial operations (120).

Plasma processing also offers several operational and cost advantages, eg, replacement of batch with flow processes. Plasma equipment often is smaller than other process hardware, resulting in associated savings in capital expenditure and floor space. Plasma units generate much smaller amounts of waste heat than most types of processing equipment and provide rapid start-up and shutdown. Plasma processing based on hydroelectrical power can be less expensive than processes that derive energy from fossil fuels. Plasma sources that are used to drive chemical reactions range from relatively low pressure and temperature, nonequilibrium glow-discharges, to dense, hotter arcs.

**Surface Modification.** Plasma surface modification can include surface cleaning, surface activation, heat treatments, and plasma polymerization. Surface cleaning and surface activation are usually performed for enhanced joining of materials (see METAL SURFACE TREATMENTS). Plasma heat treatments are not, however, limited to high temperature equilibrium plasmas on metals. Heat treatments of organic materials are also possible. Plasma polymerization crosses the boundaries between surface modification and materials production by producing materials often not available by any other method. In many cases these new materials can be applied directly to a substrate, thus modifying the substrate in a novel way.

Treating a surface with activated gas plasma is a dry process requiring no solvents or rinses of any kind. Using the correct choice of gases and process parameters, plasma cleaning can render a surface atomically clean of organic contaminants and/or activate a surface for enhanced bonding without damage to that surface (121). Figure 7 shows a continuous plasma fiber/wire treatment system that can provide surface treatments in a spool-to-spool or air-to-air mode. The plasma removal of some inorganic contaminants is possible using certain reducing, rather than oxidizing, atmospheres (see Fig. 4a). Plasma surface cleaning is done in a fully contained and controlled vacuum environment. The only effluents are the volatile reaction products of the plasma gas and the surface contaminants (see Fig. 4b). Surface drying by plasma treatment has taken on increased importance in the climate of aqueous-based wet chemical cleaning for environmentally conscious manufacturing.

Gas plasma activation of low surface energy polymers for enhanced adhesive bonding has been recognized as a viable process for many years (111,118,122–124). Plasma treatments are able to increase the surface energy of most polymers, making their surfaces more polar and wettable, thus enhancing adhesive bondability. Polar or active sites on the polymer surfaces are formed when activated species present in the plasma collide with the surface and transfer energy to it. This energy can break bonds on the surface of the material causing unsaturation, remove atoms or molecules of material, substitute atoms or molecules onto a surface, or leave partial charges where bonds have reformed in less stable ways. In some instances, oxidation of the substrate is undesirable. In these cases, activation of a surface can be accomplished using plasma gases such as argon, helium, nitrogen, or mixtures of nonoxidizing gases such as a nitrogen and hydrogen mix. Because of their inert nature, these gases do not provide the organic contaminant removal capabilities that oxygen-containing gases do (125). However, if the surface is sufficiently clean, these gases can activate the surface and provide enhanced bonding.

Direct-current arcs into which no material is introduced have many applications as heat sources. Industrial processing of metals using plasma torches has been carried out in the former USSR (126). Thermal plasmas also are used in surface and heat treatment of materials (127,128). Metals can be hardened by exposure to heat from thermal or equilibrium plasmas. Natural fibrous ma-

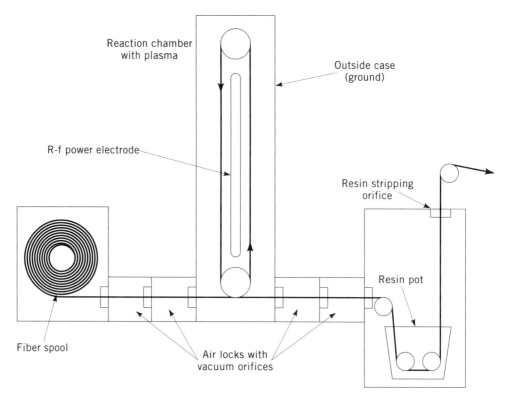

**Fig. 7.** Schematic of a self-contained plasma processing unit designed to continuously plasma-treat and impregnate with resin, reinforcing fibers for enhanced composite strength. The unit can also be used to plasma-treat wires to be coated or treated for improved adhesion. Throughput speeds of over 100 m/s are attainable for commercial use.

terials such as wool (qv) can be made shrink resistant by careful exposure to equilibrium atmospheric pressure plasmas at 150°C or less (129). Alteration of surface properties such as wettability using plasmas has been reviewed (130). The forming of materials using plasmas is widespread (131); plasma torches are employed to cut thick metal plates (22); and the use of d-c arcs for joining metals by welding (qv) is common.

**Plasma Polymerization.** Plasma polymerization is becoming a commercially viable method for modifying low cost organic substrates. Organic and other complex molecules that are exposed to low temperature nonequilibrium plasmas can be affected in terms of polymerization, rearrangements (isomerizations), surface activation, elimination of constituent parts, and total destruction of the original molecules accompanied by the generation of atoms and ions (132). The production of polymers and other heavier molecules from gaseous monomers is an attractive application of plasmas in organic chemistry (21,133,134). A wide variety of chemically inert, adherent films such as plasma-polymerized fluorocarbon films can be produced using simple, commercially available apparatus (21,133,135,136). Radio-frequency discharge polymerization of organic chemical monomers is employed to produce organosilicon films, which are useful as light guides for integrated optics (137). Semipermeable membranes for hyperfiltration also are prepared by the use of plasmas, eg, poly(vinylene carbonate) deposited on a Millipore filter. Such composite filters are dense, pinhole-free, easy to prepare, and provide high salt rejection (138). The attachment of fluorine (qv) directly onto the surface of polymers can provide fluorocarbon-like surfaces on low cost polymer substrates (139).

**Materials Production.** Substances not producible by conventional means can be made using plasmas. Plasma materials production and modification embraces processes such as production of thin coatings (qv), heat treatment, and the joining of materials. Simple organic and inorganic molecules are produced in reactions driven by plasma arcs, eg, the fixation of nitrogen to hydrogen, carbon, or oxygen (140). Numerous fluorine compounds that have been produced by plasma chemistry are tabulated in Reference 105. Reactions involving many other elements also are accomplished by plasma stimulation (79).

Inorganic small molecules also are produced in glow discharges and rf-induced plasmas by chemical vapor deposition (CVD), generally termed plasma-assisted CVD (PACVD) or plasma-enhanced CVD (PECVD). The molecules are introduced in the gaseous state and the products usually deposit on a chosen substrate. Collisional processes relevant to PECVD have been analyzed and tabulated (141–149). Amorphous silicon containing controlled concentrations of hydrogen can be produced by decomposition of $SiH_4$ in a glow discharge (150) as can layers of amorphous selenium using a hydride precursor gas, eg, $H_2S_e$ (151).

Plasma sources utilized for the production of materials and their modification are similar to those used to effect chemical reactions. Low temperature, glow-discharge, and r-f devices are employed to coat and heat the surfaces of solids and to alter them by ion bombardment sputtering. Higher temperature plasmas are used in materials processing. Plasma torches are produced by confining the heating by r-f fields or arcs to a chamber through which gas flows at high velocity. Temperatures in excess of 10,000 K are attained in the plasma, which cools as it is swept along to form a jet.

The reduction of high melting and other inorganic compounds to produce elemental solids also is achieved with plasmas. Extractive metallurgy requires breaking up ores and similar compounds (see METALLURGY, EXTRACTIVE). The decomposition of $MoS_2$ to the metal and sulfur has been demonstrated using an induction plasma torch (152). Plasma arc powered furnaces have been used for gaseous reduction of metal-containing materials (126).

Thin surface films also can be produced by plasma-sputtering deposition, without chemical reaction (see THIN FILMS). Both single- and multiple-layer materials are produced. Especially important is the use of sputtering to produce multilayered microstructures consisting of dozens of layers of two elements or compounds (153). The microstructures are used as x-ray reflectors (154). Bulk as well as thin materials can be produced, albeit at low rates, by vapor deposition. For example, PECVD can be used as a source for growth of refractory crystals such as TiN (102). Plasma deposition and etching techniques are being used to modify materials to produce fine patterns such as diffraction gratings and Fresnel lenses for use with uv and soft x-radiation, and in the >$85 billion per year semiconductor industry (155–157).

The manufacture of semiconductor chips, wafers, and devices makes extensive use of plasma processing. In the 1970s and 1980s a revolution occurred in the way that multiple layers of silicon compounds and other semiconductor materials were deposited, etched, modified, and removed, when dry plasma processing replaced the wet chemistries used previously. Significant equipment innovations in the early 1970s essentially split the fledgling plasma equipment industry into two different areas of focus: semiconductor processing and industrial applications for surface modification and cleaning. The need for higher throughput and better consistency, wafer-to-wafer and within the same wafer, required a different set of processing geometries and the equipment to handle them. The units used to process semiconductors as of the 1990s bear little resemblance to the barrel-type reactors used in industrial processes. This has led to dramatic increases in the production of devices, better quality, consistency and reliability, higher device compactness, lower costs per device, and significantly lowered environmental, health, and safety concerns regarding processing materials (158–164). Semiconductor manufacturing is a large and far reaching industry, and possesses its own distinct literature. The world market for semiconductor manufacturing equipment is in excess of $11.9 billion per year. Approximately half of that amount is plasma-related (165).

Plasmas are used extensively to melt materials for a variety of purposes. In many cases, the materials are introduced as a powder into the gas stream in a plasma torch. The molten droplets can be used to grow crystals of refractory materials such as niobium (166).

Another common and important use of plasma is spray coating of materials with plasma-melted substances (167,168). Plasma torches heated by d-c arcs can be hand-held for spray coating (22). Plasma spraying is employed to apply oxidation-resistant coatings to metals; for example, ceramic coatings of aircraft engine components and a proprietary cobalt–chromium–aluminum–yttrium coating for gas-turbine blades (169,170) (see REFRACTORY COATINGS). High temperature, self-lubricating coatings have also been applied to materials using plasma techniques (171).

In addition to coating existing structures by plasma spraying, it also is possible to build composite materials by spraying fibers or whiskers with a binding substance (see COMPOSITE MATERIALS) (167). The molten droplets produced in a plasma torch can be cooled without impacting a surface in order to obtain fine particles. Refractory metals and oxides can be made into powders of spherical particles using plasma torches. The resulting materials are used, for example, to produce filters and electrodes (126).

**Energy Production.**    The fusion or joining of two light nuclei, generally isotopes of hydrogen, leads to the formation of a heavier nucleus, eg, helium, with conversion of mass to energy. High energy is needed to overcome the mutual electrical repulsion experienced by two positive particles when they are near each other. The high velocities required for nuclear contact and reaction can be produced in plasmas if the ion temperature is greater than a few thousand electron volts. Because of the far greater number of nuclei in plasmas compared to beams of light nuclei, net energy production by nuclear fusion may occur in a plasma. Many exothermic fusion reactions are known. The easiest reaction to initiate is the deuterium–tritium reaction (see DEUTERIUM AND TRITIUM).

$$^2D + {}^3T \longrightarrow {}^4He + {}^1n + 17.6 \text{ MeV}$$

Deuterium is abundant in and easily separated from water. There is enough deuterium on earth to provide power for geological time scales. In contrast, tritium is not available in nature, but can be produced from $n$ + lithium reactions (see LITHIUM AND LITHIUM COMPOUNDS). Natural lithium is exhaustible, but sufficient tritium can be provided from it until fusion energy production is efficient enough to involve only D–D reactions:

$$^2D + {}^2D \longrightarrow {}^3T + {}^1p + 4.0 \text{ MeV} \quad \text{or} \quad {}^2D + {}^2D \longrightarrow {}^3He + {}^1n + 3.3 \text{ MeV}$$

Deuterium–deuterium reactions are harder to ignite and yield less energy than D–T reactions, but eventually should be the basis of fusion energy production (172). Research into the production of fusion power has been ongoing since the 1950s (173–177) (see FUSION ENERGY).

High temperature is an important requirement for the attainment of fusion reactions in a plasma. The conditions necessary for extracting as much energy from the plasma as went into it is the Lawson criterion, which states that the product of the ion density and the confinement or reaction time must exceed $10^{14}$ s/cm$^3$ in the most favorable cases (173). If the collisions are sufficiently violent, the Lawson criterion specifies how many of them must occur to break even. Conventional magnetic confinement involves fields of as much as 10 T ($10^5$ G) with large (1 m$^3$) plasmas of low densities ($<10^{14}$ particles/cm$^3$) and volumes and reaction times of about one second. If the magnetic flux can be compressed to values above 100 T ($10^6$ G), then a few cm$^3$ of plasma can be confined at densities of approximately $10^{19}$ particles/cm$^3$ for correspondingly shorter times ($\sim$10 $\mu$s). Inertial confinement requires compression of minute plasma ($<10^{-3}$ cm$^3$) to densities in excess of those of solids ($\sim 10^{25}$ particles/cm$^3$) for very short reaction times ($\sim$10 ps). The goal of all fusion devices is to produce high ion temperatures, which usually are in excess of the electron temperature, in contrast to low energy plasmas in which the electrons are often much hotter than the ions (see Fig. 2).

**Guns and Missiles.**   The rapid burning of powder in a gun barrel produces relatively cold plasmas which eject the projectile on a ballistic trajectory. Missiles carry a propellant which burns during flight, generating motion by high velocity ejection of mass. Modern missiles contain liquid or solid propellants having high energy densities. The chemical reactions that occur during their burning produces plasmas in the reaction chamber and exhaust nozzles.

High velocities can be reached by a variety of means (178). The highest muzzle velocities of conventional guns are about $10^6$ cm/s. Rockets can attain higher speeds and are limited by the weight of the nonfuel parts of the rocket. Laser plasma ablation can propel small masses at over $10^7$ cm/s but the technique is not likely to be useful for weaponry. A more promising plasma-based launcher is the rail gun (179), in which a plasma created by an electric discharge provides pressure to accelerate the projectile. Rail guns have the potential of speeds in excess of $10^7$ cm/s, although the motion of bullets at such speeds in the atmosphere may be problematic (see EXPLOSIVES AND PROPELLANTS).

**Impacts and Explosives.**  The collision of high velocity bullets or other projectiles with solids causes rapid conversion of kinetic to thermal energy. Plasmas result incidentally, whereas the primary effects of impact are shock and mechanical effects in the target. Impact-produced plasmas are hot enough to cause thermonuclear burn (180).

Most modern projectiles and virtually all missiles contain explosives. The plasmas that result from explosives are intrinsic to operation of warheads, bombs, mines, and related devices. Nuclear weapons and plasmas are intimately related. Plasmas are an inevitable result of the detonation of fission and fusion devices and are fundamental to the operation of fusion devices. Compressed pellets, in which a thermonuclear reaction occurs, would be useful militarily for simulation of the effects of nuclear weapons on materials and devices.

**Directed-Energy Weapons.**  Modern weapons mostly involve the propulsion of masses to inflict damage. Energetic quanta from lasers and accelerators have potential as directed-energy weapons (181). These would be based on much of the same pulsed-power technology employed to produce high temperature plasmas, and their impact on targets could produce surface plasmas. Potential laser weapons are likely to be produced within a few decades and could be useful in the atmosphere, weather permitting, and in space.

**Communications and Space Travel.**  Electromagnetic waves in and near the r-f region permit communications between points on or near earth or in space (182). The F layer of the ionosphere has the greatest electron density (see Fig. 5) and reflects radio waves of the greatest frequency. The plasma frequency, which is determined by the electron density and the angle of incidence, determines the frequency cutoff, below which reflection occurs and above which r-f waves penetrate and are partially absorbed by the ionosphere. Low frequencies (long wavelengths) which bounce off the F layer are partially absorbed in the lower D and E regions of the ionosphere.

The ionosphere is subject to sudden changes resulting from solar activity, particularly from solar eruptions or flares that are accompanied by intense x-ray emission. The absorption of the x-rays increases the electron density in the D and E layers, so that absorption of radio waves intended for F-layer reflection increases. In this manner, solar flares disrupt long-range, ionospheric bounce communications.

Earth to space (satellite) to earth communication links are relatively insensitive to ionospheric disturbances. Communications between earth and manned space vehicles are barely affected by plasmas when the spaceships are well away from the atmosphere, eg, in orbit or in a translunar trajectory. However, during reentry of a spaceship, a low temperature plasma forms around the vehicle and interrupts the communication links to it (183). Plasmas are incidental to the performance of modern rockets used to explore the solar system.

Spaceships capable of reaching stars other than the sun are expected to be more directly involved with plasmas than are contemporary spacecraft, in terms of their motion through the interstellar plasmas and their propulsion.

Very high velocities are expected to be required for travel to other stars, eg, Proxima Centauri, which is 4.3 light years distant and would require 43 years at one-tenth the speed of light.

Most schemes that have been proposed to propel starships involve plasmas. Schemes differ both in the selection of matter for propulsion and the way it is energized for ejection. Some proposals involve onboard storage of mass to be ejected, as in modern rockets, and others consider acquisition of matter from space or the picking up of pellets, and their momentum, which are accelerated from within the solar system (184,185). Energy acquisition from earth-based lasers also has been considered, but most interstellar propulsion ideas involve nuclear fusion energy; both magnetic, ie, mirror and toroidal, and inertial, ie, laser and ion-beam, fusion schemes have been considered (186–190).

## NOMENCLATURE

| Symbol | Definition | Units |
|--------|-----------|-------|
| $B$ | magnetic field strength | |
| $c$ | velocity of light | $3 \times 10^8$ m/s |
| D | particle size; region of ionosphere | |
| $d$ | average interparticle spacing | |
| E | electric field strength; region of ionosphere | |
| $E$ | oscillatory electric field | |
| $e$ | charge on electron | |
| $E_F$ | Fermi energy | |
| F | region of ionosphere | |
| $f$ | particle collision rate | |
| $h$ | Planck's constant | $6.626 \times 10^{-34}$ J·s |
| $\overline{h}$ | $h/2\pi$ | |
| $k$ | Boltzmann's constant | $1.38 \times 10^{-23}$ J/K |
| $m$ | mass | |
| $m_e$ | mass of electron | |
| $n$ | density | particles/cm$^3$ |
| $n_c$ | critical density | |
| $q$ | charge | |
| $T$ | temperature | K or eV (1 eV ~ 11,600 K) |
| $T_F$ | Fermi temperature | |
| $V$ | velocity | m/s |
| $\overline{V}$ | average velocity | |
| $V_A$ | Alfven velocity | |
| $V_\perp$ | velocity normal to magnetic field | |
| $\lambda$ | mean free path between collisions | |
| $\lambda_D$ | Debye length | |
| $\rho$ | mass density | g/cm$^3$ |
| $\omega_P$ | plasma frequency | |
| $\omega_c$ | cyclotron frequency | |

## BIBLIOGRAPHY

"Plasma Technology" in *ECT* Suppl. Vol., pp. 599–626, by D. J. Nagel, Naval Research Laboratory.

1. L. Tonks, *Am. J. Phys.* **35**, 857 (1967).
2. W. Crookes, *Radiant Matter* lecture delivered to the British Association for the Advancement of Science, Sheffield, U.K., Aug. 22, 1879.
3. W. Gilbert, *De Magnete, Magneticisque Corporibus*, Petrus Short, London, 1600.
4. N. M. Hirsh and H. J. Oskam, *Gaseous Electronics*, Academic Press, Inc., New York, 1978.
5. H. V. Boenig, *Plasma Science and Technology*, Cornell University Press, Ithaca, N.Y., 1982.
6. S. Weinberg, *The First Three Minutes*, Basic Books, Inc., New York, 1977.
7. J. L. Cecchi, in S. M. Rossnagel, J. J. Cuomo, and W. D. Westwood, eds., *Handbook of Plasma Processing Technology*, Noyes Publications, Park Ridge, N.J., 1990.
8. S. C. Brown and co-workers, *Am. J. Phys.* **31**(8), 637 (1963).
9. J. R. Hollahan and A. T. Bell, *Techniques and Applications of Plasma Chemistry*, John Wiley & Sons, Inc., New York, 1974.
10. K. Miyamoto, *Plasma Physics for Nuclear Fusion*, MIT Press, Cambridge, Mass., 1979.
11. E. Nasser, *Fundamentals of Gaseous Ionization and Plasma Electronics*, John Wiley & Sons, Inc., New York, 1971.
12. F. F. Chen, *Introduction to Plasma Physics*, Plenum Publishing Corp., New York, 1974.
13. G. Schmidt, *Physics of High Temperature Plasmas*, Academic Press, Inc., New York, 1979.
14. J. C. Ingraham, in E. U. Condon and H. Odishaw, eds., *Handbook of Physics*, 2nd ed., McGraw-Hill Book Co., Inc., New York, 1967, pp. 4, 188, 216.
15. A. Hasegawa, *Plasma Instabilities and Nonlinear Effects*, Springer-Verlag, Berlin, 1975.
16. W. H. Bostick, V. Nardi, and O. S. F. Zucker, eds., *Energy Storage, Compression and Switching*, Plenum Press, Inc., New York, 1976.
17. D. Sliney and M. L. Wolbarsht, *Safety with Lasers and Other Optical Sources*, Plenum Press, Inc., New York, 1980.
18. J. M. Meek and J. D. Craggs, *Electrical Breakdown of Gases*, John Wiley & Sons, Inc., New York, 1978.
19. A. D. McDonald, *Microwave Breakdown in Gases*, John Wiley & Sons, Inc., New York, 1966.
20. M. F. Hoyaux, *Arc Physics*, Springer-Verlag, Inc., New York, 1968.
21. H. Yasuda, *Plasma Polymerization*, Academic Press, Inc., New York, 1985.
22. B. Gross, B. Grycz, and K. Miklóssy, *Plasma Technology*, Iliffe Books Ltd., London, 1969.
23. R. A. Gross, *Nucl. Fusion* **15**, 729 (1978).
24. J. L. Vossen and W. Kern, eds., *Thin Film Processes*, Academic Press, Inc., New York, 1978.
25. H. T. Simmons, *Sciquest* **53**(7), 16 (1980).
26. K. Miyamoto, *Plasma Physics for Nuclear Fusion*, MIT Press, Cambridge, Mass., 1979.
27. D. E. Evans, ed., *Pulsed High Beta Plasmas*, Pergamon Press, Inc., Oxford, U.K., 1976.
28. H. Motz, *The Physics of Laser Fusion*, Academic Press, Inc., New York, 1969.

29. G. M. McCracken, P. E. Stott, and M. W. Thompson, eds., *Plasma Surface Interactions in Controlled Fusion Devices*, North-Holland, Amsterdam, the Netherlands, 1978.

30. M. D. Smith, *Surface Modification of High-Strength Reinforcing Fibers by Plasma Treatment*, AlliedSignal Inc., Kansas City, Mo., July 1991, p. KCP-613-4369.

31. R. H. Huddlestone and S. L. Leonard, eds., *Plasma Diagnostic Techniques*, Academic Press, Inc., New York, 1966.

32. O. Auciello and D. L. Flamm, eds., *Plasma Diagnostics*, Vol. 1, Academic Press, Inc., New York, 1989.

33. W. Lochte-Holtgreven, ed., *Plasma Diagnostics*, North-Holland, Amsterdam, the Netherlands, 1968.

34. K. Bockasten and co-workers, *Controlled Thermonuclear Fusion Research*, International Atomic Energy Agency, Vienna, Austria, 1961.

35. C. B. Wharton, in T. P. Anderson, R. W. Springer, and R. C. Warder, Jr., eds., *Physico-Chemical Diagnostics of Plasmas*, Northwestern University Press, Evanston, Ill., 1964.

36. S. W. Reeve and W. A. Weimer, *J. Vacuum Sci. Technol. A: Vacuum, Surf. Films*, **13**(2), 359 (1995).

37. K. Ashtiani, *Abstracts of the 1993 IEEE International Conference on Plasma Science*, IEEE, Piscataway, N.J., 1993, p. 6P6.

38. V. Gavrilenko and E. Oks, *Proceedings of the 1994 IEEE Conference on Plasma Science*, IEEE, Piscataway, N.J., 1994, p. 188.

39. D. Karabourniotis and E. Drakakis, *Abstracts of the 1993 IEEE International Conference on Plasma Science*, IEEE, Piscataway, N.J., 1993, p. 6P8.

40. T. P. Crowley, *IEEE Trans. Plasma Sci.* **22**(4), 291 (1994).

41. E. Oks, *Series on Atoms and Plasmas*, Vol. 9, *Plasma Spectroscopy: The Influence of Microwave and Laser Fields*, Springer-Verlag, New York, 1995.

42. K. Behringer and U. Frantz, *J. Phys. D: Appl. Phys.* **27**(10), 2128 (1994).

43. S. K. Ohorodnik and W. W. Harrison, *J. Anal. Atom. Spectrom.* **9**(9), 991 (1994).

44. D. J. Nagel, *Adv. X-Ray Anal.* **18**, 1 (1975).

45. M. D. Hester, *Gas Plasma Analysis Using an Emission Spectrometer*, AlliedSignal Inc., Kansas City, Mo., Jan. 1990, p. KCP-613-4169.

46. M. A. Uman, *Lightning*, McGraw-Hill Book Co., Inc., New York, 1969.

47. R. H. Golde, ed., *Lightning*, Academic Press, Inc., New York, 1977.

48. R. E. Orville, in R. H. Golde, ed., *Lightning*, Vol. 1, Academic Press, Inc., New York, 1977, p. 281.

49. H. Kikuchi, ed., *Dusty and Dirty Plasmas, Noise and Chaos in Space and in the Laboratory*, Plenum Publishing Corp., New York, 1994.

50. L. Kresak and P. M. Millman, eds., *Physics and Dynamics of Meteors*, D. Reidel, Dordrecht, the Netherlands, 1968.

51. A. Omholt, *The Optical Aurora*, Springer-Verlag, Inc., New York, 1971.

52. R. L. Lysak, ed., *Auroral Plasma Dynamics*, American Geophysical Union, Washington, D.C., 1993.

53. H. Risbeth and O. K. Garriot, *Introduction to Ionospheric Physics*, Academic Press, Inc., New York, 1969.

54. M. J. McEwan and L. F. Phillips, *Chemistry of the Atmosphere*, Halsted Press, a division of John Wiley & Sons, Inc., New York, 1975.

55. C. T. Russell, in K. Knott and B. Battrick, eds., *The Scientific Programme During the International Magnetospheric Study*, Academic Press, Inc., New York, 1969, p. 9.

56. E. R. Dyer, ed., *Critical Problems of Magnetospheric Physics*, National Academy of Sciences, Washington, D.C., 1972.

57. A. D. Walker, *Plasma Waves in the Magnetosphere*, Springer-Verlag, Inc., New York, 1993.

58. R. S. White, *Space Physics*, Gordon & Breach Science Publishers, Inc., New York, 1970.
59. J. L. Burch and J. H. Waite, Jr., eds., *Solar System Plasma in Space and Time*, American Geophysical Union, Washington, D.C., 1994.
60. E. N. Parker, C. F. Kennel, and L. J. Lanzerotti, eds., *Solar System Plasma Physics*, Vols. I, II, and III, North-Holland, Amsterdam, the Netherlands, 1979.
61. *Space Plasma Physics*, 3 Vols., National Academy of Sciences, Washington, D.C., 1978.
62. *Solar System Space Physics in the 1980's*, National Academy of Sciences, Washington, D.C., 1980.
63. S. Mitton, ed., *The Cambridge Encyclopedia of Astronomy*, Crown Publishers, Inc., New York, 1977.
64. H. Alfven, *Cosmic Plasma*, Kleuwer, Boston, Mass., 1981.
65. M. F. Hoyaux, *Solid State Plasmas*, Pion Ltd., London, U.K., 1970.
66. R. Bowers, *Sci. Am.* **209**(5), 46 (1963).
67. M. Glicksman, *Solid State Phys.* **26**, 275 (1971).
68. J. C. Slater, *Solid-State and Molecular Theory*, John Wiley & Sons, Inc., New York, 1975.
69. T. M. Rice, *Solid State Phys.* **32**, 1 (1977).
70. J. C. Hensel, T. G. Phillips, and G. A. Thomas, *Solid State Phys.* **32**, 87 (1977).
71. A. C. Baynham and A. D. Boardman, *Plasma Effects in Semiconductors: Helicon and Alfven Waves*, Taylor and Francis, London, 1971.
72. J. Pozhela, *Plasma and Current Instabilities in Semiconductors*, Pergamon Press, Inc., Oxford, U.K., 1981.
73. J. L. Delecroix, *Plasma Physics*, John Wiley & Sons, Inc., New York, 1965.
74. P. M. Platzman and P. A. Wolff, *Waves and Interactions in Solid State Plasmas*, Academic Press, Inc., New York, 1973.
75. D. J. Clark, *IEEE Trans. Nucl. Sci.* **NS-24**, 1064 (1977).
76. A. Bernard, in D. E. Evans, ed., *Pulsed High Beta Plasmas*, Pergamon Press, Oxford, U.K., 1976, p. 69.
77. W. T. Silfvast, L. H. Szeto, and O. R. Wood II, *Appl. Phys. Lett.* **36**(8), 617 (1980).
78. D. J. Nagel, *Naval Research Laboratory Memorandum Report* **4465** (1982).
79. R. F. Baddour and R. S. Timmins, eds., *The Application of Plasmas to Chemical Processing*, MIT Press, Cambridge, Mass., 1967.
80. *Chem Week*, 24 (Nov. 2, 1983); *Chem. Eng.*, 14 (Dec. 26, 1983).
81. H. V. Boenig, ed., *Advances in Low-Temperature Plasma Chemistry, Technology, Applications*, Vol. 1–4, Technomic Publishing Co., Lancaster, Pa., 1984, 1988, and 1991.
82. K. Upadhya, ed., *Plasma Synthesis and Processing of Materials*, The Minerals, Metals and Materials Society, Warrendale, Pa., 1993.
83. S. Veprek and M. Venugopalan, eds., *Plasma Chemistry*, Vol. 4, Springer-Verlag, Inc., New York, 1982.
84. B. M. Smirnov, ed., *Reviews of Plasma Chemistry*, Vol. 1–3, Plenum Publishing Corp., New York, 1991, 1994, and 1995.
85. V. A. Fassel, *Science* **202**, 183 (1978).
86. W. G. Schrenk, ed., *Analytical Atomic Spectroscopy*, Plenum Publishing Corp., New York, 1975; J. W. Carnahan, *Am. Lab.*, 31 (Aug. 1983).
87. C. L. Wilson, *Comprehensive Analytical Chemistry: Ultraviolet Photoelectron and Photoion Spectroscopy; Auger Electron Spectroscopy; Plasma Excitation in Spectrochemical Analysis*, Vol. 9, Elsevier Science, Inc., New York, 1979.
88. V. Loon, *Plasma Source Mass Spectroscopy*, CRC Press Inc., Boca Raton, Fla., 1994.
89. J. R. Hollahan, in Ref. 9, Chapt. 7.
90. R. S. Thomas, in Ref. 9, Chapt. 8.

91. S. Greenfield, H. McD. McGeachin, and P. B. Smith, *Talanta* **22**, 1 (1975); **22**, 553 (1975); **23**, 1 (1976).

92. P. W. Boumans, ed., *Plasma Spectrochemistry: Proceedings of the 1985 European Winter Conference on Plasma Spectrochemistry*, Elsevier Science, Inc., New York, 1985.

93. P. J. Slevin and W. W. Harrison, *Appl. Spectrosc. Rev.* **10**, 201 (1976).

94. J. P. Walers, *Science* **198**, 787 (1977).

95. C. D. Keirs and T. J. Vickers, *Appl. Spectrosc.* **31**, 273 (1977).

96. P. W. Boumans, in E. L. Grove, ed., *Analytical Emission Spectroscopy*, Part II, Marcel Dekker, New York, 1972, pp. 1–254.

97. R. K. Skogerboe and G. N. Coleman, *Anal. Chem.* **48**, 611A (1976).

98. S. A. Golden and M. W. Matthew, *NASA Technical Briefs*, Feb. 1995, pp. 51–52.

99. R. M. Measures and H. S. Kwong, *Appl. Opt.* **18**, 281 (1979).

100. F. W. Karaske, *Anal. Chem.* **43**, 1982 (1971).

101. P. H. Wieks, *Pure Appl. Chem.* **48**, 195 (1976).

102. S. Veprek, *Pure Appl. Chem.* **48**, 163 (1976).

103. P. W. Rose and E. M. Liston, *Treating Plastic Surfaces with Cold Gas Plasmas*, Plastics Engineering, Oct. 1985, pp. 41–45.

104. P. W. Rose and S. L. Kaplan, in D. Satas, ed., *Plastics Finishing and Decoration*, Van Nostrand Reinhold Co., New York, 1986, Chapt. 4.

105. S. L. Kaplan and W. P. Hansen, *Plasma—The Environmentally Safe Treatment Method*, Technical Notes, HIMONT/Plasma Science (now BOC Coating Technology), Concord, Calif., May 1991.

106. O. S. Kolluri, *Surface Cleaning with Plasma*, Technical Notes, HIMONT/Plasma Science (now BOC Coating Technology), Concord, Calif., May 1991.

107. A. Grill, *Cold Plasma in Materials Technology; From Fundamentals to Applications*, IEEE, Piscataway, N.J., 1994.

108. H. J. Oskam, ed., *Plasma Processing of Materials*, Noyes Data Corp., Park Ridge, N.J., 1985.

109. J. W. Coburn, R. A. Gottscho, and D. W. Hess, eds., *Plasma Processing*, Materials Research Society, Pittsburgh, Pa., 1986.

110. I. W. Boyd and E. F. Krimmel, eds., *Photo, Beam and Plasma Assisted Processing—Fundamentals and Device Technology*, Elsevier Science, Inc., New York, 1989.

111. O. Auciello, A. Gras-Marti, J. A. Valles-Abarca, and D. L. Flamm, eds., *Plasma–Surface Interactions and Processing of Materials*, Kluwer Academic Publishers, Norwell, Mass., 1990.

112. Staff of Panel on Plasma Processing of Materials, United States Nation Research Council, *Plasma Processing of Materials: Scientific Opportunities and Technological Challenges*, Books on Demand, Ann Arbor, Mich., 1994.

113. D. Apelian and J. Szekely, eds., *Plasma Processing and Synthesis of Materials*, Vol. 3, Materials Research Society, Pittsburgh, Pa., 1991.

114. G. S. Mathad and D. W. Hess, eds., *Proceedings of the International Symposium of Plasma Processing, 10th*, Electrochemical Society, Inc., Pennington, N.J., 1994.

115. M. A. Lieberman and A. J. Lichtenberg, *Principles of Plasma Discharges and Materials Processing*, John Wiley and Sons, Inc., New York, 1994.

116. K. Upadhya, ed., *Plasma and Laser Processing of Materials*, The Minerals, Metals and Materials Society, Warrendale, Pa., 1991.

117. J. R. Roth, *Industrial Plasma Engineering*, I.O.P. Publishing, Philadelphia, Pa., 1995.

118. M. Strobel, C. S. Lyons, and K. L. Mittal, eds., *Plasma Surface Modification of Polymers: Relevance to Adhesion*, Coronet Books, Philadelphia, Pa., 1994.

119. M. D. Smith, *Decomposition of Hazardous Organic Materials by R. F. Activated Gas Plasma*, AlliedSignal Inc., Kansas City, Mo., Mar. 1987, p. BDX-613-3687.

120. O. Auciello and D. L. Flamm, eds., *Plasma Diagnostics*, Vol. 2, Academic Press, Inc., New York, 1989.
121. M. D. Smith, *Practical Applications of Plasma Surface Modification*, AlliedSignal Inc., Kansas City, Mo., Dec. 1993, p. KCP-613-5331.
122. C. A. L. Westerdahl and co-workers, *J. Coll. Interface Sci.* **47**(3), 610 (1974).
123. R. H. Hansen and H. Schonhorn, *J. Polym. Sci.* **4**, 203 (1966).
124. J. R. Hall and co-workers, *J. Appl. Polym. Sci.* **13**, 2085 (1959).
125. M. D. Smith, *Effect of Various Gas Mixtures on Plasma Cleaned Ceramics*, AlliedSignal Inc., Kansas City, Mo., Mar. 1979, p. BDX-613-2107.
126. N. N. Rykalin, *Pure Appl. Chem.* **48**, 179 (1976).
127. M. I. Boulos, P. Fauchais, and E. Pfender, *Thermal Plasmas: Fundamentals and Applications*, Vol. 1, Plenum Publishing Corp., New York, 1994.
128. M. F. Zhukov and O. P. Solonenko, *Thermal Plasma and New Materials Technology*, Vols. 1 and 2, State Mutual Book and Periodical Service, Limited, New York, 1994.
129. A. E. Pavlath, in Ref. 9, Chapt. 4.
130. M. Hudis, in Ref. 9, Chapt. 3.
131. J. Feinman, ed., *Plasma Technology in Metallurgical Processing*, Books on Demand, Ann Arbor, Mich., 1986.
132. H. Suhr, in Ref. 9, Chapt. 2.
133. M. Millard, in Ref. 9, Chapt. 5.
134. M. Shen and A. T. Bell, *Plasma Polymerization*, American Chemical Society, Washington, D.C., 1979.
135. R. d'Agostino, ed., *Plasma Deposition, Treatment and Etching of Polymers, Plasma–Materials Interactions*, Academic Press, Inc., Orlando, Fla., 1991.
136. M. C. Shen, ed., *Plasma Chemistry of Polymers*, Books on Demand, Ann Arbor, Mich., 1976.
137. P. K. Tien, G. Smolinsky, and R. J. Martin, *Appl. Opt.* **11**, 637 (1972).
138. T. Wydeven and J. R. Hollahan, in Ref. 9, Chapt. 6.
139. M. D. Smith, *Surface Fluorination of Polymers by R. F. Activated Gas Plasma*, BKC #P-389, May 7, 1980.
140. R. S. Timmins and P. R. Ammann, in Ref. 79.
141. B. R. Bronfur, in Ref. 79, Chapt. 7.
142. M. J. Rand, *J. Vac. Sci. Tech.* **16**, 420 (1979).
143. D. W. Hess, *J. Vac. Sci. Tech.* **A8**, 1677 (1990).
144. S. Sivaram, *Principles of Chemical Vapor Deposition: Thermal Plasma Deposition of Electronic Materials*, Van Nostrand Reinhold, New York, 1995.
145. S. L. Girshick and B. W. Yu, *Proceedings of IEEE International Conference on Plasma Science 1994*, IEEE, Piscataway, N.J., 1994, p. 164.
146. Y. Nishimoto, N. Tokumasu, and K. Maeda, *Jpn. J. Appl. Phys.*, **34**(2B), 762 (1995).
147. G. Lucovsky, D. E. Ibbotson, and D. W. Hess, eds., *Characterization of Plasma-Enhanced CVD Processes: Materials Research Society Symposium Proceedings*, Vol. 165, Materials Research Society, Pittsburgh, Pa., 1990.
148. S. R. P. Silva, A. Kapoor, and G. A. J. Amaratunga, *Surf. Coat. Technol.* **73**(1–2), 132 (1995).
149. T. Baba, T. Matsuyama, T. Sawada, T. Takahama, K. Wakisaka, and S. Tsuda, *Microcrystalline and Nanocrystalline Semiconductors: Materials Research Society Symposium Proceedings*, Vol. 358, Materials Research Society, Pittsburgh, Pa., 1995, p. 895.
150. M. H. Brodsky, M. Cardone, and J. J. Cuomo, *Phys. Rev.* **16B**, 3556 (1977).
151. P. Nagles, E. Sleeckx, R. Callaerts, and L. Tichy, *Solid State Commun.* **94**(1), 49 (1995).
152. R. J. Munz and W. H. Gauvin, *AlChE J.* **21**, 1132 (1975).

153. T. W. Barbee, Jr. and D. C. Keith, *Stanford Synchrotron Radiation Laboratory Report* (78/04), Stanford, Calif., May 1978, p. III-26.

154. J. V. Gilfrich, D. J. Nagel, and T. W. Barbee, Jr., *Appl. Spectrosc.* **36**, 58 (1982).

155. J. M. Ballantyne, ed., *Proceedings of NSF Workshop on Opportunities for Microstructure Science*, National Science Foundation, Washington, D.C., 1978.

156. N. M. Ceglio, in D. T. Attwood and B. L. Henke, eds., *Low Energy X-Ray Diagnostics—1981*, American Institute of Physics, New York, 1981, pp. 210–222.

157. U.S. Department of Commerce, *U.S. Industrial Outlook 1994*, 35th ed., Government Printing Office, Pittsburgh, Pa., 1994, Sect. 15, p. 6.

158. R. W. Kirk, in Ref. 9, Chapt. 9.

159. J. J. Pouch and S. A. Alterovitz, eds., *Plasma Properties, Deposition and Etching*, Materials Science Forum Series, Vol. 140–142, L. P. S. Distribution Center, Lebanon, N.H., 1993.

160. M. Konuma, *Film Deposition by Plasma Techniques*, Atoms and Plasma Series, Vol. 10, Springer-Verlag, Inc., New York, 1992.

161. N. G. Einspruch, ed., *VLSI Electronics: Microstructure Science*, Vol. 8: *Plasma Processing for VLSI*, Academic Press, Inc., Orlando, Fla., 1984.

162. J. Mort and F. Jansen, *Plasma Deposited Thin Films*, Franklin Book Co., Inc., Elkins Park, Pa., 1986.

163. J. E. Griffiths, ed., *Monitoring and Control of Plasma-Enhanced Processing of Semiconductors*, SPIE-International Society for Optical Engineering, Bellingham, Wash., 1989.

164. D. M. Manos and D. L. Flamm, eds., *Plasma Etching: An Introduction*, Academic Press, Inc., Orlando, Fla., 1989.

165. Ref. 157, p. 21.

166. T. B. Reed, *Int. Sci. Technol.*, 42 (June 1962).

167. N. N. Rykalin and V. V. Kudinov, *Pure Appl. Chem.* **48**, 229 (1976).

168. R. Suryanarayan, *Plasma Spraying: Theory and Applications*, World Scientific Publishing Co., Inc., River Edge, N.J., 1993.

169. D. L. Ruckle, *Thin Solid Films* **64**, 327 (1979).

170. *Chem. Week* (Sept. 3, 1980).

171. H. E. Sliney, *Thin Solid Films* **64**, 211 (1979).

172. D. J. Rose and M. Clark, Jr., *Plasmas and Controlled Fusion*, MIT Press, Cambridge, Mass., 1961.

173. K. Miyamoto, *Plasma Physics for Nuclear Fusion*, MIT Press, Cambridge, Mass., 1979.

174. F. F. Chen, *The Sciences*, 6 (July/Aug. 1979).

175. T. J. Dolan, *Fusion Research*, Pergamon Press, New York, 1982.

176. S. Kuhn, K. Schopf, and R. W. Schrittwieser, *Current Research on Fusion, Laboratory, and Astrophysical Plasma*, World Scientific Publishing Co., Inc., River Edge, N.J., 1993.

177. N. J. Fisch, ed., *Advances in Plasma Physics: Proceedings of the Thomas H. Stix Symposium, American Institute of Physics*, AIP Press, Woodbury, N.Y., 1994.

178. D. J. Nagel, *IEEE Trans. Nucl. Sci.* **NS-26**, 122B (1979).

179. D. E. Thomsen, *Science News* **119**, 218 (1981).

180. A. T. Peaslee, Jr., ed., *Los Alamos Scientific Lab Report* **LA-8000C** (Aug. 1979).

181. W. J. Beane, *Naval Inst. Proc.*, 47 (Nov. 1981).

182. J. A. Ratcliff, *Sun, Earth and Radio*, Wiedenfeld and Nicolson, London, 1970.

183. J. J. Martin, *Atmospheric Reentry*, Prentice-Hall, Inc., Englewood Cliffs, N.J., 1966.

184. R. W. Bussard, *Astronaut. Acta* **6** 179 (1960).

185. E. C. Singer, *J. Brit. Interplanetary Soc.* **33**, 107 (1980).

186. G. H. Miley, *Fusion Energy Conversion*, American Nuclear Society, Washington, D.C., 1976.

187. A. A. Jackson IV and D. P. Whitmire, *J. Brit. Interplanetary Soc.* **31**, 335 (1978).
188. A. R. Martin and A. Bond, *J. Brit. Interplanetary Soc.* **32**, 283 (1979).
189. F. Winterberg, *J. Brit. Interplanetary Soc.* **32**, 403 (1979).
190. A. Martin, *J. Brit. Interplanetary Soc. Suppl.* **Sl-S192** (1978).

Mark D. Smith
AlliedSignal Aerospace Company

**PLASTIC BUILDING PRODUCTS.** See Building materials, plastic.

# PLASTICIZERS

A plasticizer is a substance the addition of which to another material makes that material softer and more flexible. This broad definition encompasses the use of water to plasticize clay for the production of pottery, and oils to plasticize pitch for caulking boats. A more precise definition of plasticizers is that they are materials which, when added to a polymer, cause an increase in the flexibility and workability, brought about by a decrease in the glass-transition temperature, $T_g$, of the polymer. The most widely plasticized polymer is poly(vinyl chloride) (PVC) due to its excellent plasticizer compatibility characteristics, and the development of plasticizers closely follows the development of this commodity polymer. However, plasticizers have also been used and remain in use with other polymer types.

The amount of plasticizer added to the polymer in question varies, depending on the magnitude of the effect required. For example, a small addition of plasticizer may be made simply to improve the workability of the polymer melt. This contrasts with larger additions made with the specific intention of completely transforming the properties of the product. For example, PVC without a plasticizer, ie, unplasticized PVC (PVC-U), is used in applications such as pipes and window profiles; with plasticizer added, articles such as PVC food film, PVC cable insulation, and sheathing and PVC floorings are formed.

There are presently ~300 plasticizers in manufacture. Of these ~100 are of commercial importance. A list of some common commodity and speciality plasticizers are given in Table 1.

## Types of Plasticizers

Two principle methods exist for softening a polymer to bring about the dramatic effects of plasticization. A rigid polymer may be internally plasticized by chemically modifying the polymer or monomer so that the flexibility of the

**Table 1. Plasticizers[a] In Common Use**

| Plasticizer | Abbreviation | Alcohol carbon number | Alcohol | CAS Registry Number | $M_w$ | Density at 20°C, $g/cm^3$ |
|---|---|---|---|---|---|---|
| | | | *Phthalates* | | | |
| diisobutyl phthalate | DIBP | 4 | isobutyl alcohol | [84-69-5] | 278.3 | 1.039 |
| dibutyl phthalate | DBP | 4 | butanol | [84-74-2] | 278.3 | 1.046 |
| diisoheptyl phthalate | DIHP | 7 | isoheptyl alcohol | [41451-28-9] [71888-89-6] | 362 | 0.991 |
| L7,9-phthalate | L79P; 79P | 7,9 | linear heptanol, nonanol | [68515-41-3] | 380 | 0.985 |
| L7,11-phthalate | L711P; 711P | 7,11 | linear heptanol, undecanol | [68648-91-9] | 414 | 0.971 |
| di-2-ethylhexyl phthalate[b] | DEHP | 8 | 2-ethylhexanol | [117-81-7] | 390 | 0.984 |
| diisooctyl phthalate | DIOP | 8 | isooctyl alcohol | [27554-26-3] | 390 | 0.983 |
| dinonyl phthalate | DNP | 9 | nonanol (3,5,5-trimethylhexanol) | [87-76-4] | 419 | 0.97 |
| diisononyl phthalate | DINP | 9 | isononyl alcohol | [28553-12-0] [68515-48-0] | 419 | 0.975 |
| diisodecyl phthalate | DIDP | 10 | isodecyl alcohol | [68515-49-1] | 447 | 0.967 |
| L9,11-phthalate | L911P; 911P | 9,11 | linear nonanol, undecanol | [68515-43-5] | 454 | 0.96 |
| diundecyl phthalate | DUP | 11 | undecanol | [3648-20-2] | 474 | 0.953 |
| diisoundecyl phthalate | DIUP | 11 | isoundecyl alcohol | [85507-79-5] | 474 | 0.962 |
| undecyldodecyl phthalate | UDP | 11,12 | undecanol, dodecanol | | 488 | 0.957 |
| diisotridecyl phthalate | DTDP | 13 | isotridecyl alcohol | [27253-26-5] [68515-47-9] | 531 | 0.952 |
| benzylbutyl phthalate | BBP | 4,7 | butanol, benzyl chloride | [85-68-7] | 312.3 | 1.119 |
| | | | *Adipates* | | | |
| di-2-ethylhexyl adipate[c] | DEHA | 8 | 2-ethylhexanol | [103-23-1] | 370.6 | 0.929 |
| diisononyl adipate | DINA | 9 | isononyl alcohol | [33703-08-1] | 398 | 0.929 |
| diisodecyl adipate | DIDA | 10 | isodecyl alcohol | [27178-16-1] | 427.1 | 0.915 |

**Table 1. (Continued)**

| Plasticizer | Abbreviation | Alcohol carbon number | Alcohol | CAS Registry Number | $M_w$ | Density at 20°C, g/cm³ |
|---|---|---|---|---|---|---|
| | | | *Trimellitates* | | | |
| tris-2-ethylhexyl trimellitate[d] | TOTM | 8 | 2-ethylhexanol | [3319-31-3] | 530 | 0.991 |
| L7,9-trimellitate | L79TM | 7,9 | linear heptanol, nonanol | [68515-60-6] | 530 | 0.996 |
| L8,10-trimellitate | L810TM | 8,10 | linear octanol, decanol | [67989-23-5] | 592 | 0.973 |
| | | | *Phosphates* | | | |
| tri-2-ethylhexyl phosphate[e] | TOP | 8 | 2-ethylhexanol | [78-42-2] | 434 | 0.926 |
| 2-ethylhexyl diphenyl phosphate | DPOP | 6,8 | phenol, 2-ethylhexanol | [1241-94-7] | 362.4 | 1.091 |
| tricresyl phosphate | TCP | 7 | methylphenol | [1330-78-5] | 368.2 | 1.165 |

[a] Esters made from the reaction of acids (or anhydrides) with alcohols. 1-Alkanols produce linear chains, hence the L designation in plasticizer names.
[b] Often dioctyl phthalate (DOP).
[c] Also dioctyl adipate (DOA).
[d] Also trioctyl trimellitate (TOTM).
[e] Also trioctyl phosphate (TOP).

260

polymer is increased. Alternatively, a rigid polymer can be externally plasticized by the addition of a suitable plasticizing agent, ie, by preparing a product consisting of a resin and a plasticizer. The external plasticizing route is the more common principally because of lower overall costs and also the fact that the use of external plasticizers allows the fabricator of the final article a certain degree of freedom in devising formulations for a range of products.

**Internal Plasticizers.** There has been much dedicated work on the possibility of internally plasticized PVC. However, in achieving this by copolymerization significant problems exist: (*1*) the affinity of the growing polymer chain for vinyl chloride rather than a comonomer implies that the incorporation of a comonomer into the chain requires significant pressure; (*2*) since the use of recovered monomer in PVC production is standard practice, contamination of vinyl chloride with comonomer in this respect creates additional problems; and (*3*) the increasing complexity of the reaction can lead to longer reaction times and hence increased costs. Thus, since standard external plasticizers are relatively cheap they are normally preferred.

**External Plasticizers.** There are two distinct groups of external plasticizers. A primary plasticizer, when added to a polymer, causes the properties of elongation and softness of the polymer to be increased. These changes are brought about by mechanisms described below. A secondary plasticizer, when added to the polymer alone, does not bring about these changes and may have limited compatibility with the polymer. However, when added to the polymer in combination with a primary plasticizer, secondary plasticizers enhance the plasticizing performance of the primary plasticizer. For this reason secondary plasticizers are also known as extenders.

**Commodity Phthalate Esters.** The family of phthalate esters are by far the most abundantly produced worldwide. Both orthophthalic and terephthalic acid and anhydrides are manufactured. The plasticizer esters are produced from these materials by reaction with an appropriate alcohol (eq. 1); terephthalate esterification for plasticizers is performed more abundantly in the United States. Phthalate esters are manufactured from methanol ($C_1$) up to $C_{17}$ alcohols, although phthalate use as PVC plasticizers is generally in the range $C_4$ to $C_{13}$. The lower molecular weight phthalates find use in nitrocellulose; the higher phthalates as synthetic lubricants for the automotive industries.

$$2 \text{ ROH } + \begin{array}{c} \text{COOH} \\ \text{COOH} \end{array} \longrightarrow \begin{array}{c} \text{COOR} \\ \text{COOR} \end{array} + 2 \text{ H}_2\text{O} \qquad (1)$$

*Di-2-Ethylhexyl Phthalate.* In Western Europe, di-2-ethylhexyl phthalate [*117-81-7*] (DEHP), also known as dioctyl phthalate (DOP), accounts for about 50% of all plasticizer usage and as such is generally considered as the industry standard. The reason for this is that it is in the mid-range of plasticizer properties. DEHP (or DOP) is the phthalate ester of 2-ethylhexanol, which is normally manufactured by the dimerization of butyraldehyde (eq. 2), the butyraldehyde itself being synthesized from propylene (see BUTYRALDEHYDES).

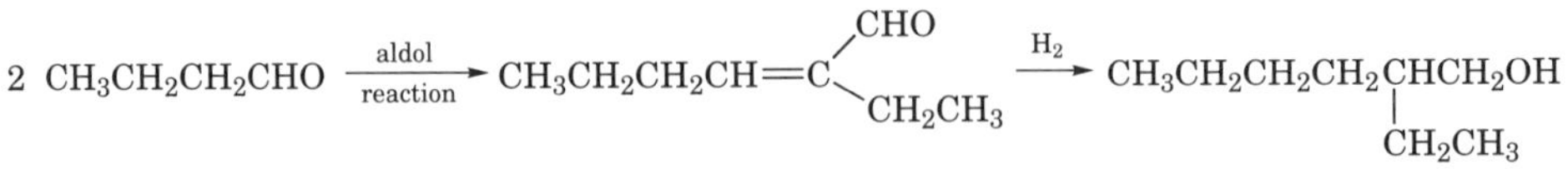

$$2 \; CH_3CH_2CH_2CHO \xrightarrow[\text{reaction}]{\text{aldol}} CH_3CH_2CH_2CH\!=\!\underset{CH_2CH_3}{\overset{CHO}{C}} \xrightarrow{H_2} CH_3CH_2CH_2CH_2\underset{\underset{CH_2CH_3}{|}}{C}HCH_2OH$$

(2)

The widespread sales of this plasticizer are a reflection of its all-around plasticizing performance and its provision of adequate properties for a great many standard products. It possesses reasonable plasticizing efficiency, fusion rate, and viscosity which, coupled with the normally competitive price, go a long way to explaining the popularity of this plasticizer. Some concerns have been periodically raised as to the possible toxicity of this material, but it can be said that these concerns are often related to the vast and widespread study of the toxicity of DEHP.

*Diisononyl Phthalate and Diisodecyl Phthalate.* These primary plasticizers are produced by esterification of oxo alcohols of carbon chain length nine and ten. The oxo alcohols are produced through the carbonylation of alkenes (olefins). The carbonylation process (eq. 3) adds a carbon unit to an alkene chain by reaction with carbon monoxide and hydrogen with heat, pressure, and catalyst. In this way a $C_8$ alkene is carbonylated to yield a $C_9$ alcohol; a $C_9$ alkene is carbonylated to produce a $C_{10}$ alcohol. Due to the distribution of the C=C double bond in the alkene and the varying effectiveness of certain catalysts, the position of the added carbon atom can vary and an isomer distribution is

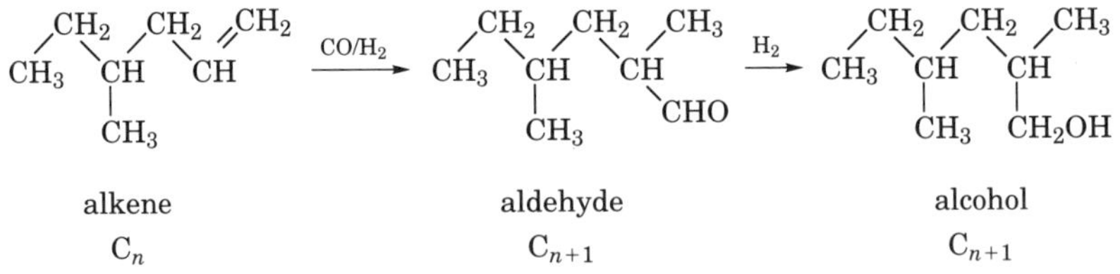

$$\text{(3)}$$

alkene $C_n$ $\xrightarrow{CO/H_2}$ aldehyde $C_{n+1}$ $\xrightarrow{H_2}$ alcohol $C_{n+1}$

generally created in such a reaction; the nature of this distribution depends on the reaction conditions. Consequently these alcohols are termed iso-alcohols and the subsequent phthalates iso-phthalates, an unfortunate designation in view of possible confusion with esters of isophthalic acid.

The $C_9$ and $C_{10}$ iso-phthalates (DINP and DIDP) generally compete with DEHP as commodity general-purpose plasticizers. Other iso-phthalates are available at opposite ends of the carbon number range (eg, diisoheptyl phthalate (DIHP), $C_7$, and diisotridecyl phthalate (DTDP), $C_{13}$), but these serve more speciality markets. The $C_8$ iso-phthalate, diisooctyl phthalate (DIOP), has also had traditional sales in the commodity plasticizer markets where it is seen as an equivalent to DEHP.

**The Specialty Plasticizers.**   For the purpose of this article, the term specialty plasticizer refers to any plasticizer other than DEHP (DOP), DIOP, DINP, or DIDP.

*Specialty Phthalates.*   These comprise the fast-fusing, low carbon number phthalates dibutyl phthalate (DBP), diisobutyl phthalate (DIBP), benzylbutyl phthalate (BBP), and diisoheptyl phthalate (DIHP); the low volatility isophthalates diisoundecyl phthalate (DIUP) and diisotridecyl phthalate (DTDP); and also the linear and semilinear phthalates for low viscosity applications, eg, L911P and L1012P. In each case these materials are the phthalate esters of alcohols of varying chain length.

*Adipate Esters.*   Alcohols of similar chain length to those used in phthalate manufacture can be esterified with adipic acid rather than phthalic anhydride to produce the family of adipate plasticizers. For example, esterification of 2-ethylhexanol with adipic acid yields di-2-ethylhexyl adipate (DEHA), also known as dioctyl adipate (DOA). The family of adipic acid esters in PVC applications has the significant properties of improved low temperature performance relative to phthalates and lower plastisol viscosities in plastisol applications, due to the lower inherent viscosities of the plasticizers themselves. Adipates used are typically in the $C_8$ to $C_{10}$ range; incompatibility problems can be encountered at higher carbon numbers, especially at high addition levels. Relative to phthalates, adipates suffer from higher volatilities and higher migration rates, and because they are a specialty for the PVC industry, higher prices. As a result, it is not uncommon to observe adipates used in blends with phthalates to produce a compromise of properties.

*Trimellitate Esters.*   These materials are produced by the esterification of a range of alcohols with trimellitic anhydride (TMA), which is similar in structure to phthalic anhydride with the exception of the third functionality (COOH) on the aromatic ring. Consequently, esters are produced in the ratio of three moles of alcohol to one mole of anhydride. Common esters in this family are tris-2-ethyhexyl trimellitate (trioctyl trimellitate, TOTM); L79TM, an ester of mixed semilinear $C_7$ and $C_9$ alcohols; and L810TM, an ester of mixed linear $C_8$ and $C_{10}$ alcohols.

The principal feature of these esters, when processed with PVC, is their low volatility, and consequently large volumes of trimellitate esters are used in high specification electrical cable insulation and sheathing. The extraction and migration resistance of these materials are also significantly improved relative to the phthalates. The low volatile loss figures also result in usage in automotive interior applications where the issue of windscreen fogging is important. In this respect they often compete with the linear high molecular weight phthalates such as 911P.

*Phosphate Esters.*   The principal advantage of phosphate esters is the improved fire retardancy relative to phthalates. The fire performance of PVC itself, relative to other polymeric materials, is very good due to its high halogen content, but the addition of plasticizers reduces this. Consequently there is a need, in certain demanding applications, to improve the fire-retardant behavior of flexible PVC.

Tris(2-ethylhexyl) phosphate shows good compatibility with PVC and also imparts good low temperature performance in addition to good fire retardancy.

2-Ethyhexyl diphenyl phosphate has widespread use in flexible PVC applications due to its combination of properties of plasticizing efficiency, low temperature performance, migration resistance, and fire retardancy.

*Sebacate and Azelate Esters.*   Esters produced from 2-ethylhexanol and higher alcohols with linear aliphatic acids are used in some demanding flexible PVC applications where superior low temperature performance is required. Di-2-ethylhexyl sebacate (DOS) and di-2-ethylhexyl azelate (DOZ) are the most commonly used members of this group, but diisodecyl sebacate (DIDS) is also encountered. They give superior low temperature performance to adipates but also command a significant premium, and their usage is generally limited to extremely demanding low temperature flexibility specifications, eg, underground cable sheathing in arctic environments.

*Polyester Plasticizers.*   These materials have found widespread use due to their exceedingly low volatility and high resistance to chemical extraction. Polyester plasticizers are based on condensation products of propanediols or butanediols with adipic acid or less commonly phthalic anhydride. The growing polymer chain may then be end-capped with an alcohol or monobasic acid, although nonend-capped polyester plasticizers can be produced by strict control of the reaction stoichiometry. Because of their higher molecular mass compared to other plasticizers, these materials have exceedingly high viscosities, ca 3–10 Pa·s (30–100 P), which can in some cases cause processing problems in dry blending and plastisol applications.

*Sulfonate Esters.*   These are marketed as efficient and easily processible plasticizers with good resistance to extraction. They are typically aryl esters of a $C_{13}$ to $C_{15}$ alkanesulfonic acid.

**Secondary Plasticizers.**   Also known as extenders, secondary plasticizers continue to play a significant role in flexible PVC formulations. They do not impart flexibility to the PVC resin alone, but when combined with a primary plasticizer act in such a way as to add flexibility to the final product. The majority of secondary plasticizers in use are chlorinated paraffins, which are hydrocarbons chlorinated to a level of 30–70%. For a given hydrocarbon chain, viscosity increases with chlorine content, as does the fire retardancy imparted to the formulation. These materials aid fire retardancy due to their chlorine content. Chlorinated paraffins of the same chlorine content may, however, have different volatilities and viscosities if they are based on different hydrocarbon chains (see CHLOROCARBONS AND CHLOROHYDROCARBONS, CHLORINATED PARAFFINS).

As well as imparting improved fire retardancy these materials may also result in volume cost savings if they can be purchased for a lower price than the commodity phthalate. Precise knowledge of the compatibility between standard plasticizers and chlorinated paraffins is required because some mixtures become incompatible with each other and the PVC resins in use at certain temperatures. Phthalate–chlorinated paraffin compatibility decreases as the molecular mass of the phthalate and the plasticizer content of the PVC formulation increase. Many compatibility graphs are available (1).

Other materials that are often referred to as secondary plasticizers include materials such as epoxidized soybean oil (ESBO) and epoxidized linseed oil (ELO) and similar materials. These can act as lubricants but also as secondary stabilizers to PVC due to their epoxy content which can remove HCl from the degrading polymer.

## The Mechanism of Plasticizer Action

This discussion refers to external plasticization only. Several theories, varying in detail and complexity, have been proposed in order to explain plasticizer action. Some theories involve detailed analysis of polarity, solubility, and interaction parameters and the thermodynamics of polymer behavior, whereas others treat plasticization as a simple lubrication of chains of polymer from each other, analogous to the lubrication of metal parts by oil. Although each theory is not exhaustive, an understanding of the plasticization process can be gained by combining ideas from each theory, and an overall theory of plasticization must include all these aspects.

The steps involved in the incorporation of a plasticizer into a PVC product can be divided into five distinct stages:

1.  Plasticizer is mixed with PVC resin.
2.  Plasticizer penetrates and swells the resin particles.
3.  Polar groups in the PVC resin are freed from each other.
4.  Plasticizer polar groups interact with the polar groups on the resin.
5.  The structure of the resin is re-established, with full retention of plasticizer.

Steps 1 and 2 can be described as physical plasticization, and the precise details of how this is carried out depends on the applications technology involved, ie, suspension or paste PVC. The rate at which step 2 occurs depends on the physical properties of plasticizer visocity, resin porosity, and particle size.

Steps 3 and 4, however, can be described as chemical plasticization since the rate at which these processes occur depends on the chemical properties of molecular polarity, molecular volume, and molecular weight. An overall mechanism of plasticizer action must give adequate explanations for this as well as the physical plasticization steps.

The importance of step 5 cannot be stressed too strongly, since no matter how rapidly and easily steps 1–4 occur, if plasticizer is not retained in the final product the product will be rendered useless. For many polymers, steps 1–4 proceed adequately but the plasticizer is not retained in the final product, leading to a product which is unacceptable.

**The Lubricity Theory.**    This is based on the assumption that the rigidity of the resin arises from intermolecular friction binding the chains together in a rigid network. On heating, these frictional forces are weakened so as to allow the plasticizer molecules between the chains. Once incorporated into the polymer bulk the plasticizer molecules shield the chains from each other, thus preventing the reformation of the rigid network. Although attractive in its simplicity, the theory does not explain the success of some plasticizers and the failure of others.

**The Gel Theory.**    This extends the lubricity theory in that it deals with the idea of the plasticizer acting by breaking the resin–resin attachments and interactions and by masking these centers of attachment from each other, preventing their reformation. Such a process may be regarded as necessary but again by itself is insufficient to explain a completely plasticized system, because although a certain concentration of plasticizer molecules provides plasticization by this process, the remainder act more in accordance with the lubricity theory, with unattached plasticizer molecules swelling the gel and facilitating the movement of plasticizer molecules, thus imparting flexibility. Molecules acting by this latter action may, on the basis of molecular size measurements, constitute the bulk of plasticizer molecules. If plasticization took place solely by this method it would not be possible to explain the ability of PVC resins to accept their own weight in plasticizer without exudation, ie, large amounts of additional space (free volume) are created which other plasticizer molecules can occupy.

**The Free Volume Theory.**    This extends the lubricity and gel theories and also allows a quantitative assessment of the plasticization process.

The free volume, $V_f$, of a polymer is described by the equation $V_f = V_t - V^0$ where $V_t$ = specific volume at a temperature $t$ and $V^0$ = specific volume of an arbitrary reference point, usually taken as zero degrees Kelvin. Free volume is a measure of the internal space available in a polymer for the movement of the polymer chain, which imparts flexibility to the resin. A rigid resin, eg, unplasticized PVC, is seen to possess very little free volume whereas resins which are flexible in their own right are seen as having relatively large amounts of free volume. Plasticizers therefore act so as to increase the free volume of the resin and also to ensure that free volume is maintained as the resin–plasticizer mixture is cooled from the melt. Combining these ideas with the gel and lubricity theories, it can be seen that plasticizer molecules not interacting with the polymer chain must simply fill free volume created by those molecules that do. These molecules may also be envisaged as providing a screening effect, preventing interactions between neighboring polymer chains thus preventing the rigid polymer network from reforming on cooling.

For the plasticized resin, free volume can arise from motion of the chain ends, side chains, or the main chain. These motions can be increased in a variety of ways, including increasing the number of end groups, increasing the length of the side chain, increasing the possibility of main group movement by the inclusion of segments of low steric hindrance and low intermolecular attraction, introduction of a lower molecular weight compound which imparts the above properties, and raising the temperature.

The introduction of a plasticizer, which is a molecule of lower molecular weight than the resin, has the ability to impart a greater free volume per volume of material because there is an increase in the proportion of end groups and the plasticizer has a glass-transition temperature, $T_g$, lower than that of the resin itself. A detailed mathematical treatment (2) of this phenomenon can be carried out to explain the success of some plasticizers and the failure of others. Clearly, the use of a given plasticizer in a certain application is a compromise between the above ideas and physical properties such as volatility, compatibility, high and low temperature performance, viscosity, etc. This choice is application dependent, ie, there is no ideal plasticizer for every application.

**Solvation–Desolvation Equilibrium.**    From the observation of migration of plasticizer from plasticized polymers it is clear that plasticizer molecules, or at least some of them, are not bound permanently to the polymer as in an internally plasticized resin, but rather an exchange–equilibrium mechanism is present. This implies that there is no stoichiometric relationship between polymer and plasticizer levels, although some quasi-stoichiometric relationships appear to exist (3,4). This idea is extended later in the discussion of specific interactions.

**Generalized Structure Theories and Antiplasticization.** In their simplest form these theories attempt to produce a visual representation of the mechanism of plasticizer action (1,5,6). The theories are based on the concept that if a small amount of plasticizer is incorporated into the polymer mass it imparts slightly more free volume and gives more opportunity for the movement of macromolecules. Many resins tend to become more ordered and compact as existing crystallites grow or new crystallites form at the expense of the more fluid parts of the amorphous material. For small additions of plasticizer, the plasticizer molecules may be totally immobilized by attachment to the resin by various forces. These tend to restrict the freedom of small portions of the polymer molecule necessary for the absorption of mechanical energy. Therefore it results in a more rigid resin with a higher tensile strength and base modulus than the base polymer itself. This phenomenon is therefore termed antiplasticization. One study used x-ray diffraction to show that small amounts of dioctyl phthalate (DOP) progressively increase the order in the PVC. Above these concentrations the order decreases and the polymer becomes plasticized.

**Interaction Parameters.** Early attempts to describe PVC–plasticizer compatibility were based on the same principles as used to describe solvation, ie, like dissolves like (2). To obtain a quantitative measure of PVC–plasticizer compatibility a number of different parameters have been used. More recently these methods have been assessed and extended by many workers (7–9). In all cases it is not possible to adequately predict the behavior of polymeric plasticizers.

*The Hildebrand Solubility Parameter.* This parameter, $d$, can be estimated (10) based on data for a set of additive constants, $F$, for the more common groups in organic molecules to account for the observed magnitude of the solubility parameter: $d = \Sigma F/V$ where $V$ represents molar volume. Solubility parameters can be used to classify plasticizers of a given family in terms of their compatibility with PVC, but they are of limited use for comparing plasticizers of different families, eg, phthalates with adipates.

*Polarity Parameter.* Despite their apparent simplicity, these parameters, $\phi$, show a good correlation with plasticizer activity for nonpolymeric plasticizers (10). The parameter is defined as $\phi = [M(A_p/P_0)]/1000$ where $M$ = molar mass of plasticizer, $A_p$ = number of carbon atoms in the plasticizer excluding aromatic and carboxylic acid carbon atoms, and $P_0$ = number of polar (eg, carbonyl) groups present. The 1000 factor is used to produce values of convenient magnitude. Polarity parameters provide useful predictions of the activity of monomeric plasticizers, but are not able to compare activity of plasticizers from different families.

*The Solid–Gel Transition Temperature.* This temperature, $T_m$, is a measure of plasticizer activity and is the temperature at which a single grain of PVC dissolves in excess plasticizer. The more efficient plasticizers show lower values of $T_m$ as a result of their higher solvating power. This can be correlated with the ease of processing of a given plasticizer, but all measurements should be conducted with a control PVC resin since clearly the choice of resin has an effect here also.

*The Flory-Huggins Interaction Parameter.* These ideas, based on a study of polymer miscibility, have been applied to plasticizers according to the following equation in which $V_1$ is the molar volume of the plasticizer, obtained from molar mass figures and density values at $T_m$, and $\chi$ represents the interaction parameter (11).

$$1/T_m = 0.002226 + 0.1351(1 - \chi)/V_1$$

*The Activity Parameter.*    Another measure ($\alpha$) of plasticizer activity that is an extension of and based on earlier work gives an indication of the ease of processing for a given plasticizer with a given resin, but does not give estimates of plasticizing performance in the final product (12). $M$ and $\chi$ are as previously defined.

$$\alpha = 1000 \frac{(1 - \chi)}{M}$$

**Specific Interactions.**    Ideas on the subject of specific interactions between PVC and a plasticizer molecule, as a basis of plasticization, can be considered a more detailed form of some of the ideas already discussed. Clearly some mechanism of attraction and interaction between PVC and plasticizer must exist for the plasticizer to be retained in the polymer after processing.

The role of specific interactions in the plasticization of PVC has been proposed from work on specific interactions of esters in solvents (eg, hydrogenated chlorocarbons) (13), work on blends of polyesters with PVC (14–19), and work on plasticized PVC itself (20–23). Modes of interaction between the carbonyl functionality of the plasticizer ester or polyester were proposed, mostly on the basis of results from Fourier transform infrared spectroscopy (ftir). Shifts in the absorption frequency of the carbonyl group of the plasticizer ester to lower wave number, indicative of a reduction in polarity (ie, some interaction between this functionality and the polymer) have been reported (20–22). Work performed with dibutyl phthalate (22) suggests an optimum concentration at which such interactions are maximized. Spectral shifts are in the range 3–8 cm$^{-1}$. Similar shifts have also been reported in blends of PVC with polyesters (14–20), again showing a concentration dependence of the shift to lower wave number of the ester carbonyl absorption frequency.

These ideas have been extended using new analytical techniques, in particular molecular modeling and solid-state nuclear magnetic resonance (nmr) spectroscopy.

*Molecular Modeling.*    The computer modeling of molecules is a rapidly growing branch of chemistry (24–26). High resolution graphics and fast computers allow the operator to build molecules in minimum energy configurations and view them in real time. This model can be constructed from crystallographic coordinates available from databases (qv) or by simple intervention from the operator. Molecular mechanics or quantum mechanics programs are then used to arrive at a likely structure.

A range of plasticizer molecule models and a model for PVC have been generated and energy minimized to observe their most stable conformations. Such models highlight the free volume increase caused by the mobility of the plasticizer alkyl chains. More detailed models have also been produced to concentrate on the polar region of the plasticizer and its possible mode of interaction with the polymer. These show the expected repulsion between areas on the polymer and plasticizer of like charge as well as attraction between the negative portions of the plasticizer and positive portions of the PVC.

*Solid-State Nuclear Magnetic Resonance Spectroscopy.* Advances in technology have made the study of solids by nmr techniques of considerably greater ease than in previous years. For the accumulation of solid-state $^{13}$C-nmr spectra, cross-polarization magic angle spinning (CPMAS) can be utilized to significantly reduce signal broadening effects present in solid state but not in the liquid state. The technique has been used to study the molecular effects of plasticization by comparing spectral shifts of PVC and plasticizer under various degrees of processing (9,27). For PVC plasticized with DIDP two different processing temperatures, 130 and 170°C were used, representing a low and high degree of plasticization, respectively. The comparison of the spectra showed no shift in the resonance frequency of the carbonyl group with processing temperature. The most significant difference in the two spectra was in the aliphatic carbon resonances. The spectra of the more plasticized sample showed resonance shifts and increased resolution of these carbon atoms. This again shows a strong dependence of successful plasticization on the conformation of the alkyl chains of the plasticizer ester (linked to the increased free volume).

It can be concluded from these theories and studies that plasticizer polarity is important in determining the gelation rate of the plasticizer but it does not explain other properties of interest in the final product. The conformation adopted by plasticizer molecules in the polymer matrix in the final product is clearly important because this determines how many PVC–PVC chain–chain interactions are screened from each other and how much free volume is created. The more recent studies have shown that although this conformation is important it is perhaps not so important in samples that have experienced high processing temperatures, since in these samples the separation of the PVC chains, and ingress of plasticizer, is controlled more by thermal energy than by plasticizer polarity. At lower processing temperatures the polarity of the plasticizer has a greater role in the attainment of acceptable physical properties of the final product.

## Plasticized Polymers: The Dominance of PVC

Well over 90% of plasticizer sales by volume are into the PVC industry. The reason for such a concentration of sales is that the benefits imparted by the plasticization of PVC are far greater than those imparted to other polymers. PVC stands alone among polymers in its ability both to accept and retain large concentrations of plasticizer. This is due in part to a morphological form comprising highly amorphous, semicrystalline, and highly crystalline regions. Without the wide range of additives available, eg, plasticizers, stabilizers, fillers, lubricants, and pigments, PVC would be of little use. The development of PVC as a commodity polymer is fundamentally linked to the development of its additives.

Different types of PVC exist on the market. The two principle types are suspension and paste-forming PVC; the latter includes the majority of emulsion PVC polymers. The plasticizer applications technologies associated with these two forms are distinctly different and are discussed separately. Details of the polymerization techniques giving rise to these two distinct polymer types can be found in many review articles (5,28) (see VINYL POLYMERS, POLY(VINYL CHLORIDE)).

## APPLICATIONS TECHNOLOGY

**Suspension PVC.** These polymers are produced by suspending vinyl chloride in water and polymerizing this monomer using a monomer-soluble initiator. PVC polymers produced via a suspension polymerization route have a relatively large particle size (typically 100–150 $\mu$m). Additionally, suspension polymers produced for the flexible sector have particles that are highly porous and are therefore able to absorb large amounts of liquid plasticizer during a formulation mixing cycle. A typical flexible PVC formulation (Table 2) using a suspension polymer is typically processed by a dry-blend cycle, during which all formulation ingredients are heated (typically 70–110°C) and intimately mixed to form a dry powder (the PVC dry blend or powder blend), ie, a powder that contains all formulation ingredients. This dry blend can be either stored or processed immediately. Processing of suspension resin formulations is performed by a variety of techniques such as extrusion, injection molding, and calendering to totally fuse the formulation ingredients and therefore produce the desired product.

The dry blend can also be extruded and the extrudate chipped to produce pellets of PVC compound which can then be subsequently reprocessed to produce the final product. This has the benefit of ease of storage of raw materials since all the formulation ingredients are contained bound in the gelled compound. Many producers of flexible PVC only purchase PVC compound, and many companies exist solely to produce PVC compound rather than a true end product such as sheet, flooring, or pipe.

**Paste-Forming Polymers.** Paste- or plastisol-forming PVC polymers differ from their suspension analogues in that after mixing with plasticizer they produce a paste or plastisol, similar in appearance to paint, rather than a dry blend. In this respect these polymers are used for flexible applications only. The plastisol can then be spread, coated, rotationally cast, or sprayed for processing. Plastisol-forming polymers are produced by microsuspension polymerization or emulsion polymerization. Microsuspension produces very fine particles of monomer to ensure small particle sizes of polymer are produced. In emulsion polymerization the vinyl chloride is polymerized using a water-soluble initiator; the vinyl chloride particles are small and stabilized using surfactants. There are also several variations of these two basic techniques.

Much lower particle size resins are produced by these routes, relative to suspension resins, but they also differ in that some residual surfactant from the

## Table 2. Typical Flexible PVC

| Ingredient | Parts by weight | Parts per hundred resin (phr) | % By weight |
|---|---|---|---|
| PVC | 75 | 100 | 50 |
| plasticizer | 45 | 60 | 30 |
| filler | 26.25 | 35 | 17.5 |
| stabilizer | 3 | 4 | 2 |
| lubricant | 0.75 | 1 | 0.5 |
| *Total* | *150* | *200* | *100* |

polymerization process is retained on the polymer. The low particle size imparts a lack of porosity to the resin and thus the mixing of formulation ingredients using a dry-blending cycle is not possible. The demands on plasticizer behavior in a plastisol tends to be more complex than that in suspension technology since choice of plasticizer is made with consideration given to the required viscosity of the plastisol and also the required rheology of the plastisol. Each paste-forming polymer shows individual characteristics with respect to particle size and particle size distribution, and therefore plastisol viscosity and rheology, and no two polymers behave the same way, thus making the choice of plasticizer for this sector somewhat complex. All plasticizer types find use in this area, the choice being governed by a compromise of properties. Even polymeric plasticizers, with their very high inherent viscosities, can find use if their addition level is very high, and it is certainly common to encounter formulations with two or three different plasticizers present.

Plastisols may also be semi-gelled for storage, ie, enough heat is imparted to convert the plastisol into a solid but without the full development of tensile properties brought about by full fusion.

Because the formulation ingredients in a plastisol are in liquid form, viscosity of the plastisol is of great importance and the intrinsic viscosity of the plasticizer contributes significantly to the plastisol viscosity, as does the precise polymerization conditions of the resin. The desired plastisol viscosity can be obtained by careful selection of polymer, plasticizer, and other formulation ingredients, but the shear rate applied to the plastisol also affects the viscosity. PVC plastisols are either (1) pseudoplastic or shear-thinning, ie, viscosity decreases with shear; (2) near-Newtonian, ie, viscosity remains nearly constant with shear; or (3) dilatant, ie, viscosity increases with shear. Example applications exist where each of these three rheological behaviors are preferred, and precise knowledge of how a given plasticizer acts with a given resin is often required.

## EFFECT OF PLASTICIZER CHOICE ON THE PROPERTIES OF FLEXIBLE PVC

A change in plasticizer affects the properties of a flexible PVC article. Certain properties are more important for some applications than others and hence some plasticizers find more extensive use in some application areas than others. The PVC technologist must ascertain the most important properties for an application and then make the correct choice of plasticizer.

**Plasticizer Efficiency.** This is a measure of the concentration of plasticizer required to impart a specified softness to PVC. Such a softness of material may be measured as a British Standard Softness (BSS) or a Shore hardness (Fig. 1). For a given acid constituent of plasticizer ester, ie, phthalate, adipate, etc, plasticizer efficiency decreases as the carbon number of the alcohol chain increases, eg, for phthalate esters efficiency decreases in the order DBP > DIHP > DOP > DINP > DIDP > DTDP. An additional six parts per hundred in PVC of DIDP rather than DOP is required to give a hardness of Shore 80 when all other formulation ingredients remain constant. The consequence of this depends on the overall formulation and product costs. In addition

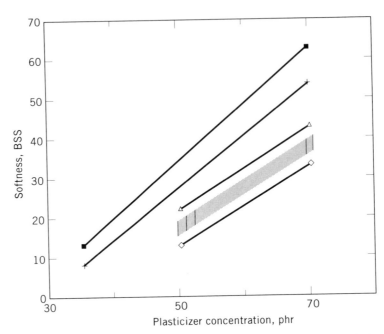

**Fig. 1.**  Relative efficiency of plasticizers where ■ represents $C_8$ phthalate; +, $C_{10}$ phthalate; △, $C_{11}$ phthalate; and ◇, $C_{13}$ phthalate; phr = parts per hundred rubber. Values for the trimellitates fall in the shaded area. BSS 35 is equivalent to a Shore A hardness of 80, test method BS2782.

to size of the carbon number of the alcohol chain, the amount of branching is also significant; the more linear isomers are of greater efficiency. Choice of the acid constituent can also be significant. For equivalent alcohol constituents, phthalate and adipate esters are approximately equivalent but both are considerably more efficient than the trimellitate equivalent.

Reasons for these trends are clearly related to the polarity of the plasticizer and its ability to impart free volume by chain separation. Differences in polarity affect the temperature at which the plasticizer can penetrate the polymer matrix and the magnitude of the interaction with the polymer chain that results.

**High Temperature Performance.**   High temperature performance in flexible PVC and its production are related to plasticizer volatilization and plasticizer degradation (Fig. 2). Plasticizer volatilization, both from the finished article during use at elevated temperatures, eg, in electrical cable insulation, and also during processing, ie, release of plasticizer fume is directly related to the volatility of the plasticizer in use. Hence the higher molecular weight plasticizers give superior performance in this area. For phthalate plasticizers, thermal stability decreases in the order DTDP > DIDP > DINP > DOP > DIHP > DBP. Higher molecular weight esters such as trimellitates are even more thermally stable and trimellitate esters find extensive use in the demanding cable specifications which have strict mass loss requirements.

Polyester plasticizers give the best performance in this area, with performance increasing with molecular weight. Additionally, branched esters have somewhat higher volatilities than their linear equivalents.

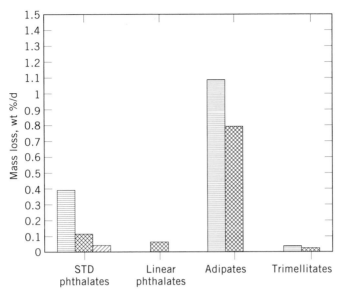

**Fig. 2**  Volatile loss of plasticizers from flexible PVC (BSS 35) where ▤ represents $C_8$; ▨, $C_{10}$; and ▧, $C_{13}$. Weight loss per day is at 100°C.

For the generation of fume in the workplace, the same structure relationships apply. Not only does excessive plasticizer volatilization have environmental consequences, but inaccuracies on formulation can be incurred since not all the plasticizer in use is entering the PVC resin, resulting in a harder material than calculated. Estimates have been made of the exact amount of plasticizer loss to the environment (29). As a result of environmental protection legislation, more end users are looking to means of recovering and re-using plasticizer fume and breakdown products, either through re-use in the processing operation or as an alternative fuel.

Plasticizer molecules can undergo thermal degradation at high temperatures. Esters based on the more branched alcohol isomers are more susceptible to such degradation. This can, however, be offset by the incorporation of an antioxidant, and plasticizer esters for cable applications frequently contain a small amount of an antioxidant such as bisphenol A.

**Low Temperature Performance.**    The ability of plasticized PVC to remain flexible at low temperatures is of great importance in certain applications, eg, external tarpaulins or underground cables. For this property the choice of the acid constituent of the plasticizer ester is also important. The linear aliphatic adipic, sebacic, and azeleic acids give excellent low temperature flexibility compared to the corresponding phthalates and trimellitates (Fig. 3).

There is also a significant contribution to low temperature performance from the alcohol portion of the ester, the greater the linearity of the plasticizer the greater the low temperature flexibility. It might be concluded that the best low temperature plasticizer would be one based on a long and linear dibasic acid with long, straight-chain alcohols. Although this is true, other requirements must be taken into account, and it is likely that such a material would have low plasticizing efficiency, poor gelation properties, and be of limited compatibility.

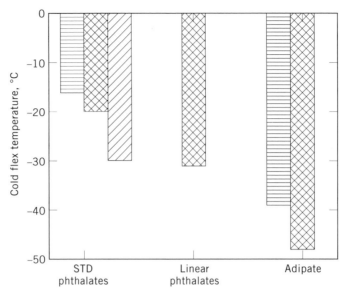

**Fig. 3.** Low temperature performance of plasticizers in flexible PVC (BSS 35) where ▤ represents $C_8$; ▨, $C_{10}$; and ▧, $C_{13}$.

**Gelation Properties.** The gelation characteristics of a plasticizer are related to its efficiency and both properties are often discussed together. The gelation characteristics are a measure of the ability of a plasticizer to fuse with the polymer so as to set up a product of maximum elongation and softness, ie, maximum plasticization properties (Fig. 4). Gelation properties are often measured

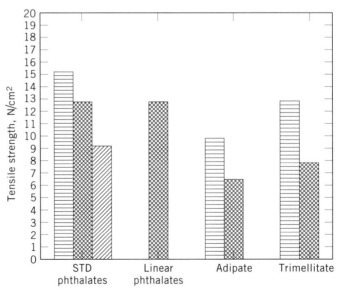

**Fig. 4.** Tensile strength of various PVC–plasticizer combinations when fused at 160°C where ▤ represents $C_8$; ▨. $C_{10}$; and ▧, $C_{13}$. To convert N/cm² to MPa, divide by 100; to convert to psi, multiply by 1.45.

either as a processing temperature, the temperature to which the plasticizer and polymer must be heated in order to obtain these properties, or as a solution temperature, the temperature at which one grain of polymer dissolves in excess plasticizer, giving a measure of the solvating power of the plasticizer. Ease of gelation is related to plasticizer polarity and molecular size. The greater the polarity of a plasticizer molecule the greater the attraction it has for the PVC polymer chain and the less additional energy, in the form of heat, is required to cause maximum plasticizer–PVC interactions. The most active plasticizers are able to bring about these effects soon after the $T_g$ of the polymer (70–80°C) is reached, whereas the less active plasticizers require temperatures on the order of 180°C in order for the maximum elongation properties to be obtained. The polarity of the plasticizer is determined by both acid type and alcohol chain length. Aromatic acids, being of greater polarity, tend to show greater ease of gelation than aliphatic acid-based esters. Molecular size also has a key contribution and explains why molecules of similar polarity can show different gelation properties. The smaller the plasticizer molecule the easier it is for it to enter the PVC matrix; larger molecules require more thermal energy to establish the desired interaction with the polymer. Because branching influences molecular size, this too has a contribution to gelation properties, with the more branched isomers showing greater activity. Thus for the phthalate esters ease of gelation rate decreases in the order BBP > DBP > DIHP > DOP > DINP > IDP > DTDP.

**Migration and Extraction.** When plasticized PVC comes into contact with other materials, plasticizer may migrate from the plasticized PVC into the other material. The rate of migration depends not only on the plasticizer employed but also on the nature of the contact material.

Plasticizer can also be extracted from PVC by a range of solvents including water. The aggressiveness of a particular solvent depends on its molecular size and its compatibility with both the plasticizer and PVC. Water extracts plasticizer very slowly, oils are slightly more aggressive, and low molecular weight solvents are the most aggressive.

The key characteristic for migration and extraction resistance is molecular size. The larger the plasticizer the less it tends to migrate or be extracted. The extreme case is seen by the use of polymeric plasticizers in applications where excellent migration and extraction resistance is required, and food packaging film applications present a relatively large market for these specialty plasticizers. There is also a contribution from the linearity of the alcohol component of the plasticizer ester. The greater the linearity of the ester the greater its migration and extraction rate in comparison to the more branched isomers.

**Plastisol Viscosity and Viscosity Stability.** After the primary contribution of the resin type in terms of its particle size and particle size distribution, for a given PVC resin, plastisol viscosity has a secondary dependence on plasticizer viscosity. The lower molecular weight and more linear esters have the lowest viscosity and hence show the lowest plastisol viscosity, ie, plastisol viscosity for a common set of other formulation ingredients increases in the sequence DBP < DIHP < DOP < DINP < DIDP. In spite of these viscosity differences, however, if plastisols are being formulated to equal softness, ie, taking into account the efficiency of the plasticizers involved, more of the less efficient plasticizer has to be employed in the plastisol so as to impart the same softness to the product being manufactured. The addition of this extra liquid to the plastisol may produce an equivalent viscosity to that of the plastisol with the less viscous plasticizer. Esters based on aliphatic acids, being of lower viscosity than the corresponding aromatic acids, show lower plastisol viscosities. Adipate esters have found widespread use in plastisol applications although due to other requirements, eg, volatility, gelation characteristics, etc, they are often employed in a blend with other esters.

Plastisols are often mixed and then stored rather than processed immediately (Fig. 5). It is of great importance in this case for the plasticizer to show little or no paste thickening action at the storage temperature, and clearly it is not advisable to use a plasticizer of too great an activity, since grain swelling, leading to plastisol viscosity increase, can occur at low temperatures for some active plasticizer systems.

**Automotive Windshield Fogging.**   The phenomenon of car windscreen fogging has been known for some time. The term fogging relates to the condensation of volatile material on the car windshield causing a decrease in visibility. Although this volatile material may arise from a variety of sources, eg, exhaust fume being sucked in through the ventilation system, material from inside the car, eg, crash pads or rear shelves, may also contribute to windscreen fogging on account of the high temperatures that can be encountered inside a car when standing in sunlight. In the case of flexible PVC such a contribution may arise from emulsifiers in the polymer, stabilizers, and plasticizers.

In each case manufacturers have studied their products in detail and recommend low fogging polymers, stabilizers, and plasticizers. Tests have been designed (eg, DIN 75 201) to assess the fogging performance of both the PVC sheet and the raw materials used in its production (Fig. 6). These tests involve heating of the sheet of raw material for a specified period at a set temperature in an enclosed apparatus with a cooled glass plate above the sheet or raw material. The reflectance of the glass plate is then compared before and after the test to

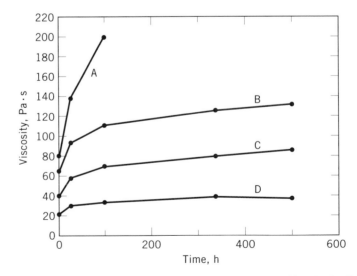

**Fig. 5.**  Viscosity aging of plastisols at 23°C where A is BBP/DIPB; B, DOP; C, F110; and D, 911P.

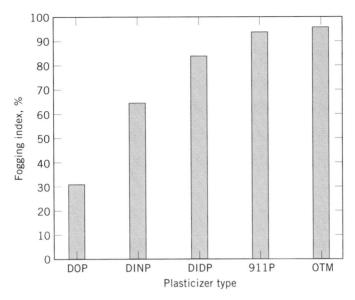

**Fig. 6.** Fogging index of various plasticizers for six hours at 90°C.

ascertain the degree of fogging. In such a test, the fogging performance of a plasticizer is related to its volatility, refractive index, and surface tension. The precise nature of refractive index and surface tension is somewhat complex (30) and attempts to improve the test are in progress (31). In the case of plasticizer volatility it is clear that a higher concentration of plasticizer leads to increased fogging in the test. The higher molecular weight and more linear plasticizers give superior performance. Trimellitate esters and 9,11-phthalates, with their high degree of linearity and consequently low viscosity for plastisol applications, have achieved widespread use as low fogging plasticizers for these applications.

**Overall Assessment of Properties.**    Table 3 shows the effect for each criterion of increasing plasticizer concentration, increasing the size of the plasticizer molecule, increasing the linearity of the plasticizer molecule, and changing the acid constituent of the ester. An I indicates improved performance for a particu-

**Table 3. Structure:Property Relationships of Plasticizer[a]**

| Property, at equal softness | Increased concentration | Increased size | Increased linearity | Acid[b] used |
|---|---|---|---|---|
| efficiency, higher temperature |  | P | I | Ph = Ad > Tr |
| high | P | I | I | Tr > Ph > Ad |
| low | I | I | I | Ad > Ph = Tr |
| gelation | (I) | P | (P) | Ph > Ad ≥ Tr |
| migration/extraction | P | I | P | Tr > Ph > Ad |
| plastisol |  |  |  |  |
| viscosity | I | (I) | I | Ad > Ph |
| aging |  | I | (I) | Ad > Ph |
| fogging | P | I | I | Tr > Ph > Ad |

[a] I = improved performance; P = poorer performance; ( ), marginal performance.
[b] Ph = phthalate; Ad = adipate; Tr = trimellitate.

lar property, a P indicates poorer performance. I and P in parentheses indicate that any changes tend to be marginal.

There is clearly no perfect plasticizer for every application. Choice depends on the performance requirements of the article being manufactured and price.

## The Plasticization of Polymers Other Than PVC

The plasticization of PVC accounts for the vast majority of plasticizer sales. However, significant amounts of plasticizers are used in non-PVC polymers and this may become increasingly important in the future. Although PVC stands alone in its ability to accept and retain large quantities of commercial plasticizer, effective plasticization of other resins using slightly modified plasticizers may be possible if certain conditions specific to the polymer of interest are met.

The first factor to be considered when looking at the plasticization of a polymer is the need; even though some polymers may be compatible with large concentrations of plasticizer, the resultant softening benefits may be of little use. Other factors are short- and long-term compatibility, ie, the ability of a polymer to accept and retain the plasticizer.

In order for a plasticizer to enter a polymer structure the polymer should be highly amorphous. Crystalline nylon retains only a small quantity of plasticizer if it retains its crystallinity. Once it has penetrated the polymer the plasticizer fills free volume and provides polymer chain lubrication, increasing rotation and movement.

The plasticizer content of a polymer may be increased by the suppression of crystallization in the polymer, but if crystallization subsequently occurs the plasticizer exudes. For highly crystalline resins, the small amounts of plasticizer allowable can change the nature of the small amorphous regions with a consequent overall change in properties.

**Acrylic Polymers.** Although considerable information on the plasticization of acrylic resins is scattered throughout journal and patent literature, the subject is complicated by the fact that acrylic resins constitute a large family of polymers rather than a single polymeric species. An infinite variation in physical properties may be obtained through copolymerization of two or more acrylic monomers selected from the available esters of acrylic and methacrylic acid (30) (see ACRYLIC ESTER POLYMERS; METHACRYLIC ACID AND DERIVATIVES).

Plasticizers, however, are used in the acrylics industry to produce tough, flexible coatings. Compatibilities with common plasticizers are up to 10 wt % although in some cases, for low molecular weight plasticizers, it can be higher. For example, a formulation of 100 phr PMMA, 150 phr DBP, 225 phr chalk, and 25 phr resorcinol has been used (6). PMMA is used in small amounts with PVC. Cast acrylics, however, require a high $T_g$ and high rigidity, hence no plasticizer application is required.

Plasticizers for acrylics include all common phthalates and adipates. There has been interest in the development of acrylic plastisols similar to those encountered with PVC. Clearly the same aspects of both plastisol viscosity and viscosity stability are important. Patents appear in the literature (32) indicating that the number of available plasticizers that show both good compatibility with acrylic resins and satisfactory long-term plastisol stability may be fewer than those showing equivalent properties with emulsion PVC resins.

Patents have appeared (33,34) which show formulations containing PMMA emulsion polymer and PMMA suspension polymer combined with benzyl butyl phthalate and octyl benzyl phthalate. It is likely that polymers of this type will require highly polar plasticizers in order to have both adequate compatibility and adequate gelation. When replacing PVC applications the use of large quantities of phosphate plasticizers is sometimes required to give equivalent fire performance.

**Nylon.** The high degree of crystallinity in nylon means that plasticization can occur only at very low levels. Plasticizers are used in nylon but are usually sulfonamide based since these are generally more compatible than phthalates. DEHP is 25 phr compatible; other phthalates less so. Sulfonamides are compatible up to 50 phr.

**Poly(ethylene terephthalate).** PET is a crystalline material and hence difficult to plasticize. Additionally, since PET is used as a high strength film and textile fiber, plasticization is not usually required although esters showing plasticizing properties with PVC may be used in small amounts as processing aids and external lubricants. Plasticizers have also been used to aid the injection molding of PET, but only at low concentrations.

The main area of interest for plasticizers in PET is in the area of dyeing. Due to its lack of hydrogen bonds PET is relatively difficult to dye. Plasticizers used in this process can increase the speed and intensity of the dyeing process. The compounds used, however, tend to be of low molecular weight since high volatility is required to enable rapid removal of plasticizer from the product (see DYE CARRIERS).

**Polyolefins.** Interest has been shown in the plasticization of polyolefins (5) but plasticizer use generally results in a reduction of physical properties (12), and compatibility can be achieved only up to 2 wt %. Most polyolefins give adequate physical properties without plasticization. There has been use of plasticizers with polypropylene to improve its elongation at break (7) although the addition of plasticizer can lower $T_g$, room temperature strength, and flow temperature. This can be overcome by simultaneous plasticization (ca 15 wt % level) and cross-linking. Plasticizers used include DOA.

**Polystyrene.** Polystyrene shows compatibility with common plasticizers but modification of properties produced is of little value. Small amounts of plasticizer (eg, DBP) are used as a processing aid.

**Fluoroplastics.** Conventional plasticizers are used as processing aids for fluoroplastics up to a level of 25% plasticizer. However, certain grades of Kel-F (chlorotrifluorethylene) contain up to 25 wt % plasticizer to improve elongation and increase softness; the plasticizers used are usually low molecular weight oily chloroethylene polymers (5).

**Rubbers.** Plasticizers have been used in rubber processing and formulations for many years (8), although phthalic and adipic esters have found little use since cheaper alternatives, eg, heavy petroleum oils, coal tars, and other predominantly hydrocarbon products, are available for many types of rubber. Esters, eg, DOA, DOP, and DOS, can be used with latex rubber to produce large reductions in $T_g$. It has been noted (9) that the more polar elastomers such as nitrile rubber and chloroprene are insufficiently compatible with hydrocarbons and require a more specialized type of plasticizer, eg, a phthalate or adipate ester. Approximately 50% of nitrile rubber used in Western Europe is plasticized at 10–15 phr (a total of 5000–6000 t/yr), and 25% of chloroprene at ca 10 phr (ca 2000 t/yr) is plasticized. Usage in other elastomers is very low although may increase due to toxicological concerns over polynuclear aromatic compounds (9).

Studies on the use of high molecular weight esters in nitrile rubber have led to further studies to compare DINP with DBP. These showed that at the 10 phr level the nitrile rubber was effectively plasticized with DINP (35).

## Economic Aspects

Worldwide consumption of plasticizers is estimated at $3.5 \times 10^6$ t (31), and is of the order of one million tons in Western Europe (Fig. 7; Table 4). The

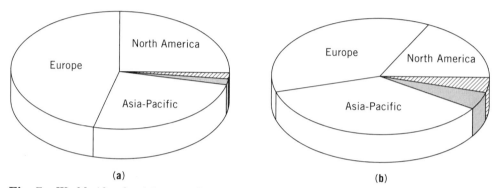

(a)                                                  (b)

**Fig. 7.**  Worldwide plasticizer production (**a**) and consumption (**b**) where ▨ represents Africa and the Near East and ▦, Latin America (31).

**Table 4.  Western European Plasticizer Market$^a$**

| Plasticizer | Market share, % |
|---|---|
| *Overall market* | |
| $C_8$ phthalates | 47 |
| $C_9$ and $C_{10}$ phthalates | 32 |
| $C_4$ phthalates | 6 |
| $C_7$ and $C_{11}$ phthalates | 2 |
| linear phthalates | 1 |
| other specialties | 12 |
| *Specialty market* | |
| epoxies$^b$ | 20 |
| adipates | 14 |
| phosphates | 13 |
| BBP | 12 |
| sulfonates | 11 |
| polymerics | 10 |
| linear phthalates | 10 |
| trimellitates | 7 |
| DTDP | 2 |
| azelates/sebacates | 1 |

$^a$Approximate split by grade types.
$^b$ESBO, etc.

distribution of plasticizers into various applications is as follows: 27%, wire and cable covering; 23%, extrusion/profile; 17%, film and sheet; 13%, coated fabrics; 12%, wall covering; and 8%, undersealing/coating.

## Health and Safety Aspects

Numerous toxicological studies have been conducted on a variety of plasticizers. However, because di-(2-ethylhexyl) phthalate (DEHP) is the most widely used plasticizer and is a well-defined single substance, it is the plasticizer that has been most thoroughly investigated in terms of its toxicology and has often been considered as a model for the other phthalates (36).

**Acute Toxicity.** Plasticizers possess an extremely low order of acute toxicity; $LD_{50}$ values are mostly in excess of 20,000 mg/kg body weight for oral, dermal, or intraperitoneal routes of exposure. In addition to their low acute toxicity, many years of practical use coupled with animal tests show that plasticizers do not irritate the skin or mucous membranes and do not cause sensitization.

**Chronic Toxicity.** The effects of repeated oral exposure to phthalates for periods ranging from a few days to two years have been studied in a number of animal species including rats, mice, hamsters, guinea pigs, ferrets, and dogs (37).

*Liver Effects.* In 1980 a two-year feeding study carried out as part of the NTP/NCI Bioassay Program in the United States (38,39) indicated that DEHP causes increased incidence of liver tumors in rats and mice and that DEHA had a similar effect in mice but not rats. In these studies the levels of plasticizers fed were very high, this being possible only because of their low acute toxicity.

A large number of more recent investigations (37) on a variety of plasticizers and different animal species have revealed the following. (*1*) Plasticizers are not genotoxic. (*2*) Oral administration of plasticizers, fats, and other chemicals including hypolipidemic drugs to rodents causes a proliferation of microbodies in the liver (peroxisomes) which may be considered to be linked to the formation of liver tumors. (*3*) Administration of plasticizers, fats, and hypolipidemic drugs to nonrodent species such as marmosets (40) and monkeys (41) does not lead to peroxisome proliferation and liver damage. Some hypolipidemic drugs which cause peroxisome proliferation in rodents have been used by humans for many years with no ill effects. (*4*) These species differences have also been observed in *in vitro* studies. Phthalates, their metabolites, and a variety of other peroxisome proliferators caused peroxisome proliferation in rat and mouse liver cells but not in those of humans, marmosets, or guinea pigs.

On the basis of these differences in species response it was concluded that phthalates do not pose a significant health hazard to humans. This view is borne out by the EU Commission decision of July 25, 1990 which states that DEHP shall not be classified or labeled as a carcinogenic or an irritant substance (42). This has been reaffirmed in a comprehensive review (43) which concludes that "peroxisome proliferators constitute a discrete class of nongenotoxic rodent hepatocarcinogens and that the relevance of their hepatocarcinogenic effects for human hazard assessment is considered to be negligible."

The International Agency for Research on Cancer (IARC) has classified DEHP (44) as "an agent possibly carcinogenic to humans." However this classification is based only on the rodent studies and does not take into account the more recent understanding of the underlying mechanisms.

*Reproductive Effects.* Phthalates have been shown to cause reproductive effects in rats and mice but primates are resistant to these effects. This may be due in part to pronounced differences in the way in which phthalates are metabolized by rodents and primates, including humans.

The reproductive toxicity of some phthalate esters has been reviewed by the Commission of the European Communities (45). This review concludes that testicular atrophy is the most sensitive indicator of reproductive impairment and that the rat is the most sensitive species.

Comparing estimates of the average human daily lifetime exposure to DEHP (0.3-6 $\mu$g/kg body weight per day) to the level at which no effects are observed in rats indicates that the margin of safety for the general public is more than 10,000. Taking into consideration the significant difference in species sensitivity between rodents and primates, the safety factor is in fact even greater.

## The Effect of Plasticizers on the Environment

About one million tons of plasticizers are used annually in Western Europe. Some 92% of this total is used to plasticize poly(vinyl chloride) (PVC) and about 95% of these PVC plasticizers are phthalate esters. In spite of the fact that there are several hundred plasticizers in commercial use in the world, only relatively few (ie, phthalates) are used in amounts that make them significant in tonnage terms, and hence in their likely environmental input and impact.

**Estimated Emissions to the Environment.** Phthalates may be emitted to the environment during their incorporation into PVC and from the finished PVC article during its use or after its final disposal. However, because their purpose is to make PVC flexible and for it to remain so over long periods of time, plasticizers are of very low volatility relative to many other commonly used products, for example solvents.

The widespread usage of phthalates in flexible PVC has resulted in many investigations being made of their concentration in the environment. Unfortunately the ubiquitous presence of phthalates in laboratory chemicals and equipment has caused problems in the analysis of very low concentrations of phthalates in environmental samples and has led to erroneously high levels being reported.

The need to identify the correct environmental concentrations of phthalates was recognized by the European Council for Plasticizers and Intermediates (ECPI). In 1990 they initiated a program of work aimed at developing rational estimates of phthalate emissions together with ecotoxicological testing to determine their environmental impact. The ECPI, whose members comprise all the important manufacturers of plasticizers in Western Europe, is one of many special interest sector groups of the European Chemical Industry Council (CEFIC).

The members of ECPI are well placed to carry out this study because in addition to their intimate knowledge of plasticizer production techniques they have in-depth technical contacts with the plasticizer consuming industries and hence access to information regarding the use and fate of plasticizers. This knowledge together with the availability of accurate statistical information has enabled ECPI to develop a much more precise model of the environmental input of plasticizers than has hitherto been possible. This long program of work has resulted in the production of soundly based emission estimates (29) (Table 5) and well-designed ecotoxicological studies.

The ECPI approach has been adopted by the European Commission in their "Technical Guidance Document on the Risk Assessment of Notified New Substances" as the model for assessment of environmental exposure from additives in plastics. It is important to note, however, that due to the effect of ultraviolet degradation and microbial attack, a significant proportion of the emissions from flexible PVC consists of plasticizer degradation products. In these instances, therefore, the level of plasticizers appearing in the environment will be significantly less than indicated by the plasticizer loss data.

*Emissions During Plasticizer Production and Distribution.* Phthalate plasticizers are produced by esterification of phthalic anhydride in closed systems hence losses to atmosphere are minimal. Inquiries of all the principal plasticizer producers indicate a maximum total emission in Western Europe of 220 t/yr, 90% of which is to the water compartment. This level is expected to decrease in the future due to increasing plant water treatment.

The transport of phthalates by road tankers and ships within Europe is carried out by international companies with sophisticated tank cleaning facilities. Wash waters from these modern facilities are passed through a series of separators to remove any residual plasticizer which is then incinerated. It is estimated that, as a result of cleaning and spillages, the maximum emission to the environment is 80 t/yr.

*Emissions During Processing.* During the production of flexible PVC products plasticizers are exposed for up to several minutes to temperatures of

**Table 5. Estimated Phthalate Emissions in Western Europe**

| Process/end use | Emissions, t/yr |
|---|---|
| production | 220 |
| distribution | 80 |
| processing | |
|   calendered film and sheet | 280 |
|   calendered flooring | 10 |
|   spread coating | 520 |
|   other plastisols | 50 |
|   extrusion/injection molding | 90 |
|   *total* | *950* |
| interior end use | |
|   flooring | |
|     evaporation | 20 |
|     water extraction | 500[a] |
|   wall covering | 20 |
|   other film, sheet, and coating | 40 |
|   wire, cable, profile, and hose | 60 |
|   *total* | *640* |
| exterior end use | 5600[a] |
| disposal | 250 |
|   *Total* | *7740* |

[a]Not well defined.

~180°C. The exact conditions depend on the processing technique employed, but it is evident that the loss of plasticizer by evaporation and degradation can be significant.

Of the various processing techniques used, injection molding and extrusion involve little or no exposure of hot product to the surrounding air, hence they give rise to no significant emission of plasticizer to the atmosphere. This is not the case in the production of sheet and film by calendering or spread coating.

Air extracted from spread coating ovens contains typically 500 mg/m$^3$ plasticizer which is present mostly as an aerosol since the saturation concentration of common plasticizers in air is low. In those installations with filtration equipment, the average phthalate concentration in the air is reduced from 500 mg/m$^3$ to below 20 mg/m$^3$. The use of incineration equipment to clean the exhaust air reduces the residual phthalate concentration to practically zero. The use of filters and incinerators on calendering and spread coating plants is steadily increasing due primarily to the need to reduce emissions of solvents and other volatile organic compounds. Consequently a large proportion (69% in 1990 and increasing) of all the phthalates in calendering and plastisol applications in Western Europe are used in processing plants equipped with incinerators or filters (Table 6).

Knowledge of the quantity of plasticizer used in each application together with the level of exhaust air treatment allows estimation of the level of plasticizer lost to atmosphere during these processes.

*Emissions During Interior End Use.*  The majority of flexible PVC is used indoors in applications such as flooring, wall covering, upholstery, wire and cable, etc. Environ Corporation, consultants to the CMA, have developed a model (46) which attempts to quantify the plasticizer losses that occur in a typical room. Using theoretical and semiexperimental approaches they have arrived at an emission rate at 25°C of $2.3 \times 10^{-4}$ mg/s/m$^2$. Using this emission rate with data on the annual production of PVC flooring, wallcovering, etc, and estimates of their lifetimes it is straightforward to calculate the plasticizer losses from all indoor flexible PVC articles.

Some products, particularly flooring, may lose plasticizer not only by evaporation but also through extraction by soapy water during cleaning. It is possible to estimate the quantity of plasticizer extracted but many assumptions have to be made including the frequency, duration, and temperature of washing and the proportion of floors cleaned in this way. Wastewater associated with the cleaning process typically goes to the municipal sewage system. Thus, the phthalates are biodegraded and do not end up in the environment.

**Table 6. Air Treatment in Flexible PVC Processing Plants in Western Europe, 1991**

| Treatment | Plasticizer usage, $10^3$ t/yr | Filter treated, % | Incinerated, % | Total treated, % |
|---|---|---|---|---|
| spread coating | 192 | 53 | 22 | 75 |
| slush, dip, and rotational molding | 17 | 26 | 6 | 32 |
| automotive underseal | 67 | | 100 | 100 |
| calendered sheet and film | 138 | 23 | 25 | 48 |
| calendered flooring | 31 | 15 | 56 | 81 |

*Emissions During Exterior End Use.*   When flexible PVC is used in exterior applications plasticizer loss may occur due to a number of processes which include evaporation, microbial attack, hydrolysis, degradation, exudation, and extraction. It is not possible, due to this wide variety of contribution processes, to assess theoretically the rate of plasticizer loss by exposure outdoors. It is necessary, therefore, to carry out actual measurements over extended periods in real life situations. Little suitable data has been published with the exception of some studies on roofing sheet (47). The data from roofing sheet has been used to estimate the plasticizer losses from all outdoor applications. This estimate may well be too high because of the extrapolation involved. Much of this extracted plasticizer does not end up in the environment because considerable degradation takes place during the extraction process.

*Emissions During Disposal and Incineration.*   The increasing use of modern incinerators to dispose of domestic waste results in complete combustion of plasticizers to carbon dioxide and water. The preponderance of plasticizer going into landfills is as plasticized PVC. Once a landfill has been capped anaerobic conditions prevail and it is biologically relatively inactive. Under these conditions the main route by which organic components are removed from the landfill contents is by ingress of water, extraction, and subsequent loss of water from the site to the environment.

In the United Kingdom there are approximately 2000 active landfill sites with an annual combined water ingress of $40 \times 106 \, m^3$ (48,49). The solubility of the most common plasticizer, DEHP, in water is difficult to measure. The true solubility is considered to be around 50 $\mu g/L$ (50,51), but various values are given in the literature. Assuming a worst case solubility of 1 ppm then the maximum quantity of plasticizer extracted is 40 t/yr. If the United Kingdom is typical of the whole of Europe, and assuming that waste is proportional to population, then a maximum of 250 t/yr plasticizer could be emitted to the environment from landfills in Western Europe. To obtain a more precise estimate of the situation, 25 effluent water samples from a variety of types of U.K.-based landfills have been analyzed. The highest concentration of DEHP detected was 30 parts per billion. If the highest figure is used instead of the assumed 1 ppm then the quantity from the whole of Europe becomes 7.5 t.

**Occurrence of Plasticizers in the Environment.**   The contamination of laboratory chemicals and equipment causes problems in the analysis of very low concentrations of phthalates in environmental samples. Strenuous efforts have been made to overcome these difficulties in recent studies but the results of many earlier investigations must be treated with caution.

*Phthalates in Air.*   Atmospheric levels of phthalates in general are very low. They vary, for DEHP, from nondetectable to 132 $ng/m^3$ (50). The latter value, measured in 1977, is the concentration found in an urban area adsorbed on airborne particulate matter and hence the biological availability is uncertain. More recent measurements (52) in both industrial and remote areas of Sweden showed DEHP concentrations varying from 0.3 to 77 $ng/m^3$ with a median value of 2 $ng/m^3$.

Atmospheric photodegradation of DEHP and DBP has been shown to be rapid (51,53) with half-life times of less than two days, hence a large proportion of phthalate emissions to the air are broken down by photodegradation.

*Phthalates in Water.* Reported levels of phthalates in natural waters are, in general, low. Concentrations found in fresh waters range from nondetectable up to 10 $\mu$g/L. Measured concentrations (54) in Swedish rivers vary from 0.3 to 3.1 $\mu$g/L. The highest values are found near industrial discharge points.

In the most recent and comprehensive study, 230 measurements from 11 sampling points along 225 km of the Rhine and adjoining rivers were made over a period of one year (1991–1992). The concentration of DEHP found varied from 0.11 to 10.3 $\mu$g/L; the latter value is unusually high as evidenced by the mean concentration of only 0.82 $\mu$g/L (55).

Plasticizer levels in surface waters are decreasing. The Netherlands National Institute of Public Health and Environmental Protection (RIVM) states that the level of phthalates in Lake Yssel fell by 75% to 0.3 $\mu$g/L over the period 1980–1988 (56).

*Phthalates in Sediments.* Phthalates are lipophilic and hence partition onto organic-rich particulate matter in water. This particulate matter on settling gives rise to sediments which contain higher levels of phthalate than the overlying water. The level of DEHP in sediment at seven points along the Rhine has been found (55) to range from 1.8 to 18.3 mg/kg dry weight. Lower levels of between 0.1 and 8.9 mg/kg dry weight have been found at 10 sampling points along the Weser. The mean concentration of DEHP in sediment in the two rivers was 5 mg/kg dry weight (dw).

By taking sections from a sediment core sample of 120 cm in depth it was found that the current phthalate level is only 15% of what it had been in 1972–1978. This is despite the fact that the total usage of plasticizers has continued to increase annually.

A thorough survey of phthalate levels in sediments has been conducted in Sweden by the Environmental Research Institute (57). Samples of sediment have been taken from 22 locations varying from isolated lakes to rivers near industrial discharge points. The concentration of DEHP in the lake and river samples varied from 0.008 to 0.79 mg/kg dry weight (dw). The only high levels were found at the discharge points from two industrial sites using phthalates. The total concentrations of all phthalates at these two discharge points were 203 mg/kg dw and 34.3 mg/kg dw. Measurements at these same points in 1983 (54) gave values of 1480 and 628 mg/kg dw, respectively. The decreasing concentrations of phthalates observed between 1983 and 1994 at these two point sources is considered to be the result of improved plant wastewater management.

*Wastewater Treatment Plants.* Numerous studies have shown that phthalates in wastewater systems are removed to a significant extent by treatment plants. The concentration of phthalates in both domestic and industrial wastewater was measured before and after treatment (55). The total level of phthalates in domestic effluent was reduced by treatment from 32.7 to 0.92 $\mu$g/L and in industrial effluent from 93.6 to 1.06 $\mu$g/L. Thus between 97 and 99% of the phthalates are removed from wastewater by treatment plants.

These data together with those from wastewater treatment plants at Darmstadt, Germany; Gothenburg and Stockholm, Sweden; and Noord-Brabant, the Netherlands, show that the concentrations of DEHP, and in some cases total phthalates, entering wastewater treatment plants vary from 1 to 167 $\mu$g/L. After treatment the concentrations range from $<1$ to 36.8 $\mu$g/L.

**Environmental Modeling.** The estimated plasticizer emissions outlined earlier have been entered into the HAZCHEM model developed by the European Centre for Ecotoxicology and Toxicology of Chemicals (ECETOC). This enables the levels of plasticizer in the various environmental compartments to be estimated. The advantage of using environmental models in conjunction with emission estimates is that they give an overview of the concentrations present in any chosen region. This is helpful in conducting realistic environmental risk assessments.

Comparison of these environmental compartment concentrations with the actual measurements made at a variety of locations show reasonable agreement but indicate that emission estimates are rather high. It is likely that the fault lies with worst case estimates for losses from outdoor applications and the washing of PVC flooring. In addition a large proportion of the phthalates lost by these routes will not enter rivers because they will be removed by wastewater treatment plants.

**Environmental Effects of Plasticizers.** Measurement of the effect of phthalates on environmental species is difficult because standard test methods are not designed to deal with poorly water-soluble substances. For this reason a number of early studies are flawed and their results should be disregarded in favor of more recent investigations where these difficulties have been overcome.

*Atmospheric Toxicity.* The only known atmospheric toxicity effect of phthalates is the phytotoxicity arising from the use of DBP plasticized glazing bars in greenhouses. However, the higher phthalates such as DEHP are not phytotoxic. General atmospheric concentrations of phthalates are extremely low and it is concluded that they pose no risk to plants or animals.

*Aquatic Toxicity.* The standard tests to measure the effect of substances on the aquatic environment are designed to deal with those that are reasonably soluble in water. Unfortunately this is a disadvantage for the primary phthalates because they have a very low water solubility (ca 50 $\mu$g/L) and this can lead to erroneous test results. The most common problem is seen in toxicity tests on daphnia where the poorly water-soluble substance forms a thin film on the water surface within which the daphnia become entrapped and die. These deaths are clearly not due to the toxicity of the substance but due to unsuitable test design.

The majority of studies on the acute and chronic toxicity of phthalates to aquatic organisms show no toxic effects at concentrations 200–1000 times the water solubility. However, there are some studies indicating higher toxicity which are believed to be due to the flotation and entrapment effects outlined above.

ECPI has commissioned further aquatic toxicity studies in order to clarify the situation and meet the requirements of Commission Directive 93/21/EEC, 18th adaptation to technical progress of 67/548/EEC which regulates the classification, labeling, and packaging of substances. The studies have been designed to ensure no entrapment problems. Some of these studies, investigating the effect of phthalates on the survival and reproduction of *Daphnia magna*, have been completed and demonstrate that DEHP and the higher phthalates tested do not show acute or chronic toxicity to daphnia at 1 mg/L. These results, coupled with data from biodegradation studies, confirm that the phthalates commonly used in the plasticization of PVC do not require classification "Dangerous for the Environment."

*Sediment Toxicity.*    Because of their low solubility in water and lipophilic nature, phthalates tend to be found in sediments. Unfortunately little work has previously been carried out on the toxicity of phthalates to sediment dwelling organisms. For this reason ECPI has commissioned some sediment toxicity studies designed to measure the effect of DEHP and DIDP in a natural river sediment on the emergence of the larvae of the midge, *Chironomus riparius.*

*Chironomus riparius* is distributed throughout North America and Europe in a wide variety of freshwater habitats. The larvae live within the sediment and after four larval stages they pupate and rise to the surface where the adult insects emerge. This 28-day study has shown that DEHP and DIDP sediment concentrations of 100, 1,000, and 10,000 mg/kg dry weight have no adverse effect on either the time to emergence or the percentage emergence of the adults. It is therefore concluded that there was no effect of either phthalate on the survival, development, and emergence of *Chironomus riparius* at any of the sediment concentrations tested. The no observed effect concentration (NOEC) of 10,000 mg/kg dry weight is far in excess of phthalate concentrations that are found even at industrial discharge points.

*Conclusions.*    The impact of plasticizers on the environment is very low and is diminishing as evidenced by analytical data showing that the levels of phthalates in surface waters and sediments are decreasing. This is despite the fact that their usage has continued to increase annually and is most likely due to improved emission controls and wastewater treatment.

## Storage and Handling

Plasticizer esters are relatively inert, thermally stable liquids with high flash points and low volatility. Consequently they can be stored safely in mild steel storage tanks or drums for extended periods of time. Exposure to high temperatures for extended periods, as encountered in drums in hot climates, is not recommended since it may lead to a deterioration in product quality with respect to color, odor, and electrical resistance.

## BIBLIOGRAPHY

"Plasticizers" in *ECT* 1st ed., Vol. 10, pp. 766–798, by A. K. Doolittle, Carbide and Carbon Chemicals Co.; in *ECT* 2nd ed., Vol. 15, pp. 720–789, by J. R. Darby and J. K. Sears, Monsanto Co.; in *ECT* 3rd ed., Vol. 18, pp. 111–183, by J. K. Sears and N. W. Touchette, Monsanto Co.

1. J. K. Sears and J. R. Darby, *The Technology of Plasticizers*, John Wiley and Sons, Inc., New York, 1982.
2. A. K. Doolittle, *J. Poly. Sci.* **2**(1), 121 (1947).
3. A. Hartmann, *Colloid Z.* **142**, 123 (1955).
4. R. S. Barshtein and G. A. Kotylarevski, *Sov. Plast.* **7**, 18 (1966).
5. D. L. Buszard, in W. V. Titow, ed., *PVC Technology*, 4th ed., Elsevier, New York, 1984.
6. W. Sommer, W. Wicke, and D. Mayer, *Ullman's Encyclopedia of Industrial Chemistry*, VCH, Weinheim, Germany, 1988, p. 350.
7. S. V. Patel and M. Gilbert, *Plast. Rubb. Proc. Appl.* **6**, 321 (1986).
8. L. Ramos de Valle and M. Gilbert, *Plast. Rubb. Proc. Appl.* **13**, 151 (1990).

9. C. J. Howick, *Plast. Rubber Compos.: Proc. Appl.* **23**, 53–60 (1995).
10. G. J. Van Veersen and A. J. Meulenberg, *SPE Tech. Paper* **18**, 314 (1972).
11. C. E. Anagnostopoulos, A. Y. Coran, and W. R. Gamrath, *J. Appl. Polym. Sci.* **4**, 181 (1960).
12. D. C. H. Bigg, *J. Appl. Polym. Sci.* **19**, 3119 (1975).
13. A. Garton, P. Cousin, and R. E. Prud'Homme, *J. Polym. Sci., Polym. Phys. Ed.* **21**, 2275 (1983).
14. M. M. Coleman and J. Zarian, *J. Polym. Sci. Polym. Phys. Ed.* **12**, 837 (1979).
15. M. M. Coleman and D. F. Varnell, *J. Polym. Sci. Polym. Phys. Ed.* **18**, 1403 (1980).
16. J. J. Schmidt, J. A. Gardella, and L. Salvati, *Macromolecules* **22**, 4489 (1989).
17. M. Aubin, Y. Bedard, M. F. Morrissette, and R. E. Prud'Homme, *J. Polym. Sci. Polym. Phys. Ed.* **21**, 233 (1983).
18. M. B. Clark, C. A. Burkhardt, and J. A. Gardella, *Macromolecules* **22**, 4495 (1989).
19. D. J. Walsh and S. Rostami, *Adv. Polym. Sci. Rev.* **70**, 119 (1985).
20. D. F. Varnell and M. M. Coleman, *Polymer* **22**, 1324 (1981).
21. D. L. Tabb and J. L. Koenig, *Macromolecules* **8**(6), 929 (1975).
22. M. Theodorou and B. Jasse, *J. Polym. Sci. Polym. Phys. Ed.* **21**, 2263 (1983).
23. E. Benedetti and co-workers, *J. Polym. Sci. Polym. Phys. Ed.* **23**, 1187 (1985).
24. *Chemistry In Britain*, Special Issue on Computational Chemistry, Nov. 1990.
25. *Chem. Ind.* (Dec. 1990).
26. J. P. Sibilia, *A Guide to Materials Characterisation and Chemical Analysis*, VCH, Weinheim, Germany, 1988.
27. N. J. Clayden and C. J. Howick, *Polymer* **34**(12), 2508 (1993).
28. M. J. Bunten, M. W. Newman, P. V. Smallwood, and R. C. Stephenson, in J. I. Kroschwitz, ed., *Encyclopedia of Polymer Science and Engineering*. Vol. 17 and Supplement, John Wiley and Sons, Inc., New York, 1989, pp. 241–392.
29. D. F. Cadogan and co-workers, *Prog. Plast. Rubb. Tech.* **10**(1), 1 (1994).
30. U.S. Pat. 3,178,386, R. J. Hickman (to General Motors).
31. R. F. Caers and A. C. Poppe, *Kunststoffe* **83**, 10 (1993).
32. S. S. Kurtz, J. S. Sweely, and W. J. Stout in P. F. Bruins, ed., *Plasticizer Technology*, Reinhold Publishing Corp., New York, 1965.
33. Bundesrepublik Deutschland Pat. Nr 38 16 710 (to Pegulan-Werke AG).
34. Bundesrepublik Deutschland Pat. Nr 39 03 669 (to Pegulan-Werke AG).
35. D. Stening, *Trwobridge Technical College Internal Report*, Wiltshire, U.K., 1991.
36. *Di-2-ethylhexyl Phthalate: A Critical Review of the Available Toxicological Information*, CEFIC, Brussels, Belgium, 1985.
37. *Environmental Health Criteria 131, Diethylhexyl Phthalate*, World Health Organization, Geneva, Switzerland, 1992.
38. *Carcinogenesis Bioassay of Di(2-ethylhexyl) Adipate in F344 Rats and B6C3F1 Mice*, Report series No. 212, National Toxicology Programme, Research Triangle Park, N.C., 1980.
39. *Carcinogenesis Bioassay of Di(2-ethylhexyl) Phthalate in F344 Rats and B6C3F1 Mice*, Report series No. 217, National Toxicology Programme, Research Triangle Park, N.C., 1982.
40. C. Rhodes and co-workers, *Environ. Health. Perspect.* **65**, 299 (1986).
41. R. D. Short and co-workers, *Toxicol. Ind. Health* **3** 185 (1987).
42. *Official Journal of the European Communities*, No. L 222/49, Office for Official Publications of the European Communities, Luxembourg, Belgium, Aug. 17, 1990.
43. J. Ashby and co-workers, *Human Experim. Toxicol.* **13** Suppl. 2 (1994).
44. *IARC Monogr.* **29**, 281 (1982).
45. F. M. Sullivan and co-workers, *The Toxicology of Chemicals, Series Two: Reproductive Toxicity*, Vol. 1, Commission of the European Communities, Brussels, Belgium, 1993.

46. Report prepared for CMA, Washington, D.C., *Indoor DEHP Air Concentrations Predicted after DEHP Volatilizes from Vinyl Products*, Environ. Corp., 1988.

47. G. Pastuska and co-workers, *Kautsch Gummi Kunststoffe* **41**, 451 (1988).

48. Private communication, Aspinwall & Co., U.K., Feb. 1991.

49. Private communication, AEA Technology, Harwell, U.K., 1991.

50. *An Assessment of the Occurrence and Effects of Dialkyl Ortho-Phthalates in the Environment*, Technical Report No. 19, European Centre for Ecotoxicology and Toxicology of Chemicals (ECETOC), Brussels, Belgium, 1985.

51. *Di-(2-Ethylhexyl)Phthalat, BUA-Stoffbericht 4*, Beratergremium fur Umweltrelevante Altstoffe (BUA) der Gesellschaft Deutscher Chemiker (Hrsg.), VCH Verlagsges, Weinheim, Germany, 1986.

52. A. Thurén and P. Larsson, *Environ. Sci. Technol.* **24**(4), 554 (1990).

53. C. Zetzsch, *Z. Umweltchem. Okotox.* **3**, 59 (1991).

54. A. Thurén, *Bull. Environ. Contam. Toxicol.* **36**, 40 (1986).

55. *Phthalate in der Aquatischen Umwelt*, Report No. 6/93, Landesamt fur Wasser und Abfall Nordrhein-Westfalen, Dusseldorf, Germany, 1993.

56. *Update of the Exploratory Report Phthalates*, Report No. 710401008, National Institute of Public Health and Environmental Protection (RIVM), Bilthoven, the Netherlands, 1991.

57. *Phthalates in Swedish Sediments*, No. 1167, The Environmental Research Institute (IVL), Stockholm Sweden, 1995.

DAVID F. CADOGAN
European Council for Plasticizers and Intermediates


CHRISTOPHER J. HOWICK
European Vinyls Corporation

# PLASTICS PROCESSING

Plastics are classified as thermoplastic or thermosetting resins, depending on the effect of heat. Thermoplastic resins, when heated during processing, soften and flow as viscous liquids; when cooled, they solidify. The heating/cooling cycle can be repeated many times with little loss in properties. Thermosetting resins liquefy when heated and solidify with continued heating; the polymer undergoes permanent cross-linking and retains its shape during subsequent cooling/heating cycles. Thus, a thermoset cannot be reheated and molded again. However, thermoplastics can be melt-reprocessed, and hence readily recycled (see RECYCLING, PLASTICS (SUPPLEMENT)).

## Thermoplastic Resins

Almost 85% of the resins produced are thermoplastics (1). Although a number of chemically different types of thermoplastics are available in the market, they can be divided into two broad classes: amorphous and crystalline. Amorphous thermoplastics shown in Table 1 (2) are characterized by their glass-transition temperature, $T_g$, a temperature above which the modulus decreases rapidly and the polymer exhibits liquid-like properties; amorphous thermoplastics are normally processed at temperatures well above their $T_g$. Semicrystalline resins shown in Table 2 (2) can have different degrees of crystallinity ranging from 50 to 95%; they are normally processed above the melting point, $T_m$, of the crystalline phase. Upon cooling, crystallization must occur quickly, ie, in a few seconds. Addition of nucleating agents increases the crystallization rate. Additional crystallization often takes place after cooling and during the first few hours following melt processing.

Over 70% of the total volume of thermoplastics is accounted for by the commodity resins: polyethylene, polypropylene, polystyrene, and poly(vinyl chloride) (PVC) (1) (see OLEFIN POLYMERS; STYRENE PLASTICS; VINYL POLYMERS). They are made in a variety of grades and because of their low cost are the first choice for a variety of applications. Next in performance and in cost are acrylics, cellulosics, and acrylonitrile–butadiene–styrene (ABS) terpolymers (see ACRYLIC ESTER POLYMERS; ACRYLONITRILE POLYMERS; CELLULOSE ESTERS). Engineering plastics (qv) such as acetal resins (qv), polyamides (qv), polycarbonate (qv), polyesters (qv), and poly(phenylene sulfide), and advanced materials such as liquid crystal polymers, polysulfone, and polyetheretherketone are used in high performance applications; they are processed at higher temperatures than their commodity counterparts (see POLYMERS CONTAINING SULFUR).

**Table 1. Glass-Transition Temperature of Amorphous Thermoplastics[a]**

| Polymer | $T_g$, °C |
|---|---|
| polyamideimide (PAI) | 295 |
| polyethersulfone (PES) | 230 |
| polyarylsulfone (PAS) | 220 |
| polyetherimide (PEI) | 218 |
| polyarylate (PAR) | 198 |
| polysulfone (PSU) | 190 |
| polyamide, amorphous (PA) | 155 |
| polycarbonate (PC) | 145 |
| styrene–maleic anhydride (SMA) | 122 |
| chlorinated PVC (CPVC) | 107 |
| poly(methyl methacrylate) (PMMA) | 105 |
| styrene–acrylonitrile (SAN) | 104 |
| polystyrene (PS) | 100 |
| acrylonitrile–butadiene–styrene (ABS) | 100 |
| poly(ethylene terephthalate) (PET) | 67 |
| poly(vinyl chloride) (PVC) | 65 |

[a]Ref. 2.

Table 2. Melting Temperature of Semicrystalline
Thermoplastics[a]

| Polymer | $T_m$, °C |
|---|---|
| polyetherketone (PEK) | 365 |
| polyetheretherketone (PEEK) | 334 |
| polytetrafluoroethylene (PTFE) | 327 |
| poly(phenylene sulfide) (PPS) | 285 |
| liquid crystal polymer (LCP) | 280 |
| nylon-6,6[b] | 265 |
| poly(ethylene terephthalate) (PET) | 260 |
| nylon-6[b] | 220 |
| nylon-6,12[b] | 212 |
| nylon-11[b] | 185 |
| nylon-12[b] | 178 |
| acetal resin[c] | 175 |
| polypropylene (PP) | 170 |
| high density polyethylene (HDPE) | 135 |
| low density polyethylene (LDPE) | 112 |

[a] Ref. 2.
[b] Nylons are polyamides (qv).
[c] Acetal resin is polyoxymethylene (POM).

With few exceptions, thermoplastics are marketed in the form of pellets. They are shipped in containers of various sizes, from 25-kg bags to railroad hopper cars. Resins are conveyed to silos for storage and from there to the processing equipment. Colored resins are available, but frequently it is more convenient and economical to buy uncolored resins and blend them with color concentrates. Using concentrates avoids handling dusty pigments and ensures uniform color distribution.

The packaging (qv) requirements for shipping and storage of thermoplastic resins depend on the moisture that can be absorbed by the resin and its effect when the material is heated to processing temperatures. Excess moisture may result in undesirable degradation during melt processing and inferior properties. Condensation polymers such as nylons and polyesters need to be specially predried to very low moisture levels (3,4), ie, less than 0.2% for nylon-6,6 and as low as 0.005% for poly(ethylene terephthalate) which hydrolyzes faster.

A variety of processing equipment and shaping methods are available to fabricate the desired thermoplastic product (5–9). Extrusion is the most popular. Approximately 50% of all commodity thermoplastics are used in extrusion process equipment to produce profiles, pipe and tubing, film, sheet, wire, and cable (1). Injection molding follows as a preferred processing method, accounting for about 15% of all commodity thermoplastics. Other common methods include blow molding, rotomolding, thermoforming, calendering, and, to some extent, compression molding. Details on the amounts of resins converted annually in the United States in terms of processes and products can be found in the following year's January issue of *Modern Plastics*. Computer-aided design software for molds and extruder screws is commercially available. These programs assist in the selection and fabrication of processing equipment, thereby saving research and development time. Modeling of polymer processing is described in specialized textbooks (10–14) (see COMPUTER-AIDED DESIGN AND MANUFACTURING (CAD/CAM); COMPUTER-AIDED ENGINEERING (CAE)).

## EXTRUSION

Extrusion is defined as continuously forcing a molten material through a shaping device. Because the viscosity of most plastic melts is high, extrusion requires the development of pressure in order to force the melt through a die. Manufacturers of plastic resins generally incorporate stabilizers and modifiers and sell the product in the form of cylindrical, spherical, or cubic pellets of about 2–3-mm in diameter. The end-product manufacturers remelt these pellets and extrude specific profiles, such as film, sheet, tubing, wire coating, or as a molten tube of resin (parison) for blow molding or into molds, as in injection molding.

To provide a homogeneous product, incorporation of any additives, such as antioxidants (qv), colorants, and fillers, requires mixing them into the plastic when it is in a molten state. This is done primarily in an extruder. The extruder, as shown in Figure 1, accepts dry solid feed (F,E,J) and melts the plastic by a combination of heat transfer through the barrel (B,C) and dissipation of work energy from the extruder drive motor (I). In the act of melting, and in subsequent sections along the barrel, the required amount of mixing is usually achieved. Venting may also be accomplished to remove undesirable volatile components, usually under vacuum through an additional deep-channel section and side vent port. The final portion of the extruder (L) is used to develop the pressure (≤50 MPa (7500 psi)) for pumping the homogenized melt through a filtering

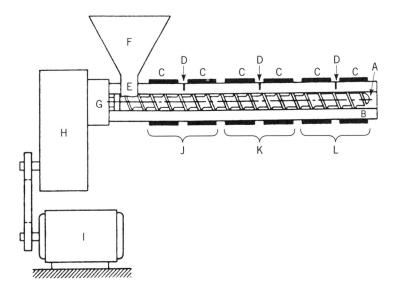

**Fig. 1.** Parts of an extruder: A, screw; B, barrel; C, heater; D, thermocouple; E, feed throat; F, hopper; G, thrust bearing; H, gear reducer; I, motor; J, deep channel feed section; K, tapered channel transition section; and L, shallow channel metering section (15).

screen (optional) and then through a shaping die attached to the end of the extruder.

Extruders are defined by their screw diameter and length, with the length expressed in terms of the length-to-diameter ratio ($L/D$). Single-screw extruders range from small laboratory size (6-mm $D$) to large commercial units (450-mm $D$) capable of processing up to 20 t/h. Melt-fed extruders run about 8 $L/D$; solids fed extruders run from 20–40 $L/D$ depending on whether intermediate venting is provided.

Varieties of twin-screw extruders are also utilized, particularly when the ingredient mixing requirements are difficult to fulfill or require multiple staging, as in reactive extrusion. Twin-screw extruders are classified as being tangential or intermeshing, and the latter as being counter- or co-rotating. These extruders are generally supplied with slip-on conveying and kneading screw elements and segmented barrels. These elements, shown in Figure 2, give the processor improved mixing and pumping versatility.

Reactive extrusion is the term used to describe the use of an extruder as a continuous reactor for polymerization or polymer modification by chemical reaction (16). Extruders are uniquely suitable for carrying out such reactions because of their ability to pump and mix highly viscous materials. Extruders readily permit multiple process steps in a single machine, including melting, metering, mixing, reacting, side-stream addition, and venting.

**Mechanism of Extruder Operation.**  An extruder employs drag flow to perform a conveying action that depends on the relative motion between the screw and the barrel. With higher friction on the barrel than on the screw, the solids are conveyed almost as solid plug in the deep-feed channel section. As the channel depth becomes shallower, the compressive action causes more frictional heat, which, combined with the conduction supplied by the barrel heaters, causes the plastic to melt. The molten plastic then enters a constant shallow depth section of the screw called the metering section, where the pumping pressure necessary for extrusion through the final shaping die is developed.

The drag flow or volumetric conveying capability, $Q_d$, for the plastic melt is dependent only on screw speed, $N$, and the geometry, $A$, of the screw:

$$Q_d = AN$$

The drag flow is most easily visualized by unwrapping the screw and dragging a flattened barrel surface diagonally across the channel (Fig. 3).

At the discharge end of the metering section, enough pressure must be generated to overcome the resistance of the transfer piping and the shaping die. This same pressure in effect also causes the plastic melt to want to flow back down the channel and possibly also over the flight tips. This pressure flow, $Q_p$, is dependent on the screw geometry, $B$, pressure driving force, $\Delta P$, filled length of the screw, $L$, and melt viscosity, $\mu$:

$$Q_p = B\Delta P/\mu L$$

The pressure flow is independent of screw speed, except as the latter affects the viscosity of the melt.

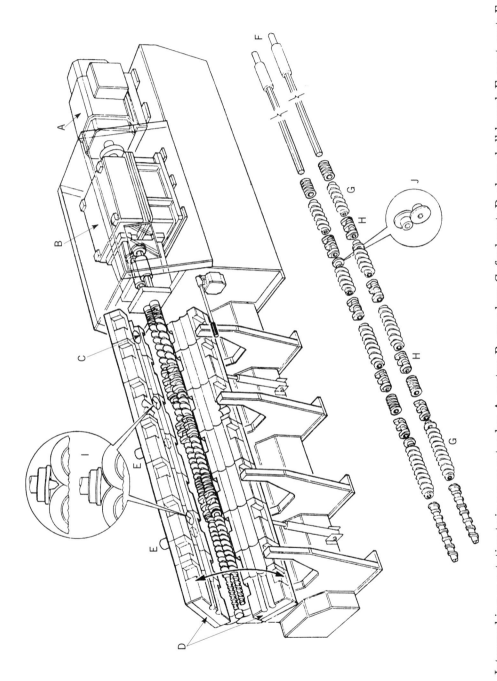

**Fig. 2.** Intermeshing co-rotating twin-screw extruder: A, motor; B, gear box; C, feed port; D, clam shell barrel; E, vent port; F, screw shafts; G, conveying screws; H, kneading paddles; I, barrel valve; and J, blister rings. Courtesy of APV Chemical Machinery Inc.

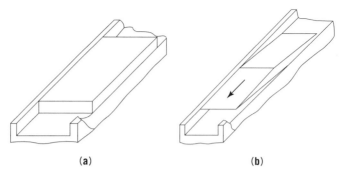

**Fig. 3.** (a) Channel with block of molten plastic; (b) channel with moved block of molten plastic; arrow indicates direction of material movement (15).

The net flow, $Q$, of an extruder is simply the difference between drag and pressure flows:

$$Q = Q_d - Q_p$$

**Mixing in Extruders.** In addition to the conveying and melting steps, extruders perform the vital task of homogenization of additives such as fillers (qv), stabilizers, and pigments (qv) into the base resin. If the cross-channel mixing provided by the drag and pressure flows in single-screw extruders is insufficient, various mixing enhancers are available. These generally assure dispersion by forced passage over a higher shear-restricting slot or an improved distribution by multiple reorientations of flow and sequences of extensional flow (see MIXING AND BLENDING).

Mixing in twin-screw extruders benefits from the additional interaction of the two screws with each other as well as with the barrel. Various arrays of kneading paddles are generally used in the most common type, namely the co-rotating, intermeshing, twin-screw extruder. The kneading paddles improve mixing by causing extensive melt reorientation, back-mixing, and elongational flow patterns (17).

Further information on extruder operation and plastic processing is available (15,18–25). Technology updates and equipment options for specific processing steps are provided annually in publications such as the *Plastics Compounding Redbook* and in *Modern Plastics Encyclopedia*.

**Pipe and Tubing.** A typical die for extruding tubular products is shown in Figure 4. It is an in-line design, ie, the center of the extruded pipe is concentric with the extruder barrel. The extrudate is formed into a tube by the male and female die parts. The male die part is supported in the center by a spider mandrel. Melt flows around legs of the mandrel and meets on the downstream side. The position of the female die part can be adjusted with bolts; adjustment is required to obtain a tube with a uniform wall thickness.

A vacuum calibrator usually is used for controlled cooling of the tube, as shown in Figure 5. It is a long, closed tank that contains cooling water and uses a vacuum to maintain a constant tube diameter and thickness.

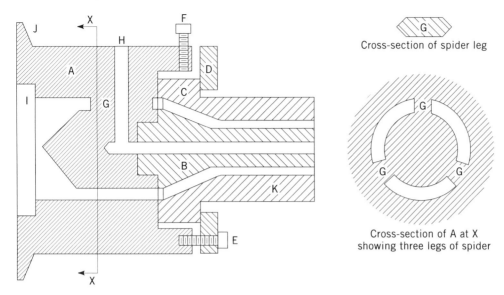

**Fig. 4.** Pipe or tubing die for in-line extrusion: A, die body; B, mandrel, pin, and male die part; C, die, die bushing, and female part; D, die-retaining ring; E, die-retaining bolt; F, die-centered bolt; G, spider leg; H, air hole; I, seat for breaker plate; J, ring for attachment to extruder; and K, die land (15).

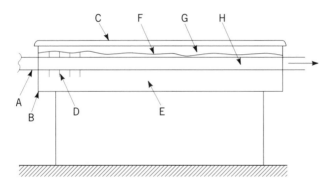

**Fig. 5.** Vacuum calibrator for pipe and tubing extrusion: A, molten tube from die; B, tank; C, hinged cover with gasket; D, sizing rings; E, circulated and temperature controlled water; F, water level; G, vacuum; and H, inside of pipe open to atmospheric pressure (15).

The molten tube enters the tank through a series of cooling rings, and atmospheric pressure inside the tube forces the melt against the rings. The rings control the expansion of the tube and establish its outside diameters. Water cools the tube and adds strength to it, allowing it to withstand expansion forces. The tube exits the tank through a rubber gasket. A variable-speed takeoff controls the rate at which the tube is pulled through the tank.

Tubing extrusion usually involves drawdown of the resin from the die, ie, the diameter and wall thickness of the tubing are less than that of the die opening. The amount of drawdown is expressed as the ratio of the cross-sectional area of the die opening to the cross-sectional area of the tube wall. Drawdowns in tubing extrusion usually range from 2:1 to 8:1. Drawdown in pipe extrusion is often less than 1.1:1. Pipe and tubing may be cut in straight lengths or may be coiled, depending on use and stiffness, eg, polyethylene pipe usually is cut to length, whereas laboratory tubing is coiled. ABS and PVC pipe are made in straight lengths, and soda straws are cut in short lengths.

Pipe extrusion is commonly used in the manufacture of thick-walled products. Die openings are large and resistance to flow is small. Cooling of the pipe is slow because plastics are poor conductors of heat. Long quench tanks and good circulation of cooling water are needed. Low melt temperatures relative to melting or softening points characterize pipe extrusion. Large die openings permit the use of viscous, high molecular weight resins, which yield tougher products. Coextruded foam-core pipe is described in Reference 26.

**Profile Extrusion.** This method is similar to pipe and tubing extrusion, except that special profile dies are used to produce a variety of asymmetrical products. Typical products include housing siding, window sash molding, decorative trim, plastic lumber, gaskets, and channels. Dies for profile extrusion are designed to allow for shape changes that occur during quenching. Usually, a water trough is used for cooling, but air or cooled metal surfaces are also used. Profile extrusion typically relies on amorphous thermoplastics, such as poly(vinyl chloride), ABS, polystyrene, thermoplastic elastomers, etc. The rapid shrinkage accompanying crystallization can result in severe distortion of the profile.

**Blown Film.** The blown film process (Fig. 6) uses a tubular die from which the extrudate expands in diameter while traveling upward to a film tower. The top of the tower has a collapsing frame followed by guide and pull rolls to transport the collapsed film to subsequent slitting and windup rolls. The tubular bubble from the die is inflated to the desired diameter by air passing through the center of the die. Although primary cooling to solidify the melt is supplied by an external air ring, chilled air may also be used internally. Polyethylene is the primary plastic used in most films, especially for packaging and trash bags. Coaxial dies can be used for manufacture of coextruded multilayer films.

The tube is characterized by its blow-up ratio, ie, a larger diameter than the die opening, which is expressed as the ratio of bubble diameter to the die diameter. Typical blow-up ratios range from 2:1 to 4:1. The final film thickness is much thinner than the die gap. Die gaps are slits of ca 0.65 mm. Typical film thicknesses are 0.007–0.125 mm. The process requires a high melt viscosity resin so that the melt can be pulled from the die in an upward direction. Since only air is used for cooling, removal of heat tends to be slow and rate-limiting. Chilled air can also be used internally to improve the efficiency of the air cooling process. The film may be treated for subsequent printing, and it can be slit into various widths and wound onto separate cores.

**Cast Film.** The cast film process provides a film with gloss and sparkle and can be used with various resins. Figure 7 is an illustration of the essential features of the extrusion equipment. The die opening is a long straight slit with an adjustable gap ca 0.4 mm wide. The die is positioned carefully with respect to the casting roll. The casting or chill roll is highly polished and plated and imparts a smooth and virtually flawless surface to the film. The roll is cooled by

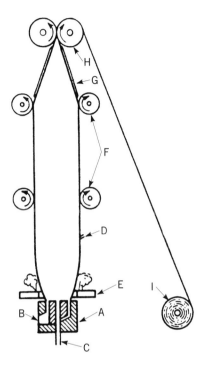

**Fig. 6.** Extrusion of blown film: A, blown-film die; B, die inlet; C, air hole and valve; D, plastic tube (bubble); E, air ring for cooling; F, guide rolls; G, collapsing frame; H, pull rolls; and I, windup roll (15).

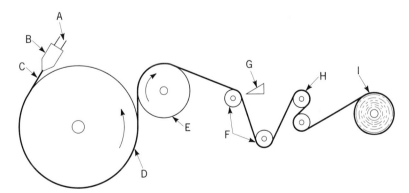

**Fig. 7.** Extrusion of roll-cast film: A, die inlet; B, cast-film die; C, air gap with molten web; D, casting roll; E, stripping roll; F, idler roll; G, edge-trim slitter; H, pull rolls; and I, windup roll (15).

rapid circulation of water. Temperature control is critical. A die somewhat longer than the width of the film is needed, because the molten web becomes narrow as it is drawn from the die; this is called neck-in. Edges of the film thicken and are mechanically removed before the film is wound on a roll. The edge trim can be reprocessed.

One of the requirements of this process is that the melt maintain good contact with the chill roll, ie, air must not pass between the film and the roll. Otherwise, air insulates the plastic and causes it to cool at a rate different from the rest of the plastic and this spoils the appearance of an otherwise satisfactory product. The melt should not emit volatiles, which condense on the chill roll, reduce heat transfer, and mar the film's appearance. The cast film process allows the use of a higher melt temperature than is characteristic of the blown film process. The higher temperature imparts better optical properties.

**Sheet.** The process used to make an extruded plastic sheet is illustrated in Figure 8. Sheeting thicknesses are 0.25–5 mm and widths are as great as 3 m. Heavier-gauge sheets are usually cut to a specified length and are stacked. Cooling is controlled by a three-roll stack. The rolls are 25–50 cm in diameter, highly polished, and chrome-plated. They are cored for cooling with circulating water, and the temperature of each roll is controlled by a circulator and temperature control unit. Often the rolls are operated at high temperature to maximize the gloss of the sheeting surface.

A web of molten plastic is pulled from the die into the nip between the top and middle rolls. At the nip, there is a very small rolling bank of melt. Pressure between the rolls is adjusted to produce sheet of the proper thickness and surface appearance. The necessary amount of pressure depends on the viscosity. For a given width, thickness depends on the balance between extruder output rate and the take-off rate of the pull rolls. A change in either the extruder screw speed or the pull-roll speed affects thickness. A constant thickness across the sheet requires a constant thickness of melt from the die. The die is equipped with bolts for adjusting the die-gap opening and with an adjustable choker bar or dam located inside the die a few centimeters behind the die opening. The choker bar restricts flow in the center of the die, helping to maintain a uniform flow rate across the entire die width.

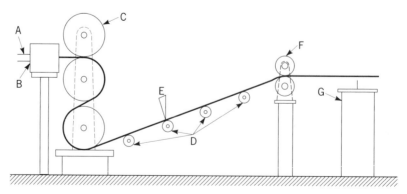

**Fig. 8.**   Sheet extrusion: A, die inlet; B, die; C, three-roll finisher; D, support rollers; E, edge-trim cutter; F, pull rolls; and G, saw or shear (15).

Drawdown from the die to nip is typically ca 10%. Sheet extrusion requires that the resin be of high melt viscosity to prevent excessive sag of the melt between the die and the nip. The melt should reach the nip before touching any other part of the middle roll to prevent uncontrolled cooling of the resin. The appearance of the lower surface of the sheet is determined by the middle roll, ie, its polish, surface temperature, and freedom from condensed materials. The upper surface is cooled by air and has a more glossy appearance. Neither the top roll nor the bottom roll greatly affect the appearance of the top surface of the sheet. Edges of the sheet are trimmed.

There usually is no need for a high melt temperature to obtain flow through a sheeting die, because die openings are large. Cooling of the sheet is slow because sheeting is thick. Most sheeting is used for thermoforming (see FILM AND SHEETING MATERIALS).

**Extrusion Coating.** A coating of an appropriate thermoplastic, such as polyethylene, may be applied to a substrate of paper, thin cardboard, or foil to provide a surface property which enables heat sealing or better barrier performance (see BARRIER POLYMERS). Figure 9 is a sketch of the extrusion-coating process. A molten web of resin is extruded downward, and the web and substrate make contact at the nip between a pressure roll and a chill roll. Typical coating thicknesses are 0.005–0.25 mm; the die opening is ca 0.5 mm. The melt web is narrower than the die, which is characteristic of a neck-in, and the edge tends to bead or thicken. Coated substrate is trimmed to the desired width. The highly polished and water-cooled chill roll determines the nature of the surface and removes the heat from the resin. The pressure roll pushes the substrate and the molten resin against the chill roll. Pressure and high melt temperatures are needed for adhesion of resin and substrate.

In contrast to most extrusion processes, extrusion coating involves a hot melt, ca 340°C. The thin web cools rapidly between the die and nip even at high linear rates. Both mechanical and chemical bonding to substrates are involved. Mechanical locking of resin around fibers contributes to the resin's adhesion to paper. Some oxidation of the melt takes place in the air gap, thereby providing sites for chemical bonding to aluminum foil. Excessive oxidation causes poor heat-sealing characteristics.

**Wire and Cable Coating.** Protective and insulating coatings can be applied continuously to wire as it is drawn through a cross-head die, as shown in

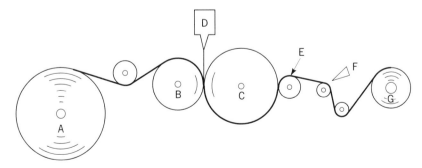

**Fig. 9.** Extrusion coating: A, unwinding of substrate; B, pressure roll; C, chill roll; D, die; E, stripper roll; F, edge trimmer; and G, windup of coated substrate (15).

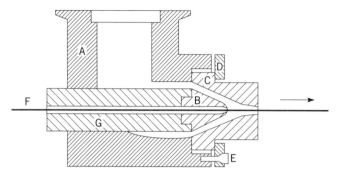

**Fig. 10.** Wire-coating pressure die: A, die body, cross-head; B, guider tip; C, die; D, die-retaining ring; E, die-retaining bolt; F, wire; and G, core tube (15).

Figure 10. A typical wire coating line consists of a wire payoff, wire preheater, extruder, die, cooling trough, capstan, and wire takeup (Fig. 11).

The melt flows from the extruder into the die where it flows around the bend and around the core tube. On the far side of the core tube, it forms a weld. Melt sticks to and is pulled by the moving wire. Details of the sizes and shapes of the die parts in contact with the melt are important in obtaining a smooth coating at high rates. The die exit usually is the same diameter as that of the coated wire and there is little drawdown. Die openings are small and pressures inside the die are high at ca 35 MPa (5000 psi). Wire takeup systems operate as high as 2000 m/min.

**Foam Extrusion.** Foamed thermoplastics provide excellent insulating properties because of their very low thermal conductivity, good shape retention, and good resistance to moisture pickup. As such, cylindrical shapes are extruded for pipe insulation, and sheets for building panel insulation. The blowing agent is dissolved and held in solution by the pressure developed in the extruder. As the molten thermoplastic exits the extruder die, the pressure release causes instantaneous foaming. The chlorinated fluorocarbons formerly used as blowing agents have been replaced with more environmentally friendly substitutes, such as HCFCs or low molecular weight hydrocarbons such as butane or isopentane. For some thermoplastics, carbon dioxide, nitrogen, or argon can be used (27). Critical to the success of most foaming extrusion operations is cooling of the melt just prior to entry to the die. Cooling is most effectively accomplished with a tandem arrangement of two extruders, as shown in Figure 12, wherein the first extruder assures complete dissolution of the blowing agent, and the second

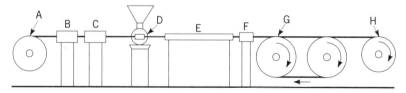

**Fig. 11.** Wire-coating line: A, unwinding of wire (payoff); B, straightener; C, preheater; D, extruder; E, water trough; F, tester; G, capstan (puller); and H, windup of coated wire (15).

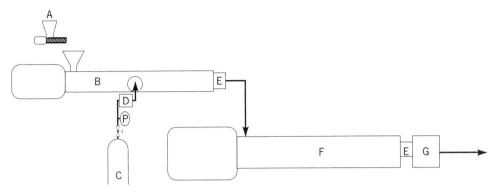

**Fig. 12.** Foam extrusion line: A, resin feed; B, extruder for melting and gas incorporation; C, gas supply; D, gas metering and compression; E, optical windows; F, cooling extruder; and G, die. P represents a pressure gauge. Courtesy of Polymer Processing Institute.

extruder is operated at slow speed for optimum cooling. Additional information on extrusion of foams is contained in Reference 28 (see FOAMED PLASTICS; INSULATION, THERMAL).

MOLDING

**Injection Molding.** In injection molding a molten thermoplastic is injected under high pressure into a steel mold. After the plastic solidifies, the mold is opened and a part in the shape of the mold cavity is removed. General descriptions of the process and related equipment are given in References 29–34.

*Equipment.* The machine consists of an injection unit and a clamp unit. The injection unit is usually a reciprocating single-screw extruder that melts the plastic and injects it into the mold or tool. The clamp unit opens, closes, and holds the mold closed against the pressure of injection. The size of an injection molding machine is described by the capacity of its clamping unit, ranging from 50 to 1000 tons. A machine with a given clamp unit can be supplied with a variety of injection units. Injection unit capacity is described by the shot size, which is the maximum volume of melt that can be injected in a single cycle. Shot size is determined by the diameter of the screw and the distance over which it is designed to reciprocate. Shot size is usually expressed as weight or as volume.

An injection molding machine is operated by hydraulic power and equipped with an electric motor and hydraulic pump. The maximum hydraulic oil pressure is ca 14 MPa (2000 psi). A hydraulic cylinder opens and closes the mold and holds the mold closed during injection; another cylinder forces the screw forward, thereby injecting the melt into the mold. A separate hydraulic motor turns the screw to plasticate, homogenize, and pressurize the melt. Control of these movements is a combined function of the hydraulic and electrical systems (35–37).

In the injection molding machine shown in Figure 13, the clamp unit is on the left and the injection unit on the right. A mold is shown in position between the platens of the clamp unit with one-half of the mold fastened to the fixed platen and the other to the movable platen. When the mold is opened,

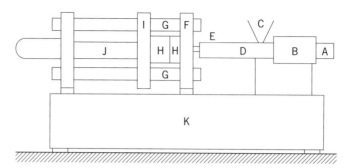

**Fig. 13.** Injection molding machine: A, hydraulic motor for turning the screw; B, hydraulic cylinder and piston allowing the screw to reciprocate about three diameters; C, hopper; D, injection cylinder (a single-screw extruder); E, nozzle; F, fixed platen; G, tie rods; H, mold; I, movable platen; J, hydraulic cylinder and piston used to move the movable platen and supply the force needed to keep the mold closed; and K, machine base.

the movable platen moves away from the fixed platen and the molded part can be removed. After part ejection, the mold is closed in preparation for the next injection cycle. In contrast to extrusion, the screw in the injection unit rotates only during part of molding cycle. When the screw turns, it pumps melt forward. The rearward movement of the screw is controlled by the placement of a limit switch, which stops the hydraulic motor. The maximum length of reciprocation is approximately three screw diameters. The shot size is directly proportional to the amount of screw reciprocation. The front of the screw is usually equipped with a check valve. While the screw rotates, the melt can move freely forward through the valve. However, the valve closes to prevent any reverse flow. The position of the screw is adjusted in such a way that there is always some melt in front of the screw when the mold is full. This pad or cushion of melt transfers pressure from the screw to the plastic in the mold. After initial filling, some additional melt flows into the mold; this is called packing. The cross section of the hydraulic cylinder, which pushes the screw, is approximately 10 times the cross section of the melt in front of the screw. Thus, 14 MPa (2000 psi) oil pressure corresponds to ca 140 MPa (20,000 psi) melt pressure.

*Molding Cycle.*   The molding cycle begins when the empty mold is closed and full clamp force develops. The screw is then pushed forward in order to inject melt into the mold. The screw holds melt under pressure in the mold for a certain time, often called the injection time, until the gate freezes. When the injection is finished (38,39), the pressure holding the screw forward is reduced to a minimum, the screw starts to rotate, and a mold-closed timer is activated. Before the time expires and the mold opens, the screw stops turning and the next shot is ready for injection. The speed of screw rotation is chosen in such a way that the screw rotation time is less than the mold-closed time. The mold remains open only long enough for the part(s) to be ejected. The length of the molding cycle depends on cooling the part sufficiently so that it may be removed from the mold without distortion. Mold cooling systems are used for cycle reduction. The length of the cycle may also be determined by the appearance of the part; a hotter mold usually imparts a more attractive finish than a cooler mold. Cycle time and quality of moldings are optimized through microprocessor-based controllers that receive inputs from temperature, pressure, and ram position sensors.

*Molds and Mold Design.* The molds are custom-machined from steel. Cavities must be polished to a very high gloss, since the plastic reproduces the surface in every detail. The cavities are hardened and frequently chrome-plated. Details on mold design are given in References 40–42. A typical shot from a mold, shown in Figure 14, consists of a sprue, runners (43), gates, and two parts. Gate design is influenced by part geometry, resin type, and processing conditions. Sprue and runners channel the melt into the cavities. After ejection they are separated from the parts, ground, and fed back into the injection unit for reprocessing. Modern mold design tends to reduce or eliminate sprue and runner scrap through a variety of techniques such as hot runners, insulated runners, or by designs that place the nozzle directly against the mold cavity (34,44).

When the mold is opened, the part should be easily removable. Cavities are made with a slight taper to reduce frictional drag of the part on the mold. The half of the mold attached to the movable platen is equipped with ejector pins, which push the part out of the cavity while the mold is being opened. When the mold is closed, the pins are flush with the cavity surface. Release agents or lubricants facilitate ejection and shorten the molding cycle. Some complex parts require that the mold open in several directions in addition to the direction of the platen movement. For a threaded part, eg, a bottle cap, part of the mold must be rotated to remove the article from the mold.

Mold designs must take into account that cooled moldings are always smaller than the cavity, owing to shrinkage. Amorphous plastics shrink less than crystalline plastics (45). In practice, mold shrinkage is expressed as linear shrinkage, rather than volumetric shrinkage. Parts are measured the day after molding and are compared with the size of the mold at room temperature. In actual molding, pressure is maintained to continue packing during cooling to minimize voids and excessive sink marks.

*Developments.* A variety of process modifications aimed at improving surface finish or weld line integrity have been described. They include gas assisted, co-injection, fusible core, multiple live feed, and push–pull injection molding (46,47). An important development includes computer-aided design (CAD)

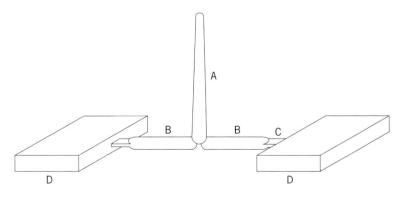

**Fig. 14.** A full shot: A, sprue; B, runner; C, gate; and D, part.

methods, wherein a proposed mold design is simulated by a computer and the melt flow through it is analyzed (48).

**Structural Foam Molding.**    Structural foam is a modified injection molding process for large articles having a cellular core and an integral solid skin with an overall 20–50% reduction in density, compared to their solid counterparts (28,49,50). It is most frequently used with polyethylene, high impact polystyrene, polypropylene, and several engineering resins. Some modifications to the resin, machine, and mold are required, and a blowing agent must be added to the resin. A chemical blowing agent, which releases gas when heated, is commonly used. The choice of blowing agents depends on the processing temperature required. The most common blowing agent is azodicarbonamide (1,1'-azobisformamide) used at 200–260°C. Upon decomposition, it releases nitrogen, carbon monoxide, and carbon dioxide (qv). Approximately 0.5 wt % of a blowing agent is normally added to the resin pellet as a surface coating or as a pelletized concentrate. Instead of a chemical blowing agent, nitrogen may be mixed with the melt while it is under pressure in an extruder and the mixture maintained under pressure until it is injected into the mold.

The injection molding machine must be equipped with a shutoff nozzle that maintains the melt under pressure while the mold is opened. The screw is retracted only part of the way needed for a full shot, and a short shot is injected into the mold. Without a blowing agent, only a section of a part, ie, a short shot, would be made; the empty space allows the blowing agent to expand the melt, forming the foam structure. Structural foam molding is limited to parts with wall thicknesses of at least 6 mm; below that thickness, reduction in part weight is usually insignificant. Parts, typically, have a dense skin and a foamed interior with various pore sizes. Compared to injection molded surfaces, the surfaces of structural-foam moldings are poor, and are characterized by a rough, swirly finish. Maximum pressure in the mold during foaming is much lower than in injection molding; also, no packing pressure needs to be maintained since the gas keeps the melt front moving. Surface appearance is improved by special techniques (51,52).

Because of low injection pressure, some cost savings are possible in mold and press construction. Molding cycles are somewhat longer than for injection molding. The part must be cooled in the mold long enough to be able to resist swelling from internal gas pressure. In structural foam parts there is almost a total absence of sink marks, even in the case of unequal section thickness. Structural foam has replaced wood, concrete, solid plastics, and metals in a variety of applications.

**Blow Molding.**    Blow molding is the most common process for making hollow thermoplastic components (53–55). In extrusion blow molding a molten tube of resin called a parison is extruded from a die into an open mold (Fig. 15**a**). In Figure 15**b** the mold is closed around the parison, and the bottom of the parison is pinched together by the mold. Air under pressure is fed through the die into the parison, which expands to fill the mold. The part is cooled as it is held under internal air pressure. Figure 15**c** shows the open mold with the part falling free.

As the parison is extruded, the melt is free to swell and sag. The process requires a viscous resin with consistent swell and sag melt properties. For a

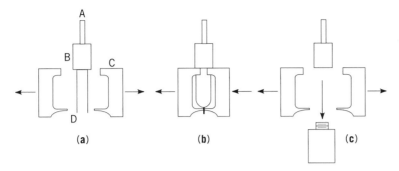

**Fig. 15.** Three stages of blow molding (**a–c**): A, air line; B, die; C, mold; and D, parison. See text.

large container the machine is usually equipped with a cylinder and a piston called an accumulator. The accumulator is filled with melt from the extruder and emptied at a much faster rate to form a large parison; this minimizes the sag of the molten tube.

With a simple parison, the large-diameter sections of the bottle have a thin wall and the small-diameter sections have a thick wall. Certain modifications of the die can control the thickness of the parison wall along its length, which results in a bottle with improved wall thickness distribution and better strength. High density polyethylene (HDPE) is the most common blow molding resin used to produce containers ranging in size from 30 cm$^3$ to 200 L.

In injection blow molding, a parison is injection molded onto a core pin; the parison is then rapidly transferred via the core pin to a blow mold, where it is blown by air into an article. This process is applied to small and intricate bottles.

Soft-drink bottles made from poly(ethylene terephthalate) (PET) are usually made by stretch-blow molding in a two-step process. First, a test-tube-shaped preform is molded, which is then reheated to just above its glass-transition temperature, stretched, and blown. Stretching the PET produces biaxial orientation, which improves transparency, strength, and toughness of the bottle (54,56). A one-step process is used for many custom containers that are injection stretch-blow molded.

**Rotational Molding.** Hollow articles and large, complex shapes are made by rotational molding, usually from polyethylene powder of relatively low viscosity (57–59). The resin is in the form of a fine powder. A measured quantity is placed inside an aluminum mold and the mold is heated in an oven and rotated at low speed. The resin sinters and fuses, coating the inside of the mold. The mold is then cooled by water spray and the part solidifies, duplicating the inside of the mold.

A rotomolding machine has three long arms extending from a central driving mechanism; each arm rotates several molds in two planes. The arms are moved from one process station to the next, ie, from unloading and loading to heating and cooling. Tooling costs are low. The molds are usually made of cast aluminum, but sheet metal is also used. The melt is forced without pressure against the mold surface during heating or cooling, resulting in uniform wall thickness, zero orientation, and high physical properties. Cycle times are long because of the heating and cooling required; they depend on wall thickness and can be as high as 15 min for a 4-mm wall thickness. Common roto-molded products include large tanks and boxes, drums, furniture, and toys. PVC plastisol, a mixture of fine PVC particles and a plasticizer, may also be processed by rotomolding. Plastisols are liquid at room temperature and are converted to soft solids when heated to ca 180°C. Playballs and toys are made from plastisols.

**Expandable Polystyrene Molding.**   Molding expandable polystyrene gives foamed products such as insulation board shapes for packaging and disposable food and cup containers. Such processes are also called bead or steam molding (60,61). Expandable polystyrene moldings are manufactured from polystyrene beads 0.25–1.5 mm in diameter containing a blowing agent, such as pentane. These beads, when exposed to heat, expand from 2–50 times their original size. The beads are pre-expanded by heating with steam or hot air to about the density of the final molded part. After pre-expansion, they are stored from 3–12 h to let the air diffuse into the cells of the foam. Shaping of the beads is done in special presses and molds, into which the preexpanded beads are transferred by air conveying systems. Steam causes further expansion and forces the surface fused particles to conform to the shape of the mold. In the manufacture of insulation board, blocks are molded into cavities up to $5.5 \times 1.2 \times 0.65$ m by injecting steam. After storage and conditioning, the blocks are cut into sheets.

## THERMOFORMING

Thermoforming is a process for converting a preform, usually an extruded plastic sheet, into an article such as a thin-wall container or a tray for packaging meat (62–65). Under vacuum the process is called vacuum forming. The sheet is clamped in a frame and exposed to radiant heaters. The sheet softens to a formable condition, is moved over a mold and sucked against the mold by vacuum. Excess plastic is trimmed and recycled. Timers control the length of the heating and cooling periods, which depend on composition and sheet thickness. Frequently, the process is improved by using mechanical aids (plug-assisted, air-cushioned plug-assisted, drape-assisted). The molds are made of wood, aluminum, steel, or epoxy (66). In some plants the vacuum forming line is run in-line with the sheeting extruder.

Amorphous resins such as styrenics, acrylics, PVC, and some modified crystalline resins are used for thermoforming. Foamed polystyrene sheet is also thermoformed. The polymers must be of fairly high molecular weight because the heated sheet must be form-stable. The significant property is melt elasticity. Sheets of these resins soften but do not sag when heated. Sagging causes thinning; a sagged sheet may have more surface area than the mold, resulting in folds and areas of uneven thickness. Thermoforming is employed for small items such as cups, plates, trays, and larger, deep-drawn moldings such as boats, bath tubs, and freezer liners. Skin packaging, which employs a flexible plastic skin drawn tightly over an article on a card backing, is made by thermoforming. The process also is used for blister packaging.

CALENDERING

Calendering is a process uniquely applied to rubbery polymers, mainly semirigid and flexible PVC, for making sheeting of uniform thickness from 0.75–0.05 mm after stretching (67–70). A calender has four heavy, large steel rolls, which are usually assembled in an inverted "L" configuration as shown in Figure 16. This design is preferred for thick sheeting because it gives a long dwell for full heating. The forces generated between the rolls are considerable and sufficient to bend the rolls, resulting in uneven sheet thickness. A common method of compensating for roll bending is by grinding an opposite contour on the roll. A two-roll mill, a Banbury mixer, or an extruder melt the resin, which is subsequently transferred to the calender. Sheet can be made 2.5 m wide and production rates can be as high as 100 m/min (71). Calendering is often followed by printing, laminating, and embossing. PVC calenders are run at temperatures approaching 200°C to produce highly oriented sheets for items such as shower curtains, rainwear, luggage, and wall paneling.

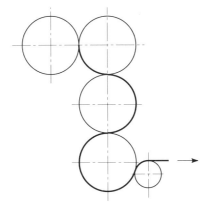

**Fig. 16.** A four-roll, inverted "L" calender.

CASTING

Casting refers to the formation of an object in a batch process by pouring a fluid monomer–polymer solution into a mold where it solidifies or by continuously pouring the liquid onto a moving belt (72). The casting process is most frequently used with acrylics. Since the 1950s, cast acrylic sheeting has been made by polymerization of methyl methacrylate in a cell assembled from two glass plates and a flexible gasket. Heating for a few hours until polymerization is complete gives sheets with excellent optical properties. Acrylic sheeting is also made by continuous casting, ie, a monomer–polymer–catalyst mixture is fed onto a stainless steel belt, on which the polymerization is completed. These products are used for glazing and thermoforming. Nylon-6 products with higher mol wt and crystallinity than extruded or molded resins are also produced by casting in the mold by a process known as *in situ* anionic polymerization.

## Thermosetting Resins

Common thermosetting resins are unsaturated polyesters, phenolic resins (qv), amino resins (qv), polyurethanes, epoxy resins (qv), and silicones (qv) (see POLYESTERS, UNSATURATED; URETHANE POLYMERS). Less common thermosets employed in specialized applications are bismaleimides, polyimides (qv), and furan resins. Thermosetting resins are usually low viscosity liquids or low mol wt solids that are formulated with suitable additives known as cross-linking agents to induce curing; curing involves permanent chemical changes resulting in infusible, insoluble products with excellent thermal and dimensional stability. Thermosetting resins are commonly used in combination with fillers or fibrous reinforcements; as a result, processing methods are often quite different from those employed for thermoplastics (73–75).

### COMPRESSION, INJECTION, AND TRANSFER MOLDING

Compression molding is the oldest process of the plastics industry. General information on the process, which is also applicable to a certain extent to thermoplastics, is found in References 76–80. It is still common, although many thermosetting materials are also injection molded (81–83). The equipment consists of a vertical hydraulic press with platens for mold attachment. The mold comprises a matched pair of male and female dies. Mold cavities are hardened and highly polished, similar to those of an injection mold. A measured quantity of thermosetting resin compound in granular, sheet, or other form is placed in the hot mold. The mold is closed and the liquified resin through pressure fills the cavity. Continued heating cures the resin within a few minutes, and the part is removed from the mold. Compression molding is commonly employed for phenol–formaldehyde, urea–formaldehyde, melamine–formaldehyde, and polyester molding compounds. To shorten the total cycle, some of these compounds are partially polymerized or preformed into various shapes that are preheated to intermediate temperature before being inserted into the mold. Laminates made of reinforcement layers impregnated with partially polymerized resins are also prepared by compression molding (see LAMINATED MATERIALS, PLASTIC).

Sheet molding compound (SMC) is used in the manufacture of large reinforced compression moldings, which are employed extensively in automotive panels. SMC is prepared as a sandwich, rolled between two polyethylene films; it contains polyester resin, filler (usually, calcium carbonate), and 20–30 wt % chopped glass fibers. Other components such as peroxides for cross-linking, thermoplastic additives for shrinkage control, and alkaline-earth oxides and hydroxides to assist maturation are present in smaller amounts. Maturation involves a thickening chemical reaction that produces within a few days after compounding a tack-free sheet with desirable flow characteristics during molding. The required amount of sheet is then cut to size for placement on the bottom of two matched mold halves to cover at least 70% of the mold area. It is molded at pressures up to 15 MPa (2100 psi) and temperatures in the range 140–180°C. Cure cycles depend on part size and complexity, pressure and heat applied, and formulation characteristics. Cure cycles of 60 s are common. The properties of cured SMC depend on the amount, length, and orientation of the glass fibers.

Bulk molding compound (BMC) is similar to SMC in formulation, but contains shorter glass fibers and higher filler levels (84). The premixed material has the consistency of dough and is supplied in bulk form or as a cylindrical extrudate. BMC is molded by conventional compression or transfer molding; it is also molded in reciprocating screw injection presses equipped with a piston stuffer attached to the barrel to force-feed the material into the screw flights. Other injection-moldable thermoset formulations in granular form (phenolics, amino, epoxy, and diallyl phthalate compounds) are fed through standard gravity hoppers. In thermoset injection molding, screws with low compression ratios are used to convey the material through a short barrel heated typically up to 120°C and decrease its viscosity; this is followed by injection into a hotter mold (typically up to 220°C) where chemical cross-linking takes place. Total cycles may range from 10–120 s, depending on part thickness and formulation characteristics.

Transfer molding is a combination of compression molding and injection molding (76,77). A measured charge is heated in a reservoir, from which it is injected or transferred through runners to the mold cavity. Forcing the polymer through the entry gate results in a rise in temperature that may promote flow but also accelerates curing. The molten polymer is held at temperature while the cross-linking reaction proceeds to completion.

## OPEN-MOLD PROCESSING

Common open-mold processes are hand or mechanized methods such as lay-up and spray-up, that use a single cavity mold and produce one finished surface (85). These techniques are used to produce fiber-reinforced structures containing the reinforcement in the form of cloth, chopped strands, mat, continuous roving, woven roving, etc. Thermosetting resins of choice are liquid unsaturated polyester or epoxy resins. In lay-up, catalyzed resin is used to impregnate the reinforcement that is preplaced on the mold. A spray-up system consists of a resin spray gun, a fiber chopper and a pumping system; the catalyst is combined with the resin just before deposition with the fibers onto the mold surface. Open-mold processes use little or no pressure during the curing cycle. Cross-linking usually takes place at room or slightly elevated temperature through a proper combination of curing agents. A variety of products such as boat hulls, automotive components, tanks, etc are made by open-mold processing (86). Types of emissions and U.S. environmental regulations for thermoset components and cure reaction by-products are included in Reference 87.

In another open-mold process, filament winding, continuous fibers such as glass, carbon, or aramid are wound on a mandrel surface in a precise geometric pattern (88). Fibers are usually preimpregnated with a low viscosity resin containing a suitable cross-linking agent for room or higher temperature curing. Most common resins are polyesters and epoxies. A tubular component with the fibers in angle-ply arrangement results after curing and the mandrel is withdrawn. The process, originally developed as a rapid method to manufacture pipes and cylindrical vessels, has expanded to allow more complicated geometries. Products include, among others, helicopter rotor blades, external fuel tanks, pressure bottles, and pipes.

## PULTRUSION

In pultrusion, reinforcing fibers in a combination of styles are wetted by a resin containing a high temperature cross-linking agent and then pulled through a forming system that positions the fibers into the desired packed structural arrangement; the material then moves through a heated die where curing takes place (89). The process can in principle produce continuous profiles of any length; rods, bars, and other more complex profiles such as I-beams and channels are mostly made from continuous glass fibers and polyester or epoxy resins (90). Glass content can be very high (50–70%) at production rates of about 1.5 m/min (91–93).

## REACTION INJECTION MOLDING

Reaction injection molding (RIM) is used for the production of solid or partially foamed polyurethane moldings by rapid injection of metered streams of polyol and isocyanate into a mold. RIM has been developed around polyurethane and polyurea chemistry; other chemical systems of lesser commercial significance can be processed by RIM (94,95). Articles are produced on a cycle of 3 min or less. The two liquid streams are mixed by impingement before entering the mold. The mold is filled at 0.35–0.70 MPa (50–100 psi), which is a very low pressure compared to that used in standard injection molding. Equipment, and in particular mixing head design, is critical. The cost of the mold and press are less for RIM than for conventional injection molding and the process is particularly suited to the production of large parts.

A wide range of formulations can be used in the RIM process. A typical polyol stream contains 75–80 wt % polyether polyol, 17–20 wt % 1,4-butane-diol, 2–6 wt % blowing agent, and amine and metal catalysts. The most commonly used isocyanates are proprietary products made from 4,4'-methylenedi(phenyl isocyanate) (MDI). Processing conditions and physical properties depend on the formulation. Often, parts are removed from the mold before curing is complete; then, additional curing is required at elevated temperatures in an oven. Glass fibers and other fillers added for reinforcement and to control shrinkage are dispersed into the polyol component. Another approach is to place continuous fiber mats directly in the mold cavity and inject a low viscosity, reactive mixture through the mat (structural RIM) (96).

## RESIN TRANSFER MOLDING

In resin transfer molding (RTM) a piston-type positive displacement pump injects a premixed resin–catalyst stream into a closed mold thoroughly impregnating a preplaced reinforcement pack. Glass, graphite, or aramid fibers in the form of continuous strand mat, woven rovings, and their combinations are typical reinforcements (97). RTM differs from structural RIM in that it uses slower-reacting formulations and injects under lower pressures components that are premixed or homogenized by passing through static mixers. Molds are made of lower cost materials and cycles are longer (10–60 min). RTM machines have been adapted for a variety of resins including polyesters, epoxies, polyurethanes, and isocyanurates (98,99).

## POLYURETHANE FOAM PROCESSING

**Flexible Foam.**　These polyurethane products have varying densities and a relatively low degree of cross-linking (100). They are made from trifunctional polyols (mol wt 3000) and TDI, a mixture of 2,4- and 2,6-toluene diisocyanates, in the presence of tin and amine catalysts, surfactants as foam stabilizers, water, and blowing agents. Physical blowing agents are Freons; $CO_2$, the reaction product of water and isocyanate, is a chemical blowing agent. Flexible foams are resilient open-cell structures with densities varying from 25–650 kg/m$^3$, depending on the choice of the raw materials. Most flexible foams are produced in the form of a slab or bun in a continuous process in widths up to 2.4 m and thicknesses up to 1.2 m. A liquid foamable mixture is pumped onto a conveyor, which moves through a tunnel where reaction and foaming occur (101). Similar mixtures can be placed in a mold and allowed to foam. This process is used in the manufacture of automobile seats (see FOAMED PLASTICS).

**Rigid Foam.**　These are closed-cell foams with excellent thermal insulation characteristics (102). Most rigid foams are made from a high functionality polyol with ca 500 mol wt and poly(methylene)–poly(phenyl isocyanate) (PMPPI), yielding a mixture with a high functionality of 2.7; hence, the foams have a higher degree of cross-linking than the flexible foams. The formulations include blowing agents, catalysts, and foam stabilizers (103,104). Rigid foams can be made as continuous slabs that are cut into panels. Other production methods include coating suitable substrates to produce laminated products used for architectural insulation; spraying, as for insulation during construction; and pouring or pumping the foamable mixture into place, as in the manufacture of refrigerators.

## BIBLIOGRAPHY

"Plastics" in *ECT* 1st ed., Vol. 10, pp. 798–818, by G. M. Kline, National Bureau of Standards, U.S. Dept. of Commerce; "Plastics Technology" in *ECT* 2nd ed., Vol. 15, pp. 790–811, by R. B. Seymour, University of Houston; "Plastics Processing" in *ECT* 3rd ed., Vol. 18, pp. 184–206, by P. N. Richardson, E. I. du Pont de Nemours & Co., Inc.

1. *Modern Plastics*, **71**(1), 73 (1994).
2. P. G. Kelleher, *Adv. Polym. Technol.* **10**(3), 219 (1990).
3. P. G. Galanty and G. A. Bujtas, in *Modern Plastics Encyclopedia '92*, Vol. 68, No. 11, McGraw-Hill Book Co., Inc., New York, 1991, p. 23.
4. P. J. Rigby in Ref. 3, p. 45.
5. J. F. Chabot, Jr., *The Development of Plastics Processing Machinery and Methods*, Society of Plastics Engineers, Brookfield, Conn., 1992.
6. R. C. Progelhof and J. L. Throne, *Polymer Engineering Principles: Properties, Processes and Tests for Design*, Hanser Publishers, Munich, Germany, 1993, Chapt. 5.
7. D. H. Morton-Jones, *Polymer Processing*, Chapman and Hall, New York, 1989.
8. E. A. Grulke, *Polymer Process Engineering*, PTR Prentice-Hall, Englewood Cliffs, N.J., 1994, Chapt. 10.
9. J.-M. Charrier, *Polymeric Materials and Processing: Plastics, Elastomers and Composites*, Hanser Publishers, New York, 1991, Chapt. 5.
10. J. F. Aggasant, P. Avenas, J. Sergent, and P. Carreau, eds., *Polymer Processing: Principles and Modeling*, Hanser Publishers, Munich, Germany, 1991.

11. A. I. Isayev, ed., *Modeling of Polymer Processing*, Hanser Publishers, Munich, Germany 1992.
12. N. S. Rao, *Computer Aided Design of Plasticating Screws*, Hanser Publishers, Munich, Germany, 1986.
13. K. T. O'Brien, *Computer Modeling for Extrusion and Other Continuous Polymer Processes*, Hanser Publishers, Munich, Germany, 1992.
14. C. L. Tucker, *Fundamentals of Computer Modeling for Polymer Processing*, Hanser Publishers, Munich, Germany, 1989.
15. P. N. Richardson, *Introduction to Extrusion*, Society of Plastics Engineers, Inc., Brookfield Center, Conn., 1974.
16. M. Xanthos, ed., *Reactive Extrusion: Principles and Practice*, Hanser Publishers, Munich, Germany, 1992.
17. D. B. Todd, *Intern. Polym. Proc.* **6**, 143 (1991).
18. E. C. Bernhardt, ed., *Processing of Thermoplastic Materials*, Reinhold Publishing Corp., New York, 1959.
19. Z. Tadmor and C. G. Gogos, *Principles of Polymer Processing*, John Wiley & Sons, Inc., New York, 1979.
20. J. R. Pearson, *Mechanics of Polymer Processing*, Elsevier Science Publishing Co., New York, 1985.
21. Ref. 7, pp. 74–125.
22. C. Rauwendaal, *Polymer Extrusion*, Hanser Publishers, Munich, Germany, 1990.
23. J. L. White, *Twin Screw Extrusion: Technology and Principles*, Hanser Publishers, Munich, Germany, 1991.
24. M. L. Berins, ed., *Plastics Engineering Handbook of the Society of Plastics Industry*, 5th ed., Van Nostrand Reinhold Co., Inc., New York, 1991, Chapt. 4.
25. W. Michaeli, *Extrusion Dies for Plastics and Rubbers*, Hanser Publishers, Munich, Germany, 1992.
26. F. R. Bush and G. C. Rollefson, *Mod. Plast.* **58**, 80 (1981).
27. C. Jacob and S. K. Dey, *Proceedings of the 52nd SPE Annual Technical Conference*, Vol. 40, Society of Plastics Engineers Inc., Brookfield Center, Conn., 1994, p. 1964.
28. D. Klemper and K. C. Frisch, eds., *Handbook of Polymeric Foams and Foam Technology*, Hanser Publishers, Munich, Germany, 1992.
29. I. I. Rubin, *Injection Molding—Theory and Practice*, John Wiley & Sons, Inc., New York, 1972.
30. N. G. McCrum, C. P. Buckley, and C. B. Bucknall, *Principles of Polymer Engineering*, Oxford University Press, New York, 1988, pp. 294–305.
31. T. Whelan and J. Goff, *Injection Molding of Thermoplastics*, Vols. 1 and 2, Society of Plastics Engineers, Brookfield Center, Conn., 1990.
32. C. L. Weir, *Introduction to Injection Molding*, Society of Plastics Engineers, Brookfield Center, Conn., 1975.
33. S. L. Kirkham, in Ref. 3, p. 267.
34. Ref. 7, pp. 146–175.
35. R. Farrell, *Plast. Technol.* **26**(4), 81 (1980).
36. B. Sanschagrin, *Polym. Eng. Sci.* **23**(8), 431 (1983).
37. D. K. Rideout and M. Kochajda, *Plast. Eng.* **41**(2), 47, (1985).
38. H. W. Cox, C. C. Mentzer and R. C. Custer, *Polym. Eng. Sci.* **24**(7), 501 (1984).
39. H. Bangert, *Kunststoffe* **75**(6), 325 (1985).
40. G. Menges and P. Mohren, *How to Make Injection Molds*, Hanser Publishers, Munich, Germany, 1993.
41. K. Stoeckhert, *Mold-Making Handbook*, Hanser Publishers, Munich, Germany, 1983.
42. J. H. Dubois and W. Pribble, *Plastics Mold Engineering Handbook*, 4th ed., Van Nostrand Reinhold, Co., Inc., New York, 1987.

43. W. B. Glenn, *Plast. Technol* **26**(4), 99 (1980).
44. Mold-Masters Ltd., in Ref. 3, p. 278.
45. R. M. Ogorkiewicz, ed., *Thermoplastics: Effects of Processing*, CRC, Cleveland, Ohio, 1969, p. 118.
46. J. Theberge, *Plast. Eng.* **47**(2), 27, 1991.
47. J. Theberge in *Opportunities for Innovation: Polymer Composites*, S. H. Munson-McGee, ed, NIST GCR 90-577-1, National Institute of Standards and Technology, Gaithersburg, Md., 1990, pp. 35–44.
48. L. T. Manzione, ed., *Applications of Computer Aided Engineering in Injection Molding*, Hanser Publishers, Munich, Germany, 1987.
49. S. Semerdjiev, *Introduction to Structural Foam*, Society of Plastics Engineers, Brookfield Center, Conn., 1982.
50. E. Hunerberg, in Ref. 3, p. 264.
51. R. B. Johnson, *Plast. Des. Forum*, **10**, 95 (1985).
52. G. Koski and D. Lessard, *Plast. Mach. Equip.* **14**, 7, 25 (1985).
53. Ref. 30, pp. 311–312.
54. D. Rosato and D. Rosato, eds., *Blow Molding Handbook*, Hanser Publishers, Munich, Germany, 1989.
55. Ref. 7, pp. 126–137; Ref. 3, pp. 222–227.
56. C. L. Kern, *Plast. Eng.* **40**(1), 37 (1984).
57. Ref. 7, pp. 237–241.
58. R. L. Fair, in Ref. 3, p. 293.
59. Ref. 24, Chapt. 14.
60. Ref. 24, pp. 593–598.
61. J. M. Joyce, in Ref. 3, p. 256.
62. Ref. 30, pp. 308–311.
63. J. L. Throne, *Thermoforming*, Hanser Publishers, Munich, Germany, 1987.
64. Ref. 7, pp. 138–145.
65. R. Whiteside, in Ref. 3, p. 294.
66. G. R. Garrison, in Ref. 3, p. 298.
67. Ref. 7, pp. 186–190.
68. Ref. 24, Chapt. 15.
69. A. Ray and A. V. Shenoy, *J. Appl. Polym. Sci.* **30**, 1 (1985).
70. J. F. Agassant and M. Espy, *Polym. Eng. Sci.* **25**(2), 118 (1985).
71. K. Marquardt, in Ref. 3, p. 228.
72. Ref. 3, pp. 230, 233.
73. R. E. Wright, *Molded Thermosets*, Hanser Publishers, Munich, Germany 1992.
74. P. Mallick and S. Newman, *Composites Materials Technology: Processes and Properties*, Hanser Publishers, Munich, Germany, 1991.
75. Ref. 30, p. 223–226.
76. R. G. Whitesides and J. C. O'Brien, in Ref. 3, p. 234.
77. Ref. 30, pp. 313–314.
78. G. N. Hartt and J. F. Wilk, Jr., in S. H. Munson-McGee, ed., *Opportunities for Innovation: Polymer Composites*, NIST GCR 90-577-1, National Institute of Standards and Technology, Gaithersburg, Md., 1990, pp. 23–34.
79. Ref. 7, pp. 176–184.
80. Ref. 24, Chapt. 9.
81. Ref. 30, pp. 305–307.
82. P. L. Leopold, in Ref. 3, p. 272.
83. Ref. 24, Chapt. 8.
84. L. N. Nunnery, in Ref. 3, p. 200.
85. L. C. McManus, in Ref. 3, p. 286.

86. E. M. Zion, in Ref. 78, pp. 1–22.
87. Radian Corp., *Plastics Processing—Technology and Health Effects*, Noyes Data Corp., Park Ridge, N.J., 1986, pp. 317–338, Append. C.
88. R. R. Roser in Ref. 3, p. 284.
89. J. Martin, in Ref. 3, p. 288.
90. J. E. Sumerak, in Ref., 78, pp. 45–58.
91. *Plast. Mach. Equip.* **9**(4), 30 (1980).
92. L. Krutchkoff, *Plast. Des. Process.* **20**(8), 37 (1980).
93. J. E. Sumerak, *Mod. Plast.* **62**(3), 58, (1985).
94. J. R. Riley, in Ref. 3, p. 281.
95. C. W. Macosko, *RIM, Fundamentals of Reaction Injection Molding*, Hanser, New York, 1989.
96. C. D. Shirrel, in Ref. 78, pp. 59–86.
97. K. A. Jacobs in Ref. 3, p. 290.
98. C. F. Johnson, in Ref. 78, pp. 87–110.
99. G. N. Hartt and J. K. Brew, in Ref. 78, pp. 111–120.
100. R. McBrayer in Ref. 3, p. 260.
101. G. Oertel, ed., *Polyurethane Handbook*, Hanser Publishers, New York, 1993.
102. Ref. 24, Chapt. 19.
103. M. J. Cartmell and B. MacNeall, *Mod. Plast.* **62**(1), 84 (1985).
104. M. Kapps, *Kunststoffe* **75**(6), 337 (1985).

M. XANTHOS
D. B. TODD
Polymer Processing Institute

# PLASTICS TESTING

Plastics testing encompasses the entire range of polymeric material characterizations, from chemical structure to material response to environmental effects. Whether the analysis or property testing is for quality control of a specific lot of plastic or for the determination of the material's response to long-term stress, a variety of test techniques is available for the researcher.

Polymer analysis has progressed significantly with improvements in instrumental techniques and computer-enhanced data analysis methods. Fourier transform techniques for infrared, nuclear magnetic resonance, and Raman spectroscopy have allowed these techniques to expand their usefulness to the micro sample region. Combinations of instrumental techniques such as pyrolysis–Fourier transform infrared (ftir), pyrolysis–gas chromatography (pyrolysis–gc), or liquid chromatography–mass spectroscopy (lc–ms) permit separation and identification of polymer components and degradation products from complex systems which were not possible previously. The use of microprocessors to capture transient signals from physical test methods, ie, high speed impact, allows the graphing of the complete impact event as a stress–strain curve versus the previously limited single-point total energy value.

## Composition and Structure

Infrared techniques, particularly Fourier transform infrared, are the most commonly used analysis methods for determining the composition of polymers. The ftir technique, which corresponds to the vibrational energies of atoms or specific groups of atoms within a molecule as well as rotational energies, identifies components by comparing the spectrum of a sample to reference spectra. The infrared spectral region is commonly from 2–50 $\mu$m, and the region of 7–15 $\mu$m is capable of differentiating similar isomers (1,2). The 7–8.5-$\mu$m region has been used in model hydrocarbon studies to show that the absorption of methylene groups is dependent on the size of the methylene sequences in compounds (3). The sequence of consecutive methylene groups develops a characteristic ir rocking vibration that can be used to distinguish between true copolymers and physical mixtures of copolymer polyolefins (4). In a similar manner, near-ir measurements at 1.69 and 1.76 $\mu$m can be used to determine propylene in ethylene–propylene copolymers down to levels of 15 mol % of propylene (5).

Other works have examined molecular mechanics of oriented polypropylene during creep and stress relaxation studies by using a stress-sensitive band at 9.75 $\mu$m and an orientation-sensitive band at 8.99 $\mu$m (6). Some examples of composition determinations using ir analysis of polymers are as follows. (1) Comparison of carbonyl absorbances due to anhydride and acid to standards has been used to determine the amount of grafted maleic anhydride on polyolefins (7). (2) The C=N stretching mode has been used for quantitative analysis of acrylonitrile in butadiene–acrylonitrile copolymers (8). (3) The relative intensity of the bands assigned to C–Cl stretching vibration at 6–7 $\mu$m is dependent on conformational and configurational structures as well as the crystallinity of the sample (9). (4) Residual catalyst support in commercial polyethylene can be detected at levels of 100 ppm (10). There are several excellent sources of spectra both in print and on computer disk for comparing and matching the fingerprints obtained from a sample (11,12).

Computer-assisted ir spectrum subtraction techniques are commonly used and are valuable in particular for comparing similar polymers (13,14). Other features of computer-assisted equipment, such as repeated scans, computer averaging, and curve smoothing, lead to high signal-to-noise ratios with both Fourier transform and newer dispersive ir equipment (15–17). Work with a prism liquid cell and an ir beam condenser has allowed monitoring at temperatures up to 200°C and 3.5 MPa (500 psi) using a reflectance measurement mode (18). Progress has been made in the development of on-line ir equipment for monitoring the production of polymers (19,20). However, the near-ir range is used rather than mid-ir with resultant lower sensitivity to molecular fragments. Work on the improvement of optical fiber sensors for handling temperature and pressure demands allows the monitoring to come closer to the midrange bands (21).

Combination techniques such as microscopy–ftir and pyrolysis–ir have helped solve some particularly difficult separations and complex identifications. Microscopy–ftir has been used to determine the composition of copolymer fibers (22); polyacrylonitrile, methyl acrylate, and a dye-receptive organic sulfonate trimer have been identified in acrylic fiber. Both normal and grazing angle modes can be used to identify components (23). Pyrolysis–ir has been used to study polymer decomposition (24) and to determine the degree of cross-linking of sulfonated divinylbenzene–styrene copolymer (25) and ethylene or propylene levels and ratios in ethylene–propylene copolymers (26).

Raman spectroscopy is an emission phenomenon as opposed to infrared absorption, and results from vibrations caused by changes in polarizability. Raman spectroscopy had limited use until the advent of lasers as the exciting source. It can be complementary to infrared and may be more useful when the sample for analysis is a filled polymer or composite containing silica, clay, other similar inorganic filler, or a piece of polymer, but particularly thermosets or rubbery products requiring no further sample preparation. Additional advantages over ir are the intensity advantage for the C=C band and the determination of thiol groups in sulfur polymers (27). Orientation of polymers, poly(methyl methacrylate) (PMMA) (28), polypropylene (PP), and polyethylene (PE) (29), as well as the crystallinity of PE (30,31) and poly(ethylene terephthalate) (PET) (32), have been investigated by using Raman. Fluorescence is a significant interference in Raman spectroscopy because it occurs simultaneously in several commercial plastic products; however, Fourier transform of Raman spectroscopy has expanded its applicability by rendering it immune to fluorescence (33).

Nuclear magnetic resonance (nmr) requires an atomic nuclei that can absorb a radio-frequency signal impinging it in a strong magnetic field to give a spectrum. The field strength at which the nucleus absorbs is a function of both the nucleus and its immediate electronic environment. The atoms normally used for nmr analysis are as follows (34): $^{1}$H, $^{19}$F, $^{31}$P, $^{11}$B, $^{14}$N, $^{29}$Si, $^{13}$C, $^{2}$H, $^{17}$O, $^{33}$S, and $^{15}$N. Of these, the most commonly used in polymer analyses are $^{1}$H and $^{13}$C; $^{13}$C-nmr in particular has been used to measure branching and branch concentrations in polyethylene (35), although chain lengths greater than six carbon branches give the same spectra as a six-carbon branch (36,37). In high density polyethylene (HDPE), $^{13}$C-nmr is useful to detect branching because only long chains are normally present (38). It is also an excellent technique for the analysis of sequence distribution and comonomer content in ethylene–propylene copolymers (39–42). Stereochemical configurations and stereochemical sequence distribution that occur in polystyrene, poly(vinyl chloride) (PVC), and chlorine-containing polymers can likewise be measured by $^{13}$C-nmr (43–45). Proton nmr is useful in determining vinylidene fluoride in copolymers with chlorotrifluoroethylene (46). In addition, nmr has been used to analyze monomers and determine polymerization mechanisms (47–50). Fourier transform has also been applied to nmr, allowing increased sensitivity on microsamples. Spectra can be obtained on as little as 10 $\mu$g (51). However, nmr is not sensitive to minor components in mixtures. Pulsed nmr techniques have been used to study molecular motions in bisphenol A polycarbonate in homopolymers and copolymers (52). It has also been used in blends, block copolymers, and composites of these materials (53). High resolution nmr may be useful in determining the composition and molecular weight of polyester urethane, polyphenylene, and poly(phenylene oxide) (54).

Other techniques that have been shown effective in the determination of particular polymer compositions employ uv methods, pyrolysis–gc, or diffraction and scattering studies. A quantitative uv method for the determination of high isocyanate, particularly aliphatic isocyanate, content polymers in copolymers, on which titration techniques typically cannot be used, was found to have good agreement with nitrogen levels by ftir analysis of urea linkages (55). Pyrolysis–gas chromatography has been used in a variety of studies on polymer composition and structure (56–58): isotacticity of polypropylene (59); structure and analysis of styrene copolymers and terpolymers, including microstructure (60,61); correlation of weight of poly(vinyl chloride) pyrolyzed and HCl measured by gas chromatography (62), as well as the presence of other compounds from the degradation of PVC, ie, benzene, toluene, ethylbenzene, and a variety of aryl chlorides; and degradation mechanisms for polybutadiene (63) and isoprene–methyl methacrylate copolymers (64). Pyrolysis–mass spectroscopy has been reported to be an improvement over pyrolysis–gas chromatography, because it allows higher molecular weight fractions of the pyrolyzate to be determined (65). Wide-angle x-ray scattering (waxs) is primarily used for examining the crystalline regions of semicrystalline polymers and composites (66–68). It is useful in estimating crystalline sizes and the degree of disorder caused by thermal history differences (69), as well as lattice expansion and unit cell parameters. By monitoring the surface structure of a polyetherketone (PEEK) composite, skin effects were observed using reflectance waxs (70). Small-angle x-ray scattering (saxs) allows the determination of morphology via domains and microstructure by varying electron density in different areas of a material or density differences between amorphous and crystalline phases (71,72).

Structure determinations in polymers have involved the use of small-angle neutron scattering (sans) techniques to evaluate not only the radius of polymer molecule gyration where light or x-ray scattering is not usable, but also phase dimensions in multicomponent polymers and suggested mechanisms of phase separation (73). Applications in polymer melts as well as concentrated and dilute solutions of polymers and solid-state samples have investigated the nature of semicrystalline polymers, polymer blends, and the development of new theories of polymer solutions and networks (74). In addition, sans studies on block copolymers have shown problems in obtaining sufficient detail to define domain morphologies (75). Although partial deuteration of all monomer and subsequent monitoring are required, it has been possible to characterize the molecular network of an epoxy thermoset system (76). Another technique finding use in structure analysis is dielectric spectroscopy. A combination of x-ray scattering and dielectric spectroscopy has been used to study the noncrystalline regions of semicrystalline polymers; this technique uses semicrystalline and amorphous polymer blends in which the amorphous polymer, although miscible in amorphous regions, is excluded from the crystal–amorphous interphase region (77). In a study on the cross-linking of poly(dimethylsiloxane), the dielectric properties were dominated by ionic conductivity up to frequencies of $10^3$ Hz; dipole orientation was the dominant factor at higher frequencies (78).

## Molecular Weight

The molecular weight and the distribution of multiple molecular weights normally found within a commercial polymer influence both the processibility of the material and its mechanical properties. For a few well-defined homopolymers, an analysis of composition and molecular weight is sufficient to define the likely mechanical properties of the polymer.

Low angle laser light scattering (turbidity) has been found to be the most accurate and reproducible technique for measuring polymer molecular weights (average molecular weight) as a primary method (79,80). Most turbidimetric techniques use polymer solubility as a function of temperature for defining molecular weights (81,82). Dynamic light scattering has been used to obtain the distribution of hydrodynamic radius, which can be converted to molecular weight distribution with appropriate calibration. This technique can be performed in corrosive solvents or with polymers that dissolve only at high temperatures (83,84). Gel-permeation chromatography (gpc), although a secondary technique requiring careful calibration and interpretation of data, is the most frequently used commercial technique because of its ease of use, low cost, short time for analysis, and generation of excellent comparative data (85). Shifting of the polymer molecular weight peak or displacement in comparison to a standard or good material can indicate the presence of a higher molecular weight species by showing a shift to early elution, or a lower molecular weight species by showing a shift to later elution. These shifts can affect the processibility and physical performance of the material. Small sharp peaks that elute late from the gpc column may indicate the presence of low molecular weight oligomers, residual monomers, additives, or even residual moisture in the polymer.

A wide variety of polymers have been analyzed by gel-permeation, or size-exclusion, chromatography (sec) to determine molecular weight distribution of the polymer and additives (86–92). Some work has been completed on expanding this technique to determine branching in certain polymers (93). Combinations of sec with pyrolysis–gc systems have been used to show that the relative composition of polystyrene or acrylonitrile–polystyrene copolymer is independent of molecule size (94). Improvements in gpc include smaller cross-linked polystyrene beads having narrow particle size distributions, which allow higher column efficiency and new families of porous hydrophilic gels to be used for aqueous gpc (95).

Temperature-rising elution fractionation (tref) is a technique for obtaining fractions based on short-chain branch content versus molecular weight (96). On account of the more than four days of sample preparation required, stepwise isothermal segregation (97) and solvated thermal analysis fractionation (98) techniques using variations of differential scanning calorimetry (dsc) techniques have been developed.

## Thermal Properties

Thermal analysis involves techniques in which a physical property of a material is measured against temperature at the same time the material is exposed to a controlled temperature program. A wide range of thermal analysis techniques have been developed since the commercial development of automated thermal equipment as listed in Table 1. Of these the best known and most often used for polymers are thermogravimetry (tg), differential thermal analysis (dta), differential scanning calorimetry (dsc), and dynamic mechanical analysis (dma).

Thermogravimetric analysis (tga), which monitors the change in the mass of a material during a controlled temperature ramp, is useful for both qualitative and quantitative analysis. It is used to help identify the types of polymers

**Table 1. Common Thermal Analysis Techniques[a]**

| Technique | Abbreviation | Property measured[b] |
|---|---|---|
| thermogravimetry | tg | mass, $m$ |
|   derivative thermogravimetry | dtg | $dm/dt$ |
| differential thermal analysis | dta | $dT/dt$ |
| differential scanning calorimetry | dsc | $dH/dt$ |
| thermomechanometry | | |
|   thermomechanical analysis | tma | $L, dL/dT$ |
|   thermal dilatometry analysis | tda | $V, dV/dT$ |
|   dynamic mechanical analysis | dma | $G \cdot (\omega)$ |
| thermoelectrometry | | |
|   thermally simulated conductivity | tsc | $di/dt$ |
|   thermally simulated polarization | tsp | |
|   thermally simulated depolarization | tsd | |
|   dielectric thermal analysis | deta | $Z \cdot (\omega)$ |
| thermophotometry | | |
|   thermoluminescence | tl | $dI/dt$ |
|   thermal microscopy | tm | |
| emanation thermal analysis | eta | radioactive gas evolution |
| thermosonimetry | ts | acoustic emission |
| evolved gas analysis | ega | gas chromatography, mass spectrometry, infrared, etc |
| thermomagnetometry | tm | magnetic effect on tg |

[a]Ref. 99. [b]$m$ = mass, $t$ = time, $T$ = temperature, $H$ = heat flow, $L$ = length, $V$ = volume, $G$ = sheer modulus, $\omega$ = angular frequency, $i$ = current, $Z$ = impedance, and $I$ = radiance.

by comparison to degradation curves (100), as well as identify and quantify additives such as carbon fillers, mineral fillers, plasticizers, antioxidants, stabilizers, uv absorbers, nucleating agents, and lubricants (101,102). Carbon black or graphite, which is commonly used as a colorant or filler in many plastics, may be analyzed using tg by running the analysis in a nitrogen atmosphere to 600°C for measurement of the polymer component and then switching to an oxygen atmosphere and heating to 800 or 900°C to oxidize the carbon (101,103). Although generally useful, tg curves are influenced by variations in the amount of material that changes temperature of decomposition, particle size of the sample, size and form of the crucible, packing of the sample, furnace atmosphere (oxidizing, inert, or vacuum), rate of heating, and position of the thermocouple in relation to the sample (104). The derivative curve of tga can be used to improve the determinations of onset and end point of decomposition of low level components or in multipolymer systems (105). Refinements in computer control and instrument design have also led to commercial instruments having smooth changes in dynamic heating rate that varies as a function of weight loss, a low mass furnace for rapid response, and horizontal purge gas flow to better remove decomposition products, among other changes (106). Early studies have shown improved resolution of vinyl acetate in ethylene–vinyl acetate copolymers and even separation of ABS decomposition into three separate weight-loss steps (107).

Differential thermal analysis (dta) monitors the temperature difference between a specimen and a reference material, whereas differential scanning calorimetry (dsc), the preferred technique for polymer analysis, heats specimen and reference materials separately and measures the heat flow between the materials. The various changes in polymers denoted by changes in enthalpy measured by dsc are shown in Figure 1 (108). Differential scanning calorimetry has become the method of choice for determinations of melting point as well as the presence and level of crystalline phase in polymers (109). The effects of cooling rates, applied strain during cooling, molecular weight, and molecular weight distribution on melting and recrystallization have been documented on a variety of polymers, eg, polyethylene (110,111), poly(phenylene sulfide) (PPS) (112–114), and poly(butylene terephthalate) (PBT) (112).

In addition, dsc has been found suitable for the determination of second-order transitions ($T_g$ = glass transition) and, in some cases, $\beta$-transitions. Procedures for the determination of $T_g$ are discussed in the literature (105,106), which includes data on the dependence of $T_g$ on number-average molecular weight ($\overline{M}_n$) for polymers. This shows a significant effect until a critical $\overline{M}_n$ is reached with little or no effect beyond. A technique employing dsc to measure specific heat of rapidly quenched polyethylene samples, which are subsequently annealed at low temperatures, shows the formation of secondary crystals that melt near the annealing temperature with small endotherms. The endotherms increase and melting points rise as time increases leading to the potential use of this technique for studying the thermal history of samples (107). Additional work on the correlation of enthalpy changes with the aging of samples has been completed on PBT (108). For amorphous polymers, changes in $T_g$ can be used to determine aging that has occurred below and near $T_g$. Accuracy is improved by matching the heating rates of dsc to cooling rates on specimen preparation (105,106). The $T_g$ response to aging relates to the continued relaxation of frozen molecular structure that continues over time below $T_g$ and results in changes in free volume

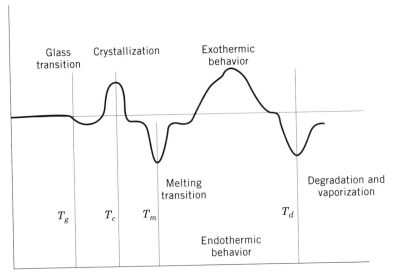

**Fig. 1.** Schematic dsc curve. Exothermic behavior includes curing, oxidation, chemical reactions, cross-linking, etc.

(118). Reduction in free volume due to aging can be correlated to reduced impact strength and improved creep response at low strain rates for polymers.

Chemical reactivity of polymers through polymerization reactions induced by thermal catalytic or radiation treatments can be shown by dsc (119); this can also be applied to hydration and dehydration reactions. Similarly, thermoset plastics exhibit exotherms during curing and cross-linking. Kinetic studies of thermoset curing that include the determination of rate constants and activation energies have been done (120). Furthermore, dsc provides quantitative data on the degree of conversion (via change in the measured $T_g$) for epoxies (121), and determination of $\Delta H$ of residual cure of a molded material compared to that of uncured resin gives the degree of conversion for several resins (122,123).

The relative effectiveness of nucleating agents in a polymer can be determined by measuring recrystallization exotherms of samples molded at different temperatures (105). The effect of catalyst concentration and filler content has been determined on unsaturated polyesters by using dynamic thermal techniques (124). Effects of formulation change on the heat of rubber vulcanization can be determined by dsc; pressurized cells may be needed to reduce volatilization during the cure process (125).

Changes in heat capacity and measurement of $T_g$ for blends have been used to determine components of copolymers and blends (126–129), although dynamic mechanical analysis has been found to give better resolution. Equations relating $T_g$ of miscible blends and ratios of components have been developed from dsc techniques, eg, the Fox equation (eq. 1), where $T_{g12}$ is the $T_g$ of the blend, $W_1$ or $W_2$ is the weight fraction of component 1 or 2, and $T_{g1}$ or $T_{g2}$ is the $T_g$ of the pure component 1 or 2 (130); the Gordon, Taylor, Wood (GTW) equation (eq. 2), where $k$ is the best fit constant for empirical data (131); and the Couchman equation (eq. 3), where $x_1$ and $x_2$ are mass fractions and $\Delta C_p$ is heat capacity changes for neat polymers (132).

$$\frac{-1}{T_{g12}} = \frac{W_1}{T_{g1}} + \frac{W_2}{T_{g2}} \tag{1}$$

$$T_{g12} = \frac{k(T_{g2} - T_g)\,W_2}{1 - W_2} + T_{g1} \tag{2}$$

$$\ln T_{g12} = \frac{x_1 \cdot \Delta C_{p1} \cdot \ln T_{g1} \cdot x_2 \cdot \Delta C_{p2} \cdot \ln T_{g2}}{x_1 \cdot \Delta C_{p1} + x_2 \cdot \Delta C_{p2}} \tag{3}$$

Thermomechanical analysis (tma) measures the dimensional changes of a fabricated polymer part at a constant heating rate. First-order ($T_m$) and second-order ($T_g$) transitions may be measured from graphing length of specimen ($L$) versus temperature ($T$). This technique may also be used to determine the coefficient of linear thermal expansion (CLTE) for a material. Thermomechanical analysis can be more sensitive to small changes in enthalpy typical of highly crystalline polymers than differential scanning calorimetry (133). This technique has also been used for evaluating dimensional changes of polymers in fluid environments. Studies have been completed on water–nylon 6 at different temperatures (134), nylon 6 at different levels of water absorption (135), chlorobenzene–polyethylene (136), and elastomers in various fluids and oils.

Dynamic mechanical analysis (dma) was originally developed as a rheological method and has been used primarily for measuring the properties of polymers beyond those determined by the thermal techniques previously discussed (see RHEOLOGICAL MEASUREMENTS). This method detects $\alpha$- (first-order or glass) transitions, $\beta$- (side chain) transitions, and $\gamma$- (crankshaft rotation of main chain segments) transitions, whereas dsc can rarely resolve $\beta$- or $\gamma$-transitions (137). In this procedure, a specimen is placed under a fixed or free oscillatory stress and the strain with which the specimen responds is recorded. Polymers are typically viscoelastic in nature, and have a dynamic response to the oscillatory stress controlled by the viscous component, which results in a strain that continues to increase until the stress is removed, but is not recoverable. The other component is the elastic response, which reacts instantaneously to the applied stress and is completely recovered when the stress is removed. Polymers respond with elements of each, resulting in a strain with the same frequency as the applied stress, but out of phase by an amount, $\delta$, dependent on the relative elastic and viscous nature of the material (see Fig. 2) (138). Samples can be tested under different stress configurations, ie, tensile, flexural, compressive, and shear, by using a constant resonance frequency under a constant temperature or constant rate of temperature change. The tangent of $\delta$ is known as the loss factor and is the ratio of $E''$, the loss modulus, over $E'$, the storage modulus; $E$ is used for tensile (extensional deformation) whereas $G$ is used for shear.

Stress levels are kept low within the elastic region of the material. Although solid-state testing is used most frequently, melt-state testing allows the examination of viscosity versus frequency at a set temperature and also shows

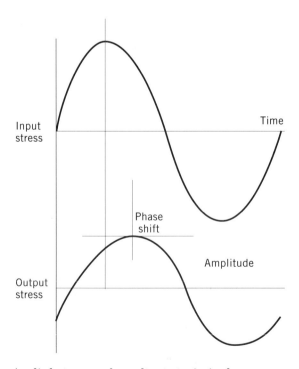

**Fig. 2.** Applied stress and resultant strain in dma measurements.

the response of the shear storage and loss modulus ($G'$ and $G''$) versus frequency. Solid-state testings of both amorphous and semicrystalline polymers by dma have shown effects on tan $\delta$ values near the $T_g$ for each type of material from differences in orientation by sample preparation method. For amorphous materials, the values of tan $\delta$ are higher in the temperature region slightly below the $T_g$ as orientation increases in the sample; the same effect is seen for crystalline materials above the $T_g$ (139). Interfacial strength has been found to have an inverse relationship with tan $\delta$ peak heights at $T_g$ for several filled polymer systems, such as glass-reinforced polyester (140), reinforced rubber (141), and epoxy–carbon fiber composites (142). The interrelationship between temperature and frequency as well as time and frequency are well established (143). The equivalence of the effects of temperature and time in viscoelastic systems results in time–temperature superpositioning (144) and is related by the Williams-Landel-Ferry (WLF) equation (eq. 4) for amorphous polymers, where $a_T$ is the shift factor, $T$ is the measurement temperature, $T_o$ is the reference temperature, and $c_1$ and $c_2$ are constants (145).

$$\log a_T = \frac{-c_1\,(T - T_o)}{c_2 + T - T_o} \qquad (4)$$

This allows the production of master curves, which can be used to estimate changes in modulus or other properties over a long period of time by shorter tests over different temperatures.

The Arrhenius relationship (eq. 5) for crystalline polymers or other transitions, where $E_a$ is the activation energy and $R$ the gas constant (8.3 J/mol), is as follows:

$$\log a_T = \frac{-E_a}{R(T - T_o)} \qquad (5)$$

Dynamic mechanical analysis of multiple frequency measurements over a range of temperatures allows the plotting of modulus versus time in similar master curves as generated by a series of short-term creep studies in a much shorter time frame. There are potential problems with these predictions, such as internal stresses in amorphous polymers interfering in superpositioning loss modulus data near $T_g$ (146), using data in temperature regions outside linear viscoelastic behavior (147), or strains outside the linear regions. In addition, dma has been found useful in studying the physical properties and morphology of blends as evidenced in a study of polycarbonate–ABS, in which the overlap of the glass-transition temperatures of the two base polymers indicates component levels and the presence of a bimodal $\beta$-transition, which can be related to immiscibility between the polycarbonate-phase and the polybutadiene of the ABS-phase (148). Other studies on blends have been documented on styrene block copolymers with poly($\alpha$-methylstyrene) (149), poly(phenylene oxide)–polystyrene blends (150), unsaturated polyesters with elastomers (151), and polycarbonate–poly(butylene terephthalate) (152). Furthermore, dma is used to study the curing of thermoset systems (120,153,154) and the cross-linking of polymers (155,156).

Dielectric analysis (dea) or dielectric thermal analysis (deta) is similar to dma. In dea, the movement of dipoles or other ionic species that can be electrically stimulated allows the monitoring of changes in ionic conductivity. Similar to dma phase angle $\delta$, in dea this is related to the lag of the polarization induced in an insulator placed in an alternating electrical field. The energy dissipated as heat is proportional to the dielectric loss, $\epsilon''$, and the energy stored is the real part of the dielectric constant, $\epsilon'$. The electrical technique is considered more sensitive to transitions below $T_g$ than dma (157). It is also more sensitive to $\beta$-transitions frequently linked to impact strength and long-term mechanical strength (158). Dielectric analysis has a broader and higher range of frequencies than dma. Changes in permittivity with time in dea measurements have been correlated with moisture content change in polymers (159) and have become an on-line technique for monitoring moisture in polymers. Thermally stimulated conductivity (tsc) uses the technique of polarizing a polymer at different temperatures, quenching the sample, and measuring the depolarizing current as the material is heated at a controlled rate (160). The rate of repolarization is related to the relaxation times of internal motions. The plot of current density or dynamic conductivity versus temperature shows the relaxations as peaks. The method claims improved characterization of polymers compared to dsc and dma (161–163). Inverse gas chromatography has been used to measure thermodynamic interactions in mixed polymer systems and with polymer–small-molecule interactions (164). It may also be useful in estimating polymer–polymer interaction parameters for immiscible polymer segments in diblock copolymers (165).

Vicat softening temperature (ASTM D1525 (166), ISO 306 (167)) uses a flat-ended needle penetration of 1 mm of the plastic specimen under controlled heating conditions to indicate short-term resistance to heat. For some plastics these curves can be extrapolated to zero force and zero rate of heat to give values in agreement with glass-transition data (168). Testing of deflection temperature under load (ASTM D648 (169), ISO 75 (170)) determines bending resistance of plastic specimen under load and at a set rate of temperature increase. This test is highly influenced by the molded-in stresses and the thickness of the specimen tested. Annealing specimen to relieve stresses prior to testing improves reproducibility of the test but also results in higher values than unannealed specimen.

## Processing Properties

The flow properties of polymers, whether the viscosity of liquid thermosets prior to gelation or of molten thermoplastics, are important parameters for the proper processing of materials. Several test methods have been developed to predict or correlate to actual processing conditions or for quality control testing of polymers. For thermoplastics, the commonly used quality control method is melt flow rate, ASTM D1238 (171). This technique can give values at melt index conditions that are inversely proportional to the molecular weight of homopolymers. However, the common addition of processing aids, lubricants, and fillers makes this correlation highly unreliable. Procedures that determine melt flow under two loads have been used primarily with polyolefins as an indication of molecular weight distribution. Melt flow rate values, although useful for quality control, are not indicative of the actual response of the material during processing, primarily due to the viscoelastic nature of polymers. Because polymers are non-Newtonian in flow, large differences in shear rate of the melt flow rate test (1–50 s$^{-1}$) versus that of processing equipment, eg, extrusion (10$^2$–10$^3$ s$^{-1}$) and injection molding (10$^3$–10$^5$ s$^{-1}$), can cause large changes in the viscosity of most polymers.

Because processing conditions cover a wide range of shear rates, tests that can simulate both temperature and shear rate conditions are more useful in predicting flow properties. Several studies have been conducted on correlating common processing techniques that use torque rheometers, capillary rheometers, and oscillating disk rheometers for both thermoplastics and thermosets (172–175). Capillary rheometers have been designed to reach shear rates of 10$^7$ s$^{-1}$ and are used to study flow behavior at high shear (176). A typical set of curves of viscosity versus shear rate is shown in Figure 3 (177). Torque rheometers having a mixing chamber capable of being controlled within 1°C and speed ranges between 10–200 rpm are useful in understanding the compounding of plastic formulations (178–181). Studies have also been completed with instrumented processing equipment (182). A study of flow testing of ABS resins showed similar flow ranking of different grades of resin by using capillary rheology (183), spiral flow molding, and on-machine rheometry (184,185), the last a technique using an injection molding machine as a rheometer. Although spiral flow molding has been established as a standard only for thermoset polymers, many resin manufacturers and molders use a similar technique for thermoplastics to profile material processing response to temperature and pressure conditions in an Archimedes spiral mold on a particular piece of injection molding equipment. The test is useful for processability ranking of materials tested at the same time

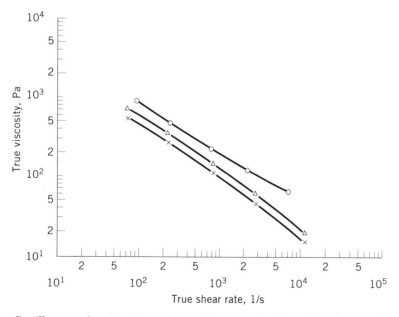

**Fig. 3.** Capillary melt viscosity, where ($\circ$) = 300°C, ($\triangle$) = 320°C, and ($\times$) = 340°C. Courtesy of The Dow Chemical Co.

and under the same conditions. Capillary rheology tests using a screw rheometer in place of a piston have been published as ASTM D5422 (186).

Flow tests designed for thermosetting resins include a cup flow mold for flash-molding of powdered phenolic and alkyd materials (187), spiral flow for low pressure thermosets (188), and torque rheometer techniques (189). In addition to these, tests for measuring time to gelation (gel) are important for processors to understand usage time and pot life. Dilute solution viscosity testing of polymers requires complete dissolution in the solvent without degradation or interaction with the solvent within the chosen temperature range of the tests. Data from these tests have been used to determine relative molecular weight of polymers (190).

New standard tests for determining pressure–volume–temperature (PVT) properties of materials using rheometers are being established to discover the correlation for the fundamental properties of plastics. These in combination with no-flow temperature, ejection temperature, and capillary rheology data are useful in predicting the behavior of molten plastic as it enters and fills a mold. Several mold-filling prediction computer programs are commercially available (191,192) to assist mold designers and plastic processors in developing plastic parts that have reduced stresses from the molding process, and in improving productivity with easier start-up of new molds and fewer rejected parts.

## Mechanical Properties

Mechanical properties of plastics can be determined by short, single-point quality control tests and longer, generally multipoint or multiple condition procedures that relate to fundamental polymer properties. Single-point tests include tensile, compressive, flexural, shear, and impact properties of plastics; creep, heat aging, creep rupture, and environmental stress-cracking tests usually result in multipoint curves or tables for comparison of the original response to post-exposure response.

Tensile properties are those of a plastic being pulled in an uniaxial direction until sufficient stress is applied to yield or break the material. Standard tests are ASTM D638 (193) and ISO 527 (194). For many materials, Hooke's law is valid for a portion of the stress–strain curve. If stress is relieved during this portion of the testing, any strain that has occurred is fully recovered. Elastomers generally do not show this linear response. Tensile curves can be used as an indication of polymer strength and toughness. Figure 4 shows the relationship normally seen for the high stress necessary for yield or break with strength, whereas high elongation beyond yield shows ductility (toughness). Similar curves can be generated for tests in comparison, flex, shear, and some forms of impact.

Mechanical properties are determined on solid polymers in arbitrary forms defined precisely by standard test methods in ISO, ASTM, or other national standards organizations. Parts are formed by either injection molding, compression molding, or milling from extruded sheet or molded plaques. Viscoelasticity of polymers dictates that the technique used to make the part must have a significant effect on the mechanical behavior of the polymer. For valid comparison of materials, they should be prepared similarly and conditioned under the same environment. Viscoelastic effects are also the reason for the rate of strain effects on the modulus values of materials under tensile, flexural, and compressive

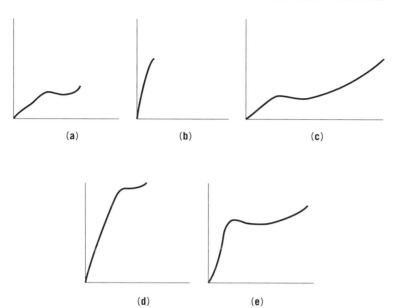

**Fig. 4.** Types of stress–strain curves: (**a**) soft and weak; (**b**) hard and brittle; (**c**) soft and tough; (**d**) hard and strong; and (**e**) hard and tough.

testing (195). Several types of impact testing have been developed to measure a plastic's response to a high rate of strain. Notched or Charpy Izod impact are pendulum impact tests essentially uniaxial in direction. Although both tests have been long established for quality control tests and material property data sheets, they have little practical value in determining the impact response of a plastic. ASTM D256 (196) lists multiple cautions on using data from these tests. Notched Izod's primary practical value in characterizing a plastic material is to establish how notch-sensitive the material is. Drop-weight impacts, either manual (197) or instrumented (198), give more practical information because these are multiaxial test procedures and closer to normal impact seen in part applications. Instrumented impact allows the recording of the full impact event as a stress–strain curve, showing similar characteristics to tensile or other modes of stress–strain tests. Instrumented impact is preferred for monitoring ductile–brittle transitions and determining the effects of polymer composition on material toughness (199). Large equipment is available for testing the actual parts with data correlated to performance (200). Newer high speed impact testing equipment (hydraulically controlled) can be used to test at speeds greater than $1.6 \times 10^4$ mm/s (201) and to simulate automotive impact conditions.

      Creep, creep rupture, and stress relaxation tests are multiple-point tests requiring long periods of time (1000 h min) to generate useful data; these are standard tests for determining more fundamental polymer properties (202,203). Data for these tests are generated under several time–temperature–stress-level conditions in either tensile, flexural, compressive, or shear modes and combined into master curves using superpositioning theory. The low stress loads applied in creep tests, and thereby the low strain rates determined, require sensitive equipment for accurate measurements. Studies that normalize the stress required to produce a given strain in a given time to the stress producing that same strain after 24 hours, have shown that plots of the normalized stress rates versus time over stress (24 h) give a single curve, which can be used with other short-term (days) creep tests at different stress levels to prepare master curves that can require up to one year for traditional data generation (204). Isometric curves can also be obtained similarly by using different temperatures for a given strain. Modulus values from these curves are used in plastic part design equations to predict part performance over the expected life of the part.

Fatigue testing of polymers may consist of static fatigue, ie, creep rupture, or dynamic fatigue. The ASTM procedure for dynamic fatigue, D671 (205), is a constant amplitude force technique in flexural mode utilizing a carefully prepared specimen that allows even stress distribution over the test span. However, the test is limited to a single frequency of 30 Hz. For many polymers, this induces an unacceptable temperature rise that can affect the results, if not lead to thermal failure in which the specimen distorts (206–208). Other modes of cyclic fatigue include tension, compression, or shear (209,210), which use sinusoidal-, square-, and sawtooth-wave forms for applying stress, although varying strain can also be used in cyclic fatigue testing (211–213). Notching of specimen and high or low temperatures have been employed to accelerate the failure via embrittlement of the materials. This can also lead to unwarranted extrapolation of the test data without additional tests (214).

Substantial work on the application of fracture mechanics techniques to plastics has occurred since the 1970s (215–222). This is based on earlier work on inorganic glasses, which showed that failure stress is proportional to the square root of the energy required to create the new surfaces as a crack grows and inversely with the square root of the crack size (223). For the use of linear elastic fracture mechanics in plastics, certain assumptions must be met (224): (1) the material is linearly elastic; (2) the flaws within the material are sharp; and (3) plane strain conditions apply in the crack front region.

An ASTM standard has been accepted for the determination of the critical stress intensity factor, $K_{IC}$ (225). Although $K_{IC}$ is temperature- and rate-dependent and the dependence is specific for any material, it is relatively independent of specimen geometry. Single-edge-notched beams are typically used for the test, but other geometries, eg, compact tension, short rod, and center-notched tension, have been used for the testing (226–229). Studies on the plastic zone in front of a propagating crack have used blunt notches and double-edge-notched tension specimens to determine crack resistance curves (230). The J-integral parameter is taken from these curves. More complex treatments of fracture incorporating nonlinear, viscoelastic effects have also been developed (231–233).

As materials deform in mechanical testing, they emit stress waves having frequencies from 10 kHz to 10 MHz called acoustic emission (AE) waves. These are detectable by piezoelectric transducers as electric signals and are used to determine the initiation and growth of cracks in materials (234). The high attenuation required for the low level of released energies limits the usefulness of AE in plastics (235–239). However, the analysis of AE signals correlating time, based on the propagation speed of AE waves in tested materials, or use of a time to amplitude converter, makes it possible to separate AE waves from mechanical noise, system noise, and other miscellaneous electromagnetic noise (240). AE waves generated during cyclic fatigue of single-edge-notched specimens of PMMA showed variation in amplitude and frequency dependent on the progression of the crack front (Fig. 5) (241). In further high pressure studies using the same type of specimen, striations found on the fracture surface corresponded with the number of AE pulses recorded, thus suggesting that the AE signal was emitted when the striation was formed. AE has also been used in detecting craze formation and crack initiation in both brittle and ductile polymers. AE signals were found throughout the tensile deformation process in amorphous polymers, but only after the yield point in crystalline polymers (242). Two types of signals were found, those prior to yield appear associated with submicrocrack formation, and those after necking (if not a brittle fracture) possibly associated with molecular chain slipping near the necking front.

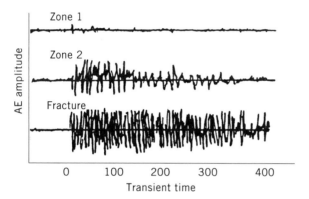

**Fig. 5.** Typical acoustic emission (AE) waves observed at different zones and during fracture of cyclic loading of PMMA. The amplitude is drawn on arbitrary scale. Courtesy of Oxford University Press.

## Environmental Effects Tests

In practical applications of plastic materials, their mechanical properties can be significantly influenced by environmental factors of chemical exposure, temperature, radiation (photon and gamma ray), biological agents, and/or combinations of several of these factors at once. Few standard tests have been developed in these areas. Mechanical properties can be determined by using ovens or cold-temperature chambers under the same procedures described in ASTM or ISO short-term tests to evaluate the effect of heat or cold. The length of exposure time, test conditions (at temperature or after being reconditioned to standard testing conditions after exposure), and reference material must all be stated to use these data. ISO 2578 (243) addresses these variables, but allows the tester to choose the characteristic for testing and the threshold value (limit of decline in performance). It recommends testing for 20,000 hours and performing the test under standard room temperature conditions after heat exposure. ASTM is developing a standard for comparative indexing of heat-aged materials to original properties, but does not indicate a minimum exposure time. Thermal expansion testing also addresses the physical dimensional changes in plastic parts exposed to temperature changes. The change must be measured in each orthogonal direction, because polymers are not generally isotropic or homogenous. ASTM D696 (244) has been used for coefficient of linear thermal expansion.

Shrinkage measurements are not related to low temperature measurements, but refer to the dimensions of post-molded parts compared to the mold dimensions (245) or to post-molding shrinkage that involves short-term, high heat exposure with post-exposure dimensional measurements (246,247). Melting points of crystalline materials may be determined by capillary tube (248) or hot plate techniques, whereas softening point (ISO 306, ASTM D1525) (249,250) by penetration or deflection under load (ISO 75, ASTM D648) (251,252) is generally used for amorphous and semicrystalline polymers.

Chemical exposure of plastics may exhibit a wide range of effects with one chemical and one type of polymer, depending on the concentration of chemical agent, temperature of exposure, molded-in stresses within the specimen, applied stress level to the part during exposure, and formulation of the specific polymer grade. Standardized tests have been developed to evaluate the effects of chemical agents on plastics with and without external stress. Tests such as ASTM D543 (253) and ISO 175 (254) propose procedures for evaluating the effects of suggested chemicals on dimensions, appearance, and mass of the polymer specimen after the specimen has been immersed for a set period of time. Mechanical properties may also be checked if the specimen has maintained sufficient integrity to test. Tests in which stress is externally applied include both stress and strain techniques. ASTM D1693 (255) uses bent strips of polyethylene with a cut, exposed to an aggressive surfactant and exposed until failure or 48 hours. Several improvements have been suggested, primarily in better defining specimen preparation (256,257) or control of the amount of strain used (258).

The method of caustics has also been used to study the formation of cracks and crazes formed by exposure of PMMA to solvents (259). ISO 4599 (260) has been developed to better control the application of stress using a jig having the curve of the arc of a circle for shaping the specimen and maintaining a set curvature during exposure to the agent. After a predetermined time the specimen is tested for tensile or flexural properties and compared to preexposure test values. ISO 4600 (261) uses the technique of impressing an oversized ball or pin into a hole drilled in the specimen to apply a strain.

Tests using a constant stress (constant load) normally by direct tension have been described in ISO 6252 (262). This test takes the specimen to failure, or a minimum time without failure, and frequently has a flaw (drilled hole or notch) to act as a stress concentrator to target the area of failure. This type of testing, as well as the constant strain techniques, requires careful control of specimen preparation and test conditions to achieve consistent results (263,264).

Weathering of plastic materials combines complex factors of temperature, radiation, oxidation, and moisture effects on a plastic part. Weathering effects vary with geography, time of year, and position of material being exposed. All of these effects make predicting the weatherability of a material extremely difficult. ISO 877 (265) and ASTM G24 (266) outline procedures for standardizing outdoor exposure of plastics under glass; ISO 4607 (267) and ASTM D1435 (268) are techniques for total exposure of plastic specimens to natural elements in racks and under specific climatic conditions. Frequently, only changes in physical appearance are measured after weathering; however, appropriate specimens can be exposed, mechanically tested, and compared to carefully stored control specimens (269,270). The long time periods required for outdoor weathering studies have led to the development of several accelerated techniques. The use of extreme

and consistent climates, eg, south Florida or Arizona, and intensifying techniques for natural sunlight have been suggested to accelerate tests (271–273).
ASTM D4364 (274) has established a method for the use of a follow-the-sun rack
with flat mirrors to reflect sunlight uniformly onto the test specimen as an intensifying technique. Because these techniques still require long exposure times
to obtain results, in-lab acceleration tests to weather materials artificially have
been developed. Various lamp sources have been used, such as low and high
pressure mercury, carbon arc, and xenon arc. Although carbon arc lamps are
still used to weather materials, xenon arc lamps are the preferred lamp for plastics weathering. These lamps simulate sunlight closely when appropriate filters
are used (Fig. 6). Various correlations have been reported between xenon testing
and outdoor weathering. For several plastics, 500 hours in xenon arc testing,
provided high humidity and water spray are used, are equivalent to one year of
outdoor exposure (275). Some equipment manufacturers indicate that the ratio
is one hour of accelerated to ten hours of outdoor exposure, whereas experi-
menters have found that the ratio can range from 1 h accelerated/5 h outdoor to
1 h accelerated/9 h outdoor (276) when evaluated by the change in plastic color.
Accelerated tests should be used to compare the relative performance of materials and ranking resistance to exposure, rather than for any direct correlation to
outdoor weathering. Both ISO and ASTM have established practices for the use
of carbon arc and xenon arc accelerated equipment (277,278).

   Fluorescent ultraviolet lamps within an apparatus that allows condensation
cycles rather than the water spray typical of xenon arc tests have been developed
for plastics testing (279). The spectral cutoff wavelength of the lamps used in
the apparatus determines the severity of the test. Ultraviolet B (UVB) 313

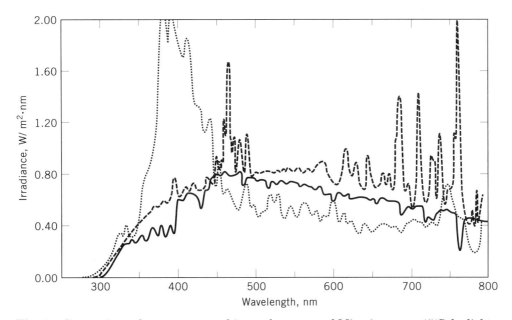

**Fig. 6.** Comparison of xenon arc, sunshine carbon arc, and Miami average 45°S daylight,
where (– – – –) represents xenon with Type S high borate filters, 0.35 W/m²·nm at 340 nm;
(·····) represents sunshine carbon arc with Corex; and (——) represents daylight. Courtesy
of Atlas Electric Devices Co.

lamps allow a significant irradiance component below 290 nm, which is normally filtered out by the earth's atmosphere. Ultraviolet A (UVA) 340 lamps have shown better correlation to the spectral irradiance of natural sunlight, although the visible light range is missing.

Other environmental testing standards have been developed for evaluating the effects of biological agents, either strains of fungi (ISO 846 (280)) or ASTM G21 (281)) or bacteria (ASTM G22 (282)) on mold growth on the plastic material, or if the plastic material has a toxic effect on the fungi. Longer-term outdoor exposure tests for resistance to microbial attack require geographic locations favorable to microbial growth or burial of the material in soil (283). ASTM E1027 (284) is used for exposure of polymers to ionizing radiation. This can be used for testing the resistance to gamma rays, electrons, neutrons, etc. These tests are important for plastics used in the medical area to determine the resistance to sterilization techniques.

## Optical Properties

Transparency, gloss, color, refractive index, and reflectance are the properties normally associated with aesthetics of plastic materials. In some areas, changes in optical properties, increases in haze after abrasion testing (285), color differences after weathering, and birefringence analysis of residual stress within a transparent part (286) are all used to measure the effects of applied stresses. Measurements of color, gloss, refractive index, and haze apply to many products beyond plastics and use similar techniques. Reference should be made to this general topic for detailed information (see COLOR).

One of the most widely accepted measurements of color in plastics is based on standards developed by the Commission Internationale de l'Eclairage (CIE) for illuminants and observers to establish tristimulus values. A common color scale used to describe color in numeric terms of lightness and hue is the $L$-$a$-$b$ tristimulus system, where $L$ is scaled vertically from a perfect white with a value of 100 to a perfect black at 0; $a$ is positive for red and negative for green; and $b$ is positive for yellow and negative for blue. Spectrophotometers are used to measure the full reflectance of a colored material over the visible range and convert to the tristimulus values via microprocessors. Color as viewed under various CIE illuminants and/or observers can be determined by a single spectrophotometric analysis as the microprocessor computes the tristimulus values for each set of conditions. Color-matching of an object can be completed by comparing spectral curves generated through the use of the spectrophotometer. However, many factors, such as gloss, texture, opacity, and changes in illuminant and observer, complicate the visual appearance of a color with the spectral curve of a color. ASTM D1729 (287) establishes procedures for visual evaluation of color and defines illuminant, viewing conditions, and use of standards for comparison; ASTM D2244 (288) describes instrumental color evaluation techniques. Specular gloss, the relative reflectance of a specimen in the specular direction, is usually measured at 60° angle of incidence for most plastics. High gloss materials may be measured at 20° incidence and low gloss specimens at 85° incidence using ASTM D523 (289). For films, ASTM D2457 (290) uses measurements at 20°, 60°, and 45° angles of incidence.

## Electrical Properties

Excellent insulating properties, along with the ability to be structural components, make plastics the ideal candidate materials for electrical applications. Although generally used as insulators, carbon black or carbon fiber can be added to make plastic materials electrically conductive, thereby expanding their usefulness in the electronics area.

Standard testing of electrical properties of plastics includes dielectric strength, permittivity, dissipation factor, surface and volume resistivity, and arc resistance. Dielectric strength is the maximum voltage required for breakdown and is determined by one of three techniques: short-time, slow-rate or -rise, and step-by-step. The two last techniques use data from the short-time test to determine a starting point (291). Dielectric strength is thickness-dependent with thin foils being used in specific space-saving devices (292).

Permittivity gives an indication of the electrical storage ability of a conductor (behaving as a capacitor). The thinner the material, the higher the permittivity or capacitance, allowing use of plastic foils in areas where high capacitance is required. However, when plastics are used as insulators, low permittivity is required. Resistivity testing can be either volume, relating to leakage of current through the body of the insulator, or surface, relating to a surface layer of moisture or contaminant leading to electrical loss (293).

Arc resistance or tracking resistance is the resistance of a material to a high voltage arc or discharge. ASTM has published three methods to test this phenomenon, depending on the conditions to be tested: D495 (294) for high voltage, low current, dry arc resistance; D2132 (295) for dust and fog tracking; and D2302 (296) for liquid-contaminant inclined-plane tracking. Tracking tests have shown high variability with standard deviations as high as 40% from average (297). A comparative tracking index (CTI) technique uses a low frequency, low level current applied between electrodes while allowing drops of aqueous solution to fall between electrodes at 30-s intervals. This technique is described in ASTM D3638 (298). However, the results or values obtained from these tests cannot be used in design studies (299).

Electromagnetic interference (EMI) testing has become more prevalent for materials that either emit or are affected by EMI. Shielding efficiency (SE) of materials is determined by measuring electric field strength between a transmitter and receiver with or without the presence of the material under test. Several researchers have suggested a correlation between volume resistivity and SE values (300,301).

Underwriters' Laboratories (UL) is an independent, nonprofit organization that develops standards for safety in electrical products. UL 746 A, B, and C describe tests and limits for materials used in electrical equipment; UL 746 D lists test requirements for the fabricated plastic parts (302).

## Flammability Properties

Plastics have become an important material in the construction industry and are used in areas of insulation, wire coating, flooring materials, and piping; they are also considered for structural components as foam panels and plastic wood. As use in the industry has grown, so has concern about the fire properties of these materials. Tests to determine ignition resistance, flame spread, heat release, and smoke or toxic gas release have been developed. However, the results of these tests are not applicable to the performance of the materials under actual fire conditions. These small-scale tests are intended for quality control and potential hazard ranking. Several organizations that go beyond testing standard bodies such as ASTM or the International Electrotechnical Commission (IEC) have developed tests and/or regulations for the use of plastics in electrical appliances or buildings. A listing of some of these tests is given in ASTM D3814 (303). General fire test terminology is given in ASTM E176 (304). Ignition temperature tests ISO 871 (305) and ASTM D1929 (306) have procedures using small external pilot flames. The limiting oxygen index test has been used in quality control primarily for its good reproducibility, but cannot be used to predict fire performance. ISO 4589 (307) and ASTM D2863 (308) determine the minimum oxygen concentration required to support flaming combustion under set conditions.

UL 94 flame testing of plastics is similar to ASTM D635 (309) for horizontal burn and ASTM D568 (310) and D3801 (311) for vertical burn. UL lists both materials and fabricated products meeting the various 94 requirements in a publication (312). Rate of heat release data is highly important in determining fire behavior, but few tests have been established. Equipment proposed by Factory Mutual has been used in fundamental fire modeling studies (313). The cone calorimeter as described by the National Institute of Standards Technology (NIST) utilizes both vertical and horizontal specimens (314). Refinements of this equipment have allowed testing in the reduced oxygen atmosphere typically found under fire conditions (315). Smoke generation tests have used changes in light absorption of a photoelectric cell within a confined chamber as the primary measurement. Results are reported as percent obscuration or smoke density. Flame or radiant panels can be used as the ignition source in these tests (316–318).

## Nondestructive Testing

Nondestructive testing (qv) can include any test that does not damage the plastic piece beyond its intended use, such as visual and, in some cases, mechanical tests. However, the term is normally used to describe x-ray, nuclear source, ultrasonics, atomic emission, as well as some optical and infrared techniques for polymers. Nondestructive testing is used to determine cracks, voids, inclusions, delamination, contamination, lack of cure, anisotropy, residual stresses, and defective bonds or welds in materials.

X-ray techniques can readily determine voids down to 1 mm in diameter and cracks (as long as the plane of the crack is close to the direction of the ray) (319). Nuclear sources and gamma and beta rays have been used to measure the thickness of extruded plastic sheet. For ultrasonics, the frequency used for plastics ranges between 0.5–5 MHz. However, isolated defects smaller than a wavelength are difficult to detect with this technique. Ultrasonics have also been used to determine elastic modulus, shear modulus, and Poisson's ratio via the relationship of sound wave velocities through a material. These relationships are given in Figure 7 (319,320). Holographic interferometry uses an expanded beam of light from a laser source divided by a beam splitter, with one portion hitting the object and the other hitting a recording screen. When both beams are combined at the screen a unique interference pattern is formed. Comparison of this generated image with another of the object under stress produces a fringe pattern for analysis. The maximum depth/diameter ratio for a detectable defect is less than 0.2 (321).

$$E = \frac{\rho \cdot V_c^2 (3V_c^2 - 4V_s^2)}{V_c^2 - V_s^2} \qquad G = \rho \cdot V_s^2$$

$$\text{(a)} \qquad\qquad\qquad \text{(b)}$$

$$\sigma = \frac{V_c^2 - 2V_s^2}{2(V_c^2 - V_s^2)} \qquad K = r\left(V_c^2 - \tfrac{4}{3}V_s^2\right)$$

$$\text{(c)} \qquad\qquad\qquad \text{(d)}$$

**Fig. 7.** Relations between elastic constants and ultrasonic wave velocities. (**a**) Young's modulus; (**b**) shear modulus; (**c**) Poisson's ratio; and (**d**) bulk modulus.

Thermal imaging is sensitive to infrared radiation that detects temperature changes over the surface of a part when heat has been applied. Thermal diffusion in a solid is affected by variation in composition or by the presence of cracks, voids, delaminations, etc; the effects are detected by surface temperature changes. Defects cannot be detected if their depth below the surface is more than two to three times their diameter. Nondestructive testing has been primarily used for composites and analysis of adhesive bonds or welds. Several studies are documented in the literature (322–327).

## Other Tests

There are tests for physical properties such as density and hardness (qv) of plastics. Microscopy (qv) is important in fracture analysis as well as in analysis of the morphology of polymer systems for an understanding of polymer blend performance.

## BIBLIOGRAPHY

"Plastics Testing" in *ECT* 2nd ed., Vol. 15, pp. 811–831, by E. Horowitz, National Bureau of Standards; in *ECT* 3rd ed., Vol. 18, pp. 207–228, by E. Horowitz, The Johns Hopkins University.

1. C. D. Craver, *Applied Polymer Science*, 2nd ed., American Chemical Society, Washington, D.C., 1985, p. 711.
2. P. C. Painter, M. M. Coleman, and J. L. Koenig, *Theory of Vibrational Spectroscopy with Application to Polymers*, John Wiley & Sons, New York, 1981.
3. G. Bucci and T. Simonazzi, *J. Polym. Sci. Part C-7*, 203 (1964).
4. J. N. Lomonte and G. A. Tirpak, *J. Polym. Sci. Part A-2*, 705 (1964).
5. C. Toni and T. Simonazzi, *Angemandte Makromolecular Chemie*, **32**, 153 (1973).

6. R. P. Wool and W. O. Stratton, *J. Polym. Sci. Polym. Phys. Ed.* **12**, 1575 (1974).

7. T. H. Kozel and R. T. Kazmierczsk, *SPE ANTEC*, **37**, 1570 (May 1991).

8. P. Biyer and M. R. Padhye, *Silk Rayon Ind. India*, **16**(5) 174 (1973).

9. T. R. Crompton, *The Analysis of Plastics*, Elsevier Science, Inc., Elmsford, N.Y., 1986.

10. D. Battiste and co-workers, *Anal. Chem.* **53**, 2232 (1981).

11. D. O. Hummel and F. Scholl, *Atlas of Polymers and Plastics Analysis*, Vol. 1, 2nd ed., Carl Hanover Verlag, Munich, 1978.

12. C. D. Craver, *Infra-Red Spectra of Plasticizers and Other Additives*, 2nd ed., The Coblentz Society, Kirkwood, Mo., 1980.

13. P. C. Painter, M. Watzek, and J. L. Koenig, *Polymer*, **18**, 1169 (1977).

14. A. Gartan, D. Carsson, and D. M. Wiles, *Appl. Spectosc.* **35**, 432 (1981).

15. H. Ishida and C. Scott, *J. Polym. Eng.* **6**, 201 (1986).

16. H. W. Seisler and K. Holland-Moritz, *Infra Red and Raman Spectroscopy of Polymers*, Marcel Dekker, Inc., New York, 1980.

17. Y. S. Yang and L. J. Lee, *Macromolecules*, **20**, 1490 (1987).

18. Y. Yang and co-workers, *SPE ANTEC*, **34**, 1062 (1988).

19. A. Bickel, *Am. Lab.* **22**(5), 94 (Oct. 1990).

20. R. Harvey, *SPE ANTEC*, **34**, 1035 (1988).

21. M. G. Hansen, *SPE ANTEC*, **37**, 840 (1991).

22. G. C. Pandey, *Analyst*, **112**(2), 231, 232 (1989).

23. F. Eng and C. Shebih, *SPE ANTEC*, **35**, 1174 (1989).

24. R. Samha and L. A. Wall, in P. H. Emmet, *Catalysis*, Vol. 6, Reinhold Publishing Corp., New York, 1958.

25. J. R. Parrish, *Anal. Chem.* **45**, 1973 (1659).

26. Ref. 10, p. 73.

27. S. K. Mukherjee, G. D. Guenther, and A. K. Bhattacharya, *Anal. Chem.* **50**, 1591 (1978).

28. D. I. Bower, *Structure and Properties of Oriented Polymers*, Applied Sciences, London, U.K., 1975.

29. P. J. Hendra and H. A. Willis, *Chem. Ind.* 2146 (1967); *Chem. Commun.* **4**, 225 (Feb. 1968).

30. G. R. Strobl and W. Hagedorn, *J. Polym. Sci. Polym. Phys. Ed.* **16**(7), 1181 (1978).

31. J. Maxfield, R. S. Stein, and M. C. Chen, *J. Polym. Sci. Polym. Phys. Ed.* **16**(1), 37 (1978).

32. J. Purvis and D. I. Bower, *J. Polym. Sci. Polym. Phys. Ed.* **14**(8), 1461 (1976).

33. H. Buijs and co-workers, *Amer. Lab.* **21**(2), 62–69 (1989).

34. Ref. 1, p. 730.

35. J. C. Randall, *J. Appl. Polym. Sci.* **22**, 585 (1978).

36. F. A. Bovey and co-workers, *Macromolecules*, **9**, 76 (1976).

37. G. N. Foster, *Polymer Preprints*, **20**, 463 (1979).

38. Ref. 35, p. 587.

39. C. J. Carmen, R. A. Harrington, and C. E. Wilkes, *Macromolecules*, **10**, 536 (1977).

40. G. J. Ray, P. E. Johnson, and J. R. Knox, *Macromolecules*, **10**, 773 (1977).

41. J. C. Randall, *Macromolecules*, **11**, 33 (1978).

42. Y. Inoue, A. Nishioka, and R. Chujo, *Makromol. Chem.* **156**, 207 (1972).

43. F. Heatley and F. A. Bovey, *Macromolecules*, **2**, 241 (1969).

44. L. Cavalli and co-workers, *Polym. Sci.* Part A-1 **8**, 801 (1970).

45. C. E. Wilkes, J. C. Westfahl, and R. H. Bachderf, *J. Polym. Sci. Part A-1* **7**, 23 (1969).

46. R. Smith and C. Smith, *SPE ANTEC*, **34**, 893 (1988).

47. A. Zambrini, *Pitture Vernici*, **54**, 169 (1978).

48. F. Mozayeni, *Appl. Spectrosc.* **33**, 520 (1979).

49. C. Caze and C. J. Loucheux, *Macromol. Sci.* **A**12, 1501 (1979).

50. K. Plochokca and H. J. Harwood, *ACS Div. Polym. Chem. Pap.* **19**, 240 (1978).
51. Ref. 1, p. 729.
52. D. Stefan and H. L. Williams, *J. Appl. Polym. Sci.* **18**, 1279 (1974).
53. Ref. 52, p. 1451.
54. F. W. Yeager and J. W. Becker, *Anal. Chem.* **49**, 722 (1977).
55. R. G. Brown and J. E. J. Glass, *Appl. Polymer Sci.* **36**(8), 1909–1923 (1988).
56. V. G. Beuzhkin, *CRC Crit. Revs. Anal. Chem.* **11**, 11 (1981).
57. Y. Sigmura, T. Nagaya, and T. Tsuge, *Macromolecules*, **14**, 520 (1981).
58. S. T. Lai and D. C. Locke, *J. Chromatog.* **314**, 283 (1984).
59. M. Toader and co-workers, *Mate. Plast. (Bucharest)*, **10**(3), 151 (1973).
60. K. G. Haeusler, E. Schroeder, and P. Muehling, *Plaste Kautsch*, **24**(8), 55 (1977).
61. T. Shimono, M. Tanaka, and T. Shoro, *Anal. Chem. Acta.* **96**, 359 (1978).
62. M. M. O'Mara, *J. Polym. Sci.* Part A-1, **8** 1887 (1970).
63. J. G. Cobler and E. P. Samsel, *SPE Transactions* **2**(2), 145 (Apr. 1962).
64. Z. Xu and co-workers, *Sepu.* **6**(4), 205–209 (1988).
65. S. Lai and J. Shen, *SPE ANTEC* **34**, 1258 (Apr. 1988).
66. J. Hay and co-workers, *Polym. Comm.* **26**(6), 175 (1984).
67. D. Rueda and co-workers, *Calliga. Polym. Comm.* **29**(9), 258 (1983).
68. N. T. Wakelyn, *J. Polym. Sci. Polym. Lett.* **25**, 25 (1987).
69. D. P. Anderson, *SPE ANTEC*, **35**, 1117 (1989).
70. P. Cebe, L. Lowry, and S. Chung, *SPE ANTEC*, **35**, 1120 (1989).
71. O. Krathy, in O. Glatter, O. Krathy, eds., *Small-Angle X-Ray Scattering*, Academic Press, Inc., New York, 1982.
72. J. Barnes and F. Mopsik, *SPE ANTEC*, **34**, 1178 (1988).
73. A. M. Jones, T. P. Russell, and D. Y. Yoon, *Proceedings of Polymer Material Science & Engineering*, San Diego, Calif., 1993, p. 394.
74. P. G. deGennes, *Scaling Concepts in Polymer Physics*, Cornell University Press, Ithaca, New York, 1979.
75. R. W. Richards and J. L. Thomason, *Macromolecules*, **16**, 982 (1983).
76. W. L. Wu, *SPE ANTEC*, **35**, 1132 (1989).
77. Ref. 73, p. 396.
78. X. Xu, V. Galiatsatos, and D. Schuele, in Ref. 73, pp. 313, 314.
79. T. Dumelow, S. R. Holding, and L. J. Maisey, *Polymer*, **24**, 307 (1983).
80. R. P. Brown, ed., *Handbook of Plastics Test Methods*, 3rd ed., John Wiley & Sons, Inc., New York, 1988, p. 84.
81. W. C. Taylor and L. H. Tung, presented at 140th Meeting of the American Chemical Society, Chicago, Ill., 1961.
82. L. W. Gamble, W. T. Nipke, and T. L. Lane, *J. Appl. Polym. Sci.* **9**, 1503 (1965).
83. N. C. Ford, in R. Pecora, ed., *Dynamic Light Scattering*, Plenum Publishing Corp., New York, 1985.
84. R. Russ, K. Guo, and L. Delong, *SPE ANTEC*, **34**, 983 (1988).
85. V. Shah, *Handbook of Plastics Testing Technology*, John Wiley & Sons, Inc., New York, 1984, p. 179.
86. P. Starch and P. Kantla, *Kem-Kemi*, **3**(2), 100 (1976).
87. J. Choi, *SPE ANTEC*, **34**, 1013 (1988).
88. D. Kranz, H. U. Pohl, and H. Baumann, *Angew. Makromol. Chem.* **26**, 67 (1972).
89. H. Matsuda and co-workers, *Kobenshi Kagahu*, **29**(6), 362 (1972).
90. J. M. Guenet and co-workers, *J. Appl. Polym. Sci.* **21**(8), 2181 (1977).
91. S. Mori, *Anal. Chem.* **61**, 1321–1325 (1989).
92. J. Janca and M. Kolinsky, *Plasty Kauc.* **13**(5), 138 (1976).
93. M. R. Ambler, R. D. Mate, and J. R. Purdon, *J. Polym. Sci.* **12**, 1759 (1974).
94. H. J. Cortes and co-workers, *Anal. Chem.* **61**, 961–965 (1989).

95. J. L. Viony and J. Lesec, in A. Abe, ed., *Advance in Polymer Science*, Springer-Verlag, Berlin, 1994.
96. L. Wold and co-workers, *J. Polym. Sci. Polym. Phys.* **20**, 441 (1982).
97. T. Kamiya and co-workers, *SPE ANTEC*, **36**, 871 (1990).
98. D. H. Parikh, B. S. Childress, and G. W. Knight, *SPE ANTEC*, **37**, 1543 (1991).
99. A. R. McGhie, in R. G. Linford, ed., *Thermal Analysis Techniques in Electrochemical Science and Technology of Polymers-2*, Elsevier Publishing Co., Inc., New York, 1990, p. 202.
100. J. Chiu, *Appl. Poly. Symp.* **2**, 25 (1966).
101. J. J. Maurer, in E. Turi, ed., *Thermal Characterization of Polymeric Materials*, Academic Press, Inc., New York, 1981, p. 585.
102. H. E. Bair, in Ref. 101, p. 845.
103. W. P. Brennan, *Thermochim. Acta.* **18**, 101 (1977).
104. T. R. Manley, *Prog. Rubber Plast. Technol.* **5**(4), 256, 257 (1989).
105. M. P. Sepe, in N. P. Cheremisinoff, ed., *Elastomer Technology Handbook*, CRC Press, Inc., Boca Raton, 1993, p. 139.
106. *Ibid.*, p. 148.
107. J. M. Criado, *Thermochim. Acta.* **28**, 307 (1979).
108. Rabeck, Wiley-Interscience, New York, 1983.
109. M. Dole, *J. Polymer Sci. Part C-18*, 57 (1967).
110. B. Wunderlich and co-workers, *J. Macromol. Sci.* **B**1, 485 (1967).
111. F. Hamadas and co-workers, *J. Phys. Chem.* **72**, 178 (1968).
112. H. Ludwig and P. Eyerer, *SPE ANTEC*, **32**, 665 (1986).
113. P. Cebe, *SPE ANTEC*, **35**, 1413 (1989).
114. J. C. Seferis, C. Ahlstrom, and S. H. Dillman, *SPE ANTEC*, **33**, 1467 (1987).
115. J. H. Hou and J. M. Bai, *SPE ANTEC*, **33**, 946 (1987).
116. B. Wunderlich, in Ref. 101, Chapts. 2, 3.
117. P. Soni, *SPE Seminars*, Chicago, Ill., June, 1991.
118. L. C. E. Struik, *Physical Aging in Amorphous Polymers and Other Materials*, Elsevier, Amsterdam, the Netherlands, 1978.
119. A. R. McGhie, in R. Linford, ed., *Electrochemical Science and Technology of Polymers-2*, Elsevier Applied Science Publishers, Ltd., London, 1990, p. 214.
120. R. B. Prime, in Ref. 101, Chapt. 5.
121. T. R. Manley, in J. Chiu, ed., *Thermal Methods of Analysis*, Marcel Dekker, Inc., New York, 1974, p. 53.
122. A. P. Grey, technical data, TAAS-2, Perkin Elmer, Norwalk, Conn., 1972.
123. R. A. Fava, *Polymer*, **9**, 137 (1968).
124. V. M. Gonzalez-Romero and N. Casillas, *SPE ANTEC*, **33**, 1119 (1987).
125. J. J. Maurer, in Ref. 101, Chapt. 6.
126. S. W. Shalaby and H. E. Bair, in Ref. 101, Chapt. 4.
127. H. E. Bair, *Polym. Eng. Sci.* **14**, 202 (1974).
128. W. M. Prest and R. S. Porter, *J. Polym. Sci.* **10**, 1639 (1972).
129. H. E. Bair, *Anal. Calorim.* **2**, 51 (1979).
130. T. G. Fox, *Bull. Am. Phys. Soc.* **1**(2), 123 (1956).
131. M. Gordon and J. S. Taylor, *J. Appl. Chem.* **2**, 493 (1952).
132. P. R. Couchman, *Macromolecules*, **11**, 1156 (1978).
133. Ref. 105, p. 157.
134. Ref. 105, p. 165.
135. G. J. Kettle, *Polymer*, **18**, 742 (1977).
136. Technical data, R. W. Thomas and M. W. Cadwallader, TA Instruments Hotline, Wilmington, Del., June, 1990.
137. Ref. 119, p. 219.

138. Ref. 119, p. 218.
139. C. Rohm and P. Herh, *SPE ANTEC*, **34**, 1135 (1988).
140. P. S. Chua, *SPI Composites Institute 42nd Annual Conference*, Session 21A, Cincinnati, Ohio, 1987.
141. C. F. Zorowski and T. Murayama, *Proceedings of 1st International Conference on Mechanical Behavior of Materials*, Vol. 5, Society of Material Scientists, Kyoto, Japan, 1972, p. 28.
142. D. D. Edie and co-workers, *SPE ANTEC*, **37**, 2248 (1991).
143. J. D. Ferry, *Viscoelastic Properties of Polymers*, 3rd ed., John Wiley & Sons, Inc., New York, 1980.
144. L. H. Sperling, *Introduction to Physical Polymer Science*, John Wiley & Sons, Inc., New York, 1986, p. 384.
145. Ref. 119, p. 266.
146. C. L. Rohn, *SPE ANTEC*, **35**, 870 (1989).
147. Ref. 105, p. 204.
148. W. Y. Chiang and D. S. Hwung, *SPE ANTEC*, **32**, 492 (1986).
149. C. D. Han and co-workers, *Macromolecules*, **22**(8), 3443–3451 (1989).
150. D. K. Yoshimura and W. D. Richards, *SPE ANTEC*, **32** 688 (1986).
151. J. Sahai, K. Nakamura, and S. Inoue, *SPE ANTEC*, **36**, 1912 (1990).
152. Ref. 105, p. 220.
153. L. Salmen and co-workers, *Composite Systems from Natural and Synthetic Polymers*, Elsevier, Amsterdam, the Netherlands, 1986.
154. W. J. Sichina and P. S. Gill, *SPE ANTEC*, **33**, 959 (1987).
155. R. J. Hawkins, *Macromolecules*, **9**, 191 (1976).
156. P. J. Phillips, W. S. Lambert, and H. D. Thomas, *SPE ANTEC*, **37**, 1575 (1991).
157. Ref. 119, p. 268.
158. T. R. Manley, J. A. Stonebanks, and P. Laggon, *Polym. Comm.*, **28**, 17 (1988).
159. D. R. Day, D. D. Sheppard, and K. J. Craven, *SPE ANTEC*, **36**, 1045 (1990).
160. A. Bernes and C. Locahanne, in S. E. Keitwath, ed., *Order in the Amorphous State of Polymers*, Plenum Publishing Corp., New York, 1987, p. 305.
161. P. Cebeillac and co-workers, *SPE ANTEC*, **35**, 1855 (1989).
162. J. P. Ibar and co-workers, *SPE ANTEC*, **37**, 1655 (1991).
163. J. P. Ibar, P. Denning, and C. DeGoys, *SPE ANTEC*, **36**, 866 (1990).
164. D. R. Lloyd, T. C. Ward, H. Schrieber, eds., *Inverse Gas Chromatography: Characterization of Polymers and Other Materials*, American Chemical Society, Washington, D.C., 1989.
165. G. DiPaola and co-workers, in Ref. 164, p. 415.
166. *ASTM D1525, Vicat Softening Temperature of Plastics*, Vol. 8.01, ASTM, Philadelphia, Pa., 1991.
167. *ISO 306, Plastics: Thermoplastic Materials-Determination of Vicat Softening Temperature*, ISO, Geneva, Switzerland, 1987.
168. D. Crofts, *RAPRA Research Report*, No. 200, Rubber and Plastics Research Association, Shrewsburg, U.K., 1972.
169. *ASTM D648, Deflection Temperature of Plastics Under Flexural Load*, Vol. 8.01, ASTM, Philadelphia, Pa., 1982 (reapproved 1988).
170. *ISO 75, Plastics and Ebonite- Determination of Temperature of Deflection Under Load*, ISO, Geneva, Switzerland, 1987.
171. *ASTM D1238, a, Determination of the Flow Rate of Thermoplastics Using an Extrusion Plastomer*, Vol. 8.01, ASTM, Philadelphia, Pa., 1993.
172. G. H. Pearson, in M. D. Baijal, ed., *Plastics Polymer Science and Technology*, John Wiley & Sons, Inc., New York, 1982.
173. J. A. Brydson, *Flow Properties of Polymer Melts*, 2nd ed., Godwin/Plastics and Rubber Institute, 1981.

174. K. F. Wissbrun, *Polymer News*, **4**(2), 55 (1977).

175. K. H. Moos, *Kunststoffe*, **75**(1), 3 (1985).

176. H. Takahashi, T. Matsuoka, and T. Kurauchi, *J. of Appl. Poly. Sci.* **30**(12), 4669 (1985).

177. Technical data, Dow Chemical Co., Midland, Mich., 1993.

178. J. Matthan, *RAPRA Research Report*, No. 165, Rubber and Plastics Research Association, Shrewsburg, U.K., 1968.

179. K. T. Paul, *RAPRA Bulletin*, Rubber and Plastics Research Association, Shrewsburg, U.K., Feb. 1972.

180. K. T. Paul, *RAPRA Mem. J.* (Nov. 1973).

181. E. O. Allen and R. F. Williams, *SPE ANTEC* **17**, 587 (1971).

182. R. L. Ballam and J. J. Brown, *Capillary Rheometer*, Instron Engineering Co., Canton, Mass.

183. *ASTM D3835, Determination of Properties of Thermoplastic Materials by Capillary Rheometer*, Vol. 8.02, ASTM, Philadelphia, Pa., 1994.

184. B. J. Furches and G. Kachin, *SPE ANTEC*, **35**, 1663 (1989).

185. R. J. Groleau, *SPE ANTEC*, **35**, 1183 (1989).

186. *ASTM D5422, Determination of Properties of Thermoplastic Materials by Screw Extrusion Capillary Rheometer*, Vol. 8.03, ASTM, Philadelphia, Pa., 1993.

187. *ASTM D731, Test Method for Molding Index of Thermosetting Molding Powder*, Vol. 8.01, ASTM, Philadelphia, Pa., 1993.

188. *ASTM D3123, Thermosetting Molding Compounds, Spiral Flow, Low Pressure*, Vol. 8.02, ASTM, Philadelphia, Pa., 1994.

189. *ASTM D3795, Test Method for Thermal Flow/Cure Properties of Thermosetting Plastics by Torque Rheometer*, Vol. 8.02, ASTM, Philadelphia, Pa., 1993.

190. Ref. 86, pp. 175–178.

191. MOLDFLOW, Moldflow Australia, Victoria, Australia.

192. C-MOLD, AC Technology, Ithaca, N.Y.; IDEAS, Structural Dynamics Research Corp., Cincinnati, Ohio.

193. *ASTM D638, Test Method for Tensile Properties of Plastics*, Vol. 8.01, ASTM, Philadelphia, Pa., 1991.

194. *ISO 527, Plastics: Determination of Tensile Properties*, ISO, Geneva, Switzerland, 1991.

195. R. J. Crawford and Y. Yigsaw, *J. Mat. Sci. Lett.* **3**(2), 171–176 (1984).

196. *ASTM D256, a, Test Methods for Impact Resistance of Plastics and Electrical Insulating Materials*, Vol. 8.01, ASTM, Philadelphia, Pa., 1993.

197. *ASTM D3029, Test Method for Impact Resistance of Rigid Plastic Sheeting or Parts by Means of a Tup (Falling Weight)*, Vol. 8.02, ASTM, Philadelphia, Pa., 1993.

198. *ASTM D3763, Test Method for High-Speed Puncture Properties of Plastics Using Load and Displacement Sensors*, Vol. 8.02, ASTM, Philadelphia, Pa., 1993.

199. E. C. Szamborski and R. J. Hutt, *SPE ANTEC*, **42**, 883–889 (1984).

200. D. R. Ireland, in R. E. Evans, ed., *Physical Testing of Plastics-Correlation with End-Use Performance*, ASTM STP 736, ASTM, Philadelphia, Pa., 1981, pp. 45–58.

201. Ref. 86, pp. 68, 69.

202. *ASTM D2990, a, Test Methods for Tensile, Compressive and Flexural Creep and Creep Rupture of Plastics*, Vol. 8.02, ASTM, Philadelphia, Pa., 1993.

203. *ISO 899 Determination of Tensile Creep of Plastic*, ISO, Geneva, Switzerland, 1981.

204. A. V. Shenoy and D. R. Saini, *Polym. Test.* **6**(1), 37–45 (1986).

205. *ASTM D671, Test Method for Flexural Fatigue of Plastics by Constant Amplitude of Force*, Vol. 8.01, ASTM, Philadelphia, Pa., 1993.

206. M. N. Riddell, G. P. Koo, and J. L. O'Toole, *Polym. Eng. Sci.* **6**, 363 (1969).

207. I. Constable, J. G. Williams, and D. J. Burns, *J. Mech. Eng. Sci.* **12**, 20 (1970).

208. S. Turner, *Mechanical Testing of Plastics*, 2nd ed., Longman Inc., New York, 1983, p. 170.
209. K. Boller, *Modern Plastics*, **41**(10), 145 (1964).
210. M. J. Owen, *J. Plast. Inst.* **35**, 353 (1967).
211. Ref. 205, p. 168.
212. R. W. Hertzberg and J. A. Manson, *Fatigue of Engineering Plastics*, Academic Press, Inc., New York, 1980.
213. K. V. Gotham, *Plast. Polym.* **37**(130), 309 (1969).
214. Ref. 205, p. 174.
215. K. Kendall, W. J. Clegg, and R. D. Gregory, *J. Mat. Sci. Lett.* **10**(11), 671–682 (1991).
216. G. Medri and R. Ricci, *Plast. Rubb. Comp. Process. Appln.* **15**(1), 47–52 (1991).
217. M. B. Jamarani, P. E. Reed, and W. R. Davies, *J. Mat. Sci.*, **23**(12), 4437–4444 (1988).
218. R. Frassine and co-workers, *J. Mat. Sci.* **23**(11), 4027–4036 (1988).
219. P. L. Fernando, *Polym. Engng. Sci.* **28**(2), 806–814 (1988).
220. R. Greco and co-workers, *Makromol. Chem., Rapid. Commun.* **9**(2), 91–95 (1988).
221. G. P. Marshall and M. W. Birch, *Proceedings of the 6th International Conference*, York, U.K., 1988, pp. 18.1–18.6.
222. R. J. Lee and D. C. Phillips, in T. Feest, ed., *Proceedings of the International Conference on Testing, Evaluation and Quality Control of Composites*, Guildford, U.K., 1983, pp. 12–26.
223. A. A. Griffiths, *Philosophical Transactions of the Royal Society*, A221, Royal Society of London, 1920, p. 163.
224. S. W. Hawley, *RAPRA Review Report* **5**(12), 14 (1992).
225. *ASTM D5045, Test Method for Plane-Strain Fracture Toughness and Strain Energy Release Rate of Plastic Materials*, Vol. 8.03, ASTM, Philadelphia, Pa., 1993.
226. J. G. Williams, *Prog. Rubber. Plast. Technol.* **6**(2), 174–200 (1990).
227. T. Watson and co-workers, *J. Mat. Sci.* **22**(4), 1249–1258 (1987).
228. R. M. Pilliar, R. Vowles, and D. F. Williams, *J. Biomed. Mat. Res.* **21**(1), 145–154 (1987).
229. C. T. Wang and R. M. Pilliar, *J. Mater. Sci.* **24**(7), 2391–2400 (1989).
230. G. C. Adams and co-workers, *Polym. Engng. Sci.* **30**(4), 24–28 (1990).
231. D. B. Barry and O. Delatycki, *J. Appl. Polym. Sci.* **38**(2), 339–350 (1989).
232. M. Kitagawa and T. Matsutani, *J. Mat. Sci.* **23**(11), 4085–4090 (1988).
233. Y.-W. Mai and P. Powell, *J. Polym. Sci. Polym. Phys.* **29**(7), 785–793 (1991).
234. K. Matsushige, in E. Baer and A. Moet, eds., *High Performance Polymers*, Hanser Publishers, Munich, Germany, 1990, p. 104.
235. A. Peterlin, *Adv. Dhem. Ser.* **174**, 651 (1979).
236. T. Nishiura and co-workers, *Polymer J.* **13**(89), 611 (1983).
237. D. Betteridge and co-workers, *Polymer*, **23**, 178 (1982).
238. K. Matsushige and co-workers, *Rept. Prog. Polym. Phys. Jpn.* **26**, 343 (1983).
239. K. Matsushige, Y. Sakurada, and K. Takahashi, *J. Mater. Sci.* **19**, 1548 (1984).
240. Ref. 228, p. 105.
241. Ref. 228, p. 113.
242. Ref. 228, pp. 114–116.
243. *ISO 2578, Determination of Time-Temperature Limits after Exposure to Prolonged Action of Heat*, ISO, Geneva, Switzerland, 1974.
244. *ASTM D696, Test Method for Coefficient of Linear Thermal Expansion of Plastics*, Vol. 8.01, ASTM, Philadelphia, Pa., 1991.
245. *ASTM D955, Test Method for Measuring Shrinkage from Mold Dimensions of Molded Plastics*, Vol. 8.01, ASTM, Philadelphia, Pa., 1989.
246. *ASTM D1299, Test Method for Shrinkage of Molded and Laminated Thermosetting Plastics at Elevated Temperatures*, Vol. 8.01, ASTM, Philadelphia, Pa., 1979.

247. *ASTM D2838, Test Method for Shrink Tension and Orientation Release Stress of Plastic Film and Thin Sheeting,* Vol. 8.02, ASTM, Philadelphia, Pa., 1989.

248. *ISO 1218, Polyamides-Determination of Melting Point,* ISO, Geneva, Switzerland, 1987.

249. *ISO 306, Plastics: Thermoplastic materials: Determination of the Vicat Softening Temperature,* ISO, Geneva, Switzerland, 1987.

250. *ASTM D1525, Test Method for Vicat Softening Temperature of Plastics,* Vol. 8.01, ASTM, Philadelphia, Pa., 1991.

251. *ISO 75, Plastic and Ebonite: Determination of Temperature of Deflection under Load,* ISO, Geneva, Switzerland, 1992.

252. *ASTM D648, ed, Test Method for Deflection Temperature of Plastics under Flexural Load,* Vol. 8.01, ASTM, Philadelphia, Pa., 1988.

253. *ASTM D543, Test Method for Resistance of Plastics to Chemical Reagents,* Vol. 8.01, ASTM, Philadelphia, Pa., 1989.

254. *ISO 175, Determination of the Effects of Liquid Chemicals, Including Water,* ISO, Geneva, Switzerland, 1992.

255. *ASTM D1693, ed, Test Method for Environmental Stress-Cracking of Ethylene Plastics,* Vol. 8.01, ASTM, Philadelphia, Pa., 1988.

256. R. J. Roe and C. Gievewski, *Polym. Eng. Sci.* **15**(6), 197 (1975).

257. H. Steinke and H. Pflasterer, *Kautschuk Gummi Kunststoffe,* **20**(9), 516 (1967).

258. A. L. Ward, X. Lu, and N. Brown, *Polym. Eng. Sci.* **30**(18), 1175–1179 (1990).

259. K. Takahashi, A. E. Abo-El-Ezz, and N. Takeda, *Proceedings of the 7th International Conference,* Cambridge, U.K., 1988, pp. 29/1–29/4.

260. *ISO 4599, Plastics: Determination of Resistance to Environmental Stress-Cracking, Bent Strip Method,* ISO, Geneva, Switzerland, 1986.

261. *ISO 4600, Plastics: Determination of Environmental Stress-Cracking, Ball or pin Impression Method,* ISO, Geneva, Switzerland, 1981.

262. *ISO 6252, Determination of Environmental Stress-Cracking: Constant Tensile,* ISO, Geneva, Switzerland, 1981.

263. Ref. 80, p. 354.

264. Ref. 253, p. 198.

265. *ISO 877, Determination of Resistance to Change upon Exposure under Glass to Daylight,* ISO, Geneva, Switzerland, 1976.

266. *ASTM G24, Practice for Conducting Natural Light Exposure under Glass,* Vol. 14.02, ASTM, Philadelphia, Pa., 1987.

267. *ISO 4607, Methods of Exposure to Natural Weathering,* ISO, Geneva, Switzerland, 1978.

268. *ASTM D1435, Recommended Practice for Outdoor Weathering of Plastics,* Vol. 8.01, ASTM, Philadelphia, Pa., 1985.

269. D. Kiessling and W. Schaaf, *Plast. u. Kaut.* **33**(8), 310–313 (1986).

270. T. Kobayashi and co-workers, *Kobunshi Ronbun,* **42**(6), 405–413 (1985).

271. M. L. Ellinger, *Paint Techn.* **27**(12), 40 (1963).

272. H. I. Garner and P. J. Papillo, *Proceedings of Division of Organization on Coatings and Plastics Chemistry,* Atlantic City, N.J., 1962, p. 110.

273. C. R. Caryl, *SPE J.* **23**(1), 49 (1976).

274. *ASTM D4364, Practice for Performing Accelerated Outdoor Weathering of Plastics Using Concentrated Natural Sunlight,* Vol. 8.03, ASTM, Philadelphia, Pa., 1984.

275. H. R. Kamal, *Coloring of Plastics 6,* RETEC, Philadelphia, Pa., 1972, p. 4.

276. G. M. Ruhnke and L. F. Biritz, *Plast. Polym.* **40**(147), 118 (1972).

277. *ASTM D1499, a, Recommended Practice for Operating Light- and Water-Exposure Apparatus (Carbon Arc Type) for Exposure of Plastics,* Vol. 8.01, ASTM, Philadelphia, Pa., 1992.

278. *ASTM D2565, a, Practice for Operating Xenon Arc-Type Light Exposure Apparatus with and without Water for Exposure of Plastics*, Vol. 8.02, ASTM, Philadelphia, Pa., 1992.

279. *ASTM D4329, Practice for Operating Light- and Water-Exposure Apparatus (Fluorescent uv–Condensation Type) for Exposure of Plastics*, Vol. 8.03, ASTM, Philadelphia, Pa., 1992.

280. *ISO 846, Behavior under the Action of Fungi and Bacteria*, ISO, Geneva, Switzerland, 1978.

281. *ASTM G21, Recommended Practice for Determining Resistance of Synthetic Polymeric Materials to Fungi*, Vol. 14.02, ASTM, Philadelphia, Pa., 1990.

282. *ASTM G22, Recommended Practice for Determining Resistance of Plastics to Bacteria*, Vol. 14.02, ASTM, Philadelphia, Pa., 1990.

283. L. A. Wienert and M. W. Hillard, *Plast. Tech.* **23**(1), 75 (1977).

284. *ASTM E1027, Practice for Exposure of Polymeric Materials to Ionizing Radiation*, Vol. 12.02, ASTM, Philadelphia, Pa., 1984.

285. *ASTM D1044, Test Method for Resistance of Transparent Plastics to Surface Abrasion*, Vol. 8.01, ASTM, Philadelphia, Pa., 1993.

286. Ref. 80, pp. 271, 272.

287. *ASTM D1729, Practice for Visual Evaluation of Color Differences of Opaque Materials*, Vol. 6.01, ASTM, Philadelphia, Pa., 1989.

288. *ASTM D2244, Method for Calculation of Color Differences from Instrumentally Measured Color Coordinates*, Vol. 6.01, ASTM, Philadelphia, Pa., 1989.

289. *ASTM D523, Test Method for Specular Gloss*, Vol. 8.01, ASTM, Philadelphia, Pa., 1989.

290. *ASTM D2457, Test Method for Specular Gloss of Plastic Films*, Vol. 8.01, ASTM, Philadelphia, Pa. 1990.

291. *ASTM D149, Test Methods for Dielectric Breakdown Voltage and Dielectric Strength of Solid Electrical Insulating Materials at Commercial Power Frequencies*, Vol. 8.01, ASTM, Philadelphia, Pa., 1987.

292. Ref. 86, p. 116.

293. *ASTM D257, Test Methods for D-C Resistance or Conductance of Insulating Materials*, Vol. 8.01, ASTM, Philadelphia, Pa., 1990.

294. *ASTM D495, Test Method for High-Voltage, Low-Current, Dry Arc Resistance of Solid Electrical Insulation*, Vol. 8.01, ASTM, Philadelphia, Pa., 1989.

295. *ASTM D2132, Test Method for Dust-and-Fog Tracking and Erosion Resistance of Electrical Insulating Materials*, Vol. 10.02, ASTM, Philadelphia, Pa., 1989.

296. *ASTM D2302, Test Method for Liquid-Contaminant, Inclined-Plane Tracking and Erosion of Insulating Materials*, Vol. 10.02, ASTM, Philadelphia, Pa., 1985.

297. R. E. Evans, in M. D. Baijal, ed., *Plastics Polymer Science and Technology*, John Wiley & Sons, Inc., New York, 1982.

298. *ASTM D3638, Test Method for Comparative Tracking Index of Electrical Insulating Materials*, Vol. 10.02, ASTM, Philadelphia, Pa., 1985.

299. Ref. 80, p. 258.

300. R. M. Simon, *Mod. Plast. Int.* **13**(9), 124 (1983).

301. D. M. Bigg, W. Mirick, and D. E. Stutz, *Polym. Testing*, **5**(3), 169 (1985).

302. UL 746 A-D, Underwriters Laboratories, Chicago, Ill.

303. *ASTM D3814, Guide for Locating Combustion Test Methods*, Vol. 8.02, ASTM, Philadelphia, Pa., 1993.

304. *ASTM E176, Terminology Relating to Fire Standards*, Vol. 4.07, ASTM, Philadelphia, Pa., 1990.

305. *ISO 871, Temperature of Evolution of Flammable Gases (Decomposition Temperature) from a Small Sample of Pulverized Material*, ISO, Geneva, Switzerland, 1994.

306. *ASTM D1929, Test Method for Ignition Properties of Plastics*, Vol. 8.01, ASTM, Philadelphia, Pa., 1991.
307. *ISO 4589, Plastics: Determination of the Oxygen Index*, ISO, Geneva, Switzerland, 1994.
308. *ASTM D2863, Test Method for Measuring the Minimum Oxygen Concentration to Support Candle-Like Combustion of Plastics (Oxygen Index)*, Vol. 8.02, ASTM, Philadelphia, Pa., 1991.
309. *ASTM D635, Test Method for Rate of Burning and/or Extent and Time of Burning of Self-Supporting Plastics in a Horizontal Position*, Vol. 8.01, ASTM, Philadelphia, Pa., 1991.
310. *ASTM D568, Test Method for Rate of Burning and/or Extent and Time of Burning of Flexible Plastics in a Vertical Position*, Vol. 8.01, ASTM, Philadelphia, Pa., 1985.
311. *ASTM D3801, Test Method for Measuring the Comparative Extinguishing Characteristics of Solid Plastics in a Vertical Position*, Vol. 8.02, ASTM, Philadelphia, Pa., 1987.
312. UL 94, Underwriters Laboratories, Chicago, Ill.
313. A. Tewarson and R. F. Pion, *Combust. Flame*, **26**, 85 (1976).
314. V. Brabraushas, NBSIR 82-2611, U.S. Bureau of Commerce, National Bureau of Standards (now NIST), Washington, D.C., 1982.
315. N. Batho, personal communication, Dark Star Research, Penley, Clwyd, U.K., Mar. 1992.
316. *ASTM D2843, Test Method for Density of Smoke from the Burning or Decomposition of Plastics*, Vol. 8.02, ASTM, Philadelphia, Pa., 1993.
317. *ASTM E162, Test Method for Surface Flammability of Materials Using a Radiant Heat Energy Source*, Vol. 4.07, ASTM, Philadelphia, Pa., 1987.
318. *ASTM D4100, ed, Test Method for Gravimetric Determination of Smoke Particulates from Combustion of Plastic Materials*, Vol. 8.02, ASTM, Philadelphia, Pa., 1989.
319. W. N. Reynolds, *RAPRA Rev. Rep.* **3**(2), 30 (1990).
320. Ref. 321, p. 30/7.
321. Ref. 321, p. 30/12.
322. J. M. Winter and R. W. Green, *Proceedings of the 11th World Conference on NDT*, Las Vegas, Nev., 1985, p. 779.
323. Y. Bar-Cohen, C. C. Yin, and A. K. Mal, *J. Adhesion*, **29**(1/4), 257–274 (1989).
324. J. Summerscales, ed., *Nondestructive Testing of Fibre-Reinforced Plastic Composites*, Vol. 1, Elsevier Applied Science Publishers, Ltd., Barking, U.K., 1987, pp. 278, 627–629.
325. R. Pepper and R. Samuels, *SPE ANTEC*, **46**, 884–888 (1988).
326. W. J. Beranek, *Proceedings of the European Workshop on NDE of Polymer and Polymer Matrix Composition*, Termar do Vimeiro, Sept. 1984, pp. 222–236.
327. W. N. Reynolds, *Mat. Des.* **5,6**(D/J), 256–270 (1985).

*General References*

T. R. Crompton, *The Analysis of Plastics*, Pergamon Press Inc., Elmsford, New York, 1986.
C. D. Craver, *Infra Red Spectra of Plasticizers and Other Additives*, 2nd ed., The Coblentz Society, Kirkwood, Mo., 1980.
D. O. Hummel and A. Solti, *Atlas of Polymer and Plastics Analysis*, Vol. 2, VCH Publishers, New York, 1988.
M. P. Sepe, in N. P. Cheremisinoff, ed., *Elastomer Technology Handbook*, CRC Press, Inc., Boca Raton, Fla., 1993, pp. 105–258.
E. Turi, ed., *Thermal Characterization of Polymeric Materials*, Academic Press, Inc., New York, 1981.

R. P. Brown, ed., *Handbook of Plastics Test Methods*, John Wiley & Sons, Inc., New York, 3rd ed., 1988.

V. Shah, *Handbook of Plastics Testing Technology*, John Wiley & Sons, Inc., New York, 1984.

ASTM Volumes 8.01–8.03, *Testing of Plastics*, ASTM, Philadelphia, Pa., published annually.

O. Olakisi, L. M. Robeson, and M. T. Shaw, *Polymer–Polymer Miscibility*, Academic Press, Inc., New York, 1979.

L. H. Sperling, *Introduction to Physical Polymer Science*, Wiley-Interscience, New York, 1986.

L. Sperling and co-workers, *SPE ANTEC*, **34**, 1219 (Apr. 1988).

BARBARA J. FURCHES
The Dow Chemical Company

**PLASTISOLS.**   See VINYL POLYMERS, VINYL CHLORIDE AND POLY(VINYL CHLORIDE).

# PLATINUM-GROUP METALS

The platinum-group metals (PGMs), which consist of six elements in Groups 8–10 (VIII) of the Periodic Table, are often found collectively in nature. They are ruthenium, Ru; rhodium, Rh; and palladium, Pd, atomic numbers 44 to 46, and osmium, Os; iridium, Ir; and platinum, Pt, atomic numbers 76 to 78. Corresponding members of each triad have similar properties, eg, palladium and platinum are both ductile metals and form active catalysts. Rhodium and iridium are both characterized by resistance to oxidation and chemical attack (see PLATINUM-GROUP METALS, COMPOUNDS).

The PGMs are of significant technological importance. They are also extremely rare, owing in part to low natural abundance and in part to the complex processes required for extraction and refining. Unlike gold (see GOLD AND GOLD COMPOUNDS), the majority of which is used for jewelry and investment purposes, the PGMs are used primarily in industrial applications.

One of the earliest recorded uses of platinum was the primitive jewelry articles found in the Esmereldas region of Ecuador (1), which date from several centuries before the Spanish conquest of South America. It is thought that pre-Colombian natives collected platinum nuggets from streams and made primitive articles of platinum or a crude platinum–gold alloy. Archaeological evidence suggests that the jewelry was made by an early form of powder metallurgy, in which small grains of platinum were mixed with gold dust, and heated to build up an homogeneous mixture (see METALLURGY).

Platinum was found in conjunction with gold after the Spanish conquest of South America. It was referred to as platina, or little silver. It was regarded as an unwanted impurity in the silver and gold, and was often discarded. However, scientific interest in platinum gradually grew and in 1741 the first samples of New World platinum were brought to England for scientific examination.

At the start of the nineteenth century, platinum was refined in a scientific manner by William Hyde Wollaston, resulting in the successful production of malleable platinum on a commercial scale. During the course of the analytical work, Wollaston discovered palladium, rhodium, iridium, and osmium. Ruthenium was not discovered until 1844, when work was conducted on the composition of platinum ores from the Ural Mountains.

The records of Faraday, Davy, and those who followed in developing knowledge of electricity show how often platinum was employed to provide a means of carrying, making, and breaking a current. The development of the electric telegraph, the incandescent lamp, and the thermionic valve all involved the use of platinum. Similarly, early internal combustion engines required platinum in the igniter tubes and magnetocontacts. In the late twentieth century, the single most widely developed application of platinum is as a catalyst in a range of chemical processes and petroleum refining, emission control systems, and in fuel cells (qv). Although industrial applications of the other five platinum-group metals came later than those of platinum itself, these too have become widely established in chemical, electrical, and electronic engineering. The most widely used unit of mass for the PGMs is the troy ounce (1 troy oz = 0.0311 kg). However, herein masses are given in metric tons.

## Properties

The relative abundance of the stable isotopes of the PGMs and their CAS Registry Numbers are shown in Table 1.

**Physical and Mechanical Properties.** Whereas there are some similarities in the physical and chemical properties between corresponding members of the PGM triads, eg, platinum and palladium, the PGMs taken as a unit exhibit a wide range of properties (2). Some of the most important are summarized in Table 2.

Ruthenium and osmium have hcp crystal structures. These metals have properties similar to the refractory metals, ie, they are hard, brittle, and have relatively poor oxidation resistance (see REFRACTORIES). Platinum and palladium have fcc structures and properties akin to gold, ie, they are soft, ductile, and have excellent resistance to oxidation and high temperature corrosion.

Hardness of the annealed metals covers a wide range. Rhodium (up to 40%), iridium (up to 30%), and ruthenium (up to 10%) are often used to harden platinum and palladium whose intrinsic hardness and tensile strength are too low for many intended applications. Many of the properties of rhodium and iridium, Group 9 metals, are intermediate between those of Group 8 and Group 10. The mechanical and many other properties of the PGMs depend on the physical form, history, and purity of a particular metal sample. For example, electrodeposited platinum is much harder than wrought metal.

Table 1. CAS Numbers and Relative Abundances of Stable Isotopes of PGMs

| Isotope | CAS Registry Number | Abundance, % |
|---|---|---|
| ruthenium | [7440-18-8] | |
| $^{96}$Ru | [15128-32-2] | 5.52 |
| $^{98}$Ru | [18393-13-0] | 1.88 |
| $^{99}$Ru | [15411-62-8] | 12.7 |
| $^{100}$Ru | [14914-60-4] | 12.6 |
| $^{101}$Ru | [14914-61-5] | 17.0 |
| $^{102}$Ru | [14914-62-6] | 31.6 |
| $^{104}$Ru | [15766-01-5] | 18.7 |
| osmium | [7440-04-2] | |
| $^{184}$Os | [14922-68-0] | 0.02 |
| $^{186}$Os | [13982-09-7] | 1.58 |
| $^{187}$Os | [15766-52-6] | 1.6 |
| $^{188}$Os | [14274-81-8] | 13.3 |
| $^{189}$Os | [15761-06-5] | 16.1 |
| $^{190}$Os | [14274-79-4] | 26.4 |
| $^{192}$Os | [15062-08-5] | 41.0 |
| rhodium-103 | [7440-16-6] | 100 |
| palladium | [7440-05-3] | |
| $^{102}$Pd | [14833-50-2] | 1.02 |
| $^{104}$Pd | [15128-18-4] | 11.14 |
| $^{105}$Pd | [15749-57-2] | 22.23 |
| $^{106}$Pd | [14914-59-1] | 27.33 |
| $^{108}$Pd | [15749-58-3] | 26.46 |
| $^{110}$Pd | [15749-60-7] | 11.72 |
| iridium | [7439-88-5] | |
| $^{191}$Ir | [13967-66-3] | 37.3 |
| $^{193}$Ir | [13967-67-4] | 62.7 |
| platinum | [7440-06-4] | |
| $^{190}$Pt | [15735-68-9] | 0.01 |
| $^{192}$Pt | [14913-85-0] | 0.79 |
| $^{194}$Pt | [14998-96-0] | 32.9 |
| $^{195}$Pt | [14191-88-9] | 33.8 |
| $^{196}$Pt | [14867-61-9] | 25.3 |
| $^{198}$Pt | [15756-63-5] | 7.2 |

**Chemical Properties and Corrosion Resistance.** Among the outstanding characteristics of the PGMs are exceptional resistance to corrosive attack by a wide range of liquid and gaseous substances, and stability at high temperatures under conditions where base and refractory metals are easily oxidized (3). This is owing to thermodynamic stability over a wide range of conditions and the formation of thin protective oxide films in aqueous media under oxidizing or anodic conditions. Electrochemical properties of the PGMs as these affect corrosion resistance are described in Reference 4. The PGMs are often used as sheaths, linings, electrodeposits, or other thin coatings on strong supporting structures. In many cases the strength, rigidity, hardness, and resistance to corrosion of the PGMs can be further improved by alloying, particularly with a second metal of the same group. Alloys of platinum with rhodium (up to 40 wt %), with iridium

**Table 2. Properties of Platinum-Group Metals**

| Parameter | Ruthenium | Rhodium | Palladium | Osmium | Iridium | Platinum |
|---|---|---|---|---|---|---|
| atomic number | 44 | 45 | 46 | 76 | 77 | 78 |
| atomic weight | 101.07 | 102.91 | 106.40 | 190.20 | 192.20 | 195.09 |
| number of stable isotopes | 7 | 1 | 6 | 7 | 2 | 6 |
| elemental abundance, ppm | $1 \times 10^{-3}$ | $1 \times 10^{-3}$ | $5 \times 10^{-3}$ | $1 \times 10^{-3}$ | $1 \times 10^{-3}$ | $1 \times 10^{-2}$ |
| usual valency | 3,4,6,8 | 3 | 2,4 | 4,6,8 | 3,4 | 2,4 |
| ionic radius, pm | 67 | 68 | 65 | 69 | 68 | 65 |
| crystal structure[a] | hcp | fcc | fcc | hcp | fcc | fcc |
| lattice constant, $a$, pm | 270.56 | 380.3 | 389 | 273.41 | 384 | 392.31 |
| color | white | silvery | steel white | bluish | yellowish white | silvery |
| reflectance, % | 63 | 79 | 54 | | 64 | 55 |
| melting point, °C | 2310 | 1960 | 1552 | 3050 | 2443 | 1769 |
| vapor pressure at mp, Pa[b] | 1.31 | 0.133 | 3.47 | 1.8 | 0.467 | 0.0187 |
| density, g/cm$^3$ | 12.45 | 12.41 | 12.02 | 22.61 | 22.65 | 21.45 |
| heat capacity at 25°C, J/(°C·mol)[c] | 24.06 | 24.98 | 25.98 | 24.70 | 24.50 | 25.85 |
| thermal expansion, °C$^{-1} \times 10^6$ | 9.1 | 8.3 | 11.11 | 6.10 | 6.8 | 9.1 |
| magnetic susceptibility, cm$^3$/g | $4.27 \times 10^{-7}$ | $9.9 \times 10^{-7}$ | $5.23 \times 10^{-6}$ | $5.2 \times 10^{-8}$ | $1.33 \times 10^{-7}$ | $9.71 \times 10^{-7}$ |

| Property | | | | | | |
|---|---|---|---|---|---|---|
| work function, eV | | 4.8 | 4.99 | | | 5.27 |
| Young's modulus, kN/m² | $4.85 \times 10^8$ | $3.86 \times 10^8$ | $1.24 \times 10^8$ | $5.56 \times 10^8$ | $5.28 \times 10^8$ | $1.71 \times 10^8$ |
| ultimate tensile strength, MPa[d] | 500–600 | 400–560 | 180–200 | | 400–500 | 120–160 |
| Poisson's ratio | 0.31 | 0.36 | 0.39 | 0.28 | 0.28 | 0.36 |
| Vickers' hardness (VPN) | 200–250 | 100–120 | 40 | 300–670 | 220 | 40 |
| elongation, % | 10 | 6.5 | 20–35 | | | 40 |
| specific strength per unit cost, MPa·cm³/(g·$)[d] | 6.66 | 0.17 | 0.26 | | 0.72 | 0.06 |
| temperature coefficient of resistance (TCR), K⁻¹ | $4.2 \times 10^{-3}$ | $4.6 \times 10^{-3}$ | $3.8 \times 10^{-3}$ | $4.2 \times 10^{-3}$ | $4.3 \times 10^{-3}$ | $3.9 \times 10^{-3}$ |
| electrical resistivity at 0°C, μΩ·cm | 6.8 | 4.33 | 9.93 | 8.12 | 4.71 | 9.85 |
| thermal conductivity, W/(m·K) | 119 | 153 | 75 | 88 | 147 | 73 |

[a]hcp = hexagonal close-packed; fcc = face-centered cubic.
[b]To convert Pa to mm Hg, multiply by 0.075.
[c]To convert J to cal, divide by 4.184.
[d]To convert MPa to psi, multiply by 145.

(up to 30 wt %), and with ruthenium (up to 10 wt %) are characterized by improved corrosion resistance and creep resistance as well as by greater hardness. However, alloys having higher alloying additions are more difficult to fabricate.

The PGMs are extremely resistant to corrosion by aqueous solutions of alkalies and salts and by dilute acids, and are generally quite resistant to more concentrated acids and halogens (Table 3) (5). In concentrated acids at high redox potentials, ie, under oxidizing conditions, there is a zone of corrosion. This accounts for the solubility of platinum in aqua regia. Rhodium is particularly resistant to chemical attack. In bulk form, rhodium is resistant even to aqua regia, although when finely divided this metal is attacked by concentrated sulfuric acid and aqua regia.

Of the nonmetals, phosphorus, arsenic, silicon, sulfur, selenium, tellurium, and carbon attack some or all of the metals at red heat, forming alloys or low melting point phases. Rhodium and iridium are resistant to fused lead oxide, silicates, molten copper, and iron at temperatures up to 1500°C. The PGMs are unaffected by most organic compounds, although the catalytic reactions that can occur on the surface of platinum and palladium result in an etched appearance of these metals. Palladium is much less resistant to chemical attack, especially by strong oxidizing acids at elevated temperatures (see Table 3). Palladium is stable in air, even at elevated temperatures, and shows no corrosion or tarnishing in hydrogen sulfide atmospheres. However, it has been reported that some discoloration owing to sulfide film formation may take place in industrial atmospheres containing sulfur dioxide (3).

**High Temperature Properties.** There are marked differences in the ability of PGMs to resist high temperature oxidation. Many technological applications, particularly in the form of platinum-based alloys, arise from the resistance of platinum, rhodium, and iridium to oxidation at high temperatures. Osmium and ruthenium are not used in oxidation-resistant applications owing to the formation of volatile oxides. High temperature oxidation behavior is summarized in Table 4.

Although platinum does not form a measurable oxide film, it is covered with a strongly held layer of adsorbed oxygen that volatilizes at an increasing rate above 1000°C, particularly in moving air or oxygen. Rhodium, iridium, and palladium exhibit oxide-film formation, as low as 600°C in the case of palladium. Palladium oxide dissociates again above 800°C, the metal appearing bright up to its melting point. However, absorption of oxygen without film formation occurs and the palladium increases in weight. Rhodium and iridium are less volatile than platinum in the temperature range of 900–1200°C, but are about equal at 1300°C.

No single metal in the platinum group possesses all the desirable characteristics necessary for industrial high temperature applications. However, most high temperature applications for platinum utilize its freedom from corrosion and oxidation, and alloying additions are thus chosen to preserve this characteristic (6). Improved high temperature strength is obtained by alloying with rhodium or iridium. Rhodium–platinum alloys show no preferential loss of either metal at high temperature and are widely used. Iridium–platinum alloys show greater loss of weight on heating in air, because of the greater rate of oxidation of iridium and the higher volatility of the oxide of this metal. Iridi-

**Table 3. Corrosion Resistance of Platinum-Group Metals**[a]

| Medium | Ruthenium | Rhodium | Palladium | Osmium | Iridium | Platinum |
|---|---|---|---|---|---|---|
| chlorine at 20°C | | | | | | |
| saturated | B | A | A | | A | A |
| moist | A | A | D | C | A | B |
| dry | A | A | C | A | A | B |
| bromine at 20°C | | | | | | |
| saturated | B | A | B | | A | A |
| moist | A | A | D | B | A | C |
| dry | A | A | D | D | A | C |
| iodine at 20°C | | | | | | |
| in alcohol | B | B | B | | A | A |
| moist | A | B | B | A | A | A |
| dry | A | A | A | B | A | A |
| HBr, 62% conc | | | | | | |
| 7 h, 100°C | A | B | C | | A | B |
| 25°C | A | B | D | A | A | B |
| HCl, 36% conc, 7 h, 25°C | A | A | A | A | A | A |
| HI, 60% conc, 4 h, 25°C | A | A | D | B | A | A |
| HF, 40% conc, 25°C | A | A | A | A | A | A |
| aqua regia, 25°C | A | A | D | D | A | D |
| H₂SO₄, 98% conc, 7 h | | | | | | |
| 25°C | A | A | A | A | A | A |
| 300°C | A | D | C | | A | B |
| H₃PO₄, 10% conc, 5 h, 100°C | A | A | B | | A | A |
| HNO₃, 95% conc, 25°C | A | A | D | D | A | A |
| selenic acid, 95% conc, 25°C | | | C | | | A |
| fused NaOH, 1 h, 330°C | C | B | B | | B | |
| fused NaCN, 1 h, 700°C | D | D | D | | C | D |

[a] A = no appreciable corrosion; B = some attack, but not enough to preclude use; C = attacked enough to preclude use; and D = rapid attack.

**Table 4. High Temperature Oxidation Behavior**

| Metal | Behavior |
|---|---|
| ruthenium | readily oxidizes to $RuO_2$ if heated in air or oxygen |
| rhodium | tarnish-resistant at room temperature; superficial oxidation when heated in air to red heat, but oxide decomposes above 1100°C; visible as dark discoloration |
| palladium | not tarnished by dry or moist air at room temperature, but at about 600°C a thin oxide film forms in air; above 800°C the superficial oxide decomposes, leaving a clean metal surface; some dissolution or oxide formation occurs again above 1000°C |
| osmium | volatile oxide $OsO_4$ is formed by action of cold air on the finely divided metal, although the compact metal must be heated before oxidation occurs |
| iridium | superficial oxidation when heated in air, but oxide dissociates above 1140°C |
| platinum | no oxide film forms; retains metallic luster on heating in air up to its melting point, even in sulfur-bearing industrial atmospheres; volatilizes above 1000°C |

um is thus lost preferentially from this alloy. Below 1100°C, alloys of platinum with rhodium or palladium are less volatile than pure platinum. However, the oxygen-absorbing properties of palladium prevent its application in alloy form at high temperatures.

Pure platinum and its alloys are subject to grain growth when operated for long periods at high temperature, resulting in mechanical failure owing to weaknesses from large intercrystal boundaries. This problem is alleviated by the incorporation of a small amount of refractory oxide, such as zirconia, to form a range of grain-stabilized materials. Zirconia grain-stabilized (ZGS) platinum and ZGS 10% rhodium–platinum are considerably stronger and have longer life at high temperatures than the corresponding pure metal or alloy.

## Sources and Production

Total known world resources of platinum-group metals have been variously estimated as between 68,000 (7) and 96,000 metric tons (8). Assuming the former estimate and 1979 levels of demand, these reserves should be sufficient to supply the Western world well into the twenty-fourth century. Reserves and relative proportions of the PGMs in the larger deposits are given in Tables 5 and 6. Relative amounts of the PGMs vary from deposit to deposit.

Almost all known deposits of platiniferous ores are related to basic igneous rocks. In many of the deposits, including those in South Africa, Canada, and the CIS, the PGM-containing ores occur in association with nickel, copper (qv), and iron sulfides. There are over 90 known minerals of the PGMs (9), of which the most numerous are the minerals of palladium, followed by platinum. However, there are few native minerals of iridium, osmium, and ruthenium. Rhodium is only known to exist as two minerals: native rhodium and hollingworthite. Rhodium is usually found as impurities dispersed throughout the other PGM

Table 5. Estimated Reserves of PGMs and Gold, t[a]

| PGM | Bushveld Igneous Complex, South Africa | | | Sudbury, Canada | Noril'sk, CIS | Stillwater, Montana |
|---|---|---|---|---|---|---|
| | Merensky Reef | UG2 | Platreef | | | |
| platinum | 10,360 | 13,590 | 4,980 | 105 | 1,550 | 220 |
| palladium | 4,380 | 11,350 | 5,440 | 110 | 4,420 | 715 |
| ruthenium | 1,400 | 3,890 | 470 | <30 | 60 | 45 |
| rhodium | 530 | 2,580 | 375 | <30 | 190 | 85 |
| iridium | 190 | 750 | 95 | <30 | | <30 |
| osmium | 150 | | 60 | <30 | | |
| gold | 560 | 220 | 400 | <35 | | <30 |
| *Total* | *17,570* | *32,380* | *11,820* | *280* | *6,220* | *1,090* |
| grade, g/t | 8.1 | 8.7 | 7–27 | 0.9 | 3.8 | 22.3 |

[a]Ref. 7.

Table 6. Relative Proportions of the PGMs in Deposits, %[a]

| PGM | Bushveld Igneous Complex, South Africa | | | Sudbury, Canada | Noril'sk, CIS | Stillwater, Montana |
|---|---|---|---|---|---|---|
| | Merensky Reef | UG2 | Platreef | | | |
| platinum | 59 | 42 | 42 | 38 | 25 | 19 |
| palladium | 25 | 35 | 46 | 40 | 71 | 66.5 |
| ruthenium | 8 | 12 | 4 | 2.9 | 1 | 4 |
| rhodium | 3 | 8 | 3 | 3.3 | 3 | 7.6 |
| iridium | 1 | 2.3 | 0.8 | 1.2 | | 2.4 |
| osmium | 0.8 | | 0.6 | 1.2 | | |
| gold | 3.2 | 0.7 | 3.4 | 13.5 | | 0.5 |

[a]Ref. 7.

minerals. The PGM minerals can be divided into three classes: (1) native platinoids and their native alloys; (2) intermetallic compounds between PGMs and other metals and semimetals such as Sn, Pb, Bi, Sb, and Te; and (3) sulfides, arsenides, and sulfoarsenides of the platinoids. The most common PGM-containing minerals occurring in significant deposits are listed in Table 7.

Table 8 shows the supply to the Western world of platinum, palladium, and rhodium. For each metal, supply increased significantly in the latter 1980s and early 1990s.

South Africa is by far the largest producer of primary (newly mined) PGMs. It supplied 76% of the platinum used in the West in 1993, and over 50% of all PGMs. South African PGMs are mined as primary product. Other metals such as nickel, copper, and cobalt are by-products. The principal PGM mining houses in South Africa are Anglo American Platinum Corporation (Amplats), Impala Platinum, Lonrho South Africa, and Northam Platinum.

In South Africa, PGM deposits are found within a geological area known as the Bushveld Igneous Complex, an area covering approximately 650,000 km$^2$ in the Central Transvaal. Commercially, the most important deposit within the Bushveld Complex is the Merensky Reef (Table 5), which has an average thick-

**Table 7. Most Common Minerals in Primary PGM Deposits[a]**

| Deposit | Mineral | Composition |
|---|---|---|
| Merensky Reef | ferroplatinum | $Pt_3Fe$ |
| | braggite | $(Pt,Pd,Ni)S$ |
| | cooperite | $PtS$ |
| | moncheite | $PtTe_2$ |
| | sperrylite | $PtAs_2$ |
| Sudbury | sperrylite | $PtAs_2$ |
| | moncheite | $PtTe_2$ |
| | michenerite | $PdBiTe$ |
| Noril'sk | ferroplatinum | $Pt_3Fe$ |
| | atokite | $Pd_3Sn$ |
| | rustenburgite | $Pt_3Sn$ |
| | zvyagintsevite | $Pd_3Pb$ |
| | paolovite | $Pd_2Sn$ |
| | plumbopalladinite | $Pd_3Pb_2$ |
| | polarite | $Pd(Pd,Bi)-Pd(Bi,Pb)$ |
| | sperrylite | $PtAs_2$ |
| | kotulskite | $PdTe$ |
| | cooperite | $PtS$ |
| | vysotskite | $(Pd,Ni,Pt)S$ |

[a]Ref. 9.

**Table 8. Supply of PGMs to the Western World, t[a]**

| Region | 1981 | 1984 | 1987 | 1990 | 1993 |
|---|---|---|---|---|---|
| | | *Platinum* | | | |
| South Africa | 56.0 | 70.9 | 78.4 | 85.8 | 104.5 |
| North America | 4.0 | 4.7 | 4.4 | 5.8 | 6.5 |
| others | 0.9 | 1.2 | 1.2 | 2.0 | 4.0 |
| CIS sales | 11.5 | 7.8 | 12.4 | 22.4 | 21.2 |
| *Total* | *72.4* | *84.6* | *96.4* | *116.0* | *136.2* |
| | | *Palladium* | | | |
| South Africa | 28.3 | 30.5 | 33.9 | 38.2 | 43.4 |
| North America | 5.0 | 5.9 | 5.9 | 11.5 | 11.5 |
| others | 2.2 | 2.8 | 2.8 | 2.2 | 2.2 |
| CIS sales | 44.5 | 52.9 | 55.7 | 58.2 | 71.5 |
| *Total* | *80.0* | *92.1* | *98.3* | *110.1* | *128.6* |
| | | *Rhodium* | | | |
| South Africa | | | 6.1 | 6.2 | 8.6 |
| North America | | | 0.5 | 0.5 | 0.5 |
| others | | | 0.0 | 0.0 | <0.1 |
| CIS sales | | | 3.1 | 4.8 | 2.5 |
| *Total* | | | *9.7* | *11.5* | *11.7* |

[a]Ref. 10.

ness of only 500 mm, but is generally continuous, having relatively little faulting. Exploitation is facilitated by the relatively shallow depth of the Merensky Reef, which in places outcrops at the surface.

There are two other known deposits of PGMs in the Bushveld Complex. The UG2 deposit has a thickness of approximately 600 mm, and is located between 36–330 m below or parallel to the Merensky Reef, which allows the site to be mined from the same access shafts. UG2 is of a higher grade than the Merensky Reef (Table 5) and contains nearly three times as much rhodium. Historically, UG2 was not exploited commercially, because its high chromite content made smelting difficult. However, refining methods have now been developed that allow the chromium to be exploited as ferrochromium. Commercial production from UG2 began in the mid-1980s, and PGMs of UG2 origin may provide up to 25% of South African metal by 1996 (11).

The Platreef deposit is located at the contact between the Bushveld Complex and the underlying rocks. It is disseminated and erratic, having a PGM-grade of 7–27 g/t. Proximity to the surface allows open-cast mining, and the high nickel content provides valuable additional revenue. Although Platreef was mined briefly in the 1920s, commercial mining did not begin until 1993.

The two other principal producers of PGMs are Canada and the CIS. In the CIS, PGMs are produced almost entirely as a by-product of nickel and copper mining at Noril'sk-Talnakh in northern Siberia. There are two distinct types of PGM-containing ore: massive sulfides, having an average PGM content of 5–15 g/t but where localized deposits contain up to 60 g/t; and disseminated ore, containing around 5 g/t PGM. The Noril'sk deposits are particularly rich in palladium (Table 6), supplying over 50% of the Western demand for this metal (Table 8). Total production capacity of these mines is 14 million metric tons of ore per year, which corresponds to between 60 and 200 tons PGM (2.2–6.7 million oz). In Canada, PGMs are present in the extensive nickel and copper sulfide deposits in Sudbury, Ontario. As in CIS, the PGMs are exploited only as a by-product. The two main producers are INCO (the bigger of the two) and Falconbridge.

The most significant PGM deposit in the United States is at Stillwater, Montana, where PGMs are mined as the primary product. The grade has been estimated as between 13 and 22 g/t, having a platinum–palladium ratio of 1:3.5. The first ore was extracted from Stillwater in 1987 and full production of 1000 t/d was reached in 1990.

Extensive PGM deposits, similar geologically to the Bushveld Complex, exist on the Great Dyke in Zimbabwe. There has been extensive exploration in the 1990s, but as of this writing (1995) exploitation has been delayed. Should the operation proceed, it has been estimated that Zimbabwe could produce up to 4.7 tons of platinum, 3.4 tons of palladium, and 0.35 tons of rhodium annually.

China is thought to contain reasonably extensive PGM deposits, in conjunction with widespread nickel deposits, around Jinchang in the north center of the republic. However, the extent to which these are exploited is not clear.

Small amounts of PGM have been produced in Australia for a number of years as a by-product of nickel recovery. There has been a significant increase in speculative exploration of the widespread known deposits in Western Australia, but as of this writing commercialization is not regarded as economically viable.

## Recovery and Refining Techniques

In order to separate the PGMs from each other and from other metals at high yield, high percentage recovery, and high purity, a multistage refining process has been developed. The actual series of processing steps that the ore undergoes varies according to the composition and grade of that ore, and the source from which it was obtained. Raw material from the Merensky Reef requires a completely different pretreatment process to PGM-containing anode slimes, produced as a by-product of nickel electrorefining. The refining processes are tailored to each particular material. Every refinery has developed its own specific reagents and separation technology. The processes described herein are intended to give an indication of the stages required, and the physical and chemical principles controlling each separation. The procedures used by any particular refinery may differ in detail from those described.

**PGM Concentration.** The ore mined from the Merensky Reef in South Africa has a maximum PGM content of 8.1 g/t, of which 50–60% is platinum, and 20–25% palladium. The PGMs are in the form of a ferroplatinum alloy, or as their sulfides, arsenides, or tellurides. The aim of the concentration process is to separate from the ore a crude metal concentrate, having a PGM content of 60%. The majority of other metals, such as nickel and copper, are separated out at this stage for further refining.

A typical concentration process used for the ore of Merensky Reef origin is depicted in Figure 1. The crude ore is crushed and pulverized, using a series of jaw crushers followed by primary and secondary ball mills. The metal sulfide particles are then separated from the gangue by froth flotation using the following steps: air is introduced into a suspension of finely divided ore in water containing a frothing agent to produce a foam and further reagents to promote hydrophobicity on the PGM and base metal minerals. These then adhere to the surface of the air bubbles and rise to the surface where they can be separated from those minerals that are hydrophilic and remain in the liquid pulp. The flotation concentrate now has a PGM content of about 100–150 g/t. Conventional gravity concentration is used to separate out free, large, and dense PGM particles (typically 30–40% of the PGM), which can go straight to the refining process. The PGM sulfide pulp is prepared for smelting by thickening, filtering, and drying. It is smelted in an electric furnace, in the presence of fluxes and other additions, to produce a matte containing principally copper, nickel, iron sulfides, and PGMs. The sulfur content of the matte is reduced by converting with oxygen, and the matte is slow-cooled (see MINERAL RECOVERY AND PROCESSING).

The resultant matte contains copper, nickel, cobalt, and PGMs, including 700–1000 g/t platinum. Nickel and copper are separated magnetically, refined, and marketed (see SEPARATION, MAGNETIC). The nonmagnetic component is pressure-leached to yield the final concentrate, which has a PGM content of approximately 60%. The gravity concentrate separated out earlier and the magnetic concentrate are then refined, using either a conventional or a solvent extraction process.

**Conventional Refining Process.** The conventional refining process is based on complex selective dissolution and precipitation techniques. The exact

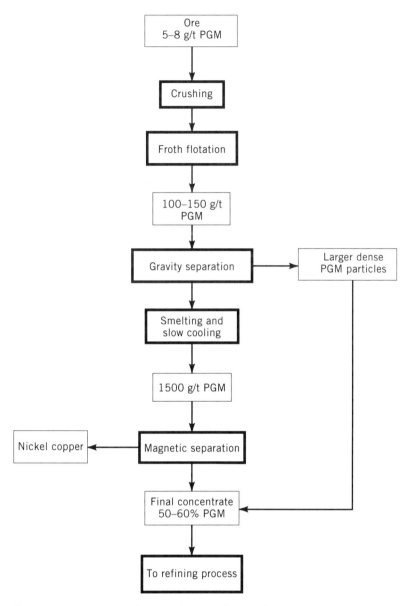

**Fig. 1.**  Concentration process for ore of Merensky Reef origin, where (□) indicate stages of the process and (□), the product obtained.

process at each refinery differs in detail (12–14), but a typical scheme is outlined in Figure 2.

The PGM concentrate is attacked with aqua regia to dissolve gold, platinum, and palladium. The more insoluble metals, iridium, rhodium, ruthenium, and osmium remain as a residue. Gold is recovered from the aqua regia solution either by reduction to the metallic form with ferrous salts or by solvent-extraction methods. The solution is then treated with ammonium chloride to produce a

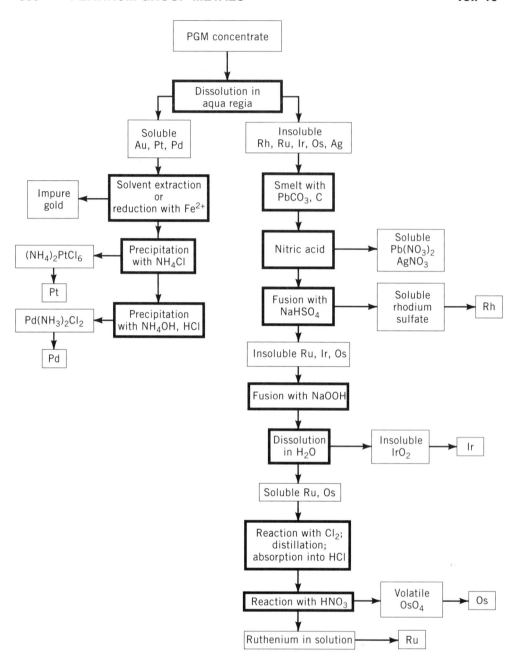

**Fig. 2.**   Conventional PGM refining process. See Figure 1 for definitions.

precipitate of ammonium hexachloroplatinate(IV), $(NH_4)_2PtCl_6$. Calcination of the precipitate gives Pt sponge, which can undergo further purification. The remaining aqua regia solution is treated with ammonium hydroxide, followed by hydrochloric acid, to precipitate dichlorodiammine-palladium. Calcination of this gives impure palladium sponge.

The Ir, Ru, Rh, and Os-containing residue from the original aqua regia dissolution is fused with lead carbonate and carbon. After solidification of the melt, the slag is removed, leaving a lead phase containing the PGMs and silver. This lead phase is melted, granulated, and treated with hot nitric acid to dissolve and separate lead and silver for recovery. The insoluble residue contains the minor PGMs. The residue is fused with sodium bisulfate; rhodium forms soluble rhodium sulfate, which can be leached out with water. A multistage dissolution and recrystallization process is then used to produce pure rhodium.

The residue, which contains Ir, Ru, and Os, is fused with sodium peroxide at 500°C, forming soluble sodium ruthenate and sodium osmate. Reaction of these salts with chlorine produces volatile tetroxides, which are separated from the reaction medium by distillation and absorbed into hydrochloric acid. The osmium can then be separated from the ruthenium by boiling the chloride solution with nitric acid. Osmium forms volatile osmium tetroxide; ruthenium remains in solution. Ruthenium and osmium can thus be separately purified and reduced to give the metals.

Insoluble iridium dioxide from the sodium peroxide fusion is dissolved in aqua regia, oxidized with nitric acid, and precipitated with ammonium chloride as impure ammonium hexachloroiridate(IV), $(NH_4)_2IrCl_6$. To purify this salt, it is necessary to redissolve the compound and precipitate out the impurities as sulfides. Nitric acid and ammonium chloride are again used to produce pure ammonium chloroiridate, which is ignited and reduced with hydrogen to give iridium metal.

The disadvantage of the conventional refining process is that any single dissolution–reprecipitation step does not give a complete separation of the metals. The separation efficiency is hence low. Coprecipitation occurs, and the structure of the precipitate means that filtrate often becomes trapped. There is a need for repeated washing and filtration cycles, resulting in a long, labor-intensive process. Precious metals are thus tied up for long periods of time. The many filtrates and residues also require recovery or retreatment.

**Solvent Extraction Technology.** The use of solvent extraction technology to replace traditional processes has been the subject of considerable research and development effort since the 1970s (12,14–21). This newer technique was being used commercially as of 1995 in at least three of the principal refineries.

The principle of solvent extraction in refining is as follows: when a dilute aqueous metal solution is contacted with a suitable extractant, often an amine or oxime, dissolved in a water-immiscible organic solvent, the metal ion is complexed by the extractant and becomes preferentially soluble in the organic phase. The organic and aqueous phases are then separated. By adding another aqueous component, the metal ions can be stripped back into the aqueous phase and hence recovered. Upon the identification of suitable extractants, and using a multistage process, solvent extraction can be used to extract individual metals from a mixture.

Solvent extraction is a relatively high cost process, owing to the specialty organic extractants required and the expenses of recovery and storage of organic solvents. However, in a precious-metal recovery operation, these costs are easily outweighed by the increased efficiency and PGM recovery as well as shortened metal-in-process time.

All solvent extraction processes are based on the chloro complexes of the PGMs, as chloride is the only effective medium by which the PGMs can be brought into solution. Selection of suitable extractants utilizes aspects of the aqueous chemistry of the metal species, such as the nature of the complex ionic species and the various redox potentials. In general, the PGMs form more stable complexes than the base metals. The most stable PGM complexes are those having heavier donor atoms. As an approximation, the overall order of stability is $S \sim C > I > Br > Cl > N > O > F$. Use of extractants forming particularly stable complexes, eg, by using sulfur ligands such as thioethers, allows good partition into the organic phase. However, the strength of the complex formed makes stripping of the metal back into the aqueous phase difficult.

The actual solvent extraction processes used, including the specific extractants and the order in which the components are separated, vary from refinery to refinery. However, a typical scheme is shown in Figure 3 (12).

The raw precious metal concentrate is totally dissolved in hydrochloric acid–chlorine solution to form the soluble chloride ions of each of the metals. Silver remains as insoluble silver chloride and can be filtered off. Gold, in the form of $[AuCl_4]$, is extracted with, eg, tributyl phosphite or methyl isobutyl ketone. Base metals are also extracted in this step, and are removed from the organic phase by scrubbing with dilute hydrochloric acid (HCl). Iron powder is then used to reduce the gold species and recover them from the organic phase.

Palladium is extracted using a beta-hydroxyoxime, which undergoes a ligand exchange reaction with $[PdCl_4]^{2-}$. The resulting organic phase is again scrubbed with dilute HCl to separate base metals. The palladium is stripped from the organic phase into aqueous ammonium hydroxide solution, and finally precipitated with hydrochloric acid. The remaining solution is neutralized, allowing the osmium and ruthenium to be separated as their volatile tetroxides. Distillation can later be used to separate these two metals. After reduction of the solution using sulfur dioxide to convert Ir(IV) to Ir(III), $[PtCl_6]^{2-}$ is extracted using a tertiary amine, tri-$n$-octylamine. Hydrochloric acid is used to strip the platinum from the organic phase, and ammonium chloride is used to precipitate it as $(NH_4)_2PtCl_6$.

Subsequently, the PGM solution is oxidized and acidified to reconvert Ir(III) to Ir(IV). Tri-$n$-octylamine is again used as the extractant, this time to extract iridium. The iridium in the organic phase is reduced to Ir(III) and recovered. The remaining element is rhodium, which is recovered from impurities in the original solution by conventional precipitation or ion exchange (qv).

**Secondary Platinum-Group Metals.**  For many PGM applications, the actual loss during use of the metal is small, and hence the ability to recover the PGM efficiently contributes greatly to the economics of PGM use. Typical sources of PGM for secondary refining include jewelry and electronics scrap, catalysts, and used equipment, eg, from the glass industry.

In the latter twentieth century, spent automotive catalysts have emerged as a significant potential source of secondary Pt, Pd, and Rh. In North America, it has been estimated that 15.5 metric tons per year of PGM from automotive catalysts are available for recycling (22). However, the low PGM loading on such catalysts and the nature of the ceramic monoliths used have required the development of specialized recovery techniques as well as the establishment of an

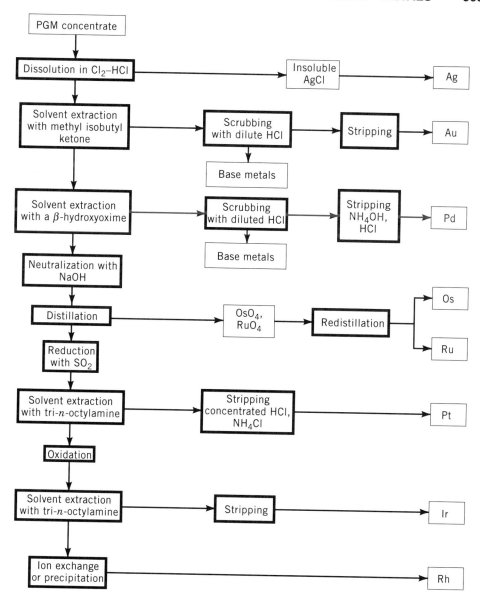

**Fig. 3.** PGM refining by solvent extraction. See Figure 1 for definitions.

infrastructure of collection centers. These factors have slowed the development of an automotive catalyst recycling industry.

Several processes are available for the recovery of platinum and palladium from spent automotive or petroleum industry catalysts. These include the following. (1) Selective dissolution of the PGM from the ceramic support in aqua regia. Soluble chloro complexes of Pt, Pd, and Rh are formed, and reduction of these gives crude PGM for further refining. (2) Dissolution of the catalyst support in sulfuric acid, in which platinum is insoluble. This technique is widely used for the treatment of γ-alumina petroleum catalysts, but is not suitable for the dissolution of cordierite honeycomb substrates often used in automotive catalysts. (*3*) A gas-phase volatilization process has also been reported in which the PGMs are selectively chlorinated to give volatile compounds, and condensed in a cooler zone (22). In principle, the ceramic substrate does not react. (*4*) The substrate containing highly dispersed PGM is melted with a flux to give a slag. The PGMs accumulate in a molten collector metal. This is known as pyrometallurgy (23,24).

## Economic Aspects

Tables 9–16 show PGM consumption by the Western world and Japan. The figures are broken down by industrial application and, for platinum and palladium, by the region in which they are used. In each case, net demand is shown, ie, total purchases of the metals by customers, minus any sales back to the market.

Demand for the PGMs has risen significantly from the mid-1980s to the mid-1990s. Demand for platinum rose by 66% over this 10-year period. The

**Table 9. Platinum Demand by the Western World, t[a]**

| Industry | 1981 | 1984 | 1987 | 1990 | 1993 |
|---|---|---|---|---|---|
| autocatalyst[b,c] | 19.9 | 26.1 (1.4) | 39.0 (3.5) | 47.7 (6.7) | 52.9 (8.9) |
| chemical | 7.8 | 8.1 | 6.1 | 6.7 | 5.7 |
| electrical | 5.7 | 5.9 | 5.6 | 6.4 | 5.4 |
| glass | 3.1 | 4.4 | 3.7 | 4.2 | 2.2 |
| investment | 6.1 | 9.9 | 15.2 | 6.2 | 9.5 |
| jewelry | 23.5 | 24.1 | 30.8 | 42.5 | 50.1 |
| petroleum | 4.4 | 0.5 | 1.7 | 4.4 | 3.1 |
| other | 5.1 | 4.2 | 3.7 | 3.7 | 5.6 |
| *Total[d]* | *75.6* | *81.8* | *102.3* | *115.1* | *125.6* |

[a]Ref. 10.
[b]Automotive emission control catalyst.
[c]Numbers in parentheses represent the quantity of platinum recovered from this industry.
[d]Total reflects the contribution of recovered platinum from the autocatalyst industry.

**Table 10. Platinum Demand by Region, t[a]**

| Region | 1981 | 1984 | 1987 | 1990 | 1993 |
|---|---|---|---|---|---|
| Western Europe | 13.0 | 12.4 | 17.4 | 21.8 | 27.8 |
| Japan | 35.8 | 35.5 | 51.3 | 57.5 | 61.0 |
| North America | 21.8 | 28.3 | 28.0 | 24.6 | 23.8 |
| rest of Western world | 5.0 | 5.6 | 5.6 | 11.2 | 13.0 |
| *Total* | *75.6* | *81.8* | *102.3* | *115.1* | *125.6* |
| Western sales to Eastern Europe and China | 0.9 | 0.9 | 0.9 | 0 | 0 |
| *Total demand* | *76.5* | *82.7* | *103.2* | *115.1* | *125.6* |

[a]Ref. 10.

**Table 11. Palladium Demand by the Western World, t[a]**

| Industry | 1981 | 1984 | 1987 | 1990 | 1993 |
|---|---|---|---|---|---|
| autocatalyst[b,c] | 8.4 | 10.0 | 8.4 (1.5) | 9.8 (2.7) | 23.5 (3.2) |
| dental | 14.9 | 28.0 | 29.7 | 31.7 | 37.6 |
| electrical | 24.9 | 38.3 | 49.0 | 52.1 | 61.1 |
| jewelry | 6.5 | 6.5 | 5.1 | 6.1 | 6.5 |
| other | 10.3 | 9.3 | 8.4 | 9.2 | 7.0 |
| *Total[d]* | *65.0* | *92.1* | *99.1* | *106.2* | *132.5* |

[a]Ref. 10.
[b]Automotive emission control catalyst.
[c]Numbers in parentheses represent the quantity of palladium recovered from this industry.
[d]Total reflects the contribution of recovered palladium from the autocatalyst industry.

**Table 12. Palladium Demand by Region, t[a]**

| Region | 1981 | 1984 | 1987 | 1990 | 1993 |
|---|---|---|---|---|---|
| Western Europe | 9.3 | 16.2 | 17.1 | 18.3 | 21.5 |
| Japan | 25.5 | 38.9 | 44.5 | 47.6 | 60.2 |
| North America | 25.5 | 30.8 | 32.2 | 33.6 | 41.8 |
| rest of Western world | 4.7 | 6.2 | 5.3 | 6.7 | 9.0 |
| *Total* | *65.0* | *92.1* | *99.1* | *106.2* | *132.5* |

[a]Ref. 10.

**Table 13. Rhodium Demand by the Western World, t[a]**

| Industry | 1984 | 1987 | 1990 | 1993 |
|---|---|---|---|---|
| autocatalyst[b,c] | 3.4 | 7.0 (0.1) | 10.4 (0.4) | 11.1 (0.9) |
| chemical | 1.5 | 0.7 | 0.8 | 0.3 |
| electrical | 0.1 | 0.4 | 0.4 | 0.2 |
| glass | 0.3 | 0.4 | 0.5 | 0.1 |
| other | 1.7 | 0.8 | 0.5 | 0.4 |
| *Total[d]* | *7.0* | *9.2* | *12.2* | *11.2* |

[a]Ref. 10.
[b]Automotive emission control catalyst.
[c]Numbers in parentheses represent the quantity of rhodium recovered from this industry.
[d]Total reflects the contribution of recovered rhodium from the autocatalyst industry.

**Table 14. Rhodium Demand by Region, t[a]**

| Region | 1984 | 1987 | 1990 | 1993 |
|---|---|---|---|---|
| Western Europe | 1.4 | 2.0 | 3.0 | 3.8 |
| Japan | >1.4 | 2.6 | 3.6 | 2.1 |
| North America | 3.9 | 4.0 | 4.7 | 4.0 |
| rest of Western world | | 0.6 | 0.9 | 1.3 |
| *Total* | *7.0* | *9.2* | *12.2* | *11.2* |

[a]Ref. 10.

**Table 15. Ruthenium Demand by the Western World, t[a]**

| Industry | 1984 | 1987 | 1990 | 1993 |
|---|---|---|---|---|
| electrical | 2.8 | 2.7 | 2.7 | 3.9 |
| chemical | 2.9 | 1.8 | 1.9 | 1.7 |
| other | 0.5 | 0.6 | 0.1 | 0.1 |
| *Total* | *6.2* | *5.1* | *4.7* | *5.7* |

[a]Ref. 10.

**Table 16. Iridium Demand by the Western World, t[a]**

| Industry | 1984 | 1987 | 1990 | 1993 |
|---|---|---|---|---|
| chemical | 0.3 | 0.3 | 0.4 | 0.6 |
| crucibles | 0.2 | 0.1 | 0.1 | 0.1 |
| other | 0.2 | 0.7 | 0.3 | 0.3 |
| *Total* | *0.7* | *1.1* | *0.8* | *1.0* |

[a]Ref. 10.

fastest growth areas include autocatalyst, a growth of 78%, and jewelry, a growth of over 100%. The palladium market grew by 44% over the period 1984–1993. Use of palladium in autocatalysts grew by over 100%, and in electronics by 60%. Demand for rhodium in autocatalysts increased most dramatically, by nearly 200% in the 10-year period prior to 1993.

In October 1995, the average prices of the PGMs were for platinum, $13.28/g; palladium, $4.44/g; rhodium, $11.96/g; iridium, $1.93/g; and ruthenium, $0.93/g.

## Analysis

Like the refining of the PGMs, the analysis is complicated by the chemical similarity of the metals. The techniques used depend on the elements present and their concentration in the sample. For some low grade samples, analysis is preceded by a concentration stage using fire assay with collection into a lead or nickel sulfide button. The individual metals can then be determined.

Colorimetric and fluorimetric techniques are sensitive and accurate but have generally been superseded by other methods in specialist analytical laboratories. Gravimetric techniques are still widely used for all the PGMs but require correct preseparation procedures. The examination of solid materials is carried out by using either x-ray fluorescence spectrometry, where the use of matrix-matched standards is essential, or dc-arc spectrography and spectrometry, where setup, sampling, and standardization are all element-specific. The most widely used method for determining low concentrations of the PGMs in solution is atomic absorption spectrometry, which is useful for single element determinations but slow and laborious for multiple elements. Plasma emission spectrometry is the method of choice for rapid multielement analysis. Detection limits are 0.1–10 ppm, depending on the element. For two-orders-of-magnitude lower detection limits, the plasma mass spectrometer is used in specialist PGM laboratories.

## Health and Safety Factors

In bulk metallic form, the PGMs are not hazardous to health. In common with many other metals, however, the PGMs in finely divided form can be hazardous to handle. For example, powdered iridium can ignite in air, palladium dust is combustible, and powdered platinum is a powerful catalyst and liable to ignite combustible materials. Platinum and rhodium metals both have a time-weighted average threshold limit value in the United States of 1 mg/m$^3$. The levels in the United Kingdom are 5 and 0.1 mg/m$^3$ for platinum and rhodium dusts, respectively. Exposure limits have not been set for the other PGMs. Fumes of ruthenium can harm the eyes and lungs, and can produce nasal ulcers. Some PGM compounds are toxic (see PLATINUM-GROUP METALS, COMPOUNDS). In particular, platinum halogeno salts are highly allergenic. On reexposure, sensitized individuals can show symptoms similar to hay fever, including asthma.

## Uses

The principal applications of the PGMs are summarized in Table 17.

**Catalytic Applications.** The PGMs are widely used as catalysts for a variety of chemical reactions, such as hydrogenation, oxidation, dehalogenation, dehydrogenation, isomerization, and cyclization. The application of the individual PGMs for particular reactions is summarized in Table 18. The main areas of commercial application are automotive emission control catalysts (autocatalysts), oil refining, ammonia oxidation, liquid-phase catalysis, and fuel cells. Autocatalyst applications consume large quantities of these metals (see Tables 9, 11, and 13).

Catalytic properties are dependent on physical form, principally the exposed surface area which is a function of particle size. Industrial PGM catalysts

**Table 17. Applications of the Platinum-Group Metals**

| Application | Pt | Pd | Ru | Rh | Ir |
|---|---|---|---|---|---|
| automotive catalysts | + | + | | + | |
| industrial emission catalysts | + | | | | |
| fuel cells | + | | | | |
| gas sensors | + | + | | | |
| jewelry | + | + | | | |
| investment | + | | | | |
| biomedical devices | + | | | | + |
| chemotherapeutics, anticancer | + | | | | + |
| dental materials | | + | | | |
| electronics | + | + | + | | + |
| electrochemical | + | | + | | + |
| chemical$^a$ | + | + | + | + | + |
| petroleum refining | + | + | | | |
| glass | + | | | + | |
| crucibles | + | | | | + |
| coatings | + | | | | |
| spark plugs | + | | | | + |

$^a$Osmium may be used as a catalyst for the chemical and pharmaceutical industries.

**Table 18. Chemical Industry Catalytic Applications of Platinum-Group Metals**

| Reaction | Pt | Pd | Ru | Rh | Ir | Os |
|---|---|---|---|---|---|---|
| hydrogenation | + | + | + | + | + | |
| oxidation | + | + | + | + | | |
| dehydrogenation | + | + | | | | |
| hydrogenolysis | + | + | + | | | |
| synthesis | | | | | | |
|   ammonia | | | + | | | |
|   methanol | | + | | | | |
|   hydrocarbons | | | + | | | |
|   acetic acid | | | | + | | |
| hydroformylation | | | + | + | + | |
| carbonylation | | | | + | | |
| cis-hydroxylation | | | | | | + |

are in the form of finely divided powder, wire, or gauze, or supported on substrates such as carbon or alumina (see CATALYSIS; CATALYSTS, SUPPORTED).

A fundamental measure of catalytic activity is heat of adsorption (qv), which measures how strongly a reactant adsorbs to a surface. If the reactant is too strongly adsorbed, no reaction with other adsorbed species occurs. Weakly adsorbed substances require only a small amount of energy to be displaced from the surface; these are desorbed so quickly that they do not have time to react. Thus, the surface is effectively noncatalytic for these reactants. Effective catalysts operate between the two extremes. Platinum is unique in that a wide spectrum of substances are adsorbed at a moderate strength, making it catalytically active for a wide range of reactions. These substances include hydrogen, oxygen, carbon monoxide, unsaturated hydrocarbons, nitrogen compounds (imines, azines, hydrazones, oximes), oxygenated compounds (acids, esters, anhydrides, aldehydes, ketones), sulfur dioxide, and ammonia. Methane is only weakly adsorbed, and platinum must be doped with other metals to improve its ability to act as a methane combustion catalyst. Platinum does not adsorb nitrogen, which makes it a poor ammonia synthesis catalyst.

*Automotive Emission Control Catalysts.* Air pollution (qv) problems caused by automotive exhaust emissions have been met in part by automotive emission control catalysts (autocatalysts) containing PGMs. In the United States, all new cars have been required to have autocatalyst systems since 1975. In 1995, systems were available for control of emissions from both petrol and diesel vehicles (see EXHAUST CONTROL, AUTOMOTIVE).

The first type of autocatalyst to be developed was an oxidation catalyst effective at oxidizing carbon monoxide (qv), CO, and hydrocarbons (qv), HC, to carbon dioxide and water. This system contains platinum, palladium, or a mixture of both metals. The ability of an oxidation catalyst to control nitrogen oxides, $NO_x$, however, is limited. As of this writing, the most widely used emission control catalyst for gasoline engines is the three-way catalyst, which serves simultaneously to oxidize CO and HC and to reduce $NO_x$. This is achieved through advanced catalyst systems, used in conjunction with accurate fuel metering to keep the air–fuel ratio near the stoichiometric point. Three-way catalysts are manufactured by dispersing PGMs onto ceramic or metal honeycomb supports.

High surface areas are achieved by coating the support with an alumina-based washcoat, into which PGMs are subsequently impregnated. Three-way catalysts are usually based on combinations of platinum, palladium, and rhodium, and are able to remove 90% of the CO, HC, and $NO_x$ pollutants from car exhausts. Platinum is the most active of the three metals for converting CO and HC, and is least susceptible to poisoning by lead, sulfur, and other impurities in the fuel. It is quick to start working from cold but does not show much ability to reduce $NO_x$. Rhodium has a high $NO_x$-reduction activity and is good at converting the other pollutants, but has a more limited availability. Palladium displays some ability to convert CO and HC and to reduce $NO_x$, but is slower to light off than platinum and is more easily poisoned.

Developments in autocatalyst technology during the 1990s include the introduction of heated catalysts designed to reduce the time taken for the catalyst to warm up from cold start. Emissions of pollutants during this interval are high for the standard three-way catalyst. Catalysts for diesel engine vehicles have also been developed. Several approaches are being considered. The first systems were based on oxidation catalysts to remove HC and CO. These catalysts contained promoters to minimize $SO_2$ oxidation. Second-generation systems are based on a combined oxidation catalyst and particulate trap to remove HC and CO, and to alleviate particulate emissions on a continuous basis. The next phase will be the development of advanced catalysts for $NO_x$ removal under oxidizing conditions. Low or zero sulfur diesel fuel will be an advantage in overall system development.

*Industrial and Stationary Source Emission Catalysts.* Catalytic incineration is used for the destruction of fumes and odors caused by the presence of volatile organic compounds (VOCs) in the exhaust gases from industrial processes. VOC emissions occur in a wide range of industries, including food processing (qv), chemicals manufacture, printing, textile treatment, tobacco, and petrochemicals. Platinum-based catalysts are particularly suitable for eliminating VOCs as these catalysts are not only highly active for the oxidation process, but also chemically and physically stable. Catalytic incineration operates at much lower temperatures and much faster rates than thermal incineration (see INCINERATORS).

PGM catalyst technology can also be applied to the control of emissions from stationary internal combustion engines and gas turbines. Catalysts have been designed to treat carbon monoxide, unburned hydrocarbons, and nitrogen oxides in the exhaust, which arise as a result of incomplete combustion. To reduce or prevent the formation of $NO_x$ in the first place, catalytic combustion technology based on platinum or palladium has been developed, which is particularly suitable for application in gas turbines. Environmental legislation enacted in many parts of the world has promoted, and is expected to continue to promote, the use of PGMs in these applications.

*Catalysts for Chemical Industry Applications.* The PGMs are widely used as catalysts in chemical manufacturing, petroleum (qv) refining, and electrochemical processing (qv). A number of the most important industrial products manufactured by using PGM catalysts are outlined herein.

*Nitric Acid.* Nitric acid (qv) is a commodity chemical manufactured by the oxidation of ammonia (qv) over platinum–rhodium gauze catalysts. Such catalysts must be able to withstand long-term use at high temperatures and pressures. Knitted gauzes were introduced in the early 1990s to replace the conventional woven gauzes. Such gauzes offer increased conversion efficiency, reduced rhodium oxide formation, greater strength, lower metal losses, and extended campaign lengths (25). Industrial ammonia oxidation plants also use palladium–gold catchment gauzes for the recovery of platinum and rhodium lost through volatilization.

*Silicones.*    Homogeneous platinum catalysts, used in silicone manufacture, are retained in the product. This has been shown to impart improved fire resistance in specialized rubbers.

*Methyl tert-Butyl Ether.*    Methyl *t*-butyl ether (MTBE) is an increasingly important fuel additive. Platinum–tin and other PGM catalysts are used for the dehydrogenation of isobutane to isobutene, an intermediate step in MTBE manufacture.

*Organic Synthesis.*    Many organic syntheses involve the use of PGM catalysts at one or more intermediate stages (see Table 18). Examples include caprolactam (qv), an intermediate in nylon production; and paraxylene, which is converted to terephthalic acid for the manufacture of polyester fibers. A range of pharmaceuticals (qv), eg, paracetamol and various antibiotics (qv), and pharmaceutical intermediates utilize palladium hydrogenation catalysts.

*n-Aldehydes.*    A rhodium-based homogeneous hydroformylation catalyst, which operates at low temperature and pressure, allows much greater selectivity for *n*-butyraldehyde than is possible using the conventional cobalt-catalyzed process (26).

$C_1$-*Chemistry.*    A great deal of research has been undertaken on the development of PGM catalysts for the manufacture of chemicals and fuels from syngas, a mixture of CO and $H_2$ obtained from coal gasification (see COAL CONVERSION PROCESSES).

*Petroleum.*    The petroleum (qv) industry uses PGMs as monometallic, bimetallic, or multimetallic catalysts in catalytic reforming, hydrogenation, and isomerization. They are also used as promoters in other catalytic reactions. The largest quantities of PGM are used in catalytic reforming, where processes such as platforming and rheniforming, ie, to reform low octane naphthas to high octane fuels and aromatics for petrochemical feedstocks, are widespread. Many downstream petrochemical processes also require PGM catalysts.

*Metal Anodes.*    Ruthenium and iridium are used as electrode materials in the chlor-alkali industry, where brine is electrolyzed to produce chlorine and sodium hydroxide. The PGMs are in the form of dimensionally stable anodes, consisting of a $RuO_2$ or $IrO_2$ coating on titania (see METAL ANODES). In the modern membrane-type electrochemical cells, iridium is increasingly favored over ruthenium. Iridium–platinum electrodes are used in the electrochemical production of sodium chlorate, which is regarded as more environmentally acceptable than chlorine for bleaching.

*Fuel Cell Catalysts.*    Fuel cells (qv) are electrochemical devices that convert the chemical energy of a fuel directly into electrical and thermal energy. The fuel cell, an environmentally clean method of power generation (qv), is more efficient than most other energy conversion systems. The main by-product is pure water.

A fuel cell contains two electrodes: the anode, which is supplied with a fuel such as hydrogen (qv), and the cathode, which is supplied with oxygen. The electrodes are separated by an electrolyte. Fuel cells are usually classified by the type of electrolyte used in the cell. Both phosphoric acid fuel cells (PAFC) and solid polymer fuel cells (SPFC) utilize catalyzed electrodes. In PAFC, phosphoric acid is the electrolyte and the electrodes are coated with platinum black or platinum supported on carbon. Platinum promotes the electrochemical reaction by weakly adsorbing hydrogen and oxygen on the metal surface, and enabling rapid formation and breaking of the metal–hydrogen and metal–oxygen bonds. Prototype and pilot-scale PAFC fuel cell plants have been built and used in trials by electricity and gas utilities as stationary power sources.

Another important potential application for fuel cells is in transportation (qv). Buses and cars powered by fuel cells or fuel cell–battery hybrids are being developed in North America and in Europe to meet zero-emission legislation introduced in California. The most promising type of fuel cell for this application is the SPFC, which uses platinum-on-carbon electrodes attached to a solid polymeric electrolyte.

**High Temperature Applications.** Exceptional mechanical as well as corrosion- and oxidation-resistant properties of PGMs are particularly pronounced at elevated temperatures. This has led to a substantial number of applications for the PGMs and their alloys in high temperature process industries. A significant application of platinum and platinum–rhodium alloys is for glass (qv) industry equipment, eg, crucibles, stirrers, tubes, furnace linings, and bushings. The latter are troughs that contain an array of nozzles in the base, through which molten glass is extruded to form glass fibers. Platinum and Pt–10% Rh are both resistant to the high temperatures and corrosive conditions encountered in molten glass. Zirconia grain-stabilized (ZGS) platinum and platinum–rhodium alloys further increase the stability and operating life.

In applications where it is not technically or economically viable to use solid PGM alloys, coatings of PGM on the refractory ceramic substrate can be used. This allows the mechanical properties of the substrate to be combined with the corrosion resistance of the PGM, and also helps to overcome the inherent brittleness and poor thermal shock characteristics of the ceramic. Johnson Matthey's proprietary coating technology (27) is being used to coat thermocouple sheaths and glass processing equipment, yielding significantly longer lifetimes than uncoated products.

Turbine blades of jet engines are coated with a protective layer of platinum aluminide to impart high temperature corrosion resistance. Platinum is electroplated onto the blade using P-salt or Q-salt electroplating solutions (28,29). The platinum is then diffusion-treated with aluminum vapor to form platinum aluminide. Standards for the inspection and maintenance of turbine blades have become more stringent. Blades are therefore being recoated several times during their lifetime.

Tips of platinum, platinum–nickel alloy, or iridium can be resistance-welded to spark-plug electrodes for improved reliability and increased lifetime. These electrodes are exposed to extremely hostile environments involving spark erosion, high temperature corrosion, thermal shock, and thermal fatigue.

Platinum and iridium crucibles are used for the growth of high purity crystals of specialty materials such as the rare earths. Examples include gadolinium gallium garnet and yttrium aluminum garnet for use in electronic memory chips and lasers (qv). The high melting point and oxidation resistance of the PGMs minimize contamination of the melt. Platinum laboratory apparatus is routinely used in chemical analysis (see CONTAMINATION CONTROL TECHNOLOGY).

**Electronic Applications.** The PGMs have a number of important and diverse applications in the electronics industry (30). The most widely used are palladium and ruthenium. Palladium or palladium–silver thick-film pastes are used in multilayer ceramic capacitors and conductor inks for hybrid integrated circuits (qv). In multilayer ceramic capacitors, the termination electrodes are silver or a silver-rich Pd–Ag alloy. The internal electrodes use a palladium-rich Pd–Ag alloy. Palladium salts are increasingly used to plate edge connectors and lead frames of semiconductors (qv), as a cost-effective alternative to gold. In 1994, 45% of total ruthenium demand was for use in ruthenium oxide resistor pastes (see ELECTRICAL CONNECTORS).

There is great interest in the use of thin-film platinum silicide in silicon-integrated circuit technology. Platinum silicide has low resistance and high stability, and can provide both ohmic low resistance contacts and rectifying contacts. Typically, silicide layers are formed *in situ* by sputtering a thin platinum layer onto the silicon surface, followed by sintering. Infrared detection is another application of platinum silicide technology.

Increasing data storage requirements for computer applications are creating demands for advanced magnetic and magnetooptic storage media. Platinum cobalt multilayers demonstrate good perpendicular magnetic anisotropy, making them excellent candidates for information storage materials (qv) (31).

**Temperature Measurement.** PGM thermocouples are widely used for high temperature measurement (qv) in the glass, steel, and semiconductor industries owing to their linear, high thermoelectric voltages. In the semiconductor and glass industries, permanently installed profiling thermocouples are used to control furnace temperatures (see FURNACES). In the steel (qv) industry, disposable thermocouples give an instantaneous temperature reading before being destroyed in the furnace. Type S (Pt vs Pt–10% Rh) and type R (Pt vs Pt–13% Rh) are used over the temperature range of 0–1450°C. Type B thermocouples (Pt–6% Rh vs Pt–30% Rh) operate from 870–1700°C. For higher temperatures, in the range 1800–2000°C, a thermocouple of Ir vs Rh–40% Ir is available.

High purity platinum wire is used in resistance thermometers because the temperature coefficient of resistance of pure platinum is linear over a wide temperature range. The platinum resistance thermometer is the recognized instrument for the interpolation of the international practical temperature scale from −259.35 to 630.74°C. Whereas such precision measurements require very high purity platinum, for most routine industrial measurements lower purity metal can be tolerated. Conventional wire-wound devices are quite fragile and this disadvantage has led to the introduction of printed resistance thermometers, which are cheap to produce and much more durable. They can be used as an inexpensive replacement for thermocouple applications in intermediate temperature applications.

A high temperature optical fiber thermometer has been developed (32,33). It consists of a sputtered iridium blackbody tip on a single crystal sapphire laser. Such a device has been shown to be accurate to within ±0.03°C at 1000°C.

**Jewelry and Investment.** Platinum is widely used in jewelry, particularly in Japan where it is preferred to gold. As pure platinum is too soft for jewelry applications, it is alloyed with other PGMs such as iridium or ruthenium to increase its hardness and wear resistance. A Pt–5% Ir alloy has excellent hardness and strength, but has a grayish color and can easily lose its surface finish and luster. These problems are overcome by the addition of palladium as a whitening agent. The 95% Pt–4.5% Pd–0.5% Ir alloy is a widely used material. Palladium is also used as a whitening agent in gold jewelry.

Platinum jewelry alloys are easily worked and readily melted and cast. However, they are difficult to machine, resulting in severe tool wear. Pure platinum (99.95%) in the form of coins and ingots is used as an investment metal.

**Medical and Dental Applications.** The most important medical use of platinum is in the anticancer drugs, carboplatin and cisplatin (see CHEMO-THERAPEUTICS, ANTICANCER; PLATINUM-GROUP METALS, COMPOUNDS). However, the PGMs are also used in metallic form in a range of biomedical devices (see PROSTHETICS AND BIOMEDICAL DEVICES). The PGMs are biocompatible as well as highly corrosion resistant and workable. Applications include platinum guidewires on balloon angioplasty catheters to allow radiographical detection of the catheter in the body; platinum coils for treating aneurysms and for emboliza-tion in vein bypass operations; and radioactive iridium, $^{192}$Ir, seeds sheathed in platinum as implants in the treatment of tumors.

Platinum and its alloys are also used as biomedical electrodes, eg, plat-inum–iridium wires for permanent and temporary pacemaker leads and de-fibrillator leads. Electrophysiology catheters, which contain platinum electrodes and marker bands, have been used to map the electrical pathways of the heart so that appropriate treatment, such as a pacemaker, can be prescribed.

One of the main uses of osmium tetroxide is as a biological staining agent for microscopic cell and tissue studies. Osmium tetroxide is unique in that it both fixes and stains biological material.

In dentistry, palladium alloys are widely used as alternatives to base metal alloys in the manufacture of crowns and bridges as well as the replacement of lost or damaged teeth (see DENTAL MATERIALS). Such alloys contain over 80% palladium, and hence offer significant cost benefits over alloys containing a high proportion of gold.

# BIBLIOGRAPHY

"Platinum-Group Metals, Alloys, and Compounds" in *ECT* 1st ed., Vol. 10, pp. 819–855, by E. M. Wise, The International Nickel Co., Inc.; "Platinum-Group Metals," in *ECT* 2nd ed., Vol. 15, pp. 833–860, by E. M. Wise, Consultant; in *ECT* 3rd ed., Vol. 18, pp. 228–253, by R. D. Lanam and E. D. Zysk, Engelhard Corp.

1. D. McDonald and L. B. Hunt, *A History of Platinum and Its Allied Metals*, Johnson Matthey, London, U.K., 1982.
2. A. S. Darling, *Internat. Met. Rev.* **18**, 91 (1973).
3. L. L. Shrier, R. A. Jarman, and G. T. Burstein, *Corrosion*, 3rd ed., Butterworth-Heinemann, Oxford, U.K., 1994, pp. 6:1–6:27.
4. M. J. N. Pourbaix, J. Van Muylder, and N. de Zoubov, *Plat. Metals Rev.* **3**(2), 47–53 (1959); **3**(3), 100–106 (1959).
5. F. L. LaQue and H. R. Copson, *Corrosion Resistance of Metals and Alloys*, 2nd ed., Reinhold, New York, 1963, pp. 601–622.
6. E. Bullock, *Research and Development of High Temperature Materials for Industry*, Elsevier Applied Science, London, U.K., 1989, pp. 95–106.
7. D. L. Buchanan, *Report of the Bureau of Mineral Studies*, No. 4, University of Wit-watersrand, South Africa, 1979.

8. *Platinum Group Metals in Mineral Commodity Summaries 1992*, U.S. Bureau of Mines, Washington, D.C., 1992.
9. Y. M. Savitskii and A. Prince, *Handbook of Precious Metals*, Hemisphere Publishing, New York, 1989.
10. *Platinum* series, Johnson Matthey, London, U.K., published annually since 1985.
11. *Platinum 1988*, Johnson Matthey, London, U.K., 1988, p. 14.
12. L. S. Benner and co-workers, eds., *Precious Metals Science and Technology*, IPMI, Allentown, Pa., 1991, pp. 375–399.
13. L. B. Hunt and F. M. Lever, *Plat. Metals Rev.* **13**(4), 126–138 (1969).
14. G. B. Harris, in R. K. Mishra, ed., *Precious Metals 1993*, IPMI, Allentown, Pa., 1993, pp. 351–374.
15. M. J. Cleare, R. A. Grant, and P. Charlesworth, in *Extraction Metallurgy '81*, London, U.K., 1981, pp. 34–41.
16. M. J. Cleare, P. Charlesworth, and D. J. Bryson, *J. Chem. Tech. Biotechnol.* **29**, 210 (1979).
17. R. A. Grant, in L. Manziek, ed., *Precious Metals Recovery and Refining*, IPMI, Allentown, Pa., 1990, p. 7.
18. P. Charlesworth, *Plat. Metals Rev.* **25**(3), 106–112 (1981).
19. L. R. P. Reavill, *Plat. Metals Rev.* **28**(1), 2–6 (1984).
20. M. B. Mooiman, in Ref. 14, pp. 411–434.
21. B. Côté, in Ref. 14, pp. 593–617.
22. R. K. Mishra, in Ref. 14, pp. 449–474.
23. U.S. Pat. 4,428,768 (1984), J. G. Day (to Johnson Matthey).
24. B. Keyworth, in M. I. El Grundy, ed., *Precious Metals 1982*, IPMI, Allentown, Pa., 1992, pp. 509–565.
25. B. T. Horner, *Plat. Metals Rev.* **37**(2), 76–85 (1993).
26. F. J. Smith, *Plat. Metals Rev.* **19**(3), 93–95 (1975).
27. D. R. Coupland, *Plat. Metals Rev.* **37**(2), 62–70 (1993).
28. P. E. Skinner, *Plat. Metals Rev.* **33**(3), 102–105 (1989).
29. M. E. Baumgärtner, *Plat. Metals Rev.* **32**(4), 188–197 (1988).
30. N. M. Davey and R. J. Seymour, *Plat. Metals Rev.* **29**(2), 2–11 (1985).
31. P. D. Gurney, *Plat. Metals Rev.* **37**(3), 130–135 (1993).
32. R. R. Dils, *J. Appl. Physics*, **54**(3), 1198 (1983).
33. R. R. Dils, J. Geist, and M. L. Reilly, *J. Appl. Physics*, **59**(4), 1005 (1986).

*General References*

References 1, 9, 10, and 12 are general references.
*Platinum Metals Review*, published quarterly since 1957 by Johnson Matthey, London, U.K.
*Precious Metals* conference proceedings, IPMI, Allentown, Pa.
E. Savitsky and co-workers, *Physical Metallurgy of Platinum Metals*, MIR Publishers, Moscow, CIS, 1978.
*Degussa Edelmetall-Taschenbuch*, Degussa, Frankfurt am Main, Germany, 1967.
R. J. Seymour, *The Metallurgist and Materials Technologist*, Nov. 1982, pp. 505–508.

RICHARD J. SEYMOUR
JULIA I. O'FARRELLY
LOUISE C. POTTER
Johnson Matthey Public Limited Company

# PLATINUM-GROUP METALS, COMPOUNDS

The platinum-group metals (qv) (PGMs) comprise the elements in the second and third row of Groups 8–10 (VIII) of the Periodic Table, ie, ruthenium, osmium, rhodium, iridium, palladium and platinum. These elements are kinetically inert relative to other transition-metal ions. However, the second-row elements ruthenium, rhodium, and palladium are more labile and more oxidizing in the higher oxidation states than are the third-row PGMs. All of these elements have several readily accessible stable oxidation states. The extreme case is that of Ru and Os, which exhibit oxidation states ranging from −2 and 0 to +8. The relative ease of conversion between oxidation states gives rise to a rich catalytic chemistry. The lower oxidation states of the PGMs tend to enable them to form complexes with soft ligands such as sulfur and phosphorus whereas higher oxidation states tend toward hard ligands such as oxygen and fluorine (see COORDINATION COMPOUNDS). The platinum-group metals form binary compounds, coordination compounds, and organometallic compounds (see ORGANOMETALLICS (SUPPLEMENT)). The latter two classes have been the most active areas of research and development in the latter part of the twentieth century. Examples of important commercial applications include several homogenous catalytic processes and platinum chemotherapeutics (see CHEMOTHERAPEUTICS, ANTICANCER). Additionally, substantial progress has been made in the areas such as catalytic enantioselective reduction and oxidation, and palladium-catalyzed carbon–carbon bond formation.

## Economic Aspects

The annual market for PGM compounds is a relatively small (<10%) but very important fraction of the overall PGM market. The total demand and average annual prices of the PGMs from 1984 to 1994 are summarized in Table 1. The high intrinsic cost of each PGM is the principal limitation on applications of these compounds. Two of the more important PGM-containing products

**Table 1.  Total Demand and Prices of PGMs, 1984–1994[a]**

| Metal | Total demand, $t^b$ | | | Price range, $/kg^c$ | | |
|---|---|---|---|---|---|---|
| | 1984 | 1989 | 1994 | 1984 | 1989 | 1994 |
| platinum | 85.5 | 107.8 | 140.26 | 12,860–9,320 | 17,850–14,690 | 13,760–12,150 |
| palladium | 90.8 | 102.9 | 150.2 | 5,240–4,150 | 5,760–4,080 | 5,180–3,950 |
| rhodium | 6.8[d] | 10.3 | 12.0 | ca 28,940–11,090 | 53,860–40,190 | 31,510–19,600 |
| ruthenium | 6.2[d] | 4.8 | 10.0 | ca 6,430–ca 1,130 | 2,090–1,960 | 870–580 |
| iridium | | 0.8 | 1.3 | 16,720–ca 9,970 | 10,130–9,650 | 2,250 |
| osmium | | | 0.2[e] | 32,150–4,500 | | 12,860 |

[a]Ref. 1.
[b]To convert metric tons to troy oz, multiply by 32,154.
[c]To convert $/kg to $/troy oz, multiply by 0.0311.
[d]Value is estimated.
[e]Data is for 1993.

are the platinum antitumor agents cisplatin and carboplatin, which produced $155 and $265 million in sales, respectively, for Bristol-Myers Squibb in 1994 (2). The worldwide market for cisplatin is even larger. Cisplatin has become a generic drug in most parts of the world. PGM catalysts, especially platinum, palladium, and rhodium, are used to produce commodity and fine chemicals and pharmaceuticals. Although large quantities of these catalysts are installed in production sites worldwide, annual demand for new PGMs is quite modest because spent catalysts are recycled (3) (see CATALYSIS; CATALYSTS, REGENERATION; RECYCLING). Ruthenium oxides employed in resistor plates for miniaturized electronic circuits account for 45% of the ruthenium demand; another 45% is required for the production of ruthenium-coated anodes for the chlor–alkali industry (3) (see ALKALI AND CHLORINE PRODUCTS; METAL ANODES). Iridium and iridium–ruthenium coatings, an alternative to ruthenium-coated anodes in the chlor–alkali industry, account for 58% of the demand for iridium (3). Worldwide demand for osmium is very small. The metal is used primarily as a biological staining agent and oxidation catalyst (3).

## Refining

PGMs are obtained from mined ores or from secondary sources such as industrial scrap and used catalysts. After initial processing to form concentrates, the PGMs are dissolved in acid under oxidizing conditions. Typically, chlorine is used (4). The overall efficiency and methods of final refining processes used to produce pure metals from solution streams have evolved considerably since the early 1980s. Historically PGMs were separated by selective precipitation of complex salts, followed by recrystallization and recycling of filtrates and washes. These processes were efficient neither in the degree of purification achieved nor in the amount of time that was required. Modern plants, such as the one opened by Johnson Matthey (Royston, Hertfordshire, U.K.) in 1983, employ solvent extraction methods to extract the individual metals (5). General principles and some details of these proprietary procedures have been published (4–7).

In sequential steps, each metal is selectively extracted from an aqueous solution into an immiscible organic solvent, from which it is then removed by reduction or back-extraction. The extractant forms a coordination compound, solvate, or ion pair with a solution species of the desired element. Thus, palladium may be extracted using oximes in the presence of organic amine, sulfur, phosphorus, or arsenic compounds which serve to accelerate the rate of formation of an extractable coordination compound (5). Extraction systems based on long-chain alkyl sulfides (8), hydroxy oximes (5), and 8-hydroxyquinolines (9) have also been reported. Provided that palladium and gold have been removed and iridium is present in the +3 oxidation state, platinum is removed by extraction of $[PtCl_6]^{2-}$ as an ion pair with tri-$n$-octyl amine (5). Ruthenium and osmium are removed by oxidation to their respective tetroxides and distilled, or extracted into carbon tetrachloride. Separation of the two oxides is based on the ability of $RuO_4$ to be reduced to Ru(IV) in the presence of $OsO_4$. Subsequent distillation separates the volatile $OsO_4$ from ruthenium (4). Alternatively, ruthenium is converted to a nitrosyl species, extracted with a tertiary amine, and eventually precipitated as ammonium hexachlororuthenate (6). Next, iridium is oxidized to Ir(IV), and $[IrCl_6]^{2-}$ is extracted in the same way as with platinum (5). Finally, rhodium is removed by conventional precipitation and isolated as impure rhodium sulfate (4–6). Rhodium may also be separated from iridium by extraction with a mono-$N$-substituted amide (10,11). Compared to earlier precipitation–dissolution methods, the solvent extraction method offers the advantages of reduced processing time, improved safety owing to enclosure of allergenic platinum solutions, improved primary yields, and greater versatility.

## Analysis

Gravimetric methods of analysis of PGM compounds have been largely superseded by spectroscopic methods, which are not only more versatile and selective, but require significantly less labor (12). The methods most often used are inductively coupled plasma-emission spectroscopy (icp-es), flame atomic absorption spectroscopy (faas), and graphite furnace atomic absorption spectroscopy (gfaas). For routine determinations where the PGM compound can be dissolved in aqueous or organic solution, icp-es is generally preferred to faas, because the former is freer from chemical interference caused by differences between the sample and standard matrix. The preferred method for trace level determinations is gfaas, which can also be used for solid samples and small (as little as 1 $\mu$L) samples (12) (see TRACE AND RESIDUE ANALYSIS). The analysis of noble metals by classical and spectroscopic methods as well as a variety of specialized spectroscopic methods has been reviewed (12–14).

## Ruthenium Compounds

The most common oxidation states and the corresponding electronic configuration of ruthenium are +2 ($d^6$) and +3 ($d^5$). Compounds are usually octahedral. Compounds in oxidations states from −2 ($d^{10}$), and 0 ($d^8$) to +8 ($d^0$) have various coordination geometries. Important applications of ruthenium compounds include oxidation of organic compounds and use in dimensionally stable anodes (DSA).

**Binary Compounds.** The ruthenium fluorides are $RuF_3$ [51621-05-7], $RuF_4$ [71500-16-8], tetrameric $(RuF_5)_4$ [14521-18-7] (15), and $RuF_6$ [13693-087-8]. The chlorides of ruthenium are $RuCl_2$ [13465-51-5], an insoluble $RuCl_3$ [10049-08-8] which exists in an $\alpha$- and $\beta$-form, ruthenium trichloride trihydrate [13815-94-6], $RuCl_3 \cdot 3H_2O$, and $RuCl_4$ [13465-52-6]. Commercial $RuCl_3 \cdot 3H_2O$ has a variable composition, consisting of a mixture of chloro, oxo, hydroxo, and often nitrosyl complexes. The overall ruthenium oxidation state is closer to +4 than +3. It is a water-soluble source of ruthenium, and is used widely as a starting material. Ruthenium forms bromides, $RuBr_2$ [59201-36-4] and $RuBr_3$ [14014-88-1], and an iodide, $RuI_3$ [13896-65-6].

The high valent ruthenium oxides are important because of their use as oxidizing agents in organic synthesis. Ruthenium(VIII) oxide [20427-56-9], $RuO_4$, the most important of these compounds, is most readily prepared by oxidation of $RuO_2$ using $NaIO_4$, followed by extraction into $CCl_4$ (16–19). Other methods of preparation have been reviewed (20,21). Ruthenium also forms a dioxide [12036-10-1], $RuO_2$. The ruthenium chalcogens $RuS_2$ [12166-20-0], $RuSe_2$ [12166-21-1], and RuTe$_2$ [12166-22-2] have all been prepared by direct reaction of ruthenium and the elemental chalcogen (22). Ruthenium also forms numerous borides, silicides, phosphides, and arsenides (23,24).

**Coordination Compounds.** Ruthenium forms a variety of complexes, particularly in oxidation states +2 and +3, with ammonia and amines. Some examples are [Ru(NH$_3$)$_6$]$^{2+}$ [19052-44-9], [Ru(NH$_3$)$_6$]$^{3+}$ [18943-33-4], and the mixed valent compounds [(NH$_3$)$_5$Ru(pyz)Ru(NH$_3$)$_5$]$^{5+}$ [26253-76-9], [(NH$_3$)$_3$-RuCl$_3$Ru(NH$_3$)$_3$]$^{2+}$ [39003-96-8], and ruthenium red [32696-80-3], [Ru$_3$O$_2$-(NH$_3$)$_{14}$]$^{6+}$. Deprotonation of [Ru(NH$_3$)$_6$]$^{3+}$ forms an amido complex [Ru(NH$_3$)$_5$-(NH$_2$)]$^{2+}$ [35273-31-5] that reacts with 2,3-butandione to form a coordinated diimine, [Ru(NH$_3$)$_4${CH$_3$C(NH)C(NH)CH$_3$}]$^{2+}$ [56370-80-0] (25,26). Basic oxidation of [Ru(NH$_3$)$_6$]$^{3+}$ produces [Ru(NH$_3$)$_5$NO]$^{2+}$ [37874-79-6] (27), whereas [Ru(en)$_3$]$^{2+}$ [21393-86-2] can be oxidized to [Ru(en)$_2$(diim)]$^{2+}$ [36216-30-5] (28). Complexes of dinitrogen such as [Ru(NH$_3$)$_5$N$_2$]$^{2+}$ [19504-40-6] and [Ru(NH$_3$)$_5$-N$_2$Ru(NH$_3$)$_5$]$^{4+}$ [25754-89-6] are also known (see NITROGEN FIXATION). Examples of compounds containing heterocyclic amine ligands include [Ru(bipy)$_3$]$^{2+}$ [50525-27-4], cis-[Ru(bipy)$_2$Cl$_2$] [19542-80-4], [Ru(terpy)(bipy)Cl]$^+$ [47690-83-5], and [RuCl(bipy)$_2$CO] [85719-81-9]. The complexes of [RuL$_3$]$^{2+}$ containing 2,2'-bipyridine, phenanthroline, and similar ligands are very stable, and the photochemistry and redox properties of this class have been the subject of intensive study (29). Ruthenium also forms complexes with nitriles, oximes, and Schiff bases. The base hydrolysis of the nitrile coordinated to Ru(NCC$_2$H$_5$)(NH$_3$)$_5^{3+}$ [58560-96-6] is accelerated by a factor of about 10$^8$ relative to the free ligand (30). Ruthenium forms macrocyclic complexes with porphyrins and phthalocyanines (31–35), and tetra and triaza macrocycles such as [RuCl$_2$(cyclam)]$^+$ [66652-53-7].

Nitric oxide has a high affinity for ruthenium, and its complexes have been reviewed (36–41). The RuNO group can occur in many coordination environments. Examples include [Ru(NO)(OH$_2$)$_2$Cl$_3$] [34416-98-3], [Ru(NO)(NH$_3$)$_5$]Cl$_3$ [37874-79-6], [Ru(NO)(S$_2$CN(C$_2$H$_5$)$_2$)$_3$] [51139-54-9], [RuCl(NO)(bipy)$_2$] [29102-12-3], [Ru(NO)Cl(P(C$_6$H$_5$)$_3$)$_2$] [38856-98-3], and the stable amido complex [Ru(NO)(NH$_3$)$_4$(NH$_2$)] [60133-06-4]. Nucleophilic attack on the coordinated NO is a characteristic reaction of the RuNO group, an example being the reaction of [RuCl(NO)(bipy)$_2$] and hydroxide to produce [RuCl(NO$_2$)(bipy$_2$)] [34398-51-1]. Ruthenium nitrosyls reduce the efficiency of nuclear fuel reprocessing because these compounds are extracted along with the uranium and plutonium, and away from other fission products, by trialkylphosphines, OPR$_3$ (42) (see NUCLEAR REACTORS, CHEMICAL REPROCESSING). Phosphines tend to stabilize oxidation states +0 and +2 of ruthenium, although other oxidation states are also accessible. Typical compounds include [Ru(P(C$_6$H$_5$)$_3$)$_3$(CO)(CN-p-C$_6$H$_4$CH$_3$)] [34830-25-6], [Ru(P(C$_6$H$_5$)$_3$)$_4$H$_2$] [19529-00-1], and one of the few known ruthenium(−2) complexes, [Ru(PF$_3$)$_4$]$^{2-}$ [26876-73-3].

Complexes of oxygen ligands include the hexaaqua ions of Ru(II) and Ru(III); ketoenolates, [Ru(acac)$_3$] [14284-93-6]; complexes of iminodiacids, [RuCl(Hedta)]$^-$ [129726-27-8]; and dicarboxylates such as [Ru(ox)$_3$]$^{3-}$ [25072-75-7]. Dinuclear and trinuclear monocarboxylates in which the metals are bridged by carboxylates, such as [Ru$_2$(OOCCH$_3$)$_4$]Cl [55598-01-1] and [Ru$_3$O-(OOCCH$_3$)$_6$(OH$_2$)$_3$]OOCCH$_3$ [38998-79-7], have also been prepared. The high oxidation state complexes of ruthenium invariably contain oxo ligands. Their properties and chemistry have been reviewed (43–45). The dark green per-ruthenate anion [14333-21-2], $[RuO_4]^-$, is the only well-characterized compound of ruthenium(VII) and is most readily prepared by oxidation of $RuCl_3$ using $KIO_4$, followed by precipitation as the tetra-$i$-propylammonium salt (46). The ruthenate anion [14333-22-3], $[RuO_4]^{2-}$, can be prepared by oxidation of $RuCl_3$ by peroxydisulfate in KOH (47–49). This orange salt is stable in basic solution but disproportionates in neutral or acidic solution (50,51). The O=Ru=O and Ru=O groups are also found in a number of coordination compounds, eg, $[Ru(VI)O_2(bipy)_2]$ [84988-24-9] and $[Ru(IV)O(bipy)_2py]$ [67202-43-1]. Ruthenium forms complexes with sulfur ligands including sulfoxides, which may be O- or S-bound (52), as in $[RuCl_2(DMSO)_4]$ [11070-19-2], as well as thioethers, thioureas, dithiolenes, dithiophosphinates, thiolates, and thiocarbamates. Very few complexes with ligands of the higher chalcogenides selenium and tellurium have been reported. Ruthenium also forms complexes of silyl, germyl, and stannyl ligands (23,29).

**Organometallic Compounds.** Ruthenium, predominately in the oxidation states 0 and +2, forms numerous mononuclear and polynuclear organometallic compounds. A few examples of compounds in both higher and lower oxidation states also exist. The chemistry of polynuclear ruthenium complexes is extensive and has been reviewed (53–59).

Numerous examples exist of complexes of carbon monoxide (qv), such as $[Ru(CO)_4]^{2-}$ [57398-60-4], $[Ru(CO)_5]$ [16406-48-7], $[Ru(CO)Cl_3]^{2-}$ [66115-12-6], $[Ru_3(CO)_{12}]$ [15243-33-1], and $[Ru_6(CO)_{18}]$ [12277-77-9] (see CARBONYLS); isocyanides, eg, $[Ru(CNCH_3)_6]^{2+}$ [49631-67-6]; carbenes, $[Ru(CNCH_3)_5\{C(NH\,CH_3)_2\}]^{2+}$ [42566-55-2]; and carbynes (60). Ruthenium alkyls and acyls are less common; examples include $[RuN(CH_3)_4]^{1-}$ [102649-23-0] and $[Ru(CO)_2(\eta^5-C_5H_5)(CO\text{-}c\text{-}C_3H_5)]$ [107769-65-3]. Ligands with $\pi$-systems form $\eta^2-\eta^6$ compounds of ruthenium. Examples include $[Ru(CO)_3(1,5\text{-cod})]$ [32874-17-2], $[Ru(\eta^6-C_7H_8)(\eta^5\text{-}C_7H_7)]$ [77357-78-9], and a large number of arene complexes such as $[Ru_2Cl_4(\eta^6\text{-}C_6H_6)_2]$ [51831-98-2]. Ruthenocene [1287-13-4], $[Ru(\eta^5\text{-}C_5H_5)_2]$, and its derivatives have chemistry similar to the ferrocenes (see ORGANOMETALLICS (SUPPLEMENT)).

**Synthesis.** The most important starting material for the synthesis of ruthenium compounds is the commercial trichloride trihydrate. Other useful starting materials include $[RuCl_5(OH_2)]^{2-}$ [3187-84-2], $[Ru_3O(OOCCH_3)_6(OH_2)_3]OOCCH_3$ [38998-79-7], and $[Ru_3(CO)_{12}]$ [15243-33-1].

**Uses.** *Oxidation.* The use of ruthenium oxo complexes as oxidants has been reviewed (44,61). The most important of these is $RuO_4$, a very strong oxidant (60,62) which reacts violently with common organic solvents such as ether and benzene (62). Ruthenium tetraoxide oxidizes alcohols to aldehydes (qv) and ketones (qv), aldehydes to acids, ethers to esters (qv) or lactones, tertiary amines to amides, and amides to imides. It generally cleaves olefins to aldehydes or ketones. It can oxidize aromatic rings and sterically hindered groups that are inert to other oxidants. Oxidations are frequently carried out catalytically in carbon tetrachloride or acetone, regenerating the $RuO_4$ with sodium hypochlorite or sodium metaperiodate. The perruthenate ion $[RuO_4]^-$, by contrast, is a much milder and more selective oxidant, able to oxidize alcohols to aldehydes and ketones in the presence of sensitive functional groups such as epoxides, silyl ethers, lactones, acetals, and allyl groups. This anion can be generated catalytically using $N$-methylmorpholine $N$-oxide (63). A variety of ruthenium complexes catalyze aerobic oxidation of alcohols (64). [$RuO_2(TPP)$] [$123051$-$66$-$1$] catalyzes the aerobic epoxidation of olefins with retention of configuration (65).

*Catalysis.* Ruthenium complexes catalyze a wide range of reactions, including hydroformylation (66) (see OXO PROCESS), carbonylation (67), water gas shift reaction (68), homogeneous hydrogenation (69), and transfer hydrogenation (64,70). Other catalyzed reactions include addition of aromatic C–H bonds to olefins (71), homologation of carboxylic acids (72,73), N-alkylation of amines by alcohols (64), N-alkyl migration of amines (74), disproportionation of aldehydes to esters (64), dehydrogenation of alcohols (75), reductive carbonylation of nitroaromatics to form carbamates (76), addition to carboxylic acids to terminal alkynes to form vinyl esters (77), photochemical reduction of carbon dioxide (qv) to formate (78,79), and formation of ethylene glycol and other products from synthesis gas in melts (80,81). A series of well-defined ruthenium carbene complexes capable of metathesizing both strained and unstrained olefins have been reported. These catalysts are tolerant to many functional groups and some are active in aqueous media (82). Enantioselective hydrogenations and transfer hydrogenations can be accomplished using chiral auxiliaries (71,83). For example, hydrogenation of itanoic acid, an $\alpha,\beta$-unsaturated carboxylic acid, occurs with an enantioselectivity of 97% in the presence of the chiral ruthenium catalyst [$RuH_2((-)binap)_2$] (70). Results are substrate- and catalyst-dependent.

*Biology and Medicine.* Biological and medicinal applications of ruthenium complexes have been reviewed (84,85). The photochemical and electrochemical properties, and kinetic stability, of ruthenium complexes have led to use as probes for the structure of deoxyribonucleic acid (DNA) (86) and for the study of the electron-transfer process in proteins (qv) (87–89). Ruthenium complexes have been developed to cleave DNA (90). Ruthenium red is used as a histological stain. The compound [$Ru_2(\mu\text{-}O)(OOCH)_2(NH_3)_8$]$^{3+}$ [$133399$-$54$-$9$] has been reported to be a potent *in vitro* noncompetitive inhibitor of biological $Ca^{2+}$ transport (91). Several compounds including *cis*-[$RuCl_2(DMSO)_4$] [$11070$-$19$-$2$] and [$RuCl_4(im)_2$]$^+$ [$103875$-$27$-$0$] have been reported to have antitumor activity (85). However, efficacy in humans has not been demonstrated.

*Miscellaneous.* Ruthenium dioxide-based thick-film resistors have been used as secondary thermometers below 1 K (92). Ruthenium dioxide-coated anodes are the most widely used anode for chlorine production (93). Ruthenium(IV) oxide and other compounds are used in the electronics industry as resistor material in applications where thick-film technology is used to print electrical circuits (94) (see ELECTRONIC MATERIALS). Ruthenium electroplate has similar properties to those of rhodium, but is much less expensive. Electrolytes used for ruthenium electroplating (95) include [$Ru_2Cl_8(OH_2)_2N$]$^{3-}$ [$55186$-$41$-$9$], $Na_2[Ru(NO_2)_4(NO)OH]$ [$13859$-$66$-$0$], and ($NH_4$)$_2$[$RuCl_5(NO)$] [$13820$-$58$-$1$]. Several photocatalytic cycles that generate $H_2$ or $O_2$ from water in the presence of a sacrificial reductant or oxidant employ a ruthenium complex, typically [$Ru(bipy)_3$]$^{2+}$, as the photon absorber (96,97). A series of mixed binuclear ruthenium complexes having a variety of bridging ligands have been the subject of numerous studies into the nature of bimolecular electron-transfer reactions and have been extensively reviewed (99–102). The first example of this system, reported in 1969 (103), is the Creutz-Taube complex [35599-57-6], $[Ru_2(pyz)(NH_3)_{10}]^{5+}$.

## Osmium Compounds

The most common oxidation states and the corresponding electronic configurations of osmium are +2 ($d^6$) and +3 ($d^5$), which are usually octahedral. Stable oxidation states that have various coordination geometries include −2 ($d^{10}$), and 0 ($d^8$), to +8 ($d^0$). The single most important application is $OsO_4$ oxidation of olefins to diols. Enantioselective oxidations have also been demonstrated.

**Binary Compounds.** Osmium forms four fluorides: $OsF_4$ [54120-05-7], $OsF_5$ [31576-40-6], $OsF_6$, and $OsF_7$ [16949-69-2]. Osmium hexafluoride [13768-38-2] is more stable than the other platinum-metal hexafluorides. Three chlorides, $OsCl_3$, $OsCl_4$ [10026-01-4], and $OsCl_5$ [71328-74-0]; two bromides, $OsBr_3$ [59201-51-3] and $OsBr_4$ [59201-52-4]; and three iodides, $OsI$ [76758-38-8], $OsI_2$ [59201-57-9], and $OsI_3$ [59201-58-0], have been described. Osmium tetroxide [20816-12-0], $OsO_4$, is a volatile, toxic liquid and is obtained by oxidizing osmium-containing solutions using $HNO_3$, or by heating the finely divided metal in air (104). The tetroxide is reduced readily by organic matter. The other stable oxide is osmium dioxide [12036-02-1], $OsO_2$. Other binary chalcogenides include $OsS_2$ [12137-61-0], $OsSe_2$ [12310-19-9], and $OsTe_2$ [12165-67-2]. Osmium also forms borides, silicides, and phosphides (24).

**Coordination Compounds.** Osmium in oxidation states from +2 to +8 forms a wide range of complexes with nitrogen ligands. Amine complexes of osmium include $[Os(NH_3)_5(CO)]^{2+}$ [70528-08-4], $[Os(NH_3)_5(OSO_2CF_3)]^{2+}$ [83781-29-7], $[Os(NH_3)_6]^{3+}$ [48016-91-7], $[OsO_2(en)_2]^{2+}$ [61202-82-2], porphyrins $[Os(TPP)CO]$ [104677-48-7] and $[Os(OEP)(O)_2]$ [59650-32-7], and the tetraamido complex $[OsO_2(o\text{-}(NH)_2C_6H_4)_2]$ [127472-00-8]. Schiff base and chelating amide complexes have also been prepared. Unlike ruthenium, Os(II) hexaammine is labile and as of this writing has not been isolated. Osmium(II) and (III) form many dinitrogen complexes, such as $[Os(NH_3)_5N_2]^{2+}$ [22840-90-0] and $[Os_2N_2(NH_3)_{10}]^{5+}$ [81988-68-3].

Like ruthenium, amines coordinated to osmium in higher oxidation states such as Os(IV) are readily deprotonated, as in $[Os(en)(NHCH_2CH_2NH_2)]^{2+}$ [111614-75-6]. This complex is subject to oxidative dehydrogenation to form an imine complex (105). An unusual Os(IV) hydride, $[OsH_2(en)_2]^{2+}$ [57345-94-5] has been isolated and characterized. The complexes of aromatic heterocyclic amines such as pyridine, bipyridine, phenanthroline, and terpyridine are similar to those of ruthenium. Examples include $[Os(bipy)_3]^{2+}$ [23648-06-8], $[Os(bipy)_2acac]$ [47691-08-7], $[Os(terpy)_2]^{3+}$ [100815-62-1], $[Os_2(pyz)(NH_3)_{10}]^{5+}$ [85282-23-1], $[Os(OSO_2CF_3)_2(bipy)_2]$ [104474-97-7], $[OsCl_4(bipy)]$ [57288-05-8], and $[Os(phen)(OH)_2(O)_2]$ [69531-97-1]. The photochemistry and redox properties of $OsL_3^{2+/3+}$ complexes, where L = heteroaromatic chelating amines, have been investigated, although less thoroughly than the corresponding ruthenium complexes.

Osmium readily forms nitrosyl complexes such as $[OsCl_5(NO)]^{2-}$ [53796-27-3], $[Os(NO)(NH_3)_5]^{3+}$ [52720-70-4], and $[Os(P(C_6H_5)_3)_2(OH)(NO)_2]^+$

[*47847-71-2*]. Coordinated NO in [Os(NO)(terpy)(bipy)]$^{2+}$ [*90219-19-5*] is subject to nucleophilic attack analogous to this same effect on ruthenium complexes. A much more unusual reaction is protonation of the NO of [OsCl(P(C$_6$H$_5$)$_3$)$_2$-(NO)(CO)] [*22186-54-5*] to produce the complex [OsCl$_2$(P(C$_6$H$_5$)$_3$)$_2$(HNO)(CO)] [*31011-34-4*], containing coordinated nitrous acid. Osmium forms a variety of imido and nitrido compounds, such as [OsO$_3$(N-t-C$_4$H$_9$)] [*50381-48-1*], [Os(N-t-C$_4$H$_9$)$_4$] [*129117-79-9*], [OsCl$_5$(N)]$^{2-}$ [*42718-64-9*], [OsCl$_4$(OH$_2$)(N)]$^-$ [*59187-86-9*], [OsCl$_2$(terpy)(N)]$^+$ [*127571-47-5*], and [Os$_3$(OH$_2$)$_6$(NH$_3$)$_8$(N)$_2$]$^{6+}$ [*76247-72-8*].

Numerous phosphine and arsine complexes have been synthesized and characterized predominately with osmium in the +2, +3, or +4 oxidation states. Examples include [OsCl$_2$(dppm)$_2$] [*108341-10-2*], [OsCl$_3$(P(CH$_3$)$_2$(C$_6$H$_5$)$_3$] [*20500-70-3*], [Os$_2$Cl$_6$(dppm)$_2$(O)] [*87883-12-3*], and [Os(AsC$_2$H$_5$(C$_6$H$_5$)$_2$)$_4$H$_2$] [*27498-19-7*]. An example of an unusually low oxidation state is the Os(−2) complex K$_2$[Os(PF$_3$)$_4$] [*26876-74-4*]. High coordination numbers and formal oxidation states are found in the phosphine hydrides, eg, [Os(P(CH$_3$)(C$_6$H$_5$)$_2$)H$_6$] [*25895-55-0*] and Os(P(C$_2$H$_5$)$_2$C$_6$H$_5$)$_3$H$_4$ [*24228-58-8*].

The oxo ligand is the most common motif in osmium chemistry. By far the single most important complex is OsO$_4$ a much milder oxidizing agent than RuO$_4$. Many complexes containing the oxo or cis or trans dioxo group (O=Os=O) are known, including [OsO$_4$(OH)$_2$]$^{2-}$ [*131725-28-5*], *trans*-[OsO$_2$(OH)$_4$]$^{2-}$ [*88993-76-4*], [OsO$_2$(OCH$_3$)$_4$]$^{2-}$ [*120169-51-9*], and [OsO$_3$N]$^-$ [*19662-38-5*]. Osmium does not form a hexaaqua complex, although the aqua ligand is known in combination with other ligands such as [Os(NH$_3$)$_5$(OH$_2$)]$^{3+}$ [*53222-99-4*]. Other complexes include [Os(acac)$_3$] [*60133-05-3*], [Os(ox)Cl$_4$]$^{2-}$ [*74325-52-3*], [Os(cat)$_3$] [*67799-34-2*], and the bridging carboxylato complex [Os$_2$Cl$_2$(O$_2$CCH$_3$)$_4$] [*81519-41-7*]. Relatively few osmium compounds having sulfur ligands have been characterized. Some complexes include sulfoxides, [OsCl$_3$(DMSO)$_3$] [*66009-49-2*]; dithiocarbamates, [Os(S$_2$CN(C$_2$H$_5$)$_2$)$_3$]; thioureas, [Os(SC(NH$_2$)$_2$)$_6$]$^{3+}$ [*12300-37-7*]; thiocarbonyls, Os(sacsac)$_3$ [*29966-35-6*]; and macrocyclic thioethers, [Os([9]aneS$_3$)$_2$]$^{2+}$ [*145312-30-7*].

**Organometallic Compounds.** Osmium forms numerous mononuclear and polynuclear organometallic complexes, primarily in lower oxidation states. There are many complexes of carbon monoxide, such as [Os(CO)$_5$] [*16406-49-8*], [Os(CO)$_4$H$_2$] [*22372-70-9*], [Os$_3$(CO)$_{11}$H$_2$] [*56398-24-4*], [Os$_6$(CO)$_{18}$] [*37216-50-5*], [Os$_3$(CO)$_{10}$(μ-O=CCH$_3$)(H)] [*65908-54-5*], and to a lesser extent isonitriles such as *trans*-[OsBr$_2$(CNCH$_3$)$_4$] [*42608-08-2*]. The most important carbon monoxide complex is the trinuclear complex [Os$_3$(CO)$_{12}$] [*15696-40-9*], which is a starting material for many mononuclear and polynuclear complexes. Osmium forms alkyl, [Os(CO)$_4$(CH$_3$)$_2$] [*22639-01-6*]; aryl, [Os(2-methylphenyl)$_4$] [*101191-32-6*]; acyl, *trans*-[Os(dppe)$_2$(CHO)(CO)]$^+$ [*89411-52-9*]; vinyl; carbene, [OsCl(P(C$_6$H$_5$)$_3$)$_2$(NO)(=CH$_2$)] [*86645-81-0*]; and carbyne complexes, [OsCl(P(C$_6$H$_5$)$_3$)$_2$(CO)(C$_6$H$_5$)] [*75346-82-6*]. Osmium is unusual in its tendency for *dihapto* binding of acyls and related ligands such as iminoacyl, and thioacyl complexes as in [Os($\eta^2$-CS(p-C$_6$H$_4$CH$_3$))(P(C$_6$H$_5$)$_3$)$_2$(O$_2$CCF$_3$(CO)] [*68111-80-8*].

Many unsaturated compounds form $\pi$-complexes with osmium including alkenes, [Os(P(C$_6$H$_5$)$_3$)$_2$(CO)$_2$($\eta$-C$_2$H$_4$)] [*79725-76-1*]; alkynes, [Os(P(C$_6$H$_5$)$_3$)$_2$(NO)(CO)($\eta$-C$_2$H$_2$)]$^+$ [*41529-78-6*]; allyls, [Os(NH$_3$)$_5$($\eta^3$-C$_3$H$_5$)]$^{3+}$ [*131617-37-3*]; dienes, [Os(CO)$_3$(1,4-$\eta$-C$_4$H$_6$)] [*75756-64-8*]; cyclopentadienes, [Os(C$_5$H$_5$)$_2$] [*1273-

81-0]; and arenes, $[Os_2Cl_3(\eta\text{-}C_6H_6)_2]^+$ [70317-01-0]. Other unusual complexes are the stable $\eta^2$-bound carbonyl, $[Os(NH_3)_5\{\eta^2\text{-}(C,O)\text{-}acetone)\}]$ [105164-47-4], and benzene adducts, $[Os(NH_3)_5(\eta^2\text{-}C_6H_6)]^{2+}$ [107202-74-4] and $[Os(NH_3)_5\text{-}(C_6H_6)Os(NH_3)_5]^{4+}$ [107202-75-5] (106,107). The chemistry of osmium clusters is replete with examples of ligands engaged in intimate and unusual bonding modes with multiple metal centers.

**Synthesis.** The most important starting material for synthesis of osmium complexes is $OsO_4$. Other important complexes are disodium hexachloroosmium [1307-81-9], $Na_2[OsCl_6]$; $[Os(NH_3)_5(O_2SCF_3)]^{2+}$; and $[Os_3(CO)_{12}]$.

**Uses.** *Oxidations.* The use of osmium for oxidations has been reviewed (108,109). The most important application is the oxidation of olefins to cis-diols using $OsO_4$. This reaction occurs via an osmate ester intermediate that is oxidatively or reductively cleaved to form the cis-diol. The reaction may be carried out stoichiometrically, but most often catalytic amounts are used in the presence of appropriate oxidants such as hydrogen peroxide (qv), ferricyanide, chloramine-T, chlorate, or N-methylmorpholine N-oxide. Osmium tetroxide can also oxidize alcohols to aldehydes, ketones, or carboxylic acids. Catalytic enantioselective dihydroxylation of achiral olefins by $OsO_4$ can be accomplished in the presence of chiral ligands (109–111). Enantiomeric excesses of 77% to >99.5% have been achieved using simple olefins in the presence of a chiral pthalazine ligand, an organic sulfonamide to accelerate the rate of osmate ester hydrolysis, and ferricyanide as the cooxidant (110).

Nitrogen-transfer reactions of osmium imido species lead to cis diamines or amino alcohols. Reaction of $OsO_2(N\text{-}t\text{-}C_4H_9)_2$ [63174-13-0] with olefins produces vicinal diamines after reductive cleavage. Catalytic oxyamination of olefins using chloramine-T or N-chlor-N-argenocarbamates yields vicinal hydroxy toluenesulfonamides (112) or carbamates (113), respectively, which may be deprotected to vicinal amino alcohols.

*Miscellaneous.* Aside from the oxidation chemistry described, only a few catalytic applications are reported, including hydrogenation of olefins (114,115), $\alpha,\beta$-unsaturated carbonyl compounds (116), and carbon monoxide (117); and the water gas shift reaction (118). This is so owing to the kinetic inertness of osmium complexes. A 1% by weight osmium tetroxide solution is used as a biological stain, particularly for preparation of samples for electron microscopy. In the presence of pyridine or other heterocyclic amines it is used as a selective reagent for single-stranded or open-form B-DNA (119) (see NUCLEIC ACIDS). Osmium tetroxide has also been used as an indicator for unsaturated fats in animal tissue. Osmium tetroxide has seen limited if controversial use in the treatment of arthritis (120,121).

## Rhodium Compounds

The most common oxidation states and corresponding electronic configurations of rhodium are $+1$ ($d^8$), which is usually square planar although some five coordinate complexes are known, and $+3$ ($d^6$) which is usually octahedral. Dimeric rhodium carboxylates are $+2$ ($d^7$) complexes. Compounds in oxidation states $-1$ ($d^{10}$) to $+6$ ($d^3$) exist. Significant industrial applications include rhodium-catalyzed carbonylation of methanol to acetic acid and acetic anhydride, and hydroformylation of propene to $n$-butyraldehyde. Enantioselective catalytic reduction has also been demonstrated.

**Binary Compounds.** Four fluorides of rhodium can be prepared. These include the most stable, RhF$_3$ [60804-25-3]; RhF$_4$ [60617-65-4]; tetrameric (RhF$_5$)$_4$ [14521-17-6]; and very reactive RhF$_6$ [13693-07-7]. The trihalides RhCl$_3$ [10049-07-7], RhBr$_3$ [15608-29-4], and RhI$_3$ [15492-38-3] are the only other well-characterized halides. These are insoluble when anhydrous. The chloride and bromide form soluble hydrates, ie, RhCl$_3$·3H$_2$O [13569-65-8] and RhBr$_3$·2H$_2$O [76758-40-2]. Rhodium forms two oxides, RhO$_2$ [12137-27-8] and Rh$_2$O$_3$ [12036-35-0]. The latter is used as a catalyst in the manufacture of HI. Rhodium(III) hydroxide [21656-02-0], Rh(OH)$_3$, has a variable water content and is likely a hydrated form of Rh$_2$O$_3$ (121). Rhodium forms a sulfide, Rh$_2$S$_3$ [12067-06-0], and a selenide, RhSe$_2$ [12038-76-5], as well as borides, silicides, phosphides, and arsenides (23,24).

**Coordination Compounds.** A large number of Rh(III) ammine and amine complexes have been isolated, including [Rh(NH$_3$)$_6$]$^{3+}$ [16786-63-3], [RhCl(NH$_3$)$_5$]$^{2+}$ [15379-09-6], [RhCl$_2$(en)$_2$]$^+$ [29701-59-5], and cis-[Rh(en)$_2$H$_2$]$^+$ [108566-47-8]. These $d^6$ complexes are substitutionally and sterochemically inert, allowing resolution of optically active complexes such as [Rh(en)$_3$]$^{3+}$ [30983-68-7]. The high degree of kinetic inertness of Rh(III) is evident in the apparent anation of [Rh(OH)$_2$(NH$_3$)$_5$]$^{3+}$ [15337-79-8], which actually occurs by electrophilic attack of NO$^+$ on coordinated hydroxyl, followed by intramolecular isomerization of the initially formed O-bound nitrito to an N-bound nitrito ligand (122). The substitution chemistry of Rh(III) amine complexes has been extensively studied and reviewed (123). The redox stability of the +3 oxidation state, combined with photolability of many amine complexes, has resulted in considerable investigation of the photochemistry of these complexes (124). Rhodium compounds containing heterocyclic amine ligands are numerous. Examples include [RhCl$_2$(py)$_4$]$^+$ [14077-03-3], mer-[RhCl$_3$(py)$_3$]$^+$ [14267-66-4], and [Rh(bipy)$_3$]$^{3+}$ [47780-17-6]. The last complex has been employed as part of an aqueous photochemical cycle that produces H$_2$ by oxidation of an organic donor (96). Other complexes contain nitriles, eg, [Rh(NCC$_6$H$_5$)-(NH$_3$)$_5$]$^{2+}$ [46343-55-9] and [RhCl$_3$(NCCH$_3$)$_3$] [15747-79-2]; oximes, eg, [Rh(OH$_2$)(DMG)$_2$CH$_3$] [26026-41-5]; Schiff bases, [Rh(salen)O$_2$] [91067-88-8]; pyrazolylborates, [RhCl$_2$(CH$_3$OH){HB(3,5-dimethylpyrazolato)$_3$}] [73117-95-0]; nitrito ligands, [Rh(NO$_2$)(NH$_3$)$_5$]$^{2+}$ [34412-13-0]; and nitrosyl ligands, [Rh(P(C$_6$H$_5$)$_3$)$_2$(O$_2$CCF$_3$)$_2$(NO)] [87652-47-9]. Basic hydrolysis of coordinated nitriles is accelerated by six orders of magnitude relative to the free ligand. However, this is two orders of magnitude slower than hydrolysis catalyzed by ruthenium. Macrocyclic complexes of rhodium include [Rh(cyclam)Cl$_2$]$^+$ [38781-23-6]; porphyrins, [RhCl(TPP)(CO)] [42892-91-1]; and phthalocyanines, [Rh(Pc)] [37099-31-3]. Phosphine complexes of rhodium are a large class of compounds that are important catalytically. Phosphines stabilize a range of oxidation states from $-1$ to $+3$, but $+1$ and $+3$ are the most important. Some phosphine complexes of rhodium are Wilkinson's catalyst [14694-95-2], [RhCl(P(C$_6$H$_5$)$_3$)$_3$]; [RhCl$_2$(P(C$_6$H$_5$)$_3$)$_2$]$^+$ [132108-94-2]; [RhCl(P(C$_6$H$_5$)$_3$)$_2$H$_2$] [12119-41-4]; [Rh(diphos)$_2$]$^+$ [65521-62-2]; and the cluster [Rh$_{55}$Cl$_{20}$(P($t$-C$_4$H$_9$)$_3$)$_{12}$] [104619-08-1]. A few arsine complexes, eg, [Rh(diars)$_2$]$^+$

[*53450-80-9*], that are similar to the corresponding phosphine complexes, are also known.

Rhodium complexes with oxygen ligands, not nearly as numerous as those with amine and phosphine complexes, do, however, exist. A variety of compounds are known, including [Rh(ox)$_3$]$^{3-}$ [*18307-26-1*], [Rh(acac)$_3$] [*14284-92-5*], the hexaaqua ion [Rh(OH$_2$)$_6$]$^{3+}$ [*16920-31-3*], and Schiff base complexes. Soluble rhodium sulfate, Rh$_2$(SO$_4$)$_3$·$x$H$_2$O, exists in a yellow form [*15274-75-6*], which probably contains [Rh(H$_2$O)$_6$]$^{3+}$, and a red form [*15274-78-9*], which contains coordinated sulfate (125). The structure of the soluble nitrate [Rh(NO$_3$)$_3$·2H$_2$O [*10139-58-9*] is also complex (126). Another class of rhodium compounds is the Rh(II) oxo-bridged dimers such as [Rh$_2$(OOCCH$_3$)$_4$L$_2$], where L is an oxygen, nitrogen, carbon, sulfur, or phosphorus donor ligand. Examples are [Rh$_2$(OOCCH$_3$)$_4$(OH$_2$)$_2$] [*28410-96-0*] and [Rh$_2$(mhp)$_4$] [*75310-08-6*]. Complexes containing sulfur ligands include dithiocarbamates, [Rh$_2$(S$_2$CN(C$_2$H$_5$)$_2$)$_5$]$^{+}$ [*15109-46-5*]; dialkyldithiophosphonates, [Rh(S$_2$P(OC$_2$H$_5$)$_2$)$_3$] [*33991-54-7*]; thioethers, [RhCl$_3$(S(C$_2$H$_5$)$_2$)$_3$] [*14784-72-6*]; thiocyanates, [Rh(SCN)$_6$]$^{3-}$ [*17731-24-7*]; sulfoxides, [RhCl$_3$(DMSO)$_3$] [*56195-04-1*]; thiocarbonyls, [Rh(sacsac)$_3$] [*26304-95-0*]; and macrocyclic complexes, [Rh((SCH$_2$CH$_2$)$_3$)$_2$] [*116115-75-4*]. A few complexes containing selenium ligands, eg, [Rh(Se$_2$C$_2$(CN)$_2$)$_3$] [*45229-91-2*], have been prepared.

**Organometallic Compounds.** The predominant oxidation states of rhodium organometallics are +1 and +3. Rhodium forms a large number of mononuclear and polynuclear carbonyl complexes, including the complexes [Rh(CO)$_4$]$^{-}$ [*44797-04-8*], *trans*-[RhI$_4$(CO)$_2$]$^{-}$ [*55634-00-9*], [Rh$_2$Cl$_2$(CO)$_4$] [*14523-22-9*], [Rh$_4$(CO)$_{12}$] [*19584-30-6*], [Rh$_6$(CO)$_{16}$] [*28407-51-4*], and the A-frame molecules, typified by [Rh$_2$Cl$_2$(dppm)$_2$(CO)$_2$] [*22427-58-3*]. An unusual dinuclear porphyrin complex is [Rh$_2$(CO)$_4$(TPP)] [*88083-36-7*], which contains two Rh(CO)$_2$ units, one on each side of the ring and each coordinated to two porphyrin nitrogens (127). Isocyanides react readily with many rhodium compounds to form complexes such as [Rh(CN-$t$-C$_4$H$_9$)$_4$]$^{+}$ [*34195-61-4*], [Rh$_2$Cl$_2$(CN-$t$-C$_4$H$_9$)$_4$] [*37017-32-6*], and the bridging complex [Rh$_2$((CNCH$_2$)$_2$CH$_2$)$_4$] [*717375-02-5*]. Carbene complexes are also known, eg, [RhCl(P(C$_6$H$_5$)$_3$)$_2$(C(NC$_2$H$_5$CH$_2$)$_2$)] [*54686-50-9*].

Rhodium forms $\sigma$-bonded alkyl, *trans*-[RhI(CNCH$_3$)$_4$CH$_3$] [*45151-89-1*] and [RhI$_2$(P(C$_6$H$_5$)$_3$)$_2$CH$_3$] [*47829-28-7*]; aryl, [Rh(P(CH$_3$)$_3$)$_3$C$_6$H$_5$] [*71110-93-5*]; and acyl, [RhI(P(C$_6$H$_5$)$_3$)(COCH$_3$)C$_5$H$_5$] [*34676-55-6*], complexes, as well as alkenyl and alkynyl compounds. Alkynes readily undergo oxidative cyclodimerization in the presence of [RhCl(P(C$_6$H$_5$)$_3$)$_2$] to form metallocyclopentadienyl complexes such as [*36463-91-9*], shown below.

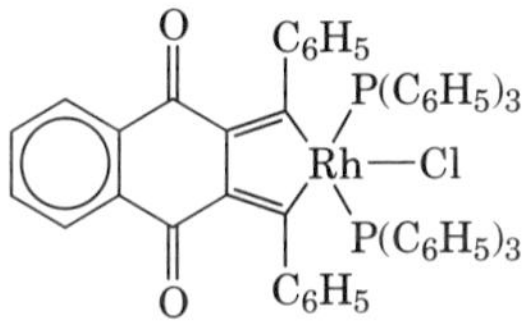

Reaction with additional alkyne releases an organic anthraquinone (128). Numerous $\pi$-complexes are known. Examples include complexes of alkenes,

[Rh$_2$Cl(CH$_2$CH$_2$)$_4$] [12081-16-2]; dienes, [Rh(cod)$_2$]$^+$ [35338-22-8]; alkynes, RhCl(P(C$_6$H$_5$)$_3$)$_2$(C$_6$H$_5$CCC$_6$H$_5$)] [12124-06-0]; allyls, [Rh(C$_3$H$_5$)$_3$] [12082-48-3]; conjugated dienes, [Rh(P(C$_6$H$_5$)$_3$)$_2$(1,4-cyclohexadiene)]$^+$ [32799-73-8]; cyclobutadienes, [Rh$_2$Cl$_2$((C$_6$H$_5$C)$_4$)$_2$] [50477-60-6]; and many others. Many $\eta^5$-cyclopentadiene and $\eta^6$-arene complexes such as [(Rh$_3$(CO)$_3$(C$_5$H$_5$)$_3$)H]$^+$ [76082-92-3] and [Rh(nbd)(C$_6$H$_6$)]$^+$ [36683-23-2] are known. Although rhodacene is unstable at room temperature (129), rhodacenium hexafluorophosphate [37205-11-1], [Rh(C$_5$H$_5$)$_2$]PF$_6$, is a colorless crystalline solid.

**Synthesis.**    The most important starting material for rhodium compounds is rhodium(III) chloride hydrate [20765-98-4], RhCl$_3$·$n$H$_2$O. Other commercially available starting materials useful for laboratory-scale synthesis include [Rh$_2$(OOCCH$_3$)$_4$] [5503-41-3], [Rh(NH$_3$)$_5$Cl]Cl$_2$ [13820-95-6], [Rh$_2$Cl$_2$(CO)$_4$] [32408-34-7], and [Rh$_2$Cl$_2$(cod)$_2$] [12092-47-6].

**Uses.**    *Catalysis.*    The readily accessible +1 and +3 oxidation states of rhodium make it a useful catalyst. There are several reviews of the catalytic properties of rhodium available (130–132). Rhodium-catalyzed methanol carbonylation (Monsanto process) accounted for 81% of worldwide acetic acid by 1988 (133). The Monsanto acetic acid process is carried out at 175°C and 1.5 MPa (200 psi). Rhodium is introduced as RhCl$_3$ but is likely reduced in a water gas shift reaction (134) to the active catalytic species [RhI$_2$(CO)$_2$]$^-$ [38255-39-9]. The basic steps of the reaction (eqs. 1–6) are reaction of methanol (qv) and HI to form methyl iodide, oxidative addition of methyl iodide, migratory insertion of the methyl to the carbonyl, coordination of CO, reductive elimination of acetyl iodide, and hydrolysis of acetyl iodide to acetic acid. The iodide promoter is important for the success of the reaction because iodide is a good leaving group, a good nucleophile, and a good ligand for rhodium.

$$CH_3OH + HI \longrightarrow CH_3I + H_2O \tag{1}$$

$$[RhI_2(CO)_2]^- + CH_3I \longrightarrow [RhI_3(CO)_2CH_3]^- \tag{2}$$

$$[RhI_3(CO)_2CH_3]^- \longrightarrow [RhI_3(COCH_3)(CO)]^- \tag{3}$$

$$[RhI_3(COCH_3)(CO)]^- + CO \longrightarrow [RhI_3(COCH_3)(CO)_2]^- \tag{4}$$

$$[RhI_3(COCH_3)(CO)_2]^- \longrightarrow [RhI_2(CO)_2]^- + CH_3COI \tag{5}$$

$$H_2O + CH_3COI \longrightarrow CH_3COOH + HI \tag{6}$$

A related but distinct rhodium-catalyzed methyl acetate carbonylation to acetic anhydride (134) was commercialized by Eastman in 1983. Anhydrous conditions necessary to the Eastman acetic anhydride process require important modifications (24) to the process, including introduction of hydrogen to maintain the active [RhI$_2$(CO)$_2$]$^-$ catalyst and addition of lithium cation to activate the alkyl methyl group of methyl acetate toward nucleophilic attack by iodide.

Homogeneous rhodium-catalyzed hydroformylation (135,136) of propene to $n$-butyraldehyde (qv) was commercialized in 1976. $n$-Butyraldehyde is a key intermediate in the synthesis of 2-ethylhexanol, an important plasticizer alcohol. Hydroformylation is carried out at <2 MPa (<290 psi) at 100°C. A large excess of triphenyl phosphine contributes to catalyst life and high selectivity for $n$-butyraldehyde ($>$10:1) yielding few side products (137). Normally, product separation from the catalyst [Rh(P(C$_6$H$_5$)$_3$)$_3$(CO)H] [$17185$-$29$-$4$] is achieved by distillation.

In 1984 Rührchemie introduced a water-soluble catalyst, [Rh(P(3-(NaO$_3$S)C$_6$H$_4$)$_3$)(CO)H] [$103823$-$25$-$2$], in a process producing 100,000 t/yr of $n$-butyraldehyde (138). Use of the water-soluble rhodium catalyst simplifies catalyst recovery and results in negligible rhodium losses, because the catalyst is not soluble in the organic product phase. Although a definitive mechanism has not been established, a plausible course based on model studies (139) includes dissociative loss of a ligand from the catalyst, coordination of propene, insertion into the metal hydride to form a rhodium propyl species, coordination of CO, migration of the propyl into the carbonyl, oxidative addition of H$_2$, and reductive elimination, forming $n$-butyraldehyde and a coordinatively unsaturated catalyst species.

Efficient enantioselective asymmetric hydrogenation of prochiral ketones and olefins has been accomplished under mild reaction conditions at low (0.01–0.001 mol %) catalyst concentrations using rhodium catalysts containing chiral ligands (140,141). Practical synthesis of several optically active natural products and medicines such as L-DOPA, ($-$)-deoxypodohyllotoxin, ($+$)-collinusin, ($R$)-rhenyephrine hydrochloride, and ($S$)-levamisole have been reported (141). The observed enantioselectivity is highly substrate- and catalyst-dependent. Much work has been devoted to developing improved chiral auxiliaries (142). The principal product enantiomer observed in the chiral bisphosphine-rhodium asymmetric hydrogenation of dehydroamino acids has been demonstrated to arise from the less stable catalyst–substrate intermediate (143).

Rhodium(II) acetate catalyzes C–H insertion, olefin addition, heteroatom-H insertion, and ylide formation of $\alpha$-diazocarbonyls via a rhodium carbenoid species (144–147). Intramolecular cyclopentane formation via C–H insertion occurs with retention of stereochemistry (143). Chiral rhodium(II) carboxamides catalyze enantioselective cyclopropanation and intramolecular C–N insertions of $\alpha$-diazoketones (148). Other reactions catalyzed by rhodium complexes include double-bond migration (140), hydrogenation of aromatic aldehydes and ketones to hydrocarbons (150), homologation of esters (151), carbonylation of formaldehyde (152) and amines (140), reductive carbonylation of dimethyl ether or methyl acetate to 1,1-diacetoxy ethane (153), decarbonylation of aldehydes (140), water gas shift reaction (69,154), C–C skeletal rearrangements (132,140), oxidation of olefins to ketones (155) and aldehydes (156), and oxidation of substituted anthracenes to anthraquinones (157). Rhodium-catalyzed hydrosilation of olefins, alkynes, carbonyls, alcohols, and imines is facile and may also be accomplished enantioselectively (140). Rhodium complexes are moderately active alkene and alkyne polymerization catalysts (140). In some cases polymer-supported versions of homogeneous rhodium catalysts have improved activity, compared to their homogenous counterparts. This is the case for the conversion of alkenes directly to alcohols under oxo conditions by rhodium–amine polymer catalysts (158).

*Miscellaneous.* Dimeric rhodium isocyanide complexes have been suggested for use to convert water to hydrogen and oxygen (159,160). Photochemical activation of some rhodium complexes leads to intermolecular oxidative addition of aliphatic and aromatic C–H bonds under ambient conditions (161,162). Rhodium complexes have been investigated for antibacterial and antitumor properties (163). However, none has been used in human clincial trials. Rhodium complexes have been used for site-specific DNA recognition and photoactivated strand scission (164). Rhodium plating has been used to deposit thin, brilliant white, tarnish-resistant, wear-resistant coatings for jewelry (165). Engineering applications utilize rhodium's good sliding contact characteristics or high stable reflectivity (165). Rhodium electroplating baths are proprietary, but are based on sulfate or phosphate salts (165).

## Iridium Compounds

The most common oxidation states, corresponding electronic configurations, and coordination geometries of iridium are $+1$ ($d^8$) usually square plane although some five-coordinate complexes are known, and $+3$ ($d^6$) and $+4$ ($d^5$), both octahedral. Compounds in every oxidation state between $-1$ ($d^{10}$) and $+6$ ($d^3$) are known. Iridium compounds are used primarily to model more active rhodium catalysts.

**Binary Compounds.**    The fluorides of iridium are IrF$_3$ [23370-59-4], IrF$_4$ [37501-24-9], the tetrameric pentafluoride (IrF$_5$)$_4$ [14568-19-5], and IrF$_6$ [7789-75-7]. Chlorides of iridium include IrCl$_3$, which exists in anhydrous [10025-83-9] $\alpha$- and $\beta$-forms, and as a soluble hydrate [14996-61-3], and IrCl$_4$ [10025-97-5]. Other halides include IrBr$_3$ [10049-24-8], which is insoluble, and the soluble tetrahydrate IrBr$_3$·4H$_2$O; IrBr$_4$ [7789-64-2]; and IrI$_3$ [7790-41-2]. Iridium forms iridium dioxide [12030-49-8], IrO$_2$; a poorly characterized sesquioxide, Ir$_2$O$_3$ [1312-46-5]; and the hydroxides, Ir(OH)$_3$ [54968-01-3] and Ir(OH)$_4$ [25141-14-4]. Other binary iridium compounds include the sulfides, IrS [12136-40-2], Ir$_2$S$_3$ [12136-42-4], IrS$_2$ [12030-51-2], and IrS$_3$ [12030-52-3], as well as various selenides and tellurides.

**Coordination Compounds.**    A large number of iridium complexes with nitrogen ligands have been isolated, particularly where Ir is in the $+3$ oxidation state. Examples of ammine complexes include [Ir(NH$_3$)$_6$]$^{3+}$ [24669-15-6], [IrCl(NH$_3$)]$^{2+}$ [29589-09-1], and trans-[Ir(O$_3$SCF$_3$)$_2$(en)$_2$]$^+$ [90065-94-4]. Compounds of N-heterocyclic ligands include trans-[IrCl$_4$(py)$_2$]$^-$ [24952-67-8], [Ir(bipy)$_3$]$^{3+}$ [16788-86-6], and an unusual C-metalated bipyridine complex, [Ir(bipy)$_2$(C$^3$, N-bipy)]$^{2+}$ [87137-18-6]. Isolation of this latter complex produced some confusion regarding the chemical and physical properties of [Ir(bipy)$_3$]$^{3+}$ (167).

Complexes of $d^6$ Ir(III) are kinetically inert and undergo octahedral substitution reactions slowly. The rate constant for aquation of [IrBr(NH$_3$)$_5$]$^{2+}$ [35884-02-7] at 298 K has been measured at $\sim 2 \times 10^{-10}$ s$^{-1}$ (168). In many cases, addition of a catalytic reducing agent such as hypophosphorous acid greatly accelerates the rate of substitution via a transient, labile Ir(II) species (169). Optical isomers can frequently be resolved, as is the case of cis-[IrCl$_2$(en)$_2$]$^+$ [15444-47-0] (170). Ir(III) amine complexes are photoactive and undergo rapid photosubstitution reactions (171). Other iridium complexes containing nitrogen ligands include complexes of nitriles, [Ir(P(C$_6$H$_5$)$_3$)$_2$(NCCH$_3$)CO]$^+$ [51540-63-7]; nitrosyl, [IrCl$_2$(P(C$_6$H$_5$)$_3$)$_2$(NO)]$^+$ [27411-12-7]; dinitrogen, [IrCl(P(C$_6$H$_5$)$_3$)$_2$N$_2$]

[*15695-36-0*]; azide, *trans*-[Ir(N$_3$)$_2$(en)$_2$]$^+$ [*26104-26-7*]; and pyrazylborate, [Ir$_2$Cl$_4$(HB(3,5-(CH$_3$)$_2$C$_3$N$_2$H$_3$)$_3$)$_2$] [*73117-58-5*]. A large number of phosphine complexes of iridium have been prepared. Examples are [Ir(P(C$_6$H$_5$)$_3$)(CO)$_3$]$^-$ [*52352-68-8*], [IrCl(P(C$_6$H$_5$)$_3$)$_2$(CO)] [*15842-08-7*], and [Ir(P(C$_6$H$_5$)$_3$)$_3$(CO)H] [*17250-25-8*], [Ir(PC$_6$H$_5$)$_3$H$_3$] [*16924-01-9*]. Fewer arsine complexes such as [Ir(As(C$_6$H$_5$)$_3$)Cl(CO)$_2$] [*59809-93-7*] and [IrCl$_2$(As(C$_6$H$_5$)$_3$)$_2$(CO)] [*15682-62-9*] are known.

Iridium complexes having oxygen ligands are not nearly as extensive as those having nitrogen. Examples include acetylacetonates [Ir(P(C$_6$H$_5$)$_3$)$_2$ (acac)H$_2$] [*64625-61-2*], aqua complexes [Ir(OH$_2$)$_6$]$^{3+}$ [*61003-29-0*], nitrato complexes [Ir(ONO$_2$)(NH$_3$)$_5$]$^{2+}$ [*42482-42-8*], and peroxides IrCl(P(C$_6$H$_5$)$_3$)$_2$(O$_2$-*t*-C$_4$H$_9$)$_2$(CO)] [*81624-11-5*]. Unlike rhodium, very few Ir(II) carboxylate-bridged dimers have been claimed and [Ir$_2$(OOCCH$_3$)$_4$] has not been reported. Some Ir(I) complexes exhibit reversible oxidative addition of O$_2$ to form Ir(III) complexes. That chemistry has been reviewed (172). The most well-known and studied is Vaska's complex [*14871-41-1*], *trans*-[IrCl(P(C$_6$H$_5$)$_3$)$_2$CO], which reversibly forms [IrCl(P(C$_6$H$_5$)$_3$)($\eta^2$-O$_2$)CO] [*29933-65-1*] in the presence of oxygen. Complexes containing sulfur ligands include dithiocarbamates [Ir{S$_2$CN(CH$_2$)$_4$}$_3$] [*39958-06-0*]; thioethers, *mer*-[IrCl$_3$(S(C$_2$H$_5$)$_2$)$_3$] [*34177-65-6*]; and carbon disulfide complexes, [IrCl(P(C$_6$H$_5$)$_2$CH$_3$)(CS$_2$)(CO)] [*81178-13-4*].

Halide ligands are found in homoleptic complexes as well as in mixed ligands systems. Halide complexes of Ir(IV) such as [IrCl$_6$]$^{2-}$ [*16918-91-5*] are readily reduced to Ir(III) species, eg [IrCl$_6$]$^{2-}$ [*14648-50-1*], in neutral or basic solution, or in the presence of reducing agents such as KI, oxalate, or photochemical activation (173).

**Organometallic Compounds.** The predominant oxidation states of iridium in organometallics are +1 and +3. Iridium forms mononuclear and polynuclear carbonyl complexes including [IrCl(P(C$_6$H$_5$)$_3$)$_2$(CO)$_2$] [*14871-41-1*], [Ir$_2$Cl$_4$-(CO)$_2$]$^-$ [*12703-90-1*], [Ir$_4$(CO)$_{12}$] [*18827-81-1*], and the conducting, polymeric [IrCl(CO)$_3$]$_n$ [*32594-40-4*]. Isonitrile and carbene complexes are also known.

Iridium forms $\sigma$-bonded alkyl, *fac*-[Ir(P(CH$_3$)$_2$C$_6$H$_5$)$_3$(CH$_3$)$_3$] [*15927-48-7*]; aryl, [IrI(Cp)(C$_6$H$_5$)(CO)] [*64867-79-4*]; acyl, [IrCl$_2$(P(C$_6$H$_5$)$_3$)$_2$OOCCH$_3$] [*36447-22-0*]; carboxylato, [IrI$_2$(bipy)(CO$_2$OCH$_3$)(CO)] [*23733-10-0*]; alkenyl; and alkynyl complexes. Many orthometallated complexes are known, such as [IrHCl(P(C$_6$-H$_5$)$_2$(C$_6$H$_4$))(P(C$_6$H$_5$)$_3$)$_2$] [*24846-80-8*]. In some cases orthometallation is reversible and gives rise to H–D exchange on the aromatic ring (172). Numerous $\pi$-complexes are known; examples include alkenes, [Ir$_2$Cl$_2$(C$_2$H$_4$)$_4$] [*39722-81-1*]; dienes, [Ir(cod)$_2$]$^+$ [*35138-23-9*]; alkynes; allyls, [Ir($\eta^3$-C$_3$H$_5$)$_3$] [*12108-64-4*]; and conjugated dienes, [IrCl($\eta^4$-1,3-butadiene)$_2$] [*39732-20-2*]. Complexes of $\eta^5$-cyclopentadiene (Cp) and $\eta^6$-arene such as [Ir(P(C$_6$H$_5$)$_3$)(CO)Cp] [*32612-68-3*], [IrBr$_2$(CO)Cp] [*64867-78-3*], and [Ir(C$_6$H$_6$)((CH$_3$)$_5$C$_5$)]$^{2+}$ [*12715-71-8*], are also known.

**Synthesis.** The principal starting material for synthesis of iridium compounds is iridium trichloride hydrate [*14996-61-3*], IrCl$_3\cdot x$H$_2$O. Another useful material for laboratory-scale reactions is [Ir$_2$Cl$_2$(cod)$_2$] [*12112-67-3*].

**Uses.** *Catalysis.* Iridium compounds do not have industrial applications as catalysts. However, these compounds have been studied to model fundamental catalytic steps (174), such as substrate binding of unsaturated molecules and dioxygen; oxidative addition of hydrogen, alkyl halides, and the carbon–hydrogen bond; reductive elimination; and important metal-centered transformations such as carbonylation, $\beta$-elimination, CO reduction, and oligomerization reactions. One of the most widely studied systems is the class of Vaska's compounds typified by trans-[IrCl(P(C$_6$H$_5$)$_3$)(CO)] which reversibly adds O$_2$ yielding [IrCl(P(C$_6$H$_5$)$_3$)$_2$(O$_2$)CO] [27628-97-3]. Oxygen complexes of this sort can catalytically oxidize olefins (175), ketones (176), NO (177), CO (178), and SO$_2$ (179). Another much-studied reaction is carbon–hydrogen bond activation (180). Complexes such as [Ir(Cp(CH$_3$)$_5$)L], where L = P(CH$_3$)$_3$ or CO, oxidatively add arene and alkane carbon–hydrogen bonds (181,182). Catalytic dehydrogenation of alkanes (183) and carbonylation of benzene (184) has also been observed. Iridium compounds have also been shown to catalyze hydrogenation (185) and isomerization of unsaturated alkanes (186), hydrogen-transfer reactions, and enantioselective hydrogenation of ketones (187) and imines (188).

*Miscellaneous.* Iridium dioxide, like RuO$_2$, is useful as an electrode material for dimensionally stable anodes (DSA) (189). Solid-state pH sensors employing IrO$_2$ electrode material are considered promising for measuring pH of geochemical fluids in nuclear waste repository sites (190). Thin films (qv) of IrO$_2$ are stable electrochromic materials (191).

## Palladium Compounds

The most common oxidation state of palladium is +2 which corresponds to a $d^8$ electronic configuration. Compounds have square planar geometry. Other important oxidation states and electronic configurations include 0 ($d^{10}$), which can have coordination numbers ranging from two to four and is important in catalytic chemistry; and +4 ($d^6$), which is octahedral and much more strongly oxidizing than platinum (IV). The chemistry of palladium is similar to that of platinum, but palladium is between $10^3$ to $5 \times 10^5$ more labile (192). A primary industrial application is palladium-catalyzed oxidation of ethylene (see OLEFIN POLYMERS) to acetaldehyde (qv). Palladium-catalyzed carbon–carbon bond formation is an important organic reaction.

**Binary Compounds.** The fluorides of palladium are PdF$_2$ [13444-96-7], PdF$_3$ [12021-58-8], and PdF$_4$ [13709-55-2]. The difluoride and tetrafluoride are unusual. The former is paramagnetic and octahedral; the latter is eight-coordinate. The trifluoride contains both Pd(II) and Pd(IV). The only chloride is palladium dichloride [7647-10-1], PdCl$_2$, which occurs in a hydrated form, and two anhydrous forms, an $\alpha$-form having a polymeric chain structure, and a $\beta$-form that consists of a Pd$_6$ core with edge-bridging chlorides. Solutions of the hydrate are readily reduced to metal by hydrogen and many organic compounds. Other halides include palladium dibromide [13444-94-5] and diiodide [7790-38-7]. Palladium monoxide [1314-08-5], PdO, is the only stable oxide of palladium. It is easily reduced to metal and used as a hydrogenation catalyst. Palladium also forms a dihydroxide, Pd(OH)$_2$ [12135-22-7]. Palladium forms sulfides, selenides, and tellurides. The best-characterized for these are PdS [12125-22-3], PdS$_2$ [12137-75-6], PdSe [12137-76-7], PdSe$_2$ [60672-19-7], PdTe [12037-94-4], and PdTe$_2$ [12037-95-5]. Borides, silicides, phosphides, and arsenides are also known (23,24).

**Coordination Compounds.** Palladium forms numerous complexes with ammonia and with simple amines. Examples are [Pd(NH$_3$)$_4$]$^{2+}$ [15974-14-

8], [PdCl(dien)]$^+$ [17549-31-4], cis-[PdCl$_2$(NH$_3$)$_2$] [15684-18-1], and trans-[PdCl$_2$(NH$_3$)$_2$] [13782-33-7]. Monoammine complexes such as [PdCl$_3$(NH$_3$)]$^-$ [15691-32-4] are stable but less common. Examples of aromatic amine complexes include trans-[PdCl$_2$(pyr)$_2$] [14052-12-1], [PdCl$_2$(bipy)] [14871-92-2], and nucleosides such as [PdCl(dien)(guanosine)]$^+$ [73601-42-0] (193). Complexes of Pd(IV) such as [PdCl$_2$(NH$_3$)$_4$]$^{2+}$ [70491-81-5] and [PdCl$_4$(bipy)] [57209-01-5] may be prepared by chlorine oxidation of the corresponding Pd(2+). The aromatic amine Pd(IV) complexes are more stable than ammine and aliphatic amine species, which are reduced to Pd(II) in water or thermally (194).

The Pd(II) compound chloro(tris[2-(dimethylamino)ethyl]amine)palladium(II) [66632-97-1] is a rare example of a Pd(II) compound having a coordination number greater than four (194,195). Palladium forms complexes with nitriles, [PdCl$_2$(NCC$_6$H$_5$)$_2$] [14220-64-5]; imines, cis-[PdCl$_2$(t-C$_4$H$_9$N=CHCH=NC$_4$H$_9$-t)] [72905-45-4]; oximes, [Pd(DMG)$_2$] [14740-97-7]; Schiff bases, [Pd(N-n-butylsalicylaldiminato)$_2$] [14363-25-8]; azides, [Pd(N$_3$)$_4$]$^{2-}$ [16843-41-7]; semicarbazones; hydrazones; and triazenes (194,196). Numerous phosphine complexes have been isolated, including [Pd$_2$Cl$_4$(P(C$_2$H$_5$)$_3$)$_2$] [15684-59-0], [PdCl$_2$-(P(CH$_3$)$_3$)$_2$] [25892-38-0], trans-[PdCl$_2$(P(C$_6$H$_5$)$_2$CH$_3$)$_2$] [26973-01-3], [Pd(PC$_2$-H$_5$)$_3$H]$^+$ [51404-97-8], and trans-[PdCl(P(C$_6$H$_5$)$_3$)H] [16841-99-9]. In solution, cis- and trans-[PdCl$_2$P$_2$] complexes, where P = phosphine, interconvert rapidly (197). Oxidation of Pd(II) phosphine and arsine complexes yields Pd(IV) complexes similar to those obtained with amines (198). Low valent Pd(0) complexes, [Pd(P(C$_6$H$_5$)$_3$)$_4$] [14221-01-3], and binuclear Pd(I) complexes, [Pd$_2$Cl$_2$-(bis(dimethylphosphino)methane)$_2$] [89178-59-6], containing bridging phosphine ligands and a metal–metal bond have been prepared. A variety of small molecules, including CO, CNR, SO$_2$, and CS$_2$, insert into the Pd–Pd bond (199) of this last species. Arsine and stibine complexes are similar to the phosphines.

Palladium coordinates oxygen ligands weakly. The aqua complexes are generally formed as intermediates in substitution reactions in water. The tetraaqua ion [Pd(OH$_2$)$_4$]$^{2+}$ [22573-07-5] is prepared by treating [PdCl$_4$]$^{2-}$ [14349-67-8] with AgClO$_4$ in water. Isolatable complexes containing oxygen ligands include [Pd(DMSO-S)$_2$(DMSO-O)$_2$] [3079-58-4]; [Pd$_2$(OH)$_2$(P(C$_6$H$_5$)$_3$)$_4$] [39151-60-5]; [Pd$_3$(OOCCH$_3$)$_6$] [29950-51-4]; [Pd(ox)$_2$]$^{2-}$ [15226-60-5]; [Pd(acac)$_2$] [14024-61-4]; and [Pd(P(C$_6$H$_5$)$_3$)$_2$(cat)] [64608-33-9]. Palladium binds to sulfur and selenium more strongly than to oxygen. Some representative complexes include the thiolates, [Pd$_6$(S-n-C$_3$H$_7$)$_{12}$] [35359-97-8]; thioethers, [Pd$_2$Br$_4$(S(CH$_3$)$_2$)$_2$] [21315-76-4]; thiocyanates [Pd(SCN)$_4$]$^{2-}$ [16057-01-5]; thioureas [Pd(SC(NH$_2$)$_2$)$_4$]$^{2+}$ [42941-76-4]; sulfoxides; sulfites, [Pd(SO$_3$)$_4$]$^{6-}$ [65466-59-3]; dithiocarbamates, [Pd(S$_2$CN(C$_2$H$_5$)$_2$)$_2$] [15170-78-2]; dithiophosphinates, [Pd(S$_2$P(C$_2$H$_5$)$_2$)$_2$] [89621-75-0]; and dithiolenes, [Pd(S$_2$C$_2$(CN)$_2$)$_2$]$^{2-}$ [19555-33-0]. Numerous other complexes with sulfur and selenium ligands are known (197).

Palladium forms halide complexes such as [PdCl$_4$]$^{2-}$ [14349-67-8] and [PdCl$_6$]$^{2-}$ [17141-41-2], and bridging complexes, [Pd$_2$Br$_6$]$^{2-}$ [33887-55-7]. Halide complexes can aquate in water. Hexachloropalladate salts release Cl$_2$ when heated as solids or in solution (200).

**Organometallic Compounds.** Mononuclear carbon monoxide complexes of palladium are relatively uncommon because of palladium's high lability, tendency to be reduced, and competing migratory insertion reactions in the presence of a Pd–C bond (201). A variety of multinuclear compounds are known (202), including [Pd$_2$Cl$_4$(CO)$_2$] [*75991-68-3*], [Pd$_3$(P(C$_6$H$_5$)$_3$)$_3$(CO)$_3$] [*36642-60-1*], and [Pd$_7$(P(CH$_3$)$_3$)$_7$($\mu_3$-CO)$_3$($\mu_2$-CO)$_4$] [*83632-51-3*]. Isonitriles and carbenes are better $\sigma$-donors, and many compounds are known (203) such as [PdI$_2$(CN-*t*-C$_4$H$_9$)$_2$] [*24917-36-0*], [Pd$_2$(CNCH$_3$)$_6$]$^{2+}$ [*56116-47-3*], [Pd$_2$Cl$_2$(CN-*t*-C$_4$H$_9$)$_4$] [*34742-93-3*], [Pd(CN-*t*-C$_4$H$_9$)$_2$] [*24859-25-4*], and [Pd(C(NHCH$_3$)$_2$)$_4$]$^{2+}$ [*45256-27-7*].

Palladium forms $\sigma$-bonded alkyl, [Pd(C$_2$H$_5$)$_2$(bipy)] [*102150-17-4*], [PdI(CH$_3$)$_3$(bipy)] [*110182-93-9*]; aryl, [Pd$_2$Cl$_2$(2-(2-pyridinyl)phenyl-C,N)$_2$] [*20832-86-4*]; vinyl, [PdCl(P(C$_6$H$_5$)$_3$)$_2$(CH=CCl$_2$)] [*31871-49-5*], acyl, [PdCl(P(C$_6$H$_5$)$_3$)$_2$(COC$_6$H$_5$)] [*50417-59-9*]; and iminoacyl, [PdI(P(C$_4$H$_9$)$_3$)$_2$(C(CH$_3$)NCC$_6$H$_{11}$)$_2$] [*42582-35-4*], complexes. Palladium alkyls and aryls can be prepared by transmetallation of Pd(II) complexes using Grignard reagants, organolithium, or other organometallic species (see GRIGNARD REACTIONS); by oxidative addition by organic halides to a Pd(0) complex; by nucleophilic attack on coordinated olefins; and by orthometallation. Palladium alkyls and aryls tend to be very reactive, a property which has been used to advantage in organic synthesis. When the alkyl is contained in a chelate ring it is more stable. Some Pd(IV) alkyls have been prepared. Palladium acyl and iminoacyl are most often prepared by migratory insertion of an alkyl into a palladium carbonyl or isonitrile.

Palladium forms $\pi$-complexes of olefins, [Pd$_2$Cl$_4$(CH$_2$=CHC$_6$H$_5$)$_2$] [*12257-72-6*]; dienes, [PdCl$_2$(1,5-cod)$_2$]$^{2+}$ [*59687-80-8*] and [Pd$_2$(dba)$_3$] [*51364-51-3*]; acetylenes, [Pd(P(C$_6$H$_5$)$_3$)$_2$((CCOOCH$_3$)$_2$)] [*15629-88-6*]; allyls, [Pd(P(C$_6$H$_5$)$_3$)$_2$-(C$_3$H$_5$)]$^+$ [*38497-96-0*]; and cyclopentadienes [Pd(CH$_3$Cp)(C$_3$H$_5$)] [*92719-24-9*]. Olefin complexes are generally very reactive and susceptible to nucleophilic attack. Diene complexes are somewhat more stable. The $\pi$-allyl ligand, the most common organometallic ligand for palladium, can be formed by deprotonation of a coordinated olefin that contains a $\beta$-hydrogen, nucleophilic attack on coordinated 1,3-dienes, hydropalladation of a diene by a palladium hydride, reaction of an alkyl halide or alcohol with a palladium salt in a protic solvent, oxidative addition of allyl halides with Pd(0) complexes, or transmetallation using an allyl Grignard. Palladacene is unknown but the Cp ligand is found in a variety of palladium complexes.

**Synthesis.** The most common starting materials for palladium complexes are PdCl$_2$ [*7647-10-1*] and [PdCl$_4$]$^{2-}$ [*14349-67-8*]. Commercially available materials useful for laboratory-scale synthesis include [Pd$_3$(OOCCH$_3$)$_6$] [*3375-31-3*], [PdCl$_2$(NCC$_6$H$_5$)] [*14220-64-5*], [Pd(acac)$_2$] [*14024-61-4*], [PdCl$_2$(cod)] [*12107-56-1*], and [Pd(P(C$_6$H$_5$)$_3$)$_4$] [*14221-01-3*].

**Uses.** *Catalysis.* The most important industrial use of a palladium catalyst is the Wacker process. The overall reaction, shown in equations 7–9, involves oxidation of ethylene to acetaldehyde by Pd(II) followed by Cu(II)-catalyzed reoxidation of the Pd(0) by oxygen (204). Regeneration of the catalyst can be carried out *in situ* or in a separate reactor after removing acetaldehyde. The acetaldehyde must be distilled to remove chlorinated by-products.

$$C_2H_4 + PdCl_4^{2-} + H_2O \longrightarrow CH_3CHO + Pd^0 + 2\ HCl + 2\ Cl^- \tag{7}$$

$$Pd^0 + 2\ CuCl_2 + 2\ Cl^- \longrightarrow PdCl_4^{2-} + 2\ CuCl \tag{8}$$

$$2\ CuCl + 1/2\ O_2 + 2\ HCl \longrightarrow 2\ CuCl_2 + H_2O \tag{9}$$

The acetaldehyde-forming step (eq. 7) involves nucleophilic attack by hydroxide or water on a coordinated Pd olefin complex followed by $\beta$-hydride elimination.

$$\underset{H_2O}{\overset{\displaystyle \begin{array}{c} H \quad H \\ \diagdown\!\!\diagup \\ C \\ \parallel\!\!-Pd \\ C \\ \diagup\!\!\diagdown \\ H \quad H \end{array}}{}} \longrightarrow \underset{}{\overset{\displaystyle HO\!-\!CH_2}{\underset{\displaystyle CH_2\!-\!Pd}{}}} \longleftarrow \overset{\displaystyle \begin{array}{c} H \qquad OH_2 \\ \diagdown\!\!\diagup \\ C\!-\!H \mid \\ \parallel\!\!-Pd \\ C \\ \diagup\!\!\diagdown \\ H \quad H \end{array}}{} \qquad (10)$$

The nature of the initial attack by the water (eq. 10) is a matter of some controversy (205,206). Stereochemical and kinetic studies of model systems have been reported that support trans addition of external water (207,208) or internal addition of cis-coordinated water (209), depending on the particular model system under study. Other palladium-catalyzed oxidations of olefins in various oxygen donor solvents produce a variety of products including aldehydes (qv), ketones (qv), vinyl acetate, acetals, and vinyl ethers (204). However the product mixtures are complex and very sensitive to conditions.

Palladium-catalyzed coupling reactions have important synthetic applications (210–212). The prototypical reaction is the Heck reaction (213) whereby an organoPd(II) undergoes coupling with an olefin (eq. 11).

$$C_6H_5\!-\!Pd\!-\!X + CH_2\!\!=\!\!CH_2 \longrightarrow C_6H_5CH\!\!=\!\!CH_2 + HX + Pd^0 \qquad (11)$$

The utility of palladium-coupling reactions arises from the fact that several mild, stereospecific methodologies compatible with a variety of functional groups have been developed for coupling organopalladium complexes and electrophiles other than olefins (214–217). The reactive organopalladium complex is usually prepared *in situ* by transmetallation of a main group organometallic, eg, an organozinc, organotin, or organomercurial, using a Pd(II) salt, or by oxidative addition of Pd(0) into an R–X bond, where X is a leaving group, often halide or triflate. Hence palladium-catalyzed cross-coupling of organotin reagents using acid chlorides, benzyl halides, allyl halides, and vinyl halides, aryl halides, and aryl triflates occurs under mild conditions in high yields (215,216). In the presence of CO, migratory insertion in the palladium alkyl intermediate leads to ketone formation (215) or ester formation (218,219). Palladium-catalyzed arylation of enol ethers (217) and cross-coupling of organic halides with organoborates (220) and organosilanes (214) are other examples of palladium-catalyzed bond formation.

Other reactions (204) catalyzed by palladium include oxidative coupling of aromatic compounds to biphenyls; aromatic acetoxylation; oxidative carbonylation of olefins, aromatics, and alcohols; air oxidation of alcohols to acetals and aldehydes; allylation of active methylene groups (221); cycloaddition reactions (222); hydrostannylation (223); copolymerization of CO and olefins (224); and reductive carbonylation of methyl acetate to 1,1-diacetoxy ethane (153). Finally, ibuprofen [51146-56-6], 4-$(CH_3CH_2CH_2)C_6H_4CH(COOH)CH_3$, is prepared commercially by Boots Hoechst Celanese using a palladium-catalyzed carbonylation.

The carbonylation of 4-$(CH_3CH_2CH_2)C_6H_4CH(OH)CH_3$ is carried out in the presence of HCl and phosphoric acid at 125°C under 5.5 MPa (800 psi) of CO in the presence of catalytic $PdCl_2(P(C_6H_5)_3)_2$ [4056-88-3] (225,226).

*Miscellaneous.*   Reduction of a palladium salt by CO is the basis of a visual test for ambient carbon monoxide (227). Palladium compounds are used as photographic sensitizers (228). The low dimensional mixed valence compound $Cs_{0.83}[Pd(S_2C_2(CN)_2)]\cdot 0.5H_2O$ behaves as a semimetal at room temperature (229). Palladium compounds isostructural with potent platinum antitumor compounds have poor antitumor activity (230).

*Plating and Coatings.*   Palladium films are used as electrical contacts and connectors and in circuit board fabrication (see ELECTRICAL CONNECTORS). Palladium, palladium–silver, and palladium–nickel alloys are used as less expensive substitutes for gold in some applications (231). Thin films also have applications as gas sensors (94). Typical electrolytes for plating baths include salts of $[Pd(NH_3)_4]^{2+}$, or $[Pd(NH_3)_2(NO_2)_2]$, called Pd–P–Salt in an ammoniacal medium. Alternatively, the Pd amine salt is prepared directly from $PdCl_2$. Palladium–silver baths can be prepared from $[Pd(NH_3)_2(NO_3)_2]$ and $AgNO_3$ in aqueous ammonia (232). Electroless plating (qv) baths contain a reducing agent such as hydrazine or hypophospite (233). Deposition of a thin, tarnish-resistant coating can be accomplished by immersion plating, using $H_2[Pd(NO_2)_2SO_4]$, known as DNS palladium solution (234). Volatile palladium compounds are used in chemical vapor deposition processes to deposit thin films of metallic palladium (94). Liquid palladium solutions employ discrete compounds such as $PdCl_2(S(n\text{-}C_4H_9)_2)_2$ [32335-75-4] or poorly defined sulforesinates.

## Platinum Compounds

The most common oxidation states and corresponding electronic configurations of platinum are +2 ($d^8$), which is square planar, and +4 ($d^6$), which is octahedral. Compounds in oxidation states between 0 ($d^{10}$) and +6 ($d^4$) exist. Platinum hydrosilation catalysts are used in the manufacture of silicone polymers. Several platinum coordination compounds are important chemotherapeutic agents used for the treatment of cancer.

**Binary Compounds.**   Three fluorides, $PtF_4$ [13455-15-7], $PtF_5$ [37782-184-8], and platinum hexafluoride [13693-05-5], $PtF_6$, are well documented. The last is a powerful oxidizing agent and can oxidize dioxygen and xenon (235). Two chlorides exist, platinum dichloride [10025-65-7], $PtCl_2$, and platinum tetrachloride [37773-49-2]. Platinum dichloride exists in an α- and β-form, the latter containing a $Pt_6$ core and edge-bridging chlorides. Platinum trichloride [25909-39-1], $PtCl_3$, contains Pt(II) and Pt(IV) centers. Other halides include two bromides, $PtBr_2$ [13455-12-4] and $PtBr_4$ [13455-11-3], and two iodides, $PtI_2$ [7790-39-8] and $PtI_4$ [7790-46-7]. The most common oxide is platinum dioxide [1314-15-4] (Adams catalyst), $PtO_2$. Three less well-defined oxides are PtO [12035-82-4], $Pt_2O_3$ [12725-92-7], and $PtO_3$ [77883-44-4]. Platinum also forms $Pt(OH)_2$ [12135-23-8]. Sulfides of platinum include PtS [12038-20-9] and $PtS_2$ [12038-21-0] (24,25,236,237). Borides, silicides, phosphides, selenides, and tellurides have been prepared (23,24).

**Coordination Compounds.** Platinum forms numerous stable ammine complexes, the most important of which is the antitumor agent cisplatin, cis-dichlorodiammineplatinum(II) [26035-31-4], cis-[PtCl$_2$(NH$_3$)$_2$]. Other examples of complexes containing ammonia or aliphatic or aromatic amines include trans-PtCl$_2$(NH$_3$)$_2$ [14913-33-8], Magnus's green salt [13820-46-7], [PtCl$_4$]-[Pt(NH$_3$)$_4$], [PtCl(dien)]$^+$ [17549-31-4], [Pt(NH$_3$)$_6$]$^{4+}$ [18536-12-4], [PtCl$_2$(bipy)] [13965-31-6], and oligonucleotide complexes such as that with deoxyguanidyl guanosine monophosphate d(GpG) [99802-17-2], cis-Pt(NH$_3$)$_2$(d(GpG))]$^+$. The p$K_a$ of ammines coordinated to Pt(IV) is reduced sufficiently to permit acetylacetone to react with [Pt(NH$_3$)$_6$]$^{4+}$ in the presence of base, forming [Pt(NH$_3$)$_4$((HNCCH$_3$)$_2$CH))]$^{4+}$ [56370-84-4] (238). Mixed valence complexes such as Wolfram's red salt [60428-75-3], [(Pt(II)(NH$_2$C$_2$H$_5$)$_4$)(Pt(IV)Cl$_2$(NH$_2$C$_2$-H$_5$)$_4$)]$^{4+}$, and platinum blues [71611-15-9], [Pt$_4$(NH$_3$)$_8$($\alpha$-pyridone)$_4$]$^{5+}$,are well known. The former consists of linear stacks of alternating platinum(II) and platinum(IV) centers bridged by chloride; the latter contains four platinum centers having an average oxidation state of 2.5 (239). Platinum forms complexes with many other nitrogen donor ligands such as nitriles, cis-[PtCl$_2$(NCC$_6$H$_5$)$_2$] [15617-19-3]; nitrite, [Pt(NO$_2$)$_4$]$^{2-}$ [22289-82-3]; azides, [Pt(N$_3$)$_4$]$^2$ [45074-06-4]; imines; oximes, [Pt(DMG)$_2$] [17632-92-7]; and imidates, trans-[PtCl$_2$((E)-HNCCH$_3$(OCH$_3$))$_2$] [15022-74-9]. Phosphorus ligands are able to stabilize oxidation states of platinum between 0 and +4, although the higher oxidation states are less common. Zero-valent complexes having coordination numbers ranging from two to four depending on the steric demands of the ligand are known. Examples include [Pt(P($t$-C$_4$H$_9$)$_3$)$_2$] [60648-70-6], [Pt(P(C$_6$H$_5$)$_3$)$_3$] [13517-35-6], and [Pt(P(C$_6$H$_5$)$_2$CH$_3$)$_4$] [27121-53-5]. Other phosphine and phosphite complexes are [PtCl$_2$(P(C$_6$H$_5$)$_3$)$_2$] [10199-34-5], [Pt$_2$(P$_2$O$_5$H$_2$)$_4$Cl$_2$]$^{4-}$ [98303-98-1], [PtCl$_4$(P(C$_6$H$_5$)$_3$)$_2$] [78309-42-9], and A-frame complexes such as [Pt$_2$Cl$_2$(dppm)$_2$] [61250-65-5], and [Pt$_2$Cl$_2$(dppm)$_2$($\mu^2$-CO)] [68851-47-8]. Arsine and stibine complexes are also known.

Oxygen ligands are relatively labile. Nonetheless, many stable complexes exist. Aqua complexes are frequently prepared as reactive intermediates for ligand substitution reactions but are generally not isolated. Hydroxide forms stable bridging complexes such as [Pt$_2$($\mu$-OH)$_2$(NH$_3$)$_4$]$^{2+}$ [62048-57-1], as well as some terminal complexes such as c,t,c-[PtCl$_2$(OH)$_2$(NH$_2$-$i$-C$_3$H$_7$] [62928-11-4] and hexahydroxyplatinic acid [52438-26-3], H$_2$Pt(OH)$_6$. Other oxygen ligands that form platinum complexes include oxygen, [Pt(P(C$_6$H$_5$)$_3$)$_2$O$_2$] [15614-67-2]; carboxylate, [Pt(CBDCA)(NH$_3$)$_2$] [41575-94-4] and [Pt$_4$(OOCCH$_3$)$_8$] [67286-38-8]; $\beta$-diketonate, [Pt(O,O'-acac)$_2$] [15170-57-5]; catecholato, [Pt(P-(C$_6$H$_5$)$_3$)$_2$(cat)] [64608-33-9]; and nitrato, cis-[Pt(NO$_3$)$_2$(NH$_3$)$_2$] [14286-03-4]. Sulfur ligands bind strongly to platinum and frequently form bridging complexes. Examples of compounds include thiocyantes, [Pt(SCN)$_4$]$^{2-}$ [38668-99-4]; thioethers, cis-[PtCl$_2$(S(CH$_3$)$_2$)$_2$ [17836-09-8] and [Pt$_2$Br$_4$($\mu$-S(C$_2$H$_5$)$_2$)$_2$] [20004-38-0]; sulfoxides, [PtCl$_3$(DMSO)] [31203-96-0]; dithiocarbamates, [Pt(S$_2$CN(C$_2$H$_5$)$_2$)$_2$] [15730-38-8]; and 1,2-dithiolenes, [Pt(S$_2$C$_2$(CN)$_2$)$_2$)]$^{2-}$ [62906-12-1]. Sulfur ligands exhibit a strong trans effect that labilizes the trans ligands toward substitution. This feature forms the basis of the Kurnakov test (240) to distinguish cis and trans isomers of [PtCl$_2$(NH$_3$)$_2$]. After substitution of the chloride by thiourea, the ammines of the cis isomer are labilized and displaced by additional thiourea, yielding $[Pt\{SC(NH_2)_2\}_4]^{2+}$ [48069-54-1], whereas the trans isomer yields trans-$[Pt\{SC(NH_2)_2\}_2(NH_3)_2]^{2+}$ [56959-55-8].

Many mixed ligand complexes contain halide ligands. Some examples of homoleptic complexes are $[PtCl_4]^{2-}$ [13965-91-8], $[Pt_2Br_6]^{2-}$ [31826-84-3], and $H_2[PtCl_6]$ [16941-12-1]. Platinum forms stable hydride complexes such as $[PtH_2(P(C_2H_5)_3)_2]$ [62945-61-3], and the A-frame complex $[Pt_2(dppm)_2H_2(\mu$-H)]$^+$ [86392-85-0] containing both terminal and bridging hydride. Complexes of silicon and tin such as $[PtH(SiCH_3(C_6H_5)_2)(P(C_6H_5)_3)_2]$ [40869-78-1], cis-$[PtCl_2(SnCl_3)_2]^{2-}$ [44967-93-3], and the unusual five-coordinate, $[Pt(SnCl_3)_5]^{3-}$ [40770-13-6], are also known and are significant to catalytic chemistry.

*Organometallic Compounds.* Organometallic complexes of platinum are usually more stable than palladium complexes. Carbon monoxide complexes of platinum are formed more readily than with palladium. Mononuclear and polynuclear complexes in oxidation states 0 to +2 exist such as $[Pt(P(C_6H_5)_3)_3CO]$ [15376-99-5], $[Pt(P(C_6H_5)_3)_2(CO)_2]$ [15377-00-1], $[Pt_3(P(C_6H_5)_3)_4(CO)_3]$ [16222-02-9], $[Pt_{19}(CO)_{22}]^{4-}$ [71966-26-2], $[Pt_2Cl_4(CO)_2]$ [25478-60-8], and trans-$[PtCl(P(C_2H_5)_3)_2(CO)]^+$ [20683-71-0]. A few in the +4 oxidation state are known, eg, $[PtBr_3(CO)H_2]^-$. Isonitrile complexes such as $[Pt_3(CN-t-C_4H_9)_6]$ [55664-26-1], $[PtBr(CN-t-C_4H_9)_3]^+$ [38317-64-5]; and carbene complexes such as $[PtCl_2(P(C_2H_5)_3)(C(OC_2H_5)(NHC_6H_5)_2)]$ [25530-58-9] are also readily prepared. Platinum forms $\sigma$-bonded alkyl, trans-$[PtCl(P(C_6H_5)(CH_3)_2)_2C_2H_5]$ [38832-88-1]; cycloalkyl, $[PtCl_2(CH_2)_3(py)_2]$ [12085-95-9]; aryl, cis-$[PtBr(P(C_6H_5)_3)_2C_6H_5]$ [57694-39-0] and $[PtI(C_6H_5)_2(CH_3)(bipy)]$ [58411-19-1], as well as alkenyl, alkynyl, and acyl complexes.

The olefin complexes of platinum are generally quite stable and have been extensively studied. Typically, compounds are prepared by ligand displacement from a Pt(0) or Pt(II) species, but a variety of methods have been developed (241). Examples of complexes include $[Pt(dba)_3]$ [11072-92-7], $[PtCl_3(CH_2CH_2)]^-$ [12275-00-2] (Zeisses salt), and $[PtCl_2(COD)]$ [12080-32-9]. Many olefin complexes yield stable nucleophilic addition products with oxygen and nitrogen nucleophiles in a step analogous to the first step of the catalytic Wacker process catalyzed by palladium. Platinum also forms $\eta^2$-complexes with dienes, allenes, acetylenes, carbonyls, and thiocarbonyls. Allyl complexes such as $[Pt(C_3H_4)_2]$ [12240-88-9] and $[Pt(P(C_6H_5)_3)_2(C_3H_4)]^+$ [12246-65-0] may be prepared by oxidative addition of allylic halides in the presence of a reducing agent, transmetallation, deprotonation of a coordinate olefin containing a $\beta$-hydrogen, or insertion of a Pt hydride into an allene or diene. Complexes of $\eta^4$-cyclobutadiene $[Pt(SnCl_3)_2(\eta^4$-$C_4(CH_3)_4)]$ and $\eta^5$-cyclopentadiene $[Pt(Cp)(CH_3)]$ [1271-07-4] are also known.

**Synthesis.** The most important starting materials for platinum compounds are potassium tetrachloroplatinate(II) [100025-98-6], $K_2[PtCl_4]$, and $H_2[PtCl_6]$ [16941-12-1]. Other useful starting materials are $PtCl_2(CCN_6H_5)_2$ [15617-19-3], $[PtCl_2(cod)]$ [12080-32-9], and $[PtCl_2(P(C_6H_5)_3)_2]$ [10199034-5].

**Uses.** *Chemotherapy.* Two platinum compounds, cisplatin, cis-$[PtCl_2(NH_3)]$, and carboplatin, $[Pt(CBDCA)(NH_3)_2]$, are approved for use in humans. Both compounds are particularly effective in the treatment of testicular tumors where disease-free status is achieved in more than 90% of patients having minimal-to-moderate disease at the start of treatment (242). Platinum compounds are also used in the treatment of ovarian, head and neck, prostate, bladder, lung, and cervical tumors (243). These compounds have been used as single-agent treatments, but are frequently employed in combination with other chemotherapeutics (see CHEMOTHERAPEUTICS, ANTICANCER). The principal dose-limiting side-effects of cisplatin, introduced in 1972, are kidney toxicity, nausea and vomiting, myleosuppression, and in 20% of patients, hearing loss (244). Carboplatin was introduced in 1981 as an analogue to cisplatin. Carboplatin has good antitumor activity and significantly reduced kidney toxicity and reduced nausea and vomiting associated with it (245). Numerous other analogues of Pt(II) and Pt(IV), active in animal models having cisplatin-resistant tumors, have been prepared. Some are in human trials (246,247). Development of an orally active platinum antitumor agent such as $c,t,c$-[PtCl$_2$(OOCCH$_3$)$_2$(NH$_2$-$c$-hexyl)(NH$_3$)] [*129580-63-8*] which as of this writing (ca 1995) is in clinical trials (247,248), is also of interest. The biochemical origins of antitumor activity and resistance have been reviewed (244,249). Platinum chemotherapeutics is an active area of research.

*Catalysis.* Platinum-catalyzed hydrosilation is used for cross-linking silicone polymers and for the preparation of functionally substituted silane monomers (250). The most widely used catalyst is chloroplatinic acid (Spier's catalyst), H$_2$PtCl$_6$. Other compounds that catalyze the reaction include Pt(II) and Pt(0) complexes such as PtCl$_2$(cod) and Pt(CH$_2$CHC$_6$H$_5$)$_3$, and a catalyst which contains predominately $\eta$-vinyl Pt(0) species, prepared from HPtCl$_6$ and 1,3-divinyltetramethyldisiloxane (251). The mechanism of the reaction is not well understood. Whereas the observation of asymmetric hydrosilation of 1-methylstyrene by a Pt(II) complex in the presence of a chiral phosphine suggests a homogenous pathway (252), other evidence suggests that in many systems, colloidal platinum, produced during an induction period, may be the active catalyst (253). Platinum halides in the presence of tin(II) chloride catalyze the hydrogenation, isomerization, and hydroformylation of alkenes in solution (254) and in melts (255,256). The water gas shift reaction is catalyzed by [Pt(P($i$-C$_3$H$_7$))$_3$] [*60648-72-8*] (257). Other compounds such as PtO$_2$ (Adam's catalyst) are reduced by hydrogen to a metallic heterogeneous hydrogenation catalyst.

*Plating and Coatings.* Thin surface coatings of platinum and platinum alloys are used as decorative finishes and in critical applications where it is necessary to provide finishes resistant to corrosion or high temperature, eg, coatings on jet-engine turbine components (258). Compounds used in the electrodeposition of platinum are based on Pt(II) and Pt(IV) and include H$_2$[PtCl$_6$] and its salts, eg, Pt–P–Salt, [Pt(NH$_3$)$_2$(NO$_2$)$_2$]; H$_2$[Pt(SO$_4$)(NO$_2$)$_2$]; Pt–Q–Salt, [Pt(NH$_3$)$_2$(HPO$_4$)]; and [Pt(OH)$_6$]$^{2-}$ (259,260). Chloride-based baths have been superseded by P-Salt-based baths, which are more stable and relatively easily prepared. Q-Salt baths offer even greater stability and produce hard, bright films of low porosity. Plating under alkaline conditions employs salts of [Pt(OH$_6$)]$^{2-}$. These baths are easily regenerated but have low stability. Platinum films have uses in the electronics industry for circuit repair, mask repair, platinum silicide production, and interconnection fabrication (94). Vapor deposition of volatile platinum compounds such as [Pt(hfacac)$_2$] and [Pt(CO)$_2$Cl$_2$]

is used to deposit thin films of metallic platinum on surfaces. Concentrated organic solutions of poorly defined platinum complexes of alkyl mercaptides or sulforesinates are used to coat ceramics and glass.

*Miscellaneous.* Chloroplatinic acid is used in the production of automobile catalysts. Platino-type prints based on reduction of Pt(II) to Pt(0) by a photosensitive reducing agent such as iron(III) oxalate are used in art photography (261,262). Infrared imaging devices based on a platinum silicide detector have been developed (263).

## Health and Safety

Compounds of most PGMs are only slightly to moderately toxic by oral ingestion (264), $LD_{50}$ (rat): $RhCl_3$, 1300 mg/kg; $Na_2[IrCl_6]$, 500 mg/kg; $Na_2[PdCl_4]$, 500 mg/kg; *cis*-$[PtCl_2(NH_3)_2]$, 20 mg/kg; $Na_2[PtCl_6]$, 25 mg/kg; $RuCl_3$, 210 mg/kg (guinea pig). The most serious acute hazard arises from exposure to volatile $RuO_4$ (bp 40°C) and $OsO_4$ (bp 131°C), which deposit black $RuO_2$ or $OsO_2$ upon contact with tissue. These substances are especially hazardous to the eyes and respiratory system (265–267). Other volatile Os compounds, eg, $OsF_6$ (bp 45.9°C) and $OsCl_4$ (bp 130°C), pose a similar hazard (268). Finely divided osmium metal can react in air to form $OsO_4$. The acceptable 8-h exposure for $OsO_4$ is 2 $\mu g/m^3$ (269).

Work using PGM compounds should be carried out in a properly functioning hood. Special care should be taken to avoid inhalation or contact of fumes with the eyes. Exposure to some anionic salts of platinum, eg, $[PtX_6]^{2-}$ and $[PtX_4]^{2-}$ where X = Cl, Br, or I, can lead to serious allergic reaction and sensitization (allergenicity Cl > Br > I) (270). Symptoms include rhinitis, conjuctivitus, asthma, urticara, and contact dermatitis. The allergic reaction may be immediate or may be delayed overnight. Whereas the allergic reaction can be life-threatening, there is no evidence of long-term effects if a sensitized individual is removed from exposure to platinum salts. Appropriate protective clothing should be worn and precautions should be taken to avoid inhalation or contact with the skin when working with these compounds. Compounds such as *cis*-$[PtCl_2(NH_3)_2]$, $[Pt(NH_3)_4]Cl_2$, and $K_2[Pt(NO_2)_4]$ are not allergenic (271). The threshold exposure limit of soluble platinum salts is 0.002 mg/m$^3$ on the basis of Pt (270). Platinum antitumor compounds are toxic. There have been no reports of adverse toxic effects of platinum compounds owing to environmental exposure arising from the use of automobile catalysts (270).

## NOMENCLATURE

| Abbreviation | Compound |
| --- | --- |
| acac | 2,4-pentanedionato(1−) |
| (−)binap | (*S*)-(−)-2,2′-bis(diphenylphosphino)-1,1′-binaphthyl |
| bipy | 2,6-bipyridine |
| cat | catecholato(2−) |
| CBDCA | cyclobutane-1,1-dicarboxylate(2−) |
| cod | 1,5-cyclooctadiene |

| Abbreviation | Compound |
|---|---|
| cyclam | 1,4,8,11-tetrazacyclotetradecane |
| dach | 1,2-diaminocyclohexane |
| dba | 1,5-diphenyl-1,4-pentadiene-2-one |
| diars | bis[diphenylarsine]ethane |
| dien | $N$-(2-aminoethyl)-1,2-ethanediamine |
| diim | ethanediimine |
| diphos | bis[diphenylphosphine]ethane |
| DMG | dimethyl glyoximato(1−) |
| DMSO | dimethyl sulfoxide |
| dppe | 1,2-bis(diphenylphosphino)ethane |
| dppm | methylenebis[diphenylphosphine] |
| EDTA | ethylenediaminetetraacetic acid anion(4−) |
| en | ethylenediamine |
| hfacac | 1,1,1,5,5,5-hexafluoroacetyl acetonate(1−) |
| im | imidazolate(1−) |
| mhp | 6-methyl-2-pyridonate(1−) |
| nbd | norbornadiene |
| OEP | 2,3,7,8,12,13,17,18-octaethyl porphine(2−) |
| ox | oxalate(2−) |
| Pc | phthalocyanine(2−) |
| pop | $[P_2O_5H_2]^{2-}$ |
| pyz | 1,4-pyrazine |
| py | pyridine |
| terpy | 2,2′:6′2″-terpyridine |
| sacsac | 2,4-pentanedithionato(1−) |
| salen | 1,2-ethylenebis(salicylideneiminato)(2−) |
| 14-tms | 1,4,8,11-tetramethyl-1,4,8,11-tetrazacyclotetradecane |
| TPP | *meso*-tetraphenylporphine(2−) |

## BIBLIOGRAPHY

"Platinum Group Metals, Alloys and Compounds" in *ECT* 1st ed., Vol. 10, pp. 819–859, by E. M. Wise, International Nickel Company, Inc., and R. Gilchrist, National Bureau of Standards, U.S. Department of Commerce; "Platinum-Group Metals, Compounds" in *ECT* 2nd ed., Vol. 15, pp. 861–878, by H. K. Straschrill and J. G. Cohn, Engelhard Minerals and Chemicals Corp.; in *ECT* 3rd ed., Vol. 18, pp. 254–277, by A. R. Amundsen and E. W. Stern, Engelhard Corp.

1. *Platinum 1985*, *Platinum 1990*, and *Platinum 1995*, Johnson-Matthey PLC, London.
2. *Financial Times*, London, Feb. 28, 1995.
3. *Platinum 1995*, Johnson Matthey PLC, London, 1995.
4. F. R. Hartley, in F. R. Hartley, ed., *Studies in Inorganic Chemistry*, Vol. 11, Elsevier Science, Inc., New York, 1991, pp. 9–31.
5. P. Charlesworth, *Plat. Met. Rev.* **25**, 106–112 (1981).
6. G. B. Harris, *Precious. Met.* **17**, 351–374 (1993).
7. R. I. Edwards, W. A. M. te Riele, and G. J. Bernfeld, in A. J. Bird, ed., *Gmelin Handbook of Inorganic Chemistry, Platinum Supplement*, Springer-Verlag, New York, 1986, Sect. A1, pp. 1–23.
8. H. Renner, *MINTEK*, Report No. M217, Dec. 24, 1985.
9. B. Côté, E. Benguerel, and G. P. Demopoulos, *Precious. Met.* **17**, 593–617 (1993).
10. R. A. Grant, *Proc. Int. Prec. Met. Inst. Sem.*, 7 (1990).

11. Eur. Pat. Appl. 210004 (Jan. 28, 1987) R. A. Grant and B. A. Murrer (to Matthey Rustenburg Refiners (Pty), Ltd., S. Africa).
12. R. R. Brooks, in R. R. Brooks, ed., *Noble Metals and Biological Systems*, CRC Press, Boca Raton, Fla., 1992, pp. 17–44.
13. J. C. Van Loon and R. R. Barefoot, *Determination of the Precious Metals*, John Wiley & Sons Ltd., Chichester, U.K., 1991.
14. F. E. Beamish and J. C. Van Loon, *Analysis of Noble Metals*, Academic Press, New York, 1977.
15. J. H. Holloway, R. D. Peacock, and R. W. H. Small, *J. Chem. Soc.*, 527 (1964).
16. D. G. Lee, D. T. Hall, and J. H. Cleeland, *Can. J. Chem.* **50**, 3741 (1972).
17. D. G. Lee and M. van den Engh, *Can. J. Chem.* **50**, 2000 (1972).
18. H. Nakata, *Tetrahedron* **19**, 1959 (1963).
19. A. H. Bowman, R. S. Evans, and A. F. Schreiner, *Chem. Phys. Lett.* **29**, 140 (1974).
20. *Gmelin Handbuch der Anorgischen Chemie*, System-Number 63, Verlag Chemie, GmbH, Berlin, 1938.
21. D. G. Lee and M. van den Engh, *Org. Chem. Ser. Monogr.* **50**, 177 (1973).
22. L. Wohler, K. Ewald and H. G. Krall, *Chem. Ber.* **66**, 1638 (1933).
23. J. E. Macintyre, F. M. Daniel, and V. M. Stirling, eds., *Dictionary of Inorganic Compounds,* Chapman and Hall, New York, 1992.
24. R. I. Edwards, W. A. M. te Riele, and G. J. Bernfeld, in Ref. 7, pp. 1–338.
25. I. P. Evans, G. W. Everett, and A. M. Sargeson, *J. Chem. Soc., Chem. Commun.*, 139 (1975).
26. I. P. Evans, G. W. Everett, and A. M. Sargeson, *J. Am. Chem. Soc.* **98**, 8041 (1976).
27. S. D. Pell and J. N. Armor, *J. Am. Chem. Soc.* **97**, 5012 (1975).
28. D. F. Mahoney and J. K. Beatti, *Inorg. Chem.* **12**, 2601 (1973).
29. E. A. Seddon and K. R. Seddon, *The Chemistry of Ruthenium*, Elsevier Science Publishing Co., Inc., New York, 1984.
30. A. W. Zanella and P. C. Ford, *Inorg. Chem.* **14**, 42 (1975).
31. D. Dolphin, B. R. James, and P. D. Smith, *Coord. Chem. Rev.* **39**, 31 (1981).
32. L. J. Boucher, in G. A. Melson, ed., *Coordination Chemistry of Macrocyclic Compounds*, Plenum Publishing Corp., New York, 1979, pp. 517–539.
33. K. M. Smith, ed., *Porphyrins and Metalloporphyrins*, Elsevier, Amsterdam, the Netherlands, 1975.
34. D. Dolphin, ed, *The Porphyrins*, Vols. I–VII, Academic Press, New York, 1978,
35. B. D. Berezin, *The Coordination Compounds of Porphyrins and Phthalocyanines*, John Wiley & Sons, Ltd., London, 1981.
36. J. A. McCleverty, *Chem. Rev.* **79**, 53 (1979).
37. J. H. Enemark and R. D. Felham, *Coord. Chem. Rev.* **13**, 339 (1974).
38. K. G. Caulton, *Coord. Chem. Rev.* **14**, 317 (1974).
39. N. G. Connelly, *Inorg. Chim Acta Rev.* **6**, 47 (1972).
40. F. Bottomley, *Coord. Chem. Rev.* **26**, 7 (1978).
41. F. Bottomley, E. V. F. Brooks, S. G. Clarkson, and S. B. Tong, *J. Chem. Soc. Chem. Commun.*, 919 (1973).
42. M. P. Dare-Edwards, J. B. Goodenough, A. Hammett, K. R. Seddon, and R. D. Wright, *Faraday Discuss. Chem. Soc.* **70**, 285 (1970).
43. D. J. Gulliver and W. Levason, *Coord. Chem. Rev.* **46**, 1 (1982).
44. W. P. Griffith, *Chem. Soc. Rev.* **21**, 179–85 (1992).
45. C-M. Che, V. W-W. YAM, *Adv. Inorg. Chem.* **39**, 233–325 (1992).
46. W. P. Griffith and co-workers, *J. Chem. Soc. Chem. Commun.*, 1625 (1987).
47. G. Green, W. P. Griffith, D. M. Hollinshead, S. V. Levy, and M. Schröder, *J. Chem. Soc. Perkin Trans. I.*, 1286 (1984).
48. A. van der Wiel, *Chem. Week.* **48**, 597 (1952).

49. M. Schröder, W. P. Griffith, *J. Chem. Soc. Chem. Commun.*, 58 (1979).
50. H. Remy, *Z. Anorg. Allg. Chem.* **126**, 185 (1923).
51. R. E. Connick, C. R. Hurley, *J. Amer. Chem. Soc.* **74**, 5012 (1952).
52. A. Mercer and J. Trotter, *J. Chem. Soc. Dalton Trans.*, 2480 (1975).
53. R. Bau, R. G. Teller, S. W. Kirtley, and T. F. Koetzle, *Accounts Chem. Res.* **12**, 176 (1979).
54. A. P. Humphries and H. D. Kasez, *Prog. Inorg. Chem.* **25**, 145 (1979).
55. J. Lewis and B. F. G. Johnson, *Pure Appl. Chem.* **44**, 43 (1975).
56. B. F. G. Johnson and Y. V. Roberts, *J. Cluster Sci.* **4**, 231–444 (1993).
57. G. Lavigne, N. Lugan, S. Rivomanana, F. Mulla, J. M. Souile, and P. Kalck, *J. Cluster Sci.* **4**, 49–58 (1993).
58. L. J. Farrugia, *J. Cluster Sci.* **3**, 361–383 (1992).
59. M. I. Bruce, M. P. Cifuentes, and M. G. Humphrey, *Polyhedron* **10**, 277–322 (1991).
60. W. R. Roper, *NATO ASI SER., SER.C.* **392**, 155–168 (1993).
61. J. L. Courtney, in *Organic Synthesis by Oxidation with Metal Compounds*, W. J. Mijs and C. R. H. I. de Jonge, eds., Plenum Press, New York, 1986, pp. 445–467.
62. C. Djerassi and R. R. Engle, *J. Am. Chem. Soc.* **75**, 3838 (1953).
63. W. P. Griffith, S. V. Ley, G. P. Whitcombe, and A. D. White, *J. Chem. Soc., Chem. Commun.*, 1625 (1987).
64. J.-E. Bäckvall, R. L. Chowdhury, U. Karlsson and G-Z. Wang, *Perspect. Coord. Chem.*, 463–486 (1992).
65. J. T. Groves and R. Quinn, *J. Am. Chem. Soc.* **107**, 5790–5792 (1985).
66. P. Kalack, Y. Peres, and J. Jenck, *Adv. Organomet. Chem.* **32**, 121–146 (1991).
67. M. M. T. Khan, *Pat. Met. Rev.* **35**, 70–82 (1991).
68. F. Peter, in S. B. Patrick, ed., *Electrochem. Electrocatal. React. Carbon Dioxide*, Elsevier, Amsterdam, the Netherlands, 1993, pp. 68–93.
69. R. A. W. Johnstone, A. H. Wilby, and I. D. Entwistle, *Chem. Rev.* **85**, 129 (1985).
70. G. Zassinovich, G. Mestroni, and S. Gladiali, *Chem. Rev.* **92**, 1051–1069 (1992).
71. S. Murai, F. Kakiuchi, S. Sekine, Y. Tanaka, A. Kamatani, M. Sonod, and N. Chatani, *Nature* **366**, 529–531 (1993).
72. J. F. Knifton. *J. Mol. Catal.* **11**, 91 (1981).
73. J. R. Zoeller, *J. Mol. Catal.* **37**, 377 (1986).
74. Y. Shvo and R. M. Laine, *J. Chem. Soc., Chem. Commun.* **16**, 753–754 (1980).
75. S. Shinoda, T. Kojuma, and Y. Saito, *J. Mol. Catal.* **18**, 99 (1983).
76. C. Cenini, C. Crotti, M. Pizzotti, and F. Porta, *J. Org. Chem.* **53**, 1243–1250 (1988).
77. C. Bruneau, M. Nueuvx, A. Kabouche, C. Rupping, and P. H. Dixneuf, *Synlett.*, 755–763 (1991).
78. J. M. Lehn and R. Ziessel, *J. Organometal. Chem.* **382**, 157 (1990).
79. J. R. Pugh, M. R. M. Bruce, B. P. Sullivan, and T. J. Meyer, *Inorg. Chem.* **30**, 86 (1991).
80. J. F. Knifton, *Plat. Met. Rev.* **29**, 63–72 (1985).
81. J. F. Knifton, in F. R. Hartley, ed., *Chemistry of the Platinum Group Metals*, Elsevier Science Inc., New York, 1991, pp. 124–146.
82. S. T. Nguyen, R. H. Grubbs, and J. W. Ziller, *J. Am. Chem. Soc.* **115**, 9858–9859 (1993); B. M. Novak and R. H. Grubbs, *J. Am. Chem. Soc.* **110**, 7542–7543 (1988).
83. H. Takaya, T. Ohata, and K. Mashima, *Advan. Chem. Ser.* **230**, 123–142 (1992).
84. M. J. Clarke, *Met. Complexes Cancer Chemother.*, 129–56 (1993).
85. B. K. Keppler, B. Stenzel, K.-G. Lipponer, R. Niebl, H. Vongerichten, and E. Vogelin, in R. R. Brooks, ed., *Noble Metals and Biological Systems*, CRC Press, Boca Raton, Fla., 1992, pp. 323–348.
86. C. J. Murphy and J. K. Barton, *Met. Enzym.* **226**, 226 (1993).
87. J. R. Winler and H. B. Gray, *Chem. Rev.* **92**, 369–379 (1992).

88. M. J. Therien, J. Chang, A. L. Raphael, B. E. Bowler, and H. B. Gray, *Struct. Bonding.* **75**, 109–129 (1991).

89. F. Millet and B. Furham, *Met. Ions Biol. Syst.* **27**, 223–264 (1991).

90. H. H. Thorp. *J. Inorg. Organomet. Polym.* **3**, 41–57 (1993).

91. W.-L. Ying, J. Emerson, M. J. Clarke, and D. R. Sandi, *Biochemistry* **30**, 4949–4952 (1991).

92. I. Bat'ko, M. Somora, D. Vanicky, and K. Flachbart, *Cryogenics* **32**, 1167–1168 (1992).

93. E. N. Balko, *Stud. Inorg. Chem.* **11**, 267–301 (1991).

94. P. D. Gurney and R. J. Seymour, in Ref. 81, pp. 594–616.

95. P. C. Hydes, *Plat. Met. Rev.* **24**, 50 (1980).

96. J. M. Lehn and J. P. Sauvage, *Nouv. J. Chem.* **1**, 449 (1977).

97. Ref. 29, pp. 1240–1260; A. Mills, in Ref. 81, pp. 302–337.

98. H. Taube, *Coord. Chem. Revs.* **26**, 33 (1978).

99. C. Creutz, *Prog. Inorg. Chem.* **30**, 1 (1978).

100. T. J. Meyer, *Adv. Chem. Ser.* **150**, 73 (1976).

101. T. J. Meyer, *Accounts Chem. Res.* **11**, 1978 (73).

102. H. Taube, *Pure Appl. Chem.* **44**, 25 (1975).

103. C. Creutz and H. Taube, *J. Am. Chem. Soc.* **91**, 3988–3989 (1969).

104. H. L. Grube, in G. Brauer, ed., *Handbook of Preparative Inorganic Chemistry*, Academic Press, Inc., New York, 1965, pp. 1601.

105. P. A. Lay, A. M. Sargeson, B. W. Skelton, and A. H. White, *J. Am. Chem. Soc.* **104**, 6161 (1984).

106. W. D. Harman, M. Gebhard, and H. Taube, *Inorg. Chem.* **29**, 567 (1990).

107. W. D. Harman and H. Taube, *J. Am. Chem. Soc.* **109**, 1883 (1987).

108. C. R. H. I. de Jonge, in W. J. Mijs and C. R. H. I. de Jonge, eds., *Organic Synthesis by Oxidation with Metal Compounds*, Plenum Press, New York, 1986, pp. 633–693.

109. H. C. Kolb, M. S. VanNiewuwenhze, and B. K. Sharpless, *Chem. Rev.* **94**, 2483–2547 (1994).

110. K. B. Sharpless and co-workers, *J. Org. Chem.* **57**, 2768–2771 (1992).

111. E. J. Corey, M. C. Noe, and S. Sarshar, *J. Am. Chem. Soc.* **115**, 3828–3829 (1993).

112. E. Herranz and K. B. Sharpless, *Org. Syn.* **61**, 85–93 (1983).

113. *Ibid.*, pp. 93–97.

114. T. J. Johnson, J. C. Huffman, K. G. Caulton, S. A. Sarah, and O. Eisenstein, *Organometallics* **8**, 2073–2074 (1989).

115. R. A. Sanchez-Delgado, A. Andriollo, J. Puga, and G. Martin, *Inorg. Chem.* **26**, 1867–1870 (1987).

116. C. P. Lau, C. Y. Ren, C. H. Yeung, and M. T. Chu, *Inorg. Chim. Acta.* **191**, 21–24 (1992).

117. P. L. Zhou, S. D. Maloney, and B. C. Gates, *J. Catal.* **129**, 315–329 (1991).

118. L. Maurizio, J. Kaspar, R. Ganzerla, A. Trovarelli, and M. Graziani, *J. Catal.* **112**, 1–11 (1988).

119. E. Palecek, and co-workers, in D. M. J. Lilley, H. Heumann, and D. Suck, eds., *Structural Tools for the Analysis of Protein-Nucleic Acid Complexes*, Birkhauser, Basel, Switzerland, 1992, pp. 1–22.

120. P. Jarvinen, R. von Essen, and M. Nissilia, *Rheumatol.* **17**, 1704–1706 (1990); H. Sheppeard and D. J. Ward, *Rheumatol. Rehabil.* **19**, 25–29 (1980).

121. C. C. Hinckley, I. A. Ali, A. F. T. Yokochi, P. D. Robinson, and J. K. Dorsey, in Ref. 85, pp. 303–321.

122. F. Basolo and G. S. Hammaker, *Inorg. Chem.* **1**, 1 (1962).

123. F. Jardine and P. S. Sheridan, in G. Wilkinson, R. D. Gillard, and J. A. McCleverty, eds., *Comprehensive Coordination Chemistry*, Vol. 4, Pergamon Press, Elmsford, N.Y., 1987, pp. 953–979.

124. F. Jardine and P. S. Sheridan, in Ref. 123, pp. 980–989.
125. H. L. Grube, in Ref. 104, p. 1589.
126. J. W. Mellor, ed., *A Comprehensive Treatise on Inorganic and Theoretical Chemistry*, Vol. 15, Longmans, Green and Co., New York, 1942, p. 589.
127. A. Takenaka, Y. Sasada, H. Ogoshi, T. Omura, and Z. I. Yoshida, *Acta Crystallogr., Sect. B* **31**, 1 (1975).
128. E. Müller, *Synthesis,* 761 (1974).
129. E. O. Fischer and H. Wawersik, *J. Organomet. Chem.* **5**, 559 (1966).
130. R. S. Dickson, *Homogenous Catalysis with Compounds of Rhodium and Iridium*, D. Reidel, Dordrecht, the Netherlands, 1985.
131. G. R. Steinmetz and J. R. Zoeller, in Ref. 81, pp. 75–105.
132. F. H. Jardine, in Ref. 81, pp. 407–469.
133. P. N. Lodal, *Chem. Ind.* **49**, 61–69 (1993).
134. J. R. Zoeller, V. H. Agreda, S. L. Cook, N. L. Lafferty, S. W. Polichnowski, and D. M. Pond, *Catal. Today* **13**, 73–91 (1992).
135. R. L. Preutt, *Adv. Organomet. Chem.* **17**, 1 (1979).
136. A. A. Oswald, D. E. Hendriksen, R. V. Kastrup, and E. J. Mozeleski, *Advan. Chem. Ser. ACS* **230**, 395–418 (1992).
137. M. J. H. Russell, *Chemie Technik* **17–18**, 21–22 (1988).
138. H. Bach, W. Gick, W. Konkol, and E. Wiebus, *Proc. Int. Congr. Catal.* **1**, 254–259 (1988).
139. P. W. N. M. van Leeuwen and G. van Koten, *Stud. Surf. Sci. Catal.* **79**, 199–248 (1993).
140. F. H. Jardine, in F. R. Hartley, ed., *The Chemistry of the Metal Carbon Bond*, Vol. 4, John Wiley & Sons, Ltd., Chichester, U.K., 1987, pp. 734–461.
141. K. Inoguchi, S. Sakuraba, and K. Achiwa, *Synlett.* **3**, 169–178 (1992).
142. J. K. Whitesell, *Chem. Rev.* **89**, 1581–1590 (1989).
143. A. S. C. Chan, J. J. Pluth, and J. Halpern, *J. Am. Chem. Soc.* **102**, 5952 (1980).
144. D. F. Taber and co-workers, *Chem. Ind. London* **40**, 43–50 (1990).
145. M. P. Doyle, *Adv. Chem. Ser.* **230**, 443–461 (1992).
146. M. P. Doyle, K. G. High, and C. L. Nesloney, *Chem. Industries* **47**, 293–305 (1992).
147. J. Adams and D. M. Spero, *Tetrahedron* **47**, 1765–1808 (1991).
148. M. P. Doyle, *ACS Symp. Ser.* **517**, 40–57 (1993).
149. F. H. Jardine, *Prog. Inorg. Chem.* **28**, 63 (1980).
150. H. A. Zahalker and H. Alper, *Organometallics* **5**, 1909 (1986).
151. E. Drent, *ACS Symp. Ser.* **328**, 154 (1987).
152. A. S. Chan, W. E. Carroll, and D. E. Willis, *J. Mol. Catal.* **19**, 377 (1983).
153. N. Rizkalla and A. Goliaszewski, *ACS Symp. Ser.* **328**, 136 (1979).
154. K. Kaneda, T. Imanaka, and S. Teranishi, *Chem. Lett.*, 1465 (1983).
155. H. Mimoun, M. M. P. Machirant, and I. S. de Roch, *J. Am. Chem. Soc.* **100**, 5437 (1978).
156. K. Yogish and N. V. S. Sastri, *Ind. Eng. Chem. Res.* **27**, 909 (1988).
157. P. Muller and C. Bobillier, *Tetrahedron Lett.* **24**, 5499 (1983).
158. C. U. Pittman, Jr., in G. Wilkinson, F. G. Stone, and E. W. Abel, eds., *Comprehensive Organomettallic Chemistry*, Vol. 8, Pergamon Press, New York, 1982, pp. 553–611.
159. H. B. Gray and co-workers, *Adv. Chem. Ser.* **168**, 44 (1978).
160. S. Fukuzumi and co-workers, *Bull. Chem. Soc. Jpn.* **55**, 2892 (1982).
161. R. G. Bergamn, *Adv. Chem. Ser.* **230**, 211–220 (1992).
162. M. Tanaka, *Chemtech* **19**, 59–64 (1989).
163. I. Haiduc and C. Silvestru, *in vivo* **3**, 285–294 (1989).
164. A. Sitlani, C. M. Dupurer, and J. K. Barton, *J. Am. Chem. Soc.* **114**, 2303 (1992).
165. R. R. Benham, *Plat. Met. Rev.* **5**, 13–18 (1961).

166. J. Fischer and D. E. Weimer, *Precious Metal Plating*, 167–200 (1964).
167. N. Serpone and M. A. Jamieson, in Ref. 123, pp. 1130–1131.
168. A. B. Lamb and L. T. Fairhill, *J. Am. Chem. Soc.* **45**, 378 (1923).
169. F. A. Cotton and G. Wilkinson, *Advanced Inorganic Chemistry*, 5th ed., John Wiley & Sons, Inc., New York, 1988, p. 911.
170. P. A. Lay, A. M. Sargeson, and H. Taube, *Inorg. Syn.* **24**, 287–289 (1986).
171. N. Serpone and M. A. Jamieson, in Ref. 123, pp. 1128–1130.
172. G. J. Leigh and R. L. Richards, in Ref. 158, Vol. 5, pp. 541–628.
173. Ref. 169, p. 915.
174. J. D. Atwood, *Coord. Chem. Rev.* **83**, 93–114 (1988).
175. W. Strohmeier and E. Eder, *J. Organomet. Chem.* **94**, C14 (1975).
176. G. D. Mercer, W. B. Beaulieu, and D. M. Roundhill, *J. Am. Chem. Soc.* **99**, 6551 (1977).
177. B. L. Haymore and J. A. Ibers, *J. Am. Chem. Soc.* **96**, 3325 (1975).
178. B. F. G. Johnson and S. Bhaduri, *J. Chem. Soc., Chem. Commun.*, 650 (1973).
179. J. Valentine, D. Valentine, and J. P. Collman, *Inorg. Chem.* **10**, 219 (1971).
180. W. D. Jones, in C. L. Hill, ed., *Activation and Functionalization of Alkanes*, John Wiley & Sons, Inc., New York, 1989, pp. 114–149.
181. R. G. Bergamn, *Science* **223**, 902 (1984).
182. J. K. Hoyano and W. A. G. Graham, *J. Am. Chem. Soc.* **104**, 3723 (1982).
183. M. J. Burk and R. H. Crabtree, *J. Am. Chem. Soc.* **109**, 8025–8032 (1987).
184. A. J. Kunin and R. Eisenberg, *J. Am. Chem. Soc.* **108**, 535 (1986).
185. L. Vaska and R. E. Rhodes, *J. Am. Chem. Soc.* **85**, 4970 (1965).
186. M. B. France, J. Feldman, and R. H. Grubbs, *J. Chem. Soc., Chem Commun.*, 1307–1308 (1994).
187. X. Zhang and co-workers, *J. Am. Chem. Soc.* **115**, 3318–3319 (1993).
188. Y. Ng Cheong Chan and J. A. Osborn, *J. Am. Chem. Soc.* **112**, 9400–9401 (1990).
189. S. Trasatti, *Croat. Chem. Acta* **63**, 313–329 (1990).
190. P. H. Huang and K. G. Kreider, *Electrochem. Eval. Solid State pH Sensors Nuclear Waste Contain.* **89**, 30 (1989).
191. N. Baba, T. Yoshino, S. Morisaki, and H. Masuda, *Mem. Fac. Technol.* **40**, 4339–4348 (1990).
192. P. M. Maitlis, P. Espinet, and M. J. H. Russell, in Ref. 158, Vol. 6, pp. 233–263.
193. F. D. Rochon, P. C. Kong, B. Coulombe, and R. Melanson, *Can. J. Chem.* **58**, 381–386 (1980).
194. C. F. Barnard and M. J. H. Russell, in Ref. 123, Vol. 5, pp. 1099–1130.
195. S. N. Battacharya and C. V. Senoff, *Inorg. Chim. Acta.* **41**, 67 (1980).
196. F. R. Hartley, in *The Chemistry of Platinum and Palladium*, John Wiley & Sons, Inc., New York, 1973.
197. A. T. Hutton and C. P. Morely, in Ref. 123, Vol. 5, pp. 1057–1170.
198. W. Leavason and co-workers, *J. Chem. Soc. Dalton Trans.*, 133 (1983).
199. M. L. Kullberg and C. P. Kubiak, *Inorg. Chem.* **25**, 16–30 (1986).
200. F. Puche, *Chem. Abstr.* **32**, 5322 (1938).
201. P. M. Maitlis, P. Espinet, and M. J. H. Russell, in Ref. 158, Vol. 6, pp. 279–342.
202. A. D. Burrows and D. Mingos, *Trans. Met. Chem.* **18**, 129–148 (1993).
203. U. Belluco and co-workers, *Inorg. Chim. Acta* **198–200**, 883-897 (1992).
204. S. F. Davison and P. M. Maitlis, in W. J. Mijs and C. R. H. I. De Jonge, eds., *Organic Synthesis by Oxidation with Metal Compounds*, Plenum Press, New York, 1986, pp. 469–502.
205. B. Akermark, B. C. Söderberg, and S. S. Hall, *Organometallics* **6**, 2608–2610 (1987).
206. J. W. Francis and P. M. Henry, *Organometallics* **11**, 2832–2836 (1992).
207. J. E. Backväll, B. Äkermark, and S. O. Ljunggren, *J. Am. Chem. Soc.* **101**, 2411–2415 (1979).

208. J. K. Stille and R. Divakaruni, *J. Am. Chem. Soc.* **100**, 1303–1304 (1978).
209. J. W. Francis and P. M. Henry, *Organometallics* **10**, 3498–3503 (1991).
210. R. F. Heck, *Palladium Reagents in Organic Synthesis*, Academic Press, Inc., New York, 1985.
211. B. M. Trost and T. R. Verhoeven, in Ref. 158, pp. 799–938.
212. G. K. Anderson, in Ref. 81, pp. 338–406.
213. R. F. Heck, *J. Am. Chem. Soc.* **90**, 5518 (1968).
214. Y. Hatanaka and T. Hiyama, *Synlett.*, 845–853 (1991).
215. J. K. Stille, *Pure Appl. Chem.* **57**, 1771–1780 (1985).
216. T. N. Mitchell, *Synthesis*, 803–815 (1992).
217. G. D. Davies, Jr., *Adv. Met.-Org. Chem.* **2**, 59–99 (1991).
218. A. Schoenberg and R. F. Keck, *J. Org. Chem.* **39**, 3327 (1974).
219. S. Cacchi, E. Morera, and G. Ortar, *Tetrahedron Lett.* **26**, 1109–1112 (1985).
220. N. Miyaura, T. Yanagi, and A. Suzuki, *Synth. Commun.* **11**, 513–519 (1981).
221. J. Tsuji, *Tetrahedron* **42**, 4361–4401 (1986).
222. B. M. Trost, *Pure Appl. Chem.* **60**, 1615–1626 (1988).
223. F. Guibe, *Main Group Met. Chem.* **12**, 437–446 (1989).
224. A. Sen, *Accounts Chem. Res.* **26**, 303–310 (1993).
225. Eur. Pat. 400,892 (Dec. 5, 1990), V. Elango, K. G. Davenport, M. A. Murphy, G. N. Mott, E. G. Zey, B. L. Smith, and G. L. Moss.
226. Eur. Pat. 284,310 (Sept. 28, 1988), V. Elango, M. A. Murphy, B. L. Smith, K. G. Davenport, G. N. Mott, G. L. Moss (to Hoechst Celanese Corp.).
227. L. Winkler, *Z. Anal. Chem.* **100**, 321 (1935).; U.S. Pat. 2,487,077 (Nov. 8, 1949), G. M. Shephard.
228. T. H. James, ed., *Theory of Photographic Process,* 4th ed., Macmillan, Inc., New York, 1977, p. 149; U.S. Pat. 4,092,171 (May 30 1978), J. H. Bigelow.
229. P. I. Clemenson, *Coord. Chem. Rev.* **106**, 171–203 (1990).
230. M. Cleare, *Rec. Results Cancer Res.* **48**, 2 (1974).
231. C. J. Raub, *Plat. Met. Rev.* **26**, 158–166 (1982).
232. B. Sturzenegger and C. Cl. Puippe, *Plat. Met. Rev.* **28**, 117–124 (1984).
233. F. A. Lowenheim, *Modern Electroplating*, John Wiley & Sons, New York, 1974, pp. 352–354.
234. J. Fischer and D. E. Weimer, *Precious Met. Plating*, 223–226 (1964).
235. N. Bartlett and D. H. Lohmann, *J. Chem. Soc.*, 5253 (1964).
236. H. Remy, in J. Kleinberg, ed., *Treatise on Inorganic Chemistry*, Vol. 2, Elsevier, Amsterdam, the Netherlands, 1956, p. 346.
237. N. V. Sidgwick, *The Chemical Elements and their Compounds*, Vol. 2, Oxford University Press, U.K., 1950, p. 1614.
238. I. P. Evans, G. W. Everett, and A. M. Sargeson, *J. Am. Chem. Soc.* **98**, 8041 (1976).
239. J. K. Barton and co-workers, *J. Am. Chem. Soc.* **100**, 3785 (1978); *ibid.* **101**, 1434 (1979); *ibid.* 7269 (1979).
240. F. R. Hartley, *The Chemistry of Platinum and Palladium*, John Wiley & Sons, Inc., New York, 1973, pp. 182–183.
241. F. R. Hartley, in Ref. 158, Vol. 6, pp. 619–640.
242. V. T. DeVita, Jr., S. Hellman, and S. A. Rosenberg, *Cancer—Principles and Practice of Oncology*, 4th ed., J. B. Lippincott, Philadelphia, Pa., 1993, pp. 1126–1151.
243. P. J. Loehrer and L. H. Einhorn, *Ann. Intern. Med.* **100**, 704 (1984).
244. E. B. Douple, in Ref. 12, pp. 349–376.
245. P. A. Bunn, R. Canetta, R. F. Ozols, and M. Rozencweig, eds., *Carboplatin (JM-8): Current Perspectives and Future Directions*, W. B. Saunders Co., Philadelphia, Pa., 1990.
246. L. R. Kelland, *Crit. Rev. Oncol. Hematol.* **15**, 191–219 (1993).

247. R. B. Weiss and M. C. Christian, *Drugs* **46**, 360–377 (1993).
248. L. R. Kelland and co-workers, *Cancer Res.* **53**, 2581–2586 (1993).
249. S. L. Bruhn, J. H. Toney, and S. J. Lippard, in S. J. Lippard, ed., *Progress in Inorganic Chemistry: Bioinorganic Chemistry*, Vol. 38, John Wiley & Sons, Inc., New York, 1990, pp. 477–516.
250. A. W. Parkins, in Ref. 81, p. 116.
251. G. Chandra, P. Y. Lo, P. B. Hitchcock, and M. F. Lappert, *Organometallics* **6**, 191–192 (1987).
252. K. Yamamoto, T. Hayashi, and M. Kumada, *J. Am. Chem. Soc.* **93**, 5301 (1971).
253. L. N. Lewis, N. Lewis, and R. J. Uriarte, *Advan. Chem. Ser.* **230**, 541–549 (1992).
254. D. M. Roundhill, in Ref. 123, Vol. 5, p. 371.
255. J. F. Knifton, in Ref. 81, Vol. 11, p. 143.
256. G. W. Parshall, *J. Am. Chem. Soc.* **94**, 8716 (1972).
257. T. Yoshida, Y. Ueda, and S. Otsuka, *J. Am. Chem. Soc.* **100**, 3941 (1978).
258. R. G. Wing and I. R. McGill, *Plat. Met. Rev.* **25**, 94–105 (1981).
259. C. J. Raub, in *Gmelin Handbuch der Angorganischen Chemie, Platinum Suppl.* Springer-Verlag, Berlin, 1987, Sect. A1, p. 137.
260. M. E. Baumgärtner and C. J. Raub, *Plat. Met. Rev.* **32**, 188–197 (1988).
261. M. J. Ware, *Brit. J. Photogr.* **133**, 1165, 1190 (1986).
262. M. J. Ware, *J. Photogr. Sci.* **34**, 12 (1986).
263. J. R. Tower, *Infrared Technol.* **25**, 103–106 (1991).
264. H. G. Seiler and H. Sigel, *Handbook on Toxicity of Inorganic Compounds*, Marcel Dekker Inc., New York, 1988.
265. R. A. Cooper, *J. Chem. Met. Soc. S.* **22**, 152 (1922).
266. S. Harris, *J. Soc. Occup. Med.* **25**, 133 (1975).
267. F. R. Brunot, *J. Industr. Hyg.* **15**, 136 (1933).
268. H. G. Seiler and H. Sigel, *Handbook on Toxicity of Inorganic Compounds*, Marcel Dekker, Inc., New York, 1988, p. 502.
269. *Threshold Limit Values for Chemical Substance in the Work Environment Adopted by ACGIH for 1983–1984*, American Conference of Governmental Hygienists, 1983.
270. Ref. 268, pp. 533–539.
271. M. J. Cleare, E. G. Hughes, B. Jacoby, and J. Pepys, *Clin. Allergy* **2**, 391 (1972).

*General References*

An excellent comprehensive review of ruthenium chemistry is E. A. Seddon and K. R. Seddon, *The Chemistry of Ruthenium*, Elsevier Science Publishing Co., Inc., New York, 1984.

Other useful references are R. F. Heck, *Palladium Reagents in Organic Synthesis* Academic Press, New York, 1985, and J. Tsuji, *Organic Synthesis with Palladium Compounds*, Springer-Verlag, New York, 1980.

There are several excellent sources of information about the platinum-group metals. The excellent reference work G. Wilkinson, R. D. Gillard, and J. A. McCleverty, eds., *Comprehensive Coordination Chemistry* Pergamon Press, Oxford, U.K., 1987, contains individual chapters devoted to descriptive chemistry of each element.

Similar treatment of the organometallic chemistry of the PGMs are found in G. Wilkinson, F. Gordon, A. Stone, and E. W. Abel, eds., *Comprehensive Organometallic Chemistry* Pergamon Press, Oxford, U.K., 1982.

Earlier descriptive chemistry is contained in F. R. Hartley, *The Chemistry of Platinum and Palladium*, John Wiley & Sons, Inc., New York, 1973, and W. P. Griffith, *The Chemistry of the Rarer Platinum Metals (Os, Ru, Ir and Rh)*, John Wiley & Sons, Ltd., Chichester, U.K., 1967.

Two works provide excellent collections of articles describing important modern applications of PGMs. These are F. H. Hartley, ed., *The Chemistry of Platinum Group Metals*, Elsevier, 1991, and *Gmelin Handbook of Inorganic Chemistry Platinum Suppl.* Vol. A1, Springer-Verlag, New York, 1986.

The *Platinum Group Metal Reviews* is a specialized review series focusing on the new developments and uses of PGMs. Each issue also provides a brief description of recent patents issued in the field.

J. E. Macintyre, F. M. Daniel, and V. M. Stiriling, eds., *Dictionary of Inorganic Compounds*, Chapman and Hall, London, 1992, and J. E. Macintyre, *The Dictionary of Organometallic Compounds*, Chapman and Hall, London, 1984, provide extensive listings of characterized compounds, with brief notes regarding uses, properties, and lead literature references.

CHRISTEN M. GIANDOMENICO
Johnson Matthey

# PLUTONIUM AND PLUTONIUM COMPOUNDS

Plutonium [7440-07-5], Pu, element number 94 in the Periodic Table, is a member of the actinide series and is metallic (see ACTINIDES AND TRANSACTINIDES). Isotopes of mass number 232 through 246 have been identified. All are radioactive. The most important isotope is plutonium-239 [15117-48-3], $^{239}$Pu; also of importance are $^{238}$Pu, $^{242}$Pu, and $^{244}$Pu.

The large energy release that accompanies the nuclear fission reaction is the most significant property of plutonium. Upon fission, one gram of plutonium-239 releases energy equivalent to that produced by combustion of three metric tons of coal (qv). The energy can be applied in electric power generating reactors or industrial or military explosives (see EXPLOSIVES AND PROPELLANTS; POWER GENERATION). Large quantities of plutonium are produced in uranium-fueled power reactors (see NUCLEAR REACTORS). Estimates of world production of plutonium are listed in Table 1 (1). Through 1990, the U.S. discharge of Pu from power reactors (176 metric tons) was estimated to be 27% of the world's discharge; by 2010, the cumulative U.S. discharge of Pu from power reactors is estimated to be 528 metric tons, which is expected to be 25% of the world's discharge. The vast majority (>98%) of this plutonium has not been separated from fuel assemblies or fission products (1).

Much of the world's separated plutonium has been used for nuclear weapons (Table 1). It is probable that 5 kg or less of $^{239}$Pu is used in most of the fission, fusion, and thermonuclear-boosted fission weapons (2). Weapons-grade plutonium requires a content of >95 wt % $^{239}$Pu for maximum efficiency. Much plutonium does not have this purity.

**Table 1. Fissile Plutonium Produced in and Separated from Military and Power Reactor Fuel,[a] t**

| Country | 1990 | | | | 2010[d] |
| | Military | Civilian[b] | Separated[c] | Total | Civilian[b] |
| --- | --- | --- | --- | --- | --- |
| U.S. | 90–105 | 176 | 1.5 | 268–283 | 528 |
| CIS[e] | 115–160 | 78 | 25 | 218–263 | 220 |
| U.K. | 8 | 53 | 50 | 111 | 103 |
| France | 6 | 78 | 49 | 133 | 296 |
| Japan | 0 | 57 | 3.6 | 61 | 228 |
| other | 3 | 251 | 1.9 | 256 | 723 |

[a]Ref. 1.
[b]Includes cumulative amounts from power reactors and spent fuel.
[c]From spent power and research reactor fuel.
[d]Predicted values.
[e]Commonwealth of Independent States (former USSR); includes Lithuania, Estonia, and Latvia.

In plutonium-fueled breeder power reactors, more plutonium is produced than is consumed (see NUCLEAR REACTORS, REACTOR TYPES). Thus the utilization of plutonium as a nuclear energy or weapon source is especially attractive to countries that do not have uranium-enrichment facilities. The cost of a chemical reprocessing plant for plutonium production is much less than that of a uranium-235 enrichment plant (see URANIUM AND URANIUM COMPOUNDS). Since the end of the Cold War, the potential surplus of $^{239}$Pu metal recovered from the dismantling of nuclear weapons has presented a large risk from a security standpoint.

The isotope plutonium-238 [13981-16-3], $^{238}$Pu, is of technical importance because of the high heat that accompanies its radioactive decay. This isotope has been and is being used as fuel in small terrestrial and space nuclear-powered sources (3,4). $^{238}$Pu-based radioisotope thermal generator systems delivered 7 W/kg and cost \$120,000/W in 1991 (3). For some time, $^{238}$Pu was considered to be the most promising power source for the radioisotope-powered artificial heart and for cardiovascular pacemakers. Usage of plutonium was discontinued, however, after it was determined that adequate elimination of penetrating radiation was uncertain (5) (see PROSTHETIC AND BIOMEDICAL DEVICES).

Plutonium was the first element to be synthesized in weighable amounts (6,7). Technetium, discovered in 1937, was not isolated until 1946 and not named until 1947 (8). Since the discovery of plutonium in 1940, production has increased from submicrogram to metric ton quantities. Because of its great importance, more is known about plutonium and its chemistry than is known about many of the more common elements. The metallurgy and chemistry are complex. Metallic plutonium exhibits seven allotropic modifications. Five different oxidation states are known to exist in compounds and in solution.

The discovery of plutonium-238, an $\alpha$-emitter having a half-life, $t_{1/2}$, of 87.7 years, by G. T. Seaborg and co-workers (9,10) was achieved by bombardment of uranium using deuterons, $^2$H (eqs. 1 and 2):

$$^{238}\text{U} + {}^2\text{H} \longrightarrow {}^{238}\text{Np} + 2\,{}^1n \tag{1}$$

$$^{238}\text{Np} \xrightarrow[\text{2.12 d}]{\beta^-} {}^{238}\text{Pu} \tag{2}$$

In early 1941, 0.5 $\mu$g of $^{239}$Pu was produced (eqs. 3 and 4) and subjected to neutron bombardment (9) demonstrating that plutonium undergoes thermal neutron-induced fission with a cross section greater than that of $^{235}$U. In 1942, a self-sustaining chain reaction was induced by fissioning $^{235}$U in a large graphite natural uranium lattice. Some of the excess neutrons from the $^{235}$U fission were captured in the abundant uranium-238, thereby producing $^{239}$Pu (eqs. 3 and 4).

$$^{238}\text{U} + n \longrightarrow {}^{239}\text{U} \tag{3}$$

$$^{239}\text{U} \xrightarrow[\text{23.5 min}]{\beta^-} {}^{239}\text{Np} \xrightarrow[\text{2.35 d}]{\beta^-} {}^{239}\text{Pu} \tag{4}$$

The first preparation, isolation, and weighing of a pure compound of plutonium, plutonium(IV) oxide [12059-95-5], $PuO_2$, was achieved in 1942 by ignition of hydrated $Pu(NO_3)_4$ [13823-27-3] (10–12). This 2.77-$\mu$g sample of $PuO_2$ was the first pure substance containing a synthetic element to be visible to the human eye. The first unequivocal preparation of metallic plutonium was in 1943 at the Metallurgical Laboratory of the University of Chicago by the reduction of plutonium tetrafluoride [13709-56-3] and barium metal in a sealed, evacuated reaction vessel at 1400°C (13,14). Fabricable amounts of metal were prepared in 1944 at Los Alamos, New Mexico (15).

## Sources

Plutonium occurs in natural ores in such small amounts that separation is impractical. The atomic ratio of plutonium to uranium in uranium ores is less than $1{:}10^{11}$; however, traces of primordial plutonium-244 have been isolated from the mineral bastnasite (16). One sample contained $1 \times 10^{-18}$ g/g ore, corresponding to a plutonium-244 [14119-34-7], $^{244}$Pu, terrestrial abundance of $7 \times 10^{-27}$ to $2.8 \times 10^{-25}$ g/g of mineral and to $<10$ g of primordial $^{244}$Pu on earth. The content of plutonium-239 [15117-48-3], $^{239}$Pu, in uranium minerals is given in Table 2.

**Table 2. Plutonium in Natural Sources**

| Ore | Uranium content, wt % | Mass ratio,[a] $^{239}$Pu/ore |
|---|---|---|
| pitchblende | | |
|   Canadian | 13.5 | $9.1 \times 10^{-12}$ |
|   Zaire | 38 | $4.8 \times 10^{-12}$ |
|   Colorado | 50 | $3.8 \times 10^{-12}$ |
|   Zaire concentrate | 45.3 | $7.0 \times 10^{-12}$ |
| monazite | | |
|   Brazilian | 0.24 | $2.1 \times 10^{-14}$ |
|   North Carolina | 1.64 | $5.9 \times 10^{-14}$ |
| fergusonite | 0.25 | $<1.0 \times 10^{-14}$ |
| carnotite | 10 | $<4.0 \times 10^{-14}$ |
| bastnasite | | $1 \times 10^{-18}{}^{b}$ |

[a]Ref. 17, unless otherwise noted.
[b]Value corresponds to a mass ratio of $^{244}$Pu/ore (16).

All of the 15 plutonium isotopes listed in Table 3 are synthetic and radioactive (see RADIOISOTOPES). The lighter isotopes decay mainly by K-electron capture, thereby forming neptunium isotopes. With the exception of mass numbers 237 [15411-93-5], 241 [14119-32-5], and 243, the nine intermediate isotopes, ie, 236–244, are transformed into uranium isotopes by $\alpha$-decay. The heaviest plu-

**Table 3. Plutonium Isotopes[a]**

| Mass number | Half-life | Mode of decay[b] | Main radiations | | Production method |
|---|---|---|---|---|---|
| | | | Particle | Energy, MeV | |
| 232 | 34 min | EC $\geq$ 80% | $\alpha$ | 6.60, 62% | $^{233}$U$(\alpha,5n)$ |
| | | $\alpha \leq$ 20% | | 6.54, 38% | |
| 233 | 20.9 min | EC 99.88% | $\alpha$ | 6.30 | $^{233}$U$(\alpha,4n)$ |
| | | $\alpha$ 0.12% | $\gamma$ | 0.235 | |
| 234 | 8.8 h | EC 94% | $\alpha$ | 6.202, 68% | $^{233}$U$(\alpha,3n)$ |
| | | $\alpha$ 6% | | 6.151, 32% | |
| 235 | 25.6 min | EC > 99% | $\alpha$ | 5.85 | $^{235}$U$(\alpha,4n)$ |
| | | $\alpha$ 3 × 10$^{-3}$% | $\gamma$ | 0.049 | $^{233}$U$(\alpha,2n)$ |
| 236 | 2.85 yr | $\alpha$ | $\alpha$ | 5.768, 69% | $^{235}$U$(\alpha,3n)$ |
| | 3.5 × 10$^9$ yr | SF | | 5.721, 31% | $^{236}$Np daughter |
| 237 | 45.4 d | EC > 99% | $\alpha$ | 5.65, 21% | $^{235}$U$(\alpha,2n)$ |
| | | $\alpha$ 3.3 × 10$^{-3}$% | | 5.36, 79% | $^{237}$Np$(d,2n)$ |
| | | | $\gamma$ | 0.059 | |
| 238 | 87.74 yr | $\alpha$ | $\alpha$ | 5.499, 70.9% | $^{242}$Cm daughter |
| | 4.8 × 10$^{10}$ yr | SF | | 5.457, 29.0% | $^{238}$Np daughter |
| 239 | 2.41 × 10$^4$ yr | $\alpha$ | $\alpha$ | 5.155 73.3% | $^{239}$Np daughter |
| | 5.5 × 10$^{15}$ yr | SF | | 5.143, 15.1% | $n$ capture |
| | | | $\gamma$ | 0.129 | |
| 240 | 6.563 × 10$^3$ yr | $\alpha$ | $\alpha$ | 5.168, 72.8% | multiple $n$ capture |
| | 1.34 × 10$^{11}$ yr | SF | | 5.123, 27.1% | |
| 241 | 14.4 yr | $\beta^-$ > 99% | $\alpha$ | 4.896, 83.2% | multiple $n$ capture |
| | | $\alpha$ 2.41 × 10$^{-3}$% | | 4.853, 21.1% | |
| | | | $\beta^-$ | 0.021 | |
| | | | $\gamma$ | 0.149 | |
| 242 | 3.76 × 10$^5$ yr | $\alpha$ | $\alpha$ | 4.901, 74% | multiple $n$ capture |
| | 6.8 × 10$^{10}$ yr | SF | | 4.857, 26% | |
| 243 | 4.956 h | $\beta^-$ | $\beta^-$ | 0.58 | multiple $n$ capture |
| | | | $\gamma$ | 0.084 | |
| 244 | 8.26 × 10$^7$ yr | $\alpha$ | $\alpha$ | 4.589, 81% | multiple $n$ capture |
| | 6.6 × 10$^{10}$ yr | SF | | 4.546, 19% | |
| 245 | 10.5 h | $\beta^-$ | $\beta^-$ | 1.28 | $^{244}$Pu$(n,\gamma)$ |
| | | | $\gamma$ | 0.327 | |
| 246 | 10.85 d | $\beta^-$ | $\beta^-$ | 0.374 | $^{245}$Pu$(n,\gamma)$ |
| | | | $\gamma$ | 0.224 | |

[a]Ref. 18.
[b]EC = K-electron capture; SF = spontaneous fission.

tonium isotopes tend to undergo $\beta$-decay, thereby forming americium. Detailed reviews of the nuclear properties have been published (18).

The technologically most important isotope, $^{239}$Pu, has been produced in large quantities since 1944 from natural or partially enriched uranium in production reactors. This isotope is characterized by a high fission reaction cross section and is useful for fission weapons, as trigger for thermonuclear weapons, and as fuel for breeder reactors. A large future source of plutonium may be from fast-neutron breeder reactors.

Commercial electric power generating reactors generally produce plutonium by irradiating uranium fuels to a total neutron exposure of more than 5000 MW·d/t. The recoverable plutonium contains a larger fraction of heavier isotopes. The rate of production and the isotopic composition depends on the reactor type and method of operation, which depend on economics. In boiling water reactors (BWRs) and pressurized water reactors (PWRs), the rates of production are 270 and 360 grams of plutonium per electrical megawatt-year of operation, respectively. The ORIGEN code has been used to calculate the growth of actinide and fission-product isotopes in a PWR operating at 30 thermal megawatts per metric ton of uranium (MTU) for 1100 days for a total burnup of 33,000 MW·d/t, at which time the $^{239}$Pu content has reached a steady state of 5190 g/MTU (19). Natural uranium gas-cooled reactors, which are used for generating electricity in the United Kingdom and France, produce 500 g plutonium per electrical megawatt-year, whereas fully enriched uranium-fueled gas-cooled plants produce no plutonium. Generally, commercial-grade plutonium (Table 4) is richer in heavy isotopes than weapons-grade plutonium from production reactors. When $^{239}$Pu is irradiated for long periods to produce $^{243}$Am and $^{244}$Cm, the converted plutonium fraction, which is isolated after irradiation, consists almost entirely of plutonium-242 [13982-10-0], $^{242}$Pu, and some $^{244}$Pu.

$^{238}$Pu has been produced in multikilogram amounts by neutron bombardment of $^{237}$Np, which was recovered in the $^{239}$Pu isolation process following the neutron capture of $^{239}$Pu to plutonium-240 [14119-33-6], $^{240}$Pu, and then $^{241}$Pu, which decays to $^{241}$Am and then to $^{237}$Np. The $^{238}$Pu has been isolated from the neptunium target material by an ion-exchange process (20). Conventional radioisotopic fuel contains ca 80 wt % $^{238}$Pu, 16.5 wt % $^{239}$Pu, 2.5 wt % $^{240}$Pu,

**Table 4. Isotopic Composition of Plutonium from Thermal Power Reactors[a–c]**

| Isotope | Reactor-grade concentrations, % of total Pu | | | |
| --- | --- | --- | --- | --- |
| | Magnox (5000 MW·d/t[d]) | CANDU (7000 MW·d/t[d]) | BWR (33,000 MW·d/t[a,d]) | PWR (33,000 MW·d/t[c,d]) |
| $^{238}$Pu | | | 1.3 | 1.8 |
| $^{239}$Pu | 68.5 | 66.6 | 56.6 | 58.4 |
| $^{240}$Pu | 25.0 | 26.6 | 23.2 | 24.3 |
| $^{241}$Pu | 5.3 | 5.3 | 13.9 | 11.5 |
| $^{242}$Pu | 1.2 | 1.5 | 4.7 | 3.9 |

[a]Ref. 25.
[b]Ref. 1.
[c]Ref. 19.
[d]MW·d/t = thermal megawatt·days per ton of uranium.

0.8 wt % $^{241}$Pu, and 0.08 wt % $^{242}$Pu. The specific power from $\alpha$-decay of this isotopic mixture is 450 mW/g (427 Btu/(g·s)), compared to 2 mW/g (1.9 Btu/(g·s)) for weapons-grade plutonium (21,22).

## Economic Aspects

**Uses of Plutonium.** The fissile isotope $^{239}$Pu had its first use in fission weapons, beginning with the Trinity test at Alamogordo, New Mexico, on July 16, 1945, followed soon thereafter by the "Little Boy" bomb dropped on Nagasaki on August 9, 1945. Its weapons use was extended as triggers for thermonuclear weapons. This isotope is produced in and consumed as fuel in breeder reactors. The short-lived isotope $^{238}$Pu has been used in radioisotope electrical generators in unmanned space satellites, lunar and interplanetary spaceships, heart pacemakers, and (as $^{238}$Pu–Be alloy) neutron sources (23).

**Cost and Value of Plutonium.** The cost of building all U.S. nuclear weapons has been estimated as \$378 billion in 1995 dollars (24). If half of this sum is attributed to U.S. weapons-grade plutonium production ($\sim$100 t), the cost is \$1.9 $\times$ 10$^6$/kg of weapons-grade Pu. Some nuclear weapons materials (Be, enriched U, $^{239}$Pu) also have value as a clandestine or terrorist commodity. The economic value of reactor-grade plutonium as a fuel for electric power-producing reactors has depended in the past on the economic value of pure $^{235}$U (see URANIUM AND URANIUM COMPOUNDS).

Plutonium has no commercial (open-market) value at the time of this writing (1995), because approximately 100 metric tons is to become available as surplus from weapons being retired by the United States and the former Soviet Union. A 1989 Organization for Economic Cooperation and Development (OECD) study (25) concluded that the only economic advantage of using plutonium as a reactor fuel was when already separated (with all costs paid) plutonium was mixed with enriched uranium as a mixed oxide (MOX) reactor fuel. The savings would be 30% for natural uranium at \$80/kg and negligible for natural uranium at \$50/kg. In fact, plutonium metal may be an economic liability. In a 1994 study it was estimated that converting excess weapons $^{239}$Pu metal into PuO$_2$ and blending it into MOX reactor fuel would increase the cost by approximately \$10,000/kg of Pu over the equivalent amount of low enriched uranium (26). Another 1994 study estimated that mixing plutonium into MOX fuel, calculated on the basis of chemical processing and fabrication of the MOX fuel in comparison with enriched UO$_2$ fuel, would save \$5000/kg of Pu (27). The uncertainty of this latter number is greater than $\pm$100%, however, because of the uncertainty in the input cost parameters. Whatever the cost, blending of excess weapons plutonium into reactor fuel continues to be advocated (28) because its intense radioactivity after irradiation into spent fuel effectively eliminates the risk of diversion or theft.

In 1992 Japan accepted 1.7 tons of Pu that had been reprocessed in France as the initial shipment of 30 tons of reprocessed Pu planned over a 10-year period. The cost of the contract, \$4 billion, for these 30 tons yields a reprocessing cost of \$133,000/kg.

## Plutonium Metal

**Preparation.**   Reductions of $PuF_4$ or $PuF_3$ on the gram or multigram scale have been carried out in centrifugal bombs and stationary bombs in order to produce coherent, pure Pu in high yield. The best reductant is calcium metal because the reduction evolves sufficient heat to melt the Pu as well as the slag. The resultant molten Pu coalesces. The direct oxide reduction (DOR) process, developed in the 1980s in the United States, reduces $PuO_2$ using calcium metal in a molten salt bath of $CaCl_2$ or $CaCl_2-CaF_2$ (29,30). Of the common refractory materials, MgO, CaO, $CaF_2$, and $Y_2O_3$ are satisfactory container materials for holding molten Pu. Tungsten and tantalum also can be used. Both, however, are measurably soluble in Pu. The solubility of many metals in molten Pu has been determined as a function of temperature (31). Melting and casting procedures for plutonium have been summarized (32–34).

**Physical Properties.**   The element exists in six allotropic modifications under ordinary pressure, and there is a high pressure allotrope (35–41) (Fig. 1). The temperatures and enthalpies of phase transitions are listed in Table 5. The high pressure $\zeta$-phase is formed reversibly at above 330°C and above 60 kPa

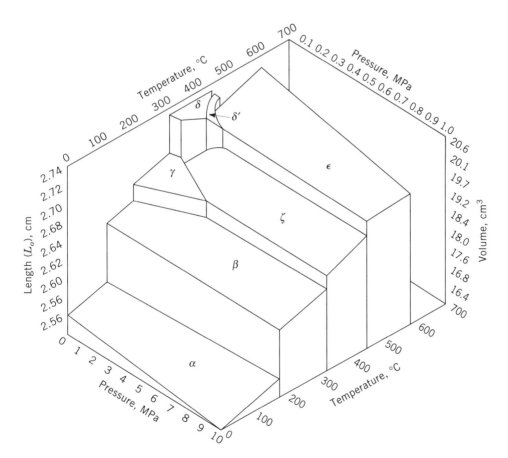

**Fig. 1.**   Pressure–volume–temperature phase diagram of plutonium metal (35). To convert MPa to atm, multiply by 9.9.

**Table 5. Thermodynamic Properties and Transition Temperatures for Plutonium Metal**[a]

| Property | Value |
|---|---|
| absolute entropy of $\alpha$-Pu at 298 K, J/(mol·K)[b] | 54.461 |
| heat capacity of $\alpha$-Pu at 298 K, J/(mol·K)[b] | 31.192 |
| heat capacity, $C_p$, J/(mol·K) | |
| $\quad\alpha$-Pu from 298–397.6 K | $18.126 + 0.0448\,T$ |
| $\quad\beta$-Pu from 397.6–487.9 K | $27.416 + 0.0131\,T$ |
| $\quad\gamma$-Pu from 487.9–593.1 K | $22.023 + 0.0299\,T$ |
| $\quad\delta$-Pu from 593.1–736 K | $28.478 + 0.0108\,T$ |
| $\quad\delta'$-Pu from 736–756 K | 35.56 |
| $\quad\epsilon$-Pu from 756–913 K | 33.72 |
| $\quad$Pu (l), >913 K | 42.248 |
| enthalpy of transition, J/mol[b] | |
| $\quad\alpha \rightarrow \beta$-Pu at 397.6 ± 1.0 K | $3703 \pm 4$ |
| $\quad\beta \rightarrow \gamma$-Pu at 487.9 ± 1.0 | $477 \pm 8$ |
| $\quad\gamma \rightarrow \delta$-Pu at 593.1 ± 1.0 K | $711 \pm 1\,3$ |
| $\quad\delta \rightarrow \delta'$-Pu at 730 ± 2 K | $84 \pm 42$ |
| $\quad\delta' \rightarrow \epsilon$-Pu at 752 ± 4 K | $1840 \pm 84$ |
| $\quad\epsilon$-Pu →liquid at 913 ± 2 K | $2825 \pm 105$ |
| enthalpy of vaporization, kJ/mol[b] | |
| $\quad$from 1200–1790 K | $345.2 \pm 4$ |
| $\quad$from 1724–2219 K[c] | $342 \pm 2$ |
| normal boiling point, K[d] | $3640 \pm 100$ |
| vapor pressure of liquid, $\log P$, in kPa[c,e] | $-(17{,}066 \pm 208)/T + 6.664 \pm 0.050$ |

[a]Ref. 41. Estimated values are in parentheses.
[b]To convert J to cal, divide by 4.184.
[c]Ref. 43.
[d]Value is extrapolated.
[e]To convert $\log P(\text{kPa})$ to $\log P(\text{atm})$, subtract 2.0056 from constant.

(8.7 psi). The crystal structures of all plutonium allotropes, with the exception of the $\zeta$-modification, are summarized in Table 6. The pivotal position of Pu in the actinide series, where itinerant (bonding, nonmagnetic) $5f$ electrons are present in metallic actinide systems through Pu, and localized (nonbonding, magnetic) $5f$ electrons are present in metallic systems containing actinides heavier than Pu, has been noted (42).

*Thermal Expansion.* Coefficients of linear thermal expansion and linear expansion during transformation are listed in Table 7. The expansion coefficient of $\alpha$-plutonium is exceptionally high for a metal, whereas those of $\delta$- and $\delta'$-plutonium are negative. The net linear increase in heating a polycrystalline rod of plutonium from room temperature to just below the melting point is 5.5%.

*Thermodynamic.* The thermodynamic properties of elemental plutonium have been reviewed (35,40,41,43–46). Thermodynamic properties of solid and liquid Pu, and of the transitions between the known phases, are given in Table 5. There are inconsistencies among some of the vapor pressure measurements of liquid Pu (40,41,43,44).

*Electrical and Magnetic.* The electrical resistivity of plutonium, as shown in Figure 2, is high in all modifications as a result of the band structure in metallic plutonium (47). There are two or more bands in Pu. The narrow $f$-band has a high

**Table 6. Phase Relationships and Crystallographic Properties of Plutonium Metal[a]**

| Allotrope | Stability range, °C | Crystal structure | Lattice parameters, pm | Reference temperature, °C | Atoms per cell, $Z$ | Density, calculated, g/cm³ | Average M–M distance, pm |
|---|---|---|---|---|---|---|---|
| $\alpha$ | <124.5 | primitive monoclinic | $a = 618.3 \pm 000.1$<br>$b = 482.2 \pm 000.1$<br>$c = 1096.3 \pm 000.1$<br>$\beta = 101.79° \pm 0.01°$ | 21 | 16 | 19.86 | 305 |
| $\beta$ | 124.5–214.8 | body-centered monoclinic | $a = 928.4 \pm 0.3$<br>$b = 1046.3 \pm 0.4$<br>$c = 785.9 \pm 0.3$<br>$\beta = 92.13° \pm 0.03°$ | 190 | 34 | 17.70 | 314 |
| $\gamma$ | 214.8–320 | face-centered orthorhombic | $a = 315.9 \pm 0.1$<br>$b = 576.8 \pm 0.1$<br>$c = 1016.2 \pm 0.2$ | 235 | 8 | 17.14 | 317 |
| $\delta$ | 320–462.9 | fcc | $a = 463.71 \pm 0.04$ | 320 | 2 | 15.92 | 328 |
| $\delta'$ | 462.9–482.6 | body-centered tetragonal | $a = 334 \pm 1$<br>$c = 444 \pm 4$ | 465 | 2 | 16.00 | 327 |
| $\epsilon$ | 482.6–640 | body-centered cubic | $a = 336.31 \pm 0.04$ | 490 | 2 | 16.51 | 318 |

[a]Refs. 30, 35, 36, and 38–40.

**Table 7. Mean Coefficients of Thermal and Transformation Expansion of a Polycrystalline Rod of Plutonium[a]**

| Phase | Temperature range, °C | Linear expansion coefficient $\times 10^6$/K | Transformation expansion, 100 $\Delta L/L$ |
|---|---|---|---|
| $\alpha$ | 21–104 | 54 | |
| $\beta$ | 93–190 | 42 | 2.9 |
| $\gamma$ | 210–310 | 34.6 ± 0.7 | 0.8 |
| $\delta$ | 320–440 | −8.6 ± 0.3 | |
| $\delta'$ | 452–480 | −65.6 ± 10.1 | −4.2 |
| $\epsilon$ | 490–550 | 36.5 ± 1.1 | −0.6 |
| liquid | 664–788 | 93 | |

[a]Refs. 35, 36, 39, and 40.

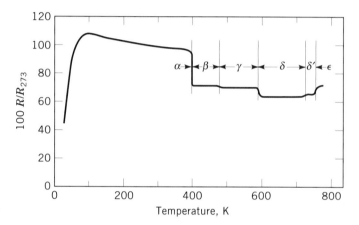

**Fig. 2.** Electrical resistivity, $R$, of plutonium metal, plotted as 100 $R/R_{273\,\text{K}}$ versus absolute temperature (47).

density of states; a broad band, probably a $d$-band, is the principal conductor. The resistivity of $\alpha$-Pu decreases uniformly from 146.45 at 273.15 K until the $\alpha \rightarrow \beta$ transition temperature is reached (48). At this point, the resistivity drops by about 20% and remains fairly constant, except for small changes associated with phase transitions. Minimum resistance is observed in the $\delta$-Pu region. The resistivity of highly oriented $\alpha$-plutonium, with current directions parallel and vertical to the 020 plane, has been measured and the resistivity curves show strong anisotropy (49).

Results from magnetic susceptibility studies have been reported (50–53). Measurements (50) obtained by the Gouy method are shown in Figure 3. These are lower than those of other investigators. However, the temperature dependences of the magnetic susceptibilities, $\chi$, for the various plutonium allotropes were similar. $\alpha$-Plutonium single crystals show a slight anisotropy of $\chi$ (54).

*Other Physical Properties.* The thermal conductivity of $^{242}$Pu is 0.084 and 0.155 W/(m·K) (0.020 and 0.037 cal/(s·cm·°C)) for the $\alpha$- and $\beta$-phase, respectively (55). Thermoelectric power for each allotrope may be found in References 30, 35, 36, 38, and 39. The surface tension of liquid plutonium is 437–475 mN/m(=dyn/cm) (56). The viscosity of molten plutonium at 645–950°C has

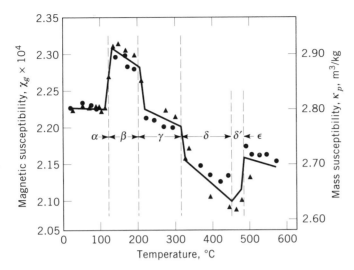

**Fig. 3.**  Magnetic susceptibility of plutonium metal as a function of temperature where ▲ represent initially high density plutonium and ●, cast plutonium (50).

been determined by the oscillating cup method (57). The temperature dependence of the viscosity, $\eta$, is given by the following equation, where $T = K \times 10^{-3}$. At the melting point, $\eta = 6.0$ mPa·s(=cP).

$$\eta(T) = 0.672/T + 0.037 \qquad (5)$$

*Radioactive Self-Heating.*  Because of $\alpha$-emission radioactivity, all plutonium isotopes generate heat. The self-heating of $^{239}$Pu metal is $(1.923 \pm 0.019) \times 10^{-3}$ W/g $(1.824 \pm 0.18$ Btu/(s·g)) (58). Large $^{239}$Pu samples have a temperature higher than that of the human body; a 30–60°C rise above ambient temperature is typical for a 100-g piece of $^{239}$Pu metal. The self-heating of $^{238}$Pu metal is 0.456 W/g (432 Btu/(s·g)) (59) for the specific power. Self-heating is sufficient to heat a multigram $^{238}$Pu metal sample to incandescence.

**Storage and Handling.**  Plutonium can be stored safely in dry air. Because of self-heating, storage accompanied by heat removal is advisable. The metal can be machined in moisture-free air containing at least 70 vol % Ar or He. Casting and foundry operations that require melting of the metal must be carried out in vacuum or inert atmospheres and special containers.

**Chemical Properties of Plutonium Metal.**  Plutonium, the fifth member of the actinide series, is a reactive metal. It can be made by the usual methods for strongly reducing metals such as electrolysis in fused salts or reduction of its halides or oxide using a stronger reducing metal such as Li, Ca, or Ba. Plutonium forms compounds with all the nonmetallic elements except the rare gases. The halogens and halogen acids form Pu halides, other chalcogens form chalcogenides, and CO forms a carbide. Nitrides (qv) are formed with $NH_3$ and $N_2$ and hydrides (qv) with $H_2$.

The metal dissolves readily in concentrated HCl, $H_3PO_4$, HI, or $HClO_4$. Nitric acid (qv) forms a protective oxide skin on the metal and can be removed by ca 0.05 $M$ HF. Dissolution of Pu metal in $HNO_3$–HF mixtures is common practice in scrap-recovery plants. The metal does not dissolve readily in $H_2SO_4$ because passivation of the metal surface occurs. The reaction of water and Pu metal is slow compared to that in HCl, HI, or $HClO_4$.

The corrosion behavior of plutonium metal has been summarized (60,61). $\alpha$-Plutonium oxidizes very slowly in dry air, typically $<10^{-2}$ mm/yr. The rate is accelerated by water vapor. Thus, a bright metal surface tarnishes rapidly in normal environments and a powdery surface soon forms. Eventually green $PuO_2$ [12059-95-9] covers the surface. Plutonium is similar to uranium with respect to corrosion characteristics. The stabilization of $\delta$-Pu confers substantial corrosion resistance to Pu in the same way that stabilization of $\gamma$-U yields a more corrosion-resistant metal. The reaction of Pu metal with liquid water produces both oxides and oxide-hydrides (62). The reaction with water vapor above 100°C also produces oxides and hydride (63).

## Properties of Atomic Plutonium and Plutonium Ions

**Thermodynamic.** Thermodynamic properties of Pu metal, gaseous species, and the aqueous ions at 298 K are given in Table 8. Thermodynamic properties of elemental Pu (44), of alloys (68), and of the gaseous ions $Pu^+$, $PuO^+$, $PuO_2^+$, and $PuO_2^-$ (67) have been reviewed, as have those of aqueous ions (64), oxides (69), halides (70), hydrides (71), and most other compounds (65).

**Spectroscopic.** The electronic configuration of Pu vapor is [Xe] $4f^{14}5d^{10}$ $5f^66s^26p^66d^07s^2$. The earliest studies of copper- and silver-spark spectra of neutral plutonium, $Pu^0$, have been reviewed (30). The singly charged ion $Pu^+$ [14700-

**Table 8. Thermochemical Properties of Plutonium Species at 298 K**[a]

| Species[b] | $\Delta H_f$, kJ/mol[c] | $\Delta G_f$, kJ/mol[c] | $S$, J/(mol·K)[c] | $C_p$, J/(mol·K)[c] |
|---|---|---|---|---|
| Pu (c) | 0 | 0 | 54.46 | 31.19 |
| Pu (g) | 342 ± 2 | 305 | 177.1 | 20.85 |
| $Pu^+$(g) | 223.7 ± 1 | | 182.822 | 20.880 |
| $PuO^+$(g) | 120 ± 13 | | 246.788 | 35.702 |
| $PuO_2^+$(g) | 119 ± 13 | | 272.929 | 52.609 |
| $PuO_2^-$(g) | | | 268.075 | 55.496 |
| $Pu^{3+}$(aq) | −592 ± 2 | −579 ± 3 | −184 ± 8 | |
| $Pu^{4+}$(aq) | −536 ± 3 | −482 ± 3 | −389 ± 20 | |
| $PuO_2^+$(aq) | −915 ± 6 | − 850± 8 | −21 ± 8 | |
| $PuO_2^{2+}$(aq) | −822 ± 7 | −756 ± 7 | −92 ± 8 | |
| $PuO_{1.5}$ (hex) | (−828) | (−790) | 81.5 ± 0.3 | 58.5 ± 0.25 |
| $PuO_{1.515}$ (bcc) | (−836 ± 11) | (−795) | 75.7 | |
| $PuO_2$ (c) | −1056.2 ± 0.7 | −998.0 ± 0.7 | 66.13 ± 0.26 | 66.25 ± 0.26 |
| $PuF_3$ (c) | −1585.7 ± 4.2 | −1516.4 | 126.11 ± 0.36 | 92.64 ± 0.28 |
| $PuF_4$ (c) | (−1846 ± 21) | (−1753 ± 21) | 147.25 ± 0.37 | 116.19 ± 0.29 |
| $PuF_6$ (c) | −1862 ± 29 | −1728 | 222 ± 21 | 167 |
| $PuF_6$ (g) | −1813 ± 20 | −1724 | 369.1 | 129.4 |
| $PuCl_3$ (c) | −959.8 ± 1.7 | 892 | (164) | 103 |

[a]Refs. 64–67. Estimated values are in parentheses.
[b](c) = crystalline; (hex) = hexagonal crystalline; (bcc) = body-centered cubic crystalline; (g) = gaseous; and (aq) = aqueous.
[c]To convert J to cal, divide by 4.184.

74-4] and the doubly charged ion $Pu^{2+}$ [17440-99-2] are produced in emission spectroscopy. A summary of energy levels, electronic configurations, and ionization potentials is found in Table 9. Summaries and comprehensive listings and assignments of spectra have been published (72,73); a comprehensive and interpretive review has also been published (74). Isotope shifts of the 238, 239, and 240 isotopes have been obtained in the spark spectrum, re-examined with hollow cathode spectral data, and supplemented by $^{241}$Pu data (75).

Hyperfine structure measurements of 75 strong lines at 4020–6200 mm were made by means of interferometer spectrograms. Doublet structures occurred in 30 lines with separations of 0.034–0.180 cm$^{-1}$ (76). Nuclear spins for $^{239}$Pu and for $^{241}$Pu have been established to be $I = 1/2$ and $I = 5/2$, respectively. The spins were based on paramagnetic resonance measurements on a RbPuO$_2$(NO$_3$)$_3$ crystal (77).

*Absorption Spectra of Aqueous Ions.* The absorption spectra of Pu(III) [22541-70-4], Pu(IV) [22541-44-2], Pu(V) [22541-69-1], and Pu(VI) [22541-41-9] in mineral acids, ie, HClO$_4$ and HNO$_3$, have been measured (78–81). The Pu(VII) [39611-88-61] spectrum, which can be measured only in strong alkali hydroxide solution, also has been reported (82). As for rare-earth ion spectra, the spectra of plutonium ions exhibit sharp lines, but have larger extinction coefficients than those of most lanthanide ions (see LANTHANIDES). The visible spectra in dilute acid solution are shown in Figure 4 and the spectrum of Pu(VII) in base is shown in Figure 5. The spectra of ions of plutonium have been interpreted in relation to all of the ions of the 5$f$ elements (83).

**Aqueous Solution Chemistry.** The aqueous solution chemistry of plutonium is complex. Plutonium can exist in acidic aqueous solutions in oxidation states III, IV, V, and VI. Additionally plutonium can be oxidized to Pu(VII) in alkaline solutions (85). Furthermore, because the formal reduction potentials between the successive oxidative states III–VI are ca 1 volt, these four states can coexist in the same solution. The least stable state is the +5, which exists only in a limited weakly acidic pH range, having greater stability at high dilution. In basic solution, however, the potentials are shifted significantly and Pu(V) is

**Table 9. Electronic Configurations and Energetics of Free Plutonium Atom and Ions**[a]

| Pu$^0$ | | Pu$^{+b}$ | |
|---|---|---|---|
| State | Energy, 10$^3$ cm$^{-1}$ | State | Energy, 10$^3$ cm$^{-1}$ |
| $5f^67s^2$ | 0 | $5f^67s^1$ | 0 |
| $5f^56d^17s^2$ | 6.314 | $5f^57s^2$ | 8.199 |
| $5f^66d^17s^1$ | 13.528 | $5f^56d^17s^1$ | 8.710 |
| $5f^56d^27s^1$ | 14.912 | $5f^66d^1$ | |
| $5f^67s^17p^1$ | 15449 | $5f^56d^2$ | |
| $5f^56d^17s^17p^1$ | 20.828 | $5f^67p^1$ | 22.039 |
| $5f^57s^17p^2$ | (35 ± 5) | $5f^57s^17p^1$ | 30.956 |
| $5f^46d^37s^1$ | (42.5 ± 1) | $5f^55d^17p^1$ | 33.793 |
| $5f^46d^27s^17p^1$ | (51 ± 1) | $5f^46d^27s^1$ | 37.641 |
| ionization potential, eV[c] | 6.06 ± 0.2 | | 11.7 ± 0.2 |

[a]Refs. 65 and 74.
[b]State for Pu$^{2+}$ = $5f^6$; for Pu$^{3+}$ = $5f^5$.
[c]For the equation Pu$^x$ → Pu$^{x+1}$ + $e^-$.

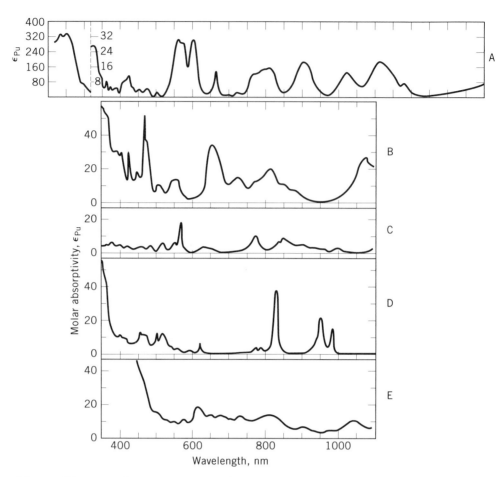

**Fig. 4.** Electronic absorption spectra of A, plutonium(III); B, Pu(IV); C, Pu(V); D, Pu(VI); and E, colloidal Pu(IV) in dilute acid at 25°C (78,79).

rather stable (86). In aqueous media such as seawater, at trace Pu concentration, Pu(V) is the dominant species (87,88). Figure **6a** shows the predominant Pu species present in aqueous solution as a function of the reduction potential vs the standard hydrogen electrode ($Eh$) and pH. Figure **6b** shows the species present in aqueous solution in the presence of carbonate.

Plutonium(III) in aqueous solution, $Pu^{3+}(aq)$, is pale blue. Aqueous plutonium(IV) is tan or brown; the nitrate complex is green. Pu(V) is pale red-violet or pink in aqueous solution and is believed to be the ion $PuO_2^+$. Pu(VI) is tan or orange in acid solution, and exists as the ion $PuO_2^{2+}$. In neutral or basic solution Pu(VI) is yellow; cationic and anionic hydrolysis complexes form. Pu(VII) has been described as blue-black. Its structure is unknown but may be the same as the six-coordinate $NpO_4(OH)_2^{3-}$ (91). Aqueous solutions of each oxidation state can be prepared by chemical oxidants or reductants (30) but the best methods appear to be electrochemical (92,93).

The chemistry of plutonium ions in solution has been thoroughly studied and reviewed (30,94–97). Thermodynamic properties of aqueous ions of Pu

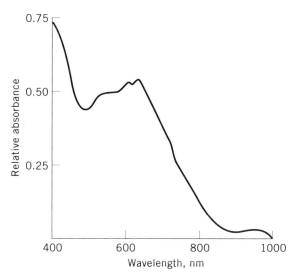

**Fig. 5.** Electronic absorption spectrum of plutonium(VII) in 1 $M$ KOH (aq) at 25°C (84).

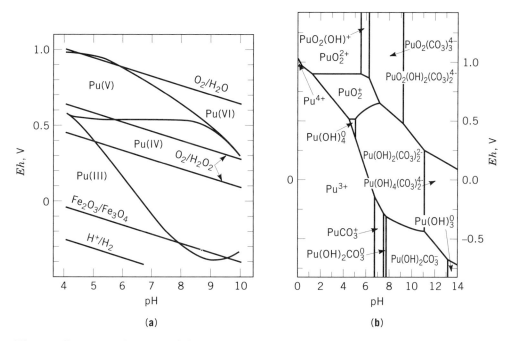

**Fig. 6.** Speciation diagram of plutonium as function of $Eh$ and pH in aqueous solution at 25°C: (**a**) carbonate-free (89); (**b**) 0.004 $M$ total carbonate (90). $Eh$ refers to the standard reduction potential in volts.

are given in Table 8 and in the literature (64–66). The formal reduction potentials in aqueous solutions of 1 $M$ $HClO_4$ or KOH at 25°C may be summarized as follows (66,86,98–100):

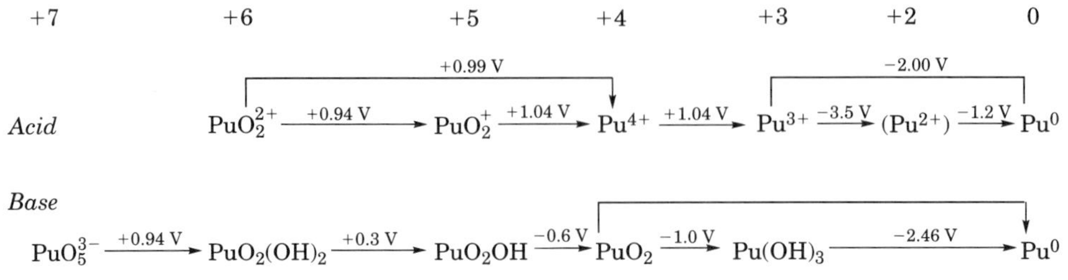

+7                    +6                    +5          +4          +3          +2          0

$$Acid \quad PuO_2^{2+} \xrightarrow{+0.94\,V} PuO_2^{+} \xrightarrow{+1.04\,V} Pu^{4+} \xrightarrow{+1.04\,V} Pu^{3+} \xrightarrow{-3.5\,V} (Pu^{2+}) \xrightarrow{-1.2\,V} Pu^0$$

with overhead arrows $+0.99\,V$ (from $PuO_2^{2+}$ to $Pu^{4+}$) and $-2.00\,V$ (from $Pu^{3+}$ to $Pu^0$).

$$Base \quad PuO_5^{3-} \xrightarrow{+0.94\,V} PuO_2(OH)_2 \xrightarrow{+0.3\,V} PuO_2OH \xrightarrow{-0.6\,V} PuO_2 \xrightarrow{-1.0\,V} Pu(OH)_3 \xrightarrow{-2.46\,V} Pu^0$$

with overhead arrow from $PuO_2$ to $Pu^0$.

The optical absorption spectra of Pu ions in aqueous solution show sharp bands in the wavelength region 400–1100 nm (Fig. 4). The maxima of some of these bands can be used to determine the concentration of Pu ions in each oxidation state (III–VI), thus quantitative determinations of oxidation–reduction equilibria and kinetics are possible. A comprehensive summary of kinetic data of oxidation–reduction reactions is available (101) as are the reduction kinetics of $Pu^{7+}$(aq) (84).

Radiolysis creates oxidants such as the $OH^{\bullet}$ radical and reductants such as $e^-$(aq), which oxidize or reduce plutonium ions. Self-radiolysis of Pu solutions has in general been found to decrease the plutonium oxidation state (78,97), but oxygen-free 0.02 $M$ $^{242}$Pu in brine was found to solution-oxidize slowly (102). Subsequent studies have confirmed that in brine Pu in lower oxidation states gradually oxidizes to Pu(VI) (103) and that Pu(VI) remains as Pu(VI) (104).

*Coordination Complexes.* The ability of the various oxidation states of Pu to form complex ions with simple hard ligands, such as oxygen, is, in order of decreasing stability, $Pu^{4+} > PuO_2^{2+} > Pu^{3+} > PuO_2^{+}$. Thus, Pu(III) forms relatively weak complexes with fluoride, chloride, nitrate, and sulfate (105), and stronger complexes with oxygen ligands (Lewis-base donors) such as carbonate, oxalate, and polycarboxylates, eg, citrate, and ethylenediaminetetraacetic acid (106). The complexation behavior of Pu(III) is quite similar to that of the light lanthanide(III) ions, particularly to Nd(III) (see COORDINATION COMPOUNDS).

Pu(IV) forms polyatomic complexes with inorganic and organic ligands. As the number of anionic ligands increases, cationic, neutral, and anionic complexes form and the sequential stability constants, $K_n$, typically decrease. For the following reaction, where M is $Pu^{x+}$ and L is a ligand,

$$q\,M + n\,L = M_qL_n$$

the overall stability constant is

$$K_1K_2\cdots K_n = \beta_{nq} = [M_qL_n]/[M]^q[L]^n$$

For example, in HCl (aq), the following chlorocomplex ions are known and stability complex constants have been determined (105).

$$Pu^{4+} + Cl^- \longrightarrow PuCl^{3+} \qquad \log \beta_{11} = -0.1$$

$$Pu^{4+} + 2\ Cl^- \longrightarrow PuCl^{2+} \qquad \log \beta_{12} = -0.3$$

$$Pu^{4+} + 3\ Cl^- \longrightarrow PuCl_3^+ \qquad \log \beta_{13} = +0.3$$

Evidence for the anionic complex $PuCl_6^{2-}$ is the precipitation of complex halides such as $Cs_2PuCl_6$ from concentrated HCl (aq). The ability of Pu(IV) to form stable nitrate complexes provides the basis for the Purex and ion-exchange (qv) process used in the chemical processing of Pu (107). Pu(VI) is similar to Pu(IV) in its ability to form complex ions. Detailed reviews of complex ion formation by aqueous plutonium are available (23,94,105).

*Hydrolysis.* Complexes formed by Pu ions with $OH^-$ represent hydrolysis reactions. There is extensive interaction between $Pu^{x+}$ and water. Pu(III) hydrolyzes at ca pH 7 (105); the first hydrolysis equilibrium is as follows:

$$Pu^{3+} + H_2O \rightleftharpoons Pu(OH)^{2+} + H^+ \qquad \log K = \log \beta_1^* \cong -7$$

The solubility product of $Pu(OH)_3$, $K_{sp} = [Pu^{3+}][OH^-]^3$, has been estimated as $2 \times 10^{-29}$ (108). The hydrolysis and complexation behavior of Pu(III) is similar to that of the rare-earth elements in the +3 oxidation state. This similarity is the result of the similarity in electronic structure, ionic radii, and electrical charge.

Hydrolysis is very extensive in Pu(IV) solutions, less so in Pu(III) and Pu(VI), and least in Pu(V). The chemical properties of Pu(IV) are somewhat similar to those of Ce(IV) and U(IV) (see CERIUM AND CERIUM COMPOUNDS). The hydrolysis thermodynamics of Pu(IV) have been assessed in perchlorate solutions (105). The first hydrolysis equilibrium is

$$Pu^{4+} + H_2O \rightleftharpoons Pu(OH)^{3+} + H^+ \qquad \log K = \log \beta_1^* \cong -1.55$$

in sodium perchlorate solution of unit ionic strength at 25°C. Thus, at an acidity of 1 $M$, 23.6% of the Pu is hydrolyzed. Complete hydrolysis of Pu(IV) results in the precipitation of green $Pu(OH)_4$, which has a calculated solubility product of $5 \times 10^{-64}$ (65). The structure and water content of this $PuO_2 \cdot nH_2O$ are not known. Pu(IV) is unstable with respect to disproportionation:

$$2\ Pu^{4+} + 2\ H_2O \longrightarrow Pu^{3+} + PuO_2^+ + 4\ H^+$$

$$Pu^{4+} + PuO_2^+ \longrightarrow Pu^{3+} + PuO_2^{2+}$$

The kinetics and equilibria involved in the disproportionation have been reviewed (30,101). An interesting but troublesome characteristic of Pu(IV) is the irreversible formation of Pu(IV) polymers. The bright green polymer (109) forms rapidly at low acidities or can form when a Pu(IV) solution is diluted with water. Depolymerization requires strong acid, high temperature, complexants, and/or oxidants (110). The polymer is of high molecular weight and can be identified from absorption spectra. Its role in the sol–gel process for preparation of $PuO_2$ ceramics (qv) has been described (109) (see SOL–GEL TECHNOLOGY).

Hydrolysis constants for Pu(IV) have been determined in aqueous solutions at many pH values and ionic strengths. In 1 $M$ $NaClO_4$ solution the first three Pu(IV) − $OH^-$ complexes and overall stability constants are (105) as follows:

$$Pu^{4+} + OH^- \longrightarrow [PuOH]^{3+} \qquad \log \beta_{11} = 12.45 \pm 0.3$$

$$Pu^{4+} + 2\ OH^- \longrightarrow [Pu(OH)_2]^{2+} \qquad \log \beta_{12} = 24.7$$

$$Pu^{4+} + 3\ OH^- \longrightarrow [Pu(OH)_3]^+ \qquad \log \beta_{13} = 35.6$$

It is also common to report hydrolysis constants as, eg, $\beta_{11}^*$, for the reaction $Pu^{4+} + H_2O \rightleftarrows PuOH^{3+} + H^+$, where $\beta_{11} K_w = \beta_{11}^*$, and $K_w = [H^+][OH^-] = 1 \times 10^{-14}$ at 25°C.

$$Pu^{4+} + H_2O \longrightarrow [PuOH]^{3+} + H^+ \qquad \log \beta_{11}^* = -1.55 \pm 0.3$$

$$Pu^{4+} + 2\ H_2O \longrightarrow [Pu(OH)_2]^{2+} + 2\ H^+ \qquad \log \beta_{12}^* = -3.3$$

$$Pu^{4+} + 3\ H_2O \longrightarrow [Pu(OH)_3]^+ + 3\ H^+ \qquad \log \beta_{13}^* = -6.4$$

Pu(V) begins to hydrolyze at ca pH 9. Pu(V) usually is unstable and disproportionates to Pu(VI) and Pu(IV) or Pu(III). The hydrolysis of Pu(VI) is slow and its stability is not well resolved, but some hydrolysis constants have been determined (105). In 0.1 $M$ $NaClO_4$ solution the first two Pu(VI)−$OH^-$ complexes and overall stability constants have been determined spectroscopically at 22°C in 0.1 $M$ $NaClO_4$ (aq) (111):

$$PuO_2^{2+} + OH^- \longrightarrow [PuO_2OH]^+ \qquad \log \beta_{11} = 8.10 \pm 0.15$$

$$PuO_2^{2+} + 2\ OH^- \longrightarrow PuO_2(OH)_2 \qquad \log \beta_{12} = 14.25 \pm 0.18$$

and the $\beta^*$ values are

$$PuO_2^{2+} + H_2O \longrightarrow [PuO_2OH]^+ + H^+ \qquad \log \beta_{11}^* = -5.9 \pm 0.15$$

$$PuO_2^{2+} + 2\ H_2O \longrightarrow PuO_2(OH)_2 + 2\ H^+ \qquad \log \beta_{12}^* = -13.75 \pm 0.18$$

The enthalpies of these hydrolysis reactions have also been determined (112). Polynuclear complexes such as $[(PuO_2)_2(OH_2]^{2+}$, $[(PuO_2)_3(OH)_5]^+$, and $[(PuO_2)_4(OH)_7]^+$ have been inferred from potentiometric titrations (105).

*Other Coordination Complexes.* Because carbonate and bicarbonate are commonly found under environmental conditions in water, and because carbonate complexes Pu readily in most oxidation states, Pu carbonato complexes have been studied extensively. The reduction potentials vs the standard hydrogen electrode of Pu(VI)/(V) shifts from 0.916 to 0.33 V and the Pu(IV)/(III) potential shifts from 1.48 to −0.50 V in 1 $M$ carbonate. These shifts indicate strong carbonate complexation. Electrochemistry, reaction kinetics, and spectroscopy of plutonium carbonates in solution have been reviewed (113). The solubility of Pu(IV) in aqueous carbonate solutions has been measured, and the stability constants of hydroxycarbonato complexes have been calculated (Fig. 6**b**) (90).

Plutonium(III), (IV), and (VI) complex stability constants have been determined for some oxygen-donor (carboxylate) (114) and a few nitrogen-donor (115,116) ligands. Complexes of plutonium with natural complexants such as humic acids have also been studied extensively (89).

## Analytical Chemistry

The analytical chemistry of plutonium has been reviewed (23,117). It is possible to analyze a plutonium-containing sample gravimetrically by precipitating it as oxalate, calcining to $PuO_2$, and weighing. The sample must contain no other metal (rare earths, barium, or other transuranics) that precipitates as an oxalate. Potentiometric (oxidation–reduction) titration of plutonium in solution is usually faster, more versatile, and potentially as accurate (as good as ±0.06%). Plutonium is reduced to $Pu^{3+}$ using $Ti^{3+}$ and then titrated with $Ce^{4+}$. For higher precision, coulometric techniques have been developed.

Only slightly less accurate (±0.3–0.5%) and more versatile in scale are other titration techniques. Plutonium may be oxidized in aqueous solution to $PuO_2^{2+}$ using AgO, and then reduced to $Pu^{4+}$ by a known excess of $Fe^{2+}$, which is back-titrated with $Ce^{4+}$. $Pu^{4+}$ may be titrated complexometrically with EDTA and a colorimetric indicator such as Arsenazo(I), even in the presence of a large excess of $UO_2^{2+}$. Solution spectrophotometry (Figs. 4 and 5) can be utilized if the plutonium oxidation state is known or controlled. The spectrophotometric method is very sensitive if a colored complex such as Arsenazo(III) is used. Analytically useful absorption maxima and molar absorption coefficients ($\epsilon$) are given in Table 10. Laser photoacoustic spectroscopy has been developed for both elemental analysis and speciation (oxidation state) at concentrations of $10^{-4}–10^{-5}$ $M$ (118). Chemical extraction can also be used to enhance this technique.

X-ray fluorescence, mass spectroscopy, emission spectrography, and ion-conductive plasma–atomic emission spectroscopy (icp–aes) are used in specialized laboratories equipped for handling radioisotopes with these instruments.

A large number of radiometric techniques have been developed for Pu analysis on tracer, biochemical, and environmental samples (119,120). In general the $\alpha$-particles of most Pu isotopes are detected by gas-proportional, surface-barrier, or scintillation detectors. When the level of $^{239}$Pu is lower than $10^{-12}$ g/g sample, radiometric techniques must be enhanced by preliminary extraction of the Pu to concentrate the Pu and separate it from other radioisotopes (121,122). Alternatively, fission–fragment track detection can detect $^{239}$Pu at a level of $10^{-13}$ g/g sample or better (123). Chemical concentration of Pu from urine, neutron irradiation in a research reactor, followed by fission track detection, can achieve a sensitivity for Pu of better than 1 mBq/L ($4 \times 10^{-19}$ g/g sample) (124).

**Separation Chemistry and Extractive Metallurgy.** Irradiated uranium fuel elements from a reactor are cooled for a period of at least 150 days, depending on the irradiation time, to allow $^{239}$Np to decay to $^{239}$Pu and to allow most of the short-lived fission products to decay to a level allowing safer handling. The technology, eg, transport, storage or cooling, dejacketing or decladding, and

**Table 10. Spectral Absorption Data for Aqueous Plutonium Ions at 25°C[a]**

| Plutonium oxidation state | Wavelength maximum, nm | Molar absorptivities, L/(mol·cm) | | | | |
|---|---|---|---|---|---|---|
| | | $\epsilon_{Pu(III)}$ | $\epsilon_{Pu(IV)}$ | $\epsilon_{Pu(V)}$ | $\epsilon_{Pu(VI)}$ | $\epsilon_{Pu(VII)}$ |
| Pu(III)[b] | 560 | 36.1 | 11.64 | 3.62 | 2.50 | |
| | 600 | 35.3 | 0.91 | 0.50 | 1.35 | |
| | 603 | 35.4 | 0.96 | 0.60 | 1.20 | |
| | 665 | 14.65 | 30.9 | 0.43 | 0.55 | |
| | 900 | 19.30 | 4.00 | 5.16 | 0.52 | |
| Pu(IV)[c] | 470[d] | 3.46 | 49.6 | 1.82 | 11.25 | |
| | 655 | 3.10 | 34.4 | 1.15 | 0.90 | |
| | 700 | 0.75 | 10.88 | 0.44 | 0.25 | |
| | 730 | 1.35 | 14.60 | 1.03 | 0.50 | |
| | 815 | 14.63 | 19.61 | 1.55 | 2.30 | |
| Pu(V)[c] | 569 | 34.3 | 5.60 | 17.10 | 1.75 | |
| | 775 | 12.40 | 11.90 | 9.87 | 2.90 | |
| Pu(VI)[b] | 833 | 5.25 | 15.5 | 4.00 | 550. | |
| | 953 | 1.20 | 0.40 | 1.76 | 19.10 | |
| | 983 | 3.15 | 1.76 | 1.18 | 8.90 | |
| Pu(VII)[e] | 635 | | | | | 530 |

[a]Refs. 78 and 79.
[b]In 0.1 $M$ HClO$_4$ (aq).
[c]In 0.5 $M$ HCl (aq).
[d]Very sharp band.
[e]In KOH (aq).

dissolving, that is involved in the handling of these materials has been discussed (125–127) (see METALLURGY, EXTRACTIVE; NUCLEAR REACTORS).

*Separation of Plutonium.*  The principal problem in the purification of metallic plutonium is the separation of a small amount of plutonium (ca 200–900 ppm) from large amounts of uranium, which contain intensely radioactive fission products. The plutonium yield or recovery must be high and the plutonium relatively pure with respect to fission products and light elements, such as lithium, beryllium, or boron. The purity required depends on the intended use for the plutonium. The high yield requirement is imposed by the price or value of the metal and by industrial health considerations, which require extremely low effluent concentrations.

The first successful production method for the separation of Pu from U and its fission products was the bismuth phosphate process, based on the carrying of Pu by a precipitate of BiPO$_4$ (126). That process has been superseded by liquid–liquid extraction (qv) and ion exchange (qv). In the liquid–liquid extraction process, an aqueous solution, usually HNO$_3$ containing U, Pu, and fission products, is extracted using an organic solution containing an organic complexing agent. The complexing agent may be diluted in a relatively inert solvent or comprise the pure solvent. The aqueous and organic phases are essentially immiscible. Conditions are adjusted for extraction of the Pu into the organic phase and later for stripping into the aqueous phase. Because liquid–liquid extraction can be adapted readily to continuous countercurrent operation, very high separation or purification factors can be achieved using simple equipment.

Many organic reagents have been used successfully in Pu separation processes. The reagents include tri-*n*-butyl phosphate (TBP); methyl isobutyl ketone; thenoyl trifluoroacetone (TTA); ethers, eg, diethyl ether, di-*n*-butyl ether, tetraethylene glycol dibutyl ether; trilaurylamine (TLA); trioctylamine (TOA); di-*n*-butyl phosphate (DBP); hexyl-di(2-ethylhexyl) phosphate (HDEHP); and many others. Of these, TBP is by far the most widely used (30,95).

The Purex process, ie, plutonium uranium reduction extraction, employs an organic phase consisting of 30 wt % TBP dissolved in a kerosene-type diluent. Purification and separation of U and Pu is achieved because of the extractability of $UO_2^{2+}$ and Pu(IV) nitrates by TBP and the relative inextractability of Pu(III) and most fission product nitrates. Plutonium nitrate and $UO_2(NO_3)_2$ are extracted into the organic phase by the formation of compounds, eg, $Pu(NO_3)_4 \cdot 2TBP$. The plutonium is reduced to Pu(III) by treatment with ferrous sulfamate, hydrazine, or hydroxylamine and is transferred to the aqueous phase; U remains in the organic phase. Further purification is achieved by oxidation of Pu(III) to Pu(IV) and re-extraction with TBP. The plutonium is transferred to an aqueous product. Plutonium recovery from the Purex process is ca 99.9 wt % (128). Decontamination factors are $10^6 - 10^8$ (97,126,129). A flow sheet of the Purex process is shown in Figure 7.

Historically, the Redox process was used to achieve the same purification as in the Purex process (97,129). The reagents were hexone (methyl isobutyl ketone) as the solvent, dichromate as an oxidant, and $Al(NO_3)_3$ as the salting agent. The chief disadvantages of hexone are its flammability and its solubility in water.

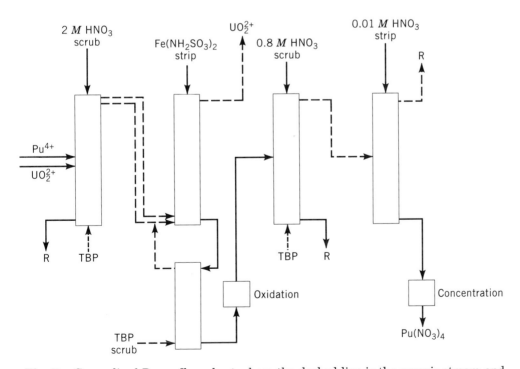

**Fig. 7.** Generalized Purex flow sheet where the dashed line is the organic stream and R = residue.

However, because $Al(NO_3)_3$ collects in the highly radioactive waste, thereby impeding the latter's further processing, the Redox process was abandoned in favor of the Purex process.

The plutonium extracted by the Purex process usually has been in the form of a concentrated nitrate solution or syrup, which must be converted to anhydrous $PuF_3$ [13842-83-6] or $PuF_4$, which are charge materials for metal production. The nitrate solution is sufficiently pure for the processing to be conducted in gloveboxes without $\beta$- or $\gamma$-shielding (130). The Pu is first precipitated as plutonium(IV) peroxide [12412-68-9], plutonium(III) oxalate [56609-10-0], plutonium(IV) oxalate [13278-81-4], or plutonium(III) fluoride. These precipitates are converted to anhydrous $PuF_3$ or $PuF_4$. The precipitation process used depends on numerous factors, eg, derived purity of product, safety considerations, ease of recovering wastes, and required process equipment. The peroxide precipitation yields the purest product and generally is the preferred route (131). The peroxide precipitate is converted to $PuF_4$ by $HF-O_2$ gas or to $PuF_3$ by $HF-H_2$ gas (31,132).

*Preparation of Plutonium Metal from Fluorides.* Plutonium fluoride, $PuF_3$ or $PuF_4$, is reduced to the metal with calcium (31). Although the reactions of Ca with both fluorides are exothermic, iodine is added to provide additional heat. The thermodynamics of the process have been described (133). The purity of production-grade Pu metal by this method is ca 99.87 wt % (134). Metal of greater than 99.99 wt % purity can be produced by electrorefining, which is applicable for Pu alloys as well as to purify Pu metal. The electrorefining has been conducted at 740°C in a NaCl–KCl electrolyte containing $PuCl_3$ [13569-62-5], $PuF_3$, or $PuF_4$. Processing was done routinely on a 4-kg Pu batch basis (135).

*Pyrometallurgical Processing.* High temperature nonaqueous chemical separations, typically by reduction of actinides to metals with more reactive metallic reducing agents or by means of molten salt electrochemistry, have been developed to separate actinide elements from fission products. These pyrometallurgical processes refine plutonium from other spent fuel wastes. The plutonium is itself contaminated with higher actinides and some fission products so that it is unsuitable for weapons and sufficiently $\gamma$-radioactive to be diversion resistant. The recovery and purification of plutonium from fast-breeder reactor fuels can be achieved most economically by pyrochemical processes (127,136). Pyrochemical separations are typically followed by direct oxide reduction (DOR) (29,30). The DOR technique is also suitable for conversion of scrap and waste materials to oxide and then to metal. In the United States, Pu metal has been produced from $PuO_2$ since the 1970s.

A more recently developed pyrometallurgical process is that of the proposed integral fast reactor, which would use metallic fuel (U–Pu–Zr alloy) and a molten salt electrorefiner as follows:

metal fuel cathode | LiCl–KCl–CdCl$_2$ eutectic electrolyte | cadmium anode

In this process, uranium metal is electrodeposited at the cathode, while plutonium and other transuranium elements remain in the molten salt as trichlorides. Plutonium is reduced in a second step at a metallic cathode to produce Cd–Pu intermetallics. The refined plutonium and uranium metals can then be refabricated into metallic fuel (137).

*Nuclear Waste Reprocessing.*   Liquid waste remaining from processing of spent reactor fuel for military plutonium production is typically acidic and contains substantial transuranic residues. The cleanup of such waste in 1996 is a higher priority than military plutonium processing. Cleanup requires removal of long-lived actinides from nitric or hydrochloric acid solutions. The transuranium extraction (Truex) process has been developed for actinide extraction and recovery (139). It modifies the Purex process using octyl(phenyl)-$N,N$-diisobutylcarbamoylmethylphosphine oxide ($O\Phi D(iB)CMPO$ or CMPO) to extract trivalent actinides, especially $Am^{3+}$, as well as tetravalent actinides $Np^{4+}$ and $Pu^{4+}$. Truex reduces the transuranic concentration in acidic Purex wastes to <3700 Bq/L ($\leq$100 nCi/L), decreasing the volume of transuranic (TRU) waste that requires burial to 1% or less. This process was at the pilot-plant stage at Hanford in 1992 (140) and is expected to reach production scale by 1997 (141). The Truex separation scheme is shown in Figure 8.

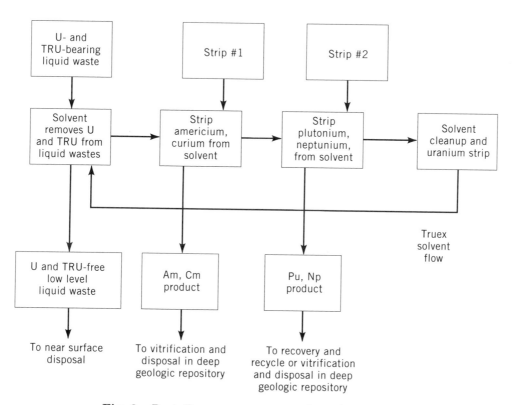

**Fig. 8.**   Basic Truex process separation scheme (139).

## Storage, Usage, and Disposal of Excess Weapons Plutonium

Beginning with the signing of arms reductions treaties in the late 1980s, and accelerating in the post-Cold War era of the 1990s, the production of plutonium for nuclear weapons in the United States and Russia appears to have ended. A vast amount, approximately 50 metric tons in each of the two countries, is being removed as metal from nuclear weapons. As of 1995, this metal was being stored in militarily secure, chemically inert environments such as at the Pantex plant in the United States. Criteria for storage facilities have been published (138).

Long-term disposition of excess weapons plutonium can be achieved by using it as a reactor fuel for electricity generation as $PuO_2$, as mixed oxide (MOX) fuel, or as a specially fabricated new fuel; by transmuting it in very large-scale accelerators; or by mixing it with high level nuclear waste and vitrifying it into glass logs which are then stored in a geologically inert repository (26). Each of these options represents an enormous commitment of secure and reliable facilities that do not exist as of this writing (1995). The U.S. National Academy of Sciences has recommended fabrication and use as nuclear fuel without reprocessing or vitrification. Other options that have been considered are subseabed burial, dispersal into the oceans, or space launch. Although these latter options are feasible from an engineering viewpoint and may be environmentally appropriate, these options violate international agreements and are not being pursued (26).

## Plutonium Compounds

Plutonium forms compounds with many of the metallic elements and all of the nonmetallic elements, except the helium-group gases (qv). The interactions of plutonium with metallic elements, plutonium alloys, and intermetallic compounds of plutonium have been reviewed (30,35–40,47,68,142–146). Alloy phase diagrams, intermetallic crystal structures, and thermodynamics of plutonium binary alloys have been reviewed in detail (30,68,143). The chemistry of plutonium hydrides, oxides, halides, carbonates, sulfates, nitrates, phosphates, carbides, silicides, nitrides, phosphides, arsenides, sulfides, tellurides, and other compounds has been summarized (30,35,36,45,46,96,97,147). Some physical properties of oxides and halides are given elsewhere in the *Encyclopedia* (see ACTINIDES AND TRANSACTINIDES). The physical and chemical properties of plutonium refractory compounds, ie, the oxides, carbides, nitrides, silicides, sulfides, hydrides, borides, and phosphides, have been thoroughly studied (30,96,97,143,147,148) because these refractory compounds are of interest as potential fast-breeder reactor fuels.

**Oxides.** The most important oxide is $PuO_2$. The high melting point (2390 ± 20°C), chemical stability, radiation stability, and similarity to $UO_2$ characterize plutonium dioxide [12059-95-9], $PuO_2$, as an attractive reactor fuel. Plutonium dioxide also is an important intermediate in processing operations such as the preparation of $PuF_4$. The ignition of plutonium or the sulfate, nitrate, chloride, fluoride, oxalate, carbonate, iodate, hydroxide, or many other Pu compounds in air results in the formation of $PuO_2$. However, for the preparation of pure crystalline $PuO_2$, ignition of Pu(III) or Pu(IV) oxalate or Pu(IV) peroxide to 1000°C is the preferred method. Dissolution of $PuO_2$ that has been calcined at high temperature, which is required in nuclear fuel reprocessing, is difficult and requires boiling with strong acid such as $HNO_3$–HF under reflux, with other oxidants (110), reductants, or catalytic oxidative electrolysis (149,150). Dissolution of high temperature calcined $PuO_2$ has been effected by repetitive heating in concentrated hydrochloric acid containing HI and boiling to dryness in nitric acid; the completeness of dissolution was monitored by coulometry (151). $PuO_2$ has the fluorite structure, a lattice parameter of 0.53960 nm, and is isomorphous with $UO_2$, which has a lattice parameter of 0.54862 nm. These dioxides form a continuous series of solid solutions, as do other dioxides ($ThO_2$, $CeO_2$) with $PuO_2$. $PuO_2$ is stoichiometric as shown in Figure 9. However, the observed O:Pu ratio is usually somewhat higher than 2.00 (30). The only higher oxides are Pu(VI) hydroxide $PuO_3 \cdot 0.8H_2O$ and complex oxides.

Plutonium dioxide can be reduced to $PuO_{2-x}$ by loss of oxygen at elevated temperatures in either an inert or reducing atmosphere or in vacuum. The best ways to prepare Pu(III) oxide $Pu_2O_3$ [*12311-78-3*] are to reduce $PuO_2$ using a stream of pure, dry $H_2$ (152), or with carbon or finely divided Pu metal in sealed vessels. The physical and chemical properties of Pu(III), Pu(IV), and mixed Pu(III,IV) oxides have been described (30) and a discussion of the Pu–O and the U–Pu–O systems has been published (46). $PuO_2$ cannot be oxidized, although some salts decompose to compositions in which the O:Pu ratio is as high as 2.09.

Pu(IV) peroxide is an important intermediate in the conversion of plutonium nitrate solution to metal. The chemical composition and the crystal structure depend on the method of precipitation (153). The two crystalline forms are hexagonal (hex) and face-centered cubic (fcc); neither form is stoichiometric. The

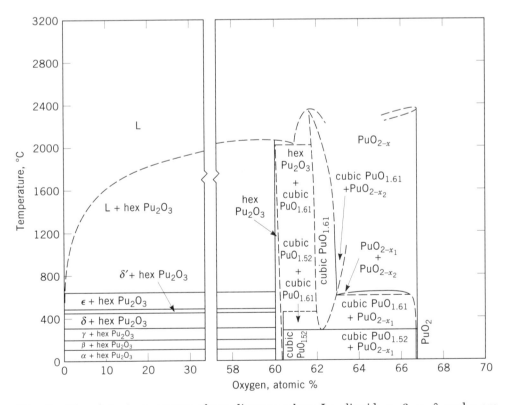

**Fig. 9.** The plutonium–oxygen phase diagram where L = liquid; $\alpha$, $\beta$, $\gamma$, $\delta$, and $\epsilon$ are metal phases; and the dashed lines represent uncertain phase boundaries (143).

cubic form contains three peroxy oxygen atoms per atom of plutonium and the hexagonal form contains 3–3.4 peroxy oxygen atoms per atom of plutonium (153). Both forms contain variable amounts of anions, eg, sulfate, nitrate, or chloride. The cubic precipitate is colloidal and is undesirable for most process applications. In low pH nitric acid solutions, the hexagonal form precipitates; the cubic form precipitates from solutions of low acidity. The recommended procedure for preparing the hexagonal form is to add 30 wt % $H_2O_2$ to a $Pu(NO_3)_4$ solution, which is 2.7 $M$ in $HNO_3$ and 0.15 $M$ in $H_2SO_4$ (153). Yellow-green plutonium(IV) hydroxide [14014-51-8] precipitates from Pu(IV) solutions on addition of ammonia. Under some conditions, a bright green polymer forms and is difficult to redissolve.

There are numerous complex (ternary and quaternary) plutonium oxides. Their properties have been reviewed (30). Plutonium oxidizes readily to Pu(VI) from binary oxides to complex oxides such as $Ba_3PuO_6$. The best way to oxidize Pu to Pu(VII) is to prepare complex oxides such as $Li_5PuO_6$ from $Li_2O$ and $PuO_2$ in flowing oxygen (85).

**Halides and Oxyhalides.** The following binary plutonium halides have been characterized: $PuF_6$ [13693-06-6], $PuF_4$, $PuF_3$, $PuCl_3$, $PuBr_3$ [15752-46-2], and $PuI_3$ [13455-01-1] (30,147,154,155). The hexafluoride, a volatile and extremely reactive brown solid, is obtained by heating $PuF_4$ in flowing $F_2$ and immediately condensing the vapor on a cold surface. The hexafluoride is essential to the fluoride volatility process for reprocessing nuclear fuel. Plutonium tetrafluoride, $PuF_4$, a pink solid, is the principal compound used in the preparation of plutonium metal, and is made by passing a mixture of HF and $O_2$ gas over $PuO_2$ at ca 600°C. Plutonium trifluoride, $PuF_3$, a blue-violet solid, is obtained from the reaction of $PuO_2$ and a mixture of HF and $H_2$ gas at about 600°C. In addition to the usual health hazards associated with plutonium, $PuF_4$ and $PuF_3$ radiate a high neutron flux from $(\alpha, n)$ reactions with fluorine. In the case of $^{238}PuF_4$, the neutron flux is so high as to require remote handling of the fluoride. In the presence of alkali fluorides, $PuF_4$ and $PuF_3$ form numerous complex compounds, which are important in molten salt reactor technology. Numerous efforts have failed to produce $PuF_5$, although a few complex Pu(V) fluorides such as $CsPuF_6$ are known and are relatively stable. Complex fluorides are also known for Pu(IV) and Pu(VI) (30).

Plutonium trichloride, $PuCl_3$, is the only known binary chloride of plutonium. No higher chloride in the solid state has been prepared so far. The trichloride is a blue-green solid and can be prepared from the reaction of Pu metal or $PuO_2$, and HCl gas at elevated temperatures, and can be purified by sublimation and zone melting. The yellow complex chloride, $Cs_2PuCl_6$, is stoichiometric, stable, and can be used as a primary analytical standard for plutonium (156).

Plutonium tribromide [15752-46-2], $PuBr_3$, and plutonium triiodide [13455-01-1], $PuI_3$, both bright green solids, have no practical applications. Comprehensive reviews of the binary and ternary halides are available (147,154,155).

Only Pu(III) oxyhalides (PuOF, PuOCl, PuOBr, and PuOI) and Pu(VI) oxyhalides ($PuO_2F_2$, $PuOF_4$, and $PuO_2Cl_2 \cdot 6H_2O$) are known. Of these the most important are PuOCl, which is the stable product of hydrolysis of $PuCl_3$ (s) with $H_2O$ (g) (157), and $PuO_2F_2$, which is the product of low temperature hydrolysis of $PuF_6$ and one of the products of room temperature hydrolysis of $PuF_6$ (158–160).

**Hydrides.** Plutonium hydrides can only be synthesized from the elements. These are important because they display unique involvement of the $5f$ electrons in bonding, thereby giving these compounds interesting magnetic and transport properties, and because they decompose to give pure and finely divided metal. The hydrides are pyrophoric and must be handled in inert atmospheres (He, Ar). $PuH_{2+x}$ exists having the $CaF_2$ (fcc) structure over the composition $0 < x < 0.7$ and having the hexagonal $LaF_3$ structure above $x = 0.9$. Syntheses, decomposition behavior, and reaction kinetics are discussed in detail (161). The phase diagram and thermodynamics of the Pu–H system have also been reviewed (71,162) (see HYDRIDES).

**Carbides and Silicides.** Plutonium carbides and silicides have received consideration as advanced reactor fuels. They can be prepared from the elements. The carbides can also be produced by reaction of $PuO_2$ or $PuH_3$ with graphite. PuC is stable in room air, oxidizes slowly at 200°C, and burns at 400°C. It reacts with hot water to form hydroxide, hydrogen, and a mixture of hydrocarbons (163) (see CARBIDES). PuSi can also be prepared by reacting $PuF_3$ with Si at 1300°C (164) and the other silicides can be prepared by reacting $PuF_3$ with $CaSi_2$ or $PuO_2$ with Si or SiC (97). The silicides are brittle and metallic but are pyrophoric and react with water at 90°C (97).

**Pnictides.** Plutonium nitride, PuN, has been studied as a possible fast-reactor fuel. It can be prepared by reaction of $PuH_3$ with $NH_3$ at 600–650°C or $N_2$ above 230°C. The pnictides are also interesting for their solid-state magnetic and electrical properties. For the latter reason they also have been prepared as single crystals (165).

**Chalcogenides.** Several sulfides are known, but only PuS has been studied in detail. It is best produced by vacuum decomposition of $PuH_3$ to finely divided metal at 400°C, followed by reaction with $H_2S$, vacuum decomposition, and re-reaction with $H_2S$ several times. It is not attacked by $H_2O$ up to 250°C (97). All of the chalcogenides can be prepared from the elements. The chalcogenides are also of interest because of their solid-state magnetic and electrical behavior.

**Oxalates.** Stable oxalates of Pu(III), Pu(IV), and Pu(VI) are known. However, only the Pu(III) and Pu(IV) oxalates are technologically important (30,147). Brilliant green plutonium(III) oxalate [56609-10-0] precipitates from nitric acid solutions containing Pu(III) ions upon addition of oxalic acid or sodium oxalate. The composition of the precipitate is $Pu_2(C_2O_4)_3 \cdot 10H_2O$. A homogeneous oxalate precipitation by hydrolysis of diethyl oxalate at 75°C minimizes plutonium loss in the filtrate and decreases filtering time (166). Heating the compound to ca 270°C in air or to 460°C in absence of air results in the formation of $PuO_2$. Yellow-green Pu(IV) oxalate [26588-74-9], $Pu(C_2O_4)_2 \cdot 6H_2O$, precipitates from acidic solutions of Pu(IV) upon addition of oxalic acid. The oxalate converts to $PuO_2$ upon heating in the presence or absence of air to ca 500°C.

**Other Compounds.** Other compounds of technological interest include deep green plutonium(IV) nitrate pentahydrate [61204-24-8], $Pu(NO_3)_4 \cdot 5H_2O$; deep red plutonyl nitrate hexahydrate [19125-90-7], $PuO_2(NO_3)_2 \cdot 6H_2O$; coral red anhydrous plutonium(IV) sulfate [13692-89-2], $Pu(SO_4)_2$; pink or red $Pu(SO_4)_2 \cdot 4H_2O$; and moss green plutonyl carbonate [39292-10-9], $PuO_2CO_3$ (30).

## Plutonium in the Environment

It has been estimated that $1.3 \times 10^{16}$ Bq of $^{239+240}$Pu has been released to the environment from atmospheric detonation of nuclear weapons; that $7.9 \times 10^{14}$ Bq of $^{238}$Pu has been released, mostly from burn-up of the nuclear powered satellite SNAP-9a; and that $3.7 \times 10^{13}$ Bq of $^{239+240}$Pu was released by the Chernobyl accident (167,168). Many studies have been done to determine the cumulative fallout on soils, plants, bodies of water, animals, and humans. For example, the cumulative $^{239+240}$Pu fallout in forest and grasslands and in the liver of elderly humans in Bavaria, Germany are approximately 50 Bq/m$^2$ and 0.02 Bq/kg wet mass, respectively (167).

## Health and Safety Factors

The principal hazards of plutonium are those posed by its radioactivity, nuclear critical potential, and chemical reactivity in the metallic state. $^{239}$Pu is primarily an $\alpha$-emitter. Thus, protection of a worker from its radiation is simple and usually no shielding is required unless very large (kilogram) quantities are handled or unless other isotopes are present.

**Protection Against Penetrating Radiation.**   The external dose at the surface of 1 kg of pure plutonium (nearly pure $^{239}$Pu) obtained from uranium irradiated at 3000 MW·d/t is ca 0.025 Gy/h (2.5 rad/h) (23). Thus, for the direct handling of kilogram quantities of Pu, lead-lined gloves (Rad-Bar) are recommended. Isotopes other than $^{239}$Pu, which are present in normal Pu, emit weak $\gamma$- and x-rays; for large amounts of these isotopes lead shielding is necessary. In plutonium compounds with light elements, especially fluorides, the $(\alpha, n)$ reaction produces significant neutron doses which require shielding to slow the neutrons to thermal energies and absorb them.

**Protection Against Internal Radiation.**   The extreme health hazard of plutonium comes from two causes. First, when Pu enters the body, it may not be excreted immediately. Once absorbed or lodged in the body its biological half-life is months or longer. In soluble form it can be metabolized and it accumulates in the blood-forming tissues of the bones and in the liver (169,170). Insoluble particles of PuO$_2$ can lodge in the alveolae of the lung. Second, the $\alpha$-particles deposit energy locally, breaking many bonds and causing much tissue damage.

Elaborate precautions must be taken to prevent the entrance of Pu into the worker's body by ingestion, inhalation, or entry through the skin, because all common Pu isotopes except for $^{241}$Pu are $\alpha$-emitters. $^{241}$Pu is a $\beta$-emitter, but it decays to $^{241}$Am, which emits both $\alpha$- and $\gamma$-rays. Acute intake of Pu, from ingestion or a wound, thus mandates prompt and aggressive medical intervention to remove as much Pu as possible before it deposits in the body. Subcutaneous deposition of plutonium from a puncture wound has been effectively controlled by prompt surgical excision followed by prolonged intravenous chelation therapy with diethylenetriaminepentaacetate (Ca$^{2+}$–DTPA) (171).

Plutonium can enter the body by the gastrointestinal tract (ingestion), by absorption through the skin, from a wound, or by inhalation. Of these possibilities, inhalation is the most likely source of entry and poses the greatest risk (172–174). Inhalation may result in insoluble particles of PuO$_2$ lodging in the respiratory tract or soluble particles migrating to bone or liver. $PuO_2$ is assigned to inhalation class Y (biological clearance time $\geq 100$ d) with fractional absorption $f_1 = 10^{-5}$ (fraction reaching body fluids following transfer to the gastrointestinal tract). Other Pu compounds are assigned to inhalation class W (biological clearance time 10–100 d) with $f_1 = 10^{-4}$ for nitrates and $f_1 = 10^{-3}$ for other compounds or mixtures. Although these $f_1$ values are small, approximately 45% of the transferred Pu deposits in liver with a retention half-time of 20 years and 45% deposits in bone with a retention half-time of 50 years (174).

The International Commission on Radiation Protection has established 20 mSv/yr (2 rads/yr) as the recommended occupational dose limit from the sum of external exposure to ionizing radiation and the 50-year committed dose from intakes of radionuclides. The annual limit of intake (ALI) is 300 Bq (0.0081 $\mu$Ci) for $^{238}$Pu, $^{239}$Pu, $^{240}$Pu, and $^{242}$Pu by inhalation. For ingestion the ALI of these isotopes is $3 \times 10^5$ Bq (8 $\mu$Ci) for $PuO_2$ and $4 \times 10^4$ Bq (1.3 $\mu$Ci) for other Pu compounds (175). Because the inhalation ALI is so low (300 Bq represents only 0.13 $\mu$g of $^{239}$Pu) most Pu laboratories are required to monitor workplace air. The maximum derived air concentration (DAC) of $^{239}$Pu for radiation workers in the United States is $2 \times 10^{-12}$ $\mu$Ci/mL (0.074 Bq/m$^3$). A worker who breathes this air for fifty 40-hour weeks can receive a maximum annual intake of 148 Bq (0.004 $\mu$Ci = 0.064 $\mu$g) of $^{239}$Pu.

**Protection Against Nuclear and Chemical Hazards.**    Safe handling of plutonium metal, its compounds, and its aqueous solutions in chemical, metallurgical, and engineering operations has been the subject of several comprehensive reports (170,176,177). In a laboratory where plutonium is processed and workers are present, a ventilation rate of 8–10 air changes per hour has been recommended (178). The design and operation of hoods, glove boxes, and laboratories for work with plutonium are specialties that have been highly developed as a result of years of experience.

Plutonium solutions that have a low activity ($\leq 3.7 \times 10^7$ Bq (1 mCi) or 10 mg of $^{239}$Pu) and that do not produce aerosols can be handled safely by a trained radiochemist in a laboratory fume hood with face velocity 125–150 linear feet per minute (38–45 m/min). Larger amounts of solutions, solutions that may produce aerosols, and plutonium compounds that are not air-sensitive are handled in glove boxes that are maintained at a slight negative pressure, ca 0.1 kPa (0.001 atm, more precisely measured as 1.0–1.2 cm (0.35–0.50 in.) differential pressure on a water column) with respect to the surrounding laboratory pressure (176,179–181). This air is exhausted through high efficiency particulate (HEPA) filters.

Plutonium metal and air-sensitive plutonium compounds, eg, hydride, must be isolated for chemical as well as nuclear safety. They are handled in airtight glove boxes containing an inert atmosphere (nitrogen, helium, or argon) which are maintained at a slight negative pressure, ca 0.1 kPa (0.001 atm, ie, 1-cm differential pressure on a water column) (182–184). These conditions are especially demanding for operations such as powder metallurgy, grinding, and machining. Plutonium-inert atmosphere boxes require extensive gas repurification because most inert atmosphere glove boxes for inorganic and organometallic chemistry are airtight but at slight positive pressure (0.1 kPa (0.001 atm)).

Although massive pieces of plutonium and its alloys are safe to handle in air, a few Pu compounds and finely powdered Pu metal are pyrophoric. The ignition and pyrophoricity of Pu metal and alloys have been thoroughly reviewed (61). Plutonium metal burns in a nitrogen atmosphere containing only 5% oxygen (185). Fire extinguishing must contain the Pu and should not create a nuclear criticality potential (177,186). The preferred method of extinguishing a Pu fire is by excluding oxygen, either by making the glove box atmosphere inert or by smothering the Pu with a powder, eg, graphite, magnesia, or sodium carbonate (187). Controlled burning of Pu metal also has been recommended as a way to convert it to an inert oxide.

**Criticality Precautions.** The presence of a critical mass of Pu in a container can result in a fission chain reaction. Lethal amounts of gamma and neutron radiation are emitted, and a large amount of heat is produced. The assembly can simmer near critical or can make repeated critical excursions. The generation of heat results eventually in an explosion which destroys the assembly. The quantity of $^{239}$Pu required for a critical mass depends on several factors: the form and concentration of the Pu, the geometry of the system, the presence of moderators (water, hydrogen-rich compounds such as polyethylene, cadmium, etc), the proximity of neutron reflectors, the presence of nuclear poisons, and the potential interaction with neighboring fissile systems (188). As little as 509 g of $^{239}$Pu(NO$_3$)$_4$ solution at a concentration Pu of 33 g/L in a spherical container, reflected by an infinite amount of water, is a critical mass (189,190). Evaluation of criticality controls is available (32,190).

**Special Precautions for $^{238}$Pu.** Because of its high heat generation rate, $^{238}$Pu samples emit a great deal of heat. Large samples may have to be handled with tongs or heat-resistant gloves drawn over the standard glove box gloves. The gloves may be penetrated by $^{238}$Pu unless Hypalon-coated gloves are used. Because the $\alpha,n$-reaction from contact of $^{238}$Pu with light elements emits large numbers of neutrons on glove boxes with 10-cm Lucite front windows and for operations with compounds, eg, $^{238}$PuF$_4$, remote-control equipment is recommended (59,191,192). Inhalation of $^{238}$Pu aerosols yield a higher liver and bone surface dose, but a lower lung dose, than $^{239}$Pu aerosols because $^{238}$PuO$_2$ appears to be more soluble *in vivo* than $^{239}$PuO$_2$ (193).

# BIBLIOGRAPHY

"Plutonium and Plutonium Compounds" in *ECT* 2nd ed., Vol. 15, pp. 879–896, by J. A. Leary and L. J. Mullins, Los Alamos Scientific Laboratory, University of California; in *ECT* 3rd ed., Vol. 18, pp. 278–301, by F. Weigel, University of Munich.

1. D. Albright, F. Berkhout, and W. Walker, *World Inventory of Plutonium and Highly Enriched Uranium, 1992*, SIPRI, Oxford University Press, U.K., 1993.
2. T. B. Cochran, W. M. Arkin, and M. M. Hoenig, *Nuclear Weapons Databook*, Ballinger Publishing Co., Cambridge, Mass., 1984–1994.
3. K. L. Biringer and co-workers, *Power System Requirements and Selection for the Space Exploration Initiative*, SAND-91-1593C, Sandia National Laboratories, N.M., 1991.
4. G. T. Seaborg and W. D. Loveland, *The Elements Beyond Uranium*, Wiley-Interscience, New York, 1990.

5. *Cardiac Pacemakers and Mechanical Hearts—A Bibliography of Radioisotope Power Sources*, TID-3336, Technical Information Center, U.S. Atomic Energy Commission, Washington, D.C., 1973.
6. B. B. Cunningham, in A. S. Coffinberry and W. N. Miner, eds., *The Metal Plutonium*, University of Chicago Press, Ill., 1961, pp. 13–17.
7. B. B. Cunningham, *Plutonium Chemistry Symposium*, The University of Chicago and Argonne National Laboratory, Chicago, Illinois, 1963.
8. *Gmelin Handbuch der Anorganischen Chemie*, Technetium, Springer-Verlag, Berlin, 1982.
9. G. T. Seaborg, in L. R. Morss and J. Fuger, eds., *Transuranium Elements: A Half Century*, American Chemical Society, Washington, D.C., 1992, pp. 10–49.
10. G. T. Seaborg, "Transuranium Elements, Products of Modern Alchemy," *Benchmark Papers in Physical Chemistry and Chemical Physics*, Vol. 1, Dowden, Hutchison & Ross, Stroudsburg, Pa., 1978.
11. B. B. Cunningham and L. B. Werner, in G. T. Seaborg, J. J. Katz, and W. M. Manning, eds., *The Transuranium Elements*, NNES-IV-14B, Vol. 1, McGraw-Hill Book Co., New York, 1949, pp. 51–78.
12. G. T. Seaborg, *The Transuranium Elements*, Addison-Wesley Publishing Co., Reading, Mass., 1958.
13. S. Fried, E. F. Westrum, Jr., H. L. Baumbach, and P. L. Kirk, *J. Inorg. Nucl. Chem.* **5**, 182 (1958).
14. S. Fried, E. F. Westrum, Jr., H. L. Baumbach, and P. L. Kirk, in Ref. 6, pp. 18–24.
15. C. S. Smith, in Ref. 6, pp. 26–35.
16. D. C. Hoffman, F. O. Lawrence, J. L. Mewherter, and F. M. Rourke, *Nature* **234**, 132 (1971).
17. C. A. Levine and G. T. Seaborg, *J. Am. Chem. Soc.* **73**, 3278 (1951).
18. E. K. Hyde, I. Perlman, and G. T. Seaborg, *The Nuclear Properties of the Heavy Elements*, Prentice-Hall, Englewood Cliffs, N.J., 1964; E. Browne, R. B. Firestone, and V. S. Shirley, eds., *Table of Radioactive Isotopes*, John Wiley & Sons, Inc., New York, 1986.
19. M. J. Bell, *ORIGEN—The ORNL Isotope Generation and Depletion Code*, ORNL-4628, Oak Ridge National Laboratory, Oak Ridge, Tenn., 1973.
20. W. W. Schulz and G. E. Benedict, *Neptunium-237 Production and Recovery*, TID-25955, U.S. Atomic Energy Commission, Washington, D.C., 1972.
21. *Plutonium-238 and Polonium-210 Data Sheets*, MLM-1441, Monsanto Co., Miamisburg, Ohio, 1967.
22. *Plutonium-238 Isotopic Fuel Form Data Sheets*, MLM-1681, Monsanto Co., Miamisburg, Ohio, 1969.
23. M. Taube, *Plutonium, A General Survey*, Verlag Chemie, Weinheim, Germany, 1974.
24. S. I. Schwartz, *Bull. Atom. Scien.* **51**, 32–52 (1995).
25. *Plutonium Fuel: An Assessment*, Organization for Economic Co-operation and Development, Nuclear Energy Agency, Paris, France, 1989.
26. *Managing and Disposition of Excess Weapons Plutonium*, U.S. National Academy of Sciences, National Academy Press, Washington, D.C., 1994.
27. *The Economics of the Nuclear Fuel Cycle*, Organization for Economic Co-operation and Development, Nuclear Energy Agency, Paris, France, 1994.
28. R. T. Kennedy and M. B. Kratzer, *Protection and Management of Plutonium*, American Nuclear Society, LaGrange Park, Ill., 1995.
29. W. T. Carnall and G. R. Choppin, eds., *Plutonium Chemistry*, American Chemical Society, Washington, D.C., 1983.
30. F. Weigel, J. J. Katz, and G. T. Seaborg, in J. J. Katz, G. T. Seaborg, and L. R. Morss, eds., *The Chemistry of the Actinide Elements*, Vol. 1, Chapman & Hall, London, 1986, pp. 499–886.

31. D. A. Orth, *Ind. Eng. Chem. Proc. Des. Dev.* **2**, 121 (1963).
32. H. C. Paxton, *Criticality Control in Operations with Fissile Material*, LA-3366 (rev.), Los Alamos Scientific Laboratory, Los Alamos, N.M., 1972.
33. J. T. Byrne, *Plutonium Management at Rocky Flats*, RFP-720, The Dow Chemical Co., Golden, Colo., 1966.
34. R. L. Rose, *Plutonium Foundry Practice at the Lawrence Livermore Laboratory*, UCRL-73740, Lawrence Livermore Laboratory, Calif., 1972.
35. *Gmelin Handbuch der Anorganischen Chemie, Transurane, Teil B1, Metalle*, Springer-Verlag, Berlin, 1976.
36. *Gmelin Handbuch der Anorganischen Chemie, Transurane, Legierungen*, Springer-Verlag, Berlin, 1976–1977.
37. U. Grison, W. B. H. Lord, and R. D. Fowler, *Plutonium 1960*, Cleaver-Hume Press, London, 1961.
38. W. N. Miner and co-workers, in C. A. Hampel, ed., *Rare Metals Handbook*, Reinhold Book Co., New York, 1961.
39. W. N. Miner and F. W. Schonfeld, in O. J. Wick, ed., *Plutonium Handbook*, Vol. I, Gordon & Breach Science Publishers, New York, 1967, Chapt. 3.
40. A. S. Coffinberry and W. N. Miner, *The Metal Plutonium*, University of Chicago Press, Chicago, Ill., 1961.
41. J. W. Ward, P. D. Kleinschmidt, and D. E. Peterson, in A. J. Freeman and C. Keller, eds., *Handbook on the Physics and Chemistry of the Actinides*, Vol. 4, Elsevier-North Holland, Amsterdam, the Netherlands, 1986, pp. 309–412.
42. R. G. Haire and J. K. Gibson, in L. R. Morss and J. Fuger, eds., *Transuranium Elements: A Half Century*, American Chemical Society, Washington, D.C., 1992, pp. 426–439.
43. M. H. Bradbury and R. W. Ohse, *J. Chem. Phys.* **70**, 2310 (1979).
44. F. L. Oetting, M. H. Rand, and R. J. Ackermann, in F. L. Oetting, ed., *The Chemical Thermodynamics of Actinide Elements and Compounds, Part 1, The Actinide Elements, STI/PUB/424/1*, IAEA, Vienna, Austria, 1976.
45. F. L. Oetting, *Chem. Rev.* **66**, 261 (1966).
46. M. H. Rand, in O. Kubaschewski, ed., *Plutonium: Physicochemical Properties of its Compounds and Alloys, Atomic Energy Review*, Vol. 4, Special Issue No. 1, IAEA, Vienna, Austria, 1966.
47. A. J. Freeman and J. B. Darby, Jr., *The Actinides, Electronic Structure and Related Properties*, 2 Vols., Academic Press, Inc., New York, 1974.
48. T. A. Sandenaw and R. B. Gibney, *J. Phys. Chem. Solids* **6**, 81 (1958).
49. R. O. Elliott, C. E. Olsen, and S. E. Bromisz, *Phys. Rev. Lett.* **12**, 276 (1954).
50. A. A. Comstock, in Ref. 40, pp. 152–156.
51. S. T. Konobeevsky, *Conference of the Academy of Sciences of the USSR on the Peaceful Uses of Atomic Energy*, Eng. Trans., Moscow, Russia, 1955.
52. J. K. Dawson, *J. Chem. Soc. (London)*, 3393 (1954).
53. J. A. Lee, R. O. A. Hall, E. King, and G. D. Meaden, in Ref. 37, pp. 39–50.
54. A. J. Arko and M. B. Brodsky, in Ref. 47, p. 364.
55. T. A. Sandenaw and R. B. Gibney, *J. Chem. Thermodyn.* **3**, 85 (1971).
56. C. E. Olsen, T. A. Sandenaw, and C. C. Herrick, *The Density of Liquid Plutonium Metal*, LA-2358, Los Alamos Scientific Laboratory, N.M., 1959.
57. L. V. Jones, D. Ofte, W. C. Rohr, and L. J. Wittenberg, *Am. Soc. Met. Trans. Quart.* **55**, 819 (1962).
58. J. W. Stout and W. A. Jones, *Phys. Rev.* **71**, 582 (1947).
59. M. W. Shupe and co-workers, *Radiation Safety in Hot Facilities, Proceedings of Symposium, Saclay, France, Oct. 13–17, 1969*, IAEA Publication STI/PUB/238, Vienna, Austria, 1970.

60. J. T. Waber and E. S. Wright, in Ref. 40, pp. 194–204.
61. J. T. Waber, in Ref. 39, Chapt. 6, pp. 145–189.
62. J. M. Haschke, A. E. Hodges III, G. E. Bixby, and R. L. Lucas, *Reaction of Pu with Water*, RFP-3416, Rockwell International, Golden, Colo., 1983.
63. J. L. Stakebake, in Ref. 42, pp. 251–259.
64. J. Fuger and F. L. Oetting, *Part 2, The Actinide Aqueous Ions*, STI/PUB/424/2, in Ref. 44, 1976.
65. L. R. Morss, in Ref. 30, Vol. 2, pp. 1278–1360.
66. J. Fuger, *J. Chem. Thermodyn.* **24**, 337 (1992).
67. D. L. Hildenbrand, L. V. Gurvich, and V. S. Yungman, in Ref. 44, *Part 13, The Gaseous Actinide Ions*, STI/PUB/424/13, 1992.
68. P. Chiotti, V. V. Akhachinskij, I. Ansara, and M. H. Rand, *Part 5, The Actinide Binary Alloys*, STI/PUB/424/5, in Ref. 44, 1981.
69. IAEA, *The Plutonium–Oxygen and Uranium–Plutonium–Oxygen Systems*, Technical Report No. 79, IAEA, Vienna, Austria, 1967.
70. J. Fuger, V. B. Parker, W. N. Hubbard, and F. L. Oetting, *Part 8, The Actinide Halides*, STI/PUB/424/8, in Ref. 44, 1983.
71. H. E. Flotow, J. M. Haschke, and S. Yamauchi, *Part 9, The Actinide Hydrides*, STI/PUB/424/9, in Ref. 44, 1984.
72. J. Blaise, J. F. Wyart, J. G. Conway, and E. F. Worden, *Phys. Scr.* **22**, 224 (1980); J. Blaise, M. Fred, W. T. Carnall, and H. W. Crosswhite, *Plutonium Chemistry*, ACS Symposium Series No. 216, American Chemical Society, Washington, D.C., 1983, pp. 173–198.
73. J. Blaise, M. Fred, and R. G. Gutmacher, *Report ANL-83-95*, Argonne National Laboratory, Argonne, Ill., 1984.
74. M. S. Fred and J. Blaise, in Ref. 30, Vol. 2, pp. 1196–1234.
75. J. G. Conway and M. Fred, *J. Opt. Soc. Am.* **43**, 216 (1953).
76. M. van den Berg, P. F. A. Klinkenberg, and P. Regnant, *Physica* **20**, 17 (1954).
77. B. Bleaney, P. M. Llewellyn, M. H. L. Pryce, and R. C. Hall, *Phil. Mag.* **45**, 991 (1954).
78. R. E. Connick, in G. T. Seaborg and J. J. Katz, eds., *The Actinide Elements, National Nuclear Energy Series, Plutonium Project Record*, Vol. Div. IV, Vol. 14A, McGraw-Hill, New York, 1954, pp. 221–300.
79. D. Cohen, *J. Inorg. Nucl. Chem.* **18**, 211 (1961).
80. W. T. Carnall and P. R. Fields, in P. R. Fields and T. Moeller, eds., *Lanthanide/Actinide Chemistry*, Advances in Chemistry Series No. 71, American Chemical Society, Washington, D.C., 1967, pp. 86–101.
81. M. N. Myers, *Absorption Spectra of Plutonium and Impurity Ions in Nitric Acid Solution*, HW-44744, General Electric Co., 1956.
82. V. I. Spitsyn and co-workers, *J. Inorg. Nucl. Chem.* **31**, 2733 (1969).
83. W. T. Carnall and H. M. Crosswhite, in Ref. 30, pp. 1235–1277.
84. P. K. Bhattacharyya, R. Veeraraghavan, and R. D. Saini, *Radiochim. Acta* **30**, 217 (1982).
85. C. Keller, in Ref. 41, Vol. 3, Chapt. 3, pp. 143–184.
86. V. F. Peretrukhin, F. David, and A. Maslennikov, *Radiochim. Acta* **65**, 161 (1994).
87. G. R. Choppin, *J. Radioanal. Nucl. Chem. Articles* **147**, 109 (1991).
88. G. L. Silver, *J. Radioanal. Nucl. Chem. Lett.* **155**, 177 (1991).
89. G. R. Choppin and B. Allard, in Ref. 41, pp. 407–429.
90. T. Yamaguchi, Y. Sakamoto, and T. Ohnuki, *Radiochim. Acta* **66/67**, 9 (1994).
91. E. H. Appelman, A. G. Kostka, and J. C. Sullivan, *Inorg. Chem.* **27**, 2002 (1988).
92. D. Cohen, *J. Inorg. Nucl. Chem.* **18**, 207 (1961).
93. T. W. Newton, D. E. Hobart, and P. D. Palmer, *The Preparation and Stability of Pure Oxidation States of Neptunium, Plutonium, and Americium*, LAUR-86-967, Los Alamos National Laboratory, Calif., 1986.

94. *Gmelin Handbuch der Anorganischen Chemie, Transurane, Teil D1, Chemie in Lösung*, Springer-Verlag, Berlin, 1975.
95. *Gmelin Handbuch der Anorganischen Chemie, Transurane, Teil D2, Chemie in Lösung*, Springer-Verlag, Berlin, 1975.
96. C. Keller, in *The Chemistry of the Transuranium Elements*, Verlag Chemie, Weinheim, Germany, 1971, Chapt. XIV.
97. J. M. Cleveland, *The Chemistry of Plutonium*, 2nd ed., American Nuclear Society, LaGrange Park, Ill., 1979.
98. L. Martinot and J. Fuger, in A. J. Bard, R. Parsons, and J. Jordan, eds., *Standard Potentials in Aqueous Solution*, Marcel Dekker, Inc., New York, 1985.
99. C. Riglet, P. Robouche, and P. Vitorge, *Radiochim. Acta* **46**, 85 (1989).
100. H. Capdevila and P. Vitorge, *Radiochim. Acta* **68**, 51 (1995).
101. T. W. Newton, *The Kinetics of the Oxidation Reduction Reactions of Uranium, Neptunium, Plutonium, and Americium in Aqueous Solution*, TID-26506, U.S. Energy, Research, and Development Administration (ERDA) Technical Information Center, Washington, D.C., 1975.
102. S. Fried and co-workers, in G. J. M. McCarthy, ed., *The Scientific Basis of Nuclear Waste Management*, Materials Research Society, Plenum Press, New York, 1979, pp. 655–664.
103. H. Nitsche and co-workers, *Radiochim. Acta* **66/67**, 3 (1994).
104. D. T. Reed, S. Okajima, and M. K. Richmann, *Radiochim. Acta* **66/67**, 95 (1994).
105. J. Fuger, I. L. Khodakovskii, V. A. Medvedev, and J. D. Navratil, in Ref. 44, *Part 12, The Actinide Aqueous Inorganic Complexes, STI/PUB/424/12*, 1992.
106. G. R. Choppin, *Radiochim. Acta* **32**, 43 (1983).
107. D. B. James, *Anion Exchange Processing of Plutonium*, LA-3499, Los Alamos Scientific Laboratory, N.M., 1966.
108. L. R. Morss and C. W. Williams, *Radiochim. Acta* **66/67**, 89 (1994).
109. M. H. Lloyd and R. G. Haire, *Radiochim. Acta* **25**, 139 (1978).
110. J. C. Sullivan and E. H. Appelman, *Radiochim. Acta* **48**, 151 (1989).
111. I. Pashalides, J. I. Kim, T. Ashida, and I. Grenthe, *Radiochim. Acta* **68**, 99 (1995).
112. E. N. Rizkalla, L. F. Rao, G. R. Choppin, and J. C. Sullivan, *Radiochim. Acta* **65**, 23 (1994).
113. T. W. Newton and J. C. Sullivan, in Ref. 41, Vol. 3, Chapt. 10, pp. 387–406.
114. S. Ahrland, in Ref. 30, pp. 1480–1546.
115. N. V. Jarvis and R. D. Hancock, *Radiochim. Acta* **64**, 15 (1994).
116. I. A. Mahamid, K. A. Becraft, and H. Nitsche, *Radiochim. Acta* **68**, 63 (1995).
117. *Gmelin Handbuch der Anorganischen Chemie, Transurane, Teil A2, Element*, Verlag Chemie, Weinheim, Germany, 1973.
118. M. P. Neu, D. C. Hoffman, K. E. Roberts, H. Nitsche, and R. J. Silva, *Radiochim. Acta* **66/67**, 251 (1994).
119. G. H. Coleman, *The Radiochemistry of Plutonium*, NAS-NS-3058, U.S. Atomic Energy Commission, Washington, D.C., 1965.
120. R. A. Roberts, G. R. Choppin, and J. F. Wild, *The Radiochemistry of Uranium, Neptunium, and Plutonium—An Updating*, NAS-NS-3063, U.S. Department of Energy, Washington, D.C., 1986.
121. *Measurement of Radionuclides in Food and the Environment; A Guidebook*, IAEA, Vienna, Austria, 1989.
122. E. P. Horwitz, M. L. Dietz, D. M. Nelson, J. J. LaRosa, and W. D. Fairman, *Anal. Chim. Acta* **238**, 263 (1990).
123. Y. T. Chuburkov and co-workers, *Radiochim. Acta* **68**, 227 (1995).
124. M. E. Wrenn, N. P. Singh, and Y. H. Xue, *Radiat. Prot. Dosim.* **53**, 81 (1994).
125. J. F. Flagg, *Chemical Processing of Reactor Fuels*, Academic Press, Inc., New York, 1961.

126. S. M. Stoller and R. B. Richards, *Reactor Handbook*, 2nd ed., Vol. II, Interscience, New York, 1961.

127. J. T. Long, *Engineering for Nuclear Fuel Processing*, American Nuclear Society, LaGrange Park, Ill., 1978.

128. J. L. Ryan and E. J. Wheelwright, *The Recovery, Purification, and Concentration of Plutonium by Anion Exchange in Nitric Acid*, HW-55893, General Electric Co., Richland, Wash., 1959.

129. J. M. Cleveland, in Ref. 39, Vol. II, 1967, Chapt. 14.

130. W. C. Hazen, *Remote Control Equipment for Plutonium Metal Production*, LA-1387, Los Alamos Scientific Laboratory, N.M., 1951.

131. J. A. Leary, A. N. Morgan, and W. N. Maraman, *Ind. Eng. Chem.* **51**, 27 (1959).

132. A. N. Morgan and co-workers, *Proceedings, 2nd International Conference on Peaceful Uses of Atomic Energy*, Geneva, Switzerland, 1958.

133. J. A. Leary and L. J. Mullins, *Practical Applications of Thermodynamics to Plutonium Process Reactions at High Temperature,* Vol. 1, *Thermodynamics,* STI/PUB/162, IAEA, Vienna, 1967, pp. 459–471.

134. M. W. Gibson and D. H. Nyman, *Recent Plutonium Metal Production Experience at Hanford*, WHC-SA-0616, Westinghouse Hanford Co., Wash., 1989.

135. L. J. Mullins and J. A. Leary, *Ind. Eng. Chem. Proc. Des. and Dev.* **4**, 394 (1965).

136. M. Levenson, J. V. C. Trice, and W. J. Mecham, *Comparative Cost Study of the Processing of Oxide, Carbide, and Metal Fast Breeder Reactor Fuels by Aqueous, Volatility and Pyrochemical Methods*, ANL-7137, Argonne National Laboratory, Argonne, Ill., 1966.

137. J. E. Battles, J. J. Laidler, C. C. McPheeters, and W. E. Miller, in B. Mishra, ed., *Actinide Processing: Methods and Materials*, The Minerals, Metals & Materials Society, 1994.

138. D. D. Wilkey, W. T. Wood, and C. D. Guenther, *Long-Term Plutonium Storage: Design Concepts*, LA-UR-94-2390, Los Alamos National Laboratory, N.M., 1994.

139. E. P. Horwitz and R. Chiarizia, in T. E. Carleson, N. A. Chipman, and C. M. Wai, eds., *Separation Techniques in Nuclear Waste Management*, CRC Press, Boca Raton, Fla., 1995, pp. 3–33.

140. G. J. Lumetta, *Pretreatment of Neutralized Cladding Removal Waste Sludge*, PNL-9747, Battelle Pacific Northwest Laboratory, Richland, Wash., 1994.

141. E. P. Horwitz and W. Schulz, in L. Cecille, M. Casarci, and L. Pietrielli, eds., *New Separation Chemistry Techniques for Radioactive Waste and Other Applications*, Elsevier Applied Science, New York, 1990, pp. 21–30.

142. M. E. Hasbrouck, *Plutonium Metallurgy Handbook*, BNWL-37, Battelle Northwest Laboratory, Richland, Wash., 1965.

143. F. H. Ellinger, W. N. Miner, D. R. O'Boyle, and F. W. Schonfeld, *Constitution of Plutonium Alloys*, LA-3870, Los Alamos Scientific Laboratory, N.M., 1968.

144. A. E. Kay and M. B. Waldron, *Plutonium 1965*, Chapman & Hall, London, 1967.

145. W. N. Miner, *Plutonium 1970 and Other Actinides*, Metallurgical Society of AIME, New York, 1970.

146. H. Blank and R. Lindner, *Plutonium 1975 and Other Actinides*, North Holland Publishing Co., American Elsevier Publishing Co., New York, 1975.

147. *Gmelin Handbuch der Anorganischen Chemie, Transurane, Teil C, Verbindungen*, Springer-Verlag, Berlin, 1972.

148. R. E. Skavdahl and T. D. Chikalla, in Ref. 39, Chapt. 80.

149. J. L. Ryan, L. A. Bray, E. J. Wheelwright, and G. H. Bryan, in Ref. 42, pp. 288–304.

150. C. Madic, P. Berger, and X. Machuron-Mandard, in Ref. 42, pp. 457–468.

151. I. S. Sklyarenko, V. V. Andriets, and T. M. Chubukova, *Radiochemistry*, **37**, 374 (343 Engl. trans.) (1995).

152. H. E. Flotow and M. Tetenbaum, *J. Chem. Phys.* **74**, 5269 (1981).
153. J. A. Leary, *Studies on the Preparation, Properties, and Composition of Plutonium Peroxide*, LA-1913, Los Alamos Scientific Laboratory, N.M., 1955.
154. D. Brown, *The Halides of the Lanthanides and Actinides*, John Wiley & Sons, Inc., New York, 1968.
155. J. J. Katz and I. Sheft, *Advances in Inorganic Chemistry and Radiochemistry*, Vol. 2, Academic Press, Inc., New York, 1960, pp. 195–236.
156. F. J. Miner, R. P. DiGrazio, and D. T. Byrne, *Anal. Chem.* **35**, 1218 (1963).
157. F. Weigel, V. Wishnevsky, and H. Hauske, *J. Less-Common Metals* **56**, 113 (1977).
158. C. J. Mandleberg and co-workers, *J. Less-Common Metals* **2**, 358 (1956).
159. A. E. Florin, I. R. Tannenbaum, and J. F. Lemons, *J. Less-Common Metals* **2**, 368 (1956).
160. B. Weinstock and J. G. Malm, *J. Less-Common Metals* **2**, 380 (1956).
161. J. M. Haschke, in G. Meyer and L. R. Morss, eds., *Synthesis of Lanthanide and Actinide Compounds*, Kluwer Academic Publishers, Dordrecht, the Netherlands, 1991, pp. 1–53.
162. J. W. Ward, in Ref. 41, Vol. 3, pp. 1–74.
163. J. L. Drummond, B. J. McDonald, H. M. Ockenden, and G. A. Welch, *J. Chem. Soc.*, 4785 (1957).
164. O. J. C. Runnalls, in Ref. 40, p. 309.
165. J. C. Spirlet, in Ref. 161, pp. 353–367.
166. S. L. Yarbro, S. B. Schreiber, S. L. Dunn, and C. W. Mills, in Ref. 42, pp. 489–493.
167. R. W. Perkins and C. W. Thomas, in W. C. Hanson, ed., *Transuranic Elements in the Environment*, U.S. Department of Energy, Springfield, Va., 1980, p. 53.
168. Y. S. Sedunov, V. A. Borzilov, and N. V. Klepikova, *Use of Mathematical Modeling to Estimate Formation of Contaminated Areas Resulting from Nuclear Accident*, IAEA-SM-306/114, IAEA, Vienna, Austria, 1990.
169. R. G. Thomas, J. W. Healy, and J. F. McInroy, *Health Physics* **46**, 839 (1984).
170. L. G. Faust, *Health Physics Manual of Good Practices for Plutonium Handling*, PNL-6534, Pacific Northwest Laboratory, Richland, Wash., 1988.
171. E. H. Carbaugh, W. A. Decker, and M. J. Swint, *Radiat. Prot. Dosim.* **26**, 345 (1989).
172. *Proceedings, Plutonium Information Meeting for an ad hoc Subcommittee of the Advisory Committee on Reactor Safeguards*, CONF–740115, Los Alamos National Laboratory, Los Alamos, N.M., 1974.
173. J. C. Nenot and J. W. Stather, *The Toxicity of Plutonium, Americium, and Curium*, Pergamon Press, Oxford, U.K., 1979.
174. *The Metabolism of Plutonium and Related Elements*, ICRP Pub. 48, Pergamon Press, Oxford, U.K., 1986.
175. *Annual Limits for Intakes of Radionuclides by Workers Based on the 1990 Recommendations*, ICRP Pub. 61, Pergamon Press, Oxford, U.K., 1991.
176. B. A. J. Lister, *Health Physics Aspects of Plutonium Handling*, AERE-L151, AERE, Harwell, U.K., 1964.
177. *Safe Handling of Plutonium, A Panel Report, Safety Series No. 38*, IAEA, Vienna, Austria, 1974.
178. C. I. Fairchild and co-workers, *Health-Related Effects of Different Ventilation Rates in Plutonium Laboratories*, LA-11948-MS, Los Alamos National Laboratory, N.M., 1991.
179. C. J. Barton, in E. S. Perry and A. Weissberger, eds., *Techniques of Chemistry*, Vol. XIII, John Wiley & Sons, Inc., New York, 1979.
180. A. Brodsky, *Radiat. Protect. Manage.* **6**, 39 (1989).
181. *Industrial Ventilation—A Manual of Recommended Practice*, American Conference of Governmental Industrial Hygienists, 21st ed., Cincinnati, Ohio, 1992.

182. L. R. Kelman, W. D. Wilkinson, A. B. Shuck, and R. C. Goertz, *The Safe Handling of Radioactive Pyrophoric Materials*, ANL-5509, Argonne National Laboratory, Argonne, Ill., 1955.

183. A. B. Shuck and R. M. Mayfield, *The Process Equipment and Protective Enclosures Designed for the Fuel Fabrication Facility, Facility No. 350*, ANL-5499, Argonne National Laboratory, Argonne, Ill., 1956.

184. L. R. Kelman, J. L. Armstrong, W. H. Livernash, and H. V. Rhude, *Proceedings of the Ninth Conference on Hot Laboratories and Equipment*, American Nuclear Society, LaGrange Park, Ill., 1963.

185. H. V. Rhude, *Fire and Explosion Tests of Pu Gloveboxes*, U.S. AEC Report TID-16826, Argonne National Laboratory, Argonne, Ill., 1962.

186. *Proceedings of the Rocky Flats Symposium on Safety in Plutonium Handling Facilities*, U.S. Atomic Energy Commission, Washington, D.C., 1971.

187. R. R. King, in Ref. 184, pp. 71–75.

188. J. J. Rehr, J. M. D. Leon, S. Zabinsky, and R. C. Albers, *Phys. Rev. B* **44**, 4146 (1991).

189. E. D. Clayton and W. A. Reardon, in Ref. 39, Vol. II, Chapt. 27.

190. *Handbook of Nuclear Criticality, Rev. 3*, RFP-4657, EG&G, Rock Flats, Golden, Colo., 1991.

191. F. D. Lonadier and J. S. Griffo, *Ind. Eng. Chem. Process Design Devel.* **3**, 336 (1964).

192. A. N. Morgan, J. L. Green, J. A. Leary, and W. J. Maraman, *Preparation of Plutonium-238 Metal*, LA-2411, Los Alamos Scientific Laboratory, N.M., 1960.

193. R. A. Guilmette, W. C. Griffith, and A. W. Hickman, *Radiat. Prot. Dosim.* **53**, 27 (1994).

*General References*

J. M. Cleveland, *The Chemistry of Plutonium*, 2nd ed., American Nuclear Society, LaGrange Park, Ill., 1979.

M. Taube, *Plutonium—A General Survey*, Verlag Chemie, Weiheim, Germany, 1974.

F. Weigel, J. J. Katz, and G. T. Seaborg, in J. J. Katz, G. T. Seaborg, and L. R. Morss, eds., *The Chemistry of the Actinide Elements*, 2nd ed., Chapman & Hall, London, 1986, Chapt. 7.

O. J. Wick, ed., *Plutonium Handbook: A Guide to the Technology*, Vols. I and II, Gordon and Breach, New York, 1967.

LESTER R. MORSS
Argonne National Laboratory

# PLYWOOD.   See WOOD-BASED COMPOSITES AND LAMINATES.

# POLISHES

Polishes are used to maintain a glossy finish on surfaces as well as to prolong the useful lives of these surfaces. Appearance enhancement provided by polishes generally results from the presence of components that leave a glossy coating, and/or materials that smooth and clean surfaces. Furniture, shoe, and most floor polishes rely on the deposition of a film. The exception in the case of floor polishes is in certain treatments for marble, which involve recrystallization of the surface rather than application of a film. In addition to providing glossy protective films, car polishes contain abrasives (qv) to remove weathered paint (qv) and soils. Metal polishes are based on either abrasive smoothing and cleaning or tarnish-removing chemicals, and sometimes deposit materials that retard future tarnishing (see COATINGS; METAL SURFACE TREATMENTS).

The terms *polish* and *wax* have been used almost interchangeably. Film formers not based on wax are used increasingly (see WAXES). Change to synthetic polymer-based films was motivated by the promise of more consistent quality, greater durability, and greater convenience in the total product experience, as well as changes in the surfaces being protected. Buffing is necessary to achieve gloss for natural wax-based formulas. Most modern polishes are self-polishing. Even buffable products have been formulated to reduce buffing effort. Some modern formulations clean and remove previous polish films as well as leave a new film, all in one operation. Thus, periodic stripping of polish film buildup is not required. Polymeric film formers, which can be synthesized to provide specific properties, have been the basis of most polish formula advancements, even though the technology involved in modifying natural waxes has also progressed. Products employing reduced levels of volatile organic compounds (VOCs) to meet environmental requirements have been developed. The nature of the surfaces that are polished has also changed over the years. Polymers are widely used as construction materials and are thence polished (see BUILDING MATERIALS). For example, whereas wood (qv) floors are popular in households, the most common hard-surface floor covering in the industrial and institutional market is vinyl polymer. Synthetic resins are commonly used to coat natural surfaces, eg, of wooden furniture.

## Furniture

Modern furniture polishes are designed for a wide variety of surfaces, eg, plastics, metals, and synthetic and natural resin coatings. Furniture polishes impart shine and provide protection from abrasion, marring, and spills. The formulations clean well in many cases. In common with most other polishes, furniture polishes are characterized by ease and speed of application and of buffing, and by either the absence of objectionable odors or the addition of pleasing ones.

Reviews of furniture polish formulations are available (1–3). Furniture polishes contain one or two classes of film-forming ingredients, solvents, and various stabilizers. Natural and synthetic waxes are the main film formers in many formulations. Most natural waxes protect plants against water loss and microorganism entry (4). Typically, these waxes are long-chain molecules that do not evaporate easily. They contain few polyunsaturated compounds that might be susceptible to atmospheric degradation, and the waxes may be immune to enzymatic breakdown by microbes. Other modern waxes include paraffins derived from petroleum; montan, a fossil vegetable wax; and synthetic materials, eg, polyethylene [9002-88-4], polypropylene [25085-53-4], and Fisher-Tropsch products. For polish use, the basis for referring to these materials as waxes is the similarity between their physical properties and those of natural waxes. Wax blends used in furniture polishes are not as hard as those used for automobile or floor polishes, but the blends must not be so soft that the resulting coating can be smeared easily.

Chemical modification of the wax can improve smear resistance (5). Silicones, which do not harm furniture finishes (6), are incorporated as film-forming ingredients in furniture polishes. The lubricant properties of silicones improve ease of application of the polish and removal of insoluble soil particles. In addition, silicones make dry films easier to buff and more water-repellent, and provide depth of gloss, ie, ability to reflect a coherent image as a result of a high refractive index (7). Wax-free polishes, which have silicones as the only film former, can be formulated to deliver smear resistance (8). Another type of film former commonly used in oil-base furniture polishes is a mineral or vegetable oil, eg, linseed oil.

Solvents facilitate oil-borne detergency, provide the solvency required for a stable formulation of the desired consistency, and control the film drying rate after its application. For furniture polishes, the nonaqueous solvents generally are aliphatic hydrocarbons (qv) that do not attack finishes. Because its use reduces cost and toxicity, and because it aids detergency against waterborne soils, water is increasingly used as a carrier. Emulsion polishes can be either of two types: oil in a continuous water phase (o/w), or water in a continuous oil phase (w/o). Generally, o/w emulsions provide better cleaning properties for water-soluble soils but poorer gloss. For o/w emulsions, ethoxylated, nonionic emulsifiers having high levels of ethoxylation are used. For w/o emulsions, sorbitan oleates [1338-43-8] commonly are used (1). The advantages of using cationic emulsifiers for w/o-emulsion furniture polishes have been described (9). It is possible to formulate shear unstable emulsions of silicones. These break when the liquid emulsion is sheared during polishing and thereby deposit a uniform silicone film (10).

The different types of furniture polishes include liquid or paste solvent waxes, clear oil polishes, emulsion oil polishes, emulsion wax polishes, and aerosol or spray polishes (3). Nonwoven wipes impregnated with polish ingredients have been targeted at consumers who do not wish to expend the time to dust before polishing (11). Compilations of representative formulas are given in References 3, 4, 12, and 13. Paste waxes contain ca 25 wt % wax, the remainder being solvent. Clear oil polishes contain 10–15 wt % oil and a small amount of wax, the rest being solvent. Aerosol or spray products may contain 2–5 wt % of a silicone polymer, 1–3 wt % wax, 0–30 wt % hydrocarbon solvent, and ca 1 wt % emulsifier. The remainder is water.

Evaluations of furniture polish application properties, gloss, uniformity, film clarity, smear and mar resistance, film healing, buffability, cleaning, water spotting, gloss retention, and dust attraction are all described in ASTM

D3751-79. Federal Specification P-P553B (1977) includes some standard test methods for the evaluation of liquid furniture polishes.

## Floor

There are two basic segments in the floor care market: the household market which has been declining as a result of changing lifestyles and no-wax floors, and the industrial and institutional (I&I) market. Whereas trends in the former have stressed convenience, the I&I market, which has been holding fairly steady, has focused on labor saving, either through increased durability or faster maintenance.

Floor polishes are subject to more mechanical abuse than other polish films and, therefore, should be resistant to abrasion, soiling, water, and detergents. Polymers, which generally are of higher molecular weight than natural waxes, can be formulated to provide films having these properties. In addition, the ability to tailor the properties of synthetic polymers has produced self-polishing aqueous formulations. These formulations result in the deposition of films having high gloss, low color, durability, little tendency to powder, and built-in mechanisms for easy removal.

Reviews of floor polish formulations are given in References 3, 13, 14, and 15. Aqueous, self-polishing, polymeric formulations generally contain two or three polymeric film formers, coalescing agents, leveling aids, plasticizers, zinc complexes, ammonia, and wetting and emulsifying agents. The polymer generally is a styrene–acrylic copolymer or an acrylic copolymer (14,15). The monomers used are styrene [100-42-5]; acrylic acid [79-10-7]; acrylic acid esters, eg, butyl acrylate [141-32-2]; methacrylic acid [79-41-4]; and methacrylic acid esters. Polyurethanes are used in some products as well (16). The polymer may be used in the formulation as an emulsion, but if sufficient acid monomer is incorporated and base is added, the emulsion particles swell and, in some cases, the polymer dissolves. This dissolved form is used in water-clear floor polishes (17,18). Polymer usually is the principal film former. Resins, which are low molecular weight and high acid content polymers, form clear dispersions of low viscosity at high solids content (14). They may be styrene–maleic anhydride copolymers, various styrene–acrylic copolymers, rosin maleates, or terpene–phenolics. In an emulsion product that otherwise might be low viscosity, resins are used to increase the application viscosity and thereby provide a thicker film.

Waxes are added to formulations in the form of emulsions. The trend of the latter twentieth century has been to use low molecular weight polyethylene, made somewhat polar by oxidation or by copolymerization with carboxylic acid monomers to make them readily emulsifiable (19). Waxes provide buffability to polish films and increase abrasion resistance. Plasticizers are used in emulsion products to lower the minimum film-forming temperature (MFT) below which the coating is applied. Below MFT, polymer emulsion particles do not coalesce to provide a coherent film. The plasticizer can be a phthalate ester, phosphate ester, glycol ether, or glycol (20). Phosphate esters, particularly tris(butoxyethyl)phosphate [78-51-3], are plasticizers but are more important as leveling agents (21). Glycol ethers are classed as coalescing agents and differ from plasticizers in that they are fugitive (22). These reduce MFT but evaporate when the film dries, yielding a hard film.

Zinc may be added as a fugitive ligand complex, eg, zinc ammonium carbonate (23). During film formation, the ligands evaporate and zinc complexes with carboxylate groups on the polymer, forming a cross-linked network (24). Zinc hardens the film and imparts detergent resistance to it. Nevertheless, the ammonia in a floor wax stripper can complex the zinc, thereby breaking the cross-link and allowing the polymer to redissolve (see ZINC COMPOUNDS). Alternatives to zinc complexes are also possible (25,26). Aqueous polymeric floor polishes contain the emulsifiers, wetting agents, and stabilizers needed to give a stable homogeneous product that wets out on the surface to which it is applied. Liquid, solvent-based products for wooden floors are solutions or dispersions of waxes in aliphatic hydrocarbons. Water and oil emulsion products for wooden floors also can be formulated using proper selection of emulsifiers and waxes (27). These are rapidly replacing solvent-based products.

Formulas for representative floor polishes are listed in References 3, 12, 13, and 25. An aqueous formula may contain 0–12 wt % polymer, 0–12 wt % resin, 0–6 wt % wax, 0.3–1.5 wt % tris(butoxyethyl)phosphate, 1–6 wt % glycol ether, and 0–1 wt % zinc, with water filling the rest. Water-clear floor finishes contain little or no wax, whereas buffable products contain relatively large amounts of wax. Sealers contain little wax and relatively large amounts of emulsion polymers (28). For industrial use, sealers are applied to porous substrates to fill the pores and prevent polishes that are used as topcoats from soaking into the floor.

One-step clean-and-shine products have become popular in the household market. These products are applied to the floor with a sponge mop and their detergent action removes and suspends soil, which collects on the mop and is removed when the mop is rinsed with water. The formulation, which remains on the floor, dries to a polish film. An earlier product of this type was dispensed from an aerosol as a foam. Formulas as of this writing (ca 1995) are applied as liquids (29,30). In one product, the dried film obtained from the formulation is soluble in the formulation, which includes low molecular weight, high acid polymers and a fairly large amount of ammonia (31). Repeated use does not contribute to a buildup of polish.

Industrial and institutional floor care demands polishes that accommodate the needs of machine-centered maintenance. The development of machines that buff or burnish with rotational speeds from 1000 to 2000 rpm has produced polish formulations that are balanced to accommodate the process (32–36).

Marble crystallizing does not involve a coating, rather it is a process whereby the stone gets a new surface. A heavy rotary machine is used with oxalic acid-based formulations to yield a highly reflective surface. The formulations can include hardening and curing agents that render the surface both harder and more stain-resistant (37). A machine-applied treatment for marble, wood, and stone floors involving silicone copolymers that penetrate and bond to the surface has been marketed (38).

Floor polishes typically are evaluated for gloss, application and leveling properties, discoloration, slip resistance, scratch resistance, heel-mark resistance, scuff resistance, damp-mopping and detergent resistance, repairability, lack of sediment, and removability (3). Recoatability and formula stability are also important. A review of test methods is available (35). More than 20 ASTM test methods for floor polishes exist. From the standpoint of product safety, slip resistance is a particularly important variable and many test methods are available (39).

## Automobiles

Automobile polishing is designed to remove road film and oxidized paint, and to lay down a continuous, glossy film, which resists removal by water and car-wash detergents (7,40–42). Much of the market is represented by one-step products which generally contain four functional ingredients. Abrasives are the principal cleaning ingredients but must not be so aggressive as to scratch the paint film. Representative types are fine grades of aluminum silicate, diatomaceous earth, and silicas (see ABRASIVES; DIATOMITE; SILICA; SILICON COMPOUNDS). Modern acrylic automobile paints, in contrast to older alkyd or nitrocellulose types, do not oxidize as rapidly and so reduce the need for highly abrasive polishes.

Straight- and branched-chain aliphatic hydrocarbons are used to facilitate detergency toward oil-based traffic soils, provide the solvency characteristics needed to produce a stable formula, and control the drying rate of the overall formulation. Other types of solvents, eg, aromatics, which attack paint films, are not employed. The solvent content of these polishes is generally being reduced (43). Waxes are used as one of the two principal film-forming ingredients. They are spread and leveled to produce a high luster by buffing. Blends of soft wax and hard wax, which provide ease of buffing and durability, respectively, generally are used. Many combinations have been used, including paraffin and microcrystalline petroleum waxes, carnauba and candelilla vegetable waxes, montan waxes derived from coal (qv), and synthetic polymer waxes, eg, oxidized polyethylene. Colored car waxes are liquids that hide minor scratches and enhance the color of the vehicle (44).

Silicones comprise the other type of film-forming ingredient. They also serve as lubricants for easy application and buffing and as release agents for dried abrasives. Silicones spread easily, providing a uniform high gloss as well as water repellency. Dimethylsilicone fluids are the most common, but amino-functional silicones are being used increasingly, particularly in premium quality products (45–47). Amino-functional silicones provide films with increased resistance to detergent removal because of the ability of the silicones to plate out on a paint surface and cross-link and bond to that surface. Furthermore, films containing amino-functional silicones in conjunction with corrosion inhibitors, eg, phenylacetic acid, inhibit surface corrosion (46). Acrylic polymers may also be used with the silicones (48). In addition to the functional ingredients, car polish formulas contain the emulsifiers, thickeners, and stabilizers needed to produce a homogeneous, stable product of the desired consistency.

Car polishes can be solid, semisolid, or liquid. They can be solvent-based or emulsions. In either case, liquid and solid forms are possible. Compilations of suggested formulas are given in References 3, 12, and 44. A representative liquid emulsion product may contain 10–15 wt % abrasive, 0–30 wt % solvent, 2–12 wt % silicone, and 0–4 wt % wax; an emulsion paste product may contain 3–15 wt %

wax with other ingredients at similar levels. Vinyl-top waxes and dressings contain no abrasives and some are similar to water-based floor polishes (49). Federal Specifications P-W-120C (1975) and P-P-546D (1975) describe laboratory tests for gloss, smear resistance, weatherability, cleaning ability, and stability. Some of the standard laboratory tests pertaining to silicone polishes are described in Reference 7. Methods for the evaluation of automotive polishes are described in ASTM D3836 and D4955. Other evaluations required in the formulating process are depth of gloss, ease of application and buffing, film stability, and resistance to water and detergent.

## Metal

In industrial metal finishing, polishing is an abrading operation involving the use of coarse abrasives, which remove significant amounts of metal from a surface and leave visible line patterns (50). Buffing is the smoothing of the resultant surface and involves the use of fine abrasives to reduce the dimensions of the polishing patterns. The only requirement for a metal surface to reflect light is that the surface roughness be small compared to the wavelength of incident light (51). Nonindustrial polishing does not remove large amounts of metal, but cleans and buffs to remove tarnish, oxides, and stubborn soils. The exposed metal is buffed to a high luster. Metal polishing is reviewed in Reference 3.

Formulated metal polishes consist of fine abrasives similar to those involved in industrial buffing operations, ie, pumice, tripoli, kaolin, rouge and crocus iron oxides, and lime. Other ingredients include surfactants (qv), eg, sodium oleate [143-19-1] or sodium dodecylbenzenesulfonate [25155-30-0]; chelating agents (qv), eg, citric acid [77-92-9]; and solvents, eg, alcohols or aliphatic hydrocarbons.

A problem associated with the use of abrasive metal polishes is that the fresh metal, which has been exposed by the cleaning, rapidly oxidizes or tarnishes. Thus, many modern polishes contain inhibitors. Sulfur compounds, eg, alkyl benzyl thiols, commonly are used, as are mercapto esters such as lauryl thioglycolate [3746-39-2] and dialkyldisulfides (52–54).

Metal polishes may contain emulsifiers and thickeners for controlling the consistency and stabilization of abrasive suspensions, and the product form can be solid, paste, or liquid. Liquid and paste products can be solvent or emulsion types; the market for the latter is growing. Formulas for metal polishes are listed in Reference 12. A representative liquid emulsion product may contain 8–25 wt % abrasive, 2–6 wt % surfactant, 0–5 wt % chelating agents, and 0–25 wt % solvent, with the remainder being water. The abrasive content in an emulsion paste product is greater than that in a solvent product.

Although abrasive polishing is the most common metal polishing operation, other forms of polishing and chemical brightening are used in industrial operations. Aluminum truck trailers often are cleaned and brightened by treatments with strong acids or alkalies, which chemically remove oxides. Chemical methods can also remove tarnish from other metals (55,56).

Federal specifications that pertain to metal polishes include P-P-566D (1964), metal polishes; A-A-105, brass polishes; P-P-580A (1975), silver polishes, cleaners, and tarnish preventers. Specification P-P-556D provides test methods for polishing capability, tarnish removal, metal weight loss resulting from polishing, chemical activity toward the metal, and gloss retention. The formulation process involves evaluation of ease of use, cleaning capability, product stability, and the efficacy of tarnish inhibitor.

## Shoe

Use of a shoe polish imparts high gloss, maintains the supple hand of the leather (qv), and increases the weather resistance of the leather (3,57–59). Three general types of polishes are produced: solvent pastes, self-polishing liquids, and emulsion creams. Solvent pastes represent ca 60% of the market (58). They are similar to the paste furniture and floor polishes, except that the former contain higher wax-to-solvent ratios and high levels of dye. The resulting film dries to a dull finish and must be buffed to a high shine. Liquid self-polishing products contain a soft polymer, which provides a more flexible coating, and coloring agents. The latter may be dyes or, in scuff-coat polishes, pigments. Shoe creams can be made in any consistency. They are emulsions of waxes, solvents, and water. Formulas for shoe polishes are listed in References 3 and 57. Silicone waxes can also be used in shoe polishes (60). The evaluation of shoe polishes is reviewed in Reference 3. The evaluation of paste shoe polishes is reviewed in Federal Specification P-P-557B (see also LEATHER; LEATHER-LIKE MATERIALS).

## Health, Safety, and Environmental Factors

Liquid polishes and waxes containing 10 wt % or more petroleum distillates must be contained in childproof packaging (61). General experience indicates that natural waxes and polyethylene waxes are nontoxic (62). Although nonsolvent floor polishes are relatively nontoxic, concern for floor waxes continues to be slip-resistance (63,64).

Under Section 183 of the Clean Air Act Amendments of 1990, the Environmental Protection Agency is required to study volatile organic compound (VOC) emissions from consumer and commercial products. The goal is to (1) determine the potential of the VOCs to contribute to ambient ozone levels; (2) establish criteria for regulating consumer and commercial products; and (3) submit a report of this study to the U.S. Congress (65). As of this writing, this report has not been submitted. Although it remains unclear what the exact nature of regulation will be, VOC reductions in polish formulations are underway. The State of California Air Resource Board has imposed limits on the content of volatile organic compounds in products sold in its jurisdiction (66). Much of the work described herein pertains to reductions in VOC content in polish formulations.

## Economic Factors

According to U.S. Census Data, the value of polishing preparations and related products shipped from U.S. factories in 1987 was $798 million in factory sales, a 23% increase from 1982 (67). These sales included $155.3 million in furniture polish, $245.2 million in floor polish, $185.1 million in automotive polish, $81.1 million in metal polish, and $42.8 million in shoe polish. Industry estimates for retail sales as of 1994 involving furniture polish were that this category remained flat at $197 million (68). Retail shoe polish sales remained constant at $200 million; household floor polish sales declined about 7% in 1993, whereas institutional floor wax and polish sales have grown at an annual rate of 2–2.5% to $370 million (69). In 1992 and 1993, automotive polish sales increased about 40% a year with the advent of colored car polishes to $240 million, but were flat in 1994.

The leading manufacturers in the polish categories are S. C. Johnson & Son, Inc., Reckitt & Coleman, Sara Lee, Scotts Liquid Gold, Turtle Wax, and Alberto Culver for furniture polishes; S. C. Johnson & Son, Inc. and L & F Products for household floor polishes; S. C. Johnson & Son, Inc., Pioneer-Eclipse, Spartan Chemical, Hillyard Chemical Co., and Butcher Co. for industrial and institutional floor polishes; Turtle Wax, Armor All, First Brands, Kit Products of Northern Labs, Meguires, Blue Coral, and Nu-Finish for automotive polishes; and Kiwi for shoe polishes (68–72).

## BIBLIOGRAPHY

"Polishes" in *ECT* 3rd ed., Vol. 18, pp. 321–328, by F. J. Randall and Seán G. Dwyer, S. C. Johnson & Son, Inc.

1. J. D. Bower, *Soap Cosmet. Chem. Spec.* **54**(5), 68 (1978).
2. L. Chalmers, *Manuf. Chem. Aerosol News*, **48**(3), 43 (1977).
3. W. J. Hackett, *Maintenance Chemical Specialties*, Chemical Publishing Co., New York, 1972.
4. N. F. Hadley, *Am. Sci.* **68**, 546 (1980).
5. Can. Pat. 1,223,296 (July 14, 1987), F. J. Steer and J. A. Ferguson (to Bristol-Meyers Co.).
6. J. C. Boomsma and E. E. Schaefer, *Chem. Times Trends*, **1**(2), 31 (1978).
7. K. A. Kasprzak, *Household Pers. Prod. Ind.* **15**(12), 67 (1978).
8. U.S. Pat. 4,936,914 (June 26, 1990), S. M. Hurley, E. J. Miller, and H. A. Rasoul (to S. C. Johnson Wax).
9. U.S. Pat. 4,163,673 (Aug. 7, 1979), R. S. Dechert (to S. C. Johnson & Son).
10. R. J. Thimineur, *Household Pers. Prod. Ind.* **16**(12), 82 (1979).
11. L. Kintish, *Soap Cosmet. Chem. Spec.* **67**(9), 58 (1991).
12. E. W. Flick, *Household and Automotive Chemical Specialties: Recent Formulations*, Noyes Data Corp., Park Ridge, N.J., 1979.
13. E. H. McMullen, *Adv. Chem. Ser.* **78**, 249 (1968).
14. G. Gregory, in J. Kroschwitz, ed., *Encyclopedia of Polymer Science and Technology*, Vol. 7, Wiley-Interscience, New York, 1987, p. 247.
15. H. W. Tiggmann, *Proceedings of the 60th Mid-Year Meeting of the Chemical Specialties Manufacturers Association*, Washington D.C., 1974, p. 174.
16. A. Tyskwicz and J. Tsirovasiles, *Soap Cosmet. Chem. Spec.* **63**(5), 37 (1987).
17. C. J. Verbrugge, *J. App. Poly. Sci.* **14**, 897 (1970).
18. *Ibid.*, p. 911.
19. E. H. Erenrich and P. G. McQuillan, *Chem. Times Trends*, **11**(1), 51 (1988).
20. R. S. Sweet, *Soap Chem. Spec.* **47**(5), 54 (1971).
21. T. R. Hopper, *Soap Chem. Spec.* **46**(10), 36 (1970).
22. S. L. Hillman, *Household Pers. Prod. Ind.* **28**(4), 96 (1991).
23. J. R. Rogers and F. J. Randall, *Am. Chem. Soc., Div. Polym. Chem.* **29**, 432 (1988).

24. U.S. Pat. 3,308,078 (Mar. 7, 1967), J. R. Rogers and L. M. Sesso (to S. C. Johnson & Son).
25. H. C. Carson, *Household Pers. Prod. Ind.* **27**(6), 78 (1990).
26. Eur. Pat. Appl. EP 438,216 (July 24, 1991), R. T. Gray, J. M. Owens, and H. S. Killam (Rohm and Haas Co.).
27. U.S. Pat. 4,055,433 (Oct. 25, 1977), R. Morones (to S. C. Johnson & Son).
28. N. Sood and I. G. Stewart, *Soap Cosmet. Chem. Spec.* **49**(10), 44 (1973).
29. Can. Pat. 871,086 (May 9, 1966), R. G. Harris and H. E. Mann (to S. C. Johnson & Son).
30. Can. Pat. 864,344 (May 9, 1966), R. G. Harris and H. E. Mann (to S. C. Johnson & Son).
31. U.S. Pat. 4,013,607 (Mar. 22, 1977), S. G. Dwyer and D. J. Hackbarth (to S. C. Johnson & Son).
32. J. Donohue, *Soap Cosmet. Chem. Spec.* **63**(7), 46 (1987).
33. J. Owens, *Household Pers. Prod. Ind.* **24**(6), 54 (1987).
34. C. R. Costin, *Soap Cosmet. Chem. Spec.* **61**(12), 46 (1985).
35. J. M. Owens, R. T. Cyrus, and C. C. Mateer, *Soap Cosmet. Chem. Spec.* **66**(6), 26 (1990).
36. H. C. Hamilton, *Household Pers. Prod. Ind.* **23**(9), 40 (1986).
37. U.S. Pat. 4,738,876 (Apr. 19, 1988), R. D. George, S. R. Pasupathikoil, and N. Stansfeld (to S. C. Johnson & Son).
38. U.S. Pat. 4,837,261 (June 6, 1989), P. Hampe and S. Kelbert (to Dyna-5, Inc.).
39. J. H. Merscher, *Chem. Times Trends*, **1**(1), 76 (1977).
40. Wax Department of American Hoechst, *Soap Cosmet. Chem. Spec.* **53**(11), 35 (1977).
41. *Chem. Week*, **123**(19), 23 (1978).
42. E. Jacobs, *Popular Sci.* **216**(4), 111 (1980).
43. R. J. Thimineur, *Household Pers. Prod. Ind.* **27**(9), 59 (1990).
44. H. C. Carson, *Household Pers. Prod. Ind.* **29**(9), 33 (1992).
45. U.S. Pat. 3,890,271 (June 17, 1975), J. G. Kakaszwa (to Dow Corning).
46. S. F. Hayes, *Soap Chem. Spec.* **47**(7), 55 (1971).
47. U.S. Pat. 3,960,575 (June 1, 1976), E. R. Martin (to SWS Silicones Corp.).
48. U.S. Pat. 4,347,333 (Aug. 31, 1982), R. H. Lohr and L. W. Morgan (to S. C. Johnson & Son).
49. R. J. Thimineur, *Household Pers. Prod. Ind.* **13**(9), 38 (1976).
50. H. L. Kellner, *Metal Finish. Guide. Directory*, **77**(13), 16 (1980).
51. L. E. Samuels, *Sci. Am.* **239**, 132 (1978).
52. U.S. Pat. 3,365,312 (Jan. 23, 1968), C. J. Nowack (to R. M. Hollingshead Corp.).
53. U.S. Pat. 3,330,672 (July 11, 1967), H. Kroll, A. R. Therrien, and P. W. Bennett (to Philip A. Hunt Corp.).
54. U.S. Pat. 3,503,883 (Mar. 31, 1970), I. A. M. Ford, B. C. Cox, and J. C. Thornton (to J. Goddard & Sons).
55. U.S. Pat. 4,640,713 (Feb. 3, 1987), R. B. Harris (to S. C. Johnson & Son).
56. U.S. Pat. 4,353,786 (Oct. 12, 1982), J. DeJager (to S. C. Johnson & Son).
57. L. Chalmers, *Manuf. Chem. Aerosol News*, **49**(7), 55 (1978).
58. *Chem. Week*, **123**(15), 43 (1978).
59. R. S. Sweet, *Soap Cosmet. Chem. Spec.* **53**(1), 48 (1977).
60. A. J. O'Lenick, Jr. and J. K. Parkinson, *Soap Cosmet. Chem. Spec.* **70**(8), 50 (1994).
61. *Code of Federal Regulations*, Title 16, Part 1700.14, Consumer Product Safety Commission, Washington, D.C., Jan. 1994, Chapt. 11.
62. J. Lange and J. Wildgruber, *Soap Cosmet. Chem. Spec.* **53**(7), 31 (1977).
63. R. J. Brungraber, *Technical Note 895*, Institute for Applied Technology, National Bureau of Standards, NBSTN-895, Washington, D.C., 1976.

64. J. Hermann, *Clean. Manage.* **73**(6), 29 (1990).
65. *Amendment to Clean Air Act*, Public Law 101-549, 101st Congress, 5.1630, Washington, D.C., Nov. 15, 1990.
66. *Soap Cosmet. Chem. Spec.* **67**(2), 10 (1991).
67. *1977 Census of Manufacturers, Soap, Cleaners, and Toilet Goods*, Bureau of the Census, U.S. Department of Commerce, MC87-1-28D, Washington, D.C., 1987.
68. J. Lawrence, *Advert. Age* **65**(12), 20 (1994).
69. T. Branna *Household Pers. Prod. Ind.* **31**(9), 61 (1994).
70. *Spray Tech. Market.* **4**(6), 16 (1994).
71. L. Kintish, *Soap Cosmet. Chem. Spec.* **70**(12), 44 (1994).
72. *1994 Waxes, Polishes, and Floor Finishes Survey by the Chemical Specialties Manufacturers Association*, CSMA, Washington, D.C.

FRANCIS J. RANDALL
SEÁN G. DWYER
S. C. Johnson & Son, Inc.

**POLLUTION.** See AIR POLLUTION; ENVIRONMENTAL IMPACT.

**POLYACETALS.** See ACETAL RESINS.

**POLYACRYLAMIDES.** See ACRYLAMIDE POLYMERS.

**POLYACRYLATES.** See ACRYLIC ESTER POLYMERS, SURVEY.

**POLYACRYLONITRILE.** See ACRYLONITRILE POLYMERS.

# POLYAMIDES

## GENERAL

Polyamides, often also referred to as nylons, are high polymers which contain the amide repeat linkage in the polymer backbone. They are generally characterized as tough, translucent, semicrystalline polymers that are moderately low cost and easily manipulated commercially by melt processing. However, significant exceptions to all these attributes occur. The regularity of the amide linkages along the polymer chain defines two classes of polyamides: AB and AABB.

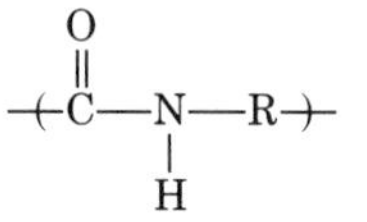

$$\begin{array}{cc}
\underset{\text{AB}}{+\!\!\overset{\displaystyle O}{\overset{\|}{C}}\!-\!\underset{\underset{H}{|}}{N}\!-\!R\!\!+} &
\underset{\text{AABB}}{+\!\!\overset{\displaystyle O}{\overset{\|}{C}}\!-\!R\!-\!\overset{\displaystyle O}{\overset{\|}{C}}\!-\!\underset{\underset{H}{|}}{N}\!-\!R'\!-\!\underset{\underset{H}{|}}{N}\!\!+}
\end{array}$$

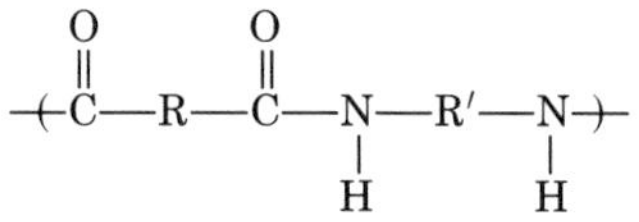

Type AB, which has all the amide linkages with the same orientation along the backbone, can be viewed as being formed in a polycondensation reaction from $\omega$-amino acids to give a polymer with the repeat unit AB. Type AABB, where the amide linkages alternate in orientation along the backbone, can be viewed as being formed from diacids and diamines in a polycondensation reaction to form a polymer with the repeat unit AABB. The R and R′ groups in these structures are hydrocarbon radicals and can be aliphatic, aromatic, or mixed. Because the chemical and physical properties of polyamides differ drastically between those which contain greater than ~15% aliphatic character, ie, nylons, and those which are predominately aromatic, ie, aramids, the aromatic polyamides are treated separately in this article.

### Nomenclature

The nomenclature (qv) of polyamides is fraught with a variety of systematic, semisystematic, and common naming systems used variously by different sources. In North America the common practice is to call type AB or type AABB polyamides nylon-x or nylon-x,x, respectively, where x refers to the number of carbon atoms between the amide nitrogens. For type AABB polyamides, the number of carbon atoms in the diamine is indicated first, followed by the number of carbon atoms in the diacid. For example, the polyamide formed from 6-aminohexanoic acid [60-32-2] is named nylon-6 [25038-54-4]; that formed from 1,6-hexanediamine [124-09-4] or hexamethylenediamine and dodecanedioic acid [693-23-2] is called nylon-6,12 [24936-74-1]. In Europe, the common practice is to use the designation "polyamide," often abbreviated PA, instead of "nylon" in the name. Thus, the two examples above become PA-6 and PA-6,12, respectively. PA is the International Union of Pure and Applied Chemistry (IUPAC) accepted abbreviation for polyamides. Occasionally abbreviations such as Ny-6 or Ny-6,12 are seen, but these are to be avoided. More complex organic radicals, eg, branched, alicyclic, or aromatic, that appear between the amide functions are generally designated by special abbreviations; many of the more common ones appear in Table 1. Copolymers are generally designated by writing the symbols for the two polymers separated by a slash, eg, nylon-6,6/6,T, with the component in the higher concentration listed first. This method is readily extended to polymers containing three or more components.

Another common naming scheme, termed source-based (or component monomer) nomenclature, is the practice of naming the polyamide after the monomer from which the polymer is made, eg, poly(caprolactam) for nylon-6. The use of poly(hexamethyleneadipamide [*9011-55-6*] for nylon-6,6 is a variation of this naming procedure, where the repeat unit is used instead of the specific starting materials. One of the significant disadvantages of source-based nomenclature is that since a given polymer can be prepared from different monomers, it can have several different names; for example, nylon-6 could also be named poly(6-aminohexanoic acid). Trivial as well as systematic names are used to designate various monomers or repeating structural portions of the polymer.

All of these methods of nomenclature are slowly falling out of use, in favor of the systematic nomenclature based on the constitutional repeat unit (CRU), advocated by the IUPAC (1,2) and used with modification as the structural repeat unit (SRU) in *Chemical Abstracts* to index polymers. In the IUPAC scheme polyamides are named as derivatives of repeating divalent, nitrogen-substituted (designated as imino-) carbon backbones. The carbonyl function is designated as a divalent oxygen (oxo-) substitution on the carbon backbone. Thus nylon-6 becomes poly[imino-(1-oxo-1,6-hexanediyl)] and nylon-6,6 becomes poly[imino(1,6-dioxo-hexanediyl)imino-hexanediyl]. The general template for naming type AB polymers is as follows: nylon-$x$ becomes poly[imimo(1-oxo-R)].

Because the rules for organic nomenclature determine the priority of naming different carbon chains from their relative lengths, the systematic names for type AABB polyamides depend on the relative length of the carbon chains between the amide nitrogens and the two carbonyl functions of the polymer: for aliphatic nylon-$x,y$, when $x < y$, the IUPAC name is poly[imino-R' imino(1,$y$-dioxo-R")]. When $x \geq y$, then the name is poly[imino(1,$y$-dioxo-R")imino-R']. Table 2 presents the alternative names of the hydrocarbon radicals for the various common polyamides (3). Although the systematic nomenclature appears complicated, and perhaps needlessly so for simple polyamides, it provides several advantages: (*1*) a unique, chemically specific name is assigned to each structure; (*2*) chemical derivatives of the linear polyamides are named in a straightforward and unequivocal manner from their linear analogues; and (*3*) the use of special, nonchemical designations is avoided. The IUPAC systematic nomenclature also provides appropriate methods for designating end groups, copolymers, and nonlinear polymers (1,4).

Tables 3 and 4 list the CAS Registry Numbers for all the common linear polyamides containing 1–12 carbon atoms in their backbone, except those derived from carbonic acid. Table 3 lists the CAS Registry Numbers for polyamides derived from diacids and diamines (or lactams); additional CAS Registry

**Table 1. Codes for Common Polyamide Monomers**

| Abbreviation or code | Chemical name | CAS Registry Number | Structure |
|---|---|---|---|
| I | isophthalic acid | [121-91-5] | 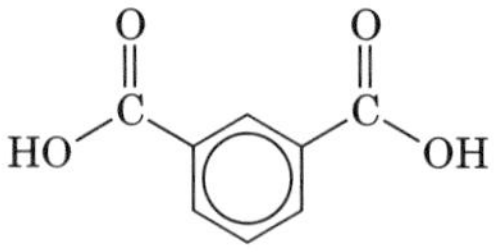 |
| T | terephthalic acid | [100-21-0] | 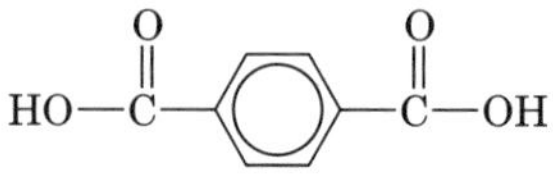 |
| N | 2,6-naphthalenedicarboxylic acid | [1141-38-4] | 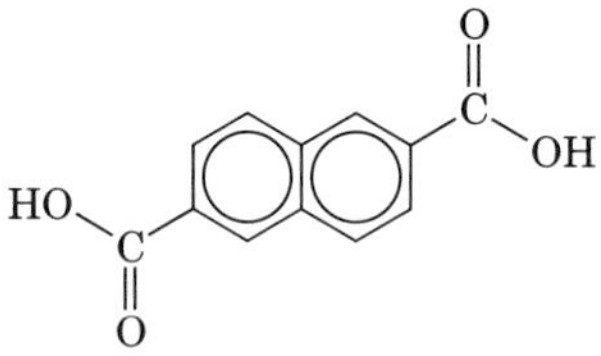 |
| MPD | m-phenylenediamine | [108-45-2] | 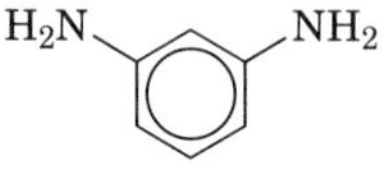 |
| PPD | p-phenylenediamine | [106-50-3] |  $H_2N$—⟨ ⟩—$NH_2$ |
| MXD | m-xylylenediamine | [1477-55-0] | 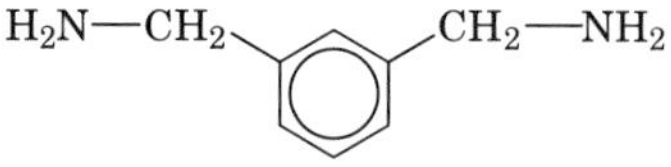 |
| Pip | piperazine | [110-85-0] |  HN ⟨ ⟩ NH |
| MPMD or D | 2-methylpentamethylenediamine or Dytek A[a] | [15520-10-2] | 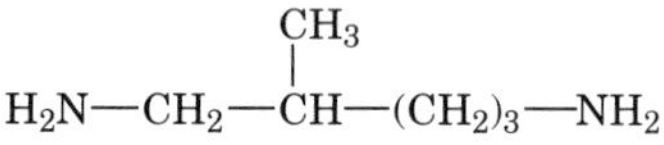 |
| TMD or 6-3 | mixture of 2,2,4- and 2,4,4-trimethylhexamethylenediamine | [3236-53-1] [3236-54-2] | 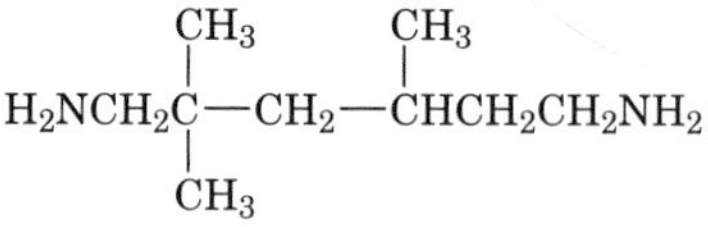 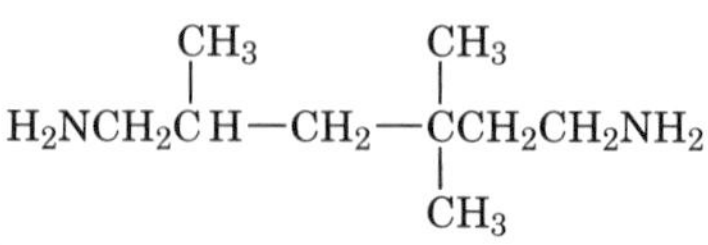 |

[a] Registered trademark of Du Pont Co.

Numbers may be available for these polymers if they are formed from other monomers such as the reaction of diamines and diacid chlorides (or ω-aminoacids). Table 4 lists the CAS Registry Numbers for the same set of polyamides based on their constitutional repeat unit. To facilitate compari-

**Table 2. Systematic Names for Radicals in the SRU Format Appearing in Chemical Abstracts and Other Sources**

| | | Associated systematic name[a] | | |
|---|---|---|---|---|
| x or y[b] | R, R', or R''[b] | ω-Amino acid replaces 1-oxo-R | Diamine radical replaces R' | Diacid radical replaces 1,y-dioxo-R'' |
| 1 | 1,1-methanediyl | 1-oxomethylene | methylene | [c] |
| 2 | 1,2-ethanediyl | 1-oxoethylene | ethylene | oxalyl |
| 3 | 1,3-propanediyl | 1-oxotrimethylene | propylene, trimethylene | malonyl[d] |
| 4 | 1,4-butanediyl | 1-oxotetramethylene | tetramethylene | succinyl |
| 5 | 1,5-pentanediyl | 1-oxopentamethylene | pentamethylene | glutaryl |
| 6 | 1,6-hexanediyl | 1-oxohexamethylene | hexamethylene | adipoyl |
| 7 | 1,7-heptanediyl | 1-oxoheptamethylene | heptamethylene | heptanedioyl, pimeloyl |
| 8 | 1,8-octanediyl | 1-oxooctamethylene | octamethylene | octanedioyl, suberoyl |
| 9 | 1,9-nonanediyl | 1-oxononamethylene | nonamethylene | nonanedioyl, azelaoyl |
| 10 | 1,10-decanediyl | 1-oxodecamethylene | decamethylene | decanedioyl, sebacoyl |
| 11 | 1,11-undecanediyl | 1-oxoundecamethylene | undecamethylene | undecanedioyl |
| 12 | 1,12-dodecanediyl | 1-oxododecamethylene | dodecamethylene | dodecanedioyl |
| 13 | 1,13-tridecanediyl | 1-oxotridecamethylene | tridecamethylene | tridecanedioyl |

[a]Not used in *Chemical Abstracts*, but appear in other references, eg, Ref. 3.
[b]In poly[imino(1-oxo-R)], poly[imino-R' imino(1,y-dioxo-R'')], and poly[imino-(1,y-dioxo-R'')imino-R'], as explained in text.
[c]Polymers based on carbonic acid and diamines are considered polyureas.
[d]The systematic name of a small radical can replace the trivial name, eg, propanedioyl for malonyl.

son, the entries are tabulated using their source-based nomenclature even in the case of the constitutional repeat unit (CRU), ie, Table 4. These tables illustrate several important points regarding use of CAS Registry Numbers to identify polymeric materials in general and polyamides in particular. First, there are at least two sets of CAS Registry Numbers for many polymers, especially polyamides; one corresponds to the source-based approach and the other to the CRU approach to the identification of the material. Secondly, depending on the identity of the starting monomer(s), there can be multiple source-based Registry Numbers. Thirdly, in searches of databases (qv) containing Registry Numbers, it should not be assumed that the various substance records are cross-referenced. Thus, if a thorough search is desired for a particular polymer, multiple Registry Numbers must generally be used. Finally, the situation can be made even more difficult in searches for information on copolymers since they have additional complexities resulting from the variety of combinations available for several monomers, and they are generally indexed in a source-based format.

## History

The first patent for the production of synthetic polyamides was issued in 1937 to Wallace H. Carothers, who was working at Du Pont Company (5). His pioneering

# Table 3. CAS Registry Numbers for Source-Based Monomers

| | $C_1$ | $C_2$ | $C_3$ | $C_4$ | $C_5$ | $C_6$ |
|---|---|---|---|---|---|---|
| **Type AB** | Methylene, $CH_6N_2$ | Ethylene, $C_2H_8N_2$ [89394-22-9] | Trimethylene, $C_3H_{10}N_2$ [27881-47-6] | Tetramethylene, $C_4H_{12}N_2$ [24968-97-6] | Pentamethylene, $C_5H_{14}N_2$ [25036-00-4] | Hexamethylene, $C_6H_{16}N_2$ [9012-16-2] |
| **Type AABB** | | | | | | |
| oxalic, $C_2H_2O_4$ | | | | | | [29322-30-3] |
| malonic, $C_3H_4O_4$ | [138251-80-6] | | | [149508-42-9] | [149508-43-0] | [36863-61-3] |
| succinic, $C_4H_6O_4$ | [138251-81-7] | [65595-82-6] | | [26098-28-2] | | [34853-53-7] |
| glutaric, $C_5H_8O_4$ | [41510-76-3] | | [56467-27-7] | | [31781-00-7] | [54215-29-1] |
| adipic, $C_6H_{10}O_4$ | [63289-13-4] | [26936-86-7] | [51025-67-3] | [50327-77-0] | [41510-67-2] | [9011-55-6] [75361-24-9] |
| pimelic, $C_7H_{12}O_4$ | | | | | [31781-01-8] | [27136-66-9] |
| suberic, $C_8H_{14}O_4$ | | [65595-76-8] | | [25916-16-9] | | [25776-74-3] |
| azelaic, $C_9H_{16}O_4$ | | [28551-99-7] | | [27136-64-7] | | [27136-65-8] |
| sebacic, $C_{10}H_{18}O_4$ | | [30585-15-0] | [138260-67-0] | [25776-75-4] | [111519-63-2] | [9011-52-3] |
| undecanedoic, $C_{11}H_{20}O_4$ | | | | | | [50733-20-5] |
| dodecanedoic, $C_{12}H_{22}O_4$ | [128743-02-2] | [99207-47-3] | | [26834-07-1] | | [26098-55-5] |
| tridecanedoic (brassylic), $C_{13}H_{24}O_4$ | | | | | | [26096-69-0] [69665-23-2] |
| isophthalic, $C_8H_6O_4$ | [30585-16-1] | [30585-16-1] | [130934-29-1] | [35885-42-8] | [130491-44-0] | [25722-07-0] |
| terephthalic, $C_8H_6O_4$ | | [29319-66-2] | [103691-99-2] | [73276-93-4] | [32761-06-1] | [24938-03-2] |

**Table 3.** (*Continued*)

| Type AB / Type AABB | C7 [25588-36-7] Heptamethylene, $C_7H_{18}N_2$ | C8 [25190-92-5] Octamethylene, $C_8H_{20}N_2$ | C9 [26779-81-7] Nonamethylene, $C_9H_{22}N_2$ | C10 [26710-03-2] Decamethylene, $C_{10}H_{24}N_2$ | C11 [25747-65-3] Undecamethylene, $C_{11}H_{26}N_2$ | C12 [25038-74-8] Dodecamethylene, $C_{12}H_{28}N_2$ |
|---|---|---|---|---|---|---|
| oxalic, $C_2H_2O_4$ | | [26763-17-7] | | [31156-16-8] | | [31778-88-8] |
| malonic, $C_3H_4O_4$ | | [149508-44-1] | | | | |
| succinic, $C_4H_6O_4$ | | [65595-77-9] | | [65595-78-0] | | |
| glutaric, $C_5H_8O_4$ | | | | | | |
| adipic, $C_6H_{10}O_4$ | [136065-07-1] | [26123-26-2] | [30848-98-7] | [26123-27-3] | [69175-20-8] | [36786-01-3] [28567-50-2] |
| pimelic, $C_7H_{12}O_4$ | [31781-02-9] | | [31781-03-0] | | | |
| suberic, $C_8H_{14}O_4$ | | [33180-14-2] | | | | |
| azelaic, $C_9H_{16}O_4$ | | | [31781-04-1] | [98260-90-3] | | |
| sebacic, $C_{10}H_{18}O_4$ | | [25988-33-4] | | [27815-37-8] | | [36683-34-8] |
| undecanedoic, $C_{11}H_{20}O_4$ | | | | | | |
| dodecanedoic, $C_{12}H_{22}O_4$ | | [55426-10-3] | [55426-11-4] | [55426-12-5] | [72928-94-0] | [36497-34-4] |
| tridecanedoic (brassylic), $C_{13}H_{24}O_4$ | | [69665-22-1] | | [60806-36-2] [69665-20-9] | | [69665-21-0] |
| isophthalic, $C_8H_6O_4$ | [29294-22-2] | [25988-35-6] | [130491-35-9] | [35885-43-9] | | [27515-23-7] |
| terephthalic, $C_8H_6O_4$ | [31017-99-9] | [73276-94-5] | [24938-10-1] | [24938-11-0] | [116220-46-3] [52733-85-4] | [26098-36-2] |

459

**Table 4. CAS Registry Numbers for Constitutional Repeat Unit**

| Type AB / Type AABB | $C_1$ [28725-43-1] Methylene, $CH_6N_2$ | $C_2$ [25734-27-4] Ethylene, $C_2H_8N_2$ | $C_3$ [24937-14-2] Trimethylene, $C_3H_{10}N_2$ | $C_4$ [24938-56-5] Tetramethylene, $C_4H_{12}N_2$ | $C_5$ [24938-57-6] Pentamethylene, $C_5H_{14}N_2$ | $C_6$ [25038-54-4] Hexamethylene, $C_6H_{16}N_2$ |
|---|---|---|---|---|---|---|
| oxalic, $C_2H_2O_4$ | [51735-52-5] | [36812-63-2] | [88007-14-1] | [91277-46-2] | | [31694-58-3] |
| malonic, $C_3H_4O_4$ | [108390-54-1] | [113040-39-4] | [41706-21-2] | [113040-38-3] | [113040-25-8] | [32027-68-2] |
| succinic, $C_4H_6O_4$ | [109672-46-0] | [27496-28-2] | [68821-49-8] | [26402-74-4] | | [24936-71-8] |
| glutaric, $C_5H_8O_4$ | [41724-64-5] | [68821-50-1] | [56467-06-2] | [50327-22-5] | [32473-29-3] | [33638-80-1] |
| adipic, $C_6H_{10}O_4$ | [138255-69-3] | [26951-61-1] | [51344-88-8] | | [41724-56-5] | [32131-17-2] |
| pimelic, $C_7H_{12}O_4$ | | | | | [32552-84-4] | [28757-64-4] |
| suberic, $C_8H_{14}O_4$ | [67937-41-1] | [41724-62-3] | | [26247-04-1] | | [24936-73-0] |
| azelaic, $C_9H_{16}O_4$ | [67937-40-0] | [28775-08-8] | | [28757-62-2] | | [28757-63-3] |
| sebacic, $C_{10}H_{18}O_4$ | [67937-42-2] | [32126-82-2] | [105063-18-1] | [26247-06-3] | [105063-19-2] | [9008-66-6] |
| undecanedoic, $C_{11}H_{20}O_4$ | | | | | | [50732-66-6] |
| dodecanedoic, $C_{12}H_{22}O_4$ | [128853-03-2] | [41724-60-1] | | [26969-09-5] | | [24936-74-1] |
| tridecanedoic (brassylic), $C_{13}H_{24}O_4$ | | | | | | [26916-49-4] |
| isophthalic, $C_8H_6O_4$ | [67937-43-3] | [32126-83-3] | [119495-33-9] | [28757-32-6] | [119495-34-0] | [25668-34-2] |
| terephthalic, $C_8H_6O_4$ | | [25722-35-4] | [35483-54-6] | [35483-53-5] | [32985-25-4] | [24938-70-3] |

**Table 4. (Continued)**

| Type AABB | Type AB: C7 [25035-01-2] Heptamethylene, C7H18N2 | C8 [25035-02-3] Octamethylene, C8H20N2 | C9 [25035-03-4] Nonamethylene, C9H22N2 | C10 [26970-31-0] Decamethylene, C10H24N2 | C11 [25035-04-5] Undecamethylene, C11H26N2 | C12 [24937-16-4] Dodecamethylene, C12H28N2 |
|---|---|---|---|---|---|---|
| oxalic, C2H2O4 | | [26854-69-3] | [36906-43-1] | [27635-11-6] | | [33182-38-6] |
| malonic, C3H4O4 | [121778-65-2] | [149474-23-7] | | [113040-26-9] | | [113040-27-0] |
| succinic, C4H6O4 | | [65586-26-7] | | [65586-27-8] | | [41724-54-3] |
| glutaric, C5H8O4 | | [77402-35-8] | | | | |
| adipic, C6H10O4 | [79569-15-6] | [26468-37-1] | [32131-10-5] | [26247-49-4] | [69221-92-7] | [36812-76-7] |
| pimelic, C7H12O4 | [32473-30-6] | | [32473-31-7] | | | |
| suberic, C8H14O4 | | [33182-53-5] | [33182-52-4] | [127340-03-8] | | [53301-89-6] |
| azelaic, C9H16O4 | | [55671-57-3] | | [98241-67-9] | | |
| sebacic, C10H18O4 | [105063-20-5] | [25950-51-0] | [51345-09-6] | [28774-87-0] | | [36731-92-7] |
| undecanedoic, C11H20O4 | | | | | | |
| dodecanedoic, C12H22O4 | | [55426-62-5] | [55426-61-4] | [55426-60-3] | [87827-35-8] | [36348-71-7] |
| tridecanedoic (brassylic), C13H24O4 | | [69662-04-0] | | [60806-42-0] | | [69662-05-1] |
| isophthalic, C8H6O4 | [32008-68-7] | [25950-52-1] | [119495-35-1] | [28883-72-9] | | [26401-02-5] |
| terephthalic, C8H6O4 | [32110-14-8] | [55445-81-3] | [24938-73-6] | [24938-74-7] | [52734-79-9] | [26009-07-4] |

461

work in the development of polymeric materials led in a few years to the commercialization of nylon-6,6 as the first synthetic fiber. In 1941 P. Schlack at I. G. Farbenindustrie in Germany was issued a patent for nylon-6 based on the polymerization of caprolactam [105-60-2] (6). Ironically, Carothers' first attempt to synthesize polyamides in 1930 was to make nylon-6 from 6-aminohexanoic acid, but for unexplained reasons he was only able to produce a low molecular weight polymer (7). At this time he and his co-workers also made other polyamides from dibasic acids and aliphatic diamines (8); however, owing to their low solubility and high melting point, this work was also abandoned for the next five years while they worked on other polymers, including neoprene and polyesters. In July 1935, nylon-6,6 was chosen by Du Pont to be the specific polyamide for commercial introduction. This choice was based on its balance of physical properties making it suitable for fiber production and the potential for a low cost source of starting materials from six-member ring carbon compounds derived from coal (qv) (9). In less than one year, Du Pont scientists and engineers built the first commercial plant in Seaford, Delaware, which began production in 1939.

The magnitude of the intellectual achievement of Carothers often overshadows the tremendous effort and success that followed in building the necessary industrial infrastructure and developing the numerous scientific and engineering innovations required to make nylon a successful commercial venture. One of the first of these was the development of a route to produce the starting materials from "coal, air, and water," and the first intermediates plant was built at Belle, West Virginia (10). Another was the invention of the autoclave polymerization process using balanced salt and acetic acid end termination to control the molecular weight of the final polymer. Because nylon-6,6 was insoluble in all common solvents, a new melt-spinning process was required to form fibers and wind them onto packages. Also, the two-step drawing process was invented to develop the full strength of the fibers. Additional inventions were required for effective downstream processing of this new synthetic fiber in order to dye and form it into finished goods. Finally, strong markets were required to support the financial investment necessary for this revolutionary product; fortunately, nylon was extremely well suited to compete in the high value silk markets.

Shortly after the commercial introduction of nylon, World War II began and most of the nylon produced was used for military purposes such as in ropes, parachutes, and tires. After the war, nylon production expanded rapidly, first into the apparel and tire markets, and then into carpets and plastic parts. During the following three decades (1950–1980), numerous technical developments were achieved to provide increased nylon capacity and cost effectiveness. These include the development of intermediates production from petroleum-based feedstocks; the invention of the continuous polymerization (CP) and solid-phase polymerization (SPP) processes; coupled draw-spinning and the invention of interlace to replace twisting of the fiber bundle; and the invention of numerous additives to improve the performance of nylon in special end uses, such as thermal and photostabilizers and rubber tougheners. This development of technology continues where high productivity–low cost, intensely competitive worldwide markets, and environmental friendliness are key factors driving the development of the polymer industry.

## Economic Aspects

Since its introduction in the 1940s as the polymer base for the first synthetic fiber, nylon has been a significant commercial polymeric material. Table 5 shows the relative percentage of the world's consumption for the principal polymers. Nylon represents approximately 4% of the total, $4.5 \times 10^6$ metric tons in 1993, of which nylon-6,6 and nylon-6 comprise 85–90%. These two polyamides are produced worldwide in the following approximate percentages: nylon-6, 60%, and nylon-6,6, 40%; their relative proportions vary among the geographic markets globally. In North America, nylon-6,6 dominates; in the Asia/Pacific area, nylon-6 is the predominant polyamide; and in Europe, the market is about equally split. The rate of growth for polyamides in the developed economies of North America and Europe is roughly equal to the overall growth rate of these economies; however, the recent rapid growth in the Asia/Pacific area is expected to continue. South America and Africa represent nascent markets with growth potential in the twenty-first century.

**Table 5. Percentage of World Consumption of Polymers by Type in 1990[a]**

| Type | Percentage |
| --- | --- |
| polyethylene | 27 |
| poly(vinyl chloride) | 18 |
| polypropylene | 13 |
| polyester | 11 |
| thermosets | 10 |
| polystyrene | 9 |
| acrylics | 4 |
| nylon | 4 |
| ABS/SAN | 3 |
| other | 1 |

[a]Calculated on a $102 \times 10^6$ metric ton basis for total production.

Table 6 shows the principal end uses for polyamides and the approximate relative percent of the total consumption, where the "other" category includes monofilaments and nonwovens. Synthetic fiber, comprising almost 75% of the

**Table 6. Percentage of World Consumption of Polyamides by End Use**

| End use or type | Percentage |
| --- | --- |
| fibers | |
|   carpet | 34 |
|   textile | 27 |
|   industrial | 13 |
| engineering resins | 15 |
| aramid | 4 |
| films/coatings | 4 |
| adhesives | 1 |
| other | 2 |

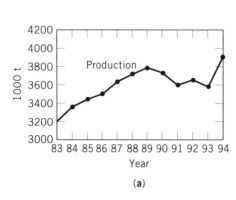
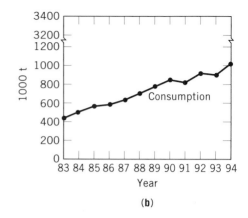

**Fig. 1.** (**a**) World production of nylon fiber (11); (**b**) world consumption of nylon resins (12).

total, still represents the principal end use for nylon polymer. As shown in Figure 1**a**, the total of nylon fiber production has declined between 1989 and 1993 (11), primarily as a result of the slowing of the world economy but also as a result of price competition from polyester and polypropylene fibers in many of its less demanding applications. In 1994, production, driven by the rebounding economies worldwide, took a sharp upturn. The long-term outlook for nylon fiber is expected to brighten as a result of the growth of the world economy and the substantial potential of the markets in Asia, especially in India and China (13).

The markets for nylon resins, on the other hand, have shown consistent growth since the mid-1980s (13), with only minor fluctuations owing to the slowing of the world's economies (Fig. 1**b**). As a result of the differences in relative growth rates of nylon plastics and fiber, production of the former as a percent of the total production is expected to continue to increase well into the twenty-first century. The position of polyamides as the lowest cost engineering resins or the highest cost (but arguably the highest performing) commodity resin has led to this continuous growth. Although the nylon resin markets, like those for fibers, are dominated by nylon-6,6 and nylon-6, there are many alternative polyamides and copolyamides in this end use. Table 7 gives the price range of

**Table 7. Range of Prices for Various Polyamides and Their Primary Product Types**

| Polymer base and product type | Price/kg |
|---|---|
| nylon-6,6 and nylon-6 | |
|     fiber | $1.98–$5.50 |
|     resin | $2.97–$4.40 |
| other nylons | |
|     resin | $4.95–$8.80 |
| *meta*-aramids | |
|     fiber and paper | $19.80–$66.00 |
| *para*-aramids | |
|     fiber and pulp | $22.00–$66.00 |
| other polyamides | |
|     adhesives and coatings | $0.66–$3.85 |

many polyamides for various product types. Specialty products and applications can command prices two times higher or greater. In addition to References 11 and 12, price and volume information concerning polyamides can be found in various public sources (14) and private databases that require a client fee for access (15).

## Physical Properties

**Crystallinity.** Linear polyamide homopolymers consist of crystalline and amorphous phases and are termed semicrystalline. Crystallinity enhances yield strength, hardness (qv), abrasion resistance, tensile strength, elastic and shear modulus, and probably resistance to thermooxidation (16), but it decreases moisture absorption and impact strength. Most commercial samples of nylon-6,6 and nylon-6 are 40–50% crystalline by weight, as determined by density measurements. A low degree of crystallinity can be achieved in these polymers by rapidly quenching them below room temperature from the melt, but this state is unstable and the sample quickly crystallizes if it is warmed, subjected to mechanical stress such as drawing, or exposed to moisture or to other plasticizers (qv). A permanent reduction in the degree of crystallinity can be achieved by chemical modification, eg, through the use of unsymmetrical monomers, copolymers, or substitution at the amide nitrogen, but then most of the desirable physical properties are lost. A few properties are improved, however, such as film clarity.

Figure 2 shows the unit cell for nylon-6,6, and Table 8 presents the crystallographic constants for several polyamides. The semicrystalline nature of

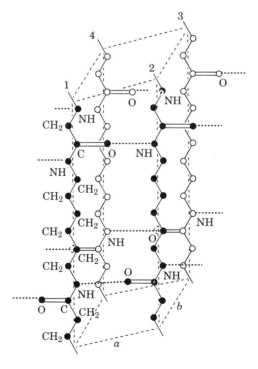

**Fig. 2.** Unit cell for the $\alpha$-triclinic form of nylon-6,6 (17).

**Table 8. Lattice Constants for Some Polyamides**[a]

| Polyamide | Crystal system | Lengths, nm | | | Angles, ° | | | $Z$[b] |
|---|---|---|---|---|---|---|---|---|
| | | $a$ | $b$ | $c$ | $\alpha$ | $\beta$ | $\gamma$ | |
| nylon-6,6 | $\alpha$-triclinic | 0.49 | 0.54 | 1.72 | 48.5 | 77 | 63.5 | 1 |
| nylon-6 | $\alpha$-monoclinic | 0.956 | 0.801 | 1.724 | 90 | 90 | 67.5 | 8 |
| | $\gamma$-monoclinic | 0.914 | 0.484 | 1.668 | 90 | 90 | 121 | 4 |
| nylon-6,10 | $\alpha$-triclinic | 0.495 | 0.54 | 2.24 | 49 | 76.5 | 63.5 | 1 |
| nylon-12 | $\alpha$-monoclinic | 0.479 | 3.19 | 0.958 | 90 | 120 | 90 | 4 |

[a]Refs. 17–19.
[b]$Z$ = number of chemical repeat units per unit cell.

polyamides and their high melting point are generally attributed to the high degree of hydrogen bonding between adjacent chains in the crystals. The apparent crystal size in polyamides has been determined by wide-angle x-ray scattering (waxs) (20) and small-angle x-ray scattering (saxs) (21). The average size depends on the mechanical and thermal history of the polymer, and in commercial samples it is typically 5–7 nm on each side, but the size distribution is very broad. Chain-folded lamellar, single crystals of polyamides have been formed from dilute solution (22) and from the melt (23). It is generally accepted that such lamellar structures are present as crystallites in isotropic samples (24), though extended chain crystals may be formed in highly oriented systems such as fibers. The crystallites can form ordered three-dimensional superstructures called spherulites which are detectable by optical microscopy using polarized light. In drawn fibers and films the crystallites as well as the polymer chains in the amorphous regions are preferentially oriented in the direction of the applied strain. The large-scale structure in bulk polyamides is usually unoriented and spherulitic, though some orientation can occur at the surface and elsewhere in the polymer where stresses were induced by flow during melt processing such as injection molding. A comprehensive review of crystallinity and structure in polyamides has been published (25), and extensive crystallographic data are also available (26).

**Solubility.** In general, the homopolymer aliphatic polyamides are insoluble in common organic solvents at room temperature. However, they are soluble in formic acid [64-18-6], phenols, chloral hydrate, minerals acids, and fluorinated alcohols such as 1,1,1,3,3,3-hexafluoroisopropyl alcohol [920-66-1] (HFIP) and 2,2,2-trifluoroethanol [75-89-8] (TFE) (27). Mixtures of TFE and methylene chloride in a 3:1 ratio are recommended as effective nmr solvents (28). Formic acid in 90% concentration, 96% sulfuric acid [7664-93-9], m-cresol [108-39-4], and HFIP are commonly used as solvents for relative viscosity measurements, and HFIP has been recommended as the preferred solvent for size exclusion chromatography (29). Also, HFIP is an excellent solvent for uv-visible spectroscopy, because it has virtually no absorption in the wavelength range 200–900 nm. At higher temperatures, lithium or calcium chloride–methanol mixtures are effective solvents, as are benzyl alcohol [100-51-6], unsaturated alcohols, alcohol–halogenated hydrocarbons, and nitro alcohols (30). Copolymers of aliphatic polyamides and polyamides with substitution on the amide nitrogen, both of which significantly reduce the degree of crystallinity, are more soluble, and methanol–chloroform mixtures can often be used. Predominately or wholly aromatic polyamides require powerful solvents, such as trifluoroacetic acid [76-05-1] (TFA) or concentrated sulfuric acid. Lower molecular weight aromatic polyamides are often soluble in basic solvents such as N,N-dimethylacetamide [127-19-5] and N-methyl-2-pyrrolidinone [872-50-4], usually with the addition of lithium chloride [7447-41-8] or calcium chloride [10043-52-4]. When polyamides are used in solution to characterize the bulk polymer, such as relative viscosity or spectroscopy, care must be taken to ensure that the very high molecular weight portion of the linear distribution or branched polymers are not left behind as small amounts of insoluble material.

**Piezoelectric Effect.** The electrical properties of piezoelectricity, ie, the ability to generate an electrical signal in response to a mechanical stress; pyroelectricity, ie, the ability to generate an electrical signal in response to a temperature change; and ferroelectricity, ie, the ability to respond repeatedly to reversing external electric fields, have been recognized in polymers for many years (31). Materials with these properties find application in microphones, tone generators, hydrophones, ir detectors, electromechanical transducers, and in numerous other devices. The odd-numbered nylons possess a strong piezoelectric effect, with their piezoelectric strain and stress coefficients, $d_{31}$ and $e_{31}$, being second in magnitude for polymeric materials only to those of poly(vinylidene fluoride) [24937-79-9] and its copolymers (32). The phenomenon has been observed in nylon-11 [25035-04-5], nylon-9 [25035-03-4], nylon-7 [25035-01-2], and nylon-5 [24138-57-6] (33), and is believed to result in part from the alignment of the high dipole moment of the amide group in the periodic array of the crystalline regions of the polymer during the process called poling; the ordered array is stabilized by hydrogen bonding between adjacent amide groups. During poling, a sample of the polymer, positioned between the plates of a capacitor, is subjected to a high electric field of up to 900 kV/cm. Heating the polymer above the glass-transition point, $T_g$, and allowing it to cool while the electric field is maintained increase the effectiveness of the poling process. Figure 3**a** shows how the amide groups in a crystal of nylon-7 are aligned in an all-trans conformation to reinforce their dipole moments perpendicular to the chain axis; this arrangement produces a microscopic polarization within the crystal. These microscopic crystalline domains must then be aligned and maintained in a mutual orientation, on average, within the polymer sample during poling in order for the macroscopic polarization to be achieved. The dipole moments are also reinforced in odd–odd nylons, as shown in Figure 3**b**. The $\alpha$-crystalline form has been associated with the development of piezoelectric properties in nylon-7 by means of [13]C and [15]N solid-state nmr (34). However, waxs studies associate the piezoelectric effects with crystal orientation (35). In nylons containing even-numbered segments, the amide groups alternate direction periodically within the crystal; thus this arrangement mutually cancels the effect of their dipole moments at the microscopic level, as shown in Figure 3**c**.

The odd-numbered nylons exhibit an additional useful property in that their piezoelectric constants increase with increasing temperature almost to the melting point of the polymer. This increase is attributed to the persistence of the

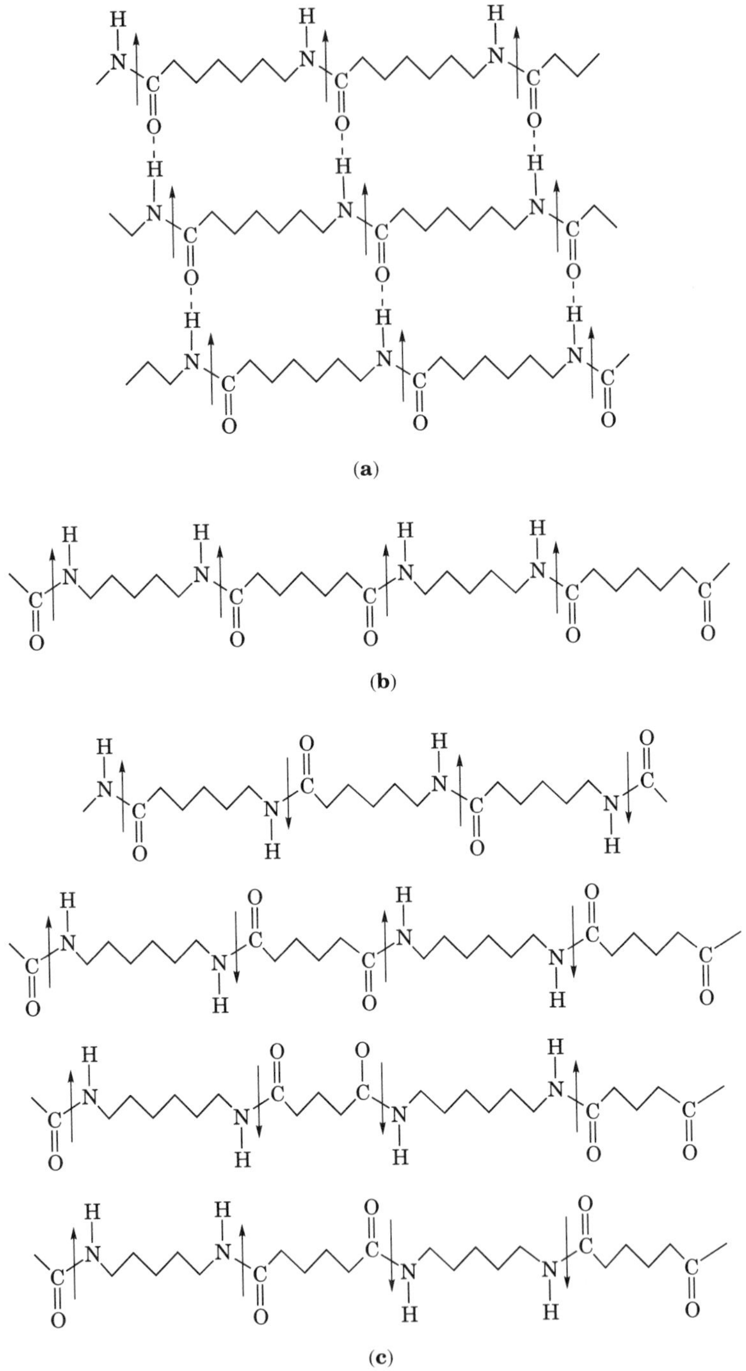

**Fig. 3.** Alignment of amide dipoles in polyamide crystals: (**a**) for a two-dimensional array of an odd nylon, nylon-7, (**b**) for a one-dimensional array of an odd–odd nylon, nylon-5,7; (**c**) for one-dimensional arrays of polyamides containing even segments: an even nylon, nylon-6; an even–even nylon, nylon-6,6; an even–odd nylon, nylon-6,5; and an odd–even nylon, nylon-5,6 (31).

hydrogen bonding in polyamides up to the melting point. This property of odd-numbered nylons is in contrast to that of poly(vinylidene fluoride), whose piezo-electric constants remain approximately unchanged with increasing temperature (36) (Fig. 4). Thus nylon could find useful application as a high temperature polymer piezoelectric material. Bilaminate films of nylon-11 and poly(vinylidene fluoride) have been prepared and appear to show enhancements of properties above those exhibited when either polymer is used individually (37).

Organic plasticizers (38) and water (39), also a plasticizer for nylons, affect the development of piezoelectric properties in nylon-11 and nylon-7. If these are applied after the samples are poled, the piezoelectric coefficients show an increase with increased plasticization. This effect is interpreted as showing that the alignment of dipoles in the amorphous region also plays a significant role in the bulk polarization. Because both the melting point and weight percent of moisture regain increase as the number of amide bonds increases, nylon-3 [25513-34-2] would be expected to demonstrate superior piezoelectric performance. To support this, the remanent polarization, ie, the polarization that remains after the sample has been poled and the applied field removed, was determined for several nylons and appears to increase linearly as the number of carbon atoms decreases from 55 mC/m$^2$ for nylon-11 to 135 mC/m$^2$ for nylon-5. This effect has been extrapolated to approximately 180 mC/m$^2$ for nylon-3 (40) (Fig. 5). Similarly, ferroelectric behavior has been observed for partially fluorinated odd–odd nylons, eg, nylon-3,5, nylon-5,5, nylon-7,5, and nylon-9,5, based on perfluoroglutaric acid (32,41). In addition, remanent polarizations have been measured for a series of polyamides containing $m$-xylylenediamine (42) and for those containing 1,3-bis(aminomethyl)cyclohexane (43). Pyroelectricity has been observed in nylon-11 (44) and nylon-5,7 (45).

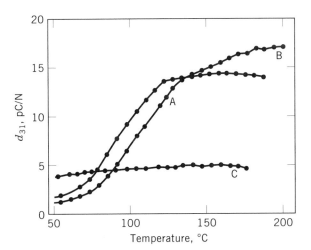

**Fig. 4.** The effect of temperature on the piezoelectric strain constant, $d_{31}$, for A, nylon-11; B, nylon-7; and C, poly(vinylidene fluoride) (PVF$_2$) films (35).

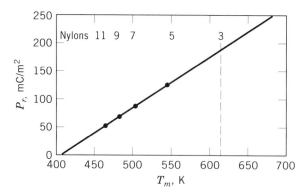

**Fig. 5.** The dependence of remanent polarization, $P_r$, on the melting point, $T_m$, in odd nylons (40).

## Chemical Properties

### PREPARATION

**Direct Amidation.** The direct reaction of amino acids to form Type AB polyamides (eq. 1) and diacids and diamines to form type AABB polyamides (eq. 2) are two of the most commonly used methods to produce polyamides. The

$$n\ \mathrm{H_2N{-}R{-}\overset{\overset{\displaystyle O}{\|}}{C}{-}OH} \rightleftharpoons \mathrm{H_2N{-}R{+}\overset{\overset{\displaystyle O}{\|}}{C}{-}NH{-}R{+}_{(n-1)}\overset{\overset{\displaystyle O}{\|}}{C}{-}OH} + (n-1)\mathrm{H_2O} \qquad (1)$$

$$n\ \mathrm{H_2N{-}R{-}NH_2} + n\ \mathrm{HO{-}\overset{\overset{\displaystyle O}{\|}}{C}{-}R'{-}\overset{\overset{\displaystyle O}{\|}}{C}{-}OH} \rightleftharpoons$$

$$\mathrm{H_2N{-}R{-}NH{+}\overset{\overset{\displaystyle O}{\|}}{C}{-}R'{-}\overset{\overset{\displaystyle O}{\|}}{C}{-}NH{-}R{-}NH{+}_{(n-1)}\overset{\overset{\displaystyle O}{\|}}{C}{-}R'{-}\overset{\overset{\displaystyle O}{\|}}{C}{-}OH} + (2n-1)\mathrm{H_2O} \qquad (2)$$

integer $n$ is called the degree of polymerization (DP). The average DP is approximately 200 for a typical nylon-6, or about 100 for nylon-6,6; thus the number-average molecular weight is approximately equal for both, since the monomer, hexamethyleneadipamide, for nylon-6,6 has twice the unit weight as the monomer, $\epsilon$-aminocaproamide, has in nylon-6. Water is released as a by-product of the reaction and depending on the conditions of the reaction can be in equilibrium with the reactants. Ideally for the amino acids, only one homologous series of linear polymers is formed, each member of which possesses one amino and one carboxyl end group, as shown in equation 2. However, for the type AABB polymers, two additional homologous series of linear polymers are possibly one with two amino end groups and one with two carboxyl end groups:

$$\mathrm{H_2NRNH{+}\overset{\overset{\displaystyle O\ \ O}{\|\ \ \|}}{CR'C}NHRNH{+}_{(n-1)}H}$$

$$\mathrm{HO{+}\overset{\overset{\displaystyle O\ \ O}{\|\ \ \|}}{CR'C}NHRNH{+}_{(n-1)}\overset{\overset{\displaystyle O}{\|}}{C}{-}R'{-}\overset{\overset{\displaystyle O}{\|}}{C}{-}OH}$$

Polymers from either of these homologous series can be made to predominate by using a small excess of the diamine or diacid, respectively. In addition to these linear polymers, cyclic oligomers are also formed, though in this case $n$

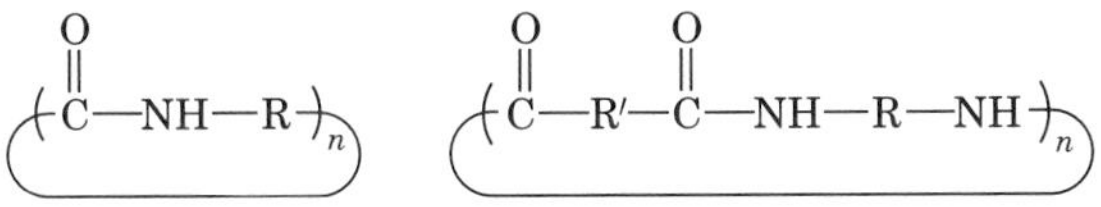

$$\left(\text{C}-\text{NH}-\text{R}\right)_n \qquad \left(\text{C}-\text{R}'-\text{C}-\text{NH}-\text{R}-\text{NH}\right)_n$$

is generally $<10$ for type AABB and $<20$ for type AB polymers. Thus, for the Type AB polyamides, direct polyamidation leads ideally to a mixture of two homologous series of polymers, one linear and one cyclic, and for the Type AABB a mixture of four is formed, three linear and one cyclic. Additional complications can arise as a result of side reactions and degradation, which can lead to different end groups, defects along the chain, or branching.

Direct amidation is generally carried out in the melt, although it can be done in an inert solvent starting from the dry salt (46). Because most aliphatic polyamides melt in the range of 200–300°C and aromatic-containing polyamides at even higher temperatures, the reactants and products must be thermally stable to be polymerized via this method.

**Acid Chloride Reaction.** In situations where the reactants are sensitive to high temperature or the polymer degrades before the melt point is reached, the acid chloride route is often used to produce the polyamide (47). The basic reaction in the presence of a base, B:, is as follows:

$$n\ H_2NRNH_2 + n\ Cl-\overset{O}{\underset{\|}{C}}-R'-\overset{O}{\underset{\|}{C}}-Cl \xrightarrow{\ B:\ }$$

$$H_2NRNH \left(\overset{O}{\underset{\|}{C}}-R'-\overset{O}{\underset{\|}{C}}-NHRNH\right)_{(n-1)}\overset{O}{\underset{\|}{C}}-R'-\overset{O}{\underset{\|}{C}}-Cl + (2n-1)B{:}HCl$$

Because almost any diacid can be readily converted to the acid chloride, this reaction is quite versatile and several variations have been developed. In the interfacial polymerization method the reaction occurs at the boundary of two phases: one contains a solution of the acid chloride in a water-immiscible solvent and the other is a solution of the diamine in water with an inorganic base and a surfactant (48). In the solution method, only one phase is present, which contains a solution of the diamine and diacid chloride. An organic base is added as an acceptor for the hydrogen chloride produced in the reaction (49). Following any of these methods of preparation, the polymer is exposed to water and the acid chloride end is converted to a carboxylic acid end. However, it is very difficult to remove all traces of chloride from the polymer, even with repeated washings with a strong base.

**Ring-Opening Polymerization.** Ring-opening polymerization is the method used to convert lactams to polyamides. There are several variations of the method, but the most commonly practiced method in industry is hydrolytic polymerization, in which lactams containing six or more carbons in the ring are heated in the presence of water above the melting point of the polyamide. The reaction begins with the hydrolytic ring opening of the lactam, which can be catalyzed by an acid or base, an amino acid, or an amine carboxylate, eg, nylon-6,6 salt.

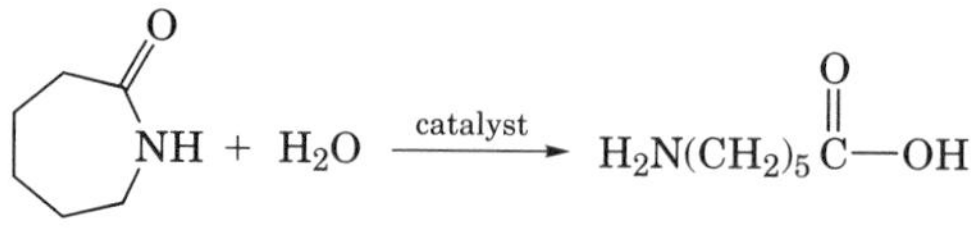

$$
\underset{}{\text{lactam}} + H_2O \xrightarrow{\text{catalyst}} H_2N(CH_2)_5\overset{O}{\overset{||}{C}}-OH
$$

The resulting amino acid then condenses in a stepwise manner to form the growing polymer chain. As in direct polymerization, cyclic oligomers are also formed; hence, caprolactam (qv) can be formed in the reverse of the reaction just shown above.

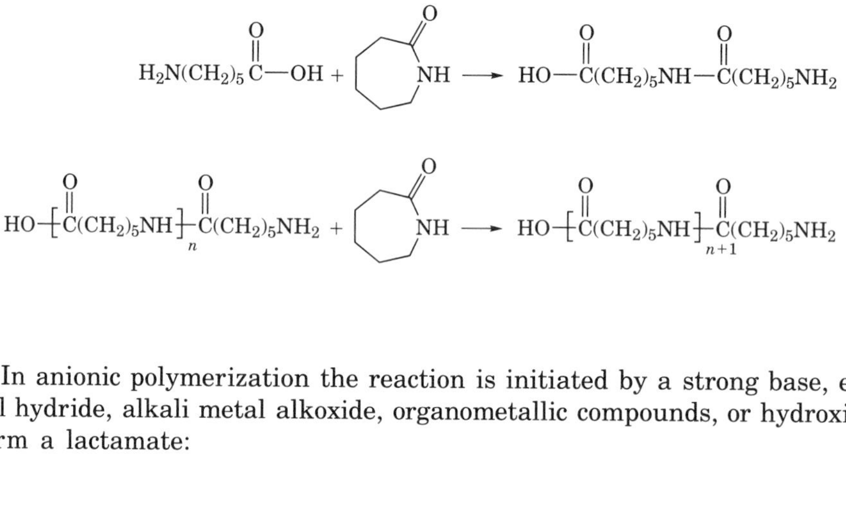

$$
H_2N(CH_2)_5\overset{O}{\overset{||}{C}}-OH + \text{lactam} \longrightarrow HO-\overset{O}{\overset{||}{C}}(CH_2)_5NH-\overset{O}{\overset{||}{C}}(CH_2)_5NH_2
$$

$$
HO-\!\!\left[\overset{O}{\overset{||}{C}}(CH_2)_5NH\right]_{\!n}\!\!\overset{O}{\overset{||}{C}}(CH_2)_5NH_2 + \text{lactam} \longrightarrow HO-\!\!\left[\overset{O}{\overset{||}{C}}(CH_2)_5NH\right]_{\!n+1}\!\!\overset{O}{\overset{||}{C}}(CH_2)_5NH_2
$$

In anionic polymerization the reaction is initiated by a strong base, eg, a metal hydride, alkali metal alkoxide, organometallic compounds, or hydroxides, to form a lactamate:

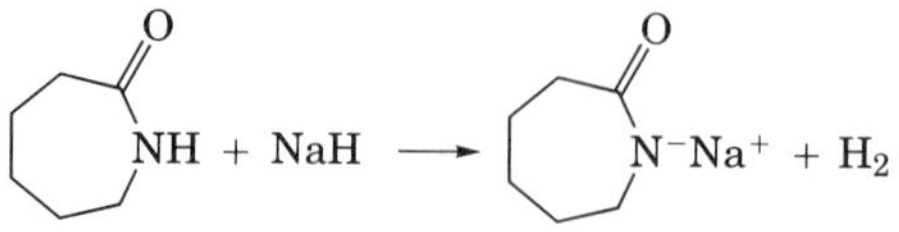

$$
\text{lactam} + NaH \longrightarrow \text{lactamate} + H_2
$$

The lactamate then initiates a two-step reaction which adds a molecule of the lactam to the polymer chain (50–52):

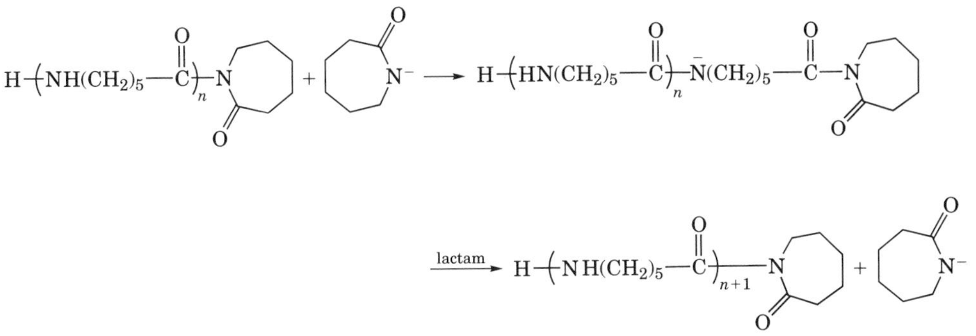

Lactams can also be polymerized under anhydrous conditions by a cationic mechanism initiated by strong protic acids, their salts, and Lewis acids, as well as amines and ammonia (51–53). The complete reaction mechanism is complex and this approach has not as yet been used successfully in a commercial process.

**Other Preparative Reactions.** Polyamidation has been an active area of research for many years, and numerous methods have been developed for polyamide formation. The synthesis of polyamides has been extensively reviewed (54). In addition, many of the methods used to prepare simple amides are applicable to polyamides (55,56). Polyamides of aromatic diamines and aliphatic diacids can also be made by the reaction of the corresponding aromatic diisocyanate and diacids (57).

## REACTIONS OF POLYAMIDES

**Acidolysis, Aminolysis, and Alcoholysis.** When heated, polyamides react with monofunctional acids, amines, or alcohols, especially above the melt temperature, to undergo rapid loss of molecular weight (58,59), eg, as in acidolysis (eq. 3) with acetic acid [*64-19-7*] or aminolysis (eq. 4) with an aliphatic amine:

$$R{-}\overset{O}{\overset{\|}{C}}{-}NH{-}R' + CH_3\overset{O}{\overset{\|}{C}}{-}OH \longrightarrow R{-}\overset{O}{\overset{\|}{C}}{-}OH + CH_3\overset{O}{\overset{\|}{C}}{-}NH{-}R' \qquad (3)$$

$$R{-}\overset{O}{\overset{\|}{C}}{-}NH{-}R' + CH_3(CH_2)_xNH_2 \longrightarrow R'{-}NH_2 + CH_3(CH_2)_xNH{-}\overset{O}{\overset{\|}{C}}{-}R \qquad (4)$$

If adipamide reacts with hexamethylenediamine, then nylon-6,6 can be prepared by aminolysis of the adipamide; this could also be viewed as reverse ammonolysis.

$$n\ H_2N{-}\overset{O}{\overset{\|}{C}}(CH_2)_4\overset{O}{\overset{\|}{C}}{-}NH_2 + n\ H_2N(CH_2)_6NH_2 \longrightarrow$$

$$H_2N(CH_2)_6NH{+}\overset{O}{\overset{\|}{C}}(CH_2)_4\overset{O}{\overset{\|}{C}}{-}NH(CH_2)_6NH{+}_{(n-1)}\overset{O}{\overset{\|}{C}}(CH_2)_4\overset{O}{\overset{\|}{C}}{-}NH_2 + (2n-1)NH_3$$

Phosphoric acid [*7664-38-2*] and its derivatives are effective catalysts for this reaction (60). Reverse alcoholysis and acidolysis can, in principle, also be used to produce polyamides, and the conversion of esters to polyamides through their reaction within diamines, reverse alcoholysis, has been demonstrated (61). In the case of reverse acidolysis, the acid by-product is usually less volatile than the diamine starting material. Thus, this route to the formation of polyamide is not likely to yield a high molecular weight polymer.

**Ammonolysis.** In a reaction closely related to aminolysis, ammonia [*7664-41-7*] reacts with polyamides, usually under pressure and at elevated temperatures (62).

$$R-\overset{\overset{\displaystyle O}{\|}}{C}-NH-R' + NH_3 \longrightarrow R-\overset{\overset{\displaystyle O}{\|}}{C}-NH_2 + H_2N-R'$$

Lewis acids, such as the halide salts of the alkaline-earth metals, Cu(I), Cu(II), zinc, Fe(III), aluminum, etc, are effective catalysts for this reaction (63). The ammonolysis of polyamides obtained from post-consumer waste has been used to cleave the polymer chain as the first step in a recycle process in which mixtures of nylon-6,6 and nylon-6 can be reconverted to diamine (64). The advantage of this approach lies in the fact that both the adipamide [*628-94-4*] and 6-aminohexanoamide can be converted to hexamethylenediamine via their respective nitriles in a conventional two-step process in the presence of the diamine formed in the original ammonolysis reaction, thus avoiding a difficult and costly separation process. In addition, the mixture of nylon-6,6 and nylon-6 appears to react faster than does either polyamide alone.

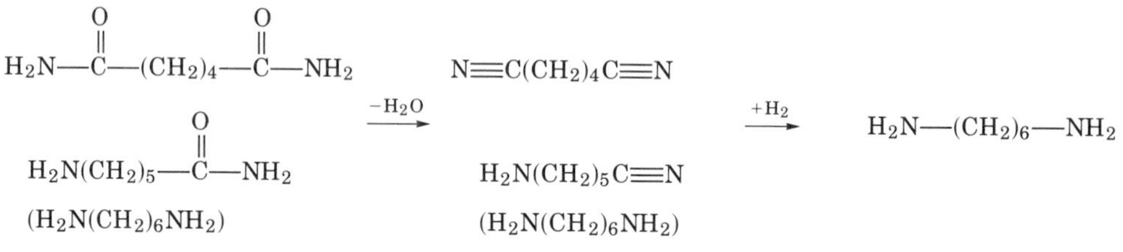

$$H_2N-\overset{\overset{\displaystyle O}{\|}}{C}-(CH_2)_4-\overset{\overset{\displaystyle O}{\|}}{C}-NH_2 \qquad N{\equiv}C(CH_2)_4C{\equiv}N$$

$$H_2N(CH_2)_5-\overset{\overset{\displaystyle O}{\|}}{C}-NH_2 \qquad \xrightarrow{-H_2O} \qquad H_2N(CH_2)_5C{\equiv}N \qquad \xrightarrow{+H_2} \qquad H_2N-(CH_2)_6-NH_2$$

$$(H_2N(CH_2)_6NH_2) \qquad\qquad (H_2N(CH_2)_6NH_2)$$

The resulting hexamethylenediamine can then be reused to produce new nylon-6,6. Impurities or contaminants from monomers of other types of polyamides can be readily removed by distillation from either the nitriles or diamine.

**Transamidation and Transesteramidation.** Transamidation is the mutual exchange of chain fragments in a polyamide, shown as follows where R, R'' and R', R''' represent polymer chain fragments of any length.

$$R-\overset{\overset{\displaystyle O}{\|}}{C}-NH-R' + R''-\overset{\overset{\displaystyle O}{\|}}{C}-NH-R''' \longrightarrow R-\overset{\overset{\displaystyle O}{\|}}{C}-NH-R''' + R''-\overset{\overset{\displaystyle O}{\|}}{C}-NH-R'$$

It is generally accepted that transamidation is not a concerted reaction, but occurs through the attack of a free end on the amide group via aminolysis (eg, eq. 4) or acidolysis (eg, eq. 3) (65). Besides those ends always present, new ends are formed by degradation processes, especially hydrolysis (eq. 5), through which the amide groups are in dynamic equilibrium with the acid and amine ends.

$$R-\overset{\overset{\displaystyle O}{\|}}{C}-NH-R' + H_2O \rightleftharpoons R-\overset{\overset{\displaystyle O}{\|}}{C}-OH + H_2N-R' \tag{5}$$

The acid and amine products of equations 3, 4, and 5 condense to form new amide end groups (eq. 6).

$$R-\overset{\overset{\displaystyle O}{\|}}{C}-OH + H_2N-R''' \rightleftharpoons R-\overset{\overset{\displaystyle O}{\|}}{C}-NH-R''' + H_2O \tag{6}$$

The step in which the free acid and amine ends recombine (eq. 6) is only accomplished statistically, since it is unlikely that any two particular ends formed in the acidolysis or aminolysis steps would find each other in the melt. Transamidation is catalyzed by both acidic and basic ends, but in general acids appear to be much more effective than bases (59,65).

Transamidation is an important process in the melt phase for polyamides because it is usually the process by which an equilibrium molecular weight distribution is reestablished and, in the case of the melt blending of two or more polyamides to form a copolymer, it is the process by which randomization of the individual monomers along the chain is effected. In the solid phase, chain mobility is restricted and equilibrium in either case often is not achieved. In the case of blending two homopolymers in the melt, eg, nylon-6,6 and nylon-6, randomization begins by forming copolymers with large blocks of nonrandomized polymer. Then, as the reaction proceeds, the size of the blocks decreases as a result of the interchange of segments between adjacent polymer chains via transamidation, until the monomeric units are distributed randomly along the polymer chain. Industrial practice indicates that the effects of incomplete randomization are usually indistinguishable from full randomization after about 15 minutes in the melt. However, laboratory studies have shown that complete randomization takes several hours (66). Equilibration of molecular weight distribution also appears to occur relatively quickly, presumably because it requires only a statistical distribution of chain lengths rather than complete randomization of monomeric units. This tendency of polyamides to randomize in the melt makes it virtually impossible to produce block copolymers via such a process.

Transesteramidation is a process similar to transamidation, except that a polyamide is mixed with a polyester rather than another polyamide (67). This is often a convenient route to produce polyesteramides.

$$R-\overset{\overset{\displaystyle O}{\|}}{C}-NH-R' + R''-\overset{\overset{\displaystyle O}{\|}}{C}-O-R''' \longrightarrow R-\overset{\overset{\displaystyle O}{\|}}{C}-O-R''' + R''-\overset{\overset{\displaystyle O}{\|}}{C}-NH-R'$$

Here, R and R' represent polyamide chain fragments, and R'' and R''' represent polyester chain fragments of any length. Polyesters are generally more

easily hydrolyzed than polyamides and thus are quite sensitive to the presence
of water in the polyamide. During transesteramidation care must be taken not
to significantly hydrolyze the polyester before it reacts. The rate constants for
randomization of copolyamides by transamidation are an order of magnitude
slower than that for copolyesters by transesterification (68).

**Grafting.** Grafting is the process of chemically bonding additional polymeric units, usually not polyamides, to the nylon polymer chain. This is most
often initiated by reaction of the grafting substrate directly with the polyamide
backbone, but grafting can also be achieved by introducing non-amide reactive
sites into the polymer chain, through the incorporation of reactive comonomers
during polymerization. In general, the polyamides are relatively inert chemically,
and a source of high energy sufficient to create free-radical sites is necessary to
initiate the grafting reaction, eg, $\alpha$-particles, electrons from linear accelerators,
$\gamma$-rays from $^{60}$Co, x-rays, glow discharges, or uv radiation. The predominant reactive intermediate is believed to be the alkyl free radical formed by the removal
of the $\alpha$-hydrogen adjacent to the amide nitrogen (69). The reaction with vinyl
monomers is typically as follows:

$$
\underset{H}{\underset{|}{R'C}}\!\!\overset{\overset{O}{\|}}{-}\!\!NCH_2R'' \xrightarrow{\text{ionizing energy}} \underset{H}{\underset{|}{R'C}}\!\!\overset{\overset{O}{\|}}{-}\!\!N\dot{C}HR'' \xrightarrow{CH_2=CHR} \underset{H}{\underset{|}{R'C}}\!\!\overset{\overset{O}{\|}}{-}\!\!NCHR''\overset{(CH_2CHR)_n}{|}
$$

Grafting can also occur in the amide nitrogen, either through an anionic-
type mechanism which is believed to operate when ethylene oxide [75-21-8] and
similar copolymers are grafted to polyamides, or through a polycondensation
mechanism when secondary amides are formed as graft copolymers (70).

Grafting can be used to change the surface properties of the final polyamide
article, especially film or fiber; its hydrophilic, antistatic, frictional, or other
characteristics are altered. Also, grafting can be made to occur in the bulk polymer; this alters, for example, the mechanical (toughening) or optical (delustering)
properties of the polyamide. Although there is a substantial amount of work
published both in the open and patent literature that discusses grafting onto
polyamides (70–74), this technology does not appear to be practiced by the primary polyamide manufacturers to any great extent and is probably reserved for
specialty end use applications, where the resulting performance changes warrant
the added cost of materials and processing.

## DEGRADATION OF POLYAMIDES

**Hydrolysis.** Hydrolysis (eq. 5) is the reverse of the amidation reaction. As
a consequence, if the water is not removed from the reaction media, the polyamidation reaction eventually approaches equilibrium and the ultimate molecular
weight of the polymer is limited (75). In many polymerization processes, a vacuum is applied to the polymer melt and the molecular weight can continue to
grow. Nonetheless, hydrolysis is important in determining the stability of the
final polymer after it has been quenched and dried. Because the equilibrium
moisture content of polyamides at room temperature at any practical relative humidity is almost always greater than the equilibrium water content in the melt,

polyamides must be dried to avoid a significant decrease in molecular weight when they are remelted. Reduction of molecular weight via hydrolysis can also occur when polyamides are in use, particularly in a high humidity environment or when they are placed in direct contact with water. The hydrolysis reaction is generally slow at room temperature, but it is accelerated at higher temperatures and when catalyzed by acids or bases. An example is the rapid loss of strength in tire cord that occurs as water comes in contact with nylon fiber in a tire carcass. The hydrolysis reaction can be used to advantage in the determination of the composition of polyamides. Most aliphatic polyamides can be completely hydrolyzed by heating in 8 $N$ HCl for 16 hours at about 110°C. The solution is then neutralized and the monomers are extracted or dried, derivatized, and analyzed by gas chromatography (gc) or high performance liquid chromatography (hplc) for the relative amounts of their constituent monomers (76).

**Thermal Degradation.**   The degradation that occurs in the absence of oxygen affects all polyamides at a sufficiently high temperature and is usually significant above 300°C. Thermooxidation reactions often occur simultaneously owing to the presence of small amounts of air, which can lead to a confusion of the two processes. The general thermal decomposition reaction in polyamides, which is the cleavage of the amide bond to eventually form an olefin and a nitrile, results in chain cleavage and thus a loss in molecular weight.

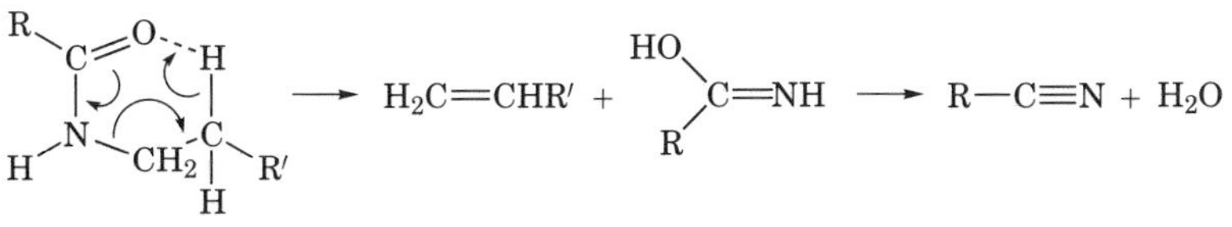

$$R-C(=O)\cdots H \quad\longrightarrow\quad H_2C=CHR' + \underset{R}{HO}C=NH \quad\longrightarrow\quad R-C\equiv N + H_2O$$

However, if there is a lower energy decomposition pathway available, then an alternative degradation reaction dominates. There is a growing body of evidence to suggest that cyclization reactions to form small stable ring compounds are one such decomposition pathway, especially for polyamides containing monomers with four to six carbon atoms (77,78) (eq. 7). The first example of this is the formation of cyclic amines, which is the principal decomposition pathway in nylon-4,6 (in eq. 7, $n = 1$ and $R = H$) (79); this has also been observed in MPMD-containing polyamides (in eq. 7, $n = 2$, $R = CH_3$) (80), as well as in nylon-6,6 (in eq. 7, $n = 3$, $R = H$).

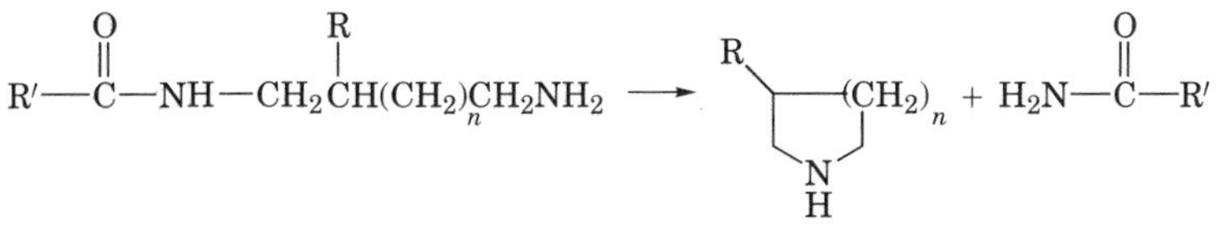

$$R'-\overset{O}{\overset{||}{C}}-NH-CH_2\overset{R}{\overset{|}{C}}H(CH_2)_nCH_2NH_2 \quad\longrightarrow\quad \underset{\underset{H}{N}}{}(CH_2)_n + H_2N-\overset{O}{\overset{||}{C}}-R' \qquad (7)$$

Another example of a cyclic product is the formation of cyclopentanone [120-92-3] as a thermal decomposition product in nylon-6,6 (81,82). The following mechanism (eqs. 8 and 9) accounts not only for the formation of the cycloketone but also for the increase in amine ends, the decrease in acid ends, and the evolution of $CO_2$ that is observed in the thermal decomposition of nylon-6,6 (82).

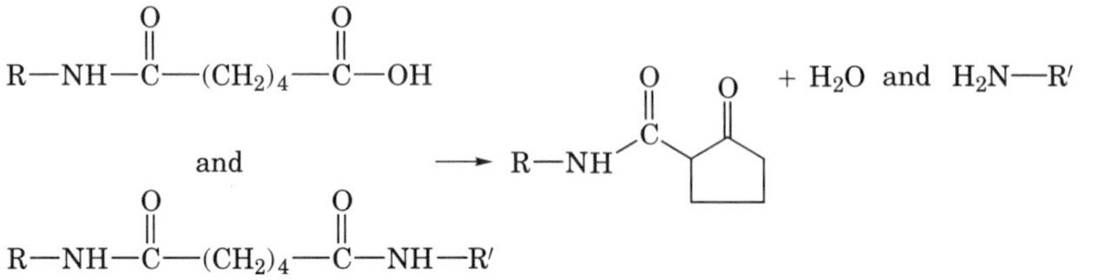

$$R-NH-\overset{\overset{\displaystyle O}{\|}}{C}-(CH_2)_4-\overset{\overset{\displaystyle O}{\|}}{C}-OH$$

and

$$R-NH-\overset{\overset{\displaystyle O}{\|}}{C}-(CH_2)_4-\overset{\overset{\displaystyle O}{\|}}{C}-NH-R' \longrightarrow R-NH \qquad + H_2O \text{ and } H_2N-R' \qquad (8)$$

$$R-NH \longrightarrow + R-N=C=O \xrightarrow{+ H_2O} H_2N-R + CO_2 \qquad (9)$$

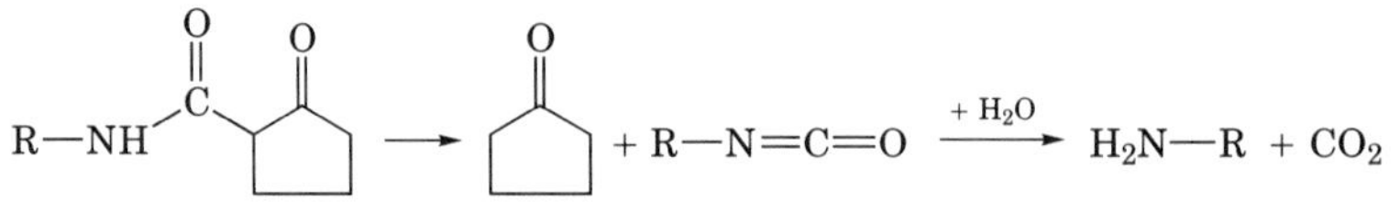

In type AB polyamides the reequilibration reaction leading to production of the starting lactam can be viewed as the decomposition of the polyamides into cyclic products (83); $n = 1, 2,$ and 3 for nylon-4, nylon-5, and nylon-6, respectively.

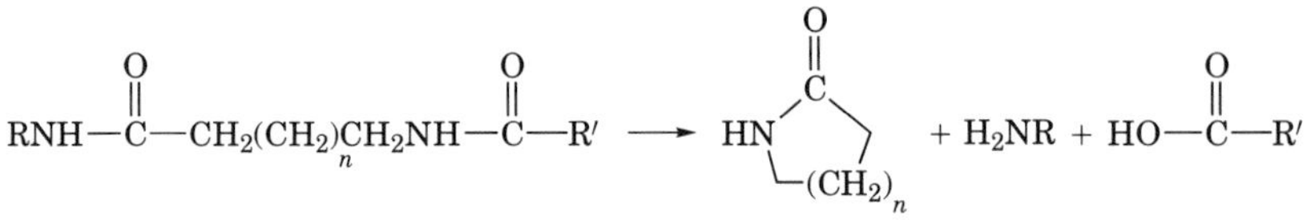

$$RNH-\overset{\overset{\displaystyle O}{\|}}{C}-CH_2(CH_2)CH_2NH-\overset{\overset{\displaystyle O}{\|}}{C}-R' \longrightarrow HN \underset{(CH_2)_n}{\overset{\overset{\displaystyle O}{\|}}{C}} + H_2NR + HO-\overset{\overset{\displaystyle O}{\|}}{C}-R'$$

Finally, when polyamides containing four or five carbon diacids, ie, succinic acid [110-15-6] and glutaric acid [110-94-1], respectively, are heated, they form cyclic imides that cap the amine ends and prevent high molecular weights from being achieved (84). For nylon-x,4, $n = 1$ and for nylon-x,5, $n = 2$.

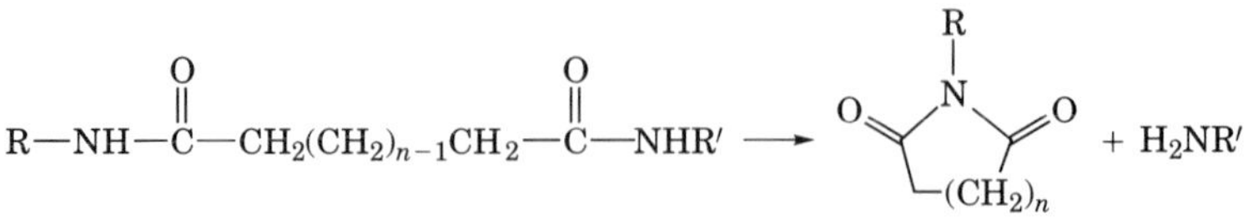

$$R-NH-\overset{\overset{\displaystyle O}{\|}}{C}-CH_2(CH_2)_{n-1}CH_2-\overset{\overset{\displaystyle O}{\|}}{C}-NHR' \longrightarrow O \underset{(CH_2)_n}{\overset{\overset{\displaystyle R}{\underset{\displaystyle N}{|}}}{}} O + H_2NR'$$

Other noncyclic reactions are observed, especially in polyamides of longer carbon chain monomers; for example, the linear analogue to the cyclic amine reaction is diamine coupling (eq. 10) to form secondary amines that can act as branch points (eq. 11).

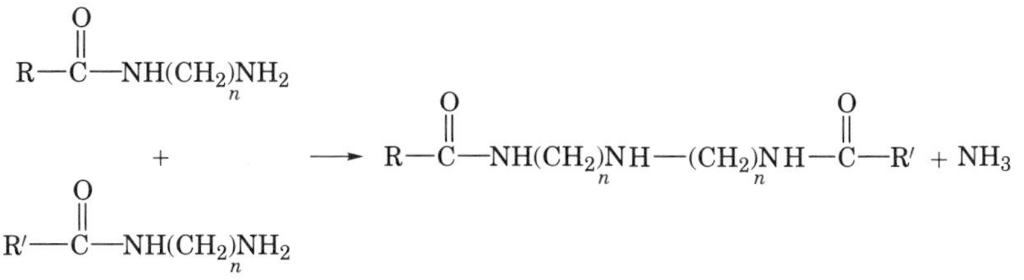

$$
\begin{array}{l}
\text{R—C—NH(CH}_2)_n\text{NH}_2 \\
\qquad\qquad\text{O} \\
\\
\qquad + \qquad\qquad \longrightarrow \quad \text{R—C—NH(CH}_2)_n\text{NH—(CH}_2)_n\text{NH—C—R}' + \text{NH}_3 \qquad (10) \\
\\
\text{R}'\text{—C—NH(CH}_2)_n\text{NH}_2
\end{array}
$$

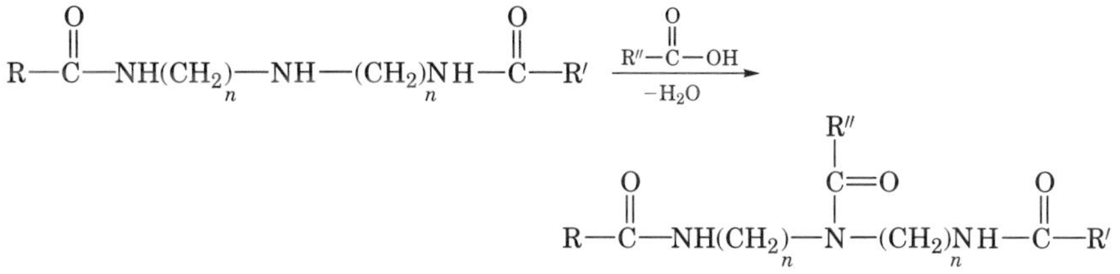

$$
\text{R—C—NH(CH}_2)_n\text{—NH—(CH}_2)_n\text{NH—C—R}' \xrightarrow[-H_2O]{\text{R}''\text{—C—OH}}
$$

$$
\text{R—C—NH(CH}_2)_n\text{—N—(CH}_2)_n\text{NH—C—R}' \qquad (11)
$$

with R'' group as C=O substituent on the central N.

In nylon-6,6, the secondary amine so formed, bis-hexamethylenetriamine [*143-23-7*] (BHMT) has been thought by some to be the source of cross-linking and gelation in this polyamide (84–86). However, nylon-6,10 and nylon-6,12 do not exhibit such gelation behavior, and the branching reaction in nylon-6,6 is generally believed to be associated with the presence of adipic acid [*124-04-9*]. Although other branch sites have been postulated (87), the nature of the specific chemical steps leading to the formation of gel in nylon-6,6 remains an open question. Several patents have appeared that claim to inhibit the gel formation in nylon-6,6, but no mechanistic details are given (86,88,89). Another type of branching reaction which has been found in aromatic polyamides is amidine formation (eq. 12) (90).

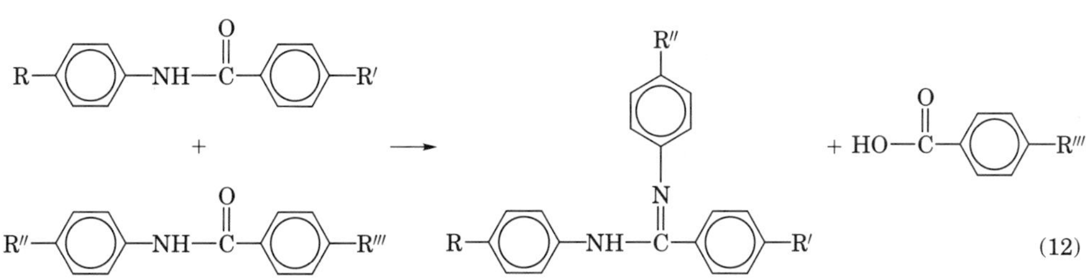

$$(12)$$

Such structures are stabilized by having the imine bond conjugated between two aromatic rings and the possibility of tautomerism between two equivalent structures:

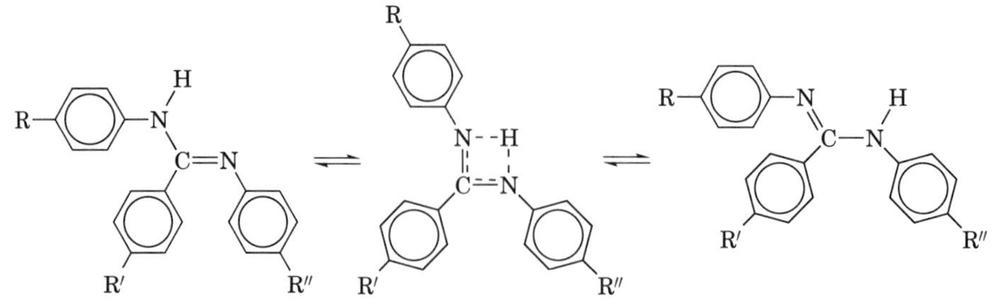

Decarbonylation of the acid ends is another reaction (eq. 13) that can occur during the thermal decomposition of polyamides, especially above 300°C. This reaction, which forms an unreactive end and thus limits the ultimate molecular weight that is attainable, is particularly troublesome during the processing of polyamides containing a high ratio of terephthalic acid; these polyamides generally have a high melting point and require high processing temperatures.

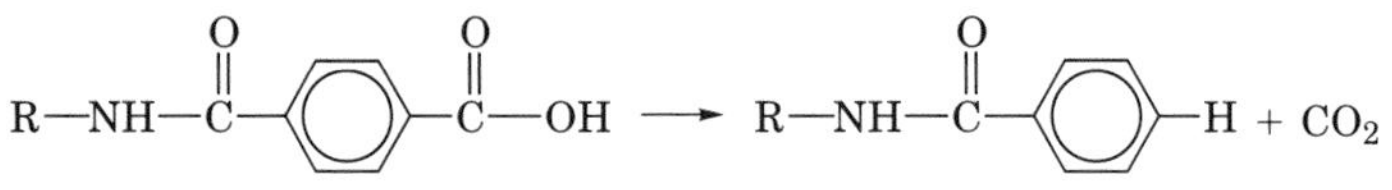

$$R-NH-\overset{\overset{\displaystyle O}{\|}}{C}-\!\!\bigcirc\!\!-\overset{\overset{\displaystyle O}{\|}}{C}-OH \longrightarrow R-NH-\overset{\overset{\displaystyle O}{\|}}{C}-\!\!\bigcirc\!\!-H + CO_2 \qquad (13)$$

**Thermooxidation.** This is an autoxidation process that occurs in all polyamides. It is significantly accelerated at elevated temperatures and can lead to carbonization of the polymer, but it also occurs during ambient temperature storage unless the polymer is protected with an antioxidant or the storage temperature is reduced. The principal effects of thermooxidation are a loss in molecular weight, increase in acid ends, decrease in amine ends, and the generation of color. Thermooxidation is the primary source of color generation in aliphatic polyamides; this is sometimes attributed incorrectly to thermal degradation. Aliphatic polyamides should be protected from air during thermal processing and in high temperature applications if the negative effects of oxidation are to be avoided. Blanketing the polymer with inert gas or the addition of antioxidants (qv) are two techniques providing good protection.

Figure 6 presents the first steps in the generally accepted mechanism for thermooxidation (and photooxidation). Isotopic labeling studies have demonstrated that the position alpha to the amide nitrogen is the predominant site for oxygen attack (91), and the corresponding alkyl radical has been observed using electron-spin resonance (esr) in polyamides exposed to ionizing radiation (92). The effectiveness of free-radical trapping agents and peroxide decomposers used as additives to inhibit oxidation in polyamides gives strong support to the remaining steps in the mechanism, which are essentially the same as those that have been thoroughly investigated for polyolefins. However, this mechanism does not account for any of the primary deleterious effects of thermooxidation in polyamides mentioned above.

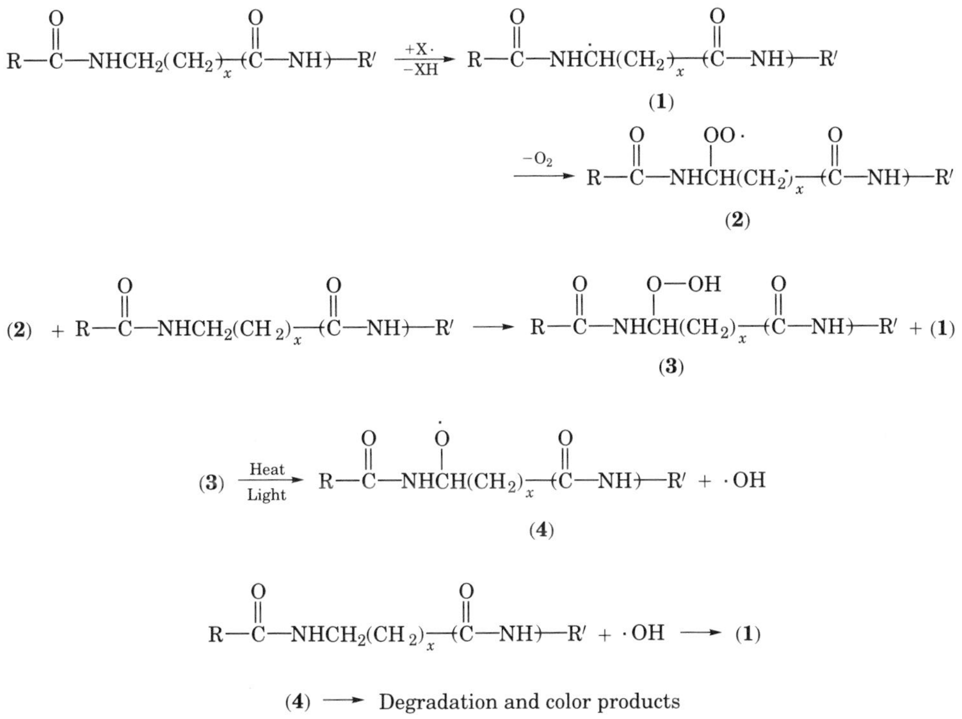

$$R-\overset{O}{\overset{\|}{C}}-NHCH_2(CH_2)_{\overline{x}}(\overset{O}{\overset{\|}{C}}-NH)-R' \xrightarrow[-XH]{+X\cdot} R-\overset{O}{\overset{\|}{C}}-NH\overset{\cdot}{C}H(CH_2)_{\overline{x}}(\overset{O}{\overset{\|}{C}}-NH)-R'$$

(1)

$$\xrightarrow{-O_2} R-\overset{O}{\overset{\|}{C}}-NH\overset{\overset{OO\cdot}{|}}{C}H(CH_2)_{\overline{x}}(\overset{O}{\overset{\|}{C}}-NH)-R'$$

(2)

$$(2) + R-\overset{O}{\overset{\|}{C}}-NHCH_2(CH_2)_{\overline{x}}(\overset{O}{\overset{\|}{C}}-NH)-R' \longrightarrow R-\overset{O}{\overset{\|}{C}}-NH\overset{\overset{O-OH}{|}}{C}H(CH_2)_{\overline{x}}(\overset{O}{\overset{\|}{C}}-NH)-R' + (1)$$

(3)

$$(3) \xrightarrow[\text{Light}]{\text{Heat}} R-\overset{O}{\overset{\|}{C}}-NH\overset{\overset{\cdot O}{|}}{C}H(CH_2)_{\overline{x}}(\overset{O}{\overset{\|}{C}}-NH)-R' + \cdot OH$$

(4)

$$R-\overset{O}{\overset{\|}{C}}-NHCH_2(CH_2)_{\overline{x}}(\overset{O}{\overset{\|}{C}}-NH)-R' + \cdot OH \longrightarrow (1)$$

$$(4) \longrightarrow \text{Degradation and color products}$$

**Fig. 6.** The initial degradation pathway for thermooxidation and photooxidation. The free radical X· is generated by the effect of heat or light on impurities, additives, and polymer.

It is usually postulated that the final product in the accepted mechanism, the alkoxyl radical (**4**), cleaves (eqs. 14 and 15) before or after hydrogen abstraction, and that this accounts for the drop in molecular weight of the

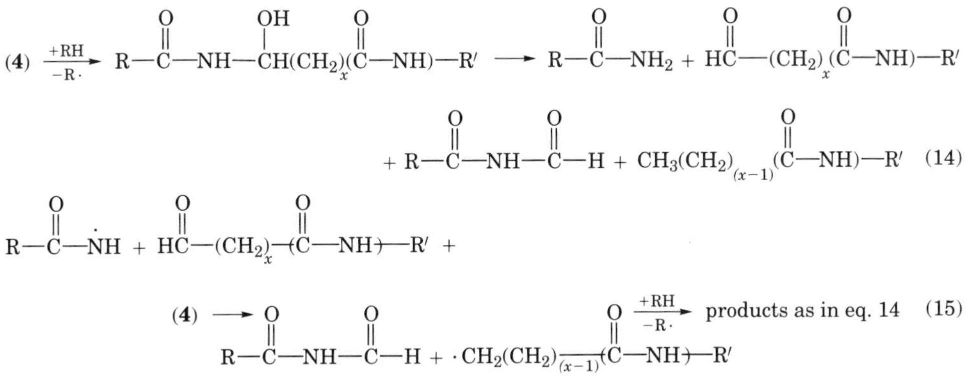

$$(4) \xrightarrow[-R\cdot]{+RH} R-\overset{O}{\overset{\|}{C}}-NH-\overset{\overset{OH}{|}}{C}H(CH_2)(\overset{O}{\overset{\|}{C}}-NH)-R' \longrightarrow R-\overset{O}{\overset{\|}{C}}-NH_2 + H\overset{O}{\overset{\|}{C}}-(CH_2)(\overset{O}{\overset{\|}{C}}-NH)-R'$$

$$+ R-\overset{O}{\overset{\|}{C}}-NH-\overset{O}{\overset{\|}{C}}-H + CH_3(CH_2)_{(x-1)}(\overset{O}{\overset{\|}{C}}-NH)-R' \quad (14)$$

$$R-\overset{O}{\overset{\|}{C}}-\overset{\cdot}{N}H + H\overset{O}{\overset{\|}{C}}-(CH_2)_{\overline{x}}(\overset{O}{\overset{\|}{C}}-NH)-R' +$$

$$(4) \longrightarrow R-\overset{O}{\overset{\|}{C}}-NH-\overset{O}{\overset{\|}{C}}-H + \cdot CH_2(CH_2)_{\overline{(x-1)}}(\overset{O}{\overset{\|}{C}}-NH)-R' \xrightarrow[-R\cdot]{+RH} \text{products as in eq. 14} \quad (15)$$

polymer. When the site of the alkyl radical is located beta or gamma to the amide nitrogen with a lower probability, as the isotopic labeling indicates, then a homologous series of products is formed. The carbonyl absorption of the aldehydes (93) and the alkyl amines formed after hydrolysis of the polymer (94) have been observed in nylon-6,6. It is also likely that further oxidation of the aldehydes leads to the increase in acid ends (eq. 16). The condensation reaction between

$$\underset{\substack{\| \\ \mathrm{O}}}{\mathrm{HC}}\!-\!(\mathrm{CH_2})_{\overline{x}}(\underset{\substack{\| \\ \mathrm{O}}}{\mathrm{C}}\!-\!\mathrm{NH})\!-\!\mathrm{R'} \xrightarrow{+\frac{1}{2}\,\mathrm{O_2}} \mathrm{HO}\!-\!\underset{\substack{\| \\ \mathrm{O}}}{\mathrm{C}}\!-\!(\mathrm{CH_2})_{\overline{x}}(\underset{\substack{\| \\ \mathrm{O}}}{\mathrm{C}}\!-\!\mathrm{NH})\!-\!\mathrm{R'} \qquad (16)$$

the acid and amine ends is probably the primary pathway for the reduction of the amine ends, particularly when the temperature is elevated.

No definitive evidence has appeared that identifies the source of the color generated during thermooxidation (95). However, two laboratories have postulated that the reactions leading to the formation of the color chromophores are aldol-type reactions, either via the reaction of aldehydes directly (96) or via imines (95,97) formed by the condensation of the aldehyde with an amine end, which could also contribute to the loss of amine ends:

$$\mathrm{RNH_2} + \underset{\substack{\| \\ \mathrm{O}}}{\mathrm{HC}}\!-\!(\mathrm{CH_2})_{\overline{x}}(\underset{\substack{\| \\ \mathrm{O}}}{\mathrm{C}}\!-\!\mathrm{NH})\!-\!\mathrm{R'} \xrightarrow{-\mathrm{H_2O}} \mathrm{RN}\!\!=\!\!\mathrm{CH}\!-\!(\mathrm{CH_2})_{\overline{x}}(\underset{\substack{\| \\ \mathrm{O}}}{\mathrm{C}}\!-\!\mathrm{NH})\!-\!\mathrm{R'}$$

These polyaldol–condensation reactions lead to a system of conjugated double bonds that, when the number of bonds is sufficient, can account for absorption at visible wavelengths, ie, yellow color. However, the color chromophores can also be formed via further reaction of these species, such as continued oxidation, cyclization, etc. Primarily because of the tendency of nylons to yellow and because of long-term strength loss, antioxidants are commonly used in commercial polymers. The copper halide system, the combination of a soluble copper(II) salt with sodium or potassium iodide, is probably the most frequently used antioxidant in polyamides (98). The mechanism for stabilization by copper halides has been reviewed and a new function for the metal ion as a peroxide decomposer has been postulated (99), as well as its accepted role as a radical scavenger (100). What is particularly attractive about this mechanism is that it offers an explanation as to why the copper halide system works so well in polyamides, whereas copper ions promote severe thermooxidation in polyolefins. In polyamides there is always a carbonyl oxygen available to coordinate the copper in a stable, six-membered ring when the hydroperoxide is formed in its most probable position, alpha to the amide nitrogen. In polyolefins, the possibility of coordination seldom occurs, and the copper ion is free to act in its usual role as an oxidation catalyst. Other antioxidants are also used in polyamides, eg, phenols, hypophosphites, and phosphites. However, the high processing temperatures, presence of moisture, and the acid–base functionality associated with polyamides significantly limit the number and type of stabilizers that can be used in polyamides. For example, most hindered amines, which have been used with great success when combined with other antioxidants for the thermo- and photostabilization of polypropylene, are thermally unstable above 200°C (101), and therefore cannot be used in a melt process for most polyamides.

**Photodegradative Processes.** Polymers can undergo two types of photodegradative processes; one in the presence of oxygen, photooxidation, and one in its absence, photodegradation. Additive-free, noncontaminated nylons appear to have only one significant chromophore in the uv-visible region, a strong, $\log \epsilon \geq 4.0$, at approximately 185 nm, which is assigned to the $\pi \rightarrow \pi^*$ transition of the amide group. There may also be a much weaker $n \rightarrow \pi^*$ at slightly longer wavelengths, but its presence is usually masked by absorption resulting from thermooxidative impurities or the carbonyl absorption of the acid ends. The strongly forbidden ground singlet state to first excited triplet state absorption, $S_0 \rightarrow T_1$, can lie as low as 285 nm, based on low temperature phosphorescence excitation–emission spectra of model alkyl-bis(hexanamides) (102). A weak, predominantly continuum absorption by the thermooxidative degradation products occurs from about 235 to at least 400 nm, where they are the primary source of yellow color in polyamides. Thermal degradation products can also show a continuum-like absorption in this region; however, nylon-6,6 shows a weak but discernible absorption peak at 290 nm (103) which has been assigned to the following chromophore (104):

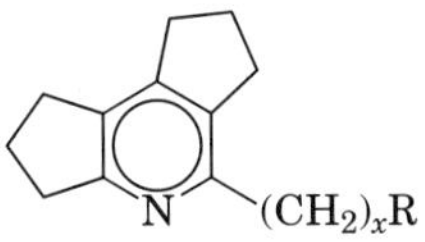

The mechanism for photodegradation at short wavelengths is generally believed to be initiated by the photolytic cleavage of the amide bond (eq. 17), which has the lowest bond strength in aliphatic polyamides (220 kJ/mol (53 kcal/mol))

$$R-CH_2-\overset{\overset{\displaystyle O}{\|}}{C}-NHCH_2R' \xrightarrow[\leq 250 \text{ nm}]{\text{light}} R-CH_2-\overset{\overset{\displaystyle O}{\|}}{C}\cdot + \cdot NHCH_2R' \qquad (17)$$

$$\hspace{5cm} \textbf{(5)} \hspace{2cm} \textbf{(6)}$$

$$\textbf{(5)} \longrightarrow R-CH_2\cdot + CO$$

$$\hspace{2cm} \textbf{(7)}$$

$$\textbf{(5)},\textbf{(6)},\textbf{(7)} + R-CH_2-\overset{\overset{\displaystyle O}{\|}}{C}-NHCH_2R' \longrightarrow R-CH_3, R'CH_2NH_2,$$

$$+ R-CH_2-\overset{\overset{\displaystyle O}{\|}}{C}-NH\overset{\displaystyle \cdot}{C}HR' + R-CH_2-\overset{\overset{\displaystyle O}{\|}}{C}H$$

(105,106). The product radicals are then consumed by recombination or by the reaction with oxygen (Fig. 6) when the sample is exposed to air.

The initial steps in the mechanism for photooxidation are generally accepted as being the same as for thermooxidation. This is supported by the facts that similar degradation products have been detected and the effectiveness of similar stabilizers, especially the copper halide system. Although there are some claims that the initial formation of free radicals at wavelengths above 300 nm is the result of the photolytic cleavage of the amide bond, an extrapolation of the logarithm of the molar extinction coefficient for the amide absorption of model amide compounds (107) to 300 nm suggests that the value of the coefficient would be on the order of 0.01 to 0.001, and thus would not be a significant source of radicals even if the quantum yield were high. A more likely source is the photolysis of additives or impurities, such as degradation products like hydroperoxides. Iron(III) halides have been suggested as a likely source of photoinitiated radicals in polyamides (108) at wavelengths above 310 nm, and iron has been shown to have significant negative impact on the photostability of nylon (109). The additive anatase titanium dioxide [1317-70-0], commonly used as a delustrant in fibers, has long been recognized as a potent photodegradant that can, however, be stabilized with the use of manganese(II) compounds (110–111). Dyes and pigments can also act as either prodegradants or as stabilizers, presumably as a result of their propensity to produce free radicals or to act as excited-state quenchers, respectively (112–114).

The generation of color during photooxidation, known as photoyellowing, has long been recognized as a source of color in nylon-6,6 and nylon-6 (115). This effect has been shown to occur in all aliphatic polyamides at wavelengths between 320 and 350 nm (116) (Fig. 7). The chemical nature of the yellow chromophore has not been identified.

**Bio-, Environmental, and Mechanical Degradation.**   Pressure on industry to reduce or remove plastic materials from waste streams is increasing. One approach to meeting this expectation is to manufacture plastic materials that degrade in the environment (117,118) (see POLYMERS, ENVIRONMENTALLY DEGRADABLE). Unfortunately, polyamides, like almost all synthetic polymers, are not directly biodegradable. However, if the polymeric material is reduced to low molecular weight oligomers, then many can be metabolized by microorganisms. Polyamide materials can be fragmented and then reduced in molecular

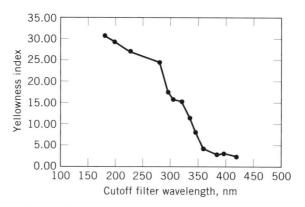

**Fig. 7.**   Photoyellowing of nylon-12 as a function of wavelength.

weight by a process of mechanical destruction, photooxidation, and hydrolysis. This process can occur in a managed waste treatment facility, but it is expensive. An alternative approach has been to incorporate naturally occurring amino acids into polyamide polymer chains to provide sites for enzymatic attack. Numerous articles describing research in this area have appeared (119–123), but it has not been commercially successful, owing to cost and the fact that at high amino acid content most of the desirable properties of synthetic polyamides are lost and at low content the polymers are not sufficiently biodegradable.

Polyamides, like other macromolecules, degrade as a result of mechanical stress either in the melt phase, in solution, or in the solid state (124). Degradation in the fluid state is usually detected via a change in viscosity or molecular weight distribution (125). However, in the solid state it is possible to observe the free radicals formed as a result of polymer chains breaking under the applied stress. If the polymer is protected from oxygen, then alkyl radicals can be observed (126). However, if the sample is exposed to air then the radicals react with oxygen in a manner similar to thermo- and photooxidation. These reactions lead to the formation of microcracks, embrittlement, and fracture, which can eventually result in failure of the fiber, film, or plastic article.

## Principal Commercial Nylons: Nylon-6,6 and Nylon-6

### NYLON-6,6

Nylon-6,6 [32131-17-2] is a tough, translucent white, semicrystalline, high melting ($T_m$ = 265°C) material. The common physical properties are shown in Table 9, and principal producers worldwide in Table 10, for nylon-6,6 and other commercial polyamides.

**Ingredients.** Nylon-6,6 is made from the reaction of adipic acid [124-04-9] and hexamethylenediamine [124-09-4]. The manufacture of intermediates for polyamides is extremely important; not only is the quality of the polymer, such as color, degree of polymerization, and linearity, strongly dependent on the ingredient quality, but also the economic success of the producer is often determined by the yields and cost of manufacture of the ingredients.

Adipic acid (qv) has a wide variety of commercial uses besides the manufacture of nylon-6,6, and thus is a common industrial chemical. Many routes to its manufacture have been developed over the years but most processes in commercial use proceed through a two-step oxidation of cyclohexane [110-83-8] or one of its derivatives. In the first step, cyclohexane is oxidized with air at elevated temperatures usually in the presence of a suitable catalyst to produce a mixture of cyclohexanone [108-94-1] and cyclohexanol [108-93-0], commonly abbreviated KA (ketone–alcohol) or KA oil:

**Table 9. Physical Constants of Commercial Aliphatic Polyamide Homopolymers[a]**

| Property | Nylon-6,6 | Nylon-6 | Nylon-4,6[b] | Nylon-6,9[b] | Nylon-6,10 | Nylon-6,12[b] | Nylon-11[b] | Nylon-12 |
|---|---|---|---|---|---|---|---|---|
| melting point, °C | 255 | 220 | 295 | 210 | 220[b] | 212 | 185 | 175 |
| at equilibrium crystalline | 270 | 231 | | | | | | |
| crystalline | | 260 | | | | | | |
| specific gravity | 1.14[b] | 1.13[b] | 1.18 | 1.07 | 1.07[b] | 1.06 | 1.05 | 1.01[b] |
| density, g/cm³ crystalline[c] | | | | | | | | |
| $\alpha$ | 1.22–1.24 T | 1.21–1.24 M | | | | | | |
| $\beta$ | 1.248 T | | | | | | | |
| $\gamma$ | | 1.13 H | | | | | | |
| $\gamma$ | | 1.17 M | | | | | | |
| amorphous | | | | | | | | |
| $\gamma$ | | 1.09 | | | | | | |
| $\alpha$ | | 1.11 | | | | | | |
| heat of fusion, kJ/kg[d] | | | | | | | | |
| crystalline form $\Delta H_m$ | 196 | 190 | 104.6 | | 215 | | | 95 |
| $\alpha$-crystalline form $\Delta H_m$ | | 240–260 | | | | | | |
| amorphous, annealed 8 h at 50°C | | 45 | | | | | | 45 |
| entropy of fusion, J/(mol·K)[d] | | | | | | | | |
| crystalline | 83–86 | 44–47.5 | | | 110–114 | | | |
| heat capacity, J/(mol·K)[d] | | | | | | | | |
| crystalline, 20°C | 374 | 204 | | | 502 | | | 354 |
| specific heat, J/(g·K)[d] | 1.67 | 1.67 | 2.09 | 1.67 | | 1.67 | | |
| heat of crystallization, kJ/kg[d] | −54 | −46.5 | | | | | | |
| heat of combustion, kJ/kg[d] | −31.4 | −31.4 | | | | | | |
| coefficient of thermal expansion | | | | | | | | |
| linear at 20°C, m/m/K × 10⁵ | 7–10 | 7–10 | 16 | 15 | 8–10 | 9 | 8.5 | 11 |
| volume at 20°C, m³/m³/K × 10⁴ | 2.8 | 2.7 | | | 3.8 | | 2.9 | 2.9 |
| thermal conductivity, W/(m·K) | | | | | | | | |
| crystalline (wet) at 30°C | 0.43 | 0.43 | 0.29 | | | | | |
| amorphous (wet) at 30°C | 0.36 | 0.36 | | | 0.35 | 0.22 | | |
| melt at ~250°C | 0.15 | 0.21 | | | | | | |
| moldings | 0.23 | 0.23 | | | 0.23 | 0.22 | | 0.25 |

**Table 9. (Continued)**

| Property | Nylon-6,6 | Nylon-6 | Nylon-4,6[b] | Nylon-6,9[b] | Nylon-6,10 | Nylon-6,12[b] | Nylon-11[b] | Nylon-12 |
|---|---|---|---|---|---|---|---|---|
| dielectric strength, kV/mm | | | | | | | | |
| dry | 120 | 100 | 26.5 | 23 | 100 | | 30 | 90 |
| dry at 100°C | 40 | 40 | 16 | | 60 | | | 25 |
| humid | 80 | 60 | | | | | | |
| volume resistivity, $\Omega \cdot cm$ | | | | | | | | |
| dry as molded | $1 \times 10^{15}$ | $1 \times 10^{15}$ | $5 \times 10^{14}$ | | $1 \times 10^{15}$ | $10^{15}$ | $10^{14}$ | $10^{15}$ |
| 50% rh at 20°C | $3 \times 10^{11}$ | $2 \times 10^{11}$ | $2 \times 10^{9}$ | | $2 \times 10^{12}$ | $10^{13}$ | | $10^{14}$ |
| dry at 100°C | $3 \times 10^{9}$ | $3 \times 10^{9}$ | | | $5 \times 10^{8}$ | | | $10^{9}$ |
| 50% rh at 100°C | $4 \times 10^{7}$ | | | | | | | |
| dielectric constant | | | | | | | | |
| 100 Hz (dry) | 3.6 | 3.5 | 3.83 | | 3.2 | 4.0 | | 4.0 |
| $10^{6}$ Hz (dry) | 3.2 | 3.3 | 3.55 | | 3.0 | 3.5 | | |
| 100 Hz (50% rh) | 7.5 | 10.9 | 11.0 | | 6.5 | 6 | | >20 |
| $10^{6}$ Hz (50% rh) | 3.7 | 3.8 | 4.5 | | 3.5 | 4 | | |
| dissipation factor | | | | | | | | |
| 100 Hz (dry) | 0.0085 | 0.0065 | 0.012 | | 0.010 | 0.02 | | 0.07 |
| $10^{6}$ Hz (dry) | 0.025 | 0.024 | 0.026 | | 0.021 | 0.02 | | |
| 100 Hz (50% rh) | 0.110 | 0.145 | 0.35 | | 0.200 | 0.15 | | >0.500 |
| $10^{6}$ Hz (50% rh) | 0.070 | 0.092 | 0.12 | | 0.080 | 0.1 | | |
| refractive index ($n_D$) | | | | | | | | |
| single crystals | | | | | | | | |
| $\alpha$ (calc) | 1.475 | | | | 1.475 | | | |
| $\beta$ (calc) | 1.565 | | | | 1.525 | | | |
| $\gamma$ (obs) | 1.58 | | | | 1.565 | | | |
| moldings | 1.53 | 1.53 | | | 1.53 | | | |

[a]All data from Ref. 127 unless indicated otherwise.
[b]From trade literature.
[c]T = triclinic; M = monoclinic; and H = hexagonal.
[d]To convert J to cal, divide by 4.184.

**Table 10. Producers[a] of Polyamides by Region**

| Producer | Manufacturing facilities[b] | Principal polymer types |
|---|---|---|
| *North and South America* | | |
| AlliedSignal | U.S. Eur | 6 and amorphous |
| Amoco | U.S. | 6,T/6,I/6,6 and 6,T/6,6 |
| Cookson Fibers | U.S. | 6 and 6,6 fibers |
| Du Pont | NA, Eur, A/P, SA | 6,6; 6,12; 6,T/D,T; aramids; 6; and amorphous |
| Fairway Filamentos[c] | SA | 6,6 |
| Hoechst-Celanese | U.S. | 6,6 resins |
| Mohawk | U.S. | 6 carpet fibers |
| Monsanto | U.S., A/P | 6,6; 6,9; and 6,10 |
| Shaw Industries | U.S. | 6 carpet fibers |
| *Europe, Middle East, and Africa* | | |
| Akzo | Eur, U.S., SA, India | 6; 6,6; and aramid |
| Aqualon | Italy | 6 |
| BASF | Eur, NA | 6,6; 6; 6,10; and 6,T/6 |
| Bayer | Eur | 6 and amorphous |
| Beaulieu | Eur, U.S., Africa | 6 carpet fiber |
| Chimvolorno (Grodno) | Belarus | 6 |
| DSM | Netherlands | 6; 6,6; 4,6; and 6/12 |
| Elf Atochem | France | 11; 12; 6; 6,6; and amorphous |
| EMS | Switzerland, U.S. | 6; 12; and amorphous |
| Hüls | Germany | 6/12; 12; and TMD,T |
| Nilit | Israel | 6,6 |
| Radici | Italy | 6 and 6,6 |
| Rhône-Poulenc | Eur, SA, A/P | 6,6; 6; and 6,10 |
| RP-Snia Joint Ventures | Eur | 6 |
| SANS | South Africa | 6,6 |
| Snia | Italy | 6 films |
| Stilon | Poland | 6 |
| *Asia/Pacific* | | |
| Asahi | A/P | 6,6 and 6 |
| Chung Shing | Taiwan | 6 |
| Formosa Chem. & Fiber Co. | Taiwan | 6,6 and 6 |
| Hualon | Taiwan | 6 |
| Kanebo | Japan | 6 |
| Kohap | Korea | 6 |
| Kolon | Korea | 6,6 |
| Liaoyiang | China | 6,6 |
| Mitsubishi Gas | Japan | MXD/6 |
| Mitsui | Japan | 6,T/6,I and 6,T/6,6 |
| PT Branta Mulia | Indonesia | 6,6 fiber |
| Shen Ma (Pingdingshan) | China | 6,6 |
| Shri Ram | India | 6 |
| Tae Kwang | Korea | 6,6 |
| Teijin | Japan | aramids |
| Teijin–Du Pont joint venture | Japan | 6 to 6,6[d] |
| Tong Yang | Korea | 6 |

**Table 10. (*Continued*)**

| Producer | Manufacturing facilities[b] | Principal polymer types |
|---|---|---|
| Toray | A/P | 6; 6,6; 6,10; and 12 |
| Toray–Du Pont joint venture | Japan | aramid |
| Toyobo | Japan | 6 |
| Ube | Japan | 6; 6,6; and 12 |
| Unitika | Japan | 6 |

[a]There are many smaller facilities producing nylon-6,6 and especially nylon-6.
[b]United States (U.S.), Europe (Eur), North America (NA), South America (SA), and Asia/Pacific (A/P).
[c]Joint venture of Hoechst-Rhodia.
[d]The polymer base is being converted from nylon-6 to nylon-6,6.

In the second step, KA is further oxidized by nitric acid [7697-37-2] to adipic acid, which is separated, purified usually through crystallization, and dried (see ADIPIC ACID).

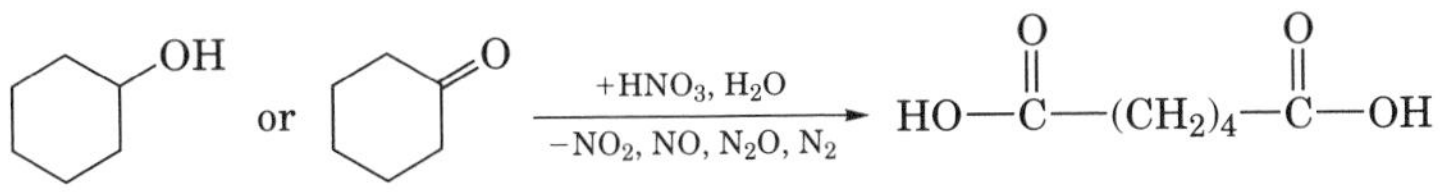

Virtually all the hexamethylenediamine manufactured is used captively in the production of nylon-6,6, with a small amount being used to produce diisocyanates. It is produced via the reduction of adiponitrile [111-69-3] (ADN):

$$N\equiv C(CH_2)_4C\equiv N + 4\,H_2 \xrightarrow{catalyst} H_2N(CH_2)_6NH_2$$

The highly exothermic reduction reaction is generally carried out in either a high or low pressure process (128). In the high pressure process, an iron or cobalt catalyst is used at 20–35 MPa (3000–5000 psi) and a temperature of 100–150°C. The low pressure process uses a Raney nickel catalyst containing iron or chromium activated by sodium or potassium hydroxide in solution and operates at a pressure of 2–3.5 MPa (300–500 psi) and a temperature of 60–100°C (Raney is a registered trademark of W. R. Grace Co.). Although the temperature and pressure ranges are substantially reduced in the low pressure process and thus the capital investment is also reduced, the productivity is lower and the catalyst costs are higher than for the high pressure processes.

There are three commercial routes to ADN in use. The first method, direct hydrocyanation of 1,3-butadiene [106-99-0], has replaced an older process, cyanation via reaction of sodium cyanide with 1,4-dichlorobutane [110-56-5] owing to the lower cost and fewer waste products of the new process. During the initial steps of the direct hydrocyanation process, a mixture of two isomers is generated, but the branched isomer is readily converted to the linear 3-pentenenitrile [4635-87-4].

$$CH_2\!=\!CH\!-\!CH\!=\!CH_2 \xrightarrow[+HCN]{catalyst} CH_2\!=\!CH\underset{\underset{\textstyle CH_3}{|}}{CH}\!-\!C\!\equiv\!N + CH_3CH\!=\!CHCH_2\!-\!C\!\equiv\!N$$

In the final step the dinitrile is formed from the anti-Markovnikov addition of hydrogen cyanide [74-90-8] at atmospheric pressure and 30–150°C in the liquid phase with a Ni(0) catalyst. The principal by-product, 2-methylglutaronitrile

$$CH_3CH\!=\!CHCH_2\!-\!C\!\equiv\!N \xrightarrow[+HCN]{catalyst} N\!\equiv\!C(CH_2)_4C\!\equiv\!N + N\!\equiv\!C\underset{\underset{\textstyle CH_3}{|}}{C}\!-\!CH(CH_2)_2C\!\equiv\!N$$

[4553-62-2], when hydrogenated using a process similar to that for the conversion of ADN to hexamethylenediamine, produces 2-methyl-1,5-pentanediamine or 2-methylpentamethylenediamine [15520-10-2] (MPMD), which is also used in the manufacture of polyamides as a comonomer.

In the second method to produce ADN, known as electrohydrodimerization, two moles of acrylonitrile [107-13-1] are combined and hydrogenated in an electrochemical cell where the two half-cells are separated by a membrane.

$$2\,CH_2\!=\!CH\!-\!C\!\equiv\!N \xrightarrow{+2\,e^-} 2\,CH_2\!=\!CH\!-\!C\!\equiv\!N^{\cdot} \xrightarrow[-2\,OH^-]{+2\,H_2O} N\!\equiv\!C(CH_2)_4C\!\equiv\!N$$

In the second-generation process a membrane is not needed to separate the two half-cell reactions; instead, a finely divided, two-phase emulsion consisting of an organic phase containing acrylonitrile is used, and ADN is suspended in the aqueous phase containing Na$_2$HPO$_4$ [13708-85-5], a tetraalkylammonium salt, and acrylonitrile (qv). The electrodes can be graphite and magnetite, or, more recently, cadmium and iron (129,130).

In the third method adipic acid is converted to ADN via dehydroamination with NH$_3$ in the gas (131) or liquid phase (132); a dehydration catalyst, usually phosphoric acid, is used.

$$HOOC(CH_2)_4COOH \xrightarrow{+2\,NH_3} H_4\overset{+}{N}\quad {}^-OOC(CH_2)_4COO^-\quad \overset{+}{NH_4} \xrightarrow[catalyst]{-2\,H_2O} H_2N\overset{\overset{\textstyle O}{\|}}{C}(CH_2)_4\overset{\overset{\textstyle O}{\|}}{C}NH_2$$

The diamide intermediate is dehydrated to form the dinitrile.

$$H_2N\overset{\overset{\textstyle O}{\|}}{C}(CH_2)_4\overset{\overset{\textstyle O}{\|}}{C}NH_2 \xrightarrow[catalyst]{-2\,H_2O} N\!\equiv\!C(CH_2)_4C\!\equiv\!N$$

This same method, followed by hydrogenation, is also used to produce the C-12 and C-13 diamines from the corresponding diacids in the production of nylon-12,12 and nylon-13,13.

**Polymer Production.** Three processes are used to produce nylon-6,6. Two of these start with nylon-6,6 salt, a combination of adipic acid and hexamethylenediamine in water; they are the batch or autoclave process and the continuous polymerization process. The third, the solid-phase polymerization process, starts with low molecular weight pellets usually made via the autoclave process, and continues to build the molecular weight of the polymer in a heated inert gas, the temperature of which never reaches the melting point of the polymer.

Nylon-6,6 salt, hexamethylenediammonium adipate [3323-53-3], is made by adding adipic acid to a solution of hexamethylenediamine in water (133). The diamine is usually stored at approximately 85% concentration in water to keep it in the liquid state for ease of handling and blanketed with an inert gas, $N_2$, to prevent oxidation and color formation. The diamine is diluted further before reaction so that the final concentration of salt is about 50%. To this solution a slight excess of diacid is added and thoroughly dissolved and neutralized; then, in a second reactor, the remaining diamine is added to reach the desired stoichiometric balance. The end point is determined very accurately via pH measurement on a diluted sample. In the past the salt solution was then filtered through activated charcoal to remove color and impurities, but in the 1990s the quality of the starting materials and the degree of process control have eliminated the need for this step. This salt solution is held under inert gas until it is sent to the polymerization process. The dilution of the ingredients and exothermic heat of reaction (110 kJ/mol (26.4 kcal/mol)) (134) generate substantial heat, which can be recovered and used in other steps in the process.

If dry salt is to be isolated from solution, it can be precipitated by the addition of methanol [67-56-1], washed, filtered, and dried, since its solubility in alcohol is low (0.4% at 25°C) (135). An alternative method has been developed to produce dry salt in a continuous process using a saturated nylon-6,6 salt solution from which the solid salt can be precipitated by adjusting the temperature and concentration (136). Adipic acid is dissolved in a recycled, saturated aqueous solution of nylon salt at approximately 60°C under nitrogen. When diamine is added as an 85% aqueous solution, additional salt is formed, most of which precipitates out of solution, and the temperature rises. The mixture is cooled, which precipitates the maximum amount of salt. This is separated by centrifuging the slurry; the isolated salt is then washed and dried. The mother liquor is reheated and recycled. Dry salt is quite stable; it can be used to transport ingredients for nylon-6,6 over long distances, and does not require an inert atmosphere to prevent degradation. However, it is never used internally for the production of nylon-6,6, because the pure diacid–diamine or salt solution has a lower manufacturing cost.

In the first step of the autoclave process, the salt solution is concentrated to about 75% by boiling in a vessel commonly referred to as an evaporator. The heat contained in the steam generated in this step can be recovered in a heat exchanger and used elsewhere in the process. After the salt is concentrated, it is fed into an autoclave, which is a large and sometimes stirred pressure vessel, where the polymerization takes place. Often additives, such as $TiO_2$, stabilizers, etc, are injected at this point in the process. During the polymerization, the temperature and pressure are adjusted to control the loss of diamine and to release steam

(137). In the first step of the autoclave cycle the vessel is closed and the temperature is raised, which allows the pressure to increase to 1.75 MPa (250 psig). During this step, most of the diamine and diacid react to form oligomers. After the boiling mass is held at pressure for a time sufficient for the reaction to come to equilibrium (138), the excess steam is vented and the polymerization reaction proceeds to build molecular weight. As a result of the equilibrium between the bound and unbound diamine, some free diamine is lost with the exiting steam owing to its low boiling point, but it can be recovered and recycled. In the final step, the temperature is raised to 275°C while the pressure is held at atmospheric pressure. The polymerization then approaches equilibrium. An additional vacuum finishing step can be added if still higher molecular weight polymer is desired; however, this process step can lead to complications such as discoloration resulting from oxidation introduced from the atmosphere via vacuum leaks, thermal degradation from the longer time at high temperature, and difficulty in draining the polymer mass from the vessel owing to the higher viscosity achieved. After polymerization is completed, the polymer is drained from the autoclave through a die which forms a series of continuous polymer strands. These are fed into a casting machine that cools and solidifies the polymer in water, cuts it into small cylindrical pellets, and separates the pellets from the water. The pellets are then dried and stored for further processing or sale. The advantages of the autoclave process are relatively low capital investment, flexibility, and relatively simple operation. The disadvantages are low throughput, high labor costs, variation of polymer properties through the batch and from batch to batch, and difficulty in obtaining high molecular weight polymer.

The continuous polymerization (CP) process was invented to solve the limitations inherent in the batch process. The process makes use of several vessels through which the polymeric material flows while it grows in molecular weight (139–141). In a modern CP unit, nylon salt is introduced into the first vessel, which functions much like an evaporator to concentrate the salt solution. It is then fed into the bottom of a pressurized column reactor, where the initial polymerization reaction takes place. The column allows the steam that is produced as a by-product to be removed without the loss of hexamethylenediamine. After a suitable residence time, higher molecular weight polymer is produced and then pumped into the flasher, a vessel designed to lower the pressure of the polymer mass while steam escapes from the polymer. Flow through the flasher is quite complex because there are two phases with decreasing pressure and increasing temperature. From the flasher the polymer enters the separator at a pressure slightly above atmospheric and at a temperature of approximately 275°C. In the separator, water vapor is separated from the polymer, which has reached about 98% of reaction. The polymer is then pumped into the vacuum finisher, which maintains the pressure at about 40 kPa (300 mm Hg). The final degree of polymerization is completed in the finisher. The final polymer is pumped from the bottom of the finisher and sent directly to a spinning machine or to a die for pelletizing. Additives addition can be made at various points along the CP process such as in the initial salt, the flasher, or after the finisher. The advantages of the CP are high, continuous throughput, uniform polymer properties, and the ability to produce high molecular weight and low cost, high volume polymer. Its disadvantages are high capital investment and complex operation and maintenance.

An alternative route to high molecular weight polymer is the solid-phase polymerization (SPP) process (142). There are two basic types of SPP process: batch and continuous. The batch process is little more than a rotary dryer operating at a higher temperature under a flowing atmosphere of controlled-moisture inert gas. In the continuous SPP process, pellets are introduced into a holding vessel where they are slowly heated to about 100°C under nitrogen. In this step, most of the atmospheric oxygen absorbed by the pellets during formation and storage is removed, and the pellets are dried to a uniform, initial moisture level. The pellets are then fed into the SPP vessel, which is a large, vertical plug-flow reactor with a counterflow of hot nitrogen gas. The moisture level is adjusted in the hot gas to control the final moisture level in the pellets and the rate of polymerization. Typical reactor conditions are a temperature of 150–200°C and a holdup time of 6–24 h. The finished pellets are removed at the bottom of the reactor and can be sent directly to an extruder for spinning or compounding, or to a cooler and then packaged or stored for future sale. An important component of an SPP reactor, whether batch or continuous, is the gas recycle loop. Because large quantities of nitrogen are needed, recycling the gas is essential. As the gas leaves the top of the reactor it contains added moisture and organic materials, mostly cyclic oligomers, which must be removed before the gas is reused. After the gas is scrubbed, the moisture level must be readjusted and the gas reheated to the level required in the process.

The SPP process has the advantage of being able to produce polymer of very high molecular weight without increasing the thermal degradation of the polymer. Although this is useful for nylon-6,6, it can be essential for other types of polyamides that cannot be processed in the melt phase owing to thermal degradation, eg, many aromatic-containing polyamides. Unfortunately, this is done at the cost of long holdup times in the continuous process or a slow processing step in batch processing. In addition, thermooxidative damage is always increased in the polymer along with the accompanying increase in yellow color, because oxygen can never be completely excluded from the reactor at a practical cost. Another disadvantage is the additional capital investment above that needed for autoclaves or CP units.

## NYLON-6

Nylon-6 [25038-54-4] was first made in 1899 by heating 6-aminohexanoic acid (143), but its commercially feasible synthesis from caprolactam was discovered by Paul Schlack at I. G. Farbenindustrie in 1938. Like nylon-6,6, it is a tough, white translucent, semicrystalline solid, but melts at a lower temperature ($T_m$ = 230°C). The physical properties and primary producers of nylon-6 are listed in Tables 9 and 10, respectively.

**Ingredients.** Nylon-6 is produced commercially from caprolactam [105-60-2], which is the most important lactam industrially. All industrial production processes for caprolactam are multistep and produce ammonium sulfate [7783-20-2] or other by-products. Approximately 95% of the world's caprolactam is produced from cyclohexanone oxime [100-64-1] via the Beckmann rearrangement (144). The starting material for cyclohexanone can be cyclohexane, phenol [108-95-2] (eq. 18), or benzene [71-43-2]. Then, through a series of reductions (for the aromatic starting materials) and oxidations, cyclohexanone is formed. This cyclic ketone then reacts with a hydroxylamine salt, usually the sulfate, to form the oxime and ammonium sulfate. The oxime is rearranged in concentrated sulfuric acid, and the resulting lactam sulfate salt is hydrolyzed to form caprolactam and more ammonium sulfate (eq. 19).

$$\text{(Ph-OH)} \xrightarrow[\text{oxidation}]{\text{reduction}} \text{(cyclohexanone)} \xrightarrow[+NH_3 \ -H_2O]{+\frac{1}{2}\ (NH_2OH)_2 \cdot H_2SO_4} \text{(cyclohexanone oxime)} + \frac{1}{2}\ (NH_4)_2SO_4 \quad (18)$$

$$\text{(cyclohexanone oxime)} \xrightarrow{H_2SO_4 \ conc} \text{(lactam} \cdot \frac{1}{2}\ H_2SO_4) \xrightarrow{NH_3} \text{(lactam)} + \frac{1}{2}\ (NH_4)_2SO_4 \quad (19)$$

An additional mole of ammonium sulfate per mole of final lactam is generated during the manufacture of hydroxylamine sulfate [*10039-54-0*] via the Raschig process, which converts ammonia, air, water, carbon dioxide, and sulfur dioxide to the hydroxylamine salt. Thus, a minimum of two moles of ammonium sulfate is produced per mole of lactam, but commercial processes can approach twice that amount. The DSM/Stamicarbon HPO process, which uses hydroxylamine phosphate [*19098-16-9*] in a recycled phosphate buffer, can reduce the amount to less than two moles per mole of lactam. Ammonium sulfate is sold as a fertilizer. However, because $H_2SO_4$ is released and acidifies the soil as the salt decomposes, it is a low grade fertilizer, and contributes only marginally to the economics of the process (145,146) (see CAPROLACTAM).

**Polymer Production.**    Commercially the ring-opening polymerization of caprolactam to nylon-6 is accomplished by both the hydrolytic and anionic mechanisms. However, the hydrolytic process is by far the most predominantly used method because it is easier to control and better adapted for large-scale production. Like nylon-6,6, the polymerization process for nylon-6 via the hydrolytic mechanism can be batch or continuous; however, the processes for the two polymers are significantly different. The hydrolytic process for nylon-6 contains the following steps: caprolactam and additives addition, hydrolysis, addition, condensation, pelletizing (for remelt processing), leaching/extraction of monomers, drying, and packaging (for pellet sales) (147). Caprolactam is usually handled as a molten liquid because its melting point is 69°C and it can be melted with hot water. Increasingly caprolactam is being shipped as a molten liquid vs a dried solid for ease of handling. When delivered as a solid, it is melted and fed into the first step of the process as a liquid.

Batch processing of nylon-6 is generally used only for the production of specialty polymers such as very high molecular weight polymer or master batch polymers for special additives. In a typical modern batch process (147–150), the caprolactam is mixed in a holding tank with the desired additives and then charged to an autoclave with a small amount (2–4%) of water. During the two-stage polymerization cycle, the temperature is raised from 80 to 260°C. In the first stage, water is held in the reactor, the pressure rises, and the hydrolysis and addition steps occur. After a predetermined time the pressure is released and the final condensation reaction step occurs. The molecular weight of the polymer can be increased by means of a vacuum finishing step, if desired. The entire process can take three to five hours. The final polymer is then drained, often with a forcing pressure of inert gas, through a die to form ribbons of polymer, which are then cooled in water and cut into pellets. Because nylon-6 has such a high monomer and oligomer content, 10–12% by weight, in the cast pellets, which would significantly reduce the quality of the final fiber or resin products, it must be extracted. This is usually done in hot water under pressure at 105–120°C for 8–20 h. Most of the caprolactam and higher oligomers that are released with the steam from the autoclave or extracted from the pellets in hot water are then recycled. The pellets must be carefully dried because excess water decreases the molecular weight of the polymer during subsequent melt processing. The final polymer processed through water extraction and drying can have an oligomer level of <0.2% and a moisture level of <0.05%. A low level of total oligomers is necessary because on remelting and further processing, the oligomers' content will increase owing to the reestablishment of the equilibrium distribution of molecular species that occurs for all condensation polymers (151). Because the approach to equilibrium progresses at a moderate rate, it is possible to utilize extracted nylon-6 in a remelt process without increasing the oligomer concentration above 2–3% and thus avoiding any significant drop in final properties.

In the continuous polymerization process for nylon-6 the three steps of polymerization can be made to take place in a series of connected vessels or in a single long, vertical, tubular reactor, sometimes referred to as the VK tube (147–150,152). There are many variations and proprietary reactor designs for the continuous process. However, the purpose of the various configurations is the same as in the batch process; first, to provide water-rich reaction media to accelerate hydrolysis and the initial coupling of the monomers; and then to provide a low water environment for the polycondensation reaction to approach equilibrium. The polymer can be cast and cut into pellets in a continuous pelletizing and drying operation similar to the batch process. However, a continuous, vacuum-stripping step can be applied to remove much of the caprolactam while the nylon-6 is still in the melt; the polymer can then be spun directly into fiber or cast into resin. The oligomer content in this case is 2–3%; and, when this level of oligomers is acceptable, the advantage of the lower cost for vacuum stripping can be attained by eliminating the water quench, extraction, drying, and remelt steps.

In order to attain higher molecular weight polymer, nylon-6 can also be polymerized in the solid phase in a manner similar to that used for nylon-6,6. The same advantages and disadvantages arise in the case of nylon-6, except that one additional difficulty occurs for nylon-6 from the presence of residual caprolactam. Since the temperature range of the SPP process is 140–170°C, the residual lactam is a liquid or vapor in the reactor. As a liquid it can collect on the surface of the pellets and reactor walls, which causes the pellets to stick together or to the walls, thus impeding their uniform flow through the SPP reactor. More vapor-phase oligomers can collect in the gas recycle loop.

Anionic polymerization is also used in commercial processing, but not for fiber production. By casting the rapidly polymerizing polymer or using reaction injection molding (RIM), stock or custom-shaped bulk polymer items can be fabricated directly (153). The stock shapes can be readily machined to the desired finished item; or, in the case of large production runs, the RIM process can produce directly the finished shape in a mold. However, much after-processing is required in the RIM-cast articles. In a typical casting process, sodium lactamide is generally used as the catalyst, and either hexamethylene diisocyanate [822-06-0] (HDI) or imidodicarbonic diamide [4035-89-6], a biuret derivative of HDI, is used as the activator. Anionic polymerization is essentially a direct monomer-to-finished item process. Two streams of caprolactam, one containing the catalyst and the other containing the activator, are mixed and then fed into a heated mold. The rate of reaction and the process temperatures are adjusted to reduce the buildup of internal stresses as the polymer reacts and cools. The finished parts can be annealed at 100–120°C for eight hours to further relieve the stresses. Anionic polymerization is usually practiced commercially below the melting point of nylon-6, at approximately 160°C. An interesting consequence of this is that the concentration of residual caprolactam is only about 2% in the final nylon-6; this is significantly below what would be expected from the extrapolated concentration as a function of temperature in the melt, ie ~7% (154). Thus, the oligomers do not need to be extracted from the bulk plastic articles, which makes this process economically feasible. Another interesting feature of anionic polymerization is its ability to produce block copolymers. Because of the two ingredient streams and the low processing temperature, it is possible to blend and react caprolactam with a prepolymer, such as modified polyols, and randomization is avoided. Thus, block copolymer plastics can be produced that might be impossible in the melt at 250°C. Random copolyamides can also be made by blending and reacting capro- and laurolactams to make nylon-6-co-nylon-12.

## COMPARISON OF NYLON-6,6 AND NYLON-6

Nylon-6,6 and nylon-6 have competed successfully in the marketplace since their respective commercial introductions in 1939 and 1941, and in the 1990s share, about equally, 90% of the total polyamide market. Their chemical and physical properties are almost identical, as the similarity of their chemical structure might suggest: the amide functions are oriented in the same direction along the polymer chain for nylon-6, but are alternating in direction for nylon-6,6.

Table 9 shows the similarity of properties; however, a few differences between the two polyamides do exist: in melting point, in ingredients preparation, and in polymer manufacturing. The melting point of nylon-6,6 is about 40°C higher than that for nylon-6. This is an advantage for nylon-6,6 in those end uses where high temperature performance is required, such as under-the-hood applications for automobiles, high speed thermal processing of fibers and films, and high temperature fatigue resistance in industrial tire cords (qv). On the other hand, the lower processing temperatures for nylon-6 result in slightly lower energy costs and could potentially permit the use of some additives which would decompose at the higher temperatures necessary for processing nylon-6,6. Nylon-6 appears to have a definite advantage in ingredients preparation because it requires the capital investment and handling costs for only one monomer, whereas nylon-6,6 requires the same for two monomers. However, the added cost of ammonium sulfate production, handling, and sales increases the cost of caprolactam production. Also hexamethylenediamine is increasingly being made from three- or four-carbon petroleum-based hydrocarbons vs higher cost six-carbon feedstocks for caprolactam (and adipic acid). In general, the nylon-6,6 ingredients are made in very large plants by a few producers, which allows for substantial economy-of-scale. This has an unexpected consequence for the relative growth of the two polymers. Although caprolactam and adipic acid are commodity chemicals, hexamethylenediamine is not, since almost its entire world production is consumed internally to produce nylon-6,6. Thus nylon-6 can be made from its monomer, purchased on the open market without the investment in an ingredients facility, whereas nylon-6,6 cannot. This probably accounts for the growth of nylon-6 in such developing areas as Asia/Pacific.

Finally, there are significant differences in polymer production. Nylon-6 requires the extraction of caprolactam and other oligomers, which increases the capital investment as well as operating costs in polymerization. On the other hand, nylon-6,6 is plagued with a propensity to branch and gel when exposed to the required higher processing temperatures for extended periods of time. Proper management of gel deposits in nylon-6,6 manufacturing and processing steps in order to maintain high polymer quality requires skill and experience, which undoubtedly adds to its overall costs. Finally, nylon-6,6 can be readily processed from ingredients to final polymer in 2 hours, whereas nylon-6 takes 12–24 hours.

The two polymers appear to be well balanced, and future competitive pressure will almost assuredly come not from each other, but rather from other polyamides and, even more likely, from other polymers, such as low cost polyolefins and polyesters or high performance engineering resins.

## Other Aliphatic Nylons

**Nylon-4,6.** In the 1980s, Dutch State Mines (DSM) introduced a new commercial aliphatic polyamide, nylon-4,6 [50327-22-5] (155). This polymer was first studied by Carothers in the 1930s but was dropped, presumably the result of difficulties in obtaining a high molecular weight and the presence of substantial color formation resulting from degradation. In order to commercialize this polymer, DSM had to overcome this difficulty and the lack of a low cost supply of the

diamine, tetramethylenediamine [110-18-9]. DSM developed a two-step process
to produce the diamine, in the first of which acrylonitrile and HCN are combined,
with triethylamine [121-44-8] (TEA) as the catalyst (156). The second step entails
hydrogenation of the dinitrile, which is similar to the process used to produce
hexamethylenediamine for nylon-6,6.

$$CH_2{=}CHC{\equiv}N + HCN \xrightarrow{\text{TEA}} N{\equiv}C(CH_2)_2C{\equiv}N \xrightarrow[+4 H_2]{\text{catalyst}} H_2N(CH_2)_4NH_2$$

The second difficulty, degradation, required the development of a two-step
polyamidation process following salt formation (157). During salt formation,
tetramethylenediammonium adipate salt is formed in water solution at approximately 50% concentration or at a higher concentration in a suspension. As in
nylon-6,6 manufacture, this salt solution, when diluted, permits easy adjustment
of the stoichiometry of the reactants by means of pH measurement.

$$H_2N(CH_2)_4NH_2 + HOOC(CH_2)_4COOH \xrightarrow{H_2O} H_3\overset{+}{N}(CH_2)_4\overset{+}{N}H_3 + {}^-OOC(CH_2)_4COO^-$$

In the first step of the polymerization process, a prepolymer is prepared
as a slurry in water. Excess diamine is added to control the degree of polymerization, eg, degree of polymerization = 6–14 (158). This prepolymerization step
is conducted at approximately 200°C under autogenous pressure for less than
90 min.

$$n\, H_3\overset{+}{N}(CH_2)_4\overset{+}{N}H_3 + n^-OOC(CH_2)_4COO \xrightarrow[-2n\,H_2O]{+H_2N(CH_2)_4NH_2}$$
$$H{-}[NH(CH_2)_4NHCO(CH_2)_4CO{-}]_n NH(CH_2)_4NH_2$$

The prepolymer is separated from the water by spray drying and then
formed into cylindrical pellets of uniform size (159). At this point additives can be
added to the porous pellets from solution or suspension. These pellets are then
placed in a solid-phase condensation reactor where they are heated to 260°C
for up to four hours under nitrogen, with a small amount of water added. The
pressure is maintained close to atmospheric pressure. At the end, $x > n$.

$$H{-}[NH(CH_2)_4NHCO(CH_2)_4CO{-}]_n NH(CH_2)_4NH_2 \longrightarrow$$
$$H{-}[NH(CH_2)_4NHCO(CH_2)_4CO{-}]_x NH(CH_2)_4NH_2$$

The use of preformed pellets of uniform size is important because the rate
of solid-phase polymerization and thus the uniformity of the degree of polymerization is dependent on particle size. This is especially important for nylon-4,6
because the polymer is not held in the melt long enough for transamidation to
establish a uniform molecular weight distribution, owing to the sensitivity of
nylon-4,6 to thermal degradation. The excess diamine is removed from the nitrogen stream and recycled. The pellets of high molecular weight nylon-4,6 exit
the solid-phase reactor and are cooled and stored under nitrogen. Although this

manufacturing process effectively minimizes any thermal degradation damage during polymer production, downstream melt-processing, such as injection molding or fiber spinning, requires increased care over what is necessary for nylon-6 or nylon-6,6, so as to avoid extended periods in the melt where further degradation damage could occur.

Nylon-4,6 has a high melting temperature ($T_m = 295°C$), high crystallinity, and a much faster crystallization rate, ie, four to eight times faster than that for nylon-6,6. As an unfilled plastic, nylon-4,6 has a high tensile strength, 80 MPa (12,000 psi), compared to 55–65 MPa (8250–9750 psi) for other polyamides; when filled with 30% glass, it has a heat deflection temperature of 285°C, vs 190–240°C for most other polyamides (160). Its dielectric properties have been well documented (161). All these properties make nylon-4,6 a good candidate for high temperature applications and end uses that require good resistance to impact and abrasion. Nylon-4,6 has been studied for fiber applications (162) and as tire cord it is claimed to be 30% better than nylon-6,6 in flat-spot index measurements, which is even better than polyester in this test (155). Besides its limited stability in the melt phase, its other significant drawback as an engineering plastic is its high moisture regain. Copolymers of nylon-4,6 with terephthalic acid, caprolactam, and nylon-6,10 have also been prepared (163–165).

**Nylon-6,9, Nylon-6,10, and Nylon-6,12.** These related polyamides are produced in a process similar to that used for nylon-6,6, where a salt of hexamethylenediamine and the appropriate diacid is formed in water. The solution is heated in an autoclave until polymerization is complete. Processing times, pressures, and temperatures are adjusted for the slightly different melting points and viscosities of these polymers. Because of the lower melting points, ie, nylon-6,9 ($T_m = 210°C$), nylon-6,10 ($T_m = 220°C$), and nylon-6,12 ($T_m = 212°C$), and the perhaps greater chemical stability of the diacids, these polymers generally experience less thermal degradation in processing than nylon-6,6. They are generally used as engineering resins for specialty applications where reduced moisture regain and chemical resistance are important. Nylon-6,12 [24936-74-1] and its copolymers are also used in the manufacture of toothbrush bristles and fishing line.

The diacids for these polymers are prepared via different processes. Azelaic acid [123-99-9] for nylon-6,9 [28757-63-3] is generally produced from naturally occurring fatty acids via oxidative cleavage of a double bond in the 9-position, eg, from oleic acid [112-80-1]:

$$CH_3(CH_2)_7CH{=}CH(CH_2)_7COOH \xrightarrow{+2\ O_2} HOOC(CH_2)_7COOH + CH_3(CH_2)_7COOH$$

The by-product of this process, pelargonic acid [112-05-0], is also an item of commerce. The usual source of sebacic acid [111-20-6] for nylon-6,10 [9008-66-6] is also from a natural product, ricinoleic acid [141-22-0] (12-hydroxyoleic acid), isolated from castor oil [8001-79-4]. The acid reacts with excess sodium or potassium hydroxide at high temperatures (250–275°C) to produce sebacic acid and 2-octanol [123-96-6] (166) by cleavage at the 9,10-unsaturated position. The manufacture of dodecanedioic acid [693-23-2] for nylon-6,12 begins with the catalytic trimerization of butadiene to make cyclododecatriene [4904-61-4], followed by reduction to cyclododecane [294-62-2] (see BUTADIENE). The cyclododecane is oxidatively cleaved to dodecanedioic acid in a process similar to that used in adipic acid production.

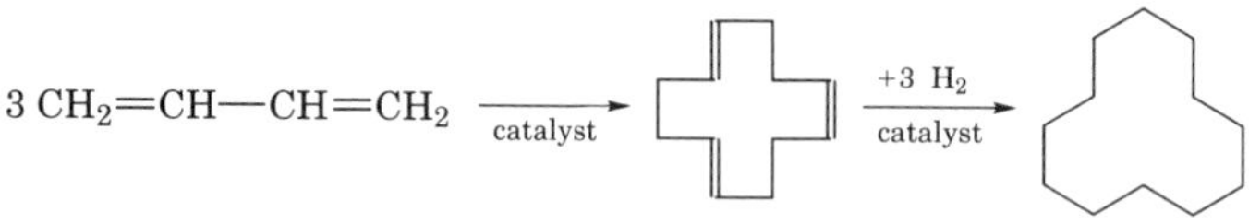

$$3\ CH_2\!\!=\!\!CH\!\!-\!\!CH\!\!=\!\!CH_2 \xrightarrow[\text{catalyst}]{} \qquad \xrightarrow[\text{catalyst}]{+3\ H_2} \qquad$$

**Nylon-11.**   Nylon-11 [25035-04-5], made by the polycondensation of 11-aminoundecanoic acid [2432-99-7], was first prepared by Carothers in 1935 but was first produced commercially in 1955 in France under the trade name Rilsan (167); Rilsan is a registered trademark of Elf Atochem Company. The polymer is prepared in a continuous process using phosphoric or hypophosphoric acid as a catalyst under inert atmosphere at ambient pressure. The total extractable content is low (0.5%) compared to nylon-6 (168). The polymer is hydrophobic, with a low melt point ($T_m$ = 190°C), and has excellent electrical insulating properties. The effect of formic acid on the swelling behavior of nylon-11 has been studied (169), and such a treatment is claimed to produce a hard elastic fiber (170).

The starting amino acid for nylon-11 is produced from methyl ricinoleate [141-24-2], which is obtained from castor oil (qv). The methyl ricinoleate is pyrolized to methyl 10-undecylenate [25339-67-7] and heptanal [111-71-7]. The unsaturated ester is hydrolyzed and then converted to the amino acid by hydrobromination, followed by ammoniation and acidification. The $\omega$-amino acid product is a soft paste containing water, which is dried in the first step of the polymerization process.

$$CH_3(CH_2)_5CHOHCH_2CH\!\!=\!\!CH(CH_2)_7COOCH_3 \xrightarrow[-\text{heptanal}]{\Delta} CH_2\!\!=\!\!CH(CH_2)_8COOCH_3 \xrightarrow[-CH_3OH]{+H_2O}$$

$$CH_2\!\!=\!\!CH(CH_2)_8COOH \xrightarrow[\text{peroxide}]{+HBr} BrCH_2CH_2(CH_2)_8COOH \xrightarrow[-NH_4Br]{+2\ NH_3\ acid} H_2N(CH_2)_{10}COOH$$

**Nylon-12.**   Laurolactam [947-04-6] is the usual commercial monomer for nylon-12 [24937-16-4] manufacture. Its production begins with the mixture of cyclododecanol and cyclododecanone which is formed in the production of dodecanedioic acid starting from butadiene. The mixture is then converted quantitatively to cyclododecanone via dehydrogenation of the alcohol at 230–245°C and atmospheric pressure. The conversion to the lactam by the rearrangement of the oxime is similar to that for caprolactam manufacture. There are several other, less widely used commercial routes to laurolactam (171).

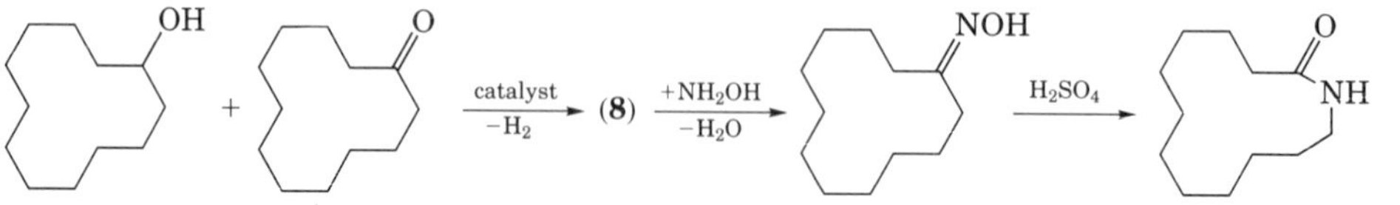

(8)

The mechanism for the production of nylon-12 from the lactam is similar to that for nylon. However, in the case of nylon-12, the ring opening is more difficult and the rate of polymerization is slower, at least in part owing to the lower solubility of the lactam in water. A catalyst such as an acid, amino acid, or nylon salt can serve as a ring-opening agent. Nylon-12 can also be produced via anionic polymerization, ie, polymerization using an anhydrous alkali catalyst. This process can be quite fast even at low temperatures, eg, a few minutes at 130°C.

The properties of nylon-12 compared to nylon-6 and nylon-6,6 include lower moisture regain, which gives it better electrical resistance and dimensional stability. It has a significantly lower melting point ($T_m = 178°C$). The dielectric and dynamical mechanical properties have recently been studied in dry and water-saturated polymer (172,173). When dry, it is less strong but retains its strength better in high relative humidity and subzero environments. It is less soluble in polar solvents, but more soluble in nonpolar ones. Because of these properties, it has found application in automotive, mechanical, and electrical parts; in food packaging (qv); and in other specialty polymer end uses.

**Nylon-12,12.** Nylon-12,12 [36497-34-4], [36348-71-7] was introduced into the marketplace by Du Pont in the late 1980s (174). This polymer possesses very low moisture absorption, high dimensional stability, and excellent chemical resistance, with a moderately high melt point ($T_m = 185°C$) (175). Its manufacture begins with the formation of dodecanedioic acid produced from the trimerization of butadiene in a process identical to that used in the manufacture of nylon-6,12. The other starting material, 1,12-dodecanediamine, is prepared in a two-step process that first converts the dodecanedioic acid to a diamide, and then continues to dehydrate the diamide to the dinitrile. In the second step, the dinitrile is then hydrogenated to the diamine with hydrogen in the presence of a suitable catalyst.

$$HOOC(CH_2)_{10}COOH \xrightarrow[-2\ H_2O]{+2\ NH_3} H_2N\overset{\overset{O}{\|}}{C}(CH_2)_{10}\overset{\overset{O}{\|}}{C}NH_2 \xrightarrow[-2\ H_2O]{} NC(CH_2)_{10}CN$$

$$\xrightarrow[\text{catalyst}]{+4\ H_2} H_2N(CH_2)_{12}NH_2$$

The polymerization process proceeds in a manner similar to that of other type AABB polyamides, such as nylon-6,6. The final resin had found application in automotive and other high performance end uses but was withdrawn from the market in 1994.

**Nylon-13,13 and Nylon-13.** The ingredients for nylon-13,13 [26796-68-9], [26796-70-3] and nylon-13 [14465-66-8], [26916-48-3] and their copolymers have become available in developmental quantities from a natural source, crambe and rapeseed oil (176). Erucic acid [112-86-7] is obtained in high yield approaching 50 wt % from the oil and oxidatively cleaved to produce the dicarboxylic acid, brassylic acid [505-55-2] and pelargonic acid:

$$CH_3(CH_2)_7CH{=}CH(CH_2)_{11}COOH \longrightarrow HOOC(CH_2)_{11}COOH + CH_3(CH_2)_7COOH$$

Then in a series of chemical transformations the diamine or lactam can be prepared from brassylic acid (177,178). The diamine is formed as described above for the 12-carbon diamine, ie, diacid → diamide → dinitrile → diamine. The lactam is made from the dinitrile as follows.

$$N\equiv C(CH_2)_{11}C\equiv N \xrightarrow[\text{catalyst}]{+2\,H_2} H_2NCH_2(CH_2)_{11}C\equiv N \xrightarrow[-NH_3]{+H_2O}$$

The pelargonic acid by-product is already a useful item of commerce, making the overall process a commercial possibility. The 13-carbon polyamides appear to have many of the properties of nylon-11, nylon-12, or nylon-12,12: toughness, moisture resistance, dimensional stability, increased resistance to hydrolysis, moderate melt point, and melt processability. Thus, these nylons could be useful in similar markets, eg, automotive parts, coatings, fibers, or films. Properties for nylon-13,13 are $T_g = 56°C$ and $T_m = 183°C$ (179).

## Copolyamides and Mixed Aliphatic–Aromatic Polyamides

Copolymers of polyamides, when prepared in the melt or in solution or when held in the melt for a sufficient time, are usually fully randomized. Such random copolyamides generally show a decrease in melting point, slower rate and lower degree of crystallization, lower modulus, and a higher solubility than that of either homopolymer. The variation in melting point with a change in the relative composition of the two polymers shows a depression in melting point similar to the eutectic behavior observed for nonpolymeric compounds (180) (Fig. 8). However, there are select combinations of polyamides for which melting point in-

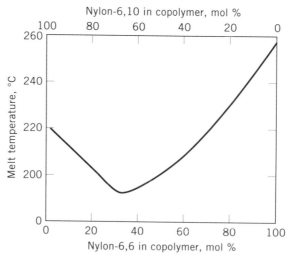

**Fig. 8.**   Reduction in melting point for nylon-6,6/6,10 copolymers.

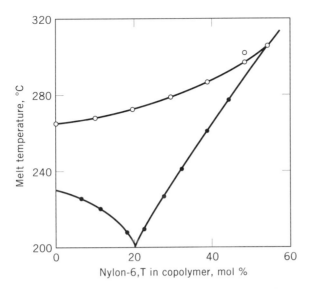

**Fig. 9.** Reduction in melting point for nylon-6,6/6,T (○) and nylon-6,10/6,T (●) copolymers (181).

creases monotonically with composition (Fig. 9). The monomers of these copolymers can replace each other isomorphically in their crystal lattices, and these copolymers are called isomorphic copolymers, ie, the monomers are sufficiently similar in their three-dimensional structure that they do not significantly distort the crystal lattice when one is replaced by the other. This effect was first observed in the nylon-6,6/6,*T* system (182) and in the nylon-6–poly(*endo*-ethylenecaprolactam) system (183).

O
‖
C
CH₂   NH          HO—C
                      CH₂
                              NH₂

*endo*-ethylenecaprolactam        4-aminocyclohexylacetic acid

Many combinations of diacids–diamines and amino acids are recognized as isomorphic pairs (184), for example, adipic acid and terephthalic acid or 6-aminohexanoic acid and 4-aminocyclohexylacetic acid. In the type AABB copolymers the effect is dependent on the structure of the other comonomer forming the polyamide; that is, adipic and terephthalic acids form an isomorphic pair with any of the linear, aliphatic C-6–C-12 diamines but not with *m*-xylylenediamine (185). It is also possible to form nonrandom combinations of two polymers, eg, physical mixtures or blends (Fig. 10), block copolymers, and strictly alternating (187–188) or sequentially ordered copolymers (189), which show a variation in properties with composition differing from those of the random copolymer. Such combinations require care in their preparation and processing to maintain their

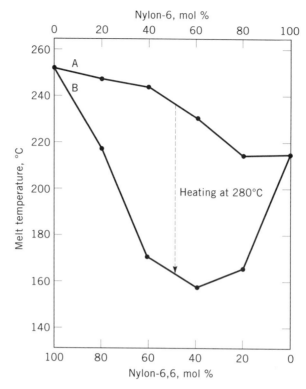

**Fig. 10.**  Effect on melting point of varying composition of nylon-6,6 and nylon-6, where A represents a physical blend, and B, an equilibrated random copolymer (186).

nonrandom structure, because transamidation introduces significant randomization in a short time above the melting point.

Because of the capacity to tailor select polymer properties by varying the ratio of two or more components, copolymers have found significant commercial application in several product areas. In fiber-spinning, ie, with copolymers such as nylon-6 in nylon-6,6 or the reverse, where the second component is present in low (<10%) concentration, as well as in other comonomers with nylon-6,6 or nylon-6, the copolymers are often used to control the effect of spherulites by decreasing their number and probably their size and the rate of crystallization (190). At higher ratios, the semicrystalline polyamides become optically clear, amorphous polymers which find applications in packaging and barrier resins markets (191).

An emerging development is the introduction of high temperature polyamide resins for automotive, under-the-hood use and in some electrical end uses, such as cores for transformer windings. At first glance, nylon-6,T appears to be an excellent candidate, because it has a very high melting point (365°C) and its components, hexamethylenediamine and terephthalic acid, are low in cost and readily available. However, its high melting point requires even higher processing temperatures, which in turn lead to substantial thermal degradation. The attempt to produce copolymers with nylon-6 or nylon-6,6 in a melt process is thwarted by the formation of macroscopic portions of high melting blocks of nylon-6,T that can act as nucleating agents for spherulite (microscopic particles) formation, as particulate contamination (visible particles), and as nonmelting reactor fouling (bulk material). In addition, when nylon-6,6 is blended with nylon-6,T, the required higher processing temperatures accelerate the rate of branching and gel formation to such an extent that the process is inoperable above 300°C. Since nylon-6,T and nylon-6,6 are isomorphic, they maintain a high degree of crystallinity in the copolymer, but when nylon-6 is used, its copolymer with nylon-6,T demonstrates the usual eutectic-like drop in melting point. Thus, a high ratio of nylon-6,T to nylon-6 is required to attain a significant increase in melting point. At high ratios, the copolymer can lose much of its crystallinity, but at a lower ratio it is still semicrystalline, and a satisfactory nylon-6,T/6 copolymer has been introduced by BASF. Amoco has introduced a proprietary process that allows the production of terephthalic acid-based copolymers which also contain isophthalic acid (192). These materials have been named polyphthalamides and show many desirable properties, such as a high melting point, high $T_g$, and low moisture regain (193).

Another approach to the production of high melting terephthalate-based copolyamides is first to make a low molecular weight prepolymer and then solid-phase the material to higher molecular weight; this process is similar in principle to that used in the manufacture of nylon-4,6. A variation of this process is used by Mitsui to produce its nylon-6,T/6,6 product, a copolymer of nylon-6,T and nylon-6,6 via a two-step process. First, an oligomer of the copolymer is made in an autoclave and spray-dried. The particles are then fed into an extruder, where the final copolymer is produced. A third approach, used by Du Pont, is to add a second diamine, 2-methylpentamethylenediamine (trade name Dytek A) rather than a second diacid to reduce the melting point (194,195). This nylon-6,T/D,T copolymer is produced via an all-melt phase process in an autoclave. Although the resulting polymer has a high melt point, the process avoids the added cost of special process equipment and handling. Table 11 presents information on most of the high temperature resins that have been introduced into the marketplace; nylon-6,6 and nylon-4,6 are included for comparison.

Two additional aromatic monomers have become commercially available for the production of polyamides: $m$-xylylenediamine and 2,6-naphthalenedicarboxylic acid. Mitsubishi Gas has introduced a nylon homopolymer,

**Table 11. High Temperature Polyamide Resins**

| Trade name[a] | Producer | Components | Melting point, °C |
|---|---|---|---|
| Amodel A-1000 | Amoco | nylon-6,T/6,I/6,6 | 310 |
| Super Amodel | Amoco | nylon-6,T/6,6 | 323 |
| Ultramid T | BASF | nylon-6,T/6 | 295 |
| Stanyl | DSM | nylon-4,6 | 295 |
| Zytel HTN | Du Pont | nylon-6,T/D,T | 300 |
| Zytel 101 | Du Pont | nylon-6,6 | 265 |
| Arlen A | Mitsui | nylon-6,T/6,I | 320 |
| Arlen C | Mitsui | nylon-6,T/6,6 | 310 |

[a]Amodel is a registered trademark of Amoco Oil Co., Ultramid is a registered trademark of BASF, Stanyl is a registered trademark of DSM, Zytel is a registered trademark of Du Pont Co., Inc., and Arlen is a registered trademark of Mitsui Petrochemical Industries, Ltd.

nylon-MXD,6, based on MXD and adipic acid under the trade name Reny (Reny is a registered trademark of Mitsubishi Gas Chemical Company, Inc.). Though the melting point of this polymer ($T_m = 243°C$) lies between those of nylon-6,6 and nylon-6, its glass-transition point is significantly higher, ie, $T_g = 102°C$ (dry), and it has lower water absorption (196). 2,6-Naphthalenedicarboxylic acid has been introduced by Amoco in development quantities for polyester and polyamide production. Several patents (197) and a publication (198) have appeared in regard to poly(hexamethylene naphthalamide) and nylon-6,N [26401-12-7] and its copolymers, but no commercial polymer has been introduced. The naphthalate aromatic system absorbs uv radiation at a longer wavelength than the terephthalate or isophthalate systems, which suggests that polymers containing the naphthalate segment would offer better uv protection for materials packaged in films containing nylon-6,N (199). If the concentration of aromatic chromophores is high enough in the polymer, it can increase the uv lifetime of the fiber, film, or plastic part itself (197) by absorbing the radiation at the surface of the polymer.

## Aromatic Polyamides

Polyamides that contain 85% or greater of the amide bonds attached to aromatic rings are classified as aramids. These aromatic polyamides have been the subject of numerous reviews (200–206) and a book (207) since 1985; and although this area of research has been very active including the development of many chemical variations (202), commercially it is still dominated by those meta- and para-aramids sold as fibrous materials under the trade names for *m*-aramids, predominately poly(*m*-phenyleneisophthalamide) [24938-60-1]: Apial, Nomex, and Tejiconex; and for *p*-aramids, predominately poly(*p*-phenyleneterephthalamide) [24938-64-5]: Kevlar, Technora, and Twaron (Apial is a registered trademark of Unitika, Ltd.; Nomex and Kevlar are registered trademarks of Du Pont Company, Inc.; Tejiconex and Technora are registered trademarks of Teijin Ltd.; and Twaron is a registered trademark of Akzo, NV) (see HIGH PERFORMANCE FIBERS).

poly(*m*-phenyleneisophthalamide)          poly(*p*-phenyleneterephthalamide)

The aramids were first introduced into the marketplace by Du Pont in 1961 with the commercialization of Nomex fiber, which is used in flame-resistant fabrics and as an industrial fiber, such as in powerhouse filtration systems. Nomex has also been processed into a variety of papers, in which form it is used primarily as electrical insulation. The *p*-aramids were also first introduced commercially by Du Pont as Kevlar fiber in 1972. They possess outstanding tensile strength and modulus and are stronger than fiber glass or steel on a performance-per-weight basis. They can be heat-treated to improve these properties even further. Their end uses include tire cord, industrial fibers, and plastic reinforcement. The *p*-aramids are also being sold as pulp for vehicle brake linings and in woven and nonwoven blends, with *m*-aramids as fire-blocking materials for aircraft seating and other end uses.

The diacid components for the manufacture of poly(*m*-phenyleneiso-phthalamide) and poly(*p*-phenyleneterephthalamide) are produced by one of two processes. In the first, the diacid chlorides are produced by the oxidation of *m*-xylene [*108-38-3*] or *p*-xylene [*106-42-3*] followed by the reaction of the diacids with phosgene [*75-44-5*]. In the second, process *m*- or *p*-xylene reacts with chlorine initiated by ultraviolet light to form the *m*- or *p*-hexachloroxylene. This then reacts with the respective aromatic dicarboxylic acid to form the diacid chloride.

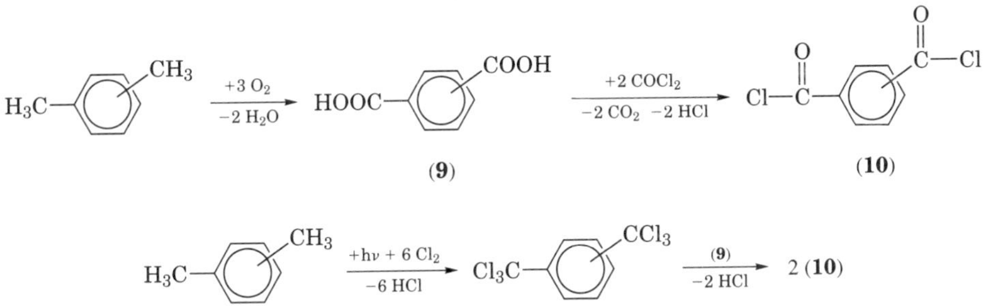

The synthesis of *m*-phenylenediamine [*108-45-2*] is also straightforward; it proceeds via the formation of *m*-dinitrobenzene [*99-65-0*] by the nitration of benzene, followed by hydrogenation to the diamine.

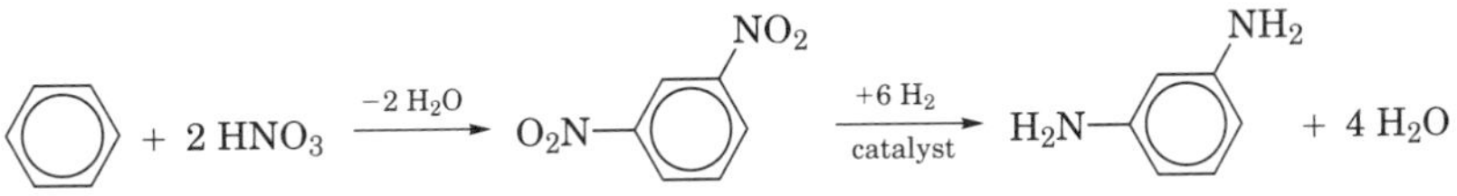

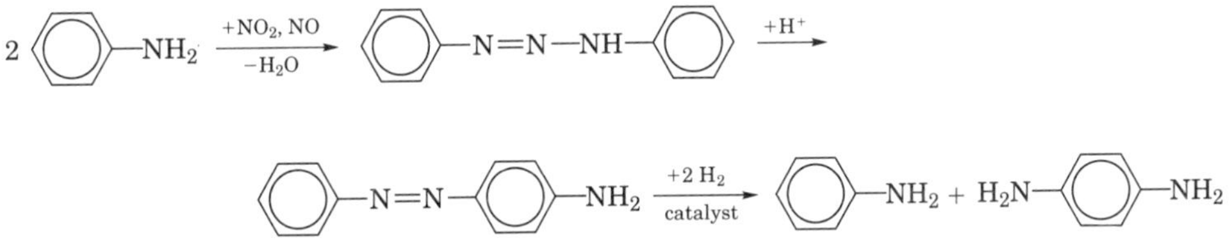

However, the production of the *p*-phenylenediamine [*106-50-3*] interme-diate is more complex, because it involves the diazotization and coupling of aniline [*65-53-3*]. Aniline reacts with nitrogen oxides, produced via the oxidation of ammonia, to form 1,3-diphenyltriazene [*136-35-6*] in the process used by Du Pont (208,209) (see AMINES, AROMATIC–ANILINE AND ITS DERIVATIVES). In the Akzo process a metal nitrite salt and acid in water is used (210). The triazene rearranges in the presence of acid and an excess of aniline to form predominately the p-aminoazobenzene [60-09-3] and a small amount of the ortho isomer, o-aminoazobenzene [2835-58-7]. The mixture of isomers is catalytically reduced to the respective diamines, and they are then separated from the aniline, which is recycled (208,209). The o-phenylenediamine [95-54-5] is used in the manufacture of herbicides (see AMINES, AROMATIC–PHENYLENEDIAMINES).

The aramids are formed in the low temperature reaction, $-10$ to $60°C$, of equimolar amounts of the diacid chloride and the diamine in an amide solvent, typically dimethyl acetamide (DMAc) or N-methyl-2-pyrrolidinone (NMP) and usually with a small amount of an alkali or alkaline-earth hydroxide and a metal salt, such as LiOH [1310-65-2], LiCl, Ca(OH)$_2$ [1305-62-0], or CaCl$_2$ added to increase the solubility of the polymer and neutralize the hydrochloric acid generated in the reaction.

The m-aramids are fully soluble at a moderate concentration in the solvent and are usually spun directly from the solution in a dry or wet spinning process, washed in water to remove excess solvent, and dried. High molecular weight poly(phenyleneterephthalamide), the polymer used to form Kevlar and Twaron aramid fiber, has a lower solubility in the amide solvents and is not typically spun directly into fiber. Rather it is precipitated from its polymerization solvent by the addition of water, neutralized, filtered, dried, and then redissolved in concentrated sulfuric acid. The acid solution is then spun through a spinnerette into a water bath after it passes through a short gap in the air that is essential to the process (211). The p-aramids form a liquid crystalline solution in the concentrated acid (212) and the spinning process further orients the stiff, rod-like polymer molecules. This high degree of orientation, which is maintained in the washed, neutralized, and dried fiber, is responsible for the extraordinary properties of these fibers. Technora aramid fiber, a poly(phenyleneterephthalamide) copolymer, is soluble in NMP or DMAc solutions with CaCl$_2$ or LiCl. It is spun from an isotropic solution, then drawn 10 times to generate its high orientation. The aramids do not appear to be sold as bulk polymer, but only as yarn or staple fiber, paper, pulp, or fabrics. Some of the basic properties of aramid fibers appear in Table 12.

## Health and Environment Aspects

**Health.**   As is the case for almost all commercial polymers, there appears to be no significant recognized health hazard for additive-free polyamides in their normal fiber, film, or bulk plastic end uses, and the same also appears to apply to their higher oligomers (213,214). However, the manufacturer's or supplier's Material Safety Data Sheet (MSDS) should always be consulted for the following reasons: (1) many commercial products contain additives which could significantly alter the health risks associated with the use of polyamides;

**Table 12. Properties of Commercial Aramid Fibers Compared to Nylon-6,6[a]**

| Fiber type | Density, g/cm$^3$ | Strength, GPa[b] | Elongation, % | Modulus, GPa[b] | Maximum use temperature, °C |
|---|---|---|---|---|---|
| Kevlar 29 | 1.43 | 2.9 | 3.6 | 70 | 250 |
| Kevlar 49 | 1.45 | 2.9 | 2.8 | 135 | 250 |
| Kevlar 119 | 1.44 | 3.1 | 4.4 | 55 | 250 |
| Kevlar 129 | 1.45 | 3.4 | 3.3 | 99 | 250 |
| Kevlar 149 | 1.47 | 2.3 | 1.5 | 143 | 250 |
| Technora | 1.39 | 3.3 | 4.3 | 70 | 250 |
| Nomex | 1.38 | 0.6 | 22 | 17 | 250 |
| Nylon-6,6[c] | 1.14 | 1.0 | 18 | 5.5 | <150 |

[a]From Ref. 207.
[b]To convert GPa to gram-force per denier (gpd), multiply by 11.33/density.
[c]Du Pont Type 728 industrial strength fiber.

(2) polyamides comprise an active area of research and changes in recognized levels of safety can occur; and (3) some potential end uses exceed the range of the manufacturer's intended applications for a given product. During incomplete combustion, polyamides can emit toxic products such as carbon monoxide [630-08-0], hydrogen cyanide, and NO$_x$, as well as other less hazardous products (213,214).

The usual starting materials for type AABB polyamides, diamines, and diacids or diacid chlorides, are hazardous materials because they are moderate-to-strong bases, acids, or highly reactive chemicals, respectively. However, there is rarely any detectable starting material in type AABB polyamides. The common starting materials for type AB polyamides, lactams, or aminoacids are generally less hazardous but can be present in the final products. The most significant case is the relatively high concentration of caprolactam in nylon-6. Although caprolactam can be an eye, skin, and respiratory irritant, this is rarely a problem in the final product where the concentrations are kept low (<2–3%) and the lactam is well below its melting point. Mechanical and thermal processing can generate higher levels of caprolactam in the air from nylon-6 or its copolymers and create dust from most polyamides, which can act as irritants and which are usually monitored and corrected in the workplace.

**Environmental Aspects.** In general, the polymerization processes for nylon-6,6 and nylon-6 generate little waste. However, because of economic advantages and governmental regulations, there has been a substantial increase in recycling of the starting materials for polyamides, especially hexamethylenediamine, caprolactam, and water; and of the energy, most often as steam. The hexamethylenediamine and caprolactam are usually emitted as vapors during the polymerization process along with steam and can be condensed, purified, and reused. The caprolactam extracted from the finished polymer, usually in aqueous solution, is also recycled in a similar manner.

During intermediates production, the generation of waste is greater; but again, significant success has been made in isolating, purifying, and selling the by-products, which previously were burned for their fuel value, in this instance as specialty chemicals into other industries. Waste streams whose contaminants are too low in concentration or in value to warrant separation are increasingly being remediated via biotreatment, rather than by more controversial methods such as burning or deep-well disposal. It has been recognized that the nitric acid oxidation of cyclohexanone–cyclohexanol mixture to adipic acid is a significant source of $N_2O$ [10024-97-2] production worldwide (215). Nitrous oxide is potentially a contributor to the greenhouse effect and contributes to catalytic ozone depletion. In 1991 an interindustry consortium was formed to share information on $N_2O$ abatement (216,217). There are three potential, commercially viable routes to solving this problem: (1) thermal combustion of $N_2O$ in a reducing flame to produce nitrogen and carbon dioxide; (2) controlled partial oxidation of $N_2O$ to NO [10102-43-9], then dissolution in $H_2O$ to form nitric acid; and (3) catalytic decomposition of $N_2O$ into nitrogen and oxygen. Du Pont has patented the third method and is making it available for implementation around the world between 1996 and 1998 to eliminate this source of nitrous oxide (218).

Polymer recycle has been practiced as part of the manufacturing process for nylon-6,6 (219) and nylon-6 (220) almost from the beginning of the industry. Acid hydrolysis by Du Pont and base hydrolysis by BASF and Rhône-Poulenc of relatively pure nylon-6,6 waste streams, followed by separation of ingredients, purification, and reuse, has been practiced for many years. Also, phosphoric acid-catalyzed hydrolysis and steam distillation of caprolactam from pure nylon-6 is still used by BASF, Rhône-Poulenc, and SNIA. However, it is the challenge of recycling post-consumer waste that has generated the greatest activity since 1990. Stimulated by more stringent governmental regulations for recycling plastic packaging and automotive plastic components in Germany and the growing landfill problem in the United States, the nylon industry has developed several technologies to address the issue of recycling post-consumer waste. One of the greatest economic challenges is the collection and separation of nylon from other wastes, including other polymers. The ultimate solution to this problem may await the development of a cost-effective waste handling infrastructure for all recycled materials, at which point relatively pure, high volume, low cost, post-consumer nylon will become available. Recycled nylon carpets constitute the largest single supply of potentially recyclable nylon. A patent has appeared for the preparation of a thermoplastic composite by remelting all the components of a nylon carpet and forming it into bulk plastic parts (221). Unfortunately, because of the thermo- and photooxidative products formed in nylon during manufacture and use, and the thermal degradation and thermooxidative products formed during further melt processing, any direct remelt processing of nylon results in a low grade product, even with the use of currently available thermal stabilizers. Other approaches have focused on depolymerization and separation of the ingredients. Several patents and articles have appeared regarding the recovery of caprolactam from post-consumer waste nylon-6 via hydrolysis (222) or polymer pyrolysis (223), and the recovery of polymer via solvent dissolution of nylon-6 from nonpolyamide contamination (224); however, these technologies are limited to waste streams that contain nylon-6 as the only polyamide. In particular, nylon-6,6 significantly interferes with these processes. Several technologies have appeared which attempt to separate nylon-6,6 and nylon-6, and convert them simultaneously to useful monomers (225,226). The most promising technology to date appears to be the ammonolysis of nylon-6,6–nylon-6 mixtures, which converts all three ingredients to hexamethylenediamine (227,228).

## BIBLIOGRAPHY

"Polyamides" in *ECT* 1st ed., Vol. 10, pp. 916–937, by F. Schulze, E. I. du Pont de Nemours & Co., Inc., and H. Wittcoff, General Mills, Inc.; "Polyamides (General)" in *ECT* 2nd ed., Vol. 16, pp. 1–46, by W. Sweeny, E. I. du Pont de Nemours & Co., Inc.; in *ECT* 3rd ed., Vol. 18, pp. 328–371, by R. E. Putscher, E. I. du Pont de Nemours & Co., Inc.

1. W. V. Metanomski, *Compendium of Macromolecular Nomenclature*, IUPAC, Macromolecular Division, Commission on Macromolecular Nomenclature, Blackwell Scientific Publications, Oxford, U.K., 1991.
2. IUPAC, Macromolecular Division, Commission on Macromolecular Nomenclature, *Pure Appl. Chem.* **48**, 375–385 (1976).
3. J. Brandrup and E. H. Immergut, eds., *Polymer Handbook*, 3rd ed., John Wiley & Sons, Inc., New York, 1989.
4. IUPAC, Macromolecular Division, Commission on Macromolecular Nomenclature, *Pure Appl. Chem.* **57**, 1427–1440 (1985).
5. U.S. Pat. 2,071,250 (Feb. 16, 1937), W. H. Carothers (to Du Pont); also see U.S. Pat. 2,071,253 (Feb. 16, 1937), W. H. Carothers (to Du Pont) and U.S. Pat. 2,130,523 (Sept. 20, 1938), W. H. Carothers (to Du Pont) filed Jan. 2, 1935 for synthesis of nylon-6,6.
6. U.S. Pat. 2,241,321 (May 6, 1941), P. Schlack (to I. G. Farbenindustrie AG) filed in Germany June 10, 1938 for synthesis of nylon-6 from caprolactam.
7. W. H. Carothers and G. J. Berchet, *J. Am. Chem. Soc.* **52**, 5289–5291 (1930); W. H. Carothers and J. W. Hill, *J. Am. Chem. Soc.* **54**, 1566–1569 (1932); H. Mark and G. S. Whitby, eds., *Collected Papers of Wallace Hume Carothers on High Polymeric Substances*, Interscience Publishers, New York, 1940.
8. W. H. Carothers, *Chem. Rev.* **8**, 353–426 (1931).
9. D. A. Hounshell and J. K. Smith, Jr., *Science and Corporate Strategy: Du Pont R & D, 1902–1980*, Cambridge University Press, Cambridge, U.K., 1988.
10. *Fortune* **22**, 56–60, 114,116 (July 1940).
11. *Fiber Organ.* **59–66**(6), (June 1988–1995).
12. *Mod. Plast. Int.* **15–24**(1) (Jan. 1985–1995).
13. A. Wood and I. Young, *Chem. Week*, 9 (July 27, 1994).
14. *Modern Plastics Encyclopedia*, McGraw-Hill Book Co., Inc., New York, published annually; S. Davies, *The Man-Made Fibre Industry in Western Europe*, Spec. Rept. 1107, The Economist Publications Ltd., London 1987; S. Davies, *The Man-Made Fibre Industry in Japan*, Spec. Rept. 1174, The Economist Publications Ltd., London 1989.
15. *Polyamide and Intermediates*, Techcon (U.K.) Ltd., London published monthly; *Nylon Intermediates and Fiber*, PCI—Fibers & Raw Materials, West Sussex, U.K. published monthly; *World Nylon 6 and 66 Supply and Demand Report*, PCI—Fibers & Raw Materials, West Sussex, U.K. published annually; *Chemical Economic Handbook*, Stanford Research Institute, Menlo Park, Calif., published and updated periodically.
16. P. Gijsman, D. Tummers, and K. Janssen, *Polym. Degr. Stab.* **49**, 121–125 (1995).
17. C. W. Bunn and E. V. Garner, *Proc. Roy. Soc. (London)* **189A**, 39–68 (1947).
18. D. R. Holmes, C. W. Bunn, and D. J. Smith, *J. Polym. Sci.* **17**, 159–177 (1955).
19. R. Pflueger, in J. Brandrup and E. H. Immergut, eds., *Polymer Handbook*, 3rd. ed., John Wiley & Sons, Inc., New York, 1989, pp. V/109–V/116.
20. J.-I. Wang and I. R. Harrison, *Meth. Exp. Phys.* **16**(B), 128–183 (1980).

21. B. S. Hsiao, A. D. Kennedy, H. Chang, and A. Biswas, Du Pont, private communication, 1995.
22. D. V. Badami and P. H. Harris, *J. Polym. Sci.* **41**, 540–541 (1959); P. H. Geil, *J. Polym. Sci.* **44**, 449–458 (1960).
23. J. H. Magill and P. H. Harris, *Polymer* **3**, 252–256 (1962).
24. P. Dreyfuss and A. Keller, *J. Macromol. Sci., Phys.* **B4**, 811–836 (1970).
25. A. Xenopoulos and E. S. Clark, in M. I. Kohan, ed., *Nylon Plastics Handbook*, Hanser Publishers, Munich, Germany, 1995, pp. 107–137.
26. R. L. Miller, in Ref. 4, pp. VI/1–VI/208.
27. Z. Tuzar, in R. Puffr and V. Kubanek, *Lactam-Based Polyamides*, Vol. 1, CRC Press, Boca Raton, Fla., 1991.
28. S. Steadman and L. J. Mathias, *Polym. Prepr.* (*Div. Polym. Chem., Am. Chem. Soc.*) **34**(1), 507–508 (1993); S. M. Aharoni, F. G. Cilurso, and J. M. Hanrahan, *J. Appl. Polymer Sci.* **30**, 2505–2525 (1995).
29. C. Jackson, H. G. Barth, and M. C. Han, *Polym. Mat. Sci. Eng. Prepr.* (*Div. Polym. Mat. Sci. Eng., Am. Chem. Soc.*) **69**, 270–271 (1993).
30. H. W. Starkweather, Jr. and G. A. Jones, *J. Polym. Sci., Polym. Phys. Ed.* **19**, 467–477 (1981).
31. T. T. Wang, J. M. Herbert, and A. M. Glass, eds., *The Applications of Ferroelectric Polymers*, Blackie & Son Ltd., Glasgow, Scotland, 1988.
32. H. S. Nalwa, in H. S. Nalwa, ed., *Ferroelectric Polymers*, Marcel Dekker, Inc., New York, 1995, pp. 281–323.
33. J. W. Lee, Y. Takase, B. A. Newman, and J. I. Scheinbeim, *J. Polym. Sci. Polym. Phys. Ed.* **29**, 273–277 (1991), *ibid.* **29**, 279–286 (1991); J. W. Lee, B. A. Newman, and J. I. Scheinbeim, *Polym. Prepr.* (*Div. Polym. Chem., Am. Chem. Soc.*) **33**(1), 274–275 (1992).
34. L. J. Mathias, C. G. Johnson, and S. J. Steadman, *Proc. SPIE Int. Soc. Optical Eng., SPIE* **1916**, 309–319 (1993).
35. B. Z. Mei, J. I. Scheinbeim, and B. A. Newman, *Ferroelectrics* **171**, 177–189 (1995).
36. B. A. Newman, J. I. Scheinbeim, J. W. Lee, and Y. Takase, *Ferroelectrics* **127**, 229–234 (1992).
37. J. Su, Z. Y. Ma, J. I. Scheinbeim, and B. A. Newman, *J. Polym. Sci. Polym. Lett. Ed.* **33**(B), 85–91 (1995).
38. S. C. Mathur, A. Sen, B. A. Newman, and J. I. Scheinbeim, *J. Mater. Sci.* **23**, 977–981 (1988).
39. B. A. Newman, K. G. Kim, and J. I. Scheinbeim, *J. Mater. Sci.* **25**, 1779–1783 (1990).
40. J. I. Scheinbeim, B. A. Newman, B. Z. Mei, and J. W. Lee, *ISAF '92: Proc. 8th IEEE Internat. Symp. Appl. Ferroelec.*, 248–249 (1992); B. Z. Mei, J. I. Scheinbeim, and B. A. Newman, *Ferroelectrics* **144**, 51–60 (1993).
41. M. Ohtani, T. Shouko, S. Tasaka, and N. Inagaki, *Polym. Prepr. Jpn.* **41**, 4559–4562 (1992).
42. Y. Murata, K. Tsunashima, N. Koizumi, K. Ogami, F. Hosokawa, and K. Yokoyama, *Jpn. J. Appl. Phys.* **32**, L849–L851 (1993).
43. Y. Murata, K. Tsunashima, and N. Koizumi, *Jpn. J. Appl. Phys.* **33**, L354–L356 (1994).
44. M. H. Litt, C. H. Hsu, and P. Basu, *J. Appl. Phys.* **48**, 2208–2212 (1977).
45. M. H. Litt and J.-C. Lin, *Ferroelectrics* **57**, 171–185 (1984); P. E. Dunn, *Piezoelectric Properties of Nylon 5,7*, Ph.D. dissertation, Northwestern University, Evanston, Ill., 1989, 178 pp.
46. C. D. Papaspyrides and E. M. Kampouris, *Polymer* **25**, 791–796 (1984); C. D. Papaspyrides, *Polymer* **31**, 490–495 (1990).
47. P. W. Morgan, *Condensation Polymers by Interfacial and Solution Methods*, John Wiley & Sons, Inc., New York, 1965.

48. F. Millich and C. E. Carraher, Jr., *Interfacial Synthesis*, Vol. 2, Marcel Dekker, Inc., New York, 1977.
49. S. L. Kwolek, *Polym. Prepr. (Div. Polym. Chem. Am. Chem. Soc.)* **21**(1), 12 (1980).
50. J. Sebenda, in Sir G. Allen and J. C. Bevengton, eds., *Comprehensive Polymer Science*, Vol. 3, Pergamon Press, Oxford, U.K., 1989, pp. 511–530.
51. H. K. Reimschuessel, *J. Polym. Sci. Macromol. Rev.* **12**, 65–139 (1977).
52. H. Sekiguchi, in K. J. Ivin and T. Sagusa, eds., *Ring-Opening Polymerization*, Vol. 2, Elsevier Applied Science Publishers, London, 1984.
53. J. Sebenda, in Ref. 27.
54. H. Sekiguchi and B. Coutin, in H. R. Kricheldorf, ed., *Handbook of Polymer Synthesis*, Part A, Marcel Dekker, New York, 1992, pp. 807–939.
55. B. C. Challis and J. A. Challis, in I. O. Sutherland, ed., *Comprehensive Organic Chemistry*, Vol. 2, Pergamon Press, Oxford, U.K., 1979, 957–1065.
56. J. Zabicky, ed., *The Chemistry of Amides*, Wiley-Interscience, New York, 1970.
57. K. Onder, eds., in W. F. Gum, W. Riese, and H. Ulrich, *Reaction Polymers*, Hanser Publishers, Munich, Germany, 1992, pp. 405–452.
58. V. V. Korshak and co-workers, *Acta Phys. USSR* **21**, 723 (1946).
59. V. V. Korshak and T. M. Frunze, *Synthetic Hetero-Chain Polyamides*, Israel Program for Scientific Translation, Jerusalem, Israel, 1968, pp. 87–93.
60. U.S. Pat. 4,543,407 (Sept. 24, 1985), B. S. Curatolo, R. C. Sentman, and G. P. Coffey (to Standard Oil Co., Ohio).
61. Brit. Pat. 1,053,100 (Oct. 17, 1963) (to BASF); U.S. Pat. 3,454,536 (July 8, 1969), G. Schade and F. Blascke (to Chemische Werke Witten GmbH); U.S. Pat. 3,607,840 (Sept. 21, 1971), S. Yura and Y. Koyake (to Honshu Chemical Industrial Co., Ltd.).
62. Brit. Pat. 1,172,997 (Dec. 3, 1969) A. Lambert and G. H. Lang (to ICI); Jpn. Pat. Appl. Publ. 54-84,525 (July 5, 1979), Y. Kobayashi, Y. Ohnishi, and G. Ishihara (to Toray).
63. U.S. Pat. 5,395,974 (Mar. 7, 1995), R. J. McKinney (to Du Pont).
64. U.S. Pat. 5,302,756 (Apr. 12, 1994), R. J. McKinney (to Du Pont).
65. I. K. Miller, *J. Polym. Sci.: Polym. Chem. Ed.* **14**, 1403–1417 (1976).
66. P. Kress, *Faserforsch. Textiltech.* **11**(8), 353 (1960); T. M. Frunze, V. V. Korshak, and V. A. Makarkin, *Vysokomol. soedin.* **1**, 500 (1959); summarized in L. B. Sokolov, *Synthesis of Polymers by Polycondensation*, Israel Program for Scientific Translation, Jerusalem, Israel, 1968, p. 70.
67. M. J. Han, H. C. Kang, and K. B. Choi, *Macromolecules* **19**, 1649–1652 (1989).
68. M. J. Han, *Macromolecules* **13**, 1009–1012 (1980); *ibid.*, **15**, 438–441 (1982); M. J. Han, H. C. Kang, and K. B. Choi, *Polymer (Korea)* **11**, 349–355 (1987).
69. E. J. Burrell, *J. Am. Chem. Soc.* **83**, 574–577 (1961).
70. J. Stehlicek and co-workers, in Ref. 27.
71. E. B. Mano and F. M. B. Coutinho, *Adv. Polym. Sci.* **19**, 97–116 (1975).
72. P. L. Nayak, *J. Macromol. Sci. Rev. Macromol. Chem.*, **C17**(2), 267–296 (1979).
73. A. K. Mukherjee and H. R. Goel, *J. Macromol. Sci., Rev. Macromol. Chem. Phys.*, **C25**(1), 99–117 (1985).
74. P. L. Nayak and S. Lenka, *J. Macromol. Sci., Rev. Macromol. Chem. Phys.*, **C31**(1), 91–116 (1991).
75. L. B. Sokolov, *Synthesis of Polymers by Polycondensation*, Israel Program for Scientific Translation, Jerusalem, Israel, 1968, pp. 60–71.
76. J. P. Sibila and co-workers, in M. I. Kohan, ed., *Nylon Plastics Handbook*, Hanser Publishers, Munich, Germany, 1995.
77. R. U. Pagilagan, in Ref. 76, pp. 49–61.
78. J. N. Weber, *Polym. Prepr. Jpn.* **44**(1), 66–68 (1995).
79. E. Roerdink and J. M. M. Warnier, *Polymer* **26**, 1582–1588 (1985).
80. Austral. Pat. A-49875/90 (Feb. 19, 1990), P.-Y. Lahary and S. Roy (to Rhône-Poulenc).

81. S. Straus and L. A. Wall, *J. Res. Nat. Bur. Stand.* **60**, 39–45 (1958); L. H. Peebles, Jr. and M. W. Huffman, *J. Polym. Sci., Pt. A-1* **9**, 1807–1822 (1971); D. H. MacKerron and R. P. Gordon, *Polym. Degr. Stab.* **12**, 277–285 (1985).

82. A. Ballistreri, D. Garozzo, M. Giuffrida, and G. Montaudo, *Macromol.* **20**, 2991–2997 (1987).

83. H. Sekiguchi, P. Tsourkas, F. Carriere, and R. Audebert, *Eur. Polym. J.* **10**, 1185–1193 (1974).

84. R. Hill, *Chem. Ind. (London)* **1954**(36), 1083–1089, (Sept. 4, 1954).

85. Jpn. Pats. Kokai 1-104652 and 1-104653 (Oct. 17, 1987), K. Kitamura and K. Kinoshita (to Asahi Chemical Industry).

86. Jpn. Pat. Kokai 1-104654 (Oct. 17, 1987), K. Kitamura and K. Kinoshita (to Asahi Chemical Industry).

87. D. D. Steppens, M. F. Doherty, and M. F. Malone, *J. Polym. Sci.* **42**, 1009–1021 (1991).

88. U.S. Pat. 5,194,578 (Mar. 16, 1993), A. Anton (to Du Pont).

89. U.S. Pats. 5,110,900 (May 5, 1992), 5,162,491 (Nov. 10, 1992), and 5,185,428 (Feb. 9, 1993), J. A. Hammond and D. N. Marks (to Du Pont).

90. J. A. Fitzgerald, R. S. Irwin, and W. Memeger, *Macromolecules* **24**, 3291–3299 (1991).

91. W. H. Sharkley and W. E. Mochel, *J. Am. Chem. Soc.* **81**, 3000–3005 (1959).

92. E. J. Burrell, *J. Am. Chem. Soc.* **83**, 574–577 (1961).

93. J. Lemaire, R. Arnaud, and J.-L. Gardette, *Polym. Degr. Stab.* **33**, 277–294 (1991).

94. E. H. Boasson, B. Kamerbeek, A. Algera, and G. H. Kroes, *Rec. Trav. Chim., Pays-Bas.* **81**, 624–634 (1962).

95. V. Rossbach and T. Karstens, *Chemiefast. Textilind.* **40/92**, 603–611, E44 (1990).

96. D. Fromageot, A. Roger, and J. Lemaire, *Angew. Makcromol. Chem.* **170**, 71–85 (1989).

97. T. Karstens and V. Rossbach, *Makromol. Chem.* **190**, 3033–3053 (1989); *ibid.*, **191**, 757–771 (1990).

98. U.S. Pat. 2,705,227 (Mar. 29, 1954), G. Stoeff (to Du Pont); Brit. Pat. 908,647 (June 22, 1961), W. Costain and A. Hill (to ICI).

99. K. Janssen, P. Gijsman, and D. Tummers, *Polym. Degr. Stab.* **49**, 127–133 (1995).

100. G. Scott, in G. Scott, ed., *Atmospheric Oxidation and Antioxidants*, 2nd. ed., Vol. 2, Elsevier Science Publishers, Amsterdam, the Netherlands, 1993, pp. 141–218.

101. F. Gugumus, in R. Gaechter and H. Mueller, eds., *Plastic Additives Handbook*, 3rd ed., Hanser Publishers, Munich, Germany, 1990, pp. 1–104.

102. J. A. Dellinger and C. W. Roberts, *J. Polym. Sci., Polym. Lett. Ed.* **14**, 167–178 (1976).

103. A. M. Liquori, A. Mele, and V. Cavelli, *J. Polym. Sci.* **10**, 510–512 (1953).

104. I. Goodman, *J. Polym. Sci.* **13**, 175–178 (1954); *ibid.*, **17**, 587–590 (1955).

105. J. F. Rabek, *Polymer Degradation: Mechanisms and Experimental Methods*, Chapman and Hall, London, 1995, pp. 296–306.

106. C. H. Do, E. M. Pearce, B. J. Bulkin, and H. K. Reimschuessel, *J. Polym. Sci., Polym. Chem. Ed.* **25**, 2301–2321 (1987).

107. L. M. Postrekov and A. L. Margolin, in V. Ya Shlyapintokh, ed., *Photochemical Conversion and Stabilization of Polymers*, Hanser Publishers, Munich, Germany, 1984, pp. 114.

108. A. T. Bett and N. Uri, *Chem. Ind. (London)*, 512–513 (1967); Y. Yokoyama, T. Suzuki, and Y. Kurita, *Bull. Chem. Soc. Japan* **59**, 2917–2919 (1986).

109. N. S. Allen, M. J. Harrison, M. Ledward, and G. W. Follows, *Polym. Degr. Stab.* **23**, 165–174 (1989).

110. G. S. Egerton, *Nature* **204**, 1153–1155 (1964).

111. H. A. Taylor, W. C. Tincher, and W. F. Hammer, *J. Appl. Polym. Sci.* **14**, 141–146 (1970); N. S. Allen, J. F. McKellar, G. O. Phillips, and C. B. Chapman, *J. Polym.*

*Sci., Polym. Lett. Ed.* **12**, 723–727 (1974); J. F. Rabek, *Mechanisms of Photophysical Processes and Photochemical Reactions in Polymers*, John Wiley & Sons, Inc., New York, 1987, pp. 581–587.

112. G. S. Egerton and A. G. Morgan, *J. Soc. Dyers Colorists* **87**, 268–277 (1971); H. S. Koenig and C. W. Roberts, *J. Appl. Polym. Sci.* **19**, 1847–1873 (1975).
113. H. S. Freeman and J. Sokolowska-Gajda, *Tex. Res. J.* **60**, 221–227 (1990); Y. Honma, N. Choji, and M. Karasawa, *Tex. Res. J.* **60**, 433–440 (1990).
114. N. S. Allen and J. F. McKellar, *Dev. Polym. Photochem.* **1**, 191–217 (1980); N. S. Allen, M. Ledward, and G. W. Follows, *Polym. Degr. Stab.* **38**, 95–105 (1992).
115. A. Anton, *J. Appl. Polym. Sci.* **9**, 1631–1639 (1965).
116. J. N. Weber, *Polym. Prepr. Jpn.* **44**(1), 66–68 (1995).
117. W. Schnabel, *Polymer Degradation*, Hanser Publishers, Munich, Germany, 1981, pp. 154–177.
118. T. Kelen, *Polymer Degradation*, Van Nostrand Reinhold Co., New York, 1983, pp. 152–156.
119. M. Nagata and T. Kiyotsukuri, *Eur. Polym. J.* **28**, 1069–1072 (1992); M. Nagata, T. Kiyotsukuri, D. Kohmoto, and N. Tsutsumi, *Eur. Polym. J.* **30**, 135–138 (1994).
120. J. L. Espartero, B. Coutin, and H. Sekiguchi, *Polym. Bull.* **30**, 495–500 (1993); S. Bechaouch, B. Coutin, and H. Sekiguchi, *Macromol. Rapid Comm.* **15**, 125–131 (1994).
121. Y. Ban, M. Rikukawa, K. Sanui, and N. Ogata, *Kobunshi Ronbunshu* **50**, 793–796 (1993).
122. P. M. Mungara and K. E. Gonsalves, *Polymer* **35**, 663–666 (1994).
123. W. J. Bailey and B. Gapud, in P. P. Klemchuk, ed., *Polymer Stabilization and Degradation*, ACS Symposium Series, Vol. 280, American Chemical Society, Washington, D.C., 1985, pp. 423–431.
124. J. Sohma, *Prog. Polym. Sci.* **14**, 451–596 (1989); Ref. 117, pp. 64–94.
125. T. Q. Nguyen and H.-H. Kausch, *Makromol. Chem.* **190**, 1389–1406 (1989).
126. T. Nagamura, *Meth. Exper. Phys.* **16**(C), 185–231 (1980); F. Szocs, J. Becht, and H. Fischer, *Eur. Polym. J.* **7** 173–179 (1971); J. Tino, J. Placek, and F. Szocs, *Eur. Polym. J.* **11**, 609–611 (1975).
127. R. Pflueger, "Physical Constants of Various Polyamides", in J. Brandrup and E. H. Immergut, eds., *Polymer Handbook*, 3rd. ed., John Wiley & Sons, Inc., New York, 1989.
128. V. D. Luedeke, in J. J. McKetta and W. A. Cunningham, *Encyclopedia of Chemical Processing and Design*, Marcel Dekker, Inc., New York, 1987, pp. 222–237.
129. K. Weissermal and H.-J. Arpe, *Industrial Organic Chemistry*, 2nd rev. ed., VCH Publishers, Weinheim, Germany, 1993, pp. 235–262.
130. D. Pletcher and F. C. Walsh, *Industrial Electrochemistry*, 2nd ed., Chapman and Hall, London, 1990, pp. 298–311.
131. V. D. Luedeke, in J. J. McKetta and W. A. Cunningham, *Encyclopedia of Chemical Processing and Design*, Marcel Dekker, Inc., New York, 1977, pp. 146–162.
132. U.S. Pat. 3,299,116 (Jan. 17, 1967), R. Romani and Fetri (to Societa Rhodiatoce).
133. I. K. Miller, in J. J. McKetta and W. A. Cunningham, eds., *Encyclopedia of Chemical Processing and Design*, Vol. 31, Marcel Dekker, Inc., New York, 1990, pp. 407–409.
134. N. Ogata, *Makromol. Chem.* **43**, 117 (1961).
135. Brit. Pat. 1,034,307 (June 29, 1966), C. Brierley (to ICI).
136. Ger. Offen. 2,403,178 (Aug. 1, 1974) (to ICI).
137. D. B. Jacobs and J. Zimmerman, in C. E. Schildknecht and I. Skeit, eds., *Polymerization Processes*, Wiley-Interscience, New York, 1977, pp. 424–467.
138. N. Ogata, *Makromol. Chem.* **42**, 52–67 (1960); *ibid.*, **43**, 117–137 (1961).
139. U.S. Pat. 3,948,862 (Apr. 6, 1976), J. M. Iwasyk (to Du Pont).
140. T. S. Chern, in *Encyclopedia of Chemical Processing and Design*, Vol. 39, Marcel Dekker, Inc., New York, 1992, pp. 401–461.

141. V. A. Augstkalns, in Ref. 76, pp. 13–32.
142. J. Zimmerman, *J. Polym. Sci., Polymer Lett. Ed.* **2**(B), 955–958 (1964); D. C. Jones and T. R. White, in G. E. Ham, ed., *Kinetics and Mechanisms of Polymerization,* Vol. 3, of D. H. Solomon, ed., *Step-Growth Polymerization,* Marcel Dekker, Inc., New York, 1975, pp. 41–94; S. Fakirov, in J. M. Schultz and S. Fakirov, eds., *Solid State Behavior of Linear Polyesters and Polyamides,* Prentice-Hall, Inc., Englewood Cliffs, N.J., 1990, pp. 1–74.
143. S. Gabriel and T. A. Maas, *Chem. Ber.* **32,** 1266–1272 (1899).
144. K. Weisermel and H.-J. Arpe, *Industrial Organic Chemistry,* 2nd rev. ed., VCH Publishers, Weinheim, Germany, 1993.
145. K. Sporka, in R. Puffr and V. Kubanek, *Lactam-Based Polyamides,* Vol. 2, CRC Press, Boca Raton, Fla., 1991, pp. 73–90.
146. V. D. Luedeke, in J. J. McKetta and W. A. Cunningham, eds., *Encyclopedia of Chemical Processing and Design,* Vol. 6, Marcel Dekker, Inc., New York, 1978, pp. 72–95.
147. M. K. K. Rao and P. G. Galanty, in Ref. 76, pp. 518–542.
148. V. Komanicky, F. Cupak, and V. Kvarda, in Ref. 145.
149. H. K. Reimschuessel, *J. Polym. Sci. Macromolec. Rev.* **12,** 65–137 (1977).
150. H. Klare, E. Fritzsche, and V. Groebe, *Synthetische Fasern aus Polyamiden,* Akademie-Verlag, Berlin, 1963.
151. S. Smith, *J. Polym. Sci.* **30,** 459–478 (1958); D. Heikens, P. H. Hermans, and S. Smith, *J. Polym. Sci.* **38,** 265–268 (1959); H. K. Reimschuessel and K. Nagasubramanian, *Polym. Eng. Sci.* **12,** 179–183 (1972).
152. P. Matthies, "Polyamides", in *Ullmanns Encyklopaedie der technischen Chemie,* Verlag Chemie, Weinheim, Germany, 1980.
153. J. Bongers and H. Mooij, in Ref. 76, pp. 236–249 and pp. 542–549.
154. O. Wichterle, *Makromol. Chem.* **35,** 174–182 (1960); W. Sweeny and J. Zimmerman, "Polyamides," in N. Bikales, ed., *Encyclopedia of Polymer Science and Technology,* Vol. 10, John Wiley & Sons, Inc., New York, 1969, pp. 483–597.
155. D. O'Sullivan, *Chem. Eng. News* **62**(21), 33–34 (May 21, 1984); *Eur. Chem. News,* 4 (Mar. 28, 1988).
156. Eur. Pat. 16,482 (Feb. 9, 1983), G. H. Suverkropp and co-workers (to Stamicarbon, BV).
157. U.S. Pat. 4,722,997 (Feb. 2, 1988), E. Roerdink and J. M. M. Warnier (to Stamicarbon, BV).
158. Eur. Pat. 534,542 (Mar. 31, 1993), J. M. M. Warnier, P. M. Hendricks-Knape, and E. De Goede (to DSM NV); E. Roerdink, P. J. de Jong, and J. Warnier, *Polym. Comm.* **25,** 194–195 (1984).
159. U.S. Pat. 4,814,356 (Mar. 21, 1989), A. J. P. Bongers and E. Roerdink.
160. J. Heard, *Chem. Eng.* **93**(1), 21, 23–24 (Jan. 1986).
161. P. A. M. Steerman and F. H. J. Maurer, *Polymer* **33,** 4236–4241 (1992).
162. K. Kudo and co-workers, *J. Appl. Polym. Sci.* **52,** 861–867 (1994).
163. R. J. Gaymans, S. Aalto, and F. H. J. Maurer, *J. Polym. Sci.,* **27**(A), 423–430 (1989).
164. Eur. Pat. 254,367 (Jan. 27, 1988), A. J. P. Bongers and E. Roerdink (to Stamicarbon BV).
165. R. G. Beaman and F. B. Cramer, *J. Polym. Sci.* **21,** 223–235 (1956).
166. R. A. Dyntham and B. C. L. Weedon, *Tetrahedron* **8,** 246 (1960); 246 (1961).
167. M. Genas, *Angew. Chem.* **74,** 535–540 (1962).
168. L. Notarbartolo, *Ind. Plastiques Mod.* **10**(2), 44, 47–48, 52 (1958).
169. H. M. Le Huy, X. Huang, and J. Rault, *Polymer* **34,** 340–345 (1993).
170. M. Dosiere, *Makromol. Chem., Macromol. Symp.* **23,** 205–211 (1989).
171. K. Weissermel and H.-J. Arpe, *Industrial Organic Chemistry,* 2nd rev. ed., VCH Publishers, Inc., New York, 1993, pp. 235–262.

172. K. Pathmanathan, J. Y. Cavaille, and G. P. Johari, *J. Polym. Sci., Polym. Phys.* **30**(B), 341–348 (1992); K. Pathmanathan and G. P. Johari, *J. Polym. Sci., Polym. Phys.* **31**(B), 265–271 (1993).

173. J. Varlet, J. Y. Caville, and J. Perez, *J. Polym. Sci., Polym. Phys.* **28**(B), 2691–2705 (1990).

174. L. Nazarenko, ed., *Modern Plast.* **64**(9), 130 (1987).

175. *Plast. World*, 14, (Aug. 1987).

176. D. L. Van Dyne, M. G. Blase, and K. D. Carlson, *Industrial Feedstocks and Products from High Erucic Acid Oil*, University of Missouri-Columbia, Columbia, Mo., 1990, 30 pp.

177. J. L. Greene, Jr. and co-workers, *J. Polym. Sci., Pt. A-1*, **5**, 391–394 (1967).

178. F. Millich and K. V. Seshadri, in K. C. Frish, ed., *Cyclic Monomers*, Wiley-Interscience, New York, 1972, p. 205.

179. L.-H. Wang, F. J. B. Calleja, T. Kanamoto, and R. S. Porter, *Polymer* **34**, 4688–4691 (1993).

180. D. E. Schwartz, in R. Vieweg and A. Mueller, eds., *Kunststoff-Handbuch*, Vol. 6, Carl Hanser Verlag, Munich, Germany, 1966, pp. 160–166; M. I. Kohan, in M. I. Kohan, ed., *Nylon Plastics*, John Wiley & Sons, Inc., New York, 1973, pp. 413–416.

181. O. B. Edgar and R. Hill, *J. Polym. Sci.* **8**, 1–22 (1952).

182. Brit. Pat. Appl. 604/49 (1949), H. Plimmer, R. J. W. Reynolds, L. Wood, and H. A. Hargreaves (to ICI Ltd.).

183. Swiss Pats. 270,546 (Aug. 20, 1949) and 280,367 (Aug. 20, 1949) (to Inventa AG).

184. M. Hewel, "Copolymerization", in Ref. 76.

185. A. J. Yu and R. D. Evans, *J. Polymer Sci.* **42**, 249–257 (1960).

186. S. J. Allen, *J. Tex. Inst.* **44**, P286–P306 (1953); U.S. Pat. 2,193,529 (Mar. 12, 1940), D. D. Coffman (to Du Pont); Brit. Pat. 540,135 (Oct. 7, 1941) (to Du Pont).

187. E. Djodeyre, F. Carriere, and H. Sekiguchi, *Eur. Polym. J.* **15**, 69–73 (1979); N.-M. Tran, F. Carriere, and H. Sekiguchi, *Makromol. Chem.* **182**, 2175–2182 (1981).

188. H. R. Kricheldorf and R. Muehlhaupt, *Angew. Makromol. Chem.* **65**, 169–182 (1977).

189. M. Ueda and co-workers, *Polymer J.* **22**, 73–743 (1990); M. Ueda and co-workers, *Macromol.* **25**, 6580–6585 (1992); M. Ueda and H. Sugiyama, *Macromol.* **27**, 240–244 (1994).

190. U.S. Pat. 4,919,874 (Apr. 24, 1990), W. T. Windley (to Du Pont); U.S. Pat. 5,422,420 (June 6, 1995), K. G. Shridharani (to Du Pont).

191. H. Dalla Torre, in Ref. 76, pp. 377–387.

192. D. Erickson, *Sci. Am.* **265**, 130–131 (1991); U.S. Pat. 4,476,280 (Oct. 9, 1984), W. Poppe and Y.-T. Chen (to Standard Oil Co.).

193. U.S. Pat. 4,603,166 (July 29, 1986), W. Poppe and co-workers (to Amoco); U.S. Pat. Re. 34,447 (Nov. 16, 1993); Technical Bulletin AM-F-50060, Amoco Performance Products, Inc., Chicago, Ill.

194. Fr. Pat. 8,902,467 (Feb. 21, 1989), P.-Y. Lahary and S. Roy (to Rhône-Poulenc).

195. U.S. Pat. 5,378,800 (Jan. 3, 1995), S. L. Mok and R. U. Pagilagan (to Du Pont).

196. *Reny Technical Bulletin*, Mitsubishi Gas Chemical Co., Inc., Tokyo, Japan.

197. Jpn. Pat. 10,224,852 (Jan. 26, 1989), M. Hasuo (to Mitsubishi Chem. Ind.); Jpn. Pat. Appl. 5,009,293 (Jan. 19, 1993), Y. Matsuki, T. Kiriyama, and T. Mita (to Teijin Ltd.); Eur. Pat. 574,297 (Dec. 15, 1993), J.-M. Sage (to Elf Atochem SA).

198. C.-P. Yang, S.-H. Hsiao, and Y.-S. Lin, *J. Appl. Polym. Sci.* **51**, 2063–2072 (1994).

199. Technical Bulletins FA-14 and GTSR-J, Amoco Chem. Co., Chicago, Ill., 1994.

200. H. H. Yang, in J. J. McKetta and W. A. Cunningham, eds., *Encyclopedia of Chemical Processing and Design*, Vol. 40, Marcel Dekker, Inc., New York, 1992, pp. 1–94.

201. L. Vollbracht, in *Comprehensive Polymer Science*, Vol. 5, Pergamon Press, Oxford, U.K., 1989, pp. 375–386.

202. H. H. Yang, *Aromatic High-Strength Fibers*, John Wiley & Sons, Inc., New York, 1989, pp. 66–289.

203. S. L. Kwolek, W. Memeger, and J. E. Van Trump, in M. Lewin, ed., *Polymers for Advanced Technologies*, VCH Publishers, Weinheim, Germany, 1988, pp. 421–454.

204. J. Preston, in J. I. Kroschwitz, ed., *Encyclopedia of Polymer Science and Engineering*, 2nd ed., Vol. 11, John Wiley & Sons, Inc., New York, 1988, pp. 381–409.

205. S. L. Kwolek, P. W. Morgan, and J. R. Schaefgen, in Ref. 204, Vol. 9, 1987, pp. 1–61.

206. M. Jaffe and R. S. Jones, in M. Lewin and J. Preston, eds., *Handbook of Fiber Science and Technology*, Vol. 3, Part A, Marcel Dekker, Inc., New York, 1985, pp. 349–392.

207. H. H. Yang, *Kevlar® Aramid Fiber*, John Wiley & Sons, Inc., New York, 1993.

208. J. March, *Advanced Organic Chemistry*, 3rd. ed., John Wiley & Sons, Inc., New York, 1985, pp. 471, 502–503.

209. U.S. Pat. 4,020,051 (Apr. 26, 1977) and 4,279,815 (July 21, 1981), F. E. Herkes (to Du Pont); U.S. Pat. 4,020,052 (Apr. 26, 1977), J. K. Detrick (to Du Pont).

210. Eur. Pat. 035815 (Apr. 3, 1981), J. de Graaf (to Akzo NV).

211. U.S. Pats. 3,767,756 (Oct. 23, 1973) and 3,869,429 (Mar. 4, 1975), H. Blades (to Du Pont).

212. U.S Pats. 3,600,350 (Aug. 17, 1971), 3,671,542 (June 20, 1972), and 3,819,587 (June 25, 1974), S. L. Kwolek (to Du Pont); S. L. Kwolek, *Polym. Prepr., Div. Polym. Chem.* **21**(1), 12 (1980).

213. *Nylon 6,6 Material Safety Data Sheet*, Du Pont, Wilmington, Del., rev. Feb. 19, 1991; technical data, Sigma-Aldrich Corp., Milwaukee, Wis.

214. *Nylon 6 Material Safety Data Sheet*, Du Pont, Wilmington, Del., rev. Nov. 13, 1995; technical data, Sigma-Aldrich Corp., Milwaukee, Wis.

215. M. H. Thiemens and W. C. Trogler, *Science* **251**, 932–934 (1991); A. K. Naj, *Wall Street J.*, A6 (Feb. 22, 1991); D. Loepp, *Plastic News*, 3 (Mar. 4, 1991).

216. R. A. Reimer and co-workers, *Environ. Prog.* **13**, 134–137 (1994).

217. R. A. Reimer and co-workers, paper presented at the *6th International Workshop on Nitrous Oxide Emissions*, Turku/Abo, Finland, June 7–9, 1994, 25 pp.

218. U.S. Pat. 5,314,673 (May 24, 1994), K. Anseth and T. A. Koch (to Du Pont).

219. U.S. Pats. 2,348,751 (May 16, 1944) and 2,364,387 (Dec. 5, 1944), W. R. Peterson (to Du Pont).

220. Ger. Pats. 887,199 (Dec. 17, 1943) and 910,056 (Jan. 1, 1944) (to BASF).

221. U.S. Pat. 5,294,384 (Mar. 15, 1994), D. J. David, J. L. Dickerson, and T. F. Sincock (to Monsanto).

222. U.S. Pat. 5,169,870 (Dec. 8, 1992), T. F. Corbin, E. A. Davis, and J. A. Dellinger (to BASF); U.S. Pat. 5,241,066 (Aug. 31, 1993), E. A. Davis and J. A. Dellinger (to BASF); U.S. Pat. 5,294,707 (Mar. 15, 1994), R. Kotek (to BASF); R. Sferrazza, A. Sarian, and G. Shore, paper presented at the *1994 Annual Chicago Section of SPE*, Schaumberg, Ill., Nov. 4, 1994, pp. 256–265.

223. R. J. Evans and co-workers, paper presented at *Recycling Plastics VII '92*, Arlington, Va., May 20–21, 1992, 15 pp.

224. R. Hagen, paper 3-6 presented at *Recycle '94*, Davos, Switzerland, Mar. 14–18, 1994, 8 pp.

225. U.S. Pats. 5,266,694 (Nov. 30, 1993) and 5,280,105 (Jan. 18, 1994), E. F. Moran (to Du Pont).

226. U.S. Pat. 5,310,905 (May 10, 1994), E. F. Moran (to Du Pont).

227. U.S. Pats. 5,302,756 (Apr. 12, 1994) and 5,395,974 (Mar. 7, 1995), R. J. McKinney (to Du Pont).

228. R. A. Smith and B. E. Gracon, paper presented at *Recycle '95*, Davos, Switzerland, May 15, 1995, 15 pp.; R. A. Smith, paper presented at the *International Symposium on Polymers and the Environment*, Tokyo, Japan, Oct. 19–20, 1995, 2 pp.

JOSEPH N. WEBER
Du Pont Company, Inc.

# FIBERS

Polyamide fibers are spun from linear thermoplastic polymers having recurring amide groups made from diamines and dicarboxylic acids (CONH—R—NHCO—R)$_n$ or lactams (RCONH)$_n$. Polyamides are generally referred to as nylons when R and R' are essentially aliphatic, alicyclic, and less than 85% of the amide linkages are attached directly to two aromatic rings. When these linkages are equal or greater than 85% aromatic, the fibers are referred to as aramids. Polyamides from the condensation of diamines and dicarboxylic acids are termed AABB type; from lactams, AB type. Aliphatic polyamides are identified by numerals that indicate the number of carbon atoms in the monomer. One number is used for the AB type and two numbers, the first designating the diamine, for the AABB type. For example, the polyamide made from caprolactam (six carbons) is nylon-6 [25038-54-4]; from hexamethylenediamine and adipic acid, nylon-6,6 [9011-56-6]. Monomers containing ring structures are coded by either a single letter such as T for terephthalic acid (nylon-6,T) or an abbreviated form such as PACM for bis(p-aminocyclohexyl)methane, eg, PACM-12. In designating a random copolyamide, the components are named in order of decreasing percentages, followed by their weight percentages in parenthesis. For example, if hexamethylenediamine, adipic acid, sebacic acid, and dodecanedioic acid are copolymerized to give 50% nylon-6,6, 40% nylon-6,10 and 10% nylon-6,12, the product can be designated as nylon-6,6/6,10/6,12 (50/40/10). A block copolyamide, which is made by melt-blending two polyamides or a yarn consisting of fibers spun from two different polymer melt streams such as nylon-6,6 and nylon-6,10, can be designated as nylon-6,6//6,10 (50//50).

Approximately $3.9 \times 10^6$ t of nylon fiber is produced worldwide; nylon-6,6 and nylon-6 account for about 98% of the total production. Nylon fibers are used for carpets, tire cord, cordage, soft-sided luggage, automotive air bags, parachutes, apparel, swimwear, and sheer hosiery. The advantages of nylon fibers over other synthetic fibers are high strength, durability, resilience, ease of dyeability, and low specific gravity.

The fiftieth anniversary of the announcement of nylon as the first synthetic organic textile fiber by the Du Pont Co. on October 27, 1938 was celebrated as a significant event by the textile industry in 1988 (1,2). The announcement was the culmination of the fundamental research efforts of W. H. Carothers and his team at Du Pont (3). Carothers synthesized diamines from $C_2$ to $C_{18}$ in order for them to react with a variety of aliphatic dicarboxylic acids to make polyamides for evaluation as fibers (4–10). Alicyclic and aromatic diamines and dicarboxylic acids were also included. Nylon-6,6 was ultimately selected for scale-up and development because of its favorable melting point (~260°C), best balance of properties, and lower manufacturing cost. The pilot plant for nylon-6,6 was completed in Wilmington, Delaware, in July, 1938, and a product was introduced on the market as Exton bristles for Dr. West's toothbrushes (2). The first nylon filament plant was built in 1939 at Seaford, Delaware, and nylon stockings went on sale on October 24, 1939 only to residents of Wilmington, and then nationally, on May 15, 1940 (2).

In Europe, I. G. Farbenindustrie decided to develop nylon-6 that had been synthesized from ε-caprolactam using an aminocaproic acid catalyst (1) and commercially introduced as Perlon L in 1940 (11,12). I. G. Farbenindustrie had evaluated over 3000 polyamide constituents without finding an improvement over nylon-6 and nylon-6,6 (13). In Italy, Societa Rhodiaceta started making nylon-6,6 in 1939. In the United Kingdom, ICI and Courtaulds formed British Nylon Spinners in 1940 and started to manufacture nylon-6,6 in 1941.

In the United States, Chemstrand Corp. started spinning nylon-6,6 in 1952, and by 1955 Allied Chemical Corporation, American Enka, and Industriale Rayon started spinning nylon-6 fibers. In the 1990s the principal producers of nylon fiber in the United States are Du Pont, Monsanto, AlliedSignal, and BASF. Outside the United States, other key fiber producers include Rhône-Poulenc (France), Snia Fibre SpA (Italy), as well as Toray Industries and Asahi Chemical Co. (both in Japan). However, the downturn in the fiber business, worldwide recession, and overcapacity in the late 1980s led to acquisitions and mergers in the early 1990s (14). Du Pont acquired ICI's nylon filament and staple business in Europe. Snia and Rhône-Poulenc started a joint business in nylon carpet yarn and staple called Novalis Fibers.

## Properties of Nylon-6 and Nylon-6,6

Both the inherent properties and those that can be engineered into the fiber and ultimately into the fabricated article account for the diverse end uses of nylon. For every end use, the fiber must offer performance and/or a perceived market value, meet mill acceptance standards, and have favorable economics.

The properties of textile fibers can be divided into three categories: geometric, physical, and chemical, which can be measured with available methods (15–17). Perceived values such as tactile aesthetics, style appearance of apparel fabrics, comfort of hosiery, as well as color, luster, and plushness of carpets are difficult to quantify and are not always associated with the properties of the fiber, but rather with the method of fabric construction and finishing.

All synthetic fibers that are useful for textile applications are linear, semicrystalline, oriented polymers, whose properties are defined by molecular structure and molecular organization. The first level of molecular organization is the chemical structure that defines the structure of the repeating unit in the base polymer and the nature of the polymeric link. This relates directly to chemical reactivity, dyeability, moisture absorption, and swelling characteristics, and indirectly to all physical properties. Macromolecular structure, the second level, describes the family of polymer molecules in terms of chain length, chain-length distribution, chain stiffness, molecular size, and molecular shape. The third level, supermolecular organization, describes the arrangement of the polymer chains in three-dimensional space. Levels two and three are directly related to the physical properties of textile fibers and have a significant bearing on some of their chemical characteristics. The chemical and macromolecular structures of polyamide fibers can be determined by chemical and instrumental analysis (18). The supermolecular organization of nylon-6 and nylon-6,6 have been defined in terms of crystallinity, crystalline and amorphous orientation, and chain folds by x-ray diffraction, fiber density, dsc, and nmr methods (15,19–23).

Molecular chains in a polyamide fiber are held in specific structural configurations by weak van der Waal and strong intermolecular hydrogen bonding forces between amide groups. An increase in the number of amide bonds raises the melting point in a homologous series as shown with AB and AABB polyamides in Figure 1. Nylon-6 and nylon-6,6 are isomers that share the same empirical formula, $C_6H_4NO$; density, 1.14 gm/cm$^3$; refractive index ($n_D$), 1.530 (unoriented fiber); and many other properties (24). However, they differ in melting point by 40°C because of differences in the alignment of molecular chains and crystallization behavior (23). These parameters are important in fiber formation in melt spinning, but are not as significant as the glass-transition temperature, $T_g$, in the downstream processing of the fiber into a specific end use article (25). The $T_g$, which signifies a transition from a glassy to a rubbery state, plays a role in the drawing, texturing, and dyeing of a fiber. By controlling melt spinning and downstream processing, the fiber properties of both nylon types can be adjusted to accommodate a variety of end use performance requirements.

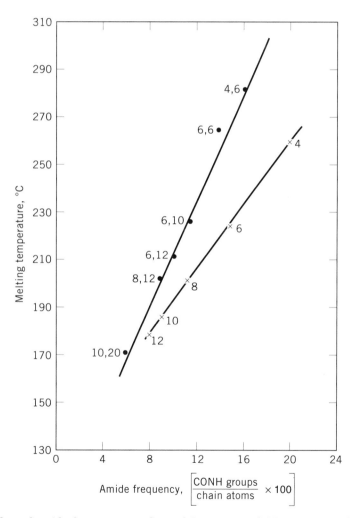

**Fig. 1.** Effect of amide frequency on the melting points of AB-type (×) and AABB-type (●) polyamides. The numbers on the curves indicate the specific nylon.

**Tensile Properties.** Tensile properties of nylon-6 and nylon-6,6 yarns shown in Table 1 are a function of polymer molecular weight, fiber spinning speed, quenching rate, and draw ratio. The degree of crystallinity and crystal and amorphous orientation obtained by modifying elements of the melt-spinning process have been related to the tenacity of nylon fiber (23,27).

Tensile properties of synthetic and natural fibers (or yarn) are measured from stress–strain curves as shown in Figure 2. These measurements are not only important in determining the suitability of a nylon fiber for a specific end use, but also in comparing it to other fiber types. The toughness of nylon is similar to polyester, but higher than silk, Nomex, wool, Kevlar, and cotton. Nylon has a lower modulus than polyester, cotton, silk, Nomex, and Kevlar, the last two produced by Du Pont. Silk has higher strength and lower elongation than standard nylon and polyester textile fiber.

Definitions of the commonly measured tensile properties are as follows: *Linear density* (tex) is the weight in grams of 1000 m of yarn. *Tenacity* is the tensile stress at break and is expressed in force-per-unit linear density of unstrained specimen, N/tex. *Knot tenacity* is the tensile stress required to rupture a single strand of yarn with an overhand knot tied in the segment of sample between the testing clamps. It is expressed as force-per-unit linear density and is an approximate measure of the brittleness of the yarn. *Loop tenacity* is the tensile stress required to rupture yarn when one strand of yarn is looped through another and broken. It is expressed as force-per-unit linear

**Table 1. Tensile Properties of Nylon-6 and Nylon-6,6[a] Continuous-Filament Yarns[b]**

| Property | Nylon-6 | | Nylon-6,6 | |
|---|---|---|---|---|
| | Regular tenacity | High tenacity | Regular tenacity | High tenacity |
| breaking tenacity, N/tex[c] | | | | |
| standard | 0.35–0.64 | 0.57–0.79 | 0.20–0.53 | 0.52–0.86 |
| wet | 0.33–0.55 | 0.51–0.72 | 0.18–0.48 | 0.45–0.71 |
| loop | 0.34–0.49 | 0.45–0.89 | 0.18–0.45 | 0.44–0.67 |
| knot | 0.34–0.48 | 0.42–0.59 | 0.18–0.45 | 0.44–0.67 |
| tensile strength, MPa[d] | 503–690 | 703–862 | 275–731 | 593–924 |
| breaking elongation, % | | | | |
| standard | 17–45 | 16–20 | 25–65 | 15–28 |
| wet | 20–47 | 19–33 | 30–70 | 18–32 |
| average modulus (stiffness), N/tex[c] | 1.6–2.0 | 2.6–4.2 | 0.44–2.1 | 1.9–5.1 |
| average toughness, N/tex[c] | 0.06–0.08 | 0.06–0.08 | 0.07–0.11 | 0.07–0.11 |
| elastic recovery, % | 98–100 at 1–10% | 99–100 at 2–8% | 88 at 3% | 89 at 3% |
| moisture regain at 21°C, % | | | | |
| 65% rh | 2.8–5.0 | 2.8–5.0 | 4.0–4.5 | 4.0–4.5 |
| 95% rh | 3.5–8.5 | 3.5–8.5 | 6.1–8.0 | 6.1–8.0 |

[a]Conditioned at 65% rh and 21°C.
[b]Ref. 26.
[c]To convert N/tex to g/den (gpd), multiply by 11.33.
[d]To convert MPa to psi, multiply by 145.

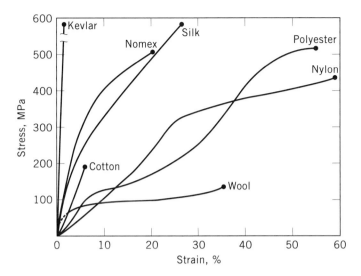

**Fig. 2.** Stress–strain curves of some textile fibers. To convert MPa to Mg/cm$^2$, divide by 100.

density. Loop tenacity is an indication of brittleness, but is not considered as sensitive as the knot test. Reported values are one-half the actual test values.

*Breaking strength* is the maximum load required to rupture a fiber. It is expressed in grams. *Tensile strength* is the maximum stress or load per unit area in units of Pa (1 MPa = 145 psi). Tensile strength (MPa) = tenacity (N/tex) × special gravity × 1005. *Elongation* at break is the increase in sample length during a tensile test and is expressed as a percentage of original length. *Tensile modulus*, ie, Young's modulus, also called initial modulus or elastic modulus, is the load required to stretch a specimen of unit cross-sectional area by a unit amount. It is expressed as the ratio of the tensile stress divided by the tensile strain in the initial straight-line portion of the stress–strain curve, extrapolated to 100% sample elongation. *Work-to-break* is the actual work required to rupture the material. It is proportional to the total area under the stress–strain curve. *Yield point*, or elastic limit, is the point on the stress–strain curve where the load and elongation cease to be directly proportional. It is the point at which the stress–strain curve deviates from the tangent drawn to the initial straight-line portion of the curve, as in tensile modulus determinations.

*Creep* is the change in shape of a material while subject to a stress; it is time-dependent. *Elasticity*, the ability of a fiber to return to its original dimensions upon release of a deformation stress, is described by the tenacity of the yield point on the stress–strain curve. *Toughness*, the ability of a fiber to absorb work before it ruptures, is evaluated from the area under the total stress–strain curve. The toughness index is defined as one-half the product of the stress and strain at break and is also in units of gram-per-unit linear density. *Creep and recovery* define the plasticity or plastic flow of a fiber in relation to load and time (28). At a critical applied load above the yield point, the fiber extends instantaneously and continues to extend at a much slower rate as time progresses, spanning from a few seconds to over a month. When the load is removed, the reverse process occurs. The fiber recovers part of its extension instantaneously and then begins to contract over a period of time (delayed recovery). The recovery is a relaxational phenomenon where the stress that generated the deformation is dissipated as a function of time, which in effect is complimentary to creep. The quantification of the plastic flow of a fiber resulting from molecular motions induced by the deformation and recovery processes are dependent on load, time, and temperature (29).

Elastic recovery is the ability of a fiber to regain its original form after being stretched at a specified percentage of its length. Shrinkage induced by hot–wet treatment of a fiber is also a form of stress relaxation. Compared to other fibers, nylon fibers have an outstanding degree of elasticity, recovering well from high loads and extension.

**Temperature and Moisture Properties.** Thermal treatment and moisture affinity significantly influence the physical properties of nylon fibers and fabrics. The absorption of water causes the fiber to swell, which alters its dimensions and in turn changes the size, shape, stiffness, and permeability of yarns and fabrics. It also alters the frictional and static behavior of yarns in mill processing and the hand and performance of fabrics in use. Water, a powerful plasticizer for nylon, preferentially penetrates the amorphous regions, readily hydrogen-bonds to the amide, and increases molecular chain mobility (30–35). As a result, the absorbed water lowers tensile modulus, the work to elongate and work recovery (36,37), and the glass-transition temperature (38). The crystallinity of nylon also increases as it absorbs water (39), which breaks up intermolecular hydrogen bonding and allows the hydrocarbon chain segments to pack better. Water also affects the heat setting properties of nylon yarns and fabrics (36,40,41). The moisture regain of nylon-6 and nylon-6,6 is shown in Table 1, the change with rh for nylon-6 and other fibers in Figure 3.

Because of water's plasticizing effect, the water content of nylon fibers and fabrics must be known and controlled when measuring physical properties. Prior

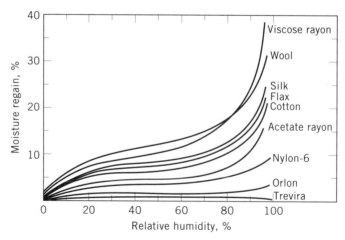

**Fig. 3.** Variation of moisture regain with rh at 20°C. Orlon is an acrylic fiber; Trevira, a polyester.

to the measurement, samples are conditioned at a specified temperature and rh for at least 24 hours.

Nylon-6 melts at 215–220°C and softens at 170°C. Nylon-6,6 melts at 255–260°C and softens at 234°C. Nylons also have thermal behavior properties between the melting and glass-transition temperatures that are important in fiber melt spinning and drawing and in fiber-to-fabric processing. These behavior properties are the result of relaxational phenomena due to the onset of mobility of the aliphatic chain segments in the amorphous and crystalline regions and occur between 70 and 180°C (35,42). Typical unrestrained nylon-6,6 fibers begin to shrink as the temperature increases from 25 to 70°C with the application of dry heat. At 70°C the fiber undergoes crystallization that tends to reduce chain mobility. From 70 to 115°C the fiber continues to shrink and at 115–125°C loses absorbed and bonded water. The crystal structure also changes from the triclinic to the hexagonal form. At 170–180°C, the hexagonal crystals begin to break up and the polymer chains take on a fluid-like mobility. Because of its higher temperature profile, nylon-6,6 requires higher fiber melt processing, texturing, heat setting (43), and end use application temperatures (eg, thermal bonding) than nylon-6. Nylon-6,6, however, is preferred by some over nylon-6 for tire construction because of its superior heat tolerance and higher strength and strength retention at elevated temperatures.

**Electrical Properties.** Nylon has low electrical conductivity (high electrical resistivity) and behaves like an insulator. Nylon-6 has a resistivity of $6 \times 10^{14}$ $\Omega \cdot$cm when dry and a resistivity of $2 \times 10^{14}$ $\Omega \cdot$cm when conditioned at 100% rh at 20°C (44); nylon-6,6 responds similarly.

Because of its insulating nature, nylon can accumulate positive or negative electrical charges on its surface when rubbed or in contact with other substances followed by separation. These charges do not dissipate readily and can cause problems in fiber processing and certain end use applications. Consequently, antistatic agents are added to spin finishes to facilitate fiber processing, conductive fibers are added to carpets to prevent static shock, and antistats are added to the fibers to prevent static cling of women's slips.

The dielectric constant of nylon-6,6 (at 1000 Hz) is 4.0 at 22°C and 18% rh, and increases to 20.0 when wet (45).

**Optical Properties.** When light falls on a nylon fiber, it can be partially transmitted, absorbed, reflected, or scattered, depending on the cross-section shape and the nature of any second substance added during polymerization or melt spinning. Additive-free nylon with a standard round cross section has a translucent high sheen luster. The inherent luster of the fiber and the color introduced in dyeing are important optical properties that relate to a fabric's visual quality and its acceptance by the customer. Fibers that need to be opaque or to cover well in certain applications such as apparel are melt-spun from polymer containing $TiO_2$ delusterant.

Optical properties also provide useful structure information about the fiber. The orientation of the molecular chains of a fiber can be estimated from differences in the refractive indexes measured with the optical microscope, using light polarized in the parallel and perpendicular directions relative to the fiber axis (46,47). The difference of the principal refractive indexes is called the birefringence, which is illustrated with typical fiber examples as follows.

Birefringence is used to monitor the orientation of nylon filament in melt spinning (48).

| | Refractive index | | Birefringence |
|---|---|---|---|
| | $n_{11}$ | $n_1$ | $(n_{11} - n_1)$ |
| nylon | 1.582 | 1.519 | 0.063 |
| polyester | 1.725 | 1.539 | 0.188 |

**Chemical Properties.**   The chemical reactivity of nylon is a function of the amide groups and the amine and carboxyl ends. The aliphatic segment of the chain is relatively stable.

*Solvolysis and Reactivity.*   Generally, nylon is insoluble in organic solvents, but soluble in formic acid, some phenolic solvents, and fluorinated alcohols such as hexafluoro-2-propanol. Nylon is inert to alkalies. Dilute solutions of strong mineral acids, such as sulfuric and hydrochloric, weaken nylon fibers and hydrolyze them at high concentration and elevated temperature. Strong oxidizing agents and mineral acids such as potassium permanganate solution and nitric acid degrade nylon. Nylon, however, can be bleached in most bleaching solutions common to household and commercial mill applications. Nylon can be hydrolyzed by water under pressure at 150°C and above. The amine end group of a nylon fiber can undergo the same reactions as primary amines, such as neutralization, acylation, and dehydrohalogenation (49).

The amine ends also react with atmospheric contaminants, such as $SO_2$ and oxides of nitrogen and ozone, under ambient storage conditions (50). This phenomenon is referred to as aging and results in reduced acid dye affinity.

*Thermochemical Properties.*   Thermal degradation of polyamides involves complex chemical reaction paths that are a function of the chemical structure, temperature, length of time exposed, and levels of moisture and oxygen (see POLYAMIDES, GENERAL). Nylon-6,6 begins to cross-link or gel after six hours at 305°C, whereas nylon-6 takes 12 days to cross-link at 280°C (51). At 305°C, both nylons drop in molecular weight and emit volatile products: water, carbon dioxide, ammonia, hexamethyleneimine, hexylamine, heptylamine, and methylamine. Nylon-6 also depolymerizes emitting caprolactam (qv). Effects of temperature and exposure time on yarn strength of nylon-6 and nylon-6,6 are illustrated in Figure 4 (52).

The preferred and also the most effective thermal stabilizers for nylon are salts and organic derivatives of copper (53–57). The copper salt is added before polymerization at concentrations that give 30–70 ppm Cu in the spun yarn. Alkali metal iodides or bromides are also added at 0.1–0.3% for synergism and stabilization of the cuprous/cupric moieties during melt processing. Organic antioxidants, such as hindered phenols alone and in combination with organic phosphites, retard thermal oxidation, but are not as effective as the copper derivatives (58). The addition of 9,9-dialkyldihydroacridine compounds before or during polymerization of nylon-6 improves yarn stability at high temperature and high loads, which is essential for tire yarn (59) (see TIRE CORDS).

*Photolytic Properties.*   The extent of photodegradation of nylon-6 and nylon-6,6 depends on the intensity and spectral distribution of the light, the length of

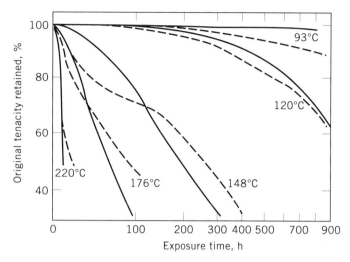

**Fig. 4.** Effects of temperature and exposure time on retained tenacity of nylon-6 (———) and nylon-6,6 (— — —).

time exposed, humidity, air quality, and the presence of photosensitizers on the surface or in the fiber. Photosensitizers include the degradation products created during melt spinning and high temperature fiber processing, $TiO_2$ (the standard fiber delusterant), iron salt contamination, as well as certain dyes and dye bath chemicals. The chemistry of the photodegradation of nylon is as complex as thermal degradation and has been studied extensively (60–63). The amide groups undergo direct photolysis when irradiated at wavelengths less than 300 nm under nitrogen, resulting in cross-linking and significant loss of tenacity and elongation (64). At 300–400 nm, the effect on mechanical properties is negligible under nitrogen, but in the presence of air (oxygen) and moisture, properties deteriorate rapidly. Under these conditions, photodegradation is initiated by a free radical created by photon absorption which proceeds to abstract the hydrogen on the carbon $\alpha$ to the amide NH. The new radical formed on the polyamide chain can react with atmospheric oxygen and moisture to form hydroperoxides. These hydroperoxides propagate other chemical reactions, resulting in chain cleavage and carbonyl by-products. The formation of hydroperoxide, the resulting fiber tenacity loss, and yellowing are all strongly dependent on irradiation wavelength in the 300–400-nm region (65).

Copper-based thermal stabilizers are also effective photostabilizers for nylon. They can be added before polymerization, or the soluble salts (eg, $CuSO_4$) can be applied to fibers as part of the finish or to fabrics as post-treatments. The effectiveness of the copper salt–alkali halide system added to prepolymer in retarding phototendering and photoyellowing of the resulting spun yarn is illustrated in Figure 5.

The $TiO_2$ delusterant is photostabilized by adding manganese salts of pyrophosphorous acids, phosphites, and phosphates (66–68) in the prepolymer, or by using $TiO_2$ coated with a manganese compound (69). The combination of manganese-coated $TiO_2$ and a cuprous halide has also been specified for improved light resistance (70).

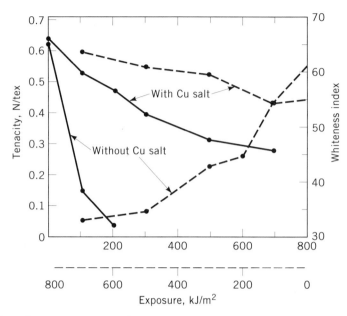

**Fig. 5.**  Effect of uv exposure on nylon-6,6 yarn tenacity and whiteness with and without copper salt and alkali halide in polymer; scoured 210-denier 34-filament yarns exposed in Xenon-arc Ci65 Weather-Ometer using radiation intensity of 0.55 W/m$^2$ at a wavelength of 340 nm. Exposure from 0 to 800 kJ/m$^2$ from left to right relates to tenacity (——), from right to left relates to whiteness (– – –) on right vertical axis. Both tenacity and whiteness decrease with exposure. To convert N/tex to g/den (gpd), multiply by 11.33.

**Staple Properties.**    In contrast to continuous filament, spun nylon staple yarns have a softer, warmer tactile hand and a more natural appearance. Nylon staple is used as the pile material in plush carpets and woven velour automotive upholstery fabrics because of its high strength, abrasion resistance, and ease of recovery from pile crush and distortion. It is also used in blends with natural fibers and regenerated cellulosics primarily to add strength in yarn spinning of both coarse and fine count yarns and for weaving apparel greige goods. For optimum reinforcement, the nylon must have higher strength at the break elongation of the natural fiber. In the case of cotton blends, a nylon staple has been engineered to exceed the strength of cotton at its break elongation of about 7% (71). Nylon staple at 5–25% is commonly used to reinforce the strength of wool yarns.

Because of its high strength and toughness, spun nylon yarns tend to form unsightly pills, ie, balls of fiber on the fabric surface containing some short lengths of fiber, in flat woven and apparel fabrics. This is also why bulked continuous-filament yarn (BCF) is preferred over spun yarns for loop pile carpets.

To accommodate the various uses in 100% form and in blends, the tenacities and elongations of the nylon staple offerings range from 0.3 to 0.6 N/tex (3–7 g/den) and from 50 to 100% elongation. Most other fiber properties of nylon staple differ little from those of the continuous filament; property characteristics of nylon-6 and nylon-6,6 are similar (see POLYAMIDES, GENERAL).

The multiplicity of nylon blends, processing systems, and uses requires a large variety of staple types. Tex per filament may be 0.1–2 (1–20 den), the cross section may be round or modified, the luster may be bright or dull, crimp may be present or absent, and the fiber may be heat-set or not, depending on its use. The staple length is about 4 cm for cotton system processing, 5–7 cm for the woolen system, 8–10 cm for the worsted system, and about 18–20 cm for carpet staple.

## Preparation of Nylon-6 and Nylon-6,6 Polymer

Hexamethylenediamine [124-09-4] and adipic acid [124-04-9] are used in the commercial production of nylon-6,6 and ε-caprolactam [105-60-2] is used for nylon-6 (see also ADIPIC ACID; CAPROLACTAM; POLYAMIDES).

Nylon-6,6 is made by the polycondensation of hexamethylene diamonium adipate salt with removal of water.

$$\overset{+}{H_3}N(CH_2)_6\overset{+}{N}H_3\overset{-}{O}\overset{\overset{O}{\|}}{C}(CH_2)_4\overset{\overset{O}{\|}}{C}O^- \longrightarrow -\!\!\left(NH(CH_2)_4NH-\overset{\overset{O}{\|}}{C}(CH_2)_4\overset{\overset{O}{\|}}{C}\right)_{\overline{n}} + 2\,H_2O$$

The diamine and diacid react in equimolar quantities to form a salt having a pH of 7–7.5 and stored as a 50–60% aqueous solution. The salt is transferred to an evaporator where it is heated at 150–160°C and concentrated to 80–85% by the removal of water. Additives such as chain terminators, dye modifiers, and antioxidants are introduced. The mixture is pumped into an autoclave purged with nitrogen and heated under pressure from 210 to 275°C until prepolymer having about 4000 molecular weight is formed. At this point, $TiO_2$ delusterant can be added. The pressure is gradually released and the temperature is maintained to remove more water in order to increase the degree of polymerization. After the molten polymer has equilibrated to the desired molecular weight, it is extruded under pressure in the form of a ribbon, which is quenched on a water-cooled casting wheel and cut into flakes of a specified size. The flake is dried in a flake conditioner and packaged.

The number-average molecular weight is adjusted in the 12,000–15,000 range for apparel fibers, >20,000 for high strength yarns for tires and industrial end uses.

Nylon-6 is the polyamide formed by the ring-opening polymerization of ε-caprolactam. The polymerization of ε-caprolactam can be initiated by acids, bases, or water. Hydrolytic polymerization initiated by water is often used in industry. The polymerization is carried out commercially in both batch and continuous processes by heating the monomer in the presence of 5–10% water to temperatures of 250–280°C for periods of 12 to more than 24 hours. The chemistry of the polymerization is shown by the following reaction sequence.

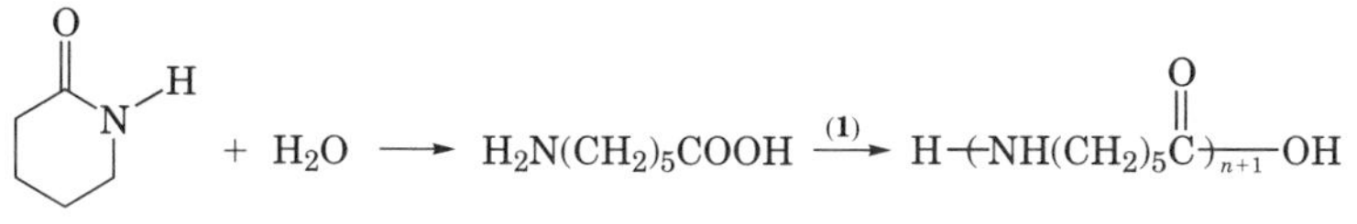

$$+ H_2O \longrightarrow H_2N(CH_2)_5COOH \xrightarrow{(1)} H\!\!-\!\!\left(NH(CH_2)_5\overset{\overset{O}{\|}}{C}\right)_{\overline{n+1}}\!\!OH$$

(1)

Depending on the final polymerization conditions, an equilibrium concentration of monomers (ca 8%) and short-chain oligomers (ca 2%) remains (72). Prior to fiber spinning, most of the residual monomer is removed. In the conventional process, the molten polymer is extruded as a strand, solidified, cut into chip, washed to remove residual monomer, and dried. In some newer continuous processes, the excess monomer is removed from the molten polymer by vacuum stripping.

The addition of small, but specific, amounts of a monofunctional acid to the polymerization is often used to control molecular weights and catalyze reactions. The polymerization is controlled to produce a number-average molecular weight of 18,000–30,000, depending on the end use.

Nylon-6 can be easily polymerized at atmospheric pressure. A continuous process was developed in 1940, called the VK process, which in German stands for *vereinfacht kontinuierlich*, or simplified continuous (73). The VK process is widely used in industry in the 1990s, whereas batch processes, being less economical, are gradually phased out of use. Procedures are also available for making gram quantities of nylon-6 and nylon-6,6 in the laboratory (74).

## Manufacture of Nylon-6 and Nylon-6,6 Fibers and Yarns

All commercial linear polyamides that melt at or below 280°C are melt- rather than solution-spun into fiber because melt spinning is more economical.

Polyamide fibers are manufactured by melt spinning (or extrusion) followed by drawing (or stretching). In spinning, the molten polymer is delivered from an extruder or a metal-grid melter, or directly from continuous polymerization located upstream of the spinning machine. The molten polymer passes through a filter and a spinnerette. The emerging molten filaments are quenched by cross-flow air in a vertical chimney, attenuated partially in the molten state to achieve the desired spun tex and partially in the solid state to develop some degree of orientation. The filaments are converged and a finish preparation is applied to the yarn to provide lubrication and static protection prior to winding on a bobbin. In a two-step process, the spun yarn packages are first lagged in a conditioned area to maintain package integrity. The yarns are then drawn to 30–600% of their original length, resulting in higher tensile strength, lower elongation, and oriented crystallization. The drawing process is controlled and modified to achieve the yarn properties for specific end use. Conventional drawing is used for moderate tenacity, apparel straight (flat) yarns and heat is applied, followed by a heat-relaxation step for the high tenacity, low shrinkage tire yarns. For carpets and some apparel applications, flat yarns are textured (crimped) to impart softness and fabric cover (bulk).

Combinations of two or more of the above steps into consecutive processes, such as spin–draw or spin–draw–texture, reduces manufacturing costs. In addition to continuous-filament yarn, nylon is also offered in staple, tow, and flock forms. Staple is made by cutting crimped continuous-filament yarn into 3–20-cm lengths. In manufacturing, tow is made by combining many yarn ends, either flat or crimped, to give a total tow size of 6–111 ktex. Flock is made by precision-cutting tow into 0.5–3-mm fiber lengths.

## Spinning Continuous-Filament Yarns

In the first commercial process for melt-spinning nylon, polymer chips were stored under nitrogen pressure in a sealed hopper from which they flowed by gravity to a pancake coil heated by a central Dowtherm system (75). The nitrogen pressure moved the molten polymer to a gear pump that forced the metered polymer stream through a sand-filled filter pack and a spinneret, both maintained at the desired temperature by the single Dowtherm system for the entire machine. The molten filaments were quenched by a cross-flow of air at ambient temperature in a chimney with side panels to prevent outside air disturbances. The quenched filaments were converged over a ceramic guide to form the single threadline per spinnerette. Finish was applied by a roll and the threadline wound on friction-driven bobbins at speeds of a few hundred meters per minute. The spun yarn was taken to another machine, the draw twister, where it was drawn, given twist for coherence, and wound on shipping packages containing about 0.5 kg of yarn.

Over the years, fiber producers have improved on the basic design and developed more efficient spinning machines that have lowered manufacturing costs and improved yarn quality. In the early 1950s, the continuous polymerization (CP) process for nylon was combined directly to the spinning machine to eliminate the use of polymer chips. The capacity of a CP direct-spinning machine can be as high as 70,000 t/yr and the machine is used in the production of apparel, carpet, industrial yarn, and staple. Extruders are also used to melt polymer chips. They can be horizontal or vertical, vented or unvented, having capacities up to 5000–7000 t/yr.

Molten polymer must be delivered to the spinnerette in precisely metered amounts or the filaments will vary in size and give unacceptable products. A metering pump is installed after the extruder to discharge exact volumes of molten polymer per unit time against pressures as high as 70 MPa (10,000 psi) at temperatures in the 300°C range. The molten polymer is also filtered under high shear to remove gel particles and particulate matter that can clog spinnerette holes. A filtration-shear device is attached directly to the distribution plate and spinneret. The typical pack is a cylindrical cavity about 3.7 cm in diameter and 3.7 cm in height, filled with layers of sand, the finest on the bottom and the coarsest on top. Fine-mesh screens in the bottom and top of the cavity keep the sand in place. Specially designed screens and sintered metal are replacing the sand. Pack designs must avoid stagnant spots where polymer can be trapped and thermally degrade, thus increasing the pressure drop through the pack and shortening its life.

Early spinnerettes were 316-stainless steel disks, ~5 cm in diameter, 0.5 cm in thickness, and with 13 round holes. Over the years, the increased productivity demands have led to as many as 500 holes for larger disk spinnerettes and 4000 holes for rectangular shape spinnerettes.

The molten filaments are extruded through the spinnerette down a vertical chimney where they are air-quenched. The filaments are then converged to form the threadline in the V shape formed by crossed ceramic pins or other similar devices. The threadline passes to the floor below where finish is applied and is wound up on the spin bobbins.

In earlier years, the windup speed was 275–375 mpm (m/min) with 0.5 kg of yarn per bobbin. In the 1990s, winders are designed for 10–25-kg bobbins, two or more bobbins per winder, operating at speeds up to 6000 mpm, and equipped with automatic bobbin changers.

A conventional spinning apparatus is shown in Figure 6 (76). Polymer chips are melted in a larger extruder which, in turn, feeds a manifold spinning line. The feed lines are made as short as possible and the polymer is distributed to a series of spin packs and chimneys where the yarn is quenched. The yarn passes to the tube conditioner, over finish rolls and take-up godet rolls, and then to winders. Spare winders switch over rapidly to a new roll during the doffing operation to avoid yarn loss. The yarns then pass on monorail conveyors to the drawing areas. In plant production, multiple yarns are spun simultaneously, each with many filaments. Efforts must be taken from polymer makeup through spinning to control the uniformity of each end of yarn and to minimize defects that can result in breaks and down time. A significant design trend in spinning has been toward modern short and ultrashort compact machines (77,78). The concept is to have one floor spinning where the maximum distance between spin

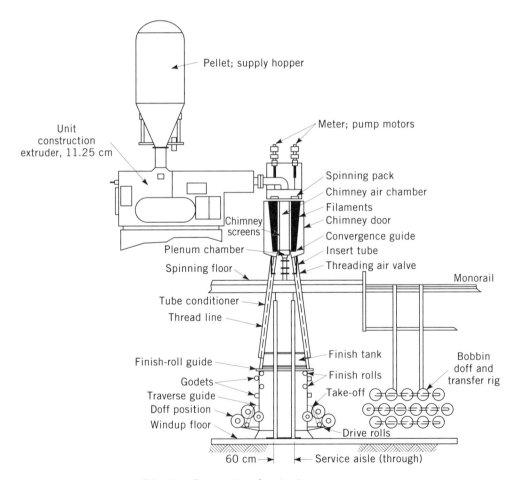

**Fig. 6.**  Conventional spinning apparatus.

die and winder is 1.3–2 m compared to conventional processes having up to 5 m between spinnerette plate and winder.

**Draw-Twisting Process.**    This is a two-step split process where yarn spun onto bobbins at 300–3000 mpm is lagged and then taken to a draw-twisting operation (Fig. 7**a**). The degree of drawing or stretching is accomplished by adjusting the speed of the draw rolls, which run faster than the feed rolls. The drawn yarn is then twisted for coherence and wound on a pirn with a standard ring-and-traveler mechanism. Yarns for apparel, hosiery, and industrial end uses have been made by this process.

**Spin-Drawing Process.**    In the early 1960s, a single-step process was commercialized in which the spinning and drawing steps were combined (Fig. 7**b,c**). This required the development of high speed drawing, winding technology, and air jet interlacing to tangle instead of twisting the yarn. For apparel and carpet yarn end uses, the process utilizes the conventional cold draw; for tire and high strength industrial yarn applications, heated draw assists such as jets, rotating pins, shoes, roll chests, and yarn relaxation steps are added (79). The spin-drawing process has also been combined with continuous polymerization. The double-sided unit in Figure 6 is commonly used where the stretch godets and draw assists are installed before the windup (80,81).

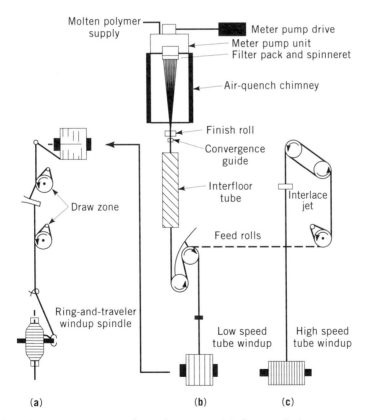

**Fig. 7.**    Spin-drawing processes for nylon yarn: (**a**) draw-twisting process, (**b**) conventional spinning process, and (**c**) coupled process.

In the coupled process (Fig. 7c), the draw ratio, which affects the tenacity and elongation, lowers with increasing spinning speed. As draw ratio is increased, tenacity and initial modulus increase and elongation decreases.

**New Developments in Spinning and Drawing.**  The continual effort by the fiber producers to increase production, lower manufacturing costs, and improve quality has led to a better understanding of the effects of the spinning and drawing process on physical and molecular structure of fibers (82–90).

The fiber property–structure relationships for nylon-6 are shown in Figures 8 and 9 (85). The earlier work cited was in the 100–2400-mpm spinning speed range, but the later effort is focused at 4000–6000 mpm (91,92). The effect of spinning speed on residual drawing under various classifications of fiber orientation is illustrated in Figure 10 (93). Low orientation yarns (LOY) have a high residual drawing, low crystallinity, and limited storage stability. These yarns are subsequently processed through the draw-twisting process. Medium oriented yarns (MOY) processed at 1800–2800 mpm are slightly more crystalline, but still have limited storage stability. Partially oriented yarns (POY) processed at 3000–4000 mpm are partially drawn (70–100% elongation) with some residual drawing, but still have low crystallinity. Partially or spin-oriented nylon feed yarns were developed in the early 1970s to meet the needs of increased false-twist texturing speeds (94). In processing, the drawing step is eliminated and the orientation required for draw texturing is controlled in spinning. POY yarns have also been used in the 1990s in the process of draw warping to make flat yarn warps for weaving and knitting (94–97). A process has been defined for making feed yarns that impart excellent uniformity with structure-sensitive large acid dye molecules in dyeing fabrics from warp-drawn yarns (98). Highly oriented yarns (HOY) spun at 4000–6000 mpm are not fully drawn (50–60% elongation). Fully oriented yarns (FOY) with elongations of 20–30% require spinning speeds of well over 6000 mpm.

Nylon-6 properties obtained at spinning speeds up to 7000 mpm under certain conditions are shown in Figure 11 (99); nylon-6,6 responds similarly.

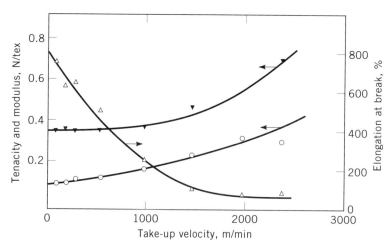

**Fig. 8.**  Tangent modulus (▼), tenacity (○), and elongation (△) as a function of take-up velocity for nylon-6. To convert N/tex to g/den (gpd), multiply by 11.33.

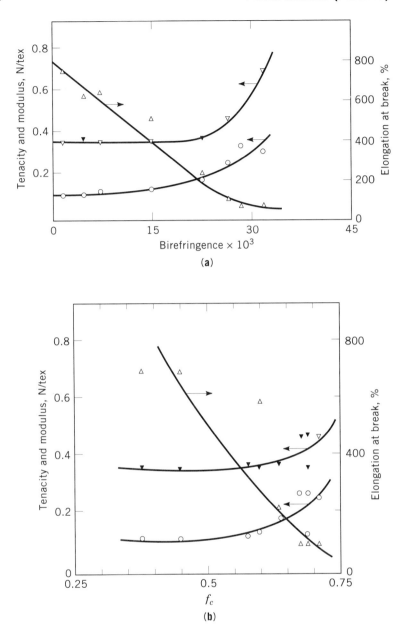

**Fig. 9.** Tangent modulus (▼), tenacity (○), and elongation (△) as a function of (**a**) birefringence and (**b**) $c$-axis crystalline orientation factor $f_c$ for nylon-6. To convert N/tex to g/den (gpd), multiply by 11.33.

Molecular weight has a strong effect on the properties of high speed melt-spun nylon. Generally, higher molecular weight leads to higher modulus and filament tenacity and lower elongation to break (100). In spinning POY nylon, the yarn bundle can be transferred directly to winders capable of 4500–6500 mpm, which is referred to as godetless spinning, or through a set of driven godets (Fig. 12)

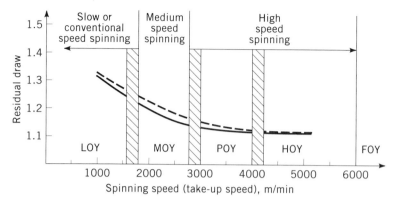

**Fig. 10.**  Dependence of the production on spinning speed (89), where LOY is low orientation yarn; MOY, medium oriented yarn; POY, partially oriented yarn; HOY, highly oriented yarn; and FOY, fully oriented yarn.

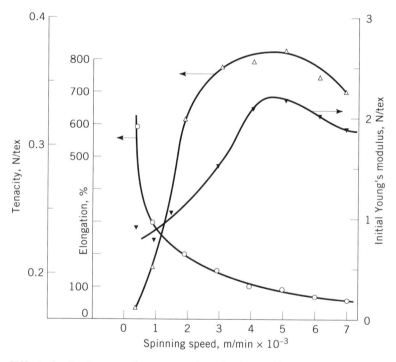

**Fig. 11.**  Effect of spinning speed on properties of nylon-6 fibers, where (▼) is the tangent modulus; (○), the tenacity; and (△), the elongation. To convert N/tex to g/den (gpd), multiply by 11.33.

(101). Improvements in the processing and properties of nylon-6,6 have been made by modifying quench and windup hardware (77,102), using higher than usual molecular weights (100,103), adding a trifunctional branching agent to increase melt viscosity (104), incorporating minor amounts of hydrogen-bonding agents in high molecular weight polymer (101), and adding pyrogenic silica to

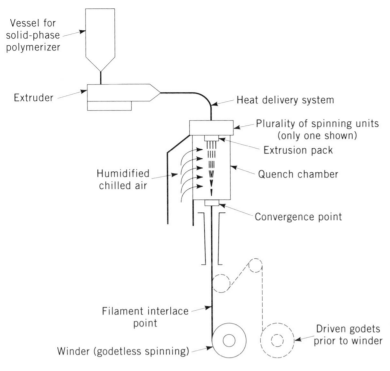

**Fig. 12.** Spinning process for partially oriented nylon yarn (97). Winder speed is between 4500 and 6500 mpm.

the polymer (105). The hydrogen-bonding agents include nylon-6 monomer or 2-methylpentamethylene diamine.

**Texturing Processes.** Texturing is the conversion of flat to crimped continuous-filament yarns to simulate properties inherent in natural and synthetic spun staple yarns, such as thermal insulation, fullness, cover (bulk), softness, and moisture transport. In texturing, the geometry and, to a degree, the surface of the filaments are mechanically deformed by bending, twisting, or compression to introduce permanent waviness (crimp) loops and coils. Prior to the texturing step, heat is applied in the range of 100–190°C to soften the filaments. After texturing, the yarn is cooled to set the crimp and the mechanical deformation is removed by tensioning the yarn during package windup. Texturing affects the tensile properties, dyeability, and the macromolecular structure of the fibers in a nonuniform way that can cause dye structure and barré problems in dyeing fabrics. Depending on how the filaments have been mechanically deformed and the temperature used, differential stresses and random disoriented molecular chains are set in specific configurations to give one of three classifications of textured yarns: stretch, modified stretch–bulk, or bulk. The textured stretch or bulk is developed when the yarn experiences a stress-relaxing environment under low or no tension, as in steaming, hot air, or dyeing. Stretch yarns have high extensibility and good recovery, but only moderate bulk compared to modified stretch and high bulk-textured yarns.

Descriptions of the various texturing processes and latest machine modifications to increase quality, speed, and efficiency have been the subject of numerous reviews (106–110). In air jet or Taslan texturing, the yarn is introduced into a venturi tube in the air jet where the turbulence from a stream of compressed air cause the filaments to entangle and form surface loops. The product has bulk, but no stretch, and the process can be varied to give a range of bulk and novelty effects (111–113). In false-twist texturing, a specially made drawn feed yarn is heated by a radiant or contact heater just below the melting point, twisted as high as 30–40 turns per centimeter by disks rotating as fast as 800,000 rpm, then untwisted and packaged. Draw-texturing is a continuous process where partially oriented yarns are drawn and false-twist-textured (107). False-twist-textured filaments are coiled, have good bulk, but are regarded primarily as stretch yarns.

In another form, bulked continuous-filament (BCF) yarns are spin-drawn and textured with a fluid jet in a continuous one-step process. After drawing, the yarn is heated over rolls in a hot chest and passed through a hot air or superheated steam jet under controlled conditions. In some designs, texturing is achieved by impinging the yarn onto a baffle plate, a screen, or the walls of a specially shaped chamber (114–117). BCF yarns have bulk with some stretch and each filament has a random, three-dimensional curvilinear crimp form. Edge crimping is accomplished by drawing a heated yarn over a knife edge that compresses (flattens) the edge side and stretches the outside of the filament. The metastable stresses so imposed are relaxed during heating, causing the yarn to crimp and coil to give a stretch yarn (118–120). Gear crimping is adaptable to draw-twisting or spin–draw–texture operation and generally used to texture high tex yarns. Stuffer-box crimping is used for both low and high tex yarns and for crimping two for staple. Most of the crimped nylon made in the United States is stuffer-boxed, gear, false-twist, and air- or fluid-jet-crimped.

Commercial textured yarns cover a range of yarn sizes and end uses. The 1.7–3.3-tex (15–30 denier) yarns for hosiery are false-twist-textured, as is the 4.4–22.2-tex (40–200 denier) yarns for apparel. The 56–333-tex (500–3000 denier) yarns for carpets, upholstery, and soft-sided luggage are air- and fluid-jet-textured.

Most of the textured apparel and industrial yarns are woven or knitted directly into fabric. The carpet BCF yarns can be tufted directly off package into loop pile or velvet constructions. For the textured saxony constructions, the BCF and the spun staple yarns must be ply-twisted and heat-set. The heat-setting temperature for nylon-6 and nylon-6,6 is 180–220°C in hot–dry atmosphere, and 120–140°C in saturated steam. The yarns are twist-set in pressurized autoclaves or continuously on the Superba and Suesson machines (121). Before setting the twist, the yarn is heated and relaxed for predevelopment of the bulk.

**Staple Fiber.** Staple manufacturing consists of spinning, drawing, crimping, cutting, and baling (122). The spun yarn from individual spinnerettes, ranging from several hundred to a few thousand filaments, are piddled into a can, lagged, the ends of which combined into a tow, and drawn. Finish is applied in spinning to help drawing, and is usually applied again after drawing to assist in the crimping and cutting operations. In crimping, multiple tow ends are forced continuously into a constricted stuffer box by preset feed rolls. The box is sealed by an adjustable hinged and weighted gate. The tow ends filling the chamber are arranged in uniform folds by compression. When the pressure in the stuffer box exceeds the pressure of the hinged gate, the gate rises to permit discharge of the crimped tow, and then falls, thereby repeating the process (123). Prior to cutting, water is applied to the crimped tow to reduce static and maintain wear life of the knives. The compact spinning systems, approximately 6-m high by 6-m wide and developed for nylon, polyester, and polypropylene, can be integrated into accommodating, designed downstream staple process operations (crimping, heat-setting, cutting, baling) to reduce space and cost as well as to expand product flexibility and improve quality (124). This process is a spin–draw type where tow is fed directly to drawing and then through the staple-making operations. The spinning and drawing speed must be coordinated for proper performance. When utilizing the conventional low speed staple process, the spinning and drawing speeds must be reduced substantially. To compensate for the loss in throughput capacity per position, spinnerets having high number of holes (8,000–75,000) are used. In another process where spinning and drawing are maintained close to conventional spinning, the stuffer-box crimper must be designed to process large tow titers, and to prevent the sticking of filaments on the roller and escaping of filaments that break loose from the bundle. Developed by Takehara, Japan, a high speed crimper capable of 6000 m/min avoids the problems encountered with the conventional roller (125). The Takehara Type T-K-K has a large-diameter ring roller that rotates in one direction, and an inner roller that lies on the inner surface of the ring roller and rotates with it in the same direction and speed while holding the tow to be crimped (126).

Staple is used directly in the manufacturing of nonwoven fabrics (qv) (127) and spun into yarn through the cotton, worsted, and woolen systems in 100% form or in blends with other synthetic or natural fibers (128,129).

**Flock.** Flocking is the mechanical and/or electrostatic application of finer fiber particles to adhesive-coated fabrics, paper, yarns, plastic, or metal objects (130). Flocking offers a soft velvety surface for decorative and visual appeal and has a variety of functions: sound dampening, thermal insulation, friction reduction of sliding surfaces, increased surface exposure for evaporation and filtering, buffing and polishing, liquid retention or dispersal, and cushioning of heavy objects.

Flock is made by precision-cutting drawn, uncrimped tow using a rotary or guillotine cutter into 0.3–6-mm lengths, depending on the denier per filament (dpf). The higher the dpf, the longer the tolerable length before curvature of the cut fiber becomes a problem in flock preparation. The tow presented to the cutter can be a combination of tow with lower total tow ktex creeled directly out of shipping boxes or a tow that had been scoured, pad-dyed, rinsed, and treated with special finish on a continuous range. The tow size can be from 111 to 3889 ktex (1,000,000–35,000,000 denier) at the cutter.

Whether the process is continuous or batch, the flock must be similarly treated for electrostatic application. The spin finish is removed by scouring and then a finish mixture is applied to impart the following properties to the flock (131): (*1*) good siftability; (*2*) low interfiber adhesion needed for proper coverage; (*3*) good movement and orientation in the electrostatic field needed to get the flock to go straight into the adhesive coating; (*4*) correct level of conductivity appropriate to the strength of the field and compatible with an ac- or dc-applied voltage; and (5) good wetting-out properties in the adhesive. Fiber length has a great effect on flock properties; increasing the fiber length decreases the number of fibers sifted, the number of fibers flocked, and the percentage of maximum flock density (132).

Nylon is the preferred fiber for flocking because of its good chemical bonding to a wide range of adhesives, its toughness, and its ease of dyeability and printability. Nevertheless, the tow must be manufactured with the proper ktex, cohesion, and spin finish to be readily converted to flock (133).

**Finishes.**    Fiber finishes are designed to provide fiber cohesion, lubricity, and static-free operability at low and high traverse speeds over a variety of metallic and ceramic surfaces encountered in fiber plant and mill operations (134–136). Fiber cohesion is important in that loose or protruding filaments can catch on processing equipment and cause snags and breaks, or become entrapped within a windup package, causing unevenness in texturing, knitting, and weaving. Finishes, consisting primarily of lubricants, emulsifiers, and antistatic agents, are generally applied as aqueous emulsions at concentrations giving a finish-on-yarn level of 0.3–1% after the evaporation of water. Lubricants can consist of mineral, vegetable, and animal oils or waxes (triglycerides), or of such synthetic types as esters, polyethers, silicones, and ethoxylated esters. Depending on lubricant compatibility, the degree of fiber wetting desired, and the required oil–water balance, emulsifiers can be anionic (eg, sulfated vegetable oil), cationic (eg, quaternary ammonium compounds), nonionic (polyglycols), or amphoteric (eg, betaines). Finish antistats fall into the same categories. Other functional chemicals, such as biocides, antioxidants, dyeing assists, and rubber adhesion promoters for tire manufacturing, are added on an as-needed basis.

Over time, finish components tend to separate and migrate within the fiber and throughout the yarn package. With nylon, the ionic emulsifiers and antistats tend toward the core of the fiber whereas the hydrocarbon lubricants remain on the surface. It is, therefore, essential to scour yarns and fabrics at neutral to basic pH to reemulsify the lubricant and remove the finish emulsifier prior to dyeing. In formulating any new finish, environmental issues such as biodegradability, water and air pollution must be considered (137).

In application, the aqueous-finish emulsion is pumped to a holding or storage tank from which it is circulated through a system that feeds the spinning machine. The finish is applied by passing the yarn either through the circulating finish across a constantly revolving "kiss" roll that is partially immersed in the emulsion, or through a slotted pin or guide in which the emulsion stream is metered through an orifice.

**Modified Cross Sections.**    Nylon filaments are spun in a variety of cross-section shapes that include the conventional round to irregular solid and hollow shapes (Fig. 13). The cross-section shape is an important variant in designing the functionality and luster of fibers. The round cross section is used for strength in industrial applications and for subdued luster in apparel and upholstery. The multilobal cross sections are used to enhance bulk and for bright luster in both BCF and spun staple yarns for carpets and upholstery. The grooves in the multilobal shapes also enhance moisture transport by wicking water through capillary action. Flat-sided ribbon-like cross sections provide cover in apparel applications.

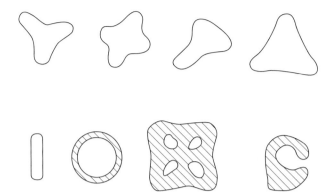

**Fig. 13.**  Typical nylon fiber cross-section shapes.

Bright (no delusterant) round and hollow fiber cross sections have a smooth sheen-like luster, whereas multilobal cross sections vary from bright sheen to bright glitter luster, depending on how they are designed to converge and transmit internally reflected light (138).

In spinning the different fiber cross sections, spinnerette orifice shape and dimensions must be made to exacting tolerances, and changes in polymer melt viscosity, block temperature, and air quenching conditions must be anticipated to assure the desired shape, spinning continuity, and quality yarn. In spinnerette manufacture, the final orifice is produced after counterboring all but the thin sections of the spinnerette blank. Noncircular spinnerette orifice shapes are made by electron-beam milling and electrodischarge machining. For the solid, non-round shapes, a single orifice design is used. Various specific orifice shapes have been developed and fiber cross sections have been defined using mathematical relationships (139,140).

In making hollow fibers, both precoalescent and post-coalescent spinnerettes and spinning techniques can be used. The former requires injection of a gas through the capillary to create the void (141,142), whereas the latter involves entrapment of gas (air, nitrogen) by coalescence of the molten polymer spun exiting from a segmented orifice designed to give the desired number of cross section holes, shape, and percentage void (143,144).

The open hollow fiber shape shown in Figure 13 is made by a unique process requiring bicomponent yarn technology (145). A yarn is spun with a water-soluble copolyester core and nylon sheath where the core is dissolved out with an alkali treatment in fabric dyeing.

**Additives.**  Additives can be introduced in salt preparation, polymerization, or the molten polymer stream prior to extrusion of the filaments. Additives are used to alter the basic characteristics and performance of the fiber and can be classified by function.

Delusterants reduce the transparency, increase the whiteness, and alter the fiber's reflectance of light. $TiO_2$ is the delusterant of choice for all melt-spun fiber types because of its high cover, whiteness, and chemical and thermal stability. Nylon is translucent and requires $TiO_2$ in most textile and home furnishing applications for the cover and to reduce objectionable sheen and gloss in fabrics. Semidull nylon yarns contain ca 0.3% $TiO_2$; middull, ca 1.0%; and full-dull, ca 1.5–2.0%.

TiO$_2$ requires careful slurry preparation and introduction to the nylon to minimize the formation of agglomerates and to maintain an average particle size of 0.1–0.5 $\mu$m in fiber. Otherwise, filter packs will clog and require frequent replacement. Large TiO$_2$ agglomerates can also cause filaments to break while still molten, creating drips at the face of the spinnerets, erratic spinning, and drawing. The anatase, not the rutile form of TiO$_2$, is preferred because it is softer and contributes less to abrasion wear of machine guides and rolls. TiO$_2$ is added as an aqueous dispersion to the prepolymer in the manufacturing of nylon-6,6, and to the caprolactam in the manufacturing of nylon-6, frequently as a master batch containing up to 30% TiO$_2$ in low molecular weight polymer (146). Nonmelt-compatible polymeric materials such as polyoxyethylenes and polystyrene have also been used as delusterants (147).

Colorants can be introduced into the fiber by adding dyes and pigments in salt preparation, during polymerization, or into the molten polymer just before spinning (148) (see COLORANTS FOR PLASTICS; DYES, APPLICATION AND EVALUATION). Pigmented fibers are referred to as mass-dyed, dope-dyed, solution-dyed, or producer-colored. Inorganic pigments are used more than the organics, especially where high color, light-, and crock fastness are required, such as in upholstery and carpet for automotive interiors. The organic pigments have higher chroma, but are not as colorfast to heat and light (see PIGMENTS). Nylon-6 can accommodate more pigments than nylon-6,6 because of its lower melt processing temperatures. Like TiO$_2$, the pigments must be dispersible and heat stable in polymer and fiber manufacturing (149). A common and efficient approach to adding pigment to the base polymer in spinning is first to disperse the pigment as a concentrate in a carrier polymer, usually a lower melting copolymer. The concentrates range from 25 to 50% pigment content and are offered as a single color or a compounded color blend in pellet or flake form. The flake can be blended with the base polymer flake at a specified loading and charged to the feed hopper of the spinning process or remelted in a vessel that allows it to be metered directly into the molten polymer prior to spinning. Quality pigmented fibers are spun with processes equipped with mixing screws and volumetric or gravimetric automatic feeders. Typical pigments for nylon are carbon black, red iron oxide, aluminum cobalt blue, and phthalocyanine blue and green. Carbon black enhances nylon's resistance to photodegradation. A number of light-stable pigments that are also environmentally friendly are available (150).

Antioxidants (qv) are used to prevent thermal and oxidative degradation of nylon in manufacturing, post-fiber and fabric processing, and final use.

Antistats such as polyoxyethylenes (151,152) and $N$-alkyl polycarbonamide (153) are added to nylon to reduce static charge and improve moisture transport and soil release in fabrics. These additives also alter the luster of fiber spun from bright polymer. Static reduction in carpets is achieved primarily by the use of fibers modified with conductive carbon black (see ANTISTATIC AGENTS; CARBON, CARBON BLACK).

Antimicrobial agents are used where there is a need to inhibit bacterial and fungal growth. The additives can consist of copper, germanium, zinc and zinc compounds, metal oxides or sulfides, metal zeolites, as well as silver and copper oxide-coated inorganic core particles (154–159) (see INDUSTRIAL ANTIMICROBIAL AGENTS).

Flame retardants designated for nylon include halogenated organic compounds, phosphorous derivatives, and melamine cyanurate (160–163). Generally, flame retardants are difficult to spin in nylon because of the high loading required for effectiveness and their adverse effects on melt viscosity and fiber physical properties.

Both fiber producers and fabric mills have realized that many of the performance variants that are difficult to incorporate into fiber melt spinning can be accomplished by post-treating yarns or fabrics. Mills in the 1990s can apply flame retardants, softeners, dye-fade inhibitors, and stain- and soil-resisting agents as part of the finishing of a fabric.

## Dyeability

Because of its physical and chemical structure, nylon has an affinity for every dye class: disperse, direct, vat, fiber reactive, chrome, acid, premetallized, and cationic dyes with special nylons (164) (see DYES, APPLICATION AND EVALUATION). Commonly used dyes are the disperse, acid, and premetallized. Disperse dyes are easy to apply, but have marginal wash and lightfastness on nylon. Acid dyes, unlike the disperse dyes, are water-soluble, react with the amine ends, and, as expected, have better wash and lightfastness than the disperse dyes. There are two categories of acid dyes. Those containing one sulfonate group on the dye molecule are referred to as leveling acid dyes; those containing two or more sulfonates, nonleveling or reserving acid dyes. Premetallized dyes are large metal complex molecules. They have the best wash and lightfastness and are used for high uv-exposure end uses such as automotive upholstery. Leveling acid dyes are used in piece-dyeing and wet roller and screen printing of carpets, upholstery, and apparel. Disperse dyes are used to dye carpets and for transfer printing primarily of flocked upholstery and carpets. In transfer printing, dyes printed on special papers are transferred to the fabric under heat and pressure (165–167).

**Effect of Fiber Properties.** Acid dyes are attracted to the accessible amine ends of the nylon chains located in the amorphous regions of the fiber. Acid dye affinity of nylon can be adjusted by adding excess diamine or diacid to the polymer salt or by changing the molecular weight in polymerization. A light acid-dyeable nylon-6,6 is spun with 15–20 amine ends, expressed in terms of gram equivalents per $10^6$ g of polymer. A medium or regular acid-dyeable nylon has 35–45 and a deep acid-dyeable nylon has 60–70 amine ends. Ultradeep acid-dyeable nylons have 80 amine ends or more and are made by adding basic compounds such as N-(2-aminoethyl)piperazine in polymerization to enhance the total available dye sites (168). Nylon-6,6 can also be made basic-dyeable by increasing the level of carboxyl ends (169) or by introducing sulfonic acid groups through the addition of a copolymerizable dicarboxylic acid derivative to a polymer salt, such as the sodium salt of 3,5-dicarboxybenzene sulfonate (170). By using the proper dyes, dye-bath auxiliaries, and pH control, light, medium, and deep acid- and basic-dyeable yarns can be dyed in combination in a single bath to give multitonal and two-color styling effects (171). This concept of differential-dyeing yarns is unique to nylon and is used extensively in the piece dyeing of carpet, apparel, and upholstery fabrics.

Macromolecular structure and supermolecular organization also affect dye affinity. Drawn (oriented) nylon-6 has more of a random open structure than nylon-6,6 (172). Nylon-6, therefore, dyes more rapidly than nylon-6,6, but is also more susceptible to color crocking, especially with disperse dyes. High dye rates can be achieved with nylon-6,6 by adding a minor random copolyamide component in polymerization. Also, any melt-spinning process adjustment or post-fiber and fabric treatment that decreases orientation in the amorphous domain increases dyeability. Examples are lowering the draw ratio, applying steam or swelling chemicals (eg, phenol), and texturing particularly with pre- or post-heat relaxation (173–176). Nonuniform drawing or heat application can cause streaks with structure-sensitive nonleveling and premetallized dyes. Variations in fiber geometry and cross-section shape and yarn crimp level in texturing cause optically related color streaks.

Prior to dyeing, the fiber or fabric is scoured to remove spinning and mill-processing oils and waxes (177). Nylon is atmospherically dyed in staple form (stock-dyed), yarn form, and fabric form (piece-dyed) in batch systems (becks, jet dyers, kettles) or continuously on a pad-dye, steam (fixation), and rinse range. Uniformity is best achieved in batch systems where equilibrium can be established between dye diffusing in and out of the fiber and the dye remaining in the bath. The pH adjustment of the dye bath is important for controlling dye strike and leveling. The nitrogen atoms in nylon absorb hydrogen ions increasingly as the pH of the dye bath is lowered from 9. The amine ends fix acid dyes at relatively high pH. Acid dyes fixed at high pH are only slightly ionized (weakly acidic). At a pH of less than 6.5, the fiber becomes strongly cationic and absorbs dye anions much more rapidly. Under such conditions, the dye is rapidly exhausted on the most readily accessible dye sites, and variations in dyeing appear as streaks in both warp knit and woven fabrics from continuous flat or textured filament yarn. The streaks are caused by a single or groups of yarns that take up dye differently from neighboring groups of yarns. The yarns can differ in heat history, amine-end sites, and/or processing tensions, which can either open or close the molecular structure.

The appearance of streaks with leveling or nonleveling acid and premetallized dye can be subdued by increasing the dye-bath pH from 5.5 to 6–7, at a sacrifice in dye exhaust, by adding chemical agents that retard the dye strike or, more effectively, by metering all or a portion of the dye in a concentrated solution at or near the dyeing temperature of the fiber (87.8–104.4°C) instead of at the usual 26.7–48.9°C practiced by the trade (178).

For end uses demanding high dye lightfastness such as automotive interior fabrics, select premetallized dyes and uv inhibitors are applied through the dye bath (179–182). Nylon can also be codyed with polyurethane elastomeric fibers, wool, acrylics, polyesters, and cellulosics (183).

## Modified Nylon-6 and Nylon-6,6 Fibers

**Bicomponent and Biconstituent Fibers.** Bicomponent fibers consist of two polymers of the same generic class, eg, nylon-6 and nylon-6,10; biconstituent fibers consist of two dissimilar generic polymers, eg, nylon-6,6 and polyester. Both fiber types are made by melt-spinning separately the two dif-

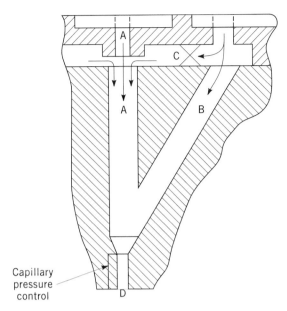

**Fig. 14.** Bicomponent spinneret for sheath-core fiber, where A represents copolymer; B, homopolymer; and D, bicomponent fiber capillary exit. If side-by-side fiber is desired, the interconnecting channel, C, is sealed at ✕.

ferent polymers through a common, specially designed spinneret such as the one in Figure 14. The spinneret hole and block channels can be designed so that the two polymers emerge side by side, as sheath-core or conjugate fibers. The fibers are processed through conventional drawing or spin-drawing operations. The original intent of the side-by-side bicomponent yarn was to impart stretch in tricot knit and hosiery applications. Nylon-6 and nylon-6,6 homopolymers were paired to a copolyamide on the basis of shrinkage difference as measured on the individually spun polymers. Depending on the differential shrinkage, the bicomponent yarn would assume a level of helical crimp in dyeing or steaming. The most notable example was Du Pont's Cantrece for hosiery (184). Some of the polymer components used in earlier bicomponent stretch yarns are shown in Table 2. Stretch yarn for hosiery can also be made by spinning side by side two nylon-6,6 polymers having a relative viscosity difference of at least 15 units (185). Monsanto's Monvelle, which

**Table 2. Commercial Bicomponent Fibers**

| Type | Copolymer component | | Homopolymer component |
| | Copolymer | Composition | |
|---|---|---|---|
| sheath core | 6,6–6,10 | 50:50 | 6,6 |
| | 6–6,T[a] | 40:60 | 6 |
| side by side | 6,6–6,I[b] | 80:20 | 6,6 |
| | 6,6–6 | 80:20 | 6,6 |

[a] 6,T = hexamethyleneterephthalamide.
[b] 6,I = hexamethyleneisophthalamide.

is no longer marketed, is an example of a biconstituent yarn (nylon-6 and melt-spinnable polyurethane) for stretch hosiery (186,187). Most of the stretch yarns made for hosiery today are textured POY because of lower cost.

Bicomponent technology has been used to introduce functional and novelty effects other than stretch to nylon fibers. For instance, antistatic yarns are made by spinning a conductive carbon-black polymer dispersion as a core with a sheath of nylon (188) and as a side-by-side configuration (189). At 0.1–1.0% implants, these conductive filaments give durable static resistance to nylon carpets without interfering with dye coloration. Conductive materials such as carbon black or metals as a sheath around a core of nylon interfere with color, especially light shades.

Dye selectivity can be altered by spinning a high amine-end nylon core with a sulfonate-containing nylon sheath (190). In standard dyeing conditions, the core accepts leveling acid dyes, but not reserving or basic dyes; the sheath accepts leveling and nonleveling basic dyes. The effect creates a third color when dyed in combination with acid and basic dyeable yarns.

Microdenier fibers, ie, very fine fibers, can be made by spinning biconstituent conjugate fibers such as those illustrated in Figure 15. The technology was developed by Kanebo and marketed under the fabric names Belseta and Glacem. The concept, referred to as alkalization, is to spin a readily alkali-soluble copolyester with nylon and then to dissolve away the polyester in the fabric stage. For example, the wedge-like nylon portion of the round 1.7-decitex (1.5-d) filament in Figure 15 can split into eight components, each 0.20 decitex (0.18 d) (191). The hexagonal cross-section fiber is of the "islands-in-the-sea" type where the orange wedge-shaped nylon "islands" in the copolyester "sea" connect slightly at the matrix center (192).

Microdenier nylon and polyester were a significant spinning breakthrough when demonstrated in 1985. The finer-than-silk fibers added a new dimension to fabric aesthetics, comfort, and performance. Microdenier nylons are used in weaving, warp knits, and weft knits for sports-, leisure-, and fashion-wear. Polyester can be melt-spun into 0.55 decitex/filament, but the finest

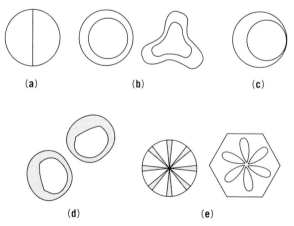

(a)                    (b)                    (c)

(d)                    (e)

**Fig. 15.** Bicomponent cross-section forms: (**a**) side by side, (**b**) concentric sheath core, (**c**) eccentric sheath core, (**d**) kidney-shaped sheath core, and (**e**) conjugate.

nylon is spun at 1.1 decitex/filament. The ultramicro staple nylons made by the polyester alkalization procedure are used to manufacture suede and water-resistant fabrics.

**Copolymers.**    There are two forms of copolymers, block and random. A nylon block copolymer can be made by combining two or more homopolymers in the melt, by reaction of a preformed polymer with diacid or diamine monomer; by reaction of a complex molecule, eg, a bisoxazolone, with a diamine to produce a wide range of multiple amide sequences along the chain; and by reaction of a diisocyanate and a dicarboxylic acid (193). In all routes, the composition of the melt is a function of temperature and more so of time. Two homopolyamides in a moisture-equilibrated molten state undergo amide interchange where amine ends react with the amide groups.

As time progresses, the two homopolyamides in the melt form a block and eventually a random copolymer as a result of amide interchange (Fig. 16). Block copolymerization is a way of introducing a new variant into a base polymer without grossly affecting the spinning performance and physical properties of the yarn. The process requires careful control, however, to maintain the desired composition in order to ensure uniform product and spinning continuity. Examples of this technology are the block copolymers of poly(4,7-dioxadecamethylene adipamide) and nylon-6 (194), and the block copolymers of nylon-6 and polyethylene oxide diamine (195) for hydrophilic nylons.

Random copolymers are made by combining three or more monomers in the polymerization process. The melting temperature of random copolymers are lowered as the regularity with which the monomer groups are spaced along the

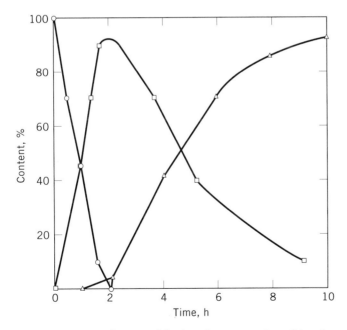

**Fig. 16.**   Typical composition change with time for conversion of two homopolymers (○), first to a block (□), and then to a random (△) copolymer.

backbone is reduced. Hydrogen bonding between the amides is also reduced. This is illustrated in the melting point–composition relationship for nylon-6,6/6 and nylon-6,6/6,10 in Figure 17. Crystallinity and orientation are also reduced with increasing randomization. However, in the case of nylon-6,6/6,T, the crystallinity is not reduced because the copolyamide segment is similar in size to that which is replaced.

The general properties of random copolyamides are high dyeability (especially with nonleveling large dye molecules), lower melting and softening points, reduced dry and wet strength properties, high creep failure, and high shrinkage.

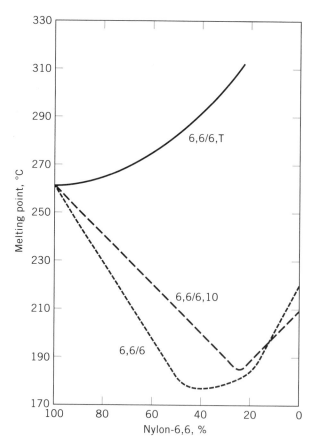

**Fig. 17.**  Melting point–composition curves for random copolyamides of nylon-6,6.

The last property led to the use of the copolyamides in bicomponent self-crimping yarns. Copolyamides are also suitable for thermal bonding of fiber and fabrics because of their low softening temperature. Random copolyamides are also used to modify the dyeability of nylon-6,6. The addition of 0.5–2% comonomer increases the dyeing rate to almost that of nylon-6. Copolymerization with 5-sulfoisophthalic acid gives cationic dyeable fiber and copolymerization with N-(2-aminoethyl)piperazine a deep acid-dyeing fiber. Comonomers are also used as assists in the high speed godetless spinning of draw-texturing feed yarns. Nylon-6,6 copolyamides with ethyltetramethyleneadipamide, pentamethyleneadipamide, or 2-methylpentamethyleneadipamide units lower the gelation rate of nylon-6,6 at the spinning temperature of nylon-6,6 (196–198).

**Graft Polymers.** The grafting of polymers on nylon fiber surfaces as a means of altering chemical or physical properties has been explored and demonstrated over the years. However, as of 1996, a process has not yet been found that is both economical and efficient for grafting yarns at conventional spinning speeds. Nevertheless, efforts continue because the property changes are dramatic, permanent, and the process is more conducive to slower speed fabric finishing processes. Vinyl monomers can be grafted to almost any nylon by ionizing radiation usually from a Co-60 source, high energy ultraviolet radiation, or chemical means. Radiation grafting of acrylonitrile on nylon-6 gives basic dyeability; radiation grafting of styrene, hydrophobicity (199). Nylon-6,6 can also be made basic-dyeable by chemically grafting acrylic or methacrylic acid with a water-soluble formaldehyde sulfoxylate salt (200). Stain-resistant nylon fibers for carpets can be made by attaching stain-blocking compounds, phenyl–vinyl, and ether–maleic anhydride copolymers to the fiber surface using agents grafted to the nylon with uv light and a photoactivator (201).

# Other Nylons

Nylon-4,6 [24936-71-8], introduced as Stanyl by Dutch State Mines, is synthesized from 1,4-tetramethylenediamine and adipic acid (202). Stanyl has a high melting temperature (295°C), improved chemical resistance, better dimensional stability, and higher modulus than nylon-6 and nylon-6,6; it is therefore highly suited for industrial yarn applications, including tire cord.

Qiana, introduced by Du Pont in 1968 but later withdrawn from the market, was made from bis(4-aminocyclohexyl)methane and dodecanedioic acid. This diamine exists in several cis–trans and trans–trans isomeric forms that influence fiber properties such as shrinkage. The product offered silk-like hand and luster, dimensional stability, and wrinkle resistance similar to polyester. The yarn melted at 280°C, had a high wet glass-transition temperature of ~85°C and a density of 1.03 g/cm$^3$, the last was lower than that of nylon-6 and nylon-6,6. Qiana required a carrier for effective dyeing (see DYE CARRIERS).

Other polyamides having higher moduli and $T_g$ than nylon-6 and nylon-6,6 have been evaluated in an effort to reduce wrinkle resistance and eliminate flat-spotting of bias and bias-belted tires (Table 3). Nylons have also been tested extensively over the years for apparel and carpets (Table 4).

**Table 3. Dimensionally Stable Experimental Nylons**

| Diamine[a] | Diacid | CAS Registry Number | Mp, °C | $T_g^b$, °C | Tenacity, N/tex[c] | Elongation, % |
|---|---|---|---|---|---|---|
| PACM[d] | dodecanedioic | [25035-12-5] | 298 | ~220 | 0.78 | 13 |
| CBMA[e] | dodecanedioic | [53830-65-2] | 280 | | 0.8 | 13–16 |
| | sebacic | [53830-66-3] | 291 | 175 | 0.6 | 14 |
| CBEA | dodecanedioic, | [52277-94-8] | 254 | | 0.35–0.44 | 20–30 |
| | sebacic | | | 135 | | |
| MDA[f] | dodecanedioic, | | 270 | | 0.48 | 11 |
| | sebacic | | | 160 | | |
| PXD[e] | sebacic | [31711-07-6] | 290 | | 0.44–0.53 | 12–15 |
| HMD[g] | terephthalic | [24938-70-3] | 370[h] | 158 | 0.22–0.44 | 30–45 |
| | terephthalic/ | | 310 | 180 | 0.3–0.66 | 8–40 |
| | isophthalic[i] | | | 165 | | |

[a]PACM = bis(p-aminocyclohexyl)methane; CBMA = p-cyclohexanebis(methylamine); CBEA = p-cyclohexanebis(ethylamine); HMD = hexamethylenediamine; MDA = 4,4'-methylenedianiline (bis(p-aminophenyl)methane); PXD = p-xylylenediamine.
[b]Dry; $T_g$ measured by Vibron analysis.
[c]To convert N/tex to g/den (gpd), multiply by 11.33.
[d]Ref. 203.
[e]Ref. 204.
[f]Ref. 205.
[g]Ref. 206.
[h]Melting point is too high for melt spinning.
[i]Ref. 207; T/I = 65/35.

**Table 4. Miscellaneous Nylons**

| Polyamide | CAS Registry Number | Mp, °C | $T_g$, °C | Tenacity, N/tex[a] | Moisture regain, % |
|---|---|---|---|---|---|
| poly(m-xylylene-adipamide)[b] | [25805-74-7] | 243 | 90 | 0.26–0.44 | 4.0 |
| nylon-4[c] | [24938-56-2] | 262 | 72 | 0.22–0.40 | 7.3 |
| nylon-7[d] | [25035-01-2] | 235 | 60 | 0.26–0.40 | 2.8 |
| nylon-11[d] | [25035-04-5] | 192 | 57 | 0.26–0.44 | 1.3 |
| nylon-12[d,e] | [24937-16-4] | 175 | 40 | 0.22–0.40 | 2.0 |

[a]To convert N/tex to g/den (gpd), multiply by 11.33.
[b]Refs. 208 and 209; hot–wet properties and light stability inferior to nylon-6 and nylon-6,6.
[c]Refs. 209 and 210; excessive property loss when wet.
[d]Ref. 209.
[e]Melting point is too low.

## Applications

The world fiber mill consumption of nylon has been forecasted to increase from 3.72 to 4.56 million tons from the year 1990 to 2000 (211). During the same period, polyester consumption is expected to grow from 8.41 to 12.88 million tons, which is the largest forecasted increase of synthetic fibers. Polyester is expected to take market share from nylon, even though nylon continues to grow, especially in applications where its intrinsic functional properties are highly desired. Coupled with consumer demand and the innovative endeavors of fiber producers and downstream processors to offer timely and economically competitive technology and products, the nylon fiber industry has sustained a growth pattern since its creation in 1938. Nylon fiber was cited as an example of how the life cycle of a maturing business can be extended by employing sound marketing and technical principles to position products effectively for continual growth (212,213). Technical, product, and marketing breakthroughs led to new uses in the first 25 years that postponed the total time of market maturity for nylon fibers. New end uses introduced during this period were circular knits in 1943; broadwovens in 1944; tricot warp knits in 1945; carpet staple, automotive upholstery, and sheer hosiery in 1947; tire cord in 1948; texturized yarns in 1955; and carpet BCF in 1959. In the 1990s, most of the end uses for nylon-6 and nylon-6,6 filament, tow, staple, and flock are categorized into four major application areas, in descending order of market size: carpets, industrial, apparel, and home furnishings. Carpet applications include residential, commercial, automotive, and rugs.

In carpets, nylon continues to be the dominant fiber, but it has lost some of its market share to polyester and especially polypropylene since the 1980s. From 1983 to 1993, the total carpet fiber shipments in the United States had grown from 0.65 to 1.3 million tons, but nylon's share decreased from 78% to 66% (214). Nevertheless, nylon remains the preferred fiber for carpets because of its excellent wear resistance, appearance retention, and ease of coloration. For a given appearance retention and wear life, particularly in plushes, less nylon (in terms of weight per area unit) is required than polyester and polypropylene. Technical innovations such as soil and stain resistance and shock-free carpeting have contributed significantly to the growth of nylon in this industry. During the mid-1980s, fiber producers introduced carpet performance guarantees, new styling, and high performance pigmented yarns to compete for market share. Du Pont had its Stainmaster certification program, Monsanto its Stain Blocker with a wear-dated guarantee, Allied its Anso Worry-Free program, and BASF its Zeftron product. With trade emphasis on stain resistance, producers of inherently stain-resistant polyester and polypropylene fibers have benefited and been able to gain market share.

In the category of industrial applications, nylon is the predominant fiber used in the carcass of bias truck, racing car, and airplane tires because of its excellent strength, adhesion to rubber, and fatigue resistance. Nylon is used less in the carcass of radial tires for automobiles and in replacement bias and bias-belted tires because of the development of temporary flat spots. For this reason, nylon has lost most of this market to polyester.

A ruling issued by the U.S. Department of Transportation in 1984 that all new cars sold in the United States must be equipped with automatic crash-restraints for driver and passenger by 1990 led to the adoption of the air bag concept which created a new fabric opportunity for nylon industrial yarns. Air bags are expected to become a 50-million-yard business by 1997. The air bag fabric is woven with high tenacity nylon-6 or nylon-6,6 industrial yarns, ranging from 47 to 94 tex (420–840 denier), in plain or rip-stop construction, and then scoured, heat-set, and coated with a neoprene formulation to protect the fabric from environmental pollutants and changes in temperature and humidity. In addition to automotive uses such as tire cords and air bags, other industrial applications include cordage, hose, webbing, sporting goods, coated fabrics, belting, printer ribbon tapes, personal flotation devices, luggage, upholstery (automotive, contract), and paper maker felts (staple).

In apparel, nylon has lost market share to polyester in broadwoven applications requiring easy care and wrinkle resistance, but continues to be the principal fiber in knit applications such as hosiery, pantyhose, and lingerie. A wrapped polyurethane-based elastic fiber (spandex) introduced by Du Pont under the trademark Lycra was used in the development of stretch nylon hosiery and swimwear (see FIBERS, ELASTOMERIC). The introduction of 0.88–1.12-dtex (0.8–1-dpf) fiber technology greatly impacted the nylon apparel market. Du Pont's introduction of 1.12-dtex filament Supplex nylon in 1985 penetrated the skiwear and activewear markets because of its durability, competitive cottony hand, and resistance to abrasion, water, and wind. In 1991, BASF introduced its 3.3 tex (30-denier) 37-filament microdenier nylon, Silky Touch. Nylon staple continues to be recognized as a reinforcing fiber for both cotton and wool blend fabrics used for military uniforms and women's sportswear. Other apparel uses are in shell cloth, work clothes, sacks, and interliners.

Home furnishings, which is primarily dominated by cotton and polyester, includes drapes, curtains, bedding, table cloth, bathroom fabrics, and upholstery. In the United States, nearly all of the nylon upholstery velours are flocked, printed fabrics. Nylon-flocked blankets are also made in the United States.

## Recycling

The recycling of fibers, fabrics, and end use articles to conserve materials and minimize landfilling has become an important technical and business issue in the 1990s. The commitment to recycle and protect the environment in compliance to regulatory legislation will have a profound effect on how fiber producers design their offerings for the future. For example, a car seat with upholstery fabric and cushioning material of the same polymer type would be easier to recycle than a unit with heterogeneous polymeric materials. Recycling can occur in many forms. A fabric or article of commerce can be reconditioned for reuse, mechanically converted to a new use (eg, shredded for filler material), burned for fuel value, or converted to its original raw materials (see RECYCLING).

Producers of carpet fiber have initiated carpet reclamation programs (209). The approaches of these programs differ, but all aim at reducing the 1.5 × $10^6$ t of carpets annually disposed through incineration or landfilling in the United States. Du Pont is reclaiming post-consumer carpets through carpet mills and dealers. The carpets are collected, assorted, packaged, and shipped to a processing center where the surface yarn is sheared from the backing and then converted into new products for carpet cushion and plastic reinforcement. AlliedSignal and BASF are collecting only nylon-6 carpets; the companies then remove the surface yarn and convert it back to caprolactam for repolymerization and melt spinning. Nylon-6 depolymerization processes have been designed for carpet reclamation, conversion of manufacturing waste fiber to caprolactam (215–217), and separation of nylon-6 and nylon-6,6 fiber mixtures (218).

## BIBLIOGRAPHY

"Polyamides" in *ECT* 1st ed., Vol. 10, pp. 916–937, by F. Schulze, E. I. du Pont de Nemours & Co., Inc., and H. Wittcoff, General Mills, Inc.; "Polyamide Fibers" in *ECT* 2nd ed., Vol. 16, pp. 46–87, by O. E. Snider and R. J. Richardson, Allied Chemical Corp.; in *ECT* 3rd ed., Vol. 18, pp. 372–405, by J. H. Saunders, Monsanto Co.

1. M. Thomas, *Text. World*, **138**(3), 61 (1988).
2. G. B. Kauffman, *Chem. Tech.* **17**(12), 725 (1988).
3. H. Mark and G. S. Whiby, eds., *Collected Papers of Wallace Hume Carothers on High Polymeric Substances, High Polymers*, Vol. I, Interscience Publishers, New York, 1940.
4. E. K. Bolton, *Ind. Eng. Chem.* **34**, 53 (1942).
5. C. S. Marvel and C. E. Carraher, Jr., *Chem. Tech.* **14**, 716 (1984).
6. O. L. Shealy, *Fiber Prod.* **12**(2), 32 (1984).
7. U.S. Pat. 2,071,250 (Feb. 16, 1937), W. H. Carothers (to E. I. du Pont de Nemours & Co., Inc.).
8. U.S. Pat. 2,071,253 (Feb. 16, 1937), W. H. Carothers (to E. I. du Pont de Nemours & Co., Inc.).
9. U.S. Pat. 2,130,523 (Sept. 20, 1938), W. H. Carothers (to E. I. du Pont de Nemours & Co., Inc.).
10. U.S. Pat. 2,130,948 (Sept. 20, 1938), W. H. Carothers (to E. I. du Pont de Nemours & Co., Inc.).
11. P. A. Schlack, *Chemiefasern/Textilindus.* **17**, 961 (1967).
12. H. Klare, *Technologie und Chemie der Synthetischem Fasern aus Polyamiden*, Verlag Technik, Berlin, Germany, 1954, p. 13.
13. H. Hupff, *Synthetic Fiber Development in Germany*, Part II, H. M. Stationery Office, London, U.K., 1945, p. 101.
14. M. S. Reisch, *C&EN*, **31**(71), 12 (Aug. 2, 1993).
15. W. E. Morton and J. W. S. Hearle, *Physical Properties of Textile Fibers*, Textile Institute, Manchester, U.K., 1993.
16. H. D. W. Smith, *Rayon Text. Mar.* **26**(6), 271 (1945).
17. R. Meredith, *Mechanical Properties of Textile Fibers*, Interscience Publishers, New York, 1956, Chapt. 16.
18. F. D. Snell and L. S. Ettre, eds., *Encyclopedia of Industrial Chemical Analysts*, Vol. 17, John Wiley & Sons, Inc., New York, 1973, pp. 275–305.
19. C. W. Bunn and E. V. Garner, *Proc. Royal Soc. (London)*, **A189**, 39 (1947).
20. P. F. Dismore and W. O. Statton, *Polym. Sci.*, **C13**, 133 (1966).
21. H. F. Mark, S. M. Atlas, and E. Cermia, eds., *Man-Made Fibers: Science and Technology*, Vol. 1, Wiley-Interscience, New York, 1967, pp. 24–374.
22. J. H. Dumbleton, D. R. Buchanan, and B. B. Bowles, *J. Applied Polym. Sci.* **12**, 2067 (1968).
23. R. F. Stepaniak and co-workers, *J. Appl. Polym. Sci.* **23**, 1747 (1979).
24. J. Brandrup and E. H. Immergut, eds., *Polymer Handbook*, Wiley-Interscience, New York, 1989, pp. V-109, VI-258.
25. R. A. F. Moore, *Text. Chem. Color.* **21**(2), 19 (1989).
26. M. Isaccs, III, ed., *Textile World Manmade Fiber Chart*, Maclean Hunter Publication Co., Chicago, Ill., 1992.
27. P. G. Simpson, J. H. Southern, and R. L. Ballman, *Text. Res. J.* **51**, 97 (1981).

28. Ref. 15, pp. 341–356.
29. D. G. Hunt and D. W. Darlington, *Polymer*, **20**(2), 241 (1979).
30. H. H. Starkweather, Jr. and co-workers, *J. Polym. Sci.* **21**, 189 (1956).
31. R. Puffr and J. Sebenda, *J. Polym. Sci. Part C*, (16), 79 (1967).
32. H. G. Olf and A. Peterlin, *J. Polym. Sci. Part A-2*, **11**, 2033 (1971).
33. E. G. Smith, *Polymer*, **17**(9), 7751 (1976).
34. G. Weber, *Angew. Makromol. Chem.* **74**, 187 (1978).
35. A. Anton, *Text. Res. J.* **43**(9), 524 (1973).
36. W. T. Holfeld and M. S. Shepard, *Can. Tex. J.* **94**, 80 (1977).
37. J. Freitag, *Chemiefasern/Textilindus., Man-Made Fiber Yearbook*, 58, 1989.
38. G. J. Kettle, *Polymer*, **18**(7), 742 (1977).
39. G. Hinrickson, *Colloid Polym. Sci.* **256**, 9 (1978).
40. J. L. Rush and J. C. Miller, *Am. Dyest. Rep.* **24**(2), 37 (1969).
41. M. E. Gibson, Jr. and co-workers, *Text. Chem. Color.* **18**(9), 55 (1980).
42. N. S. Murty and co-workers, *Macromolecules*, **24**, 3215 (1991).
43. H. W. Schmidlin, *Preparation and Dyeing of Synthetic Fibers*, Reinhold Publishing Corp., New York, 1963, p. 45.
44. H. F. Mark, S. M. Atlas, and E. Cernia, *Man-Made Fibers*, Vol. 2, John Wiley & Sons, New York, 1968, p. 280.
45. M. Dole and B. Wunderlich, *Macromol. Chem.* **34**, 29 (1969).
46. *AATCC Technical Manual*, American Association of Textile Chemists and Colorists, Research Triangle Park, N.C., 1993, p. 50.
47. Ref. 15, p. 564.
48. Ref. 44, p. 259.
49. A. Anton, *Text. Chem. Col.* **13**(2), 46 (1981).
50. M. Makansi, *Text. Chem. Col.* **18**, 27 (1986).
51. N. B. Neiman, *Aging and Stabilization of Polymers*, Consultants Bureau, New York, 1965, p. 236.
52. R. W. Moncrief, *Man-Made Text.* **41**(481), 136 (1964).
53. Ref. 51, p. 249.
54. U.S. Pat. 2,705,227 (Mar. 29, 1955), G. S. Stanatoff (to E. I. du Pont de Nemours & Co., Inc.).
55. U.S. Pat. 3,113,120 (Dec. 3, 1963), P. V. Papero and R. L. Morter (to Allied Chemical Corp.).
56. U.S. Pat. 3,272,773 (Aug. 13, 1966), D. H. Edison (to E. I. du Pont de Nemours & Co., Inc.).
57. U.S. Pat. 3,457,325 (July 22, 1969), A. Anton (to E. I. du Pont de Nemours & Co., Inc.).
58. *CIBA-GEIGY Technical Bulletins, Irganox 1098* (Apr. 1974); *Irgafos 168 in Combination with Irganox Antioxidants* (Oct. 1989), CIBA-GEIGY Corp., Ardsley, N.Y.
59. U.S. Pat. 3,003,995 (Oct. 10, 1961), E. C. Shule (to Allied Chemical Corp.).
60. R. F. Moore, *Polymer*, **4**, 493 (1963).
61. W. H. Sharkey and W. E. Mochel, *J. Am. Chem. Soc.* **81**, 3000 (1959).
62. C. V. Stephenson, B. C. Moses, and W. S. Wilcox, *J. Polym. Sci.* **55**, 477 (1961).
63. G. Reinert and F. Thommen, *Tex. Chem. Col.* **23**, 31 (1991).
64. Ref. 51, p. 253.
65. A. Anton, *J. Appl. Polym. Sci.* **9**, 1631 (1965).
66. U.S. Pat. 2,887,462 (May 19, 1955), J. G. Van Oot (to E. I. du Pont de Nemours & Co., Inc.).
67. U.S. Pat. 3,002,947 (Oct. 3, 1961), D. E. Maple (to E. I. du Pont de Nemours & Co., Inc.).
68. U.S. Pat. 3,242,134 (Mar. 22, 1966), P. V. Papero (to Allied Chemical Corp.).

69. U.S. Pat. 4,710,535 (Dec. 1, 1987) P. Perrot and G. Vuillemey (to Rhone-Poulenc Fibre).

70. Jpn. Pat. 62106959-A (May 5, 1987), O. Shinonome, T. Kitahara, and S. Murakam (to Unitika KK).

71. Du Pont Technical Information Bulletin, *NY-11*, E. I. du Pont de Nemours & Co., Inc., Wilmington, Del., June 1980.

72. H. K. Reimschuessel, *J. Polym. Sci. Macromol. Rev.* **12**, 65 (1977).

73. H. Ludewig, *Faserforsh. Textiltech.* **2**, 341 (1951); obituary, *Acta Polym.* **35**, 113 (1984).

74. W. R. Sorenson and T. W. Campbell, *Preparative Methods of Polymer Chemistry*, Interscience Publishers, New York, 1961, pp. 60, 236.

75. F. S. Fiordan, Jr. and J. H. Saunders, eds., *Forty Years of Melt Spinning*, American Chemical Society, Washington, D.C., 1981.

76. A. Alexander, *Man-Made Fiber Processing*, Noyes Data Corp., Parkridge, N.J., 1966, p. 82.

77. U.S. Pat. 4,804,508 (Feb. 14, 1989), Jean-Pierre Double and Cecile Lecluse (to Rhone-Poulenc Fibres).

78. I. E. Lenk, *Chemifasern/Textilindus. Man-Made Fiber Yearbook*, 1994, p. 62.

79. U.S. Pat. 3,311,691 (Mar. 29, 1967), A. N. Good (to E. I. du Pont de Nemours & Co., Inc.).

80. H. Klare, *Synthetic Fibers from Polyamides*, Akademie-Verlag GmbH, Berlin, Germany, 1963.

81. F. Fourne, *Synthetische Fasern*, Wissenschaftlicher Verlag, Stuttgart, Germany, 1964, p. 39.

82. A. Ziabicki, *Fundamentals of Fiber Formation*, John Wiley & Sons, Inc., New York, 1976.

83. A. Wasiak and A. Ziabicki, *Appl. Polym. Symp.* **27**, 111 (1975).

84. J. E. Spruiell and J. L. White, *Appl. Polym. Symp.* **27**, 121 (1975).

85. J. G. Bankar, J. E. Spruiell, and J. L. White, *J. Appl. Polym. Sci.* **21**, 2341 (1977).

86. M. D. Danford, J. E. Spruiell, and J. L. White, *J. Appl. Polym. Sci.* **22**, 3351 (1978).

87. J. Gianchandani, J. E. Spruiell, and E. S. Clark, *J. Appl. Polym. Sci.* **27**, 3527 (1982).

88. W. O. Statton, *J. Polym. Sci. Part A-2*, **10**, 1587 (1972).

89. J. P. Bell, *Tex. Res. J.* **42**, 292 (1972).

90. J. Elad and J. M. Schultz, *J. Polym. Sci. Polym. Phys.* **22**, 781 (1984).

91. G. Perez and E. Jung, *ISF-85*, Hakone, Japan, 1985, p. 20.

92. H. Brever and co-workers, *Chemiefasern/Textilindus.* **42**, 87 (1992).

93. H. Treptow, *Chemiefasern/Textilindus. Man-Made Fiber Yearbook*, 1986, p. 6.

94. R. T. Maier, *Chemiefasern/Textilindus. Man-Made Fiber Yearbook*, 1986, p. 106.

95. R. C. Mears, *Chemiefasern/Textilindus. Man-Made Fiber Yearbook*, 1986, p. 108.

96. J. F. Hagewood and Monroe, *Chemiefasern/Textilindus. Man-Made Fiber Yearbook*, 1988, p. 102.

97. I. F. Maag, *Chemiefasern/Textilindus. Man-Made Fiber Yearbook*, 1990, p. 79.

98. U.S. Pat. 5,219,503 (June 15, 1993), R. L. Boles, Jr. and co-workers (to E. I. du Pont de Nemours & Co., Inc.).

99. J. Shimizu and co-workers, *J. Soc. Fiber Sci. Technol. Jpn.* **37**(4), T143 (1981).

100. K. Koyanna, J. Suryadevara, and J. E. Spruiell, *J. Appl. Polym. Sci.* **31**, 2203 (1986).

101. U.S. Pat. 5,202,182 (Apr. 13, 1993), B. H. Knox and co-workers (to E. I. du Pont de Nemours & Co., Inc.).

102. U.S. Pat. 5,182,068 (Jan. 26, 1993), J. Richardson (to Imperial Chemical Industries PLC).

103. U.S. Pat. Re 33,059 (U.S. Pat. 4,583,357) (Sep. 19, 1989), J. M. Chamberlin and co-workers (to Monsanto Co.).

104. U.S. Pat. 4,721,650 (Jan. 26, 1988), W. J. Nunning and co-workers (to Monsanto Co.).
105. U.S. Pat. 5,238,637 (Aug. 24, 1993), O. Chaubet and co-workers (to Rhone-Poulenc Fibers).
106. B. L. Hathorne, *Woven Stretch and Textured Fabrics*, Wiley-Interscience, New York, 1964.
107. H. M. El-Behery, Draw Texturing Technical Conference, Clemson, S.C., Apr. 10, 11, 1973.
108. *Texturing Today*, Shirley Institute Conference, Shirley Institute Publication S. 46, Manchester, U.K., 1982.
109. H. Schellenberg, *Chemiefasern/Textilindus. Man-Made Fibers Yearbook*, 1986, p. 96.
110. T. Ishida, *JTN*, **1990**, 98 (1990).
111. U.S. Pat. 2,958,112 (Nov. 1, 1960), J. N. Hall.
112. U.S. Pat. 2,884,756 (May 5, 1959), W. I. Head (to Eastman Kodak, Co.).
113. U.S. Pats. 2,783,609 (Mar. 5, 1957), 2,852,906 (Sep. 23, 1958), and 2,869,967 (Jan. 20, 1959), A. L. Breen (to E. I. du Pont de Nemours & Co., Inc.).
114. U.S. Pat. 3,005,251 (Oct. 24, 1961), C. E. Hallden, Jr. and K. Murenbeeld (to E. I. du Pont de Nemours & Co., Inc.).
115. U.S. Pat. 2,942,402 (June 28, 1960), C. W. Palm (to Celanese Corp.).
116. U.S. Pat. 2,884,756 (May 5, 1959), W. I. Head (to Eastman Kodak Co.).
117. U.S. Pat. 2,852,906 (June 23, 1958), A. L. Breen (to E. I. du Pont de Nemours & Co., Inc.).
118. U.S. Pat. 2,875,502 (May 3, 1959), H. W. Matthews and W. K. Wyatt (to Turbo Machine Co.).
119. U.S. Pat. 2,245,874 (June 17, 1941), W. S. Robinson.
120. Brit. Pat. 801147 (Oct. 12, 1955), J. A. Place (to British Nylon Spinners Ltd.).
121. F. Werny and co-workers, *Ullmann's Encyclopedia of Industrial Chemistry*, VCH Verlagsgesellschaft, Weinheim, Germany, 1988, p. 263.
122. A. Alexander, *Manmade Fiber Processing*, Noyes Development Corp., Park Ridge, N.J., 1966, pp. 149, 150.
123. U.S. Pat. 2,575,781 (Nov. 20, 1951), J. L. Barach (to Alexander Smith).
124. S. D. Fredericks, *Int. Fiber J.* **5**(4) 49–62 (1990).
125. *Chemiefasern/Textilindus.* **43/95**, 499 (E77) (1993).
126. U.S. Pat. 4,908,920 (May 20, 1990), Katsuomi Takehara (to Kabushiki Kaisha Takehara Kikai Kenkyusho).
127. M. Grayson, ed., *Encyclopedia of Textiles, Fibers and Nonwoven Fabrics*, Wiley-Interscience, 1989, pp. 252, 284.
128. E. Oxtoby, *Spun Yarn Technology*, Butterworths, London, U.K., 1987.
129. P. R. Lord, *The Economics, Science and Technology of Yarn Production*, The Textile Institute, Manchester, U.K., 1981, pp. 30–192.
130. *Design with Flock in Mind*, American Flock Association, Boston, Mass., 1992.
131. W. Irganells and N. Ranadan, *J. Soc. Dyers Colour.* **108**(5/6), 270 (1992).
132. R. L. Coldwell and H. P. Solomon, *IEEE Transactions on Industry Applications*, **1A-14**(2), 175 (Mar./Apr., 1978).
133. Eur. Pat. 93420253.2 (June 17, 1993), C. Cavalie (to Novalis Fibers).
134. M. J. Schick, ed., *Surface Characteristics of Fibers and Textiles*, Part 1, Marcel Dekker, Inc., New York, 1975, pp. 1–65.
135. H. R. Billica, *Fiber Prod.* **12**(2), 21 (1984).
136. H. R. Billica, *Fiber Prod.* **12**(3), 24 (1984).
137. S. E. Ross, *IFJ*, **8**(5), 78 (Oct. 1993).
138. A. Anton, K. A. Johnson, and P. A. Jansson, *Tex. Res. J.* **48**(5), 247 (1978).
139. U.S. Pats. 2,939,201 and 2,939,202 (June 7, 1960), M. C. Holland (to E. I. du Pont de Nemours & Co., Inc.).

140. U.S. Pats. 3,109,220 (Nov. 5, 1963), A. H. McKinney and H. E. Stanley (to E. I. du Pont de Nemours & Co., Inc.).
141. U.S. Pat. 3,075,242 (Jan. 29, 1963), E. Grafried (to W. C. Heraeus).
142. U.S. Pat. 3,081,490 (Mar. 19, 1963), W. Heymen and W. Martin (to Vereingte Glanzstoff-Fabriken AG).
143. U.S. Pat. 3,745,061 (July 10, 1973), A. J. Champaneria and M. R. Lindbeck (to E. I. du Pont de Nemours & Co., Inc.).
144. U.S. Pat. 5,190,821 (Mar. 2, 1993), M. T. Goodall, C. A. Jackson, and P. H. Lin (to E. I. du Pont de Nemours & Co., Inc.).
145. *JTN*, **459**(2), 44 (1993).
146. U.S. Pat. 2,689,332 (Aug. 8, 1958), G. A. Nesty (to Allied Chemical Corp.).
147. U.S. Pat. 3,903,348 (Sep. 2, 1975), B. B. Esteridge and B. M. Lyon (to Akzona Inc.).
148. P. Schaeffer, *Chem. Tech.* **12**, 242 (1960).
149. S. L. Handen and J. R. Graff, *Fiber Prod.* **8**(4), 35 (1980).
150. S. S. Parikh and co-workers, *Material Processes and Products in Interior Trim, Seating & Headliners*, Society of Automotive Engineers, Inc., Detroit, Mich., 1991, p. 47.
151. U.S. Pat. 3,329,557 (July 4, 1967), E. E. Magat and D. Tanner (to E. I. du Pont de Nemours & Co., Inc.).
152. U.S. Pat. 3,388,104 (June 11, 1968), L. W. Crojatt, Jr. (to Monsanto Co.).
153. U.S. Pat. 3,900,676 (Aug. 19, 1975), T. Alderson (to E. I. du Pont de Nemours & Co., Inc.).
154. Jpn. Pat. 03275603-A (9204) (Dec. 6, 1991), T. Kawachi (to Teisan Seiyaku KK).
155. World Pat. 09207037-A (9220) (Mar. 30, 1992), A. K. Philpott (to Aboe Pty. Ltd.).
156. Jpn. Pat. 04240205-A (9241) (Aug. 27, 1991), Y. Takeda and I. Yuaja (to Kuraray Co., Ltd.).
157. Jpn. Pat. 04248943-A (9242) (Sept. 4, 1992), Y. Mizukami, H. Tamemasa, and Y. Kimuka (to Kanebo Ltd.).
158. Eur. Pat. 505638-A1 (9240) (Sept. 30, 1992), Z. Hagiwara and M. Okubo (to Hagiwara Giken KK).
159. U.S. Pat. 5,180,585 (Jan. 19, 1993), H. W. Jacobson, S. L. Samuels, and M. H. Scholla (to E. I. du Pont de Nemours & Co., Inc.).
160. H. E. Stepniczka, *Ind. Eng. Chem. Prod. Res. Dev.* **12**, 29 (1975).
161. J. W. Stoddard and co-workers, *Tex. Res. J.* **45**, 474 (1975).
162. U.S. Pat. 5,238,982 (Sep. 24, 1992), A. Adhya and R. L. Lilly (to BASF Corp.).
163. U.S. Pat. 4,298,518 (Nov. 3, 1981), Y. Ohmura, Y. Murakani, and R. Hidaka (to Mitsubishi Chemical Industries, Ltd.).
164. H. U. Schanidlin, *Preparation and Dyeing of Synthetic Fibers*, Van Nostrand Reinhold Co., Inc., New York, 1963, p. 172.
165. A. Hebeish, M. A. ElKashoutl, and I. Abo-El-Thalouth, *Am. Dyest. Rep.* (2), 24 (1983).
166. *Chemiefasern/Textilindus.* **34/86**, 354 (E50) (1984).
167. R. Christmann and co-workers, *Chemiefasern/Textilindus.* **42/94**, E128 (1992).
168. U.S. Pat. 3,375,651 (Apr. 2, 1968), L. B. Greeson, Jr. (Monsanto Co.).
169. U.S. Pat. 4,017,255 (Apr. 12, 1977), M. C. Cobb (Imperial Chemical Industries).
170. U.S. Pat. 3,184,436 (May 18, 1965), E. E. Magat (E. I. du Pont de Nemours & Co., Inc.).
171. M. D. Chantler, G. A. Partlett, and J. A. B. Whiteside, *JSDC*, **85**(12), 621 (1969).
172. G. A. Nesty, *Text. Res. J.* **29**, 765 (1944).
173. J. L. Rush and J. C. Miller, *Am. Dyest. Rep.* **58**, 37 (1969).
174. R. A. F. Moore and H. D. Weigman, *Text. Res. J.* **56**, 180 (1986).
175. L. Han. T. Wakida, and T. Takagishi, *Text. Res. J.* **57**, 519 (1987).
176. H. A. Davis, *Text. Chem. Color.* **24**(6), 19 (1992).

177. *Dyeing and Finishing of DuPont Textile Nylon*, DuPont Technical Information Bulletin, *NY-12*, E. I. du Pont de Nemours & Co., Inc., Wilmington, Del., 1981.
178. U.S. Pat. 5,230,709 (July 27, 1993), W. T. Holfeld and D. E. Mancuso (E. I. du Pont de Nemours & Co., Inc.).
179. A. Anton, *Text. Chem. Color.* **4**(10), 32 (1982).
180. *Cibafast N-2, CIBA-GEIGY Technical Bulletin 10*, CIBA-GEIGY Corp., Greensboro, N.C.
181. *Cibafast W, CIBA-GEIGY Technical Bulletin 7*, CIBA-GEIGY Corp., Greensboro, N.C.
182. U.S. Pat. 4,902,299 (Feb. 20, 1990), A. Anton (E. I. du Pont de Nemours & Co., Inc.).
183. J. R. Aspland, *Text. Chem. Color.* **25**(9), 79 (1993).
184. E. A. Tippetts, *Tex. Res. J.* **37**, 524 (1967).
185. U.S. Pat. 4740339 (Apr. 26, 1988), H. C. Bach and W. B. Black (to Monsanto Co.).
186. A. H. Bruner, N. W. Boe, and P. Bryne, *J. Elastoplast.* **5**, 201 (1973).
187. J. H. Saunders and co-workers, *J. Appl. Polym. Sci.* **19**, 1387 (1975).
188. E. E. Magat and R. E. Morrison, *J. Polym. Sci Polym. Symp.* **51**, 203 (1975).
189. U.S. Pat. 3,969,559 (May 27, 1975), N. W. Boe (to Monsanto Co.).
190. U.S. Pat. 4,075,378 (Feb. 21, 1978), A. Anton and J. A. B. Nolin (to E. I. du Pont de Nemours & Co., Inc.).
191. S. Shiomura, *Text. Asia*, **22**(9), 140 (1991).
192. U.S. Pat. 5,047,189 (Sep. 10, 1991), Chen-Ling Lin (to Nan Ya Plastics Corp.).
193. "Block Copolymers," in J. I. Kroschwitz, ed., *Encyclopedia of Polymer Science and Engineering*, Vol. 2, Wiley-Interscience, New York, 1985, pp. 324–434.
194. U.S. Pat. 4,113,794 (Sep. 12, 1978), R. M. Thompson and R. S. Stearns (to Sun Ventures Inc.).
195. R. A. Lofquist and co-workers, *Text. Res. J.* **55**(6), 325 (1985).
196. U.S. Pat. 5,110,900 (May 5, 1992), J. A. Hammond, Jr. and D. N. Marks (to E. I. du Pont de Nemours & Co., Inc.).
197. U.S. Pat. 5,185,428 (Feb. 9, 1993), J. A. Hammond, Jr. and D. N. Marks (to E. I. du Pont de Nemours & Co., Inc.).
198. U.S. Pat. 5,194,578 (May 16, 1993), A. Anton (to E. I. du Pont de Nemours & Co., Inc.).
199. K. El-Salmawi and co-workers, *Am. Dyest. Rep.* **82**(5), 47 (1993).
200. U.S. Pat. 3,394,985 (July 30, 1968), H. H. Froehlich, (to E. I. du Pont de Nemours & Co., Inc.).
201. U.S. Pat. 5,236,464 (Aug. 17, 1993), D. K. Barnes and co-workers (to Allied-Signal, Inc.).
202. E. Roerdink and J. M. M. Warnier, *Polymer*, **26**(10), 1582 (1985).
203. J. W. Hannell, *Polym. News*, **1**(1) 9 (1970).
204. A. Bell, J. G. Smith, and C. J. Kibler, *J. Polym. Sci. Part A*, **3**(1), 19 (1965).
205. D. A. Holmer, O. A. Pickett, Jr., and C. J. Kibler, *J. Polym. Sci. Part A-1*, **10**, 1547 (1972).
206. B. S. Sprague and R. W. Singleton, *Text. Res. J.* **35**, 999 (1965).
207. R. D. Chapman and co-workers, *Text. Res. J.* **51**, 564 (1981).
208. E. F. Carlston and F. G. Lum, *Ind. Eng. Chem.* **49**, 1239 (1957).
209. R. W. Longbottom, *Mod. Text.* **49**(12), 19 (1968).
210. E. M. Peters and J. A. Geruasi, *Chem. Tech.* **2**, 16 (1972).
211. P. S. Collishaw, A. D. Cunningham, and D. P. Lindsay, *Proceedings from International Conference and Exhibition, AATCC*, Atlanta, Ga., 1992, p. 63.
212. J. P. Yale, *Mod. Text. Mag.* **44**(2), 33 (1964).
213. T. Levitt, *The Marketing Imagination*, Free Press, New York, 1962, pp. 173–199.
214. A. H. Snider, ed., *Int. Fiber J.* **9**(3), 35 (1994).

215. U.S. Pat. 5,169,870 (Dec. 8, 1992), T. F. Corbin, E. A. Davis, and J. A. Dellinger (to BASF Corp.).

216. R. N. Goel and K. N. Seetha, *Syn. Fibres,* **11**, 10–19 (1982).

217. *Int. Fiber J.,* **7**(5), 12, 34 (1992).

218. U.S. Pat. 5,280,105 (Jan. 18, 1994), E. F. Moran (to E. I. du Pont de Nemours & Co.).

*General References*

E. M. Hicks, Jr. and co-workers, "The Production of Synthetic-Polymer Fibers," *Text. Prog.* **3**(1), 1 (1971).

A. J. Hughes and J. E. McIntyre, "Nylon Fibers," *Text. Prog.* **8**, 18 (1976).

H. F. Mark, S. M. Atlas, and E. Cernia, eds., *Man-Made Fibers Science and Technology,* 3 Vols., Wiley-Interscience, New York, 1967–1968.

Z. K. Walczak, *Formation of Synthetic Fibers,* Gordon & Breach, New York, 1977.

A. Ziabicki, *Fundamentals of Fiber Formation,* John Wiley & Sons, Inc., New York, 1976.

M. Lewin and E. M. Pearce, *Handbook of Fiber Science and Technology,* Vol. IV, Marcel Dekker, Inc., New York, 1985, pp. 75–169.

D. A. Clancy, *Flooring,* **99**(1), 14 (1993).

ANTHONY ANTON
E. I. du Pont de Nemours & Company, Inc.

# PLASTICS

Polyamides can claim to have been the first engineering plastics as a result of their excellent combination of mechanical and thermal properties. Despite being introduced as long ago as the 1930s, these materials have retained their vitality and new applications, and indeed new types of nylon continue to be developed.

Amide groups along the backbone of long-chain molecules link the monomeric units in polyamides (1). Polyamide is generally referred to as nylon and was first introduced in 1938 by Du Pont after its invention in the company's laboratories in 1935 (2). Nylons comprise a range of materials, depending on the monomers employed, and can be prepared by either a condensation reaction between a diacid and a diamine or by a ring-opening addition of a lactam. The type of nylon is designated by numbers or letters after the name. For example, nylon-6,6, which was the first nylon commercialized, is produced from hexamethylenediamine and adipic acid; the designation 6,6 reflects the number of carbon atoms in the diamine and the diacid, 6 and 6. Nylon-6, on the other hand, is the second of the two most important polyamides, first successfully produced in Germany in 1938 (3). Nylon-6 is produced from the single lactam monomer caprolactam containing six carbon atoms in the lactam ring. The names of polyamides based on aromatic monomers generally use letters to indicate the structure, eg, T for terephthalic acid and MXD for *m*-xylylenediamine.

Nylon-6,6 [32131-17-2] and nylon-6 [25038-54-4] continue to be the most popular types, accounting for approximately 90% of nylon use. There are a number of different nylons commercially available; Table 1 gives a summary of the properties of the more common types. In the 1990s there has been a spurt of new polyamide introductions designed for higher temperatures, better stiffness and strength, and/or lower moisture uptake.

**Table 1. Properties of the More Common Nylons, Dry as Molded**

| Property | Nylon-6,6[a] | Nylon-6[b] | Nylon-11[c] | Nylon-12[d] | Nylon-6,9[e] | Nylon-6,12[a] | ASTM test method |
|---|---|---|---|---|---|---|---|
| CAS Registry Number | [32131-17-2] | [25038-54-4] | [25035-04-5] | [24937-16-4] | [28757-63-3] | [24936-74-1] | |
| specific gravity | 1.14 | 1.13 | 1.04 | 1.02 | 1.09 | 1.07 | D792 |
| water absorption, wt % | | | | | | | |
| 24 h | 1.2 | 1.6 | 0.3 | 0.25 | 0.5 | 0.25 | |
| equilibrium at 50% rh | 2.5 | 2.7 | 0.8 | 0.7 | 1.8 | 1.4 | |
| saturation | 8.5 | 9.5 | 1.9 | 1.5 | 4.5 | 3.0 | |
| melting point, °C | 255 | 215 | 194 | 179 | 205 | 212 | D2117 |
| tensile yield strength, MPa[f] | 83 | 81 | 55 | 55 | 55 | 61 | D638 |
| elongation at break, % | 60–90 | 50–150 | 200 | 200 | 125 | 150 | D638 |
| flexural modulus, MPa[f] | 2800 | 2800 | 1200 | 1100 | 2000 | 2000 | D790 |
| Izod impact strength, J/m[g] | 53–64 | 55–65 | 40–68 | 95 | 58 | 53 | D256 |
| Rockwell hardness, R scale | 121 | 119 | 108 | 107 | 111 | 114 | D785 |
| deflection temperature under load, °C | | | | | | | D648 |
| at 0.5 MPa[f] | 235 | 185 | 150 | 150 | 150 | 180 | |
| 1.8 MPa[f] | 90 | 75 | 55 | 55 | 55 | 90 | |
| dielectric strength, kV/mm | | | | | | | D149 |
| short time | 24 | 17 | 16.7 | 18 | 24 | 16 | |
| step by step | 11 | 15 | | 16 | 20 | | |
| dielectric constant | | | | | | | D150 |
| at 60 Hz | 4.0 | 3.8 | 3.7 | 4.2 | 3.7 | 4.0 | |
| $10^3$ Hz | 3.9 | 3.7 | 3.7 | 3.8 | 3.6 | 4.0 | |
| $10^6$ Hz | 3.6 | 3.4 | 3.1 | 3.1 | 3.3 | 3.5 | |
| starting acid[h] or lactam | adipic acid[h] | caprolactam | 11-amino-undecanoic acid | dodecano-lactam | azaleic acid[h] | dodecanedioic acid[h] | |

[a] Ref. 4.
[b] Ref. 5.
[c] Ref. 6.
[d] Ref. 7.
[e] Ref. 8.
[f] To convert MPa to psi, multiply by 145.
[g] To convert J/m to ft·lbf/in., divide by 53.38.
[h] The starting amine is hexamethylenediamine for nylon-6,6, nylon-6,9, and nylon-6,12.

## Properties

When nylon-6,6 was first introduced, its main attraction was as a fiber-forming material the strength, elasticity, and high dye uptake of which were considered the most important properties, along with the ability to withstand ironing temperatures. It soon became apparent, however, that the properties of the material held many advantages for use as a plastic. In particular, the relatively high tensile strength and stiffness, together with good toughness, high melting point (and therefore temperature stability), and good chemical resistance, all combined to allow a wide range of applications. The material soon came to be seen as an engineering plastic that could be used for metal replacement in structural or semistructural end uses. These properties are present to a greater or lesser extent in the entire semicrystalline polyamide family, together with the "Achilles heel" of nylon, ie, the hygroscopic nature that leads to moisture uptake, change of properties, and the potential for hydrolysis.

Appropriate choice of monomer can provide a balance of properties to meet particular types of applications. In general, the effects of different monomers and therefore the property balance of different types of nylon can be summarized as follows: lengthening the aliphatic segments between the amide groups results in lower moisture absorption, reduced strength and stiffness, and lower melting point (eg, nylon-11 or nylon-12 compared to nylon-6,6). The introduction of aromatic groups increases stiffness and strength but reduces moisture uptake and, to a lesser extent, impact strength. Some semiaromatic polyamides also have an increased melting point.

In addition to the semicrystalline nylons, which comprise the vast majority of commercial resins, nylon is also available in an amorphous form that gives rise to transparency and improved toughness at the expense of high temperature properties and chemical stress crack resistance. Table 2 shows the properties of some different polyamide types.

### PHYSICAL PROPERTIES

**Crystallinity.** The presence of the polar amide groups allows hydrogen bonding between the carbonyl and NH groups in adjacent sections of the polyamide chains. For common nylons such as nylon-6,6 and nylon-6, the regular spatial alignment of amide groups allows a high degree of hydrogen bonding to be developed when chains are aligned together, giving rise to a crystalline structure in that region. These nylons are semicrystalline materials that can be thought of as a combination of ordered crystalline regions and more random amorphous areas having a much lower concentration of hydrogen bonding. This semicrystalline structure gives rise to the good balance of properties. The crystalline regions contribute to the stiffness, strength, chemical resistance, creep resistance, temperature stability, and electrical properties; the amorphous areas contribute to the impact resistance and high elongation. The crystallinity can be disrupted by substituents on the chains that interfere with the alignment process. Amorphous nylons are produced by deliberately engineering this effect, eg, nylon-NDT/INDT (also known as PA-6-3-T or PA-TMDT), which uses trimethyl-substituted hexamethylenediamine isomers combined with terephthalic acid.

**Table 2. Properties of Other Nylons, Dry as Molded**

| Property | Nylon-4,6 | Nylon-MXD,6 | Nylon-NDT/INDT | Polyphthalamide (PPA) |
|---|---|---|---|---|
| CAS Registry Number | [50327-22-5] | [25805-74-7] | [9071-17-4] | |
| water absorption, % | | | | 0.81 |
| 24 h | 2.0 | 0.31 | | |
| 50% rh | 3.4 | | 3.0 | |
| saturation | 13.0 | 5.5 | 7.0 | |
| melting point, °C | 295 | 243 | amorphous | 310 |
| glass-transition temperature, $T_g$, °C | ~85 | 102 | 149 | 123–135 |
| tensile strength, MPa[a] | 95 | 103 | 85 | 104 |
| flexural modulus, MPa[a] | 3100 | 4500 | 2900 | 3300 |
| elongation at break, % | 50 | 2.3 | 70 | 6.4 |
| notched Izod impact strength, J/m[b] | 110 | 20 | ~160 | 53 |
| DTUL[c] at 1.8 MPa[a], °C | 160 | 96 | 130 | 120 |
| starting materials | | | | |
| amine | diamino-butane | m-xylylene-diamine | trimethylhexa-methylene-diamine | hexamethylene-diamine |
| acid | adipic acid | adipic acid | terephthalic acid | adipic acid, iso/tere-phthalic acids |
| Reference | 9 | 10 | 11 | 12 |

[a]To convert MPa to psi, multiply by 145.
[b]To convert J/m to ft·lbf/in., divide by 53.38.
[c]Deflection temperature under load.

**Thermal Properties.** The high melting point of polyamides such as nylon-6,6 is a function of the strong hydrogen bonding between the chains and the crystal structure. This also allows the materials to retain significant stiffness above the glass-transition temperature ($T_g$) and almost up to the melting point. The effect is further increased when reinforcements such as glass fiber are added, giving a high deflection temperature under load even at high loading. The effect also results in the sharp melting points of nylon as the majority of the hydrogen bonding rapidly breaks down at that temperature, giving a low viscosity, water-like melt. The melting point is mainly related to the degree of hydrogen bonding between the chains, which depends on the density of amide groups. The melting point therefore drops as the length of aliphatic groups between the amide links increases, eg, nylon-6,6 melting at 264°C, compared to nylon-6,12 at 212°C. The influence of structure on the melting point is further complicated by factors that affect the ease of crystallization. For even–even nylons such as nylon-6,6 and nylon-6,12, the monomers have a center of symmetry and the amide groups easily align to form hydrogen bonds in whichever direction the chains are facing when placed on top of one another. For even nylons, such as nylon-6, that have no center of symmetry, the amide groups are in the correct positions only if the chains are aligned in one particular direction (antiparallel). For this reason, nylon-6 has a melting point more than 40°C lower than nylon-6,6, despite having the same density of amide groups. It also has a slower crystallization rate and therefore wider processing window. Other types of nylon, such as even–odd and odd nylons, also differ from the above types for similar reasons of crystallization and crystal packing (see POLYAMIDES, GENERAL). In addition, crystallization is impeded and melting point reduced by copolymerization and substituents on the chains, although in certain cases isomorphism of comonomers avoids this effect, eg, terephthalic acid increases the melting point of nylon-6,6.

**Moisture Absorption.** A characteristic property of nylon is the ability to absorb significant amounts of water (13) (Fig. 1). This again is related to the polar amide groups around which water molecules can become coordinated. Water absorption is generally concentrated in the amorphous regions of the polymer where it has the effect of plasticizing the material by interrupting the polymer hydrogen bonding, making it more flexible (with lower tensile strength) and increasing the impact strength. The $T_g$ is also reduced. Moisture absorption, determined by both the degree of crystallinity and the density of amide groups, is, as with the melting point, reduced with increasing length of aliphatic groups in the chain. Aromatic monomers also reduce the moisture absorption. Nylon-6 has a higher moisture absorption than nylon-6,6 because of its lower crystallinity. The effect of moisture absorption on the mechanical properties of nylon-6,6 is included in Table 3.

**Electrical Properties.** Nylons are frequently used in electrical applications mainly for their combination of mechanical, thermal, chemical, and electrical properties. They are reasonably good insulators at low temperatures and humidities and are generally suitable for low frequency, moderate voltage applications. The relatively high dissipation factor of nylon causes problems under conditions of high electrical stress, particularly when moist, because of the likelihood of overheating. Dry nylon has volume resistivities in the $10^{14}$–$10^{15}$ ohm·cm region, but this decreases with increasing moisture and temperature. Dielectric constant displays large increases with moisture and temperature. For moist nylon, however, the value decreases with increasing frequency, as the water molecules are less able to respond at higher frequencies.

Nylons have excellent arc and tracking resistance. Arc resistance is not affected by moisture or temperature up to about 100°C. Comparative tracking resistance of most unmodified nylons is greater than 600 V. Incorporation of additives such as flame retardants often reduces the electrical properties because of the introduction of ionic species.

**Flammability.** Most nylons are classified V-2 by the Underwriters' Laboratory UL-94 test, which means that these nylons are self-extinguishing within a certain time-scale under the conditions of the test. They achieve this performance by means of giving off burning drips. Inclusion of reinforcement such as glass fiber converts this behavior to HB, where the sample continues to burn as

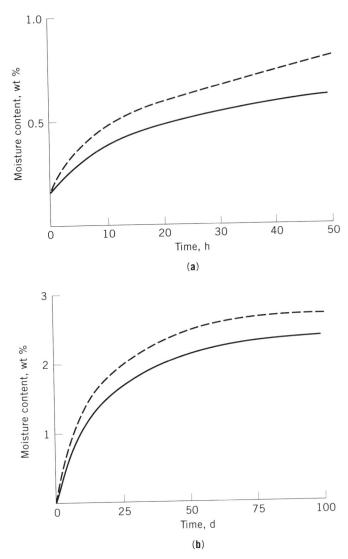

**Fig. 1.**  Rate of moisture pickup for nylon-6,6 (——) and nylon-6 (–––) granules at 50% rh and 23°C: (**a**) 0–48 hours; (**b**) 0–100 days.

a result of the reinforcement holding it together. The flammability performance can be improved by adding flame-retardant additives that can eliminate burning drips and produce nylon that meets the most stringent UL-94 test (V-0), even with materials containing glass fibers.

## MECHANICAL PROPERTIES

The semicrystalline structure of most commercial nylons imparts a high strength (tensile, flexural, compressive, and shear) as a result of the crystallinity and good toughness (impact strength) due mainly to the amorphous region. The properties of nylon are affected by the type of nylon (including copolymerization),

**Table 3. Effect of Additives on Nylon-6,6 and Nylon-12**

| Property | Nylon-6,6[a] | Nylon-6,6 + | | Nylon-12[b] | Nylon-12 + plasticizer[b] |
|---|---|---|---|---|---|
| | | 30% Glass fiber (wt/wt)[a] | Impact modifier[a] | | |
| tensile strength, MPa[c] | | | | | |
| dry | 83 | 193 | 52 | 50–55 | 27 |
| 50% rh[d] | 77 | 130 | 41 | | |
| flexural modulus, MPa[c] | | | | | |
| dry | 2800 | 9300 | 1800 | 1500 | 330 |
| 50% rh[d] | 1200 | 6600 | 900 | | |
| elongation at break, % | | | | | |
| dry | 60 | 3 | 60 | 200 | 300 |
| 50% rh[d] | >300 | 5 | 210 | | |
| notched Izod impact strength, J/m[e] | | | | | |
| dry | 53 | 110 | 910 | 60 | no break |
| 50% rh[d] | 112 | 133 | 1070 | | |
| deflection temperature under load at 1.8 MPa[c], °C | 90 | 254 | 83 | 50 | 55 |

[a]Ref. 4.
[b]Ref. 7.
[c]To convert MPa to psi, multiply by 145.
[d]50% rh = conditioned to 50% relative humidity at 23°C.
[e]To convert J/m to ft·lbf/in., divide by 53.38.

molecular weight, moisture content, temperature, and the presence of additives. Strength and modulus (stiffness) are increased by increasing density of amide groups and crystallinity in aliphatic nylons; impact strength and elongation, however, are decreased. Nylon-6 having a lower crystallinity than nylon-6,6 has a higher impact strength and slightly lower tensile strength. Nylons containing aromatic monomers tend to have increased stiffness and strength by virtue of the greater rigidity of the chains. Increasing molecular weight gives increased impact strength without having a significant effect on tensile strength. Moisture content affects the properties of nylon-6 and nylon-6,6; the effect is similar to that of temperature. Increasing moisture content reduces the $T_g$ above which the modulus and tensile strength drop significantly; however, some polyamides with a high $T_g$, such as those containing aromatic monomers, have little change in properties with changing moisture as the $T_g$ remains above room temperature. Increasing moisture for nylon-6 and nylon-6,6 also gives a steady increase in impact strength as a result of increasing plasticization, although at very low temperatures moisture can embrittle nylon. For nylons that absorb lower amounts of water, the effects on properties are less.

The effect of temperature on properties can be seen in Figure 2, which shows the effect on modulus of increasing temperature of unmodified and glass-reinforced nylon-6,6. Impact strength, however, shows a steady increase with temperature as it does with moisture.

Generally, nylon is notch-sensitive and the unnotched impact strength is dramatically reduced when a notch or flaw is introduced into the material. This needs to be considered when designing parts so that sharp angles are avoided where possible. This notch sensitivity can be considerably reduced by incorporating impact modifiers. For the most effective of these materials, the notched impact strength approaches the unnotched impact performance of the unmodified resin. The increased ductility of the material that accompanies impact modification does, however, reduce stiffness and strength. Moisture conditioning of moldings is often used to increase impact strength and flexibility before such operations as snap fitting or assembling cable ties, which can be avoided in some cases by using impact-modified resins. The effect of impact modifier on the properties of nylon-6,6 is shown in Table 3.

Properties such as stiffness and strength can be considerably increased by adding a reinforcing agent to the polymer, particularly glass or carbon fiber. Inclusion of a filler or reinforcement forces the material to fail in a brittle rather than ductile fashion. As a result, the unnotched impact strength and elongation are reduced, although the notched impact strength may be increased. These materials maintain their mechanical integrity under a high load almost up to the melting point of the nylon, eg, deflection temperature under load (see Table 3). Mechanical properties can also be modified by the inclusion of plasticizers (qv), which have a similar effect to that of water in breaking down hydrogen bonding in the amorphous region and increasing ductility, flexibility, and impact strength. Table 3 also shows the effect of glass and plasticizers on nylon-6,6 and nylon-12 properties, respectively.

As with most plastics the properties of nylons are time-dependent. The strain in a molding constantly under load increases with time (creep); equally,

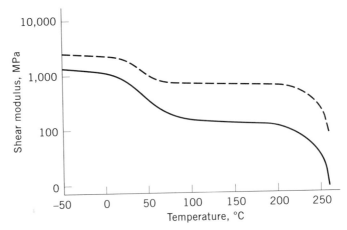

**Fig. 2.**   Effect of temperature on the shear modulus of dry nylon-6,6 (——) and nylon-6,6 plus 30% glass fiber (— — —). To covert MPa to psi, multiply by 145.

the load or stress required to maintain a constant deformation decays with time (stress relaxation). Some creep curves are given in Figure 3. Glass fiber reinforcement considerably improves the creep performance. Nylons have good resistance to dynamic fatigue, ie, the application of cyclic loads. This is influenced both by the frequency and wave form imposed as well as by moisture, temperature, and the presence of notches. Again, glass fiber reinforcement considerably improves the number of cycles that can be withstood. Nylon-6,6 has much better fatigue resistance than nylon-6, and it is claimed that nylon-4,6 is much better still (9). Nylon is also particularly resistant to damage from repeated impacts; much better, for example, than some metals that have a high impact resistance to a single blow.

Two more properties for which nylon shows particular advantages are abrasion resistance and coefficient of friction. These properties make the material suitable for use in, for example, unlubricated bearings and intermeshing gears; nylon has been used in such applications from an early stage in its development.

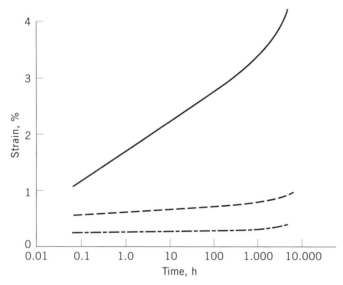

**Fig. 3.**   Tensile creep of nylon-6,6 at 50% rh, 23°C, and 15 MPa (2175 psi) (——), reinforced by 15 wt % glass (– – –), and 33 wt % glass (–·–·–).

### CHEMICAL PROPERTIES

**Hydrolysis and Polycondensation.** One of the key properties of polyamides relates to the chemical equilibrium set up when the material is polymerized. The polymerization of nylon is a reversible process and the material can either hydrolyze or polymerize further, depending on the conditions.

In the melt the material is in a dynamic situation and only at a certain (equilibrium) moisture content does the rate of hydrolysis equal the rate of polymerization. This equilibrium moisture content (in a sealed system) depends on the polymer, the temperature, the molecular weight, and the end group balance of the polymer. Below this moisture content, the melt increases in viscosity (polymerizes) and above it hydrolysis occurs with reduction in viscosity and molecular weight. For nylon-6,6, the equilibrium moisture content is close to 0.15% for most standard injection-molding resins; however, the figure is less for reinforced materials as less nylon is present per unit weight. For high molecular weight nylons used for extrusion applications, the equilibrium moisture content is less as the concentration of end groups is less; therefore, these materials need to be processed at lower moisture contents to avoid lowering the molecular weight. Nylons also polymerize in the solid form (solid-phase or solid-state polymerization) if heated significantly above 100°C in the absence of water. The equilibrium also means that nylons can hydrolyze when parts are exposed to aqueous environments for long periods at high temperatures, leading to loss of properties. However, this depends on the conditions of exposure. Nylon-6,6 has long been used successfully for automobile radiator end tanks and is used in washing machine valves. Nylons that absorb lower amounts of moisture have improved hydrolysis resistance.

**Thermal Degradation.**   Although nylons have good thermal stability, they tend to degrade in the melt when held for long periods of time or at high temperatures. This is particularly the case for nylons containing adipic acid such as nylon-6,6. The adipic acid segment can cyclize, leading to chain scission and the production of cyclopentanone and derivatives and evolution of carbon dioxide and ammonia. As well as reduction of molecular weight, cross-linking also occurs, and the material eventually sets into an intractable gel. This is normally not a problem with plastics processing operations where the residence time is relatively short, but loss of molecular weight during injection molding can occur as a result of this, particularly at high temperatures (over 300°C) or where the shot size is a small proportion of the machine capacity. Significant evolution of carbon dioxide also occurs. Processing machines should not be left containing molten nylon for any length of time but should be either emptied or purged out with, for example, a polyolefin.

**Oxidation.**   All polyamides are susceptible to oxidation. This involves the initial formation of a free radical on the carbon alpha to the NH group, which reacts to form a peroxy radical with subsequent chain reactions leading to chain scission and yellowing. As soon as molten nylon is exposed to air it starts to discolor and continues to oxidize until it is cooled to below 60°C. It is important, therefore, to minimize the exposure of hot nylon to air to avoid discoloration or loss of molecular weight. Similarly, nylon parts exposed to high temperature in air lose their properties with time as a result of oxidation. This process can be minimized by using material containing stabilizer additives.

**Ultraviolet Aging.**   Nylon parts exposed to sunlight and uv rays undergo a similar free-radical aging process. Again, this can be reduced with appropriately stabilized materials.

**Effect of Chemicals and Solvents.**   Nylons have excellent resistance to many chemicals, although the effect varies depending on the nature of the nylon. Generally, polyamides tend to be particularly resistant to nonpolar materials such as hydrocarbons. Resistance is least to strong acids and phenols which are most effective at disrupting the hydrogen bonding and which can sometimes dissolve the nylon. Highly polar materials such as alcohols are absorbed and sometimes dissolve the nylons containing lower concentrations of amide groups. Certain metal salts can attack nylon causing stress cracking, eg, zinc chloride, or even dissolve the material in alcoholic solution, eg, lithium chloride.

## Manufacture

**Nylon-6,6.** This nylon is a condensation product of hexamethylenediamine [124-09-4] and adipic acid [124-04-9] (14). The raw materials are derived from crude oil (see ADIPIC ACID; AMINES). The monomers are stoichiometrically combined into an aqueous salt solution, the pH of which can be closely controlled. Additives may be added to the salt, such as a monoacid (eg, acetic acid), for molecular weight control. The polymerization can be carried out by either a batch process in an autoclave or by a continuous process involving the same basic steps. The salt solution is first concentrated by evaporation under conditions to minimize hexamethylenediamine loss. In the batch process this is typically carried out in a separate vessel before charging to the autoclave. The progress of the polymerization is illustrated in Figure 4. In the first stage, the pressure and temperature are steadily increased, starting the polymerization. Subsequently, the solution is held at 1.7 MPa (250 psi) while steam is bled off and the temperature is increased to 275°C. The pressure is then dropped to atmospheric and finally held at temperature (under partial vacuum if necessary) for the time required to achieve the final molecular weight before casting and pelletizing.

**Nylon-6.** This nylon is produced from caprolactam (qv) in the presence of water. The reaction is initiated by a hydrolytic ring opening to aminocaproic acid followed by reaction of the amine end with caprolactam [105-60-2], giving ring opening and further reaction. The polymerization, which takes place at 240–280°C, can be carried out at atmospheric pressure; a continuous process (the VK tube process) was developed as early as 1940 (15,16). Nylon-6 is almost always produced by continuous means in the 1990s. Figure 5 shows one such

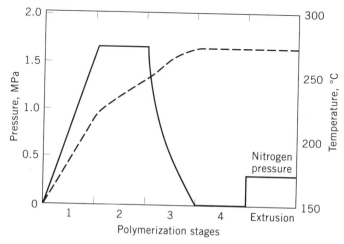

**Fig. 4.** Typical nylon-6,6 autoclave polymerization cycle showing the changes in pressure (——) and temperature (———). To convert MPa to psi, multiply by 145.

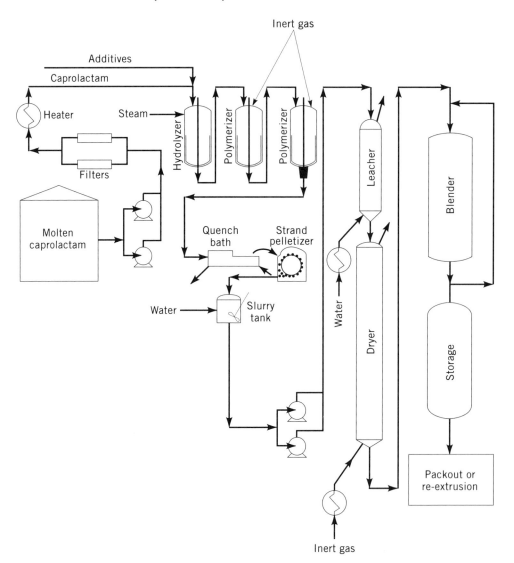

**Fig. 5.** Polymerization of nylon-6 for pellet production.

system, which involves three vessels for hydrolysis and polymerization; the single VK tube for these operations is still also widely used. The equilibrium with caprolactam established in the polymerization leaves about 10% unreacted monomer in the product. This needs to be removed for most plastics operations, unless plasticized material is required. The removal is normally done by water extraction followed by drying, although vacuum removal from the melt is also employed. The whole process is carried out under an inert gas such as nitrogen to avoid discoloration.

Nylon-6 can also be produced from molten caprolactam using strong bases as catalysts (anionic polymerization); this is used as the basis of monomer casting and reaction injection molding (RIM). Anionic polymerization proceeds much faster than the hydrolytic route but products retain catalysts which may need to be extracted.

**Nylon-6,9, Nylon-6,10, and Nylon-6,12.** These nylons are all produced from hexamethylenediamine and either azelaic, sebacic, or dodecanoic acid. They are produced by a process similar to that for nylon-6,6, usually using batch autoclaves.

**Nylon-11.** This nylon is produced from 11-aminoundecanoic acid, which is derived from castor oil. The acid is polymerized by heating to 200°C with continuous removal of water. Catalysts such as phosphoric acid are frequently used. There is no appreciable amount of unreacted monomer left in the product.

**Nylon-12.** This nylon is produced by a similar process to that for nylon-6. The monomer is dodecanolactam, which is made from butadiene. The process uses a catalyst and higher temperatures than nylon-6 (300–350°C) on account of the stability of the 13-membered ring. Again, there is little residual unreacted monomer.

**Nylon-4,6.** This nylon is produced from diaminobutane and adipic acid. The process is similar to that for nylon-6,6, but the amine has a high tendency to cyclize and the temperatures are therefore kept low. This results in a low molecular weight polymer, which is subsequently increased in viscosity by solid-state polymerization.

**Other Nylons.** A number of other nylons are produced, including copolymers of the above types by variations of the same processes. However, some semiaromatic nylons can give problems as a result of the high melt viscosity. A process for producing polymers of hexamethylenediamine, adipic acid, terephthalic acid, and isophthalic acid has been developed, which involves vaporizing the salt mixture in a high temperature flash reactor followed by molecular weight increase in a twin-screw extruder with efficient moisture removal (17).

A number of high melting point semiaromatic nylons, introduced in the 1990s, have lower moisture absorption and increased stiffness and strength. Apart from nylon-6/6,T (copolymer of 6 and 6,T), the exact structure of these is usually proprietary and they are identified by trade names. Examples include Zytel HTN (Du Pont); Amodel, referred to as polyphthalamide or PPA (Amoco); and Arlen (Mitsui Petrochemical). Properties for polyphthalamide are given in Table 2. A polyphthalamide has been defined by ASTM as "a polyamide in which the residues of terephthalic acid or isophthalic acid or a combination of the two comprise at least 60 molar percent of the dicarboxylic acid portion of the repeating structural units in the polymer chain" (18).

**Solid-State Polymerization.** There is a limit to the molecular weight that can be obtained in a melt polymerization process as a result of degradation with long melt residence times and limits to moisture removal from high viscosity melts. In order to produce high viscosity resins to be used in, for example, extrusion operations, it is necessary to polymerize further by heating in the solid state. This is carried out under vacuum, steam, or inert gas. Nylon-6,6, for example, can be solid-state-polymerized in the temperature range of 150–240°C. Below the melting point, the hydrolysis rate is negligible compared to the reaction time and degradation reactions are also much reduced.

**Compounding.** Although low levels of additives can often be incorporated during the polymerization process, in order to add the required higher levels of, eg, glass fiber reinforcement, impact modifiers, or flame retardants, it is necessary for the nylon to undergo a second manufacturing step to melt-incorporate the additives. This is carried out by extrusion compounding, which usually consists of a gravimetric feed of additives and nylon granules to a single- or twin-screw extruder. The archimedian screw conveys the mixture along a heated barrel and, by the action of shear on the polymer, melts the material and mixes in the ingredients. Twin-screw extruders usually have intermeshing corotating screws that are more effective at mixing the material than single screws, and can be built up from modular components, allowing flexibility in selecting the average and localized shear/mixing regime along the screw. Additives such as glass fiber can be added along the barrel and into the melt in such designs; screw-side feeders can be used to do this. A vacuum line attached to a vent in the barrel allows the removal of moisture and volatiles. Material is extruded through a die into laces, which are water-cooled and pelletized.

## Processing

Nylons need to be processed dry to avoid molecular weight loss and processing problems. Figure 6 indicates both the usable moisture range for nylon-6,6 and nylon-6 in injection molding and the relationship with temperature (19). Extrusion applications require lower moisture contents (max 0.1–0.15% for nylon-6,6 and -6) as do some other nylon types (eg, max 0.1% for nylon-11 and -12; 0.05% for nylon-4,6). The materials are normally supplied dry by the manufacturer in moistureproof packaging such as foil-lined 25-kg bags or lined one-ton boxes.

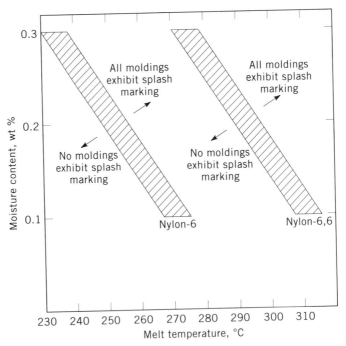

**Fig. 6.** Relationship between moisture content and temperature for an unvented cylinder.

Once opened, the material should be used within a few hours or resealed. Material that has absorbed some moisture can be redried using a vacuum oven at 80°C or a dehumidifier hopper drier.

Material should not be processed at too high a temperature, eg, above 310°C for nylon-6,6 or 290°C for nylon-6, in order to avoid degradation. Residence times at the higher temperatures should be kept to a minimum. Molten nylon should not be left in an idle machine for more than 30 min maximum. Exposure of molten and hot nylon to air should also be minimized to avoid discoloration.

Generally, nylon scrap or regrind can be reused satisfactorily, provided that it is dry. The level allowed depends on the amount of degradation and the specification of the final products.

**Injection Molding.** This is the largest single processing route for nylon taking more than 60% of the material produced. This technique is generally carried out using a screw preplasticizing injection-molding machine (19–22) (see PLASTICS PROCESSING). Material is heated and plasticized on a screw similar to that described above for compounding extruders. As melt accumulates at the screw front, the screw is forced back against a residual oil pressure (screw back pressure). When enough melt has accumulated, the screw stops rotating and moves forward, injecting material into the mold. A back-flow stop valve on the tip of the screw prevents the material from flowing back. Nylons can often be processed on general-purpose screws having a constant decreasing flight depth along the screw. Better performance can be achieved, however, by a nylon-type screw having a sharp reduction in flight depth within a few flights in the compression or transition zone, ie, between the first, conveying, and final metering sections. This allows for the sharp melting point of semicrystalline nylons, which results in a rapid reduction in bulk density.

The sharp melting point and the low melt viscosity also mean that nylon can give problems with nozzle drool and/or premature freeze-off. For this reason, it is normally necessary to use either a reverse-taper nozzle (fitted with a heater to avoid freeze-off), a mechanical shut-off nozzle, or melt decompression. Melt decompression, or suck-back, involves the screw retracting slightly after the screw retraction time, pulling the material back and preventing drool. The design of molds should also take into account this low viscosity and sharp melting point. Mold surfaces should be well-mated to avoid flash. Runners and gates should be of an adequate size; 1.4–1.6 times the component wall thickness for runners and 0.5 times for gates are recommended, although significantly smaller sizes can be used with experience. Runners should be round in section or, more commonly, trapezoidal and as short as possible. The use of hot runners eliminates waste material from sprues and runners, and allows smaller gates so that the material can be kept molten right up to the molding. Careful temperature control is important in hot runner systems. Multicavity tools, used for small components, should be balanced for even filling, these are often of three-plate design for automatic stripping. A mold temperature control unit should be used to ensure consistent components; high mold temperatures are normally used for nylon. Computer-aided design (CAD) techniques are frequently used for part and mold design. These techniques allow the creation of the design, mechanical analysis of the part using finite element analysis, and an evaluation of tooling performance and material flow behavior.

**Table 4. Injection-Molding Temperatures for Different Polyamides**[a]

| Polyamide | Melt temperature, °C[b] | Mold temperature, °C |
|---|---|---|
| nylon-6 | 250–290 | 80 |
| nylon-6,6 | 290–300 | 80–100[b] |
| nylon-4,6 | 315 | 120 |
| nylon-6,9 and nylon-6,10 | 270 | 80 |
| nylon-6,12 | 240–270 | 80 |
| nylon-11 | 210–260 | 80 |
| nylon-12 | 200–250 | 70–80[c] |
| nylon-MXD,6 | 250–280 | 130 |
| nylon-6(3),T | 280–300 | 80 |
| polyphthalamide[d] | 327–332 | 135–150 |

[a]Ref. 23.
[b]The higher temperatures are used for glass-filled or high viscosity materials and the lower temperatures for low viscosity or plasticized resins.
[c]The lower mold temperature is used if >5% plasticizer.
[d]Ref. 18.

Table 4 gives injection molding temperatures for various nylons. Typical molding conditions for polyamides are as follows: average injection velocity, 100–300 mm/s; hold pressure, 70–90 MPa; hold pressure time, 20–30 s; maximum injection pressure, 70–90 MPa; cooling time, ≤10 s; and total cycle time, ≤50 s. Nylons generally require a fast injection speed and are particularly good for achieving fast cycle times because of the low viscosity and rapid setting-up.

Significant mold shrinkage occurs with nylon mainly on account of the increase in density with crystallization. This can give rise to voiding and sink marks in moldings. The dimensional stability can be much improved by using reinforcing agents or nucleated materials.

Developments in injection molding that have become increasingly important in the 1990s include gas injection and fusible-core molding. Gas injection enables a saving in part weight and avoidance of sink marks in thick sections. Fusible-core technology involves overmolding a low melting metal alloy core that is subsequently melted out in an oil bath. This technique allows much more complex moldings containing internal passageways to be produced. The method has become important for making automobile air inlet manifolds. The molding of the core, nylon overmolding, melt-out, and alloy recycling can all be automated as a total operation.

**Extrusion.** Extrusion accounts for about 30% of nylon produced and is used in various processes (24). Nylons can be extruded on conventional equipment having the following characteristics. The extruder drive should be capable of continuous variation over a range of screw speeds. Nylon often requires a high torque at low screw speeds; typical power requirements would be a 7.5-kW motor for a 30-mm machine or 25-kW for 60-mm. A nylon screw is necessary and should not be cooled. Recommended compression ratios are between 3.5:1 and 4:1 for nylon-6,6 and nylon-6; between 3:1 and 3.5:1 for nylon-11 and nylon-12. The length-to-diameter ratio, $L/D$, should be greater than 15:1; at least 20:1 is recommended for nylon-6,6, and 25:1 for nylon-12.

Nylon screws are usually of single-start design and the pitch, ie, the distance between the flights, is equal to the screw diameter. The feed zone having a compression ratio of 1:1 can be up to two-thirds of the length of the screw and acts as a conveyor of pellets from the feed throat area through the heated barrel. The compression zone compacts the melt and can vary between half a turn (suitable for nylon-6,6) and six turns (suitable for nylon-12); values in between are usually used. The metering zone having a shallow flight depth then homogenizes the melt and acts as a melt pump to give a constant pressure at the die. The metering zone is normally about one-third of the screw length and should have a constant cross section as molten nylon is not very compressible. A range of screw sizes can be used. Typical dimensions are, for a 30–40-mm barrel diameter, feed zone channel depth of 7.9 mm and metering zone depth of 2.0 mm; for 85–95-mm diameter, 14.7 mm and 4.0 mm, respectively. Typical operating temperatures are shown in Table 5. Most extrusion operations require high viscosity (high molecular weight) nylon in order to give a high melt strength to maintain the shape of the extrudate. In the following sections, reference is made to three broad viscosity ranges that correspond to these approximate number-average molecular weights when nylon-6,6 is used: high viscosity nylon ($M_n$ = 30,000–40,000), medium viscosity (20,000–30,000), and standard viscosity (15,000–18,000, as used for injection molding).

*Film.* Nylon film can be produced as either tubular or cast film. In tubular film, melt is extruded through a screen pack and a tubular die, and a bubble is formed with air pressure. Total drawdown (extension of the melt) on the order of 10:1 to 20:1 is achieved. High viscosity nylon is required for this operation. A relatively stiff, hazy film is produced; nylon-6 has a lower haze level than nylon-6,6. Cast film is produced from medium viscosity nylon by extruding through a straight slot die and then rapidly quenching on highly polished rolls at a controlled temperature; speed of the rolls affects drawdown as above. Nylon film is also produced as coextruded multilayer structures mainly with olefin-type polymers. Nylon film has a low permeability to oxygen, nitrogen, and carbon dioxide, but a high permeability to water vapor (see BARRIER POLYMERS).

*Tubing and Pipe.* Medium to high molecular weight polymers are used for tube extrusion. Small bore tube (up to 10 mm) can be extruded through a conventional die and cooled in an open water bath. Air is injected into the center of the tube using a torpedo to prevent the extrudate from collapsing and to adjust the wall thickness. For tubing larger than 10 mm, a pressure sizing die

**Table 5. Typical Temperatures for Nylon Extrusion, °C**

| Location | Nylon-6,6[a] | Nylon-6[a] | Nylon-12[b] |
|---|---|---|---|
| screw | | | |
| feed zone | 265–275 | 220–230 | 185–205 |
| compression zone | 275–285 | 235–250 | 190–215 |
| metering zone | 285–295 | 245–265 | 190–220 |
| head | 285–295 | 250–270 | 200–230 |
| die | 285–295 | 250–270 | 195–230 |

[a] Ref. 24.
[b] Ref. 25.

is used and tube is drawn into the water bath through a series of sizing plates in the end of the bath, 25–50 mm away from the die. A drawdown of about 2:1 is usually used, but this depends on the molecular weight and line speed. High molecular weight nylons require less drawdown than the less stiff lower molecular weight materials. Similar methods are used for large diameter pipe, for which high viscosity material is normally used, particularly with high wall thicknesses. As this is a slow operation, melt temperatures are kept up to 30°C above the normal melt temperature.

*Monofilament.* Standard and medium viscosity nylons are used. Close control of diameter is important and a gear pump is used before the die (after filtering through a fine filter pack) to minimize pressure variations. The die hole diameter is normally 1.5–2 times the diameter of the undrawn filament. The filaments are drawn through a quench tank at approximately 40°C, after which they are separated by a comb-type guide and passed through two sets of Godet rolls (pull rolls) separated by an electrically heated chamber. The second set of rolls operates at four or five times the speed of the first set. This drawing process enhances the properties by orienting the chains. The drawn filament then needs to be set in the oriented form by being drawn through a heated conditioning chamber by a third set of rolls operating at a similar speed as the second set. The filaments are then wound onto separate bobbins.

*Rod and Profiles.* Medium to high viscosity grades are used. Accurate temperature control is important. For rods of diameter greater than 3 mm, slow solidification is essential to avoid voids and cracks caused by nonuniform shrinkage. Two processes are used. For rods up to 150 mm, a forming box is used whereby nylon is extruded under pressure through a water-cooled cylindrical tube. The tube is isolated from the die by a nonmetallic gasket, the rod is pulled off at a constant rate by a haul-off. The second process, which can be used for more complex shapes, involves extruding the nylon into a series of interconnecting open-ended molds.

*Wire and Cable Coating.* Nylons are widely used for wire covering and sheathing cable. In the latter application, nylon is usually overcoated onto insulation of another material, such as PVC, in order to provide cut and abrasion resistance. In a specialized application, nylons-11 and -12 are used to provide termite resistance. For cable sheathing, two extruders are normally used; the primary coated cable is fed through the center of a tubular die on the nylon extruder and the nylon is drawn down onto the cable using a vacuum applied through the torpedo. Standard viscosity polymer is used. Wire coating is carried out by feeding the wire at a constant tension to the torpedo, which guides the wire centrally through the die orifice. Thickness of the covering is controlled by the diameter and the haul-off speed.

**Blow Molding.** Blow molding of nylons has become more important as a means of making large hollow moldings. As well as high molecular weight nylon, resins modified to increase melt strength and containing glass fiber have been introduced. Blow molding of nylons is usually carried out by extrusion blow molding, whereby melt is produced in an extruder and formed into a tube called a parison (26). The molten parison is captured in a mold that pinches and seals the ends, and the rigid hollow part is formed by inflation under air pressure. Intermittent extrusion blow molding is most common, which involves the storing of melt in an accumulator die head until required to form the next parison. Melt storage with a reciprocating screw or accumulator pot with ram is also used, as is continuous blow molding. Injection blow molding can also be carried out for small parts. This involves the molding of a preform around a core, which is transferred to another mold for inflating or blowing to the final part shape. The design of the manifold and accumulator head should be such as to avoid material hold-up locations (giving degradation and gel formation), and should also incorporate even, carefully controlled heating. Adequate venting of the head is necessary to avoid buildup of gas.

**Rotomolding.** Nylon-6, nylon-11, and nylon-12 can be used in rotomolding and are generally supplied for these applications as a powder or with a small pellet size. The process involves tumbling the resin in a heated mold to form large, thin-walled moldings. Nylon-11 and nylon-12 use mold temperatures of 230–280°C and nylon-6 is processed at over 300°C. An inert gas atmosphere is preferred to avoid oxidation.

**Reaction Injection Molding.** RIM uses the anionic polymerization of nylon-6 to carry out polymerization in the mold. A commercial process involves the production of block copolymers of nylon-6 and a polyether by mixing molten caprolactam, catalyst, and polyether prepolymer, and reacting in a mold (27,28).

**Powder Coating.** Nylon-11 and nylon-12 are used in powder form for anti-corrosion coating of metals. Dip coating and electrostatic and flame spraying are used. Dip coating, which involves immersing a preheated article into fluidized nylon powder, is most suitable for automation.

**Additives and Modifications.** For plastics uses, nylon is only rarely employed as the pure polymer, and is almost always modified to some extent even if only with the addition of a small amount of lubricant. There has been a dramatic increase in the range and number of combinations of additives used to modify nylons, resulting in a huge expansion in the number of commercial grades available and the uses to which they can be put. It is not unusual to find formulations that contain less than 50% nylon and half a dozen or more additives.

*Lubricants.* Lubricants are used to improve the melt flow, screw feeding, and mold release of nylons. Long-chain acids, esters, and amides are used together with metal salts, eg, metal stearates. Improved melt flow is mainly a function of molecular weight reduction during molding. Mold release is improved by waxes of limited compatibility with nylon, which migrate to and lubricate the mold surface.

*Nucleants.* Although nylons crystallize quickly, it is often an advantage, particularly for small parts, to accelerate this process to reduce cycle time and increase productivity. Nylon-6, which crystallizes more slowly than nylon-6,6, also benefits from nucleation in unreinforced formulations. Nucleants are generally fine-particle-size solids or materials that crystallize as fine particles before the nylon. The materials, eg, finely dispersed silicas or talc, seed the molten nylon and result in a higher density of small uniformly sized spherulites; in nylon-6 the crystalline form is also changed. Nucleation increases tensile strength and stiffness but makes the material more brittle. Mold shrinkage is lower for nucleated resins.

*Stabilizers.* Stabilizers are often added to slow down the rate of oxidation and uv aging. Heat stabilizers can be organic antioxidants (such as hindered phenols or aromatic amines), hydroperoxide decomposers, or metal salts. The latter are most commonly used in the form of copper halide mixtures. This system, which has the side effect of discoloring the nylon, acts as a regenerative free-radical suppressor. Above about 120°C, the copper halide system is by far the most effective and allows the use of glass-reinforced nylon-6,6 in high temperature automotive underhood applications. Ultraviolet (uv) stabilizers can be free-radical acceptors, uv absorbers, or hindered amine light stabilizers. The most common uv stabilizer, however, continues to be finely dispersed carbon-black pigment.

*Impact Modifiers.* Notched impact strength and ductility can be improved with the incorporation of impact modifiers, which also can lower the brittle–ductile transition temperature and give much improved low temperature toughness. Impact modifiers are rubbers (often olefin copolymers) that are either modified or contain functional groups to make them more compatible with the nylon matrix. Dispersion of the rubber into small (micrometer size) particles is important in order to obtain effective toughening (29). Impact modifiers can be combined with other additives, such as glass fiber and minerals, in order to obtain a particular balance of stiffness and toughness.

*Flame Retardants.* Flame retardants are added to nylon to eliminate burning drips and to obtain short self-extinguishing times. Halogenated organics, together with catalysts such as antimony trioxide, are commonly used to give free-radical suppression in the vapor phase, thus inhibiting the combustion process. Some common additives are decabromodiphenyl oxide, brominated polystyrene, and chlorinated dodecahydrodimethanodibenzocyclooctene. In Europe, red phosphorus is widely used. It is effective at much lower levels and promotes char formation as well as inhibiting combustion, but is a more hazardous raw material to handle and the compounds are not available in light colors. Concern about the possibility of dioxin formation from resins containing halogenated organics, particularly those containing phenoxy groups, has led to increasing interest in nonhalogen, nonphosphorus flame retardants (qv). Melamine derivatives have been used and magnesium hydroxide has been promoted for this type of application (30,31).

*Plasticizers.* Plasticizers are used to increase the flexibility of nylon and improve impact strength. They are most commonly used in nylon-11 and nylon-12 for such applications as flexible fuel hoses for automobiles. Unextracted nylon-6 is also used with the caprolactam acting as the plasticizer. Other common plasticizers are long-chain diols and sulfonamides.

*Reduced-Moisture Nylon.* A modified nylon-6 has been commercialized that has approximately 30% less moisture uptake (22). Patented compositions use various amine and phenolic additives to obtain such a reduced moisture uptake effect (33,34).

*Reinforcement.* Nylon is particularly suitable for reinforcement and the melt incorporation of short glass fibers has long been practiced, being developed around 1960 by ICI in England (35) and Fiberfil Inc. in the United States (36). The tensile strength of nylon-6,6 is increased by more than 2.5 times and stiffness by almost 4 times by adding 30% glass fiber. Glass fiber also improves dimensional stability, notched impact strength, and long-term creep, and is normally used in the 15–60% (wt/wt) range. The glass fibers used need to be treated with a specific sizing to enable bonding with the nylon and dispersion in the melt; the size formulations are proprietary but often contain an aminosilane coupling agent and a polyurethane or acrylic binder.

A disadvantage of glass fibers is the warpage of flat moldings. This is the result of differential shrinkage caused by anisotropy from the glass fibers, which tend to align with the direction of melt flow. Mineral reinforcements are used to obviate this by stiffening the material with a more isotropic mixture. Properties are lower than for glass fiber. Minerals having a higher aspect ratio such as talc and mica tend to have higher stiffness and strength; surface-treated kaolin and wollastonite have better impact properties. Combinations with glass are also used. Other reinforcements include mineral fibers, carbon fiber, and para-aramid fibers (Kevlar).

**Polymer Blends.**  Commercial blends of nylon with other polymers have also been produced in order to obtain a balance of the properties of the two materials or to reduce moisture uptake. Blends of nylon-6,6 with poly(phenylene oxide) have been most successful, but blends of nylon-6,6 and nylon-6 with polypropylene have also been introduced.

## Economic Aspects

The principal worldwide manufacturers of nylon resins are given in Table 6. Total sales of nylon plastics in the United States and Canada in 1993 were 331,000 metric tons (37). West European sales were 352,000 t and Japanese sales 220,000 t (37). Figure 7 shows how sales in the United States have steadily increased since 1967 (38) and also how the price of nylon-6,6 has changed (39). The effect of the oil price rises, the boom of the mid-1980s, as well as the oil price reduction and the recession that followed are clearly evident. Table 7 shows the variation of price across different polyamide types.

## Specifications, Standards, and Quality Control

Raw material specifications may be agreed between the supplier and the molder or end user, or they may be defined as requirements by an external body. The standard ASTM D4066 (41) identifies how to classify and specify nylon materials and gives details of tests and test methods that may be used, as well as required values; ASTM D5336 for polyphthalamides and ISO 1874-1/2 are also useful. The tests include mechanical, thermal, electrical, and flammability properties as appropriate. In addition to these, it is normally necessary to specify the viscosity of the material, maximum moisture content, and ash content (if reinforced). Viscosity is generally measured as solution viscosity that corresponds directly to molecular weight, rather than melt viscosity, which is moisture-dependent. In the United States, solution viscosity is generally measured as relative viscosity (ASTM D789) or inherent viscosity (ASTM D2857), normally in 90% formic acid or $m$-cresol solvent. Elsewhere, viscosity measurements are moving to the internationally agreed viscosity number (ISO 307) in formic acid, sulfuric acid, or *meta*-cresol. Moisture content is determined according to ASTM D789. The test employs a coulometric Karl Fischer titration technique using a nitrogen-flushed heated chamber to drive moisture into the solution. Commercial equipment is

**Table 6. Manufacturers of Polyamide Plastics**

| Manufacturer | Nylon type | Trade name |
|---|---|---|
| United States | | |
| E. I. du Pont de Nemours & Co., Inc. | 6,6, 6,12, and others | Zytel and Minlon[a] |
| Hoechst Celanese | 6,6 | Celanese |
| Monsanto Co. | 6,6, 6,9 | Vydyne |
| Allied Corp. | 6 | Capron |
| Badische Corp. | 6 | Ultramid |
| Emser Industries | 6 | Grilon |
| Nylon Corp. of America | 6 | Nycoa |
| Rilsan Corp. | 11 | Rilsan |
| Amoco | PPA | Amodel |
| France | | |
| ATO Chimie | 6 | Orgamide |
| | 11,12 | Rilsan |
| Nyltech France | 6,6, 6,10 | Technyl |
| Germany | | |
| BASF AG | 6, 6,6, 6,10 | Ultramid |
| Bayer AG | 6 | Durethan |
| Du Pont de Nemours | 6,6 | Zytel and Minlon[a] |
| Hüls AG | 12 | Vestamid |
| | 6(3),T | Trogamid |
| Italy | | |
| Nyltech | 6, 6,6 | Sniamid |
| Radici Novacips SpA | 6, 6,6 | Radilon |
| Netherlands | | |
| DSM | 6, 6,6 | Akulon |
| | 4,6 | Stanyl |
| Japan | | |
| Asahi Chemical Industries, Ltd. | 6,6 | Leona |
| Mitsubishi Chemical Industries, Ltd. | 6 | Novamid |
| Mitsubishi Gas Chem. Co. | MXD,6 | Reny |
| Toray Industries, Inc. | 6, 6,6, 12 | Amilan |
| Ube Industries, Ltd. | 6, 6,6, 12 | Ubepol |
| Unitika, Ltd. | 6 | Unitika |
| Switzerland | | |
| EMS Chemie AG | 6 | Grilon |
| | 12 | Grilamid |
| United Kingdom | | |
| BIP Chemicals Ltd. | 6 | Beetle |

[a]Mineral-filled.

available to carry out this test. Manufacturer's quality control tests normally include tests for contamination, color, moisture, ash, viscosity, packaging (qv), and other properties as appropriate.

## Uses

More than 60% of nylon is used in injection-molding applications. About 55% of this use is in the transportation industries, and most of this use is concerned

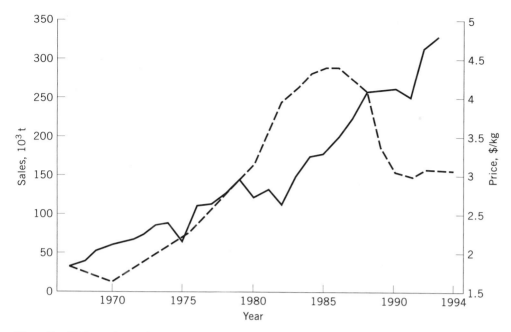

**Fig. 7.** U.S. nylon plastics consumption (——) and nylon-6,6 price (———) from 1967–1994. 1992 and 1993 consumption figures include Canada.

**Table 7. Prices of Polyamides, 1994**[a]

| Type | Price, $/kg |
|---|---|
| nylon-6 | 2.84–2.93 |
|    mineral filled | 2.69–2.87 |
|    30% glass fiber | 3.44–3.53 |
| nylon-6,6 | 2.99–3.15 |
|    mineral filled | 2.71–2.80 |
|    30% glass fiber | 3.64–3.86 |
| nylon-6,9 | 5.51–6.08 |
| nylon-6,10 | 6.30–6.90 |
| nylon-6,12 | 6.31–7.21 |
| nylon-4,6 | 6.50 |
| nylon-11 | 7.25–7.52 |
| nylon-12 | 7.01–7.52 |
| transparent amorphous | 5.45–7.94 |

[a]Ref. 40.

with automobile production. Nylons are used throughout automobiles and the amount used is continually increasing as a result of the drive to reduce weight. Some exterior applications are wheel covers, door mirrors, handles, lamp housings, gas tank filler covers, and decorative louvers. In the interior, applications include parts of the window, chair and seat belt mechanisms, handles, light housings, cable ties and gears in the speedometer, and windshield wipers. One of the fastest growing areas of automotive use is underhood applications, including radiator end tanks, filter housings, valve covers, fuse blocks, fuel hoses, and

**Table 8. Polyamide Consumption in North America, 1993**[a]

| Market | Consumption, $10^3$ t |
|---|---|
| extruded products | |
|   filaments | 12 |
|   film | 37 |
|   sheet, rod, tube | 13 |
|   wire and cable | 21 |
|     *total* | *83* |
| injection molding | |
|   transportation | 96 |
|   electrical and electronics | 28 |
|   industrial | 22 |
|   consumer products | 18 |
|   appliances and power tools | 10 |
|     *total* | *174* |
| export | 56 |
| other | 18 |
|   *Total* | *331* |

[a]Ref. 37.

fuel rails and air inlet manifolds for injection engines. Table 8 shows the split of applications for which nylon was used in the United States and Canada in 1993.

About 16% of injection-molded nylon has been used in the electronic and electrical industries in the 1990s, mainly in cable ties and various electrical fittings such as switchgear, contactors, and a range of plugs, sockets, connectors, and switches. The domestic appliance and power tool markets represent another 6%, including uses in casings and handles for various power drills, chain saws, etc, as well as parts of equipment ranging from hair driers to washing machines and lawn mowers. The consumer products area (10%) includes such things as cigarette lighters, combs, racquet frames, kitchen utensils, buttons, and ski boots. Finally, approximately 13% of injection-molding applications are industrial, a broad area that is difficult to generalize but which includes a wide range of gears, bearings, pulleys, screws, cams, sprockets, as well as larger structural parts. In the extrusion area (about 30% of nylon use), monofilaments are used for fishing lines and nets as well as toothbrushes and brushes of other sorts. Film is mainly used for packaging in the food and medical industries, including cook-in-the-bag food envelopes and sausage skins.

## BIBLIOGRAPHY

"Polyamides, Plastics" in *ECT* 2nd ed., Vol. 16, pp. 88–105, by E. C. Schule, Allied Chemical Corp.; in *ECT* 3rd ed., Vol. 18, pp. 406–425, by R. J. Welgos, Allied Corp.

1. M. I. Kohan, *Nylon Plastics*, John Wiley & Sons, Inc., New York, 1973.
2. U.S. Pat. 2,130,523 (Sep. 20, 1938), W. H. Carothers (to E. I. du Pont de Nemours & Co., Inc.).
3. U.S. Pat. 2,241,321 (May 6, 1941), P. Schlack (to I. G. Farbenindustrie).
4. *Zytel Nylon Resins, Product Guide and Properties*, E-96368, E. I. du Pont de Nemours & Co., Inc., Wilmington, Del., Sept. 1991.

5. *Capron Nylon Homopolymers for Molding and Extrusion*, CL-100-884, Allied Corp., Morristown, N.J.

6. *Design Guide to a Versatile Engineering Plastic*, Rilsan Corp., France, 1984.

7. *Vestamid, Range of Grades*, 1129e, Hüls AG, Marl, Germany, Oct. 1988.

8. *Vydyne Engineering Plastic Composite Data Sheet*, 6313-E, Monsanto Co., St. Louis, Mo.

9. *Stanyl, General Information*, DSM, Heerlen, the Netherlands.

10. *New Engineering Plastics, Reny*, 60612-4R, Mitsubishi Gas Chemical Co. Inc., Tokyo, Japan.

11. *Technical Information, Trogamid-T*, KR 85/5.87/5000/5457, Hüls AG, Marl, Germany.

12. *Plast. Technol.* **37**, 45 (Apr. 1991).

13. *Maranyl Nylon Compounds, Data for Design*, N101/GB/1/7/90, Du Pont de Nemours Int. SA, Geneva, Switzerland.

14. D. B. Jacobs and J. Zimmerman, in C. E. Schildknecht and I. Skeist, eds., *Polymerisation Processes, High Polymers*, Vol. XXIX. Wiley-Interscience, New York, 1977, pp. 424, 467. A very detailed review of nylon-6,6 polymerization.

15. H. Ludewig, *Faserforschg. Textiltechn.* **2**, 341 (1951).

16. P. Matthies and W. F. Seydl, in R. B. Seymour and G. S. Kirehenbaum, eds., *High Performance Polymers: Their Origin and Development*, Elsevier Science Publishing Co., Inc., New York 1986, pp. 39, 53. A good review of nylon-6 development.

17. U.S. Pat. 4,831,108 (May 16 1987), J. A. Richardson, W. Poppe, B. A. Bolton, and E. E. Paschke (to Amoco).

18. *ASTM D5336-93, Standard Specification for Polyphthalamide (PPA) Injection-Molding Materials*, ASTM, Philadelphia, Pa., 1993.

19. *Maranyl Nylon Compounds, Injection Moulding*, N102, 11th ed., Du Pont de Numours Int. SA, Geneva, Switzerland, 1987.

20. *Moulding of Zytel 6,6 Nylon Resins*, TRZ 30, E-39416, Du Pont de Nemours Int. SA, Geneva, Switzerland, May 1992.

21. J. B. Dym, *Injection Molds and Molding*, Van Nostrand Reinhold Co., Inc., New York, 1979.

22. I. I. Rubin, *Injection Molding Theory and Practice*, John Wiley & Sons, Inc., New York, 1972.

23. *ISO/DIS 1874-2, Plastics Polyamide (PA) Moulding and Extrusion Materials, Part 2: Preparation of Test Specimens and Determination of Properties*, International Standards Organization, Geneva, Switzerland, 1994.

24. *Maranyl Nylon Compounds, Extrusion*, N103, Du Pont De Nemours Int. SA, Geneva, Switzerland, 1987.

25. *Rilsan Technical Bulletin*, TB-11-76-1006, Rilsan Corp., France.

26. *Blow Molding Processing Guide, Zytel Nylon Resins*, H06024, E. I. du Pont de Nemours & Co., Wilmington, Del., Oct. 1988.

27. *Nyrim Nylon Block Copolymers*, NY-01-01, Monsanto Co., St. Louis, Mo.

28. U.S. Pat. 4,031,164 (June 21, 1977), R. M. Hedrick and J. D. Gabbert (to Monsanto Co.).

29. U.S. Pat. 4,174,358 (Nov. 13, 1979), B. N. Epstein (to E. I. du Pont de Nemours & Co., Inc.).

30. *Plast. Technol.* **38**, 27 (Mar. 1992).

31. *Plast. Technol.* **39**, 45 (Sept. 1993).

32. J. Döring and co-workers, *Kunststoffe*, **79**(10), 989–992 (1989).

33. U.S. Pat. 4,616,053, (Oct. 7, 1986), K. D. Schultz and co-workers (to Bayer AG).

34. U.S. Pat. 4,628,069, (Dec. 9, 1986), R. V. Meyer and co-workers (to Bayer AG).

35. Brit. Pat. 950,656, (Feb. 26, 1964), J. Maxwell and A. Rutherford (to ICI).

36. *Chem. Week*, 109 (Jan. 25, 1964).

37. *Mod. Plast. Int.*, 47 (Jan. 1994).
38. "Resin Statistics," *Mod. Plast.* Yearly, in Jan.
39. "Pricing Update," *Plast Technol.* Monthly.
40. *Plast. Technol.* **40**, 160 (May, 1994).
41. *ASTM D4066-93, Standard Specification for Nylon Injection and Extrusion Materials (PA)*, ASTM, Philadelphia, Pa., 1993.

*General References*

R. S. Williams and T. Daniels, *Rapra Rev. Rep.* **3**(3), 33/1–33/116 (1990). Useful review of polyamide plastics with 473 refs.
M. I. Kohan, *Nylon Plastics Handbook*, Carl Hanser Verlag, Munich, Germany, 1995. An excellent reference book covering all aspects of nylon technology.

ROBERT J. PALMER
Du Pont de Nemours Int. SA

## POLYBENZIMIDAZOLES.    See HIGH PERFORMANCE FIBERS.

## POLYBLENDS.    See POLYMER BLENDS.

## POLYBUTADIENE.    See ELASTOMERS, SYNTHETIC-POLYBUTADIENE.

## POLY(BUTYLENE TEREPHTHALATE).    See POLYESTERS, THERMO-PLASTIC.

# POLYCARBONATES

Polycarbonates are an unusual and extremely useful class of polymers. The vast majority of polycarbonates are based on bisphenol A [*80-05-7*] (BPA) and sold under the trade names Lexan (GE), Makrolon (Bayer), Calibre (Dow), and Panlite (Idemitsu). BPA polycarbonates [*25037-45-0*], having glass-transition temperatures in the range of 145–155°C, are widely regarded for optical clarity and exceptional impact resistance and ductility at room temperature and below. Other properties, such as modulus, dielectric strength, or tensile strength are comparable to other amorphous thermoplastics at similar temperatures below their respective glass-transition temperatures, $T_g$. Whereas below their $T_g$s most amorphous polymers are stiff and brittle, polycarbonates retain their ductility.

Polycarbonates are prepared commercially by two processes: Schotten-Baumann reaction of phosgene (qv) and an aromatic diol in an amine-catalyzed interfacial condensation reaction; or via base-catalyzed transesterification of a bisphenol with a monomeric carbonate. Important products are also based on polycarbonate in blends with other materials, copolymers, branched resins, flame-retardant compositions, foams (qv), and other materials (see FLAME RETAR-DANTS). Polycarbonate is produced globally by several companies. Total manufacture is over one million tons annually. Polycarbonate is also the object of academic research studies, owing to its widespread utility and unusual properties. Interest in polycarbonates has steadily increased since 1984. Over 4500 publications and over 9000 patents have appeared on polycarbonate. Japan has issued 5654 polycarbonate patents since 1984; Europe, 1348; United States, 777; Germany, 623; France, 30; and other countries, 231.

## Historical Development

The first polycarbonates were prepared in the late 1890s via reaction of hydroquinone or resorcinol with phosgene in pyridine (1). A few years later, the same materials were prepared via transesterification using diphenyl carbonate (2). The hydroquinone polymer is brittle, crystalline, insoluble in most solvents, and infusible. The polymer from resorcinol is glassy and brittle, although it crystallizes from solution, and melts at about 200°C. Attempts to prepare the polycarbonate of catechol led only to the cyclic five-membered carbonate. Both of these polymers were apparently of low molecular weight and owing to difficulties of processing and characterization, were not developed further.

In the early 1930s, the preparation of aliphatic carbonates was studied during the investigation of the preparation and properties of polyesters (qv) in general (3). Because the reactions of aliphatic alcohols and phosgene proceed more slowly than those of phenols, two other methods were used to prepare the aliphatic polycarbonates: direct transesterification reactions, and ring-opening polymerization of low molecular weight cyclic polycarbonates prepared by a distillative transesterification/depolymerization. Further work was carried out in the 1940s. 1,6-Hexanediol polycarbonates were prepared via transesterification using dibutyl carbonate (4). The aliphatic polycarbonates had low melting points and did not prove interesting commercially.

In 1941, the Pittsburgh Plate Glass Company (PPG) introduced a liquid casting resin designated as CR-39 (5). This material, formally a polycarbonate, was a cross-linked resin prepared by a peroxide-initiated radical polymerization of the bisallyl carbonate of diethylene glycol (see GLYCOLS). The starting material was prepared from allyl alcohol and diethylene glycol bischloroformate (eq. 1). Once polymerized, CR-39 was a colorless, transparent, scratch-resistant plastic which was used in optical applications. Although its nature and chemical makeup are completely different from modern polycarbonates, CR-39 was the first commercially available polycarbonate.

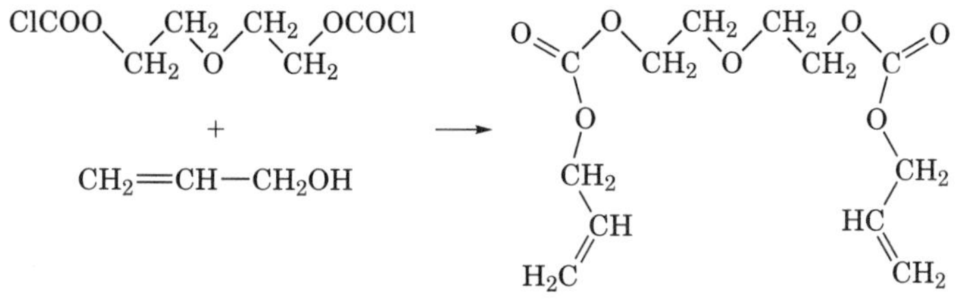

$$\text{(1)}$$

A reexamination of polycarbonate chemistry was carried out about 50 years after the first aromatic polycarbonates of resorcinol and hydroquinone were discovered. In independent investigations at Bayer AG and General Electric, it was discovered that the polycarbonates of BPA could be prepared (eq. 2). Unlike the aliphatic polycarbonates prepared earlier, which were either liquids or low melting solids, the aromatic polycarbonates were amorphous solids having elevated glass-transition temperatures.

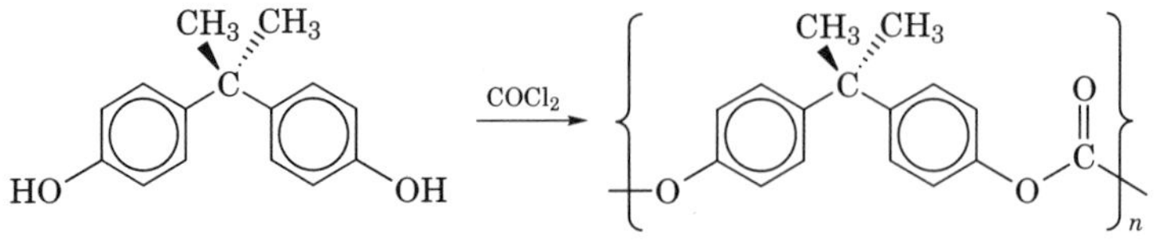

$$(2)$$

$$(1)$$

Owing to the unusual properties of the BPA polycarbonate (**1**), ie, toughness, transparency, and thermal stability, each company began development programs. Bayer AG, the first to report the properties of a series of polycarbonates (6), had patents issuing as early as 1954 (7). Commercial production of polycarbonate by Bayer AG began in Germany in 1958, and in the United States in 1960. General Electric (GE) started U.S. commercial production in 1960. After a period of litigation, U.S. patents were issued to Bayer AG, which claimed an interfacial process for preparation of polycarbonates, and had multiple claims to various polycarbonates (8). The basic GE patent claimed the transesterification process and the polycarbonate product so formed (9). Since that time extensive research has been carried out on polycarbonates. Several manufacturers have developed niches of new products, blends, or processes for production of these materials. Although GE and Bayer AG remain the principal producers, at least 50 companies have patented some aspect of polycarbonate chemistry. Over a dozen producers exist worldwide.

## Properties and Characterization

**Solubility and Solvent Resistance.**  The majority of polycarbonates are prepared in methylene chloride solution. Chloroform, cis-1,2-dichloroethylene, sym-tetrachloroethane, and methylene chloride are the preferred solvents for polycarbonates. The polymer is soluble in chlorobenzene or o-dichlorobenzene when warm, but crystallization may occur at lower temperatures. Methylene chloride is most commonly used because of the high solubility of the polymer (350 g/L at 25°C), and because this solvent has low flammability and toxicity. Nonhalogenated solvents include tetrahydrofuran, dioxane, pyridine, and cresols. Hydrocarbons (qv) and aliphatic alcohols, esters (see ESTERS, ORGANIC), or ketones (qv) do not dissolve polycarbonates. Acetone (qv) promotes rapid crystallization of the normally amorphous polymer, and causes catastrophic failure of stressed polycarbonate parts.

In general, polycarbonate resins have fair chemical resistance to aqueous solutions of acids or bases, as well as to fats and oils. Chemical attack by amines or ammonium hydroxide occurs, however, and aliphatic and aromatic hydrocarbons promote crazing of stressed molded samples. For these reasons, care must be exercised in the choice of solvents for painting and coating operations. For sheet applications, polycarbonate is commonly coated with a silicone–silicate hardcoat which provides abrasion resistance as well as increased solvent resistance. Coated films are also available.

Certain blends and copolymers of polycarbonate demonstrate dramatically improved solvent resistance. The blend of polycarbonate and poly(butylene terephthalate), eg, Xenoy (GE) or Makroblend (Bayer), combines the toughness of polycarbonate with the solvent resistance of the semicrystalline polyester. Hydroquinone polycarbonates were reinvestigated in the late 1980s (10). Several binary and ternary copolycarbonates were prepared using monomers such as hydroquinone, biphenol, and substituted hydroquinones. No thermal transitions other than the $T_g$ were noted, however, and the copolymer with hydroquinone had a very low molecular weight ($\eta_{\text{inh}} = 0.09$–$0.10$; inh = inherent). Difficulty in preparation of hydroquinone polycarbonates, owing to the insolubility of the oligomers, had been noted in the 1950s (11).

Copolycarbonates of BPA and hydroquinone (HQ) can be prepared via the intermediacy of cyclic oligomeric cocyclics (eq. 3) (12). Although hydroquinone linear oligomers having degrees of polymerization greater than two are insoluble in $CH_2Cl_2$, the cyclic analogues remain soluble when randomly cyclized with BPA. Polymerization of the hydroquinone–BPA cocyclics via anionically initiated ring-opening polymerization leads to high molecular weight semicrystalline polymers. Using this methodology, hydroquinone can be incorporated into the polycarbonate in levels up to 60%. The copolycarbonates show dramatically increased solvent resistance, and are insoluble in all common polycarbonate solvents such as methylene chloride or tetrahydrofuran. Furthermore, when molded bars of the polycarbonate are exposed to gasoline while under stress, impact properties are retained, whereas standard polycarbonate grades fail.

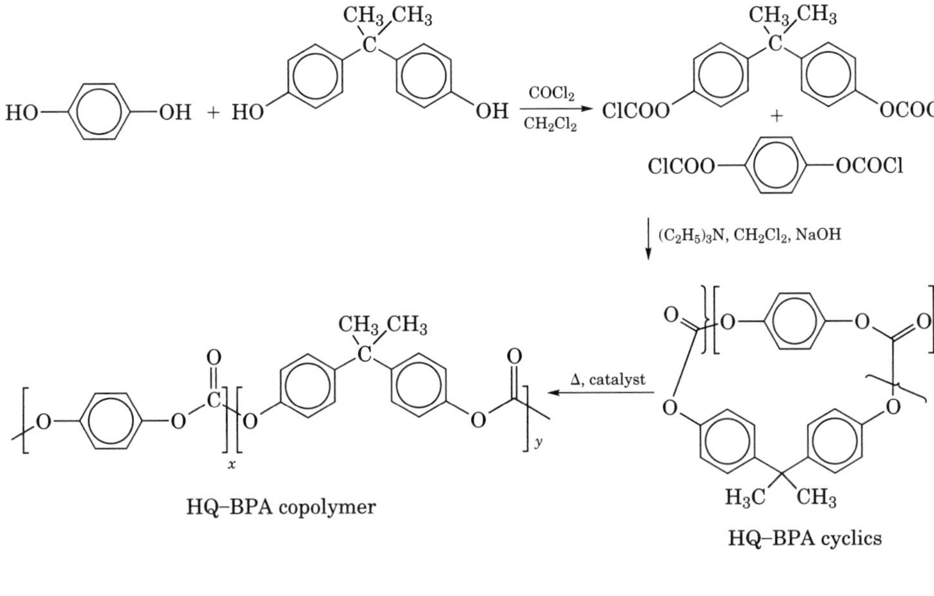

HQ–BPA copolymer

HQ–BPA cyclics        (3)

BPA polycarbonate has excellent resistance to hydrolysis. Contact with water at 60°C or intermittent contact at 100°C has little effect on polycarbonate, but extended contact can lead eventually to embrittlement. For example, exposure of polycarbonate film to steam at 101 kPa (1 atm) at 100 and 150°C showed failure of the film after 700 and 200 hours, respectively (13). Although acids have little effect, aqueous base can lead to etching. Hydrolysis occurs at the surface. The hydrolytic stability can be attributed to the low water solubility in the resin, which leads to essentially no swelling, and to the high glass-transition temperature of the resin. Heating at elevated temperatures, eg, during molding, however, can lead to degradation owing to hydrolysis. Drying of all grades of polycarbonate is recommended prior to molding to avoid hydrolysis to lower molecular weight materials.

**Molecular Weight and Viscosity.**    BPA polycarbonates are commercially available in a wide range of molecular weights. As the molecular weight increases, melt and solution viscosities increase proportionally. Molecular weights may be determined or inferred by several means, including gel-permeation chromatography, light-scattering chromatography, measurement of intrinsic or inherent viscosity, and measurement of melt viscosity and flow. Correlation of intrinsic viscosity (IV), $[\eta]$, with weight-average mol wt ($M_w$) has been carried out on carefully characterized polycarbonate samples (14). The following relationship exists when $[\eta]$ is in mL/g.

$$[\eta] = 41.2 \times 10^{-3} \cdot M_w^{0.69}$$

For chemical studies the chromatographic methods or solution viscosities are methods of preference, but for practical applications, melt flow is most important. Standard injection-molding grades of polycarbonate have intrinsic viscosities in the range of 0.50–0.55 dL/g in chloroform at 30°C, with $M_w =$ 50,000–55,000 and number-average mol wt ($M_n$) = 20,000–24,000 determined by gel-permeation chromatography (gpc) using polystyrene standards, or $M_w =$ 21,000–30,000 as determined by light scattering. The polydispersity ratio, $M_w/M_n$, of polycarbonate is about 2.5–2.7. The range of molecular weights and viscosities available is shown in Table 1.

The mechanical properties of polycarbonate, eg, tensile strength, impact resistance, flexural strength, elongation, etc, improve dramatically with increasing polymer intrinsic viscosity up to a value of about 0.45 dL/g. After that point, slight increases in mechanical properties are seen with increasing molecular weight, but melt viscosity continues to climb. At IV values greater than 0.6 dL/g, the melt viscosity becomes so high that processing is very difficult. Because some compromise between polymers having high molecular weight and good mechanical properties must balance the processibility of the resin, newer formulations having increased melt flow are being marketed. Lexan SP copolyestercarbonates (GE), for example, demonstrate enhanced flow rates as measured by melt flow rate (10–22 g/min) compared to standard grades of resin (6–16 g/min), yet retain excellent impact resistance, eg, notched Izod of 642–910 J/m (12–17 ft·lb/in.). Ultrahigh molecular weight polycarbonates can be prepared via ring-opening polymerization of cyclic aromatic oligomeric car-

**Table 1. Molecular Weight and Viscosity of Lexan Resins**

| Grade | Description | $MFI^a$ | $IV^b$ | $M_w{}^c$ | $M_n{}^c$ | $PDI^d$ | $M_w{}^e$ |
|-------|-------------|---------|--------|-----------|-----------|---------|-----------|
| 131 | ultrahigh viscosity | 3.1 | 0.629 | 72,600 | 28,100 | 2.58 | 35,500 |
| 1881 | very high viscosity | 4.9 | 0.581 | 66,100 | 25,400 | 2.6 | 32,000 |
| 101 | high viscosity | 6.5 | 0.551 | 62,000 | 25,400 | 2.44 | 29,000 |
| 161 | medium, high viscosity | 7.4 | 0.538 | 60,600 | 24,400 | 2.48 | 27,900 |
| 141 | medium viscosity | 9.2 | 0.510 | 57,000 | 23,900 | 2.38 | 26,300 |
| 141L | medium low viscosity | 11.2 | 0.493 | 54,500 | 22,700 | 2.40 | 27,400 |
| 121 | low viscosity | 16.2 | 0.454 | 49,800 | 20,400 | 2.44 | 21,200 |
| HF1110 | high flow | 20.9 | 0.434 | 46,900 | 18,400 | 2.55 | 22,700 |
| SP1110 | superior flow | 22 | 0.53 | 60,000 | 24,000 | 2.50 | 27,500 |
| OQ1020 | optical quality | 78 | 0.35 | 35,800 | 13,900 | 2.57 | 16,600 |

$^a$MFI = melt flow index.
$^b$IV = intrinsic viscosity in $CH_2Cl_2$ at 25°C.
$^c$From gel-permeation chromatography using polystyrene standards.
$^d$PDI = polydispersivity ratio, $M_w/M_n$.
$^e$Molecular weight from light scattering.

bonates. These materials, which are not commercially available, can lead to polycarbonates having intrinsic viscosities > 1.0 dL/g, and molecular weights of 300,000–500,000 (15).

**Spectroscopy and Analysis.** Polycarbonates have a strong $C{=}O$ stretching band at 1770 cm$^{-1}$, and strong $C{-}O$ stretching bands at 1220 and 1235 cm$^{-1}$, distinguishing them from polyesters. The amount of phenol end groups can be determined from the $O{-}H$ absorption at 3595 cm$^{-1}$. Proton nmr spectroscopy shows a symmetrical $A_2B_2$ aromatic pattern at 7.16 and 7.24 ppm, and absorption for the *gem*-dimethyl at 1.68 ppm, relative to $(CH_3)_4Si$. The $^{13}C$-nmr shows seven distinct absorptions: 152.1, attributable to $C{=}O$; 148.9; 148.2; 127.9; 120.3; 42.5; and 30.9. X-ray spectroscopy shows a weak absorption at 725 cm$^{-1}$, related to crystallinity, and a band at 917 cm$^{-1}$, independent of crystallinity.

Differential scanning calorimetry reveals a glass-transition temperature at around 154°C, shifting slightly with molecular weight or level of branching. End group and impurity analysis is best revealed by hydrolysis of the polycarbonate using KOH–methanol in tetrahydrofuran under nitrogen, followed by reversed-phase hplc analysis or by spectroscopic techniques. Trace levels of impurities, such as methylene chloride, amine, chloride, and sodium, are determined by standard analytical techniques, eg, atomic absorption or titration.

**Structure and Crystallinity.** The mechanical–optical properties of polycarbonates are those common to amorphous polymers. The polymer may be crystallized to some degree by prolonged heating at elevated temperature (8 d at 180°C) (16), or by immersion in acetone (qv). Powdered amorphous powder appears to dissolve partially in acetone, initially becoming sticky, then hardening and becoming much less soluble as it crystallizes. Enhanced crystallization of polycarbonate can also be caused by the presence of sodium phenoxide end groups (17).

Film or fibers derived from low molecular weight polymer tend to embrittle on immersion in acetone; those based on higher molecular weight polymer (>0.60 dL/g) become opaque, dilated, and elastomeric. When a dilated sample is stretched and dried, it retains orientation and is crystalline, exhibiting enhanced tensile strength. The tensile heat-distortion temperature of the crystalline film is increased by about 20°C, and the gas permeability and resistance to solvent attack is increased.

Thermotropic polycarbonates have been prepared from mixtures of 4,4'-dihydroxybiphenyl and various diphenols (10). Nematic melts were found for copolycarbonates prepared from methylhydroquinone, chlorohydroquinone, 4,4'-dihydroxydiphenyl ether, and 4,4'-dihydroxybenzophenone. Slightly crystalline polycarbonates have been prepared from mixtures of hydroquinone and BPA ($T_g = 154$°C, $T_m = 313$°C, $\Delta H_m = 11.0$ J/g (2.63 cal/g)), and from methylhydroquinone ($T_g = 155$°C, $T_m = 289$°C, $\Delta H_m = 31.0$ J/g (7.41 cal/g)) (12).

Experimental and theoretical studies on the structure of BPA polycarbonates have been the object of considerable interest since the work of Williams and Flory in the late 1960s (18). Because of the low conformational barriers to rotation, phenyl ring-flipping and cis–trans isomerization about the carbonate group have been invoked as mechanisms for energy absorption providing polycarbonates with low temperature impact strength. Crystal structures of diphenyl carbonate, described in detail, have been published (19,20). The crystal structures of a model carbonate (the bisphenyl carbonate of BPA) has also appeared (20). All of the published structures indicate that the thermodynamically preferred backbone conformation about the carbonate functionality is a trans–trans conformation. In this form, the dihedral angles of the aromatic rings with the carbonyl oxygen, C–O–C(=O)–O, (174.8 and 176.5°) are nearly eclipsed, and as a consequence the planes of the carbonyl groups are skewed from the planes of the aromatic rings by 59.1 and 53.2°. A cis–trans relationship about the carbonyl group was first seen in a complex of the bisphenyl carbonate of BPA with two molecules of a thiopyrilium salt (22). The cyclic dimer carbonate of BPA also shows a cis–trans relationship of aromatic rings about the carbonyl (20). The crystal structure of cyclic tetramer has also been described (20), showing two distinct types of aromatic conformations about the carbonyl. Several mathematical (23) and physical (24) methods have been used to analyze the conformational features of BPA polycarbonate. Estimations of the energy differences between conformations have been investigated by a variety of techniques, including *ab initio* calculations (25), nmr spectroscopy (26), and infrared spectroscopy (27). Molecular simulation studies on the conformation of cyclic oligomers have also appeared (28).

**Glass-Transition Temperature and Melt Behavior.** The $T_g$ of BPA polycarbonate is around 150°C, which is unusually high compared to other thermoplastics such as polystyrene (100°C), poly(ethylene terephthalate) (69°C), nylon-6,6 (45°C), or polyethylene (−45°C). The high glass-transition temperature can be attributed to the bulky structure of the polymer, which restricts conformational changes, and to the fact that the monomer has a higher molecular weight than the monomer of most polymers. The high $T_g$ is important for the utility of polycarbonate in many applications, because, as the point which marks the onset of molecular mobility, it determines many of the polymer's properties such as dimensional stability, resistance to creep, and ultimate use temperature. Polycarbonates of different structures may have significantly higher or lower glass-transition temperatures.

BPA polycarbonate becomes plastic at temperatures around 220°C. The viscosity decreases as the temperature increases, exhibiting Newtonian behavior, with the melt viscosity essentially independent of the shear rate. At the normal injection molding temperature of 270–315°C, the melt viscosity drops from 1,100 to 360 Pa·s (11,000 to 3,600 poise), which is about five times the viscosity of poly(ethylene terephthalate) of similar molecular weight over the same temperature range. Because the viscosity of polycarbonate can only be reduced by increasing the temperature, the ultimate limit on molecular weight is controlled by the processing conditions and the thermal stability of the polymer. Branched polycarbonates can be prepared by incorporation of small amounts of tri- or tetrafunctional phenols or carboxylic acids. The rheological properties of the branched resin are different from those of linear resins. The branched resins demonstrate non-Newtonian behavior, and viscosity depends on shear rate. The melt viscosity of branched resins decreases with increasing shear, allowing extrusion at lower temperatures of materials with exceptional melt strength for blow molding applications (Fig. 1).

**Thermal, Flame-Retardant, and Hydrolytic Behavior.** BPA polycarbonate exhibits excellent thermal stability, especially in the absence of oxygen and water. The dry polymer may be heated at 320°C for several hours, or for short times as high as 330–350°C with minimal degradation. At these high temperatures, thermal–oxidative degradation leads to slight yellowing, requiring color stabilization. Low levels (<0.5%) of stabilizers (phosphites, phosphonites, phosphines, epoxide compounds, and organosilicon compounds) are usually added during processing. At temperatures above 400°C, rapid decomposition and cracking occur. BPA has an oxygen index of 26 according to ASTM D2863-70, indicating that under test conditions, an atmosphere of 26% oxygen is required for combustion. Owing to thermal–oxidative stability, polycarbonate has some inherent flame resistant properties and can be classified as V-2 according to UL94 of the Underwriters Laboratory. Several polycarbonate grades have additives to increase the flame-retardant properties, and to decrease smoke. Flame-retardant agents include brominated oligomers from tetrabromoBPA;

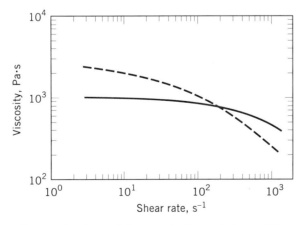

**Fig. 1.** Melt viscosity as a function of shear rate for (——) linear BPA polycarbonate and (– – –) branched polycarbonate. To convert Pa·s to poise, multiply by 10.

alkali metal salts, usually alkali sulfonates; and polytetrafluoroethylene (PTFE). The flame-retardant grades are classed as V-0 according to UL94 testing procedures, using a wall thickness of 3.2 mm (see FLAME RETARDANTS).

Because of the low solubility of water in the resin, BPA polycarbonates are inherently resistant to aqueous acid and base, although strong nucleophilic bases can cause hydrolysis. Prolonged contact with water at 100°C can lead to reduced impact strength as a result of microcrack formation. Hydrolytic stability can be increased using additives, especially epoxides, but other additives such as phosphites or ammonium salts can reduce the hydrolytic stability. Polycarbonates should be dried prior to melt processing, because the low levels of water present in the resin can lead to partial hydrolysis and lowering of the molecular weight. The equilibrium water content at 60% relative humidity is 0.2% at ambient temperature, 0.36% in water, and 0.9% in water at 100°C.

**Optical Properties.** Polycarbonate is a transparent colorless polymer, making it attractive for glass replacement. Visible light transmission is about 90%, and haze is minimal (1–2%). Absorption in the ultraviolet region is essentially complete. Polycarbonate's high (1.584) refractive index and light weight relative to glass make it attractive for eyewear. Exposure of non-uv stabilized polycarbonate to strong uv or outdoor conditions can lead to photoyellowing of the surface. Uv-screens are typically combined with hard coats to protect the surface of polycarbonate, and also provide resistance to solvent attack in sheet products used for glazing applications.

Special polycarbonate grades have been developed for the rapidly growing optical information storage market (see INFORMATION STORAGE MATERIALS). Compact disks (CDs) utilize the transparency, low birefringence, mechanical and dynamic stability, and good heat distortion properties of polycarbonate. The use of $p$-alkylphenols with branched alkyl groups containing eight or nine carbon atoms results in extremely high melt flow polycarbonates having sustained impact strength (29). Use of high flow resins decreases molding cycle time to about 7 s, and also reduces birefringence.

**Mechanical Properties.** Mechanical and other properties of BPA polycarbonate are listed in Table 2. The room temperature modulus and tensile strength are similar to those of other amorphous thermoplastics, but the impact strength and ductility are unusually high. Whereas most amorphous polymers are glasslike and brittle below their glass-transition temperatures, polycarbonate remains ductile to about −10°C. The stress–strain curve in uniaxial tension (Fig. 2) is typical of ductile materials, having an initial Hookean region, followed by shear-induced yielding and plastic deformation. Breakage occurs at about 120% elongation. The area under the stress–strain curve is a measure of energy absorption per unit volume, and is about 65 J/m$^3$ (15.5 cal/m$^3$). The value is about 8.5 times higher than the energy absorption of cast aluminum, and 60% that of carbon steel, placing polycarbonate in an ideal position for use as a metal replacement. Weight savings as a metal replacement are substantial, because polycarbonate is only 44% as dense as aluminum and one-sixth as dense as steel.

Impact strength can be measured by a variety of methods, including notched Izod, tensile impact, and falling dart impact. The notched Izod test is most widely used to measure the toughness of a resin. This test involves the energy absorption under impact conditions of a sample in which a notch has been cut

## Table 2. Bisphenol A Polycarbonate Properties

| Property | Lexan 141 | Lexan 3414[a] | ASTM method |
|---|---|---|---|
| *Physical properties* | | | |
| specific gravity | 1.52 | 1.20 | D792 |
| water absorption, 23°C, % | | | |
|   24 h | 0.15 | 0.12 | D570 |
|   equil | 0.35 | 0.23 | D570 |
| melt flow rate, 300°C, | 10.5 | | D1238 |
|   1.2 kgf, g/10 min | | | |
| mold shrinkage, 3.2-mm part, % | 0.1–0.2 | 0.5–0.7 | D955 |
| light transmittance, 550 nm, % | 86–89 | | D1003 |
| haze, % | 1–1.5 | | D1003 |
| refractive index | 1.586 | | |
| *Mechanical properties* | | | |
| tensile strength, Type I, MPa[b] | | | |
|   yield | 60 | | D638 |
|   break | 70 | 160 | D638 |
| tensile elongation, break, Type I, % | 130 | 3.0 | D638 |
| flexural strength, MPa[b] | 97 | 190 | D790 |
| flexural modulus, MPa[b] | 2,300 | 9,600 | D790 |
| compressive strength, MPa[b] | 86 | 140 | D695 |
| compressive modulus, MPa[b] | 2,400 | 10,300 | D695 |
| shear strength, MPa[b] | | | |
|   yield | 41 | 75 | D732 |
|   break | 695 | | D732 |
| shear modulus, MPa[b] | 785 | 2,200 | D732 |
| hardness, Rockwell R | 118 | 119 | D785 |
| fatigue limit, $2.5 \times 10^6$ cycles, MPa[b] | 6.9 | 50 | D671 |
| deformation under load, 27 MPa[b], % | | | |
|   at 23°C | 0.2 | 0.1 | D621 |
|   70°C | 0.5 | 0.2 | D621 |
| *Impact properties* | | | |
| Izod impact, J/m[c] | | | |
|   notched | 801 | 133 | D256 |
|   unnotched | no break | 1300 | D256 |
| tensile impact, Type S, kJ/m$^{2}$ [c] | 578 | 67 | D1822 |
| *Thermal properties* | | | |
| softening temperature, Vicat, °C | 154 | 166 | D1525 |
| heat deflection, °C | | | |
|   at 0.45 MPa[b] | 138 | 154 | D648 |
|   1.8 MPa[b] | 134 | 146 | D648 |
| specific heat, J/(g·°C)[c] | 1.25 | 1.0 | C351 |
| coefficient of thermal expansion, | $6.75 \times 10^{-3}$ | $1.67 \times 10^{-3}$ | E831 |
|   −40 to 95°C, % | | | |
| thermal conductivity, W/(m·°C) | 0.19 | 0.22 | C177 |
| brittle temperature, °C | −129 | | D746 |
| continuous use temperature, °C | 121 | | |
| *Electrical properties* | | | |
| dielectrical strength, mV/m | 15 | 17.7 | D149 |

**Table 2.** (*Continued*)

| Property | Lexan 141 | Lexan 3414[a] | ASTM method |
|---|---|---|---|
| dielectric constant | | | |
|   60 Hz | 3.17 | 3.53 | D150 |
|   1 MHz | 2.96 | 3.48 | D150 |
| dissipation factor | | | |
|   60 Hz | 0.0009 | 0.0013 | D150 |
|   1 MHz | 0.010 | 0.0067 | D150 |
| *Flame classifications* | | | |
| 100 Series 94, mm | | | |
|   V-0 rating | 6.10 | 3.05 | UL 94 |
|   V-2 rating | 1.14 | | UL 94 |
| oxygen index, % | 26 | 30 | D2863 |

[a]This polycarbonate is 40% glass reinforced.
[b]To convert MPa to psi, multiply by 145.
[c]To convert J to cal, divide by 4.184.

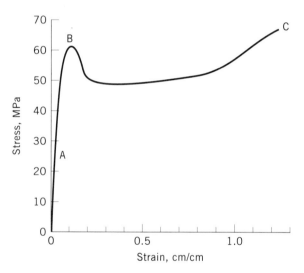

**Fig. 2.**  Stress–strain curve for standard polycarbonate resin at 23°C where the points A, B, and C correspond to the proportional limit (27.6 MPa), the yield point (62 MPa), and the ultimate strength (65.5 MPa), respectively. To convert MPa to psi, multiply by 145.

to serve as a stress concentrator. Polycarbonates are among the highest rated engineering polymers for impact resistance, and are the toughest transparent materials known (see ENGINEERING PLASTICS). Under high strain rate conditions (impact), polycarbonates are sensitive to the nature of the stress concentrator, and different results can be obtained depending on how the notch is cut and the thickness of the sample. At room temperature, decreasing the radius of the notch tip from 0.038 to 0.20 mm reduces the notched impact strength from 960 to 130 J/m (230–31 cal/m). A similar transition is noted when samples are tested over a range of temperatures, showing a transition from ductile to brittle behavior at temperatures around −20°C. As the thickness of the specimen is increased from 3.18 to 6.35 mm, a similar transition occurs, with brittle failure of thicker samples. The critical thickness phenomenon, as well as low temperature embrittlement, can be alleviated by use of copolymers or blends.

Glass-reinforced polycarbonates are sold as high modulus materials having properties approaching those of metals, while retaining the basic plastic attributes of low cost processing, dielectric character, resistance to corrosion, and inherent color. As the level of glass is increased, tensile and flexural strength and modulus increase to almost double their original values. Compressive strength increases by about 65%. Fatigue endurance increases dramatically to seven times the value for neat resin; deformation under load drops to 0.1% under 27 MPa (6800 psi) at 23°C. Both the coefficient of thermal expansion and mold shrinkage drop in glass-reinforced resins. As expected for a stiffer resin, the impact properties drop as well to a notched Izod impact value of 133 J/m (32 cal/m). Properties of a 40% glass-reinforced resin are compared to a standard injection molding resin in Table 2.

## Preparation

The historical direct reaction route, which utilized phosgenation of a solution of BPA in pyridine, proved inefficient commercially because of the need for massive pyridine recycle. Calcium hydroxide was used as an HCl scavenger for a period of time. In the historical transesterification process, BPA and diphenyl carbonate are heated in the melt in the presence of a catalyst, driving off by-product phenol, which is recycled to diphenyl carbonate. Using a series of reactors providing higher heat and vacuum, the product polymer was eventually produced as a neat melt.

**Interfacial Polymerization.** Most BPA polycarbonate is produced by an interfacial polymerization process utilizing phosgene [75-44-5]. The interfacial process for polycarbonate preparation involves stirring a slurry or solution of BPA and 1–3% of a chain stopper, such as phenol, $p$-$t$-butylphenol, or $p$-cumylphenol, in a mixture of methylene chloride and water, while adding phosgene (qv) in the presence of a tertiary amine catalyst (eq. 4). Sodium hydroxide solution is added concurrently to maintain the appropriate reaction pH. Efficient mixing is important, to ensure contact between the four phases present: solid BPA, gaseous phosgene, methylene chloride (into which the polymer dissolves), and the aqueous phase, in which the by-product sodium chloride is concentrated.

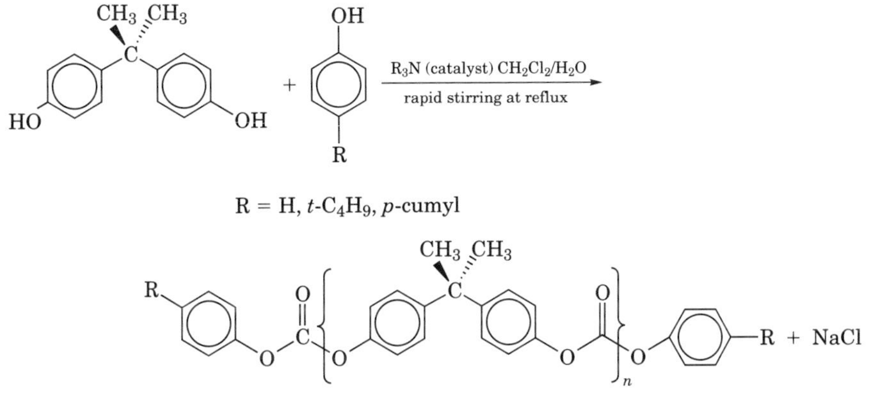

$$R = H, t\text{-}C_4H_9, p\text{-cumyl}$$

$$n = 35\text{--}60$$

Phosgene addition is continued until all the phenolic groups are converted to carbonate functionalities. Some hydrolysis of phosgene to sodium carbonate occurs incidentally. When the reaction is complete, the methylene chloride solution of polymer is washed first with acid to remove residual base and amine, then with water. To complete the process, the aqueous sodium chloride stream can be reclaimed in a chlor-alkali plant, ultimately regenerating phosgene. Many variations of this polycarbonate process have been patented, including use of many different types of catalysts, continuous or semicontinuous processes, methods which rely on formation of bischloroformate oligomers followed by polycondensation, etc.

The mechanism of the polycarbonate formation reaction can be broken down into several steps. In the first step, the bisphenol reacts with phosgene to form a chloroformate (eq. 5, Ar = aryl group). Although this reaction can proceed in the absence of catalyst, it is accelerated both by phase-transfer and by nucleophilic catalysts. The reaction occurs quickly enough to compete effectively with phosgene hydrolysis, even at the reaction pH of 10–12. Reaction of a chloroformate functionality and a phenol to form the carbonate functionality requires a catalyst to proceed effectively (eq. 6). In the absence of catalyst the reaction takes several hours (30). Tertiary amines provide the greatest rate acceleration, but phase-transfer catalysts are also effective. The amines function by nucleophilic catalysis, forming an acyl ammonium salt, which is more susceptible toward attack by a phenol. The rates of formation of acyl ammonium salts, as well as the rates of their condensation reactions, have been measured using stop-flow ftir techniques (30). Normally, polycarbonate reactions are over-phosgenated, to ensure complete conversion of bisphenol to product, forming slight stoichiometric excesses of chloroformate. Hydrolysis of the chloroformate so formed is then necessary to complete the reaction, converting the chloroformate first to a phenoxide, so that conversion to a carbonate can occur (eq. 7).

$$\text{ArOH} + \text{COCl}_2 \xrightarrow[\text{NaOH}]{\text{catalyst}} \text{ArO}-\underset{\underset{O}{\parallel}}{C}-\text{Cl} + \text{NaCl} \qquad (5)$$

$$\text{ArO}-\underset{\underset{O}{\parallel}}{C}-\text{Cl} \xrightarrow{\text{R}_3\text{N}} \text{ArO}-\underset{\underset{O}{\parallel}}{C}-\text{N}^+\text{R}_3\text{Cl}^- \xrightarrow{\text{ArOH}} \text{ArO}-\underset{\underset{O}{\parallel}}{C}-\text{OAr} \qquad (6)$$

$$\text{ArO}-\underset{\underset{O}{\parallel}}{C}-\text{Cl} \xrightarrow{\text{R}_3\text{N}} \text{ArO}-\underset{\underset{O}{\parallel}}{C}-\text{N}^+\text{R}_3\text{Cl}^- \xrightarrow[\text{H}_2\text{O}]{\text{NaOH}} \text{ArO}^-\text{Na}^+ \qquad (7)$$

Methods for isolation of the product polycarbonate remain trade secrets. Feasible methods for polymer isolation include antisolvent precipitation, removal of solvent in boiling water, spray drying, and melt devolatization using a wiped film evaporator. Regardless of the technique, the polymer must be isolated dry, to avoid hydrolysis, and essentially be devoid of methylene chloride. Most polycarbonate is extruded, at which point stabilizers and colors may be added, and sold as pellets.

**Transesterification.** There has been renewed interest in the transesterification process for preparation of polycarbonate because of the desire to transition technology to environmentally friendly processes. The transesterification process utilizes no solvent during polymerization, producing neat polymer directly and thus chlorinated solvents may be entirely eliminated. General Electric operates a polycarbonate plant in Chiba, Japan which produces BPA polycarbonate via this melt process.

The polymerization is carried out in several stages. In the first stage, diphenyl carbonate and BPA are combined with small amounts (<0.01% molar) of basic catalysts (Na, Li, K, or tetraalkylammonium hydroxides or carbonates) in a melt reactor, in which phenol begins to be liberated according to the equilibrium shown in equation 8. As the raw materials pass from reactor to reactor, the temperature of the reaction is increased, and higher and higher vacuum is applied, so that phenol can be removed from the equilibrium, driving the process toward polymer. In the later stages of reaction, the polymer melt becomes viscous and specialized equipment such as wiped film evaporators, helicone reactors, or multiply vacuum-vented extruders are necessary to expedite mass transfer of by-product phenol via good surface renewal of the melt.

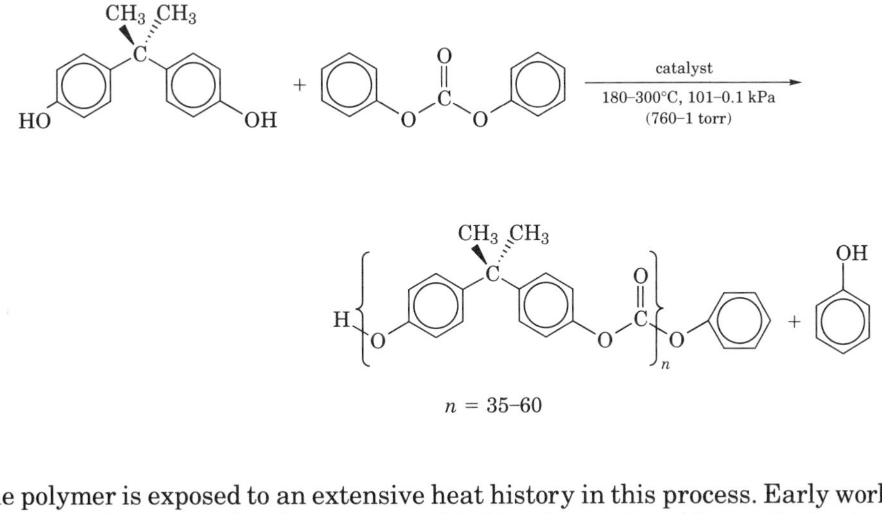

$$n = 35\text{--}60 \tag{8}$$

The polymer is exposed to an extensive heat history in this process. Early work on transesterification technology was troubled by thermal–oxidative limitations of the polymer, especially in the presence of the catalyst. More recent work on catalyst systems, more reactive carbonates, and modified processes have improved the process to the point where color and decomposition can be suppressed. One of the key requirements for the transesterification process is the use of clean starting materials. Methods for purification of both BPA and diphenyl carbonate have been developed.

Initially, diphenyl carbonate to be used for the transesterification process was prepared by reaction with phosgene. The phosgenation can be carried out in water, avoiding use of organic solvents in the polycarbonate process altogether

(31). More recently, methods for preparation of diphenyl carbonate have been developed which avoid the use of phosgene (qv). Enichem has a commercial plant for the preparation of diphenyl carbonate which utilizes a two-step process. In the first step (eq. 9) methanol is carbonylated using carbon monoxide in the presence of a catalyst to produce dimethyl carbonate. Reaction of dimethyl carbonate and phenol under specific transesterification conditions produces diphenyl carbonate. The by-product is methanol which can be recycled. Because the transesterification reaction is thermodynamically an unfavorable equilibrium process, the reaction is normally carried out in two stages: first, formation of a mixed carbonate (eq. 10), then disproportionation to dimethyl carbonate and diphenyl carbonate (eq. 11). An alternative method for formation of diphenyl carbonate is the direct carbonylation of phenol, using palladium catalysts (32). Although this latter chemistry works well for formation of diphenyl carbonate, it is apparently not quite efficient enough for the formation of high molecular weight polycarbonate directly from BPA.

$$CH_3OH + CO \xrightarrow{\text{catalyst}} H_3C{\diagdown}O{\diagup}\overset{\overset{O}{\|}}{C}{\diagdown}O{\diagup}CH_3 \qquad (9)$$

$$H_3C{\diagdown}O{\diagup}\overset{\overset{O}{\|}}{C}{\diagdown}O{\diagup}CH_3 + \text{PhOH} \underset{\phantom{xx}}{\overset{Ti(OR)_4}{\rightleftharpoons}} H_3C{\diagdown}O{\diagup}\overset{\overset{O}{\|}}{C}{\diagdown}O{\diagup}CH_3 + H_3C{\diagdown}O{\diagup}\overset{\overset{O}{\|}}{C}{\diagdown}O{\diagup}C_6H_5 \qquad (10)$$

$$C_6H_5{\diagdown}O{\diagup}\overset{\overset{O}{\|}}{C}{\diagdown}O{\diagup}CH_3 \overset{Ti(OR)_4}{\rightleftharpoons} H_3CO\overset{O}{\overset{\|}{C}}OCH_3 + H_3CO\overset{O}{\overset{\|}{C}}OC_6H_5 + C_6H_5O\overset{O}{\overset{\|}{C}}OC_6H_5 \qquad (11)$$

An analogue of the transesterification process has also been demonstrated, in which the diacetate of BPA is transesterified with dimethyl carbonate, producing polycarbonate and methyl acetate (33). Removal of the methyl acetate from the equilibrium drives the reaction to completion. Methanol carbonylation, transesterification using phenol to diphenyl carbonate, and polymerization using BPA is commercially viable. The GE plant is the first to produce polycarbonate via a solventless and phosgene-free process.

Another technique for the preparation of polycarbonates which can lead to novel applications is the ring-opening polymerization of cyclic oligomers. Although discrete cyclic oligomers of BPA and other bisphenols were isolated in the early 1960s, exploitation of this chemistry was not possible owing to the difficulty of preparation and purification (34). Furthermore, the cyclic tetramer of BPA, which could be prepared in about 25% yield, was a crystalline high melting (375°C) solid, which made melt polymerization studies difficult.

A newer process for preparation of mixtures of cyclic oligomers has been discovered (15,35). This process utilizes a triethylamine-catalyzed hydrolysis condensation of BPA bischloroformate in a pseudo-dilution reaction. This efficiently provides 85–90% yield of a mixture of cyclic oligomers having a degree of polymerization of 2 to about 12 (eq. 12).

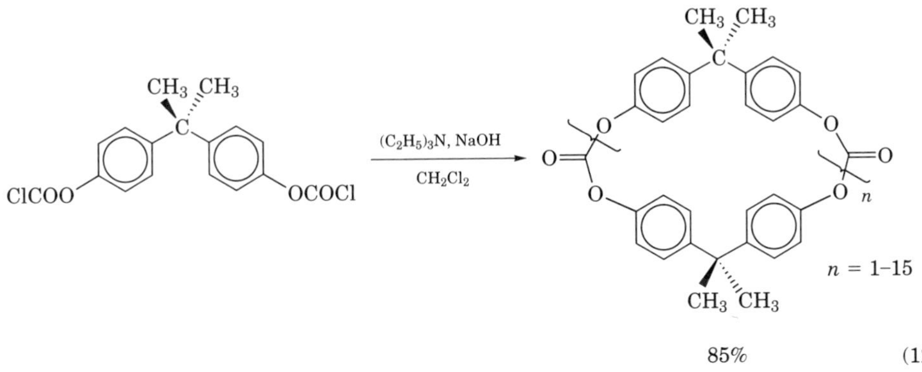

$$n = 1\text{--}15$$

$$85\% \qquad\qquad (12)$$

Two important features distinguish this work from the previous preparation of cyclic tetramer. First, the reaction is extremely selective toward formation of cyclic oligomers. Less than 0.01% linear oligomers are formed. Second, a mixture of cyclics is formed. Because the linear oligomers incorporate into the polymer via chain-transfer reactions during ring-opening polymerization, their presence limits ultimate molecular weight, and it is important to limit their level. The fact that a broad mixture of cyclics is formed results in depression of the melting point of the cyclics, allowing a clear melt to form at 200–210°C and allowing melt processing. The melting point of the cyclic mixture can be depressed further by the incorporation of long-chain bisphenols (36). Melt polymerization of the cyclic oligomers can be achieved at 200–300°C, and solution polymerization is possible at ambient temperature (37). The ring-opening polymerization is driven almost entirely by entropy, in a reaction with almost no exotherm, leading to the highest molecular weight polycarbonates achievable by any process. Because the viscosity of the cyclic oligomers is about $10^5$ lower than the ultimate polymer, a variety of processing techniques become possible, including pultrusion, resin-transfer molding, and other methods for fabrication of composites. The use of the cyclic oligomer technique also makes it possible to prepare polycarbonates which had been unattainable by other means, eg, hydroquinone–BPA copolymers. The cyclic oligomer approach has not been commercialized.

**Processing.** Polycarbonates may be fabricated by all conventional thermoplastic processing operations, of which injection molding is the most common. Recommended operating conditions are stock temperatures of 275–325°C and molding pressures of 69–138 MPa (10,000–20,000 psi). Thorough predrying is necessary to prevent hydrolysis of the polymer by dissolved water at the high processing temperatures. Inadequate drying may lead to surface-streaked parts and loss of molecular weight. Extrusion produces film, sheet, and stock shapes. Molecular weights higher than those used for injection molding are preferred. Temperatures for extrusion vary between 285 and 315°C. Predrying is also important for extrusion.

Structural foam molding is also a valuable commercial technique. The molding compound is charged with an inert gas under pressure and at the same time thermally plasticized in an extruder. Nitrogen gas may be pumped into the melt or the molding compound may contain a chemical blowing agent that decomposes with gas generation at the plasticizing temperature. The pressurized plasticized melt is injected into a low pressure mold with a volume change taking place causing the polymer to expand and fill the mold cavity. Very large parts having relatively low clamping pressures can be produced with this process.

Injection blow molding of polycarbonates produces an assortment of containers from 20-L water bottles and 0.25-L milk bottles to outdoor lighting protective globes. The polymer normally contains a small amount of polyfunctional monomer that serves as a branching agent. Low level branching enhances the melt strength, as well as inducing non-Newtonian flow behavior, ie, sensitivity of melt viscosity to shear.

Conventional thermoforming of sheet and film is applicable to the production of skylights, radomes, signs, curved windshields, prototype production of body parts for automobiles, skimobiles, boats, etc. Because BPA polycarbonate is malleable, it can be cold-formed like metal, and may be cold-rolled, stamped, or forged.

**Production.** Production of polycarbonate has steadily increased since the opening of the first commercial plants in the early 1960s. Worldwide capacity reached 665,000 t in 1992 (38). Since the mid-1980s, production has increased at 8–13% per year. Expected capacity in 1995 is over one million metric tons. Plants have opened in several countries, including Japan, China, Korea, and Brazil, although as of this writing (1995) the United States remains the primary producer of polycarbonate. U.S. production is about 47% of that worldwide.

General Electric is the largest polycarbonate producer, having close to 45% of worldwide capacity (350,000 t). Bayer AG has the capacity to produce about 230,000 t/yr. Other producers include Dow Chemical (86,000 t), Mitsubishi Gas (50,000 t), and Teijin (50,000 t). Because most polycarbonate is used in the automotive and construction industries, consumption depends greatly on the state of the economy, and growth forecasts vary widely according to region. The fastest growing region is in the Far East, ie, Japan, China, and Korea. Growth projections there are 11–15% into the twenty-first century, followed by the United States, at about 5–10%/yr, and Western Europe, at 0–5%/yr net growth projected. Normal grades of polycarbonates sold in 1995 for $1.65–2.25/lb ($3.30–$5.00/kg). Many specialty grades demanded significantly higher prices.

## Health and Safety Factors

BPA polycarbonate is an amorphous solid, supplied as extruded pellets averaging 2.5 mm in diameter and 3.2 mm long, or as a powder. Polycarbonate is considered a slight or nonexistent fire hazard. The ignition temperatures are quite high; flash ignition is at 449°C and autoignition at 632°C, according to ASTM D1929. Odor and volatiles are negligible. Processing fumes, which include water, carbon dioxide, diphenyl carbonate, methylene chloride, and phenol, are not formed in levels considered to be hazardous. Nonetheless, good housekeeping and industrial hygiene (qv) techniques should be followed. Polycarbonate has very low acute oral and dermal toxicity, is not a primary skin irritant, and does not cause systemic or local sensitization. In a finely divided form, polycarbonate is a mild eye irritant, consistent with the abrasive nature of the ground resin particle. Polycarbonate does not degrade during storage, and no heating or cooling requirements are necessary. For transportation purposes, polycarbonate is not classified as a hazardous material by U.S. Department of Transportation code, Title 49.

## Uses

Extreme toughness, transparency, low color, resistance to burning, and maintenance of engineering properties over a wide thermal range are the outstanding properties of polycarbonate that make it useful for a variety of applications. Glazing and sheet are the largest markets for polycarbonate resins. About 50,000 t are used in these markets annually in the United States, and 110,000 t worldwide. Clarity, and an impact resistance 250 times greater than glass and 30 times greater than acrylic sheet, makes polycarbonate the natural choice for window replacement in areas where breakage is common. Windows in airplanes, trains, and schools commonly use polycarbonate. Exotic applications include military use, for example in high speed aircraft canopies, where tests have shown polycarbonate to withstand impact with fowl at Mach 2. Polycarbonate is also used for security applications as laminates with glass or other materials. Polycarbonate offers unsurpassed projectile-stopping capability, as the material softens upon impact with a bullet, absorbing the projectile's energy.

Automotive applications account for about 116,000 t of worldwide consumption annually, with applications for various components including headlamp assemblies, interior instrument panels, bumpers, etc. Many automotive applications use blends of polycarbonate with acrylonitrile–butadiene–styrene (ABS) or with poly(butylene terephthalate) (PBT) (see ACRYLONITRILE POLYMERS). Both large and small appliances also account for large markets for polycarbonate. Consumption is about 54,000 t annually. Polycarbonate is attractive to use in light appliances, including houseware items and power tools, because of its heat resistance and good electrical properties, combined with superior impact resistance.

Packaging (qv) is a growing segment of the polycarbonate market, accounting for about 20,000 t consumption annually. Polycarbonate has been the preferred choice for large returnable, refillable 20-L water bottles because of its light weight and impact resistance. GE Plastics has introduced returnable, refillable 0.25-L milk bottles for use in schools. These bottles, which can be refilled 50 to 100 times, are attractive alternatives to nonreusable paper milk cartons in communities which are concerned about waste (see RECYCLING, PLASTICS).

Electrical, electronic, and technical applications use polycarbonates for a variety of purposes. The worldwide market is about 156,000 t annually. Because of excellent electrical properties (dielectric strength, volume resistivity), and resistance to heat and humidity, polycarbonate is used for electrical connectors (qv), telephone network devices, outlet boxes, etc. Polycarbonate had been popular for use in computer and business machine housings, but the use of neat resin has been largely supplanted by blends of polycarbonate with ABS. Overall, however, the total use of polycarbonate continues to increase. A total of 15,000–17,000 t of resin is used annually. Polycarbonate also has many technical uses in instrument panels and devices, especially for membrane switches and insulators. Optical quality polycarbonate is the only suitable material for the compact disk market. Since their introduction in 1983, compact disks have shown explosive growth in the consumption of polycarbonate, with utility for audio, video, and computer applications. Consumption of optical quality resin more than doubled between 1988 and 1992, and as of 1995 accounted for about 20,000 t of annual production.

Medical and health-care related applications consume about 21,000 t of polycarbonate annually. Polycarbonate is popular because of its clarity, impact strength, and low level of extractable impurities. Special grades have been developed to maintain clarity and resistance to yellowing upon gamma radiation sterilization (qv) processes. Leisure and safety applications are many and varied, accounting for about 22,000 t of consumption annually. The largest markets are for protective headgear such as football and motorcycle helmets, and safety helmets for firefighters and construction workers. Protective eyewear also uses polycarbonate, because of its clarity and impact resistance.

## Other Polycarbonates, Blends, and Copolymers

During the early development of polycarbonates, many bisphenols were investigated for potential useful products. Some of these monomers and polymers are listed in Table 3. Despite this intensive search, however, no homopolycarbonates other than that of BPA have been produced. Copolymers and blends, on the other hand, have been quite successful. Blends of polycarbonate with ABS and with poly(butylene terephthalate) (PBT) in particular have shown significant growth since the mid-1980s.

**Copolymers.** The copolymer of tetrabromoBPA and BPA was one of the first commercially successful copolymers. Low levels of the brominated comonomer lead to increased flame resistance (V-0 rating by UL 94) while having little effect on other properties. The polycarbonate of bis(4-hydroxyphenyl)-1,1-dichloroethylene, prepared from chloral and phenol, followed by dehydrohalogenation, was investigated as another flame-resistant polymer which retained good impact properties.

For utility in compact disks and other laser-readable data storage systems, polycarbonates of low birefringence are advantageous. Birefringence can be reduced by using bulky polarizable side groups, or eliminated entirely in structures such as the spirobiindane bisphenol (SBI) shown in Table 3. Birefringence can also be reduced by eliminating molded-in stresses via the use of high flow resins.

Polyester carbonates can be prepared by the copolymerization of BPA with diacyl chlorides such as iso- or terephthaloyl chloride (eq. 13). In some cases, a diacid can be used directly, because reaction with phosgene converts it into an acid chloride prior to reaction with the bisphenol. The polyester carbonates prepared using aromatic diacids are useful as high heat materials; although the polyarylate esters are crystalline as homopolymers, copolymerization as copolyestercarbonates allows formation of amorphous resins which have impact strength similar to BPA polycarbonate, but have the glass-transition temperature elevated to about 190°C. Incorporation of aliphatic esters into polycarbonate

## Table 3.  Aromatic Polycarbonates Derived from Bisphenols

| Monomer | | Polymer | | |
|---|---|---|---|---|
| R | Mp, °C | $T_g$, °C | Melt range, °C | Refs. |
| *When R′ = R″ = H* | | | | |
| —CH₂— | 163[a] | 147 | >300 | 6,39,40 |
| —O— | 161 | 145 | 230–235 | 6,39,40 |
| —S— | 152 | 113 | 220–240 | 6,39,40 |
| O=S=O | 249 | | 200–210 | 6,39 |
| —C(=O)— | 213–215 | | | 41 |
| Cl–C–Cl (—C—) | 214–215[b] | 168 | 230–260 | 42,43 |
| H, CH₃ (—C—) | 122 | 130 | 185–195 | 6,39,40 |
| H, CH₂CH₂CH₃ (—C—) | 129 | 123 | 150–170 | 6,39,40 |
| H, CH(CH₃)₂ (—C—) | 155 | 149 | 170–180 | 6,39,40 |
| H, cyclohexyl (—C—) | 224 | 190 | | 44 |
| H, phenyl (—C—) | 161 | 121 | 200–215 | 44 |
| CH₃, CH₃ (—C—) | 157[c] | 149 | 215–230 | 6,39,40 |
| CH₃, CH₂CH₂CH₃ (—C—) | 149 | 137 | 200–220 | 6,39,40 |
| CH₃, phenyl (—C—) | 188 | 176 | 210–230 | 6,39,40 |
| phenyl, phenyl (—C—) | 295 | 220 | 240 | 6,39,40 |
| CF₃, CF₃ (—C—) | 161[d] | 149 | | 45 |
| CN, CH₃ (—C—) | 170 | 200 | | 46 |

**Table 3. (*Continued*)**

| Monomer | | Polymer | | |
| --- | --- | --- | --- | --- |
| Monomer, R | Mp, °C | $T_g$, °C | Melt range, °C | Refs. |
| NC–C–CN (structure) | 147 | 186 | | 46 |
| (cyclohexylidene structure) | 190$^e$ | 179 | 250–260 | 6,39,40 |
| (cyclobutylidene structure) | 157 | 167 | 240–250 | 6,39,40 |
| (fluorenylidene structure) | 224 | 275 | 390 | 47,48 |
| H₃C (trimethylcyclohexyl, CH₃, CH₃) structure | 239 | 239 | | 49 |
| CH₃–C(CH₃)–phenyl–C(CH₃)–CH₃ structure | 304 | 228 | | 50 |
| *When R′ = R″ = CH₃* | | | | |
| C(CH₃)₂ | 165$^f$ | 207 | | 51 |
| *When R′ = R″ = Br* | | | | |
| C(CH₃)₂ | 178–180$^g$ | 265 | | 6,39,40 |
| *Other monomers* | | | | |
| HO– (SBI, CH₃ CH₃ / H₃C CH₃) –OH structure | $h$ | 230 | | 52 |
| HO– (phenyl, CH₃) –OH structure | 128–130$^i$ | 155 | | 12 |

$^a$Crystalline and insoluble.  $^b$Bisphenol C; flame retardant.  $^c$Bisphenol A.  $^d$Flame retardant.
$^e$Bisphenol Z.  $^f$TMBPA; hydrolytically stable.  $^g$TBBPA; flame retardant.  $^h$SBI; low birefringence.
$^i T_m = 289°C$; crystalline.

leads to elastomeric materials, or copolymers with somewhat decreased $T_g$, but dramatically increased melt flow (eq. 14). Some of these block copolymers also have improved low temperature impact strength and higher stress–crack resistance than neat BPA polycarbonate.

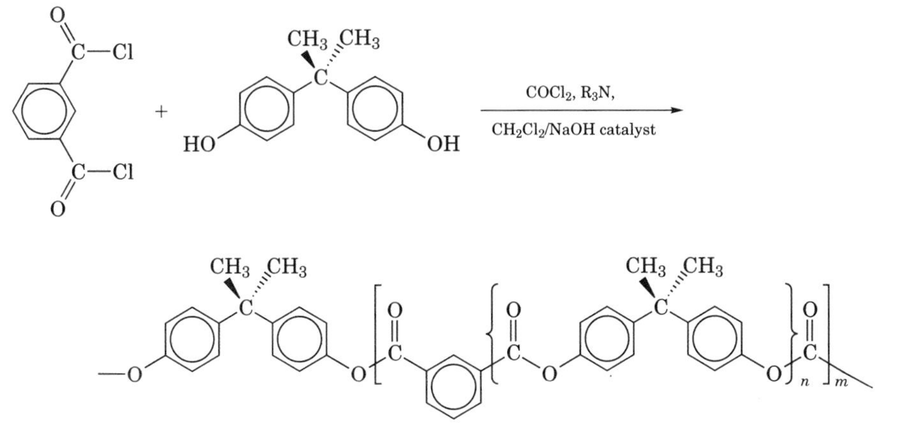

(13)

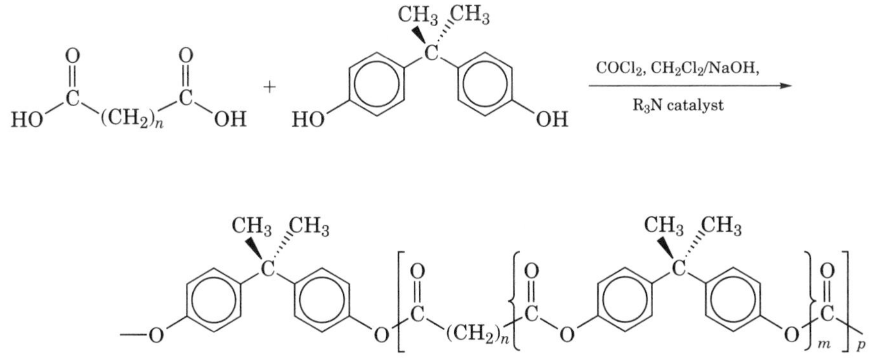

(14)

A variety of methods have been developed for the preparation of poly(dimethylsiloxane-carbonates). Phosgenation of silanol-terminated siloxane oligomers in the presence of BPA, or reaction of BPA sodium salt with chlorosilane-stopped siloxane oligomers, followed by phosgenation with BPA, are two of the described procedures (53). Use of cyclic oligomers for the preparation of these copolymers has also been described (54). These materials display a wide range of properties depending on the block length of the individual polymers and on the weight level of silicone present. Some of the block copolymers having low levels of silicone have outstanding low temperature impact strength; those having higher levels of silicone are thermoplastic elastomers (see ELASTOMERS, SYNTHETIC)

In 1991 Bayer reported the preparation of a family of polycarbonates (Apec) using a monomer based on the condensation of a hydrogenated isophorone with phenol (49). Because of the bulky structure of the linking unit, the polycarbonate retains the low temperature loss modulus of BPA, which affords low temperature impact strength, yet has a high glass-transition temperature owing to rigidity derived from hindered rotational freedom. The homopolymer has a $T_g$ of 239°C, and copolymers with BPA provide intermediate glass-transition temperatures.

**Blends.** The concept of blending two or more commercially available materials to create a new material having properties different from either starting material has generated a great deal of interest. One of the fastest growing market segments within the polycarbonate industry is this arena of blends. Polycarbonate blends are used to tailor performance and price to specific markets. Despite its strengths, polycarbonate suffers from shortcomings with regard to flow characteristics and solvent resistance. The two principal blends are produced to improve those characteristics. Polycarbonate–ABS blends (GE, Cycoloy; Bayer, Bayblend; Dow, PULSE; and Monsanto, Triax 2000) have improved flow and lower cost, and polycarbonate–polyesters which use either PBT or PET (GE, Xenoy; Bayer, Makroblend; Dow, SABRE; and Hoechst Celanese, Vandar) utilize semicrystalline polymers to improve chemical resistance, especially for automotive applications.

Fundamental studies of blends of polycarbonate with ABS indicate that the presence of ABS greatly decreases the melt viscosity in the blend, enhancing processibility. Even as little as 20% ABS provides a decrease in melt viscosity of a factor of 4–5 (55). A synergistic improvement of the notched impact strength at low temperature is also seen for polycarbonate–ABS blends (Fig. 3). About 35,000 t of these blends are produced annually, and significant growth in the market has been seen, with a yearly growth rate of 12–13%. More than half of the blends produced are used in the automotive market, mostly on instrument panels. The other large use for polycarbonate blends is in office and business machines such as computer housings.

Polycarbonate–polyester blends were introduced in 1980, and have steadily increased sales to a volume of about 70,000 t. This blend, which is used on exterior parts for the automotive industry, accounting for 85% of the volume, combines the toughness and impact strength of polycarbonate with the crystallinity and inherent solvent resistance of PBT, PET, and other polyesters. Although not quite miscible, polycarbonate and PBT form a fine-grained blend, which upon analysis shows the glass-transition temperature of the polycarbonate and the melting point of the polyester.

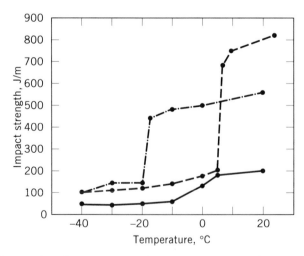

**Fig. 3.** Dependence of Izod impact strength on temperature for (—––—) polycarbonate, (——) ABS, and (—··—) polycarbonate–ABS blends. To convert J to cal, divide by 4.184.

Other blends of polycarbonate have limited markets so far. The most significant blends are with polyurethanes, polyetherimides, acrylate–styrene–acrylonitrile (ASA), acrylonitrile–ethylene–styrene (AES), and styrene–maleic anhydride (SMA).

## BIBLIOGRAPHY

"Polycarbonates" in *ECT* 2nd ed., Vol. 16, pp. 106–116, by H. Schnell, Farbinfabriken Bayer AG; in *ECT* 3rd ed., Vol. 18, pp. 479–494, by D. W. Fox, General Electric Co.

1. A. Einhorn, *Liebigs Ann. Chem.* **300**, 135 (1898).
2. C. A. Bischoff and A. von Hedenstroem, *Ber.* **35**, 3431 (1902).
3. W. H. Carothers and F. J. Van Natta, *J. Am. Chem. Soc.* **52**, 314 (1930); J. W. Hill and W. H. Carothers, *J. Am. Chem. Soc.* **55**, 5031 (1933); E. W. Spanagel and W. H. Carothers, *J. Am. Chem. Soc.* **57**, 929 (1935).
4. U.S. Pat. 2,210,817 (1940), W. R. Peterson (to E. I. duPont de Nemours & Co., Inc.).
5. R. L. Wakeman, *The Chemistry of Commercial Plastics*, Reinhold Publishing Corp., New York, 1947, p. 518.
6. H. Schnell, *Ang. Chem.* **68**, 633 (1956).
7. Belg. Pat. 532,543 (1954); H. Schnell, L. Bottenbruch, and H. Krimm (to Bayer AG).
8. U.S. Pat. 3,028,365 (1962), H. Schnell, L. Bottenbruch, and G. Grimm (to Bayer AG).
9. U.S. Pat. 3,153,008 (1964), D. W. Fox (to General Electric).
10. H. R. Kricheldorf and D. Lübbers, *Makromol. Chem. Rapid Commun.* **10**, 383 (1989); H. R. Kricheldorf and D. Lübbers, *Macromolecules* **23**, 2656 (1990).
11. H. Schnell, *Angew. Chem.* **68**, 633 (1956).
12. D. J. Brunelle, H. O. Krabbenhoft, and D. K. Bonauto, *Polym. Prep.* **34**(1), 73 (1993); D. J. Brunelle, H. O. Krabbenhoft, and D. K. Bonauto, *Makromolecular Chemie* **194**, 1249 (1993).
13. R. Myers and J. Longs, eds., *Treatise on Coatings*, Marcel Dekker, New York, 1972, p. 283.
14. C. Bailly and co-workers, *Polymer* **27**(9), 1410 (1986).
15. D. J. Brunelle and T. G. Shannon, *Macromolecules* **24**, 3035 (1991).
16. G. Kampf, *Kolloid* Z. **172**, 50 (1960).
17. Ch. Bailly, R. Legras, and J. P. Mercier, *Polym. Prep.* **26**(2), 170 (1985).
18. A. D. Williams and P. J. Flory, *J. Polym. Sci., Polym. Phys. Ed.* **6**, 1945 (1968).
19. B. Erman, D. C. Marvin, P. A. Irvine, and P. J. Flory, *Macromolecules* **15**, 664 (1982).
20. D. J. Brunelle and M. A. Garbauskas, *Macromolecules* **26**, 2724 (1993); J. A. King and G. L. Bryant, Jr., *Acta. Cryst.* **C49**, 550 (1993).
21. S. Perez and R. P. Scaringe, *Macromolecules* **20**, 68 (1987); P. M. Henricks, H. R. Luss, and R. P. Scaringe, *Macromolecules* **22**, 2731 (1989).
22. W. J. Dulmage and co-workers, *J. Appl. Phys.* **49**, 5543 (1978).
23. M. Hutnik, A. S. Argon, and U. Suter, *Macromolecules* **24**, 5870 (1991); M. Hutnik, F. T. Gentile, P. J. Ludovice, U. W. Suter, and A. S. Argon, *Macromolecules* **24**, 5962 (1991); M. Hutnik, A. S. Argon, and U. W. Suter, *Macromolecules* **24**, 5956 (1991); M. A. Mora, M. Rubio, and C. A. Cruz-Ramos, *J. Polym. Sci., Part B: Polym. Phys.* **24**, 239 (1986).
24. F. J. Hoerth, K. J. Kuhn, J. Mertes, and G. P. Hellmann, *Polymer* **33**, 1223 (1992); L. B. Liu, A. F. Yee, and, D. W. Gidley, *J. Polym. Sci., Part B: Polym. Phys.* **30**, 221 (1992); L. B. Liu, A. F. Yee, J. C. Lewis, X. Li, and D. W. Gidley, *Mater. Res. Soc. Symp. Proc.* **61** (1991).
25. B. C. Laskowski, D. Y. Yoon, D. McLean, and R. L. Jaffe, *Macromolecules* **21**, 1629 (1988).

26. P. M. Henrichs and H. R. Luss, *Macromolecules* **21**, 860 (1988); P. M. Henrichs and V. A. Nicely, *Macromolecules* **23**, 3193 (1990).
27. P. Schmidt, J. Dybal, E. Turska, and A. Kulczycki, *Polymer* **32**, 1862 (1991).
28. K. R. Stewart, *Polym. Prep.* **30**(2), 140 (1989).
29. Ger. Pat. 2,842,005 (1980), D. Freitag, W. Nouvertne, K. Burkhardt, and F. Kleiner.
30. E. Aquino, W. J. Brittain, and D. J. Brunelle, *Polym. Int.* **33**, 161 (1994); E. Aquino, W. J. Brittain, and D. J. Brunelle, *J. Polym. Sci., Part A: Polym. Chem.* **32**, 741 (1994).
31. Eur. Pat. Appl. 228,672 (1986); M. Janatpour and S. J. Shafer (to General Electric).
32. J. E. Hallgren and R. O. Matthews, *J. Organomet. Chem.* **175**, 135 (1979); J. E. Hallgren, G. M. Lucas, and R. O. Matthews, *J. Organomet. Chem.* **204**, 135 (1981).
33. U.S. Pat. 4,452,968 (1983), D. A. Bolon and J. E. Hallgren (to General Electric).
34. H. Schnell and L. Bottenbruch, *Makromol. Chem.* **57**, 1 (1962).
35. D. J. Brunelle, E. P. Boden, and T. G. Shannon, *J. Am. Chem. Soc.* **112**, 2399 (1990).
36. D. J. Brunelle, *Makromol. Chem., Macromol. Symp.* **64**, 65–74 (1992).
37. T. L. Evans, C. B. Berman, J. C. Carpenter, D. Y. Choi, and D. A. Williams, *Polym. Preprints* **30**, (2), 573 (1989).
38. C. S. Read, *CEH Marketing Research Report*, SRI International, 1993. Latest production figures available are for 1992.
39. H. Schnell, *Polym. Rev.* **9** (1964).
40. H. Schnell, *Ind. Eng. Chem.* **51**, 157 (1959).
41. Ger. Pat. 2,746,141 (1979), S. Aldelmann, D. Margotte, and H. J. Rosenkranz (to Bayer AG).
42. Z. Sobiczewski and Z. Wielgosz, *Plaste. Kautsch* **15** 176 (1968).
43. A. Factor and C. M. Orlando, *J. Polym. Sci. Polym. Chem. Ed.* **18**, 579 (1980); M. R. MacLaury, A. D. Chan, A. M. Colley, A. Saracino, and A. M. Toothaker, *J. Polym. Sci. Polym. Chem. Ed.* **18**, 2501 (1980).
44. W. J. Jackson, Jr. and J. R. Caldwell, *Ind. Eng. Chem. Prod. Res. Dev.* **2**, 246 (1963).
45. WO Pat. 82/02402 (1982), V. Mark and C. V. Hedges (to General Electric).
46. E. G. Banucci, *J. Polym. Sci. Polym. Chem. Ed.* **11**, 2947 (1973).
47. P. W. Morgan, *Macromolecules* **3**, 536 (1970).
48. R. P. Kambour, J. E. Corn, S. Miller, and G. E. Niznik, *J. Appl. Polym. Sci.* **20**, 3275 (1976); R. P. Kambour, W. V. Ligon, and R. R. Russell, *J. Polym. Sci. Polym. Lett. Ed.* **16**, 327 (1978).
49. D. Freitag, G. Fengler, and L. Morbitzer, *Angew Chem. Int. Ed. Engl.* **30**, 1598 (1991).
50. Ger. Pat. 2,746,141 (1975), D. Neuray, E. Tresper, and D. Freitag (to Bayer AG).
51. V. Serini, D. Freitag, and H. Vernaleken, *Angew. Makromol. Chem.* **55**, 175 (1976).
52. K. C. Stueben, *J. Polym. Sci., Part A: Gen. Pap.* **3**, 3209 (1965); Eur. Pat. Appl. 287887 (1988), G. R. Faler and J. C. Lynch (to General Electric).
53. U.S. Pat. 3,189,662 (1961), H. A. Vaughn (to General Electric); D. W. Dwight, J. E. McGrath, A. R. Beck, and J. S. Riffle, *Polym. Prep.* **29**, 702 (1979).
54. T. L. Evans and J. C. Carpenter, *Makromol. Chem., Macromol. Symp.* **42/43**, 177 (1991).
55. J-S. Wu, S-C. Shen, and F-C. Chang, *J. Appl. Polym. Sci.* **50**, 1379 (1993).

DANIEL J. BRUNELLE
General Electric

# POLYCHLORINATED BIPHENYLS. See CHLOROCARBONS AND
CHLOROHYDROCARBONS, TOXIC AROMATICS.

**POLYCHLOROPRENE.**    See Elastomers, synthetic-polychloroprene.


**POLYCHLOROTRIFLUOROETHYLENE**    See Fluorine compounds, organic.


**POLYELECTROLYTES.**    See Ionomers.


**POLYENE ANTIBIOTICS.**    See Antiparasitic agents, antimycotics.


**POLYESTER FIBERS.**    See Fibers, polyester.


# POLYESTERS, THERMOPLASTIC

Taken collectively, thermoplastic polyesters constitute a significant item of commerce, entering into almost every imaginable end use: fibers, textiles, industrial yarns, tire cord (qv), ropes, molded items, consumer goods, medical accessories, automotive and electronic items, photographic film, magnetic tape filmbase for audio and video recording, packaging materials, bottles, containers, etc. The broad span of their applications illustrates the wide utility of such materials. This article considers thermoplastic polyesters as materials for injection molding and similar applications and is confined to semicrystalline thermoplastic polyesters. Thermoset, unsaturated polyesters, extruded or melt-spun polyester films and fibers, and thermoformed sheets are covered elsewhere (see Fibers, polyester; Films and sheeting; Polyesters, unsaturated). Blow-molded polyester containers are considered. This is a rapidly growing business which already represents a huge market for thermoplastic polyesters, almost exclusively poly(ethylene terephthalate).

The principal polymers to be described are poly(butylene terephthalate) [26062-94-2] (PBT); poly(ethylene terephthalate [25038-59-9] (PET); poly(cyclohexanedimethylene terephthalate) [24936-69-4] (CHDMT), and mention will be made of poly(ethylene naphthalene-2,6-dicarboxylate) [24968-11-4] (PEN). This article also covers the increasingly commercially important high performance liquid crystalline all-aromatic polyesters, eg, VECTRA [70679-92-4], [82538-13-4] and XYDAR [31072-56-7].

Noncrystalline aromatic polycarbonates (qv) and polyesters (polyarylates) and alloys of polycarbonate with other thermoplastics are considered elsewhere, as are aliphatic polyesters derived from natural or biological sources such as poly(3-hydroxybutyrate), poly(glycolide), or poly(lactide); these, too, are separately covered (see Polymers, environmentally degradable; Sutures). Thermoplastic elastomers derived from poly(ester–ether) block copolymers such as PBT/PTMEG-T [82662-36-0] and known by commercial names such as HYTREL and RITEFLEX are included here in the section on poly(butylene terephthalate). Specific polymers are dealt with largely in order of volume, which puts PET first by virtue of its enormous market volume in bottle resin.

## History of Commercial Polyesters

Polyesters are linear polymeric molecules containing in-chain ester groups, formally derived by condensation of a diacid with a diol. The topic has been reviewed in depth (1). During World War I, such polyesters as cellulose acetate and glycerol–phthalic anhydride (glyptal) resins were developed as nonflammable aircraft dopes, varnishes, and coatings (qv), but not until the classic studies of W. H. Carothers at Du Pont were linear polyesters examined in a systematic fashion (2). Although he studied a wide range of aliphatic polyesters, Carothers, in his pioneering work during the 1930s, did not pursue polyesters derived from aromatic diacids and alkylene diols. The aliphatic polymers he studied were low melting and mostly soluble in common organic solvents and thus had little utility as practical textile fibers; thus Carothers turned his attention to polyamides. The first successful synthesis of satisfactory high molecular weight poly(ethylene terephthalate) was made in the United Kingdom during the early days of World War II by J. Rex Whinfield and W. Dickson in 1942 while working at the Calico Printers' Association (3,4). The material was quickly recognized as a basis for a valuable melt-spinnable synthetic fiber but no serious commercialization could be carried out until the end of the World War II. In the United Kingdom, fiber melt-spun from the new polyester was manufactured by Imperial Chemical Industries Ltd. under the trade name TERYLENE; Du Pont introduced it to the United States in 1953 under the trade name DACRON.

Work had gone on in both the United States and Europe, notably at Du Pont and ICI Ltd. in the United Kingdom, on exploring the whole series of alkylene terephthalate polymers in connection with new synthetic fibers. Poly(1,4-butylene terephthalate) (PBT) was investigated in detail, as it had very attractive fiber properties. Notably, it was white and resisted photooxidative yellowing much better than nylon, and it accepted disperse dyes more easily and had better resilience and elastic recovery properties than PET. These characteristics made it attractive for such end uses as women's wear, hosiery, and carpets. Despite this, PBT fiber did not meet the twin criteria of pleat retention and crease resistance in apparel fabrics, which were commercially important at that time. Later it was realized that PBT polymer was both highly crystalline and had a high rate of crystallization, two qualities well suited to an injection molding resin. Unlike nylon, it had a low moisture uptake and was much more dimensionally stable to changes in atmospheric humidity. Molding-grade resins were introduced by Celanese Corporation in 1970.

Some time earlier, Eastman-Kodak has been working on a novel polyester as an entry into the important polyester fiber market and had devised a new alicyclic diol, 1,4-cyclohexanedimethanol [105-08-5], effectively made by exhaustive hydrogenation of dimethyl terephthalate. Reaction of the new diol with dimethyl terephthalate gave a crystalline polyester with a higher melting point than PET and it was introduced in the United States in 1954 as a new polyester fiber under the trade name KODEL (5). Much later the same polyester, now called PCT, and a cyclohexanedimethanol–terephthalate/isophthalate copolymer were introduced as molding resins and thermoforming materials (6). More recently still, copolymers of PET with CHDM units have been introduced for blow molded bottle resins (7).

Notwithstanding its very good physical properties and chemical stability, PET itself was not outstandingly successful as an injection molding resin due to its low rate of crystallization in a cold mold. The mold has to be heated to 130–140°C, well above the PET glass–rubber-transition temperature, $T_g$, to obtain adequate crystallization rates. However, satisfactory moldings were still not obtained due to the uncontrolled crystal morphology. During the mid-1960s and in later years, fast-crystallizing grades of PET were developed which gave uniform and controlled morphology because of the presence of specific additives acting as nucleating agents (8–10). Akzo and Du Pont were early entrants in the field with the ARNITE and RYNITE range of PET molding resins; other manufacturers followed and PET molding resins are now widely used, particularly in the auto industry. However, the biggest single market for moldable PET in the 1990s is in blow-molded bottles, which exceeds every other single end use for PET polymer except polyester fibers. In 1990 the annual world production of PET fibers was about 9 million metric tons, whereas the annual production of PET bottle resin in 1990 was 1.2 million metric tons and growing rapidly.

For many years polymer chemists were aware of the desirable properties of films and fibers (Fiber-Q) made from poly(ethylene 2,6-naphthalenedicarboxylate), usually abbreviated PEN. This was first synthesized by ICI workers in 1948 (11). The polyester was not a commercial possibility until an economical route to naphthalene-2,6-dicarboxylic acid [1141-38-4] (NDA) was in existence. However, two firms have brought commercial plants on-stream for the production of NDA and its dimethyl ester (DMNDA) ready for 1996. These are Amoco in the United States (Decatur, Alabama), and Mitsubishi Gas Chemical (Japan) (12).

PEN film for audio- and videotape and various electronic applications and blow molded PEN containers for hot-fill applications are already being marketed in Japan. NDA is unlikely to ever become as inexpensive as terephthalic acid but novel NDA-based polyesters will become available if a market need exists. One example could be the experimental polyester PBN (Celanese Corp.); this is the NDA analogue of PBT, poly(1,4-butylene naphthalene-2,6-dicarboxylate) [28779-82-0]. It has a high rate of crystallization, faster even than that of PBT, and its combination of physical properties is well-suited for injection molding.

Liquid crystalline thermotropic polyesters were discovered independently in several places during the late 1960s and early 1970s. Notably work at Carborundum Corporation (13–15), Eastman (16,17), and Du Pont (18–21) led to the crucial discovery that certain aromatic copolyesters derived either from aromatic hydroxyacids or by reaction of a diacid with an aromatic diol, could exist in a mesophase or liquid crystal state well below their thermal decomposition temperature. In fact, work on all-aromatic high melting polyesters was carried out at ICI Fibres (Harrogate, England) during 1962–1963; thermotropic behavior (birefringent melt) was actually observed in 1963, but at this early date its true significance was not appreciated (22). In the mid-1970s, during work on a melt-spun high modulus all-aromatic polyester fiber at Celanese Corporation, the unique properties of all-aromatic polyesters derived from 6-hydroxy-2-naphthoic acid [16712-64-4] (HNA) were discovered (23). These materials were readily processible and liquid crystalline, yet had moderate melting points, ie, a solid-nematic mesophase transition, in the region 240–300°C. The copolyester of HNA and 4-hydroxybenzoic acid [99-96-7] (HBA) became the basis of the VECTRA series of liquid crystal polymers (LCP). Meanwhile, the original Carborundum work had led via licensing arrangements to commercialization of other high melting thermotropic polyesters, firstly by Sumitomo (Japan) and later by Dart Corporation (now Amoco) in the United States under the trade name XYDAR. Du Pont has entered the field with the ZENITE range of LCPs based on naphthalene-2,6-dicarboxylic acid.

Liquid crystal polymers have a unique combination of high thermal stability, outstanding chemical and solvent resistance, very good melt-flow properties, and low shrinkage, all ideally suited to the precision injection molding of complex shapes such as those widely used in electronic components. During the early 1990s the business has seen rapid growth, both in the United States and the Far East. Because the commercial success of LCP molded plastics is well established, commercial development of fibers and monofils (eg, VECTRAN) from the same materials is now underway in Japan.

## Manufacture of Raw Materials and Monomers

**Terephthalic Acid and Dimethyl Terephthalate.** PET and PBT can both be made from terephthalic acid or its dimethyl ester (see PHTHALIC ACIDS AND OTHER BENZENEPOLYCARBOXYLIC ACIDS). Terephthalic acid [100-21-0] (TA) is made by the air oxidation of p-xylene [106-42-3] in acetic acid under moderate pressure in the presence of catalysts such as divalent cobalt and manganese bromides (24). p-Xylene is the highest melting of the three isomeric dimethylbenzenes and is separated by fractional crystallization from the C-8 aromatic fraction (including ethylbenzene) during petroleum refining (25). Alternatively, it may be separated by selective adsorption on a zeolite bed combined with an isomerization process (26). For PET fiber production, very pure TA is required and in the early days, the oxidation of p-xylene was achieved with 50% aqueous nitric acid under pressure, a process which left many undesirable by-products in the crude TA. Due to the insolubility of TA, it was not itself easily purified and accordingly was converted to its dimethyl ester, dimethyl terephthalate [120-61-6] (DMT), and the DMT was purified by redistillation under reduced pressure and finally recrystallized. PET was made by reaction of DMT with excess ethylene glycol in the presence of catalysts to promote ester interchange and polymerization. During the late 1960s and early 1970s, pure TA became available in large quantities (27) due to improvements in the xylene oxidation process and the purification of crude TA. Notably, impurities such as p-toluic acid and 4-carboxybenzaldehyde (4-CBA) were eliminated by recrystallization from aqueous solution under pressure with concomitant hydrogenation over a fixed catalyst bed. Under these conditions, p-toluic acid is easily soluble in water and any 4-CBA present is hydrogenated to toluic acid, thus removing it (28). The availability of pure TA caused significant changes in the production of PET for fibers as direct esterification (DE) processes superseded the former ester-interchange (EI) process based on DMT. Interestingly, DMT has of late made a partial comeback due to recycling waste PET (29,30). Distillation of DMT derived from the methanolysis–glycolysis of PET waste is being carried out on a significant scale.

**Naphthalene-2,6-Dicarboxylic Acid.** NDA is more complicated to synthesize on an industrial scale. The preferred starting material is 2,6-dimethylnaphthalene [581-42-0]. This is synthesized by a three-step reaction (31–33).

Toluene reacts with carbon monoxide and butene-1 under pressure in the presence of hydrogen fluoride and boron trifluoride to give 4-methyl-*sec*-butyrophenone which is reduced to the carbinol and dehydrated to the olefin. The latter is cyclized and dehydrogenated over a special alumina-supported catalyst to give pure 2,6-dimethylnaphthalene, free from isomers. It is also possible to isomerize various dimethylnaphthalenes to the 2,6-isomer in the presence of hydrogen fluoride (34). The 2,6-dimethylnaphthalene is air oxidized to NDA (35) under similar conditions to those used for terephthalic acid. However, there are additional difficulties with NDA oxidation. Undesirable oxidation by-products are trimellitic acid [528-44-9] and 6-formyl-2-naphthoic acid.

Another problem arises from brominated aromatic species derived from inorganic bromides used as oxidation cocatalysts. As a result, the crude NDA is converted to its dimethyl ester, DMNDA [840-65-3], and solvent recrystallized to give a high purity diester (36–38). A process for purifying NDA directly by hydrogenation (pure TA process) has also been described (39).

**Diol Components.** Ethylene glycol (ethane 1,2-diol) is made from ethylene by direct air oxidation to ethylene oxide and ring opening with water to give 1,2-diol (40) (see GLYCOLS). Butane-1,4-diol is still made by the Reppe process; acetylene reacts with formaldehyde in the presence of catalyst to give 2-butyne-1,4-diol which is hydrogenated to butanediol (see ACETYLENE-DERIVED CHEMICALS). The ethynylation step depends on a special cuprous acetylide–bismuth salt catalyst which minimizes side reactions (41). The hydrogenation step is best done in two stages over special catalysts (42). An alternative butanediol route also starts from butadiene (qv). In this process, due to Mitsubishi Chemical Industries, butadiene reacts with acetic acid and oxygen in the presence of a palladium catalyst to give 1,4-diacetoxy-but-2-ene (43) and the latter is hydrogenated over a special catalyst (44) and finally hydrolyzed to 1,4-butanediol. The other butanediol process relies on the hydroformylation of allyl alcohol (qv) (made from propylene oxide) over a rhodium catalyst (45) to give 4-hydroxybutyraldehyde. This is reduced to butanediol by again using a two-stage hydrogenation route (46) to minimize side reactions. The ethynylation route is still the most important, but Arco has commercialized a process based on hydroformylation. Butanediol freezes at 20°C, hence its shipping and handling facilities need to be heat-traced.

Cyclohexanedimethanol (47) starts from dimethyl terephthalate. The aromatic ring is hydrogenated in methanol to dimethyl cyclohexane-1,4-dicarboxylate (hexahydro-DMT) and the ester groups are further reduced under high pressure to the bis primary alcohol, usually as a 68/32 mixture of trans and cis forms. The mixed diol is a sticky low melting solid, mp 45–50°C. It is of interest that waste PET polymer may be directly hydrogenated in methanol to cyclohexanedimethanol (48).

Another novel polyester resin has been announced (49) by Shell Chemical Company which plans to introduce it during 1996 for carpet fibers. This polymer is poly(trimethylene terephthalate) (PPT), sometimes called 3GT [26546-03-2], which is derived from terephthalic acid and propane-1,3-diol [504-63-2]. This is another polyester which has been known for many years but was not considered commercially feasible until recently. The change is a new process (50) for the economic production of propane-1,3-diol by the hydroformylation of ethylene oxide to 3-hydroxypropanal using a special cobalt–ruthenium catalyst and its reduction to propandiol. The commercial target of the new polyester is carpet fiber, which exploits its exceptional resilience (51). Little has been published on the properties of PPT as a molding resin but it appears to crystallize readily and has a similar melting point to PBT. Its crystal structure has been determined by two groups of workers (52,53). One source (54) suggests that the new process may make propane-1,3-diol significantly cheaper than butane-1,4-diol and only a little more expensive than ethylene glycol. If this proves to be true, PPT may be re-evaluated in nonfiber end uses.

**Liquid Crystal Polyesters.** These high performance, high added-value products are derived from all-aromatic precursors and the raw materials are inevitably more expensive. 4-Hydroxybenzoic acid (HBA) and 6-hydroxy-2-naphthoic acid (HNA) are both made by the Kolbe-Schmitt carboxylation reaction (55). In this reaction the solid potassium phenoxide is heated under pressure with carbon dioxide. A principal process improvement uses an inert hydrocarbon oil as a heat-transfer fluid (56). Using phenol and 2-naphthol, respectively, HBA and HNA are made, although reaction conditions differ. 2-Naphthol can give both the 2,3- and 2,6-isomers depending on the alkali metal and the reaction conditions (57). Other LCP comonomers are 4,4'-dihydroxybiphenyl [92-88-6], hydroquinone [123-31-9] (HQ), terephthalic acid, NDA, and 4-acetamidophenol [103-90-2]. The latter is used in minor amounts in certain liquid crystal polyesteramides, eg, VECTRA B [82538-13-4] (58). Hydroquinone is manufactured on a large scale, much of it for photographic chemicals and the synthesis of antioxidants (qv). There are several routes to hydroquinone (see HYDROQUINONE, RESORCINOL, AND CATECHOL). One is the alkylation of benzene or cumene with propene or 2-propanol in the liquid phase with a zeolite catalyst to a mixture of 1,3- and 1,4-diisopropylbenzenes (59). These are separated and air oxidized to the hydroperoxides and decomposed into acetone and either HQ or resorcinol. The acetone can be reduced to 2-propanol and recycled via the alkylation stage (60). The latest process for HQ uses the direct oxidation of phenol with hydrogen peroxide and a strongly acid catalyst, such as trifluoromethanesulfonic acid, to a mixture of HQ and catechol (61). Another process uses a special titanium zeolite with hydrogen peroxide to achieve the same result (62).

An important intermediate for liquid crystal polyesters, particularly XY-DAR, is 4,4'-biphenol [92-88-6] or 4,4'-dihydroxybiphenyl. One route starts from 2,6-di-t-butylphenol [128-39-2] which is oxidatively coupled (63) to 3,3',5,5'-tetra-t-butyl-4,4'-biphenol [128-38-1]. This intermediate is dealkylated at high temperatures giving 4,4'-biphenol and isobutene. The isobutene is recycled to make more di-t-butylphenol (64). An alternative route relies on sulfonation of biphenyl to the 4,4'-disulfonic acid and caustic fusion of the alkali metal sulfonate to biphenol. 4-Aminophenol and 4-acetamidophenol, the analgesic acetaminophen, have been made for many years by two routes. One starts with nitrobenzene (qv) [98-95-3] which is reduced to phenylhydroxylamine and rearranged in sulfuric acid to 4-nitrosophenol. The latter is reduced to 4-aminophenol. The other route starts with 1-chloro-4-nitrobenzene [100-00-5] which is hydrolyzed to 4-nitrophenol and then hydrogenated to 4-aminophenol. Because 4-aminophenol itself is very susceptible to oxidation, it is more convenient to use the N-acetyl derivative which is chemically equivalent during liquid crystal polymerization. An acetaminophen process developed by Hoechst Celanese Corporation starts from phenyl acetate [122-79-2] which undergoes a Fries rearrangement in anhydrous hydrogen fluoride to 4-hydroxyacetophenone [99-93-4] (4-HAP). Beckmann rearrangement of 4-HAP oxime gives acetaminophen directly (65).

## Manufacture of Polyesters and Polymerization Processes

Thermoplastic polyesters are step-growth polymers that need to be made to high molecular weight (12,000–50,000) to be useful (66). The first stage is an esterification (qv) or ester-exchange stage where the diacid or its dimethyl ester reacts with the appropriate diol to give the bis(hydroxyalkyl)ester and some linear oligomers. Water or methanol is evolved at this stage and is removed by fractional distillation, often under reduced pressure at the conclusion of the cycle. For ester interchange, weakly basic metallic salt catalysts are used: the list is extremely long and many recipes are proprietary, but such salts as calcium, zinc, and manganese acetates, tin compounds, and titanium alkoxides have been widely used (67–69). Certain ester-interchange (EI) catalysts have the undesirable effect of promoting thermal degradation at high temperatures (70) encountered during the latter stages of high polymerization. This is particularly true of PET and PEN. To overcome this, the EI catalysts are sequestered at the end of the ester-interchange stage, frequently by adding phosphorus compounds such as triphenyl phosphite, triphenyl phosphate, or polyphosphoric acid in very small amounts (71). Again, such recipes are often proprietary. Titanium and tin compounds act as universal catalysts for both EI and polymerization reactions and are left unchanged. For the manufacture of poly(butylene terephthalate), ester exchange using DMT and a titanium alkoxide catalyst is the route of choice because butanediol readily cyclizes to tetrahydrofuran in the presence of acids (72). Nevertheless, there has been work done on the direct polycondensation of butanediol with terephthalic acid using special reaction catalysts and conditions to minimize THF formation (73–76).

The final polymerization stage is usually done in an autoclave fitted with a powerful mechanical stirrer to handle the viscous melt under high vacuum at a temperature above the melting point of the final polymer. During this critical stage it is important to eliminate oxygen and the process is blanketed with inert gas (nitrogen or argon). During the polycondensation stage, the linear oligomers and the bishydroxyalkyl terephthalate esters undergo a succession of EI reactions, eliminating the diol which is removed under high vacuum, and thus molecular weight increases steadily. A polymerization catalyst is needed: tin and titanium compounds are suitable for both EI and polymerization, but for PET, antimony trioxide is the usual polymerization catalyst (77). It only becomes active at high temperatures and thus can be added at the start of the EI stage along with the other catalysts. There has been a move away from heavy metals (eg, antimony), particularly in Europe, where they are viewed with increasing disfavor on environmental grounds. Even in the United States problems can arise with heavy-metal contaminants (including antimony) in waste glycolysis still-bottoms. These cannot be landfilled for environmental reasons and their safe disposal causes added expense. A less toxic metal is clearly advantageous. However, alternatives are not universally satisfactory. Titanium alkoxides cause unacceptable yellowing of PET, apparently due to reaction with vinyl ends. This does not occur with PBT. Germanium compounds, either as the dioxide, tetraalkoxide, or glycol oxide, are good catalysts, nontoxic, and give very white polymers. However, they are considered too expensive due to the scarcity of the metal. Germanium or germanium–titanium mixtures have been disclosed in a patent relating to PET bottles (78). Antimony trioxide is a robust polymerization catalyst; however, in PET it is susceptible to reduction to metallic antimony, which can cause a grayish blue color in the final polymer.

As the polymer molecular weight increases, so does the melt viscosity, and the power to the stirrer drive is monitored so that an end point can be determined for each batch. When the desired melt viscosity is reached, the molten polymer is discharged through a bottom valve, often under positive pressure of the blanketing gas, and extruded as a ribbon or as thick strands which are water-quenched and chopped continuously by a set of mechanical knives. Large amounts of PET are also made by continuous polymerization processes. PBT is made both by batch and continuous polymerization processes (79–81).

The polymer is then dried thoroughly and stored for subsequent processing. Whenever a polyester is made by melt polycondensation, a small amount of cyclic oligomer is formed which is in equilibrium with the polymer. This can be extracted with solvents from solid polymer but when the extracted polymer is remelted, more oligomer forms until the equilibrium is re-established. The level of such oligomers is about 1.4–1.8% by weight for both PET and PBT, thus it is not possible to completely remove cyclic oligomers from any melt-processed polyesters. In the case of PET, the main oligomer is a cyclic trimer (82,83), whereas in the case of PBT, the oligomers comprise roughly an equal mixture of cyclic dimer and trimer, together with much smaller amounts of higher oligomers (84). The presence of such oligomers usually does no harm, but under certain conditions the oligomers can exude to the surface. This is usually more of a problem with polyester fibers with their very high surface/volume ratio which can interfere with the fiber dyeing process, for example. However, thermoplastic polyester molded articles are not immune to oligomer effects on the polymer surface, which can sometimes give trouble during such processes as electroless metal plating.

It is often necessary to make polymer of much higher molecular weight than would be practicable in the melt, either by reason of excessively high melt viscosity or because degradation reactions would begin to overtake the rate of polymerization and limit molecular weight. For this reason, solid-state polymerization (SSP) is frequently used. Dried polymer chip of moderate molecular weight is heated at a temperature roughly $20°C$ below its softening point, either in a high vacuum or in a stream of hot inert gas in a device which agitates the solid. Typical devices might be a twin-cone rotary vacuum drier or a fluidized-bed unit. There are many types of commercial SSP units available.

One important practical consideration with PET is that the polymer chips must be fully crystallized by careful annealing before the solid-phase polymerization process begins. Usually the polymer has been water quenched before cutting and as such it is largely amorphous. If not precrystallized, the chips may sinter together on attempted solid-phase polymerization. The difficulties caused by several tons of polymer setting to a solid mass are obvious. It is possible by careful annealing to raise the $T_m$ of the PET considerably above the usual figure (85), thus allowing the solid-phase polymerization to take place at higher temperatures and shorter reaction times. Various agitation devices (86) and polymer chip treatments (87) have been described to prevent sticking. An integrated crystallizing and solid-state polymerization process has been described in the patent literature (88).

In the solid-state process, the volatile by-products of the polycondensation reaction (traces of water, methanol, excess diol, etc) escape by vapor diffusion through the solid chip and are rapidly removed from the chip surface instead of being limited by viscous diffusion through a bulk melt. Esters of aliphatic diols and aromatic diacids begin to decompose thermally even in an inert atmosphere at about 250°C (89). In the polymer melt, local mechanical heating by agitating the viscous melt can exceed this and thus cause thermal degradation. SSP eliminates this and the molecular weight rises in a few hours to the desired figure. The polymer chip is also further crystallized. For certain applications it may be necessary to use chain extension to increase the molecular weight of a polyester. This can be done by adding a highly reactive diester, such as diphenyl terephthalate, which couples together hydroxyl-ended chains as terephthalate esters evolving phenol. This reaction goes very readily at high temperatures (90), but is unsuitable for products, eg, food containers, where traces of phenol are objectionable.

During the polymerization reaction, various by-products are formed. In the case of PBT the principal one is tetrahydrofuran (THF) formed by dehydration of butanediol or by internal cyclization of C-4 ester units. This by-product is largely harmless because it is nonreactive under polymerization conditions and, being volatile, is quickly removed. PET by contrast has two troublesome by-products. One is the generation of diethylene glycol (DEG) units in the chain by dehydration of 2-hydroxyethyl ester chain ends to form an ether link. This process cannot be prevented entirely and the concentration of DEG units is minimized by restricting time–temperature combinations that tend to favor DEG formation. DEG content is related to the softening point by the empirical relation $\Delta T_m = (-2.2)\, m\,°C$, where $m$ is the molal concentration of DEG (91). As well as depressing melting point, DEG units have an adverse effect on the crystallinity of the polymer, reducing the strength of both fibers and oriented films and increasing the susceptibility of the polymer to chemical attack and aqueous hydrolysis.

The other significant by-product is acetaldehyde which is produced by thermal degradation of the PET unit. Random oxygen–alkyl scission of ester units leaves a vinyl ester end and a carboxyl-ended chain. The vinyl ester reacts with a polymer end group to form a new polymer link and expel acetaldehyde, the tautomer of vinyl alcohol (92). The vinyl ester end can also thermally polymerize to give chain-branched and cross-linked products and gel particles, and further thermal degradation of these polyvinyl units gives rise to colored polyenes (93,94). Although acetaldehyde is highly volatile, its presence is particularly objectionable in PET resin used for soda bottles. Its presence cannot exceed 3 ppm in the final container if used for potable substances as it imparts an off-taste to popular cola drinks (95). Every time PET is melted during its processing, more acetaldehyde is generated; one reason for bottle resin undergoing a final solid-phase polymerization before stretch blow molding is to remove the last traces of residual acetaldehyde. Only the tiny amount of fresh acetaldehyde produced during the actual bottle molding is present in the final article and this is within specification.

Liquid crystal polyesters are made by a different route. Because they are phenolic esters, they cannot be made by direct ester exchange between a diphenol and a lower dialkyl ester due to unfavorable reactivities. The usual method is the so-called reverse ester exchange or acidolysis reaction (96) where the phenolic hydroxyl groups are acylated with a lower aliphatic acid anhydride, eg, acetic or propionic anhydride, and the acetate or propionate ester is heated with an aromatic dicarboxylic acid, sometimes in the presence of a catalyst. The phenolic polyester forms readily as the volatile lower acid distills from the reaction mixture. Many liquid crystal polymers are derived formally from hydroxyacids (97,98) and their acetates readily undergo self-condensation in the melt, stoichiometric balance being automatically obtained.

The phenolic acetates frequently have convenient melting points and are readily purified as well-crystallized materials. If the melting point is inconveniently high, the propionate is sometimes used instead. It is not necessary to isolate the acetates; reaction of a mixture of diphenol, diacid, and/or hydroxyacid and acetic anhydride generates the acetates *in situ*, which then polymerize evolving acetic acid (99). In favorable cases (100) it is possible to directly polycondense acids and phenols evolving water at temperatures around 300°C. This process works well when catalyzed with compounds of Group IV or V metals. Tin salts are suitable, notably dialkyltin dialkanoates or oxides (101). The process is less suitable for copolymers derived from 4-hydroxybenzoic acid which undergoes decarboxylation above 200°C. This does not affect the polymerization as the phenol formed simply volatilizes away, but the polymer is left with less than the theoretical amount of HBA units. Hydroxynaphthoic acid, by contrast, is not decarboxylated under such conditions and copolymerization with non-HBA comonomers is possible by using direct esterification (102).

## PET Blow-Molded Bottles

One of the largest uses of PET resin and certainly the most dramatic in growth during the 1980s and 1990s is the stretch blow-molded PET soda bottle, the annual consumption of which runs into billions of units in the United States alone with corresponding rapid expansion on a worldwide scale. The advantages of blow-molded thermoplastic soda bottles are self-evident: these bottles are lightweight, shatterproof, and potentially reusable. The early market drive was reduction in weight, which reduced transportation costs at a time (early 1970s) when the United States was suffering from a severe energy shortage. Improved product safety associated with shatterproof plastic bottles was another factor. Recycling (qv) did not become a factor until somewhat later.

The principal technical problem facing any thermoplastic bottle manufacturer is the permeability of the bottle wall to oxygen and carbon dioxide, which affects the shelf-life of the contents. The average 2-L soda bottle maintains an internal pressure of roughly 500 kPa (5 atm) of $CO_2$. To stop the product from going flat, the carbon dioxide pressure must be retained during storage before unsealing for several weeks. Likewise, oxygen from the air must not diffuse in through the bottle walls to cause oxidation of the contents, thus spoiling the flavor of the product. PET is semipermeable, particularly when undrawn, and the first bottles experimentally manufactured used styrene–acrylonitrile (SAN) copolymers, not PET (103). These SAN bottles were banned by the U.S. Food and Drug Administration in September 1977 due to the presence of traces of acrylonitrile monomer in the bottle walls which allegedly could be leached out. Some early PET bottles had coatings of impermeable polymers such as poly(vinylidene chloride) or its copolymer with acrylonitrile applied to the outside. However, the process involved extra steps and the product was not readily recyclable (104). The most desirable bottle material is plain PET with a reduced permeability.

One extremely important development was the invention of the stretch blow-molded PET bottle process in 1973 (105,106). In this process, the polymer bottle wall is subjected to a rapid biaxial drawing stage which greatly increases its molecular orientation. Not only does this increase the mechanical strength of the bottle but it also reduces the permeability of the walls to carbon dioxide diffusing out (107). The stretch blow molding process has been highly successful, and following its introduction the economic growth of the product has seen very high rates. Although the 2- and 3-L soda bottles have satisfactory shelf-lives of several months, this is not always true of the smaller 1- and 0.5-L bottles, due to their less favorable surface/volume ratios. The criterion for shelf-life is the time to 15% loss of carbonation (107).

**The Blow-Molding Process.** Blow-molding thermoplastic hollow articles is a highly specialized process (108). A brief outline of the process is given to demonstrate the reasons for various polymer properties. Of the various processes in use the most popular is the two-stage (reheat) blow-molding process. A bottle preform is molded from PET by a conventional injection molding process. The preform looks like a large test tube with thick walls and the screw-cap threads and neck molded in place. Multiple cavity dies are used to increase productivity. In the second stage, the preforms are heated in a mold cavity to a carefully controlled temperature above the glass–rubber-transition temperature, typically to 90–100°C. The inside of the mold cavity is the size and shape of the finished bottle and the preform is subjected to a biaxial stretching process by a combination of mechanical deformation and air pressure inflation. A hollow metal mandrel passes into the preform and partially elongates it in the axial direction; simultaneously dry air at about 0.35–0.7 MPa (50–100 psi) is applied to blow the walls of the softened preform outward to fill the mold, thus giving radial stretching. The mold opens to allow the bottle to cool. This combined process results in both radial and axial drawing of the bottle walls causing stress-induced crystallization and giving a container with superior strength, clarity, and freedom from environmental stress cracking. The kinematics of stretch blow-molding PET bottles have been investigated (109), and it was found that depending on the mold temperature and inflation pressure, the preform inflated in either of two ways. In one mode the preform ballooned out at the neck and the "bubble" traveled down the mold to fill it. In the other mode, the bubble formed halfway along the preform and expanded in two directions simultaneously to fill the mold. The high rates of deformation involved in this process can produce additional temperature rises of 15°C due to both adiabatic heating and stress-induced crystallization. The morphology of the PET bottle walls in relation to blow molding conditions has also been examined (110).

The usual 2-L bottle preforms weigh about 50 g and the final blown bottle has a wall thickness of about 0.38 mm (0.015 in.) (111). In view of the enormous number of bottles produced annually, the processes are constantly being modified to raise throughput rates and are all highly automated. Several large machinery manufacturers, eg, Cincinnati-Milacron and Groupe Sidel, specialize in building the complex blow molding equipment. Typical cycle times for the actual blow molding are about six seconds, although the latest technology has almost halved this time. The process results in a bottle with a hemispherical base which is clearly not suitable for standing upright on a shelf. For several years PET soda bottles were fitted with separate flat bottomed basecaps, usually molded from high density polyethylene (HDPE) and secured with a hot-melt adhesive. This meant extra cost due to extra material and processing steps and also interfered with recycling (112). The invention by the Continental Group of the so-called petaloid base bottles with usually a five- or six-lobed pattern base was a significant advance. It made possible a one-piece, fully biaxially oriented bottle which could stand up by itself (113).

**Properties of Bottle Resin.** Stretch blow molding is a mechanically severe operation with deformation rates up to 25,000%/min (109,110). The consumer wants a glass-clear bottle and any opacity caused by stress-induced crystallization is highly undesirable. PET bottle resins are usually made to high molecular weight (IV 0.75–0.90 dL/g) so that the preforms can be blow molded without problems. Such an intrinsic viscosity (IV) is too high for melt polymerization and solid-stage polymerization is required. The process has been reviewed (114). The base polymer is made in a continuous melt polymerization plant using either direct esterification of ethylene glycol with terephthalic acid or by ester interchange using purified dimethyl terephthalate (DMT), often recovered from recycled PET bottles. The illustration shows a plant using the terephthalic acid process.

First, the acid and glycol are thoroughly mixed to a paste and catalyst and stabilizers are added. The paste is pumped to the esterifiers where water is driven off. The molten mixture of low polymer and oligomers passes through various polymerization stages: a prepolymerizer; an intermediate polymerizer, which on large plants can sometimes be divided into low and high polymer stages; and finally the melt arrives at the high polymerizer or finishing stage where the IV is about 0.65 dL/g. During the various successive stages the melt grows increasingly viscous and high vacuum is applied at the finisher to complete the reaction. The agitators used in the polymerizers are highly specialized and proprietary designs are built by specialist suppliers. The basic design principle is maximum agitation to disengage volatiles without high local shear rates and absence of dead spots where polymer melt could stagnate and undergo color-forming degradation reactions.

At the exit from the high polymerizer the molten polymer is extruded as multiple strands into a water bath and the quenched strands are continuously diced to small chips about 2.5–3.5 mm across. The chips are then thoroughly dried and passed to the crystallizers, which sometimes have several stages, where they are annealed in the solid state above $T_g$, gradually rising to a temperature close to the point of maximum crystallization rate (~170°C). They are slowly agitated to prevent sintering together as they move to the final solid-state polymerizer. During the crystallization stage the chip density increases from 1.333 to ~1.400 g/cm$^3$. Finally the chips pass into the solid phasing towers, where they descend slowly under gravity in a plug flow mode through a long hot zone under a countercurrent flow of inert gas to sweep away the volatile by-products. The speed of descent is controlled so that optimum time–temperature profiles are maintained for the desired final IV. The scale of operation is impressive; typical melt continuous polymerizers run at 10 t/h which implies a capacity of approximately 70,500 t of polymer per year. A large manufacturer may have several such units and each one provides enough polymer for several billion 2-L soda bottles.

The number-average molecular weight for typical bottle resin is between 24,000 and 31,000 daltons/mol. As has been stated, one of the most objectionable by-products of PET polymerization is acetaldehyde which affects the taste of cola drinks at concentrations as low as 60 ppb. The specification for acetaldehyde in the final product must not exceed 3 $\mu$g of acetaldehyde per liter of headspace. The bulk of the acetaldehyde produced in the polymerization process is removed during the final SSP stage. Because blow molding is carried out well below 200°C, only minute amounts of aldehyde are formed at the last stage. Originally, bottle preforms weighed 60–70 g but this has been reduced to about 50 g. The lighter weight bottles have thinner walls so that during biaxial drawing excessive stress crystallization (opacity) again becomes a problem. As a result some manufacturers (115,116) have introduced copolymers of PET containing minor amounts (2–5 mol %) of such comonomers as isophthalic acid or cyclohexanedimethanol to reduce the polymer melting point by about 4–12°C with a correspondingly lower tendency to crystallize. This solves unwanted opacity problems. Another problem which may arise is bottle pearlescence due to microvoids in the bottle wall. These result from mechanical degradation due to excessive deformation rates during blow molding. Polyesters hydrolyze very rapidly at 280°C in the melt and rigorous polymer drying to a chip moisture content below 50 ppm moisture is necessary before any melt processing, such as the injection molding of bottle preforms. Some customers require colored bottles for their specific products, eg, certain popular noncola sodas are packed in pale green bottles. Melt-dyed polymers using U.S. FDA approved dyes are used to mold the preforms for colored bottles.

**Hot-Fill Applications.** A growing market for blow-molded containers is the so-called hot-fill market. This covers such items as tomato ketchup, pasteurized fruit juices, and salad dressings that are packaged while still hot. The PET blow-molded bottle suffers from its relatively low $T_g$ (70°C) which results in severe bottle distortion if the temperature of the contents approaches $T_g$ (the

maximum fill temperature is ~60–65°C). A brief heat treatment under constrained conditions at 160–220°C, analogous to the heat-setting stage during
the drawing of PET fibers, cures this problem but introduces another process
step (117). Also, it can cause another problem: although the oriented regions in
the bottle remain clear, poorly oriented or unoriented regions crystallize randomly and become opaque. This can be partly solved by the actual design of the
bottle shape, but the best solution to the whole problem is to use a polyester with
a higher $T_g$. PEN blow-molded bottles are being introduced in Japan (118). They
have many advantages despite their considerably greater cost. PEN has better
barrier properties than PET as well as a higher $T_g$ and greater stiffness. It absorbs uv-light and this is exploited in packaging light-sensitive materials such
as multivitamins. DMNDA plants are coming on-stream in several parts of the
world and it is hoped that PEN polymer will become more cost-competitive. Some
manufacturers are examining PET/PEN copolymers. The $T_g$ of PET/PEN copolymers rises linearly with NDA content from the PET value (70°C) to that of pure
PEN (125°C). However, only certain compositions (below 15 mol % NDA or over
85% NDA) are crystalline. In the intermediate amorphous range, the molded
bottles have poor barrier properties. Blends of the two homopolymers are complicated by their immiscibility; high shear mixing gives clear blends but partial
ester exchange and randomization occurs (119). Other approaches described use
NDA polyesters or copolyesters made with other diols such as butane-1,4-diol
(120), cyclohexane-1,4-dimethanol (121), or diethylene glycol (122) and alternative diacids such as *trans*-cyclohexane-1,4-dicarboxylic acid (123). These compositions are said to reduce oxygen and carbon dioxide permeability and to improve
resistance to distortion in boiling water. A considerable effort is in progress both
in the United States and in Europe to solve these problems.

**Economic Aspects.**    The total world market for PET bottle resin is growing at a rapid rate. One survey indicates that the annual growth rate worldwide
exceeded 15% per annum over the years 1990–1995. Another article (124) states
that the global consumption of PET grew at 19% during 1992 to a world total
of 1,720,000 t. Much of this growth is due to the demand for PET bottle resin.
Table 1 shows the world PET consumption by end use in 1992.

The world total devoted to carbonated beverages (qv) is astonishing and
the United States is by far the largest consumer, with Europe next. There
are many interesting variations, notably the strong demand for packaging for
mineral waters in Europe and the large hot-fill market in Japan, compared with
the United States. Rapid growth in demand is projected up to the year 2000
(124). Due to the economic downturn during 1990–1992 in the United States,
there was little investment in new PET capacity. As a result prices were driven
upward during the recovery in 1992–1993 due to shortfalls in polymer supply.
The growth of PET appears assured; its balance of properties and recyclability
are extremely strong factors. It is expected that it will continue to displace
glass, PVC, and polyolefins in food containers. The projected figures for world
regional PET consumption is shown in Table 2. These figures may already be
underestimates; in November 1995, Hoechst Celanese Corporation announced
plans to increase its global PET bottle resin capacity to 1.4 million metric tons
per year by the year 2000. The U.S. market price for PET bottle resin in rail
hopper car lots as of January 1996 was $1.47–$1.54/kg (125).

**Table 1. World PET Consumption by End Use[a] 1992, $10^3$ t**

| Country | Carbonated soft drink containers | Hot-fill containers | Returnable containers | Mineral water containers | Edible-oil containers | Other[b] |
|---|---|---|---|---|---|---|
| Europe | 244 | | 21 | 111 | 22 | 92 |
| United States | 413 | 45 | | 13 | 18 | 213 |
| Canada | 25 | | | 4 | | 3 |
| South America | 14 | 1 | 17 | 4 | 5 | 3 |
| Japan | 35 | 61 | | 4 | 19 | 29 |
| Korea | 28 | 10 | | 2 | 5 | 11 |
| Taiwan | 9 | 1 | | 2 | 3 | 2 |
| China[c] | 40 | | | 13 | 2 | 1 |
| Middle East | 20 | | | 10 | 3 | 9 |
| Africa | 10 | | 2 | 1 | 1 | 2 |
| rest of world | 54 | 1 | 10 | 19 | 21 | 12 |
| Total | 892 | 119 | 50 | 183 | 99 | 377 |

[a]Ref. 124.
[b]Includes other drink containers, nonfood packaging, and all other uses.
[c]Includes Hong Kong.

**Table 2. Projected PET Consumption by Region, 1990–2000, $10^3$ t**

| Region | 1990 | 1991 | 1992 | 1993 | 1994 | 1995 | 1996 | 2000 |
|---|---|---|---|---|---|---|---|---|
| North America | 564 | 644 | 759 | 847 | 950 | 1032 | 1113 | 1398 |
| Europe | 367 | 404 | 490 | 567 | 641 | 710 | 765 | 999 |
| Far East | 205 | 276 | 318 | 383 | 468 | 567 | 634 | 938 |
| rest of world | 89 | 119 | 153 | 188 | 225 | 266 | 304 | 457 |
| Total | 1225 | 1443 | 1720 | 1985 | 2284 | 2575 | 2816 | 3792 |

## PET Molding Resins

It is difficult to establish the exact breakdown of the world demand for PET as a molding resin from the open literature. The total tonnage is small compared with the vast amounts used for fibers and bottle resin. Some estimates are given of total compounded thermoplastic polyester use in the United States, but this includes both PET and PBT (126). The bulk of the U.S. market for PET molding resin is the automotive industry. The heat distortion temperature and $T_m$ of PET are higher than the corresponding values for PBT and its low moisture uptake and dimensional stability with respect to changes in humidity make it superior to nylon. Both PET and PBT engineering resins have good resistance to chemicals, and because they are crystalline do not suffer from the solvent stress cracking problems that plague amorphous materials. Polyesters in general are only attacked by severe chemicals such as powerful acidic or phenolic solvents; hot, strong aqueous alkali; and certain bases, such as hydrazine.

In the unfilled state, PET is not a good molding resin and all commercial grades are filled with either chopped glass strand, 3–4-mm long (1/8 in.);

mineral fillers, usually mica (qv); or a mixture of the two. Some grades also have longer glass fibers. Various proprietary nucleating agents are added; these are often sodium salts of various organic carboxylic acids. Some manufacturers supply fire-retardant polymer grades as well. Such formulations often involve a synergistic mixture of an aryl halide with antimony oxide (see FLAME RETARDANTS). One difficulty with flame-retardant (FR) PET polymer is that recipes which contain antimony trioxide can suffer severe polymer degradation at molding temperatures around 280–290°C.

As has been discussed, antimony trioxide (a polymerization catalyst) can also act as a prodegradant at high temperatures. This is a much less serious problem with FR grades of PBT due to the lower processing temperatures involved (240–250°C). To mitigate the problem with PET, pentavalent antimony as sodium antimonate is used in some FR formulations. The halogenated species is often a ring-brominated polystyrene (127–132). These materials are approved for Underwriters Laboratory (UL) V-0 classification and because they are polymeric, are nonfugitive and less likely to cause problems due to migration and blooming than small-molecule additives such as decabromodiphenyl oxide.

In the United States there has been pressure on manufacturers from the automotive industry, led notably by Ford Motor Company, to use at least 25% by weight of post-consumer recycled (PCR) material in their resins rather than 100% virgin polymer. In the case of PET, this is readily achievable due to the large volume of recovered PET bottle polymer chip available in the United States and Europe. During 1993, as much as 190,000 t, 41% of all the PET soda bottle used in the United States, was recycled (133) (see RECYCLING, PLASTICS). Such material certainly counts as PCR polymer, and some suppliers use 100% recycled polymer in their compounded PET resins. The PCR recovery process and subsequent melt-compounding and reformulation reduce the initial IV of the original PET bottle resin from around 0.75–0.85 dL/g to values more like 0.62–0.65 dL/g, but this is still in the range of virgin melt-polymerized resin, as opposed to SSP resin. In the auto industry the majority of components are pigmented black so differences in feedstock PCR chip color are less important. If pale-colored resin is needed for a specific end use, then virgin polymer may need to be used.

In the early days PET moldings were used in small components, notably electrical connections and covers for switchboxes or fuses. This is important in the 1990s, even more so with the growing complexity of the modern automobile with its numerous on-board electronics. The trend, however, has been to use PET moldings more and more for nonloadbearing structural parts, such as radiator grille supports and headlamp mountings. Glass–mineral-filled PET moldings do not have a smooth enough surface (Class A finish) for exterior body parts that would show. However, they are very suitable for internal structural components. Because a reduction in automobile weight improves gas mileage, this trend will continue. Some of the moldings are dimensionally quite large, weighing 4.5 kg (10 lbs) per shot. Improvements in mold design and better understanding of flow behavior in molds, brought about by increasing use of computer-aided design and flow-simulation programs, have all helped to make these large moldings possible on a routine production basis.

Another factor is the increasing use of robots in molding shops. Large structures are removed hot from the mold by robots and clamped to a jig to cool, ensuring that no distortion or warpage takes place during the cooling cycle (see ROBOTICS, MANUFACTURING).

**Properties of PET Molding Resins.** The full crystal structure of poly(ethylene terephthalate) has been established by x-ray diffraction (134–137). It forms triclinic crystals with one polymer chain per unit cell. The original cell parameters were established in 1954 (134) and numerous groups have re-examined it over the years. Cell parameters are $a = 0.444$ nm, $b = 0.591$ nm, and $c = 1.067$ nm; $\alpha = 100°$, $\beta = 117°$, and $\gamma = 112°$; and density $= 1.52$ g/cm$^2$. One difficulty is determining when crystallinity is fully developed. PET has been annealed at up to 290°C for two years (137).

Thermochemical data depend on the degree of crystallinity in the polymer and a very highly annealed polymer sample can have $T_m = 280°$C, much higher than the usual value of 260–265°C (138). The heat of fusion (139) is about 140 J/g, (33.5 cal/g). The glass–rubber-transition temperature ($T_g$) depends both on the method of measurement and the state of the polymer. A solid chip sample as measured by differential scanning calorimetry (dsc) gives a value around 78°C (140), but a highly oriented crystalline drawn fiber measured by the dynamic loss method gives values as high as 120°C (141). Likewise, the specific gravity of undrawn amorphous PET is 1.33, whereas crystalline drawn fiber has a value of 1.39 (142).

As a step-growth polymer made under equilibrium conditions, PET has a molecular weight distribution very close to the theoretical value of 2.0. The Mark-Houwink equation relates the intrinsic solution viscosity to the molecular weight; $M_v$ is the viscosity average molecular weight and $K$ and $\alpha$ are the Mark-

$$[\eta] = K \cdot M_v^\alpha$$

Houwink constants determined experimentally for individual solvents. The usual solvents historically used for PET are 60/40 wt/wt phenol–tetrachlorethane (P/TCE) and 2-chlorophenol (OCP). Neither solvent system is entirely satisfactory and better results have been obtained using either hexafluoroisopropyl alcohol (HFIP) or a 50/50 by volume mixture of HFIP and pentafluorophenol at 25°C (143). Although expensive, these acidic fluorinated solvents readily dissolve even highly crystalline samples of PET at moderate temperatures, thus avoiding the degradation problems commonly encountered with the older solvents at high temperatures. Degradation under normal conditions is minimal. The Mark-Houwink constants for PET for a range of solvent systems are shown in Table 3.

PET does not crystallize well in the unoriented state even in a hot mold unless nucleating agents and/or plasticizers are added. Commercial PET molding-grade polymers are nearly always filled. Typical compounded polymer properties are shown in Table 4.

**Table 3. Mark-Houwink Constants for PET in Various Solvents at 25°C**

| Solvent | $K$, dL/g | $\alpha$ |
|---|---|---|
| OCP | $6.31 \times 10^{-4}$ | 0.658 |
| P/TCE | $7.44 \times 10^{-4}$ | 0.648 |
| HFIP | $5.20 \times 10^{-4}$ | 0.723 |
| PFP/HFIP | $4.50 \times 10^{-4}$ | 0.705 |

**Table 4. Typical Properties of PET Molding Resins[a]**

| Property | ASTM method | Glass | | | |
|---|---|---|---|---|---|
| | | 30% | 45% | 35% min | |
| specific gravity | D792 | 1.58 | 1.70 | 1.60 | 1.60 |
| tensile strength, MPa[b] | D638 | 166 | 197 | 97 | 103 |
| elongation at break, % | D638 | 2.0 | 2.0 | 2.2 | 2.1 |
| flexural strength at 5%, MPa[b] | D790 | 245 | 310 | 148 | 152 |
| flexural modulus, GPa[c] | D790 | 9.66 | 14.5 | 9.66 | 9.66 |
| notched Izod, J/m[d] | D256 | 80.1 | 107 | 58.7 | 58.7 |
| heat deflection temperature at 1.82 MPa[b], °C | D648 | 224 | 229 | 202 | 216 |
| flammability[e] | UL-94 | HB | HB | HB | HB |
| dielectric strength, V/25 $\mu$m | D149 | | | | |
| 5.2 mm | | 565 | 540 | 500 | 450 |
| 1.6 mm | | 904 | 631 | 550 | 575 |
| 0.8 mm | | 975 | 951 | 810 | 860 |
| volume resistivity at 23°C, 50% rh, $\Omega \cdot$cm | D257 | $3.0 \times 10^{15}$ | | 1.0 | 1.0 |
| dielectric constant, $\epsilon$ | D150 | | | | |
| $10^3$ Hz | | 3.2 | 3.5 | 3.8 | 3.8 |
| $10^5$ Hz | | 3.1 | 3.4 | 3.6 | 3.7 |

[a]Ref. 144.
[b]To convert MPa to psi, multiply by 145.
[c]To convert GPa to psi, multiply by 145,000.
[d]To convert J/m to ft·lbf/in., divide by 53.38.
[e]HB = Brinell hardness.

**Economic Aspects.** The first company to introduce nucleated PET molding resins was Akzo Plastics BV with their ARNITE range. This business has been taken over by DSM NV as part of a portfolio exchange between the two companies. Du Pont introduced their RYNITE fast crystallizing materials in 1978, and other manufacturers followed.

The present North American market is dominated by Du Pont, no doubt by virtue of their early lead; other suppliers are AlliedSignal with PETRA and Hoechst Celanese Corporation with IMPET. Eastman also markets PET injection molding grades under the trade name THERMX. In Europe the situation with PET is confined primarily to two suppliers, DSM NV and Du Pont; in Japan the main suppliers are Teijin, Toray (Du Pont), and Toyobo.

The market prices for PET molding resins in January 1996 in the United States were $2.90–$3.15/kg for 30% glass-filled PET, $3.26–$3.41/kg for 55% glass-filled, and $3.23–$3.45/kg for 30% glass-filled flame-retardant-grade PET (125).

**Safety and Environmental Factors.** PET polymer is safe and poses no threat to animals or humans. PET fibers have been in use since the 1950s and PET has U.S. Food and Drug Administration (FDA) approval for use as a food packaging (qv) material. PET fibers have been used in internal arterial prostheses. The only significant hazard in handling PET resins is dust associated with mineral or glass fillers during chip grinding or compounding operations. Appropriate protective equipment must be worn. All extruders or machinery handling molten polymer should be properly ventilated to remove harmful fumes from the stray decomposition of molten polymer. Molten PET causes serious thermal burns in contact with bare flesh; adequate thermal protection, eg, heat-proof gloves, should always be worn when handling hot polymer.

## PCT Molding Resins

The latest prominent thermoplastic polyester, poly(cyclohexyldimethylene terephthalate) (PCT), was first produced by Eastman Kodak in the 1950s as a polyester fiber (5). PCT was introduced as a molding resin in 1987 in glass-filled and flame-retarded grades with specific end uses (7). Eastman is the sole polymer supplier from the plant in Kingsport, Tennessee, although some polymer is believed to have been supplied to General Electric, who marketed it as part of their VALOX range of thermoplastic polyesters. The targeted end uses appear to be connectors for both the electronic and automotive markets. A development in the electronic market associated with miniaturization is the so-called surface mount technology (SMT) (145). This has already taken over about 50% of the electronic connector market and the trend is expected to continue. There is a much lower level of penetration of this technique in the automotive connector market but its use is also expected to grow. The SMT process uses a solvent vapor-heated reflow soldering process in which the whole electronic component is immersed in hot vapor to melt the solder alloy and allow it to flow. The vapor temperature is usually about 215°C (420°F) for 60/40 tin–lead solder. This puts greater thermal demands on the thermoplastic parts and as a result the blistering temperature of these components has to be above the soldering bath temperature. This is beyond the capability of PBT and most nylons, except nylon-4,6. Interestingly, the heat-deflection temperature (HDT) is not a very precise guide to performance in the blistering test, although obviously the higher the HDT, the better. PCT has a HDT around 260°C (500°F) and its blistering performance under vapor reflow soldering conditions is very good. PCT is significantly less expensive than the ultrahigh performance liquid crystalline polymer (LCP) engineering resins. One specific advantage of PCT is that it has similar flow characteristics (although at higher temperatures) during molding to both PET and PBT which means that extensive mold redesign is not necessary; which makes it attractive to molders.

PCT has low moisture uptake and is not affected by changes in humidity as is nylon. However, as with all high melting polyesters, care must be taken to dry the polymer chip thoroughly before melting. PCT has both a high melting point (285–290°C) and a high melt viscosity. Processing temperatures must be over 300°C, which is high for non-LCP polyesters. The combination of properties can lead to thermal and thermooxidative degradation and severe IV drop during molding, giving brittle parts (146). The presence in the molecule of two tertiary hydrogen atoms both in a $\beta$-position to the ester group is possibly a source of structural weakness here. Several patents have appeared on the use of extra additives, antioxidants, and other polymer stabilizers to overcome this problem

(147–149). Another patent describes copolymers (150) with better crystallization rates and molding properties, and two patents have appeared on the use of terephthalic acid in place of dimethyl terephthalate in the production of the polymer, both in batchwise (151) and continuous polymerization processes (152). The new family of PCT resins was originally called EKTAR, but has been re-designated THERMX since the reorganization of Eastman Chemicals into an independent corporation.

   **Properties of PCT.**   Typical properties of PCT are shown in Tables 5 and 6 (154). Table 5 shows the standard grades and Table 6 the FR grades. This

**Table 5. Properties of Standard Grades (THERMX PCT)**

| Polymer | CG011 | CG007 | CG041 |
|---|---|---|---|
| filler, % glass fiber | 15 | 30 | 40 |
| specific gravity | 1.32 | 1.43 | 1.53 |
| tensile strength, MPa$^a$ | 98 | 130 | 139 |
| elongation at break, % | 3.3 | 2.6 | 2.0 |
| flexural strength, MPa$^a$ | 153 | 192 | 207 |
| flexural modulus, GPa$^b$ | 4.9 | 8.3 | 11.1 |
| notched Izod, J/m$^c$ | 42.7 | 74.7 | 80.1 |
| HDT at 1.82 MPa$^a$, °C | 245 | 263 | 265 |

$^a$To convert MPa to psi, multiply by 145.
$^b$To convert GPa to psi, multiply by 145,000.
$^c$To convert J/m to ft·lbf/in., divide by 53.38.

**Table 6. Properties of Flame-Retardant Grades (THERMX PCT)**

| Polymer | CG912 | CG922 | CG932 | CG942 |
|---|---|---|---|---|
| filler, % glass | 15 | 20 | 30 | 40 |
| specific gravity | 1.50 | 1.54 | 1.61 | 1.69 |
| tensile strength at break, MPa$^a$ | 91 | 110 | 130 | 137 |
| elongation, % | 2.2 | 2.4 | 2.2 | 1.6 |
| flexural strength, MPa$^a$ | 142 | 162 | 184 | 203 |
| flexural modulus, GPa$^b$ | 5.1 | 6.3 | 8.6 | 12.0 |
| notched Izod, J/m$^c$ | 48 | 58.7 | 74.7 | 74.7 |
| HDT at 1.82 MPa$^a$, °C$^d$ | 224 | 239 | 241 | 251 |

$^a$To convert MPa to psi, multiply by 145.
$^b$To convert GPa to psi, multiply by 145,000.
$^c$To convert J/m to ft·lbf/in., divide by 53.38.
$^d$The UL flammability rating is 94V-0.

material is in effect a copolymer of terephthalic acid with two diols, the cis and trans forms of cyclohexanedimethanol. These two isomers do not form a eutectic copolymer with a minimum melting point, but instead the melting point rises monotonically between the two $T_m$ values of the pure isomeric homopolymers (153). The pure cis polymer melts at ~250°C and the pure trans polymer at ~305°C. The $T_g$ likewise rises from about 60°C (cis) to about 90°C (trans). There is little difference in the degrees of crystallinity or crystallization rates over the composition range which means that variations in isomer ratio do not have a marked effect on the physical properties of the polymer. The usual commercial form of PCT has an isomer ratio of approximately 70/30 trans/cis, this being governed by the isomer ratio of the diol (CHDM) produced at the hydrogenation stage. PCT does not crystallize as rapidly as PBT or nylon; its crystallization behavior is similar to that of PET and nucleants, and plasticizers are often added to improve molding performance. THERMX PCT comes in various glass-filled and flame-retardant (FR) grades.

## PBT Molding Resins

Poly(butylene terephthalate) is the oldest of the crystalline thermoplastic polyester molding resins, having been introduced by Celanese Corporation in 1970 under the trade name CELANEX. General Electric Company then brought out their own version (VALOX) and in 1996 there are numerous suppliers including BASF (ULTRADUR), Bayer AG (POCAN), Du Pont (CRASTIN; Du Pont acquired Ciba-Geigy's PBT business in 1993) (155), and DSM NV (ARNITE). In Japan, the principal manufacturers are Polyplastics, Toray, Teijin, and Mitsubishi. PBT has a unique ability to crystallize rapidly (as rapidly as nylon-6,6), even in a cold mold, to give tough, distortion-free moldings without special additives or nucleants. Although the unmodified polymer has very good flow properties and is used in electrical connectors and fiber optical cable buffer tubes, it performs even better if reinforced with inorganic fillers, notably chopped glass fiber, usually about 3.2 mm (1/8 in.) long. Additional mineral fillers are incorporated for special applications where high heat deflection temperature and stiffness are important. These mineral fillers include mica, talc, wollastonite, and even barium sulfate, the last for special applications such as countertops.

The filled grades of PBT are tougher, stiffer, and stronger materials and they also have improved notched Izod impact strength, since unfilled PBT is notch-sensitive (156,157). Even when unfilled, the plastic has good strength, rigidity, and toughness, low creep and minimal moisture absorbance, and does not undergo dimensional changes with fluctuations in humidity. It is characterized by excellent electrical and dielectric properties and high surface finish. As a result, it has found wide acceptance in a variety of end uses where precision molding and a high quality finish are required. Typical applications are in the electrical and electronic markets where it is widely used in such parts as connectors, plugs, switches, typewriter and computer keyboard components, plug-in printed circuit boards, and small electric motor components. PBT is widely used in the automotive industry for electrical and ignition system components such as distributor caps, bobbins, coil-formers, rotors, windshield wiper arms, headlight mountings, and other fittings. In the auto market, under-the-hood components have to maintain their dimensional stability at elevated temperatures as well as stay resistant to various automotive fluids. Other uses for PBT are home appliances, such as food mixers, hair driers, coffee makers, toasters, and camera parts. PBT is also used in industrial machinery, for example, in molded conveyor-belt links. An appreciable quantity is used in polymer alloys and blends with other polymers. PBT is marketed in both standard and flame-retardant grades, the latter being essential to meet Underwriters' Laboratory 94V-0 standards in thin-walled sections. Glass-filled PBT is a mature product; to sustain growth the tendency is for manufacturers and compounders to make more specialty grades with tailored properties to suit specific end-use demands.

**Physical Properties of PBT.**   Unlike PET, the polymer PBT exists in two polymorphs called the $\alpha$- and $\beta$-forms, which have distinctly different crystal structures. The two forms are interconvertible under mechanical stress (158,159). Both crystal forms are triclinic and the crystal parameters are shown in Table 7.

The melting point of PBT is 222–224°C depending on the degree of crystallization and annealing conditions. The heat of fusion is about 140 J/g (161) and the $T_g$ is usually quoted at about 45°C, although this depends on the physical nature of the sample (162). PBT crystallizes readily from the melt and it is difficult to obtain a truly amorphous sample. Its crystallization kinetics are not easy to determine (163). The density of the annealed crystalline unfilled polymer is 1.33 and 1.26 for the amorphous material (164). Like PET, PBT is made to various molecular weights; $M_n$ values are in the 20,000–50,000 range. Intrinsic viscosities are usually measured in $o$-chlorophenol (OCP) or a phenol–tetrachlorethane mixture. The Mark-Houwink parameters are $K = 1.17 \times 10^{-4}$, $\alpha = 0.87$ for P/TCE at 30°C, and $K = 6.62 \times 10^{-5}$, $\alpha = 0.915$ for OCP at 25°C (165,166).

Flame-retardant recipes used in PBT–FR grades usually consist of synergistic mixtures of antimony trioxide with various halogenated (brominated) aromatic compounds. A typical recipe for PBT might be 10% wt/wt decabromodiphenyl oxide and 5% wt/wt antimony oxide. The trend has been to use polymeric or oligomeric brominated additives. A typical additive is an end-capped polycarbonate derived from tetrabromobisphenol A [94334-64-2]; another is a mixture of epoxy oligomers derived from the diglycidyl ether of tetrabromobisphenol A [68928-70-1]. Brominated polystyrenes have limited use in PBT because they have a low compatibility (167). During the years 1993–1994 the market price of antimony metal and the trioxide rose very steeply. In fact the oxide tripled in price (168) which caused some FR-grade PBT producers and compounders to seek cheaper formulations. During 1995, antimony trioxide prices partially retreated from their 1994 peak (169) but were still at twice their previous level.

**Chemical Properties.**   PBT is highly crystalline and, as in the case of PET, does not suffer from solvent stress corrosion cracking as do amorphous materials. It is resistant at room temperature to most common chemicals and solvents, lubricants, greases, and automotive fluids. Ketones attack PBT at

**Table 7. Crystal Parameters[a] for the Two Forms of PBT**

| Parameter | $\alpha$-Form, unstretched | $\beta$-Form, stretched |
|---|---|---|
| $a$, nm | 0.482 | 0.469 |
| $b$, nm | 0.593 | 0.580 |
| $c$, nm | 1.174, 1.165[b] | 1.300 |
| $\alpha$, deg | 100, 98.9[b] | 102 |
| $\beta$, deg | 115.5, 116.6[b] | 120.5 |
| $\gamma$, deg | 111 | 105 |
| volume, nm$^3$ | 0.260 | 0.267 |
| density, g/cm$^3$ | 1.41 | 1.37 |

[a]Refs. 158 and 159, except as noted.
[b]Ref. 160.

temperatures above ambient. Parts made from PBT are dishwasher safe, but do not withstand steam autoclaving repeatedly. PBT is attacked by aqueous alkali and other strong bases and by dilute acids, particularly at elevated temperatures. PBT has very good resistance to weathering, and black pigmented grades with uv-stabilizers have excellent outdoor stability. As with all polyesters, PBT is susceptible to hydrolytic attack by moisture in the melt. Injection molding screw temperatures are usually about 250°C and IV drop is very rapid unless the polymer chip is dried to below 50 ppm moisture content and kept dry in the hopper. Inadequate drying is probably the cause of more molding problems than any other single cause.

    **Mechanical Properties.**  Properties of typical grades of PBT, either as unfilled neat resin, glass-fiber filled, and FR-grades, are set out in Table 8. This table also includes impact-modified grades which incorporate dispersions of elastomeric particles inside the semicrystalline polyester matrix. These dispersions act as effective toughening agents which greatly improve impact properties. The mechanisms are not fully understood in all cases. The subject has been discussed in detail (171) and the particular case of impact-modified polyesters such as PBT has also been discussed (172,173).

    **Processing.**  PBT is one of the easiest thermoplastics to injection mold provided the polymer is thoroughly dried before melting. Producers recommend 3–4 h at 121–138°C predrying. Owing to its good flow properties and extremely high rate of crystallization in the mold, cycle times are short (5–45 s). For typical glass-filled PBT resin the melt temperatures (240–250°C) may be reduced for FR grades; the melt temperature should not exceed 270°C.

    PBT resins are very fluid in the melt and sometimes drooling from the injection nozzle can be a problem in machines that do not have melt decompres-

**Table 8. Mechanical Properties of PBT**[a]

| Property | Unfilled, grade low mol wt | 30% Glass | | |
| --- | --- | --- | --- | --- |
| | | General purpose | Flame retardant | High impact |
| specific gravity | 1.31 | 1.54 | 1.66 | 1.53 |
| tensile strength, MPa[b] | 57 | 135 | 135 | 97 |
| tensile modulus, GPa[c] | 2.5 | 9.7 | 11.7 | 8.3 |
| elongation, % | 5 | 2 | 1.5 | 3.1 |
| flexural strength, MPa[b] | 85.5 | 193 | 193 | 152 |
| flexural modulus, GPa[c] | 2.5 | 8.3 | 10.3 | 6.9 |
| notched Izod, J/m[d] | 37.4 | 90.7 | 69.4 | 160 |
| unnotched Izod, J/m[d] | 1228 | 240 | 214 | 641 |
| HDT at 1.82 MPa[b], °C | 51 | 206 | 208 | 191 |
| volume resistivity, $\Omega \cdot$cm | $10^{15}$ | $10^{16}$ | $5 \times 10^{15}$ | $4 \times 10^{14}$ |
| dielectric strength, V/25 $\mu$m | 420 | 560 | 490 | 500 |
| dielectric constant, $\epsilon$, 100 Hz | 3.2 | 3.7 | 3.9 | 4.3 |
| flammability UL94, at 0.8 mm | HB | HB | V0 | HB |

[a]Ref. 170.
[b]To convert MPa to psi, multiply by 145.
[c]To convert GPa to psi, multiply by 145,000.
[d]To convert J/m to ft·lbf/in., divide by 53.38.

sion. A simple free-flow nozzle can be used satisfactorily to minimize drooling if its temperature can be adjusted by a separate heater band. PBT does not drool as badly as nylons.

**Health and Safety Aspects.**  PBT resins are not harmful or hazardous when handled at room temperature under normal conditions according to their Materials Safety Data Sheets. No problem regarding contact with the pellets has been encountered under normal conditions. Glass fines can, however, cause skin irritation, and if glass-filled resins are being ground or reground, due precautions must be taken. Inhalation of dust must be guarded against, as is true for grinding any glass-filled resin. During molding the temperature must not exceed 271°C and never 287°C as decomposition with the evolution of harmful vapors can occur. As with all thermoplastics, adequate ventilation must be provided around injection molding machines, etc.

**Economic Aspects.**  PBT is one of the fastest growing commercial thermoplastics. In 1993 the North American market alone exceeded 90,000 t, a 15% increase over 1992 (174). This rapid growth was accounted for mainly by the electrical and electronic and automotive markets, which together accounted for over half the PBT used. The principal manufacturers of PBT in North America, Europe, and Japan are shown in Table 9.

The economic growth prospects for PBT are promising. It is still displacing thermosets such as phenolics from some markets and its versatility, excellent flow properties, and ease of molding will assure it a prominent place in the thermoplastic world for many years to come. Quoted prices (November 1995) for PBT resins are $3.61–$3.85/kg for unfilled resin, $3.74–$4.18/kg for 30% FR-grade glass-filled, and $4.29–$4.51/kg for high impact grades (175). The U.S. market for all thermoplastic polyesters, PBT plus PET, and smaller volume products like PCT is shown in Table 10 (126).

**Table 9. Principal World Manufacturers of PBT by Region**

| United States | Europe | Japan |
|---|---|---|
| General Electric | Bayer AG | Polyplastics |
| Hoechst Celanese | BASF | Toray |
| Du Pont (Ciba-Geigy) | Du Pont | General Electric |
| BASF | DSM NV | Mitsubishi Engineering Plastics |
| | General Electric | Teijin |
| | Hoechst AG | Toyobo |
| | Hüls | |

## Thermoplastic Copolyester Elastomers

Thermoplastic copolyester elastomers are generally block copolymers produced from short-chain aliphatic diols, aromatic diacids, and polyalkylene ether-diols. They are often called polyesterether or polyester elastomers. The most significant commercial product is the copolymer from butane-1,4-diol, dimethyl terephthalate, and polytetramethylene ether glycol [25190-06-1], which produces a segmented block copolyesterether with the following structure.

**Table 10. Total U.S. Market[a] for Thermoplastic Polyesters by End Use, $10^3$ t**

| End use | 1986 | 1988 | 1990 | 1992 | 1994 |
|---|---|---|---|---|---|
| consumer recreation | 3.2 | 4.1 | 5.5 | 4.5 | 7.7 |
| transport | 17.3 | 22.7 | 21.8 | 28.1 | 37.7 |
| electrical, electronic | 14.5 | 20.5 | 19.5 | 23.6 | 28.6 |
| appliances | 2.7 | 3.2 | 3.6 | 4.1 | 6.4 |
| industrial | 10.9 | 12.3 | 13.6 | 12.7 | 13.2 |
| exports | | 6.4 | 8.2 | 9.5 | 11.4 |
| other | 7.7 | 17.3 | 17.7 | 14.1 | 12.3 |
| *Total including exports* | *56.3* | *86.5* | *89.9* | *96.6* | *117.3* |
| *Total domestic* | *56.3* | *80.1* | *81.7* | *87.1* | *105.9* |

[a]On a compounded basis.

The polymer with this structure has "hard" crystallizable segments of poly(tetramethylene terephthalate) (PBT or 4GT) (176–178). The "soft" segment phase is poly(tetramethylene ether glycol terephthalate), called PTMEG-T. The ratio of soft to hard segments determines the elastomeric nature of the copolymer. As the percentage of soft segments increases, so does the elasticity of the copolymer. The 4GT hard segment imparts crystallinity and chemical and thermal stability to the copolymer whereas the soft segment contributes toughness and elasticity.

Copolyesterether elastomers are considered to be high performance elastomers. In general they are used in applications which involve some type of repeated mechanical movement such as bending, flexing, pushing, rotating, pulsing, impacting, or recoiling. Such elastomers are used for one or all of the following reasons: (1) they have an excellent combination of strength, toughness, flexibility, and recovery from deformation; (2) they have excellent chemical resistance; and (3) they are capable of being used at high temperatures. Several of grades of such elastomers are supplied depending on the amount of PTMEG-T units in the final polymer. The polymers are classified, as are other elastomers and rubbers, by their Shore A or D hardness. The copolymers are generally available in hardnesses ranging from Shore D35 to Shore D82.

As with all thermoplastic elastomers, the copolyesterethers can be processed as thermoplastics. They are linear polymers and contain no chemical cross-links, thus the vulcanization step needed for thermosetting elastomers is eliminated and scrap elastomer can be re-used in the same process as virgin material (176–180).

**Physical Properties and Morphology.** Elastomers require a system of cross-bonding which links the flexible molecular chains to each other to give a network structure. Without this network, the elastomer would not have the properties to make it commercially useful. Thermoset elastomers accomplish this cross-bonding by forming molecular cross-links, irreversible chemical bonds formed during the vulcanization step. Thermoplastic elastomers do not undergo a vulcanization step and the network is produced by reversible physical bonds between the polymer chains (181). Copolyesterether elastomers are cross-bonded through a crystallization process. These block copolymers contain a crystalline phase which physically locks the flexible soft segments into an elastic network below the crystalline melting point of this phase.

In these structures, the crystalline 4GT units provide ties to the soft amorphous regions (181) which are made up of both PTMEG-T and amorphous 4GT units. From a structural viewpoint, this crystallization of the 4GT units is a fundamental determinant of the attributes of the elastomer. Morphological studies of such copolymers suggest that the 4GT crystallites are dispersed within the soft amorphous regions. The properties are related to this two-phase morphology; apparently the extent of hard segment/soft segment mixing depends on hard segment content, both segment lengths, and the affinity of one segment for the other (182).

Because the crystalline cross-links are reversible, the morphology and hence the utility of the elastomer is temperature dependent. Once the 4GT units have melted there is no more structure. This melting point depends on the percentage 4GT content in the copolymer. As the melt cools, the crystalline regions reform and the morphological structure re-establishes itself. 4GT crystals grow which tie together the amorphous segments. There are in fact two models for the two-phase structure in the polyesterether elastomers. The first model (181) considers the structure to be that of an interpenetrating network (Fig. 1) in which it is suggested that " . . . continuous and interpenetrating crystalline and amorphous regions exist where the randomly oriented and interconnected lamellar hard segments serve to anchor elastomeric portions of the molecule and thus provide physical cross-linking . . . " (183). The second model (184) considers a spherulitic structure (Fig. 2), consisting of " . . . 4GT radial lamellae with inter-radial amor-

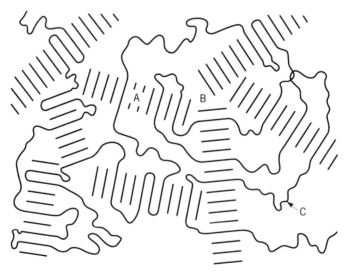

**Fig. 1.** Interpenetrating network morphology of thermoplastic elastomer where A = the crystalline domain, B = the junction of crystalline lamellae, and C = the noncrystalline 4GT segments (181).

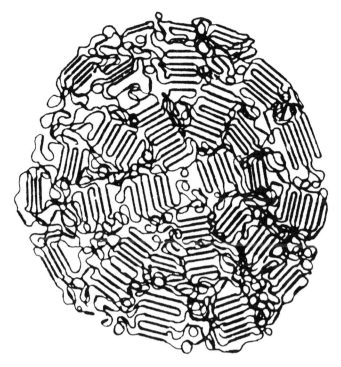

**Fig. 2.** Spherulitic morphology of thermoplastic elastomer where the heavy lines represent polyester segments (184).

phous regions which are a mixture of PTMEG-T soft segments and uncrystallized 4GT hard segments . . . ". In either case the crystallized 4GT segments provide a superstructure to tie together the amorphous 4GT and the PTMEG-T soft segments, with the latter two forming a single phase. It should be noted that several papers have been published that provide evidence for an analogy between copolyesterether elastomer spherulitic morphology and semicrystalline thermoplastics and copolymers (183–188). Like semicrystalline copolymers, the 4GT polyesterether elastomers exhibit only one glass–rubber-transition temperature and one crystalline melting point temperature as measured by dsc or dma.

Other morphological features of interest are as follows. Increasing the concentration of the amorphous region decreases the melt temperature and lowers the $T_g$ to the same values as those of the PTMEG-T homopolymer (183). There can be different states of order within the amorphous area. This is evidenced by a broadening of the glass-transition peak (188). Sample preparation affects the details of the morphological characterization. In an investigation of the chain conformation of 4GT–PTMEG-T elastomers (189) it was found that the PTMEG-T segments have a slightly expanded coil conformation, whereas the 4GT segments exhibit a chain-folded structure. Three types of long spacing were observed in x-ray scattering in samples studied with and without stress (190). Scattering behavior of drawn annealed bristles was studied by small-angle x-ray scattering for the copolyesterether elastomer based on 4GT and poly(ethylene glycol terephthalate) (191). Table 11 lists some of the characteristics of a series of 4GT–PTMEG-T elastomers.

**Mechanical Properties.** Commercial grades of copolyesterether elastomers generally range from Shore Hardness (SH) 35 to SH-82. These products are mechanically durable with high tensile strength and high load-bearing capabilities for an elastomer. They are very resilient, have low hysteresis, and excellent creep resistance. In addition they have high tear strength, good abrasion resistance, and a long flex life. Of special importance are the high temperature properties coupled with very good chemical resistance. Table 12 shows the properties of selected grades of polyesterether elastomers (192). These grades are representative of commercially available polyesterether elastomers. Table 12 shows several key properties, notably Vicat softening point, tear strength, flex fatigue, stress–strain characteristics, low temperature properties, and resistance to chemicals.

Polyester elastomers are resistant to a variety of common solvents including aqueous acids or bases. The chemical resistance of copolyesterether elastomers is shown in Table 13 (193) which gives examples of solvent resistance and is not inclusive.

Copolyesterether elastomers have excellent resistance to flex fatigue. For example, in the Ross flex test (ASTM D1052), all of the samples listed in Table 13 resisted cut-growth over 300,000 cycles.

Typical stress–strain curves are shown in Figure 3 (181). The stress–strain curve has three regions. At low strains, below about 10%, these materials are considered to be essentially elastic. At strains up to 300%, orientation occurs which degrades the crystalline regions causing substantial permanent set. At strains over 300% the stress occurs mostly in the amorphous regions up to the point where the sample breaks. All of the grades exhibit permanent set, and the curves of grades with a Shore Hardness of 55 and higher exhibit a yield point. This means that parts have to be designed for low strains to stay within the area of elastic recovery. Special grades of elastomer are available to provide hydrolysis resistance (194), improved heat aging (195), and improved uv-stability (196).

**Manufacture.** Polyesterether elastomers are made by a polycondensation reaction, either batchwise or by continuous polymerization processes. The reac-

**Table 11. Polyesterether Elastomer Characterization**[a]

| 4GT content, % | 4GT block length, nm | $T_m$, °C | Heat of fusion[b], J/g[c] | Crystallinity, % | $E''$ peak[d], °C | Tan peak[d], °C |
|---|---|---|---|---|---|---|
| 33 | 0.264 | 163 | 16.3 | 11.5 | −63 | −51 |
| 50 | 0.495 | 189 | 32.6 | 22.9 | −58 | −42 |
| 57 | 0.643 | 196 | 41.0 | 28.6 | −53 | −34 |
| 63 | 0.814 | 200 | 47.7 | 33.3 | −48 | −27 |
| 76 | 1.48 | 209 | 58.1 | 40.7 | −30 | 10 |
| 84 | 2.43 | 214 | 61.1 | 42.8 | −4 | 30 |

[a] Ref. 183.
[b] From dsc.
[c] To convert J/g to cal/g, divide by 4.184.
[d] From dma.

**Table 12. Selected Mechanical Properties of 4GT–PTMEP-T Elastomers**

| Properties | ASTM test | 35D | 40D | 55D | 72D |
|---|---|---|---|---|---|
| hardness, Shore D | D2240 | 35 | 40 | 55 | 72 |
| specific gravity | D792 | 1.14 | 1.15 | 1.19 | 1.26 |
| tensile strength, MPa[a] | D638 | | | | |
|   type I | | | | | |
|     5% yield | | 2.62 | 3.87 | 8.90 | 19.0 |
|     10% yield | | 4.19 | 6.29 | 13.0 | 25.0 |
|     type IV at break | | 13.2 | 20.7 | 25.2 | 29.2 |
| elongation at break, % | D638 | 470 | 480 | 430 | 300 |
| flexural modulus, MPa[a] | D790 | | | | |
|   at −40°C | | 84.8 | 154.0 | 677.0 | 2370.0 |
|   23°C | | 60.7 | 84.8 | 210.0 | 512.0 |
|   100°C | | 32.1 | 51.7 | 111.0 | 222.0 |
| Izod impact, J/m[b] | D256 | | | | |
|   at −40°C | | no break | no break | 144.0 | 37.4 |
|   23°C | | no break | no break | no break | no break |
| tear strength, kN/m[c] | D1004 | 101 | 110 | 147 | 217 |
| Tabor abrasion, H-18, mg/1000 cycles | D1044 | 121 | 90 | 85 | 30 |
| melting point, °C | D3418 | 164 | 180 | 200 | 214 |
| Vicat temperature, °C | D1525 | 107 | 136 | 181 | 204 |

[a]To convert MPa to psi, multiply by 145.
[b]To convert J/m to ft·lbf/in., divide by 53.38.
[c]To convert kN/m to ppi, divide by 0.175.

**Table 13. Chemical Resistance of Polyesterether Elastomers**

| Chemical | Rating[a] | Chemical | Rating[a] |
|---|---|---|---|
| acetone | G | mineral oil | E |
| ASTM oil No. 1 | E | nitric acid 10% | G |
| ASTM oil No. 3 | E | nitric acid >30% | NR |
| ASTM ref fuel A | E | SAE #10 oil | E |
| ASTM ref fuel C | E | seawater | E |
| ethanol | E | Skydrol 500B | E |
| formic acid | G | soap solution | E |
| 2-propanol | E | water (70°C) | G |
| methyl ethyl ketone | G (35-55D) | water (boiling) | E |
| methyl ethyl ketone | E (72-82D) | zinc chloride solution | E |

[a]E = no adverse reaction, little or no absorption, little or no effect on mechanical properties; G = some effect, some absorption causing slight swelling, and reduction in mechanical properties; and NR = not recommended, material adversely affected in a short time.

tion proceeds in two steps, the first stage being an ester interchange followed by a polycondensation step, exactly analogous to the steps used in the manufacture of PET or PBT. The manufacture generally proceeds in the following manner. Dimethyl terephthalate (DMT), poly(tetramethylene oxide diol) (PTMEG), and excess butane-1,4-diol react in the presence of a titanium catalyst and the methanol evolved is distilled out of the reaction mixture (178). When the ester

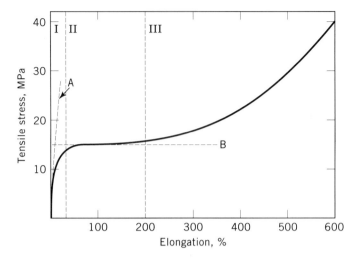

**Fig. 3.** Stress–strain curve of typical polyesterether elastomer showing the three main regions (I, II, and III) (181), where A is the slope (Young's modulus) and B is the yield stress. To convert MPa to psi, multiply by 145.

exchange is complete the reaction mixture is subjected to gradually diminished pressure to distill out the excess butanediol and drive the polymerization to completion, eventually under full vacuum. The batch temperature must be kept below 250°C to avoid thermal degradation, and generally an antioxidant is added to protect the polyether segment from oxidative degradation (see ANTIOXIDANTS). As the batch builds molecular weight, the melt viscosity increases and the power consumed by the agitator motor is taken as a measure of the degree of polymerization. When the desired end point is reached, the vacuum is released with inert gas and the batch discharged under pressure through a die plate into molten strands which are cooled in a water bath and chopped into pellets for drying and bagging. The distilled methanol and butanediol can then be redistilled for recovery.

Other comonomers can be incorporated at the polymerization stage. Some typical comonomers that have been described are but-2-en-1,4-diol (197), dimethyl sebacate (176,178), dimethyl *o*-phthalate (176,178), dimethyl iso-phthalate (176,178), dimethyl *m*-t erphenyl-4,4''-dicarboxylate (176,178), and the dimethyl ester of cyclohexane-1,4-dicarboxylic acid (187). In some cases the butanediol is replaced partly or completely by other diols such as ethylene glycol (176,178) or cyclohexanedimethanol (198). Other polyether diols can be used in place of PTMEG. The most common are poly(ethylene glycol) and poly(propylene glycol), derived respectively from ethylene oxide and propylene oxide, and polyether diols that are random or block copolymers from ethylene oxide and propylene oxide (see POLYETHERS). Elastomers made from these polyether diols have been compared (199). Elastomeric copolymers can also be made by replacing the PTMEG with a hydrocarbon soft segment such as those derived from dimer acids (200,201). The dimethyl terephthalate can be replaced totally or partially by the dimethyl ester of naphthalene-2,6-dicarboxylic acid (NDA) (202).

**Processing.**    Injection molding is probably the most important process for shaping polyesterether elastomer parts. However, extrusion and blow molding are also important. The polymer pellets must be thoroughly dried before melt processing. The polymer has an equilibrium moisture content of 0.2% but this must be dried to below 0.05% if hydrolytic degradation is to be avoided. Recommended conditions for drying are 4 h at 105–110°C (180). Molding conditions depend on elastomer grade but using the same grades as in Table 12, the temperatures as set out in Table 14 are suitable. For all grades, injection pressure is low–medium; injection speed is fast; screw speed is 60–125 rpm; back-pressure is 0–345 kPa (0–50 psi); and cushion/pad is 5–70 mm.

Processing conditions for blow molding and extrusion are similar to those for injection molding, but cooler temperatures are required to achieve higher viscosities for extrusion and blow molding. Thus the front end temperatures are at the colder end of the ranges in Table 14. Other methods of fabrication are rotational molding, film extrusion, and melt casting. The processing of polyesterether elastomers has been compared with the processing of other thermoplastic elastomers such as styrenics and urethanes (203).

**Table 14. Molding Temperatures for Polyesterether Elastomers**

| Region[b] | Polymer grade[a] | | | |
|---|---|---|---|---|
| | 35D | 40D | 55D | 72D |
| rear zone | 155–170 | 162–182 | 200–215 | 200–215 |
| mid-zone | 170–182 | 182–200 | 215–232 | 215–232 |
| front zone | 170–182 | 182–205 | 215–238 | 215–238 |
| nozzle | 170–188 | 182–205 | 215–238 | 215–238 |
| melt | 170–188 | 182–205 | 220–238 | 220–238 |
| mold | 25–52 | 25–52 | 25–52 | 25–52 |

[a]Hardness, Shore D.
[b]Temperatures in °C.

**Health and Safety Issues.** Polyesterether elastomers derived from dimethyl terephthalate, butanediol, and PTMEG are not hazardous according to the published Materials Safety Data Sheets (MSDS) for this elastomer. Polymers of a similar structure containing isophthalic acid are also not considered hazardous. For other copolymer elastomers, the MSDS put out by suppliers should be consulted by potential users before evaluation. One environmental advantage of thermoplastic elastomers of this type is that they are melt-reprocessible and thus scrap and off-specification material and even obsolete parts can be easily recycled. Up to 25% by weight of recycled material can be incorporated (see RECYCLING, PLASTICS).

**Uses.** The market for polyesterether elastomers is expected to grow to 60,800 t by 1996 (204). A projected world market situation is shown in Table 15.

**Table 15. Polyesterether Elastomer World Market[a]**

| Region | 1996 Projection, $10^3$ t | Growth rate, %[b] |
|---|---|---|
| United States | 34,000 | 10.1 |
| Western Europe | 13,600 | 11.6 |
| Japan | 13,200 | 14.0 |

[a]Ref. 204.
[b]For the years 1986–1996.

The principal market segments for thermoplastic elastomers in 1987 were reported to have the various market shares as shown in Table 16 (204).

**Table 16.  Market Share of Polyesterether Elastomers in 1987[a], %**

| Market | United States | Western Europe | Japan |
|---|---|---|---|
| wire and cable | 13 | 10 | 11 |
| automotive | 13 | 25 | 44 |
| footware | 4 | 12 | |
| hose and tubing | 13 | 14 | 11 |
| other industrial | 57 | 39 | 34 |

[a]Ref. 204.

Specific applications for polyesterether elastomers are too numerous to mention in detail, but a list of applications with references where appropriate is as follows. This is by no means a complete list, but is merely representative of the vast range of potential applications.

hydraulic hosing (180)
release binders (180)
door lock bumpers (205)
headphones (206)
jacketing (206)
diaphragms for railroad cars (177)
railroad draft gear (177)
CVJ boots (207)
conductive rubbers (208)
energy management devices (207)
compression spring pads (206)
gun holsters (177)
run-flat tire inserts (207)

rail car couplers (209)
auto vacuum control tubing
railroad car shock absorbers (177)
specialty fibers, films, and sheets
automotive shock absorbers
corrugated plastic tubing
auto electric window drive tapes
recreational footware
wire coatings (180)
telephone handset cords (205)
wire clamps (207)
drive belts (177)
medical films (199)

**Manufacturers.**  The main suppliers of polyesterether elastomers and the trade names of their products are as follows. An important end use for these elastomers is in blends with PBT to make polymer alloys such as VANDAR or BEXLOY. These alloys are very tough and find uses in such applications as automobile fascias (210), wheelcovers, and air-bag housings.

| Manufacturer | Trade name |
|---|---|
| Du Pont | HYTREL (BEXLOY V) |
| General Electric | LoMod |
| Hoechst Celanese | RITEFLEX (VANDAR) |
| Eastman | ECDEL |
| Akzo | ARNITEL |

## Liquid Crystal Polymers

Most crystalline substances undergo a single transition from a solid state to an isotropic liquid phase. However, some molecules, both small molecules and polymers, can exist in an ordered liquid phase or mesophase. Such substances are called liquid crystals (see LIQUID CRYSTALLINE MATERIALS). The mesophase state is intermediate between a structured solid and an isotropic or disordered liquid. A striking manifestation of this structure is that a liquid crystal melt is often birefringent between crossed polarizers and displays vivid colors when sheared. There are many types of mesophase, but they all fall into two main categories: those which possess some measure of three-dimensional order and those which have only rotational order and are therefore fluid. It is this dual property of behaving like a liquid and a solid that gives rise to many interesting properties.

Most polymeric liquid crystals are based on stiff rod-like molecular units which are called calamitic mesogens. There are some unusual polymers (which are not discussed here) that contain flat disk-like molecular units called discotic mesogens in which the disks form columnar arrays like stacks of poker chips.

Another distinction is that between mesophases which exist only in solution and are called lyotropic mesophases, and those that can exist by altering the temperature, called thermotropic mesophases. Lyotropic mesophases typically form from solutions of stiff rod-like polymers in solvents above a certain critical concentration (211); an excellent example is the polyamide Kevlar which was first commercialized in the 1960s by Du Pont. Other examples are peptides such as poly($\gamma$-benzyl-L-glutamate) and proteins such as tobacco mosaic virus. Liquid crystals can also be classified, according to their molecular symmetry, as nematic, cholesteric, and smectic. In nematic liquid crystals there exists long-range orientational order and the molecules tend to align themselves parallel to each other; there is no long-range correlation of molecular position. Cholesteric mesophase order is only found in chiral nematic molecules and here layers or sheets of nematic order exist but superimposed upon this is a spiral ordering like a twisted deck of cards. The smectic order is the most complex and many smectic states exist. It is also possible to have chiral smectic states. All are characterized by both orientational and positional order to varying degrees. Due to the positional order of smectic phases they are much more viscous than nematic polymers. An interesting anecdote concerns unpublished work at ICI Fibers Division (Harrogate, U.K.), during 1959–1961 concerning the polyester from 4,4'-bibenzoic acid and hexane-1,6-diol. Despite much effort, this material could not be melt-spun into fibers even though its melting point was below its decomposition point. No coherent threadline could be maintained and the extrudate flew off the windup as short, brittle, crystalline lengths. Not until many years later did other workers show that this polymer on cooling exhibits a mesophase transition directly from the isotropic melt to a smectic A phase. Good sources of information on liquid crystals and liquid crystal polymers are available (212–216).

Polymers have been described exhibiting all types of mesophase characters but the only ones to have practical commercial utility as structural materials are rod-like polymers which exist in nematic mesophases. These polymers are aromatic polyesters (or polyesteramides) which frequently contain units derived from 4-hydroxybenzoic acid. The literature on liquid crystal polyesters is voluminous. The abbreviation LCP will hereafter refer to thermotropic liquid crystalline aromatic polyesters. The history and outline polymerization chemistry of LCPs has been discussed. In commercial terms, Dartco licensed the Carborundum technology and built a plant in the United States to manufacture XYDAR for Tupperware. This plant was sold to Amoco in 1987. Meanwhile Celanese Corporation had introduced its VECTRA range of LCP resins as an advanced engineering plastic in 1985 and a new plant was built at Shelby in North Carolina. Du Pont has begun to produce their ZENITE range in the United States. Sumitomo, Unitika, and Polyplastics in Japan have made commercial LCPs; in Europe, ICI began, then abandoned, a LCP product. BASF and Bayer AG also began to produce commercial material, but in 1996 the LCP market is dominated by Hoechst Celanese and Amoco products.

**Molecular Structure of LCPs.** There are no rigorous criteria for predicting the liquid crystal nature of any polymeric structure but some generalizations can be made. Firstly, the persistence length of LCPs is greater by a factor of 2–3 than that of random coil polymers. Generally, lyotropic polymers have a greater persistence length than thermotropics. An axial ratio of 6.42 is required for a polymer to be stiff enough to form a thermotropic melt (217–219). Chain stiffness and interaction between the neighboring chains are the key parameters that determine whether or not a polymer forms a mesophase. Most LCPs have a large polarizability along the rigid chain axis compared to that in a transverse direction. 4-Hydroxybenzoic acid (HBA) is a key ingredient in the majority of commercial LCPs. When copolymerized into polyester structures it imparts the correct amount of polymer chain stiffness and degree of interaction to lead to the formation of a mesophase state. It is also available and reasonably inexpensive. The homopolymer has an exceedingly high melting transition (220) and copolymers are made to lower the solid nematic transition (melting point) to make the polymer tractable and useful.

During thermoplastic processing (eg, extrusion or injection molding) the polymer flows in the liquid crystal (nematic) phase. The relatively rigid rodlike polymer chains align themselves in the flow direction and there is little or no entanglement, giving a very fluid melt. As a result, such polymers are ideally suited to molding intricate and finely detailed components. After cooling, the aligned chains are frozen with some net alignment remaining in the flow direction, called the frozen liquid crystal state. Due to this orientation effect, LCPs typically exhibit anisotropic properties.

**LCPs as Molding Resins.** LCPs are in many ways ideally suited for injection molding. Their balance of properties is particularly well suited to the electrical connector and electronics industries. Some of the principal advantages of LCPs as molding resins include (1) extremely low shrinkage and warpage and exceptional dimensional repeatability, (2) high melt strength for versatility of fabrication, (3) low melt viscosity for high flow rates in thin sections and intricate molds, (4) low heat of fusion for very fast cooling, short cycle times, (5) low flash due to low injection pressures and shear-sensitive viscosity, and (6) high stiffness at high temperatures which allows parts to be ejected while hot.

In addition, LCPs offer, by their very chemical nature as all-aromatic crystalline polyesters, a combination of physical and chemical properties that no other single class of polymers can demonstrate. They have excellent thermal stability, UL relative thermal index up to 240°C, depending on grade of LCP. LCPs exhibit high rigidity and strength; typical values are modulus 10–24 GPa ($1.4$–$3.5 \times 10^6$ psi), tensile strength 125–225 MPa (18,000–37,000 psi), and notched Izod values from 85.4–294 J/m (1.6–5.5 ft·lb/in.). LCPs retain 70% impact value even down to $-270$°C. Most other plastics are extremely brittle at cryogenic temperatures. They are inert to a wide range of chemical agents, acids, bleaches, chlorinated solvents, fuels, and alcohols. Because of their close-knit molecular structure, LCPs have high barrier properties and very low permeation rates. They have very low levels of ionic contamination, well below those needed for corrosion-free environments in electronic integrated circuitry. LCPs are inherently flame retardant, produce low smoke density, and their combustion products are relatively nontoxic. Regrind polymer maintains nearly all properties of virgin resin.

**Synthesis.**    The outline of LCP synthesis has already been discussed: nearly all are made by the acidolysis route. The *in situ* acetylation polymerization route, which starts from the free hydroxy compounds and carboxylic acids rather than preformed acetylated monomers, is increasingly used because larger polymer batches can be made in the same reactor vessel. Polymerizations can be run batchwise or continuously and the volatile acetic acid is removed by distillation under vacuum in the conventional way. Because of the corrosive nature of this by-product, the reactors must be made of a corrosion-resistant alloy. Stainless steel is generally adequate but some manufacturers prefer to use Hastelloy. Several unique problems arise in LCP polymerization due to the extreme shear sensitivity of the LCP melt. One concerns agitator design: it is possible under certain circumstances for the whole batch to revolve with the agitator shaft as a quasi-solid lump, due to the high rate of shear thinning at the walls of the reactor. Under such conditions no effective melt agitation would be taking place. Such problems can be solved by correct equipment design. In the mesophase nematic state the polymer melt is opaque and cloudy, unlike the clear melt of a normal isotropic polymer.

For very high melting polymers, sometimes a slurry or dispersion polymerization route is used. In this case the reaction is run in a high boiling inert solvent as a reaction medium. The starting materials are usually soluble at the commencement but as the acetic acid distills out and the polymer molecular weight grows, the product separates out as a slurry or dispersion of fine particles, hence the name. The low molecular weight prepolymers initially formed presumably undergo a form of solid-state polymerization in the later stages of the process. This method enables high melting polymers to be made under more normal conditions and produces the polymer in a finely divided powder form. Disadvantages are that the presence of the inert solvent reduces the size of the polymer batch, and the product usually has to be solvent washed to remove and recover the inert reaction medium. Both of these factors add to the process cost; however, the technique has been described in the patent literature several times (221–224).

The two principal commercial LCP polymers are VECTRA and XYDAR. Both are based on 4-hydroxybenzoic acid (HBA), which is copolymerized with 6-hydroxy-2-naphthoic acid in the case of VECTRA (23) and with a mixture of 4,4'-biphenol and terephthalic acid (some compositions are believed to also include isophthalic acid) in the case of XYDAR (220). Both types of copolymers lead to lower melting points, but the $T_m$ of XYDAR is about 420°C which is considerably above that of most thermoplastic polyesters. This gives XYDAR outstanding high temperature performance at the cost of some added difficulties during processing. VECTRA comes in various grades and compositions (225–228), some of which are intended for high temperature end uses. VECTRA A polymer melts at 280°C, which is much more in the range of normal processing conditions. The VECTRA B series of resins is particularly suitable for the extrusion market. This polymer is a polyesteramide (58,229–232) and has exceptional stiffness and high modulus. This appears to be due to a favorable molecular alignment brought on during elongational flow during extrusion, which is more effective for this composition than for normal LCPs (233,234). VECTRA B resins have a melting point very close to that of the VECTRA A series.

Du Pont's ZENITE range of LCPs have been reintroduced. These materials are also LCP polyesters based on HBA copolymers with, it is believed, various amounts of 4,4'-biphenol, terephthalic acid, and naphthalene-2,6-dicarboxylic acid (NDA) as comonomers. Their compositions are thus similar to those of XYDAR with added NDA to further reduce the melting point (235,236).

**Injection-Molding LCPs.** VECTRA and XYDAR are also the two leading commercial LCP materials for injection molding. The high degree of molecular alignment in the flow direction during the molding process gives the finished part an anisotropic quality; the part is stronger in the direction of flow than across it, a phenomenon observed in the early days of LCP technology (16). Owing to this molecular alignment the shrinkage properties are very low or zero in the direction of alignment, but across the flow direction the shrinkage is close to that of a normal isotropic polymer. Typically, along the flow direction the shrinkage is 0.00–0.001 cm/cm, and across the flow, 0.05 cm/cm.

LCPs are remarkable for combinations of properties unmatched by any other thermoplastic. They have low melt viscosity, low flash, high heat resistance, and high dimensional stability. The rigid (or semirigid) rod-like nature of the polymer chains in LCPs results in a profile of molecular orientation that resembles the physical orientation of fibers in a fiber-reinforced thermoplastic. During injection molding the skin is oriented in the flow direction and may comprise 15–30% of the thickness of the total part. Generally as the section becomes thinner, the proportion of skin increases. The skin has high strength and modulus and good tensile performance, but the very high level of orientation can cause the surface of unfilled parts to fibrillate or peel off as fine fibers under abrasion. This is prevented by filling the polymer with an inorganic filler such as glass, wollastonite, etc (see FILLERS). Typically LCPs are sold as glass-filled compounded resins. As the percentage of glass increases the stiffness and strength go up, and impact properties such as notched Izod values come down. Heat-deflection temperature (HDT) also increases with glass content and the degree of anisotropy decreases in the final molded part.

Weld-line effects are common to all injection moldings where two polymer melt fronts meet as in multiple gate molds. This effect is particularly noticeable in the case of LCPs because of the highly anisotropic liquid crystalline nature of the melt and the domain microstructures associated with it. Because of these, the two melt fronts do not readily interpenetrate and the final molding has poor weld-line strength (237). This problem can be partially solved by using glass-filled resins, but manufacturers recommend that weld-line problems be allowed for during the initial mold design and gates be positioned accordingly.

**Properties of Typical Commercial LCPs.** Table 17 presents data on three significant, commercially available LCPs (238). Elongation values have been quoted for HX-2000 as 0.6%, for XYDAR G-930 as 1.6%, and for XYDAR G-540 as 1.5%. Further details can be obtained from the manufacturers.

**Rheology of LCPs.** Most conventional thermoplastics show evidence of shear thinning at high shear rates (1000 reciprocal seconds or more). LCPs show evidence of shear thinning at much lower shear rates (16,239–241). Typically, low injection pressures are needed in LCP molding compared with those for conventional thermoplastics. A very good overview of the rheology of LCPs has been published (242). The long relaxation time of molten LCPs in the mesophase state results in very low die swell. It has been demonstrated that significantly improved tensile properties result from extrusion through a die with either zero "land" or very minimal land length (243). Injection-molded LCP parts have a distinct fibrous morphology that can only be obtained in an extensional flow

**Table 17. Mechanical Properties of LCPs**

| Resin[a] | Filler[b] | Density, g/cm$^3$ | Tensile strength, MPa[c] | Flexural modulus, GPa[d] | Notched Izod, J/m[e] | HDT at 1.82 MPa[c], °C |
|---|---|---|---|---|---|---|
| VECTRA | | | | | | |
| A950 | none | 1.4 | | 8.97 | 534 | 180 |
| A130 | 30% CG | 1.61 | 207 | 14.5 | 149 | 230 |
| A150 | 50% CG | 1.79 | 179 | 20 | 90.7 | 232 |
| A410 | 50% GF/MIN | 1.24 | 169 | 18.6 | 85.4 | 280 |
| B950 | none | 1.4 | 186 | 15.2 | 427 | 200 |
| C130 | 30% CG | 1.61 | 152 | 13.1 | 128 | 240 |
| C150 | 50% CG | 1.79 | 166 | 19.3 | 85.4 | 252 |
| ZENITE | | | | | | |
| 6130 | 30% CG | 1.61 | 145 | 12.4 | 139 | 250 |
| 7130 | 30% CG | 1.61 | 145 | 13.1 | 139 | 279 |
| HX-2000 | none | 1.27 | 117 | 17.9 | 198 | 185 |
| XYDAR | | | | | | |
| G-930 | 30% GF | 1.6 | 135 | 13.4 | 587 | 271 |
| G-540 | 40% GF | 1.7 | 146 | 15.7 | | 241 |

[a]Every sample has a UL-94 rating of V-0.
[b]CG = chopped glass, GF = glass fiber, and MIN = mineral filler.
[c]To convert MPa to psi, multiply by 145.
[d]To convert GPa to psi, multiply by 145,000.
[e]To convert J/m to ft·lbf/in., divide by 53.38.

process. Key processing and property advantages of LCPs in injection molding are the ability to fill long flow paths, thin walls, multiple cavities, and multiple inserts; fast mold cycles; reduced injection and clamp pressures; low shrinkage and warpage; and low thermal expansion coefficients in the machine direction due to the orientation of the molecules.

**Chemical Resistance of LCPs.**   Certain liquid crystal polymers (eg, VEC-TRA) have extremely high chemical resistance to a variety of aggressive chemicals and solvents. Table 18 shows the chemical stability of VECTRA test-bars to various agents (244).

**Applications of LCPs.**   The most advanced application for LCPs is as injection-molding compounds for electrical interconnect devices (245). In the electrical and electronics area, surface mount components, connectors, chip carriers, ceramic replacements, bobbins, electric motor insulation, fiber optic components, closures, and fuse-holders are some of the applications. Other applications are bearings, bushings, gears, cams, microwave components, under-the-hood automotive components (including fuel systems and electronic and electrical systems), aerospace applications, and aircraft components (eg, secondary structures, interior components for passenger aircraft, mechanical linkages, and structural components). Future applications will include biomedical devices, printed circuit boards, pump, valves, etc. Because of their great chemical resistance and excellent electrical properties, additional ease of fabrication, and the ability to replace complex assembly and machining operations with a single molded thermoplastic component, which may well perform better, the utility of LCPs is particularly advantageous. The whole applications area for LCPs has been reviewed (246); another useful review of commercial LCP materials in general is available (247).

**LCP Economics.**   LCPs are expensive materials. Prices in January 1996 (248) ranged from $15.20/kg for mineral-filled resin, $15.73–$23.43/kg for glass-filled resin, and up to $48.40/kg for unfilled extrusion-grade polymer. One of the basic reasons is the fundamental high cost of monomers and intermediates which is a consequence of low volume.

**Table 18. Chemical Resistance of VECTRA After 30-d Exposure**

| Agent | Temperature, °C | Rating[a] |
|---|---|---|
| 80% formic acid | 216 | A |
| 70% chromic acid | 190 | B |
| 37% hydrochloric acid | 190 | A |
| 70% nitric acid | 190 | B |
| caustic soda | | |
| 10% | 190 | A |
| 30% | 190 | C |
| acetone | 133 | A |
| methylene chloride | 148 | A |
| ethyl acetate | 171 | A |
| gasoline | 250 | A |
| methanol fuel | 250 | B |
| Skydrol | 160 | A |

[a]A = essentially no effect, B = some change, and C = not recommended.

The exact size of the world LCP market is difficult to determine from the open literature. It is a highly fragmented and specialized area and both suppliers and customers closely protect their proprietory sales information. The total world sales based on compounded resin is estimated at between 2270–4545 t/yr, and in view of the close association with the electrical/electronics industry, it is expected that a significant area of this business is in the Far East. The LCP market has only been in existence since the 1980s and only since the early 1990s has it begun to make a noticeable impact. However, the future looks bright for materials that offer so many combined property advantages if the specialized end use can justify the price.

## BIBLIOGRAPHY

"Polyesters, Thermoplastic" in *ECT* 3rd ed., Vol. 18, pp. 549–574, by D. B. G. Jaquiss, W. F. H. Borman, and R. W. Campbell, General Electric Co.

1. I. Goodman, in J. I. Kroschwitz, ed., *Encyclopedia of Polymer Science and Engineering*, 2nd ed., Vol. 12, John Wiley & Sons, Inc., New York, 1988, pp. 1–75.
2. H. Mark and G. S. Whitby, eds., *The Collected Works of W. H. Carothers on High Polymeric Substances*, Interscience Publishers, Inc., New York, 1940.
3. Brit. Pat. 578,079 (June 14, 1946), J. R. Whinfield and J. T. Dickson (to ICI Ltd.).
4. U.S. Pat. 2,465,319 (Mar. 22, 1949), J. R. Whinfield and J. T. Dickson (to Du Pont).
5. E. V. Martin and C. J. Kibler, in H. F. Mark, S. M. Atlas, and E. Cernia, eds., *Science and Technology of Man-Made Fibers*, Vol. 3, New York, 1967.
6. *Mod. Plast. Int.* **20**(4), 8–10 (Apr. 1990).
7. P. A. Aspy and E. E. Denison, *Mod. Plast.*, 74–76 (Aug. 1983).
8. Brit. Pat. 1,239,751 (July 21, 1971), A. J. Dijkstra, J. A. W. Reid, and I. Goodman (to ICI Ltd.).
9. U.S. Pat. 4,393,178 (July 12, 1983), R. M. H. Legras and co-workers (to Imperial Chemical Industries plc).
10. U.S. Pat. 4,380,621 (Apr. 19, 1983), E. Nield, D. E. Higgins, and M. W. Young (to Imperial Chemical Industries plc).
11. Brit. Pat. 604,073 (1948), J. G. Cook, H. P. W. Hugill, and A. R. Low.
12. *Plast. Technol.*, 55 (May 1995).
13. Ger. Offen. 2,052,971 (Dec. 3, 1970), S. Cottis, J. Economy, and B. E. Nowak (to Carborundum Corp.).
14. U.S. Pat. 3,829,406 (Oct. 1, 1971), S. Cottis, J. Economy, and B. E. Nowak (to Carborunum Corp.).
15. Ger. Offen. 2,507,066 (Sept. 2, 1976), S. Cottis, J. Economy, and L. C. Wohrere (to Carborundum Corp.).
16. W. J. Jackson and H. F. Kuhfuss, *J. Polym. Sci. Polym. Chem.* **14**, 2043–2058 (1976).
17. U.S. Pat. 4,181,792 (Jan. 1, 1980), W. J. Jackson and J. C. Morris (to Eastman Kodak Co.).
18. U.S. Pat. 3,991,013 (Nov. 9, 1976), T. C. Pletcher (to Du Pont).
19. U.S. Pat. 3,991,014 (Nov. 9, 1976), J. J. Kleinschuster (to Du Pont).
20. U.S. Pat. 4,066,620 (Jan. 3, 1978), T. C. Pletcher and J. J. Kleinschuster (to Du Pont).
21. U.S. Pat. 4,118,372 (Oct. 3, 1978), J. R. Schaefgen (to Du Pont).
22. Brit. Pat. 993,272 (May 26, 1965), I. Goodman, J. E. McIntyre, and D. H. Aldred (to ICI Ltd.).
23. U.S. Pat. 4,161,470 (July 17, 1979), G. W. Calundann (to Celanese Corp.).
24. W. S. Witts, *Chem. Process. Eng.* **51**, 55 (1970).

25. U.S. Pat. 3,467,724 (Sept. 16, 1969), S. A. Laurich (to Chevron Research Co.).
26. U.S. Pat. 3,636,180 (Jan. 18, 1972), D. B. Broughton (to Universal Oil Products Co.).
27. Brit. Pat. 994,769 (June 10, 1965), (to Standard Oil Co.).
28. U.S. Pat. 3,584,039 (June 8, 1971), D. H. Meyer (to Standard Oil Co.).
29. J. Milgrom, *Outlook for Plastics Recycling in the United States*, Decision Resources, Inc., 1993, pp. 25–27.
30. K. A. Domeshek, in D. Brunnschweiler and J. Hearle, eds., *Tomorrow's Ideas and Profits: Polyester: Fifty Years of Achievement*, The Textile Institute, Manchester, U.K., 1993, pp. 288–290.
31. U.S. Pat. 5,008,479 (Apr. 16, 1991), T. Abe and co-workers (to Mitsubishi Gas Chemical Co.).
32. U.S. Pat. 5,276,230 (Jan. 4, 1994), K. Inamasa, N. Fushimi, and M. Takagawa (to Mitsubishi G. C. Co.).
33. U.S. Pat. 5,321,178 (June 14, 1994), K. Inamasa, N. Fushimi, and M. Takagawa (to Mitsubishi G. C. Co.).
34. U.S. Pat. 5,254,769 (Oct. 19, 1993), M. Takagawa, K. Yamagishi, and K. Nagagata (to Mitsubishi G. C. Co.).
35. U.S. Pat. 5,183,933 (Feb. 2, 1993), Harper and co-workers (to Amoco Corp.).
36. U.S. Pat. 5,169,977 (Dec. 8, 1992), T. Tanaka and co-workers (to Mitsubishi Gas Chemical Co.).
37. U.S. Pat. 5,254,719 (Oct. 19, 1993), J. K. Holzhauer and D. A. Young (to Amoco Corp.).
38. U.S. Pat. 5,262,560 (Nov. 16, 1993), J. K. Holzhauer and co-workers (to Amoco Corp.).
39. U.S. Pat. 5,256,817 (Oct. 26, 1993), D. L. Sikkenga and S. V. Hoover (to Amoco Corp.).
40. G. O. Curme and F. Johnston, *Glycols*, ACE Monograph No. 114, Rheinhold Publishing Corp., New York, 1952, Chapt. 2.
41. U.S. Pat. 3,920,759 (Nov. 18, 1975), E. V. Hort (to GAF Corp.).
42. U.S. Pat. 3,950,441 (Apr. 13, 1976), S. Rudolf and W. DeThomas (to GAF Corp.).
43. U.S. Pat. 3,922,300 (Nov. 25, 1975), T. Onoda and co-workers (to Mitsubishi Chemical Industries Ltd.).
44. U.S. Pat. 4,010,197 (Mar. 1, 1977), J. Toriya and K. Shiraga (to Mitsubishi Chemical Industries Ltd.).
45. U.S. Pat. 4,064,145 (Dec. 20, 1977), P. D. Taylor (to Celanese Corp.).
46. U.S. Pat. 4,356,125 (Oct. 26, 1982), N. A. de Munck and J. F. Scholten (to Stamicarbon BV).
47. Ger. Offen. 2,526,312 and 2,526,775 (Dec. 30, 1977), H. Moella and co-workers (to Henkel and Cie GmbH).
48. Japan Kokkai 75,142,537 (Nov. 17, 1975), Mizumoto and Kamatani (to Toyobo Co., Ltd.).
49. *Chem. Mark. Rep.*, 5 (May 15, 1995).
50. U.S. Pat. 5,304,691 (Apr. 19, 1994), J. P. Arhancet and L. H. Slaugh (to Shell Oil Co).
51. R. Jakeways, I. M. Ward, and co-workers, *J. Polym. Sci. Polym. Phys.* **13**, 799–813 (1975).
52. S. Poulin-Dandurand and co-workers, *Polymer* **20**, 419–426 (1979).
53. J. Desborough, I. H. Hall, and J. Z. Neisser, *Polymer* **20**, 545–552 (1979).
54. A. M. Brownstein, *CHEMTECH*, 5 (Jan. 1995).
55. A. S. Lindsey and H. Jeskey, *Chem. Rev.* **57**, 583 (1957).
56. WO 86 02,924 (May 22, 1986), R. Ueno and co-workers (to Kabushiki Kaisha Ueno Seiyaku Oyo Kenkyujo).
57. Eur. Pat. Appl. EP 53,824 (June 16, 1982), R. Ueno and co-workers (to Kabushiki Kaisha Ueno Seiyaku Oyo Kenkyusho).
58. U.S. Pat. 4,330,457 (May 18, 1982), A. J. East, L. F. Charbonneau, and G. W. Calundann (to Celanese Corp.).

59. Eur. Pat. 148,584 (Jan. 4, 1984), W. W. Kaeding (to Mobil Oil).
60. Eur. Pat. 271,623 (Dec. 19, 1986), S. Koskimies and T. Haimela (to Nesté Oy).
61. Fr. Pat. 2,071,464 (Dec. 30, 1969), F. Bourdin and co-workers (to Rhône Poulenc).
62. Eur. Pat. 480,800 (Oct. 8, 1990), M. Costandini and M. Jouffret (to Rhône Poulenc).
63. U.S. Pat. 3,306,874 (Feb. 28, 1967), A. S. Hay (to General Electric Co.).
64. U.S. Pat. 4,205,187 (May 27, 1980), J. N. Cardenas and W. T. Reichle (to Union Carbide Corp.).
65. U.S. Pat. 4,524,217 (June 18, 1985), K. G. Davenport and C. B. Hilton (to Celanese Corp.).
66. W. L. Hergenrother and C. J. Nelson, *J. Poly. Sci. Chem. Ed.*, **12**, 2905 (1974).
67. R. E. Wilfong, *J. Polym. Sci.* **54**, 388 (1961).
68. Fr. Pat. 1,169,659 (1959), (to ICI Ltd.).
69. Can. Pat. 573,301 (1959), W. K. Easley and co-workers (to Chemstrand Corp.).
70. H. Zimmerman and N. T. Kim, *Polym. Sci. Eng.* **20**(10), 680–683 1980.
71. U.S. Pat. 4,501,878 (Feb. 26, 1985) L. J. Adams (to Eastman Kodak Co.).
72. U.S. Pat. 2,822,348 (1958), J. H. Haslam (to E. I. du Pont de Nemours & Co., Inc.).
73. U.S. Pat. 4,439,597 (Mar. 27, 1984), H. K. Hall and A. B. Padias (to Celanese Corp.).
74. U.S. Pat. 4,824,930 (Apr. 25, 1989). M. L. Doerr (to Celanese Fibers Inc.).
75. U.S. Pat. 4,511,708 (Apr. 16, 1985), T. Kasuga and co-workers (to Polyplastics Co. Ltd.).
76. U.S. Pat. 4,656,241 (Apr. 7, 1987), H. Iida, K. Azuma, and M. Hayashi (to Toray Industries Inc.).
77. C. M. Fontana, *J. Polym. Sci.*, **6**(A-1), 2343–2358 (1968); Ref. 10 lists numerous U.S. patents relating to antimony oxide as a polymerization catalyst.
78. U.S. Pat. 4,820,795 (Apr. 11, 1989), S. Hirata and Y. Watanabe (to Toyo Seikan Kaisha).
79. U.S. Pat. 4,499,261 (Feb. 12, 1985), H. Heinze and co-workers (to Davy McKee AG).
80. U.S. Pat. 4,973,655 (Nov. 27, 1990), G. Pipper and co-workers (to BASF AG).
81. U.S. Pat. 5,064,935 (Nov. 12, 1991), R. Jackson, D. J. Lowe, and C. A. Stewart (to E. I. du Pont de Nemours and Co., Inc.).
82. I. Goodman and B. F. Nesbitt, *J. Poly. Sci.* **48**, 423–433 (1960).
83. S. Shiono; *J. Poly. Sci. Poly. Chem. Ed.* **17**, 4123 (1979).
84. G. C. East and A. M. Girshab, *Polymer* **23**, 323–324 (1982).
85. U.S. Pat. 3,718,621 (Feb. 27, 1973), W. K. Wilson (to Goodyear Tire & Rubber Co.).
86. U.S. Pat. 3,816,377 (June 11, 1974), Y. Okuzumi (to Goodyear Tire & Rubber Co.).
87. U.S. Pat. 4,064,112 (Dec. 20, 1977), H. J. Rothe and co-workers (to Zimmer AG).
88. U.S. Pat. 4,161,578 (July 17, 1979), D. J. Herron (to Bepex Corp.).
89. L. H. Buxbaum, *Angew. Chem., Int. Ed.* **7**, 182–190 (1968).
90. U.S. Pat. 4,137,278 (Jan. 30, 1979), A. L. Lemper and J. C. Rosenfeld (to Hooker Chemicals and Plastics Corp.).
91. I. Goodman, Ref. 1, p. 22.
92. H. Zimmerman and N. T. Kim, *Polym. Eng. Sci.* **20**(10), 680–683 (1980).
93. E. P. Goodings, *Soc. Chem. Ind. (London), Chem. Eng. Group*, monograph 13, Society of Chemical Industry, London, 1961, p. 2112.
94. K. Tomita, *Polymer* **18**, 295 (1977).
95. A. von Hassell, *Plast. Technol.*, 70 (Jan. 1979).
96. U.S. Pat. 3,549,593 (Dec. 22, 1970), T. Takekoshi (to General Electric Co.).
97. U.S. Pat. 4,161,470 (July 17, 1979), G. W. Calundann (to Celanese Corp.).
98. U.S. Pat. 4,431,770 (Feb. 14, 1984), A. J. East and G. W. Calundann (to Celanese Corp.).
99. U.S. Pat. 4,429,105 (Jan. 31, 1984), L. F. Charbonneau (to Celanese Corp.).
100. U.S. Pat. 4,393,191 (July 12, 1983), A. J. East (to Celanese Corp.).

101. U.S. Pat. 4,093,595 (June 6, 1978), S. P. Elliot (to Du Pont).
102. U.S. Pat. 4,421,908 (Dec. 20, 1983), A. J. East (to Celanese Corp.).
103. *Plast. Technol.*, 55 (Apr. 1977), *Ibid.*, 43 (Sept. 1977).
104. P. T. DeLassus and co-workers, *Mod. Plast.* 86–88 (Jan. 1983).
105. U.S. Pat. 3,733,309 (May 15, 1973), N. C. Wyeth and R. N. Roseveare.
106. U.S. Pat. 3,778,214 (Dec. 11, 1973), N. C. Wyeth and co-workers.
107. R. W. Tock and co-workers, *Adv. Polym. Technol.* **4**, 307–315 (1984).
108. E. A. Muccio, *Plastics Processing Technology*, Materials Information Society, 1994, Chapt. 5.
109. M. Cakmak, J. L. White, and J. E. Spruiell, *J. Appl. Poly. Sci.* **30**, 3679–3695 (1985).
110. M. Cakmak, J. E. Spruiell, and J. L. White, *Polym. Eng. Sci.* **24**(18), 1390–1395 (1984).
111. A. von Hassell, *Plast. Technol.* 74 (Jan. 1979).
112. *Ibid.*, pp. 69–76.
113. *Ibid.*, p. 76.
114. D. G. Callander, *Polym. Eng. Sci.* **25**(8), 453–457 (1985).
115. U.S. Pat. 4,578,437 (Mar. 25, 1986), R. R. Light (to Eastman Kodak Co.).
116. P. A. Aspy and E. E. Denison, *Mod. Plast.*, 74–76 (Aug. 1983).
117. D. G. Callander, Ref. 117, p. 456.
118. L. M. Sherman, *Plast. Technol.*, 55 (May 1995).
119. L. M. Sherman, *Plast. Technol.*, 57–58 (May 1995).
120. *Res. Discl.* **283**, 735–737 (1987).
121. *Ibid.*, pp. 705–710.
122. *Ibid.*, pp. 711–716.
123. *Ibid.*, pp. 699–703.
124. A. K. Mitchell, *Mod. Plast.* **71**(12), B48–B50 (1994).
125. L. M. Sherman, *Plast. Technol.*, 77 (May 1995).
126. R. Seymour, *Mod. Plast.* (1982–1994).
127. U.S. Pat. 4,074,032 (Feb. 14, 1978), H. Naarman and co-workers (to BASF AG).
128. U.S. Pat. 4,143,221 (Mar. 6, 1979), H. Naarman and co-workers (to BASF AG).
129. U.S. Pat. 4,200,703 (Apr. 29, 1980), K. Diebel and co-workers (to Chemische Werke Hüls AG).
130. U.S. Pat. 4,360,455 (Nov. 23, 1982), G. Lindenschmidt and co-workers (to BASF AG).
131. U.S. Pat. 4,352,909 (Oct. 5, 1982), H. J. Barda and S. L. Gray (to Ferro Corp.).
132. U.S. Pat. 4,879,353 (Nov. 7, 1989), D. C. Sanders and co-workers (to Great Lakes Chemical Corp.).
133. *Mod. Plast.* **71**(12), A40 (1994).
134. R. de P. Daubeny and C. W. Bunn, *Proc. Royal Soc.* (*London*) **A226**, 531 (1954).
135. S. Fakirov, E. W. Fischer, and G. F. Schmidt, *Makromol. Chem.* **176**, 2459 (1975).
136. E. Bornschegl and R. Bonart, *Coll. Polym. Sci.* **258**, 319 (1980).
137. Y. Kitano, Y. Kinoshita, and T. Ashida, *Polymer* **36**, 10 (1995).
138. M. Droscher, *Makromol. Chem.* **181**, 789 (1980).
139. H. W. Starkweather, P. Zoller, and G. A. Jones, *J. Poly. Sci. Poly. Phys. Ed.* **21**, 295 (1983).
140. J. Menczel and B. Wunderlich, *J. Poly. Sci., Poly. Lett. Ed.* **19**, 261 (1981).
141. K. H. Illers and H. Breuer, *J. Coll. Sci.* **18**, 1–31 (1963); J. H. Dumbleton, T. Murayama, and J. P. Bell, *Kolloid Z. Z. Polym.* **228**, 54–58 (1968).
142. J. Y. Jadhav and S. W. Kantor, in J. I. Kroschwitz, ed., *Encyclopedia of Polymer Science and Engineering*, 2nd ed., Vol. 12, John Wiley & Sons, Inc., New York, 1988, pp. 217–256.
143. S. Berkowitz, *J. Appl. Poly. Sci.* **29**, 4353–4361 (1984).
144. Technical data, *IMPET Resins*, Hoechst-Celanese Corp., Summit, N. J., 1994.

145. M. Grossman, *Electron. Des.* **37**(7), 19–24 (1989); C. Lodge, *Plast. World* **47**(1), 48–50 (1989).
146. A. B. Auerbach and J. W. Sell, *Poly. Sci. Eng.* **30**(17), 1041–1050 (1990).
147. WO Pat. 9,117,209 (Nov. 14, 1991), M. L. Cassell and co-workers (to Eastman Kodak Co.).
148. EP 273,149 (July 6, 1988), M. G. Minnick (to General Electric Co.).
149. Jpn. Pat. 3,009,948 (June 6, 1989), (to Toray Industries).
150. U.S. Pat. 5,124,435 (June 23, 1992), H. Mori and co-workers (to Mitsubishi Rayon Co. Ltd).
151. U.S. Pat. 5,194,573 (Mar. 16, 1993), W. Schmidt (to Zimmer AG).
152. U.S. Pat. 5,198,530 (Mar. 30, 1993), M. Kyber and co-workers, (to Zimmer AG).
153. R. M. Schulken, R. E. Boy, and R. H. Cox, *J. Poly. Sci., Part C*, 17–25 (1964).
154. Eastman Publication PPD-115B, Nov. 1994.
155. M. C. Gabriele, *Plast. Technol.*, 43 (May 1995).
156. Ger. Offen. 2,042,447 (Apr. 15, 1971) D. D. Zimmerman and R. B. Isaacson; U.S. Appl. (Aug. 29, 1969), (to Celanese Corp.).
157. Ger. Offen. 2,533,358 (Feb. 26, 1976) F. M. Berardinelli; U.S. Appl. 496,607 (Aug. 12, 1974), (to Celanese Corp.).
158. E. Bornschegl and R. Bonart, *Coll. Poly. Sci.* **258**, 319 (1980).
159. I. H. Hall and M. G. Pass, *Polymer* **17**, 807 (1976).
160. R. Jakeways, I. M. Ward, and M. A. Wilding, *J. Polym. Sci., Poly. Phys.* **13**, 807–809 (1973).
161. K. H. Illers, *Coll. Polym Sci.* **258**(2), 117 (1980).
162. S. Z. D. Cheng, R. Pan, and B. Wunderlich, *Makromol. Chem.* **189**, 2443 (1988).
163. C. F. Pratt and S. Y. Hobbs, *Polymer* **17**, 12 (1976).
164. K. Tashiro and co-workers, *Macromolecules* **13**, 137 (1980).
165. W. F. H. Borman, *J. Appl. Poly. Sci.* **22**, 2119 (1978).
166. H-N. Sung, personal communication, Hoechst Celanese Corp., Summit, N.J., Aug. 1995.
167. J. C. Gill, *Plast. Compound.*, 77–81 (Sept.–Oct. 1989).
168. L. M. Sherman, *Plast. Technol.*, 32–35 (May 1994); B. Miller, *Plast. World*, 38–40 (Jan. 1995).
169. L. M. Sherman, *Plast. Technol.*, 27–28 (July 1995).
170. *CELANEX*, data sheet CX-1A, Hoechst Celanese Corp., Summit, N.J., 1992.
171. C. B. Bucknall, *Toughened Plastics*, Applied Science, London, 1977.
172. C. B. Bucknall and C. J. Page, *J. Mater. Sci.*, 808 (1982).
173. D. J. Hourston, S. Lane, and H. X. Zhang, *Polymer* **32**(12), 2215–2220 (1991).
174. S. Ostrowski de Kiewiet, *Mod. Plast.* **71**(12), B67–68 (1994).
175. L. M. Sherman and J. DeGaspari, *Plast. Technol.*, 71 (Nov. 1995).
176. N. R. Legge, G. Golden, and H. E. Schroeder, eds., *Thermoplastic Elastomers*, Hanser, Munich, Germany, pp. 163–196, 1987.
177. B. M. Walker and C. P. Rader, eds., *Handbook of Thermoplastic Elastomers*, 2nd ed., Van Nostrand Rheinhold Co., New York, 1988 pp. 181–223.
178. W. Witsiepe, *ACS Adv. Chem. Ser.* **129**, 39 (1973).
179. W. H. Buck, R. J. Cella, E. Gladding, and J. Wolfe, Jr., *J. Poly. Sci. (C)* **48**, 47–60 (1974).
180. S. Nelsen, S. Gromelski, and J. J. Charles, *J. Elastom. Plast.* **15**, 256–264 (Oct. 1983).
181. R. J. Cella, *J. Poly. Sci., Symp. Ser.* **42**, 727–740 (1973).
182. T. Hasketh, J. Van Bogart and S. Cooper; *Poly. Eng. Sci.* **20**(3), 190–197 (Feb. 1980).
183. A. Liliaonitkul, J. West, and S. Cooper, *J. Macromol. Sci.-Phys.* **B-12**(4), 563–597 (1976).

184. R. W. Seymour, J. R. Overton, and L. S. Corley, *Macromolecules* **8**, 331 (1975).

185. M. Shen, U. Mehra, M. Niiromi, J. T. Koberstein, and S. L. Cooper, *J. Appl. Phys.* **45**, 4182 (1974).

186. J. West, A. Lilaonitkul, S. Cooper, U. Mehra, and M. Shen, *ACS Polym. Prepr.* **15**(2), 191 (1974).

187. A. Lilaonitkul and S. Cooper, *Rubber Chem. Technol.* **50**, 1–23 (1977).

188. N. K. Kaloglou, *J. Appl. Poly. Sci.* **21**, 543–554 (1977).

189. J. Miller, J. McKenna, G. Pruckmayr, J. E. Epperson, and S. L. Cooper, *Macromolecules* **18**, 1727–1736 (1985).

190. A. A. Apostolov and S. Fakirov, *J. Macromol. Sci. Phys.*, 329–355 (1992).

191. S. Fakirov and co-workers, *Colloid Polym. Sci.*, 811–823 (1993).

192. *Riteflex Thermoplastic Polyester Elastomer*, Bulletin RF-4, Hoechst Celanese Corp., Summit, N.J., 1993.

193. *Riteflex Thermoplastic Polyester Elastomer Chemical Resistance Table*, Material Monograph MRF-002, Hoechst-Celanese Corp., Summit, N.J.

194. *HYTREL 10MS*, Bulletin HYT-114, Du Pont, Wilmington, Del.

195. U.S. Pat. 4,520,149 (May 28, 1985), M. D. Golder (to GAF Corp.).

196. M. D. Golder and Mulholland, *Thermoplastic Elastomers*, Vol. 3, Rapra Technology Ltd., 1991, pp. 20–23.

197. U.S. Pat. 4,405,749 (Sept. 20, 1983), S. B. Nelsen (to GAF Corp.).

198. U.S. Pat. 3,261,812 (July 19, 1966), A. Bell, C. J. Kibler, and J. G. Smith (to Eastman Kodak Co.).

199. J. R. Wolfe, *Rubber Chem. Technol.* **50**, 688 (1977).

200. G. K. Hoeschele, *Angew. Makromol. Chemie.* **58–59**, 299 (1977).

201. U.S. Pat. 3,954,689 (May 4, 1976), G. K. Hoeschele (to E. I. du Pont de Nemours and Co., Inc.).

202. J. Wolfe, *ACS Adv. Chem. Ser.* **176**, 129 (1979).

203. J. Hudson, *Rubber World*, 23–26 (1991).

204. H. Blum, *1989 TPE Conference*, Orlando, Fla., Mar. 1989.

205. P. Stahl, Paper No. 57, ACS Rubber Division, Cincinnati, Ohio, Oct. 1988.

206. *Mod. Plast.*, 42–45 (Mar. 1985).

207. R. Peffer, in Ref. 204.

208. Harding and Gerteisen, *SPE ANTEC*, 1991, pp. 2409–2013.

209. M. Dolde, *1991 TPE Conference*, Orlando, Fla., Feb. 1991.

210. M. D. Golder, *1990 TPE Conference*, Dearborn, Mich., Mar. 1990.

211. U.S. Pat. 3,819,587 (June 25, 1974), S. L. Kwoleck (to E. I. du Pont de Nemours and Co., Inc.).

212. E. B. Priestley, P. J. Wojtowicz, and P. Sheng, *Introduction to Liquid Crystals*, Plenum Press, New York, 1975.

213. C. Carfagna, *Liquid Crystal Polymers*, Permagon Press, Elmsford, N.Y., 1994.

214. M. G. Dobb and J. E. McIntyre, *Adv. Polym. Sci.* **60–61**, 63–98 (1984).

215. G. W. Calundann and M. Jaffe, "Anisotropic Polymers—Their Synthesis and Properties," *Proceedings of the Robert A. Welch Conference on Chemical Research XXVI*, Nov. 1982, Houston, Tex., p. 247.

216. T. S. Chung, G. W. Calundann, and A. J. East, in N. P. Cheremisinoff, ed., *Handbook of Polymer Science and Technology*, Marcel Dekker, New York, 1989, pp. 625–675.

217. P. J. Flory, *Proc. Royal Soc. (London)*, Ser. A **234**, 60 (1956).

218. P. J. Flory and G. Ronca, *Mol. Cryst. Liq. Cryst.* **54**, 289 (1979).

219. P. J. Flory and M. Gordon, *Advances in Polymer Science Series*, Vol. 59, Springer-Verlag, New York, 1984, p. 1.

220. U.S. Pat. 3,637,597 (Jan. 25, 1972), S. G. Cottis, J. Economy, and B. E. Nowak.

221. U.S. Pat. 4,067,852 (Jan. 10, 1978), G. W. Calundann (to Celanese Corp.).

222. U.S. Pat. 4,083,829 (Apr. 11, 1978), G. W. Calundann, H. L. Davis, F. J. Gorman, and R. M. Mininni (to Celanese Corp.).

223. U.S. Pat. 4,902,772 (Feb. 20, 1990), N. Carter, B. P. Griffin, W. A. McDonald, and T. G. Ryan (to Imperial Chemical Industries plc).
224. A. Bunn, B. P. Griffin, W. A. McDonald, and D. G. Rance, *Polymer* **33**(14), 3066–3072 (1992).
225. U.S. Pat. 4,184,996 (Jan. 22, 1980), G. W. Calundann (to Celanese Corp.).
226. U.S. Pat. 4,219,461 (Aug. 26, 1980), G. W. Calundann (to Celanese Corp.).
227. U.S. Pat. 4,256,624 (Mar. 17, 1981), G. W. Calundann (to Celanese Corp.).
228. U.S. Pat. 4,473,682 (Sept. 25, 1984), G. W. Calundann, L. F. Charbonneau, and B. C. Benicewicz (to Celanese Corp.).
229. U.S. Pat. 4,341,688 (July 27, 1982), L. F. Charbonneau, G. W. Calundann, and A. J. East (to Celanese Corp.).
230. U.S. Pat. 4,351,917 (Sept. 28, 1982), G. W. Calundann, L. F. Charbonneau, and A. J. East (to Celanese Corp.).
231. U.S. Pat. 4,351,918 (Sept. 28, 1982), L. F. Charbonneau, A. J. East, and G. W. Calundann (to Celanese Corp.).
232. U.S. Pat. 4,355,132 (Oct. 19, 1982), A. J. East, G. W. Calundann, and L. F. Charbonneau (to Celanese Corp.).
233. A. J. East, L. F. Charbonneau, and G. W. Calundann, *Mol. Cryst. Liq. Cryst.* **157**, 615–637 (1988).
234. S. Kenig, *Polym. Eng. Sci.* **27**(12), 87–892 (1987).
235. U.S. Pat. 5,110,896 (May 5, 1992), M. G. Waggoner and M. R. Samuels (to E. I. du Pont de Nemours and Co., Inc.).
236. U.S. Pat. 5,250,654 (Oct. 5, 1993), G. R. Alms, M. R. Samuels, and M. G. Waggoner (to E. I. du Pont de Nemours and Co., Inc.).
237. G. Kiss, *Polym. Eng. Sci.* **27**(6), 410–423 (1987).
238. *Plastics Technology, Manufacturing Handbook and Buyers' Guide*, Aug. 1995, pp. 490–492.
239. P. Zhuang, T. Kyu, and J. L. White, *Polym. Eng. Sci.* **28**(17), 1095–1106.
240. F. N. Cogswell, in L. L. Chapoy, *Recent Advances in Liquid Crystalline Polymers*, Elsevier, New York, 1985.
241. D. W. Giles and M. M. Denn, *J. Rheol.* **38**, 617–637 (1994).
242. J. M. Dealy and K. Wissbrun, *Melt Rheology and Its Role in Plastics Processing*, Van Nostrand Reinhold, New York, 1990, pp. 424–440.
243. U.S. Pat. 4,468,364 (Aug. 28, 1984), Y. Ide (to Celanese Corp.).
244. C. E. McChesney and J. R. Dole, *Mod. Plast.*, 112–117 (Jan. 1988).
245. F. C. Jaarsma, *SPE ANTEC 1990 Preprints*, pp. 406–407.
246. A. Kaslusky, *Mod. Plast.*, B69–B70 (Nov. 1994).
247. L. K. English, *Mater. Eng.* **106**(6), 29 (June 1989).
248. L. M. Sherman, *Plast. Technol.*, 75 (May 1995).

*General References*

L. M. Sherman, *Plast. Technol.*, 62 (Mar. 1996).
*Chem. Eng. News*, 10 (Apr. 1996).

ANTHONY J. EAST
MICHAEL GOLDEN
Hoechst Celanese Corporation


SUBHASH MAKHIJA
Consultant

# POLYESTERS, UNSATURATED

Low molecular weight polyester polymers derived from unsaturated dibasic acids (or anhydrides) dissolved in unsaturated vinyl monomers, comprise a versatile family of thermosetting materials known generally as unsaturated polyester resins. Prototype resins produced in the early 1940s were used in conjunction with glass fiber reinforcements to produce the first composite plastics. The high strength and radiotransparency of this material expanded its role in radome covers employed in large quantities toward the end of World War II.

Markets for the resins have expanded rapidly; the dominant applications are still in conjunction with glass fiber reinforcement to form laminar composites known generically as fiber glass-reinforced plastic (FRP) in the United States and glass fiber-reinforced plastic (GRP) in Europe and elsewhere. Resins have also evolved for use in casting processes, which usually contain high loadings of fillers or mineral aggregate and are defined as one form of polymer concrete. By 1990, global production of unsaturated polyester resins had exceeded 1,500,000 metric tons, with the United States being the largest market, at over 600,000 metric tons.

## Raw Materials

The properties of polymers formed by the step growth esterification (1) of glycols and dibasic acids can be manipulated widely by the choice of coreactant raw materials (Table 1) (2). The reactivity fundamental to the majority of commercial resins is derived from maleic anhydride [108-31-6] (MAN) as the unsaturated component in the polymer, and styrene as the coreactant monomer. Propylene glycol [57-55-6] (PG) is the principal glycol used in most compositions, and (ortho)-phthalic anhydride (PA) is the principal dibasic acid incorporated to moderate the reactivity and performance of the final resins.

Isophthalic (m-phthalic) acid [121-91-5] (IPA) is selected to enhance thermal endurance as well as to produce stronger, more resilient cross-linked plastics that demonstrate improved resistance to chemical attack. Terephthalic (p-phthalic) acid [100-21-0] (TA) provides somewhat similar properties as isophthalic acid but is only used in selective formulations due to the limited solubility of these polyester polymers in styrene [100-42-5] (see PHTHALIC ACID AND OTHER BENZENEPOLYCARBOXYLIC ACIDS).

Other glycols (qv) can be used to impart selective properties to these simple compositions. Ethylene glycol [107-21-1] (EG) is used to a limited degree to reduce cost, whereas diethylene glycol [111-46-6] (DEG) produces a more flexible polymer that can resist cracking when impacted. Neopentyl glycol [126-30-7] (NPG) is used in most commercial products to improve uv and water resistance. Alkoxylated derivatives of bisphenol A (BPA) not only impart a high degree of resistance to strong acidic and alkaline environments, but also provide resistance to deformation and creep at elevated temperatures.

Long-chain aliphatic acids such as adipic acid (qv) [124-04-9] are generally used to improve flexibility and enhance impact properties, demonstrating subtle

**Table 1. Ingredients of Polyester Resin Formulations in Descending Order of Commercial Significance**

| Glycol | Dibasic acid or anhydride | Unsaturated acid or anhydride | Unsaturated monomer |
|---|---|---|---|
| propylene glycol | phthalic anhydride | maleic anhydride | styrene |
| diethylene glycol | dicyclopentadiene– | fumaric acid | vinyltoluene |
| ethylene glycol | maleic anhydride[a] | methacrylic acid | methyl methacrylate |
| neopentyl glycol | isophthalic acid | acrylic acid | diallyl phthalate |
| dipropylene glycol | adipic acid | itaconic acid | $\alpha$-methylstyrene |
| dibromoneopentyl | chlorendic | | triallyl cyanurate |
| glycol | anhydride[b] | | divinylbenzene |
| bisphenol A diglycidyl | tetrabromophthalic | | |
| ether | anhydride | | |
| bisphenol A dipropoxy | tetrahydrophthalic | | |
| ether | anhydride | | |
| tetrabromobisphenol | terephthalic acid | | |
| diethoxy ether | tetrachlorophthalic | | |
| propylene oxide | anhydride | | |
| 1,4-butanediol | | | |

[a] Acid addition product, formed *in situ*.
[b] 1,4,5,6,7,7-Hexachlorobicyclo(2,2,1)-5-heptene-2,3-dicarboxylic anhydride [115-27-5].

improvements over resins modified with the ether glycols (diethylene glycol) and polyether glycols (polypropylene glycol) (see POLYETHERS).

Novel polyester compositions have also been derived from dicyclopentadiene [77-73-6] (DCPD), which can enter into two distinct reactions with maleic anhydride to modify properties for lower cost. These compositions have effectively displaced o-phthalic resins in marine and bathtub laminating applications.

Recycled poly(ethylene terephthalate) (PET), which offers excellent properties at potentially lower cost, is finding wider use as a raw material component and meeting increasing demands for environmentally compatible resins (see POLYESTERS, THERMOPLASTIC; RECYCLING, PLASTICS).

Other minor raw materials are used for specific needs. Fumaric acid [110-17-8], the geometric isomer of maleic acid, is selected to maximize thermal or corrosion performance and is the sole acid esterified with bisphenol A diol derivatives to obtain optimum polymer performance. Cycloaliphatics such as hydrogenated bisphenol A (HBPA) and cyclohexanedimethanol (CHDM) are used in selective formulations for electrical applications. Tetrahydrophthalic anhydride [85-43-8] (THPA) can be used to improve resilience and impart useful air-drying properties to polyester resins intended for coating or lining applications.

Halogenated intermediates, dibromoneopentyl glycol [3296-90-0] (DB-NPG), and alkoxylated derivatives of tetrabromobisphenol A are used extensively in flame-retardant applications. Similar properties can be derived from halogenated dibasic acids, chlorendic anhydride [115-27-5] (CAN), and tetrabromophthalic anhydride [632-79-1] (TBPA). Processes can be used to produce brominated products by the *in situ* bromination of polymers derived from tetrahydrophthalic anhydride.

Bisphenol A diglycidyl ether [1675-54-3] reacts readily with methacrylic acid [71-49-4] in the presence of benzyldimethylamine catalyst to produce bisphenol epoxy dimethacrylate resins known commercially as vinyl esters. The resins display beneficial tensile properties that provide enhanced structural performance, especially in filament-wound glass-reinforced composites. The resins can be modified extensively to alter properties by extending the diepoxide with bisphenol A, phenol novolak, or carboxyl-terminated rubbers.

Monomers such as methyl methacrylate [80-62-6] are often used in combination with styrene to modify refractive index and improve uv resistance. Vinyltoluene [25013-15-4] and diallyl phthalate [131-17-9] are employed as monomers in selective molding compositions for thermal improvements.

## Process Equipment

The polyesterification reaction is normally carried out in stainless steel vessels ranging from 8,000–20,000 liters, heated and cooled through internal coils (Fig. 1). Blade agitators revolving at 70–200 rpm are effective in stirring the low viscosity mobile reactants, which are maintained under an inert atmosphere of nitrogen or carbon dioxide during the reaction at temperatures up to 240°C.

Weigh tanks or meters measure the liquid glycols into the reactor. Solids are usually added from 25-kg bags or 1000-kg supersacks. Silo and auger are used widely for isophthalic acid, whereas phthalic anhydride [85-44-9] and maleic anhydride are metered in molten form. A packed condenser efficiently separates water from glycol. The glycol is refluxed back into the reactor and the remaining

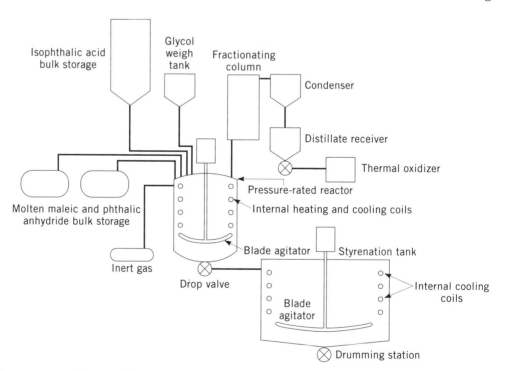

**Fig. 1.**   Reactor system for the manufacture of polyester resins.

condensate is incinerated on-site using a thermal oxidizer. Once the polymer is formed, it is cooled to below 180°C and drained into a cooled blend tank containing styrene monomer under high agitation.

Glass-lined reactor systems are used occasionally for halogenated resins to prevent corrosion of the reactor components. Copper and brass fitting should be avoided due to the significant influence on resin cure characteristics.

## Polyesterification

The reaction of glycols with dibasic acid anhydrides, such as phthalic and maleic anhydrides, proceeds at above 100°C, ending with the exothermic formation of the acid half-ester produced by the opening of the anhydride ring. The reaction exotherm effectively raises the temperature of the reactants to over 150°C, at which point the half-esters condense into polymers with the evolution of by-product water. As the reaction proceeds, the viscosity of the reactants increases, restricting the release of water, so that the temperatures must be gradually increased to 200°C to maintain a steady evolution of condensate water. Resins normally lose between 8–12% of initial charge weight as condensate.

The polyesterification reaction is reversible because it is influenced by the presence of condensate water in equilibrium with the reactants and the polymer. The removal of water in the latter part of the reaction process is essential for the development of optimum molecular weight, on which the ultimate structural performance depends.

The polyesterification reaction is carried out in the presence of an inert gas, such as nitrogen or carbon dioxide, to prevent discoloration. Usually, the sparge rate of the inert gas is increased in the final stages of polyesterification to assist the removal of residual water. Although the removal of water can be facilitated by processing under vacuum, this is rarely used on a commercial scale.

The esterification rate can be accelerated by acid catalysts (3) such as *para*-toluenesulfonic acid and tetrabutyl titanate, but tin salts such as hydrated monobutyl tin oxide are preferred for reasons of product stability during storage. The polyesterification process can be reversed by injecting steam into the reactants; this can be used to control the final molecular weight attained by the polymer. Transesterification during polymer formation also occurs, which leads to changes in the distribution of polymers in the final product. Transesterification is employed on a commercial scale to recycle poly(ethylene terephthalate) waste into useful resins by digesting an equivalent amount of glycol in the presence of catalysts such as tetraisopropyl titanate or zinc acetate. The terephthalate esters subsequently react with maleic anhydride to produce unsaturated terephthalate polyester polymers, which have properties similar to their isophthalic homologues.

The viscosity of the final polymer melt usually limits the progress of molecular weight development, and number-average values ($M_n$) between 1800–2500 are normally observed. Other side reactions also modify molecular weight growth. Side reactions are influenced by the choice of reactants. Ethylene glycol can form cyclic esters with phthalic anhydride; maleic anhydride can produce addition products (4) with lower glycols, forming trifunctional succinate derivatives that lead to high molecular weight branched polymers and the possibility of gelation during esterification. To maintain maximum unsaturation levels during esterification, fumaric acid is often used in place of maleic anhydride. Fumaric acid is also used widely in formulations involving alkoxylated bisphenol A to obtain optimum thermal and corrosion performance.

Polyesterification involving insoluble reactants such as isophthalic acid is normally carried out in two-stage reactions, in which isophthalic acid reacts first with the glycol to form a clear melt. The balance of the reactants, including maleic anhydride, is then added to complete the polyester polymer, thus avoiding longer cycle times and some discoloration.

Polyester resins can also be rapidly formed by the reaction of propylene oxide (5) with phthalic and maleic anhydride. The reaction is initiated with a small fraction of glycol initiator containing a basic catalyst such as lithium carbonate. Molecular weight development is controlled by the concentration of initiator, and the highly exothermic reaction proceeds without the evolution of any condensate water. Although this technique provides many process benefits, the low extent of maleate isomerization achieved during the rapid formation of the polymer limits the reactivity and ultimate performance of these resins.

Comparable process techniques involving the reaction of bisphenol A diglycidyl ether with methacrylic acid produce the corresponding hydroxy esters rapidly. Because of the monofunctional nature of methacrylic acid, the resins formed are not strictly polyesters but oligomer diesters. Their reactions with styrene have characteristics similar to polyesters.

**Maleic Isomerization.** Polyester polymers are formulated using maleic anhydride (**1**) as the common unsaturated moiety. During the course of the polyesterification reaction at 200°C, the *cis*-maleate ester (**2**) isomerizes to the *trans*-fumarate (**3**). The fundamental reactivity of the final polyester with styrene is directly proportional to the degree of isomerization and the level of fumarate polymers formed during the course of esterification. Maleate polymers in the cis form create strain across the double bond, causing some displacement from a planar configuration; fumarate polymers in the trans configuration are influenced less by steric effects and the double bond can assume a planar configuration conducive to addition copolymerization reactions with styrene. Branched asymmetric reactants such as propylene glycol and bulky aromatic dibasic acids such as isophthalic acid create sufficient steric interference to promote isomerization to the *trans*-fumarate polymers, whereas linear glycols such as ethylene and diethylene glycol and linear dibasic acids such as adipic acid produce resins that have higher levels of the maleate polymer.

(1)     (2)     (3)

The temperature of esterification has a significant influence on isomerization rate, which does not proceed above 50% at reaction temperatures below 150°C. In resins produced rapidly by using propylene oxide and mixed phthalic and maleic anhydrides at 150°C, the polyester polymers, which can be formed almost exclusively in the maleate conformation, show low cross-linking reaction rates with styrene.

Isomerization is facilitated by esterification at temperatures above 200°C or by using catalysts, such as piperidine and morpholine (6), that are effective in raising isomerization of fumarate to 95% completion. Resins made by using fumaric acid are exclusively fumarate polymers, demonstrate higher reactivity rates with styrene, and lead to a complete cross-linking reaction.

**End-Group Analysis.** The reaction of balanced stoichiometric glycol–dibasic acid ingredients theoretically produces polymers with equivalent acid and hydroxyl end groups. Commercial processes, however, normally include an excess of glycol to offset distillation losses. Depending on column efficiency, the polymer can have a higher relative hydroxyl value and a much lower acid value (Fig. 2), and this controls the ultimate molecular size. Most commercial laminating resins have acid values ranging from 25–30, whereas higher molecular weight isophthalic resins have values that fall between 10–15. Epoxy dimethacrylates, produced in a slight excess of methacrylic acid, have a final acid value of 3–5.

**Molecular Weight.** Unsaturated polyester resins are relatively low in molecular weight, and are formulated to achieve low working viscosities when dissolved in styrene. The $M_n$ normally falls between 1800–2500, although dicyclopentadiene and orthophthalic resins can be useful below this range. The molecular weight follows a Gaussian distribution curve (Fig. 3), the shape of which influences final solution viscosities. In phthalic resins, the presence of a

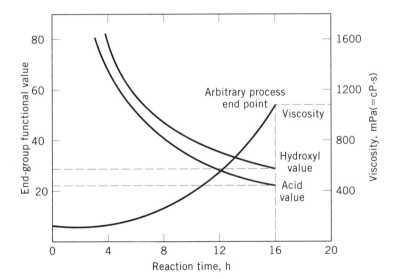

**Fig. 2.** Functional end groups and solution viscosity during polyester resin manufacture. Acid value is defined as the milligrams of KOH required to neutralize 1 g of polymer; hydroxyl value is defined as the milligrams of acid equivalent required to neutralize 1 g of polymer. Solution viscosities are determined at 60 wt % in styrene.

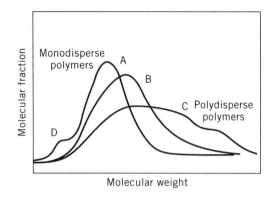

**Fig. 3.**  Molecular weight distribution curves as determined by gel-permeation chromatography. A represents *ortho*-phthalic resins; B, highest molecular weight isophthalic resins; C, high molecular weight polydisperse polymers having high maleic addition reactions; and D, low molecular weight fraction seen in most phthalic resins.

low molecular weight fraction is usually observed, whereas in high maleic resins, a high molecular weight shoulder is observed. This indicates significant glycol addition across the double bond to form a trifunctional reactant. Polymers having different chain lengths and compositions exist as a compatible mixture, the ratio of weight-average molecular weight $(M_w)$ and number-average molecular weight $(M_n)$ is defined as the polydispersity, $D$. For $o$-phthalic and isophthalic resins, $D$ is just over two, but higher molecular weight resins and resins containing high maleic levels are more polydisperse and the distribution curve becomes flatter as the resin viscosity increases.

ortho-Phthalic resins consist of chains having an average of 15 ester groups and a narrow distribution curve. The molecular weight development during the manufacturing process is normally followed by using solution viscosity that tends to increase exponentially in the final stages of the reaction. Arbitrary end points are usually established well before this stage to avoid gelation in the reactor (see Fig. 2).

## Formulation

*ortho*-**Phthalic Resins.**    Resins based on *ortho*-phthalic anhydride (Table 2) comprise the largest group of polyester resins and are used in a variety of commercially significant applications, including marine craft, translucent glazing, simulated-marble vanity sets, and buttons. Most laminating and casting processes also rely on both colored and clear gel coats to provide some level of surface protection. The glycol generally controls required performance; the phthalic–maleic anhydride ratio is adjusted to modify the reactivity according to fabrication needs.

The glycol is charged to the reactor and heated to 50°C under an inert-gas sparge. The lower melting maleic anhydride is metered in, allowing the temperature to rise to 100°C, followed by the phthalic anhydride. A clear solution forms and the temperature is increased to 120°C to initiate the exothermic reaction between the glycol and anhydride forming the half-ester. The temperature increases rapidly to 150°C as the exotherm heat is released, at which point the

**Table 2. Molar Component Ratio Used in *ortho*-Phthalic Formulations**

| Resin | Glycol | | | Acid | | Styrene monomer |
|---|---|---|---|---|---|---|
| | PG | DEG | NPG | PA | MAN | |
| marine | 1.0 | | | 0.5 | 0.5 | 1.0 |
| marble | 0.8 | 0.2 | | 0.65 | 0.35 | 1.0 |
| gel-coat | 0.5 | | 0.5 | 0.5 | 0.5 | 1.0 |
| button | 0.85 | 0.15 | | 0.60 | 0.4 | 1.0 |

half-ester begins to condense into low molecular weight polymers. The increasing viscosity of the melt tends to restrict the release of water vapor, and the reactants can foam into the condenser unless the reaction is controlled through cooling. The temperature is gradually increased to 190°C to maintain a constant removal of the condensate, but falls off as the reaction nears completion, usually in 12 hours. Solid phthalic anhydride sublimes out in the latter stages of the reaction, requiring close control of the inert-gas flow and temperature. Some processors have used xylene as an azeotropic solvent to assist water removal while suppressing the sublimation of the phthalic anhydride, but this is not widely applied.

Phthalic resins are usually processed to an acid number of 25–35, yielding a polymer with an average $M_n$ of 1800–2000. The solution viscosity of the polymer is usually followed to ascertain the polymer end point. The resin is cooled to 150°C and hydroquinone stabilizer (150 ppm) is added to prevent premature gelation during the subsequent blending process with styrene at 80°C. The final polymer solution is cooled to 25°C before a final quality check and drumming out for shipment.

**Isophthalic Resins.** Isophthalic acid (IPA) can be substituted for phthalic anhydride to enhance mechanical and thermal performance and improve resistance to corrosive environments. Significant products include underground gasoline storage tanks and large diameter sewer and water pipe. Although phthalic resins find wide application in ambient fabrication processes, isophthalic resins (Table 3) are more widely used in products employing high temperature forming processes such as pultruded profile and electrical-grade laminate.

Isophthalic resins are manufactured by two-stage processing to facilitate the dissolution of the isophthalic acid. In the first stage, the glycol and isophthalic acid react at temperatures of 240°C under an inert atmosphere to produce a clear melt. High pressure processing (207 kPa (30 psi)) and esterification catalysts such as hydrated monobutyl tin oxide are also widely employed to reduce the cycle times of two-stage processing. Maleic anhydride is added in the

**Table 3. Molar Component Ratio Used in Isophthalic Formulations**

| Resin | Glycol | | Acid | | Styrene monomer |
|---|---|---|---|---|---|
| | PG | DEG | IPA | MAN | |
| tank | 1.0 | | 0.5 | 0.5 | 1.2 |
| pipe | 0.5 | 0.5 | 0.4 | 0.6 | 1.2 |
| pultrusion | 0.3 | 0.7 | 0.4 | 0.6 | 1.0 |
| electrical | 1.0 | | 0.3 | 0.7 | 1.0 |

second stage and the final resin completed at 210°C to control color and molecular weight development. Isophthalic resins intended for corrosion application are processed to $M_n$ (2200–2500); the reaction cycle is about 24 hours.

The melting point of the higher molecular weight isophthalic polymer is much higher than *ortho*-phthalic resins, and blending temperatures in styrene must be increased to avoid freezing the polymer out. Stabilizers such as toluhydroquinone and benzoquinone are more effective gelation inhibitors for the higher styrene-blending temperatures used for these resins.

**Dicylopentadiene Resins.** Dicyclopentadiene (DCPD) can be used as a reactive component in polyester resins in two distinct reactions with maleic anhydride (7). The addition reaction of maleic anhydride in the presence of an equivalent of water produces a dicyclopentadiene acid maleate that can condense with ethylene or diethylene glycol to form low molecular weight, highly reactive resins. These resins, introduced commercially in 1980, have largely displaced

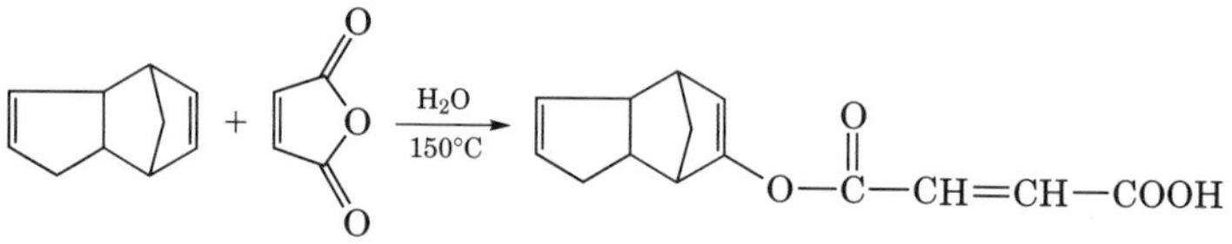

*ortho*-phthalic resins in marine applications because of beneficial shrinkage properties that reduce surface profile. The inherent low viscosity of these polymers also allows for the use of high levels of fillers, such as alumina trihydrate, to extend the resin-enhancing, flame-retardant properties for application in bathtub products (Table 4).

**Table 4. Molar Component Ratio Used in Dicyclopentadiene Formulations**

| Resin | Glycol | | | DCPD | MAN | Styrene monomer |
| | PG | EG | DEG | | | |
| --- | --- | --- | --- | --- | --- | --- |
| marine |  | 1.0 |  | 2.0 | 2.0 | 3.0 |
| molding | 1.0 |  |  | 0.5 | 1.0 | 1.0 |
| bathtub |  |  | 1.0 | 2.0 | 2.0 | 3.0 |

The cleavage of dicyclopentadiene into cyclopentadiene can be accomplished at temperatures above 160°C, producing the heterocyclic Diels-Alder maleic addition product, which opens to the diacid. This product can be esterified with propylene glycol to produce resins that demonstrate enhanced resilience and thermooxidative resistance suitable for molded electrical components.

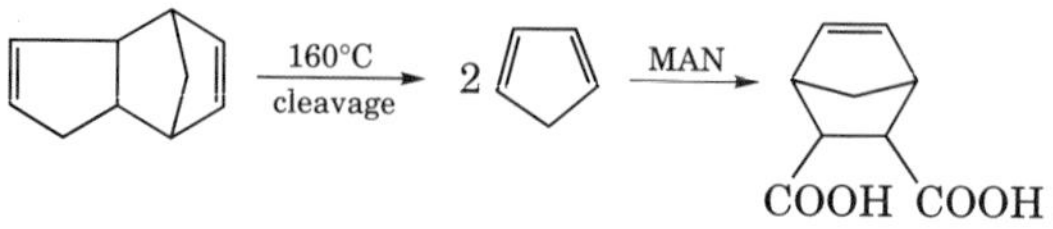

**Flame-Retardant Resins.** Flame-retardant resins are formulated to conform to fire safety specifications developed for construction as well as marine and electrical applications. Resins produced from halogenated intermediates (Table 5) are usually processed at lower temperatures (180°C) to prevent excessive discoloration. Dibromoneopentyl glycol [3296-90-0] (DBNPG) also requires glass-lined equipment due to its corrosive nature. Tetrabromophthalic anhydride (TBPA) and chlorendic anhydride (8) are formulated with ethylene glycols to maximize flame-retardant properties; reaction cycle times are about 12 hours. Resins are also produced commercially by the *in situ* bromination of polyester resins derived from tetrahydrophthalic anhydride (9).

Methyl methacrylate is often used in combination with styrene to improve light transmission and uv stability in flame-retardant glazing applications. Phosphate ester (triethyl phosphate) additives are also included to supplement flame-retardant efficiency; benzophenone uv stabilizers are required to prevent yellowing of these uv-sensitive resins.

**Table 5. Molar Component Ratio Used in Flame-Retardant Formulations**

| Resin | Glycol | | Acid | | | | Styrene monomer |
|---|---|---|---|---|---|---|---|
| | EG | DBNPG | TBPA | PA | CAN | MAN | |
| building | 1.0 | | | | 0.6 | 0.4 | 1.5 |
| glazing | | 1.0 | | 0.5 | | 0.5 | 2.0 |
| marine | 1.0 | | 0.15 | 0.35 | | 0.5 | 1.0 |

**Bisphenol Resins.** Derivatives of bisphenol A form the basis for two distinct resin groups that demonstrate superior thermal and corrosion resistance. The addition product of propylene oxide [75-56-9] and bisphenol A, reacted with fumaric acid and dissolved in styrene monomer, has established commercial significance in applications involving extreme corrosive environments. The resins known generically as bisphenol fumarates (10) have been used since 1955 in the fabrication of tanks and piping used in the chloralkali and pulp and papermaking industries. This resin is unique among polyester compositions in resisting strong alkaline solutions that readily decompose *ortho*-phthalic and isophthalic resins.

Bisphenol A diglycidyl ether reacts readily with methacrylic acid in the presence of basic catalysts, such as benzyldimethylamine and triphenyl phosphine, to form the corresponding methacrylate hydroxy esters. Unlike polyesters that rely on fumarate unsaturation distributed within the polymer chain to form the cross-linked network, the bisphenol A epoxy dimethacrylates (vinyl esters) have two reactive sites on the extremities of the polymer, and the low polydispersity defines a reactive polymer having almost uniform molecular weight. This provides some control of the cross-linked network, which expands by a regulated mechanism and results in products that have high flexural properties and inherent resiliency characterized by high tensile elongation. The resins can be modified extensively by extending the diepoxide with bisphenol A or polyphenols, such as the novolaks that allow for branched polymers having higher reactivity and enhanced thermal and solvent resistance. Novel polymers with higher resilience have been produced by extending the diepoxides with carboxy-terminated polybutadiene rubber. In view of their high strength and resilience, vinyl esters are used extensively in filament-wound products such as pipe and tanks.

**Stabilizers.** Hydroquinone [123-31-9] (**4**) is widely used in commercial resins to provide stability during the dissolution of the hot polyester resin in styrene during the manufacturing process. Aeration of the styrene with oxygen (air) is required to activate the stabilizer, which is converted to an equilibrium mixture of quinone and the quinhydrone (**5**) (11). At levels of 150 ppm, a shelf life of over six months can be expected at ambient temperatures.

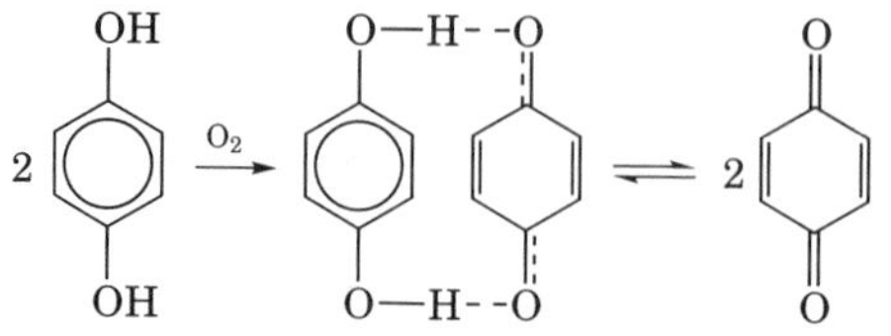

$$2 \quad \underset{\text{OH}}{\overset{\text{OH}}{\bigcirc}} \quad \xrightarrow{\text{O}_2} \quad \underset{\text{O—H--O}}{\overset{\text{O—H--O}}{\bigcirc}} \quad \rightleftharpoons \quad 2 \quad \underset{\text{O}}{\overset{\text{O}}{\bigcirc}}$$

(**4**)         (**5**)

Toluhydroquinone and methyl *tert*-butylhydroquinone provide improved resin color retention; 2,5-di-*t*-butylhydroquinone also moderates the cure rate of the resin. Quaternary ammonium compounds, such as benzyl trimethylammonium hydroxide, are effective stabilizers in combination with hydroquinones and also produce beneficial improvements in color when promoted with cobalt octoate. Copper naphthenate is an active stabilizer at levels of 10 ppm; at higher levels (150 ppm) it influences the cure rate. Tertiary butylcatechol (TBC) is a popular stabilizer used by fabricators to adjust room temperature gelation characteristics (see STABILIZERS).

## Cross-Linking Mechanism

The reaction rate of fumarate polyester polymers with styrene is 20 times that of similar maleate polymers. Commercial phthalic and isophthalic resins usually have fumarate levels in excess of 95% and demonstrate full hardness and property development when catalyzed and cured. The addition polymerization reaction between the fumarate polyester polymer and styrene monomer is initiated by free-radical catalysts, commercially usually benzoyl peroxide (BPO) and methyl ethyl ketone peroxide (MEKP), which can be dissociated by heat or redox metal activators into peroxy and hydroperoxy free radicals.

The free radicals initially formed are neutralized by the quinone stabilizers, temporarily delaying the cross-linking reaction between the styrene and the fumarate sites in the polyester polymer. This temporary induction period between catalysis and the change to a semisolid gelatinous mass is referred to as gelation time and can be controlled precisely between 1–60 min by varying stabilizer and catalyst levels.

As the quinone stabilizer is consumed, the peroxy radicals initiate the addition chain propagation reactions through the formation of styryl radicals. In dilute solutions, the reaction between styrene and fumarate ester follows an alternating sequence. However, in concentrated resin solutions, the alternating addition reaction is impeded at the onset of the physical gel. The liquid resin forms an intractable gel when only 2% of the fumarate unsaturation is cross-linked with styrene. The gel is initiated through small micelles (12) that form the nuclei for the expansion of the cross-linked network. The free styrene monomer is restrained within the gel and further reaction with fumarate groups is determined by the spacial arrangement; the styrene polymerizes in homopolymer blocks as it intercepts fumarate reaction sites. As individual micelles expand and deplete available fumarate sites in the short polymer chains, the remaining styrene forms homopolymer blocks that terminate at the boundaries between overlapping micelles (Fig. 4).

As the micelles expand, the soft gel is transformed into a hard, rubbery transition stage that demonstrates low physical strength before the onset of the exotherm. The temperature increases exponentially as the micelle expands. As the temperature subsides, the resulting cross-linked thermoset solid develops superior properties characteristic of the polymer. Generally, optimum strength characteristics are obtained in resins having a styrene/fumarate molecular ratio of 2:1. Most resins are formulated with a styrene content consistent with this relationship (Tables 2–5). In resins having equivalent molar ratios of dibasic and unsaturated acid, this equates to resin polymer solutions containing around 40% styrene. However, this imposes some limitation on viscosity and most commercial resins contain between 40–45% styrene to achieve lower application viscosities.

In resins having low isomerization levels (80%), the fumarate–styrene reactions run to completion, leaving many unreacted maleate groups within the cross-linked structure. This results in an excess of styrene that inevitably forms larger homopolymer blocks between the intersecting micelles. The performance of such resins is characterized by lower softening temperatures and lower physical properties due to the additional plasticizing effects of the maleate ester group.

The cross-linking reaction mechanism is also influenced by the presence of other monomers. Methyl methacrylate is often used to improve the uv resistance of styrene-based resins. However, the disparate reaction rates of styrene and methacrylate monomer with the fumarate unsaturation not only preclude the use of more than 8% of the methacrylate monomer due to the significant slowing of the cross-linking reaction but also result in undercured products.

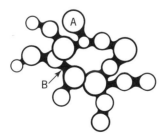

**Fig. 4.** Micellular gelation mechanism. A shows micelle nuclei, highly cross-linked; B, boundary where micelle growth terminates in styrene block polymers. Styryl free radicals simultaneously initiate micelle nuclei at points of high fumarate concentration. The micelles continue to expand, interacting with free styrene until the fumarate groups are depleted. The micelles eventually overlap at the boundaries that contain higher levels of terminal styrene homopolymer blocks.

Methacrylate monomers are most effective with derivatives of bisphenol A epoxy dimethacrylates, in which the methacrylate–methacrylate cross-linking reaction proceeds at a much faster pace than with styrene monomer. This proves beneficial in some fabrication processes requiring faster cure, such as pultrusion and resin-transfer molding (RTM).

**Catalyst Selection.**   The low resin viscosity and ambient temperature cure systems developed from peroxides have facilitated the expansion of polyester resins on a commercial scale, using relatively simple fabrication techniques in open molds at ambient temperatures. The dominant catalyst systems used for ambient fabrication processes are based on metal (redox) promoters used in combination with hydroperoxides and peroxides commonly found in commercial MEKP and related perketones (13). Promoters such as styrene-soluble cobalt octoate undergo controlled reduction–oxidation (redox) reactions with MEKP that generate peroxy free radicals to initiate a controlled cross-linking reaction.

$$Co^{3+} + ROOH \longrightarrow ROO\cdot + H^+ + Co^{2+}$$

$$Co^{2+} + ROOH \longrightarrow RO\cdot + OH^- + Co^{3+}$$

This catalyst system is temperature-sensitive and does not function effectively at temperatures below 10°C; but at temperatures over 35°C the generation of free radicals can be too prolific, giving rise to incomplete cross-linking formation. Redox systems are preferred for fabrication at temperatures ranging from 20–30°C (Fig. 5).

Some fabrication processes, such as continuous panel processes, are run at elevated temperatures to improve productivity. Dual-catalyst systems are commonly used to initiate a controlled rapid gel and then a fast cure to complete

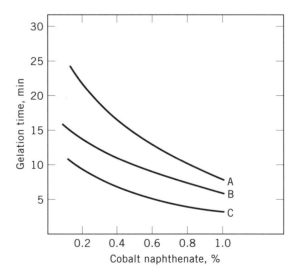

**Fig. 5.**  Influence of catalyst systems on cure rate; gelation time is at 25°C as a function of the initiator concentration. A represents MEKP (1.0%); B, MEKP (1.0%) and dimethylaniline (DMA) (0.05%); and C, MEKP (2.0%).

the cross-linking reaction. Cumene hydroperoxide initiated at 50°C with benzyl trimethylammonium hydroxide and copper naphthenate in combination with *tert*-butyl octoate are preferred for panel products. Other heat-initiated catalysts, such as lauroyl peroxide and *tert*-butyl perbenzoate, are optional systems. For higher temperature molding processes such as pultrusion or matched metal die molding at temperatures of 150°C, dual-catalyst systems are usually employed based on *t*-butyl perbenzoate and 2,5-dimethyl-2,5-di-2-ethylhexanoylperoxyhexane (Table 6).

The action of redox metal promoters with MEKP appears to be highly specific. Cobalt salts appear to be a unique component of commercial redox systems, although vanadium appears to provide similar activity with MEKP. Cobalt activity can be supplemented by potassium and zinc naphthenates in systems requiring low cured resin color; lithium and lead naphthenates also act in a similar role. Quaternary ammonium salts (14) and tertiary amines accelerate the reaction rate of redox catalyst systems. The tertiary amines form beneficial complexes with the cobalt promoters, facilitating the transition to the lower oxidation state. Copper naphthenate exerts a unique influence over cure rate in redox systems and is used widely to delay cure and reduce exotherm development during the cross-linking reaction.

Another unique redox system used for extending gel times consists of cumene hydroperoxide and manganese naphthenate, which provides consistent gel times of between two and four hours over a temperature range of 25–50°C.

For application temperatures below 10°C or for acceleration of cure rates at room temperature, nonredox systems such as benzoyl peroxide initiated by tertiary amines such as dimethylaniline (DMA) have been applied widely. Even more efficient cures can be achieved using dimethyl-*p*-toluidine (DMPT), whereas moderated cures can be achieved with diethylaniline (DEA).

Tertiary amines are also effective as accelerators in cobalt redox systems to advance the cure rate (Fig. 6). Hardness development measured by Shore D or Barcol D634-1 penetrometer can be used to demonstrate this benefit, which is useful in increasing mold turnover at ambient temperatures.

**Table 6. Optimal Temperature Range of Conventional Catalyst Systems for Unsaturated Polyesters**

| Catalyst | CAS Registry Number | Activator | Processing temperature, °C |
|---|---|---|---|
| benzoyl peroxide | [94-36-0] | dimethylaniline | 0–25 |
| methyl ethyl ketone peroxides (MEKP) | [1338-23-4] | cobalt octoate | 20–25 |
| cumene hydroperoxide | [80-15-9] | manganese naphthenate | 25–50 |
| lauroyl peroxide | [105-74-8] | heat | 50–80 |
| *t*-butyl peroctoate | [13467-82-8] | heat | 80–120 |
| benzoyl peroxide | [94-36-0] | heat | 80–140 |
| 2,5-dimethyl-2,5-di-2-ethyl-hexanoylperoxyhexane | [13052-09-0] | heat | 93–150 |
| *t*-butyl perbenzoate | [614-45-9] | heat | 105–150 |

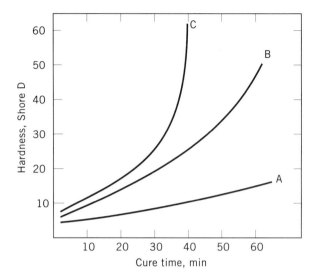

**Fig. 6.**  Influence of catalyst systems on cure rate; effect of dimethylaniline (DMA) on cure rate of cast polymer resin at 25°C. Initiator system contains cobalt naphthenate (0.5%), MEKP (1.0%), and one of the following: A, DMA (0.0%); B, DMA (0.05%); or C, DMA (0.1%).

**Cure Exotherm.**  The cross-linking reaction between the unsaturated polymer and styrene results in a spontaneous change from liquid to a solid state with the onset of the exotherm. The exothermic heat generated is proportional to the fumarate level in the polymer, but increasing styrene levels can enhance it further. Although some exotherms can be tolerated in molding processes, these can lead to excessive shrinkage, warpage, and cracking in large moldings or glass fiber-reinforced laminations in excess of 9-mm thick. The cure exotherm can be suppressed in a number of ways to afford a more controllable fabrication system, without adversely affecting the final cure or structural performance.

Copper naphthenate added to the resin at levels between 100–200 ppm effectively extends gel and cure characteristics, resulting in a reduction in exothermic heat (Fig. 7). Copper additives are used widely in commercial laminating resins to modify process exothermic effects. $\alpha$-Methylstyrene [98-83-9] substituted for styrene at levels of 5–8% has also been used effectively in resins cured at above ambient temperatures. The inhibitor 2,5-di-$t$-butylhydroquinone exerts significant exotherm suppression at levels of 200–400 ppm and is useful in high temperature molding processes.

**Shrinkage.**  Polyester resins that undergo cross-linking reactions have as a result a net contraction in volume. Resins dissolved in styrene demonstrate a volumetric shrinkage of 17%. Such high shrinkage can lead to cracking or warping of castings or laminates. Monomers such as vinyltoluene or diallyl phthalate reduce shrinkage; these monomers are often used in high temperature molding compositions to avert warpage or internal voids. With vinyltoluene, volumetric shrinkage during cure is 12.6% and with diallyl phthalate, 11.8%. The effects of shrinkage can also be modified by incorporating soluble poly(vinyl acetate) and related thermoplastic additives into the liquid resins, which phase

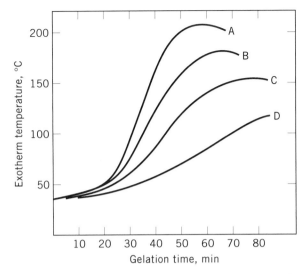

**Fig. 7.** Influence of copper naphthenate on exotherm temperature. Composition in pph; isophthalic laminating resin is cobalt naphthenate (0.20), dimethylaniline (0.05), and copper naphthenate, A (0.00); B (0.01); C (0.015); or D (0.02).

out during the cross-linking reaction, thus significantly reducing shrinkage. Additional benefits can be obtained by incorporating high filler loadings into resins used in casting or high temperature molding processes.

**Air Inhibition.**   Polyester resins are widely used in open-mold lamination or casting and quite often develop a tacky surface feel after curing. The free-radical polymerization process is sensitive to oxygen, which interferes with the surface cross-linking mechanism and impedes the cure. Lower molecular weight phthalic resins display a pronounced effect, but higher molecular weight isophthalic resins have less tack. Resins based on dicyclopentadiene react with oxygen through an air-drying mechanism and give rise to tack-free surfaces; resins derived from tetrahydrophthalic anhydride have similar air-drying qualities useful for polyester coatings and linings.

Paraffin wax additives are effective in overcoming surface inhibition by forming a monomolecular wax layer at the curing surface. Although effective in excluding oxygen, this waxy layer must be removed for subsequent lamination or bonding processes (see WAXES).

## Performance Characteristics

Polyester resins undergo a rapid transformation from a viscous liquid to a solid plastic state that comprises a three-dimensional cross-linked polymer structure. The level of polyester polymer unsaturation determines essential performance characteristics (Table 7), although polymer components can influence subtle features that affect thermal, electrical, and mechanical performance as defined by ASTM procedures.

The cross-linked polymers form a thermoset plastic which cannot be changed or returned to its original condition by heating as it can with ther-

**Table 7. Standard Test Methods for Polyester Resins and Compounds**

| Properties | ASTM designation |
|---|---|
| specific gravity | D792 |
| Barcol hardness | D2583 |
| heat distortion | D648 |
| flexural strength modulus | D790 |
| tensile strength modulus, elongation | D638 |
| tensile impact | D1822 |
| compressive strength modulus | D695 |
| Izod impact | D256 |
| shear | D732 |
| Taber abrasion | D1044 |
| flammability | E84, E162 |
| dielectric strength | D149 |
| dissipation factor | D150 |
| arc resistance | D495 |
| water absorption | D570 |
| water vapor transmission | C355 |
| chemical resistance | C581 |

moplastics. This thermoset characteristic is beneficial in providing high temperature properties, good solvent and chemical resistance, and high flexural modulus. Cross-linked polyester resins are rigid materials and are highly sensitive to brittle fracture. Reinforcing with glass fiber produces a composite plastic, which has exceptional strength characteristics suitable for replacing conventional fabricating materials such as wood, steel, and concrete. Aggregates and fillers such as ground limestone also improve the strength characteristics of polyester resins and are used widely in cast objects such as bathroom vanity sets and building components. Polyester resins are used in an unfilled condition in cast objects such as bowling balls and buttons, in thin films for gel coats, and, to a lesser extent, as clear wood coatings.

**Mechanical Properties.** The performance of various polyester resin compositions can be distinguished by comparing the mechanical properties of thin castings (3 mm) of the neat resin defined in ASTM testing procedures (15). This technique is used widely to characterize subtle changes in flexural, tensile, and compressive properties that are generally overshadowed in highly filled or reinforced laminates.

Resins of higher molecular weight demonstrate higher tensile strength, whereas high fumarate resins have higher flexural modulus. Formulations containing diethylene glycol and adipic acid produce resins that have higher resilience reflected in enhanced tensile and flexural strength but lower flexural modulus. Isophthalic resins provide better mechanical properties than *ortho*-phthalic resins and are consequently preferred in laminate applications requiring higher structural performance. Resins that have exceptional tensile performance are formulated from the bisphenol A epoxy dimethacrylates. These resins demonstrate superior fracture resistance properties beneficial in filament-wound structures such as tanks and pipes, which are subjected to cyclic stresses that can result in fatigue failure. The strength of all polyester resins is enhanced significantly by glass and other fibrous reinforcements. Laminates are usually fabricated from glass fiber mat, having individual fibers 5 cm in length. The structural properties increase in proportion to the glass fiber content, which can be varied from 25–40%. Increased reinforcement levels can be achieved by using woven glass roving in alternating plies with chopped strand glass mat. Higher strength can be realized from continuous glass fiber rovings used in filament-wound structures in the form of pipes or tanks. Helical wind angles are varied to achieve design requirements in hoop or axial directions; a wind angle of 55° can generate twice the strength in the hoop direction. Continuous glass fiber roving used at reinforcement levels of 65% in pultruded products provides composites that have the highest flexural and tensile properties. Carbon fiber can be useful in developing composites that have higher modulus characteristics, but economics have reduced their wider attractiveness in combinations with polyester resins. Kevlar cloth can be used in combination with bisphenol epoxy dimethacrylates to produce light weight, high strength composite plastics for sporting equipment such as kayaks and skis.

Resins filled with ground limestone to levels of 80% by weight are useful in solid cast products. The fillers reduce sensitivity to brittle fracture and improve modulus, but have little effect on general strength properties (Table 8).

**Thermomechanical Properties.** The highly cross-linked structure of cured unsaturated polyester resins produces thermoset characteristics in which the resistance to softening and deformation is greatly enhanced at elevated temperatures. The cross-linked network undergoes a structural transition during heating, in which the rigid crystalline state transforms to a softer amorphous condition at the glass-transition temperature, $T_g$, accompanied by a small expansion in volume, thus facilitating some relaxation along the amorphous micellular boundaries and allowing deformation to take place. Aromatic constituents enhance the $T_g$, as do high fumarate and high styrene levels, whereas aliphatic derivatives (adipic) and reduced fumarate levels lower $T_g$ values. Resins containing high adipic acid levels display rubbery or elastomeric properties at below ambient temperatures. Glass-reinforced products using these resins have exceptional impact properties and demonstrate high tolerance to low temperature cryogenic applications. Deformation at higher temperatures is moderated

**Table 8. Strength Characteristics of Isophthalic Resin and Composite Derivatives**

| Characteristic | Cast resin | Filled resin | Glass-reinforced laminate | Filament-wound laminate | Pultruded profile |
|---|---|---|---|---|---|
| glass fiber content, % | 0 | 0 | 30 | 50 | 60 |
| flexural strength, MPa[a] | 110 | 82 | 193 | 296 | 448 |
| flexural modulus, GPa[b] | 3.44 | 4.68 | 5.86 | 13.7 | 20.6 |
| tensile strength, MPa[a] | 68.9 | 44.8 | 110 | 193 | 241 |
| tensile modulus, GPa[b] | 3.1 | 2.6 | 5.51 | 12.4 | 15.1 |
| tensile elongation, % | 2.5 | 0.5 | 1.6 | 1.6 | 1.5 |
| compressive strength, MPa[a] | 103 | 110 | 137 | 193 | 200 |

[a]To convert MPa to psi, multiply by 145.
[b]To convert GPa to psi, multiply by 145,000.

by fibrous reinforcements. However, as the temperature exceeds the $T_g$ of the cross-linked polymer, laminate properties fall off considerably (Fig. 8).

Although reinforcements can improve the structural behavior of the composite at elevated temperature, the polymer, irrespective of its composition, begins to disassociate chemically in the presence of oxygen. *ortho*-Phthalic resins having the weaker ester bonds depolymerize readily at temperatures over 200°C and form sublimed phthalic anhydride on the laminate surface. Determination of laminate weight loss or flexural property retention indicates that alicyclic diols, including hydrogenated bisphenol A and cyclohexanedimethanol, perform better than lower glycols such as propylene, and neopentyl glycol imparts exceptional thermal stability at temperatures up to 200°C. However, all resins depolymerize spontaneously at around 300°C as the styrene disassociates from the polymer network, producing by-products that include styrene, benzaldehyde, and benzene. Vinyltoluene provides enhanced thermal performance on account of the increased bond strength resulting from the inductive influence of the methyl group para to the vinyl unsaturation. Oxidative disassociation at elevated temperature can be suppressed with antioxidants, but these eventually interfere with the catalyst's activity, thus reducing cure rate.

Halogenated intermediates based on chlorendic anhydride and alkoxylated, brominated bisphenol are quite stable and are used extensively in flame-retarded high temperature compositions, but brominated alicyclics, such as dibromotetrahydrophthalic resin, are rapidly dehydrohalogenated at lower temperatures.

Cross-linked polyester composites have a relatively low coefficient of thermal conductivity that can provide beneficial property retention in thick laminates at high temperatures as well as remove the need for secondary insulation.

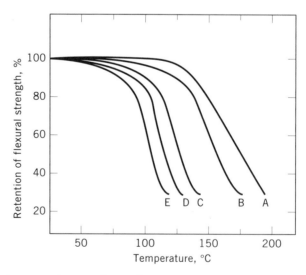

**Fig. 8.** Flexural properties at elevated temperatures. Laminates constructed from alternating plies of 46.7-g (1.5-oz) mat and 746-g/m$^2$ (24-oz/yd$^2$) woven roving at a nominal glass content of 45%. A represents bisphenol fumarate ($T_g = 130°C$); B, novolak epoxy methacrylate ($T_g = 130°C$); C, epoxy dimethacrylate ($T_g = 100°C$); D, isophthalic resin ($T_g = 100°C$); and E, *ortho*-phthalic resin ($T_g = 80°C$).

The coefficient of thermal expansion of glass-reinforced composites is similar to aluminum but higher than most common metals.

**Dielectric Properties.** Polyester resins are nonconductors, have relatively low dipolar characteristics, and provide high dielectric strength and surface resistivity. At high voltage or high current, however, the cross-linked plastics fail due to carbon arcing or tracking caused by the charring of the polymer surface into a conductive carbonaceous residue. Hydrated fillers such as alumina trihydrate (16) suppress surface char formation and are used extensively with glass reinforcement in molding compounds intended for electrical applications. High temperature electrical applications usually specify bisphenol fumarate resins or high molecular weight isophthalic resins having high fumarate reactivity. Dicyclopentadiene resins provide enhanced oxidative resistance, whereas vinyltoluene monomer is used for superior thermal properties. Specialized granulated molding compounds are also formulated by using diallyl phthalate monomers, and are used for injection molding smaller electrical components.

**Chemical Properties.** The three-dimensional cross-linked network resists penetration and attack by most corrosive chemicals and nonpolar solvents, although weak alkalies and especially polar solvents such as lower ketones, chlorinated aliphatics, and aromatics readily attack *ortho*-phthalic, isophthalic, and dicyclopentadiene resins. Water has wide ranging effects on different resin compositions as it penetrates into the plastic network. Cross-linking density and the presence of steric constituents local to the ester groups can enhance water resistance. Isophthalic resins have better water absorption characteristics than corresponding *ortho*-phthalic resins. Neopentyl glycol and alkoxylated bisphenol constituents provide maximum performance in aqueous media. Ethylene and diethylene glycols demonstrate high water absorption that leads to the loss of mechanical strength. The presence of glass reinforcements within the cross-linked plastic are also affected by chemical attack. Water absorbed onto surface fibers is carried into the plastic matrix, decoupling the resin–glass fiber interfacial bond in the process. Alkalies compliment this attack on the glass reinforcement and lead to surface blistering and rapid loss of laminate strength. Chemical attack is usually minimized by incorporating surface protection in the form of corrosion-resistant (C glass) surface veils and high surface resin content in laminates intended for chemical service. Reinforced polyester composites or fiber-reinforced plastics (FRP) perform well in concentrated hydrochloric and phosphoric acids; concentrated sulfuric and nitric acids oxidize and degrade the polymer rapidly. High performance resins based on bisphenol A fumarate or bisphenol A vinyl esters and their brominated homologues have featured prominently in FRP piping, tanks, chemical ducting, and scrubbers used to handle corrosive chemicals and emissions in pulp and paper plants, mining and ore classification, and the chloralkali industries. FRP has also emerged as a strategic solution for designing equipment intended for high temperature applications in fossil fuel electric power generating utilities such as chimney stack liners and scrubbers for the control of sulfur dioxide emissions.

Corrosion attack on the polymer is influenced by permeation rate, as well as internal stresses or fatigue, that distorts or fractures the resin glass fiber interface. Localized corrosion failure in large tanks or piping systems can sometimes be explained by stress-induced corrosion failure. FRP composites normally contain air voids that distort into elliptical voids adjacent to the reinforcing fiber. Water and chemicals penetrating these voids set up osmotic cells that plasticize and soften the surrounding polymer, initiating brittle fracture along the fibers and progressively opening up new surface area for corrosion attack in areas under high stress. Nonpolar solvents have little effect on polyester resins and have been used extensively for underground storage of gasoline. However, with the reformulation to unleaded gasoline, the higher aromatic fuels require resins that have higher cross-linking density. This is provided by high fumarate isophthalic derivatives or novalak-modified bisphenol dimethacrylates.

**Flammability.** Polyester resin products ignite and burn by emitting sooty smoke. Flammability can be reduced significantly through halogen-modified components either formulated into the polyester polymer as chlorendic anhydride, tetrabromophthalic anhydride, or dibromoneopentyl glycol, or as part of the monomer system, ie, dibromostyrene. Additives such as phosphate esters are frequently used to enhance flame retardance, whereas antimony trioxide at levels of 5% on resin provides optimum retardance in combination with halogenated intermediates. The flame-retardant mechanism depends on the thermal disassociation of the organohalogen component, which, upon exposure to the heat of the flame, releases hydrogen chloride or hydrogen bromide gases that become active as radical transfer agents in suppressing the oxidation–combustion process (17). The activity of the halogen gases is synergized through the formation of antimony oxy halides and trihalides; arsenic and molybdenum salts offer comparable activity. Organophosphorus compounds behave in a supplementary role by forming polyphosphoric acids that induce char at the burning surface. However, phosphates plasticize the polymer and reduce performance at levels exceeding 5%, thus offering marginal benefits. Organoferrous compounds such as ferrocene demonstrate unique synergism with halogenated resins, reducing both flame spread and smoke generation. Also, ferrous oxide (18) is active in chlorendic resins in a similar capacity. The formation of ferric chloride promotes aromatization and enhances char formation and the subsequent reduction of smoke emissions. The exothermic nature of the combustion process can provide sufficient energy for sustained burning, unless halogen modification is used. Nonhalogenated resin systems, based on hydrated fillers such as alumina trihydrate, have emerged. Alumina trihydrate undergoes endothermic disassociation at above 300°C, and can offset the combustion heat and reduce surface temperatures sufficiently to interrupt the high energy oxidation mechanism. Resins incorporating over 40% by weight of aluminum trihydrate not only qualify for most construction and electrical flammability specifications but also provide significant reduction in smoke emissions. Useful bench-scale tests, such as ASTM E162 and ASTM D2863, have been developed for characterizing flammability and smoke emissions, but the dynamic nature of the combustion process requires large or full-scale fire testing, such as ASTM E84, to be performed for applications intended for confined areas. Halogenated resins are sensitive to heat and ultraviolet radiation; resins based on tetrabromophthalic anhydride yellow after a few days in direct sunlight. Benzophenone and benzotriazole stabilizers used at levels of 1% are highly effective with chlorendic, dibromoneopentyl, and dibromotetrahydrophthalic derivatives, but resins based on brominated aromatics such as tetrabromophthalic anhydride and tetrabromobisphenol eventually discolor in exterior applications.

**Weathering.**   Polyester resins in the form of laminates, coatings (gel coats), and castings perform well in outdoor exposures; marine craft, tanks, pipes, and architectural facia produced in the 1960s are still in service. Polyesters undergo some change in their surface features in direct sunlight, discoloration or yellowing being most obvious. Dulling and microcrazing occur only in products not appropriately formulated for the exposure. Long-term surface erosion of laminates exposes the glass fibers, which, unless corrected, can lead to rapid loss of structural integrity. Absorption of water influences the interfacial bond between the resin matrix and either the fibrous or aggregate reinforcement, leading to some loss of mechanical properties over 30 years. However, observations indicate that most applications reach this equilibrium after three–five years and do not show much significant change thereafter. Most FRP products are designed with protective and decorative gel coats (19) formulated from neopentyl glycol, which, in combination with some methyl methacrylate monomer and benzophenone uv stabilizers, provides improved weather resistance. Platelet fillers such as talcs and clays as well as high pigment levels are incorporated into gel coats to obscure the underlying laminate from the effects of direct sunlight. Surface oxidation and discoloration can also be controlled by using uv-resistant lacquers based on polymethacrylate resins or by cladding with poly(vinyl fluoride) (PVF) or acrylic and poly(ethylene terephthalate) (PET) films.

## Application Processes

**Open-Mold Processes.**   Polyester resins are fabricated easily in open molds at room temperature. Such processes account for over 80% of production volume, the remaining being fabricated using matched metal dies in high temperature semiautomated processes.

The hand lay-up or spray-up process, used universally for the production of laminar composites incorporating glass fiber reinforcement, is most efficient for the manufacture of large parts, such as boats, bathtubs, tanks, architectural shapes, and recreational accessories. Resins intended for spray-up processes are usually modified with thixotropic additives, such as fumed silica (1%), to reduce the risk of drainage when applied over large vertical mold surfaces. Molds are also made from FRP for short-run products usually surfaced with a tooling gel coat to provide consistent surface quality and appearance.

Gel coats are pigmented polyester coatings applied to the mold surface and are an integral part of the finished laminate. Gel coats are used widely on hand lay-up and spray-up parts to enhance surface aesthetics and coloration as well as to provide an abrasion-resistant waterproof surface that protects the underlying glass-reinforced structure.

Thermoformed acrylic sheet is displacing gel coats in some bathtub applications; spas have converted almost exclusively to formed acrylic sheet reinforced with glass-reinforced polyester laminations due to higher temperatures and higher structural requirements.

Products of axial symmetry such as pipes and tanks can be produced by filament-winding resin-impregnated glass rovings over a rotating mandrel. This process provides for some versatility in adjusting the winding angle to meet design cost considerations. A winding angle of 55° to the axis produces pipes that have twice the strength in the hoop direction, suitable for conveying high pressure liquids that may be corrosive to metals. Large-diameter piping systems (1–3 m) are also produced by centrifugal casting techniques, by which high levels of silica sand can be used to improve pipe stiffness. FRP pipes have been used extensively as slip linings for deteriorated concrete sewer pipes that avoid costly excavation.

Resins containing fillers and aggregates can be usefully cast into attractive components that simulate marble and granite. Bathroom vanity sets and kitchen countertop components have emerged as a significant commercial outlet for filled resins cast in open molds that often employ gel coats having superior hydrolytic properties to enhance resistance to thermal shock and crazing. Aggregate-filled bisphenol dimethacrylate resins have emerged in high strength, cast polymer concrete structures used in corrosion-resistant electrolytic cells for the purification of copper. Specially formulated isophthalic resins in admixture with sand have been developed for surfacing worn concrete roads and bridges. Polyester resins are also emerging in filled pigmented coatings used as permanent road marking, as some U.S. states displace solvent-based alkyd marking materials. Other products such as buttons and bowling balls are cast from resins containing low filler levels.

**Closed-Mold Processes.**    In an effort to improve the productivity of the hand lay-up process, closed-mold systems containing two mating dies have evolved. In resin-transfer molding (RTM), glass reinforcement is placed in the open mold; once the molds are in place, precatalyzed resin is injected into the cavity under pressure. The process has been further modified to use a vacuum that ensures complete air removal and faster mold filling. The process is adaptable to large components and can be used to encapsulate foam, aluminum, and wood components into the structure. The process is also beneficial in reducing styrene emissions that are regulated under Title III of the U.S. Clean Air Act passed in 1990.

High temperature compression molding has grown rapidly between 1985 and 1995 as applications for glass-reinforced composites have expanded in the automotive body panel market. Molding compounds incorporating resins, catalysts, fillers, pigments, and glass fiber reinforcements have evolved as bulk molding compounds (BMC) and sheet molding compounds (SMC) to meet requirements in the electrical, business machine, as well as structural automotive markets.

Matched die molding is the most efficient process to produce high volumes of relatively large parts. The mold cycle is under two minutes at 150°C and the process can be incorporated into an automated manufacturing system, reducing labor and scrap while improving quality. Resins have been uniquely formulated to reduce shrinkage and provide composite surfaces that can be primed and coated by using conventional baked enamels. The polyester technology providing these low profile, low shrink resins is based on the action of thermoplastic additives (20) incorporated into unique polyester resin formulations based on propylene glycol fumarate.

Poly(methyl methacrylate) and poly(vinyl acetate) precipitate from the resin solution as it cures. This mechanism offsets the contraction in volume as the polyester resin cross-links, resulting in a nonshrinking thermoset. Other polymer additives such as poly(butylene adipate) provide similar shrinkage control. The precipitation of the polymer produces a whitening of the composite and results in nonuniform coloration in pigmented products. Polystyrene additives used in bulk molding compound (BMC) can be formulated into nonshrinking, pigmentable compounds suitable for colored electrical products and kitchen utensils.

Injection molding of BMC and TMC is expanding to improve the efficiency of the matched die process. TMC is a compound intermediate between BMC and sheet molding compound (SMC) in which the glass fiber strand integrity is retained in the molded product, thus greatly enhancing strength and impact properties.

High strength composites with linear symmetry can be produced by the pultrusion process (21) using continuous glass fiber reinforcements in the form of rovings. Special continuous glass mats have also evolved to meet the process requirements of this technology. Glass fiber levels of 65% can be formed into polyester composites that demonstrate exceptional flexural performance characteristics required in such applications as flag poles and automotive springs. The process can also form hollow structures and profiles compatible with extruded aluminum and PVC useful in window and portal construction.

## BIBLIOGRAPHY

"Unsaturated Polyesters" in *ECT* 2nd ed., Vol. 20, pp. 791–839, by R. M. Nowak and L. C. Rubens, The Dow Chemical Co.; "Polyesters, Unsaturated" in *ECT* 3rd ed., Vol. 18, pp. 575–594, by J. Makhlouf, PPG Industries.

1. W. H. Carothers, *J. Am. Chem. Soc.* **51**, 2548 (1929).
2. H. Boenig, *Unsaturated Polyesters Structures and Properties*, Elsevier Science, Inc., New York, 1964.
3. U.S. Pat. 3,345,339 (Oct. 3, 1967), E. E. Parker and J. G. Baker (to PPG Industries).
4. E. E. Parker, *Ind. Eng. Chem.* **58**, 53 (1966).
5. U.S. Pat. 3,355,434 (Nov. 28, 1967), J. G. Milligan and H. G. Waddill (to Jefferson Chemical Corp.).
6. U.S. Pat. 3,576,909 (Apr. 27, 1971), C. J. Schmidle and A. E. Schmucker (to General Tire and Rubber).
7. D. L. Nelson, *Soc. Plast. Ind. Conf.* **34**, 1G (1979).
8. U.S. Pat. 2,909,501 (Oct. 20, 1959), P. Robitschek and J. L. Olmstead (to Hooker Chemical Corp.).
9. U.S. Pat. 3,536,782 (Oct. 27, 1970), U. Toggwerler and F. Rosselli (to Diamond Shamrock Corp.).
10. U.S. Pat. 2,634,251 (Apr. 7, 1953), P. Kass (to Atlas Powder Co.).
11. C. C. Price and D. H. Read, *J. Polym. Sci.* **1**, 44 (1946).
12. Y.-S. Yang and L. Suspene, *Polym. Eng. Sci. Mar.* **31**(5), 321 (1991).
13. J. Harrison, O. Mageli, and S. Stengel, *Rein. Plast. Conf.* **15**(14) (1960).
14. U.S. Pat. 3,886,113 (May 27, 1975), M. W. Uffner (to Air Products and Chemicals, Inc.).

15. *The American Standards and Testing Methods Manual*, The American Society for Testing and Materials, Philadelphia, Pa., 1992.
16. T. D. Bautista, *Polym. Plast. Tech. Eng.* **18**, 179 (1982).
17. J. E. Selley, *Soc. Plast. Ind. Conf.* **33**, 2E (1978).
18. U.S. Pat. 4,152,368 (May 1, 1979), E. Dorfman, R. R. Hindersinn, and W. T. Schwartz (to Hooker Chemical Corp.).
19. J. H. Davis and S. L. Hillman, *Soc. Plast. Ind. Conf.* **26**, 12C (1971).
20. U.S. Patent 3,701,748 (Oct. 31, 1972), C. H. Kroekel (to Rohm & Haass Co.).
21. J. E. Sumerak, *Soc. Plast. Ind. Conf.* **40**, 2B (1985).

JEFFREY SELLEY
Consultant

## POLYETHER ANTIBIOTICS. See ANTIBIOTICS, POLYETHERS.

## POLYETHER ELASTOMERS. See ELASTOMERS, SYNTHETIC–POLYETHERS.

# POLYETHERS

## AROMATIC

Aromatic polyethers are best characterized by their thermal and chemical stabilities and mechanical properties. The aromatic portion of the polyether contributes to the thermal stability and mechanical properties, and the ether functionality facilitates processing but still possesses both oxidative and thermal stability. With these characteristic properties as well as the ability to be processed as molding materials, many of the aromatic polyethers can be classified as engineering thermoplastics (see ENGINEERING PLASTICS).

One class of aromatic polyethers consists of polymers with only aromatic rings and ether linkages in the backbone; poly(phenylene oxide)s are examples and are the principal emphasis of this article. A second type contains a wide variety of other functional groups in the backbone, in addition to the aromatic units and ether linkages. Many of these polymers are covered in other articles, based on the other functionality (see POLYMERS CONTAINING SULFUR, POLYSULFONES).

## Poly(phenylene oxide)s

The poly(phenylene oxide)s are also referred to as polyoxyphenylenes and poly(phenylene ether)s. Variations in the configuration of the ether group, ie, ortho, meta, or para, and in the extent and type of substitution, eg, alkyl, halo, etc, on the aromatic backbone give rise to a large number of possible homo- and copolymers. The polymers with para-oriented ethers have been studied most extensively and several have significant utility. Poly(2,6-dimethyl-1,4-phenylene oxide) [25134-01-4] (DMPPO), prepared by General Electric by the oxidative coupling polymerization of 2,6-dimethylphenol, is marketed as PPO resin. Blends of DMPPO with polystyrene and additives is marketed under the trade name of Noryl thermoplastic resin. Blends of DMPPO with nylon plus additives are sold as Noryl GTX resin. Other poly(phenylene oxide)s are also of commercial interest. For example, low molecular weight, unsubstituted poly(phenylene oxide)s are of use as heat-transfer fluids and liquid phases for gas chromatography. Poly(2,6-diphenyl-1,4-phenylene oxide) is sold as a gas absorbent and for gas chromatography.

Many newer poly(phenylene oxide)s have been reported in the early 1990s. For example, a number of poly(2,6-diphenyl-1,4-phenylene oxide)s were prepared with substituents in the 4-positions of the pendent phenyl groups. Of particular interest is the 4-fluoro substituent, which imparts a lower melting point, enhanced solubility, and a lesser tendency to crystallize than has been found for the parent material (1).

**Physical Properties.** The glass-transition temperatures and melting points of a variety of poly(1,4-phenylene oxide)s are summarized in Table 1. The very high glass-transition temperatures of DMPPO and poly(2,6-dichloro-1,4-phenylene oxide) relative to their melting points provide polymers with excellent high temperature physical properties. Crystallization of several polymers can be induced with certain liquids, eg, $\alpha$-pinene, decalin, and toluene for DMPPO, and tetrachloroethane for poly(2,6-diphenyl-1,4-phenylene oxide) (2). Although DMPPO dissolves readily in methylene chloride, a complex of polymer and methylene chloride soon precipitates from the solution (3). Although precipitation of the polymer as the complex can be quantitative, the complex is unstable when separated from the liquid methylene chloride phase and dissociates rapidly to leave the pure polymer. Low molecular weight polymers with very narrow molecular weight distribution ($M_w/M_n = 1.1$) can be prepared by cooling an equilibrating mixture of low molecular weight oligomers in methylene chloride. As the molecular weights of some of the oligomers increase owing to redistribution, these oligomers reach the critical size for complexation and precipitate from solution. The molecular weight at which this occurs is approximately 2000 (4).

Mechanical and electrical properties for DMPPO are given in Table 2.

**Chemical Properties.** The phenolic end groups in poly(phenylene oxides)s react with oxidizing agents in a variety of ways; the type of product depends in part on other reagents that may be present. Thus, in the presence of other

**Table 1. Thermal Properties of Poly(Phenylene Oxides)s**

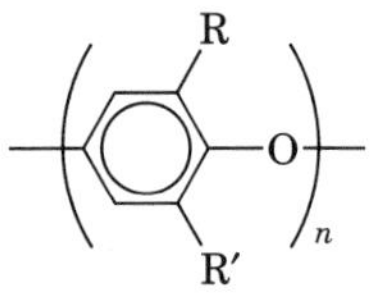

| Name | CAS Registry Number | R | R′ | $T_g$, °C | Ref. |
|---|---|---|---|---|---|
| poly(1,4-phenylene oxide)[a] | [25667-40-7] | H | H | 82 | 5 |
| poly(2,6-dimethyl-1,4-phenylene oxide)[b] | [25134-01-4] | $CH_3$ | $CH_3$ | 211 | 6 |
| poly(2-methyl-6-phenyl-1,4-phenylene oxide) | [25805-39-4] | $CH_3$ | $C_6H_5$ | 169 | 7 |
| poly(2-benzyl-6-methyl-1,4-phenylene oxide) | [26545-37-9] | $C_6H_5CH_2$ | $CH_3$ | 99 | 8 |
| poly(2-isopropyl-6-methyl-1,4-phenylene oxide) | [31985-12-3] | $(CH_3)_2CH$ | $CH_3$ | 144 | 9 |
| poly(2,6-dimethoxy-1,4-phenylene oxide) | [25667-13-4] | $CH_3O$ | $CH_3O$ | 167 | 10 |
| poly(2,6-dichloro-1,4-phenylene oxide)[c] | [26023-26-7] | Cl | Cl | 228 | 11 |
| poly(2,6-diphenyl-1,4-phenylene oxide)[d] | [24938-68-9] | $C_6H_5$ | $C_6H_5$ | 230 | 12 |
| poly(2-m-tolyl-6-phenyl-1,4-phenylene oxide) | [79569-12-3] | m-tolyl | $C_6H_5$ | 219 | 13 |
| poly(2-p-tolyl-6-phenyl-1,4-phenylene oxide) | [79569-09-8] | p-tolyl | $C_6H_5$ | 218 | 13 |
| poly[2-(4-t-butyl)phenyl-6-phenyl-1,4-phenylene oxide] | [79569-10-1] | 4-t-butylphenyl | $C_6H_5$ | 240 | 13 |
| poly(2-naphthyl-6-phenyl-1,4-phenylene oxide) | [79569-11-2] | 2-naphthyl | $C_6H_5$ | 234 | 13 |

[a]Mp = 298°C.
[b]Mp = 268°C.
[c]Mp = 269°C.
[d]Mp = 480°C.

phenols, a catalytic amount of oxidizing agent generates aroxy radicals and the ensuing coredistribution produces low molecular weight products (eq. 1, where $x = 1, 2, 3 \ldots$ ) (15). In the absence of the phenol HOAr, and with an equivalent

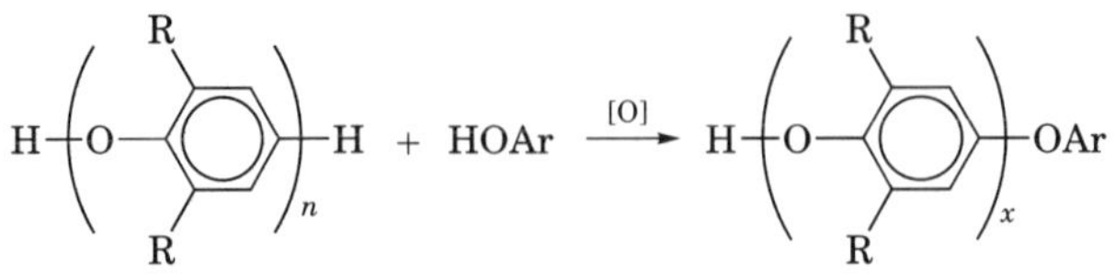

$$H\left(O-\underset{R}{\overset{R}{\bigcirc}}\right)_n H + HOAr \xrightarrow{[O]} H\left(O-\underset{R}{\overset{R}{\bigcirc}}\right)_x OAr \qquad (1)$$

of tetrasubstituted diphenoquinone as an oxidizing agent, biphenyl units are introduced into the polymer at the terminal position and then, through

**Table 2. Properties of DMPPO[a]**

| Property | Value |
| --- | --- |
| tensile yield (at 23°C), MPa[b] | 80 |
| tensile modulus, MPa[b] | 2,690 |
| flexural strength, MPa[b] | 114 |
| flexural modulus, MPa[b] | 2,590 |
| Izod impact (notched), J/m[c] | 64–96 |
| oxygen index | 28–29 |
| density, g/cm$^3$ | 1.06 |
| dielectric constant (at 60 Hz) | 2.54 |
| dielectric strength (thickness, 3.18 mm), V/mm | 20,000 |

[a]Ref. 14.
[b]To convert MPa to psi, multiply by 145.
[c]To convert J/m to ft·lb/in., divide by 53.38 (ASTM D256).

redistribution, internally (16) (eq. 2, where $a = 0, 1, 2, 3$, etc). Under acidic conditions with excess oxidizing agent, benzoquinones are generated (eq. 3) (17).

The preceding reactions do not occur if the terminal hydroxyl group in the polymer is no longer present, eg, if it has been end-capped by acetylation.

Side reactions during polymerization can lead to abnormal end-group formation and branching. For example, when secondary amines are present during a copper-catalyzed polymerization, they are often covalently bound to the methyl group of a head end group (18). With a manganese catalyst system, secondary amines are bound to the polymer analogously if they are present, but if they are absent polymer chains are bound at this site through the oxygen of a head end group to form a benzyl ether linkage. Additional polymerization attaches more units to produce a branched molecule (19), as follows.

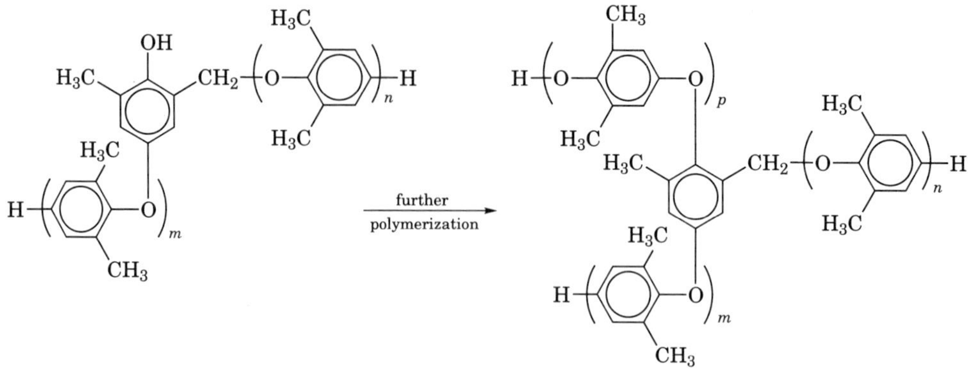

The backbone of poly(phenylene oxide)s is cleaved under certain extreme reaction conditions. Lithium biphenyl reduces DMPPO to low molecular weight products in the dimer and trimer molecular weight range (20) and converts poly(2,6-diphenyl-1,4-phenylene oxide) to 3,5-diphenylphenol in 85% yield (21) (eq. 4).

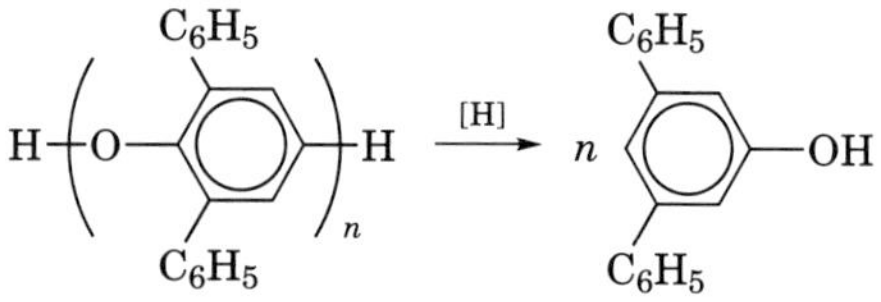

$$\qquad\qquad\qquad\qquad\qquad\qquad\qquad\qquad\qquad\qquad (4)$$

At temperatures near 370°C, DMPPO undergoes a rearrangement (eq. 5).

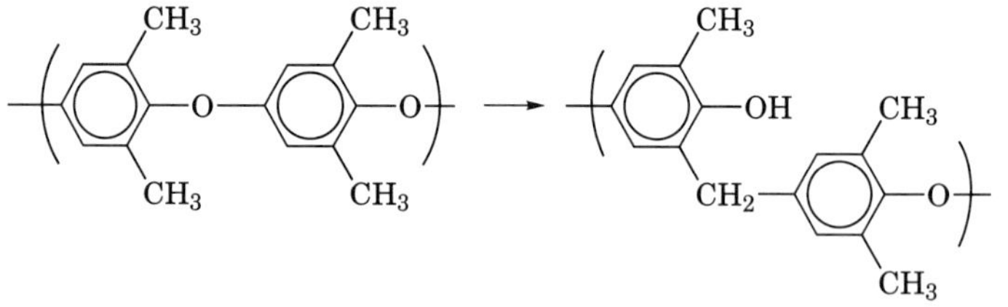

$$\qquad\qquad\qquad\qquad\qquad\qquad\qquad\qquad\qquad\qquad (5)$$

This appears from model studies to be a radical-chain process (22). Abstraction of a hydrogen from a methyl group, followed by migration of the adjacent phenyl group, generates a phenoxy radical that continues the chain process by abstracting a hydrogen from another methyl group.

Ultraviolet radiation causes cleavage of the aryl ether linkage (23). DMPPO undergoes oxidation when exposed to ultraviolet light and oxygen by direct attack on the aromatic ring to produce a variety of ring-cleaved and quinoidal structures (24).

Poly(phenylene oxide)s undergo many substitution reactions (25). Reactions involving the aromatic rings and the methyl groups of DMPPO include bromination (26), displacement of the resultant bromine with phosphorus or amines (27), lithiation (28), and maleic anhydride grafting (29). Additional reactions at the open 3-position on the ring include nitration, alkylation (30), and amidation with isocyanates (31).

SYNTHESIS

**Oxidative Coupling.** Many poly(1,4-phenylene oxide)s have been prepared in a one-step polymerization of 2,6-disubstituted phenols by oxidative coupling (32). The scope and mechanism of polymerization has been studied in detail by

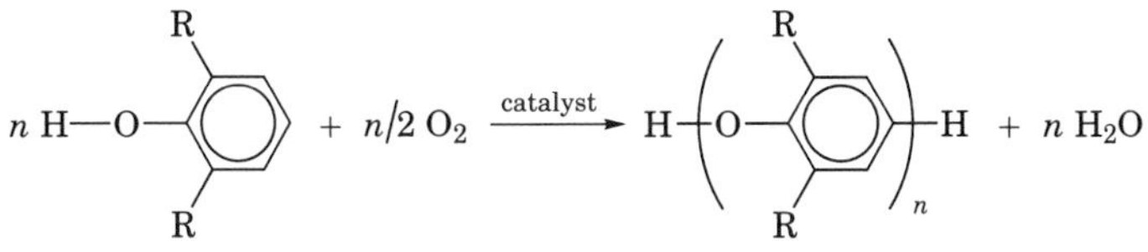

$$n \text{ H}-\text{O}-\!\!\!\!\bigcirc\!\!\!\!- + \; n/2 \; O_2 \;\; \xrightarrow{\text{catalyst}} \;\; \text{H}\!\!-\!\!\left(\!\text{O}-\!\!\!\!\bigcirc\!\!\!\!-\right)_{\!n}\!\!\text{H} \; + \; n \; H_2O$$

many investigators (33). For the preparation of DMPPO, oxygen is passed through a vigorously stirred solution of 2,6-dimethylphenol and a catalyst at ca 25–50°C. A typical catalyst is composed of a copper halide and one or more aliphatic amines or pyridine. If the catalyst is susceptible to hydrolysis, a desiccant, eg, anhydrous magnesium sulfate, is added to remove the water coproduct. With an amine such as dibutylamine, the drying agent is not required. The polymerization is exothermic and often requires cooling to maintain catalytic activity and attain a high molecular weight product. Catalysts based on manganese and cobalt salts also are effective (34). Noncatalytic oxidizing agents, eg, silver oxide, lead dioxide, and manganese dioxide, when used in excess of the stoichiometric amount, also polymerize many 2,6-disubstituted phenols (35).

Oxidative coupling polymerizations represent a general reaction for the preparation of high molecular weight linear polymers from many 2,6-di- and 2,3,6-trisubstituted phenols. When the ortho substituents on the phenols are relatively unhindered alkyl or aryl groups, the poly(phenylene oxide) is the chief product (36,37). Bulky ortho substituents, eg, *tert*-butyl, lead to the formation of the 3,3',5,5'-tetrasubstituted 4,4'-diphenoquinone (36).

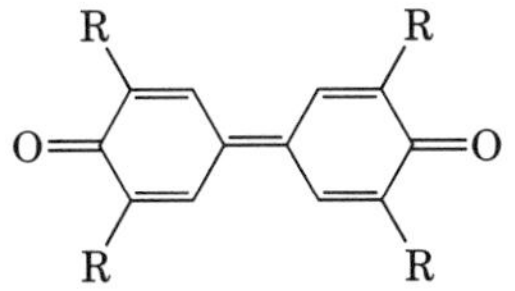

$$\text{O}=\!\!\!\bigcirc\!\!\!=\!\!\!\bigcirc\!\!\!=\text{O}$$

Significant quantities of the diphenoquinone are also produced if the ortho substituents are methoxy groups (36). Phenols with less than two ortho substituents produce branched and colored products from the reactions that occur at the open ortho sites. It is possible to minimize such side reactions in the case of *o*-cresol oxidation by using a bulky ligand on the copper catalyst to block the open ortho position (38).

The selectivity of the oxidation of 2,6-disubstituted phenols depends on the type of oxidizing agent. For example, with a series of cobalt-containing catalysts of the salcomine type, oxidation of 2,6-dimethylphenol produces three products: the poly(phenylene oxide), the diphenoquinone, and 2,6-dimethylbenzoquinone. The product ratio changes with variation in the nature of the ligands (39). The formation of the benzoquinone suggests that ionic processes as well as free-radical reactions take place. Oxidations with reagents that proceed mainly through ionic pathways produce such nonpolymeric products as benzoquinones and cyclohexadienones. Thus, 2,6-dimethylbenzoquinone and its 3-substituted derivatives form when 2,6-dimethylphenol is oxidized with peracids or metal-catalyzed hydrogen peroxide, whereas dimers of 6-hydroxy-2,6-dimethyl-2,4-cyclohexadien-1-one are produced from periodate oxidations (40).

*Polymerization Mechanism.*   The mechanism that accounts for the experimental observations of oxidative coupling of 2,6-disubstituted phenols involves an initial formation of aryloxy radicals from oxidation of the phenol with the oxidized form of the copper–amine complex or other catalytic agent. The aryloxy radicals couple to form cyclohexadienones, which undergo enolization and redistribution steps (32). The initial steps of the polymerization scheme for 2,6-dimethylphenol are as in equation 6.

In equation 6, the dimer (**2**) is formed by enolization of the quinol ether (**1**). Oxidation of the dimer produces the dimer radical, which can couple with other radicals that are present. If it couples with a 2,6-dimethylphenoxy radical to form a quinol ether, enolization can produce the trimer in a reaction analogous to equation 6. Similarly, oxidation of the trimer followed by coupling with a 2,6-dimethylphenoxy radical and enolization produces tetramer. By this process, each oligomeric product can form the next higher oligomer, and eventually high molecular weights are attained. However, the polymerization does not proceed only by addition of one unit at a time to the growing chain, because coupling also can occur between other combinations of oligomers. Thus, this is a step-growth polymerization rather than a chain-growth polymerization. As an example, dimer radical also can couple with another dimer radical to form a quinone ketal (**3**) (eq. 7) or with a larger oligomeric radical to form an analogous ketal. The reaction proceeds further by dissociation of the ketal to form new oligomeric radicals.

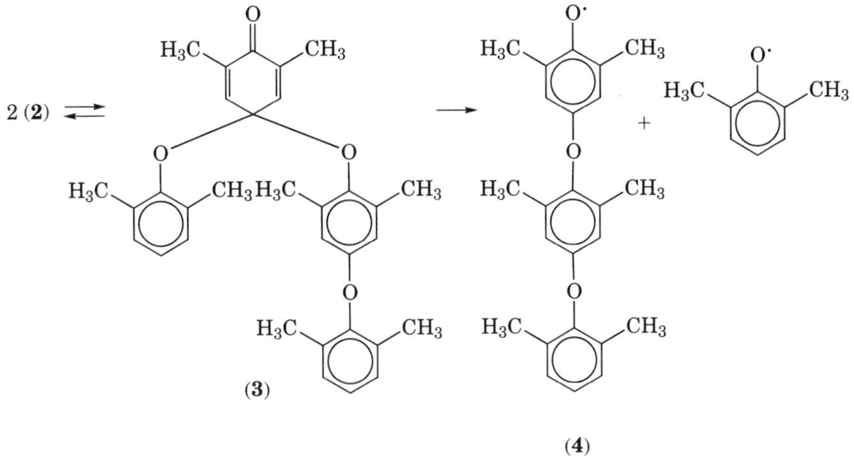

(7)

In equation 7, trimer radical (4) is produced when (3) dissociates. Whenever (4) couples with the other product of equation 7, ie, the 2,6-dimethylphenoxy radical, the tetramer is produced as described. These redistribution reactions of oligomers that proceed by ketal formation and subsequent dissociation ultimately generate terminal quinol ethers which enolize to the more stable terminal phenol (eq. 8).

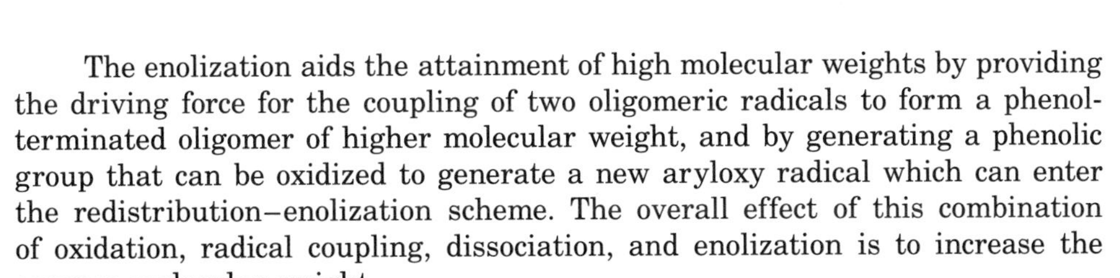

(8)

The enolization aids the attainment of high molecular weights by providing the driving force for the coupling of two oligomeric radicals to form a phenol-terminated oligomer of higher molecular weight, and by generating a phenolic group that can be oxidized to generate a new aryloxy radical which can enter the redistribution–enolization scheme. The overall effect of this combination of oxidation, radical coupling, dissociation, and enolization is to increase the average molecular weight.

A second process that occurs concurrently with the dissociation–redistribution process is an intermolecular rearrangement by which cyclohexadienone groups move along a polymer chain. The reaction may be represented as two electrocyclic reactions analogous to a double Fries rearrangement. When the cyclohexadienone reaches a terminal position, the intermediate is the same as in equation 8, and enolization converts it to the phenol (eq. 9).

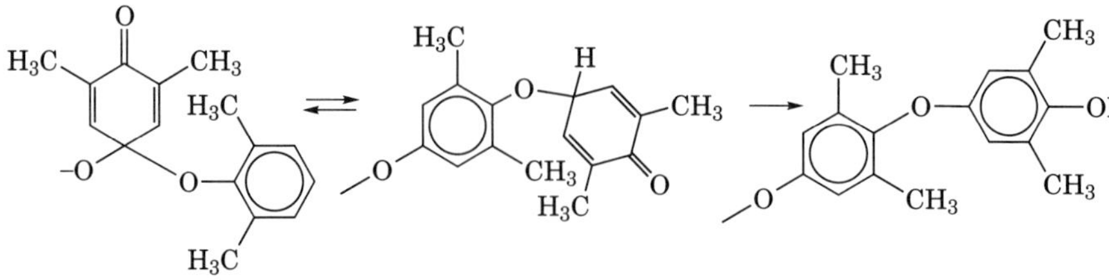

$$(9)$$

This process, which predominates at low temperatures, causes migration of internal ketal structures along a chain but does not involve the dissociation to separate aryloxy radicals that occurs during the redistribution process.

Considerable evidence supports these routes and rules out several alternative mechanisms (15,41–47).

**Halogen Displacement.**    Poly(phenylene oxide)s can also be prepared from 4-halo-2,6-disubstituted phenols by displacement of the halogen to form the ether linkage (48). A trace of an oxidizing agent or free radical initiates the displacement reaction. With 4-bromo-2,6-dimethylphenol, the reaction can be represented as in equation 10:

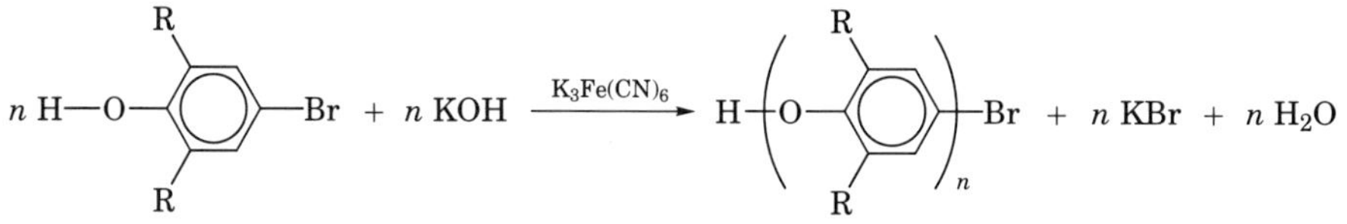

$$(10)$$

High molecular weight DMPPO is formed in high yield at room temperature. The reaction is applicable to certain 2,3,5-tri- and 2,3,5,6-tetrasubstituted-4-halophenols. When such compounds have halogens in the 3,4- or 3,4,5-positions only the one in the 4-position is displaced. When a mixture of 3,4-dibromo-2,6-dimethylphenol and 3,4,5-tribromo-2,6-dimethylphenol is polymerized, a random copolymer is formed, where $a$ and $b = 1$, 2, etc (49). The two homopolymers

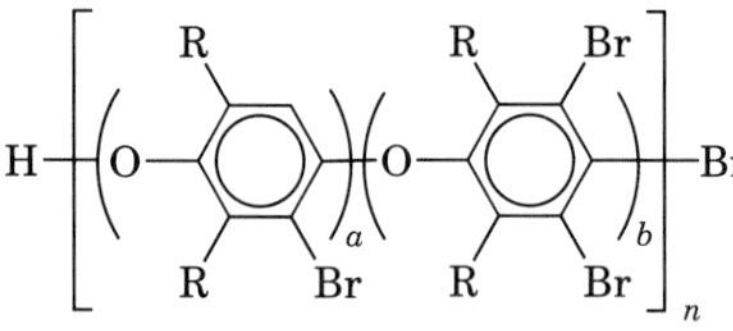

from polymerization of these phenols by themselves are sufficiently insoluble in halogenated and aromatic solvents that they precipitate from the reaction mixture with low molecular weight. The copolymer made from an equimolar mixture of the two monomers is soluble and can be prepared with high molecular weight.

The halogen displacement polymerization proceeds by a combination of the redistribution steps described for oxidative coupling polymerization and a sequence in which a phenoxide ion couples with a phenoxy radical (eq. 11) and then expels a bromide ion. The resultant phenoxy radical can couple with another phenoxide in a manner that is analogous to equation 11 or it can redistribute with other aryloxy radicals in a process analogous to equations 7 and 8.

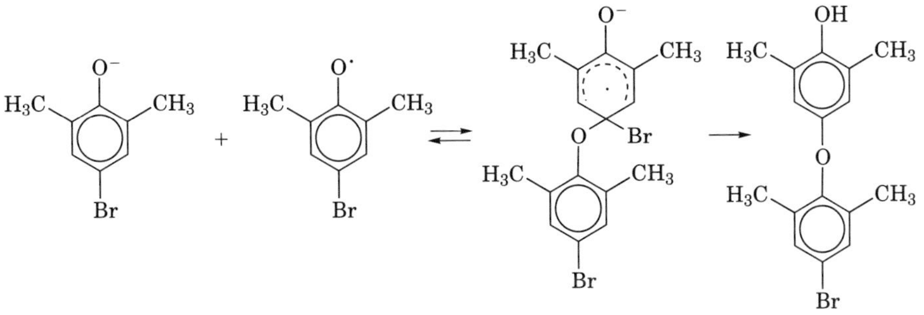

(11)

2,4,6-Trihalophenols can be converted to poly(dihalophenylene oxide)s by a reaction that resembles radical-initiated displacement polymerization. In one procedure, either a copper or silver complex of the phenol is heated to produce a branched product (50). In another procedure, a catalytic quantity of an oxidizing agent and the dry sodium salt in dimethyl sulfoxide produces linear poly(2,6-dichloro-1,4-polyphenylene oxide) (51). The polymer can also be prepared by direct oxidation with a copper–amine catalyst, although branching in the ortho positions is indicated by chlorine analyses (52).

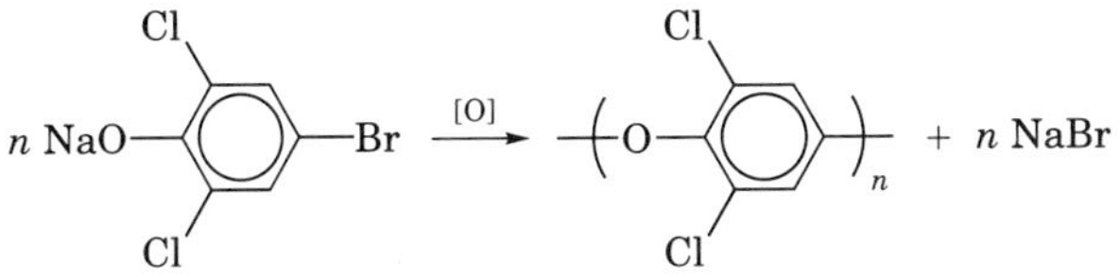

4-Halophenols without 2,6-disubstitution do not polymerize under oxidative displacement conditions. Oxidative side reactions at the ortho position may consume the initiator or interrupt the propagation step of the chain process. To prepare poly(phenylene oxide)s from unsubstituted 4-halophenols, it is necessary to employ the more drastic conditions of the Ullmann ether synthesis. A cuprous chloride–pyridine complex in 1,4-dimethoxybenzene at 200°C converts the sodium salt of 4-bromophenol to poly(phenylene oxide) (1):

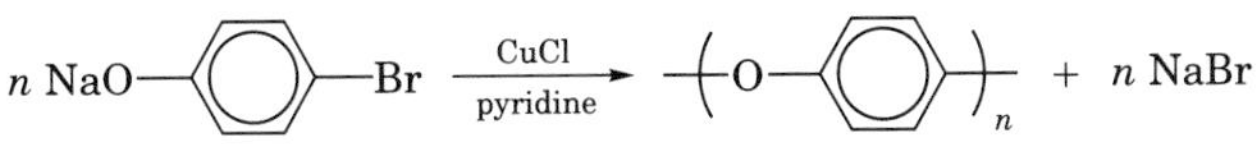

Low molecular weight poly(1,3-phenylene oxide) [25190-64-1] has been prepared from the sodium salt of $m$-chlorophenol with copper as a catalyst (53).

**2,4,6-Trimethylphenol Polymerization.** An unusual aspect of the oxidative coupling of substituted phenols is the formation of DMPPO from 2,4,6-trimethylphenol. Oxidative cleavage of the 4-methyl group produces

$$n \text{ HO}\!-\!\!\bigcirc\!\!-\!\text{CH}_3 \xrightarrow{[O]} \text{H}\!\!\left(\!\text{O}\!-\!\!\bigcirc\!\!\right)_{\!n}\!\!\text{CH}_3 + n/3\,(\text{CH}_2\text{O})_3$$

paraformaldehyde and relatively low molecular weight ($M_n = 1000-3000$) DMPPO with manganese dioxide (54). DMPPO can also be prepared from 2,4,6-trimethylphenol with oxygen as the oxidizing agent and a catalytic amount of copper halide and an amine (55).

**Copolymers.** Copolymers of poly(phenylene oxide)s can be prepared in several ways. Oxidative coupling of mixtures of phenols, eg, combinations of 2,6-dimethylphenol, and 2-methyl-6-phenylphenol usually provide random copolymers. With a pair of phenols that have different oxidation potentials or that coredistribute at different rates, such as a mixture of 2,6-dimethylphenol and 2,6-diphenylphenol, block copolymers can form (56). Another route is the oxidation of mixed dimers which forms random copolymers (57). Copolymers can also be produced by allowing only some of the rings to undergo reaction in a substitution reaction, eg, bromination of only a fraction of the aromatic rings in DMPPO converts normal units to 3-bromo units (26). Block copolymers have been prepared by condensing the phenolic end groups of DMPPO with other polymers bearing reactive leaving groups on their end groups. For example, block copolymers were formed from aromatic polyesters with $\alpha,\omega$-bis(chloroformyl) end groups by using a phase-transfer catalyst and sodium hydroxide to effect the displacement of the halogen (58). These reactions can be carried out with high yields, thereby enabling the formation of ABA types of block copolymers. In the case of low molecular weight di- or triacyl halides, coupled linear or branched polymers with double or triple the original molecular weight are formed.

### POLYETHER BLENDS

DMPPO and polystyrene form compatible blends. The two components are miscible in all proportions (59). Reported dynamic–mechanical results that indicate the presence of two phases in some blends apparently are caused by incomplete mixing (60). Transition behavior of thoroughly mixed blends indicates that the polymers are truly compatible on a segmental level (61). Compatibility may be attributed to a $\pi-\pi$ interaction between the aromatic rings of the two polymers sufficient to produce a negative heat of mixing. However, the forces are very small, ie, $\Delta H_{\text{mix}} = \text{ca } 40$ J/mol (9.6 cal/g), and any change in the substitution pattern of either polymer usually makes them incompatible (62). Tensile strength and modulus of blends of DMPPO and crystal polystyrene reach a maximum with a composition containing about 80 wt % DMPPO, but most properties of blends are close to the weighted average for the two polymers (63). Blends with rubber-modified polystyrene, ie, high impact polystyrene (HIPS), also have intermediate property values, but the ductile PPO matrix is toughened more effectively by rubber than is the brittle polystyrene. Therefore, blends of DMPPO with HIPS have much higher impact strength than either material alone (64). This characteristic makes it possible to prepare a family of tough plastics or modified poly(phenylene oxide)s in which some of the high temperature capability of the DMPPO is combined with the easy processibility of polystyrene.

Blends with good mechanical properties can be made from DMPPO and polymers with which DMPPO is incompatible if an appropriate additive, compatibilizing agent, or treatment is used to increase the dispersion of the two phases. Such blends include mixtures of DMPPO with nylon, polycarbonate, polyester, ABS, and poly(phenylene sulfide).

Sulfonation has been used to change some characteristics of blends. Poly(2,6-diphenyl-1,4-phenylene oxide) and polystyrene are immiscible. However, when the polymers were functionalized by sulfonation, even though they remained immiscible when blended, the functionalization increased interfacial interactions and resulted in improved properties (65). In the case of DMPPO and poly(ethyl acrylate) the originally immiscible blends showed increased miscibility with sulfonation (66).

Blends have also been prepared by dissolving DMPPO in a monomer and then polymerizing the monomer. An example is an epoxy–DMPPO blend prepared by curing a solution of DMPPO in Epon 828 at 85°C with an aluminum–tetramethylguanidine catalyst. Some copolymer formation is observed. The solutions can be applied to glass cloth before curing to produce prepregs for composites in applications such as printed circuit boards (67).

Interpenetrating networks of DMPPO and polymers such as polystyrene, polybutadiene, poly(urethane acrylate), and poly(methyl methacrylate) have been prepared by cross-linking solutions of DMPPO containing bromomethyl groups with ethylenediamine in the presence of the other polymer (68).

**Noryl.** Noryl engineering thermoplastics are polymer blends formed by melt-blending DMPPO and HIPS or other polymers such as nylon with proprietary stabilizers, flame retardants, impact modifiers, and other additives (69). Because the rubber characteristics that are required for optimum performance in DMPPO–polystyrene blends are not the same as for polystyrene alone, most of the HIPS that is used in DMPPO blends is designed specifically for this use (70). Noryl is produced as sheet and for vacuum forming, but by far the greatest use is in pellets for injection molding.

Noryl is a rigid dimensionally stable material. Dimensional stability results from a combination of low mold shrinkage, low coefficient of thermal expansion ($5.9 \times 10^{-5}$ per °C), good creep resistance (0.6–0.8% in 300 h at 13.8 MPa (2000 psi)), and the lowest water absorption rate of any of the engineering thermoplastics (0.07% in 24 h at room temperature). Noryl resins are completely stable to hydrolysis. They are not affected by aqueous acids or bases and have good resistance to some organic solvents, but they are attacked by aromatic or chlorinated aliphatic compounds.

Noryl has good impact strength in unfilled grades: Gardner impact strength of 200 to >400 J (150–300 ft·lbf), and notched Izod 0.27–0.53 J/mm (5–10 ftlb·f/in) at room temperature and 0.13 J/mm (2.5 ftlb·f/in.) at 140°C. Other mechanical properties are typical of amorphous engineering thermoplastics: tensile strength 50–70 MPa (8,000–11,000 psi), flexural strength 60–100 MPa (8,000–15,000 psi), and flexural modulus ca 2500 MPa (350,000 psi). Glass- or mineral-filled products have flexural moduli as high as 7600 MPa ($1.1 \times 10^6$ psi). Dielectric properties are good: dielectric constant is 2.64, dielectric strength is 22 kV/mm at a thickness of 3.18 mm, and it is insensitive to changes in temperature or humidity and to frequencies of $60-10^6$ Hz.

DMPPO–polystyrene blends, because of the inherent flame resistance of the DMPPO component (oxygen index ca 29.5), can be made flame retardant without the use of halogenated additives that tend to lower impact strength and melt stability in other polymers. Approximately one-half of total Noryl sales volume is in flame-retarded grades, ie, V0 or V1 in a 1.6-mm section (UL-94).

Properties of DMPPO–polystyrene blends, especially flammability, deflection temperature under load (DTUL) (ASTM D648), and melt viscosity, can be varied over a wide range by changing the ratio of DMPPO to HIPS, the amount and type of additives, and the characteristics of the HIPS. Noryl originally was introduced to fill a price–performance gap between ABS and polycarbonate, but the property range has been extended steadily and 50 standard grades are available with heat-distortion temperatures of 88–150°C; these include products specially designed for foaming, profile extrusion, and electroplating. In addition to the regular grades, approximately equal numbers of special products are available and each is designed primarily for a single application, eg, an easy-flow V0 product for television cabinets, a high heat Noryl for electrical connectors and food packaging for microwave reheating, Noryl GTX blends for automobile parts, etc.

Principal application areas are in water distribution, electrical–electronic applications, business machines, and automobiles. Water distribution applications include pump housings, impellers and filters, shower heads, faucets, valve handles, and other plumbing fixtures. Among the applications in the electrical–electronic areas are cabinets and internal television and radio parts, lighting fixtures, connectors, junction boxes, wiring ducts, cable covers, protective devices, and motor housings. Noryl is used in many small appliances and personal care products, especially those involving exposure to heat and moisture (food processors, hair dryers, steam irons, hot combs, etc) and in console and decorative trim in dryers, washing machines, and other large appliances. Frames, bases and supports for copiers, display stations, computers, typewriters, printers, and other business machines are commonly made of Noryl. The low temperature impact strength of Noryl and Noryl GTX contribute to its use in a variety of external automotive applications (wheel covers, grilles, fenders, mirror housings steering-column shrouds), instrument panels, electrical connectors, and filler panels.

General Electric is the only U.S. producer of Noryl resin and also has facilities in Japan and Europe. DMPPO is also produced in Japan by Ashahi, Mitsubishi Gas Chemical, and Sumitomo.

**Health and Safety Factors.**   Animal-feeding studies of DMPPO itself have shown it to be nontoxic on ingestion. The solvents, catalyst, and monomers that are used to prepare the polymers, however, should be handled with caution. For example, for the preparation of DMPPO, the amines used as part of the catalyst are flammable; toxic on ingestion, absorption, and inhalation; and are also severe skin and respiratory irritants (see AMINES). Toluene, a solvent for DMPPO, is not a highly toxic material in inhalation testing; the TLV (71) is set at 375 mg/m$^3$, and the lowest toxic concentration is reported to be 100–200 ppm (72). Toxicity of 2,6-dimethylphenol is typical of alkylphenols (qv), eg, for mice, the acute dermal toxicity is $LD_{50}$, 4000 mg/kg, whereas the acute oral toxicity is $LD_{50}$, 980 mg/kg (73). The Noryl blends of DMPPO and polystyrene have FDA approval for reuse food applications.

## Polyethersulfones

The aromatic sulfone polymers are a group of high performance plastics, many of which have relatively closely related structures and similar properties (see POLYMERS CONTAINING SULFUR, POLYSULFONES). Chemically, all are polyethersulfones, ie, they have both aryl ether (ArOAr) and aryl sulfone (ArSO$_2$Ar) linkages in the polymer backbone. The simplest polyethersulfone (5) consists of aromatic rings linked alternately by ether and sulfone groups.

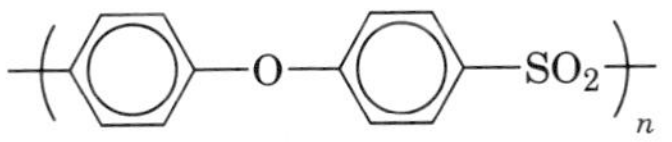

(5)

**Effect of Structure on Properties.**   The effect of structure on chain flexibility in the series of polymers having the following common structure has been

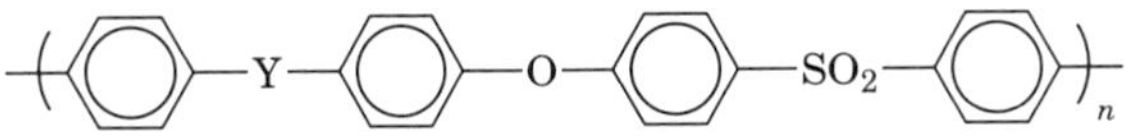

studied by varying the linking group Y (74). The glass-transition temperature changes in a predictable manner with the nature of Y. For example, when Y is a flexible ether, thio, or methylene group, the glass-transition temperature is ca 180°C. Replacement of the hydrogens of the methylene group by alkyl or aryl groups decreases the flexibility and increases the glass-transition temperature; the larger the substituent, the higher the value of $T_g$. Polar groups, eg, carbonyl and sulfonyl, that are capable of conjugation with the aromatic ring increase the glass-transition temperature, eg, to 245°C for the sulfonyl group, as in (5).

   **Synthesis.**   Aromatic polyethersulfones can be prepared by two different routes. In polyetherification, the sulfone group is present in one of the monomers and the ether linkage is formed in the polymerization step. In polysulfonation, the alternative approach is used and the aryl ethers are coupled through a reaction that forms the sulfone linkage. Both processes have been developed commercially. Unlike oxidative coupling polymerization, the mechanisms of both reactions are relatively straightforward, ie, both are polycondensations corresponding to types of reactions that are well-established in organic chemistry. In general, the two processes are used to produce different types of polymers but some structures, including the basic structure (5), are available by either route. Polyetherification and polysulfonation have been reviewed extensively (74,75).

   *Polyetherification.*   Aromatic polyethersulfones are formed by reaction of the salts of dihydroxyaromatic compounds with di(haloaryl)sulfones. Reactions of this type, ie, nucleophilic aromatic displacement of halogen in aromatic compounds activated by sulfonyl or other strongly electron-withdrawing groups, are familiar in organic chemistry, but successful application to formation of high molecular weight polymers depends on the accelerating effects of certain dipolar aprotic solvents, which are believed to solvate the cations selectively, thus increasing the reactivity of the phenolate anions. The polar compounds must be stable under the conditions of the reaction and be capable of dissolving both the reactants and the polymer. Dimethyl sulfoxide (DMSO) is commonly employed, and other solvents, eg, tetramethylene sulfone (sulfolane), sometimes are useful when high temperatures are required (74).

The first aromatic sulfone polymer produced commercially was introduced as Bakelite polysulfone but now is sold by Union Carbide under the trade name Udel. It is made by reaction of the disodium salt of bisphenol A (BPA) with 4,4'-dichlorodiphenyl sulfone in a mixed solvent of chlorobenzene and dimethyl sulfoxide (eq. 12).

Polyetherification is similar to a polycondensation process: formation of high molecular weight polymer requires precise adjustment of composition to approximately 1:1 ratio of bisphenol to dihalosulfone. Trace amounts of water greatly reduce the molecular weight attainable owing to side reactions that unbalance the stoichiometry (76). The reactivity of the halosulfone is in the order expected for two-step nucleophilic aromatic displacement reactions: F > Cl > Br. In addition to being the most reactive of the halides, fluorosulfones are the least affected by traces of water. For economic reasons, only the chlorosulfone is used commercially. The reactivity of bisphenols in polyetherification decreases with increasing acidity of the phenol, but aromatic polyethersulfones can be prepared from almost all dihydric phenols under sufficiently vigorous conditions. The hydroxyl group and the active halogen can be combined in the same molecule; for example, (5) is produced by the self-condensation of 4-chlorophenyl-4'-hydroxyphenyl sulfone (77). The dichlorosulfone in equation 12 may be replaced

(5)

by other dihalogen compounds having strongly electron-withdrawing groups ortho or para to the halogens, such as carbonyl, azo, sulfamido, and nitro groups (78).

*Copolymers.* Copolymers from mixtures of different bisphenols or from mixtures of dichlorosulfone and dichlorobenzophenone have been reported in the patent literature. Bifunctional hydroxyl-terminated polyethersulfone oligomers are prepared readily by the polyetherification reaction simply by providing a suitable excess of the bisphenol. Block copolymers are obtained by reaction of the oligomers with other polymers having end groups capable of reacting with the phenol. Multiblock copolymers of BPA-polysulfone with polysiloxane have been made in this way by reaction with dimethylamino-terminated polydimethylsiloxane; the products are effective impact modifiers for the polyethersulfone (79). Block copolymers with nylon-6 are obtained when chlorine-terminated oligomers, which are prepared by polyetherification with excess dihalosulfone, are used as initiators for polymerization of caprolactam (80).

*Polysulfonylation.* The polysulfonylation route to aromatic sulfone polymers was developed independently by Minnesota Mining and Manufacturing (3M) and by Imperial Chemical Industries (ICI) at about the same time (81). In the polymerization step, sulfone links are formed by reaction of an aromatic sulfonyl chloride with a second aromatic ring. The reaction is similar to the Friedel-Crafts acylation reaction. The key to development of sulfonylation as a polymerization process was the discovery that, unlike the acylation reaction which requires equimolar amounts of aluminum chloride or other strong Lewis acids, sulfonylation can be accomplished with only catalytic amounts of certain halides, eg, $FeCl_3$, $SbCl_5$, and $InCl_3$. The reaction is a typical electrophilic substitution by an arylsulfonium cation (eq. 13).

$$ArSO_2Cl + MCl_n \rightleftharpoons ArSO_2^+ M(Cl)_{n+1} \xrightarrow[-MCl_n]{Ar'H} ArSO_2Ar' + HCl \qquad (13)$$

Reaction of bis(sulfonyl chloride)s with diaryl ether produces polyethersulfones. For example, condensation of diphenyl ether with the disulfonylchloride of diphenyl ether yields polyethersulfone (**5**):

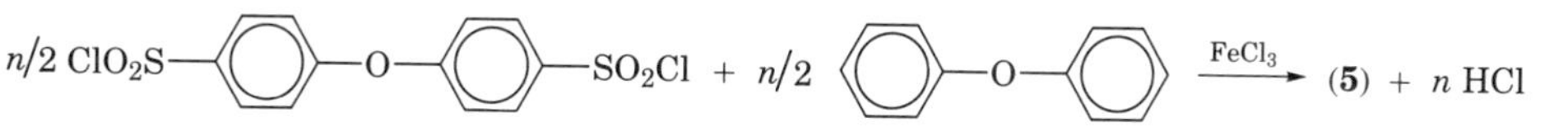

The reaction is carried out either in the melt or in a suitable inert solvent such as acetonitrile; ferric chloride is a preferred catalyst. Successful application of the polysulfonylation reaction to polymer synthesis requires a high degree of para coupling. Chain branching or any substantial degree of ortho coupling without branching results in brittle products. Specificity is improved if the sulfonyl group and the reactive ring are in the same molecule. The process described results in 10–20% of ortho coupling, but self-coupling of the monosulfonyl chloride is without detectable ortho coupling and the product is identical with that obtained by polyetherification. The polysulfonylation reaction is not limited to aryl ethers; it can be applied to most compounds having two or more independent aromatic rings, eg, biphenyl, terphenyl, and naphthalene, but the products are not polyethersulfones unless at least one of the reactants contains the ether linkage. Copolymers of (**5**) in which some of the diphenyl ether units are replaced by other groups are readily available by the sulfonation reaction. Incorporation of rigid biphenyl units by copolymerization of the sulfonyl chlorides of biphenyl and diphenyl ether increases the glass-transition temperature of the polyethersulfones; the Astrel polyethersulfones are of this type (82).

## Polyetherketones

The polyetherification route to polyethersulfones can be adapted to the synthesis of polyethers containing strongly electron-withdrawing groups other than sulfone groups. Poly(1,4-oxyphenylenecarbonyl-1,4-phenylene) [27380-27-4] (**6**) is produced by condensation of 4,4'-dihydroxybenzophenone or by the self-condensation of 4-chloro-4'-hydroxybenzophenone. It has a melting point of 367°C and a glass-transition temperature of 154°C (83).

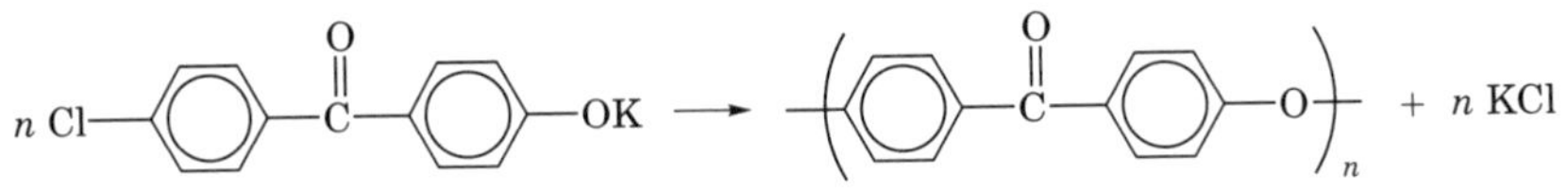

$$n \; Cl\!-\!\langle\bigcirc\rangle\!-\!\overset{\overset{O}{\|}}{C}\!-\!\langle\bigcirc\rangle\!-\!OK \; \longrightarrow \; \left(\!\langle\bigcirc\rangle\!-\!\overset{\overset{O}{\|}}{C}\!-\!\langle\bigcirc\rangle\!-\!O\!\right)_{\!n} \; + \; n \; KCl$$

(**6**)

A polyetheretherketone (PEEK) (**7**) was introduced by Imperial Chemical Industries on a developmental scale in 1978. It is crystalline with a melting point

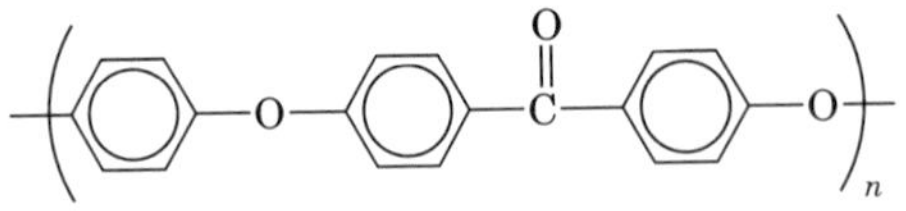

$$\left(\!\langle\bigcirc\rangle\!-\!O\!-\!\langle\bigcirc\rangle\!-\!\overset{\overset{O}{\|}}{C}\!-\!\langle\bigcirc\rangle\!-\!O\!\right)_{\!n}$$

(**7**)

of 334°C and a glass-transition temperature of ca 145°C (84). PEEK can be molded and is used in applications such as liquid chromatography fittings. It is also used for coatings, in electrical insulation for high temperature service, and in composites. Care must be taken on molding composites since the surface of the mold can affect the surface properties of the composite (85). Differential scanning calorimetry and microscopic techniques have been used to determine the morphology of molded PEEK graphite composites (86).

Stilan 1000, introduced by RayChem in 1974 and later withdrawn, is a polyetherketone made by a different route:

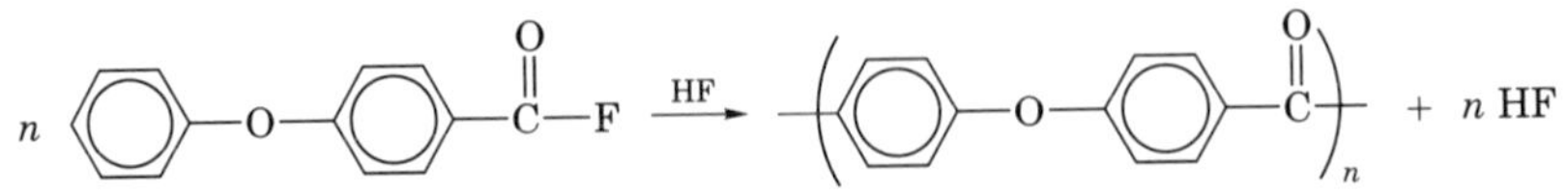

$$n \; \langle\bigcirc\rangle\!-\!O\!-\!\langle\bigcirc\rangle\!-\!\overset{\overset{O}{\|}}{C}\!-\!F \; \xrightarrow{\text{HF}} \; \left(\!\langle\bigcirc\rangle\!-\!O\!-\!\langle\bigcirc\rangle\!-\!\overset{\overset{O}{\|}}{C}\!\right)_{\!n} \; + \; n \; HF$$

Cyclic aryl ether ketones have been prepared from 1,2-bis(4-fluorobenzoyl)-benzene and bisphenols under pseudo high dilution conditions. These materials undergo ring-opening polymerization in the presence of an anionic catalyst (87).

## Polyetherimides

A variety of polyetherimides have been described in reviews on polyimides (qv) (88). Many more recent materials have additional heterocyclic units such as quinoxaline and benzimidazole units, besides the ether and imide function-alities (89).

An all aromatic polyetherimide is made by Du Pont from reaction of py-romellitic dianhydride and 4,4'-oxydianiline and is sold as Kapton. It possesses

excellent thermal stability, mechanical characteristics, and electrical properties, as indicated in Table 3. The high heat-deflection temperature of the resin lim-its its processibility. Kapton is available as general-purpose film and used in applications such as washers and gaskets. Often the resin is not used directly; rather, the more tractable polyamide acid intermediate is applied in solution to a surface and then is thermally imidized as the solvent evaporates.

Not all polyetherimides are limited by their tractability, however. Certain aromatic polyetherimides are characterized by a combination of properties that makes them potential engineering thermoplastics (90). One of these polymers contains an isopropylidene unit in the backbone to enhance the solubility. It is a molding material introduced by General Electric in 1981 and sold as Ultem resin. Attractive features include high temperature stability, flame resistance without

**Table 3. Selected Properties of Kapton**[a]

| Property | Value |
|---|---|
| CAS Registry Number | [25036-53-7] |
| heat deflection temperature, °C | 360 |
| tensile yield (at 23°C), MPa[b] | 86 |
| tensile modulus, MPa[b] | 3110 |
| flexural strength, MPa[b] | 131 |
| flexural modulus, MPa[b] | 3110 |
| oxygen index | 36.5 |
| density, g/cm$^3$ | 1.43 |
| dielectric strength, V/m | 58 |

[a]Ref. 83.
[b]To convert MPa to psi, multiply by 145.

added halogen or phosphorus, high strength, solvent resistance, hydrolytic stability, and injection moldability.

**Syntheses.** The presence of the ether and imide functionalities provides two general approaches for synthesis. Polyetherimides can be prepared by a nucleophilic displacement polymerization similar to the halide displacement in polysulfone synthesis or by a condensation of dianhydrides and diamines that is similar to normal polyimide synthesis (see POLYIMIDES).

*Nitro-Displacement Polymerization.* The facile nucleophilic displacement of a nitro group on a phthalimide by an oxyanion has been used to prepare polyetherimides by heating bisphenoxides with bisnitrophthalimides (91). For example with 4,4'-dinitro monomers, a polymer with the Ultem backbone is prepared as follows (92). Because of the high reactivity of the nitro phthalimides, the

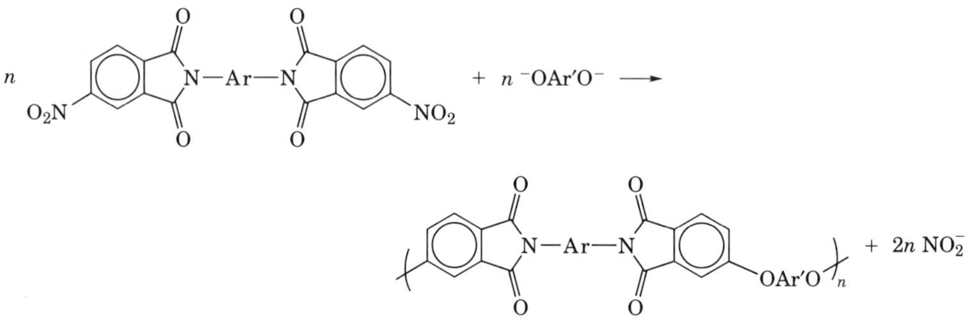

polymerization can be carried out at temperatures below 75°C. Relative reactivities are nitro compounds over halogens, *N*-aryl imides over *N*-alkyl imides, and 3-substituents over 4-substituents. Solvents are usually dipolar aprotic liquids such as dimethyl sulfoxide, and sometimes an aromatic liquid is used, in addition.

Bisnitrophthalimides can be prepared in high yields and high purity from nitrophthalic anhydrides and diamines under acidic conditions (93).

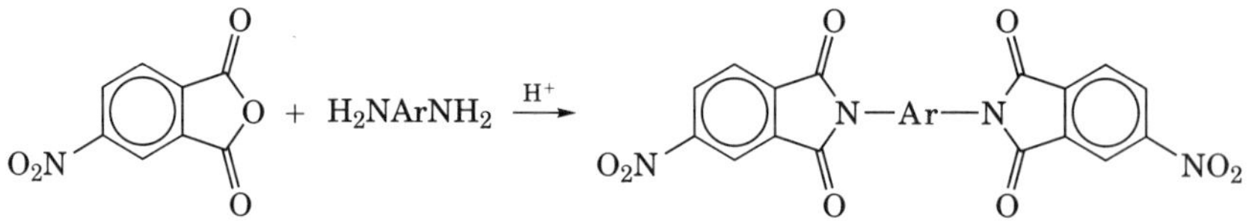

*Condensation of Dianhydrides with Diamines.* The preparation of polyetherimides by the reaction of a diamine with a dianhydride has advantages over nitro-displacement polymerization: sodium nitrite is not a by-product and thus does not have to be removed from the polymer, and a dipolar aprotic solvent is not required, which makes solvent-free melt polymerization a possibility. Aromatic dianhydride monomers (8) can be prepared from *N*-substituted nitrophthalimides by a three-step sequence that utilizes the nitro-displacement reaction in the first step, followed by hydrolysis and then ring closure. For the 4-nitro compounds, the procedure is as follows.

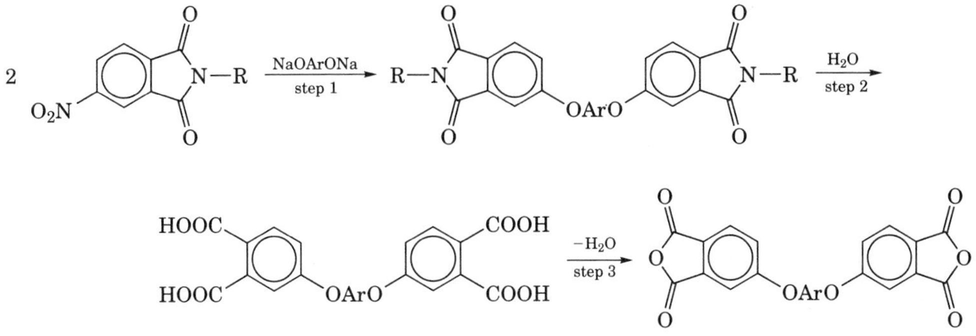

(8)

Polymerization of the dianhydride and diamine proceeds through an intermediate poly(amide acid) stage before ring closure converts the adjacent acid and amide groups to the polyetherimide (94). The polymerization can be carried directly to the polyetherimide as a single-step process, or first to an amide–acid-containing prepolymer, which can be isolated, and then to the polyetherimide.

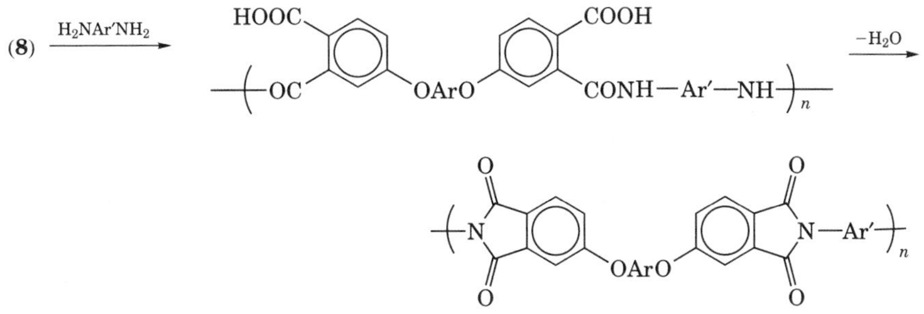

**Properties.** The mechanical properties and oxygen index of Ultem polyetherimide are listed in Table 4. In flammability testing, the oxygen index of many polyetherimides is high and they are self-extinguishing (V-0), non-dripping, and generate little smoke (95). Many polyetherimides with additional chemical functional groups have been reported. Such materials and a variety of polyetherheterocyclics have received only limited commercial attention.

### Table 4. Ultem Polyetherimide Properties

| Property | Value |
| --- | --- |
| glass-transition temperature, °C | 217 |
| heat-distortion temperature, °C | ca 200 |
| tensile yield (at 23°C), MPa[a] | 103 |
| tensile modulus, MPa[a] | 3000 |
| flexural strength, MPa[a] | 145 |
| flexural modulus, MPa[a] | 3300 |
| oxygen index | 47 |
| density, g/cm$^3$ | 1.28 |

[a]To convert MPa to psi, multiply by 145.

Ultem polyetherimides have applications in areas where high strength, dimensional stability, creep resistance, and chemical stability at elevated temperatures are important. Uses include electrical connectors, wave guides and printed circuit boards for electronic equipment, food applications (microwaveable containers, utensils, and films), aircraft interior materials, and sterilizable medical equipment.

A large variety of newer poly(ether imide)s has been described. Included among these are perfluorinated polymers (96), poly(ester ether imide)s (97), poly(ether imide)s derived from $N,N'$-diamino-1,4,5,8-naphthalenetetracarboxylic bisimide (98), and poly(arylene ether imide ketone)s (99). In addition, many other heterocylic groups have been introduced into polyether systems, eg, poly(pyrazole ether)s (100) and poly(aryl ether phenylquinoxaline)s (101); poly(aryl ether oxazole)s with trifluoromethyl groups (102); and polyethers with other heterolinkages, eg, poly(arylether azine)s (103).

## BIBLIOGRAPHY

"Phenolic Ethers" in *ECT* 1st ed., Vol. 10, pp. 325–334, by W. R. Brookes, Chemical Division, General Electric Co.; in *ECT* 2nd ed., Vol. 15, pp. 165–175, by J. E. Cantrill, General Electric Co.; "Polyethers–Aromatic Polyethers" in *ECT* 3rd ed., Vol. 18, pp. 594–615, by D. M. White and G. D. Cooper, General Electric Co.

1. H. Yang and A. S. Hay, *J. Polym. Sci. Polym. Chem. Ed.* **31**, 1261, 2015 (1993).
2. W. A. Butte, C. C. Price, and R. E. Hughes, *J. Polym. Sci.* **61**, 528 (1961); S. Horikiri, *J. Polym. Sci., Part A-2* **10**, 1167 (1972); A. R. Schultz, and C. R. McCullough, *J. Polym. Sci., Part A-2* **10**, 307 (1972); D. M. Koehen, C. A. Smolders and M. Gordon, *J. Poly Sci. Polym. Symp.* **61**, 93 (1977); E. Turski and H. Janeczed, *Polymer* **19**, 81 (1978).
3. A. Factor, D. E. Heinsohn, and L. H. Vogt, *Polym. Lett.* **7**, 205 (1969).
4. D. M. White, *Macromolecules* **12**, 1008 (1979).
5. H. M. van Dort and co-workers, *Eur. Polym. J.* **4**, 275 (1968).
6. A. R. Schultz and B. M. Gendron, *J. Appl. Polym. Sci.* **16**, 461 (1972).
7. D. M. White and H. J. Klopfer, *J. Polym. Sci., Part A-1* **10**, 1565 (1972).
8. A. R. Schultz and B. M. Gendron, *J. Polym. Sci. Polym. Symp.* **43**, 89 (1973).
9. Technical data, General Electric Co., Selkivk, N.Y., 1976.
10. H. M. van Dort and co-workers, *J. Polym. Sci., Part C* **22**, 431 (1968).
11. G. D. Cooper and A. Katchman, *Addition and Condensation Polymerization Processes, Advances in Chemistry Series No. 91*, American Chemical Society, Washington, D.C., 1969, p. 660.
12. A. S. Hay, *Macromolecules* **2**, 107 (1969).
13. A. S. Hay and R. F. Clark, *Macromolecules* **3**, 533 (1970).
14. *Poly(phenylene oxide)s, Properties Comparison Chart*, General Electric Co., Pittsfield, Mass., 1969.
15. G. D. Cooper, A. R. Gilbert, and H. L. Finkbeiner, *Polym. Prep.* **7**, 166 (1966); D. A. Bolon, *J. Org. Chem.* **32**, 1584 (1967); D. M. White, *J. Org. Chem.* **34**, 297 (1969).
16. D. M. White, *J. Polym. Sci. Polym. Chem. Ed.* **19**, 1008 (1979); U.S. Pat. 4,140,675 (Feb. 20, 1979), D. M. White (to General Electric Co.).
17. H. L. Finkbeiner and A. M. Toothaker, *J. Org. Chem.* **33**, 4347 (1968).
18. D. M. White and S. A. Nye, *Macromolecules* **23**, 1318 (1990).
19. D. M. White and L. A. Socha, *Preprints of the 3rd Pacific Polymer Conference, Gold Coast, Australia*, Dec. 1993.

20. U.S. Pat. 3,442,858 (May 6, 1969), A. S. Hay (to General Electric Co.).
21. P. C. Juliano, J. V. Crivello, and D. E. Floryan, *Polym. Prep.* **15**, 210 (1974).
22. A. Factor, H. Finkbeiner, R. A. Jerussi, and D. M. White, *J. Org. Chem.* **35**, 57 (1970); A. Factor, *J. Polym. Sci., Part A-1* **7**, 363 (1969).
23. H. J. Hageman, W. L. Louwerse, and W. J. Mijs, *Tetrahedron* **26**, 2045 (1970); H. J. Hageman and W. G. B. Huysmans, *Rec. Trav. Chim. Pays-Bas* **91**, 528 (1972).
24. J. E. Pickett, *5th International Conference on Advances in Stabilization and Controlled Degradation of Polymers*, Zurich, Switzerland, 1983; *ACS Symp. Ser* **280**, 313, 1985. J. Peeling and D. T. Clark, *J. Appl. Polym. Sci.* **26**, 3761 (1981); A. Dilks, and D. T. Clark, *J. Polym. Sci. Polym. Chem. Ed.* **19**, 2487 (1981).
25. J. Liska and E. Borsig, Chem. Listy **86**(12), 900 (1992). A review of substitution reactions as well as degradation reactions.
26. D. M. White and C. M. Orlando, *Amer. Chem. Soc. Symp. Ser.* **6**, 179 (1975).
27. I. Cabasso, J. Jagur-Grodzinski, and D. Vofsi, *J. Polym. Sci., Part A-1* **12**, 1141 (1974).
28. U.S. Pat. 3,378,505 (Apr. 16, 1968), A. S. Hay (to General Electric Co.).
29. J. H. Glans and M. K. Akkapeddi, *Macromolecules*, **24**, 384 (1991).
30. A. S. Hay and A. J. Chalk, *Polym. Lett.* **6**, 105 (1668); *J. Polym. Sci., Part A-1* **7**, 691 (1969).
31. S. S. Mahajan, B. D. Sarwade, and P. P. Wadgaonkar, *Polym. Bull.* **20**, 153 (1988).
32. A. S. Hay, H. S. Blanchard, G. F. Endres, and J. W. Eustance, *J. Amer. Chem. Soc.* **81**, 6335 (1959).
33. H. L. Finkbeiner, A. S. Hay, and D. M. White, in C. E. Schildknecht and I. S. Skeist, eds., *Polymerization Processes, High Polymers*, Vol. 29, John Wiley & Sons, Inc., New York, 1977, p. 537.
34. U.S. Pat. 3,965,069 (June 22, 1976), W. K. Olander (to General Electric Co.); U.S. Pat. 3,573,257 (Mar. 30, 1971), S. Nakashio and I. Nakagawa (to Sumitomo Chemical Co.).
35. B. O. Lindgren, *Acta Chem. Scand.* **14**, 1203 (1960); E. J. McNelis, *J. Org. Chem.* **31**, 1255 (1966); M. van Dort, *J. Polym. Sci. Part C* **22**, 431 (1968).
36. A. S. Hay, *J. Polym. Sci.* **58**, 581 (1962).
37. D. M. White and H. J. Klopfer, *J. Polym. Sci Part A-1* **8**, 1427 (1970).
38. A. S. Hay and G. F. Endres, *Polym. Lett.* **3**, 887 (1965).
39. L. H. Vogt, J. G. Wirth, and H. L. Finkbeiner, *J. Org. Chem.* **34**, 273 (1969); D. L. Tomaja, L. H. Vogt, and J. G. Wirth, *J. Org. Chem.* **35**, 2029 (1970).
40. J. D. McClure, *J. Org. Chem.* **28**, 69 (1963); R. G. R. Bacon and A. R. Izzat, *J. Chem. Soc. C*, 791 (1966).
41. G. F. Endres and J. Kwiatek, *J. Polym. Sci.* **58**, 593 (1962); G. F. Endres, A. S. Hay, and J. W. Eustance, *J. Org. Chem.* **28**, 1300 (1963).
42. G. D. Cooper, H. S. Blanchard, G. F. Endres, and H. L. Finkbeiner, *J. Amer. Chem. Soc.* **87**, 3996 (1965).
43. D. M. White, *Polym. Prepr.* **9**(1), 663 (1967).
44. W. J. Mijs, O. E. van Lohuizen, J. Bussink, and L. Volbracht, *Tetrahedron* **23**, 2253 (1967).
45. D. J. Williams and R. Kreilick, *J. Amer. Chem. Soc.* **89** 3408 (1967); *J. Amer. Chem. Soc.* **90**, 2775 (1968).
46. H. L. Finkbeiner, G. F. Endres, H. S. Blanchard, and J. W. Eustance, *Polym. Prep.* **2**, 340 (1961).
47. G. D. Cooper and J. G. Bennett, *J. Org. Chem.* **37**, 317 (1972).
48. G. D. Staffin and C. C. Price, *J. Amer. Chem. Soc.* **82**, 3632 (1960); U.S. Pat. 3,257,358 (June 21, 1966), G. S. Stamatoff (to E. I. du Pont de Nemours & Co., Inc.).
49. D. M. White, *Polym. Prep.* **15**, 210 (1974); D. M. White and H. J. Klopfer, in E. J. Vandenberg, ed., *Polyethers*, ACS Symposium Series, American Chemical Society, Washington, D.C., 1975, p. 169.

50. H. S. Blanchard, H. L. Finkbeiner, and G. A. Russell, *J. Polym. Sci.* **58**, 469 (1962).
51. U.S. Pat. 3,236,807 (Feb. 22, 1966), G. S. Stamatoff (to E. I. du Pont de Nemours & Co., Inc.).
52. U.S. Pat. 3,431,238 (Mar. 4, 1969), W. Borman (to General Electric Co.).
53. J. H. Beeson and R. E. Pecsar, *Anal. Chem.* **41**, 1678 (1969).
54. E. McNelis, *J. Amer. Chem. Soc.* **88**, 1074 (1968).
55. U.S. Pat. 3,749,693 (July 31, 1973), G. D. Cooper (to General Electric Co.).
56. G. D. Cooper, J. G. Bennett, Jr., and A. Factor, *Polym. Prep.* **13**, 551 (1972).
57. G. D. Cooper, J. G. Bennett, Jr., and A. Factor, *Polymerization Kinetics and Technology, Advances in Chemistry Series*, No. 128, American Chemical Society, Washington, D.C., 1973, p. 230.
58. D. M. White and G. R. Loucks, *Polym. Preprints*, **25**(1), 129 (1984).
59. A. R. Schultz and B. M. Beach, *Macromolecules* **7**, 902 (1974).
60. S. Krause, *J. Macromol. Sci. Rev. Macromol. Chem.* **7**, 251 (972).
61. J. Stoelting, F. E. Karasz, and W. J. McKnight, *Polym. Eng. Sci.* **10**(3), 133 (1970); A. F. Yee, *Polym. Eng. Sci.* **17**, 213 (1977).
62. W. J. McKnight, F. E. Karasz, and J. R. Fried, *Polymer Blends*, Vol. 1, Academic Press, Inc., New York, 1978, p. 185.
63. L. W. Kleiner, F. E. Karasz, and W. J. McKnight, *Polym. Eng. Sci.* **19**, 519 (1979).
64. U.S. Pat. 3,383,435 (May 14, 1968), E. P. Cizek (to General Electric Co.).
65. D. T. Hseih and D. G. Peiffer, *J. Appl. Polym. Sci.* **44**, 2003 (1992).
66. R. Murali, A. Eisenberg, R. K. Gupta, and F. W. Harris, in D. E. Bergbreiter and C. R. Martin, eds., *Functional Polymers 6th IUCCP Symposium on Functional Polymers*, Plenum Publishing Corp., New York, 1988, p. 37.
67. H. S. Chao and J. M. Whalen, *J. Appl. Polym. Sci.* **49**, 1537 (1993).
68. H. L. Fricsh, D. Klemperner, H. K. Yoon, and K. C. Frisch, *Macromolecules* **13**, 1016 (1980); H. L. Frisch and Y. Hua, *Macromolecules* **22**, 91 (1989); P. C. Mengnjoh and H. L. Frisch, *J. Polym. Sci. Polym. Chem. Ed.* **27**, 3363 (1989); P. C. Mengnjoh and H. L. Frisch, *J. Polym. Sci., Part C, Polym. Lett.* **127**, 285 (1989); S. Singh, H. Ghiradella, and H. L. Frisch, *Macromolecules* **23**, 375 (1990).
69. M. Kramer, *Appl. Polym. Symp.* **15**, 227 (1971).
70. G. D. Cooper, G. F. Lee, A. Katchman, and C. P. Shank, *Mater. Technol.* 12 (1981).
71. J. M. Nielsen, *Solvents and Safety*, General Electric Co., Schenectady, N.Y., 1977, p. 31.
72. *Registry of Toxic Effects of Chemical Substances, 1979 Edition*, U.S. Government Printing Office, Washington, D.C., 1980, p. 621.
73. *Ibid.*, p. 709.
74. R. N. Johnson, A. G. Farnham, R. A. Clendenning, W. F. Hale, and C. N. Merriam, *J. Polym. Sci., Part A-1* **5**, 2375 (1967).
75. M. E. A. Cudby, R. G. Feasy, B. E. Jennings, M. E. B. Jones, and J. B. Rose, *Polymer* **6**, 589 (1965); M. E. A. Cudby, R. G. Feasy, S. Gaskin, V. Kendall, and J. B. Rose, *Polymer* **9**, 265 (1968); K. J. Ivin and J. B. Rose, in W. M. Pasika, ed., *Advances in Macromolecular Chemistry*, Vol. 1, Academic Press, Inc., New York, 1968, p. 366; V. J. Leslie, G. O. Rudkin, Jr., J. B. Rose, and J. Feltzin, *Amer. Chem. Soc. Symp. Ser.* **4**, 63 (1974).
76. R. N. Johnson and A. G. Farnham, *J. Polym. Sci., Part A-1* **7**, 2415 (1967).
77. U.S. Pat. 4,105,636 (Aug. 8, 1978), I. C. Taylor (to Imperial Chemical Industries, Ltd.).
78. U.S. Pat. 4,108,837 (Aug. 22, 1978), R. N. Johnson and A. G. Farnham (to Union Carbide Corp.).
79. A. Noshay, M. Matzner, B. P. Barth, and K. W. Watson, *J. Polym. Sci., Part A 1* **9**, 3147 (1970).

80. S. E. McGrath, L. M. Robeson, and M. Matzner, in L. M. Sperling, ed., *Recent Advances in Polymer Blends, Grafts and Blocks*, Plenum Press, New York, 1974, p. 195.
81. U.S. Pat. 3,321,449 (May 23, 1967), H. A. Vogel (to Minnesota Mining and Mfg. Co.); Brit. Pat. 1,016,245 (1965), M. E. B. Jones (to Imperial Chemical Industries, Ltd.).
82. T. E. Attwood and co-workers, *Polym. Prep.* **20**(1), 191 (1979).
83. *Plastics for Electronics*, Cordura Publications, Cordura, Calif., 1979.
84. *Plast. World*, 11 (May 1980); I. Schmidt and K. J. Binder, *J. Phys. (Paris)* **46**, 1631 (1985).
85. B. R. Prime, *J. Polym. Sci.* **25**, 641, 1986; R. D. McElhaney, D. G. Castner, and B. D. Ratner, in E. Sacher, J. J. Perreaux, and S. P. Kowalczyk, eds., *Metallization of Polymers*, ACS Symp. Series No. 440, American Chemical Society, Washington, D.C., 1990, p. 370.
86. A. Lustiger, F. S. Uralil, and F. M. Newaz, *Polym. Composites* **11**, 65, 1990.
87. K. P. Chan, Y. Wang, and A. S. Hay, *Macromolecules* **28**, 653 (1995).
88. T. Takekoshi, *Advances in Polymer Science*, Vol. 94. Springer-Verlag, Berlin, 1990; C. E. Sroog, *Prog. Polym. Sci.* **16**, 561 (1991).
89. P. M. Hergenrother and S. J. Havens, *Macromolecules* **27**, 4659 (1994).
90. U.S. Pat. 3,838,097 (Sept. 24, 1974), J. G. Wirth and D. R. Heath (to General Electric Co.); U.S. Pat. 3,803,085 (Apr. 9, 1974), T. Takekoshi and J. E. Kochanowski (to General Electric Co.).
91. L. R. Caswell and T. L. Kao, *J. Heterocycl. Chem.* **3**, 333 (1965); F. J. Williams and P. E. Donahue, *J. Org. Chem.* **42**, 3414 (1979).
92. D. M. White and co-workers, *J. Polym. Sci. Polym. Chem. Ed.* **19**, 1635 (1981).
93. T. Takekoshi, J. E. Kochanowski, J. S. Manello, and M. J. Webber, *NASA Contract Report No. CR-145 007*, NASA, Langley, Va., 1976.
94. T. Takekoshi, D. R. Heath, J. E. Kochanowski, J. S. Manello, and M. J. Webber, *Polym. Prep.* **20**(1), 179 (1979).
95. D. E. Floryan and G. L. Nelson, *J. Fire Flammabil.* **11**, 284 (1980).
96. S. Ando, T. Matsuura, and S. Sasaki, *Macromolecules* **25**, 5858 (1992).
97. D. Venkatesan and M. Srinivasan, *J. M. S.—Pure Appl. Chem.* **A30**, 801 (1993).
98. H. Ghassemi and A. S. Hay, *Macromolecules* **27**, 3116 (1994).
99. S. Matsuo and K. Mitsuhashi, *J. Polym. Sci. Polym. Chem. Ed.* **32**, 1969 (1994).
100. J. A. Moore and P. G. Mehta, *Macromolecules* **28**, 444 (1995).
101. J. Hedrick, R. Tweig, T. Matray, and K. Carter, *Macromolecules* **26**, 4833 (1993).
102. G. Maier, R. Hecht, O. Nuyken, K. Borger, and B. Helmreich, *Macromolecules* **26**, 2583 (1993).
103. K. R. Carter and J. L. Hedrick, *Macromolecules* **27**, 3426 (1994).

Dwain M. White
General Electric Company

# ETHYLENE OXIDE POLYMERS

Poly(ethylene oxide) [*25322-68-3*] (PEO) is a water-soluble, thermoplastic polymer produced by the heterogeneous polymerization of ethylene oxide. The white, free-flowing resins are characterized by the following structural formula:

$$-(\text{CH}_2\text{CH}_2\text{O})_{\overline{n}}$$

The resins are available in a broad range of molecular weight grades, from as low as 100,000 to over $7 \times 10^6$. Although most commonly known as poly(ethylene oxide) resins, they are occasionally referred to as poly(ethylene glycol) or poly(oxyethylene) resins. The CAS Registry Number of these resins is also used for low molecular weight oligomers of ethylene oxide, eg, tetraethylene glycol.

## Physical Properties

**Crystallinity.** At molecular weights of $10^5$–$10^7$, poly(ethylene oxide) forms a highly ordered structure. This has been confirmed by nmr and x-ray diffraction patterns and by the sharpness of the crystalline melting point (62–67°C). However, the highest degree of crystallinity (ca 95%) is obtained at a molecular weight of 6000. The polymer chain contains seven structural units per fiber identity period (1.93 nm) (1). The diffraction pattern of the monoclinic unit cell of poly(ethylene oxide) contains four molecular chains, in which $a = 0.796$ nm, $b = 1.311$ nm, $c = 1.939$ nm, and $\beta = 124°48'$. Infrared studies show that the oxygen atoms of the crystalline polymer are in the gauche configuration. Because of this arrangement, the intermolecular dipole forces are oriented along the axis of the helix and ca 15 monomer units are involved within a single repeat unit. The heat of fusion of the polymer is 8.3 kJ (1980 cal) per structural unit (2). The high molecular weight poly(ethylene oxide) resins are of the spherulitic structure (3). Proper annealing of a melt-cast film produces a distinct lamellar structure. The molecular conformation of poly(ethylene oxide), as determined by the use of x-ray diffraction, ir, and Raman spectroscopic methods, is shown in Figure 1.

**Density.** Although the polymer unit cell dimensions imply a calculated density of 1.33 g/cm³ at 20°C, and extrapolation of melt density data indicates a density of 1.13 g/cm³ at 20°C for the amorphous phase, the density actually measured is 1.15–1.26 g/cm³, which indicates the presence of numerous voids in the structure.

**Glass-Transition Temperature.** The glass-transition temperature, $T_g$, of poly(ethylene oxide) has been measured over the molecular weight range of $10^2$–$10^7$ (5,6). The $T_g$–molecular weight relationship is shown in Figure 2. These data indicate a rapid rise in the transition temperature to a maximum of $-17$°C for a molecular weight of 6000. The highest percentage of crystalline character develops at that molecular weight, and it is at that point that $T_g$ is the highest. Beyond this point, chain entanglement reduces crystallinity.

**Solubility.** Poly(ethylene oxide) is completely soluble in water at room temperature. However, at elevated temperatures (>98°C) the solubility decreases. It is also soluble in several organic solvents, particularly chlorinated hydrocarbons (see WATER-SOLUBLE POLYMERS). Aromatic hydrocarbons are bet-

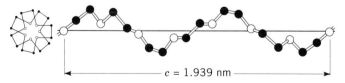

$c = 1.939$ nm

**Fig. 1.** Molecular conformation of poly(ethylene oxide).

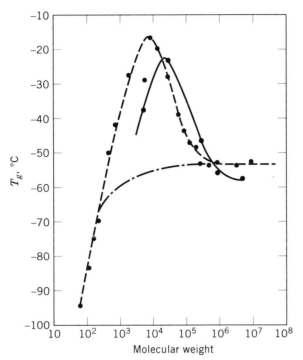

**Fig. 2.** Glass-transition temperature–molecular weight relationship for poly(ethylene oxide): ( ———·——— ) represents classical $T_g$–mol wt relationship; (———), data from Ref. 4; and (— — — —), data from Ref. 5.

ter solvents for poly(ethylene oxide) at elevated temperatures. Solubility characteristics are listed in Table 1.

Aqueous poly(ethylene oxide) solutions of higher molecular weight (ca $10^6$) become stringy at polymer concentrations less than 1 wt %. At concentrations of 20 wt %, solutions become nontacky elastic gels; above this concentration, the solutions appear to be hard, tough, water-plasticized polymers.

*Concentration and Molecular Weight Effects.* The viscosity of aqueous solutions of poly(ethylene oxide) depends on the concentration of the polymer solute, the molecular weight, the solution temperature, concentration of dissolved inorganic salts, and the shear rate. Viscosity increases with concentration and this dependence becomes more pronounced with increasing molecular weight. This combined effect is shown in Figure 3, in which solution viscosity is presented as a function of concentration for various molecular weight polymers.

The dependence of the intrinsic viscosity $[\eta]$ on molecular weight $M$ for these polymers can be expressed by the Mark-Houwink relationship:

$$[\eta] = KM^a \tag{1}$$

The constants $K$ and $a$ for high molecular weight poly(ethylene oxide) in several solvents at various temperatures are summarized in Table 2.

**Table 1. Solubility of Poly(Ethylene Oxide)[a] in Several Solvents[b]**

| Solvent[c,d] | Temperature | |
| --- | --- | --- |
| | Resin dissolves on heating >25°C, °C | Resin precipitates on cooling, °C |
| dissolves at room temperature in | | |
| water | | <0 |
| carbon tetrachloride | | <0 |
| acetonitrile | | <0 |
| ethylene dichloride | | <0 |
| trichloroethylene | | <0 |
| methylene dichloride | | <0 |
| benzene | | 2 |
| 2-propanol (91 wt %) | | 2 |
| dimethylformamide | | 14 |
| tetrahydrofuran | | 18 |
| methanol | | 20 |
| methyl ethyl ketone | | 20 |
| dissolves with heating in[e] | | |
| toluene | 30 | 20 |
| xylene | 30 | 20 |
| acetone | 35 | 20 |
| Cellosolve[f] acetate | 35 | 25 |
| anisole | 40 | 0 |
| 1,4-dioxane | 40 | 4 |
| ethyl acetate | 40 | 25 |
| ethylenediamine | 40 | 26 |
| dimethyl Cellosolve[f] | 40 | 27 |
| Cellosolve[f] solvent | 45 | 28 |
| ethanol (dry) | 45 | 31 |
| Carbitol[f] solvent | 50 | 32 |
| n-butanol | 50 | 33 |
| butyl Cellosolve[f] | 50 | 33 |
| n-butyl acetate | 50 | 34 |
| 2-propanol (dry) | 50 | 36 |
| methyl Cellosolve[f] | 50 | 46 |

[a]Mol wt $(1–50) \times 10^5$.
[b]Ref. 4.
[c]Solution concentration = ca 1 wt %.
[d]All solvents except 2-propanol (91 wt %) were carefully dried before testing.
[e]The polymer was insoluble in 1,3-butanediol, ethylene glycol, diethylene glycol, and glycerol at all temperatures.
[f]Registered trademark of Union Carbide Corp.

*Temperature Effect.* Near the boiling point of water, the solubility–temperature relationship undergoes an abrupt inversion. Over a narrow temperature range, solutions become cloudy and the polymer precipitates; the polymer cannot dissolve in water above this precipitation temperature. In Figure 4, this limit or cloud point is shown as a function of polymer concentration for poly(ethylene oxide) of $2 \times 10^6$ molecular weight.

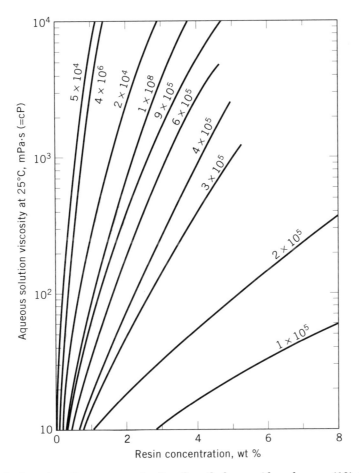

**Fig. 3.** Solution viscosity vs concentration for ethylene oxide polymers (10). The molecular weight of the polymer is indicated on each curve.

**Table 2. Mark-Houwink Constants for Poly(ethylene oxide)**

| Solvent | Temperature, °C | $K \times 10^5$ | $a$ | Approximate mol wt, $M$ | Reference |
|---|---|---|---|---|---|
| water | 25 | 11.92 | 0.76 | $(5-40) \times 10^5$ | 7 |
| | 35 | 6.4 | 0.82 | $10^4-10^7$ | 8 |
| | 45 | 6.9 | 0.91 | $10^4-10^7$ | 8 |
| 0.45 $M$ K$_2$SO$_4$ (aq) | 35 | 130 | 0.5 | $10^4-10^7$ | 8 |
| 0.39 $M$ MgSO$_4$ (aq) | 45 | 100 | 0.5 | $10^4-10^7$ | 8 |
| benzene | 25 | 39.7 | 0.686 | $(8-500) \times 10^4$ | 9 |
| | 30 | 61.4 | 0.64 | $(3-20) \times 10^5$ | 10 |

The viscosity of the aqueous solution is also significantly affected by temperature. In polymers of molecular weights $(1-50) \times 10^5$, the solution viscosity may decrease by one order of magnitude as the temperature of measurement is increased from 10 to 90°C. Figure 5 shows this effect.

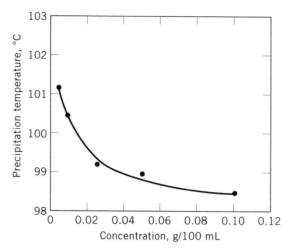

**Fig. 4.** Upper temperature limit for solubility of poly(ethylene oxide) in water. Molecular weight is $2 \times 10^6$ (3).

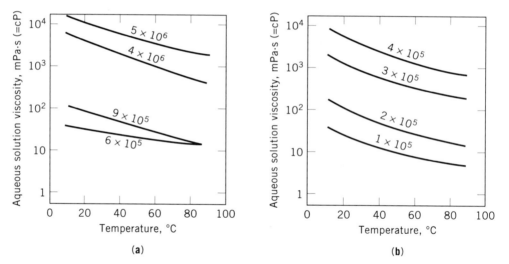

(a)                                                            (b)

**Fig. 5.** Solution viscosity vs temperature: (**a**) 1.0 wt % solution, (**b**) 5.0 wt % solution (11), for polymers of various molecular weights, indicated on the curves

*Effects of Salts.* The presence of inorganic salts in aqueous solutions of poly(ethylene oxide) reduces the upper temperature limit of solubility and viscosity. The upper temperature limit of solubility decreases in proportion to the concentration and valence of the ionic species present. The size of the ions is also important, eg, smaller hydrated ions have the greatest effect. The effect of a number of inorganic salts on the upper temperature limit of solubility of poly(ethylene oxide) in water is illustrated in Figure 6. The decrease in temperature is nearly a linear function of salt concentration. However, this salting-out effect cannot always be correlated with the ionic strength principle. For example, potassium and magnesium sulfate have approximately the same ef-

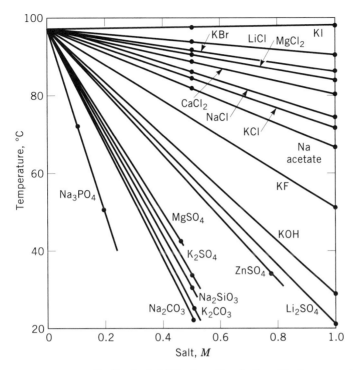

**Fig. 6.** Upper temperature limit of solubility in salt solution. Resin concentration is 5.0 wt % (10).

fect but potassium halides are widely different. Thus, it appears that the anion has the greater effect on the upper temperature limit of solubility. The effectiveness of anions to reduce the $\theta$ temperature of aqueous poly(ethylene oxide) solution decreases in the following order: $PO_4^{3-} > HPO_4^{2-} > S_2O_3^{2-} > H_2PO_4^- > F^- > HCO_2^- > CH_3CO_2^- > Br^- > I^-$. The order for cations was found to be $K^+ \approx Rb^+ \approx Na^+ \approx Cs^+ > Sr^{2+} > Ba^{2+} \approx Ca^{2+} > NH_4^+ > Li^+$ (12).

The presence of inorganic salts in solutions of poly(ethylene oxide) also can reduce the hydrodynamic volume of the polymer, with attendant reduction in intrinsic viscosity; this effect is shown in Figure 7.

*Effect of Shear.* Concentrated aqueous solutions of poly(ethylene oxide) are pseudoplastic. The degree of pseudoplasticity increases as the molecular weight increases. Therefore, the viscosity of a given aqueous solution is a function of the shear rate used for the measurement. This relationship between viscosity and shear rate for solutions of various molecular weight poly(ethylene oxide) resins is presented in Figure 8.

**Thermoplasticity.** High molecular weight poly(ethylene oxide) can be molded, extruded, or calendered by means of conventional thermoplastic processing equipment (13). Films of poly(ethylene oxide) can be produced by the blown-film extrusion process and, in addition to complete water solubility, have the typical physical properties shown in Table 3. Films of poly(ethylene oxide) tend to orient under stress, resulting in high strength in the draw direction.

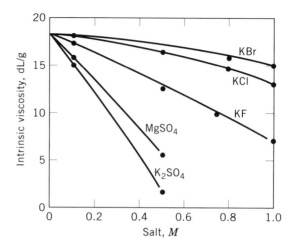

**Fig. 7.** Effects of salts on the intrinsic viscosity of poly(ethylene oxide) at 30°C. Molecular weight is $5.5 \times 10^6$ (3).

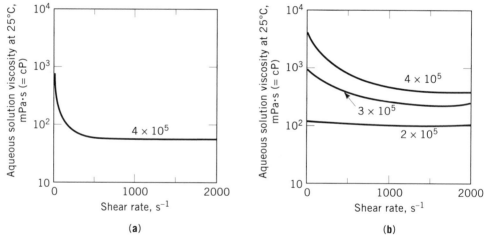

**Fig. 8.** Effect of shear on aqueous solution viscosities of poly(ethylene oxide) resins: (**a**) 1.0 wt % solution, (**b**) 5.0 wt % solution (10). Each curve represents a different molecular weight.

The physical properties, melting behavior, and crystallinity of drawn films have been studied by several researchers (14–17).

At 100–150°C above the melting point, the melt viscosities of these polymers may exceed 10 kPa·s ($10^5$ P) (Fig. 9). These high melt viscosities indicate extremely high molecular weight. Melt viscosities are relatively unaffected by temperature changes but are directly proportional to the molecular weight of the polymer. Thus, polymers with molecular weights of $(1–3) \times 10^5$ usually are used for applications involving thermoplastic forming processes.

**Polymer Blends.** The miscibility of poly(ethylene oxide) with a number of other polymers has been studied, eg, with poly(methyl methacrylate) (18–23),

**Table 3. Typical Physical Properties of Poly(Ethylene Oxide) Film**

| Property | Value |
|---|---|
| specific gravity | 1.2 |
| tensile strength, MPa[a] | |
| machine direction | 16 |
| transverse direction | 13 |
| secant modulus, MPa[a] | |
| machine direction | 290 |
| transverse direction | 480 |
| elongation, % | |
| machine direction | 550 |
| transverse direction | 650 |
| tear strength, kN/m[b] | |
| machine direction | 100 |
| transverse direction | 240 |
| dart impact at 50% failure, kN/m[b] | 80 |
| release time in water, s | 15 |
| $O_2$ transmission, $\mu mol/(m \cdot s \cdot GPa)$[c] | 85.8 |
| melting point, °C | 67 |
| heat-sealing capability | excellent[d] |
| heat-sealing temperature, °C | 71–107 |
| cold-crack resistance, °C | −46 |

[a]To convert MPa to psi, multiply by 145.
[b]To convert kN/m to lbf/in., multiply by 57.14.
[c]To convert $\mu mol/(m \cdot s \cdot GPa)$ to $cm^3 \cdot mil/(in.^2 \cdot d \cdot atm)$, multiply by 5.
[d]Equal to low density polyethylene.

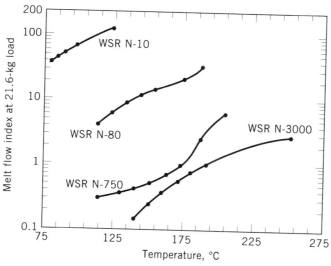

**Fig. 9.** Melt flow index as a function of temperature for varying molecular weights of poly(ethylene oxide). WSR = Polyox water-soluble resins.

poly(vinyl acetate) (24–27), polyvinylpyrrolidinone (28), nylon (29), poly(vinyl alcohol) (30), phenoxy resins (31), cellulose (32), cellulose ethers (33), poly(vinyl chloride) (34), poly(lactic acid) (35), poly(hydroxybutyrate) (36), poly(acrylic acid) (37), polypropylene (38), and polyethylene (39).

## Chemical Properties

**Association Complexes.**   The unshared electron pairs of the ether oxygens, which give the polymer strong hydrogen bonding affinity, can also take part in association reactions with a variety of monomeric and polymeric electron acceptors (40,41). These include poly(acrylic acid), poly(methacrylic acid), copolymers of maleic and acrylic acids, tannic acid, naphtholic and phenolic compounds, as well as urea and thiourea (42–47).

When equal amounts of solutions of poly(ethylene oxide) and poly(acrylic acid) are mixed, a precipitate, which appears to be an association product of the two polymers, forms immediately. This association reaction is influenced by hydrogen-ion concentration. Below ca pH 4, the complex precipitates from solution. Above ca pH 12, precipitation also occurs, but probably only poly(ethylene oxide) precipitates. If solution viscosity is used as an indication of the degree of association, it appears that association becomes more pronounced as the pH is reduced toward a lower limit of about four. The highest yield of insoluble complex usually occurs at an equimolar ratio of ether and carboxyl groups. Studies of the poly(ethylene oxide)–poly(methacrylic acid) complexes indicate a stoichiometric ratio of three monomeric units of ethylene oxide for each methacrylic acid unit.

These association reactions can be controlled. Acetone or acetonylacetone added to the solution of the polymeric electron acceptor prevents insolubilization, which takes place immediately upon the removal of the ketone. A second method of insolubilization control consists of blocking the carboxyl groups with inorganic cations, ie, the formation of the sodium or ammonium salt of poly(acrylic acid). Mixtures of poly(ethylene oxide) solutions with solutions of such salts can be precipitated by acidification.

Poly(ethylene oxide) associates in solution with certain electrolytes (48–52). For example, high molecular weight species of poly(ethylene oxide) readily dissolve in methanol that contains 0.5 wt % KI, although the resin does not remain in methanol solution at room temperature. This salting-in effect has been attributed to ion binding, which prevents coagulation in the nonsolvent. Complexes with electrolytes, in particular lithium salts, have received widespread attention on account of the potential for using these materials in a polymeric battery. The performance of solid electrolytes based on poly(ethylene oxide) in terms of ion transport and conductivity has been discussed (53–58). The use of complexes of poly(ethylene oxide) in analytical chemistry has also been reviewed (59).

**Oxidation.**   Because of the presence of weak C–O bonds in the backbone, high molecular weight polymers of ethylene oxide are susceptible to oxidative degradation in bulk, during thermoplastic processing, or in solution. The mechanistic aspects of poly(ethylene oxide) oxidation have been reviewed (60). During thermoplastic processing at elevated temperature, oxidative degradation is manifested by a rapid decrease in melt viscosity with time. In aqueous solution at ambient temperatures, the decay of solution viscosity also is an indication of oxidative degradation, and the rate of decay is increased by the presence of traces of chlorine, peroxides, permanganate, or persulfate and certain transition-metal ions such as $Cu^+$, $Cu^{2+}$, $Cu^{3+}$, $Fe^{3+}$, and $Ni^{2+}$. A combination of these agents can lead to severe viscosity losses.

Several stabilizers are useful in minimizing oxidative degradation during thermoplastic processing or in the bulk solid. Phenothiazine, hindered phenolic antioxidants such as butylated hydroxytoluene, butylated hydroxyanisole, and secondary aromatic amines in concentrations of 0.01–0.5% based on the weight of polymer, are effective.

Aqueous solutions can be stabilized against viscosity loss by addition of 5–10 wt % anhydrous isopropyl alcohol, ethanol, ethylene glycol, or propylene glycol. The manganous ion ($Mn^{2+}$) also is an effective stabilizer at concentrations of $10^{-5}$–$10^{-2}$ wt % of the solution.

## Manufacture and Processing

**Heterogeneous Catalytic Polymerization.** The preparation of polymers of ethylene oxide with molecular weights greater than 100,000 was first reported in 1933. The polymer was produced by placing ethylene oxide in contact with an alkaline-earth oxide for extended periods (61). In the 1950s, the low yield and low polymerization rates of the early work were improved upon by the use of alkaline-earth carbonates as the catalysts (62). Further improvements in reaction rates and polymerization control have led to the commercial availability of poly(ethylene oxide) of varying molecular weights.

The polymerization of ethylene oxide to produce high molecular weight polymer involves heterogeneous reaction with propagation at the catalyst surface. The polymerization can involve anionic or cationic reactions of ethylene oxide that generally produce lower molecular weight products. The mechanism for production of extremely high molecular weight polymers is thought to involve a coordinate anionic reaction where ethylene oxide is coordinated with a metal atom of the catalyst and is then attacked by an anion. The various polymerization mechanisms have been described (63).

Catalysts capable of polymerizing ethylene oxide to high molecular weight polymers include many metal compounds. Among those reported are alkaline-earth carbonates and oxides (64), alkyl zinc compounds (65), alkyl aluminum compounds and alkoxides (66,67), and hydrates of ferric chloride, bromide, and acetate (68,69). Other catalysts include various alkyls and alkoxides of aluminum, zinc, magnesium, and calcium, and mixtures of these materials with various other inorganic salts. The preparation and utilization of the various catalysts have also been described (70,71). The molecular weight of the polymer appears to be controlled by the catalyst systems as well as by polymerization conditions. Rigid control of catalyst preparation and raw material quality appear to be mandatory for successful laboratory preparation of high molecular weight poly(ethylene oxide).

**Polymer Suspensions.** Poly(ethylene oxide) resins are commercially available as fine granular solids. However, the polymer can be dispersed in a nonsolvent to provide better metering into various systems. Production processes involve the use of high shear mixers to disperse the solids in a nonsolvent vehicle (72–74).

**Thermoplastic Processing.** Poly(ethylene oxide) resins can be thermoplastically formed into solid products, eg, films, tapes, plugs, retainers, and fillers (qv). Through the use of plasticizers (qv), poly(ethylene oxide) can be extruded, molded, and calendered on conventional thermoplastic processing equipment. Sheets and films of this resin are heat sealable.

**Irradiation and Cross-Linking.** Exposure of poly(ethylene oxide) to ionizable radiation (gamma irradiation, electron beam, or ultraviolet light) can result in molecular weight breakdown or cross-linking, depending on the environmental conditions. If oxygen is present, hydroperoxides are formed and chain scission leads to an overall decrease in molecular weight (75). However, in the absence of oxygen, cross-linking becomes the preferred reaction (76–78). The resulting polymer network exhibits hydrogel properties of high water capacity (79,80).

Studies of the cross-linking mechanism and structure of the cross-linked polymer indicate that a complex network of cross-linked chains of varying lengths is present (81–84). When the cross-linking is performed in solution, the cross-links can be both intermolecular and intramolecular; the overall structure of the cross-linked polymer is the combined result of chain scission, intramolecular bonding, and intermolecular bonding. Under conditions of high aggregation in solution, for example, high concentration, intermolecular cross-linking is preferred and a continuous gel is formed. When the polymer is not aggregated in solution, intramolecular cross-linking predominates, and microgels, rather than a cohesive gel network, are formed.

## Economic Aspects

Only Japan and the United States have significant commercial facilities for the production of poly(ethylene oxide) resins. In Japan, Meisei Chemical Works Ltd. produces Alkox and Sumitomo Seika Kagaky Co., Ltd., PEO. In the United States, Union Carbide Corp. produces Polyox. Precise figures have not been released on capacities or annual production.

## Specifications, Standards, and Quality Control

Of the three worldwide manufacturers of poly(ethylene oxide) resins, Union Carbide Corp. offers the broadest range of products. The primary quality control measure for these resins is the concentrated aqueous solution viscosity, which is related to molecular weight. Specifications for Polyox are summarized in Table 4. Additional product specifications frequently include moisture content, particle size distribution, and residual catalyst by-product level.

## Analytical and Test Methods

**Molecular Weight.** Measurement of intrinsic viscosity in water is the most commonly used method to determine the molecular weight of poly(ethylene oxide) resins. However, there are several problems associated with these measurements

**Table 4. Aqueous Solution Viscosity Specifications for Polyox Resins[a]**

| Polyox resin-grade | Approx. mol wt | Nominal resin concentration, wt % | Brookfield viscometer spindle number | Speed, rpm | Viscosity at 25°C, Pa·s[b] |
|---|---|---|---|---|---|
| WSR-308 | $8 \times 10^6$ | 1.0 | 2 | 2 | 10,000–15,000 |
| WSR-303 | $7 \times 10^6$ | 1.0 | 2 | 2 | 7,500–10,000 |
| WSR | $5 \times 10^6$ | 1.0 | 2 | 2 | 5,500–7,500 |
| WSR-301 | $4 \times 10^6$ | 1.0 | 1 | 2 | 1,650–5,500 |
| WSR-N-60K | $2 \times 10^6$ | 2.0 | 3 | 10 | 2,000–4,000 |
| WSR-N-12K | $1 \times 10^6$ | 2.0 | 1 | 10 | 400–800 |
| WSR-1105 | $9 \times 10^5$ | 5.0 | 2 | 2 | 8,800–17,600 |
| WSR-205 | $6 \times 10^5$ | 5.0 | 2 | 2 | 4,500–8,800 |
| WSR-3333 | $4 \times 10^5$ | 5.0 | 1 | 2 | 2,250–3,350 |
| WSR-N-3000 | $4 \times 10^5$ | 5.0 | 1 | 2 | 2,250–4,500 |
| WSR-N-750 | $3 \times 10^5$ | 5.0 | 1 | 10 | 600–1,000 |
| WSR-N-80 | $2 \times 10^5$ | 5.0 | 1 | 50 | 65–115 |
| WSR-N-10 | $1 \times 10^5$ | 5.0 | 1 | 50 | 12–50 |

[a]Ref. 85.
[b]To convert Pa·s to P, multiply by 10.

(86,87). The dissolved polymer is susceptible to oxidative and shear degradation, which is accelerated by filtration or dialysis. If the solution is purified by centrifugation, precipitation of the highest molecular weight polymers can occur and the presence of residual catalyst by-products, which remain as dispersed, insoluble solids, further complicates purification.

A number of techniques, including static and dynamic light scattering (88), viscometry (89), and gpc with low angle laser light-scattering detection (90), have been used to study the behavior of poly(ethylene oxide) in solution. Dynamic light scattering (91) has also been used to determine the molecular weight distribution of poly(ethylene oxide) and to study crystallization from dilute solutions (92,93).

Attempts to measure average mol wt and molecular weight distribution of poly(ethylene oxide) for molecular weights above one million, using gel-permeation chromatography (gpc), are extremely difficult because of the effect of shear on the high molecular weight polymer molecules in the gpc column and the lack of adequate standards for calibration of the columns. However, one group has been successful in using high speed gel filtration to fractionate high molecular weight poly(ethylene oxide) and provide materials with narrow molecular weight distributions suitable for use as standards for gpc (94). An alternative method for average molecular weight determination is cloud point titration (95).

**Aqueous Solution Viscosity.** A special solution preparation method is used for one type of measurement of aqueous solution viscosity (96). The appropriate amount of poly(ethylene oxide) resin is dispersed in 125 mL of anhydrous isopropyl alcohol by vigorous stirring. Because the resin is insoluble in anhydrous isopropyl alcohol, a slurry forms and the alcohol wets the resin particles. An appropriate amount of water is added and stirring is slowed to about 100 rpm to avoid shear degradation of the polymer. In Table 4, the nominal resin concentration reported is based on the amount of water present and ignores the isopropyl alcohol.

**Analysis for Poly(Ethylene Oxide).**    Another special analytical method takes advantage of the fact that poly(ethylene oxide) forms a water-insoluble association compound with poly(acrylic acid). This reaction can be used in the analysis of the concentration of poly(ethylene oxide) in a dilute aqueous solution. Freshly prepared poly(acrylic acid) is added to a solution of unknown poly(ethylene oxide) concentration. A precipitate forms, and its concentration can be measured turbidimetrically. Using appropriate calibration standards, the precipitate concentration can then be converted to concentration of poly(ethylene oxide). The optimum resin concentration in the unknown sample is 0.2–0.4 ppm. Therefore, it is necessary to dilute more concentrated solutions to this range before analysis (97). Low concentrations of poly(ethylene oxide) in water may also be determined by viscometry (98) or by complexation with $KI_3$ and then titration with $Na_2S_2O_3$ (99).

## Health and Safety Factors, Toxicology

Poly(ethylene oxide) resins are safely used in numerous pharmaceutical and personal-care applications. Poly(ethylene oxide) resins show a low order toxicity in animal studies by all routes of exposure. Because of their high molecular weight, they are poorly adsorbed from the gastrointestinal tract and completely and rapidly eliminated. The resins are not skin irritants or sensitizers, nor do they cause eye irritation.

Considerable interest has been shown in poly(ethylene oxide) for diverse applications in food, drug, and cosmetic products. Such uses fall within the scope of the Federal Food, Drug, and Cosmetic Act. The U.S. FDA has recognized and approved the use of poly(ethylene oxide) for specific food and food packaging uses. USP/NF-grades of Polyox water-soluble resins (Union Carbide Corp.) are available for pharmaceutical applications.

## Uses

Significant use properties of poly(ethylene oxide) are complete water solubility, low toxicity, unique solution rheology, complexation with organic acids, low ash content, and thermoplasticity.

**Pharmaceutical and Biomedical Applications.**    On account of its low toxicity and unique properties, poly(ethylene oxide) is utilized in a variety of pharmaceutical and biomedical applications.

*Denture Adhesives.*    Fast hydration and gel-forming properties are ideally mated to produce a thick, cushioning fluid between the dentures and gums (100). The biologically inert nature of poly(ethylene oxide) helps reduce unpleasant odors and taste in this type of personal-care product (see DENTAL MATERIALS).

*Mucoadhesives.*    Poly(ethylene oxide) has good adhesive properties to mucosal surfaces because of its high molecular weight, linear molecules, and fast hydration properties. *In vivo* results have shown that the duration of adhesion increases with molecular weight up to 400,000. Further increase in molecular weight results in a concomitant decrease in adhesive properties, most likely due to the swelling of the resulting hydrogel (101). The mucoadhesive properties have been utilized in the design of buccal-sustained drug delivery systems (102,103).

*Ophthalmic Solutions.* The viscoelastic properties of poly(ethylene oxide) produce unique benefits for vitreous fluid substitution for ophthalmic surgery. Solutions of high molecular weight poly(ethylene oxide) have been used as vehicles for therapeutics for the eye (104) and as a contact lens fluid for hard or gel-type lenses (105) (see CONTACT LENSES). A treated lens appears to have a high viscosity layer at the low shear rates that occur on the inside surface. This provides a thick, comfortable cushioning layer. At high shear rates caused by the blinking eyelid, the apparent viscosity is much lower. This allows the lid to move smoothly and effortlessly over the outside surface of the lens. Unlike the cellulosics, poly(ethylene oxide) does not support bacterial growth. Lens solutions are easier to keep sterile.

*Wound Dressings.* Cross-linked poly(ethylene oxide) solutions form hydrogels which contain about 90–97% of water. These hydrogels are clear, transparent, permeable to gases, and absorb 5–100 times their weight in water. Such characteristics make these hydrogels interesting materials for wound dressings. Compared to other occlusive dressings, these hydrogels have shown the promotion of rapid healing (106). Release of therapeutic substances from these dressings has been demonstrated (107,108).

*Oral Drug Release.* The dissolution rate of tableted poly(ethylene oxide) depends on the molecular weight and particle size distribution. High molecular weight resin provides an excellent tablet binder for sustained drug release from matrix tablets. The good flow properties and compressibility of poly(ethylene oxide) powder can be advantageously exploited in preparing tablets by direct compression. The high swelling capacity of high molecular weight poly(ethylene oxide) tablets when exposed to intestinal fluids has been successfully used in osmotic delivery systems for water-insoluble drugs (109–112). A zero-order drug release has been reported from films produced from poly(ethylene oxide) and polycaprolactone. The change in drug release profile with respect to film composition, thickness, and morphology has been described (113).

*Biomaterials with Low Thrombogenicity.* Poly(ethylene oxide) exhibits extraordinary inertness toward most proteins and biological macromolecules. The polymer is therefore used in bulk and surface modification of biomaterials to develop antithrombogenic surfaces for blood contacting materials. Such modified surfaces result in reduced concentrations of cell adhesion and protein adsorption when compared to the nonmodified surfaces.

*Lubricious Coatings for Biomaterials.* Coatings of poly(ethylene oxide) when dry are tactile. If brought into contact with water, the poly(ethylene oxide) hydrates rapidly and forms a lubricious coating. This type of technology is of great interest for biomedical devices introduced into the human body, such as catheters and endotracheal tubes, and for sutures (114–117).

**Industrial Applications.** Poly(ethylene oxide)s also have numerous industrial uses.

*Flocculation.* Poly(ethylene oxide)s of molecular weights greater than four million have been used as specialty flocculants. The ability of the PEO molecule to hydrogen-bond with the surface hydroxyl layers of silica, kaolinites, and other mineral oxides leads to adsorption of the polymer on the substrate. The balance between the hydrophilicity of the ether oxygen moiety and the hydrogen-bonding forces on the solid substrate results in a loop-tail conformation essential to flocculation. Some of the end uses for PEO as a flocculant are as a fines retention aid in the paper industry, a low pH flocculant of silica in beryllium, uranium, and copper mines that use acid leaching, and as a dewatering aid in industrial waste treatment (see FLOCCULATING AGENTS).

In the paper industry, PEO is widely used as a retention aid and pitch control agent in the newsprint industry (118–135). Typically, a phenol formaldehyde-type resin is added to the substrate before the addition of PEO. The chemical that is added before PEO has been referred to as an enhancer. Recent publications on designing enhancers that work with PEO have resulted in expanding the use of PEO in flocculation of several substrates (128,129).

Several technical articles suggest that the use of PEO increases the dewatering efficiency of mineral sludges significantly (136–151). In the mining industry, PEO is used to flocculate siliceous substrates at pH <2 during the acid leaching operations. The low pH stability provides PEO polymers with unique advantages in this application.

*Drag Reduction.* The addition of 0.03% of high molecular weight PEO (greater than four million) to aqueous solutions has resulted in a 100% increase in the flow rate at fixed pump pressures (152). The significant reduction in friction as a result of the addition of PEO has been attributed to supermolecular structure formation of PEO (153) and to expansion and orientation of the polymer in rotation-free draining flow (154). Drag reduction properties have been demonstrated by trials using fire hoses, which show that water travels 50% further due to the addition of small quantities of PEO (63). Investigations on the effect of Reynold's number, orifice size of the pumping device, and polymer blends have led to a better understanding for suitable drag-reducing systems that may use PEO (152–163). Degradation of PEO and point of addition of the polymer appear to play a significant role in successful manipulation of this property in end uses. Some references also suggest the use of this property in reducing arterial pressure in medical applications (162).

*Binders in Ceramics, Powder Metallurgy, and Water-Based Coatings of Fluorescent Lamps.* In coatings and ceramics applications, the suspension rheology needs to be modified to obtain a uniform dispersion of fine particles in the finished product. When PEO is used as a binder in aqueous suspensions, it is possible to remove PEO completely in less than five minutes by baking at temperatures of 400°C. This property has been successfully commercialized in several ceramic applications, in powder metallurgy, and in water-based coatings of fluorescent lamps (164–168).

*Detergents and Lotions.* The addition of PEO provides a silky feel to solid and liquid products. This unique lubricious property has been successfully exploited in formulation of razor strips (169) and in shampoos, detergents, and other personal-care applications. Formulations are available from the manufacturers.

*Adhesives.* High concentration (>10%) solutions of poly(ethylene oxide) exhibit wet tack properties that are used in several adhesive applications. The tackiness disappears when the polymer dries and this property can be success-fully utilized in applications that require adhesion only in moist conditions. PEO is also known to form solution complexes with several phenolic and phenoxy resins. Solution blends of PEO and phenoxy resins are known to exhibit syner-gistic effects, leading to high adhesion strength on aluminum surfaces. Adhesive formulations are available from the manufacturers.

*Acid Cleaners.*   The addition of PEO can significantly increase the viscosity of acid solutions. Highly viscous acid solutions are used in cleaning formulations for glass, ceramic, and metal surfaces. The increase in viscosity increases the contact time of the cleaning solution when it is sprayed on vertical surfaces. Some acids that can be thickened by using PEO are hydrochloric, sulfuric, phosphoric, and oxalic. The order of addition of the polymer and oxidative stabilizers appears to play an important role in formulating highly viscous acid solutions. The manufacturers provide several formulations to thicken different acids.

*Jet Cutting and Drift Control.*   The pseudoplastic properties of PEO solutions reduce mist formation during spraying of aqueous solutions that contain PEO. This property is used in metal-working fluids to lower worker exposure to mists from the cutting and grinding aids. PEO may also be used to focus the spraying area of herbicides (qv) and water-based coatings. Another significant application that utilizes the unique solution properties of PEO is jet cutting. Patent literature shows that the coherent needle-like stream of a PEO solution can also be successfully used to cut semisoft solids such as cardboard, leather, cheese, and partially cured asbestos brake shoes (170,171).

*Construction.*   The addition of PEO to concrete has been a subject of several investigations (172). Research studies and patent literature suggests that PEO can be used as a pumping aid to concrete where the lubricity of PEO allows concrete to be pumped to longer distances (173–176). In addition, PEO is also used to disperse the water more uniformly in the concrete mixture that promotes better uniformity of the concrete mixture. Formulations in the construction industry are proprietary and not easily available.

*Batteries.*   Polymer electrolytes based on PEO have been widely reviewed (53–58,177). The prospect of using a thin-layer, flexible battery for applications ranging from cellular phones to electric vehicles has led to several patents and research papers in this field. Typically, a salt such as potassium iodide, lithium triflate, or lithium perchlorate is complexed with PEO in a methylene chloride solvent. The solution complex is cast into thin films and the solvent is evaporated. The complex has been characterized; it is believed that the 7C2 helical structure of PEO allows an ideal structure for ion transport and leads to effective use as a battery. The dissociation of the anion–cation pair in the PEO salt complex has been attributed to the oxygen atoms, which form a cage around the cation and lead to ionic conductivity. The crystallinity of PEO at room temperature has limited the use of this technology to batteries that are used at temperatures higher than 65°C, the melting point of PEO. Research in the 1990s focuses on modifying the complex or the PEO molecule to overcome the crystallinity problem (see BATTERIES).

*Other Applications.*   PEO has also been used as an antistat additive (178,179), water-soluble packaging material of seeds and fertilizers (180), and rheology modifier in aqueous flexographic printing inks (qv) (181).

## BIBLIOGRAPHY

"Polyethers, Ethylene Oxide Polymers" in *ECT* 3rd ed., Vol. 18, pp. 616–632, by D. B. Braun and D. J. DeLong, Union Carbide Corp.

1. F. E. Bailey, Jr. and J. V. Koleske, *Poly(Ethylene Oxide)*, Academic Press, Inc., New York, 1976, p. 105.
2. P. J. Flory, *Principles of Polymer Chemistry*, Cornell University Press, Ithaca, N.Y., 1953, pp. 266–313.
3. S. Z. Cheng, J. S. Barley, and E. D. Von Meerwall, *J. Polym. Sci. Part B: Polym. Phys.* **29**(5), 515–525 (1991).
4. F. W. Stone and J. J. Stratta, in N. Bikales, ed., *Encyclopedia of Polymer Science and Technology*, Vol. 6, Wiley-Interscience, New York, 1967, pp. 103–145.
5. B. E. Read, *Polymer*, **3**, 529 (1962).
6. J. A. Faucher and co-workers, *J. Appl. Phys.* **37**, 3962 (1966).
7. G. Nabi, *J. Sci.* **20**(3), 136 (1968).
8. F. E. Bailey, Jr. and R. W. Callard, *J. Appl. Polym. Sci.* **1**, 56 (1959).
9. G. Allen and co-workers, *Polymer*, **8**, 391 (1967).
10. T. A. Ritscher and H. G. Elias, *Makromol. Chem.* **30**, 48 (1959).
11. *Union Carbide Technical Bulletin*, F44029C, (1981).
12. M. Ataman, *Colloid Polym. Sci.* **265**(1), 19–25 (1987).
13. *Union Carbide Technical Bulletin*, UC-420 (1994).
14. B. Kim and R. Porter, *Macromolecules*, **18**, 1214 (1985).
15. D. Mitchell and R. Porter, *Macromolecules*, **18**, 1218 (1985).
16. B. Bogdanov and M. Mikhailov, *J. Therm. Anal.* **30**, 551 (1985).
17. B. Bogdanov and M. Mikhailov, *Acta Polym.* **35**, 469 (1984).
18. C. Marco and co-workers, *Macromolecules*, **23**, 2183 (1990).
19. G. Weill, *J. Macromol. Sci. Chem.* **A27**, 1769 (1990).
20. S. Liberman, A. Gomes, and E. Macchi, *J. Polym. Sci. Polym. Chem. Ed.* **22**, 2809 (1984).
21. E. Martuscelli and co-workers, *Makromol. Chem.* **187**, 1557 (1986).
22. S. Cimmino, *Makromol. Chem.* **191**, 2447 (1990).
23. S. Cimmino, E. Martuscelli, and C. Silvestre, *Journ. Calorim. Anal. Therm. Thermodym. Chim.* **17**, 188 (1986).
24. E. Munoz and co-workers, *Polym. Bull. (Berlin)*, **7**, 295 (1982).
25. E. Martuscelli, C. Silvestre, and C. Gismondi, *Makromol. Chem.* **186**, 2161 (1985).
26. E. Martuscelli, L. Vicini, and A. Seves, *Makromol. Chem.* **188**, 607 (1987).
27. C. Han, H. Chung, and J. Kim, *Polymer*, **33**, 546 (1992).
28. J. Horrion, R. Jerome, and P. Teyssie, *J. Polym. Sci. Part A: Polym. Chem.* **28**, 153 (1990).
29. M. Coleman, J. H. C. Serman, and P. Painter, *Polym. Mater. Sci. Eng.* **59**, 321 (1988).
30. J. Quintana and co-workers, *Polymer*, **32**, 2793 (1991).
31. M. Iriarte and co-workers, *Macromolecules*, **24**, 5546 (1991).
32. Y. Nishio, N. Hirose, and T. Takahashi, *Polym. J. (Tokyo)*, **21**, 347 (1989).
33. P. Sundararajan, P. Bluhm, and Y. Piche, *Polym. Bull. (Berlin)*, **27**, 345 (1991).
34. A. Margaritia and N. Kalfoglou, *J. Polym. Sci. Part B: Polym. Phys.* **26**, 1595 (1988).
35. C. Nakafuku and M. Sakoda, *Polym. J. (Tokyo)*, **25**, 909 (1993).
36. M. Avella and E. Martuscelli, *Polymer*, **29**, 1731 (1988).
37. P. Sanjay and E. Goethals, *New Polym. Mater.* **4**, 1 (1993).
38. T. Tang and B. Huang, *J. Polym. Sci. Part B: Polym. Phys.* **32**, 1991 (1994).
39. B. Pukanszky and F. Tudos, *Makromol. Chem. Macromol. Symp*, **38**, 205 (1990).
40. Ref. 1, p. 87.
41. *Union Carbide Technical Bulletin*, SC-1170 (1991).

42. C. M. Pradip and co-workers, *Langmuir*, **7**, 2108 (1991).
43. K. L. Smith, A. E. Winslow, and D. E. Peterson, *Ind. Eng. Chem.* **51**, 1361 (1959).
44. A. Guner and O. Guven, *Makromol. Chem.* **179**, 2789 (1978).
45. G. E. Baker and H. J. Ronsanto, *J. Am. Oil Chem. Soc.* **32**, 249 (1955).
46. F. E. Bailey, Jr. and H. G. Brance, *J. Polym. Sci.* **49**, 397 (1961).
47. A. Chenite and F. Brisse, *Macromolecules*, **24**, 2221 (1991).
48. H. Daoust and B. Cloutier, *Makromol. Chem. Macromol. Symp.* **20/21**, 221 (1988).
49. P. Wright, *Polymer*, **30**, 1179 (1989).
50. H. Honda and K. Ono, *Macomolecules*, **23**, 4950 (1990).
51. R. Sartori and co-workers, *Macromelecules*, **23**, 3878 (1990).
52. K. Ono, H. Honda, and K. Murakami, *Macromol. Sci. Chem.* **A26**, 567 (1989).
53. R. Huq and G. Farrington, *Solid State Ionics 1987*, **28–30**, 990 (1988).
54. L. Yang and co-workers, *Solid State Ionics 1987*, **28–30**, 1029, (1988).
55. M. Munshi and B. Owens, *Solid State Ionics*, **26**, 41 (1988).
56. A. Wendajoe and H. Yang, *Int. Symp. Polym. Electrolytes*, **2**, 225 (1990).
57. H. Yang and G. Farrington, *Report TR-1992-22 from Gov. Rep. Announce. Index* (*U.S.*), **92**, 23 (1992).
58. L. Jean, E. Mintz, and I. Khan, *Polym. Prepr.* (*Am Chem. Soc., Div. Polym. Chem.*), **34**, 612 (1993).
59. T. Okada, *Analyst*, **118**, 959 (1993).
60. M. Donbrow, *Surfactant Sci. Ser*, **23**, 1011 (1987).
61. H. Staudinger and H. Lohmann, *Ann. Chemie*, **505**, 41 (1933).
62. F. N. Hill, F. E. Bailey, Jr., and J. T. Fitzpatrick, *Ind. Eng. Chem.* **50**, 5 (1958).
63. Ref. 1, p. 20.
64. F. N. Hill, F. E. Bailey, Jr., and J. T. Fitzpatrick, *Ind. Eng. Chem.* **50**, 5 (1958).
65. F. E. Bailey, Jr., G. M. Powell, III, and K. L. Smith, *Ind. Eng. Chem.* **50**, 8 (1958).
66. R. A. Miller and C. C. Price, *J. Polym. Sci.* **34**, 161 (1959).
67. Y. Zhang, X. Chen, and Z. Shen, *Inor. Chim. Acta*, **155**(2), 263 (1989).
68. U.S. Pat. 2,706,181 (Apr. 12, 1955), M. E. Pruitt and J. M. Baggett (to Dow Chemical Co.).
69. U.S. Pat. 2,706,189 (Apr. 12, 1955), M. E. Pruitt and J. M. Baggett (to Dow Chemical Co.).
70. W. R. Sorenson and T. W. Cambell, *Preparative Methods of Polymer Chemistry*, Interscience Publishers, New York, 1961.
71. U.S. Pat. 3,470,078 (Sep. 30, 1969), P. A. King (to Union Carbide Corp.).
72. U.S. Pat. 3,736,288 (May 29, 1973), J. J. Stratta, C. W. Frank, and J. A. Barrere (to Union Carbide Corp.).
73. U.S. Pat. 3,843,589 (Oct. 22, 1974), L. H. Wartman (to Union Carbide Corp.).
74. U.S. Pat. 3,900,043 (Aug. 19, 1975), J. H. Bowen and K. J. Sollman (to Union Carbide Corp.).
75. U.S. Pat. 3,470,078 (Sep. 30, 1969), P. A. King (to Union Carbide Corp.).
76. U.S. Pat. 3,336,129 (Aug. 15, 1967), R. A. Herrett and P. A. King (to Union Carbide Corp.).
77. U.S. Pat. 3,734,876 (May 22, 1973), N. S. Chu (to Union Carbide Corp.).
78. U.S. Pat. 3,900,378 (Aug. 19, 1975), S. N. Yen and F. D. Osterholtz (to Union Carbide Corp.).
79. E. Nedkov and S. Tsvetkova, *Radiat. Phys. Chem.* **44**, 257 (1994).
80. E. Merrill, K. Dennison, and C. Sung, *Biomaterials*, **14**, 1117 (1993).
81. E. Nedkov and S. Tsvetkova, *Radiat. Phys. Chem.* **44**, 251 (1994).
82. L. Minkova and co-workers, *J. Polym. Sci.* **27**, 621 (1989).
83. L. Zhang and co-workers, *Radiat. Phys. Chem.* **40**, 501 (1992).
84. E. Nedkov and S. Tsvetkova, *Radiat. Phys. Chem.* **43**, 397 (1994).

85. *Union Carbide Technical Bulletin*, F44029C, (1981).
86. E. Bartel and A. Kochanowski, *Makromol. Chem. Rapid Commun.* **1**, 205 (1980).
87. C. Wolfe, *Can. J. Chem. Eng.* **58**, 634 (1980).
88. K. Devanand and J. C. Selser, *Macromolecules*, **24**, 5943 (1991).
89. D. Woodley and co-workers, *Macromolecules*, **25**, 5283 (1992).
90. A. Eshuis and P. F. Mijnlieff, *Polymer*, **27**, 1951 (1986).
91. M. Mettille and R. Hester, *ACS Symp. Ser.* **467**, 276 (1991).
92. N. Ding and co-workers, *Polymer*, **29**, 2121 (1988).
93. N. Ding, R. Salovey, and E. Amis, *Polym. Prepr. (Am. Chem. Soc., Div. Polym. Chem.)*, **29**, 385 (1988).
94. Y. Kato and co-workers, *J. Chromatogr.* **190**, 297 (1980).
95. E. Bortel and A. Kochanowski, *J. Appl. Polym. Sci.* **28**, 2445 (1983).
96. *Union Carbide Technical Bulletin*, SC-1169A (Apr. 1993).
97. *Union Carbide Technical Bulletin*, F44029C (1981).
98. B. Moudgil, B. Shah, and S. Jayanti, *Colloids Surf.* **20**, 101 (1986).
99. G. Perina and co-workers, *Khim. Tekhnol. Svoista Primen. Plastmass*, 159 (1990).
100. U.S. Pat. 2,978,812 (Apr. 11, 1961), M. W. Rosenthal and H. A. Cohen (to Block Drug Co., Inc.).
101. P. Bottenberg and co-workers, *J. Pharm. Pharmacol.* **43**, 457–464 (1991).
102. U.S. Pat. 4,713,243 (Dec. 15, 1987), M. Shiraldi and co-workers (to Johnson and Johnson Products).
103. U.S. Pat. 4,764,378 A. Keith and W. Snipes (to Zetachron, Inc.).
104. E. Dimitrova and co-workers, *Int. J. Pharmaceutics*, **93**, 21–26 (1993).
105. Brit. Pat. 1,340,516 (Dec. 12, 1973), B. F. Rankin (to Burton Parsons Chemicals, Inc.).
106. S. Mandy, *J. Dermatol. Surg. Oncol.* **9**(2), 153–155 (1983).
107. P. Mertz and co-workers, *Arch. Dermatol.* **122**(10) 1133–1138 (1983).
108. J. Fulton, Jr., *J. Dermatol. Surg. Oncol.* **16**(5) 460–467 (1990).
109. A. Apicella and co-workers, *Biomaterials*, **14**(2) (1993).
110. U.S. Pat. 4,859,470 (Aug. 22, 1989), R. Cortese and co-workers (to Alza Corp.).
111. U.S. Pat. 5,330,762 (July 19, 1994), A. Ayer and D. Ridzon (to Alza Corp.).
112. U.S. Pat. 4,404,183 (Sept. 13, 1983), K. Hiroitsu and co-workers (to Yamanouchi Parmaceutical Co.).
113. K. C. Yoo and C. Y. Chung, *Pollimo*, **18**(1) 103–112 (1994).
114. U.S. Pat. 5,077,352 (Dec. 31, 1991), R. Elton (to C. R. Bard Co.).
115. U.S. Pat. 5,041,100 (Aug. 20, 1991), S. Rowland and R. Wright (to Cordis Corp.).
116. U.S. Pat. 4,487,808 (Dec. 11, 1984), H. Lambert (to Astra Meditec Aktiebolag).
117. U.S. Pat. 4,649,920 (Mar. 17, 1987), J. Rhum (to Pfizer Hospital Products Group, Inc.).
118. J. Koppelman and I. K. Migliorini, *Tappi J.* **69**(11), 74–78 (1986).
119. R. Waagberg and T. Lindstroem, *Colloids Surfaces*, **27**(1–3), 29–42 (1987).
120. P. K. Leung and E. D. Goddard, *Tappi J.* **70**(7), 115–118 (1987).
121. L. A. Stack and K. Noel, *Appita J.* **43**(2), 125–129 (1990).
122. D. Coghill, *Appita J.* **42**(5), 373–375; (1989).
123. WO Pat. 9322243-A1 (Nov. 11, 1993), A. O. P. Anderson and co-workers (to Eka Nobel AB).
124. WO Pat. 9315271-A1 (Aug. 5, 1993), M. Owens (to Kemba Kemi AB).
125. Brine, R. M. Brauer, and N. Wiseman, *Appita J.* **45**(2), 118–120 (1992).
126. Jpn. Pat. 04240296-A (Jan. 22, 1991), S. Owari (to OJI Paper Co.).
127. Can. Pat. 2042907-A (Nov. 1, 1992) (to Radu G.).
128. Can. Pat. 2040967-A (Oct. 24, 1992) (to Radu G.).
129. L. A. Stack and N. K. Roberts, *J. Wood Chem. Technol.* **13**(2), 283–308; (1993).

130. D. A. Braun and K. Ehms, *Tappi J.* **67**(9), 110–114 (1984).
131. A. Bokstroem and L. Oedberg, *Res. Discl.* **237**, 5 (1984).
132. G. Lindstroem and Glad-Nordmark, *J. Colloid Interface Sci.* **97**(1), 62–67, (1984).
133. G. Lindstroem and Glad-Nordmark, *J. Colloid Interface Sci.* **4**(2), 404–411 (1983).
134. S. Tay and T. A. Cauley, *Papermakers Conf.* 205–210 (1982).
135. N. Furusawa and Y. Watanabe, *Kobunshi Ronbunshu*, **41**(12), 727–732 (1984).
136. R. H. Shen and co-workers, *Trans. Soc. Min. Metall. Explor.* 286, (1989).
137. T. Tadros, *Polym. J.* **23**(5), 683–696 (1991).
138. F. Liu and R. Audebert, *Colloid Polym. Sci.* **272**(2), 196–203 (1994).
139. M. A. Cohen-Stuart and G. J. Fleer, *ACS Symp. Ser.* **532**, 14–22 (1993).
140. S. Behl and B. Moudgil, *J. Colloid Interface Sci.* **161**(2), 422–429 (1993).
141. S. Behl and B. Moudgil, *J. Colloid Interface Sci.* **161**(2), 443–449 (1993).
142. B. Moudgil and S. Behl, *Miner. Metall. Process*, **10**(2), 62–65 (1993).
143. B. J. Scheiner and G. M. Wilemon, *Process Technol. Proc.*, 175–185 (1987).
144. P. M. Brown, D. A. Stanley, and B. J. Scheiner, *Miner. Metall. Process*, **6**(4), 196–200 (1989).
145. B. J. Scheiner and D. A. Stanley, *Miner. Metall. Process*, **6**(4), 206–210 (1989).
146. D. A. Stanley, B. J. Scheiner, and P. M. Brown, *Miner. Metall. Process*, **7**(2), 114–117 (1990).
147. B. J. Scheiner, *Coal Sci. Technol.* **9**, 135 (1985).
148. B. J. Scheiner and A. G. Smelley, *Rep. Invest. 9131 U.S. Bur. Mines*, (1982).
149. B. J. Scheiner, A. G. Smelley, and D. R. Brooks, *Proceedings of the Flocculation and Dewatering Symposium*, New York, 1988, pp. 272–277.
150. B. J. Scheiner and A. G. Smelley, *Report Gov. Rep. Announce. Index (U.S.)* **93**(8) (1993).
151. E. Koksal and co-workers, *Powder Technol.* **62**(3), 253–259 (1990).
152. *Union Carbide Technical Bulletin*, UC-728, (1995).
153. K. Kazimierz, G. Piotr, *Chem. Inz. Chem*, **16**, 29–39, 1986.
154. H. Usui, K. Matsuru, S. Yuji, *J. Chem. Eng. (Japan)*, **21**(2), 132–140, (1988).
155. V. N. Kalshnikov and M. G. Tsiklauri, *Inzh.-Fiz. Zh*, **58**(1), 1990.
156. O. K. Kim, L. S. Choi, T. D. Long, and T. H. Yoon, *Polym. Prepr. (Am. Chem. Soc. Div. Polym. Chem.)*, **28**(2), 68, 69, (1987).
157. A. Abdelhak, J. Patrice, G. Etienne *Acad. Sci. Ser. 2*, **303**(13), 1161–1164 (1986).
158. G. H. Sedahmed and M. A. Fawzy, *Br. Cooros. J.*, **21**(4), 225–227, (1986).
159. C. R. Phillips, *SPE Reservoir Eng.* **5**(4), 481–486 (1990).
160. L. Ralph and co-workers, *Ind. Eng. Chem. Res.* **30**(2), 403–407, (1991).
161. A. S. Golub and co-workers, *Dokl. Acad. Nauk. SSSR*, **295**(4), 813–816, (1987).
162. P. I. Polimeni and B. T. Ottenbreit, *J. Cardiovasc. Pharmacol*, **14**(3), 374–380 (1989).
163. U.S. Pat. 5,045,588 (Sept. 3, 1991), B. A. Alexander and S. Madre (to U.S. Sec. of Navy).
164. Jpn. Pat. JP 63199043 and JP 92029451 (Aug. 17, 1988 and May 19, 1992), H. Uba and co-workers (to UBE Industries KK).
165. *Union Carbide Technical Bulletin*, BBTL 9051A (1994).
166. Jpn. Pat. JO 3083-805 (Sept. 4, 1991), K. Saida and S. Fujiwara (to Sumitomo Chem.).
167. Jpn. Pat. JP 62226874 (May 10, 1987) (to Tagaitt.).
168. Jpn. Pat. 63317581 (Dec. 26, 1988), M. Magai (to Hitachi KK).
169. Eur. Pat. 276-094-A (July 27, 1988), B. B. Braun and co-workers (to Warner-Lambert Co.).
170. U.S. Pat. 3524367 (Aug. 18, 1970) (to N. C. Franz).
171. U.S. Pat. 3568926 (Mar. 9, 1971), R. E. Bowles (to Bowles Engineering Corp.).
172. B. D. Saucier, *Concrete Int. Design Construct.* **9**(5), 42–47, (1987).

173. Jpn. Pat. J6 3050-356-A (Aug. 21, 1986), O. Hatakeyama (to Hazama Gumi).
174. Jpn. Pat. J6 3288 934-A (Nov. 15, 1988), T. Kidda (to Denki Kagaku Kogyo KK).
175. U.S. Pat. GB2211-183-A (Oct. 18, 1988) (to Courtaulds Pls.).
176. Jpn. Pat. JO 1160-853-A (Dec. 18, 1987) (to Tokyo Kensetsu KK).
177. B. Scrosati, ed., *Second International Symposium on Polymer Electrolytes*, Elsevier Science Publishing Co. Inc., New York, 1990.
178. U.S. Pat. 3425981 (Feb. 12, 1965) (to Union Carbide Corp.).
179. Jpn. Pat. JO 3188-142-A (Dec. 15, 1989) (to Mitsui Toatsu Chem Inc.).
180. U.S. Pat. 5206278 (Apr. 27, 1993) (to Air Products & Chemicals Inc.).
181. U.S. Pat. 4014833 (Mar. 29, 1977) (to Owens-Illinois, Inc.).

DARLENE M. BACK
ELKE M. CLARK
RAMESH RAMACHANDRAN
Union Carbide Corporation

# PROPYLENE OXIDE POLYMERS

Propylene oxide and other epoxides undergo homopolymerization to form polyethers. In industry the polymerization is started with multifunctional compounds to give a polyether structure having hydroxyl end groups. The hydroxyl end groups are utilized in a polyurethane forming reaction. This article is mainly concerned with propylene oxide (PO) and its various homopolymers that are used in the urethane industry.

Poly(propylene oxide) [25322-69-4] may be abbreviated PPO and copolymers of PO and ethylene oxide (EO) are referred to as EOPO. Diol poly(propylene oxide) is commonly referred to by the common name poly(propylene glycol) (PPG). Propylene oxide [75-56-9] and poly(propylene oxide) and its copolymers, with ethylene oxide, have by far the largest volume and importance in the polyurethane (PUR) and surfactant industry compared to all other polyepoxides. Articles reviewing propylene oxide (1), poly(propylene oxide) (2–4), other poly(alkylene oxides) (4), and polyurethanes (5–7) are cited to lead the interested reader to additional detail not in the scope of this article.

Homopolymers of PO and other epoxides are named a number of ways: after the monomer, eg, poly(propylene oxide) (PPO) or polymethyloxirane; from a structural point of view, polyoxypropylene or poly(propylene glycol); or from the *Chemical Abstracts* (CA) name, poly[oxy(methyl-1,2-ethanediyl)], $\alpha$-hydro-$\omega$-hydroxy-. Common names are used extensively in the literature and in this article.

**History.**    Propylene oxide was discovered in 1860 in Wurz's laboratory (8). It became an important commercial industrial chemical after World War II when its importance in polyurethanes was recognized. As a general guide, polyethers give softer, more resilient foams (qv) with better hydrolytic resistance than polyesters, whereas the polyester-based foams have greater tensile strength and better resistance to oils, solvents, and oxidation. Bayer (9) received one of the first patents which was applied for in 1951 and granted in 1960. This patent used polyethers but was limited to poly(ethylene oxide). A patent for copolymers of PO and EO, which were used as surfactants (qv), was granted in 1954 (10). Flexible foam systems based on 80:20 (2,4- and 2,6-isomer ratio) toluene diisocyanate (TDI) and PO polymers or EOPO copolymers [9003-11-6] were introduced in the United States in 1957. That year Mobay offered commercial prepolymers based on these ingredients.

**Uses of Poly(propylene oxide).** The vast majority of uses of PPO and EOPO copolymers are in polyurethanes, surfactants, and the medical area. Many but by no means all other applications follow. Taking advantage of their surfactant properties (especially EOPO copolymers), they have been used as lubricants (10–17), dispersants (18–21), antistatic agents (22–24), foam control agents (25–28), in printing inks (29,30), in printing processes (31), and as a solubilizers (32). PPO and EOPO copolymers have been used in aqueous hydraulic fluids (33–37) and in coolant compositions (38). They are used in secondary oil recovery operations (39,40), as plastic additives (41,42), in nonpolyurethane adhesives (43), and in propellant compositions (44). In the medical field they find applications as protective bandages (45–49), in drug delivery systems (qv) (50–53), in organ preservation (54), in dental compositions (55), and as a fat substitute (56).

## Propylene Oxide Monomer

**Synthesis.** The total annual production of PO in the United States in 1993 was 1.77 billion kg (57) and is expected to climb to 1.95 billion kg with the addition of the Texaco plant (Table 1). There are two principal processes for producing PO, the chlorohydrin process favored by The Dow Chemical Company and indirect oxidation used by Arco and soon Texaco. Molybdenum catalysts are used commercially in indirect oxidation (58–61). Capacity data for PO production are shown in Table 1 (see PROPYLENE OXIDE).

Miscellaneous synthesis methods for propylene oxide include the following. Arco received a patent (62) for producing PO and other epoxides in an integrated process involving air oxidation of a secondary alcohol in the presence of a titanium silicate catalyst. A 1993 Olin patent (63) described the production of PO by a noncatalytic, gas-phase oxidation process. In this process a gaseous mixture of propylene, oxygen, and acetaldehyde is allowed to react at a temperature in the range of 200 to ~350°C and at a pressure up to about 6.8 MPa (1000 psig). Propylene oxide is obtained in 42% selectivity at low propylene conversion. This process has potential because of its simplicity and could be realized if the selectivity and conversion are improved. Several groups have reported that PO and

**Table 1. U.S. Capacity for Propylene Oxide**[a]

| Producer | Location | Capacity, $10^3$ t | Process |
|---|---|---|---|
| Arco Chemical | Bayport, Tex.; Channelview, Tex. | 1.05 | indirect oxidation |
| Dow | Freeport, Tex. | 0.5 | chlorohydrin |
| | Plaquemine, La. | 0.22 | chlorohydrin |
| Texaco | Port Neches, Tex. | 0.18 | indirect oxidation |
| *Total* | | *1.95* | |

[a]Ref. 57.

other epoxides can be microbiologically synthesized from alkenes (64–69), some even producing chiral epoxides (70–76). PO can be prepared electrochemically and pilot-scale reactors have been described (77–79). There have been several reports of the direct oxidation of propene with hydrogen peroxide (80,81).

**Purification.**    A three-phase distillation for producing high purity PO has been reported (82). PO can be purified in the laboratory by refluxing with a drying agent, such as calcium hydride, then fractionally distilling (83). Texaco has reported that PO can be purified by extractive distillation (84–89).

**Optically Active PO.**    The synthesis of optically pure PO has been accomplished by microbial asymmetric reduction of chloroacetone [78-95-5] (90). (S)-2-Methyloxirane [16088-62-3] (PO) can be prepared in 90% optical purity from ethyl (S)-lactate in 44% overall yield (91). This method gives good optical purity from inexpensive reagents without the need for chromatography or a fermentation step. (S)-PO is available from Aldrich Chemical Company, having a specific rotation $[\alpha]_D^{20} -7.2$ ($c = 1$, $CHCl_3$).

**Physical Properties.**    The physical properties of PO are shown in Table 2.

**Table 2. Physical Properties of Propylene Oxide**[a]

| Property | Value |
|---|---|
| molecular weight | 58.08 |
| boiling point, °C | 33.9 |
| $dp/dt$, kPa/°C[b] | 3.70 |
| vapor pressure constants[c] | |
| A | 6.095 |
| B | 1065.27 |
| C | 226.283 |
| freezing point, °C | −104.4 |
| coefficient of thermal expansion $\alpha$[d] | $0.00151^{20}$ |
| refractive index $n_D$ | |
| at 20°C | 1.36603 |
| 25°C | 1.36322 |
| viscosity $\eta$, mPa·s(=cP) | |
| at 0°C | 0.41 |
| 25°C | 0.28 |
| heat of vaporization $\Delta H_v$, kJ/mol[e] | $28.75^{34}$ |
| heat of combustion $\Delta H_c$, kJ/mol[e] | −1917 |
| heat capacity, $C_p$, J/(mol·K)[e] | 120.37 |
| critical temperature, °C | 209.1 |
| solubility in water, at 20°C, wt % | 40.5 |
| solubility of water in, at 20°C, wt % | 12.8 |
| flash point, closed cup, °C | −35.0 |
| explosive limits in air, wt % | |
| upper | 21.5 |
| lower | 2.1 |
| specific gravity at 20°C | 0.830 |

[a]Refs. 4, 92, and 93.
[b]To convert kPa to mm Hg, multiply by 7.5.
[c]For log $P = A − (B/T)$ or the Antoine equation log $P = A − [B/(T + C)]$.
[d]$\alpha = 1/V(\partial V/\partial T)_P = (d_1/d_2) − 1/(T_2 − T_1)$. Superscript is temperature, °C.
[e]To convert J to cal, divide by 4.184.

## Propylene Oxide Polymers

Propylene oxide and other epoxides polymerize by ring opening to form polyether structures. Either the methine, CH–O, or the methylene, $CH_2$–O, bonds are broken in this reaction. If the epoxide is unsymmetrical (as is PO) then three regioisomers are possible: head-to-tail (H–T), head-to-head (H–H), and tail-to-tail (T–T) dyads, ie, two monomer units shown as a sequence. The anionic and

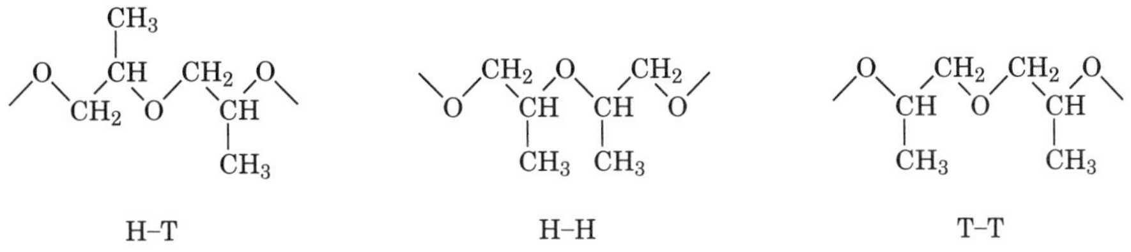

coordination polymerization of PO results in nearly all (95–98%) H–T sequences because the $S_N2$ attack occurs on the least substituted carbon atom, the methylene carbon. A small amount of the other sequences are also found and can be identified by nmr. Tacticity describes stereoregular polymers and indicates the orderliness of the succession of configurational repeating units in the main chain of the polymer. PPO from anionic polymerization always results in an atactic polymer which is a regular polymer (H–T), the molecules of which have equal numbers of possible configurational base units in a random sequential distribution. On the other hand, certain coordination catalysts produce a stereoregular isotactic polymer which can be described in terms of only one species of configurational unit, having chiral or prochiral atoms in the main chain, in a single sequential unit. A triad, a sequence of three monomer units, is necessary to show this behavior (Fig. 1). PPO would have an excess of one enantiomer only if optically active PO were used as starting material.

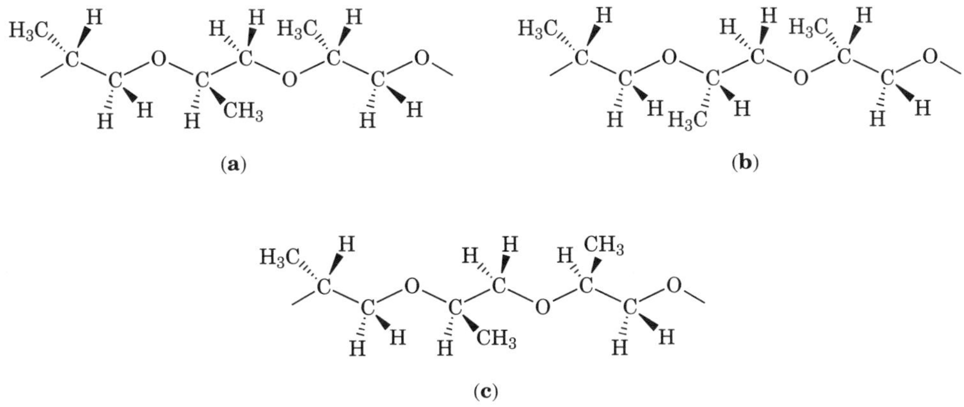

**Fig. 1.** Triad sequences for stereoregular poly(propylene oxide) where (**a**) shows isotactic (RRR or SSS), (**b**) syndiotactic (RSR or SRS), and (**c**) heterotactic (RRS or SSR, or SRR or RSS) units.

BASE-CATALYZED POLYMERIZATION OF PROPYLENE OXIDE

**The Reaction.** Most polyether polyols used commercially for urethanes and surfactants are produced by anionic polymerization. The bases of choice are potassium hydroxide or sodium hydroxide. The sequence of reactions leading to polymer are shown in Figure 2. Reaction 1 is the formation of alkoxide ion with the formation of water which is normally removed. ROH (ROK) is commonly referred to as the starter. Reaction 2 is the propagation reaction and the rate depends on catalyst and PO concentration. Reaction 3 is a proton-transfer reaction which is very fast and gives rise to the narrow molecular weight distribution normally seen in commercial polyether polyols. Reaction 4 shows the generation of allyl alkoxide which in turn polymerizes with PO to form monofunctional polyetherol (Reaction 5). Another reaction which is not shown is the isomerization of allyl alkoxide to propenyl alkoxide. The unsaturation present at the end of anionic polymerization is nearly all allyl but isomerization to propenyl occurs in unneutralized polymerizates (94). In measurements of the rate of isomerization (allyl → propenyl) simple second-order kinetics were found; the rate $= k_2[allyl][base]$ (94). The activation energy is 116 kJ/mol (27.7 kcal/mol) over a temperature range 90–130°C.

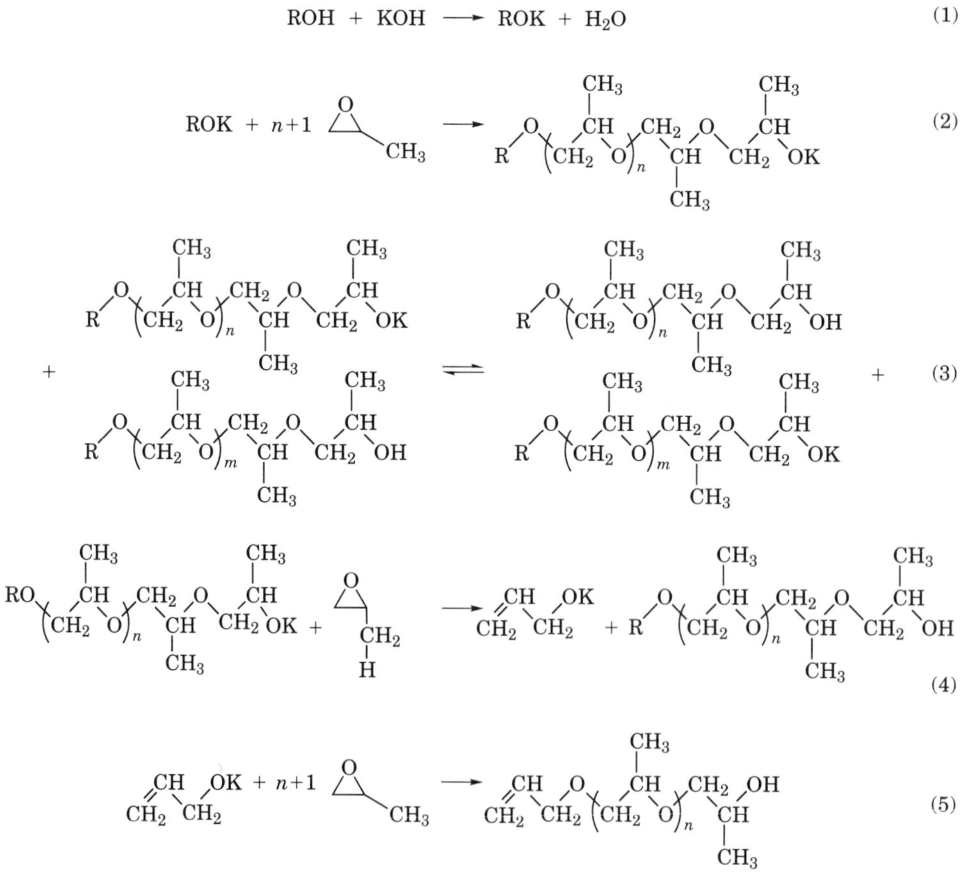

$$\text{ROH} + \text{KOH} \longrightarrow \text{ROK} + \text{H}_2\text{O} \tag{1}$$

$$\text{ROK} + n+1 \; \triangle\text{CH}_3 \longrightarrow \text{(polyether alkoxide)} \tag{2}$$

$$\text{(alkoxide)} + \text{(hydroxyl polymer)} \rightleftharpoons \text{(hydroxyl polymer)} + \text{(alkoxide)} \tag{3}$$

$$\text{(polyether alkoxide)} + \text{(propylene oxide/allyl)} \longrightarrow \text{(allyl alkoxide)} + \text{(polyetherol)} \tag{4}$$

$$\text{CH}_2{=}\text{CH--CH}_2\text{OK} + n+1 \; \triangle\text{CH}_3 \longrightarrow \text{(monofunctional polyetherol)} \tag{5}$$

**Fig. 2.** Reaction scheme for the anionic polymerization of propylene oxide.

Tetrabutylammonium benzoate has been used as a catalyst for the polymerization of PO over the temperature range 40–108°C and the yield of polymer was typically low (2–78%); a large amount of unsaturation was present due to chain transfer (95). When synthetic hydrotalcite, $Mg_6Al_2(OH)_{16}CO_3 \cdot 4H_2O$, is used to polymerize PO and is activated by calcining at 450°C, a quantitative yield of PPO is obtained at 50°C in two hours (96). At Olin, POLY-L polyols have been produced with reduced unsaturation, but the catalyst used to produce them has not been disclosed (97). The use of zinc hexacyanocolbaltate to prepare low unsaturation polyols has been reported (98).

**General Procedure of Base Catalysis.**   This yields a 3000 number-average molecular weight triol. In order to make this polyol in a reasonable amount of time, high temperature and consequently high pressure are required; therefore a stainless steel autoclave reactor is employed instead of a glass apparatus. The reactor is nominally 3.78 L (1 gal) in size, and has the following features: an oxide addition tube which extends to the bottom of the vessel and is pointed toward a high speed stirrer; a means of adding the oxide at a constant rate such as a pump or a flow controller; an inlet for vacuum or inert gas; a means to monitor the temperature and pressure (the oxide feed rate should also be monitored to give reproducible results); a charge port to add starter and catalyst; a water and steam inlet and outlet for cooling and heat; a high speed stirrer; a water jacket to help control the temperature; and a discharge port.

The charges for this polyol are shown in Table 3. Glycerol and 90% KOH are charged to the autoclave which is then purged with nitrogen. The charge of glycerol is only 3% of the total charge and may not be enough material for efficient stirring. A 4 or 5 mol PO adduct of glycerol can be made and used as the starter. The reactor is pressurized with nitrogen to 450 kPa (50 psig), where it is held for 15 minutes to check for leaks. The pressure is relieved and the reactor is heated to 105°C. Then the reactor is evacuated to 8 kPa (60 mm Hg), and the required amount of water is removed by stripping. The oxide is then added at a constant rate (600–900 g/oxide per mole initiator) in five hours at 105°C. The pressure is not allowed to exceed 722 kPa (90 psig) during the addition.

**Table 3. Charges for a 3000 Molecular Weight Glycerol-Initiated PPO Triol**

| Charges | Wt, g | Moles of hydroxyl | Hydroxyl equivalents | Hydroxyl number contribution |
|---|---|---|---|---|
| glycerol | 178.0 | 1.93 | 5.80 | 54.23 |
| potassium hydroxide, 90% | 36.0 | 0.58 | 0.58 | |
| water from KOH | −14.0 | | | |
| propylene oxide (PO) | 5822 | | | |
| water from PO | 0.6 | 0.03 | 0.06 | 0.56 |
| unsaturation[a] | | 0.14 | 0.14 | 1.35 |
| −K + H[b] | −22.0 | −0.58 | −0.58 | |
| *Total[c]* | *6000* | *2.10* | *6.00* | *56.14* |

[a]Unsaturation is normally expressed in meq/g but it is convenient to convert it to hydroxyl units for charge calculation.
[b]The replacement of K by H in the equation ROK + $H_2O$ ⟶ ROH (polyol) + KOH.
[c]The functionality can be calculated from the hydroxyl equivalents and hydroxyl moles: $f = 6.00/2.10 = 2.86$.

The mixture is kept for three hours at 105°C after the oxide addition is complete. By this time, the pressure should become constant. The mixture is then cooled to 50°C and discharged into a nitrogen-filled bottle. The catalyst is removed by absorbent (magnesium silicate) treatment followed by filtration or solvent extraction with hexane. In the laboratory, solvent extraction is convenient and effective, since polyethers with a molecular weight above about 700 are insoluble in water. Equal volumes of polyether, water, and hexane are combined and shaken in a separatory funnel. The top layer (polyether and hexane) is stripped free of hexane and residual water. The hydroxyl number, water, unsaturation value, and residual catalyst are determined by standard titration methods.

**Hydroxyl Number.** The molecular weight of polyether polyols for urethanes is usually expressed as its hydroxyl number or percent hydroxyl. When KOH (56,100 meg/mol) is the base, the hydroxyl number is defined as 56,100/equivalent weight (eq wt). Writing the equation as eq wt = 56,100/OH No. allows one to calculate the equivalents of polyol used in a urethane formulation, and then the amount of isocyanate required. The molecular weight can be calculated from these equations if the functionality, $f$, is known: mol wt = $f^*$eq wt.

The hydroxyl number can be determined in a number of ways such as acetylation, phthalation, reaction with phenyl isocyanate, and ir and nmr methods. An imidazole-catalyzed phthalation has been used to measure the hydroxyl number for a number of commercial polyether polyols and compared (favorably) to ASTM D2849 (uncatalyzed phthalation) (99). The uncatalyzed method requires two hours at 98°C compared to 15 minutes at the same temperature.

**Starters.** Nearly any compound having an active hydrogen can be used as starter (initiator) for the polymerization of PO. The common types are alcohols, amines, and thiols. Thus in Figure 2 ROH could be $RNH_2$ or RSH. The functionality is derived from the starter, thus glycerol results in a triol. Some common starters are shown in Table 4. The term starter is preferred over the commonly used term initiator because the latter has a slightly different connotation in polymer chemistry. Table 5 lists some homopolymer and copolymer products from various starters.

**Unsaturation Value.** The reaction temperature, catalyst concentration, and type of counterion of the alkoxide affect the degree of unsaturation. The tendency for rearrangement of PO to allyl alcohol is greatest with lithium hy-

**Table 4. Common Starters for Polyurethane Polyols**

| Name | Abbreviation | CAS Registry Number | $f$ | Mol wt | OH No. |
|------|-------------|-------------------|-----|--------|--------|
| water | | | 2 | 18 | 6233.3 |
| propylene glycol | PG | [57-55-6] | 2 | 76.1 | 1474.4 |
| dipropylene glycol | DPG | [110-98-5] | 2 | 134.2 | 836.1 |
| glycerol | Gly | [56-81-5] | 3 | 92.1 | 1827.4 |
| trimethylolpropane | TMP | [77-99-6] | 3 | 134.2 | 1254.1 |
| pentaerythritol | PE | [115-88-5] | 4 | 136.2 | 1647.6 |
| ethylenediamine | EDA | [107-15-3] | 4 | 60.1 | 3733.8 |
| toluenediamine | TDA | [25376-45-8] | 4 | 122.2 | 1836.3 |
| sorbitol | Sorb | [50-70-4] | 6 | 182.2 | 1847.4 |
| sucrose | Suc | [57-50-1] | 8 | 342.3 | 1311.1 |

**Table 5. CAS Numbers for Some Polyols Listed in the TSCA Inventory**

| Homopolymer | | Copolymer | |
| --- | --- | --- | --- |
| Composition[a] | CAS Registry Number | Composition[a] | CAS Registry Number |
| PG–PO | [25322-69-4] | PO–EO | [9003-11-6] |
| Gly–PO | [25791-96-2] | PG–PO–EO | [53637-25-5] |
| TMP–PO | [25723-16-4] | Gly–PO–EO | [9082-00-2] |
| PE–PO | [9051-49-4] | TMP–PO–EO | [52624-57-4] |
| EDA–PO | [25214-63-5] | PE–PO–EO | [30374-35-7] |
| TDA–PO | [63641-63-4] | EDA–PO–EO | [26316-40-5] |
| Sorb–PO | [52625-13-5] | TDA–PO–EO | [67800-94-6] |
| Suc–PO | [9049-71-2] | Suc–PO–EO | [26301-10-0] |

[a]The abbreviation for the composition has the form initiator–1st oxide–2nd oxide. The abbreviations for the initiators are shown in Table 4. PO and EO are propylene oxide and ethylene oxide, respectively.

droxide and decreases in the following order (100): $Li^+ > Na^+ > K^+ > Cs^+$. The amount of unsaturation also increases with number-average molecular weight ($M_n$) suggesting that the rate of polymerization decreases relative to the rate of isomerization (chain transfer). The maximum molecular weight of base-catalyzed PPO, limited to 6000 by the ratio of polymerization, $k_p$, to transfer, $k_{tr}$, is about $k_p/k_{tr} \approx 100$. A theoretical upper limit to the molecular weight of PPO has been calculated.

The unsaturation present at the end of the polyether chain acts as a chain terminator in the polyurethane reaction and reduces some of the desired physical properties. Much work has been done in industry to reduce unsaturation while continuing to use the same reactors and holding down the cost. In a study (102) using 18-crown-6 ether with potassium hydroxide to polymerize PO, a rate enhancement of approximately 10 was found at 110°C and slightly higher at lower temperature. The activation energy for this process was found to be 65 kJ/mol (mol ratio, $r = 1.5$ crown ether/KOH) compared to 78 kJ/mol for the KOH-catalyzed polymerization of PO. It was also feasible to prepare a PPO with $M_n \sim 10,000$ having narrow distribution at 40°C with added crown ether ($r = 1.5$) (103). The polymerization rate under these conditions is about the same as that without crown ether at 80°C.

Unsaturation value can be determined by the reaction of the allyl or propenyl end group with mercuric acetate in a methanolic solution to give ace-toxymercuric methoxy compounds and acetic acid (ASTM D4671-87). The amount of acetic acid released in this equimolar reaction is determined by titration with standard alcoholic potassium hydroxide. Sodium bromide is normally added to convert the insoluble mercuric oxide (a titration interference) to mercuric bromide. The value is usually expressed as meq KOH/g polyol which can be converted to OH No. units using multiplication by 56.1 or to percentage of vinyl using multiplication by 2.7.

## ACID CATALYSIS

The ring-opening polymerization of PO using acid catalysts has been extensively studied. The products range from isomerization of PO to low molecular weight oligomers (104–111). Measurement of the kinetics of hydration of PO to PG using an ion-exchange resin catalyst (acid) showed that the order of reaction for PO was 0.43 for homogenous reaction and 0.55 under heterogenous conditions (112). The activation energies obtained for the homogenous and heterogenous reactions were 51.5 and 53.4 kJ/mol, respectively. Studies (113) of the reaction of PO with boron trifluoride etherate catalyst showed varying results with different solvents. In dioxane, the isomerization to propionaldehyde proceeded smoothly and selectively. In THF, copolymerization of PO and THF was observed. In benzene the primary product was low mol wt PPO. In another study (111) the oligomerization PO using $BF_3(C_2H_5)_2O$ stopped before the monomer was completely exhausted. Substituted vinyl bromides were studied as photoinitiators in combination with onium salts for cationic polymerization of cyclohexene oxide, CHO (110). The product had a mol wt range of 200–2000 and a rate enhancement of approximately 10 was observed for irradiated samples over onium salt alone. The vinyl radicals generated by irradiation were oxidized by onium ions and the vinyl cation thus formed initiated the cationic polymerization of monomers such as cyclohexene oxide. Phosphoric acid has been used as a catalyst for the oligomerization of EO (109).

Chemicals responsible for odor in some PUR foams were synthesized by polymerization of PO in $CH_2Cl_2$ with $BF_3(C_2H_5)_2O$ catalyst (114). The yield was 25% volatile material and 75% polymeric material. The 25% fraction consisted of dimethyldioxane isomers, dioxolane isomers, DPG, TPG, crown ethers, tetramers, pentamers, etc, and 2-ethyl-4,7-dimethyl-1,3,6-trioxacane (acetal of DPG and propionaldehyde). The latter compound is mainly responsible for the musty odor found in some PUR foams. This material is not formed under basic conditions but probably arises during the workup when acidic clays are used for catalyst removal.

## COORDINATION POLYMERIZATION OF PO

A variety of ring-opening polymerization catalysts, called coordination catalysts, have been reported. Among them are organoaluminum and organozinc compounds that have been modified with alcohols, ketones, phenols, and others. These polymerizations are characterized by controlled molecular weight with narrow molecular weight distribution and result in some amount of stereoregular polymer. The process is described as living polymerization, defined as consisting only of initiation and propagation reactions with no termination or chain-transfer reactions. Another way of saying this is that polymerization can be stopped by cutting off the flow of monomer and can be restarted by adding a new monomer. The end of the reaction is reached when a reagent such as acetic acid to hydrolyze the metal species is added. The term immortal polymerization has been used (115). In immortal polymerization the mixture continues to initiate polymerization until the reaction is specifically quenched. Many groups have studied the mechanism of immortal polymerization. It was found that the propagation and chain transfer could be accelerated by use of a Lewis acid such as methylaluminum bis(2,6-di-*tert*-butyl-4-methylphenolate) (115). Living polymerization especially with metalloporphyrins has been reviewed (116). An organozinc complex has been used to polymerize *tert*-butylethylene oxide and the polymerization compared to that of PO (117). Partially stereoregular PPO has been separated on glass beads using isooctane as the eluent and controlling the temperature based on desired stereoregularity and mol wt (118,119).

**Procedure for Porphyrin–(C₂H₅)₂AlCl Catalyst.** The following procedure is considered typical (120). The reaction of tetraphenylporphyrin with pyrophoric diethylaluminum chloride was carried out in a Pyrex flask fitted with a three-way stopcock. The flask containing the porphyrin (1 mmol) was purged with dry nitrogen, and dichloromethane (20 mL) was added to dissolve the porphyrin. To this solution was added $(C_2H_5)_2AlCl$ (1.2 mmol) in 20% excess to the porphyrin. After about four hours, the volatile materials were removed under reduced pressure from above the reaction mixture to leave crystalline materials, which were used as the polymerization catalyst. Dichloromethane (20 mL) was added to the catalyst and the mixture was cooled with liquid nitrogen. Purified PO (200 mmol) was added by trap-to-trap distillation to the cooled catalyst. Polymerization was carried out at room temperature for several days. A large amount of methanol was added to stop the polymerization. The volatile compounds were removed under reduced pressure. This procedure gave a 100% yield of polymer having mol wt of 10,000 and polydispersity of 1.13. The catalyst was removed by dissolving the residue in THF and filtering off the insoluble catalyst residue.

## AUTOXIDATION OF PPO

PPO is prone to oxidation, as are short-chain aliphatic ethers. The mechanism of oxidation has been studied both with and without added stabilizers. The oxidation is initiated by the formation of a radical on the carbon (usually secondary) $\alpha$ to the ether oxygen. The radical is then trapped by oxygen to form an $\alpha$-alkoxy hydroperoxide (121). The hydroperoxide decomposes unimolecularly to give an oxy radical and a hydroxyl radical (eq. 6). The oxy radical (**1**) can follow two paths to products: the first occurs from split of the C–C bond (indicated by arrow) and leads to an acetate end group (and acetic acid), a formate end group, one chain scission, and two moles of water formed per two moles of oxygen consumed; the second path occurs from split of the C–O bond (indicated by arrow) and leads to formation of methyl ketone and an alcohol group for one mole of oxygen consumed (122).

(6)

(**1**)

The autoxidation of PPO is characterized by having an induction period which becomes longer upon addition of increasing amounts of an antioxidant such as 2,6-di-*tert*-butyl-4-methylphenol, more commonly referred to as butylated hydroxy toluene (BHT) (123). Both the induction period and rate of reaction are sensitive to temperature. The rate of autoxidation is independent of molecular weight and therefore the autoxidation must occur randomly along the chain and not on the end groups. The formation of phenyl urethane end groups increases the induction period and decreases the rate of oxygen consumption.

The tendency of aliphatic ethers toward oxidation requires the use of antioxidants such as hindered phenolics (eg, BHT), secondary aromatic amines, and phosphites. This is especially true in polyether polyols used in making polyurethanes (PUR) because they may become discolored and the increase in acid number affects PUR production. The antioxidants also reduce oxidation during PUR production where the temperature could reach 230°C. A number of new antioxidant products and combinations have become available (115,120,124–139) (see ANTIOXIDANTS).

## Manufacture of Polyols

A list of polyol producers is shown in Table 6. Each producer has a varied line of PPO and EOPO copolymers for polyurethane use. Polyols are usually produced in a semibatch mode in stainless steel autoclaves using basic catalysis. Autoclaves in use range from one gallon (3.785 L) size in research facilities to 20,000 gallon (75.7 m³) commercial vessels. In semibatch operation, starter and catalyst are charged to the reactor and the water formed is removed under vacuum. Sometimes an intermediate is made and stored because a 30–100 dilution of starter with PO would require an extraordinary reactor to provide adequate stirring. PO and/or EO are added continuously until the desired OH

**Table 6. Polyether[a] Producers for Urethane Applications**

| Company | Location | Trademark | Annual capacity, $10^3$ t |
|---|---|---|---|
| Arco[b] Chemical Co. | Channelview, Tex. | Arcol | 77 |
| | Conroe, Tex. | | 45 |
| AC West Virginia Polyols Co. | | | 200 |
| BASF Corp. | Geismar, La. | Pluracol | 82 |
| | Wyandotte, Mich. | | 36 |
| | Washington, N.J. | | 77 |
| Bayer[c] | Baytown, Tex. | Multranol | 70 |
| | New Martinville, W. Va. | | 32 |
| E. R. Carpenter Co., Inc. | Bayport, Tex. | Carpol | 125 |
| Dow Chemical U.S.A. | Freeport, Tex. | Voranol | 227 |
| Olin Corp. | Brandenburg, Ky. | Poly-G | 114[d] |
| Pelron Corp. | Lyons, Ill. | PEL-RIG | 1.4 |
| | | PEL-FLEX | |
| | | PEL-PPG | |
| Stepan Co. | Milsdale, Ill. | Stepanol | 3.6 |
| ICI Corp. | Geismar, La. | | |
| *Total* | | | *1090* |

[a]Each company has a full line of products including PPG (diols) and glycerol adducts, as well as other initiator adducts.
[b]Texaco sold its polyol product line and a tolling agreement to Arco in 1987.
[c]Formerly Miles (formerly Mobay).
[d]Plus 18,000 t non-PUR product.

No. is reached; the reaction is stopped and the catalyst is removed. A uniform addition rate and temperature profile is required to keep unsaturation the same from batch to batch. The KOH catalyst can be removed by absorbent treatment (140), extraction into water (141), neutralization and/or crystallization of the salt (142–147), and ion exchange (148–150).

## Characterization and Properties of Polyethers

**Viscosity.** In the molecular weight range of 200–6000, PPO polyols are liquids. The viscosity depends on the functionality. Polyols with higher functionality have higher viscosity at a given equivalent weight. At low equivalent weight the viscosity depends strongly on the initiator. Monols and diols decrease the viscosity of a triol. PPO polyethers used for flexible foam have a viscosity of approximately 600 mPa·s(=cP). Polyethers for rigid foam have short chains and are therefore viscous. Some polyethers initiated by toluenediamine, which gives a tetrol, and used for rigid foams have a viscosity range of 10,000–100,000 mPa·s. Some high functionality sucrose-initiated polyols have viscosity as high as $10^6$ mPa·s. Many sucrose-initiated polyols are co-initiated with lower functionality materials such as glycerol to reduce the viscosity of the final product. The viscosity of mixtures of diols and triols is intermediate and can be estimated from the weight fraction of the components.

In measurements (151) of the viscosity–temperature–mol wt relationship for PPO diols and triols the viscosity of PPG diols was found to be independent of shear rate, that is they are Newtonian fluids. A plot of viscosity vs reciprocal temperature ($\eta - 1/T$) of a series of PPO diols or triols gives a family of straight lines; higher mol wt gives higher viscosity. A plot of viscosity vs mol wt is a straight line but breaks at a limiting viscosity corresponding to a mol wt of ca 600 where no further decrease is observed. This means that a Mark-Houwink relationship should fail for mol wt less than 600. Equation 7 gives a polydispersity-corrected limiting viscosity number–molecular weight relationship for PPO in benzene solution at 25°C (152).

$$[\eta] = 0.000246\ M^{0.71}\ (\text{dL/g}) \tag{7}$$

This value differs by a factor of 2 from that of the often quoted relationship (153) shown by equation 8.

$$[\eta] = 0.000129\ M_v^{0.75}\ (\text{dL/g}) \tag{8}$$

**Nmr Studies.** [1]H- and [13]C-nmr has been valuable in elucidating the structure of PPO and copolymers of EO and PO, especially since high field nmr has become widely available.

The primary and secondary hydroxyl content in polyethers has been determined by high field (360 MHz) [1]H-nmr of trichloroacetyl isocyanate (TAIC)-modified polymers (154). Methylene or methine protons in the $\alpha$-position are shifted and both are well resolved for integration. When the molar range of secondary hydroxyl groups is 10–90%, 5–8% accuracy is claimed (154). The resonance due to allylic end groups confounds the signal at 4.43 but is separable. $^1$H-nmr (300 and 500 MHz) has been used to determine the number-average molar masses and molar ratio of the double-bond content of anionically polymerized PO over a range of conversions (136). Triad sequences of statistical and block copolymers of EO and PO were measured using resolution enhancement and subspectrum editing techniques (155). $^{13}$C-nmr has been used to differentiate between random and block copolymers, and study persistence ratio (a measure of the deviation from fully random statistics), mean sequence length of EOPO sequences, triad probabilities, and starter and end groups (156). In $^1$H and $^{13}$C studies (157), to characterize the end groups in PPO polymers, it was found that the peaks corresponding to propenyl end groups were only observed in freshly prepared solutions and disappeared in about two days due to hydrolysis by trace DCl in the DCCl$_3$, producing propanol and a CH$_2$OH end group. Dipropylene glycol (DPG) and tripropylene glycol (TPG) prepared from chiral PO using an aluminum complex has been studied by means of a $J$-modulated spin echo technique $^{13}$C-nmr (158). Findings were applied to a racemic oligomer (DP = 11) and it was possible to identify central and terminal units. End-group carbon atoms present large configurational effects compared to internal carbon atoms. Measurement of relaxation times of PPO complexed with sodium trifluoromethanesulfonate using $^{13}$C-nmr demonstrated the influence of polyether–salt interactions on local segmental motion of the polymer chains (159). Relaxation time, $T_1$, decreases with decreasing temperature for uncomplexed PPO but increases with decreasing temperature for complexes. Increasing the salt concentration to create more virtual cross-links causes PPO lifetime, $\tau_0$, to increase rapidly. This behavior is unusual.

$^{13}$C-nmr was used to analyze the stereoregularity of PPO prepared with the diphenylzinc–water system at various (H$_2$O/(C$_6$H$_5$)$_2$Zn) ratios (160). Two nmr methods were used to determine the primary hydroxyl content of EOPO copolymers (161). The first method was integration of the $^{13}$C-nmr resonance from the carbon bearing the hydroxyl group. The second method used $^{19}$F-nmr of trifluoroacetic acid derivatives. The $^{13}$C method had good accuracy and easy sample preparation but poor sensitivity and precision. The $^{19}$F method had good sensitivity and precision but poor accuracy (probably because of workup during derivatization) and difficult sample preparation. These tests are called ASTM 4273-83 methods A and B. Two-dimensional $J$-resolved spectroscopy was used to separate overlapping multiplets in atactic PPO (162). Proton chemical shift is sensitive to triad sequences but the homonuclear coupling constants for head-to-head monomer units are the same regardless of stereosequence.

A $^{13}$C-nmr study of regioisomers (eg, H–T, H–H, T–T) of oligomeric poly(propylene glycol) has been done (163). A method for determining the number-average functionality and functionality distribution of polyether polyols based on measured intensities of relevant end groups observed by $^{13}$C-nmr has been described (164). Chromium(III) acetylacetonate, Cr(acac)$_3$, was used as a relaxation reagent and allows analysis to be speeded up by decreasing relaxation times. Chemical shift reagent Eu(DPM)$_3$ has been used to measure the molecular weight of PPO (165). The molecular weight values found were within a few percent of the values reported by suppliers. This method is not practical due to the ease of use and accuracy of gpc compared to the expense of the shift reagents. Multipulse $^{13}$C-nmr INEPT and DEPT techniques have been used for the determination of PPO microstructure (166). The $^{13}$C-nmr chemical shifts were also calculated using the $\gamma$-gauche effect for various dyad sequences (166).

**Refractive Index.**   The effect of mol wt (1400–4000) on the refractive index (RI) increment of PPG in benzene has been measured (167). The RI increments of polyglycols containing aliphatic ether moieties are negative: $d\eta/dc(\mathrm{mL/g}) = -0.055$. A plot of RI vs $1/M$ is linear and approaches the value for PO itself (109). The RI, density, and viscosity of PPG–salt complexes, which may be useful as polymer electrolytes in batteries and fuel cells have been measured (168). The variation of RI with temperature and salt concentration was measured for complexes formed with PPG and some sodium and lithium salts. Generally, the RI decreases with temperature, with the rate of change increasing as the concentration increases.

**Infrared Spectroscopy.**   The following bands are seen in the ir spectrum of PPG: 2970, 2940, 2880 cm$^{-1}$ (C–H stretch, m); 1460, 1375 cm$^{-1}$ (C–H bend, m); 1100, 1015 cm$^{-1}$ (C–O stretch, m) of which the 2940 and 1015 band are specific. The latter are also present in copolymers of EO and PO. Absorptions due to unsaturated end groups are found at 1650 cm$^{-1}$ (allyl ether) and 1672 cm$^{-1}$ (1-propenyl ether). The O–H stretching band at 3470 cm$^{-1}$ shows the greatest variation for different hydroxyl number polyols and has been used to estimate the hydroxyl number (169).

**Chromatography.**   One gpc study (170) of low molecular weight polyethers used two systems: THF solvent and PLgel columns and water with TSK gel column sets. In THF the elution volume depends predominantly on chain length, whereas in water the composition as well as chain length influences the elution volume. THF is a good solvent for PPO homopolymer and EOPO copolymers. Gpc calibration is typically done with poly(ethylene glycol) (PEG) or polystyrene standards but the latter tend to overestimate the mol wt of PPO. In some cases unsaturated polyethers can be resolved as a shoulder on the low mol wt side of the main peak. In many cases it is possible to see additives in polyols, but quantitation is difficult owing to the low sensitivity of the RI detector. Additives are better determined by gc or hplc. An on-line gpc analyzer for the detection of PO and EO in the polymerization reactor has been described (171). The system consists of a PL aquagel column, a RI detector, and a sampling valve that allows on-stream measurement. This system was evaluated over the range 8% to <0.3% PO and compared favorably with that calculated from the vapor pressure. Another study used a gpc system consisting of Fractogel TSK columns, water–acetonitrile as the eluent, and PEG–PEO calibration standards.

The composition of PPG–PEG blends has been determined using gpc with coupled density and RI detectors. PEG and PPG have different response factors for the density and RI detectors which were exploited (173). An hplc system with CHROMPAC RP-18C$_{18}$ column at 298°C and acetonitrile–water or methanol–water as the mobile phase has been used to gather information about the functionality of PPO (174).

Reversed-phase hplc has been used to separate PPG into its components using evaporative light scattering and uv detection of their 3,5-dinitrobenzoyl derivatives. Acetonitrile–water or methanol–water mixtures effected the separation (175). Polymer glycols in PUR elastomers have been identified (176) by pyrolysis-gc. The pyrolysis was carried out at 600°C and produced a small amount of ethane, $CO_2$, propane, and mostly propylene, CO, and $CH_4$. The species responsible for a musty odor present in some PUR foam was separated and identified by gc (Supelco SP-2100 capillary column) (114). Unsaturated oligomers up to allyl-penta PG from a pentane extraction of polyol have also been separated and identified (114).

**Solubility.** PPO polyols with a molecular weight below 700 are water soluble. The triol is slightly more water soluble than the diol. The solubility in water decreases with increasing temperature. This inverse solubility causes a cloud point which is important in characterizing copolymers of propylene oxide and ethylene oxide.

Polyethers prepared from propylene oxide are soluble in most organic solvents. The products with the highest hydroxyl number (lowest molecular weight) are soluble in water, not in nonpolar solvents such as hexane. The solubility of 3000 molecular weight triols is high enough in solvents such as toluene, hexane, and methylene chloride that the triols can be purified by a solvent extraction process.

The following components of solubility parameters for PPO have been obtained (177): $\delta_d = 16.3 \pm 1$, $\delta_p = 4.7 \pm 0.5$, $\delta_h = 7.4 \pm 0.5$, and $\delta_o = 18.5 \pm 1.2$ with units $(J/mL)^{1/2}$. The determination was based on the use of three mixtures of solvents. For each mixture, the point of maximum interaction between the mixture and the polyol was obtained from the maximum value of the intrinsic viscosity. The parameter $\delta_d$ measures dispersion; $\delta_p$, polar bonding; $\delta_h$, hydrogen bonding; and $\delta_o$ is the Hildebrand solubility parameter which is the radius vector of the other orthogonal solubility parameters. Water solubility of PPO has been determined using turbidimetric titration (178) (Table 7).

**Mass Spectrometry.** Field desorption mass spectrometry has been used to analyze PPO (179). Average molecular weight parameters ($M_n$ and $M_w$) could be determined using either protonated ($MH^+$) or cation attachment ($MNa^+$) ions. Good agreement was found between fdms and data supplied by the manufacturer, usually less than 5% difference in all cases up to about 3000 amu. Laser desorption Fourier transform mass spectrometry was used to measure PPG ion and it was claimed that ions up to m/z 9700 (PEG) can be analyzed by this method (180).

**Density.** At low equivalent weight, the specific gravity (density) of polyethers depends on the initiator and at high equivalent weight it depends on the alkylene oxide. The effects of molecular weight on specific gravity are

**Table 7. Water Solubility of PO Polyols[a]**

| | Solubility, wt % | |
| --- | --- | --- |
| Polyol description | at 23°C | at 50°C |
| 2000 mol wt diol | 5.3 | 3.6 |
| 2000 mol wt diamine | 7.3 | 5.4 |
| 3000 mol wt triol | 4.8 | 4.1 |
| 5000 mol wt triol | 2.8 | |

[a] Ref. 178.

minor. Most commercial products have a specific gravity of 1.005–1.020 at 25°C unless they are of low hydroxyl equivalent weight. The specific gravity of PPO diol can be calculated from the hydroxyl number by the linear equation sp gr (25°C) = 0.000127 × (hydroxyl number) + 0.9976 over the molecular weight range 2000–7000.

The specific volume (mL/g) of PPO is inversely proportional to the molecular weight (181) and is described by sp vol = $1.0013 - 2.524/M_n$. The temperature dependence of the specific volume is given by $V_{sp} = 1.000 + 0.0007576\,(T - 25)$ over the temperature range 25–80°C (94).

**Other Properties.** The glass-transition temperature for PPO is $T_g \sim$ 190 K and varies little with molecular weight (182). The temperature dependence of the diffusion coefficient of PPO in the undiluted state has been measured (182).

The thermal conductivity of PPO is approximately 0.16 W/(m·K) for a 3000 mol wt polyol and 0.15 W/(m·K) for a 5000 mol wt polyol. The thermal conductivity is relatively insensitive to the temperature. The specific heat of PPO varies with temperature but not with the molecular weight. At 25°C the specific heat is 1950 J/(kg·°C) and at 150°C it is 2300 J/(kg·°C). Intermediate values can be interpolated with the following equation:

$$\text{specific heat (J/(kg·°C))} = 2.84 \times \text{temperature (°C)} + 1875$$

## Health and Safety

Propylene oxide is highly reactive. It reacts exothermically with any substance that has labile hydrogen such as water, alcohols, amines, and organic acids; acids, alkalies, and some salts act as catalysts.

Propylene oxide is a primary irritant, a mild protoplasmic poison, and a mild depressant of the central nervous system. Skin contact, even in dilute solution (1%), may cause irritation to the eyes, respiratory tract, and lungs. Propylene oxide is a suspected carcinogen in animals. The $LC_{50}$ (lowest lethal concentration by inhalation in rats) is 4000 mg/kg body weight. The $LD_{50}$ (oral) is 930 mg/kg. The $LD_{50}$ (dermal) is 1500 mg/kg. The TWA (8-h exposure) is 100 ppm and the STEL (15-min exposure) is 150 ppm.

PPO and EOPO copolymers are low hazard–low vapor pressure liquids. Contact with skin, eyes, or inhalation cause irritation. There are no known acute or chronic affects associated with polyols. First aid for contact with polyols involves washing the affected area with water. The flash point of PPO is greater than 93°C.

## BIBLIOGRAPHY

"Propylene Oxide Polymers and Higher 1,2-Epoxide Polymers" under "Polyethers" in *ECT* 3rd ed., Vol. 18, pp. 633–645, by R. A. Newton, Dow Chemical USA.

1. D. Kahlich, U. Wiechern, and J. Lindner, in B. Elvers, S. Hawkins, and G. Schulz, eds., *Ullmann's Encyclopedia of Industrial Chemistry*, 5th ed., Vol. A22; VCH Publishers, Inc., New York, 1993, pp. 239–260.

2. S. D. Gagnon, pp. 273–307; and N. Clinton, and P. Matlock, pp. 225–273, in J. I. Kroschwitz, ed., *Encyclopedia of Polymer Science and Engineering*, 2nd ed., Vol. 6, John Wiley and Sons, Inc., New York, 1986.
3. F. E. Bailey, in Ref. 1, Vol. A21; 1992, pp. 579–589.
4. L. C. Pizzini, and J. T. Patton, Jr., pp. 145–167; F. W. Stone, and J. J. Stratta, pp. 103–145; J. Furukawa, and T. Saegusa, pp. 175–195; and L. C. Pizzini, J. T. Patton, Jr., pp. 168–175, in H. F. Mark, N. G. Gaylord, and N. M. Bikales, eds., *Encyclopedia of Polymer Science and Technology*, Vol. 6, John Wiley and Sons, Inc., New York, 1967.
5. R. W. Body, and V. L. Kyllingstad, in Ref. 2, pp. 307–322.
6. D. Dieterich, and K. Uhlig, in Ref. 3, pp. 665–716.
7. R. A. Briggs, and E. E. Gruber, in Ref. 4, pp. 195–209.
8. B. Osner, *Bull. Soc. Chim. Fr.*, 235 (1860).
9. U.S. Pat. 2,948,691 (Aug. 9, 1960), E. Windemuth, H. Schnell, and O. Bayer, (to Bayer).
10. U.S. Pat. 2,674,619 (Apr. 6, 1954), L. G. Lundsted, (to Wyandotte Chemical Corp.).
11. Eur. Pat. 460,317 (Dec. 11, 1991), D. K. Walters, and R. I. Barber (to Ethyl Petroleum Additives, Ltd., U.K.).
12. Eur. Pat. 244,733 (Nov. 11, 1987), G. C. Weitz (to American Polywater Corp.).
13. U.S. Pat. 4,555,549 (Nov. 26, 1985), R. L. Camp, E. M. Dexheimer, and M. J. Anchor, (to BASF Corp.).
14. U.S. Pat. 4,452,711 (June 5, 1984), J. T. Laemmie (to Aluminum Co. of America, USA).
15. U.S. Pat. 4,414,121 (Nov. 8 1983), R. P. Aiello.
16. U.S. Pat. 4,402,839 (Sept. 6, 1983), R. H. Davis, and A. B. Piotrowski (to Mobil Oil Corp., USA).
17. H. S. Koenig and G. M. Bryant, *Text. Res. J.* **50**, 1 (1980).
18. U.S. Pat. 4,560,482 (Dec. 24, 1985), G. P. Canevari (to Exxon Research and Engineering Co., USA).
19. U.S. Pat. 4,505,716 (Mar. 19, 1985), E. W. Sawyer, Jr. (to ITT Corp., USA).
20. U.S. Pat. 4,441,889 (Apr. 10, 1984), S. Mark (to Gulf and Western Industries, Inc., USA).
21. Eur. Pat. 158,996 (Oct. 23, 1985), W. McCormick (to Adamantech, Inc., USA).
22. WO Pat. 9,205,220 (Apr. 2, 1992), S.H.-P Yu and T. R. Mass (to B.F. Goodrich Co., USA).
23. U.S. Pat. 4,542,095 (Sept. 17, 1985), D. J. Steklenski and J. E. Littman (to Eastman Kodak Co., USA).
24. U.S. Pat. 4,304,562 (Dec. 8, 1981), J. A. Bolan and M. A. Grimmer (to Drackett Co., USA).
25. WO Pat. 9,101,171 (Feb. 7, 1991), M. S. Dahanayake (to GAF Chemicals Corp., USA).
26. U.S. Pat. 4,836,951 (June 6, 1989), G. E. Totten, and G. C. Johnson (to Union Carbide Corp., USA).
27. U.S. Pat. 4,510,067 (Apr. 9, 1985), A. C. Ozmeral (to BASF Corp.).
28. U.S. Pat. 4,411,810 (Oct. 25, 1983), D. R. Dutton, E. J. Parker, R. A. Ott, and J. G. Otten (to BASF Corp.).
29. U.S. Pat. 5,098,478 (Mar. 24, 1992), R. Krishnan, R. W. Bassemir, and T. C. Vogel (to Sun Chemical Co.).
30. Eur. Pat. 397,431 (Nov. 14, 1990), H. Tomita and Y. Sonoda (to Kabushiki Laisha Kako Co., Ltd.).
31. Eur. Pat. 469,724 (Feb. 5, 1992), C. R. Frisby (to McGean-Rohco, Inc.).
32. U.S. Pat. 4,528,075 (July 9, 1985), M. J. Anchor and R. L. Camp (to BASF Corp.).
33. Eur. Pat. 359,071 (Mar. 21, 1990), L. Levrero, F. Granata, and R. Latorrata (to BP Chemicals Ltd., U.K.).

34. U.S. Pat. 4,548,726 (Oct. 22, 1985), B. J. Morris-Sherwood, E. C. Brink, Ir., D. R. McCoy, and E. E. McEntire (to Texaco).
35. U.S. Pat. 4,552,686 (Nov. 12, 1985), B. J. Morris-Sherwood and E. C. Brink Jr. (to Texaco Inc., USA).
36. U.S. Pat. 4,543,199 (Sept. 24, 1985), L. F. Kuntschik and C. L. Dowe (to Texaco Inc., USA).
37. U.S. Pat. 4,481,125 (Nov. 6, 1984), R. V. Holgado (to E. F. Houghton and Co., USA).
38. Can Pat. 2,027,303 (Apr. 12, 1991), S. T. Hirozawa and D. E. Coker (to BASF Corp.).
39. U.S. Pat. 5,057,234 (Oct. 15, 1991), R. G. Bland and D. K. Clapper (to Baker Hughes, Inc., USA).
40. U.S. Pat. 4,780,220 (Oct. 25, 1988), T. E. Peterson (to Hydra Fluids, Inc., USA).
41. WO Pat. 9,115,542 (Oct. 17, 1991), J. S. Peanasky, J. M. Long, and R. P. Wool (to AGRI-Tech Industries, Inc., USA).
42. U.S. Pat. 4,857,593 (Aug. 15, 1989), P. S. Leung, E. D. Goddard, and F. H. Ancker (to Union Carbide Corp., USA).
43. WO Pat. 9,114,727 (Oct. 3, 1991), G. M. Vanhaeren (to Exxon Chemical Ltd., U.K.).
44. U.S. Pat. 4,343,664 (Aug. 10, 1982), S. Iyer (to U.S. Dept. of the Army).
45. U.S. Pat. 5,160,328 (Nov. 3, 1992), J. V. Cartmell and W. R. Sturtevant (to NDM Aquisition Corp., USA).
46. Eur. Pat. 481,600 (Apr. 22, 1992), N. I. Payne, M. Gibson, and P. M. Taylor (to American Cyanamid Co.).
47. U.S. Pat. 4,879,109 (Nov. 7, 1989), R. L. Hunter (to Emory University, USA).
48. D. Attwood, J. H. Collett, and C. J. Tait, *Int. J. Pharm.* **26**, 25 (1985).
49. R. M. Nalbandian, R. L. Henry, K. W. Balko, D. V. Adams, and N. R. Neuman, *J. Biomed. Mater. Res.* **21**, 1135 (1987).
50. Eur. Pat. 441,307 (Aug. 14, 1991), Y. Kawasaki and Y. Suzuki (to Showa Yakuhin Kako Co., Ltd.).
51. Eur. Pat. 122,799 (Oct. 24, 1984), R. C. Harrison (to Amersham International PLC, U.K.).
52. S. Miyazaki, and co-workers, *Chem Pharm. Bull.* **34**, 1801 (1986).
53. T. K. Law, T. L. Whateley, and A. T. Florence, *J. Controlled Release* **3**, 279 (1986).
54. WO Pat. 9,013,307 (Nov. 15, 1990), G. Collins and W. Wicomb.
55. Eur. Pat. 464,545 (Jan. 8, 1992), M. Shibuya and S. Ishii (to Showa Yakuhin Kako Co., Ltd.).
56. Eur. Pat. 481,717 (Apr. 22, 1992), C. F. Cooper (to ARCO Chemical Technology, Inc.).
57. *Chem. Mark. Rep.*, (Apr. 1993).
58. Y.-C. Yen and C.-S. Liu, *Ethylene Oxide/Propylene Oxide, Process Economics Report 2D*, Stanford Research Institute, Menlo Park, Calif., 1985.
59. U.S. Pat. 5,101,052 (Mar. 31, 1992), R. A. Meyer and E. T. Marquis (to Texaco Chemical Co.).
60. U.S. Pat. 5,107,067 (Apr. 21, 1992), E. T. Marquis, K. P. Keating, J. R. Sanderson, and W. A. Smith (to Texaco Inc.).
61. U.S. Pat. 5,093,509 (Mar. 3, 1992), R. A. Meyer and E. T. Marquis (to Texaco Chemical Co.).
62. U.S. Pat. 5,214,168 (May 25, 1993), J. G. Zajacek and G. L. Crocco (to Arco Chemical Technology, L.P.).
63. U.S. Pat. 5,241,088 (Aug. 31, 1993), J. L. Meyer, B. T. Pennington, and M. C. Fullington, (to Olin Corp.).
64. A. Q. H. Habets-Cruetzen and J. A. M. De Bont, *Appl. Microbiol. Biotechnol.* **26**, 434 (1987).
65. T. Imai, and co-workers, *Appl. Environ. Microbiol.* **52**, 1403 (1986).
66. V. Subramanian, *J. Ind. Microbiol.* **1**, 119 (1986).

67. L. E. S. Brink and J. Tramper, *Enzyme Microb. Technol.* **8**, 281 (1986).
68. N. R. Woods and J. C. Murrell, *Bitechnol. Lett.* **12**, 409 (1990).
69. G. A. Kovalenko and V. D. Sokolovskii, *React. Kinet. Catal. Lett.* **48**, 447 (1992).
70. C. A. G. M. Weijers, E. J. T. M. Leenen, N. Klijn, and J. A. M. De Bont, *Meded. Fac. Landbouwwet., Rijksuniv. Gent.* **53**(4b), 2098 (1988).
71. M. Mahmoudian and A. Michael, *J. Ind. Microbiol.* **11**, 29 (1992).
72. S. H. Stanley and H. Dalton, *Biocatalysis* **6**(3), 163 (1992).
73. S. Li, C. Gao, and A. Liu, *Chin. Chem. Lett.* **2**(4), 303 (1991).
74. R. L. Kelley, D. E. Hoefer, J. R. Conrad, V. J. Srivastava, and C. Akin, in C. Akin and J. I. Smith, eds., *Gas, Oil, Coal, Environ. Biotechnol.* 2nd ed., Gas Technology, Chicago, Ill., 1990, pp. 433–455.
75. M. Shimoda, Y. Seki, and I. Okura, *J. Mol. Catal.* **78**(2), L27 (1993).
76. S. H. Stanley, A. O'L. Richards, M. Suzuki, and H. Dalton, *Biocatalysis* **6**(3), 177 (1992).
77. K. Scott, C. Odouza, and W. Hui, *Chem. Eng. Sci.* **47**(9–11), 2957 (1992).
78. L. Franke, A. Zimmer, and K. Seibig, *Chem. Ing. Tech.* **64**(7), 652 (1992).
79. J. D. Lisius and P. W. Hart, *J. Electrochem. Soc.* **138**(12), 3678 (1991).
80. U.S. Pat. 5,214,168 (May 25, 1993), J. Zajacek (to Arco Chemical).
81. R. L. Burwell, Jr., *Chemtracts: Inorg. Chem.* **3**(6), 344 (1991).
82. A. Zhou, B. Zhu, and H. Chen, in X. I. A. Hou, ed., *Proceedings of the International Conference on Petroleum Refining and Petrochemical Processing*, Vol. 3; Beijing, China, 1991, pp. 1176–1182.
83. J. A. Riddick, W. B. Bunger, and T. K. Sakano, *Organic Solvents, Techniques of Chemistry*, 4th ed., Vol. II, John Wiley and Sons, Inc., New York, 1986.
84. U.S. Pat. 5,154,803 (Oct. 13, 1992), E. T. Marquis, G. P. Speranza, Y. H. E. Sheu, W. K. Culbreth, III, and D. G. Pottratz (to Texaco Chemical Co.).
85. U.S. Pat. 5,154,804 (Oct. 13, 1992), E. T. Marquis, G. P. Speranza, Y. H. E. Sheu, W. K. Culbreth, III, and D. G. Pottratz (to Texaco Chemical Co.).
86. U.S. Pat. 5,139,622 (Aug. 18, 1992), E. T. Marquis, G. P. Speranza, Y. H. E. Sheu, W. K. Culbreth, III, and D. G. Pottratz (to Texaco Chemical Co.).
87. U.S. Pat 5,116,466 (May 26, 1992), E. T. Maarquis, G. P. Speranza, Y. H. E. Sheu, W. K. Culbreth, III, and D. G. Pottratz, (to Texaco Chemical Co.).
88. U.S. Pat. 5,116,467 (May 26, 1992), E. T. Marquis, G. P. Speranza, Y. H. E. Sheu, W. K. Culbreth, III, and D. G. Pottratz, (to Texaco Chemical Co.).
89. U.S. Pat. 5,116,465 (May 26, 1992), E. L. Yeakey and E. T. Marquis (to Texaco Chemical Co.).
90. C. A. G. M. Weijers, M. J. J. Litjens, and J. A. M. de Bont, *Appl. Microbiol. Biotechnol.* **38**(3), 297 (1992).
91. R. G. Ghirardelli, *J. Am. Chem. Soc.* **95**, 4987 (1973).
92. *Propylene Oxide*, Technical bulletin, Arco. Chemical Co., Philadelphia, Pa.
93. J. A. Riddick, and W. B. Bunger, *Organic Solvents, Techniques of Chemistry*, 3rd ed., Vol. II, Wiley-Interscience, New York, 1970.
94. G. Yu, F. Heatley, C. Booth, and T. G. Blease, *J. Polym. Sci., Part A: Polym. Chem.* **32**, 1131 (1994).
95. X. Chen and M. Van De Mark, *J. Appl Polym. Sci.* **50**, 1923 (1993).
96. S. Kohjiya, T. Sato, T. Nakayama, and S. Yamashita, *Makromol. Chem., Rapid Commun.* **2**, 231 (1981).
97. A. T. Chen, and co-workers, "Comparison of the Dynamic Properties of Polyurethane Elastomers Based on Low Unsaturation Polyoxypropylene Glycols and Poly(tetramethylene oxide) Glycols," *Polyurethanes World Congress 1993*, Vancouver, B.C., Canada, Oct. 10–13, 1993.

98. N. Barksby and G. L. Allen, "Low Monol Polyols and Their Effects in Urethane Systems," *Polyurethanes World Congress 1993*, Vancouver, B.C., Canada, Oct. 10–13, 1993.

99. M. A. Carey, S. L. Wellons, and D. K. Elder, *J. Cell. Plast.*, 42 (1984).

100. J. Furukawa and T. Saegusa, *Polymerization of Aldehydes and Oxides*; Wiley-Interscience, New York, 1963.

101. A. Penati, C. Maffezzoni, and E. Moretti, *J. Appl. Polym. Sci.* **26**, 1059 (1981).

102. J. Ding, C. Price, and C. Booth, *Eur. Polym. J.* **27**, 891 (1991).

103. J. Ding, F. Heatley, C. Price, and C. Booth, *Eur. Polym. J.* **27**, 895 (1991).

104. U.S. Pat. 4,223,160 (Sept. 16, 1980), L. G. Hess (to Union Carbide Corp., USA).

105. P. Brüzga, J. V. Grazulevicius, R. Kavallünas, R. Kublickas, and I. Liutviniene, *Polym. Bull.* **30**, 509 (1993).

106. I. I. Abu-Abdoun, *Eur. Polym. J.* **28**, 73 (1992).

107. I. I. Abu-Abdoun, *Eur. Polym. J.* **29**, 1445 (1993).

108. A. Martinez, F. Mijangos, and L. M. Leon, *Eur. Polym. J.* **22**, 243 (1986).

109. T. Biela and P. Kubisa, *Makromol. Chem.* **192**, 473 (1991).

110. N. Johnen, S. Kobayashi, Y. Yagci, and W. Schnabel, *Polym. Bull.* **30**, 279 (1993).

111. S. S. Ivanchev, and co-workers, *J. Polym. Sci., Polym. Chem. Ed.* **18**, 2051 (1980).

112. R. Jaganathan, R. V. Chaudhari, and P. A. Ramachandran, *AIChe J.* **30**, 1 (1984).

113. S. Sugiyama, S. Ohigashi, K. Sato, S. Fukunaga, and H. Hayashi, *Bull. Chem. Soc. Jpn.* **62**, 3757 (1989).

114. S. H. Harris, P. E. Kreter, and C. W. Polley, "Characterization of Polyurethane Foam Odor Bodies," *Polyurethanes World Congress 1987*, Aachen, Germany.

115. M. Akatsuka, T. Aida, and S. Inoue, *Macromolecules* **27**, 2820 (1994).

116. S. Inoue and T. Alda, *Chemtech*, 28 (1994).

117. T. Tsuruta and Y. Hasebe, *Macromol. Chem. Phys.* **195**, 427 (1994).

118. K. Alyürük and K. Hartani, *Polymer* **30**, 2328 (1989).

119. K. Alyürük, T. Özden, and N. Çolak, *Polymer* **27**, 2009 (1986).

120. T. Aida and S. Inoue, *Macromolecules* **14**, 1166 (1981).

121. U. Hähner, W. D. Habicher, and K. Schwetlick, *Polym. Degrad. Stabil.* **34**, 111 (1991).

122. P. J. F. Griffiths, J. G. Hughes, and G. S. Park, *Eur. Polym. J.* **29**, 437 (1993).

123. U. Hähner, W. D. Habicher, and K. Schwetlick, *Polym. Degrad. Stabil.* **34**, 119 (1991).

124. R. G. Skorpenske, A. K. Schrock, and G. E. Beal, "Antioxidant Behavior in Flexible Polyurethane Foam," *33rd Annual Polyurethane Technical/Marketing Conference*, Orlando, Fla., Sept. 30–Oct. 3, 1990.

125. L. B. Barry and M. C. Richardson, "Novel AO System for Polyether Polyol Stabilization," *33rd Annual Polyurethane Technical/Marketing Conference*, Orlando, Fla., Sept. 30–Oct. 3, 1990.

126. L. B. Barry and M. C. Richardson, "A Less Volatile Performance Equivalent to BHT for Polyether Polyol Stabilization," *Polyurethane World Congress 1993*, Vancouver, B.C., Canada, Oct. 10–13, 1993.

127. L. B. Barry and M. C. Richardson, "Recent Developments in Polyol Stabilization Systems," *34th Annual Polyurethane Technical/Marketing Conference*, New Orleans, La., Oct. 21–24, 1992.

128. R. A. Hill, "New Antioxidant Package for Polyether Polyols, with Reduced Fogging Behaviour," *34th Annual Polyurethane Technical/Marketing Conference*, New Orleans, La., Oct. 21–24, 1992.

129. U.S. Pat. 4,904,745 (Feb. 27, 1990), S. Inoue and T. Aida (to Kanegafuchi Kagaka Kogyo).

130. S. Asano, T. Aida, and S. Inoue, *J. Chem. Soc., Chem. Commun.*, 1148 (1985).

131. T. Yasuda, T. Aida, and S. Inoue, *Bull. Chem. Soc. Jpn.* **59**, 3931 (1986).

132. Y.-S. Gal, B. Jung, W.-C. Lee, and S.-K. Choi, *J. M. S.-Pure Appl. Chem.* **30**, 531 (1993).

133. S. Inoue, T. Aida, Y. Watanabe, and K.-I. Kawaguchi, *Makromol. Chem., Macromol. Symp.* **42**, 365 (1991).

134. Y. Watanabe, T. Aida, and S. Inoue, *Macromolecules* **23**, 2612 (1990).

135. S. Murouchi, Y. Hasebe, and T. Tsuruta, *Makromol. Chem., Rapid Commun.* **11**, 129 (1990).

136. Ga-E. Yu, A. J. Masters, F. Heatley, C. Booth, and T. G. Blease, *Macromol. Chem. Phys.* **195**, 1517 (1994).

137. Eur. Pat. Appl. 18,609 (Nov. 12, 1980), H. Thurow, (to Hoechst A-G).

138. M. P. Calcagno, F. Lopez, J. M. Contreras, M. Ramirez, and F. M. Rabagliati, *Eur. Polym. J.* **27**(8), 751 (1991).

139. Z. Oktem, A. Sari, K. Alyuruk, *Eur. Polym. J.* **29**(5), 637 (1993).

140. Eur. Pat. Appl. 418,533 (Mar. 27, 1991), A. Penati, E. Moretti, C. Maffezzoni, G. Agopian, and B. Mazdrakov (to Pressindustria SpA, Verila State Enterprises, Italy).

141. Ger. Offen. 3,016,113 (Oct. 29, 1981), H. Hetzel, P. Gupta, R. Nast, H. Echterhof, and U. Brocker (to Bayer A-G).

142. Ger. Offen. 3,907,911 (Sept. 13, 1990), P. Gupta, H. J. Sandhagen, and H. J. Rosenbaum (to Bayer A-G).

143. Eur. Pat. Appl. 383,333 (Aug. 22, 1990), T. Watabe, H. Takeyasu, T. Doi, and N. Kunii (to Asahi Glass Co., Ltd.).

144. Eur. Pat. Appl. 414070 (Feb. 27, 1991), P. Gupta, H. J. Sanndhagen, and H. J. Rosenbaum (to Bayer A-G).

145. U.S. Pat. 4,521,548 (June 4, 1985), J. D. Christen, H. B. Taylor, III (to Dow Chemical Co.).

146. Ger. Offen. 3,229,216 (Feb. 9, 1984), W. Straechle, R. Denni, and M. Marx (BASF A-G).

147. U.S. Pat. 4,306,943 (Dec. 22, 1981) (to Daiichi Kogyo Seiyaku Co., Ltd.).

148. U.S. Pat. 4,985,551 (Jan. 15, 1991), J. G. Perry and W. A. Spelyng (to BASF Corp.).

149. U.S. Pat. 4,994,627 (Feb. 19, 1991), M. Cuscurida and A. J. Faske (to Texaco Chemical Co.).

150. U.S. Pat. 5,182,025 (Jan. 26, 1993), R. G. Duranleau, M. J. Plishka, and M. Cuscurida (to Texaco Chemical Co.).

151. C. Vervloet and D. E. Knibbe, *Cellular Polym.* **1**, 15 (1982).

152. P. Szewczyk, *J. Appl. Polym. Sci.* **31**, 1151 (1986).

153. G. Allen, C. Booth, and M. N. Jones, *Polymer* **5**, 195 (1964).

154. J. Loccufier, M. Van Bos, and E. Schacht, *Polym. Bull.* **27**, 201 (1991).

155. F. Heatley, Y.-Z. Luo, J.-F. Ding, R. H. Mobbs, and C. Booth, *Macromolecules* **21**, 2713 (1988).

156. W. Gronski, *Makromol. Chem.* **192**, 591 (1991).

157. F. Heatley, J. Ding, G. Yu, and C. Booth, *Makromol. Chem., Rapid Commun.* **14**, 819 (1993).

158. A. Le Borgne, M. Moreau, and V. Vincens, *Macromol. Chem. Phys.* **195**, 375 (1994).

159. J. P. Manning, C. B. Frech, B. M. Fung, and R. E. Frech, *Polymer* **32**, 2939 (1991).

160. F. M. Rabagliati and F. López, *Makromol. Chem., Rapid Commun.* **6**, 141 (1985).

161. C. L. LeBas and P. A. Turley, *J. Cell. Plast.*, 194 (1984).

162. M. D. Bruch, F. A. Bovey, and R. E. Cais, *Macromolecules* **18**, 1253 (1985).

163. C. Campbell, F. Heatley, G. Holcroft, and C. Booth, *Eur. Polym. J.* **25**, 831 (1989).

164. R. H. Carr, J. Hernalsteen, and J. Devos, *J. Appl Polym. Sci.* **52**, 1015 (1994).

165. F. F. Ho, *Polym. Lett.* **9**, 491 (1971).

166. F. C. Schilling and A. E. Tonelli, *Macromolecules* **19**, 1337 (1986).

167. N. Binboga, D. Kisakürek, and B. M. Baysal, *J. Polym. Sci. Polym. Phys.* **23**, 925 (1985).

168. W. Wixwat, Y. Fu, and J. R. Stevens, *Polymer* **32**, 1181 (1991).
169. J. Loertscher and F. Weesner, *Determination of Hydroxyl Number in Polyols by Mid-Infrared Spectroscopy*, Nicolet FT-IR Application Note (AN-9146), Nicolet Instrument Corp.
170. Y. Luo, and co-workers, *Eur. Polym. J.* **24**, 607 (1988).
171. C. W. Amoss, R. W. Slack, and L. R. Taylor, *J. Liq. Chromat.* **10**, 583 (1987).
172. J. Van Dam, P. Daenens, and R. Busson, *J. Appl. Polym. Sci.* **50**, 2115 (1993).
173. B. Trathnigg, *J. Liq. Chromat.* **13**, 1731 (1990).
174. A. V. Gorshkov and co-workers, *J. Chromatogr.* **523**, 91 (1990).
175. K. Rissler, H.-P. Künzi, and H.-J. Grether, *J. Chromatogr.* **635**, 89 (1993).
176. M. Furukawa, N. Yoshitake, and T. Yokoyama, *J. Chromatogr.* **435**, 219 (1988).
177. R. Mieczkowski, *Eur. Polym. J.* **27**, 377 (1991).
178. R. L. Tabor, K. J. Hinze, R. D. J. Priester, and R. B. Turner, "The Compatibility of Water with Polyols," *34th Annual Polyurethane Technical/Marketing Conference*, New Orleans, La., Oct. 21–24, 1992.
179. R. P. Lattimer and G. E. Hansen, *Macromolecules* **14**, 776 (1981).
180. C. F. Ijames and C. L. Wilkins, *J. Am. Chem. Soc.* **110**, 2687 (1988).
181. L. S. Sandell and D. A. I. Goring, *Macromolecules* **3**, 50 (1970).
182. S. J. Mumby, B. A. Smith, E. T. Samulski, Li-P. Yu, and M. A. Winnik, *Polymer* **27**, 1826 (1986).

STEVEN D. GAGNON
BASF Corporation

# TETRAHYDROFURAN AND OXETANE POLYMERS

The polymerizations of tetrahydrofuran [*1693-74-9*] (THF) and of oxetane [*503-30-0*] (OX) are classic examples of cationic ring-opening polymerizations. Under ideal conditions, the polymerization of the five-membered tetrahydrofuran ring is a reversible equilibrium polymerization, whereas the polymerization of the strained four-membered oxetane ring is irreversible (1,2).

The polymerization of tetrahydrofuran was first studied in the late 1930s (3,4). In 1960, this work was summarized (4), and the literature on tetrahydrofuran polymers and polymerization has been growing ever since. Polytetrahydrofuran with hydroxy end groups has become a large-scale commercial product, used mainly as the flexible polyether segment in elastomeric polyurethanes and polyesters. It is commercially available under the trade names Terathane (Du Pont), Polymeg (QO Chemicals), and PolyTHF (BASF). Comprehensive review articles and monographs have been published (2,5–8).

Tetrahydrofuran [*1693-74-9*] (tetramethylene oxide, 1,4-epoxybutane, oxolane) is a cyclic ether containing four methylene groups; the end groups are connected by an oxygen bridge. It has some internal strain resulting from repulsion of the eclipsed hydrogen atoms, and this small strain is sufficient to cause ring-opening polymerization under appropriate conditions.

$$\langle\text{\_}\text{O} \rangle \longrightarrow \text{---}(CH_2CH_2CH_2CH_2O)_{\overline{n}}$$

The resulting linear polymer is called polytetrahydrofuran (PTHF), poly(tetra-methylene oxide) (PTMO), poly(oxytetramethylene), or poly(tetramethylene ether) (PTME). It is filed under two CAS Registry Numbers, as poly(oxy-1,4-butanediyl) [*26913-43-9*] for the repeat-unit-based formula $(C_4H_8O)_n$ and under the monomer-based formula as tetrahydrofuran homopolymer [*24979-97-3*]. The most important difunctional derivative is the $\alpha,\omega$-dihydroxy-terminated polymer. It is called poly(tetramethylene ether) glycol (PTMEG, PTMG, PTG), poly(tetramethylene oxide) glycol, poly(oxytetramethylene) glycol (or diol), poly(1,4-oxybutylene) diol, or $\alpha,\omega$-dihydroxy poly(1,4-butanediyl), and is listed in *Chemical Abstracts* (CA) as $\alpha$-hydro-$\omega$-hydroxy poly(oxy-1,4-butanediyl [*25190-06-1*]. Poly(tetramethylene oxide) with other known end groups may be filed under still other names. For example, diamino poly(tetramethylene oxide) is indexed by CA as $\alpha$-4-aminobutyl-$\omega$-4-aminobutoxy poly(oxy-1,4-butanediyl) [*27417-83-0*].

The four-membered oxetane ring (trimethylene oxide [*503-30-0*]) has much higher ring strain, and irreversible ring-opening polymerization can occur rapidly to form polyoxetane [*25722-06-9*]:

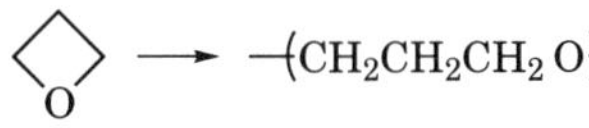

$$\langle\rangle_O \longrightarrow -\!(CH_2CH_2CH_2O)_{\overline{n}}$$

Because of the high ring strain of the four-membered ring, even substituted oxetanes polymerize readily, in contrast to substituted tetrahydrofurans, which have little tendency to undergo ring-opening homopolymerization (5).

Polymers of tetrahydrofuran are relatively low melting, crystallizable materials, characterized by a low glass-transition temperature, $T_g$. It is possible to make THF polymers of almost any molecular weight. Above the melting point (ca 55°C), the high molecular weight homopolymer is a tough, somewhat elastomeric material. However, only the low molecular weight hydroxy-terminated oligomers have commercial applications. These materials, when incorporated as flexible soft segments into a polyurethane or polyester chain, impart useful elastomeric properties.

No oxetane polymer, either substituted or unsubstituted, has as of 1996 any commercial significance.

## TETRAHYDROFURAN POLYMERS

### Physical Properties

Some typical physical properties of high molecular weight poly(tetrahydrofuran) are summarized in Table 1. The type of end groups is not specified because in high molecular weight polymers, the effect of end groups becomes negligible. Tetrahydrofuran polymers crystallize readily near ambient temperature. Moderately high molecular weight polymers turn into waxy solids, whereas high molecular weight polymers display thermoplastic behavior. Typical mechanical properties of high molecular weight crystalline polytetrahydrofuran, as well as elastomeric properties of amorphous high molecular weight polymer, are also given in Table 1. Infrared and Raman spectroscopic studies have aided determination of the structure of polytetrahydrofuran (23–26), and nmr spectroscopy

**Table 1. Typical Properties of High Molecular Weight Polytetrahydrofuran**

| Property | Value | Reference |
|---|---|---|
| melting temperature, $T_m$, °C | 43,58–60[a] | 4,9,10 |
| glass-transition temperature, $T_g$, °C | −84 | 4,9–12 |
| density at 25°C, g/cm³ | | |
|   amorphous | 0.975 | 9 |
|   crystalline | 1.07–1.08 | 13 |
| 300% modulus, MPa[b,c] | 1.6–14.3[d] | 14 |
| tensile strength, MPa[b,e] | 29.0 | 15 |
| elongation, %[e] | 820 | 15 |
| modulus of elasticity, MPa[b] | 97.0 | 15 |
| Shore A hardness | 95 | 16–18 |
| thermal expansion coefficient[f], K$^{-1}$ | $(4–7) \times 10^{-4}$ | 19,20 |
| compressibility[g], kPa$^{-1}$[b] | $(4–10) \times 10^{-7}$ | 19,20 |
| internal pressure, $P_i$, MPa[b] | 281 | 19,20 |
| heat capacity, $C_p$ at $T_g$, J/(mol·K)[h] | | |
|   rapidly cooled | 19.4 | 21 |
|   annealed | 15.8 | 21 |
| coefficient of expansion[i], cm³/(g·K) | $7.3 \times 10^{-4}$ | 9 |
| refractive index at 20°C | 1.48 | 16–18 |
| dielectric constant at 25°C | 5.0 | 16–18 |
| solubility parameter, $\delta^p$, (J/cm³)$^{1/2}$[j] | 17.3–17.6 | 22 |
| unit cell (oriented) | | |
|   monoclinic | $C_{2/c}–C_{2n}^6$ | 23,24 |
| | $C_{2/c}$ | 25 |
|   $a$, nm | 0.548–0.561 | 23–25 |
|   $b$, nm | 0.873–0.892 | 23–25 |
|   $c$, nm | 1.297–1.225 | 23–25 |
|   $\beta$, degrees | 134.2–134.5 | 23–25 |

[a]The melting temperature most often reported for PTHF (see also Tables 2 and 3 for oligomeric glycols) is about 43°C, although these higher values have been observed after annealing.
[b]To convert MPa to psi, multiply by 145.
[c]Low to high molecular weight.
[d]Varies with the molecular weight.
[e]High molecular weight.
[f]$\alpha = (1/V)(\delta V/\delta T)_P$.
[g]$\beta = (1/V)(\delta V/\delta T)_T$.
[h]To convert J to cal, divide by 4.184.
[i]$(dV/dT)$.
[j]To convert (J/cm³)$^{1/2}$ to (cal/cm³)$^{1/2}$, divide by 2.045.

has been used in studies of the kinetics of polymerization and in end-group analysis (27–31). The principal absorptions and chemical shifts are well known, and many of them have been summarized (6).

The only THF polymers of commercial importance in the 1990s are diprimary low molecular weight poly(tetramethylene ether) glycols or their derivatives. These materials generally are waxy solids when crystallized and colorless viscous fluids when melted. Number-average molecular weights of the most widely used PTME glycols are 1000 and 2000. Polymer glycols are soluble in many polar organic solvents such as alcohols, esters, ketones, and aromatic hydrocarbons. The high molecular weight fractions are slightly soluble in aliphatic hydrocarbons; the low molecular weight fractions are slightly soluble in water. The polymeric glycols normally contain ca 300–1000 ppm of butylated hydroxytoluene (BHT) to prevent peroxide formation. They should be stored under nitrogen to prevent oxidation and moisture pickup, which could adversely affect later reactions and detract from the properties of the final product. The ether oxygens of PTHF greatly increase the solubility of PTHF in polar solvents as compared to hydrocarbon polymers. Solubility varies with molecular weight, but there are solvents, such as THF, that dissolve PTHF of all molecular weights. Lower molecular weight polymers dissolve more readily in more solvents and form less viscous solutions than higher molecular weight materials. Typical solvents for nearly all molecular weights include THF, toluene, methylene chloride, ethyl acetate, and other esters. Ethyl ether, alcohols, acetone, and water are solvents for lower molecular weight THF polymers. Nonsolvents are aliphatic hydrocarbons, eg, pentane, petroleum ether, and hexane. Narrow molecular weight distribution samples of PTHF can therefore be prepared by fractional precipitation from methanol–water (32), or solvent fractionation using a hydrocarbon (such as toluene or cyclohexane), methanol, and water (33–38). PTMEG of narrow molecular weight distribution can also be prepared by direct polymerization or depolymerization.

Intrinsic viscosity is often used to characterize tetrahydrofuran polymers. Intrinsic viscosities in a variety of solvents and Mark-Houwink constants for the equation $[\eta] = KM^{\alpha}$ have been determined for a wide variety of solvents (39–45), where $[\eta]$ is the intrinsic viscosity, $M$ is molecular weight, and $K$ and $\alpha$ are constants; many of the constants have been summarized and tabulated (6).

The phase structure of PTME has been studied by $^{13}$C-nmr (46–48); the effects of chain entanglement (49,50) and the PTME chain dimensions in model polyurethanes have been measured by small-angle neutron scattering using perdeuterated PTMEG (51,52). Molecular interactions and cohesive energy densities of PTMEG have been determined (53), molecular orientation has been measured by x-ray diffraction (54), and molecular motion of PTMEG in solution and in melts has been studied by using coupled spin relaxation $^{13}$C-nmr (47,48,55). Thermal properties (56,57), dielectric relaxation (58), and pressure–volume–temperature data have been measured, as well as surface tension (59). Plastic deformation of PTMEG has been determined, the influence of molecular weight distribution on crystallinity (60) and on equilibrium melting points has been studied (61), and thermal properties of phthalate end-capped PTMEG have been measured (62). Thermal crystallization of polytetrahydrofuran chains in cross-linked networks and the influence of chain length and polydispersity on crystallization has been examined (63). Among other published reports are the segmental adsorption energies of PTMEG on silica and alumina surfaces (64) and measurements of the crystallite core size (65).

Number-average molecular weights, $M_n$, of THF polymers have been determined by standard techniques. For polymers of low molecular weight, $M_n$ has been calculated primarily from data obtained by end-group analysis. Depending on the nature of the end group, the methods used include titration (39), nmr (6,31,66,67), ir (39), uv (39,68,69), elemental analysis (38,70), liquid scintillation counting of C-labeled end groups (71), and fluorescence spectroscopy (70). Ebulliometry and osmometry have also been used (39,71,72). For polymers with $M_n > 15,000$, osmometry has been used frequently, with toluene as the most common solvent (39,45,72–74).

Weight-average molecular weights of high molecular weight THF polymers have been determined by light scattering in a variety of solvents (42–45,75). Refractive index increments, $dn/dc$, and refractive indexes, $n_D$, for typical solvents have been calculated (42–45,75–78). Molecular weight distribution of THF polymers have been determined from $M_w/M_n$ ratios derived from light scattering and osmometry (78), turbidimetric titration (78), sedimentation velocity in a theta-solvent (42,79), and gel-permeation chromatography (gpc) (80–83). THF is used as the preferred solvent for measurements of the molecular weight distribution by gpc. Usual detectors are differential refractometers or viscometers (84,85). High pressure liquid chromatography (hplc) with gradient elution separates PTMEG into individual oligomer fractions and gives information about molecular weights and molecular weight distributions without requiring calibration standards (86).

## Chemical Properties

The most important tetrahydrofuran polymers are the hydroxy-terminated polymers, that is, the $\alpha,\omega$-poly(tetramethylene ether) glycols used commercially to manufacture polyurethanes and polyesters (see URETHANE POLYMERS; POLYESTERS, THERMOPLASTIC).

**End-Group Reactions.** PTME glycols are normally the primary THF polymerization products, but THF polymers can be prepared with other end groups, either by direct polymerization of THF, or by chemical transformation of the hydroxy groups of the preformed polymer. For example, polymerization of THF in presence of carboxylic acid anhydrides results in PTME chains with the corresponding ester end groups, such as acetate, acrylate, or methacrylate groups (87–89). Such diesters can also be prepared by conventional esterification of the polymeric diols (90–93). Diols are also the starting materials for a number of other reported PTME derivatives, such as dialkylsulfonates (94), dithiols via disulfonates (95) or dihalides (96), alkyl ethers via alkylsulfates (97), and unsaturated alkenyl ethers such as diallyl ethers via alkenyl chlorides or sulfones (98,99). Unsaturated ether end groups have also been introduced by catalytic vinylation of PTMEG with acetylene (100). Such unsaturated polyether macromers are claimed as comonomers for olefin polymerizations (99). Among other PTME derivatives are diglycidyl ethers (101), dialdehydes (102), and diacids (103,104). Reaction of PTME diols with difunctional reagents results in coupling and formation of longer PTME chains with formaldehyde (105), short-chain diols (106), or carbonate links (107).

A number of different routes have been described for the preparation and manufacture of amine-terminated tetrahydrofuran polymers. The simplest route to PTMEG diamines appears to be esterification with $p$-aminobenzoic acid. Such a product is commercially available under the tradename Versalink (Air Products Corporation). Amines have also been prepared by reductive amination of the corresponding polymer glycols (108,109) or diacetates (110). PTME diamines can be prepared by the addition of acrylonitrile to the hydroxy groups, followed by hydrogenation. The resulting diamines are bis(1,3-aminopropyl)poly(tetramethylene oxides) [72088-96-1]. Amine-terminated PTMEs have also been prepared from PTMEG disulfonates (94) and alkylene diamines (111). The resulting polymer is a mixture of primary, secondary, and tertiary amines. PTME diamines are claimed by reacting PTMEG with diisocyanates, such as methylene-bis(phenylene isocyanate) (MDI) or toluene diisocyanate (TDI), followed by hydrolysis (112). Finally, PTME diamines can also be prepared directly by polymerizing THF with triflic anhydride, $(CF_3SO_2)_2O$, and quenching the resulting dioxonium ions with hexamethylenetetramine. Exclusive formation of primary amines is claimed (113), but quenching even with an excess of ammonia results in a mixture of primary and secondary amines (114–116). A similar method has been used to put a variety of different end groups on PTME (116), for instance dimethylamino groups (117) or amino phenoxy groups (118). PTME with aminophenoxy end groups has also been prepared by hydrogenation of the corresponding nitro compounds (119). The basicity of different aromatic PTME diamines has been studied (120). In most cases these reactions do not result in complete amination of the PTME chain. Nevertheless, PTME diamines have been chain-extended with diisocyanates to give polyureas in fast reactions, or with common epoxy resins to give unusual elastomeric composites (121).

Commercially, the most important reaction of PTME glycols is the reaction with diisocyanates. Reaction with an excess of diisocyanate yields a prepolymer having isocyanate end groups, which can further react with short-chain diols or amines to give high molecular weight polyurethanes or polyurethane ureas (122,123). Some of these prepolymers are also available commercially. Typical examples are Adiprene and Vibrathane from Uniroyal, Airthane from Air Products Corporation, Conathane from Conap, and Hyprene from Mitsui Toatsu. These prepolymers can be chain-extended (cured) to high molecular weight polyurethanes with aliphatic diols or to polyureas with diamines, or they can be cross-linked by reaction with moisture (124). High molecular weight, fully cured PTMEG polyurethanes are also commercially available as thermoplastic polyurethanes (TPUs). Examples are certain grades of Pellethane (Dow Chemical Corporation), Estane (BF Goodrich), and Texin (Bayer/Miles Corporation). Thermoplastic polyesters based on PTMEG are other types of important commercial elastomer. They are generally based on PTMEG soft segments and terephthalate hard segments with aliphatic short-chain diols, such as ethylene glycol, 1,4-butandiol, or cycloaliphatic diols (123–139). Examples of such polyester elastomers are Hytrel (Du Pont), Pelprene (Toyobo), Ecdel (Eastman Chemicals), Lomod (General Electric), and Arnitel (DSM). PTME is also the soft segment in an elastomeric polyamide, which contains nylon-type hard segments (Pebax, by Elf Atochem). The chemical and physical properties of thermoplastic elastomers have been reviewed (140).

**Main-Chain Reactions.**    The backbone of PTME (PTHF) consists of a series of linear aliphatic ether sequences. Like monomeric ethers, it is subject to oxidation to hydroperoxides and subsequent thermal degradation. The addition of common antioxidants, such as amines or hindered phenols, inhibits these reactions and thereby imparts adequate stability to the polymer (141,142). In vacuum in the absence of acidic impurities, temperatures of over 200°C are needed to degrade the polymer completely. THF polymers are also subject to degradation by ionizing radiation. They undergo both chain scission and cross-linking with evolution of hydrogen and traces of other gases (143). The ratio of main-chain scission to cross-linking was found to be 0.37. PTHF is quite stable to attack by bases but can be degraded by strong acids. Because the synthesis of

PTHF often involves the use of strong acid initiators, it is important to remove these before storage.

The PTHF chain is also subject to attack and reaction during the polymerization process. These are reactions of the active end group of the growing polymer chain in the polymerizing mixture and an oxygen atom in the main chain. The nucleophilicity of the oxygen atom decreases when it is transformed from a monomeric THF oxygen to a polymeric ether oxygen, but detectable reactivity remains. These polymeric oxygen atoms can react with the THF oxonium ion to produce a branched ion. Further reaction with THF can occur at any of

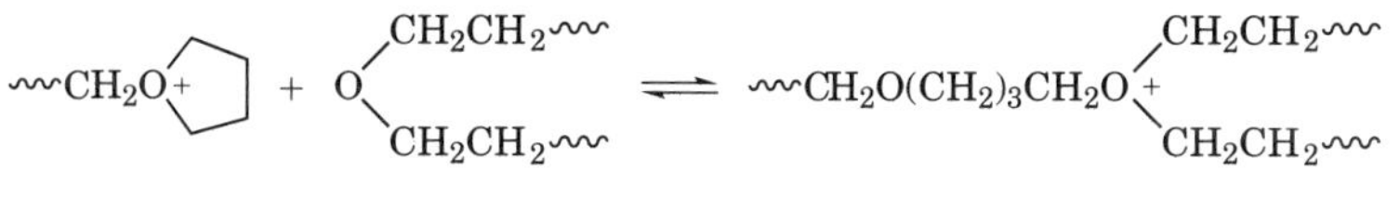

the three carbon atoms $\alpha$ to the acyclic polymeric oxonium ion. The result is a randomization of the molecular weight distribution of PTHF. This process ensures a polydispersity $M_w/M_n$ of about 2.0 for high conversion polymerizations. Under certain conditions, lower polydispersity may be achieved by limiting polymerization to low conversions or, especially for low molecular weight (500–3000) materials, by post-treatments designed to remove short-chain fractions (86). Depropagation with expulsion of monomer occurs on intramolecular reaction of the penultimate oxygen atom with the acyclic carbon atom $\alpha$ to the oxonium ion. If the oxygen atom that reacts is a few monomer units removed from the active end, intramolecular reaction with polymer oxygen atom results in a macrocyclic oxonium ion (144). Further reaction with THF may lead to the formation of macrocycles. The proportions of cyclic oligomers depend on the initiator and the polymerization conditions. The concentration of the macrocycles generally is less than 1% of the total product formed at 25°C.

**Polymerization.**   The THF ring contains an oxygen atom with two unshared pairs of electrons. THF is therefore a nucleophilic monomer having little steric interference toward potential electron acceptors. The free energy of polymerization is only about $-4.2$ kJ/mol ($-1$ kcal/mol) at ambient temperatures. Consequently, a polymerization characterized by modest rates and heats of polymerization is not surprising. Simply substituted tetrahydrofurans, in general, do not polymerize, although oligomers have been reported for monomethyl derivatives (6). Copolymerization of these substituted THFs is possible and has been realized. Purity of reagents, dryness of apparatus, choice of initiator, atmosphere, monomer concentration, and solvent greatly influence the course of the polymerization.

Cationic ring-opening polymerization is the only polymerization mechanism available to tetrahydrofuran (5,6,8). The propagating species is a tertiary oxonium ion associated with a negatively charged counterion:

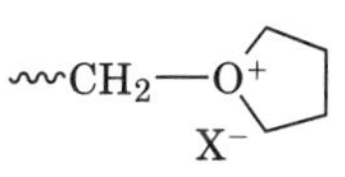

For continuing polymerization to occur, the ion pair must display reasonable stability. Strongly nucleophilic anions, such as $Cl^-$, are not suitable, because the ion pair is unstable with respect to THF and the alkyl halide. A counterion of relatively low nucleophilicity is required to achieve a controlled and continuing polymerization. Examples of anions of suitably low nucleophilicity are complex ions such as $SbF_6^-$, $AsF_6^-$, $PF_6^-$, $SbCl_6^-$, $BF_4^-$, or other anions that can reversibly collapse to a covalent ester species: $CF_3SO_3^-$, $FSO_3^-$, and $ClO_4^-$. In order to achieve reproducible and predictable results in the cationic polymerization of THF, it is necessary to use pure, dry reagents and dry conditions. High vacuum techniques are required for theoretical studies. Careful work in an inert atmosphere, such as dry nitrogen, is satisfactory for many purposes, including commercial synthesis.

The polymerization of THF is an equilibrium polymerization. It fits the equation that relates the enthalpy of polymerization, $\Delta H_p$, and entropy of polymerization at 1 $M$, $\Delta S_p^0$, to the equilibrium monomer concentration, $[M]_e$, as a function of the absolute temperature, $T$, where $R$ is the gas constant (6,145).

$$\ln[M]_e = \frac{\Delta H_p}{RT} - \frac{\Delta S_p^0}{R} \tag{1}$$

Thus, to obtain significant conversions to polymer, it is essential both to choose the polymerization temperature carefully and to limit the amount of solvent. The polymerization of bulk THF has a relatively low ceiling temperature, $T_c$, of $83 \pm 2°C$. At $T_c$ or above, no polymer is formed. At temperatures below $T_c$, there is an equilibrium monomer concentration below which no further polymerization occurs. As indicated by equation 1, the conversion to PTHF at any given temperature depends on the monomer concentration and is independent of polymer concentration. Thus the use of a solvent (inert diluent) reduces the conversion to polymer at that temperature. When a solvent is used, the $M_e$ characteristic of a given temperature and the monomer−solvent mixture can be achieved only at the expense of polymer conversion (146–148). The $T_c$ of a system diluted by solvent is substantially lower than that observed for bulk monomer and drops in direct proportion to the amount of solvent employed. For example, in THF polymerization carried to equilibrium conversion at 30°C, polymerization in bulk monomer gives 72% polymer, whereas polymerization in a solution containing $THF/CH_2Cl_2$ in a 5/3 ratio by volume gives only 27% polymer. Largely because of these considerations, THF polymerizations are generally carried out near or below room temperature, especially if an inert diluent is used. Below ca −20°C, the rate of polymerization of THF becomes prohibitively slow unless the initiator concentration is raised substantially. The enthalpy and entropy data on THF polymerization have been obtained from data relating $M_e$ with temperature. Values reported for the enthalpy of polymerization fall in the range of −18.0 to −23.4 kJ/mol (−4.3 to −5.6 kcal/mol); for the entropy of polymerization, −71 to −87 J/(mol·K) (−17 to −21 cal/(mol·K)) (13). The values are somewhat dependent on the counterion and solvent. The reasons for these differences are related to structural factors and to solvation (149).

It is possible to balance all of these thermodynamic, kinetic, and mechanistic considerations and to prepare well-defined PTHF. Living oxonium ion polymerizations, ie, polymerizations that are free from transfer and termination reactions, are possible. PTHF of any desired molecular weight and with controlled end groups can be prepared.

*Initiation.* The basic requirement for polymerization is that a THF tertiary oxonium ion must be formed by some mechanism. If a suitable counterion is present, polymerization follows. The requisite tertiary oxonium ion can be formed in any of several ways.

Direct alkylation or acylation of the oxygen of THF by exchange or addition occurs with the use of trialkyloxonium salts, carboxonium salts, super-acid esters or anhydrides, acylium salts, and sometimes carbenium salts.

*Exchange*

$$(C_2H_5)_3O^+X^- + \text{(THF)} \longrightarrow C_2H_5{-}O^+\text{(ring)} \ X^- + (C_2H_5)_2O$$

$$R{-}\text{(cyclic)}^+ \ X^- + \text{(THF)} \longrightarrow \overset{O}{\overset{\|}{R}C}OCH_2CH_2{-}O^+\text{(ring)} \ X^-$$

$$CF_3SO_2OR + \text{(THF)} \longrightarrow RO^+\text{(ring)} \ CF_3SO_2O^-$$

*Addition*

$$\overset{O}{\overset{\|}{R}C}{}^+X^- + \text{(THF)} \longrightarrow \overset{O}{\overset{\|}{R}C}{-}O^+\text{(ring)} \ X^-$$

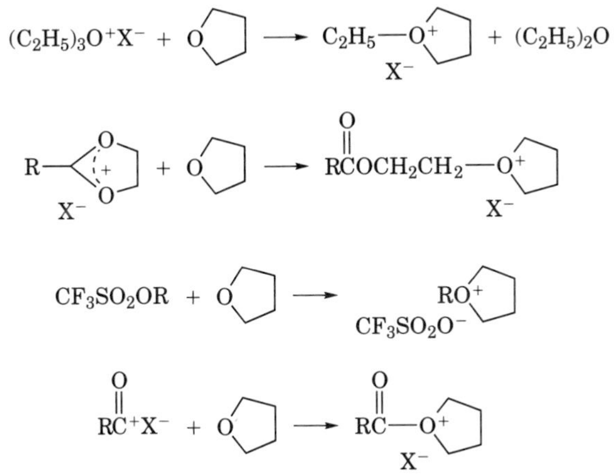

The first three equations illustrate exchange reactions in which an alkyl group is exchanged from the initiator to the THF molecule and the last equation illustrates the addition mechanism.

Often the requisite THF oxonium ion is generated *in situ* by using a combination of reagents based on the Meerwein syntheses of trialkyl oxonium salts (150). These combinations include epichlorohydrin or a reactive halide with a Lewis acid, a reactive halide with a metal salt, or sometimes just a Lewis acid alone. The epoxide portion is often referred to as a promoter.

A protonic acid derived from a suitable or desired anion would seem to be an ideal initiator, especially if the desired end product is a poly(tetramethylene oxide) glycol. There are, however, a number of drawbacks. The protonated THF,

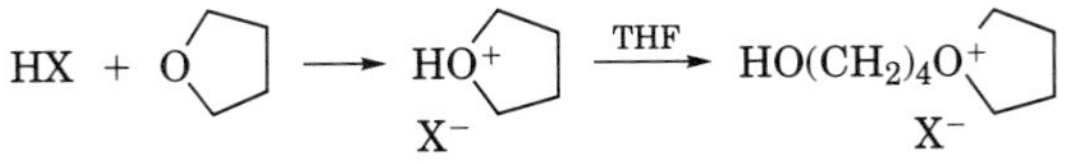

$$HX + \text{(THF)} \longrightarrow HO^+\text{(ring)} \ X^- \xrightarrow{\text{THF}} HO(CH_2)_4O^+\text{(ring)} \ X^-$$

ie, the secondary oxonium ion, is less reactive than the propagating tertiary oxonium ion. This results in a slow initiation process. Also, in the case of several of the readily available acids, eg, $CF_3SO_3H$, $FSO_3H$, $HClO_4$, and $H_2SO_4$, there is an ion–ester equilibrium with the counterion, which further reduces the concentration of the much more reactive ionic species. The reaction is illustrated for $CF_3SO_3^-$ counterion as follows:

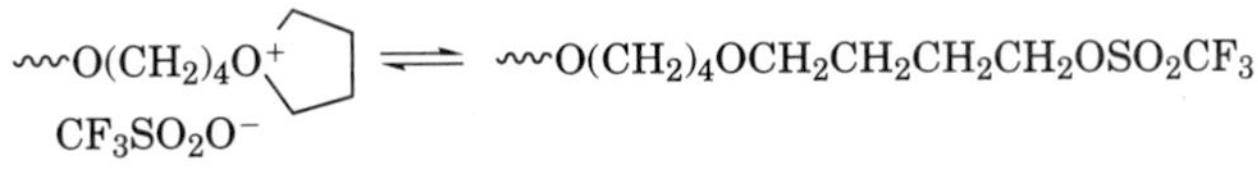

$$\text{\textsf{www}}O(CH_2)_4O^+\text{(ring)} \ CF_3SO_2O^- \rightleftharpoons \text{\textsf{www}}O(CH_2)_4OCH_2CH_2CH_2CH_2OSO_2CF_3$$

The reverse reaction (ion formation) can occur in two ways: internally, by attack of the penultimate polymer oxygen atom, or externally, by attack of a monomer oxygen atom (chain growth). The external process is about 10 times slower than the internal process in bulk THF (1). Since ion formation is a slow process compared to ion chain growth, chain growth by external attack of monomer on covalent ester makes a negligible contribution to the polymerization process.

The hydroxyl end group is a reactive functional group, the interaction of which with the propagating tertiary oxonium ion can lead to chain coupling and a rapid increase of molecular weight. In these cases chain length and molecular weight are not simple functions of initiator concentration. Covalent coupling is more important with nonhydrolyzable anions, such as $CF_3SO_3^-$, than with hydrolyzable anions, such as $FSO_3^-$ or $ClSO_3^-$. Long reaction time with the latter anions results in equilibration closer to the predicted molecular weight (1,6).

Protonic initiation is also the end result of a large number of other initiating systems. Strong acids are generated *in situ* by a variety of different chemistries (6). These include initiation by carbenium ions, eg, trityl or diazonium salts (151); by an electric current in the presence of a quartenary ammonium salt (152); by halonium, triaryl sulfonium, and triaryl selenonium salts with uv irradiation (153–155); by mercuric perchlorate, nitrosyl hexafluorophosphate, or nitryl hexafluorophosphate (156); and by interaction of free radicals with certain metal salts (157). Reports of "new" initiating systems are often the result of such secondary reactions. Other reports suggest standard polymerization processes with perhaps novel anions. These latter include $(Tf)_4Al^-$ (158); heteropoly acids, eg, tungstophosphate anion (159,160); transition-metal-based systems, eg, Pt (161) or rare earths (162); and numerous systems based on triflic acid (158,163–166). Coordination polymerization of THF may be in a different class (167).

*Propagation.* The tertiary THF oxonium ion undergoes propagation by an $S_N2$ mechanism as a result of a bimolecular collision with THF monomer. Only collisions at the ring $\alpha$-carbon atoms of the oxonium ion result in chain growth. Depropagation results from an intramolecular nucleophilic attack of the penultimate chain oxygen atom at the exocyclic $\alpha$-carbon atom of the oxonium ion, followed by expulsion of a monomer molecule.

Studies have shown that, in marked contrast to carbanionic polymerization, the reactivity of the free oxonium ion is of the same order of magnitude as that of its ion pair with the counterion (6). On the other hand, in the case of those counterions that can undergo an equilibrium with the corresponding covalent ester species, the reactivity of the ionic species is so much greater than that of the ester that chain growth by external attack of monomer on covalent ester makes a negligible contribution to the polymerization process. The relative concentration of the two species depends on the dielectric constant of the polymerization medium, ie, on the choice of solvent.

*Termination.* THF can be polymerized in the virtual absence of termination and transfer reactions. Under these conditions a living polymerization results and the number-average molecular weight of the polymer produced can be calculated from the number of active sites introduced and the amount of polymer produced at equilibrium. For many initiators, the number of active sites corresponds directly to the number of moles of initiator used. In other cases, a number of methods are available that allows analytical determination of the number of active sites (6). In order to eliminate all termination and transfer reactions, it is necessary to carry out the polymerization while carefully avoiding any adventitious impurities such as air or water. This is generally most easily accomplished by working under high vacuum. In living polymerizations, the head group of the polymer is generally determined by the initiator, whereas the end group is a function of the method of termination chosen. Any strong nucleophile can be used for chain termination. For example, water leads to hydroxyl end groups; ammonia gives amine end groups. Some counterions, such as the halides, are strong enough nucleophiles to prevent polymerization. They can be used to terminate a polymerization by addition at the desired time. Some counterions, such as $BF_4^-$ or $SbCl_6^-$, are weak nucleophiles and their use results in slow chain termination during the course of the polymerization, especially if used at room temperature or above. In the case of $SbCl_6^-$, the $SbCl_5$ that forms upon termination can itself initiate a new active center. The net result is one of chain transfer. Termination by some nucleophiles allows the determination of the number of active sites that were present at the time of termination. Thus, the use of sodium phenoxide is followed by a uv analysis of phenoxide end groups (69), and triphenylphosphine permits analysis with $^{31}$P-nmr (31,66). Mercaptans or sulfides are also effective terminating agents. If a polymerizable cyclic sulfide is employed, a block polymer of PTHF and the cyclic sulfide results (168). No further polymerization of the THF occurs. The cyclic ether polymerization is effectively terminated because sulfur is a much stronger nucleophile than oxygen.

*Chain Transfer.* A number of materials act as true transfer agents in THF polymerization; notable examples are dialkyl ethers and orthoformates. In low concentrations, water behaves as a transfer agent, and hydroxyl end groups are produced. The oxygen of dialkyl ethers are rather poor nucleophiles compared to THF and are therefore not very effective as transfer agents. On the other hand, orthoformates are effective transfer agents and can be used to produce alkoxy-ended PTHFs of any desired molecular weight (169).

$$\text{CH}_2\text{O}^+ \text{(ring)} + \text{HC(OCH}_3)_3 \longrightarrow \text{CH}_2\text{O(CH}_2)_4\text{OCH}_3 + \text{HC} \begin{matrix} \text{OCH}_3 \\ \\ \text{OCH}_3 \end{matrix} + \xrightarrow{\text{THF}}$$

$$\underset{\text{O}}{\overset{\text{O}}{\text{HCOCH}_3}} + \text{CH}_3\text{O}^+ \text{(ring)}$$

Acetic and other acid anhydrides were at first thought to behave as transfer agents. Acetate end groups and low molecular weight polymers can be obtained, but proton initiation and rather long reaction times are required, and suggest a more complex process (1,170,171). Transfer to polymer ether oxygen does occur. Depropagation is an example. Back-biting and formation of cyclic oligomers exemplify intramolecular transfer to polymer ether oxygen. The analogous intermolecular reaction can also occur.

*Kinetics.* Details of the kinetics of polymerization of THF have been reviewed (6,148). There are five main conclusions. (*1*) Macroions are the principal propagating species in all systems. (*2*) With stable complex anions, such as $PF_6^-$, $SbF_6^-$, and $AsF_6^-$, the polymerization is living under normal polymerization conditions. When initiation is fast, kinetics of polymerizations in bulk can be closely approximated by equation 2, where $k_p$ is the specific rate constant of propagation; $t$ is time; $[I]_0$ is the initiator concentration at $t = 0$; and $[M]_0$, $[M]_e$, and $[M]_t$ are the monomer concentrations at $t = 0$, at equilibrium, and at time $t$, respectively.

$$k_p t = \left[ \frac{1}{[I]_0} \right] \left[ \ln \frac{[M]_0 - [M]_e}{[M]_t - [M]_e} \right] \qquad (2)$$

The rate of ion propagation, $k_{ip}$, is independent of the counterion and has been found to be about $46 \times 10^{-2}$ in all cases for $CF_3SO_3^-$, $AsF_6^-$, $SbF_6^-$, $SbCl_6^-$, $PF_6^-$, and $BF_4^-$ counterions. Conditions were the same for all counterions, ie, 8.0 $M$ of monomer in $CCl_4$ solvent and 25°C polymerization temperature. With less stable counterions such as $SbCl_6^-$ and $BF_4^-$ at most temperatures, the influence of transfer and termination reactions must be taken into account (71).

(*3*) PTHF does not behave ideally in solution and the equilibrium monomer concentration varies with both solvent and temperature. Kinetics of THF polymerizations fit equation 2, provided that the equilibrium monomer concentration is determined for the conditions used.

(*4*) For counterions that can form esters with the growing oxonium ions, the kinetics of propagation are dominated by the rate of propagation of the macroions. For any given counterion, the proportion of macroions compared to macroesters varies with the solvent–monomer mixture and must be determined independently before a kinetic analysis can be made. The macroesters can be considered to be in a state of temporary termination. When the proportion of macroions is known and initiation is sufficiently fast, equation 2 is satisfied.

(*5*) When the initiation is slow, the number of growing centers as a function of time must be determined in a separate step before the kinetic analysis can be carried out. Several different methods are available (6,31,66,69–71).

**Copolymerization.** THF copolymerizations are of interest for several reasons. Random copolymerization provides a way of reducing the melting temperature of THF polymers to room temperature and below. The crystallization tendency of THF segments in products is thereby reduced and mechanical properties of the products are often improved. Copolymers also provide a way of introducing unsaturated units for vulcanization purposes (172) or active groups such as $-CH_2N_2$, $-NO_2$, and $-CH_2OCH_2C(NO_2)_2CH_3$ units for energetic polymers (173).

Many copolymerization studies have been made. A detailed discussion and critique of the results has been published (1) and the breadth of the comonomers

studied has been summarized (6). Among the comonomers used are oxiranes, oxetanes, 1,3-dioxolane, substituted tetrahydrofurans, 1,4-dioxane, $\epsilon$-caprolactone,
and cyclopentadiene. Random copolymerization with cyclic sulfides and most
vinyl compounds does not occur. Some monomers that do not homopolymerize do
copolymerize. In addition to chemical functionality, relative basicity, ring strain,
and ring size all seem to be important factors in determining the reactivity of different monomers in copolymerizations. Many attempts have been made to apply
the usual copolymer equation (174) and associated methods of analyses (175,176)
to determine relative reactivity ratios for the comonomers. Numbers are readily
obtained, but often they are meaningless because some of the assumptions on
which the derivation of the equation is based are not obeyed, especially at low
concentrations of THF and polymerization temperatures near room temperature.
In particular, THF polymerizations have a significant rate of reversibility, which
markedly affects the composition of the polymer obtained and sometimes makes
THF appear less reactive than it is. Also, THF oxonium ion copolymerizations
are sometimes complicated by the possibility of interconversion of active species
without propagation, by redistribution of molecular weights of initially formed
copolymers, and by significant effects of the penultimate or even antipenultimate
monomer unit in copolymer chains. Other theoretical treatments have been published (177–179) to account for some of these factors, especially reversibility, but
the solutions are more complex, and more constants need to be determined or
assigned than for the original copolymer equation.

*Block, Graft, and Star Copolymers.* A host of copolymers of these types
have been prepared. They include block copolymers from $\epsilon$-caprolactam and
PTMEG as well as block copolymers from PTHF and other cationically polymerizable heterocycles, including 3,3-bis(chloromethyl)oxetane (180), 7-oxabi-
cyclo[2.2.1]heptane, 1,3-dioxolane, pivalolactone, and ethylene oxide. Block
copolymers from polystyrene and PTHF have been prepared, including AB,
ABA, and $(AB)_n$ copolymers, where A and B can be either polystyrene or PTHF
(6,180–183). One-, two-, three-, and four-arm stars have been prepared with
PTHF arms (184). Graft copolymers with PTHF branches have been made from
a variety of hydrocarbon backbones (11,185–188), and graft copolymers with
poly(vinyl chloride) branches have been prepared from PTHF backbones (118).
The properties determined for some of these copolymers are sometimes unusual,
but as of 1996 none had achieved commercial importance.

In addition to the primary application of PTMEG in polyurethanes,
polyureas, and polyesters, a considerable number of reports of other block and
graft polymers highlighting PTME units have appeared. Methods have been
developed that allow the conversion of a cationically polymerizing system to an
anionic one or vice versa (6,182).

Cationic polymerization followed by carefully defined free-radical polymerization has also been designed for preparing novel block polymers. The
latter often involves the preparation of PTHF containing an azo or peroxy linkage, which is then used to initiate free-radical polymerization of standard vinyl
or acrylate monomers (189–191). The preparation of PTHF block and graft
polymers has been reviewed (192,193). Thus, THF has been grafted from cellulose derivatives (194), from silane polymers (195), from ethylene propylene
diene monomer (EPDM) rubber (185,196), from butyl rubber (185,197), and from

poly(phenylmethylsilylene) (198). Newer block systems that have been prepared include blocks with azetidine (199), with p-chlorostyrene (200), with oxazolines (201,202), with polyether sulfones (203), and with methyl methacrylate (MMA) and styrene (204). Availability of various block and graft THF polymers has also led to their use as compatibilizers in blending studies (205).

## Manufacture and Processing

THF can be polymerized by many strongly acidic catalysts, but not all of them produce the required bifunctional polyether glycol with a minimum of by-products. Several large-scale commercial polymerization processes are based on fluorosulfonic acid, $HFSO_3$, catalysis, which meets all these requirements. The catalyst is added to THF at low temperatures and an exothermic polymerization occurs readily. The polymerization products are poly(tetramethylene ether) chains with sulfate ester groups (8).

In a subsequent product work-up, the sulfates are hydrolyzed and the acid is removed by water extraction (206,207). In the extraction step, most water-soluble short polyether chains are also removed, and the molecular weight distribution becomes narrower, from close to the theoretical $M_w/M_n$ value of 2 to 1.6 or less, depending on the particular molecular weight (86,208). The lower molecular weight grades become narrower than the higher molecular weight grades because they contain a larger concentration of short, water-soluble chains which are partly removed in this step. After neutralization and vacuum drying, the product is filtered hot and loaded into drums or tank trucks under nitrogen.

Many other polymerization processes have been patented, but only some of them appear to be developed or under development in 1996. One large-scale process uses an acid montmorrillonite clay and acetic anhydride (209); another process uses strong perfluorosulfonic acid resin catalysts (170,210). The polymerization product in these processes is a poly(tetramethylene ether) with acetate end groups, which have to be removed by alkaline hydrolysis (211) or hydrogenolysis (212). If necessary, the product is then neutralized, eg, with phosphoric acid (213), and the salts removed by filtration. Instead of montmorrillonite clay, other acidic catalysts can be used, such as Fuller's earth or zeolites (214–216).

The primary polymerization product in these processes has a relatively wide molecular weight distribution, and a separate step is often used to narrow the polydispersity. Such a narrowing step may consist of high vacuum stripping to remove volatile polymer chains, often followed by a solvent fractionation step (35,36), sometimes a solvent fractionation step alone (37,38), or a fractional precipitation from organic solvent (32). The molecular weight distribution can also be narrowed by depolymerization at elevated temperatures in the presence of a depolymerization catalyst (217–220).

A more recent catalyst system introduced by Asahi Chemical Industries (221) involves the use of heteropoly acids (HPA) such as phosphomolybdic or phosphotungstic acids (222,223). These catalysts appear to work in the presence of small amounts of water (224) and lead directly to the polymeric glycols (59). In the presence of alcohols instead of water, the products are PTMEG monoalkyl ethers (225). The process is carried out by mixing two liquid phases: a THF phase and an aqueous heteropoly acid phase. Control of water concentration is critical for control of molecular weight and conversion (226). Asahi is operating a small PTMEG plant based on this technology.

A number of papers and patents describe polymerization processes to poly(tetramethylene ether) glycols having a narrow molecular weight distribution ($M_w/M_n$ = 1.2–1.4). In principle, this can be achieved by having all chains grow quickly at one time, either by high temperature initiation (33) followed by low temperature propagation, or by a temperature-cycling process, which can give product of narrow distribution (227). On a larger scale, temperature-cycling processes to narrow distribution PTMEG can be carried out with advantage in pipeline reactors. PTMEG of narrow molecular weight distribution can also be obtained by rapid initiation with an active catalyst system, such as fluorosulfonic acid/acetic anhydride or perchloric acid/acetic anhydride (228–230). If such polymerizations are quenched after short polymerization times, products with narrower than normal distributions can be obtained. Rapid initiation and chain growth can also be achieved by initiation with fluorosulfonic acid anhydride ($FSO_2)_2O$ or trifluoromethane sulfonic anhydride ($CF_3SO_2)_2O$ (231,232). In these cases, chain growth occurs at both chain ends simultaneously, and short polymerization times can lead to products having narrow molecular weight distribution.

Many patents claim PTMEG of narrow polydispersity by using special catalyst blends or polymerization conditions (233–239). Whereas some of these processes may lead to a certain narrowing of the polydispersity, others give product with marginal narrowing. Polyurethanes made from PTMEG of narrow molecular weight distribution are claimed to have some improved properties, but other properties are adversely affected (240).

Other THF polymerization processes that have been disclosed in papers and patents, but which do not appear to be in commercial use in the 1990s, include catalysis by boron trifluoride complexes in combination with other cocatalysts (241–245), modified montmorrillonite clay (246–248) or modified metal oxide composites (249), rare-earth catalysts (250), triflate salts (164), and sulfuric acid or fuming sulfuric acid with cocatalysts (237,251–255).

A number of patents claim THF copolymers by direct copolymerization of THF and other cyclic ethers (168,256–259). Although samples of THF copolyethers are available occasionally, none had any industrial importance as of 1996.

## Storage and Handling

PTMEG is normally available in 20-kg steel pails, 200-kg steel drums, and in stainless steel tank cars or tank trucks, which are insulated and equipped with heating coils. Shipping temperature for tank trucks is normally 80°C (175°F) to maintain the product in liquid state. If subjected to low temperatures, it has to be reheated and melted, eg, with low pressure steam (350 kPa ≈ 50 psig). Product transfer from tank trucks requires special precaution; the procedure has been described in detail in product bulletins (260). Drums should be stored in a warm room to prevent freezing. Frozen drums can be thawed by loosening the bung and storing for one day at 70°C (160°F). The content should be mixed thoroughly before use, because fractionation and stratification by molecular weight may occur during freezing. Centrifugal or gear pumps, and butyl rubber- or neoprene-lined hose or stainless steel flexible hose, are used for product transfer. All piping should be heat-traced (260).

PTMEG is a polymeric ether susceptible to both thermal and oxidative degradation. It usually contains 300–1000 ppm of an antioxidant such as 2,6-di-*tert*-butyl-4-hydroxytoluene (BHT) to prevent oxidation under normal storage and handling conditions. Thermal decomposition in an inert atmosphere starts at 210–220°C (410–430°F) with the formation of highly flammable THF. In the presence of acidic impurities, the decomposition temperature can be significantly reduced; contact with acids should therefore be avoided, and storage temperatures have to be controlled to prevent decomposition to THF (261).

Oxidative degradation occurs when PTMEG is heated in contact with air (262). Under these conditions, decomposition can start at temperatures as low as 100°C, along with evolution of THF and formation of aldehydes and ketones. Degradation occurs particularly readily when PTMEG is in contact with high surface area materials, eg, pipe insulation. In such cases the heat of oxidative degradation is normally sufficient to sustain the degradation reaction, and spontaneous combustion can occur. Contaminated fibrous insulation should therefore immediately be cooled and discarded. PTMEG is also hygroscopic and should be stored in completely enclosed tanks under a dry nitrogen blanket. Mild steel tanks, tanks lined with phenolic resin, or stainless steel (type 304) tanks are usually satisfactory. The storage tank must be provided with external or internal heating to maintain a temperature of about 50°C. The storage tanks should be pressure-tested and equipped with a vacuum pressure-relief conservation vent and an emergency vent. Vents should be heat-traced and insulated to prevent plugging by solidified polymer. New equipment should be cleaned and dried carefully to avoid contamination. If necessary, PTMEG can be dried, eg, by removing water by azeotropic distillation with toluene, by heating (120°C) under reduced pressure (263), or by sparging the hot polyglycol with dry nitrogen.

PTMEG is not regulated as a hazardous material by the U.S. Department of Transportation. Liquid spills may be absorbed with a material such as vermiculite and handled as nontoxic waste. Larger spills, if fluid, may be pumped into drums or, if solid, shoveled into drums for later recovery or disposal by burning under controlled conditions. PTMEG has a flash point above 150°C, and little fire hazard exists under normal conditions. When handled above 150°C, particularly when in contact with high surface area material, ignition may occur by flames, sparks, or hot surfaces. PTMEG fires can be extinguished with water spray, foam, dry chemical, or $CO_2$ extinguishers (263).

## Specifications, Standards, and Analysis

The standard commercial molecular weight grades for polytetramethylene ether glycol are 650, 1000, 1800, and 2000, but other molecular weight grades, such as 1400 and 2900, are available for special applications. Commercial poly(tetramethylene ether) glycols are waxy, white solids that melt over a temperature range near room temperature to clear, colorless, viscous liquids. The viscosities of PTMEG 650, 1000, and 2000, up to a temperature of 80°C, are shown in Figure 1, and the corresponding densities in Figure 2.

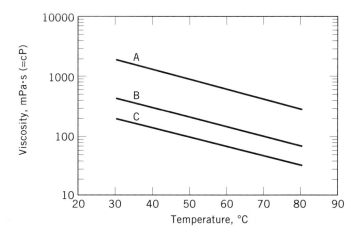

**Fig. 1.** Viscosities (Brookfield) of poly(tetramethylene ether) glycol of the following molecular weights: A, 2000 (Terathane 2000); B, 1000 (Terathane 1000); and C, 650 (Terathane 650).

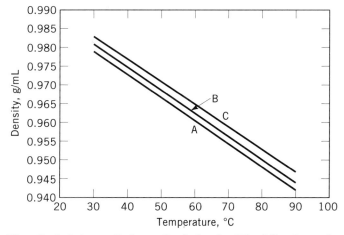

**Fig. 2.** Densities of poly(tetramethylene ether) glycols of the following molecular weights: A, 2000 (Terathane 2000); B, 1000 (Terathane 1000); and C, 650 (Terathane 650).

There are no generally accepted specifications; manufacturers set their own specifications for hydroxyl number range, melt viscosity, water content, etc. Typical properties of Du Pont Terathane PTMEG are listed in Table 2, and selected thermal properties are given in Table 3 (260). For all grades, water content is <0.015 wt %; ash, <0.001 wt %; iron, <1 ppm; peroxide, <5 ppm as $H_2O_2$; and flash point TOC is >163°C.

**Analytical and Test Methods.**   General analytical procedures are applicable in most cases, although a number of specific test methods have been developed for the analysis of polyether glycols. One of the most important tests is

**Table 2. Properties[a] of Terathane Polyether Glycols**

| Property | Terathane 650 | Terathane 1000 | Terathane 1400 | Terathane 2000 |
|---|---|---|---|---|
| molecular weight | 625–675 | 950–1050 | 1350–1450 | 1900–2100 |
| hydroxyl number | 166–180 | 107–118 | 77–83 | 53–59 |
| viscosity at 40°C, mPa·s(=cP) | 100–200 | 260–320 | 525–600 | 950–1450 |
| density at 40°C, g/mL | 0.978 | 0.974 | 0.973 | 0.972 |
| melting point, °C | 11–19 | 25–33 | 27–35 | 28–40 |
| color, APHA[b] | <20 | <10 | <10 | <10 |
| refractive index, $n_D^{25}$ | 1.462 | 1.463–1.465 | 1.464 | 1.464 |
| alkalinity number, meq·KOH/ kg × 30 | −2.0 to 1.0 | −2.0 to 1.0 | −2.0 to 1.0 | −3.0 to 1.0 |
| heat of fusion, kJ/kg[c] | | 90.4 | | ~109 |

[a]Property values = specifications unless otherwise noted.
[b]Color specification is <40 for Terathane 1000, 1400, and 2000; <50 for Terathane 650.
[c]To convert kJ to kcal, divide by 4.184.

**Table 3. Selected Thermal Properties of Poly(Tetramethylene Ether) Glycols[a]**

| Property | Value |
|---|---|
| heat of fusion, kJ/kg[b] | 90.4 |
| | 109[c] |
| heat of combustion, MJ/kg[b] | 35 |
| | 35[c] |
| thermal conductivity, W/(m²·K) | |
| at 20°C | 0.62 |
| at 60°C | 0.57 |
| specific heat, J/(g·K)[b] | 0.52 |
| | 0.55[d] |
| | 0.50[c] |
| enthalpy of combustion[e], MJ/(kg·mol) | −29.8 |

[a]Molecular weight = 1000 unless otherwise noted; Ref. 260.
[b]To convert J to cal, divide by 4.184.
[c]Molecular weight = 2000.
[d]Molecular weight = 650.
[e]Liquid at 25°C to all gases, calculated.

the determination of the hydroxyl number, ie, the number of milligrams of KOH (formula weight = 56.1) equivalent to the hydroxyl content of 1 g of the polymer diol sample (264). Because all the PTME chains are strictly difunctional, the number-average molecular weight is calculated from the hydroxyl number according to the following relation:

$$\text{molecular weight} = 56.1 \times 1000 \times 2/\text{hydroxyl number}$$

Hydroxyl number and molecular weight are normally determined by end-group analysis, by titration with acetic, phthalic, or pyromellitic anhydride (264). For lower molecular weights (higher hydroxyl numbers), $^{19}$F- and $^{13}$C-nmr methods have been developed (265). Molecular weight determinations based on colligative properties, eg, vapor-phase osmometry, or on molecular size, eg, size exclusion chromatography, are less useful because they do not measure the hydroxyl content.

Other important tests are for acid and alkalinity number and for water content (266), because water content and alkalinity of the polyether glycol can influence the reaction with isocyanates. The standard ASTM test for acid and alkalinity number, ASTM D4662 (267), is not sensitive enough for the low acidity and alkalinity numbers of PTMEG, and special methods have been developed. A useful alkalinity number (AN) has been defined as milliequivalents KOH per 30 kg of PTMEG, as titrated in methanol solution with 0.005 $N$ HCl (268). Other useful nonstandard tests are for heavy metals, sulfated ash, and peroxide. The peroxides formed initially in oxidations are quickly transformed into carbonyl groups, which are detectable by infrared spectroscopy. On oxidation, a small C=O peak develops at 1726 cm$^{-1}$ and can be detected in thick (0.5-mm) films. A relative ratio of this peak against an internal standard peak at 2075 cm$^{-1}$ is sometimes defined as the carbonyl ratio.

APHA color (269) is usually one of the specifications of PTMEG, sometimes viscosity is another (270). Melt viscosity at 40°C is often used as a rough measure of the molecular weight distribution within a narrow molecular weight range. Sometimes an empirical molecular weight ratio, MWR = $M_v'/M_n$, is used, where $M_v'$ is a melt-viscosity-average molecular weight defined by the equation $\overline{M}_v' = $ antilog (0.493 log $\eta$ + 3.5576), where $\eta$ is the viscosity in Pa·s at 40°C (240). Commercially available PTMEG-grades usually have a molecular weight ratio in the range of 2.0–2.1. The molecular weight distribution is generally characterized by the dispersity index or polydispersity MWD = $M_w/M_n$. This ratio is lower than the ratio $M_v'/M_n$. For a commercial-grade PTMEG 1000, it is approximately 1.6. The molecular weight distribution can be determined by size exclusion/gel-permeation chromatography (gpc).

The normally used polystyrene or poly(methyl methacrylate) standards may give erroneous gpc results, and primary PTMEG standards have to be used for calibration. Such primary standards can be defined by high performance liquid chromatography (hplc), which effectively separates every molecular weight fraction. Molecular weight and molecular weight distribution can be calculated directly from such a chromatogram, and no calibration is required (86). A bar-graph of Terathane 1000, based on an hplc chromatogram, is shown in Figure 3.

The most important general test methods are issued as ASTM Test Methods and are periodically updated by the Polyurethane Raw Materials Analysis

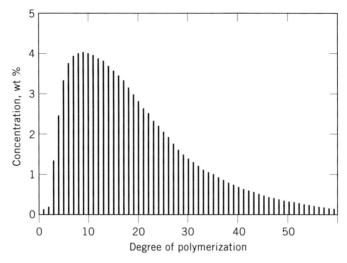

**Fig. 3.** Graph based on an hplc chromatogram of a commercial PTMEG of molecular weight $M_n = 1000$. The bars represent the weight percentage of the individual oligomer fractions. The degree of polymerization is the number of repeating monomer units per polymer chain.

Committee (PURMAC) of the Society for the Plastics Industry (SPI). PURMAC has collected all pertinent analytical methods in a manual (271).

## Health and Safety Factors

Poly(tetramethylene ether) glycols were found to have low oral toxicity in animal tests. The approximate lethal oral dose, $LD_{50}$, for Terathane 1000 has been found to be greater than 11,000 mg/kg (272). No adverse effects on inhalation have been observed. The polymer glycols are mild skin and eye irritants, and contact with skin, eyes, and clothing should be avoided. Goggles and gloves are recommended. In case of contact with the skin, wash thoroughly with water and soap. If swallowed, no specific intervention is indicated, because the compounds are not hazardous. However, a physician should be consulted (260).

Tests with bacterial or mammalian cell cultures demonstrated no mutagenic activity. PTMEG is not listed as a carcinogen by the Occupational Safety and Health Administration (OSHA), International Agency for Research on Cancer (IARC), National Toxicology Program (NTP), or American Conference of Governmental Industrial Hygienists (ACGIH), and exposure limits have not been established by either OSHA or ACGIH. Poly(tetramethylene ether) glycols are cleared as indirect food additives under 21 CFR 175.105 (adhesives) and 21 CFR 177.1680 (polyurethane resins). In addition, elastomers made from PTMEG may be acceptable under 21 CFR 177.1590 (polyester elastomers) and 21 CFR 177.2600 (rubber articles for repeated use).

Various medical devices based on Terathane have been approved by the U.S. FDA, including those used within the body. Formulators are cautioned, however, that FDA approval is not given generically for these devices; it must be applied for separately by each manufacturer for each device. Additional data on safety of PTMEG may be found in the material safety and data sheets provided by the manufacturers.

## Economic Aspects

Poly(tetramethylene ether) glycols of molecular weights 1000, 1800, and 2000 represent the bulk of commercially produced and used PTMEG. Polyglycols of molecular weights 650, 1400, and 2900 are produced only on a smaller scale. Other molecular weight grades are sometimes manufactured as specialty products. The 1996 prices range from about \$3.50/kg to about \$4.20/kg. Du Pont is the world capacity leader, with the largest PTMEG plants both in the United States and in Europe and a total capacity in 1993 of about 60,000 metric tons, 30,000 tons of which were produced each in the United States and in Europe. Du Pont has announced plans for construction of another PTMEG plant based on an internally developed catalyst system to come onstream in 1997 (273,274). Du Pont has a large internal demand for PTMEG; the rest is sold on the merchant market under the trade name Terathane. BASF, the second largest producer of PTMEG, has an estimated capacity of about 16,000 metric tons in the United States and 12,000 metric tons in Japan. Construction of a new PTMEG facility in Germany, with a projected capacity of 16,000 metric tons, has been announced (275). BASF also has a substantial internal demand. In 1993, QO Chemicals, a subsidiary of Great Lakes Chemicals, was the third largest producer of PTMEG, with a capacity of about 8000 metric tons. There are several manufacturers in Japan that produce PTMEG on a smaller scale, each one of them using a different process. The largest are Hodogaya, with about 6000 tons capacity, and Mitsubishi (MKC), with an estimated capacity of about 4000 tons. Sanyo and Asahi Chemicals each produce about 1000 t/yr. An estimated world capacity of PTMEG is shown in Figure 4.

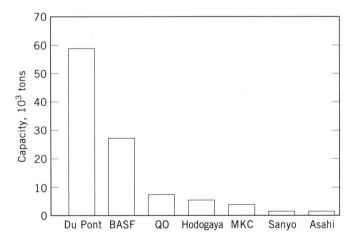

**Fig. 4.** Estimated world PTMEG capacity in 1993.

## Uses

The most important use area for poly(tetramethylene ether) glycols is polyurethane technology. Polyurethanes based on PTMEG have some outstanding properties that set them apart from polyurethanes based on other soft segments. They have excellent hydrolytic stability, high abrasion resistance, and excellent elastomeric properties, such as low hysteresis, high rebound, and high flexibility and impact resistance, even at low temperatures. Other benefits include strength, toughness, durability, low compression set, and high water vapor permeability. These premium properties warrant their use in areas where polyurethanes with soft segments based on polyesters or poly(propylene ether) glycols are inadequate (276).

The largest polyurethane end use areas are in spandex fibers for apparel, which in 1993 accounted for about 50% of the PTMEG market (277). Examples of PTMEG spandex fibers are Lycra (Du Pont), Dorlastan (Bayer), Opelon (Du Pont–TDC), Roica (Asahi), and Acelan (Taekwang). (see FIBERS, ELASTOMERIC). Next are thermoplastic polyurethanes (TPUs) and castable polyurethanes, accounting for about 15 and 14% of the market, respectively. Thermoplastic polyurethanes are fully cured polyurethanes suitable for conventional thermoplastics processing, such as injection molding and extrusion. Examples of TPUs are Pellethane (Dow), Estane (BFGoodrich), and Texin (Miles/Bayer). Castable urethanes are made by mixing and casting reactive liquid components. End use areas of TPUs and castables include wheels, high speed rolls, automotive parts, bushings, specialty hose, cable sheathing and coating, and pipeline liners (278–280). PTMEG is also used in millable gums as rubber substitutes for specialty applications, under conditions where natural rubber would degrade or could not be used for other reasons. PTMEG-based polyurethanes can be manufactured as clear, colorless films, and as such find use in security glazing as glass laminates for aircraft and in bulletproof glass (281).

Elastomers based on PTMEG have excellent microbial and fungus resistance. Their hydrolytic stability make these elastomers prime candidates for use in ground-contact applications, for example, as jacketing material for buried cables. Because of their good biocompatibility, they have also found uses in medical applications, such as catheter tubing.

A smaller but rapidly growing area is the use of PTMEG in thermoplastic polyester elastomers. Formation of such polyesters involves the reaction of PTMEG with diacids or diesters. The diols become soft segments in the resulting elastomeric materials. Examples of elastomeric PTMEG polyesters include Hytrel (Du Pont) and Ecdel (Eastman Chemicals).

For a total of 80,000 t, the estimated world end use markets for PTMEG in 1993 were spandex fibers, 50%; TPU, 15%; castable polyurethane, 14%; other thermoplastic elastomers, 10%; and other uses, 11%. Most of the PTMEG-grades used commercially have molecular weights between 1000–2000. High molecular weight PTMEG ($M_w > 5000$) finds no significant commercial applications as of 1996.

## OXETANE POLYMERS

### Physical Properties

In contrast to THF, substituted oxetanes polymerize readily as a result of the added ring strain associated with the smaller ring size. A large number of polyoxetanes have been prepared and partially characterized (282). The properties of the polymers vary greatly with the symmetry, bulk, and polarity of the substituents on the chain (283). The polymers range from totally amorphous liquids to highly crystalline, high melting solids. The unsubstituted oxetane polymer (POX) has a melting temperature, $T_m$, of 35°C, not far above ambient temperature. A single methyl or azido substituent at either the second or the third position gives an amorphous polymer with stereoisomeric possibilities (282,283). The 3,3-dimethyl derivative (PDMOX) is a crystalline polymer having a $T_m$ of 47°C, whereas 3,3-bis(halomethyl)oxetane polymers have a $T_m$ of 135, 180, 220, and 290°C for fluoro-, chloro-, bromo-, and iodo-derivatives, respectively (282). For the 3,3-dinitro derivative (PDNOX) and the 3,3-diazidomethyl derivative (PDAMO), the melting temperatures are 200–202°C and 75–78°C, respectively. As usual, the effect of copolymerization is to lower the $T_m$. For example, the random, high molecular weight copolymer of 3,3-bis(chloromethyl)oxetane [78-71-7] (BCMO) and THF is a tough, amorphous rubber (74).

Other crystallization parameters have been determined for some of the polymers. The dependence of the melting temperature on the crystallization temperature for the orthorhombic form of POX ($T_m$ = 323 K) and both monoclinic ($T_m$ = 348 K) and orthorhombic ($T_m$ = 329 K) modifications of PDMOX has been determined (284). The enthalpy of fusion, $\Delta H_m$, for the same polymers has been determined by the polymer diluent method and by calorimetry at different levels of crystallinity (284). $\Delta H_m$ for POX was found to be 150.9 J/g (36.1 cal/g); for the dimethyl derivative, it ranged from 85.6 to 107.0 J/g (20.5–25.6 cal/g). Numerous crystal structure studies have been made (285–292). Isothermal crystallization rates of POX from the melt have been determined from 19 to −50°C (293,294). Similar studies have been made for PDMOX from 22 to 44°C (295,296).

The glass-transition temperatures, $T_g$, of several polyoxetanes with alkyl substituents in position three were determined by differential scanning calorimetry (dsc). The results showed a regular increase of $T_g$ according to the size of the substituents, except for the dibutyl derivative. Intramolecular interactions had a greater effect on $T_g$ than intermolecular interactions (297,298). The pressure dependence of the $T_g$ of diazidomethyl oxetane–THF copolymers has been determined by high pressure (≤850 MPa = 8.5 kbar) differential thermal analysis (dta) and dielectric methods (299). Glass-transition temperature increases with increasing pressure. The results fit a modified Gibbs-DiMarzio theory. The effect of copolymer composition on $T_g$ (300) and on dipole moments (301) has been studied over the whole composition range for DMOX–THF copolymer. The copolymers exhibit a $T_g$ intermediate between those of the parent homopolymers.

Solubility parameters of 19.3, 16.2, and 16.2 $(J/cm^3)^{1/2}$ (7.9 $(cal/cm^3)^{1/2}$) have been determined for polyoxetane, poly(3,3-dimethyloxetane), and poly(3,3-diethyloxetane), respectively, by measuring solution viscosities (302). Heat capacities have been determined for POX and compared to those of other polyethers and polyethylene (303,304). The thermal decomposition behavior of poly[3,3-bis(ethoxymethyl)oxetane] has been examined (305).

Extensive physical property measurements have been reported for only PBCMO (283,306). This polymer was commercially available in the United States from Hercules, Inc. for about 15 years under the trade name Penton, but it is not currently (1996) produced for commercial sale. It has been studied extensively in the former USSR (307), where it was called Pentaplast, and also in Japan. The comparatively high heat distortion temperature, combined with a low water absorption value, makes PBCMO suitable for articles that require sterilization. It also has excellent electrical properties, a high degree of chemical resistance, and, because of the chlorine on the side chain, is self-extinguishing. PBCMO is resistant to most solvents, specifically to water, ketones, aldehydes, esters, aromatic hydrocarbons, weak organic bases such as aniline and ammonia, weak acids, and weak and strong alkalies. Like most polyethers, PBCMO is attacked by strong acids (306). Most other crystalline oxetane polymers are also insoluble in common organic solvents (286). The amorphous and low melting oxetane polymers, in contrast, are soluble in a wide variety of organic solvents.

Properties have been determined for a series of block copolymers based on poly[3,3-bis(ethoxymethyl)oxetane] and poly{[3,3-bis(methoxymethyl)oxetane]-co-tetrahydrofuran}. The block copolymers had properties suggestive of a thermoplastic elastomer (308). POX was a good main chain for a well-developed smectic liquid crystalline state when cyano- or fluorine-substituted biphenyls were used as mesogenic groups attached through a four-methylene spacer (309,310). Other side-chain liquid crystalline polyoxetanes were observed with a spacer-separated azo moiety (311) and with laterally attached mesogenic groups (312).

## Chemical Properties

The chemistry of polymerization of the oxetanes is much the same as for THF polymerization. The ring-opening polymerization of oxetanes is primarily accomplished by cationic polymerization methods (283,313–318), but because of the added ring strain, other polymerization techniques, eg, insertion polymerization (319), anionic polymerization (320), and free-radical ring-opening polymerization (321), have been successful with certain special oxetanes. With appropriately substituted oxetanes, aluminum-based initiators (321) impose a degree of microstructural control on the substituted polyoxetane structure that is not obtainable with a pure cationic system. A polymer having largely the structure of poly(3-hydroxyoxetane) has been obtained from an anionic rearrangement polymerization of glycidol or its trimethylsilyl ether, both oxirane monomers (322). Polymerization-induced epitaxy can produce ultrathin films of highly oriented POX molecules on, for instance, graphite (323). Theoretical studies on the cationic polymerization mechanism of oxetanes have been made (324–326).

Other differences between THF and oxetane polymerizations are noteworthy. The added ring strain in the four-membered ring oxetane results in much faster polymerization rates and higher heats of polymerization than are observed for THF polymerization. At near ambient temperature and moderately high initiator concentrations, high conversions to PTHF usually require hours to achieve, whereas most oxetane polymerizations are near 100% conversion in seconds or minutes. Typical heat of polymerization values, $\Delta H_p$, are 84.6, 80.9, 67.9, and 25 kJ/mol (20.2–6 kcal/mol) for BCMO, OX, DMOX, and THF, respectively (6,282). Nominally, the polymerization of oxetanes is reversible. However, the formation of a four-membered ring under the usual polymerization conditions is energetically so disfavored that the depolymerization reaction can be ignored; the formation of cyclic oligomers becomes much more significant in oxetane polymerizations. The amount of cyclic oligomer formed depends on the particular oxetane, the polymerization conditions, and the initiating system, polymerization temperature, and solvent (327–331). Super acids and their esters have been found to be of little value in oxetane polymerizations. The equilibrium between ester and the oxonium ion in a BCMO polymerization, for example, is far toward the ester form, which is virtually unreactive to the cyclic ether and does not participate in the propagation reaction. Moreover, the active oxonium ion, when formed, collapses immediately to the corresponding ester (332). Thus a suitable mechanism for rapid polymer formation with this group of initiators does not exist for BCMO, and probably not for other oxetanes.

**Kinetics.** Results of the earliest studies of the kinetics of polymerization of OX and DMOX were reported in 1956 (327). Subsequent analyses led to the conclusion that all oxetane polymerizations, regardless of the initiator used, could be described by the kinetic expression $-d[M]/dt = k_p[I]_0 \cdot [M]$, where $k_p$ is the propagation rate constant; $[I]_0$ is the concentration of a fast, 100% efficient initiator or the concentration of active sites as measured directly; and $[M]$ is the monomer concentration (67,133). Later studies revealed that not all systems are that simple. For example, under some conditions the rate of initiation with trialkyl oxonium ions is slow compared to the rate of polymerization; because a competing fast reaction of the growing oxonium ion with polymer oxygen forms a polymeric acyclic oxonium ion, this reaction effectively leads to dormant or sleeping ions by a process often referred to as temporary termination (334). Also, the formation of cyclic oligomers by competing reactions, when they occur, needs to be considered.

## Manufacture and Processing

The only commercially important products in this area are the low molecular weight polyethers with hydroxyl end groups resulting from THF polymerization. As the next lower homologue, and having increased potential for functional or modifying ring substitution, oxetane offers an enticing extension of this chemistry. Indeed it is possible to control the rapid polymerizations and to prepare low molecular weight products with the necessary end-group functionality. Japanese patents (335,336) describe the preparation of low molecular weight oxetane–THF copolymers using methods that appear similar to those used to prepare PTHF glycols. The preparation of polyoxetanes and oxetane–THF copolymers where the oxetane moieties are substituted with energetic groups, such as azido or nitro groups, has also been described (337–339). The latter polymerizations are similar to oxirane polymerizations occurring by the activated monomer mechanism (340,341) and generally involve polymerization in the presence of relatively large amounts of organic diols, eg, 1,4-butanediol, or the use of preformed Lewis acid–diol mixtures as polymerization initiators (342–344). Low molecular weight glycols have also been obtained by degrading high molecular weight polymer (342,345). PBCMO, the only oxetane polymer that was once commercialized, is no longer available. Some details about its processing can be found in the literature (283,306). Macroglycols from oxetanes or from mixtures of oxetanes with other monomers have been converted to polyurethanes using technology similar to that for the conversion of macroglycols from THF.

**Economic Aspects.** Oxetanes are expensive monomers and are not readily available in commercial quantities. Commercial production of PBCMO has been discontinued; its end uses were not able to support its comparatively high selling price. Energetic polymers prepared from appropriately substituted oxetanes have opened a new market for their use to prepare solid rocket propellants and explosives. Should this specialty market result in the large-scale production of these oxetanes even at current (1996) high prices and/or in a cheap synthetic route to oxetanes, this economic picture could change.

**Analytical and Test Methods.** Most of the analytical and test methods described for THF and PTHF are applicable to OX and POX with only minor modifications (346). Infrared and nmr are useful aids in the characterization of oxetanes and their polymers. The oxetane ring shows absorption between 960 and 980 $cm^{-1}$, regardless of substituents on the ring (282). Dinitro oxetane (DNOX) has its absorption at 1000 $cm^{-1}$. In addition, $^1$H-nmr chemical shifts for $CH_2$ groups in OX and POX are typically at 4.0–4.8 $\delta$ and 3.5–4.7 $\delta$, respectively (6,347,348); $^{13}$C-nmr is especially useful for characterizing the microstructure of polyoxetanes.

## Health and Safety Factors

Toxicity and hazards of handling oxetanes and their polymers are influenced markedly by their substituents. For many of the monomers, these factors are described in detail in material safety data sheets. Special monomers such as those designed to yield energetic polymers require special care in handling because they are sensitive explosives. Under no circumstances should an energetic monomer be distilled; rather, purification via column chromatography is suggested (348–350). Polymers derived from energetic monomers burn with a large amount of smoke. Special procedures have been developed for disposal of these energetic polymers.

## Uses

A large number of uses have been explored worldwide for high performance PBCMO (283), but none of these uses were adequate in the marketplace to support its high price. The energetic polymers derived from appropriately substituted oxetanes were specifically designed to increase the energy content of explosives and solid fuels that use polymeric binders (349). In this the oxetanes have advantages over THF and the oxiranes, which may ultimately help the oxetanes find continuing demand. Substituted oxetanes homopolymerize and copolymerize well with THF. The technology developed for converting PTHFs to useful products is readily adaptable to polyoxetanes. The formation of cyclic oligomers can be essentially eliminated in oxetane polymerizations by proper choice of the initiating system, counterion, and temperature of polymerization. Research on these applications continues (351–356).

## BIBLIOGRAPHY

"Polyethers, Tetrahydrofuran and Oxetene Polymers" in *ECT* 3rd ed., Vol. 18, pp. 645–670, by P. Dreyfuss, University of Akron, and M. P. Dreyfuss, BF Goodrich Co.
    1. S. Penczek, P. Kubisa, and K. Matyjaszewski, *Adv. Polym. Sci.* **68,69** (1985).

2. S. Inoue, T. Aida, in K. J. Ivin and T. Saegusa, eds., *Ring-Opening Polymerization*, Elsevier, London, U.K., 1984; and T. Saegusa in O. Vogl and J. Furukawa, eds., *Polymerization of Heterocycles*, Marcel Dekker, Inc., New York, 1973.
3. Ger. Pat. 741,478 (June 21, 1939), H. Meerwein (to I. G. Farben).
4. A. C. Farthing, in N. G. Gaylord, ed., *High Polymers*, Vol. XIII, Part I, Interscience Publishers, New York, 1963, Chapt. 5, p. 310.
5. S. Penczek, P. Kubisa, and K. Matyjasewski, *Adv. Polym. Sci.* **37** (1980).
6. P. Dreyfuss, *Poly(tetrahydrofuran)*, Gordon & Breach, New York, 1982.
7. P. Dreyfuss, M. P. Dreyfuss, and G. Pruckmayr, in J. I. Kroschwitz, ed., *Encyclopedia of Polymer Science and Engineering*, 2nd ed., Vol. 16, Wiley-Interscience, New York, 1989, pp. 649–681.
8. G. Pruckmayr and T. K. Wu, *Macromolecules*, **11**, 662 (1978).
9. G. S. Trick and J. M. Ryan, *J. Polym. Sci. Part C*, **18**, 93 (1967).
10. R. E. Wetton and G. Williams, *Trans. Faraday Soc.* **61**, 2132 (1965).
11. J. Lehmann and P. Dreyfuss, *Adv. Chem. Ser.* **176**, 587 (1979).
12. A. H. Wilbourn, *Trans. Faraday Soc.* **54**, 717 (1958).
13. P. Dreyfuss and M. P. Dreyfuss, *Adv. Polym. Sci.* **4**, 528 (1967).
14. Brit. Pat. 1,006,316 (Sept. 29, 1965), D. Heppert and L. Joginder (to Goodyear Tire and Rubber Co.).
15. R. C. Burrows, *ACS Polym. Prepr.* **6**, 600 (1965).
16. G. G. Winspear, *The Vanderbilt Rubber Handbook*, R. T. Vanderbilt Co., Inc., New York, 1958.
17. *Compounders Technical Guide*, Vol. 1, CTG No. 49, The BF Goodrich Co., Research Center, Brecksville, Ohio.
18. C. D. Hodgman, *Handbook of Chemistry and Physics*, 61st ed., The Chemical Rubber Publishing Co., Boca Raton, Fla., 1981.
19. Y. Tsujita, T. Nose, and T. Hata, *Polym. J.* **5**, 201 (1973).
20. *Ibid.*, **6**, 51 (1974).
21. S. Yoshida, H. Suga, and S. Seki, *Polym. J.* **5**, 25 (1973).
22. M. B. Huglin and D. J. Pass, *J. Appl. Polym. Sci.* **12**, 473 (1968).
23. K. Imada and co-workers, *Makromol. Chem.* **83**, 113 (1965).
24. S. Kobayashi and co-workers, *J. Polym. Sci., Polym. Lett. Ed.* **14**, 591 (1976).
25. M. Cesari, G. Perego, and A. Muzzi, *Makromol. Chem.* **83**, 106 (1965).
26. D. Makino, M. Kobayashi, and H. Tadokoro, *Spectrochim. Acta, Part A*, **31A**(9,10), 1481 (1975).
27. T. K. Wu and G. Pruckmayr, *Macromolecules*, **8**, 77 (1975).
28. G. Pruckmayr and T. K. Wu, *Macromolecules*, **8**, 954 (1975).
29. K. Matyjaszewski and S. Penczek, *J. Polym. Sci. Polym. Chem. Ed.* **15**, 247 (1977).
30. K. Matyjaszewski, T. Diem, and S. Penczek, *Makromol. Chem.* **180**, 1817 (1979).
31. Y. Eckstein and co-workers, *J. Polym. Sci. Polym. Chem. Ed.* **18**, 2021 (1980).
32. Jpn. Pat. 60,108,424 (1985), (to Asahi Chem. Ind. Co.).
33. T. G. Croucher and R. E. Wetton, *Polymer*, **17**, 205 (1976).
34. U.S. Pat. 3,478,109 (Nov. 11, 1969), M. Wayne (to Eastman Kodak).
35. U.S. Pat. 4,762,951 (Aug. 9, 1988), H. Mueller (to BASF).
36. U.S. Pat. 4,933,503 (June 12, 1990), H. Mueller (to BASF).
37. Jpn. Pat. 60,042,421 (July 16, 1985), T. Sueyoshi and M. Shirato (to Mitsubishi Chem. Ind.).
38. U.S. Pat. 5,053,553 (Oct. 1, 1991), S. Dorai (to E. I. du Pont de Nemours & Co., Inc.).
39. D. Sims, *Makromol. Chem.* **98**, 235 (1966).
40. M. B Huglin and D. H. Whitehurst, *J. Appl. Polym. Sci.* **12**, 1889 (1968).
41. R. M. Bell, C. Fitzsimmons, and A. Ledwith, *Polymer*, **6**, 661 (1965).
42. M. Kurata, H. Uttiyama, and K. Kamada, *Makromol. Chem.* **88**, 281 (1965).
43. J. M. Evans and M. B. Huglin, *Makromol. Chem.* **127**, 141 (1969).

44. J. M. Evans, M. B. Huglin, and R. F. T. Stepto, *Makromol. Chem.* **146**, 91 (1971).
45. H. G. Elias and G. Adank, *Makromol. Chem.* **103**, 230 (1967).
46. L. Mandelkern, *Macromol. Chem.* **1**(4), 301 (1990).
47. A. Hirai and co-workers, *Macromolecules*, **23**(11), 2913 (1990).
48. H. Weber and R. Kimmich, *Macromolecules*, **26**(10), 2597 (1993).
49. J. S. Higgins and J. E. Roots, *J. Chem. Soc. Farad. Trans.* **2**, 81, 757 (1985).
50. J. S. Higgins, in J. Mark and B. Erman, eds., *Elastomeric Polymer Networks*, Prentice-Hall, Inc., Englewood Cliffs, N.J., 1992.
51. J. A. Miller and co-workers, *Polymer*, **26**, 1915 (1985).
52. S. Visser, G. Pruckmayr, and S. L. Cooper, *Macromolecules*, **24**, 6769 (1991).
53. G. Marchionni and co-workers, *Macromolecules*, **26**(7), 1751 (1993).
54. M. Kretz and co-workers, *J. Polym. Sci. Part B, Polym. Phys.* **26**(8), 1553 (1988).
55. M. M. Fuson and J. B. Miller, *Macromolecules*, **26**, 3218 (1993).
56. S. Ishikawa, *Eur. Polym. J.* **29**(12), 1621 (1993).
57. E. S. Domalski and E. D. Hearing, *J. Phys. Chem. Ref. Data*, **19**(4), 881 (1990).
58. H. S. Faruque and C. Lacabanne, *J. Mater. Sci. Lett.* **8**(1), 58 (1989).
59. B. B. Sauer and G. T. Dee, *J. Colloid Interface Sci.* **162**(1), 25 (1994).
60. M. Kretz and co-workers, *J. Polym. Sci. Part B, Polym. Phys.* **26**(3), 663 (1988).
61. F. Schultze-Gebhardt, *Acta Polym.* **38**(1), 36 (1987).
62. C. P. Lillya and co-workers, *Macromolecules*, **25**, 2076 (1992).
63. C. M. Roland and G. S. Buckley, *Rubber Chem. Tech.* **64**(1), 74 (1991).
64. G. P. Van der Beek and co-workers, *Macromolecules*, **24**, 6600 (1991).
65. S. Ishikawa, K. Ishizu, and T. Fukutomi, *Makromol. Chem.* **192**(5), 1177 (1991).
66. K. Brzezinska and co-workers, *Makromol. Chem.* **178**, 2491 (1977).
67. A. Hase and T. Hase, *Analyst*, **97**, 998 (1972).
68. T. Saegusa and S. Kobayashi, *Prog. Polym. Sci. Jap.* **6**, 107 (1973).
69. T. Saegusa and S. Matsumoto, *J. Polym. Sci A-1*, **6**, 1559 (1968).
70. Y. Eckstein and P. Dreyfuss, *Anal. Chem.* **52**, 537 (1980).
71. D. Vofsi and A. V. Tobolsky, *J. Polym. Sci. Part A*, **3**, 3261 (1965).
72. U.S. Pat. 4,972,701 (Nov. 11, 1990), W. Yau (to E. I. du Pont de Nemours & Co., Inc.).
73. S. M. Ali and M. B. Huglin, *Makromol. Chem.* **84**, 117 (1965).
74. M. P. Dreyfuss and P. Dreyfuss, *J. Polym. Sci. A-1*, **4**, 2179 (1966).
75. J. M. Evans and M. B. Huglin, *Eur. Polym. J.* **6**, 1161 (1970).
76. N. V. Makletsova and co-workers, *Polym. Sci. USSR*, **7**, 73 (1965).
77. G. Adank, *Thesis No. 3703*, E. T. H., Zürich, 1965, Clearinghouse Publication N66–36762, Springfield, Va.
78. B. A. Rozenberg and co-workers, *Polym. Sci. USSR*, **7**, 1163 (1965).
79. T. Fujimoto and co-workers, *Polym. J.* **11**, 193 (1979).
80. E. A. Ofstead, *ACS Polym. Prepr.* **6**, 674 (1965).
81. D. H. Richards, S. B. Kingston, and T. Souel, *Polymer*, **19**, 68 (1978).
82. *Ibid.*, 806 (1978).
83. M. P. Dreyfuss and P. Dreyfuss, *ACS Polym. Prepr.* **11**, 203 (1970).
84. J. M. Goldwasser, *Chromatog. Polym. ACS Symp. Ser.* **521**, 243 (1993).
85. B. Trathnigg, *J. Chromat.* **552**(1,2), 507 (1991).
86. G. D. Andrews, A. Vatvars, G. Pruckmayr, *Macromolecules*, **15**, 1580 (1982).
87. W. Stix and W. Heitz, *Makromol. Chem.* **180**, 1367 (1979).
88. Ger. Pat. DE 300428 (1981), W. Heitz, H. J. Kress, and W. Stix (to Bayer).
89. N. G. Matseeva, L. N. Turovska, and A. I. Kutsaev, *Acta Polym.* **34**(11,12), 735 (1983).
90. U.S. Pat. 5,034,559 (July 23, 1991), E. Hickmann (to BASF).
91. Eur. Pat. 411,415 (1991).
92. J. Sierra-Vargas and co-workers, *Polym. Bull.* **3**, 83 (1980).
93. A. A. Berlin, N. G. Matseeva, and E. S. Pankova, *Vysokomol. Soedin, Ser. A*, **9**(6), 1325 (1967).

94. Eur. Pat. 363,769 (Apr. 18, 1990), U. Koehler, W. Schoenleben, and H. Siegel (to BASF).
95. Ger. Pat. 3,840,204 (May 31, 1990), Hickmann and co-workers (to BASF).
96. Eur. Pat. 370,445 (May 30, 1990), J. P. Wolf and F. Setiabudi (to Ciba-Geigy).
97. Eur. Pat. 302,487 (Feb. 8, 1989), S. Birnbach and co-workers (to BASF).
98. Eur. Pat. 400,436 (Dec. 5, 1990), E. Hickmann (to BASF).
99. U.S. Pat. 5,124,488 (June 23, 1990), E. Hickmann (BASF).
100. U.S. Pat. 4,967,015 (Oct. 30, 1990), M. Karcher, H. Eckhardt, and J. Henkelmann (to BASF).
101. Eur. Pat. 91836 (Oct. 19, 1983), K. Yasuda and co-workers (to Japan Synth. Rubber Corp.).
102. E. Yoshida, T. Takata, and T. Endo, *Makromol. Chem.* **194**, 2507 (1993).
103. Ger. Pat. 3,700,709 (July 21, 1988), M. Steiniger and K. Halbritter (to BASF).
104. S. Kobayashi and co-workers, *Macromolecules*, **23**, 2861 (1990).
105. U.S. Pat. 5,254,744 (Jan. 19, 1993), J. Neumer (to E. I. du Pont de Nemours & Co., Inc.).
106. U.S. Pat. 5,284,980 (Feb. 8, 1994), G. Pruckmayr and R. Osborn (to E. I. du Pont de Nemours & Co., Inc.).
107. Eur. Pat. 442,402 (Aug. 21, 1991), K. Bott, W. Straehle, and U. Abel (to BASF); P. Groll, *Proceedings of the SPI Polyurethane World Congress*, Nice, France, 1991, p. 858.
108. U.S. Pat. 5,003,107 (Mar. 26, 1991), R. Zimmerman and J. Larkin, (to Texaco).
109. Ger. Pat. 3,827,119 (May 12, 1990), W. Schoenleben and H. Mueller (to BASF).
110. Eur. Pat. 541,252 (May 12, 1993), D. R. Hollingsworth and R. L. Zimmerman (to Texaco).
111. Ger. Pat. 3,835,040 (Apr. 18, 1990), U. Koehler, E. Hickmann, and H. Siegel (to BASF).
112. Ger. Pat. 3,920,928 (Jan. 3, 1991), H. Schaper (to Phoenix AG).
113. Eur. Pat. 296,852 (Dec. 28, 1988), C. Leir, J. Hoffmann, and J. Stark (to 3M Co.).
114. Brit. Pat. 1,120,304 (Feb. 23, 1966), S. Smith and A. J. Hubin (to 3M Co.).
115. U.S. Pat. 3,436,359 (Apr. 1, 1969), A. J. Hubin and S. Smith (to 3M Co.).
116. S. Smith and A. J. Hubin, *J. Macromol. Sci. Chem.* **A-7**, 1399 (1973).
117. S. Kohija and co-workers, *Bull. Chem. Soc. Jpn.* **63**(7), 2089 (1990).
118. Jpn. Pat. 02,242822 (Sept. 27, 1990), S. Kobayashi and co-workers (to Hodogaya).
119. Ger. Pat. 4,022,931 (Jan. 23, 1992), U. Thiery (to Bayer).
120. C. Bougherara, B. Boutevin, and J. J. Robin, *Makromol. Chem.* **194**(4), 1225 (1993).
121. S. Smith and co-workers, Contribution No. 43, Div. Rubb. Chem., ACS, Montreal, Canada, 1967.
122. C. Hepburn, *Polyurethane Elastomers*, Applied Science Ltd, London, U.K., 1982.
123. *Uses for Du Pont Glycols in Polyurethanes*, Du Pont Publication E-70687, E. I. du Pont de Nemours & Co., Inc., Wilmington, Del., 1985.
124. G. Oertel, *Polyurethane Handbook*, Hanser Publishers, New York, 1985.
125. M. Brown and W. K. Witsiepe, *Rubber Age*, **104**, 35 (Mar. 1972).
126. W. K. Witsiepe, *Adv. Chem. Ser.* **129**, 39 (1973).
127. J. R. Wolfe, Jr., *Rubber Chem. Tech.* **50**, 688 (1977).
128. J. R. Wolfe, Jr., *Adv. Chem. Ser.* **176**, 129 (1979).
129. Neth. Pat. 6,508,295 (1965), (to Eastman Kodak Co.).
130. G. K. Hoeschele and W. K. Witsiepe, *Angew. Makromol. Chem.* **29/30**, 267 (1973).
131. *Chem. Week*, 20 (July 6, 1977).
132. A. Ghaffer, I. Goodman, and I. H. Hall, *Brit. Polym. J.* **5**, 315 (1973).
133. A. Ghaffer, I. Goodman, and R. H. Peters, *Brit. Polym. J.* **10**, 115 (1978).
134. A. Ghaffer and co-workers, *Brit. Polym. J.* **10**, 123 (1978).
135. H. W. Hasslin, M. Droscher, and G. Wegner, *Makromol. Chem.* **179**, 1373 (1978).

136. R. J. Cella, *J. Polym. Sci. Polym. Symp.* **42**, 727 (1973).
137. A. Lilaonitkul and S. L. Cooper, *Rubb. Chem. Tech.* **50**, 1 (1977).
138. A. Lilaonitkul, J. C. West, and S. L. Cooper, *J. Macromol. Sci. Phys.* **B12**, 563 (1976).
139. K. Matsuda, *Chemtech*, **4**, 744 (Dec. 1974).
140. R. K. Adams and G. K. Hoeschele, in N. R. Legge, G. Holden, H. E. Schroeder, eds., *Thermoplastic Elastomers*, Hanser Publishers, Munich, Germany, 1987.
141. P. A. Okunev, *Sb. Nauch. Tr. Ivanov. Energ. Inst.* **14**, 130 (1972).
142. A. Davis and J. H. Golden, *Makromol. Chem.* **81**, 38 (1965).
143. J. H. Golden, *Makromol. Chem.* **81**, 51 (1965).
144. G. Pruckmayr and T. K. Wu, *Macromolecules*, **11**, 265 (1978).
145. F. S. Dainton and K. Ivin, *Q. Rev. Chem. Soc.* **12**, 61 (1958).
146. K. Matyjaszewski, S. Słomkowski, and S. Penczek, *J. Polym. Sci. Polym. Chem. Ed.* **17**, 2413 (1979).
147. P. Dreyfuss and J. P. Kennedy, *J. Appl. Polym. Sci. Appl. Polym. Symp.* **30**, 165 (1977).
148. S. Penczek, *Makromol. Chem. Suppl.* **3**, 17 (1979).
149. A. M. Buyle, K. Matyjaszewski, and S. Penczek, *Macromolecules*, **10**, 269 (1977).
150. H. Meerwein, *Org. Syn. Coll.* **5**, 1085,1096 (1973).
151. M. P. Dreyfuss, J. C. Westfahl, and P. Dreyfuss, *Macromolecules*, **1**, 437 (1968).
152. S. Nakahama, S. Hino, and N. Yamazaki, *Polym. J.* **2**, 56 (1971).
153. J. V. Crivello and J. H. W. Lam, *Macromolecules*, **10**, 1307 (1977).
154. J. V. Crivello and J. H. W. Lam, *J. Polym. Sci. Polym. Letts. Ed.* **16**, 563 (1978).
155. J. V. Crivello and J. H. W. Lam, *J. Polym. Sci. Polym. Chem. Ed.* **17**, 977,1047 (1979).
156. Y. Eckstein and P. Dreyfuss, *J. Polym. Sci. Polym. Chem. Ed.* **17**, 4115 (1979).
157. F. A. M. Abdul-Rasoul, A. Ledwith, and Y. Yagci, *Polym. Bull.* **1**, 1 (1978).
158. Eur. Pat. 442,635 (1991), O. Farooq (to 3M Co.).
159. A. Aoshima, S. Tonomura, and S. Yamamatsu, *Polym. Adv. Tech.* **1**(2), 127–132 (1990).
160. M. Bednarek and co-workers, *Makromol. Chem.* **190**, 929–938 (1989).
161. J. V. Crivello and M. Fan, *J. Polym. Sci. Part A: Polym. Chem.* **29**, 1853–1863 (1991).
162. F. Li and co-workers, *J. Appl. Polym. Sci.* **50**, 2017–2020 (1993).
163. G. A. Olah and co-workers, *J. Appl. Polym. Sci.* **45**, 1355–1360 (1992).
164. Ger. Pat. 2,364,859 (July 4, 1974), K. Matsuda, Y. Tanaka, and T. Sakai (to Kao Soap Co., Ltd.).
165. J. S. Hrkach and K. Matyjaszewski, *Macromolecules*, **23**, 4042–4046 (1990).
166. S. Slomkowski and co-workers, *Macromolecules*, **22**, 503 (1989).
167. E. J. Vandenberg and J. C. Mullis, *J. Polym. Sci. Part A: Polym. Chem.* **29**, 1421–1438 (1991).
168. J. L. Lambert and E. J. Goethals, *Makromol. Chem.* **133**, 289 (1970).
169. P. Dreyfuss, *Chemtech*, **3**, 356 (1973).
170. U.S. Pat. 4,163,115 (July 31, 1979), G. E. Heinsohn and co-workers (to E. I. du Pont de Nemours & Co., Inc.).
171. H. J. Kress and W. Heitz, *Makromol. Chem. Rapid Commun.* **2**, 427 (1981).
172. M. P. Dreyfuss and J. H. Macey, *ACS Polym. Prepr.* **20**(2), 324 (1979).
173. U.S. Pat. 4,393,199 (July 12, 1983), G. E. Manser (to SRI International).
174. F. R. Mayo and F. M. Lewis, *J. Am. Chem. Soc.* **66**, 1594 (1944).
175. M. Fineman and S. D. Ross, *J. Polym. Sci.* **5**, 259 (1950).
176. T. Kelen and T. Tüdös, *J. Macromol. Sci.-Chem.* **A9**, 1 (1975); **A10**, 151 (1976); **A16**, 1283 (1981).
177. Y. Yamashita and co-workers, *Makromol. Chem.* **117**, 242 (1968).
178. P. Kubisa and S. Penczek, *J. Macromol. Sci.-Chem.* **A7**, 1509 (1973).
179. S. G. Entelis and G. V. Korovina, *Makromol. Chem.* **175**, 1253 (1974).
180. T. Saegusa, S. Matsumoto, and Y. Hashimoto, *Macromolecules*, **3**, 377 (1970).

181. G. Berger, M. Levy, and D. Vofsi, *Polym. Lett.* **4**, 183 (1966).
182. F. J. Burgess and co-workers, *Polymer*, **18**, 733 (1977).
183. M. L. Hallensleben, *Makromol. Chem.* **178**, 2125 (1977).
184. J. Lehmann and co-workers, *Abstracts to First International Symposium on Polymerization of Heterocycles (Ring-Opening)*, Warsaw-Jablonna, Poland, June 23–25, 1975, p. 141.
185. P. Dreyfuss and J. P. Kennedy, *J. Polym. Sci. Polym. Symp.* **56**, 129 (1976).
186. E. Franta and co-workers, *J. Polym. Sci. Polym. Symp.* **56**, 139 (1976).
187. P. Dreyfuss and co-workers, *ACS Polym. Prepr.* **20**(2), 94 (1979).
188. U.S. Pat. 3,696,173 (Oct. 3, 1972), T. Sakamura and co-workers (to Toyo Soda Co.).
189. G. Hizal and J. Yagci, *Polymer*, **30**, 722–725 (1989).
190. B. Hazer, *Eur. Polym. J.* **26**, 1167–1170 (1990).
191. G. Galli and co-workers, *Makromol. Chem. Rapid Commun.* **14**, 185–193 (1993).
192. E. Franta and L. Reibel, *Makromol. Chem. Macromol. Symp.* **47**, 141–150 (1990).
193. S. Kobayashi, *ACS Polym. Prepr.* **29**(2), 40–41 (1988).
194. N. Aoki, K. Furuhata, and M. Sakamoto, *J. Appl. Polym. Sci.* **51**, 721–730 (1994).
195. K. Matyjaszewski and J. S. Hrkach, Report, TR-323, Order No. AD-A251086, *Gov. Rep. Announce. Index (US)*, **92**(18), Abstr. No. 249057 (1992).
196. Y. Sun, W. Jin, and A. C. L. Aaron, *ACS Polym. Prepr.* **32**(1), 561–562 (1991).
197. G. Cameron and K. Sarmouk, *Makromol. Chem.* **191**, 17–23 (1990).
198. J. Hrkach, K. Ruehl, and K. Matyjaszewski, *ACS Polym. Prepr.* **29**(2) 112–113 (1988).
199. Eur. Pat. 454,226 (Oct. 30, 1991), E. J. Goethals and F. P. F. Hosteaux (to Stamicarbon BV).
200. M. Zsuga and J. P. Kennedy, *J. Polym. Sci. Part A: Polym. Chem.* **29**, 875–880 (1991).
201. S. Kobayashi and co-workers, *Macromolecules*, **23**, 1586–1589 (1990).
202. Eur. Pat. 450,724 (Oct. 9, 1991), F. P. Hosteaux and P. Francine (to Stamicarbon BV).
203. L. Zhao and co-workers, *Polym. Mater. Sci. Eng.* **66**, 308–309 (1992).
204. G. C. Eastmond and J. Woo, *Polymer*, **31**, 358–361 (1990).
205. W. L. Wu and L. Jong, *Polymer*, **34**, 2357–2362 (1993).
206. Jpn. Pat. 74 015 074 (Apr. 12, 1974), K. Matsusuwa and K. Oya (to Mitsubishi Chem. Ind.).
207. U.S. Pat. 3,935,252 (Sept. 14, 1976), H. Tomomatsu (to Quaker Oats Co.).
208. Jpn. Pat. 52 32678 (1977), M. Matsui and co-workers (to Mitsubishi Chem. Ind.).
209. U.S. Pat 4,189,566 (Feb. 19, 1980), H. Mueller, O. Huchler, and H. Hofmann (to BASF).
210. U.S. Pat. 5,118,869 (June 2, 1992), S. Dorai and co-workers (to E. I. du Pont de Nemours & Co., Inc.).
211. U.S. Pat. 4,230,892 (Oct. 28, 1980), G. Pruckmayr (to E. I. du Pont de Nemours & Co., Inc.).
212. U.S. Pat. 4,608,422 (Aug. 26, 1986), H. Mueller (to BASF).
213. Eur. Pat. 59368 (Sept. 8, 1982), H. Mueller (to BASF).
214. Eur. Pat. 239,787 (Oct. 7, 1991), H. Mueller (to BASF).
215. U.S. Pat. 4,803,299 (Feb. 7, 1989), H. Mueller (to BASF).
216. U.S. Pat. 4,303,782 (Dec. 1, 1981), G. Bendoraitis and W. D. Michale (to Mobil Oil Corp.).
217. U.S. Pat. 4,670,519 (June 2, 1987), H. Mueller (to BASF).
218. U.S. Pat. 3,925,484 (Dec. 9, 1975), M. C. Baker (to E. I. du Pont de Nemours & Co., Inc.).
219. Jpn. Pat. 62 240 319 (1987), M. Shirato (to Mitsubishi Chem. Ind.).
220. Jpn. Pat. 93 031 572 (May 12, 1993), (to Asahi Chem. Ind.).
221. A. Aoshima and S. Tonomura, *Polym. Adv. Technol.* **1**(2), 127 (1990).

222. U.S. Pat. 4,568,775 (Feb. 4, 1986) A. Aoshima and S. Tonomura (to Asahi Chem. Ind.).
223. U.S. Pat. 5,099,074 (Mar. 24, 1992), H. Mueller, G. Jeschke, and R. Fischer (to BASF).
224. Eur. Pat. 126,471 (Nov. 28, 1984) A. Aoshima, S. Tonomura, and R. Mitsui (to Asahi Chem. Ind.).
225. Eur. Pat. 503,392 (Sept. 16, 1992), H. Mueller (to BASF).
226. A. Aoshima, *Shokubai*, **33**(1), 34 (1991).
227. U.S. Pat. 4,510,333 (Apr. 9, 1985), G. Pruckmayr (to E. I. du Pont de Nemours & Co., Inc.).
228. Jpn. Pat. 54 3718 (1976), H. Kondo, N. Okabe, and Y. Nakanishi (to Hodogaya).
229. S.U. Pat. 1,696,439 (Dec. 7, 1990), A. E. Baltser, V. M. Komarov, and L. S. Lokai (to Appl. Chem. Inst.).
230. Jpn. Pat. 51 084 896 (July 24, 1976), S. Maeda, A. Kondo, K. Okabe, and Y. Nakanishi (to Hodogaya).
231. U.S. Pat. 3,644,567 (Feb. 22, 1972), A. J. Hubin and S. Smith (to 3M Co.).
232. Jpn. Pat. 61141 729 (June 28, 1986), K. Masaoka, O. Kishiro, and M. Shirato (to Mitsubishi Chem. Ind.).
233. Jpn. Pat. 91 041 490 (June 24, 1991) (to Mitsubishi Chem. Ind.).
234. U.S. Pat. 5,208,385 (May 4, 1993), A. P. Kahn, R. G. Gastinger, and R. Pichai, (to ARCO).
235. Ger. Pat. DE 4,205,984 (May 6, 1993), B. Vollmert.
236. Jpn. Pat. 58 204 026 (Nov. 28, 1983), T. Sueyoshi and M. Shirato (to Mitsubishi Chem. Ind.).
237. Jpn. Pat. 52 022 097 (Feb. 19, 1977), S. Maeda, A. Kondo, and K. Okabe (to Hodogaya).
238. Jpn. Pat. 01 098 624 (Apr. 17, 1989) R. Ooshima and co-workers (to Sanyo Chem. Ind.).
239. Jpn. Pat. 61 162 522 (July 23, 1986), T. Sueyoshi and co-workers (to Mitsubishi Chem. Ind.).
240. E. Pechhold and G. Pruckmayr, *Rubber Chem. Tech.* **55**, 76 (1982).
241. Jpn. Pat. 56 118 025 (Sept. 16, 1981), S. Matsumoto and co-workers (to Japan Synth. Rubber).
242. U.S. Pat. 4,988,797 (Jan. 29, 1989), R. B. Wardle and J. C. Hinshaw (to Thiokol Chem. Corp.).
243. Ger. (DD) Pat. 291,562 (July 4, 1991), R. Grunert and co-workers (to Schiller Univ., Jena).
244. Jpn. Pat. 58 134,120 (Aug. 10, 1983), M. Iwafuji, H. Kawarasaki, and A. Saitoh (to Sanyo Chem. Ind.).
245. Jpn. Pat. 52 32 797 (1977), A. Saitoh and T. Saidai (to Sanyo Chem. Ind.).
246. U.S. Pat. 5,262,562 (Nov. 16, 1992), D. R. Hollingsworth and J. F. Kalfton (to Texaco).
247. Jpn. Pat. 04 253 724 (Sept. 9, 1992), S. Yokota and co-workers (to Daicel Chem. Ind.).
248. Jpn. Pat. 06 016 805 (Jan. 25, 1994), S. Yokota and co-workers (to Daicel Chem. Ind.).
249. Jpn. Pat. 04 277 522 (Oct. 2, 1992), S. Yokota (to Daicel Chem. Ind.); Jpn. Pat. 04 306 228 (1992), S. Yokota, M. Mori, and K. Tagawa (to Daicel Chem. Ind.).
250. F. Li and co-workers, *J. Appl. Polym. Sci.* **50**(11), 2017 (1993).
251. Jpn. Pat. 53 028 195 (Mar. 16, 1975), S. Kanai, O. Maruyama, and T. Aoki (to Sanyo Chem. Ind.).
252. Jpn. Pat. 88 061 964 (Nov. 30, 1988) (to Japan Synth. Rubber Corp.).
253. U.S. Pat. 5,001,277 (Mar. 19, 1991), J. L. Schuchardt (to Arco Chem. Tech. Inc.).
254. U.S. Pat. 3,778,480 (Dec. 11, 1973), K. Matsuda, T. Sakai, and Y. Tanaka (to Kao Soap).
255. Jpn. Pat. 73 01 100 (Jan. 9, 1973), K. Matsuzawa, Y. Suzuki, and K. Ohya (to Mitsubishi Chem. Ind.).

256. U.S. Pat. 5,218,141 (June 8, 1993), H. Mueller (to BASF).
257. Ger. Pat. 2,827,510 (U.S. 4,153,786), (Nov. 17, 1980) G. Pruckmayr (to E. I. du Pont de Nemours & Co., Inc.).
258. Eur. Pat. 382,931 (Aug. 16, 1990), M. Stehr, H. W. Voges, and H. Voges (to Huels AG).
259. Ger. Pat. 3,606,479 (Sept. 3, 1987), H. Mueller (to BASF).
260. *Terathane Polyether Glycol*, Du Pont Product Bulletin H-37161, E. I. du Pont de Nemours & Co., Inc., Wilmington, Del., 1991.
261. *Tetrahydrofuran, Properties, Uses, Storage and Handling*, Du Pont Bulletin E-85109, E. I. du Pont de Nemours & Co., Inc., Wilmington, Del., 1986.
262. C. Thomassin and J. Marchal, *Makromol. Chem.* **178**, 1327 (1977).
263. *Fire Hazard Properties of Flammable Liquids, Gases, Volatile Solids*, Bull. No. 325M, National Fire Protection Association, Boston, Mass., 1977.
264. *Annual Book of ASTM Standards*, ASTM D4274, The American Society for Testing and Materials, Philadelphia, Pa.
265. Ref. 264, ASTM D4273.
266. Ref. 264, ASTM D4672.
267. Ref. 264, ASTM D4662.
268. Alkalinity number, Du Pont Test Method T2550.005.03, E. I. du Pont de Nemours & Co., Inc., Wilmington, Del.
269. Ref. 264, ASTM D4890.
270. Ref. 264, ASTM D4878.
271. *Test Methods for Polyurethane Raw Materials*, 2nd ed., The Society of the Plastics Industry, New York, 1992.
272. *Terathane Polyether Glycols, Material Safety and Data Sheet*, E. I. du Pont de Nemours & Co., Inc., Wilmington, Del., 1991.
273. A. Wood, *Chem. Week*, 9 (Oct. 14, 1992).
274. *Chem. Eng. News*, 15 (July 25, 1994).
275. *Chem. Mark. Report.* (July 19, 1993).
276. J. R. Harrison, *J. of Elastomers Plast.* **17**, 1 (1985).
277. Technical data, E. I. du Pont de Nemours & Co., Inc., Wilmington, Del., 1993.
278. *Du Pont Magazine*, (Dec. 1984).
279. *Du Pont Magazine*, European ed., **78**, 4 (1984).
280. *Terathane Bulletin*, E-84616, E. I. du Pont de Nemours & Co., Inc., Wilmington, Del.
281. *Uses for Du Pont Glycols in Polyurethanes, Terathane Product Bulletin*, E-70687, E. I. du Pont de Nemours & Co., Inc., Wilmington, Del.
282. E. J. Goethals, *Ind. Chim. Belg.* **30**, 559 (1965).
283. M. P. Dreyfuss and P. Dreyfuss, in Ref. 7, Vol. 10, 1987, pp. 653–670.
284. E. Perez, J. G. Fatuo, and A. Bello, *Eur. Polym. J.* **23**(6), 469–474 (1987).
285. P. Geil, *Polymer Single Crystals*, Wiley-Interscience, New York, 1963, p. 526.
286. R. L. Miller, in J. Brandrup and E. G. Immergut, eds., *Polymer Handbook*, 2nd ed., John Wiley & Sons, Inc., New York, 1975, Chapt. III, p. 1.
287. Y. Takahashi, Y. Osaki, and H. Tadokoro, *J. Polym. Sci. Polym. Phys. Ed.* **18**, 1863 (1980).
288. *Ibid.*, **19**, 1153 (1981).
289. R. Gilardi, C. George, and J. Karle, *Off. Nav. U.S. Res. Rep.* No. LSM 81-1 (Oct. 15, 1981).
290. B. Moss and D. L. Dorset, *J. Polym. Sci. Polym. Phys. Ed.* **20**, 1789 (1982).
291. E. Perez and co-workers, *Colloid Polym. Sci.* **261**, 571 (1983).
292. K. E. Hardenstine and co-workers, *Gov. Rep. Announce (U.S.)* No. AD-A148271/0/GAR 85, 1985, p. 172.
293. E. Perez, A. Bello, and J. G. Fatuo, *An. Quim. Ser.* **A80**, 509 (1984).
294. E. Perez, A. Bello, and J. G. Fatuo, *Colloid Polym. Sci.* **262**, 605 (1984).
295. E. Perez, A. Bello, and J. G. Fatuo, *Colloid Polym. Sci.* **262**, 933 (1984).

296. E. Perez, A. Bello, and J. G. Fatuo, *Makromol. Chem.* **186**, 439 (1985).
297. A. Bello and E. Perez, *Thermochim. Acta*, **134**, 155–159 (1988).
298. E. Perez, A. Bello, and J. M. Perena, *Polym. Bull. (Berlin)* **20**(3), 291–296 (1988).
299. K. D. Pae, C. L. Tang, and E. S. Shin, *J. Appl. Phys.* **56**, 2426 (1984).
300. L. Garrido, E. Riande, and J. Guzman, *Makromol. Chem. Rapid Comm.* **4**, 725 (1983).
301. L. Garrido, E. Riande, and J. Guzman, *J. Polym. Sci. Polym. Phys. Ed.* **21**, 1493 (1983).
302. E. Perez and co-workers, *J. Appl. Polym. Sci.* **27**, 3721 (1982).
303. U. Guar and B. Wunderlich, *ACS Polym. Prepr.* **20**(2), 429 (1979).
304. U. Guar and B. Wunderlich, *J. Phys. Chem. Ref. Data*, **10**, 1001 (1981).
305. R. B. Jones and co-workers, *J. Appl. Polym. Sci.* **30**, 95 (1985).
306. D. C. Miles and J. H. Briston, *Polymer Technology*, Chemical Publishing Co., Inc., New York, 1979.
307. N. M. Ishmuratova and co-workers, *Plast. Massy*, **3**, 7–10 (1989).
308. K. E. Hardestine and co-workers, *J. Appl. Polym. Sci.* **30**, 2051 (1985).
309. M. Motoi and co-workers, *Bull. Chem. Soc. Japan*, **66**(6), 1778–1789 (1993).
310. Y. Kawakami and co-workers, *Polym. Int.* **31**(1), 35–40 (1993).
311. Y. Kawakami, K. Takahashi, and K. Hifino, *Macromolecules*, **24**, 4531–4537 (1991).
312. Y. Kawakami and K. Takahashi, *Polym. Bull. (Berlin)*, **25**(4), 439–442 (1991).
313. S. Penczek and P. Kubisa, *Makromol. Chem.* **130**, 186 (1969).
314. F. Andruzzi and co-workers, *Eur. Polym. J.* **18**, 685 (1982).
315. H. Hayashi and S. Okamura, *Macromol. Chem.* **47**, 230 (1961).
316. K. Takakura, K. Hayashi, and S. Okamura, *J. Polym. Sci. Part A-1*, **4**, 1747 (1966).
317. H. Sasaki and J. V. Crivello, *J. Macromol. Sci. Pure Appl. Chem.* **A29**, 915–930 (1992).
318. B. Xu, C. P. Lillya, and J. C. W. Chien, *J. Polym. Sci. Part A Polym. Chem.* **30**, 1899–1909 (1992).
319. E. J. Vandenberg and A. E. Robinson, in E. J. Vandenberg, ed., *ACS Symp. Ser.* **6**, 101–119 (1975).
320. T. Hiramo, S. Nakayama, and T. Tsuruta, *Makromol. Chem.* **176**, 1897 (1975).
321. N. L. Sidney, S. E. Shaffer, and W. J. Bailey, *ACS Polym. Prepr.* **22**(2), 373 (1981).
322. E. J. Vandenberg, *J. Polym. Sci. Polym. Chem. Ed.* **23**, 915 (1985).
323. M. Sano, D. Y. Sasaki, and T. Kunitake, *Macromolecules*, **25**, 6961–6969 (1992).
324. F. L. Tobin, P. C. Hariharan, and J. J. Kaufman, *ACS Polym. Prepr.* **22**(1), 244–245 (1981).
325. J. J. Kaufman, P. C. Hariharan, and P. B. Keegstra, *Int. J. Quantum Chem. Symp.* **21**, 623–643 (1987).
326. J. G. Cheun, J. T. Kim, and S. K. Park, *J. Korean Chem. Soc.* **35**, 636–644 (1991).
327. J. B. Rose, *J. Chem. Soc.*, 542 (1956).
328. E. J. Goethals, *Adv. Polym. Sci.* **23**, 103 (1977).
329. P. Dreyfuss and M. P. Dreyfuss, *Polym. J.* **8**, 81 (1976).
330. Y. Arimatsu, *J. Polym. Sci. A-1*, **4**, 728 (1966).
331. J. Dale, *Tetrahedron*, **49**, 8707–8725 (1993).
332. T. Saegusa and S. Kobasyashi, in Ref. 319, pp. 150–168.
333. E. Riande and co-workers, *Macromolecules*, **17**, 1431 (1984).
334. P. E. Black and D. J. Worsfold, *Can. J. Chem.* **54**, 3325 (1976).
335. Jpn. Kokai Tokkyo Koho 58 125 718 (July 26, 1983) (to Daicel Chemical Ind.).
336. Ger. Offen. 3,326,178 (Oct. 18, 1984), Y. Toga, I. Okamoto, and T. Kanno (to Daicel Chem. Ind.).
337. M. B. Frankel and co-workers, *JANNAF Propulsion Committee Meeting*, No.Ad-A103 844, CPIA Publ. No. 340, Chemical Propulsion Information Agency, Johns Hopkins University, Baltimore, Md., 1981, p. 39.

338. R. S. Miller and co-workers, *Off. Nav. Res. Rev.*, 21 (spring 1981).
339. *Ind. Chem. News*, cover (Nov. 1981).
340. P. Kubisa and S. Penczek, *ACS Polym. Prepr.* **31**(1), 89–90 (1990).
341. M. Wojtania, P. Kubisa, and S. Penczek, *Makromol. Chem. Symp.* **6**, 201–206 (1986).
342. S. V. Conjeevaram, R. S. Benson, and D. J. Lyman, *J. Polym. Sci. Polym. Chem. Ed.* **23**, 429 (1985).
343. H. Desae and co-workers, *Polymer*, **34**, 642–645 (1993).
344. H. Cheradame, J. P. Andreolety, and E. Rousset, *Makromol. Chem.* **192**, 901–918 (1991).
345. E. J. Vandenberg, *J. Polym. Sci. Polym. Chem. Ed.* **10**, 2887 (1972).
346. R. T. Keen, *Anal. Chem.* **29**, 1041 (1957).
347. K. Matyjaszewski and S. Penczek, *Macromolecules*, **7**, 173 (1974).
348. K. Baum and co-workers, *J. Org. Chem.* **48**, 2953 (1983).
349. U.S. Pat. 4,483,978 (Nov. 20, 1984), G. E. Manser (to SRI International).
350. G. E. Manser, "Nitrate Ester Polyether Glycol Prepolymer," *JANNAF Propulsion Committee Meeting*, New Orleans, La., Chemical Information Agency, John Hopkins University, Baltimore, Md., 1984.
351. U.S. Pat. 4,707,540 (Nov. 17, 1987), G. E. Manser (to Morton Thiokol, Inc.).
352. U.S. Pat. 5,210,153 (May 11, 1993), G. E. Manser and R. S. Miller (to Morton Thiokol, Inc.).
353. Jpn. Pat. 05 051 444 (1993), K. Bando and co-workers (to Daicel Chem. Ind.).
354. M. A. H. Talukder, *Makromol. Chem. Makromol. Symp.* **42/43**, 501–511 (1991).
355. Jpn. Pat. 05 78 442 (1993), K. Bando (to Daicel Chem. Ind.).
356. H. Cheradame and E. Gojon, *Makromol. Chem.* **192**, 919–933 (1991).

*General References*

References 1,5,6,7, and 283.
K. J. Ivin and T. Saegusa, eds., *Ring-Opening Polymerization*, Vols. 1 and 2, Elsevier, New York, 1984.

GERFRIED PRUCKMAYR
E. I. du Pont de Nemours & Co., Inc.

P. DREYFUSS
M. P. DREYFUSS
Consultants

# POLYETHYLENE. See OLEFIN POLYMERS.

# POLY(ETHYLENE OXIDE). See POLYETHERS, ETHYLENE OXIDE POLYMERS.

# POLYFLUOROSILICONES. See FLUORINE COMPOUNDS, ORGANIC.

# POLYFORMALDEHYDE. See ACETAL RESINS.

# (POLYHYDROXY)BENZENES

Polyhydric phenols with more than two hydroxy groups (ie, the three positional isomers of benzenetriol, the three isomeric benzenetetrols, benzenepentol [4270-96-6], and benzenehexol [608-80-0]) are discussed in this article. The benzenediols are catechol, resorcinol, and hydroquinone (see HYDROQUINONE, RESORCINOL, AND CATECHOL).

The following names of the benzenetriols have been used.

| Common (trivial) name | CAS Registry Number | *Chemical Abstracts* | Other usage |
|---|---|---|---|
| pyrogallol (pyrogallic acid) | [87-66-1] | 1,2,3-benzenetriol | 1,2,3-trihydroxybenzene |
| hydroxyhydro-quinone | [533-73-3] | 1,2,4-benzenetriol | 1,2,4-trihydroxybenzene |
| phloroglucinol | [108-73-6] | 1,3,5-benzenetriol | 1,3,5-trihydroxybenzene |

The benzenetetrols, -pentol, and -hexol do not have trivial names, except for 1,2,3,4-benzenetetrol [642-96-6], which was named apionol in some of the older literature. The other two benzenetetrol isomers are 1,2,3,5- [634-94-6] and 1,2,4,5-benzenetetrol [636-32-8].

Derivatives of these compounds or their corresponding quinones are of widespread occurrence in nature. They are abundant in plants and fruits as glucosides, chromones, coumarin derivatives, flavonoids, essential oils, lignins, tannins, and alkaloids (see ALKALOIDS; COUMARIN; LIGNIN; OILS, ESSENTIAL). They also occur in microorganisms and animals. Many of these compounds have distinct properties and uses, eg, antibiotics (qv), plant-growth factors, insecticides, astringents, antioxidants (qv), toxins, sweeteners (qv), pigments (qv) and dyes, drugs, and many others (see DYES AND DYE INTERMEDIATES; INSECT CONTROL TECHNOLOGY; PHARMACEUTICALS). Developing uses for the benzenepolyols and derivatives appear particularly valuable in the pharmaceutical and agricultural chemical areas. The most recent applications of these compounds are as components of photosensitive compounds in high resolution heat-resistant photoresist compositions.

Identification, isolation, and removal of (polyhydroxy)benzenes from the environment have received increased attention throughout the 1980s and 1990s. The biochemical activity of the benzenepolyols is at least in part based on their oxidation–reduction potential. Many biochemical studies of these compounds have been made, eg, of enzymic glycoside formation, enzymic hydroxylation and oxidation, biological interactions with biochemically important compounds such as the catecholamines, and humic acid formation. The range of biochemical function of these compounds and their derivatives is not yet fully understood.

## Pyrogallol

Pyrogallol (**1**) was first observed by Scheele in 1786 as a product of the dry distillation of gallic acid [*149-91-7*] (3,4,5-trihydroxybenzoic acid). Pyrogallol, which is of widespread occurrence in nature, is incorporated in tannins, anthocyanins, flavones, and alkaloids (1).

**Properties.** Pyrogallol (mp 133–134°C) forms colorless needles or leaflets which gray on contact with air or light. Its boiling point at atmospheric pressure with partial decomposition is 309°C; at 13.3 kPa (100 mm Hg), 232°C; and at 1.3 kPa (10 mm Hg), 168°C. When heated slowly, pyrogallol sublimes without decomposition; sp gr at 4°C, 1.453; heat of combustion, 2.673 MJ/mol (638.9 kcal/mol); solubility in parts per 100 parts solvent: 40 in water at 13°C, 62.5 in water at 25°C, 100 in alcohol at 25°C, 83.3 in ether at 25°C, slightly soluble in benzene, chloroform, and carbon disulfide. Pyrogallol is the strongest reducing agent among the benzenepolyols. Therefore, it is oxidized rapidly in air; its aqueous alkaline solution absorbs oxygen from the air and darkens rapidly. Sodium sulfite retards such oxidation.

Pyrogallol oxidized, which is obtained by the action of air and ammonia on pyrogallol, is a brownish black to black lustrous powder and is almost insoluble in water, alcohol, or ether but is soluble in alkalies. Hexahydroxybiphenyl [*4371-20-4*] (diphenylhexol), $(HO)_3C_6H_2C_6H_2(OH)_3$, is formed by mixing pyrogallol with barium hydroxide solution while air is passed through the reaction mixture. In solutions of hydrogen peroxide, pyrogallol oxidizes rapidly in the presence of catalysts, eg, colloidal suspensions of metals or metallic oxides, and luminescence occurs.

Purpurogallin (**5**), a red-brown to black mordant dye, forms from electrolytic and other mild oxidations of pyrogallol (**1**). The reaction is believed to proceed through 3-hydroxy-*o*-benzoquinone (**2**) and 3-hydroxy-6-(3,4,5-trihydroxyphenyl)-*o*-benzoquinone (**3**). The last, in the form of its tautomeric triketonic structure, represents the vinylogue of a $\beta$-diketone. Acid hydrolysis leads to the formation of (**4**), followed by cyclization and loss of formic acid to yield purpurogallin.

Methylation of (**1**) with methyl iodide or dimethyl sulfate in the presence of alkali gives 3-methoxy-1,2-benzenediol, 2,3-dimethoxyphenol, or 1,2,3-trimethoxybenzene (**2**). On heating with aqueous potassium bicarbonate, pyrogallol-4-carboxylic acid (**6**) and a lesser amount of gallic acid (**7**) are formed (**3**). Reaction of pyrogallol with phosgene (qv) in the presence of pyridine gives pyrogallol carbonate (**8**) (**4**). Bromination of this carbonate yields both

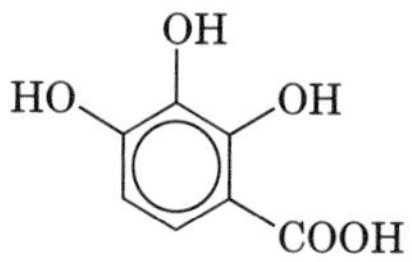

(**6**) 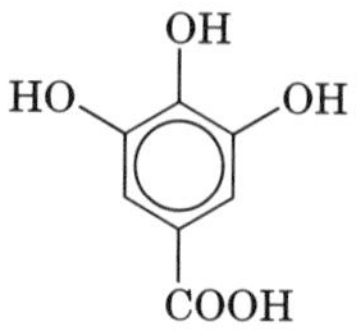          (**7**) 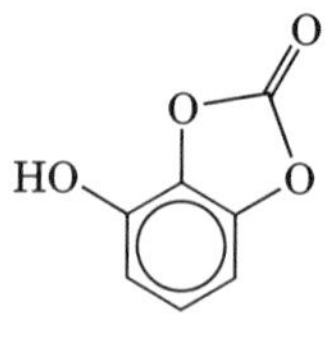          (**8**)

4-bromopyrogallol and 4,6-dibromopyrogallol. The direct bromination of pyrogallol in carbon tetrachloride produces 4,5,6-tribromopyrogallol, and with more bromine, 1,2,6,6-tetrabromocyclohexene-3,4,5-trione (**9**) is obtained (**5**).

The formation of (**9**) is evidence for the ability of pyrogallol to react in keto forms. However, in contrast to phloroglucinol, pyrogallol does not react as a ketone with hydroxylamine.

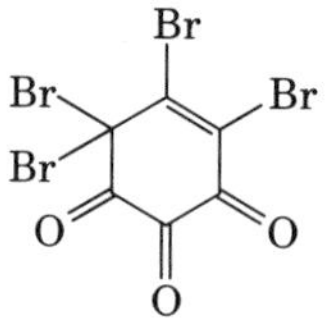

(**9**) 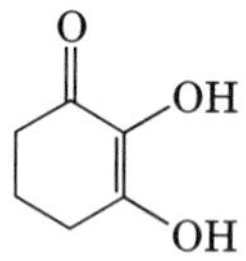          (**10**)

The hydrogenation of pyrogallol in ethanol at 100°C and 17 MPa (2400 psig), using a moist Raney nickel catalyst, results in a 60% yield of the γ-isomer, ie, *cis,cis*-1,2,3-cyclohexanetriol with only minor amounts of the cis,trans-, ie, β-isomer, and the trans,trans-, ie, α-isomer (**6**). The hydrogenation of pyrogallol in water in the presence of 1 mol of alkali at 60°C and 7 MPa (1000 psig), with a Raney nickel catalyst, gives dihydropyrogallol [4337-36-4] (**10**) (**7**). Pyrogallol forms salts or chelates with many metals, some of which are used for identification in analysis, as pigments or lakes, eg, in inks (qv), or for medicinal purposes.

**Manufacture and Synthesis.** The commercial manufacturing process is based on Scheele's original procedure starting with crude gallic acid, which is extracted from nutgalls or tara powder. It proceeds according to the following equation: $C_6H_2(OH)_3COOH \rightarrow C_6H_3(OH)_3 + CO_2$.

Gallic acid is heated with about half its weight of water in a copper autoclave until the pressure reaches 1.2 MPa (12 atm) and the temperature is 175°C.

Steam and carbon dioxide are released but sufficient water is retained to maintain the pyrogallol as a liquid. The cooled solution is decolorized with animal charcoal and is then evaporated until the volatile pyrogallol distills into iron receivers. The solidified material is purified by repeated distillation, sublimation, or vacuum distillation at 200°C in the presence of dialkyl phthalates (8).

In 1981 Mallinckrodt was the only U.S. manufacturer of pyrogallol via decarboxylation of plant-derived gallic acid, but it has since ceased production. Harshaw Chemical (Europe) makes pyrogallol by the same process. Because of the continuing uncertainties of supply of plant materials for gallic acid–pyrogallol manufacture, and because of valuable uses for pyrogallol, there is much interest in the development of synthetic processes. Gallic acid (7) can be made, presumably in good yield, by the sodium alkoxide-catalyzed condensation of a tricarballylic ester with an acetal of mesoxalic ester, eg, dialkyl(ethylenedioxy)malonate, via substituted cyclohexane-1,2,3-triones (9).

Resorcinol can be hydroxylated with 50 wt % hydrogen peroxide in the presence of hexafluoroacetone at ca 60°C to give a mixture of pyrogallol and 1,2,4-trihydroxybenzene (10). The hydrolysis of 2,6-diamino-4-$t$-butylphenol with aqueous hydrochloric acid at 250°C for 8 h in a pressure reactor provides a 48% yield of pyrogallol and a 9% yield of 5-$t$-butylpyrogallol (11). Pyrogallol or 5-alkylpyrogallols can be prepared from 2,6-dibromophenol or 4-alkyl-2,6-dibromophenol by treatment with sodium methoxide. The 2,6-dimethoxyphenols produced are subjected to ether cleavage with dealkylation in the case of 4-$t$-butyl-2,6-dimethoxyphenol with aqueous 48 wt % hydrobromic acid to give pyrogallol in good yield (12).

2,2,6,6-Tetrachlorocyclohexanone (11) can be hydrolyzed with a base, eg, sodium acetate, to give pyrogallol in high yield and purity (13). The preparation

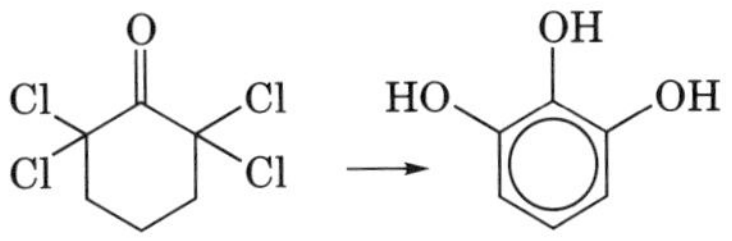

(11)  (1)

of the starting material (11) by chlorination of cyclohexanone in the presence of collidine as the catalyst has been patented (14). The sodium hydride-catalyzed condensation of dialkyl glutarate (12) with a dialkyl dialkoxymalonate (13) to pyrogallol-4,6-dicarboxylic acid (14) by means of intermediate (15) has been patented (15); the dicarboxylic acid (14) is decarboxylated in methanol at 200°C under pressure to give pyrogallol.

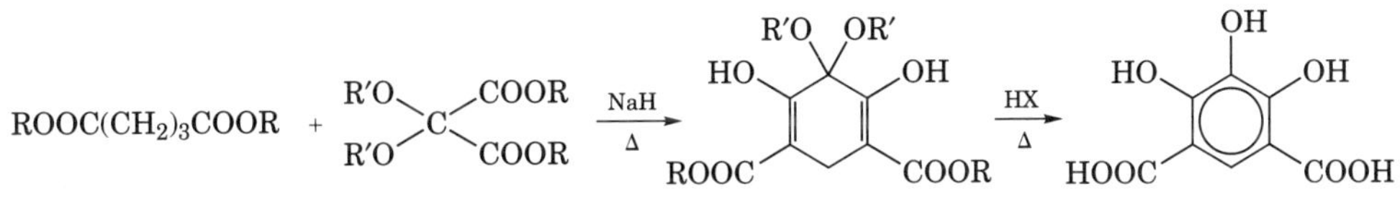

ROOC(CH₂)₃COOR +

(12)  (13)  (15)  (14)

Another synthesis of pyrogallol is hydrolysis of cyclohexane-1,2,3-trione-1,3-dioxime derived from cyclohexanone and sodium nitrite (16). The dehydrogenation of cyclohexane-1,2,3-triol over platinum-group metal catalysts has been reported (17) (see PLATINUM-GROUP METALS). Other catalysts, such as nickel, rhenium, and silver, have also been claimed for this reaction (18).

The first synthetic pyrogallol plant using hydrolysis of chlorinated cyclohexanol (2,2,6,6-tetrachlorocyclohexanone) was built by BFC Chemicals, Inc. (Muskegon, Michigan) and has been producing pyrogallol for the carbamate insecticide Beniocarb since 1982 (8,19). Société Française Hoechst offers pyrogallol for sale in the United States (American Hoechst Corp.), and Japan is also a source of this chemical.

**Grades and Specifications.** Harshaw Chemical sells material from their overseas operation in four grades of pyrogallic acid: pure crystal, pure powder, resublimed, and technical grade. Depending on grade, pyrogallic acid is coarse and white to slightly yellow, having lustrous crystals with some smaller crystals. It may contain some black or brown specks, and has a characteristic odor. Its melting point is 131.5–135°C and maximum residue on ignition is 0.1%.

**Analysis.** Freshly prepared ferrous sulfate test solution produces a blue color in an aqueous solution of pyrogallol (20). Pyrogallol can be detected in amounts of ca 0.5 $\mu$g by the violet to orange color that results from the addition of phloroglucinol to ammoniacal pyrogallol. Pyrogallol reacts with osmium tetroxide [20816-12-0] to form a compound that is reddish violet in dilute solution and almost black in concentrated solution. This reaction is extremely sensitive and can be used to detect as little as one part pyrogallol in $2 \times 10^6$ parts water (21). Various other color tests for pyrogallol have been reported (22,23). Derivatives used for the identification of pyrogallol are tris(phenylurethane), mp 173°C; tris-(3,5-dinitrobenzoate), mp 205°C; and tribenzoate, mp 90°C. Thin-layer chromatography is applicable to the detection of pyrogallol (24). Of the modern instrumental methods of analysis, liquid chromatography is particularly well-suited to the analysis of pyrogallol.

**Health and Safety Factors.** Pyrogallol is extremely toxic. Extensive exposure of the skin may cause discoloration, local irritation, eczema, or death if it is absorbed. Repeated contact with the skin may lead to sensitization. The principal symptom of poisoning attributable to pyrogallol is its effect on the red blood corpuscles which break down and lose their hemoglobin. The tremendous affinity of pyrogallol for oxygen of the blood has been shown in experimental animals, where complete removal of oxygen from the blood occurred as well as fragmentation and destruction of the erythrocytes. Severe pyrogallol poisoning also leads to degeneration of the liver and kidneys, and symptoms exhibited in such cases include urinary disturbance, headache, cyanosis, chills, vomiting, and diarrhea (25). A yeast test has proved useful in checking acute toxicity of a number of chemicals including pyrogallol (26). The effect of pyrogallol on algae has also been studied (27). Acute toxicity data include oral $LD_{50}$ (rat) = 789 mg/kg, intraperitoneal $LD_{50}$ (mouse) = 400 mg/kg, and oral $LD_{50}$ (rabbit) = 1600 mg/kg (28).

**Uses.** The main commercial applications of pyrogallol are in pharmaceuticals (qv) and pesticides (qv). Pyrogallol is the oldest and one of the more versatile of the photographic developing agents in use (see PHOTOGRAPHY). Strong contrasts are possible with concentrated solutions, and soft delicate shades are achieved with more dilute solutions and lower alkali concentrations. However, because pyrogallol oxidizes readily, the yellow oxidation product stains the gelatin so that the useful life of a pyrogallol developer is short. Pyrogallol is relatively costly, quite toxic, and water soluble, therefore its removal from the environment has become important. Hydrogen peroxide-mediated photodegradation has been studied by flash photolysis/hplc techniques (29), as well as oxidation as pretreatment in wastewater discharge (30). Electrochemical oxidation has also received scrutiny (31). The ease of oxidation of pyrogallol is the basis for its use in fur and hair dyeing and as a chemical reagent for the estimation of oxygen. Pyrogallol has been used as part of analytical procedures, thus it is a component for a "screw-cap test" for fat-soluble vitamins (qv) (32). Pyrogallol is used to demonstrate chemiluminescence (33) and traces of chromium(III) are determined with a pyrogallol chemiluminescence system (34) (see LUMINESCENT MATERIALS, CHEMILUMINESCENCE).

Scale-preventing coatings for polymerization reactors may contain pyrogallol or hydroxyhydroquinone condensed with amines (35–37). The use of pyrogallol and certain derivatives as stabilizers for photographic silver halide recording materials to improve storage stability and to reduce fogging has been patented (38). An improved synthesis of 5-bromopyrogallol-1,3-dimethyl ether by bromination of pyrogallol-1,3-dimethyl ether with $N$-bromosuccinimide in chloroform–ethanol has been disclosed; the product is an intermediate for photographic optical filter agents (39) (see OPTICAL FILTERS (SUPPLEMENT)).

An adhesive with good peel strength and soldering tolerance for copper in printed circuits is based on a mixture of poly(vinyl butyral) and a modified melamine resin containing pyrogallol (40). A rubberized pyrogallol–formaldehyde adhesive improves the adhesion of rubber to nylon (41). Pyrogallol 1-methyl, 3-propyl or allyl ethers are useful in natural smoke-aroma compositions for food or tobacco (42). Pyrogallol-1,3-dimethyl ether imparts a bonito-like aroma to dried fish (43). Zinc or chrome-plated steel is treated with aqueous pyrogallol or gallic acid to improve the adhesion of a final alkyd or melamine resin coating (44). Glass fiber for reinforcement of cementitious products is protected from corrosion by a pyrogallol dip (45).

Certain natural products containing a pyrogallol moiety, eg, myricetin, are effective inhibitors of radical-chain reactions (46). Pyrogallol, as an antioxidant, is useful for preventing decomposition of alkali cellulose (47). A mixture of defatted rice bran and alkalized pyrogallol is useful in protecting foodstuffs from oxygen (48). Phosphite esters of 4,6-dialkylpyrogallol are effective heat- and light-stabilizers for plastics (49). The formation of deposits on reactor walls during vinyl chloride polymerization can be prevented by treating the reactor surface with a compound comprised in part of chelate-forming and free-radical chain-inhibiting groups, eg, pyrogallol (50). Pyrogallol can be used as a stabilizer to inhibit peroxide formation in dicyclopentadiene (51).

Many patents have been issued on the use of pyrogallol derivatives as pharmaceuticals. Pyrogallol has been used externally in the form of an ointment or a solution in the treatment of skin diseases, eg, psoriasis, ringworm, and lupus erythematosus. Gallamine triethiodide (**16**) is an important muscle relaxant in surgery; it also is used in convulsive-shock therapy.

**(16)**

Trimethoprim (2,4-diamino-5-(3,4,5-trimethoxybenzyl)pyrimidine) is an antimicrobial and is a component of Bactrin and Septra. Trimetazidine (1-(2,3,4-trimethoxybenzyl)piperazine; Vastarel; Yosimilon) is used as a coronary vasodilator. 1,2,3,4-Tetrahydro-6-methoxy-1-(3,4,5-trimethoxyphenyl)-9$H$-pyrido[3,4-$b$]indole hydrochloride is useful as a tranquilizer (52) (see HYPNOTICS, SEDATIVES, ANTICONVULSANTS, AND ANXIOLYTICS). Substituted indanones made from pyrogallol trimethyl ether depress the central nervous system (CNS) (53). Tyrosine- and glycine(2,3,4-trihydroxybenzyl)hydrazides are characterized by antidepressant and anti-Parkinson activity (54). 2-($\omega$-Dialkylaminoalkoxy)-3',4',5'-trimethoxychalcones are effective as antihypertensives (55). $\beta$-(3,4,5-Trimethoxyphenyl)propionitrile was patented as a bactericide and a fungicide (56). Numerous 3,4,5-trialkoxycinnamamides are claimed to have therapeutic properties (57). Many $N$-alkyl, $N$-aminoalkyl-3,4,5-, or 2,3,4-trimethoxybenzylamines are useful therapeutic agents (58). Bendiocarb (2,2-dimethyl-1,3-benzodioxol-4-yl $N$-methylcarbamate; Ficam) is used for the control of cockroaches, crickets, carpet beetles, earwigs, ants, silverfish, wasps, fleas, and bedbugs in foodstores and houses. Related benzodioxolyl and benzodioxepinyl carbamates have been patented as insecticides (59).

Pyrogallol has been cited for use in photosensitive compositions. It is used in the form of sulfonate esters of quinonediazides which hydrolyze when exposed to actinic light to liberate the acid which, in turn, catalyzes further reaction of novolak resins (60).

The synthesis and phase structure of a three-arms—nine-chain liquid crystal (**17**) based on pyrogallol and phloroglucinol has been reported (61), as has a complexing agent for amino acids in water (**18**) (62) (Fig. 1).

**Derivatives.** Gallic acid (**7**) is the most important derivative of pyrogallol. It is a colorless solid that crystallizes from water as the monohydrate and begins to dehydrate at ca 100°C. The anhydrous compound melts at 253°C with decomposition, its sp gr at 4°C is 1.694, and its dissociation constant at 25°C is $3.8 \times 10^{-5}$. It is soluble in alcohol and acetone, sparingly soluble in water, and insoluble in chloroform and benzene. Gallic acid darkens on exposure to light. It is manufactured by the chemical or enzymic hydrolysis of tannin from nutgalls (ie, gallnuts), Aleppo galls, or tara powder, ie, the ground seed pod of the Peruvian tree, *Coulteria tinctoria*. Gallic acid is sold as the monohydrate or in anhydrous grades; bulk price in March of 1995 was listed as $28.67/kg.

Gallic acid has been used medicinally as a urinary astringent and internal antihemorrhageant and in veterinary medicine for the treatment of diarrhea (see VETERINARY DRUGS). It is also used to manufacture pyrogallol by decarboxylation and as a plant-growth regulator (see GROWTH REGULATORS, PLANT). The rates of

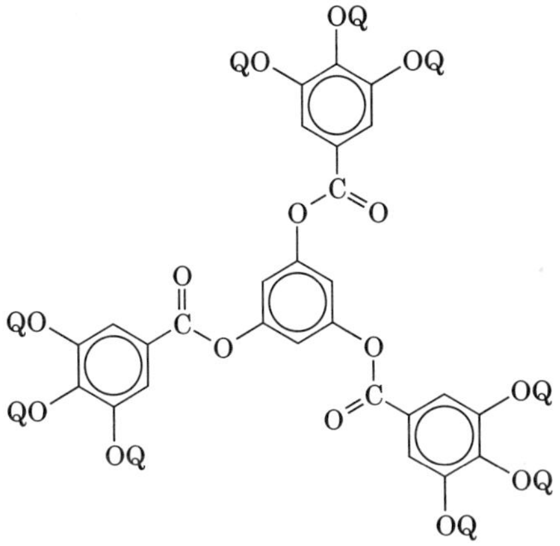

**Fig. 1.** Structures based on pyrogallol and phloroglucinol where $Q = C_{12}H_{25}$.

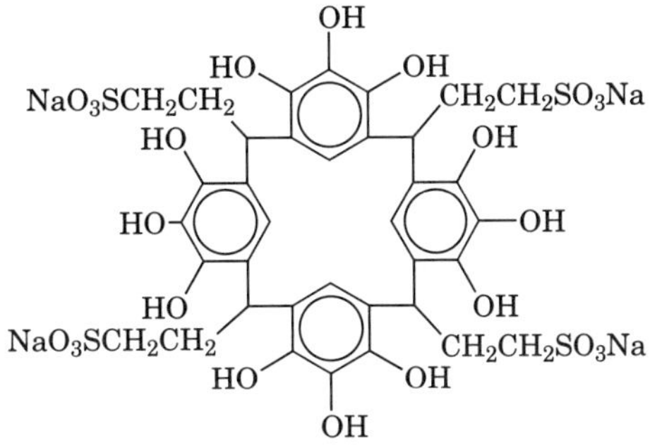

decarboxylation of gallic acid in pyrogallol, catechol, and resorcinol have been studied. The reaction is first order and rate constants decrease in the above solvent order (63).

Gallic acid has traditionally been used with ferrous sulfate to make various types of inks, particularly the blue-black permanent-type writing inks. It is used in photothermographic reproduction processes, as a process chemical in engraving and lithography, as a developer in photography, and in tanning and fur- and hair-dyeing preparations (64). Miscellaneous applications of gallic acid include its use as a deflocculating, thickening, and sizing agent in the manufacture of wallboard; as a mordant in the manufacture of colored paper and fiberboard; as an analytical reagent for alkaloids, metals, and mineral acids; and in the manufacture of alizarin, thioflavine, and gallocyanine dyes.

Propyl gallate [121-79-9] is a food antioxidant made in the United States by Eastman Chemical Products, Inc.; it is used in synergistic combination with the antioxidants 4-methyl-2,6-di-*tert*-butylphenol (BHT) and butylated hydroxyanisole (BHA). The December 1995 price was ca $35.85/kg. Bismuth subgallate (gallic acid, bismuth basic salt) (**19**) is used as a dusting powder in dermatology. It is an ingredient in Bongast. Hexobendin (N,N'-dimethyl-N,N'-bis-[3-(3′,4′,5′-trimethoxybenzoyloxy)propyl]ethylenediamine; Reoxyl) (**20**) is a coronary vasodilator. Trioxazin (N-(3,4,5-trimethoxybenzoyl)morpholine), has been used as a tranquilizer. There has been considerable interest in amides of gallic acid trialkyl ethers as therapeutic agents (65).

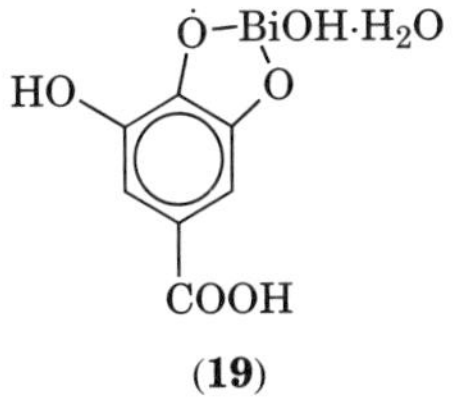

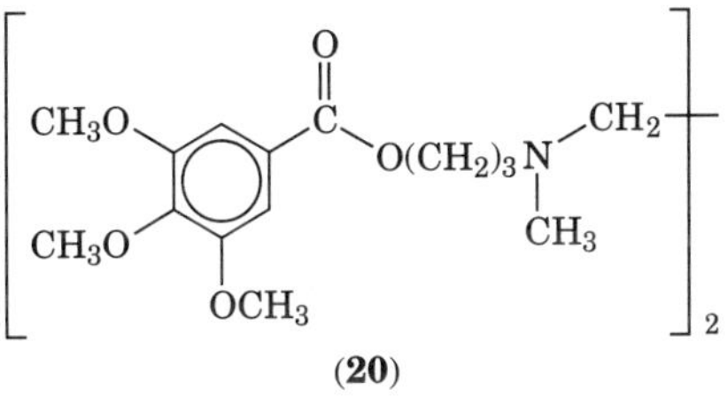

An ink composition containing a mixture of zinc diethyldithiocarbamate and bismuth subgallate changes color from white/yellowish to green when steam sterilized; thus the mixture is useful as a sterilization indicator in medicine (66) (see CHROMOGENIC MATERIALS; STERILIZATION TECHNIQUES). Alkyl gallates and gallic acid amides are useful stabilizers for aromatic amines (67). Gallic acid and its esters are adhesion promoters for cyanoacrylate adhesives (68) (see ACRYLIC ESTER POLYMERS). The addition of 0.01–0.04 wt % gallic acid to gypsum-based building-panel formulations increases the resistance of these panels to collapse (69). Carbon steel can be made corrosion resistant by surface treatment with a gallic acid–chromite solution (70) (see CORROSION AND CORROSION CONTROL). The addition of small amounts of gallic esters to tin-ore flotation systems results in increased collection of tin (71) (see FLOTATION). Photographic silver halide emulsions can be stabilized with gallic acid or an alkyl gallate to give long storage stability and to eliminate fogging (72). Alkyl gallates are useful components in latent-image printing inks which are developed with iron salts (73). Carbamate esters of gallic acid are useful fungicides (74).

The alkaloid reserpine [50-55-5], which is isolated from the roots of *Rauwolfia serpentina L.*, contains a gallate trimethyl ether moiety. Reserpine is used as an antihypertensive and a tranquilizer. A vinylogue of reserpine, rescinnamine [24815-24-5], is also an antihypersensitive (75). Methoxsalen [298-81-7] (8-methoxypsoralen; 7H-9-methoxy-furo[3,2-g][1]benzopyran-7-one) (21), a furocoumarin that occurs in plants, eg, *Leguminosae* and *Umbelliferae*, is used in the treatment of vitiligo, as a suntanning promoter, and as a sunburn protectant. It is also available by synthesis (76).

(21)                              (22)

Gallein [2103-64-2] (pyrogallolphthalein, 4,5-dihydroxyfluorescein, tetrahydroxyfluoran; CI 45445) (22) forms greenish yellow crystals when anhydrous and red crystals in a 1:1.5 ratio with water. It is obtained by heating one part of phthalic anhydride with two parts of pyrogallol or gallic acid. Gallein is used as a sensitive indicator for acids, alkali hydroxides, and ammonia but not for carbonates. A dilute solution of gallein in 50% alcohol is used as a colorimetric reagent for determining phosphates in urine. Monophosphates give a yellow color, dibasic phosphates give red, and tribasic phosphates give violet.

Gallacetophenone [528-21-2] (4-acetylpyrogallol, 2,3,4-trihydroxyacetophenone; Alizarin Yellow C) forms gray leaf crystals or a yellow or brown powder, mp 173°C. It is slightly soluble in water, soluble in alcohol and in ether, and very slightly soluble in benzene. It is used medicinally as an antiseptic for skin diseases. It and other 4-acylpyrogallols are useful protective agents against harmful radiation (77) (see RADIOPROTECTIVE AGENTS).

Mescaline [54-04-6] (2-(3,4,5-trimethoxyphenyl)ethylamine) is the active ingredient in mescal buttons (peyotl or peyote), which are the dried tops of the Mexican dumpling cactus *Lophopora williamsi*. Mescaline produces visual hallucinations on ingestion. Its possible use as a psychotomimetic drug in the field of mental health has been studied (see ALKALOIDS; PSYCHOPHARMACOLOGICAL AGENTS).

Colchicine (**23**) is a toxic substance occurring in *Colchicum autumnale*; it contains the nucleus of pyrogallol trimethyl ether. Colchicine has been used in the treatment of acute gout, and in plant genetics research to effect doubling of chromosomes.

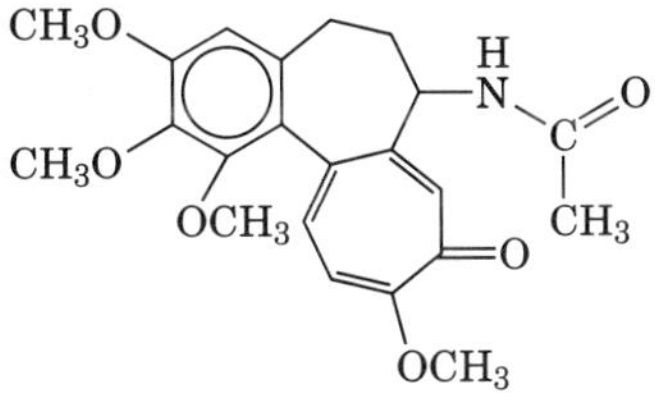

(**23**)

## Hydroxyhydroquinone

Hydroxyhydroquinone (**24**) forms colorless plates from diethyl ether when freshly prepared. It occurs in many plants and trees in the form of ethers, quinonoid pigments, coumarin derivatives, and complex compounds. Sponges from the coastal waters of Florida have been found to contain small amounts of 1,2,4-trihydroxybenzene and traces of 2,2′,4,4′,6,6′-hexahydroxybiphenyl (**25**) (78). The benzenetriol has also been isolated from tobacco leaves and tar from tobacco smoke (79). Hydroxyhydroquinone has strong reducing properties. Applications have been suggested in the synthesis of agricultural and photographic chemicals, drugs, and stabilizers (qv).

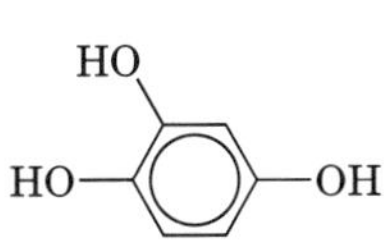

(**24**) 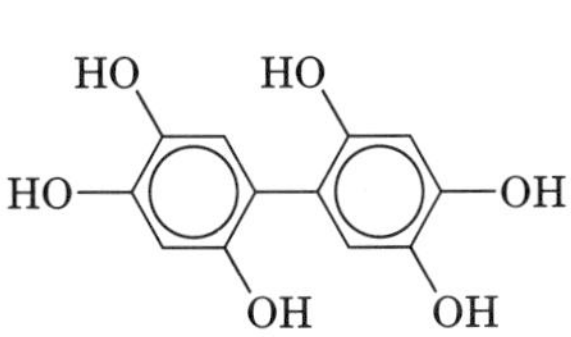                    (**25**)

**Properties.**   Hydroxyhydroquinone forms platelets or prisms (mp 140.5°C). The compound is easily soluble in water, ethanol, diethyl ether, and ethyl acetate and is very sparingly soluble in chloroform, carbon disulfide, benzene, and ligroin.

Hydroxyhydroquinone reacts as a typical oxidizable polyhydric phenol, but also undergoes certain keto-group reactions. In aqueous alkaline solution, it absorbs oxygen as effectively as pyrogallol. These solutions darken rapidly in the presence of oxygen, hydrogen peroxide, or potassium peroxysulfate, and produce a dark, humic acid-type precipitate. Mixing with excess bromine results in the formation of 2-hydroxy-3,5,6-tribromo-1,4-benzoquinone (80). Reduction with sodium amalgam produces dihydroresorcinol (**26**) (81). Condensation with aldehydes in the presence of sulfuric acid leads to the formation of 9-substituted 2,6,7-trihydroxyfluorones (**27**) (82).

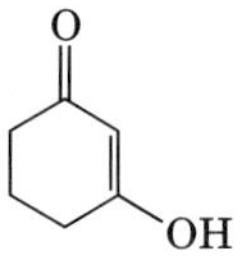

(**26**) 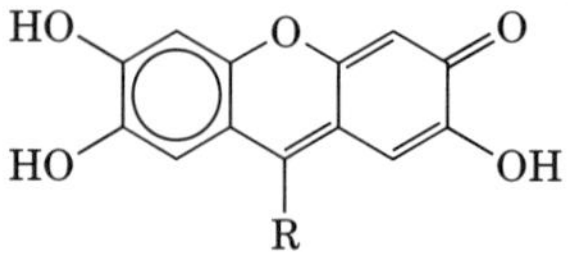                               (**27**)

Condensation of hydroxyhydroquinone with ethyl acetoacetate gives 6,7-dihydroxy-4-methylcoumarin (83). Condensation with phthalic anhydride gives hydroxyhydroquinone phthalein (**28**) (84). Mild oxidants, eg, silver oxide, produce 2-hydroxy-*p*-benzoquinone. Reaction with ammonia or amines in aqueous solution at room temperature in the absence of air yields the corresponding 2,4-dihydroxyanilines. After prolonged heating with sodium bisulfite, an adduct containing 2 mol of sulfite is formed (85). Hydrogenation of hydroxyhydroquinone with a nickel catalyst in water or alcohol gives a mixture of stereoisomeric 1,2,4-cyclohexanetriols. Ethers or esters can be formed in the usual manner with alkylating and acylating agents.

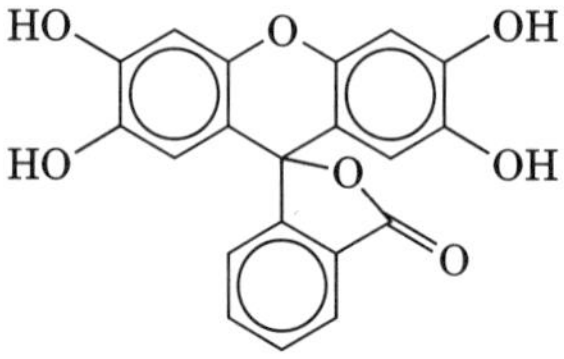

(**28**) 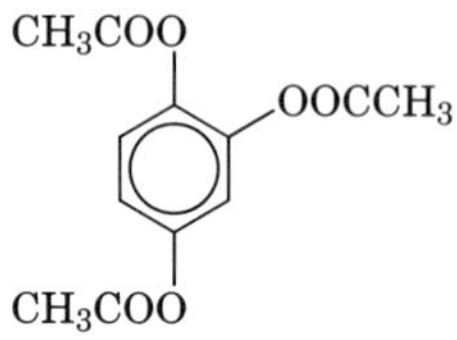                               (**29**)

**Synthesis.**  Hydroxyhydroquinone is not produced on a large scale, but many uses for it are being developed. The most convenient preparation of hydroxyhydroquinone is the reaction of *p*-benzoquinone with acetic anhydride in the presence of sulfuric acid or phosphoric acid. The resultant triacetate (**29**) can be hydrolyzed to hydroxyhydroquinone (86).

Hydroxyhydroquinone was first synthesized by the caustic fusion of hydroquinone (80,87). The oxidation of aqueous alkaline solutions of 2,4- or 3,4-dihydroxybenzaldehyde or 2,4- or 3,4-dihydroxyacetophenone with hydrogen peroxide yields hydroxyhydroquinone (88). The oxidation of vanillin (qv) in this manner gives 2-methoxyhydroquinone. 5-*tert*-Alkyl-2-hydroxy-1,4-benzoquinone can be obtained in good yield by the oxidation of 4-*tert*-alkylcatechol with oxygen in methanolic potassium hydroxide (89). Reduction of the quinone yields the corresponding alkylhydroxyhydroquinone. Conversion of dilute aqueous solutions of 5-hydroxymethyl-2-furaldehyde and D-fructose under near-critical conditions leads to 1,2,3-benzenetriol in good yields (90). It is also a product, along with catechol and hydroquinone, in the oxidation of phenol (qv) in the presence of horseradish peroxidase (91). The acid-catalyzed oxidation of phenol with hydrogen peroxide in acetonitrile has yielded 88 mol %, based on $H_2O_2$ of 1,2,3-benzenetriol (92).

**Analysis.** Dilute aqueous solutions of hydroxyhydroquinone turn blue-green temporarily when mixed with ferric chloride. The solutions darken upon addition of small amounts, and turn red with additions of larger amounts of sodium carbonate. Derivatives used for identification are the picrate, which forms orange-red needles (mp of 96°C), and the triacetate (mp 96–97°C). Thin-layer chromatography and liquid chromatography are well suited for the qualitative and quantitative estimation of hydroxyhydroquinone (93,94).

**Health and Safety Factors.** The $LD_{50}$ of 1,2,4-trihydroxybenzene in mice after intracutaneous injection is 371 $\mu g/g$ (95). Contact with hydroxyhydroquinone may blacken skin and fingernails. 1,2,3-Benzenetriol and other polyhydroxybenzenes have been found in water sources; when these react with chlorine and nitrite, derivatives having higher mutagenic potentials than their parent compounds, are formed. The mutagenicity of 1,2,3-benzenetriol is surpassed by that of hydroquinone but is greater than that of pyrogallol. The other di- and trihydroxybenzenes have been found to be nonmutagenic (96).

**Uses.** Hydroxyhydroquinone has been used in hair and mordant dyes, for healing plant wounds, and in corrosion inhibitors and adhesives.

Sesamex [51-14-9] (Sesoxane) (**30**) is a synergist of low toxicity, acute oral $LD_{50}$ (rat) = 2000–2270 mg/kg, for pyrethrins and allethrin. 6,7-Dihydroxy 4-methylcoumarin has been offered as an antioxidant for phenolics and polymers, and as an anthelmintic. 2,4,5-Trihydroxybutyrophenone has been available as an antioxidant and light stabilizer for polyolefins, waxes, and foods. Isoflavones, eg (**31**), have been patented as components of antioxidant compositions for foods and cosmetics (qv) (97).

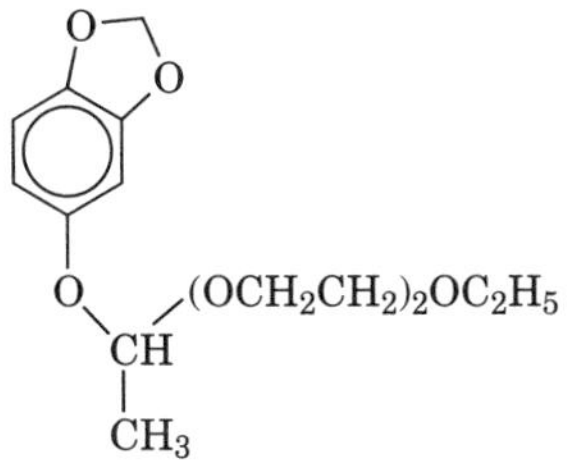

(**30**) 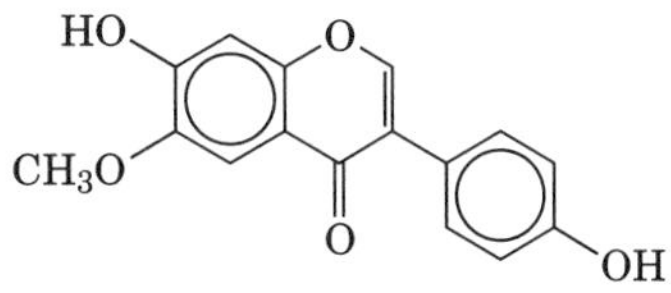                                          (**31**)

Hydroxyhydroquinone and pyrogallol can be used for lining reactors for vinyl chloride suspension polymerization to prevent formation of polymer deposits on the reactor walls (98). Hydroxyhydroquinone and certain of its derivatives are useful as auxiliary developers for silver halide emulsions in photographic material; their action is based on the dye diffusion-transfer process. The transferred picture has good contrast and stain-free highlights

(99). 5-Acylhydroxyhydroquinones are useful as stabilizer components for poly(alkylene oxide)s (100).

4-Methylesculetol-6,7-dinicotinate (**32**) is useful as an antiinflammatory and vasodilator of low toxicity (101). The synthesis of asarone [5353-15-1] (2,4,5-trimethoxy-1-propenylbenzene), which is used as a tranquilizer, has been patented (102). It occurs in calamus root, *Acorus calamus L.*, and is a chemosterilant for insects (103). 6,7-Dihydroxycoumarin-4-methylsulfonic acid and its salts are useful in the treatment of capillary permeability and fragility and for protecting oxidizable metabolites and drugs against biooxidation (104). Certain chromones derived from hydroxyhydroquinone, eg (**33**), and its salts, esters, and amides are valuable in the prophylactic treatment of asthma (105) (see ANTIASTHMATIC AGENTS). 2-Methoxy-6-multiprenyl-1,4-benzoquinones are intermediates in the microbiological synthesis of coenzyme Q compounds (106).

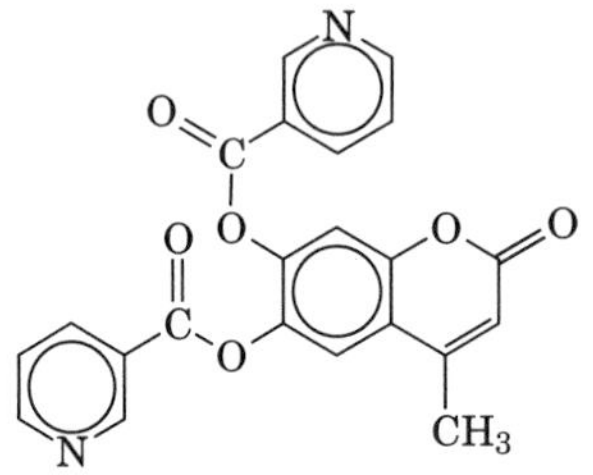

(**32**) 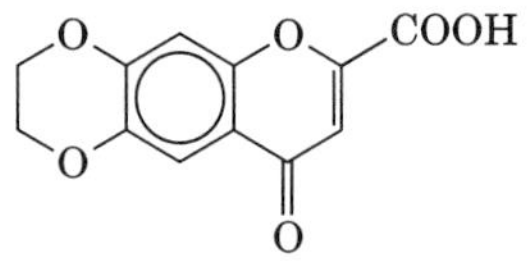          (**33**)

Dihydrochalcones (**34**, where R is D-glycosyl or the disaccharide neohesperidosyl) in conjunction with stevioside, a diterpenoid glycoside, are effective sweeteners (qv) for foods (107). Another series of trihydroxybenzene derivatives, namely 4-aryl esters and 4-carbonates with methoxy groups in the 1-position, are also sweeteners (108). 1,2,4-Triphenoxybenzene is an excellent high temperature heat-transfer agent, lubricant, hydraulic fluid, diffusion pump oil, and processing fluid (109) (see HEAT-EXCHANGE TECHNOLOGY; LUBRICATION AND LUBRICANTS; HYDRAULIC FLUIDS). 2,4-Dialkoxyphenols are antioxidants for a broad range of polymers (110).

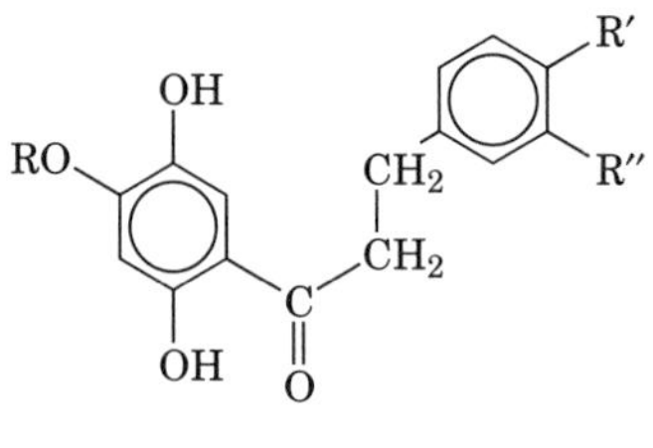

(**34**)

Cinnamyl–sesamol ethers, eg (**35**), are useful as insect chemosterilants (111). 3,4-Methylenedioxyphenyl-3-halo-2-propynyl ethers (**36**, X = halogen) are synergists for carbamate insecticides (112). Haloalkyl or haloalkenyl ethers, eg (**37**), show acaricidal and insect juvenile hormone activity (113). The first total synthesis of gibberellic acid was from 2-methoxy-6-alkoxyethyl-1,4-benzoquinone, a derivative of hydroxyhydroquinone (114).

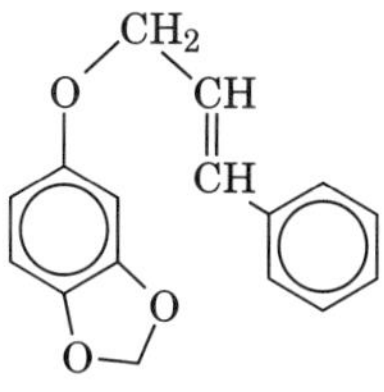

(**35**) 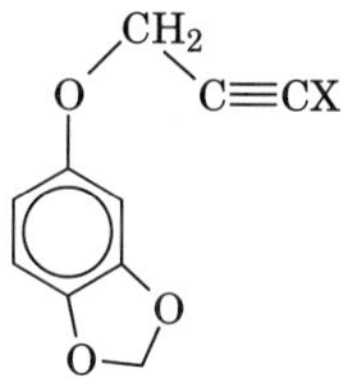     (**36**) 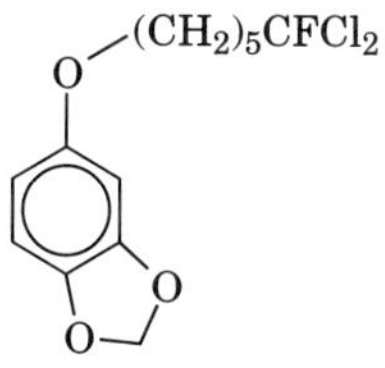     (**37**)

**Derivatives.** Scopoletin [*92-61-5*] (6-methoxyumbelliferone) (**38**) occurs, for example, in *Solanaceae* as a growth factor in plants. Primin [*15121-94-5*] (2-methoxy-6-pentyl-1,4-benzoquinone) is a skin irritant isolated from *Primula obconica*. Versicolin [*4389-44-0*] (1,2,4-trihydroxy-3-methylbenzene) is an anti-fungal antibiotic isolated from the cultural filtrate of a strain of *Aspergillum versicolor*. An epoxygeranyl ether (**39**) of 3,4-methylenedioxyphenol is an insect hormonomimetic. Rotenone [*83-79-4*] (**40**) occurs in many leguminous plants of the tropics and contains a hydroxyhydroquinone nucleus; it is used as an insecticide but is toxic to humans. Precocene-2 (2,2-dimethyl-6,7-dimethoxy-2*H*-chromene) is a chromene isolated from the bedding plant *Ageratum houstonianum*. It is an antijuvenile hormone and induces precocious metamorphosis in insects. It is expected to be developed as an insecticide (115) and a synthetic route has been reported (116). Maesanin (**41**), isolated from berries of the *Maesa lanceolata* bush, has been claimed to kill gram-negative bacteria (117).

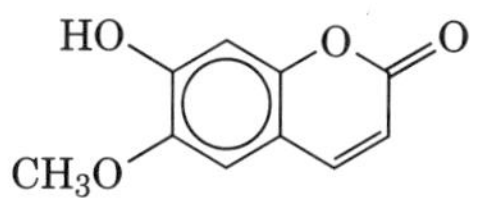

(**38**) 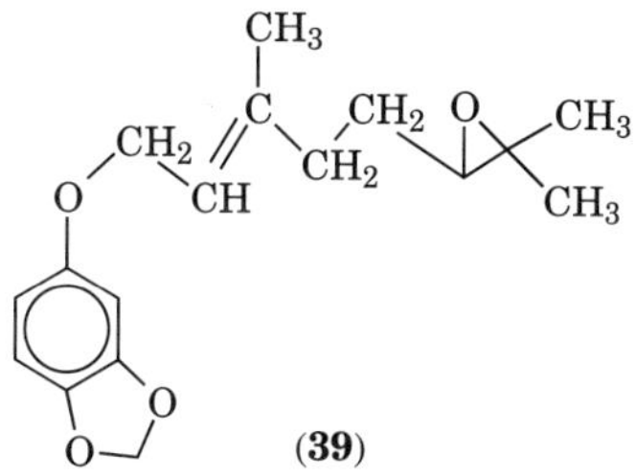     (**39**)

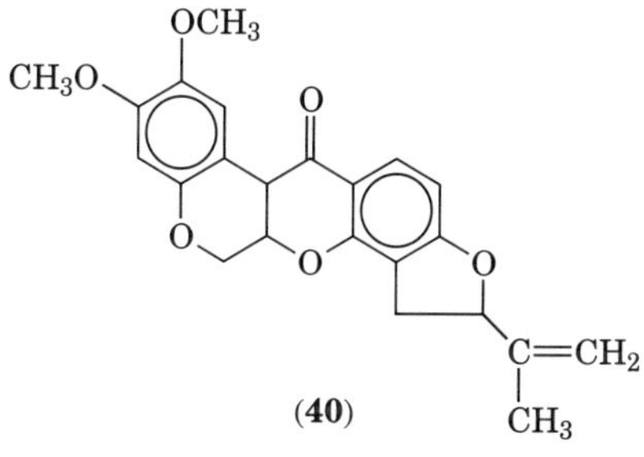

(**40**) 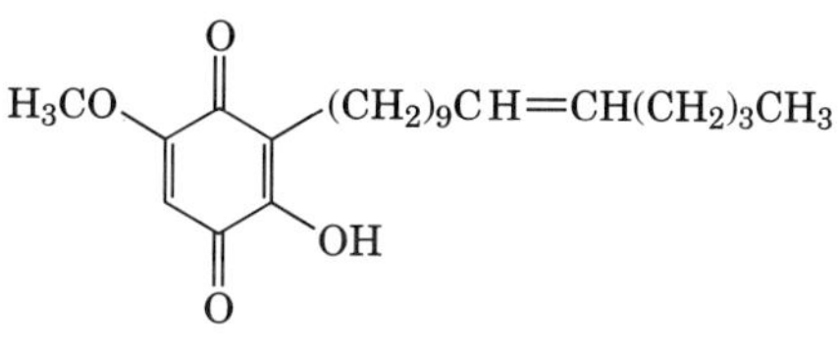     (**41**)

## Phloroglucinol

Phloroglucinol (**42**) is a colorless and odorless solid which is only sparingly soluble in cold water (82). It was discovered in 1855 in the hydrolysis products of the glucoside phloretin, which was obtained from the bark of fruit trees. Phloroglucinol occurs in many other natural products in the form of derivatives such as flavones, catechins, coumarin derivatives, anthocyanidins, xanthins, and glucosides.

There has been much interest in improved synthetic processes for phloroglucinol and in natural product-derived food sweeteners, each of which are characterized by a phloroglucinol nucleus in the structure. Phloroglucinol is of low toxicity, but complex natural products containing a phloroglucinol moiety range in biological properties from antibiotic and antimitotic to potently carcinogenic. Dose–effect relationships of these natural products can be beneficial or harmful. Some of the applications of phloroglucinol derivatives are patterned after the natural product model.

**Properties.**   Phloroglucinol forms odorless, colorless, sweet-tasting, rhombic crystals which tend to discolor on exposure to air or light. The dihydrate loses its water of crystallization at about 110°C (mp 113–116°C on quick heating); the anhydrous material melts at 217–219°C when heated rapidly. Phloroglucinol sublimes at higher temperatures with partial decomposition. The heat of combustion of phloroglucinol is 2.59 MJ/mol (618 kcal/mol); $K_1 = 4.5 \times 10^{-10}$ at 25°C. It is soluble to the extent of 1 part in 100 parts water at 25°C, 10 parts in 100 parts ethanol at 25°C, and 296 parts in 100 parts pyridine; it also is soluble in ether. Phloroglucinol is a mild reducing agent, eg, it reduces Fehling's solution. In aqueous alkali, it is slowly oxidized by air.

Although most of the physical and chemical properties of phloroglucinol characterize it as a polyhydric phenol, in many cases it reacts in a tautomeric keto form or as the $\beta$-triketone, 1,3,5-cyclohexanetrione. This tautomeric triketone has never been isolated; however, phloroglucinol dianion has been shown by $^1$H-nmr spectroscopy to exist as the ketone (**43**) (118). The rapid hydrogen–deuterium exchange of phloroglucinol in weakly alkaline solutions may be evidence for this ketone-enolate tautomerism (119).

Based on this tautomerism, certain addition and replacement reactions at the hydroxyl or keto groups can be effected easily as with certain reactions of resorcinol and 2-naphthalenol. Thus, phloroglucinol forms a trioxime with hydroxylamine (120); it forms mono-, di-, and triaddition compounds with sodium bisulfite (121); it undergoes the Bucherer reaction with ammonia at room temperature (122) to give at first phloramine (5-amino-1,3-dihydroxybenzene) and eventually 3,5-diaminophenol. Displacement of the hydroxyl groups with aromatic amines is possible at higher temperatures (123). Alkylation with a methyl halide in alkaline media leads to the formation of 2,2,4,4,6,6-hexamethylcyclohexane-1,3,5-trione (124). Cyanoethylation in the presence of sodium methylate gives 2-(2-cyanoethyl)phloroglucinol (125). Sodium borohydride reduces phloroglucinol to resorcinol (126). Halogenation in anhydrous solvents yields halogenated cyclohexane-1,3,5-triones. The reaction of phloroglucinol with potassium cyanide in the presence of sulfuric acid yields the cyanohydrin of dihydrophloroglucinol which gives $\gamma$-resorcyclic acid [99-10-5] (3,5-dihydroxybenzoic acid) on treatment with concentrated hydrochloric acid (127).

Etherification with diazomethane gives phloroglucinol trimethyl ether (128). With dimethyl sulfate at pH 8–9, the mono-, di-, or trimethyl ether can be obtained (129). Friedel-Crafts acylation with acid chlorides and aluminum chloride in carbon disulfide gives the nuclear monoacylated phloroglucinols in good yield (130) (see FRIEDEL-CRAFTS REACTIONS). The reaction of phloroglucinol with excess acetyl chloride yields phloroglucinol triacetate. The Gatterman reaction of phloroglucinol with hydrogen cyanide and hydrochloric acid gives 2,4,6-trihydroxybenzaldehyde; similarly, with zinc cyanide the dialdehyde of phloroglucinol can be formed (131). The Hoesch reaction of phloroglucinol with nitriles yields the corresponding ketones; eg, reaction with benzonitrile in the presence of hydrochloric acid yields phlorobenzophenone. Phloroglucinol couples readily with aryldiazonium salts to give di- and triazo compounds. Phloroglucinol reacts readily in the presence of alkaline or acid catalysts with aliphatic and aromatic aldehydes to give various condensation products, which often are colored. The reaction of phloroglucinol with phthalic anhydride gives phloroglucinol phthalein (132). The Perkin condensation of 2,4,6-trihydroxybenzaldehyde with sodium acetate and acetic anhydride, and the Pechmann reaction of phloroglucinol with ethyl acetoacetate in the presence of sulfuric acid yield coumarin derivatives. The Lewis acid-catalyzed reaction of phloroglucinol with olefins or alkyl halides gives nuclear alkylation products; eg, reaction with ethylene and catalyzed by $FeF_2 \cdot BF_3$ and HF gives triethylphloroglucinol [2437-88-9] (133). Alkylphloroglucinols can be prepared by Clemmensen reduction of acylphloroglucinols. The catalytic hydrogenation of phloroglucinol gives a mixture of the stereoisomeric cyclohexane-1,3,5-triols. The hydrogenation of phloroglucinol in the presence of a rhodium-on-alumina catalyst gives cyclohexane-1$\beta$,3$\beta$,5$\beta$-triol in good yield (134). Aqueous alkali bicarbonate or carbonate reacts with phloroglucinol at 20°C to give phloroglucinolcarboxylic acid (135). 2,4,6-Trinitrosophloroglucinol is obtained by reaction of phloroglucinol with nitrous acid in acetic acid (135).

**Manufacture and Synthesis.** The only commercial process in use in the United States through the 1970s involved the oxidation of 2,4,6-trinitrotoluene (TNT) with dichromate in sulfuric acid to 2,4,6-trinitrobenzoic acid. This was followed by the reduction of the nitro groups to amino groups with iron and hydrochloric acid with simultaneous decarboxylation to give 1,3,5-triaminobenzene. Acid hydrolysis at ca 108°C gave phloroglucinol in ca 75% yield (136). The process involved some explosion hazard in the initial stages. Phloroglucinol is no longer made in the United States because of the problem with waste disposal involving acid liquors and iron, chromium, and ammonium salts. The largest producer using the process based on TNT is Océ-Andeno BV (the Netherlands) (137).

An improved version of the amine hydrolysis process involves catalytic hydrogenation of 1,3,5-trinitrobenzene or 2,4,6-trinitrobenzoic acid in acetone solvent (138). Acid hydrolysis of 2,4,6-triaminobenzoic acid has been improved by addition of copper catalyst and gives phloroglucinol in 80% yield (139).

The reaction of 1,3,5-tribromobenzene with excess sodium methoxide in methanol–$N,N$-dimethylformamide and in the presence of a catalytic amount of cuprous iodide gives ca 90% yield phloroglucinol trimethyl ether (1,3,5-trimethoxybenzene). The latter is hydrolyzed with 35 wt % hydrochloric acid at room temperature to give a 90% yield of phloroglucinol (140–142).

Preparation of phloroglucinol or its monomethyl ether by reaction of a halogenated phenol with an alkali metal hydroxide in an inert organic medium by means of a benzyne intermediate has been patented (142). For example, 4-chlororesorcinol reacts with excess potassium hydroxide under nitrogen in refluxing pseudocumene (1,2,4-trimethylbenzene) with the consequent formation of pure phloroglucinol in 68% yield. In a version of this process, the solvent is omitted but a small amount of water is employed (143).

Phloroglucinol can be obtained from 1,3,5-triacetylbenzene by conversion to the tris-oxime which is subjected to a Beckman rearrangement in trifluoroacetic acid to give 1,3,5-triacetamidobenzene. The latter undergoes hydrolysis in aqueous hydrochloric acid to give phloroglucinol in 88% yield (144). A modification of this process has also been patented (145). Another patented process starts with the chlorination of benzene-1,3,5-tricarboxylic acid triamide to the tri-$N$-chloroamide, which reacts with ammonia to produce 1,3,5-triureidobenzene (**44**); the latter is hydrolyzed with aqueous mineral acid to give phloroglucinol in 94% overall yield (146). Phloroglucinol also has been prepared in high yield from hexachlorobenzene by reaction with sodium isopropylate in an aprotic solvent to give trichlorophloroglucinol triisopropyl ether (**45**). The latter is dechlorinated with sodium to phloroglucinol triisopropyl ether, followed by ether cleavage (147). Another process involves hydrolysis of an ether of triacetic acid $\delta$-lactone (**46**) with aqueous hydrochloric acid to give phloroglucinol in good yield (148).

(**44**)                              (**45**)                              (**46**)

Much work on the hydroperoxidation of triisopropylbenzene to make phloroglucinol, similar to the process of phenol from cumene, has been reported (149–155). The shortest route is based on readily available 4-chlororesorcinol. World production of phloroglucinol is estimated to be in excess of 200 metric tons annually (156).

**Grades and Specifications.**   Two grades of phloroglucinol, ie, grades 2 and pure, are offered in the United States by Haake, Inc., who resell material made by Fisons (Table 1). The product discolors slowly on exposure to light.

**Table 1. Grades and Specifications of Phloroglucinol Dihydrate**[a]

| Property | Grade 2 | Pure |
|---|---|---|
| mp, °C | 215–219 | 217–221 |
| appearance | off-white to buff powder | off-white to cream powder |
| loss at 105°C, wt % | 20–24 | 20–24 |
| sulfated ash, wt % | 0.20 max | 0.1 max |

[a]CAS Registry Number [6099-90-7].

**Analysis.** The following analyses for reagent-grade phloroglucinol are suggested: insolubles in alcohol, dissolve 1 g in 20 mL alcohol and a clear and complete solution results, mp 215–219°C; residue on ignition, ignite 1 g with 0.5 mL sulfuric acid, resulting in a residue which weighs not more than 1 mg (0.1 wt %); diresorcinol, heat a solution of 100 mg in 10 mL acetic anhydride to bp, cool the solution, and superimpose it on 10 mL sulfuric acid. No violet color appears at the zone of contact of the liquids.

With ferric chloride, phloroglucinol in aqueous solution gives a bluish violet color, which reddens on addition of a few drops of ammonia. With furfuryl alcohol and hydrochloric acid, phloroglucinol gives a greenish black precipitate. Derivatives of phloroglucinol that are used for identification are the tris(phenylurethane) (mp 190–191°C), tris(3,5-dinitrobenzoate) (mp 162°C), tribenzoate (mp 173–174°C), and picrate (mp 101–103°C). The instrumental methods of analysis are applicable, especially gas chromatography, with possible derivatization, and liquid chromatography.

**Health and Safety Factors.** Phloroglucinol has low toxicity by ingestion. Prolonged severe overexposure may disrupt the thyroid function. High dust concentration may cause respiratory irritation; the product is irritating to eyes and skin. Toxicity data include $LD_{50}$ oral (rat) = 5800 mg/kg; $LD_{50}$ percutaneous (rat) = 2600 mg/kg; $TC_{50}$ for 48 h (rainbow trout) = >2000 mg/L; Ames test = negative.

**Uses.** Two of the principal commercial applications of phloroglucinol, ie, in the diazotype copying process and textile dyeing processes, are based on the ability of each mole of phloroglucinol to couple rapidly with 3 mol of diazo compound. The azo dyes (qv), which are produced, give fast superior black shades. Phloroglucinol also is used in resins and adhesives, as a plastics component or additive, as an intermediate for hydraulic fluids, as a rubber additive, as a photographic chemical, and as a starting material for priming compositions.

Phloroglucinol is listed in the *Colour Index* as CI Developer 19. It is particularly valuable in the dyeing of acetate fiber but also has been used as a coupler for azoic colors in viscose, Orlon, cotton (qv), rayon (qv), or nylon fibers, or in union fabrics containing these fibers (157). For example, cellulose acetate fabric is treated with an aromatic amine such as *o*-dianisidine or a disperse dye such as *p*-hydroxyphenylazo-2-naphthylamine and the amine diazotizes on the fiber; the fabric is then rinsed, freed of excess nitrite, and the azo color is developed in a phloroglucinol bath at pH 5–7. Depending on the diazo precursor used, intense blue to jet-black shades can be obtained with excellent light-, bleach-, and rubfastness.

The condensation on the fabric of 1-amino-3-iminoisoindolenines or 2-amino-5-iminopyrrolenines with phloroglucinol, preferably in the presence of metal salts and solvents, yields fast dyeings in brown shades (158). Metallized azo dyes derived from phloroglucinol yield fast dyeings on leather (qv) or silk (qv) (159).

The diazotype duplicating and copying (white printing) processes are methods for making positive, direct copies of written, drawn, or typed material (160). A light-sensitive diazonium compound is coated onto a sheet of supporting material such as paper. The tracing to be copied is placed on this light-sensitive sheet and exposure to a suitable light source is made. Where not protected by the lines of tracing or drawing, the diazonium compound decomposes with loss of nitrogen and loss of the ability to form azo compounds by coupling. The copy develops by coupling of the remaining diazonium compound with, eg, phloroglucinol. Both wet and dry processes are used. In the wet process, the sensitized paper contains only the diazo compound and, after exposure, development is effected by means of a dilute aqueous solution of phloroglucinol. Dry processes are more complicated and employ ammonia, heat, or ir radiation for development (161,162).

The use of phloroglucinol and its derivatives as developer for light-sensitive planographic plates and for other photographic purposes has been described (163–167).

Cyclohexane-$1\alpha,3\alpha,5\alpha$-triol (*cis*-hexahydrophloroglucinol, $\alpha$-phloroglucite) is a starting material in Woodward's synthesis of prostaglandin $F_{2\alpha}$ and $F_{3\alpha}$ (168) (see PROSTAGLANDINS). $C_1$–$C_3$ alkyl ethers of phloroglucinol are administered as urethral and gastrointestinal antispasmodics (169). 5-Methoxypsoralen (**47**) is made from phloroglucinol and is effective in the treatment of psoriasis (170). 3-Pyrrolidinobutyrylphloroglucinol increases the blood flow of the femoral artery in dogs and inhibits blood platelet aggregation in rats (171). *O*-Substituted (+)-cyanidan-3-ols (**48**, R = alkyl, acyl, or sulfonyl) have been

(**47**)

(**48**)

patented as antihepatitis agents (172). Acylated phloroglucinols are bactericides (173). Other acylated phloroglucinols are useful as fungicides (174). Substituted 1-(2,4,6-trihydroxyphenyl)-1,2-propanediones prevent liver damage in mice (175). Phloroglucinol-3,5-dimethyl-1-(2-amino-3-hydroxybutyryl)ether is characterized by antiarrhythmic activity (176). 2,4-Diacylphloroglucinols were patented as compounds with pronounced anthelmintic activity (177). Phloroglucinol mono- and di-(2-chloroethyl) ethers have antispasmodic or tranquilizing activities (178). 2-(3,5-Dialkoxyphenoxy)ethylamines have antispasmodic, choloretic, sedative, and vasodilating effects (179). 2-Dimethylaminoethyl-2,4,6-trimethoxybenzoates and similar esters are useful as spasmolytics and in relieving indigestion (180). Bis-chromonyl compounds derived from phloroglucinol are valuable in the treatment of asthma (see ANTIASTHMATIC AGENTS) (181).

2,4,6-Trihydroxypropiophenone is useful in cosmetics, as it protects the skin from sunlight (182). Diphenylated acylphloroglucinols and their preparation have been patented as intermediates for bitter-flavoring agents for beer (183). There has been considerable development of dihydrochalcone glycosides and their metal salts as artificial sweeteners (qv) for foods (184) (Fig. 2). Quaternary ammonium salts of these dihydrochalcones are used as sweet-tasting bactericides, especially in dental compositions (185) (see QUATERNARY AMMONIUM COMPOUNDS). The parent dihydrochalcone (**49**, R, R′ = H) is useful as an artificial sweetener (186). Another dihydrochalcone such as (**49**), wherein R′ is H (ie, a phloroglucinol ring) and the other ring is a 4-substituted pyrogallol, is an antiulcer agent (187).

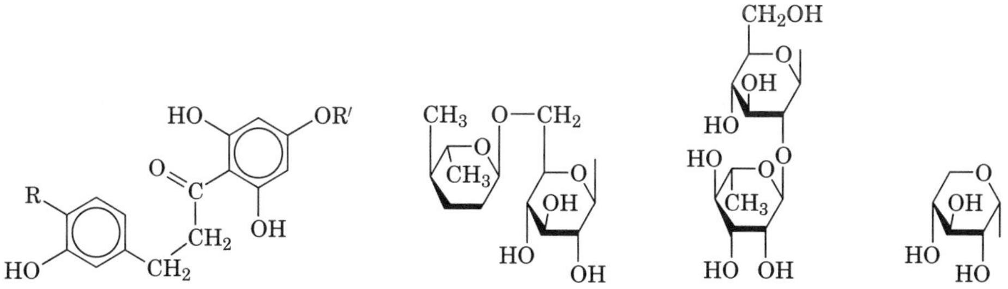

**Fig. 2.** Dihydrochalcone glycoside (**49**), where R = H, OH, or O–alkyl and R′ = glucosyl, rutinosyl, neohesperidosyl, or xylosyl.

Diacetylphloroglucinol and its homologues have been prepared and found to be inhibitors of the herpes virus (188). Syzygiol (**50**), a skin tumor promotion inhibitor, has been prepared from phloroglucinol (189). The first natural mor-phogen (cell-differentiation agent) (**51**) has also been identified as a phloroglu-cinol derivative (190).

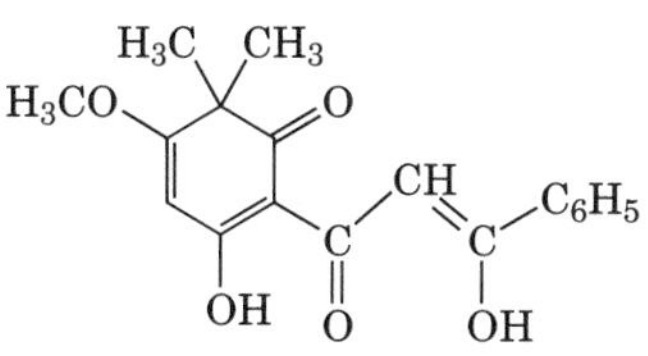

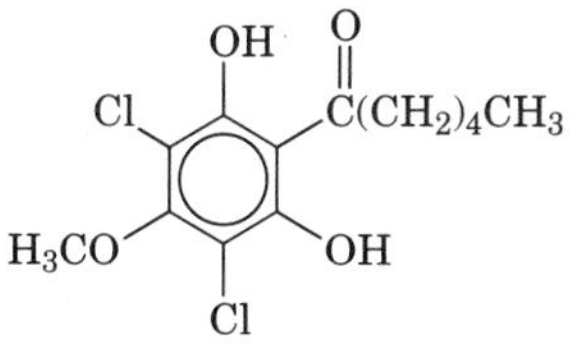

Phloroglucinol or certain of its simple derivatives in conjunction with an organic phosphite are improved heat stabilizers for vinyl chloride polymers and copolymers (190). 1,3,5-Tris(benzoyloxy)benzene was patented as an uv light stabilizer for polyolefins (191). Thermoplastic polycarbonates are produced by the polycondensation of a dihydric phenol and a carbonyl halide in the presence of phloroglucinol (192). Polybutadienes containing two terminal allylic halide groups per unit can be vulcanized in the presence of an inorganic base with phloroglucinol (193).

Alkanoyl esters of phloroglucinol, eg, phloroglucinol trisheptanoate, are high temperature-resistant lubricants and high performance fluids (194). An aqueous solution of phloroglucinol (or of several of its simple derivatives) is used as a corrosion-resistant coating on galvanized sheet (195). The alkali or ammonium salts of 2,4,6-trihydroxy-1,3,5-benzenetricarboxylic acid and 2,4,6-trihydroxy-1,3,5-benzenetrisulfonic acid are sequestering agents useful for synthetic detergent formulations (196,197).

Phloroglucinol can be used in place of silver iodide for cloud seeding to modify weather conditions (198). A nutrient medium containing cytokinin, auxin, and phloroglucinol improves rooting of cuttings from woody plant material (199). Reagent-grade phloroglucinol is used as a sensitive analytical reagent for the detection and estimation of aliphatic and aromatic aldehydes; carbohydrates, eg, pentoses, pentosans, glycuronic acids, galactoses, and galactans; lignin (qv); and hydrochloric acid.

**Derivatives.**   Many derivatives of acylated phloroglucinols that bear a benzene ring substituent or an ether or glycoside linkage occur in nature. Examples are cotoin [479-21-0] (**52**) in coto bark and conglomerone [480-25-1] (**53**) in *Eucalyptus conglomerata.*

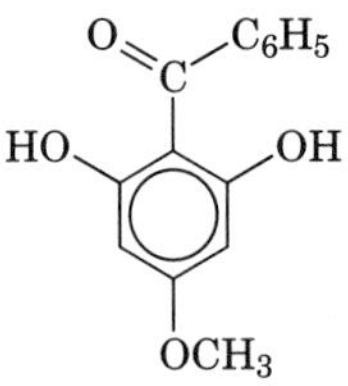

(**52**) 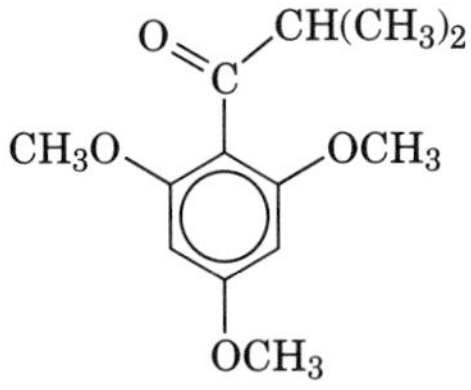            (**53**)

Griseofulvin [126-07-8] (**54**) contains the phloroglucinol nucleus. It is an important oral antifungal agent in humans and animals, elaborated by certain strains of *Penicillium*. One synthesis of griseofulvin is based on the appropriately substituted phloroglucinol (196). Uvaretin [58449-06-2] (**55**), which is extracted from *Uvaria acuminata*, inhibits lymphocytic leukemia (200).

Aflatoxins B are fungal metabolites and are produced by *Aspergillus flavus*. There are several related products; all contain a phloroglucinol segment in their structure and all are extremely toxic and carcinogenic, eg, aflatoxin B (**56**) (201).

The bioflavanoids (vitamin P complex) are substances which maintain the small blood vessel walls. The substances are widely distributed among plants, eg, all citrus fruits, and have been used medicinally to decrease capillary permeability and fragility.

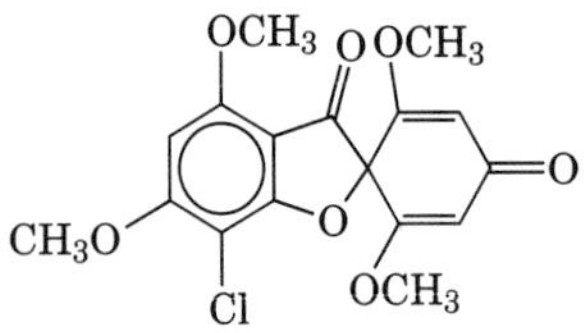

(**54**) 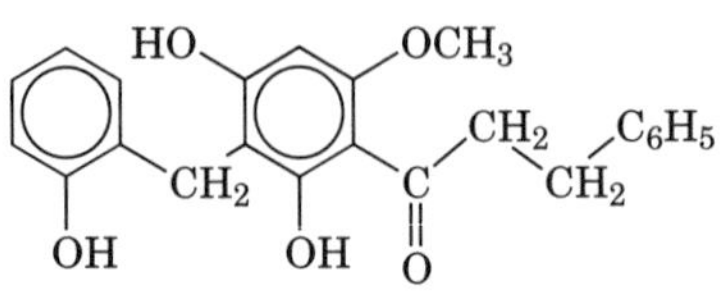                    (**55**) 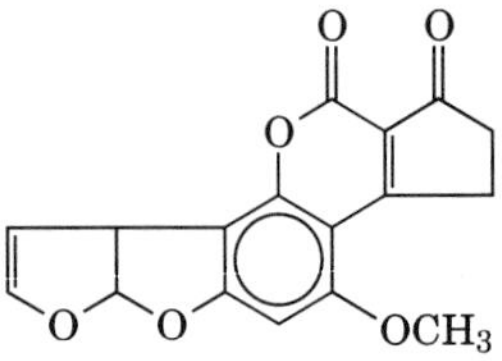                    (**56**)

Hesperidin [*520-26-3*] (hesperetin(7-rhamnoglucoside or 7-rutinoside)) (**57**) contains a core structure of phloroglucinol. A relative of this series is troxerutin [*7085-55-4*], which is a component of Paroven (**58**) and is used in the treatment of venous problems.

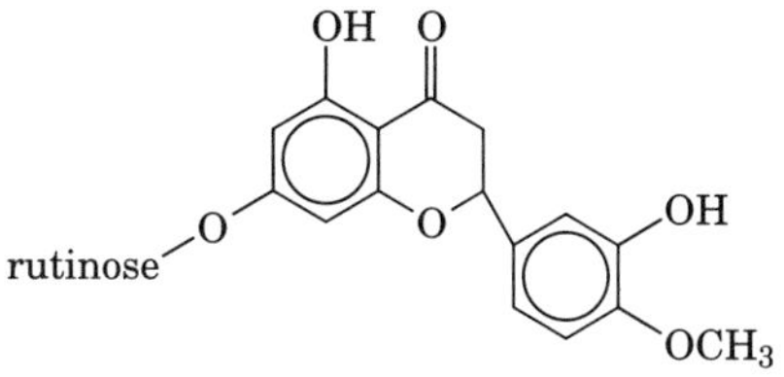

(**57**) 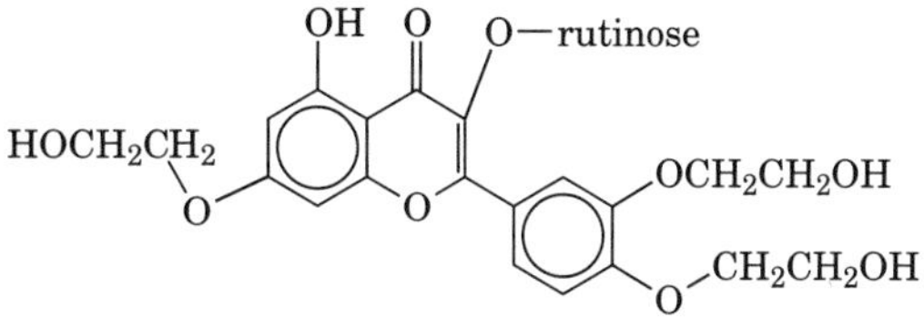                    (**58**)

## Benzenetetrols

**1,2,3,4-Benzenetetrol.** 1,2,3,4-Tetrahydroxybenzene or apionol (**59**) forms needles from benzene (mp 161°C). It is easily soluble in water, diethyl ether, ethanol, and glacial acetic acid and is sparingly soluble in benzene. It has been identified as one of the many constituents of wood–vinegar distillate (202).

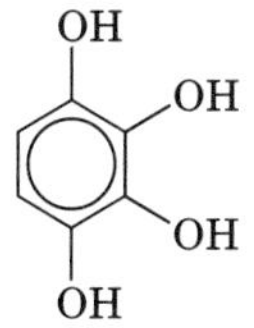

(**59**) 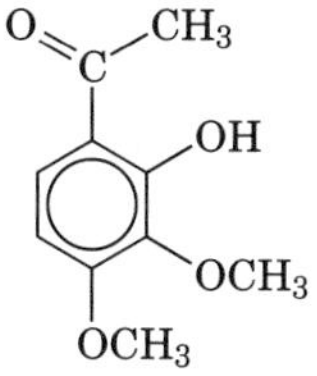          (**60**)

1,2,3,4-Benzenetetrol is best prepared by the hydrolysis of 4-amino-pyrogallol hydrochloride (203). Its 1,2-dimethyl ether (bp 160–170°C at 203 kPa (2 atm)) can be prepared by the oxidation of gallacetophenone-3,4-dimethyl ether (**60**) with hydrogen peroxide or with potassium peroxysulfate (204,205). The oxidation of pyrogallol-1,2-dimethyl ether with potassium peroxysulfate gives the 2,3-dimethyl ether of 1,2,3,4-benzenetetrol, ie, 1,4-dihydroxy-2,3-dimethoxybenzene (mp 84–85°C) (206). Similarly, the oxidation of 2,3,4-trimethoxybenzaldehyde with peracetic acid affords 2,3,4-trimethoxyphenol in 95% yield (207). Formylation of pyrogallol-1,2-dimethyl ether by methyl chloromethyl ether in the presence of titanium tetrachloride followed by reduction and then oxidation by nitrosodisulfonate gives 2,3-dimethoxy-5-methyl-1,4-benzoquinone (208). This product also can be obtained by formylation of 3,4,5-trimethoxytoluene with dimethylformamide and phosphorus oxychloride, treatment with hydrogen peroxide, and oxidation (209). A procedure based on gallic acid has also been reported (210). γ-Irradiation of gallic acid in aqueous solution in the presence of hydrogen peroxide and oxygen gives 2,3,4,5-tetrahydroxybenzoic acid in good yield, and similar treatment of 5-nitropyrogallol gives 2,3,4,5-tetrahydroxynitrobenzene (211).

*Derivatives.* The most important derivatives of 1,2,3,4-benzenetetrol are the ubiquinones, eg, coenzyme Q, which are dimethoxytoluquinones with polyisoprenoid side chains (**61**). They occur in plants and animals. Mice with hereditary muscular dystrophy have a deficiency of coenzyme Q in their heart and hind leg muscles. Therapeutic administration of coenzyme Q [*1339-63-5*] produces physical improvement and a significantly prolonged lifespan (212). Coenzyme Q also has been used to treat deafness when administered either orally or parenterally (213).

The preparation of coenzyme Q usually involves either 2,3-dimethoxy-5-methylbenzoquinone or hydroquinone as the starting material. Treatment of the hydroquinone with geranyl bromide followed by oxidation affords (**61**, $n = 2$) (214). A facile and efficient preparation of ubiquinone-10 (**61**, $n = 10$) has been developed (215).

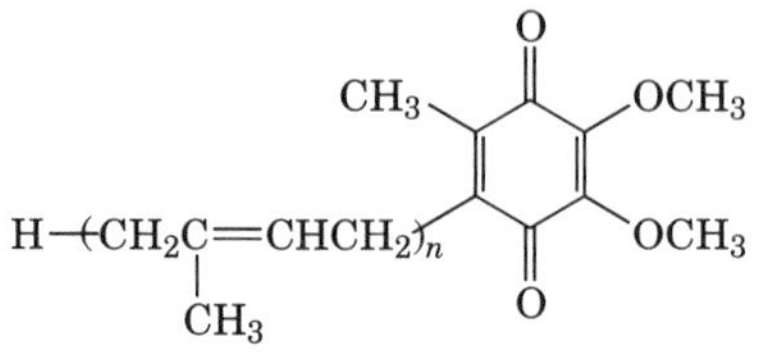

(**61**)

Fumigatin [*484-89-9*] (3-hydroxy-2-methoxy-5-methyl-*p*-benzoquinone) is isolated from metabolism of *Aspergillus fumigatus* and is used as an antimicrobial. 5-Allyl-1,6-dimethoxy-2,3-methylenedioxybenzene (dillapiole) (**62**) is a synergist for pyrethrum. Derivatives have been prepared and evaluated (216).

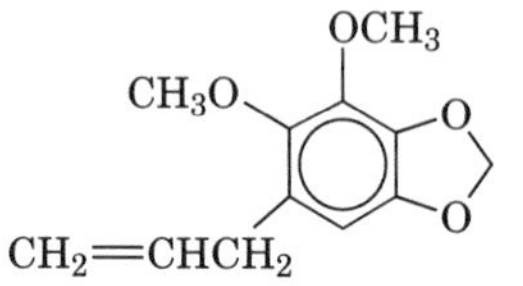

(**62**) 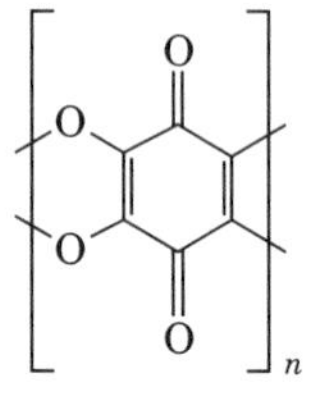          (**63**)

Derivatives of ubiquinones are antioxidants for foodstuffs and vitamins (qv) (217,218). Ubichromenol phosphates show antiinflammatory activity (219). Chromanol compounds inhibit oxidation of fats and can be used in treatment of macrocytic anemias (220). Monosulfate salts of 2,3-dimethoxy-5-methyl-6-substituted hydroquinone have been reported to be inhibitors of lipid oxidation in rats (221). Polymers based on chloranilic and bromanilic acid have been prepared and contain oxygenated quinones (**63**), which are derived from 1,2,3,4-benzenetetrol (222).

**1,2,3,5-Benzenetetrol.** 1,2,3,5-Tetrahydroxybenzene (**64**) forms needles (mp 165°C) from water. The compound is easily soluble in water, alcohol, and ethyl acetate and is insoluble in chloroform and benzene. In aqueous potassium bicarbonate solution sparged with carbon dioxide, 1,2,3,5-benzenetetrol yields 2,3,4,6-tetrahydroxybenzoic acid (mp 308–310°C dec).

1,2,3,5-Benzenetetrol has been prepared by the hydrolysis of 2,4,6-tri-aminophenol with dilute hydrochloric acid and by heating aqueous solutions of <0.2 $M$ 2,4,6-triaminophenol at >130°C (223–225). The acid hydrolysis is improved by copper (226). 1,2,3,5-Benzenetetrol also has been prepared in 46% overall yield by the nitration of hydroquinone diacetate at low temperature to 2,6-dinitrohydroquinone acetate, followed by reduction to the corresponding diamine hydrochloride with tin and hydrochloric acid. The diamine hydrochloride is hydrolyzed to the tetrol with 1 wt % hydrochloric acid at 155–160°C (224). Hydrogenation of 2,6-dibenzoyloxy-$p$-benzoquinone over Pd–C gives a 90% yield of 1,2,3,5-tetrahydroxybenzene (227).

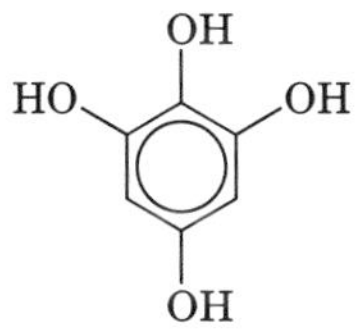

(64) 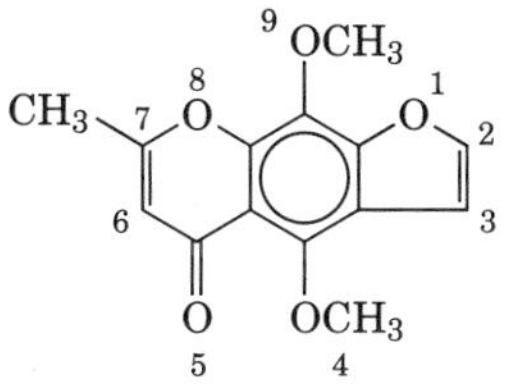                              (65)

*Derivatives.* Oxidation of pyrogallol trimethyl ether with nitric acid, followed by reduction in acetic anhydride and treatment of the product with aluminum chloride, affords 3,6-dihydroxy-2,4-dimethoxyacetophenone (228). 3,4,5-Trimethoxyphenol (antiarol) has been prepared by treatment of 3,4,5-trimethoxyacetophenone with peracetic acid and in 75% yield from 3,4,5-trimethoxybenzoic acid by conversion to the azide, decomposition of the azide, and hydrolysis of the resulting amine (229,230). In contrast, treatment of 3,4,5-trimethoxybenzaldehyde with peracetic acid affords 2,6-dimethoxybenzoquinone as does oxidation of 4-hydroxy-3,5-dimethoxybenzaldehyde with peroxides (231,232).

Many 1,2,3,5-benzenetetrol derivatives are used medicinally. For example, khellin [82-02-0] (65), which is a naturally occurring benzopyranone, is used as a coronary vasodilator and bronchodilator (233). Derivatives of khellin are effective local anesthetics and antiarrythmics (234). Similarly, amine derivatives (68) that are prepared from khellinone oxime (66) exhibit hypnotic, sedative, anticonvulsant, antiinflammatory, cardiac analeptic, diuretic, and antiulcerous activity (235) (see ANALGESICS, ANTIPYRETICS, AND ANTIINFLAMMATORY AGENTS).

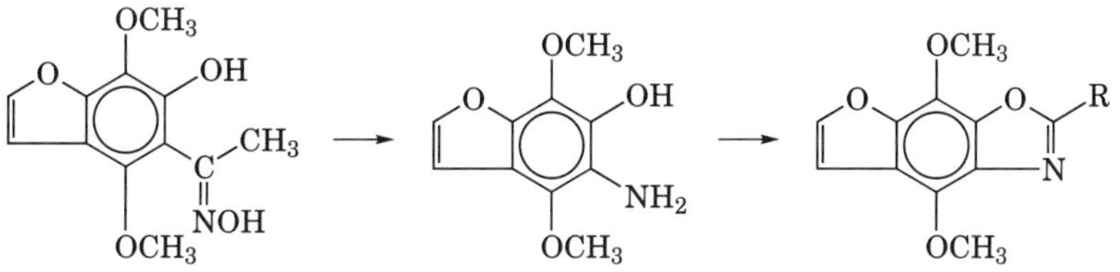

(66)                     (67)                      (68)

Eupatin (69, R = H) and Eupatoretin (69, R = CH$_3$), which are isolated from thistle perennials, show moderate cytotoxicity against human carcinoma of the nasopharynx (236). Baicalein (**70**) salts exhibit antiallergic and antiinflammatory activity. 3,4,5-Trimethoxyphenoxyacetamides are hypotensives and diuretics and are useful for controlling arrhythmia during anesthesia (237).

2,6-Dimethoxy-$p$-benzoquinone is a naturally occurring antiinflammatory (238). 3-Alkyl derivatives of it have also been prepared (239).

(**69**)                              (**70**)

**1,2,4,5-Benzenetetrol.** 1,2,4,5-Tetrahydroxybenzene (**71**) forms leaflets from glacial acetic acid (mp 215–220°C). It is easily soluble in water, ethanol, and diethyl ether but is not quite as soluble in concentrated hydrochloric acid and glacial acetic acid. Ferric chloride produces a precipitate of 2,5-dihydroxy-1,4-benzoquinone (137). The same compound also is produced by aeration of the alkaline solution. Aeration of its acid solutions precipitates a black quinhydrone.

1,2,4,5-Benzenetetrol is obtained by the reduction of 2,5-dihydroxyl-1,4-benzoquinone, which is readily made by oxidation of hydroquinone dissolved in strong aqueous sodium hydroxide with hydrogen peroxide, with stannous chloride and hydrochloric acid or by catalytic hydrogenation (240). Etherification with methyl iodide in the presence of base gives the tetramethyl ether of 1,2,4,5-benzenetetrol (mp 103°C). Several partial ethers, halo, and amino derivatives of 1,2,4,5-benzenetetrol are obtained by reduction of the appropriately substituted 1,4-benzoquinones. For example, 4,5-dimethoxy-1,2-benzoquinone is obtained by reaction of pyrocatechol with lead dioxide and sodium methoxide in methanol (241).

The saturated derivative of maesarin (**41**), dihydromaesarin, has been synthesized and showed activity as a bacteriocide and antitumor agent (242).

Phosphorus derivatives of 1,2,4,5-benzenetetrol (**71**) are effective antiwear and antioxidant additives for lubricating oils and also have flame-retardant properties (see FLAME RETARDANTS; LUBRICATION AND LUBRICANTS) (243). Bis(cyclic acetals) derived from 1,2,4,5-benzenetetrol are used in perfume and fragrance compositions (244) (see PERFUMES). Polyamides for use as tire cord are stabilized against thermal degradation by incorporation of 2,5-dihydroxybenzoquinone (245). Reaction of 2,5-dialkoxy-$p$-benzoquinones with diamines give polyaminoquinones of good heat and chemical stability (246). Intermediates in dyestuff manufacture and especially dioxazine dyestuffs and auxiliaries are prepared from (**72**, R = $C_1$–$C_5$ alkyl or allyl) and its derivatives (247). Derivatives of (**72**) are reported to be useful as antifogging and stabilizing agents for photographic silver halide emulsions (250). Compounds, eg (**73**), possess moderate activity against Walker carcinosarcoma and leukemia (248). Redox polymers have been prepared from $p$-benzoquinonediols and diisocyanates (249).

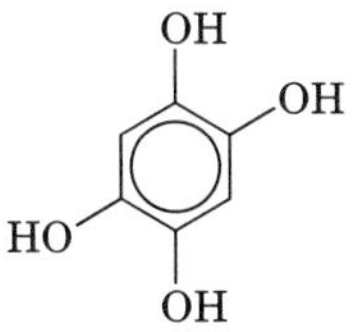

(71) 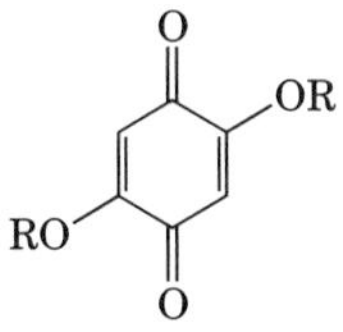     (72) 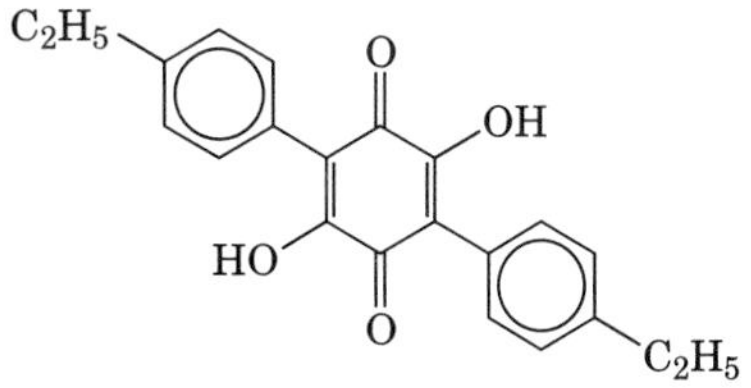     (73)

## Benzenepentol

Benzenepentol [4270-96-6] (pentahydroxybenzene) (74) has been prepared by boiling 2,4,6-triaminoresorcinol diethyl ether with water, followed by ether cleavage with HI (251). The product is very soluble in water but sparingly soluble in organic solvents. Benzenepentol prepared by hydrolysis of 4,6-diaminopyrogallol hydrochloride is sparingly soluble in water, easily soluble in diethyl ether, ethanol, and ethyl acetate, and insoluble in benzene (252).

Ethers of benzenepentol have been obtained by Dakin oxidation of the appropriately substituted acetophenone. Thus, the oxidation of 2-hydroxy-3,4,6-trimethoxyacetophenone and 2-hydroxy-3,4,5-trimethoxyacetophenone with hydrogen peroxide in the presence of alkali gives 1,2-dihydroxy-3,4,6-trimethoxybenzene and 1,2-dihydroxy-3,4,5-trimethoxybenzene, respectively; further methylation of these ethers yields the pentamethyl ether of benzenepentol (mp 58–59°C) (253). The one-step aromatization of myoinositol to produce esters of pentahydroxybenzene is achieved by treatment with carboxylic acid anhydrides in DMSO and in the presence of pyridine (254) (see VITAMINS). 6-Alkyl- or alkenyl-2,3-dimethoxy-5-hydroxy-1,4-hydroquinones (75) and benzoquinones (76), where R = $C_{10}-C_{50}$ alkyl or alkenyl, are coenzyme Q antagonists and antioxidants for fats and oils (255). Several naturally occurring flavenoids derived from pentahydroxybenzene have been synthesized (256–258).

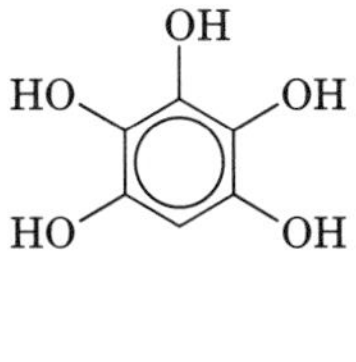

(74) 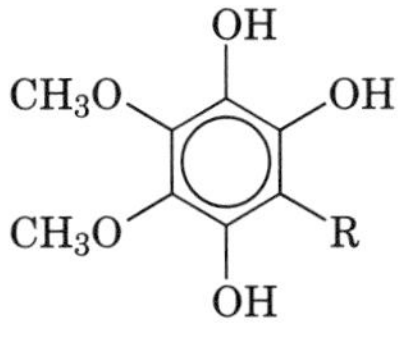     (75) 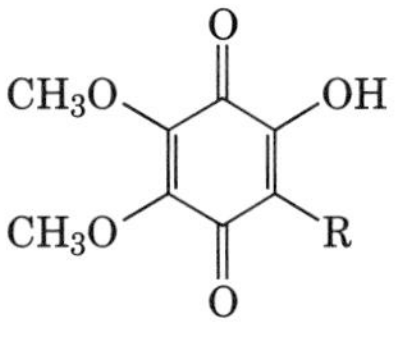     (76)

## Benzenehexol

**Properties.** Benzenehexol [608-80-0] (hexahydroxybenzene) (77) forms snow-white crystals when freshly prepared and collected in an inert atmosphere. Benzenehexol of good purity does not melt up to at least 310°C. It is sparingly soluble in water, ethanol, diethyl ether, and benzene. It readily reduces silver nitrate solution and is oxidized by air in sodium carbonate solution to tetrahydroxy-p-benzoquinone. Triquinoyl (78) is obtained from oxidation with concentrated nitric acid. Catalytic hydrogenation gives inositols, ie, stereoisomeric cyclohexanehexols, and quercitols, ie, cyclohexanepentols, although the hydrogenation of benzenehexol with platinum oxide catalyst at 50–55°C yields phloroglucinol (**42**) (259,260). On evaporation of benzenehexol with potassium carbonate, the potassium salt of croconic acid (**79**) forms by a benzilic acid-type rearrangement from quinonoid intermediates.

**Synthesis.** Benzenehexol is available only from laboratory reagent suppliers. The simplest laboratory preparation involves the aeration of the glyoxal–bisulfite addition product in sodium carbonate solution at 40–80°C, isolation of the sodium salt of tetrahydroxybenzoquinone, followed by acidification to obtain the free tetrahydroxy-*p*-benzoquinone in about 8% yield; the latter is reduced with stannous chloride in boiling dilute hydrochloric acid solution to benzenehexol (**77**) in 77% yield (261). A similar procedure affords dipotassium rhodizonate (**80**) in good yield (262).

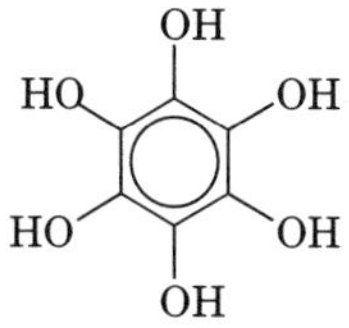

(**77**) 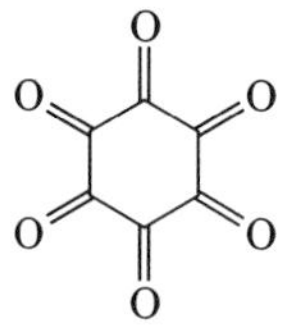          (**78**) 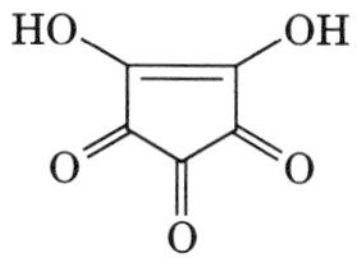          (**79**) 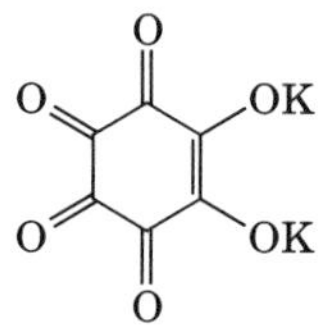          (**80**)

The oldest method of preparation of benzenehexol involves the reaction of molten potassium with carbon monoxide to give the potassium salt of the hexol; the free phenol is obtained by neutralization of the salt with dilute acid (263). This reaction has been reinvestigated and improved (264).

A simple synthesis of tetrahydroxybenzoquinone by methoxylation–hydrolysis of chloranil has been reported (265). Similarly, tetraaryloxybenzoquinones have been prepared from chloranil and alkali salts of phenols (266).

**Analysis.** Benzenehexol gives a violet color with ferric chloride. Derivatives which can be used for its identification are the hexaacetate (mp 205°C) and the hexabenzoate (mp 313°C).

**Derivatives.** A considerable number of compounds that contain the benzenehexol structure possess therapeutic activity. Esterification of benzenehexol with a pyridinecarbonyl chloride gives the corresponding hexaesters, which are antiatherogenics (267). Tetroquinone [*319-89-1*] (tetrahydroxy-*p*-benzoquinone) is administered orally for the treatment of keloids. The dipotassium salt of rhodizonic acid (**80**) is useful as a remedy for diabetes mellitus (262). Compounds, eg (**81**) and (**82**), which are derived from rhodizonic acid, are useful as antiinflammatory agents and diuretics (qv) (268).

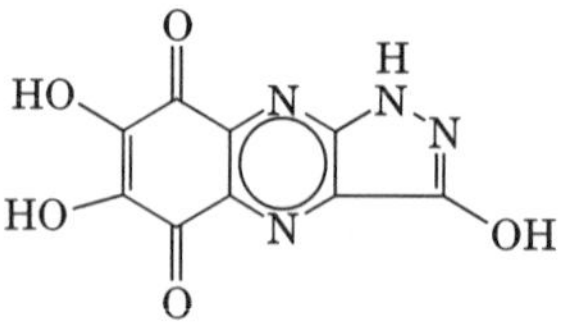

(**81**) 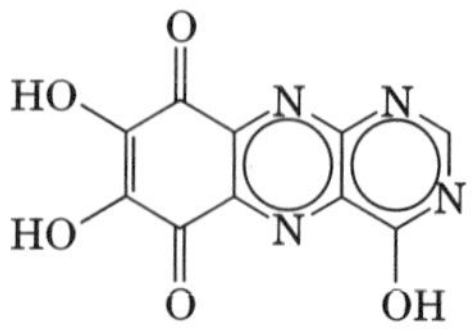                    (**82**)

Inositols, ie, hexahydrobenzenehexols, are sugars that have received increasing study and are useful in the treatment of a wide variety of human disorders, including vascular disease, cancer, cirrhosis of the liver, frostbite, and muscular dystrophy (269). Myoinositol esters prepared by reaction with lower fatty acid anhydrides are useful as liver medicines and nonionic surfactants; the aluminum and ammonium salts of inositol hexasulfate are useful anticancer agents (270). Tetraaryloxybenzoquinones are intermediates in the preparation of dioxazine dyes (266,271). The synthesis of hexakis(aryloxy)benzenes has also been published (272).

## BIBLIOGRAPHY

"Phloroglucinol" in *ECT* 1st ed., Vol. 10, pp. 386–391, by J. F. Kaplan, The Edwal Laboratories, Inc.; "Pyrogallol" in *ECT* 1st ed., Vol. 11, pp. 315–320, by D. M. C. Reilly, Midwest Research Institute; "(Polyhydroxy)benzenes" in *ECT* 2nd ed., Vol. 16, pp. 190–218, by H. Dressler, Koppers Co., Inc.; in *ECT* 3rd ed., pp. 670–704, by H. Dressler and S. N. Holter, Koppers Co., Inc.

1. A. Critchlow, R. D. Haworth, and P. L. Pauson, *J. Chem. Soc.*, 1318 (1951).
2. E. Chapman, A. G. Perkin, and R. Robinson, *J. Chem. Soc.*, 3028 (1927).
3. O. Widmer, *Z. Phys. Chem. (Leipzig)* **140A**, 175 (1929).
4. A. Einhorn and J. Cobliner, *Ber.* **37**, 106 (1904).
5. F. J. Moore and R. M. Thomas, *J. Am. Chem. Soc.* **39**, 987 (1917).
6. W. R. Christian, C. J. Gogek, and C. B. Purves, *Can. J. Chem.* **29**, 911 (1951).
7. B. Peacherer, L. M. Jampolsky, and H. M. Wuest, *J. Am. Chem. Soc.* **70**, 2587 (1948).
8. Pol. Pat. 83,989 (May 20, 1976), C. Osnowski (to Przedsiebiorstwo Przemyslowo-Handlowe "Polskie Odczynniki Chemiczne").
9. U.S. Pat. 3,560,569 (Feb. 2, 1971), C. D. Hurd (to Commercial Solvents Corp.).
10. Jpn. Kokai 75 151,832 (Dec. 6, 1975), Y. Suzuki and T. Maki (to Mitsubishi Chem. Inc.).
11. Ger. Offen. 2,445,336 (Apr. 10, 1975), H. Obara, J. Onodera, A. Matukuma, and K. Yoshida (to Mitsubishi Chem. Ind.).
12. U.S. Pat. 4,172,960 (Oct. 30, 1979), D. Baldwin and P. S. Gates (to Fisons Ltd.).
13. Ger. Offen. 2,653,446 (June 8, 1977), J. F. Harris and co-workers (to Fisons Ltd.).
14. Brit. Pat. 1,358,700 (Mar. 7, 1974) (to Quaker Oats Co.).
15. U.S. Pat. 4,092,351 (May 30, 1978), M. T. Shipchandler (to IMC Chemical Group, Inc.); U.S. Pat. 4,046,877 (Sept. 6, 1977), M. T. Shipchandler (to IMC Chemical Group, Inc.).
16. U.S. Pat. 4,275,247 (June 23, 1981), J. F. Harris (to Fisons Ltd.).
17. Eur. Pat. 0 031 530 Al (Dec. 12, 1980), T. Maki and K. Murayama (to Mitsubishi Chemical Industries).
18. Jpn. Pat. 5,7002-228 (Jan. 7, 1992), (Mitsubishi Chemical Industries KK).
19. *Chem. Eng.*, 109 (Feb. 6, 1981).
20. *The United States Pharmacopeia XX* (USPXX–NFXV), The United States Pharmacopeial Convention, Rockville, Md., 1980, p. 1107.
21. M. B. Jacobs, *Analytical Chemistry of Industrial Poisons, Hazards and Solvents*, 2nd ed., Interscience Publishers, Inc., New York, 1949, pp. 707–708.
22. F. Feigl, *Qualitative Analysis by Spot Tests*, 3rd ed., Elsevier, New York, 1946, pp. 329–332.
23. L. S. Malowan, *Mikrochem. Mikrochim. Acta* **38**, 212 (1951).
24. K. Randerath, *Thin-Layer Chromatography*, Academic Press, Inc., New York, 1963; S. S. Timofeeva and D. I. Storm, *Zh. Anal. Khim.* **31**, 198 (1976).

25. G. D. Clayton and F. E. Clayton, eds., *Patty's Industrial Hygiene and Toxicology*, 3rd ed., rev. Wiley-Interscience, New York, 1982.
26. H. P. Koch and co-workers, *Methods Find Exp. Clin. Pharmacol.* **15**(3), 141–152 (1993).
27. B. Rakowska, *Bromatol. Chem. Toksykol.* **24**(3–4), 273–277 (1991).
28. *Registry of Toxic Effects of Chemical Substances*, NIOSH, Washington, D.C., 1976.
29. E. Lipczynska-Kochany, *Environ. Pollut.* **80**(2), 147–152 (1993).
30. A. R. Bower and co-workers, *Hazard. Ind. Wastes* **24**,135–140 (1992).
31. C. P. Huang and C. S. Chu, *Chem. Oxid. Proc. Int. Symp.* **1**, 239–253 (1992).
32. Jpn. Kokai Tokkyo Koho 04,325,067 (Nov. 13, 1992), N. Mataura and co-workers.
33. B. Z. Shakhashiri, *Chemical Demonstrations*, Vol. 1, The University of Wisconsin Press, Madison, 1983, p. 175.
34. S. Nakano and co-workers, *Talanta* **40**(1), 75–80 (1993).
35. Jpn. Kokai Tokkyo Koho 05 59,105 (Mar. 9, 1993), T. Shimizu and M. Watanabe (to Shinetsu Chem. Ind. Co.).
36. JP 05 59,104 (Mar. 9, 1993), T. Shimizu and M. Watanabe (to Shinetsu Chem. Ind. Co.).
37. JP 04,314,766 (Nov. 5, 1992), T. Shimizu and M. Watanabe (to Shinetsu Chem. Ind. Co.).
38. Ger. Offen. 2,914,510 (Oct. 18, 1979) (to Konishiroku Photo KK).
39. U.S. Pat. 4,182,912 (Jan. 8, 1980), J. W. Foley (to Polaroid Corp.).
40. Jpn. Pat. 49 99,637 (Sept. 20, 1974) (to Hitachi Chemical Ltd.).
41. Jpn. 75 1,909 (Jan. 22, 1975) (to Asahi Chem. Ind. Co.).
42. Belg. Pat. 785,924 (Nov. 3, 1972) (to Bush Boake Allen Ltd.).
43. Jpn. Pat. 74 48,508 (Dec. 21, 1974) (to Takeda Chem. Ind.).
44. Jpn. Pat. 76 42,131 (Nov. 13, 1976) (to Nippon Steel Corp.).
45. Brit. Pat. 1,465,059 (Feb. 23, 1977) (to Pilkington Bros. Ltd.).
46. V. V. Polyakov, T. K. Chumbalov, L. T. Pashimina, and N. A. Zakharova, *Zh. Obsh. Khim.* **42**, 1601 (1972).
47. Ger. Offen. 2,000,082 (July 15, 1971), L. Langmarck (to Wolff Walsrode A.-G.).
48. Jpn. Pat. 52 102,446 (Feb. 18, 1976) (to Dia Tokkyo Project).
49. Ger. Offen. 2,552,796 (Nov. 25, 1976) (to Hoechst AG).
50. Belg. Pat. 859,630 (Apr. 12, 1978) (to Kanegafuchi Kagaku).
51. Ind. Pat. 138,878 (Apr. 10, 1976), D. Chodhury, K. C. Sah, and R. Kapoor (to Union Carbide India Ltd.).
52. U.S. Pat. 3,345,376 (Oct. 3, 1967), J. B. Hester, Jr. and J. Szmuszkovicz (to Upjohn Co.).
53. U.S. Pat. 3,454,565 (July 28, 1969), S. R. Safir and R. P. Williams (to American Cyanamid Co.).
54. Belg. Pat. 737,418 (Feb. 13, 1970) (to Hoffmann-LaRoche and Co.); S. Afr. Pat. 6,905,530 (Mar. 3, 1970), G. Bartholini and B. Heredus (to Hoffmann LaRoche and Co.).
55. Jpn. Pat. 73 37,027 (Nov. 8, 1973), H. Igasa, M. Tsukamoto, and J. Uno (to Dainippon Pharmaceutical Co.).
56. Ger. Offen. 2,331,969 (Jan. 10, 1974), T. Suzuki, S. Himoto, and K. Nakagawa (to Nisshin Flour Milling Co.).
57. Ger. Offen. 2,360,545 (June 12, 1974), H. Offermanns and K. Posselt (to Deutsche Gold and Silber-Scheidenanstalt vorm. Roessler); Fr. Demande 2,244,518 (Apr. 18, 1975), C. Fawan, M. Furin, J. F. Ancher, G. Raynaud, and J. Thomas (to Delalande S. A.); Fr. Demande 2,262,521 (Sept. 26, 1975), C. Fauran and co-workers (to Degussa).
58. Ger. Offen. 2,518,534 (May 4, 1974), J. Granados and co-workers (to Laboratorios Made SA); Jpn. Kokai 78 112,886 (Oct. 2, 1978), H. Murai and Y. Aoyagi (to Nippon

Shinyaku Co.); Jpn. Kokai 78 28,137 (Mar. 16, 1978), H. Tada and co-workers (to Sato Pharmaceutical Co.).

59. Ger. Offen. 2,611,042 (Oct. 7, 1976), M. S. Chodnekar, P. Loeliger, U. Schwieter, A. Pfiffner, and M. Suchy (to Hoffmann-LaRoche and Co.); Ger. Offen. 2,607,655 (Sept. 16, 1976), M. S. Chodnekar (to Hoffmann-LaRoche and Co.); Ger. Offen. 2,732,453 (Jan. 26, 1978), M. S. Chodnekar and co-workers (to Hoffmann-LaRoche and Co.).
60. T. Ueno and co-workers, *Polym. Eng. Sci.* **32**(20), 1511–1515 (1992).
61. B. Liu and co-workers, *Huaxue Tongbao* (9), 39–41 (1992).
62. K. Kobayashi and co-workers, *Tetrahedron Lett. (Eng.)* **34**(32), 5121–5124 (1993).
63. M. A. Haleem and co-workers, *J. Chin. Chem. Soc. (Taipei)* **29**(2), 139–142 (1992).
64. Ger. Pat. 1,086,719 (Apr. 1, 1957), C. S. Miller and C. A. Kuhrmeyer (to Minnesota Mining and Manufacturing Co.).
65. Belg. Pat. 698,796 and 698,797 (Nov. 3, 1967) (to Instituto Chemioterapico Italiano); U.S. Pat. 3,383,407 (May 14, 1968), J. Nordmann and H. B. Swierkot (to Etablissements Kuhlmann); U.S. Pats. 3,370,066 (Feb. 20, 1968), 3,423,512 (Jan. 21, 1969), and 3,495,008 (Feb. 10, 1970), M. L. Thorimet and E. L. Engelhardt (to Societe d'Etudes Scientifiques et Industrielles de l'Ile-de-France); Jpn. Kokai 73 68,541 (Sept. 18, 1973), G. Ootani, N. Nara, and M. Hirata (to Kowa Co.); Span. Pats. 403,313, 403,314 and 410,505 (Dec. 16, 1975) (to Laboratorio Farmaceutico Quimico-Lafarquim, S.A.); Span. Pat. 422,190 (Apr. 16, 1976) (to Zambeletti Espana S.A.); Span. Pat. 436,591 (Apr. 1, 1977) (to Instituto Chemioterapico Italiano (SpA).
66. U.S. Pat. 3,386,807 (June 4, 1968), M. I. Edenbaum (to Johnson & Johnson).
67. Jpn. Pat. 74 15,252 (Apr. 13, 1974) (to Mitsui Petrochemical Ind.).
68. U.S. Pat. 4,139,693 (Feb. 13, 1979), J. E. Schoenberg (to National Starch and Chemical Corp.).
69. Ger. Offen. 2,450,366 (Apr. 28, 1977) (to BPB Industries Ltd.).
70. U.S. Pat. 3,578,508 (May 11, 1971) (to M. B. Pearlman).
71. USSR Pat. 624,653 (Aug. 8, 1978) (to V. M. Golov and co-workers).
72. U.S. Pat. 3,457,079 (July 22, 1969), K. Koda, S. Sato, M. Shoono, and H. Hori (to Komishiroku Photo Ind.).
73. U.S. Pat. 3,850,649 (Nov. 26, 1974), D. D. Buerkley and H. E. Lange (to Minnesota Mining and Manufacturing Co.).
74. Ger. Offen. 2,529,648 (Jan. 20, 1977), W. Daum, W. Brandes, and P. E. Frohberger (to Bayer A-G).
75. S. Budavari, ed., *The Merck Index*, 11th ed., Merck & Co., Rahway, N.J., 1989, p. 1295.
76. U.S. Pat. 4,129,576 (Dec. 12, 1978), L. J. Glunz and D. E. Dickson (to Thomas C. Elder, Inc.).
77. Fr. Pat. 1,204,793 (Jan. 28, 1960) (to J. C. Seailles).
78. S. J. Wotten and J. Meinwold, *Experimentia* **37**, 3 (1981).
79. Jpn. Kokai Tokkyo Koho 04,210,643 (July 31, 1992), K. Hayashi and co-workers.
80. L. Barth and J. Schreder, *Monatsh.* **5**, 595 (1884).
81. J. Thiele and K. Jaeger, *Ber.* **34**, 2837 (1901).
82. C. Liebermann and S. Lindenbaum, *Ber.* **37**, 1176 (1904).
83. E. v. Pechmann and E. v. Krafft, *Ber.* **34**, 423 (1901).
84. W. Fuerstein and M. Dutoit, *Ber.* **34**, 2637 (1901).
85. W. Fuchs and B. Elsner, *Ber.* **57**, 1228 (1924).
86. J. Thiele, *Ber.* **31**, 1248 (1898); Ger. Pats. 101,607 (Dec. 31, 1897) and 107,508 (Dec. 31, 1897) (to Farbenfabriken Bayer); U.S. Pat. 2,118,141 (May 24, 1938), F. R. Bean (to Eastman Kodak Co.); E. B. Vliet, *Organic Syntheses*, Coll. Vol. I, John Wiley & Sons, Inc., New York, 1958, p. 317.
87. L. Barth and J. Schreder, *Monatsh.* **4**, 176 (1883).

88. H. Dakin, *Am. Chem. J.* **42**, 495 (1909); W. Baker, *J. Chem. Soc.*, 1684 (1934).

89. J. Pospisil and V. Ettel, *Chem. Listy* **52**, 939 (1958); *Coll. Czech. Chem. Commun.* **24**, 729 (1959).

90. G. C. Luijkx and co-workers, *Recl. Trav. Chim. Pays-Bas* **110**(7–8), 343–344 (1991).

91. H. Durliat and co-workers, *J. Mol. Catal.* **75**(3) (1992).

92. Jpn. Kokai Tokkyo Koho 02 78,641 (Mar. 19, 1990), K. Yorozu and co-workers.

93. S. S. Timofeeva and D. I. Stom, *Zh. Anal. Khim.* **31**(1), 198 (1976).

94. K. Takimoto, K. Sato, and S. Tsuda, *Bunseki Kagaku* **27**, 514 (1978).

95. P. Marquardt, R. Koch, and J. P. Aubert, *Z. Ges. Inn. Med. Ihre Grenzgebiete* **2**, 333 (1947).

96. J. K. Lin and S. F. Lee, *Mutat. Res.* **269**(2), 217–224 (1992).

97. U.S. Pat. 4,157,984 (June 12, 1979), F. W. Zilliken (to Z-L Ltd.).

98. Belg. Pat. 849,176 (June 8, 1977), and U.S. Pat. 4,076,951 (Feb. 28, 1978), K. Katayama and co-workers (to Kanegafuchi KK).

99. Ger. Offen. 2,459,059 (June 26, 1975) (to Fuji Photo Film KK).

100. Jpn. 7,011,434 (Apr. 24, 1970) (to Farbwerke Hoechst AG).

101. Fr. M. 5,383 (Oct. 23, 1967) (to Roussel-UCLAF).

102. Fr. M. 6,894 (June 4, 1969) (to J. F. Gauthier).

103. B. P. Saxena, O. Koul, K. Tikkiu, and K. Atal, *Nature* **270**, 512 (1977).

104. U.S. Pat. 3,438,988 (Apr. 15, 1969), D. P. R. L. Giudicelli and H. Najer (to Les Laboratoires Dausse).

105. U.S. Pat. 3,551,572 (Dec. 29, 1970), A. H. Wragg.

106. U.S. Pats. 3,564,024 and 3,564,025 (Feb. 16, 1971), K. Folkers and G. D. Daves, Jr. (to Merck & Co.).

107. Jpn. Pat. 51,022,862 (Feb. 23, 1976) (to Toray Industries KK).

108. U.S. Pats. 4,544,566 (Oct. 1, 1985), 4,545,999 (Oct. 8, 1985), 4,546,000 (Oct. 8, 1985), and 4,547,584 (Oct. 15, 1985) (to General Foods).

109. U.S. Dept. of Def. Publ., U.S. Pat. Off. 883,008 (Feb. 2, 1971), G. Irick, Jr., L. P. Foster, and R. W. Kennedy.

110. U.S. Pat. 3,816,542 (June 11, 1974), E. F. Zaweski (to Ethyl Corp.).

111. U.S. Pat. 3,968,234 (July 6, 1976), L. Jurd (to U.S. Dept. of Agriculture).

112. U.S. Pats. 3,423,428 (Jan. 21, 1969), and 3,524,915 (Aug. 18, 1970), J. Fellig and E. I. Rachlin (to Hoffmann-LaRoche Inc.).

113. Ger. Offen. 2,716,241 (Oct. 27, 1977), P. Piccardi, P. Massardo, and A. Longoni (to Montedison SpA).

114. E. J. Corey and co-workers, *J. Am. Chem. Soc.* **100**, 8031, 8034 (1978).

115. W. J. Bowers, T. Ohta, J. S. Cleere, and P. A. Marsello, *Science* **193**, 542 (1976).

116. M. Miranda and co-workers, *Heterocycles* **32**(6), 1150–1166 (1991).

117. *Chem. Brit.* 598 (July 1984).

118. R. J. Highet and T. J. Batterham, *J. Org. Chem.* **29**, 475 (1964).

119. E. S. Hand and R. M. Horowitz, *J. Am. Chem. Soc.* **86**, 2084 (1964).

120. A. Bayer, *Ber.* **19**, 159 (1886).

121. W. Fuchs, *Ber.* **54**, 245 (1921).

122. J. Pollak, *Monatsh.* **14**, 419 (1893).

123. G. Minunni, *Ber.* **21**, 1984 (1888).

124. A. R. Stein, *Can. J. Chem.* **43**, 1493 (1965).

125. G. S. Misra and R. S. Asthana, *Ann.* **609**, 240 (1957).

126. G. I. Fray, *Tetrahedron* **3**, 316 (1958).

127. W. T. Gradwell and A. M. McGookin, *Chem. Ind.* (*London*), 377 (1956).

128. J. Herzig and F. Wenzel, *Monatsh.* **27**, 785 (1906).

129. H. Bredereck, I. Henning, and W. Rau, *Ber.* **86**, 1085 (1953).

130. Ger. Pat. 941,372 (Apr. 12, 1956), W. Riedl.

131. W. Gruber, *Ber.* **75**, 29 (1942).
132. G. Link, *Ber.* **13**, 1652 (1880).
133. Ger. Pat. 1,144,727 (Mar. 7, 1963), C. B. Linn (to Universal Oil Products Co.).
134. P. N. Strong and J. F. W. Keana, *J. Org. Chem.* **40**, 956 (1975).
135. R. Mayer and A. Melhorn, *Z. Chem.* **3**, 390 (1963).
136. A. G. Perkin, *J. Chem. Soc.* **71**, 1154 (1897).
137. M. L. Kastens and J. F. Kaplan, *Ind. Eng. Chem.* **42**, 402 (1950); U.S. Pat. 2,614,126 (Oct. 14, 1952), J. Krueger (to Edwal Laboratories).
138. A. Bruggink, Research Group Océ-Andeno BV, private communication.
139. Brit. Pat. 1,106,088 (Mar. 13, 1968), E. Vero and J. N. Vickers (to Fisons Ind. Chem.).
140. Ger. Pat. 1,195,327 (June 24, 1965), S. Pietsch (to Kalle AG).
141. A. McKillop, B. D. Howarth, and R. J. Kobylecki, *Synth. Commun.* **4**(1), 35 (1974).
142. Brit. Pat. 1,431,501 (Apr. 7, 1976) (to Océ-Andeno BV); Ger. Offen. 2,458,191 (June 19, 1975), B. D. Howarth and R. J. Kobylecki (to Océ-Andeno BV).
143. U.S. Pat. 3,959,388 (May 25, 1976), N. A. deHaij and A. J. J. Hendrickx (to Andeno BV).
144. U.S. Pat. 3,904,695 (Sept. 9, 1975), A. J. J. Hendrickx and N. A. deHaij (to Adeno NV).
145. U.S. Pat. 4,115,451 (Sept. 19, 1978), H.-G. Zengel and M. Bergfeld (to Akzona, Inc.).
146. U.S. Pat. 4,157,450 (June 6, 1979) (to Akzona, Inc.).
147. U.S. Pat. 4,057,588 (Nov. 8, 1977), H.-G. Zengel and M. Bergfeld (to Akzona, Inc.).
148. Ger. Offen. 2,705,874 (Aug. 18, 1977), V. Huber (to L. Givaudan et Cie., S.A.); U.S. Pat. 4,112,003 (Sept. 5, 1978), V. Huber (to Givaudan).
149. Jpn. Pat. 5,857,736 (Mar. 11, 1982) (to Missui Petrochemical Ind.).
150. Jpn. Pat. 58,150,529 (Sept. 7, 1983) (to Mitsui Petrochemical Ind.).
151. Jpn. Pat. 58,150,530 (Sept. 7, 1983) (to Mitsui Petrochemical Ind.).
152. Jpn. Pat. 6,0036-433 (Sept. 8, 1983) (to Mitsui Petrochemical Ind.).
153. U.S. Pat. 4,463,199 (July 31, 1984) (to Sumitomo Chem. Ind.).
154. Brit. Pat. 2,104,892 (Sept. 4, 1985) (to Sumitomo Chem. Ind.).
155. Brit. Pat. 2,110,679 (Sept. 11, 1985) (to Sumitomo Chem. Ind.).
156. *Chem. Mark. Rep.*, 18 (May 3, 1982).
157. U.S. Pat. 2,546,861 (Mar. 27, 1951), C. E. Maher, P. F. Pascoe, *Chem. Prod.* **18**, 454 (1955); Ger. Pat. 917,991 (Sept. 16, 1954), R. Fleischhauer (to Cassella Farbwerke Mainkur A-G); Ger. Pat. 946,976 (Oct. 9, 1956) (to Farbwerke Hoechst A-G); Brit. Pat. 823,446 (Nov. 11, 1959), J. G. Kennedy (to Whiffen & Sons, Ltd.).
158. Ger. Pat. 1,012,406 (July 18, 1957), A. Tartter and O. Weissbarth (to Badische Anilin- und Soda-Fabrik A-G); Ger. Pat. 1,051,242 (Sept. 3, 1959), H. A. Dortmann, P. Schmitz, and J. Eibl (to Farbenfabriken Bayer A-G).
159. Brit. Pat. 668,474 (Mar. 19, 1952), J. R. Atkinson and D. A. Plant (to Imperial Chemical Industries, Ltd.); Ger. Pat. 760,951 (Mar. 30, 1953), E. Fellmer (to I. G. Farbenindustrie A-G).
160. J. Kosar, *Light Sensitive Systems*, John Wiley & Sons, Inc., New York, 1965.
161. U.S. Pat. 3,113,865 (Dec. 10, 1963), J. J. Sagura and J. A. Van Allen (to Eastman Kodak Co.).
162. Belg. Pat. 615,436 (Sept. 24, 1962) (to Ozalid Co., Ltd.).
163. U.S. Pat. 3,607,271 (Jan. 9, 1969), A. H. J. H. Helden and P. J. H. Tummers (to Van Der Grinten NV).
164. Ger. Offen. 1,939,808 (Aug. 5, 1969) (to Ricoh KK).
165. Brit. Pat. 1,208,395 (Mar. 25, 1969) (to Ricoh KK); Brit. Pat. 1,284,760 (Aug. 9, 1972) (to Ricoh KK).
166. U.S. Pat. 3,770,833 (Nov. 6, 1973), H. Bader and E. G. Jahngen (to Polaroid Corp.).
167. Jpn. Kokai 76 1,113 (Jan. 7, 1976), M. Sasaki and co-workers (to Ricoh Co., Ltd.).

168. R. B. Woodward and co-workers, *J. Am. Chem. Soc.* **95**, 6852 (1973); U.S. Pat. 3,898,248 (Aug. 5, 1975), R. B. Woodward (to Ciba-Geigy Corp.).

169. Belg. Pat. 732,900 (Mar. 2, 1970) (to Orsymonde SA).

170. Belg. Pat. 871,429 (Feb. 15, 1979), J. J. Goupil.

171. Ger. Offen. 2,841,702 (Apr. 5, 1979), L. Lafon (to Laboratoire L. Lafon SA).

172. Jpn. Kokai 79 81,274 (June 28, 1979) (to Zyma SA).

173. Brit. Pat. 1,184,731 (Mar. 18, 1970), J. F. Davies (to Unilever Ltd.); Jpn. Kokai 79 3,030 (Jan. 11, 1979), S. Mizobuchi (to Kirin Brewery Co.).

174. Jpn. Kokai 80 15,443 (Feb. 2, 1980), S. Mizobuchi (to Kirin Brewery Co.).

175. Ger. Offen. 2,428,680 (Jan. 2, 1976), R. Madaus, G. Halbach, and W. Trost (to Dr. Madaus & Co.).

176. Fr. Demande 2,208,653 (June 28, 1974), L. Lafon (to Orsymonde SA).

177. U.S. Pat. 3,467,715 (Sept. 16, 1969), J. L. Broadbent, K. Bowden, and W. J. Ross (to Smith, Kline & French Labs.).

178. Belg. Pat. 737,960 (Sept. 9, 1968) (to Orsymonde SA).

179. Ger. Offen. 2,020,464 (Nov. 12, 1970), L. Lafon (to Orsymonde SA).

180. Ger. Offen. 2,035,341 (Dec. 3, 1970), M. Vaille (to Orsymonde SA).

181. U.S. Pat. 3,519,652 (July 7, 1970), C. Fitzmaurice and T. B. Lee (to Fisons Pharmaceuticals Ltd.).

182. Jpn. Pat. 51 101,138 (Sept. 7, 1976) (to Ichimaru Boeki KK).

183. U.S. Pat. 4,088,688 (Feb. 13, 1976), T. Sigg-Gruetter and J. Wild (to Givaudan & Cie.); Ger. Offen. 2,519,990 (Apr. 5, 1975) (to Atlantic Research Corp.); Brit. Pat. 1,355,236 (June 5, 1974), E. Collins and P. Vivian (to Brewing Patents Ltd.); U.S. Pat. 4,101,585 (Aug. 30, 1976), V. Burckhardt, L. Werthemann, and R. J. Troxner (to Ciba-Geigy Corp.).

184. Ger. Offen. 2,455,373 (July 10, 1975) (to L. Givaudan & Cie.); Jpn. Kokai 78 37,646 (Apr. 6, 1978), S. Kamiya, S. Ezaki, F. Konishi, and T. Watanabe (to Meiji Seika Kaisha, Ltd.); Brit. Pat. 1,347,202 (Mar. 23, 1972) (to Nutrilite Products, Inc.); Brit. Pat. 1,310,329 (Mar. 21, 1973) (to Warner Lambert Co.); U.S. Pat. 3,890,298 (June 17, 1975), R. M. Horowitz and B. Gentili (to U.S. Dept. of Agriculture).

185. Ger. Offen. 2,400,955 (July 25, 1974) (to Unilever, NV).

186. U.S. Pat. 3,855,301 (Dec. 17, 1974), G. P. Rizzi (to Procter & Gamble Co.).

187. Jpn. Kokai 01,242,540 (Sept. 27, 1989), T. Watanabe and co-workers (to Tsumura and Co.).

188. Jpn. Kokai 04,124,129 (Apr. 24, 1992), Y. Kuribayashi and A. Kanamori (to Fujirebio, Inc.).

189. S. Sato and co-workers, *Bull. Chem. Soc. Jpn. (Eng.)* **65**(9), 2552–2554 (1992).

190. U.S. Pat. 3,998,782 (Dec. 21, 1976), R. E. Hutton, B. R. Iles, and V. Oakes (to Akzo, NV).

191. Jpn. Pat. 70 18,967 (June 29, 1970), H. Seki and M. Funata (to Sumitomo Chemical Co.).

192. Jpn. Pat. 72 23,918 (July 3, 1972) (to Farbenfabriken Bayer A-G); Belg. Pat. 686,236 (Oct. 1967) (to General Electric Co.).

193. Can. Pat. 830,012 (Dec. 16, 1969), P. Dolezal (to Polymer Corp.).

194. U.S. Pat. 3,336,349 (Aug. 15, 1967), W. H. Voris (to Koppers Co.).

195. Jpn. Pat. 50 54,539 (Sept. 14, 1973) (to Nippon Steel Corp. KK).

196. U.S. Pat. 3,699,159 (Oct. 17, 1972), D. S. Connor and H. K. Krummel (to the Procter and Gamble Co.).

197. U.S. Pat. 3,812,044 (May 21, 1974), D. S. Connor and H. K. Krummel (to the Procter & Gamble Co.).

198. T. E. Hoffer and M. L. Ogne, *J. Geophys. Res.* **70**, 3857 (1965); V. V. Piotrovich, *Meterol. Gidrol.*, 111 (1975).

199. Neth. Appl. 7,707,944 (Jan. 18, 1978) (to National Seed Development Organisation Ltd., East Malling Research Station).
200. J. R. Cole, S. J. Torrence, R. M. Wiedhopf, S. K. Arora, and R. B. Bates, *J. Org. Chem.* **41**, 1852 (1976).
201. L. A. Goldblatt, *Pure Appl. Chem.* **21**, 331 (1970).
202. T. Matsui and co-workers, *Kogakubu Kenkyu Hokoku (Miyazaki Daigaku)* **38**, 91–100 (1992).
203. A. Einhorn, J. Cobliner, and H. Pfeiffer, *Ber.* **37**, 119 (1904).
204. W. Baker, E. H. T. Jukes, and C. A. Subrahmanyam, *J. Chem. Soc.*, 1681 (1934).
205. G. Bargellini, *Gazz. Chim. Ital.* **46**, 249 (1916).
206. W. Baker and R. I. Savage, *J. Chem. Soc.*, 1604 (1938).
207. Jpn. Kokai 78 46,926 (Apr. 27, 1978), I. Yamatsu and K. Minami (to Eisai Co., Ltd.).
208. USSR Pat. 197,598 (June 9, 1967), E. A. Obol'rikova and co-workers (to All-Union Scientific Research Vitamin Institute).
209. Jpn. Pat. 9,080,031 (Dec. 7, 1972) (to Wakamoto Pharm. Co., Ltd.).
210. Jpn. Pat. 72 18,740 (May 30, 1972) (to Takeda Chemicals Inds. Ltd.).
211. Ger. Pat. 1,228,258 (Nov. 10, 1966), F. Merger and D. Graesslin (to Gesellschaft für Kernforschung GmbH).
212. *Chem. Eng. News* **48**, 19 (1970).
213. U.S. Pat. 4,073,883 (Mar. 2, 1977) (to Eisai).
214. Jpn. Kokai 79 79,240 (June 25, 1979), S. Aoyagi and co-workers (to Wakamoto Pharmaceutical Co., Ltd.).
215. S. Terao and co-workers, *J. Org. Chem.* **44**, 868 (1979).
216. Y. P. Talwar, J. B. Srivastava, and M. C. Nigam, *Indian Perfum* **10**, 43 (1966).
217. Jpn. Kokai 73 72,149 (Sept. 29, 1973), T. Seki and co-workers (to Taisho Pharmaceutical Co., Ltd.).
218. U.S. Pat. 3,517,070 (June 23, 1970), U. Gloor, R. Ruegg, and U. Schwieter (to Hoffmann-LaRoche Inc.).
219. Jpn. Pat. 72 43,555 (Nov. 2, 1972) (to Taisho Pharmaceutical Co., Ltd.).
220. U.S. Pat. 3,364,234 (Jan. 16, 1968), E. F. Schoenewaldt (to Merck and Co., Inc.).
221. Eur. Pat. Appl. 124,379 (Nov. 7, 1984), I. Imada (to Takeda Chem. Ind.).
222. A. A. Berlin and co-workers, *Vysokomol. Soedin Ser.* **A9**, 532 (1967).
223. K. Oettinger, *Monatsh.* **11**, 248 (1895); M. Nierenstein, *J. Chem. Soc.* **111**, 5 (1917).
224. G. Zemplén and J. Schwartz, *Acta Chim. Acad. Sci. Hung.* **3**, 487 (1953).
225. Jpn. Pat. 76 11,102 (Apr. 8, 1976), H. Obara and J. Onodera (to Mitsubishi Chemical Industries Co., Ltd.).
226. Ger. Pat. 1,195,327 (Mar. 10, 1966), S. Pietzsch (to Kalle A-G).
227. R. A. Baxter and J. P. Brown, *Chem. Ind. (London)*, 1171 (1967).
228. S. Matsuura and co-workers, *Gifu Yakka Daigaku Kiyo*, 1 (1974).
229. J. Andrieux and G. Emptoz, *C.R. Acad. Sci. Paris Ser. C* **265**, 1294 (1967).
230. J.-P. Brouard, A. Michaillides, and A. Resplandy, *Chem. Ther.* **8**, 113 (1973).
231. J. Andrieux and G. Emptoz, *C.R. Acad. Sci., Paris Ser. C* **265**, 681 (1967).
232. Jpn. Pat. 78 82,730 (July 21, 1978), Y. Nakamura and T. Higuchi (to Sanyo Kokusaku Pulp Co., Ltd.).
233. M. Windholz, ed., *Merck Index*, 9th ed., Merck & Co., Inc., Rahway, N.J., 1976, p. 5156.
234. Fr. Demande 2,232,311 (Jan. 3, 1975) (to Laboratories Sobio SA).
235. U.S. Pat. 3,878,207 (Apr. 15, 1975), C. P. Fauran and co-workers (to Delalande SA).
236. S. M. Kupchan and co-workers, *J. Org. Chem.* **34**, 1460 (1969).
237. Belg. Pat. 696,899 (Apr. 11, 1967).
238. M. Matsumoto and H. Kobayashi, *Synth. Comm.* **15**(6), 515–520 (1985).
239. L. Gu and Y. Zhong, *Youje Huaxe*, **9**(3), 239–241 (1989).

240. U.S. Pat. 3,780,114 (Dec. 18, 1973) (to S. A. Texaco Belgium NV).
241. USSR Pat. 638,537 (Dec. 25, 1978), A. M. Zvonok and co-workers (to Belorussian State University).
242. O. Reinaud and co-workers, *Tetrahedron Lett. (Fr.)* **26**(33), 3993–3996 (1985).
243. U.S. Pat. 3,819,748 (June 25, 1974), L. G. Dulog and S. A. R. Dewaele (to S. A. Texaco Belgium NV).
244. U.S. Pat. 4,092,331 (May 25, 1976), S. A. R. Dewaele (to Texaco Belgium NV).
245. Jpn. Pat. 74 31,109 (Mar. 12, 1970) (to Toyo Spinning Co., Ltd.).
246. Jpn. Pat. 72 00,897 (Jan. 11, 1972) (to Toyo Spinning Co., Ltd.).
247. Ger. Pat. 1,935,131 (Jan. 29, 1970), S. Hari (to Ciba Ltd.).
248. K.-Y. Zee-Cheng and C. C. Cheng, *J. Med. Chem.* **13**, 264 (1970).
249. G. Wegner and co-workers, *J. Polym. Sci. A-1* **6**, 3151 (1968).
250. U.S. Pat. 3,396,022 (Aug. 6, 1968), F. Dersch and S. L. Paniccia (to GAF Corp.).
251. F. Wenzel and H. Weidel, *Chem. Zentr.* (II), 829 (1903).
252. A. Einhorn, J. Cobliner, and F. Pfeiffer, *Ber.* **37**, 132 (1904).
253. W. Baker, *J. Chem. Soc.*, 662 (1941).
254. A. J. Fatiadi, *J. Chem. Eng. Data* **14**, 118 (1969).
255. U.S. Pat. 3,644,435 (Feb. 2, 1972), K. Folkers, J. C. Catlin, and G. D. Danes, Jr.
256. P. K. Dutta and co-workers, *Ind. J. Chem. Sect. B* **21B**(11) 1037–1038 (1982).
257. D. K. Bhardwaj and co-workers, *Proc. Indian Natl. Sci. Acad. Part A* **56**(2), 161–162 (1990).
258. N. R. Ayyanger and co-workers, *Tetrahedron Lett.* **29**(19), 2347–2348 (1988).
259. R. Kuhn, G. Quadbeck, and E. Rohm, *Ann.* **565**, 1 (1949).
260. H. Wieland and R. S. Wishart, *Ber.* **47**, 2082 (1914); S. J. Avgyol and D. S. McHugh, *Chem. Ind. (London)*, 947 (1955).
261. A. J. Fatiadi and W. F. Sager, *Org. Syn.* **42**, 66, 91 (1962).
262. Jpn. Pat. 67 12,413 (July 14, 1967), E. Ochiai and co-workers (to Shionogi E. Co., Ltd. and Pharmacological Research Foundation).
263. J. Liebig, *Ann.* **11**, 182 (1834); R. Nietski and T. Benkiser, *Ber.* **18**, 1834 (1885).
264. W. Buechner and W. Weiss, *Helv. Chim. Acta* **47**, 1415 (1964); U.S. Pat. 2,736,752 (Feb. 28, 1956), U. Hoffman, O. Schweitzer, and K. Rinn (to Deutsche Gold- und Silber-Scheideanstalt).
265. H. Junek, B. Unterweger, and R. Peltzmann, *Z. Naturforsch. B. Anorg. Chem. Org. Chem.* **33B**, 1201 (1978).
266. Brit. Pat. 1,375,334 (Nov. 27, 1974) and Ger. Pat. 2,322,927 (May 7, 1973) (to Ciba-Geigy AG).
267. U.S. Pat. 3,479,364 (Nov. 18, 1969), C. P. Krimmel (to G. D. Searle and Co.).
268. U.S. Pats. 3,431,262 (Mar. 4, 1969), and 3,498,983 (Mar. 3, 1970), G. R. Wendt and K. W. Ledig (to American Home Products Corp.).
269. R. Bernhard, *Sci. Res.*, 34 (Oct. 28, 1968).
270. Jpn. Kokai 79 19,942 (Feb. 15, 1979) and 79 61,153 (May 17, 1979), M. Ikuro, Y. Yamada, and T. Umemoto (to Mitsui Toatsu Chemicals, Inc.).
271. U.S. Pat. 3,907,839 (Sept. 23, 1975), K. Burdeska (to Ciba-Geigy Corp.).
272. C. J. Christopher and co-workers, *Tetrahedron Lett.* **24**(31), 3269–3272 (1983).

GERD LESTON
Consultant

# POLY(HYDROXYBENZOIC ACID).    See HEAT-RESISTANT POLYMERS.

# POLYIMIDES

Polyimides (PI) are polycondensation products (**1**) prepared from derivatives of tetracarboxylic acids and primary diamines (1–5). Descriptions of self-polycondensation polymers (**2**) based on aminodicarboxylic acid derivatives are also found in the literature (6–9).

(**1**)                              (**2**)

The main chain of these polymers contains, as the principal component, five- or six-membered heteroaromatic rings, ie, imides, which are usually present as condensed aromatic systems, such as with benzene (phthalimides, **3**) and naphthalene (naphthalimides, **4**) rings.

(**3**)                              (**4**)

Among imide-containing polymers, polyimides derived from aromatic tetracarboxylic acids and aromatic diamines are of primary importance and represent typical high performance specialty polymers that are commercially employed in various applications. In structures (**1–4**), the oxidatively unstable amino group has been converted to an imino group that is stabilized by two carbonyl groups directly attached to it. The powerful electron-withdrawing effect of the carbonyl groups results in the formation of heteroaromatic systems that are low in electron density. Because oxidation is an electron-abstracting process, such heteroaromatic systems are generally resistant to oxidation. Aromatic polyimides exhibit outstanding mechanical and electrical properties as well as high thermoxidative and chemical resistance. Polyimides are widely used in critical components in aerospace, automotive, electrical, electronic, film, and coating applications.

## Synthetic Methods

Since successful commercialization of Kapton by Du Pont Company in the 1960s (10), numerous compositions of polyimide and various new methods of syntheses have been described in the literature (1–5). A successful result for each method depends on the nature of the chemical components involved in the system, including monomers, intermediates, solvents, and the polyimide products, as well as on physical conditions during the synthesis. Properties such as monomer reactivity and solubility, and the glass-transition temperature, $T_g$, crystallinity, $T_m$, and melt viscosity of the polyimide products ultimately determine the effectiveness of each process. Accordingly, proper selection of synthetic method is often critical for preparation of polyimides of a given chemical composition.

### TWO-STEP POLY(AMIC ACID) METHOD

The two-step poly(amic acid) process is the most commonly practiced procedure. In this process, a dianhydride and a diamine react at ambient temperature in a dipolar aprotic solvent such as $N,N$-dimethylacetamide [127-19-5] (DMAc) or $N$-methylpyrrolidinone [872-50-4] (NMP) to form a poly(amic acid), which is then cyclized into the polyimide product. The reaction of pyromellitic dianhydride [26265-89-4] (PMDA) and 4,4'-oxydianiline [101-80-4] (ODA) proceeds rapidly at room temperature to form a viscous solution of poly(amic acid) (**5**), which is an ortho-carboxylated aromatic polyamide.

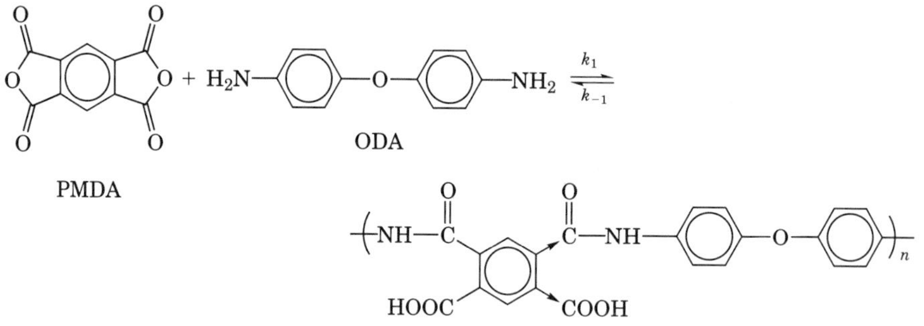

(1)

(**5**)

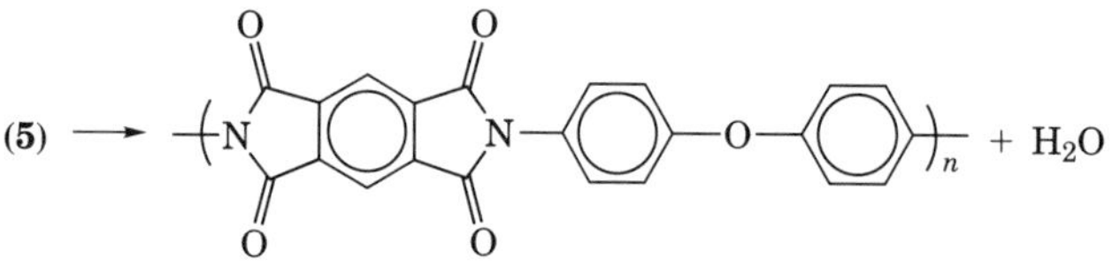

(2)

Kapton polyimide

Poly(amic acid)s are shaped into articles such as films and fibers by removal of the solvent. The shaped poly(amic acid) films, for example, are thermally or chemically converted to the final polyimide products (eq. 2). The conversion produces water as a by-product. Because the water must be removed during *in situ* imidization, the process in generally suitable only for the preparation of thin objects such as films and coatings. The arrows in poly(amic acid) structure (**5**) denote isomerism, indicating that the main chain is composed of a mixture of 1,3- and 1,4-phenylenebisamide linkages. The structural randomness may contribute to keep the polymer from crystallizing. Most of the structurally regular fully aromatic polyamides such as Nomex and Kevlar are highly crystalline and difficult to dissolve.

An additional important aspect of this polymerization is that it is an equilibrium reaction, and therefore the molecular weight of the poly(amic acid) is dependent on the reactivity of the monomers, more accurately the ratio of forward ($k_1$) and reverse reaction ($k_{-1}$) rates. One unique feature of the reaction system is that the dianhydride is neutral and the diamine is a weak base in the Brønsted sense, yet the product is a relatively strong carboxylic acid. One of the driving forces to promoting reaction is strong interaction of the product amic acid with basic amide solvents, such as DMAc, which are good proton acceptors. More weakly basic ether solvents such as tetrahydrofuran and diethylene glycol dimethyl ether (diglyme) are also reported to be good solvents for certain polyamic acids (11).

**Monomer Reactivity.** The poly(amic acid) groups are formed by nucleophilic substitution by an amino group at a carbonyl carbon of an anhydride group. Therefore, the electrophilicity of the dianhydride is expected to be one of the most important parameters used to determine the reaction rate. There is a close relationship between the reaction rates and the electron affinities, $E_a$, of dianhydrides (12). These $E_a$ were independently determined by polarography. Structures and electron affinities of various dianhydrides are shown in Table 1.

PMDA is composed of a single central benzene ring to which four strongly electron-withdrawing carbonyl groups are attached in a planar conformation. Therefore, PMDA is the most electron-deficient dianhydride among those listed, showing the largest $E_a$ value. Comparison between several dianhydrides of similar structures shows that electron-withdrawing groups such as C=O, SO$_2$, and CF$_3$ substituents enhance the $E_a$ value and reactivity, whereas electron-donating substituents such as ether groups reduce reactivity.

Similarly, nucleophilicity or basicity of diamines enhances their reactivity toward a given dianhydride. Electron-donating substituents such as ether groups increase the reaction rate but electron-withdrawing substituents such as C=O particularly located at the para position to the amino group drastically reduce reactivity. The reactivity of diamines correlates well with its basicity (p$K_a$) in a Hammett's relation (13). Because the molecular weight of their poly(amic acid)s is mainly governed by the equilibrium reaction shown previously, relatively low molecular weight polyamic acids may be obtained from diamines with C=O and SO$_2$ substituents. However, it may still be possible to prepare high molecular weight polyimides from these low molecular weight poly(amic acid)s as long as sufficient molecular mobility can be maintained to allow equilibration during the entire imidization process to its completion. This may be achieved, for example, by high temperature solution polycondensation for soluble polyimide systems.

**Effect of Solvents.** The most commonly used solvents in poly(amic acid) preparation are dipolar amide solvents such as DMAc and NMP. The strong acid–base interaction between the amide solvents and poly(amic acid)s is the principal source of exothermicity of the reaction, and one of the important driving forces. It is also expected that the reaction rate is generally faster in more basic solvents. Rate measurements for a model reaction, phthalic anhydride and 4-phenoxyaniline, showed an increasing order of reaction rate in solvents in the order THF < acetonitrile < DMAc (14). However, the rate was even faster in acidic solvents such as $m$-cresol than in DMAc. This confusing observation was found to be the result of acid catalysis. In basic solvents such as dipolar

**Table 1. Electron Affinity of Dianhydrides**

| Dianhydride | Name | $E_a$, eV |
|---|---|---|
| | pyromellitic dianhydride (PMDA) | 1.90 |
| | diphenylsulfone-3,3',-4,4'-tetracarboxylic dianhydride (DSDA) | 1.57 |
| | benzophenone-3,3',4,4'-tetracarboxylic dianhydride (BTDA) | 1.55 |
| | biphenyl-3,3',4,4'-tetra-carboxylic dianhy-dride (BPDA) | 1.38 |
| | diphenyl ether 3,3',4,4'-tetracarboxylic dian-hydride or oxydi-(phthalic anhydride) (ODPA) | 1.30 |
| | 1,4-bis(3,4-dicarboxy-phenoxy)benzene dianhydride or hydroquinone di(phthalic anhy-dride) (HQDA) | 1.19 |
| | bisphenol A di(phthalic anhydride)[a] (BPADA) | 1.12 |
| | 1,2-bis(3,4-dicarboxy-phenoxy)ethane dianhydride or ethylene glycol bis(phthalic anhydride) (EBPA) | 1.10 |

[a] 2,2-Bis[4-(3,4-dicarboxyphenoxy)phenyl]propane dianhydride.

amide solvents, the amic acid formed is buffered by strong hydrogen bonding with the solvent. In less basic solvents such as ether solvents, autocatalytic kinetics were observed. The amic acid formed during the reaction is strong enough to catalyze the reaction (14,15). The reaction of benzophenone-3,3′,4,4′-tetracarboxylic dianhydride [2421-28-5] (BTDA) and weakly basic diamines such as 3,3′- and 4,4′-diaminodiphenylsulfone [80-08-0] affords high molecular weight poly(amic acid)s in THF or diglyme (11). The molecular weights are twice as high as those obtained in dipolar amide solvents. This result is explained by the specific interaction of ether oxygen with hydrogen atoms of the amino groups, the basicity of which is enhanced as a result.

**Effect of Stoichiometry and Side Reactions.** Like any other typical polycondensation reactions, the molecular weight of poly(amic acid)s is primarily determined by the stoichiometry of the monomers, diamines, and dianhydrides. However, the molecular weight is also affected by several other parameters. In the poly(amic acid) equilibrium, the forward reaction is a bimolecular reaction and the reverse reaction monomolecular. As a result, the higher the concentration, the higher the molecular weight of poly(amic acid) formed. The forward reaction is influenced by the nature of solvents and also greatly accelerated by use of monomers with higher reactivity, resulting in higher molecular weight.

In addition, however, several minor but important side reactions concurrently proceed with the main reaction. These side reactions may become significant under certain conditions, particularly when the main reaction is slow because of low monomer reactivities or low concentrations. The principal pathways involved in the formation of poly(amic acid) are as shown in Figure 1.

The reverse reaction is an intramolecular acidolysis of amide group by the o-carboxylic acid to reform anhydride and amine. This unique feature is the result of an ortho neighboring effect. In contrast, the acylation of an amine with benzoic anhydride is an irreversible reaction under the same reaction conditions.

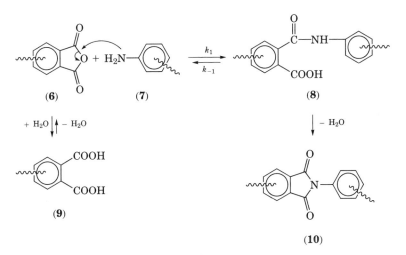

**Fig. 1.** Pathways to formation of poly(amic acid).

The poly(amic acid) structure (**8**) can be considered as a class of polyamides. Aromatic polyamides which lack ortho carboxylic groups are very stable. In contrast, poly(amic acid)s are known to undergo hydrolytic degradation even at ambient temperatures (1–3). When poly(amic acid)s are in solution, a minute but significant amount of anhydride (**6**) always exists as an end group in an equilibrium concentration. When water is present, the anhydride is hydrolyzed to *ortho*-dicarboxylic acid (**9**). The reaction is driven by enhanced nucleophilicity of water in the dipolar solvent and by the strong acid–base interaction of the product with the solvent. The formation of the dicarboxylic acid group is irreversible under these conditions and it remains as an inactive end group. The effect of water on the molecular weight of poly(amic acid)s during their formation and the effect of added water after the formation of poly(amic acid) are well documented. Although the most common source of water is in solvents, water is also formed *in situ* by imidization of amic acid groups. Although the rate of imidization, and therefore formation of water, is slow at ambient temperatures, it is still sufficient to cause a gradual reduction of molecular weight over a period of several weeks. In an aging study of an 11% DMAc solution of PMDA–ODA poly(amic acid) at 35°C after 21 d, 20% of the amic acid was transformed to the imide, generating water which was equivalent to having 0.19 wt % water in the solvent used (16). When long-term storage is necessary, poly(amic acid) solutions should be kept refrigerated.

The mode of addition of monomers during the preparation of poly(amic acid)s is another important parameter that significantly influences the result. The effect of excess amount of monomers has been studied by addition of diamine or dianhydride to the preformed high molecular weight poly(amic acid) solution (16). As expected, the addition of excess of a diamine causes reequilibration of the main equilibrium reaction, resulting in formation of amine end groups and lowering of the molecular weight. However, subsequent addition of the equivalent amount of dianhydride to compensate for the excess diamine restores the original molecular weight of the poly(amic acid). The addition of excess of a dianhydride to the original poly(amic acid) solution also results in a reduction of the molecular weight. In this case, however, the molecular weight cannot be restored by addition of the equivalent amount of diamine to compensate for the excess dianhydride. Reactive anhydrides (**6**) such as PMDA are strong dehydration agents and react with amic acid, converting it to imide (**10**) and itself to the ortho diacid (**9**). The ortho diacid group (**9**) is inactive under the conditions under which poly(amic acid)s are prepared. When both anhydride groups of the added dianhydride are hydrolyzed, it is converted to the corresponding tetraacid, and when only one is hydrolyzed, it is incorporated into the polymer as an end group, lowering the molecular weight. Therefore, the preferred mode of monomer addition is to add the dianhydride to the diamine solution. In this manner, it is best assured that no part of the reaction mixture contains excess dianhydride at any time. In particular, when diamine is only partially soluble the addition of dianhydride must be carried out slowly.

**Thermal Imidization of Poly(amic acid)s.**    Transformation of poly(amic acid)s to the corresponding polyimides is most commonly practiced thermally in the solid state. The method is suitable for preparation of thin objects such as films, coatings, and powders in order to allow the diffusion of by-products and solvent without forming bristles and voids in the final polyimide products. The cast films are dried and heated gradually up to 250–350°C depending on the stability and $T_g$ of the polyimide. When an $N,N$-dimethylacetamide solution of poly(amic acid) is cast and dried at ambient temperature to a nontacky state, the resulting film still contains a substantial amount of the solvent, typically up to 25% by weight depending on the drying conditions. Upon subsequent heating, the imidization reaction takes place not in a true solid state but rather in concentrated viscous solution, at least during the initial and intermediate stages of thermal imidization. The presence of residual solvent plays an important role. The imidization proceeds faster in the presence of dipolar amide solvents, possibly because of a plasticizing effect of the solvent increasing the mobility of the reacting functional groups. The specific effect of amide solvents also suggests that amide basicity may be responsible for the ability to accept protons. The proton of the carboxylic group is strongly hydrogen-bonded to the carbonyl group of the amide solvent. The cyclization of the $o$-carboxyamide group results in destruction of hydrogen bonds and release of the solvent molecule along with water of condensation. The process is very complex, and it has not been successfully described by a simple kinetic expression.

The imidization process involves several interrelated elementary reactions and dynamically changing physical properties, such as diffusion rate, chain mobility, solvation, and acidity. Such complex concurrent events make it difficult to analyze the system in simple terms. Generally, imidization proceeds rapidly at the initial stage and tapers off at a plateau, typical of diffusion-limited kinetics. As the degree of imidization increases, the $T_g$ or stiffness of the polymer chain increases. When the $T_g$ approaches the reaction temperature, the imidization rate markedly slows down. At a higher temperature, a higher degree of imidization is achieved. The initial rapid stage of imidization is attributed (17) to the ring closure of amic acid in the favorable conformation (**8a**). The slower rate in the later stage of imidization is attributed to the unfavorable conformation (**8b**) which has to rearrange to conformation (**8a**) before ring closure. Such a conformational rearrangement requires rotational motion of the adjoining polymer chains and strongly bound solvent molecules with it. The effect of the conformation of amic acid on the imidization rate is also consistent with the observation that the thermal cyclization of model compounds, N-substituted phthalamic acids, is strongly influenced by the steric effect imposed by N-substituents (18).

Analytical methods useful for determining the degree of chemical transformations in the solid state are limited. The most commonly used method is infrared spectroscopy. There are several characteristic absorption bands used for the quantification of five-membered aromatic imides (3,19,20). The strongest absorption occurs at 1720 cm$^{-1}$ (C=O symmetric stretching). However, this band partially overlaps with the strong carboxylic acid band (1700 cm$^{-1}$, C=O) of poly(amic acid). The more useful bands of imide groups are 1780 cm$^{-1}$ (C=O asymmetric stretching), 1380 cm$^{-1}$ (C–N stretching), and 725 cm$^{-1}$ (C=O bending). The absorption of anhydrides also occurs at 1780 and around 720 cm$^{-1}$, requiring proper correction. The carboxylic acid bands, 1700 cm$^{-1}$ (C=O) and 2800–3200 cm$^{-1}$ (OH), and amide bands, 1660 cm$^{-1}$ (C=O, amide I), 1550 cm$^{-1}$ (C–NH, amide II), and bands at 3200–3300 cm$^{-1}$ (NH) often appear as broad peaks particularly when they are strongly associated by hydrogen bonding. They are useful only for qualitative assessment during the imidization process.

Formation of poly(amic acid)s from dianhydrides and diamines is an exothermic reaction. On heating, therefore, the equilibrium shifts toward the left in equation 1. The reversion should temporarily result in a higher level of anhydride and amino groups and a lower molecular weight. Using a soluble polyimide system, 6FDA and 2,2-bis(4(4-aminophenoxy)phenyl)hexafluoropropane [69563-88-8], such a decrease in molecular weight has been observed at intermediate temperatures (80–150°C) (19). The molecular weight gradually increases thereafter up to 325°C. Examination of a poly(amic acid) system based on 6FDA and 4,4′-bis(4-aminophenoxy)biphenyl shows formation of anhydride groups by ftir (20). In a ramped temperature imidization study, the anhydride formation maximizes at around 220°C. The degree of anhydride formation is also dependent on the solvent used, in the increasing order of 2-methoxyethyl ether (diglyme) < NMP < 2-(2-ethoxy)ethoxyethanol (ethyl carbitol). The formation of anhydride group at first peaks at around 135°C when ether solvents are used, and at 156°C in the presence of NMP. The anhydride formation peaks again to a maximum at around 225°C for all three samples prepared from different solvents. At the lower temperature range, anhydride is mainly formed via the reversion of solvated amic acid. Since amic acid is more stabilized in the more basic and polar NMP, the process requires a higher temperature in the presence of NMP. The disappearance of the lower temperature peak indicates the consumption of amic acid which occurs at about 225°C. During the cure, the water formed by imidization reacts with anhydride to form o-dicarboxylic acid. The second peak at around 219–227°C is attributed to the formation of anhydride by thermal dehydrocyclization of the o-dicarboxylic acid. The anhydride and amino groups eventually react with each other as the temperature increases toward 300°C and the chain mobility is regained. However, for more rigid polyimides with very high $T_g$s, the polymer chain may not regain enough mobility to attain complete imidization. Instead, the unreacted anhydride and amino groups may decompose thermally or oxidatively. A study (21) on the thermal stability of anhydride-terminated oligoimides showed that the anhydride group completely disappeared during the ramping temperature of 200–316°C at 2°C/min. Diffuse reflectance Fourier transfer ir (ftir) spectra exhibit a close relationship between the anhydride disappearance and the $CO_2$ formation, indicating decarboxylation is the main mode of decomposition. When an excess amount of diamine is employed it may cause branching of the resulting polyimide via imide–imine formation. Using model reactions, this side reaction was demonstrated and the imide–imine by-product (**11**) was characterized (22).

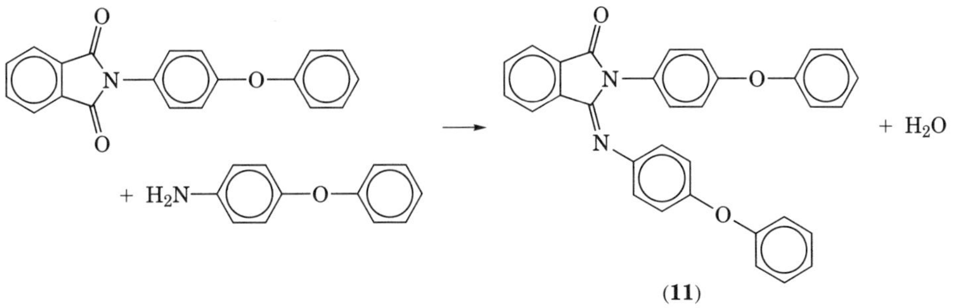

(11)

**Chemical Imidization and Polyisoimides.** Imidization of poly(amic acid)s can also be performed by means of chemical dehydration at ambient temperatures. Commonly used reagents are acid anhydrides in dipolar aprotic solvents or in the presence of tertiary amines (23–27). Acetic anhydride, trifluoroacetic anhydride, etc, were among the dehydrating agents used. Among the amine catalysts were pyridine, trialkylamines, etc. The outcome of the reaction could be very different depending on the type of dehydrating reagents, monomer components of poly(amic acid)s, reaction temperature, solvent, etc. For example, the use of acetyl chloride as a dehydrating agent results in isoimides (28,29). Similarly, the use of $N,N'$-dicyclohexylcarbodiimide (DCC) also leads to essentially quantitative conversion of amic acids to isoimides, rather than imides (30,31). Combinations of trifluoroacetic anhydride–triethylamine and ethyl chloroformate–triethylamine also result in high yields of isoimides (30). A kinetic study on model compounds has revealed that isoimides and imides are formed via a mixed anhydride intermediate (**12**) that is formed by the acylation of the carboxylic group of amic acid (**8**).

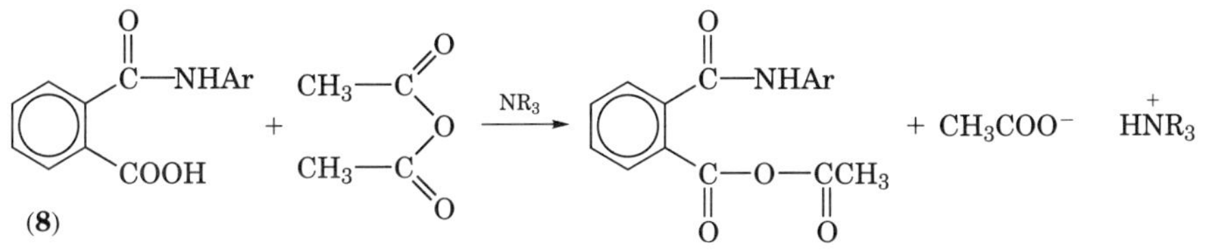

(12)

The presence of the intermediate mixed anhydride was detected by ir as well as proton nmr (31). Convincing evidence was also observed in cases where a difunctional acid chloride such as sebacoyl chloride was used instead of acetyl chloride for cyclization of poly(amic acid) (31). The solution viscosity temporarily increased during the reaction because of the interchain mixed anhydride formation. The viscosity gradually decreased back to the normal level as the cyclization proceeded. Imide (**10**) is formed by intramolecular nucleophilic substitution at the anhydride carbonyl by the amide nitrogen atom via intermediate (**13**), whereas isoimide (**14**) is formed as a result of substitution by the amide oxygen via intermediate (**15**) (Fig. 2). The cyclization of $N$-phenylphthalamic acids with acetic anhydride proceeds smoothly at room temperature in DMAc in the presence of

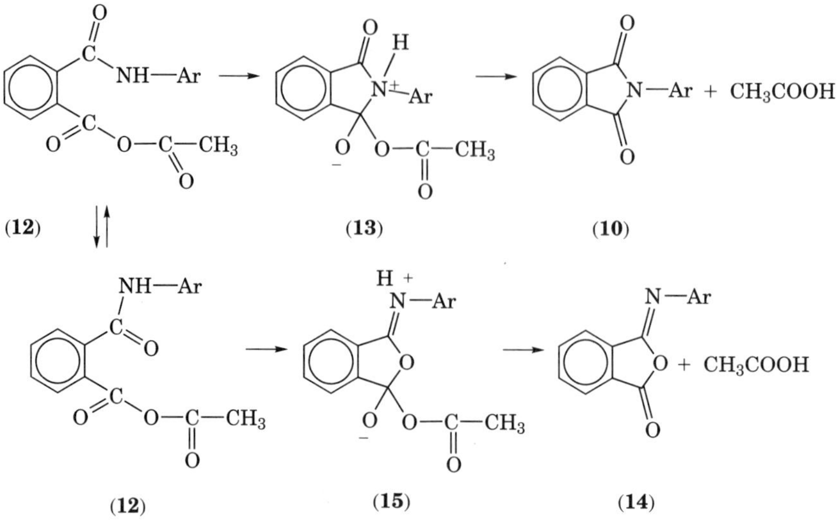

**Fig. 2.** Cyclization of amic acid to imides or isoimides via (**12**). Formation of the mixed anhydride intermediate (**12**) is shown in text.

a tertiary amine. The amine acts as a catalyst as well as an acid acceptor. The isoimide should not be considered as the intermediate to the imide. However, after completion of the cyclization the isoimide slowly rearranges to the imide. The rearrangement reaction was found to be efficiently catalyzed by acetate ion (30). Use of trifluoroacetic anhydride or $N,N'$-dicyclohexylcarbodiimide as the dehydrating agent affords exclusively isoimides for both model compounds and poly(amic acid)s. The presence of isoimide structure is readily identified by ir spectroscopy. A strong absorption at 1750–1820 cm$^{-1}$ is characteristic carbonyl absorption of imino lactone structure. In addition, isoimides show a strong absorption band at 921–934 cm$^{-1}$. Isoimides generally exhibit intense, brightly yellow or yellowish orange color, arising from a strong absorption at around 350–400 nm in the uv–visible spectrum.

## ONE-STEP METHODS

**High Temperature Solution Polymerization.** A single-stage homogeneous solution polymerization technique can be employed for polyimides that are soluble in organic solvents at the polymerization temperature. In this process, a stoichiometric mixture of monomers is heated in a high boiling solvent or a mixture of solvents at a temperature of 140–250°C, where the imidization reaction proceeds rapidly. Commonly used solvents are benzonitrile, $\alpha$-chloronaphthalene, $o$-dichlorobenzene, trichlorobenzenes, and phenolic solvents such as $m$-cresol, and chlorophenols as well as dipolar aprotic amide solvents such as NMP. Toluene is often used as a co-solvent in order to facilitate the removal of the water of condensation as an azeotrope. Importantly, preparation of high molecular weight poly(amic acid) is not necessary in this procedure, although imidization still proceeds via amic acid intermediates. However, the concentration of amic acid groups is small at any given time during the polymerization because amic acids are unstable at high temperatures and rapidly imidize or revert to amine and anhydride. Because water is formed by the formation of the imide, the anhydride group is present in an equilibrium with its hydrolyzed form, o-dicarboxylic acid. When a mixture composed of diamine, dianhydride, and a solvent is heated, a viscous solution is formed at intermediate temperatures of approximately 30–100°C. At this stage, the composition of the product is mainly poly(amic acid), and phase separation is usually observed in nonpolar solvents such as chlorinated aromatic hydrocarbons. However, on raising the temperature to 120–160°C, a vigorous evolution of water occurs and the reaction mixture becomes homogeneous. At this stage, product is essentially a low molecular weight polyimide having o-dicarboxy and amino end groups. Thereafter, a slow, stepwise polycondensation takes place according to the reaction between the end groups. The rate is slower in basic aprotic amide solvents, and faster in acidic solvents such as m-cresol. In general, the imidization reaction has been shown to be catalyzed by acid (14,32,33). Thermal imidization of poly(amic acid)s is catalyzed by tertiary amines (34). High temperature solution polymerization in m-cresol is often performed in the presence of high boiling tertiary amines such as quinoline as catalyst. Dialkylaminopyridines and other tertiary amines are effective catalysts in neutral solvents such as dichlorobenzene (35). Alkali metal and zinc salts of carboxylic acids (36) and salts of certain organophosphorus compounds (37) are also very efficient catalysts in one-step polycondensation of polyimides.

When use of one-step solution polymerization is an available option, generally it is the superior method to synthesize structurally pure polyimides because complete imidization and quantitative capping of end groups can be readily achieved in solution.

## TETRACARBOXYLIC ACIDS WITH DIAMINES

Early work (38) on polyimide synthesis included the methods that used tetracarboxylic acids and their ester derivatives. The carboxylic acids were combined with diamines to form salts in analogy to the synthesis of nylon via nylon salts. The salts were thermally imidized to form polyimides. For intractable polyimides such as those based on pyromellitic acid and aromatic diamines, high molecular weight poly(amic acid) intermediates have to be made which require the use of dianhydrides as monomer. For more tractable polyimides, however, one-step solution or melt polymerization can be employed as long as the reaction system can be maintained in solution or above the glass-transition temperature in the melt. Although the majority of literature reports that dianhydrides and diamines are used for such step-growth high temperature processes, the tetracarboxylic acids can be used in place of dianhydrides without significant differences in the overall results. The ortho dicarboxylic acid group is thermally converted *in situ* to the anhydride and then immediately reacts with amino groups.

Various high molecular weight linear polyetherimides were prepared from tetracarboxylic acid and diamines by melt polymerization (39) and by a high temperature solution method (40). In melt polymerization, it is advantageous to use tetracarboxylic acids because high molecular weight poly(amic acid) intermediates of very high melt viscosities are not formed during the initial heating-up stage.

OTHER SYNTHETIC METHODS

**Diesters of Tetracarboxylic Acids With Diamines.** Dianhydrides react with alcohols at a moderate temperature, producing the tetracarboxylic acid diesters. Benzophenonetetracarboxylic acid diester (BTTA diester) is obtained from BTDA as a mixture of three positional isomers. Partly because it is an isomeric mixture, the diester of BTTA is soluble in methanol at high concentrations. A state-of-the-art thermoset polyimide, PMR-15, is based on this unique property of the diester. Processed according to the polymerization of monomeric reactants (with molecular weight of 1500), whence its name, PMR-15 was originally developed by NASA at Lewis Research Center (41–43). The chemistry of PMR-15 is illustrated in Figure 3. End-capping cross-linker norbornenedicarboxylic anhydride (NA) and BTDA are first converted to BTTA dimethyl ester (BTDE) and norbornenedicarboxylic acid monomethyl ester (NE) (**16**) in methanol, and then methylenedianiline (MDA) (**17**) is added. The tetracarboxylic acid diester can be considered a deactivated dianhydride, and the mixture does not react at ambient temperatures. Therefore, this monomeric solution possesses a low viscosity at high concentrations and can be readily impregnated into graphite fiber cloths. In addition, methanol is readily removable, making it highly desirable for composite processing. On heating at an intermediate temperature range of 100–150°C, the ortho carboxy ester groups revert to anhydride and methanol, the anhydride then reacts with amine to form amic acid groups, which in turn are converted to imide groups at higher temperatures. The final curing proceeds via a retro-Diels-Alder reaction of nadimide groups to form maleimide end groups and cyclopentadiene, which in turn undergo addition copolymerization. PMR-15 is extensively used as matrix resin for graphite fiber-based high temperature composites.

**Polyisoimides as Precursors.** In general, polyisoimides are significantly more soluble and possess lower melt viscosities and lower glass-transition temperatures than the corresponding polyimides, mainly because of their structural irregularity arising from *p*- and *m*-imino group linkages along the chain.

Also, the R groups on the imino N may have syn and anti configurations. Taking advantage of these properties, acetylene-terminated oligoisoimides were developed as readily processible adhesives and matrix resins for graphite fiber composites (44–49). Thermid IP-600 resin is an acetylene-terminated oligoisoimide produced by National Starch Company. The structure is shown in Figure 4a. After shaping such composites, the cross-linked polyisoimides are thermally converted *in situ* to the corresponding polyimides at higher temperatures. The process does not produce gaseous condensation by-products, allowing production of void-free composites.

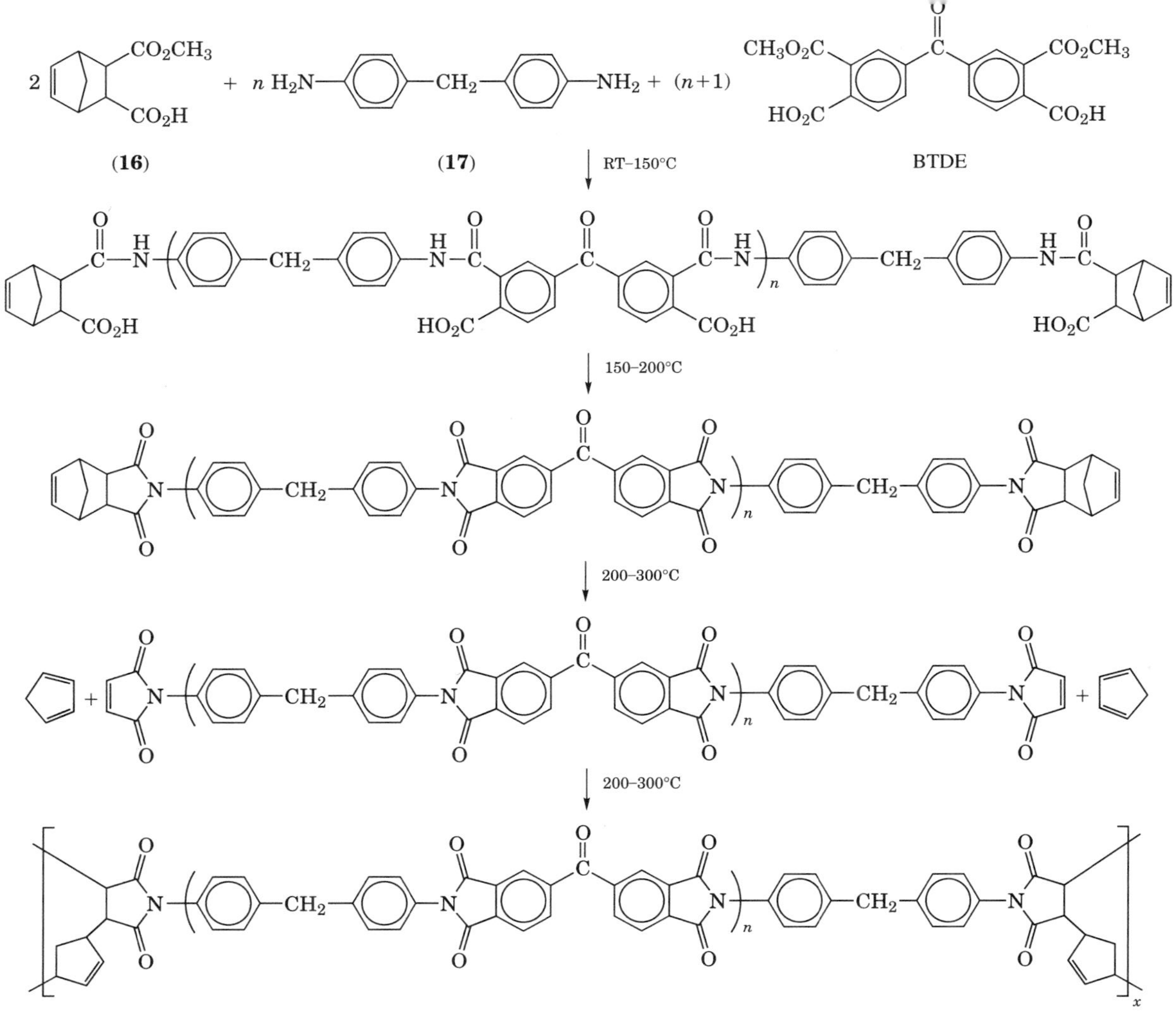

**Fig. 3.** PMR-15 thermoset polyimide resin.

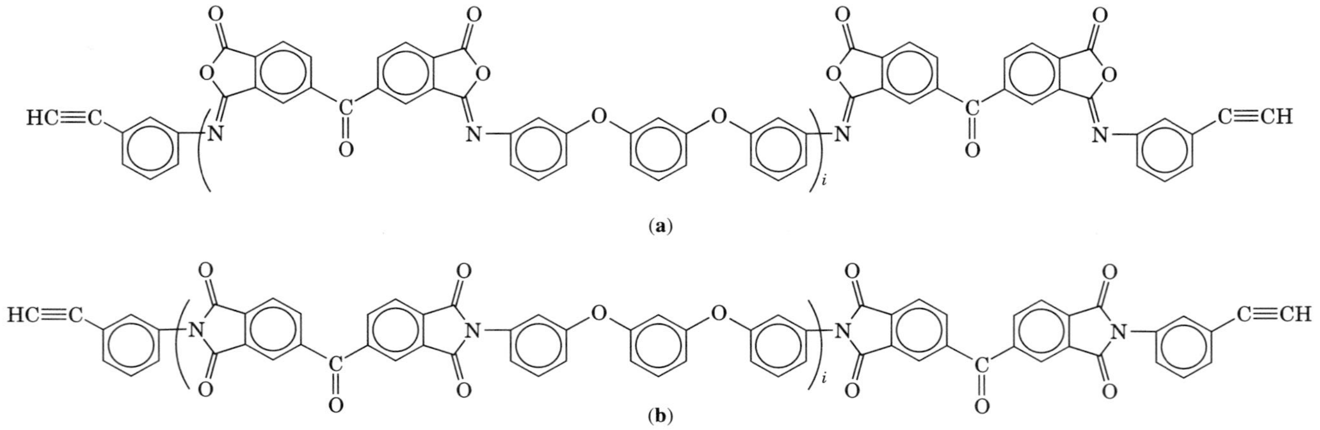

**Fig. 4.** (a) Thermid thermoset oligoisoimide; (b) acetylene-terminated oligoimide.

Intractable, rigid, rod-like polyimides have been synthesized by way of poly-isoimides (50). A large number of high molecular weight polyisoimides were also synthesized and characterized for the purpose of preparing semiinterpenetrating (SIPN) polyimide matrices and adhesives (51–53).

**Ester Derivatives of Poly(amic acid)s.** Poly(amic acid)s in solution are inherently unstable and must be stored under refrigeration. However, their ester derivatives are significantly more stable; the inability of the esters to form carboxylate anions prohibits the reverse reaction. They can be isolated, for example, by precipitation, or stored for practically indefinite periods. Such stability is highly desirable to obtain reliable and reproducible results in some applications, such as microelectronics. In addition, the ester derivatives are generally more soluble in a wider range of solvents, including conventional solvents. Conversion of esters of poly(amic acid) to polyimides proceeds thermally, but at slower rate, and generally requires a temperature significantly higher than 200°C. Alkyl esters of poly(amic acid)s are prepared by the reaction of diamines with diester diacid chlorides, which are synthesized by the chlorination of diesters of tetracarboxylic acids (54–56).

$N,N'$-Bis(trimethylsilyl)diamines react rapidly at room temperature with various aromatic dianhydrides to form high molecular weight silyl esters of poly(amic acid)s (57). Poly(amic silylester)s are soluble in solvents that are less polar than dipolar solvents. A homogeneous solution of poly(amic silylester) of PMDA–ODA analogue is obtained in THF. Homogeneous solutions of various other poly(amic silylester)s have also been prepared in aromatic hydrocarbon solvents containing a minor amount of dipolar solvent. Like alkyl esters of poly(amic acid)s the resulting silyl esters are substantially more stable than the poly(amic acid)s. The cyclization of the silyl esters to the polyimides requires temperatures higher than 200°C and proceeds with elimination of trimethylsilanol. Similarly, polyimides have been prepared by the reaction of PMDA and $N,N,N',N'$-tetratrimethylsilyldiamines; in this case the by-product is hexamethyldisiloxane (58).

**Polymerization via Nucleophilic Substitution Reaction.** Halo- and nitrogroups attached to phthalimide groups are strongly activated toward nucleophilic substitution reactions. Thus polyetherimides are synthesized by the nucleophilic substitution reaction of bishaloimides (59,60) and bisnitroimides (61,62) with anhydrous bisphenol salts in dipolar aprotic solvents.

In general, high molecular weight polyetherimides are readily obtained by nitrosubstitution polymerization from bisphenols such as bisphenol A and 4,4'-dihydroxydiphenyl sulfide, in which each hydroxy group is located in a different aromatic ring. However, lower molecular weight polymers are obtained when benzenediols are employed. An electron-transfer redox reaction occurs between electron-rich benzenediol dianions and highly electron-deficient nitrophthalimides. Alternatively, ether-containing dianhydrides, bis(ether anhydride)s, are prepared by a nitrosubstitution reaction, followed by hydrolysis of the product bis(etherimide)s according to the following (63).

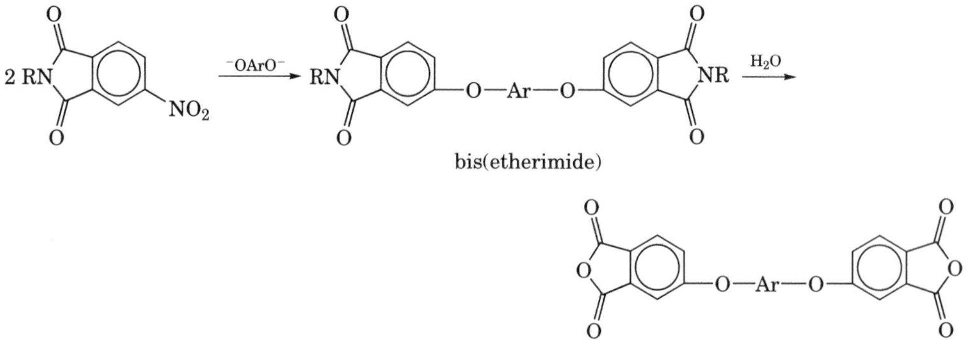

bis(etherimide)

bis(ether anhydride)

Unlike common commercial dianhydrides such as PMDA and BTDA, bis(ether anhydride)s possess moderate reactivity toward nucleophiles because of the electron-donating property of the ether groups. Bis(ether anhydride)s are hardly affected by atmospheric moisture. The stability and generally better solubility of bis(ether anhydride)s provide significant advantages in manufacturing operations.

The majority of polyetherimides are noncrystalline, and their glass-transition temperatures are in the range of 200–280°C (64). Polyetherimides are injection-moldable and exhibit high strength and high modulus, good ductility, and excellent thermal stability and flame resistance, as well as good electrical properties. ULTEM 1000 resin, manufactured by General Electric Company, is derived from bisphenol A [80-05-7] and m-phenylenediamine [108-45-2].

**Polymerization by Transimidization Reaction.** Exchange polymerization via equilibrium reactions is commonly practiced for the preparation of polyesters and polycarbonates. The two-step transimidization polymerization of polyimides was described in an early patent (65). The reaction of pyromellitic diimide with diamines in dipolar solvents resulted in poly(amic amide)s that were thermally converted to the polyimides. High molecular weight polyimides were obtained by employing a more reactive bisimide system (66). The intermediate poly(amic ethylcarboamide) was converted to the polyimide at 240°C.

The majority of polyetherimides are tractable and their polymerization can be performed in solution or in the melt. High molecular weight polyetherimides have been synthesized via one-step imide–amine exchange reaction between bis(etherimide)s and diamine (67) according to the following:

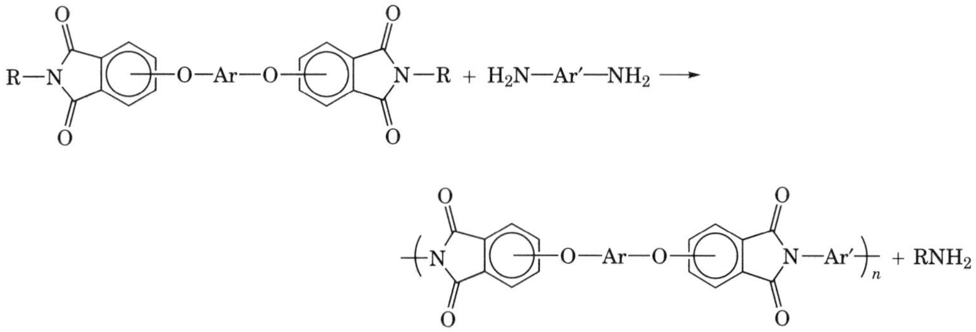

$$R-N \cdots O-Ar-O \cdots N-R + H_2N-Ar'-NH_2 \longrightarrow$$

$$\left(-N \cdots O-Ar-O \cdots N-Ar'\right)_n + RNH_2$$

This scheme eliminates the process of converting bis(etherimide)s to bis(ether anhydride)s. When polyetherimides are fusible the polymerization is performed in the melt, allowing the monamine to distill off. It is advantageous if the amino groups of diamines are more basic or nucleophilic than the by-product monoamine. Bisimides derived from heteroaromatic amines such as 2-aminopyridine are readily exchanged by common aromatic diamines (68,69). High molecular weight polyetherimides have been synthesized from various $N,N'$-bis(heteroaryl)bis(etherimide)s.

**Polymerization of Dianhydrides and Diisocyanates.** Phthalic anhydride does not react readily with an aromatic isocyanate (70). $N$-Phenylphthalimide is obtained from phthalic anhydride and phenylisocyanate only in refluxing pyridine, suggesting the reaction may be catalyzed by base. The reaction proceeds in dipolar solvents at moderate temperatures and protic compounds such as water, alcohols, and amines serve as catalysts (71). High molecular weight polyimides have been obtained in dipolar solvents by several investigators (72–76). High molecular weight polyetherimides are synthesized from bis(ether anhydride)s and diisocyanates by melt polymerization or high temperature solution polymerization in nonpolar aromatic solvents (77). In the absence of polar solvent, the complex side reactions are minimal but the reaction rate is slow. However, in the presence of a stable base such as alkali carbonates, the polymerization proceeds efficiently at temperatures above 200°C. PMDA and an aromatic diisocyanate with bulky substituents have been polymerized (78). High molecular weight polyimides have formed also in high boiling nonpolar aromatic solvents such as nitrobenzene, benzonitrile, and anisole, whereas polymers with only moderate molecular weights were obtained in dipolar solvents such as NMP.

**Polymerization by C–C Coupling.** An aromatic carbon–carbon coupling reaction has been employed for the synthesis of rigid rod-like polyimides from imide-containing dibromo compounds and aromatic diboronic acids in the presence of palladium catalyst, $Pd[P(C_6H_5)_3]_4$ (79,80).

**Polymerization by Cycloaddition.** Bisimides and oligoimides capped with reactive unsaturations such as maleimide, acetylene, and xylylene groups, can be chain-extended by a cycloaddition reaction with proper bisdienes.

Polyimides have been synthesized by Diels-Alder cycloaddition of bismaleimides and substituted biscyclopentadienones (81,82). The intermediate tricyclic ketone structure spontaneously expell carbon monoxide to form dihydrophthalimide rings, which are readily oxidized to imides in the presence of nitrobenzene.

The benzocyclobutene ring structure is a latent reactive diene because it tautomerizes to o-xylylene on heating according to the following:

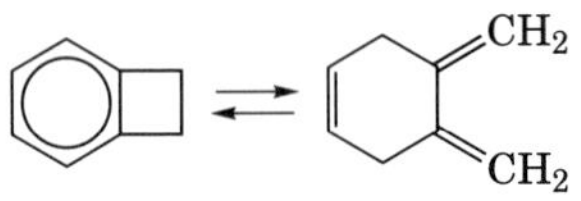

The diene undergoes Diels-Alder cycloaddition with dienophiles. Polyimides have been synthesized from various imide-containing benzocyclobutenes and dienophiles (83–85).

## Thermoset Polyimides

Various thermoset polyimide resins are used as matrix resins for advanced composites (see COMPOSITE MATERIALS, POLYMER-MATRIX). PMR-15, acetylene-terminated isoimide resins, and benzocyclobutene resins have been discussed. In general, low molecular weight oligomers with functional end groups (cross-linkable groups) are used in impregnating the fiber fabrics. Such resins possess low viscosities which facilitate flow of resins and good wetting of the fibers. The resulting preimpregnated fabrics (prepregs) are consolidated in the mold and thermally cross-linked (cured). The impregnation and consolidation have to take place before the cross-linking reaction predominates in order to attain void-free composites. The temperature at which the cross-linking reaction takes place sets the primary limitation on the chemical structures of the oligomers to be used. Thermoset polyimides produce creep-resistant composites of high rigidity. One of the more widely used thermoset polyimides is bismaleimide resin (BMI). A commonly used BMI component is a bismaleimide, 4,4'-bis(maleimido)diphenylmethane [13676-54-5] (**18**), derived from the lowest cost aromatic diamine, 4,4'diaminodiphenylmethane [101-77-9] or 4,4'-methylenedianiline (MDA).

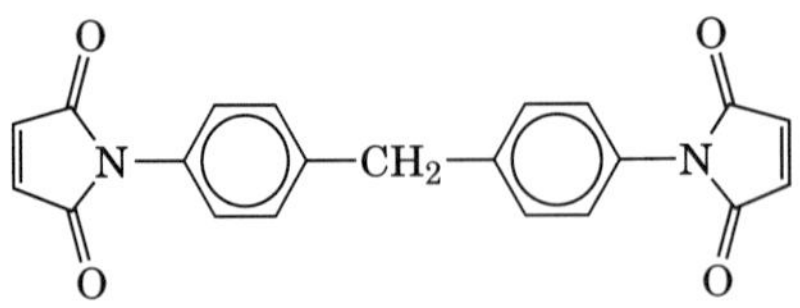

(18)

The maleimide groups cure via an addition reaction in the temperature range of 150–250°C without evolution of volatiles. The BMI resins are capable of performing at temperatures up to approximately 230°C. Because of relatively low cure temperatures, efforts are made to produce BMI of a low melting range by formulating mixtures of several components (86). The curing reaction and chain extension take place in Rhône-Poulenc's Kerimid 601 via Michael addition of amino functions to the maleimide double bonds. A great many modifications have been applied to BMIs in order to improve toughness and strength as well as processability (87). Acetylene-terminated oligoimides have been reported as useful thermoset matrix resins and adhesives (see Fig. 4b) (88). One problem associated with this system is a relatively low curing temperature of 200–250°C. The oligoimides which can be employed in this system must have low $T_g$s. More recently, oligoimides end-capped with disubstituted acetylene groups (**19**) and (**20**) have been reported to have higher curing temperatures, ie, 300–350°C. The system enables the use of oligoimides with higher $T_g$s, resulting in the formation of composites of higher temperature capability (89–91).

(**19**)                                    (**20**)

## Properties and Applications of Polyimides

Because a wide range of properties are realized with various compositions, polyimides are used in diverse areas of application.

Aromatic high temperature polyimides based on PMDA, BPDA, and other dianhydrides are commercially produced, mostly in the form of films. They possess high $T_g$ and $T_m$ and therefore the films are produced via the two-stage poly(amic acid) process. Kapton films derived from PMDA and ODA are the earliest commercial polyimides available in various thicknesses and also as surface-modified forms and others containing various inorganic fillers. Because of their outstanding thermal, mechanical, and electrical properties as well as radiation resistance, these films are widely used as insulation materials in aerospace, electric, and electronic components. For example, wires and conductors in traction motors and generators are often insulated with Kapton tapes. Another common application is flexible circuits made from copper–polyimide laminates. Similar high temperature films marketed as Upilex are based on BPDA and manufactured by UBE Industries. Upilex R is produced from BPDA and ODA. It has a significantly lower $T_g$ of 285°C, but higher tensile strength and modulus at its use temperatures. It is also superior in hydrolytic stability. Upilex S is a polyimide produced from BPDA, with $p$-phenylenediamine (PPD) as the primary diamine component, and it possesses much higher modulus and strength. Its $T_g$ is above 500°C. Some of the properties of Kapton and Upilex films are given in Table 2.

High temperature aromatic polyimides are difficult to shape thermally because of their high $T_g$ and $T_m$. However, simple shaped raw stock moldings are made by a high pressure compression sintering of polyimide powders. Moldings containing various fillers are also available. High performance parts resistant to extreme environmental conditions are made by machining these moldings. Vespel and Upimol, manufactured by Du Pont and UBE, respectively, are examples of such products.

**Table 2. Properties of High Temperature Polyimide Films**

| Properties | Kapton H | Upilex R | Upilex S |
|---|---|---|---|
| tensile strength, MPa[a] | 172 | 241 | 393 |
| tensile modulus, GPa[b] | 3.0 | 3.7 | 8.8 |
| elongation, % | 70 | 130 | 30 |
| glass-transition, $T_g$, °C | ~390 | 285 | >500 |
| limited oxygen index, % | 37 | 55 | 66 |
| CTE[c] $\times 10^5$, °C$^{-1}$ | 2 | 1.5 | 0.8 |
| water absorption, % | | | |
|   at 23.5°C | 1.3 | | |
|   at 50°C | | 1.1 | 0.9 |
| specific gravity | 1.42 | 1.39 | 1.47 |
| dielectric constant, 1 kHz | 3.5 | 3.5 | 3.5 |
| dissipation factor, 1 kHz | 0.003 | 0.0014 | 0.0013 |

[a]To convert MPa to psi, multiply by 145.
[b]To convert GPa to psi, multiply by 145,000.
[c]Coefficient of thermal expansion.

Various melt-fusible polyimides whose $T_g$s range from 250 to 350°C are available as molding powders or pellets. LARC-TPI was developed by NASA and derived from BTDA and 3,3′-diaminobenzophenone. It has a $T_g$ of 260°C. The polymer is also reported to have good adhesive properties. Du Pont's Avimid N is a m-phenylenediamine (MPD)–p-phenylenediamine (PPD) copolyimide of 6FDA and has a high $T_g$ of 340–370°C. Avimid N is used as a high temperature graphite composite matrix resin and also as an adhesive. Polyimide 2080 was developed by Upjohn Company and produced from BTDA and a mixture of toluenediisocyanate and 4,4′-diphenylmethanediisocyanate. It is an amorphous polyimide with a $T_g$ of 310°C. Some of the properties of Avimid N and PI 2080 are listed in Table 3.

Polyetherimides are composed of ether-containing dianhydrides. Ultem 1000 is manufactured by General Electric Company from the ether BPADA (bisphenol A di(phthalic anhydride)) and MPD. Ultem 1000 has a $T_g$ of 217°C. Because of flexible ether linkages in the main chain, Ultem 1000 possesses ex-

**Table 3. Properties of Thermoplastic Polyimides**

| Properties | Avimid N | LARC TPI |
|---|---|---|
| tensile strength, MPa[a] | 110 | 136 |
| tensile modulus, GPa[b] | 4.13 (flexural) | 3.72 |
| elongation, % | 6 | 4.8 |
| notched Izod impact strength, J/m[c] | 42 | |
| $T_g$ °C | 370 | 250 |
| CTE[d] $\times 10^5$, °C$^{-1}$ | 5.6 | |
| dielectric constant, $10^3$ Hz | 2.7 | |
| specific gravity | 1.44 | 1.33 |

[a]To convert MPa to psi, multiply by 145.
[b]To convert GPa to psi, multiply by 145,000.
[c]To convert J/m to ftlbf/in., divide by 53.38.
[d]Coefficient of thermal expansion.

cellent flow characteristics and melt stability. It is used as a general-purpose high temperature injection molding thermoplastic. In addition to high flexural strength and high modulus, Ultem 1000 possesses extraordinary nonflammability and low smoke-generating characteristics. These properties of polyetherimides provide a halogen-free, flame-retardant material for aircraft interiors and reliable electrical insulation. Some of the principal properties of polyetherimides are shown in Table 4.

Relatively few processible polyimides, particularly at a reasonable cost and in reliable supply, are available commercially. Users of polyimides may have to produce intractable polyimides by themselves *in situ* according to methods discussed earlier, or synthesize polyimides of unique compositions in order to meet property requirements such as thermal and thermoxidative stabilities, mechanical and electrical properties, physical properties such as glass-transition temperature, crystalline melting temperature, density, solubility, optical properties, etc. It is, therefore, essential to thoroughly understand the structure–property relationships of polyimide systems, and excellent review articles are available (1–5,92).

Because of the high functional values that polyimides can provide, a small-scale custom synthesis by users or toll producers is often economically viable despite high cost, especially for aerospace and microelectronic applications. For the majority of industrial applications, the yellow color generally associated with polyimides is quite acceptable. However, transparency or low absorbance is an essential requirement in some applications such as multilayer thermal insulation blankets for satellites and protective coatings for solar cells and other space components (93). For interlayer dielectric applications in semiconductor devices, polyimides having low and controlled thermal expansion coefficients

## Table 4. Properties of Polyetherimides

| Properties | Ultem 1000 | Ultem 6000 |
|---|---|---|
| tensile strength, MPa[a] | 105 | 103 |
| tensile modulus, GPa[b] | 3.0 | |
| ultimate elongation, % | 60 | 30 |
| flexural strength, MPa[a] | 150 | 145 |
| flexural modulus, GPa[b] | 3.3 | 3.0 |
| Izod impact strength, J/m[c] | | |
| notched | 50 | 40 |
| unnotched | 1300 | 1300 |
| heat deflection temperature at 1.82 MPa,[a] 6.4 mm, °C | 200 | 215 |
| limited oxygen index, % | 47 | 44 |
| CTE[d] $\times 10^5$, °C$^{-1}$ | 5.6 | 5.1 |
| thermal conductivity, W/(m·K) | 0.12 | |
| specific gravity | 1.27 | 1.29 |
| dielectric constant, 1 kHz, 50% rh | 3.15 | 3.0 |
| dissipation factor, 1 kHz, 50% rh | 0.0013 | 0.001 |

[a]To convert MPa to psi, multiply by 145.
[b]To convert GPa to psi, multiply by 145,000.
[c]To convert J/m to ftlbf/in., divide by 53.38.
[d]Coefficient of thermal expansion.

are required to match those of substrate materials such as metals, ceramics, and semiconductors used in those devices (94).

In assembling high temperature composites and composites with other materials such as ceramics and metals, high temperature polyimide adhesives are becoming increasingly important. An excellent review is available (5,95). Another interesting application of polyimides is gas-separation membranes, because of their high strength and high temperature capability. Despite good selectivity in gas permeability, the majority of polyimides exhibit low permeability. However, structural modifications such as alkyl and fluoroalkyl substitutions often increase permeability. There have been numerous literature and patent citations which have been excellently reviewed (5).

Polyimides containing C–F bonds have been receiving strong attention (96–98). Fluorine-containing polyimides possess lower dielectric constant and dielectric loss because of reduced water absorption and lower electronic polarization of C–F bonds vs the corresponding C–H bonds. Fluorine-containing polyimides are often more soluble and readily processible without sacrificing thermal stabilities. The materials are applied primarily in microelectronics as passivation layers and protective coatings.

## Health and Safety Factors

Handling of monomers, solvents, and solutions required in the preparation of polyimides should be practiced under standard safe environmental conditions for chemical processes. Solid powders of monomers are irritants. Inhalation and skin contact should be avoided by use of appropriate protective devices and in a well-ventilated atmosphere. Some of the aromatic diamines are suspected carcinogens. Accordingly, handling of these materials must meet OSHA guidelines. The Material Safety Data Sheet (MSDS) of each reagent should always be consulted before use. Ether-containing solvents such as THF and diglyme and solutions containing them may build up peroxides. They should be stored under an inert atmosphere and should be periodically tested for peroxide.

## BIBLIOGRAPHY

"Polyimides" in *ECT* 2nd ed., Suppl. Vol., pp. 746–773, by J. Preston, Chemstrand Research Center, Inc., Monsanto Co.; in *ECT* 3rd ed., Vol. 18, pp. 704–719, by P. E. Cassidy, Southwest Texas State University, and N. C. Fawcett, University of Southern Mississippi.

1. C. E. Sroog, *J. Polym. Sci. Macromol. Rev.* **11**, 161 (1976).
2. M. I. Bessonov, M. M. Koton, V. V. Kudryavtsev, and L. A. Laius, *Polyimides, Thermally Stable Polymers*, Plenum Press, New York, 1987.
3. D. Wilson, H. D. Stenzenberger, and P. M. Hergenrother, eds., *Polyimides*, Chapman and Hall, New York, 1990.
4. K. L. Mittal, *Polyimides*, Vols. 1 and 2, Plenum Press, New York, 1982.
5. C. E. Sroog, *Prog. Polym. Sci.* **16**, 561–694 (1991).
6. M. T. Bogert and R. R. Renshaw, *J. Am. Chem. Soc.* **28**, 617 (1906).
7. M. T. Bogert and R. R. Renshaw, *J. Am. Chem. Soc.* **30**, 1135 (1908).
8. G. I. Nosova and co-workers, *Vysokomol. Soedin., Ser. A* **26**(5), 998 (1984).

9. R. G. Bryant and T. L. St. Clair, *Abstracts of the 4th International Conference on Polyimides*, Oct.–Nov. 1991, Ellenville, N.Y., p. II-69.

10. C. E. Sroog and co-workers, *J. Polym. Sci., Part A* **3**, 1373 (1965).

11. A. H. Egli and T. L. St. Clair, in W. D. Weber and M. R. Gupta, eds., *Recent Advances in Polyimide Science and Technology*, Society of Plastics Engineers, Pouchkeepsie, N.Y., 1985, p. 57.

12. V. M. Svetlichnyi, K. K. Kalnin'sh, V. V. Kudryatsev, and M. M. Koton, *Dokl. Acad. Nauk SSSR* **237**(3), 612 (1977); English trans., **237**(3), 693 (1977).

13. V. A. Zubkov, M. M. Koton, V. V. Kudryatsev, and V. M. Svetlichnyi, *Zh. Org. Khim.* **17**(8), 1682 (1981); English trans., **17**(8), 1501 (1982).

14. V. A. Solomin and co-workers, *Dokl. Akad. Nauk USSR*, **236**(1), 139 (1977); English trans., **236**(1), 510 (1977).

15. R. L. Kaas, *J. Polym. Sci. Polym. Chem. Ed.* **19**, 2255 (1981).

16. L. W. Frost and I. Kesse, *J. Appl. Polym. Sci.* **8**, 1039 (1964).

17. L. A. Laius and co-workers, *Polym. Sci. USSR*, **A9**(10) 2470 (1967).

18. J. W. Verbicky, Jr. and L. Williams, *J. Org. Chem.* **46**, 175 (1981).

19. P. R. Young, J. R. Davis, A. C. Chang, and J. N. Richardson, *J. Polym. Sci. Polym. Chem. Ed.*, **28**, 3107 (1990).

20. D. E. Fjare and R. T. Roginski, in C. Feger and co-workers, eds., *Advances in Polyimide Science and Technology*, Technomics, Lancaster, Pa., 1991, p. 326.

21. R. Yunk and C. Watson in Ref. 9, p. II-21.

22. A. K. Saini, C. M. Carlin, and H. H. Patterson, *J. Polym. Sci. Polym. Chem. Ed.* **31**, 2751 (1993).

23. U.S. Pat. 3,179,630 (1965), A. L. Endrey (to Du Pont).

24. U.S. Pat. 3,179,631 (1965), A. L. Endrey (to Du Pont).

25. U.S. Pat. 3,179,632 (1965), W. R. Hendrix (to Du Pont).

26. U.S. Pat. 3,179,633 (1965), A. L. Endrey (to Du Pont).

27. U.S. Pat. 3,282,898 (1966), R. J. Angelo (to Du Pont).

28. S. Hoogewerff and W. A. van Dorp, *Recl. Trav. Chim. Pays Bas* **13**, 93 (1894).

29. P. H. van der Muellen, *Recl. Trav. Chim. Pays Bas*, **15**, 282 (1896).

30. R. J. Cotter, C. K. Sauers, and J. M. Whelan, *J. Org. Chem.* **26**, 10 (1961).

31. R. J. Angelo, R. C. Golike, W. E. Tatum, and J. A. Kreuz in Ref. 11, p. 67.

32. S. V. Lavrov, A. Y. Ardashnikov, I. Y. Kardash, and A. N. Pravednikov, *Vysokomol. Soyed.* **A19**(5), 1052 (1977); English trans. *Polym. Sci. USSR* **19**, 1212 (1977).

33. S. V. Lavrov and co-workers, *Vysokomol. Soyed.* **A22**(8), 1886 (1980); English trans., *Polym. Sci. USSR* **22**(8), 2069 (1980).

34. J. A. Kreuz, A. L. Endrey, F. P. Gay, and C. E. Sroog, *J. Polym. Sci., Part A-1*, **4**, 2607 (1966).

35. U.S. Pat. 4,324,884; 4,324,885; and 4,330,666 (1982), D. M. White (to General Electric).

36. U.S. Pat. 4,293,683 (1981), T. Takekoshi and H. J. Klopfer (to General Electric).

37. U.S. Pat. 4,324,882 (1982), T. Takekoshi (to General Electric).

38. U.S. Pat. 2,710,853 (1955) and 2,867,609 (1959), W. M. Edwards and I. M. Robinson (to Du Pont).

39. U.S. Pat. 3,833,546 (1974), T. Takekoshi and J. E. Kochanowski (to General Electric).

40. U.S. Pat. 3,905,942 (1975), T. Takekoshi and J. E. Kochanowski (to General Electric).

41. T. T. Serafini, P. Delvigs, and G. R. Lightsey, *NASA TN* D-6611 (1972).

42. T. T. Serafini, P. Delvigs, and G. R. Lightsey, *J. Appl. Polym. Sci.* **16**, 905 (1972).

43. T. T. Serafini, P. Delvigs, and W. B. Elston, *PMR-15 Polyimides Review and Update*, NASA TM-82821, Washington, D.C., 1982.

44. A. L. Landis and A. B. Naselow, *14th Natl. SAMPE Tech. Conf.* **14**, 236 (1982).

45. R. H. Bott, L. T. Taylor and T. C. Ward, *Amer. Chem. Soc. Polym. Prepr.*, **27**(2), 72 (1982).

46. T. Murray and N. Tessier, *31st Intl. SAMPE Symp.*, p. 693 (1986).
47. F. W. Harris, A. Pamidimukolas, R. Gupta, S. Das, T. Wu, and G. Mock, *J. Macromol. Sci. Chem.* **A21**, 1117 (1984).
48. M. R. Unroe, B. A. Reinhardt, and F. E. Arnold, *Amer. Chem. Soc. Polym. Prepr.* **26**, 136 (1985).
49. D. J. Capo and J. Schoenberg, *18th Natl. SAMPE Tech. Conf.*, 710 (1986).
50. J. S. Wallace, F. E. Arnold and L. S. Tan, *Am. Chem. Soc. Polym. Prepr.* **28**(2), 316 (1987).
51. A. L. Landis, A. W. Chow, R. D. Hamlin, and K. S. Y. Lau in Ref. 20, p. 84.
52. *Ibid.*, p. 110.
53. A. L. Landis, A. W. Chow, R. D. Hamlin, K. S. Y. Lau, and R. H. Boschan in Ref. 20, p. 128.
54. V. L. Bell and R. A. Jewell, *J. Polym. Sci., Part A-1* **5**, 3043 (1967).
55. V. V. Korshak, S. V. Vinogradova, Y. S. Vygodskii, and Z. V. Gerashchenko, *Polym. Sci. USSR* **13**, 1341 (1971).
56. F. M. Houlihan, B. J. Backman, C. W. Wikins, and C. A. Pryde, *Macromolecules* **22**, 4477 (1989).
57. U.S. Pat. 3,303,157 (1967), E. M. Boldebuck and J. F. Klebe (to General Electric).
58. G. Greber, *Angew. Chem Int. Ed. Engl.* **8**, 899 (1969).
59. U.S. Pat. 3,787,364 (1974), J. G. Wirth and D. R. Heath (to General Electric).
60. U.S. Pat. 3,847,869 (1974), F. J. Williams (to General Electric).
61. T. Takekoshi, J. G. Wirth, D. R. Heath, J. E. Kochanowski, J. S. Manello, and M. J. Webber, *J. Polym. Sci. Polym. Chem. Ed.* **18**, 3069 (1980).
62. D. M. White and co-workers, *J. Polym. Sci. Polym. Chem. Ed.* **19**, 1635 (1981).
63. T. Takekoshi, J. E. Kochanowski, J. S. Manello, and M. J. Webber, *J. Polym. Sci. Polym. Chem. Ed.* **23**, 1759 (1985).
64. T. Takekoshi, J. E. Kochanowski, J. S. Manello, and M. J. Webber, *J. Polym. Sci., Polym. Symp.* **74**, 93 (1986).
65. Neth. Pat. Appl. 6,413,552 (1965) (to Du Pont).
66. Y. Imai, *J. Polym Sci.* **B8**, 555 (1970).
67. U.S. Pat. 3,847,870 (1974), T. Takekoshi (to General Electric).
68. Eur. Pat. Appl. EP 132,547 (1985), J. L. Webb (to General Electric).
69. T. Takekoshi, J. L. Webb, P. P. Anderson, and C. E. Olsen, *Abst. IUPAC 32nd Intl. Symp. Macromol.*, 464 (1988).
70. C. D. Hurd and A. G. Prapas, *J. Org. Chem.* **24**, 338 (1959).
71. W. J. Farrissey, J. S. Rose, and P. S. Carleton, *ACS Polym. Prepr.* **9**, 1581 (1968); *J. Appl. Polym. Sci.* **14**, 1093 (1970).
72. R. A. Meyers, *J. Polym. Sci. Part A* **7**, 2757 (1969).
73. P. S. Carleton, W. J. Farrissey, and J. S. Rose, *J. Appl. Polym. Sci.* **16**, 2983 (1972).
74. W. M. Alvino and L. E. Edelman, *J. Appl. Polym. Sci.* **19**, 2961 (1975).
75. N. D. Ghatge and U. P. Mulik, *J. Polym. Sci. Polym. Chem. Ed.* **18**, 1905 (1980).
76. G. D. Khune, *J. Macromol. Sci. Chem.* **A14**(5), 687 (1980).
77. U.S. Pat. 3,833,544 (1974), T. Takekoshi and J. S. Manello (to General Electric).
78. M. Kakimoto, R. Akiyama, S. Negi, and Y. Imai, *J. Polym. Sci., Polym. Chem. Ed.* **26**, 99 (1988).
79. F. Helmer-Metzmann, M. Rehahn, L. Schmitz, M. Ballauff, and G. Wegner, *Makromol. Chem.* **193**, 1847 (1992).
80. L. Schmitz, M. Rehahn, and M. Ballauff, *Polymer* **34**, 646 (1993).
81. F. W. Harris, and S. O. Norris, *J. Polym. Sci. Polym. Chem. Ed.* **11**, 2143 (1973).
82. J. K. Stille, F. W. Harris, H. Mukamal, R. O. Rakutis, C. L. Schilling, G. K. Noren, and J. A. Reeds, *Am. Chem. Soc. Adv. Chem. Ser.* **91**, 628 (1969).
83. L. Tan, E. J. Soloski, and F. E. Arnold, *ACS Polym. Prepr.* **27**(2), 240 (1986).

84. L. Tan, E. J. Soloski, and F. E. Arnold, *ACS Polym. Prepr.* **27**(1), 453 (1986).
85. L. Tan and F. E. Arnold, *ACS Polym. Prepr.* **29**(1), 194 (1988).
86. H. D. Stenzenberger, *Appl. Polym. Symp.* **310**, 91 (1977).
87. D. Wilson, in Ref. 3, pp. 187–226.
88. H. D. Stenzeneberger, in Ref. 3, pp. 108–112.
89. C. W. Paul, R. A. Schultz, and S. P. Fenelli, in Ref. 51, p. 220.
90. J. A. Johnston, F. M. Li, F. W. Harris, and T. Takekoshi, *Polymer* **35**(22), 4865 (1994).
91. T. Takekoshi and J. M. Terry, *Polymer* **35**(22), 4874 (1994).
92. T. L. St. Clair, in Ref. 3, pp. 58–78.
93. A. K. St. Clair, T. L. St. Clair, and W. S. Slemp, in W. D. Weber and M. R. Gupta, eds., *Recent Advances in Polyimide Science and Technology*, Society of Plastics Engineers, Poughkeepsie, N.Y., 1987, p. 16.
94. S. Numata, N. Kinjo and D. Makino, *Polym. Eng. Sci.* **28**(14), 906 (1988).
95. P. M. Hergenrother, in Ref. 3, pp. 158–186.
96. S. Trofimenko, in Ref. 20, p. 3.
97. B. C. Auman, in Ref. 20, p. 15.
98. G. Hougham, G. Tesoro, and J. Shaw, *Macromolecules* **27**, 3642 (1994).

TOHRU TAKEKOSHI
General Electric Company

**POLYISOBUTYLENE.** See ELASTOMERS, SYNTHETIC–BUTYL RUBBER.

**POLYISOPRENE.** See ELASTOMERS, SYNTHETIC–POLYISOPRENE.

# POLYMER BLENDS

Mixing of two or more polymers of different chemical composition offers a powerful way of tailoring performance and economic relationships using existing materials. As a result, the area of polymer blends or alloys has become an important one for both scientific investigation and commercial product development. Fundamental issues that affect the properties of blends include equilibrium phase and interfacial behavior, physical and chemical interactions between the components, phase morphology, and rheology, all of which relate to pragmatic issues of compatibility. One of the most important examples of polymer blends is the judicious incorporation of an elastomeric phase in a rigid matrix to enhance mechanical toughness.

## Equilibrium-Phase Behavior

A mixture of two amorphous polymers may form a single phase of intimately mixed segments of the two macromolecular components or separate into two distinct phases consisting primarily of the individual components; which occurs at equilibrium is dictated by the principles of solution thermodynamics. The older polymer literature often refers to the molecularly mixed blends as compatible, but the modern literature uses the more scientifically precise term miscible (1) to avoid confusion with other uses of the former term. The issues involved in liquid–liquid phase behavior of polymer blends are discussed briefly in terms of basic thermodynamics and theories used in the 1990s. However, it is also important to note that amorphous polymers form glasses on sufficient cooling and that a homogeneous (or miscible) polymer blend exhibits a single, composition-dependent glass-transition temperature, $T_g$, whereas an immiscible blend has separate glass transitions associated with each phase (2). If one of the polymers in a blend can crystallize, a separate crystalline phase of that component can form even when the two polymers are miscible in the melt.

**Basic Thermodynamics.**    Equilibrium-phase behavior of mixtures is governed by the free energy of mixing and how this quantity, consisting of enthalpic

$$\Delta G_{\text{mix}} = \Delta H_{\text{mix}} - T \Delta S_{\text{mix}} \tag{1}$$

($\Delta H$) and entropic ($\Delta S$) parts, is affected by concentration, temperature $T$, and pressure $P$, or volume $V$. To have miscibility (2–4), $\Delta G_{\text{mix}}$ must be negative and satisfy the additional following requirement which ensures stability against

$$\left( \frac{\partial^2 \Delta G_{\text{mix}}}{\partial \phi_i^2} \right)_{T,P} = \left( \frac{\partial^2 G}{\partial \phi_i^2} \right)_V + \left( \frac{\partial V}{\partial p} \right)_{T,\phi_i} \left( \frac{\partial^2 G}{\partial \phi_i \partial V} \right)^2 > 0 \tag{2}$$

phase segregation. Volume fraction of component $i$, $\phi_i$, is employed here, but any other measure of mixture concentration could also be used. Figure 1c shows the dependence of the free energy of a mixture on composition at three temperatures. At $T_1$, the above two conditions are fully satisfied, so that miscible, single-phase mixtures occur for all compositions. At $T_2$, equation 2 is not satisfied for all compositions, and mixtures between the points B and B′ separate into two phases having these compositions as this results in a lower total free energy than that of the homogeneous phase. The curve for an intermediate temperature, $T_c$, has been drawn in a manner satisfying the conditions of a critical point (4) at C. In Figure 1a, $T_1 > T_2$, so that $T_c$ is an upper critical solution temperature (UCST), whereas in Figure 1b, $T_2 > T_1$, so that $T_c$ is a lower critical solution temperature (LCST). More complex liquid–liquid phase diagrams, including those with both UCST and LCST, are possible. The envelope dividing the single-phase and two-phase regions, ie, the locus of all points B and B′, is called the binodal curve. The inflection points S and S′ on the free-energy curve define the spinodal curve, shown as dashed lines in Figures 1a and 1b. The spinodal has significance with respect to the mechanisms and kinetics of phase-separation processes (1,5). The critical point where the binodal and spinodal curves touch may not always lie at the extreme limit of the binodal.

**Flory-Huggins Theory.**    The simplest quantitative model for $\Delta G_{\text{mix}}$ that includes the most essential elements needed for polymer blends is the Flory-Huggins theory, originally developed for polymer solutions (3,4). It assumes the

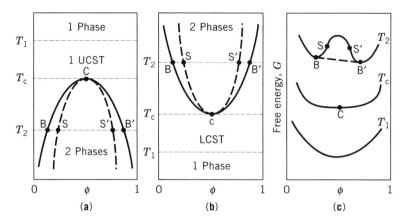

**Fig. 1.** Phase diagram for mixtures: (**a**) upper critical solution temperature (UCST); (**b**) lower critical solution temperature (LCST); (**c**) composition dependence of the free energy of the mixture (on an arbitrary scale) for temperatures above and below the critical value.

only contribution to the entropy of mixing is combinatorial in origin and is given by equation 3, for a unit volume of a mixture of polymers $A$ and $B$. Here, $\rho_i$ and

$$\Delta S_{\text{mix}} = -R\left[\frac{\rho_A \phi_A}{M_A} \ln \phi_A + \frac{\rho_B \phi_B}{M_B} \ln \phi_B\right] \tag{3}$$

$M_i$ refer to the density and molecular weight of $i$, and $R$ is the gas constant. For simplicity, we assume each component to be monodisperse; more complex expressions result when polydispersity is considered (6). This model also assumes the heat of mixing per unit volume follows a van Laar-type relation where $B$ is

$$\Delta H_{\text{mix}} = B \phi_A \phi_B \tag{4}$$

an interaction energy density for mixing segments of the two components and can be alternately expressed in terms of a dimensionless parameter $\chi = B V_{\text{ref}}/RT$ through the use of an arbitrary reference volume, as often seen in the polymer literature. Application of the thermodynamic requirements for phase stability to this model reveals that miscibility of two polymers occurs only when $B$ is less than a critical value given by equation 5.

$$B_c = \frac{RT}{2}\left(\left(\frac{\rho_A}{M_A} + \frac{\rho_B}{M_B}\right)^{1/2}\right)^2 \tag{5}$$

Some important conclusions can be learned from this simple model. First, it shows that $\Delta H_{\text{mix}}$ does not depend on polymer molecular weight, whereas $\Delta S_{\text{mix}}$ does. The combinatorial entropy of mixing becomes progressively smaller as the molecular weights of the components increase and becomes zero as they approach infinity. Endothermic mixing, where $B > 0$, does not favor miscibility. Thus, forming a homogeneous mixture requires that molecular weights must be low enough that the favorable entropic contribution offsets the unfavorable enthalpic effect. For exothermic mixing, where $B < 0$, this theory predicts that the conditions for miscibility will be satisfied no matter how large the molecular weights are. Thus, miscibility of high molecular weight polymers is only assured when mixing is exothermic.

When the interaction energy density is positive, equation 5 defines a critical temperature of the UCST type (Fig. 1a) that is a function of component molecular weights. The LCST-type phase diagram, quite common for polymer blends, is not predicted by this simple theory unless $B$ is temperature-dependent, in which case basic thermodynamic relations reveal that this quantity is not strictly an enthalpic parameter as defined in equation 4, but also contains a noncombinatorial entropic contribution (3). There is substantial literature on extending the simple theory to allow for cases where the interaction energy depends on temperature and composition (7–11).

**Equation-of-State Theories.**    The Flory-Huggins theory does not account for the compressible nature of the mixture or any changes in volume on mixing, which, of course, is only an approximation of how real systems behave. Theories allowing for these volumetric issues, called equation-of-state theories, have received much attention in the polymer literature with a recent focus on blends (9–18). These theories expand the terms of the Flory-Huggins theory by introducing contributions to the entropy and enthalpy of mixing resulting from volume considerations. The latter are expressed in terms of reduced temperature, volume, and pressure by an equation-of-state of the corresponding state type. The resulting equations are quite complex and are not reproduced here; instead some general conclusions are summarized.

The stability condition for a binary mixture can be expanded into the constant volume and volume-dependent terms shown in equation 2 (19). Roughly speaking, the first term to the right of the equal sign is, in effect, what the Flory-Huggins-type approach yields. The second term contains the effects of the compressible nature and is the additional contribution estimated by the equation-of-state theories; it is always negative. These equation-of-state effects always favor phase separation and often grow as the temperature is raised. Thus, one important aspect of these theories is that they can, unlike the basic Flory-Huggins theory, predict LCST type behavior (9) even when the binary interaction parameter is not a function of temperature. Of course, LCST type behavior may indeed occur because the interaction parameter does vary with temperature.

Equation-of-state theories employ characteristic volume, temperature, and pressure parameters that must be derived from volumetric data for the pure components. Owing to the availability of commercial instruments for such measurements, there is a growing data source for use in these theories (9,11,20). Like the simpler Flory-Huggins theory, these theories contain an interaction parameter that is the principal factor in determining phase behavior in blends of high molecular weight polymers.

**Ternary Blends.**    Discussion of polymer blends is typically limited to those containing only two different components. Of course, inclusion of additional components may be useful in formulating commercial products. The recent literature describes the theoretical treatment and experimental studies of the phase behavior of ternary blends (10,21). The most commonly studied ternary mixtures are those where two of the binary pairs are miscible, but the third pair is not. There are limited regions where such ternary mixtures exhibit one phase. A few cases have been examined where all three binary pairs are miscible; however, theoretically this does not always ensure homogeneous ternary mixtures (10,21).

**Crystalline Phases.**   The ternary blends discussion deals with liquid–liquid phase behavior; however, sometimes one or both components of the blend can crystallize. For a polymer pair that is miscible in the melt state, cooling well below the melting point of the pure crystallizable component leads to a thermodynamic driving force to create a pure crystalline phase of that component, as may be seen by simple thermodynamic calculations using typical values for the energetics of mixing and of crystallization (22). Far below the melting point, the free energy of crystallization is considerably larger than that of mixing. At equilibrium, a separate crystalline phase forms, although the kinetics of this process may be greatly affected by the presence of the other polymer (22,23). Because polymers never become 100% crystalline, the pure crystals coexist with a mixed amorphous phase consisting of the material that did not crystallize, and it has a composition different than that of the overall blend (22,24). Evidence for an interphase of different composition has been described for some systems, eg, blends of poly(vinylidene fluoride), $PVF_2$, with poly(methyl methacrylate) (PMMA) (25,26). Cocrystallization of the two polymers is unlikely but has been reported in some cases, such as mixtures of different poly(arylether ketones) (27).

## Experimental Techniques for Characterizing Phase Behavior

A variety of experimental techniques have been used to prepare and characterize polymer blends; some of the more important ones for establishing the equilibrium-phase behavior and the energetic interactions between chain segments are described here (3,5,28,29).

**Blend Preparation.**   The most common techniques for preparing blends are melt mixing and solution casting (30). For immiscible pairs, the details of the mixing process determine the morphology of the resulting composite. To determine whether the components are miscible or not, considerable care must be exercised in the preparation stage, to assure that a physical equilibrium has in fact been achieved. For condensation-type polymers, interchange reactions may occur during melt mixing, giving a copolymer; a single-phase structure can result, which may give the impression that the two polymers are physically miscible, although they may not be (31–33). Solution methods, often used for polymers not amenable to melt processing or when only small amounts of the components are available, offer other possibilities for artifacts. For example, two polymers which are indeed miscible may form two-phase mixtures when cast from certain solvents (34) because of differences in solvency power for the two polymers, leading in some cases to a closed-loop, two-phase region in the ternary phase diagram (35,36). When in question, several different solvents should be tried (34) or the polymers precipitated by a nonsolvent (37). On the other hand, two immiscible polymers form a single-phase solution when diluted enough by solvent, and rapid solvent removal can trap the polymers in a nonequilibrium homogeneous state (38,39). When heated above $T_g$, phase separation may be relatively fast (40) or very slow (39,41).

**Transition Behavior.**   Immiscibility of a blend is usually readily apparent, because phase separation causes light scattering or limited transparency. However, simple visual inspection may not be reliable since the domains may be small relative to the wavelength of light or may have similar refractive indexes, thus limiting the extent of light scattering. Conversely, a miscible blend may contain a crystallizable component, as described above, which scatters light and reduces transparency.

A simple and usually reliable approach for determining whether a blend system is miscible or not is to examine its glass-transition behavior using thermal, mechanical, or dielectric techniques. Miscible blends show a single, composition-dependent glass transition reflecting the mixed environment of the blend (Fig. 2), whereas two-phase blends show two $T_g$s characteristic of each phase. This method is of limited usefulness when the glass transitions of the two polymers are very close together and cannot be adequately resolved. Similarly, crystallinity may render the $T_g$ hard to detect by thermal analysis. When two glass transi-

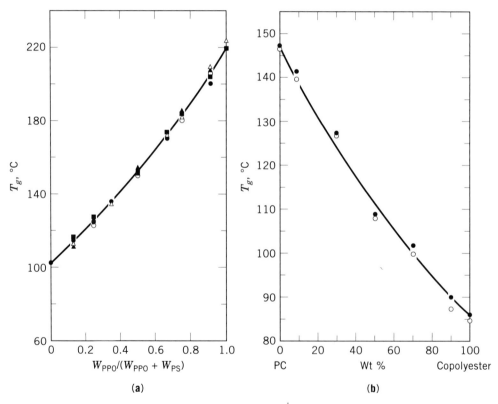

**Fig. 2.**   Glass-transition temperature, $T_g$, for two commercially available, miscible blend systems: (**a**) poly(phenylene oxide) (PPO) and polystyrene (PS) (42); (**b**) polycarbonate (PC) and a copolyester based on cyclohexane dimethanol and a mixture of terephthalic and isophthalic acids (43). The solid lines are calculated from the equation $T_g^{-1} = W_1(T_g)_1^{-1} + W_2(T_g)_2^{-1}$. Points in (**a**) are experimental data. In (**b**), samples were quenched from the melt and points correspond to first (•) and second (○) heats in dsc scans. Courtesy of The American Chemical Society.

tions are observed, they may not be identical to those of the pure polymers if there is partial miscibility of the components in these phases.

For miscible systems containing a crystallizable component, a single $T_g$ is observed as a result of the remaining mixed amorphous phase; however, the composition of the amorphous phase differs from that of the overall blend by an amount that depends on how much of the crystallizable component exists in the separate crystalline phase (24). This in turn influences the relations between $T_g$ and overall blend composition (24). The degree of crystallinity observed depends on prior history, owing to the effect of blending on crystallization kinetics (22). Information about the interaction energy parameter $B$ can, in principle, be obtained by appropriate analysis of the melting point depression of the crystallizable component as a function of blend composition (44–48). Whereas this approach is relatively simple to use, attention must be paid to the effect of crystallite size on the observed melting point (46,47). The results of this technique are sometimes at variance with those obtained using other methods (48).

Miscible blends of high molecular weight polymers often exhibit LCST behavior (3); blends that are miscible only because of relatively low molecular weights may show UCST behavior (11). The cloud-point temperatures associated with liquid–liquid phase separation can often be adequately determined by simple visual observations (39); nevertheless, instrumented light transmission or scattering measurements frequently are used (49). The cloud point observed may be a sensitive function of the rate of temperature change used, owing to the kinetics of the phase-separation process (39).

**Sorption.** Measurement of the equilibrium sorption of vapors or gases in miscible blends can, in principle, give information about the interaction energy parameter, and considerable literature on this subject has appeared (50–52). The concept of inverse gas chromatography has been one of the most popular methods of obtaining measurements (50–52). Despite its apparent simplicity, the latter requires the exercise of considerable skill and care to be reliable in terms of the results obtained. The interpretation of the data is usually based on a ternary version of the Flory-Huggins theory for the limit of very dilute concentrations of the probe molecule, but the results are not simple enough to regard this as a routine technique.

**Analogue Calorimetry.** The heat of mixing for two high molecular weight polymers cannot be measured directly. Basically, the interactions between polymer segments are the same as those between lower molecular weight compounds of similar molecular structures, and the heats of mixing for these liquid analogues can be measured by direct calorimetry. This approach has been used successfully in a number of instances (53–62) in an attempt to understand the relationship between polymer molecular structure and blend phase behavior. The method is limited by the fact that suitable analogues may not be available, or they may be too volatile, too viscous, or solid.

For example, this technique has been used in the study of blends of poly(methyl methacrylate) with polyethylene chlorinated to various degrees (56,57). These polymers are completely miscible when the chlorinated polyethylene contains 50% by weight or more of chlorine. The energetics of mixing for this system were modeled using an oligomer of methyl methacrylate (degree of polymerization of about 4) and octadecane chlorinated to varying degrees. The heat of mixing is positive when the chlorination level is below 50% and negative when above, in good accord with the observations on blend miscibility. Data from analogue calorimetry have been used to simulate the LCST behavior of various blends using an equation-of-state theory (63).

**Spectroscopy.** A variety of spectroscopic techniques, including nmr, eximer fluorescence, and nonradiative energy transfer, have been used to obtain information about blend phase structure and the underlying molecular interactions (30,64–67). However, Fourier transform infrared (ftir) has been most extensively used to learn about mechanisms of specific interactions involved in blend miscibility (68–75). Much attention has been devoted to systems which involve hydrogen bond formation, and the sophistication of spectral analyses varies considerably in the reported studies. Simple observation of band shifts is often used to identify the interacting chemical groups. Many miscible blends involve polymers having ester moieties in the chain or part of the pendent group, and significant shifts in the carbonyl band are observed in these cases. A diminution in this shift with temperature has been observed just prior to phase separation by an LCST mechanism (74). One study used ftir spectroscopy as the source of quantitative information about hydrogen bonding for use in thermodynamic models (76).

**Scattering.** Use of the classical techniques of small-angle scattering of light and x-rays to study polymer blends has been reviewed (77). Scattering techniques provide a sensitive method for determining whether blends are homogeneous or heterogeneous and to follow the kinetics of phase-separation processes (78). Small-angle neutron scattering (sans) and reflection have emerged as powerful tools for investigating many aspects of polymer blends (76,79–91). The sans technique can be used to obtain thermodynamic interaction energies for miscible blends; however, one of its more unusual aspects is the ability to determine conformational information about the components. For example, the radius of gyration of a styrene–acrylonitrile (AN) copolymer containing 19% AN by weight as a function of its weight average molecular weight when blended with a deuterated poly(methyl methacrylate) has been determined (79,83). The chain is slightly more expanded in the blend than the unperturbed chain dissolved in an ideal or $\theta$ solvent, as would be expected for a solvent which is somewhat better than an ideal one. Thus, it is confirmed that the copolymer is molecularly dissolved in PMMA in the same sense as polymers are in low molecular weight solvents; this requires mixing at the segmental level for the blend (79,80,83). The sans technique has become an important tool in elucidating the phase behavior and quantifying the interactions in polyolefin blends (88–91). A significant issue in the use of sans for these purposes is that one of the samples generally needs to be deuterated to produce scattering contrast. This can pose some synthesis challenges. Of more serious concern is the fact that deuterated polymers appear to have measurably different interactions with other molecules than their hydrogenous counterparts (86–90).

**Critical Molecular Weight.** Two polymers that form immiscible blends at molecular weight levels typical of commercial materials may show miscibility if the molecular weights of one or both components can be reduced enough. Quantitative determination of these critical molecular weights can then be used to compute the interaction energy for this pair using an appropriate mixing theory (11,92–95). This technique is most useful for polymer pairs where the interaction energy is positive but small and the components are available in a wide range of molecular weights, preferably monodisperse. In some cases, end-group effects can be important (11,94).

Blends of low molecular weight PMMA and polystyrene (PS) show that UCST behavior and the interaction energy, calculated from the observed phase diagram using the Flory-Huggins theory, agree with values obtained by neutron scattering (86) and other techniques (96). On the other hand, blends of low molecular weight PMMA with poly($\alpha$-methyl styrene) (P$\alpha$MS), show LCST behavior owing to the rather different PVT properties of P$\alpha$MS than those found in PMMA and PS, and calculation of an interaction energy from this type of phase diagram has been made using an equation-of-state theory (11).

If available molecular weight combinations do not lead to observable phase-diagram boundaries of either the UCST or LCST type, then the interaction energy can only be estimated to lie within upper and lower bounds using this technique (93).

## Effect of Molecular Structure on Polymer–Polymer Interactions

The literature contains extensive reports on investigations of the equilibrium-phase behavior for an enormous number of polymer–polymer pairs (1,97). The number of blends known to be miscible has grown so rapidly since the mid-1980s that it is more instructive to attempt to understand these observations in terms of the molecular structures of the components rather than to catalog them.

**Dispersive Interactions.** For pairs of nonpolar polymers, the intermolecular forces are primarily of the dispersive type, and in such cases the energy of interaction between unlike segments is expected to be closely approximated by the geometric mean of the energies of interaction between the two like pairs (98). In this case, the Flory-Huggins interaction energy between this polymer pair can be expressed in terms of the solubility parameters $\delta$ of the pure components.

$$B = (\delta_A - \delta_B)^2 \qquad (6)$$

Obviously, $B$ can never be negative in these cases; to the extent that equation 5 is valid, miscibility can only be driven by combinatorial entropy, and this possibility is maximized by matching the values of $\delta_A$ and $\delta_B$ as nearly as possible. In general, high molecular weight, nonpolar polymers are rarely miscible with each other.

**Hydrogen Bonding.** Miscibility caused by exothermic mixing, ie, $B < 0$, can occur among pairs of polar polymers (1). The obvious possibilities involve those pairs which contain groups capable of relatively strong specific interactions (75). A great deal of attention has been devoted to blends where one component contains monomer units with strong hydrogen bond donors and the other component contains hydrogen bond acceptors such as ester, carbonyl, or ether units. There are many examples of miscibility among such pairs (70,75,99). Examples of hydrogen bond donors are as follows:

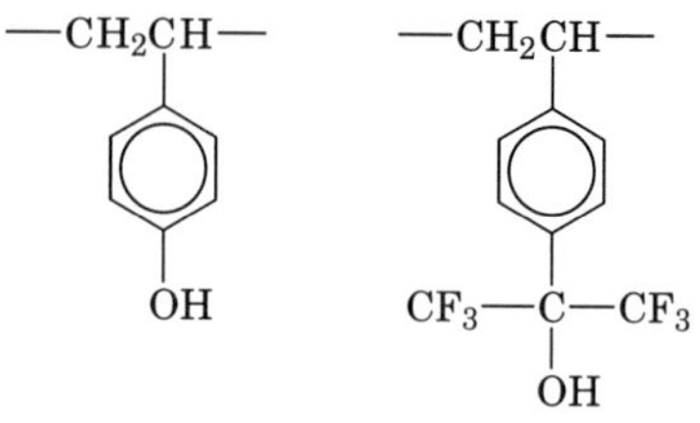

$$-CH_2CH- \qquad\qquad -CH_2CH-$$

(chemical structures)

OH $\qquad\qquad$ CF$_3$—C—CF$_3$

$\qquad\qquad\qquad\qquad$ OH

One proposed approach (75) to modeling the phase behavior for hydrogen bonding pairs uses the following expression for the free energy of mixing (eq. 7).

$$\Delta G_{\mathrm{mix}} = \Delta G_{\mathrm{FH}} + \Delta G_{\mathrm{QC}} \tag{7}$$

Other formulations based on more rigorous statistical mechanics approaches are also available (12). The first term on the right in equation 7 is the standard Flory-Huggins (FH) expression, except that the interaction term, $B$ or $\chi$, is evaluated from solubility parameters, computed from a group contribution method as described (75) using equation 6 or its equivalent. The second term is a quasichemical (QC)-type formulation (98) that treats hydrogen bond formation in analogy with chemical reaction equilibria. The temperature-dependent equilibrium constants have been approximated from infrared spectroscopy observations (75). Of course, the possibility of hydrogen bond formation between two polymers does not guarantee their miscibility. The self-association that must be disrupted within the pure components may outweigh the favorable intercomponent hydrogen bonds formed in the mixture and the overall contribution to the free energy is thus unfavorable. Likewise unfavorable nonhydrogen bond interactions may outweigh any favorable contribution from hydrogen bonding. Other potentially strong specific interaction mechanisms, including ionic interactions, have been proposed for producing miscible blends (1,101–105). In all cases, whether miscibility occurs or not depends on a delicate balance of issues with the net free energy effect rarely being as great as might be imagined by a simplistic accounting of only a strong specific interaction. For one thing, any favorable gain in the energy of mixing is accompanied by an unfavorable noncombinatorial entropy effect (106,107).

**Mean Field Approximations.** The two cases outlined above may be regarded as the extreme limits of polymer–polymer interactions. There is an important intermediate situation for somewhat polar polymer pairs where equation 6 is not valid, but the interactions are far from specific enough to be regarded in quasichemical terms. These cases may lead to exothermic interactions as clearly illustrated by the following well-known examples: poly(phenylene oxide)–polystyrene (PPO–PS) (76); tetramethyl bisphenol polycarbonate–PS (9); bisphenol A polycarbonate–polyesters (43,54,108,109); PS–poly(vinyl methyl ether) (PVME) (110); etc. As of this writing (ca 1995), there is no method for predicting *a priori* the interaction energies for such systems. However, patterns emerge from the literature which can help to guide the selection of pairs that are likely to be miscible. For example, Table 1 summarizes a variety of observations

**Table 1. Phase Behavior of Halogenated Polymers With Families of Polymers Containing Ester Units[a,b]**

| Structure | Aliphatic polyesters | Polymethacrylates | Polyacrylates |
|---|---|---|---|
| $-CH_2-CH-$ <br> $\quad\quad\mid$ <br> $\quad\quad F$ | not miscible[c] | [d] | [d] |
| $-CH_2-CF_2-$ | not miscible[c] | miscible | miscible |
| $-CH_2-CH-$ <br> $\quad\quad\mid$ <br> $\quad\quad Cl$ | miscible | miscible | miscible |
| $-CH_2-CCl_2-$ [e] | miscible | miscible | miscible |
| $-OCH_2-CH-$ <br> $\quad\quad\quad\mid$ <br> $\quad\quad CH_2Cl$ | miscible | miscible | miscible |
| $-CH_2-CH-$ <br> $\quad\quad\mid$ <br> $\quad\quad Br$ | miscible | [d] | [d] |

[a]See Ref. 111 for literature citations.
[b]The designation miscible indicates that some (not all) members of the family have been reported miscible with the indicated halogen-containing polymer.
[c]No member of this family, eg, aliphatic polyesters, has been reported miscible with the indicated halogen-containing polymer.
[d]No information available in the literature on these systems.
[e]Because of the intractability of poly(vinylidene chloride), these studies have used copolymers containing various comonomers; however, the results are believed to be indicative of the interactions with vinylidene chloride units and not primarily those of the comonomer.

(111) that suggest a high incidence of miscibility of polymers containing halogen units with members of three families of polymers with ester units in their chains or pendent groups; other examples of this kind also exist (112,113). Some authors suggest that systems like those in Table 1 involve hydrogen bonding, via hydrogen atoms that are either $\alpha$ or $\beta$ to the halogen (1,75). Small shifts in the infrared carbonyl bands have been noted in some systems of this type (61). The exact nature of the interactions involved in systems like those illustrated and how specific they may be are far from being certain at this time. In any case, there is considerable evidence that simple mean field-type treatments of relatively weak favorable interactions like those involved in most systems of interest, using standard Flory-Huggins or equation-of-state theories, can provide good descriptions of the thermodynamic phase behavior, and that quasichemical type formulations are not necessary. On this basis, interaction energies have been evaluated (9,11,58,61,92–95,110–132) for a number of systems, including ones with either negative or positive values, using a variety of experimental techniques.

**Copolymer Models.** A significant advance dating from the 1980s is the use of a simple, mean field, binary interaction energy approach to describe or to analyze the phase behavior of blends involving statistical copolymers (133–135). This model relates the net interaction energy $B$ (or $\chi$) for mixing two multicomponent polymers, $A$ and $B$, in terms of interaction energies $B_{ij}$ between their monomer unit pairs, $i$ and $j$, regardless of what these units are bonded to:

$$B = \Sigma\Sigma B_{ij}(\phi_i' - \phi_j'')(\phi_i'' - \phi_j') \tag{8}$$

where $\phi_i'$ and $\phi_i''$ are the volume fractions of $i$ units in $A$ and $B$, respectively, with $B_{ii} = 0$ and no double counting in the sum. Similar expressions apply for corresponding quantities in equation-of-state theories (9,128) or in the Flory-Huggins $\chi$ notation, provided care is exercised in dealing with arbitrary reference volumes (117). The positive terms in equation 8 stem from $i$ and $j$ interactions between copolymers $A$ and $B$, whereas the negative terms reflect intramolecular interactions within $A$ or $B$. It is important to note that $B$ can be negative even when all $B_{ij}$ are positive.

There are many examples known where a random copolymer $A$, comprised of monomers 1 and 2, is miscible with a homopolymer $B$, comprised of monomer 3, even though neither homopolymer 1 or 2 is miscible with homopolymer 3, as illustrated by Table 2. The binary interaction model offers a relatively simple explanation for the increased likelihood of random copolymers forming miscible blends with other polymers. The overall interaction parameter for such blends can be shown (eg, by simplifying eq. 8) to have the form of equation 9 (133–134).

$$B = B_{13}\phi_1' + B_{23}\phi_2' - B_{12}\phi_1'\phi_2' \tag{9}$$

For this case, both $B_{12}$ and $B_{23}$ would be positive. If $B_{12}$ is a large enough positive value, then $B$ can be negative over a certain range of copolymer compositions. The mathematical condition for such a region is given in equation 10 (133).

$$B_{12} > \left( (B_{13} + B_{23})^{1/2} \right)^2 \tag{10}$$

Physically, this occurs because addition of polymer 3 to the copolymer dilutes the highly unfavorable interactions between 1 and 2, causing a net favorable mixing condition even though no individual binary interaction is favorable. Experimental confirmation of this principle has been demonstrated (62,142) using direct calorimetry of liquids to simulate the copolymer–homopolymer

**Table 2. Examples of Random Copolymers Which Form Miscible Blends With Other Polymers When Corresponding Homopolymers Do Not**

| Copolymer | Homopolymer | References |
|---|---|---|
| styrene–acrylonitrile | poly(methyl methacrylate) and poly(ethyl methacrylate) | 136 |
| | polyethyloxazoline | 137 |
| | poly($\epsilon$-caprolactone) and other aliphatic polyesters | 138,139 |
| ethylene–vinyl acetate | poly(vinyl chloride) | 140 |
| butadiene–acrylonitrile | poly(vinyl chloride) | 141 |
| $\alpha$-methylstyrene–acrylonitrile | poly(methyl methacrylate) | 96 |
| o-chlorostyrene–p-chlorostyrene | poly(phenylene oxide) | 114 |
| styrene–maleic anhydride | poly(methyl methacrylate) and poly(ethyl methacrylate) | 129 |

situation. Thus, intrachain repulsion can be a driving force for miscibility in the absence of specific interactions. However, $B$ cannot be negative if all $B_{ij}$ values are represented by equation 6, ie, the geometric mean relation must be violated for at least one $i,j$ pair. The homopolymer is miscible with the copolymer over any copolymer composition region, often called a miscibility window, where $B$, given by equation 9, is less than $B_c$, given by equation 5. Precisely defining the copolymer composition limits of the miscibility window gives two pieces of information that can be used in the evaluation of the three $B_{ij}$ at the temperature of observation. One other piece of information would be needed to deduce all three $B_{ij}$.

Blends of poly(vinyl chloride) (PVC) and $\alpha$-methylstyrene–acrylonitrile copolymers ($\alpha$-MSAN) exhibit a miscibility window that stems from an LCST-type phase diagram. Figure 3 shows how the phase-separation temperature of 50% PVC blends varies with the AN content of the copolymer (96). This behavior can be described by an appropriate equation-of-state theory and interaction energy of the form given by equation 9.

When copolymer $A$ contains monomers 1 and 2 and copolymer $B$ contains monomers 3 and 4, equation 8 gives equation 11. Clearly, there is a double

$$B = B_{13}\phi_1'\phi_3'' + B_{14}\phi_1'\phi_4'' + B_{23}\phi_2'\phi_3'' + B_{24}\phi_2'\phi_4'' - B_{12}\phi_1'\phi_2' - B_{34}\phi_3''\phi_4'' \quad (11)$$

opportunity for miscibility caused by intramolecular repulsion stemming from the last two terms even when all $B_{ij}$ are positive (143). There are many possibilities in this situation which can be conveniently represented as isothermal

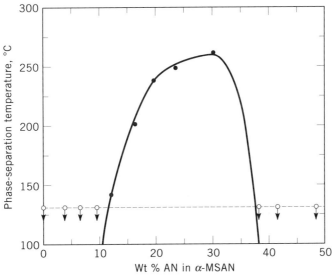

**Fig. 3.** Phase-separation temperatures for 50:50 PVC–$\alpha$-MSAN blends, where (---) represents the blend drying temperature, ($\circ$) are experimentally determined phase-separation temperatures, (⚲) correspond to blends with phase-separation temperatures below the drying temperature, and (——) was calculated using the Sanchez-Lacombe theory combined with the binary interaction model (96). Courtesy of Butterworth-Heinemann, Ltd.

plots of composition of copolymer $A$ versus the composition of copolymer $B$ for a fixed proportion of $A$ to $B$. On this map, regions of miscibility occur for compositions where $B < B_c$. For fixed molecular weights of $A$ and $B$, these regions have the shape of a conic section, ie, parabola, ellipse, or hyperbola. The theoretical conditions for various possibilities have been described and many experimental examples given in the literature (94,120,123,124,143). These diagrams can be predicted if the required $B_{ij}$ are known or the procedure can be reversed to extract information about these parameters.

A commercially important example of the special case where one monomer is the same in both copolymers is blends of styrene–acrylonitrile, 1 + 2, or SAN copolymers with styrene–maleic anhydride, 1 + 3, or SMA copolymers. The SAN and SMA copolymers are miscible (128,133,144) so long as the fractions of AN and MA are nearly matched, as shown in Figure 4. This suggests that miscibility is caused by a weak exothermic interaction between AN and MA units (128,133) since miscibility by intramolecular repulsion occurs in regions where $\phi_2' \neq \phi_3''$, as can be shown (143) by equation 11.

A simple but important case involves blends of two copolymers made from the same monomers, 1 and 2, but differing in composition where equation 11 reduces to equation 12 (133).

$$B = B_{12}(\phi_2' - \phi_2'')^2 \tag{12}$$

Assuming $M_A = M_B = M$ and $\rho_A = \rho_B = \rho$, equation 12 can be equated to $B_c$ from equation 5 to obtain equation 13. This formula defines the critical

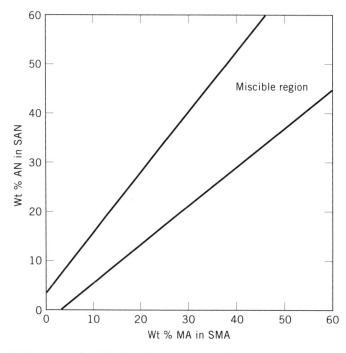

**Fig. 4.** Miscibility map for blends of styrene–acrylonitrile copolymers (SAN), with styrene–maleic anhydride copolymers (SMA).

$$B_{12}(\phi_2' - \phi_2'')^2 = \frac{2\rho RT}{M} \qquad (13)$$

composition difference distinguishing miscible mixtures from immiscible ones. The lines in Figure 5 were computed from equation 13 using equation 6 to estimate $B_{12}$, having solubility parameters of 18.6 $(J/cm^3)^{1/2}$ (9.1 $(cal/cm^3)^{1/2}$) for polystyrene and 26 $(J/cm^3)^{1/2}$ (12.7 $(cal/cm^3)^{1/2}$) for polyacrylonitrile, and compare well with published data (145) for blends of SANs of different compositions with a SAN containing 19 wt % AN at different levels of $M$. Note that as $M$ becomes very large, the tolerable difference in copolymer compositions for miscibility approaches zero, because the entropic driving force varies as $1/M$. The formalism of equation 8 has been successfully used to describe the behavior of binary blends of terpolymers with other terpolymers, copolymers, and the homopolymer (117,122). It has also been extended to chemical groups (eg, $CH_2$, —⬡—, COO, and NHCO) smaller than monomer units for a number of systems (55,131–133,139); but this use requires defining a rather large number of parameters except in a few simple cases. The value of these parameters may be rather sensitive to the nature of neighboring units in the chain because of alteration of the electronic structure of small groups.

The binary interaction model used here assumes that mixtures may be modeled by interactions between groups without regard to details about the spatial placements of these groups on the molecules, ie, a kind of molecular "soup" of interacting units. Clearly, this approximation is not adequate for all

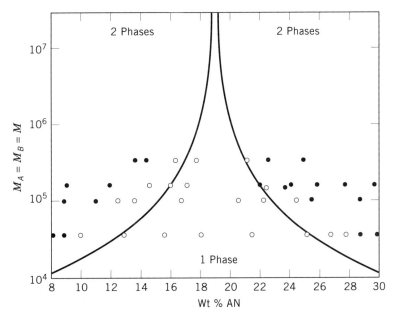

**Fig. 5.** Phase behavior of blends of a styrene–acrylonitrile copolymer containing 19 wt % of acrylonitrile with other SAN copolymers of varying AN content and as a function of the molecular weight of the two copolymers: (○) one-phase mixture; (●) two-phase mixtures as judged by optical clarity. Curve computed as described in text (145). Courtesy of Hüthig & Wepf Verlag, Zug, Switzerland.

situations. In some cases, sequence distribution in copolymers could complicate the description (146), particularly when strong inductive effects are possible. Stereo-regularity has been shown to have an effect on the phase behavior of some, but not all, blends (147), which is a fact that is not accounted for in this approach. Aliphatic polyesters with branched units in their repeat structures differ very little from corresponding ones without branches in terms of miscibility with PVC (148,149), whereas similar branching tends to promote immiscibility in other cases (53,54,150,151); the effects may include shielding of certain groups from interactions, inductive effects, increased intermolecular spacing, or changes in equation-of-state contributions to the free energy.

## Property Relationships

The relationship between the physical properties of a blend and those of its components can depend on the thermodynamic interaction between the components and many other factors. Some generalizations are possible, but exceptions are common and fundamental understanding for some properties remains incomplete in spite of the central importance of this issue in blend technology.

Fully miscible blends generally represent the simplest case. In the absence of crystallinity, most properties follow some additive relationship, as suggested in Figure 6; miscible blends are similar to random copolymers in this regard. The glass-transition temperature, and hence the softening point, is generally a monotonic function of composition, as shown in Figure 2, but maxima have been observed in strongly interacting systems (70). As a rule, most mechanical properties, permeation (logarithmic scale) to small molecules, etc, follow nearly

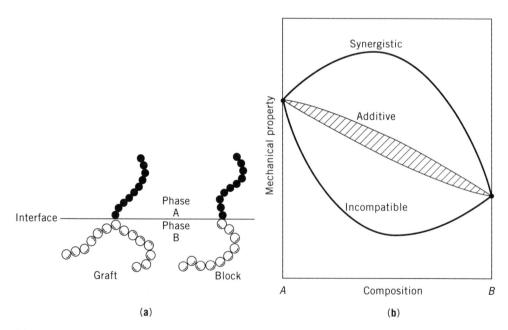

**Fig. 6.**   Illustration of (**a**) compatibilization of immiscible blends of polymers $A$ and $B$ by block or graft copolymers and (**b**) the subsequent modification of property responses.

linear relations with composition in such systems, but exceptions are known (109,152–155). Of course, if the blend converts from rubbery to glassy with composition, the usual changes in properties at the glass transition become superimposed on this relationship. Crystallization of one of the components changes properties in similar ways, as would be expected for a single-component polymer (156,157).

For blends where the components form separate phases, properties depend on the arrangement of these phases in space and the nature of the interface between the phases. Immiscible blends behave like composite materials (qv) in many respects. Properties like softening temperature, modulus, permeation, etc, are dominated by the properties of the component that forms the continuous phase (153). Failure properties, especially those related to ductility, eg, elongation at break and impact strength, often depend on the dimensions of the phases and the degree of interfacial adhesion between the components. The nature of the interface is related to the thermodynamic interaction between the components (95), which then governs morphology generation in the melt via interfacial tension, and adhesion in the solid state via the thickness of the interfacial zone and the degree to which chains from the two phases entangle. When the thermodynamic affinity is very low, the blend can exhibit a degree of toughness well below that of either component (lower curve in Fig. 6) and is regarded as incompatible. Improvement of this property response is often called compatibilization and approaches to this important aspect of blend technology are described later. When the interaction between phases is good enough, additive properties may be obtained without compatibilization. The PC–SAN interface (95) seems to be such a case and forms the basis of commercial PC–ABS blends.

For some blends, properties exceeding additive or those of either component can be achieved (109,152,158). The curve labeled synergistic in Figure 6 illustrates this case.

## Compatibilization

Incompatible blends often have little commercial value because of the deficiencies in ductility-related properties such as impact resistance, elongation at break, and strength at break. This overall response is usually related to the lack of interfacial adhesion between the phases in immiscible systems and to poor control of morphology. Two general methods are used to remedy these problems: copolymer addition and reactive compatibilization. As illustrated in Figure 6, both methods seek to place a block or graft copolymer at these interfaces. This has the effect of joining the phases by the covalent bonds that form the backbones of the compatibilizing copolymers. It also lowers interfacial tension and retards coalescence of domains, both of which improve dispersion and reduce domain size. Retarding coalescence leads to a more stable morphology and lessens the dependence of blend performance on processing history.

**Copolymer Addition.** Addition of block or graft copolymers to improve the mechanical properties of immiscible polymer blends has been used since the mid-1970s with varying degrees of success. Studies have shown that interfacial adhesion (159–162) and mechanical compatibility (159,163–166) can be improved by the addition of appropriate block and graft copolymers. In their simplest forms the copolymers have block or graft segments that are chemically identical to those in the respective phases, although nonidentical segments, which are miscible or partially miscible in the respective phases, are also effective. For example, the compatibilization of immiscible blends of PE and PS by various hydrogenated butadiene–styrene block copolymers has been investigated (163,166). As little as 0.5–2.0% of diblock (164,167) can be sufficient to achieve a uniform and stable phase dispersion with most of the copolymer at the interface, as depicted in Figure 6. Diblocks are thought to be the most efficient interfacial agents, but triblock and graft copolymers are also useful. Blends of PE and PS containing such compatibilizers have attracted commercial interest for packaging applications.

The additive approach to compatibilization is limited by the fact that there is a lack of economically viable routes for the synthesis of suitable block and graft copolymers for each system of interest. The compatibilizer market is often too specific and too small to justify a special synthetic effort. Moreover, commercially available triblock copolymers designed to be thermoplastic elastomers, not compatibilizers, are often used in lieu of the more appealing diblock materials. Since the mid-1980s, the generation of block or graft copolymers *in situ* during blend preparation (158,168–176), called reactive compatibilization, has emerged as an alternative approach and has received considerable commercial attention.

**Reactive Compatibilization.**  Polymers functionalized with anhydride, carboxylic acid, amine, hydroxyl, epoxide, etc, groups have been utilized in reactive compatibilization (168). These groups can react by means of condensation chemistry to give block or graft copolymers whose constituent chains are joined by ester, amide, and imide linkages during the melt-blending process. In the specific case of rubber-toughened polyamides, the naturally occurring functional end groups, eg, the primary amine ends, of the polyamide provide the needed reaction capability. The other functionalized polymer might be a maleic anhydride-modified ethylene–propylene rubber (EPR). During melt mixing, reactions at the interface produce a nylon-grafted rubber phase with significantly reduced dispersed rubber particle size. A variety of other polymers, including poly(phenylene oxide) (PPO) (173,177), styrene-based block copolymers (178), and PS (179,180) can be modified by anhydride functionality prior to reaction-blending them with nylon. Alternatively, small amounts of polymers containing reactive functionality can be added to one of the phases to produce the reaction (158). Various polymers miscible with SAN but reactive with amine groups have been added to ABS to effect reactive compatibilization of polyamide–ABS blends (158,174,181,182). Systems of this type can show reductions in domain size of up to a factor of 100 as a result of such reactions.

Blends that contain no nylon can also be prepared by reactive compatibilization. However, interest in these systems has been limited somewhat by lack of control of the reaction pathways. For polyester-based systems, epoxide functionality appears to be an effective chemistry, involving reaction of the polyester chain ends (183,184).

Grafting reactions are often employed to chemically modify polymer melts so as to achieve a particular desired functionality. Polyolefins such as PP, PE, and EPR are functionalized with MA, typically by adding a free-radical initiator and the functional monomer to the polymer melt (185–189) in specially configured twin-screw extruders to provide the necessary degree of control over reactant residence time and temperature (190–192). Grafting with MA has an added advantage in that it self-polymerizes very slowly. As a result, unused monomer can be easily removed near the end of the extruder barrel, and the polymer product primarily contains the MA graft. Grafting with monomers that can self-polymerize, such as styrene, leads to more complex product mixtures of homopolymer polyolefin, polystyrene, and polyolefin-$g$-polystyrene compatibilizer. Such systems are somewhat difficult to predict and control, but can be commercially quite interesting, as illustrated by the debut of such materials in 1994 (193).

## Blend Morphology

In many instances, phase-separated blends are the preferred means of achieving useful results. For example, such polymer–polymer composites yield materials whose stiffness can be adjusted, in principle, to any value between those of the component polymers. However, tailoring blends to achieve this or any other characteristic requires, among other things, control over the spatial arrangement or morphology of the phases, and some degree of stability once they are formed. These arrangements may consist of one phase dispersed as simple spheres in a matrix of the other polymer, as shown in Figure 7**a**. On the other hand, the dispersed phase may take the form of platelets or fibrils with varying aspect

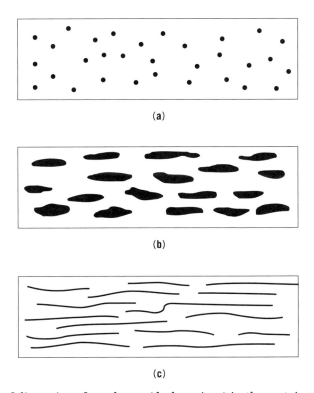

(a)

(b)

(c)

**Fig. 7.** Types of dispersion of a polymer (dark regions) in the matrix of an immiscible polymer. The spherical droplets (**a**) are progressively extended into platelets (biaxial) or fibrils (uniaxial); (**b**) and (**c**) by deformation.

ratios, as shown in Figure 7**b** and 7**c**, respectively. Another distinct morphology consists of both phases simultaneously having a continuous character (194) or an interpenetrating network of phases (195). The typical dimensions of the phases are important in all these morphologies.

**Morphology Generation and Control.**   Homogeneous mixtures of two polymers may phase-separate upon a change in the temperature or removal of the solvent (196–203). In most commercially significant blends, however, the component polymers are immiscible, and the morphology of the dispersed phase is often generated by either added block and graft copolymer compatibilizers or chemical reactivity of the two principal components. Such blends are typically prepared by extrusion-melt blending.

The morphology generated during mixing depends on the interfacial tension between the phases and the viscosity and elasticity (2,204–209). The component occupying the most space tends to assume the role of the continuous phase. The size of the dispersed phase is determined by the balance between drop breakup (210) and coalescence processes (211), which in turn are governed by the deformation field imposed by the mixing device, interfacial tension, and the rheological characteristics of the components. The shape of the dispersed phase may be deformed from spheres into fibrils by uniaxial elongational flow, eg, extrusion through a die, or into platelets by biaxial stretching, eg, blow molding, as shown in Figure 7.

The component with the lower viscosity tends to encapsulate the more viscous (or more elastic) component (207) during mixing, because this reduces the rate of energy dissipation. Thus the viscosities may be used to offset the effect of the proportions of the components to control which phase is continuous (2,209). Frequently, there is an intermediate situation where a cocontinuous or interpenetrating network of phases can be generated by careful control of composition, microrheology, and processing conditions. Rubbery thermoplastic blends have been produced by this route (212).

The morphology created during processing is a dynamic structure that may be subject to further changes during subsequent processing steps, as can be observed in a number of blends, eg, polyethylene–PS systems (166,213,214). In these blends, appropriate polystyrene-containing block copolymers improve the morphology and its stability. Another example relates to the morphology of near-miscible PC–SAN (matrix of ABS) blends, but they have poor morphological stability at melt temperatures (215).

Compatibilization with appropriate graft or block copolymers or via chemical reactivity during melt blending should give rise to a more stable morphology. Melt-blended nylon-6–elastomer compositions that are chemically reactive or are compatibilized with a reactive third component are very tough and typically have a more stable phase morphology. Other unique features of morphology development are involved in formulating toughened PVC, PP, PPO, and polyesters. *In situ* polymerization and the use of the core-shell impact modifiers are additional approaches to forming heterophase blends.

**Characterization.**   Electron microscopy has become one of the most widely used techniques for characterizing blend morphology (216). Scanning electron microscopy (sem) offers the simplest procedure (217). Because it only reveals surface features, the internal structure of blends is investigated by viewing fracture surfaces created at ambient or cryogenic temperatures. At low temperature, yielding is suppressed during the fracture process. Photomicrographs of this type often give information about the extent of adhesion between phases. To aid the identification of phases and enhance morphological features, one of the phases may be selectively extracted by a solvent, leaving the other phase to be viewed in the microscope. Further, by solvent extraction the matrix polymer can be selectively removed in materials such as high impact polystyrene (HIPS), leaving the particulate rubber phase intact for viewing by sem (218), as shown in Figure 8.

Transmission electron microscopy (tem) requires viewing thin sections of material. Normally, tem is not directly useful for polymer blends because of insufficient contrast between most organic polymers in the electron beam; hence, staining methods to enhance contrast are often used. Osmium tetroxide, $OsO_4$, is a useful stain for rubber-modified polymers (219). The rubber phase, through its unsaturation, reacts with $OsO_4$, whereas the matrix phase does not. This can produce sufficient contrast between them owing to the high electron density of the former phase. The use of $OsO_4$ is primarily limited to polymers with

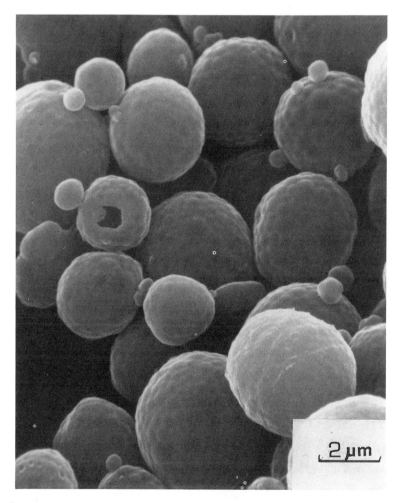

**Fig. 8.**   Scanning electron photomicrograph of rubber particles extracted from HIPS (218). Courtesy of Butterworth-Heinemann Publishers, Ltd.

double bonds or other reactive sites. For modified PVC and other saturated polymers, chemical etching methods are utilized (220). Reaction with $OsO_4$ causes hardening, which facilitates microtoming of ultrathin sections for viewing. In addition, this stain makes the craze structure visible in stress-whitened polymers. Figure 9 is an example of $OsO_4$ staining for tem contrast.

Since 1980, ruthenium tetroxide, $RuO_4$, has been used for staining a number of heterophase polymers for tem (221); it seems to be a more versatile staining agent than $OsO_4$. For instance, in SAN modified with acrylate rubber, where the rubber phase is fully saturated, an excellent contrast between the rubber and the matrix has been achieved (222). Crystalline polymers have been stained with $RuO_4$ (223), and excellent craze structures have been revealed (221). The stain

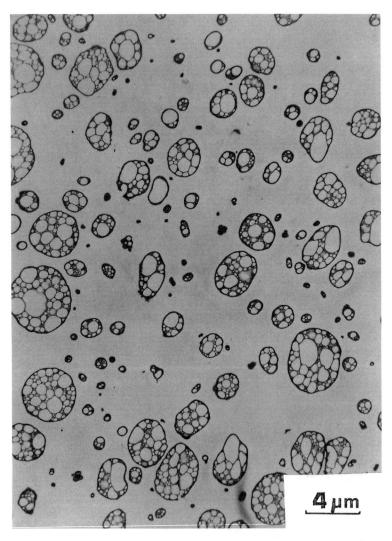

**Fig. 9.** Transmission electron photomicrographs of HIPS where the dark phase is $OsO_4$-stained rubber (218). Courtesy of Butterworth-Heinemann Publishers, Ltd.

may be prepared by dissolving $RuCl_3 \cdot 3H_2O$ in aqueous sodium hypochlorite for immediate use (224).

As more complex multicomponent blends are being developed for commercial applications, new approaches are needed for morphology characterization. Often, the use of $RuO_4$ staining is effective, as it is sensitive to small variations in the chemical composition of the component polymers. For instance PS, PC, and styrene–ethylene/butylene–styrene block copolymers (SEBS) are readily stained, SAN is stained to a lesser degree, and PBT and nylons are not stained (158,225–228).

Because $RuO_4$ is very reactive, staining times of ultrathin slices should be kept below 15 min to prevent overstaining and development of artifacts in the form of small extra-stained domains. Multiple staining techniques can be especially useful. Osmium tetroxide and $RuO_4$ used in combination are effective for discerning the morphology of PC–ABS (229) and nylon-6–ABS (158) blends. The blends containing polyamides are effectively characterized by the use of phosphotungstic acid stain (2% solution in water), in combination with some other staining agent such as $OsO_4$ or $RuO_4$. Phosphotungstic acid stain renders nylons dark without affecting the other typical toughening agents. It is especially useful for revealing any occlusions of polyamide material inside the rubber particles (228). In polyamide blends with PPO, other approaches such as plasma etching (for sem) and bromine staining of PPO have been used effectively for morphology characterization (227).

Some of the most difficult heterophase systems to characterize are those based on hydrocarbon polymers such as rubber-toughened polypropylene or other blends of rubbers and polyolefins. Because of its selectivity, $RuO_4$ staining has been found to be useful in these cases (221,222,230). Also, $OsO_4$ staining of the amorphous blend components has been reported after sorption of double-bond-containing molecules such as 1,7-octadiene (231) or styrene (232). In these cases, the solvent is preferentially sorbed into the amorphous phase, and the reaction with $OsO_4$ renders contrast between the phases.

## Toughened Polymers

Improvement of toughness is frequently a reason for blending (233–235). Usually, this is accomplished by addition of a small amount of an elastomer as a discrete particulate second phase. When it is properly done, this can result in a significant improvement of impact strength, with only a small reduction of modulus and tensile strength. The improvement of toughness is caused by plastic deformation processes, eg, crazing or shear-yielding mechanisms, triggered by the rubber phase.

The matrix polymers can be divided into brittle or ductile categories, each having specific requirements for achieving toughness (Table 3). Numerous variations are possible. For instance, often rubber particles that vary in both size and kind are desirable for optimum performance. In these cases, the requirements of the rubber phase and the toughening mechanisms are complex.

Commercial growth of most polymers has been significant, illustrated by Table 4 (220). Many of these polymers may be toughened by the addition of a dispersed elastomer phase, and the growth of the rubber-toughened versions of

**Table 3. Rubber Toughening of Plastics**

| Brittle matrix | Ductile matrix |
|---|---|
| PS, SAN, PMMA, epoxy | PC, polyamide, PP, PVC, PBT, PPO |

*Requirements*

| | |
|---|---|
| requires strong interfacial adhesion | strong interfacial adhesion may not always be critical |
| optimum particle size 0.1–3 $\mu$m, depending on matrix polymer; often bimodal particle size distribution is desirable | optimum particle size <0.5 $\mu$m; lower limit depends on the matrix |
| chemical or physical cross-linking of the rubber | cross-linking not required but may improve blend stability |
| minimum volume fraction | sometimes small amounts of rubber may have large effects, eg, PVC |

*Toughening mechanisms*

| | |
|---|---|
| multiple craze formation in PS and SAN matrix; shear yielding may occur in some cases; cavitation of rubber phase has been observed in some cases | shear yielding initiated by cavitation of rubber particles is the principal mechanism of matrix deformation, although crazing has been cited in some systems |

**Table 4. U.S. Sales of Commercial Plastics, $10^6$ t[a]**

| | Year | | | |
|---|---|---|---|---|
| Polymer | 1981 | 1985 | 1989 | 1993 |
| LDPE | 3.4 | 4.1 | 4.8 | 6.0 |
| PVC | 2.6 | 3.1 | 3.8 | 4.7 |
| HDPE | 2.2 | 2.9 | 3.7 | 4.8 |
| PP | 1.8 | 2.3 | 3.3 | 4.1 |
| PS | 1.7 | 1.8 | 2.4 | 2.5 |
| thermoplastic polyesters | 0.56 | 0.65 | 0.96 | 1.3 |
| ABS | 0.44 | 0.48 | 0.57 | 0.62 |
| thermoplastic elastomers | 0.17 | 0.22 | 0.25 | 0.32 |
| nylon | 0.13 | 0.18 | 0.27 | 0.33 |
| epoxy | 0.15 | 0.16 | 0.22 | 0.25 |
| PC | 0.11 | 0.13 | 0.28 | 0.28 |
| PPO and copolymers | 0.06 | 0.08 | 0.08 | 0.10 |

[a]Ref. 236.

these polymers has been large as well. The polymers that are most frequently rubber-modified include PS, PPO, PVC, and ABS. Also, a significant growth of modified nylons and PC (via maleated rubbers and ABS) has occurred since the early 1980s. More detailed information on the commercially available products, including trade names, properties, and applications, is given in Reference 237.

**High Impact Polystyrene.** The toughening of brittle, glassy PS was introduced in the late 1940s. High impact polystyrene (HIPS) grew rapidly because of its improved toughness and satisfactory rigidity, and in the 1990s it accounts for about 50% of all commercial polystyrene production. It is manufactured by the solution polymerization of styrene in the presence of ca 5–10% dissolved polybutadiene (PBD) or a copolymer rubber (233,238–240). Early in the polymerization, phase separation begins at ca 2% conversion because of the immiscibility of the rubber with the polystyrene being formed and the depletion of the solvent (styrene); grafting of PBD with PS takes place (220,241,242). As the volume ratio of the PS–PBD solutions in styrene approaches one, phase inversion takes place. The presence of some PS-grafted PBD at this stage is crucial for the stabilization of the polymeric oil-in-oil emulsion. Shearing agitation of the reaction mixture is essential to control the rubber droplet size. The PS–styrene content of the rubber droplet is affected by the shear stresses imposed on the system. Accordingly, the PS occlusions within the rubber particle and the rubber-phase volume fraction are strongly influenced by the agitation during and shortly after phase inversion. During the latter stage, shearing agitation is no longer crucial. In fact, some processes use suspension polymerization for finishing. Near the completion of polymerization and during the removal of residual monomer between 180 and 240°C, cross-linking of the rubber phase readily takes place. During the earlier stages of polymerization, only grafting occurs. Cross-linking is crucial to maintaining the particle morphology and toughness during heat fabrication, as uncross-linked particles are prone to disintegrate during melt shearing. Typical rubber particles of HIPS are shown in Figure 9. The light portions within the rubber particles represent the occluded or entrapped PS. Rubber particles separated from the PS matrix are shown in Figure 8.

The toughness as well as the other mechanical and rheological properties of HIPS are strongly affected by the nature of the rubber phase. Some of the variables that control HIPS performance include rubber composition and concentration, rubber-phase volume, particle-size distribution, degree of grafting and cross-linking, and molecular weight and molecular weight distribution of the matrix PS. Antioxidants (qv), plasticizers (qv), flame retardants (qv), and other additives also influence performance (220).

Adhesion between the matrix and rubber particles is necessary for stress transfer in order to prevent premature crack development from crazes. The PS graft layer on the particles causes the rubber particles to adhere to the PS matrix. For optimum toughening, rubber particles of 1–3 $\mu$m are required. However, to achieve adequate toughness and superior molded surface appearance (gloss) dual-size particle populations (eg, 2 $\mu$m and 0.2 $\mu$m) are desirable (240). In addition to the control of the particle size by shearing, initiators, chain-transfer agents, and diluents play an important role (220), especially the type and the concentration of peroxides for the control of grafting and molecular weight. Excessive grafting may readily occur and give rise to a phase morphology that can no longer be controlled. In such cases, a spontaneous disintegration of the rubber phase into fragments may take place without shearing (243). If small particles are desired for reducing opaqueness or for blending, particle size may be controlled by careful adjustment of initiator concentration. However, for optimum toughening, HIPS should have a volume fraction of the rubber phase above 30%. The control of grafting and PS occlusions is crucial, as most HIPS products contain only 5–7% rubber by weight.

HIPS may be modified by mechanical blending with elastomeric polymers such as styrene–butadiene–styrene block copolymers, for the improvement of toughness and stress-crack resistance. Block copolymers resulting in stringy or particulate morphologies are useful; these morphologies coexist with the parent HIPS morphology (244). High styrene content block copolymers have been used in blends with PS for transparent packaging (qv) applications.

Mechanical properties of HIPS differ from those of PS (237,241). Although modulus and tensile strength are lower, impact strength and elongation at failure are significantly higher for HIPS, whereas heat-deformation temperature and melt viscosity are similar. Stress whitening of HIPS starts at ca 1% strain and becomes more intense at higher strains and impact conditions. Stress whitening is caused by extensive dilational processes, such as cavitation within the rubber particles and crazing in the PS matrix. Studies by the use of real-time small-angle x-ray scattering reveal the sequence of these two mechanisms and their relative contributions (245).

**Acrylonitrile–Butadiene–Styrene.**    ABS is an important commercial polymer, with numerous applications. In the late 1950s, ABS was produced by emulsion grafting of styrene–acrylonitrile copolymers onto polybutadiene latex particles. This method continues to be the basis for a considerable volume of ABS manufacture. More recently, ABS has also been produced by continuous mass and mass-suspension processes (237). The various products may be mechanically blended for optimizing properties and cost. Brittle SAN, toughened by SAN-grafted ethylene–propylene and acrylate rubbers, is used in outdoor applications. Flame retardancy of ABS is improved by chlorinated PE and other flame-retarding additives (237).

Emulsion polymerization of ABS (241) gives a rubber-phase particle morphology which is mostly determined by the rubber-seed latex. Since the rubber particle size, polydispersity, and cross-linking are established before the preparation, the main variables relate to grafting, molecular weight distribution of SAN, and the final cross-linking of the rubber phase. The internal structure of the rubber phase depends on the initiators. Oil-soluble initiators give the highest amounts of SAN inclusions within the rubber particle. A study of rubber-seed latices with a bimodal particle distribution as well as of high and low grafting revealed that the structure of the graft shell of the larger particles is particularly important. Low grafting in large particles, combined with highly grafted small particles, gives optimum toughening (246). Other studies on the relationship of rubber-phase structure to properties have yielded similar conclusions. At low grafting, some coalescence of particles takes place, which may result in further improvement in toughening (247–250). The particle size in an emulsion-made ABS ranges from 0.1 to <1 $\mu$m for optimum performance.

In mass (bulk) and solution polymerization of ABS, the morphology of the rubber phase depends on the same variables described previously for the HIPS process (220). In addition, the fact that the styrene and acrylonitrile mixture is a poorer solvent for the rubber than styrene influences the final rubber-phase morphology. A study of the grafting kinetics of ABS in solution reveals that the grafted and the free SAN differ in composition because of the preferential solvation of the rubber by styrene, resulting in a 1–2% higher AN content in the graft-free SAN (251). Because of these differences, mass- and solution-polymerized ABS have rubber particles that resemble those of HIPS but are usually smaller (ca 1 $\mu$m) and have fewer rigid-phase occlusions.

Rubber particle cavitation, shear yielding, and crazing all contribute to the toughness of ABS (244,252). In an emulsion-made ABS, shear yielding is the predominant mechanism at low extension rates, whereas at higher rates cavitation and crazing also become important (251,253). In mass-made ABS, dilatational processes remain the important mechanisms of energy absorption. A study of the micromechanics of a commercial ABS, containing both small and large (1.5 $\mu$m) rubber particles, shows that crazing (nucleated at the large particles), shear yielding, and cavitation (associated with both small and large rubber particles) all take place (245,254). Studies on the fracture phenomena (255), including fracture mechanics (256) and fatigue damage (253) of ABS illustrate its complex response to stress.

**Poly(vinyl chloride).** PVC is one of the most important and versatile commodity polymers (Table 4). It is inherently flame retardant and chemically resistant and has found numerous and varied applications, principally because of its low price and capacity for being modified. Without modification, processibility, heat stability, impact strength, and appearance all are poor. Thermal stabilizers, lubricants, plasticizers, impact modifiers, and other additives transform PVC into a very versatile polymer (257,258).

The improvement of its toughness by melt blending with impact modifiers is of considerable importance. The most frequently used impact modifiers are chlorinated polyethylenes, styrene–acrylonitrile-grafted elastomers, methyl methacrylate–styrene-grafted elastomers, ethylene–vinyl acetate copolymers, polyacrylates, and nitrile rubbers. Numerous elastomers have been grafted with vinyl chloride for PVC modification (259). Further opportunities are provided by monomers giving PVC-miscible grafts (260). These modifiers are usually melt-blended with PVC along with stabilizers, flow modifiers, fillers, and pigments to give the final composition for a specific application. The method used and the conditions of blending are extremely important to the resulting morphology and performance.

PVC is insoluble in vinyl chloride and precipitates during polymerization. Because of this insolubility, primary (1–2 $\mu$m) and secondary (50–250 $\mu$m) grain structure is formed in mass or suspension polymerization (261). For optimum performance of a toughened PVC, the modifier must be thoroughly dispersed in the interstitial spaces between the primary grains (Fig. 10) (220). Two conditions must be avoided: inadequate mixing, and a too high overall or local temperature during melt blending, the latter causing the melting of the primary grains and resulting in a too finely dispersed modifier throughout the PVC matrix (262). Because of the unusual distribution of the impact modifier in the interstitial spaces, a small amount of modifier is required. In some cases, ca 2–10% of modifier may be sufficient.

Rubber-modified PVC is toughened principally by a shear-yielding mechanism (255). Stress whitening, which occurs in some toughened compositions, is usually a result of the internal rupture of elastomer particles, not crazing (263). Stress dilatometry and microscopy have further revealed that short microcracks may form in the interstitial spaces between the primary PVC grains. The stressed sample remains clear, as the microcracks are smaller (20 nm) than the wavelength of light (264). Upon subsequent stressing, the microcracks facilitate shear band formation and do not readily grow into catastrophic cracks because

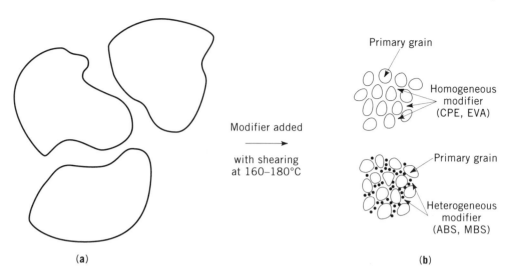

**Fig. 10.** Preparation and morphology of toughened PVC: (**a**) secondary PVC grain (50–250 $\mu$m); (**b**) modified PVC with coherent primary grain (ca 1 $\mu$m) (220). CPE = chlorinated polyethylene; EVA = ethylene–vinyl acetate copolymers; ABS = acrylonitrile–butadiene–styrene; MBS = methyl methacrylate–butadiene–styrene copolymer. Courtesy of MMI Press.

of the blunting by the shear bands. As a result of the various contributions to energy absorption, toughened PVC without stress whitening can be made with a number of modifiers.

**Poly(2,6-dimethyl-1,4-phenylene oxide).** It is well established that PS and PPO are miscible in all proportions and that the rubber particles from HIPS are distributed uniformly throughout the new mixed matrix.

Because of the miscibility and the significantly different properties of component polymers, production of a large number of products with well-balanced properties is possible. Blends containing large amounts of PPO are particularly useful because of their high heat-deformation temperature; those containing less PPO are less expensive. They have an excellent combination of strength and toughness and are easily processed (265).

As in other heterophase blends, morphology of the dispersed phase is of critical importance in determining toughness of the blend. In this case, the rubber-particle size is established in HIPS. Rubber particles smaller than those found in most commercial HIPS (ca 1 $\mu$m) are most effective in toughening the PS–PPO matrix. Because of the larger particle size in commercial HIPS, specialty grades of HIPS are used. It is also apparent from the extensive patent literature on PPO blends that PPO is quite amenable to impact modification with a series of block and graft copolymers containing a rubbery phase. Polyethylene is also useful as a flow promoter (266).

Both crazing and shear yielding of the matrix can contribute to the energy absorption of PPO–HIPS blends (267); at high deformation rates, shear yielding is of increasing importance. The HIPS rubber particles fail by rupture, in both cyclic fatigue testing as well as fast fracture, indicating strong adhesion to the PS–PPO matrix (266).

For high performance applications in the automotive industry, nylon–PPO blends with impact modifiers have been introduced (173,177).

**Polypropylene.** PP is a versatile polymer, use of which continues to grow rapidly because of its excellent performance characteristics and improvements in its production economics, eg, through new high efficiency catalysts for gas-phase processes. New PP-blend formulations exhibit improved toughness, particularly at low temperatures. PP has been blended mechanically with various elastomers from a time early in its commercialization to reduce low temperature brittleness.

PP is also modified by copolymerization. A two-phase rubber-toughened product is obtained by adding ethylene during the later stages of propylene polymerization (241,268). Both PP and the two-phase copolymer may be blended with a number of elastomers, and styrene–butadiene rubber (SBR) (269–271), ethylene–propylene–diene rubber (EPDM) (270,271), ethylene–propylene rubber (270,271), PBD (271), thermoplastic rubber (272,273), natural rubber (274), polyisobutylene (275,276), and PE (268,277) have been used to modify the physical and mechanical behavior of PP. In choosing the rubber modifier, consideration should be given to its compatibility, adhesion, and dispersibility. Even after an appropriate choice, the varied shearing conditions prevailing in extrusion blending give rise to blends with different morphologies. Control of the rubber-phase morphology in PP is complicated, as it is affected by the rubber composition, molecular weight, cross-linking, and rheological properties and by the crystallization behavior of the PP matrix. Further changes in morphology may take place upon subsequent heat fabrication. Low cross-linking of the EPDM rubbers gives melt-blended products with PP which have more stable morphologies and excellent properties and which are not affected by further melt shearing (278,279). Chemical modification of the PP interface has been reported (280).

Introduction of a dispersed rubber phase into PP reduces modulus and tensile strength as well as the heat-deformation temperature and hardness. These changes, however, have a minor effect on rigidity. The changes in tensile properties and toughness that result from the rubber phase are well-documented (270,272,278,281). Rubber particles of ca 0.5–2 $\mu$m seem to give optimum overall balance of rigidity and toughness (268–270,282). In general, PP blends containing smaller particles are more ductile and impact-resistant than those with larger particles (268,270).

Molding conditions influence the mechanical behavior of PP to a large extent (272). During injection molding, high anisotropy develops (282,283) and a highly oriented skin is formed, whereas the interior remains isotropic. Compression moldings remain isotropic throughout the molding (282). Annealing of injection moldings further affects the morphology and mechanical properties of PP blends. Annealing PP near its melting point reduces impact strength, whereas annealing of a rubber-modified PP can improve both rigidity and impact strength (281).

Introduction of the rubber phase into PP causes significant changes in its response to stress. In unmodified PP, the dominant mechanism of creep is shear-yielding and there is no detectable dilatational response (284). High deformation rates and low temperatures, however, favor crazing (285). Rubber-modified PPs show a higher propensity for craze formation, even at low strains (269,270,284). Void formation within rubber particles adds to the dilatational processes. Injection molding of rubber-modified PP further complicates the stress response. The highly oriented skin layer of an injection-molded specimen after stressing is free from crazing, in contrast to the core, which has a high craze concentration (283).

**Nylon.**  Nylons comprise a large family of polyamides with a variety of chemical compositions (234,286,287). They have excellent mechanical properties, as well as abrasion and chemical resistance. However, because of the need for improved performance, many commercial nylon resins are modified by additives so as to improve toughness, heat fabrication, stability, flame retardancy, and other properties.

Nylons are tough under many impact conditions but may be notch-sensitive and brittle, particularly at low temperatures. In rubber modification, an elastomeric second phase is incorporated, usually by mechanical blending with a variety of chemically modified elastomers (eg, maleic anhydride) to effect interaction between the nylon matrix and the elastomer phase. Importantly, the morphology of the elastomer phase is influenced by the degree of chemical reaction between the two phases. The control of rubber-phase morphology is critical in determining the mechanical properties of the modified nylons (234,287). Significant improvements in ductility are accompanied by a corresponding drop in flexural modulus and tensile strength. Nylons may be toughened by the solution polymerization of caprolactam in the presence of ethylene−propylene elastomers (288). Impact properties of these heterophase polymers are often comparable to those prepared by melt-blending.

Most elastomers that are used for nylon modification contain a small amount of maleic anhydride (0.3 to 2%). In the melt blending process, these elastomers react with the primary amine end groups in nylon, giving rise to nylon grafted elastomers. These grafts reduce the interfacial tension between the phases and provide steric stabilization for the dispersed rubber phase. Typically, thermally stable, saturated rubbers such as EPR, EPDM, and styrene−ethylene/butylene−styrene (SEBS) are used.

In addition to the strong influence of the compatibilizers in rubber-particle dispersion in nylon, the nature of the polyamide also plays an important role in rubber-phase morphology development (178,228,289). The monofunctional nylons (eg, nylon-6) that comprise molecules with only one amine end group are readily toughened. Nylon-6 is formed by a ring-opening reaction of $\epsilon$-caprolactam. In the blends of nylon-6 and maleic anhydride-modified rubbers, only single-end grafts are formed and these are effective for controlling the development of particle size and stability. Difunctional nylons (eg, nylon-6,6) may have molecules with both ends containing amine end groups. In this case, cross-linking as well as grafting may take place during melt blending, and this usually makes it more difficult to disperse the rubber and control the rubber-phase morphology (289). Accordingly, blends based on nylon-6,6 require more rigorous melt-blending conditions.

Numerous approaches to super-tough nylons have been described (234,287). Rubber-phase particle size and distribution are critical in determining the toughness of the blends. Particles in the range of 0.2−0.4 $\mu$m in diameter appear to give the optimum performance. A sharp brittle−tough transition occurs at a critical particle size. This critical particle size increases with increasing volume

fraction of the rubber phase, and it has been suggested that the former is related to a critical interparticle distance (290). However, it has also been reported
that there is a lower limit of particle size, below which poor toughening is observed (291).

Emulsion-made core-shell polymers (eg, MBS and ABS) are also used for
impact modification of nylons. In the latter case, a third polymer component is
often used to compatibilize the impact modifier and nylon. For example, nylon-
6–ABS blends may be compatibilized with imidized acrylic polymers (157), as
well as with styrenic copolymers containing maleic anhydride (292,293). Here the
compatibilizing copolymer (158,293) is miscible with the SAN matrix in ABS. The
resulting blend has a well-defined ABS dispersion in the nylon-6 matrix owing
to the grafting reaction between maleic anhydride of the compatibilizer and the
primary amine end group in nylon-6 (158). Super-tough blends are prepared
using this approach.

The primary deformation mechanism in toughened nylons is shear-yielding,
but some dilatational processes such as crazing, rubber particle cavitation and
debonding may occur simultaneously (294–296). Subsurface deformation and
debonding processes contribute to energy dissipation during fracture. The latter
may play an important role in reducing notch sensitivity in toughened nylons.

**Polycarbonates.**   Bisphenol A polycarbonate (PC) is used in a wide variety of applications because of its excellent balance of properties, including
optical clarity, high heat-deformation temperature, toughness, and electrical
properties. However, PC has some characteristics that deter its use in some
areas. The exceptional toughness it is noted for is not retained in thick molded
sections, when there are sharp notches, at low temperatures, and after physical
aging. In addition, it has poor radiation, solvent, and hydrolysis resistance. A
number of approaches have been used to enhance these end-use properties. For
example, polycarbonate blends with acrylonitrile–butadiene–styrene materials
are commercially important plastics that have been found to be useful in many
molding applications, particularly in the automotive industry. These blends are
successful because they economically combine some of the best properties of the
components, eg, excellent impact strength (including improved notch sensitivity
and thick section toughness), high heat distortion temperature, and relatively
low melt viscosity for improved processability.

There is extensive literature on PC blends with ABS, and blends of PC
with related materials such as SAN, methacrylate–butadiene–styrene (MBS)
emulsion-made core-shell rubber modifiers (297–299), and other impact modifiers. One report reviews some of these approaches and compares PC blends
based on emulsion vs bulk ABS (229). In PC–ABS blends, no additional compatibilizer is used, because of the near-miscibility of the SAN matrix of ABS
and PC.

The use of PC–ABS blends has grown significantly in the early 1990s.
These blends exhibit excellent properties, particularly low temperature ductility,
reduced notch sensitivity, and ease of melt fabrication. The blend morphology
(229), ABS composition, thermal history (215), PC content and molecular weight
(300), processing conditions, etc, all affect the mechanical behavior of PC–ABS
blends. These blends have been most frequently used in automotive and other
engineering applications.

**Polyesters.**  A variety of polyester blends are commercially available, and most of these are rubber-toughened, typically by the incorporation of core-shell impact modifiers. A toughened version of poly(butylene terephthalate) (PBT) may serve as an illustration to reveal one of the approaches used to achieve excellent toughness of such a blend. In a blend of a core-shell (PMMA) modifier with PBT, the emulsion-made impact modifier particles are not well-dispersed in the PBT matrix. The addition of a few percent of PC to the blend improves the rubber phase dispersion significantly and renders the system super-tough. In this system, PC acts as a dispersing aid for the core-shell impact modifier (301,302). Blends containing comparable amounts of PC and PBT give excellent low temperature toughness. Such blends have become commercially significant, particularly in the automotive industry, where toughness, high heat deformation temperature, solvent resistance, and surface quality of the moldings are important.

Also, PBT is blended with poly(ethylene terephthalate) (PET), polysulfone, and SMA (303). PET may also be blended with a number of other engineering polymers, such as PC and impact modifiers.

**Epoxy Resins.**  The possibility of toughening epoxy resins (qv) by an elastomeric second phase was first proposed in 1968 (304). Since then, low molecular weight liquid carboxy-terminated butadiene–acrylonitrile rubbers have become widely used for modification of epoxy resins. Toughened epoxy resins are typically prepared *in situ* by quiescent bulk polymerization of epoxy in the presence of dissolved rubber. To control the rubber particle morphology and the final properties, the composition and concentration of both rubber and hardener and the curing temperature must be considered. During curing, the rubber phase separates directly as droplets without passing through a phase inversion. Before the gel point, further nucleation and growth of the rubber phase may take place. Occlusions of epoxy resins are often formed during the mass polymerization of cure. Matrix–rubber particle adhesion develops because of the early chain-extension reaction between the carboxyl groups in the rubber and the epoxy resin. The cured resin usually contains dissolved and solid rubber particles (305).

Studies of the particle–epoxy interface and particle composition have been helpful in understanding the rubber-particle formation in epoxy resins (306). Based on extensive dynamic mechanical studies of epoxy resin cure, a mechanism was proposed for the development of a heterophase morphology in rubber-modified epoxy resins (307). Other functionalized rubbers, such as amine-terminated butadiene–acrylonitrile copolymers (308) and *n*-butyl acrylate–acrylic acid copolymers (309), have been used for toughening epoxy resins.

Mechanical properties of rubber-modified epoxy resins depend on the extent of rubber-phase separation and on the morphological features of the rubber phase. Dissolved rubber causes plastic deformation and necking at low strains, but does not result in impact toughening. The presence of rubber particles is a necessary but not sufficient condition for achieving impact resistance. Optimum properties are obtained with materials comprising both dissolved and phase-separated rubber (305).

A proposed mechanism for toughening of rubber-modified epoxies based on the microstructure and fracture characteristics (310–312) involves rubber cavitation and matrix shear-yielding. A quantitative expression describes the fracture toughness values over a wide range of temperatures and rates.

## Commercial Blends and Applications

The commercial development of polymer blends is strongly influenced by a set of more favorable economics than those affecting the more conventional chemical routes to new material systems. One crucial factor underlying this contrast is that blend systems, comprising pre-existing materials, can be developed much more quickly than newer polymers, thus allowing manufacturers to respond more rapidly, and at reduced cost, to new market requirements (313). Because the properties of an existing blend system are functions of the blend composition, an existing blend can also be easily and quickly modified to meet different performance objectives relative to cost required by new or changing markets. Discounting research and development costs, blends are a particularly attractive choice when one of the principal blend components is much less expensive than the others, because this permits the blend to be produced at a cost that is potentially lower than that of the higher cost ingredient (314–316). Even where all ingredients are comparably valued, blends are commercially attractive if the blend possesses unique improvements in processability or performance.

Tables 5 and 6 summarize key properties and applications for miscible and immiscible blends which are either commercial as of 1996 or were commercialized

**Table 5. Commercial Blends with Miscible Principal Components**

| Principle components | Property advantages[a] | Applications | References |
|---|---|---|---|
| PVC–nitrile rubber | permanent plasticization of PVC, improved processibility | wire and cable insulation, food contact service | 141,317,318 |
| PVC–related co- and terpolymers | permanent plasticization and improved processibility | pond liners, auto interiors, shoes | 140,317,319 |
| PVC–$\alpha$-MeSAN | increased HDT and improved processibility | rigid house siding | 27,320 |
| PPO–HIPS | better processibility and toughness than PPO, better HDT than PS | appliance components, business machine housings | 321,322 |
| ABS–$\alpha$-MeSAN | better HDT than ABS | appliance components | 323,324 |
| ABS–SMA | increased HDT | automotive | 144 |
| PMMA–PVF$_2$, PEMA–PVF$_2$ | better chemical and uv resistance than PMMA or PEMA, better clarity than PVF$_2$ | outdoor film, decorative striping on automobiles | 45,325,44 |
| PBT–PET | lower cost than PBT, better gloss and flexibility than PET | electrical/electronic, brake and fuel lines | 326,327 |
| PC–copolyester | better processibility, hydrolytic, and gamma radiation resistance than PC | sterilizable applications | 43,109 |

[a]HDT = heat-distortion temperature.

**Table 6. Commercial Blends with Nonmiscible Principal Components**

| Principle components | Property advantages[a] | Applications | References |
|---|---|---|---|
| PVC–ABS and PVC–acrylic | better processibility and toughness than PVC, better fire retardancy than ABS or acrylic | mass transit interiors, appliance housings | 314,328 329,330 |
| PVC–chlorinated PE (<41% Cl) | better impact than PVC | pipe and house siding | 48,331,332 |
| PC–ABS | better toughness and HDT than ABS, better processibility and lower cost than PC | appliance and business machine housings, automotive components | 313,333 |
| PSF–ABS | similar to PC–ABS, composition can be electroplated due to ABS, lower cost than polysulfone | plumbing fixtures, food service trays, appliances | 334 |
| PC–PET | better chemical resistance, processibility, and lower cost than PC | tubing, auto bumpers, and business machine housings | 335,336, 313 |
| PC–PBT | better solvent resistance and processibility than PC | tubing, auto bumpers, and business machine housings | 313,315 |
| PC–SMA | impact modified, better toughness and ductility than SMA, better retention of properties on aging at high temperature, lower cost than PC | automotive, appliances, and cookware | 315 |
| nylon–PPO | high HDT, solvent resistance toughness | automotive | 173,177 |
| polyolefin blends | better impact and toughness than either PE or PP components | wire and cable insulation, auto bumpers, hose and gaskets, film | 337–341 |

[a]HDT = heat-distortion temperature.

in the past (2,314–316,342,343). Most of the listed blends contain only two primary components, although many are compatibilized and impact-modified. Consequently, an immiscible system consisting of two primary components or phases may contain impact modifiers for each phase and a compatibilizer copolymer, for a total of five or more components.

Tables 5 and 6 suggest that a particular property advantage can be obtained by appropriately forming either a miscible or immiscible blend. For example, the heat distortion temperature (HDT) of ABS can be enhanced by blending it with either a miscible SMA component (144) or an immiscible PC component. Similarly, the processibility of PVC can be improved by addition of a miscible plasticizer such as EVA copolymer or an immiscible flow aid such as ABS. The choice of blend type typically depends on the entire balance of mechanical, thermal, and rheological properties. Specifically, the mechanical properties of immiscible mixtures can depend strongly on the phase morphology that is generated by the flow process (344), especially if interfacial agents are not present, whereas those of miscible mixtures are no more affected than those of comparable homopolymers. Weld lines in immiscible blends can be quite weak (345), and great care must be taken to assure that weld lines are placed in nonstressed regions of the part. On the other hand, immiscible mixtures of nylons or polyesters with amorphous high $T_g$ materials such as PPO or PC can be quite useful if properly modified and morphologically stabilized, because of the improved flow and solvent resistance provided by the semicrystalline component and the dimensional stability and rigidity provided by the amorphous component (173,177,225).

Because most polymer pairs are immiscible, it should not be surprising that a majority of the blend products which have emerged to date are formed from such mixtures. Continued growth in products of this type can be expected, fueled principally by the need to combine high performance characteristics, such as strength and toughness, with resistance to heat and aggressive chemical environments, and this process will be aided by careful control of morphology and the interface between phases by reactive compatibilization techniques. Because of advances in understanding of the role of molecular structure on polymer–polymer interactions, design of copolymers and terpolymers to achieve specific miscibility, interfacial, or compatibility effects in commercial products will become more common. Concern about the recycling or disposal of products after their useful life will no doubt play an increased role in the selection or design of blends that may be used in commercial applications.

Although the concept of polymer blends is sometimes a route for avoiding the development of new polymers, it often has been an integral part of the utilization of new polymer chemistry, eg, the commercial success of PPO hinged on the advantages of its blends with PS.

## Specialty Blends

There has been a great deal of interest in blends of rigid rod polymers, which form liquid crystals either in solution or in the melt state with flexible polymers. One of the principal reasons for this interest stems from the concept known as molecular composites (346). Because of their rigid chains, liquid crystal polymers (LCPs) have a very high modulus in their axial direction, and the concept of molecular composites envisions the use of these chains more or less like glass or carbon fibers to form composites on the molecular scale. In practice, the rigid chains aggregate so that true molecular composites are not achieved; however, the formation of microfibrillar structures can give some improvements in stiffness and strength (346,347). Even if this fundamental issue were to be resolved, the types of systems proposed for molecular composites involve solution processing and do not appear to be practical. Simple melt blends of commercial

LCP materials with a variety of other polymers have been studied extensively (348) but few, if any, products have emerged from this effort, largely because of issues of incompatibility and cost vs benefits.

The notion of lightweight materials as electrical conductors has fueled an intensive research effort to develop conducting polymers. However, these polymers are frequently intractable (ie, they do not dissolve or melt, so they cannot be easily formed into the desired shape), quite brittle, and normally not stable in the ambient environment because of sensitivity to oxygen. Reports have suggested that these problems may be lessened at least to some degree by *in situ* formation of the conducting polymer in the matrix of another polymer to produce a blend (155,349,350). For example, the catalyst for polymerizing acetylene has been dispersed in polymers like polybutadiene by an appropriate solvent technique (349). Next, this polymer is formed into the desired shape, eg, a thin film. This film is exposed to acetylene gas, which diffuses into the polymer and polymerizes *in situ*. The polyacetylene forms a separate phase within the host polymer, but apparently under appropriate circumstances cocontinuous morphology is created so that quite high levels of electrical conductivity are achieved. Polyaniline in conductive form has more recently been melt blended into thermoplastic matrices at levels where the percolation threshold is exceeded, with the result being a truly melt-processable conducting polymer (351,352). Commercial applications for it are anticipated.

## BIBLIOGRAPHY

"Polyblends" in *ECT* 3rd ed., Vol. 18, pp. 443–478, by O. Olabisi, Union Carbide Corp.

1. O. Olabisi, L. M. Robeson, and M. T. Shaw, *Polymer–Polymer Miscibility*, Academic Press, Inc., New York, 1979.
2. D. R. Paul and J. W. Barlow, *J. Macromol. Sci. Rev. Macromol. Chem.* **18**, 109 (1980).
3. D. R. Paul in D. J. Walsh, J. S. Higgins, and A. Maconnachie, eds., *Polymer Blends and Mixtures*, NATO ASI Series, Series E, Applied Sciences, No. 89, Martinus Nijhoff Publishers, Dordrecht, the Netherlands, 1985, p. 1.
4. P. J. Flory, *Principles of Polymer Chemistry*, Cornell University Press, Ithaca, N.Y., 1953.
5. D. J. Walsh in C. Booth and C. Price, eds., *Comprehensive Polymer Science*, Vol. 2, Pergamon Press, Oxford, U.K. 1989, p. 135.
6. R. Koningsveld, H. A. G. Chermin, and M. Gordon, *Proc. R. Soc. London Series A* **319**, 331 (1970).
7. R. Koningsveld and L. A. Kleintjens, in Ref. 3, p. 89.
8. I. C. Sanchez, *Polymer* **30**, 471 (1989).
9. C. K. Kim and D. R. Paul, *Polymer* **33**, 1630, 2089, and 4941 (1992).
10. G. R. Brannock and D. R. Paul, *Macromolecules* **23**, 5240 (1990).
11. T. A. Callaghan and D. R. Paul, *Macromolecules* **26**, 2439 (1993).
12. C. Panayiotou and I. C. Sanchez, *Macromolecules* **24**, 6231 (1991).
13. P. J. Flory, *Discuss. Faraday Soc.* **49**, 7 (1970).
14. D. Patterson and A. Robard, *Macromolecules* **11**, 690 (1978).
15. I. C. Sanchez and R. H. Lacombe, *J. Phys. Chem.* **80**, 2352, 2568 (1976).
16. I. C. Sanchez and R. H. Lacombe, *Macromolecules* **11**, 1145 (1978).
17. I. C. Sanchez, *Ann. Rev. Mater. Sci.* **13**, 387 (1983).
18. D. J. Walsh and S. Rostami, *Macromolecules* **18**, 216 (1985).

19. I. C. Sanchez in K. Solc, ed., *Polymer Compatibility and Incompatibility: Principles and Practice*, MMI Press Symposium Series, Vol. 2, Harwood Academic Publishers GmbH, New York, 1982, p. 59.
20. P. A. Rodgers, *J. Appl. Polym. Sci.* **48**, 1061 (1993).
21. C. K. Kim and D. R. Paul, *Polym. Eng. Sci.* **34**, 24 (1994).
22. D. R. Paul and J. W. Barlow, *Polym. Sci. Technol.* **11**, 239 (1980).
23. T. T. Wang and T. Nishi, *Macromolecules* **10**, 421 (1977).
24. L. M. Robeson and A. B. Furtek, *J. Appl. Polym. Sci.* **23**, 645 (1979).
25. D. Y. Yoon, Y. Ando, S. Rojstaczer, S. K. Kumar, and G. C. Alfonso, *Makromol. Chem., Macromol. Symp.* **50**, 183 (1991).
26. J. P. Runt, C. A. Barron, X. F. Zhang, and S. K. Kumar, *Macromolecules* **24**, 3468 (1991).
27. J. E. Harris and L. M. Robeson, *J. Polym. Sci.: Part B: Polym. Phys.* **25**, 311 (1987).
28. L. M. Robeson, in Ref. 19, p. 177.
29. F. E. Karasz, in Ref. 3, p. 25.
30. M. T. Shaw, in Ref. 3, pp. 37 and 57.
31. J. Devaux, P. Godard, and J. P. Mercier, *Polym. Eng. Sci.* **22**, 229 (1982).
32. L. M. Robeson, *J. Appl. Polym. Sci.* **30**, 4081 (1985).
33. Y. Takeda and D. R. Paul, *Polymer* **32**, 2771 (1991) and **33**, 3899 (1992).
34. A. C. Fernandes, J. W. Barlow, and D. R. Paul, *J. Appl. Polym. Sci.* **32**, 5481 (1986).
35. A. K. Nandi, B. M. Mandal, and S. N. Bhattacharyya, *Macromolecules* **18**, 1454 (1985).
36. L. Zeman and D. Patterson, *Macromolecules* **5**, 513 (1972).
37. E. M. Woo, J. W. Barlow, and D. R. Paul, *J. Polym. Sci. Polym. Symp.* **71**, 137 (1984).
38. A. R. Shultz and A. L. Young, *Macromolecules* **13**, 663 (1980).
39. M. Nishimoto, H. Keskkula, and D. R. Paul, *Polymer* **32**, 272 (1991).
40. S. Ichihara, A. Komatsu, and T. Hata, *Polym. J.* **2**, 640 (1971).
41. J. Maruta, T. Ougizawa, and T. Inoue, *Polymer* **29**, 2056 (1988).
42. P. S. Tucker, J. W. Barlow, and D. R. Paul, *Macromolecules* **21**, 2794 (1988).
43. R. N. Mohn, D. R. Paul, J. W. Barlow, and C. A. Cruz, *J. Appl. Polym. Sci.* **23**, 575 (1979).
44. R. L. Imken, D. R. Paul, and J. W. Barlow, *Polym. Eng. Sci.* **16**, 593 (1976).
45. T. Nishi and T. T. Wang, *Macromolecules* **8**, 909 (1975).
46. R. S. Stein, *J. Polym. Sci. Polym. Phys. Ed.* **19**, 1281 (1981).
47. B. S. Morra and R. S. Stein, *J. Polym. Sci. Polym. Phys. Ed.* **20**, 2243 (1982).
48. P. P. Gan and D. R. Paul, *J. Polym. Sci.: Part B: Polym. Phys.* **33**, 1693 (1995).
49. C. P. Doube and D. J. Walsh, *Polymer* **20**, 1115 (1979).
50. G. DiPaola-Biranyi and P. Degre, *Macromolecules* **14**, 1456 (1981).
51. Z. Y. Al-Saigh and P. Munk, *Macromolecules* **17**, 803 (1984).
52. M. J. El-Hibri, W. Cheng, and P. Munk, *Macromolecules* **21**, 3458 (1988).
53. J. E. Harris, D. R. Paul, and J. W. Barlow, *Adv. Chem. Ser.* **206**, 43 (1984).
54. C. A. Cruz, J. W. Barlow, and D. R. Paul, *Macromolecules* **12**, 726 (1979).
55. E. M. Woo, J. W. Barlow, and D. R. Paul, *Polymer* **26**, 763 (1985).
56. D. J. Walsh, J. S. Higgins, and C. Zhikuan, *Polymer Commun.* **23**, 336 (1982).
57. C. Zhikuan, S. Ruona, D. J. Walsh, and J. S. Higgins, *Polymer* **24**, 263 (1983).
58. C. H. Lai, D. R. Paul, and J. W. Barlow, *Macromolecules* **21**, 2492 (1988); **22**, 374 (1989).
59. C. J. T. Landry and D. M. Teegarden, *Macromolecules* **24**, 4310 (1991).
60. C. J. T. Landry and B. K. Coltrain, D. M. Teegarden, and W. T. Ferrar, *Macromolecules* **26**, 5543 (1993).
61. E. Espi and J. J. Iruin, *Macromolecules* **24**, 6458 (1991).
62. C. A. Cruz-Ramos and D. R. Paul, *Macromolecules* **22**, 1289 (1989).
63. S. Rostami and D. J. Walsh, *Macromolecules* **18**, 1228 (1985).

64. K. Fukumori, N. Sato, and T. Kurauchi, *Rubber Chem. Technol.* **64**, 522 (1991).

65. Y. H. Chin, P. T. Inglefield, and A. A. Jones, *Macromolecules* **26**, 5372 (1993).

66. M. A. Gashgari and C. W. Frank, *Macromolecules* **21**, 2782 (1988).

67. F. Mikes, H. Morawetz, and K. S. Dennis, *Macromolecules* **17**, 60 (1984).

68. M. M. Coleman and J. Zarian, *J. Polym. Sci. Polym. Phys. Ed.* **17**, 837 (1979).

69. S. T. Wellinghoft, J. C. Koenig, and E. Bair, *J. Polym. Sci. Polym. Phys. Ed.* **15**, 1913 (1977).

70. S. P. Ting, E. M. Pearce, and T. K. Kwei, *J. Polym. Sci. Polym. Lett. Ed.* **18**, 201 (1980).

71. A. Garton, M. Aubin, and R. E. Prud'homme, *J. Polym. Sci. Polym. Lett. Ed.* **21**, 45 (1983).

72. D. Garcia, *J. Polym. Sci. Polym. Phys. Ed.* **22**, 107 (1984).

73. E. J. Lu, E. Benedetti, and S. L. Hsu, *Macromolecules* **16**, 1525 (1983).

74. M. M. Coleman, E. J. Moskala, P. C. Painter, D. J. Walsh, and S. Rostami, *Polymer* **24**, 1410 (1983).

75. M. M. Coleman, J. F. Graf, and P. C. Painter, *Specific Interactions and the Miscibility of Polymer Blends,* Technomic, Lancaster, Pa., 1991.

76. R. J. Composto, J. W. Mayer, E. J. Kramer, and D. M. White, *Phys. Rev. Lett.* **57**, 1312 (1986).

77. R. S. Stein in D. R. Paul, and S. Newman, eds., *Polymer Blends*, Vols. I and II, Academic Press, Inc., New York, 1978, Chapt. 9.

78. H. L. Snyder and P. Meakin, *J. Polym. Sci. Polym. Symp.* **73**, 217 (1985).

79. W. A. Kruse, R. G. Kirste, J. Haas, B. J. Schmitt, and D. J. Stein, *Makromol. Chem.* **117**, 1145 (1976).

80. D. G. H. Ballard, M. G. Rayner, and J. Schelten, *Polymer* **17**, 640 (1976).

81. R. P. Kambour, R. C. Bopp, A. Maconnachie, and W. J. MacKnight, *Polymer* **21**, 133 (1980).

82. G. Hadziioannou and R. S. Stein, *Macromolecules* **17**, 567, 1059 (1984).

83. B. J. Schmitt, *Angew. Chem. Int. Ed. Engl.* **18**, 273 (1979).

84. J. S. Higgins, *Makromol. Chem., Macromol. Symp.* **15**, 201 (1988).

85. M. C. Fernandez, J. S. Higgins, and J. Penfold, *Makromol. Chem., Macromol. Symp.* **62**, 103 (1992).

86. T. P. Russell, R. P. Hjelm, and P. A. Seeger, *Macromolecules* **23**, 890 (1990).

87. P. Russell, *Macromolecules* **26**, 5819 (1993).

88. G. D. Wignall, in J. I. Kroschwitz, ed., *Encyclopedia of Polymer Science and Engineering*, Vol. 10, Wiley-Interscience, New York, 1987, p. 112.

89. J. C. Nicholson, T. M. Finerman, and B. Crist, *Polymers* **31**, 2287 (1990).

90. J. Rhee and B. Crist, *J. Chem. Phys.* **98**, 4174 (1993).

91. R. Krishnamoorti, W. N. Graessley, N. P. Balsara, and D. J. Loshe, *Macromolecules* **27**, 3073 (1994).

92. R. P. Kambour, P. E. Gundlach, I. C. W. Wang, D. M. White, and G. N. Yeager, *Polym. Comm.* **29**, 170 (1988).

93. T. A. Callaghan and D. R. Paul, *J. Poly. Sci.: Part B: Polym. Phys.* **32**, 1813 and 1847 (1994).

94. K. Takakuwa, S. Gupta, and D. R. Paul, *J. Polym. Sci.: Part B: Polym. Phys.* **32**, 1719 (1994).

95. T. A. Callaghan, K. Takakuwa, D. R. Paul, and A. R. Padwa, *Polymer* **34**, 3796 (1993).

96. P. P. Gan, D. R. Paul and A. R. Padwa, *Polymer* **35**, 1487 (1994).

97. S. Krause in J. Brandrup and E. H. Immergut, eds., *Polymer Handbook*, 3rd ed., Wiley-Interscience, New York, 1989, p. VI/347.

98. A. F. M. Barton, *Handbook of Solubility Parameters and Other Cohesion Parameters*, CRC Press, Boca Raton, Fla., 1983.

99. E. M. Pearce, T. K. Kwei, and B. Y. Min, *J. Macromol. Sci. Chem.* **21**, 1181 (1984).

100. E. A. Guggenheim, *Mixtures*, Clarenden, Oxford, U.K., 1952.
101. A. Eisenberg, P. Smith, and Z. L. Zhou, *Polym. Eng. Sci.* **22**, 1117 (1982).
102. Z. L. Zhou and A. Eisenberg, *J. Appl. Polym. Sci.* **27**, 657 (1982).
103. Z. L. Zhou and A. Eisenberg, *J. Polym. Sci. Polym. Phys. Ed.* **21**, 595 (1983).
104. X. Lu and R. A. Weiss, *Macromolecules* **25**, 3242 and 6185 (1992).
105. A. Molnar and A. Eisenberg, *Polymer* **34**, 1918 (1993).
106. I. C. Sanchez and A. C. Balazs, *Macromolecules* **22**, 2325 (1989).
107. A. Sikora and F. E. Karasz, *Macromolecules* **26**, 3438 (1993).
108. R. S. Barnum, J. W. Barlow, and D. R. Paul, *J. Appl. Polym. Sci.* **27**, 4065 (1982).
109. E. A. Joseph, M. D. Lorenz, J. W. Barlow, and D. R. Paul, *Polymer* **23**, 112 (1982).
110. C. C. Han and co-workers, *Polymer* **29**, 2002 (1988).
111. E. M. Woo, J. W. Barlow, and D. R. Paul, *J. Appl. Polym. Sci.* **30**, 4243 (1985).
112. C. K. Kim and D. R. Paul, *Macromolecules* **25**, 3097 (1992).
113. C. K. Kim and D. R. Paul, *Polymer* **33**, 4929 (1992).
114. P. Alexandrovich, F. E. Karasz, and W. J. MacKnight, *Polymer* **18**, 1022 (1977).
115. D. R. Paul, *Pure Appl. Chem.* **67**, 977 (1995).
116. M. Nishimoto, H. Keskkula, and D. R. Paul, *Polymer* **32**, 1274 (1991).
117. M. Nishimoto, H. Keskkula, and D. R. Paul, *Polymer* **30**, 1279 (1989).
118. M. Nishimoto, H. Keskkula, and D. R. Paul, *Macromolecules* **23**, 3633 (1990).
119. J. M. G. Cowie, *Macromol. Symp.* **78**, 15 (1994).
120. J. M. G. Cowie, *Makromol. Chem. Macromol. Symp.* **58**, 63 (1992).
121. S. Y. Bell, J. M. G. Cowie, and I. J. McEwen, *Polymer* **35**, 786 (1994).
122. J. M. G. Cowie, R. Ferguson, I. J. McEwen, and J. C. M. Reid, *Polymer* **35**, 1473 (1994).
123. T. Shiomi, F. E. Karasz, and W. J. MacKnight, *Macromolecules* **19**, 2274 (1986).
124. C. Zhikuan and F. E. Karasz, *Macromolecules* **25**, 4716 (1992).
125. S. Djadoun and F. E. Karasz, *Macromol. Symp.* **78**, 155 (1994).
126. F. E. Karasz, *Makromol. Chem. Macromol. Symp.* **28**, 1 (1989).
127. T. Shiomi and K. Imai, *Polymer* **32**, 73 (1991).
128. J. H. Kim, J. W. Barlow, and D. R. Paul, *J. Polym. Sci.: Part B: Polym. Phys.* **27**, 223 (1989).
129. G. R. Brannock, J. W. Barlow, and D. R. Paul, J. Polym. Sci.: *Part B: Polym. Phys.* **29**, 413 (1991).
130. P. P. Gan and D. R. Paul, *J. Appl. Polym. Sci.* **54**, 317 (1994).
131. T. S. Ellis, *Polymer* **33**, 1469 (1992).
132. T. S. Ellis, J. Polym Sci.: *Part B: Polym. Phys.* **31**, 1109 (1993).
133. D. R. Paul and J. W. Barlow, *Polymer* **25**, 487 (1984).
134. R. P. Kambour, J. T. Bendler, and R. C. Bopp, *Macromolecules* **16**, 753 (1983).
135. G. ten Brinke, F. E. Karasz, and W. J. MacKnight, *Macromolecules* **16**, 1827 (1983).
136. M. E. Fowler, J. W. Barlow, and D. R. Paul, *Polymer* **28**, 1177 (1987).
137. H. Keskkula and D. R. Paul, *J. Appl. Polym. Sci.* **31**, 1189 (1986).
138. S. C. Chiu and T. G. Smith, *J. Appl. Polym. Sci.* **29**, 1781, 1797 (1984).
139. A. C. Fernandes, J. W. Barlow, and D. R. Paul, *J. Appl. Polym. Sci.* **32**, 5357 (1986).
140. C. F. Hammer, *Macromolecules* **4**, 69 (1971).
141. G. A. Zakrzewski, *Polymer* **14**, 348 (1973).
142. J. L. G. Pfennig, H. Keskkula, J. W. Barlow, and D. R. Paul, *Macromolecules* **18**, 1937 (1985).
143. F. E. Karasz and W. J. MacKnight, *Adv. Chem. Ser.* **211**, 67 (1986).
144. W. J. Hall, R. L. Kruse, R. A. Mendelson, and Q. A. Trementozzi, *ACS Symp. Ser.* **229**, 49 (1983).
145. B. J. Schmitt, R. G. Kirste, and J. Jelenic, *Makromol. Chem.* **181**, 1655 (1980).
146. A. C. Balazs, I. C. Sanchez, I. R. Epstein, F. E. Karasz, and W. J. MacKnight, *Macromolecules* **18**, 2188 (1985).

147. E. Roerdink and G. Challa, *Polymer* **21**, 1161 (1980).
148. J. J. Ziska, J. W. Barlow, and D. R. Paul, *Polymer* **22**, 918 (1981).
149. M. Aubin and R. E. Prud'homme, *Macromolecules* **13**, 365 (1980).
150. J. E. Harris, S. H. Goh, D. R. Paul, and J. W. Barlow, *J. Appl. Polym. Sci.* **27**, 839 (1982).
151. C. A. Cruz, J. W. Barlow, and D. R. Paul, *J. Appl. Polym. Sci.* **24**, 2399 (1979).
152. A. F. Yee, *Polym. Eng. Sci.* **17**, 213 (1977).
153. H. B. Hopfenberg and D. R. Paul in Ref. 77, Chapt. 10.
154. Y. Maeda and D. R. Paul, *Polymer* **26**, 2055 (1985).
155. R. P. Kambour and S. A. Smith, *J. Polym. Sci. Polym. Phys. Ed.* **20**, 2069 (1982).
156. S. R. Murff, J. W. Barlow, and D. R. Paul, *Adv. Chem. Ser.* **211**, 313 (1986).
157. W. E. Preston, J. W. Barlow, and D. R. Paul, *J. Appl. Polym. Sci.* **29**, 2251 (1984).
158. B. Majumdar, H. Keskkula, and D. R. Paul, *Polymer* **35**, 1386 (1994).
159. J. W. Barlow and D. R. Paul, *Polym. Eng. Sci.* **24**, 525 (1984).
160. P. F. Green and E. J. Kramer, *Macromolecules* **19**, 1108 (1986).
161. K. H. Dai, L. J. Norton, E. J. Kramer, *Macromolecules* **27**, 1949 (1994).
162. J. Washiyama, E. J. Kramer, C. F. Creton, and C.-Y. Hui, *Macromolecules* **27**, 2019 (1994).
163. R. Fayt, R. Jerome, and Ph. Teyssie, *J. Polym. Sci. Polym. Phys. Ed.* **20**, 2209 (1982).
164. R. Fayt, R. Jerome, and Ph. Teyssie, *Makromol. Chem.* **187**, 837 (1986).
165. D. W. Bartlett, J. W. Barlow, and D. R. Paul, *SPE Tech. Pap.* **27**, 487 (1981); *Mod. Plast.* **58**(12), 60 (1981).
166. R. Fayt, R. Jerome, and P. Teyssie, *J. Polym. Sci. Polym. Lett. Ed.* **19**, 79 (1981).
167. R. Fayt, R. Jerome, and Ph. Teyssie, *J. Polym. Sci. Polym. Lett. Ed.* **24**, 25 (1986).
168. N. C. Liu and W. E. Baker, *Adv. Polym. Technol.* **11**, 249 (1992).
169. M. Xanthos and S. S. Dagli, *Polym. Eng. Sci.* **31**, 929 (1991).
170. S. Cimmino, *Polym. Eng. Sci.* **24**, 48 (1984).
171. U.S. Pat. 4,174,358 (Oct. 30, 1979), B. N. Epstein (to du Pont).
172. R. J. M. Borggreve and R. Gaymans, *Polymer* **30**, 63 (1989).
173. S. Y. Hobbs and M. E. J. Dekkers, *J. Mater. Sci.* **24**, 1316 (1989).
174. V. J. Triacca, S. Ziaee, J. W. Barlow, H. Keskkula, and D. R. Paul, *Polymer* **32**, 1401 (1991).
175. D. R. Paul, *Adv. Polym. Blends Alloys Technol.* **4**, 80 (1993).
176. M. K. Akkapeddi, B. Van Buskirk, and J. H. Glans, *Adv. Polym. Blends Alloys Technol.* **4**, 87 (1993).
177. H.-J. Sue and A. F. Yee, *J. Mater. Sci.* **24**, 1447 (1989).
178. A. J. Oshinski, H. Keskkula, and D. R. Paul, *Polymer* **33**, 268 (1992).
179. I. Park, J. W. Barlow, and D. R. Paul, *J. Polym. Sci.: Part B: Polym. Phys.* **30**, 1021 (1992).
180. I. Park, J. W. Barlow, and D. R. Paul, *J. Polym. Sci. Polym. Chem. Ed.* **29**, 1329 (1991).
181. Y. Takeda and D. R. Paul, *J. Polym. Sci.: Part B: Polym. Phys.* **30**, 1273 (1992).
182. A. R. Padwa and R. E. Lavengood, *ACS Symp. Ser.* **33**, 600 (1992).
183. M. E. Stewart, S. E. George, R. L. Miller, and D. R. Paul, *Polym. Eng. Sci.* **33**, 675 (1993).
184. C.-T. Maa and F.-C. Chang, *J. Appl. Polym. Sci.* **49**, 913 (1993).
185. Eur. Pat. Appl. 81300266.4 (Jan. 21, 1981), W. H. Staas (to Rohm & Haas Co.).
186. Eur. Pat. Appl. 82306789.7 (Dec. 20, 1982), M. G. Waggoner (to E. I. du Pont de Nemours & Co., Inc.).
187. F. Ide and A. Hasegawa, *J. Appl. Polym. Sci.* **18**, 963 (1974).
188. U.S. Pats. 4,087,588 (May 2, 1975) and 4,087,587 (May 2, 1978), M. Shida, J. Barrington, Jr., S. Schaumburg, and R. Zeitlin (to Chemplex Corp.).
189. U.S. Pat. 3,873,643 (Mar. 25, 1985), W. Wu, L. Krebaum, and J. Machonis, Jr. (to Chemplex Co.).

190. L. Wielgolinski and J. Nangeroni, *Adv. Polym. Technol.* **3**(2), 99 (1983).
191. K. Eise, *Adv. Polym. Technol* **3**(2), 113 (1983).
192. M. Xanthos, ed., *Reactive Extrusion: Principles and Practice*, SPE, Brookfield, Conn., 1992.
193. P. Galli, *Macromol. Symp.* **7B**, 269 (1994).
194. D. R. Paul, in Ref. 77, Chapt. 1.
195. W. P. Gergen, S. Davison, and R. G. Lutz, in N. R. Legge, G. Holden, and H. E. Schroeder, eds., *Thermoplastic Elastomers—Research and Development*, Carl Hanser Verlag, Munich, Germany, 1986.
196. L. P. McMaster, *Adv. Chem. Ser.* **142**, 43 (1975).
197. T. K. Kwei and T. T. Wang, in Ref. 77, Chapt. 4.
198. P. G. DeGennes, *J. Chem. Phys.* **72**, 4756 (1980).
199. T. Hashimoto, J. Kumaki, and H. Kawai, *Macromolecules* **16**, 641 (1983).
200. T. Hashimoto, K. Sasaki, and H. Kawai, *Macromolecules* **17**, 2812 (1984).
201. K. Sasaki and T. Hashimoto, *Macromolecules* **17**, 2818 (1984).
202. G. R. Strobl, *Macromolecules* **18**, 558 (1985).
203. G. Ronca and T. P. Russell, *Macromolecules* **18**, 665 (1985).
204. H. van Oene, in Ref. 77, Chapt. 7.
205. A. P. Plochocki, in Ref. 77, Chapt. 21.
206. J. L. White, in Ref. 19, p. 413.
207. J. L. White, in Ref. 3, p. 413.
208. C. D. Han, *Multiphase Flow in Polymer Processing*, Academic Press, Inc., New York, 1981.
209. C. J. Nelson, G. N. Avgeropoulos, F. C. Weissert, and G. G. A. Bohm, *Angew. Makromol. Chem.* **60/61**, 49 (1977).
210. J. J. Elmendorp and R. J. Maalcke, *Polym. Eng. Sci.* **25**, 1041 (1985).
211. C. M. Roland and G. G. A. Bohm, *J. Polym. Sci. Polym. Phys. Ed.* **22**, 79 (1984).
212. E. N. Kresge, in Ref. 77, Chapt. 20.
213. D. R. Paul, in Ref. 77, Chapt. 12.
214. R. Fayt, P. Hadjiandreou, and P. Teyssie, *J. Polym. Sci. Polym. Chem. Ed.* **23**, 337 (1985).
215. T. W. Cheng, H. Keskkula, and D. R. Paul, *J. Appl. Polym. Sci.* **45**, 1245 (1992).
216. M. T. Shaw, in Ref. 3, p. 37.
217. J. R. White and E. L. Thomas, *Rubber Chem. Technol.* **57**, 457 (1984).
218. H. Keskkula and P. A. Traylor, *Polymer* **19**, 465 (1978).
219. K. Kato, *J. Electron Microsc. Jpn.* **14**, 219 (1965).
220. H. Keskkula, in Ref. 19, p. 328.
221. J. S. Trent, J. I. Scheinbeim, and P. R. Couchman, *Macromolecules* **16**, 589 (1983).
222. R. Vitali and E. Montani, *Polymer* **21**, 1220 (1980).
223. D. E. Morel and D. T. Grubb, *Polym. Commun.* **25**, 68 (1984).
224. D. Montezinos, B. G. Wells, and J. L. Burns, *J. Polym. Sci. Polym. Lett. Ed.* **23**, 421 (1985).
225. M. E. J. Dekkers, S. Y. Hobbs, and V. H. Watkins, *Polymer* **32**, 2150 (1991).
226. S. Y. Hobbs, M. E. J. Dekkers, and V. H. Watkins, *Polymer* **29**, 1598 (1988).
227. S. Y. Hobbs, M. E. J. Dekkers, and V. H. Watkins, *J. Mat. Sci.* **24**, 2025 (1989).
228. B. Majumdar, H. Keskkula, and D. R. Paul, *Polymer* **35**, 1386 (1994).
229. B. S. Lombardo, H. Keskkula, and D. R. Paul, *J. Appl. Polym. Sci.* **54**, 1697 (1994).
230. Y. Trevoort-Engelen, and J. Gisbergen, *Polym. Comm.* **32**, 261 (1991).
231. T. K. Kwei, H. L. Frisch, W. Radigan, and S. Vogel, *Macromolecules* **10**, 157 (1977).
232. H.-J. Sue, B. L. Garcia-Meitin, and C. C. Garrison, *J. Appl. Polym. Sci.* **29**, 1623 (1991).
233. A. Echte in C. K. Riew, ed., *Adv. Chem. Ser.* **222**, Am. Chem. Soc., 15 (1989).

234. H. Keskkula and D. R. Paul, in A. A. Collyer, ed., *Rubber-Toughened Engineering Polymers*, Chapman and Hall, London, 1994, pp. 136–209.

235. A. Yee, in *Toughening of Plastics II*, Paper No. 19, Plastics and Rubber Institute, London, 1985.

236. *Mod. Plast.* **59**, 77 (Jan. 1982); **63**, 59 (Jan. 1986); **67**, 99 (Jan. 1990); **71**, 73 (Jan. 1994).

237. G. Graft, ed., *Modern Plastics Encyclopedia '94*, Vol. 70, No. 12, McGraw-Hill Book Co., Inc., New York, 1993.

238. Z. Kromolicki, in Ref. 235, Paper No. 20.

239. H. Keskkula, D. R. Paul, and A. E. Platt, in J. J. McKetta, ed., *Encyclopedia of Chemical Processing and Design*, Vol. 41, Marcel Dekker, New York, 1992, pp. 1–31.

240. M. E. Soderquist and R. P. Dion, in J. I. Kroschwitz, ed., *Encyclopedia of Polymer Science and Engineering*, Wiley-Interscience, New York, 1989, pp. 88–97.

241. C. B. Bucknall, *Toughened Plastics*, Applied Science Publishers, Ltd., London, 1977.

242. J. P. Fischer, *Angew. Makromol. Chem.* **33**, 35 (1973).

243. H. Keskkula, *Plast. Rubber Mater. Appl.* **16**(5), 66 (1979).

244. A. L. Bull and G. Holden, *J. Elastomers Plast.* **9**, 281 (1977).

245. R. A. Bubeck, D. J. Buckley, Jr., E. J. Kramer, and H. R. Brown, *J. Mat. Sci.* **26**, 6249 (1991).

246. L. Morbitzer, D. Kranz, G. Humme, and K. H. Ott, *J. Appl. Polym. Sci.* **20**, 2691 (1976).

247. Y. Aoki, *Macromolecules* **20**, 2208 (1987).

248. D. Kranz, L. Morbitzer, K. H. Ott, and R. Casper, *Angew. Makromol. Chem.* 58/59, 213 (1977).

249. T. Ricco, A. Pavan, and F. Danusso, *Polymer* **16**, 685 (1975).

250. M. Rink, T. Ricco, W. Lubert, and A. Pavan, *J. Appl. Polym. Sci.* **22**, 429 (1978).

251. G. Riess and J. L. Locatelli, *Adv. Chem. Ser.* **142**, 186 (1975).

252. H. Breuer, J. Stabenow, and F. Haaf, *Toughening of Plastics*, Plastics and Rubber Institute, London, 1978, Paper No. 13.

253. C. B. Bucknall and W. W. Stevens, *J. Mater. Sci.* **15**, 2950 (1980).

254. A. M. Donald and E. J. Kramer, *J. Mater. Sci.* **17**, 1765 (1982).

255. C. B. Bucknall, in Ref. 77, Chapt. 14.

256. L. V. Newmann and J. G. Williams, *J. Mater. Sci.* **15**, 773 (1980).

257. J. T. Lutz, Jr., and D. L. Dunkelberger, *Impact Modifiers for PVC: The History and Practice*, John Wiley & Sons, Inc., New York, 1992.

258. J. F. Gabbett, in R. Juran, ed., *Modern Plastics Encyclopedia '89*, McGraw-Hill Book Co., Inc., Vol. 65, No. 11, (1988), p. 118.

259. D. Hardt in R. Ceresa, ed., *Block and Graft Polymerization*, Vol. 2, John Wiley & Sons, Inc., New York, 1972, pp. 315–373.

260. C. Tremblay and R. E. Prud'homme, *J. Polym. Sci. Polym. Phys. Ed.* **22**, 1857 (1984).

261. P. G. Faulkner, *J. Macromol. Sci. Phys.* **11**, 251 (1975).

262. E. B. Rabinovitch and J. W. Summers, *SPE Ann. Tech. Conf.* **38**, 403 (1980).

263. H. Breuer, F. Haaf, and J. Stabenow, *J. Macromol. Sci. Phys.* **14**, 387 (1977).

264. K. P. Richter and G. Goldbach in Ref. 235, Paper No. 23.

265. G. D. Cooper, G. F. Lee, A. Katchman, and C. P. Shank, *Mater. Tech.*, 12 (Spring 1981).

266. C. M. Rimnac, R. W. Hertzberg, and J. A. Manson, *Polymer* **23**, 1977 (1982).

267. C. B. Bucknall, D. Clayton, and W. E. Keast, *J. Mater. Sci.* **7**, 1443 (1972).

268. F. C. Stehling, T. Huff, C. S. Speed, and G. Wissler, *J. Appl. Polym. Sci.* **26**, 2693 (1981).

269. B. Z. Jang, D. R. Uhlmann, and J. B. Vander Sande, *Polym. Eng. Sci.* **25**, 643 (1985).

270. B. Z. Jang, D. R. Uhlmann, and J. B. Vander Sande, *J. Appl. Polym. Sci.* **30**, 2485 (1985).

271. D. Yang, B. Zhang, Y. Yang, Z. Fang, G. Sun, and Z. Feng, *Polym. Eng. Sci.* **24**, 612 (1984).
272. A. K. Gupta and S. N. Purwar, *J. Appl. Polym. Sci.* **29**, 3513 (1984).
273. A. Ghijsels, N. Groesbeek, and C. W. Yip, *Polymer* **23**, 1913 (1982).
274. A. J. Tinker, *Polymer Comm.* **25**, 325 (1984).
275. E. Martuscelli, C. Silvestre, and L. Bianchi, *Polymer* **24**, 1458 (1983).
276. L. Bianchi, S. Cimmino, A. Forte, R. Greco, E. Martuscelli, F. Riva, and C. Silvestre, *J. Mater. Sci.* **20**, 895 (1985).
277. F. Ramsteiner, G. Kanig, W. Heckmann, and W. Gruber, *Polymer* **24**, 365 (1983).
278. K. C. Dao, *J. Appl. Polym. Sci.* **27**, 4799 (1982).
279. K. C. Dao, *Polymer* **25**, 1527 (1984).
280. A. Y. Coran, R. Patel, and D. Williams-Headd, *Rubber Chem. Tech.* **58**, 1014 (1985).
281. J. Ito, K. Matani, and Y. Mizutani, *J. Appl. Polym. Sci.* **29**, 75 (1984).
282. W.-J. Ho and R. Salovey, *Polym. Eng. Sci.* **21**, 839 (1981).
283. B. Z. Jang, D. R. Uhlmann, and J. B. Vander Sande, *J. Appl. Polym. Sci.* **29**, 4377 (1984).
284. C. B. Bucknall and C. J. Page, *J. Mater. Sci.* **17**, 808 (1982).
285. Ref. 269, p. 98.
286. M. L. Kohan, *Nylon Plastics*, SPE Monograph, John Wiley & Sons, Inc., New York, 1973.
287. H. Keskkula and D. R. Paul, in M. I. Kohan, ed., *Nylon Plastics*, Carl Hanser Verlag, Munich, Germany, 1995, pp. 412–434.
288. C. Cimmino and co-workers, *Polym. Eng. Sci.* **25**, 193 (1985).
289. A. J. Oshinski, H. Keskkula, and D. R. Paul, *Polymer* **33**, 284 (1992).
290. S. Wu, *Polymer* **26**, 1855 (1985).
291. A. J. Oostenbrink, K. Dijkstra, A. van de Wal, and R. J. Gaymans, *Proceedings of the PRI Conference on Deformation, Yield and Fracture of Polymers*, Plastics & Rubber Institute, Cambridge, U.K., Apr. 1991, pp. 50/1–50/4.
292. R. E. Lavengood and F. M. Silver, *Soc. Plast. Eng., ANTEC*, **33**, 1369 (1987).
293. U.S. Pat. 4,713,415 (Dec. 15, 1987), R. E. Lavengood, A. R. Padwa, and A. F. Harris (to Monsanto).
294. S. Wu, *J. Polym. Sci. Polym. Phys. Ed.* **21**, 699 (1983).
295. D. Neuray and K.-H. Ott, *Angew. Makromol. Chem.* **98**, 213 (1981).
296. F. Ramsteiner and W. Heckman, *Polym. Comm.* **26**, 199 (1985).
297. F.-C. Chang, J.-S. Wu, and L.-H. Chu, *J. Appl. Polym. Sci.* **44**, 491 (1992).
298. T. W. Cheng, H. Keskkula, and D. R. Paul, *J. Appl. Polym. Sci.* **45**, 531 (1992).
299. T.-W. Cheng, H. Keskkula, and D. R. Paul, *Polymer* **33**, 1606 (1992).
300. J.-S. Wu, S.-C. Shen, and F.-C. Chang, *J. Appl. Polym. Sci.* **50**, 1379 (1993).
301. A. J. Brady, H. Keskkula, and D. R. Paul, *Polymer* **35**, 3665 (1994).
302. S. Y. Hobbs, and M. E. J. Dekkers, V. H. Watkins, *J. Mater. Sci.* **23**, 1219 (1988).
303. R. J. McCready, *Soc. Plast. Eng., ANTEC* **35**, 1828 (1989).
304. F. J. McGarry and A. M. Willner, *Research Report R68-8*, School of Engineering, Massachusetts Institute of Technology (MIT), Cambridge, Mass., 1968.
305. L. T. Manzione, J. K. Gillham, and C. A. McPherson, *J. Appl. Polym. Sci.* **26**, 907 (1981).
306. J. A. Sayre, R. A. Assink, and R. R. Lagasse, *Polymer* **22**, 87 (1981).
307. J. K. Gillham, *SPE Ann. Tech. Conf.* **38**, 268 (1980).
308. G. Levita, A. Marchetti, and E. Butta, *Polymer* **26**, 1110 (1985).
309. M. Ochi and J. P. Bell, *J. Appl. Polym. Sci.* **29**, 1381 (1984).
310. A. J. Kinloch, S. J. Shaw, D. A. Tod, and D. L. Hunston, *Polymer* **24**, 1341 (1983).
311. A. J. Kinloch, S. J. Shaw, and D. L. Hunston, *Polymer* **24**, 1355 (1983).
312. H.-J. Sue, *Polym. Eng. Sci.* **31**, 270, 1991.

313. *Chem. Week*, **132**(9), 72–76 (Mar. 2, 1983).
314. M. T. Shaw, *Polym. Eng. Sci.* **22**, 115 (1982).
315. L. M. Robeson, *Polym. Eng. Sci.* **24**, 587 (1984).
316. L. A. Utracki, *Polym. Eng. Sci.* **22**, 1166 (1982).
317. G. H. Hofmann, in Ref. 3, p. 117.
318. R. A. Emmet, *Ind. Eng. Chem.* **36**, 730 (1944).
319. E. W. Anderson and co-workers, *Adv. Chem. Ser.* **176**, 413 (1979).
320. U.S. Pat. 3,644,577 (Feb. 22, 1972), Y. C. Lee and Q. A. Trementozzi (to Monsanto Co.).
321. M. B. Djordjevic and R. S. Porter, *Polym. Eng. Sci.* **23**, 650 (1983).
322. U.S. Pat. 3,383,435 (May 14, 1968), E. P. Cizek (to General Electric Co.).
323. U.S. Pat. 3,010,936 (Nov. 28, 1961), H. H. Irwin (to Borg-Warner Corp.).
324. R. J. Slocomb, *J. Polym. Sci.* **26**, 9 (1957).
325. U.S. Pat. 3,253,060 (May 25, 1966), F. F. Koblitz, R. G. Petrella, A. A. Dukert, and A. Christofas (to Pennwalt Chemical Co.).
326. A. S. Wood, *Mod. Plast.* **56**, 44 (1979).
327. H. M. Li and A. H. Wong, in Ref. 19, p. 395.
328. Y. J. Shur and B. G. Ranby, *J. Appl. Polym. Sci.* **20**, 3121 (1976).
329. D. G. Walsh and J. G. McKeown, *Polymer* **21**, 1330 (1980).
330. G. R. Forger, *Mater. Eng.* **85**, 44 (1977).
331. S. Krause, in Ref. 77, Chapt. 2.
332. C. F. Hammer, in Ref. 77, Chapt. 17.
333. J. D. Keitz, J. W. Barlow, and D. R. Paul, *J. Appl. Polym. Sci.* **29**, 3131 (1984).
334. U.S. Pat. 3,636,140 (Jan. 18, 1972), A. F. Ingulli and H. L. Alter (to Uniroyal, Inc.).
335. S. R. Murff, J. W. Barlow, and D. R. Paul, *J. Appl. Polym. Sci.* **29**, 3231 (1984).
336. F. Pilati, E. Marianucci, and C. Berti, *J. Appl. Polym. Sci.* **30**, 1267 (1985).
337. L. D'Orazio, C. Greco, C. Mancarella, E. Martuscelli, G. Ragosta, and C. Silvestre, *Polym. Eng. Sci.* **22**, 536 (1982).
338. W. K. Fisher, *Mod. Plast.* **51**, 116 (1974).
339. G. R. Forger, *Plast. World* **40**, 28 (1982).
340. R. Martino, *Mod. Plast.* **58**, 42 (1981).
341. A. Y. Coran and R. Patel, *Rubber Chem. Technol.* **53**, 141 (1980).
342. L. A. Utracki, *Polym. Plast. Technol. Eng.* **22**(1), 27 (1984).
343. J. W. Barlow and D. R. Paul, *Polym. Eng. Sci.* **21**, 985 (1981).
344. J. R. Stell, J. W. Barlow, and D. R. Paul, *Polym. Eng. Sci.* **16**, 496 (1976).
345. C. R. Lindsey, J. W. Barlow, and D. R. Paul, *J. Appl. Polym. Sci.* **26**, 1 (1981).
346. W. F. Hwang, D. R. Wiff, C. Verschoore, G. E. Price, T. E. Helminiak, and W. W. Adams, *Polym. Eng. Sci.* **23**, 784 (1983).
347. M. Takayanagi and K. Goto, *Polym. Bull.* **13**, 35 (1985).
348. C. Carfagna, ed., *Liquid Crystalline Polymers*, Pergamon Press, Oxford, U.K., 1994.
349. O. Niwa, M. Hikita, and T. Tamamura, *Makromol. Chem. Rapid Comm.* **6**, 375 (1985).
350. M. E. Galvin and G. E. Wnek, *Polym. Prepr. Am. Chem. Soc. Div Polym. Chem.* **25**, 229 (1984).
351. L. W. Shacklette, C. C. Han, and M. H. Luly, *Synth. Met.* **57**, 3532 (1993).
352. J. E. Osterholm and co-workers, *ACS Polym. Preprints* **35**, 244 (1994).

*General References*

J. A. Manson and L. H. Sperling, *Polymer Blends and Composites*, Plenum Press, New York, 1976.
D. R. Paul and S. Newman, eds., Polymer Blends, Vols. I and II, Academic Press, Inc., New York, 1978.

O. Olabisi, L. M. Robeson, and M. T. Shaw, *Polymer–Polymer Miscibility*, Academic Press, Inc., New York, 1979.

K. Solc, ed., *Polymer Compatibility and Incompatibility: Principles and Practice*, MMI Press Symposium Series, Vol. 2, Harwood Academic Publishers GmbH, New York, 1982.

D. J. Walsh, J. S. Higgins, and A. Maconnachie, eds., *Polymer Blends and Mixtures*, NATO ASI Series, Series E, Applied Sciences, No. 89, Martinus Nijhoff Publishers, Dordrecht, the Netherlands, 1985.

L. A. Utracki, *Polymer Alloys and Blends*, Hanser, Munich, Germany, 1989.

B. M. Culbertson, ed., *Multiphase Macromolecular Systems*, Contemporary Topics in *Polymer Science*, Vol. 6, Plenum Publishing Corp., New York, 1989.

M. M. Coleman, J. F. Graf, and P. C. Painter, *Specific Interactions and the Miscibility of Polymer Blends*, Technomic, Lancaster, Pa., 1991.

I. S. Miles and S. Rostami, eds., *Multicomponent Polymer Systems*, Longman, Essex, U.K., 1992.

A. A. Collyer, ed., *Rubber-Toughened Engineering Plastics*, Chapman and Hall, London, 1994.

H. KESKKULA
D. R. PAUL
J. W. BARLOW
University of Texas at Austin

# POLYMERS

Polymers are very large molecules made by covalently binding many smaller molecules. The word polymer is derived from the Greek *poly* (many) and *meros* (part). The size of polymer molecules imparts many interesting and useful properties not shared by low molecular weight materials. Polymers are the fundamental materials of plastics, rubbers and most fibers, and surface coatings and adhesives, and as such are essential to modern society. Also, many important constituents of living organisms, eg, proteins (qv) and cellulose (qv), are biopolymers (qv).

## Classification and Nomenclature

Polymers were initially classified according to their response to temperature. Those that are softened (plasticized) reversibly by heat are known as thermoplastics. By analogy, wax behaves as a thermoplastic. Others, though they might initially be liquid or soften once upon heating, undergo a curing (setting) reaction that solidifies them, and further heating leads only to degradation. These are known as thermosets. Again by analogy, an egg behaves as a thermoset. The ability of polymers to soften and flow at least once is one of their most valuable assets, as it allows them to be formed into complex shapes easily and inexpensively.

In general, polymers are formed by two types of reactions: condensation and addition. The formation of a polyester by polycondensation may be illustrated as follows.

$$x \text{ HOROH} + x \text{ HOOCR}'\text{COOH} \longrightarrow \text{H} \text{-}(\text{ORO} - \overset{\overset{\text{O}}{\|}}{\text{C}}\text{R}'\overset{\overset{\text{O}}{\|}}{\text{C}}\text{-})_{\overline{x}}\text{OH} + (2x - 1) \text{ H}_2\text{O}$$

diol                  diacid                              polyester

In the polyester formula shown, parentheses enclose the repeating unit. The quantity $x$ is the degree of polymerization, sometimes also called the chain length, the number of repeating units strung together like identical beads on a string. Neglecting the ends of the molecule, which is usually justified for large $x$, the molecular weight $M$ of the polymer molecule is given by $M = mx$, where $m$ is the molecular weight of the repeating unit. Since $x$ can easily be in the thousands, it is not surprising that the term macromolecules is also used to describe these materials.

The ester linkage in the repeating unit characterizes polyesters. R and R' represent portions of the monomer molecule that do not participate in the polymerization. They may vary widely, giving rise to many different polyesters. Poly(ethylene terephthalate) (PET), made from ethylene glycol (R = (CH$_2$)$_2$) and terephthalic acid (R' = $-\bigcirc-$ ), is familiar in the form of soda bottles, recording tape, and polyester fiber (see FIBERS, POLYESTER).

Another common polycondensation involves reaction of diamines and diacids to form polyamides, commonly called nylons:

$$x \text{ H}_2\text{NRNH}_2 + x \text{ HOOCR}'\text{COOH} \longrightarrow \text{H} \text{-}(\text{NHRNH} - \overset{\overset{\text{O}}{\|}}{\text{C}}\text{R}'\overset{\overset{\text{O}}{\|}}{\text{C}}\text{-})_{\overline{x}}\text{OH} + (2x - 1) \text{ H}_2\text{O}$$

diamine                diacid                             polyamide

The amide linkage characterizes nylons. In the first commercial nylon, nylon-6,6, R = (CH$_2$)$_6$ and R' = (CH$_2$)$_4$. Nylon-6,6 is familiar as a textile fiber (nylon stockings) and a molded plastic (see POLYAMIDES).

The two complementary functional groups that react to form condensation polymers may also occur in a single monomer, eg, a hydroxy acid, HO–R–COOH, or an amino acid, H$_2$N–R–COOH. In some cases, such monomers self-condense to a cyclic structure, which is what actually polymerizes. For example, $\epsilon$-caprolactam (**1**) can be thought of as the self-condensation product of an amino acid. Caprolactam undergoes a ring-opening polymerization to form another

$$\text{H}_2\text{N} \text{-}(\text{CH}_2\text{-})_{\overline{5}}\text{COOH} \longrightarrow \quad \overset{\displaystyle \text{HN-C}{\overset{\text{O}}{\diagup}}}{\bigcirc}$$

(**1**)

important nylon, nylon-6 (see Caprolactam). Even though no water is eliminated in the actual polymerization step, the polymer is usually considered a condensation polymer. Table 1 illustrates some other important condensation polymers.

$$x\ (\mathbf{1}) \longrightarrow \ -\!\!\!\left(\text{NH(CH}_2)_5\overset{\overset{\displaystyle O}{\displaystyle \|}}{\text{C}}\right)_{\!\!\overline{x}}$$

Addition or chain-growth polymerization involves the opening of a double bond to form new bonds with adjacent monomers, as typified by the polymerization of ethylene to polyethylene:

$$x\ \text{H}_2\text{C}\!=\!\text{CH}_2 \longrightarrow \ -\!\!\!\left(\text{CH}_2\!-\!\text{CH}_2\right)_{\!\!\overline{x}}$$

Because no molecule is split out, the molecular weight of the repeating unit is identical to that of the monomer. Vinyl monomers, $\text{H}_2\text{C}\!=\!\text{CHR}$ (Table 2) undergo addition polymerization to form many important and familiar polymers. Diene (two double bonds) monomers also undergo addition polymerization. Normally, one double bond remains, leaving an unsaturated polymer, with one double bond per repeating unit. These double bonds provide sites for subsequent reaction, eg, vulcanization.

In terms of molecular structure, there are three principal categories of polymers, illustrated schematically in Figure 1. If each monomer is difunctional, that is, can react with other monomers at two points, a linear polymer is formed. All the examples given above are linear polymers. Polymers that contain two different repeating units, say A and B, are known as copolymers (qv). A linear polymer with a random (AABBABAAABABB) arrangement of the repeating units is a random or statistical copolymer, or just copolymer. It is termed poly($A-co-B$), with the primary constituent listed first. A molecule in which the two repeating units are arranged in long, contiguous blocks ($[A]_x[B]_y$) is a block ($b$) copolymer, poly($A-b-B$).

A few points of tri- or higher functionality introduced along the polymer chains, either intentionally or through side reactions, give a branched polymer. A branched structure with the backbone consisting of one repeating unit (A) and the branches of another (B), is a graft ($g$) copolymer, poly($A-g-B$). Dendrimers are a more recent development. They are molecules that branch repeatedly as they grow outward from a central core (1–3).

As the length and frequency of branches increase, they may ultimately reach from chain to chain. If all the chains are connected together, a cross-linked or network polymer is formed. Cross-links may be built in during the polymerization reaction by incorporation of sufficient tri- or higher functional monomers, or may be created chemically or by radiation between previously formed linear or branched molecules (curing or vulcanization). For example, a liquid epoxy (Table 1) oligomer (low molecular weight polymer) with $x \approx 6-8$ is cured to a cross-linked solid by reaction of the hydroxyl and terminal epoxide groups with a diamine or acid anhydride. In a fully cross-linked polymer, all the atoms are connected to one another by covalent bonds, so the entire macroscopic polymer mass is literally a single molecule. Thus, the cross-linked polyester in a bowling ball has a molecular weight on the order of $10^{27}$ g/mol.

**Table 1. Some Commercial Condensation Polymers**

| Polymer | Monomers | Repeating unit | Eliminated molecule[a] |
|---|---|---|---|
| polycarbonate | 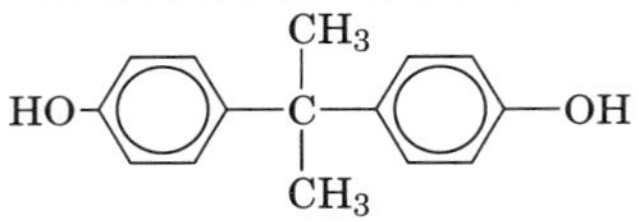 bisphenol A[b] | 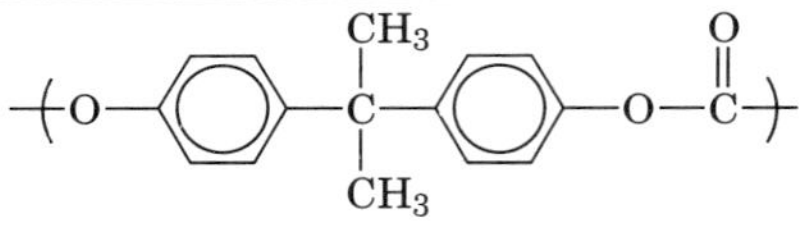 | HCl |
| epoxy resin | 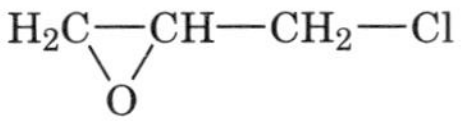 epichlorohydrin[c] | 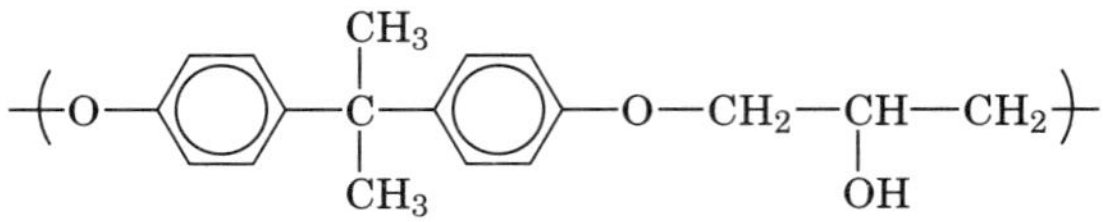 | HCl |
| polydimethylsiloxane (silicone rubber) | 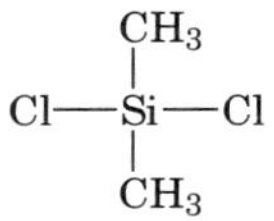 dimethyl dichlorosilane[d] | 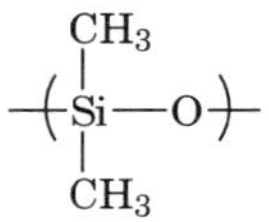 | HCl |

| Polymer | Monomer structure | Repeat unit structure | Byproduct |
|---|---|---|---|
| poly(phenylene sulfide) | Cl—⟨⟩—Cl + Na₂S | —⟨⟩—S— | NaCl |
| polysulfone[e] | 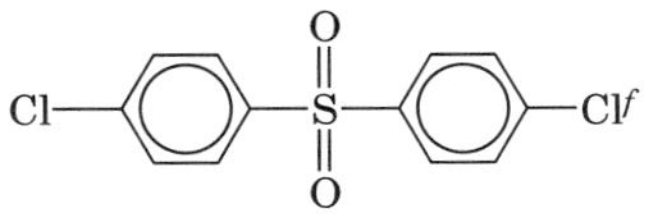 | 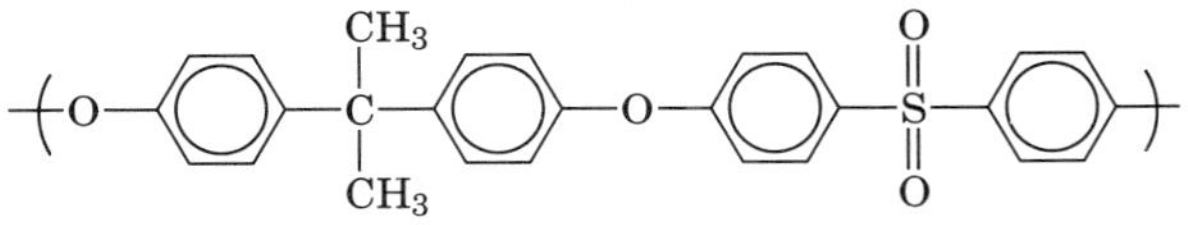 | NaCl |
| polyimide | dianhydride[g] | | H₂O |

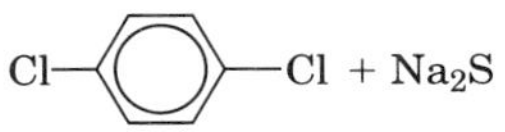

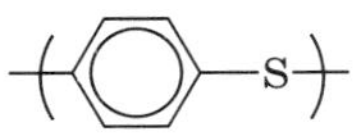

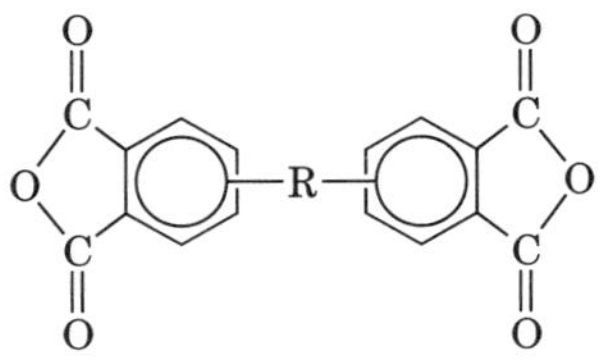

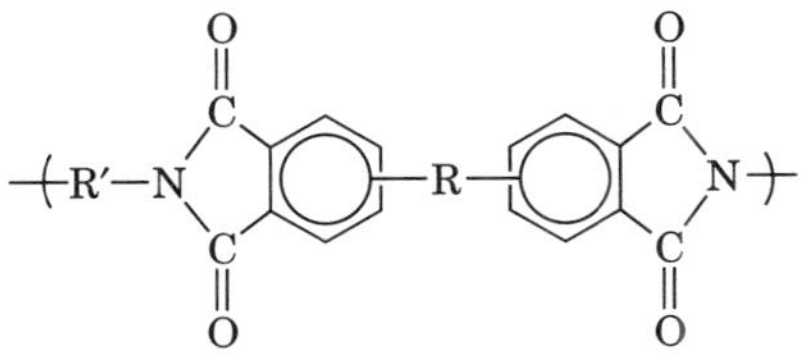

885

**Table 2. Some Vinyl Monomers, CH$_2$=CHR**

| R | Common name | Polymer repeating unit | *Encyclopedia* reference |
|---|---|---|---|
| Cl | vinyl chloride | $+CH_2-CHCl+$ | VINYL POLYMERS, VINYL CHLORIDE AND PVC |
| C$_6$H$_5$ | styrene | $+CH_2-CH(C_6H_5)+$ | STYRENE PLASTICS |
| CH$_3$ | propylene | $+CH_2-CH(CH_3)+$ | OLEFIN POLYMERS, POLYPROPYLENE |
| C≡N | acrylonitrile | $+CH_2-CH_2CN+$ | ACRYLONITRILE POLYMERS |
| COOH | acrylic acid | $+CH_2-CH+$ <br> $\quad\quad\;$ COOH | ACRYLIC ACID AND DERIVATIVES |
| OOCCH$_3$ | vinyl acetate | $+CH_2-CH+$ <br> CH$_3$COO | VINYL POLYMERS, POLY(VINYL ACETATE) |
| CONH$_2$ | acrylamide | $+CH_2-CH+$ <br> $\quad\;$ O=C—NH$_2$ | ACRYLAMIDE POLYMERS |

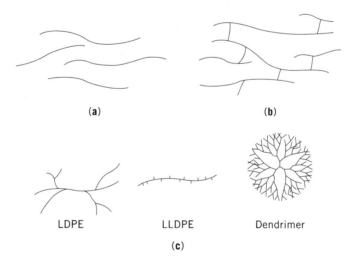

**Fig. 1.** Schematic diagram of polymer structures: (**a**) linear; (**b**) cross-linked; and (**c**) branched, where LDPE = low density polyethylene and LLDPE = linear low density polyethylene.

## Structure and Properties

Various levels of structure ultimately determine the properties of a polymer. The characterization of structure and how it influences properties is outlined in the following.

**Molecular Weights.** With the exception of some naturally occurring polymers, all linear and branched polymers consist of molecules with a distribution of molecular weights. Two average molecular weights are commonly defined; the number-average, $\overline{M}_n$, and the weight-average, $\overline{M}_w$:

$$\overline{M}_n = \frac{W}{N} = \frac{\sum\limits_{x=1}^{\infty} n_x M_x}{\sum\limits_{x=1}^{\infty} n_x} \qquad\qquad \overline{M}_w = \frac{\sum w_x M_x}{\sum w_x} = \frac{\sum n_x M_x^2}{\sum n_x M_x}$$

where $W$ = total sample weight = $\sum\limits_{x=1}^{\infty} w_x = \sum\limits_{x=1}^{\infty} n_x M_x$

$w_x$ = total weight of $x$-mer

$N$ = total number of moles in the sample (of all sizes) = $\sum\limits_{x=1}^{\infty} n_x$

$n_x$ = number of moles of $x$-mer

$M_x$ = molecular weight of $x$-mer

It may be shown that $\overline{M}_w \geq \overline{M}_n$. The two are equal only for a monodisperse material, in which all molecules are the same size. The ratio $\overline{M}_w/\overline{M}_n$ is known as the polydispersity index and is a measure of the breadth of the molecular weight distribution. Values range from about 1.02 for carefully fractionated samples or certain polymers produced by anionic polymerization, to 20 or more for some commercial polyethylenes.

Colligative property techniques that measure the moles of polymer in solution give $\overline{M}_n$. These include membrane and vapor-pressure osmometry and freezing-point depression. Procedures which, in effect, determine the mass of polymer at each size level give $\overline{M}_w$. These include light scattering and sedimentation in an ultracentrifuge. Another technique, intrinsic viscosity, makes use of the large increase in viscosity caused by relatively small amounts of polymeric solute. Intrinsic viscosity gives yet another average molecular weight, $\overline{M}_v$ (viscosity-average), which is between $\overline{M}_n$ and $\overline{M}_w$, but is closer to the latter. Unlike the other techniques mentioned, it must be calibrated with monodisperse samples of known molecular weight. Still, because much calibration data are available in the literature (4), the necessary equipment is inexpensive, and the measurements are precise, straightforward, and rapid, it is used frequently.

In the 1990s, most molecular weight characterization is done by size-exclusion chromatography (sec) (5), also known as gel-permeation chromatography (gpc). Sec makes use of a column packed with particles of a porous substrate, most commonly a cross-linked polystyrene gel (hence gpc). Solvent is pumped through the column at a constant rate. A small amount of polymer solution is injected ahead of the column. The solvent flow carries the polymer through the column. The smaller molecules have easy access to the substrate pores and diffuse in and out, following a circuitous route as they progress through the column. Larger molecules cannot fit into the smaller pores and are swept more directly through the interstices between the substrate particles. Thus, a separation is obtained, the largest molecules leaving the column first, followed by successively smaller ones. A detector, for example a differential refractometer, records the concentration of the solution leaving the column. Until recently, calibration with monodisperse samples of known molecular weight was necessary. However, as of 1995, a laser light-scattering photometer may be added to the differential re-fractometer, giving an on-line $\overline{M}_w$ determination and eliminating the need for calibration.

Size-exclusion chromatography (sec) easily and rapidly gives the complete molecular weight distribution and any desired average (6). Thus, it has become the technique of choice for determining molecular weights despite its relatively high initial cost.

**Secondary Bonding.**   The atoms in a polymer molecule are held together by primary covalent bonds. Linear and branched chains are held together by secondary bonds: hydrogen bonds, dipole interactions, and dispersion or van der Waal's forces. By copolymerization with minor amounts of acrylic ($CH_2{=}CHCOOH$) or methacrylic acid followed by neutralization, ionic bond-ing can also be introduced between chains. Such polymers are known as iono-mers (qv).

$$\sim\!COOH + M(OH)_2 + HOOC\sim \longrightarrow \sim COO]^{-+}[M]^{+-}[OOC\sim + 2\,H_2O$$

Secondary bonds are considerably weaker than the primary covalent bonds. When a linear or branched polymer is heated, the dissociation energies of the secondary bonds are exceeded long before the primary covalent bonds are broken, freeing up the individual chains to flow under stress. When the material is cooled, the secondary bonds reform. Thus, linear and branched polymers are generally thermoplastic. On the other hand, cross-links contain primary covalent bonds like those that bond the atoms in the main chains. When a cross-linked polymer is heated sufficiently, these primary covalent bonds fail randomly, and the material degrades. Therefore, cross-linked polymers are thermosets. There are a few exceptions such as cellulose and polyacrylonitrile. Though linear, these polymers are not thermoplastic because the extensive secondary bonds make up for in quantity what they lack in quality.

Similarly, polymers dissolve when a solvent penetrates the mass and re-places the interchain secondary bonds with chain-solvent secondary bonds, sep-arating the individual chains. This cannot happen when the chains are held together by primary covalent cross-links. Thus, linear and branched polymers dissolve in appropriate solvents, whereas cross-linked polymers are insoluble, although they may be swelled considerably by absorbed solvent. The extent of swelling is inversely related to the degree of cross-linking.

**Stereoisomerism.**   Vinyl monomers, $CH_2{=}CHR$, generally polymerize in a head-to-tail fashion, placing the R group on every other carbon atom in the chain backbone. If a chain is conceptually stretched out, the carbon atoms in the backbone will lie in a plane (Fig. 2). The arrangement in which the R groups are all on one side of that plane is the isotactic stereoisomer. Regular alternation of the R groups from side to side is the syndiotactic form. Random placement of the R groups is the atactic (without order) polymer. Stereoisomers are formed during polymerization, and cannot be altered subsequently by rotation about the bonds. It is important to note that isotactic and syndiotactic chains are regular, whereas atactic chains are irregular (7).

**Crystallinity.**   Crystals are an ordered, regular arrangement of units in a repeating, three-dimensional lattice structure. Small molecules, which in the

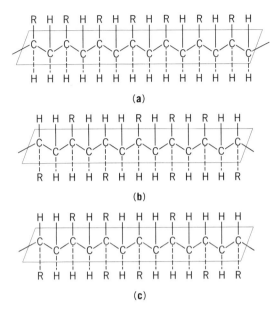

**Fig. 2.** Stereoisomerism in vinyl polymers: (**a**) isotactic; (**b**) syndiotactic; and (**c**) atactic.

liquid state have three-dimensional mobility, crystallize readily when cooled. It is not so easy for polymers, because a repeating unit cannot move independently of its neighbors in the chain. Nevertheless, some polymers can and do crystallize, though never completely. Not surprisingly, a regular chain structure is required if the chains are to fit into a regular crystal lattice. Thus, isotactic and syndiotactic polypropylenes crystallize, but atactic polypropylene does not. Similarly, branches protruding from a chain sterically inhibit crystallization in their vicinity.

Polymer crystals most commonly take the form of folded-chain lamellae. Figure 3 sketches single polymer crystals grown from dilute solution and illustrates two possible modes of chain re-entry. Similar structures exist in bulk-crystallized polymers, although the lamellae are usually thicker. Individual lamellae are held together by tie molecules that pass irregularly between lamellae. This explains why it is difficult to obtain a completely crystalline polymer.

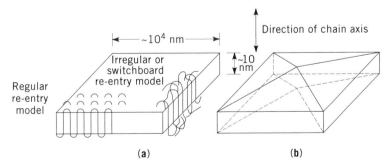

**Fig. 3.** Polymer single crystals: (**a**) flat lamellae; and (**b**) pyramidal lamellae. Two concepts of chain re-entry are illustrated (6).

Tie molecules and material in the folds at the lamellae surfaces cannot readily fit into a lattice.

In bulk-crystallized polymers, lamellae are often organized into spherulites, spherical structures which grow outward from a point of nucleation, typically to about 0.01 mm in diameter. Spherulites are in some ways similar to the grain structure in metals. They can make a polymer brittle and also reduce transparency.

Linear polyethylene fibers containing extended-chain crystals, in which the chains are arranged parallel to one another over great distances with a minimum of folding, are produced by using a flow field to align the molecules in solution just prior to crystallization. The resulting fibers are then drawn (stretched) to enhance orientation further. These fibers (Spectra, by AlliedSignal) are impressively strong, particularly on a strength-to-weight basis, because the chains are oriented most efficiently, in the fiber direction, to resist stress (see FIBERS, OLEFIN; OLEFIN POLYMERS). Table 3 compares some properties of extended-chain linear polyethylene (8) with quiescently crystallized linear polyethylene.

**Table 3. Comparison of Some Properties of PE$^a$ and Steel**

| Property | Extended-chain PE | Ordinary PE | Steel |
|---|---|---|---|
| tensile modulus, GPa$^b$ | 44 | 1.2 | 206 |
| tensile strength, GPa$^b$ | 1.8 | 0.04 | 0.8 |
| density, g/cm$^3$ | 0.97 | 0.97 | 7.6 |

$^a$Extended-chain linear PE and ordinary linear PE.
$^b$To convert GPa to psi, multiply by 145,000.

Similarly, liquid-crystal polymers exhibit considerable order in the liquid state, either in solution (lyotropic) or melt (thermotropic). When crystallized from solution or melt, they have a high degree of extended-chain crystallinity, and thus have superior mechanical properties. Kevlar (Du Pont) is an aromatic polyamide (aramid) with the repeating unit designated as (**2**). It is spun into

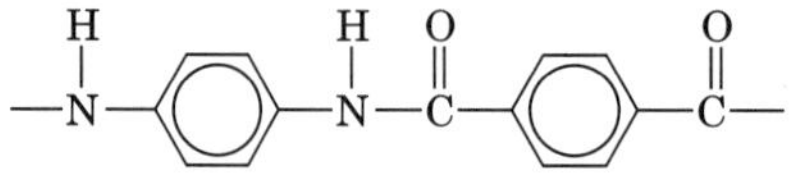

(**2**)

fibers from a lyotropic liquid-crystal solution in concentrated $H_2SO_4$. The solution is extruded through small holes into a bath which leaches out the acid, forming the fibers (wet spinning). Because the molecules are oriented prior to crystallization, the fibers maintain a high degree of extended-chain crystalline order in the fiber direction, imparting remarkable strength. The fibers are used in antiballistic armor, as a tire cord (see TIRE CORDS), and as a reinforcement in high strength composites (see POLYAMIDES, FIBERS).

The thermotropic liquid-crystal polyester Vectra A (Hoechst-Celanese) is reported to have the repeating units (**3**) and (**4**) and Xydar (Dartco) the repeating units (**5**) and (**3**).

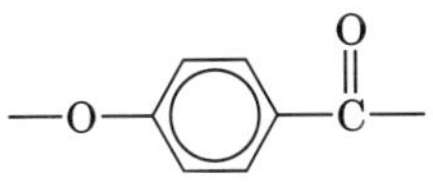

The macroscopic orientation of their extended-chain crystals depends on the orientation imparted by flow during molding. Because of the fibrous nature of the extended-chain crystals, these materials behave as self-reinforcing composites, with excellent mechanical properties.

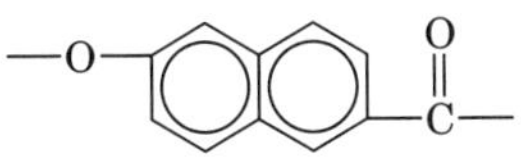

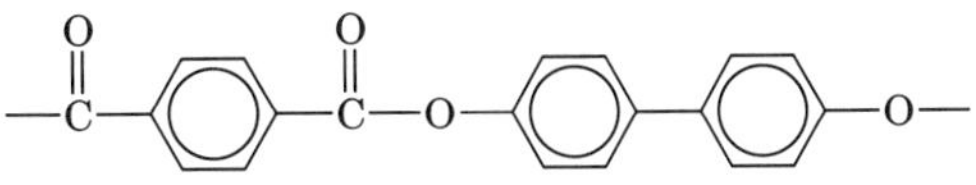

The crystalline melting point is governed by $\Delta G = \Delta H - T\Delta S$. At the equilibrium crystalline melting point, $T_m$, $\Delta G = 0$, and $T_m = \Delta H_m / \Delta S_m$, where $\Delta H_m$ is the heat (enthalpy) of melting and $\Delta S_m$ is the entropy of melting. $\Delta H_m$ represents the secondary bond energy holding the chains in the crystal lattice. Thus, polymers which form strong and/or extensive secondary bonds, eg, hydrogen bonds, between chains have high crystalline melting points, other things being equal. $\Delta S_m$ characterizes the degree of randomness the chains gain upon melting. The highly aromatic liquid crystalline polymers shown above have stiff, rigid chains, and stackable, nearly planar aromatic rings, causing them to maintain a high degree of order in the liquid state. This results in a low $\Delta S_m$ and therefore a high $T_m$ in addition to good mechanical strength.

Because shorter chains randomize more upon melting, they have a larger $\Delta S_m$ than longer chains. Therefore, $T_m$ decreases as the molecular weight is reduced, though this effect becomes important only at relatively low molecular weights.

**The Amorphous Phase and $T_g$.** Not all polymers crystallize, and even those that do are not completely crystalline. Noncrystalline polymer is termed amorphous. Four types of molecular motion have been identified in amorphous polymers. Listed in order of decreasing activation energy, they are (1) translational motion of entire molecules, (2) coiling and uncoiling of 40–50 C-atom segments of chains, (3) motion of a few (five to six) atoms along the main chain or on side groups, and (4) vibrations of individual atoms. Type 1 motions are responsible for flow. Type 2 motions give rise to rubber elasticity. The temperature below which type 1 and 2 motions are frozen out is known as the glass-transition temperature, $T_g$. Below its $T_g$, an amorphous polymer is a glass; hard, rigid, and often brittle. Above $T_g$, it becomes rubbery, and at still higher temperatures, if it is not cross-linked, it flows easily. Consider the demonstration in which a rubber ($T_g = -70°C$) ball bounces at room temperature, but when cooled in liquid nitrogen ($-196°C$), shatters when dropped. Thus amorphous polymers intended for use as plastics, which require mechanical rigidity, must be below their $T_g$ at use temperature. Rubber polymers must be amorphous, at least in their unstretched state, because crystallinity inhibits type 1 and 2 motions. They also must be above their $T_g$ to permit type 2 motions, and are usually lightly cross-linked (vulcanized) to eliminate the type 1 motions which lead to irrecoverable deformation.

In general, things that impede type 1 and 2 motions raise $T_g$. These include strong secondary forces between chains and steric hindrance to rotation about main-chain bonds. Chemists take advantage of these to design plastics that maintain rigidity at higher temperatures. Too high a $T_g$ may make a polymer difficult to process, however. A good way of viewing linear polymers that are not thermoplastic, eg, cellulose and polyacrylonitrile, is that their $T_g$ and/or $T_m$ exceeds their thermal decomposition temperature. Similarly, even light cross-linking eliminates type 1 motions. Increasing degrees of cross-linking begin to inhibit type 2 motions as well, raising $T_g$. Highly cross-linked polymers maintain their rigidity until they degrade, ie, their $T_g$ is above their decomposition temperature.

The $T_g$ of random copolymers varies with composition between the $T_g$ of the corresponding homopolymers. Thus, random copolymerization with a monomer whose homopolymer has a low $T_g$ provides internal plasticization. Addition of an external plasticizer, a liquid with a molecular weight on the order of several hundred, displaces the secondary bonds between polymer chains, reducing the $T_g$ and softening the polymer mass. This is most often practiced with poly(vinyl chloride) (PVC) which has $T_g = 86°C$. At room temperature, nearly unplasticized PVC is a rigid material used for pipe, window frames, siding, etc. Addition of a plasticizer, commonly di-2-ethylhexyl phthalate (DEHP), also known as dioctyl phthalate (DOP), reduces its $T_g$ and converts it into a material suitable for flexible tubing (Tygon, by Norton), wire insulation, refrigerator door gaskets, and a leather-like upholstery material (Naugahyde, by Uniroyal Engineered Products) (see LEATHER-LIKE MATERIALS; PLASTICIZERS).

**Effects of Crystallinity on Properties.** In polymers that can crystallize, the ratio of crystalline to amorphous material has a profound effect on properties. Because the chains are packed more tightly and efficiently in the crystalline areas than in the amorphous, the crystalline phase has a higher density and greater mechanical strength. In fact, density is a common measure of degree of crystallinity. The mechanical differences are greater if the amorphous phase is above its $T_g$, giving rigid crystallites in a rubbery matrix, as is the case for polyethylene ($T_m = 135°C$, $T_g = -120°C$) at room temperature. Table 4 illustrates how the strength and stiffness increase with the degree of crystallinity for several polyethylenes. In polyethylenes, crystallinity is sterically reduced by branching (see Fig. 1). High density PE (HDPE) (eg, milk jugs) has little branching. Side reactions during polymerization give rise to long, branched branches which lead successively to low density polyethylene (LDPE) (eg, squeeze bottles). In the newer linear, low density polyethylenes (LLDPE), short branches are introduced by random copolymerization with $\alpha$-olefins, $CH_2 = CHC_nH_m$, such as butene, hexene, or octene. Isotactic polypropylene ($T_m = 165°C$, $T_g = -30°C$) is highly crystalline and an important plastic. The atactic form does not crystallize, however, and is above $T_g$ at room temperature. It is used as a caulking compound, pressure-sensitive adhesive, etc.

Similarly, the random introduction by copolymerization of sterically incompatible repeating unit B into chains of crystalline A reduces the crystalline melt-

**Table 4. Influence of Crystallinity on Properties of Polyethylene[a,b]**

| Commercial product | Low density | Medium density | High density |
|---|---|---|---|
| density range, g/cm$^3$ | 0.910–0.925 | 0.926–0.940 | 0.941–0.965 |
| approximate crystallinity, % | 42–53 | 54–63 | 64–80 |
| branching, equivalent CH$_3$ groups/1000 carbon atoms | 15–30 | 5–15 | 1–5 |
| crystalline melting point, °C | 110–120 | 120–130 | 130–136 |
| hardness, Shore D | 41–46 | 50–60 | 60–70 |
| tensile modulus, MPa$^c$ | 97–260 | 170–380 | 410–1240 |
| tensile strength, MPa$^c$ | 4.1–16 | 8.3–24 | 21–38 |
| flexural modulus, MPa$^c$ | 34–410 | 410–790 | 690–1800 |

[a]Ref. 6 (see OLEFIN POLYMERS, POLYETHYLENE).
[b]It must be kept in mind that mechanical properties are influenced by factors other than the degree of crystallinity (molecular weight, in particular).
[c]To convert MPa (N/mm$^2$) to psi, multiply by 145.

ing point and degree of crystallinity. If $T_m$ is reduced to $T_g$, crystals cannot form. Isotactic polypropylene and linear polyethylene homopolymers are each highly crystalline plastics. However, a random 65% ethylene–35% propylene copolymer of the two, poly(ethylene-*co*-propylene) is a completely amorphous ethylene–propylene rubber (EPR). On the other hand, block copolymers of the two, poly(ethylene-*b*-propylene) of the same overall composition, are highly crystalline. X-ray studies of these materials reveal both the polyethylene lattice and the isotactic polypropylene lattice, as the different blocks crystallize in their own lattices.

The stiffest polymers are both crystalline and have a glassy amorphous phase, ie, are below both $T_m$ and $T_g$. They are often useful as engineering (structural) plastics (qv). Nylon-6,6 and crystalline poly(ethylene terephthalate) are examples of this class of materials, as is the newly developed syndiotactic polystyrene. Traditional polystyrene is atactic, and therefore cannot crystallize. It has $T_g = 100°C$ and so is rigid at room temperature, but is brittle and not particularly strong, limiting its structural applications (see STYRENE PLASTICS). The syndiotactic form does crystallize, providing greater strength and a higher heat-distortion temperature.

Crystalline polymers undergo a discontinuous decrease in volume when cooled through $T_m$ (Fig. 4). This can lead to nonuniform shrinkage and warping in molded objects. On the other hand, it also causes the polymer to "lock on" to reinforcing fibers, eg, glass (qv), so that crystalline thermoplastics benefit much more than amorphous thermoplastics from fiber reinforcement.

Crystallinity also influences optical properties. In a crystalline polymer, the denser crystalline areas generally have a higher refractive index than the amorphous areas and have dimensions comparable to the wavelength of visible light (0.4–0.7 $\mu$m) or greater. Thus, light is scattered as it passes between phases, so crystalline polymers have a characteristic white, translucent-to-opaque appearance, depending on the thickness of the specimen, its degree of crystallinity, and the size of the crystallites. Large spherulites scatter considerable light, so nucleating agents are used to promote smaller spherulites and enhance clarity. Isotactic poly(4-methylpentene) is the only polymer known which

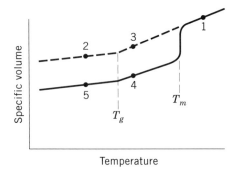

**Fig. 4.** Specific volume vs temperature for a crystallizable polymer. The dashed line represents amorphous material (6). Numbers refer to individual states. See text.

is both transparent and highly crystalline, because the refractive indexes of the crystalline and amorphous phases are almost identical (see OLEFIN POLYMERS, POLYMERS OF HIGHER OLEFINS).

Pure amorphous polymers, being homogeneous materials, are transparent. Atactic polystyrene is a good example. The crystalline syndiotactic form is not transparent. A lack of transparency does not necessarily indicate crystallinity, however. It can also be caused by inorganic fillers, pigments, gas bubbles (as in a foam), a second polymer phase, etc.

Even though a polymer is capable of crystallizing, it may not. Long, entangled polymer chains cannot rearrange instantaneously from the disordered amorphous state into a regular crystal lattice. Some polymers, if cooled rapidly enough from above $T_m$ to below $T_g$, where the chains lose the mobility to rearrange (Fig. 4, state 2), persist indefinitely in a metastable amorphous state as long as they are maintained below $T_g$. If heated above $T_g$ (state 3) for a while, such a material anneals to the crystalline form (state 4), and goes from transparent to opaque in the process. The metastable amorphous state (state 2) is easily obtained with poly(ethylene terephthalate) ($T_m = 267°C$, $T_g = 69°C$), with its bulky, aromatic repeating unit. It has important commercial applications (eg, soda bottles) in state 2 and also in crystalline state 5, eg, fibers, molded plastic (see POLYESTER, THERMOPLASTIC; FIBERS, POLYESTER).

Transitions such as $T_g$ and $T_m$ are rapidly and conveniently studied using differential scanning calorimetry (dsc). This technique monitors changes in enthalpy that accompany the transitions as a sample is heated at a constant rate.

**Solubility.** Cross-linking eliminates polymer solubility. Crystallinity sometimes acts like cross-linking because it ties individual chains together, at least well below $T_m$. Thus, there are no solvents for linear polyethylene at room temperature, but as it is heated toward its $T_m$ (135°C), it dissolves in a variety of aliphatic, aromatic, and chlorinated hydrocarbons. A rough guide to solubility is that like dissolves like, ie, polar solvents tend to dissolve polar polymers and nonpolar solvent dissolve nonpolar polymers.

Solubility is thermodynamically feasible if the sign of the Gibbs free energy ($\Delta G = \Delta H - T\Delta S$) for the process is negative. Distributing the solute molecules throughout the solvent molecules represents a decrease in order (positive $\Delta S$), so the entropy term favors solubility. The difference between polymeric and low molecular weight solutes lies in the magnitude of the entropy term. Because a repeating unit in a polymer chain is confined between its neighbors in the chain whereas low molecular weight solutes are free to move anywhere, the entropy of solution for high molecular weight polymers is much smaller than for equivalent masses or volumes of monomeric solutes. The higher the molecular weight, the lower the entropy of solution, which accounts for the observed decrease in solubility with molecular weight.

Because the magnitude of the entropy term is so small for high molecular weight polymers, one of the schemes for predicting polymer solubility concentrates only on the enthalpy term. If $\Delta H$ is negative, solubility is assured. Negative $\Delta H$ arises from strong polymer–solvent secondary bonds, eg, hydrogen bonds. In the more common situation, where $\Delta H$ is positive, it must be quite small, smaller than $-T\Delta S$ for solution to occur. Regular solution theory predicts equation 1 where the subscripts 1 and 2 refer to solvent and solute, respectively;

$$\Delta H \approx \phi_1\phi_2(\delta_1 - \delta_2)^2 \tag{1}$$

$\phi$ represents volume fractions; and $\delta$ represents solubility parameters as defined by equation 2, where $\Delta E_v$ is the energy of vaporization and $v$ the molar volume.

$$\delta = (\Delta E_v/v)^{1/2} \tag{2}$$

The solubility parameter is therefore a measure of the energy density holding the molecules in the liquid state. Note that regular solution theory can only predict positive $\Delta H$. Thus, with this approach, prediction of solubility involves matching the solute and solvent solubility parameters as closely as possible to minimize $\Delta H$. As a very rough rule of thumb $|\delta_1 - \delta_2|$ must be less than 2 $(\text{J/cm}^3)^{1/2}$ for solubility.

The practical utility of the solubility parameter approach was enhanced considerably by Hansen (9–11), who reasoned that $\Delta E_v$ was made up of hydrogen bonding, $h$; permanent dipole interaction, $p$; and dispersion, $d$, contributions, so that equation 3 holds where $\delta_j = (\Delta E_j/v)^{1/2}$; $j = d, p, h$. This equation is

$$\delta^2 = \delta_d^2 + \delta_p^2 + \delta_h^2 \tag{3}$$

reminiscent of the relation between the magnitude of a vector and its components. Hansen observed empirically that if the solvent components $\delta_{j1}$ and the solute components $\delta_{j2}$ are plotted in three dimensions, using a $\delta_d$ scale twice the size of the others, solvents that fall within a sphere of radius $R$ from the polymer point dissolve the polymer. Thus, equation 4 states the solubility condition:

$$\left[(\delta_{p1} - \delta_{p2})^2 + (\delta_{h1} - \delta_{h2})^2 + 4(\delta_{d1} - \delta_{d2})^2\right]^{1/2} < R \quad \text{for solubility} \tag{4}$$

Because values of $\delta_d$ cover a rather small range, the three-dimensional scheme is often reduced to two dimensions, with polymers and solvents represented on $\delta_h$–$\delta_p$ coordinates with a solubility circle of radius $R$.

Values of solubility parameters, their three-dimensional components, and $R$ are available (9,10,12). Not surprisingly, the three-dimensional scheme, with more constants to adjust, does a better job of predicting solubility.

More fundamental treatments of polymer solubility go back to the lattice theory developed independently and almost simultaneously by Flory (13) and Huggins (14) in 1942. By imagining the solvent molecules and polymer chain segments to be distributed on a lattice, they statistically evaluated the entropy of solution. The enthalpy of solution was characterized by $\chi$, the Flory-Huggins interaction parameter, which is related to solubility parameters by equation 5. For high molecular weight polymers in monomeric solvents, the Flory-Huggins solubility criterion is $\chi \leq 0.5$.

$$\chi = \frac{v(\delta_1 - \delta_2)^2}{RT} \tag{5}$$

Experimental values of $\chi$ have been tabulated for a number of polymer-solvent systems (4,12). Unfortunately, they often turn out to be concentration and molecular weight dependent, reducing their practical utility. The Flory-Huggins theory qualitatively predicts several phenomena observed in solutions of polymers, including molecular weight effects, but it rarely provides a good quantitative fit of data. Considerable work has been done subsequently to modify and improve the theory (15,16).

A newer approach uses group-contribution methods to predict solubility. It has been remarkably successful when applied to nonpolymer solutions and there are indications that it will be equally successful for treating polymer solutions (17).

Polymer solutions are often characterized by their high viscosities compared to solutions of nonpolymeric solutes at similar mass concentrations. This is due to the mechanical entanglements formed between polymer chains. In fact, where entanglements dominate flow, the (zero-shear) viscosity of polymer melts and solutions varies with the 3.4 power of weight-average molecular weight.

## Polymer Synthesis

### STEP-GROWTH POLYMERIZATION

Step-growth polymerization is characterized by the fact that chains always maintain their terminal reactivity and continue to react together to form longer chains as the reaction proceeds, ie, $x$-mer + $y$-mer $\rightarrow$ $(x + y)$-mer. Because there are reactions that follow this mechanism but do not produce a molecule of condensation, eg, the formation of polyurethanes from diols and diisocyanates (eq. 6), the terms step-growth and polycondensation are not exactly synonymous (6,18,19).

$$x \ \text{HOROH} + x \ \text{O}{=}\text{C}{=}\text{NR}'\text{N}{=}\text{C}{=}\text{O} \longrightarrow \ {\leftarrow}\text{ORO}{-}\overset{\overset{\text{O}}{\|}}{\text{C}}{-}\text{NHR}'\text{NH}{-}\overset{\overset{\text{O}}{\|}}{\text{C}}{\rightarrow}_x \tag{6}$$

diol                    diisocyanate                                        polyurethane

For linear step-growth, the number-average degree of polymeriza-tion, $\bar{x}_n$, is given by equation 7, where $r = N_A/N_B$, the stoichiometric ratio of the

$$\bar{x}_n = \frac{\overline{M}_n}{m} = \frac{1 + r}{1 + r - 2\,rp} \tag{7}$$

complementary functional groups $A$ and $B$, and $p$ is the fractional conversion of the limiting functional group, $A$. Here, $\bar{x}_n$ refers to the number of monomer residues rather than repeating units in the chain. This equation reveals that even with perfect stoichiometry ($r = 1$), a conversion $p = 0.99$ is required to achieve an $\bar{x}_n = 100$ (a modest value). A stoichiometric imbalance further lowers $\bar{x}_n$ at any value of $p$. Thus, producing high molecular weight, linear condensation polymers requires both high conversions and a close approach to stoichiometric equivalence. On the other hand, molecular weights can be limited when desired by a deliberate stoichiometric imbalance.

High conversions are promoted by efficient removal of a molecule of con-densation. This may be done with a vacuum if the molecule of condensation is sufficiently volatile, by carrying out the reaction in a refluxing organic solvent, condensing the vapors and decanting the water of condensation before returning the solvent to the reactor, by having a molecule of condensation that is insoluble in the reaction medium, or by a secondary reaction that consumes the mole-cule of condensation as it is formed. An example of the last is the production of a polycarbonate from bisphenol A and phosgene (see Table 1). By carrying out this reaction in a basic medium, the HCl is consumed as it is formed (see POLYCARBONATES).

If a step-growth polymerization batch includes some monomer with a functionality greater than two, and if the reaction is carried to a high enough conversion, a cross-linked network or gel may be formed. In the production of thermosetting molding compounds, the reaction must be terminated short of the conversion at which a gel is formed, or the material could not be molded (cross-linking is later completed in the mold). Flory worked out the statistics of gelation, relating the gel-point conversion to the stoichiometry of the reactor charge (6,20).

## CHAIN-GROWTH POLYMERIZATION

Chain-growth polymerizations are characterized by chains that propagate by adding one monomer molecule at a time, ie, $x$-mer + monomer $\rightarrow$ $(x + 1)$-mer. There are, however, several mechanisms by which this occurs.

**Free-Radical Addition.** In free-radical addition polymerization, the prop-agating species is a free radical. The free radicals, R·, are most commonly gener-ated by the thermal decomposition of a peroxide or azo initiator, I (see INITIATORS, FREE-RADICAL):

*Decomposition*                          $\mathrm{I} \xrightarrow{k_d} 2\,\mathrm{R}\cdot$

The radicals then initiate chain growth by adding an unsaturated monomer molecule, M, to form growing chain $\mathrm{P}_1^{\cdot}$.

*Addition*                          $\mathrm{R}\cdot + \mathrm{M} \xrightarrow{k_a} \mathrm{P}_1^{\cdot}$

$P_1^{\cdot}$ is a growing polymer chain with one monomer unit. The chain end remains an active radical and propagates by the sequential addition of monomer.

*Propagation*
$$P_x^{\cdot} + M \xrightarrow{k_p} P_{(x+1)}^{\cdot}$$

Chains terminate by either of two mechanisms: combination or disproportionation. Two chain radicals may combine to form a single bond between them:

*Combination*
$$P_x^{\cdot} + P_y^{\cdot} \xrightarrow{k_{tc}} P_{(x+y)}$$

Alternatively one chain radical may abstract a proton from the penultimate carbon atom of the other, giving one saturated and one unsaturated dead chain:

*Disproportionation*
$$P_x^{\cdot} + P_y^{\cdot} \xrightarrow{k_{td}} P_x + P_y$$

Chain transfer may also occur:

*Transfer*
$$P_x^{\cdot} + R'X \xrightarrow{k_{tr}} P_x + R'^{\cdot}$$

*Addition*
$$R'^{\cdot} + M \xrightarrow{k_{a'}} P_1^{\cdot}$$

The compound R'X is a chain-transfer agent, with X usually H or Cl. The net effect of chain transfer is to kill a growing chain and start a new one in its place, thus shortening the chains. Mercaptan chain-transfer agents are often used to limit molecular weight, but under appropriate conditions, almost anything in the reaction mass (solvent, dead polymer, initiator) can act as a chain-transfer agent to a certain extent.

The above mechanism, together with the assumptions that initiator decomposition is rate controlling and that a steady state in chain radicals exists, results in the classical expressions (eqs. 8 and 9) for polymerization rate, $r_p$, and number-average degree of polymerization, $\bar{x}_n$, in a homogeneous, free-radical polymerization (6).

$$r_p = k_p \left( \frac{fk_d}{k_t} \right)^{1/2} [I]^{1/2}[M] \tag{8}$$

$$\bar{x}_n = \frac{k_p[M]}{\zeta(fk_dk_t[I])^{1/2} + k_{tr}[R'X]} \tag{9}$$

In equations 8 and 9, $f$ is the initiator efficiency, the fraction of initiator radicals that actually initiates chain growth, $k_t = k_{tc} + k_{td}$, and $\zeta$ is the number of dead chains formed per termination reaction; $\zeta = 2$ for disproportionation, $\zeta = 1$ for combination.

Unlike step-growth polymerization, free-radical chains do not continue to grow as the reaction proceeds. Typically, the monomer half-life is many hours, but the average lifetime of a growing chain, from initiation to termination, is less than a second. Thus, high molecular weight polymer is produced right from the beginning; high conversions are not required.

The minimum polydispersity index from a free-radical polymerization is 1.5 if termination is by combination, or 2.0 if chains are terminated by disproportionation and/or transfer. Changes in concentrations and temperature during the reaction can lead to much greater polydispersities, however. These concepts of polymerization reaction engineering have been introduced in more detail elsewhere (6).

**Polymerization Processes.** Free-radical polymerization is carried out in a variety of ways. One of the practical problems that must be dealt with is runaway reactions which can result from autoacceleration, an increase in rate of polymerization caused by diffusion-limited termination (reduced $k_t$) at high conversions. The effects of autoacceleration are compounded by large, exothermic heats of polymerization, typically 42–84 kJ/mol ($-10$ to $-20$ kcal/mol), and the great increase in polymerization rate with temperature. Also, the low heat capacities and thermal conductivities of organic reaction masses and the very high viscosities of the reaction mass at high conversions limit heat removal and make temperature control difficult.

*Bulk Polymerization.* This involves only monomer, initiator, and perhaps chain-transfer agent. It gives the greatest polymer yield per unit of reactor volume and a very pure polymer. However, in large-scale batch form, it must be run slowly or in continuous form with a lot of heat-transfer area per unit of conversion to avoid runaway. Objects are conveniently cast to shape using batch bulk polymerization. Poly(methyl methacrylate) glazing sheets are produced by batch bulk polymerization between glass plates. They are also made by continuous bulk polymerization between polished stainless steel belts. Polystyrene and other thermoplastic molding compounds are produced by continuous bulk polymerization processes (6).

*Solution Polymerization.* In this process an inert solvent is added to the reaction mass. The solvent adds its heat capacity and reduces the viscosity, facilitating convective heat transfer. The solvent can also be refluxed to remove heat. On the other hand, the solvent wastes reactor space and reduces both rate and molecular weight as compared to bulk polymerization. Additional technology is needed to separate the polymer product and to recover and store the solvent. Both batch and continuous processes are used.

*Suspension Polymerization.* In this process the organic reaction mass is dispersed in the form of droplets 0.01–1 mm in diameter in a continuous aqueous phase. Each droplet is a tiny bulk reactor. Heat is readily transferred from the droplets to the water, which has a large heat capacity and a low viscosity, facilitating heat removal through a cooling jacket.

Agitation is critical to carrying these reactions out successfully. It is used, along with suspending agents in the aqueous phase, to control droplet size and maintain dispersion. If it fails or weakens before the particles become sufficiently rigid (the polymer must be below its $T_g$), the particles coalesce, leading to a runaway bulk reaction, which not only can be dangerous, but may require the use of mining tools to clear a commercial reactor. For this reason also, it has not been possible to run suspension polymerization continuously, since any flow system has some stagnant corners where polymer would accumulate, nor can it be used to produce rubbers. Commercially, the reactions are carried out in jacketed, stainless steel or glass-lined stirred tanks of up to 75.5 m$^3$ (20,000 gal) capacity.

The product of a successful suspension polymerization is small, uniform polymer spheres. For certain applications, they are used directly, eg, as the precursors for ion-exchange resins or bead foams. For others, they may be extruded and chopped to form larger, more easily handled molding pellets.

*Emulsion Polymerization.* When the U.S. supply of natural rubber from the Far East was cut off in World War II, the emulsion polymerization process was developed to produce synthetic rubber. In this complex process, the organic monomer is emulsified with soap in an aqueous continuous phase. Because of the much smaller ($\leq 0.1$ $\mu$m) dispersed particles than in suspension polymerization and the stabilizing action of the soap, a proper emulsion is stable, so agitation is not as critical. In classical emulsion polymerization, a water-soluble initiator is used. This, together with the small particle size, gives rise to very different kinetics (6,21–23).

The product of an emulsion polymerization is a latex; ie, polymer particles on the order of 0.5–0.15 $\mu$m stabilized by the soap. These form the basis for the popular latex paints. Solid rubber is recovered by coagulating the latex with ionic salts and acids (see LATEX TECHNOLOGY).

The original wartime process was run batchwise in reactors similar to those used for suspension polymerization. Since then, in many plants, the reactors have been hooked together as a series of continuous stirred tanks.

**Ionic Polymerization.** *Anionic Polymerization.* Addition polymerization may also be initiated and propagated by anions (23–26), eg, in the polymerization of styrene with *n*-butyllithium. The Li$^+$ gegen ion, held electrostatically in

$$[n\text{-}C_4H_9\text{:}]^-Li^+ + CH_2 \!=\! CHC_6H_5 \longrightarrow [n\text{-}C_4H_9 \!-\! CH_2 \!-\! CHC_6H_5\text{:}]^-Li^+$$

$$[n\text{-}C_4H_9\!\!-\!\!(CH_2CHC_6H_5)_x\!\!-\!\!CH_2CHC_6H_5\text{:}]^-Li^+ + CH_2 \!=\! CHC_6H_5 \longrightarrow$$

$$[n\text{-}C_4H_9\!\!-\!\!(CH_2CHC_6H_5)_{x+1}CH_2CHC_6H_5\text{:}]^-Li^+$$

the vicinity of the propagating chain end, can exert a steric influence on the addition of monomer molecules to the chain. Growing chains are terminated immediately by proton-donating impurities, eg, water:

$$[n\text{-}C_4H_9\!\!-\!\!(CH_2CHC_6H_5)_x\!\!-\!\!CH_2CHC_6H_5\text{:}]^-Li^+ + H_2O \longrightarrow$$

$$n\text{-}C_4H_9\!\!-\!\!(CH_2CHC_6H_5)_x\!\!-\!\!CH_2CH_2C_6H_5 + LiOH$$

There are some important differences between anionic and free-radical addition. First, unlike free-radical initiators, which decompose and start chains randomly throughout the course of the reaction, anionic initiators ionize readily in fairly polar organic solvents (eg, tetrahydrofuran) or at low concentrations ($< 10^{-4}$ molar) in hydrocarbons, and chains are started immediately, one for each molecule of initiator. Second, in the absence of impurities, there is no termination. Thus, in a batch reactor, all chains are initiated at the same time, the number of propagating chains remains constant, and the propagating chains continue to grow, competing for monomer on an even basis. This gives rise to an essentially monodisperse polymer, with $x = M_oX/I_o$, where $M_o$ is the moles monomer charged, $I_o$ the moles initiator charged, and $X$ the fractional conversion of monomer to polymer. In practice, polydispersity indexes of less than 1.1 can be achieved with reasonable care.

When the initial monomer supply is exhausted, the anionic chain ends retain their activity. Thus, these anionic chains have been termed living polymers. If more monomer is added, they resume propagation. If it is a second monomer, the result is a block copolymer.

Anionic polymerization provided, for the first time, a convenient means of synthesizing nearly monodisperse polymers and block copolymers. It also can provide chains with useful terminal functionality. For example, bubbling $CO_2$ through a batch of living chains, followed by exposure to water, gives carboxyl-terminated chains. Substituting ethylene oxide for the $CO_2$ gives hydroxyl-terminated chains. By using multifunctional initiators or reagents which couple the living chains, star polymers can be produced.

*Cationic Polymerization.* For decades cationic polymerization has been used commercially to polymerize isobutylene and alkyl vinyl ethers, which do not respond to free-radical or anionic addition (see ELASTOMERS, SYNTHETIC–BUTYL RUBBER). More recently, development has led to the point where living cationic chains can be made, with many of the advantages described above for anionic polymerization (27,28).

*Group-Transfer Polymerization.* Du Pont has patented (29) a technique known as group-transfer polymerization and applied it primarily to the polymerization of acrylates and methacrylates. It is mechanistically similar to anionic polymerization, giving living chains, except that chain transfer can occur (30).

Ionic polymerizations are almost exclusively solution processes. To produce monodisperse polymers or block copolymers, they must be run batchwise, so that all chains grow for the same length of time under identical conditions.

## HETEROGENEOUS STEREOSPECIFIC POLYMERIZATION

In the early 1950s, Ziegler observed that certain heterogeneous catalysts based on transition metals polymerized ethylene to a linear, high density material at modest pressures and temperatures. Natta showed that these catalysts also could produce highly stereospecific poly-$\alpha$-olefins, notably isotactic polypropylene, and polydienes. They shared the 1963 Nobel Prize in chemistry for their work.

A typical Ziegler-Natta catalyst might be made from $TiCl_4$ or $TiCl_3$ and $Al(C_2H_5)_3$. Vanadium and cobalt chlorides are also used, as is $Al(C_2H_5)_2Cl$. When these substances are mixed in an inert solvent, a crystalline solid is obtained. Early catalysts consisted of the finely divided solid alone, but in modern catalysts, it is often supported on $SiO_2$ or $MgCl_2$.

Polymerization occurs at active sites formed by interaction of the metal alkyl with metal chloride on the surface of the metal chloride crystals. Monomer is chemisorbed at the site, thus accounting for its orientation when added to the chain, and propagation occurs by insertion of the chemisorbed monomer into the metal–chain bond at the active site. The chain thus grows out from the surface (31). Hydrogen is used as a chain-transfer agent. Chain transfer with the metal alkyl also occurs.

$$\text{site} \overset{\frown}{-} P_x + M \longrightarrow \text{site} - P_{x+1}$$

$$\text{site} - P_x + H_2 \longrightarrow \text{site} - H + H - P_x$$

$$\text{site} - P_x + Al(R)_3 \longrightarrow \text{site} - R + Al(R)_2 - P_x$$

Because initiation and transfer occur randomly in these systems, they have much more in common with free-radical addition than they have with anionic addition. Based on this randomness, a polydispersity index of two would be expected. Much greater values, up to 20 or more, are observed in practice. This results mainly from sites of widely differing activity on the catalyst surface (32). More recently, single-site metallocene catalysts have been introduced which can give polydispersity indexes approaching the theoretical minimum of two, allowing closer control over polymer properties (33). They also have been used to produce two promising new polymers, syndiotactic (crystalline) polystyrene and syndiotactic polypropylene, and they are reported to give a more uniform distribution of comonomer in the production of linear low density polyethylenes (LLDPE) (34–36) (see METALLOCENE CATALYSTS (SUPPLEMENT)).

Modern catalysts produce a much higher percentage of isotactic polypropylene than in the past, eliminating the need for a costly extraction step to remove an atactic fraction. Yields are high enough ($>10,000$ g polymer/g catalyst) so that a catalyst removal (de-ashing) step is no longer required.

These heterogeneous catalysts are used either in solution or gas-phase bulk processes. When a rubber is produced in a solution process, it remains in solution. Crystalline polymers such as polyethylene and isotactic polypropylene may be made at temperatures high enough so that they remain in solution. More commonly, the temperatures are low enough so that solid polymer coats the catalyst particle as it forms. These are known as slurry processes. In the popular Unipol gas-phase process (37), polymer grows on the catalyst particles which are suspended in a fluidized bed by circulating monomer. The heat of polymerization is removed from the monomer in external heat exhangers, and polymer particles are discharged from the reactor looking like laundry detergent powder.

Great care must be exercised in the preparation and use of Ziegler-Natta catalysts. They are easily poisoned by moisture, among other things. They are pyrophoric and are used in conjunction with large amounts of flammable monomer and solvent, and so can present a significant safety hazard.

## BIBLIOGRAPHY

"Polymers" in *ECT* 1st ed., Vol. 10, pp. 957–971, by H. F. Mark, Polytechnic Institute of Brooklyn; in *ECT* 2nd ed., Vol. 16, pp. 242–253, by H. F. Mark, Polytechnic Institute of Brooklyn; in *ECT* 3rd ed., Vol. 18, pp. 745–755, by F. W. Billmeyer, Jr., Rensselaer Polytechnic Institute.

1. D. A. Tomalia and co-workers, *Polym. J.* **17**, 1, 117 (1985).
2. T. M. Miller and co-workers, *J. Am. Chem. Soc.* **115**, 356 (1993).
3. D. A. Tomalia, *Aldrichimica Acta* **26**, 4, 91 (1993).
4. J. Brandrup and E. H. Immergut, eds., *Polymer Handbook*, 3rd ed., Wiley-Interscience, New York, 1989.
5. T. Provder, ed., *Chromatography of Polymers: Characterization by SEC and FFF*, ACS Symposium Series No. 521, American Chemical Society, Washington, D.C., 1993.
6. S. L. Rosen, *Fundamental Principles of Polymeric Materials*, 2nd ed., Wiley-Interscience, New York, 1993.
7. G. Natta, *Sci. Am.* **205**(2), 33 (1961).
8. U.S. Pat. 4,413,110 (Nov. 1, 1983), S. Kavesh and D. C. Prevorsek (to Allied Corp.).
9. C. M. Hansen, *J. Paint Technol.* **39**, 505, 104 (1967).
10. C. M. Hansen, *The Three-Dimensional Solubility Parameter and Solvent Diffusion Coefficient*, Danish Technical Press, Copenhagen, Denmark, 1967.
11. C. M. Hansen, *Ind. Eng. Chem. Prod. Res. Dev.* **8**, 2 (1969).
12. A. F. M. Barton, *Handbook of Polymer–Liquid Interaction Parameters and Solubility Parameters*, CRC Press, Boca Raton, Fla., 1990.
13. P. J. Flory, *J. Chem. Phys.* **10**, 51 (1942).
14. M. L. Huggins, *Ann. N.Y. Acad. Sci.* **43**, 1 (1942).
15. I. C. Sanchez and R. H. Lacombe, *Macromolecules* **11**(6), 1145 (1978).
16. A. I. Pesci and K. F. Freed, *J. Chem. Phys.* **90**(3), 2017 (1989).
17. J. Holten-Andersen, P. Rasmussen, and A. Fredenslund, *Ind. Eng. Chem. Res.* **26**, 1382 (1987).
18. G. Odian, *Principles of Polymerization*, 3rd ed., Wiley-Interscience, New York, 1991.
19. S. R. Sandler and W. Karo, *Polymer Synthesis*, Vol. 1, 2nd ed., Academic Press, Inc., San Diego, Calif., 1992.
20. P. J. Flory, *Principles of Polymer Chemistry*, Cornell UP, Ithaca, N.Y., 1953.
21. D. C. Blackley, *Emulsion Polymerization*, John Wiley & Sons, Inc., New York, 1975.
22. D. R. Basset and A. E. Hamielec, eds., *Emulsion Polymers and Emulsion Polymerization*, ACS Symposium Series 165, American Chemical Society, Washington, D.C., 1981.
23. I. Piirma, ed., *Emulsion Polymerization*, Academic Press, Inc., New York, 1982.
24. M. Morton, *Anionic Polymerization: Principles and Practice*, Academic Press, Inc., New York, 1983.
25. M. Szwarc, *Living Polymers and Mechanisms of Anionic Polymerization*, Springer-Verlag, Berlin, 1983.
26. M. Szwarc and M. Van Beylen, *Ionic Polymerization and Living Polymers*, Chapman & Hall, New York, 1993.
27. J. P. Kennedy and B. Ivan, *Designed Polymers by Carbocationic Macromolecular Engineering*, Carl Hanser Verlag, New York, 1992.
28. M. Sawamoto, *Trends in Polymer Sci.* **1**, 4, 111 (1993).
29. U.S. Pat. 4,414,372 (Nov. 8, 1983), W. B. Farnham and D. Y. Sogah (to E. I. du Pont de Nemours & Co., Inc.); U.S. Pat. 4,417,034 (Nov. 22, 1983), O. W. Webster (to E. I. du Pont de Nemours & Co., Inc.).
30. S. C. Stinson, *Chem. Eng. News*, 43 (Apr. 27, 1983).
31. P. J. T. Tait, *CHEMTECH* **5**, 688 (1975).
32. S. Floyd and co-workers, *J. Appl. Polym. Sci.* **33**, 1021 (1987).
33. V. Wigotsky, *Plastics Eng.* **50**, 7, 14 (1994).
34. J. Chowdhury and S. Moore, *Chem. Eng.*, 34 (Apr. 1993).
35. D. Schwank, *Modern Plast.*, 49 (Aug. 1993).

36. A. A. Montagna and J. C. Floyd, *Hydrocarbon Process.*, 57 (Mar. 1994).
37. F. J. Karol, *CHEMTECH* **13**(4), 222 (1983).

STEPHEN L. ROSEN
University of Missouri−Rolla

**POLYMERS, CONDUCTIVE.**   See   ELECTRICALLY   CONDUCTIVE   POLYMERS; PHOTOCONDUCTIVE POLYMERS.

# POLYMERS CONTAINING SULFUR

Poly(phenylene sulfide), **904**
Polysulfides, **933**
Polysulfones, **945**

## POLY(PHENYLENE SULFIDE)

Poly(*p*-phenylene sulfide) (PPS), (**1**) (poly(thiophenylene) [*9016-75-5*] or poly-(thio-1,4-phenylene) [*25212-74-2*]) has a remarkably long history in the chemical literature, which has been reviewed in detail (1−4). The structure of PPS consists of alternating para-disubstituted aromatic rings (*p*-phenylene moieties) and divalent sulfur atoms (sulfide linkages). PPS is a semicrystalline polymer possessing a desirable combination of properties that include good mechanical properties, excellent electrical and thermal properties, as well as inherent flame resistance. Combined with the ease of molding (5), PPS plays an important role in the class of materials known as engineering thermoplastics. Designers have been exploiting the properties of PPS in end uses such as coatings, injection molding, film, fiber, pipe, and advanced composite materials (qv).

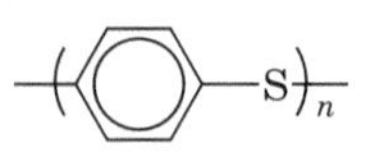

(**1**)

The earliest reported reference describing the synthesis of phenylene sulfide structures is that of Friedel and Crafts in 1888 (6). The electrophilic reactions studied were based on reactions of benzene and various sulfur sources. These electrophilic substitution reactions were characterized by low yields

(50–80%) of rather poorly characterized products by the standards of 1990s. Products contained many by-products, such as thianthrene. Results of self-condensation of thiophenol, catalyzed by aluminum chloride and sulfuric acid (7), were analogous to those of Friedel and Crafts.

The first reported assignment of the PPS structure to reaction products prepared from benzene and sulfur in the presence of aluminum chloride was made by Genvresse in 1897 (8). These products were oligomeric and contained too much sulfur to be pure PPS. Genvresse isolated thianthrene and an amorphous, insoluble material that melted at 295°C. These early synthetic efforts have been reviewed (9–11).

The electrophilic substitution reactions of benzene and sulfur, catalyzed by aluminum chloride, were reinvestigated in 1984 in the laboratories of Phillips Petroleum Company (11) and the findings of Genvresse and others were confirmed. Using improved spectroscopic and chromatographic analytical techniques, products containing a variety of structures, including phenanthrylene linkages, phenylene sulfide linkages, and polysulfides, were identified. Further, it was demonstrated that the desired phenylene sulfide structures were unstable toward the electrophilic substitution reaction conditions. The lack of reaction selectivity and the instability of the desired product contributed to the low molecular weight of the products. The molecular weight of polymers from electrophilic reactions of benzene and sulfur is typically only ca 3500, which is too low to have useful mechanical properties.

The recognition that PPS had significant commercial potential as an advanced material came in the late 1940s (12). Macallum's PPS process is based on the reaction of elemental sulfur, $p$-dichlorobenzene, and sodium carbonate in sealed vessels at 275–300°C (12). Typical products produced by the Macallum process contain more than one sulfur per repeating unit ($x = 1.2$–2.3):

$$\text{Cl}-\!\!\langle\bigcirc\rangle\!\!-\text{Cl} + \text{S} + \text{Na}_2\text{CO}_3 \longrightarrow -\!\!\left(\langle\bigcirc\rangle-\text{S}_x\right)_{\!n}$$

The reaction is highly exothermic, difficult to control, and incapable of being scaled up, but developmental products showed outstanding thermal stability. Also, Macallum's high molecular weight polymers could be molded into parts having good tensile strength and toughness (13). This recognition, among others, led to a growing interest in PPS in the industrial sector.

Dow Chemical Company purchased the rights to Macallum's patents (14), initiated a detailed study of the process and other improved syntheses of PPS in the 1950s and early 1960s, and published the results of their investigation (9,15,16). Clearly, alternative routes to PPS were desirable and the most promising of these involved the nucleophilic self-condensation of cuprous $p$-bromothiophenoxide, carried out at 200–250°C in the solid state or in the presence of pyridine (16):

$$\text{Br}-\!\!\langle\bigcirc\rangle\!\!-\text{SCu} \longrightarrow -\!\!\left(\langle\bigcirc\rangle-\text{S}\right)_{\!n} + \text{CuBr}$$

The main benefit of the Dow process was control of the polymer architecture. The polymer from the self-condensation process possessed a linear structure, but there were other difficulties. The monomer was costly and removal of the cuprous bromide by-product was difficult (17); ultimately, scale-up difficulties terminated the Dow PPS development. However, there was a growing recognition that PPS was an attractive polymer with an excellent combination of properties.

Researchers at Phillips Petroleum Company developed a commercially viable process for the synthesis of PPS involving the polymerization of p-dichlorobenzene and a sodium sulfide source in a polar organic compound at elevated temperature and pressure. This Phillips process was patented in 1967 (18). Between 1967 and 1973, Phillips built and operated a pilot plant, estab-

$$\text{Cl}-\!\!\bigcirc\!\!-\text{Cl} + Na_2S \longrightarrow -\!\!\left(\bigcirc\!\!-S\right)_{\!n} + 2\,NaCl$$

lished market demand, and constructed a full-scale commercial plant. In 1973, the world's first PPS plant came on-stream in Phillips' facility in Borger, Texas.

The expiration of Phillips' basic PPS patent in 1984 ushered in a large interest from the industrial sector. Companies, based largely in Europe and Japan, began acquiring patents worldwide for both the synthesis of PPS and a wide variety of applications, including compounds, blends, alloys, fiber, film, advanced composite materials, as well as end use products.

The first and largest commercial supplier of PPS, with a capacity of approximately 8000 t of neat resin per year, is Phillips Petroleum Company. Their product slate ranges from low to high molecular weight linear PPS, as well as a variety of cured and long-chain branched polymers. Phillips commercial trademark is Ryton poly(phenylene sulfide). Kureha Chemical Industry Company in Japan developed an alternative synthesis of PPS and commercialized their PPS process in Japan. Kureha markets PPS in combination with their compounding partner, Polyplastics, under the trade name Fortron PPS. Kureha has also formed a joint operation with Hoechst Celanese Corporation. Hoechst Celanese has constructed a new PPS commercial production facility at its Cape Industries plant in Wilmington, North Carolina. This is the second U.S. domestic PPS plant, increasing the U.S. neat resin capacity by approximately 2700 t/yr. This facility went on-stream in 1994.

Bayer AG in Germany has been active in PPS research since the mid-1980s, culminating in the construction of a commercial PPS plant in Belgium. Bayer marketed PPS compounds in the United States under the trade name Tedur, but the company has exited the PPS business. PPS is also marketed in the United States by GE Plastics, whose source of neat resin is Tosoh Corporation of Japan. GE Plastics markets PPS under the trade name Supec PPS. Patent activity by Tennessee Eastman describes an alternative process for the production of poly(phenylene sulfide/disulfide), although samples of such product have not appeared as of early 1996. Both Phillips and Hoechst Celanese have announced plans to debottleneck their existing U.S. facilities in order to meet anticipated market growth.

The roster of PPS suppliers in Japan is much larger than in the United States. Multiple market presences in Japan include TOPPS (Toray PPS, formerly TO–PP, which was a joint venture between Toray and Phillips); Dainippon Ink and Chemicals, Inc.; Tosoh; Tohpren; Kureha/Polyplastics; and Idemitsu. PPS marketed by Toray is sold under the Torelina trademark. Production capacity in Japan was estimated in 1995 at 11,400 t of neat resin per year. At the time that this capacity was created, the situation in Japan was characterized by overcapacity and underutilization. Additionally, further PPS capacity was brought on by Sunkyong in Korea, which is marketing a low cost PPS product. Although excess PPS capacity still exists in Japan, market growth has narrowed the gap between supply and demand.

## Polymerization Processes

The neat resin preparation for PPS is quite complicated, despite the fact that the overall polymerization reaction appears to be simple. Several commercial PPS polymerization processes that feature some steps in common have been described (1,2). At least three different mechanisms have been published in an attempt to describe the basic reaction of a sodium sulfide equivalent and *p*-dichlorobenzene; these are $S_N Ar$ (13,16,19), radical cation (20,21), and Bunnett's (22) $S_{RN} 1$ radical anion (23–25) mechanisms. The benzyne mechanism was ruled out (16) based on the observation that the para-substitution pattern of the monomer, *p*-dichlorobenzene, is retained in the repeating unit of the polymer. Demonstration that the step-growth polymerization of sodium sulfide and *p*-dichlorobenzene proceeds via the $S_N Ar$ mechanism is fairly recent (1991) (26). Further complexity in the polymerization is the incorporation of comonomers that alter the polymer structure, thereby modifying the properties of the polymer. Additionally, post-polymerization treatments can be utilized, which modify the properties of the polymer. Preparation of the neat resin is an area of significant latitude and extreme importance for the end user.

There are two commercial PPS processes being practiced worldwide: the Phillips process and the Kureha process. Although these processes contain some common steps, there are distinguishing features, most notably in the reagents used to facilitate the synthesis of high molecular weight linear PPS.

The first commercial PPS process by Phillips synthesized a low molecular weight linear PPS that had modest mechanical properties. It was useful in coatings and as a feedstock for a variety of cured injection-molding resins. The Phillips process for preparing low molecular weight linear PPS consists of a series of nucleophilic displacement reactions that have differing reactivities (26).

(1)

(2)

Equation 1 is an oversimplification of the actual process. The polymerizable sulfur source for the PPS polymerization consists of a dehydrated product of $N$-methyl-2-pyrrolidinone [872-50-4] (NMP) and aqueous sodium sulfide feedstocks. During the course of this dehydration, one equivalent of NMP is hydrolyzed to form sodium $N$-methyl-4-aminobutanoate (SMAB) (eq. 3).

$$\text{\raisebox{1em}{}} \quad N-CH_3 + Na_2S + H_2O \longrightarrow CH_3NH(CH_2)_3COO^-Na^+ + NaSH$$

$$(3)$$

The process implications of equation 3 go beyond the well-known properties (27–29) of NMP to facilitate $S_NAr$ processes. The function of the aminocarboxylate is also to help solubilize the sulfur source; anhydrous sodium sulfide and anhydrous sodium hydrogen sulfide are virtually insoluble in NMP (26). It also provides a necessary proton acceptor to convert thiophenol intermediates into more nucleophilic thiophenoxides. A block diagram for the Phillips low molecular weight linear PPS process is shown in Figure 1.

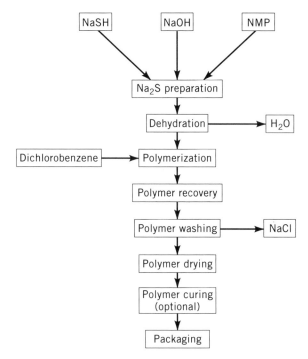

**Fig. 1.** The key steps for the Phillips PPS process are (*1*) production of aqueous sodium sulfide from aqueous sodium hydrogen sulfide (or hydrogen sulfide) and aqueous sodium hydroxide; (*2*) dehydration of the aqueous sodium sulfide and NMP feedstocks; (*3*) polymerization of the dehydrated sulfur source with *p*-dichlorobenzene to yield a slurry of PPS and by-product sodium chloride in the solvent; (*4*) polymer recovery; (*5*) polymer washing for the removal of by-product salt and residual solvent; (*6*) polymer drying; (*7*) optional curing, depending on the application; and (*8*) packaging.

The polymer produced by the Phillips low molecular weight linear PPS process is an off-white powder, having approximately 150–200 repeating units. There are no known solvents for PPS below 200°C, which complicates traditional molecular weight characterization methods. However, dilute solution light-scattering experiments and gel-permeation chromatography in 1-chloronaphthalene at 220°C show that the weight-average molecular weight is approximately 18,000 (30,31). The inherent viscosity, as measured at 206°C in 1-chloronaphthalene solution, of PPS having this molecular weight is approximately 0.16 dL/g.

Low molecular weight linear PPS possesses only modest mechanical properties; however, it can be converted to a much tougher material by an oxidative heat treatment. It was recognized early in the development of PPS that the polymer had the property of undergoing change during thermal treatments and could favorably alter the usable properties of PPS (32). This ability to undergo change by appropriate thermal treatment is not to be confused with traditionally thermosetting polymers. PPS behaves as a true thermoplastic material during normal processing conditions. It can be repeatedly melted and reprocessed with only minor changes in its rheology. However, under more extensive oxidative heat treatments, PPS properties can be favorably altered and this behavior has been termed curing. Curing is a deliberate process step in which the polymer is heated in an oxidizing environment (often air) or after an oxidizing treatment for an extended period of time. Several changes in the polymer take place during curing: toughness increases, melt viscosity increases, rate and extent of crystallization decrease, and color changes from white to tan/brown/black. The extent to which these changes occur depends on the extent of cure, which is a function of both the time and temperature of curing. The reactions, which take place when PPS cures, are not well characterized, in part, because of the difficulties in analyzing a polymer that only dissolves at elevated temperatures.

The complex curing reactions have been studied (33–39) and the findings have been reviewed (1,3). Evidence for thermally induced homolysis of carbon–sulfur bonds and the resultant reactions of thiyl and aryl radicals to produce cross-linking via biphenyl-type structures have been reported (17). Chain scission, cross-linking, and oxidation products are found during PPS curing (33). Curing of PPS also involves the loss of hydrogen from the aromatic moiety, resulting in cross-linking and the formation of new carbon–sulfur bonds and there is evolution of low molecular weight fragments, such as phenyl sulfide (33). Evidence for the evolution of low molecular weight species, such as the dimer and trimer of PPS and other sulfur compounds, has been found (34). Studies of the effects of variables such as temperature and atmosphere showed that the curing reactions are complex and result in polymers having increased viscosity and decreased crystallizability (35). The latter finding was the same as that found in a previous study (36), which showed that PPS cured below its crystalline melting point has no reduction in the percentage of crystallinity, but upon melting and annealing shows dramatically reduced crystallinity. The extent to which the crystallinity is decreased is a direct function of the extent of cure. Curing is dramatically faster in air than it is in nitrogen; oxygen is important in hydrogen atom abstraction during free-radical reactions forming biphenyl structures (35). Quinones can also function to abstract hydrogen atoms during PPS curing (37). Curing of PPS oligomers ($n = 6-8$) comprised a complex mixture of chain extension reactions, oxidative cross-linking reactions, thermally induced cross-linking reactions, oxygen uptake, reactions involving the extrusion of sulfur dioxide (presumably from sulfone moieties created by oxidation of PPS sulfide bonds), and arylthio metathesis (38). The rheological properties of PPS cured in the solid state, ie, below the crystalline melting point of the polymer, differ from those of PPS cured in the melt, ie, at temperatures above the crystalline melting point (39). Melt-cured PPS shows less shear rate sensitivity, ie, more Newtonian behavior, than does solid-state cured PPS.

PPS can be cured either in the solid state or in the melt. Melt curing (40) requires the polymer to be heated above the crystalline melting point of the polymer in the presence of air. As the melt cure progresses, the melt viscosity of the polymer increases. Extended curing results in continued increases in melt viscosity, gelation, and eventual vitrification, yielding a dark infusible solid. Solid-state curing of PPS (41) is a convenient process for curing large batches of powder, generally carried out at 175–280°C. The extent of cure is easily followed by measuring changes in the flow rate of the polymer in a modified version (316°C, 5 kg driving weight) of ASTM D1238 (42). The rate of solid-state curing is a strong function of the cure temperature, as Figure 2 shows. More rapid decreases in flow rate are observed when PPS is cured at higher temperatures.

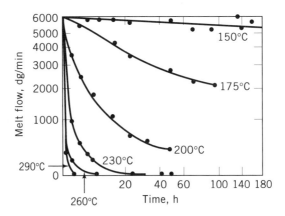

**Fig. 2.**   Cure rate of PPS as a function of solid-state cure temperature.

Solid-state curing can be carried out effectively in the laboratory in an air-circulating oven. It is advisable to have good air–polymer contact for the purpose of exposing the polymer to air and maintaining good thermal control of the process. Solid-state curing is exothermic; if care is not given to maintaining good temperature control, localized hot spots may develop, resulting in fused or aggregated polymer. Ultimately, the result of such hot spots is uneven curing throughout the batch of polymer. One solution to this potential problem is to use an agitated bed of polymer in which air passes through the polymer.

One of the conveniences afforded by curing PPS is that a single uncured feedstock can give rise to an entire family of cured polymers. The flow rates, ie, the extent of cure, of the cured polymers are optimized for specific applications.

Table 1. Typical Flow Rates for PPS Resins

| PPS polymer | Flow rate, g/10 min |
|---|---|
| uncured PPS | 3000–8000 |
| powder coating PPS | 1000 |
| PPS for mineral- or glass-filled compounds | 600 |
| PPS for glass-filled compounds | 60–150 |
| compression-molding PPS | 0 |

Table 1 shows typical melt flow values of cured PPS polymers for various types of applications.

Advances in the understanding of PPS polymerization chemistry in the Phillips Petroleum Company laboratories have made possible the synthesis of high molecular weight linear PPS directly in the polymerization vessel (30,43). Although curable, high molecular weight linear PPS does not require curing prior to end use. The process for the Phillips high molecular weight linear PPS process is essentially the same as that shown in Figure 1 for lower molecular weight Ryton PPS. In the high molecular weight process, alkali metal carboxylate, a polymerization modifier, is added in the $Na_2S$ preparation step; curing is optional. Such high molecular weight polymers are useful as extrusion resins for the production of film, fiber, and extruded profiles, such as pipe, and as feedstocks for injection-molding compounds and advanced composite materials.

The polymers produced by the alkali metal carboxylate-modified Phillips process are off-white polymers having a linear structure. The molecular weight of high molecular weight linear PPS has been reported to be approximately 35,000 (44). More recent measurements (45), however, indicate that weight-average molecular weight for linear, uncured PPS having a flow rate (42) of approximately 150 g/10 min is in the range of 50,000–55,000. High molecular weight linear PPS was commercialized in 1979 by Phillips.

Comonomers can be used to create a variety of polymer structures that can impart desirable properties. For example, even higher molecular weight PPS polymers can be produced by the copolymerization of a tri- or tetrafunctional comonomer (18). The resultant polymer molecules can have long-chain branching, which can be used to tailor the rheological response of the polymer to the application.

The second PPS process practiced commercially was developed by Kureha Chemical Industry Company. Kureha has built a commercial PPS plant in Nishiki, Fukushima (46), and has formed a joint venture, Fortron Industries, with Hoechst Celanese (47). Fortron Industries has completed a commercial PPS plant at Hoechst Celanese's plant in Wilmington, North Carolina. Fortron Industries represents the only other PPS producer in North America. Figure 3 shows a flow diagram for the Kureha PPS process.

Although examples in the Kureha patent literature indicate latitude in selecting hold times for the low and high temperature polymerization periods, the highest molecular weight polymers seem to be obtained for long polymerization times. The addition of water to PPS polymerizations has been reported to effect polymer stabilization (49), to improve molecular weight (50,51), to cause or enhance the formation of a second liquid phase in the reaction mixture (52),

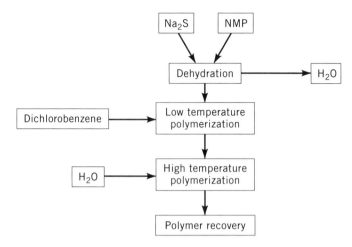

**Fig. 3.** The key steps of the Kureha process, as disclosed in the patent literature (48), are (1) dehydration of aqueous feedstocks (sodium sulfide or its functional equivalent) in the presence of N-methyl-2-pyrrolidinone; (2) polymerization of the dehydrated sodium sulfide with p-dichlorobenzene at a low temperature to form a prepolymer; (3) addition of water to the prepolymer; (4) a second, higher temperature polymerization step; and (5) polymer recovery.

and to help reprecipitate PPS from NMP solution (51). It has also been reported that water can be added under pressure in the form of steam (53).

Other PPS processes have been described in the patent literature by Bayer (54–60). Distinguishing features of the Bayer PPS processes are the use of N-methylcaprolactam, which is a higher boiling analogue to the polar organic compound, NMP, reported by Phillips and Kureha. Many of Bayer's patent examples indicate the use of a branching comonomer, eg, 1,2,4-trichlorobenzene. Other polymerization modifiers reported by Bayer include carboxylic amides (54), carboxylic esters (56), carboxylic anhydrides (56), amino acids (57), and a variety of inorganic salts (58,59). Bayer has also disclosed control over the end group chemistry via chain termination with monofunctional phenols (60) for the purposes of improved control of the polymer melt viscosity and also improved melt stability. Bayer built and operated a commercial PPS plant in Antwerp, Belgium (61), but chose to exit the PPS business in early 1992 (62).

Several polymerization processes for PPS or PPS-like structures have been reported but are not of commercial significance. However, they are of chemical interest. For example, Idemitsu has reported a process for the synthesis of PPS (63) in which an alkali metal sulfide and a dihaloaromatic compound react in a two-phase solvent system comprising high molecular weight ethylene glycol and water. A lower molecular weight poly(ethylene glycol) serves as a phase-transfer catalyst for the sodium sulfide monomer and the sodium chloride by-product. The polymer is reported to be isolated from the poly(ethylene glycol) phase.

A low temperature catalytic process has been reported (64). The process involves the divalent nickel- or zero-valent palladium-catalyzed self-condensation of halothiophenols in an alcohol solvent. The preferred halothiophenol is p-bromothiophenol. The relatively poor solubility of PPS under the mild reaction conditions results in the synthesis of only low molecular weight PPS. An advantage afforded by the mild reaction conditions is that of making telechelic PPS with functional groups that may not survive typical PPS polymerization conditions.

The Eastman Chemical Company has published extensively in the patent literature (65–74) and the scientific literature (75–77) on processes for making poly(phenylene sulfide)-co-(phenylene disulfide), and related copolymers. The Eastman process involves the reaction of elemental sulfur with p-diiodobenzene to yield a phenylene sulfide polymer that also contains phenylene disulfide repeating units in the polymer. The fraction of repeating groups containing

$$\text{I—Ar—I} + \text{S} \longrightarrow \text{+Ar—S}\overline{)_{1-x}}\text{(Ar—S—S}\overline{)_{x}} + \text{I}_2$$

disulfides, $x$ in the equation shown, can be controlled by varying the monomer stoichiometry. Increasing the amount of sulfur reportedly increases the fraction of disulfides found in the polymer (65). Disulfide fractions of 0.001–0.5 are reported for the Eastman process and, surprisingly, have not been found to influence polymer properties such as chemical resistance and thermal stability (65–68,76). Crystallization kinetics, however, are strongly influenced by the disulfide content. When the disulfide fraction is in the range of 0.2–0.5, the resultant polymers are reportedly amorphous or, at best, difficult to crystallize. Polymers having disulfide fractions lower than 0.2 show a clear trend toward faster kinetics of crystallization. Eastman has also reported that polymers made by this process can be further polymerized by a solid-state high temperature treatment in an inert atmosphere (72).

The synthesis of poly(arylene sulfide)s via the thermolysis of bis(4-iodophenyl) disulfide has been reported (78). The process leads to the formation of PPS and elemental iodine. This process presumably occurs analogously to that reported by Eastman Chemical Company.

$$\text{I—Ar—S—S—Ar—I} \xrightarrow{\Delta} \text{+Ar—S}\overline{)_n} + \text{I}_2$$

Alternative synthetic routes to poly(arylene sulfide)s have been published (79–82). The general theme explored is the oxidative polymerization of diphenyl disulfide and its substituted analogues by using molecular oxygen as the oxidant, often catalyzed by a variety of reagents:

$$\text{Ar—S—S—Ar} + \text{O}_2 \xrightarrow{\text{catalyst(s)}} \text{+Ar—S}\overline{)_n} + \text{O}_2^{2-}$$

Other PPS polymerizations involving the electrooxidative polymerization of thiophenols have also been reported (83). Polymerizations described in the early work by this research group were characterized by low molecular weight due to the limited solubility of the product. Although aromatic disulfide-containing monomers are costly, the mild and selective reaction conditions allowed the synthesis of many functionalized PPS analogues. Work by this group has resulted in many patents and patent applications (84–86). Idemitsu researchers have also published synthetic routes to PPS via oxidative polymerization of diphenyl disulfides or thiophenols in the presence of acids, catalysts, and oxygen (87). More recent (1993–1994) work describes the synthesis of other poly(arylene sulfide)s (88) as well as PPS (89,90) via soluble polysulfonium cationic polymer precursors. The route for PPS is based on the self-condensation of methyl 4-phenylthiophenyl sulfide under oxidizing conditions in methane sulfonic acid. The oxidation is catalyzed by cerium(IV) salt. The resultant polysulfonium cation, a soluble precursor for PPS, is demethylated, yielding PPS qualitatively.

Mechanistic studies on the formation of PPS from polymerization of copper(I) 4-bromobenzenethiolate in quinoline under inert atmosphere at 200°C have been published (91). PPS synthesized by this synthetic procedure is characterized by high molar mass at low conversions and esr signals consistent with a single-electron-transfer mechanism, the $S_{RN}1$-type mechanism described earlier (22).

In another process for the synthesis of PPS, as well as other poly(arylene sulfide)s and poly(arylene oxide)s, a pentamethylcyclopentadienylruthenium(I) $\pi$-complex is used to activate $p$-dichlorobenzene toward displacement by a variety of nucleophilic comonomers (92). Important facets of this approach, which allow the polymerization to proceed under mild conditions, are the tremendous activation afforded by the $\pi$-coordinated transition-metal group and the improved solubility of the resultant organometallic derivative of PPS. Decomplexation of the organometallic derivative polymers may, however, be complicated by precipitation of the polymer after partial decomplexation.

## Properties of PPS Neat Resins

Highly desirable properties of PPS include excellent chemical resistance, high temperature thermal stability, inherent flame resistance, good inherent electrical insulating properties, and good mechanical properties.

**Thermal Properties.**    Thermodynamic stability of the chemical bonds comprising the PPS backbone is quite high. The bond dissociation energies (at 25°C) for the carbon–carbon, carbon–hydrogen, and carbon–sulfur bonds found in PPS are as follows: C–C, 477 kJ/mol (114 kcal/mol); C–H, 414 kJ/mol (99 kcal/mol); and C–S, 276 kJ/mol (66 kcal/mol). The large expenditure of energy required to dissociate these bonds (and therefore initiate thermal degradation) implies that PPS should have excellent thermal stability. Thermogravimetric analysis (tga) is used to measure the percentage weight loss of a sample versus temperature. The material is typically heated in a controlled atmosphere at a prescribed heating rate. Figure 4 shows tga results in a nitrogen atmosphere of PPS in comparison to several known polymers. PPS retained approximately 40% of its original weight even after reaching 1000°C.

Tga has been used to calculate apparent activation energies for the thermal degradation of commercially available PPS samples (94) and the activation energies used to predict 10-year maximum use temperatures for Ryton PPS (grade P4) and Fortron PPS (grade 0205B4). This temperature for Ryton PPS was calculated to be between 198 and 235°C, whereas that for Fortron PPS was between 180 and 222°C. Correlations of weight loss data to useful property retention should be supplemented with actual lifetime property measurements. Pyroly-

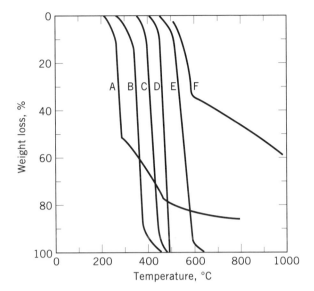

**Fig. 4.** Comparative thermogravimetric analyses of polymers in nitrogen: A, poly(vinyl chloride); B, poly(methyl methacrylate); C, polystyrene; D, polyethylene; E, polytetrafluoroethylene; and F, PPS (93).

sis–gc/ms analysis of volatile compounds obtained from thermal decomposition of PPS showed predominantly random chain scission and cyclization reactions at lower pyrolysis temperatures (up to 550°C) (95). Temperatures higher than 550°C resulted in depolymerization and production of compounds such as benzene and thiophenol.

Other thermal measurements are more relevant to the melt processing of PPS into molded parts. PPS is a semicrystalline polymer comprising a crystalline fraction and an amorphous fraction. The crystalline fraction displays a crystalline melting point, $T_m$, which for PPS is a broad endothermic transition at approximately 285°C as measured by differential scanning calorimetry (dsc). The amorphous region displays a modest glass-transition temperature, $T_g$, at approximately 85°C. A typical dsc thermogram is shown in Figure 5. The crystallization behavior of PPS has been extensively investigated (13,36,96–110) and reviewed (2).

At temperatures between $T_g$ and $T_m$, PPS crystallizes readily. However, because the rate of crystallization is slow, rapid cooling from the melt can result in a molded part that is not fully crystallized. If the cooling is sufficiently rapid, a nearly amorphous part is obtained. Amorphous PPS crystallizes readily when heated to temperatures above $T_g$, as indicated by the exothermic crystallization peak located at approximately 120–130°C (Fig. 5). PPS also crystallizes readily when cooled from the melt. PPS displays a melt crystallization exotherm located between 160 and 250°C, depending on variables such as polymer structure (98–100), end-group chemistry (99), level of cure (36,111), molecular weight (103,105), the presence or absence of materials that function as nucleating agents (109), blends with other polymers (104), and plasticizing agents such as sorbed gases (110). The implication of these studies is that processing conditions can

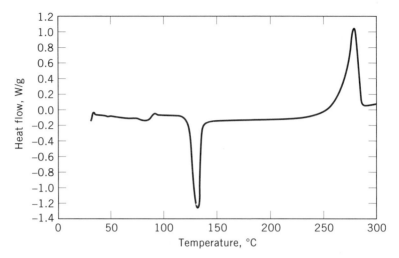

**Fig. 5.** Differential scanning calorimetry thermogram. Amorphous PPS is heated from room temperature to 325°C at 20°C/min.

determine the level of crystallinity in molded PPS parts. Proper choice of molding and/or annealing conditions allows the molder to obtain crystalline parts having both dimensional stability and good mechanical properties.

The value for the heat of fusion of PPS, extrapolated to a hypothetical 100% crystalline state, is not agreed upon in the literature. Reported values range from approximately 80 J/g (19 cal/g) (36,96,101) to 146 J/g (35 cal/g) (102), with one intermediate value of 105 J/g (25 cal/g) (20). The lower value, 80 J/g, was originally measured by thermal analysis and then correlated with a measure of crystallinity determined by x-ray diffraction (36). The value of 146 J/g was determined independently on uniaxially oriented PPS film samples by thermal analysis, density measurement via density-gradient column, and the use of a calculated density for 100% crystalline PPS to arrive at a heat of fusion for 100% crystalline PPS (102). The value of 105 J/g was obtained by measuring the heats of fusion of well-characterized linear oligomers of PPS and extrapolation to infinite molecular weight.

**Melt and Solution Properties.**    Melt rheology of PPS is complicated by the high temperatures needed and the requirement that air be excluded to avoid adventitious oxidative curing. One study (44) of the melt rheology of three commercially important classes of PPS, ie, linear and uncured, branched and uncured, and linear and cured, showed the unexpected result that the Newtonian viscosity of linear, uncured PPS increased with the 4.9 power of the weight-average molecular weight, rather than the expected 3.4 power. A homologous series of samples containing increasing amounts of long-chain branching showed increasingly non-Newtonian behavior. This was rationalized on the basis of the branched polymers having higher weight-average molecular weight and broader molecular weight distributions.

The melt rheology of products of the polymerization of *p*-diiodobenzene and sulfur have been described (112). The properties of PPS made by this synthetic route are complex and different from those made from sodium sulfide (or the chemical equivalent) and $p$-dichlorobenzene. During rheological testing, melt viscosities of PPS made from $p$-diiodobenzene and sulfur showed a significant increase, probably a result of continued polymerization reactions involving disulfide moieties. In contrast to these results (112), PPS synthesized from sodium sulfide and $p$-dichlorobenzene shows nearly constant or slightly decreasing dynamic viscosity as a function of time during measurements at temperatures from 316 to 360°C for up to two hours (113).

Other rheological studies have been directed toward understanding the effect of processing conditions on the properties of PPS neat resin and PPS reinforced with carbon fiber (114). The expected curing effects of processing PPS in air versus nitrogen were found. Similarly, the effects of melt curing on the rheological properties of PPS have been investigated; melt curing of PPS resulted in a polymer still possessing Newtonian behavior, indicating maintenance of a predominantly linear structure (39).

PPS is well-recognized for its exceptional chemical resistance. There are no known solvents for PPS below 200°C. A comprehensive survey of solvents for PPS has been published (115). Extreme conditions are required to dissolve PPS in both common and exotic solvents. Solution viscosity measurements are made difficult by this high temperature requirement. Inherent viscosity measurements are performed in 1-chloronaphthalene at 206°C at a concentration of 0.4 g of polymer per deciliter of solution. The inherent viscosity of PPS solutions shows a useful response to increasing molecular weight. Table 2 shows a correlation of inherent viscosity measurements with melt flow measurements.

Molecular weight distribution for a polymer affects polymer properties. Hence size-exclusion chromatography (sec) has become an essential analytical tool for the synthetic polymer chemist striving to enhance the performance characteristics of a polymer. The molecular weight distribution of PPS was first measured by using a custom-built, high temperature size-exclusion chromatograph (45) utilizing a viscometric detector. The methodology was based on combining the infinite dilution approximation with a universal calibration (116) that allows the use of well-characterized standard polymer samples having different chain structures. This method requires knowledge of the Mark-Houwink-Sakurada parameters, $K$ and $a$, for the particular polymer–solvent system being studied. $K$ was determined as $8.91 \times 10^{-5}$ and $a$ as 0.747 for PPS in 1-chloronaphthalene at 208°C (45). Although the concept of high temperature sec for PPS was clearly proven, viscometric detection of low molecular weight species was found to have low sensitivity. To improve the detection of low molecular weight species, a flame ionization detector (FID) was employed as a concentration detector (117). Concern about complete removal of the carrier solvent, 1-chloronaphthalene, from

**Table 2. Inherent Viscosities of PPS**[a]

| PPS polymerization process | Melt flow, g/10 min | Inherent viscosity, dL/g |
|---|---|---|
| unmodified | >6000 | 0.18 |
| carboxylate-modified | 1295 | 0.24 |
| | 665 | 0.28 |
| | 93 | 0.35 |

[a]Ref. 43.

the moving quartz belt of the FID led to the speculation that this method, too, might suffer from detector nonuniformities for low molecular weight species (118), but this concern was later shown to be unfounded in a study using dual detector sec (119) in which virtually identical chromatograms were obtained using FID and uv–vis detectors. Use of a nonaromatic carrier solvent, 1-cyclohexyl-2-pyrrolidinone, has extended the window for uv–vis detection, allowing polystyrene standards to be used in an sec experiment with PPS (120). Suggestions of the utility of sec for PPS have also been found in the patent literature (121,122); however, workable details of the analytical technique are not disclosed.

## Properties of PPS Injection-Molding Compounds

PPS injection-molding compounds are distinguished among other thermoplastic compounds for their high performance properties, including excellent high temperature resistance, dimensional stability, and chemical resistance; inherent flame resistance; high electrical resistance properties; and a good balance of mechanical properties.

**Thermal Properties.** The inherent thermal stability of PPS translates into high temperature resistance for short- as well as long-term exposure and retention of properties at elevated temperature.

Highest thermal performance with PPS compounds requires that parts be molded under conditions leading to a high level of crystallinity. Glass-filled PPS compounds can be molded so that crystalline or amorphous parts are obtained. Mold temperature influences the crystallinity of PPS parts. Mold temperatures below approximately 93°C produce parts with low crystallinity and those above approximately 135°C produce highly crystalline parts. Mold temperatures between 93 and 135°C yield parts with an intermediate level of crystallinity. Part thickness may also influence the level of crystallinity. Thinner parts are more responsive to mold temperature. Thicker parts may have skin-core effects. When thick parts are molded in a cold mold the skin may not develop much crystallinity. The interior of the part, which remains hot for a longer period of time, may develop higher levels of crystallinity.

A useful measure of an engineering material's resistance to short-term exposure to heat is the heat deflection temperature (HDT). This test (ASTM D648) indicates the temperature at which a molded test specimen deflects 0.254 mm under a stress (1.82 MPa for engineering plastics) when heated at a rate of 2°C/min. This test does not predict long-term thermal performance of a plastic. The heat deflection temperatures for a variety of fiber glass-reinforced engineering plastics are shown in Table 3. Glass-filled PPS compounds have high heat deflection temperatures, indicating excellent short-term retention of properties at high temperatures. This is important when a part is exposed to high temperatures for short periods of time, such as in soldering operations where temperatures are typically 224°C, but can reach 260°C, for short periods (usually <1 min).

Long-term exposure to high temperature is best described by the Underwriters' Laboratory (UL) temperature index. The UL temperature index is determined by oven-aging test specimens at several temperatures, usually considerably higher than the expected use temperature. Samples are removed at

**Table 3. Heat Deflection Temperatures of Various Fiber Glass-Reinforced Engineering Materials**[a]

| Polymer | Heat deflection temperature, °C |
|---|---|
| liquid crystal polymer (LCP) | 241 |
| poly(phenylene sulfide) (PPS) | 252–268 |
| poly(etherimide) (PEI) | 209–215 |
| poly(cyclohexyldimethylene terephthalate) (PCT) | 249 |
| poly(ethylene terephthalate) (PET) | 224 |
| polyphthalamide (PPA) | 275 |
| phenolic resin | 182–315 |
| polycarbonate (PC) | 142–152 |

[a]Ref. 123.

regular intervals and tested (usually tensile strength) at room temperature. A plot is made for each temperature to determine the time required for 50% property reduction. These times are used to make a plot of time to failure versus temperature. Extrapolation of the data to an arbitrary time (usually 100,000 h) results in the temperature index. Because this method measures long-term polymer degradation as a function of time, it is a good measure of a material's long-term performance at elevated temperature. UL temperature indexes are usually reported as ranges to account for different properties tested. The UL temperature indexes for a variety of engineering plastics are in Table 4. Glass-filled PPS compounds exhibit a high UL temperature index, which indicates excellent retention of properties for long-term exposure to high temperature. Other fillers, eg, mineral (talc), may also be used.

**Table 4. UL Index of Various Glass-Filled Engineering Materials**[a]

| Polymer | UL temperature index, °C |
|---|---|
| liquid crystal polymer (LCP) | 220–260 |
| poly(phenylene sulfide) (PPS) | 200–240 |
| poly(etherimide) (PEI) | 170–180 |
| phenolic resin | 150–180 |
| poly(ethylene terephthalate) (PET) | 120–150 |
| poly(cyclohexyldimethylene terephthalate) (PCT) | 120–140 |
| polycarbonate (PC) | 100–125 |

[a]Ref. 124.

For semicrystalline polymers such as PPS, the relationship between a strength property and the temperature is not linear. As the test specimen goes through the glass-transition temperature (85–95°C for PPS), the strength of the material falls off fairly rapidly. However, with semicrystalline materials, a significant amount of mechanical strength is retained due to the crystalline portion of the polymer matrix. The effect of temperature on flexural strength of a 40% glass-filled PPS is shown in Figure 6. This figure demonstrates that glass-filled PPS compounds exhibit good strength up to 200°C.

**Dimensional Stability.** Plastics, in general, are subject to dimensional change at elevated temperature. One important change is the expansion of

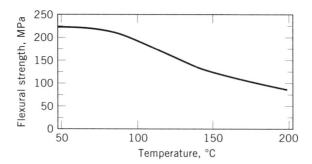

**Fig. 6.**  Flexural strength vs temperature for Ryton PPS R-4 02XT. To convert MPa to psi, multiply by 145.

plastics with increasing temperature, a process that is also reversible. However, the coefficient of thermal expansion (CTE), measured according to ASTM E831, frequently is not linear with temperature and may vary depending on the direction in which the sample is tested, that is, samples may not be isotropic (Fig. 7).

Long-term irreversible dimensional change is described by creep modulus, which provides a numerical value to predict the load-bearing capability of a material as a function of temperature and time. Apparent creep modulus, measured according to ASTM D2990, is determined by applying a constant tensile load and measuring the percentage of linear strain as a function of time. At constant load, any change in measured strain can be considered tensile creep. A calculation of the tensile modulus at this point in time can be made by dividing the applied stress by the measured strain. Changing test conditions such as temperature and stress level provide data for specific applications. The tensile creep modulus of a 40% glass-filled PPS at various stress levels and temperatures is shown in Figure 8. Glass-filled PPS exhibits low creep even at high load and temperature.

**Chemical Resistance.**  The chemical resistance of PPS compounds is outstanding, even at elevated temperatures, but as an organic polymer, PPS can be

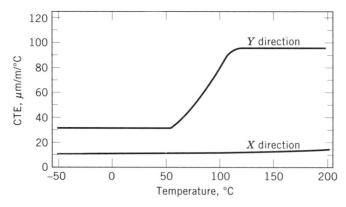

**Fig. 7.**  CTE of a 40% glass-filled PPS as functions of both temperature and direction. CTE values are measured in micrometers of dimensional change per meter of sample per degree C.

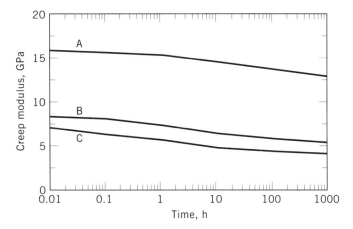

**Fig. 8.**  Tensile creep of Ryton PPS R-4XT. Conditions: A, 65°C and 34.5 MPa (5000 psi) stress; B, 121°C, 34.5 MPa; and C, 121°C, 70 MPa (10,000 psi). To convert GPa to psi, multiply by 10,000.

affected by some chemicals under certain conditions. Time and temperature are critical factors which must be considered when determining the level of chemical resistance required for a specific application. The effective chemical resistance should be evaluated on the basis of how well the material performs over time to chemical exposure relative to the required performance level. In a comparison of chemical resistance of various plastics (125), five materials were exposed to 127 different reagents for 24 h at 93°C. A passing grade was assigned if the material retained at least 75% of its initial tensile strength. PPS was the best performer, passing on 120 reagents. Phenolic resins were closest at ∼109, followed by nylon at 70, PPO at 65, and PC at 50. More detailed discussions of specific chemical exposure have been published (1,3). To characterize the chemical resistance of PPS compounds, tensile specimens of 40% glass-filled PPS were immersed in various chemicals at 93°C and tested periodically for their retention of tensile strength. Based on these data, a rating was established as a general guide to illustrate the degrees of chemical resistance one might expect PPS compounds to exhibit within a chemical class (125).

**Flame Resistance.**  Because of its aromatic structure and its tendency to char when exposed to an external flame, PPS is inherently flame-resistant. There are several tests that measure the flame resistance of plastics, including ignition temperature, flash point, UL flammability rating, oxygen index, and smoke obscuration time. The ignition temperature is a UL test which determines the minimum temperature that induces the material to burn. For 40% glass-filled PPS, the minimum reported combustion temperature is 540°C. The flash point for 40% glass-filled PPS was determined by ASTM D1929 and is above 499°C. In the standard UL 94 laboratory flammability tests, 40% glass-filled PPS is classified as either V-0 or 5VA, depending on the thickness of the test specimen. This rating indicates that the compound neither supports prolonged combustion nor drips flaming particles. The oxygen index is a relative indication of flammability under a specific set of conditions. The oxygen index is the minimum concentration of oxygen required to maintain continued burning of the material. The oxygen index of glass-filled PPS is greater than 46%, which demonstrates its excellent flame resistance. The NIST smoke test yields the time to reach a critical smoke density called the obscuration time. This test is designed to provide an estimate of the time available before typical occupants in typical rooms find their vision obscured by smoke that hinders escape. The obscuration time for glass-filled PPS is 15.5 min smoldering and 3.2 min flaming.

**Electrical Properties.**  PPS resins and compounds possess good overall electrical properties which indicate that they are good insulators. The importance of any particular electrical property depends on the application (either high or low voltage use). For high voltage applications, the key properties are dielectric strength, arc resistance, and comparative tracking index (CTI). The dielectric strength of a material, measured by ASTM D149, is the voltage required for electrical breakdown through the material. It is obtained by dividing the total breakdown voltage by the thickness of the specimen, and is reported in kilovolts per millimeter. The arc resistance of a material is a measure of the ability of a material to resist the formation of a permanent conductive path or track when exposed to arcing. Arc resistance measurements, tested by ASTM D495, determine the resistance of specimens to surface tracking by a high voltage arc. The value reported is the average number of seconds required to initiate surface tracking. CTI is also a measure of arc resistance; however, it is measured in a wet environment with an electrolyte present. Table 5 summarizes these high voltage properties for both 40% glass-filled PPS and a glass–mineral-filled PPS. PPS compounds that contain minerals in addition to glass fiber tend to exhibit better high voltage properties (see also Table 4).

For low voltage applications, the key properties are dielectric constant, dissipation factor, volume resistivity, and insulation resistance. The dielectric constant, measured by ASTM D150, determines the extent to which a material polarizes when placed in an electric field. It can be significant because it affects the amount of energy stored in dielectric components of an electrical circuit. Although a small added capacitance may not cause a problem, it is usually preferable to minimize the dielectric constant rather than to compensate for it via circuitry design. Thus, low dielectric constant is particularly desirable for

**Table 5. Electrical Properties of PPS[a]**

| Electrical property | Ryton PPS R-4XT, 40% glass-filled | Ryton PPS R-7, glass–mineral-filled |
|---|---|---|
| *High voltage* | | |
| dielectric strength, kV/mm | 19.7 | 17.7 |
| arc resistance, s | 130 | 167 |
| comparative tracking index, V | 130 | 225 |
| *Low voltage* | | |
| dielectric constant at 1 MHz | 3.9 | 4.0 |
| dissipation factor at 1 MHz | 0.004 | 0.0088 |
| volume resistivity, $\Omega \cdot$cm | $1 \times 10^{16}$ | $1 \times 10^{15}$ |
| insulation resistance, $\Omega$ at 90°C, 95% rh, 48 h | $1 \times 10^{11}$ | $1 \times 10^{12}$ |

[a]Ref. 126.

communications and electronic circuits that rely on rapid switching and low loss transmission. Dissipation factor, measured by ASTM D150, is the ratio of energy dissipated as heat compared to the energy stored in the system. The energy dissipated as heat in many applications remains small in terms of temperature rise; however, the effect on the signals being transmitted can be substantial. Volume resistivity, measured by ASTM D257, is the inherent resistance of a material to current flow through its volume. It is generally an indication of a material's insulation characteristics. Insulation resistance is a measure of surface or volume resistance at use conditions, combining high temperature and high humidity. Table 5 summarizes the low voltage electrical properties for both 40% glass-filled PPS and glass–mineral-filled PPS. Glass-filled PPS compounds exhibit the lowest dielectric constant and dissipation factor, but they all possess a good combination of low voltage electrical properties.

**Mechanical Properties.** Articles molded from PPS compounds are generally characterized by high strength, high stiffness, and moderate impact. In addition to standard grades, PPS can be custom-tailored with fillers, additives, and fibers to suit specific applications. The mechanical properties of 40% glass-filled and glass–mineral-filled PPS is shown in Table 6. PPS compounds possess a good combination of mechanical properties.

**Table 6. Mechanical Properties of PPS Compounds[a]**

| Property | Ryton PPS R-4XT, 40% glass-filled | Ryton PPS R-7, glass–mineral-filled |
|---|---|---|
| tensile strength, MPa[b] | 193 | 128 |
| elongation, % | 1.6 | 0.8 |
| flexural strength, MPa[b] | 276 | 186 |
| flexural modulus, GPa[c] | 14.5 | 17.2 |
| modulus of elasticity, GPa[c] | 15.1 | 17.2 |
| compressive strength, MPa[b] | 235 | 172 |
| Izod impact, J/m[d] | | |
| notched | 85 | 70 |
| unnotched | 560 | 240 |

[a]Ref. 126.
[b]To convert MPa to psi, multiply by 145.
[c]To convert GPa to psi, multiply by 145,000.
[d]To convert J/m to lbf·ft/in., divide by 53.38.

## Applications

PPS resin and compounds combine a unique and useful combination of properties, facile processing conditions, and an attractive price for the level of performance afforded by the polymer. Inherent properties of PPS that determine the utility of PPS resins and compounds include excellent electrical insulation, excellent long- and short-term thermal stability, inherent flame resistance, easy moldability, outstanding chemical resistance, good mechanical properties, and dimensional stability of molded parts. PPS finds acceptance in a wide variety of applications, including coatings, injection-molding compounds, fiber, film, composites, blends, and alloys. Market segments utilizing PPS resins and compounds comprise electrical, electronic, automotive, appliance, and industrial.

**Coatings.**   PPS coatings can be applied to a variety of substrates, usually metal, by a variety of techniques and have been reviewed (127,128). Corrosion-resistant, pinhole-free, thermally stable coatings of PPS having good release characteristics can be applied to steel, aluminum, and other metals from aqueous slurries of PPS, by electrostatic powder coating, fluidized-bed coating, and powder spraying and flocking. Cured coatings of PPS with proper additives may be used in cookware applications (129).

Low molecular weight linear PPS is conventionally used in coatings. A curing cycle is required to melt the polymer and cure the fused coating into a tough, insoluble coating. Cure time and temperature must be selected so that the polymer melts and forms a contiguous coating, and that appropriate curing takes place to yield a durable coating. Undercured coatings are brittle as a result of low molecular weight and polymer crystallization. Overcured coatings are likewise brittle because the cross-link density is too high. Optimum curing conditions have been described (127).

Substrate surface preparation such as grit blasting, heat treating, and primer application can enhance adhesion of PPS to steel. Adhesion to aluminum is good and does not require special pretreatments. Downhole oil well applications are challenging adhesion tests for any coating. Special low molecular weight grades of PPS have been developed (130) which provide good adhesion in these demanding environments.

**Injection-Molding Compounds.**   PPS injection-molding compounds typically contain reinforcing fibers such as glass or carbon fiber, often with coupling agents, and/or mineral fillers to enhance specific properties. Filled compounds also have good processibility and can be used for molding of intricate parts. Typical barrel temperatures range between 300 and 360°C. High temperature increases flow and reduces equipment wear. Typical molding conditions utilize high injection pressure, slow injection rate, medium screw speed, and maximum clamping pressure.

Mold temperature can vary widely. Typical temperatures range from 40 to 150°C. Higher mold temperatures favor polymer crystallization and result in more dimensionally stable parts. Crystallinity can be developed in parts molded in cold molds by annealing at approximately 200°C.

Although PPS compounds do not absorb much water, drying is recommended to minimize any effects of hygroscopic fillers. Typical drying conditions are 150°C for six hours.

**Fiber.**   High molecular weight linear PPS is well-suited for fiber applications. The inherent properties of PPS (flame resistance, chemical resistance, and thermal stability) make PPS fiber highly desirable in textile applications (128). PPS fiber has been designated by the U.S. Federal Trade Commission as a new generic class of materials called sulfar. Typical fiber properties are listed in Table 7 (see HIGH PERFORMANCE FIBERS).

PPS fiber has excellent chemical resistance. Only strong oxidizing agents cause degradation. As expected from inherent resin properties, PPS fiber is flame-resistant and has an autoignition temperature of 590°C as determined in tests at the Textile Research Institute. PPS fiber is an excellent electrical

**Table 7. Properties of PPS Fiber**

| Property | Value |
|---|---|
| tenacity, N/tex[a] | 0.31 |
| elongation, % | 40 |
| modulus, N/tex[a] at 10% extension | 1.4 |
| elastic recovery, % | |
|     2% extension | 100 |
|     5% extension | 96 |
|     10% extension | 86 |
| dry shrinkage, % at 130°C | 4 |
| moisture regain, % | 0.6 |
| specific gravity | 1.37 |

[a]To convert N/tex to gf/den, divide by 0.08826.

insulator; it finds application in hostile environments such as filter bags for filtration of flue gas from coal-fired furnaces, filter media for gas and liquid filtration, electrolysis membranes, protective clothing, and composites.

**Composites.** High molecular weight PPS can be combined with long (0.6 cm to continuous) fiber to produce advanced composite materials (131). Such materials having PPS as the polymer matrix have been developed by using a variety of reinforcements, including glass, carbon, and Kevlar fibers as mat, fabric, and unidirectional reinforcements. Thermoplastic composites based on PPS have found application in the aircraft, aerospace, automotive, appliance, and recreation markets (see COMPOSITES, POLYMER-MATRIX).

The preparation of composite materials has been described extensively in the patent literature (132–142). Stampable sheet composites are prepared by combining PPS with chopped or continuous fiber mat. Long reinforcing fibers result in dramatic increases in toughness. Mat-reinforced composite materials can be rapidly compression-molded into parts. Unidirectional (132,139) and woven fabric prepreg materials (137) are produced using proprietary processes. Comingling of PPS fiber and long reinforcing fibers produce yarns that, when heated above the melting point of PPS, produce long-fiber-reinforced PPS composite materials (138). PPS resins have been custom-synthesized for composite applications (133–135) that result in improved composite properties. PPS prepreg materials have unlimited shelf life, are tack-free, and, unlike many of the thermosetting prepreg materials, do not require refrigerated storage. PPS prepregs can be easily laid up and held into place by spot welding. They are converted into laminates by heating the prepregs above the crystalline melting point of the polymer and applying pressure of about 345–1380 kPa (50–200 psig). PPS laminates and prepregs can be further shaped in subsequent melt processing operations (139). PPS prepreg material can be formed into useful shapes using a heated filament winding technique (140). Subsequent operations can be carried out to form ribs and other reinforcing shapes on already formed laminates (141,142). PPS composite materials are characterized by exceptional flexibility in manufacturing technology. Typical properties of PPS composites are shown in Table 8.

**Table 8. Properties of PPS Composites**

| Property | Stampable sheet | Prepreg or laminate |
|---|---|---|
| fiber reinforcement | glass, carbon | glass, carbon, aramid |
| fiber form | random mat | unidirectional, fabric |
| fiber loading, wt % | 20–40 | 40–70 |
| strength, MPa[a] | 90–207 | 207–1724 |
| modulus, GPa[b] | 6.2–13.8 | 13.8–138 |
| impact, J/m[c] | 534–1335 | 64–1600 |

[a]To convert MPa to psi, multiply by 145.
[b]To convert GPa to psi, multiply by 145,000.
[c]To convert J/m to lbf·ft/in., divide by 53.38.

**Film.**   High molecular weight PPS is suitable for film-making applications (143). PPS film is amenable to biaxial orientation (144). Biaxially oriented film is manufactured by extruding sheet that must be amorphous to accommodate the drawing operation. After biaxial stretching, the drawn film is heat-set to allow polymer crystallization. Biaxial orientation of amorphous PPS results in some strain-induced crystallinity as indicated by a modest increase in density. The level of strain-induced crystallinity is likely to be influenced by process variables such as the rate and temperature of drawing, as well as polymer variables such as the molecular weight, molecular weight distribution, and the distribution of structural features (eg, long-chain branching) within the molecular weight distribution. Amorphous PPS film has a density of $1.321–1.323$ g/cm$^3$, which increases to approximately $1.323–1.325$ g/cm$^3$ after biaxial orientation. Subsequent heat setting increases the density to approximately $1.36$ g/cm$^3$.

Biaxially oriented PPS film is transparent and nearly colorless. It has low permeability to water vapor, carbon dioxide, and oxygen. PPS film has a low coefficient of hygroscopic expansion and a low dissipation factor, making it a candidate material for information storage devices and for thin-film capacitors. Chemical and thermal stability of PPS film derives from inherent resin properties. PPS films exposed to toluene or chloroform for eight weeks retain 75% of their original strength. The UL temperature index rating of PPS film is 160°C for mechanical applications and 180°C for electrical applications. Table 9 summarizes the properties of PPS film.

**Blends.**   Several research groups have investigated the properties of blends comprising PPS and thermotropic liquid crystalline copolyesters. These blends have been found to be incompatible, resulting in a dispersed LCP-phase in a continuous PPS matrix (145). The dispersed LCP-phase possesses a processing-dependent fibrilar morphology. Films of these blends were drawn, causing orientation of the LCP fibrils and thereby producing dramatic improvements in the torsional modulus. Injection-molded specimens of PPS–LCP blends were studied to determine mechanical properties and morphology (146). Tensile and impact properties of the blends were improved over those of pure PPS. PPS was found to be incompatible with the LCP. Blends of PPS and LCP were extruded into monofilaments and the mechanical properties of drawn fibers studied (147). The strength and modulus of fibers of PPS–LCP

**Table 9. Properties of Biaxially Oriented PPS Film**

| | |
|---|---|
| tensile yield, MPa[a] | 90–110 |
| tensile strength, MPa[a] | 117–158 |
| elongation, % | 40–60 |
| tear strength, N/mm[b] | 1.57–3.14 |
| haze, % | 2–10 |
| shrinkage | 2–5 |
| thermal expansion coefficient, cm/cm/°C | $2.2 \times 10^{-5}$ |
| hygroscopic expansion coefficient, cm/cm/%rh | $0.18 \times 10^{-5}$ |
| dielectric constant at 1.0 kHz | 3.0 |
| dissipation factor at 1.1 kHz | 0.0005 |
| volume resistivity, $\Omega \cdot$cm | $1.3 \times 10^{-17}$ |
| permeability, nmol/(m·s·GPa)[c] | |
| $\quad H_2O$ | 1.44 |
| $\quad CO_2$ | 84 |
| $\quad O_2$ | 20 |

[a]To convert MPa to psi, multiply by 145.

[b]To convert N/mm to ppi, divide by 0.175; to convert to Elmendorf tear (gf/mil), multiply by 2.549.

[c]To convert nmol/(m·s·GPa) to (cc·mil)/(100 in. $^2$·d·atm), divide by 2.

blends were higher than those of pure PPS. A study of the morphology and crystallization kinetics of blends of PPS with thermotropic liquid crystalline copolyesters showed that PPS–LCP blends are incompatible (148); however, the crystalline morphology of the PPS showed considerably smaller spherulite size. The linear growth rate of the PPS spherulites was shown by optical microscopy to be virtually unaffected by the presence of the LCP. The bulk kinetics of crystallization, however, were enhanced by the presence of the LCP, leading to the conclusion that the LCP functioned as a nucleating agent for PPS.

The effect of a second polymer blended with PPS which causes enhanced nucleation of PPS has been previously observed. It was found that low concentrations (1–2 wt %) of poly(phenylene sulfide ketone) and poly(ether ether ketone), when melt-blended with PPS, function effectively to increase the nucleation density of PPS (149).

The crystallization kinetics of blends of PPS and poly(ethylene terephthalate) have been studied (150–152). Isothermal crystallization kinetics of PET–PPS blends showed that the crystallization of the PPS was accelerated over that of pure PPS. The nucleation density for PPS was increased in the PPS–PET blend as determined by polarized light microscopy, resulting in smaller spherulite size. PPS blends with a variety of nylons have been reported (153,154). Blends comprising aromatic amorphous appeared to impart greater toughness than corresponding blends made with aliphatic nylons. Blends of PPS and polyethylene (155) were prepared by melt-mixing and displayed immiscibility at all blend ratios. Blends of PPS and polyetherimide have been prepared by reactive extrusion (156). Compatibilization of PPS and PEI, presumably accomplished by block copolymers of PPS and PEI being synthesized during reactive extrusion, resulted in a dramatically reduced PEI domain size.

## Health and Safety Considerations

Information on health and safety considerations cited herein for Ryton PPS powders and pellets can be found in Reference 157. Ryton PPS [26125-40-6] is listed in the Toxic Substance Control Act (TSCA) Inventory of Chemicals.

For personal protection when using PPS, employ adequate ventilation to control airborne powder concentration and off-gases from molding and extruding processes. Local exhaust may be needed to control off-gases. No respiratory protection is generally required unless needed to control respiratory irritation from dust or off-gases. To control off-gases during molding, use a NIOSH- or Mine Safety Health Administration (MSHA)-approved air purifying respirator equipped with an organic vapor cartridge and face mask. For eye protection, use safety glasses with side shields and provide eyewash stations in the work area. No special garments are required for skin protection; avoid unnecessary skin contamination with material. Use heat-resistant gloves when handling hot or molten material. When cleaning thermal decomposition off-gas condensate from equipment, use full-body, long-sleeved garments to prevent skin contact. Molten polymer may cause severe thermal burns. The interior of molten masses may remain hot for some time because of the low thermal conductivity of the polymer.

PPS dust should be treated as a nuisance particulate. The OSHA permissible exposure limit for respirable dust is 5 mg/m$^3$ for dust containing no asbestos and less than 1% silica. The principal decomposition products released during molding of PPS and their permissible exposure limits are given in Table 10. Sulfur dioxide and carbonyl sulfide are the most significant off-gases for production of mucous membrane irritation.

Table 10. Exposure Limits for PPS Decomposition Products

| Decomposition product | OSHA PEL, ppm |
| --- | --- |
| carbon dioxide | 10,000 |
| carbon monoxide | 35 |
| sulfur dioxide | 2 |
| carbonyl sulfide | not established |

Acute effects of overexposure are as follows. Exposure to dust may cause mechanical irritation of the eye. PPS is essentially nonirritating to the skin, although freshly molded material may occasionally cause dermatitis. Inhalation of PPS dust may cause mechanical irritation to mucous membranes of nose, throat, and upper respiratory tract.

Subchronic effects of overexposure have been studied in feeding tests of PPS powder at dietary levels of up to 5%. No detrimental effects in laboratory animals were observed (157).

The flash point of PPS, as measured by ASTM D1929, is greater than 500°C. Combustion products of PPS include carbon, sulfur oxides, and carbonyl sulfide. Specific hazards are defined by the OSHA Hazard Communication Standard (158). Based on information in 1995, PPS does not meet any of the hazard definitions of this standard.

# BIBLIOGRAPHY

"Polymers Containing Sulfur, Poly(Phenylene Sulfide)" in *ECT* 3rd ed., Vol. 18, pp. 793–814, by H. W. Hill, Jr., and D. G. Brady, Phillips Petroleum Co.

1. J. F. Geibel and R. W. Campbell, in G. C. Eastmond, and co-workers, eds., *Comprehensive Polymer Science*, Vol. 5, Pergamon Press, Oxford, U.K. 1989, pp. 543–560.
2. L. C. Lopez and G. L. Wilkes, *J. Macromol. Sci., Rev. Macromol. Chem. Phys.* C29(1) 83–151 (1989).
3. J. F. Geibel and R. W. Campbell, in J. J. McKetta and W. A. Cunningham, eds., *Encyclopedia of Chemical Processing and Design*, Marcel Dekker, Inc., New York, 1992, pp. 94–125.
4. R. S. Shue, *Dev. Plast. Tech.* **2**, 259–295 (1985).
5. M. C. Gabriele, *Plastics Tech.* **38**, 59–62 (1992).
6. C. Friedel and J. M. Crafts, *Ann. Chim. Phys.* **14**(6) 433–472 (1888).
7. N. G. Gaylord, *Polyethers*, Interscience, New York, 1962, p. 31.
8. P. Genvresse, *Bull. Soc. Chim. Fr.* **17**, 599 (1897).
9. R. W. Lenz and W. K. Carrington, *J. Polym. Sci.* **41**, 333–358 (1959).
10. H. A. Smith, *Encycl. Polym. Sci. Tech.* **10**, 653–659 (1969).
11. J. W. Cleary, in B. M. Culbertson and J. E. McGrath, eds., *Advances in Polymers Synthesis*, Plenum Press, New York, 1985, pp. 159–172.
12. A. D. Macallum, *J. Org. Chem.* **13**, 154–159 (1948).
13. D. G. Brady, *J. Appl. Polym. Sci., Appl. Polym. Symp.* **36**, 231–239 (1981).
14. U.S. Pat. 2,513,188 (June 27, 1950) and 2,538,941 (Jan. 23, 1951), A. D. Macallum.
15. R. W. Lenz and C. E. Handlovits, *J. Polym. Sci.* **43**, 167–181 (1960).
16. R. W. Lenz, C. E. Handlovits, and H. A. Smith, *J. Polym. Sci.* **58**, 351–367 (1962).
17. H. A. Smith and C. E. Handlovits, *Report on Conference on High Temperature Polymer and Fluid Research, Part II, Phenylene Sulfide Polymers*, ASD-TDR-62-322, Dayton, Ohio, 1962, pp. 18, 19.
18. U.S. Pat. 3,354,129 (Nov. 21, 1967), J. T. Edmonds and H. W. Hill (to Phillips Petroleum Co.).
19. V. A. Sergeev and V. I. Nedelkin, *Makromol. Chem. Macromol. Symp.* **26**, 333–346 (1989).
20. W. Koch and W. Heitz, *Makromol. Chem.* **184**, 779–792 (1983).
21. W. Koch, W. Risse, and W. Heitz, *Makromol. Chem. Suppl.* **12**, 105–123 (1985).
22. J. F. Bunnett, *Acc. Chem. Res.* **11**, 413–420 (1978).
23. V. Z. Annenkova and co-workers, *Dokl. Akad. Nauk SSSR*, **286**, 1400–1403 (1986).
24. V. Z. Annenkova and co-workers, *Vysokomol. Soed. Ser. B*, **28**, 137–140 (1986).
25. M. Novi, G. Petrillo, and M. L. Sartirana, *Tetrahedron Lett.* **27**, 6129–6132 (1986).
26. D. R. Fahey and C. E. Ash, *Macromolecules*, **24**, 4242–4249 (1991).
27. J. Miller and A. J. Parker, *J. Am. Chem. Soc.* **83**, 117–123 (1961).
28. B. G. Cox and A. J. Parker, *J. Am. Chem. Soc.* **95**, 408–410 (1973).
29. J. R. Campbell, *J. Org. Chem.* **29**, 1830–1833 (1964).
30. H. W. Hill, *I. EC Prod. Res. Dev.* **18**, 252–253 (1979).
31. C. J. Stacy, *Polym. Prepr. Am. Chem. Soc. Polym. Div. Chem.* **26**(1), 180, 181 (1985).
32. Ref. 17, p. 123.
33. R. M. Black, C. F. List, and R. J. Wells, *J. Appl. Chem.* **17**, 269–275 (1967).
34. G. F. L. Ehlers, K. R. Fish, and W. R. Powell, *J. Polym. Sci. A-1*, **7**, 2955–2967 (1969).
35. A. B. Port, and R. H. Still, *Poly Deg. Stab.* **2**, 1–22 (1980).
36. D. G. Brady, *J. Appl. Polym. Sci.* **20**, 2541–2551 (1976).
37. M. Wejchan-Judek, B. Perkowska, and B. Karska, *J. Mat. Sci. Lett.* **12**, 433–435 (1993).
38. R. T. Hawkins, *Macromolecules*, **9**(2), 189–194 (1976).

39. J. J. Scobbo, *J. Appl. Polym. Sci.* **48**, 2055–2061 (1993).

40. U.S. Pat. 3,524,835 (Aug. 18, 1970), J. T. Edmonds, Jr., and H. W. Hill (to Phillips Petroleum Co.).

41. U.S. Pat. 3,717,620 (Feb. 20, 1973), R. G. Rolfing (to Phillips Petroleum Co.).

42. ASTM D1238, American Society for Testing Materials, Philadelphia, Pa.

43. U.S. Pat. 3,919,177 (Nov. 11, 1975), R. W. Campbell (to Phillips Petroleum Co.).

44. G. Kraus and W. M. Whitte, *Proceedings from the 28th Macromolecular Symposium of the IUPAC*, Amherst, Mass., July 12, 1982.

45. C. J. Stacy, *J. Appl. Polym. Sci.* **32**, 3959–3969 (1986).

46. *Chem. Week*, **140**, 30 (Apr. 15, 1987).

47. *Chem. Eng. News*, **70**, 7, 8 (June 15, 1992).

48. U.S. Pat. 4,645,826 (Feb. 24, 1987), Y. Iizuka and co-workers (to Kureha Kagaku Kogyo Kabushiki Kaisha).

49. U.S. Pat. 4,071,509 (Jan. 31, 1978), J. T. Edmonds, Jr. (to Phillips Petroleum Co.).

50. U.S. Pat. 4,116,947 (Sept. 26, 1978), J. T. Edmonds, Jr., and L. E. Scoggins (to Phillips Petroleum Co.).

51. U.S. Pat. 4,748,231 (May 31, 1988), A. M. Nesheiwat (to Phillips Petroleum Co.).

52. U.S. Pat. 4,415,729 (Nov. 15, 1983), L. E. Scoggins and B. L. Munro (to Phillips Petroleum Co.).

53. U.S. Pat. 4,963,651 (Oct. 16, 1990), A. M. Nesheiwat (to Phillips Petroleum Co.).

54. U.S. Pat. 4,433,138 (Feb. 21, 1984), K. Idel, D. Freitag, and L. Bottenbruch (to Bayer AG).

55. U.S. Pat. 4,663,430 (May 5, 1987), E. Ostlinning and K. Idel (to Bayer AG).

56. Ger. Pat. DE 3,428,986 (Feb. 20, 1986), K. Idel, E. Ostlinning, and D. Freitag (to Bayer AG).

57. Ger. Pat. DE 3,428,984 (Feb. 20, 1986), K. Idel and co-workers (to Bayer AG).

58. Ger. Pat. DE 3,205,996 (Sep. 1, 1983), K. Idel and co-workers (to Bayer AG).

59. Ger. Pat. DE 3,019,732 (Dec. 3, 1981), K. Idel and co-workers (to Bayer AG).

60. U.S. Pat. 4,771,120 (Sep. 13, 1988), W. Alewelt and co-workers (to Bayer AG).

61. M. Roberts, *Chem. Week*, **147**(9), 9 (Aug. 8, 1990).

62. M. S. Reisch, *Chem. Eng. News*, **71**(35), 24–37 (Aug. 30, 1993).

63. Jpn. Pat. 61,145,226 (July 2, 1986), R. G. Sinclair, H. B. Benekay, and S. Sowell (to Idemitsu Petrochemicals Co.).

64. U.S. Pat. 4,841,018 (June 20, 1988), R. G. Gaughan (to Phillips Petroleum Co.).

65. U.S. Pat. 4,786,713 (Nov. 22, 1988), M. Rule and co-workers (to Eastman Kodak Co.).

66. U.S. Pat. 4,792,600 (Dec. 20, 1988), M. Rule, D. R. Fagerburg, and J. J. Watkins (to Eastman Kodak Co.).

67. U.S. Pat. 4,826,956 (May 2, 1989), D. R. Fagerburg, P. B. Watkins, and M. Rule (to Eastman Kodak Co.).

68. U.S. Pat. 4,855,393 (Aug. 8, 1989), M. Rule and co-workers (to Eastman Kodak Co.).

69. U.S. Pat. 4,857,629 (Aug. 15, 1989), M. Rule, D. R. Fagerburg, and J. J. Watkins (to Eastman Kodak Co.).

70. U.S. Pat. 4,859,762 (Aug. 22, 1989), M. Rule, D. R. Fagerburg, and J. J. Watkins (to Eastman Kodak Co.).

71. U.S. Pat. 4,877,851 (Oct. 31, 1989), D. R. Fagerburg, J. J. Watkins, and P. B. Lawrence (to Eastman Kodak Co.).

72. U.S. Pat. 4,877,862 (Oct. 31, 1989), D. R. Fagerburg and J. J. Watkins (to Eastman Kodak Co.).

73. U.S. Pat. 5,241,038 (Aug. 31, 1993), D. M. Teegarden and co-workers (to Eastman Kodak Co.).

74. U.S. Pat. 5,258,489 (Nov. 2, 1993), D. R. Fagerburg, J. J. Watkins, and P. B. Lawrence (to Eastman Chemical Co.).

75. M. Rule and co-workers, *Makromol. Chem. Rapid Comm.* **12**, 221–226 (1991).
76. M. Rule and co-workers, *Makromol. Chem. Symp.* **54/55**, 233–246 (1992).
77. D. R. Fagerburg and D. E. Van Sickle, *J. Appl. Polym. Sci.* **51**, 989–997 (1994).
78. Z. Y. Wang and A. S. Hay, *Macromolecules*, **24**, 333–335 (1991).
79. E. Tsuchida and co-workers, *Macromolecules*, **20**, 2030,2031 (1987).
80. E. Tsuchida and co-workers, *Macromolecules*, **20**, 2315,2316 (1987).
81. E. Tsuchida and co-workers, *Macromolecules*, **22**, 4138–4140 (1989).
82. E. Tsuchida and co-workers, *Macromolecules*, **23**, 930–934 (1990).
83. K. Yamamoto and co-workers, *Eur. Polym. J.* **28**, 341–346 (1992).
84. U.S. Pat. 4,983,720 (Jan. 8, 1991), E. Tsuchida and co-workers (to Idemitsu Petrochemical Co., Ltd.).
85. U.S. Pat. 5,250,657 (Oct. 5, 1993), E. Tsuchida and co-workers (to Seisan Kaihatsu Kagaku Kenkyusho).
86. U.S. Pat. 5,290,911 (Mar. 1, 1994), E. Tsuchida and co-workers (to Research Institute for Production Development and Idemitsu Petrochemical Co., Ltd.).
87. Jpn. Pat. Appl. 04,055,433 (Feb. 24, 1992), N. Ogata (to Idemitsu Sekiyu Kagaku K.K.).
88. E. Tsuchida, K. Yamamoto, and E. Shouji, *Macromolecules*, **26**, 7389, 7390 (1993).
89. K. Yamamoto and co-workers, *J. Am. Chem. Soc.* **115**, 5819, 5820 (1993).
90. E. Tsuchida and co-workers, *Macromolecules*, **27**, 1057–1060 (1994).
91. A. C. Archer and P. A. Lovell, *Makromol. Chem. Macromol. Symp.* **54/55**, 257–274 (1992).
92. A. A. Dembek, P. J. Fagan, and M. Marsi, *Macromolecules*, **26**, 2992–2994 (1993).
93. J. N. Short and H. W. Hill, *Chemtech*, **2**, 481–485 (1972).
94. M. Day and D. R. Budgell, *Thermochimica Acta*, **203**, 465–474 (1992).
95. D. R. Budgell, M. Day, and J. D. Cooney, *Poly Deg. Stab.* **43**, 109–115 (1994).
96. A. J. Lovinger, D. D. Davis, and F. J. Padden, Jr., *Polymer*, **26**, 1595–1604 (1985).
97. J. P. Jog and V. M. Nadkarni, *J. Appl. Polym. Sci.* **30**, 997–1009 (1985).
98. L. C. Lopez and G. L. Wilkes, *Polymer*, **29**, 106–113 (1988).
99. L. C. Lopez, G. L. Wilkes, and J. F. Geibel, *Polymer*, **30**, 147–155 (1989).
100. L. D. Lopez and G. L. Wilkes, *Polymer*, **30**, 882–887 (1989).
101. S. Z. D. Cheng, Z. Q. Wu, and B. Wunderlich, *Macromolecules*, **20**, 2802–2810 (1987).
102. S. S. Song, J. L. White, and M. Cakmak, *Polym. Eng. Sci.* **30**, 944–949 (1990).
103. D. R. Fagerburg, J. J. Watkins, and P. B. Lawrence, *Macromolecules*, **26**, 114–118 (1993).
104. K. H. Seo and co-workers, *Polymer*, **34**, 2524–2527 (1993).
105. J. D. Menczel and G. L. Collins, *Polym. Eng. Sci.*, **32**, 1264–1269 (1992).
106. G. L. Collins and J. D. Menczel, *Polym. Eng. Sci.* **32**, 1270–1277 (1992).
107. C. Auer and co-workers, *J. Appl. Polym. Sci.* **51**, 407–413 (1994).
108. J. S. Chung and P. Cebe, *Polymer*, **33**, 2312–2324 (1992).
109. J. S. Chung and P. Cebe, *Polymer*, **33**, 2325–2333 (1992).
110. J. D. Schultze, M. Boehning, and J. Springer, *Makromol. Chem.* **194**, 339–351 (1993).
111. S. G. Joshi, *Thin Solid Films*, **142**, 213–226 (1986).
112. D. R. Fagerburg, J. J. Watkins, and P. B. Lawrence, *J. Appl. Polym. Sci.* **50**, 1903–1907 (1993).
113. C. C. M. Ma and co-workers, *J. Appl. Polym. Sci.* **39**, 1399–1415 (1990).
114. C. C. M. Ma and co-workers, *Polym. Composites*, **8**, 256–264 (1987).
115. H. N. Beck, *J. Appl. Polym. Sci.* **45**, 1361–1366 (1992).
116. Z. Grubisic, P. Rempp, and H. Benoit, *J. Polym. Sci.* **B-5**, 753–759 (1967).
117. A. Kinugawa, *Kobunshi Ronbunshu*, **44**, 139–141 (1987).
118. T. Housaki and K. Satoh, *Polym. J.* **20**, 1163–1166 (1988).
119. T. Housaki, *J. Appl. Polym. Sci.* **48**, 75–83 (1991).

120. S. Mayeda and M. Nagata, *J. Appl. Polym. Sci.* **52**, 173–181 (1993).
121. U.S. Pat. 4,645,825 (Feb. 24, 1987), K. Idel and B. Willenberg (to Bayer AK).
122. Eur. Pat. Appl. 0,171,021 (July 29, 1985), K. Idel and co-workers (to Bayer AK).
123. G. Graff, ed., *Modern Plastics Encyclopedia*, Vol. 70, No. 12, McGraw-Hill Book Co., Inc., New York, 1994.
124. *Recognized Component Directory*, Vol. 2, Underwriters Laboratories Inc., Northbrook, Ill., 1994.
125. *Chemical Resistance Guide*, Phillips Petroleum Co., Bartlesville, Okla., 1994.
126. *Engineering Properties Guide*, Phillips Petroleum Co., Bartlesville, Okla., 1993.
127. H. W. Hill and D. G. Brady, *J. Coatings Tech.* **49**, 33–37 (1977).
128. J. G. Scruggs and J. O. Reed, *Proceedings of Textile Research Institute 52nd Annual Research and Technical Conference*, Charlotte, N.C., 1982.
129. See U.S. Food and Drug Administration Regulation 177.2490(d).
130. L. R. Kallenbach and M. R. Lindstrom, *NACE Corrosion*, Paper No. 233, 1987.
131. J. E. O'Connor, C. C. Ma, and A. Y. Lou, *SPI Reinforced Plastics/Composites Symposium Proceedings*, **39**, 11E, 1984.
132. U.S. Pat. 4,680,224 (July 14, 1987), J. E. O'Connor (to Phillips Petroleum Co.).
133. Eur. Pat. Appl. 0,418,455 A2 (Mar. 27, 1991), R. L. Hagenson and co-workers (to Phillips Petroleum Co.).
134. U.S. Pat. 5,334,701 (Aug. 2, 1994), C. E. Ash (to Phillips Petroleum Co.).
135. U.S. Pat. 5,039,572 (Aug. 13, 1991), R. L. Bobsein, S. D. Mills, and M. L. Stone (to Phillips Petroleum Co.).
136. U.S. Pat. 5,019,427 (May 28, 1991), D. A. Soules (to Phillips Petroleum Co.).
137. U.S. Pat. 4,925,729 (May 15, 1990), J. E. O'Connor (to Phillips Petroleum Co.).
138. U.S. Pat. 4,800,113 (Jan. 24, 1989), J. E. O'Connor (to Phillips Petroleum Co.).
139. U.S. Pat. 5,026,447 (June 25, 1991), J. E. O'Connor (to Phillips Petroleum Co.).
140. U.S. Pat. 4,848,745 (July 18, 1989), J. R. Bohannan, W. H. Beever, and J. A. Stirling (to Phillips Petroleum Co.).
141. U.S. Pat. 5,053,263 (Oct. 1, 1991), J. R. Krone and J. H. Barber (to Phillips Petroleum Co.).
142. U.S. Pat. 5,139,405 (Aug. 18, 1992), J. R. Krone and J. H. Barber (to Phillips Petroleum Co.).
143. H. W. Hill, *ACS Symp. Ser.* **95**, 183–197 (1979).
144. K. Iwakura, Y. D. Wang, and M. Cakmak, *Int. Polym. Process.* **7**, 327–333 (1992).
145. D. G. Baird and T. Sun, *ACS Symp. Ser.* **435**, 416–438 (1990).
146. P. R. Subramanian and A. I. Isayev, *Polymer*, **32**, 1961–1969 (1991).
147. M. T. Heino and J. V. Seppala, *J. Appl. Polym. Sci.* **44**, 2185–2195 (1992).
148. M. Pracella, P. Magagnini, and L. Minkova, *Polym. Networks Blends*, **2**, 225–231 (1992).
149. U.S. Pat. 4,690,972 (Sept. 1, 1987), T. W. Johnson, W. H. Beever, and J. P. Blackwell, (to Phillips Petroleum Co.).
150. V. M. Nadkarni, V. L. Shingankuli, and J. P. Jog, *J. Appl. Polym. Sci.* **46**, 339–351 (1992).
151. J. P. Jog, V. L. Singankuli, and V. M. Nadkarni, *Polymer*, **34**, 1966–1969 (1993).
152. V. L. Skingankuli, J. P. Jog, and V. M. Nadkarni, *J. Appl. Polym. Sci.* **51**, 1463–1477 (1994).
153. S. Akhtar and J. L. White, *Polym. Eng. Sci.* **32**, 690–698 (1992).
154. T. Takaki, Y. Naganuma, and K. Nakashima, *Kobunshi Ronbunshu*, **50**, 199–204 (1993).
155. T. H. Chen and A. C. Su, *Polymer*, **34**, 4826–4831 (1993).
156. J. J. Scobbo, *Annu. Tech. Conf. Soc. Plast. Eng.* **50**, 605–608 (1992).
157. *Material Safety Data Sheet for Ryton PPS Powders and Pellets*, Phillips Petroleum Co., Bartlesville, Okla., Aug. 31, 1993.

158. OSHA Hazard Communication Standard, 29 CFR Section 1910.1200, OSHA, Washington, D.C., 1983.

*General Reference*

J. P. Blackwell, D. G. Brady, and H. W. Hill, *J. Coatings Tech.* **50**, 62–66 (1978).

JON GEIBEL
JOHN LELAND
Phillips Petroleum Company

## POLYSULFIDES

Polysulfide polymers have the following general structure:

$$HS \text{---} ( R \text{---} S_x \text{---})_n SH$$

**(1)**

where $x$ is referred to as the rank and represents the average number of sulfur atoms in the polysulfide unit. This article is limited to polymers of this type where R is an aliphatic group and $x > 1$. The rank, $x$, usually ranges from slightly less than two to about four. A recent monograph (1) provides extensive information on the history, properties, and uses of aliphatic polysulfides. Other sulfur-containing polymers have been reviewed (2).

The history of polysulfides began over 150 years ago. In 1838 chemists in Switzerland reported that the reaction of chloraetherin (1,2-dichloroethane) with potassium polysulfide gaveambivalent a rubbery, intractable, high sulfur semisolid. Subsequently there were reports of similar products obtained by various methods, but the first useful products were developed from studies in the late 1920s. This led to the formation of Thiokol Corp. which began production of the ethylene tetrasulfide polymer Thiokol A in 1928, the first synthetic elastomer manufactured commercially in the United States. One of the first successful applications of Thiokol A [*14807-96-6*] was for seals where its resistance to solvents justified its relatively high price.

These new synthetic rubbers were accessible from potentially low cost raw materials and generated considerable worldwide interest. For a time, it was hoped that the polysulfide rubbers could substitute for natural rubber in automobile tires. Unfortunately, these original polymers were difficult to process, evolved irritating fumes during compounding, and properties such as compression set, extension, and abrasion characteristics were not suitable for this application.

During the 1930s gradual improvements in the product and processing overcame some of the drawbacks of this material. Nonetheless, the applications were limited and Thiokol Corp. struggled to remain solvent. The first year Thiokol reported a profit was in 1941, 13 years after its foundation. This was realized when the U.S. Air Force discovered that the aliphatic polysulfides were useful as a fuel-resistant sealant for aircraft tanks and hoses. Polysulfides also began to be used as sealants for boat hulls and decks.

The most significant improvement came in the early 1940s when a method for preparing thiol-terminated liquid polysulfides was developed. Cure of the liquid polysulfides could be accomplished by oxidative coupling. Thus, in effect, a rubber could be compounded without the need of heavy mixing equipment. One of the first large-scale applications of the liquid polysulfides was as a binder for solid rocket fuel. From about 1946 until 1958, these binders were used in various rocket systems and the aliphatic polysulfides achieved commercial success. The switch to predominately liquid-fueled rockets in 1958 ended this phase of the polysulfide business.

Since then, uses have shifted more toward civilian applications. Polysulfides have unusually good resistance to solvents and to the environment and good low temperature properties. This makes them particularly useful in a variety of sealant applications. For example, the outstanding resistance of polysulfides to petroleum (qv) products has made them the standard sealant for virtually all aircraft integral fuel tanks and bodies. Another important application is in insulating glass window sealants (qv). Sealants based on liquid polysulfides have had an excellent record since the 1950s and are the worldwide market leader in this application.

Polysulfides also have a long record as construction sealants. In 1953, the Lever House in New York, New York, was one of the first high rise buildings to abandon the traditional structural masonry for the attractive curtain wall construction. Originally, the joints were sealed with a typical oil-based caulk. Within six months, there were serious leakage problems through virtually every joint. The original sealant was removed and replaced with a polysulfide-based elastomeric sealant that could expand and compress with the movements of the panels. Much of this original sealant is still performing in the 1990s. Polysulfides became the first high performance elastomeric sealants to be used in building construction and have been applied successfully to many large-scale projects around the world (see BUILDING MATERIALS).

Some of the early Thiokol solid rubbers are still made and used in printing rolls, solvent-resistant spray hose, gaskets, and gas-meter diaphragms. Many of the polysulfide products have been in use since the 1940s with an excellent track record. Continuing improvements in technology keep these products competitive.

## Physical Properties

The commercial polysulfides are made from bis-chloroethylformal (formal) as shown later in equation 11. In some products trichloropropane is added as a branching agent. Table 1 shows typical properties of polysulfides available from Morton International. These products were acquired from Thiokol Corp. in 1983.

The solid polysulfide products are light brown millable rubbers. Thiokol ST [9065-29-6] (2) is made as in equation 11 with 2% branching agent added. Its

$$HS(C_2H_4OCH_2OC_2H_4SS)_nC_2H_4OCH_2OC_2H_4SH \qquad ClCH_2CH_2OCH_2OCH_2CH_2Cl$$

$$(2) \qquad\qquad\qquad\qquad\qquad\qquad (3)$$

**Table 1. Properties of LP Liquid Polysulfide Polymers[a]**

| Property | LP-31 | LP-2 | LP-32 | LP-12 | LP-3 | LP-33 | LP-977 | LP-980 |
|---|---|---|---|---|---|---|---|---|
| *Specification requirements* | | | | | | | | |
| viscosity at 25°C, Pa·s[b] | 95–155 | 41–52.5 | 41–52.5 | 41–52.5 | 0.94–1.44 | 1.5–2 | 10–15 | 10–15 |
| moisture content, % | 0.12–0.22 | 0.3 max | 0.27 max | 0.27 max | 0.1 max | 0.1 max | 0.26 max | 0.26 max |
| mercaptan content, % | 1.0–1.5 | 1.50–2.00 | 1.50–2.00 | 1.50–2.00 | 5.9–7.7 | 5.0–6.5 | 2.5–3.5 | 2.5–3.5 |
| *General properties* | | | | | | | | |
| average molecular weight | 8,000 | 4,000 | 4,000 | 4,000 | 1,000 | 1,000 | 2,500 | 2,500 |
| pour point, °C | 10 | 7 | 7 | 7 | −26 | −23 | 4 | 4 |
| branching agent, % | 0.5 | 2.0 | 0.5 | 0.2 | 2.0 | 0.5 | 2.0 | 0.5 |
| average viscosity | | | | | | | | |
| at 4°C, Pa·s[b] | 740 | 380 | 380 | 380 | 9 | 16.5 | 77 | 77 |
| at 65°C, mPa·s(=cP) | 14,000 | 6,500 | 6,500 | 6,500 | 150 | 210 | 1,100 | 1,100 |
| low temperature flex, °C[c] | | | | | | | | |
| at 69 MPa[d] | −54 | −54 | −54 | −54 | −54 | −54 | −54 | −54 |

[a]All products listed have flash point (PMCC) > 177°C. The specific gravity at 25°C ranges from 1.27 to 1.31. CAS Registry Number for all the LPs listed is [68611-50-7], ie, they are copolymers made from (**3**) and 1,2,3-trichloropropane, and sodium polysulfide.
[b]To convert Pa·s to P, multiply by 10.
[c]Cured compound.
[d]To convert MPa to psi, multiply by 145.

935

Mooney viscosity (ML 1 + 3 at 100°C) ranges from 28 to 38. Thiokol FA [68611-48-3] is a copolymer made from formal (3) and ethylene chloride with sodium polysulfide; Mooney viscosity (ML 1 + 4 at 121°C) ranges from 60 to 112. Both rubbers have excellent resistance to a wide range of chemicals. They also have low permeability to gases, water, and organic liquids, excellent low temperature flexibility, and superior resistance to the effects of sunlight, ozone (qv), aging, and weathering.

## Chemical Properties

**Oxidative Curing.** The rich chemistry of the thiol end group provides versatility in modifying and curing polysulfide polymers. The most common means of curing polysulfides is by chain extension with oxidizing agents, eg, equation 1, where R–SH represents liquid polysulfide; O, oxidizing agent; and the product is a disulfide.

$$2\,R\text{—}SH + O \longrightarrow R\text{—}S\text{—}S\text{—}R + H_2O \tag{1}$$

Because thiols are easily oxidized, a host of organic and inorganic oxidants may be used. Mild oxidants such as oximes, nitro compounds, or air can be effective. Various oxidants have been used in special applications, but only a few are used in large-scale applications.

For a long time, lead(IV) oxide, $PbO_2$, was the most widely used oxidizing agent for the high molecular weight liquid polysulfides (mol wt > 2500). It was not suitable for lower molecular weight polymers because of the difficulty in controlling the strongly exothermic reaction. Since the early 1970s manganese dioxide, $MnO_2$, has become the predominate oxidizing agent. Manganese dioxide [11129-60-5] has several advantages over lead, such as reduced toxicity, better pot life stability, and better light resistance, elasticity, and recovery for the cured rubber.

The newest curing system is sodium perborate monohydrate [10332-33-9], $NaBO_2H_2O_2H_2O$, a well-known bleaching agent used in certain laundry cleaning formulations. Therefore, it is produced on a large scale at a reasonable price. Sodium perborate offers several advantages as a curing agent for building sealants. One is that it has a light color rather than the dark color of manganese or lead oxides. Thus, it is nonstaining and sealants can be manufactured in a variety of colors. Sealants cured with sodium perborate have low modulus, excellent elasticity, and outstanding resistance to water, weather, uv light, and mold, even without additives. They have good adhesion to most surfaces and are environmentally friendly.

*One-Part Oxidative Curing Systems.* The inorganic peroxide curing agents for liquid polysulfides are activated by water. By formulating and packaging polysulfides under anhydrous conditions, one-part sealants are prepared. These cure when exposed to atmospheric moisture and are used in construction sealant applications. The curing agents most commonly used have been calcium peroxide, zinc peroxide/amine, zinc/lithium peroxide, or manganese dioxide. Sodium perborate is also effective in preparing light-colored, fast curing, one-part sealants and is growing in importance.

*Epoxy Resins.* Polysulfides may also be cured by reaction with epoxy resins (qv) according to the reaction in equation 2. Amines or other catalysts are used and often primary or secondary amine resins are cured together with the polysulfide.

$$R-CH-CH_2 + HS-R'-SH \longrightarrow R-CH-CH_2-S-R'-S-CH_2-CH-R \quad (2)$$
$$\phantom{R-CH-}\underset{O}{\diagdown\diagup}\phantom{CH_2} \qquad\qquad\qquad \underset{OH}{|}\phantom{-CH_2-S-R'-S-CH_2-}\underset{OH}{|}$$

LP-3, the lowest molecular weight liquid polysulfide, is used as a reactive diluent to lower the viscosity of the formulation and to facilitate mixing and application of the resin. The liquid polysulfide also acts as a flexibilizer. The addition of LP-3 to epoxy formulations gives products with good flexibility, high impact strength, excellent chemical resistance, and good adhesion. ELP-3 [*117527-71-6*] (**4**) is an epoxy-terminated polysulfide derived from LP-3. An advantage of this product is its low odor, especially in comparison with the thiol-terminated LP-3.

$$CH_2-CHCH_2-S\!-\!(C_2H_4-OCH_2O-C_2H_4-S-S)_{\overline{x}}C_2H_4OCH_2O-C_2H_4-S-CH_2CH-CH_2$$
$$\underset{O}{\diagdown\diagup}\phantom{CHCH_2-S-(C_2H_4-OCH_2O-C_2H_4-S-S)_xC_2H_4OCH_2O-C_2H_4-S-CH_2CH}\underset{O}{\diagdown\diagup}$$

(**4**)

**Diisocyanates or Polyisocyanates.** The thiol end groups of the liquid polysulfides are quite reactive with isocyanates (eq. 3). Typical diisocyanates such as 1,3-toluene diisocyanate (m-TDI) or diphenylmethane-4,4'-diisocyanate (MDI) are effective in curing liquid polysulfides. Using liquid polysulfides in-

$$2\,RSH + O{=}C{=}N-R'-N{=}C{=}O \longrightarrow RSC-NH-R'-NH-CSR \quad (3)$$
$$\phantom{2\,RSH + O=C=N-R'-N=C=O \longrightarrow RS}\underset{O}{\|}\phantom{-NH-R'-NH-}\underset{O}{\|}$$

stead of the common hydroxy-terminated polymers brings the advantages of the polysulfide to the cured product. Thus, good chemical and solvent resistance, weatherability, adhesion, etc, can be attained. The isocyanate-cured systems have some advantages over oxidative-cured systems, such as improved adhesion to plastic substrates. There are also drawbacks to this method of curing. For example, water must be eliminated from the formulation to prevent foaming. This requires additional steps to dry all of the formulation ingredients (fillers, plasticizers, liquid polysulfide, etc) and to protect them from atmospheric moisture. Also, the mix ratio must be tightly controlled to give close to stoichiometric amounts of polysulfide and isocyanate. Actually, a slight excess of isocyanate is usually used. For the oxidative-cured formulations, a significant excess of oxidant is used and the ratio is more forgiving. Furthermore, the isocyanates are more hazardous to work with than the commonly used oxidizing curing agents.

**Phenolic Resins.** At elevated temperatures, phenolic resins are cured with polysulfide resins through a condensation reaction. The product may be considered a block copolymer of the rigid phenolic resin and the flexible polysulfide. Thus, the polysulfide acts to flexibilize the resulting polymer.

**Miscellaneous Curing Reactions.**   Other functional groups can react with the thiol terminal groups of the polysulfides to cross-link the polymer chains and build molecular weight. For example, aldehydes can form thioacetals and water. Organic and inorganic acids or esters can form thioesters. Active dienes such as diacrylates can add to the thiols (3). Examples of these have been mentioned in the literature, but none have achieved commercial significance.

**Reactions of the Disulfide Group.**   Besides the thiol end groups, the disulfide bonds also have a marked influence on both the chemical and physical properties of the polysulfide polymers. One of the key reactions of disulfides is nucleophilic attack on sulfur (eq. 4). The order of reactivity for various thiophiles has been reported as $(C_2H_5O)_3P > R^-, HS^-, C_2H_5S^- > C_6H_5S^- > C_6H_5P, CN^- > SO_3^{2-} > OH^- > 2,4\text{-}(NO_2)_2C_6H_3S^- > N_3^- > SCN^-, I^-, C_6H_5NH_2$ (4). These thiophiles are capable of splitting the disulfide bond and thus reducing the molecular weight of the polymers.

$$RS-SR + X^- \longrightarrow RS-X + RS^- \qquad (4)$$

An important aspect of this is the splitting of the polymer chain with thiol (eq. 5) or mercaptide ion (thiol + base catalyst). In fact, sodium sulfide or organic monothiols, eg, mercaptoethanol or decylmercaptan, are utilized to lower the molecular weight of polysulfides or to limit the extent of curing reactions.

$$RSSR + R'SH \rightleftharpoons R'SSR + RSH \qquad (5)$$

This reaction also plays a role in the degradation of polysulfides. A back-biting mechanism as shown in equation 6 results in formation of the cyclic disulfide (**5**). Steam distillation of polysulfides results in continuous gradual collection of (**5**). There is an equilibrium between the linear polysulfide polymer and the cyclic disulfide. Although the linear polymer is favored and only small amounts of the cyclic compound are normally present, conditions such as steam distillation, which remove (**5**), drive the equilibrium process toward depolymerization.

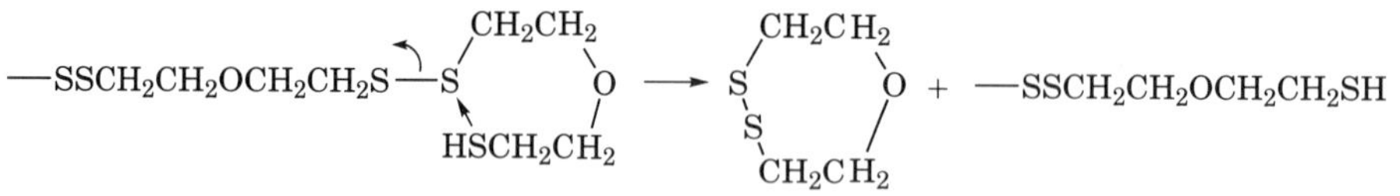

$$(5)$$

Another aspect of the splitting reaction (eq. 5) is that it allows for the potential recycle of cured polysulfide. Oxidatively cured polysulfides can be broken down by adding low molecular weight liquid polysulfide with mixing and heating. The resulting liquid material can be reworked and cured again (5). A related reaction is disulfide interchange (eq. 7). This process leads to redistribution of the polymers. For example, if thiol-terminated polymers made from different monomers are combined, they redistribute to form a random polymer. Thus, block copolymers of polysulfides are difficult to obtain. One way to make use of this phenomenon is shown in equation 8. Here a high molecular weight thiol-terminated polysulfide polymer is reduced to a lower molecular weight polymer with hydroxyl terminals. In this way a different functionality can be introduced into the end group.

$$\text{RSSR} + \text{R}'\text{SSR}' \longrightarrow 2\,\text{RSSR}' \tag{7}$$

$$\text{RSSR} + \text{HOCH}_2\text{CH}_2\text{SSCH}_2\text{CH}_2\text{OH} \longrightarrow 2\,\text{RSSCH}_2\text{CH}_2\text{OH} \tag{8}$$

Disulfide interchange also effects the physical properties of the cured polysulfide polymers. Polysulfide polymers undergo stress relaxation in a manner markedly different from conventional rubbers. Stress applied to stretch a sample of polysulfide rubber rapidly falls to zero. There is no change in the chemical and physical properties of the polymer recovered after the tests. The polysulfide polymer can be repeatedly recycled through the relaxation process. With vulcanized hydrocarbon rubbers, the stress decay takes place much more slowly and the activation energy for the relaxation is higher. Studies have attributed the behavior of the polysulfides to interchange between the polysulfide linkages of adjacent polymer chains (6). Addition of free sulfur or free thiol groups dramatically increases the rate of relaxation. Small amounts of free thiol can increase the rate several hundredfold (6).

Disulfides are susceptible to attack by strong oxidizing agents and this can result in decomposition of polysulfides. For example, nitric acid causes violent decomposition of polysulfide polymers.

## Manufacture and Processing

Polysulfide polymers are made commercially according to the reactions shown in equations 9–12. Details of the process and alternative approaches have been described (1,7). Although other dihalides can be used, its favorable economics, minimal competition with ring formation, and the desirable physical properties of the resulting polymer have made bis-chloroethylformal the monomer of choice. Only occasionally are other dihalides used in special applications. 1,2,3-Trichloropropane [96-18-4] is sometimes added as a branching agent.

Many of the reagents used are hazardous and require special equipment and handling. At Morton International, there has been ongoing effort to minimize waste from the processes. In equation 9 excess HCl is used to avoid formation of higher homologues from reaction of ethylene oxide with itself. Excess HCl is recovered and recycled. Similarly, excess 2-chloroethanol (eq. 10) is removed from the product by distillation and recycled. Little waste is generated in these steps. In the polymerization reaction (eq. 11), excess sodium polysulfide is required to drive the reaction to completion. Specialized equipment is used for washing and collecting the high molecular weight solid polymer in order to optimize its recovery. The wash water contains NaCl, the excess sodium polysulfide, and only small amounts of organics. The amount of excess $Na_2S_x$ is minimized and efforts are underway to recover the sulfur species from the aqueous stream. High molecular weight solid (**2**) is converted to liquid polysulfide (**2**) by reaction with NaSH and $Na_2SO_3$.

$$CH_2\!\!-\!\!CH_2 + HCl \longrightarrow HOCH_2CH_2Cl \qquad (9)$$
$$\underset{O}{\diagdown\diagup}$$

$$2\ HOCH_2CH_2Cl + CH_2O \longrightarrow \mathbf{(3)} \qquad (10)$$

$$\mathbf{(3)} + NaS_x \longrightarrow \mathbf{(2)} \qquad (11)$$

**Formulation.**  Polysulfide-based sealants are formulated with appropriate ingredients to obtain the desired properties for a particular application. A typical formulation contains liquid polysulfide polymer, curing agent, cure accelerators (bases) or retarders (acids), fillers, plasticizers, thixotropes, and adhesion promoters.

For a two-part (A and B) sealant, Part A consists of liquid polysulfide, filler, plasticizer, thixotrope, and adhesion promoter. Part B contains the curing agent, plasticizer, a small amount of filler, and the accelerator or retarder. Part A can be mixed on any type of mixer that will ensure thorough dispersion. Generally, the equipment should utilize a double-arm kneading action mixer or a combination kneading action and high speed dispersing blade. Representative types are Change Can, Ross Planetary or Power Mixer, Meyers Mixer, Kneader Extruder, and Hockmeyer Mixer. Optimum dispersion of the fillers is obtained on a paint mill where the fillers are subjected to a grinding action which breaks down the agglomerates. If the mixer itself provides sufficient grinding of the pigments, then milling can be eliminated. The curing paste (Part B) can be prepared either by ball milling or three-roll paint milling. Paint milling is generally preferred, however, solvent-based pastes are best prepared in a ball mill to minimize solvent loss. In order to obtain the most uniform results, pastes containing the accelerators or retarders should be aged for at least two weeks at room temperature prior to use.

The one-part sealant bases require drying the ingredients. Predrying the ingredients before manufacturing is cumbersome and expensive. Other alternatives include vacuum drying during manufacture, azeotropic distillation to remove water before moisture-sensitive components are added, and use of desiccants (qv). The most common method is through the use of the desiccant barium oxide (8). The mixing is done on equipment as described except the processes are carried out under an atmosphere of dry nitrogen and the final step is to combine the pastes into a single mix.

## Specifications and Testing

Typical specifications for the polysulfide polymers are summarized in Table 1. Specifications for the sealants vary widely depending on the specific application and the needs of the applicators. Standards for sealant testing vary in different countries. Ultimately, the tests should simulate the environment the sealants will be exposed to and measure their performance under these conditions.

Examples of typical tests for sealants include those for stability, curing characteristics, and cured sealants. Stability varies with formulation and storage temperature, and may range from three months to over one year. Curing characteristics include tack-free time, which varies with relative humidity and can range from hours to days; work life; and cure time. Cured sealant tests are adhesion to appropriate substrates; measurement of physical and mechanical properties such as tensile strength, elongation, modulus, hardness, compression and recovery, and low temperature flexibility; and environmental effects, eg, uv light, water or special fluids, and high and low temperature.

## Health and Safety

Because they are sulfur-containing polymers, the polysulfides have a characteristic odor. Although the odor is somewhat objectionable it probably arises from small amounts of lower molecular weight compounds. In fact, the lower molecular weight polymers exhibit the strongest odor. Toxicity tests conducted on a representative LP brand liquid polysulfide used in Morton sealant applications indicate that the polymers are not eye irritants and have a low order of oral toxicity ($LD_{50} > 5$ g/kg). Tests on the lower molecular weight liquid polysulfide products show similar findings. These materials are not eye or skin irritants, do not cause allergic skin reactions, and are not toxic when administered orally ($LD_{50} > 3.4$ g/kg). Rats exposed to a low molecular weight LP at a level of 230 mg/L for four hours developed apparent eye and respiratory tract irritation. However, this level far exceeds that which would be encountered in an industrial setting and is not indicative of real life exposure. Under the criteria set forth under OSHA's Hazard Communication Standard (29 CFR 1910.1200), Morton classifies liquid polysulfide products as nonhazardous. When used in accordance with prescribed procedures, they do not pose a health hazard (9).

When considering sealants or other formulated products, the health and safety considerations relating to the other ingredients should be taken into account.

## Economic Aspects

Worldwide production capacity of liquid polysulfides is about 33,000 t with manufacturing sites in the United States, Japan, and Germany. Total consumption is about 28,600 t. Approximately 50% is for insulating glass sealants, 30% for construction applications, and ~10% for aircraft sealants. In addition, ~909 t of the solid polysulfide rubbers are sold each year.

Polysulfide sealants have ~83% of the market share of insulating glass sealants in Europe, 35% in the United States, and a total of ~60% worldwide. Some growth in this area is expected as polysulfides are gaining market share in the United States and sales of insulating glass windows is expected to increase overall. Polysulfides have only a modest share of the building sealant market. The perborate-cured sealants offer advantages and some increase in sales is expected in this area. Polysulfides are virtually the only sealant specified for aircraft integral fuel tanks worldwide. In addition, they comprise about 90% of other aircraft sealant applications. The total volume used in the aircraft sealant market fluctuates with aircraft construction and repair. This segment is expected to remain fairly constant (see SEALANTS).

Most of the polysulfide sales are in Europe, the United States, and Japan. Over the next few years, there should be some expansion into other countries.

## Uses

**Insulating Glass Sealants.**   One of the largest scale applications of polysulfide polymers is as a sealant for insulating glass windows. The window consists of two panes of glass separated by a hollow spacer that is filled with desiccant to remove moisture or volatiles from the air space. This prevents condensation and fogging of the window at low temperature. Dual seal windows contain a primary seal along the edge of the spacer that is usually made from polyisobutylene (PIB). PIB has good resistance to moisture vapor transmission and extends the life of the unit. The outer edge of the unit is closed off with the secondary seal of polysulfide or other sealant material. The intermediate air space is sometimes filled with an inert gas such as argon to improve insulation and reduce noise. Polysulfide sealants have low permeability for water vapor or inert gas and can be used either in dual- or single-seal units. In the dual-seal units, if there are small leaks in the primary seal, the polysulfide provides a good backup. Other sealants such as silicones have high permeability, and can only be used in dual-seal units that rely on the integrity of the primary seal.

Besides water vapor and gas diffusion, other requirements for good edge sealants are water resistance, uv resistance, heat/cold resistance ($-40$ to $80°C$), adhesion to glass and metal, and good characteristics for application. Polysulfide sealants have maintained an excellent record in use since the 1950s. Development of new polysulfides and sealant formulations continues in order to meet market needs of the 1990s.

**Aircraft Sealants.**   Polysulfides have been used for sealing fuel tanks and aircraft structural components since the 1940s. There are stringent requirements for these sealants. They must have outstanding resistance to fuels and excellent adhesion to many different materials such as various metal alloys and protective coatings used in aircraft construction. The sealants must also perform in extremely variable weather conditions (10). The polysulfides have an excellent performance record.

**Construction Sealants.**   In the 1950s, curtain wall construction became dominant in high rise buildings. This technique requires sealing of the joints against wind and weather. Sealants used up until that time were not suitable to withstand the temperature changes, sun, wind, rain, vibration, etc. Polysulfide-based sealants were developed for this application and have served in many large-scale projects since that time. Polysulfide construction sealants are used to seal glass in aluminum frames, concrete moving joints, steel/stone joints, and in other applications. For earlier sealants, in order to obtain good adhesion to the variety of substrates, primers were usually used. Besides the disadvantage of requiring application of the primer, there was the risk that the primer would not be properly or uniformly applied. More recently, polysulfide sealants have been developed that incorporate adhesion promoters in the formulation and eliminate the need for primers. These sealants have been tested against ASTM C920 as a class 25 building sealant on glass, anodized aluminum, and concrete and meet all requirements (11).

Other advancements in construction sealants are the improvements in one-part curing of polysulfide sealants. In the past, one-part polysulfide sealants had a reputation for being very slow to cure through. In the 1990s, one-part sealants have been developed that cure much faster. These use either manganese dioxide or sodium perborate as the curing agent. Sodium perborate has several advantages over other curing agents. This is especially true in construction sealants, where the neutral color, low modulus, good mold resistance, and low physiological effects make it attractive for both one- and two-part sealants.

**Below-Ground Sealants.**   Immersion in water for long periods or continuous exposure to high humidity are especially difficult conditions for organic-based materials to withstand. Sealants for use in water purification plants or wastewater treatment plants have special demands for physical, chemical, and microbiological properties. Polysulfide based sealants have proved themselves useful in this area. Special consideration must also be given to the other ingredients, eg, fillers, plasticizers, adhesion promoters, or curing agents, in formulating for these applications (12).

The chemical and fuel resistance of polysulfides makes them useful as sealants and coatings for secondary containment areas, where they prevent chemicals, solvents, fuels, etc, from seeping into the ground in the event of spillage or a storage tank leak. Polysulfide-based coatings and sealants are also used for bridges, air fields, and road construction.

**Epoxy Flexibilizers.**   Polysulfides are useful as flexibilizers in epoxy resin formulations. Compounders can target the properties desired for a particular application through the selection and balance of the epoxy, the liquid polysulfide–epoxy ratio, curing agent, and filler. Most of the compounds are two-component: one containing the epoxy resin, the second containing polymer and curing agent. Probably the most important factor is the ratio of polysulfide to epoxy resin. The most effective ratio ranges from 1:2 to 2:1 liquid polysulfide to epoxy. Liquid polysulfide–epoxy formulations are used for a variety of applications including adhesives, protective coatings, barrier coatings, electrical potting compounds, resilient plastic tooling materials, aggregate liquid polysulfide–epoxy mortars, and surface sealers.

**Water Dispersions.**   Polysulfide products are offered as aqueous dispersions (Thiokol WD-6). These are useful for applying protective coatings to line fuel tanks, and for concrete, wood, and in some cases fabrics, felt, leather (qv), and paper (qv). It has been found that a stable emulsion can be made that contains both LP and manganese oxide curing agent. The emulsion can be thinned and applied as a spray coating. After it is applied, water evaporates and the LP cures to form a solid rubber (13).

**Rubber Articles.**   The solid polysulfide rubber products, Thiokol FA and Thiokol ST, are used in the manufacture of printing rollers, printing blankets, gas meter diaphragms, O-rings, and specialty molded items.

## BIBLIOGRAPHY

"Polymers Containing Sulfur, Polysulfides" in *ECT* 2nd ed., Vol. 16, pp. 253–272, by M. B. Berenbaum, Thiokol Chemical Corp.; "Polysulfide Resins" under "Polymers Containing Sulfur" in *ECT* 3rd ed., Vol. 18, pp. 814–831, by S. M. Ellerstein and E. R. Bertozzi, Thiokol Corp.

1. H. Lucke, *ALIPS Aliphatische Polysulfide*, Hüthig & Wepf, Basel, Switzerland, 1992; English transl. in press.

2. A. Duda and S. Penczec, in J. I. Kroschwitz, ed., *Encyclopedia of Polymer Science and Engineering*, Wiley and Sons, Inc., New York, 1989, pp. 246–368.
3. S. J. Hobbs, *Polym. Mater. Sci.* **67**, 415 (1992).
4. W. A. Pryor, *Mechanisms of Sulfur Reactions*, McGraw-Hill Book Co., Inc., New York, 1962, pp. 59–64.
5. Ger. Pat. Appl. DE 4142500 (1991), R. J. Hecktor, F. Specht, R. Theobald, and G. Unger (to Metallgesellschaft AG).
6. A. V. Tobolsky and W. J. MacKnight, *Polymeric Sulfur and Related Polymers*, Interscience, New York, 1965.
7. D. E. Vietti, *Comprehensive Polymer Science*, Vol. 5, Pergamon Press, Oxford, U.K, 1989, pp. 533–542.
8. U.S. Pat. 3,912,696 (1975), J. I. Doughty (to Minnesota Mining and Manufacturing Co.).
9. D. Wienckowski, personal communication, Morton International, Inc., Woodstock, Ill., 1994.
10. D. B. Paul, P. J. Hanhela, and R. H. E. Huang, *Polymer Science and Technology*, Plenum Press, New York, 1988, pp. 269–280.
11. A. R. Fiorillo and J. R. Harding, in D. H. Nicastro, ed., *Science and Technology of Building Seals, Sealants, Glazing and Waterproofing*, Vol. 4, ASTM STP 1243, ASTM, Philadelphia, Pa., 1994.
12. T. Lee, T. Rees, and A. Wilford, in C. J. Parise, ed., *Science and Technology of Building Seals, Sealants, Glazing and Waterproofing*, STP 1168 ASTM, Philadelphia, Pa., 1992, pp. 47–56.
13. U.S. Pat. 5,073,577 (1991), P. H. Anderson (to Morton International, Inc.).

*General References*

M. B. Berenbaum, in N. G. Gaylord, ed., *Polyethers*, Part III, Vol. 13, Interscience, New York, 1962, p. 43.
E. R. Bertozzi, *Rubber Chem. Technol.*, **41**, 114 (1968).
S. M. Ellerstein, in J. I. Kroschwitz, ed., *Encyclopedia of Polymer Science and Engineering*, Vol. 13, John Wiley & Sons, Inc., New York, 1988, p. 186.
E. M. Fettes, in N. Kharasch, ed., *Organic Sulfur Compounds*, Vol. 1, Pergamon, London, 1961, p. 266.
H. Lucke, *ALIPS Aliphatische Polysulfide*, Hüthig & Wepf, Basel, Switzerland, 1992; English transl. in press.
J. R. Panek, in N. G. Gaylord, ed., *Polyethers*, Part III, Vol. 13, Interscience, New York, 1962, p. 115.
A. V. Tobolsky and W. J. MacKnight, *Polymeric Sulfur and Related Polymers*, John Wiley & Sons, Inc., New York, 1965.
A. V. Tobolsky, *The Chemistry of Sulfides*, Interscience, New York, 1968.
D. E. Vietti, *Comprehensive Polymer Science*, Vol. 5, Pergamon, London, 1989, p. 533.

DAVID VIETTI
MICHEAL SCHERRER
Morton International, Inc.

## POLYSULFONES

A polysulfone is characterized by the presence of the sulfone group as part of its repeating unit. Polysulfones may be aliphatic or aromatic. Aliphatic polysulfones

(R and R' are alkyl groups) were synthesized by radical-induced copolymerization of olefins and sulfur dioxide and characterized many years ago. However, they never demonstrated significant practical utility due to their relatively unattractive physical properties, not withstanding the low cost of their raw materials (1,2). The polysulfones discussed in this article are those based on an aromatic backbone structure. The term polysulfones is used almost exclusively to denote aromatic polysulfones.

Polysulfones are a class of amorphous thermoplastic polymers characterized by high glass-transition temperatures, good mechanical strength and stiffness, and outstanding thermal and oxidative resistance. These polymers are characterized by the presence of the para-linked diphenylsulfone group (**2**) as part of their backbone repeat units. By virtue of their mechanical, thermal, and other desirable characteristics, these polymers enjoy an increasingly wide and diversified range of commercial applications. The basic repeat unit of any polysulfone always contains sulfone, aryl, and ether units as part of the main backbone structure and are thus often referred to in the polymer literature as poly(arylethersulfone)s. Other names include polysulfones, polyethersulfones, and polyarylsulfones and the lack of standardization in nomenclature has often resulted in confusion as to the meaning of a designation. The different designations are somewhat arbitrary and all refer to the same class of polymers.

In addition to sulfone, phenyl units, and ether moieties, the main backbone of polysulfones can contain a number of other connecting units. The most notable such connecting group is the isopropylidene linkage which is part of the repeat unit of the well-known bisphenol A-based polysulfone. It is difficult to clearly describe the chemical makeup of polysulfones without reference to the chemistry used to synthesize them. There are several routes for the synthesis of polysulfones, but the one which has proved to be most practical and versatile over the years is by aromatic nucleophilic substitution. This polycondensation route is based on reaction of essentially equimolar quantities of 4,4'-dihalodiphenylsulfone (usually dichlorodiphenylsulfone (DCDPS)) with a bisphenol in the presence of base thereby forming the aromatic ether bonds and eliminating an alkali salt as a by-product. This route is employed almost exclusively for the manufacture of polysulfones on a commercial scale.

The diphenylsulfone group is supplied to the repeat unit of all polysulfones by DCDPS; the differentiating species between various polysulfones comes from the choice of bisphenol. There are three commercially important polysulfones referred to generically by the common names polysulfone (PSF), polyethersulfone (PES), and polyphenylsulfone (PPSF). The repeat units of these polymers are shown in Table 1.

**Table 1. Chemical Structures and Glass-Transition Temperatures, $T_g$, of PSF, PES, and PPSF**

| Polymer | Repeat unit structure | $T_g$, °C |
|---|---|---|
| polysulfone | | 185 |
| polyethersulfone[b,c] | | 220 |
| polyphenylsulfone[d] | | 220 |

[a]Bisphenol A polysulfone [25135-51-7].

[b]PES repeat unit structure can alternatively be drawn as

[c]Victrex polyethersulfone [25667-42-9].

[d]RADEL R polyphenylsulfone [25608-64-4].

## Polymerization

**Nucleophilic Substitution Route.** Commercial synthesis of poly(aryleth-ersulfone)s is accomplished almost exclusively via the nucleophilic substitution polycondensation route. This synthesis route, discovered at Union Carbide in the early 1960s (3,4), involves reaction of the bisphenol of choice with 4,4′-dichlorodiphenylsulfone in a dipolar aprotic solvent in the presence of an alkali base. Examples of dipolar aprotic solvents include N-methyl-2-pyrrolidinone (NMP), dimethyl acetamide (DMAc), sulfolane, and dimethyl sulfoxide (DMSO). Examples of suitable bases are sodium hydroxide, potassium hydroxide, and potassium carbonate. In the case of polysulfone (PSF) synthesis, the reaction is a two-step process in which the dialkali metal salt of bisphenol A (**1**) is first formed *in situ* from bisphenol A [80-05-7] by reaction with the base (eg, two molar equivalents of NaOH), followed by the nucleophilic substitution reaction of (**1**) with 4,4′-dichlorodiphenylsulfone [80-07-9] (**2**). Polysulfone is typically prepared as described in Reference 5 according to equation 1 where $n = 40–60$. The minimum degree of polymerization is dictated by the minimum molecular weight required to achieve useful mechanical properties from the polymer. For polysulfone, this corresponds to a reduced viscosity of about 0.35 dL/g (in chloroform at 25°C and 0.2 g/dL concentration).

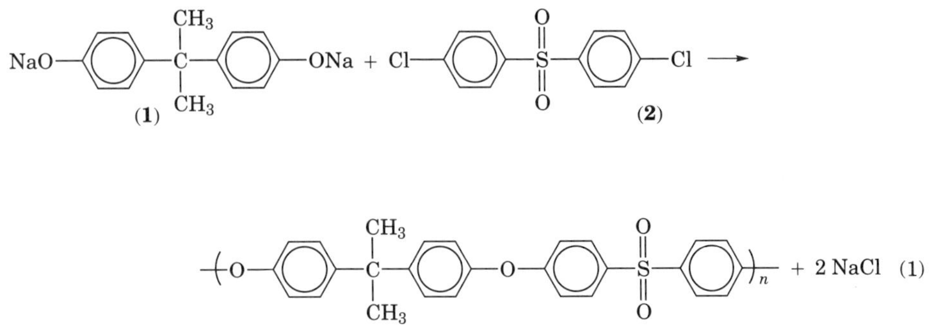

The rate of polymerization in this type of reaction depends on both the basicity of the bisphenol salt and the electron-withdrawing capacity of the activating group (in this case sulfone) in the dihalide monomer. The difluoride-based sulfone monomer is more reactive than the dichloride and thus gives higher polymerization rates, but the latter is by far the more economical raw material. Another important consideration in reactivity and the attainment of high molecular weight is the purity of the 4,4′-isomer in the dichlorodiphenyl sulfone monomer. Chlorines in the meta position on the phenyl ring are not activated and do not undergo nucleophilic displacement. Hence, a significant presence of the 3,4′-isomer of the sulfone monomer can act as a monofunctional terminating species for the polycondensation and prevent the attainment of target molecular weights. Factors governing rates of reaction have been discussed (6).

DMSO is an effective solvent for the polymerization as it affords good solubility for both the polymer and disodium bisphenol A [2444-90-8]. Typical polymerization temperatures for polysulfone are in the range 130–160°C. At temperatures below 130°C, the polymerization slows down considerably due to poor solubility of the disodium bisphenol A salt.

The reaction of NaOH with bisphenol A generates water. This water must be thoroughly removed from the system to allow the reaction to be driven to completion, and more importantly, to preclude any residual water in the system from hydrolyzing part of the DCDPS monomer (2). Before the introduction of DCDPS for the polymerization step, all but traces of water must be removed. Failure to do so results in regeneration of NaOH, which rapidly reacts with DCDPS to form the monosodium salt of 4-chloro-4′-hydroxydiphenylsulfone [18995-09-0] (3) (6).

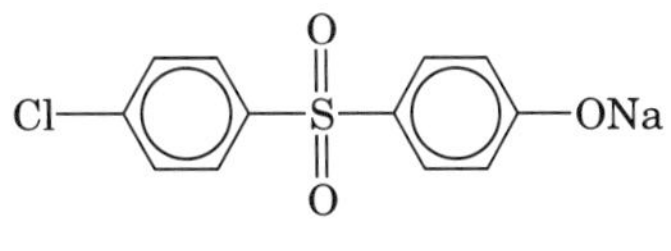

(3)

With as little as 0.5% hydrolysis of the sulfone monomer, the polymerization stoichiometric balance is sufficiently upset to prevent high molecular weight polymer from being achieved. The dependence of maximum attainable PSF molecular weight on water content during polymerization can be inferred from Figure 1.

Molecular weight control for nucleophilic condensation polymerizations of this type is achieved by one of several methods: (1) the addition of a monohalogen compound such as chloromethane once target molecular weight is achieved; ie, such compounds terminate chain growth by reacting with available sodium phenate end groups forming nonreactive methoxyl end groups on the chains; (2) excess DCDPS monomer can be used to limit the maximum attainable molecular weight via stoichiometric imbalance; (3) monohydric phenols or monohalo-activated aromatic compounds can also be used as chain terminators. Generally, the first or the second method (or a combination of them) is used. In the case of termination methods (2) or (3), the terminator is typically added at the beginning of the reaction with the monomer charge. The actual effect of any such terminator depends on its relative reactivity as well as the amount added. The terminator unbalances the stoichiometry of the two monomers thereby placing a ceiling on the maximum theoretically attainable molecular degree of polymerization according to Carothers' principle of functionality. A mole ratio of unity for the two parent monomers results in the highest attainable molecular weight as illustrated in Figure 2. The most favorable reaction rates are also achieved by maintaining the monomer ratio at or very near unity.

The polymerizations of polyethersulfone (PES) and polyphenylsulfone (PPSF) are analogous to that of PSF, except that in the case of these two polymers, solvents which are higher boiling than DMSO are needed due to the higher reaction temperatures required. Diphenyl sulfone, sulfolane, and NMP are examples of suitable solvents for PES and PPSF polymerizations. Chlorobenzene or toluene are used as cosolvents at low concentrations. These cosolvents form an

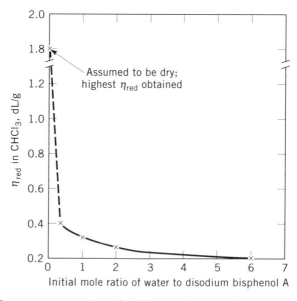

**Fig. 1.** Effect of water presence in polysulfone polymerization on maximum attainable polymer reduced ($\eta_{red}$) viscosity.

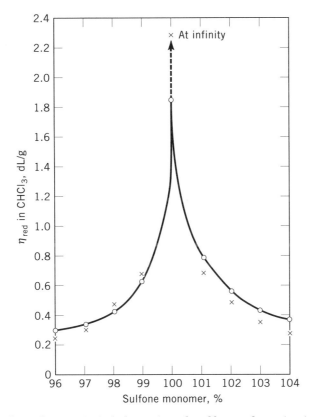

**Fig. 2.** Effect of stoichiometric imbalance in polysulfone polymerization on maximum attainable polymer reduced ($\eta_{red}$) viscosity where ($\times$) is theoretical, and ($\circ$), experimental.

azeotrope with water as they distill out of the reaction mixture, thereby keeping the polymerization medium dehydrated. Potassium carbonate is a suitable choice for base. The synthesis of PES and PPSF differ from the PSF case in that the reaction is carried out in a single-step process. In other words, the formation of the dipotassium salt of the bisphenol is not completed in a separate first step. Equations 2 and 3 represent polymerizations based on the dipotassium salts of bisphenol S and biphenol to make PES and PPSF, respectively.

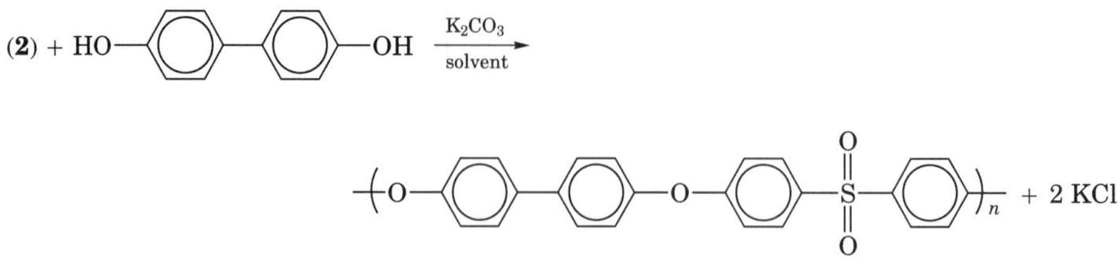

$$(2) + HO\text{---}\langle\text{---}\rangle\text{---}\langle\text{---}\rangle\text{---}OH \xrightarrow[\text{solvent}]{K_2CO_3}$$

$$\left[O\text{---}\langle\rangle\text{---}\langle\rangle\text{---}O\text{---}\langle\rangle\text{---}\overset{\overset{O}{\|}}{\underset{\underset{O}{\|}}{S}}\text{---}\langle\rangle\right]_n + 2\ KCl \quad (3)$$

An alternative synthesis route for PES involves the partial hydrolysis of dichlorodiphenyl sulfone (**2**) with base to produce 4-chloro-4′-hydroxydiphenylsulfone [7402-67-7] (**3**) followed by the polycondensation of this difunctional monomer in the presence of potassium hydroxide or potassium carbonate (7).

As a variation on the base-catalyzed nucleophilic displacement chemistry described, polysulfones and other polyarylethers have been prepared by cuprous chloride-catalyzed polycondensation of aromatic dihydroxy compounds with aromatic dibromo compounds. The advantage of this route is that it does not require that the aromatic dibromo compound be activated by an electron-withdrawing group such as the sulfone group. Details of this polymerization method, known as the Ullmann synthesis, have been described (8).

A method for the polymerization of polysulfones in nondipolar aprotic solvents has been developed and reported (9,10). The method relies on phase-transfer catalysis. Polysulfone is made in chlorobenzene as solvent with (2.2.2)cryptand as catalyst (9). Less reactive crown ethers require dichlorobenzene as solvent (10). High molecular weight polyphenylsulfone can also be made by this route in dichlorobenzene; however, only low molecular weight PES is achievable by this method. Cross-linked polystyrene-bound (2.2.2)cryptand is found to be effective in these polymerizations which allow simple recovery and reuse of the catalyst.

**Other Synthesis Routes.** Several alternative routes to the nucleophilic substitution synthesis of polysulfones are possible. Polyethersulfone can be synthesized by the electrophilic Friedel-Crafts reaction of bis(4-chlorosulfonylphenyl)ether [121-63-1] with diphenyl ether [101-84-8] (11–13).

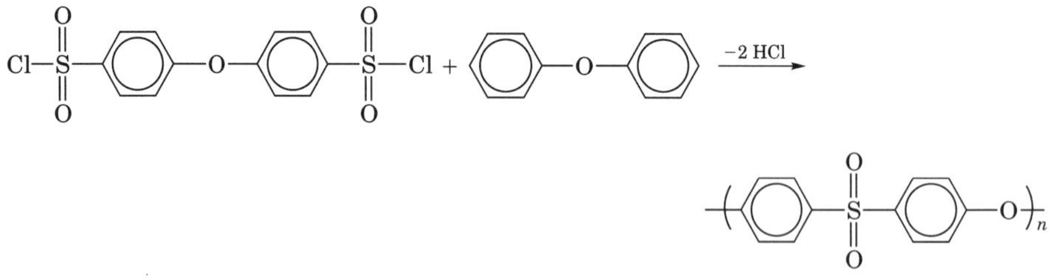

$$(4)$$

The same reaction can be carried out using 4-chlorosulfonyldiphenyl ether [1623-92-3] as a single monomer:

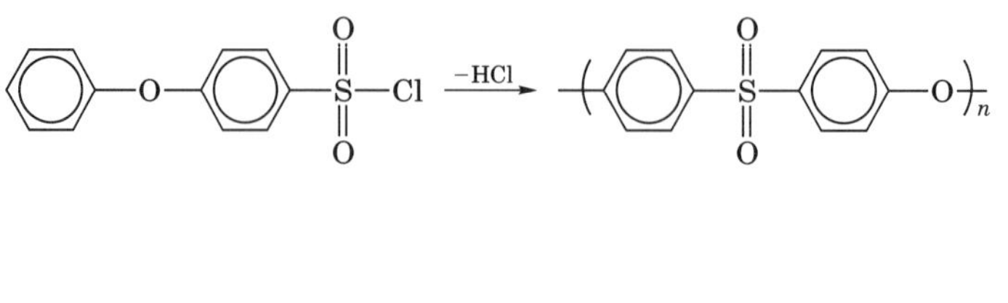

$$(5)$$

The single-monomer route (eq. 5) is preferred as it proves to give more linear and para-linked repeat unit structures than the two-monomer route. Other sulfone-based polymers can be similarly produced from sulfonyl halides with aromatic hydrocarbons. The key step in these polymerizations is the formation of the carbon–sulfur bond. High polymers are achievable via this synthesis route although the resulting polymers are not always completely linear.

An elegant synthesis method which is specific to sulfone polymers containing phenyl–phenyl linkages (such as PPSF) is the nickel-catalyzed coupling of aryl dihalides. The scheme for this synthesis involves a two-step process. First, an aromatic dihalide intermediate is formed which carries the backbone features of the desired polymer. This aromatic dihalide intermediate is then self-coupled in the presence of zero-valent nickel, triphenylphosphine, and excess zinc to form the biphenyl- or terphenyl-containing polymer. Application of this two-step scheme to PPSF can be depicted as follows:

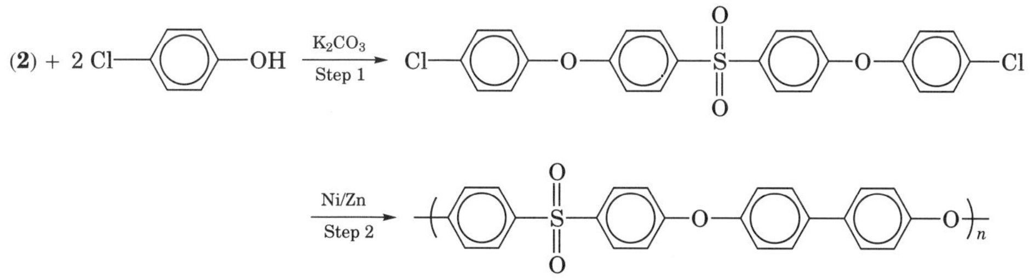

The first step in this scheme is a classical aromatic nucleophilic substitution. Details of the method have been expounded (14–17). References 14 and 15 are concerned with the synthesis of the diaryl halide intermediate whereas References 16 and 17 discuss the synthesis of the polymers, with emphasis on the polymerization of PPSF by this route.

Oxidative coupling of aromatic compounds via the Scholl reaction has been applied successfully to synthesize a polyarylethersulfone (18). High molecular weight polymer was obtained upon treating 4,4'-di(1-naphthoxy)diphenylsulfone and 4,4'-di(1-naphthoxy)benzophenone with ferric chloride. Equimolar amounts of the Lewis acid are required and the method is limited to naphthoxy-based monomers and other systems that can undergo the Scholl reaction.

## Properties

**Structure–Property Relationships.**   The characteristic feature of each of the polymers in Table 1 is the highly resonant diaryl sulfone grouping. As a consequence of the sulfur atom being in its highest state of oxidation and the enhanced resonance of the sulfone group being in the para position, these resins offer outstanding thermal stability and resistance to thermal oxidation. The thermal stability is further augmented by the high bond dissociation energies inherent in the aromatic backbone structure. As a result, these polymers can be melt fabricated at temperatures of up to 400°C with no adverse consequences. The high degree of oxidative stability also allows for prolonged or continuous exposure to temperatures of anywhere between 150–190°C, depending on polymer, formulation, and use conditions. The ether linkages in these polymers contribute to chain flexibility leading to mechanical toughness and favorable melt rheological properties. The relatively inert ether and sulfone backbone functionalities contribute to resistance against hydrolysis and chemical attack by acids and bases. Medical and food contact applications for polysulfones are possible in part because of this characteristic.

The high glass-transition temperature, $T_g$, of polysulfones is attributed to the rigid phenyl rings in the backbone and also the sulfone group which increases $T_g$ by providing strong dipole interactions and restricting rotation of the aromatic units relative to other connecting groups. The ether groups in these polymers are the main flexibilizing units and the $T_g$ of polysulfones is inversely related to the ether content in the backbone on a repeat unit weight basis. Other connecting groups can contribute either an increase or decrease in chain rigidity and $T_g$ depending on the conformational freedom and polarity of those units. The most convenient way to tailor properties of a polysulfone to specific needs is through the selection of the bisphenol. The $T_g$s of polysulfones based on the polycondensation of DCDPS with different bisphenols are listed in Table 2.

The three polysulfones in Table 1 exhibit several important common attributes, but there are distinguishing features for each member of this group. Examination of the repeat unit formulas shows that a primary distinguishing feature of PES is that it contains almost double the sulfone group content on a weight basis compared to PSF. Another differentiation is that the flexibilizing isopropylidene unit is absent in PES. As a consequence of these structural changes, the $T_g$ of PES is 35°C higher than that of PSF. On the other hand, because the sulfone group is the most hygroscopic moiety in the backbone of these polymers and the isopropylidene is hydrophobic, the moisture uptake at equilibrium is significantly lower for PSF than it is for PES (Table 3). The ability to maintain water absorption of a plastic at a minimum is desirable for most engineering applications. Thus backbone structure of a polysulfone is tied to physical properties and ultimately to performance in various applications.

Mechanical properties of aromatic polysulfones are intimately tied to backbone structure. For the achievement of good strength and toughness together with favorable melt processing characteristics the first and foremost requirement is a linear (unbranched) and para-linked structure for the aryl groups in the backbone. The nature of the permanent deformation mechanism is not the same in all polysulfones. Rather, it is dependent on subtle backbone structural features, the best illustration of which is the step improvement in impact and toughness properties of PPSF over those of PES and PSF. The dependence of polysulfone mechanical toughness on backbone structural features has been discussed in detail (19).

**Physical, Chemical, and Optical Properties.**   Aromatic polysulfones possess several common key attributes including high glass-transition temperatures (generally >170°C) and a high degree of thermal oxidative stability (Table 3). Thermal oxidative stability of PSF, PES, and PPSF can be inferred from the thermogravimetric data shown in Figure 3. Because of their fully amorphous nature, these resins exhibit optical transparency. The glass-transition temperature of polysulfones produced via nucleophilic polycondensation can be tailored by the choice of the bisphenol as illustrated in Table 2. By virtue of the chemi-

**Table 2. Glass-Transition Temperatures of Polysulfones Produced from the Polycondensation of Dichlorodiphenylsulfone with Various Bisphenols**[a]

| Bisphenol | Structure | $T_g,$[b] °C |
|---|---|---|
| 4,4'-dihydroxydiphenyl oxide | HO—C₆H₄—O—C₆H₄—OH | 170 |
| 4,4'-dihydroxydiphenyl sulfide | HO—C₆H₄—S—C₆H₄—OH | 175 |
| 4,4'-dihydroxydiphenyl methane | HO—C₆H₄—CH₂—C₆H₄—OH | 180 |
| 2,2-bis(4-hydroxyphenyl)-propane | HO—C₆H₄—C(CH₃)₂—C₆H₄—OH | 185 |
| hydroquinone | HO—C₆H₄—OH | 200 |
| 2,2-bis(4-hydroxyphenyl)-perfluoropropane | HO—C₆H₄—C(CF₃)₂—C₆H₄—OH | 205 |
| 4,4'-dihydroxybenzo-phenone | HO—C₆H₄—C(=O)—C₆H₄—OH | 205 |
| 4,4'-dihydroxydiphenyl sulfone | HO—C₆H₄—S(=O)₂—C₆H₄—OH | 220 |
| 4,4'-dihydroxydiphenyl | HO—C₆H₄—C₆H₄—OH | 220 |
| 1,4-bis(4-hydroxyphenyl)-benzene | HO—C₆H₄—C₆H₄—C₆H₄—OH | 250 |
| 4,4'-bis(4''-hydroxyben-zenesulfonyl)diphenyl | HO—C₆H₄—S(=O)₂—C₆H₄—C₆H₄—S(=O)₂—C₆H₄—OH | 265 |

[a] Ref. 17.
[b] Glass-transition values reported rounded to nearest 5°C.

cally nonlabile aromatic ether backbone, these polymers exhibit superb resistance to hydrolysis in hot water and steam environments. Furthermore, they can withstand acidic and alkali media over a wide range of concentrations and temperatures.

In addition to conferring transparency on these polymers, the amorphous noncrystallizable nature of polysulfones assures minimal shrinkage during fab-

**Table 3. Physical and Thermal Properties of PSF, PES, and PPSF**

| Property | ASTM test method | PSF | PES | PPSF |
|---|---|---|---|---|
| color | | light yellow | light amber | light amber |
| haze[a], % | D1004 | <7 | <7 | <7 |
| light transmittance[b], % | | 80 | 70 | 70 |
| refractive index | D1505 | 1.63 | 1.65 | 1.67 |
| density, g/cm$^3$ | D1505 | 1.24 | 1.37 | 1.29 |
| glass-transition tempera-ture[c], °C | | 185 | 220 | 220 |
| heat deflection tempera-ture[d], °C | D648 | 174 | 204 | 207 |
| continuous service tempera-ture[e], °C | | 160 | 180 | 180 |
| coefficient of linear thermal expansion | D696 | $5.1 \times 10^{-5}$ | $5.5 \times 10^{-5}$ | $5.5 \times 10^{-5}$ |
| specific heat at 23°C, J/(g·K)[f] | | 1.00 | 1.12 | 1.17 |
| thermal conductivity, W/(m·K)[g] | C177 | 0.26 | 0.18 | 0.35 |
| water absorption, % | | | | |
| in 24 h | D570 | 0.22 | 0.61 | 0.37 |
| at equilibrium | D570 | 0.62 | 2.1 | 1.1 |
| mold shrinkage, cm/cm | D955 | 0.005 | 0.006 | 0.006 |
| temperature at 10% weight loss (tga)[h] | | | | |
| in nitrogen | | 512 | 547 | 550 |
| in air | | 507 | 515 | 541 |

[a]As measured on 3.1-mm thick specimens.
[b]Typical values; varies with color. All three resins are transparent.
[c]Onset value as measured by differential scanning calorimetry.
[d]As measured on 3.1-mm thick ASTM specimens under a load of 1.82 MPa (264 psi).
[e]Practical maximum long-term use temperatures for PSF and PES based on UL 746 thermal rating data; value for PPSF is estimated.
[f]To convert J/(g·K) to Btu/(lb·°F), divide by 4.184.
[g]To convert W/(m·K) to Btu/(h·ft·°F), multiply by 1.874.
[h]Thermogravimetric analysis (tga) run at heating rate of 10°C/min and 20 mL/min gas (nitrogen or air) flow rate.

rication of the resins into finished parts. The absence of crystallinity also assures dimensional stability during the service life of the parts where high use temperatures are encountered. Good dimensional stability is important to many structural and engineering applications.

Although commercially available polysulfones are transparent in their natural form, they show a slight yellow–amber tinge. This color is related to the high melt processing temperatures required during resin manufacture and finishing steps. Great progress has been made over the years in controlling and minimizing color (yellowness) in polysulfones. Among the three commercially available polysulfones, the bisphenol A-based polymer (PSF) exhibits the lowest color and highest transmittance of visible light. The light transmittance value

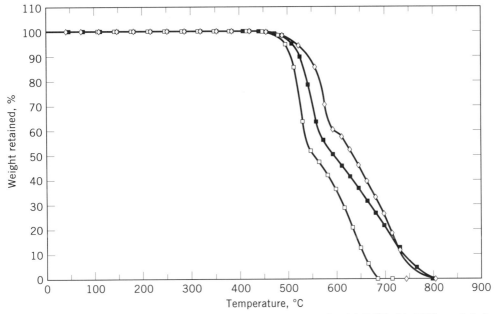

**Fig. 3.** Thermogravimetric analysis (tga) curves in air for (□) PSF, (■) PES, and (◇) PPSF. Tga in nitrogen at a heating rate of 10°C/min (17).

ranges given in Table 3 are based on the color ranges that are typical for commercially available grades.

Because of the presence of the sulfone moiety, polysulfones are slightly hygroscopic. Absorbed water expands linear dimensions of a polysulfone of the order of 0.010–0.012% for every 0.1 weight percent of moisture absorbed. This dimensional change is relatively small, but it can be important in applications where very close dimensional tolerances are a requirement. In addition to effecting small dimensional changes, absorbed moisture causes a slight plasticization of the plastic, contributing to some lowering of stiffness and strength in hot, wet environments. This effect is again small and can be easily compensated for by judicious part design and engineering.

**Mechanical Properties.** Polysulfones are rigid and tough with practical engineering strength and stiffness properties even without reinforcement. Their strength and stiffness at room temperature are high compared to traditional aliphatic backbone amorphous plastics. The polymers exhibit ductile yielding over a wide range of temperatures and deformation rates. High unnotched impact resistance has been tied to a second-order ($\beta$) transition which is observed in these polymers under dynamic mechanical thermal analysis. The $\beta$-transition occurs at around −100°C and is believed to be due to two mechanisms. The first involves 180 degree flips of aromatic units about the ether bond (20). The second has been proposed to be a concerted motion of the sulfone group with complexed absorbed water (21). The presence of moisture increases the magnitude of the $\beta$-transition peak although it is not necessary for its existence. The effect of polysulfone backbone structure on the sub-$T_g$ relaxations has been the subject of detailed study (22).

The room temperature mechanical properties of bisphenol A, bisphenol S, and biphenol-based polysulfones are given in Table 4. The elastic limit (yield) elongation among these polymers is highest for PPSF; PES offers slightly higher tensile strength than the other two polymers. Otherwise the tensile and flexural properties for these three polymers are quite comparable. The main distinguishing feature in PPSF mechanical properties is its very high notched impact strength and the ability of the resin to retain a high degree of ductility after prolonged heat exposure. The data in Table 4 represent short-term mechanical properties under simple loading conditions. These values should only be regarded as typical values for generic polysulfone, polyethersulfone, and polyphenylsulfone of practical molecular weights. Mechanical performance assessment of an engineering polymer for consideration in a specific end use should include long-term aspects such as creep and fatigue properties where applicable. Temperature and environmental factors should also be taken into consideration.

The effect of temperature on PSF tensile stress–strain behavior is depicted in Figure 4. The resin continues to exhibit useful mechanical properties at temperatures up to 160°C under prolonged or repeated thermal exposure. PES and PPSF extend this temperature limit to about 180°C. The dependence of flexural moduli on temperature for polysulfones is shown in Figure 5 with comparison to other engineering thermoplastics.

The tensile and flexural properties as well as resistance to cracking in chemical environments can be substantially enhanced by the addition of fibrous reinforcements such as chopped glass fiber. Mechanical properties at room temperature for glass fiber-reinforced polysulfone and polyethersulfone are shown in Table 5.

**Table 4. Room Temperature Mechanical Properties of PSF, PES, and PPSF**

| Property | ASTM test method | PSF | PES | PPSF |
|---|---|---|---|---|
| tensile[a] (yield) strength, MPa[b] | D638 | 70.3 | 83.0 | 70.0 |
| tensile modulus, GPa[c] | D638 | 2.48 | 2.60 | 2.30 |
| elongation at yield, % | D638 | 5.7 | 6.5 | 7.2 |
| elongation at break, % | D638 | 75 | 40 | 90 |
| flexural strength, MPa[b] | D790 | 106 | 111 | 91 |
| flexural modulus, GPa[c] | D790 | 2.69 | 2.90 | 2.40 |
| compressive strength, MPa[b] | D695 | 96 | 100 | 99 |
| compressive modulus, GPa[c] | D695 | 2.58 | 2.68 | 1.73 |
| shear (yield) strength, MPa[b] | D732 | 41.4 | 50 | 62 |
| notched Izod impact, J/m[d,e] | D256 | 69 | 85 | 694 |
| tensile impact, kJ/m$^{2f}$ | D1822 | 420 | 340 | 400 |
| Poisson ratio, at 0.5% strain | | 0.37 | 0.39 | 0.42 |
| Rockwell hardness | D785 | M69 | M88 | M86 |
| abrasion resistance,[g] mg/1000 cycles | D1044 | 20 | 19 | 20 |

[a]Tensile, flexural, and impact properties based on 3.1-mm thick ASTM specimens.
[b]To convert MPa to psi, multiply by 145.
[c]To convert GPa to psi, multiply by 145,000.
[d]To convert J/m to ft·lbf/in., divide by 53.38.
[e]No break for unnotched samples.
[f]To convert kJ/m$^2$ to ft·lbf/in$^2$, divide by 2.10.
[g]Taber abrasion test using CS-17 wheel and 1000-g load for 1000 cycles.

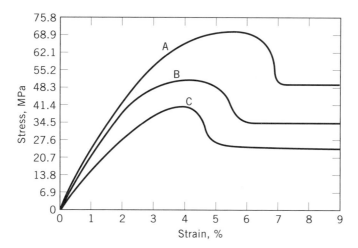

**Fig. 4.** Tensile stress–strain curves for polysulfone showing yield behavior at A, 20°C; B, 99°C; and C, 149°C. To convert MPa to psi, multiply by 145.

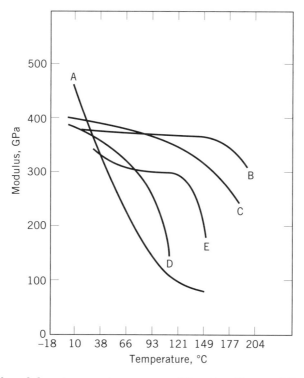

**Fig. 5.** Flexural modulus–temperature curves of C, polysulfone and B, polyethersulfone compared to the moduli curves of A, polyacetal; D, heat-resistant ABS; and E, polycarbonate. To convert GPa to psi, multiply by 145,000.

**957**

**Table 5. Properties of Glass Fiber-Reinforced (GR) Polysulfone and Polyethersulfone**

| Property | ASTM test method | Polysulfone, % GR | | | Polyethersulfone, % GR | | |
|---|---|---|---|---|---|---|---|
| | | 10 | 20 | 30 | 10 | 20 | 30 |
| tensile strength, MPa[a] | D638 | 77.9 | 96.5 | 108 | 86 | 105 | 126 |
| tensile modulus, GPa[b] | D638 | 3.65 | 5.17 | 7.38 | 3.8 | 5.7 | 8.6 |
| tensile elongation, % | D638 | 4.1 | 3.2 | 2.0 | 5.8 | 3.2 | 1.9 |
| flexural strength, MPa[a] | D790 | 128 | 148 | 154 | 145 | 162 | 179 |
| flexural modulus, GPa[b] | D790 | 3.79 | 5.52 | 7.58 | 4.1 | 5.2 | 8.1 |
| Izod impact, J/m[c] | D256 | 64 | 69 | 74 | 48 | 59 | 75 |
| tensile impact, kJ/m[2d] | D1822 | 101 | 114 | 109 | 59 | 65 | 71 |
| heat deflection temperature, °C | D648 | 179 | 180 | 181 | 211 | 214 | 216 |
| CLTE[e], mm/mm·°C | D696 | 3.6 | 2.5 | 2.0 | 3.6 | 3.1 | 3.1 |
| specific gravity | D1505 | 1.33 | 1.40 | 1.49 | 1.43 | 1.51 | 1.58 |
| mold shrinkage, mm/mm | D955 | 0.004 | 0.003 | 0.002 | 0.005 | 0.004 | 0.003 |

[a]To convert MPa to psi, multiply by 145.
[b]To convert GPa to psi, multiply by 145,000.
[c]To convert J/m to ft·lbf/in., divide by 53.38.
[d]To convert kJ/m$^2$ to ft·lbf/in$^2$, divide by 2.10.
[e]Coefficient of linear thermal expansion.

**Flammability.**   An important consideration in the selection of thermoplastic resins involves the need for good flame retardancy characteristics (see FLAME RETARDANTS). Polysulfones exhibit excellent inherent burning resistance characteristics compared to many engineering thermoplastics. The wholly aromatic polysulfones such as PES and PPSF possess particularly outstanding flame retardancy and very low smoke release characteristics. The flammability of PSF is also good but can be enhanced further by the addition of a small proportion of a nonhalogen-based flame retardant. This forms the basis of the UDEL P-1720 commercial polysulfone grade from Amoco Corporation. Flammability properties of various polysulfones are given in Table 6.

**Table 6. Flammability and Burning Behavior of PSF, PES, and PPSF**

| Property | Test method | PSF | PES | PPSF |
|---|---|---|---|---|
| flammability rating | UL 94 | V0 at 6.1 mm | V0 at 0.8 mm | V0 at 0.8 mm |
| limiting oxygen index | ASTM D286 | 30.0 | 38.0 | 38.0 |
| smoke density[a] | ASTM E662 | 90 at 1.5 mm | 35 at 6.2 mm | 30 at 6.2 mm |
| self-ignition temperature, °C | ASTM D 1929 | 621[b] | 502[b] | |

[a]Specific optical density, $D_m$, flaming condition.
[b]Thickness.

**Electrical Properties.**   Polysulfones offer excellent electrical insulative capabilities and other electrical properties as can be seen from the data in Table 7. The resins exhibit low dielectric constants and dissipation factors even in the GHz (microwave) frequency range. This performance is retained over a wide temperature range and has permitted applications such as printed wiring board substrates, electronic connectors, lighting sockets, business machine components, and automotive fuse housings, to name a few. The desirable electrical properties along with the inherent flame retardancy of polysulfones make these polymers prime candidates in many high temperature electrical and electronic applications.

**Table 7. Electrical Properties of PSF, PES, and PPSF**

| Property | ASTM test method | PSF | PES | PPSF |
|---|---|---|---|---|
| dielectric strength[a], kV/mm | D149 | 16.6 | 15.5 | 14.6 |
| volume resistivity, Ω·cm | D257 | $7 \times 10^{16}$ | $9 \times 10^{16}$ | $9 \times 10^{15}$ |
| dielectric constant, Hz | D150 | | | |
| at 60 | | 3.18 | 3.65 | 3.44 |
| $10^3$ | | 3.17 | 3.65 | 3.45 |
| $10^6$ | | 3.19 | 3.52 | 3.45 |
| dissipation factor, Hz | D150 | | | |
| at 60 | | 0.0008 | 0.0019 | 0.0006 |
| $10^3$ | | 0.0008 | 0.0023 | |
| $10^6$ | | 0.0051 | 0.0048 | 0.0076 |

[a]Thickness = 3.2 mm.

**Resistance to Chemical Environments and Solubility.**   As a rule, amorphous plastics are susceptible, to various degrees, to cracking by certain chemical environments when the plastic material is placed under stress. The phenomenon is referred to as environmental stress cracking (ESC) and the resistance of the polymer to failure by this mode is known as environmental stress cracking resistance (ESCR). The tendency of a polymer to undergo ESC depends on several factors, the most important of which are applied stress, temperature, and the concentration of the aggressive species. Polysulfones, being completely amorphous, exhibit susceptibility to stress cracking by some organic environments. The potency of the stress cracking agent is generally related to the match between the solubility parameter of the solvent with that of the polymer. For example, PSF, which has solubility parameter $\delta = 21.8$ $(J/cm^3)^{1/2}$, resists aliphatic hydrocarbons (lower $\delta$) and most alcohols (higher $\delta$), but readily undergoes stress cracking by ketones which have $\delta$-values sufficiently close to that of the polymer.

The exact mechanism of environmental stress cracking of a polymer is still not completely understood, but in essence it involves a weakening of the secondary intermolecular forces between polymeric segments due to the solubility of the chemical in the polymer. A number of crazes are generated at stressed polymer surfaces as a result, and if the number of crazes is large the stress level is moderated and cracking is prevented, or at least delayed. If only a few crazes form, on the other hand, they tend to grow readily and propagate as cracks resulting in rupture.

The ESCR performance of PSF, PES, and PPSF is summarized in Table 8. These polymers are highly resistant to hydrolysis by hot aqueous media, including boiling water, high pressure steam, mineral acids and alkalies, and salt solutions. This resistance is usually a key reason behind the selection of polysulfones over the other engineering plastics like polycarbonates, polyesters, polyamides, and polyetherimides (23). The resistance to stress cracking by organic solvents varies according to the one-dimensional solubility parameter concept described. The most problematic chemical families are aromatics, chlorinated hydrocarbons, ketones, and esters. The resistance of the three polysulfones to these and other environments in general follows the ascending order PSF < PES < PPSF as illustrated in Table 8. When the aggressive environment does not actually dissolve or swell the polymer, the ESCR problem can usually be overcome by the use of a glass fiber-reinforced grade of the polysulfone resin of interest. Glass fiber or any other effective reinforcing agent usually provides enough added load-bearing capability to compensate for the ESCR deficiency of the polymer.

Solubility of the three commercial polysulfones follows the order PSF > PES > PPSF. At room temperature, all three of these polysulfones as well as the vast majority of other aromatic sulfone-based polymers can be readily dissolved in a few highly polar solvents to form stable solutions. These solvents include NMP, DMAc, pyridine, and aniline. 1,1,2-Trichloroethane and 1,1,2,2-tetrachloroethane are also suitable solvents but are less desirable because of their potentially harmful health effects. PSF is also readily soluble in a host of less polar solvents by virtue of its lower solubility parameter. These solvents include tetrahydrofuran (THF), 1,4-dioxane, chloroform, dichloromethane, and chlorobenzene. The relatively broad solubility characteristics of PSF have been key in the development of solution-based hollow-fiber spinning processes

**Table 8. Resistance**[a,b] **of Natural (Unreinforced) PSF, PES, and PPSF Resins to Various Chemical Environments**[c]

| | Ranking | | |
|---|---|---|---|
| Environment | PSF | PES | PPSF |
| hydrocarbons | | | |
|   aliphatic | 1 | 1 | 1 |
|   aromatic | 10 | 9 | 7 |
|   chlorinated | 10 | 8 | 8 |
| alcohols/glycols | 2 | 2 | 2 |
| esters | 9 | 8 | 6 |
| ketones | 10 | 8 | 7 |
| amines | | | |
|   aliphatic | 2 | 1 | 1 |
|   aromatic | 8 | 7 | 6 |
| electrolyte solutions | 1 | 1 | 1 |
| acids[d] | 3 | 2 | 1 |
| bases | 3 | 2 | 1 |
| surfactants | | | |
|   ionic | 1 | 1 | 1 |
|   nonionic | 5 | 4 | 3 |

[a]Ranking codes: 1= excellent; 5 = good; 10 = very poor.
[b]Data are for comparative purposes only; actual resistance depends on many factors including stress, temperature, concentration, and exposure duration.
[c]Stress cracking resistance is substantially enhanced in the presence of reinforcement such as glass fibers.
[d]Nonoxidizing.

in the manufacture of polysulfone asymmetric membranes (see HOLLOW-FIBER MEMBRANES). The solvent list for PES and PPSF is short because of the propensity of these polymers to undergo solvent-induced crystallization in many solvents. When the PES structure contains a small proportion of a second bisphenol comonomer, as in the case of RADEL A (Amoco Corp.) polyethersulfone, solution stability is much improved over that of PES homopolymer.

**Radiation Resistance.** Polysulfones exhibit resistance to many electromagnetic frequencies of practical significance, including microwave, visible, and infrared. Especially notable is the excellent resistance to microwave radiation, which has contributed to the excellent fit of polysulfones in cookware applications. Polysulfone also shows good resistance to x-rays, electron beam (24), and gamma (25,26) radiation under many practical application conditions.

Like the majority of aromatic polymers, polysulfones exhibit poor resistance to ultraviolet light. Polysulfones absorb in the uv region, with attendant discoloration and losses in mechanical properties due to polymer degradation at and directly beneath the exposed surface. Prolonged exposure to sunlight is therefore not recommended for these polymers in neat form. Retention of mechanical integrity is improved with pigmentation and/or reinforcement of the resin, and full resistance to uv is achieved when the resins are pigmented with carbon black. Painting is another option that can be used to avoid the effects of uv light. The retention of mechanical integrity of PPSF upon exposure to uv light is somewhat better than it is for either PSF or PES, especially when the resin is in an opaque form. This is a consequence of the more ductile nature of PPSF compared to both PSF and PES.

## Fabrication

As with most linear amorphous polymers, polysulfones are fully thermoplastic materials and readily flow at temperatures ≥150°C above their respective glass-transition temperatures. The backbone structure is extremely thermally stable during melt processing, remaining unchanged even when subjected to several melt fabrication cycles. Polysulfones can be melt-processed on conventional equipment used for thermoplastics fabrication. Typical melt viscosity behavior as a function of shear rate is shown in Figure 6 for polysulfones and some other polymers. As illustrated from these plots, the shear thinning characteristics of polysulfones are much more muted than they are for aliphatic backbone polymers such as polyethylene and polystyrene. The rheological behavior of polysulfones is fundamentally more similar to that of bisphenol A polycarbonate.

Injection molding is the most common fabrication technique. Melt temperatures for PSF injection molding can be in the range 325–400°C depending on part thickness, length, and complexity. PES and PPSF are generally molded using temperatures in the range 360–400°C. Mold temperatures suitable for PSF injection molding are in the range 100–170°C; for PES and PPSF this range is 120–190°C. Lower mold temperatures can be used but are not recommended as they can result in unacceptably high levels of molded-in residual stresses in the parts. Because of the relatively high viscosities of polysulfones and the limited

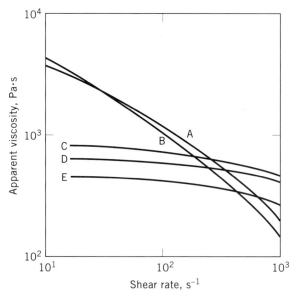

**Fig. 6.** Melt viscosity dependence on shear rate for various polymers: A, low density polyethylene at 210°C; B, polystyrene at 200°C; C, UDEL P-1700 polysulfone at 360°C; D, LEXAN 104 polycarbonate at 315°C; and E, RADEL A-300 polyethersulfone at 380°C.

shear-thinning they exhibit over practical shear rate ranges, the generous sizing of runners and gates in mold design is recommended. The insensitivity to shear has the beneficial effect of yielding parts that are low in molecular orientation and hence much more nearly isotropic than those molded from many other thermoplastics.

Polysulfones are easily processible by other thermoplastic fabrication techniques, including extrusion, thermoforming, and blow molding. Extrusion into film, sheet, tubing, or profile can be accomplished on conventional extrusion equipment with a metering screw of moderate depth having a 2.5:1 compression ratio. Stock temperatures during extrusion of PSF, PES, and PPSF are in the range 315–375°C depending on resin viscosity grade and type of product being produced. A common fabrication technique is sheet extrusion followed by thermoforming. Prior to thermoforming, the sheet must be dry to prevent foaming. The surface temperatures required to produce sag are in the range 230–260°C for PSF and 275–305°C for PES and PPSF. Blow molding is possible on most commercial blow molding equipment provided that equipment is capable of maintaining melt temperatures in the 300–360°C range.

Once formed, parts made of polysulfones (particularly those produced by injection molding) can be annealed to reduce molded-in stress. Increased part stiffness, dimensional stability, and resistance to creep during service life are generally enhanced by reducing molded-in stresses. More importantly, the resistance of the part to environmental stress cracking or crazing is usually improved if molded-in stresses are minimized. Annealing can be easily accomplished either in an air oven or a glycerol bath. Typical time–temperature conditions for properly annealing polysulfones in an air oven are 1 h/170°C for PSF and 1 h/200°C for PES and PPSF. Annealing at similar temperatures by immersion in hot glycerol can be accomplished in 1–5 min. Because over-annealing may result in a reduction of impact toughness the use of conservative molding practices, most notably a hot mold, to produce low stress parts is preferred whenever contact with aggressive chemical environments is anticipated during the service life of the component.

Prior to melt processing, the resin must be dried to reduce the level of absorbed atmospheric moisture, which can be as high as 0.8 wt %, to below 0.05 wt %. The presence of significant amounts of moisture during melt processing causes structural and appearance defects in the fabricated parts due to bubbling and foaming of the trapped moisture. Drying to the target 0.05 wt % moisture can be easily achieved in a circulating hot air oven or a dehumidifying hopper dryer in 3–4 h at 135–165°C. Drying temperatures can be increased up to 180°C for PES and PPSF, if needed to cut down on drying time.

## Blends and Alloys

The blending of two or more polymers to achieve unique property combinations of the parent polymers continues to be an attractive way to tailor existing commercial polymers to specific end use requirements. The blending of polysulfones with other polymers presents opportunities, but at the same time poses some significant technical challenges. Miscibility of PSF or PES with any nonsulfone-based polymer is extremely rare. Some examples of limited miscibility have been discussed (27–31). Only one of these cases involves stable one-phase behavior in the melt state (31). The inert nature of the phenyl, sulfone, and ether moieties precludes interchain reactions, such as ester–ester interchange, which can facilitate the achievement of miscibility in otherwise immiscible systems. As a result, blends of PSF or PES with other nonsulfone-based polymers generally rely on interfacial adhesion and good shear mixing during compounding to produce an intimately mixed blend with good mechanical compatibility. None of the binary blends comprising PSF, PES, and PPSF are miscible, although their blends form mechanically compatible mixtures with relatively stable phase morphologies. Properties of these blends are available (32–34). Blends comprising two or more polysulfones can provide a convenient way to achieve specific properties or simply to reduce cost.

One of the factors precluding blending of polysulfones and commodity polymers such as styrenics and polyolefins is the fact that these polymers do not possess the requisite thermal stability to endure melt processing in the temperature range that is required for polysulfones. As a result, there are few commercially available polymer classes that make good candidates for blending with polysulfones. These are, in general, other engineering thermoplastics such as polycarbonates, some polyester, polyimides (qv), polyaryletherketones, and poly(phenylene sulfide) (PPS). Polymer blends involving polysulfones are discussed in numerous patents and journal articles, eg, a study (35) dealing with the interaction energies of polysulfones with several polymers as they relate to miscibility. Properties of PSF–PPS blends are well documented (36,37). Another study (38) describes the mechanical and rheological properties of PSF when blended with a thermotropic liquid crystalline polyester.

For reasons that are not fully understood, PPSF exhibits generally improved compatibility characteristics over either PSF or PES in a number of systems. An example of this is blends of PPSF with polyaryletherketones (39,40). These blends form extremely finely dispersed systems with synergistic strength, impact, and environmental stress cracking resistance properties. Blends of PPSF with either PSF or PES are synergistic in the sense that they exhibit the super-toughness characteristic of PPSF at PSF or PES contents of up to 35 wt % (33,34). The miscibility of PPSF with a special class of polyimides has been discovered and documented (41). The miscibility profile of PPSF with high temperature ($T_g > 230°C$) polysulfones has been reported (42).

Proprietary blend formulations based on polysulfone, polyethersulfone, and polyphenylsulfone are sold commercially by Amoco Corporation to meet various end use requirements. The blends based on polysulfone are sold under the MINDEL trademark. A glass fiber-reinforced blend based on PES is offered under the trade name RADEL AG-360. This offers most of the performance characteristics of 30% glass fiber-reinforced polyethersulfone but at a lower cost. Two blend product lines are offered based on PPSF. These are designated as the RADEL R-4000 and R-7000 series of products. The former is a lower cost alternative to RADEL R PPSF homopolymer offering most of the performance attributes unique to PPSF. The R-7000 series of resins have been formulated for use in aircraft interiors for civil air transport. They exhibit a very high degree of resistance to flammability and smoke release.

## Health and Safety

Polysulfones are chemically inert polymers for the most part and to date (ca 1996) have no known negative health effects. These polymers have been used for many years in applications where safety is of the utmost importance. Numerous grades comply with U.S. and international governmental regulations for direct food contact. UDEL (Amoco Corp.) polysulfone has been in use for food processing (qv), plumbing, and medical and prosthetic device applications since the 1960s. Various grades of UDEL polysulfone and RADEL A polyethersulfone meet U.S. Food and Drug Administration (FDA) requirements for direct food contact. Similar approvals are given by the U.S. National Sanitation Foundation under Standard 51 permitting use of certain polysulfone and polyethersulfone resins in food processing equipment. RADEL R polyphenylsulfone complies with *U.S. Pharmacopeia* Class VI requirements for use in medical device components.

The thermally and oxidatively stable backbones of polysulfones preclude development of any significant amount of toxic volatile degradation by-products when the resins are heated during melt processing. The polymers remain essentially odorless when injection-molded at stock temperatures of up to 380°C. At temperatures above 380°C, trace amounts of sulfur dioxide, methane, and other organic compounds and residual solvents begin to be evolved. As with other plastic materials, adequate ventilation of the molding area is recommended when injection-molding polysulfones.

## Uses

Polysulfones are used in a wide variety of applications that take advantage of hydrolytic and acid/caustic stability, clarity, and high heat deflection temperatures. These application areas include consumer items (ie, cookware and appliances), electrical and electronic packaging and substrates, automotive, aerospace, and a host of industrial and plumbing uses. The resistance of polysulfones to chemical attack has resulted in their use in chemical processing equipment. Examples of components in this area are corrosion-resistant pipe, pumps, filter modules, support plates, and tower packing. Glass-reinforced grades can be used in very severe chemical environments for enhanced resistance and long service life.

Polysulfones also offer desirable properties for cookware applications, eg, microwave transparency and environmental resistance to most common detergents. Resistance to various sterilizing media (eg, steam, disinfectants, and gamma radiation) makes polysulfones the resin family of choice for many medical devices. Uses in the electrical and electronic industry include printed circuit boards, circuit breaker components, connectors, sockets, and business machine parts, to mention a few. The good clarity of PSF makes it attractive for food service and food processing uses. Examples of applications in this area include coffee decanters and automated dairy processing components.

One unique application area for PSF is in membrane separation uses. Asymmetric PSF membranes are used in ultrafiltration, reverse osmosis, and ambulatory hemodialysis (artificial kidney) units. Gas-separation membrane technology was developed in the 1970s based on a polysulfone coating applied to a hollow-fiber support. The PRISM (Monsanto) gas-separation system based on this concept has been a significant breakthrough in gas-separation technology (see MEMBRANE TECHNOLOGY). Additional details are available on the use of polysulfone in membrane separations (43), as well as gas transport properties of polysulfone and polyethersulfone (44–48).

Polysulfone, polyethersulfone, and polyphenylsulfone may be used interchangeably in many applications. In general, polysulfone is selected because of its lower cost. PES is selected over PSF in applications demanding higher temperatures and/or additional environmental resistance. PPSF represents a step improvement over PES in hydrolytic stability, impact, and chemical resistance with similar temperature capabilities to those of PES. It is selected only when both PSF and PES fail to meet engineering performance requirements. Examples of PPSF uses include steam-autoclaveable surgical sterilization trays, transformer magnet wire coatings, and aircraft interior parts.

## Economic Aspects

There are three commercial suppliers that manufacture polysulfones: Amoco Corporation in the United States, BASF Corporation in Germany, and Sumitomo Chemical Company in Japan. A listing of the resins supplied by each of these companies along with the trade names particular to each of these suppliers is shown in Table 9. All three companies supply a polyethersulfone-type product. Polysulfone, on the other hand, is supplied by Amoco and BASF, and Amoco is the sole supplier of polyphenylsulfone.

As of November 1, 1995, the price range of UDEL polysulfone resins in large quantities was \$9.10–\$12.03/kg depending on grade and color. RADEL A polyethersulfone grades sold for \$10.46–\$13.26/kg and the prices of RADEL R polyphenylsulfone resins were in the range \$15.75–\$25.95/kg. MINDEL resins based on polysulfone had list prices between \$6.72–\$7.42/kg.

**Table 9. Manufacturers and Trade Names of Commercially Available Polysulfones[a]**

| Supplier | Polysulfone types offered | Trade name |
|---|---|---|
| Amoco Corp. | PSF | UDEL |
| | PES | RADEL A |
| | PPSF | RADEL R |
| BASF Corp. | PSF | ULTRASON S |
| | PES | ULTRASON E |
| Sumitomo Chemical Co. | PES | SUMIKAEXCEL |

[a] As of November 1, 1995.

## BIBLIOGRAPHY

"Polysulfone Resins" under "Polymers Containing Sulfur" in *ECT* 2nd ed., Vol. 16, pp. 272–281, by R. N. Johnson, Union Carbide Corp.; in *ECT* 3rd ed., Vol. 18, pp. 832–848, by N. J. Ballintyn, Union Carbide Corp.

1. E. J. Goethals, in N. M. Bikales, ed., *Encyclopedia of Polymer Science and Technology*, Vol. 13, Interscience Publishers, a division of John Wiley & Sons, Inc., New York, 1969, pp. 448–477.
2. R. N. Johnson, in Ref. 1, Vol. 11, pp. 447–463.
3. U.S. Pat. 4,108,837 (Aug. 22, 1978), R. N. Johnson and A. G. Farnham (to Union Carbide Corp.).
4. R. N. Johnson, A. G. Farnham, R. A. Clendinning, W. F. Hale, and C. N. Merriam, *J. Polym. Sci., Part A-1* **5**, 2375 (1967).
5. R. N. Johnson and J. E. Harris, in J. I. Kroschwitz, ed., *Encyclopedia of Polymer Science and Engineering*, 2nd ed., Vol. 13, John Wiley & Sons, Inc., New York, 1988, pp. 196–211.
6. S. R. Shulze, *155th National American Chemical Society Meeting*, San Francisco, Calif., Mar.–Apr. 1968, p. L-090.
7. Brit. Pat. 1,153,035 (Sept. 24, 1965), D. A. Barr and J. B. Rose (to Imperial Chemical Industries).
8. U.S. Pat. 3,332,909 (July 25, 1967), A. G. Farnham and R. N. Johnson (to Union Carbide Corp.).
9. U.S. Pat. 5,239,043 (Aug. 24, 1993), S. Savariar (to Amoco Corp.).
10. U.S. Pat. 5,235,019 (Aug. 10, 1993), S. Savariar (to Amoco Corp.).
11. U.S. Pat. 4,008,203 (Feb. 15, 1977), M. E. B. Jones (to Imperial Chemical Industries).
12. B. E. Jennings, M. E. B. Jones, and J. B. Rose, *J. Polym. Sci., Part C: Polym. Lett.* **16**, 715 (1967).
13. J. B. Rose, *Polymer* **15**, 456 (1974).
14. I. Colon and D. R. Kelsey, *J. Org. Chem.* **51**, 2627 (1986).
15. U.S. Pat. 4,263,466 (Apr. 21, 1981), I. Colon, L. M. Maresca, and G. T. Kwiatkowski (to Union Carbide Corp.).
16. U.S. Pat. 4,400,499 (Aug. 23, 1983), I. Colon (to Union Carbide Corp.).
17. G. T. Kwiatkowski, I. Colon, M. J. El-Hibri, and M. Matzner, *Makromol. Chem., Macromol. Symp.* **54/55**, 199–224 (1992).
18. V. Percec and H. Nava, *J. Polym. Sci., Part A: Polym. Chem.* **26**, 783 (1988).
19. T. E. Attwood, M. B. Cinderey, and J. B. Rose, *Polymer* **34**, 1322 (1993).
20. J. J. Dumias, A. L. Cholli, L. W. Jelinski, J. L. Hendrick, and J. E. McGrath, *Macromolecules* **19**, 1884 (1986).
21. L. M. Robeson, A. G. Farnham, and J. E. McGrath, *Appl. Polym. Symp.* **26**, 373 (1975).
22. C. L. Aitken, W. J. Koros, and D. R. Paul, *Macromolecules* **25**, 3424 (1992).
23. L. M. Robeson and S. T. Crisafulli, *J. Appl. Polym. Sci.* **28**, 2925 (1983).
24. A. Davis, M. H. Gleaves, J. H. Golden, and M. B. Huglin, *Makromol. Chem.* **129**, 63 (1969).
25. J. R. Brown and J. H. O'Donnell, *Polym. Lett.* **8**, 121 (1970).
26. A. R. Lyons, M. C. R. Symons, and J. K. Yandel, *Makromol. Chem.* **157**, 103 (1972).
27. D. J. Walsh and V. B. Singh, *Makromol. Chem.* **185**, 1979 (1984).
28. D. J. Walsh, S. Rostami, and V. B. Singh, *Makromol. Chem.* **186**, 145 (1985).
29. H. Nakamura, J. Maruta, T. Ohnaga, and T. Inoue, *Polymer* **31**, 303 (1990).
30. K. Jeremic, F. E. Karasz, and W. J. Macknight, *New Polym. Mater.* **3**, 163 (1992).
31. U.S. Pat. 5,191,035 (Mar. 2, 1993), M. J. El-Hibri, J. E. Harris, and J. L. Melquist (to Amoco Corp.).
32. U.S. Pat. 4,743,645 (May 10, 1988), J. E. Harris and L. M. Robeson (to Amoco Corp.).
33. U.S. Pat. 5,086,130 (Feb. 4, 1992), B. L. Dickinson, M. J. El-Hibri, and M. E. Sauers (to Amoco Corp.).
34. U.S. Pat. 5,164,466 (Nov. 17, 1992), M. J. El-Hibri, B. L. Dickinson, and M. E. Sauers (to Amoco Corp.).
35. T. A. Callaghan and D. R. Paul, *J. Polym. Sci., Part B: Polym. Phys.* **32**, 1847 (1994).

36. M.-F. Cheung, A. Golovoy, H. K. Plummer, and H. van Oene, *Polymer* **31**, 2299 (1990).
37. M.-F. Cheung, A. Golovoy, and H. van Oene, *Polymer* **31**, 2307 (1990).
38. S. M. Hong, B. C. Kim, K. U. Kim, and I. J. Chung, *Polym. J.* **23**, 1347 (1991).
39. U.S. Pat. 4,804,724 (Feb. 14, 1989), J. E. Harris and L. M. Robeson (to Amoco Corp.).
40. U.S. Pat. 4,713,426 (Dec. 15, 1987), J. E. Harris and L. M. Robeson (to Amoco Corp.).
41. U.S. Pat. 5,037,902 (Aug. 6, 1991), J. E. Harris and G. T. Brooks (to Amoco Corp.).
42. U.S. Pat. 4,804,723 (Feb. 14, 1989), J. E. Harris and L. M. Robeson (to Amoco Corp.).
43. W. J. Koros and R. T. Chern, in R. W. Rousseau, ed., *Handbook of Separation Process Technology*, John Wiley & Sons, Inc., New York, 1987.
44. A. J. Erb and D. R. Paul, *J. Membr. Sci.* **8**, 11 (1981).
45. K. Ghosal and R. T. Chern, *J. Membr. Sci.* **72**, 91 (1992).
46. J. S. McHattie, W. J. Koros, and D. R. Paul, *Polymer* **32**, 840 (1991).
47. *Ibid.*, p. 2618.
48. K. Ghosal, R. T. Chern, and B. D. Freeman, *J. Polym. Sci., Part B: Polym. Phys.* **31**, 891 (1993).

M. JAMAL EL-HIBRI
Amoco Polymers, Inc.

# POLYMERS, ENVIRONMENTALLY DEGRADABLE

Interest in environmentally degradable plastics began in the early 1960s with the recognition that the common packaging plastics such as polyethylene (PE), polypropylene (PP), polystyrene (PS), poly(vinyl chloride) (PVC), and poly(ethylene terephthalate) (PET) were accumulating in the environment. Their designed and well-established resistance to environmental degradation was observed to be contributing to landfill depletion and litter problems resulting from careless disposal after use. At that time, the principal focus was on replacing these synthetic packaging plastics in all applications with environmentally degradable substitutes.

More recently, it has become increasingly apparent that in addition to the primary synthetic plastics, water-soluble and other specialty polymers and plastics such as poly(acrylic acid), polyacrylamide, poly(vinyl alcohol) and poly(alkylene oxides), and even some modified natural polymers, eg, cellulosics, also potentially contribute to environmental problems and thus become targets for environmentally degradable substitutes. These polymers are widely used as coatings additives, pigment dispersants, temporary coatings, and detergents, in mining, water treatment, etc. Therefore, this article covers polymers in a general sense rather than focusing on commodity plastics. The term polymer will refer to both water-soluble polymers and plastics, unless there is a need to differentiate the two principal polymer types that are the subject of most of the attention given to environmentally degradable polymers. The discussion covers all polymeric materials, natural, synthetic, and modified natural, designed to be degradable in the environment by any of the accepted degradation pathways: photodegradation, biodegradation, and chemical degradation which is hydrolytic or oxidative degradation. Of these pathways, biodegradation is recognized as by far the most important, insofar as it is the only one that can lead to complete removal from the environment, and accordingly it has received the greatest attention. The other degradation pathways are more appropriately described as biodeterioration or biodisintegration because their products are left in the environment unless they are biodegradable. There have been several excellent reviews on environmentally degradable polymers covering all aspects of the subject (1–18).

Concurrent with the movement toward a broadened focus of research to cover all polymeric materials, the initially high expectation for environmentally degradable polymers as a total solution to the polymer and plastic waste management issue was gradually accepted as being no longer tenable. Alternative technologies such as recycling of plastic materials, including recycling of plastics; recycling of plastics to monomers and subsequent repolymerization to the same or new polymers; recycling to olefinic feedstocks by pyrolysis; continued burial in landfill sites; and incineration are recognized as viable options, along with environmental degradation (see RECYCLING, PLASTICS). Each disposal method has a part to play in polymer waste management; preference among them depends on many factors, including available processing facilities, collection of waste material, cost of new polymers, property requirements, etc. Therefore, research in 1996 on environmentally degradable polymeric materials is aimed at developing polymers for applications where they offer unique advantages over the competitive alternative: for example, in agricultural film where cost savings are possible with use of photo- or biodegradable polymers over collection and recycle of current products; compostable plastics for fast food wrappers which will eliminate the need for separation of food and plastic; difficult-to-recover water-soluble polymers and plastics; and personal and industrial hygiene products for diapers, feminine products, and hospital disposables. These products offer challenging synthesis opportunities for environmentally degradable polymers.

Environmentally degradable polymers face many issues and challenges not apparent or recognized in the early years of research. Their development requires a multidisciplinary approach, involving polymer synthesis chemists, analytical chemists, environmental scientists for establishing testing protocols for laboratory simulation of disposal environments, and microbiologists for evaluating the environmental fate and effects of the degradable polymers, their degradation products, and any residues left in the environment. In addition to the scientific issues, there are issues related to perceptions among the public in which strong emotions can play a part, and any new polymer developed as an environmentally degradable product will be scrutinized by consumers of the products it goes into and by national and international legislative bodies for confirmation that it is free from real or perceived adverse environmental effects, before global acceptance can become a reality. Thus, environmentally degradable polymers and plastics must meet very stringent guidelines for acceptance by a wide-ranging panel of reviewers. The importance of meeting this requirement is reflected in the search for acceptable definitions for environmentally degradable polymers and new, more meaningful laboratory testing protocols for quantitatively measuring degradation and environmental fate and effects, and correlating the results of these experiments with real-world exposures. Consequently, definitions and test methods are addressed early in this article, prior to describing the important synthetic approaches under evaluation for environmentally degradable plastics and polymers and identifying some current and potential commercial products.

## DEFINITIONS

There have been numerous definitions proposed for environmentally degradable plastics and polymers. ASTM D1566 defines a polymer as "a macromolecular material formed by the chemical combination of monomers having either the same or different chemical composition"; a plastic, as defined by ASTM D1695, is "a material that contains as an essential ingredient an organic substance of large molecular weight, is solid in its finished state, and, at some stage in its manufacture or in its processing into finished articles, can be shaped by flow." Definitions are important because they are indicative of expectations for the acceptance of environmentally degradable polymers and of the types of testing protocols that are needed to establish the acceptability of the polymers in the environment. The definitions developed by the American Society for Testing and Materials, ASTM D883-93, for degradable, biodegradable, hydrolytically degradable, and oxidatively degradable plastics and given here are probably the most widely accepted, either as written or in some slightly modified form. They are equally applicable to polymers, in general.

*Degradable plastic* is a plastic designed to undergo a significant change in its chemical structure under specific environmental conditions, resulting in a loss of some properties that may vary as measured by standard test methods appropriate to the plastic and the application in a particular period of time that determines its classification.

*Biodegradable plastic* is a degradable plastic in which the degradation results from the action of naturally occurring microorganisms such as bacteria, fungi, and algae.

*Hydrolytically degradable plastic* is a degradable plastic in which the degradation results from hydrolysis.

*Oxidatively degradable plastic* is a degradable plastic in which the degradation results from oxidation.

*Photodegradable plastic* is a degradable plastic in which the degradation results from the action of natural daylight.

The definitions do not quantify the extent of degradation by any of the pathways and indicate only the mechanism that is operating to promote degradation. Although this is acceptable in a scientific sense in order to define the chemical process, it does not really go far enough to satisfy the requirements for environmentally acceptable polymers (19), which in the minds of legislators and lay people is the key issue. If environmentally degradable polymers and plastics are to be acceptable as a waste management option, definitions must be more practical and descriptive in conveying the assurance that no harmful residues are left in the environment after degradation has occurred. The ASTM definitions, therefore, require elaboration in order to address this deficiency (20). The environmental degradation processes are interrelated, as shown schematically in Figure 1.

All four degradation pathways, ie, biodegradation, oxidation, hydrolysis, and photodegradation, initially give intermediate products or fragments that may biodegrade further to some other residue, biodegrade completely and be removed from the environment entirely (being converted into biomass and carbon dioxide and/or methane, depending on whether it is an aerobic or anaerobic system) and ultimately mineralized, or remain unchanged in the environment. Mineralization, usually a slow process for polymeric materials and fragments, refers to complete conversion of a polymer (or any organic compound) to carbon dioxide or methane, water, and salts; it is used here loosely to indicate complete or total removal from the environment to carbon dioxide or methane, water, and biomass. In the cases where residues remain in the environment, they must be established as harmless by suitably rigorous fate and effect evaluations. Only biodegradation has the potential to remove plastic and polymers completely from the environment. Thus, when developing and designing polymers and plastics for degradation in the environment by other pathways, the final stage preferably should be complete biodegradation and removal from the environment, with ultimate mineralization. In this way, the polymers are essentially recycled through nature into microbial cells, plants, and higher animals (21), and thence back into chemical feedstocks. Therefore, an environmentally acceptable degradable plastic or polymer may be defined as one which degrades by any of the above-defined mechanisms, ie, biodegradation, photodegradation, oxidation, or hydrolysis, to leave no harmful residues in the environment (20). This definition has the advantage of not limiting the degree of degradation for a particular polymer but requiring sufficient testing for fragments and degradation products that are incompletely removed from the environment to ensure that no long-term damage or adverse effects to the ecological system remain a possibility. Polymers and plastics meeting this definition should be completely acceptable for disposal in the appropriate environment anywhere in the world.

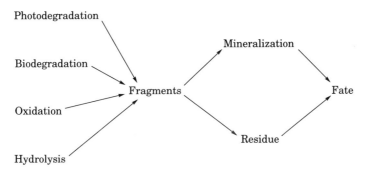

**Fig. 1.** Interrelationships of processes for environmentally degradable polymers.

## OPPORTUNITIES FOR ENVIRONMENTALLY DEGRADABLE
## PLASTICS AND POLYMERS

Opportunities play a significant part in driving research into the property requirements for environmentally degradable plastics and polymers and for the development of laboratory testing protocols. The conditions providing a significant stimulus for the development of environmentally degradable polymers and plastics are the same ones that govern waste management programs and the restriction of the uses for nondegradable polymers and plastics in the environment. Figures 2 and 3 illustrate the disposal pathways of water-soluble polymers and plastics, respectively, two significant environmental waste-management problem areas.

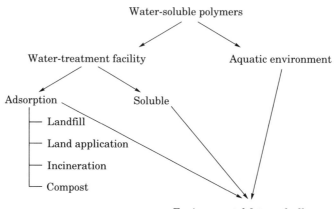

**Fig. 2.**   Environmental disposal of water-soluble polymers.

Water-soluble polymers (qv) after use are usually very dilute solutions and are preferably disposed of through wastewater treatment facilities or sometimes directly into the aquatic environment. On entering a wastewater treatment plant, a polymer may pass straight through it into streams, rivers, lakes, and other aquatic environments, or it may be adsorbed onto the suspended solids in the treatment plant. If it passes straight through, it is no different from direct disposal into those same aqueous environments, and the two disposal methods raise similar questions as to the fate and effects of the polymers. On the other

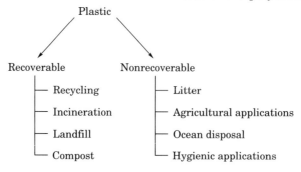

**Fig. 3.**   Environmental disposal of plastics.

hand, adsorption of a polymer onto sewage sludge results in the possibility of the polymer being landfilled, incinerated, composted, or land-applied as fertilizer or for soil amendment, depending on the local options available. However, here again, environmental fate and effects need to be addressed; it must be established how these polymers move in their new environmental compartments, and what are their incineration products. For water-soluble polymers, therefore, there is an obvious and distinct advantage for environmentally acceptable degradable polymers. There is a preference for biodegradable water-soluble polymers, as it is unlikely that the other degradation paths would be applicable in dilute aqueous solutions.

Water-soluble polymers should be designed to be completely biodegradable and removed in the disposal environment, generally the sewage treatment facility, because they may move rapidly and without difficulty throughout the environment. If biodegradation is not complete in the disposal environment, it must be assessed in subsequent environmental compartments that it enters. Once the biodegradation of a polymer has been confirmed, no uncertainty remains as to its fate and effects in any subsequent environmental compartments. The advantage of complete biodegradation is that it can be established with a high degree of certainty with the appropriate test methods, whereas the assessment of environmental fate and effect of any residue is always a risk assessment based on a limited study with a few aquatic species for all environments with which a water-soluble polymer may come into contact as it moves through the environment.

Plastic materials, as indicated, offer more disposal options than water-soluble polymers because they are usually solid, handleable materials and are recoverable in most cases after use for several disposal options, including landfilling, recycling, incineration, and composting. Composting, which is predominantly biodegradation with the possibility of oxidation and hydrolysis, is an opportunity for environmentally degradable plastics which are used in food applications, such as wrappers and utensils. In these uses, plastics are contaminated with food residues and the mix is ideally suitable for composting without separation. Where recovery of current plastics is not economically feasible, viable, controllable, or attractive, the plastics remain as litter; may be discarded at sea from naval vessels; may be used in farm and agricultural applications such as pre-emergence plant protection with sheets and mulch; or in hygienic applications such as diapers, sanitary napkins, and hospital garments and swabs, etc. These each constitute opportunities or a stimulus for the development of environmentally acceptable degradable plastics and polymers.

For water-soluble polymers, there is a well-established disposal infrastructure, with the widely available wastewater treatment plants, whereas plastics being developed for composting require large-scale implementation of a composting infrastructure. This fact will certainly influence the rate of their acceptance.

## TEST METHODS FOR ENVIRONMENTALLY DEGRADABLE POLYMERS

Among the degradation categories, by far the most significant research and development activity has been in biodegradation, followed by photodegradation, and very much less in oxidation and hydrolysis. However, it should be remembered that hydrolysis is frequently followed by biodegradation of the degradation products, and that oxidation and photodegradation often operate in tandem.

Test methods are usually laboratory evaluation of environmental degradation under the simulated real-world conditions to which a particular polymer or plastic will be exposed on disposal; however, it is sometimes desirable to do the testing under the most favorable conditions, to establish whether degradation is possible. Results of such tests only indicate the rate and extent of degradation under the test conditions and must be correlated with exposure in the real-world environment. The ultimate goal of laboratory test methods is predictability of environmental response to new polymers and plastics. This goal is probably more difficult for biodegradation than for photodegradation, because the environments for biodegradation differ widely in microbial composition, pH, temperature, moisture, etc, and are not readily reproduced in the laboratory. Once an environment has been sampled and placed within the confinements of a laboratory vessel, it can no longer interact with the greater environment in response to an added xenobiotic; thus results are frequently difficult to reproduce and may not always be representative of the response in real-world exposures. Photodegradation can be measured more readily.

## Test Methods

**Photodegradation.** Test methods for measuring photodegradation are usually a combination of an exposure to some form of radiation and subsequent property loss measurement in another test, for example a tensile strength loss (ASTM D882-83), impact resistance loss (ASTM D1709-85), tear strength loss (ASTM D1922-67), molecular weight loss, friability, disintegration, brittle point, etc. Several standard test practices have been developed within ASTM (22) for plastic exposure and are listed in Table 1. Subsequent property testing by the standard ASTM test methods, including those already named, may be done at various time intervals throughout the exposure to assess the rate of degradation. In ASTM terminology (ASTM D883-93), a test practice is a procedure which does not lead to a result in itself, that is, it is a conditioning for a standard test method which measures the change(s) that may have occurred.

**Biodegradation.** In the early years of biodegradation testing for polymers and plastics, the only tests conducted to establish biodegradability were related to microbial growth, weight loss, tensile changes, and other physical property losses. These are all indirect measurements of biodegradation and often led to results that were difficult to reproduce from laboratory to laboratory, giving rise to confusion on the susceptibility to biodegradation of a given polymer. The tests

**Table 1. ASTM Standard Practices for Photodegradation**

| Test number | Standard test practice |
|---|---|
| D3826-91 | degradation end points using a tensile test |
| D5071-91 | operation of a xenon arc ARC-type exposure apparatus |
| D5208-91 | operation of a fluorescent ultraviolet (uv) and condensation apparatus |
| D5272-92 | outdoor exposure testing of photodegradable plastics |

and their results have been reviewed in many articles (4,22–24). More recent test methodology development has stressed the proper selection of environment to reflect probable disposal sites for a given polymer or plastic and the need for quantitative testing as the most important aspect of assessing acceptability for environmental disposal. Qualitative tests, however, are still recognized as important for indicating the rate of disintegration of plastics, which has a bearing on such disposal methods as composting for compaction and volume reduction of the compost.

Some of the early tests include Growth Ratings based on ASTM Tests G21-70 and ASTM G22-76, which actually were tests developed for assessing the resistance of plastics to fungal and bacterial growth, respectively. Fungal organisms such as *Aspergillus niger*, *Aspergillus flavus*, *Chaetomium globosum*, and *Penicillium funiculosum* and bacterial standards such as *Pseudomonas aeruginosa* are suggested in the test protocol, though it is not limited to these, and are evaluated for growth on suitable plastics. After a suitable time period, growth is assessed by a subjective numerical rating in which higher numbers are considered to correlate with the susceptibility of the plastic to biodegradation:

| Rating | Growth |
|--------|--------|
| 0 | no visible growth |
| 1 | <10% of surface with growth |
| 2 | 10–30% surface with growth |
| 3 | 30–60% surface with growth |
| 4 | 60–100% surface with growth |

The advantage of this test is that it is quick and easy to do, and gives an indication of biodegradation potential. However, the test is not definitive, because any impurities in the plastic, such as plasticizers and solvents, may interfere with the test by promoting growth, and thus give false positive results. Other simple tests include the soil burial test used to demonstrate the biodegradability of polycaprolactone (25), following its disappearance as a function of time, and the clear zone method which indicates biodegradation by the formation of a clear zone in an agar medium of the test polymer or plastic as it is consumed (26). The burial test is still used as a confirmatory test method in the real-world environment after quantitative laboratory methods indicate biodegradation.

The need to develop better tests, both qualitative and quantitative, has become apparent and is generating a great deal of activity in the areas of both plastics and water-soluble polymers. Many of the quantitative test protocols are based on the tests developed over the last 25 years in the detergent industry for water-soluble organic compounds, particularly surfactants, which have been scrutinized for biodegradability since the late 1960s. Many of these tests have been summarized (27) and are formalized in publications by the U.S. Environmental Protection Agency (EPA) (28) and the Organization for Economic and Cooperative Development (OECD) (29). This earlier work showed the value of choosing the environment applicable to the disposal method with detergents, the wastewater treatment plant, rivers, aquifers, and soil; the conditions for running laboratory tests with temperature control; and addressing toxicity issues, acclimation potential, etc. Above all is recognition of the value of quantitative measurements of the products of biodegradation as the only reasonable means of assessing biodegradation (20).

Biodegradation in aerobic and anaerobic environments may be described by chemical equations, assuming a hydrocarbon polymer as is done in equations 1 (aerobic) and 2 (anaerobic). The equations are readily modified to include other elements that may be present in a particular polymer, and appear in the oxidized or reduced form, depending on the type of environment, aerobic or anaerobic, respectively. Most of the testing reported in the literature over the years has been with aerobic biodegradation conditions, probably because this is easier to do in the laboratory and because most disposal of polymers and plastics is into these environments. However, anaerobic degradation, which is particularly pertinent to water-soluble polymers that may enter anaerobic digestors in sewage treatment facilities, is the object of growing interest, and more information will undoubtedly be developed in the future on this condition. In this article, unless otherwise specified, biodegradation should be understood to be in an aerobic environment.

$$\text{polymer} + O_2 \longrightarrow CO_2 + H_2O + \text{biomass} + \text{residue} \tag{1}$$

$$\text{polymer} \longrightarrow CO_2 + CH_4 + H_2O + \text{biomass} + \text{residue} \tag{2}$$

To assess the degree of biodegradation quantitatively, analytical techniques are needed for any or all of the reactants and products; the polymer; oxygen uptake, known as biochemical oxygen demand (BOD); and the residue. The more rigorous the analysis, the more reliable the measurement of the extent of biodegradation and, of course, the greater the acceptability of the data and the conclusions drawn. For total biodegradation, there should be no residue remaining in the environment.

Qualitative assessment of biodegradation where changes such as weight loss, tensile strength loss, disintegration, etc, are measured is useful in that it indicates the loss of properties as a guideline to the physical breakdown of the plastic and its decomposition in various environments such as compost, landfill, etc, and particularly in the intended disposal environment. To differentiate biodegradation and abiotic degradation such as oxidation and hydrolysis, it is usually necessary to do a simultaneous control test with a killed inoculum (cyanide or mercury salts are acceptable) in which no degradation should be observed if only biodegradation is involved in the plastic degradation. This test would also serve as a test method for hydrolytic or oxidative degradation, should degradation occur in the abiotic environment.

There have been numerous communications on the subject of biodegradation test methods, including aerobic compost (30), anaerobic bioreactor (31), general methodology and future directions (32–34), and a fine review article (24). ASTM (22) and MITI (35) have also set forth standard testing protocols for plastics, as shown in Table 2, whereas OECD test methods (29) are more suited to water-soluble polymers.

The current state of testing (ca 1996) is greatly improved over that which prevailed during the early days of research. However, it should be recognized

**Table 2. Biodegradation Test Protocols**

| Test number | Environment | Measurement |
|---|---|---|
| ASTM D5209-92 | aerobic sewage sludge | $CO_2$ |
| ASTM D5210-92 | anaerobic sewage sludge | $CO_2$ and $CH_4$ |
| ASTM D5247-92 | aerobic specific microorganisms | molecular weight |
| ASTM D5271-93 | aerobic activated sewage sludge | $O_2$ and $CO_2$ |
| ASTM D5338-92 | aerobic controlled composting | $CO_2$ |
| ASTM D5437-93 | marine floating conditions | physical properties |
| MITI Test | mixed microbial | $O_2$ |

that the biodegradation test protocols discussed here are only screening tests for readily biodegradable polymers and plastics. Failure in these tests does not exclude the possibility of biodegradation; it merely indicates that under the environmental conditions evaluated there is no biodegradation. Repeated tests, particularly in other environments, are recommended before acceptance of non-biodegradability. The possibility of toxicity of the polymer and the need for lower concentration testing should also be explored. In the latter case it may be necessary to resort to isotopic labeling in order to measure biodegradation by monitoring low concentrations of carbon dioxide evolution. In many cases with synthetic polymers, it may also be important to allow acclimation to occur so that enzymes may be induced that will biodegrade them. In addition to attention to improved biodegradation test method development, there is a need to establish the fate and effect of residues and degradation fragments in the environment where biodegradation is incomplete, whether it stems from biodegradable polymers and plastics or from any of the other environmental degradation pathways. Some efforts are being made in this area, eg, work on water-soluble polymers (36) and ASTM Standard Practice D5152-91 (22) for extracting aqueous solubles from the solid fragments produced by the environmental degradation of plastics for testing in standard aquatic toxicity protocols. Biodegradation has no value in itself; it is a part of fate and effects to be utilized in the important Environmental Safety Assessment (34).

## Degradation Mechanisms

### PHOTODEGRADATION

Photodegradable polymers degrade in the environment by chain scission promoted by natural daylight and usually oxygen to yield low molecular weight fragments that are more susceptible to biodegradation than the original high molecular weight polymer. The polymers are generally structurally similar to currently used environmentally stable polymers but have been modified during synthesis or post-treatment to insert photochemically active groups. The addition of carbonyl functionality (37–40) into the polymer main or side chain, or of external photosensitizers and pro-oxidants such as metal salts (41,42), benzophenone (43), ketones (44), ethers (44), mercaptans (44), and polyunsaturated compounds (45) are representative examples. The concept and mechanism of fragmentation is well established and there are commercial products based on ethylene–carbon monoxide copolymers for use in six-pack holders and agricultural film. The ultimate fate of the fragments produced is not yet fully established; in most cases the argument is put forward that if the molecular weight of the degradation products is low enough, then they will biodegrade.

The synthesis of copolymers of olefins with carbon monoxide (eq. 3) or ketones (eq. 4) leads to backbone or side-chain carbonyl functionality, respectively. When $R = C_6H_5$, the monomer is styrene, and when $R = H$, the monomer is ethylene (46–48).

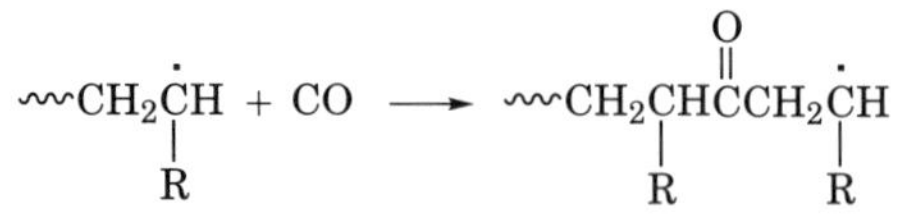

$$\sim\!\!\!\sim\!CH_2\overset{\cdot}{C}H + CO \longrightarrow \sim\!\!\!\sim\!CH_2CH\overset{O}{\overset{\|}{C}}CH_2\overset{\cdot}{C}H \qquad (3)$$
$$\underset{R}{|} \qquad\qquad\qquad \underset{R}{|}\quad\underset{R}{|}$$

$$\sim\!\!\!\sim\!CH_2\overset{\cdot}{C}H + \underset{\underset{R'}{\diagdown}C=O}{\overset{CH_2=CH}{|}} \longrightarrow \sim\!\!\!\sim\!CH_2CHCH_2CHCH_2\overset{\cdot}{C}H \qquad (4)$$
$$\underset{R}{|}\qquad \underset{R}{|}\quad\underset{\underset{R'}{\diagup}C=O}{|}\;\underset{R}{|}$$

When the polymers are exposed to ultraviolet radiation, the activated ketone functionalities can fragment by two different mechanisms, known as Norrish types I and II. The degradation of polymers with the carbonyl functionality in the backbone of the polymer results in chain cleavage by both mechanisms, but when the carbonyl is in the polymer side chain, only Norris type II degradation produces main-chain scission (37,49). A Norrish type I reaction for backbone carbonyl functionality is shown by equation 5, and a Norrish type II reaction for backbone carbonyl functionality is equation 6.

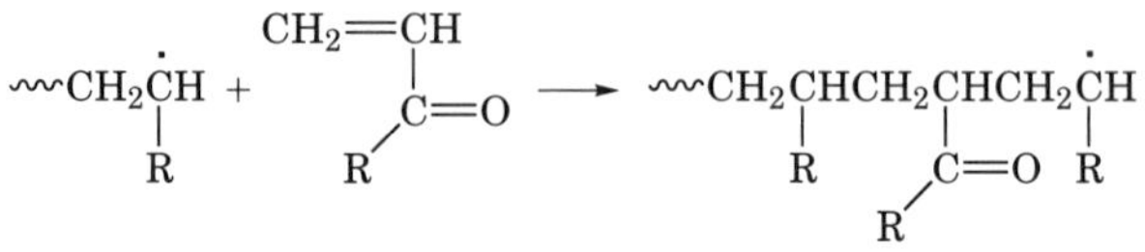

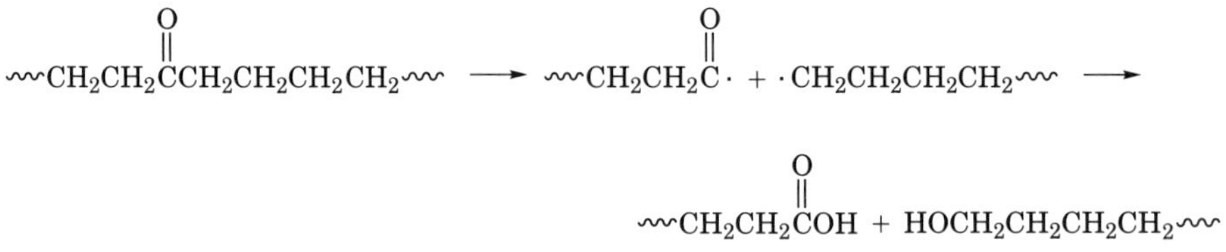

$$\sim\!\!\!\sim\!CH_2CH_2\overset{O}{\overset{\|}{C}}CH_2CH_2CH_2CH_2\!\sim\!\!\!\sim \longrightarrow \sim\!\!\!\sim\!CH_2CH_2\overset{O}{\overset{\|}{C}}\cdot + \cdot CH_2CH_2CH_2CH_2\!\sim\!\!\!\sim \longrightarrow$$

$$\sim\!\!\!\sim\!CH_2CH_2\overset{O}{\overset{\|}{C}}OH + HOCH_2CH_2CH_2CH_2\!\sim\!\!\!\sim \quad (5)$$

$$\sim\!\!\!\sim\!CH_2CH_2\overset{O}{\overset{\|}{C}}CH_2CH_2CH_2CH_2\!\sim\!\!\!\sim \longrightarrow \sim\!\!\!\sim\!CH_2CH_2\overset{O}{\overset{\|}{C}}CH_3 + H_2C\!=\!CHCH_2\!\sim\!\!\!\sim \quad (6)$$

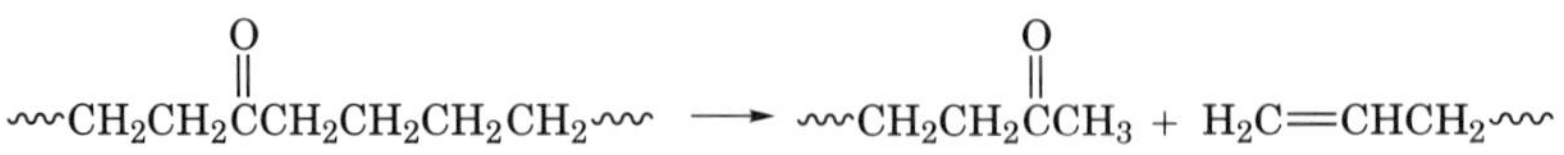

A Norrish type I reaction for side-chain carbonyl functionality is equation 7, and a Norrish type II reaction for side-chain carbonyl functionality is equation 8.

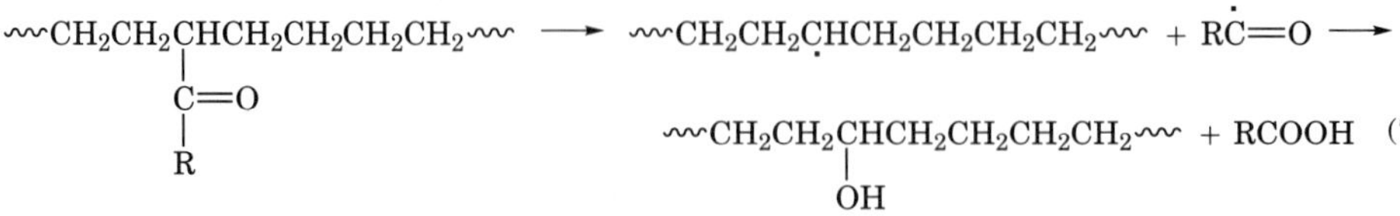

$$\sim\!\!\!\sim\!CH_2CH_2CHCH_2CH_2CH_2CH_2\!\sim\!\!\!\sim \longrightarrow \sim\!\!\!\sim\!CH_2CH_2\overset{\cdot}{C}HCH_2CH_2CH_2CH_2\!\sim\!\!\!\sim + R\overset{\cdot}{C}\!=\!O \longrightarrow$$
$$\underset{\underset{R}{|}}{\overset{|}{C=O}}$$

$$\sim\!\!\!\sim\!CH_2CH_2CHCH_2CH_2CH_2CH_2\!\sim\!\!\!\sim + RCOOH \quad (7)$$
$$\underset{OH}{|}$$

$$\sim CH_2CH_2\underset{\underset{R}{\overset{|}{\underset{C=O}{|}}}{\overset{|}{CH}}CH_2CH_2CH_2CH_2\sim \longrightarrow \sim CH_2CH_2CH_2 + CH_2\!\!=\!\!CH\underset{\underset{R}{\overset{|}{\underset{C=O}{|}}}{}}{}CH_2CH_2\sim \quad (8)$$

Norrish type I chemistry is claimed to be responsible for about 15% of the chain scission of ethylene–carbon monoxide polymers at room temperature, whereas at 120°C it promotes 59% of the degradation. Norrish I reactions are independent of temperature and oxygen concentration at temperatures above the $T_g$ of the polymer (50).

Degradation of polyolefins such as polyethylene, polypropylene, polybutylene, and polybutadiene promoted by metals and other oxidants occurs via an oxidation and a photo-oxidative mechanism, the two being difficult to separate in environmental degradation. The general mechanism common to all these reactions is that shown in equation 9. The reactant radical may be produced by any suitable mechanism from the interaction of air or oxygen with polyolefins (42) to form peroxides, which are subsequently decomposed by ultraviolet radiation. These reaction intermediates abstract more hydrogen atoms from the polymer backbone, which is ultimately converted into a polymer with ketone functionalities and degraded by the Norrish mechanisms (eq. 5–8).

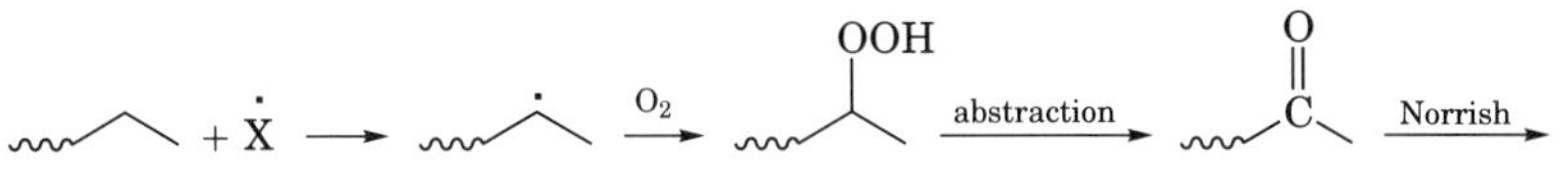

$$\sim\!\!\wedge\!\!\wedge + \overset{\cdot}{X} \longrightarrow \sim\!\!\wedge\!\!\wedge \overset{O_2}{\longrightarrow} \sim\!\!\wedge\!\!\overset{\overset{OOH}{|}}{\wedge} \overset{abstraction}{\longrightarrow} \sim\!\!\wedge\!\!\overset{\overset{O}{\|}}{C}\!\!\wedge \overset{Norrish}{\longrightarrow} \quad (9)$$

Research on photodegradable polymers is still very active as an end in itself and for their use as additives for biodegradable polymers to encourage more rapid biodegradation by decreasing molecular weight. Exxon (51) and Quantum (52) have received patents for polyolefin–polyester–carbon monoxide compositions. The Exxon route is by copolymerization of ethylene, carbon monoxide, and 2-methylene-1,3-dioxapane. The use of dioxapane is based on a clever invention by W. J. Bailey to introduce an ester linkage into polyolefins during free-radical polymerization such that they become susceptible to biodegradation. Thus Exxon scientists combine Bailey's idea with a known photodegradable product to enhance biodegradation of the fragments. The polymer has the structural elements shown in equation 10 in concentrations related to the degradation response required and controlled by the synthesis variables.

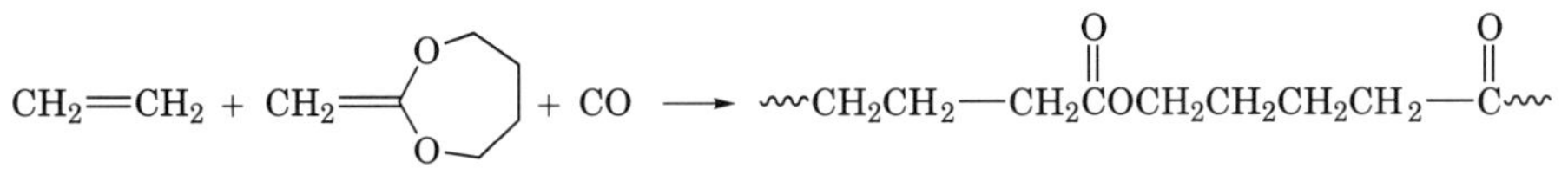

$$CH_2\!\!=\!\!CH_2 + CH_2\!\!=\!\!\!\!\left\langle\!\!\begin{array}{c}O\!\!\smallfrown\\ \\ O\!\!\smallsmile\end{array}\!\!\right\rangle + CO \longrightarrow \sim\!CH_2CH_2\!-\!CH_2\overset{\overset{O}{\|}}{C}OCH_2CH_2CH_2CH_2\!-\!\overset{\overset{O}{\|}}{C}\!\sim \quad (10)$$

Quantum, by contrast, converted an ethylene–carbon monoxide polymer into a polyester-containing terpolymer by treatment with acidic hydrogen peroxide, the Baeyer-Villiger reaction (eq. 11). Depending on the degree of conversion to polyester, the polymer is totally or partially degraded by a biological mechanism.

$$\text{\sim\sim CH}_2\text{CH}_2\text{CH}_2\text{CH}_2\text{\sim\sim}\overset{\overset{\displaystyle O}{\|}}{C}\text{CH}_2\text{CH}_2\text{\sim\sim}\xrightarrow[\text{H}^+]{\text{H}_2\text{O}_2} \text{\sim\sim CH}_2\text{CH}_2\text{CH}_2\text{CH}_2\text{\sim\sim}\overset{\overset{\displaystyle O}{\|}}{C}\text{OCH}_2\text{CH}_2\text{\sim\sim} \qquad (11)$$

Other patents include copolymers of vinyl ketones with acrylates, methacrylates, and styrene (53); an ethylene–carbon monoxide (1–7 wt %) blend as a photo-initiator in polycaprolactone–polyethylene blends (54); ethylene–carbon monoxide for degradable golf tees (55); a vinyl ketone analogue of Exxon's carbon monoxide–dioxapane–ethylene (56); a photodegradable food wrapper based on blends of a polyolefin–starch and photoactivators for polyolefin degradation (57); and a carboxylated polyethylene–carbon monoxide–norbornene-2,3-dicarboxylic acid (58).

Photodegradation chemistry has evolved to a highly practical state; commercial products are in existence and others are being evaluated. The degradation mechanisms are understood to the point of property loss for the polymers. Complete environmental acceptability is still lacking. It is not sufficient to expect low molecular weight fragments to be biodegradable; this must be demonstrated.

## BIODEGRADATION

Biodegradable polymers and plastics are readily divided into three broad classifications: (*1*) natural, (*2*) synthetic, and (*3*) modified natural. These classes may be further subdivided for ease of discussion, as follows: (*1*) natural polymers; (*2*) synthetic polymers may have carbon chain backbones or heteroatom chain backbones; and (*3*) modified natural may be blends and grafts or involve chemical modifications, oxidation, esterification, etc.

**Natural Polymers.**    Natural polymers, or biopolymers, are produced in nature by all living organisms. As a class they represent truly renewable resources since they are biodegradable, even if slowly in some cases. Because they are produced in nature there is no concern about this slow rate of biodegradation, contrary to concerns about synthetic polymers. Biopolymers are considered environmentally acceptable degradable polymers. The most widespread natural polymers are the polysaccharides, such as cellulose and starch. Other important classes include polyesters such as polyhydroxyalkanoates, proteins like silk and poly($\gamma$-glutamic acid), and hydrocarbons such as natural rubber. Usually the natural polymers exist in an optically active form, biopolymers with asymmetric centers being always isotactic. An excellent description of many biopolymers has been given in an edited book (59) with chapters on silk proteins, collagen, polyhydroxyalkanoates, microbial polysaccharides, microbial cellulose, hyaluronic acid, alginates, and other miscellaneous biomaterials. Other sources include a review on biodegradable polymers (17) and the proceedings of the NATO Advanced Research Workshop on New Biodegradable Microbial Polymers (60). With few exceptions, natural polymers are not suitable polymers for practical applications as of 1996, either because they lack the property requirements or because they are too expensive for other than specialty, high value-niche markets such as biomedical applications. There are opportunities in blends and for less costly chemically modified polymers such as starch and celluloses.

Polysaccharides are largely limited to starch and cellulose derivatives for practical applications either in plastics or as water-soluble polymers. Both these polymers are composed of D-glycopyranoside repeating units to very high molecular weight, thousands of units. They differ in that starch is poly(1,4-$\alpha$-D-glucopyranoside) (1) and cellulose is poly(1,4-$\beta$-D-glucopyranoside) (2). This difference in structure controls biodegradation rates and properties of the polymers.

(1)                    (2)

Complex carbohydrates (qv) such as microbially produced xanthan, curdlan, pullulan, hyaluronic acid, alginates, carageenan, and guar are accepted as biodegradable and are finding uses where cost is not an impediment. Xanthan is the predominant microbial polysaccharide on the market, ca 10,000 t worldwide (59), and finds use in the food industry and as a thickener in many industrial applications (see GUMS). It is foreseeable that other polysaccharides will gain acceptance in specialty areas where biodegradability is essential.

Proteins (qv) have not found widespread use as plastic materials because they are not soluble or fusible without decomposition; hence, they must be used as found in nature. They are of course widely used as fibers. Examples are wool, silk, and gelatin (collagen), which is used as an encapsulant in the pharmaceutical and food industries. The structure of proteins is an extended chain of amino acids joined through amide linkages which are readily degraded by enzymes, particularly proteases. Recent activity in poly($\gamma$-glutamic acid) (60) with control of stereochemistry by the inclusion of manganese ions may be important for future developments in biodegradable water-soluble polymers with carboxyl functionality, which is an intensely researched and desirable goal for detergent applications.

Polyesters are known to be produced by many bacteria as intracellular reserve materials for use as a food source during periods of environmental stress. They have received a great deal of attention since the 1970s because they are biodegradable, can be processed as plastic materials, are produced from renewable resources, and can be produced by many bacteria in a range of compositions. The thermoplastic polymers have properties that vary from soft elastomers to rigid brittle plastics in accordance with the structure of the pendent side-chain of the polyester. The general structure of this class of compounds is shown by (**3**), where $R = +CH_2\overline{)_m}CH_3$, $n = >100$, and $m = 0-8$.

$$-\!\!+\!OCHCH_2\overset{\displaystyle \overset{O}{\|}}{C}\!\!\overline{)_n}\!-$$
$$\underset{\displaystyle R}{|}$$

(**3**)

These polyesters have been comprehensively reviewed (61–63). All the polyesters are 100% optically pure and are 100% isotactic. When R is $CH_3$, ie, poly($\beta$-hydroxybutyrate) (PHB) the polymer is highly crystalline, with a melting point of 180°C and a glass-transition temperature, $T_g$, of 5°C (64). This combination of high $T_g$ and high crystallinity makes polymer films and plastics very brittle and plasticization is preferred to improve the properties. This is accomplished in the commercially available BIOPOL from Zeneca (formerly ICI) by using bacteria to produce a copolymer containing $\beta$-hydroxyvalerate ($R = C_2H_5$). By feeding the bacteria, *Alcaligenes eutrophus*, a mixed feed of propionic acid and glucose (65) a random copolymer is produced with some control over the composition. These polymers have better mechanical properties and are produced on a relatively large scale for this new polymer fermentation technology (a few hundred tons per year). Their acceptance would be more widespread if the price were closer to the synthetic commodities with which they compete, since they are produced from renewable resources and are of natural origin and thus environmentally appealing. Attempts to reduce costs are underway in a number of laboratories: the cost of processing with the current polymers is excessive due to the isolation steps which include clean-up of bacterial debris. If these polymers could be produced by isolated enzymatic processes this would be avoided. Another intriguing possibility is to produce these natural polymers in plants rather than bacteria by transferring the bacterial genes to suitable plants. Some work (66) in mustard plants has succeeded in the production of minute quantities. The day when poly(hydroxyalkanoate)s are produced in the volumes and at the price of starch may herald a new age for plastics and polymers, but it is not yet here. Some of the opportunities for these materials have been discussed (67).

The longer side-chain polyesters (**3**), where $m$ is 3–6, are produced by a variety of bacteria, usually as copolymers and with low crystallinity, low melting points, and low glass-transition temperatures. These polyesters are elastomeric and have excellent toughness and strength (63). They are inherently biodegradable but as the chain length is increased the biodegradation rate is greatly reduced, indicating that the hydrophilic–hydrophobic balance of the polymer plays a large role in biodegradation (68). Other biodegradation studies are underway to evaluate mechanisms (69) using $^{13}C$-labeled poly(hydroxybutyrate) and effect of environment (70) on the rate of biodegradation in lake Lugano, Switzerland.

All these polyesters are produced by bacteria in some stressed conditions in which they are deprived of some essential component for their normal metabolic processes. Under normal conditions of balanced growth the bacteria utilizes any substrate for energy and growth, whereas under stressed conditions bacteria utilize any suitable substrate to produce polyesters as reserve material. When the bacteria can no longer subsist on the organic substrate as a result of depletion, they consume the reserve for energy and food for survival; or upon removal of the stress, the reserve is consumed and normal activities resumed. This cycle is utilized to produce the polymers which are harvested at maximum cell yield. This process has been treated in more detail in a paper (71) on the mechanism of biosynthesis of poly(hydroxyalkanoate)s.

**Synthetic Polymers.** Synthetic polymers are well established in many applications where their environmental resistance properties are highly valued. Evolving environmental awareness and waste-disposal problems have led to increasing activity to develop biodegradable synthetic analogues of these polymers, particularly water-soluble polymers and plastics, which are used in packaging and other areas of opportunity. Natural polymers are ultimately degraded and consumed in nature in a continuous recycling of resources, but since synthetic polymers did not evolve naturally, the plethora of enzymes available in nature for degrading natural polymers are not in the main useful for synthetic polymers. The search for synthetic polymeric structures that can be biodegraded has progressed from minor modification of the nondegradables in use to structures that mimic nature, for which more success has been achieved.

Despite the fact that biodegradation testing until the late 1980s has been very unreliable and the interpretation of many reported data in the literature is questionable, some guidelines based on polymer structure, polymer physical properties, and environmental conditions at the exposure site have emerged for predicting the biodegradability of synthetic polymers (9,72,73). In considering polymer structure, the following generalizations can be made:

A higher hydrophilic/hydrophobic ratio is better for biodegradation.

Carbon chain polymers are unlikely to biodegrade.

Chain branching is deleterious to biodegradation.

Condensation polymers are more likely to biodegrade.

Lower molecular weight polymers are more susceptible to biodegradation.

Crystallinity slows biodegradation.

Favorable polymer physical properties include water solubility and sample purity. Environmental conditions to consider in evaluating biodegradability are temperature, pH, moisture, oxygen, nutrients, suitable microbial population (fungal, algae, bacterial), concentration, and test duration.

*Carbon Chain Backbone Polymers.* These polymers may be represented by (**4**) and considered derivatives of polyethylene, where $n$ is the degree of polymerization and R is (an alkyl group or) a functional group: hydrogen (polyethylene), methyl (polypropylene), carboxyl (poly(acrylic acid)), chlorine (poly(vinyl chloride)), phenyl (polystyrene) hydroxyl (poly(vinyl alcohol)), ester (poly(vinyl acetate)), nitrile (polyacrylonitrile), vinyl (polybutadiene), etc. The functional groups and the molecular weight of the polymers, control their properties which vary in hydrophobicity, solubility characteristics, glass-transition temperature, and crystallinity.

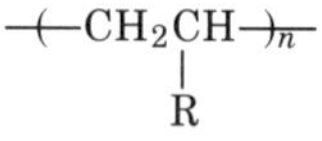

$$-\!\!\!\!\;(\!-CH_2CH\!-\!)_{\!n}-$$
$$\mid$$
$$R$$

(**4**)

Polyethylene with molecular weights into the oligomer or hydrocarbon range have been studied very extensively (4,23,25) and also at the higher range (74–82). The latter experiments are long term with radiolabeled polymer running for many years. The earlier work was based on fungal and bacterial growth tests and indicated that polyethylene and other high molecular weight carbon chain polymers did not support growth. A few anomalous results were attributed to plasticizer or low molecular weight impurities which had already been shown to be biodegradable in similar tests with pyrolyzed polyethylene and simple hydrocarbons. In the same work (4,23,25) it was also found that branching of hydrocarbon chains limits biodegradation and the molecular weight cut-off for linear molecules appears to be in the 500 dalton range, but more rigorous testing is needed for confirmation. Some of these data are shown in Tables 3 and 4.

The increase in degradation with lower molecular weight may be a result of many factors. For example, transportation of polymer across cell walls is more likely at lower molecular weight, or it may be the mechanism of biodegradation or because of random or chain-end cleavage prior to entering the cell. Chain-end *exo*-biodegradation by $\beta$-oxidation is indicated schematically in Figure 4 for a simple hydrocarbon. The individual oxidation steps are enzyme-catalyzed with a final hydrolysis to a two-carbon diminished chain and acetic acid, which is then presumably biodegraded. Such a mechanism would explain the occurrence of slower biodegradation at higher molecular weight, where there would be fewer chain ends. The terminal groups found in oxidized and photodegraded

**Table 3. Hydrocarbon Branching and Molecular Weight Effects on Biodegradability**

| Compound | Mol wt | Number of branches | Growth test[a] |
|---|---|---|---|
| dodecane | 170 | 0 | 4 |
| 2,6,11-trimethyldodecane | 212 | 3 | 0 |
| hexadecane | 226 | 0 | 4 |
| 2,6,11,15-hexadecane | 282 | 4 | 0 |
| tetracosane | 338 | 0 | 4 |
| squalene | 422 | 6 | 0 |
| dotriacontane | 450 | 0 | 4 |
| hexatriacontane | 506 | 0 | 0 |
| tetracontane | 562 | 0 | 0 |
| tetratetracontane | 618 | 0 | 0 |

[a]ASTM tests G21-70 and G22-76. Higher numbers are considered to correlate with the susceptibility of the plastic to biodegradation.

**Table 4. Biodegradability of Low Mol Wt Pyrolysis Products of HDPE and LDPE**

| Pyrolysis temperature, °C | Mol wt | Growth rating[a] |
|---|---|---|
| control (HDPE) | 123,000 | 0 |
| 400 | 16,000 | 1 |
| 450 | 8,000 | 1 |
| 500 | 3,200 | 3 |
| 535 | 1,000 | 3 |
| control (LDPE) | 56,000 | 0 |
| 400 | 19,000 | 1 |
| 450 | 12,000 | 1 |
| 500 | 2,100 | 2 |
| 535 | 1,000 | 3 |

[a]ASTM tests G21-70 and G22-76. Higher numbers are considered to correlate with the susceptibility of the plastic to biodegradation.

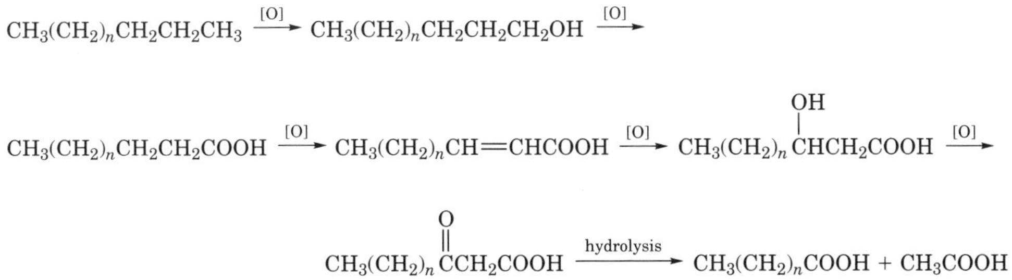

$$CH_3(CH_2)_nCH_2CH_2CH_3 \xrightarrow{[O]} CH_3(CH_2)_nCH_2CH_2CH_2OH \xrightarrow{[O]}$$

$$CH_3(CH_2)_nCH_2CH_2COOH \xrightarrow{[O]} CH_3(CH_2)_nCH{=}CHCOOH \xrightarrow{[O]} CH_3(CH_2)_n \overset{OH}{\underset{|}{C}}HCH_2COOH \xrightarrow{[O]}$$

$$CH_3(CH_2)_n \overset{O}{\overset{\|}{C}}CH_2COOH \xrightarrow{hydrolysis} CH_3(CH_2)_nCOOH + CH_3COOH$$

**Fig. 4.** $\beta$-Oxidative degradation of hydrocarbons.

polyethylene are oxygen-containing and these should expedite biodegradation via a $\beta$-oxidation mechanism, but proof of this is preferable to prediction.

Results of experiments (74–82) that are, as of 1996, several years along and continuing suggest that polyethylene is slowly biodegraded and is slightly accelerated by pretreatment with surfactants or an oxidation process. This meticulous work is identifying some of the myriad of degradation products, many of which are oxygen-containing, as is expected if the degradation process for polyethylene is, as projected, initially oxidation and then enzymatic.

Other high molecular weight hydrocarbon polymers are not biodegradable, but oligomers of *cis*-1,4-isoprene (83), butadiene (84), and styrene (85), are degradable. And there has been further confirmation of biodegradation of oligomeric ethylene (86).

Functional derivatives of polyethylene, particularly poly(vinyl alcohol) and poly(acrylic acid) and derivatives, have received attention because of their water-solubility and disposal into the aqueous environment. Poly(vinyl alcohol) is used in a wide variety of applications, including textiles, paper, plastic films, etc, and poly(acrylic acid) is widely used in detergents as a builder, a super-absorbent for diapers and feminine hygiene products, for water treatment, in thickeners, as pigment dispersant, etc (see VINYL POLYMERS, POLY(VINYL ALCOHOL)).

Poly(vinyl alcohol), obtained by the hydrolysis of poly(vinyl acetate) is probably the only carbon chain polymer to be fully biodegradable. The biodegradation is a random chain cleavage of 1,3-diketones formed by an enzyme-catalyzed oxidation of the secondary alcohol functional groups in the polymer backbone. The biodegradation was first observed as a reduction in aqueous viscosity of the polymer in the presence of soil bacteria (87). Subsequently, a *Pseudomonas* species was identified as the soil bacterium responsible for the degradation over a degree of polymerization range of 500–2000 (88). Utilizing the polymer as a sole carbon source in an aqueous polymer solution at a concentration of 2700 ppm, it was reduced to 250–300 ppm concentration in 7–10 d at pH 7.5–8.5 and 35–45°C. An oxidative endo mechanism was proposed and later substantiated (89) by quantifying the oxygen uptake at one mole for every mole of hydrogen peroxide produced and identifying the degradation products as ketones and carboxylic acids, as shown in the following (eq. 12). An alternative mechanism (90,91), in which the products were identified as an alcohol and a carboxylic acid was subsequently proved to be in error, and the mechanism shown is widely accepted. It is also supported by the rapid biodegradation of chemically oxidized poly(vinyl alcohol) (92–94).

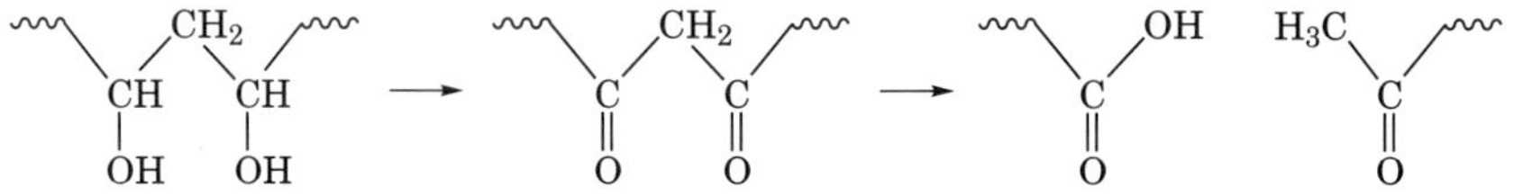

$$\text{(see structure)} \quad (12)$$

Other bacterial strains identified as biodegrading poly(vinyl alcohol) include *Flavobacterium* (95) and *Acinetobacter* (96) and many others, as well as fungi, molds, and yeasts (97). Industrial evaluations at Du Pont (98) and Air Products (99) indicate that over 90% of poly(vinyl alcohol) entering wastewater treatment plants is removed, and hence no environmental pollution is likely.

Poly(vinyl acetate), the precursor of poly(vinyl alcohol), hydrolyzed to less than 70%, is claimed to be nonbiodegradable under conditions similar to those that biodegrade the fully hydrolyzed polymer (100) (see VINYL POLYMERS, POLY(VINYL ACETATE)).

Carboxylate derivatives of poly(vinyl alcohol) are biodegradable and function in detergents as cobuilders, although they are too costly to be practical as of 1996. Vinyloxyacetic acid has been polymerized (5) (101,102), and Lever has patented polymers, eg, poly(vinyloxyaspartic acid) (6), based on vinyl carbamates obtained from the reaction of vinyl chloroformates and amino acids such as aspartic and glutamic acids (103). Both hydrolyze to poly(vinyl alcohol) and then biodegrade.

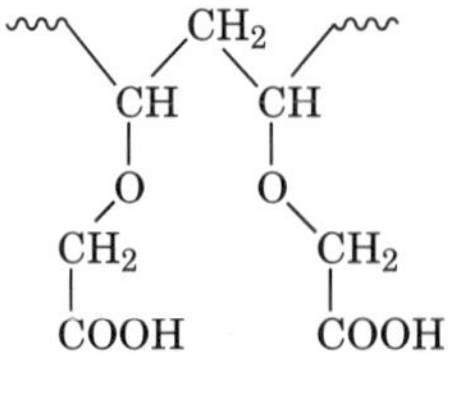

(5) 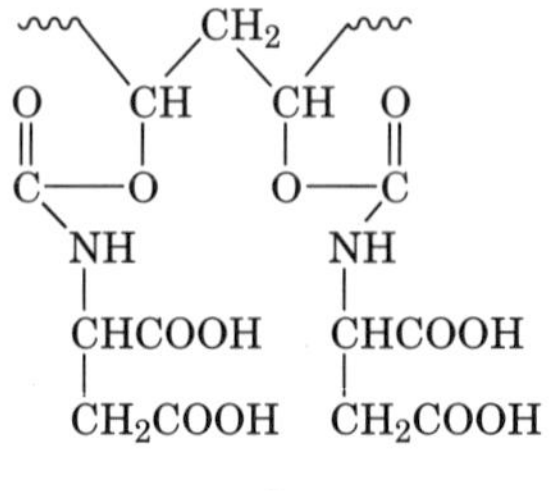                              (6)

Copolymers of vinyl alcohol with acrylic or maleic acid have been evaluated in detergents as potentially biodegradable cobuilders by a number of laboratories (104–106), but the results were not encouraging for balancing biodegradation and performance. Higher than 80 mol % of vinyl alcohol is required for high levels of biodegradation, and less than 20 mol % for acceptable performance.

The use of poly(carboxylic acids) in detergents is well established and has been well reviewed (107). Their lack of biodegradability at preferred performance molecular weights, ca 70 (5,000 dalton) for poly(acrylic acid) and 1,000 (70,000 dalton) for copoly(acrylic–maleic acids), even though there are ample data to indicate no harmful environmental effects, has resulted in a massive search for degradable replacements. After many efforts to copolymerize, by a free-radical mechanism, acrylic and maleic acids to biodegradable polymers with a whole range of vinyl monomers (108–113) and graft substrates, including polysaccharides (114), it is recognized that earlier work (115) with functional polymers and oligomeric hydrocarbons (4,23,25) is correct and only low molecular weight oligomeric carbon chain polymers are likely to be biodegradable, regardless of functionality. More recent confirmation has come from Japan (116,117).

High molecular weight poly(acrylic acid), polyacrylamide, and poly-(vinylpyrrolidinone) ozonized to oligomers with molecular weights less than 14 (1000 daltons) showed a marked increase in their biodegradability, with the exception of polyacrylamide (115). Later work (116,117) was based on oligomers of acrylic acid obtained by chromatographic separation from low molecular weight polymers. The results all indicate that poly(acrylic acids) are not completely biodegradable above about a degree of polymerization of 6–8 (400–600 daltons).

Other efforts to use radical polymerization to synthesize biodegradable carboxylated polymers have been based on combining low molecular weight oligomers through degradable linkages and by introducing weak links into the polymer backbone. BASF (118) and NSKK (119) have patented acrylic oligomers chain-branched with degradable linkages designated X in (7). Grillo Werke has patented copolymers of acrylic acid and enol sugars (120). The degradability of these polymers has not been clearly established, but the branching is likely to be a problem.

(**7**)

Several miscellaneous carbon chain backbone polymers have been claimed as biodegradable without clear evidence, including copolymers of methyl methacrylate and vinyl pyridinium salts (121), where the pyridinium salt is hypothesized to act as a magnet for bacteria which then cleave the chain into small fragments that biodegrade completely. An ethylene–vinyl alcohol copolymer has been converted into a terpolymer ester by a Baeyer Villiger reaction (122). Also, polymers of $\alpha$-hydroxyacrylic acid have been made (123). The terpolymer ester is probably biodegradable, but the polyacid fails to differentiate adsorption on solids in a sewage sludge test from biodegradation.

*Heteroatom Chain Backbone Polymers.* This class of polymers includes polyesters, which have been widely studied from the initial period of research on biodegradable polymers, polyamides, polyethers, polyacetals, and other condensation polymers. Their linkages are quite frequently found in nature and these polymers are more likely to biodegrade than hydrocarbon-based polymers.

Low melting, low molecular weight aliphatic polyesters were shown to be readily biodegradable using soil burial tests (4) and ASTM bacterial and fungal growth methods (ASTM G21-70 and G22-76). From this work, polycaprolactone was recognized as one of a select few commercially available synthetic polymers that is beyond a doubt biodegradable. Since that time many other workers have confirmed the biodegradability of aliphatic polyesters using other test protocols (124–126) such as lipase hydrolysis with measurement of the rate of production of water-soluble oligomers. Results also indicate that as the aliphatic polyesters become more hydrophobic, either from acid or alcohol chain length extension, the biodegradation rate is slowed. Amorphous regions of polyesters are more readily biodegradable than crystalline regions (127).

During this early period, a very ingenious free-radical route to polyesters was used to introduce weak linkages into the backbones of hydrocarbon polymers and render them susceptible to biodegradability (128–131). Copolymerization of ketene acetals with vinyl monomers incorporates an ester linkage into the polymer backbone by rearrangement of the ketene acetal radical as illustrated in equation 13. The ester is a potential site for biological attack. The chemistry has been demonstrated with ethylene (128–131), acrylic acid (132), and styrene (133).

$$\text{CH}_2{=}\text{C}\begin{smallmatrix}H\\\\R\end{smallmatrix} + \text{CH}_2{=}\underset{O}{\overset{O}{\diamondsuit}} \longrightarrow \left(\!\!\left(\begin{smallmatrix}\overset{R}{|}\\CH\\CH_2\end{smallmatrix}\right)_{\!n}\!\!\left(\begin{smallmatrix}\overset{O}{\|}\\C\\CH_2\end{smallmatrix}\!\!-\!\!O\!\!-\!\!\begin{smallmatrix}CH_2\\CH_2\end{smallmatrix}\right)\!\!\right)_{\!m} \tag{13}$$

Interest in biodegradable polyesters has been predominantly focussed on aliphatic polyester structures and includes ring opening of 1,5-dioxepan-2-one (134), polyesters of aliphatic acids with glycols, but also some terephthalic acid and sulfoterephthalic acid as a compostable diaper (135). Perhaps the biggest advances in the synthetic polyester area are the close match for BIOPOL, the expensive bacterial polyester (136) which has the chirality but lacks the molecular weight of the natural polymer; and the new product BIONOLLE from Showa

High Polymer (137–139) that is supposedly a biodegradable polyester. Patents (137) suggest that BIONOLLE is an aliphatic ester coupled with a polyisocyanate and that ease of biodegradation and properties are inversely related. Both these inventions indicate that progress is being made on meeting the property requirements for biodegradable polymers.

Aliphatic polyesters are also available by the chemical reaction of carbon monoxide and formaldehyde (140), carbon dioxide and epoxy compounds (141), and bisepoxies and biscarboxylic acids (142).

Poly(lactic acid) [50-21-5] has until very recently been known only in the medical field as an expensive polymer for the manufacture of sutures (qv) and other biomaterials. With cheap lactic acid becoming available from the fermentation of waste agricultural products, there has been a surge of activity to develop new polymerization methods and to find commercial outlets for the products. Though still more expensive than common commodity polymers, the introduction of products for niche markets such as agricultural films and mulch and the fast-food industry is anticipated to occur by the mid- to late 1990s. There are many patents issuing in the homo- and copolymer synthesis and process areas, principally from just a few sources, ie, Cargill (143), Du Pont (144), Mitsui Toatsu (145), and Battelle Memorial Laboratories (146). Others active in this area include Shimadzu and Argonne National Laboratories.

Polyamides have received some attention and the results indicate that the stereochemistry of the groups close to the amide linkages and the hydrophilic nature control biodegradability (147–148). A more general study on polyesters, polyureas, polyurethanes, and polyamides is a good fundamental early study (151) guiding the later work in this general area. It has been demonstrated that polyesteramides are difficult to hydrolyze chemically yet can be biodegraded rapidly at ambient conditions in the right environment (150). Support for this observation comes from Japan (151,152). Although polymers of nylon-6 are considered nonbiodegradable, oligomers and low molecular weight polymers of less than 157 (11,000 daltons) will biodegrade (153).

Water-soluble polyesters and polyamides containing carboxyl functionality are reported to be biodegradable detergent polymers by BASF and may be obtained by condensation polymerization of monomeric poly(carboxylic acid)s such as citric acid, butane-1,2,3,4-tetracarboxylic acid, tartaric acid, and malic acid with polyols (154); with amino compounds, including amino acids (155); and with polysaccharides (156). Early work (157,158) demonstrated the self-condensation of malic acid to biodegradable polyesters regardless of the ester, $\alpha$- or $\beta$-linkage formed. Procter and Gamble has patented succinylated poly(vinyl alcohol) (159).

Polyanionics are available also from poly(amino acid)s based on poly-(carboxyamino acid)s such as glutamic acid and aspartic acid. Though both are known and claimed as biodegradable homopolymers, aspartic acid is more amenable to a practical industrial synthesis by thermal polymerization, since it has no tendency to form an internal $N$-anhydride. An alternative synthesis is from ammonia and maleic acid. Only the acid-catalyzed condensation of L-aspartic acid yields an authenticated biodegradable polymer (160). The noncatalyzed process and the ammonia–maleic acid processes give partially (ca 30 wt % residue remains in the Sturm test for $CO_2$ evolution, ASTM D5209-92) biodegradable polymers, owing to the molecules being branched and resistant to enzymatic attack. The structures are shown in Figure 5. Starting from aspartic acid, pathway A is acid-catalyzed thermal condensation, and B is noncatalyzed thermal condensation. The polysuccinimides shown hydrolyze at the point indicated to give mixtures of α- and β-poly(aspartic acid) salts. Regardless of the stereochemistry of the starting aspartic acid, L or D, the final product is the DL racemate.

Many synthesis patents and publications from aspartic acid (161–166) and ammonia–maleic (167,168) processes have issued and the product is expected to find use in many applications, including dispersants (169,170) and detergents (171,172). BASF has an aspartic acid copolymer patent with carbohydrates and polyols (173), and Procter and Gamble (174) has a patent for poly(glutamic acid) [25513-46-6], both for biodegradable detergent cobuilders. There is one patent for poly(methyl γ-glutamate) as a transparent plastic with excellent strength and biodegradability (175).

Polyethers have been investigated since about 1962, especially poly-(ethylene glycol), which is water soluble and is widely used in detergents and as a synthesis intermediate in polyurethanes. It has been established that the nature of the degradation of poly(ethylene glycols) with molecular weights higher than 6000 daltons is symbiotic, but below a molecular weight of 1000 daltons the polymer is biodegraded by many individual bacteria (176). The enzymatic *exo*-degradation pathway described in Reference 176 is shown schematically as equation 14. The first stage is dehydrogenation (oxidation), the second stage is oxidation, the third stage is an oxidation that is followed by a hydrolysis to remove a two-carbon fragment as glyoxylic acid. Degradation of poly(ethylene glycols) with molecular weights of 20,000 daltons has been reported.

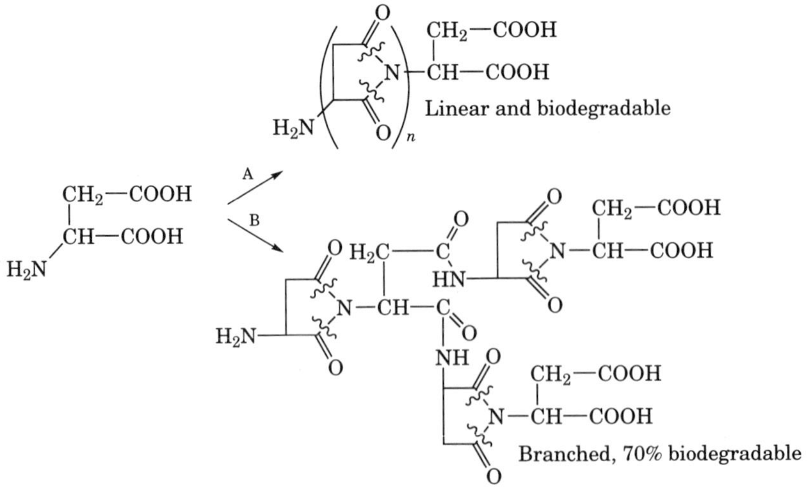

**Fig. 5.** Condensation polymers of aspartic acid.

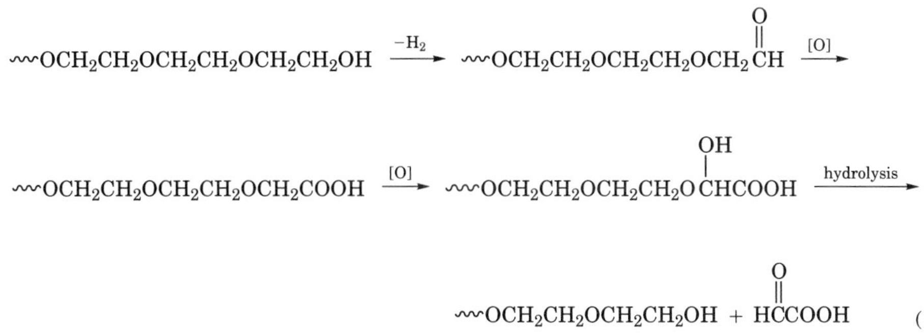

$$\text{\raise1pt\hbox{$\sim$}}OCH_2CH_2OCH_2CH_2OCH_2CH_2OH \xrightarrow{-H_2} \text{\raise1pt\hbox{$\sim$}}OCH_2CH_2OCH_2CH_2OCH_2\overset{\displaystyle O}{\overset{\displaystyle \|}{C}}H \xrightarrow{[O]}$$

$$\text{\raise1pt\hbox{$\sim$}}OCH_2CH_2OCH_2CH_2OCH_2COOH \xrightarrow{[O]} \text{\raise1pt\hbox{$\sim$}}OCH_2CH_2OCH_2CH_2O\overset{\displaystyle OH}{\overset{\displaystyle |}{C}}HCOOH \xrightarrow{hydrolysis}$$

$$\text{\raise1pt\hbox{$\sim$}}OCH_2CH_2OCH_2CH_2OH + H\overset{\displaystyle O}{\overset{\displaystyle \|}{C}}COOH \qquad (14)$$

Anaerobically, poly(ethylene glycol) degrades slowly, although molecular weights up to 2000 daltons have been reported (177,178) to biodegrade.

The biodegradation of poly(alkylene glycols) is hindered by their lack of water solubility, and only the low oligomers of poly(propylene glycol) are biodegradable with any certainty (179–181), as are those of poly(tetramethylene glycol) (182). A similar *exo*-oxidation mechanism to that reported for poly-(ethylene glycol) has been proposed.

Polyether carboxylates have been evaluated as biodegradable detergent polymers (183–185). They all fit the general structure of the series made in an extensive evaluation of anionic and cationic polymerized epoxy compounds in the molecular weight range of several hundred to a few thousand, where X or Y may be carboxyl functionality and X or Y may be hydrogen or a substituent bearing a carboxyl functionality (185). Biodegradability, based on biochemical oxygen demand (BOD), is structure-dependent.

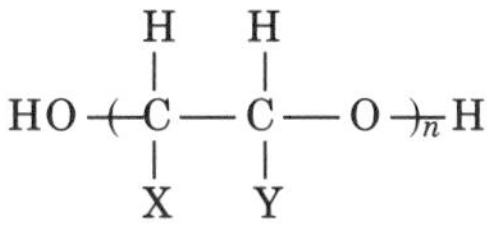

$$HO\left(\!\!\begin{array}{c} H \\ | \\ C \\ | \\ X \end{array}\!\!-\!\!\begin{array}{c} H \\ | \\ C \\ | \\ Y \end{array}\!\!-\!\!O\right)_{\!\!n}\!\!H$$

Water-soluble biodegradable polycarboxylates with an acetal or ketal weak link were inventions of Monsanto scientists in the course of their search for biodegradable detergent polymers. However, the polymers were prevented by economics from reaching commercial status. The polymers are based on the anionic or cationic polymerization of glyoxylic esters at low temperature (molecular weight is inversely proportional to the polymerization temperature) and subsequent hydrolysis to the salt form of the polyacid, which is a hemiacetal (R = H) or ketal (R = CH_3) if methylglyoxylic acid is used, and stable under basic conditions.

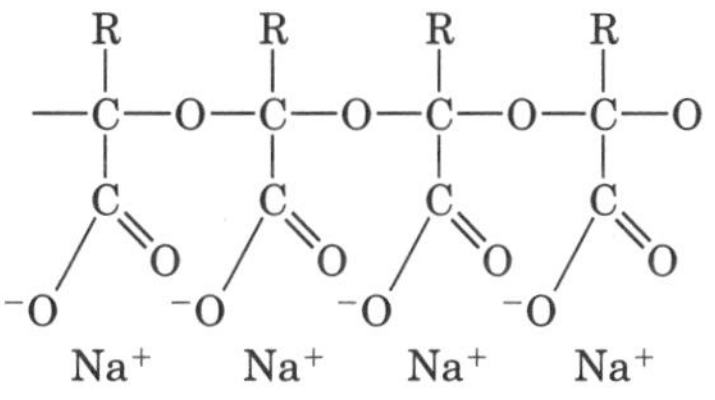

$$\begin{array}{cccccccc} & R & & R & & R & & R \\ & | & & | & & | & & | \\ -C & -O- & C & -O- & C & -O- & C & -O \\ | & & | & & | & & | \\ C & & C & & C & & C \\ \diagup\!\!\diagdown & & \diagup\!\!\diagdown & & \diagup\!\!\diagdown & & \diagup\!\!\diagdown \\ ^-O \quad O & & ^-O \quad O & & ^-O \quad O & & ^-O \quad O \\ Na^+ & & Na^+ & & Na^+ & & Na^+ \end{array}$$

Biodegradation results from the pH drop such a detergent polymer experiences as it leaves the alkaline laundry environment (pH ca 10) and enters the sewage or ground water environment (pH close to neutral); the polymer (now a polyacid rather than a salt) is unstable and hydrolyzes to monomer which rapidly biodegrades. The chemistry has been reported in many patents (186) and several publications (187,188).

Similar polyacetals were prepared by BASF scientists from $\omega$-aldehydic aliphatic carboxylic acids (189,190) and by the addition of poly(hydroxycarboxylic acid)s such as tartaric acid to divinyl ethers (191) as biodegradable detergent polymers.

**Modified Natural Polymers.** Modifying natural polymers offers a way of capitalizing on their well-accepted biodegradability in the development of polymers that might be environmentally acceptable. The modification must not interfere with the biodegradation process and the product must meet guidelines for environmental acceptability, ie, they must be either demonstrated to be totally biodegraded and removed from the environment, or be biodegradable to the extent that no environmentally harmful residues remain. With this requirement in mind, the approaches that have received the most attention include blends with other natural and synthetic polymers, grafting of another polymeric composition, and chemical modification to introduce some desirable functional group by oxidation or some other simple chemical reaction, such as esterification or etherification.

Starch is made thermoplastic at elevated temperatures in the presence of water as a plasticizer, allowing melt processing alone or in blends with other thermoplastics (192–194). Good solvents such as water lower the melt-transition temperature of amylose, the crystalline component of starch, so that processing can be done well below the decomposition–degradation temperature.

The most important commercial application has been the blending of polyethylene with starch in the presence or absence of other additives to promote compatibility. The interest in this approach goes back to the 1970s (195), and there is continuing activity with commercial products from several companies. There are a great many other contributors to this field. One pioneer developed starch–polyethylene compatibilized with ethylene–acrylic acid copolymers (196) and ethylene–vinyl alcohol polymers (197). Later work with polyethylene capitalizing on this early research includes Novamont (also known as Butterfly in some of its patents) with starch blends containing hydroxy acids, urethanes, polyamides, and polyvinyls (198), Warner-Lambert (Novon) (199,200), the U.S. Army with cellulose acetate (201), Henkel with alkyds (202), Iowa State University with proteins and oxidized polyethylene (203,204), ADM (205), Solvay with polycaprolactone (206), and Agritech with a starch minimum of 30 wt % (207).

Biodegradation studies of starch blends have not been conclusive where a nondegradable synthetic polymer has been the blend component; probably biodisintegration would be a better term to describe these polymers. The principal deficiencies of products based on this chemistry, aside from the incomplete biodegradation, are water-sensitivity of manufactured articles, and the balance of this and biodegradation with the starch level in the product.

Other blends such as polyhydroxyalkanoates (PHA) with cellulose acetate (208), PHA with polycaprolactone (209), poly(lactic acid) with poly(ethylene glycol) (210), chitosan and cellulose (211), poly(lactic acid) with inorganic fillers (212), and PHA and aliphatic polyesters with inorganics (213) are receiving attention. The different blending compositions seem to be limited only by the number of polymers available and the compatibility of the components. The latter blends, with all natural or biodegradable components, appear to afford the best approach for future research as property balance and biodegradability is attempted. Starch and additives have been evaluated in detail from the perspective of structure and compatibility with starch (214).

Starch has also been a substrate of choice for biodegradable polymers, by grafting with synthetic polymers to achieve property improvement and new properties such as carboxyl functionality, not available in starch, with retention of as much biodegradability as possible. Obtaining thermoplastic polymers from the ionic grafting of styrene to starch has been demonstrated (215), and radical grafting of acrylate esters (216) has also been reported. The latter was recommended as a mulch, as it rapidly decomposed in the presence of fungi. The extent of biodegradation of both these materials must be questioned as the acrylic and styrene components are known for their resistance to biodegradation.

Other grafts to natural materials are exemplified by work in which polyesters are produced from sugars and polycarboxylates by enzyme catalysis (217). These polymers and the method of synthesis may well be one of the future directions of renewable resource chemistry. This method is similar to some very early research on cellulose condensation with polyfunctional isocyanates and optionally propylene glycol (218); some degradation was claimed. The utility and potential for lignin grafted with styrene has been shown (219), and it is claimed that the product is totally biodegradable, owing to the potency of white rot Basidiomycetes, a lignin degrader. Further proof is required, but this is a promising lead.

Natural polymers have also received attention as graft sites for carboxylic monomers to produce detergent polymers (DP), though without great success. The synthetic portion of the graft is not usually biodegradable, although in some cases attempts were made to meet molecular weight limitations (less than DP of ca 6–8). Acrylic grafts onto polysaccharides in the presence of alcohol chain-transfer agents (220) were not completely biodegradable, nor were the ones based on initiation with $Ce^{4+}$ (221) and mercaptan (222,223). Protein substrates (224) are expected to be similar to the starch grafts; the fundamental problem is the need to control acrylic acid polymerization to the oligomer range, as indicated earlier, in order to have complete biodegradability.

Simple chemical reactions on natural polymers are widely known to produce polymers such as hydroxyethylcellulose, hydroxypropylcellulose, carboxymethyl cellulose, cellulose acetates and propionates, and many others that have been in commerce for many years (see CELLULOSE ETHERS; CELLULOSE ESTERS, ORGANIC). Their biodegradability is not at all well established. Carboxymethylcellulose, for example, has been claimed to be biodegradable below a degree of substitution of about 2, which is similar to that of cellulose acetate. More recently, there have been attempts to more rigorously quantify biodegradation of the cellulose acetates (218,225) and to establish a property–biodegradation relationship. Rhône-Poulenc also indicates that cellulose acetate with a degree of substitution of about 2 is biodegradable, in agreement with the earlier references (226). Cellulose has been discussed as a renewable resource (227). The ampholytic product of the reaction of chitosan with citric acid is claimed to be biodegradable (228).

Carboxylated natural polymers have been known for many years, with the introduction of carboxymethylcellulose. This product has wide use in detergents and household cleaning formulations, although it is of questionable biodegradability at the level of substitution on cellulose required for performance. Nevertheless, carboxylated polysaccharides are a desirable choice for many applications, and the balance of biodegradation with performance that is achievable has been recognized as an attractive possible goal met in this case with a high probability of success. Three approaches have been employed: esterification, oxidation, or Michael addition of the hydroxyl groups to unsaturated carboxylic acids such as maleic and acrylic, with some attempts to react specifically at the primary or secondary sites.

Esterification with poly(carboxylic anhydride)s can be controlled to minimize diesterification and cross-linking to produce carboxylated cellulosic esters. An Eastman Kodak patent claimed the succinylation of cellulose to different degrees, 1 per 3 anhydroglucose rings (229) and 1 per 2 rings (230). Henkel (231) also has a patent for a surfactant by the esterification of cellulose with alkenylsuccinic anhydride; presumably, substitution governs the hydrophile–hydrophobe balance of the product.

Oxidation of polysaccharides is a far more attractive route to polycarboxylates, potentially cleaner and less costly than esterification. Selectivity at the 2,3-secondary hydroxyls and the 6-primary is possible. Total biodegradation with acceptable property balance has not yet been achieved. For the most part, oxidations have been with hypochlorite–periodate under alkaline conditions. In the 1990s, catalytic oxidation has appeared as a possibility, and chemical oxidations have also been developed that are specific for the 6-hydroxyl oxidation.

A wide range of polysaccharides, starch, xyloses, amyloses, pectins, etc, with hypochlorite–periodate have been oxidized (232–235). The products are either biodegradable at low oxidation levels or functional at high oxidation levels; the balance has not yet been established. From Delft University have come contributions in the search to control the hypochlorite–periodate liquid-phase oxidations of starches (236–237) and research for finding catalytic processes to speed up the oxidation with hypochlorite. Hypobromite offers one solution and is generated *in situ* from the inexpensive hypochlorite and bromide ion (238,239). At the same time, a method has been published for oxidizing specifically the 6-hydroxyl group (primary) of starch by using TEMPO and bromide–hypochlorite (240), as shown in Figure 6.

Chemical oxidation with strong acid is reportedly selective at the 6-hydroxyl, either with nitric acid–sulfuric acid–vanadium salts (241) which is claimed as specific for the 6-hydroxyl up to 40% conversion, or with dinitrogen tetroxide in carbon tetrachloride, with similar specificity up to 25% conversion (242).

Catalytic oxidation in the presence of metals is claimed as both nonspecific and specific for the 6-hydoxyl depending on the metals used and the conditions employed for the oxidation. Nonspecific oxidation is achieved with silver or copper and oxygen (243), and noble metals with bismuth and oxygen (244). Specific oxidation is claimed with platinum at pH 6–10 in water in the presence of oxygen

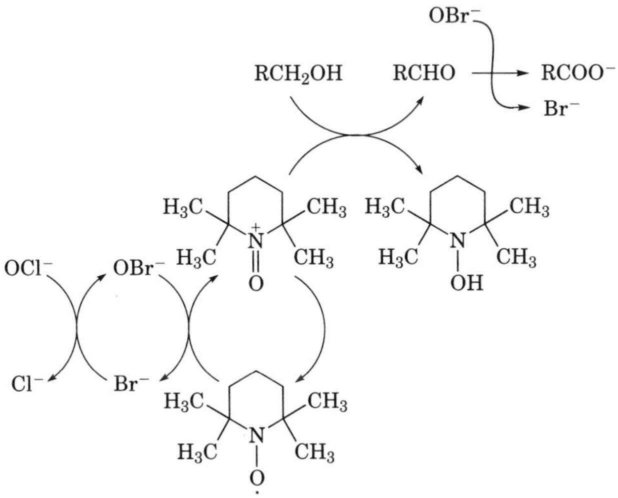

**Fig. 6.** Specific oxidation of the 6-hydroxyl of starch using bromide–hypochlorite and tetramethylpiperidine oxide (TEMPO).

(245). Related patents to water-soluble carboxylated derivatives of starch are Hoechst's on the oxidation of ethoxylated starch and another on the oxidation of sucrose to a tricarboxylic acid. All the oxidations are specific to primary hydroxyls and are with a platinum catalyst at pH near neutrality in the presence of oxygen (246,247). Polysaccharides as raw materials in the detergent industry have been reviewed (248).

## PRODUCTION OF ENVIRONMENTALLY DEGRADABLE POLYMERS

There are signs that the use of environmentally degradable polymers and plastics is expanding. As the market begins to become aware of the availability of these new materials, it is expected that they will move into niche opportunities. When this occurs, production will increase, and costs, the biggest barrier to acceptance, should begin to come down. Some of the polymers in production at some scale larger than laboratory are shown in Table 5.

The only product with substantial sales in 1996 is a photodegradable, environmentally degradable ethylene–carbon monoxide polymer used for six-pack holders, which are often carelessly thrown away so as to litter the environment. This litter will slowly disappear as it degrades into fine fragments. The U.S. EPA has let it be known that although this is acceptable for now, future products should degrade by a combination of photodegradation and biodegradation to ensure complete removal from the environment.

International agreement is close as of 1996 on what an acceptable environmentally degradable polymer should do in the environment; succinctly put, it must not harm the environment. There has been much progress in the early 1990s on this issue; standard protocols are available to determine degradation in the environment of disposal, and definitions are understood and accepted in a broad sense, if not in detail. Fate and effects issues for these new polymers are being addressed, and these will be resolved and appropriate tests developed.

**Table 5. Environmentally Degradable Polymers**

| Polymer | Developer | Degradation |
| --- | --- | --- |
| poly(lactic acid) | Cargill, Ecochem, Biopak, Mitsui Toatsu | hydrolysis and biodegradation |
| cellophane | Flexel | biodegradation |
| PHBV | Zeneca | biodegradation |
| starch-based | Novamont | biodegradation |
| polycaprolactone | Union Carbide, Solvay | biodegradation |
| starch-activator | Ecostar | photo- or biodegradation |
| starch foam | National Starch | biodegradation |
| polyolefin-activator | Plastigone | photodegradation |
| polyester | Showa High Polymer | biodegradable |
| poly(ethylene–CO) | Dow | photodegradation |
| poly(vinyl alcohol) | Rhône-Poulenc, Air Products Kuraray, Hoechst | biodegradation |
| poly(ethylene glycol) | Union Carbide, Dow | biodegradation |
| cellulosics | Rhône-Poulenc, Eastman | biodegradation |
| poly(aspartic acid) | Rohm and Haas | biodegradable |

Perhaps the most significant remaining issue is the cost of these polymers; most are at least three to five times that of the current products that must be displaced. There must be a recognition that a clean environment has a price. Regardless of cost, the interested parties in this area realize that environmentally degradable polymers are unlikely to be a significant force in the waste management of polymers. They will be used where they are best suited and offer advantages over the competing technologies, such as recycle, landfilling, and incineration. Some obvious areas of opportunity are in fast food, agriculture, sanitary articles for plastics, and water-soluble polymers which are not recoverable after use. A significant advantage for environmentally degradable water-soluble polymers is the ready availability of a disposal infrastructure in municipal wastewater treatment plants. Composting, a promising disposal avenue for plastics, on the other hand, is in need of development, and no plans are in place to do this, a fact that is hindering the growth of these plastics.

## BIBLIOGRAPHY

"Plastics, Environmentally Degradable" in *ECT* 3rd ed., Suppl. Vol., pp. 626–668, by J. E. Potts, Union Carbide Corp.

1. W. M. Heap and S. H. Morell, *J. Appl. Chem.* **18**, 189 (1968).
2. F. Rodriquez, *Chemtech.* **1**, 409 (1970).
3. W. Coscarelli, in W. L. Hawkins, ed., *Polymer Stabilization*, John Wiley & Sons, Inc., New York, 1972.
4. J. E. Potts, in H. H. J. Jellinek, ed., *Aspects of Degradation and Stabilization of Polymers*, Elsevier, Amsterdam, The Netherlands, 1978, p. 617.
5. E. Kuster, *J. Appl. Polym. Sci. Appl. Polym. Symp.* **35**, 395 (1979).
6. G. S. Kumar and co-workers, *J. Macromol. Sci. Rev., Macromol. Chem. Phys.* **C22**(2), 225 (1981–1983); G. S. Kumar, *Biodegradable Polymers*, Marcel Dekker, Inc., New York, 1987.
7. T. Jopski, *Kunststoffe* **83**(10), 248–251 (1993).

8. C. David, C. DeKersel, F. LeFevre, and M. Wieland, *Angew. Makromol. Chem.* **216**, 21–35 (1994).
9. G. Swift, *Polym. Mater. Sci. Prepr. Amer. Chem. Soc.*, 846–852 (1990).
10. Jan-Chan Huang, A. S. Shetty, and M. S. Wang, *Adv. Polym. Technol.* **10**(1), 23–30 (1990).
11. T. M. Amirabhavi, R. H. Balundgi, and P. E. Cassidy, *Polym. Plast. Technol. Eng.* **29**(3), 235–262 (1990).
12. A. Yabbana and R. Bartha, *Soil Biol. Biochem.* **25**(11), 1469–1475 (1993).
13. R. Narayan, *Biotechnology*, NIST GCR 93–633 (1993).
14. S. Nakamura, Report No. 10, Update No. 1, Kansai Research Institute (KRI), Osaka, Japan, 1994.
15. K. Udipi and A. M. Zolotor, *J. Polym. Sci.* **75**, 109–117 (1993).
16. D. Satyananaya and P. R. Chatterji, *J. Macromol. Sci.; Rev. Macromol. Chem. Phys.* **C33**(3), 349–368 (1993).
17. R. W. Lenz, *Adv. Polym. Sci.* **107**, 1–40 (1993).
18. G. Swift, *ACS Symp. Ser.* **433**, 1–12 (1990).
19. G. Swift, *FEMS Microbiol. Revs.* **103**, 339–346 (1992).
20. G. Swift, *Accounts Chem. Res.* **26**, 105–110 (1993).
21. R. Narayan, *Ann. Mtg. Air and Waste Mgmt. Assoc., June 24–29*, 40 (1990).
22. *ASTM Standards on Environmentally Degradable Plastics*, ASTM Publication Code Number (PCN): 03-420093-19, American Society for Testing and Materials, Philadelphia, Pa., 1993.
23. J. E. Potts and co-workers, *Polymers and Ecological Problems*, Plenum Press, New York, 1973; EPA Contract, CPE-70-124 U.S. Environmental Protection Agency, Washington, D.C., 1972.
24. A. L. Andrady, *J. Mater. Sci. Rev. Macromol. Chem. Phys.* **C34**(1), 25–76 (1994).
25. R. A. Clendinning, S. Cohen, and J. E. Potts, *Great Plains Agric. Council Publ.*, **68**, 244 (1974).
26. R. D. Fields, F. Rodriquez, and R. K. Finn, *J. Am. Chem. Soc., Div. Poly., Chem.*, **14**, 2411 (1973).
27. R. D. Swisher, *Surfactant Biodegradation*, 2nd ed., Marcel Dekker, Inc., New York, 1987.
28. *Chemical Fate Testing Guidelines*, NTIS No. PB82-233008, U.S. Environmental Protection Agency, Washington, D.C., 1982.
29. *OECD Guidelines for Testing Methods*, Degradation and Accumulation Section, Nos. 301A–E, 302A–C, 303A, and 304A, OECD, Washington, D.C., 1981.
30. R. J. Tanna, R. Gross, and S. P. McCarthy, *Polym. Mater. Sci. Eng.* **67**, 230–231 (1992).
31. J. D. Gu, S. P. McCarthy, G. P. Smith, D. Eberiel, and R. Gross, *Polym. Mater. Sci. Eng.* **67**, 294–295 (1992).
32. A. C. Palmisano and C. A. Pettigrew, *Bioscience* **42**(9), 680–685 (1992).
33. G. Swift, *Polym. News* **19**, 102–106 (1994).
34. G. Swift, *Expectations for Biodegradation Testing Methods*, International Biodegradable Polymer Workshop, Osaka, Japan, Nov. 1993.
35. T. Masuda, *Technol. Japan* **24**, 56 (1991).
36. N. Scholz, *Tenside Surfact. Deterg.* **28**, 277–281 (1991).
37. F. J. Golemba and J. E. Guillet, *SPE J.* **26**, 88 (1970).
38. G. H. Hartley and J. E. Guillet, *Macromolecules* **1**, 165 (1968).
39. M. Heskins and J. E. Guillet, *Macromolecules* **1**, 97 (1968).
40. Y. Americk and J. E. Guillet, *Macromolecules* **4**, 375 (1971).
41. G. J. C. Griffin, *J. Amer. Chem. Soc., Divn. Org. Coat. Plast. Chem.* **33**(2), 88 (1973); *J. Polym. Sci.*, **57**, 281 (1976).
42. J. D. Hodsworth, G. Scott, and D. Williams, *J. Chem. Soc.*, 4692 (1964); A. J. Sipinen and D. R. Rutherford, *J. Environ. Polym. Degrad.* **1**(3), 193–203 (1993).

43. U.S. Pat. 3,888,804 (June 10, 1975), C. E. Swanholm (to Biodegradable Plastics).
44. U.S. Pat. 4,056,499 (Nov. 1, 1977), L. J. Taylor (to Owens-Illinois).
45. U.S. Pat. 3,847,852 (Nov. 12, 1974), R. A. White (to DE Bill and Richardson).
46. G. Scott, D. C. Mellor, and A. B. Moir, *European Polym. J.* **9**, 219 (1973).
47. C. G. Cooney, J. D. Carlsson, and D. J. Wiles, *J. Appl. Poly. Sci.* **26**, 509 (1981).
48. J. E. Guillet, *Degradable Materials: Perspective, Issues and Opportunities*, CRC Press, Boca Raton, Fla., 1990, pp. 55–97.
49. M. Kato and Y. Yoneshige, *Macromol. Chem.* **164**, 159, (1973).
50. L. Reich and S. S. Stivala, *Elements of Polymer Degradation*, McGraw Hill Book Co., Inc., New York, 1971, pp. 32–35.
51. U.S. Pat. 5,281,681 and WO 9212185-A2 (July 23, 1992), R. G. Austin (to Exxon).
52. U.S. Pat. 5,064,932 (Nov. 12, 1991), B. H. Chang and L. Y. Lee (to Quantum).
53. U.S. Pat. 5,194,527 (Mar. 16, 1993), J. J. O'Brien and co-workers (to Dow).
54. U. S. Pat. 5,147,712 (Sept. 15, 1992), K. Hirsoe (to Nippon Unicar).
55. Jpn. Pat. 04058962 (Feb. 25, 1992), I. Akimoto and co-workers (to Nippon Unicar).
56. U.S. Pat. 5,115,058 (June 19, 1992), D. B. Priddy and K. D. Sikkema (to Dow).
57. Jpn. Pat. 04173869 (June 22, 1992), T. Kurata and co-workers (to Japan Synthetic Rubber).
58. U.S. Pat. 5,059,676 (Oct. 22, 1991) E. Dent (to Shell).
59. D. Byrom, ed., *Biopolymers: Novel Materials from Biological Sources*, Stockton Press, New York, 1991.
60. E. Dawes, ed., *Novel Biodegradable Microbial Polymers*, NATO ASI Series E: Applied Sciences, Vol. 186, Kluwer Academic Publishers, Dordrecht, Germany, 1990; M. Vert, J. Feijin, A. Albertsson, G. Scott, and E. Chiellini, eds., *Proceedings of the 2nd International Scientific Workshop on Biodegradable Polymers and Plastic, Montpelier, France, Nov. 25–27, 1991*, The Royal Society of Chemistry, London, 1992.
61. Y. Doi, *Microbial Polyesters*, VCH Publishers, New York, 1990.
62. A. J. Anderson and E. A. Dawes, *Microbiol. Rev.* **54**, 450 (1990).
63. H. Brandl, R. A. Gross, R. W. Lenz, and R. C. Fuller, *Adv. Biochem. Eng. Biotech.* **41**, 77 (1990).
64. R. H. Marchessault and co-workers, *Makromol. Chem. Makromol. Symp.* **19**, 235 (1988).
65. M. K. Cox, *Spec. Publ. R. Soc. Chem.* **109**, 95 (1992).
66. World Pat. 9302187–A1 (Feb. 4, 1993), D. C. Sommerville and co-workers (to University of Michigan); see also *Adv. Mater.* **5**(1), 30–36 (1993).
67. D. Byrom, *Int. Biodeter. Biodegrad.* **31**(3), 199–208 (1993).
68. D. F. Gilmore, S. Antoria, R. W. Lenz, and R. C. Fuller, *J. Environ. Polym. Degrad.* **1**(4), 269–274 (1993).
69. T. Saito, M. Shiraki, and M. Tatsumichi, *Kobunshi Robun* **50**(10), 781–783 (1993).
70. H. Brandl and P. Pucchner, *Biodegradation* **2**(4), 237–243 (1992).
71. A. Steinbuchel, *Acta Biotechnol.* **11**(5), 419–427 (1991).
72. T. F. Cooke, *J. Polym. Eng.* **9**, 171 (1990).
73. S. J. Huang, *Comprehensive Polymer Science*, Vol. 6, Pergamon Press, Elmsford, N.Y., 1989, p. 597.
74. A. C. Albertsson, *Proc. Int. Biodeg. Symp.*, 743 (1976).
75. A. C. Albertsson, *J. Appl. Polym. Sci., Polym. Sci. Symp.* **35**, 423 (1979).
76. A. C. Albertsson, *Eur. Polym. J.* **16**, 123 (1978).
77. A. C. Albertsson, *J. Appl. Poly. Sci.* **22**(11), 3419, 3435 (1978).
78. A. C. Albertsson, *J. Appl. Poly. Sci.* **25**(12), 1655 (1980).
79. A. C. Albertsson and S. Karlsson, Ref. 9, pp. 60–64.
80. A. C. Albertsson, S. O. Andersson, and S. Karlsson, *Polym. Deg. Stab.* **17**, 73 (1987).
81. A. C. Albertsson and S. Karlsson, *J. Appl. Polym. Sci.* **35**, 1289 (1988).
82. A. C. Albertsson, *J. Makromol. Sci. Pure Appl. Chem.*, **A30**(9,10), 757–765 (1993).

83. T. Suzuki and co-workers, *Agric. Biol. Chem.* **43**(12), 2441 (1979).
84. T. Suzuki and co-workers, *Agric. Biol. Chem.*, **42**(6), 1217 (1978).
85. T. Suzuki and co-workers, *Agric. Biol. Chem.* **41**(12), 2417 (1977).
86. T. Suzuki and co-workers, *Report of the Fermentation Institute*, Tsukuba, Japan, 1980.
87. T. Yamamoto, H. Inagaki, J. Yagu, and T. Osumi, *Abst. Ann. Mtg. Agric. Chem. Soc. Jpn.*, 133 (1966).
88. T. Suzuki, Y. Ichihara, M. Yamada, and K. Tonomura, *Agric. Biol. Chem.* **34**(4), 747–756 (1973).
89. T. Suzuki, Y. Ichihara, M. Yamada, and K. Tonomura, *J. Appl. Polym. Sci., Appl. Polym. Symp.* **35**, 431–437 (1979).
90. Y. Watanabe, M. Morita, N. Hamada, and Y. Tsujisaka, *Agric. Biol. Chem.* **39**(12), 2447–2448 (1975).
91. Y. Watanabe, M. Morita, N. Hamada, and Y. Tsujisaka, *Arch. Biochem. Biophys.* **174**, 575 (1976).
92. S. J. Huang, E. Quingua, and I. F. Wang, *Org. Coat. Appl. Polym. Sci. Proc.* **46**, 345 (1982).
93. J. C. Huang, A. S. Shetty, and M. S. Wang, *Adv. Polym. Tech.* **10**, 23 (1990).
94. Jpn. Pats. 03263406 and 03263407 (both Nov. 22, 1991), T. Endo (to Kuraray).
95. Jpn. Kokai 7794471 (Aug. 9, 1977) to F. Fukunaga and co-workers.
96. Japan Kokai, 76125786 (Nov. 2, 1976) to F. Fukunaga and co-workers.
97. M. Shimao, N. Kato, *Int. Symp. Biodeg. Polym. Abstr.*, 80 (1990).
98. J. P. Casey and D. G. Manley, *Proc. 3rd Int. Biodeg. Symp. Appl. Sci. Publ.*, 731–741 (1976).
99. O. D. Wheatley, C. F. Baines, *Text. Chem. Color* **8**(2), 28–33 (1976).
100. S. Matsumura, S. Maeda, J. Takahashi, and S. Yoshikawa, *Kobunshi Robunshu*, **45**(4), 317, (1988).
101. S. Matsumura, J. Takahashi, S. Maeda, and S. Yoshikawa, *Macromol. Chem. Rapid Commun.* **9**(1), 1–5 (1988).
102. S. Matsumura, J. Takahashi, S. Maeda, and S. Yoshikawa, *Kobunshi Ronbunshi* **45**(4), 325–331 (1988).
103. U.S. Pat. 5,062,995 (Nov. 5, 1991), A. Garafalo and S. R. Wu (to Lever).
104. S. Matsumura and co-workers, *J. Amer. Oil Chem. Soc.* **70**, 659–665 (1993).
105. S. Matsumura and co-workers, *Makromol. Chem.* **194**(12), 3237–3246 (1993).
106. U.S. Pat. 5,191,048 (Mar. 2, 1993), G. Swift and B. Weinstein (to Rohm and Haas).
107. M. Hunter, D. M. L. daMotta Marques, J. N. Lester, and R. Perry, *Environ. Technol. Lett.* **9**, 1–22 (1988).
108. S. Matsumura and co-workers, *Yukagaku* **33**, 211, 228 (1984).
109. S. Matsumura and co-workers, *Yukagaku* **34**, 202, 456 (1985).
110. S. Matsumura and co-workers, *Yukagaku* **35**, 167 (1985).
111. S. Matsumura and co-workers, *Yukagaku* **35**, 937 (1986).
112. S. Matsumura and co-workers, *Yukagaku* **30**, 31 (1980).
113. S. Matsumura and co-workers, *Yukagaku* **30**, 757 (1981).
114. Jpn. Pat. 6131498 (Feb. 13, 1986), K. Murai and A. Oota (to Sanyo).
115. J. Suzuki, K. Hukushima, and S. Suzuki, Environ. Sci. Technol. **12**(10), 1180–1183 (1978).
116. Jpn. Pat. 05237200-A (1991), F. Kawai (to Rohm and Haas Co.); F. Kawai, *Appl. Microbiol. Biotech.* **39**(3), 382–385 (1993).
117. Y. Tani and co-workers, *Appl. Environ. Microbiol.*, 1555–1559 (1993).
118. Ger. Pat. DE 3716543/4A (Nov. 24, 1988) and Eur. Pat. 291808A (Nov. 23, 1988); and Eur. Pat. 292766A (Nov. 30, 1988), R. Baur and co-workers; Eur. Pat. 289787/8A (Nov. 9, 1988); Eur. Pat. 289827A (Nov. 9, 1988) and Ger. Pat. DE 3733480A (Nov. 9, 1988), D. Boeckh and co-workers; (all to BASF).

119. Eur. Pat. 529910A (Mar. 3, 1993), Y. Irie (to NSKK).
120. Eur. Pat. 289895A (Nov. 9, 1988), K. Driemel, K. Bunthoff, and H. Nies (to Grillo Werke).
121. N. Kawabata, *Nippon Gomu Kyokaishi*, **66**(2), 80–87 (1993).
122. U.S. Pat. 5,219,930 (June 15, 1993), T. S. Brima, B. Chang, and J. Kwiatek (to Quantum Chemicals).
123. J. Mulders and J. Gilain, *Water Res.* **11**(7), 571–574 (1977).
124. T. Suzuki and Y. Tokiwa, *Agric. Biol. Chem.* **50**(5), 1323 (1986).
125. T. Suzuki and Y. Tokiwa, *Agric. Biol. Chem.* **42**(5), 1071 (1978).
126. T. Suzuki and Y. Tokiwa, *Nature* **270**(5632), 76 (1977).
127. J. P. Kendricks, *Diss. Abstr.*, 82329391.
128. W. J. Bailey and co-workers, *Contemp. Topics Polym. Sci.* **3**, 29 (1979).
129. W. J. Bailey and co-workers, *J. Polym. Sci. Polym. Lett. Ed.* **13**, 193 (1975).
130. W. J. Bailey, W. J. Gu, Y. Lin, and Z. Zheng, *Makromol. Chem. Makromol. Symp.* **42/42**, 195 (1991).
131. W. J. Bailey and B. Gapud, *Polym. Stab. Deg.* **280**, 423 (1985).
132. U.S. Pat. 4,923,941 (May 8, 1990), W. J. Bailey (to American Cyanamid).
133. Y. Tokiwa, personal communication, 1993.
134. A. C. Albertsson, *J. Makromol. Sci. Pure Appl. Chem.*, **A30**(2), 919–931 (1993).
135. U.S. Pat. 5,171,308 (Dec. 15, 1992), 5171309 (Dec. 15, 1992), and 5219646 (June 15, 1993), F. G. Gallagher (to Procter and Gamble).
136. Kobayashi, *Macromol. Chem. Rapid Commun.* **14**, 785–790 (1993).
137. Jpn. Pat. 05070543, -566, -571, -572, -574, -575, -576, -577, -579 (all Mar. 23, 1993), E. Takiyama; and Eur. Pat. 572682-A1 (Dec. 8, 1993) and U.S. Pat. 5,310,782 (May 10, 1994), T. Fujimaka and co-workers (all to Showa High Polymer).
138. W. Taniguchi and co-workers, *Polym. Prepr. Jpn.* **42**, 3787 (1993).
139. E. Takeyama and T. Fujimaki, *Plastics* **43**, 87 (1992).
140. T. Masuda, *Kagaku to Kogyo* **44**, 1737 (1991).
141. Y. Yoshida, *Fine Chem.* **21**, 12 (1992).
142. N. Yamamoto, *Polym. Prepr. Jpn.* **41**, 2240 (1991).
143. U.S. Pats. 5,142,023 (Aug. 25, 1992), 5,247,058 (Sept. 21, 1993), 5,247,059 (Sept. 21, 1993), and 5,258,488 (Nov. 2, 1993), R. D. Benson and co-workers (to Cargill).
144. U.S. Pats. 5,210,108 (May 1, 1993), 5,097,005 (Mar. 17, 1992), World Pat. 9204410 (Mar. 19, 1992), T. M. Ford, S. Hyvnkook, and H. E. Bellis (to Du Pont).
145. Jpn. Pat. 05339557-A1 (Dec. 21, 1993); Eur. Pat. 5726750-A1 (Dec. 8, 1993); Eur. Pat. 510999-A1 (Oct. 28, 1992), T. Kitamura and co-workers (to Mitsui Toatsu).
146. World Pat. 9204413-A (Mar. 19, 1992), J. R. Preston and R. G. Sinclair (to Battelle Memorial Institute).
147. W. J. Bailey, in Ref. 98, p. 765.
148. S. J. Huang, in Ref. 98, p. 731.
149. S. J. Huang and co-workers, *Polym. Prepr. Jpn.* **18**(1), 438–441 (1977).
150. K. Gonsalves and co-workers, *J. Appl. Polym. Sci.* **50**, 1999–2006 (1993).
151. Y. Tokiwa, *J. Appl. Polym. Sci.* **24**, 1701 (1979).
152. N. Yamamoto, *Prepr. Ann. Mtg. Kagaku Kogaku-Kai*, 34 (1992).
153. V. S. Andreoni, G. Baggi, C. Guaita and P. Manflin, Int. Biodeter. Biodegrad. **31**(1), 41–53 (1993).
154. U.S. Pat. 5,217,642 (June 8, 1993); World Pat. 9216493-A1 (Oct. 1, 1992); Ger. Pat. DE 4108626-A1 (Sept. 17, 1992), R. Baur and co-workers (to BASF).
155. Ger. Pat. DE 4225620-A1 (Feb. 10, 1994), Ger. Pat. DE 4213282-A1 (Oct. 28, 1993), R. Baur and co-workers (to BASF).
156. Ger. Pat. DE 4108626-A1 (Sept. 17, 1992), Ger. Pat. DE 4034334-A1 (Apr. 30, 1992), R. Baur and co-workers (to BASF).

157. Y. Abe, S. Matsumura, and K. Imai, *Yukagaku* **35**(11), 937–944 (1986).
158. R. W. Lenz, M. Vert, *ACS Polym. Prepr.* **20**, 608 (1978).
159. U.S. Pat. 5,093,170 (Mar. 3, 1992), C. R. Degenhardt and B. A. Kozikowski; and U.S. Pat. 5,093,170 (June 25, 1991), L. Schechtman (to Procter and Gamble).
160. G. Swift, M. B. Freeman, Y. H. Paik, S. Wolk, and K. M. Yocom, *ACS Biotech. Secetariat Abstr. (San Diego)* (Spring 1994); *6th International Conference on Polymer Supported Reactions in Organic Chemistry (POC), Venice, Italy, June 19–23, 1994, Abstr.*, p. 1.13; and *35th IUPAC International Symposium on Macromolecules, Akron, Ohio, July 11–15, 1994, Abstr. 0-4.4-13*, p. 615.
161. U.S. Pat. 5,219,952 (June 15, 1993), L. P. Koskin and A. Meah (to Donlar).
162. Eur. Pat. 578448-A1 (Jan. 12, 1994), Y. H. Paik, E. S. Simon, and G. Swift (to Rohm and Haas).
163. Eur. Pat. 511037-A1 (Oct. 28, 1992), A. Ponce and F. Tournilhac (to Rhône-Poulenc).
164. U.S. Pat. 5,221,733 (June 22, 1993), L. P. Koskin and A. M. Atencio (to Donlar Corp.).
165. V. S. Rao, *Makromol. Chem.* **194**, 1095–1104 (1993).
166. World Pat. 9214753 (Sept. 9, 1992), A. M. Atencio (to Donlar Corp.).
167. U.S. Pat. 5,288,783 (Feb. 22, 1994), L. L. Wood (to SR Chem).
168. World Pat. 9,323,452 (Nov. 25, 1993), L. L. Wood (to SR Chem).
169. U.S. Pat. 5,260,272 (Nov. 9, 1993), J. Donachy and S. Sikes (to University of South Alabama).
170. U.S. Pat. 5,116,513 (May 26, 1993), L. P. Koskan and K. C. Low (to Donlar Corp.).
171. Eur. Pat. 454126 (Oct. 30, 1991), T. Imanaka and co-workers (to Montedipe).
172. Eur. Pat. 561464 and 561452 (both Sept. 22, 1993), A. P. A. Rocourt (to Lever).
173. Ger. Pat. DE 4221875-A1 (Jan. 5, 1994), R. Baur and co-workers (to BASF).
174. World Pat. 93 06202 (April 1, 1993), R. Hall, A. D. Wiley, and R. G. Hall (to Procter and Gamble).
175. Eur. Pat. 445923 (Sept. 11, 1991), T. Endo and co-workers (to Meija-Seika K.K.).
176. F. Kawai, CRC Crit. Revs Biotechnol. **6**, 273 (1987).
177. B. Schink and H. Strab, *Appl. Environ. Microbiol.* **45**, 1905 (1983).
178. B. Schink and H. Strab, *Appl. Microbiol. Biotech.* **25**, 37 (1986).
179. F. Kawai, K. Hanada, Y. Tani, and K. Ogata, *J. Ferment. Technol.* **55**, 89 (1977).
180. F. Kawai, T. Okamoto, and T. Suzuki, *J. Ferment. Technol.* **63**, 239 (1985).
181. F. Kawai, *J. Kobe Univ. Commerce*, **18**(1–2), 23 (1982).
182. F. Kawai and H. Yamanaka, *Ann. Mtg. Agric. Chem. Soc. Jpn.* Kyoto, 1986.
183. M. M. Crutchfield, *J. Amer. Oil Chem. Soc.* **55**, 58 (1978).
184. U.S. Pats. 4,654,159 (Mar. 31, 1987), J. A. Henderson and co-workers, 4,663,071 (May 5, 1987), R. D. Bush and co-workers; 4,689,167 (Aug. 25, 1987), G. L. Spadini and co-workers; Eur. Pats. 192441 (Aug. 27, 1986), G. L. Spadini; 192442 (Aug. 27, 1986), J. H. Collins and co-workers; 236007 (Sept. 9, 1987), R. D. Bush and co-workers; 264977 (Apr. 27, 1988), I. Herbots and co-workers (all to Procter and Gamble).
185. S. Matsumura, K. Hashimoto, and S. Yashikawa, *Yukagaku* **36**(110), 874–881 (1987).
186. U.S. Pats. 4,144,226 (Mar. 13, 1979), 4,146,495 (Mar. 27, 1979), 4,204,052 (May 20, 1980), to M. M. Crutchfield and co-workers, 4,233,422 and 4,233,423 (Nov. 11, 1980), V. D. Papanu (all to Monsanto).
187. W. E. Gledhill and V. W. Saeger, *J. Ind. Microbiol.* **2**(2), 97 (1987).
188. W. E. Gledhill, *Appl. Environ. Microbiol.* **12**, 591 (1978).
189. Ger. Pat. DE 4204808-A1 (Aug. 19, 1993), D. Koeffer, P. M. Lorz, and M. Roeper (to BASF).
190. Ger. Pat. DE 4106354-A1 (Sept. 3, 1992), World Pat. 9215629-A1 (Sept. 17, 1992), R. Baur and co-workers (to BASF).
191. Ger. Pat. DE 4142130-A1 (June 24, 1993), R. Baur (to BASF).
192. H. F. Zobel, in R. L. Whistler, J. N. Bemiller, E. F. Paschall, eds., *Starch: Chemistry and Technology*, 2nd ed., Academic Press, New York, 1984, p. 285.

193. R. F. T. Stepto and I. Tomka, *Chimia*, 41 (1987).
194. U.S. Pat. 4,673,438, F. Whitmer and I. Thomka.
195. G. J. L. Griffin, *Adv. Chem. Ser.* **134**, 159 (1971); U.S. Pat. 4,016,117 (Apr. 5, 1977); Brit. Pats. 1485833 (Sept. 14, 1979) and 1487050 (Sept. 28, 1977) to G. J. L. Griffin; G. J. L. Griffin, ed., *The Chemistry and Technology of Biodegradable Polymers*, Blackie Academic and Professional Press, Chapman and Hall, London, 1994, p. 18.
196. F. H. Otey, R. P. Westoff, W. M. Doane, *Ind. Eng. Chem. Res.* **26**, 1659 (1987); *Org. Coat. Plast. Chem.* **37**(2), 297 (1977); C. L. Swanson, R. L. Shogren, G. F. Fanta, and S. H. Imam, *J. Environ. Polym. Degrad.*, **1**(2), 155–166 (1993).
197. U.S. Pat. 3,949,145 (April 6, 1976), F. H. Otey and A. M. Mark (to U.S. Department of Agriculture).
198. World Pat. 9219680-A1 (Nov. 12, 1992); U.S. Pats. 5,262,458 (Nov. 16, 1993), 5,286,770 (Feb. 15, 1994), and 5,288,765 (Feb. 22, 1994); and World Pats. 9214782-A1 (Sept. 3, 1992) and WO9202363 (Feb. 20, 1992), C. Bastioli (to Novamont); *C. Bastioli, Spec. Publ. R. Soc. Chem.* **109**, 101 (1992); C. Bastioli, V. Bellotti, L. Del Guidice, and G. Gilli, *J. Environ. Polym. Degrad.*, **1**(3), 181–192 (1993).
199. World Pat. 9314911-A1 (Aug. 5, 1993); Brit. Pat. 2218994 (Nov. 29, 1989), F. C. Wexler, and R. Stepto, and J. Silbiger (to Warner Lambert).
200. B. Miller, *Plast. World*, **48**(3), 12 (1990).
201. U.S. Pat. 5288318 (Feb. 22, 1994), G. R. Elion and J. M. Meyer (to U.S. Army, Natick Laboratory).
202. Ger. Pat. DE 4209095-A (Sept. 23, 1993), M. Beck and W. Ritter (to Henkel).
203. World Pat. 9319125-A1 (Sept. 30, 1993), J. Jane and S. Lim (to Iowa State University).
204. U.S. Pat. 5,115,000 (May 19, 1992), J. Jane and co-workers (to Iowa State University).
205. U.S. Pat. 5,271,766 (Dec. 21, 1993), G. Koutlakis, C. C. Lane, and R. P. Lenz (to Archer Daniels Midland).
206. Eur. Pat. 580032-A1 (Jan. 26, 1994), C. Dehennau and T. Depireux (to Solvay SA).
207. World Pat. 9115542 (Oct. 17, 1991), J. S. Peanasky, J. M. Long, and R. P. Wool (to Agritech).
208. C. M. Buchanan and co-workers, *Macromolecules* **25**, 7381 (1992).
209. Jpn. Pat. 0429261-A, (to Mita Industry).
210. Eur. Pats. 520888 and 520889 (Dec. 30, 1992), D. Bazile (to Rhône Poulenc).
211. Jpn. Pat. 06001881-A (Jan. 11, 1994), M. Nishiyama and co-workers (to Agency Institute of Science and Technology (AIST)).
212. Jpn. Pat. 05237180-A (Sept. 17, 1993), Y. Shikinami and co-workers (to Takiron).
213. Jpn. Pats. 04146952-A, 04146953-A, 04146929-A (May 20, 1992), Y. Tokiwa and M. Koyama (to AIST).
214. H. Poteate, A. Ruecker, B. Nartop, *Starch* **46**(20), 52–59 (1994).
215. N. Stacy, Z. J. Lu, Z. X. Chen, and R. Narayan, *Antec '89, Confer. Proc.* 1362–1364 (1989).
216. R. J. Dennenberg, R. J. Bothast, and T. P. Abbot, *J. Appl. Polm. Sci.* **22**(2), 459–465 (1978).
217. World Pat. 9221765 (Dec. 10, 1992), J. Dordick, D. R. Patil, and D. G. Rethwisch (to University of Iowa, State Research Foundation).
218. S. Kim, V. T. Stannett, and R. D. Gilbert, *J. Polym. Sci. Polym. Lett. Ed.* **11**(12), 731–735 (1973).
219. O. Milstein, R. Gersonde, A. Hutterman, M. J. Chen, and J. J. Meister, *Environ. Microbiol.*, **58**(10), 3225–3237 (1992).
220. World Pat. 9302118-A1 (Feb. 4, 1993), Y. W. Kim, I. H. Park, and T. S. Park (to Taechang Moolsan Co. Ltd.).
221. Eur. Pat. 465286 (Jan. 8, 1992), C. Vidal and S. Vaslin; and Eur. Pat. 465287, P. Jost and F. Tournilhac (to Rhone-Poulenc).
222. Ger. Pat. DE 4003172 (Aug. 8, 1991), R. Baur and co-workers (to BASF).

223. World Pat. 9401476-A1 (Jan. 20, 1994), H. Klimmek and F. Krause (to Stockhausen).
224. Ger. Pat. DE 4029348 (Mar. 19, 1992), M. Kroner and co-workers (to BASF).
225. C. M. Buchanan, R. Komanek, D. Dorschel, C. Boggs, and A. W. White, *J. Appl. Polym. Sci.* **52**(10), 1477–1488 (1994); J. D. Gu, D. T. Ebereiel, S. P. McCarthy, and R. A. Gross, *J. Environ. Polym. Degrad.* **1**(2), 143–155 (1994).
226. Rhône-Poulenc Announcement, *Eur. Plast. News* **20**, 16 (1993).
227. A. Arch, *J. Macromol. Sci.*, **A30**(9–10), 733–740 (1993).
228. W. A. Monal and C. P. Covac, *Macromol. Chem. Rapid Commun.* **14**, 735–740 (1993).
229. World Pat. 9210521-A1 (Sept. 22, 1993), J. W. H. Faber (to Eastman Kodak).
230. Eur. Pat. 560891-A1 (Sept. 22, 1993), J. W. H. Faber (to Eastman Kodak).
231. Eur. Pat. 254025-B (Jan. 27, 1988), K. Engelskirc and co-workers (to Henkel).
232. S. Matsumura and co-workers, *Angew. Makromol. Chem.* **205**, 117–129 (1993).
233. S. Matsumura, S. Maeda, and S. Yoshikawa, *Macromol. Chem.* **191**(6), 1269–1275 (1990).
234. S. Matsumura and co-workers, *Polym. Prepr. Jpn.* **41**(7), 2394–2396 (1992); S. Matsumura, K. Amaya, S. Yoshikawa, *J. Environ. Polym. Degrad.* **1**(1), 23–31 (1993).
235. S. Matsumura, K. Aoki, and K. Toshima, *J. Amer. Oil. Chem. Soc.* **71**(7), 749–755 (1994).
236. H. van Bekkum and co-workers, *Prog. Biotech.* **3**, 157 (1987).
237. H. van Bekkum, *Starch Starke*, 192 (1988).
238. A. C. Bessemer, Ph.D. dissertation, University of Delft, the Netherlands, 1993; Eur. Pat. 4273459-A2, World Pat. 9117189.
239. A. C. Bessemer and H. van Bekkum, *Starch-Starke* **46**, 95–100 (1994); **46**, 101–106 (1994).
240. A. E. J. deNooy, A. C. Bessemer, and H. van Bekkum, *Rec. Trav. Chim.* **113**(3), 165–166 (1994).
241. Eur. Pat. 542496-A1 (May 19, 1993), J. S. Dupont and S. W. Heinzman (to Procter and Gamble).
242. Ger. Pat. DE 4203923-A1 (Aug. 12, 1993) and World Pat. 9308251-A1 (Apr. 29, 1993), K. Engelskirchen and co-workers (to Henkel).
243. Eur. Pat. 548399-A1 (June 30, 1979) and World Pat. 9218542-A1 (Oct. 29, 1993), A. M. Sakharov, I. P. Skibida, and G. Brussani (to Novamont).
244. Eur. Pat. 455522-A (Nov. 6, 1991), S. Gosset and D. Videau; U.S. Pat. 4,985,553 (Jan. 15, 1991), G. Fleche and P. Fuertes (to Roquette Frères).
245. Jpn. Pat. 05017502-A (Jan. 26, 1993), Watanabe and Y. Tsuchiyama (to Mercian Corp.).
246. U.S. Pat. 5,223,642 (June 29, 1993) and World Pat. 9102712 (Mar. 7, 1991), F. J. Dany and co-workers (to Hoechst).
247. U.S. Pat. 5,238,597 (Aug. 24, 1993), W. Fritschela and co-workers (to Hoechst).
248. G. Swift, Y. H. Paik, and E. S. Simon, *ACS Polym. Mater. Sci. and Eng. Abstr.* **69**, 496 (1993).

*General References*

*Biodegradable Plastics and Polymers*, Y. Doi and K. Fukuda, eds., *Studies in Polymer Science*, Vol. 12, Elsevier, Amsterdam, the Netherlands, 1994.
C. G. Gebelein, ed., *Biotechnological Polymers*, Technomic Publishing Co., Lancaster, Pa., 1993.
W. McGucken, *Biodegradable Detergents and the Environment*, Texas A and M University Press, College Station, 1994.
M. Veret, L. Feijen, A-C. Albertsson, G. Scott, and E. Chiellini, eds., *Proceedings of the International Conference on Biodegradable Polymers and Plastics*, Nov. 25–27, 1991, Royal Society of Chemistry, London, 1992.

E. A. Dawes, *Novel Biodegradable Microbial Polymers, Proceedings of an Advanced Work-shop on New Biosynthetic Biodegradable Polymers of Industrial Interest*, Sitges, Spain, 1990, NATO ASI Series E: Applied Science, Vol. 186, Kluwer Academic Publishing, London, 1990.

H. G. Schlegel and A. Steinbuchel, eds., *International Symposium on Biodegradable PHAs*, FEMS Microbiological Reviews, 1993.

C. G. Gebelein and C. E. Carraher, eds., *Biotechnology and Bioactive Polymers*, Plenum Press, New York, 1994.

*Polym. Degrad. Stabil.* **45**(2) (1994), papers from the *33rd IUPAC International Sympo-sium on Macromolecules, Montreal, Canada.*

*J. Macromol. Sci. Pure Appl. Chem.*, **A32**(4) (1995), selected papers from the *International Workshop on Controlled Life Cycle of Polymeric Materials, Stockholm, Sweden, Apr. 21–24, 1994.*

GRAHAM SWIFT
Rohm and Haas Company

## POLYMERS OF HIGHER OLEFINS.    See OLEFIN POLYMERS.

## POLYMETHACRYLATES.    See METHACRYLIC POLYMERS.

# POLYMETHINE DYES

Polymethine dyes (PMD) represent a large class of organic colored compounds that contain a chain of methine groups as the basic constitutive elements. According to Daehne's triad theory (1), polymethines, together with polyenes and aromatics, are the three main types of conjugated systems. The term polymethine dyes was introduced by W. Koenig in 1922 (2). The formula of PMDs having the stable, closed electron shell, in its general form, can be written as two resonance structures, where $n$ is the number of vinylene groups in the polymethine chain,

$$[G_1^+ \!\!-\!\!(CH\!=\!CH)_n\!\!CH\!=\!G_2]X^{\mp} \longleftrightarrow [G_1\!=\!CH\!\!-\!\!(CH\!=\!CH)_n\!\!G_2^+]X^{\mp}$$

$$(1)$$

and $G_1$ and $G_2$ are terminal or end groups of various chemical structure, ie, acyclic, carbo-, or heterocyclic residues. Polymethine chains of symmetrical dyes ($G_1 = G_2$) consist of an odd number of carbon atoms, and these systems carry charges, ie, they exist as either cations or anions. In formula (**1**), end groups differ by a number of $\pi$-electrons, and thus one is written in electron donor, and the other in electron acceptor, form.

The simplest PMDs include the polymethines (**2**), streptocyanines (**3**), oxonols (**4**), and merocyanines (**5**).

$$H_2C = CH + CH = CH \rightarrow_{\overline{n}} CH_2^{\pm} \qquad R_2N^+ = CH + CH = CH \rightarrow_{\overline{n}} NR_2$$

(**2**) $\qquad\qquad\qquad\qquad\qquad$ (**3**)

$$O = CH + CH = CH \rightarrow_{\overline{n}} O^- \qquad R_2N + CH = CH \rightarrow_{\overline{n}} CH = O$$

(**4**) $\qquad\qquad\qquad\qquad\qquad$ (**5**)

A great number of different heterocyclic residues have been used as the terminal groups of PMDs. Examples appear throughout this article. PMDs containing residues with quaternary nitrogen atoms are traditionally called cyanine dyes (qv).

Polymethines with branched polymethine chains also exist. Among these, PMD with symmetrically branched chains are the best known; they are referred to as trinuclear polymethine dyes (TPMD) (**6**).

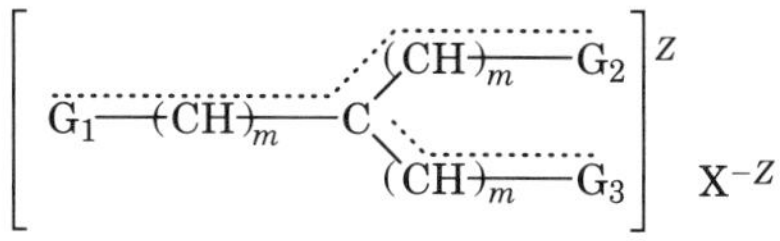

$$\left[ G_1 + (CH)_{\overline{m}} C \begin{array}{c} (CH)_{\overline{m}} G_2 \\ (CH)_{\overline{m}} G_3 \end{array} \right]^Z \quad X^{-Z}$$

(**6**)

PMDs demonstrate pronounced absorption and contain fluorescence bands that are relatively narrow and highly intense, which arise from electron transitions occurring within the polymethine chromophore $+(CH-CH)_{\overline{n}}CH+$ (1–10). These spectral properties give rise to the wide range of applications of PMDs as silver halide sensitizers (11), laser media components (12,13), polymerization initiators (14), etc.

According to one classification (15,16), symmetrical dinuclear PMDs can be divided into two classes, A and B, with respect to the symmetry of the frontier molecular orbital (MO). Thus, the lowest unoccupied MO (LUMO) of class-A dyes is antisymmetrical and the highest occupied MO (HOMO) is symmetrical, and the $\pi$-system contains an odd number of $\pi$-electron pairs. On the other hand, the frontier MO symmetry of class-B dyes is the opposite, and the molecule has an even number of $\pi$-electron pairs.

For convenience, unsymmetrical PMDs should be considered as the derivatives of the corresponding symmetrical polymethines, commonly called parent dyes (7,9). In contrast to symmetrical compounds, unsymmetrical PMDs can contain an even number of methine groups in the polymethine chains, for example, styryls (**7**), where $X = S, O, NCH_3, C(CH_3)_2$, or $CH = CH$; and (**8**), where $Y = O, S, Se$ or $NCH_3$.

(7)                                                            (8)

Unsubstituted PMDs (**2**) or dyes containing odd alternate hydrocarbon residues as end groups can exist in two relatively stable forms distinguished by a $\pi$-electron pair, eg, $\alpha,\omega$-diphenylpolymethines (**9**).

(9a)                                      (9b)

It has been proposed to designate polymethine cations (**9a**) in an electron-deficient $(N - 1)$ form, and anion (**9b**) in an electron-excessive $(N + 1)$ form (17). As a rule, typical PMDs have only one stable form, in contrast to related polyenes with the same terminal residues which have two or even three relatively stable forms (17,18).

Neutral or charged PMD radicals that have open electron shells are derived by chemically or polarographically reducing or oxidizing the corresponding dyes having closed electron shells. Pyrylocyanine radicals and their heteroanalogues, represented by (**10**), where X = O or S, and $n = 0$, 1, or 2 (19,20), are examples.

(10)

The literature on polymethine dyes has been reviewed (3,4,7,9–11,21). Reference 3 is the best among recent (1995) sources on the chemistry and spectral properties of this dye class; it also contains a large bibliography.

## Electron Structure

A considerable number of experiments have shown that symmetrical PMDs in the ground state have an all-trans configuration and are nearly planar with practically equalized carbon–carbon bonds and slightly alternating valence angles within the polymethine chain (1,3,5,22,23). This is caused by some significant features of the PMD electron structure.

**Frontier Level Positions.**   As the polymethine chain lengthens, the energies of frontier MO, LUMO, and HOMO tend regularly toward the energy of the Fermi level, or energy of the nonbonding $\pi$-electron (Fig. 1), in contrast to related polyenes in which there is a nonzero energy gap due to the alternation of the bond lengths (1,3).

The nonbonding level of unsubstituted polymethines is the lowest vacant one in cations and the highest occupied one in anions. The nonbonding MO modes fall on odd atoms, and the other frontier MO has its modes on even atoms. As a rule, an attachment of end groups causes the frontier level shifts. A parameter, $\varphi_0$, called electron donor ability, has been proposed for quantitative estimation of the position of the frontier levels (16):

$$\varphi_0 = 90°(\epsilon_e - \alpha)/(\epsilon_e - \epsilon_g) \qquad (1)$$

where $\epsilon_e$ and $\epsilon_g$ are the LUMO and the HOMO energies, and $\alpha$ is the Fermi energy ($\alpha \cong -5.60$ eV). If the energy gap is constant, an increase in $\varphi_0$ is accompanied by a decrease in the ionization and oxidation potentials, and an increase in the reduction potential. For polymethines having the stable, closed

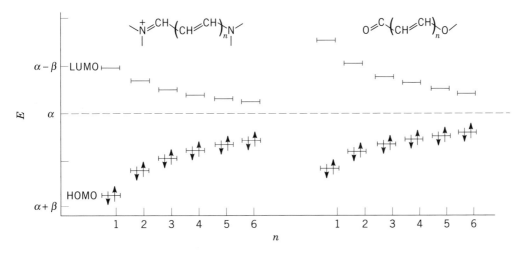

**Fig. 1.**   Frontier electron levels for polymethines containing the simplest end groups with low basicity ($\alpha$, $\Phi_0 = 33°$) and high basicity ($\beta$, $\Phi_0 = 63°$). $E$ = energy.

electron shell, $\varphi_0$ has only positive values and is within the interval $0° \leq \varphi_0 \leq 90°$. Taking into account the linear relation between the redox potentials or the transition energy and the calculated level energies or the energy gap, the electron donor ability can be determined from experimental data (10,17,24).

Estimation of the influence of end groups by using their topological indexes, $\Phi_0$ and $L$, has been proposed. The first parameter, $\Phi_0$, characterizes the shift of the MO modes and the level positions of a PMD containing end groups relative to unsubstituted polymethines. Thus it corresponds to the end-group basicity concept (7). The parameter $\Phi_0$ was found to be related directly to the electron donor ability: $\Phi_0 = \lim_{n \to \infty} \varphi_0$. The other index, $L$, quantitatively estimates the lengthening of polymethine chromophore at the expense of the conjugated system of terminal groups.

A large body of quantum chemical data shows the alternating positive and negative charges at the carbon atoms within the polymethine chain in both ground and excited states. The charge sign at a polymethine chain atom changes on excitation. If $\Phi_0 > 45°$, the negative charges exceed the positive ones, but if $\Phi_0 < 45°$, the positive charge values are greater. If $\Phi_0 = 45°$, the absolute charge values at the neighboring atoms are equal (10,25). Considerable alternation of the electron density has been confirmed experimentally by means of nmr spectroscopy (3).

The bond orders in the polymethine chain are equalized in the ground and excited states. If $\Phi_0 = 45°$, the bond equalization is maximum. This is the ideal polymethine state (1) of the polymethine chain. Any deviation from this state (ie, $\Phi_0$ is greater than or less than $45°$) causes the bond to alternate from the polymethine chain center to its ends. The alternation amplitude is found to be proportional to the absolute value $|45° - \Phi_0|$.

The perturbation of the PMD symmetry is accompanied by a decrease in the charge alternation and by the appearance of bond alternation from one end group to another. The bond alternation amplitude has been revealed to be proportional to the asymmetry degree, which can be calculated as the difference of topological indexes: $\Delta\Phi_{12} = \Phi_{01} - \Phi_{02}$. The effect is maximum if $\Phi_{01} > 45°$ and $\Phi_{02} < 45°$. If $\Delta\Phi_{12} = 90°$, the ideal polyene state is reached. In the excited state of the asymmetrical PMD, bond orders are essentially equalized.

In PMD radicals, the bond orders are the same as those in the polymethines with the closed electron shell, insofar as the single occupied MO with its modes near atoms does not contribute to the bond orders. Also, an unpaired electron leads the electron density distribution to equalize. PMD radicals are characterized by a considerable alternation of spin density, which is confirmed by epr spectroscopy data (3,19,20).

## Electron Transitions

PMD color or the nature of the electron transitions produces the widest application for PMDs. Depending on the polymethine chain length, the end-group topology, and the electron shell occupation, polymethines can absorb light in uv, visible, and near-ir spectral regions.

**Polymethine and Local Transitions.** The long wavelength absorption band corresponds to the transition of the $\pi$-electron from the HOMO to the LUMO.

As a rule, in typical PMDs the frontier level is delocalized along the whole polymethine chromophore. Therefore it is convenient to call this transition the polymethine electron transition (PMET) (26). Because the bond length change upon excitation is small, the absorption bands corresponding to PMETs are relatively narrow, as for example in thiapyrylocyanines (**11**) (27). On the other hand, the considerable redistribution of the electron density upon excitation causes a large transition moment and, as a result, high intensity of the corresponding absorption band (Fig. 2).

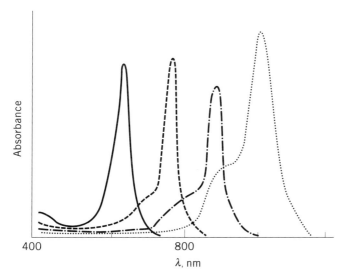

If the terminal groups contain a local level lying near the frontier polymethine level, then a new transition type involving this level, ie, the local or quasilocal electron transition (LET), can appear. It has been proposed to consider the terminal groups as local chromophores (26). LET can have lower energy than PMET, for example, in such dyes as (**12**) (28). In this case the absorption spectrum changes essentially (Fig. 3) and the short-wavelength band is narrower and appears to be a typical polymethine one.

**Vinylene Shift.** The extension of the chromophore in symmetrical PMDs by the successive attachment of vinylene groups shifts absorption maxima to

**Fig. 2.** Absorption spectra of dyes (**11**) showing polymethine electron transition where for (——), $n = 1$; (– – –), 2; (—·—), 3; and (···), 4 (27).

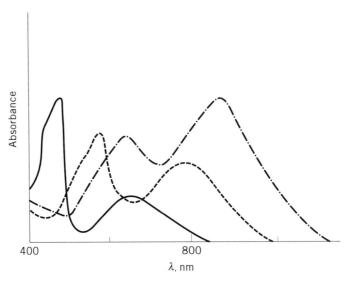

**Fig. 3.**   Absorption spectra of dyes (**12**) showing local electron transition (28). See Fig. 2 for curve descriptions.

longer wavelengths; the spectral shift per one vinylene unit, traditionally called a vinylene shift, is near 100 nm (29) (see Fig. 2). Various relations between the band maximum wavelength and the number of vinylene groups in the polymethine chains have been proposed (1,3,5,10). In general form, they can be written as in equation 2, where $V$ is a vinylene shift value and $L$ the topological index. As a rule, $V$ is independent of the end-group constitution. Nevertheless,

$$\lambda_{\mathrm{max},n} = Vn + \lambda_0 = V(n + L) \tag{2}$$

if the topological index $\Phi_0$ tends to its extreme values, ie, $0°$ or $90°$, then the vinylene shift becomes smaller and falls down to 60–70 nm, as is the case for $\alpha,\omega$-diphenylpolymethine cations (30). The vinylene shifts of the absorption bands originating from the LETs are much smaller (26).

   **Topology of End Groups.**   For the same polymethine chain length, the PMD absorption region depends on end-group topology. For instance, the carbocyanines (**13**) absorb light in the region of 400–600 nm (3), whereas the corresponding pyrylocyanines (**14**) have their band maxima at longer wavelengths, ie, they are shifted to 100–200 nm longer wavelength (27). In (**13**), $\mathrm{X = NCH_3, C(CH_3)_2, CH{=}CH, O, S,}$ or Se; in (**14**), $\mathrm{Y = NCH_3, O, S,}$ or Se.

(**13**)                                              (**14**)

This topology phenomenon can be treated as an additional lengthening of the polymethine chromophore by attachment of the terminal groups. It is convenient to estimate this chromophore lengthening by topological index $L$, the effective length of the end group (see eq. 2). Parameter $L$ is calculated theoretically and reflects significant topological features such as linear or cyclic constitution, the number and sizes of rings, the position and nature of heteroatoms, the position of the atom bound to the polymethine chain, etc (10,17).

   **Asymmetry.**   As a quantitative characteristic of the PMD asymmetry degree, deviation, $D$, is used. The parameter $D$ is calculated from equation 3, where $\lambda_a$ is a band maximum of asymmetrical dye; $\lambda_1$ and $\lambda_2$ are maxima of the symmetrical parent dyes.

$$D = (\lambda_1 + \lambda_2)/2 - \lambda_a \qquad (3)$$

   It has been found (7,9) that $D > 0$, ie, the maximum of an asymmetrical dye, $\lambda_a$, is shifted to the short-wavelength region with respect to the arithmetical mean of the parent dye maxima. The phenomenon has been named a deviation (7). The positive deviations in PMDs are explained by the bond order alternation within the polymethine chain caused by different contributions of both end groups to the dye energetic stability (7,9,10). The deviation reaches its maximum at $\Phi_{01} > 45°$ and $\Phi_{01} < 45°$; if the end groups have $\Phi_{01} > 45°$ and $\Phi_{02} > 45°$, or $\Phi_{01} < 45°$ and $\Phi_{02} < 45°$, then the deviation is negligible. It should be mentioned that the deviation can also be negative if the parent dyes are distinguished considerably by their absorption regions.

   Studies show polymethine chain lengthening in highly asymmetrical dyes to be accompanied by strong quadratic increases in deviations (3,7,9,10,30,31). In contrast to polymethines, the deviations of the related asymmetrical polyenes are negative as the break of the symmetry leads to a decrease in bond alternation (32).

   **Chromophore Branching.**   Among PMDs that have the branched chromophores, dyes with the symmetrically branched polymethine chain, ie, trinuclear dyes (TPMD) (**6**), have been investigated in most detail. If end groups are of the same constitution, the ground state of the dye cation is of the $C_3$ symmetry (33–35). As a result, the antisymmetrical MOs appear to be degenerate. The degeneration of one of the frontier MOs leads the electron-transition energies to also be degenerate. Therefore, two PMETs correspond to the one long-wavelength band observed spectroscopically. The bandwidth of a TPMD has been found to be greater than that of the corresponding PMD having two terminal groups (36). The parallel lengthening of each polymethine chain branch in the TPMD (**6**) by one methine group accompanied by a regular shift of the absorption maximum so that $\lambda_m = T(m + L)$, where $T = \lambda_m - \lambda_{m-1}$, is known as the trimethine shift (36,37). Whereas the vinylene shift, $V$, is practically independent of the polymethine chain length, the parameter $T$ is sensitive to the parity of the methine group number $m$, and $T = V + (-1)^m \Delta$, where $\Delta \cong 10$ nm (38).

   If one terminal group differs from the other two, the $C_3$ symmetry is broken, and the level and electron-transition degeneration vanishes. Level splitting has been found to be proportional to the difference of the effective lengths of the nonidentical end groups, $\Delta L$. Experimentally, transition splitting of the asymmetrical TPMD manifests itself by an increase in the bandwidths; if the value of $\Delta L$ is great enough, the transitions manifest themselves as two separate maxima (36).

**Polymethine Radicals.**   Two transitions occur with low energy involving the single occupied MO (SOMO) in the PMD radical: (1) the electron transition from the highest double occupied MO to the SOMO; and (2) the transition from the SOMO to the LUMO.

In the first approximation, the energy of one of the transitions should be equal to the transition energy of the original PMD, and the energy of another transition should be close to that of the original. However, because both electron transitions in the radical involve the same MO (SOMO), they interact significantly. As a result, the energy and the moment of the former transition decrease but the energy and the moment of the first one increase. Two bands are then observed in the spectra of radicals (Fig. 4) (36).

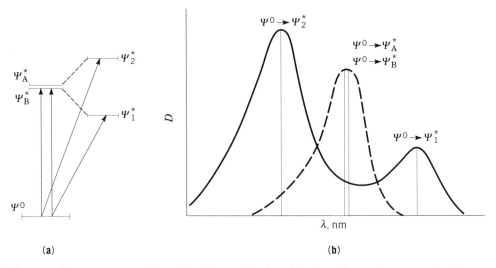

(a)                                                 (b)

**Fig. 4.**  Electron transitions in dye radicals. (**a**) A schematic spectral picture showing the ground ($\Psi^0$) state and two excited ($\Psi_1^*, \Psi_2^*$) states obtained after (**b**), the interaction of two initial quasidegenerate electron configurations, $\Psi_A^*$ and $\Psi_B^*$. $\Psi_1^* = C_1\Psi_A^* + C_2\Psi_B^*; \Psi_2^* = C_1\Psi_A^* - C_2\Psi_B^*$.

**Chromophore Constitution Variations.**   In addition to the main topological factors considered previously, there are many relatively slight constitution variations affecting the PMET energy. These can be identified by the substitution of methine groups by heteroatoms, introduction of conjugated or unconjugated substituents, acetylenic bonds, cyclizations, spatial effects, etc. Compare —CH=CH—CH=CH— to the following:

If the perturbations thus caused are relatively slight, the accepted perturbation theory can be used to interpret observed spectral changes (3,10,39). The spectral effect is calculated as the difference of the long-wavelength band positions for the perturbed and the initial dyes. In a general form, the band maximum shift, $\Delta\lambda$, can be derived from equation 4 analogous to the well-known Hammett equation. Here $\rho$ is a characteristic of an unperturbed molecule, eg, the electron density or bond order change on excitation or the difference between the frontier level and the level of the substitution. The other parameter, $\sigma$, is an estimate of the perturbation.

$$\Delta\lambda = \rho\sigma \qquad (4)$$

The direction of the long-wavelength maximum shift caused by the heterosubstitution or the introduction of substituents is determined by the Forster-Dewar-Knott rule (40–42). Spatial hindrances within the symmetrical PMDs cause bathochromic effects (39,43), whereas the introduction of an acetylenic bond is accompanied by the maximum shift to the short-wavelength spectral region (44).

**Chromophore Interaction.** A great number of systems containing several polymethine chromophores exist. As the simplest objects of this sort, dyes with coupled chromophores, or bis-dyes, can be considered (3,10) as follows (**15**), where M is either a conjugated or a nonconjugated bridge.

$$G_1^+ \!\!+\!\! CH\!\!=\!\!CH\!\!\to_{\overline{n}} CH\!\!=\!\!G_2 \!-\!M\!-\!G_1^+ \!\!+\!\! CH\!\!=\!\!CH\!\!\to_{\overline{n}} CH\!\!=\!\!G_2$$

**(15)**

Unlike PMDs having a single chromophore, the absorption band of a bis-dye splits into two components, so that one maximum is shifted bathochromically and the other hypsochromically with respect to the absorption maximum of the parent dye. The distance between bis-dye maxima depends on the magnitude of the chromophore interaction. Interaction of this kind has been discovered to be universal (45,46).

In the dipole–dipole approximation, the intensities of the long-wavelength, $I-$, and short-wavelength, $I+$, bands are related to the angle between the interacting chromophores, $\delta$ (47):

$$I-/I+ = tg^2(\delta/2) \qquad (5)$$

If both chromophores are in line, the long-wavelength maximum appears only; its intensity is much greater than that of the parent dye. Conversely, if chromophores are parallel, only the highly intensive short-wavelength band is observed.

**Dye Aggregation.** Another important kind of chromophore interaction is the dye aggregation. A number of reviews exist that treat the structure and application of polymethine dye aggregates (3,48–51) (see also CYANINE DYES).

The phenomenon of dye aggregation was discovered in the 1930s (52,53). Polymethine dyes were found to form molecular complexes in solution or on certain crystal surfaces. Molecules within aggregates are bound together by nonvalence bonds, ie, resonance interactions exist between them.

There are two main kinds of dye aggregates, characterized by their typical spectral properties: J-aggregates and H-aggregates. The absorption band maximum (J-band) of the J-aggregates is shifted bathochromically with respect to that of an isolated molecule (M-band); the absorption maximum of the H-aggregates is shifted hypsochromically (H-band). The dyes can also form dimers with a shorter absorption wavelength (D-band).

An aggregate model in which the molecules are packed plane-to-plane with the short axis perpendicular to the aggregate axis has been elaborated (52,54,55). Different structures of J- and H-aggregates were proposed that correspond to the slip angles, $\alpha$, between the long molecular axis and the aggregate axis. J-aggregates were found to be characterized by a small angle ($\alpha < 45°$) and H-aggregates by a greater slip angle. According to the chromophore interaction theory, a different orientation of aggregate molecules leads J-aggregates to form highly intensive and narrower long-wavelength bands than H-aggregates. Solvents, surfaces of ionic crystals, and other factors can also play a significant part in dye aggregation (3,48,50).

**Fluorescence Spectra.**   The main distinction between absorption and fluorescence spectra consists in the equilibrium geometry of the initial state from which the electron transition occurs (3,10). Thus, light absorption occurs at the geometry of the ground state according to the Frank-Condon rule, from the equilibrium ground state $E^{0(0)}$ to the nonequilibrium excited state $E^{*(0)}$ having the same geometry. In contrast, the fluorescence spectra correspond to electron transition from the equilibrium excited state $E^{*(*)}$ to the unstable ground state $E^{0(*)}$ with the same geometry. As a rule, the transition energy for light emission is lower than the energy of the absorbed quanta, and hence the fluorescence maximum is shifted to longer wavelengths. The difference, $\Delta\nu_S = \nu_{max}^{ab} - \nu_{max}^{fl}$, is called the Stokes shift. The Stokes shifts should be proportional to bond order or length change on excitation. Because all bonds within the polymethine chain of symmetrical PMDs are significantly equalized and change slightly on excitation, relatively small Stokes shifts ($500-600$ cm$^{-1}$) are observed in their spectra. In unsymmetrical PMDs, the essential bond alternation exists in the ground state. However, bond orders in the excited state are found to be insensitive to the symmetry perturbation. As a result, the deviations of fluorescence maxima, $D^{fl}$, are much lower than those of absorption maxima, $D^{ab}$ (3,10,56–58). The vinylene shifts of fluorescence maxima of unsymmetrical PMDs are practically constant and equal to 100 nm (57).

**Solvent Influence.**   Solvent nature has been found to influence absorption spectra, but fluorescence is substantially less sensitive (9,58). Sensitivity to solvent media is one of the main characteristics of unsymmetrical dyes, especially the merocyanines (59). Some dyes manifest positive solvatochromic effects (60); the band maximum is bathochromically shifted as solvent polarity increases. Other dyes, eg, highly unsymmetrical ones, exhibit negative solvatochromicity, and the absorption band is blue-shifted on passing from nonpolar to highly polar solvent (59). In addition, solvents can lead to changes in intensity and shape of spectral bands (58).

As a rule, the fluorosolvatochromic effects are less as the dipole moment decreases on excitation, but the media environment can considerably influence quantum yield (61).

## Chemical Reactivity

As conjugated systems with alternating $\pi$-charges, the polymethine dyes are comparatively highly reactive compounds (3). Substitution rather than addition occurs to the equalized $\pi$-bond. If the nucleophilic and electrophilic reactions are charge-controlled, reactants can attack regiospecifically.

**Protonation.** As expected from $\pi$-electron distribution, the proton attacks the negatively charged odd positions in the polymethine chain, eg,

$$G^+\text{—CH}=\text{CH—CH}=\text{CH—CH}=\text{G} \xrightarrow{\text{H}^+} G^+\text{—CH}_2\text{—CH}=\text{CH —CH}=\text{CH—G}^+$$

The destruction of the total $\pi$-system causes the color to vanish; the protonated molecule absorbs light in the uv region. Protonation has been proved by H/D exchange to be reversible (62,63).

The p$K_a$ values of polymethine dyes depend on terminal group basicity (64); thus the protonation ability diminishes if the basic properties of the residues decrease, passing from benzimidazole, quinoline, benzothiazole, to indolenine. On the other hand, the p$K_a$ of higher homologues increases with chain lengthening. The rate constant of protonation is sensitive to other features, for example, substituents and rings in the chain and steric hindrance for short-chain dyes.

**Other Electrophilic Reactants.** Reversibility of the electrophilic reactions enables substituted dye derivatives to be obtained. Thus, the halogenation of cyanines, oxonoles, and merocyanines has been studied (3,65,66). Halogen atoms are mobile in the polymethine chain, and the derivatives themselves can function as halogenation reagents.

The dye can be formulated by means of a phosgene (qv) (67,68). Chloromethylation leads to the alkylated polymethines; nitration of cyanines results in mononitro-substituted compounds (66).

**Nucleophilic Reagents.** In contrast to electrophilic reactions, nucleophiles attack positively charged, even carbons in the chain. The reactions lead to the exchanging of substituents or terminal residues. Thus, SR and OR groups, or halogen atoms can be exchanged by other suitable nucleophiles (4,69,70), for example, by aniline:

$$G^+\text{—CH}=\text{CH—}\underset{\underset{\text{SCH}_3}{|}}{\text{C}}=\text{CH—CH}=\text{G} \xrightarrow[-\text{HSCH}_3]{+\text{C}_6\text{H}_5\text{NH}_2} G^+\text{—CH}=\text{CH—}\underset{\underset{\text{NHC}_6\text{H}_5}{|}}{\text{C}}=\text{CH—CH}=\text{G}$$

$$\textbf{(16)} \qquad\qquad\qquad\qquad \textbf{(17)}$$

If the dye contains no mobile substituents in the chain, nucleophiles attack primarily the end carbon atoms (changing of terminal residues). Streptocyanines can be hydrolyzed in aqueous alkaline solution to form the corresponding

$$\textbf{(3)} \xrightarrow[(+\text{H}_2\text{O})]{+\text{OH}^-} \textbf{(5)} \longrightarrow \textbf{(4)}$$

merocyanines and then the oxonoles (71,72). These processes are reversible. Nucleophilic reactions with the methylene bases of the corresponding heterocycles result in polymethines containing new end groups (Fig. 5).

The dianilinopolycarbocyanines, $C_6H_5HN=CH+CH=CH)_{\overline{n}}NHC_6H_5$, are useful synthetic intermediates for polymethine dyes, offering needed spectral and other properties (3,4,73,74). The asymmetrical polymethines appear to be ambivalent systems, and the number of possible reaction paths increases considerably as a result (75,76).

**Reactions with Parting of Radicals.**   The one-electron oxidation of cationic dyes yields a corresponding radical dication. The stability of the radicals depends on the molecular structure and concentration of the radical particles. They are susceptible to radical–radical dimerization at unsubstituted, even-membered methine carbon atoms (77) (Fig. 6).

**Photochemistry.**   The most important photochemical processes that proceed from the excited state are geometrical isomerization and photochromic reactions.

Photoisomerization of polymethines is a reversible trans–cis transfer first observed in 1953 (78). The cis-isomer absorbs at longer wavelength with a smaller intensity than the trans-isomer. Extensive investigations of this phenomenon began in 1970s (79,80). In the first time-resolved spectroscopic study of the photoisomerization of dyes with various end groups, the lifetime of photoisomers was found to be in the ≥1 ms range. The molecular structure of the photoisomers is relatively unknown as of 1996. For some dyes, two or more photoisomers have

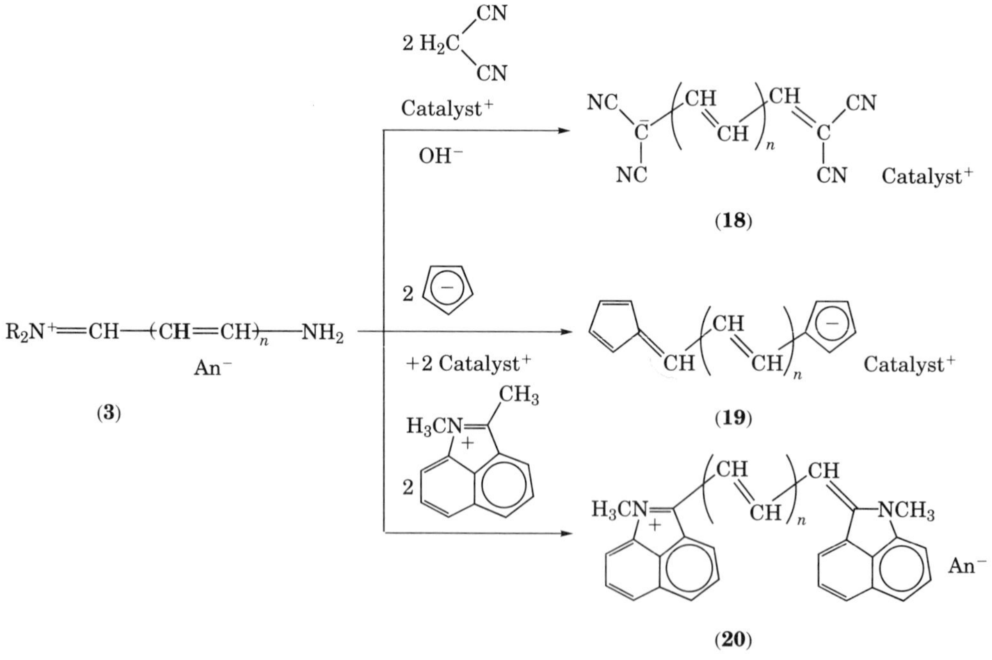

**Fig. 5.**   Alteration of end groups by reaction with methylene bases.

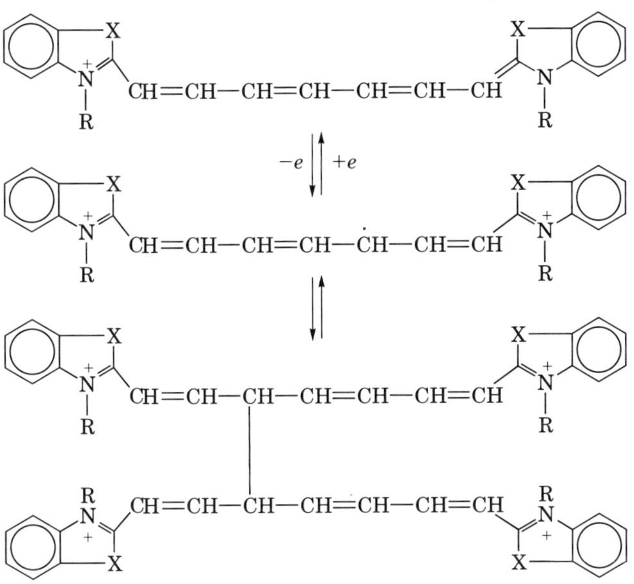

**Fig. 6.** One-electron oxidation and dimerization where (**21a**) is a dye, (**21b**) a radical cation, and (**21c**) a dimer.

been found to exist, especially in the case of polymethines with longer chains (81–83).

A kinetic scheme and a potential energy curve picture in the ground state and the first excited state have been developed to explain photochemical trans–cis isomerization (80). Further investigations have concluded that the activation energy of photoisomerization amounts to about 20 kJ/mol (4.8 kcal/mol) or less, and the potential barrier of the reaction back to the most stable trans-isomer is about 50–60 kJ/mol (3).

In addition to photoisomerization, there are reversible photochemical reactions of special types for asymmetrical polymethines, producing spiropyranes (84–86) as in equation 6, where X = NR, S, or $C(CH_3)_2$.

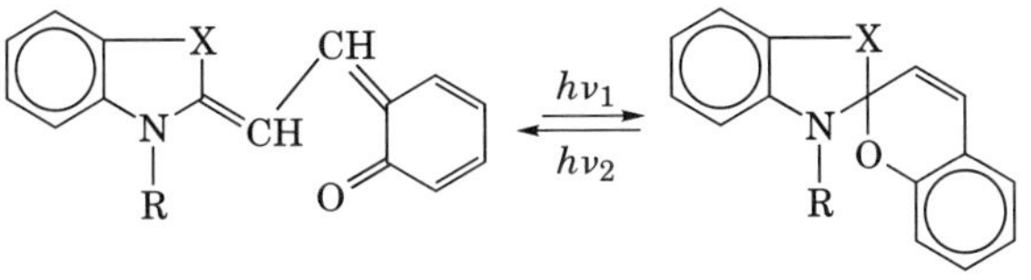

The reverse reaction, the photochemical ring opening of spiropyranes (**22b**), takes place by absorption in the short-wave uv region of the spectrum and the merocyanine isomer (**22a**) is obtained. The electron transition of (**22a**) is in the visible spectral region, whereas (**22b**) is colorless. As a result, the dye solution can change from colorless to a colored solution (87,88). These photochromic reactions can be used for technical applications (89).

## Applications of Polymethines

The most important reason for the large number of technical applications of polymethine dyes is their relatively low electron-transition energies and their highly intense and narrow spectral bands. Indeed, polymethines display strong light absorption and emission, between 300 and 1600 nm. In the 1990s, these dyes are mainly used as photographic sensitizers and desensitizers (11,90), as laser dyes (12,13,91), as probes of membrane potentials (14), and in other applications where the theoretical aspects of polymethines are useful.

**Spectral Sensitization.**  Photographic silver halide emulsions are active with light only up to about 500 nm. However, their sensitivity can be extended within the whole visible and near-ir spectral region up to about 1200–1300 nm. This is reached by the addition of deeply colored dyes that transfer excited electrons or excitation energy to the silver halide.

According to the electron-transfer mechanism of spectral sensitization (92,93), the transfer of an electron from the excited sensitizer molecule to the silver halide and the injection of photoelectrons into the conduction band are the primary processes. Thus, the lowest vacant level of the sensitizer dye is situated higher than the bottom of the conduction band. The regeneration of the sensitizer is possible by reactions of the positive hole to form radical dications (94). If the highest filled level of the dye is situated below the top of the valence band, desensitization occurs because of hole production.

Based on correlations between energy level positions and electrochemical redox potentials, it has been established that polymethine dyes with reduction potentials less than $-1.0$ V (vs SCE) can provide good spectral sensitization (95). On the other hand, dyes with oxidation potentials lower than $+0.2$ V are strong desensitizers.

Improvement of spectral sensitization can be accomplished by dye combinations. The effect has been found to often be greater than the predicted additive sensitivity increase. This phenomenon is called supersensitization (94), which is applied most effectively to polymethine aggregates (96).

The opposite phenomenon, a decrease of sensitivity, is known as desensitization. The main reasons for densensitization are the results of relative electron level positions as well as the secondary processes of the photoelectrons, for example (97),

$$2\ \text{dye}^+ + O_2 + 2\ H^+ \longrightarrow 2\ \text{dye}^{2+} + H_2O_2$$

$$2\ \text{dye}^{2+} \longrightarrow \text{dimer (desensitizer)}$$

**Quantum Electronics and Laser Dyes.**  In quantum electronics, PMDs are usually applied as mode-locking compounds in passive mode-locked lasers as well as active laser media (12,13,91). A solution of cryptocyanine ($\lambda = 704$ nm) was first used for passive mode-locking in the ruby laser ($\lambda_{\text{gen}} = 694.3$ nm) in 1964 (98). A gigantic or ultrashort pulse with the period of $\Delta t = 10^{-8}$ s and peak capacity of 20 MW was generated. Using PMDs as active laser media allows the generation wave to be varied in a wide region, from 200 to 1200 nm and more. In 1967, generation in ethanolic cryptocyanine solution was realized upon excitation by the gigantic 30–40-ns pulse of a ruby laser with $\lambda = 694$ nm (99). Generation occurred in the spectral region, $\lambda = 808.5$ nm of width $\Delta\nu = 40$ cm$^{-1}$.

The required characteristics of dyes used as passive mode-locking agents and as active laser media differ in essential ways. For passive mode-locking dyes, short excited-state relaxation times are needed; dyes of this kind are characterized by low fluorescence quantum efficiencies caused by the highly probable nonradiant processes. On the other hand, the polymethines to be applied as active laser media are supposed to have much higher quantum efficiencies, approximating a value of one (91).

If extreme pumping capacities are provided, superluminescence can occur. For a cyanine dye solution, this phenomenon was observed and described for the first time in 1969 (100).

**Photopolymerization.** In many cases polymerization is initiated by irradiation of a sensitizer with ultraviolet or visible light. The excited state of the sensitizer may dissociate directly to form active free radicals, or it may first undergo a bimolecular electron-transfer reaction, the products of which initiate polymerization (14). Triphenylalkylborate salts of polymethines such as (**23**) are photoinitiators of free-radical polymerization. The sensitivity of these salts throughout the entire visible spectral region is the result of an intra-ion pair electron-transfer reaction (101).

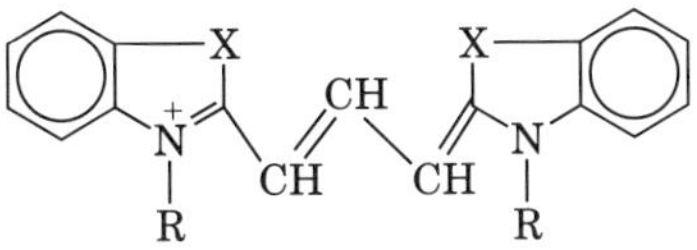

(**23**)

Single-electron transfer from a borate anion particle to the excited polymethine cation generates a dye radical and an alkylphenylboranyl radical. The alkylphenylboranyl radical fragments to form an active alkyl radical. It is the alkyl radical particles that initiate the polymerization reactions (101).

$$(\text{dye})^+[n\text{-}C_4H_9B(C_6H_5)_3]^- \xrightarrow{h\nu} (\text{dye}^*)^+[n\text{-}C_4H_9B(C_6H_5)_3]^- \longrightarrow (\text{dye})^\cdot + [n\text{-}C_4H_9B(C_6H_5)_3]^\cdot$$

$$[n\text{-}C_4H_9B(C_6H_5)_3]^\cdot \longrightarrow n\text{-}C_4H_9^\cdot + B(C_6H_5)_3$$

## Synthesis

By varying the molecular structure, it is possible to synthesize dye initiators with the required characteristics. The synthesis of polymethine dyes with different chain length, end groups, and substituents, or with other variations of the chromophore, has been summarized (3,4,9,21,73,74) (see also CYANINE DYES).

Polymethine dyes consist of three main structural elements: two identical or different end groups and a conjugated chain containing an odd number of methine groups. There are many possibilities for changing the chromophore constitution: using new heterocyclic systems for the end groups, introducing specific substituents in either the chain or in the residues, branching of the polymethine chromophore, replacement of the methine groups by heteroatoms, and cyclization of the chain by conjugated or unconjugated bridges.

**General Aspects.** As a rule, the end-group synthones have the following reactive centers: an activated methyl or methylene group with high CH acidity, a functional group (OR, SR, X(halide), $NR_2$) leaving as an anion in the reaction, and a carbonyl or heteroanalogous group as a leaving group. Complementary reactive centers are needed in the chain synthones in the $\alpha$- and $\omega$-positions. In particular, derivatives of formic acid are used to prepare monomethine dyes; for dyes with longer chromophores, the application of vinylogous aminals or $\omega$-methylpolyenals are preferred. In the most famous reactions, the orthoformate was originally proposed (2), and vinylogous amidines that can be regarded as dianils of conjugated dialdehydes are widely used. A general reaction is shown in equation 7 (3), where X = OR, SR, halide, or $NR_2$; and Y = O, S, NR, or $N^+R_2$;

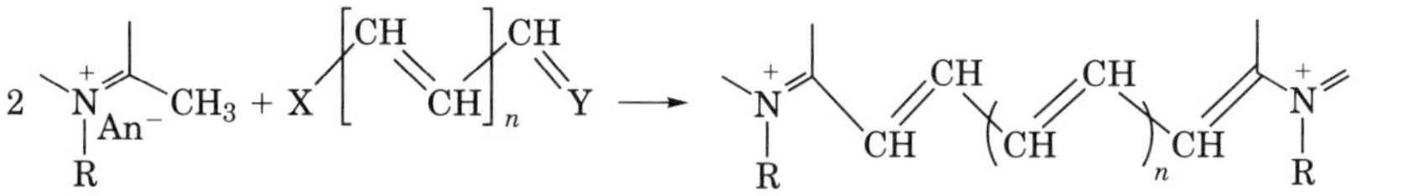

(7)

$H_2Y$, HAn, and HX also form. An alternative method of polymethine synthesis is by ring closure; the residues and chain are formed simultaneously (3), as in, for example, equation 8.

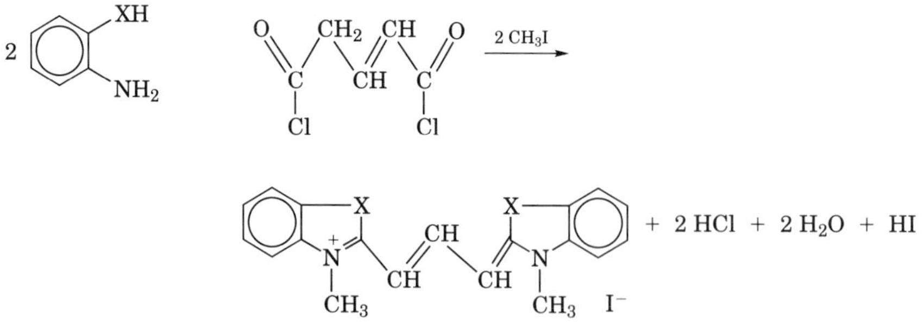

(8)

(24)

Another synthesis reaction is the hydrolytic cleavage of a suitable heterocyclic ring, for example, an N-methylpyridine salt, as shown in equation 9.

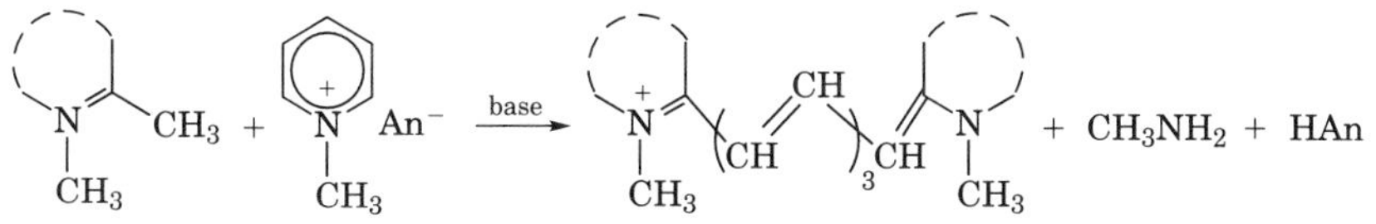

$$2 \begin{array}{c} & CH \\ N & CH_3 + N & An^- \xrightarrow{\text{base}} & N & CH & N \\ CH_3 & CH_3 & & CH_3 & CH_3 \end{array} + CH_3NH_2 + HAn \qquad (9)$$

The methods described above are used widely in the synthesis of dyes with new heterocyclic residues such as tropylium (**25**) [*76430-96-1*] (102), 2,4-diphenylpyrylium and its heteroanalogues (**26**) with a high effective length (103,104), or exotic systems (**27**) with long-wavelength local electron transitions (105) (Fig. 7).

Another route to long-wavelength dyes is through the synthesis of polymethines with a long chain, for example, tetra- or pentacarbocyanines (73). As chain-forming synthones, the alkoxyacetals of heptadiene- and nonatrienedials have been used, as in (**28**), where $n = 0$ or 1. Derivatives (**28**) interact with *N*-methylaniline in acid media to form the corresponding salts of di(*N*-methyl)anyls (**29**).

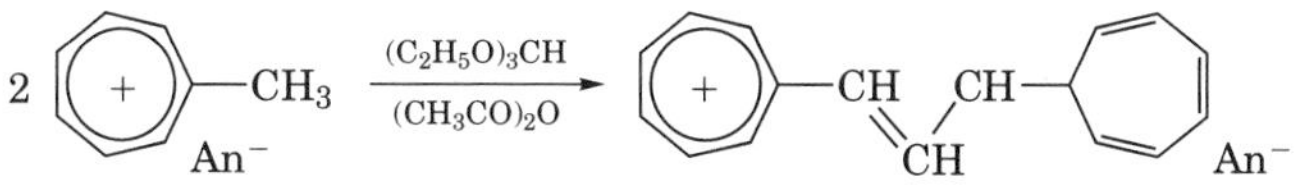

$$2 \quad \overset{+}{\bigcirc}-CH_3 \quad \xrightarrow[(CH_3CO)_2O]{(C_2H_5O)_3CH} \quad \overset{+}{\bigcirc}-CH \quad CH-\bigcirc \quad An^-$$

**(25)**

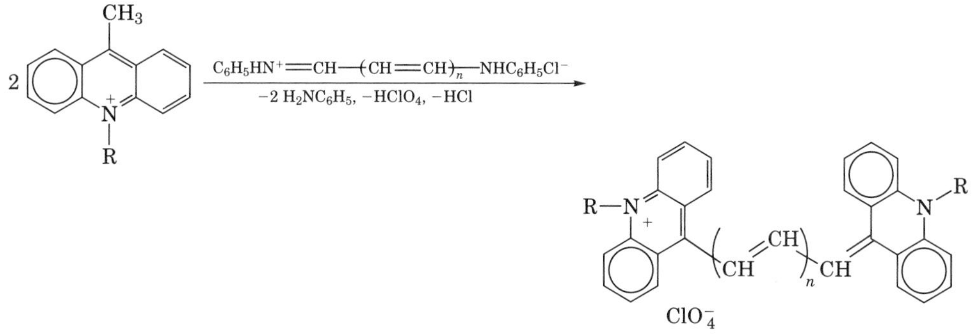

**(26)**

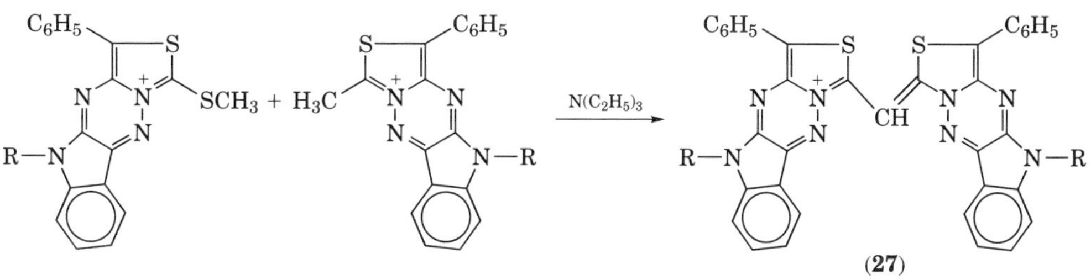

**(27)**

**Fig. 7.** Synthesis of a number of polymethine dyes.

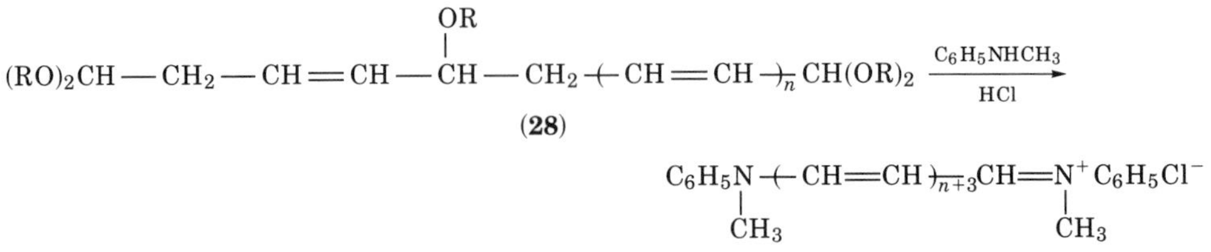

$$\underset{\textbf{(28)}}{(RO)_2CH-CH_2-CH=CH-\overset{\overset{\displaystyle OR}{|}}{CH}-CH_2 \!\!+\!\! CH=CH \!\!\xrightarrow{}_{\!\!n} CH(OR)_2} \xrightarrow[\text{HCl}]{C_6H_5NHCH_3}$$

$$C_6H_5\underset{\overset{|}{CH_3}}{N} \!\!+\!\! CH=CH \!\!\xrightarrow{}_{\!\!\overline{n+3}} CH=\overset{+}{N} C_6H_5Cl^-$$

$$\underset{\textbf{(29)}}{\phantom{x}}$$

The salts (**29**) condense with quartenary salts of heterocyclic bases containing an activated methyl group to yield the polycarbocyanines (**30**), where $n = 4$ or 5 (73). Higher vinylogous dyes (**30**), hexa- and heptacarbocyanines ($n = 6$ or 7), have been synthesized by analogous methods (106).

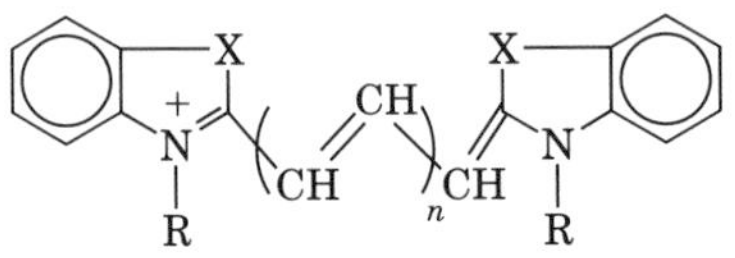

(**30**)

Substituents in the polymethine chromophore are mainly introduced by choosing the correspondingly substituted chain synthon (4,73,74,107).

**Dyes with Bridges in the Chromophore.** An important modification of dye constitution consists of bridging the chain by conjugated or unconjugated rings. This causes rigidization of the polymethines, enhances the stability, and increases fluorescence quantum efficiency.

The synthesis of polymethines with dimethylene bridges starts with alicyclic ketones (**31**), ketals (**32**), and enamines (X = $NR_2$), or enol ethers (X = OR) (**33**). They possess two activated centers, methine or methylene groups, which react with Vilsmeier's reagent to produce the corresponding dyes.

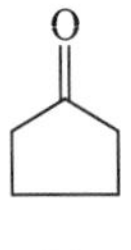

(**31**) 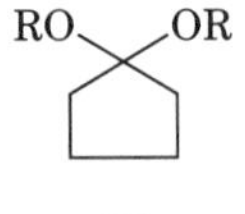              (**32**) 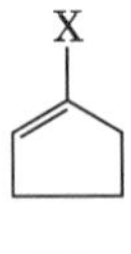              (**33**)

In the next step, the polymethine chain can be lengthened by interaction with the methylene bases of appropriate heterocyclic compounds (3).

Polymethines with trimethylene bridges have been obtained by interaction of the corresponding salts with 2,6-dimethoxy-1,4-dihydrobenzene (**34**) (74), for example, as follows, where Y = O, S, or Se.

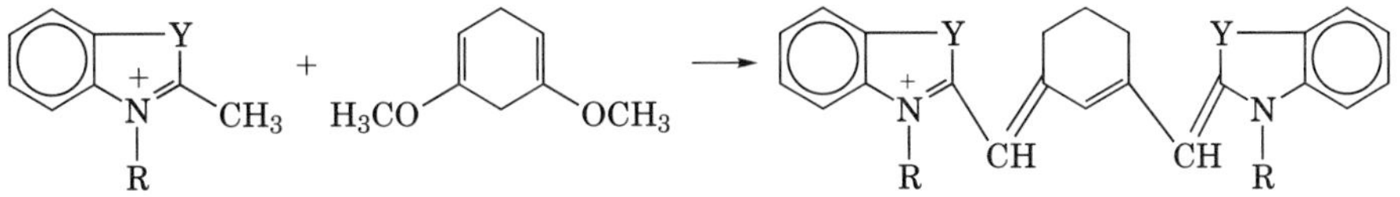

(**34**)

The reaction of methoxy-substituted 1,4-dihydroaromatic systems is a general one. Other condensed systems react in a similar manner, for example, 3,6-dimethoxy-1,4,5,8-tetrahydronaphthalene and derivatives of anthracene (**35**) and xanthene (**36**) (74). The proposed method enables synthesis of the tri- and tetracarbocyanines where the whole chromophore is integrated into a rigidizing skeleton. Asymmetrical polymethines can also be obtained similarly.

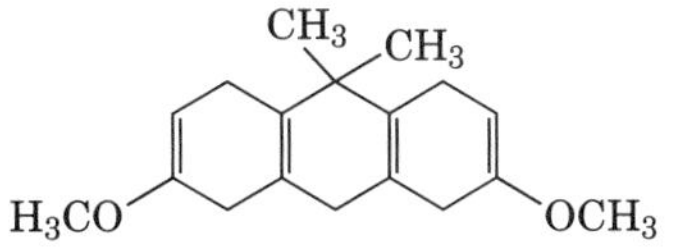

(**35**) 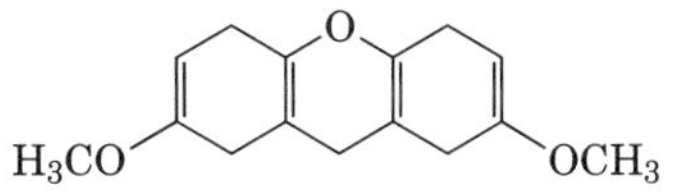                 (**36**)

**Dyes with Branched Chains.** Symmetrical trinuclear dyes with the shortest chain branches, [1.1.1]tetramethines (**37**), have been prepared by reaction of the corresponding heterocyclic bases with tetrachloro- or tetrabromomethane (33).

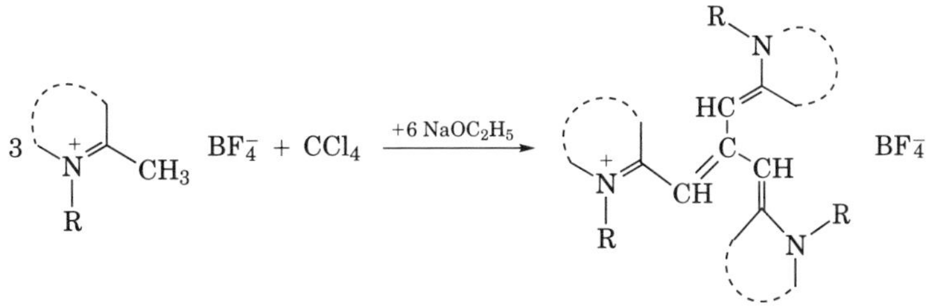

(**37**)

Higher methinylogous trinuclear symmetrical [2.2.2]heptamethinecyanines (**38**) are synthesized by the interaction of appropriate residue-forming synthons, which contain an active center with triformylmethane as a branched chain-forming synthon (33,36):

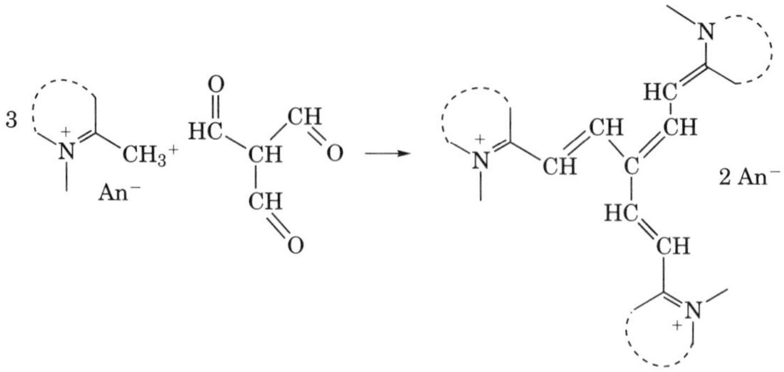

(38)

Asymmetrical dyes can be obtained by condensation of the γ-formyl-substituted dinuclear polymethine with the corresponding heterocyclic salt.

Reaction of Fisher's base and its quaternary salt with the trisacetal leads to the only synthesized [3.3.3]decamethine dye (33). Higher methinylogous trinuclear dyes have not been obtained.

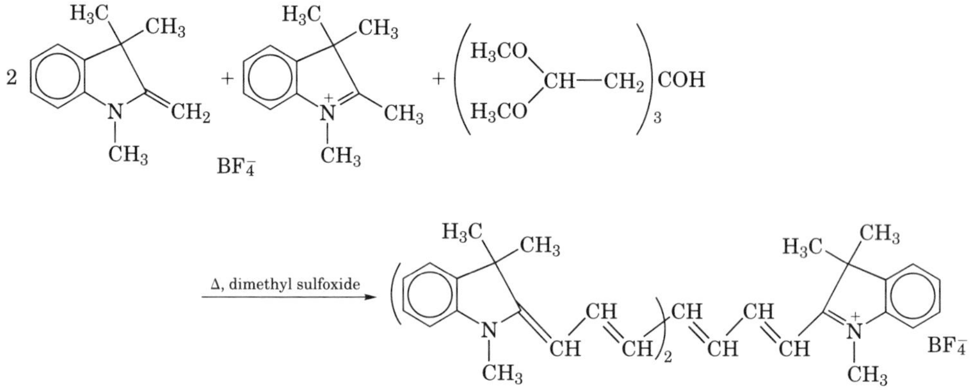

(39)

## Molecular Design

Extension of the applications of polymethine dyes has required special spectral and other characteristics. As a rule, the search for and synthesis of promising new compounds having desired properties imply the preliminary estimation of their most important parameters on the basis of elaborated theoretical conceptions. Thus, an effective way of governing electron properties consists of the variation of molecular topology of polymethines.

The first encouraging results for engineered dyes having desired ground- and excited-state properties have been achieved and summarized (108). For example, to design effective spectral sensitizers, it is necessary to engineer dyes having special positions of their frontier levels, desired wavelengths of the absorption band, and special thermodynamic stability after light excitation.

Also, using dyes as laser media or passive mode-locked compounds requires numerous special parameters, the most important of which are the band position and bandwidth of absorption and fluorescence, the luminiscence quantum efficiency, the Stokes shift, the possibility of photoisomerization, chemical stability, and photostability. Applications of PMDs in other technical or scientific areas have additional special requirements.

**Electron-Transition Energies.**  According to equation 2, the desired wavelength of absorption or fluorescence band maximum can be provided mainly in two ways: variation of the length of the chain, and introduction of end groups with appropriate effective length $L$. However, chain lengthening of infrared dyes tends to decrease the stability of the molecule because of the twisting of molecular fragments and of photoisomerization. Hence, a more practical way of deepening the color of polymethines is to construct dyes with residues that have high effective length $L$, for example, benz[$c,d$]indole ($L = 6.50$) (109), 2,6-diphenylpyrylium ($L = 6.07$) or its S analogue ($L = 7.24$) and Se analogue ($L = 7.94$) (27), and thiazolopyridinium ($L = 6.89$) and its derivates (110). Also, an absorption band maximum can be shifted bathochromically if the introduced end groups have their own deep color; then interaction of local and polymethine chromophores can lead to an additional shift as observed in polymethine derivatives of azaazulene residues (111). Comparatively smaller spectral effects can be reached by introducing substituents in different positions of both chain and end groups (3).

**Electron Level Position.**  One essential condition of spectral sensitization by electron transfer is that the LUMO of the dye be positioned above the bottom of the conduction band, eg, $\epsilon_e > |-3.23 \text{ eV}|$ in AgBr or $\epsilon_e > |-4.25 \text{ eV}|$ in ZnO (108). To provide the desired frontier level position respectively to the valence and conduction bands of the semiconductor, it is necessary to use a polymethine with suitable electron-donor ability $\varphi_0$. Increasing the parameter $\varphi_0$ leads to the frontier level shift up, and vice versa. Chain lengthening is known to be accompanied by a decrease of LUMO energy and hence by a decrease of sensitization properties. As a result, it is necessary to use dyes with high electron-donor ability for sensitization in the near-ir. The desired value of $\varphi_0$ can be provided by end groups with the needed topological index $\Phi_0$ or suitable substituents (112).

**Stability.**  The most stable molecules among dyes having the same absorption region are those with middle electron donor ability: $\varphi_0 = 45°$. Then all carbon−carbon bonds within the chain are completely equalized in both the ground and the excited states. Introducing terminal groups with high $\Phi_0$ leads to an increase of the negative charges at odd-carbon atoms within the chain; those dyes are sensitive to electrophile reactions. On the other hand, polymethines containing end groups with low index $\Phi_0$ are sensitive to nucleophiles. The stability of long dye molecules can also be increased by introducing polymethylene bridges.

**Other Characteristics.**  Other spectral and chemical parameters of dyes can be optimized by using suitable molecular fragments. Thus, the band

**Table 1. Commercial Prices of Representative Polymethine Dyes**[a]

| Name[b] | CAS Registry Number | Price, 1994–1995 |
|---|---|---|
| 1,1,3,3,3,3-hexamethylindotricarbocyanine iodide[c] | [19764-96-6] | $14.30/100 mg |
| | | $45.80/500 mg |
| 3,3′-diethyloxadicarbocyanine iodide[d] (DODCI) | [14806-50-9] | $28.10/500 mg |
| Stains-all[e] | [7423-31-6] | $11.85/250 mg |
| | | $32.75/1 g |
| IR-786[f] | [102185-03-5] | $11.90/100 mg |
| | | $40.00/500 mg |
| IR-27[g] | [83592-28-3] | $20.55/25 mg |
| | | $56.20/100 mg |
| IR-140[h] | [53655-17-7] | $24.45/100 mg |
| | | $69.95/500 mg |
| 5-cyano-2-(3-(5-cyano-1,3-diethyl-1,3-dihydro-2H-benzimidazol-2-ulidene)-1-propenyl)-1-ethyl-3-(4-sulfobutyl)-1H-benzimidazolium hydroxide (inner salt) | [32634-36-9] | $20.55/100 mg |
| 1-(3-sulfopropyl)-2-(2-((1-(3-sulfopropyl)-naphtho(1,2-d)thiazol-2(1H)-ylidene)methyl)-1-butenylnaphtho(1,2-d)thiazolium hydroxide (inner salt, triethylammonium salt) | [23216-67-3] | $44.30/250 mg |
| | | $18.40/100 mg |
| 2,4-di-3-guaiazulenyl-1,3-dihydroxycyclobutenediyilium dihydroxide (bis(inner salt)) | [72939-79-8] | $21.60/250 mg |
| | | $63.75/1 g |

[a]Ref. 113.
[b]Common names or trade names are given if necessary or appear as footnotes. Systematic names are used if necessary or appear as footnotes.
[c]2-(7-(1,3-Dihydro-1,3,3-trimethyl-2H-indol-2-ylidene)-1,3,5-heptotrienyl)-1,3,3-trimethyl-3H-indolium iodide.
[d](3-Ethyl-2(5-(3-ethyl-2-benzoxazolinylidene)-1,3-pentadienyl)benzoxazolium iodide.
[e]1-Ethyl-2-(3-(1-ethylnaphtho(1,2-d)thiazolin-2-ylidene)-2-methylpropenyl)naphtho-(1,2-d)thiazolium bromide.
[f]2-(2-Chloro-3-((1,3-dihydro-1,3,3-trimethyl-2H-indol-2-ylidene)ethylidene)-1-cyclohexen-1-ylethenyl)-1,3,3-trimethylindolium perchlorate.
[g]4-(2-(2-Chloro-3-(2-phenyl-4H-1-benzopyran-4-ylidene)ethylidene)-1-cyclohexen-1-yl)-2-phenyl-1-benzopyrylium perchlorate.
[h]5,5-Dichloro-11-diphenylamino-3,3-diethyl-10,12-ethylenethiatricarbocyanine perchlorate.

absorption width and Stokes shift appear to depend on the difference of $|45° - \Phi_0|$ in symmetrical dyes and the difference of $|\Phi_{01} - \Phi_{02}|$ in asymmetrical polymethines (10). Additional color deepening can be achieved by introducing end groups with long-wavelength local electron transitions. Short-wavelength absorptions also depend on local chromophores.

## Economic Aspects

Numerous patents and papers demonstrate a wide spectrum of applications for the polymethine dyes (3,6,11–14,34,58,87–92), particularly in special purposes such as in active laser media or passive mode-locking components, photographic sensitizers, and initiators of polymerization reactions. As a rule, polymethine manufacture is needed in significantly smaller quantities than dyes of other classes. On the other hand, special applications demand a high degree of polymethine dye purity. Also, typical syntheses include multistep, complicated reactions with low yields. Dyes may also be toxic; some solvents can provide transport of dye molecules through skin. Consequently, polymethine dyes are comparatively high in price. Some important representatives are given in Table 1.

The main suppliers and manufacturers of polymethine dyes are Aldrich Chemical Company, Eastman Organic Chemicals (U.S.), Japanese Institute for Photosensitizing Dyes, NK Dyes (Japan), Riedel deHaen (Germany), Institute of Organic Chemistry of the National Academy of Sciences (Ukraine), and NIIKhim-FotoProekt (Russia).

## BIBLIOGRAPHY

"Polymethine Dyes" in *ECT* 2nd ed., Vol. 16, pp. 282–300, by A. J. Cofrancesco and L. C. Hensley, GAF Corp.; in *ECT* 3rd ed., Vol. 18, pp. 848–874, by D. M. Sturmer and D. R. Diehl, Eastman Kodak Co.

1. S. Daehne, *Science*, **199**, 1163 (1978).
2. W. Koenig, *Ber. Dtsch. Chem. Ges.* **55**, 3306 (1922).
3. N. Tyutyulkov and co-workers, *Polymethine Dyes: Structure and Properties*, St. Kliment Ohridski University Press, Sofia, Bulgaria, 1991.
4. F. M. Hamer, *The Cyanine Dyes and Related Compounds*, Interscience Publishers, New York, 1964.
5. S. F. Mason, in K. Venkataraman, ed., *The Chemistry of Synthetic Dyes*, Vol. 3, Academic Press, Inc., New York, 1970, p. 169.
6. J. Fabian and H. Hartmann, *Light Absorption of Organic Colorants*, Springer-Verlag, Berlin, Germany, 1980.
7. L. G. S. Brooker, *Rev. Mod. Phys.* **25**, 275 (1942).
8. J. Griffiths, *Colour and Constitution of Organic Molecules*, Academic Press, Inc., London, 1976.
9. A. I. Kiprianov, *Colour and Constitution of Cyanine Dyes*, Naukova Dumka, Kiev, CIS, 1979.
10. A. D. Kachkovski, *Colour and Constitution of Polymethine Dyes*, Naukova dumka, Kiev, CIS, 1989.
11. L. G. S. Brooker, in C. E. K. Mees and T. H. James, ed., *The Theory of the Photographic Process*, 3rd ed., Macmillan Co., New York, 1972, p. 198.

12. M. Maeda, *Laser Dyes: Properties of Organic Compounds for Dye Laser*, Academic Press, Inc., New York, 1984.
13. F. J. Duarte and L. W. Hillmann, eds., *Dye Laser Principles with Applications*, Academic Press, Inc., New York, 1990.
14. J. C. Bevington, *Radical Polymerization*, J. Wiley & Sons, Inc., New York, 1966.
15. G. G. Dyadyusha, *Ukrain. Khim. Zurn.* **30**, 929 (1964).
16. A. D. Kachkovski, G. G. Dyadyusha, and M. L. Dekhtyar, *Dyes Pigments*, **15**, 191 (1991).
17. A. D. Kachkovski and M. L. Dekhtyar, *Dyes Pigments*, **22**, 83 (1993).
18. S. Hunig and H. Berneth, *Topics Curr. Chem.* **92**, 1 (1980).
19. F. Baer and H. Oehling, *Org. Magn. Reson.* **6**, 421 (1974).
20. A. V. Shpakov and co-workers, *Khimia Geterocyklich. Soed.*, 1638 (1993).
21. G. E. Ficken, in K. Venkataraman, ed., *The Chemistry of Synthetic Dyes*, Vol. 4, Academic Press, Inc., New York, 1971, p. 211.
22. S. Kulpe and B. Schulz, *Kristall. Technik*, **11**, 707 (1976).
23. S. Kulpe and co-workers, *Cryst. Res. Techn.* **22**, 375 (1987).
24. A. D. Kachkovski and co-workers, *Inf. J. Rec. Mater.* **20**, 569 (1993).
25. G. G. Dyadyusha, I. V. Repyakh, and A. D. Kachkovski, *Teoret. Eksperim. Khimia.* **20**, 398 (1984).
26. A. D. Kachkovski, *Dyes Pigments*, **24**, 171 (1994).
27. A. D. Kachkovski and co-workers, *Dyes Pigments*, **16**, 137 (1991).
28. Y. P. Kovtun, N. N. Romanov, and A. D. Kachkovski, *Khimia Geterotsicl. Soed.*, 985 (1988).
29. W. Koenig, *Z. Prakt. Chem.* **112**, 1 (1925).
30. S. Hunig, *Optische Anregung Organischer Systeme*, II Intern. Farbensymp., Verlag Chemie, Weinheim, Germany, 1966, p. 184.
31. G. G. Dyadyusha and co-workers, *Teoret. Eksperim. Khimia.* **15**, 412 (1979).
32. Y. A. Nesterenko and co-workers, *Khimia Geterocyclic. Soed.*, 252 (1990).
33. C. Reinhardt, *Studies Org. Chem.* **31**, 3 (1987).
34. R. Steiger and J. F. Reber, *Photo. Sci. Eng.* **25**, 127 (1981).
35. W. Grahn, *Tetrahedron*, **32**, 1931 (1976).
36. Y. L. Bricks and co-workers, *Dyes Pigments*, **8**, 353 (1987).
37. C. Reinhardt and co-workers, *Chem. Ber.* **116**, 1982 (1983).
38. G. G. Dyadyusha, A. D. Kachkovski, and Y. L. Bricks, *Ukrain. Khim. Zhurn.* **57**, 1152 (1991).
39. H. Hartmann, *J. Signal AM*, **7**, 101 (1979).
40. M. J. S. Dewar, *J. Chem. Soc.*, 2339 (1950).
41. T. Foerster, *Z. Phys. Chem.* **48**, (1940) 12.
42. E. B. Knott, *J. Chem. Soc.*, 1024 (1951).
43. A. I. Kiprianov, G. G. Dyadyusha, and F. A. Mikhaylenko, *Uspekhi Khimii* **35**, 823 (1966).
44. D. Mee, *J. Am. Chem. Soc.* **96**, 4712 (1974).
45. A. I. Kiprianov and I. L. Mushkalo, *Zhurn. Organ. Khimii*, **1**, 744 (1965).
46. A. I. Kiprianov, *Uspekhi Khimii*, **40**, 1283 (1971).
47. A. I. Kiprianov and G. G. Dyadyusha, *Ukrain. Khim. Zhurn.* **35**, 608 (1969).
48. D. I. Smith, *Photo. Sci. Eng.* **18**, 309 (1974).
49. W. West and B. H. Carroll, in C. E. K. Mees and T. H. James, eds., *The Theory of the Photographic Process*, Macmillan Co., New York, 1969, p. 233.
50. F. Dietz, *J. Signal AM*, **1**, 237 (1973).
51. V. I. Yuzhakov, *Uspekhi Khimii*, **61**, 1114 (1992).
52. G. Scheibe, *Angew. Chem.* **49**, 563 (1936).
53. E. Jelly, *Nature*, **139**, 1009 (1936).

54. G. Scheibe and I. Kandler, *Naturwiss.* **26**, 412 (1938).

55. G. S. Levinson, W. T. Simpson, and W. Curtis, *J. Am. Chem. Soc.* **79**, 4314 (1957).

56. M. V. Bondar and co-workers, *Kvant. Elektronika,* **11**, 464 (1984).

57. G. G. Dyadyusha and co-workers, *Ukrain. Khim. Zhurn.* **51**, 298 (1985).

58. A. A. Ishchenko, *Uspekhi Khimii,* **60**, 1708 (1991).

59. C. Reinhardt, *Solvent Effects in Organic Chemistry*, Verlag Chemie, Weinheim, Germany, 1979.

60. A. P. Marchetti and M. Scozzafava, *J. Chem. Phys.* **65**, 2382 (1976).

61. E. Lippert and co-workers, *Angew. Chem.* **73**, 695 (1961).

62. G. Scheibe and co-workers, *Angew. Chem.* **76**, 270 (1964).

63. R. Radeglia, *J. Prakt. Chem.* **320**, 539 (1978).

64. A. I. Kiprianov, S. G. Fridman, and L. S. Pupko, in Ref. 9, p. 501.

65. R. Steiger, *Photo. Sci. Eng.* **25**, 10 (1981).

66. J. Kuchera and Z. Arnold, *Collect. Czechosl. Chem. Commun.* **32**, 1707 (1967).

67. Z. Arnold, *Collect. Czechosl. Chem. Commun.* **25**, 1308 (1960).

68. J. Becher, *Synthesis*, 589 (1980).

69. A. F. Vompe and co-workers, *Zhurn. Organ. Khimii* **21**, 584 (1985).

70. D. Lloyd and co-workers, *Angew. Chem.* **93**, 193 (1981).

71. Z. Arnold, *Collect. Czechosl. Chem. Commun.* **24**, 3051 (1959).

72. S. S. Malhotra and M. C. Whiting, *J. Chem. Soc.*, 3812 (1960).

73. A. F. Vompe, *Uspekhi Nauchn. Fotografii* **22**, 5 (1984).

74. A. I. Tolmachev, Y. L. Slominski, and M. A. Kudinova, *Uspekhi Nauchn. Fotografii* **22**, 12 (1984).

75. J. Liebsher and H. Hartmann, *Synthesis*, 241 (1979).

76. V. A. Usov, L. V. Timokhina, and M. G. Voronkov, *Uspekhi Khimii*, **55**, 1761 (1986).

77. J. R. Lenhardt, B. R. Hein, and A. A. Muenter, *J. Phys. Chem.* **97**, 8269 (1993).

78. L. Zechmeister and J. H. Pinkard, *Experomentia*, **9**, 16 (1953).

79. V. A. Kusmin, *Uspekhi Nauchn. Fotografii*, **22**, 90 (1984).

80. C. Ruliere, *Chem. Phys. Lett.* **30**, 352 (1975).

81. S. K. Rentsch, *Chem. Phys.* **69**, 81 (1982).

82. E. N. Kalitievskaja, T. K. Razumova, and G. M. Rubanova, *Opt. Spektrosk.* **60**, 272 (1986).

83. G. R. Fleming and co-workers, *Chem. Phys. Lett.* **49**, 1 (1977).

84. G. H. Brown, *Photochromism*, Vol. 3, John Wiley & Sons, Inc., New York, 1981.

85. D. Gaude and co-workers, *Bull. Soc. Chim. France*, **2**, 489 (1981).

86. F. Zerbetto, S. Monti, and G. Orlandi, *J. Chem. Soc. Faraday 2* **80**, 1513 (1980).

87. C. Lenoble and R. S. Becker, *J. Photochem.* **34**, 83 (1984).

88. S. A. Kyrsanov and M. V. Alfimov, *Laser Chem.* **4**, 129 (1984).

89. E. Fisher, *Topics Curr. Chem.* **73**, 125 (1978).

90. B. I. Shapiro, *Zh. Nauchn. Prikl. Fotogr. Kinematogr.* **22**, 143 (1977).

91. V. I. Malyshev, *Uspekhi Nauchn. Fotografii* **22**, 177 (1984).

92. R. W. Gurney and N. F. Mott, *Proc. Roy. Soc.* **A164**, 151 (1938).

93. T. Tani, S. Kikuchi, and K. Honda, *Photo. Sci. Eng.* **12**, 80 (1968).

94. S. H. Ehrlich, *Photo. Sci. Eng.* **18**, 179 (1974).

95. T. Tani, *Photo. Sci. Eng.* **15**, 161 (1971).

96. W. West and B. H. Carroll, *J. Chem. Phys.* **19**, 417 (1951).

97. B. I. Shapiro and co-workers, *J. Inf. Rec. Mater.* **20**, 265 (1992).

98. B. Soffer, *J. Appl. Phys.* **35**, 255 (1964).

99. B. I. Stepanov, A. N. Rubinov, and V. A. Mostovnikov, *Pisma v Zh. Eksperim. Teoret. Fisiki* **5**, 144 (1967).

100. M. Mack, *Appl. Phys. Lett.* **15**, 166 (1969).

101. S. Chatterjiee and co-workers, *J. Am. Chem. Soc.* **110**, 2326 (1988).

102. K. Hofner, H. W. Riedel, and M. Danielish, *Angew. Chem. Int. Ed. Engl.* **2**, 215 (1969).

103. R. Wizinger and U. Arni, *Ber.* **92**, 2309 (1959).

104. A. I. Tolmachev, N. A. Derevjanko, and M. A. Kudinova, *Khimia Geterocyclich. Soed.*, 617 (1975).

105. Y. P. Kovtun and N. N. Romanov, *Khimia Geterocyclich. Soed.*, 1547 (1988).

106. S. M. Makin and A. I. Pomogaev, *Uspekhi Nauchn. Fotografii* **22**, 27 (1982).

107. S. M. Makin, R. I. Kruglinova, and O. V. Kharitonova, *Zh. Org. Khimii* **24**, 1987 (1988).

108. S. Daehne, *Proc. Indian Acad. Sci. (Chem. Sci.)* **104**, 311 (1992).

109. F. A. Mikhailenko and co-workers, *Zhurn. Organ. Khimii* **18**, 435 (1982).

110. G. G. Dyadyusha and co-workers, *Dyes Pigments* **4**, 179 (1983).

111. Y. L. Bricks and N. N. Romanov, *Khimia Geterotsicl. Soed.*, 1432 (1991).

112. B. I. Shapiro and co-workers, *J. Inf. Rec. Mater.* **20**, 265 (1992).

113. *Catalog Handbook of Fine Chemicals, 1994-1995*, Aldrich Chemical Co. Inc., Milwaukee, Wis., 1994.

*General Reference*

J. Fabian and S. Daehne, *J. Mol. Structure (Theochem.)* **92**, 217 (1983).

ALEXY D. KACHKOVSKI
Institute of Organic Chemistry
National Academy of Sciences of the Ukraine

# POLYMETHYLBENZENES

Polymethylbenzenes (PMBs) are aromatic compounds that contain a benzene ring and three to six methyl group substituents (for the lower homologues see BENZENE; TOLUENE; XYLENES AND ETHYLBENZENE). Included are the trimethylbenzenes, $C_9H_{12}$ (mesitylene (**1**), pseudocumene (**2**), and hemimellitene (**3**)), the tetramethylbenzenes, $C_{10}H_{14}$ (durene (**4**), isodurene (**5**), and prehnitene (**6**)), pentamethylbenzene, $C_{11}H_{16}$ (**7**), and hexamethylbenzene, $C_{12}H_{18}$ (**8**). The PMBs are primarily basic building blocks for more complex chemical intermediates.

## Physical Properties

The structures of the eight PMBs are shown here and their physical and thermodynamic properties are given in Table 1.

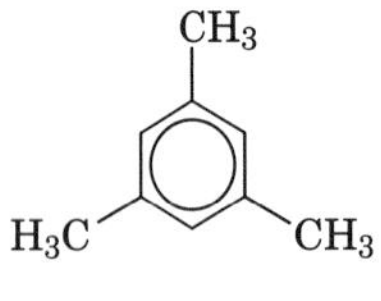

|   (1)   |   (2)   |   (3)   |   (4)   |

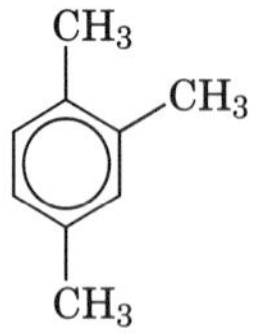

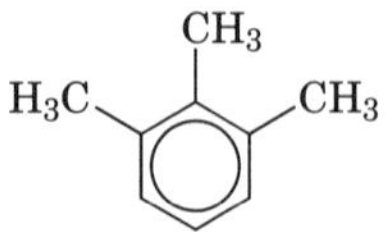

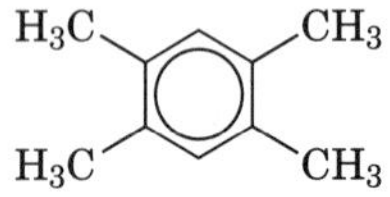

**Table 1. Physical and Thermodynamic Properties of Polymethylbenzenes**[a]

| Property | 1,3,5-Tri-methyl-benzene | 1,2,4-Tri-methyl-benzene | 1,2,3-Tri-methyl-benzene | 1,2,4,5-Tetra-methyl-benzene | 1,2,3,5-Tetra-methyl-benzene | 1,2,3,4-Tetra-methyl-benzene | Penta-methyl-benzene | Hexa-methyl-benzene |
|---|---|---|---|---|---|---|---|---|
| | | | | Systematic (benzene) name | | | | |
| CAS Registry Number | [108-67-8] | [95-63-6] | [526-73-8] | [95-93-2] | [527-53-7] | [488-23-3] | [700-12-9] | [87-85-4] |
| mol wt | 120.194 | 120.194 | 120.194 | 134.221 | 134.221 | 134.221 | 148.248 | 162.275 |
| bp, °C | 164.74 | 169.38 | 176.12 | 196.80 | 198.00 | 205.04 | 231.9 | 263.8 |
| flash point, °C | 43.0 | 46.0 | 51.0 | 67.0 | 68.0 | 73.0 | | |
| density, $g/cm^3$ | | | | | | | | |
| at 20°C | 0.8651 | 0.8758 | 0.8944 | $0.8875^b$ | 0.8903 | 0.9052 | $0.917^c$ | solid |
| 25°C | 0.8611 | 0.8718 | 0.8905 | $0.8837^b$ | 0.8865 | 0.9015 | $0.913^c$ | solid |
| freezing point, °C in air at 101.3 $kPa^c$ | −44.694 | −43.881 | −25.344 | 79.240 | −23.689 | −6.229 | 54.35 | 165.7 |
| refractive index, $n_D$ at 25°C | 1.49684 | 1.50237 | 1.51150 | $1.5093^b$ | 1.5107 | 1.5181 | $1.525^b$ | solid |
| surface tension, mN/m (=dyn/cm), at 20°C | 28.84 | 29.72 | 31.28 | solid | 33.51 | 35.81 | solid | solid |
| critical temperature, °C | 364.20 | 376.02 | 391.32 | 401.85 | 405.85 | 416.55 | | |
| critical pressure, $kPa^c$ | 3127 | 3232 | 3454 | 2940 | 2860 | 2860 | | |
| critical volume, $cm^3/mol$ | 427 | 427 | 427 | 482 | 482 | 482 | | |
| heat of vaporization at bp, $kJ/mol^d$ | 39.0 | 39.2 | 40.0 | 45.52 | 43.81 | 45.02 | 45.1 | 48.2 |
| heat of formation at 25°C, liquid, $kJ/mol^d$ | −63.4 | −61.8 | −58.5 | $−119.87^e$ | −96.35 | −90.20 | $−135.1^e$ | $−171.5^e$ |
| heat of combustion, $kJ/mol^d$ at 25°C | 5193.1 | 5194.8 | 5198.0 | $5816.0^b$ | 5839.6 | 5845.7 | $6490.8^b$ | |
| dielectric constant, at 20°C | | 2.383 | 2.636 | | | | | |
| specific heat, $C_p$, liquid, at 25°C, $J/(mol \cdot K)^d$ | 200.5 | 214.9 | 216.4 | | 240.7 | 238.3 | | |

[a] Refs. 1–3.
[b] Supercooled liquid.
[c] To convert kPa to atm, divide by 101.3.
[d] To convert J to cal, divide by 4.184.
[e] Crystal.

(5)               (6)               (7)               (8)

## Chemical Properties

The PMBs, when treated with electrophilic reagents, show much higher reaction rates than the five lower molecular weight homologues (benzene, toluene, *o*-, *m*-, and *p*-xylene), because the benzene nucleus is highly activated by the attached methyl groups (Table 2). The PMBs have reaction rates for electrophilic substitution ranging from 7.6 times faster (sulfonylation of durene) to ca 607,000 times faster (nuclear chlorination of durene) than benzene. With rare exception, the PMBs react faster than toluene and the three isomeric dimethylbenzenes (xylenes).

The methyl groups direct the entering group primarily to the ortho and para positions (Table 3). The preferred site of attack by an electrophile on pseudocumene and hemimellitene is shown as follows (13); however, steric hindrance

**Table 2. Relative Reaction Rates for PMBs**

| PMB | Sulfonylation[a] | Acylation[b] | Chloromethylation[c] | Ring chlorination[d] |
|---|---|---|---|---|
| | | *Reaction conditions* | | |
| reagent | benzenesulfonyl chloride | benzoyl chloride | chloromethyl methyl ether[e] | chlorine |
| solvent | nitrobenzene | nitrobenzene | acetic acid | acetic acid |
| catalyst | $AlCl_3$ | $AlCl_3$ | none | none |
| temperature, °C | 25 | 25 | 65 | 25 |
| | | *Relative rate* | | |
| benzene | 1.0 | 1.0 | 1.0 | 1.0 |
| toluene | 9 | 150 | | 95 |
| *o*-xylene | 18 | 1,360 | | 1,140 |
| *m*-xylene | 22 | 3,910 | 24 | 48,700 |
| *p*-xylene | 19 | 140 | | 570 |
| pseudocumene | 31 | 7,600 | 46 | |
| mesitylene | 30 | 125,000 | 600 | |
| hemimellitene | 27 | 13,300 | | |
| durene | 8 | 11,000 | | 607,200 |
| isodurene | | 212,000 | | |
| prehnitene | | 35,500 | 850 | |
| pentamethyl- benzene | 25 | 139,000 | | |

[a]Ref. 4.   [b]Ref. 5.   [c]Refs. 6 and 7.   [d]Ref. 8.   [e]Aqueous formaldehyde or paraformaldehyde in aqueous HCl is preferred.

**Table 3. PMB Reactions and Isomer Distribution for Monosubstitution, %[a]**

| Reaction | Toluene | | o-Xylene | | m-Xylene | | Pseudocumene | | Hemimellitene | |
|---|---|---|---|---|---|---|---|---|---|---|
| orientation[b] | 1,4 | 1,2 | 1,2,4 | 1,2,3 | 1,3,4 | 1,2,3 | 1,2,4,5 | 1,2,3,4 | 1,2,3,4 | 1,2,3,5 |
| chloromethyl- | 54 | 43 | 72 | 28 | 89 | 11 | 80 | 18 | 95 | 5 |
| ation |  |  |  |  |  |  |  |  |  |  |
| nitration | 38 | 58 |  |  |  |  | mainly |  |  |  |
| chlorination | 42 | 58 |  |  | mainly |  |  |  |  |  |
| sulfonation | 62 | 32 |  |  |  |  | mainly |  |  |  |
| bromination | 63 | 37 |  |  | mainly |  |  |  |  |  |

[a]Refs. 9–12.
[b]The bold number shows the position on the parent (poly)methylbenzene where the incoming group will substitute.

(2)                          (3)

can cause a shift such as a *tert*-butylation of hemimellitene predominantly at the 5-position. Mesitylene, all three tetramethylbenzenes, and pentamethylbenzene can only form one mono- and one disubstituted isomer (except pentamethylbenzene). Hence, high purity derivatives are possible.

The relative basicity of PMBs is summarized in Table 4, which shows that basicity increases with increasing number of methyl groups, and that methyl groups in the 1,3-position to one another greatly influence basicity.

**Table 4. Relative Basicity of the PMBs Toward HF–BF$_3$[a]**

| Hydrocarbon | Positions of methyl group | Number of pairs in 1,3-positions | Basicity[b] |
|---|---|---|---|
| toluene | 1- | 0 | ca 0.01 |
| o-xylene | 1,2- | 0 | 3 |
| p-xylene | 1,4- | 0 | 1 |
| m-xylene | 1,3- | 1 | 9 |
| hemimellitene | 1,2,3- | 1 | ca 18 |
| pseudocumene | 1,2,4- | 1 | 18 |
| mesitylene | 1,3,5- | 3 | 1,400 |
| prehnitene | 1,2,3,4- | 2 | 85 |
| durene | 1,2,4,5- | 2 | 60 |
| isodurene | 1,2,3,5- | 3 | 2,800 |
| pentamethylbenzene | 1,2,3,4,5- | 4 | 4,350 |
| hexamethylbenzene | 1,2,3,4,5,6- | 6 | 44,500 |

[a]Ref. 14.
[b]Relative to *p*-xylene.

## Manufacture

High purity mesitylene, hemimellitene, and durene are often produced syn-thetically, whereas pseudocumene is obtained from extracted $C_9$ reformate by superfractionation. The composition of a typical extracted $C_9$ reformate and the boiling points of the nine $C_9$ isomers present are shown in Table 5. Pseudo-cumene is separated in high purity (>98%) by superfractionation alone, whereas mesitylene, hemimellitene, and durene cannot be cleanly separated because of the presence of close boiling compounds, eg, 2-ethyltoluene, indane, and iso-durene, respectively.

Exploiting the relative basicity of the xylene isomers, commercial units employ superacids, typically $HF-BF_3$, as the acid complexing agent for the separation of $m$-xylene (feedstock for isophthalic acid) (15). Amoco produces high purity $m$-xylene at its Texas City facility using the $HF-BF_3$ process (see BTX PROCESSING). Similar processes can be used for the separation of high purity mesitylene and isodurene from their $C_9$ and $C_{10}$ isomers, respectively.

Selective absorption of durene from heavy gasoline (bp 150–225°C) is possible using a version of UOP's Sorbex technology where the X zeolite is made selective for durene by replacing the exchangeable sodium cations with lithium ions (16).

Mesitylene can be synthesized from acetone by catalytic dehydrocyclization (17). Similarly, cyclotrimerization of acetylenes has produced PMBs such as hexamethylbenzene (18). Durene has been recovered from Methanex's methanol-to-gasoline (MTG) plant in New Zealand (19).

Koch Chemical Company is the only U.S. supplier of all PMBs (except hexamethylbenzene). Its process has the flexibility of producing isodurene, prehnitene, and pentamethylbenzene, should a market develop. Koch's primary

**Table 5. Distribution of $C_9$ Aromatic Compounds in Reformate[a]**

| Compound | CAS Registry Number | Boiling point, °C | Freezing point, °C | Wt %[b] | Wt %[c] |
|---|---|---|---|---|---|
| isopropylbenzene (cumene) | [98-82-8] | 152 | −96 | 0.6 | 2.2 |
| $n$-propylbenzene | [103-65-1] | 159 | −99 | 5.2 | 6.2 |
| 3-ethyltoluene | [620-14-4] | 161 | −95 | 17.4 | 18.4 |
| 4-ethyltoluene | [622-96-8] | 162 | −62 | 8.6 | 7.9 |
| 1,3,5-trimethylbenzene (mesitylene) | [108-67-8] | 165 | −45 | 7.6 | 8.9 |
| 2-ethyltoluene | [611-14-3] | 165 | −81 | 9.1 | 7.5 |
| 1,2,4-trimethylbenzene (pseudocumene) | [95-63-6] | 169 | −44 | 41.3 | 38.1 |
| 1,2,3-trimethylbenzene (hemimellitene) | [526-73-8] | 176 | −25 | 8.2 | 8.8 |
| indane | [496-11-7] | 178 | −51 | 2.0 | 2.0 |

[a]Based on the total $C_9$ aromatic compounds.
[b]Ref. 11. Solvent extracted to remove nonaromatics.
[c]Severe reforming (two refinery average, 1994).

process (20) is based on isomerization, alkylation, and disproportionation conducted in the presence of a Friedel-Crafts catalyst. For the synthesis of mesitylene and hemimellitene, pseudocumene is isomerized. If durene, isodurene, or prehnitene and pentamethylbenzene are desired, pseudocumene is alkylated with methyl chloride (see ALKYLATION; FRIEDEL-CRAFTS REACTIONS).

The thermodynamic equilibria are illustrated in Figures 1 and 2. Figure 1 shows the resulting composition after pure pseudocumene or a recycle mixture of $C_9$ PMBs is disproportionated with a strong Friedel-Crafts catalyst. At 127°C (400 K), the reactor effluent contains approximately 3% toluene, 21% xylenes, 44% $C_9$ PMBs, 29% $C_{10}$ PMBs, and 3% pentamethylbenzene. The equilibrium composition of the 44% $C_9$ PMB isomers is shown in Figure 2. Based on the values at 127°C, the distribution is 29.5% mesitylene, 66.0% pseudocumene, and 4.5% hemimellitene (Fig. 2). After separating mesitylene and hemimellitene by fractionation, toluene, xylenes, pseudocumene (recycle plus fresh), $C_{10}$ PMBs, and pentamethylbenzene are recycled to extinction.

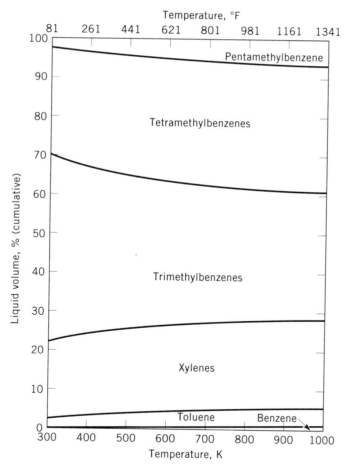

**Fig. 1.** Disproportionation of trimethylbenzenes. Composition cited in text derived by differences between curves at given temperature.

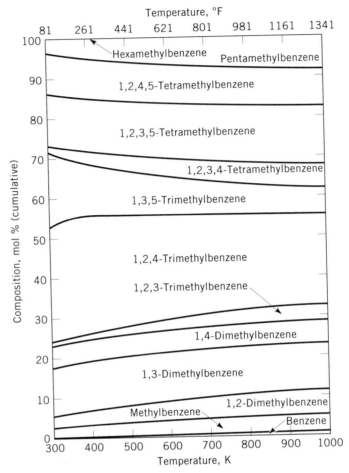

**Fig. 2.** Calculated equilibria among benzene and the $C_7$–$C_{12}$ methylated benzenes. Composition cited in text derived by differences between curves at given temperature.

If only durene is desired, pseudocumene is alkylated with methyl chloride to obtain a $C_{10}$ alkylate having a composition similar to that shown in Figure 3. The composition (at 127°C) is approximately 2% xylenes, 17% $C_9$ PMBs, 55% $C_{10}$ PMBs, 25.5% pentamethylbenzene, and 0.5% hexamethylbenzene. The isomer breakdown of the 55% $C_{10}$ PMBs is similar to the ratios shown in Figure 2 (37% durene, 52% isodurene, and 11% prehnitene). Because isodurene and durene boil very closely together, durene cannot be recovered in high purity by fractionation. However, the freezing points of durene and isodurene are 79 and −23.6°C, respectively. Consequently, the reactor effluent ($C_{10}$ PMBs alkylate) is distilled to give essentially a binary mixture of durene and isodurene which is fractionally crystallized to yield 96+% durene. The lighter and heavier PMBs present in the alkylate are separated by distillation (qv) and recycled to the reactor along with fresh pseudocumene and methyl chloride feed and the $C_{10}$ PMBs filtrate, which is rich in isodurene, from the crystallization unit.

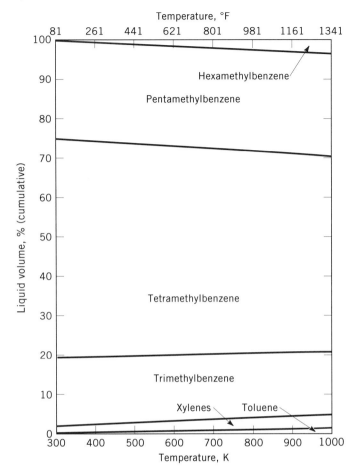

**Fig. 3.** Disproportionation of tetramethylbenzenes. Composition cited in text derived by differences between curves at given temperature.

Koch's process also permits recovery of prehnitene, isodurene, and pentamethylbenzene by fractionation. In the case of isodurene, only 85% purity is achieved when the filtrate from the durene crystallization unit is redistilled; durene is the principal impurity. Although hexamethylbenzene can be synthesized via the trimethylation of the readily available pseudocumene, its high melting point (105.5°C) presents more of a processing challenge.

## Production and Shipping

The only significant U.S. producer of PMBs is Koch Chemical Company. The only PMBs produced in any sizable quantity are pseudocumene, durene, mesitylene, and hemimellitene. Koch's production capability for pseudocumene is ca 35,000 t/yr; the combined production rate of the other PMBs is less than 5,000 t/yr.

Pseudocumene is shipped in barges, tank cars, tank trucks, isocontainers, and drums. Mesitylene is shipped in tank trucks, isocontainers, and drums, whereas durene is shipped molten in heated tank trucks, isocontainers, and occasionally as a cast solid in drums. Mesitylene, pseudocumene, and hemimellitene are classified as flammable liquids; the higher homologues are classified as combustible. The higher melting PMBs require additional precautions when handled in the molten state to avoid thermal burns. Detailed shipping and handling procedures are described in manufacturers' material safety data sheets (MSDS).

## Specifications

The typical properties of commercial PMBs are shown in Table 6.

**Table 6. Typical Properties of Commercial PMBs[a]**

|  | (1) | (2) | (3) | (4) | (5) | (6) |
|---|---|---|---|---|---|---|
| | | | *Composition, wt %* | | | |
| mesitylene | 98.5 | 0.5 | | | | |
| pseudocumene | 1.0 | 97.7 | 4.0 | | | |
| hemimellitene | nil | 0.5 | 90.5 | | | |
| durene | | | | 96.3 | 16.5 | 0.5 |
| isodurene | | | | 3.0 | 82.5 | 1.5 |
| prehnitene | | | | | | 90.5 |
| other aromatics | 0.5 | 1.2 | 5.5 | 1.5 | 1.0 | 2.5 |
| | | | *Distillation, °C* | | | |
| initial boiling point, °C | 164.5 | 168.9 | 175.0 | 194.0 | 197.0 | 203.0 |
| dry point | 166.5 | 170.6 | 176.8 | 197.5 | 200 | 206 |
| freezing point, °C | | | | 77.0 | | |
| color, Saybolt | +30 | +30 | +30 | snow white | +30 | +30 |
| sp gr 15/15°C | 0.8690 | 0.8800 | 0.8950 | | 0.8956 | 0.9042 |

[a]The composition of pentamethylbenzene is 95.5 wt % (**7**) and 4.5% other aromatics; its fp is 51°C and its color is snow white.

## Health and Safety Factors

The PMBs, as higher homologues of toluene and xylenes, are handled in a similar manner, even though their flash points are higher (see Table 1). Containers are tightly closed and use areas should be ventilated. Breathing vapors and contact with the skin should be avoided. Toxicity and primary irritation data are given in Table 7.

## Environmental Considerations

All the PMBs are listed on the U.S. EPA's Toxic Substances Control Act Non-Confidential Chemical Substances Inventory (Table 8). In the early to mid-1980s, pseudocumene, mesitylene, hemimellitene, and trimethylbenzene were covered by TSCA Section 8(a) Preliminary Assessment Information Rule (PAIR) reporting requirements (22) and by TSCA Section 8(d) for health and safety data (23). Mesitylene is the subject of a test rule; subacute oral toxicity and subchronic oral toxicity in rats were underway in 1994 (24). The Safe Drinking Water Act

**Table 7. Toxicological Properties**[a]

| Compound | Structure number | ACGIH TLV–TWA,[b] ppm | $LD_{50}$[c], mg/kg | Skin irritation, mg/24 h | Eye irritation, mg/24 h |
|---|---|---|---|---|---|
| mesitylene | (1) | 25 | | 20, moderate | 500, mild |
| pseudocumene | (2) | 25 | 5000 | | |
| hemimellitene | (3) | 25 | | | |
| trimethylbenzene[d] | | 25 | 8970 | 500, mild | 500, mild |
| durene | (4) | NE | 6989 | | |
| isodurene | (5) | NE | 5157 | | |
| prehnitene | (6) | NE | 6408 | 100, mild | |
| pentamethylbenzene | (7) | NE | | | |
| hexamethylbenzene | (8) | NE | | | |

[a]Ref. 21.
[b]NE = none established.
[c]Oral (rat).
[d]Trimethylbenzene is a generic listing, ie, mixed isomers.

**Table 8. PMB Regulatory Status**[a,b]

| PMB | CAS Registry Number | TSCA[a] listed | EINECS[a] number | Canadian status[b] |
|---|---|---|---|---|
| mesitylene | [108-67-8] | yes | 202-604-4 | DSL |
| pseudocumene | [95-63-6] | yes | 202-436-9 | DSL |
| hemimellitene | [526-73-8] | yes | 208-394-8 | DSL |
| trimethylbenzene[c] | [25551-13-7] | yes | 247-099-9 | DSL |
| durene | [95-93-2] | yes | 202-465-7 | DSL |
| isodurene | [527-53-7] | yes | 208-417-1 | DSL |
| prehnitene | [488-23-3] | yes | 207-673-1 | DSL |
| pentamethylbenzene | [700-12-9] | yes | 211-837-8 | NDSL |
| hexamethylbenzene | [87-85-4] | yes | 201-777-0 | DSL |

[a]All eight PMBs are listed on the U.S. EPA's TSCA Inventory (May 1, 1994 ed.) and on the European Communities' EINECS inventory (June 15, 1990).
[b]All but pentamethylbenzene are listed on the Canadian Domestic Substances List (DSL) (Apr. 6, 1994). Pentamethylbenzene is listed on the Canadian Nondomestic Substances List (NDSL) (Apr. 6, 1994) and as of July 1, 1994 was subject to New Substance Notification if manufactured in or imported into Canada.
[c]Trimethylbenzene is a generic listing, ie, mixed isomers.

(SDWA) allows monitoring for pseudocumene and mesitylene at the discretion of the State (25). Of the PMBs, only pseudocumene is subject to SARA Title III section 313 annual release reporting (26).

Other evaluations (27,28) of environmental considerations report vapor pressure, water solubility, and the octanol/water partition coefficient for trimethylbenzenes.

## Uses

Pseudocumene is used as a component in liquid scintillation cocktails for clinical analyses. Prehnitene, isodurene, pentamethylbenzene, and hexamethylbenzene have no significant commercial uses, however, there is a great deal of intriguing patent activity. The higher polymethylbenzenes show potential as highly regiospecific methylation agents for methylation of 4-alkylbiphenyls to form 4,4′-alkylmethylbiphenyls (29,30) which can be oxidized to the monomer 4,4′-biphenyldicarboxylic acid [787-70-2] (see LIQUID CRYSTALLINE MATERIALS).

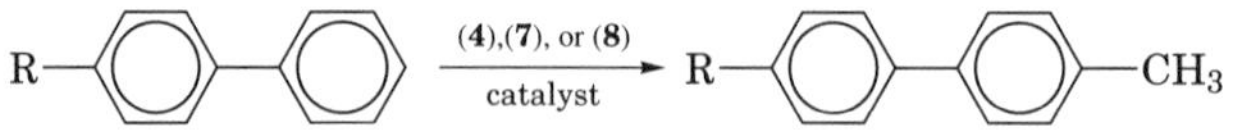

$$ R-\!\bigcirc\!-\!\bigcirc \xrightarrow[\text{catalyst}]{\textbf{(4),(7), or (8)}} R-\!\bigcirc\!-\!\bigcirc\!-CH_3 $$

Similar regiospecific syntheses employing PMBs and dialkylnaphthalene have been used to prepare 2,6-dimethylnaphthalene [581-42-0] which is oxidized to the corresponding 2,6-naphthalenedicarboxylic acid [1141-38-4], also used as a monomer for liquid crystal polymers.

## Derivatives

Little interest has been shown in prehnitene and pentamethylbenzene, even though dozens of derivatives have been prepared from each (13).

**Mesitylene.** One of the principal derivatives of mesitylene is the sterically hindered phenol of the structure shown in Figure 4. Its trade name is Ethanox 330 and it is produced by Albemarle Corporation (formerly Ethyl Corporation) (31). Ethanox 330 is an important noncoloring antioxidant and thermal stabilizer for plastics, adhesives, rubber, and waxes (qv) (32,33) (see ANTIOXIDANTS). The oral toxicity of Antioxidant 330 is extremely low (oral $LD_{50}$ in rats $>15$ g/kg) since its large size, $C_{54}H_{78}O_3$, effectively eliminates absorption from the gastrointestinal tract.

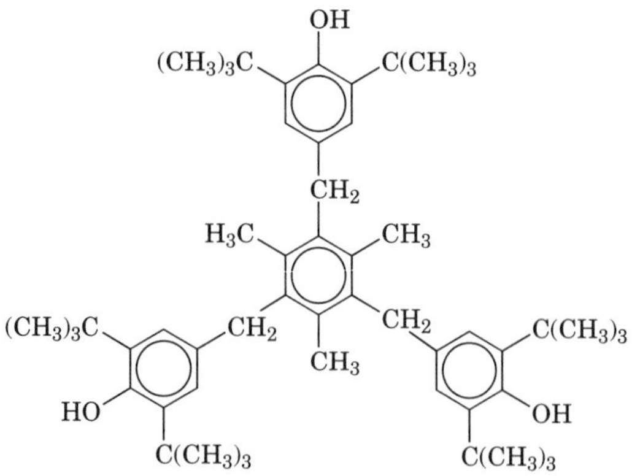

**Fig. 4.** 1,3,5-Trimethyl-2,4,6-tris(3,5-di-*tert*-butyl-4-hydroxybenzyl)benzene [1709-70-2] derived from mesitylene.

Another significant use for mesitylene is in the production of mesitaldehyde, [487-68-3], 2,4,6-trimethylbenzaldehyde (**9**) (AlliedSignal Inc.), an early intermediate in Zeneca's commercial synthesis of certain 5-(2,4,6-trimethylphenyl)cyclohexane-1,3-dione derivatives used as plant growth regulators (34) (see GROWTH REGULATORS, PLANTS). The Gattermann-Koch formylation reaction of mesitylene with carbon monoxide yields mesitaldehyde when carried out in the presence of a superacid catalyst typically derived from combination of a Brønsted acid and a Lewis acid, eg, $HF–BF_3$, $HCl–AlCl_3$ (35), or $HF–SbF_5$, although certain Brønsted acids such as trifluoromethanesulfonic acid [*1493-13-6*] (triflic acid) alone are strong enough to catalyze the formylation of mesitylene (36) (see FRIEDEL-CRAFTS REACTIONS). Mesitaldehyde can also be prepared from mesitylene by chloromethylation to $\alpha^2$-chloroisodurene [*1585-16-6*] followed by oxidation.

plant growth regulators

Liquid-phase air oxidation of mesitylene with Co, Mn, and Br catalysis produces 1,3,5-benzenetricarboxylic acid [*554-95-0*] (trimesic acid) (**10**) (37) as does the oxidation with dilute nitric acid (qv). Amoco has oxidized mesitylene to trimesic acid on a small scale (see PHTHALIC ACID AND OTHER BENZENECARBOXYLIC ACIDS). Less vigorous stepwise oxidation of mesitylene can yield 3,5-dimethylbenzoic acid [*499-06-9*] (**11**) and 5-methylisophthalic acid [*499-49-0*] (**12**).

Mesitylene is converted to a dye intermediate, 2,4,6-trimethylaniline [*88-05-1*] (mesidine), via nitration to 1,3,5-trimethyl-2-nitrobenzene [*603-71-4*] followed by reduction, eg, catalytic hydrogenation (38). Trinitromesitylene has been prepared for use in high temperature tolerant explosives (39). The use of mesitylene to scavenge contaminant NO from an effluent gas stream has been patented (40).

Mesitylene continues to be of considerable interest as a research chemical. Its trilateral structure, extremely high reactivity for electrophilic reactions, and the fact that large supplies are available point to good growth potential.

**Hemimellitene.** In addition to some specialized solvent applications, hemimellitene can be converted to musk Tibetine [*145-39-1*], 1-*tert*-butyl-3,4,5-

trimethyl-2,6-dinitrobenzene (**14**) via 5-*tert*-butyl-1,2,3-trimethylbenzene [*98-23-7*] (**13**).

Important derivatives are obtained from PMBs by oxidation (see PHTHALIC ACID AND OTHER BENZENECARBOXYLIC ACIDS). For example, hemimellitene is oxidized to hemimellitic acid dihydrate [*36362-97-7*] or hemimellitic acid [*569-51-7*] which can be dehydrated to hemimellitic anhydride [*3786-39-8*].

**Pseudocumene.** The liquid-phase air oxidation of pseudocumene yields trimellitic acid [*528-44-9*] (**15**) which is dehydrated to trimellitic anhydride [*552-30-7*] (TMA) (**16**). Amoco Chemical is the sole U.S. producer with a plant at Joliet, Illinois, having a capacity of ca 47,000 t/yr. Lonza in Italy is the sole European TMA producer.

Trimellitic anhydride is converted to PVC plasticizers, polyesters, water-soluble alkyd coatings, and polyamide–imide resins. The trimellitate plasticizers have a lower volatility than those derived from phthalic anhydride (see PLASTICIZERS).

**Durene.** The oxidation of durene (**4**) yields pyromellitic acid [*89-05-4*] (**17**) or pyromellitic dianhydride [*89-32-7*] (**18**) directly. The oxidation can be carried out with dilute nitric acid in solution, with air and catalyst either in the vapor phase over a solid vanadium pentoxide-based catalyst or in the liquid phase with a soluble catalyst system based on Co, Mn, and Br. PMDA (**18**) is also made from pseudocumene (**2**) by carbonylation to 2,4,5-trimethylbenzaldehyde [*5779-72-6*] (**19**) using HF–BF$_3$ followed by oxidation to pyromellitic acid. PMDA can also be converted to plasticizers; however, the main application is for the production of polyimides resistant to high temperature (see POLYIMIDES).

## BIBLIOGRAPHY

"Polymethylbenzenes" in *ECT* 1st ed., Suppl. 2, pp. 610–655, by H. W. Earhart, Humble Oil and Refining Co.; "(Polymethyl)benzenes" in *ECT* 2nd ed., Vol. 16, pp. 300–305; "Polymethylbenzenes" in *ECT* 3rd ed., Vol. 18, pp. 874–887, by H. W. Earhart, Koch Chemical Co.

1. TRCTHERMO database, *TRC Thermodynamics Tables: Hydrocarbons and Non-Hydrocarbons*, Thermodynamics Research Center, Texas A&M University System, College Station, Tex., July 1993.
2. *Beilstein On-line*, Beilstein Information Systems GmbH, Springer Verlag, Heidelberg, Germany.
3. D. R. Lide and H. V. Kehiaian, *CRC Handbook of Thermophysical and Thermochemical Data*, CRC Press Inc., Boca Raton, Fla., 1994.
4. F. R. Jensen and H. C. Brown, *J. Am. Chem. Soc.* **80**, 4038 (1958).
5. H. C. Brown, B. H. Bolto, and F. R. Jensen, *J. Org. Chem.* **23**, 417 (1958).
6. G. Vavon, J. Bolle, and J. Calin, *Bull. Soc. Chim.* (5), 6, 1025 (1939).
7. C. D. Shacklett and H. A. Smith, *J. Am. Chem. Soc.* **73**, 766 (1951).
8. R. M. Keefer and L. J. Andrews, *J. Am. Chem. Soc.* **79**, 4348 (1957).
9. R. C. Fuson, *Advanced Organic Chemistry*, John Wiley & Sons, Inc., New York, 1950.
10. E. E. Gilbert, *The Chemistry of Petroleum Hydrocarbons*, Vol. III, Reinhold Publishing Corp., New York, 1955, Chapt. 58.
11. H. W. Earhart and H. E. Cier, *Advances in Petroleum Chemistry and Refining*, Wiley-Interscience, New York, 1964, Chapt. 6.
12. K. L. Nelson and H. C. Brown, in Ref. 10, Chapt. 56.
13. H. W. Earhart, *The Polymethylbenzenes*, Noyes Development Corp., Park Ridge, N.J., 1969.
14. D. A. McCaulay and P. P. Lien, *J. Am. Chem. Soc.* **73**, 2018 (1951).
15. U.S. Pat. 2,528,892 (Nov. 7, 1950), D. A. McCaulay and A. P. Lein (to Standard Oil Co. of Indiana).
16. U.S. Pat. 5,223,589 (June 29, 1993), S. Kulprathipanja, K. K. Kuhnle, and M. S. Patton (to UOP).
17. U.S. Pat. 5,087,781 (Feb. 11, 1992), L. A. Cullo and A. A. Schutz (to Aristech Chemical Corp.) vapor phase over niobium on silica; U.S. Pat. 3,267,165 (Aug. 16, 1966) J. B. Braunwarth and R. C. Kimble (to Union Oil) with sulfuric acid and polyphosphoric acid; U.S. Pat. 2,917,561 (Dec. 15, 1959), L. T. Eby (to Exxon) vapor phase over tantalum.

18. U.S. Pat. 3,082,269 (Mar. 19, 1963), J. B. Armitage (to E. I. Du Pont de Nemours and Co., Inc.).

19. U.S. Pats. 4,524,227, 4,524,228, and 4,524,231 (June 18, 1985), P. F. Fowles and T-Y Yan (to Mobil Corp.).

20. U.S. Pat. 3,542,890 (Nov. 24, 1970), H. W. Earhart and G. Sugerman (Sun to Koch Industries Inc.).

21. National Library of Medicine, *NIOSH's Registry of Toxic Effects Chemical Substances (RTECS) database*, Nov. 1994 rev.; the data have not been critically evaluated.

22. *Code of Federal Regulations*, Title 40, Part 712.30, U.S. Environmental Protection Agency, Washington, D.C.

23. Ref. 22, Part 716.120, Feb. 13, 1994.

24. *Fed. Reg.* **58** (216), 59667–59682 (Nov. 10, 1993).

25. Ref. 22, Parts 141.40(j)(1) and (8), respectively.

26. Ref. 22, Part 372.65.

27. *Chemical Evaluation Search and Retrieval System (CESARS)*, CCinfo CD ROM A2, Canadian Centre for Occupational Health and Safety, Ontario, Canada, 1994.

28. G. M. Shaul, *U.S. EPA Risk Reduction Engineering Laboratory (RREL) Treatability Database*, rev. 5.0, RREL, Environmental Protection Agency, Cincinnati, Ohio, Aug. 3, 1994.

29. U.S. Pat. 5,177,286 (Jan. 5, 1993), G. P. Hagen and D. T. Hung (to Amoco Corp.).

30. Jpn. Kokai Tokkyo Koho JP 06 87,768 (94 87,768) (Mar. 29, 1994), K. Matsuzawa (to Japan Enajii Kk).

31. U.S. Pat. 5,292,969 (Mar. 8, 1994), B Berris (to Ethyl Corp., now Albemarle Corp.); U.S. Pat. 4,898,994 (Feb. 6, 1990), J. Borger, G. L. Livingston, and G. L. Mina (to Ethyl Corp., now Albemarle Corp.); U.S. Pat. 4,754,077 (June 28, 1988), G. L. Mina (to Ethyl Corp., now Albemarle Corp.); U.S. Pat. 4,340,767 (July 20, 1982), G. L. Mina (to Ethyl Corp., now Albemarle Corp.); U.S. Pat. 3,925,488 (Dec. 9, 1975), K. H. Shun (to Ethyl Corp., now Albemarle Corp.); U.S. Pat. 4,259,534 (Oct. 31, 1981), R. A. Filipova and co-workers.

32. *Antioxidant 330 Brochure*, Specialty and Ag Intermediates, Albemarle Corp., Baton Rouge, La., Dec. 1994.

33. *Chem. Mark. Rep.*, 32 (Feb. 25, 1980).

34. Brit. Pat. Appl. GB 2,124,198A (Feb. 15, 1984), R. B. Warner and J. W. Slater (to ICI, now Zeneca).

35. U.S. Pat. 4,195,040 (Mar. 25, 1980), C. A. Renner (to E. I. Du Pont de Nemours and Co., Inc.).

36. G. A. Olah, K. Laali, and O. Farooq, *J. Org. Chem.* **50**, 1483 (1985).

37. U.S. Pat. 5,107,020 (Apr. 21, 1992), A. C. Reeve (to Amoco Corp.).

38. U.S. Pat. 3,906,045 (Sept. 9, 1975), J. F. Knifton and R. M. Suggitt (to Texaco Inc.).

39. U.S. Pat. 3,515,604 (June 2, 1970), J. T. Hamrick.

40. U.S. Pat. 3,894,141 (July 8, 1975), W. R. Moser (to Exxon).

H. W. EARHART
Consultant


ANDREW P. KOMIN
Koch Chemical Company

**POLYMYXIN.** See ANTIBIOTICS, PEPTIDES.

**POLYOLEFINS.** See OLEFIN POLYMERS.

**POLYOXETANE.** See POLYETHERS, TETRAHYDROFURAN AND OXETANE POLY-MERS.

**POLYOXYMETHYLENE.** See ACETAL RESINS.

**POLYPEPTIDE ANTIBIOTICS.** See ANTIBIOTICS, PEPTIDES.

**POLYPEPTIDES.** See PROTEINS.

**POLY(PHENYLENE OXIDE).** See POLYETHERS, AROMATIC.

**POLY(PHENYLENE SULFIDE).** See HIGH PERFORMANCE FIBERS; POLYMERS CONTAINING SULFUR, POLY(PHENYLENE SULFIDE).

**POLYPHOSPHAZENES.** See ELASTOMERS, SYNTHETIC–PHOSPHAZENES; INORGANIC HIGH POLYMERS.

**POLYPROPYLENE.** See OLEFIN POLYMERS.

**POLYPROPYLENE FIBERS.** See FIBERS, OLEFIN.

**POLY(PROPYLENE OXIDE).** See POLYETHERS, PROPYLENE OXIDE POLY-MERS.

**POLYSACCHARIDES.** See CARBOHYDRATES; MICROBIAL POLYSACCHA-RIDES; STARCH.

**POLYSTYRENE.** See STYRENE PLASTICS.

**POLYSULFIDES.** See POLYMERS CONTAINING SULFUR, POLYSULFIDES.

**POLYSULFONES.** See POLYMERS CONTAINING SULFUR, POLYSULFONES.

**POLYTETRAFLUOROETHYLENE.**   See FLUORINE COMPOUNDS, OR-
GANIC—POLYTETRAFLUOROETHYLENE.

**POLYTETRAHYDROFURAN.**   See POLYETHERS, TETRAHYDROFURAN
AND OXETANE POLYMERS.

**POLYURETHANES.**   See URETHANE POLYMERS.

**POLY(VINYL ACETALS).**   See VINYL POLYMERS, POLY(VINYL ACETALS).

**POLY(VINYL ACETATE).**   See VINYL POLYMERS, POLY(VINYL ACETATE).

**POLY(VINYL ALCOHOL).**   See VINYL POLYMERS, POLY(VINYL ALCOHOL).

**POLY(VINYL CHLORIDE).**   See VINYL POLYMERS, VINYL CHLORIDE AND
POLY(VINYL CHLORIDE).

**POLY(VINYL ETHERS).**   See VINYL POLYMERS, VINYL ETHER MONOMERS
AND POLYMERS.

**POLY(VINYL FLUORIDE).**   See FLUORINE COMPOUNDS, ORGANIC—
POLY(VINYL) FLUORIDE.

**POLY(VINYLIDENE CHLORIDE).**   See VINYLIDENE CHLORIDE AND
POLY(VINYLIDENE CHLORIDE).

**POLY(VINYLIDENE FLUORIDE).**   See FLUORINE COMPOUNDS, OR-
GANIC—POLY(VINYLIDENE FLUORIDE).

**POLY(*N*-VINYLPYRROLIDINONE).**   See VINYL POLYMERS, *N*-VINYL
MONOMERS AND POLYMERS.

**POLY(XYLYENE).**   See XYLYENE POLYMERS.

**PORCELAIN.**   See CERAMICS; DENTAL MATERIALS; ENAMELS, PORCELAIN OR
VITREOUS.

**POROMERIC MATERIALS.**   See LEATHER-LIKE MATERIALS.

# POTASSIUM

Potassium [7440-09-7], K, is the third element in the alkali metal series. The name designation for the element is derived from potash, a potassium mineral; the symbol from the German name *kalium*, which comes from the Arabic *qili*, a plant. The ashes of these plants (*al qili*) were the historical source of potash for preparing fertilizers (qv) or gun powder. Potassium ions, essential to plants and animals, play a key role in carbohydrate metabolism in plants. In animals, potassium ions promote glycolysis, lipolysis, tissue respiration, and the synthesis of proteins (qv) and acetylcholine. Potassium ions are also believed to function in regulating blood pressure.

Potassium and sodium share the position of the seventh most abundant element on earth. Common minerals such as alums, feldspars, and micas are rich in potassium. Potassium metal, a powerful reducing agent, does not exist in nature.

## Physical Properties

Potassium, a soft, low density, silver-colored metal, has high thermal and electrical conductivities, and very low ionization energy. One useful physical property of potassium is that it forms liquid alloys with other alkali metals such as Na, Rb, and Cs. These alloys have very low vapor pressures and melting points.

Potassium has three naturally occurring isotopes: $^{39}$K (93.08%), $^{40}$K (0.01%), and $^{41}$K (6.91%). The radioactive decay of $^{40}$K to argon ($^{40}$Ar), half-life of $10^9$ years, makes it a useful tool for geological dating. Some physical properties of potassium are summarized in Table 1 (1–3).

**Table 1. Physical Properties of Potassium**

| Property | Value |
|---|---|
| atomic weight | 39.102 |
| atomic radius, pm | 235 |
| ionic radius, pm | 133 |
| Pauling electronegativity | 0.8 |
| crystal lattice | body-centered cubic |
| analytical spectral line, nm | 766.4 |
| melting point, °C | 63.7 |
| boiling point, °C | 760 |
| density, at 20°C, g/cm$^3$ | 0.86 |
| specific heat, J/(g·K)$^a$ | 0.741 |
| heat of fusion, J/g$^a$ | 59.591 |
| heat of vaporization, kJ/g$^a$ | 2.075 |
| electrical conductance, at 20°C, $\mu$S | 0.23 |
| surface tension, at 100°C, mN/m(=dyn/cm) | 86 |
| thermal conductivity, at 200°C, W/(m·K) | 44.77 |

$^a$To convert J to cal, divide by 4.184.

## Chemical Properties

Potassium has an electron configuration of $1s^2 2s^2 2p^6 3s^2 3p^6 4s^1$. All the alkali metals (Li, Na, K, Rb, Cs) are good reducing agents, because of the strong tendency to attain inert gas electron configuration. Reducing power increases from lithium to cesium. Cesium and rubidium are too reactive for safe handling and are not commercially available in large quantities. Potassium is the most electropositive reducing agent used in industry.

The alkali metals share many common features, yet differences in size, atomic number, ionization potential, and solvation energy leads to each element maintaining individual chemical characteristics. Among K, Na, and Li compounds, potassium compounds are more ionic and more nucleophilic. Potassium ions form loose or solvent-separated ion pairs with counteranions in polar solvents. Large potassium cations tend to stabilize delocalized (soft) anions in transition states. In contrast, lithium compounds are more covalent, more soluble in nonpolar solvents, usually existing as aggregates (tetramers and hexamers) in the form of tight ion pairs. Small lithium cations stabilize localized (hard) counteranions (see LITHIUM AND LITHIUM COMPOUNDS). Sodium chemistry is intermediate between that of potassium and lithium (see SODIUM AND SODIUM ALLOYS).

The superb reducing power of potassium metal is clearly demonstrated by its facile displacement of protons in the weakly acidic hydrocarbons (qv), amines, and alcohols (Table 2). Reactions with inorganics and gaseous elements are summarized in Table 3.

**Table 2. Potassium Products from Hydrocarbons, Amines, and Alcohols**

| Starting material | $pK_a{}^a$ | Product |
|---|---|---|
| 1,3-diaminopropane | 35 | $KNH(CH_2)_3NH_2$ |
| ammonia | 35 | $KNH_2$ |
| hexamethyldisilazane | 28 | $KN[Si(CH_3)_3]_2$ |
| aniline | 27 | $KNHC_6H_5$ |
| acetylene | 25 | $KC\equiv CH$ |
| tert-butyl alcohol | 18 | $KOC(CH_3)_3$ |
| methanol | 16 | $KOCH_3$ |
| phenol | 10 | $KOC_6H_5$ |

$^a$Values correspond to the $pK_a$ of the conjugate acid of the potassium base.

## Preparation and Manufacture

On the laboratory scale, potassium can be prepared by the following reactions, however, these reactions are not easily adaptable to a commercial scale.

$$K_2CO_3 + 2\ C \xrightarrow{\text{heat}} 3\ CO + 2\ K^0 \tag{1}$$

$$2\ KCl + CaC_2 \xrightarrow{\text{heat}} CaCl_2 + 2\ C + 2\ K^0 \tag{2}$$

$$2\ KN_3 \xrightarrow{\text{heat}} 3\ N_2 \uparrow + 2\ K^0 \tag{3}$$

In industry, chemical reduction is preferred over electrolytic processes for potassium production. Application of the Down's electrolytic sodium process to

**Table 3. Chemical Reactions of Potassium**

| Reactant | Reaction | Product |
|---|---|---|
| $H_2$ | begins slowly at ca 200°C; rapid above 300°C | KH |
| $O_2$ | begins slowly with solid; fairly rapid with liquid | $K_2O$, $K_2O_2$, $KO_2$ |
| $H_2O$ | extremely vigorous and frequently results in hydrogen–air explosions | KOH, $H_2$ |
| $C_{(graphite)}$ | 150–400°C | $KC_4$, $KC_8$, $KC_{24}$ |
| CO | forms unstable carbonyls | (KCO) |
| $NH_3$ | dissolves as K; iron, nickel, and other metals catalyze in gas and liquid phase | $KNH_2$ |
| S | molten state in liquid ammonia | $K_2S$, $K_2S_2$, $K_2S_4$ |
| $F_2$, $Cl_2$, $Br_2$ | violent to explosive | KF, KCl, KBr |
| $I_2$ | ignition | KI |
| $CO_2$ | occurs readily, but is sometimes explosive | CO, C, $K_2CO_3$ |

produce potassium has not been successful. Potassium–sodium alloy is easily prepared by the reaction of sodium with molten KCl, KOH, or solid $K_2CO_3$ powder (see SODIUM AND SODIUM ALLOYS).

Mine Safety Appliances Company, USA (MSA) developed a reduction process using sodium and KCl to produce potassium metal in the 1950s (4):

$$KCl + Na^0 \rightleftharpoons K^0 + NaCl \qquad\qquad (4)$$

The technology is based on the rapid equilibrium established between the reactants and products at high temperatures. The equilibrium shifts to the product side when potassium is removed continuously by distillation through a packed column. This process can produce high purity potassium metal. Appropriate adjustments of conditions give a wide range of potassium–sodium alloys of specified compositions.

The commercial production equipment consists of a furnace, heat-exchanger tubes, a fractionating column packed with Rachig rings, a KCl feed, a waste removal system, and a vapor condensing system (Fig. 1).

During operation, KCl is melted and introduced through a trap to the column. Molten sodium is fed to the bottom of the column. The lower portion of the column serves as a reactor, the upper portion as a fractionator. Potassium vapor is fractionated and condensed in an air-cooled condenser with the reflux pumped back to the top of the column. Waste sodium chloride is continuously removed from the bottom of the column through a trap.

## Shipping

Potassium of 98–99.5% purity is supplied in carbon steel or stainless steel drums and cylinders. Potassium–sodium alloy is shipped in carbon or stainless steel containers (3, 10, 25, 200, 750 lbs (1.36, 4.54, 11.3, 90.7, and 340-kg)) having dip tubes and valves.

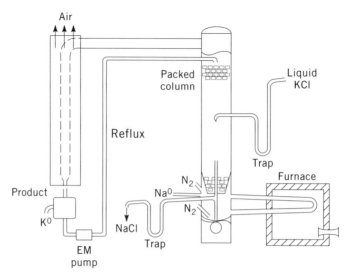

**Fig. 1.** Schematic diagram for the commercial production of potassium from sodium and potassium chloride. EM = electromagnetic.

Transport regulations are as follows (5):

| Metal/alloy | Identification number |
| --- | --- |
| potassium | UN 2257 |
| potassium–sodium alloys | UN 1422 |

Potassium and the alloys are classified as Hazard Class 4.3, "substances which in contact with water may emit flammable gases, solids," by U.S. Department of Transportation (DOT) regulations. A blue background label with a "Flame" pictogram, the words "Dangerous When Wet" and the number "4" is required. Quantities less than 500 g may be shipped via United Parcel Service (UPS) packaging in exemption specified in the October 1992 UPS "Guide for Shipping Hazardous Materials." Larger quantities are shipped in DOT Specification 4BW240 cylinders via common carrier, regulated under 49 CFR. Air shipments are restricted to cargo air craft only, 15 kg (max), and packaged under packing instruction 412 of the International Air Transport Association (IATA) "Dangerous Goods Regulations" (1994) or the International Civil Aviation Organization (ICAO) "Technical Instructions for the Safe Transport of Dangerous Goods by Air" (1993–1994).

## Economic Aspects

Total U.S. production of potassium metal is less than 500 t/yr. There are few commercial producers worldwide, although some companies produce potassium captively. The more prominent producers are Callery Chemical Company (a division of Mine Safety Appliances Company) in the United States and the People's Republic of China. Potassium may be manufactured in Russia as well. Strem Chemicals (U.S.) supplies small quantities in ampuls.

     Prices in 1994 were \$30–40/kg for bulk potassium and \$15.50–22/kg for NaK (78% K).

     Potassium up to 99.99% purity can be produced by zone refining (qv) or further distillation (qv) of commercial potassium. Technical-grade potassium is minimum 99% and is packaged under nitrogen.

     Smaller quantities of potassium and NaK are available in glass or metal ampuls (<100 g), or in stainless cylinders containing 12–13 kg. Potassium is also available in 84- or 209-L drums (containing 75 and 135 kg, respectively). NaK is available in cylinders containing 91 or 340 kg.

## Analytical and Test Methods

The principal impurity in potassium metal is sodium. Potassium's purity can be accurately determined by a melting point test (Fig. 2) or atomic absorption if necessary after quenching with alcohol and water. Traces of nonmetallic impurities such as oxygen, carbon, and hydrogen can be determined by various chemical and physical methods (7,8).

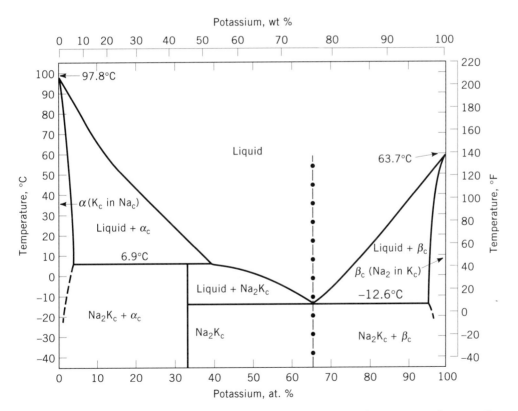

**Fig. 2.** Sodium–potassium equilibrium diagram where (—•—) represents the eutectic condition: 67.3 at. %, 77.8 wt % (6). $\alpha$ and $\beta$ represent distinct solid phases; c = crystalline.

## Health and Safety Factors

Reactions of potassium with water and oxygen are hazardous and safe handling is a concern. Potassium oxidizes slowly in air at room temperature, and it usually ignites if it sprays hot into the air. The peroxide and superoxide products may explode in contact with free potassium metal or organic materials including hydrocarbons. Thus, packaging (qv) under oils is less desirable than packaging under an inert cover gas or in a vacuum. Potassium can react with entrapped air in oils to form the superoxide. The encrustation of potassium with superoxide (as a yellow crust) developed during storage has been known to detonate by friction from cutting. Potassium encrusted with a peroxide and superoxide layer should be destroyed immediately by careful, controlled disposal (9).

Potassium forms corrosive potassium hydroxide and liberates explosive hydrogen gas upon reaction with water and moisture. Airborne potassium dusts or potassium combustion products attack mucous membranes and skin causing burns and skin cauterization. Inhalation and skin contact must be avoided. Safety goggles, full face shields, respirators, leather gloves, fire-resistant clothing, and a leather apron are considered minimum safety equipment.

Steam (qv) may be used to destroy potassium residues. It is imperative that equipment be designed to permit substantially complete drainage of potassium metal prior to steam cleaning. Precautions must also be taken to avoid hydrogen–air explosions. Small quantities of the metal can be safely destroyed under nitrogen by adding t-butyl alcohol or high boiling alcohols, followed by methanol, and then water. The caustic aqueous potassium waste should be disposed of in compliance with local regulations.

**Fire Fighting.** Potassium metal reacts violently with water releasing flammable, explosive hydrogen gas. Dry soda ash, dry sodium chloride, or Ansul's Met-L-X must be used for potassium or potassium alloy fires. Water, foam, carbon dioxide, or dry chemical fire extinguishers should never be employed. A NIOSH/MSHA approved self-contained breathing apparatus with full facepiece operated in a positive pressure mode and full protective clothing should be worn when removing spills and fighting fire (8).

## Uses

Historically, potassium metal was used by the Mine Safety Appliances Company (parent company of Callery Chemical Company) to develop potassium superoxide [12030-88-5], $KO_2$, for use as an oxygen source in self-contained breathing equipment (see OXYGEN-GENERATION SYSTEMS). Greater understanding of potassium chemistry since the 1950s and 1960s has led to numerous additional and important industrial applications (10–17). Potassium, potassium–sodium alloys, and potassium derivatives such as alkoxides, amides, and the hydride are extensively used, both industrially and academically, to synthesize organic and inorganic materials.

**Potassium–Sodium Alloys.** Potassium–sodium alloys, NaK, represent a use for potassium metal. Physical properties such as a wide range of compositions in liquid–metal forms (see Fig. 2), combined with excellent thermal and electrical conductivities, render these alloys ideal for use as heat-exchange fluids (18–20), cooling liquids in hollow valve stems (21), contact liquids for high temperature thermostats and homopolar generators, and hydraulic fluids (qv) (22) (see HEAT-EXCHANGE TECHNOLOGY). Potassium and potassium–sodium alloys were also used as working fluids on power plant topping cycles during the 1940s and 1950s.

A eutectic at 77.2% potassium has a freezing point of −12.56°C (see Fig. 2). A ternary eutectic (3% Na–24% K–73% Cs) melts at −76°C. The physical properties of NaK for 78 wt % K, 22 wt % Na are summarized in Table 4.

**Table 4. Physical Properties of NaK[a]**

| Parameter | Value |
|---|---|
| melting point, °C | −12.6 |
| boiling point[b], °C | 785 |
| electrical resistance, $\mu\Omega \cdot cm$ | |
|     at −12.6°C | 33.5 |
|     20°C | 38.0 |
| density at 20°C, $g/cm^3$ | 0.867 |
| thermal conductivity, $W/(m \cdot K)$ | |
|     at 20°C | $2.18 \times 10^{-3}$ |
|     550°C | $2.62 \times 10^{-3}$ |
| specific heat, $J/(g \cdot K)^c$ | |
|     at −12.6°C | 0.975 |
|     100°C | 0.937 |

[a]78 wt % K, 22 wt % Na.
[b]Value is estimated.
[c]To convert J to cal, divide by 4.184.

Chemically the liquid NaK alloy, usually used as a dispersion and on an inert support, provides more reactive surface area than either potassium or sodium metal alone, thus enhancing the reducing reactivity and permitting reactions to proceed at lower (eg, −12°C) temperatures. NaK alloys are suitable for chemical reactions involving unstable intermediates such as carbanions and free radicals.

NaK alloys have been used successfully in the following applications: side-chain alkylation of toluenes (see TOLUENE) and xylenes (see XYLENES AND ETHYLBENZENE) (23); isomerization of $\alpha$-olefins to internal olefins (see OLEFINS, HIGHER) (24,25); free-radical (26) and condensation (27–30) polymerization; reduction of metal halides to active metal powder (31); reduction of organic functional groups such as arenes, ketones (qv), aldehydes (qv), and alkyl halides (32,33); cleavage of functional groups containing C–X, C–O, and O–S bonds (34–38); interesterification of tallow with cottonseed oil to improve pour and clarification temperatures (39,40) (see COTTON); and impurity scavenging of acetylenic and allenic contaminants (41).

**Potassium Bases.** Potassium metal is used to prepare potassium bases from reactions of the metal with alcohols and amines.

*Potassium Alkoxides.* The most widely used potassium bases are potassium *tert*-butoxide [865-47-4] (KTB) and potassium *tert*-amylate [41233-93-6] (KTA). These strong alkoxide bases offer such advantages as base strength

($pK_a = 18$), solubility (Table 5), regio/stereoselectivity because of bulky *tert*-alkyl groups, and stability because of the lack of $\alpha$-protons. On storage, KTB and KTA have long shelf lives under inert atmosphere (see ALKOXIDES, METAL).

KTB and KTA are superior to alkali metal hydrides for deprotonation reactions because of the good solubilities, and because no hydrogen is produced or oil residue left upon reaction. Furthermore, reactions of KTA and KTB can be performed in hydrocarbon solvents as sometimes required for mild and nonpolar reaction conditions. Potassium alkoxides are used in large quantities for addition, esterification, transesterification, isomerization, and alkoxylation reactions.

*Potassium Amides.* The strong, extremely soluble, stable, and nonnucleophilic potassium amide base (42), potassium hexamethyldisilazane [*40949-94-8*] (KHMDS), KN[Si(CH$_3$)$_2$], $pK_a = 28$, has been developed and commercialized. KHMDS, ideal for regio/stereospecific deprotonation and enolization reactions for less acidic compounds, is available in both THF and toluene solutions. It has demonstrated benefits for reactions involving kinetic enolates (43), alkylation and acylation (44), Wittig reaction (45), epoxidation (46), Ireland-Claison rearrangement (47,48), isomerization (49,50), Darzen reaction (51), Dieckmann condensation (52), cyclization (53), chain and ring expansion (54,55), and elimination (56).

Potassium 3-aminopropylamide [*56038-00-7*] (KAPA), KNHCH$_2$CH$_2$CH$_2$-NH$_2$, $pK_a = 35$, can be prepared by the reaction of 1,3-diaminopropane and potassium metal or potassium hydride [*7693-26-7*] (57–59). KAPA powder has been known to explode during storage under nitrogen in a drybox, and is therefore made *in situ*. KAPA is extremely effective in converting an internal acetylene or allene group to a terminal acetylene (60) (see ACETYLENE-DERIVED CHEMICALS).

*Potassium Graphite.* Potassium, rubidium, and cesium react with graphite and activated charcoal to form intercalation compounds: C$_8$M, C$_{24}$M, C$_{36}$M, C$_{48}$M, and C$_{60}$M (61,62). Potassium graphite [*12081-88-8*], C$_8$K, made up of gold-colored flakes, is prepared by mixing molten potassium with graphite at 120–150°C.

Potassium graphite is a powerful solid reducing agent because of the free electrons from the ionized potassium trapped inside the graphite lattice. In packed beds, it can effectively eliminate undesired organic contaminants such as halogenated hydrocarbon impurities from gas streams and liquid solutions. The scope of its reducing reactions includes (61) Birch-type reactions of $\alpha,\beta$-unsaturated ketones, carboxylic acids, and Schiff's bases; reductive cleavage of vinylic and allylic sulfones; selective alkylation of aliphatic esters, imines, and

**Table 5. Solubilities of KTA and KTB at 25°C, wt %**

| Solvent | KTA | KTB |
|---|---|---|
| tetrahydrofuran | >50 | 44 |
| 2-methoxyethyl ether | 47 | 42 |
| *tert*-butyl methyl ether | ~30 | 2.0 |
| toluene | 36 | 2.3 |
| cyclohexane | 27 | 0.4 |
| hexane | 30 | 0.3 |

nitriles; and reductive decyanation of nitriles. Potassium graphite has been used to prepare active metals supported on graphite (62): Zn–Gr [69704-06-9] for Reformatsky reactions, Sn–Gr [89248-61-3] for allylic organometallics, Ti–Gr [6970-05-8] for coupling carbonyl compounds, Fe–Gr [55957-21-6] for debromination, Pd–Gr [59873-73-3] for vinylic and allylic substitution, and Pd–Gr, Ni–Gr [59873-69-7] for hydrogenation, etc. Potassium graphite has also been applied in polymerization reactions (see CARBON AND ARTIFICIAL GRAPHITE).

**Potassium Hydride.** Potassium hydride [7693-26-7], KH, made from reaction of molten potassium metal with hydrogen at ca 200°C, is supplied in an oil dispersion. Pressure Chemical Company (U.S.) is a principal supplier. KH is much more effective than NaH or LiH for enolization reactions (63,64). Use of KH as a base and nucleophile has been reviewed (65).

A noteworthy development is the use of KH for complexing alkylboranes and alkoxyboranes to form various boron hydrides used as reducing agents in the pharmaceutical industry. Potassium tri-sec-butylborohydride [54575-50-7], $KB(CH(CH_3)C_2H_5)_3H$, and potassium trisiamylborohydride [67966-25-0], $KB(CH(CH_3)CH(CH_3)_2)_3H$, are useful for the stereoselective reduction of ketones (66) and for the conjugate reduction and alkylation of $\alpha,\beta$-unsaturated ketones (67).

**Potassium Superoxide.** Potassium, rubidium, and cesium form superoxides, $MO_2$, upon oxidation by oxygen or air. Sodium yields the peroxide, $Na_2O_2$; lithium yields the oxide, $Li_2O$, when oxidized under comparable conditions. Potassium superoxide [12030-88-5], $KO_2$, liberates oxygen in contact with moisture and carbon dioxide (qv). This important property enables $KO_2$ to serve as an oxygen source in self-contained breathing equipment.

The unique chemical behavior of $KO_2$ is a result of its dual character as a radical anion and a strong oxidizing agent (68). The reactivity and solubility of $KO_2$ is greatly enhanced by a crown ether (69). Its usefulness in furnishing oxygen anions is demonstrated by its applications in $SN_2$-type reactions to displace methanesulfonate and bromine groups (70,71), the oxidation of benzylic methylene compounds to ketones (72), and the syntheses of $\alpha$-hydroxyketones from ketones (73).

# BIBLIOGRAPHY

"Potassium" under "Alkali Metals" in *ECT* 1st ed., Vol. 1, pp. 447–451, by E. Burkey, J. A. Morrow, and M. S. Andrew, E. I. du Pont de Nemour & Co., Inc.; "Potassium" in *ECT* 2nd ed., Vol. 16, pp. 361–369, by M. Blum, Atomergic Chemetals Co., and J. H. Madaus, MSA Research Corp.; in *ECT* 3rd ed., Vol. 18, pp. 912–920, by J. S. Greer, J. H. Madaus, and J. W. Mausteller, MSA Research Corp.

1. J. W. Mellor, *Supplement III to Mellor's Comprehensive Treatise on Inorganic and Theoretical Chemistry*, Vol. II, John Wiley & Sons, Inc., New York, 1963.
2. O. J. Foust, ed., *Sodium and Sodium–Potassium Engineering Handbook*, Vol. I, Gordon and Breach, New York, 1972, Chapt. 2.
3. Landolt-Börnstein, **4**(2C), 351 (1965).
4. U.S. Pat. 2,480,655 (1949), C. B. Jackson and R. C. Werner (to Mine Safety Appliances Co.).
5. *Fed. Reg.* **55**, 52402–52729 (Dec. 21, 1990); *Fed. Reg.* **56**, 66124–66287 (Dec. 20, 1991).

6. W. Mialki, *Metall. (Berlin)* **3**, 174 (1959).
7. R. E. Lee and S. L. Walter, *Techniques of Sampling and Analyzing Hot Flowing Sodium–Potassium Alloy TR-4*, MSA Research Corp., Evans City, Pa., 1950.
8. J. W. Mausteller, F. Tepper, and S. J. Rodgers, *Alkali Metal Handling and Systems Operating Techniques*, Gordon and Breach Scientific Publishers, Inc., New York, 1967.
9. M. A. Armour, L. M. Browne, and G. L. Weir, *Hazardous Chemical Information and Disposal Guide*, 2nd ed., University of Alberta, Edmonton, Alberta, Canada, 1984, p. 202.
10. A. A. Morton, *Solid Organoalkali Metal Reagents*, Gordon & Breach, New York, 1964.
11. H. Pines, W. M. Stalick, and H. B. Injovanovich, ed., *Base-Catalyzed Reactions of Hydrocarbons and Related Compounds*, Academic Press, Inc., New York, 1977.
12. M. Szware, *Carbanions, Living Polymers, and Electron Transfer Processes*, Interscience Publishers, New York, 1968.
13. E. Grovenstein, Jr., K. W. Chiu, and B. B. Patil, *J. Am. Chem. Soc.* **102**, 5848 (1980); E. Grovenstein, *Adv. Organomet. Chem.* **16**, 167 (1977).
14. L. Lochmann, J. Pospisil, and D. Lim, *Tetrahedron Lett.*, 257–262 (1966).
15. L. Lochmann and J. Petranek, *Tetrahedron Lett.* **32**, 1483 (1991).
16. L. Lochmann and J. Trekoval, *J. Organomet. Chem.* **326**, 1–7 (1987).
17. M. Schlosser and S. Strank, *Tetrahedron Lett.* **25**, 741–744 (1984).
18. *Chem. Eng. News* **33**, 648 (1955).
19. M. A. Turchin and co-workers, *Appl. Therm. Sci.* **1**(1), 39–43 (1988).
20. W. G. Anderson and co-workers, *Proceedings of the 25th Intersociety of Energy Conversion Engineering Conference* **25**(5), 268–273 (1990).
21. R. N. Lyon, ed., *Liquid-Metals Handbook*, 2nd ed., U.S. Government Printing Office, Washington, D.C., 1952, p. 5.
22. D. A. Wallace and co-workers, *Proceedings of the Intersociety of Energy Conversion Engineering Conference* **26**(5), 349–354 (1991).
23. U.S. Pat. 4,929,783 (1990), R. S. Smith (to Ethyl Corp.).
24. U.S. Pat. 4,720,601 (1989), G. S. Osaka and M. F. Shiga (to Sumitomo Chem.).
25. EP-A 211,448 (1987), G. Suzukamo, M. Fukao, M. Minobe, and A. Sakamoto (to Sumitomo Chem.).
26. Brit. Pat. 820,263 (1959), Z. Karl.
27. C. W. Carlson and R. West, *Organometallics* **2**(12), 1972–1977 (1983).
28. P. A. Bianconi and T. W. Weidman, *J. Am. Chem. Soc.* **110**(7), 2342–2344 (1988).
29. H. Y. Qui and Z. D. Du, *Goadeng Xucxiao Huaxue Xuebao* **10**(4), 423–425 (1989).
30. U.S. Pat. 4,800,221 (1989), O. W. Marko (to Dow Corning Corp.).
31. R. D. Rieke, *Acc. Chem. Res.* **10**, 301 (1988).
32. R. M. Schramm and C. E. Langlois, *Prepr. Div. Petr. Chem. Am. Chem. Soc.* **4**(4), B-53 (1959).
33. PCT Int. Appl. WO 8,607,097 (1986), (to W. R. B. Martin).
34. T. Ohsawa, T. Takagaki, A. Haneda, and T. Oishi, *Tetrahedron Lett.* **22**(27), 2583–2586 (1981).
35. T. Ohsawa and co-workers, *Chem. Phar. Bull.* **30**(9), 3178–3186 (1982).
36. C. A. Ogle, T. E. Wilson, and J. A. Stowe, *Synthesis*, (6), 495–496 (1989).
37. T. A. Thornton and co-workers, *J. Am. Chem. Soc.* **111**(7), 2434–2440 (1989).
38. Jpn. Kokai Tokyo Koho, JP 63,218,705 (1988), M. Ohata and co-workers.
39. Brit. Pat. 785,147 (1957), (to Thomas Hedley & Co. Ltd.).
40. D. Kazimierz, *Tluszcze i Srodki Piorace* **5**, 143–149 (1961).
41. Ger. Pat. 235,184 (1986), G. Heublein, D. Stadermann, H. Hartung, and W. Muller.
42. U.S. Pat. 5,025,096 (1991), K. W. Chiu, L. C. Yu, and J. R. Strickler.
43. J. Tsuji, I. Minami, and I. Shimizu, *Tetrahedron Lett.* **24**, 1973 (1983).
44. G. M. Coppola, *Synth. Common.* **15**, 135 (1985).

45. E. J. Corey and M. M. Mehrotra, *Tetrahedron Lett.* **27**, 5173 (1986).
46. W. C. Still and V. J. Novack, *J. Am. Chem. Soc.* **103**, 1282 (1982).
47. A. B. Smith and co-workers, *J. Am. Chem. Soc.* **104**, 4105 (1982).
48. J. Kallmerter and T. J. Gould, *J. Org. Chem.* **51**, 1155 (1986).
49. Y. Ikeda and H. Yamamoto, *Tetrahedron Lett.* **25**, 5581 (1984).
50. Y. Ikeda, J. Ukai, N. Ikeda, and H. Yamamoto, *Tetrahedron Lett.* **25**, 5177 (1984).
51. Eur. Pat. 0,076,643 (1982), P. Weeks (to Pfizer Corp.).
52. D. Heissler and J. Riehl, *Tetrahedron Lett.* **21**, 4707 (1980).
53. K. Takeda and co-workers, *J. Org. Chem.* **51**, 4735 (1986).
54. D. Horne, J. Gaudino, and W. J. Thompson, *Tetrahedron Lett.* **25**, 3529 (1984).
55. T. Tekahashi, S. Hashiguchi, K. Kasuga, and J. Tsuji, *J. Am. Chem. Soc.* **100**, 7424 (1978).
56. S. Hanessiam, A. Ugolini, P. J. Hodges, and D. Dube, *Tetrahedron Lett.* **27**, 2699 (1986).
57. C. Heathcock and J. A. Stafford, *J. Org. Chem.* **57**, 2566 (1992).
58. C. Almonsa, A. Moyano, M. A. Pericas, and F. Serratosa, *Synthesis*, 707 (1988).
59. J. H. Wotiz, P. M. Barelski, and D. F. Koster, *J. Org. Chem.* **38**, 489 (1973).
60. C. A. Brown and A. Yamashita, *J. Am. Chem. Soc.* **97**, 891 (1975).
61. R. Csuk, B. I. Glanzer, and A. Furstner, *Adv. Organomet. Chem.* **28**, 85 (1988).
62. W. Rudorff, in J. Emcleus and A. G. Sharpe, eds., *Graphite Intercalation Compounds, Advances in Inorganic Chemistry and Radiochemistry*, Vol. 1, Academic Press, Inc., New York, 1959, pp. 224–264; G. R. Henning, in F. A. Cotton, ed., *Interstitial Compounds of Graphite, Progress in Inorganic Chemistry*, Vol. 1, Interscience Publishers, New York, 1959, pp. 125–205.
63. C. A. Brown, *J. Org. Chem.* **39**, 3913 (1974).
64. C. A. Brown, *Synthesis*, 326 (1975).
65. H. W. Pinnick, *Org. Prep. Proced. Int.* **15**, 199 (1983).
66. C. A. Brown, *J. Am. Chem. Soc.* **59**, 4100 (1973).
67. B. Ganem, *J. Org. Chem.* **40**, 146 (1975).
68. D. T. Sawyer and J. S. Valentine, *Acc. Chem. Res.* **14**(12), 393 (1981); E. Lee-Ruff, *Chem. Soc. Rev.* **6**, 195 (1977).
69. T. Itoh, K. Nagata, M. Okada, and A. Ohsawa, *Tetrahedron Lett.* **31**, 7193 (1990).
70. R. A. Johnson, E. G. Nidy, L. Baczynski, and R. R. Gorman, *J. Am. Chem. Soc.* **99**, 7738 (1977).
71. N. A. Porter, J. D. Byers, R. C. Mebane, D. W. Gilmore, and J. R. Nixon, *J. Org. Chem.* **43**, 2088 (1978).
72. Y. H. Kim, K. S. Kim, and H. K. Lee, *Tetrahedron Lett.* **30**(46), 6357–6360 (1989).
73. C. Betancor, C. G. Francisco, R. Freire, and E. Suarez, *J. Chem. Soc., Chem. Commun.*, 947 (1988).

KUEN-WAI CHIU
Callery Chemical Company

# POTASSIUM COMPOUNDS

Potassium as a mineral is essential for life. As a consequence, potassium compounds or potash have been regarded as articles of commerce since the days of antiquity. Historically, the term potash denoted an impure form of potassium carbonate mixed with other potassium salts obtained by leaching wood ashes. As of the late twentieth century, within the English-speaking world, potash refers to the whole group of water-soluble potassium salts that are easily accessed by growing crops (1). The majority of the world's potash-bearing salt deposits likely originated from the evaporation of seawater or of mixtures of brines and seawater (see CHEMICALS FROM BRINE) (2). Potassium carbonate, obtained as crystals from the evaporation of potash solutions in large iron pots, served as the main building block for the production of other potassium compounds. In Pompeii, potassium carbonate was converted to potassium hydroxide by reaction of lime and was subsequently used to make soap (qv).

Potash manufacture was the first chemical industry in North America (3) and, until the latter part of the nineteenth century, potash was essential to the manufacture of soap, glass (qv), dyed fabrics, and gunpowder. As of the mid-1990s, approximately 90% of Canadian and U.S. potash production shipped to U.S. destinations is used in agriculture. Fertilizer potash is potassium chloride, potassium sulfate–magnesium sulfate, and potassium permanganate (see FERTILIZERS). The remaining 10% of the potash goes to the manufacture of glass, medicines, food, and soaps, as well as the polymerization of synthetic rubbers and plastics, and to export, mainly as potassium sulfate (4). About 1% of nonagricultural potash is potassium sulfate, used in gypsum board manufacture (see CALCIUM COMPOUNDS). The remaining 99% is potassium chloride, about 94% of which is diverted to the potassium hydroxide market (4). In total potash, the United States annually consumes more than it produces by nearly 3.9 million metric tons (5).

The very first patent issued by the U.S. government was for improving the "making of Pot ash and Pearl ash [a purer form of potassium carbonate] by a new Apparatus and Process." The said patent went to Samuel Hopkins on July 31, 1790, and was signed by both President George Washington and Attorney General William Randolph (3). Over 90,000 standard short tons of potash, selling for $200–300 a short ton, were exported from the United States during the 14-year life of that patent, at a time when coal sold for $8 a short ton. However, the production of less than one ton of potash required the consumption of approximately 400–500 standard tons of wood, and world demand was outstripping the supply of readily accessible forests. Close to the time when the Hopkins patent was issued, the invention of the sodium sulfate reduction process led to the general substitution of sodium carbonate for potassium carbonate as an alkali source.

Potassium nitrate, essential in the manufacture of black gun powder, was produced by the Chinese, who had developed gun powder by the tenth century AD. The process involved the leaching of soil in which nitrogen from urine had combined with mineral potassium. By the early 1800s, potassium nitrate had become a strategic military chemical and was still produced, primarily in India, by using the ancient Chinese method. The caliche deposits in Chile are the only natural source of potassium nitrate (2). These deposits are not a rich source of potassium nitrate, purifying only to about 14% as $K_2O$.

In 1840, potassium was recognized as an essential element for plant growth (6). This discovery and the invention in 1861 of a process to recover potassium chloride from rubbish salt, a waste in German salt mines, started the modern potassium chemical industry (5). Potassium compounds produced throughout the world in 1993 amounted to ca 22 million metric tons as $K_2O$ equivalent (4), down from ca 24 million t in 1992, having fallen annually from 32 million t in 1989 (2). Estimated production capacity was between 29 and 32 million t in 1992 (2).

Between 1869 and the beginning of World War I, most of the world's supply of potassium salts came from the Stassfurt deposits in Germany. During World War I, U.S. production, measured as $K_2O$, rose from 1000 metric tons in 1914 to 41,500 t in 1919. Following the end of World War I, U.S. production declined as imports increased. By the time the United States entered World War II, however, production had expanded enough to meet domestic needs. Since then, production has fluctuated, but has fallen below consumption as of the mid-1990s. Total annual U.S. demand peaked at $6.9 \times 10^6$ t in 1979 and has leveled off at approximately 5.1 to 5.5 million t. Canada is the principal potash exporter.

## Occurrence

Potassium, one of the mineral nutrients (qv), is the seventh most abundant element, constituting about 2.4% by weight of the earth's crust. Commercial production of potassium compounds is generally limited to the extraction of ores from underground deposits containing significant concentrations of soluble potassium salts. Exceptions include commercial potassium chemical operations on the Dead Sea and the Great Salt Lake, where potassium sulfate production capacity has been increasing (see CHEMICALS FROM BRINE). These natural brine refining operations exploit the highly favorable climatic conditions at those locations for the utilization of solar energy (qv). Estimated world reserves and the production of soluble potassium salts are presented in Table 1.

Canada leads the world in both reserves and production, operating at about 57% of estimated production capacity (4). Mines as of the mid-1990s were operating at about 40% of capacity (5). New Mexico is the primary source of potash in the United States, accounting for more than 80% of total marketable salts produced in 1993 (5). The average ore grade has declined from 20.7% in 1946 to 12.7% in 1992. A 9% decrease occurred from 1991 to 1992 (2). At 1995 prices and resulting production rates, most mines have 20 to 100 years reserves (4). North American agricultural potash is delivered to every State in the Union. Illinois is the destination for more potash than any other state. More North American produced nonagricultural potash is shipped to Alabama than to any other state. Alabama receives approximately 2.5 times the amount received by Ohio, the second-place destination (2).

Four minerals are the principal commercial sources of potash (Table 2). In all ores, sodium chloride is the principal soluble contaminant. Extraneous water-insoluble material, eg, clay and silica, is a significant contaminant in some of the evaporates being mined from underground deposits. Some European potassium ores contain relatively large amounts of the mineral kieserite, $MgSO_4 \cdot H_2O$. It is

**Table 1. World Reserves and Production of Soluble Potassium Salts, t × 10³ K₂O**

| Country | Reserves[a,b] | Production 1992[c] | 1993[b,c] |
|---|---|---|---|
| Belarus | | 3,311 | 2,900 |
| Brazil | | 85 | 100 |
| Canada | 4,536,000 | 7,327 | 7,200 |
| Chile | | 35 | 35 |
| China | | 25 | 25 |
| CIS | | 3,454 | 2,800 |
| France | 91,000 | 1,141 | 1,050 |
| Germany | | 3,525 | 3,300 |
| Israel | | 1,296 | 1,300 |
| Italy | 9,000 | 86 | 100 |
| Jordan | | 808 | 820 |
| Spain | 72,000 | 594 | 600 |
| Ukraine | | 182 | 200 |
| United Kingdom | | 530 | 530 |
| United States | 187,000 | 1,705 | 1,450 |
| *Total* | *~4,895,000* | *24,018* | *22,410* |

[a]Ref. 7.
[b]Estimated.
[c]Ref. 4.

**Table 2. Principal Commercial Potash Minerals and Contaminants**

| Mineral | CAS Registry Number | Chemical formula | Pure mineral K, % | K₂O, % |
|---|---|---|---|---|
| | | *Ores* | | |
| sylvite | [14336-88-0] | KCl | 52.44 | 63.17 |
| carnallite | [1318-27-0] | KCl·MgCl₂·6H₂O | 14.07 | 16.95 |
| kainite | [1318-72-5] | KCl·MgSO₄·2.75H₂O | 16.00 | 19.27 |
| langbeinite | [67320-08-5] | K₂SO₄·2MgSO₄ | 18.84 | 22.69 |
| | | *Contaminants* | | |
| halite | [14336-88-0] | NaCl | | |
| polyhalite | | K₂SO₄·MgSO₄·2CaSO₄·H₂O | | 15.62 |
| leonite | | K₂SO₄·MgSO₄·4H₂O | | 25.68 |
| kieserite | | MgSO₄·H₂O | | |
| anhydrite | | CaSO₄ | | |

recovered for captive use to produce potassium sulfate compounds or is marketed in relatively pure form as a water-soluble magnesium fertilizer.

There are two basic classes of potash-containing evaporites (2), those deposits that are rich in magnesium sulfate, such as polyhalite and kieserite, and those that are poor in magnesium sulfate. The primary source of potash in the magnesium-rich deposits is carnallite, whereas both carnallite and sylvite are found in the magnesium-poor deposits.

Approximately 98% of the potassium recovered in primary ore and natural brine refining operations is recovered as potassium chloride. The remaining

2% consists of potassium recovered from a variety of sources. Potassium produced from these sources occurs as potassium sulfate combined with magnesium sulfate. From a practical point of view, the basic raw material for all of the potassium compounds discussed in this article, except potassium tartrate, is potassium chloride. Physical properties of selected potassium compounds are listed in Table 3, solubilities in Table 4.

## Potassium Chloride

Potassium chloride [7447-40-7], or muriate of potash (MOP) as it is known in the fertilizer industry (at about 97% purity), is the world's most commonly used potash (5). Chemical-grade potassium chloride (99.9%) is the basis for manufactured production of most potassium salts (10).

**Mining.** Potassium chloride is produced mostly from solid ores occurring in underground deposits 300–1700-m deep. All New Mexico potash producers mine underground bedded ore zones and all except one mine sylvinite ore, a mixture of sylvite and halite, from which most muriate of potash originates (1). Conventional mining methods adapted from coal (qv) and hard-rock mining operations and machine mining are employed to mine ores to a depth of ca 1100 m. Solution mining is employed to recover potassium chloride from deposits exceeding ca 1100 m or from deposits that cannot be mined conventionally because of geological anomalies (see MINERAL RECOVERY AND PROCESSING).

Although it is a relatively small source in terms of total world production of potassium chloride, significant tonnages of potassium chloride, ca 3% of world production, are produced by solar evaporation of natural brines. These sources are mined by pumping the brines to evaporating ponds for enrichment under controlled conditions. Natural ores being mined for potassium chloride are either a simple mixture of potassium chloride and sodium chloride (sylvinite [12174-64-0]) or complex salt mixtures that contain langbeinite [67320-08-5] and kainite [1318-72-5]. All ores contain sodium chloride as a principle contaminant. Extensive studies of the geology of potash deposits have been made in countries where extensive deposits are known to occur (2,11,12).

Descriptions of contemporary methods employed in potash salt mining are available (2,10,13). Because the evaporate beds are generally regular, tabular bodies, underground potash mining is a relatively clean and straightforward operation having few complications. One rare hazard is the existence of occasional pockets of gas under pressure that can cause explosions when the pressure is released. Water inflow problems have become a primary concern at some Canadian mine sites, prohibiting continued conventional mining. For the most part, mining methods have been adapted from coal and salt mining and the equipment used is similar.

The room-and-pillar configuration has been used predominantly in underground mines in the United States and Canada and widely in the CIS and Europe. This results in a checkered pattern of mining areas separated by pillars. Timbering is not often used, but roof-bolting is commonly practiced by most operators. A funnel system, similar to sublevel stopping with waste fill, is used in many mines in Germany along with mechanical drilling and blasting, and particularly on the steeply dipping ore bodies of the Hannover district. In New Mexico,

**Table 3. Physical Constants of Potassium Compounds**[a]

| Potassium compound | CAS Registry Number | Formula | Mol wt | Form | Specific gravity | Melting point, °C |
|---|---|---|---|---|---|---|
| acetate | [127-08-2] | $KC_2H_3O_2$ | 98.14 | white powder | 1.57 | 292 |
| bromide | [7758-02-3] | KBr | 119.01 | cubic | 2.75 | 734 |
| carbonate | [584-08-7] | $K_2CO_3$ | 138.20 | monoclinic | 2.428 | 891 |
| bicarbonate | [298-14-6] | $KHCO_3$ | 100.11 | monoclinic | 2.17 | 100–200 dec |
| chlorate | [3811-04-9] | $KClO_3$ | 122.55 | monoclinic | 2.32 | 356 |
| chloride | [7447-40-7] | KCl | 74.55 | cubic | 1.984 | 770 |
| formate | [590-29-4] | $KHCO_2$ | 84.11 | rhombic | 1.91 | 167.5 |
| hydroxide | [1310-58-3] | KOH | 56.10 | rhombic | 2.044 | 360.4 ± 0.7 |
| iodide | [7681-11-0] | KI | 166.02 | cubic | 3.13 | 681 |
| nitrate | [7757-79-1] | $KNO_3$ | 101.10 | rhombic, trigonal | 2.109 | 334 |
| nitrite | [7758-09-0] | $KNO_2$ | 85.10 | colorless, prism | 1.915 | 440 |
| orthophosphates | | | | | | |
| normal phosphate | [7758-53-2] | $K_3PO_4$ | 212.27 | rhombic | 2.564 | 1340 |
| monohydrogen phosphate | [7758-11-4] | $K_2HPO_4$ | 174.18 | amorphous | | dec |
| dihydrogen phosphate | [7778-77-0] | $KH_2PO_4$ | 136.09 | tetragonal | 2.338 | 252.6 |
| sulfate | [7778-80-5] | $K_2SO_4$ | 174.26 | rhombic or hexagonal | 2.662 | 1069 |
| bisulfate | [7646-93-7] | $KHSO_4$ | 136.17 | monoclinic | 2.24–2.61 | 210 |
| | | | | rhombic | 2.322 | 214 |
| sulfite | [10117-38-1] | $K_2SO_3$ | 194.29 | monoclinic | | dec |
| acid tartrate | | $KHC_4H_4O_6$ | 188.1 | rhombic | 1.984 | |

[a]Ref. 8.

**Table 4. Aqueous Solubility of Potassium Compounds, wt %[a,b]**

| Potassium compound | Temperature, °C | | |
|---|---|---|---|
| | 30 | 50 | 100 |
| acetate | 73.9 | 77.1 | |
| bromide | 41.4 | 44.7 | 51.0 |
| carbonate | 53.2 | 54.8 | 60.9 |
| bicarbonate | 28.1 | 34.2 | |
| chlorate | 9.2 | 15.0 | 36.0 |
| chloride | 27.1 | 30.0 | 36.0 |
| formate | | 80.7 | |
| hydroxide | 55.8 | 58.3 | 65.2 |
| iodide | 60.4 | 62.8 | 67.4 |
| nitrate | 31.3 | 46.0 | 71.0 |
| nitrite | 76.1 | 77.3 | 80.2 |
| sulfate | 11.5 | 14.2 | 19.4 |
| sulfite | 51.8 | | 55.5 |
| acid tartrate | 0.762 | 1.8[c] | 6.5[c] |

[a]Ref. 9.
[b]Values may be converted to solubilities, $S$, expressed as potassium salt g/100 g water, by $S = v/(100 - v) \times 100$, where $v$ is the solubility in g as listed.
[c]Value may be too high.

after the intended limit of the mining area is reached, the pillars are mined and the overlying strata are allowed to subside as mining retreats toward the shaft. Shaft depths are 233 to 575 m. Slight subsidence is noted on the surface. A total recovery of well over 80% is obtained. In other areas, notably Saskatchewan, the overlying strata are water-bearing and removal of pillars floods the mine. Recovery of ore therefore is only ca 35%. The lesser problem of water-bearing strata in New Mexico is controlled by grouting and concrete lining of the shafts.

A portion of the mining in New Mexico is conducted by conventional mining methods, that is, a cycle of undercutting, drilling, and blasting with an ammonium nitrate–fuel oil mixture and loaders, use of shuttle cars, and belt haulage. Such conventional methods are used, although not exclusively, in the CIS, Germany, France, Spain, and the United Kingdom.

Sylvinite ore is relatively soft and easily broken. Thus, continuous miners of the boring- or rotating drum-type can be used. These machines are modified continuous coal miners. This method is used in some of the mines in New Mexico, in all Canadian mines except the solution mines, and in many European mines. Ore is removed from the face by a number of methods, eg, with extensible belt conveyor ore-loading machines and shuttle cars to the main haulage line. To obtain maximum benefits from the continuous miners, the ore transport system must be as nearly continuous as possible.

Potash is removed by long-wall mining in the CIS and European countries such as France and Spain. One U.S. company has considered usage in its Canadian and domestic operations. In long-wall operation, material is removed in slices up to 76 cm thick and up to 183 m or more long. Upon completion of a cut, the shear-cutting system, the conveyor system, and the powered roof-support system are advanced a distance equivalent to one-cut thickness, thereby allowing a portion of the roof behind the system to fall. The rock structure must fit certain strength and fracture requirements to allow the use of the long-wall system. The main advantages of the system over room-and-pillar methods are essentially complete removal of the ore and the potential for more rapid removal of the ore.

The long-wall method is used to remove about half of the ore from the mines owned by the Byeloruskali combine. In one mine, two 1-m thick sylviniti beds, separated by a lone meter-thick sylvinite bed, and the upper face are mined ca 5.5 m in advance of the lower. This increased recovery from 45 to 70% and increased the $K_2O$ content of the ore from 26 to 36 wt % (21.6–29.9 wt % K). Approximately one half of French ore production is by long-wall mining and occurs mainly in the Marie Louise mine. Long-wall mining is also used in the Potasas de Navarra mine in Spain.

Solution mining is used in North America to extract sylvinite from strata underlying the earth's surface at relatively great depths. Three producers in Utah evaporate saline brines, and the muriate and sulfate are produced at Searles Lake in California via solar evaporation and selective crystallization from underground brines. In the early 1970s, one firm in Utah abandoned conventional mining because of unfavorable geologic conditions (14). The mine was sealed, flooded with water, and the resulting saturated brine pumped to the surface where the sylvinite was recrystallized by solar evaporation. An injection well was used to add makeup water. Moab Salt, Inc. uses two-well solution mining and solar evaporation to recover muriate of potash in Utah. In Saskatchewan, Canada, which has the world's largest high grade sylvinite ore reserves (5), sylvinite is extracted by solution mining from an ore zone too deep to be mined conventionally (15). Brine recovered from the wells is processed in multiple-effect evaporators to recover the potassium chloride values dissolved in the brine.

**Refining.** Process selection for the separation of potassium chloride as a relatively pure product from other constituents is based on the physical and chemical characteristic of a given ore and the unique mineralogy of each potash mining area. Ores amenable to treatment by the physical separation methods commonly used in other nonmetallic minerals processing industries generally are chosen to recover the potassium chloride. These methods include heavy-media and froth-flotation separations. Physical separation processes are less energy-intensive than fractional crystallization (qv), which is the traditional method of producing potassium chloride. Mainly because of the cost of energy, conventional fractional crystallization is generally confined to ores that are not amenable to processing by physical separation methods. However, pond crystallization methods are receiving increased consideration owing to favorable energy costs. Fractional crystallization is included as a secondary operation in some physical separation plants for recovery of potassium chloride crystals that are trapped in the waste streams exciting the primary ore beneficiation plant.

Relatively pure potassium chloride is a product containing not less than 60 wt % $K_2O$ (49.8 wt % K or 95 wt % KCl equivalent). Contaminants making up the remaining 5 wt % are not objectionable. Some, such as magnesium sulfate salts, may enhance the agricultural value of the agricultural-grade muriate (chloride) products. In Europe, where the principal agricultural areas are near the mines and thus significantly reduce the importance of transportation as a cost element, potassium chloride products contain 40 wt % $K_2O$ (33.2 wt % K). Except for the presence of minor by-products, all production in North America must meet the 60 wt % $K_2O$ (49.8 wt % K) specification.

*Froth Flotation.* Froth flotation (qv) of potassium chloride from sylvinite ores accounts for ca 80% of the potassium chloride produced in North America and about 50% of the potassium chloride in Europe and the CIS. Fractional crystallization and heavy-media processing account for the remaining amounts produced. Froth flotation has been described (6,16,17).

Unlike some ores mined in Europe, sylvinite mined in North America consists of coarse crystals of potassium chloride and sodium chloride. When the mine-run ore is crushed to a grain size suitable for application of the froth flotation to separate the salts, the crystals of each must be reduced to individual grains (see SIZE REDUCTION). This grain size is called the liberation grain size of the ore. Sylvinite ores in New Mexico are liberated when crushed through a ca 2.4-mm (8-mesh) screen. Maximum grain size for efficient froth flotation is about this size. Consequently, liberation grain size and froth flotation are compatible for processing New Mexican ores. In contrast, sylvinite ores mined in Saskatchewan are composed of much coarser grains. These ores are liberated in a ca 9-mm screen. In conventional mining of sylvinite ore, the ore is crushed through a ca 2.3-mm screen before it is sent to the flotation plant. However, some Canadian mine operators produce a coarse crystalline product, called granular muriate of potash, by crushing the ore through a 9-mm screen followed by screening the crushed ore at 1.7 mm. Screen undersize (<1.7 mm ($-$12 mesh)) is sent to froth flotation. Screen oversize, ie, the 9 × 1.7-mm fraction, is processed by heavy-medium separation to produce granular 60 wt % $K_2O$ muriate for the agricultural market. Basic unit operations of the flotation process are shown in Figure 1.

*Crushing.* The ore is crushed through a ca 12.7-cm screen in the mine to facilitate conveying and hoisting. It is further reduced to an appropriate size for milling by multiple-stage dry crushing or by initial dry crushing in a rod mill. Gentle crushing is required because sylvinite ores are friable. Severe crushing causes excessive formation of fine particles. Because of mechanical entrainment, sodium chloride grains less than ca 0.2 mm in diameter tend to entrain with the KCl product recovered in the flotation cells. Contamination of the KCl in this manner interferes with product quality and potassium chloride recovery.

*Scrubbing and Desliming.* Sylvinite ores in North America contain 1–6 wt % water-insoluble clays. A significant portion of these clays is less than 0.002 mm in diameter. If not removed or controlled in some manner, clay bodies that are dispersed in the flotation solution, ie, brine saturated with KCl and NaCl, absorb the amine collector, which is added to effect flotation separation, and the collector is rendered ineffective. Clay is the most troublesome impurity encountered in the processing of sylvinite ore.

In North America flotation plants, 90–95 wt % of the clay and other water-insoluble material, eg, silica particles, are removed by hydraulic desliming. This operation entails scrubbing the ore mixed with process brine in agitated tanks at a solids-to-brine weight ratio of ca 1:1. Residence time is ca 3 min. The scrubbers are frequently operated in series to ensure complete dispersion of the clay. Amorphous clay agglomerates and clay adhering to the crystal surfaces disperse

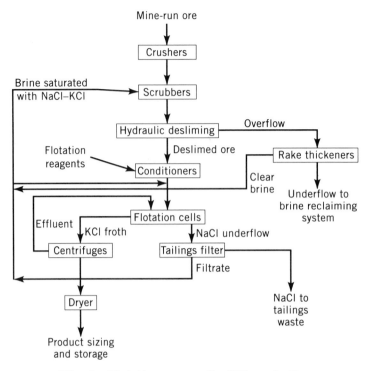

**Fig. 1.** Flotation process for KCl production.

into the brine as finely divided particles. Pulp from the scrubbers, which consists of ore grains, disseminated clay particles, and process brine, is separated by wet screening or mechanical classification into a coarse fraction (>0.8 mm (ca 20 mesh)) and a fine fraction (<0.8 mm). Each fraction is separately deslimed. The coarse fraction is mixed with clear brine to wash out entrained clay particles and is rescreened or reclassified to separate the wash ore from the muddy brine. The muddy brine stream is returned to the hydroseparators, which separate the clay particles from the fine fraction. Sylvanite ores containing more than ca 2 wt % clay require multiple-stage scrubbing for essentially complete removal of clay from the coarse fraction of the ore. Deslimed coarse ore advances to the coarse reagent conditioners. After desliming in a complex hydrocyclone–hydroseparator circuit for thorough cleansing, the fine fraction flows to the fine-ore reagent conditioners.

Clay exits the desliming circuit along with ore particles that are less than ca 0.07 mm (200 mesh) in diameter. Brine saturated with KCl and NaCl is the carrying medium. The solids content of the suspension is 3–5 wt %. Because of the KCl content (typically 10 wt %), and for water-conservation reasons, most of the process brine is recovered in conventional thickeners for recycling to the process as clarified brine. Underflow from the primary slime thickeners contains substantial quantities of process brine. A typical ratio is one part of solids per four parts of brine. Several methods are used for secondary recovery of brine. These include prolonged settling in sealed lagoons, multiple-stage countercurrent decantation, and filtration with sodium chloride residues as a filter aid (18).

*Reagent Conditioning.* This unit operation is carried out in two stages. Optimum pulp density in both stages is 50–60 wt %. The first stage involves agitating the deslimed ore with a clay depressant or slime blinder to deactivate clay particles that are entrained in the washed ore flowing from the hydraulic desliming operations. Suitable depressants include starches, guar gum, carboxymethyl cellulose, and polyacrylamides. Depending on the amount of entrained clay and the type of depressant used, dosage rates are 50–500 g of depressant per metric ton of ore processed. Cost-effectiveness is usually the sole criterion in selecting a depressant. After one or two minutes of conditioning, the pulp enters the second-stage conditioner.

Reagents, including a collector, required for the flotation separation are added in the second stage. Specially developed flotation oils called extenders are added if the ore contains relatively large percentages of coarse KCl crystals, ie, >15 wt % of 0.8–1.7-mm crystals. An extender makes the amine-coated surfaces more hydrophobic, thereby enabling the particles to float readily in the flotation cells. Retention time in second-stage conditioning is 1–2 min.

Potassium chloride collectors are primary amines derived from beef tallow. Commercially available amine is a homogeneous mixture of palmityl-, stearyl-, and oleylamines. Several years after the discovery that normal tallow amine has a specific affinity for KCl crystals only, it was realized that hydrogenation of normal amine greatly increases overall effectiveness of the collector. Hydrogenation converts the oleylamine to stearylamine, thereby bringing all alkyl groups to near saturation. Free-base amines are converted to salts with acetic or hydrochloric acid, the latter preferred by most operators. Consistent amine quality is mandatory. Some companies specify amine containing not less than 97 wt % amine. Others find that the normal commercial-grade, ie, not less than 96 wt % primary amine, is satisfactory. For homogeneous distribution of the salt with the crushed ore, the amine salt is prepared as an aqueous 4–5 wt % solution.

Longer-chain amines, ie, arachidyl–behenyl ($C_{20}$ to $C_{22}$) amines, are used in special cases in which brine temperatures exceed 35°C. At temperatures higher than ambient, normal tallow amine tends to dissolve and therefore is unavailable to coat the surfaces of the potassium chloride crystals. Amine consumption is from 50 g/t (ca 40 wt % KCl) of high grade ore, to 150 g/t (ca 20 wt % KCl) of low grade ore.

Pulp exiting the conditioners is diluted by using process brine to a solids content of 30–35 wt % for use as feed for the flotation cells. In some plants, the coarse- and fine-fraction flows are floated separately. In most plants, the two fractions are recombined and the flotation is conducted in a common operation.

*Flotation.* Tallow amines contain small amounts of short-chain compounds, eg, the octyl, lauryl, and myristyl ($C_{14}$) groups. These amines usually produce enough froth for the flotation. Small quantities of specialized synthetic alcohols, eg, Dowfroth P, are frequently added to the flotation system to supplement the natural frothers. The quantities of the reagents used in potash flotation plants are listed in Table 5.

Separation of amine-coated potassium chloride grains from sodium chloride grains requires about three minutes of retention time in the first bank of flotation cells and is referred to as the rougher operation. Froth impregnated with KCl grains and a small amount of entrained fine NaCl grains is screened at ca

**Table 5. Reagents Used In Potash Flotation**

| Reagent type | Ore, g/t |
|---|---|
| clay depressants | |
| starches | 500–900 |
| gums and others | 75–125 |
| amine collectors | 50–150 |
| amine extenders | 75–200 |
| frothers | 25–50 |

0.8 mm (20 mesh). Oversized material meets agricultural-grade potassium spe-cifications. Consequently, it is passed directly to the product debrining and dry-ing operations. Undersize material, which is contaminated with sodium chloride, is sent to a second bank of flotation cells, called cleaner cells, for final separation of potassium chloride from sodium chloride grains. Froth from the cleaner cells flows to the product debrining and drying operations. Process brine saturated with sodium chloride and potassium chloride contains ca two parts of sodium chloride per one part of potassium chloride. Thus, product crystals from the de-brining operation must contain less than ca 10 wt % process brine to meet the minimum standard $K_2O$ content of 60 wt %.

Underflow from the rougher cells consists of sodium chloride suspended in process brine and other salts in the ore. This waste stream or tailing is passed to debrining devices for separation of the solid material from the process brine. The separation must be efficient because it is essential that the brine is recovered for recycling to minimize potassium losses. Tailing salt from the debrining operation is rejected from the process. Underflow from the cleaner cells contains considerable quantities of potassium chloride grains that fail to float, and is recycled to the rougher cells.

*Product Debrining and Drying.*   Solid-bowl and pusher centrifuges are used to separate the product crystals from the process brine. The latter is recycled to the flotation circuit as dilution brine. Damp centrifuge cake is sent to the product dryers. Dryer temperature and the final moisture levels in the product vary. Potassium chloride having higher moisture content tends to cake during storage, especially when shipped to locations that are subject to high humidity. Amine remaining on the product is beneficial in that it acts as an anticaking reagent, thereby preventing pile set. Anticaking, which is added to the final product, is usually a mixture of oil (0.1–0.3 kg/t) and amine (0–6%) used for agricultural products. A straight-chain amine (0.1–0.3 kg/t) is used for high grade industrial products. Some industrial customers for the potassium chloride, eg, the chlor-alkali industries (see ALKALI AND CHLORINE PRODUCTS), require that no amine be present. In this case, granulation using compactors or loading into sparger cars is carried out.

*Sizing.*   In most flotation plants, flotation concentrates, after being dried, are sized into three fractions and each serves a specific agricultural market. The fractions are coarse-, standard-, and suspension-grades of muriate of potash. Typical screen analyses are presented in Table 6; other physical characteristics are summarized in Table 7.

**Table 6. Screen Analysis of Muriate of Potash[a]**

| Coarse-grade | | Standard-grade | | Suspension-grade | |
|---|---|---|---|---|---|
| Screen aperture, mm (mesh) | Cumulative, %[b] | Screen aperture, mm (mesh) | Cumulative, %[b] | Screen aperture, mm (mesh) | Cumulative, %[b] |
| 2.38 (8) | 16 | 1.19 (16) | 4 | 0.42 (40) | 12 |
| 1.68 (12) | 45 | 0.84 (20) | 23 | 0.30 (50) | 35 |
| 1.19 (16) | 80 | 0.59 (30) | 55 | 0.21 (70) | 70 |
| 0.84 (20) | 95 | 0.42 (40) | 76 | 0.15 (100) | 86 |
| 0.59 (30) | 98 | 0.30 (50) | 90 | 0.11 (140) | 95 |
| | | 0.21 (70) | 96 | | |

[a]Ref. 19.
[b]Percentage of cumulative equals the percentage of MOP that is guaranteed not to pass through a given screen aperture.

**Table 7. Typical Characteristics of Muriate of Potash[a]**

| Grade | MOP, wt % | Density, loose bulk, g/cm$^3$ | Angle of repose, deg |
|---|---|---|---|
| coarse | 96.7 | 1.11–1.14 | 31–33 |
| standard | 96.3 | 1.07–1.15 | 29–31 |
| suspension | 95.9 | 1.07–1.12 | 28–30 |

[a]Ref. 20.

*Products.* Significant quantities of undersized potassium chloride particles, which make up 10–15% of total production, are generated in the flotation
processes. These particles are collected by air-emission control devices (see AIR
POLLUTION CONTROL METHODS) and are processed into completely water-soluble
grades of muriate by single or double recrystallization. Potassium chloride produced by a single recrystallization is marketed in the fertilizer industry as
a source of completely water-soluble potassium chloride. It is used to prepare
specialty water-soluble fertilizer solutions containing nitrogen, phosphorus, and
potassium plus other elements required for a specific agricultural use. If impurity concentrations of the process streams are controlled during single recrystallization, a grade of potassium chloride is produced of sufficient quality to be
used as feedstock for the production of potassium chemicals. Refined potassium
chloride, which is prepared by recrystallizing a portion of the first recrystallized
potassium chloride crystal crop, is sold for applications requiring exceptionally
pure potassium chloride.

An increasingly important segment of the complete fertilizer-formulation
industry is called bulk blending and requires particles of potassium chloride
that are compatible in size with other plant nutrients used to formulate the
fertilizers. Size guide number (SGN) and uniformity index (UI) are two measurements employed when dealing with blending applications. These blended fertilizers usually contain the three primary elements essential for plant growth, ie,
nitrogen, phosphorus, and potassium. Secondary elements, ie, magnesium, sulfur, zinc, iron, etc, that are deficient in the soil are also included in the blends.
Potassium chloride destined for this important market is called granular muriate
of potash. It is produced by compacting normal flotation concentrates.

Screen analyses vary but typical examples for granular and water-soluble agricultural-grade potassium chloride products are given in Table 8. Typical loose bulk densities of granular and agricultural white MOP are $1.056-1.088$ g/cm$^3$ and $1.104-1.152$ g/cm$^3$, respectively. The angles of repose for these two grades are $32-34°$ and $24-26°$, respectively. Chemical specifications are 49.8 wt % potassium for granular MOP and 51.5 wt % potassium for white soluble MOP.

Unlike agricultural potassium chloride products, grain-size distribution is not important for chemical grades of potassium chloride, as long as the products are free flowing and do not generate dust in materials handling operations. Chemical purity is of paramount importance. Typical specifications for chemical grades of KCl are given in Table 9. Typical bulk densities of the industrial and refined grades are $1.216-1.264$ g/cm$^3$ and $1.28$ g/cm$^3$, respectively. The angle of repose is the same for each, $24-26°$.

**Compaction.** A compacting machine consists of a set of two powered, inward-turning rolls (6). One roll has a fixed bearing, the other is floating in a slide and is arranged so that it can be forced toward the fixed roll by hydraulically

**Table 8. Screen Analyses for Agricultural-Grade Potassium Chloride Products[a]**

| Granular MOP | | Agriculural white MOP | |
|---|---|---|---|
| Screen aperture, mm (mesh) | Cumulative, %[b] | Screen aperture mm (mesh) | Cumulative, %[b] |
| 3.36 (6) | 7 | 0.420 (40) | 6 |
| 2.38 (8) | 45 | 0.297 (50) | 33 |
| 1.68 (12) | 80 | 0.210 (70) | 65 |
| 1.19 (16) | 96 | 0.149 (100) | 87 |
| 0.84 (20) | 98 | | |

[a]Ref. 19.
[b]Percentage of cumulative equals the percentage of MOP that is guaranteed not to pass through a given screen aperture.

**Table 9. Specifications for Chemical Grades of KCl[a]**

| Component | Guaranteed composition, wt % | |
|---|---|---|
| | Industrial | Refined[b] |
| KCl | 99.5 | 99.9[c] |
| Na | 0.18[b] | 0.0150 |
| Br | 0.09[b] | 0.0600 |
| SO$_4$ | 0.001[b] | 0.0010 |
| Ca | 0.0075 | |
| Ca + Mg | 0.0180 | 0.0030 |
| Pb | 0.0003 | 0.0003 |
| Fe | 0.0005 | 0.0005 |
| Cu, Ni | 0.0001 | 0.00005 |
| Cr, Mo, V, Ti | 0.00004 | 0.00001 |

[a]Ref. 19.
[b]Values given are maximum unless otherwise noted.
[c]Value given is minimum.

actuated pistons acting on the bearing blocks. The potassium chloride is fed continuously into the nip of the rolls from above. High pressure can be exerted on the material as it is forced or drawn between the rolls where a momentary phase change to a plastic flow condition takes place. The potassium chloride crystals are compressed into an almost continuous sheet of product, which is ejected beneath the rolls. The specific gravity of the sheet approaches the specific gravity of pure potassium chloride, 1.99 g/cm$^3$. The sheet thickness is ca 3.18–12.70 mm, depending on the grain size of the feed and the pressure applied to the rolls. A simplified flow sheet for a compaction plant is shown in Figure 2.

*Feed Grain Size, Quality, and Temperature.* The optimum particle-size distribution of the new feed to a compaction circuit is the size range within which ca 92% of the particles pass through a 1.2-mm (16-mesh) screen and less than 8% pass through a 0.07-mm (200-mesh) screen. The actual feed includes <1.4-mm(−14-mesh) material, which is recycled as undersize grains from the compacted sheet-crushing and sheet-screening operations. Excessively coarse feed results in a friable and grainy sheet. Excessive quantities of <0.07-mm (−200-mesh) particles decrease the specific gravity of the sheet and cause severe mechanical stresses on the drive gears of the compacting machine because of chattering, ie, vibration and slipping of the rolls under load.

Residual film of organic reagents on the grains inhibits fusing and results in poorly textured sheet. On crushing for granular muriate production, the flake tends to degrade to the original grain size of the compactor feed. Optimum feed temperature is 49–66°C. Higher temperatures can be employed at the expense of increased plant maintenance costs. Low temperatures result in poor quality sheet.

*Heavy-Media Separation.* Heavy-media separation, depicted in Figure 3, can only be used for relatively rich sylvinite ores that consist of large crystals of KCl and NaCl, such as those mined in Saskatchewan (6,20). Crystals of the two salts in the Saskatchewan deposits are 6–9 mm in diameter. Mine ore that is crushed sufficiently to pass through 6–9-mm screens in this size range results in a mixture consisting of discrete grains of each salt.

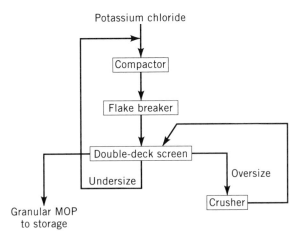

**Fig. 2.** Simplified compaction plant operation.

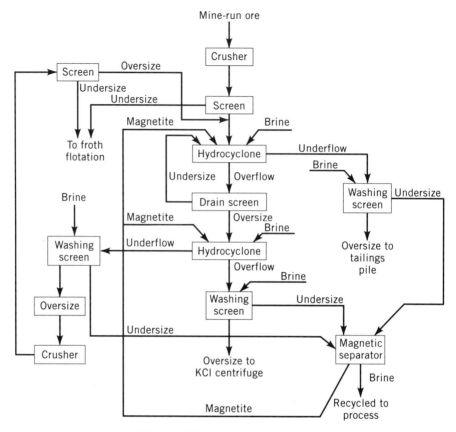

**Fig. 3.**  Heavy-medium process.

The specific gravity of potassium chloride is 1.99; of sodium chloride, 2.17. A separation of the two species can be made by placing a mixture of the salts in a liquid or medium having an intermediate specific gravity of 2.08. The result is that the KCl crystals float to the surface of the medium and the NaCl crystals sink. In laboratory determinations of whether an ore is amenable to processing by gravity separation, brominated hydrocarbons of appropriate specific gravities are employed as the media. However, these liquids are not economical for commercial operation. Consequently, a mixture of pulverized magnetite (sp gr = 5.17) and saturated brine (sp gr = 1.25) is adjusted to an appropriate specific gravity and is used to make the separation. Time is a critical variable because the magnetite settles if the mixture is held too long in a static condition. Use of hydrocyclones makes possible the separation in a fraction of a minute (21). These devices impart centrifugal force to the system, thereby permitting a separation to be made at a specific gravity less than that required in static, heavy-liquid separations.

Heavy-medium separation of potash ores is restricted to processing of the coarse fraction of the crushed ore (1.7 × 8 mm). After heavy-media separation, ashing the two products of the separation free of magnetite requires a relatively simple and inexpensive wet-screening operation. Permanent magnets are used to separate the magnetite recovered in this manner and it is recycled within the process. In contrast to grains of coarse ore and after separation into the two products, fine-ore particles (<1.7 mm) are difficult to wash free of magnetite. This is the main reason fine-ore products are excluded from the process. This processing barrier is reported to have been surmounted in coal (qv) processing (22). Mine-run ore, after being crushed through a 9-mm screen, is screened at 1.7 mm. Particles passing through the screen are sent to flotation. Coarse-ore particles that are retained on the screen are processed by heavy-media separation. A satisfactory separation cannot be made in a single-pass operation because a small quantity of salt is trapped as misplaced particles in the flowing stream of the other salt. In a sylvinite heavy-medium separation plant, an initial separation is made at an effective specific gravity of ca 2.10 to minimize the quantity of misplaced KCl particles in the sodium chloride fraction. A sufficient number of NaCl particles report to the KCl fraction as misplaced particles to cause the KCl fraction to contain less than 60 wt % $K_2O$.

After being washed to remove the magnetite, hydrocyclone underflow containing most of the NaCl is rejected from the process. Overflow containing KCl grains and misplaced NaCl grains is processed in a second stage of hydrocyclones at an effective specific gravity of 2.05. After being washed to remove the magnetite, the cyclone overflow contains KCl at ca 60 wt % $K_2O$. Underflow containing the misplaced NaCl grains and a small amount of misplaced KCl grains is recycled to the process feed.

A variation of the process is operated in New Mexico to separate a potash ore mixture consisting of langbeinite, $K_2SO_4 \cdot 2MgSO_4$; KCl; and NaCl. In this operation, langbeinite (sp gr = 2.38) is recovered as a product in the first stage of hydrocycloning. The chloride salt mixture is separated in the second-stage hydrocyclones into a sodium chloride waste and a potassium chloride–sodium chloride fraction, which is sufficiently concentrated with respect to KCl for recovery of the KCl by froth flotation.

*Fractional Crystallization.* Fractional crystallization is the traditional method for the refining of potassium chloride from underground ore deposits that contain KCl as the principal salt (Fig. 4) (23). It is a much more energy-intensive process compared with froth flotation and its use is often confined to ores that cannot be processed by froth flotation. Ores in this classification may have a fine grain structure, contain excessive amounts of clay, or have the potassium chloride chemically combined with magnesium chloride as the double salt carnallite, $KCl \cdot MgCl_2 \cdot 6H_2O$. Fractional crystallization is also used to produce high grade industrial or refined products in which the processing may involve double recrystallization and product leaching, as well as control of circuit brine chemistries to achieve the required product grade.

Fractional crystallization is based on favorable solubility relationships. Potassium chloride is much more soluble at elevated temperatures than at ambient temperatures in solutions that are saturated with sodium and potassium chlorides. Sodium chloride is slightly less soluble at elevated temperatures than at ambient temperatures in solutions that are saturated with KCl and NaCl. Working process temperatures are usually 30–110°C. The system, $KCl–NaCl–H_2O$, within this range is presented in Table 10.

Some of the ores being mined, especially from the German deposits, contain soluble salt contaminants other than NaCl, eg, magnesium chloride, which occurs as carnallite. Process brines are bled from the process and contain quantities of

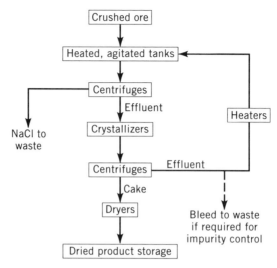

**Fig. 4.**  Fractional crystallization of potassium chloride ore.

**Table 10.  Solubility Relationships KCl–NaCl–H$_2$O$^a$**

| Temperature, °C | Quantity, wt % | | | Quantity, t/100 t water | | |
|---|---|---|---|---|---|---|
| | KCl | NaCl | H$_2$O | KCl | NaCl | Total tonnage$^b$ |
| 30 | 11.70 | 20.25 | 68.05 | 17.19 | 29.76 | 146.95 |
| 40 | 13.16 | 19.66 | 67.18 | 19.59 | 29.26 | 148.85 |
| 50 | 14.70 | 19.02 | 66.28 | 22.18 | 28.70 | 150.88 |
| 60 | 16.07 | 18.57 | 65.36 | 24.59 | 28.42 | 153.01 |
| 70 | 17.59 | 18.05 | 64.36 | 27.33 | 28.04 | 155.37 |
| 80 | 19.03 | 17.59 | 63.38 | 30.02 | 27.76 | 157.78 |
| 90 | 20.32 | 17.24 | 62.44 | 32.54 | 27.61 | 160.15 |
| 100 | 21.68 | 16.90 | 61.42 | 35.29 | 27.51 | 162.80 |

$^a$Ref. 9.
$^b$Total = $t_{KCl}$ + $t_{NaCl}$ + 100 $t_{H_2O}$.

the contaminants equal to the quantities in the ore. The bleed or waste streams are retained in waste-storage lagoons or are pumped into deep, underground porous strata. The latter method was commonly used in the former FRG for reject brine disposal. Potash production in Germany as of the mid-1990s has fallen more than 50% when compared to the years preceding reunification.

The fractional crystallization process is outlined in Figure 4. Essential steps are (1) heating the process-saturated solution remaining after separation of the product potassium chloride crystals from 30 to 100°C; (2) mixing the hot solution with ore containing sufficient KCl to saturate the solution with potassium chloride (sodium chloride is insoluble); (3) separating the sodium chloride from the hot saturated solution and rejecting it from the process; (4) cooling the solution to 30°C to recrystallize the potassium chloride dissolved from the ore; and (5) separating the potassium chloride crystals from the cool solution. The crystals are dried and, after screening into the specified products, sent to product storage. The cool solution is recycled to the ore dissolvers, ie, step (1). Appropriate adjustments are made to maintain a process water balance and to control the concentrations of water-soluble impurities.

To avoid generation of waste brines and the associated serious problem of brine disposal, the potash industry in the former FRG began converting some operations to electrostatic separation, a dry process for separating potassium salts from other soluble salts (24,25).

**Economic Aspects.**    Tables 11 and 12 illustrate the relative importance of the various grades of potassium chloride in the U.S. domestic market.

**Table 11. Sales of North American MOP to U.S. Customers, 1000 t K$_2$O$^a$**

| Grade | 1989 | 1990 | 1991 | 1992 |
|---|---|---|---|---|
| | *Agricultural* | | | |
| standard | 310 | 263 | 251 | 271 |
| coarse | 2036 | 1882 | 1862 | 1988 |
| granular | 1658 | 1658 | 1482 | 1651 |
| soluble | 342 | 334 | 349 | 383 |
| *total* | *4346* | *4137* | *3944* | *4293* |
| | *Nonagricultural* | | | |
| soluble | 116 | 131 | 85 | 107 |
| other | 305 | 314 | 387 | 365 |
| *total* | *421* | *445* | *473* | *472* |
| *Total MOP* | *4767* | *4582* | *4417* | *4765* |
| as pure KCl equivalent | 7546 | 7254 | 6992 | 7543 |

$^a$Ref. 1.

**Table 12. Bulk Prices of U.S. Potash, $/t K$_2$O$^{a,b}$**

| Muriate, 60% K$_2$O min | 1990 | | 1991 | | 1992 | |
|---|---|---|---|---|---|---|
| | Jan.–June | July–Dec. | Jan.–June | July–Dec. | Jan.–June | July–Dec. |
| standard | 129.84 | 120.52 | 126.21 | 132.27 | 132.78 | 137.05 |
| coarse | 137.32 | 129.07 | 143.36 | 129.17 | 128.72 | 136.38 |
| granular | 135.36 | 131.60 | 135.63 | 130.29 | 134.64 | 138.99 |
| all muriate$^c$ | 133.25 | 124.38 | 133.25 | 130.84 | 133.48 | 138.04 |
| sulfate, 50% K$_2$O min | 326.34 | 307.47 | 315.67 | 324.32 | 325.87 | 334.03 |

$^a$Ref. 1.
$^b$Average price, freight on board (fob) mine, based on sales.
$^c$Excluding soluble and chemical muriates.

## Potassium Sulfate

Compared to potassium chloride, potassium sulfate [7778-80-5], K$_2$SO$_4$, and its complexes with magnesium sulfate play a minor role as sources of potassium in agriculture. For example, during 1992, U.S. sales and use of agricultural-grade KCl and K$_2$SO$_4$ compounds by U.S. producers amounted to $1.552 \times 10^6$ t (expressed as $K_2O$), of which only $2.67 \times 10^5$ t consisted of potassium sulfate compounds (1). However, despite this relative difference in sales, potassium sulfate in simple form or combined with $MgSO_4$ as a double salt is an essential source of potassium for crops that are chloride-sensitive. In arid parts of the world and places where saline water is used for irrigation, potassium sulfate must be used to provide potassium in order to avoid chloride toxicity. This undesirable characteristic of chloride and other undesirable soluble elements is called the salt index. Any soluble element not utilized by a growing plant contributes to the salt index.

Potassium sulfate is also a well-established source of soluble sulfur, an essential element for plant growth. Complexes with magnesium sulfate supply water-soluble magnesium, an agronomically essential element. Although much less significant than potassium in terms of tonnage consumption, magnesium and sulfur are becoming increasingly important as essential fertilizer elements (see FERTILIZERS). U.S. production and consumption data for potassium sulfate are given in Table 13.

Quantities of potassium sulfate produced and consumed as potassium magnesium sulfate [13826-56-7], $K_2SO_4 \cdot 2MgSO_4$, are omitted in the U.S. Department of the Interior reports as classified information. Consumption data for potassium compounds identified as other potassium salts imply that the amount of potassium magnesium sulfate consumed in the United States is about double that of $K_2SO_4$. This gap is expected to widen as soils become more depleted of natural magnesium- and sulfur-containing minerals.

Several types of chemical processes are used to produce potassium sulfate. The traditional Mannheim process is used in countries that produce KCl but lack a natural source of sulfate salts for converting the KCl to $K_2SO_4$. In this process, KCl reacts with sulfuric acid to yield $K_2SO_4$ and HCl as a co-product.

$$2\,KCl + H_2SO_4 \longrightarrow K_2SO_4 + 2\,HCl$$

**Table 13. U.S. Production and Consumption of Sulfate of Potash[a,b]**

| Parameter | 1989 | 1990 | 1991 | 1992 |
|---|---|---|---|---|
| production, t $\times 10^3$ | 166 | 219 | 230 | 243 |
| sales by producers, t $\times 10^3$ | 147 | 221 | 211 | 267 |
| value, fob mine, $ $\times 10^3$ | 47,355 | 70,226 | 67,432 | 87,884 |
| exports, t $\times 10^3$ | 78 | 124[c] | 104 | 158 |
| imports, t $\times 10^3$ | 32 | 26 | 29 | 35 |
| value, CIF[d] to U.S. port, $ $\times 10^3$ | 11,700 | 11,000 | 11,800 | 13,600 |
| apparent consumption[e], t $\times 10^3$ | 101 | 123 | 136 | 146 |
| year-end producers' stocks, t $\times 10^3$ | 42 | 39 | 58 | 34 |

[a]Ref. 1.
[b]Potassium magnesium sulfate not included.
[c]Valued at $43,300,000.
[d]CIF = cost, insurance, and freight.
[e]Apparent consumption = sales + imports − exports.

This process or a variation called the Hargreaves process is also used in areas where sulfuric acid is available as a by-product or where co-product HCl is needed for the production of other chemicals.

Potassium sulfate is produced in Sicily by controlled decomposition of the natural mineral kainite, $KCl \cdot MgSO_4 \cdot 2.75H_2O$ (26). This salt is first converted to schoenite in an aqueous solution from a potassium sulfate conversion step. A similar process is used in the United States. Kainite is obtained as the potassium feedstock by stage evaporation of Great Salt Lake bitterns (see CHEMICALS FROM BRINES).

$$4(KCl \cdot MgSO_4 \cdot 2.75H_2O)(aq) + H_2O \longrightarrow 2(K_2SO_4 \cdot MgSO_4 \cdot 6H_2O) + 2\ MgCl_2$$

$$K_2SO_4 \cdot MgSO_4 \cdot 6H_2O \longrightarrow K_2SO_4 + MgSO_4 + 6\ H_2O$$

Water for the kainite conversion comes from the hydrated $MgSO_4$. This solution is saturated with $K_2SO_4$. Use of potassium sulfate mother liquor as a source of water for the reaction lowers the $K_2SO_4$ lost in the $MgCl_2$ solution, which is rejected as a waste stream from the process. It also is a solvent for sodium chloride that enters the process as a contaminant in kainite.

In Canada, ion-exchange (qv) technology has been used to produce potassium sulfate (4). Ion-exchange resins remove sulfate ions from lake water containing sodium sulfate. This is followed by a wash with aqueous solutions prepared from lower grade muriate of potash. High purity potassium sulfate is collected from the crystallizers into which the wash runs.

Approximately one-half of the world production of $K_2SO_4$ is obtained by the reaction of potassium chloride with magnesium sulfate as the double salt schoenite, $K_2SO_4 \cdot MgSO_4 \cdot 6H_2O$ (27), or double salt langbeinite, $K_2SO_4 \cdot 2MgSO_4$. In Germany, kieserite, $MgSO_4 \cdot H_2O$, is the predominant soluble magnesium sulfate salt. This salt is converted to schoenite by the reaction of kieserite with the KCl that remains dissolved in the solution from the $K_2SO_4$ reaction step. Schoenite, after separation from the final $MgCl_2$ solution, reacts with KCl. The basic reactions are as follows:

$$2(MgSO_4 \cdot H_2O) + 4\ H_2O + 2\ KCl \longrightarrow K_2SO_4 \cdot MgSO_4 \cdot 6H_2O + MgCl_2$$

$$K_2SO_4 \cdot MgSO_4 \cdot 6H_2O + 2\ KCl \longrightarrow 2\ K_2SO_4 + MgCl_2\ (aq)$$

In contrast to sodium chloride, langbeinite has an extremely slow rate of solution. Upon control of agitation time, essentially all the sodium chloride dissolves but most of the langbeinite remains as a solid. Langbeinite is separated from the brine, dried, and then screened into granular, standard, and special-standard particle sizes. These fractions are marketed directly. In one plant, the unsalable fines are used as the source of sulfate reactant for the production of potassium sulfate.

Kieserite is not present in U.S. potassium salt deposits in commercial quantities. Langbeinite is the predominant U.S. magnesium sulfate salt. The latter, a raw material for the production of potassium sulfate in New Mexico, reacts directly with potassium chloride:

$$K_2SO_4 \cdot 2MgSO_4 + 4\ KCl \longrightarrow 3\ K_2SO_4 + 2\ MgCl_2$$

A simplified flow sheet depicting the process used in New Mexico is presented in Figure 5.

The value of langbeinite as a fertilizer is enhanced because, in pure form, it contains 18.8 wt % potassium, 11.7 wt % magnesium, and 23.0 wt % sulfur. All three elements are essential nutrients for plant growth. Commercial grades contain ca 97 wt % mineral; the remaining 3 wt % consists of water-insoluble clays and residual sodium chloride.

Processes used to refine ores containing langbeinite are selected on the basis of the amount of langbeinite in the ores and the quantity of recoverable potassium chloride in the ores. In one potassium salt mineralized layer, identified as USGS (U.S. Geological Survey) Fourth Ore Zone, the ore contains relatively large amounts of langbeinite (30–45 wt %) and is generally lean in KCl. Except in areas where there are packets of ore containing significant amounts of KCl, ore from the Fourth Ore Zone is processed in a simple, freshwater washing process (29). Ore crushed through a 6-mm screen is agitated with fresh water in a ratio so as to produce an almost saturated NaCl brine.

Potassium ore in the USGS Fifth Ore Zone and pockets of ore in the Fourth Ore Zone contain commercially recoverable quantities of both langbeinite and KCl. Processing mixed KCl–langbeinite ore in the washing process to purify the langbeinite results in the solution and eventual loss of potassium chloride values. Operations to recover the potassium chloride must be conducted in brine

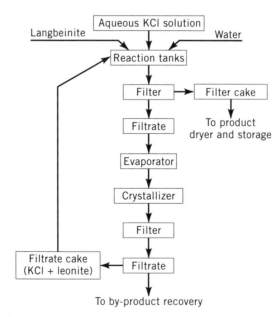

**Fig. 5.**   Process for the production of potassium sulfate (28).

saturated with potassium chloride. In commercial processing for KCl production, the process brine is also saturated with sodium chloride. A complex process used in New Mexico to recover KCl and langbeinite separately involves crushing the mine-run ore through a 6-mm screen (30). It is then wet-screened at 1 mm and oversize material is processed by heavy media to separate langbeinite from the chloride salts, KCl and NaCl. The recovered langbeinite is washed with brine to remove fine particles of magnetite. The washed langbeinite is then passed into the langbeinite product dryers. Undersize from the screens is processed by magnetic separation to recover the magnetite, which is recycled to the process.

Further upgrading of the potassium chloride content of the chloride salts recovered from the initial heavy-medium separation takes place in a second heavy-medium separation at a somewhat lower specific gravity than the first separation. Sodium chloride is discarded as a waste; the enriched KCl fraction is sent to a flotation process where a final separation of KCl from NaCl is made. Mine-run ore less than 1 mm that is not amenable to heavy-medium separation is processed by amine flotation to separate KCl from NaCl and langbeinite. The NaCl and langbeinite mixture, as tailings from the KCl flotation process, is treated by anionic flotation to recover the langbeinite.

Chemical fertilizer is the predominant market for langbeinite. Comparatively small but increasing amounts of langbeinite are used by the animal feed ingredient industry (see FEEDS AND FEED ADDITIVES). Producers who supply this market must take special precautions to be sure that any langbeinite intended as an animal feed ingredient meets all USDA specifications for toxic heavy metals and other impurities.

Potassium sulfate and langbeinite may be screened into three different size ranges to serve all segments of the fertilizer industry. Typical screen analyses of the size products are presented in Table 14. Chemical analyses for the products are shown in Table 15.

**Table 14. Typical Screen Analyses of Potassium Sulfate and Langbeinite[a]**

| | Cumulative, %[b] | | |
|---|---|---|---|
| Screen aperture, mm (mesh) | Granular | Standard | Special standard |
| 3.36 (6) | 2 | | |
| 2.38 (8) | 20 | | |
| 1.68 (12) | 60 | | |
| 1.19 (16) | 90 | 10 | |
| 0.84 (20) | 97 | 35 | |
| 0.59 (30) | | 63 | |
| 0.42 (40) | | 90 | 8 |
| 0.30 (50) | | 95 | 40 |
| 0.21 (70) | | 98 | 60 |
| 0.15 (100) | | | 76 |
| 0.07 (200) | | | 94 |

[a]Ref. 19.
[b]Percentage of cumulative equals the percentage of MOP that is guaranteed not to pass through a given screen aperture.

**Table 15. Chemical Analyses of Potassium Sulfate and Langbeinite[a], wt %**

| Component | $K_2SO_4$ | Langbeinite-grade | |
| | | Granular and standard | Special standard |
| --- | --- | --- | --- |
| K | 41.9 | 18.5 | 18.0 |
| Mg | 0.9 | 10.9 | 10.9 |
| Ca | 0.3 | 0.2 | 0.2 |
| Na | 0.2 | 0.8 | 0.8 |
| Cl | 1.5 | 1.5 | 1.5 |
| $SO_4$ | 51.5 | 67.4 | 67.4 |
| insoluble[b] | 1.0 | 0.3 | 0.3 |
| $H_2O$ | 0.1 | 0.1 | 0.1 |
| S | 17.5 | 22.4 | 22.4 |

[a]Ref. 19.
[b]For instance, clay.

## Other Potassium Compounds

**Potassium Acetate.** Potassium acetate [127-08-2], $KC_2H_3O_2$, is usually made from carbonate and acetic acid as follows:

$$K_2CO_3 + 2\,HC_2H_3O_2 \longrightarrow 2\,KC_2H_3O_2 + H_2O + CO_2$$

It is also made industrially by the simple reaction of acetic acid and potassium hydroxide:

$$HC_2H_3O_2 + KOH \longrightarrow KC_2H_3O_2 + H_2O$$

Potassium acetate is very soluble and is used in the manufacture of glass (qv), as a buffer (see HYDROGEN-ION ACTIVITY) or a dehydrating agent, and in medicine as a diuretic (see DIURETICS). It is deliquescent and is used as a softening agent for papers and textiles.

**Potassium Bromide.** Potassium bromide [7758-02-3], KBr, can be prepared by a variation of the process by which bromine is absorbed from ocean water. Potassium carbonate is used instead of sodium carbonate:

$$3\,K_2CO_3 + 3\,Br_2 \longrightarrow KBrO_3 + 3\,CO_2 + 5\,KBr$$

Potassium bromate [7758-01-7] is much less soluble than the bromide and can mostly be removed by filtration; the remaining bromate is reduced with iron. After filtration of the iron oxide, the KBr is crystallized.

An alternative method, which does not involve the formation of bromate, is treatment of iron turnings with a 35 wt % aqueous solution of bromine. Ferroso-ferric bromide, $Fe_3Br_8\cdot16H_2O$, forms and can be crystallized. The iron bromide is heated to a boil with a slight excess of 15 wt % potassium carbonate solution.

$$Fe_3Br_8\cdot16H_2O + 4\,K_2CO_3 \longrightarrow 8\,KBr + 4\,CO_2 + Fe_3O_4 + 16\,H_2O$$

Ferrosoferric bromide effects formation of a precipitate that is readily filtered in the second step. No final purification should be necessary.

Potassium bromide is extensively used in photography (qv) and engraving. It is the usual source of bromine in organic synthesis. In medicine, it is a classic sedative.

**Potassium Carbonate.** Except for small amounts produced by obsolete processes, eg, the leaching of wood ashes and the Engel-Precht process, potassium carbonate is produced by the carbonation, ie, via reaction with carbon dioxide, of potassium hydroxide. Potassium carbonate is available commercially as a concentrated solution containing ca 47 wt % $K_2CO_3$ or in granular crystalline form containing 99.5 wt % $K_2CO_3$. Impurities are small amounts of sodium and chloride plus trace amounts (<2 ppm) of heavy metals such as lead. Heavy metals are a concern because potassium carbonate is used in the production of chocolate intended for human consumption.

A phase diagram depicting the solubility of potassium carbonate in water is shown in Figure 6. Solubilities in the system $KOH-K_2CO_3-KHCO_3-H_2O$ are shown in Figure 7.

In many heavy-chemical manufacturing operations requiring an intermediate alkaline metal carbonate reactant, potassium carbonate and sodium

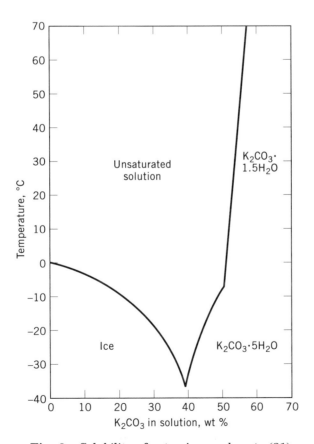

**Fig. 6.** Solubility of potassium carbonate (31).

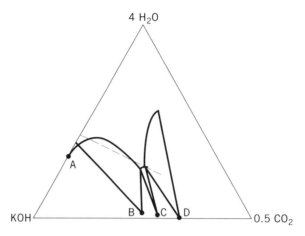

**Fig. 7.** Solubilities in the system $KOH-K_2CO_3-KHCO_3-H_2O$ at 25°C: A, $KOH \cdot 2H_2O$; B, $K_2CO_3 \cdot 1.5H_2O$; C, $K_2CO_3 \cdot 2KHCO_3 \cdot 1.5H_2O$; and D, $KHCO_3$.

carbonate can be used with equal effectiveness. The cost of producing potassium carbonate is four to five times greater than that of producing sodium carbonate. Thus, sodium carbonate is almost always used for applications in which the two carbonates are equivalent chemically. Potassium carbonate possesses properties for some applications that preclude the substitution of sodium carbonate, eg, in television glass. Television glass accounts for a substantial portion of the consumption of potassium carbonate because the potassium salt is more compatible with the lead, barium, and strontium oxides contained in these glasses than is sodium carbonate. The properties of these specialty glasses include high electrical resistivity, high index of refraction, good brilliance or luster, low softening point, and wide temperature working range. In addition, potassium carbonate produces improved behavior of colorant in glass. Uses of potassium carbonate other than glass include applications in ceramics (qv), engraving processes, finishing leather (qv), chemical dyes and pigments, foods, cleaners, and gas purification, eg, $CO_2$ and $H_2S$ (31).

**Potassium Bicarbonate.** Potassium bicarbonate, $KHCO_3$, is made by absorption of $CO_2$ in a carbonate solution, ie, potassium hydroxide is carbonated to $K_2CO_3$, which in turn is carbonated to $KHCO_3$. The changes of solubility during carbonation are presented in Figure 7. Usually the carbonate is crystallized, washed, and redissolved before carbonation and crystallization of the bicarbonate. The solutions can also be filtered. Thus, the purest carbonate is made by calcining the bicarbonate. Potassium bicarbonate is more stable than $NaHCO_3$ at normal temperatures, but $KHCO_3$ decomposes at ca 190°C.

Potassium bicarbonate is used in foods and medicine. It is approximately twice as effective as $NaHCO_3$ in dry-powder fire extinguishers, perhaps because the potassium affects the free-radical mechanism of flame propagation. However, the material does not have good handling characteristics.

**Potassium Formate.** Potassium formate [590-29-4], HCOOK, is made by the following reaction (32):

$$CO + KOH \longrightarrow HCOOK$$

Carbon monoxide (qv), eg, by-product CO from phosphorus manufacture or extracted from synthesis gas, is freed of acidic gases and absorbed in 50–80 wt % KOH at 100–200°C at a partial pressure of $P_{CO} > 690$ kPa (>100 psi). The reaction is fairly slow.

Potassium formate melts at 167°C and decomposes almost entirely to oxalate [583-52-8] at ca 360°C. Above and below this temperature, different decomposition products form. Most of the formate produced is converted to oxalate.

**Potassium Hydroxide.** Potassium hydroxide [1310-58-3] is produced industrially by the electrolysis of potassium chloride. The electrolysis cells are similar to those used for $NaOH–Cl_2$ production (see ELECTROCHEMICAL PROCESSING, INORGANIC). In diaphragm cells, the product liquor contains 10–15 wt % KOH and ca 10 wt % KCl. Most of the KCl crystallizes during concentration by evaporation and subsequent cooling, which results in purification of the KOH solution. The concentrated 50 wt % KOH solution contains ca 0.6 wt % KCl. A phase diagram of the system $KOH–H_2O$ is given in Figure 8.

For purity and energy conservation reasons, mercury cells are used to produce most of the KOH in the United States. Feed to the cells consists of brine saturated with KCl at a moderate temperature. After purification to remove assorted metal impurities, the brine is fed into the cells, which operate on dc. These cells can be visualized as large batteries containing positive and

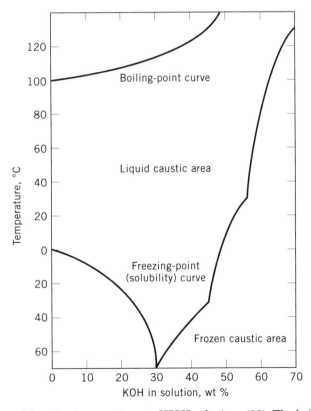

**Fig. 8.** Boiling and freezing temperatures of KOH solutions (33). The boiling point curve assumes a pressure of 101.3 kPa (760 mm Hg).

negative terminals. The positive terminal or anode is constructed of titanium; the negative terminal or cathode is a flowing layer of metallic mercury. A potassium–mercury amalgam forms from the reaction of the potassium in the brine with the flowing layer of mercury, and the amalgam flows into the denuder. Water is added and reacts with the elemental potassium in the amalgam to form potassium hydroxide and hydrogen. The stripped mercury is recycled into the cells. Potassium is recovered as 50 wt % KOH solution. After purification to remove the mercury, the hydrogen (qv) is compressed to liquid hydrogen or is used to produce hydrochloric acid (see HYDROGEN CHLORIDE).

Environmental awareness is a prime concern in all KOH plants. Safety precautions required in KOH and chlorine operations are well documented in operating manuals and sales brochures published by all commercial producers. Discharges of waste effluents containing mercury are strictly forbidden.

Principal uses of KOH include chemicals, particularly the production of potassium carbonate and potassium permanganate, pesticides (qv), fertilizers (qv), and other agricultural products; soaps and detergents; scrubbing and cleaning operations, eg, industrial gases; dyes and colorants; and rubber chemicals (qv) (10,34).

**Potassium Iodide.** Some potassium iodide [7681-11-0], KI, is made by the iron and carbonate process similar to that used for the bromide. However, most U.S. production involves absorption of iodine in KOH.

$$3 I_2 + 6 KOH \longrightarrow 5 KI + KIO_3 + 3 H_2O$$

Approximately 80 wt % of the potassium iodate [7758-05-6], $KIO_3$, crystallizes from the reaction mixture and is separated for sale. Of the remainder, 90 wt % is removed by evaporation, fusion, and heating to ca 600°C.

$$2 KIO_3 \longrightarrow 2 KI + 3 O_2$$

The iodate is a poison; potassium iodide, however, is used in foodstuffs. Thus the iodate must be completely removed frequently by a final reduction with carbon. After re-solution in water, further purification is carried out before recrystallization. Iron, barium, carbonate, and hydrogen sulfide are used to effect precipitation of sulfates and heavy metals.

Approximately half of the iodine consumed is used to make potassium iodide (see IODINE AND IODINE COMPOUNDS). Production of KI is almost 1000 t/yr. Its main uses are in animal and human food, particularly in iodized salt, pharmaceuticals (qv), and photography (qv).

**Potassium Nitrate.** Potassium nitrate [7757-79-1], $KNO_3$, is produced commercially in the United States based on the reaction of potassium chloride and nitric acid (qv) (35). Ammonia (qv) oxidation is the source for the nitric acid and the reaction is manipulated chemically to yield chlorine as a co-product. The process is operated at an elevated temperature to drive the reaction to completion according to the following equation:

$$3 KCl + 4 HNO_3 \longrightarrow 3 KNO_3 + Cl_2 + NOCl + 2 H_2O$$

Nitrosyl chloride, a product of the basic reaction, has no commercial value and is converted to salable chlorine and to nitric acid for recycling.

$$2 \text{ NOCl} + 4 \text{ HNO}_3 \longrightarrow 6 \text{ NO}_2 + \text{Cl}_2 + 2 \text{ H}_2\text{O}$$

$$4 \text{ NO}_2 + \text{O}_2 + 2 \text{ H}_2\text{O} \longrightarrow 4 \text{ HNO}_3$$

A simplified flow diagram of the process is presented in Figure 9.

Physicochemical relationships are such that solid potassium chloride can be converted to solid potassium nitrate in a one-stage operation of the simplest kind. The conversion takes place in a stirred reaction system (Fig. 10). The overall separation is analogous to a rectification and stripping operation in a distillation process.

A methathesis reaction between hot aqueous sodium nitrate and solid potassium chloride generates aqueous potassium nitrate and solid sodium chloride.

$$\text{NaNO}_3 \text{ (aq)} + \text{KCl (s)} \longrightarrow \text{NaCl (s)} + \text{KNO}_3 \text{ (aq)}$$

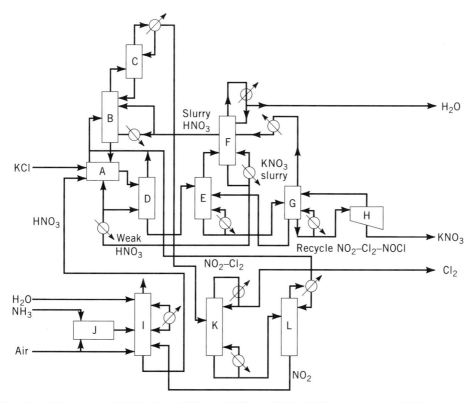

**Fig. 9.** Schematic of $KNO_2$ from $NH_2$ and KCl: A, $KCl$–$HNO_2$ reactor; B, NOCl oxidizer; C, acid eliminator; D, gas stripper; E, water stripper; F, $H_2O$–$HNO_2$ fractionator; G, evaporator–crystallizer; H, centrifuge; I, NO–$NO_2$ absorber; J, $NH_2$ burner; K, $Cl_2$ fractionator; and L, $NO_2$ fractionator.

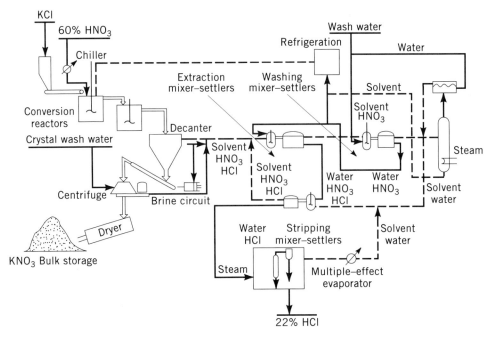

**Fig. 10.** Flow sheet for the production of KNO₃–IMI process, where (——) represents the product stream; (– – –) the reaction solvent; and (— — —) aqueous streams (36).

The hot potassium nitrate solution is drawn through the crystalline NaCl and the KNO₃ is isolated. Solubility data for KNO₃ in water are presented in Table 16.

Solvent extraction has been employed in Israel to produce KNO₃ according to the following equation (37).

$$KCl + HNO_3 \longrightarrow HCl + KNO_3$$

Among the properties sought in the solvent are low cost, availability, stability, low volatility at ambient temperature, limited miscibility in aqueous systems

**Table 16. Solubility of KNO₃ in Water[a]**

| Temperature, °C | KNO₃, g/100 g saturated solution | Temperature, °C | KNO₃, g/100 g saturated solution |
|---|---|---|---|
| 0 | 11.7 | 60 | 52.2 |
| 10 | 24.0 | 70 | 57.8 |
| 20 | 24.0 | 80 | 62.8 |
| 30 | 31.3 | 90 | 67.0 |
| 40 | 39.0 | 100 | 71.0 |
| 50 | 46.0 | 110 | 74.8 |

[a]Ref. 36.

present in the process, no solvent capacity for the salts, good solvent capacity for the acids, and sufficient difference in distribution coefficient of the two acids to permit their separation in the solvent-extraction operation. Practical solvents are $C_4$, $C_5$, and $C_6$ alcohols. For industrial process, $C_5$ alcohols are the best choice (see AMYL ALCOHOLS). Small quantities of potassium nitrate continue to be produced from natural sources, eg, the caliche deposits in Chile.

Although no longer in great demand for the preparation of black gunpowder, potassium nitrate is used in the preparation of smokeless powder and slow-burning fuses (10). The main uses of potassium nitrate include fertilizer, fireworks, steel (qv) making, the food industry, and in a eutectic mixture with $NaNO_2$, service as a heat-transfer agent (see HEAT-EXCHANGE TECHNOLOGY) (10).

**Potassium Phosphates.** Phosphoric acid is the source of phosphate for the production of potassium phosphates (see PHOSPORIC ACID AND PHOSPHATES). Orthophosphates, ranging from monopotassium phosphate [7758-11-4], $KH_2PO_3$, to tripotassium phosphate [7758-53-2], $K_3PO_4$, are made by simple neutralization with KOH. Similar to the sodium analogues, potassium orthophosphates have buffering, complexation, and cleaning capabilities. Because of their significantly lower cost, the sodium compounds are much more widely used than potassium phosphates. The latter are generally limited to special applications where solubility is important, eg, the removal of $H_2S$ from gases in a process similar to that involving amines. In this case, the high solubility minimizes the heat requirements. A growing market has developed for specialty liquid fertilizers containing potassium orthophosphates, usually a mixture of $KH_2PO_4$ and $K_2HPO_4$, for pH control. Although these are expensive sources of phosphorus and potassium, the fertilizers do not increase the salt index of soil for house plants or the media used in greenhouse horticulture.

Devising an economical method of producing agricultural-grade potassium phosphates from potassium chloride and wet-process phosphoric acid has been the subject of intense agricultural–chemical research (37–39). Limited quantities have been produced industrially. The impact on the overall quantities of phosphorus and potassium compounds consumed by the fertilizer industry is small. Because potassium phosphates are an excellent source of two essential fertilizer elements, this research is expected to continue.

Condensed potassium phosphates have been used as builders in liquid detergents. The compound tetrapotassium pyrophosphate [7320-34-5], which forms by dehydrating $K_2HPO_4$ at 400°C, was used.

$$2\ K_2HPO_4 \longrightarrow K_4P_2O_7 + H_2O$$

Potassium tripolyphosphate [13845-36-8], $K_5P_3O_{10}$, also used in detergents, is made by dehydrating an equimolar mixture of mono- and dipotassium phosphates.

$$KH_2PO_4 + 2\ K_2HPO_4 \longrightarrow K_5P_3O_{10} + 2\ H_2O$$

**Potassium Acid Tartrate.**  The monopotassium salt of tartaric acid, $KHC_4H_4O_6$, also called potassium acid tartrate or cream of tartar, is a common ingredient in baking powders and leavening systems (see BAKERY PROCESSES AND LEAVENING AGENTS). It has limited solubility in cold water and therefore does not prematurely activate the leavening reactions during dough mixing processes. These reactions are accelerated during the baking process, partially owing to an increase in the solubility of cream of tartar at the higher temperature. Cream of tartar, tartaric acid, and its Rochelle salt are manufactured from the by-products of the wine (qv) industry, ie, press cakes, lees, and argols.

## Biological Role of Potassium

Potassium ions are essential to both plants and animals (see MINERAL NUTRIENTS). Within cells, potassium serves the critical role as counterion for various carboxylates, phosphates, and sulfates, and stabilizes macromolecular structures. $K^+$ is the principal cation mediating the osmotic balance of the extra cellular fluids, and it is accumulated in cells in concert with the expulsion of $Na^+$ (40). If a cell dies or has its metabolism blocked, the concentration gradient for intracellular potassium to extracellular potassium disappears as potassium ions slowly diffuse out across the cell membrane. This implies that metabolic energy is expended in maintaining the gradient. The term sodium pump is used for the mechanism of active transport in rejection of sodium and accumulation of potassium by a cell. Energy for the sodium pump is provided by adenosine triphosphate (ATP) hydrolysis. This $K^+/Na^+$ separation has allowed the evolution of the reversible transmembrane electrical potentials essential for nerve and muscle action in animals. The $K^+/Na^+$ ratio for excitable membranes requires precise homeostatic control over internal electrolyte concentrations. Alkali metal ion levels in the circulating fluids are adjusted by the effect of thirst and salt-craving on intake and the influence of antidiuretic hormone and aldosterone on reabsorption in the kidney tubules.

Potassium transport through the hydrophobic interior of a membrane can be facilitated by a number of natural compounds that form lipid-soluble alkali metal cation complexes (41). For example, nonactin and valinomycin act as antibiotics by disrupting the essential $K^+/Na^+$ concentration gradient of bacteria. The natural ionophores are grouped as nigericins, macrotetrolides, depsipeptides, and cyclic peptides. Many of these organic ligands have a remarkably higher specificity for forming complexes with potassium over sodium. Selection of $K^+$ over $Na^+$ may be a size effect or may be the result of $Na^+$ being less easily stripped of coordinated water molecules. Ionophores have been synthesized primarily as cyclic polyethers generally called crown compounds, and as bicyclic polyether–diamine compounds generally known as kryptates.

Potassium is required for enzyme activity in a few special cases, the most widely studied example of which is the enzyme pyruvate kinase. In plants it is required for protein and starch synthesis. Potassium is also involved in water and nutrient transport within and into the plant, and has a role in photosynthesis. Although sodium and potassium are similar in their inorganic chemical behavior, these ions are different in their physiological activities. In fact, their functions are often mutually antagonistic. For example, $K^+$ increases both the respiration rate in muscle tissue and the rate of protein synthesis, whereas $Na^+$ inhibits both processes (42).

Information about a food's potassium content is required on the nutrition facts panel only if the food contains added potassium as a nutrient or if claims about it as a nutrient appear on the label. In all other cases, it is voluntary. The recommended daily value for potassium is 3500 mg. The following labels have been designated for foods: high potassium (700 mg or more per serving); good source of potassium (350–665 mg per serving); more or added potassium (at least 350 mg more per serving than the reference food) (43).

## Analytical Methods

Potassium is analyzed in chemicals that are used in the fertilizer industry and in finished fertilizers by flame photometric methods (44) or volumetric sodium tertraphenylboron methods (45) as approved by the AOAC. Gravimetric determination of potassium as $K_2PtCl_6$, known as the Lindo-Gladding method (46), and the wet-digestion determination of potassium (47) have been declared surplus methods by the AOAC. Other methods used for control purposes and special analyses include atomic absorption spectrophotometry, inductively coupled plasma (icp) emission spectrophotometry, and a radiometric method based on measuring the radioactivity of the minute amount of the $^{40}K$ isotope present in all potassium compounds (48).

## Health and Safety Factors

Potash mining and refining operations in the United States are strictly regulated by appropriate federal and state agencies. Field studies conducted by NIOSH failed to disclose any evidence of predisposition of undergound miners to any of the diseases evaluated, including lung cancer (49). Exposure to dust, ie, of sodium and potassium salts, commonly encountered in the potash industry did not influence mortality for various medical conditions that are believed to be significant statistically in some nonmetallic mining and processing industries.

Potassium compounds listed as hazardous substances by the U.S. EPA are given in Table 17. The U.S. Department of Transportation (DOT) maintains a Hazardous Materials Table that designates the listed materials as hazardous for the purpose of transportation, packaging, and labeling (50). Potassium compound DOT hazard classifications are also listed in Table 17.

Potassium compounds commonly used in fertilizers, eg, KCl and $K_2SO_4$, are not considered to be hazardous substances. Detailed information concerning health and safety precautions recommended for a specific, industrially produced potassium chemical can be obtained by contacting a manufacturer directly. Principal potassium chemical producers are listed in buyers' guides published annually by chemical trade magazines (52).

**Table 17. Hazardous Potassium Chemicals**

| DOT shipping name[a] | CAS Registry Number | DOT classification[a] | EPA reportable quantity[b], kg |
|---|---|---|---|
| potassium arsenate | [7784-41-0] | poison B | 0.454 |
| potassium arsenite | [10124-50-2] | poison B | 0.454 |
| potassium bromate | [7758-01-2] | oxidizer | |
| potassium chromate | [7789-00-6] | ORM-E[c] | 4.54 |
| potassium cyanide | [151-50-8] | poison B | 4.54 |
| potassium cyanide solution | [151-50-8] | poison B | 4.54 |
| potassium dichloro-s-triazinetrione | [2244-21-5] | oxidizer | |
| potassium dichromate | [7778-50-9] | ORM-A[c] | 4.5410 |
| potassium fluoride | [7789-23-3] | ORM-B[c] | |
| potassium fluoride solution | [7789-23-3] | corrosive material | |
| potassium hydrogen fluoride solution | [7789-29-9] | corrosive material | |
| potassium hydrogen sulfate | [7646-93-7] | ORM-B[c] | |
| potassium hydroxide | | | |
| dry solid | [1310-58-3] | corrosive material | 454 |
| liquid or solution | [1310-58-3] | corrosive material | |
| potassium metabisulfite | [16731-55-8] | ORM-B[c] | |
| potassium metal or metallic | [7440-09-7] | flammable solid | |
| potassium metal, liquid alloy | [7440-09-7] | flammable solid | |
| potassium nitrate | [7757-79-1] | oxidizer | |
| potassium perchlorate | [7778-74-7] | oxidizer | |
| potassium permanganate | [7722-64-7] | oxidizer | 45.4 |
| potassium peroxide | [17014-71-0] | oxidizer | |
| potassium superoxide | [12030-88-5] | oxidizer | |
| potassium sulfide | [1312-73-8] | flammable solid | |

[a]Ref. 50.
[b]Ref. 51.
[c]ORM = other regulated material.

# BIBLIOGRAPHY

"Potassium Compounds" in *ECT* 1st ed., Vol. 11, pp. 12–29, by W. A. Cunningham, University of Texas; in *ECT* 2nd ed., Vol. 16, 369–400, by J. J. Jacobs, Jacobs Engineering Co.; in *ECT* 3rd ed., Vol. 18, 920–950, by W. B. Dancy, International Minerals & Chemical Corp.

1. J. Searls, *Mining Eng.* **45**(6), 579–582 (June 1993).
2. J. P. Searls, *Minerals Yearbook*, 1007–1033 (1992).
3. H. M. Paynter, *Chemtech* **21**(7), 391–393 (1991).
4. J. P. Searls, *Eng. Mining J.* **195**, 53–55 (1994).
5. S. Williams-Stroud, *Mining Eng.* **46**(6), 540–542 (1994).
6. V. A. Zandon, E. A. Schoeld, and J. McManus, *Minerals Processing Handbook*, 3rd ed., Society of Mining Engineers of AIME, Denver, Colo., 1982, Section 22.
7. R. C. Weisner, J. F. Lemons, Jr., and L. V. Coppa, *Valuation of Potash Occurrence Within the Nuclear Waste Isolation Pilot Plant Site in Southeastern New Mexico*, U.S. Department of the Interior, Bureau of Mines, Washington, D.C., 1980, p. 6.

8. D. R. Lide, ed., *Handbook of Chemistry and Physics*, 75th ed., CRC Press, Inc., Boca Raton, Fla., 1995.

9. W. F. Linke and A. Seidell, *Solubilities of Inorganic and Metal-Organic Compounds*, 4th ed., The American Chemical Society, Washington, D.C., 1965.

10. G. T. Austin, ed., *Shreve's Chemical Process Industries*, 5th ed., McGraw-Hill Book Co., Inc., New York, 1984, pp. 288–302.

11. O. Braitsch, *Salt Deposits, Their Origin and Composition*, Springer-Verlag, New York, 1971.

12. J. D'Ans, *Die Losungsgleichgewichte Der Salze Ozeanischer Salzablagerung*, Kaliforschungs-Anstalt GmbH, Berlin, Germany, 1933.

13. R. H. Singleton, *Potash Mineral Commodity*, Profiles MCP11, U.S. Department of the Interior, Bureau of Mines, Washington, D.C., Feb. 1978, pp. 12–13.

14. D. Jackson, Jr., *Eng. Min. J.* **174**, 59 (July 1973).

15. U.S. Pat. 3,058,729 (Oct. 16, 1962), J. B. Dahms and B. P. Edmonds (to Pittsburgh Plate Glass Co.); Can. Pat. 672,308 (Oct. 15, 1963), J. B. Dahms (to Pittsburgh Plate Glass Co.); U.S. Pat. 4,007,964 (Feb. 15, 1977), E. L. Goldsmith (to Pittsburgh Plate Glass Co., Canada, Ltd.); U.S. Pat. 3,262,741 (July 26, 1966), B. P. Edmonds and J. B. Dahms (to Pittsburgh Plate Glass Co.); U.S. Pat. 3,433,530 (Mar. 18, 1969), J. B. Dahms and B. P. Edmonds (to Pittsburgh Plate Glass Co.).

16. B. S. Crocker, J. T. Dew, and R. J. Roach, *CIM Bull.* **62**(688), 729 (July 1969).

17. H. R. Armstrong, *A Modern Canadian Potash Plant*, Wellman/Lord, Inc., Lakeland, Fla., 1970.

18. U.S. Pat. 3,904,520 (Sept. 9, 1975), W. B. Dancy (to International Minerals and Chemical Corp.).

19. *Particle Size for Agricultural Grade KCl Products*, published specifications, International Minerals and Chemical Corp., Northbrook, Ill., Mar. 1979.

20. Can. Pat. 792,819 (Aug. 10, 1968), W. B. Dancy (to International Minerals and Chemical Corp.).

21. H. H. Dreissen and F. J. Fontein, *Assoc. Min. Eng.* **226**, 101 (Mar. 1963).

22. J. Mengelers and J. H. Absil, *Coal Min. Process*, 62 (May 1976).

23. O. Krull, *Das Kali II*, **8**, 178 (1928).

24. H. Autenrieth, *Kali Steinsalz* **8**, 171 (June 1969).

25. A. Singewald, *Kali Steinsalz* **8**, 2 (Mar. 1980).

26. U.S. Pat. 2,902,344 (Sept. 1, 1959), G. Cevidalli, J. Marchi, and P. Saccardo (to Sincat Societa Industriale Contanese).

27. O. Krull, *Das Kali II*, **8**, 315 (1928).

28. U.S. Pat. 2,684,285 (July 20, 1954), W. B. Dancy (to International Minerals and Chemical Corp.).

29. G. T. Harley and G. E. Atwood, *Langbeinite: Mining and Processing*, International Minerals and Chemical Corp., Carlsbad, N.M., Jan. 1947.

30. U.S. Pat. 3,538,791 (Feb. 1, 1972), M. H. Harrison (to International Minerals and Chemical Corp.).

31. *Diamond Shamrock Technical Brochure*, Col. 16, 2nd ed., p. 5.

32. U.S. Pat. 1,930,146 (Oct. 25, 1933), D. F. Othmer (to Eastman Kodak).

33. *KOH Caustic Potash*, technical data, International Minerals and Chemical Corp., Northbrook, Ill., Oct. 1976, p. 25.

34. *Welcome to Ashtabula*, technical data, International Minerals and Chemical Corp., Northbrook, Ill., Dec. 1976.

35. M. L. Spilman, *Chem. Eng.* **72**(23), 198 (1965).

36. Y. Araten, A. Baniel, and R. Blumberg, *Potassium Nitrate*, Proceedings No. 99, The Fertiliser Society, Alembic House, London, Oct. 1967.

37. J. M. Potts, W. C. Scott, and J. F. Anderson, Jr., *Proceedings from the American Chemical Society Meeting, Chicago, Ill., Sept. 10, 1958*, ACS, Washington, D.C.

38. *Phosphorus Potassium* (54), 44 (July/Aug. 1971).
39. E. K. Drechsel, *Proceedings from the American Chemical Society Meeting, Aug. 28, 1973*, ACS, Washington, D.C.
40. C. F. Stevens, *Nature*, **349**, 657 (1991).
41. G. Eisenman, J. Aqvist, and O. Alvarez, *J. Chem. Soc. Faraday Trans.* **87**(13), 2099–2109 (1991).
42. J. J. R. Frausto da Silva and R. J. P. Williams, *The Biological Chemistry of the Elements: the Inorganic Chemistry of Life*, Oxford University Press, Oxford, U.K., 1991, pp. 223–242.
43. P. Kurtzweil, *FDA Consumer* **28**(7), 18 (Sept. 1994).
44. *Official Methods of Analysis of the Association of Official Analytical Chemists*, 15th ed., Association of Agricultural Chemists, Washington, D.C., 1990, Section 983.02.
45. *Methods of the Association of Official Analytical Chemists*, 11th ed., Association of Agricultural Chemists, Washington, D.C., 1970, Sections 958.02 and 969.04.
46. Ref. 45, Section 935.02.
47. Ref. 45, Section 949.01.
48. I. M. Korenman, *Analytical Chemistry of Potassium Academy of Sciences of the USSR*, Vernadskii Institute of Geochemistry and Analytical Chemistry, Israel Program for Scientific Translations, Jerusalem, 1965.
49. R. J. Waxweiler, J. K. Wagoner, and V. E. Archer, *J. Occup. Med.* **15**, 486 (June 1973).
50. DOT Hazardous Materials Transportation Table, *Chemical Regulation Reporter*, The Bureau of National Affairs, Inc., Washington, D.C., 1980.
51. *Code of Federal Regulations*, Title 40, Part 302.4, Washington, D.C., July 1, 1993, p. 251.
52. *82nd Annual OPD Chemical Buyers Directory*, Schnell Publishing Co., New York, 1995.

MARK B. FREILICH
RICHARD L. PETERSEN
The University of Memphis

# POWDER COATINGS.    See COATING PROCESSES, POWDER TECHNOLOGY.

# POWDER METALLURGY.    See METALLURGY, POWDER.

# POWDERS, HANDLING

## DISPERSION OF POWDERS IN LIQUIDS

Suspensions (dispersions or slurries) of powders in liquids (vehicles) are involved in many commercial processes and products, eg, inks, paints, asphalt, and pharmaceuticals (qv). Surface forces control the flow and sedimentation characteristics of particles whose diameters are 0.02–200 $\mu$m. The preparation of a dispersion is not a trivial matter; special equipment and chemicals have been developed to facilitate the steps of dispersion, ie, wetting the external surface of the particles, breaking up clumps, and preventing reagglomeration. A surfactant (surface-active agent) is a chemical that concentrates in the interfacial region between the solid and the liquid. A surfactant that creates an interparticle repulsion large enough to overcome normal interparticle attractions may be called a dispersant for that particular solid–liquid pair.

Several factors have combined to improve the capabilities of dispersion science. Improved methods for polymer synthesis have led to a proliferation of highly effective dispersants; larger and faster computers have facilitated modeling details of interactions between particles, dispersants, and the liquid; and highly sophisticated instruments are available to measure particle–particle and particle–dispersant interactions. This article describes the technical bases and practical aspects of the three stages of dispersing a powder in a liquid: (1) wetting the powder into the liquid, (2) deagglomerating the wetted clumps, and (3) preventing reagglomeration (flocculation). Modification of the surface and use of additives (dispersants) to prevent reagglomeration are also described (1–7) (see COLLOIDS; DISPERSANTS; EMULSIONS; SURFACTANTS).

### Wetting Powders Into Liquids

The atoms and molecules at the interface between a liquid (or solid) and a vacuum are attracted more strongly toward the interior than toward the vacuum. The material parameter used to characterize this imbalance is the interfacial energy density $\gamma$, usually called surface tension. It is highest for metals ($\leq$1 J/m$^2$) (1 J/m$^2$ = N/m), moderate for metal oxides ($\leq$0.1 J/m$^2$), and lowest for hydrocarbons and fluorocarbons (0.02 J/m$^2$ minimum) (4). The International Standards Organization describes well-established methods for determining surface tension, eg, ISO 304 for liquids containing surfactants and ISO 6889 for two-liquid systems containing surfactants.

**Fundamentals.** *Contact Angle.* When a drop of liquid is placed on a flat, horizontal, solid surface it spreads out until it attains an equilibrium shape with a fixed angle from the solid surface to the gas surface (through the liquid). This is the contact angle. Figure 1 shows how the contact angle is related to the balance of surface tension forces at the drop perimeter where the three phases meet the shoreline. In cases where neither gas nor solid dissolves in the liquid

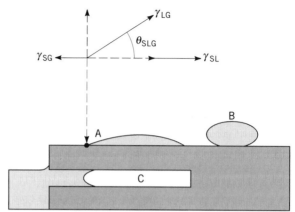

**Fig. 1.** Contact angles. The shapes of drops that are A, wetting and B, nonwetting with respect to the solid (■), and C, penetration of a wetting liquid into a pore to compress the gas (□) inside. The vector diagram shows the balance of forces at the perimeter of the liquid drop on the solid plate.

and there are no surface-active solutes present, the contact angle is related to the interfacial energy densities at the solid–gas (SG), solid–liquid (SL), and liquid–gas (LG) interfaces by the Young-DuPre equation (8):

$$\gamma_{LG} \cos\theta_{SLG} + \gamma_{SL} = \gamma_{SG} + \Pi \tag{1}$$

The surface film or spreading pressure, $\Pi$, is used to account for the change in gas–solid interaction caused by adsorption of vapor evaporated from the liquid. A liquid is called wetting if the contact angle from solid to liquid through gas, $\theta_{SLG}$, $<90°$ and nonwetting if $\theta_{SLG} > 90°$. Because it is easier to measure the contact angle than to measure $\gamma_{SL}$, the Young-DuPre equation is usually used to determine $\gamma_{SL}$ from a knowledge of the other parameters. This expression was derived for pure liquid and solid phases, so more complex expressions must be used if the solid is even slightly soluble in the liquid, if the solid has soluble impurities, or if the liquid contains surfactants or other solutes.

*Wetting Single Particles.* When a particle is submerged in a liquid, the work of wetting a surface, $w$, is the change in interfacial energy density times area, $a$, as the solid–gas and liquid–gas interfaces are replaced by a solid–liquid interface.

$$w = a[\gamma_{SL} - \gamma_{SG} - \gamma_{LG}] \tag{2}$$

A buoyant spherical particle of diameter $D$ floating in a liquid which has no tendency to wet, ie, pull the shoreline up, or to reject, ie, push the shoreline down, the solid submerges to a depth $d$ at which the downward force of gravity on the sphere equals the upward buoyancy force of the displaced liquid. This occurs when

$$\rho_S V_{SPH} = \rho_L V_{SUBM} \tag{3}$$

where the volume, $V$, of the sphere is $V_{\text{SPH}} = (\pi/6)D^3$. For $d < D/2$, the volume of the submerged portion is as follows:

$$V_{\text{SUBM}} = \frac{\pi d}{6D}[D^3 - (D - 2d)^3] \tag{4}$$

For $d > D/2$ the volume of the submerged portion is as follows:

$$V_{\text{SUBM}} = V_{\text{SPH}} - \frac{\pi}{3}\left(1 - \frac{d}{D}\right)[D^3 - (2d - D)^3] \tag{5}$$

If the liquid wets the solid, surface tension pulls the solid down farther into the liquid. Centrifugal force does not affect the depth of submergence because it increases the buoyancy force as much as it increases the settling force. A non-buoyant particle submerges fully unless the liquid is sufficiently nonwetting, eg, 0.2-mm polytetrafluoroethene [9002-84-0] particles in water. Here centrifugal force can be used to overcome the rejection and force submergence. The force of surface tension along the microdepressions in a rough surface causes wicking of a drop of wetting liquid so that it spreads farther than it would on a smooth horizontal surface. Rejection of a nonwetting liquid from the microdepressions keeps it from spreading as far as it would on a smooth surface (9).

*Wetting Clumps.* The density of a clump, $\rho_{\text{CLUMP}}$, is related to the fraction of void space within the perimeter of the clump (porosity), $\epsilon_{\text{PORE}}$, by

$$\rho_{\text{CLUMP}} = (1 - \epsilon_{\text{PORE}})\rho_S + \epsilon_{\text{PORE}}\rho_G \tag{6}$$

If the liquid is nonwetting it does not enter the pores to displace the gas unless the hydrostatic pressure, $P_{\text{HYDRO}}$, at the bottom of the submerged clump, $P_{\text{HYDRO}} = g\rho_L d$, exceeds the pressure required to drive it into the pores, which for a cylindrical pore is as follows:

$$P_{\text{ANTI}} = \frac{-4}{D_{\text{PORE}}}\gamma_{\text{LG}}\cos\theta_{\text{SLG}} \tag{7}$$

Gravity, $g$, or centrifugation rarely provide enough hydrostatic pressure to force liquid into nonwetting pores. If the liquid wets the solid the clump density increases as gas is displaced from the interior. It is best if submersion does not occur until the liquid has completely displaced gas from the pores (Fig. 2). For wetting liquids and pores with average diameter, $D_{\text{PORE}}$, and tortuosity, $f_{\text{TORT}}$, the length, $L_{\text{WET}}$, to which liquid is pulled into a bed of powder by capillary action depends on the viscous drag exerted by the increasing length of pore through which it is flowing as time, $t$, in seconds increases (10):

$$L_{\text{WET}} = \frac{1}{f_{\text{TORT}}}\left(\frac{D_{\text{PORE}}t\gamma_{\text{LG}}\cos\theta_{\text{SLG}}}{8\eta}\right)^{1/2} \tag{8}$$

$D_{\text{PORE}}$ is related to $\epsilon_{\text{PORE}}$, which is in turn related to the closeness of packing of the powder. The number of particles adjacent to a given particle

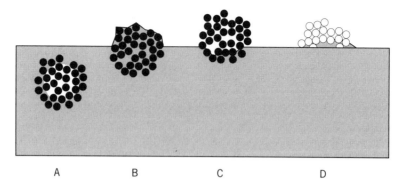

**Fig. 2.** Problems in wetting: A, liquids that wet the exterior before displacing gas from pores leave gas trapped in the submerged clump; B, fully wetted clumps of buoyant particles do not sink; C, nonwetting liquids do not penetrate and displace gas from pores, so clump remains buoyant and cannot submerge; and D, foam produced from air is drawn under the surface, sheared into small bubbles, and stabilized by the wetting agent.

is represented by $n_{ADJ}$. The maximum packing density for monosize spheres occurs at hexagonal close packing, where $n_{ADJ} = 12$ and $\epsilon_{PORE} = 0.2595$; for cubic packing of monosize spheres, $n_{ADJ} = 6$ and $\epsilon_{PORE} = 0.4764$. In a loose network $n_{ADJ}$ approaches 2 and $\epsilon_{PORE}$ approaches unity (11). Packing densities higher than hexagonally close-packed spheres ($\epsilon_{PORE}$ approaching unity) can be achieved by using spheres that are not monosize or by using nonspherical particles.

If the clump submerges in a wetting liquid before all the gas has escaped, the liquid continues to advance into the pores, compressing the trapped gas, until the gas pressure balances the force pulling the liquid into the pores:

$$P_{TRAP} = \frac{4}{D_{PORE}} \gamma_{LG} \cos\theta_{SLG} \qquad (9)$$

Complete wetting cannot occur until either the clump is broken up to let the gas escape or the trapped gas dissolves in the liquid. A sudden decrease in hydrostatic pressure can help remove gas trapped in a submerged clump by expanding the bubble volume to break up the clump or extend the bubble past the clump's exterior so that it may escape.

If a wetting agent is present in the liquid, rather than as a pretreatment on the solid, it acts by adsorbing on the solid surface. As the liquid advances through a pore it becomes depleted in wetting agent so that the rate of advance of the meniscus depends on diffusion of fresh wetting agent from the bulk liquid outside the pore. A low molar mass, $M$, for the wetting agent (WA) gives the highest diffusion coefficient and hence the fastest rate of wet-in, as shown by the diffusion constant, $F$, for a spherical molecule where $N_{AVO}$ represents the Avogadro constant, $k$, and $T$ is the Boltzmann constant, absolute temperature in K.

$$F_{WA} = \frac{kT}{6\pi\eta_L}\left(\frac{4\pi N_{AVO}}{3M_{WA}}\right)^{1/3} \qquad (10)$$

*Flushing a Dispersion Into a Second Liquid.* If a dispersion in one liquid is mixed with a second liquid which is immiscible with the first but which preferentially wets the powder, the powder transfers into and becomes dispersed in the second liquid. This process, called flushing, is sometimes used to prepare an oil-based suspension directly from a water-based dispersion. A dispersant may be needed to promote preferential wetting in the second phase. For example, stearic acid [57-11-4] dissolved in dodecane [112-40-3] transfers across a dodecane–water interface to adsorb on the surface of iron(III) oxide [1309-37-1] in the aqueous phase. Adsorption occurs with the acid group anchored on the particle and the alkane chain extending outward, making the coated particle hydrophobic and making its transfer into the dodecane phase energetically favorable (12).

**Practical Matters and Process Equipment.** The rate of wet-in may be enhanced by adding a wetting agent to the liquid or by exposing the powder, treated or not, to water or some other vapor. Some powders sold commercially have already been given surface treatments so they readily wet into the liquid used in a particular market application. Selection of a wetting agent can be difficult, because although many surfactants may be effective wetting agents in the laboratory for a given solid–liquid pair, few or none may be acceptable in a commercial formulation for that pair. The wetting agent may interfere with end use behavior, perhaps by altering the surface tension of a film during application or drying.

Gas that is drawn beneath the liquid surface, through a vortex or plunging jet, and enters a high shear region is dispersed into small bubbles which may be stabilized by a surfactant and rise to the surface to form a foam which is hard to dissipate. This is a particular problem when excessive wetting agent is present. Foam can be avoided even in the presence of a foaming agent by preventing air from being drawn beneath the liquid surface during the wet-in process (see FOAMS).

Figure 3 shows typical design features of commercial equipment to wet powders into liquids. These exploit the factors that lead to rapid wet-in: low gas pressure, high centrifugal force, distribution of powder in a thin and deagglomerated layer on the liquid, and liquid surface refreshing to expose a large area for powder addition. Provision for clear liquid flow down the walls helps prevent scale and encrustation. Scale forms above the liquid level when airborne dust sticks to a wall wetted either by vapor condensation or liquid drawn up the wall by capillary action through previously deposited scale. Encrustation may form near the shoreline in a tank as it alternately wets and recedes due to hydraulic pulses during agitation, then dries near that level.

## Deagglomerating Wetted Clumps

**Fundamentals.** Individual particles in a dry powder are held together in clumps by polarizability attraction of adjoining particles, surface tension at wetted contacts, or moderately strong sintering or precipitation (particle or soluble salt) bonds. These bonds are formed at the contact points during drying or because of exposure to humidity or temperature cycles. Breaking these bonds to fully disperse the submerged and wetted clumps into individual particles may require high shear stress or mechanical impact (12–14). In contrast to milling,

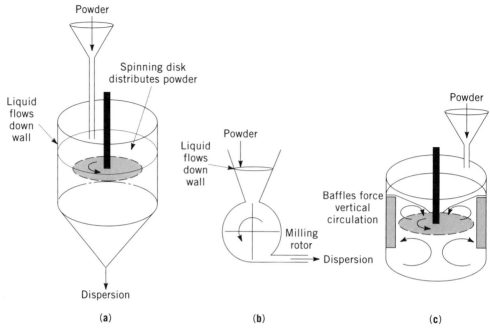

**Fig. 3.** Typical design elements for wet-in: (**a**) a spinning disk deagglomerates powder just prior to wet-in; (**b**) a rotor breaks clumps as they wet-in, and centrifugal force helps submerge nonbuoyant powders; and (**c**) a disk impeller provides a rapidly refreshed liquid surface. In (**a**) and (**c**), the system may be under vacuum to speed pore penetration.

deagglomeration generates only a small amount of new surface area and thus uses only moderate energy per unit mass of material. The surface area newly created by deagglomeration may have higher adsorptivity and reactivity than the conditioned surface of the clump. ISO 8780 (parts 1–6) describes methods of deagglomeration for pigments (qv) and extenders.

High shear can be achieved in high viscosity systems, eg, molten plastic, using slowly turning mixing paddles. For intermediate viscosity systems, eg, highly loaded inks, the suspension may be passed through the gap between two mill rolls rotating with different surface velocities. For low viscosity systems, eg, mineral ore in water, suspensions may be circulated through the small gaps between a slotted stator and a close-fitting slotted rotor moving at a high rpm or forced at high pressure through a very small clearance in a spring-loaded valve.

Impact can be achieved using grinding media beads set in motion by stirring with bars or disks, by vibration, or by cascading within a cylindrical container rotating on a horizontal axle. The cascading force can be enhanced by mounting several such rotating containers on a centrifuge with the axles parallel to the axis of centrifugation. Standard practice calls for using media approximately 20 times the diameter of the ultimate particle size; however, with high energy input size reduction can produce particles of submicrometer size. In autogenous milling, particles are ground by directing one high velocity jet of suspension against another or by vigorously stirring a highly loaded suspension (>40 vol

% solids). Because it is difficult to achieve high energy impacts in a viscous medium, impact methods are restricted to low viscosity systems.

Ultrasonic probes, which create alternating pulses of vacuum (that pull the liquid apart to form evacuated cavities) and pressure (that cause the cavities to collapse) at up to millions of cycles per second, are commonly used to prepare laboratory dispersions. The cavities tend to form next to surfaces, so they are well placed to shatter bonds holding clumps together. Although large ultrasonic generators are available, industrial applications are limited. Cavitation occurs only in the volume near the surface of the probe and it is difficult to pass all the suspension through this zone and have sufficient residence time to achieve full dispersion without overheating the suspension, which diminishes the intensity of cavity collapse.

**Practical Matters and Process Equipment.**   The strength of bonds holding clumps together may increase with storage time because of surface migration of material into the high energy contact region. Migration results from solubilization and mobilization caused by adsorbed vapors or surface-treatment chemicals. Some powders sold commercially have been coated with a solid or liquid to decrease the likelihood of such changes and to keep the clumps soft or easy to deagglomerate.

Figure 4 shows several typical commercial devices used to deagglomerate clumps in low viscosity suspensions. These use slotted rotors, multiple teeth, and clearances that decrease as the suspension is pumped by centrifugal force through the annular gap. High shear stress is produced in systems with high viscosity, which can be achieved through high solids loading, close clearances between the shearing elements, and high rotational speed of the moving element. The heat generated by the shear may limit throughput.

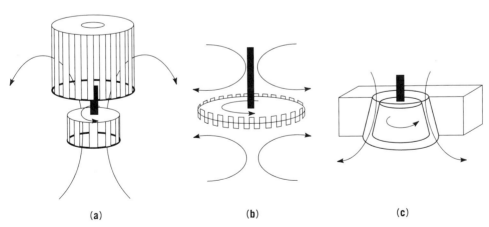

(a)                    (b)                    (c)

**Fig. 4.**  Typical design elements for wet deagglomeration in low viscosity systems: (**a**) a high rpm rotor (shown below its normal position within stator) produces turbulence and cavitation as blades pass each other; (**b**) a rotating disk creates a deep vortex to rapidly refresh the surface, and up- and downturned teeth at the edge cause impact, turbulence, and sometimes cavitation; and (**c**) the clearance of a high rpm rotor can be reduced as the batch becomes more deagglomerated.

## Preventing Reagglomeration

**Factors Affecting Dispersion Stability.** *Polarizability Attraction.* All matter is composed of electrical charges which move in response to (become electrically polarized in) an external field. This field can be created by the distribution and motion of charges in nearby matter. The Hamaker constant for interaction energy, $A$, is a measure of this polarizability. As a first approximation it may be computed from the dielectric permittivity, $\epsilon$, and the refractive index, $n$, of the material (15), where $\nu_e$ is the frequency of the principal electronic absorption

$$A = \frac{3}{4} kT \left[ \frac{\epsilon - 1}{\epsilon + 1} \right]^2 + \frac{3 h \nu_e (n^2 - 1)^2}{16 \sqrt{2} \, (n^2 + 1)^{1.5}} \tag{11}$$

band and $h$ is Planck's constant. For two identical spheres of diameter, $D$, whose surfaces are separated in a liquid by a distance $s < D$ the attractive energy, $u$, due to polarizability is $u_{\text{POLAR}} = -D/12s \, ((A_S - A_L)^{1/2})^2$, which becomes infinite at $s = 0$. A strongly adsorbed ($u_{\text{ADS}} > kT$) layer of liquid (vicinal fluid) or other solute limits the minimum separation to $s_{\text{MIN}}$, and the maximum attractive energy to $u_{\text{PMAX}}$; $s_{\text{MIN}}$ may be set equal to twice the molecular diameter of the liquid (16).

$$s_{\text{MIN}} = 2 \left[ \frac{M_L}{N_{\text{AVO}} \rho_L} \right]^{1/3} \tag{12}$$

For example, water has $A_L = 3.7 \times 10^{-20}$ J, $M = 0.018$ kg/mol, $\rho = 1000$ kg/m$^3$, and $s_{\text{MIN}} = 0.62$ nm. Two silica [7631-86-9] particles with $A_S = 6.6 \times 10^{-20}$ J and $D = 50$ nm have $u_{\text{PMAX}} = -2.8 \times 10^{-20}$ J (17).

*Thermal Jostling.* The thermally driven random motion of molecules jostles particles to provide a one-dimensional translational energy which averages $kT/2$ over several seconds. However, it is conventional to use $u_{\text{THERMAL}} = kT$ as a measure of thermal energy. At 298 K, $u_{\text{THERMAL}} = 4.11 \times 10^{-21}$ J. At any instant many particles have less energy than this, and some have several times as much. If $u_{\text{PMAX}} < u_{\text{THERMAL}}$, thermal jostling disrupts the agglomerates sufficiently often to keep the particles dispersed. In water, 4-nm silica particles have a maximum attractive energy only 55% of $u_{\text{THERMAL}}$.

*Sedimentation.* Gravity makes all particles that are more dense than the suspending liquid move downward, and also causes beds of particles to compress toward the state of minimum-included liquid. The sedimentation force, $f_{\text{SED}}$, for a single spherical particle (not part of a bed) submerged in a liquid is as follows:

$$f_{\text{SED}} = \frac{\pi}{6} g (\rho_S - \rho_L) D^3 \tag{13}$$

A larger sedimentation force can be developed by centrifuging the suspension. At an angular velocity of $\omega$ radians per second the centrifugal acceleration, $a_{\text{CENT}}$, at a distance $R$ from the axis is as follows:

$$a_{\text{CENT}} = \omega^2 R \qquad \omega = 2\pi \frac{\text{rpm}}{60} \tag{14}$$

If the principal force is centrifugal rather than gravitational, $a_{\text{CENT}}$ can be substituted for $g$ in the above equations. If $u_{\text{THERMAL}} > f_{\text{SED}}H$, where $H$ is the height of a container, then thermal jostling keeps the particles well distributed throughout the container and prevents them from settling into a dense bed at the bottom (or top) of the container.

*Viscous Drag.* The velocity, $v$, with which a particle can move through a liquid in response to an external force is limited by the viscosity, $\eta$, of the liquid. At low velocity or creeping flow ($N_{Re} < 1$), the viscous drag force is $f_{\text{DRAG}} = 3\pi\eta_{\text{L}}Dv$. The Reynolds number ($Re$) is determined from $N_{Re} = \rho_{\text{L}}Dv/\eta_{\text{L}}$. The balance between the forces due to gravity and viscous drag determines the terminal sedimentation velocity, which is upward for particles less dense than the liquid.

$$v_{\text{SED}} = \frac{g}{18\eta}(\rho_{\text{S}} - \rho_{\text{L}})D^2 \tag{15}$$

*Electrostatic Interaction.* Similarly charged particles repel one another. The charges on a particle surface may be due to hydrolysis of surface groups or adsorption of ions from solution. The surface charge density can be converted to an effective surface potential, $\psi$, when the potential is $<30$ mV, using the following equation, where $N_{\text{Far}}$ represents the Faraday constant and $N_{\text{gas}}$, the gas law constant.

$$\psi_0 = \frac{\sigma}{N_{\text{Far}}}\left(\frac{N_{\text{gas}}T}{2\epsilon_0\epsilon_{\text{L}}}\right)^{1/2} \tag{16}$$

The layer of solution immediately adjacent to the surface that contains counterions not part of the solid structure, but bound so tightly to the surface that they never exchange with the solution, is the Stern layer. The plane separating this layer from the next is the Stern plane. The potential at the Stern plane is smaller than that at the surface.

Figure 5 shows the enhanced concentration of oppositely charged ions near the charged surface, and the depleted concentration of similarly charged ions near the charged surface due to electrostatic attractions and repulsions. Both factors reduce the effective potential, $\psi$, as the distance from the surface, $x$, increases. The distance at which $\psi$ drops to $1/e$ (37%) of its value at the Stern plane is called the counterion atmosphere decay distance, $t_c$:

$$t_c = \frac{\epsilon_0\epsilon_{\text{L}}N_{\text{gas}}T}{4\pi N_{\text{Far}}^2 I} \tag{17}$$

where the ionic strength, $I$, depends on both the charge, $z_j$, and the concentration, $C_j$, of the ions, $I = 1/2 \, \Sigma z_j^2 C_j$.

At the shear plane, fluid motion relative to the particle surface is zero. For particles with no adsorbed surfactant or ionic atmosphere, this plane is at the particle surface. Adsorbed surfactant or ions that are strongly attracted to the particle, with their accompanying solvent, prevent liquid motion close to

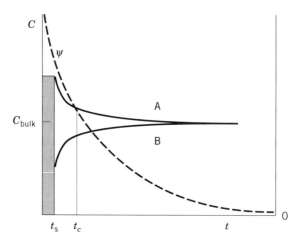

**Fig. 5.** Variation of ion concentrations, $C$ (——), and effective surface potential, $\psi$ (– – –), with distance from a positively charged surface. The layer of strongly bound anions, A, is the Stern layer, and beyond this anions are attracted into, whereas cations, B, are driven out of, the region near the surface (■); $t$ = distance from the surface, m; $t_c$ = the counterion atmosphere decay distance; and $t_s$ = thickness of the Stern layer.

the particle, thus moving the shear plane away from the particle surface. The effective potential at the shear plane is called the zeta potential, $\zeta$. It is smaller than the potential at the surface, but because it is difficult to determine $t_{\mathrm{SHEAR}}$ or $\psi_0$, the usual assumption is that $\psi_0$ is effectively equal to $\zeta$, which can be measured by observing the velocity, $v$, with which the charged particle moves in response to an electric (ELEC) field gradient, $E$:

$$v_{\mathrm{ELEC}} = \frac{2\epsilon_0\epsilon_{\mathrm{L}}}{3\eta_{\mathrm{L}}} E\zeta \tag{18}$$

In an aqueous system with large particles well-separated by a distance, $s$, ($D$ and $s > t_c$) the electrostatic repulsion energy between two identical charged spheres may be approximated (1):

$$u_{\mathrm{ELEC}} = \pi\epsilon_0\epsilon_{\mathrm{L}}D\zeta^2 \ln[1 + \exp(-s/t_c)] \tag{19}$$

If this repulsion exceeds the polarizability attraction at the distance of closest approach, determined by vicinal fluid, the particles can be prevented from agglomerating. For example, water with $\epsilon_{\mathrm{L}} = 81$, containing NaCl at $I = 10^{-7}$ mol/m$^3$, creates a counterion layer with $t_c = 0.15$ nm. The maximum repulsion in water between two silica particles with $D = 50$ nm and $\zeta = 50$ mV is $u_{\mathrm{EMAX}} = 2.86 \times 10^{-19}$ J. Because $u_{\mathrm{EMAX}} > u_{\mathrm{PMAX}}$, electrostatic repulsion prevents the surfaces from coming into contact unless the particles are in a bed of sediment and are being pressed together by the weight of it. Particles are more commonly characterized by surface potential than by surface charge density (18).

In a nonaqueous system with small closely spaced particles ($s < t_c$) the electrostatic repulsion energy between two identical charged spheres may be approximated (1):

$$u_{\mathrm{ELEC}} = \pi\epsilon_0\epsilon_{\mathrm{L}}\zeta^2\beta\frac{D^2}{s + D}\,\exp\!\left(\frac{-s}{t_c}\right) \tag{20}$$

where $\beta$ is a numerical factor between 0.6 and 1 and depends on the relative values of parameters in differential equations. Most surfactants have small solvation energies in organic liquids and consequently do not dissociate into ions in the clear liquid. However, when a particle with a strong tendency to donate protons is present, a surfactant with a strong tendency to accept protons can adsorb on the particle, accept transfer of a proton from the surface, and desorb, leaving a charged particle with the surfactant forming the counterion atmosphere.

*Polymer Chain Interactions.* If the surface of each particle is covered with links to polymer chains, whose segments are soluble in the surrounding liquid, then particles will be hindered from coming close together. The chains extend out a distance, $\delta$. In this context the liquid is referred to as the solvent. There are two phenomena involved: (*1*) mixing energy, based on the enthalpy change due to increased segment–segment interaction and decreased segment–solvent interaction as the solvent is squeezed out of the region between the particles, and (*2*) elastic energy, based on the entropy change as the number of possible chain configurations becomes restricted when the particle surfaces come closer together (19). Unlike the polarizability and electrostatic interactions which extend out to large distances, the steric repulsive interactions act only when $s < 2\delta$. As a first approximation the joint effect of these two contributions to the interparticle potential is a single term:

$$u_{\mathrm{STERIC}} = kT\frac{\rho_{\mathrm{L}}N_{\mathrm{Avo}}}{M_{\mathrm{L}}}\phi^2(0.5 - \chi)\pi D(2\delta - s)^2 \tag{21}$$

where $\phi$ is the volume fraction of polymer in the adsorbed layer and $\chi$ is the Flory polymer–solvent interaction parameter. The value of $\chi$ is both solvent- and temperature-dependent; when the segment–segment interaction equals the segment–solvent interaction, $\chi = 0.5$ and the system is said to be at the theta point. Lower values of $\chi$ indicate a better solvent for the segments and cause a more extended configuration for the polymer chains; larger values lead to a more compact configuration.

*Total Interaction Energy.* Figure 6**a** shows the dependence of individual energy components of two-particle interaction on the separation between their surfaces; Figure 6**b** shows the shapes that arise when these are combined in typical situations. The suspension can be considered stabilized against reagglomeration if the maximum attractive (negative) energy is smaller than the thermal energy. Note from the component energy equations that whereas thermal energy

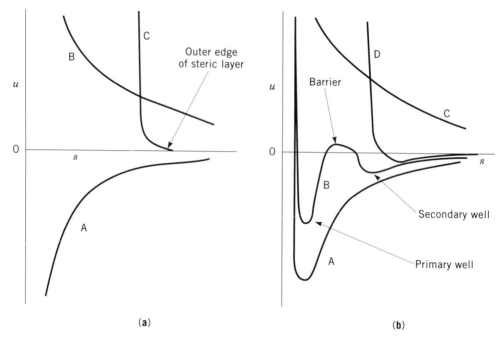

(a)                                                                (b)

**Fig. 6.**  (**a**) Components of potential energy and (**b**) typical total potential energies as a function of separation between two particles, where $s$ is the separation of particle surfaces and $u$ is the energy of interaction. In (**a**), A represents polarizability attraction; B, electrostatic repulsion, and C, steric exclusion. In (**b**), A, polarizability attraction dominates (vertical portion at small $s$ is due to vicinal fluid); B, electrostatic repulsion nearly balances polarizability attraction providing a shallow secondary minimum that leads to weak flocs, an energy barrier that can prevent weak flocs from becoming strong if it is several times larger than $kT$, and a deep primary minimum that can bind particles into a strong floc; C, electrostatic repulsion dominates; and D, polarizability attraction is almost completely offset by steric exclusion.

is independent of particle diameter $D$, the polarizability attraction, electrostatic repulsion, and the steric repulsion energies increase with $D$.

**Surface Modification.**   Reaction or adsorption at the solid surface can alter its properties and lead to a surface charge or steric stabilization (20,21).

*Hydrolysis.*   The surfaces of metal oxides and hydroxides can take up or release $H^+$ or $OH^-$ ions and become charged. Potentials as high as $\pm 100$ mV may be sustained in aqueous solutions. For aqueous solutions this is a function of the pH; the zeta potential for the particle is positive if the solution pH is below the particle's isoelectric pH ($pH_{iso}$), and negative if the pH is above $pH_{iso}$. Isoelectric points for metal oxides are presented in several publications (22,23). Reactions of hydroxyl groups at a surface, Q, with acid and base may be written as follows:

$$Q\text{---}AlOOH + H^+ \longrightarrow Q\text{---}Al(OH)_2^+$$

$$Q\text{---}AlOOH + OH^- \longrightarrow Q\text{---}AlOO^- + H_2O \qquad (22)$$

For boehmite [1318-23-6], AlOOH, $pK_{iso} = 7.6$ (22). In the region where the potential varies rapidly with pH, the slope is $-59$ mV per pH. Figure 7 shows

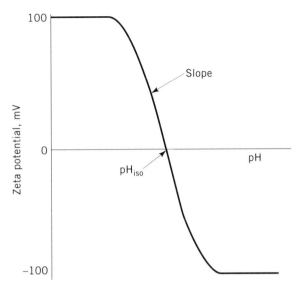

**Fig. 7.** Dependence of zeta potential on pH for a typical metal hydroxide particle in water. The isoelectric pH ( pH$_{iso}$) is at low pH for acidic hydroxides and high pH for basic hydroxides. Maximum slope = $-59$ mV/pH.

how the zeta potential varies with pH for a typical metal oxide. In many cases a zeta potential greater than 30 mV (positive or negative) is sufficient to keep the particles well dispersed. Lower values might be sufficient for particles with weak attractive forces (Hamaker constant close to the liquid), whereas higher values may be needed for particles with strong attractions, eg, magnetic particles.

*Differential Dissolution/Complexation.* For ionic salts, one of the ions making up the lattice may be complexed or hydrated more readily than the other resulting in an imbalance in particle charge. In the case of silver iodide [7783-96-2], dispersion in water leads to a net positive particle charge as the iodide ion is hydrated to a greater extent and thus goes into solution more readily than the silver ion (24). The surface charge of the particle may change with solution composition as the degree of complexation is affected by additives that adsorb on the surface, complex with the ion or the solvent, or affect the ionic strength and thus the distance over which the particle potential is effective.

*Chemical Grafting.* Polymer chains which are soluble in the suspending liquid may be grafted to the particle surface to provide steric stabilization. The most common technique is the reaction of an organic silyl chloride or an organic titanate with surface hydroxyl groups in a nonaqueous solvent. For typical interparticle potentials and a particle diameter of 10 μm, steric stabilization can be provided by a soluble polymer layer having a thickness of ~10 nm. This can be provided by a polymer tail with a molar mass of 10 kg/mol (25) (see DISPERSANTS).

*Surface Coating.* A dense surface coating (encapsulation) that contains no occluded solvent decreases interparticle attraction provided that the coating has a Hamaker constant intermediate between the particle and the liquid. This is called semisteric stabilization (ST). The energy of interaction between coated spheres is as follows (26):

$$u_{\text{COATED}} = \frac{-1}{12}\left[A_{\text{LT}}X_{\text{T}} + A_{\text{ST}}X_{\text{S}} + 2X_{\text{ST}}(A_{\text{LT}}A_{\text{ST}})^{1/2}\right] \tag{23}$$

where

$$A_{\text{LT}} = \left[(A_{\text{L}} - A_{\text{T}})^{1/2}\right]^2 \qquad A_{\text{ST}} = \left[(A_{\text{S}} - A_{\text{T}})^{1/2}\right]^2 \tag{24}$$

and for each subscript T, S, or ST

$$X = \frac{y}{Y_{\text{N}}} + \frac{y}{Y_{\text{D}}} + 2\ln\left(\frac{Y_{\text{N}}}{Y_{\text{D}}}\right)$$
$$Y_{\text{N}} = x^2 + yx + x \qquad Y_{\text{D}} = Y_{\text{N}} + y \tag{25}$$

For subscript T, $y = 1$, $x = s/(D + 2\delta)$; for subscript S, $y = 1$, $x = (s + 2\delta)/D$; and for subscript ST, $y = 1 + 2\delta/D$, $x = (s + 2\delta)/D$, where $\delta$ represents the thickness of the steric layer. For example, a 10-nm coating of polytetrafluoroethene ($A = 3.8 \times 10^{-20}$ J) reduces the attraction between 1-$\mu$m particles of silica ($A = 6.6 \times 10^{-20}$ J) in water ($A = 3.7 \times 10^{-20}$ J, $s_{\text{MIN}} = 0.62$ nm) from 53 to 1.6 times the thermal energy (17).

**Additives Used To Stabilize Dispersions.**  A dispersant for a particular solid–liquid pair consists of molecules having one region that is soluble in the liquid and a second region that is not very soluble. The insolubility of the second region causes it to adsorb on the particles and, at concentrations above the critical micelle concentration (CMC), to force the dispersant molecules into multimolecular clusters with the soluble regions on the outside. The adsorbed dispersant may prevent agglomeration by keeping the particles from coming close enough to be strongly attracted (steric stabilization) or by creating a large enough charge to overwhelm interparticle attraction (electrostatic stabilization). Polarizability interactions attract dispersant molecules from solution onto the particle surface.

If an adsorbed chemical group (anchor) is more strongly bound to the surface than a solvent molecule would be at that site, an equilibrium expression may be written for the displacement of solvent by adsorbate. Adsorption is particularly strong if the chemical nature of the adsorbed group is similar to that of the particle surface; for example, in aqueous systems perfluoroalkane groups adsorb well on polytetrafluoroethene particles and aromatic polyethene oxides adsorb well on polystyrene.

In the simplest case, for which all adsorption sites are equivalent and do not interact with each other, the fraction of surface covered by adsorbate, $f_{\text{COVERED}}$, is related to the surfactant concentration, $C_{\text{SURF}}$, and the adsorption constant, $K_{\text{ADS}}$, in mol/m$^3$, by the Langmuir adsorption isotherm:

$$f_{\text{COVERED}} = \frac{K_{\text{ADS}}C_{\text{SURF}}}{1 + K_{\text{ADS}}C_{\text{SURF}}} \tag{26}$$

Near-monolayer coverage ($f_{\text{COVERED}} > 0.95$) can be achieved only at $K_{\text{ADS}}C_{\text{SURF}} > 19$. This can pose problems for surfactants, since solution concentration can be no higher than the CMC (20,27,28).

*Inorganic Ions.* Because of electrostatic attraction, positive ions are attracted to negatively charged surfaces and have a higher concentration near the surface than in the bulk. Negative ions are repelled from the negative surface and have a lower concentration near that surface. Ions which are very strongly bound ($\mu_{\text{ADS}} > kT$) are in the Stern layer, whereas those that can move into and out of the ionic atmosphere ($u_{\text{ADS}} \leq kT$) are in the Helmholtz layer. The effect of ionic attraction or repulsion from the surface is to enhance or reduce the nonionic adsorption coefficient:

$$K_{\text{ION}} = K_{\text{NONION}} \exp\left[\frac{z N_{\text{Far}} \psi_0}{N_{\text{gas}} T}\right] \qquad (27)$$

Multiply charged ions such as $Ba^{2+}$ adsorb on an ionic surface more strongly than singly charged ions such as $Cl^-$. The addition of a solution of $BaCl_2$ to a suspension of AgI in water can provide sufficient charge to stabilize the dispersion at low ionic strength.

*Short-Chain Organics.* Adsorption of an organic dispersant can reduce polarizability attraction between particles, ie, provide semisteric stabilization, if $A_L < A_T < A_s$ or $A_s < A_T < A_L$ (T = dispersant) and the adsorption layer is thick. Adsorption in aqueous systems generally does not follow the simple Langmuir profile because the organic tails on adsorbed molecules at adjacent sites attract each other strongly.

Ionizable organic dispersants can lose or gain protons ($H^+$) in aqueous solution if the pH is in the right region for that particular dispersant. Organic acids (anionic surfactants) lose a $H^+$ ion at pH > $pK_a$ to attain a negative charge; organic amines (cationic surfactants) add a $H^+$ ion at pH < $pK_b$ to attain a positive charge. Amino acids (zwitterionic surfactants) become positively charged at pH < $pK_b$, are neutral at $pK_b$ < pH < $pK_a$, and become negatively charged at pH > $pK_a$. The ionic species may adsorb on an oppositely charged (typically inorganic) surface and provide semisteric stabilization or may adsorb on an uncharged (typically organic) surface to create a surface potential that can stabilize the suspension.

An ionizable organic dispersant can provide electrostatic stabilization for a typically inorganic particle in an organic liquid even though it does not ionize appreciably in the absence of such particles. The mechanism for ionization is adsorption of the nonionized dispersant molecule, followed by proton ($H^+$) transfer to or from the particle, or a film of water adsorbed on the particle, and desorption of the organic ion, creating a charged particle with the dispersant forming the counterion atmosphere (29,30).

*Polymeric Nonionics.* Some polymeric units can be hydrated, such as polyethylene oxide, and thus are water soluble or adsorb strongly to highly polar materials. Other polymeric units, such as alkane or fluorinated alkane, are oil soluble and adsorb strongly on nonpolar materials. Langmuir adsorption theory is not applicable for polymeric adsorbates because adsorption at one site strongly influences the adsorption at a neighboring site. Figure 8 depicts the structures of several significant types of polymer. Homopolymers consist of a series of identical segments. If the segments adsorb strongly on the particle the polymer layer will have a compressed conformation (little occluded solvent) and provide only semisteric stabilization. Less strongly adsorbed homopolymers provide a coating whose off-surface sequences may be anchored to the surface at both ends (loops) or at only one end (tails) to provide a steric layer. Because the desorption and readsorption of an adsorbed polymer segment is severely restricted by the location of nearby adsorbed segments, attainment of conformational equilibrium for the adsorbed molecule may be so slow that the configuration will effectively remain what it was when the dissolved polymer first made contact with the surface.

Random copolymers consist of random length sequences of soluble segments interspersed with less soluble segments. The less soluble segments adsorb on the particle and the more soluble sequences form loops or tails extending into the liquid. The provision of separate anchor segments and soluble segments provides better control over adsorbed layer depth and desorption than can be achieved using a homopolymer. The movement of the loops and tails through the solution is driven by thermal energy and provides a steric barrier whose thickness in a good solvent is roughly one-third the fully extended length of the loop or tail. Steric stabilization can often be achieved with a steric barrier thickness of about 10 nm, which can be provided by tails having a molar mass of about 10 kg/mol (mol wt 10,000 g/mol) (31).

Diblock copolymers consist of one sequence of anchor segments and a second sequence of backbone segments. The relative lengths of the two sequences can be controlled to provide a wide variety of adsorption and barrier characteristics. Typical commercial dispersants may use alkane $-(CH_2)$, ester $-(C_5H_{10}-(CO)-O)$, methyl methacrylate [80-62-6], $-(C(CH_3)(COOCH_3)-CH_2)$, propylene oxide $-(CH_2-CH(CH_3)-O)$, or ethylene oxide [75-21-8], $-(CH_2-CH_2-O)$, segments. Comb polymers have a soluble backbone with a number of side chains, each of which contains an anchor group (see POLYMERS).

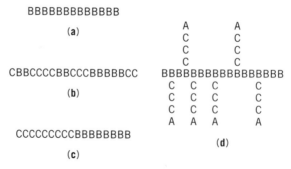

**Fig. 8.** General structures of polymeric dispersants: (**a**) homopolymer, (**b**) random copolymer, (**c**) diblock copolymer, and (**d**) comb polymer, where A = anchor group, B = soluble repeat unit, and C = repeat unit with solubility different from B. The repeat units may occur in sequences hundreds of units long that fill space as random coils thrashing due to thermal jostling.

*Selecting A Dispersant.* To select a dispersant for a system that includes a third component (such as a binder), the following steps may be used so that only a few candidates remain for the testing required at the final step (32,33). (*1*) Evaluate the dispersant solubility in the liquid with the third component, but not the powder, present and eliminate from further consideration any dispersant not readily soluble. (*2*) Measure the zeta potential of the particles in the liquid with the third component present. If the zeta potential is positive, eliminate from further consideration all cationic dispersants; if negative, eliminate all anionic dispersants. (*3*) Measure the viscosity of a highly loaded (>40 vol % solids) suspension containing only the liquid and the powder. Eliminate from further consideration any dispersant that has poor flow characteristics (high viscosity). (*4*) Observe sedimentation behavior of a moderately loaded (5–10 vol % solids) suspension containing only the liquid and powder. For small powders use centrifugation to force settling in a reasonable time. Eliminate from further consideration any dispersant that gives a large sediment volume. (*5*) Check the quality of the complete formulation, ie, liquid, powder, dispersant, third component, and other components, in whatever end-use quality tests are important and eliminate any dispersants which are incompatible with other components or cause failure in the end-use tests.

Reference 4 and several standards provide guidelines for selecting dispersants: ASTM standard B821-92 for metal powders (and some carbides and sulfides) in water; British Standard (BS) 3466 (part 4) for oxides; and ISO TC24/SC4/WG11 (1997) for a wide range of powder–liquid combinations.

**Associative Thickeners.** Although low viscosity is convenient for mixing and pumping operations during manufacture, high viscosity is desirable to deter the formation of a dense sediment bed and possible agglomeration during transportation and storage. Associative thickeners are polymers which have insoluble (with respect to the suspending liquid) end groups and a long soluble backbone (34). The insoluble end groups form weakly bonded clusters which are kept apart by the backbones. In some cases the end groups may adsorb on small particles in the system and incorporate these in the network. The three-dimensional network of linked micelles forms quickly in the absence of shear to prevent particles from settling when the suspension is quiescent, but breaks up readily when subjected to low shear forces. Short-term disruption of the network (and consequent small-scale sedimentation) may occur if the system is jolted or vibrated during transportation or storage.

**Independent Floc Network.** A space-filling structure similar to that created by an associative thickener can be provided by a loose network of weakly bonded particles. For example, at pH < 7 and low salt concentration, kaolinite [*1318-74-7*] clay platelets have positively charged faces and negatively charged edges which floc to form a weakly bonded space-filling house of cards structure (35). The addition of clay to a suspension of particles that are not stable to reagglomeration can prevent the particles from coming into contact with each other and forming large flocs.

**Adsorption of Microparticles.** Microparticles, having diameters less than one-tenth that of the particles to be dispersed, can be attached to the larger particles by sintering prior to wet-in, polarizability attraction if the Hamaker constant is midway between the liquid and larger particle, or electrostatic attraction if the microparticles have a charge opposite that of the larger particle. The attached microparticles prevent the surfaces of the larger particles from coming close together, thus reducing the maximum attractive energy in essentially the same way that encapsulation does.

## Evaluating Dispersions

Several publications discuss in detail a variety of methods for evaluating dispersions (36) and the chemical character of the particle surface (37,38). Numerous standards provide the details of well-established methods for evaluating dispersions, eg, ASTM D185-84 for measuring coarse particles in paints (qv), ASTM D869 for evaluating settling and ease of remixing in paints, ASTM D1210-79 for fineness of grind in coatings (qv), ASTM D1316-93 for measuring fineness of grind in inks (qv), ASTM D2066-91 for evaluating tinting strength in inks, ASTM D2067-92 for determining coarse particles in inks, ASTM D3015-72 for microscopic evaluation of the dispersion of pigments in plastics, ASTM E20-85 for particle size analysis by optical microscopy, ASTM F20-843 for color strength in ink, and ISO 8781 (parts 1–3) for evaluating tinting strength, fineness of grind, and gloss.

**Degree of Deagglomeration.** *Viscosity.* Because a clump of particles contains occluded liquid, the effective volume fraction of a suspension of clumps is larger than the volume fraction of the individual particles; that is, there is less free liquid available to facilitate the flow than if the clumps were deagglomerated. The viscosity of a suspension containing clumps decreases as the system becomes deagglomerated. This method is not very sensitive in the final stages of deagglomeration when there are only a few small clumps left.

*Particle Size Distribution.* Much of the progress of deagglomeration may be monitored using either sieving (particles >20 $\mu$m) or instrumental particle size analysis (smaller particles). The instrumental analysis techniques cannot reliably detect part per million levels of large particles (grit) unless the larger particles are concentrated by removing smaller particles. This can be done by letting the smaller particles escape from the suspension through a screen, an array of parallel capillary tubes, or a filter with openings of well-defined diameter in a series of partial filtrations followed by dilutions with fresh liquid.

*Tinting Strength.* When a fixed mass of particles is well-dispersed, it blocks more light than when the mass is present as clumps. To follow the progress of deagglomeration of a suspension, a series of samples may be taken as deagglomeration proceeds. A fixed quantity of each sample is mixed with a fixed quantity of a well-dispersed contrasting color pigment and a thin film is painted on a flat sheet. A sensitive colorimeter can quantify changes in color as a function of improving deagglomeration. Ultraviolet or infrared light may be used for samples that are transparent in the visible. The method does not work if the contrasting pigment is agglomerated by the sample or if the sample and contrasting pigment settle out. The colorimeter can be eliminated if the samples can be placed side by side for direct visual comparison (12).

*Grind Gauge.* This gauge consists of a metal bar with a rectangular milled depression the depth of which gradually decreases from 25 $\mu$m to zero as a function of length from one end to the other. A sample of paste, more than enough to fill the depression, is placed in the depression and the excess is scraped off by drawing a straight-edged drawdown bar toward the shallow end in a single pass. Particles having a diameter nearly equal to or larger than the depth of the depression at any given point rise up through the surface or cause a streak in the surface as the bar passes by. Although this method makes evident the presence of a few large particles among millions of smaller ones, the results are subject to considerable variability (12).

**Stability to Reagglomeration.** *Examination with a Microscope.* Examination of a dilute (<1 mass % solids) suspension using an optical microscope at a magnification of 100–400 can be helpful for evaluating the dispersion stability of particles 1–10 $\mu$m in diameter. If the dispersion is stable, thermal jostling can be seen and particles avoid coming close to one another or separate soon after collision. If the dispersion is unstable, particles remain attached after collision. The motion of particles as small as 0.1 $\mu$m may be observed with an ultramicroscope which uses side lighting and a dark field. Even in suspensions having an average diameter larger than 10 $\mu$m there are usually some smaller particles that can indicate whether or not the suspension is stable.

*Sediment Volume.* If the dispersion is unstable, the sediment bed will be quite deep and sedimenting particles will stick together where they first strike the sediment bed, thus forming an open structure with considerable occluded liquid. If the dispersion is stable to reagglomeration, the particles will move freely past one another to avoid contact as long as possible. The result is a thin sediment bed with maximum solids packing and minimum occluded liquid (12). Since dispersed particles settle more slowly than flocs, centrifugation may be needed to force sedimentation of small particles within a reasonable analysis time.

*Response to Electric and Acoustic Fields.* If the stabilization of a suspension is primarily due to electrostatic repulsion, measurement of the zeta potential, $\zeta$, can detect whether there is adequate electrostatic repulsion to overcome polarizability attraction. A common guideline is that the dispersion should be stable if $\zeta > 30$ mV. In electrophoresis the applied electric field is held constant and particle velocity is monitored using a microscope and video camera. In the electrosonic amplitude technique the electric field is pulsed, and the sudden motion of the charged particles relative to their counterion atmospheres generates an acoustic pulse which can be related to the charge on the particles and the concentration of ions in solution (18).

The attenuation of ultrasound (acoustic spectroscopy) or high frequency electrical current (dielectric spectroscopy) as it passes through a suspension is different for well-dispersed individual particles than for flocs of those particles because the flocs adsorb energy by breakup and reformation as pressure or electrical waves jostle them. The degree of attenuation varies with frequency in a manner related to floc breakup and reformation rate constants, which depend on the strength of the interparticle attraction, size, and density (inertia) of the particles, and viscosity of the liquid.

*Spectroscopy.* The spin-lattice coupling in nuclear magnetic resonance (nmr) is sensitive to the environment of the proton, so it can be used to determine whether the soluble parts of a polymeric dispersant are well solvated (desired situation for steric stabilization) or lying in a compact grouping near the particle surface (inadequate solvency) (39). Electron spin resonance (esr) or fluorescence spectroscopy can provide a more location-specific analysis, but these techniques require grafting special chemical probe groups onto the polymer chains. One negative factor is that the presence of the probe groups may alter the segment–solvent interaction parameters from the unsubstituted values (38).

   *Drift In Tinting Strength.*   Just as tinting strength can be used to follow deagglomeration, it may also be used to detect reagglomeration of fine powders. Sensitive colorimeters can follow changes in the color of a mixture of the sample suspension and a contrasting pigment. A drift in color indicates flocculation. The method does not work if the contrasting pigment is agglomerated by the sample or if the sample or contrast pigment settle out. The colorimeter can be eliminated if the agglomerates (flocs) can be broken up by passing a brush through part of the film so that a freshly deagglomerated region is adjacent to the agglomerated region (12).

## BIBLIOGRAPHY

"Dispersion of Powders in Liquids" in *ECT* 3rd ed., Suppl. Vol., pp. 339–371, by G. D. Parfitt, Carnegie-Mellon University.

1. R. J. Hunter, *Introduction to Modern Colloid Science*, Oxford University Press, Oxford, U.K., 1993.
2. W. B. Russel, D. A. Saville, and W. R. Showalter, *Colloidal Dispersions*, paperback ed., Cambridge University Press, Cambridge, U.K., 1989.
3. H. Lyklema, *Fundamentals of Interface and Colloid Science*, Academic Press, London, Vol. 1, 1991; Vol. 2, 1993; further volumes in press.
4. R. D. Nelson, *Dispersing Powders in Liquids*, Elsevier, Amsterdam, the Netherlands, 1988.
5. J. C. Berg, ed., *Wettability*, Marcel Dekker, New York, 1993.
6. R. J. Pugh, "Dispersion and Stability of Ceramic Powders in Liquids," in R. J. Pugh and L. Bergstrom, eds., *Surface and Colloid Chemistry in Advanced Ceramic Processing*, Marcel Dekker, New York, 1994.
7. P. C. Hiemenz, *Principles of Colloid and Surface Chemistry*, 2nd ed., Marcel Dekker, New York, 1986.
8. S. Ross and I. D. Morrison, *Colloidal Systems and Interfaces*, John Wiley & Sons, Inc., New York, 1988.
9. T. D. Blake, "Wetting," in Th. F. Tadros, ed., *Surfactants*, Academic Press, London, 1984.
10. E. D. Washburn, *Phys. Rev.* **17**, 374 (1921).
11. D. J. Cumberland and R. J. Crawford, "The Packing of Particles," in J. C. Williams and T. Allen, eds., *The Handbook of Powder Technology*, Vol. 6, Elsevier, Amsterdam, the Netherlands, 1987.
12. T. C. Patton, *Paint Flow and Pigment Dispersion*, 2nd ed., Wiley-Interscience, New York, 1979.
13. E. Jarvis, *Chem. Eng. (London)* **387**, 477–481 (1980).
14. I. A. Manas-Zloczower, A. Nir, and Z. Tadmor, *Rubber Chem. Technol.* **57**, 583–620 (1984).
15. J. N. Israelachvili, *Intermolecular and Surface Forces*, Academic Press, London, 1985, p. 145.
16. Ref. 4, p. 179.
17. Ref. 2, p. 148.

18. R. J. Hunter, *Zeta Potential in Colloid Science*, Academic Press, Inc., New York, 1981.

19. D. H. Napper, *Polymeric Stabilization of Colloidal Suspensions*, Academic Press, London, 1983; Ref. 6, p. 169.

20. M. J. Rosen, *Surfactants and Interfacial Phenomena*, 2nd ed., John Wiley & Sons, Inc., New York, 1989.

21. M. J. Schick and F. M. Fowkes, eds., *Surfactant Science Series*, Marcel Dekker, Inc., New York. Over 50 special topic volumes covering many aspects of surfactants; started in 1967.

22. G. A. Parks, *Chem. Revs.* **65**, 177–198 (1965). A comprehensive list of isoelectric points for solid oxides and hydroxides.

23. R. O. James, *Adv. Ceram.* **21**, 349–410 (1987).

24. B. Tezak, *Faraday Soc. Disc.* **42**, 175–186 (1966).

25. M. Green and co-workers, *Adv. Ceram.* **21**, 449–465 (1987).

26. M. J. Vold, *J. Colloid Sci.* **16**, 1 (1961); Ref. 2, p. 53.

27. *McCutcheon's Emulsifiers & Detergents—North American Edition*, MC Publishing, Glen Rock, N.J., published annually.

28. E. Matijevic, ed., *Surface and Colloid Science*, John Wiley & Sons, Inc., New York. A series of volumes containing review articles; started in 1968.

29. F. M. Fowkes, *Adv. Ceram.* **21**, 411–421 (1987).

30. M. E. Labib, *Colloids Surfaces* **29**, 293–304 (1988).

31. H. L. Jakubauskas, *J. Coat. Techn.* **58**, 71–82 (1986).

32. K. Mikeska and W. R. Cannon, *Adv. Ceram.* **9**, 164–183 (1984).

33. S. J. Schneider, Jr., ed., *Engineered Materials Handbook*, Vol. 4, ASM International, Metals Park, Ohio, 1991, p. 117.

34. D. J. Lundberg and J. E. Glass, *J. Coat. Technol.* **64**, 53–61 (1992).

35. A. S. Michaels and J. C. Bolger, *Ind. Eng. Fund.* **3**, 14–20 (1964).

36. R. A. Williams, C. R. Bragg, and W. P. C. Amarsinghe, *Colloids Surfaces* **45**, 1–32 (1990).

37. L. Bergstrom, "Surface Characterization of Ceramic Powders," in Ref. 6.

38. M. J. Kelley, *CHEMTECH* (Jan.–Oct. 1987). Voted best *CHEMTECH* papers of 1987.

39. F. Blum, *Colloids Surfaces* **45**, 361–376 (1990).

RALPH D. NELSON, JR.
E. I. du Pont de Nemours & Co., Inc.

# BULK POWDERS

Work on the development of the theory of bulk solids flow began in the early 1950s. Until that time, there had not been a recorded organized scientific approach to the analysis and design of devices to store and discharge bulk solids. Design of bins, hoppers, and feeders for powder flow was based on experiment and previous experience. Bulk solids were thought to behave much like liquids, and thus were expected to flow easily from bins. Powder flow is not always reliable, however, and attempts to approach the field of bulk solids flow using the techniques of fluid mechanics (qv) were not successful.

During the late 1950s and early 1960s Jenike employed a soil mechanics continuum approach to powder handling, developing a logical, theoretical basis to bulk solids flow (1). Testing equipment and methods were developed along with design techniques. The basic concepts of bulk solids flow have since been expanded to allow design of bins, hoppers, and feeders (2–5).

## Definitions

The following definitions are particular to bulk powders handling.

| | |
|---|---|
| Bin or silo | container for bulk solids having one or more outlets for withdrawal of solids either by gravity alone or by flow-promoting devices which assist gravity |
| Bulk solid | material consisting of discrete solid particles, which are submicrometer to several centimeters in size, handled in bulk form, as opposed to unit handling |
| Cylinder | vertical part of a bin (constant cross-sectional area) |
| Discharger | device used to enhance material flow from a bin but which is not capable of controlling the rate of withdrawal |
| Feeder | device for controlling the rate of withdrawal of bulk solid from a bin |
| Flow channel | space in a bin through which a bulk solid is actually flowing during withdrawal |
| Hopper | converging part of a bin (changing cross-sectional area) |

## Flow Problems

Bulk solids do not always discharge reliably. Unreliable flow, which can occur with some frequency, can be expensive in terms of inefficient processes, wasted product, and operational complications. Predictable flow is often impeded by the formation of an arch or rathole, or fine powders may flood uncontrollably.

Solids flow problems can occur individually or in combination. Five common flow problems, which occur when handling bulk solids follow.

*No-flow.* The problem of no flow may occur when an attempt is made to initiate flow such as by opening a gate or starting a feeder. One of two things may happen. In the first, an arch (bridge, dome) forms over the outlet (Fig. 1a). Sometimes the only way to initiate flow is to use force greater than that of gravity to overcome the arch and force material flow. Devices such as sledge hammers, vibrators, and/or air blasters are commonly used. The second form of a no-flow problem is commonly referred to as a stable rathole (pipe, core) (Fig. 1b). In this case, some material discharges as the feeder or gate is operated. However, because of a material's cohesive strength, as the flow channel empties out, the resulting hole becomes stable, and material stops flowing. Extreme measures are often required to reinstate flow.

*Erratic flow.* A combination of the two no-flow conditions can lead to erratic flow. If flow has been initiated but a stable rathole develops, then when the rathole collapses using vibration, the material may arch as it impacts the outlet. Flow may be restarted by vibration and maintained for a short time until the rathole forms again. Erratic flow can be a serious problem when handling bulk solids owing to fluctuating flow rates and bulk densities, and unreliable

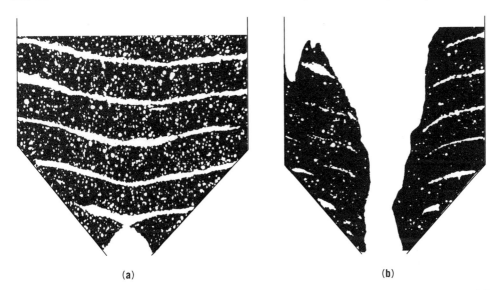

(a)                                                    (b)

**Fig. 1.** Examples of no-flow situations where the darkened areas represent material within the bin: (**a**) cohesive arch at the outlet of a bin, and (**b**) stable rathole formed within bin.

discharge. It can also jeopardize the structural integrity of the bin when a rathole collapses.

*Flooding*. When a stable rathole forms in a bin and fresh material is added, or when material falls into the channel from above, a flood can occur if the bulk solid is a fine powder. As the powder falls into the channel, it becomes entrained in the air in the channel and becomes fluidized (aerated). When this fluidized material reaches the outlet, it is likely to flood from the bin, because most feeders are designed to handle solids, not fluids (see FLUIDIZATION).

*Limited discharge rate*. Bulk solids, especially fine powders, sometimes flow at a rate lower than required for a process. This flow rate limitation is often a function of the material's air or gas permeability. Simply increasing the speed of the feeder does not solve the problem. There is a limit to how fast material flows through a certain sized opening. For fine powders the limit is strongly influenced by the counterflow of gas through the outlet.

*Segregation*. The problem of segregation occurs when a bulk solid composed of different particle sizes or densities separates. The result can be quite serious if uniform density or mixed material is required for a process.

## Flow Patterns

**Funnel Flow.** A funnel flow pattern occurs when some of the material in a bin or hopper moves toward the outlet while the rest remains stationary. This happens when the walls of the hopper section at the bottom of the bin are not sufficiently steep or smooth to cause the material to flow along them. In other words, the friction which develops between the hopper and material is great enough to inhibit flow at the interface. As a result, the material flows only in a narrow channel, usually directly over the outlet. The size of the flow channel approximates the largest dimension of the outlet. It is equal to the diameter of a circular outlet or the diagonal of a square or rectangular outlet. Using relatively free-flowing materials, the funnel flow channel may expand to the vertical walls of the bin if the bin is tall enough.

A funnel flow bin typically exhibits a first-in/last-out type of flow sequence. If the material has sufficient cohesive strength, it may bridge over the outlet. Also, if the narrow flow channel empties out, a stable rathole may form. This stable rathole decreases the bin's live or usable capacity, causes materials to cake or spoil, and/or enhances segregation problems. Collapsing ratholes may impose loads on the structure that it was not designed to withstand.

For fine powders, funnel flow bins often exhibit high discharge rates, thus controlling flow rate is always a challenge. A funnel flow channel is often unstable. The actual size and shape of the stagnant region is neither well defined nor constant. This channel can change its size radically or collapse, creating flow rates that range from no-flow conditions to complete flooding, also affecting the bulk density at the outlet.

Funnel flow bins are only suitable for bulk solids that are coarse, free flowing, and do not degrade, and for use when segregation is not important. For such materials, the principal benefits of funnel flow bins are reduced headroom and lower initial cost for the bin (excluding feeders or dischargers). Examples of funnel flow bins are shown in Figure 2.

**Mass Flow.**   A material is considered to flow in mass flow when all of it is in motion whenever any is withdrawn. This means that material flows along the walls. The walls of the hopper section are thus required to be steep and smooth (Fig. 3). As long as the outlet is large enough to prevent arching, all the material starts to move as discharge begins keeping the contents of the bin fully live. Stable ratholes cannot form.

Because of the first-in/first-out sequence of flow, mass flow bins can usually handle bulk solids that are cohesive or degrade with time. The bin's smooth, steep hopper section promotes uniform flow, preventing stagnation. A material's discharge bulk density is almost independent of the head of material in the bin. Segregation of particles is minimized as the fines and coarse particles are re-united at the outlet. In some instances, mass flow bins having special velocity profiles are used for in-bin blending. One possible disadvantage of mass flow bins is the headroom required for the steep hopper section.

Fine powders are easily handled in a mass flow bin where the flow channel is stable and predictable. However, the maximum flow rate of a fine powder through the outlet of a mass flow bin is low compared to that of a coarse, granular solid. For fine materials, the expansion and contraction of voids during flow can create an upward air pressure gradient at the outlet of a mass flow bin. During discharge, this upward gradient acts against gravity, reducing the discharge rate. Such gradients do not usually form with coarser particle materials. The latter are more permeable and thus allow air to flow freely into and out of the voids as they expand and contract.

**Expanded Flow.**   Expanded flow uses the best aspects of funnel flow and mass flow by attaching a mass flow hopper section below one that exhibits funnel

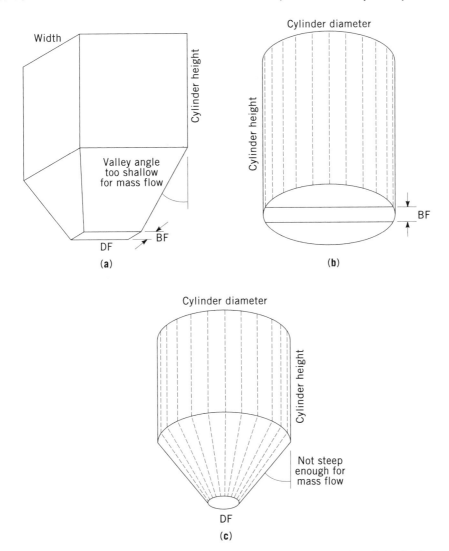

**Fig. 2.** Examples of funnel flow bins where BF is the outlet width and DF is the outlet diameter or length: (**a**) pyramidal hopper; (**b**) flat bottom; and (**c**) conical hopper.

flow. The flow pattern expands sufficiently at the top of the mass flow hopper to prevent a stable rathole from forming in the funnel flow hopper above it. In this way, the flow channel is expanded, material flow is uniform, and the bin height is limited.

Expanded flow bins, shown in Figure 4, are recommended for storing large quantities of nondegrading solids. This design is sometimes useful when modifying existing funnel flow bins. An expanded flow bin should be considered for applications that require bin diameters greater than ca 6 m. Bins of smaller diameter should usually be designed for full mass flow, if flow along the walls is required.

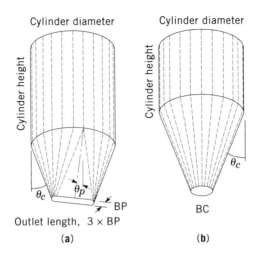

**Fig. 3.** Examples of mass flow bins: (**a**) transition hopper, where $\theta_c$ is the conical end-wall angle, $\theta_p$ is the side-wall angle, and BP is the outlet width; (**b**) conical hopper, where $\theta_c$ is the hopper angle and BC is the outlet diameter.

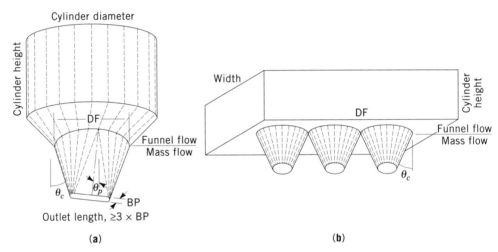

**Fig. 4.** Examples of expanded flow bins showing where funnel flow becomes mass flow: (**a**) funnel flow cone modified with mass flow transition hopper, where BP = outlet width, DF = diameter, $\theta_c$ = end-wall angle, and $\theta_p$ = side-wall angle; (**b**) long funnel flow slot modified with mass flow cones, where DF = slot length and $\theta_c$ = hopper angle.

## Measurement of Flow Properties

In order to develop the proper flow pattern, knowledge of a material's flow properties is essential. Standard test equipment and procedures for evaluating solids flow properties are available (6). Direct shear tests, run to measure a material's friction and cohesive properties, allow determination of hopper wall angles for mass flow and the opening size required to prevent arching. Other devices available to evaluate solids flowability include biaxial and rotary shear testers.

**Wall Friction Angles.**   Wall friction values, important when characterizing the flow properties of a bulk solid, are expressed as the wall friction angle or coefficient of sliding friction. The lower the friction, the less steep the hopper walls need to be to achieve mass flow.

Wall friction values can be measured by sliding a sample of material across a stationary wall surface. This test is performed by placing a retaining ring, as shown in Figure 5a, on a flat piece of wall material. Then, using weights and a cover, forces are applied to the material in a direction perpendicular to the wall surface. Material in the ring is forced to slide along the stationary wall material. The resulting shear force is measured as a function of the applied normal force.

Figure 5b shows the results of a typical wall friction test. The wall friction angle, designated as $\phi'$, is defined as the angle formed by a line drawn from the origin to a point on the shear stress–normal pressure curve. The point is selected based on the pressue level at which the angle is to be determined. As the location in a hopper changes, so does the pressure acting on the hopper walls. The tangent of the wall friction angle is the coefficient of sliding friction. For a given bulk material and wall surface, wall friction angle is not necessarily a constant but often varies with normal pressure, usually decreasing as normal pressure increases. Once the wall friction angles have been determined, hopper angles for mass flow can be chosen.

The following variables can affect wall friction values of a bulk solid. (1) Pressure: as the pressure acting normal to the wall increases, the coefficient of sliding friction often decreases. (2) Moisture content: as moisture increases, many bulk solids become more frictional. (3) Particle size and shape: typically, fine materials are somewhat more frictional than coarse materials. Angular particles tend to dig into a wall surface, thereby creating more friction. (4) Temperature: for many materials, higher temperatures cause particles to become more frictional. (5) Time of storage at rest: if allowed to remain in contact with a wall surface, many solids experience an increase in friction between the particles and the wall surface. (6) Wall surface: smoother wall surfaces are typically less frictional. Corrosion of the surface obviously can affect the ability of the material to slide on it.

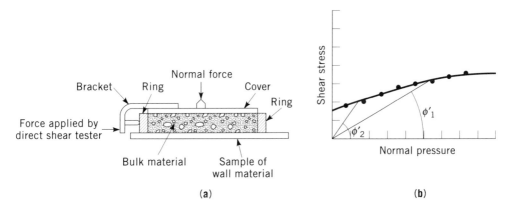

(a)                                          (b)

**Fig. 5.**   Wall friction test: (**a**) apparatus for measurement, and (**b**) results, where $\phi'$ is the wall friction angle.

**Flow Function.**    Another important consideration in bin design is the opening size required to prevent formation of arches and ratholes. A flow obstruction can occur in a bin caused by cohesive arching. Particles can bond together physically, chemically, or electrically. This tendency to bond is termed a material's cohesive strength. Many bulk solids flow like a liquid when poured from a bag. Under these conditions, the material has no cohesive strength. However, when squeezed in the palm of a hand, the material may gain enough cohesive strength to retain the shape of the palm once the hand is opened.

In order to characterize this bonding tendency, the flow function of a material must be determined. Data on flow function can be generated in a testing laboratory by measuring the cohesive strength of the bulk solid as a function of consolidation pressure applied to it. Such strength is directly related to the ability of the material to form arches and ratholes in bins and hoppers.

A material's flow function is usually measured on the same tester as the wall friction angle, although the cell arrangement is somewhat different (Fig. 6). Consolidation values are easily controlled, and the cohesive strength of the bulk solid is determined by measuring interparticle shear stresses while some predetermined normal pressure is being applied.

The cohesiveness of a bulk solid is often a function of the following variables. (1) Moisture: typically, cohesiveness rises as moisture content increases. Hygroscopic materials can experience significant increases in moisture when exposed to humid air. (2) Particle size and shape: there is no direct correlation between particle size, shape, and cohesiveness. Even so, in most cases, as a bulk solid becomes finer, it also becomes more cohesive and difficult to handle. (3) Temperature: a bulk solid's temperature can affect its cohesiveness. For example, many chemicals and plastic powders become more difficult to handle as the temperature rises. Some materials have more strength at constant temperature; others gain cohesive strength as the temperature changes during heating or cooling. (4) Time of storage at rest: when a solid resides in a bin or hopper for a period without moving, it can become more cohesive and difficult to handle. (5) Chemical additives: in some cases, a small amount of a chemical additive such as fumed silica can cause a cohesive solid to flow more easily.

**Compressibility.**    The bulk density of a solid is an essential value used in the analysis of its flow properties, such as when calculating mass flow hopper angles, opening sizes, bin loads, etc. Loose and/or packed density values are not sufficient. Bulk solids exhibit a range of densities that vary as a function of

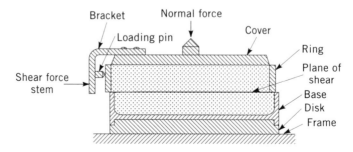

**Fig. 6.**    Test apparatus for measurement of cohesive strength.

consolidating pressure. This range of densities, called the compressibility of the solid, can often be expressed on a log–log plot as a line or relationship.

The following variables can affect a material's bulk density. (1) Moisture: higher moisture content often makes a material more compressible. (2) Particle size and shape: often, the finer the bulk solid, the more compressible it is. The shape of the particles can affect how they fit together and their tendency to break while being compacted. (3) Temperature: some materials become more compressible as their temperature increases. This could be due, for example, to softening of the particles. (4) Particle elasticity: elastic materials tend to deform significantly when they are compressed.

**Permeability.** Two-phase (gas–solids) interactions can be analyzed by considering how gas flows through a bed of powder in the presence of a pressure differential across the bed. When flow through the bed is laminar, Darcy's law, which can be used to relate gas velocities to gas pressure gradients within or across the bed, can be written as in equation 1, where $K$ = permeability factor

$$u = -K\left(\frac{dp/dx}{\gamma}\right) \tag{1}$$

of the bulk solid; $u$ = superficial relative gas velocity through the bed of solids; $\gamma$ = bulk density of the solid in the bed; and $dp/dx$ = gas pressure gradient acting at the point in the bed of solids where the velocity is being calculated.

The permeability factor, $K$, has units of velocity and is inversely proportional to the viscosity of the gas. A permeability test is run by passing air through a representative column of solids. The pressure across the bed is regulated, and the rate at which the gas flows is measured. This approach allows the permeability of the bulk solid to be determined as a function of its bulk density. Typically a linear log–log plot results.

Because mass flow bins have stable flow patterns that mimic the shape of the bin, permeability values can be used to calculate critical, steady-state discharge rates from mass flow hoppers. Permeability values can also be used to calculate the time required for fine powders to settle in bins and silos. In general, permeability is affected by particle size and shape, ie, permeability decreases as particle size decreases and the better the fit between individual particles, the lower the permeability; moisture content, ie, as moisture content increases, many materials tend to agglomerate which increases permeability; and temperature, ie, because the permeability factor, $K$, is inversely proportional to the viscosity of the air or gas in the void spaces, heating causes the gas to become more viscous, making the solid less permeable.

## Equipment Design

**Hopper Angles for Mass Flow.** The wall friction angle for a given bulk material/wall surface combination can be calculated from the results of a wall friction test. This angle, the tangent of which is the coefficient of sliding friction, often varies with the pressure acting normal to the wall surface. From this angle the hopper angles compatible with mass flow can be determined. This is most easily done for hoppers which are either conical or wedge-shaped. There is no magic angle, because mass flow is dependent on both the smoothness and steepness of the hopper wall and the properties of the bulk material involved.

*Conical Hoppers.* Design charts for conical hoppers typically are plots of wall friction angles, $\phi'$, vs hopper angle, $\theta_c$. Charts such as that in Figure 7**a** may be used in several ways. For example, if a 20° from vertical (70° from horizontal) corroded carbon steel hopper is experiencing ratholing, indicative of a funnel flow pattern, and a smooth stainless steel liner to convert the flow pattern to mass flow is suggested, rather than taking a try-it-and-see approach, a wall friction test is run. If the resulting wall friction angle is 15°, this angle combined with the 20° hopper angle is within the mass flow region of the design chart. The stainless liner should be acceptable. If, on the other hand, the wall friction angle on the proposed liner were 30°, the funnel flow pattern would remain.

Another way to use the chart in Figure 7**a** is during the design of a new hopper. The goal is often to determine the least steep hopper angle that allows mass flow for a certain wall material. A wall friction test using the proposed material of construction and actual bulk material can be run. Then, using the measured value of the wall friction angle, the corresponding mass flow region and the appropriate hopper angle can be determined. The edge of the mass flow region yields the shallowest recommended angle for mass flow for this application.

There is an uncertain region which lies between funnel flow and mass flow. Whereas the flow pattern within this region should theoretically be mass flow, slight differences in material properties or hopper angle can result in funnel flow. In effect, this region represents a margin of safety to prevent using a design too close to the funnel flow line. A switch back and forth between mass flow and funnel flow can cause bin vibrations and other problems.

*Wedge Hoppers.* Different design charts are used for wedge hoppers (Fig. 7**b**) than for conical hoppers. Values of hopper angle $\theta_p$ (measured from vertical) appear on the horizontal axis; wall friction angles $\phi'$ are on the ver-

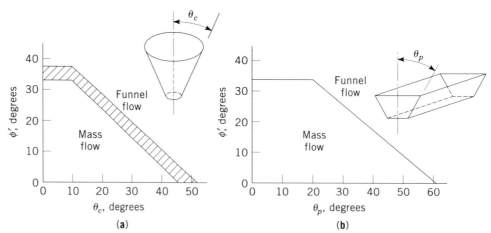

**Fig. 7.** Design chart to determine mass flow for hopper wall angles for (**a**) conical design, where the ▨ area designates uncertainty and (**b**) wedge design. See Figures 3 and 5 for definitions of $\theta_c$, $\theta_p$, and $\phi'$.

tical axis. There is no uncertain region in this chart as on the conical hopper design chart, because there is no sharp boundary line between mass flow and funnel flow in wedge-shaped hoppers. In fact, mass flow can at times occur to the right of the design line, in the region labeled funnel flow. Thus, a wedge geometry is more forgiving than a conical one and therefore more capable of handling materials having a wider range of flowability.

Wedge hopper design charts are used in the same way as the conical design charts. If $\phi'$ is 15°, the resulting maximum wedge hopper angle for mass flow is 40° from vertical. This is 12° less steep than the required conical hopper angle. Mass flow wedge-shaped configurations require significantly less headroom than conical hoppers. The side walls of transition- and chisel-shaped hoppers can be designed based on $\theta_p$ provided the outlet length-to-width ratio is at least 3:1.

*Other Designs.* Mass flow can also be achieved by the use of inserts. One popular type, called a BINSERT (Jenike & Johanson, Inc.), consists of a hopper-within-a-hopper (Fig. 8) (7). Material flows through both hoppers. By choosing appropriate angles and materials of construction, the inner hopper can force flow along the walls of the outer hopper at a less steep angle than would be possible if the inner hopper were not present. By careful choice of the hopper design, a BINSERT can provide in-bin blending or, alternatively, a uniform velocity profile which is excellent in preventing segregation.

**Outlet Size Determination.** The second consideration for proper design of a mass flow bin is the size of the outlet required to prevent arching and to achieve the required discharge rate.

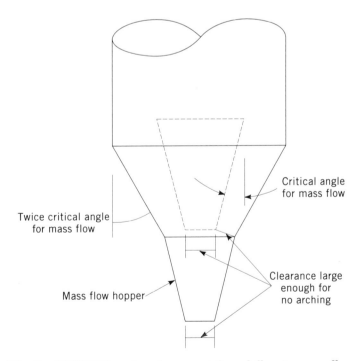

**Fig. 8.** BINSERT system to convert funnel flow to mass flow.

There are two mechanisms by which arching can occur: particle interlocking and cohesive strength. The minimum outlet size required to prevent mechanical interlocking of particles is directly related to the size of the particles. The diameter of a circular outlet must be at least six to eight times the particle size, and the width of a slotted outlet must be at least three to four times the particle size. These ratios normally only govern the outlet size of mass flow hoppers if the particles are at least 0.6 cm or larger.

If most of the particles are less than ca 0.6 cm in size, flow obstructions can occur by physical, chemical, or electrical bonds between particles. This cohesiveness is characterized by the bulk material's flow function. The forces acting to overcome a cohesive arch and cause flow are described by a hopper's flow factor, which can be obtained from the design charts (see Fig. 7). The minimum opening size required to prevent a cohesive arch from forming can be calculated from the comparison of the flow factor and flow function.

As for the particle interlocking mechanism, the minimum diameter to prevent cohesive arching for a circular outlet is about twice the minimum width of a slotted outlet. For example, if a 30-cm diameter opening is required to prevent arching in a cone, arching in a slotted outlet can be prevented using as little as a 15-cm wide opening. Note that the length of a slotted outlet should be at least three times its width.

Sizing an outlet to achieve the required discharge rate is more difficult but no less important, particularly for fine powders. All bulk materials have a maximum rate at which they discharge through a hopper opening of a given size. For coarse, free-flowing bulk materials, a good approximation of this rate from a mass flow hopper is shown in equation 2, where $Q$ = maximum steady dis-

$$Q = \gamma A (Bg/2(1 + m) \tan \theta)^{1/2} \tag{2}$$

charge rate, $\gamma$ = bulk density, $A$ = cross-sectional area of outlet, $B$ = outlet diameter or width, $g$ = acceleration owing to gravity, $m$ = 1 for circular opening and 0 for slotted opening, and $\theta$ = hopper angle (measured from vertical) in degrees. A modification of this equation takes particle size into account. This modification is only important if the particle size is a significant fraction of the outlet size (8).

Usually the rate, $Q$, is far in excess of the required rate, especially if the bulk material consists primarily of coarse particles. Slowing down the discharge rate requires a feeder. Fine powders, on the other hand, have considerably lower maximum discharge rates when exiting from a mass flow bin, because of the interaction between air (or gas) and solid particles as reflected in the permeability of the material.

Sizing the outlet of a funnel flow bin involves consideration of both arching and ratholing. Minimum dimensions to overcome both can be calculated from the material's flow function.

**Structural Considerations.** Silos, bins, and hoppers fail, in one way or another, each year. The causes of silo failures are many and varied (9). Such failures can range from a complete and dramatic structural collapse, to cracking in a concrete wall, or denting of a steel shell. This last is often a danger signal indicating that corrective measures are required.

*Design.* Silo design requires knowledge of the material's flow properties, flow channel geometry, flow and static pressure development, and dynamic effects. Problems like ratholing and vibration have to be prevented, while assuring reliable discharge at the required rate. Nonuniform loads, thermal loads, and the effects of nonstandard fabrication details must be considered. Above all the designer must know when to be cautious in the face of incomplete or misleading information or recommendations that come from handbooks.

The designer must have a full appreciation of load combinations, load paths, primary and secondary effects on structural elements, and the relative flexibility of the elements. Special attention must be given to how the most critical details in the structure are to be constructed so that the full requirements and intent of the design can be realized (10).

Some flow-related loading conditions which many designers fail to anticipate include bending of circular walls caused by eccentric withdrawal; nonsymmetric pressures caused by some types of inserts; and self-induced vibrations.

*Construction.* In the construction phase there are two ways problems can arise. The more common of these is poor workmanship. Uneven foundation settlement and faulty construction, eg, use of the wrong materials or insufficient quantity of rebars, are but two examples. Only qualified builders, close inspection during construction, and enforcement of a tightly written specification are necessary factors.

The other cause of construction problems is the introduction of badly chosen, or even unauthorized, changes during construction in order to expedite the work. Any changes in details, material specifications, or erection procedure must be given careful consideration by both the builder and silo designer.

*Silo Usage.* If a bulk material other than the one for which the silo was designed is placed in the silo, the flow pattern and loads may be completely different. The load distribution can be radically changed if alterations to the outlet geometry are made, if a side outlet is put in a center discharge silo, or if a flow controlling insert or constriction is added. Some of the problems which can occur include collapse of large voids resulting in immense impact forces; development of mass flow in silos designed structurally for funnel flow; drastic means of flow promotion, eg, use of explosives or air cannons; buckling of an unsupported cylindrical wall below an arch of stored bulk material; metal fatigue caused by externally mounted bin vibrators; and dust explosions.

*Maintenance.* There are two types of maintenance work required (11). The first is regular preventative work, such as the periodic inspection and repair of a liner used to promote flow, protect the structure, or both. Loss of a liner may be unavoidable for an abrasive or corrosive product, yet maintaining a liner in proper working condition is necessary if the bin or silo is to operate as designed. The second area of maintenance involves looking for signs of distress, eg, cracks, wall distortion, or tilting of the structure. If evidence of a problem appears, expert help should be summoned immediately. An inappropriate response to a sign that something is going wrong can precipitate a failure even faster than leaving it alone. The common, instinctive response to discharge material so as to lower the fill level is often inappropriate.

Wear owing to corrosion and/or erosion can be particularly dangerous. For example, as carbon steel corrodes, the reduced wall thickness can eventually lead to a structural failure. This problem can be compounded through erosive wear of the silo wall.

## Feeders

Most flow problems can be overcome by using a mass flow design if the mass flow pattern developed by the bin is not disturbed. Thus a properly designed feeder or discharger must be employed. A feeder is used whenever there is a requirement to transfer solids at a controlled rate from the bin to a process or a truck. A discharger is used when there is a need to discharge solids, not control the rate of discharge.

To be consistent with a mass flow pattern in the bin above it, a feeder must be designed to maintain uniform flow across the entire cross-sectional area of the hopper outlet. In addition, the loads applied to a feeder by the bulk solid must be minimized. Accuracy and control over discharge rate are critical as well. Knowledge of the bulk solid's flow properties is essential.

There are several types of feeders available to handle bulk solids, but these can be divided into two categories: volumetric and gravimetric. A volumetric feeder discharges a certain volume of material as a function of time. This type of discharge is adequate for many solids feeding applications. For mass flow designs, feed uniformities in the range of $\pm 2$ to 5% on a minute-to-minute basis are easily achieved using most volumetric designs. A disadvantage of volumetric feeding is that the feeder does not recognize bulk density changes.

A gravimetric feeder relies on weighing the material to achieve the required rate. Feed accuracies of $\pm 1/4\%$ are obtainable using a properly designed gravimetric feeder. A disadvantage of such a feeder is that it is usually more expensive than a volumetric device.

**Volumetric Feeders.** Examples of volumetric feeders are screws, belts, rotary valves, louvered, and vibratory.

*Screw Feeders.* Screws are primarily used when feed over a slotted outlet is required. Screws are a good choice when an enclosed feeder is required, when space is restricted, when handling dusty or toxic materials, or when attrition (particle breakage) is not a problem. A screw is composed of a series of flights that are wound around a common shaft. The flights have a particular diameter and pitch (the distance between flights). Some screws have constant pitch flights; others vary. The screw shaft has to be sized to prevent deflection (12).

Many applications use screws with constant pitch to feed material from a slotted opening. The configuration shown in Figure 9a shows a constant pitch and constant diameter causing a preferential flow channel to form at the back (over the first flight) of the screw. This type of flow destroys the mass flow pattern and potentially allows some or all of the problems discussed about funnel flow.

The key to proper feeder design is to allow material to be withdrawn over the entire cross-sectional area of the outlet. In order to ensure this, the capacity of the feed device must increase as material is transferred to the discharge end. Tapering the diameter theoretically increases capacity; however, material

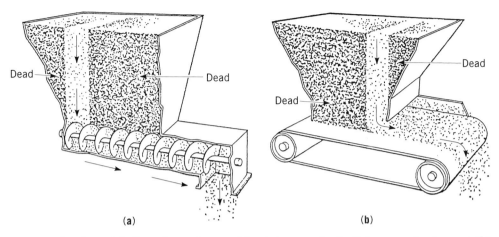

**Fig. 9.** Preferential flow channel caused by (**a**) a constant pitch screw feeder and (**b**) a belt feeder.

is likely to bridge over the narrowed end flights, creating even more problems. Increasing the pitch from no less than one-half pitch to full has many successful applications. However, the length of the slot over which one can feed reliably is limited to about three times the diameter ($3d$) of the screw owing to tolerances of fabrication. Another approach is to use a tapered diameter shaft where the screw's capacity increases in the discharge direction. This approach is limited as well by fabrication tolerances of the screw to $3d$.

A combination of tapered shaft diameter and increasing pitch is shown in Figure 10**a**. This allows a length-to-diameter ratio of about 6:1 instead of 3:1. A half pitch screw is used over the tapered diameter. This approach results in an excellent mass flow pattern provided that the hopper to which it attaches is also designed for mass flow.

A stepped diameter shaft screw (Fig. 10**b**) has also been developed (13) allowing even longer (up to as much as 12:1) length-to-diameter ratios. Other appropriate applications for screws are as sealing devices, multiple screws for

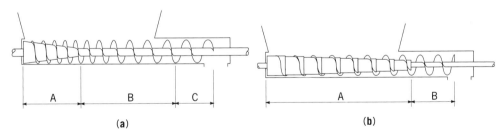

**Fig. 10.** Mass flow screw feeder designs. (**a**) Combined tapered shaft and variable pitch screw feeder where A represents a conical shaft and constant pitch (feed section); B, constant shaft and increasing pitch (feed section); and C, constant shaft and constant pitch (conveying section). (**b**) Stepped shaft screw feeder where A represents a stepped diameter shaft and constant pitch (feed section) and B, constant shaft and constant pitch (conveying section).

larger openings, multiple discharge points, and as blending and cooling (or heating) devices.

*Belt Feeders.*    Belts are used to feed over long slotted openings. Typically, belt feeders are used to handle friable, coarse, fibrous, elastic, sticky, or very cohesive solids. Because belts are available in widths up to 2.74 m and unrestricted lengths, such feeders can be designed for very large outlets.

An improperly designed interface to the belt can cause solids compaction, abrasive wear of the belt, and excessive power required to move the belt. The preferential flow channel shown in Figure 9b withdraws material from one end of the outlet. Depending on the gate opening, this could be at the back or front. Some methods of providing increased capacity over the slot length are shown schematically in Figure 11. The taper in plan is not recommended for most applications because the narrow opening at the back is prone to bridging. The taper in elevation is useful but will likely withdraw material preferentially from the front.

The optimum design of a belt feeder interface is shown in Figure 12. The nose is slanted to provide stress relief as material is transferred to the discharge end. Either flat or troughed idlers may be used to train the belt. Idlers should

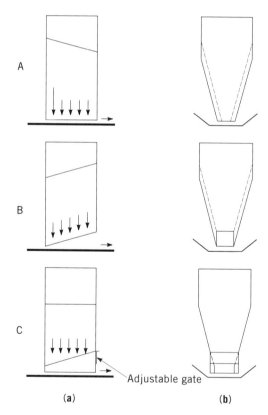

**Fig. 11.**  Schematics of belt feeder interfaces to provide increasing capacity, where (**a**) shows side elevation and (———) is the side view of the belt feeder on its centerline, and (**b**) shows end elevation: A, slot tapers in plan; B, slot tapers in elevation; and C, compound taper with vertical side walls.

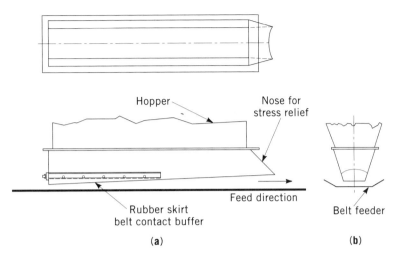

**Fig. 12.** Optimum belt feeder interface: (**a**) elevation views, where (——) is the belt feeder on its centerline, and (**b**) plan view.

be closely spaced so as to prevent belt sag. Skirts are also used to prevent spillage. The skirts should expand slightly in the direction of belt travel so as not to interfere with material flow. A large enough opening to prevent bridging and sloping side walls, which are steep and smooth enough for mass flow, are necessary.

*Rotary Valve Feeders.* Devices known as rotary valve feeders are commonly used for circular or square configured outlets. These are particularly useful when discharging materials to a pneumatic conveying system where a seal is required to prevent air flow through the hopper outlet. The discharge rate is set by the speed of rotation of the vanes or pockets of the valve.

A potential problem for rotary valve usage is that they tend to pull material preferentially from the upside of the valve, which can affect the mass flow pattern. Another problem is that once solid drops from the vane, the air or gas that replaces it is often pumped back up into the bin. In addition, air can leak around the valve rotor. Such air flows can decrease the solids flow rates and/or cause flooding problems. A vertical section shown in Figure 13 can alleviate the preferential flow problem because the flow channel expands in this area, usually opening up to the full outlet. To rectify the countercurrent air flow problem, a vent line helps to take the air away to a dust collector or at least back into the top of the bin.

*Louvered Feeders.* Louvered feeders, designed to withdraw material uniformly across the entire outlet cross section, can control discharge from a circular, square, or rectangular cross section using the material's natural angle of repose. When the drive is energized, the material's angle of repose is overcome and material discharges usually very evenly and accurately. When the drive is stopped, the material stops flowing. The design is simple, robust, economical, and has many applications.

*Vibratory Pan Feeders.* Vibratory pan devices act much in the same way as louvered feeders in that they can feed material gently and accurately. Unfortunately, they are limited primarily to applications involving round or square

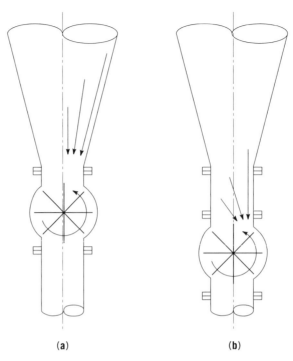

(a)                                              (b)

**Fig. 13.** (**a**) Preferential flow through rotary valve and (**b**) correction from installation of a vertical spool piece.

outlets. If a long rectangular outlet is used, the feeder must operate across the short dimension of the slot.

There are several other types of volumetric feeders that rely on external or internal agitation to initiate or maintain flow. Other types include devices that aerate material or use flexible walls that are agitated to maintain flow. These devices are all useful when used properly. However, they use something other than gravity to maintain flow and, as such, can be maintenance intensive. Knowledge of the solid's flow properties is essential to properly design the feeder. Bin opening size and hopper angles are required to ensure mass flow to the feeder. The feeder then must be capable of controlling discharge rate.

**Gravimetric Feeders.** Examples of gravimetric feeders are weigh belts, loss-in-weight systems, and gain-in-weight systems. Gravimetric feeders rely on weighing the material to achieve the required discharge rate. A gravimetric feeder would be used when accuracies of less than 5% are required, particularly over short time periods; when the material's bulk density varies; or when the weight of material used for a particular process needs to be recorded. There are basically two systems: continuous and batching. A continuous gravimetric system controls the weight/unit time. A batch system simply controls the weight of material discharged, such as ~40 kg of material to a mixer.

*Weigh Belt Feeders.* This type of gravimetric feeder, shown in Figure 14, typically is used in continuous feeding applications as opposed to batches. A belt feeder can be used as a weigh belt by locating weigh idlers under the belt downstream of the outlet. Load cells weigh the material crossing them and send

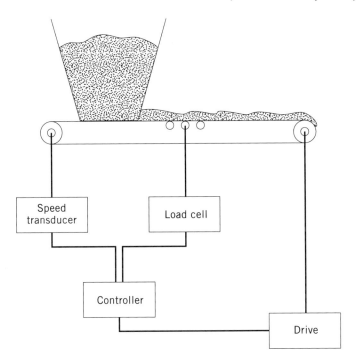

**Fig. 14.**  Weigh belt feeder.

a signal to a controller where it is integrated with the belt speed and compared to a set point. The speed of the drive is adjusted accordingly to regulate the discharge rate.

Some drawbacks to the weigh belt are that it is a zero-reference device and thus needs frequent calibration (re-zeroing). Buildup on the belt and rollers affects accuracy and operation, as does belt tension and dusty or floodable materials. Flexible connections are required to isolate the feeder from upstream and downstream equipment, unless the belt feeder/weigh idler concept is used.

*Loss-in-Weight Feeders.*   The loss-in-weight (LIW) gravimetric feeder is used when feed accuracy is essential. It measures the loss in weight of material discharged from the system. As such, it can be used both in continuous and batching systems, and may be used for liquids by replacing the feeder with a pump. Load cells are attached to the bin or hopper that are capable of weighing the bin, feeder, and the contents. These load cells sense changes that take place as material is discharged and send a signal to a controller to speed up or slow down the feeder.

A disadvantage of LIW systems is that they cannot weigh while being filled. A typical LIW system switches to volumetric feeding while the filling process occurs, and when the bin is filled, it switches back to gravimetric. Some systems are available with no-freeze designs in which one bin and feeder discharges while another is being filled, and these are easily switched back and forth. Screws, belts, rotary valves, and louvered or vibratory pan feeders can be used to control discharge.

*Gain-in-Weight Feeders.*   These types of feeders are used only for batching applications. The receiving container rests on a scale or on load cells and the system controls the discharge from the filling bin, which can use a volumetric feeder to control rate. A batch accuracy of $\pm 1/4\%$ at two standard deviations is not unusual.

## Special Considerations

**Particle Segregation Mechanisms.**   Segregation is the process by which an assembly of solid particles separates as it is being handled. This often results in costly quality control problems due to the waste of raw or finished materials, lost production, increased maintenance, and capital costs required to retrofit existing facilities.

Segregation problems occur in a wide range of industries handling materials as diverse as coal (qv) and pharmaceuticals (qv). The cost implications can be great, even when handling small quantities of material. In the batch processing of pharmaceutical tablets, it is not unusual for the value of a single batch of ingredients to be in excess of $100,000. Strict U.S. quality control standards dictate that some or all of the batch may have to be discarded if it is found that the amount of active ingredient or total weight of just five tablets in a batch varies outside narrow limits.

The primary mechanisms responsible for most particle segregation problems are sifting, particle velocity, air entrainment, particle entrainment, and dynamic effects (14).

*Sifting.*   The movement of smaller particles through a mixture of larger ones (sifting) is the most common mechanism by which particles segregate. In order for this mechanism to occur, all of the following conditions must be present: (*1*) a difference in particle size, ie, a difference smaller than 2:1 can cause sifting segregation; (*2*) a sufficiently large mean particle size, ie, sifting segregation is more common in mixtures having a mean size greater than 100–200 $\mu$m; (*3*) free-flowing material, ie, the lesser the tendency for particles to adhere to each other, the more likely they are to segregate; and (*4*) interparticle motion, eg, due to formation of a pile or movement of material on a conveyor.

Sifting may occur while a pile is being formed. Fine particles are concentrated in the center under the fill point. However, as the pile is formed, the slope stability is such that layers of finite thickness intermittently move from the central fill point, carrying some of the finer particles with them.

*Particle Velocity on a Surface.*   Smaller particles, those that are more irregular in shape and/or those that have a higher surface roughness, typically have a higher frictional drag on a hopper or chute surface.

On a chute, higher drag results in lower particle velocity which can be accentuated by stratification on the chute surface because of the sifting mechanism. Concentrations of smaller particles close to the chute surface and larger particles at the top of the bed of material, combined with the typically higher frictional drag of finer particles, often result in a concentration of fine particles close to the end of the chute, and coarse particles farther away. This can be particularly detrimental if portions of the pile go to different processing points, as is often the case with multiple outlet bins or bins with vertical partitions.

*Air Entrainment.* Fine particles generally have a lower permeability than coarse particles, and therefore tend to retain air longer in void spaces. Heavier particles settle more quickly in a fluidized mixture than lighter particles. Thus, when a mixture of particles is charged into a bin, it is not uncommon to find a vertical segregation pattern, where the coarser, heavier particles concentrate at the bottom of the bed and the finer, lighter particles concentrate near the top.

*Entrainment of Particles in an Air Stream.* The lighter the particle and the finer its size, the longer it may remain suspended in an airstream such as upon filling of a bin. Secondary air currents can carry airborne particles away from a fill point into outer areas of a bin, scattering them in a way that bears no resemblance to the calculated trajectory.

Particles may also be affected by air resistance as they fall, resulting in finer particles having a lower free-fall terminal velocity than coarser particles and thereby not traveling as far horizontally when they exit a chute. However, a stream of particles drags a stream of air with it, preventing the full drag force from being felt on individual particles. Thus, only particles on the edge of the stream are affected by air drag.

*Dynamic Effects.* Particles often differ in their resilience, inertia, and other dynamic characteristics which can cause them to segregate, particularly when they are forming a pile such as when charged into a bin or discharged from a chute.

**Correcting Particle Segregation.** The main techniques to consider when segregation problems are present are to change the material, change the process, or change the design of the equipment.

*Properties of the Material.* A common characteristic of most highly segregating materials is that they are free flowing, and therefore the particles easily separate from each other. Thus, one way to decrease segregation tendencies of a material is to increase its cohesiveness by, for example, adding water or oil. If overdone, other flow problems such as arching or ratholing may replace segregation, potentially resulting in an even greater disruption to the process.

Another technique is to change the particle size distribution. There are, however, disadvantages. If segregation is occurring by the sifting mechanism, the particles must be almost identical in size before sifting is prevented. Alternatively, the mean particle size can be reduced below 100 $\mu$m, but this size reduction (qv) increases the probability of segregation by the too fine powder mechanisms.

*The Process.* If a mixture is handled that consists of several ingredients, which are more or less uniform in themselves but vary distinctly from one to another, each ingredient should be handled separately up to the final processing step, then proportioned and mixed just before this step. When pneumatically conveying a fine fluidizable powder into a bin, a tangential entry into the side of the bin should be used rather than going in at 90° either to the side wall or the top.

*The Design of the Equipment.* If materials have segregated from side to side while filling a bin, a mass flow pattern tends to minimize segregation upon discharge, whereas a funnel flow pattern makes the segregation worse. Increasing the height-to-diameter ratio of the cylinder section of a mass flow bin above 1.0 usually results in a uniform velocity pattern across the top surface.

This lessens the tendency for segregation compared to using a short cylinder section or only a converging hopper.

When segregation is a concern, a single-outlet, symmetric bin should be used. However, if a multiple-outlet bin is required, the hoppers and outlets should be located symmetrically with respect to the fill point to prevent concentrations of fines in one hopper section which could cause pluggage or downstream quality control problems.

An alternative to traditional mass flow bin design is to use a patented BIN-SERT, which consists of a hopper-within-a-hopper below which is a single-hopper section (Fig. 15). The velocity pattern in such a unit is controlled by the position of the bottom hopper. A completely uniform velocity profile can be achieved which results in an absolute minimum level of segregation. Alternatively, by changing the geometry at the bottom of the hopper, a velocity profile can be developed in which the center section moves faster than the outside, thus providing in-bin blending of the materials (7).

**Fine Powder Flow Phenomena.** A fine powder is a material where the flow behavior is significantly influenced by the effects of entrained gas in its void spaces. This is in contrast to a coarse material such as plastic pellets in which the voidage is so great that gas effects can be neglected. Thus, for example, the limiting steady flow rate through an orifice can be described by equation 2 in which interstitial gas effects are ignored. For a fine powder such effects are extremely important, causing, for example, the maximum flow rate of such a material through a mass flow bin outlet to be several orders of magnitude lower than predicted by equation 2.

When handling a fine powder in a bin, funnel flow, mass flow, and fluidized handling can be considered (15). Funnel flow is seldom recommended because discharge is so unpredictable, ie, from no flow owing to arching or ratholing, to uncontrolled flow owing to flooding. With mass flow, discharge is controlled and predictable, but the rate is limited. Fluidized handling is generally only practical if the powder is easy-flowing and has relatively low permeability. Although high rates of discharge are possible, the powder's bulk density is much lower and more variable than if gravity-driven mass flow is used.

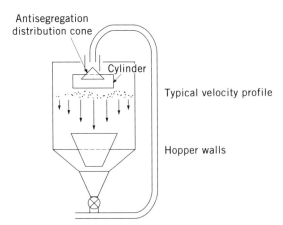

**Fig. 15.** BINSERT in-bin blender. The hopper walls are chosen to cause a significant velocity differential center to the outside.

For a mass flow bin, one of three flow-rate dependent modes of discharge can occur. (*1*) Steady gravity flow of partially deaerated material controlled by a feeder is most desirable. The limiting condition occurs when compaction in the cylinder section of the bin forces too much gas out through the material top surface. Because a bulk solid expands while flowing through the converging portion of a bin, a slight vacuum develops in its voids which causes a gas counterflow through the bin outlet that forces the solids contact pressure to drop to zero and limits the steady solids flow rate. (*2*) This mode occurs at flow rates somewhat greater than the limiting rate and is characterized by an erratic, partially fluidized powder discharging from the bin which can be controlled by some types of feeders. Even better, a steady rate can often be achieved by using an air permeation system at an intermediate point in the bin to replace the lost gas, increasing the outlet and corresponding feeder size, adding a standpipe between the bin outlet and feeder, and/or lowering the bin fill level. (*3*) This mode of flow occurs when the flow rate is too high to allow much, if any, gas to escape from the void spaces. In this extreme, the material may be completely fluidized and flood through the outlet unless the feeder can control fluidized powder.

Testers are available to measure the permeability and compressibility of powders and other bulk solids (6). From such tests critical, steady-state flow rates through various outlet sizes in mass flow bins can be calculated. With this information, an engineer can determine the need for changing the outlet size and/or installing an air permeation system to increase the flow rate. Furthermore, the optimum number and location of air permeation levels can be determined, along with an estimate of air flow requirements.

*Fluidized Handling of Powders.*   Sometimes it is more practical to handle fine powders in a fluidized state rather than to deaerate the particles and handle by gravity alone. Through the use of air pads, air slides, and/or air nozzles, some or all of the contents of a bin can be fluidized. This overcomes the common no-flow problems of arching and ratholing, and allows discharge rates through bin outlets several orders of magnitude faster than if the powder is deaerated.

When evaluating fluidized handling several factors must be considered. (*1*) The powder must have low cohesive strength, otherwise channels are likely to form which prevent uniform fluidization. In addition, it is helpful if the powder has low permeability so as to limit the amount of gas needed for fluidization as well as allow retention of the gas when it is turned off. (*2*) The bulk density of the powder as it exits the bin will be low and nonuniform, thus it may not be possible to fit the required mass of material into a downstream container, eg, a truck or rail car. If the powder is being fed into a process and close control of flow rate is important, the nonuniformity of bulk density may create significant control problems. (*3*) The cost of energy required to fluidize the bin may be significant, particularly if dry air must be used because the powder is hygroscopic. (*4*) Particle segregation may be made worse. Simply putting an air pad or an air slide into a bin is likely to segregate a powder, causing a vertical striation pattern with fine, light particles on top and dense, large particles on the bottom. (*5*) The amount of bin volume necessary to be fluidized must be evaluated. If possible, the entire contents should be fluidized, but this may be neither practical nor necessary, particularly if the bin is relatively large. If only a portion of a bin is fluidized, the potential for stagnant material supporting a rathole may be a concern. Void space must be provided for the material to expand. (6) What to do with the fluidizing air when there is no powder discharge taking place must be considered. If the powder becomes cohesive when it is deaerated, refluidization is difficult. Intermittent fluidization during periods of no discharge may be necessary.

*Purge and Conditioning Vessels.*    Vessels designed for processing solids are often adaptations of conventional storage bins modified to achieve the desired process activity. A wide variety of solids including chemicals, plastics, and sugar (qv) are processed in this way. Some of the processes carried out include heating, cooling, polymeric phase transformation, drying, curing, and suppressing or enhancing a particular chemical reaction.

Purge and conditioning vessels often exhibit nonuniform purge of conditioning which may be the result of nonuniform solids or fluid (gas or liquid) velocity profiles causing the solid's exposure time to the fluid to be nonuniform; cross-contamination, particularly after grade changeovers, which can easily happen unless the solids flow sequence is first-in/first-out; no flow due to formation of a stable obstruction at the outlet of the vessel or a rathole; erratic flow which involves fluctuations in either flow rate of bulk density; or segregation which creates quality control problems in the final product.

The key to solving these problems is to design the vessel for a mass flow pattern. This involves consideration of both the hopper angle and surface finish, the effect of inserts used to introduce gas and control the solids flow pattern, and sizing the outlet valve to avoid arching and discharge rate limitations. In addition, the gas or liquid must be injected such that the solid particles are uniformly exposed to it, and flow instabilities such as fluidization in localized regions are avoided.

**Mixing and Blending.**    Quality control is more important in the 1990s competitive global economy than ever before, and blending is a useful technique by which quality is improved. Improving uniformity of a raw material stream which enters a process, or the output stream from a process, can be extremely helpful. Static in-bin blenders are sometimes used to dampen upsets either going into or coming from a process, while batch and tumble blenders are sometimes used for close, well-defined quality control (see MIXING AND BLENDING).

*Requirements for Static In-Bin Blenders.*    An in-bin blender consists of basically a storage bin or silo which doubles as a blender (16). Requirements to make such a blender effective include no stagnant regions, ie, mass flow design; large velocity gradients throughout the blender, ie, the differential between the time it takes a particle in the fastest flowing region to exit the blender compared to a particle in the slowest flowing region should be as large as possible, and particles in the fastest flowing region should start to discharge as soon as possible after entering the blender; minimum need for recirculation to provide blending; blending uniformity independent of the blender's fill and discharge rates as well as level of material; the ability to blend a wide variety of materials, eg, fine and coarse particles, free-flowing, and cohesive materials; the ability to prevent segregating materials from demixing as they discharge from the blender; and cost effectiveness.

*Multitube Blenders.*    These blenders, eg, the Phillips and Young type, are widely used to blend materials that are free-flowing, uniform in size, and have low angles of internal friction.

Some problem applications for multitube blenders include blending in a small amount of additive; materials that contain a wide range of particle sizes, because sifting segregation can cause fines to percolate through the coarse particles while flowing toward a tube opening causing the fines to discharge last; materials that have high angles of internal friction, because flow problems in the tubes and steep flow channels outside the tubes can result and this can severely limit blending; cohesive, ie, nonfree flowing materials that may cause pluggage to tube openings; and fine powders which tend to aerate and flood through the tubes because of brief retention times.

*BINSERT Blenders.*    The design of a BINSERT blender consists of a hopper-within-a-hopper, both of which are usually conical in shape (Fig. 15). Particles flow through the inner hopper as well as through the annulus between the inner and outer hoppers. By varying the relative position of these two hoppers as well as the configuration of the outlet geometry, it is possible to achieve between a 5:1 and 10:1 velocity differential between particles in the inner hopper compared to particles in the outer annular region (7,17).

Although a BINSERT blender often requires more recirculation to achieve an acceptable blend than a multitube blender, it has a number of potential advantages including ease of cleaning since all of the internal parts are exposed and accessible; blending cohesive, ie, nonfree-flowing materials since the outlet can be sized as large as necessary for flow; blending materials with high angles of internal friction and materials which are highly segregating, eg, containing a wide range of particle sizes; low headroom requirements since the walls of the outer hopper can be made relatively shallow; no mechanical moving parts other than perhaps a feeder, eg, belt, screw, or rotary valve used at the outlet to control the discharge rate, and a recirculation system; and it can often be retrofitted to an existing storage bin causing it to act as an in-bin blender.

*Batch Blending.*    A typical batch blending system is shown in Figure 16. The basic components are a blender, one or more portable or stationary containers, and a chute to a process, eg, a tabletting press. A typical operation consists of mixing the ingredients in the blender, A, discharging the blend into a container, B, which is then moved into position over the chute, C, and gradually discharging the blend into the process, D.

Solid–solid blending can be accomplished by a number of techniques. Some of the most common include mechanical agitation which includes devices such as ribbon blenders, impellers, paddle mixers, orbiting screws, etc; a rotary fixed container which includes twin-shell (Vee) and double-cone blenders; and fluidization, in which air is used to blend some fine powders.

An important aspect of the performance of these blenders is the sequence of loading the batch to achieve optimum performance. As material discharges from a blender into a portable or stationary container, demixing or segregation often occurs. Typically, segregation due to sifting, fluidization, or entrainment of fine particles in the air takes place depending on the rate of discharge and the particle size distribution of the blend.

The flow pattern that develops as the material discharges from the portable or stationary container strongly influences what effect, if any, the filling segregation has. In many systems the container promotes a funnel flow pattern. Upon discharge, the material in the center, which is often predominantly fines,

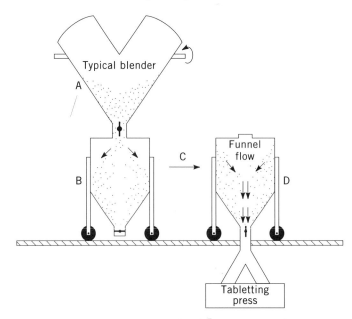

**Fig. 16.**  A typical batch blending operation. See text.

is discharged first. Mass flow eliminates this problem; however, if the blender discharge is segregated, mass flow locks in the sequence of fines and coarse product as they exit the blender.

*Tumble Blending.*    Tumble blending has a number of advantages over other common blending techniques. First, it eliminates having to discharge the blender into a portable or stationary container. Thus, at the end of the blending cycle, the material in the container is blended avoiding segregation that can take place during filling. Second, tumble blending eliminates the costly step of blender cleaning between batches. All the material is confined within the portable container, so when cleaning is required only the container needs to be cleaned, not the stationary blender. Third, tumble blending eliminates downtime required to fill and empty a blender, because the tumble station is ready to accept a new container as soon as the previous one has finished tumbling. Other advantages include maintaining strict batch integrity, minimizing dust, and providing more throughput since one tumble station can service different materials.

Problems of segregation, no-flow/erratic flow, flooding, etc, must be minimized if not eliminated, therefore containers designed for mass flow as opposed to funnel flow are preferred. The BINSERT has significant advantages over standard mass flow designs, which are particularly useful in tumble blending applications (Fig. 17) (18).

**Flow Aids.**    Flow aids are devices used to assist in discharging materials from a bin or other storage container. The best use of such a device is when gravity alone is insufficient or impractical to provide reliable discharge. However, in many instances, flow aid devices are overused in applications in which they are either unnecessary or create new problems.

Flow aids can generally be divided into three categories: mechanical, eg, those that rely on vibration or agitation of the material, such as vibrating

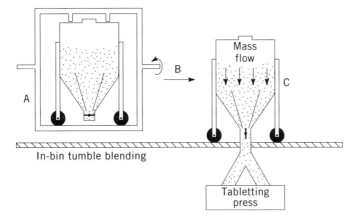

**Fig. 17.** Tumble blend operation using portable BINSERT containers where A represents the blend stage; B, transport; and C, discharge.

dischargers, external vibrators, rotating arms, vibrating panels, etc; introduction of air, eg, air cannons, air slides, or air nozzles; and chemical, eg, fumed silica. Some flow aids rely on a combination of these types.

*Vibrating Dischargers.* These mechanical devices, sometimes called bin activators, consist of an outer shell to which one or more motors with eccentric weight are mounted. By hanging the unit from the bin and using a rubber gasket to prevent spillage, the unit vibrates in a horizontal motion to assist in discharging material. The vibration is transmitted to material inside the unit through the use of a central baffle, which is either dome-shaped or conical, and supported by beams which are welded to the outer vibrating shell.

The area of influence of a vibrating discharger is limited to a cylinder, the diameter of which is roughly equal to the top diameter of the discharger. Hence, if a vibrating discharger is mounted onto a conical hopper section, flow is confined predominantly to a central flow channel located directly above the discharger. This is true unless the slope and smoothness of the static cone meet requirements for mass flow, or the diameter of the flow channel exceeds the critical rathole diameter for the material.

A vibrating discharger is incapable of controlling the rate of material discharged from it, ie, it is not a feeder. Therefore, if feed rate control is required, some type of feeder, eg, screw, rotary valve, or vibrating pan, must be mounted below the discharger's outlet. To avoid overcompacting the material by trying to force it to flow at a rate higher than that of the feeder, it is often necessary to cycle the discharger on and off using a field-adjusted timer. It is important that a vibrating discharger be operated on a regular basis, not just when complete discharge of the bin is required, otherwise the flow channel may locate preferentially on one side of the bin, which can cause significant structural problems resulting from nonuniform loading on the bin walls. If the material being handled is pressure sensitive or if segregation is important, the use of a vibrating discharger may not be advisable.

*External Vibrators.* Air- and electrically operated mechanical vibrators are sometimes placed on the exterior of hoppers and chutes. The type of vibration transmitted can vary from a high frequency, low amplitude mode, to a low frequency, high amplitude thumping condition. In general, such devices are better suited to cleaning off chutes than for use on hoppers to be filled with bulk solids. Disadvantages include noise pollution (see SUPPLEMENT) and possible fatigue damage to the structure.

*Air Blasters.*    Air cannons, or air blasters, consist of a cylinder of compressed air that has been pressurized, typically to plant air conditions 0.55–0.69 MPa (80–100 psi). When the unit is fired, a quick-acting valve is opened allowing the air to quickly exit the blaster and enter the bin. Such devices can be effective in collapsing a stable arch, but are usually far less effective in breaking up a stable rathole. If the material is severely caked, firing an air blaster into it may cause the material to shatter into large chunks, making subsequent firing of the air blaster ineffective because the air has a convenient path by which to dissipate its energy.

The effective range of a typical air blaster is on the order of 1.8–2.4 m. If a single air blaster does not provide sufficient energy to break an arch, multiple units can be used. These are fired simultaneously if at the same elevation, then start sequentially from lower elevations in the hopper and work upward.

When firing an air blaster, the pressure in the chamber is initially felt at nearly its maximum value in a localized area of the bin wall around the blaster nozzle which may require reinforcement to take the pressure.

*Mechanical Agitation.*    Devices consisting of a horizontal or vertical shaft with arms may be used to break up material and thereby cause it to flow. However, if the force resisting movement of such devices is large enough it can render the device useless either because of insufficient power to turn it or enough power to cause it to self-destruct. In general, such devices should only be used in relatively small bins and hoppers.

Other types of mechanical agitation consist of vibrating screens or expanded metal panels. However, if the device fails to perform for any reason, discharging material from the bin will be much more difficult than if the device were not present.

Uniform, reliable flow of bulk solids can allow the production of quality products with a minimum of waste, control dust and noise, and extend the life of a plant and maximize its productivity and output. By conducting laboratory tests and utilizing experts with experience in applying solids flow data, plant start-up delays that can impact schedule and cost can be eliminated.

Existing facilities present daily bulk solids flow problems; solids flow testing and analysis saves many hours of expensive downtime and thousands of dollars, thus moving from the complication of quick-fix solutions that are not satisfactory, to proven engineering solutions that work every time.

## BIBLIOGRAPHY

1. A. W. Jenike, *Storage and Flow of Solids*, Bulletin No. 123, University of Utah Engineering Experiment Station, Salt Lake City, Nov. 1964.
2. R. Kulwiec, ed., *Materials Handling Handbook*, John Wiley and Sons, Inc., New York, 1985.

3. C. R. Woodcock and J. S. Mason, *Bulk Solids Handling*, Chapman and Hall, London, 1987.
4. M. E. Fayed and L. Otten, *Handbook of Powder Science and Technology*, Van Nostrand Reinhold Co., New York, 1984.
5. P. A. Shamlou, *Handling of Bulk Solids—Theory and Practice*, Butterworths, Kent, U.K., 1988.
6. J. W. Carson and J. Marinelli, "Characterize Bulk Solids to Ensure Smooth Flow," *Chem. Eng.* (Apr. 1994).
7. D. S. Dick and R. J. Hossfeld, "Versatile BINSERT System Solves Wide Range of Flow Problems," presented at *The Powder and Bulk Solids 12th Annual Conference*, Rosemont, Ill., May 1987.
8. W. A. Beverloo, H. A. Leniger, and J. Van de Velde, *Chem. Eng. Sci.* **15**, 260 (1961).
9. R. T. Jenkyn and D. J. Goodwill, "Silo Failures: Lessons to be Learned," *Eng. Digest* (Sept. 1987).
10. J. W. Carson and R. T. Jenkyn, "Load Development and Structural Considerations in Silo Design," presented at *Reliable Flow of Particulate Solids II*, Oslo, Norway, Aug. 1993.
11. J. W. Carson and R. T. Jenkyn, "How to Prevent Silo Failures with Routine Inspections and Proper Repair," *Powder Bulk Eng.* **4**(1) (Jan. 1990).
12. J. Marinelli and J. W. Carson, "Use Screw Feeders Effectively," *Chem. Eng. Prog.* (Dec. 1992).
13. U.S. Pat. 5,101,961.
14. J. W. Carson, T. A. Royal, and D. J. Goodwill, *Bulk Solids Handl.* **6**, 139–144 (Feb. 1986).
15. T. A. Royal and J. W. Carson, "How to Avoid Flooding in Powder Handling Systems," *Powder Handl. Proc.* **1** (Mar. 1993).
16. J. W. Carson and T. A. Royal, "In-Bin Blending Improves Process Control," *Powder Handl. Proc.* (Sept. 1992).
17. U.S. Pat. 4,268,883.
18. J. W. Carson, T. A. Royal, and R. J. Hossfeld, "Tumble Blending with Mass Flow Containers Improves Productivity and Quality," *Powder Handl. Proc.* (Nov. 1994).

JOSEPH MARINELLI
Peabody SolidsFlow


JOHN W. CARSON
Jenike & Johanson, Inc.